Electric Circuits

James S. Kang
California State Polytechnic University, Pomona

CENGAGE Learning®

Australia • Brazil • Mexico • Singapore • United Kingdom • United States

CENGAGE Learning®

***Electric Circuits*, First Edition**
James S. Kang

Product Director, Global Engineering: Timothy L. Anderson
Associate Media Content Developer: Ashley Kaupert
Product Assistant: Alexander Sham
Marketing Manager: Kristin Stine
Director, Higher Education Production: Sharon L. Smith
Senior Content Project Manager: Kim Kusnerak
Production Service: MPS Limited
Senior Art Director: Michelle Kunkler
Cover/Internal Designer: Grannan Graphic Design Ltd.
Cover Image: Dabarti CGI/Shutterstock.com
Internal Images:
©Daumantas Liekis/Shutterstock.com;
©iStockPhoto.com/NesneJkraM;
©iStockPhoto.com/Denis Dryashkin;
©iStockPhoto.com/Zorandimzr
Intellectual Property
Analyst: Christine Myaskovsky
Project Manager: Sarah Shainwald
Text and Image Permissions Researcher: Kristiina Paul
Manufacturing Planner: Doug Wilke

Library of Congress Control Number: 2016955676

ISBN: 978-1-305-63521-0

Cengage Learning
20 Channel Center Street
Boston, MA 02210
USA

Cengage Learning is a leading provider of customized learning solutions with employees residing in nearly 40 different countries and sales in more than 125 countries around the world. Find your local representative at **www.cengage.com**.

Cengage Learning products are represented in Canada by Nelson Education Ltd.

To learn more about Cengage Learning Solutions, visit **www.cengage.com/engineering**.

Purchase any of our products at your local college store or at our preferred online store **www.cengagebrain.com**.

Printed in the United States of America
Print Number: 1 Print Year: 2016

Contents

CHAPTER 9

PHASORS AND IMPEDANCES 615

CHAPTER 10

ANALYSIS OF PHASOR-TRANSFORMED CIRCUITS 668

CHAPTER 11

AC POWER 733

CHAPTER 12

THREE-PHASE SYSTEMS 778

CHAPTER 13

MAGNETICALLY COUPLED CIRCUITS 829

CHAPTER 14

THE LAPLACE TRANSFORM 886

CHAPTER 15

CIRCUIT ANALYSIS IN THE *s*-DOMAIN 954

CHAPTER 16

FIRST- AND SECOND-ORDER ANALOG FILTERS 1074

CHAPTER 17

ANALOG FILTER DESIGN 1166

CHAPTER 18

FOURIER SERIES 1259

CHAPTER 19

FOURIER TRANSFORM 1399

CHAPTER 20

TWO-PORT CIRCUITS 1457

Preface

This book is intended to be an introductory text on the subject of electric circuits. It provides simple explanations of the basic concepts, followed by simple examples and exercises. When necessary, detailed derivations for the main topics and examples are given to help readers understand the main ideas. MATLAB is a tool that can be used effectively in Electric Circuits courses. In this text, MATLAB is integrated into selected examples to illustrate its use in solving circuit problems. MATLAB can be used to check the answers or solve more complex circuit problems. This text is written for a two-semester sequence or a three-quarters sequence on electric circuits.

Suggested Course Outlines

The following is a list of topics covered in a typical Electric Circuits courses, with suggested course outlines.

ONE-SEMESTER OR -QUARTER COURSE

If Electric Circuits is offered as a one-semester or one-quarter course, Chapters 1 through 12 can be taught without covering, or only lightly covering, sections 1.6, 2.10, 2.11, 3.6, 4.7, 5.6, 5.7, 5.8, 6.7, 7.6, 7.7, 8.8, 8.9, 9.9, 9.10, 10.12, 11.7, 12.5, 12.6, and 12.7.

TWO-SEMESTER OR -QUARTER COURSES

For two-semester Electric Circuit courses, Chapters 1 through 8, which cover dc circuits, op amps, and the responses of first-order and second-order circuits, can be taught in the first semester. Chapters 9 through 20, which cover alternating current (ac) circuits, Laplace transforms, circuit analysis in the *s*-domain, two-port circuits, analog filter design and implementation, Fourier series, and Fourier transform, can then be taught in the second semester.

THREE-QUARTER COURSES

For three-quarter Electric Circuit courses, Chapters 1 through 5, which cover dc circuits and op amps, can be taught in the first quarter; Chapters 6 through 13, which cover the responses of first-order and second-order circuits and ac circuits, can be taught in the second quarter, and Chapters 14 through 20, which cover Laplace transforms, circuit analysis in the *s*-domain, two-port circuits, analog filter design and implementation, Fourier series, and Fourier transform, can be taught in the third quarter.

Depending on the catalog description and the course outlines, instructors can pick and choose the topics covered in the courses that they teach. Several features of this text are listed next.

Features

After a topic is presented, examples and exercises follow. Examples are chosen to expand and elaborate the main concept of the topic. In a step-by-step approach, details are worked out to help students understand the main ideas.

In addition to analyzing RC, RL, and RLC circuits connected in series or parallel in the time domain and the frequency domain, analyses of circuits different from RC, RL, and RLC circuits and connected other than in series and parallel are provided. Also, general input signals that are different from unit step functions are included in the analyses.

In the analog filter design, the specifications of the filter are translated into its transfer function in cascade form. From the transfer function, each section can be designed with appropriate op amp circuits. The normalized component values for each section are found by adopting a simplification method (equal R equal C or unity gain). Then, magnitude scaling and frequency scaling are used to find the final component values. The entire design procedure, from the specifications to the circuit design, is detailed, including the PSpice simulation used to verify the design.

Before the discussion of Fourier series, orthogonal functions and the representation of square integrable functions as a linear combination of a set of orthogonal functions are introduced. The set of orthogonal functions for Fourier series representation consists of cosines and sines. The Fourier coefficients for the square pulse train, triangular pulse train, sawtooth pulse train, and rectified sines and cosines are derived. The Fourier coefficients of any variation of these waveforms can be found by applying the time-shifting property and finding the dc component.

MATLAB can be an effective tool in solving problems in electric circuits. Simple functions such as calculating the equivalent resistance or impedance of parallel connection of resistors, capacitors, and inductors; conversion from Cartesian coordinates to polar coordinates; conversion from polar coordinates to Cartesian coordinates; conversion from the wye configuration to delta configuration; and conversion from delta configuration to wye configuration provide accurate answers in less time. These simple functions can be part of scripts that enable us to find solutions to typical circuit problems.

The complexity of taking the inverse Laplace transforms increases as the order increases. MATLAB can be used to solve equations and to find integrals, transforms, inverse transforms, and transfer functions. The application of MATLAB to circuit analysis is demonstrated throughout the text when appropriate. For example, after finding inverse Laplace transforms by hand using partial fraction expansion, answers from MATLAB are provided as a comparison.

Examples of circuit simulation using OrCAD PSpice and Simulink are given at the end of each chapter. Simulink is a tool that can be used to perform circuit simulations. In Simulink, physical signals can be converted to Simulink signals and vice versa. Simscapes include many blocks that are related to electric circuits. Simulink can be used in computer assignments or laboratory experiments.

The Instructor's Solution Manual for the exercises and end-of-chapter problems is available for instructors. This manual includes MATLAB scripts for selected problems as a check on the accuracy of the solutions by hand.

Overview of Chapters

In **Chapter 1**, definitions of voltage, current, power, and energy are given. Also, independent voltage source and current source are introduced, along with dependent voltage sources and current sources.

In **Chapter 2**, nodes, branches, meshes, and loops are introduced. Ohm's law is explained. Kirchhoff's current law (KCL), Kirchhoff's voltage law (KVL), the voltage divider rule, and the current divider rule are explained with examples.

In **Chapter 3**, nodal analysis and mesh analysis are discussed in depth. The nodal analysis and mesh analysis are used extensively in the rest of the text.

Chapter 4 introduces circuit theorems that are useful in analyzing electric circuits and electronic circuits. The circuit theorems discussed in this chapter are the superposition principle, source transformations, Thévenin's theorem, Norton's theorem, and maximum power transfer.

Chapter 5 introduces op amp circuits. Op amp is a versatile integrated circuit (IC) chip that has wide-ranging applications in circuit design. The concept of the ideal op amp model is explained, along with applications in sum and difference, instrumentation amplifier, and current amplifier. Detailed analysis of inverting configuration and noninverting configuration is provided.

In **Chapter 6**, the energy storage elements called *capacitors* and *inductors* are discussed. The current voltage relation of capacitors and inductors are derived. The energy stored on the capacitors and inductors are presented.

In **Chapter 7**, the transformation of RC and RL circuits to differential equations and solutions of the first-order differential equations to get the responses of the circuits are presented. In the general first-order circuits, the input signal can be dc, ramp signal, exponential signal, or sinusoidal signal.

In **Chapter 8**, the transformation of series RLC and parallel RLC circuits to the second-order differential equations, as well as solving the second-order differential equations to get the responses of the circuits are presented. In the general second-order circuits, the input signal can be dc, ramp signal, exponential signal, or sinusoidal signal.

Chapter 9 introduces sinusoidal signals, phasors, impedances, and admittances. Also, transforming ac circuits to phasor-transformed circuits is presented, along with analyzing phasor transformed circuits using KCL, KVL, equivalent impedances, delta-wye transformation, and wye-delta transformation.

The analysis of phasor-transformed circuits is continued in **Chapter 10** with the introduction of the voltage divider rule, current divider rule, nodal analysis, mesh analysis, superposition principle, source transformation, Thévenin equivalent circuit, Norton equivalent circuit, and transfer function. This analysis is similar to the one for resistive circuits with the use of impedances.

Chapter 11 presents information on ac power. The definitions of instantaneous power, average power, reactive power, complex power, apparent power, and power factor are also given, and power factor correction is explained with examples.

As an extension of ac power, the three-phase system is presented in **Chapter 12**. The connection of balanced sources (wye-connected or delta-connected) to balanced loads (wye-connected or delta connected) are presented, both with and without wire impedances.

Magnetically coupled circuits, which are related to ac power, are discussed in **Chapter 13**. Mutual inductance, induced voltage, dot convention, linear transformers, and ideal transformers are introduced.

The Laplace transform is introduced in **Chapter 14**. The definition of the transform, region of convergence, transform, and inverse transform are explained with examples. Various properties of Laplace transform are also presented with examples.

The discussion on Laplace transform is continued in **Chapter 15**. Electric circuits can be transformed into an *s*-domain by replacing voltage sources and current sources to the *s*-domain and replacing capacitors and inductors to impedances. The circuit laws and theorems that apply to resistive circuits also apply to *s*-domain circuits. The time domain signal can be obtained by taking the inverse Laplace transform of the *s*-domain representation. The differential equations in the time domain are transformed to algebraic equations in the *s*-domain. The transfer function in the *s*-domain is defined as the ratio of the output signal in the s-domain to the input signal in the s-domain. The concept of convolution is introduced with a number of examples. Also, finding the convolution using Laplace transforms are illustrated in the same examples. Plotting the magnitude response and phase response of a circuit or a system using the Bode diagram is introduced.

The first-order and the second-order analog filters that are building blocks for the higher-order filters are presented in **Chapter 16**. The filters can be implemented by interconnecting passive elements consisting of resistors, capacitors, and inductors. Alternatively, filters can be implemented utilizing op amp circuits. Sallen and Key circuits for implementing second-order filters are discussed as well, along with design examples.

The discussion on analog filter design is extended in **Chapter 17**. A filter is designed to meet the specifications of the filter. The transfer function that satisfies the specification is found. From the transfer function, the corner frequency and Q value can be found. Then, the normalized component values and scaled component values are found. PSpice simulations can be used to verify the design.

Orthogonal functions and the representation of signals as a linear combination of a set of orthogonal functions are introduced in **Chapter 18**. If the set of orthogonal functions consists of harmonically related sinusoids or exponential functions, the representation is called the *Fourier series.* Fourier series representation of common signals, including the square pulse train, triangular pulse train, sawtooth waveform, and rectified cosine and sine, are presented in detail, with examples. The derivation and application of the time-shifting property of Fourier coefficients are provided. In addition, the application of the Fourier series representation in solving circuit problems are presented, along with examples.

As the period of a periodic signal is increased to infinity, the signal becomes nonperiodic, the discrete line spectrums become a continuous spectrum, and multiplying the Fourier coefficients by the period produces the Fourier transform, as explained in **Chapter 19**. Important properties of the Fourier transform, including time shifting, frequency shifting, symmetry, modulation, convolution, and multiplication, are introduced, along with interpretation and examples.

Two-port circuits are defined and analyzed in **Chapter 20**. Depending on which of the parameters are selected as independent variables, there are six different representations for two-port circuits. The coefficients of the representations are called *parameters*. The six parameters ($z, y, h, g, ABCD, b$) for two-port circuits are presented along with examples. The conversion between the parameters and the interconnection of parameters are provided in this chapter.

Instructor Resources

Cengage Learning's secure, password-protected Instructor Resource Center contains helpful resources for instructors who adopt this text. These resources include Lecture Note Microsoft PowerPoint slides, test banks, and an Instructor's Solution Manual, with detailed solutions to all the problems from the text. The Instructor Resource Center can be accessed at https://login.cengage.com.

MindTap Online Course

Electric Circuits is also available through **MindTap**, Cengage Learning's digital course platform. The carefully crafted pedagogy and exercises in this textbook are made even more effective by an interactive, customizable eBook, automatically graded assessments, and a full suite of study tools.

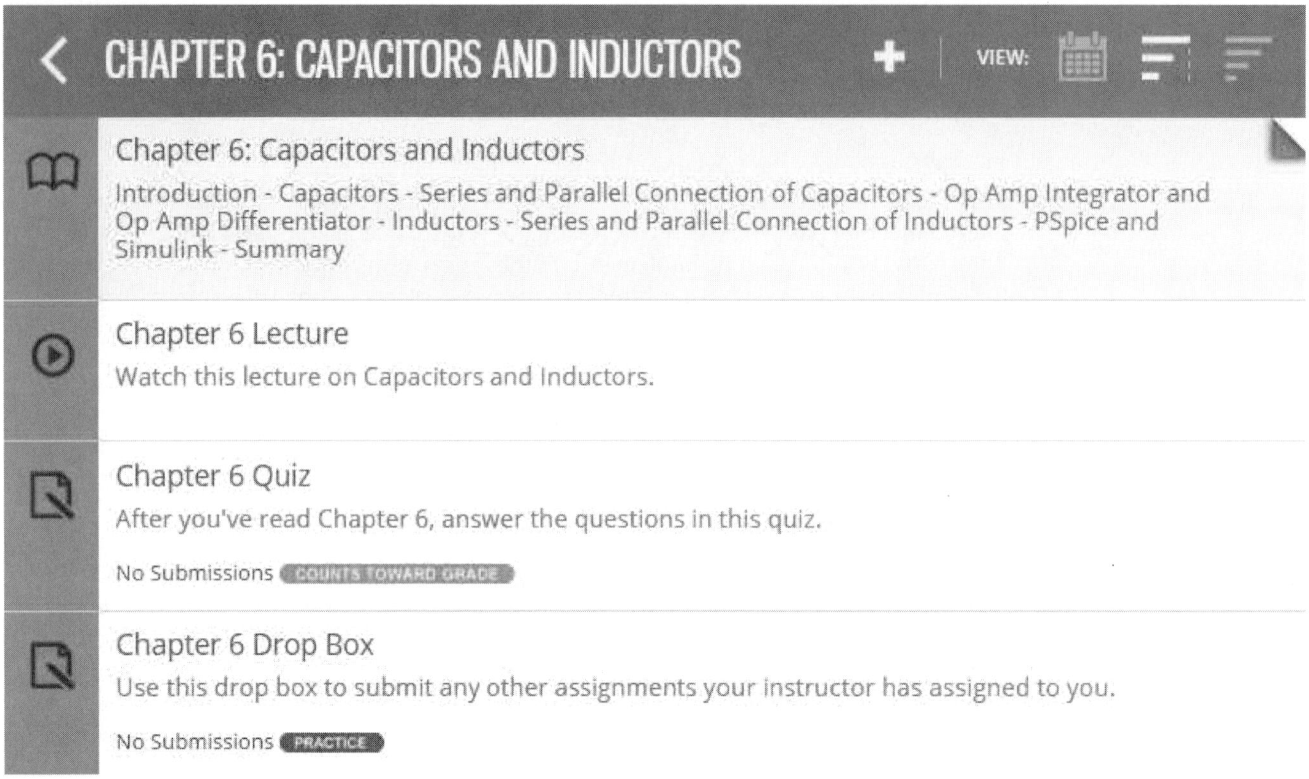

As an instructor using MindTap, you have at your fingertips the full text and a unique set of tools, all in an interface designed to save you time. MindTap makes it easy for instructors to build and customize their course so that they can focus on the most relevant material while also lowering costs for students. Stay connected and informed through real-time student tracking that provides the opportunity to adjust your course as needed based on analytics of interactivity and performance. **End-of-chapter assessments** test students' knowledge of topics in each chapter. In addition, a **curated collection of lecture videos** helps students better understand key concepts as they progress through the course.

HOW DOES MINDTAP BENEFIT INSTRUCTORS?

- Instructors can build and personalize their courses by integrating their own content into the **MindTap Reader** (like lecture notes or problem sets to download) or pull from sources such as Really Simple Syndication (RSS) feeds, YouTube videos, websites, and more. Control what content students see with a built-in learning path that can be customized to your syllabus.
- MindTap saves time by providing instructors and their students with **automatically graded assignments and quizzes**. These problems include immediate, specific feedback so students know exactly where they need more practice.
- The **Message Center** helps instructors to contact students quickly and easily from MindTap. Messages are communicated directly to each student via the communication medium (email, social media, or even text messages) designated by the student.
- **StudyHub** is a valuable tool that allows instructors to deliver important information and empowers students to personalize their experience. Instructors can choose to annotate the text with **notes** and **highlights**, share content from the MindTap Reader, and create **flashcards** to help their students focus and succeed.
- The **Progress App** lets instructors know exactly how their students are doing (and where they might be struggling) with live analytics. They can see overall class engagement and drill down into individual student performance, enabling them to adjust their course to maximize student success.

HOW DOES MINDTAP BENEFIT YOUR STUDENTS?

- The **MindTap Reader** adds the ability to have content read aloud, to print from the MindTap Reader, and to take notes and highlight text, while also capturing them within the linked **StudyHub App**.
- The **MindTap Mobile App** keeps students connected with alerts and notifications, while also providing them with on-the-go study tools like flashcards and quizzing, helping them manage their time efficiently.
- **Flashcards** are prepopulated to provide a jump start on studying, and students and instructors also can create customized cards as they move through the course.
- The **Progress App** allows students to monitor their individual grades, as well as their performance level compared to the class average. This not only helps them stay on track in the course, but also motivates them to do more, and ultimately to do better.
- The unique **StudyHub** is a powerful, single-destination studying tool that empowers students to personalize their experience. They can quickly and easily access all notes and highlights marked in the MindTap Reader, locate bookmarked pages, review notes and flashcards shared by their instructor, and create custom study guides.

For more information about MindTap for Engineering, or to schedule a demonstration, please call (800) 354-9706 or email higheredcs@cengage.com. For instructors outside the United States, visit http://www.cengage.com/contact/ to locate your regional office.

Acknowledgments

I wish to acknowledge and thank the Global Engineering team at Cengage Learning for their dedication to this new book: Timothy Anderson, Product Director; Ashley Kaupert, Associate Media Content Developer; Kim Kusnerak, Senior Content Project Manager; Kristin Stine, Marketing Manager; Elizabeth Brown and Brittany Burden, Learning Solutions Specialists; and Alexander Sham, Product Assistant. They have skillfully guided every aspect of this text's development and production to successful completion. I also would like to express my appreciation to the following reviewers, whose helpful comments and suggestions improved the manuscript:

Elizabeth Brauer, Northern Arizona University
Mario Edgardo Magana, Oregon State University
Malik Elbuluk, The University of Akron
Timothy A. Little, Dalhousie University
Ahmad Nafisi, California Polytechnic State University—San Luis Obispo
Scott Norr, University of Minnesota—Duluth
Nadipuram Prasad, New Mexico State University
Vignesh Rajamani, Oklahoma State University
Pradeepa Yahampath, University of Manitoba

About the Author

Dr. James S. Kang is a professor of electrical and computer engineering at the California State Polytechnic University, Pomona, commonly known as Cal Poly Pomona. Cal Poly Pomona is famous for its laboratory-oriented, hands-on approach to engineering education. Most of the electrical and computer engineering courses offered there include a companion laboratory course. Students design, build, and test practical circuits in the laboratory based on the theory that they learned in the lecture course. This book, *Electric Circuits,* incorporates this philosophy.

Chapter 1

Voltage, Current, Power, and Sources

1.1 Introduction

The seven base units of the International System of Units (SI), along with derived units relevant to electrical and computer engineering, are presented in this chapter. The definitions of the terms *voltage, current*, and *power* are given as well.

A voltage source with voltage V_s provides a constant potential difference to the circuit connected between the positive terminal and the negative terminal. A current source with current I_s provides a constant current of I_s amperes to the circuit connected to the two terminals. If the voltage from the voltage source is constant with time, the voltage source is called the *direct current (dc) source*. Likewise, if the current from the current source is constant with time, the current source is called the *dc source*. If the voltage from the voltage source is a sinusoid, the voltage source is called *alternating current (ac) voltage source*. Likewise, if the current from the current source is a sinusoid, the current source is called the *ac current source*.

The voltage or current on the dependent sources depends solely on the controlling voltage or controlling current. Dependent sources are introduced along with circuit symbols.

The elementary signals that are useful throughout the text are introduced next. The elementary signals are *Dirac delta function, step function, ramp function, rectangular pulse, triangular pulse*, and *exponential decay*.

1.2 International System of Units

The International System of Units (SI) is the modern form of the metric system derived from the meter-kilogram-second (MKS) system. The SI system is founded on seven base units for the seven quantities assumed to be mutually independent. Tables 1.1–1.6, which give information on the SI system, come from the NIST Reference on Constants, Units, and Uncertainty (http://physics.nist.gov/cuu/Units/units.html), the official reference of the National Institute of Standards and Technology.

A **meter** is defined as the length of a path traveled by light in a vacuum during a time interval of 1/299,792,458 [$(\approx 1/(3 \times 10^8))$] of a second.

A **kilogram** is equal to the mass of the international prototype of the kilogram.

TABLE 1.1 SI Base Units.

Base Quantity	Name	Symbol
Length	meter	m
Mass	kilogram	kg
Time	second	s
Electric current	ampere	A
Thermodynamic temperature	kelvin	K
Amount of a substance	mole	mol
Luminous intensity	candela	cd

TABLE 1.2 Examples of SI Derived Units.

Derived Quantity	Name	Symbol
Area	square meter	m^2
Volume	cubic meter	m^3
Speed, velocity	meter per second	m/s
Acceleration	meter per second squared	m/s^2
Wave number	reciprocal meter	m^{-1}
Mass density	kilogram per cubic meter	kg/m^3
Specific volume	cubic meter per kilogram	m^3/kg
Current density	ampere per square meter	A/m^2
Magnetic field strength	ampere per meter	A/m
Luminance	candela per square meter	cd/m^2

TABLE 1.3 SI Derived Units with Special Names and Symbols.

Derived Quantity	Name	Symbol	Expression in terms of other SI units
Plane angle	radian	rad	—
Solid angle	steradian	sr	—
Frequency	hertz	Hz	—
Force	newton	N	—
Pressure, stress	pascal	Pa	N/m^2
Energy, work, quantity of heat	joule	J	$N \cdot m$
Power, radiant flux	watt	W	J/s
Electric charge, quantity of electricity	coulomb	C	—
Electric potential difference, electromotive force	volt	V	W/A
Capacitance	farad	F	C/V
Electric resistance	ohm	Ω	V/A
Electric conductance	siemens	S	A/V
Magnetic flux	weber	Wb	$V \cdot s$
Magnetic flux density	tesla	T	Wb/m^2
Inductance	henry	H	Wb/A
Celsius temperature	degrees Celsius	°C	—
Luminous flux	lumen	lm	$cd \cdot sr$
Illuminance	lux	lx	lm/m^2

TABLE 1.4 Examples of SI Derived Units with Names and Symbols (Including Special Names and Symbols.)

Derived Quantity	Name	Symbol
Dynamic viscosity	Pascal second	Pa · s
Moment of force	newton meter	N · m
Surface tension	newton per meter	N/m
Angular velocity	radian per second	rad/s
Angular acceleration	radian per second squared	rad/s^2
Heat flux density, irradiance	watt per square meter	W/m^2
Thermal conductivity	watt per meter kelvin	W/(m · K)
Energy density	joule per cubic meter	J/m^3
Electric field strength	volt per meter	V/m
Electric charge density	coulomb per cubic meter	C/m^3
Electric flux density	coulomb per square meter	C/m^2
Permittivity	farad per meter	F/m
Permeability	henry per meter	H/m
Exposure (X- and γ-rays)	coulomb per kilogram	C/kg

TABLE 1.5 Metric Prefixes.

Prefix	Symbol	Magnitude
yocto	y	10^{-24}
zepto	z	10^{-21}
atto	a	10^{-18}
femto	f	10^{-15}
pico	p	10^{-12}
nano	n	10^{-9}
micro	μ	10^{-6}
milli	m	10^{-3}
centi	c	10^{-2}
deci	d	10^{-1}
deka	da	10^{1}
hecto	h	10^{2}
kilo	k	10^{3}
mega	M	10^{6}
giga	G	10^{9}
tera	T	10^{12}
peta	P	10^{15}
exa	E	10^{18}
zetta	Z	10^{21}
yotta	Y	10^{24}

TABLE 1.6 Units Outside the SI That Are Accepted for Use with the SI System.

Name	Symbol	Value in SI Units
Minute (time)	min	1 min = 60 s
Hour	h	1 h = 60 min = 3600 s
Day	d	1 d = 24 h = 86,400 s
Degree (angle)	°	$1° = (\pi/180)$ rad
Minute (angle)	′	$1' = (1/60)° = (\pi/10{,}800)$ rad
Second (angle)	″	$1'' = (1/60)' = (\pi/648{,}000)$ rad
Liter	L	$1\ L = 1\ dm^3 = 10^{-3}\ m^3$
Metric ton	t	1 t = 1000 kg
Neper	Np	$1\ Np = 20 \log_{10}(e)\ dB = 20/\ln(10)\ dB$
Bel	B	$1\ B = (1/2)\ln(10)\ Np$, 1 dB = 0.1 B
Electronvolt	eV	$1\ eV = 1.60218 \times 10^{-19}\ J$
Unified atomic mass unit	u	$1\ u = 1.66054 \times 10^{-27}\ kg$
Astronomical unit	ua	$1\ ua = 1.49598 \times 10^{11}\ m$

A **second** is the duration of 9,192,631,770 periods of the radiation corresponding to the transition between the two hyperfine levels of the ground state of the cesium 133 atom.

An **ampere** is the constant current which, if maintained in two straight parallel conductors of infinite length, of negligible circular cross section, and placed 1 meter apart in vacuum, would produce between these conductors a force equal to 2×10^{-7} newtons per meter of length.

A **kelvin**, is 1/273.16 of the thermodynamic temperature of the triple point of water.

A **mole** is the amount of substance of a system that contains as many elementary entities as there are atoms in 0.012 kilogram of carbon 12; its symbol is **mol**. When the mole is used, the elementary entities must be specified; they may be atoms, molecules, ion, electrons, other particles, or specified groups of such particles.

The **candela** is the luminous intensity, in a given direction, of a source that emits monochromatic radiation of frequency 540×10^{12} hertz (Hz) and that has the radiant intensity in that direction of 1/683 watt per steradian.

1.3 Charge, Voltage, Current, and Power

1.3.1 ELECTRIC CHARGE

Atoms are the basic building blocks of matter. The nucleus of atoms consists of protons and neutrons. Electrons orbit around the nucleus. Protons are positively charged, and electrons are negatively charged, while neutrons are electrically neutral. The amount of charge on the proton is given by

$$e = 1.60217662 \times 10^{-19}\ C$$

Here, the unit for charge is in coulombs (C).

$$-e = -1.60217662 \times 10^{-19}\ C$$

Notice that the charge is quantized as the integral multiple of e. Since there are equal numbers of protons and electrons in an atom, it is electrically neutral. When a plastic is rubbed by fur, some electrons from the fur are transferred to the plastic. Since the fur lost electrons and the plastic gained them, the former is positively charged and the latter negatively charged. When the fur and the plastic are placed close together, they attract each other. Opposite charges attract, and like charges repel. However, since the electrons and protons are not destroyed, the total amount of charge remains the same. This is called the *conservation of charge*.

1.3.2 ELECTRIC FIELD

According to Coulomb's law, the magnitude of force between two charged bodies is proportional to the charges Q and q and inversely proportional to the distance squared; that is,

$$F = \frac{1}{4\pi\varepsilon}\frac{Qq}{r^2} \tag{1.1}$$

Here, ε is permittivity of the medium. The permittivity of free space, ε_0, is given by

$$\varepsilon_0 = \frac{1}{4\pi c^2 10^{-7}}\,(\text{F/m}) = 8.8541878176 \times 10^{-12}\,(\text{F/m}) \tag{1.2}$$

Here, c is the speed of light in the vacuum, given by $c = 299{,}792{,}458$ m/s $\approx 3 \times 10^8$ m/s. The unit for permittivity is farads per meter (F/m). The direction of the force coincides with the line connecting the two bodies. If the charges have the same polarity, the two bodies repel each other. On the other hand, if the charges have the opposite polarity, they attract each other.

If a positive test charge with magnitude q is brought close to a positive point charge with magnitude Q, the test charge will have a repulsive force. The magnitude of the force is inversely proportional to the distance squared between the point charge and the test charge. The presence of the point charge creates a field around it, where charged particles experience force. This is called an **electric field**, which is defined as the force on a test charge q as the charge q decreases to zero; that is,

$$\boldsymbol{E} = \lim_{q \to 0} \frac{\boldsymbol{F}}{q} \quad \text{(V/m)} \tag{1.3}$$

The electric field is a force per unit charge. The electric field $\boldsymbol{E}$ is a vector quantity whose direction is the same as that of the force. Figure 1.1 shows the electric field for a positive point charge and charged parallel plates.

FIGURE 1.1

Electric field for (a) a point charge and (b) parallel plates.

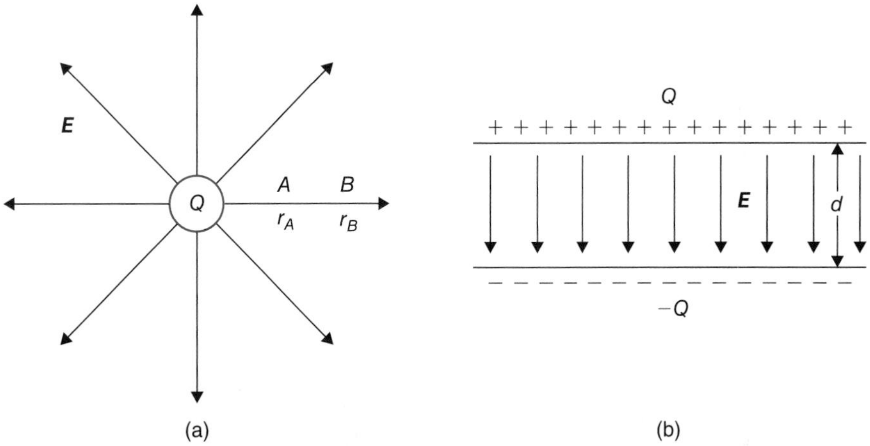

If an object with charge q is placed in the presence of electric field $\boldsymbol{E}$, the object will experience a force as follows:

$$\boldsymbol{F} = q\boldsymbol{E} \tag{1.4}$$

For a positive point charge Q, the electric field is given by

$$\boldsymbol{E} = \frac{1}{4\pi\varepsilon} \frac{Q}{r^2} \boldsymbol{a}_r \tag{1.5}$$

where $\boldsymbol{a}_r$ is a unit vector in the radial direction from the positive point charge Q. For parallel plates with area S per plate, distance d between the plates, the electric field is constant within the plates and the magnitude of the electric field is given by

$$E = \frac{Q}{\varepsilon S} \tag{1.6}$$

The direction of the field is from the plate with positive charges to the plate with negative charges, as shown in Figure 1.1(b).

1.3.3 VOLTAGE

If a positive test charge dq is moved against the electric field created by a positive charge, an external agent must apply work to the test charge. Let dw_{AB} be the amount of the work needed to move the test charge from B (initial) to A (final). Here, dw_{AB} is the potential energy in joules. Then, the potential difference between points A and B is defined as the work done per unit charge against the force; that is,

$$v_{AB} = v_A - v_B = \frac{dw_{AB}}{dq} \quad \text{(J/C)} \tag{1.7}$$

The unit for the potential difference is joules per coulomb, which is also called a *volt* (V):

$$1\text{ V} = 1\text{ J/C}$$

The potential difference between A and B is called *voltage*. The potential difference between points A and B is given by

$$v_{AB} = v_A - v_B = -\int_B^A \boldsymbol{E} \cdot d\ell \tag{1.8}$$

The negative sign implies that moving against the electric field increases the potential. For a positive point charge Q at origin with an electric field given by Equation (1.5), the potential difference between two points A and B with distances r_A and r_B, respectively, from Q is given by

$$v_{AB} = v_A - v_B = -\int_{r_B}^{r_A} \frac{1}{4\pi\varepsilon}\frac{Q}{r^2}dr = -\frac{Q}{4\pi\varepsilon}\left(\frac{-1}{r}\right)\Bigg|_{r_B}^{r_A} = \frac{Q}{4\pi\varepsilon}\left(\frac{1}{r_A} - \frac{1}{r_B}\right)\text{ V} \tag{1.9}$$

Notice that the integral of $1/r^2$ is $-1/r$. If r_B is infinity, the potential difference is

$$v_{AB} = v_A - v_B = v_A = \frac{Q}{4\pi\varepsilon r_A}\text{ V} \tag{1.10}$$

The potential is zero at infinity. This is a reference potential. For the parallel plates shown in Figure 1.1(b), the potential difference between A and B is

$$v = Ed = \frac{Q}{\varepsilon S}d \tag{1.11}$$

If the potential at B is set at zero ($v_B = 0$), the potential at point A is given by

$$v_A = \frac{dw_A}{dq} \quad \text{(J/C)} \tag{1.12}$$

or simply

$$v = \frac{dw}{dq} \quad \text{(J/C)} \tag{1.13}$$

The potential difference v is called *voltage*. A **battery** is a device that converts chemical energy to electrical energy. When a positive charge is moved from the negative terminal to the positive terminal through the 12-V battery, the battery does 12 joules of work on each unit charge. The potential energy of the charge increases by 12 joules. The battery provides energy to the rest of the circuit.

1.3.4 CURRENT

In the absence of an electric field, the free electrons in the conduction band of conductors such as copper wire make random movements. The number of electrons crossing a cross-sectional area of the copper wire from left to right will equal the number of electrons crossing the same cross-sectional area from right to left. The net number of electrons crossing this area will be zero. When an electric field is applied along the copper wire, the negatively charged electrons will move toward the direction of higher potential. The current is defined as the total amount of charge q passing through a cross-sectional area in t seconds; that is,

$$I = \frac{q}{t} \tag{1.14}$$

The unit for the current is coulombs per second (C/s) or amperes (A). If the amount of charge crossing the area changes with time, the current is defined as

$$i(t) = \frac{dq(t)}{dt} \tag{1.15}$$

The direction of current is defined as the direction of positive charges. Since the charge carriers inside the conductors are electrons, the direction of electrons is opposite to the direction of the current. Figure 1.2 shows the directions of the electric field, current, and electron inside a conductor.

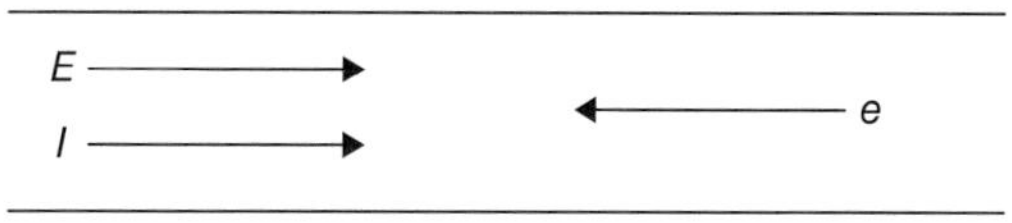

FIGURE 1.2

The directions of E, I, and e.

The charge transferred between time t_1 and t_2 can be obtained by integrating the current from t_1 and t_2; that is,

$$q = \int_{t_1}^{t_2} i(\lambda)d\lambda \tag{1.16}$$

EXAMPLE 1.1

The charge flowing into a circuit element for $t \geq 0$ is given by

$$q(t) = 2 \times 10^{-3}(1 - e^{-1000t}) \text{ coulomb}$$

Find the current flowing into the element for $t \geq 0$.

$$i(t) = \frac{dq(t)}{dt} = 2 \times 10^{-3} \times 1000e^{-1000t} \ A = 2e^{-1000t} \ A \text{ for } t \geq 0$$

Exercise 1.1

The charge flowing into a circuit element for $t \geq 0$ is given by

$$q(t) = 4 \times 10^{-3} e^{-2000t} \text{ coulomb}$$

Find the current flowing into the element for $t \geq 0$.

Answer:

$$i(t) = \frac{dq(t)}{dt} = -8e^{-2000t} \ A \text{ for } t \geq 0$$

EXAMPLE 1.2

The current flowing into a circuit element is given by

$$i(t) = 5\sin(2\pi 10t) \text{ mA}$$

for $t \geq 0$. Find the charge flowing into the device for $t \geq 0$. Also, find the total charge entered into the device at $t = 0.05$ s.

$$q(t) = \int_0^t i(\lambda)d\lambda = \frac{5 \times 10^{-3}}{2\pi 10}[1 - \cos(2\pi 10t)]$$
$$= 7.9577 \times 10^{-5}[1 - \cos(2\pi 10t)] \text{ coulomb}$$

At $t = 0.05$ s, we have

$$q(0.05) = 1.5915 \times 10^{-4}[1 - \cos(2\pi 10 \times 0.05)] = 1.5915 \times 10^{-4} \text{ coulombs}$$

Exercise 1.2

The current flowing into a circuit element is given by

$$i(t) = 5\cos(2\pi 10t) \text{ mA}$$

for $t \geq 0$. Find the charge flowing into the device for $t \geq 0$. Also, find the total charge entered into the device at $t = 0.0125$ s.

Answer:

$$q(t) = \int_0^t i(\lambda)d\lambda = \frac{5 \times 10^{-3}}{2\pi 10}\sin(2\pi 10t) = 7.9577 \times 10^{-5}\sin(2\pi 10t) \text{ coulombs}$$

$$q(0.0125) = 7.9577 \times 10^{-5}\sin(2\pi 10 \times 0.0125) = 5.6270 \times 10^{-5} \text{ coulombs}$$

1.3.5 POWER

The battery provides a constant potential difference (voltage) of v volts from the negative terminal to the positive terminal. When a positive charge dq is moved from the negative terminal to the positive terminal through the battery, the potential energy is increased by $dq\,v = dw$. When the positive charge dq moves through the rest of the circuit from the positive terminal to the negative terminal, the potential energy is decreased by the same amount ($dq\,v$). The rate of potential energy loss is given by

$$p = \frac{dw}{dt} = \frac{dq\,v}{dt} = iv \quad \textbf{(1.17)}$$

The rate of energy loss is defined as *power*. Equation (1.17) can be rewritten as

$$dw = dq\,v = p\,dt \quad \textbf{(1.18)}$$

The energy is the product of power and time. If Equation (1.18) is integrated as a function of time, we get

$$w(t) = \int_{-\infty}^{t} p(\lambda)d\lambda \quad \textbf{(1.19)}$$

According to Equation (1.19), the energy is the integral of power. As shown in Equation (1.17), power is the derivative of energy. Taking the derivative of Equation (1.19), we obtain

$$p(t) = \frac{dw(t)}{dt} \quad \textbf{(1.20)}$$

If the voltage and the current are time-varying, the power is also time-varying. If the voltage and current are expressed as a function of time, Equation (1.17) can be written as

$$p(t) = i(t)v(t) \quad \textbf{(1.21)}$$

FIGURE 1.3

(a) Power is positive.
(b) Power is negative.

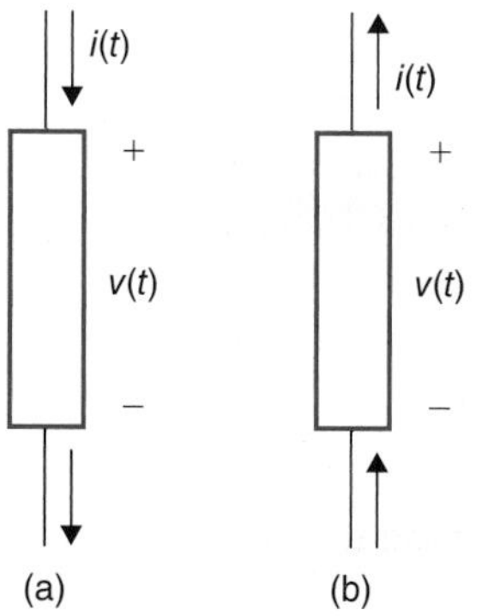

The power given by Equation (1.21) is called *instantaneous power*. According to Equation (1.21), instantaneous power is the product of current and voltage as a function of time. In the **passive sign convention**, if the direction of current is from the positive terminal of a device, through the device, and to the negative terminal of the device [as shown in Figure 1.3(a)], the power is positive. On the other hand, if the current leaves the positive terminal of a device, flows through the rest of the circuit, and enters the negative terminal of the device [as shown in Figure 1.3(b)], the power is negative.

If power is positive [i.e., $p(t) > 0$], the element is absorbing power. On the other hand, if power is negative, the element is delivering (supplying) power. In a given circuit, the total absorbed power equals the total delivered or supplied power. This is called *conservation of power*.

EXAMPLE 1.3

Let the voltage across an element be $v(t) = 100\cos(2\pi 60t)$ V, and the current though the element from positive terminal to negative terminal be $i(t) = 5\cos(2\pi 60t)$ A for $t \geq 0$. Find the instantaneous power $p(t)$ and plot $p(t)$.

continued

Example 1.3 continued

$$p(t) = i(t)\,v(t) = 5\cos(2\pi 60t) \times 100\cos(2\pi 60t) = 500\cos^2(2\pi 60t)$$
$$= 250 + 250\cos(2\pi \times 120t)\text{ W}$$

The power $p(t)$ is shown in Figure 1.4. Since $p(t) \geq 0$ for all t, the element is not delivering power any time. On average, the element absorbs 250 W of power.

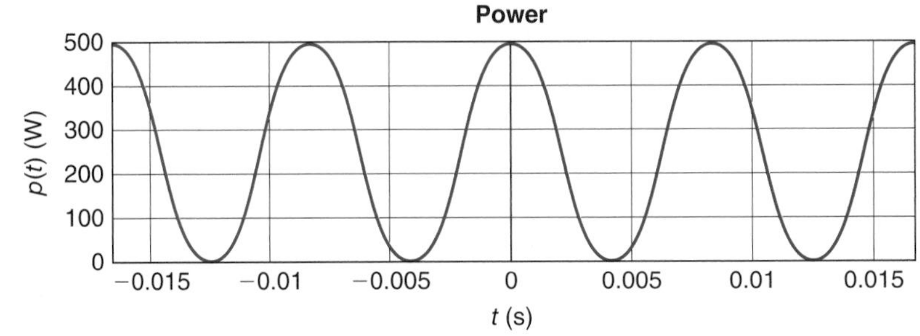

FIGURE 1.4

Plot of $p(t)$.

Exercise 1.3

Let the voltage across an element be $v(t) = 100\cos(2\pi 60t)$ V and the current though the element from positive terminal to negative terminal be $i(t) = 6\sin(2\pi 60t)$ A for $t \geq 0$. Find the instantaneous power $p(t)$ and plot $p(t)$.

$$p(t) = i(t)\,v(t) = 6\sin(2\pi 60t) \times 100\cos(2\pi 60t) = 300\sin(2\pi 120t)\text{ W}.$$

The power $p(t)$ is shown in Figure 1.5. Since $p(t) > 0$ half of the time and $p(t) < 0$ the other half of the time, the element absorbs power for $1/240$ s, then delivers power for the next $1/240$ s, and then repeats the cycle. On average, the element does not absorb any power.

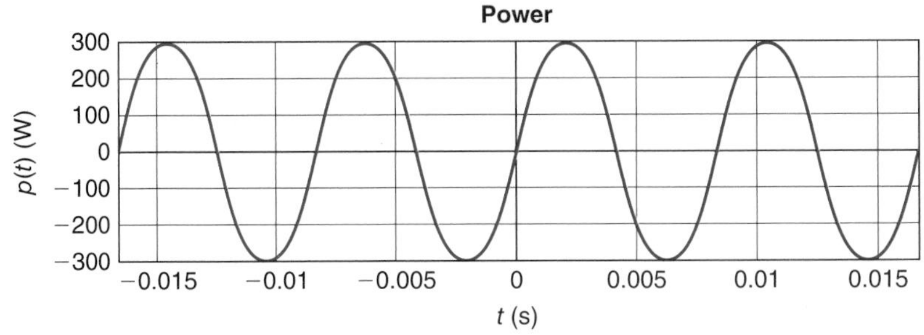

FIGURE 1.5

Power $p(t)$.

1.4 Independent Sources

FIGURE 1.6

Circuit symbols for voltage sources.

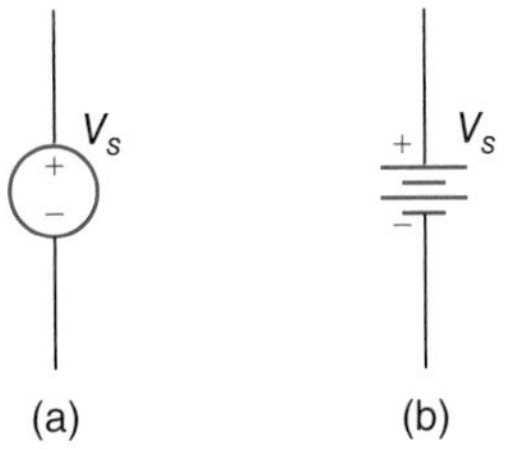

A voltage source with voltage V_s provides a constant potential difference to the circuit connected between the positive terminal and the negative terminal. The circuit notations for the voltage source are shown in Figure 1.6.

If a positive charge Δq is moved from the negative terminal to the positive terminal through the voltage source, the potential energy of the charge is increased by $\Delta q V_s$. If a negative charge with magnitude Δq is moved from the positive terminal to the negative terminal through the voltage source, the potential energy of the charge is increased by $\Delta q V_s$. A battery is an example of a voltage source.

FIGURE 1.7

A circuit symbol for the current source.

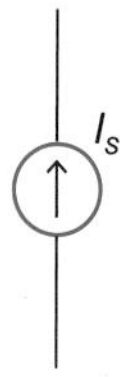

A current source with current I_s provides a constant current of I_s amperes to the circuit connected to the two terminals. The circuit notation for the current source is shown in Figure 1.7.

1.4.1 DIRECT CURRENT SOURCES AND ALTERNATING CURRENT SOURCES

If the voltage from the voltage source is constant with time, the voltage source is called the *direct current (dc) source.* Likewise, if the current from the current source is constant with time, the current source is called the *direct current (dc) source.*

If the voltage from the voltage source is a sinusoid, as shown in Figure 1.8, the voltage source is called *alternating current (ac) voltage source.* Likewise, if the current from the current source is a sinusoid, the current source is called *alternating current (ac) current source.* A detailed discussion of ac signals is given in Chapter 9. The circuit notation for an ac voltage source and ac current source are shown in Figure 1.9. The phase is given in degrees. The circuit notation for dc voltage shown in Figure 1.6(a) and the circuit notation for dc current shown in Figure 1.7 are also used for ac voltage and ac current, respectively.

FIGURE 1.8

Plot of a cosine wave with period T, amplitude V_m, and phase zero.

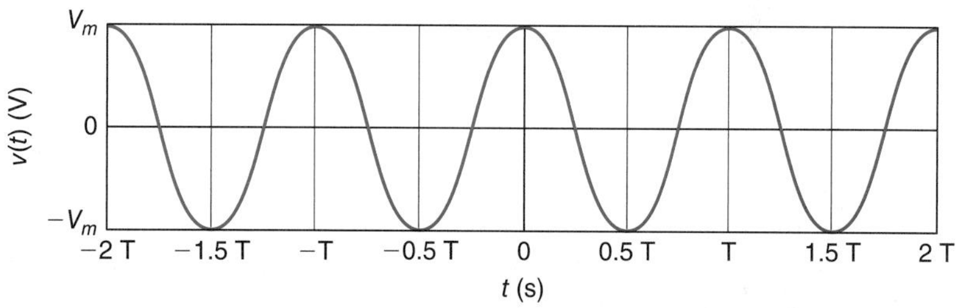

FIGURE 1.9

Circuit symbols for (a) ac voltage source; (b) ac current source.

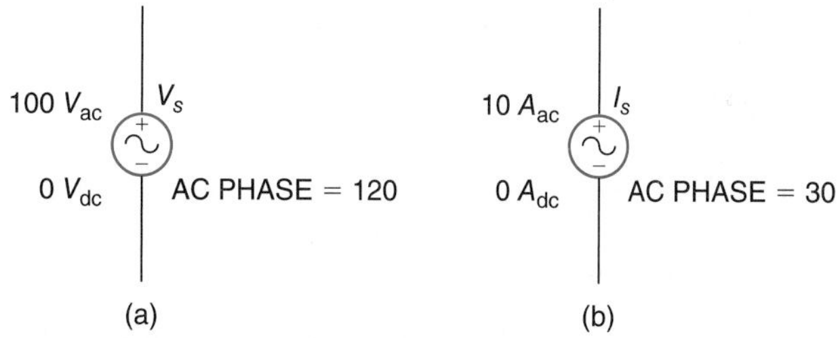

FIGURE 1.10

An equivalent voltage source.

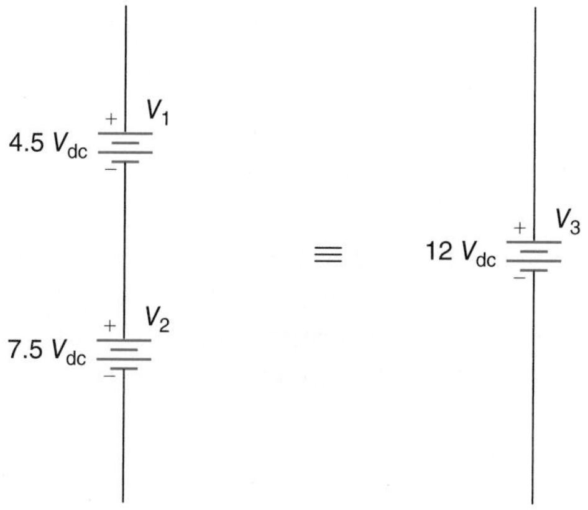

When dc voltage sources are connected in series, they can be combined into a single equivalent dc voltage source, as shown in Figure 1.10, where $V_3 = V_1 + V_2 = 4.5\text{ V} + 7.5\text{ V} = 12\text{ V}$. If there are other components, such as the resistors between V_1 and V_2 in the circuit shown in Figure 1.10, the voltage sources can be combined, so long as all the components are connected in series. Resistors are discussed further in Chapter 2.

When dc current sources are connected in parallel, they can be combined into a single equivalent dc current source, as shown in Figure 1.11, where $I_3 = I_1 + I_2 = 3\text{ A} + 5\text{ A} = 8\text{ A}$. If other components such as resistors are connected in parallel to I_1 and I_2 in the circuit shown in Figure 1.11, the current sources can be combined, so long as all the components are connected in parallel between the same points.

FIGURE 1.11

An equivalent current source.

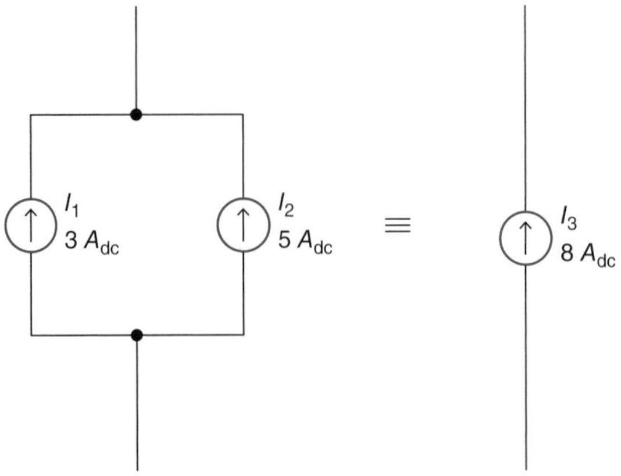

EXAMPLE 1.4

Redraw the circuit shown in Figure 1.12 with one voltage source and one current source, without affecting the voltages across and currents through the resistors in the circuit.

FIGURE 1.12

Circuit for EXAMPLE 1.4.

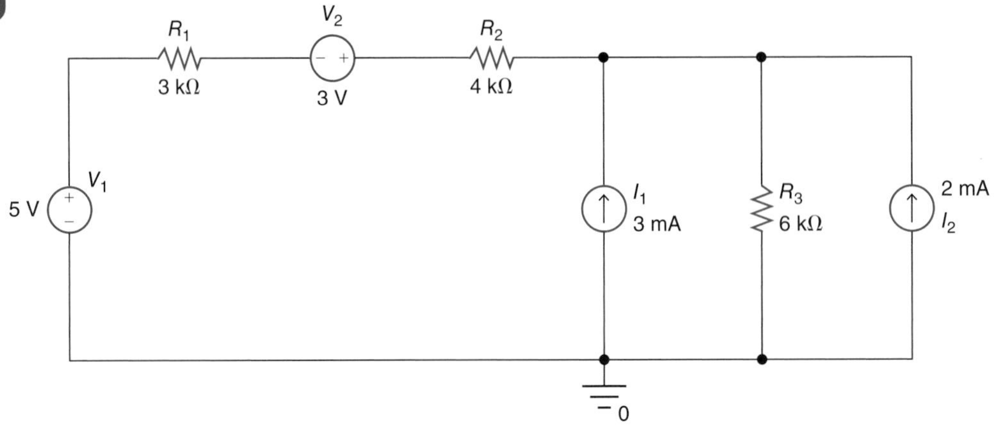

Since V_1 and V_2 are part of a single wire, they can be combined into the single voltage source V_3. Since V_2 has the same polarity as V_1, the value of V_3 is given by

$$V_3 = V_1 + V_2 = 5\text{ V} + 3\text{ V} = 8\text{ V}$$

Since I_1 and I_2 are connected between the same points in the circuit, they can be combined into the single current source I_3. Since I_2 has the same polarity as I_1, the value of I_3 is given by

$$I_3 = I_1 + I_2 = 3\text{ mA} + 2\text{ mA} = 5\text{ mA}$$

The equivalent circuit, with one voltage source and one current source, is shown in Figure 1.13.

continued

Example 1.4 continued

FIGURE 1.13

A circuit with one current source and one voltage source.

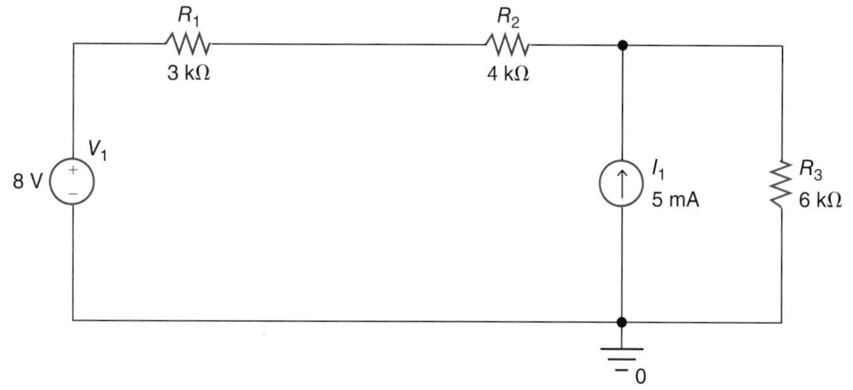

Exercise 1.4

Redraw the circuit shown in Figure 1.14 with one voltage source and one current source, without affecting the voltages across and currents through the resistors in the circuit.

FIGURE 1.14

Circuit for EXERCISE 1.4.

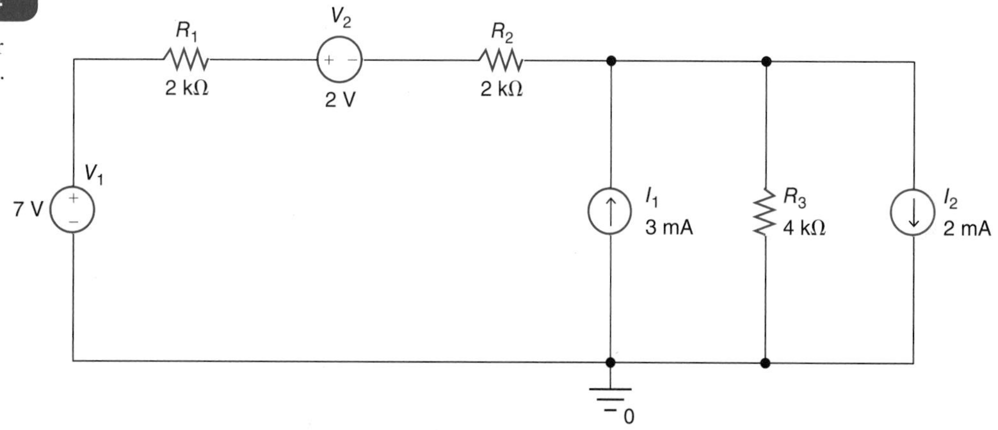

Answer:

The equivalent circuit with one voltage source and one current source is shown in Figure 1.15.

FIGURE 1.15

A circuit with one current source and one voltage source.

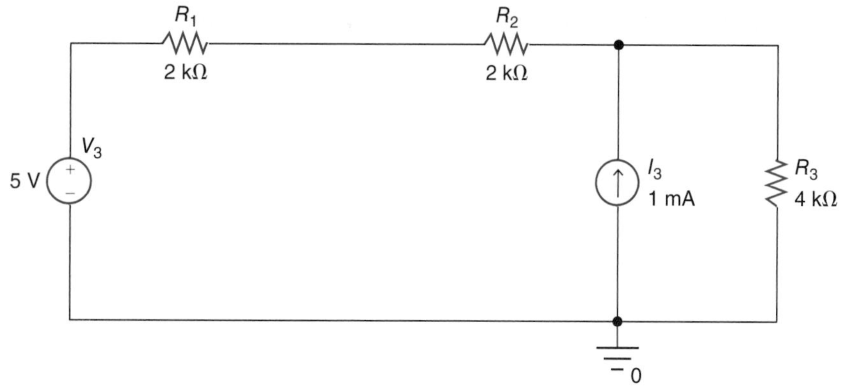

An ac voltage waveform can be represented as

$$v(t) = V_m \cos\left(\frac{2\pi t}{T} + \phi\right) \text{V} \tag{1.22}$$

Here, V_m is the amplitude (peak value) of the cosine wave, T is the period of the cosine wave, and ϕ is the phase of the cosine wave. The peak-to-peak amplitude is $2V_m$. The cosine wave repeats itself every T seconds. The number of periods per second, called *frequency* and denoted by f, is given by

$$f = \frac{1}{T} \text{ Hz} \tag{1.23}$$

The unit for the frequency is 1/s and is called *hertz* (Hz). In terms of the frequency in hertz, the ac voltage waveform can be written as

$$v(t) = V_m \cos(2\pi ft + \phi) \text{ V} \tag{1.24}$$

Since the angle changes by 2π radians in one period, and there are f periods in 1 second, the changes in angle in 1 second is given by

$$\omega = 2\pi f = \frac{2\pi}{T} \tag{1.25}$$

The parameter ω is called the *angular velocity* of the cosine wave and has a unit of radians per second (rad/s). In terms of radian frequency ω, the cosine wave becomes

$$v(t) = V_m \cos(\omega t + \phi) \tag{1.26}$$

The ac current waveform can be written as

$$i(t) = I_m \cos\left(\frac{2\pi t}{T} + \phi\right) = I_m \cos(2\pi ft + \phi) = I_m \cos(\omega t + \phi) \text{ A} \tag{1.27}$$

If the cosine wave shown in Figure 1.8 is shifted to the right by $T/4$, we get

$$\begin{aligned} v(t) &= V_m \cos\left[\frac{2\pi\left(t - \frac{T}{4}\right)}{T}\right] = V_m \cos\left(\frac{2\pi}{T}t - \frac{\pi}{2}\right) = V_m \sin\left(\frac{2\pi}{T}t\right) \\ &= V_m \sin(2\pi ft) = V_m \sin(\omega t) \end{aligned} \tag{1.28}$$

The sine wave given by Equation (1.28) is shown in Figure 1.16.

FIGURE 1.16

Plot of a sine wave with period T and amplitude V_m.

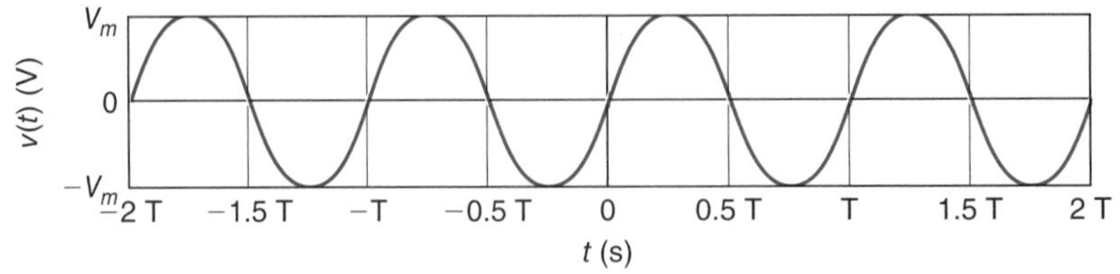

Notice that if the phase of the cosine is shifted by −90 degrees, the cosine wave becomes a sine wave.

EXAMPLE 1.5

Find the equation of the sinusoidal signal shown in Figure 1.17.

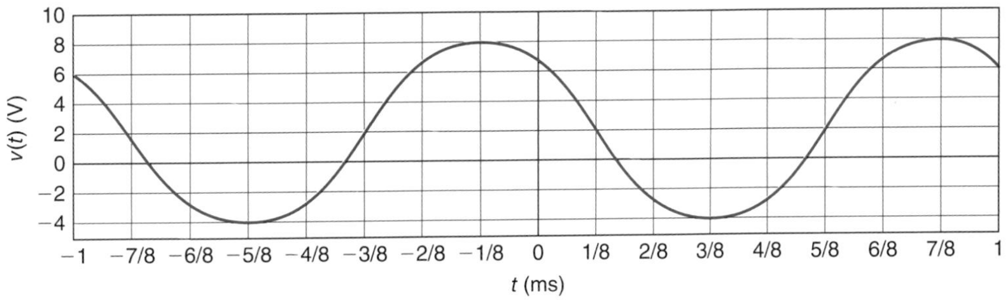

FIGURE 1.17 A sinusoid for EXAMPLE 1.5.

Since the period is $T = 1$ ms, the frequency is $f = 1/T = 1/1\text{ms} = 1000$ Hz $= 1$ kHz. The radian frequency is $2\pi f = 2\pi 1000 = 6283.1853$ rad/s. The difference between the maximum and minimum is $8 - (-4) = 12$ V, which is the peak-to-peak amplitude. The peak value of the amplitude is $V_m = 12\text{V}/2 = 6$ V. The average amplitude is $(8 - 4)/2 = 2$ V, which is the dc component. The cosine wave is shifted to the left by $T/8$ ms, which is $\pi/4$ rad $= 45°$. Therefore, the equation is given by

$$v(t) = 2 + 6\cos(2\pi 1000t + 45°) \text{ V}$$

Exercise 1.5

Plot $v(t) = 4 + 2\cos(2\pi 2000t - 72°)$ V.

Answer:
The signal $v(t)$ is shown in Figure 1.18.

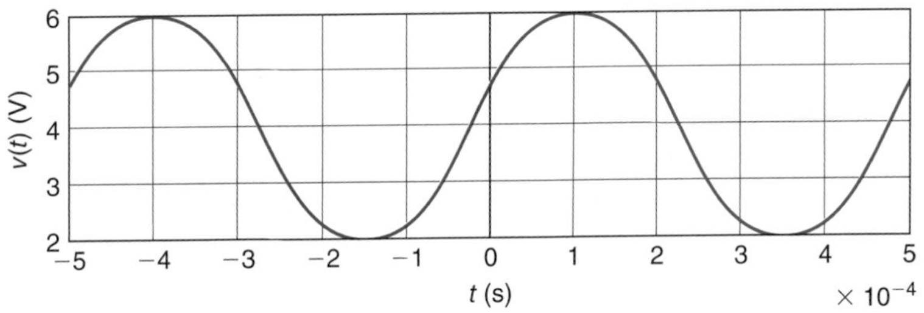

FIGURE 1.18 The plot of $v(t)$.

1.5 Dependent Sources

The voltage sources and current sources discussed previously are called **independent sources** because they are stand-alone sources that provide power to the external circuit connected to the sources. Usually, the independent sources convert one form of energy to electrical energy. For example, a battery converts chemical energy into electrical energy.

A solar cell converts energy from the Sun into electrical energy. A wind turbine converts wind energy into electrical energy. The amount of energy supplied from the source to the circuit per unit time is the power of the source. Dependent sources do not have the ability to convert one form of energy into electrical energy. The voltage or current of the dependent sources depend solely on the controlling voltage or controlling current. The dependent sources are used to model integrated circuit (IC) devices.

Depending on whether the dependent source is a voltage source or a current source, and whether the dependent source is controlled by a voltage or a current, there are four different dependent sources. The four types of dependent sources are:

Voltage-controlled voltage source (VCVS)
Voltage-controlled current source (VCCS)
Current-controlled voltage source (CCVS)
Current-controlled current source (CCCS)

These four types of dependent sources are discussed next.

FIGURE 1.19

Circuit symbol for VCVS.

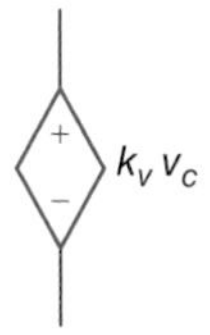

1.5.1 VOLTAGE-CONTROLLED VOLTAGE SOURCE (VCVS)

The voltage on the VCVS is proportional to the controlling voltage, which is the voltage in another part of the circuit. For example, the controlling voltage can be the voltage across a circuit element in another part of the circuit. Let v_d be the controlled voltage and v_c be the controlling voltage. Then, we have

$$v_d = k_v v_c$$

where k_v is the unitless (V/V) proportionality constant. Figure 1.19 shows the circuit symbol for VCVS.

FIGURE 1.20

Circuit symbol for VCCS.

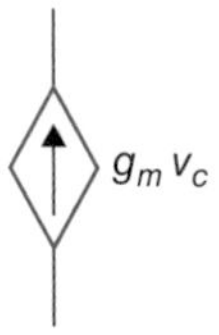

1.5.2 VOLTAGE-CONTROLLED CURRENT SOURCE (VCCS)

The current on the VCCS is proportional to the controlling voltage. Let i_d be the controlled current and v_c be the controlling voltage. Then, we have

$$i_d = g_m v_c$$

where g_m is the conductance in siemens (S). Figure 1.20 shows the circuit symbol for VCCS.

FIGURE 1.21

Circuit symbol for CCVS.

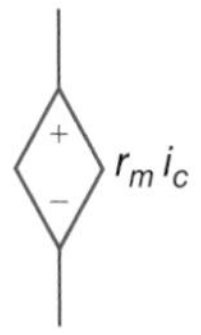

1.5.3 CURRENT-CONTROLLED VOLTAGE SOURCE (CCVS)

The voltage on the CCVS is proportional to the controlling current, the current in another part of the circuit. For example, the controlling current can be the current through a circuit element in another part of the circuit. Let v_d be the controlled voltage and i_c be the controlling current. Then, we have

$$v_d = r_m i_c$$

where r_m is the resistance in ohms (Ω). Figure 1.21 shows the circuit symbol for CCVS.

FIGURE 1.22

Circuit symbol for CCCS.

1.5.4 CURRENT-CONTROLLED CURRENT SOURCE (CCCS)

The current on the CCCS is proportional to the controlling current. Let i_d be the controlled current and i_c be the controlling current. Then, we have

$$i_d = k_i i_c$$

where k_i is the unitless (A/A) proportionality constant. Figure 1.22 shows the circuit symbol for CCCS.

EXAMPLE 1.6

In the circuit shown in Figure 1.23, the controlling voltage, which is the voltage across R_2, is $v_a = 0.9851$ V. Find the controlled current through the VCCS.

FIGURE 1.23

Circuit for EXAMPLE 1.6.

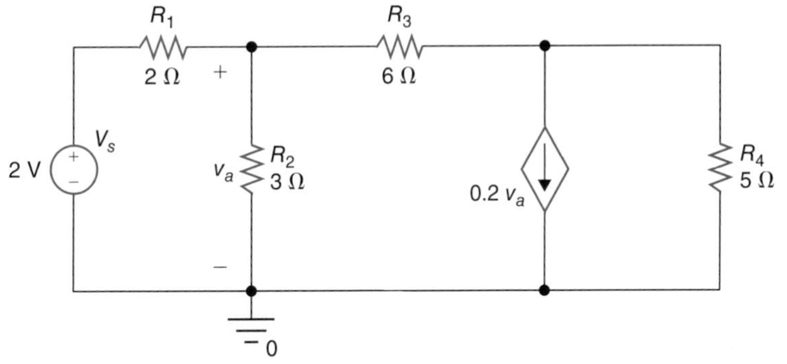

The current through the VCCS in the direction indicated in Figure 1.23 (↓) is

$$0.2\ v_a = 0.2\ (\text{A/V}) \times 0.9851\ \text{V} = 0.1970\ \text{A}$$

Exercise 1.6

In the circuit shown in Figure 1.24, the controlling current, which is the current through R_2, is $i_a = 0.5625$ A. Find the controlled voltage across the CCVS.

FIGURE 1.24

Circuit for EXERCISE 1.6.

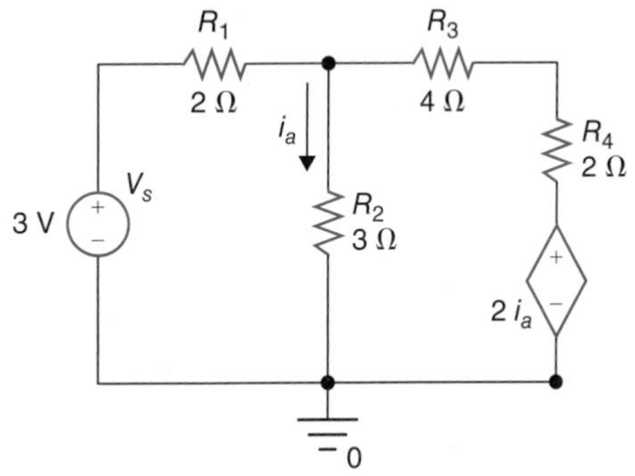

Answer:

$2\ i_a = 1.125$ V.

1.6 Elementary Signals

Several elementary signals that will be useful in later chapters are presented in this section.

1.6.1 DIRAC DELTA FUNCTION

A rectangular pulse with height $1/\tau$ and width τ is shown in Figure 1.25. The pulse is centered at $-\tau/2$ and the area of the pulse is 1. The rectangular pulse can be written as

$$f(t) = \frac{1}{\tau} rect\left(\frac{t + \frac{\tau}{2}}{\tau}\right) \quad \textbf{(1.29)}$$

If the pulse width τ is decreased to zero, the height of the pulse is increased to infinity while maintaining the area at 1. The limiting form of a rectangular pulse shown in Figure 1.25 as $\tau \to 0$ is defined as the *Dirac delta function* (or *delta function*) and is denoted by $\delta(t)$; that is,

FIGURE 1.25

A rectangular pulse.

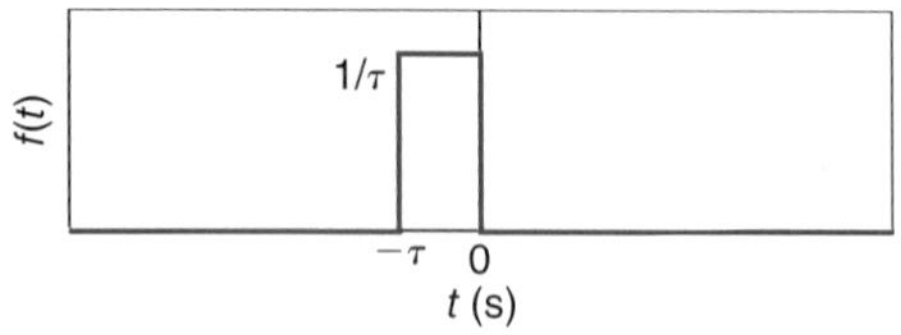

$$\delta(t) = \lim_{\tau \to 0} \frac{1}{\tau} rect\left(\frac{t + \frac{\tau}{2}}{\tau}\right) \quad \textbf{(1.30)}$$

The mathematical symbol for the Dirac delta function is shown in Figure 1.26.

FIGURE 1.26

Symbol for the Dirac delta function.

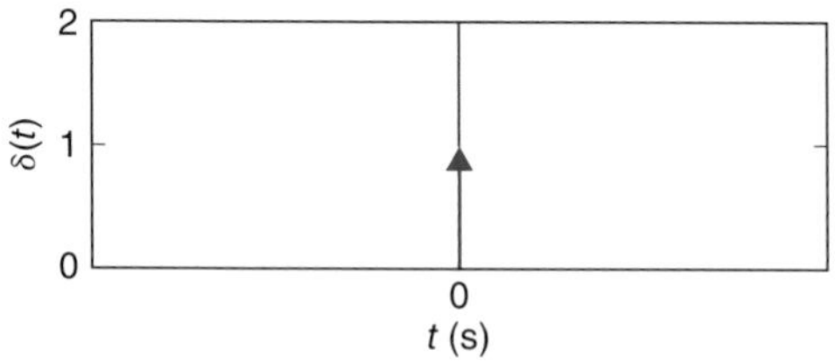

If the rectangular pulse given by Equation (1.29) is shifted to the right by $\tau/2$, it becomes

$$g(t) = \frac{1}{\tau} rect\left(\frac{t}{\tau}\right) \quad \textbf{(1.31)}$$

The Dirac delta function can also be defined as

$$\delta(t) = \lim_{\tau \to 0} \frac{1}{\tau} rect\left(\frac{t}{\tau}\right) \quad \textbf{(1.32)}$$

EXAMPLE 1.7

Plot $f(t) = 4\,\delta(t - 1)$.

The Dirac delta function is located at $t = 1$ and has an area of 4. The signal $f(t)$ is shown in Figure 1.27.

FIGURE 1.27

Plot of $f(t)$.

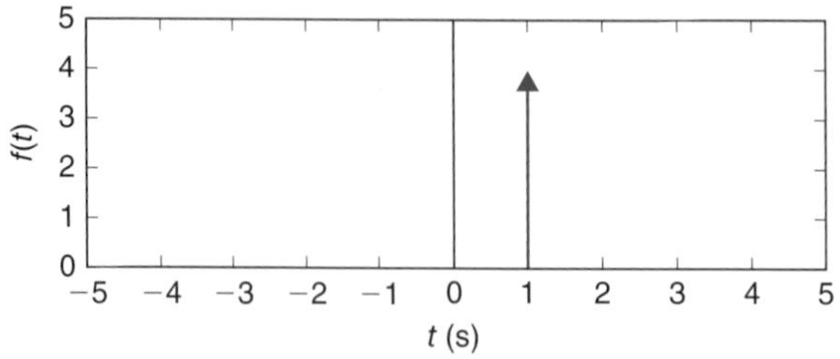

Exercise 1.7

Plot $f(t) = -2\,\delta(t+3)$.

Answer:
$f(t)$ is shown in Figure 1.28.

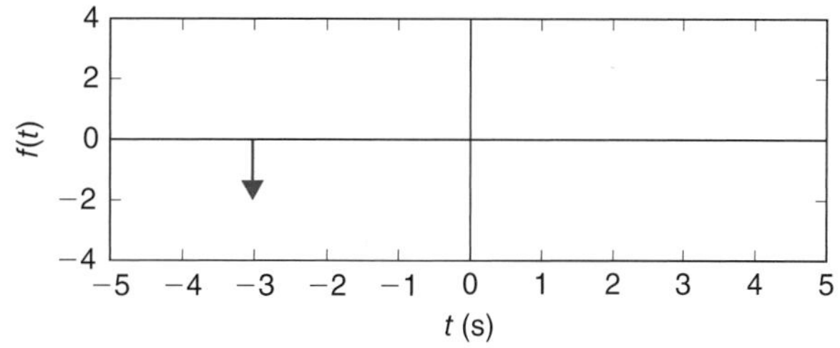

FIGURE 1.28
Plot of $f(t)$.

When a continuous signal $f(t)$ is multiplied by $\delta(t - a)$ and integrated from $-\infty$ to ∞, we obtain $f(a)$; that is,

$$\int_{-\infty}^{\infty} f(t)\delta(t-a)dt = f(a) \tag{1.33}$$

This result is called the **sifting property** of the delta function because it sifts out a single value of $f(t)$, $f(a)$, at the location of the delta function ($t = a$). To prove the sifting property, we replace $\delta(t - a)$ with

$$\delta(t-a) = \lim_{\tau\to 0} \frac{1}{\tau} rect\left(\frac{t-a}{\tau}\right)$$

Then, the integral becomes

$$\int_{-\infty}^{\infty} f(t)\delta(t-a)dt = \lim_{\tau\to 0}\frac{1}{\tau}\int_{-\infty}^{\infty} f(t)rect\left(\frac{t-a}{\tau}\right)dt = \lim_{\tau\to 0}\frac{1}{\tau}\int_{a-\frac{\tau}{2}}^{a+\frac{\tau}{2}} f(t)rect\left(\frac{t-a}{\tau}\right)dt$$

As $\tau \to 0$, $f(t) \to f(a)$ for $(a - \tau/2) < t < (a + \tau/2)$. Thus, the integral becomes

$$\int_{-\infty}^{\infty} f(t)\delta(t-a)dt = \lim_{\tau\to 0}\frac{1}{\tau}\int_{a-\frac{\tau}{2}}^{a+\frac{\tau}{2}} f(a)\times 1\,dt = \lim_{\tau\to 0}\frac{1}{\tau}f(a)\tau = f(a)$$

1.6.2 STEP FUNCTION

The unit step function $u(t)$ is the integral of the Dirac delta function $\delta(t)$. If Equation (1.29) is integrated, we obtain

$$\int_{-\infty}^{t} f(\lambda)d\lambda = \frac{1}{\tau}\int_{-\infty}^{t} rect\left(\frac{\lambda+\frac{\tau}{2}}{\tau}\right)d\lambda = \begin{cases} 0, & t < -\tau \\ \frac{t}{\tau}+1, & -\tau \le t < 0 \\ 1, & 0 \le t \end{cases} \tag{1.34}$$

The unit step function is defined as the limiting form of Equation (1.34). In the limit as $\tau \to 0$, Equation (1.34) becomes

$$u(t) = \begin{cases} 0, & t < 0 \\ 1, & 0 \le t \end{cases} \tag{1.35}$$

Notice that at $t = 0$, $u(t) = 1$. The unit step function defined by Equation (1.35) is shown in Figure 1.29.

FIGURE 1.29

A unit step function.

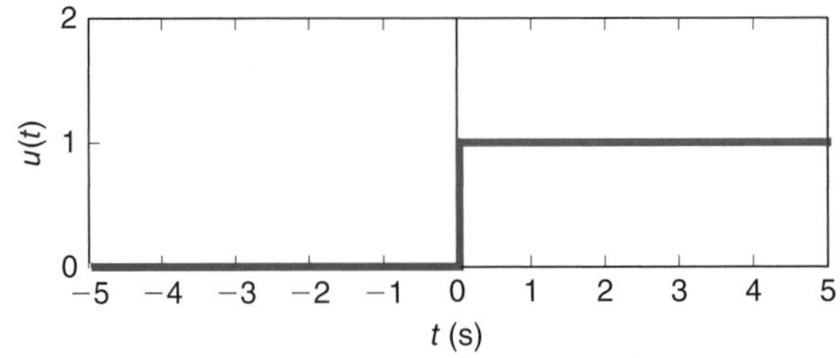

EXAMPLE 1.8

Plot $f(t) = u(t) - u(t-2)$.

Notice that $u(t) = 1$ for $t \ge 0$ and zero for $t < 0$, and $u(t-2) = 1$ for $t \ge 2$ and zero for $t < 2$. Thus, $u(t) - u(t-2) = 0$ for $t \ge 2$, and $u(t) - u(t-2) = 1$ for $0 \le t < 2$, and zero for $t < 0$. The signal $f(t)$ is shown in Figure 1.30.

FIGURE 1.30

Plot of $f(t)$.

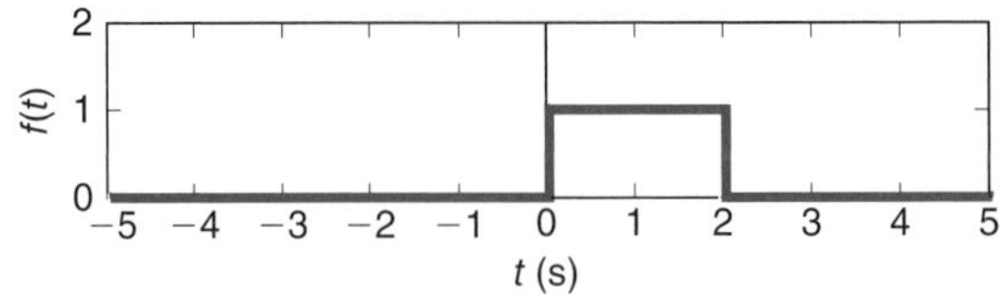

Exercise 1.8

Plot $f(t) = -2\,u(t+3)$.

Answer:

$f(t)$ is shown in Figure 1.31.

FIGURE 1.31

Plot of $f(t)$.

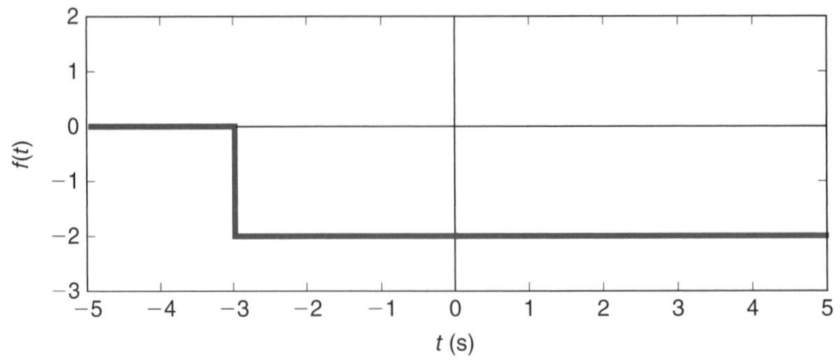

If Equation (1.31) is integrated, we obtain

$$\int_{-\infty}^{t} g(\lambda)d\lambda = \frac{1}{\tau}\int_{-\infty}^{t} rect\left(\frac{\lambda}{\tau}\right)d\lambda = \begin{cases} 0, & t < \frac{-\tau}{2} \\ \frac{t}{\tau} + \frac{1}{2}, & \frac{-\tau}{2} \le t < \frac{\tau}{2} \\ 1, & \frac{\tau}{2} \le t \end{cases} \quad (1.36)$$

If $u(t)$ is defined as the limiting form of Equation (1.36) as $\tau \to 0$, we obtain

$$u(t) = \begin{cases} 0, & t < 0 \\ \frac{1}{2}, & t = 0 \\ 1, & 0 \le t \end{cases} \quad (1.37)$$

In this text, the definition of $u(t)$ given by Equation (1.35) is used. Since $u(0) = 1$, it does include voltages and currents at $t = 0$.

1.6.3 RAMP FUNCTION

A unit ramp function is defined by

$$r(t) = t\,u(t) \quad (1.38)$$

The unit ramp function is shown in Figure 1.32. The unit ramp function is the integral of the unit step function:

$$r(t) = \int_{-\infty}^{t} u(\lambda)d\lambda \quad (1.39)$$

The derivative of the unit ramp function is the unit step function.

$$u(t) = \frac{dr(t)}{dt} \quad (1.40)$$

FIGURE 1.32

A unit ramp function.

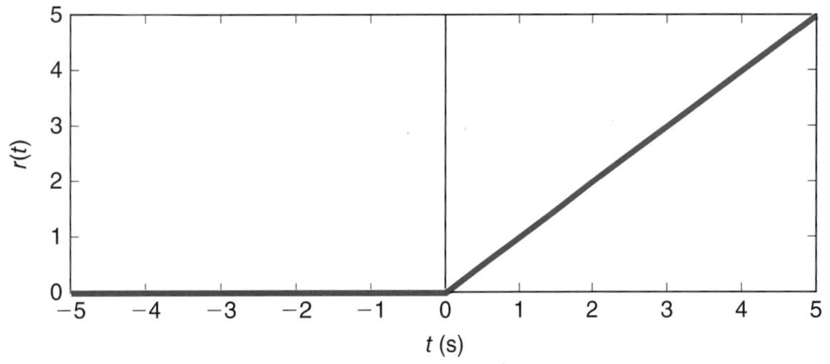

EXAMPLE 1.9

Plot $f(t) = 2tu(t) - 4(t-1)u(t-1) + 4(t-3)u(t-3) - 4(t-5)u(t-5) + 2(t-6)u(t-6)$.

For $t < 0, f(t) = 0$.
For $0 \le t < 1, f(t)$ is a linear line with slope of 2.
For $1 \le t < 3, f(t)$ is a linear line with slope of -2.
For $3 \le t < 5, f(t)$ is a linear line with slope of 2.
For $5 \le t < 6, f(t)$ is a linear line with slope of -2.
For $6 \le t, f(t) = 0$.
The waveform $f(t)$ is shown in Figure 1.33.

FIGURE 1.33

Waveform $f(t)$.

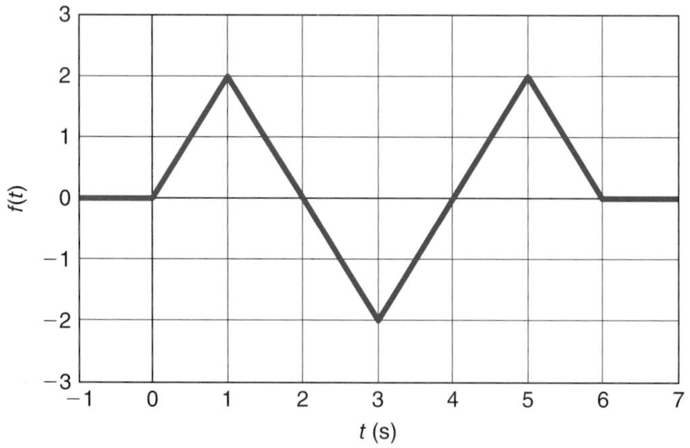

Exercise 1.9

Plot $f(t) = tu(t) - (t-1)u(t-1) - (t-3)u(t-3) + (t-4)u(t-4)$.

Answer:
The waveform is shown in Figure 1.34.

FIGURE 1.34

Waveform $f(t)$.

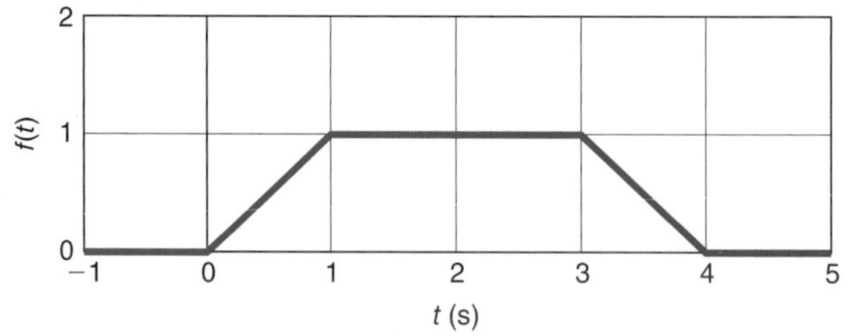

EXAMPLE 1.10

Find the equation of the waveform shown in Figure 1.35.

continued

Example 1.10 continued

For $t < 0$, $f(t) = 0$.

For $0 \le t < 1$, $f(t)$ is a linear line with slope of 3. Thus, $f(t) = 3tu(t)$.

For $1 \le t < 3$, $f(t)$ is a linear line with slope of -3. To change the slope from 3 to -3, we need to add $-6(t-1)u(t-1)$. At this point, we have $f(t) = 3tu(t) - 6(t-1)u(t-1)$.

For $3 \le t < 6$, $f(t)$ is a linear line with slope of 1. To change the slope from -3 to 1, we need to add $4(t-3)u(t-3)$. At this point, we have $f(t) = 3tu(t) - 6(t-1)u(t-1) + 4(t-3)u(t-3)$.

FIGURE 1.35

Waveform $f(t)$.

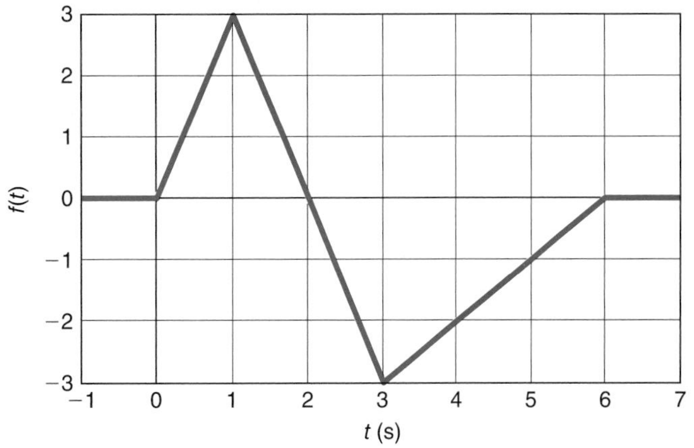

For $6 \le t$, $f(t) = 0$. To change the slope from 1 to 0, we need to add $-(t-6)u(t-6)$. Thus, we have the final equation given by

$$f(t) = 3tu(t) - 6(t-1)u(t-1) + 4(t-3)u(t-3) - (t-6)u(t-6).$$

Exercise 1.10

Find the equation of the waveform shown in Figure 1.36.

FIGURE 1.36

Waveform for EXERCISE 1.10.

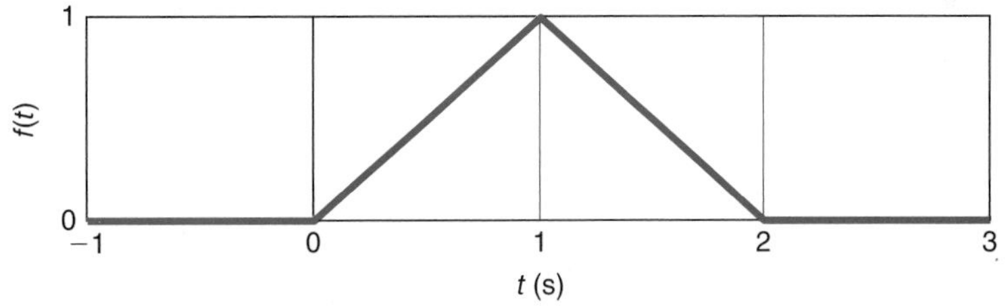

Answer:
$f(t) = tu(t) - 2(t-1)u(t-1) + (t-2)u(t-2)$.

1.6.4 EXPONENTIAL DECAY

A signal that decays exponentially can be written as

$$f(t) = e^{-at}u(t), \quad a > 0 \qquad \textbf{(1.41)}$$

The signal $f(t)$ for $a = 0.5$ is shown in Figure 1.37.

FIGURE 1.37

$f(t) = e^{-at}u(t), a = 0.5.$

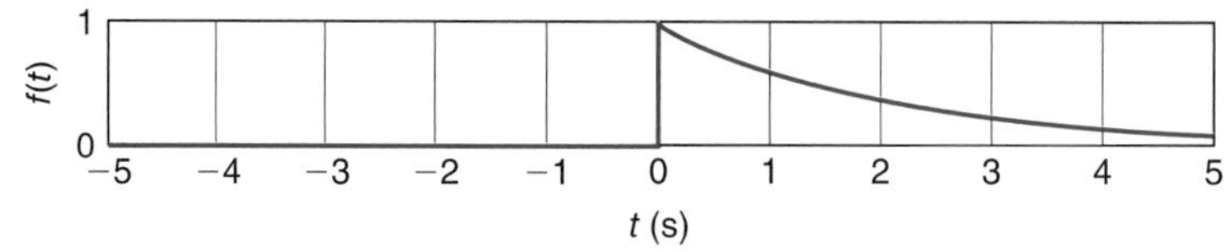

A damped cosine and damped sine, respectively, can be written as

$$f(t) = e^{-at}\cos(bt)u(t), \quad a > 0 \tag{1.42}$$

$$f(t) = e^{-at}\sin(bt)u(t), \quad a > 0 \tag{1.43}$$

A damped cosine signal is shown in Figure 1.38 for $a = 0.5$ and $b = 4$.

FIGURE 1.38

A damped cosine signal.

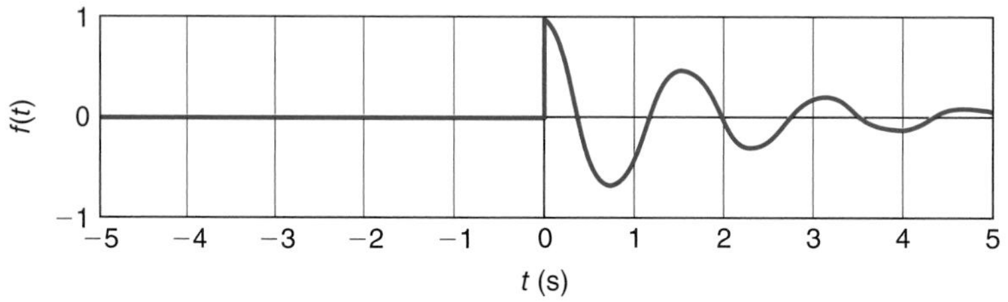

1.6.5 RECTANGULAR PULSE AND TRIANGULAR PULSE

A rectangular pulse with amplitude A and pulse width τ is shown in Figure 1.39. The center of the pulse is at $t = 0$.

FIGURE 1.39

A rectangular pulse.

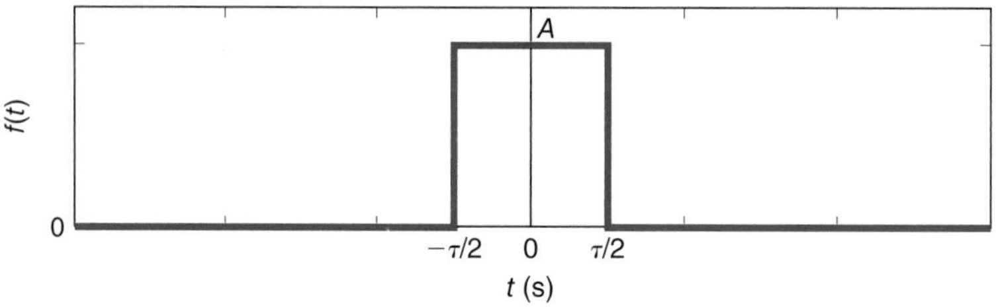

The rectangular pulse shown in Figure 1.39 is denoted by

$$f(t) = A\,rect\left(\frac{t}{\tau}\right)$$

EXAMPLE 1.11

Plot $f(t) = rect\left(\frac{t+1}{2}\right) + 3\,rect\left(\frac{t-1}{2}\right) - 2\,rect\left(\frac{t-3.5}{3}\right)$.

The first rectangle is centered at $t = -1$ and has a height of 1 and width of 2. The second rectangle is centered at $t = 1$ and has a height of 3 and width of 2. The third rectangle is centered at $t = 3.5$ and has a height of -2 and width of 3. The waveform $f(t)$ is shown in Figure 1.40.

continued

Example 1.11 continued

FIGURE 1.40

Waveform $f(t)$.

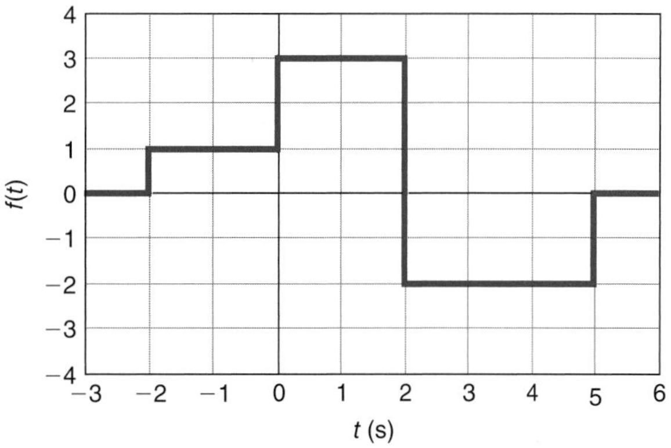

Exercise 1.11

Plot $f(t) = -3\,rect\left(\frac{t+5}{2}\right) + 2\,rect\left(\frac{t-5}{4}\right)$.

Answer:
The waveform $f(t)$ is shown in Figure 1.41.

FIGURE 1.41

Waveform $f(t)$.

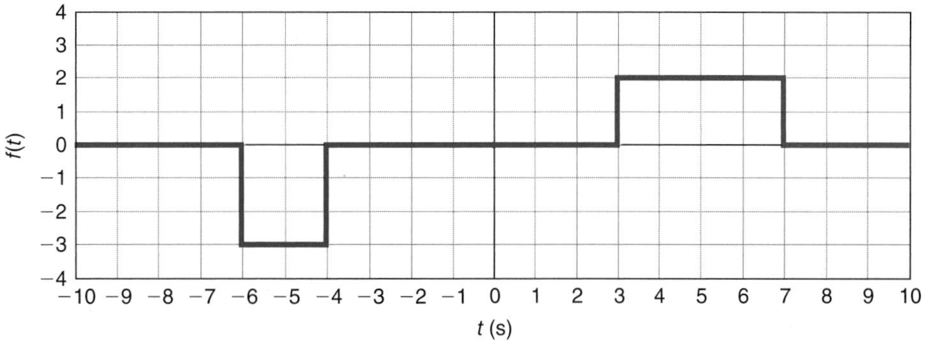

A triangular pulse with amplitude A and base 2τ is shown in Figure 1.42. The center of the pulse is at $t = 0$.

FIGURE 1.42

A triangular pulse.

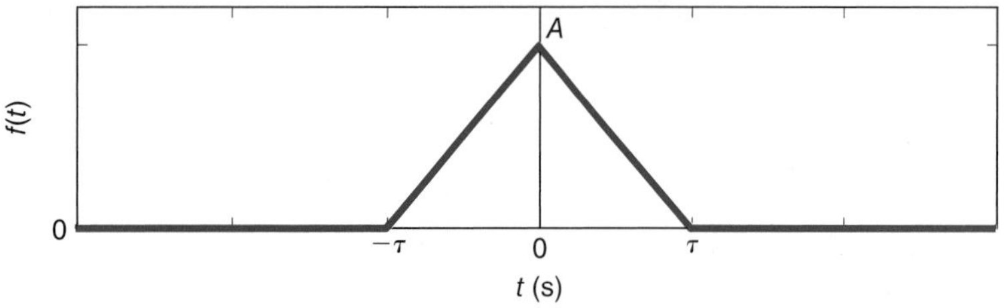

The triangular pulse shown in Figure 1.42 is denoted by

$$f(t) = A\,tri\left(\frac{t}{\tau}\right)$$

EXAMPLE 1.12

Plot $f(t) = 2\,tri\left(\frac{t+2}{2}\right) - 2\,tri\left(\frac{t}{2}\right) + 2\,tri\left(\frac{t-2}{2}\right)$.

The first triangle is centered at $t = -2$ and has a height of 2 and base of 4. The second triangle is centered at $t = 0$ and has a height of -2 and base of 4. The third triangle is centered at $t = 2$ and has a height of 2 and base of 4. The waveform $f(t)$ is shown in Figure 1.43.

FIGURE 1.43

Waveform $f(t)$.

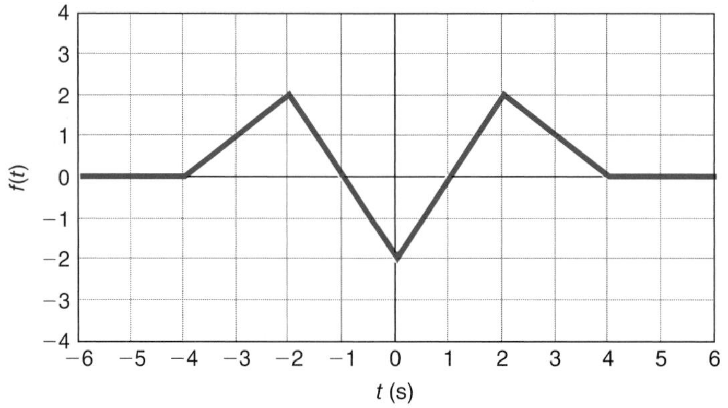

Exercise 1.12

Find the equation of the waveform $f(t)$ shown in Figure 1.44.

FIGURE 1.44

Waveform for EXERCISE 1.12.

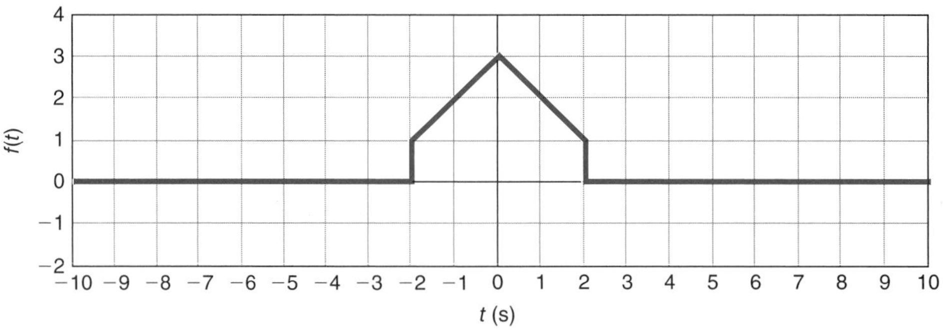

Answer:

$$f(t) = rect\left(\frac{t}{4}\right) + 2tri\left(\frac{t}{2}\right)$$

SUMMARY

In this chapter, the seven base units of the International System of Units (SI), along with derived units relevant to electrical and computer engineering, are presented. The definitions of *voltage, current,* and *power*, among other terms, are given. The potential difference per unit charge between A and B is called *voltage between A and B*, $v_{AB} = w_{AB}/q$, where w_{AB} is the amount of the work needed to move the test charge from B to A.

Current is defined as the rate of change of charge:

$$i(t) = \frac{dq(t)}{dt}$$

Power is the product of current and voltage:

$$p(t) = i(t)v(t)$$

Energy is the integral of power:

$$w(t) = \int_{-\infty}^{t} p(\lambda)d\lambda$$

The four types of dependent sources are:

Voltage-controlled voltage source (VCVS)
Voltage-controlled current source (VCCS)
Current-controlled voltage source (CCVS)
Current-controlled current source (CCCS)

PROBLEMS

1.1 **Find the current flowing through an element if the charge flowing through the element is given by**

$$q(t) = \begin{cases} 0.002t, \text{ C} & t \geq 0 \\ 0, & t < 0 \end{cases}$$

1.2 **Find the current flowing through an element if the charge flowing through the element is given by**

$$q(t) = \begin{cases} 5e^{-0.2t}, \text{ C} & t \geq 0 \\ 0, & t < 0 \end{cases}$$

1.3 **Find the current flowing through an element if the charge flowing through the element is given by**

$$q(t) = \begin{cases} 8(1 - e^{-0.003t}), \text{ C} & t \geq 0 \\ 0, & t < 0 \end{cases}$$

1.4 **Find the current flowing through an element if the charge flowing through the element is given by**

$$q(t) = \begin{cases} 7te^{-0.003t}, \text{ C} & t \geq 0 \\ 0, & t < 0 \end{cases}$$

1.5 **Find the current flowing through an element if the charge flowing through the element is given by**

$$q(t) = \begin{cases} 8 \times 10^{-6} \sin(2\pi \times 1000t), \text{ C} & t \geq 0 \\ 0, & t < 0 \end{cases}$$

1.6 **The charge entering an element is shown in Figure P1.6. Plot the current through the element for $0 \leq t < 7$ s.**

FIGURE P1.6

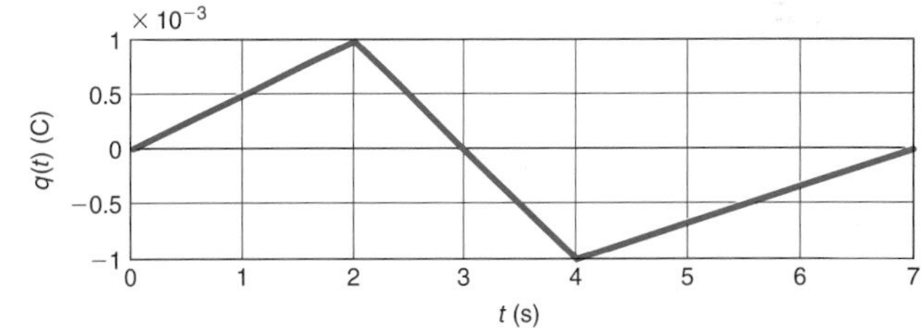

1.7 **Find the total charge passing through an element at one cross section over the time interval $0 \leq t \leq 5$ s if the current through the same cross section is given by**

$$i(t) = 5 \text{ mA}$$

1.8 Find the total charge passing through an element at one cross section over the time interval $0 \le t \le 5$ s if the current through the same cross section is given by

$$i(t) = \begin{cases} 5e^{-0.2t} \ \mu A, & t \ge 0 \\ 0, & t < 0 \end{cases}$$

1.9 Find the total charge passing through an element at one cross section over the time interval $0 \le t \le 5$ s if the current through the same cross section is given by

$$i(t) = \begin{cases} 3(1 - e^{-0.5t}) \ A, & t \ge 0 \\ 0, & t < 0 \end{cases}$$

1.10 Find the total charge passing through an element at one cross section over the time interval $0 \le t \le 5$ s if the current through the same cross section is given by

$$i(t) = \begin{cases} 2te^{-3t} \ A, & t \ge 0 \\ 0, & t < 0 \end{cases}$$

1.11 Find the total charge passing through an element at one cross section over the time interval $0 \le t \le 5$ s if the current through the same cross section is given by

$$i(t) = \begin{cases} 7 \sin(\pi t/5) \ A, & t \ge 0 \\ 0, & t < 0 \end{cases}$$

1.12 Find the power in the circuit element shown in Figure P1.12 and state whether the element is absorbing power or delivering power.

FIGURE P1.12

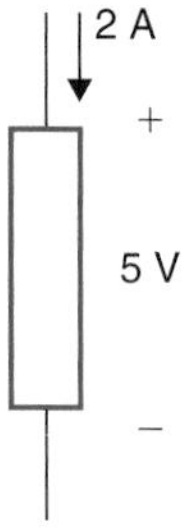

1.13 Find the power in the circuit element shown in Figure P1.13 and state whether the element is absorbing power or delivering power.

FIGURE P1.13

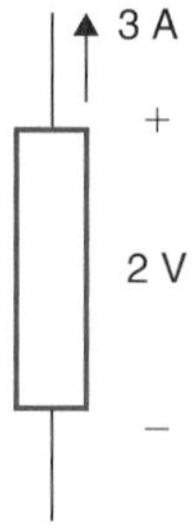

1.14 Find the power in the circuit element shown in Figure P1.14 and state whether the element is absorbing power or delivering power.

FIGURE P1.14

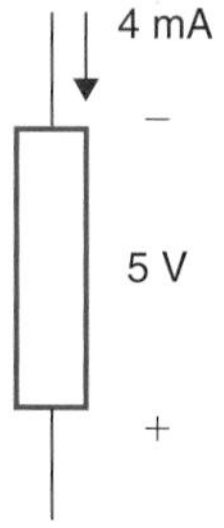

1.15 Find the power in the circuit element shown in Figure P1.15 and state whether the element is absorbing power or delivering power.

FIGURE P1.15

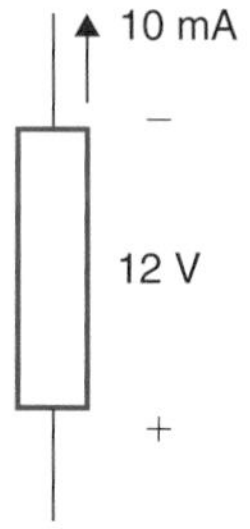

1.16 Find the power $p(t)$ on the element when the current through the element $i(t)$ from positive terminal to negative terminal and voltage $v(t)$ across the element are given by

$$i(t) = 2 \text{ mA}, \quad v(t) = 5 \text{ V}$$

1.17 Find the power $p(t)$ on the element when the current through the element $i(t)$ from positive terminal to negative terminal and voltage $v(t)$ across the element are given by

$$i(t) = 25 \cos(2\pi 1000t) \text{ mA},$$

$$v(t) = 5 \sin(2\pi 1000t) \text{ V}$$

1.18 Find the power $p(t)$ on the element when the current through the element $i(t)$ from positive terminal to negative terminal and voltage $v(t)$ across the element are given by

$$i(t) = 60\ e^{-0.07t}\ u(t)\ \text{mA},$$

$$v(t) = 7\ e^{-0.08t}\ u(t)\ \text{V}$$

1.19 Find the power $p(t)$ on the element when the current through the element $i(t)$ from positive terminal to negative terminal and voltage $v(t)$ across the element are given by

$$i(t) = 8\cos(2\pi 100t)\ \text{mA},$$

$$v(t) = 3\cos(2\pi 100t)\ \text{V}$$

1.20 Find the power $p(t)$ on the element when the current through the element $i(t)$ from positive terminal to negative terminal and voltage $v(t)$ across the element are given by

$$i(t) = 6\sin(2\pi 100t)\ \text{mA},$$

$$v(t) = 2\sin(2\pi 100t)\ \text{V}$$

1.21 Redraw the circuit shown in Figure P1.21 with one voltage source and one current source, without affecting the voltages across and currents through the resistors in the circuit.

FIGURE P1.21

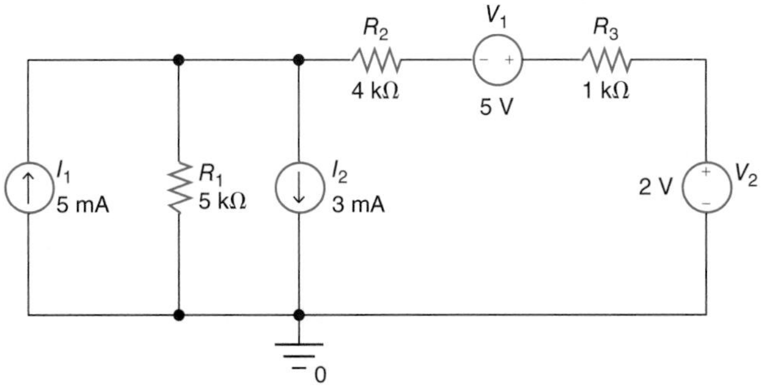

1.22 Redraw the circuit shown in Figure P1.22 with one voltage source and one current source, without affecting the voltages across and currents through the resistors in the circuit.

FIGURE P1.22

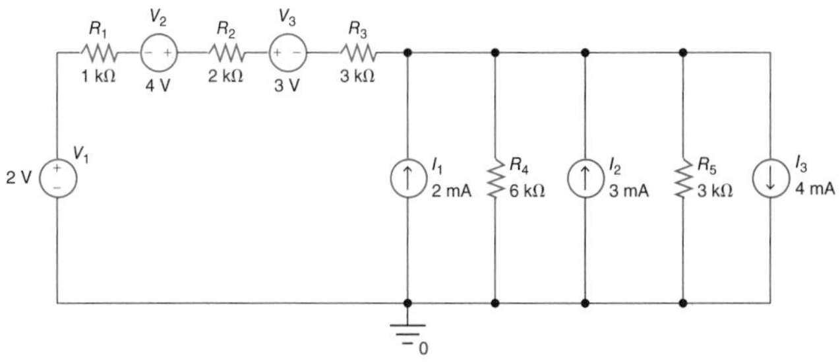

1.23 Plot $v(t) = -2 + 6\cos(2\pi 5000t - 90°)$ V.

1.24 Find the equation of the sinusoid shown in Figure P1.24.

FIGURE P1.24

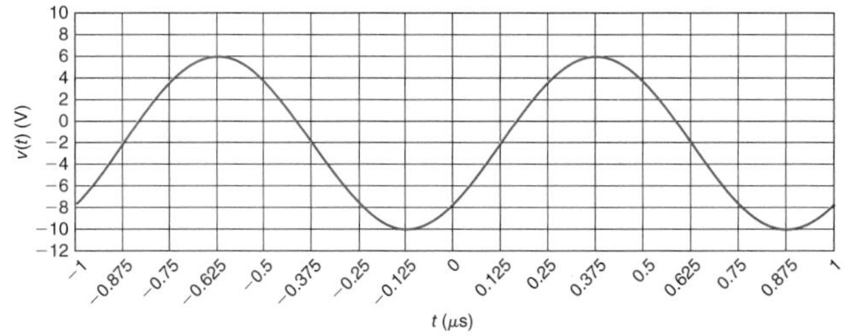

1.25 In the circuit shown in Figure P1.25, the controlling voltage, which is the voltage across R_2, is $v_a = 1.2908$ V. Find the controlled voltage across the VCVS and the controlled current through the VCCS.

FIGURE P1.25

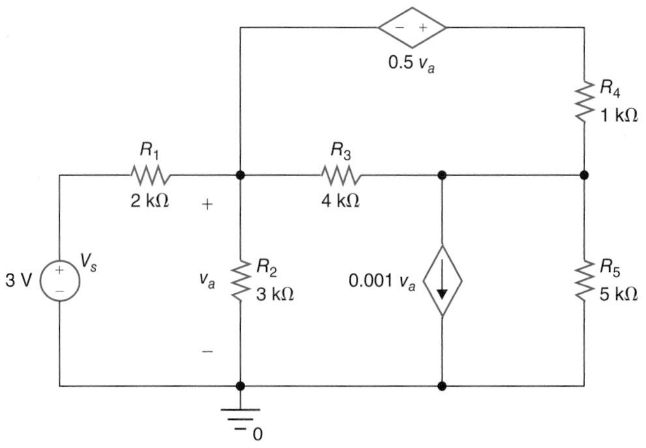

1.26 **In the circuit shown in Figure P1.26, the controlling current, which is the current through R_1, is $i_a = 0.8714$ mA. Find the controlled voltage across the CCVS and the controlled current through the CCCS.**

FIGURE P1.26

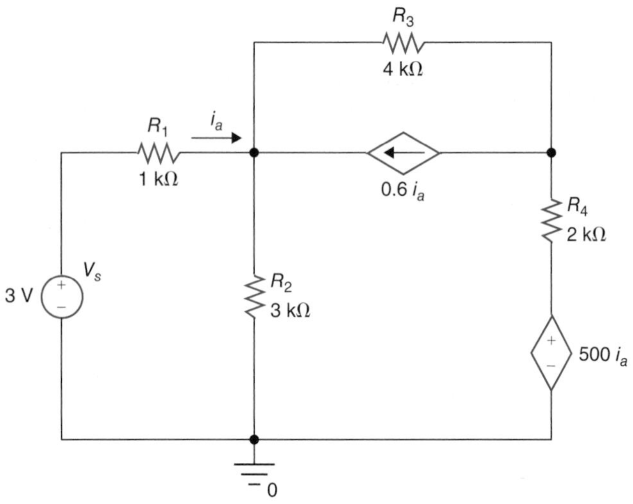

1.27 Plot

$$f(t) = u(t) - 3\,u(t-2) + 6\,u(t-5) - 4\,u(t-8)$$

1.28 Plot

$$f(t) = -5\,\delta(t+2) + 7\,\delta(t-6)$$

1.29 Plot

$$f(t) = 2t\,u(t) - 4(t-1)\,u(t-1) + 3\,(t-3) \times u(t-3) - (t-5)\,u(t-5)$$

1.30 Plot

$$f(t) = -2t\,u(t) + 6(t-2)\,u(t-2) - 5(t-3) \times u(t-3) + (t-5)\,u(t-5)$$

1.31 Plot

$$f(t) = 2\,rect\left(\frac{t+3}{4}\right)$$

1.32 Plot

$$f(t) = 2\,tri\left(\frac{t+4}{2}\right)$$

Chapter 2

Circuit Laws

2.1 Introduction

Nodes, branches, loops, and meshes are defined in this chapter. The equation of resistance of a conductor is expressed as a function of conductivity (or resistivity), and the dimension of the conductor. Ohm's law is introduced.

Kirchhoff's current law (KCL) and Kirchhoff's voltage law (KVL) are presented in this chapter. These two Kirchhoff's laws provide the theoretical basis for the nodal analysis and mesh analysis discussed in the next chapter and applied in circuit analysis in the rest of the chapters.

Finding the equivalent resistance of series and parallel connection of resistors are discussed. Simple circuit rules can be applied to analyze circuits after simplifying the circuits using equivalent resistances.

The voltage divider rule and the current divider rule are useful tools to analyze circuits without too much effort.

If a circuit contains resistors in wye (Y) shape, it can be changed to delta (Δ) shape. On the other hand, if a circuit contains resistors in delta shape, it can be changed to wye shape. The transformation from wye to delta and delta to wye may make it easier to simplify the circuit.

2.2 Circuit

A **circuit** is an interconnection of elements, which can be voltage sources, current sources, resistors, capacitors, inductors, coupled coils, transformers, op amps, etc. A **node** is a point in a circuit where two or more elements are joined. A **simple node** is a node that connects two elements. A **path** in a circuit is a series of connected elements from a node to another node that does not go to the same node more than once. A **branch** is a path in a circuit consisting of a single element. The voltage of a node measured with respect to a **reference node** is called **node voltage**. The **ground node** where the voltage is at ground level is usually taken to be the reference node. A **loop** of a circuit is a closed path starting from a node and returning to the same node. The loop must have minimum of two branches in its closed path. A **mesh** is a loop that does not contain another loop inside it.

EXAMPLE 2.1

Find all the nodes, loops, and meshes for the circuit shown in Figure 2.1.

FIGURE 2.1

Circuit for EXAMPLE 2.1.

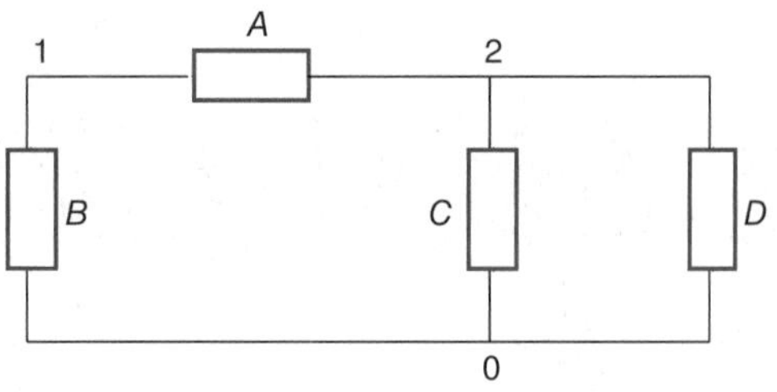

In the circuit shown in Figure 2.1, there are three nodes: 0, 1, and 2. Elements B, C, and D are joined at node 0. Node 1 connects elements A and B, and node 2 connects elements A, C, and D. If node 0 is the ground node, then the potential is set to zero at node 0. The voltages at node 1 and node 2 are measured with respect to node 0.

There are four branches in the circuit shown in Figure 2.1: A, B, C, and D. There are three loops in the circuit shown in Figure 2.1:

0-B-1-A-2-D-0
0-B-1-A-2-C-0
0-C-2-D-0

There are two meshes in the circuit shown in Figure 2.1:

0-B-1-A-2-C-0
0-C-2-D-0

The loop 0-B-1-A-2-D-0 contains two meshes: 0-B-1-A-2-C-0 and 0-C-2-D-0.

Exercise 2.1

Find all the nodes, meshes, and loops for the circuit shown in Figure 2.2.

FIGURE 2.2

Circuit for EXERCISE 2.1.

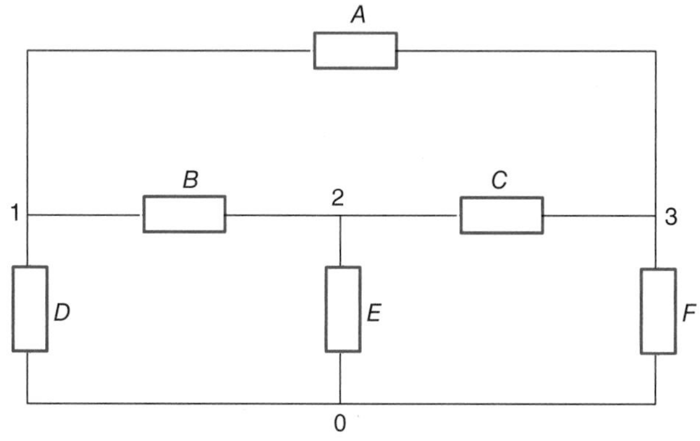

Answer:
Nodes: 0, 1, 2, 3
Meshes:
1-A-3-C-2-B-1
0-D-1-B-2-E-0
0-E-2-C-3-F-0
Loops: In addition to the meshes, we have
0-D-1-A-3-F-0
0-D-1-B-2-C-3-F-0
0-D-1-A-3-C-2-E-0
0-E-2-B-1-A-3-F-0

2.3 Resistor

A *resistor* is a circuit component that regulates the flow of current. The resistance of a resistor measures its ability to limit the current. When the resistance value is large, the amount of current flow through the resistor is small. On the other hand, if the resistance value is small, the amount of current flow through the resistor is large. The resistance value of a resistor is determined by the resistivity of the material used to make it, as well as its dimensions.

Low-power resistors can be made from carbon composition material made of fine granulated graphite mixed with clay. For high power, wire-wound resistors can be used. The wire-wound resistors are constructed by twisting a wire made of nichrome or similar material around a ceramic core. The circuit symbol for a resistor is shown in Figure 2.3.

FIGURE 2.3

Circuit symbol for a resistor.

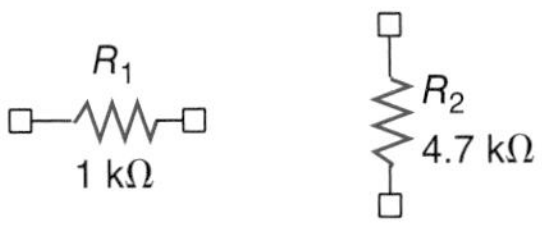

The *current density* is defined as the amount of current through the unit area. If A is the cross-sectional area of a conductor that carries a constant current I, the current density is given by

$$J = \frac{I}{A} \tag{2.1}$$

The current is obtained by integrating the current density through the area. Thus, we have

$$I = \int \mathbf{J} \cdot d\mathbf{A} \tag{2.2}$$

It can be shown that the current density is proportional to the electric field intensity; that is,

$$J = \sigma E \tag{2.3}$$

where σ is the **conductivity** of the material. The unit for conductivity is siemens per meter (S/m). Equation (2.3) is called the *microscopic Ohm's law.*

Let the length of a cylindrical conductor with the cross-sectional area A be ℓ. Let the potential difference between the ends of the conductor be V. This potential difference generates a constant electric field E inside the conductor. The potential difference V is related to the electric field through

$$V = E\ell \tag{2.4}$$

Thus, we have $E = V/\ell$. Substituting this result into Equation (2.3), we get

$$J = \sigma E = \sigma\left(\frac{V}{\ell}\right) \tag{2.5}$$

Since $J = I/A$, Equation (2.5) becomes

$$\frac{I}{A} = \sigma\left(\frac{V}{\ell}\right) \tag{2.6}$$

Solving for V in Equation (2.6), we get

$$V = \frac{\ell}{\sigma A} I = RI \tag{2.7}$$

where R is defined as the resistance of the conductor. This is called the macroscopic Ohm's law. The SI unit for the resistance R is the ohm (Ω = V/A). The resistance of a material is given by

$$R = \frac{\ell}{\sigma A} \tag{2.8}$$

The resistance is proportional to the length and inversely proportional to the conductivity and the cross-sectional area. The **resistivity** ρ of a material is defined as the reciprocal of the conductivity σ. Thus, we have

$$\rho = \frac{1}{\sigma} \tag{2.9}$$

The SI unit for resistivity is $\Omega \cdot \mathrm{m}$. In terms of the resistivity, the resistance R can be written as

$$R = \frac{\rho \ell}{A} \tag{2.10}$$

The resistance is proportional to the resistivity and length and inversely proportional to the cross-sectional area. The inverse of resistance is called **conductance** and is denoted by G. The unit for conductance is S (siemens). Notice that $\mathrm{S} = \Omega^{-1} = \mathrm{A/V}$.

$$G = \frac{1}{R} \tag{2.11}$$

With resistivity (ρ) and cross-sectional area (A) fixed, let R_1 be the resistance when the length is ℓ_1, R_2 be the resistance when the length is ℓ_2, and R be the resistance when the length is $\ell = \ell_1 + \ell_2$. Then, we have

$$R = \frac{\rho \ell}{A} = \frac{\rho(\ell_1 + \ell_2)}{A} = \frac{\rho \ell_1}{A} + \frac{\rho \ell_2}{A} = R_1 + R_2 \tag{2.12}$$

Equation (2.12) says that when two resistors are connected in series, the equivalent resistance is the sum of the two resistances. In general, as shown in section 2.7 later in this chapter, if n resistors, $R_1, R_2, \ldots, R_n$, are connected in series, the equivalent resistance is given by

$$R_{eq} = R_1 + R_2 + \cdots + R_n \tag{2.13}$$

With resistivity (ρ) and length (ℓ) fixed, let R_1 be the resistance when the cross-sectional area is A_1, R_2 be the resistance when the cross-sectional area is A_2, and R be the resistance when the cross-sectional area is $A = A_1 + A_2$. Then, we have

$$R = \frac{\rho \ell}{A} = \frac{\rho \ell}{A_1 + A_2} = \frac{1}{\dfrac{A_1}{\rho \ell} + \dfrac{A_2}{\rho \ell}} = \frac{1}{\dfrac{1}{\frac{\rho \ell}{A_1}} + \dfrac{1}{\frac{\rho \ell}{A_2}}} = \frac{1}{\dfrac{1}{R_1} + \dfrac{1}{R_2}} = \frac{R_1 R_2}{R_1 + R_2} \tag{2.14}$$

Equation (2.14) says that when two resistors are connected in parallel, the equivalent resistance is given by $R_1R_2/(R_1+R_2)$. In general, as shown in section 2.7 later in this chapter, if n resistors, $R_1, R_2, \ldots, R_n$, are connected in parallel, the equivalent resistance is given by

$$R_{eq} = \frac{1}{\dfrac{1}{R_1} + \dfrac{1}{R_2} + \cdots + \dfrac{1}{R_n}} \tag{2.15}$$

2.4 Ohm's Law

As shown in Equation (2.7), the voltage-current relation across the resistor is given by

$$V = RI \tag{2.16}$$

Equation (2.16) states that the voltage across the resistor is proportional to the current flowing through the resistor. The proportionality constant of this linear equation is resistance R. Equation (2.16) can be rewritten as

$$I = \frac{V}{R} \tag{2.17}$$

Equation (2.17) states that the current through the resistor is proportional to the voltage applied across the resistor and inversely proportional to resistance R.

Equation (2.16) can be rewritten as

$$R = \frac{V}{I} \tag{2.18}$$

Equation (2.18) states that the resistance is the ratio of voltage to current. Notice that the SI unit for voltage is volt (V), the SI unit for current is ampere (A), and the SI unit for resistance is ohm (Ω), as discussed in Chapter 1. According to Equations (2.16)–(2.18), we have

$$\begin{aligned} V &= \Omega \cdot A \\ A &= V/\Omega \\ \Omega &= V/A \end{aligned}$$

Figure 2.4 shows a circuit consisting of a voltage source and a resistor with resistance 1 $k\Omega$. Figure 2.5 shows the current I through the resistor as the voltage is swept from 0 V to 10 V. The current increases linearly from 0 mA to 10 mA. The slope of the I-V characteristic is given by 10 mA/10 V = 0.001 S. The slope is the conductance. The inverse of the conductance is the resistance, given by

$$R = \frac{10\text{ V}}{10\text{ mA}} = \frac{10\text{ V}}{0.01\text{ A}} = 1000\,\Omega = 1\,k\Omega$$

The power absorbed by a resistor is given by the product of current and voltage. Thus, we have

$$P = IV = VI\text{ (W)} \tag{2.19}$$

Substitution of V by RI from Equation (2.16) into Equation (2.19) yields

$$P = RI \times I = RI^2\text{ (W)} \tag{2.20}$$

Substitution of I by V/R from Equation (2.17) into Equation (2.19) yields

$$P = \frac{V^2}{R}\text{ (W)} \tag{2.21}$$

The power on the voltage source in Figure 2.4 is

$$P_s = V(-I) = -IV \tag{2.22}$$

FIGURE 2.4

A circuit with a voltage source and a resistor.

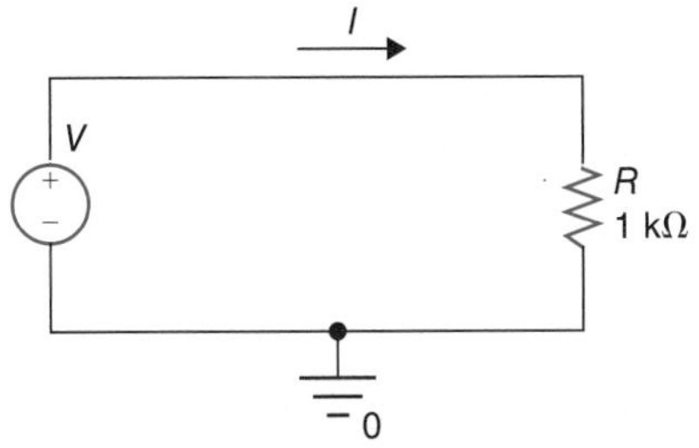

FIGURE 2.5

I-V characteristic of a resistor.

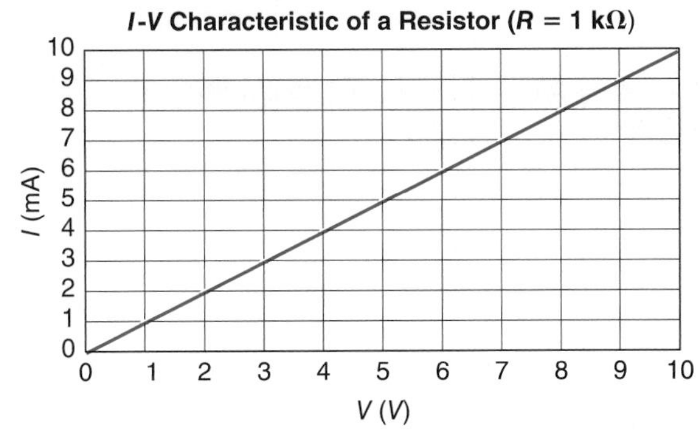

The direction of the current through the voltage source from the positive terminal to the negative terminal is negative. When the power is negative, the circuit element (in this case the voltage source) is delivering power to the rest of the circuit. The power absorbed by the resistor in Figure 2.4 is IV. The total power on the circuit shown in Figure 2.4 is

$$-IV + IV = 0$$

In general, the total power delivered equals the total power absorbed. This is called *conservation of power.* The energy is the integral of the power. If the power is constant, the energy is the power times the time. The energy spent on the resistor in Figure 2.4 in T seconds is

$$W = PT = IVT = I^2RT = \frac{V^2}{R}T \tag{2.23}$$

FIGURE 2.6

Circuit with two resistors and a voltage source.

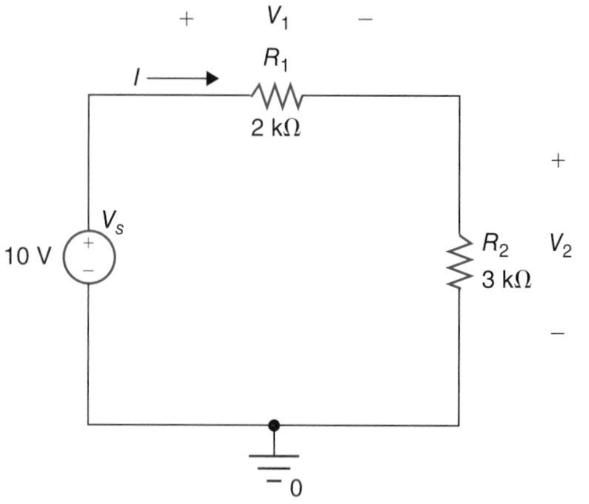

Consider a circuit shown in Figure 2.6. The current I around the mesh in the circuit shown in Figure 2.6 is given by $I = 2$ mA. We are interested in finding the voltage V_1 across R_1, voltage V_2 across R_2, and powers absorbed by R_1, R_2, and delivered by V_s.

Since the current through R_1 is 2 mA, according to Ohm's law, the voltage across R_1 is given by

$$V_1 = R_1 \times I = 2000\ \Omega \times 0.002\ A = 4\ \text{V}$$

Similarly, the voltage across R_2 is given by

$$V_2 = R_2 \times I = 3000\ \Omega \times 0.002\ A = 6\ \text{V}$$

The power absorbed by R_1 is given by

$$P_{R_1} = I \times V_1 = 0.002 \times 4 = 0.008\ \text{W} = 8\ \text{mW}.$$

Similarly, the power absorbed by R_2 is given by

$$P_{R_2} = I \times V_2 = 0.002 \times 6 = 0.012\ \text{W} = 12\ \text{mW}$$

The power from V_s is given by

$$P_{Vs} = (-I) \times V_s = -0.002 \times 10 = -0.020\ \text{W} = -20\ \text{mW}$$

According to passive sign convention, the power is defined as the product of the voltage and the current through the device. Since the current through the voltage source is −2 mA, the power is negative. When the power is negative, the device delivers power to the rest of the circuit. Since the power is positive for R_1 and R_2, the power is absorbed by the resistors. The total power absorbed is 8 mW + 12 mW = 20 mW. The power absorbed is equal to the power delivered by the source, which is 20 mW. This confirms the principle of conservation of power.

EXAMPLE 2.2

Assume that the voltage across R_2, which is also the voltage across R_3, is given by $V_2 = 9$ V in the circuit shown in Figure 2.7. Find I_2, I_3, V_1, I_1, and powers absorbed or delivered by R_1, R_2, R_3, and V_s.

continued

Example 2.2 continued

FIGURE 2.7

Circuit for EXAMPLE 2.2.

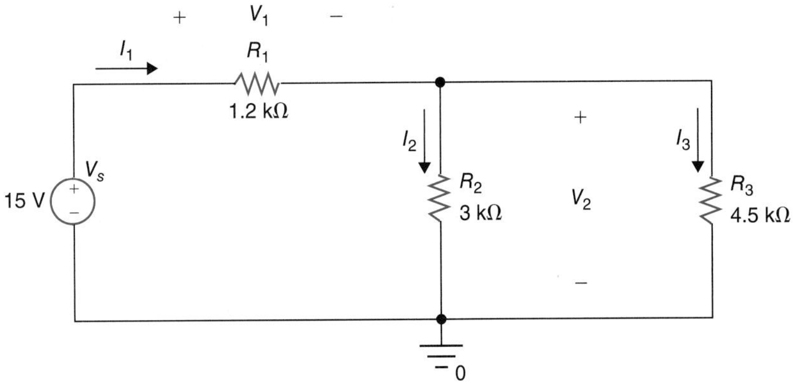

The current through R_2 can be found from Equation (2.17):

$$I_2 = \frac{V_2}{R_2} = \frac{9\text{ V}}{3\ k\Omega} = \frac{9\text{ V}}{3 \times 10^3\ \Omega} = 3 \times 10^{-3} A = 3\text{ mA}$$

Similarly, the current through R_3 is given by

$$I_3 = \frac{V_2}{R_3} = \frac{9\text{ V}}{4.5\ k\Omega} = \frac{9\text{ V}}{4.5 \times 10^3\ \Omega} = 2 \times 10^{-3} A = 2\text{ mA}$$

The potential difference across R_1 is given by

$$V_1 = V_s - V_2 = 15\text{ V} - 9\text{ V} = 6\text{ V}$$

The current through R_1 is given by

$$I_1 = \frac{V_1}{R_1} = \frac{6\text{ V}}{1.2\ k\Omega} = \frac{6\text{ V}}{1.2 \times 10^3\ \Omega} = 5 \times 10^{-3} A = 5\text{ mA}$$

The power absorbed by R_1, R_2, and R_3, respectively, are given by

$$P_{R_1} = I_1 V_1 = 5\text{ mA} \times 6\text{ V} = 5 \times 10^{-3} A \times 6\text{ V} = 30 \times 10^{-3} W = 30\text{ mW}$$

$$P_{R_2} = I_2 V_2 = 3\text{ mA} \times 9\text{ V} = 3 \times 10^{-3} A \times 9\text{ V} = 27 \times 10^{-3} W = 27\text{ mW}$$

$$P_{R_3} = I_3 V_2 = 2\text{ mA} \times 9\text{ V} = 2 \times 10^{-3} A \times 9\text{ V} = 18 \times 10^{-3} W = 18\text{ mW}$$

The power from the voltage source is

$$P_s = (-I_1)V_s = -5\text{ mA} \times 15\text{ V} = -5 \times 10^{-3} A \times 15\text{ V} = -75 \times 10^{-3} W = -75\text{ mW}$$

The current from the positive terminal to the negative terminal of the voltage source is $-I_1$. Since the power is negative, the voltage source delivers 75 mW of power to the rest of the circuit. The sum of powers absorbed by three resistors is

$$P_{R_1} + P_{R_2} + P_{R_3} = 30\text{ mW} + 27\text{ mW} + 18\text{ mW} = 75\text{ mW}$$

matching the power supplied by the voltage source.

Exercise 2.2

The voltage V_1 in the circuit shown in Figure 2.8 is given by 6 V. Find I_1, I_2, and powers on R_1, R_2, and I_s.

FIGURE 2.8

Circuit for EXERCISE 2.2.

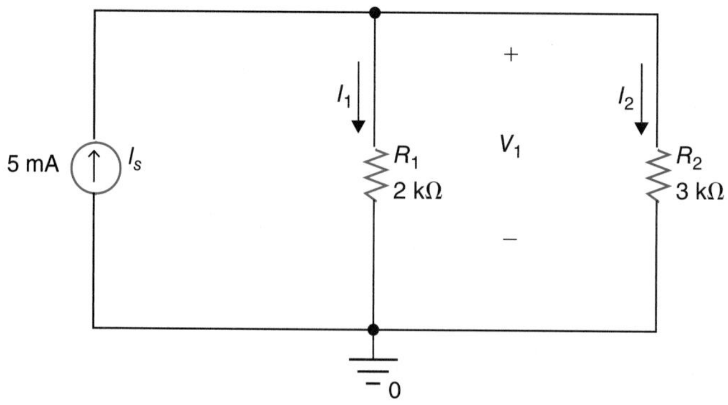

Answer:

$I_1 = 3$ mA, $I_2 = 2$ mA, $P_{R_1} = 18$ mW, $P_{R_2} = 12$ mW, $P_{I_s} = -30$ mW.

2.5 Kirchhoff's Current Law (KCL)

As defined in section 2.2, a node is a point in a circuit where two or more elements are connected. It is part of wires that interconnect elements. A node cannot store or destroy electric charges. What comes into a node must leave the same node. From this fact, we have the following theorem, called **Kirchhoff's current law (KCL):**

The sum of currents entering a node equals the sum of currents leaving the same node.

Another way to describe KCL is:

The sum of currents leaving a node is zero.

Notice that for this statement to be true, at least one of the currents leaving the node must be negative (meaning that the current actually enters the node).

Still another way to describe KCL is:

The sum of currents entering a node is zero.

Notice that for this statement to be true, at least one of the currents entering the node must be negative (meaning that the current actually leaves the node).

Consider a circuit shown in Figure 2.9. We are interested in finding currents I_1, I_2, I_3, and I_4.

FIGURE 2.9

Circuit with three resistors and a voltage source.

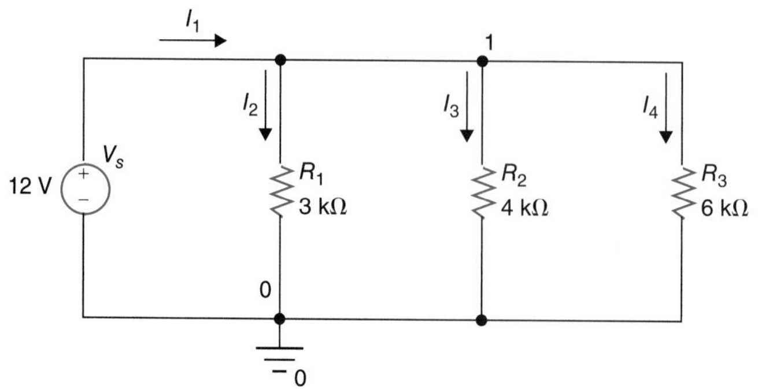

The voltage at the ground node (node 0) is 0 V. The voltage at node 1, connecting V_s, R_1, R_2, and R_3 is 12 V. Thus, the voltages across R_1, R_2, and R_3 are all 12 V. Applying Ohm's law to each resistor, we have

$$I_2 = \frac{V_s}{R_1} = \frac{12\ \text{V}}{3\ k\Omega} = 4\ \text{mA}$$

$$I_3 = \frac{V_s}{R_2} = \frac{12\ \text{V}}{4\ k\Omega} = 3\ \text{mA}$$

$$I_4 = \frac{V_s}{R_3} = \frac{12\ \text{V}}{6\ k\Omega} = 2\ \text{mA}$$

According to KCL, the current entering node 1 (I_1) must equal the currents leaving node 1 ($I_2 + I_3 + I_4$). Thus, we have

$$I_1 = I_2 + I_3 + I_4 = 4\ \text{mA} + 3\ \text{mA} + 2\ \text{mA} = 9\ \text{mA}$$

Notice that the sum of the currents leaving node 1 is zero:

$$-I_1 + I_2 + I_3 + I_4 = -9\ \text{mA} + 4\ \text{mA} + 3\ \text{mA} + 2\ \text{mA} = 0\ \text{mA}$$

Notice that the sum of the current entering node 1 is also zero:

$$I_1 - I_2 - I_3 - I_4 = 9\ \text{mA} - 4\ \text{mA} - 3\ \text{mA} - 2\ \text{mA} = 0\ \text{mA}$$

In addition, notice that the initial assignment of the direction of currents can be arbitrary. If the actual current flows in the opposite direction, the value of the current will be negative. Figure 2.10 shows the circuit shown in Figure 2.9, but with a different current assignment. Applying Ohm's law to each resistor, we have

FIGURE 2.10

Circuit shown in Figure 2.9 with a different current assignment.

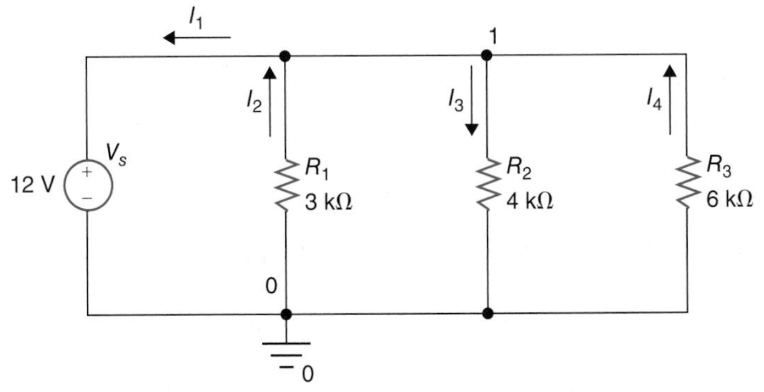

$$I_2 = \frac{0\ \text{V} - 12\ \text{V}}{3\ k\Omega} = -4\ \text{mA}$$

$$I_3 = \frac{12\ \text{V} - 0\ \text{V}}{4\ k\Omega} = 3\ \text{mA}$$

$$I_4 = \frac{0\ \text{V} - 12\ \text{V}}{6\ k\Omega} = -2\ \text{mA}$$

According to KCL, the sum of currents entering node 1 ($I_2 + I_4$) must equal the sum of currents leaving node 1 ($I_1 + I_3$). Thus,

$$I_2 + I_4 = I_1 + I_3$$

Substituting $I_2 = -4$ mA, $I_3 = 3$ mA, $I_4 = -2$ mA, we have

$$-4\ \text{mA} - 2\ \text{mA} = I_1 + 3\ \text{mA}$$

Thus, $I_1 = -9$ mA.
Notice that the sum of currents leaving node 1 is zero:

$$I_1 - I_2 + I_3 - I_4 = -9\ \text{mA} + 4\ \text{mA} + 3\ \text{mA} + 2\ \text{mA} = 0\ \text{mA}$$

Notice that the sum of current entering node 1 is also zero:

$$-I_1 + I_2 - I_3 + I_4 = 9\text{ mA} - 4\text{ mA} - 3\text{ mA} - 2\text{ mA} = 0\text{ mA}$$

EXAMPLE 2.3

Assume that the current through R_3 in the circuit shown in Figure 2.11 is given by $I_3 = 3$ mA. Find V_3, I_4, I_2, V_2, I_1, and V_1.

FIGURE 2.11

Circuit for EXAMPLE 2.3.

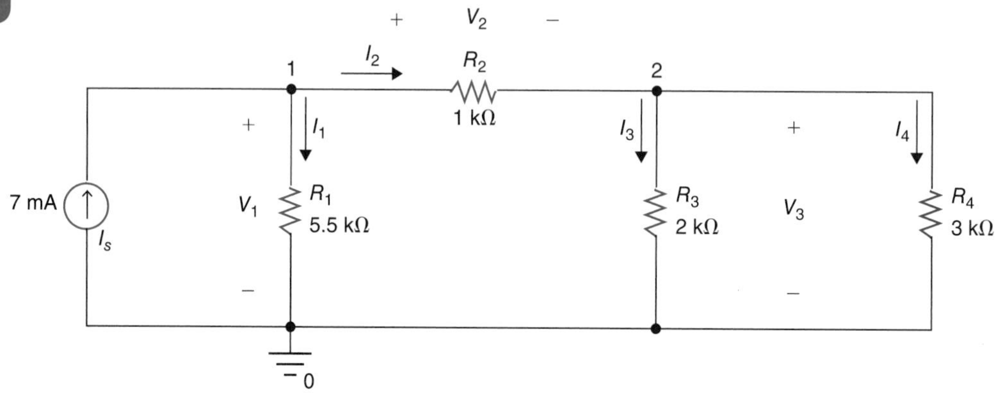

According to Ohm's law, the voltage across R_3 is given by

$$V_3 = R_3 I_3 = 2\ k\Omega \times 3\text{ mA} = 2000\ \Omega \times 0.003\ A = 6\text{ V}$$

Since R_3 and R_4 share the same nodes, the voltage across R_4 is V_3. According to Ohm's law, the current through R_4 is given by

$$I_4 = \frac{V_3}{R_4} = \frac{6\text{ V}}{3\ k\Omega} = 2\text{ mA}$$

According to KCL, the current entering node 2, I_2, equals the sum of currents leaving node 2, $I_3 + I_4$. Therefore, we have

$$I_2 = I_3 + I_4 = 3\text{ mA} + 2\text{ mA} = 5\text{ mA}$$

According to Ohm's law, the voltage across R_2 is given by

$$V_2 = R_2 I_2 = 1\ k\Omega \times 5\text{ mA} = 5\text{ V}$$

According to KCL, the current entering node 1, I_s, equals to the sum of currents leaving node 1, $I_1 + I_2$. Therefore, we have

$$I_s = I_1 + I_2$$

Thus, the current through R_1 is given by

$$I_1 = I_s - I_2 = 7\text{ mA} - 5\text{ mA} = 2\text{ mA}$$

continued

Example 2.3 continued

According to Ohm's law, the voltage across R_1, which is also the voltage across I_s, is given by

$$V_1 = R_1 I_1 = 5.5\ k\Omega \times 2\ \text{mA} = 11\ \text{V}$$

Exercise 2.3

Find I_2, I_3, I_1, I_4, and I_5 in the circuit shown in Figure 2.12.

FIGURE 2.12

Circuit for EXERCISE 2.3.

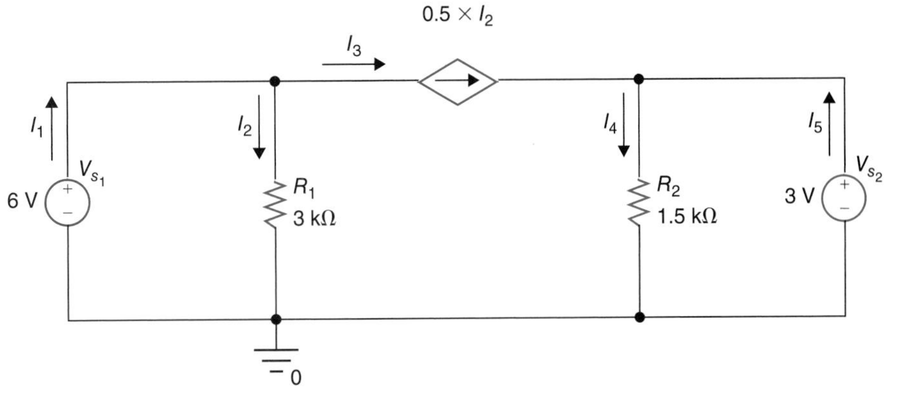

Answer:
$I_2 = 2$ mA, $I_3 = 1$ mA, $I_1 = 3$ mA, $I_4 = 2$ mA, and $I_5 = 1$ mA.

EXAMPLE 2.4

In the circuit shown in Figure 2.13, find the voltage V_1 and currents I_1, I_2, and I_3.

FIGURE 2.13

Circuit for EXAMPLE 2.4.

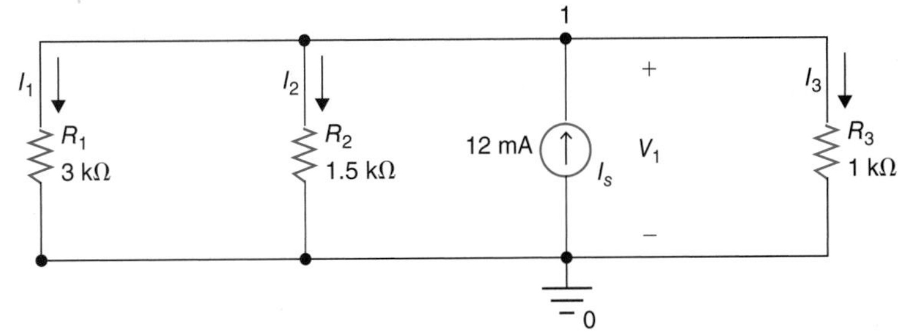

According to KCL, the current entering node 1 must equal the sum of currents leaving node 1; that is,

$$I_s = I_1 + I_2 + I_3 \tag{2.24}$$

The current entering node 1 from the current source is $I_s = 12$ mA. From Ohm's law, the current I_1 through R_1 is given by

$$I_1 = \frac{V_1}{R_1}, \tag{2.25}$$

continued

Example 2.4 continued

where V_1 is the voltage across R_1, as shown in Figure 2.13. Notice that V_1 is also the voltage across I_s, R_2, and R_3. Similarly, the currents I_2 and I_3 are given, respectively, by

$$I_2 = \frac{V_1}{R_2} \tag{2.26}$$

$$I_3 = \frac{V_1}{R_3} \tag{2.27}$$

Substitution of Equations (2.25)–(2.27) into Equation (2.24) yields

$$I_s = \frac{V_1}{R_1} + \frac{V_1}{R_2} + \frac{V_1}{R_3} = \left(\frac{1}{R_1} + \frac{1}{R_2} + \frac{1}{R_3}\right)V_1 \tag{2.28}$$

Solving Equation (2.28) for V_1, we obtain

$$V_1 = \frac{I_s}{\frac{1}{R_1} + \frac{1}{R_2} + \frac{1}{R_3}} = \frac{12 \times 10^{-3}}{\frac{1}{3000} + \frac{1}{1500} + \frac{1}{1000}}\,\text{V} = \frac{36}{1+2+3}\,\text{V} = 6\,\text{V} \tag{2.29}$$

Substitution of $V_1 = 6$ V into Equations (2.25)−(2.27) results in

$$I_1 = \frac{V_1}{R_1} = \frac{6\text{ V}}{3000\,\Omega} = 2\text{ mA} \tag{2.30}$$

$$I_2 = \frac{V}{R_2} = \frac{6\text{ V}}{1500\,\Omega} = 4\text{ mA} \tag{2.31}$$

$$I_3 = \frac{V}{R_3} = \frac{6\text{ V}}{1000\,\Omega} = 6\text{ mA} \tag{2.32}$$

Exercise 2.4

Find v, I_1, and I_2 in the circuit shown in Figure 2.14.

FIGURE 2.14

Circuit for EXERCISE 2.4.

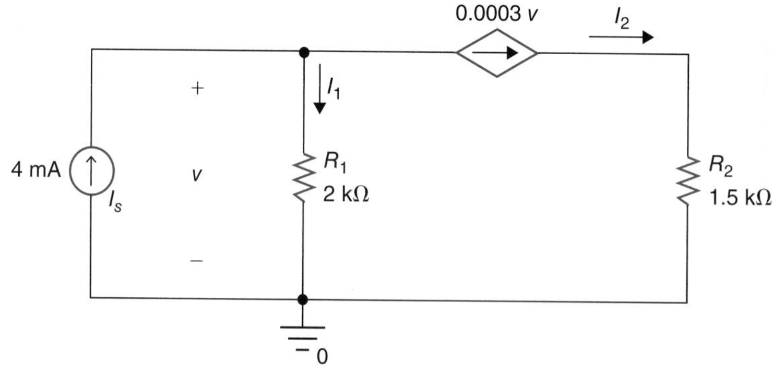

Answer:
$v = 5$ V, $I_1 = 2.5$ mA, $I_2 = 1.5$ mA.

EXAMPLE 2.5

In the circuit shown in Figure 2.15, let $I_1 = 3$ A, $I_3 = 10$ A, and $I_6 = -8$ A. Find I_2, I_4, and I_5.

FIGURE 2.15

Circuit for EXAMPLE 2.5.

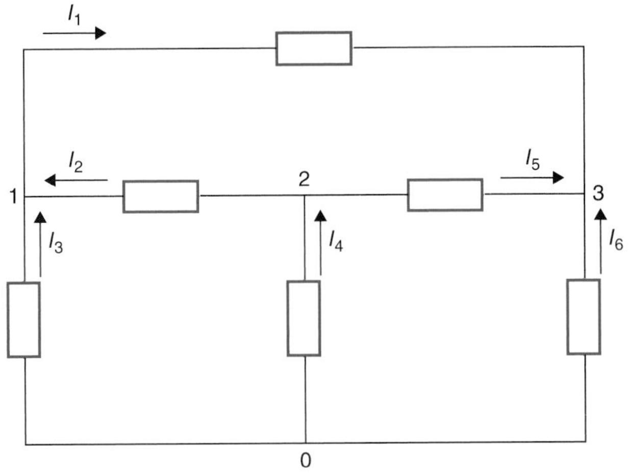

Summing the currents leaving node 0, we obtain

$$I_3 + I_4 + I_6 = 0$$

Solving for I_4, we get

$$I_4 = -I_3 - I_6 = -10 - (-8) = -10 + 8 = -2 \text{ A}$$

Summing the currents leaving node 1, we obtain

$$I_1 - I_2 - I_3 = 0$$

Solving for I_2, we get

$$I_2 = I_1 - I_3 = 3 - 10 = -7 \text{ A}$$

Summing the currents leaving node 3, we obtain

$$-I_1 - I_5 - I_6 = 0$$

Solving for I_5, we get

$$I_5 = -I_1 - I_6 = -3 - (-8) = -3 + 8 = 5 \text{ A}$$

When the current is negative, the current actually flows in the opposite direction. Figure 2.16 shows the circuit with the actual direction (positive direction) of current.

FIGURE 2.16

Circuit for EXAMPLE 2.5 with positive current direction.

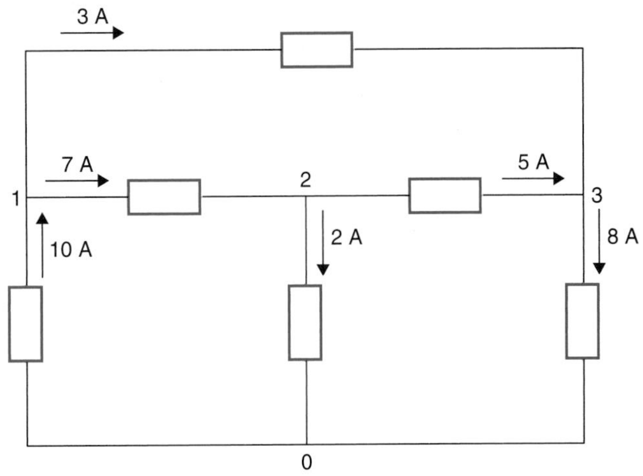

Exercise 2.5

Let $I_1 = 9$ mA, $I_2 = 6$ mA, $I_4 = 2$ mA in the circuit shown in Figure 2.17. Find I_3, I_5, and I_6.

FIGURE 2.17

Circuit for EXERCISE 2.5.

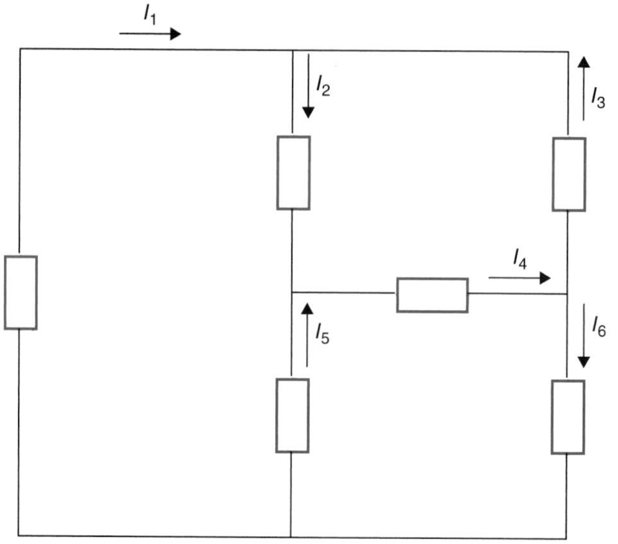

Answer:
$I_3 = -3$ mA, $I_5 = -4$ mA, $I_6 = 5$ mA.

EXAMPLE 2.6

Find I_1, I_2, I_3, I_4, I_5, and I_6 in the circuit shown in Figure 2.18.

FIGURE 2.18

Circuit for EXAMPLE 2.6.

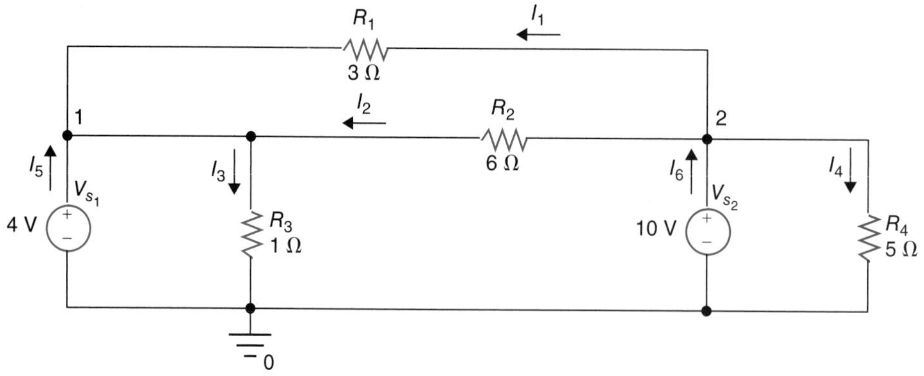

The voltage across R_1 is $V_{S_2} - V_{S_1} = 10 - 4 = 6$ V. The current I_1 through R_1 is given by

$$I_1 = \frac{V_{S_2} - V_{S_1}}{R_1} = \frac{10 - 4}{3} = \frac{6\text{ V}}{3\,\Omega} = 2\,A$$

continued

Example 2.6 continued

from Ohm's law. Similarly, the current I_2 through R_2 is given by

$$I_2 = \frac{V_{S_2} - V_{S_1}}{R_2} = \frac{10 - 4}{6} = \frac{6\text{ V}}{6\ \Omega} = 1\,A$$

from Ohm's law. Application of Ohm's law on R_3 and R_4 yields, respectively,

$$I_3 = \frac{V_{S_1}}{R_3} = \frac{4\text{ V}}{1\ \Omega} = 4\,A$$

$$I_4 = \frac{V_{S_2}}{R_4} = \frac{10\text{ V}}{5\ \Omega} = 2\,A$$

Summing the current leaving node 1, we obtain

$$-I_1 - I_2 + I_3 - I_5 = -2 - 1 + 4 - I_5 = 0$$

from KCL. Solving for I_5, we obtain

$$I_5 = -2 - 1 + 4 = 1\,A$$

Summing the current leaving node 2, we obtain

$$I_1 + I_2 + I_4 - I_6 = 2 + 1 + 2 - I_6 = 0$$

from KCL. Solving for I_6, we obtain

$$I_6 = 2 + 1 + 2 = 5\,A$$

Exercise 2.6

Find I_1, I_2, I_3, I_4, I_5, and I_6 in the circuit shown in Figure 2.19.

FIGURE 2.19

Circuit for EXERCISE 2.6.

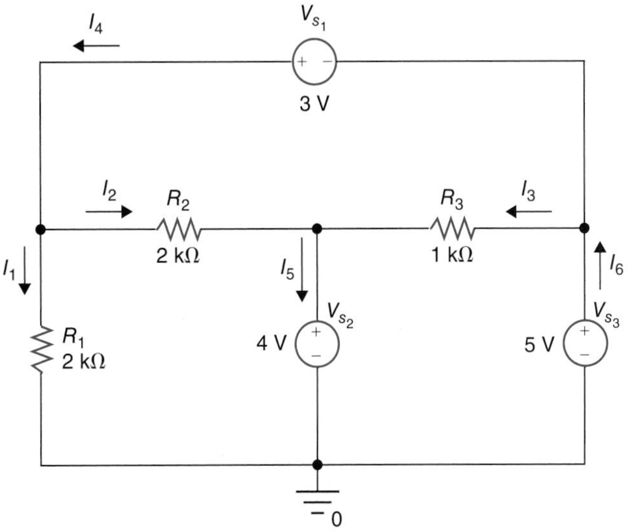

continued

Exercise 2.6 continued

Answer:
$I_1 = 4$ mA, $I_2 = 2$ mA, $I_3 = 1$ mA, $I_4 = 6$ mA, $I_5 = 3$ mA, $I_6 = 7$ mA.

2.6 Kirchhoff's Voltage Law (KVL)

As defined in section 2.2, a node is a point in a circuit where two or more elements are connected. It is part of wires that interconnect elements. The voltage of a node must be unique, and the voltage for any node cannot have two different values. We have the following theorem, called **Kirchhoff's voltage law (KVL):**

The sum of voltage drops around a loop equals the sum of voltage rises of the same loop.

Another way to describe KVL is:

The sum of voltage drops around a loop is zero.

Notice that for this statement to be true, at least one of the voltage drops around the loop must be negative (meaning that the voltage actually rises on the branch).

Still another way to describe KVL is:

The sum of voltage rises around a loop is zero.

Notice that at least one of the voltage rises around the loop must be negative (meaning that the voltage actually drops on the branch) for this statement to be true. Since a mesh is also a loop, the KVL applies to mesh as well.

Consider a circuit shown in Figure 2.20. We are interested in finding the voltages across the resistors R_1, R_2, and R_3 and the current through them.

FIGURE 2.20

A circuit with three resistors and a voltage source.

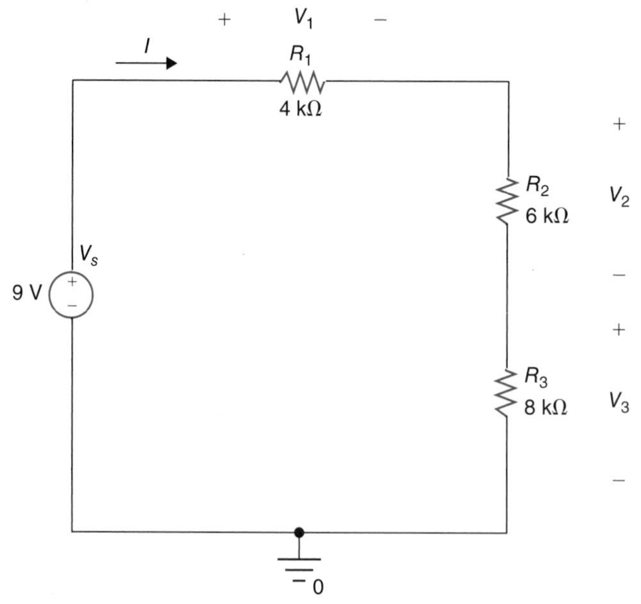

Let the current through the circuit be I. Let the voltage across R_1, R_2, and R_3 be V_1, V_2, and V_3, respectively. According to Ohm's law, the voltage V_1 across R_1 is given by

$$V_1 = R_1 I$$

Similarly, the voltages V_2 and V_3 are given respectively by

$$V_2 = R_2 I$$

$$V_3 = R_3 I$$

According to KVL, the sum of voltage drops around the circuit equals zero. When the circuit is traversed in the clockwise direction starting from the negative terminal of the voltage source, through the voltage source to the positive terminal of the voltage source and through the resistors R_1, R_2, and R_3 in that order, and coming back to the negative terminal of the voltage source, the sum of voltage drops is zero:

$$-V_s + R_1 I + R_2 I + R_3 I = 0$$

Solving this equation for I, we have

$$I = \frac{V_s}{R_1 + R_2 + R_3} = \frac{9\text{ V}}{18\ k\Omega} = 0.5\text{ mA}$$

From this current, we can find the voltages V_1, V_2, and V_3 using Ohm's law. Thus,

$$V_1 = R_1 I = 4\ k\Omega \times 0.5\text{ mA} = 2\text{ V}$$

$$V_2 = R_2 I = 6\ k\Omega \times 0.5\text{ mA} = 3\text{ V}$$

$$V_3 = R_3 I = 8\ k\Omega \times 0.5\text{ mA} = 4\text{ V}$$

Notice that multiplication of $k\Omega$ and mA results in V; that is,

$$1\ k\Omega \times 1\text{ mA} = 10^3\ \Omega \times 10^{-3}\ A = 1\text{ V}$$

Also, division of V by $k\Omega$ yields mA, and division of V by mA yields $k\Omega$, as shown here:

$$\frac{1\text{ V}}{1\ k\Omega} = \frac{1\text{ V}}{10^3\ \Omega} = 10^{-3} A = 1\text{ mA}$$

$$\frac{1\text{ V}}{1\text{ mA}} = \frac{1\text{ V}}{10^{-3}\ A} = 10^3\ \Omega = 1\ k\Omega$$

For more details on prefixes, see Chapter 1.

EXAMPLE 2.7

Suppose that $V_2 = 6$ V in the circuit shown in Figure 2.21. Find I_2, I_3, I_4, V_4, I_1, V_1, and V_s.

continued

Example 2.7 continued

FIGURE 2.21

Circuit for EXAMPLE 2.7.

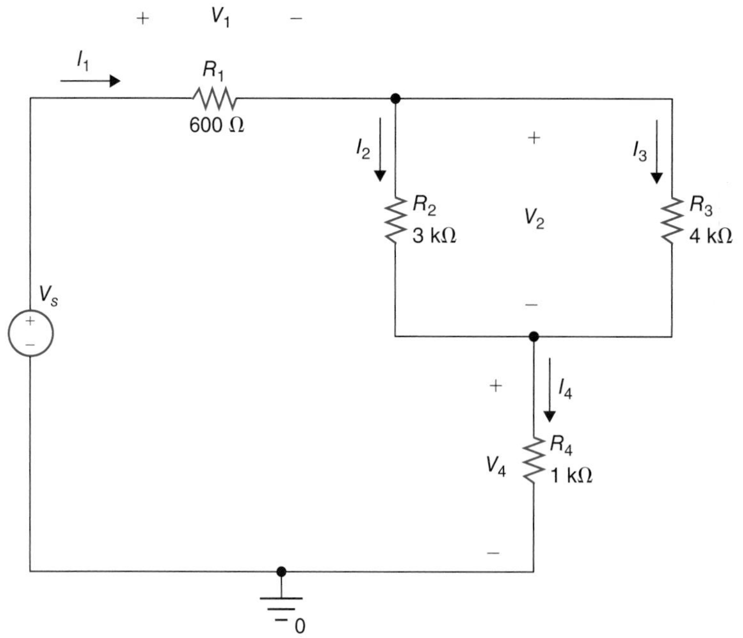

According to Ohm's law, the current through R_2 is given by

$$I_2 = \frac{V_2}{R_2} = \frac{6\text{ V}}{3\ k\Omega} = 2\text{ mA}$$

Similarly, the current through R_3 is given by

$$I_3 = \frac{V_2}{R_3} = \frac{6\text{ V}}{4\ k\Omega} = 1.5\text{ mA}$$

From KCL, the current through R_4 is the sum of I_2 and I_3. Thus, we have

$$I_4 = I_2 + I_3 = 2\text{ mA} + 1.5\text{ mA} = 3.5\text{ mA}$$

Notice that the current through R_1 is also the sum of I_2 and I_3. Thus, we have

$$I_1 = I_2 + I_3 = 2\text{ mA} + 1.5\text{ mA} = 3.5\text{ mA}$$

According to Ohm's law, the voltage across R_4 is given by

$$V_4 = R_4 I_4 = 1\ k\Omega \times 3.5\text{ mA} = 1 \times 10^3\ \Omega \times 3.5 \times 10^{-3} A = 3.5\text{ V}$$

Similarly, the voltage across R_1 is given by

$$V_1 = R_1 I_1 = 600\ \Omega \times 3.5\text{ mA} = 600\ \Omega \times 3.5 \times 10^{-3} A = 2.1\text{ V}$$

According to KVL, the sum of voltage drops around the mesh on the left side equals zero. Summing the voltage drops starting from the voltage source in the clockwise direction, we obtain

$$-V_s + V_1 + V_2 + V_4 = 0$$

continued

Example 2.7 continued

Solving for V_s, we get

$$V_s = V_1 + V_2 + V_4 = 2.1\text{ V} + 6\text{ V} + 3.5\text{ V} = 11.6\text{ V}$$

Exercise 2.7

Find the voltages V_1, V_2, and V_3, and currents I_1, I_2, and I_3 in the circuit shown in Figure 2.22.

FIGURE 2.22

Circuit for EXERCISE 2.7.

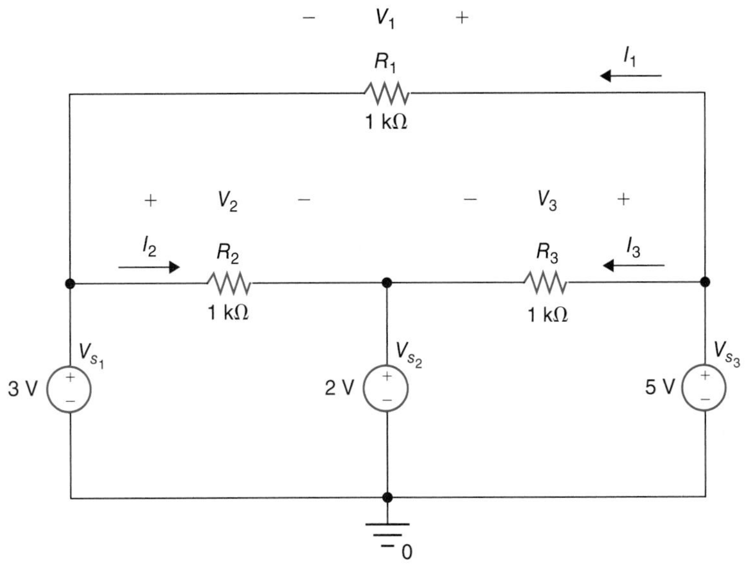

Answer:
$V_1 = 2\text{ V}, V_2 = 1\text{ V}, V_3 = 3\text{ V}, I_1 = 2\text{ mA}, I_2 = 1\text{ mA}, I_3 = 3\text{ mA}.$

EXAMPLE 2.8

Let $V_1 = 6$ V, $V_5 = 5$ V, $V_6 = 3$ V, and $V_7 = 7$ V in the circuit shown in Figure 2.23. Find V_2, V_3, V_4, and V_8.

FIGURE 2.23

Circuit for EXAMPLE 2.8.

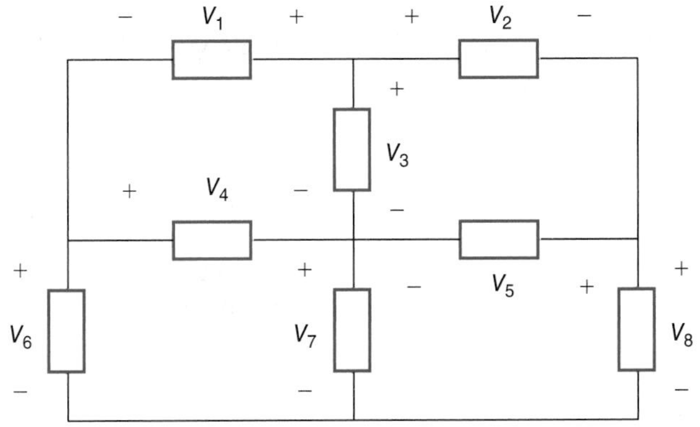

Collecting the voltage drops around the mesh at the lower left of the circuit in the clockwise direction (KVL), we obtain

$$-V_6 + V_4 + V_7 = 0$$

Thus,

$$V_4 = V_6 - V_7 = 3 - 7 = -4\text{ V}$$

Collecting the voltage drops around the mesh at the upper left of the circuit in the clockwise direction (KVL), we obtain

$$-V_1 + V_3 - V_4 = 0$$

continued

Example 2.8 continued

Thus,

$$V_3 = V_1 + V_4 = 6 - 4 = 2\text{ V}$$

Collecting the voltage drops around the mesh at the upper right of the circuit in the clockwise direction (KVL), we obtain

$$-V_3 + V_2 + V_5 = 0$$

Thus,

$$V_2 = V_3 - V_5 = 2 - 5 = -3\text{ V}$$

Collecting the voltage drops around the mesh at the lower right of the circuit in the clockwise direction (KVL), we obtain

$$-V_7 - V_5 + V_8 = 0$$

Thus,

$$V_8 = V_7 + V_5 = 7 + 5 = 12\text{ V}$$

Exercise 2.8

Let $V_1 = 7$ V, $V_3 = 4$ V, and $V_5 = 5$ V in the circuit shown in Figure 2.24. Find V_2, V_4, and V_6.

FIGURE 2.24

Circuit for EXERCISE 2.8.

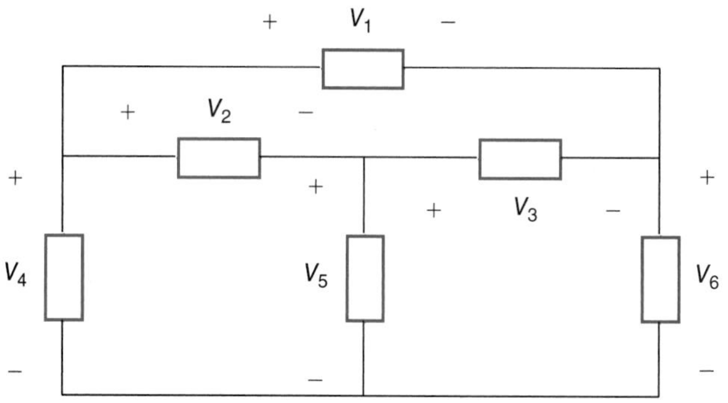

Answer:
$V_2 = 3$ V, $V_4 = 8$ V, $V_6 = 1$ V.

EXAMPLE 2.9

Find i, V_1, V_2, I_2, I_3, I_4, I_5, I_6 in the circuit shown in Figure 2.25.

continued

Example 2.9 continued

FIGURE 2.25

Circuit for EXAMPLE 2.9.

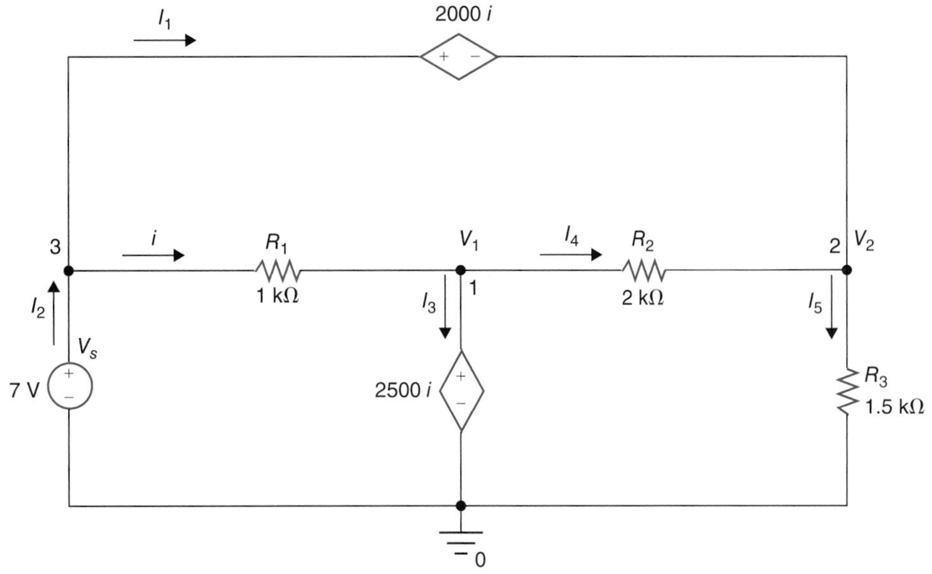

Collecting the voltage drops around the mesh (KVL) in the lower-left mesh of the circuit, we obtain

$$-7 + 1000i + 2500i = 0$$

Solving for i, we get

$$i = \frac{7}{3500} A = 0.002\ A = 2\text{ mA}$$

The voltage drop across R_1 is given by

$$V_{R_1} = iR_1 = 2\text{ mA} \times 1\ k\Omega = 2 \times 10^{-3} \times 1000\text{ V} = 2\text{ V}$$

The voltage V_1 is given by

$$V_1 = V_s - V_{R_1} = 7 - 2 = 5\text{ V}$$

The voltage V_2 is given by

$$V_2 = V_s - 2000i = 7 - 2000 \times 0.002 = 7 - 4 = 3\text{ V}$$

The current through R_2 is given by

$$I_4 = \frac{V_1 - V_2}{R_2} = \frac{5\text{ V} - 3\text{ V}}{2000\ \Omega} = \frac{2\text{ V}}{2000\ \Omega} = 1\text{ mA}$$

The current through R_3 is given by

$$I_5 = \frac{V_2}{R_3} = \frac{3\text{ V}}{1500\ \Omega} = 2\text{ mA}$$

continued

Example 2.9 continued

Summing the currents leaving node 1 (KCL), we obtain

$$-i + I_3 + I_4 = -2\text{ mA} + I_3 + 1\text{ mA} = 0$$

Thus, we have

$$I_3 = 1\text{ mA}$$

Summing the currents leaving node 2 (KCL), we obtain

$$-I_1 - I_4 + I_5 = -I_1 - 1\text{ mA} + 2\text{ mA} = 0$$

Thus, we have

$$I_1 = 1\text{ mA}$$

Summing the currents leaving node 3 (KCL), we obtain

$$i + I_1 - I_2 = 2\text{ mA} + 1\text{ mA} - I_2 = 0$$

Thus, we have

$$I_2 = 3\text{ mA}$$

Exercise 2.9

Find I_2, V_4, V_1, V_2, V_3, I_1, and I_3 in the circuit shown in Figure 2.26.

FIGURE 2.26

Circuit for EXERCISE 2.9.

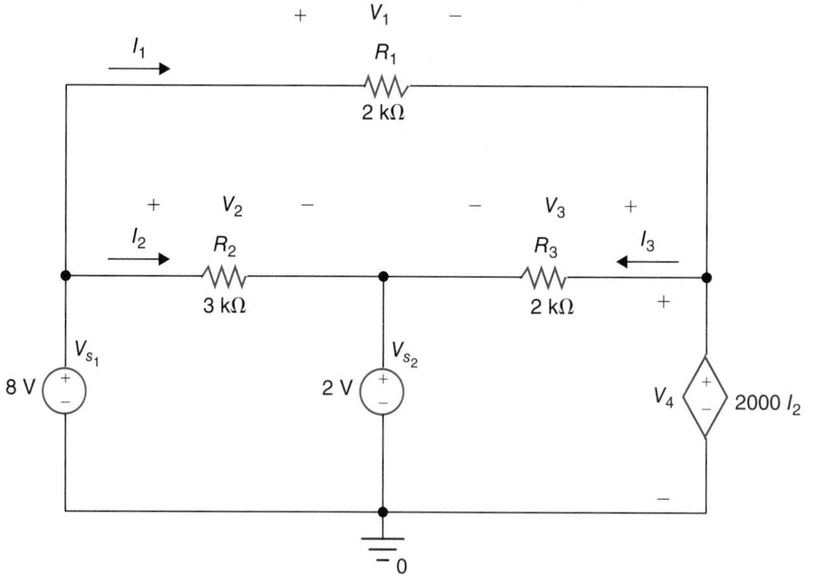

Answer:
$I_2 = 2\text{ mA}$, $V_4 = 4\text{ V}$, $V_1 = 4\text{ V}$, $V_2 = 6\text{ V}$, $V_3 = 2\text{ V}$, $I_1 = 2\text{ mA}$, $I_3 = 1\text{ mA}$.

2.7 Series and Parallel Connection of Resistors

In this section, the equivalent resistance is found when the resistors are connected in series and in parallel. Reducing the circuit with fewer resistors makes it easier to analyze the circuit. Also, the nonstandard resistor values can be obtained by combining two or more resistors.

2.7.1 SERIES CONNECTION OF RESISTORS

Consider a circuit shown in Figure 2.27(a). In this circuit, two resistors with resistances R_1 and R_2 are connected in series. Let the voltage across R_1 be V_1, the voltage across R_2 be V_2, voltage across both R_1 and R_2 be V, and the current through the resistors be I. Then, the voltage across R_1 is

$$V_1 = R_1 I$$

and the voltage across R_2 is

$$V_2 = R_2 I$$

since the current through R_1 and the current through R_2 are identical. There cannot be two different currents on one wire. Thus, we have

$$V = V_1 + V_2 = R_1 I + R_2 I = (R_1 + R_2)I = R_{eq} I$$

FIGURE 2.27

(a) Series connection of two resistors. (b) Equivalent resistor.

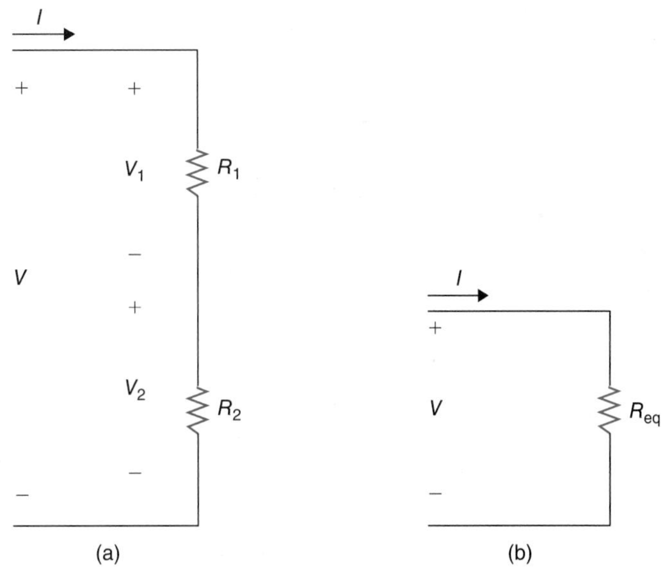

where

$$R_{eq} = R_1 + R_2$$

is the **equivalent resistance** of the series connection of R_1 and R_2. We can replace the series connection of R_1 and R_2 by a single resistor with resistance

$$R_{eq} = R_1 + R_2 \qquad \textbf{(2.33)}$$

as shown in Figure 2.27(b), without changing the voltage V and the current I.

If n resistors with resistances $R_1, R_2, \ldots, R_n$ are connected in series, as shown in Figure 2.28(a), the equivalent resistance R_{eq} is

$$R_{eq} = R_1 + R_2 + \cdots + R_n \tag{2.34}$$

Let $V_1, V_2, \ldots, V_n$ be the voltages across $R_1, R_2, \ldots, R_n$, respectively, and V be the voltage across all the resistors, and I be the current through all the resistors. Then, from Ohm's law, we have

$$V_1 = R_1 I, V_2 = R_2 I, \ldots, V_n = R_n I$$

and

$$V = V_1 + V_2 + \cdots + V_n = R_1 I + R_2 I + \cdots + R_n I$$
$$= (R_1 + R_2 + \cdots + R_n)I = R_{eq} I$$

where

$$R_{eq} = R_1 + R_2 + \cdots + R_n$$

is the **equivalent resistance** of the series connection of the n resistors. We can replace the series connection of the n resistors by a single resistor with resistance

$$R_{eq} = R_1 + R_2 + \cdots + R_n$$

as shown in Figure 2.28(b) without changing the voltage V and the current I.

FIGURE 2.28

(a) Series connection of n resistors.
(b) Equivalent resistor.

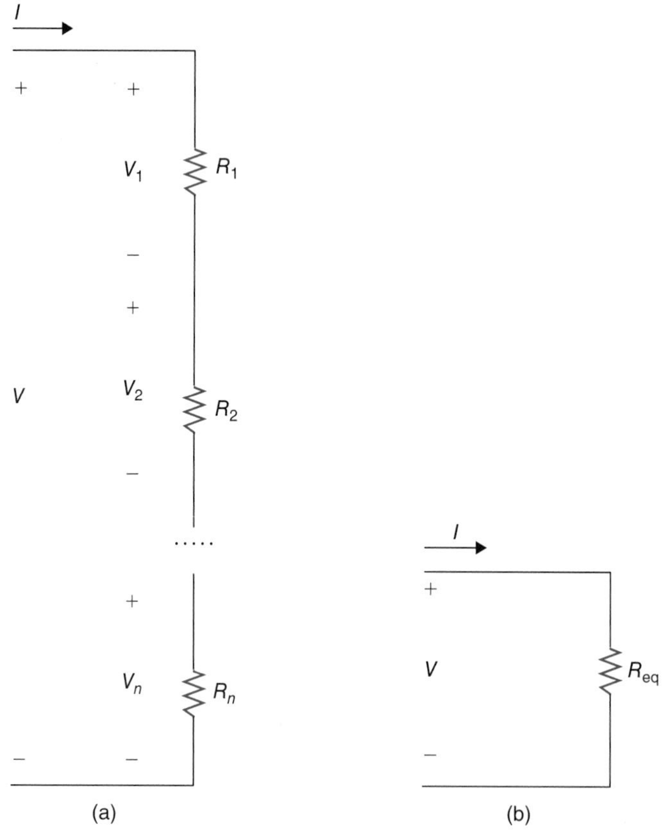

Since the conductance is the inverse of the resistance ($G = 1/R$), the equivalent conductance of two resistors connected in series is, from $R_{eq} = R_1 + R_2$,

$$\frac{1}{G_{eq}} = \frac{1}{G_1} + \frac{1}{G_2}$$

Solving for G_{eq}, we have

$$G_{eq} = \frac{1}{\dfrac{1}{G_1} + \dfrac{1}{G_2}} \tag{2.35}$$

or

$$G_{eq} = \frac{G_1 G_2}{G_1 + G_2}$$

The equivalent resistance is given by

$$R_{eq} = \frac{1}{G_1} + \frac{1}{G_2} = \frac{G_1 + G_2}{G_1 G_2} \tag{2.36}$$

In terms of R_1 and R_2, the equivalent conductance is given by

$$G_{eq} = \frac{1}{R_1 + R_2} \tag{2.37}$$

In general, when n resistors are connected in series, the equivalent conductance is given by

$$G_{eq} = \frac{1}{\dfrac{1}{G_1} + \dfrac{1}{G_2} + \cdots + \dfrac{1}{G_n}} = \frac{1}{R_1 + R_2 + \cdots + R_n} \tag{2.38}$$

Consider a circuit shown in Figure 2.29. We are interested in finding I, V_1, and V_2.

FIGURE 2.29

Circuit with series connection of resistors.

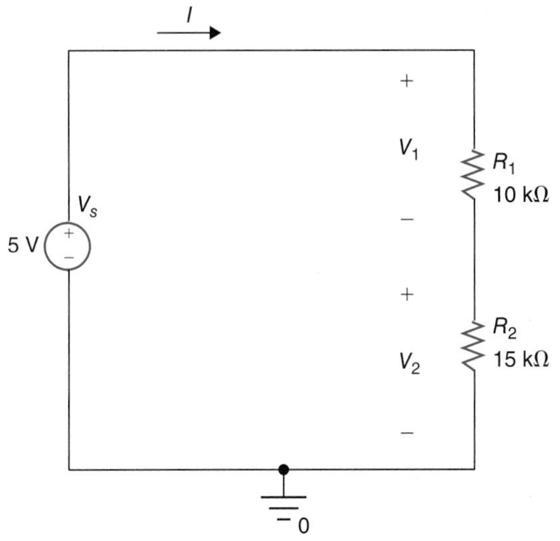

The equivalent resistance of R_1 and R_2 is given by

$$R_{eq} = R_1 + R_2 = 10\ k\Omega + 15\ k\Omega = 25\ k\Omega$$

The circuit with the equivalent resistance is shown in Figure 2.30.

The voltage across the equivalent resistance is the same as the voltage of the voltage source since they share the same nodes. The current I through the equivalent resistance is given by

$$I = \frac{V}{R_{eq}} = \frac{5\ \text{V}}{25\ k\Omega} = 0.2\ \text{mA}$$

According to Ohm's law, the voltage across R_1 is given by

$$V_1 = R_1 I = 10\ k\Omega \times 0.2\ \text{mA} = 2\ \text{V}$$

Similarly, the voltage across R_2 is given by

$$V_2 = R_2 I = 15\ k\Omega \times 0.2\ \text{mA} = 3\ \text{V}$$

FIGURE 2.30

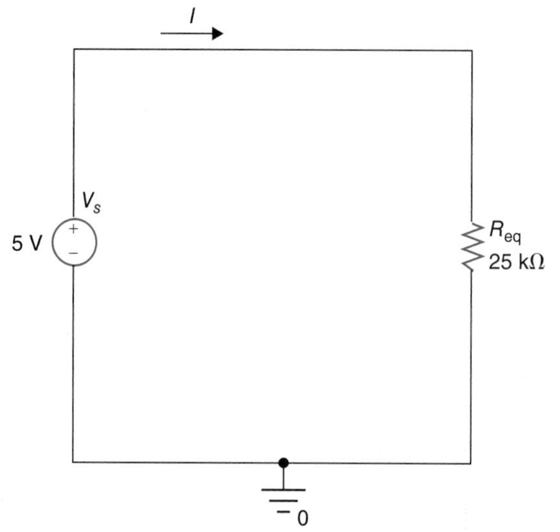

Circuit with the equivalent resistance.

EXAMPLE 2.10

Suppose that $I_3 = 750\ \mu A$ in the circuit shown in Figure 2.31. Find V_a, I_2, I_4, I_1, and V_s.

FIGURE 2.31

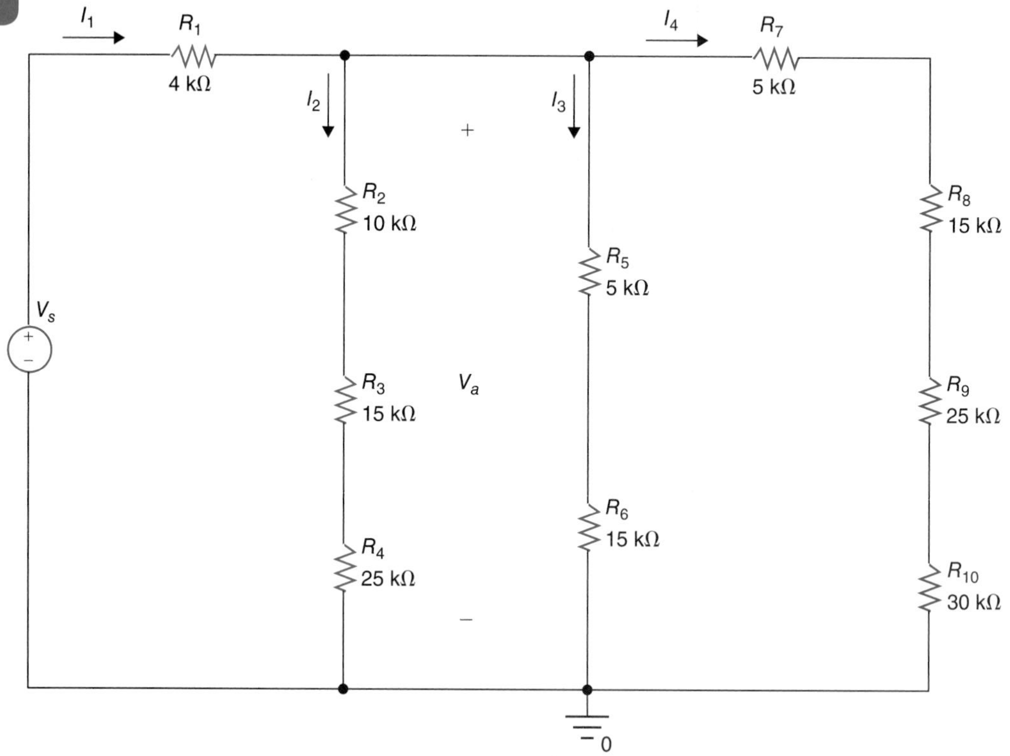

Circuit for EXAMPLE 2.10.

The equivalent resistance of the series connection of R_2, R_3, and R_4 is given by

$$R_a = R_2 + R_3 + R_4 = 50\ k\Omega$$

continued

Example 2.10 continued

The equivalent resistance of the series connection of R_5 and R_6 is given by

$$R_b = R_5 + R_6 = 20\ k\Omega$$

The equivalent resistance of the series connection of R_7, R_8, R_9, and R_{10} is given by

$$R_c = R_7 + R_8 + R_9 + R_{10} = 75\ k\Omega$$

The circuit shown in Figure 2.32 is equivalent to the circuit shown in Figure 2.31.

FIGURE 2.32

Circuit with equivalent resistances.

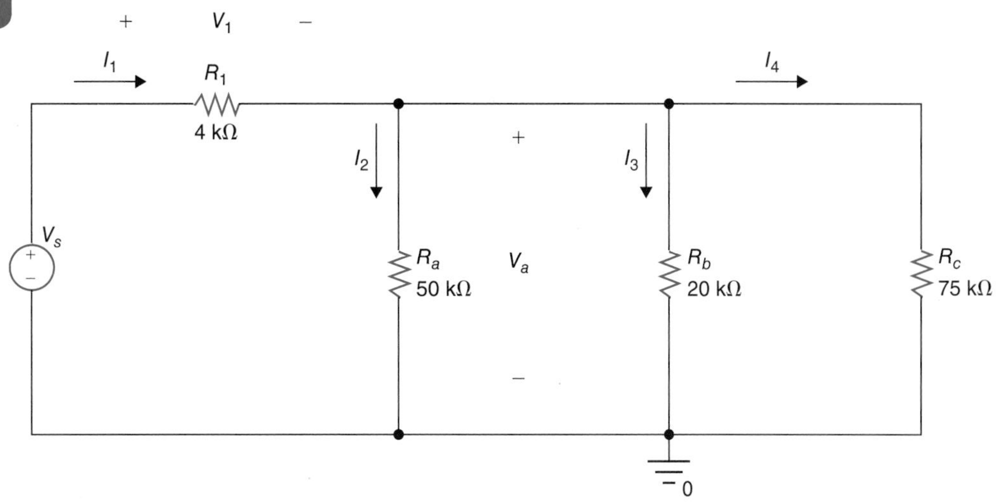

According to Ohm's law, the voltage across R_b is given by

$$V_a = R_b I_3 = 20\ k\Omega \times 0.75\ \text{mA} = 15\ \text{V}$$

The voltage across R_a and R_c is also V_a. From Ohm's law, the current I_2 is given by

$$I_2 = \frac{V_a}{R_a} = \frac{15\ \text{V}}{50\ k\Omega} = 300\ \mu\text{A}$$

Similarly, the current through R_c is given by

$$I_4 = \frac{V_a}{R_c} = \frac{15\ \text{V}}{75\ k\Omega} = 200\ \mu\text{A}$$

From KCL, we have

$$I_1 = I_2 + I_3 + I_4 = 300\ \mu\text{A} + 750\ \mu\text{A} + 200\ \mu\text{A} = 1250\ \mu\text{A} = 1.25\ \text{mA}$$

The voltage across R_1 is

$$V_1 = R_1 I_1 = 4\ k\Omega \times 1.25\ \text{mA} = 5\ \text{V}$$

From KVL, we have

$$V_s = V_1 + V_a = 5\ \text{V} + 15\ \text{V} = 20\ \text{V}$$

Exercise 2.10

Find I, V_1, V_2, V_3, and V_4 in the circuit shown in Figure 2.33.

FIGURE 2.33

Circuit for EXERCISE 2.10

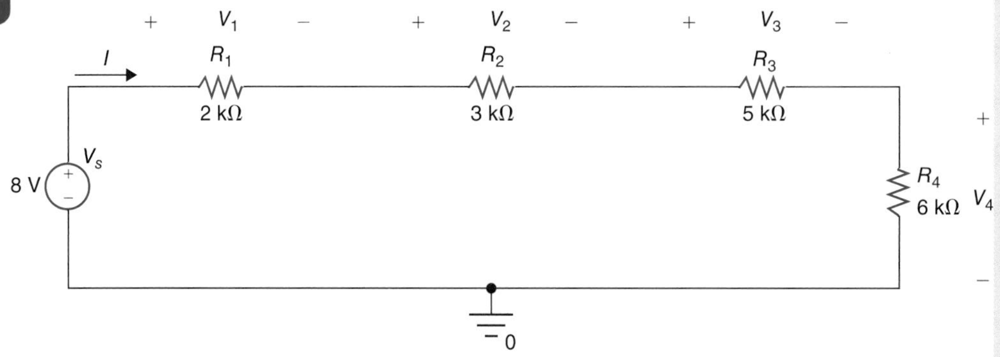

Answer:
$I = 0.5$ mA, $V_1 = 1$ V, $V_2 = 1.5$ V, $V_3 = 2.5$ V, $V_4 = 3$ V.

2.7.2 PARALLEL CONNECTION OF RESISTORS

Consider a parallel connection of two resistors with resistances R_1 and R_2, respectively, as shown in Figure 2.34(a).

FIGURE 2.34

(a) Parallel connection of two resistors. (b) Equivalent resistance.

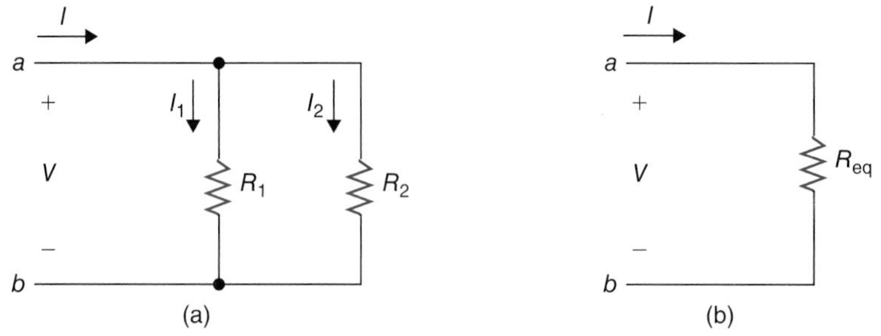

Let the voltage across the resistors be V, the current through the resistor R_1 be I_1, the current through the resistor R_2 be I_2, the current from a to b be I. Then, from Ohm's law, the current through R_1 is given by

$$I_1 = \frac{V}{R_1}$$

and the current through R_2 is given by

$$I_2 = \frac{V}{R_2}$$

Since the resistors are connected between the same points, the voltage across R_1 is identical to the voltage across R_2. The current I is given by the sum of I_1 and I_2:

$$I = I_1 + I_2 = \frac{V}{R_1} + \frac{V}{R_2} = \left(\frac{1}{R_1} + \frac{1}{R_2}\right)V = \frac{V}{R_{eq}} \tag{2.39}$$

where R_{eq} is the equivalent resistance of R_1 and R_2. From the last two terms of the Equation (2.39), we obtain

$$\frac{1}{R_{eq}} = \frac{1}{R_1} + \frac{1}{R_2}$$

or

$$R_{eq} = \frac{1}{\frac{1}{R_1} + \frac{1}{R_2}} = \frac{R_1 R_2}{R_1 + R_2} \quad \textbf{(2.40)}$$

Thus, the equivalent resistance of R_1 connected in parallel to R_2 is given by

$$R_{eq} = \frac{R_1 R_2}{R_1 + R_2}$$

We can replace the parallel connection of R_1 and R_2 by a single resistor with resistance

$$R_{eq} = \frac{R_1 R_2}{R_1 + R_2}$$

as shown in Figure 2.34(b) without changing the voltage V and the current I. The equivalent resistance of two resistors in parallel is denoted by $R_1 \| R_2$. Thus, we have

$$R_{eq} = R_1 \| R_2 = \frac{1}{\frac{1}{R_1} + \frac{1}{R_2}} = \frac{R_1 R_2}{R_1 + R_2}$$

Notice that $\frac{R_2}{R_1 + R_2} < 1$. Multiplication of R_1 on both sides yields $\frac{R_1 R_2}{R_1 + R_2} < R_1$. Thus, we have $R_{eq} < R_1$. Similarly, multiplication of R_2 on both sides of $\frac{R_1}{R_1 + R_2} < 1$ yields $\frac{R_1 R_2}{R_1 + R_2} < R_2$. Thus, we have $R_{eq} < R_2$. This result shows that the equivalent resistance R_{eq} of the parallel connection of two resistors R_1 and R_2 is smaller than R_1 and smaller than R_2.

If $R_1 = 0$ and $R_2 \neq 0$, $R_{eq} = R_1 \| R_2 = 0$. If a resistor is connected in parallel with a short circuit, all currents flow through the short circuit.

If $R_1 < \infty$ and $R_2 = \infty$, $R_{eq} = R_1 \| R_2 = R_1$.

If $R_1 \ll R_2$, $R_{eq} = R_1 \| R_2 \cong R_1$. If R_2 is significantly greater than R_1, the equivalent resistance R_{eq} of the parallel connection of two resistors R_1 and R_2 is close to R_1 (slightly smaller). For example, the equivalent resistance R_{eq} of two resistors $R_1 = 1\ k\Omega$ and $R_2 = 1\ \text{M}\Omega$ is $999.001\ \Omega = 0.999001\ k\Omega$. If we take $R_1 = 1\ k\Omega$ as the equivalent resistance, the error is 0.1%, which is within the tolerance of most resistors.

If $R_1 < R_2$, more currents flow through R_1 than R_2 ($V/R_1 > V/R_2$).

If n resistors with resistances $R_1, R_2, \ldots, R_n$, respectively, are connected in parallel, as shown in Figure 2.35(a), the equivalent resistance R_{eq} satisfies the equation

$$\frac{1}{R_{eq}} = \frac{1}{R_1} + \frac{1}{R_2} + \cdots + \frac{1}{R_n}$$

FIGURE 2.35

(a) Parallel connection of n resistors. (b) Equivalent resistance.

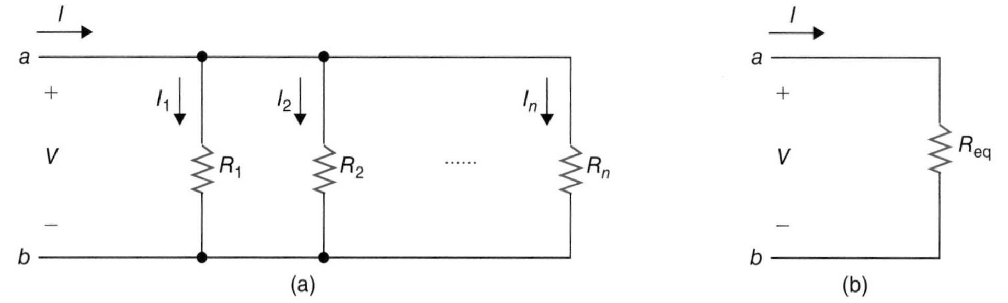

from which we have

$$R_{eq} = \frac{1}{\dfrac{1}{R_1} + \dfrac{1}{R_2} + \cdots + \dfrac{1}{R_n}} \quad \textbf{(2.41)}$$

The proof of this result is similar to the one for two resistors. Let the current through R_i, $i = 1, 2, \ldots, n$, be I_i, the voltage across the resistors be V, and the current from a to b be I. Then, from Ohm's law, we have

$$I_1 = \frac{V}{R_1}, I_2 = \frac{V}{R_2}, \ldots, I_n = \frac{V}{R_n}$$

$$I = I_1 + I_2 + \cdots + I_n = \frac{V}{R_1} + \frac{V}{R_2} + \cdots + \frac{V}{R_n}$$

$$= \left(\frac{1}{R_1} + \frac{1}{R_2} + \cdots + \frac{1}{R_n}\right) V = \frac{V}{R_{eq}}$$

Thus,

$$\frac{1}{R_{eq}} = \frac{1}{R_1} + \frac{1}{R_2} + \cdots + \frac{1}{R_n}$$

from which we obtain Equation (2.41). The equivalent circuit is shown in Figure 2.35(b).

Since conductance is the inverse of resistance ($G = 1/R$), the equivalent conductance of two resistors connected in parallel is, from $\dfrac{1}{R_{eq}} = \dfrac{1}{R_1} + \dfrac{1}{R_2}$,

$$G_{eq} = G_1 + G_2 \quad \textbf{(2.42)}$$

The equivalent resistance R_{eq} is given by

$$R_{eq} = \frac{1}{G_{eq}} = \frac{1}{G_1 + G_2} \quad \textbf{(2.43)}$$

In general, when n resistors are connected in parallel, the equivalent conductance is given by

$$G_{eq} = G_1 + G_2 + \cdots + G_n \quad \textbf{(2.44)}$$

The equivalent resistance is given by

$$R_{eq} = \frac{1}{G_{eq}} = \frac{1}{G_1 + G_2 + \cdots + G_n} \quad \textbf{(2.45)}$$

Consider a circuit shown in Figure 2.36. We are interested in finding I_1, V_1, V_2, I_2, I_3, V_3, I_4, I_5, and I_6.

FIGURE 2.36

Circuit to be analyzed.

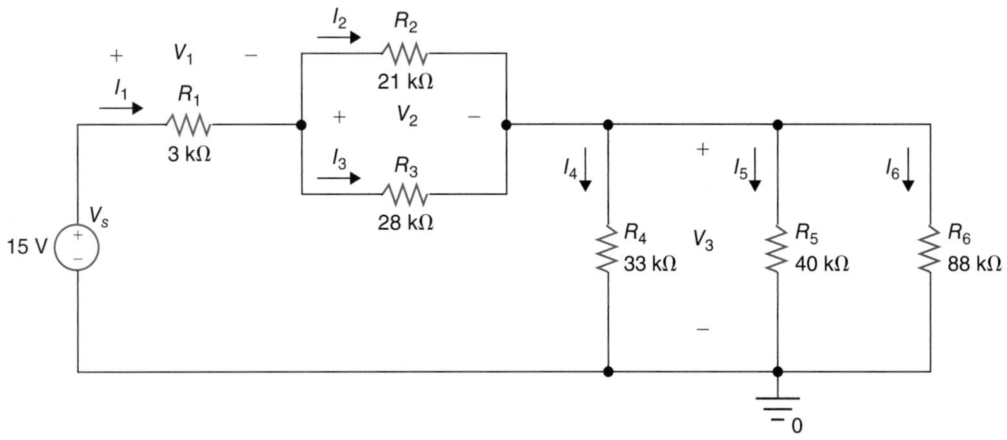

The equivalent resistance of the parallel connection of R_2 and R_3 is given by

$$R_a = R_2 \| R_3 = \frac{21\ k\Omega \times 28\ k\Omega}{21\ k\Omega + 28\ k\Omega} = \frac{588}{49}\ k\Omega = 12\ k\Omega$$

The equivalent resistance of the parallel connection of R_4, R_5, and R_6 is given by

$$R_b = R_4 \| R_5 \| R_6 = \frac{1}{\frac{1}{33} + \frac{1}{40} + \frac{1}{88}}\ k\Omega = \frac{1}{0.0666667}\ k\Omega = 15\ k\Omega$$

The circuit shown in Figure 2.36 can be redrawn using R_1, R_a, and R_b, as shown in Figure 2.37.

Resistors R_1, R_a, and R_b are connected in series. The total resistance seen from the voltage source is

$$R_{eq} = R_1 + R_a + R_b = 3\ k\Omega + 12\ k\Omega + 15\ k\Omega = 30\ k\Omega$$

FIGURE 2.37

Equivalent circuit with R_1, R_a, R_b.

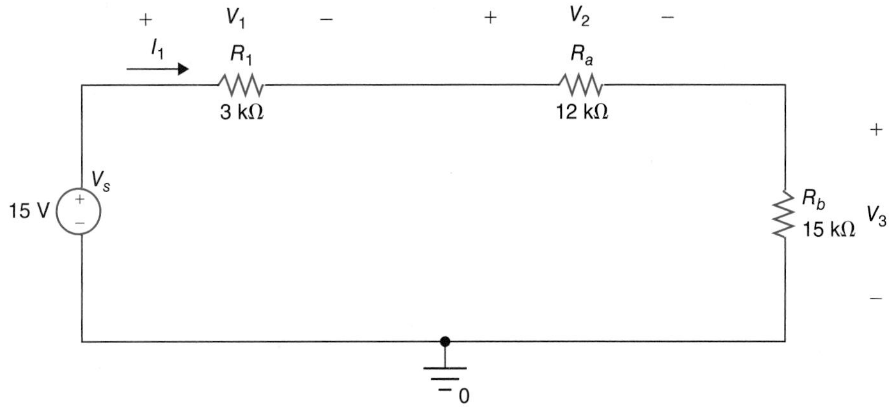

The current I_1 flowing out of the positive terminal of the voltage source is given by

$$I_1 = \frac{V_s}{R_{eq}} = \frac{15 \text{ V}}{30 \ k\Omega} = 0.5 \text{ mA}$$

Notice that the current through R_a and R_b is also I_1. The voltage drop across R_1 is given by

$$V_1 = R_1 I_1 = 3 \ k\Omega \times 0.5 \text{ mA} = 1.5 \text{ V}$$

The voltage drop across R_a (also across R_2 and R_3) is given by

$$V_2 = R_a I_1 = 12 \ k\Omega \times 0.5 \text{ mA} = 6 \text{ V}$$

Similarly, the voltage drop across R_b (also across R_4, R_5, and R_6) is given by

$$V_3 = R_b I_1 = 15 \ k\Omega \times 0.5 \text{ mA} = 7.5 \text{ V}$$

From Ohm's law, the currents I_2, I_3, I_4, I_5, and I_6 are given, respectively, by

$$I_2 = \frac{V_2}{R_2} = \frac{6 \text{ V}}{21 \ k\Omega} = 0.2857 \text{ mA}$$

$$I_3 = \frac{V_2}{R_3} = \frac{6 \text{ V}}{28 \ k\Omega} = 0.2143 \text{ mA}$$

$$I_4 = \frac{V_3}{R_4} = \frac{7.5 \text{ V}}{33 \ k\Omega} = 0.2273 \text{ mA}$$

$$I_5 = \frac{V_3}{R_5} = \frac{7.5 \text{ V}}{40 \ k\Omega} = 0.1875 \text{ mA}$$

$$I_6 = \frac{V_3}{R_6} = \frac{7.5 \text{ V}}{88 \ k\Omega} = 0.08523 \text{ mA}$$

Notice that $I_2 + I_3 = I_1 = 0.5$ mA and $I_4 + I_5 + I_6 = I_1 = 0.5$ mA.

EXAMPLE 2.11

Find the equivalent resistance between terminals *a* and *b* for the circuit shown in Figure 2.38.

FIGURE 2.38

Circuit for EXAMPLE 2.11.

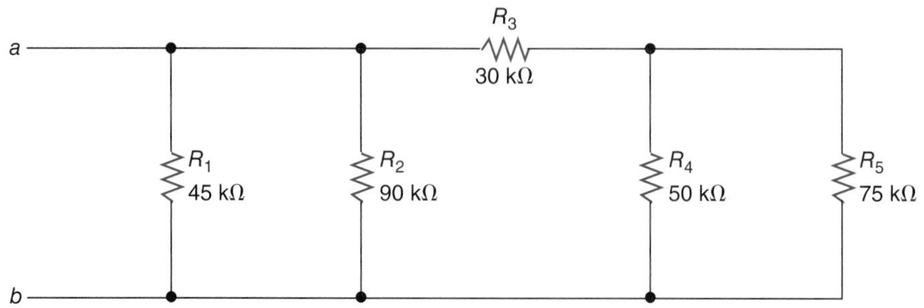

continued

Example 2.11 continued

Let R_6 be the equivalent resistance of the parallel connection of R_4 and R_5. Then, from Equation (2.40), we have

$$R_6 = \frac{R_4 R_5}{R_4 + R_5} = \frac{50\,k\Omega \times 75\,k\Omega}{50\,k\Omega + 75\,k\Omega} = \frac{3750}{125}\,k\Omega = 30\,k\Omega$$

Let R_7 be the equivalent resistance of the series connection of R_3 and R_6. Then, from Equation (2.33), we have

$$R_7 = R_3 + R_6 = 30\,k\Omega + 30\,k\Omega = 60\,k\Omega$$

The equivalent resistance R_{eq} between a and b is the equivalent resistance of the parallel connection of R_1, R_2, and R_7. Application of Equation (2.41) yields

$$R_{eq} = \frac{1}{\frac{1}{R_1} + \frac{1}{R_2} + \frac{1}{R_7}} = \frac{1}{\frac{1}{45\,k\Omega} + \frac{1}{90\,k\Omega} + \frac{1}{60\,k\Omega}}$$

$$= \frac{180}{\frac{180}{45} + \frac{180}{90} + \frac{180}{60}}\,k\Omega = \frac{180}{9}\,k\Omega = 20\,k\Omega$$

MATLAB® can be used to facilitate the calculation of the equivalent resistance. Create a function named P given here, and save the file in the current folder. The function P calculates the equivalent resistance of resistors connected in parallel.

```
function [Req] = P(x)
% Equivalent Resistance of Parallel Connection.
% Req = 1/(1/R1 + 1/R2 + ... + 1/Rn).
Req=1/sum(1./x);
end
```

The input is a vector of resistance values. As an example, we can find the equivalent resistance of parallel connection of 3 $k\Omega$ resistor and 6 $k\Omega$ resistor by calling the function P:

```
>>Req=P([3000,6000])
Req =
        2000
```

The answer is $3 \times 6/(3 + 6)\,k\Omega = 18/9\,k\Omega = 2\,k\Omega$. The equivalent resistance for the circuit shown in Figure 2.38 can be found using MATLAB:

```
>>R1=45000;R2=90000;R3=30000;R4=50000;R5=75000;
>> Req=P([R1,R2,R3+P([R4,R5])])
Req =
        20000
```

Exercise 2.11

Find the equivalent resistance between terminals *a* and *b* for the circuit shown in Figure 2.39.

FIGURE 2.39

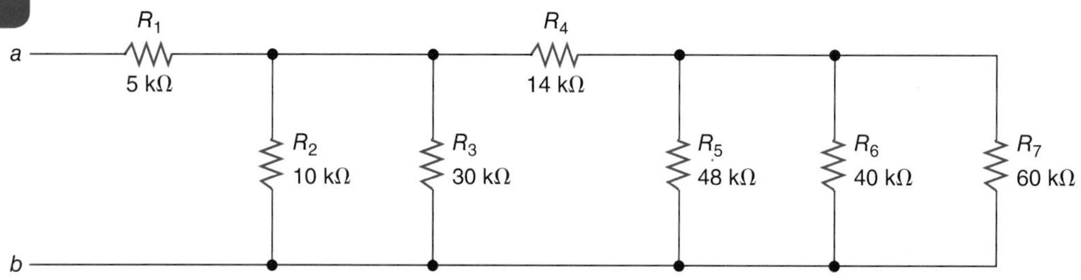

Circuit for EXERCISE 2.11.

Answer:

$R_{eq} = 11\ k\Omega.$

```
>> R1=5000;R2=10000;R3=30000;R4=14000;R5=48000;R6=40000;R7=60000;
>> Req=R1+P([R2,R3,R4+P([R5,R6,R7])])
Req =
        11000
```

EXAMPLE 2.12

For the circuit shown in Figure 2.40, find the equivalent resistance of R_1, R_2, R_3, and R_4. Also, find the currents I, I_1, and I_2, and voltages across R_1, R_2, R_3, and R_4 and powers absorbed by R_1, R_2, R_3, and R_4.

FIGURE 2.40

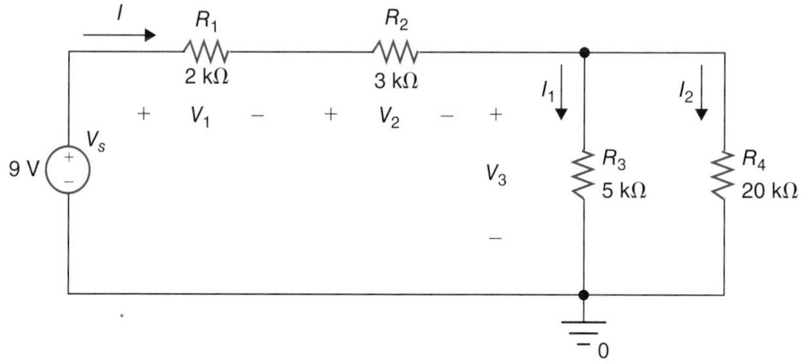

Circuit for EXAMPLE 2.12.

The equivalent resistance of R_1, R_2, R_3, and R_4 is given by

$$R_{eq} = R_1 + R_2 + \frac{R_3 R_4}{R_3 + R_4} = 2\ k\Omega + 3\ k\Omega + \frac{5\ k\Omega \times 20\ k\Omega}{5\ k\Omega + 20\ k\Omega} = 9\ k\Omega$$

With the equivalent resistance, the circuit shown in Figure 2.40 can be redrawn as the one shown in Figure 2.41.

continued

Example 2.12 continued

FIGURE 2.41

Circuit with the equivalent resistance.

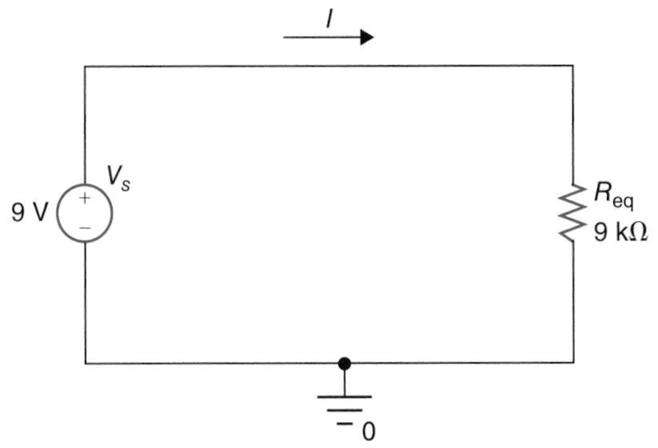

The current I from the positive terminal of the voltage source into the equivalent resistor is

$$I = \frac{V_s}{R_{eq}} = \frac{9\ \text{V}}{9\ k\Omega} = 1\,\text{mA}$$

From Ohm's law, the voltage drop across R_1 is given by

$$V_1 = IR_1 = 1\ \text{mA} \times 2\ k\Omega = 2\ \text{V}$$

The polarity of V_1 is shown in Figure 2.40. Similarly, the voltage drop across R_2 is given by

$$V_2 = IR_2 = 1\ \text{mA} \times 3\ k\Omega = 3\ \text{V}$$

From KVL, the sum of voltage drops around the circuit is zero. Starting from the negative terminal of the voltage source, the sum of voltage drops is

$$-9 + V_1 + V_2 + V_3 = -9 + 2 + 3 + V_3 = 0$$

Thus, $V_3 = 4$ V. This result can also be obtained by simply subtracting V_1 and V_2 from V_s; that is,

$$V_3 = 9 - V_1 - V_2 = 9 - 2 - 3 = 4\ \text{V}$$

From Ohm's law, the current through R_3 is given by

$$I_1 = \frac{V_3}{R_3} = \frac{4\ \text{V}}{5\ k\Omega} = 0.8\ \text{mA}$$

Similarly, the current through R_4 is given by

$$I_2 = \frac{V_3}{R_4} = \frac{4\ \text{V}}{20\ k\Omega} = 0.2\ \text{mA}$$

Notice that I_2 can also be found from KCL. Since the sum of the currents leaving a node (connecting R_2, R_3, and R_4) is zero, we have

$$-1\ \text{mA} + 0.8\ \text{mA} + I_2 = 0$$

or $I_2 = 0.2$ mA. The power absorbed by R_1 is found to be

$$P_1 = I^2R_1 = (1\ \text{mA})^2 \times 2\ k\Omega = 2\ \text{mW}$$

Similarly, the power absorbed by R_2, R_3, and R_4 is given, respectively, by

$$P_2 = I^2R_2 = (1\ \text{mA})^2 \times 3\ k\Omega = 3\ \text{mW}$$

$$P_3 = I_1^2R_3 = (0.8\ \text{mA})^2 \times 5\ k\Omega = 3.2\ \text{mW}$$

$$P_4 = I_2^2R_4 = (0.2\ \text{mA})^2 \times 20\ k\Omega = 0.8\ \text{mW}$$

continued

Example 2.12 continued

The sum of powers absorbed by the four resistors R_1, R_2, R_3, and R_4 is given by

$$P_1 + P_2 + P_3 + P_4 = 2\text{ mW} + 3\text{ mW} + 3.2\text{ mW} + 0.8\text{ mW} = 9\text{ mW}$$

Notice that the power delivered by the voltage source is

$$P_s = -IV_s = (-1\text{ mA}) \times 9\text{ V} = -9\text{ mW}$$

The current through the voltage source V_s is $-I$, which is the current from the positive terminal of the voltage source, through the voltage source, to the negative terminal of the voltage source. The negative power implies that the voltage source delivers power to the rest of the circuit. The sum of all the powers on the circuit is

$$P_s + P_1 + P_2 + P_3 + P_4 = 0 \quad (2.46)$$

This result is called *conservation of power.* Equation (2.46) can be rewritten as

$$-P_s = P_1 + P_2 + P_3 + P_4 \quad (2.47)$$

The power on the left side is the power supplied by the source, and the power on the right side is the power absorbed by all the resistors. In general, the power supplied equals the power absorbed in a circuit. This is another description of the conservation of power.

Instead of using $P = I^2R$ to calculate the power absorbed by each resistor, we can use $P = IV$ or $P = \dfrac{V^2}{R}$. The result will be identical to those given previously.

Exercise 2.12

For the circuit shown in Figure 2.42, find the equivalent resistance R_{eq} seen from the current source. Also, find the currents I_1, I_2, I_3, and I_4, voltages V_1 and V_2, voltage across R_1, and powers absorbed or supplied by R_1, R_2, R_3, R_4, and I_s.

FIGURE 2.42

Circuit for EXERCISE 2.12.

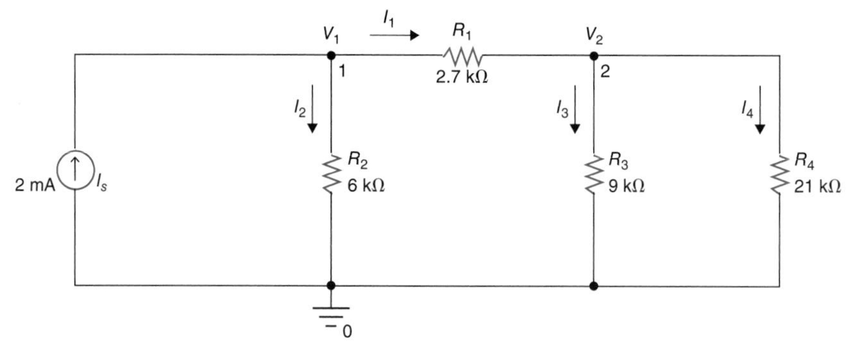

Answer:

$R_{eq} = 3.6\ k\Omega$, $I_1 = 0.8$ mA, $I_2 = 1.2$ mA, $I_3 = 0.56$ mA, $I_4 = 0.24$ mA, $V_1 = 7.2$ V, $V_2 = 5.04$ V, $V_{R_1} = 2.16$ V, $P_{R_1} = 1.728$ mW, $P_{R_2} = 8.64$ mW, $P_{R_3} = 2.8224$ mW, $P_{R_4} = 1.2096$ mW, $P_{I_s} = -14.4$ mW, $P_{R_1} + P_{R_2} + P_{R_3} + P_{R_4} = 14.4$ mW.

EXAMPLE 2.13

For the circuit shown in Figure 2.43, find the equivalent resistance R_{eq} seen from the voltage source. Also, find the currents I_1, I_2, I_3, I_4, and I_5, voltages V_1 and V_2 and voltages across R_1, R_2, R_3, R_4, and R_5 and powers absorbed or supplied by R_1, R_2, R_3, R_4, R_5, and V_s.

FIGURE 2.43

Circuit for EXAMPLE 2.13.

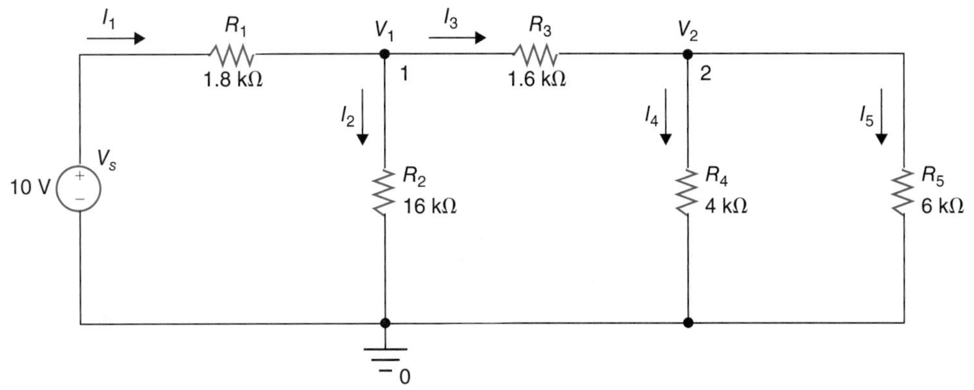

Let R_a be the equivalent resistance of the parallel connection of R_4 and R_5. Then, we have

$$R_a = R_4 \| R_5 = \frac{R_4 \times R_5}{R_4 + R_5} = \frac{4\,k\Omega \times 6\,k\Omega}{4\,k\Omega + 6\,k\Omega} = \frac{4 \times 6}{4 + 6} k\Omega = \frac{24}{10} k\Omega = 2.4\,k\Omega$$

Let R_b be the sum of R_3 and R_a. Then, we have

$$R_b = R_3 + R_a = 1.6\,k\Omega + 2.4\,k\Omega = 4\,k\Omega$$

Let R_c be the equivalent resistance of the parallel connection of R_2 and R_b. Then, we have

$$R_c = R_2 \| R_b = \frac{R_2 \times R_b}{R_2 + R_b} = \frac{16\,k\Omega \times 4\,k\Omega}{16\,k\Omega + 4\,k\Omega} = \frac{16 \times 4}{16 + 4} k\Omega = \frac{64}{20} k\Omega = 3.2\,k\Omega$$

FIGURE 2.44

The circuit with equivalent resistance.

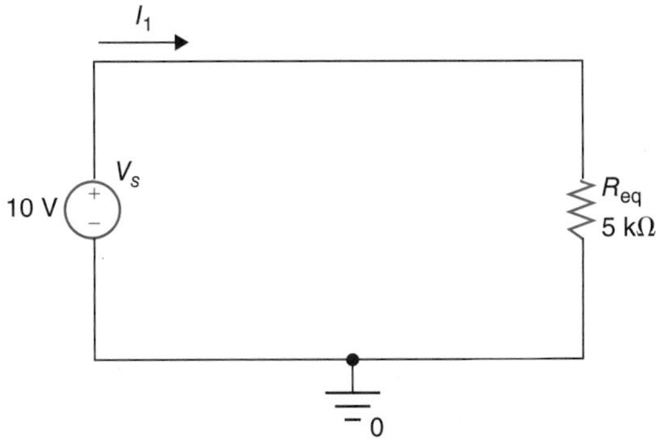

The equivalent resistance R_{eq} is the sum of R_1 and R_c. Thus, we get

$$R_{eq} = R_1 + R_c = 1.8\,k\Omega + 3.2\,k\Omega = 5\,k\Omega$$

Figure 2.44 shows the circuit with the voltage source and the equivalent resistance.

The current I_1, flowing out of the positive terminal of the voltage source, is given by

$$I_1 = \frac{V_s}{R_{eq}} = \frac{10\text{ V}}{5\,k\Omega} = 2\text{ mA}$$

The voltage drop across R_1 is given by (from Ohm's law)

$$V_{R_1} = R_1 I_1 = 1.8\,k\Omega \times 2\text{ mA} = 1.8 \times 10^3 \times 2 \times 10^{-3}\text{ V} = 3.6\text{ V}$$

continued

Example 2.13 continued

The voltage V_1 at node 1 is obtained by subtracting V_{R_1} from V_s:

$$V_1 = V_s - V_{R_1} = 10\text{ V} - 3.6\text{ V} = 6.4\text{ V}$$

The voltage across R_2 is V_1. Thus,

$$V_{R_2} = V_1 = 6.4\text{ V}$$

The current I_2 through R_2 is given by

$$I_2 = \frac{V_1}{R_2} = \frac{6.4\text{ V}}{16\ k\Omega} = 0.4\text{ mA}$$

Application of KCL at node 1 results in

$$I_3 = I_1 - I_2 = 2\text{ mA} - 0.4\text{ mA} = 1.6\text{ mA}$$

The voltage drop across R_3 is given by (from Ohm's law)

$$V_{R_3} = R_3 I_3 = 1.6\ k\Omega \times 1.6\text{ mA} = 1.6 \times 10^3 \times 1.6 \times 10^{-3}\text{ V} = 2.56\text{ V}$$

The voltage V_2 at node 2 is obtained by subtracting V_{R_3} from V_1:

$$V_2 = V_1 - V_{R_3} = 6.4\text{ V} - 2.56\text{ V} = 3.84\text{ V}$$

The voltage across R_4 and R_5 is V_2. Thus,

$$V_{R_4} = V_2 = 3.84\text{ V}$$

$$V_{R_5} = V_2 = 3.84\text{ V}$$

The current I_4 through R_4 is given by

$$I_4 = \frac{V_2}{R_4} = \frac{3.84\text{ V}}{4\ k\Omega} = 0.96\text{ mA}$$

The current I_5 through R_5 is given by

$$I_5 = \frac{V_2}{R_5} = \frac{3.84\text{ V}}{6\ k\Omega} = 0.64\text{ mA}$$

Notice that the sum of I_2, I_4, and I_5 is 2 mA, confirming KCL at node 0. The powers at the resistors and V_s are:

$$P_{R_1} = I_1 \times V_{R_1} = 2\text{ mA} \times 3.6\text{ V} = 7.2\text{ mW}$$

$$P_{R_2} = I_2 \times V_{R_2} = 0.4\text{ mA} \times 6.4\text{ V} = 2.56\text{ mW}$$

$$P_{R_3} = I_3 \times V_{R_3} = 2\text{ mA} \times 2.56\text{ V} = 4.096\text{ mW}$$

$$P_{R_4} = I_4 \times V_{R_4} = 0.96\text{ mA} \times 3.84\text{ V} = 3.6864\text{ mW}$$

continued

Example 2.13 continued

$$P_{R_5} = I_5 \times V_{R_5} = 0.64 \text{ mA} \times 3.84 \text{ V} = 2.4576 \text{ mW}$$

$$P_{Vs} = -I_1 \times V_s = -2 \text{ mA} \times 10 \text{ V} = -20 \text{ mW}$$

The power from the voltage source is negative, indicating that power is delivered from the voltage source. The total absorbed power (20 mW) by five resistors equals the power delivered (−20 W) by the voltage source.

MATLAB

```
%EXAMPLE 2.13
%Function P.m should be in the same folder as this file.
clear all;format long;
Vs=10;
R1=1800;R2=16000;R3=1600;R4=4000;R5=6000;
Ra=P([R4,R5])
Rb=R3+Ra
Rc=P([R2,Rb])
Req=R1+Rc
I1=Vs/Req
VR1=R1*I1
V1=Vs-VR1
I2=V1/R2
I3=I1-I2
VR3=R3*I3
V2=V1-VR3
I4=V2/R4
I5=V2/R5
PR1=I1^2*R1
PR2=I2^2*R2
PR3=I3^2*R3
PR4=I4^2*R4
PR5=I5^2*R5
PVs=-I1*Vs
PSum=PR1+PR2+PR3+PR4+PR5+PVs

Answers:
Ra =
     2400
Rb =
     4000
Rc =
     3200
Req =
       5000
I1 =
   0.002000000000000
VR1 =
   3.600000000000000
V1 =
   6.400000000000000
I2 =
     4.000000000000000e-04
I3 =
   0.001600000000000
```

continued

Example 2.13 continued

MATLAB continued

```
VR3 =
   2.560000000000000
V2 =
   3.840000000000000
I4 =
     9.600000000000000e-04
I5 =
     6.400000000000001e-04
PR1 =
   0.007200000000000
PR2 =
   0.002560000000000
PR3 =
   0.004096000000000
PR4 =
   0.003686400000000
PR5 =
   0.002457600000000
PVs =
  -0.020000000000000
PSum =
     0
```

Exercise 2.13

For the circuit shown in Figure 2.45, find the equivalent resistance R_{eq} seen from the voltage source. Also, find the currents I_1, I_2, I_3, I_4, and I_5, voltages V_1 and V_2 and voltages across R_1 and R_3, and powers absorbed or supplied by R_1, R_2, R_3, R_4, R_5, and V_s.

FIGURE 2.45

Circuit for EXERCISE 2.13.

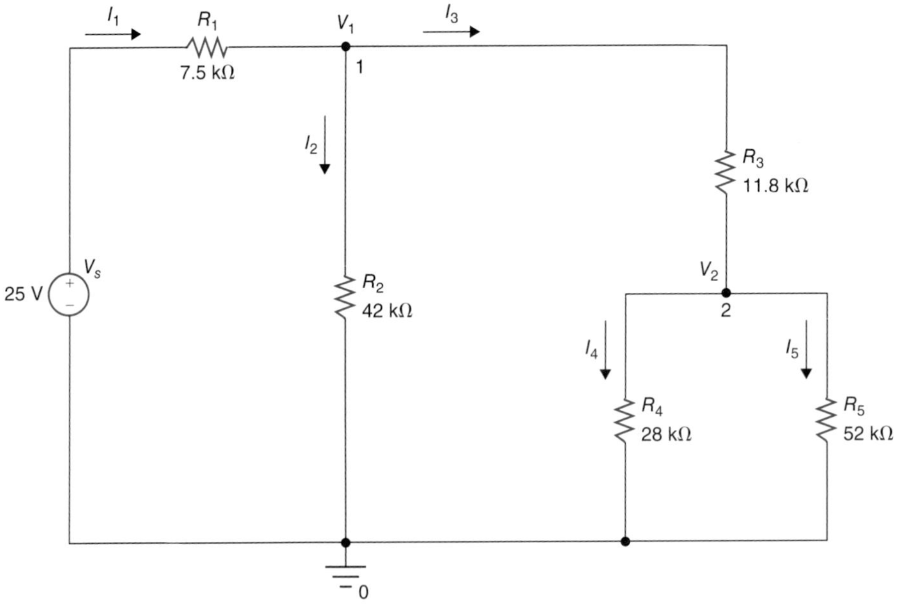

continued

Exercise 2.13 continued

Answer:
$R_{eq} = 25\ k\Omega, I_1 = 1$ mA, $I_2 = 0.416667$ mA, $I_3 = 0.583333$ mA, $I_4 = 0.37916667$ mA, $I_5 = 0.20416667$ mA, $V_1 = 17.5$ V, $V_2 = 10.6167$ V, $V_{R_1} = 7.5$ V, $V_{R_3} = 6.8833$ V, $P_{R_1} = 7.5$ mW, $P_{R_2} = 7.2916667$ mW, $P_{R_3} = 4.01528$ mW, $P_{R_4} = 4.0254861$ mW, $P_{R_5} = 2.167569$, $P_{Vs} = -25$ mW.

EXAMPLE 2.14

For the circuit shown in Figure 2.46, find the equivalent resistance R_{eq} seen from the current source. Also, find the currents I_a, I_1, I_2, I_3, I_4, I_5, I_6, and I_7, and voltages V_1, V_2, and V_3.

FIGURE 2.46

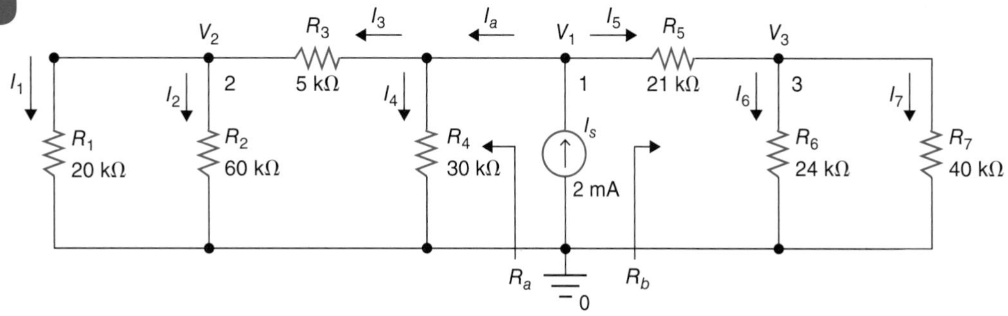

Circuit for EXAMPLE 2.14.

Let R_a be the equivalent resistance to the left of the current source, and R_b be the equivalent resistance to the right of the current source. Let R_8 be the equivalent resistance of the parallel connection of R_1 and R_2. Then, we have

$$R_8 = \frac{R_1 \times R_2}{R_1 + R_2} = \frac{20\ k\Omega \times 60\ k\Omega}{20\ k\Omega + 60\ k\Omega} = 15\ k\Omega$$

Let R_9 be the sum of R_3 and R_8. Then we have

$$R_9 = R_3 + R_8 = 5\ k\Omega + 15\ k\Omega = 20\ k\Omega$$

The equivalent resistance R_a is the parallel connection of R_9 and R_4. Thus, we get

$$R_a = \frac{R_9 \times R_4}{R_9 + R_4} = \frac{20\ k\Omega \times 30\ k\Omega}{20\ k\Omega + 30\ k\Omega} = 12\ k\Omega$$

The equivalent resistance R_b is given by

$$R_b = R_5 + (R_6 \| R_7) = R_5 + \frac{R_6 \times R_7}{R_6 + R_7}$$
$$= 21\ k\Omega + \frac{24\ k\Omega \times 40\ k\Omega}{24\ k\Omega + 40\ k\Omega} = 21\ k\Omega + 15\ k\Omega = 36\ k\Omega$$

continued

Example 2.14 continued

The equivalent resistance R_{eq} seen from the current source is the parallel connection of R_a and R_b. Therefore,

$$R_{eq} = \frac{R_a \times R_b}{R_a + R_b} = \frac{12\ k\Omega \times 36\ k\Omega}{12\ k\Omega + 36\ k\Omega} = 9\ k\Omega$$

The circuit with the current source and the equivalent resistance is shown in Figure 2.47. The voltage across the equivalent resistance, which is also the voltage across the current source, is given by

$$V_1 = I_s \times R_{eq} = 2\text{ mA} \times 9\ k\Omega = 2 \times 10^{-3}\text{A} \times 9000\ \Omega = 18\text{ V}$$

FIGURE 2.47

The current source and the equivalent resistance.

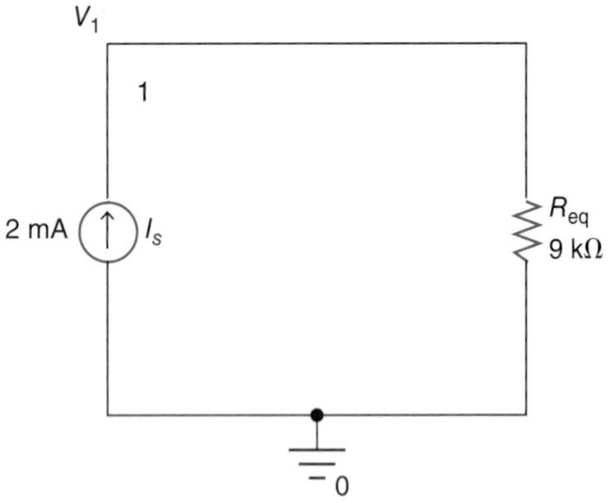

The current to the left side of the current source is given by

$$I_a = \frac{V_1}{R_a} = \frac{18\text{ V}}{12\ k\Omega} = 1.5\text{ mA}$$

The current to the right side of the current source is given by

$$I_5 = \frac{V_1}{R_b} = \frac{18\text{ V}}{36\ k\Omega} = 0.5\text{ mA}$$

The current I_5 can also be obtained from $I_5 = I_s - I_a$. The current through R_4 is

$$I_4 = \frac{V_1}{R_4} = \frac{18\text{ V}}{30\ k\Omega} = 0.6\text{ mA}$$

The current through R_3 is

$$I_3 = I_a - I_4 = 1.5\text{ mA} - 0.6\text{ mA} = 0.9\text{ mA}$$

The voltage across R_3 is

$$V_{R_3} = R_3 \times I_3 = 5\ k\Omega \times 0.9\text{ mA} = 4.5\text{ V}$$

The voltage V_2 is given by

$$V_2 = V_1 - V_{R_3} = 18\text{ V} - 4.5\text{ V} = 13.5\text{ V}$$

The current through R_1 can be found by applying Ohm's law:

$$I_1 = \frac{V_2}{R_1} = \frac{13.5\text{ V}}{20\ k\Omega} = 0.675\text{ mA}$$

Similarly, the current through R_2 is found to be

$$I_2 = \frac{V_2}{R_2} = \frac{13.5\text{ V}}{60\ k\Omega} = 0.225\text{ mA}$$

continued

Example 2.14 continued

The voltage across R_5 is given by

$$V_{R_5} = R_5 \times I_5 = 21\ k\Omega \times 0.5\ \text{mA} = 10.5\ \text{V}$$

The voltage V_3 is

$$V_3 = V_1 - V_{R_5} = 18\ \text{V} - 10.5\ \text{V} = 7.5\ \text{V}.$$

The current through R_6 is given by

$$I_6 = \frac{V_3}{R_6} = \frac{7.5\ \text{V}}{24\ k\Omega} = 0.3125\ \text{mA}$$

The current through R_7 is given by

$$I_7 = \frac{V_3}{R_7} = \frac{7.5\ \text{V}}{40\ k\Omega} = 0.1875\ \text{mA}$$

Notice that

$$I_1 + I_2 + I_4 + I_6 + I_7 = 2\ \text{mA} = I_s$$

Exercise 2.14

Find R_{eq} seen from the current source, V_1, V_2, I_1, I_2, I_3, I_4, I_5, and I_6 in the circuit shown in Figure 2.48.

FIGURE 2.48

Circuit for EXERCISE 2.14.

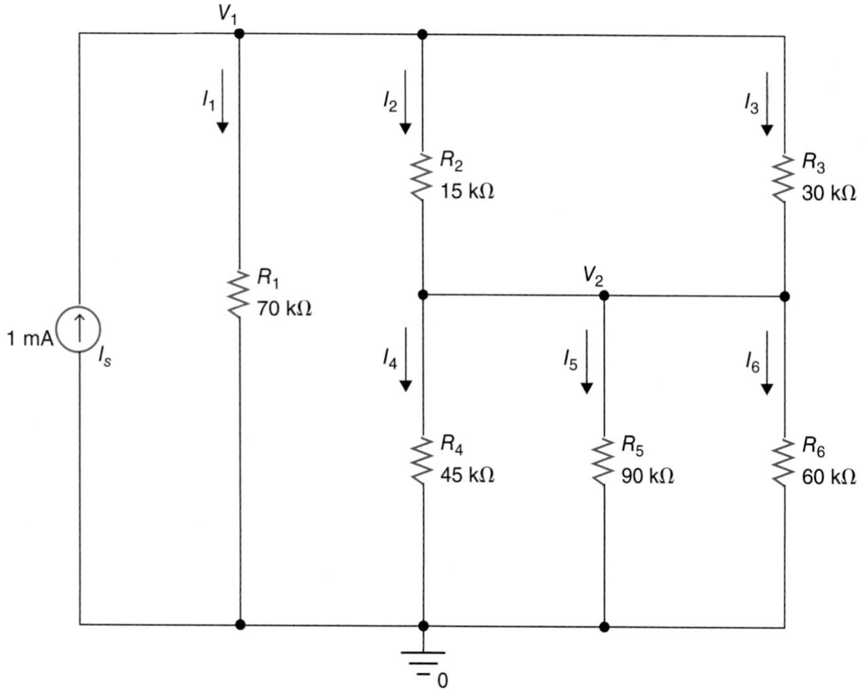

Answer:

$R_{eq} = 21\ k\Omega$, $V_1 = 21$ V, $V_2 = 14$ V, $I_1 = 0.3$ mA, $I_2 = 0.46667$ mA, $I_3 = 0.23333$ mA, $I_4 = 0.31111$ mA, $I_5 = 0.15556$ mA, $I_6 = 0.23333$ mA

2.8 Voltage Divider Rule

Suppose that two resistors with resistances R_1 and R_2, respectively, are connected in series to a voltage source with voltage V_s volts, as shown in Figure 2.49. The equivalent resistance is $R_1 + R_2$. According to Ohm's law, the current through the mesh is

$$I = \frac{V_s}{R_1 + R_2}$$

Thus, the voltage V_1 across the first resistor with resistance R_1 is

$$V_1 = R_1 I = R_1 \frac{V_s}{R_1 + R_2} = \frac{R_1}{R_1 + R_2} V_s \tag{2.48}$$

and the voltage V_2 across the second resistor with resistance R_2 is

$$V_2 = R_2 I = R_2 \frac{V_s}{R_1 + R_2} = \frac{R_2}{R_1 + R_2} V_s \tag{2.49}$$

FIGURE 2.49

A circuit with two resistors in series.

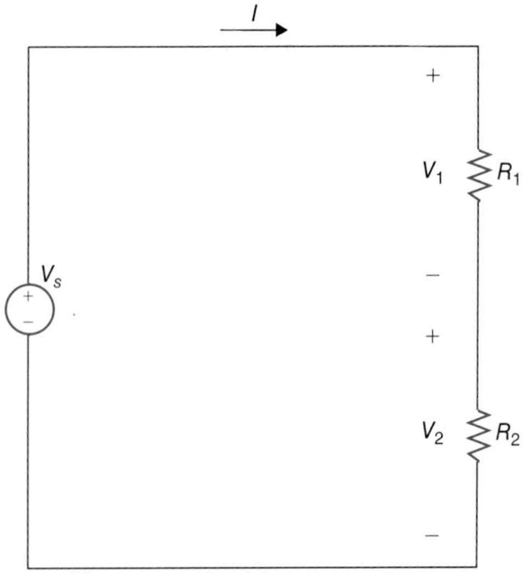

This result shows that when two resistors are connected in series to a voltage source with voltage V_s volts, the total voltage (V_s volts) is divided between V_1 and V_2; that is,

$$V_s = V_1 + V_2$$

in proportion to resistance values R_1 and R_2. This is referred to as the **voltage divider rule**. If R_1 is greater than R_2, the voltage across R_1 is more than the voltage across R_2. On the other hand, if R_1 is smaller than R_2, the voltage across R_1 is less than the voltage across R_2. In other words, the voltage drop across the resistor with a larger resistance value is more than the voltage drop across the resistor with a smaller resistance value. If $R_1 \gg R_2$, $V_1 \approx V_s$, and $V_2 \approx 0$. If $R_1 \ll R_2$, $V_1 \approx 0$, and $V_2 \approx V_s$.

The voltage divider rule can be generalized to include more than two resistors. If a voltage source with voltage V_s is connected to a series connection of n resistors with resistances $R_1, R_2, \ldots, R_n$, respectively, the voltage across each resistor R_i, $1 \le i \le n$, is given by

$$V_i = \frac{R_i}{R_1 + R_2 + \cdots + R_n} V_s \tag{2.50}$$

The voltage divider rule can be described by using conductance rather than resistance. Suppose that two resistors with resistances R_1 and R_2, respectively, are connected in series to a voltage source with voltage V_s volts. The voltage across R_1 is given by

$$V_1 = \frac{R_1}{R_1 + R_2} V_s = \frac{\frac{1}{G_1}}{\frac{1}{G_1} + \frac{1}{G_2}} V_s = \frac{G_2}{G_1 + G_2} V_s$$

Similarly, the voltage across R_2 is given by

$$V_2 = \frac{R_2}{R_1 + R_2} V_s = \frac{\frac{1}{G_2}}{\frac{1}{G_1} + \frac{1}{G_2}} V_s = \frac{G_1}{G_1 + G_2} V_s$$

Let G be the conductance of a series connection of R_1 and R_2. Then,

$$G = \frac{1}{R_1 + R_2}$$

The current through the resistors is $I = GV_s$. The voltage across the resistor R_1 is

$$V_1 = R_1 I = R_1 G V_s = \frac{G}{G_1} V_s$$

Similarly, the voltage across the resistor R_2 is given by

$$V_2 = R_2 I = R_2 G V_s = \frac{G}{G_2} V_s$$

If a voltage source with voltage V_s is connected to a series connection of n resistors with resistances $R_1, R_2, \ldots, R_n$, respectively, the voltage across each resistor R_i, $1 \le i \le n$, is given by

$$V_i = \frac{R_i}{R_1 + R_2 + \cdots + R_n} V_s = \frac{\dfrac{1}{R_1 + R_2 + \cdots + R_n}}{\dfrac{1}{R_i}} V_s = \frac{G}{G_i} V_s$$

where

$$G = \frac{1}{\dfrac{1}{G_1} + \dfrac{1}{G_2} + \cdots + \dfrac{1}{G_n}} = \frac{1}{R_1 + R_2 + \cdots + R_n}$$

Consider a circuit shown in Figure 2.50. We are interested in finding V_1, V_2, and V_3.

FIGURE 2.50

Circuit to be analyzed by the voltage divider rule.

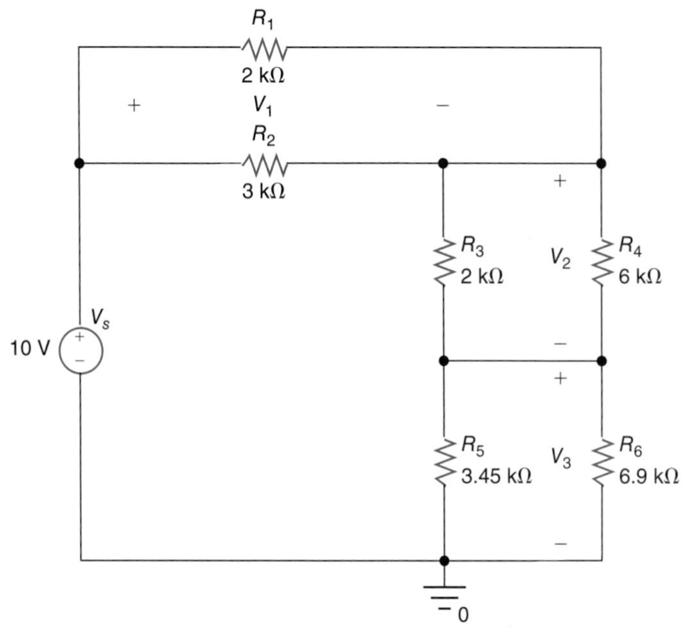

Let $R_a = R_1 \| R_2$, $R_b = R_3 \| R_4$, $R_c = R_5 \| R_6$. Then, we have

$$R_a = R_1 \| R_2 = \frac{R_1 \times R_2}{R_1 + R_2} = \frac{6}{5} k\Omega = 1.2\ k\Omega$$

$$R_b = R_3 \| R_4 = \frac{R_3 \times R_4}{R_3 + R_4} = \frac{12}{8} k\Omega = 1.5\ k\Omega$$

$$R_c = R_5 \| R_6 = \frac{R_5 \times R_6}{R_5 + R_6} = \frac{23.805}{10.35} k\Omega = 2.3\ k\Omega$$

The circuit shown in Figure 2.51 is equivalent to the circuit shown in Figure 2.50.

FIGURE 2.51

A circuit with equivalent resistances.

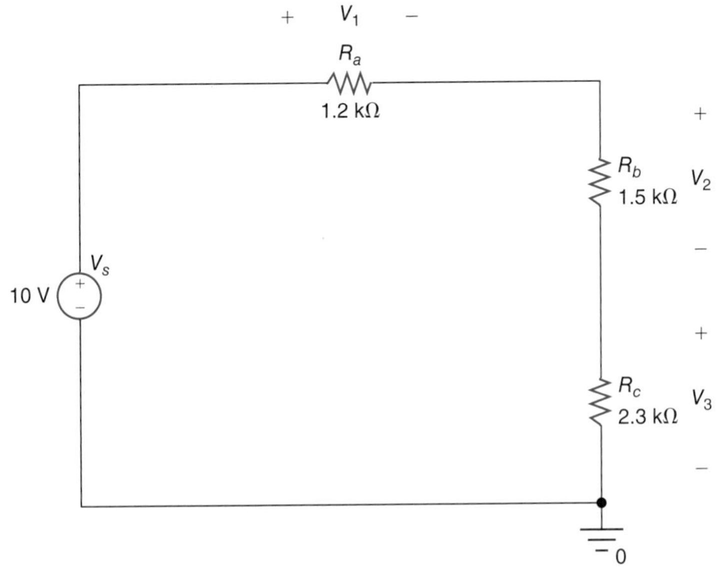

Resistors R_a, R_b, and R_c are connected in series. The total equivalent resistance seen from the voltage source is given by

$$R_{eq} = R_a + R_b + R_c = 1.2\ k\Omega + 1.5\ k\Omega + 2.3\ k\Omega = 5\ k\Omega$$

Application of the voltage divider rule to the circuit shown in Figure 2.51 yields

$$V_1 = \frac{R_a}{R_a + R_b + R_c} V_s = \frac{1.2\ k\Omega}{5\ k\Omega} \times 10\ \text{V} = 2.4\ \text{V}$$

$$V_2 = \frac{R_b}{R_a + R_b + R_c} V_s = \frac{1.5\ k\Omega}{5\ k\Omega} \times 10\ \text{V} = 3\ \text{V}$$

$$V_3 = \frac{R_c}{R_a + R_b + R_c} V_s = \frac{2.3\ k\Omega}{5\ k\Omega} \times 10\ \text{V} = 4.6\ \text{V}$$

Notice that

$$V_1 + V_2 + V_3 = 2.4\ \text{V} + 3\ \text{V} + 4.6\ \text{V} = 10\ \text{V} = V_s$$

confirming KVL.

EXAMPLE 2.15

Use the voltage divider rule to find the voltages V_1, V_2, and V_3 for the circuit shown in Figure 2.52.

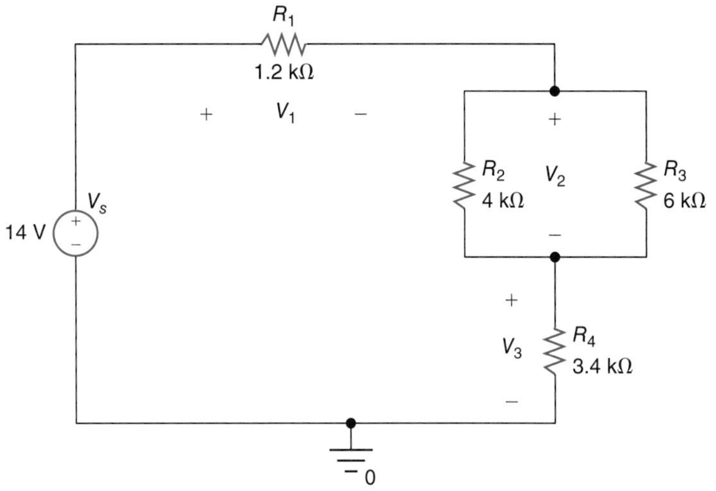

FIGURE 2.52
Circuit for EXAMPLE 2.15.

The equivalent resistance of the parallel connection of R_2 and R_3 is

$$R_2 \| R_3 = \frac{R_2 R_3}{R_2 + R_3} = \frac{4 \times 6}{4 + 6} k\Omega = 2.4\, k\Omega$$

The total resistance of the circuit seen from the voltage source V_s is

$$R_1 + (R_2 \| R_3) + R_4 = 1.2\, k\Omega + 2.4\, k\Omega + 3.4\, k\Omega = 7\, k\Omega$$

Thus, from the voltage divider rule, the voltages V_1, V_2, and V_3 are given, respectively, by

$$V_1 = \frac{R_1}{R_1 + (R_2 \| R_3) + R_4} V = \frac{1.2\, k\Omega}{7\, k\Omega} \times 14 \text{ V} = 2.4 \text{ V}$$

$$V_2 = \frac{R_2 \| R_3}{R_1 + (R_2 \| R_3) + R_4} V = \frac{2.4\, k\Omega}{7\, k\Omega} \times 14 \text{ V} = 4.8 \text{ V}$$

$$V_3 = \frac{R_4}{R_1 + (R_2 \| R_3) + R_4} V = \frac{3.4\, k\Omega}{7\, k\Omega} \times 14 \text{ V} = 6.8 \text{ V}$$

Notice that the sum of V_1, V_2, and V_3 equals the voltage from the voltage source, V_s, confirming KVL.

Exercise 2.15

Find V_1, V_2, V_3, and V_4 in the circuit shown in Figure 2.53.

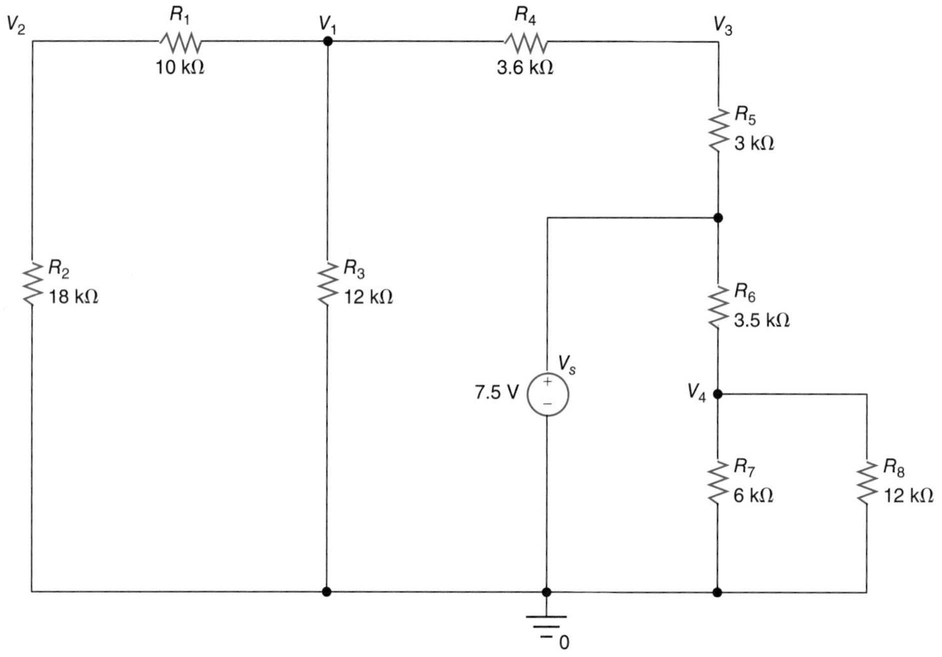

FIGURE 2.53 Circuit for EXERCISE 2.15.

Answer:
$V_1 = 4.2$ V, $V_2 = 2.7$ V, $V_3 = 6$ V, and $V_4 = 4$ V.

EXAMPLE 2.16

Find V_1, V_2, and V_3 in the circuit shown in Figure 2.54.

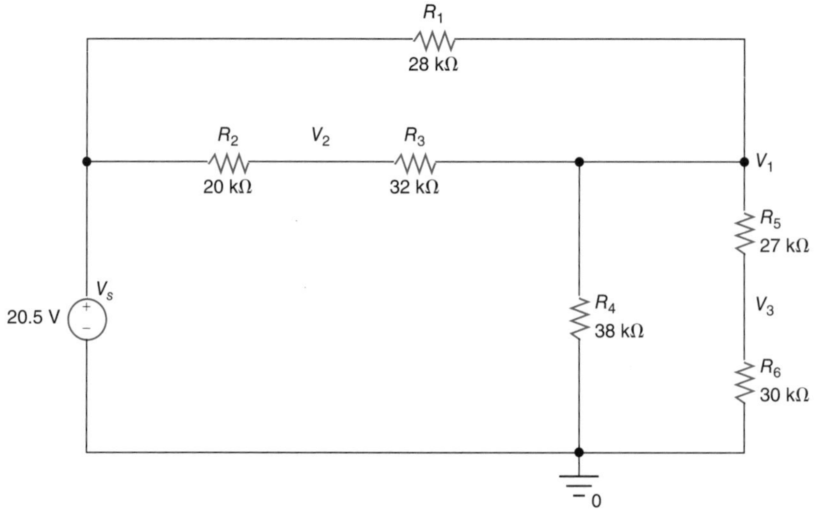

FIGURE 2.54 Circuit for EXAMPLE 2.16.

continued

Example 2.16 continued

Let R_a be the equivalent resistance of R_1, R_2, and R_3, and R_b be the equivalent resistance of R_4, R_5, and R_6. Then we have

$$R_a = R_1 \| (R_2 + R_3) = 28\ k\Omega \| 52\ k\Omega = \frac{28\ k\Omega \times 52\ k\Omega}{28\ k\Omega + 52\ k\Omega} = \frac{1456}{80}\ k\Omega = 18.2\ k\Omega$$

$$R_b = R_4 \| (R_5 + R_6) = 38\ k\Omega \| 57\ k\Omega = \frac{38\ k\Omega \times 57\ k\Omega}{38\ k\Omega + 57\ k\Omega} = \frac{2166}{95}\ k\Omega = 22.8\ k\Omega$$

Since R_a and R_b are connected in series, the voltage V_1 across R_b is given as follows (from the voltage divider rule):

$$V_1 = V_s \times \frac{R_b}{R_a + R_b} = 20.5\ \text{V} \times \frac{22.8\ k\Omega}{18.2\ k\Omega + 22.8\ k\Omega} = 20.5\ \text{V} \times \frac{22.8}{41} = 11.4\ \text{V}$$

The voltage across the R_2–R_3 path is $V_s - V_1 = 20.5\ \text{V} - 11.4\ \text{V} = 9.1\ \text{V}$. This voltage is split across R_2 and R_3 in proportion to the resistance values. Thus, we have

$$V_2 = V_1 + (V_s - V_1) \times \frac{R_3}{R_2 + R_3} = 11.4\ \text{V} + (20.5\ \text{V} - 11.4\ \text{V}) \times \frac{32\,k\Omega}{20\,k\Omega + 32\,k\Omega} = 17\ \text{V}$$

The voltage across the $R_5 - R_6$ path is $V_1 = 11.4$ V. This voltage is split across R_5 and R_6 in proportion to the resistance values. Thus, we have

$$V_3 = V_1 \times \frac{R_6}{R_5 + R_6} = 11.4\ \text{V} \times \frac{30\ k\Omega}{27\ k\Omega + 30\ k\Omega} = 6\ \text{V}$$

MATLAB

```
%EXAMPLE 2.16
%Function P.m should be in the same folder as this file.
clear all;format long;
Vs=20.5;R1=28000;R2=20000;R3=32000;R4=38000;R5=27000;R6=30000;
Ra=P([R1,R2+R3])
Rb=P([R4,R5+R6])
V1=Vs*Rb/(Ra+Rb)
VR1=Vs-V1
V2=V1+VR1*R3/(R2+R3)
V3=V1*R6/(R5+R6)

Answers:
Ra =
       18200
Rb =
     2.280000000000000e+04
V1 =
  11.399999999999999
VR1 =
   9.100000000000001
V2 =
    17
V3 =
   5.999999999999999
```

Exercise 2.16

Find V_1, V_2, V_3, V_4, and V_5 in the circuit shown in Figure 2.55.

FIGURE 2.55

Circuit for EXERCISE 2.16.

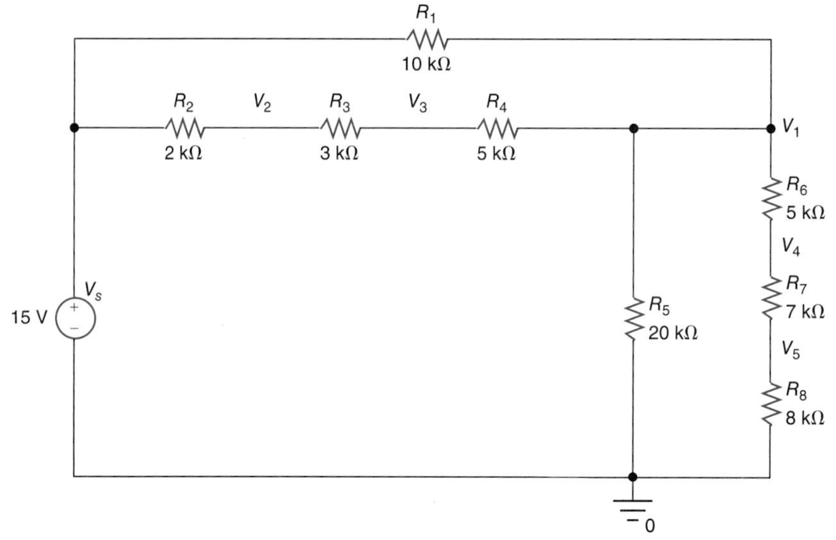

Answer:
$V_1 = 10$ V, $V_2 = 14$ V, $V_3 = 12.5$ V, $V_4 = 7.5$ V, and $V_5 = 4$ V.

FIGURE 2.56

A Wheatstone bridge circuit.

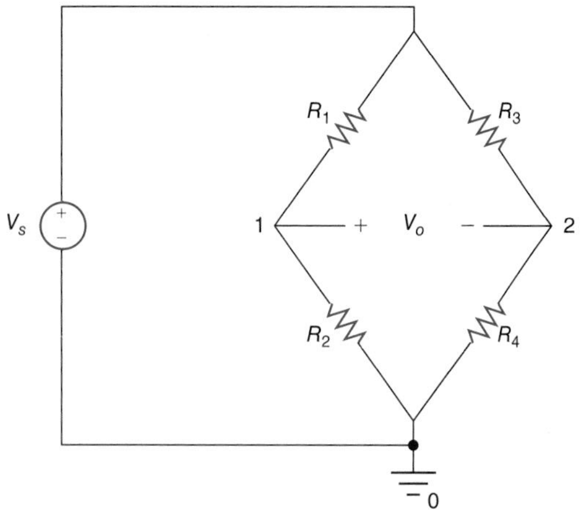

2.8.1 WHEATSTONE BRIDGE

A Wheatstone bridge circuit is shown in Figure 2.56.

According to the voltage divider rule, the voltage V_s is divided between R_1 and R_2 in proportion to the resistance values. Thus, the voltage at node 1, which is the voltage across R_2, is given by

$$V_1 = \frac{R_2}{R_1 + R_2} \times V_s$$

Similarly, voltage V_s is divided between R_3 and R_4 in proportion to the resistance values. Thus, the voltage at node 2, which is the voltage across R_4, is given by

$$V_2 = \frac{R_4}{R_3 + R_4} \times V_s$$

The output voltage V_o is the difference between voltage V_1 and voltage V_2. Therefore, we have

$$V_o = V_1 - V_2 = \frac{R_2}{R_1 + R_2} \times V_s - \frac{R_4}{R_3 + R_4} \times V_s = \left(\frac{R_2}{R_1 + R_2} - \frac{R_4}{R_3 + R_4}\right) \times V_s \quad \textbf{(2.51)}$$

The bridge is called **balanced** if V_o is zero. If the bridge is balanced, we have

$$\frac{R_2}{R_1 + R_2} = \frac{R_4}{R_3 + R_4}$$

or

$$R_2R_3 + R_2R_4 = R_1R_4 + R_2R_4$$

or

$$R_1R_4 = R_2R_3 \tag{2.52}$$

FIGURE 2.57

A Wheatstone bridge with a voltmeter.

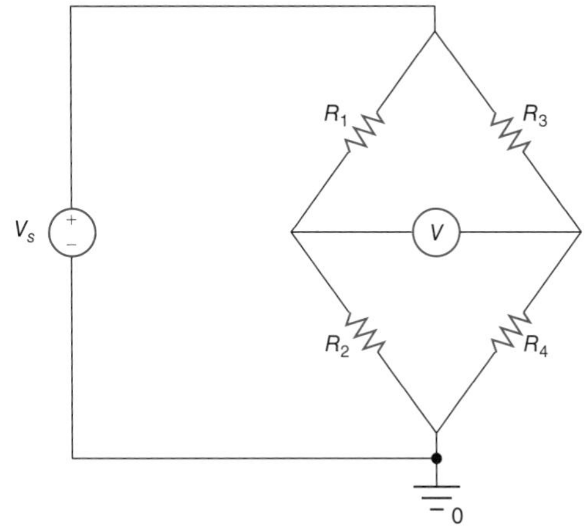

If three of the four resistance values are known, we can find the fourth resistance value. The Wheatstone bridge can be used to find the value of unknown resistance. A voltmeter is placed at the output, as shown in Figure 2.57. Let R_4 be the unknown resistance to be determined. The resistance values of R_1 and R_2 are fixed, and R_3 is a precision variable resistor. R_3 is varied until the voltmeter reading is zero, which is called the *null condition.* Then, the unknown resistance value is given by

$$R_4 = \frac{R_2R_3}{R_1} \tag{2.53}$$

The Wheatstone bridge can be used to measure small changes in resistance due to changes in the physical condition of sensors, such as temperature sensors, light sensors, and strain gauges. In this application, resistor R_4 is replaced by a sensor and R_4 represents the resistance value under certain conditions. For example, R_4 may be the resistance when no force is applied to a strain gauge, or R_4 may be the resistance when no light is applied to a light sensor. R_3 is adjusted to create balanced condition with R_4. When the physical condition changes, R_4 changes by a small amount. Let the amount of change be $\pm\,\Delta R$. Then, R_4 becomes $R_4 \pm \Delta R$, and the bridge is no longer balanced and V_o is no longer zero. The output voltage given by Equation (2.51) becomes

$$\begin{aligned}
V_o = V_1 - V_2 &= \left(\frac{R_2}{R_1 + R_2} - \frac{R_4 \pm \Delta R}{R_3 + R_4 \pm \Delta R}\right) \times V_s \\
&= \left[\frac{R_2}{R_1 + R_2} - \frac{R_4\left(1 \pm \dfrac{\Delta R}{R_4}\right)}{(R_3 + R_4)\left(1 \pm \dfrac{\Delta R}{R_3 + R_4}\right)}\right] \times V_s \\
&= \frac{R_2}{R_1 + R_2}\left(1 - \frac{1 \pm \dfrac{\Delta R}{R_4}}{1 \pm \dfrac{\Delta R}{R_3 + R_4}}\right) \times V_s = \frac{R_2}{R_1 + R_2}\left(\frac{1 \pm \dfrac{\Delta R}{R_3 + R_4} - 1 \mp \dfrac{\Delta R}{R_4}}{1 \pm \dfrac{\Delta R}{R_3 + R_4}}\right) \times V_s \\
&= \frac{R_2}{R_1 + R_2}\left(\frac{\pm\Delta R\left(\dfrac{1}{R_3 + R_4} - \dfrac{1}{R_4}\right)}{1 \pm \dfrac{\Delta R}{R_3 + R_4}}\right) \times V_s \\
&= \frac{R_2}{R_1 + R_2} \times \frac{R_3}{(R_3 + R_4)R_4}\left(\frac{\mp\Delta R}{1 \pm \dfrac{\Delta R}{R_3 + R_4}}\right) \times V_s
\end{aligned} \tag{2.54}$$

Assume $\Delta R \ll (R_3 + R_4)$. Then, $\Delta R/(R_3 + R_4) \approx 0$ and Equation (2.54) can be approximated by

$$V_o = \frac{R_2}{R_1 + R_2} \times \frac{R_3}{(R_3 + R_4)R_4} \times (\mp\,\Delta R)V_s \tag{2.55}$$

If $R_1 = R_2 = R_3 = R_4 = R$, Equation (2.55) becomes

$$V_o = \mp\frac{\Delta R}{4R}V_s \tag{2.56}$$

The output voltage of the bridge is proportional to the change of the resistance ΔR with sign inversion. Notice that $-V_o = V_2 - V_1 = \pm \Delta R \times V_s/(4R)$. This signal can be amplified by an instrumentation amplifier, as discussed in Chapter 5.

2.9 Current Divider Rule

FIGURE 2.58

A circuit with two resistors in parallel.

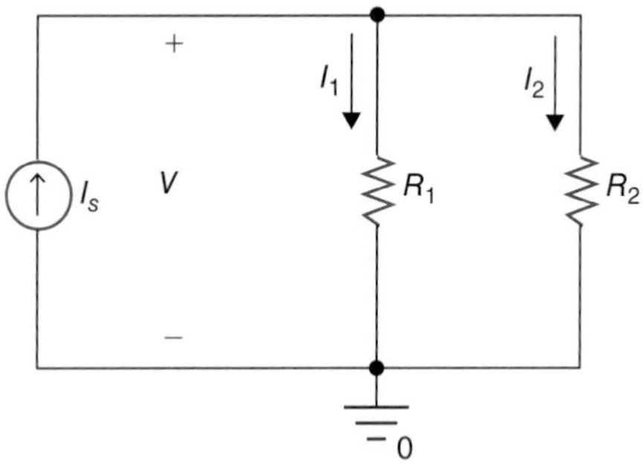

Suppose that a current source with I_s amperes of current is connected in parallel to a pair of resistors with resistances R_1 and R_2, respectively, as shown in Figure 2.58.

The equivalent resistance value is

$$R = R_1 \| R_2 = \frac{1}{\frac{1}{R_1} + \frac{1}{R_2}} = \frac{R_1R_2}{R_1 + R_2}$$

Thus, the voltage across the resistors is, from Ohm's law,

$$V = RI_s = \frac{I_s}{\frac{1}{R_1} + \frac{1}{R_2}} = \frac{R_1R_2}{R_1 + R_2}I_s$$

The current I_1 through R_1 is

$$I_1 = \frac{V}{R_1} = \frac{\frac{1}{\frac{1}{R_1} + \frac{1}{R_2}}}{R_1}I_s = \frac{\frac{1}{R_1}}{\frac{1}{R_1} + \frac{1}{R_2}}I_s = \frac{R_2}{R_1 + R_2}I_s \tag{2.57}$$

and the current through R_2 is

$$I_2 = \frac{V}{R_2} = \frac{\frac{1}{\frac{1}{R_1} + \frac{1}{R_2}}}{R_2}I_s = \frac{\frac{1}{R_2}}{\frac{1}{R_1} + \frac{1}{R_2}}I_s = \frac{R_1}{R_1 + R_2}I_s \tag{2.58}$$

From KCL, the sum of I_1 and I_2 is I_s; that is,

$$I_s = I_1 + I_2$$

This result suggests that the current through R_1 is given by I_s times the ratio of R_2 to $R_1 + R_2$, and the current through R_2 is given by I_s times the ratio of R_1 to $R_1 + R_2$. If R_1 is greater than R_2, the current through R_1 is smaller than the current through R_2. On the other hand, if R_1 is smaller than R_2, the current through R_1 is more than the current through R_2. In other words, more current flows through the resistor with smaller resistance value than through the resistor with larger resistance value. This is referred to as the **current divider rule**. If $R_1 \gg R_2$, $I_1 \approx 0$ and $I_2 \approx I_s$. On the other hand, if $R_1 \ll R_2$, $I_1 \approx I_s$ and $I_2 \approx 0$. If $R_1 = 0$ and $R_2 \neq 0$, $I_1 = I_s$ and $I_2 = 0$. If $R_1 \neq 0$ and $R_2 = 0$, $I_1 = 0$ and $I_2 = I_s$.

The current divider rule can be generalized to include more than two resistors. If a current source with current I_s is connected to a parallel connection of n resistors with resistances $R_1, R_2, \ldots, R_n$, respectively, then the voltage across the resistors is given by

$$V = R_{eq} I_s = \frac{1}{\frac{1}{R_1} + \frac{1}{R_2} + \cdots + \frac{1}{R_n}} I_s$$

The current through resistor with resistance R_i, $i = 1, 2, \ldots, n$, is given by

$$I_i = \frac{V}{R_i} = \frac{\frac{1}{R_i}}{\frac{1}{R_1} + \frac{1}{R_2} + \cdots + \frac{1}{R_n}} I_s \tag{2.59}$$

The current divider rule can be described by using conductance rather than resistance. Suppose that two resistors with resistances R_1 and R_2, respectively, are connected in parallel to a current source with current I_s amperes. The current through R_1 is given by

$$I_1 = \frac{\frac{1}{R_1}}{\frac{1}{R_1} + \frac{1}{R_2}} I_s = \frac{G_1}{G_1 + G_2} I_s$$

Similarly, the current through R_2 is given by

$$I_2 = \frac{\frac{1}{R_2}}{\frac{1}{R_1} + \frac{1}{R_2}} I_s = \frac{G_2}{G_1 + G_2} I_s$$

Let G_{eq} be the equivalent conductance of the parallel connection of R_1 and R_2. Then,

$$G_{eq} = G_1 + G_2$$

The voltage across the resistors is $V = I_s/G_{eq}$. The current through the resistor R_1 is

$$I_1 = \frac{V}{R_1} = \frac{I_s}{G_{eq} R_1} = \frac{G_1}{G_{eq}} I_s$$

Similarly, the current through the resistor R_2 is given by

$$I_2 = \frac{V}{R_2} = \frac{I_s}{G_{eq} R_2} = \frac{G_2}{G_{eq}} I_s$$

If a current source with current I_s is connected to a parallel connection of n resistors with resistances $R_1, R_2, \ldots, R_n$, respectively, the current through each resistor R_i, $1 \le i \le n$, is given by

$$I_i = \frac{\frac{1}{R_i}}{\frac{1}{R_1} + \frac{1}{R_2} + \cdots + \frac{1}{R_n}} I_s = \frac{G_i}{G_1 + G_2 + \cdots + G_n} I_s$$

where $G_i = 1/R_i$.

Consider a circuit shown in Figure 2.59. We are interested in finding I_1, I_2, and I_3.

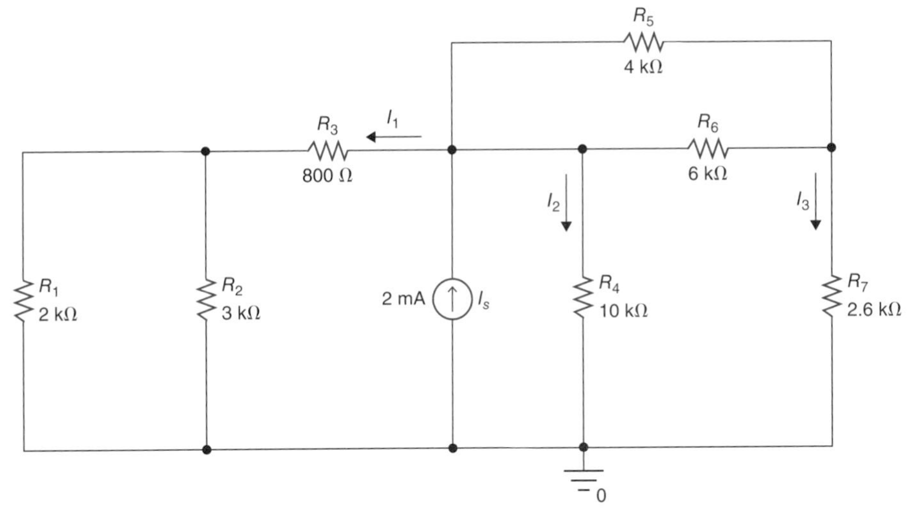

FIGURE 2.59 A circuit to be analyzed by the current divider rule.

Let $R_a = R_3 + (R_1 \| R_2)$, $R_b = R_7 + (R_5 \| R_6)$. Then, we have

$$R_a = R_3 + (R_1 \| R_2) = R_3 + \frac{R_1 \times R_2}{R_1 + R_2} = 0.8\,k\Omega + \frac{6}{5}\,k\Omega = 2\,k\Omega$$

$$R_b = R_7 + (R_5 \| R_6) = R_7 + \frac{R_5 \times R_6}{R_5 + R_6} = 2.6\,k\Omega + \frac{24}{10}\,k\Omega = 5\,k\Omega$$

The circuit shown in Figure 2.60 is equivalent to the circuit shown in Figure 2.59.

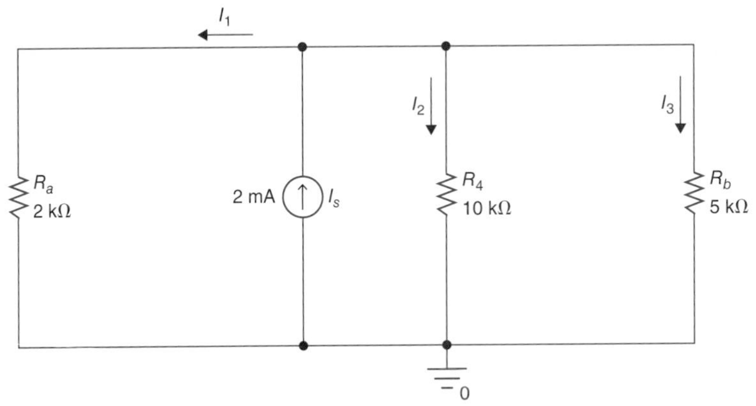

FIGURE 2.60 A circuit with equivalent resistances.

Notice that R_a, R_b, and R_4 are connected in parallel. Application of the current divider rule to the circuit shown in Figure 2.60 yields

$$I_1 = \frac{\frac{1}{R_a}}{\frac{1}{R_a} + \frac{1}{R_b} + \frac{1}{R_4}} I_s = \frac{\frac{1}{2}}{\frac{1}{2} + \frac{1}{5} + \frac{1}{10}} \times 2\text{ mA} = \frac{5}{5 + 2 + 1}$$

$$\times 2\text{ mA} = \frac{5}{8} \times 2\text{ mA} = 1.25\text{ mA}$$

$$I_2 = \frac{\frac{1}{R_4}}{\frac{1}{R_a} + \frac{1}{R_b} + \frac{1}{R_4}} I_s = \frac{\frac{1}{10}}{\frac{1}{2} + \frac{1}{5} + \frac{1}{10}} \times 2\text{ mA} = \frac{1}{5 + 2 + 1} \times 2\text{ mA}$$

$$= \frac{1}{8} \times 2\text{ mA} = 0.25\text{ mA}$$

$$I_3 = \frac{\frac{1}{R_b}}{\frac{1}{R_a} + \frac{1}{R_b} + \frac{1}{R_4}} I_s = \frac{\frac{1}{5}}{\frac{1}{2} + \frac{1}{5} + \frac{1}{10}} \times 2\text{ mA} = \frac{2}{5 + 2 + 1} \times 2\text{ mA}$$

$$= \frac{2}{8} \times 2\text{ mA} = 0.5\text{ mA}$$

Notice that

$$I_1 + I_2 + I_3 = 1.25\text{ mA} + 0.25\text{ mA} + 0.5\text{ mA} = 2\text{ mA} = I_s$$

EXAMPLE 2.17

In the circuit shown in Figure 2.61, use the current divider rule to find the currents I_1, I_2, I_3, I_4, and I_5.

FIGURE 2.61

Circuit for EXAMPLE 2.17.

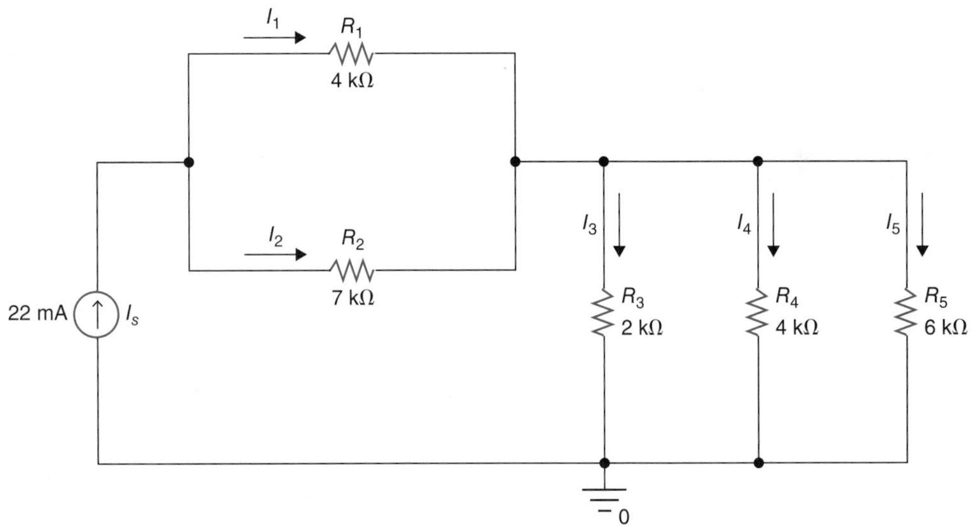

The current from the current source I_s is split between I_1 and I_2. From the current divider rule, we have

$$I_1 = \frac{R_2}{R_1 + R_2} I_s = \frac{7}{4 + 7} \times 22\text{ mA} = 14\text{ mA}$$

$$I_2 = \frac{R_1}{R_1 + R_2} I_s = \frac{4}{4 + 7} \times 22\text{ mA} = 8\text{ mA}$$

continued

Example 2.17 continued

Notice that once I_1 is found, I_2 can be obtained using KCL. The current from the current source I_s is split between I_3, I_4, and I_5. From the current divider rule, we have

$$I_3 = \frac{\frac{1}{R_3}}{\frac{1}{R_3}+\frac{1}{R_4}+\frac{1}{R_5}} I_s = \frac{\frac{1}{2000}}{\frac{1}{2000}+\frac{1}{4000}+\frac{1}{6000}} \times 22 \text{ mA}$$

$$= \frac{\frac{12{,}000}{2000}}{\frac{12{,}000}{2000}+\frac{12{,}000}{4000}+\frac{12{,}000}{6000}} \times 22 \text{ mA} = \frac{6}{6+3+2} \times 22 \text{ mA} = 12 \text{ mA}$$

$$I_4 = \frac{\frac{1}{R_4}}{\frac{1}{R_3}+\frac{1}{R_4}+\frac{1}{R_5}} I_s = \frac{\frac{1}{4000}}{\frac{1}{2000}+\frac{1}{4000}+\frac{1}{6000}}$$

$$\times 22 \text{ mA} = \frac{3}{6+3+2} \times 22 \text{ mA} = 6 \text{ mA}$$

$$I_5 = \frac{\frac{1}{R_5}}{\frac{1}{R_3}+\frac{1}{R_4}+\frac{1}{R_5}} I_s = \frac{\frac{1}{6000}}{\frac{1}{2000}+\frac{1}{4000}+\frac{1}{6000}}$$

$$\times 22 \text{ mA} = \frac{2}{6+3+2} \times 22 \text{ mA} = 4 \text{ mA}$$

Exercise 2.17

Find I_1, I_2, I_3, and I_4 in the circuit shown in Figure 2.62.

FIGURE 2.62

Circuit for EXERCISE 2.17.

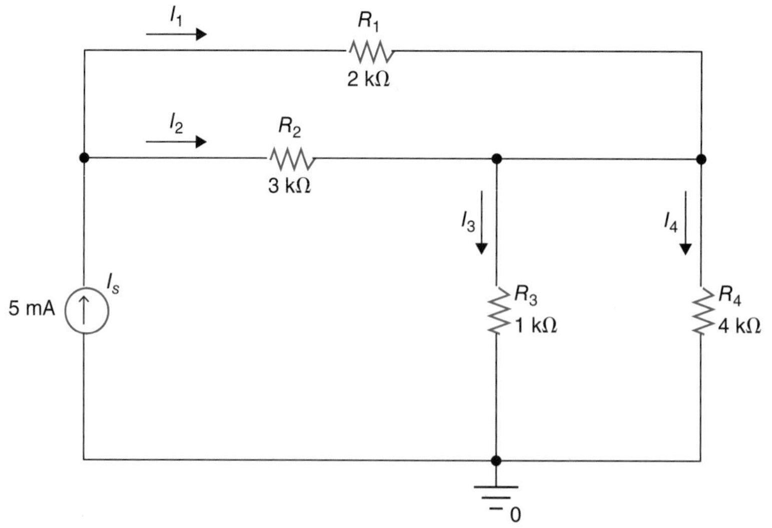

Answer:

$I_1 = 3$ mA, $I_2 = 2$ mA, $I_3 = 4$ mA, and $I_4 = 1$ mA.

EXAMPLE 2.18

Find I_1, I_2, I_3, I_4, I_5, I_6, I_7, I_8, I_9, V_1, V_2, and V_3 in the circuit shown in Figure 2.63.

FIGURE 2.63

Circuit for EXAMPLE 2.18.

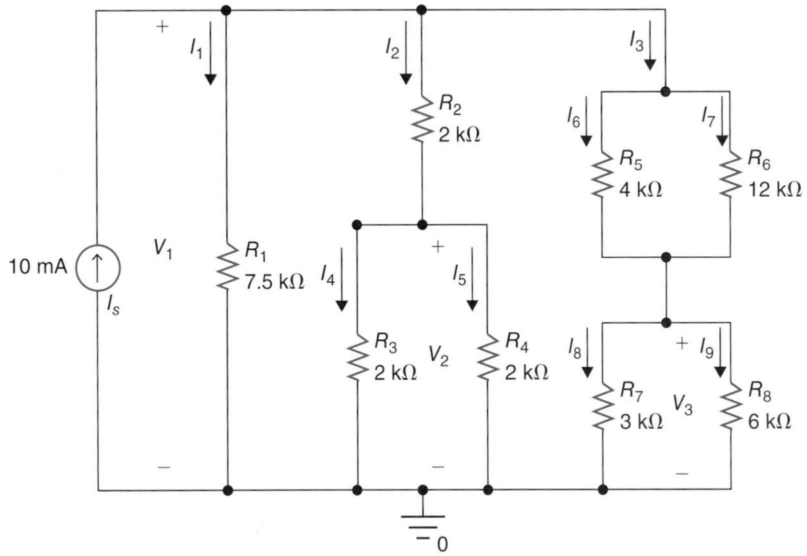

Let R_a be the equivalent resistance of R_2, R_3, and R_4. Then, we have

$$R_a = R_2 + (R_3 \| R_4) = R_2 + \frac{R_3 \times R_4}{R_3 + R_4} = 2\,k\Omega + \frac{2\,k\Omega \times 2\,k\Omega}{2\,k\Omega + 2\,k\Omega}$$
$$= 2\,k\Omega + 1\,k\Omega = 3\,k\Omega$$

Let R_b be the equivalent resistance of R_5, R_6, R_7, and R_8. Then, we have

$$R_b = (R_5 \| R_6) + (R_7 \| R_8) = \frac{R_5 \times R_6}{R_5 + R_6} + \frac{R_7 \times R_8}{R_7 + R_8} = \frac{4 \times 12}{4 + 12} k\Omega + \frac{3 \times 6}{3 + 6} k\Omega$$
$$= 3\,k\Omega + 2\,k\Omega = 5\,k\Omega$$

Application of the current divider rule results in

$$I_1 = I_s \frac{\frac{1}{R_1}}{\frac{1}{R_1} + \frac{1}{R_a} + \frac{1}{R_b}} = 10\text{ mA} \times \frac{\frac{1}{7500}}{\frac{1}{7500} + \frac{1}{3000} + \frac{1}{5000}}$$

$$= 10\text{ mA} \times \frac{\frac{15{,}000}{7500}}{\frac{15{,}000}{7500} + \frac{15{,}000}{3000} + \frac{15{,}000}{5000}} = 10\text{ mA} \times \frac{2}{2 + 5 + 3} = 2\text{ mA}$$

continued

Example 2.18 continued

$$I_2 = I_s \frac{\frac{1}{R_a}}{\frac{1}{R_1} + \frac{1}{R_a} + \frac{1}{R_b}} = 10 \text{ mA} \times \frac{\frac{1}{3000}}{\frac{1}{7500} + \frac{1}{3000} + \frac{1}{5000}}$$

$$= 10 \text{ mA} \times \frac{\frac{15{,}000}{3000}}{\frac{15{,}000}{7500} + \frac{15{,}000}{3000} + \frac{15{,}000}{5000}} = 10 \text{ mA} \times \frac{5}{2 + 5 + 3} = 5 \text{ mA}$$

$$I_3 = I_s \frac{\frac{1}{R_b}}{\frac{1}{R_1} + \frac{1}{R_a} + \frac{1}{R_b}} = 10 \text{ mA} \times \frac{\frac{1}{5000}}{\frac{1}{7500} + \frac{1}{3000} + \frac{1}{5000}}$$

$$= 10 \text{ mA} \times \frac{\frac{15{,}000}{5000}}{\frac{15{,}000}{7500} + \frac{15{,}000}{3000} + \frac{15{,}000}{5000}} = 10 \text{ mA} \times \frac{3}{2 + 5 + 3} = 3 \text{ mA}$$

Similarly, application of the current divider rule results in

$$I_4 = I_2 \frac{\frac{1}{R_3}}{\frac{1}{R_3} + \frac{1}{R_4}} = 5 \text{ mA} \times \frac{\frac{1}{2000}}{\frac{1}{2000} + \frac{1}{2000}} = 5 \text{ mA} \times \frac{\frac{2000}{2000}}{\frac{2000}{2000} + \frac{2000}{2000}}$$

$$= 5 \text{ mA} \times \frac{1}{2} = 2.5 \text{ mA}$$

$$I_5 = I_2 \frac{\frac{1}{R_4}}{\frac{1}{R_3} + \frac{1}{R_4}} = 5 \text{ mA} \times \frac{\frac{1}{2000}}{\frac{1}{2000} + \frac{1}{2000}} = 5 \text{ mA} \times \frac{\frac{2000}{2000}}{\frac{2000}{2000} + \frac{2000}{2000}}$$

$$= 5 \text{ mA} \times \frac{1}{2} = 2.5 \text{ mA}$$

$$I_6 = I_3 \frac{\frac{1}{R_5}}{\frac{1}{R_5} + \frac{1}{R_6}} = 3 \text{ mA} \times \frac{\frac{1}{4000}}{\frac{1}{4000} + \frac{1}{12{,}000}} = 3 \text{ mA} \times \frac{\frac{12{,}000}{4000}}{\frac{12{,}000}{4000} + \frac{12{,}000}{12{,}000}}$$

$$= 3 \text{ mA} \times \frac{3}{4} = 2.25 \text{ mA}$$

$$I_7 = I_3 \frac{\frac{1}{R_6}}{\frac{1}{R_5} + \frac{1}{R_6}} = 3 \text{ mA} \times \frac{\frac{1}{12{,}000}}{\frac{1}{4000} + \frac{1}{12{,}000}} = 3 \text{ mA} \times \frac{\frac{12{,}000}{12{,}000}}{\frac{12{,}000}{4000} + \frac{12{,}000}{12{,}000}}$$

$$= 3 \text{ mA} \times \frac{1}{4} = 0.75 \text{ mA}$$

continued

Example 2.18 continued

$$I_8 = I_3\frac{\frac{1}{R_7}}{\frac{1}{R_7}+\frac{1}{R_8}} = 3\text{ mA}\times\frac{\frac{1}{3000}}{\frac{1}{3000}+\frac{1}{6000}} = 3\text{ mA}\times\frac{\frac{6000}{3000}}{\frac{6000}{3000}+\frac{6000}{6000}}$$

$$= 3\text{ mA}\times\frac{2}{3} = 2\text{ mA}$$

$$I_9 = I_3\frac{\frac{1}{R_8}}{\frac{1}{R_7}+\frac{1}{R_8}} = 3\text{ mA}\times\frac{\frac{1}{6000}}{\frac{1}{3000}+\frac{1}{6000}} = 3\text{ mA}\times\frac{\frac{6000}{6000}}{\frac{6000}{3000}+\frac{6000}{6000}}$$

$$= 3\text{ mA}\times\frac{1}{3} = 1\text{ mA}$$

Notice that

$$I_1 + I_4 + I_5 + I_8 + I_9 = 2\text{ mA} + 2.5\text{ mA} + 2.5\text{ mA} + 2\text{ mA} + 1\text{ mA} = 10\text{ mA} = I_s$$

The voltages are given by

$$V_1 = R_1I_1 = 15\text{ V}$$

$$V_2 = R_3I_4 = 5\text{ V}$$

$$V_3 = R_7I_8 = 6\text{ V}$$

MATLAB

```
%EXAMPLE 2.18
%Function P.m should be in the same folder as this file.
clear all;format long;
Is=10e-3;
R1=7500;R2=2000;R3=2000;R4=2000;R5=4000;R6=12000;R7=3000;R8=6000;
Ra=R2+P([R3,R4])
Rb=P([R5,R6])+P([R7,R8])
I1=Is*(1/R1)/(1/R1+1/Ra+1/Rb)
I2=Is*(1/Ra)/(1/R1+1/Ra+1/Rb)
I3=Is*(1/Rb)/(1/R1+1/Ra+1/Rb)
I4=I2*R4/(R3+R4)
I5=I2*R3/(R3+R4)
I6=I3*R6/(R5+R6)
I7=I3*R5/(R5+R6)
I8=I3*R8/(R7+R8)
I9=I3*R7/(R7+R8)
V1=R1*I1
V2=R3*I4
V3=R7*I8

Answers:
Ra =
        3000
```

continued

Example 2.18 continued
MATLAB continued

```
Rb =
         5000
I1 =
   0.002000000000000
I2 =
   0.005000000000000
I3 =
   0.003000000000000
I4 =
   0.002500000000000
I5 =
   0.002500000000000
I6 =
   0.002250000000000
I7 =
     7.500000000000001e-04
I8 =
   0.002000000000000
I9 =
     1.000000000000000e-03
V1 =
    15
V2 =
     5
V3 =
   6.000000000000002
```

Exercise 2.18

Find I_1, I_2, I_3, and I_4 in the circuit shown in Figure 2.64.

FIGURE 2.64

Circuit for EXERCISE 2.18.

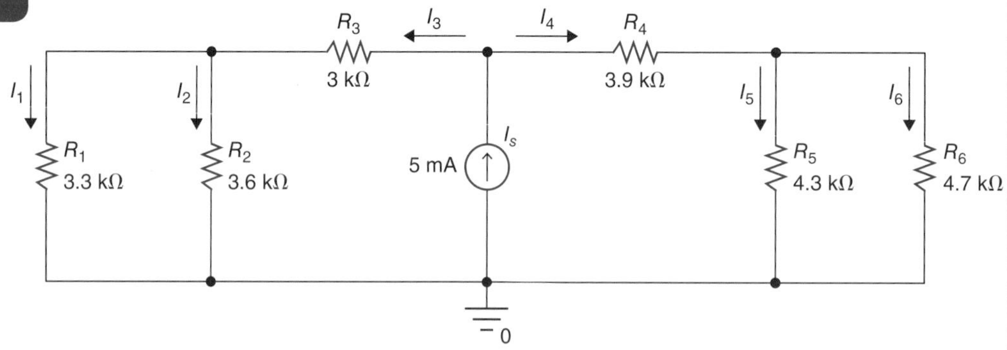

Answer:
I_1 = 1.4752 mA, I_2 = 1.3523 mA, I_3 = 2.8275 mA, I_4 = 2.1725 mA, I_5 = 1.1345 mA, and I_6 = 1.0380 mA.

2.10 Delta-Wye (Δ-Y) Transformation and Wye-Delta (Y-Δ) Transformation

FIGURE 2.65

(a) delta configuration. (b) wye configuration.

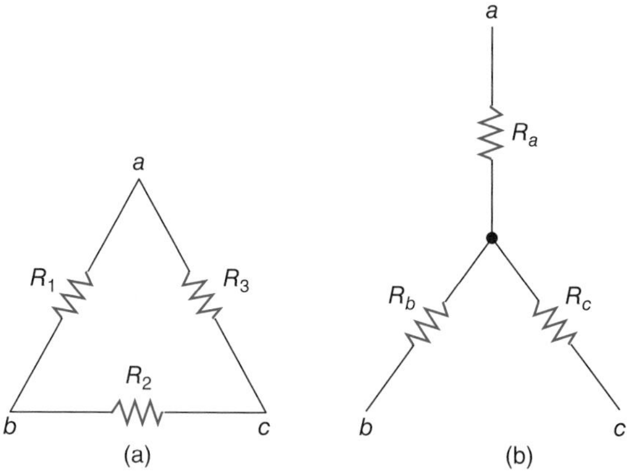

The three resistors R_1, R_2, and R_3 in Figure 2.65(a) are configured in triangular shape. This configuration is called *delta* (Δ). If a delta configuration appears in a circuit, it may be difficult to reduce the circuit directly using series or parallel connection of resistors. It may be easier to work with the wye (Y) configuration shown in Figure 2.65(b), where the three resistors R_a, R_b, and R_c are arranged in a *Y*-shape. In other cases, the delta form may be easier to work with than the *Y* form.

The transformations from delta to wye and wye to delta can be found by equating the resistances seen from the two terminals of wye and delta. Let us find the resistance for terminals *a* and *b*, R_{ab}, with terminal *c* open. From the wye circuit, we have $R_{ab} = R_a + R_b$. From the delta circuit, we have $R_{ab} = R_1 \| (R_2 + R_3)$. These two are equal. Thus, we have

$$R_{ab} = R_a + R_b = \frac{R_1(R_2 + R_3)}{R_1 + R_2 + R_3} \tag{2.60}$$

Similarly, R_{bc} and R_{ca} are given, respectively, by

$$R_{bc} = R_b + R_c = \frac{R_2(R_1 + R_3)}{R_2 + R_1 + R_3} \tag{2.61}$$

$$R_{ca} = R_c + R_a = \frac{R_3(R_1 + R_2)}{R_3 + R_1 + R_2} \tag{2.62}$$

When Equation (2.61) is subtracted from the sum of Equations (2.60) and (2.62), we have

$$R_a = \frac{R_1 R_3}{R_1 + R_2 + R_3} \tag{2.63}$$

When Equation (2.62) is subtracted from the sum of Equations (2.60) and (2.61), we have

$$R_b = \frac{R_1 R_2}{R_1 + R_2 + R_3} \tag{2.64}$$

When Equation (2.60) is subtracted from the sum of Equations (2.61) and (2.62), we have

$$R_c = \frac{R_2 R_3}{R_1 + R_2 + R_3} \tag{2.65}$$

Equations (2.63)–(2.65) provide resistance values for a wye configuration from a delta configuration.

The sum of the products of Equations (2.63) and (2.64), (2.64) and (2.65), and (2.63) and (2.65) [(2.63) × (2.64) + (2.64) × (2.65) + (2.63) × (2.65)] results in

$$R_aR_b + R_bR_c + R_aR_c = \frac{R_1R_2R_3}{R_1 + R_2 + R_3} \quad (2.66)$$

When Equation (2.66) is divided by Equation (2.65), we get

$$R_1 = \frac{R_aR_b + R_bR_c + R_aR_c}{R_c} \quad (2.67)$$

When Equation (2.66) is divided by Equation (2.63), we get

$$R_2 = \frac{R_aR_b + R_bR_c + R_aR_c}{R_a} \quad (2.68)$$

When Equation (2.66) is divided by Equation (2.64), we get

$$R_3 = \frac{R_aR_b + R_bR_c + R_aR_c}{R_b} \quad (2.69)$$

Equations (2.67)–(2.69) provide resistance values for a delta configuration from a wye configuration.

EXAMPLE 2.19

Find the voltage V_c in the circuit shown in Figure 2.66.

FIGURE 2.66

Circuit for EXAMPLE 2.19.

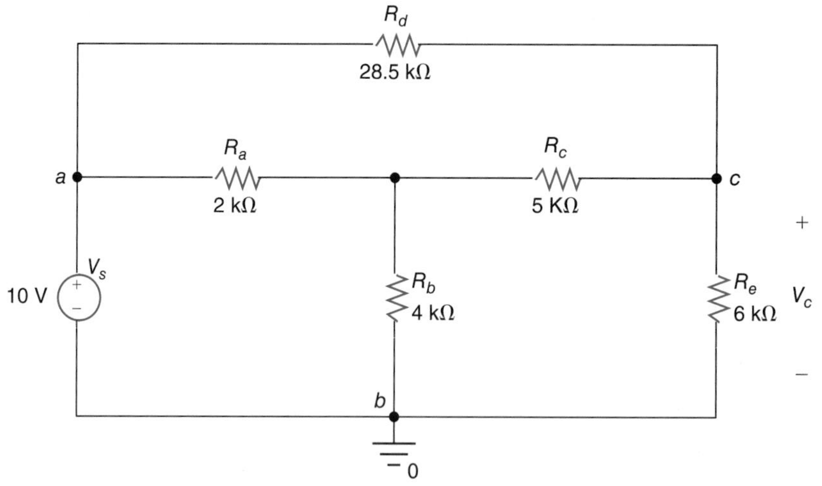

The three resistors R_a, R_b, and R_c form a wye configuration. The wye configuration can be transformed to a delta configuration, as shown in Figure 2.67.

continued

Example 2.19 continued

FIGURE 2.67

Wye to delta transformation.

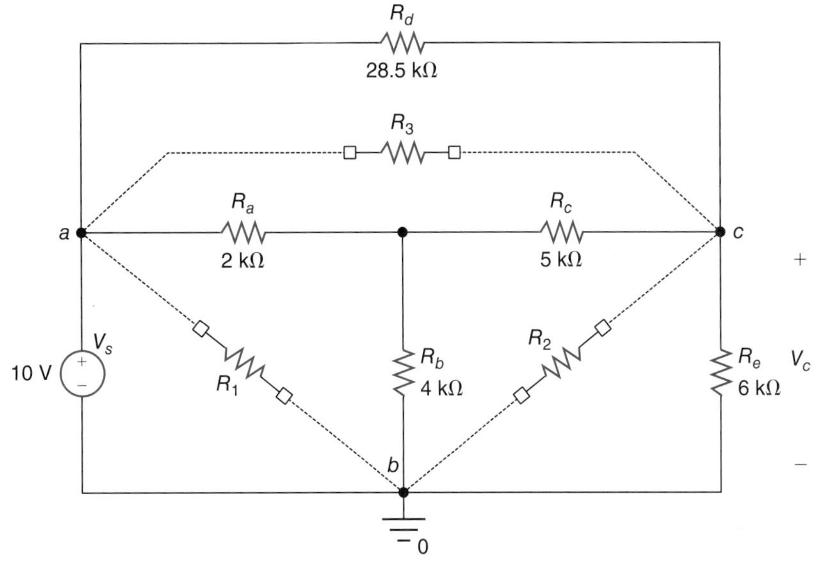

The resistor values for a delta configuration are given by

$$R_1 = \frac{R_aR_b + R_bR_c + R_aR_c}{R_c} = \frac{2\,k\Omega \times 4\,k\Omega + 4\,k\Omega \times 5\,k\Omega + 2\,k\Omega \times 5\,k\Omega}{5\,k\Omega} = 7.6\,k\Omega \quad \textbf{(2.70)}$$

$$R_2 = \frac{R_aR_b + R_bR_c + R_aR_c}{R_a} = \frac{2\,k\Omega \times 4\,k\Omega + 4\,k\Omega \times 5\,k\Omega + 2\,k\Omega \times 5\,k\Omega}{2\,k\Omega} = 19\,k\Omega \quad \textbf{(2.71)}$$

$$R_3 = \frac{R_aR_b + R_bR_c + R_aR_c}{R_b} = \frac{2\,k\Omega \times 4\,k\Omega + 4\,k\Omega \times 5\,k\Omega + 2\,k\Omega \times 5\,k\Omega}{4\,k\Omega} = 9.5\,k\Omega \quad \textbf{(2.72)}$$

After the conversion to the delta configuration, R_d and R_3 are in parallel, and R_e and R_2 are in parallel. Let $R_f = R_d \| R_3$ and $R_g = R_e \| R_2$. Then, we have

$$R_f = R_d \| R_3 = \frac{R_d \times R_3}{R_d + R_3} = \frac{28.5\,k\Omega \times 9.5\,k\Omega}{28.5\,k\Omega + 9.5\,k\Omega} = 7.125\,k\Omega$$

$$R_g = R_e \| R_2 = \frac{R_e \times R_2}{R_e + R_2} = \frac{6\,k\Omega \times 19\,k\Omega}{6\,k\Omega + 19\,k\Omega} = 4.56\,k\Omega$$

continued

Example 2.19 continued

The circuit reduces to the one shown in Figure 2.68.

FIGURE 2.68

The circuit after simplification.

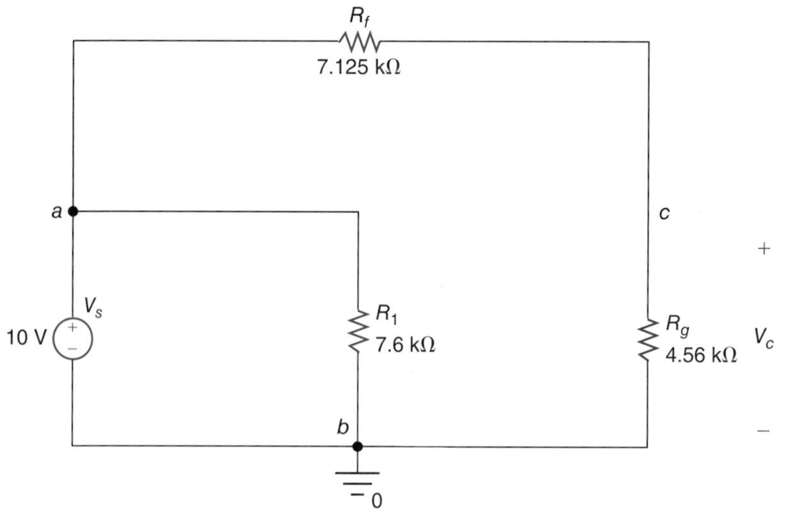

The voltage V_c is obtained by applying the voltage divider rule:

$$V_c = V_s \times \frac{R_g}{R_f + R_g} = 10\ \text{V} \times \frac{4.56\ k\Omega}{7.125\ k\Omega\ +\ 4.56\ k\Omega} = 3.9024\ \text{V}$$

A MATLAB function Y2D changes a wye configuration to a delta configuration:

```
% Wye to delta conversion
function [R1 R2 R3]=Y2D(x)
R1=(x(1)*x(2)+x(2)*x(3)+x(1)*x(3))/x(3);
R2=(x(1)*x(2)+x(2)*x(3)+x(1)*x(3))/x(1);
R3=(x(1)*x(2)+x(2)*x(3)+x(1)*x(3))/x(2);
end
```

The transformation of a wye configuration with $R_a = 2\ k\Omega$, $R_b = 4\ k\Omega$, and $R_c = 5\ k\Omega$ (shown in Figure 2.66) to a delta configuration with $R_1 = 7.6\ k\Omega$, $R_2 = 19\ k\Omega$, and $R_3 = 9.5\ k\Omega$ can be achieved using Y2D:

```
>> [R1 R2 R3]=Y2D([2000,4000,5000])
R1 =
   7600
R2 =
  19000
R3 =
   9500
```

Exercise 2.19

Find V_a in the circuit shown in Figure 2.69.

FIGURE 2.69

Circuit for EXERCISE 2.19.

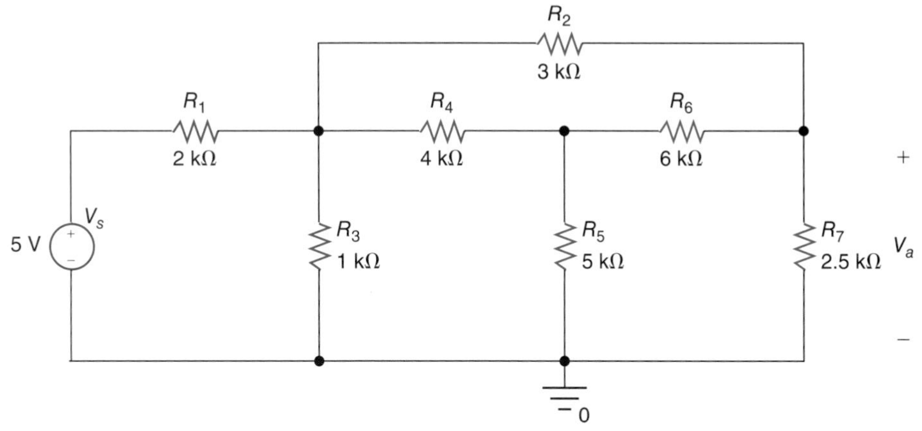

Answer:

$V_a = 0.6535$ V.

EXAMPLE 2.20

Find the voltages V_1, V_2, V_3, V_4, and V_5, and currents I_1, I_2, I_3, I_4, I_5, and I in the circuit shown in Figure 2.70.

FIGURE 2.70

Circuit for EXAMPLE 2.20.

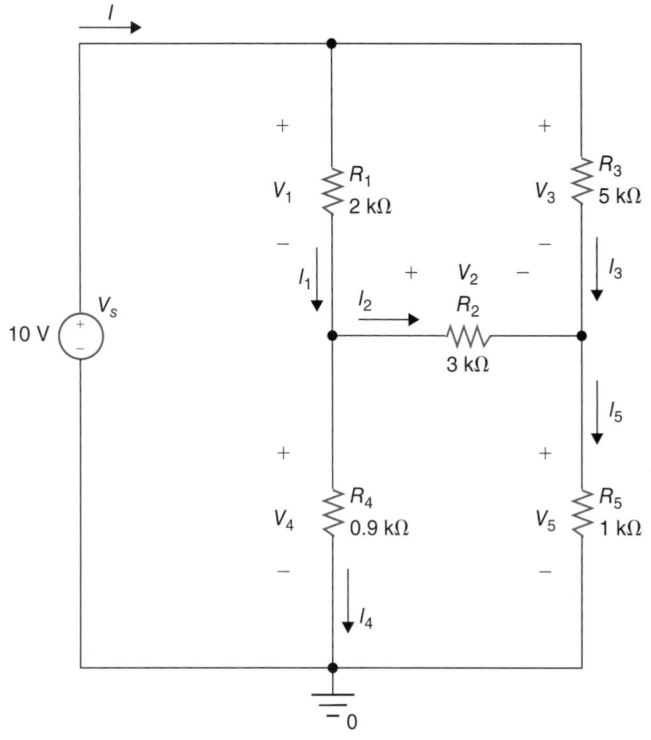

continued

Example 2.20 continued

When the delta consisting of R_1, R_2, and R_3 is transformed to a wye consisting of R_a, R_b, and R_c, we obtain the circuit shown in Figure 2.71.

FIGURE 2.71

The circuit after the delta to wye conversion.

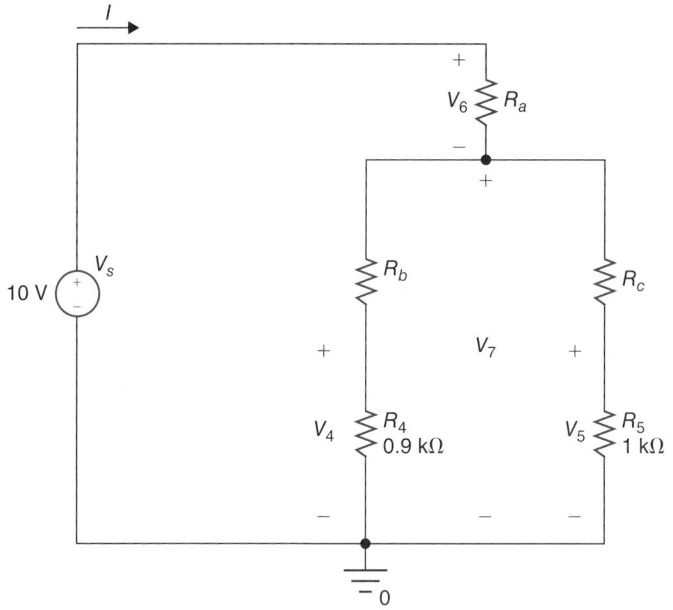

The values of R_a, R_b, and R_c are found from Equations (2.63)–(2.65):

$$R_a = \frac{R_1R_3}{R_1 + R_2 + R_3} = \frac{2\,k\Omega \times 5\,k\Omega}{2\,k\Omega + 3\,k\Omega + 5\,k\Omega} = 1\,k\Omega$$

$$R_b = \frac{R_1R_2}{R_1 + R_2 + R_3} = \frac{2\,k\Omega \times 3\,k\Omega}{2\,k\Omega + 3\,k\Omega + 5\,k\Omega} = 0.6\,k\Omega$$

$$R_c = \frac{R_2R_3}{R_1 + R_2 + R_3} = \frac{3\,k\Omega \times 5\,k\Omega}{2\,k\Omega + 3\,k\Omega + 5\,k\Omega} = 1.5\,k\Omega$$

The equivalent resistance seen from the voltage source is given by

$$R_{\text{eq}} = R_a + (R_b + R_4)\|(R_c + R_5) = R_a + \frac{(R_b + R_4)(R_c + R_5)}{R_b + R_4 + R_c + R_5}$$
$$= 1\,k\Omega + \frac{(0.6\,k\Omega + 0.9\,k\Omega)(1.5\,k\Omega + 1\,k\Omega)}{0.6\,k\Omega + 0.9\,k\Omega + 1.5\,k\Omega + 1\,k\Omega} = 1.9375\,k\Omega$$

The current I from the voltage source is given by

$$I = \frac{V_s}{R_{\text{eq}}} = \frac{10\text{ V}}{1.9375\,k\Omega} = 5.16129\text{ mA}$$

From Ohm's law, the voltage drop V_6 across R_a is given by

$$V_6 = R_aI = 1\,k\Omega \times 5.16129\text{ mA} = 5.16129\text{ V}$$

continued

Example 2.20 continued

The voltage V_7 in the middle of the wye is given by

$$V_7 = V_s - V_6 = 10\,\text{V} - 5.016129\,\text{V} = 4.83871\,\text{V}$$

Applying the voltage divider rule, we get

$$V_4 = \frac{R_4}{R_b + R_4} V_7 = \frac{0.9\,k\Omega}{0.6\,k\Omega + 0.9\,k\Omega} 4.8387\,\text{V} = 2.903226\,\text{V}$$

$$V_5 = \frac{R_5}{R_c + R_5} V_7 = \frac{1\,k\Omega}{1.5\,k\Omega + 1\,k\Omega} 4.8387\,\text{V} = 1.935484\,\text{V}$$

The voltage V_1 is obtained by subtracting V_4 from V_s; that is,

$$V_1 = V_s - V_4 = 10\,\text{V} - 2.903226\,\text{V} = 7.096774\,\text{V}$$

Similarly, the voltage V_3 is obtained by subtracting V_5 from V_s; that is,

$$V_3 = V_s - V_5 = 10\,\text{V} - 1.935484\,\text{V} = 8.064516\,\text{V}$$

The voltage V_2 is obtained by subtracting V_5 from V_4; that is,

$$V_2 = V_4 - V_5 = 2.903226\,\text{V} - 1.935484\,\text{V} = 0.967742\,\text{V}$$

The currents I_1, I_2, I_3, I_4, and I_5 are obtained by applying Ohm's law to each branch:

$$I_1 = \frac{V_1}{R_1} = \frac{7.096774\,\text{V}}{2\,k\Omega} = 3.548387\,\text{mA}$$

$$I_2 = \frac{V_2}{R_2} = \frac{0.967742\,\text{V}}{3\,k\Omega} = 0.322581\,\text{mA}$$

$$I_3 = \frac{V_3}{R_3} = \frac{8.064516\,\text{V}}{5\,k\Omega} = 1.612903\,\text{mA}$$

$$I_4 = \frac{V_4}{R_4} = \frac{2.903226\,\text{V}}{0.9\,k\Omega} = 3.225806\,\text{mA}$$

$$I_5 = \frac{V_5}{R_5} = \frac{1.935484\,\text{V}}{1\,k\Omega} = 1.935484\,\text{mA}$$

A MATLAB function D2Y changes a delta configuration to a wye configuration:

```
%Delta to wye conversion
function [Ra Rb Rc]=D2Y(x)
Ra=x(1)*x(3)/sum(x);
Rb=x(1)*x(2)/sum(x);
Rc=x(2)*x(3)/sum(x);
end
```

continued

Example 2.20 continued

The transformation of the delta configuration with $R_1 = 2\ k\Omega$, $R_2 = 3\ k\Omega$, and $R_3 = 5\ k\Omega$ (shown in Figure 2.70) to the wye configuration with $R_a = 1\ k\Omega$, $R_b = 600\ \Omega$, and $R_c = 1.5\ k\Omega$ (shown in Figure 2.71) can be achieved using D2Y:

```
>> [Ra Rb Rc]=D2Y([2000,3000,5000])
Ra =
   1000
Rb =
600
Rc =
   1500
```

MATLAB

```
%EXAMPLE 2.20
%Functions D2Y.m and P.m should be in the same folder as this file.
clear all;format long;
Vs=10;
R1=2000;R2=3000;R3=5000;R4=900;R5=1000;
[Ra,Rb,Rc]=D2Y([R1,R2,R3])
Req=Ra+P([Rb+R4,Rc+R5])
I=Vs/Req
V6=Ra*I
V7=Vs-V6
V4=V7*R4/(Rb+R4)
V5=V7*R5/(Rc+R5)
V1=Vs-V4
V3=Vs-V5
V2=V4-V5
I1=V1/R1
I2=V2/R2
I3=V3/R3
I4=V4/R4
I5=V5/R5

Answers:
Ra =
        1000
Rb =
   600
Rc =
        1500
Req =
      1.937500000000000e+03
I =
   0.005161290322581
V6 =
   5.161290322580645
V7 =
   4.838709677419355
V4 =
   2.903225806451613
```

continued

Example 2.20 continued

MATLAB continued

```
V5 =
    1.935483870967742
V1 =
    7.096774193548387
V3 =
    8.064516129032258
V2 =
    0.967741935483871
I1 =
    0.003548387096774
I2 =
      3.225806451612903e-04
I3 =
    0.001612903225806
I4 =
    0.003225806451613
I5 =
    0.001935483870968
```

Exercise 2.20

Find V_a in the circuit shown in Figure 2.72.

FIGURE 2.72

Circuit for EXERCISE 2.20.

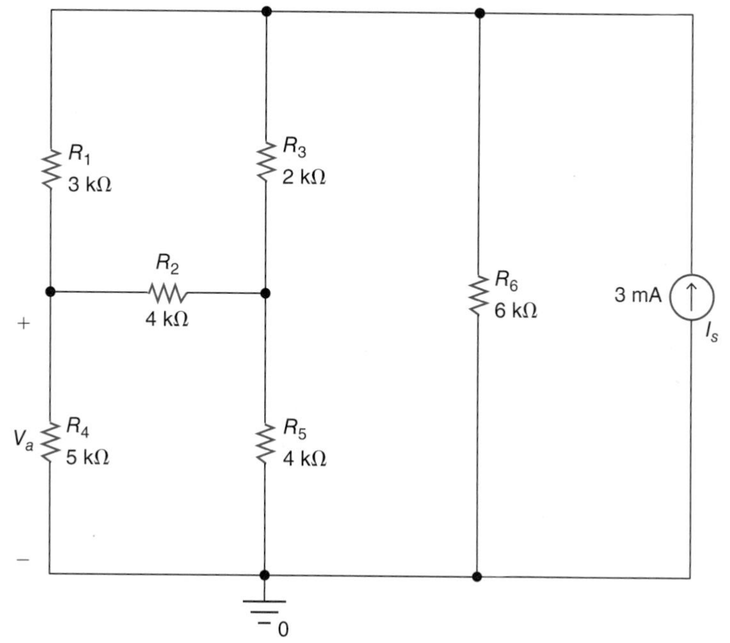

Answer:

$V_a = 4.1597$ V.

2.11 PSpice and Simulink

PSpice® is an application for PCs that can be used to simulate circuits. In this text, PSpice is introduced using the OrCAD demo software. Click on the Place Part icon to start. If not already done, click on Add Library and add the following libraries:

analog, analog_p, breakout, eval, source, special

Other libraries can be added if needed. Select all the libraries and enter *vdc* inside Part. Click Enter and move the cursor to the left side of the window and place the dc voltage source by clicking. Right-click and choose the End mode to stop placing dc voltage sources. Double-click on 0Vdc and change the voltage to 12Vdc, as shown in Figure 2.73.

Enter R inside Part. Place a resistor R_1 as shown in Figure 2.74. Click Ctrl + R to rotate the resistor in the vertical direction and place R_2 to the right side, as shown in Figure 2.74. Right-click and choose the End mode. Change the values of the resistors to $2k$ and $4k$.

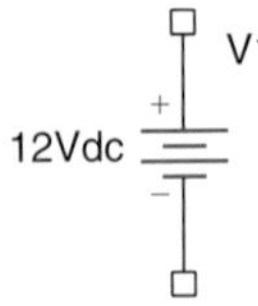

FIGURE 2.73

Placement of a voltage source.

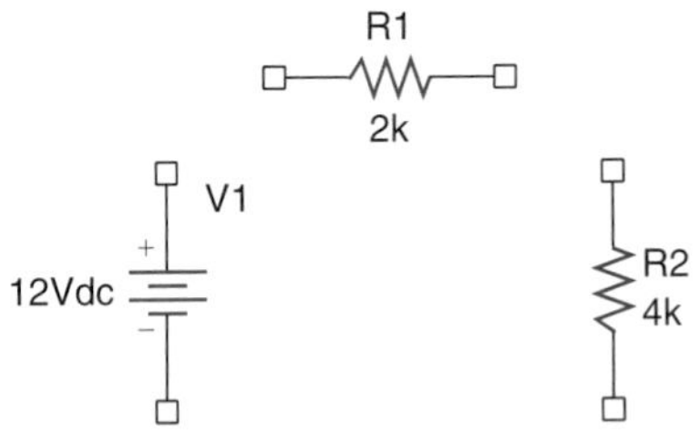

FIGURE 2.74

Placement of resistors.

Click on the Place Ground icon. Click on 0/SOURCE and OK. Place the ground at the bottom of the circuit, as shown in Figure 2.75.

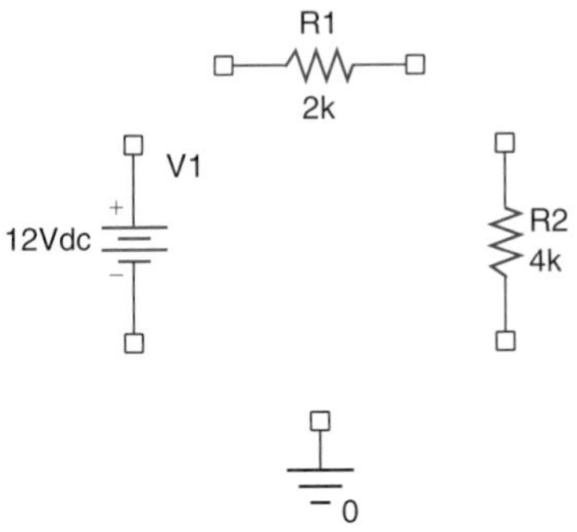

FIGURE 2.75

Placement of ground.

Click on the Place Wire icon. Place the cursor at the beginning of the wire, drag the mouse to the end of the wire, and click. To change direction, click and move in a different direction. Repeat this procedure to complete the wiring. When finished, right-click and choose End Wire. Figure 2.76 shows the finished schematic.

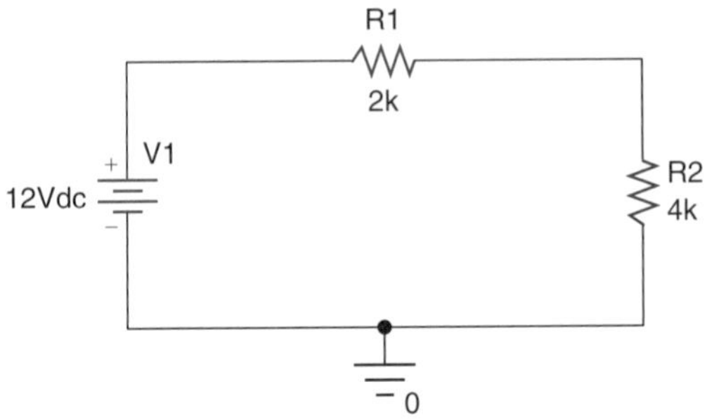

FIGURE 2.76

The wiring schematic.

Click on New Simulation Profile icon or select PSpice → New Simulation Profile. Enter a name and select Bias Point as the Analysis type. Click on Run PSpice icon or select PSpice → Run. Click on Enable Bias Voltage Display (V). The positive voltage for each element is displayed as shown in Figure 2.77.

FIGURE 2.77

Voltage display.

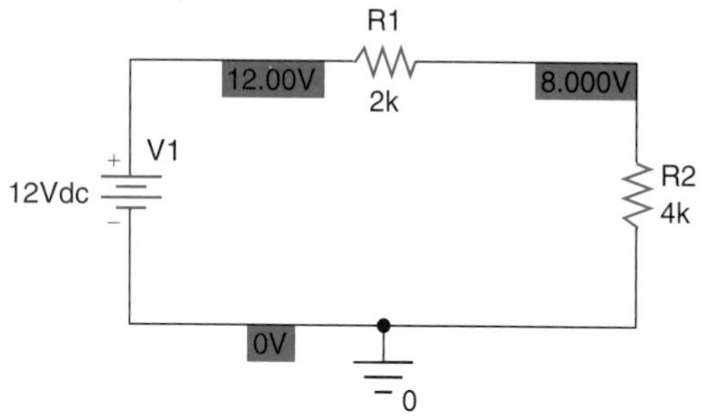

You can move the voltage display by clicking on it and moving it to the location that you want. Click on Enable Bias Current Display (I). The current values are displayed as shown in Figure 2.78. The current entering each element, 2 mA, is displayed.

FIGURE 2.78

Current display.

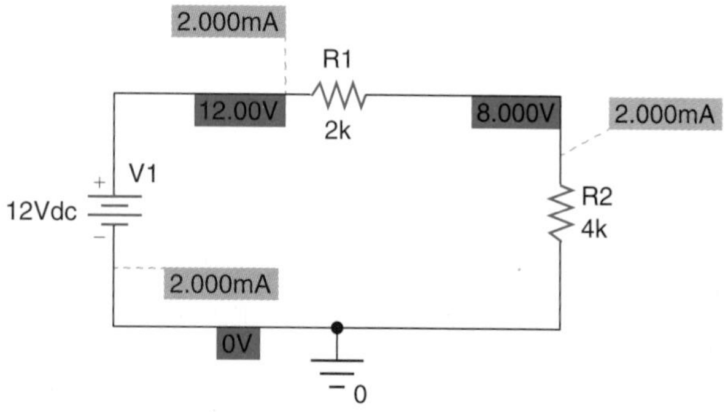

Click on Enable Bias Power Display (W). The power is displayed for each element as shown in Figure 2.79. The power on the voltage source is −24 mW, the power on R_1 is 8 mW, and the power on R_2 is 16 mW. The power delivered, 24 mW, is equal to the power absorbed: 8 mW + 16 mW = 24 mW.

FIGURE 2.79

Power display.

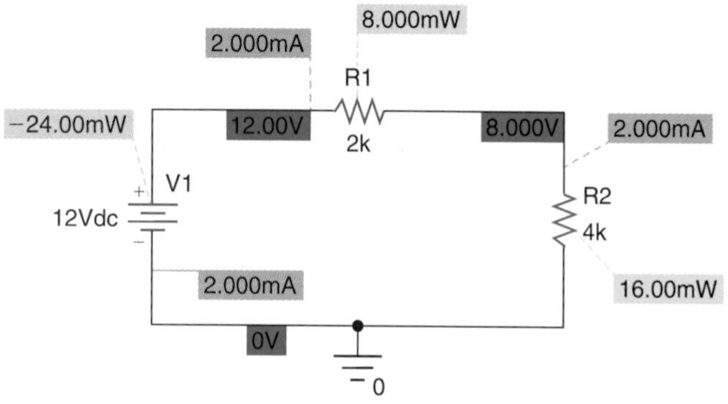

More detailed information on the simulation is available in the output file (to view, select PSpice→View Output File or click on the View Simulation Output File icon).

EXAMPLE 2.21

Build the circuit shown in Figure 2.80 and simulate it using PSpice. Find the voltages and currents everywhere.

The part name for voltage source is VSRC and the part name for current source is ISRC. For both sources, enter the values for dc only. The results of the simulation are shown in Figure 2.81.

FIGURE 2.80

Circuit for EXAMPLE 2.21.

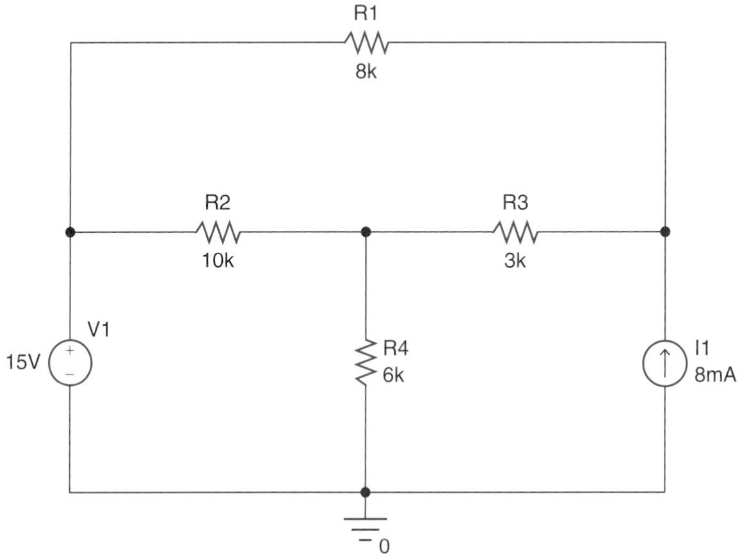

FIGURE 2.81

Results of the simulation.

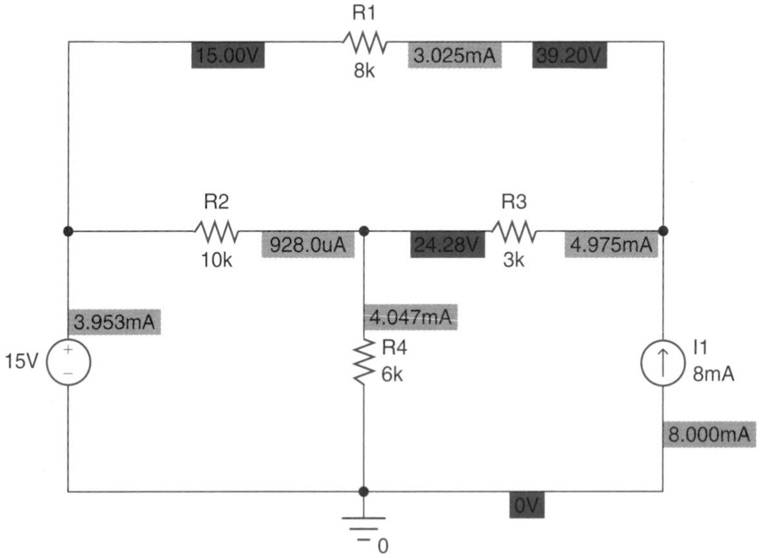

Exercise 2.21

Build the circuit shown in Figure 2.82 and simulate it using PSpice. Find the voltages at every node in the circuit.

FIGURE 2.82

Circuit for EXERCISE 2.21.

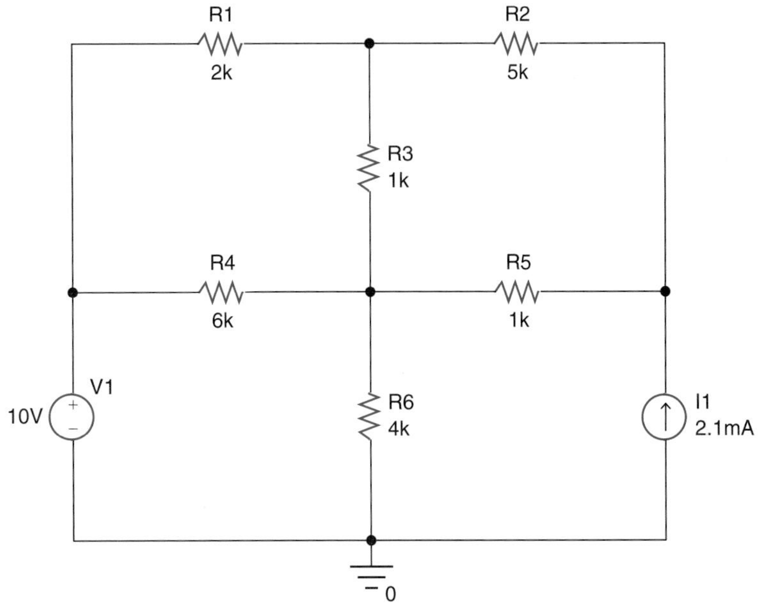

Answer:

Figure 2.83 shows the result of the simulation.

FIGURE 2.83

Circuit with node voltages.

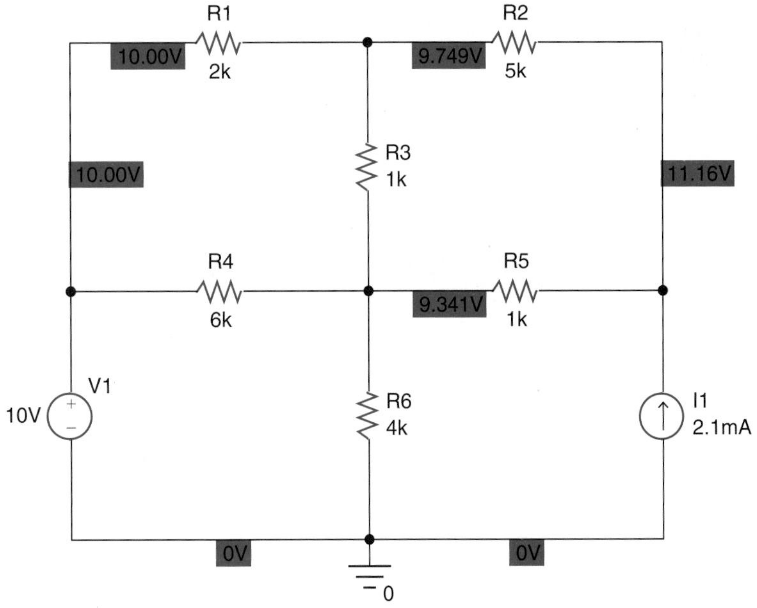

2.11.1 SIMULINK

Simscapes of Simulink® make it possible to simulate electric circuits, as shown in Figure 2.84. The blocks are available in the Simscape Library. PS-Simulink Converter converts physical signals to Simulink output signals for further processing in Simulink. Figure 2.85 displays the voltage across the $2k\Omega$ resistor on the scope.

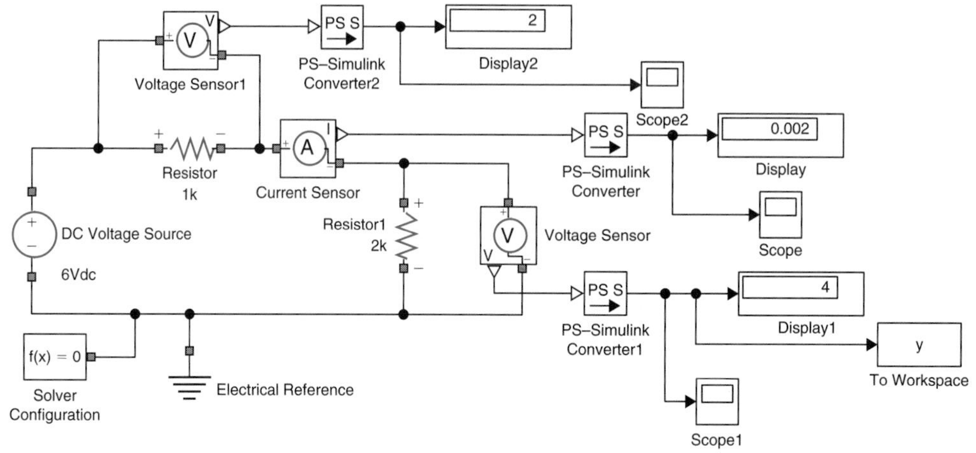

FIGURE 2.84
A Simulink model for a circuit with dc voltage source and two resistors.

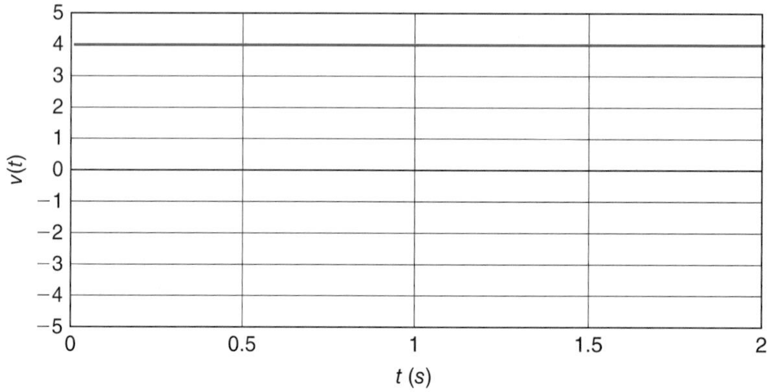

FIGURE 2.85
The voltage across the 2-$k\Omega$ resistor.

SUMMARY

According to Ohm's law, the voltage across the resistor is proportional to the current through the resistor. The proportionality constant is called the *resistance*. For the given voltage, the amount of current flowing through the resistor decreases as the resistance value increases.

KCL says that the sum of currents entering a node equals the sum of currents leaving the same node. Stated another way, the sum of currents leaving a node is zero. KCL is the basis in setting up node equations that lead to node voltages.

KVL says that the sum of voltage drops around a mesh equals the sum of voltage rises around the same mesh. Stated another way, the sum of voltage drops around a mesh is zero. KVL is the basis in setting up mesh equations that lead to mesh currents.

The voltage divider rule says that if resistors are connected to a voltage source in series, the voltage from the voltage source is divided among the resistors in proportion to the resistance value. For certain circuits, the voltage divider rule makes it possible to calculate voltages across resistors with minimal effort.

The current divider rule says that if resistors are connected to a current source in parallel, the current from the current source is divided among the resistors in proportion to the conductance value. For certain circuits, the current divider rule makes it possible to calculate currents through resistors with minimal effort.

PROBLEMS

Ohm's Law

2.1 In the circuit shown in Figure P2.1, the voltage across resistors R_1 and R_2 is 6 V. Find the current I_1 through R_1, and the current I_2 through R_2.

FIGURE P2.1

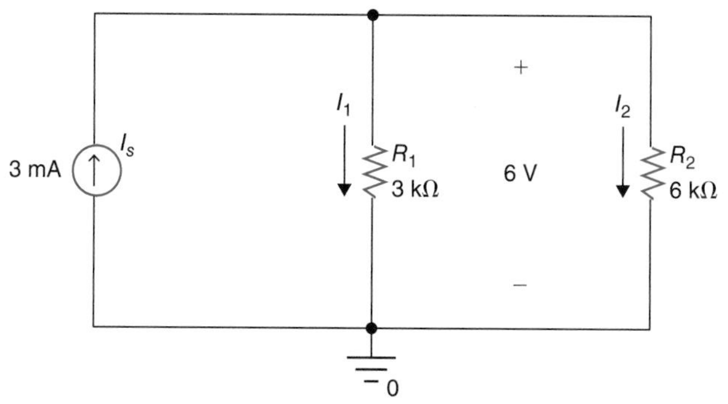

2.2 In the circuit shown in Figure P2.2, the voltage across the resistor R_1 is 2.4 V, and the voltage across the resistors R_2 and R_3 is 3.6 V. Find the currents I_1, I_2, and I_3.

FIGURE P2.2

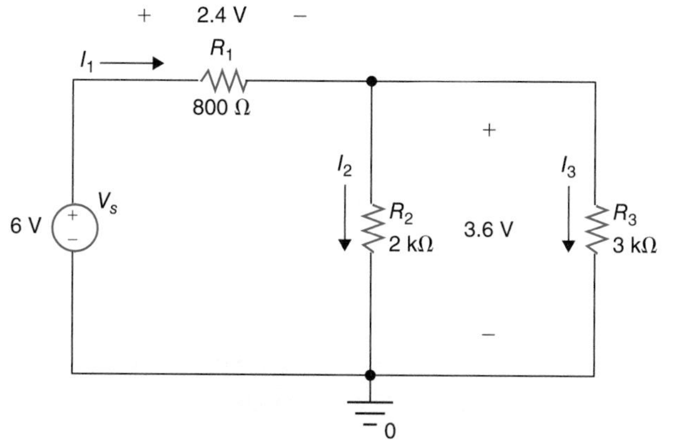

2.3 In the circuit shown in Figure P2.3, the voltage across the resistor R_1 and R_2 is 2.4 V, and the voltage across the resistors R_3, R_4, and R_5 is 1.2 V. Find the currents I_1, I_2, I_3, I_4, and I_5.

FIGURE P2.3

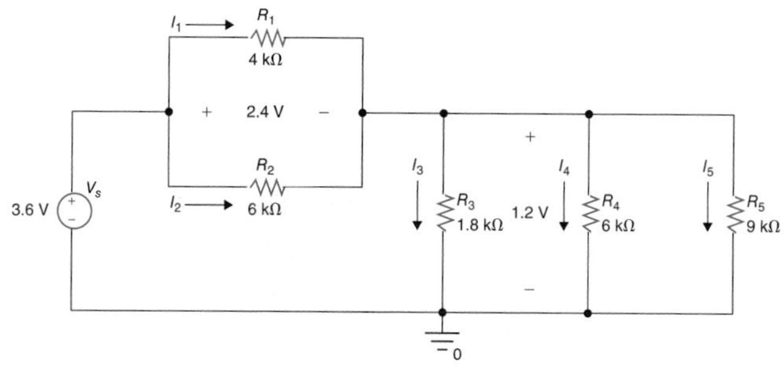

2.4 In the circuit shown in Figure P2.4, the current through the resistor R_2 is $I_2 = 1.2$ mA, and the voltage across the resistor R_1 is 2.8 V. Find the voltage V_o and currents I_1, and I_3.

FIGURE P2.4

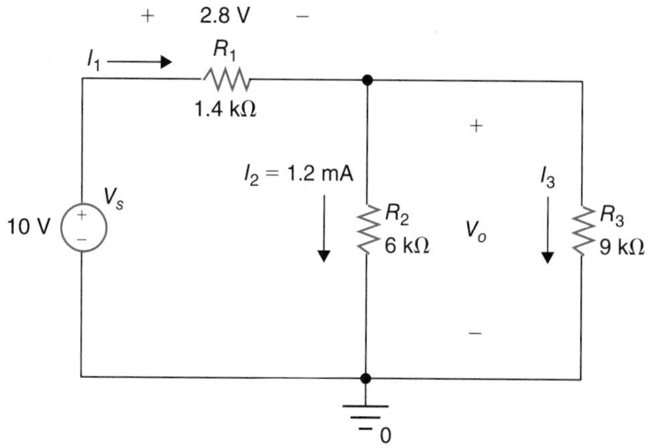

2.5 In the circuit shown in Figure P2.5, the current through the resistor R_4 is $I_4 = 0.2$ mA. Find the voltage V_o and current I_3.

FIGURE P2.5

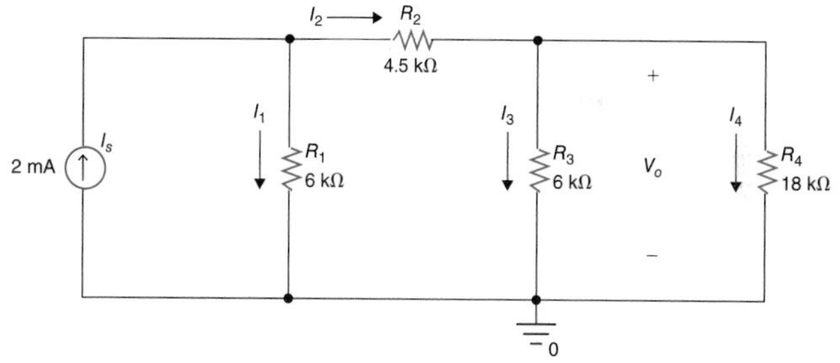

2.6 In the circuit shown in Figure P2.6, the current through the resistor R_4 is $I_4 = 0.4$ mA. Find the voltage V_o and currents I_2 and I_3.

FIGURE P2.6

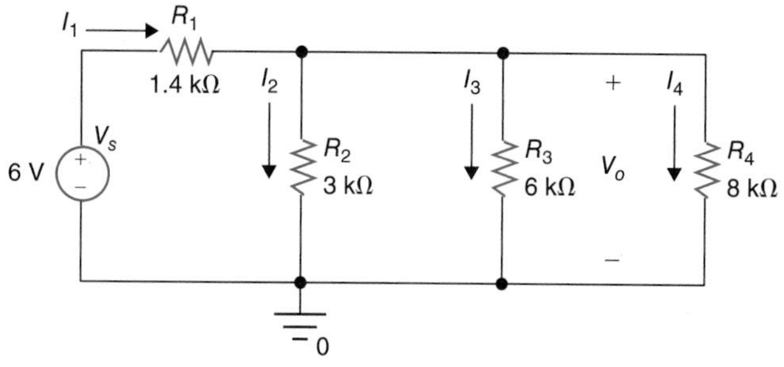

2.7 In the circuit shown in Figure P2.7, the current through the resistor R_2 is $I_2 = 7/60$ mA, and the current through the resistor R_3 is $I_3 = 1/12$ mA. Find the voltage V_o and the resistance value of R_2.

FIGURE P2.7

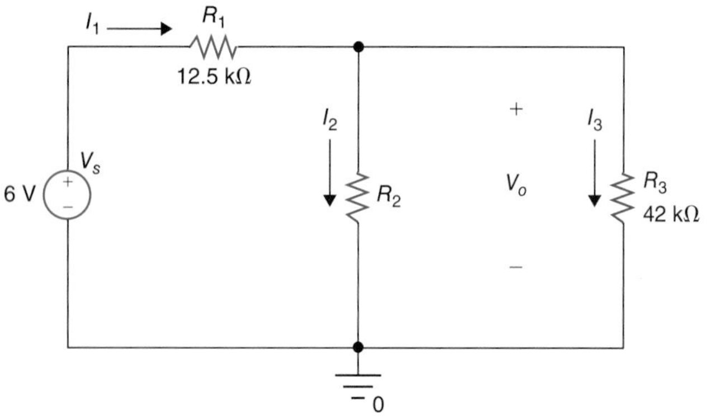

2.8 In the circuit shown in Figure P2.8, the current through the mesh is $I = 2$ mA. Find the power values on R_1, R_2, and V_s and state whether the power is absorbed or delivered.

FIGURE P2.8

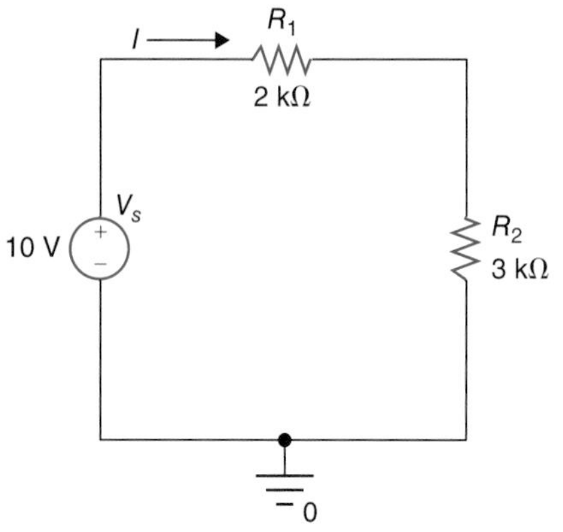

2.9 In the circuit shown in Figure P2.9, the voltage across I_s, R_1, and R_2 is $V_o = 4.8$ V. Find the power values on R_1, R_2, and I_s and state whether the power is absorbed or delivered.

FIGURE P2.9

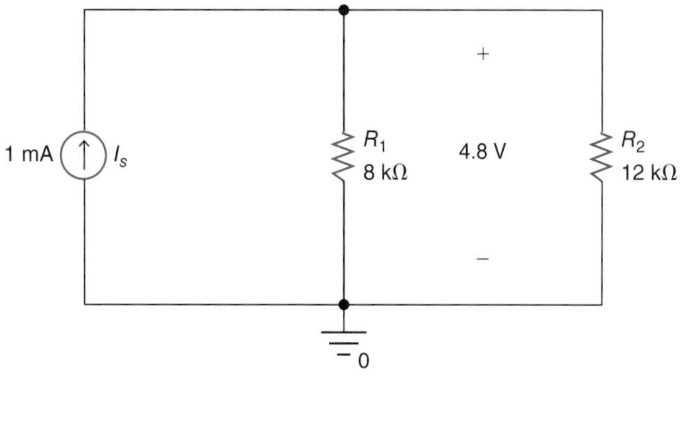

2.10 In the circuit shown in Figure P2.10, all the node voltages are given. Find the values of currents I_1, I_2, I_3, and I_4.

FIGURE P2.10

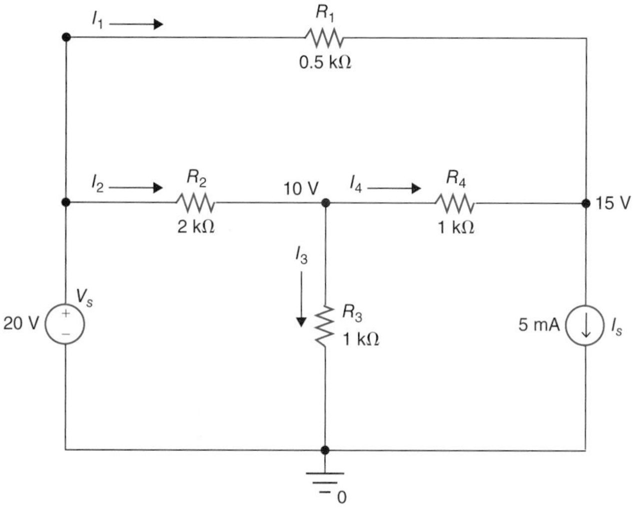

2.11 In the circuit shown in Figure P2.11, find i, I_1, I_2, I_3, I_4, and I_5.

FIGURE P2.11

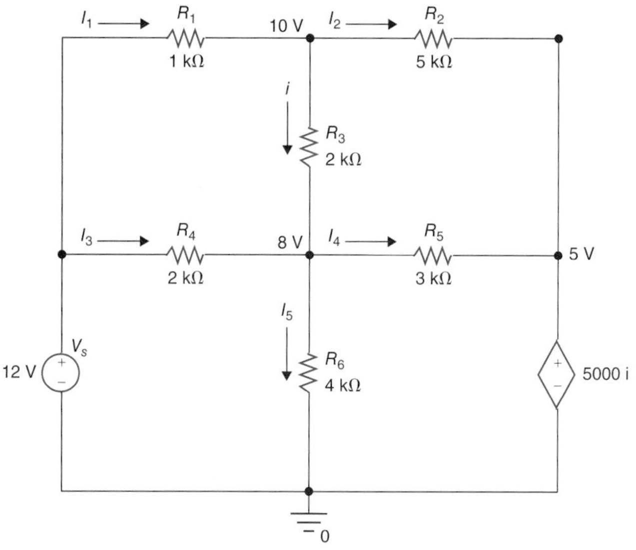

2.12 In the circuit shown in Figure P2.12, find I_1, I_2, I_3, I_4, I_5, I_6, and I_7.

FIGURE P2.12

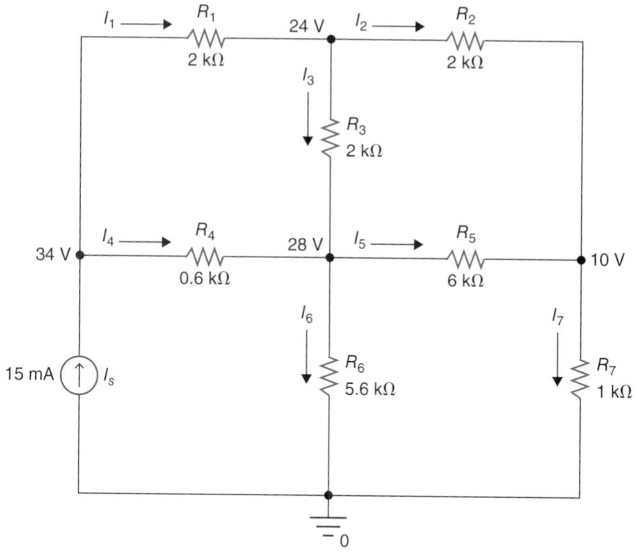

2.13 Find I, V_1, V_2, V_3, and V_4 in the circuit shown in Figure P2.13.

FIGURE P2.13

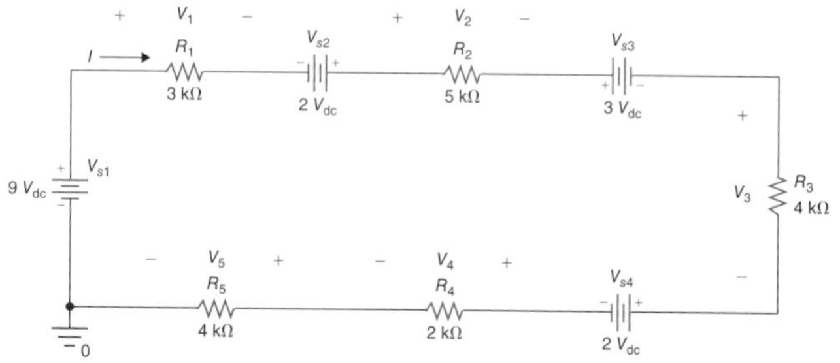

2.14 The diameter of a 26 AWG copper (conductivity $= \sigma = 5.69 \times 10^7$ S/m) wire is 0.405 mm. Find the resistance of the copper wire when the length of the wire is

a. $\ell = 20$ m
b. $\ell = 200$ m
c. $\ell = 2$ km
d. $\ell = 20$ km

Kirchhoff's Current Law (KCL)

2.15 In the circuit shown in Figure P2.15, the current through R_2 is given as $I_2 = 3$ mA. Find V_2, I_3, I_1, and V_1.

FIGURE P2.15

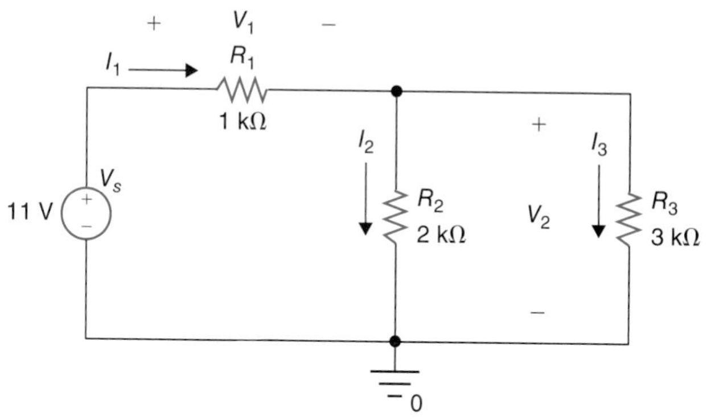

2.16 In the circuit shown in Figure P2.16, voltage V_2 is given as $V_2 = 6$ V. Find I_2, I_3, I_4, I_1, and V_1.

FIGURE P2.16

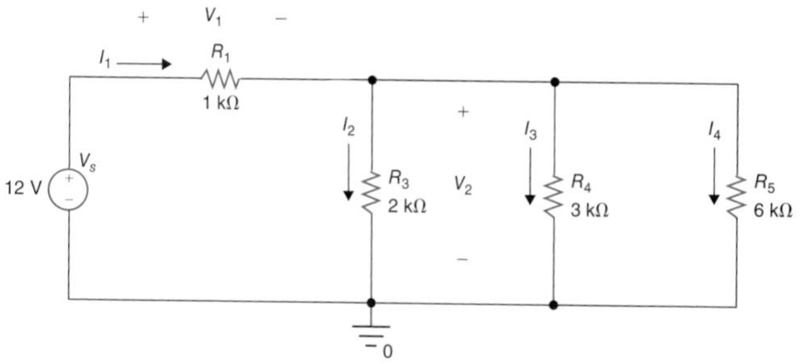

2.17 In the circuit shown in Figure P2.17, the current I_4 is given as $I_4 = 1$ mA. Find V_2, I_3, I_2, I_1, and V_1.

FIGURE P2.17

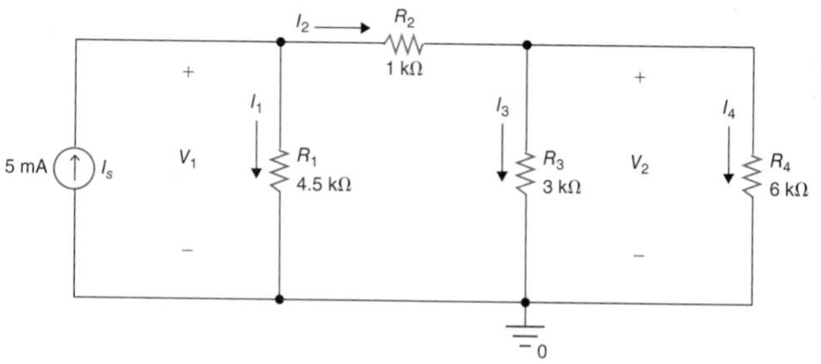

2.18 In the circuit shown in Figure P2.18, the voltage V_o is given as $V_o = 8$ V. Find I_3, I_4, I_1, and I_2.

FIGURE P2.18

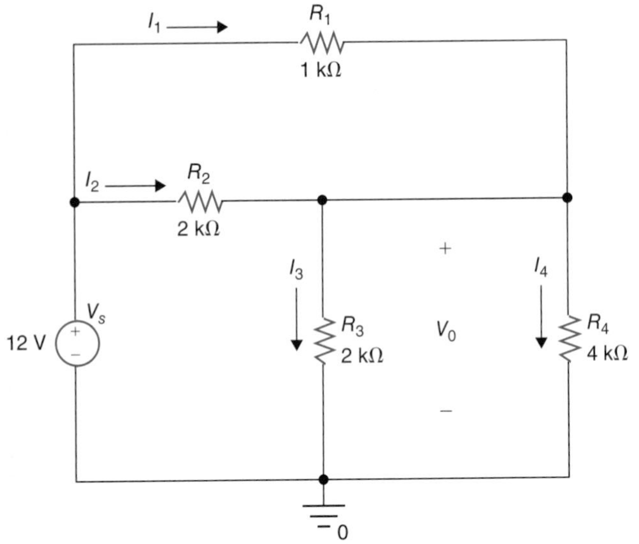

2.19 In the circuit shown in Figure P2.19, the voltage V_4 is given as $V_4 = 5$ V. Find I_3, V_3, V_2, I_2, I_1 and V_1.

FIGURE P2.19

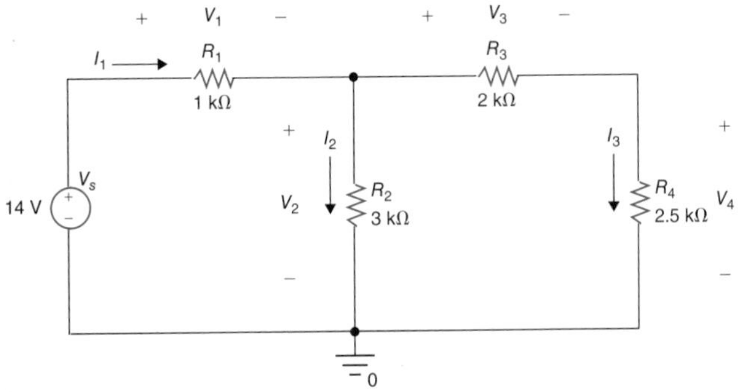

2.20 Let $I_1 = 5$ mA, $I_3 = 2$ mA, and $I_4 = 2$ mA in the circuit shown in Figure P2.20. Find I_2 and I_5.

FIGURE P2.20

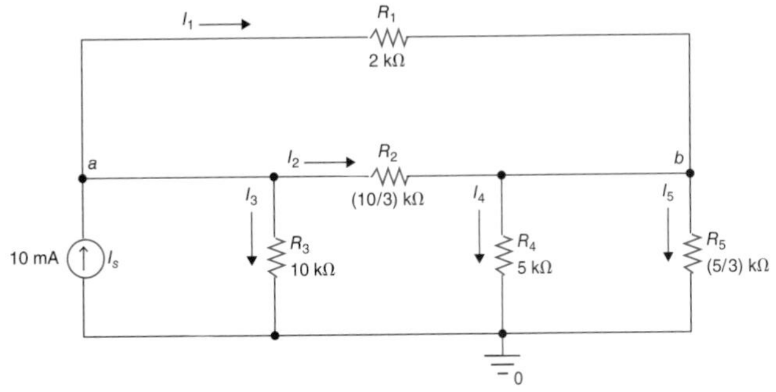

2.21 Let $I_1 = 2$ mA, and $I_3 = 10$ mA in the circuit shown in Figure P2.21. Find I_2, I_4, and I_5.

FIGURE P2.21

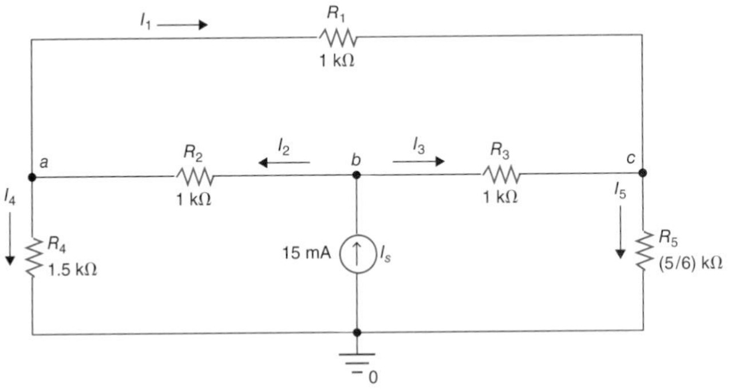

2.22 Let $I_3 = 5$ mA, $I_4 = 10$ mA, and $I_5 = 5$ mA in the circuit shown in Figure P2.22. Find I_1, I_2, I_6, and I_7.

FIGURE P2.22

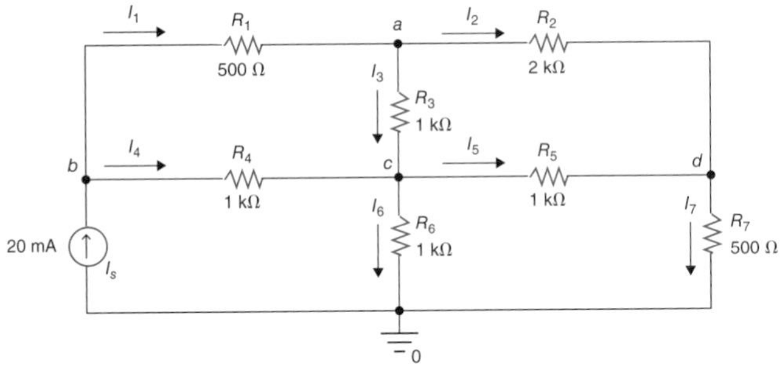

2.23 Find I_1, I_2, I_3, I_4, and I_5 in the circuit shown in Figure P2.23.

FIGURE P2.23

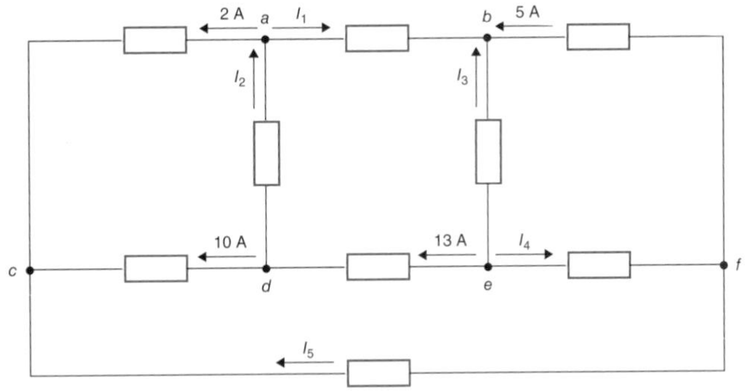

Kirchhoff's Voltage Law (KVL)

2.24 Let $V_{R_1} = 10$ V and $V_{R_4} = 15$ V in the circuit shown in Figure P2.24. Find V_{R_3} and V_{R_2}.

FIGURE P2.24

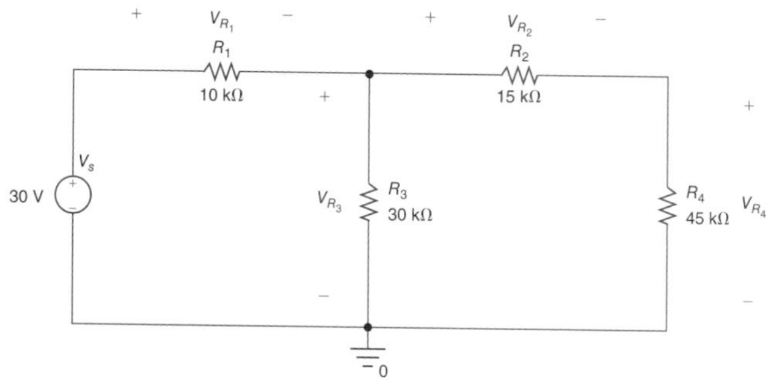

2.25 Find V_{R_2}, V_{R_4}, and V_{R_5} in the circuit shown in Figure P2.25.

FIGURE P2.25

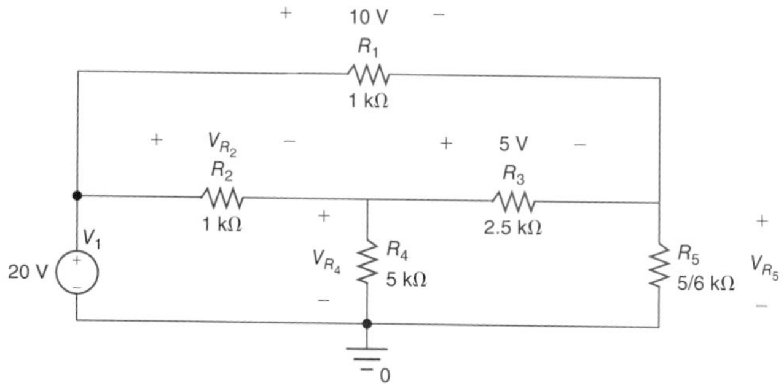

2.26 Find V_{R_2}, V_{R_4}, V_{R_6}, and V_{R_7} in the circuit shown in Figure P2.26.

FIGURE P2.26

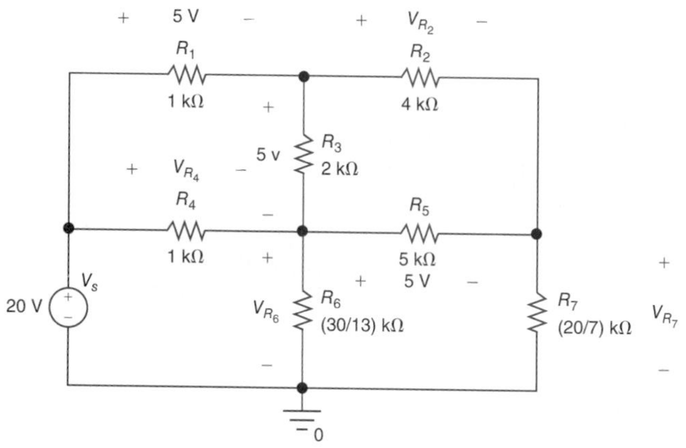

2.27 In the circuit shown in Figure P2.27, let V_5 be 6 V. Find I_5, I_1, I_3, V_4, I_4, and I_2.

FIGURE P2.27

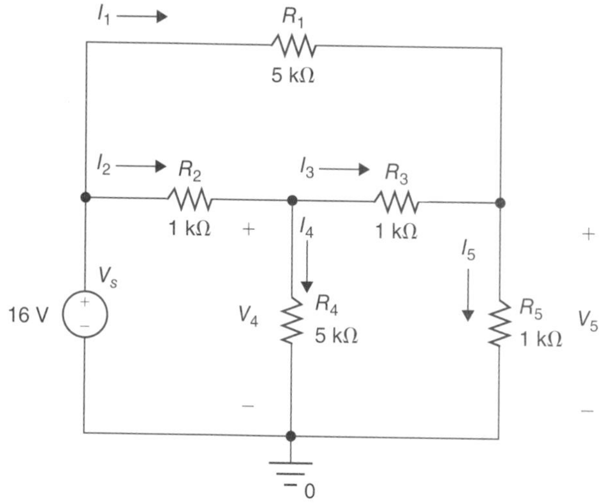

2.28 In the circuit shown in Figure P2.28, the current through R_3 is $I_3 = 2$ mA. Find V_2, I_4, I_2, I_1, and V_1.

FIGURE P2.28

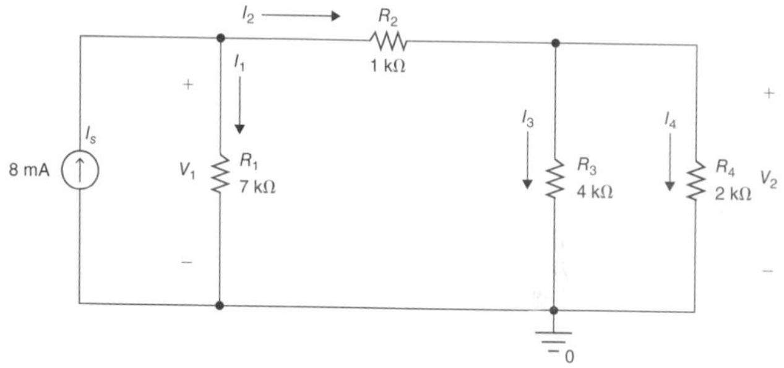

2.29 In the circuit shown in Figure P2.29, the current through R_1 is $I_1 = 1$ mA. Find V_1, I_2, V_2, I_3, and I_4.

FIGURE P2.29

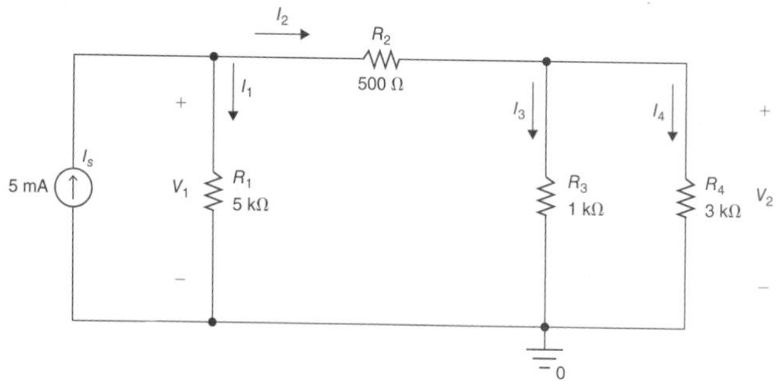

2.30 Find V_1, V_2, V_3, V_4, and V_5 in the circuit shown in Figure P2.30.

FIGURE P2.30

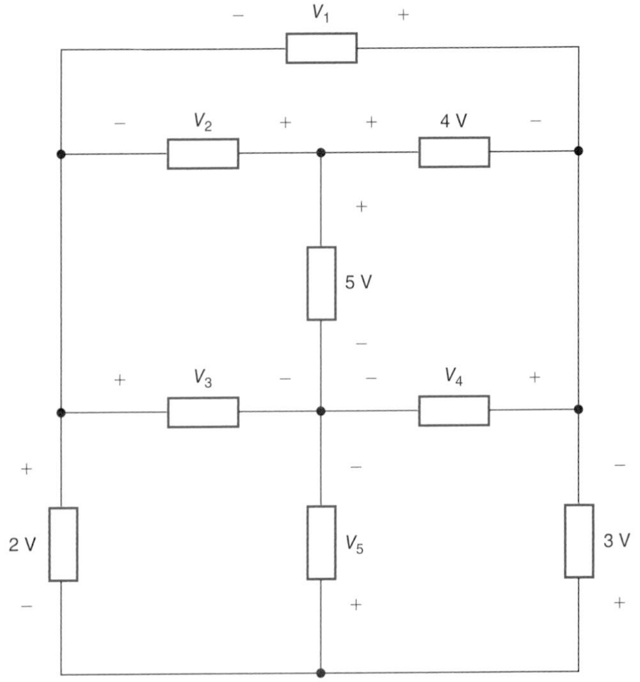

2.31 Find V_1, V_2, V_5, I_1, and I_4 in the circuit shown in Figure P2.31.

FIGURE P2.31

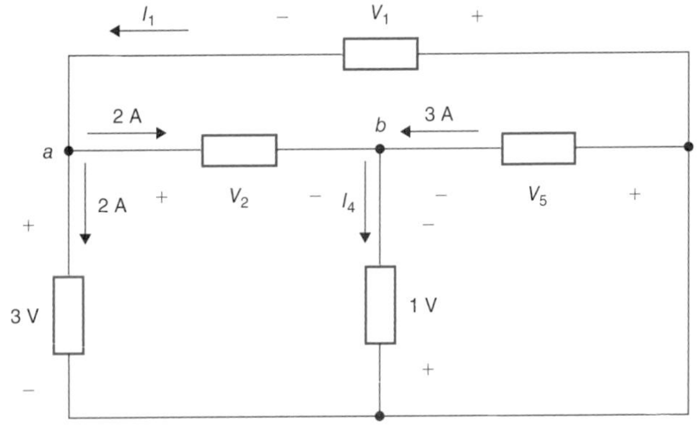

Equivalent Resistance

2.32 Find the equivalent resistance R_{eq} of the circuit shown in Figure P2.32.

FIGURE P2.32

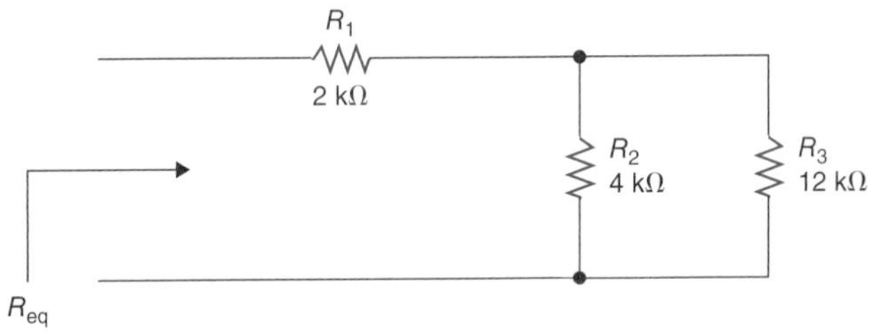

2.33 Find the equivalent resistance R_{eq} of the circuit shown in Figure P2.33.

FIGURE P2.33

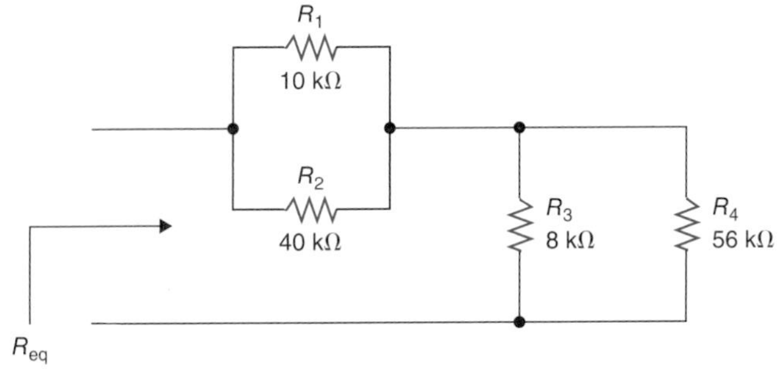

2.34 Find the equivalent resistance R_{eq} of the circuit shown in Figure P2.34.

FIGURE P2.34

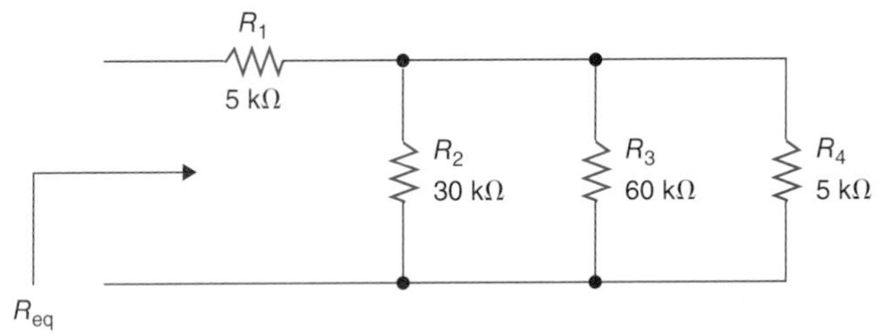

2.35 Find the equivalent resistance R_{eq} of the circuit shown in Figure P2.35.

FIGURE P2.35

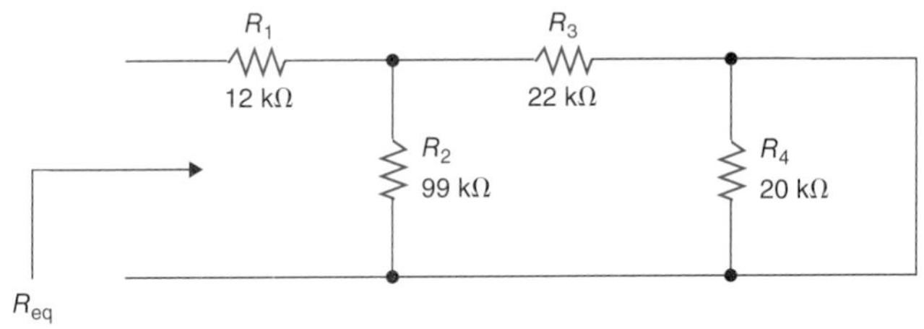

2.36 Find the equivalent resistance R_{eq} of the circuit shown in Figure P2.36.

FIGURE P2.36

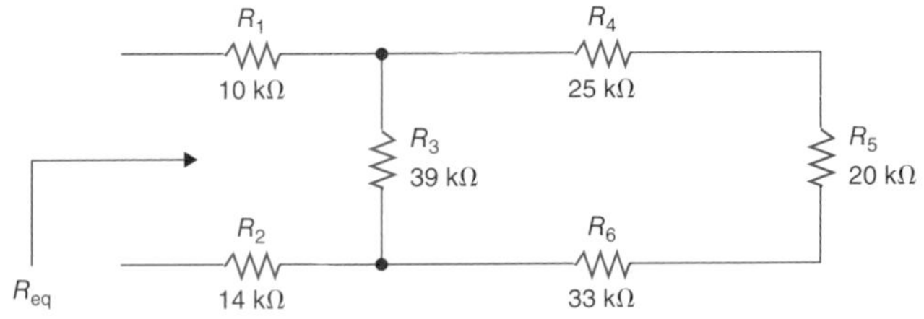

2.37 Find the equivalent resistance R_{eq} of the circuit shown in Figure P2.37.

FIGURE P2.37

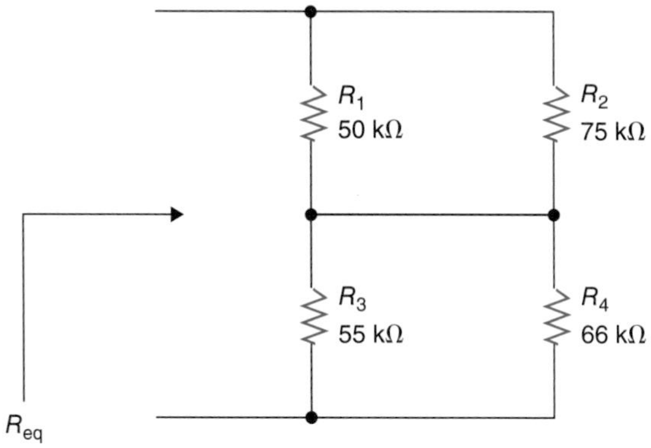

2.38 Find the equivalent resistance R_{eq} of the circuit shown in Figure P2.38.

FIGURE P2.38

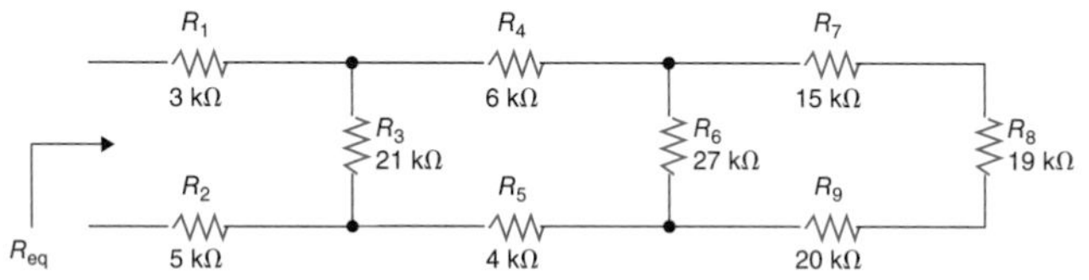

2.39 Find the equivalent resistance R_{eq} of the circuit shown in Figure P2.39.

FIGURE P2.39

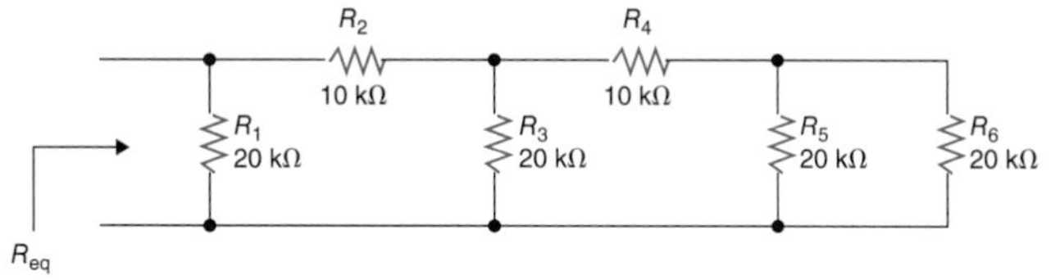

2.40 Find the equivalent resistance R_{eq} of the circuit shown in Figure P2.40.

FIGURE P2.40

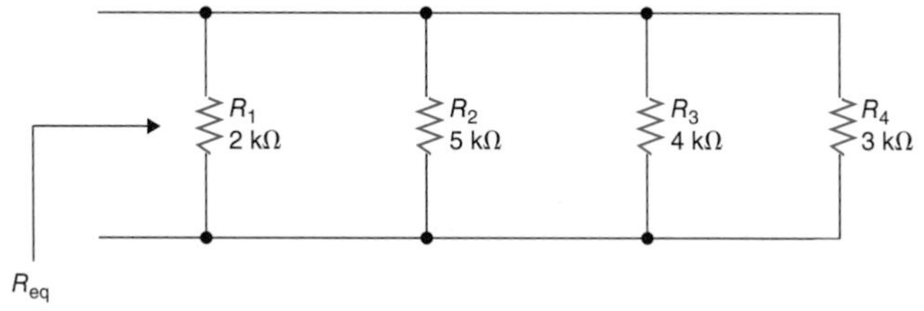

2.41 Find the equivalent resistance R_{eq} of the circuit shown in Figure P2.41.

FIGURE P2.41

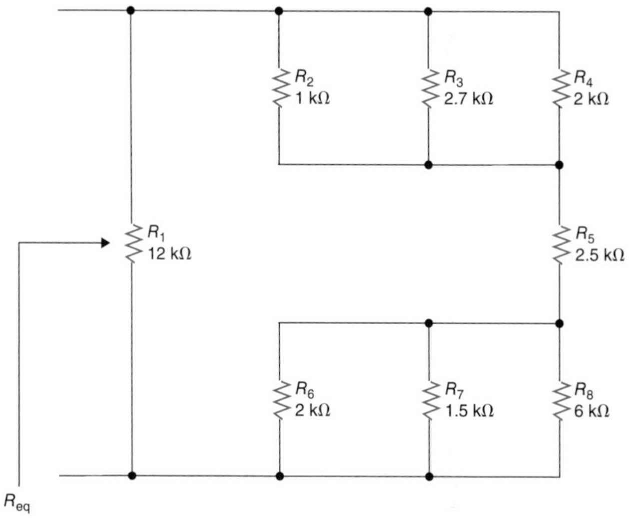

2.42 Find the equivalent resistance R_{eq} of the circuit shown in Figure P2.42.

FIGURE P2.42

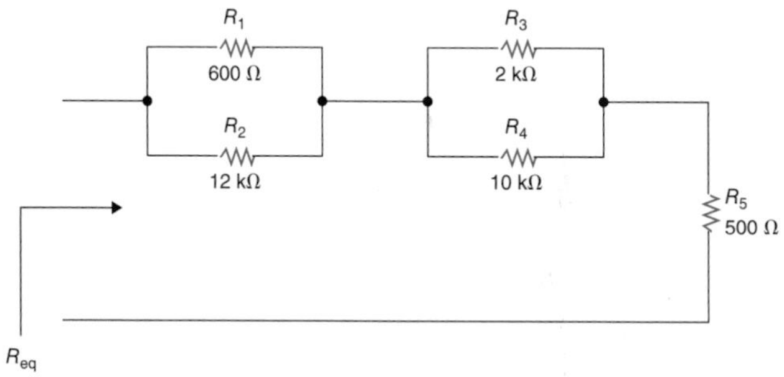

2.43 Find the equivalent resistance R_{eq} of the circuit shown in Figure P2.43.

FIGURE P2.43

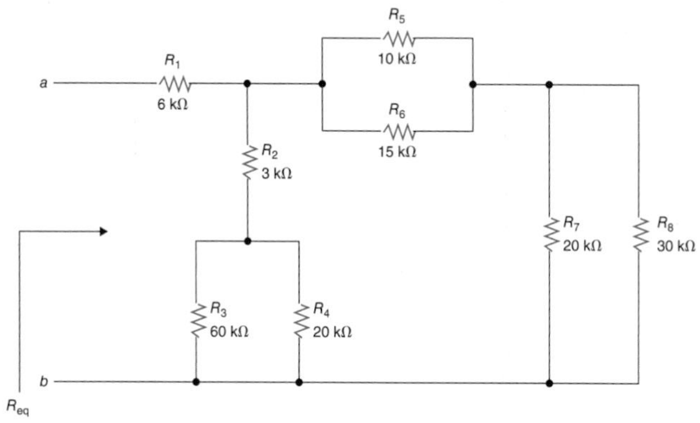

2.44 Find the equivalent resistance R_{eq} of the circuit shown in Figure P2.44.

FIGURE P2.44

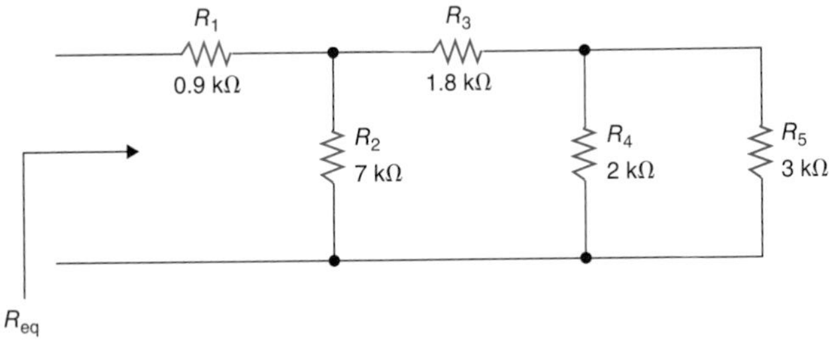

2.45 Find the equivalent resistance R_{eq} of the circuit shown in Figure P2.45.

FIGURE P2.45

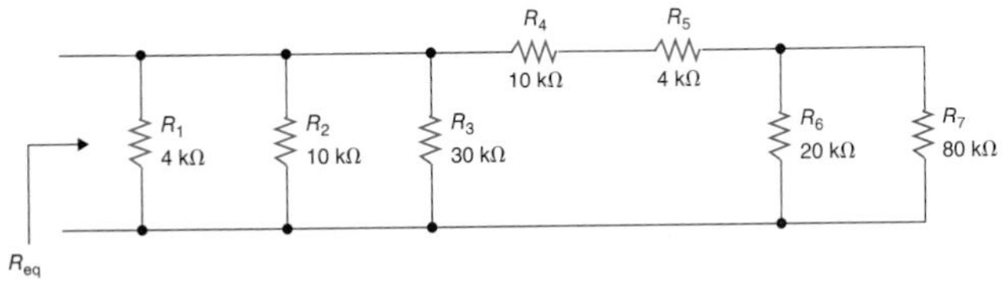

2.46 Find the equivalent resistance R_{eq} of the circuit shown in Figure P2.46.

FIGURE P2.46

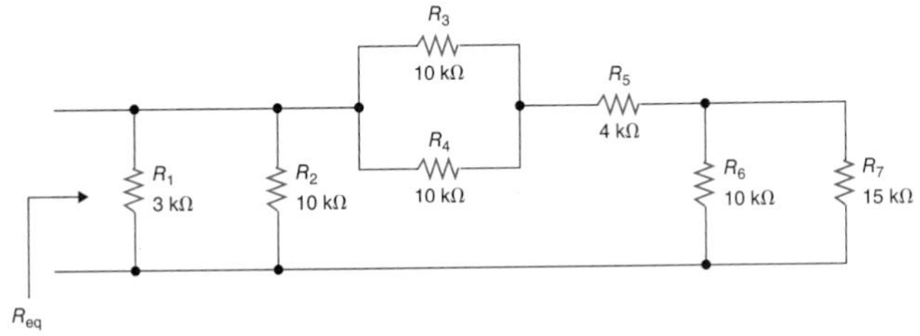

Voltage Divider Rule

2.47 Use the voltage divider rule to find voltages V_1 and V_2 in the circuit shown in Figure P2.47.

FIGURE P2.47

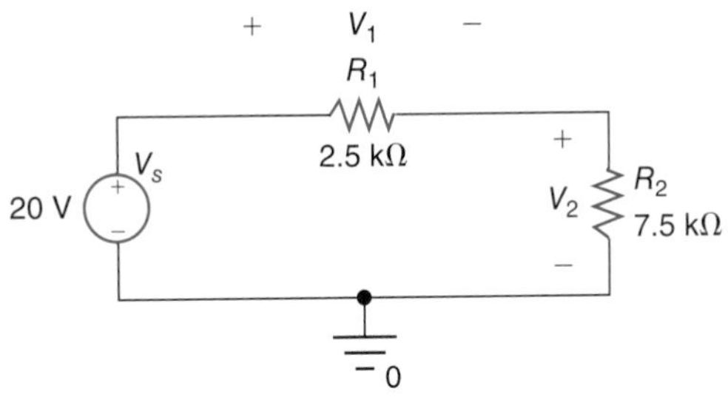

2.48 Use the voltage divider rule to find voltages V_1 and V_2 in the circuit shown in Figure P2.48.

FIGURE P2.48

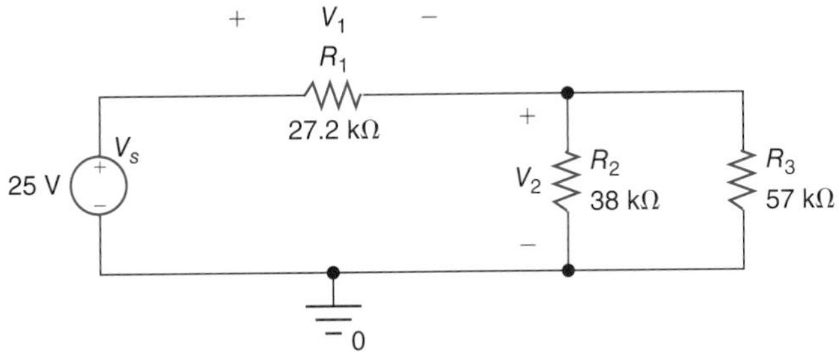

2.49 Use the voltage divider rule to find voltages V_1 and V_2 in the circuit shown in Figure P2.49.

FIGURE P2.49

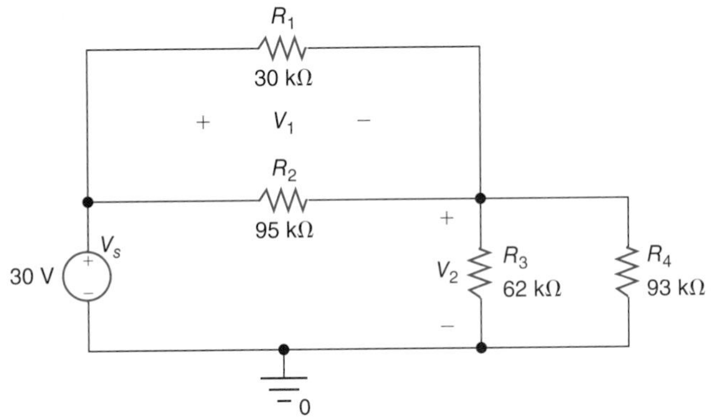

2.50 Use the voltage divider rule to find voltages V_1 and V_2 in the circuit shown in Figure P2.50.

FIGURE P2.50

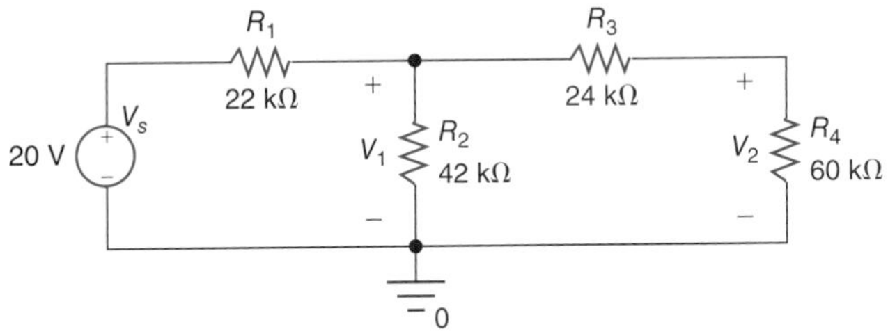

2.51 Use the voltage divider rule to find voltages V_1 and V_2 in the circuit shown in Figure P2.51.

FIGURE P2.51

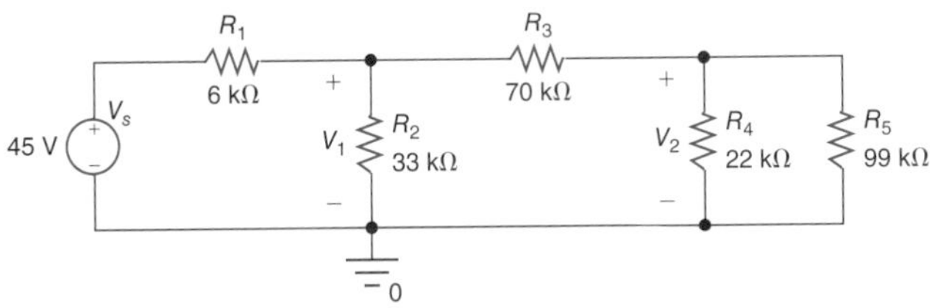

2.52 Use the voltage divider rule to find voltages V_1, V_2, and V_3 in the circuit shown in Figure P2.52.

FIGURE P2.52

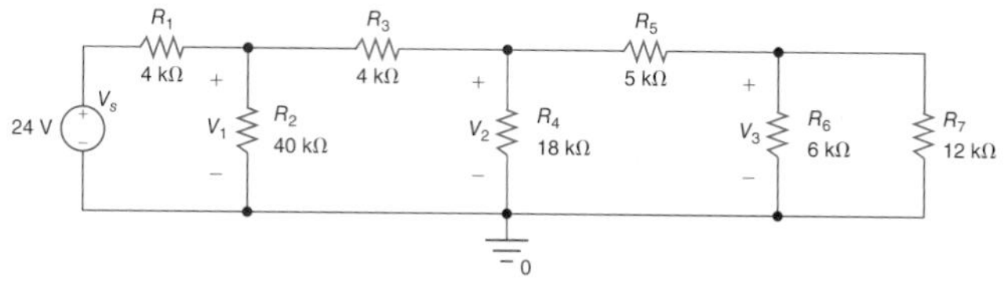

2.53 Use the voltage divider rule to find voltages V_1 and V_2 in the circuit shown in Figure P2.53.

FIGURE P2.53

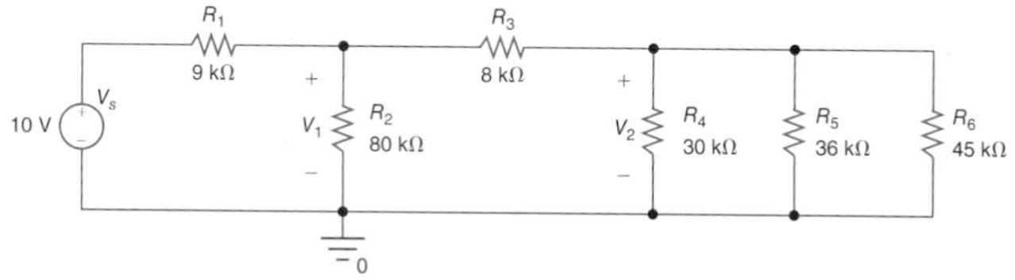

2.54 Use the voltage divider rule to find voltages V_1, V_2, and V_3 in the circuit shown in Figure P2.54.

FIGURE P2.54

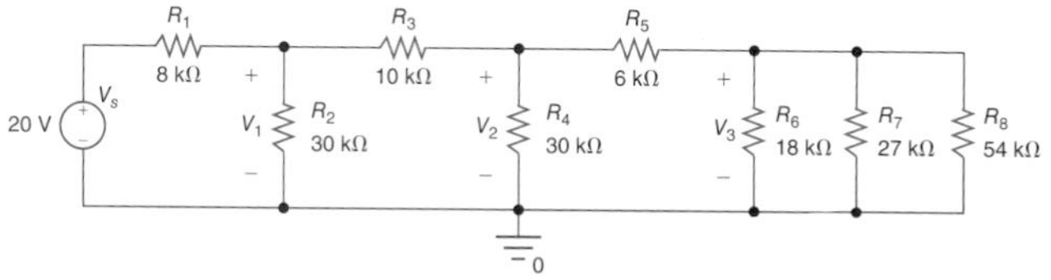

2.55 Use the voltage divider rule to find voltages V_1 and V_2 in the circuit shown in Figure P2.55.

FIGURE P2.55

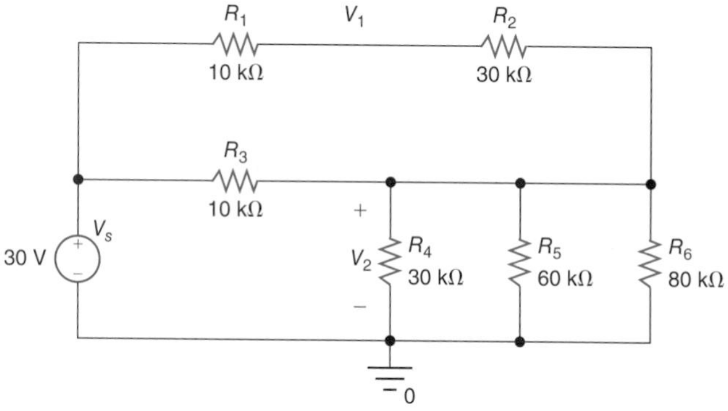

2.56 Use the voltage divider rule to find voltage V_{ab} in the circuit shown in Figure P2.56.

FIGURE P2.56

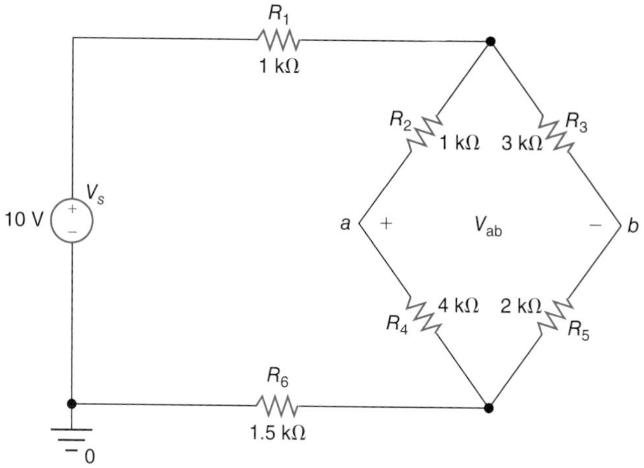

2.57 Use the voltage divider rule to find I_1, I_2, I_3, I_4, I_5, V_1, V_2, and V_3 in the circuit shown in Figure P2.57.

FIGURE P2.57

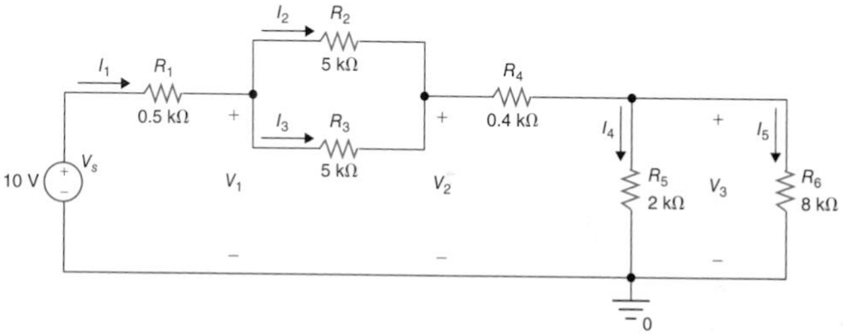

Current Divider Rule

2.58 Use the current divider rule to find currents I_1 and I_2 in the circuit shown in Figure P2.58.

FIGURE P2.58

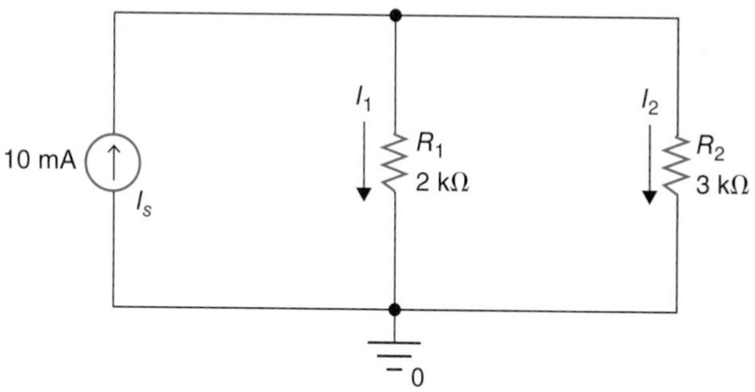

2.59 Use the current divider rule to find currents I_1, I_2, and I_3 in the circuit shown in Figure P2.59.

FIGURE P2.59

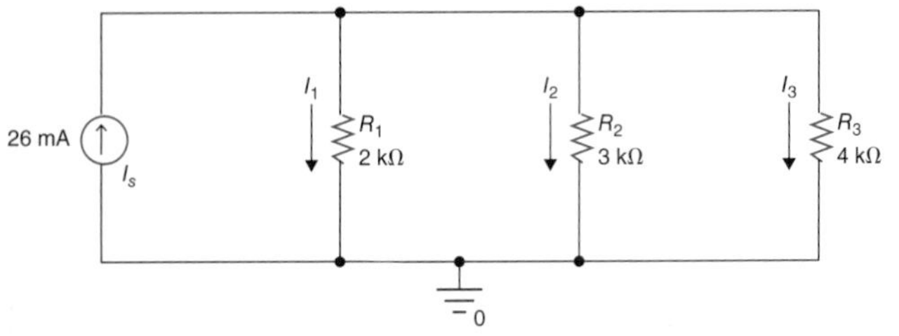

2.60 Use the current divider rule to find I_1, I_2, I_3, I_4, and I_5 in the circuit shown in Figure P2.60.

FIGURE P2.60

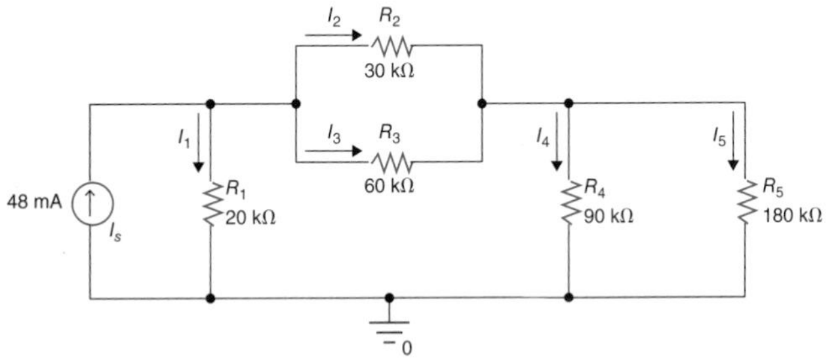

2.61 Use the current divider rule to find I_1, I_2, I_3, I_4, V_1, and V_2 in the circuit shown in Figure P2.61.

FIGURE P2.61

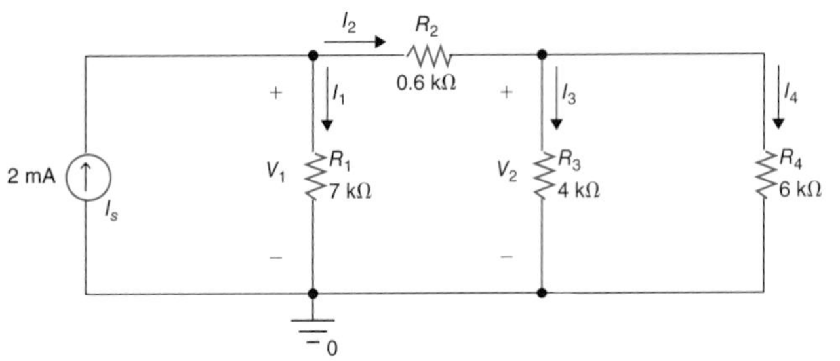

2.62 Find I_1, I_2, V_1, and V_2 in the circuit shown in Figure P2.62.

FIGURE P2.62

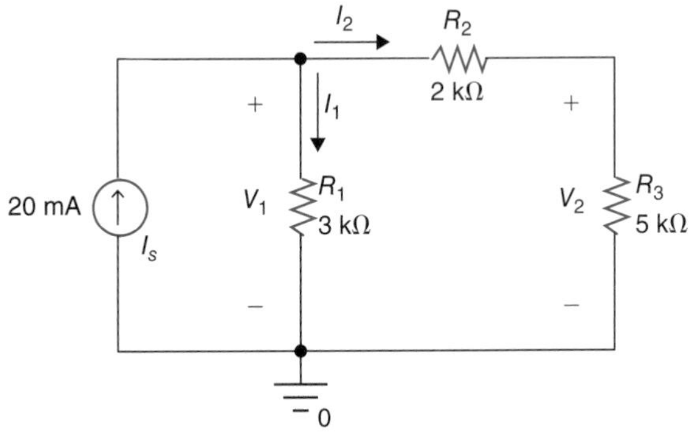

2.63 Find V_1, I_1, I_2, and I_3 in the circuit shown in Figure P2.63.

FIGURE P2.63

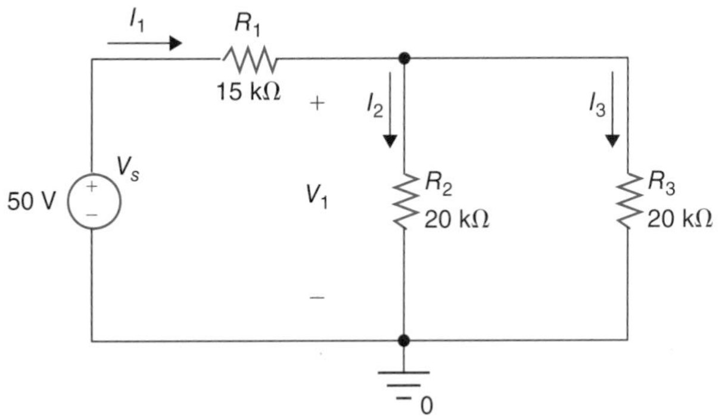

2.64 Use the current divider rule to find I_1, I_2, I_3, I_4, I_5, I_6, and I_7 in the circuit shown in Figure P2.64.

FIGURE P2.64

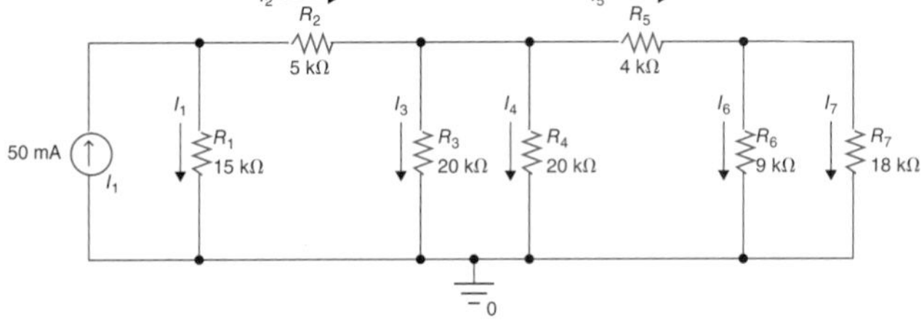

2.65 Find I_1, I_2, I_3, I_4, I_5, V_1, and V_2 in the circuit shown in Figure P2.65.

FIGURE P2.65

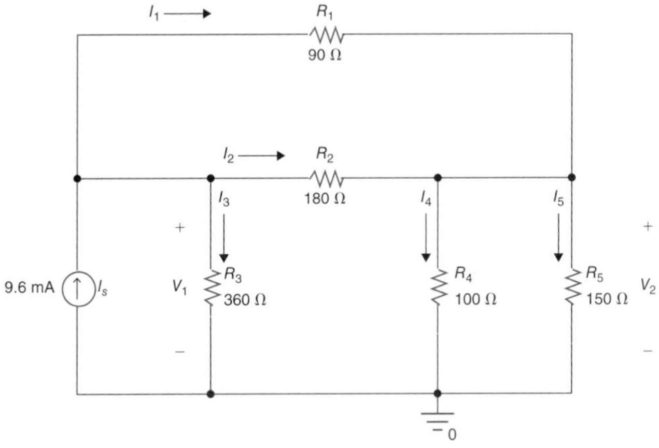

2.66 **Find I_1, V_1, V_2, I_2, I_3, V_3, V_4, I_4, I_5, V_5, and V_6 in the circuit shown in Figure P2.66.**

FIGURE P2.66

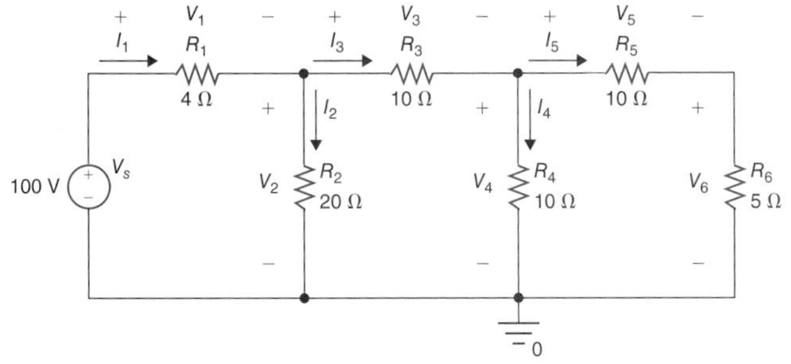

2.67 **Find I in the circuit shown in Figure P2.67.**

FIGURE P2.67

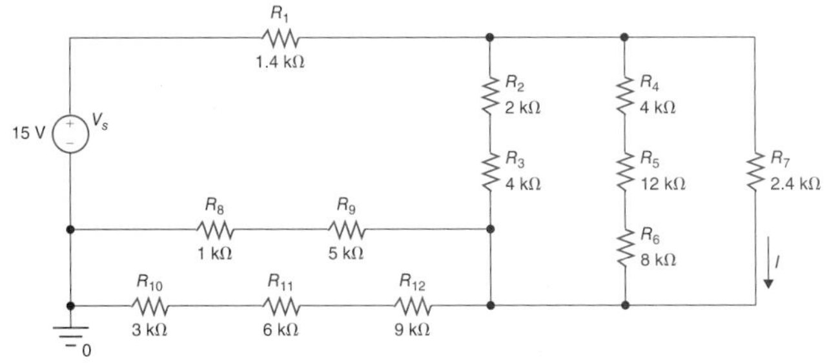

Delta-Wye and Wye-Delta

2.68 **Find the equivalent resistance R_{eq} of the circuit shown in Figure P2.68.**

FIGURE P2.68

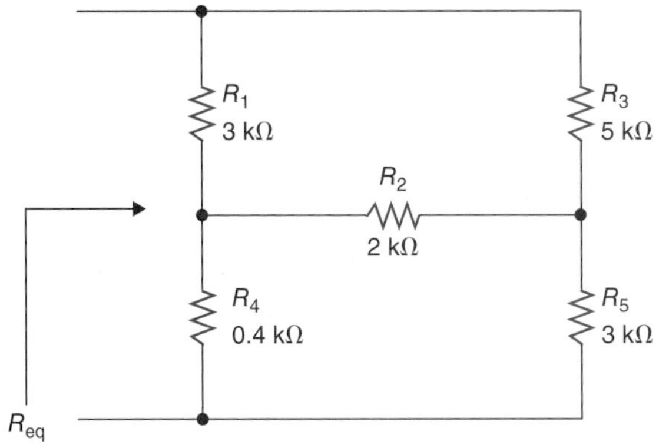

2.69 **Find the equivalent resistance R_{eq} of the circuit shown in Figure P2.69.**

FIGURE P2.69

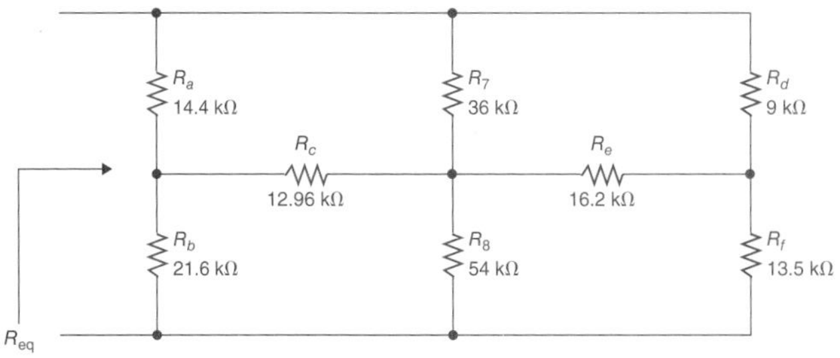

2.70 **Find V_o in the circuit shown in Figure P2.70.**

FIGURE P2.70

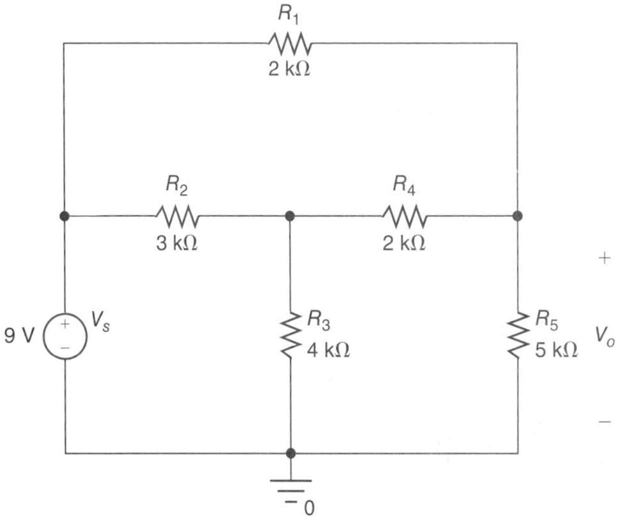

2.71 **Find V_1 and V_2 in the circuit shown in Figure P2.71.**

FIGURE P2.71

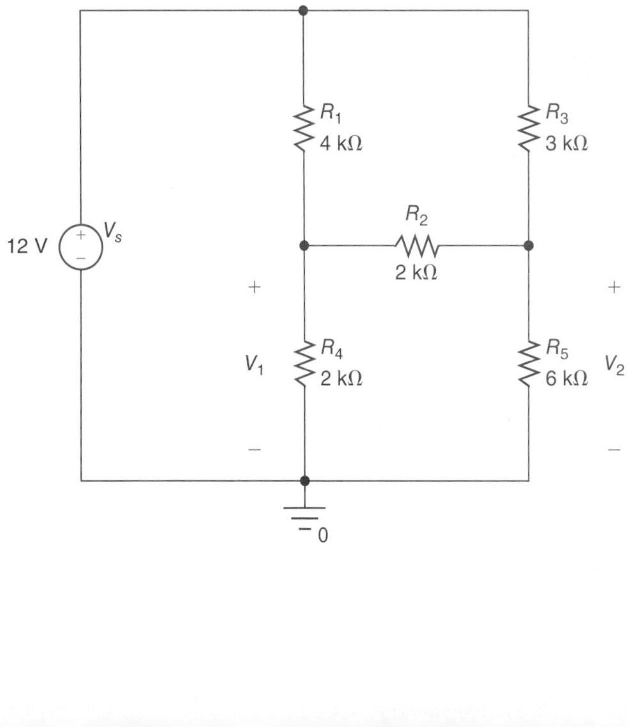

2.72 Find V_1 in the circuit shown in Figure P2.72.

FIGURE P2.72

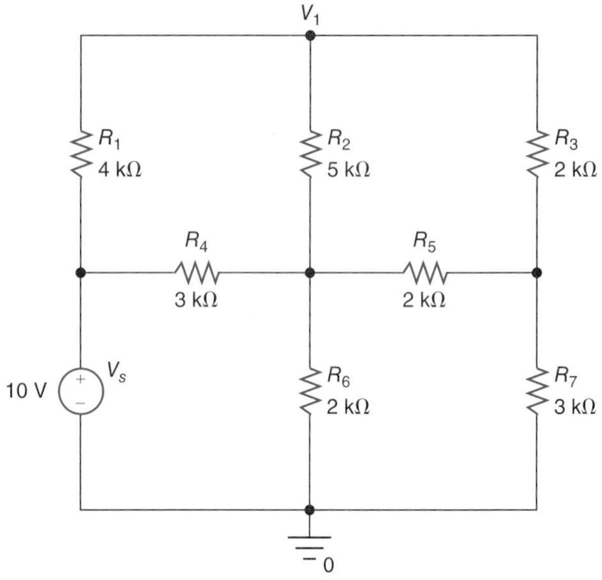

Circuit Analysis Methods

3.1 Introduction

In this chapter, systematic approaches to analyze electric circuits are presented. The approaches refer to nodal analysis and mesh analysis. These two analysis methods can be universally applied to solve circuit problems.

Nodal analysis is a method of finding all the unknown node voltages of a circuit. The method is based on Kirchhoff's current law (KCL). Nodes can be labeled 1, 2, 3, . . . , or a, b, c, . . . (0 can be used for the reference node), and voltages on these nodes can be labeled $V_1, V_2, V_3, \ldots$, or $V_a, V_b, V_c, \ldots$. The node voltage of a reference node (0 V) and nodes with specified voltage sources to a reference node are known. As such, for each node whose voltage is unknown, we can write a node-voltage equation by summing the currents leaving (entering, or some entering and the rest leaving) the node. This is tantamount to writing KCL at each node. The currents leaving the node through resistors can be found by applying Ohm's law. A solution to the node voltages is obtained by solving the set of node-voltage equations. Once all the node voltages are computed, the current in each branch can be found using Ohm's law. In a special case, if there is a voltage source connecting two nodes, we can form a "supernode" by first excluding the voltage source and then writing the sum of currents that are leaving its two node-voltage terminals. In writing such an equation instead of two linearly independent equations, the result is one linear equation with two variables. One additional equation, known as the **constraint equation**, therefore, is needed; it is obtained by relating the potential difference across the two node-voltage terminals.

A *mesh* is a fundamental loop formed by circuit elements in which no other closed path exists. Mesh analysis is a method of finding all unknown mesh currents of a circuit, and it is based on Kirchhoff's voltage law (KVL). Like nodes, meshes can be labeled 1, 2, 3, . . . , or a, b, c, . . . ; and mesh currents can be labeled $I_1, I_2, I_3, \ldots$, or $I_a, I_b, I_c, \ldots$. If there is a current source along the path of a labeled mesh current that is not shared by other meshes, then that mesh current is known. For each mesh whose mesh current is unknown, we can write an equation by summing the voltage drops (or rises) around the mesh. The voltage drops on resistors can be found by applying Ohm's law on each branch that is part of the mesh.

The mesh currents are found by solving these equations. Once all the mesh currents are found, the current through each branch can be found by taking the difference of the two mesh currents. Once all the currents through branches are known, the voltages across all branches and nodes can be found. If there is a current source between two meshes, we can form a supermesh consisting of two meshes. One additional equation is obtained by relating the current from the current source to the two mesh currents.

3.2 Nodal Analysis

If the voltages are known for all the nodes in a circuit, the voltage across each branch can be evaluated by subtracting the node voltage on one side of the branch from the node voltage on the other side. Once the voltage across the branch is found, the current through the branch can be evaluated by applying Ohm's law. Once the voltage and current are known for each branch, the power absorbed or supplied by the branch can be found by multiplying the voltage and the current. Thus, a circuit can be analyzed completely by finding the voltage on every node in the given circuit. The nodal analysis provides us voltages at all the nodes of the given circuit. If a voltage source is connected between a node and the ground, the node voltage is already known, and we do not need to find the node voltage on this node.

Excluding the nodes whose voltages are known from the voltage sources, we assign variables such as $V_1, V_2, \ldots, V_n$ to each unknown node voltage. For each node with an assigned variable, write a node equation by applying KCL. Any of the three interpretations of KCL given in Chapter 2 can be used. For example, the sum of the currents leaving a node must be zero. If there are n unknown node voltages, we get n equations in n unknowns. Thus, we can solve these n systems of linear equations with constant coefficients to find the unique solution for the unknown node voltages $V_1, V_2, \ldots, V_n$. If n is a small number, substitution method can be used to solve the system of n linear equations with constant coefficients. Cramer's rule can also be used to solve the system of n linear equations with constant coefficients. MATLAB is useful in finding the solution of an n system of linear equations with constant coefficients.

Labeling a circuit for nodal analysis requires identifying essential nodes where KCL equations can be written. The term *essential node* refers to a node where three or more elements are connected. As an example, consider the circuit shown in Figure 3.1.

FIGURE 3.1

A circuit with two unknown node voltages.

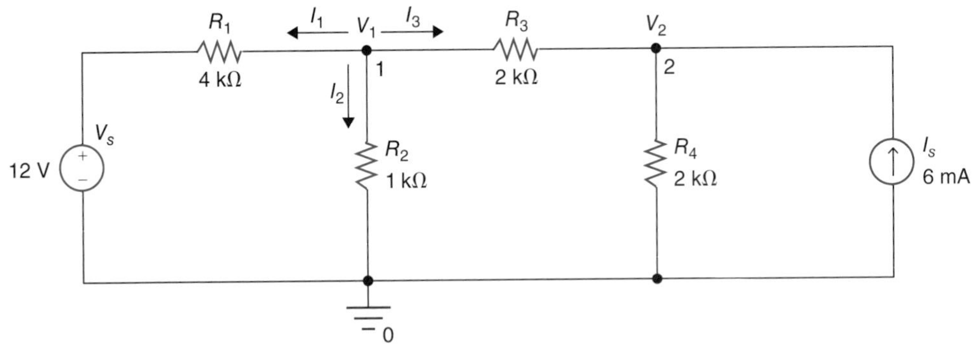

There are only three essential nodes in Figure 3.1. The node formed by resistors R_1, R_2, and R_3 is labeled node 1, and the node formed by resistors R_3, R_4, and the current source is labeled node 2. The node formed by the voltage source, resistors R_2 and R_3, and the current source is labeled node 0; this is the *ground node.* Recognize that with n essential nodes, there can be only $(n - 1)$ KCL equations, and one of the n essential nodes is selected as the ground node.

According to KCL, the sum of the currents leaving a node equals zero. Three branches are connected to node 1. The voltage across R_1 is $(V_1 - V_s)$. The current I_1 through R_1, from right to left, is given by $(V_1 - V_s)/R_1$ (Ohm's law). Similarly, the current I_2 through R_2, from top to bottom, is given by $(V_1 - 0)/R_2$, and the current I_3 through R_3, from left to right, is given by $(V_1 - V_2)/R_3$. The three currents I_1, I_2, and I_3 are shown in Figure 3.1. When these three currents leaving node 1 are added, it should be zero; that is,

$$I_1 + I_2 + I_3 = 0$$

or

$$\frac{V_1 - V_s}{R_1} + \frac{V_1 - 0}{R_2} + \frac{V_1 - V_2}{R_3} = 0 \quad (3.1)$$

Substituting the values of V_s, R_1, R_2, and R_3, we obtain

$$\frac{V_1 - 12}{4000} + \frac{V_1 - 0}{1000} + \frac{V_1 - V_2}{2000} = 0$$

Multiplication of this equation by 4000 yields

$$V_1 - 12 + 4V_1 + 2V_1 - 2V_2 = 0$$

which can be simplified to

$$7V_1 - 2V_2 = 12 \quad (3.2)$$

Instead of summing the currents leaving (away from, out of) the node, we can sum the currents entering (into) the node. The direction of current on each branch is arbitrary. If the actual current flows in the opposite direction, it will be negative.

Three branches are connected to node 2. In Figure 3.2, the current I_5 through R_3, from right to left, is given by $(V_2 - V_1)/R_3$ (Ohm's law). Notice that I_5 is identical to $-I_3$ in Figure 3.1. Similarly, the current I_4 through R_4, from top to bottom, is given by $(V_2 - 0)/R_4$, and the current I_6 is given by $I_6 = -I_s = -6$ mA. The current through the current source, from top to bottom, is given by $-I_s = -6$ mA. When these three currents leaving node 2 are added, it should be zero; that is,

$$I_5 + I_4 + I_6 = 0$$

or

$$\frac{V_2 - V_1}{R_3} + \frac{V_2 - 0}{R_4} - I_s = 0 \quad (3.3)$$

Substituting the values of I_s, R_3, and R_4, we obtain

$$\frac{V_2 - V_1}{2000} + \frac{V_2 - 0}{2000} - 6 \times 10^{-3} = 0$$

FIGURE 3.2

Currents leaving node 2.

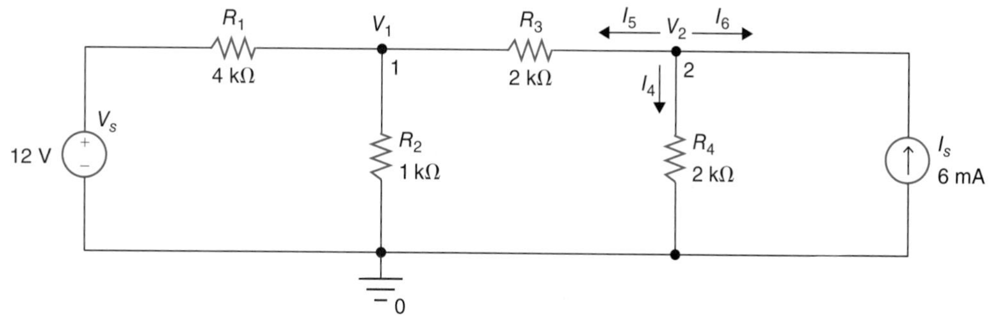

Multiplication of this equation by 2000 yields

$$V_2 - V_1 + V_2 = 12$$

which can be rewritten as

$$-V_1 + 2V_2 = 12 \tag{3.4}$$

The unknown node voltages are found by solving Equations (3.2) and (3.4) or Equations (3.1) and (3.3). Several methods can be used to solve a system of linear equations. For two or three unknowns, we can solve the equations by substitution method. If the number of unknowns is greater than or equal to 2, Cramer's rule can be used to find the solution. MATLAB can be used to find the solution to a system of linear equations. We start with the substitution method. From Equation (3.4), we have

$$V_1 = 2V_2 - 12 \tag{3.5}$$

Substitution of Equation (3.5) into Equation (3.2) yields

$$7(2V_2 - 12) - 2V_2 = 12$$

which can be simplified to

$$12V_2 = 96$$

Thus, we have $V_2 = 8$ V. Substituting this into Equation (3.5), we obtain

$$V_1 = 2V_2 - 12 = 16 - 12 = 4\text{ V}$$

Equations (3.2) and (3.4) can be put into matrix form as

$$\begin{bmatrix} 7 & -2 \\ -1 & 2 \end{bmatrix}\begin{bmatrix} V_1 \\ V_2 \end{bmatrix} = \begin{bmatrix} 12 \\ 12 \end{bmatrix} \tag{3.6}$$

Let

$$A = \begin{bmatrix} 7 & -2 \\ -1 & 2 \end{bmatrix}, \quad V = \begin{bmatrix} V_1 \\ V_2 \end{bmatrix}, \quad b = \begin{bmatrix} 12 \\ 12 \end{bmatrix}$$

Then, Equation (3.6) becomes

$$AV = b \tag{3.7}$$

Using Cramer's rule on Equations (3.2) and (3.4) or Equation (3.6), we have

$$V_1 = \frac{\begin{vmatrix} 12 & -2 \\ 12 & 2 \end{vmatrix}}{\begin{vmatrix} 7 & -2 \\ -1 & 2 \end{vmatrix}} = \frac{12 \times 2 - (-2) \times 12}{7 \times 2 - (-2) \times (-1)} = \frac{24 + 24}{14 - 2} = \frac{48}{12} = 4\text{ V}$$

Notice that the denominator is the determinant of matrix A. Let the determinant of A be Δ. Then, we have

$$\Delta = \begin{vmatrix} 7 & -2 \\ -1 & 2 \end{vmatrix} = 7 \times 2 - (-2) \times (-1) = 14 - 2 = 12$$

When the first column of matrix A is replaced by b, we obtain

$$A_1 = \begin{bmatrix} 12 & -2 \\ 12 & 2 \end{bmatrix}$$

The numerator is the determinant of A_1. Let the determinant of A_1 be Δ_1. Then, we have

$$\Delta_1 = \begin{vmatrix} 12 & -2 \\ 12 & 2 \end{vmatrix} = 12 \times 2 - (-2) \times (12) = 24 + 24 = 48$$

The voltage V_1 is the ratio of Δ_1 to Δ; that is,

$$V_1 = \frac{\Delta_1}{\Delta} = \frac{48}{12} = 4 \text{ V}$$

The voltage V_2 is computed as

$$V_2 = \frac{\begin{vmatrix} 7 & 12 \\ -1 & 12 \end{vmatrix}}{\Delta} = \frac{7 \times 12 - 12 \times (-1)}{12} = \frac{84 + 12}{12} = \frac{96}{12} = 8 \text{ V}$$

Notice that the denominator is the determinant of matrix A (that is, Δ). When the second column of matrix A is replaced by b, we obtain

$$A_2 = \begin{bmatrix} 7 & 12 \\ -1 & 12 \end{bmatrix}$$

The numerator is the determinant of A_2. Let the determinant of A_2 be Δ_2. Then, we have

$$\Delta_2 = \begin{vmatrix} 7 & 12 \\ -1 & 12 \end{vmatrix} = 7 \times 12 - (12) \times (-1) = 84 + 12 = 96$$

The voltage V_2 is the ratio of Δ_2 to Δ; that is,

$$V_2 = \frac{\Delta_2}{\Delta} = \frac{96}{12} = 8 \text{ V}$$

Equations (3.2) and (3.4) or Equation (3.6) can be solved using MATLAB, as shown here:

```
clear all;
A=[7 -2;-1 2];b=[12;12];
V=A\b

Answer:
V =
         4.0000
         8.0000
```

Equations (3.1) and (3.3) can be rearranged, respectively, as

$$\left(\frac{1}{R_1} + \frac{1}{R_2} + \frac{1}{R_3}\right) V_1 - \frac{1}{R_3} V_2 = \frac{V_s}{R_1}$$

and

$$-\frac{1}{R_3}V_1 + \left(\frac{1}{R_3} + \frac{1}{R_4}\right)V_2 = I_s$$

These equations can be put into matrix form as

$$\begin{bmatrix} \frac{1}{R_1} + \frac{1}{R_2} + \frac{1}{R_3} & -\frac{1}{R_3} \\ -\frac{1}{R_3} & \frac{1}{R_3} + \frac{1}{R_4} \end{bmatrix} \begin{bmatrix} V_1 \\ V_2 \end{bmatrix} = \begin{bmatrix} \frac{V_s}{R_1} \\ I_s \end{bmatrix}$$

This equation can be solved using MATLAB, as shown here:

```
clear all;
Vs=12;Is=6e-3;R1=4000;R2=1000;R3=2000;R4=2000;
A=[1/R1+1/R2+1/R3,-1/R3;-1/R3,1/R3+1/R4]
b=[Vs/R1;Is]
V=A\b

Answer:
A =
     0.001750000000000   -0.000500000000000
    -0.000500000000000    0.001000000000000
b =
     0.003000000000000
     0.006000000000000
V =
         4
         8
```

This approach does not require numerical values for the elements of A and b to find V_1 and V_2.

One other method of solving Equations (3.1) and (3.3) is to use the MATLAB function **solve**, as shown here:

```
clear all;
syms V1 V2
Vs=12;Is=6e-3;R1=4000;R2=1000;R3=2000;R4=2000;
[V1,V2]=solve((V1-Vs)/R1+V1/R2+(V1-V2)/R3==0,(V2-V1)/R3+V2/R4-Is==0)
V1=vpa(V1,6)
V2=vpa(V2,6)

Answer:
V1 =
4
V2 =
8
V1 =
4.0
V2 =
8.0
```

This method provides a solution by entering equations. The equations can be entered without ==0; that is, the line with **solve** can be replaced by

```
[V1,V2]=solve((V1-Vs)/R1+V1/R2+(V1-V2)/R3,(V2-V1)/R3+V2/R4-Is)
```

Once V_1 and V_2 are known, we can find the current through each element. The current I_1 through R_1, from right to left, is given by $I_1 = (V_1 - V_s)/R_1 = (4 - 12)/4 = -2$ mA. The current I_2 through R_2, from top to bottom, is $I_2 = V_1/R_2 = 4/1 = 4$ mA. The current I_3 through R_3, from left to right, is $I_3 = (V_1 - V_2)/R_3 = (4 - 8)/2 = -2$ mA. The current I_4 through R_4, from top to bottom, is $I_4 = V_2/R_4 = 8/2 = 4$ mA. The current I_5 through R_3, from right to left, is $I_5 = (V_2 - V_1)/R_3 = (8 - 4)/2 = 2$ mA. Notice that $I_5 = -I_3$. Figure 3.3 shows the circuit shown in Figure 3.1 with voltages and currents specified. This figure is obtained using PSpice®, a circuit simulation package. The direction of the current through each element is shown in Figure 3.3 by an arrow. In PSpice, the label of current is connected to the terminal of the part where current enters the part. For example, for R_1, label 2.000 mA is connected to the left side of R_1. This implies that current of 2 mA enters R_1 from the left and flows through R_1 from left to right.

FIGURE 3.3

Voltages and currents for the circuit shown in Figure 3.1.

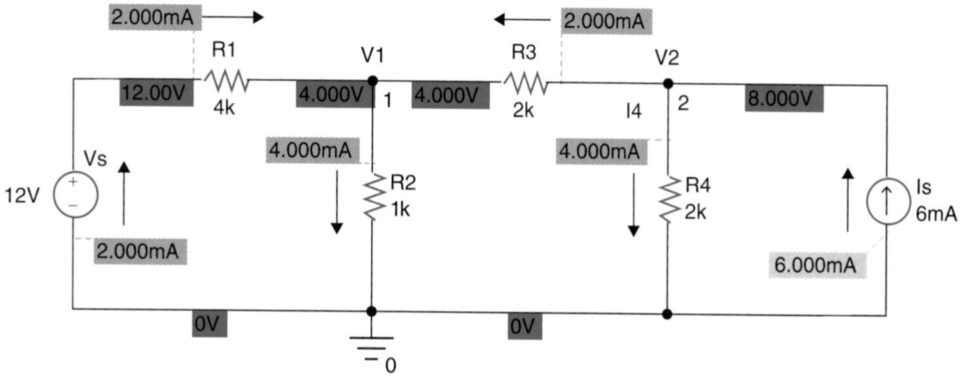

EXAMPLE 3.1

Find the node voltages V_1 and V_2 in the circuit shown in Figure 3.4.

FIGURE 3.4

The circuit for EXAMPLE 3.1.

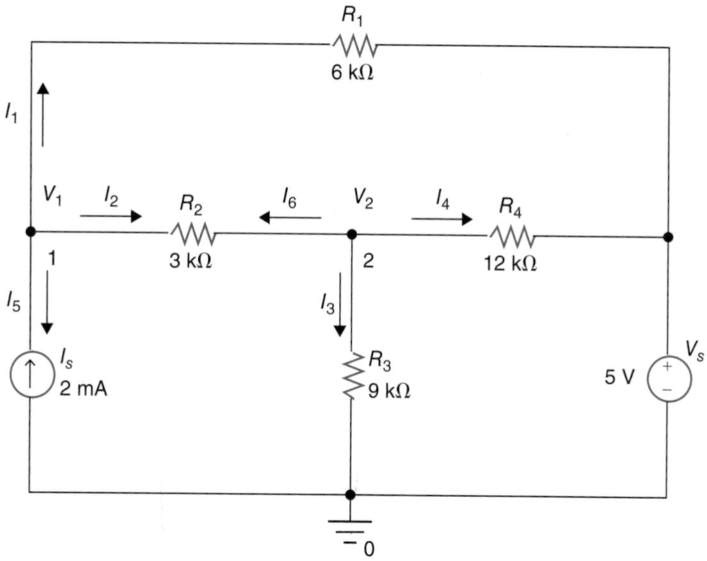

continued

Example 3.1 continued

Two essential nodes where KCL equations can be written are labeled as node 1 and node 2 in Figure 3.4. According to KCL, the sum of the currents leaving node 1 equals zero; that is, $I_1 + I_2 + I_5 = 0$. The voltage across R_1 is $V_1 - V_s$. From Ohm's law, we have $I_1 = (V_1 - V_s)/R_1$. Similarly, the current I_2 through R_2 is given by $I_2 = (V_1 - V_2)/R_2$. The current I_5 is $-I_s$. Thus, the node equation $I_1 + I_2 + I_5 = 0$ becomes

$$\frac{V_1 - V_s}{R_1} + \frac{V_1 - V_2}{R_2} - I_s = 0$$

Substituting the values of V_s, I_s, R_1, and R_2, we obtain

$$\frac{V_1 - 5}{6000} + \frac{V_1 - V_2}{3000} - 2 \times 10^3 = 0$$

Multiplication by 6000 yields

$$V_1 - 5 + 2V_1 - 2V_2 - 12 = 0$$

which can be simplified to

$$3V_1 - 2V_2 = 17 \tag{3.8}$$

According to KCL, the sum of currents leaving node 2 equals zero; that is, $I_6 + I_4 + I_3 = 0$. The voltage across R_2 is $V_2 - V_1$. From Ohm's law, we have $I_6 = (V_2 - V_1)/R_2$. Notice that $I_6 = -I_2$. The voltage across R_4 is $V_2 - V_s$. From Ohm's law, we have $I_4 = (V_2 - V_s)/R_4$. The voltage across R_3 is V_2. From Ohm's law, we have $I_3 = V_2/R_3$. Thus, the node equation $I_6 + I_4 + I_3 = 0$ becomes

$$\frac{V_2 - V_1}{R_2} + \frac{V_2 - V_s}{R_4} + \frac{V_2}{R_3} = 0$$

Substituting the values of V_s, R_2, R_3, and R_4, we obtain

$$\frac{V_2 - V_1}{3000} + \frac{V_2 - 5}{12{,}000} + \frac{V_2}{9000} = 0$$

Multiplication by 36,000 yields

$$12V_2 - 12V_1 + 3V_2 - 15 + 4V_2 = 0$$

which can be simplified to

$$-12V_1 + 19V_2 = 15 \tag{3.9}$$

Equations (3.8) and (3.9) can be solved using the substitution method:

Solving Equation (3.8) for V_2, we obtain

$$V_2 = 1.5V_1 - 8.5$$

Substituting V_2 into Equation (3.9), we get

$$-12V_1 + 19(1.5V_1 - 8.5) = 15$$

continued

Example 3.1 continued

or

$$16.5V_1 = 176.5$$

Thus, we have

$$V_1 = 10.697 \text{ V}$$

and

$$V_2 = 1.5V_1 - 8.5 = 7.5455 \text{ V}$$

Alternatively, Equations (3.8) and (3.9) can be solved using Cramer's rule:

$$\Delta = \begin{vmatrix} 3 & -2 \\ -12 & 19 \end{vmatrix} = 3 \times 19 - (-2) \times (-12) = 57 - 24 = 33$$

$$V_1 = \frac{\begin{vmatrix} 17 & -2 \\ 15 & 19 \end{vmatrix}}{\Delta} = \frac{17 \times 19 - (-2) \times 15}{33} = \frac{353}{33} = 10.6970 \text{ V}$$

$$V_2 = \frac{\begin{vmatrix} 3 & 17 \\ -12 & 15 \end{vmatrix}}{\Delta} = \frac{3 \times 15 - 17 \times (-12)}{33} = \frac{249}{33} = 7.5455 \text{ V}$$

MATLAB

```
>> A=[3 -2;-12 19];b=[17;15];
>> V=A\b
V =
    10.6970
     7.5455
```

Exercise 3.1

Find V_1 and V_2 in the circuit shown in Figure 3.5.

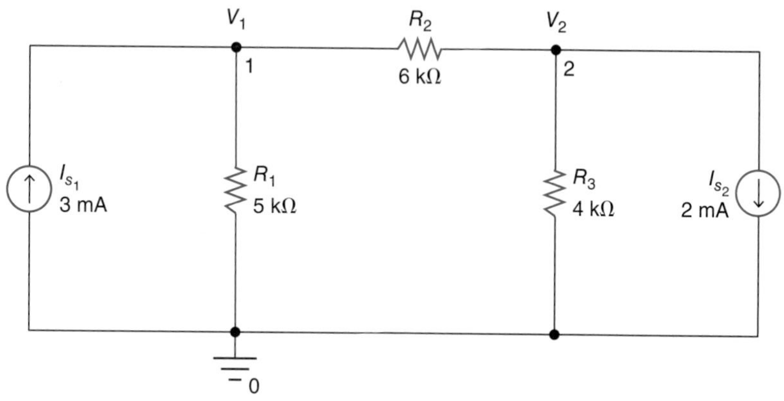

FIGURE 3.5
The circuit for EXERCISE 3.1.

Answer:
$V_1 = 7.3333$ V, $V_2 = -1.8667$ V.

EXAMPLE 3.2

For the circuit shown in Figure 3.6, use nodal analysis to find the node voltages at nodes 1 and 2.

FIGURE 3.6

The circuit for EXAMPLE 3.2.

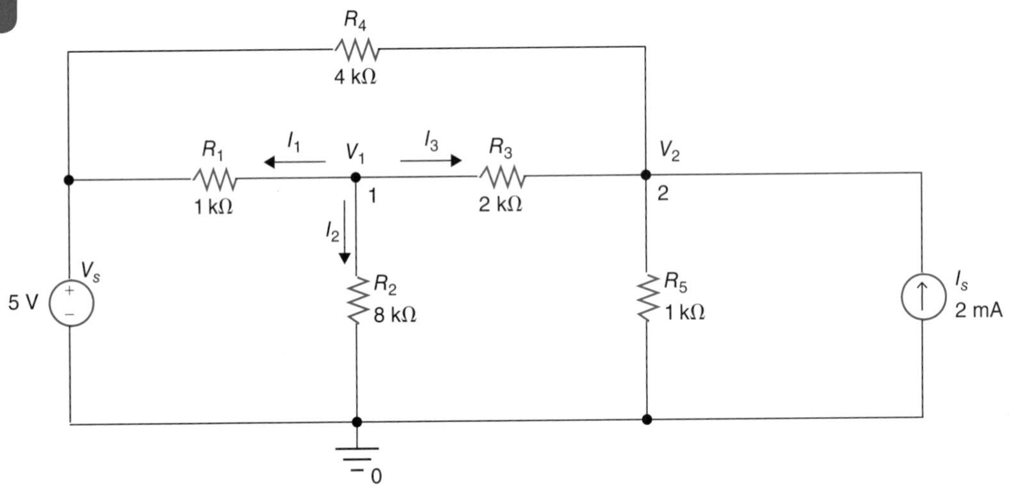

There are four nodes in the circuit shown in Figure 3.6. The voltage at the ground node (node 0) is zero. The voltage at the node connecting the positive terminal of the voltage source V_s to resistors R_1 and R_4 is 5 V. The voltage V_1 at the node (node 1) connecting R_1, R_2, and R_3 and the voltage V_2 at the node (node 2) connecting the current source I_s and resistors R_3, R_4, and R_5 are unknown. Application of KCL leads us to find these unknown voltages.

According to KCL, the sum of currents leaving a node equals zero. Three branches are connected to node 1. The current I_1 through R_1, from right to left, is given by $(V_1 - V_s)/R_1$ (Ohm's law). Similarly, the current I_2 through R_2, from top to bottom, is given by $(V_1 - 0)/R_2$, and the current I_3 through R_3, from left to right, is given by $(V_1 - V_2)/R_3$. The three currents I_1, I_2, and I_3 are shown in Figure 3.6. When these three currents leaving node 1 are added, it should be zero; that is,

$$I_1 + I_2 + I_3 = 0$$

or

$$\frac{V_1 - V_s}{R_1} + \frac{V_1}{R_2} + \frac{V_1 - V_2}{R_3} = 0$$

Substituting the values of V_s, R_1, R_2, and R_3, we obtain

$$\frac{V_1 - 5}{1000} + \frac{V_1}{8000} + \frac{V_1 - V_2}{2000} = 0$$

Multiplication by 8000 yields

$$8V_1 - 40 + V_1 + 4V_1 - 4V_2 = 0$$

which can be simplified to

$$13V_1 - 4V_2 = 40 \tag{3.10}$$

continued

Example 3.2 continued

Four branches are connected to node 2. The current I_6 through R_3, from right to left, is given by $(V_2 - V_1)/R_3$ (Ohm's law). Notice that I_6 is identical to $-I_3$. Similarly, the current I_4 through R_4, from right to left, is given by $(V_2 - V_s)/R_4$, and the current I_5 through R_5, from top to bottom, is given by $(V_2 - 0)/R_5$. The current I_7 is in the opposite direction to I_s. Thus, we have $I_7 = -I_s = -2$ mA. The direction of currents I_4, I_5, I_6, and I_7 are shown in Figure 3.7. When these four currents leaving node 2 are added, it should be zero; that is,

$$I_6 + I_4 + I_5 + I_7 = 0$$

or

$$\frac{V_2 - V_1}{R_3} + \frac{V_2 - V_s}{R_4} + \frac{V_2}{R_5} - I_s = 0$$

FIGURE 3.7

Currents leaving node 2.

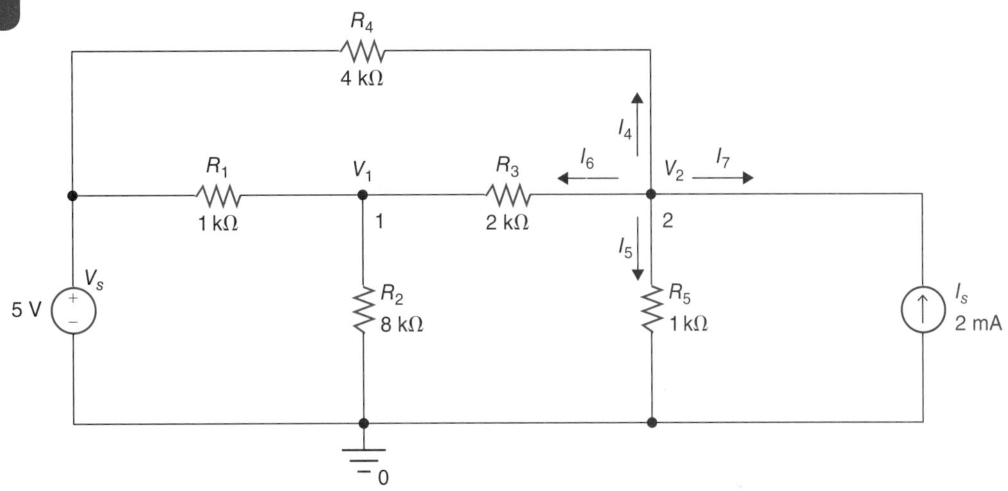

Substituting the values of V_s, I_s, R_3, R_4, and R_5, we obtain

$$\frac{V_2 - V_1}{2000} + \frac{V_2 - 5}{4000} + \frac{V_2}{1000} - 2 \times 10^{-3} = 0$$

Multiplication by 4000 yields

$$2V_2 - 2V_1 + V_2 - 5 + 4V_2 - 8 = 0$$

which can be simplified to

$$-2V_1 + 7V_2 = 13 \qquad \textbf{(3.11)}$$

Equations (3.10) and (3.11) can be solved using the substitution method:

Solving Equation (3.11) for V_1, we obtain

$$V_1 = 3.5V_2 - 6.5$$

Substituting V_1 into Equation (3.10), we get

$$13(3.5V_2 - 6.5) - 4V_2 = 40$$

or

$$41.5V_2 = 124.5$$

continued

Example 3.2 continued

Thus, we have

$$V_2 = 3\text{ V}$$

and

$$V_1 = 3.5V_2 - 6.5 = 4\text{ V}$$

Alternatively, Equations (3.10) and (3.11) can be solved using Cramer's rule:

$$\Delta = \begin{vmatrix} 13 & -4 \\ -2 & 7 \end{vmatrix} = 13 \times 7 - (-4) \times (-2) = 83$$

$$V_1 = \frac{\begin{vmatrix} 40 & -4 \\ 13 & 7 \end{vmatrix}}{\Delta} = \frac{40 \times 7 - (-4) \times 13}{83} = \frac{332}{83} = 4\text{ V}$$

$$V_2 = \frac{\begin{vmatrix} 13 & 40 \\ -2 & 13 \end{vmatrix}}{\Delta} = \frac{13 \times 13 - 40 \times (-2)}{83} = \frac{249}{83} = 3\text{ V}$$

MATLAB

```
clear all;
A=[13 -4;-2 7];b=[40;13];
V=A\b
Answer:
V =
    4.0000
    3.0000
```

Exercise 3.2

Find V_1 and V_2 in the circuit shown in Figure 3.8.

FIGURE 3.8

The circuit for EXERCISE 3.2.

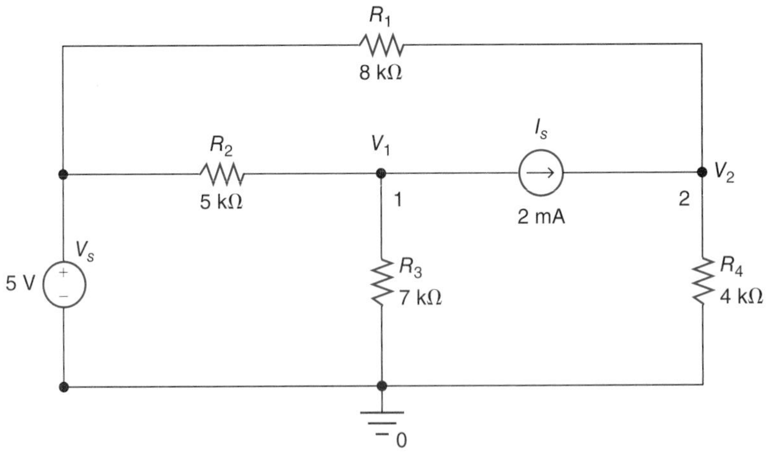

Answer:
$V_1 = -2.9167$ V, $V_2 = 7$ V.

EXAMPLE 3.3

Find V_1, V_2, and V_3 in the circuit shown in Figure 3.9.

FIGURE 3.9

The circuit for EXAMPLE 3.3.

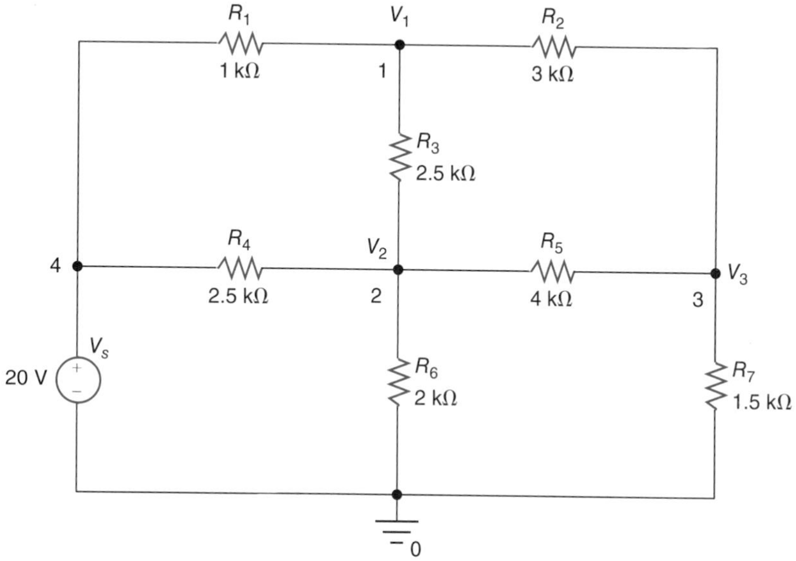

There are five nodes (0, 1, 2, 3, 4) in the circuit shown in Figure 3.9. The voltage at node 0 is zero (reference or ground). The voltage at node 4 is $V_s = 20$ V. The voltages at nodes 1, 2, and 3 are unknown at this point. Let the voltage at node 1 be V_1; at node 2, it is V_2, and at node 3, it is V_3. Summing the currents away from node 1 using Ohm's law, we obtain

$$\frac{V_1 - V_s}{R_1} + \frac{V_1 - V_2}{R_3} + \frac{V_1 - V_3}{R_2} = 0 \qquad (3.12)$$

Substituting the resistance and voltage values, we get

$$\frac{V_1 - 20}{1000} + \frac{V_1 - V_2}{2500} + \frac{V_1 - V_3}{3000} = 0$$

Multiplication by 15,000 results in

$$15(V_1 - 20) + 6(V_1 - V_2) + 5(V_1 - V_3) = 0$$

which can be simplified to

$$26V_1 - 6V_2 - 5V_3 = 300 \qquad (3.13)$$

Summing the currents away from node 2, using Ohm's law, we obtain

$$\frac{V_2 - V_1}{R_3} + \frac{V_2 - V_s}{R_4} + \frac{V_2}{R_6} + \frac{V_2 - V_3}{R_5} = 0 \qquad (3.14)$$

continued

Example 3.3 continued

Substituting the resistance and voltage values, we get

$$\frac{V_2 - V_1}{2500} + \frac{V_2 - 20}{2500} + \frac{V_2}{2000} + \frac{V_2 - V_3}{4000} = 0$$

Multiplication by 20,000 results in

$$8(V_2 - V_1) + 8(V_2 - 20) + 10V_2 + 5(V_2 - V_3) = 0$$

which can be simplified to

$$-8V_1 + 31V_2 - 5V_3 = 160 \quad \textbf{(3.15)}$$

Summing the currents away from node 3 using Ohm's law, we obtain

$$\frac{V_3 - V_1}{R_2} + \frac{V_3 - V_2}{R_5} + \frac{V_3}{R_7} = 0 \quad \textbf{(3.16)}$$

Substituting the resistance values, we get

$$\frac{V_3 - V_1}{3000} + \frac{V_3 - V_2}{4000} + \frac{V_3}{1500} = 0$$

Multiplication by 12,000 results in

$$4(V_3 - V_1) + 3(V_3 - V_2) + 8V_3 = 0$$

which can be simplified to

$$-4V_1 - 3V_2 + 15V_3 = 0 \quad \textbf{(3.17)}$$

Equations (3.13), (3.15), and (3.17) can be solved using the substitution method: Multiplying Equation (3.15) by 3, we obtain

$$-24V_1 + 93V_2 - 15V_3 = 480$$

Adding this equation and Equation (3.17), we get

$$-28V_1 + 90V_2 = 480$$

Solving for V_2, we obtain

$$V_2 = \frac{28}{90}V_1 + \frac{48}{9}$$

Subtracting Equation (3.15) from Equation (3.13), we get

$$34V_1 - 37V_2 = 140$$

Substituting $V_2 = \frac{28}{90}V_1 + \frac{48}{9}$, we obtain

$$34V_1 - 37\left(\frac{28}{90}V_1 + \frac{48}{9}\right) = 140$$

continued

Example 3.3 continued

Thus, we have

$$V_1 = \frac{140 + 37\dfrac{48}{9}}{34 - 37\dfrac{28}{90}} = 15 \text{ V}$$

$$V_2 = \frac{28}{90}V_1 + \frac{48}{9} = 10 \text{ V}$$

From Equation (3.17), we get

$$V_3 = \frac{4}{15}V_1 + \frac{3}{15}V_2 = 6 \text{ V}$$

Alternatively, Equations (3.13), (3.15), and (3.17) can be solved using Cramer's rule:

$$\begin{bmatrix} 26 & -6 & -5 \\ -8 & 31 & -5 \\ -4 & -3 & 15 \end{bmatrix} \begin{bmatrix} V_1 \\ V_2 \\ V_3 \end{bmatrix} = \begin{bmatrix} 300 \\ 160 \\ 0 \end{bmatrix} \qquad \textbf{(3.18)}$$

Let

$$A = \begin{bmatrix} 26 & -6 & -5 \\ -8 & 31 & -5 \\ -4 & -3 & 15 \end{bmatrix}, \quad V = \begin{bmatrix} V_1 \\ V_2 \\ V_3 \end{bmatrix}, \quad b = \begin{bmatrix} 300 \\ 160 \\ 0 \end{bmatrix}$$

Then, Equation (3.18) can be written as

$$AV = b.$$

Let Δ be the determinant of matrix A. Then, Δ can be written as

$$\Delta = |A| = \begin{vmatrix} 26 & -6 & -5 \\ -8 & 31 & -5 \\ -4 & -3 & 15 \end{vmatrix}$$

One method of computing the determinant of the 3×3 matrix A is to copy the first two columns and place them at the end as the fourth and fifth columns, and then multiply the numbers in the diagonals. The determinant is the difference of the sum of the products to the right and the sum of the products to the left:

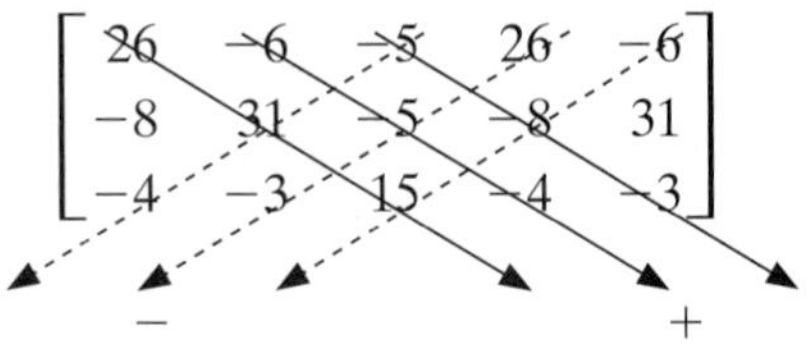

continued

Example 3.3 continued

Thus, we have

$$\begin{aligned}\Delta &= [(26)\times(31)\times(15) + (-6)\times(-5)\times(-4) + (-5)\times(-8)\times(-3)]\\ &\quad - [(-5)\times(31)\times(-4) + (26)\times(-5)\times(-3) + (-6)\times(-8)\times(15)]\\ &= [12{,}090 + (-120) + (-120)] - [620 + 390 + 720] = 11{,}850 - 1730 = 10{,}120\end{aligned}$$

In MATLAB, the determinant of A is found by det(A):

```
>> A=[26 -6 -5;-8 31 -5;-4 -3 15]
A =
     26    -6    -5
     -8    31    -5
     -4    -3    15
>> det(A)
ans =
        10,120
```

The determinant of an $n \times n$ matrix can also be found as a linear combination of determinants of matrices with dimension $(n-1)\times(n-1)$. Choose a row or a column of a matrix. For the 3×3 matrix A given previously, we select the first row. A minor M_{ij} is a determinant of a matrix with the ith row and jth column removed. For matrix A, we have

$$M_{11} = \begin{vmatrix} 26 & -6 & -5 \\ -8 & 31 & -5 \\ -4 & -3 & 15 \end{vmatrix} = \begin{vmatrix} 31 & -5 \\ -3 & 15 \end{vmatrix} = (31)\times(15) - (-5)\times(-3)$$
$$= 465 - 15 = 450$$

$$M_{12} = \begin{vmatrix} 26 & -6 & -5 \\ -8 & 31 & -5 \\ -4 & -3 & 15 \end{vmatrix} = \begin{vmatrix} -8 & -5 \\ -4 & 15 \end{vmatrix} = (-8)\times(15) - (-5)\times(-4)$$
$$= -120 - 20 = -140$$

$$M_{13} = \begin{vmatrix} 26 & -6 & -5 \\ -8 & 31 & -5 \\ -4 & -3 & 15 \end{vmatrix} = \begin{vmatrix} -8 & 31 \\ -4 & -3 \end{vmatrix} = (-8)\times(-3) - (31)\times(-4)$$
$$= 24 - (-124) = 148$$

A cofactor C_{ij} is the product of the minor M_{ij} and $(-1)^{i+j}$; that is,

$$C_{ij} = (-1)^{i+j} M_{ij}$$

If the sum of row index i and column index j is even, $(-1)^{i+j} = 1$, and if $i + j$ is odd, $(-1)^{i+j} = -1$. For the 3×3 matrix A given previously, the cofactors are given by

$$C_{11} = (-1)^{1+1}M_{11} = M_{11} = 450,\quad C_{12} = (-1)^{1+2}M_{12} = -M_{12} = 140,$$
$$C_{13} = (-1)^{1+3}M_{13} = M_{13} = 148$$

continued

Example 3.3 continued

The determinant of a matrix can be expanded as a sum of cofactors. The coefficients of this expansion are the elements of a chosen row or column. For the 3×3 matrix A given previously, with the first row chosen, we have

$$\Delta = |A| = 26C_{11} - 6C_{12} - 5C_{13} = 26 \times 450 - 6 \times 140 - 5 \times 148$$
$$= 11{,}700 - 840 - 740 = 10{,}120$$

The voltages V_1, V_2, and V_3 are given, respectively, as

$$V_1 = \frac{\begin{vmatrix} 300 & -6 & -5 \\ 160 & 31 & -5 \\ 0 & -3 & 15 \end{vmatrix}}{\begin{vmatrix} 26 & -6 & -5 \\ -8 & 31 & -5 \\ -4 & -3 & 15 \end{vmatrix}}, \quad V_2 = \frac{\begin{vmatrix} 26 & 300 & -5 \\ -8 & 160 & -5 \\ -4 & 0 & 15 \end{vmatrix}}{\begin{vmatrix} 26 & -6 & -5 \\ -8 & 31 & -5 \\ -4 & -3 & 15 \end{vmatrix}},$$

$$V_3 = \frac{\begin{vmatrix} 26 & -6 & 300 \\ -8 & 31 & 160 \\ -4 & -3 & 0 \end{vmatrix}}{\begin{vmatrix} 26 & -6 & -5 \\ -8 & 31 & -5 \\ -4 & -3 & 15 \end{vmatrix}}$$

The determinants in the numerators are given, respectively, by

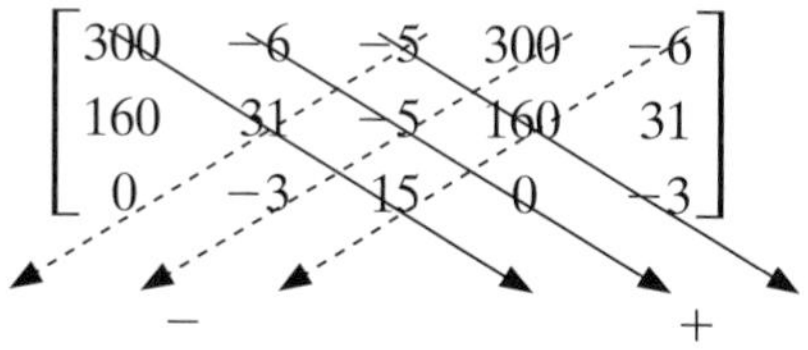

$$\Delta_1 = [(300) \times (31) \times (15) + (-6) \times (-5) \times (0) + (-5) \times (160) \times (-3)]$$
$$- [(-5) \times (31) \times (0) + (300) \times (-5) \times (-3) + (-6) \times (160) \times (15)]$$
$$= [139{,}500 + 0 + 2400] - [0 + 4500 - 14{,}400] = 141{,}900 + 9900 = 151{,}800$$

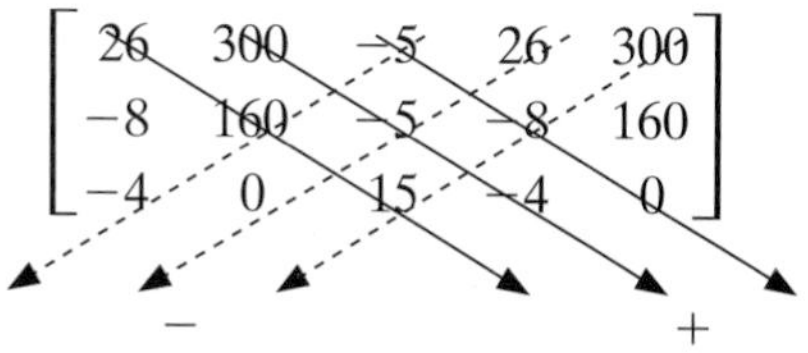

$$\Delta_2 = [(26) \times (160) \times (15) + (300) \times (-5) \times (-4) + (-5) \times (-8) \times (0)]$$
$$- [(-5) \times (160) \times (-4) + (26) \times (-5) \times (0) + (300) \times (-8) \times (15)]$$
$$= [62{,}400 + 6000 + 0] - [3{,}200 + 0 - 36{,}000] = 68{,}400 + 32{,}800 = 101{,}200$$

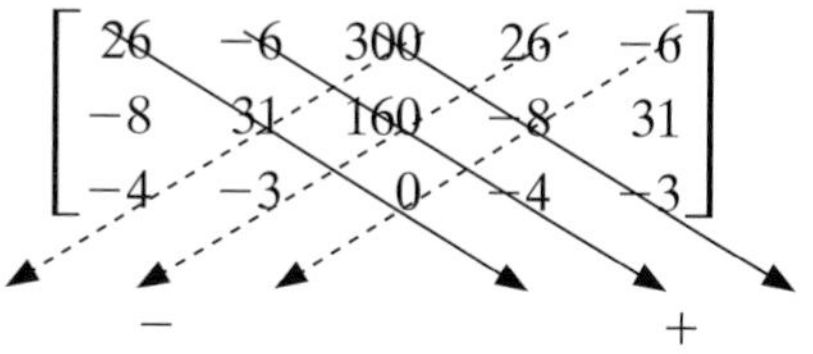

continued

Example 3.3 continued

$$\Delta_3 = [(26) \times (31) \times (0) + (-6) \times (160) \times (-4) + (300) \times (-8) \times (-3)]$$
$$- [(300) \times (31) \times (-4) + (26) \times (160) \times (-3) + (-6) \times (-8) \times (0)]$$
$$= [0 + 3{,}840 + 7{,}200] - [-37{,}200 - 12{,}480 + 0] = 11{,}040 + 49{,}680 = 60{,}720$$

The voltages V_1, V_2, and V_3 are given, respectively, as

$$V_1 = \frac{\Delta_1}{\Delta} = \frac{151{,}800}{10{,}120} = 15 \text{ V}, \quad V_2 = \frac{\Delta_2}{\Delta} = \frac{101{,}200}{10{,}120} = 10 \text{ V},$$

$$V_3 = \frac{\Delta_3}{\Delta} = \frac{60{,}720}{10{,}120} = 6 \text{ V}$$

MATLAB

```
A=[26 -6 -5;-8 31 -5;-4 -3 15];b=[300;160;0];
V=A\b

Answer:
V =
    15
    10
     6
```

The MATLAB function **solve** can be used to solve Equations (3.12), (3.14), and (3.16). The ellipsis (...) is used to continue long statements in multiple lines:

```
%EXAMPLE 3.3
clear all;
Vs=20;R1=1000;R2=3000;R3=2500;R4=2500;R5=4000;R6=2000;R7=1500;
syms V1 V2 V3
[V1,V2,V3]=solve((V1-Vs)/R1+(V1-V2)/R3+(V1-V3)/R2,...
(V2-V1)/R3+(V2-Vs)/R4+V2/R6+(V2-V3)/R5,...
(V3-V1)/R2+(V3-V2)/R5+V3/R7,V1,V2,V3);
V1=vpa(V1,6)
V2=vpa(V2,6)
V3=vpa(V3,6)

Answers:
V1 =
15.0
V2 =
10.0
V3 =
6.0
```

Exercise 3.3

Find V_1 and V_2 in the circuit shown in Figure 3.10.

continued

Exercise 3.3 continued

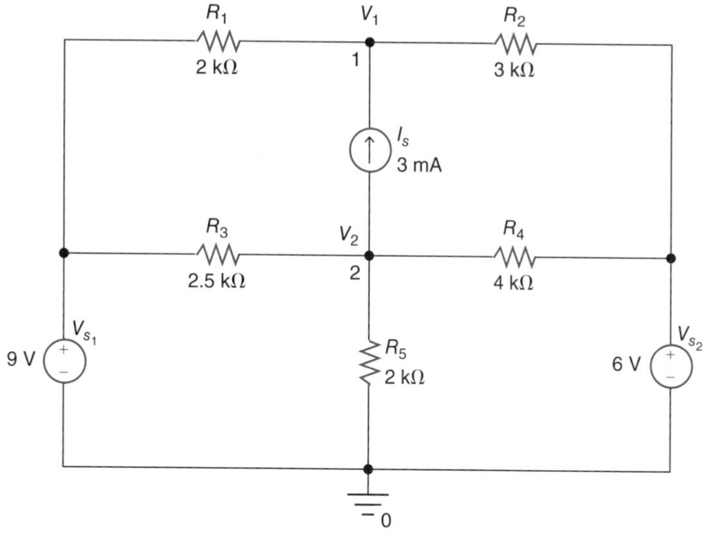

FIGURE 3.10 The circuit for EXERCISE 3.3.

Answer:
$V_1 = 11.4$ V, $V_2 = 1.8261$ V.

EXAMPLE 3.4

Find V_1 and V_2 in the circuit shown in Figure 3.11. Find currents I_1, I_2, I_3, and I_4, and the current through the voltage-controlled current source.

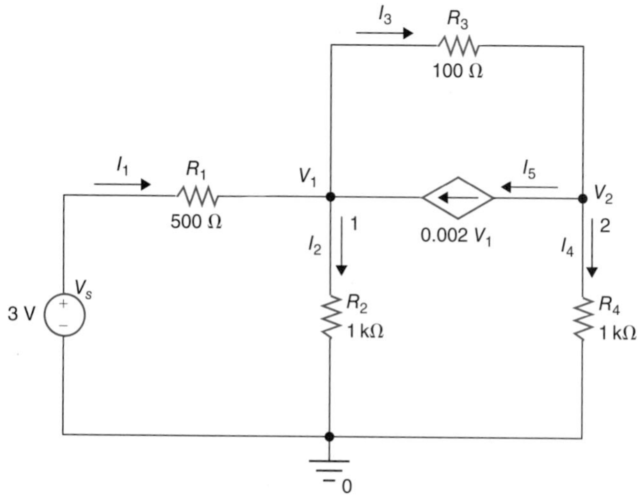

FIGURE 3.11 The circuit for EXAMPLE 3.4.

Summing the currents leaving node 1, we have

$$\frac{V_1 - V_s}{R_1} + \frac{V_1}{R_2} + \frac{V_1 - V_2}{R_3} - 0.002V_1 = 0$$

Substituting the resistance and voltage values, we get

$$\frac{V_1 - 3}{500} + \frac{V_1}{1000} + \frac{V_1 - V_2}{100} - 0.002V_1 = 0$$

Multiplication by 1000 yields

$$2V_1 - 6 + V_1 + 10V_1 - 10V_2 - 2V_1 = 0$$

which can be simplified to

$$11V_1 - 10V_2 = 6 \quad \textbf{(3.19)}$$

Summing the currents leaving node 2, we have

$$\frac{V_2 - V_1}{R_3} + \frac{V_2}{R_4} + 0.002V_1 = 0$$

continued

Example 3.4 continued

Substituting the resistance values, we obtain

$$\frac{V_2 - V_1}{100} + \frac{V_2}{1000} + 0.002V_1 = 0$$

Multiplication by 1000 yields

$$10V_2 - 10V_1 + V_2 + 2V_1 = 0$$

which can be simplified to

$$-8V_1 + 11V_2 = 0 \tag{3.20}$$

Equations (3.19) and (3.20) can be solved using the substitution method:

Solving Equation (3.19) for V_2, we obtain

$$V_2 = 1.1V_1 - 0.6$$

Substituting V_2 into Equation (3.20), we get

$$-8V_1 + 11(1.1V_1 - 0.6) = 0$$

or

$$4.1V_1 = 6.6$$

Thus, we obtain

$$V_1 = \frac{66}{41} = 1.6098 \text{ V}$$

$$V_2 = 1.1V_1 - 0.6 = \frac{48}{41} = 1.1707 \text{ V}$$

Alternatively, Equations (3.19) and (3.20) can be solved using Cramer's rule:

$$\Delta = \begin{vmatrix} 11 & -10 \\ -8 & 11 \end{vmatrix} = 11 \times 11 - (-10) \times (-8) = 121 - 80 = 41$$

$$V_1 = \frac{\begin{vmatrix} 6 & -10 \\ 0 & 11 \end{vmatrix}}{\Delta} = \frac{6 \times 11 - (-10) \times 0}{41} = \frac{66}{41} = 1.6098 \text{ V}$$

$$V_2 = \frac{\begin{vmatrix} 11 & 6 \\ -8 & 0 \end{vmatrix}}{\Delta} = \frac{11 \times 0 - 6 \times (-8)}{41} = \frac{48}{41} = 1.1707 \text{ V}$$

The currents are given as follows:

$$I_1 = \frac{V_s - V_1}{R_1} = 2.7805 \text{ mA}, \; I_2 = \frac{V_1}{R_2} = 1.6098 \text{ mA}, \; I_3 = \frac{V_1 - V_2}{R_3} = 4.3902 \text{ mA},$$

continued

Example 3.4 continued

$$I_4 = \frac{V_2}{R_4} = 1.1707 \text{ mA}, \quad I_5 = 0.002V_1 = 3.2195 \text{ mA}$$

Notice that the sum of currents leaving node 1 and node 2 are zero; that is,

$$-I_1 + I_2 + I_3 - I_5 = 0, \quad -I_3 + I_4 + I_5 = 0$$

Exercise 3.4

Find V_1 and V_2 in the circuit shown in Figure 3.12.

FIGURE 3.12

The circuit for EXERCISE 3.4.

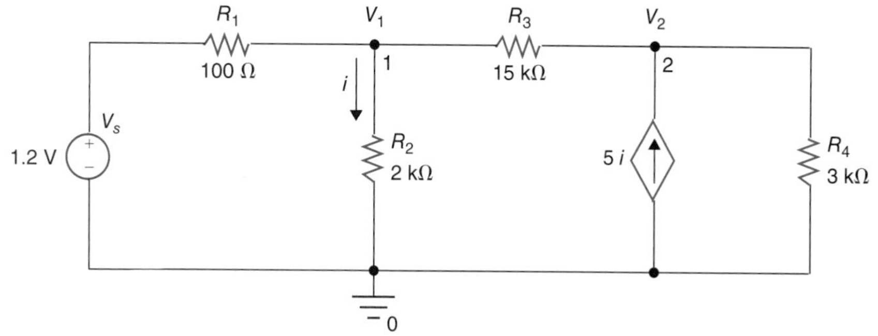

Answer:
$V_1 = 1.1836$ V, $V_2 = 7.5945$ V.

EXAMPLE 3.5

Find V_1 and V_2 in the circuit shown in Figure 3.13.

FIGURE 3.13

The circuit for EXAMPLE 3.5.

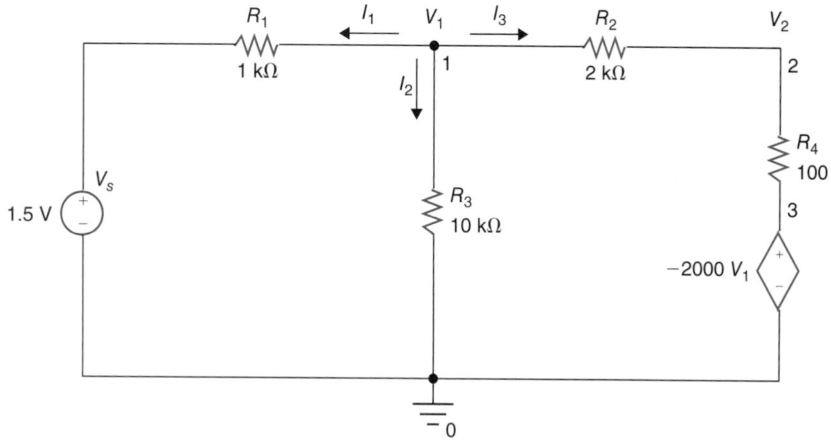

This circuit has only one node-voltage, labeled node 1. Summing the currents leaving the node gives

$$I_1 + I_2 + I_3 = 0$$

continued

Example 3.5 continued

or

$$\frac{V_1 - V_s}{R_1} + \frac{V_1}{R_3} + \frac{V_1 - (-2000V_1)}{R_2 + R_4} = 0$$

Substituting the resistance and voltage values, we get

$$\frac{V_1 - 1.5}{1000} + \frac{V_1}{10{,}000} + \frac{V_1 - (-2000V_1)}{2100} = 0 \tag{3.21}$$

Equation (3.21) can be rearranged as

$$\left(\frac{1}{1000} + \frac{1}{10{,}000} + \frac{2001}{2100}\right)V_1 = \frac{1.5}{1000} \tag{3.22}$$

Solving Equation (3.22) for V_1, we obtain

$$V_1 = \frac{\frac{1.5}{1000}}{\frac{1}{1000} + \frac{1}{10{,}000} + \frac{2001}{2100}} = 0.001572397681837 \text{ V} = 1.5724 \text{ mV}$$

The current I_3 is given by

$$I_3 = \frac{V_1 - (-2000V_1)}{R_2 + R_4} = \frac{2001V_1}{2100} = 0.00149827036255 \text{ A}$$

The voltage V_2 is given by

$$\begin{aligned} V_2 &= V_1 - R_2 I_3 \\ &= 1.572397681837 \times 10^{-3} - 2000 \times 0.00149827036255 = -2.994968 \text{ V} \end{aligned}$$

MATLAB

```
%EXAMPLE 3.5
clear all;format long;
Vs=1.5;R1=1000;R2=2000;R3=10000;R4=100;
syms V1 V2
[V1,V2]=solve((V1-Vs)/R1+V1/R3+(V1-V2)/R2, ...
(V2-V1)/R2+(V2+2000*V1)/R4,V1,V2);
I3=(V1-(-2000*V1))/(R2+R4);
V2b=V1-R2*I3;
V1=vpa(V1,15)
V2=vpa(V2,15)
I3=vpa(I3,15)
V2b=vpa(V2,15)

Answers:
V1 =
0.00157239768183656
V2 =
-2.99496832741812
```

continued

Example 3.5 continued

MATLAB continued

```
I3 =
0.00149827036254998
V2b =
-2.99496832741812
```

Exercise 3.5

Find V_1 and V_2 in the circuit shown in Figure 3.14.

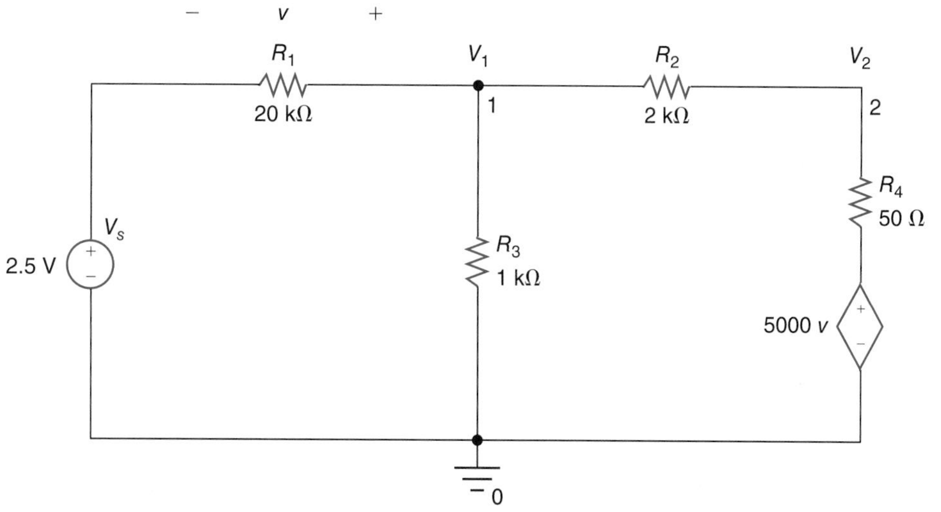

FIGURE 3.14 The circuit for EXERCISE 3.5.

Answer:
$V_1 = 2.501526$ V, $V_2 = 7.504730$ V.

EXAMPLE 3.6

Find V_1 and V_2 in the circuit shown in Figure 3.15.

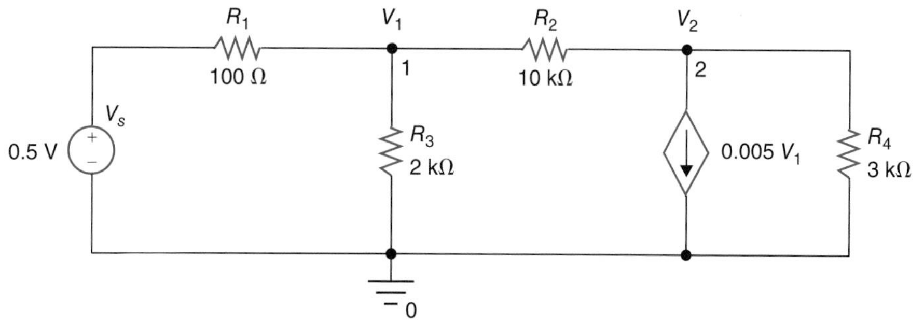

FIGURE 3.15 The circuit for EXAMPLE 3.6.

continued

Example 3.6 continued

Summing the currents leaving node 1, we obtain

$$\frac{V_1 - V_s}{R_1} + \frac{V_1}{R_3} + \frac{V_1 - V_2}{R_2} = 0$$

Substituting the resistance and voltage values, we get

$$\frac{V_1 - 0.5}{100} + \frac{V_1}{2000} + \frac{V_1 - V_2}{10{,}000} = 0$$

Multiplication by 10,000 yields

$$100(V_1 - 0.5) + 5V_1 + V_1 - V_2 = 0$$

which can be simplified to

$$106V_1 - V_2 = 50 \quad \textbf{(3.23)}$$

Summing the currents leaving node 2, we obtain

$$\frac{V_2 - V_1}{R_2} + 0.005V_1 + \frac{V_2}{R_4} = 0$$

Substituting the resistance values, we get

$$\frac{V_2 - V_1}{10{,}000} + 0.005V_1 + \frac{V_2}{3000} = 0$$

Multiplication by 30,000 yields

$$3V_2 - 3V_1 + 150V_1 + 10V_2 = 0$$

which can be simplified to

$$147V_1 + 13V_2 = 0 \quad \textbf{(3.24)}$$

Equations (3.23) and (3.24) can be solved using the substitution method:

Solving Equation (3.23) for V_2, we obtain

$$V_2 = 106V_1 - 50$$

Substituting V_2 into Equation (3.24), we get

$$147V_1 + 13(106V_1 - 50) = 0$$

or

$$1525V_1 = 650$$

Thus, we obtain

$$V_1 = \frac{650}{1525} = \frac{26}{61} = 0.42622951 \text{ V}$$

$$V_2 = 106V_1 - 50 = -4.8197 \text{ V}$$

continued

Example 3.6 continued

Alternatively, Equations (3.23) and (3.24) can be solved using Cramer's rule:

$$\Delta = \begin{vmatrix} 106 & -1 \\ 147 & 13 \end{vmatrix} = 106 \times 13 - (-1) \times 147 = 1378 - (-147) = 1525$$

$$V_1 = \frac{\begin{vmatrix} 50 & -1 \\ 0 & 13 \end{vmatrix}}{\Delta} = \frac{50 \times 13 - (-1) \times 0}{1525} = \frac{650}{1525} = 0.4262 \text{ V}$$

$$V_2 = \frac{\begin{vmatrix} 106 & 50 \\ 147 & 0 \end{vmatrix}}{\Delta} = \frac{106 \times 0 - 50 \times 147}{1525} = \frac{-7350}{1525} = -4.8197 \text{ V}$$

MATLAB

```
%EXAMPLE 3.6
clear all;format long;
Vs=0.5;R1=100;R2=10000;R3=2000;R4=3000;
syms V1 V2
[V1,V2]=solve((V1-Vs)/R1+V1/R3+(V1-V2)/R2, ...
(V2-V1)/R2+0.005*V1+V2/R4,V1,V2);
V1=vpa(V1,10)
V2=vpa(V2,10)

Answers:
V1 =
0.4262295082
V2 =
-4.819672131
```

Exercise 3.6

Find V_1 and V_2 in the circuit shown in Figure 3.16.

FIGURE 3.16

The circuit for EXERCISE 3.6.

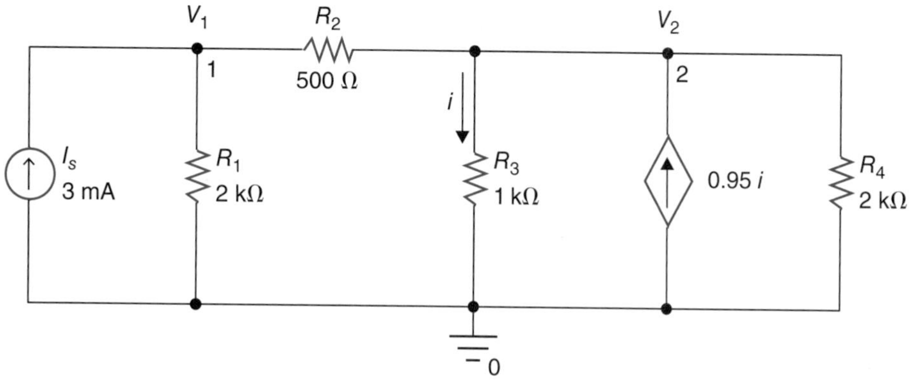

Answer:
$V_1 = 3.2211$ V, $V_2 = 2.5263$ V.

3.3 Supernode

If there is a voltage source in a circuit between two nodes whose voltages are unknown, we do not know the current through the voltage source, and it is not possible to write the node equations for the two nodes that include the voltage source. In this case, combine the two nodes to form a **supernode**. Then, we can write the node equation for this supernode. One additional equation, commonly referred to as a *constraint equation* relating the two node voltages, can be obtained by representing the voltage source as a potential drop or as a potential rise between the two nodes.

As an example, consider the circuit shown in Figure 3.17. Let I_5 be the current through the voltage source V_{s_2} from the negative terminal to the positive terminal (→), and let I_6 be the current through the voltage source V_{s_2} from the positive terminal to the negative terminal (←). Then, $I_6 = -I_5$. Application of KCL to node 1 yields

$$I_1 + I_2 + I_5 = 0 \tag{3.25}$$

FIGURE 3.17

A circuit with a supernode.

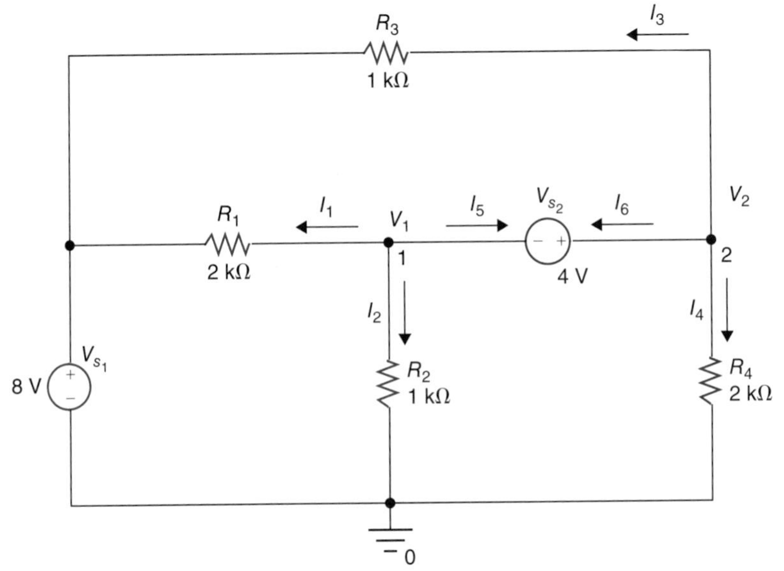

Similarly, application of KCL to node 2 yields

$$I_3 + I_4 + I_6 = 0 \tag{3.26}$$

When an ideal voltage source is present between two essential nodes, it is impossible to determine the current through the source because the internal resistance of an ideal voltage source is zero. We cannot apply Ohm's law. However, to proceed with applying KCL, we arbitrarily define a current out of the node. By writing both equations and summing, the arbitrary current vanishes, giving rise to an equation that represents the sum of all currents flowing out of both nodes 1 and 2. Adding Equations (3.25) and (3.26), we obtain

$$I_1 + I_2 + I_3 + I_4 + I_5 + I_6 = 0 \tag{3.27}$$

Since $I_6 = -I_5$, Equation (3.27) becomes

$$I_1 + I_2 + I_3 + I_4 = 0 \tag{3.28}$$

Equation (3.28) suggests that, ignoring the currents I_5 and I_6, the sum of currents leaving nodes 1 and 2 is zero. Nodes 1 and 2 can be treated as a supernode. The supernode for the circuit shown in Figure 3.17 is shown in Figure 3.18. Notice that Equation (3.28) is the sum of currents leaving the supernode. Applying Ohm's law, we can rewrite Equation (3.28) as

$$\frac{V_1 - V_{S_1}}{R_1} + \frac{V_1}{R_2} + \frac{V_2 - V_{S_1}}{R_3} + \frac{V_2}{R_4} = 0$$

Substituting the resistance and voltage values, we get

$$\frac{V_1 - 8}{2000} + \frac{V_1}{1000} + \frac{V_2 - 8}{1000} + \frac{V_2}{2000} = 0$$

FIGURE 3.18

The supernode includes node 1 and node 2.

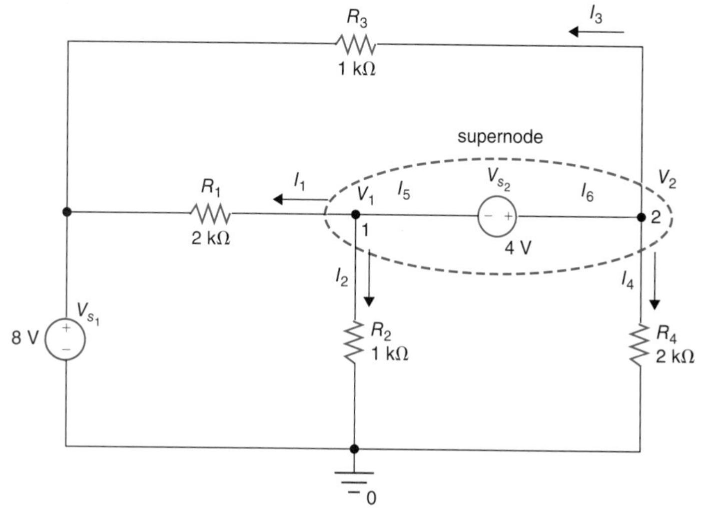

Multiplication by 2000 yields

$$V_1 - 8 + 2V_1 + 2V_2 - 16 + V_2 = 0$$

which can be simplified to $3V_1 + 3V_2 = 24$. Thus, we have

$$V_1 + V_2 = 8 \qquad (3.29)$$

The voltage at node 2, V_2, is higher than the voltage at node 1, V_1, by $V_{S_2} = 4$ V; that is,

$$V_2 - V_1 = 4$$

or

$$V_1 - V_2 = -4 \qquad (3.30)$$

Summing Equations (3.29) and (3.30) yields

$$2V_1 = 4$$

Thus, we have $V_1 = 2$ V. Since V_2 is 4 V higher than V_1, we have $V_2 = 6$ V.

EXAMPLE 3.7

Find V_1, V_2, V_3 in the circuit shown in Figure 3.19.

continued

Example 3.7 continued

FIGURE 3.19

The circuit for EXAMPLE 3.7.

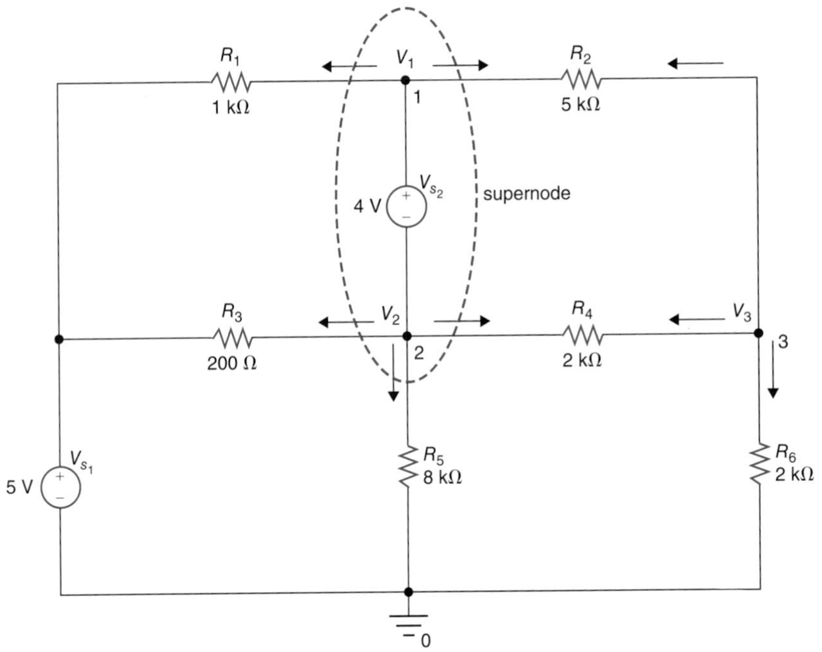

Notice the presence of an ideal voltage source connected between nodes 1 and 2. This makes nodes 1 and 2 part of a supernode. A constraint equation describing this source as a potential drop from node 1 to node 2 is

$$V_1 - V_2 = 4 \tag{3.31}$$

Since the current from node 1 to node 2 is the negative of the current from node 2 to node 1, they cancel each other out. Summing the currents leaving supernode, we obtain

$$\frac{V_1 - V_{s_1}}{R_1} + \frac{V_1 - V_3}{R_2} + \frac{V_2 - V_{s_1}}{R_3} + \frac{V_2}{R_5} + \frac{V_2 - V_3}{R_4} = 0$$

Substituting the resistance and voltage values, we get

$$\frac{V_1 - 5}{1000} + \frac{V_1 - V_3}{5000} + \frac{V_2 - 5}{200} + \frac{V_2}{8000} + \frac{V_2 - V_3}{2000} = 0$$

Multiplication by 40,000 results in

$$40(V_1 - 5) + 8(V_1 - V_3) + 200(V_2 - 5) + 5V_2 + 20(V_2 - V_3) = 0$$

which can be simplified to

$$48V_1 + 225V_2 - 28V_3 = 1200 \tag{3.32}$$

Substitution of Equation (3.31) into Equation (3.32) yields

$$273V_2 - 28V_3 = 1008 \tag{3.33}$$

Summing the currents leaving node 3, we obtain

$$\frac{V_3 - V_1}{R_2} + \frac{V_3 - V_2}{R_4} + \frac{V_3}{R_6} = 0$$

continued

Example 3.7 continued

Substituting the resistance values, we get

$$\frac{V_3 - V_1}{5000} + \frac{V_3 - V_2}{2000} + \frac{V_3}{2000} = 0$$

Multiplication by 10,000 results in

$$2(V_3 - V_1) + 5(V_3 - V_2) + 5V_3 = 0$$

which can be simplified to

$$-2V_1 - 5V_2 + 12V_3 = 0 \tag{3.34}$$

Substitution of Equation (3.31) into Equation (3.34) yields

$$-7V_2 + 12V_3 = 8 \tag{3.35}$$

Equations (3.33) and (3.35) can be solved using the substitution method:

Solving Equation (3.35) for V_2, we obtain

$$V_2 = \frac{12}{7}V_3 - \frac{8}{7}$$

Substituting V_2 into Equation (3.33), we get

$$273\left(\frac{12}{7}V_3 - \frac{8}{7}\right) - 28V_3 = 1008$$

or

$$440V_3 = 1320$$

Thus, we obtain

$$V_3 = \frac{132}{44} = \frac{12}{4} = 3\text{ V}$$

$$V_2 = \frac{12}{7}V_3 - \frac{8}{7} = 4\text{ V}$$

$$V_1 = V_2 + 4 = 4 + 4 = 8\text{ V}$$

Alternatively, Equations (3.33) and (3.35) can be solved using Cramer's rule:

$$\Delta = \begin{vmatrix} 273 & -28 \\ -7 & 12 \end{vmatrix} = 273 \times 12 - (-28) \times (-7) = 3276 - 196 = 3080$$

$$V_2 = \frac{\begin{vmatrix} 1008 & -28 \\ 8 & 12 \end{vmatrix}}{\Delta} = \frac{1008 \times 12 - (-28) \times 8}{3080} = \frac{12{,}320}{3080} = 4\text{ V}$$

$$V_3 = \frac{\begin{vmatrix} 273 & 1008 \\ -7 & 8 \end{vmatrix}}{\Delta} = \frac{273 \times 8 - 1008 \times (-7)}{3080} = \frac{9240}{3080} = 3\text{ V}$$

continued

Example 3.7 continued

From Equation (3.31), we have

$$V_1 = V_2 + 4 = 4 + 4 = 8\text{ V}$$

MATLAB

```
%EXAMPLE 3.7
clear all;format long;
Vs1=5;Vs2=4;
R1=1000;R2=5000;R3=200;R4=2000;R5=8000;R6=2000;
syms V1 V2 V3
[V1,V2,V3]=solve(V1-V2==Vs2, ...
(V1-Vs1)/R1+(V1-V3)/R2+(V2-Vs1)/R3+V2/R5+(V2-V3)/R4, ...
(V3-V1)/R2+(V3-V2)/R4+V3/R6,V1,V2,V3);
V1=vpa(V1,15)
V2=vpa(V2,15)
V3=vpa(V3,15)

Answers:
V1 =
8.0
V2 =
4.0
V3 =
3.0
```

Exercise 3.7

Find V_1, V_2, and V_3 in the circuit shown in Figure 3.20.

FIGURE 3.20

The circuit for EXERCISE 3.7.

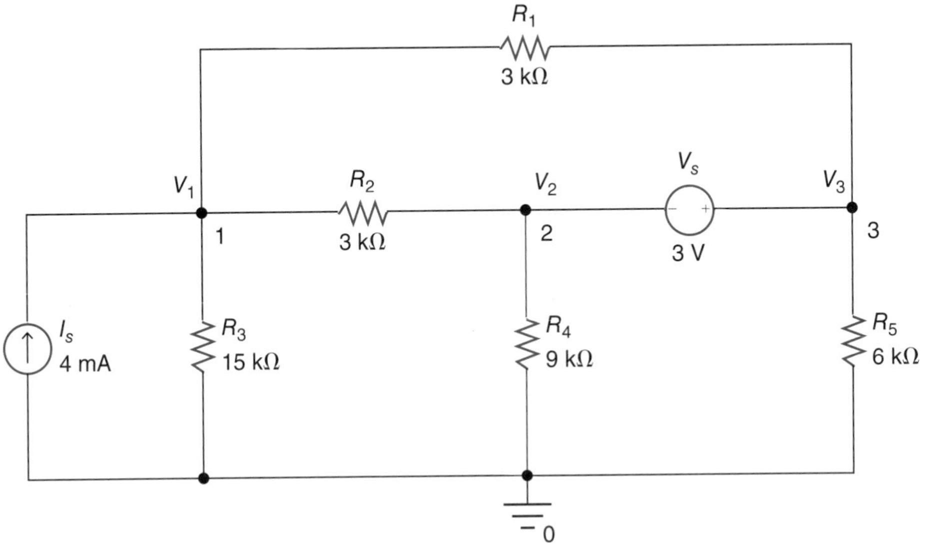

Answer:
$V_1 = 15\text{ V}, V_2 = 9\text{ V}, V_3 = 12\text{ V}.$

EXAMPLE 3.8

Find V_1, V_2, and V_3 in the circuit shown in Figure 3.21.

FIGURE 3.21

The circuit for EXAMPLE 3.8.

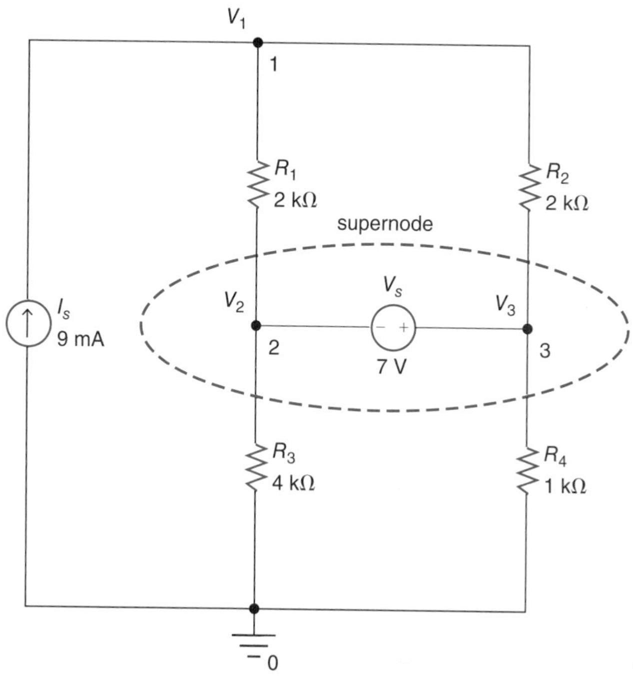

The source voltage V_s between nodes 2 and 3 represents either a potential rise from node 2 to node 3 or a potential drop from node 3 to node 2. It is good practice to write out equations with the variables listed in the same order. This makes it easy for collecting coefficients to form a set of matrix-vector equations and eliminates the possibilities for errors in algebraic signs. Considering this effect as a potential rise gives the constraint equation

$$V_2 - V_3 = -7 \tag{3.36}$$

Summing the currents leaving the supernode, circled in Figure 3.21, we obtain

$$\frac{V_2 - V_1}{R_1} + \frac{V_2}{R_3} + \frac{V_3 - V_1}{R_2} + \frac{V_3}{R_4} = 0$$

Substituting the resistance values, we get

$$\frac{V_2 - V_1}{2000} + \frac{V_2}{4000} + \frac{V_3 - V_1}{2000} + \frac{V_3}{1000} = 0$$

Multiplication by 4000 yields

$$2(V_2 - V_1) + V_2 + 2(V_3 - V_1) + 4V_3 = 0$$

continued

Example 3.8 continued

which can be simplified to

$$-4V_1 + 3V_2 + 6V_3 = 0 \tag{3.37}$$

Substitution of Equation (3.36) into Equation (3.37) yields

$$-4V_1 + 9V_2 = -42 \tag{3.38}$$

Summing the currents leaving node 1, we have

$$-I_s + \frac{V_1 - V_2}{R_1} + \frac{V_1 - V_3}{R_2} = 0$$

Substituting the resistance and current values, we get

$$-9 \times 10^{-3} + \frac{V_1 - V_2}{2000} + \frac{V_1 - V_3}{2000} = 0$$

Multiplication by 2000 yields

$$-18 + V_1 - V_2 + V_1 - V_3 = 0$$

which can be simplified to

$$2V_1 - V_2 - V_3 = 18 \tag{3.39}$$

Substitution of Equation (3.36) into Equation (3.39) yields

$$2V_1 - 2V_2 = 25 \tag{3.40}$$

Equations (3.38) and (3.40) can be solved using the substitution method: Solving Equation (3.40) for V_2, we obtain

$$V_2 = V_1 - 12.5$$

Substituting V_2 into Equation (3.38), we get

$$-4V_1 + 9(V_1 - 12.5) = -42$$

or

$$5V_1 = 70.5$$

Thus, we obtain

$$V_1 = \frac{70.5}{5} = 14.1 \text{ V}$$

$$V_2 = V_1 - 12.5 = 1.6 \text{ V}$$

$$V_3 = V_2 + 7 = 1.6 + 7 = 8.6 \text{ V}$$

continued

Example 3.8 continued

Alternatively, Equations (3.38) and (3.40) can be solved using Cramer's rule:

$$\Delta = \begin{vmatrix} -4 & 9 \\ 2 & -2 \end{vmatrix} = (-4) \times (-2) - 9 \times 2 = 8 - 18 = -10$$

$$V_1 = \frac{\begin{vmatrix} -42 & 9 \\ 25 & -2 \end{vmatrix}}{\Delta} = \frac{(-42) \times (-2) - 9 \times 25}{-10} = \frac{-141}{-10} = 14.1 \text{ V}$$

$$V_2 = \frac{\begin{vmatrix} -4 & -42 \\ 2 & 25 \end{vmatrix}}{\Delta} = \frac{(-4) \times 25 - (-42) \times 2}{-10} = \frac{-16}{-10} = 1.6 \text{ V}$$

From Equation (3.36), we have

$$V_3 = V_2 + 7 = 1.6 + 7 = 8.6 \text{ V}$$

MATLAB

```
A=[-4 9;2 -2];b=[-42;25];
V=A\b
V =
   14.1000
    1.6000
```

Exercise 3.8

Find the voltages V_1, V_2, and V_3 for the circuit shown in Figure 3.22.

FIGURE 3.22

The circuit for EXERCISE 3.8.

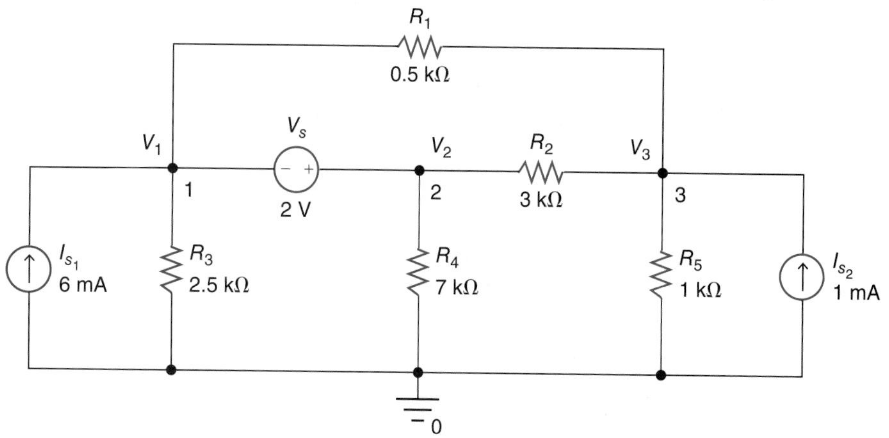

Answer:
$V_1 = 5$ V, $V_2 = 7$ V, $V_3 = 4$ V.

EXAMPLE 3.9

Find V_1, V_2, and V_3 for the circuit shown in Figure 3.23.

FIGURE 3.23

The circuit with VCVS.

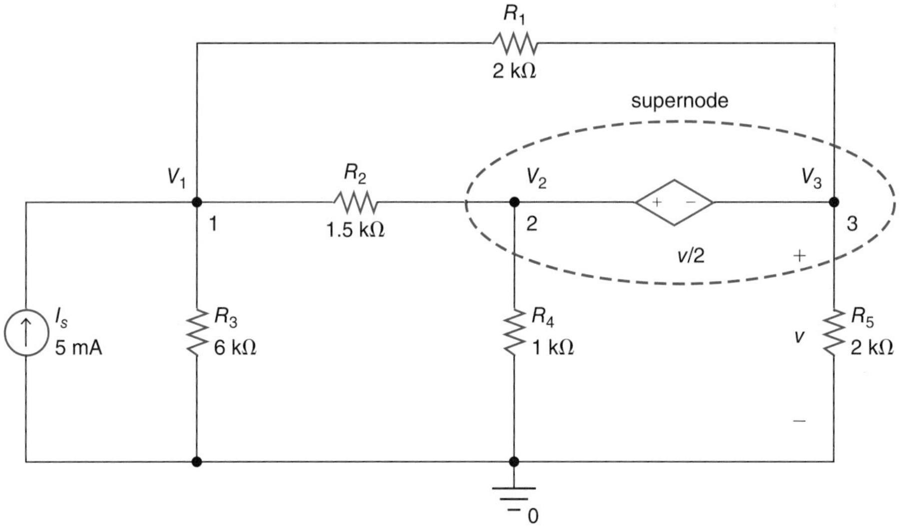

The presence of a dependent source across two essential nodes is no different than an ideal voltage source across two essential nodes for application of nodal analysis. The key is to use the dependent variable in terms of the appropriate node voltage variable. Notice that $v = V_3$. Also notice that

$$V_2 = V_3 + v/2 = V_3 + V_3/2 = 1.5V_3 \quad \textbf{(3.41)}$$

Summing the currents leaving the supernode consisting of nodes 2 and 3, we obtain

$$\frac{V_2 - V_1}{R_2} + \frac{V_2}{R_4} + \frac{V_3 - V_1}{R_1} + \frac{V_3}{R_5} = 0$$

Substituting the resistance values, we get

$$\frac{V_2 - V_1}{1500} + \frac{V_2}{1000} + \frac{V_3 - V_1}{2000} + \frac{V_3}{2000} = 0$$

Multiplication by 6000 yields

$$4V_2 - 4V_1 + 6V_2 + 3V_3 - 3V_1 + 3V_3 = 0$$

which can be simplified to

$$-7V_1 + 10V_2 + 6V_3 = 0 \quad \textbf{(3.42)}$$

Substitution of Equation (3.41) into Equation (3.42) yields

$$-7V_1 + 21V_3 = 0 \quad \textbf{(3.43)}$$

continued

Example 3.9 continued

Summing the currents leaving node 1, we obtain

$$-I_s + \frac{V_1 - V_3}{R_1} + \frac{V_1}{R_3} + \frac{V_1 - V_2}{R_2} = 0$$

Substituting the resistance and current values, we get

$$-5 \times 10^{-3} + \frac{V_1 - V_3}{2000} + \frac{V_1}{6000} + \frac{V_1 - V_2}{1500} = 0$$

Multiplication by 6000 yields

$$-30 + 3V_1 - 3V_3 + V_1 + 4V_1 - 4V_2 = 0$$

which can be simplified to

$$8V_1 - 4V_2 - 3V_3 = 30 \tag{3.44}$$

Substitution of Equation (3.41) into Equation (3.44) yields

$$8V_1 - 9V_3 = 30 \tag{3.45}$$

Equations (3.43) and (3.45) can be solved using the substitution method:

Solving Equation (3.43) for V_1, we obtain

$$V_1 = 3V_3$$

Substituting V_1 into Equation (3.45), we get

$$8(3V_3) - 9V_3 = 30$$

or

$$15V_3 = 30$$

Thus, we obtain

$$V_3 = 2\text{ V}$$

$$V_1 = 3V_3 = 6\text{ V}$$

$$V_2 = 1.5V_3 = 3\text{ V}$$

Alternatively, Equations (3.43) and (3.45) can be solved using Cramer's rule:

$$\Delta = \begin{vmatrix} -7 & 21 \\ 8 & -9 \end{vmatrix} = (-7) \times (-9) - 21 \times 8 = 63 - 168 = -105$$

$$V_1 = \frac{\begin{vmatrix} 0 & 21 \\ 30 & -9 \end{vmatrix}}{\Delta} = \frac{0 \times (-9) - 21 \times 30}{-105} = \frac{-630}{-105} = 6\text{ V}$$

continued

Example 3.9 continued

$$V_3 = \frac{\begin{vmatrix} -7 & 0 \\ 8 & 30 \end{vmatrix}}{\Delta} = \frac{(-7)\times 30 - 0 \times 8}{-105} = \frac{-210}{-105} = 2\text{ V}$$

From Equation (3.41), we obtain

$$V_2 = 1.5V_3 = 3\text{ V}$$

MATLAB

```
%EXAMPLE 3.9
clear all;
Is1=5e-3;R1=2000;R2=1500;R3=6000;R4=1000;R5=2000;R6=2000;R7=1500;
syms V1 V2 V3
[V1,V2,V3]=solve(-Is1+(V1-V3)/R1+V1/R3+(V1-V2)/R2, ...
V2==V3+V3/2, ...
(V2-V1)/R2+V2/R4+(V3-V1)/R1+V3/R5,V1,V2,V3);
V1=vpa(V1,6)
V2=vpa(V2,6)
V3=vpa(V3,6)

Answers:
V1 =
6.0
V2 =
3.0
V3 =
2.0
```

Exercise 3.9

Find V_1 and V_2 for the circuit shown in Figure 3.24.

FIGURE 3.24

The circuit for EXERCISE 3.9.

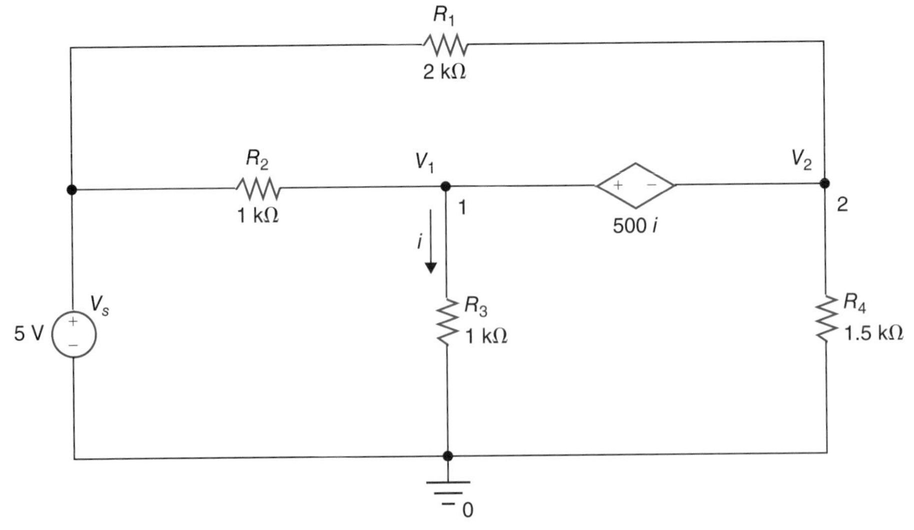

Answer:
$V_1 = 2.9032\text{ V}, V_2 = 1.4516\text{ V}.$

3.4 Mesh Analysis

A **loop** in a circuit is a closed path in which no node is encountered more than once. A **mesh** is a loop that does not contain any other loops. A circuit is **planar** if it can be drawn in a two-dimensional space without wires crossing over other wires. We are interested in finding mesh currents in a planar circuit. From the mesh currents, we can find the current through every branch of the circuit. If the current through every branch of a circuit is known, the voltage across every branch can be found by applying Ohm's law. The power on every branch can be found by applying $P = IV$. Thus, if we can find the current through every branch, we can find the voltages and powers everywhere. Mesh analysis is based on KVL. We assign mesh current variables such as $I_1, I_2, I_3, \ldots, I_n$ in the meshes whose currents are unknown. Then, for each mesh with unknown mesh current, we apply KVL. Specifically, we sum the voltage drops around the mesh and let that equal zero. If a mesh contains a current source, the mesh current is the same as the current from the current source if they point in the same direction. If the direction is opposite, the mesh current is the negative of the current from the current source.

If there are n unknown mesh currents, we get n equations in n unknowns. Thus, we can solve these n system of linear equations with constant coefficients to find the unique solution for the unknown mesh currents $I_1, I_2, I_3, \ldots, I_n$. A simple algorithm to find the solution of the n system of linear equations with constant coefficients is Cramer's rule. MATLAB is useful in finding the solution of an n system of linear equations with constant coefficients. The presence of dependent sources in the circuit requires extra equations describing the controlling voltages or currents, and controlled voltages and currents in terms of mesh currents.

As an example, consider the circuit shown in Figure 3.25.

FIGURE 3.25

Circuit with three meshes.

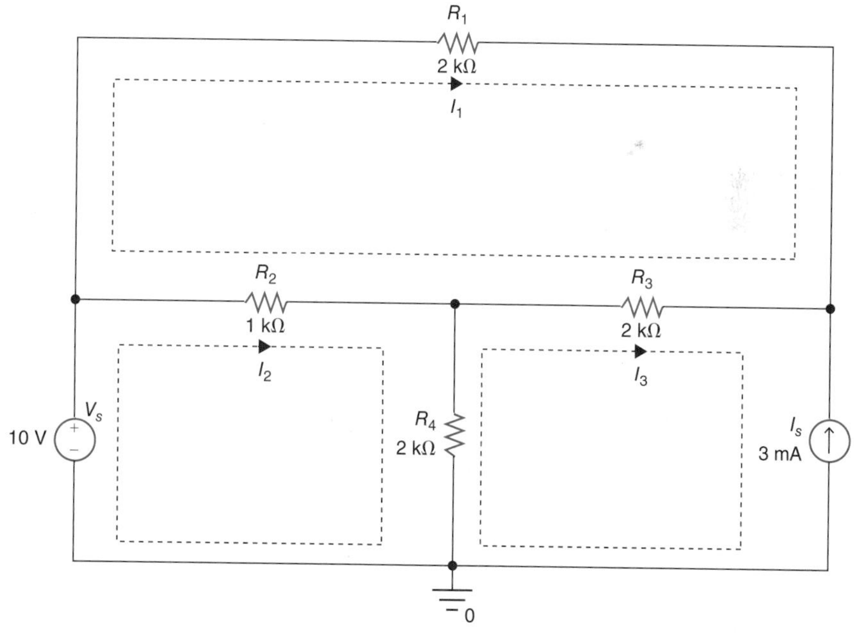

The circuit shown in Figure 3.25 has three meshes. We assign mesh currents for these three meshes. At the outset, it is good practice to assign all mesh currents in the same direction. It is common practice to choose all three mesh currents in the clockwise direction. Figure 3.25 shows the circuit with mesh currents I_1, I_2, and I_3. Notice that the mesh current I_3 is in the opposite direction of the current from current source I_s. Thus, $I_3 = -I_s = -3$ mA. This leaves two unknown mesh currents, I_1 and I_2.

Three resistors, R_1, R_3, and R_2, form mesh 1 with mesh current I_1. Starting from the left terminal of R_1, we add voltage drops across each resistor using Ohm's law. The voltage drops are measured in the clockwise direction following the direction specified by the arrow. According to KVL, the sum of voltage drops in any closed loop must be zero. The voltage drop across R_1, from left to right, is R_1I_1. The directions of I_1 and I_3 are opposite on R_3. The net current through R_3, from right to left, is $I_1 - I_3$. Thus, the voltage drop across R_3, from right to left, is $R_3(I_1 - I_3)$. Similarly, the voltage drop across R_2, from right to left, is $R_2(I_1 - I_2)$. Summing all three voltage drops in the clockwise direction, we have the mesh equation for mesh 1:

$$R_1I_1 + R_3(I_1 - I_3) + R_2(I_1 - I_2) = 0 \tag{3.46}$$

Substituting the component values, we have

$$2000I_1 + 2000(I_1 - I_3) + 1000(I_1 - I_2) = 0 \tag{3.47}$$

Since $I_3 = -3$ mA $= -0.003$ A, Equation (3.47) becomes

$$5000I_1 - 1000I_2 = 2000 \times (-0.003) = -6 \tag{3.48}$$

Summing voltage drops around mesh 2 with mesh current I_2 starting from the negative terminal of the voltage source V_s, we obtain the mesh equation for mesh 2:

$$-V_s + R_2(I_2 - I_1) + R_4(I_2 - I_3) = 0 \tag{3.49}$$

Substituting the component values, we have

$$-10 + 1000(I_2 - I_1) + 2000(I_2 - I_3) = 0 \tag{3.50}$$

Since $I_3 = -3$ mA $= -0.003$ A, Equation (3.50) becomes

$$-1000I_1 + 3000I_2 = 10 + 2000 \times (-0.003) = 4 \tag{3.51}$$

Multiplying Equation (3.48) by 3, we obtain

$$15{,}000I_1 - 3000I_2 = -18$$

Adding this equation and Equation (3.51), we get

$$14{,}000I_1 = -14$$

Thus, we have

$$I_1 = -0.001 \text{ A} = -1 \text{ mA}$$

Substituting I_1 into Equation (3.48), we obtain

$$I_2 = 0.001 \text{ A} = 1 \text{ mA}$$

Alternatively, the two mesh currents I_1 and I_2 can be found by applying Cramer's rule to Equations (3.48) and (3.51):

$$\Delta = \begin{vmatrix} 5000 & -1000 \\ -1000 & 3000 \end{vmatrix} = 1.5 \times 10^7 - 0.1 \times 10^7 = 1.4 \times 10^7$$

$$I_1 = \frac{\begin{vmatrix} -6 & -1000 \\ 4 & 3000 \end{vmatrix}}{\Delta} A = \frac{-1.8 \times 10^4 + 0.4 \times 10^4}{1.4 \times 10^7} A = \frac{-1.4 \times 10^4}{1.4 \times 10^7} A = -1 \text{ mA}$$

$$I_2 = \frac{\begin{vmatrix} 5000 & -6 \\ -1000 & 4 \end{vmatrix}}{\Delta} A = \frac{2.0 \times 10^4 - 0.6 \times 10^4}{1.4 \times 10^7} A = \frac{1.4 \times 10^4}{1.4 \times 10^7} A = 1 \text{ mA}$$

MATLAB

```
A=[5000 -1000;-1000 3000];b=[-6;4];
I=A\b
I =
   1.0e-03 *
   -1.0000
    1.0000
```

Equations (3.46) and (3.49), respectively, can be rearranged as

$$(R_1 + R_3 + R_2)I_1 - R_2I_2 - R_3I_3 = 0$$

and

$$-R_2I_1 + (R_2 + R_4)I_2 - R_4I_3 = V_s$$

Also, we have

$$I_3 = -I_s$$

These equations can be put into matrix form as

$$\begin{bmatrix} R_1 + R_2 + R_3 & -R_2 & -R_3 \\ -R_2 & R_2 + R_4 & -R_4 \\ 0 & 0 & 1 \end{bmatrix} \begin{bmatrix} I_1 \\ I_2 \\ I_3 \end{bmatrix} = \begin{bmatrix} 0 \\ V_s \\ -I_s \end{bmatrix}$$

This equation can be solved using MATLAB, as shown here:

```
clear all;
Vs=10;Is=3e-3;R1=2000;R2=1000;R3=2000;R4=2000;
A=[R1+R2+R3,-R2,-R3;-R2,R2+R4,-R4;0 0 1]
b=[0;Vs;-Is]
I=A\b

Answer:
A =
        5000        -1000        -2000
       -1000         3000        -2000
           0            0            1
b =
          0
   10.0000
   -0.0030
I =
   -0.0010
    0.0010
   -0.0030
```

This approach does not require numerical values for the elements of A and b to find I_1 and I_2.

One other method of solving Equations (3.46) and (3.49) is to use the MATLAB function **solve**, as shown here:

```
clear all;
syms I1 I2 I3
Vs=10;Is=3e-3;R1=2000;R2=1000;R3=2000;R4=2000;
[I1,I2,I3]=solve(R1*I1+R3*(I1-I3)+R2*(I1-I2),...
-Vs+R2*(I2-I1)+R4*(I2-I3),I3==-Is)
I1=vpa(I1,7)
I2=vpa(I2,7)
I3=vpa(I3,7)
```

```
Answers:
I1 =
-1/1000
I2 =
1/1000
I3 =
-3/1000
I1 =
-0.001
I2 =
0.001
I3 =
-0.003
```

This method provides the solution by entering equations.

Let I_{R_1} be the current through R_1, from right to left, as shown in Figure 3.26. Then, $I_{R_1} = -I_1$; that is, $I_{R_1} = -I_1 = 1$ mA. The voltage across R_1, from right to left, is $V_{R_1} = R_1 \times I_{R_1} = 2\ k\Omega \times 1$ mA $= 2$ V. Let I_{R_2} be the current through R_2, from left to right, as shown in Figure 3.26. Then, I_{R_2} can be represented as $I_{R_2} = I_2 - I_1 = 1$ mA $-$ $(-1$ mA$)$ $=$ 2 mA. The voltage across R_2, from left to right, is $V_{R_2} = R_2 \times I_{R_2} = 1\ k\Omega \times 2$ mA $= 2$ V. Let I_{R_3} be the current through R_3, from right to left, as shown in Figure 3.26. Then, $I_{R_3} = I_1 - I_3 = -1$ mA $-$ $(-3$ mA$)$ $= 2$ mA. The voltage across R_3, from right to left, is $V_{R_3} = R_3 \times I_{R_3} = 2\ k\Omega \times 2$ mA$= 4$ V. Let I_{R_4} be the current through R_4, from top to bottom, as shown in Figure 3.26. Then, $I_{R_4} = I_2 - I_3 = 1$ mA $-$ $(-3$ mA$)$ $= 4$ mA. The voltage across R_4, from top to bottom, is $V_{R_4} = R_4 \times I_{R_4} = 2\ k\Omega \times 4$ mA $= 8$ V. Notice that the voltage at node 1 (V_1 in Figure 3.26) is given by $V_1 = V_{R_4} = 8$ V, and the voltage at node 2 is given by $V_2 = V_1 + V_{R_3} = 8$ V $+ 4$ V $= 12$ V. The current through the voltage source, from bottom to top, is given by $I_{V_s} = I_{R_4} - I_s = 1$ mA.

FIGURE 3.26

The circuit shown in Figure 3.25 with current and voltage directions on the branches.

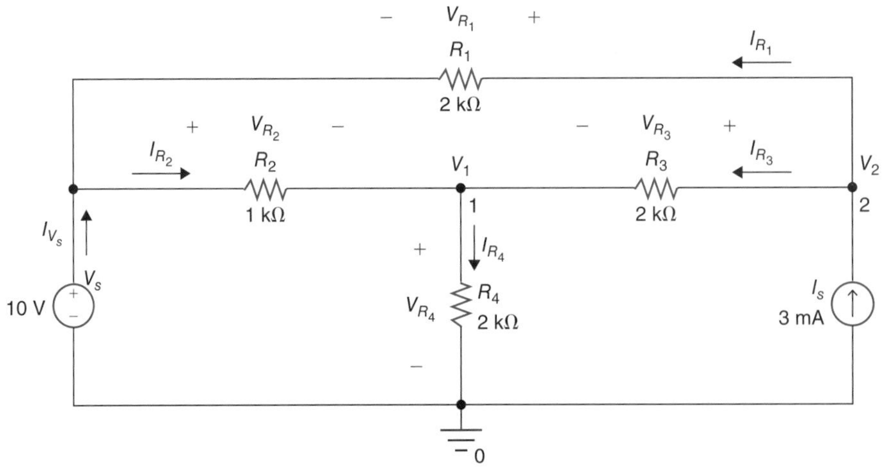

EXAMPLE 3.10

Find the mesh currents I_1 and I_2 in the circuit shown in Figure 3.27. Also, find V_1, V_2, V_3 and the power absorbed or supplied by all elements.

FIGURE 3.27

The circuit for EXAMPLE 3.10.

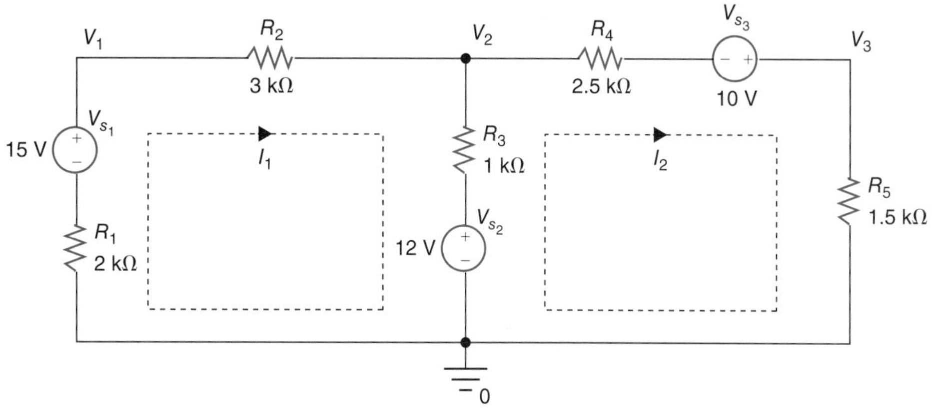

Summing the voltage drops around the mesh 1 (left side), we obtain

$$R_1I_1 - V_{s_1} + R_2I_1 + R_3(I_1 - I_2) + V_{s_2} = 0$$

Substituting the resistance and voltage values, we have

$$2000I_1 - 15 + 3000I_1 + 1000(I_1 - I_2) + 12 = 0$$

which can be simplified to

$$6000I_1 - 1000I_2 = 3 \quad \textbf{(3.52)}$$

Summing the voltage drops around mesh 2 (on the right side), we obtain

$$-V_{s_2} + R_3(I_2 - I_1) + R_4I_2 - V_{s_3} + R_5I_2 = 0$$

Substituting the resistance and voltage values, we have

$$-12 + 1000(I_2 - I_1) + 2500I_2 - 10 + 1500I_2 = 0$$

which can be simplified to

$$-1000I_1 + 5000I_2 = 22 \quad \textbf{(3.53)}$$

Equations (3.52) and (3.53) can be solved using the substitution method:

Multiplying Equation (3.52) by 5, we obtain

$$30{,}000I_1 - 5000I_2 = 15$$

Adding this equation and Equation (3.53), we get

$$29{,}000I_1 = 37$$

continued

Example 3.10 continued

Thus, we obtain

$$I_1 = \frac{37}{29{,}000} = 1.2759 \text{ mA}$$

Solving Equation (3.52) for I_2, we get

$$I_2 = 6I_1 - \frac{3}{1000} = \frac{135}{29{,}000} = 4.6552 \text{ mA}$$

Alternatively, Equations (3.52) and (3.53) can be solved using Cramer's rule:

$$\Delta = \begin{vmatrix} 6000 & -1000 \\ -1000 & 5000 \end{vmatrix} = 6000 \times 5000 - (-1000) \times (-1000) = 2.9 \times 10^7$$

$$I_1 = \frac{\begin{vmatrix} 3 & -1000 \\ 22 & 5000 \end{vmatrix}}{\Delta} A = \frac{3 \times 5000 - (-1000) \times 22}{2.9 \times 10^7} A = \frac{37{,}000}{2.9 \times 10^7} A = 1.2759 \text{ mA}$$

$$I_2 = \frac{\begin{vmatrix} 6000 & 3 \\ -1000 & 22 \end{vmatrix}}{\Delta} A = \frac{6000 \times 22 - 3 \times (-1000)}{2.9 \times 10^7} A = \frac{135{,}000}{2.9 \times 10^7} A = 4.6552 \text{ mA}$$

The voltages are given by

$$V_1 = V_{s_1} - R_1 \times I_1 = (15 - 2000 \times 1.2759 \times 10^{-3})\text{V} = 12.4483 \text{ V}$$

$$V_2 = R_3 \times (I_1 - I_2) + V_{s_2} = [1000 \times (1.2759 \times 10^{-3} - 4.6552 \times 10^{-3}) + 12] \text{ V}$$
$$= 8.6207 \text{ V}$$

$$V_3 = R_5 \times I_2 = 1500 \times 4.6552 \times 10^{-3} \text{ V} = 6.9828 \text{ V}$$

The powers absorbed or supplied by all elements are given by

$$P_{R_1} = I_1^2 R_1 = 3.2556 \text{ mW}$$

$$P_{R_2} = I_1^2 R_2 = 4.8835 \text{ mW}$$

$$P_{R_3} = (I_1 - I_2)^2 R_3 = 11.4197 \text{ mW}$$

$$P_{R_4} = I_2^2 R_4 = 54.1766 \text{ mW}$$

$$P_{R_5} = I_2^2 R_5 = 32.5059 \text{ mW}$$

$$P_{Vs_1} = (-I_1) V_{s_1} = -19.1379 \text{ mW}$$

$$P_{Vs_2} = (I_1 - I_2) V_{s_2} = -40.5517 \text{ mW}$$

$$P_{Vs_3} = (-I_2) V_{s_3} = -46.5517 \text{ mW}$$

Notice that $P_{R_1} + P_{R_2} + P_{R_3} + P_{R_4} + P_{R_5} + P_{Vs_1} + P_{Vs_2} + P_{Vs_3} = 0$.

continued

Example 3.10 continued

MATLAB

```
%EXAMPLE 3.10
clear all;format long;
Vs1=15;Vs2=12;Vs3=10;
R1=2000;R2=3000;R3=1000;R4=2500;R5=1500;
syms I1 I2
[I1,I2]=solve(R1*I1-Vs1+R2*I1+R3*(I1-I2)+Vs2, ...
-Vs2+R3*(I2-I1)+R4*I2-Vs3+R5*I2,I1,I2);
V1=Vs1-R1*I1;
V2=R3*(I1-I2)+Vs2;
V3=R5*I2;
PR1=I1^2*R1;
PR2=I1^2*R2;
PR3=(I1-I2)^2*R3;
PR4=I2^2*R4;
PR5=I2^2*R5;
Ps1=-I1*Vs1;
Ps2=(I1-I2)*Vs2;
Ps3=-I2*Vs3;
SumP=PR1+PR2+PR3+PR4+PR5+Ps1+Ps2+Ps3;
I1=vpa(I1,7)
I2=vpa(I2,7)
V1=vpa(V1,7)
V2=vpa(V2,7)
V3=vpa(V3,7)
PR1=vpa(PR1,7)
PR2=vpa(PR2,7)
PR3=vpa(PR3,7)
PR4=vpa(PR4,7)
PR5=vpa(PR5,7)
Ps1=vpa(Ps1,7)
Ps2=vpa(Ps2,7)
Ps3=vpa(Ps3,7)
SumP=vpa(SumP,7)

Answers:
I1 =
0.001275862
I2 =
0.004655172
V1 =
12.44828
V2 =
8.62069
V3 =
6.982759
PR1 =
0.003255648
PR2 =
0.004883472
PR3 =
0.01141974
PR4 =
0.05417658
```

continued

Example 3.10 continued

MATLAB continued

```
PR5 =
0.03250595
Ps1 =
-0.01913793
Ps2 =
-0.04055172
Ps3 =
-0.04655172
SumP =
0.0
```

Exercise 3.10

Find the mesh currents I_1, I_2, voltages V_1, V_2, and current I_{R_2} in the circuit shown in Figure 3.28.

FIGURE 3.28

The circuit for EXERCISE 3.10.

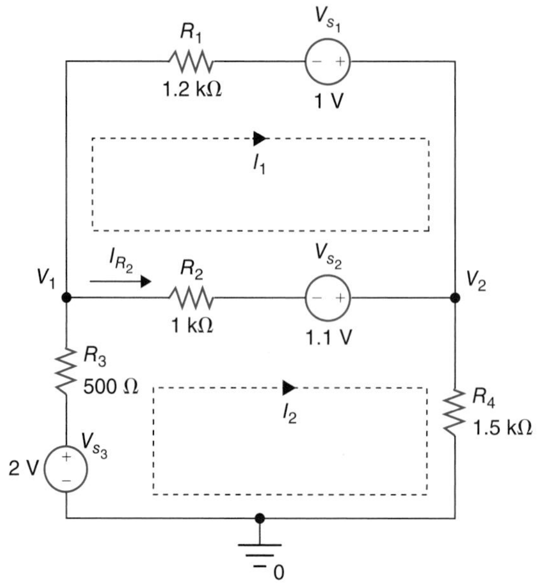

Answer:
$I_1 = 0.5$ mA, $I_2 = 1.2$ mA, $V_1 = 1.4$ V, $V_2 = 1.8$ V, $I_{R_2} = 0.7$ mA.

EXAMPLE 3.11

Find the mesh currents I_1, I_2, and I_3, voltages V_1 and V_2, and current I_{R_2} in the circuit shown in Figure 3.29.

continued

Example 3.11 continued

FIGURE 3.29

The circuit for EXAMPLE 3.11.

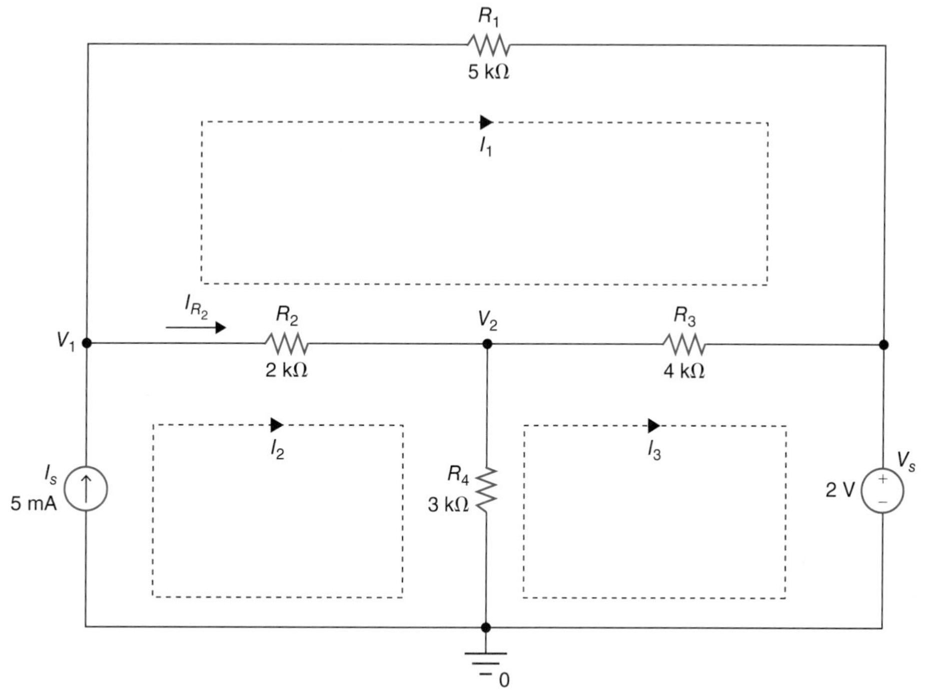

The mesh current I_2 is I_s. Thus, we have

$$I_2 = 5 \text{ mA}$$

Summing the voltage drops around mesh 1 (on the upper side), we obtain

$$R_1 I_1 + R_3(I_1 - I_3) + R_2(I_1 - I_2) = 0$$

Substituting the resistance values, we have

$$5000I_1 + 4000(I_1 - I_3) + 2000(I_1 - I_2) = 0$$

which can be simplified to

$$11{,}000I_1 - 2000I_2 - 4000I_3 = 0$$

Substitution of $I_2 = 5 \text{ mA} = 5 \times 10^{-3}$ A and dividing by 1000 yields

$$11I_1 - 4I_3 = 0.01 \qquad \textbf{(3.54)}$$

Summing the voltage drops around mesh 3 (on the lower right side), we obtain

$$R_4(I_3 - I_2) + R_3(I_3 - I_1) + V_s = 0$$

Substituting the resistance and voltage values, we have

$$3000(I_3 - I_2) + 4000(I_3 - I_1) + 2 = 0$$

which can be simplified to

$$-4000I_1 - 3000I_2 + 7000I_3 = -2$$

continued

Example 3.11 continued

Substitution of $I_2 = 5\text{ mA} = 5 \times 10^{-3}\text{ A}$ yields

$$-4000I_1 + 7000I_3 = 13 \quad \textbf{(3.55)}$$

Equations (3.54) and (3.55) can be solved using the substitution method: Solving Equation (3.54) for I_3, we obtain

$$I_3 = \frac{11}{4}I_1 - \frac{1}{400}$$

Substituting I_3 into Equation (3.55), we get

$$-4000I_1 + 7000\left(\frac{11}{4}I_1 - \frac{1}{400}\right) = 13$$

which can be rearranged as

$$\frac{61{,}000}{4}I_1 = \frac{12{,}200}{400}$$

Thus, we obtain

$$I_1 = \frac{122}{61{,}000} = 2\text{ mA}$$

$$I_3 = \frac{11}{4}I_1 - \frac{1}{400} = \frac{11}{4} \times 0.002 - \frac{1}{400} = 3\text{ mA}$$

Alternatively, Equations (3.54) and (3.55) can be solved using Cramer's rule:

$$\Delta = \begin{vmatrix} 11{,}000 & -4000 \\ -4000 & 7000 \end{vmatrix} = 11{,}000 \times 7000 - (-4000) \times (-4000) = 6.1 \times 10^7$$

$$I_1 = \frac{\begin{vmatrix} 10 & -4000 \\ 13 & 7000 \end{vmatrix}}{\Delta}A = \frac{10 \times 7000 - (-4000) \times 13}{6.1 \times 10^7}A = \frac{122{,}000}{6.1 \times 10^7}A = 2\text{ mA}$$

$$I_3 = \frac{\begin{vmatrix} 11{,}000 & 10 \\ -4000 & 13 \end{vmatrix}}{\Delta}A = \frac{11{,}000 \times 13 - 10 \times (-4000)}{6.1 \times 10^7}A = \frac{183{,}000}{6.1 \times 10^7}A = 3\text{ mA}$$

The voltages are given by

$$V_1 = V_s + R_1 \times I_1 = (2 + 5000 \times 2 \times 10^{-3})\text{V} = 12\text{ V}$$

$$V_2 = R_4 \times (I_2 - I_3) = 3000 \times (5 \times 10^{-3} - 3 \times 10^{-3})\text{V} = 6\text{ V}$$

The current through R_2 is given by

$$I_{R_2} = I_2 - I_1 = 5\text{ mA} - 2\text{ mA} = 3\text{ mA}$$

Exercise 3.11

Find the mesh currents I_1, I_2, I_3, voltages V_1, V_2, and current I_{R_1} in the circuit shown in Figure 3.30.

FIGURE 3.30

The circuit for EXERCISE 3.11.

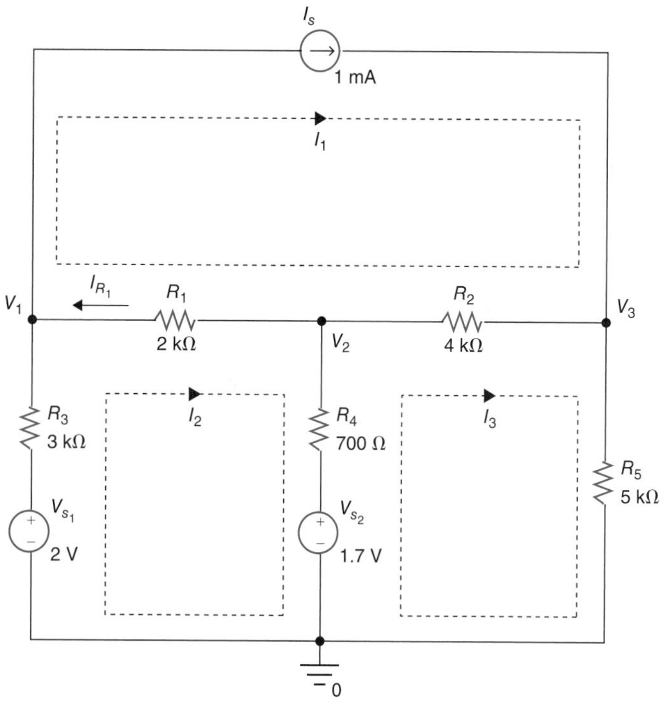

Answer:

$I_1 = 1$ mA, $I_2 = 0.4799$ mA, $I_3 = 0.6223$ mA, $V_1 = 0.5602$ V, $V_2 = 1.6004$ V, $V_3 = 3.1113$ V, $I_{R_1} = 0.5201$ mA.

EXAMPLE 3.12

Use mesh analysis to find the voltages and currents everywhere for the circuit shown in Figure 3.31.

FIGURE 3.31

The circuit for EXAMPLE 3.12.

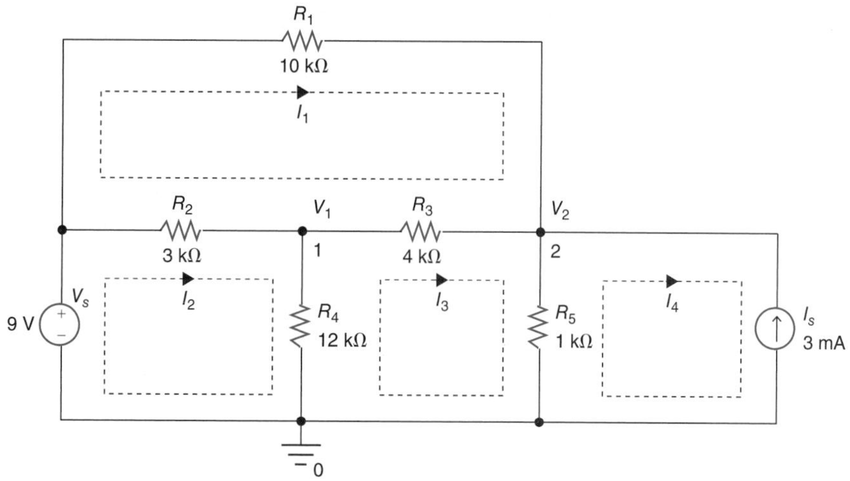

continued

Example 3.12 continued

The circuit shown in Figure 3.31 has four meshes. We assign mesh currents for these four meshes. Figure 3.31 shows the circuit with mesh currents.

The mesh with mesh current I_4 includes the current source I_s. Since the mesh current I_4 is in the opposite direction of the current from the current source, we have $I_4 = -I_s = -3$ mA. Since the mesh current I_4 is already known, we only need to write mesh equations for the other three meshes. Three resistors (R_1, R_3, and R_2) form mesh 1 with mesh current I_1. Starting from the left terminal of R_1, we add voltage drops on each resistor using Ohm's law. The voltage drops are measured in the clockwise direction following the direction specified by the arrow. According to KVL, the sum of voltage drops in any closed mesh must be zero. The voltage drop across R_1, from left to right, is R_1I_1. The directions of I_1 and I_3 are opposite on R_3. The net current through R_3, from right to left, is $I_1 - I_3$. Thus, the voltage drop across R_3, from right to left, is $R_3(I_1 - I_3)$. Similarly, the voltage drop across R_2, from right to left, is $R_2(I_1 - I_2)$. Summing all three voltage drops in the clockwise direction, we have a mesh equation for mesh 1:

$$R_1I_1 + R_3(I_1 - I_3) + R_2(I_1 - I_2) = 0 \quad \textbf{(3.56)}$$

Substituting resistance values, we get

$$10{,}000I_1 + 4000(I_1 - I_3) + 3000(I_1 - I_2) = 0 \quad \textbf{(3.57)}$$

Equation (3.57) can be simplified to

$$17{,}000I_1 - 3000I_2 - 4000I_3 = 0$$

Dividing by 1000, we get

$$17I_1 - 3I_2 - 4I_3 = 0 \quad \textbf{(3.58)}$$

Summing voltage drops around mesh 2 with mesh current I_2 starting from the negative terminal of the voltage source V_s, we obtain the mesh equation for mesh 2:

$$-V_s + R_2(I_2 - I_1) + R_4(I_2 - I_3) = 0 \quad \textbf{(3.59)}$$

Substitution of the component values yields

$$-9 + 3000(I_2 - I_1) + 12{,}000(I_2 - I_3) = 0 \quad \textbf{(3.60)}$$

Equation (3.60) can be simplified to

$$-3000I_1 + 15{,}000I_2 - 12{,}000I_3 = 9$$

Dividing by 3000, we get

$$-I_1 + 5I_2 - 4I_3 = 0.003 \quad \textbf{(3.61)}$$

Summing voltage drops around mesh 3 with mesh current I_3 starting from the bottom terminal of the resistor R_4, we obtain the mesh equation for mesh 3:

$$R_4(I_3 - I_2) + R_3(I_3 - I_1) + R_5(I_3 - I_4) = 0 \quad \textbf{(3.62)}$$

Substitution of the component values yields

$$12{,}000(I_3 - I_2) + 4000(I_3 - I_1) + 1000(I_3 + 0.003) = 0 \quad \textbf{(3.63)}$$

Equation (3.63) can be simplified to

$$-4000I_1 - 12{,}000I_2 + 17{,}000I_3 = -3$$

continued

Example 3.12 continued

Dividing by 1000, we get

$$-4I_1 - 12I_2 + 17I_3 = -0.003 \quad \textbf{(3.64)}$$

Equations (3.58), (3.61), and (3.64) can be solved using the substitution method: Solving Equation (3.61) for I_1, we obtain

$$I_1 = 5I_2 - 4I_3 - 0.003$$

Substituting I_1 into Equations (3.58) and (3.64) respectively, we get

$$17(5I_2 - 4I_3 - 0.003) - 3I_2 - 4I_3 = 0$$

$$-4(5I_2 - 4I_3 - 0.003) - 12I_2 + 17I_3 = -0.003$$

These two equations can be simplified to

$$82I_2 - 72I_3 = 0.051$$

$$-32I_2 + 33I_3 = -0.015$$

Solving $-32I_2 + 33I_3 = -0.015$ for I_2, we obtain

$$I_2 = \frac{33}{32}I_3 + \frac{0.015}{32}$$

Substituting I_2 into $82I_2 - 72I_3 = 0.051$, we get

$$82\left(\frac{33}{32}I_3 + \frac{0.015}{32}\right) - 72I_3 = 0.051$$

Thus, we have

$$I_3 = \frac{0.051 - \frac{82}{32} \times 0.015}{82 \times \frac{33}{32} - 72} = 1 \text{ mA}$$

$$I_2 = \frac{33}{32}I_3 + \frac{0.015}{32} = 1.5 \text{ mA}$$

$$I_1 = 5I_2 - 4I_3 - 0.003 = 0.5 \text{ mA}$$

Alternatively, the three mesh currents I_1, I_2, and I_3 can be found by applying Cramer's rule to Equations (3.58), (3.61), and (3.64):

$$\Delta = \begin{vmatrix} 17 & -3 & -4 \\ -1 & 5 & -4 \\ -4 & -12 & 17 \end{vmatrix} = 17\begin{vmatrix} 5 & -4 \\ -12 & 17 \end{vmatrix} + 3\begin{vmatrix} -1 & -4 \\ -4 & 17 \end{vmatrix} - 4\begin{vmatrix} -1 & 5 \\ -4 & -12 \end{vmatrix}$$

$$= 17[5 \times 17 - (-4)(-12)] + 3[(-1)(17) - (-4)(-4)] - 4[(-1)(-12) - (5)(-4)]$$

$$= 17[85 - 48] + 3[-17 - 16] - 4[12 + 20] = 17 \times 37 - 3 \times 33 - 4 \times 32 = 402$$

continued

Example 3.12 continued

$$I_1 = \frac{\begin{vmatrix} 0 & -3 & -4 \\ 0.003 & 5 & -4 \\ -0.003 & -12 & 17 \end{vmatrix}}{\Delta} = \frac{0\begin{vmatrix} 5 & -4 \\ -17 & 17 \end{vmatrix} + 3\begin{vmatrix} 0.003 & -4 \\ -0.003 & 17 \end{vmatrix} - 4\begin{vmatrix} 0.003 & 5 \\ -0.003 & -12 \end{vmatrix}}{\Delta}$$

$$= \frac{0.201}{402} = 0.0005\,\text{A} = 0.5\,\text{mA}$$

$$I_2 = \frac{\begin{vmatrix} 17 & 0 & -4 \\ -1 & 0.003 & -4 \\ -4 & -0.003 & 17 \end{vmatrix}}{\Delta} = \frac{17\begin{vmatrix} 0.003 & -4 \\ -0.003 & 17 \end{vmatrix} + 0\begin{vmatrix} -1 & -4 \\ -4 & 17 \end{vmatrix} - 4\begin{vmatrix} -1 & 0.003 \\ -4 & -0.003 \end{vmatrix}}{\Delta}$$

$$= \frac{0.603}{402} = 0.0015\,\text{A} = 1.5\,\text{mA}$$

$$I_3 = \frac{\begin{vmatrix} 17 & -3 & 0 \\ -1 & 5 & 0.003 \\ -4 & -12 & -0.003 \end{vmatrix}}{\Delta} = \frac{17\begin{vmatrix} 5 & 0.003 \\ -12 & -0.003 \end{vmatrix} + 3\begin{vmatrix} -1 & 0.003 \\ -4 & -0.003 \end{vmatrix} + 0\begin{vmatrix} -1 & 5 \\ -4 & -12 \end{vmatrix}}{\Delta}$$

$$= \frac{0.402}{402} = 0.001\,\text{A} = 1\,\text{mA}$$

MATLAB

```
%EXAMPLE 3.12
clear all;
syms I1 I2 I3 I4
Vs=9;Is=3e-3;
R1=10000;R2=3000;R3=4000;R4=12000;R5=1000;
[I1,I2,I3,I4]=solve(R1*I1+R3*(I1-I3)+R2*(I1-I2),...
I4==-Is,...
-Vs+R2*(I2-I1)+R4*(I2-I3),...
R4*(I3-I2)+R3*(I3-I1)+R5*(I3-I4),I1,I2,I3,I4);
IR1=I1;
IR2=I2-I1;
IR3=I3-I1;
IR4=I2-I3;
IR5=I3-I4;
V1=R4*IR4;
V2=R5*IR5;
I1=vpa(I1,7)
I2=vpa(I2,7)
I3=vpa(I3,7)
I4=vpa(I4,7)
IR1=vpa(IR1,7)
IR2=vpa(IR2,7)
IR3=vpa(IR3,7)
IR4=vpa(IR4,7)
IR5=vpa(IR5,7)
V1=vpa(V1,7)
V2=vpa(V2,7)
```

continued

Example 3.12 continued
MATLAB continued

```
Answers:
I1 =
0.0005
I2 =
0.0015
I3 =
0.001
I4 =
-0.003
IR1 =
0.0005
IR2 =
0.001
IR3 =
0.0005
IR4 =
0.0005
IR5 =
0.004
V1 =
6.0
V2 =
4.0
```

Let I_{R_4} be the current through R_4, from top to bottom, as shown in Figure 3.32. Then, I_{R_4} can be represented as $I_{R_4} = I_2 - I_3 = 1.5\text{ mA} - 1\text{ mA} = 0.5\text{ mA}$. The voltage across R_4 is given by $V_{R_4} = R_4 I_{R_4} = 12\ k\Omega \times 0.5\text{ mA} = 6\text{ V}$. Notice that $V_1 = V_{R_4} = 6\text{ V}$. Let I_{R_5} be the current through R_5, from top to bottom, as shown in Figure 3.32. Then, I_{R_5} can be represented as $I_{R_5} = I_3 - I_4 = 1\text{ mA} - (-3\text{ mA}) = 4\text{ mA}$. The voltage across R_5 is given by $V_{R_5} = R_5 I_{R_5} = 1\ k\Omega \times 4\text{ mA} = 4\text{ V}$. Notice that $V_2 = V_{R_5} = 4\text{ V}$. Figure 3.32 shows currents through all branches. Also, we have $V_{R_1} = V_s - V_2 = 5\text{ V}$, $V_{R_2} = V_s - V_1 = 3\text{ V}$, and $V_{R_3} = V_1 - V_2 = 2\text{ V}$.

FIGURE 3.32

A circuit with currents.

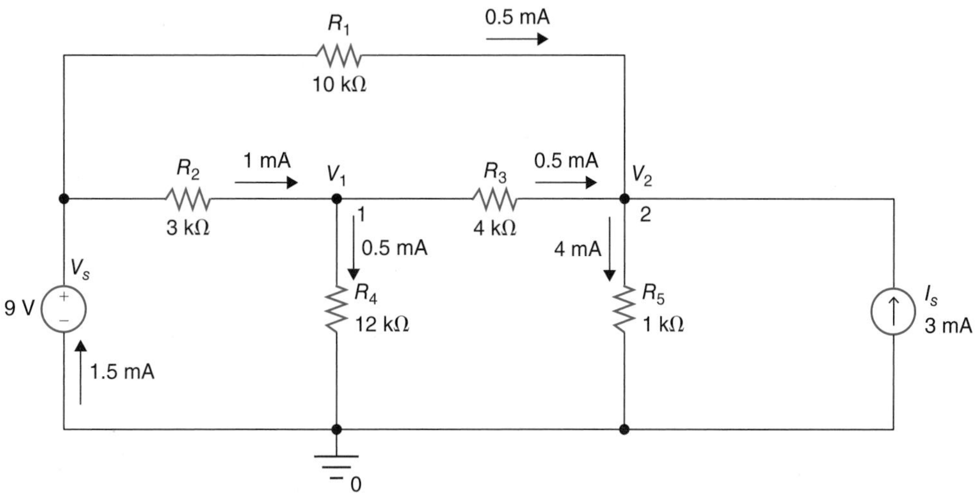

Exercise 3.12

Find the mesh currents I_1, I_2, and I_3, voltages V_1 and V_2, and current I_{R_2} in the circuit shown in Figure 3.33.

FIGURE 3.33

The circuit for EXERCISE 3.12.

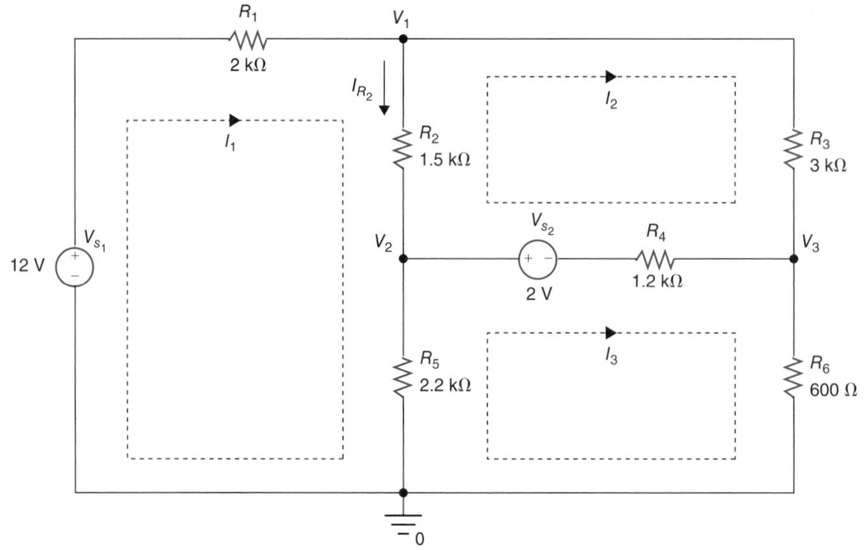

Answer:
$I_1 = 3.1707$ mA, $I_2 = 1.5447$ mA, $I_3 = 1.7073$ mA, $V_1 = 5.6585$ V, $V_2 = 3.2195$ V, $V_3 = 1.02439$ V, $I_{R_2} = 1.6260$ mA.

EXAMPLE 3.13

Find the mesh currents I_1, I_2, and I_3, voltages V_1 and V_2, and powers on all elements in the circuit shown in Figure 3.34.

FIGURE 3.34

The circuit for EXAMPLE 3.13.

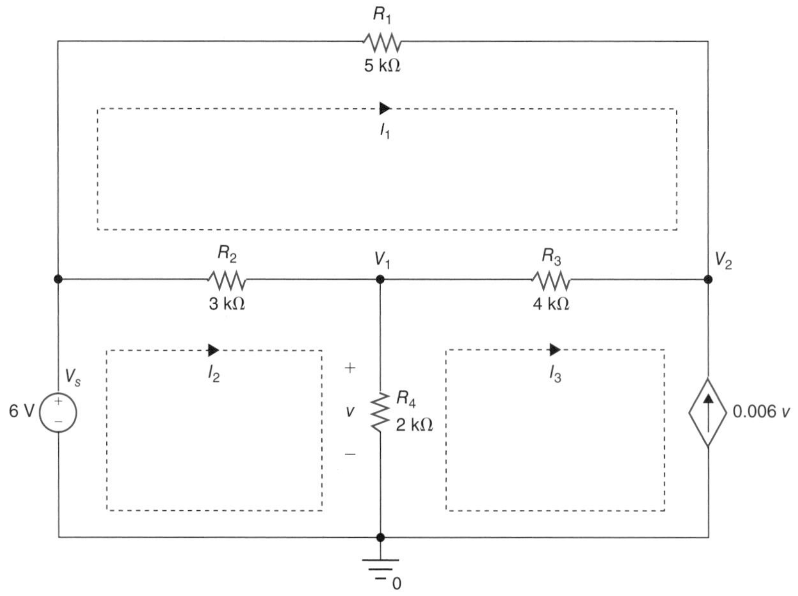

continued

Example 3.13 continued

Here, the voltage-dependent current source produces current that is in the opposite direction of the mesh current I_3; that is, $I_3 = -0.006v$, where $v = R_4(I_2 - I_3)$. Hence,

$$I_3 = -0.006v = -0.006R_4(I_2 - I_3) = -0.006 \times 2000(I_2 - I_3) = -12(I_2 - I_3)$$

Thus,

$$I_3 = \frac{12}{11} I_2 = 1.0909 I_2 \quad \textbf{(3.65)}$$

Summing the voltage drops around mesh 1 (on the upper side), we obtain

$$R_1 I_1 + R_3(I_1 - I_3) + R_2(I_1 - I_2) = 0$$

Substituting the resistance values, we get

$$5000 I_1 + 4000(I_1 - I_3) + 3000(I_1 - I_2) = 0$$

which can be simplified to

$$12{,}000 I_1 - 3000 I_2 - 4000 I_3 = 0 \quad \textbf{(3.66)}$$

Substitution of Equation (3.65) into Equation (3.66) yields

$$12{,}000 I_1 - 7363.6364 I_2 = 0 \quad \textbf{(3.67)}$$

Summing the voltage drops around mesh 2 (on the lower left side), we obtain

$$-V_s + R_2(I_2 - I_1) + R_4(I_2 - I_3) = 0$$

Substituting the resistance and voltage values, we get

$$-6 + 3000(I_2 - I_1) + 2000(I_2 - I_3) = 0$$

which can be simplified to

$$-3000 I_1 + 5000 I_2 - 2000 I_3 = 6 \quad \textbf{(3.68)}$$

Substitution of Equation (3.65) into Equation (3.68) yields

$$-3000 I_1 + 2818.1818 I_2 = 6 \quad \textbf{(3.69)}$$

Equations (3.67) and (3.69) can be solved using the substitution method:

Multiplying Equation (3.69) by 4, we obtain

$$-12{,}000 I_1 + 11{,}272.7272 I_2 = 24$$

Adding this equation and Equation (3.67), we get

$$3909.0908 I_2 = 24$$

Thus, we obtain

$$I_2 = \frac{24}{3909.0908} = 6.139535 \text{ mA}$$

continued

Example 3.13 continued

$$I_1 = \frac{7363.6364 I_2}{12{,}000} = 3.767442 \text{ mA}$$

$$I_3 = \frac{12}{11} I_2 = 1.0909 I_2 = 6.69767 \text{ mA}$$

Alternatively, Equations (3.67) and (3.69) can be solved using Cramer's rule:

$$\Delta = \begin{vmatrix} 12{,}000 & -7363.6364 \\ -3000 & 2818.1818 \end{vmatrix} = 1.172727 \times 10^7$$

$$I_1 = \frac{\begin{vmatrix} 0 & -7363.6364 \\ 6 & 2818.1818 \end{vmatrix}}{\Delta} A = \frac{44{,}181.8182}{1.172727 \times 10^7} A = 3.767442 \text{ mA}$$

$$I_2 = \frac{\begin{vmatrix} 12{,}000 & 0 \\ -3000 & 6 \end{vmatrix}}{\Delta} A = \frac{72{,}000}{1.172727 \times 10^7} A = 6.139535 \text{ mA}$$

$$I_3 = \frac{12}{11} I_2 = 1.0909 I_2 = 1.0909 \times 6.139535 \text{ mA} = 6.69767 \text{ mA}$$

The voltages are given by

$$V_1 = R_4 \times (I_2 - I_3) = 2000 \times (6.139535 \times 10^{-3} - 6.69767442 \times 10^{-3})\text{V} = -1.11628 \text{ V}$$

$$V_2 = V_s - R_1 \times I_1 = (6 - 5000 \times 3.767442 \times 10^{-3})\text{V} = -12.83721 \text{ V}$$

The power absorbed or supplied by all elements are computed as:

$$P_{R_1} = I_1^2 R_1 = 70.9681 \text{ mW}$$

$$P_{R_2} = (I_1 - I_2)^2 R_2 = 16.8805 \text{ mW}$$

$$P_{R_3} = (I_1 - I_3)^2 R_3 = 34.3451 \text{ mW}$$

$$P_{R_4} = (I_2 - I_3)^2 R_4 = 0.6230 \text{ mW}$$

$$P_{Vs} = (-I_2) V_s = -36.83721 \text{ mW}$$

$$P_{VCCS} = I_3 V_2 = -85.97945 \text{ mW}$$

Notice that $P_{R_1} + P_{R_2} + P_{R_3} + P_{R_4} + P_{Vs} + P_{VCCS} = 0$.

MATLAB

```
%EXAMPLE 3.13
clear all;format long;
Vs=6;
R1=5000;R2=3000;R3=4000;R4=2000;
syms I1 I2 I3
[I1,I2,I3]=solve(R1*I1+R3*(I1-I3)+R2*(I1-I2), ...
-Vs+R2*(I2-I1)+R4*(I2-I3), ...
I3==-0.006*R4*(I2-I3),I1,I2,I3);
V1=R4*(I2-I3);
V2=R3*(I1-I3)+V1;
PR1=R1*I1^2;
PR2=R2*(I1-I2)^2;
```

continued

Example 3.13 continued

MATLAB continued

```
PR3=R3*(I1-I3)^2;
PR4=R4*(I2-I3)^2;
PVCCS=I3*V2;
PVs=-I2*Vs;
PSum=PR1+PR2+PR3+PR4+PVCCS+PVs;
I1=vpa(I1,7)
I2=vpa(I2,7)
I3=vpa(I3,7)
V1=vpa(V1,7)
V2=vpa(V2,7)
PR1=vpa(PR1,7)
PR2=vpa(PR2,7)
PR3=vpa(PR3,7)
PR4=vpa(PR4,7)
PVCCS=vpa(PVCCS,7)
PVs=vpa(PVs,7)
PSum=vpa(PSum,7)

Answers:
I1 =
0.003767442
I2 =
0.006139535
I3 =
0.006697674
V1 =
-1.116279
V2 =
-12.83721
PR1 =
0.07096809
PR2 =
0.01688048
PR3 =
0.03434505
PR4 =
0.0006230395
PVCCS =
-0.08597945
PVs =
-0.03683721
PSum =
0.0
```

Exercise 3.13

Find the mesh currents I_1, I_2 and voltages V_1, V_2, and V_3 in the circuit shown in Figure 3.35.

continued

Exercise 3.13 continued

FIGURE 3.35

The circuit for EXERCISE 3.13.

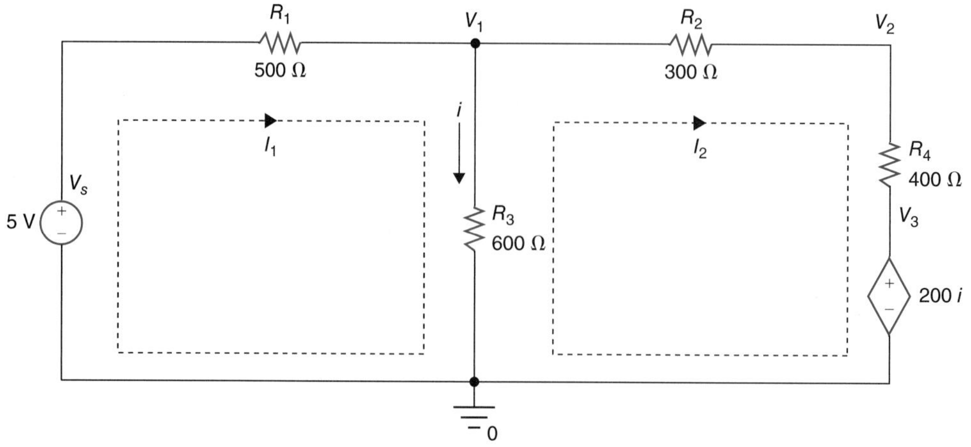

Answer:
$I_1 = 5.6701$ mA, $I_2 = 2.0619$ mA, $V_1 = 2.16495$ V, $V_2 = 1.5464$ V, $V_3 = 0.72165$ V.

EXAMPLE 3.14

Find the mesh currents I_1, I_2, and I_3 and voltages V_1, V_2, and V_3 in the circuit shown in Figure 3.36.

FIGURE 3.36

A circuit with dependent sources.

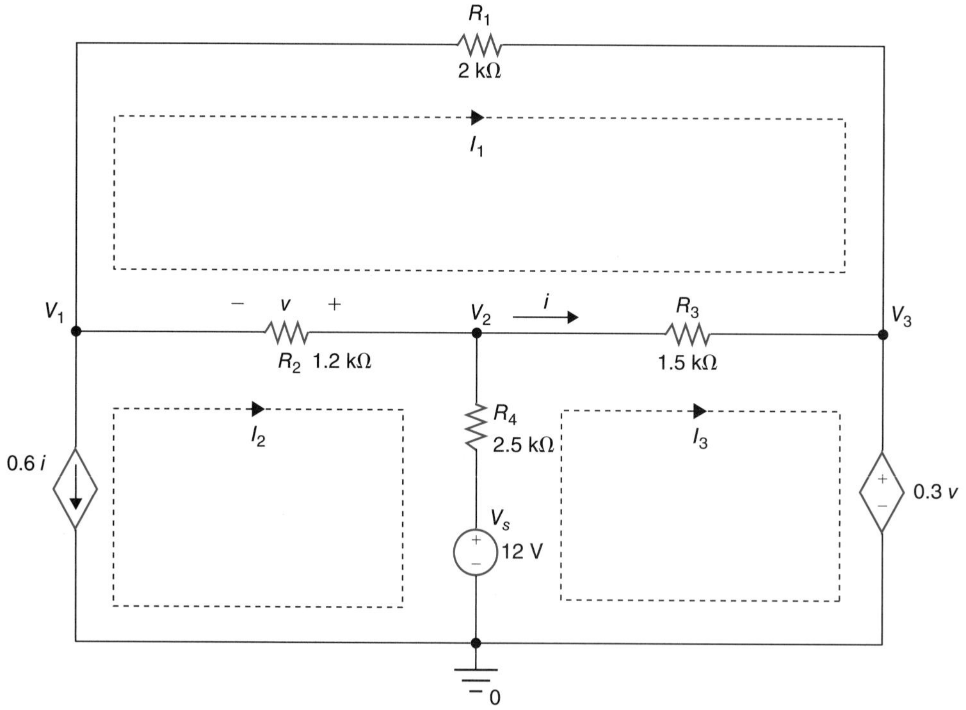

continued

Example 3.14 continued

In this circuit, the current-dependent current source has a direction opposite that of the direction chosen for mesh current I_2; that is,

$$I_2 = -0.6i = -0.6(I_3 - I_1) \tag{3.70}$$

Summing the voltage drops around mesh 1 (on the upper side), we obtain

$$R_1I_1 + R_3(I_1 - I_3) + R_2(I_1 - I_2) = 0$$

Substituting the resistance values, we get

$$2000I_1 + 1500(I_1 - I_3) + 1200(I_1 - I_2) = 0$$

which can be simplified to

$$4700I_1 - 1200I_2 - 1500I_3 = 0 \tag{3.71}$$

Substitution of Equation (3.70) into Equation (3.71) yields

$$3980I_1 - 780I_3 = 0 \tag{3.72}$$

Summing the voltage drops around mesh 3 (on the lower right side), we obtain

$$-V_s + R_4(I_3 - I_2) + R_3(I_3 - I_1) + 0.3R_2(I_1 - I_2) = 0$$

Substituting the resistance and voltage values, we get

$$-12 + 2500(I_3 - I_2) + 1500(I_3 - I_1) + 0.3 \times 1200 \times (I_1 - I_2) = 0$$

which can be simplified to

$$-1140I_1 - 2860I_2 + 4000I_3 = 12 \tag{3.73}$$

Substitution of Equation (3.70) into Equation (3.73) yields

$$-2856I_1 + 5716I_3 = 12 \tag{3.74}$$

Equations (3.72) and (3.74) can be solved using the substitution method:

Solving Equation (3.72) for I_3, we obtain

$$I_3 = \frac{398}{78}I_1$$

Substituting I_3 into Equation (3.74), we get

$$-2856I_1 + 5716 \times \frac{398}{78}I_1 = 12$$

Thus, we obtain

$$I_1 = \frac{12}{-2856 + 5716 \times \dfrac{398}{78}} = 456.0959\ \mu\text{A}$$

continued

Example 3.14 continued

$$I_3 = \frac{398}{78} I_1 = 2.3273 \text{ mA}$$

$$I_2 = -0.6(I_3 - I_1) = -1.1227 \text{ mA}$$

Alternatively, Equations (3.72) and (3.74) can be solved using Cramer's rule:

$$\Delta = \begin{vmatrix} 3980 & -780 \\ -2856 & 5716 \end{vmatrix} = 3980 \times 5716 - (-780) \times (-2856) = 2.0522 \times 10^7$$

$$I_1 = \frac{\begin{vmatrix} 0 & -780 \\ 12 & 5715 \end{vmatrix}}{\Delta} A = \frac{9360}{2.0522 \times 10^7} A = 456.0959 \ \mu\text{A}$$

$$I_3 = \frac{\begin{vmatrix} 3980 & 0 \\ -2856 & 12 \end{vmatrix}}{\Delta} A = \frac{47{,}760}{2.0522 \times 10^7} A = 2.3273 \text{ mA}$$

$$I_2 = -0.6(I_3 - I_1) = -1.1227 \text{ mA}$$

The voltages are given by

$$V_2 = V_s - R_4 \times (I_3 - I_2) = 3.3751 \text{ V}$$

$$V_1 = V_2 - R_2 \times (I_1 - I_2) = 1.4806 \text{ V}$$

$$V_3 = V_1 - R_1 \times I_1 = 0.5684 \text{ V}$$

MATLAB

```
%EXAMPLE 3.14
clear all;format long;
Vs=12;R1=2000;R2=1200;R3=1500;R4=2500;
syms I1 I2 I3
[I1,I2,I3]=solve(R1*I1+R3*(I1-I3)+R2*(I1-I2), ...
I2==-0.6*(I3-I1), ...
-Vs+R4*(I3-I2)+R3*(I3-I1)+0.3*R2*(I1-I2),I1,I2,I3);
V2=Vs-R4*(I3-I2);
V1=V2-R2*(I1-I2);
V3=V1-R1*I1;
I1=vpa(I1,10)
I2=vpa(I2,10)
I3=vpa(I3,10)
V1=vpa(V1,10)
V2=vpa(V2,10)
V3=vpa(V3,10)

Answers:
I1 =
0.0004560958971
I2 =
-0.001122697593
I3 =
0.002327258552
```

continued

Example 3.14 continued
MATLAB continued

```
V1 =
1.480557451
V2 =
3.375109638
V3 =
0.5683656564
```

Exercise 3.14

Find the mesh currents I_1, I_2, and I_3 and voltages V_1, V_2, and V_3 in the circuit shown in Figure 3.37.

FIGURE 3.37

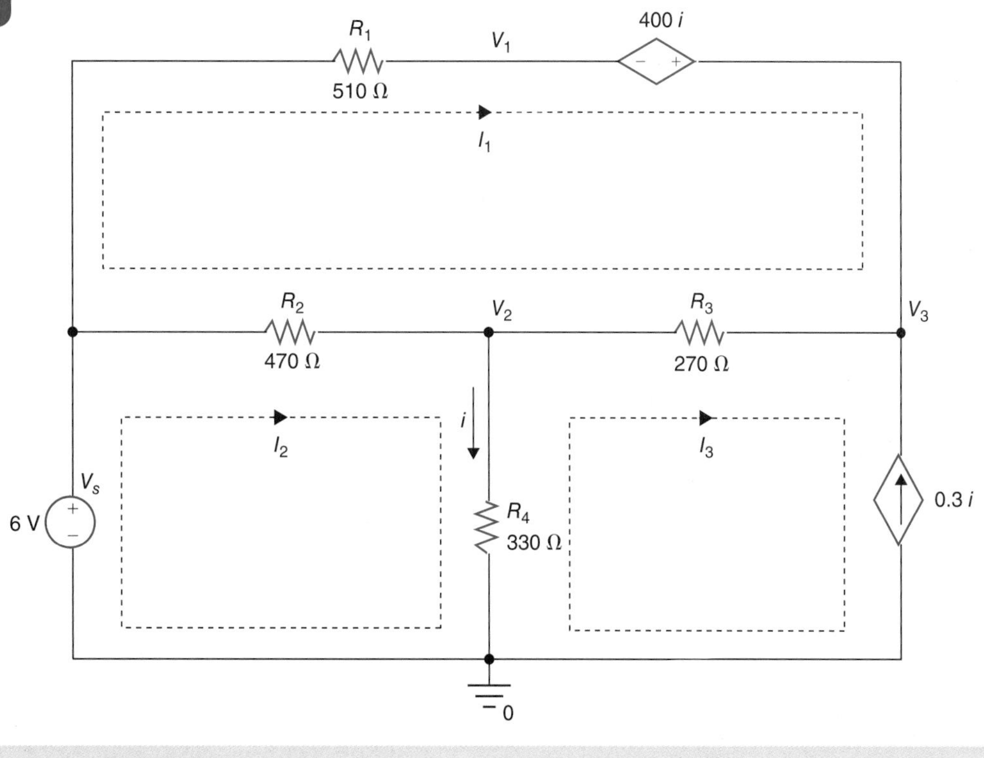

The circuit for EXERCISE 3.14.

Answer:
$I_1 = 7.4886$ mA, $I_2 = 10.1119$ mA, $I_3 = -4.3337$ mA, $V_1 = 2.1808$ V, $V_2 = 4.7670$ V, $V_3 = 7.9591$ V.

3.5 Supermesh

Recognizing that we use KVL in mesh analysis, if there is a current source that is a common branch between two different meshes, we do not know the voltage drop across the current source. Let the unknown voltage across the current source be included in the loop equation. Ohm's law cannot be applied because of the unknown internal resistance of the current source. However, to overcome this, and to complete the loop equation, let the unknown voltage across the current source be v. Then, write the mesh equation for each mesh and add the two equations to remove the unknown voltage v. The voltage v

is removed when added because in one equation, the voltage drop across the current source is v and in the other equation, the voltage drop across the current source is $-v$. Alternatively, the sum of the two equations can be obtained directly by defining a **supermesh** consisting of the two meshes, excluding the current source. When the voltage drops around the supermesh are added, we get the same equation that we obtain by adding the two equations. The extra equation needed to find the mesh currents is obtained by representing the current of the current source by the difference of the two mesh currents. The current of the current source is obtained by subtracting the mesh current pointing in the opposite direction from the mesh current pointing in the same direction as the current source.

EXAMPLE 3.15

Use mesh analysis to find the voltage v across the current source in the circuit shown in Figure 3.38.

FIGURE 3.38

The circuit for EXAMPLE 3.15.

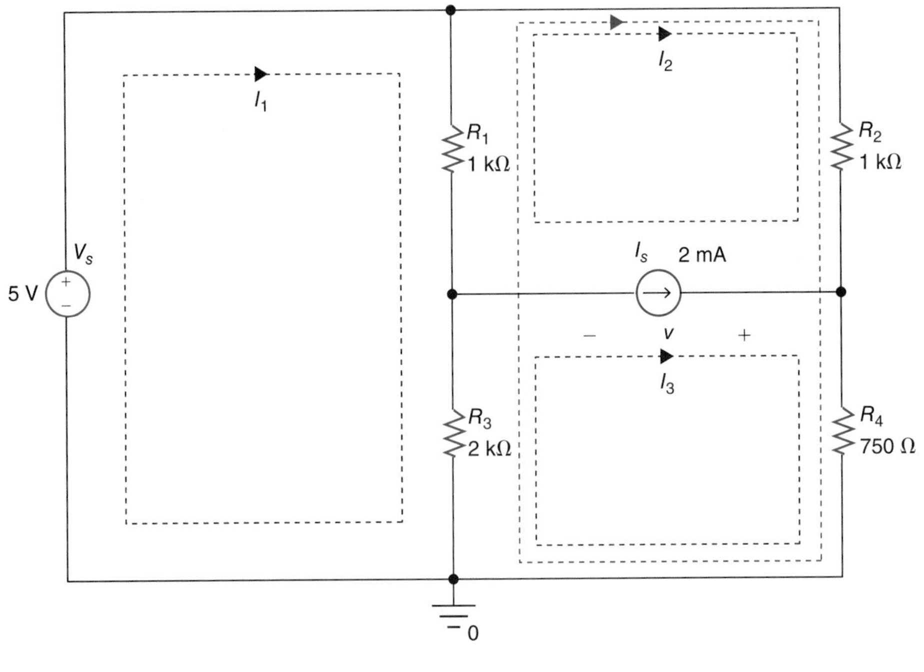

Summing the voltage drops around mesh 1 in the clockwise direction, we obtain

$$-V_s + R_1(I_1 - I_2) + R_3(I_1 - I_3) = 0$$

Substituting the resistance and voltage values, we get

$$-5 + 1000(I_1 - I_2) + 2000(I_1 - I_3) = 0$$

which can be rearranged as

$$3000I_1 - 1000I_2 - 2000I_3 = 5 \quad \textbf{(3.75)}$$

Summing the voltage drops around mesh 2 in the clockwise direction, we obtain

$$R_1(I_2 - I_1) + R_2I_2 + v = 0$$

continued

Example 3.15 continued

Substituting the resistance values, we get

$$1000(I_2 - I_1) + 1000I_2 + v = 0 \tag{3.76}$$

Summing the voltage drops around mesh 3 in the clockwise direction, we obtain

$$R_3(I_3 - I_1) - v + R_4I_3 = 0$$

Substituting the resistance values, we get

$$2000(I_3 - I_1) - v + 750I_3 = 0 \tag{3.77}$$

The unknown voltage v across the current source can be removed by adding Equations (3.76) and (3.77):

$$1000(I_2 - I_1) + 1000I_2 + 750I_3 + 2000(I_3 - I_1) = 0 \tag{3.78}$$

which can be rearranged as

$$-3000I_1 + 2000I_2 + 2750I_3 = 0 \tag{3.79}$$

Notice that Equation (3.78) can be directly obtained by summing the voltage drops around the supermesh consisting of mesh 2 and mesh 3, shown in blue in Figure 3.38. The current through the current source $I_s = 2$ mA is given by

$$I_3 - I_2 = I_s = 2 \text{ mA}$$

which can be rearranged as

$$-I_2 + I_3 = 0.002 \tag{3.80}$$

Equations (3.75), (3.79), and (3.80) can be solved using the substitution method: Solving Equation (3.80) for I_3, we obtain

$$I_3 = I_2 + 0.002$$

Substituting I_3 into Equations (3.75) and (3.79), we obtain

$$3000\, I_1 - 3000\, I_2 = 9$$

$$-3000\, I_1 + 4750\, I_2 = -5.5$$

Adding these two equations, we obtain

$$1750\, I_2 = 3.5$$

Thus, we get

$$I_2 = \frac{3.5}{1750} = 0.002 \text{ A} = 2 \text{ mA}$$

$$I_3 = I_2 + 0.002 = 0.004 \text{ A} = 4 \text{ mA}$$

continued

Example 3.15 continued

From Equation (3.79), we get

$$I_1 = \frac{2000}{3000}I_2 + \frac{2750}{3000}I_3 = 0.005 \text{ A} = 5 \text{ mA}$$

Alternatively, Equations (3.75), (3.79), and (3.80) can be solved using Cramer's rule:

Let

$$\Delta = \begin{vmatrix} 3000 & -1000 & -2000 \\ -3000 & 2000 & 2750 \\ 0 & -1 & 1 \end{vmatrix} = 5{,}250{,}000$$

$$\Delta_1 = \begin{vmatrix} 5 & -1000 & -2000 \\ 0 & 2000 & 2750 \\ 0.002 & -1 & 1 \end{vmatrix} = 26{,}250$$

$$\Delta_2 = \begin{vmatrix} 3000 & 5 & -2000 \\ -3000 & 0 & 2750 \\ 0 & 0.002 & 1 \end{vmatrix} = 10{,}500$$

$$\Delta_3 = \begin{vmatrix} 3000 & -1000 & 5 \\ -3000 & 2000 & 0 \\ 0 & -1 & 0.002 \end{vmatrix} = 21{,}000$$

Then, we have

$$I_1 = \frac{\Delta_1}{\Delta} = 5 \text{ mA}, \quad I_2 = \frac{\Delta_2}{\Delta} = 2 \text{ mA}, \quad I_3 = \frac{\Delta_3}{\Delta} = 4 \text{ mA}$$

MATLAB

```
A=[3000 -1000 -2000;-3000 2000 2750;0 -1 1];
b=[5;0;0.002];
I=A\b

Answer:
I =
    0.005000000000000
    0.002000000000000
    0.004000000000000
```

The voltage across R_3 is given by

$$V_{R_3} = R_3(I_1 - I_3) = 2000(0.005 - 0.004) = 2 \text{ V}$$

The voltage across R_4 is given by

$$V_{R_4} = R_4 I_3 = 750 \times 0.004 = 3 \text{ V}$$

Thus, the voltage v across the current source is given by

$$v = V_{R_4} - V_{R_3} = 3 \text{ V} - 2 \text{ V} = 1 \text{ V}$$

Exercise 3.15

Use mesh analysis to find V_1 in the circuit shown in Figure 3.39.

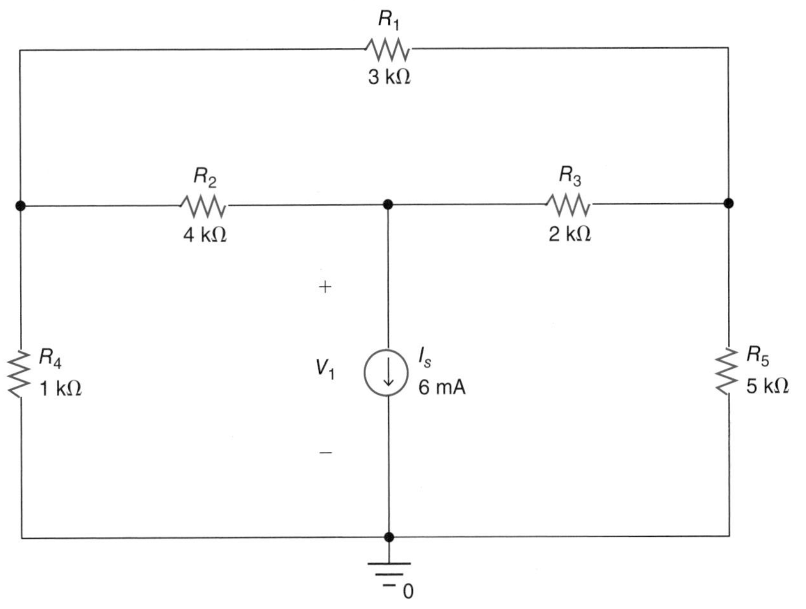

FIGURE 3.39

The circuit for EXERCISE 3.15.

Answer:
$V_1 = -15.25$ V.

EXAMPLE 3.16

Find the mesh currents I_1, I_2, I_3, and I_4 for the circuit shown in Figure 3.40.

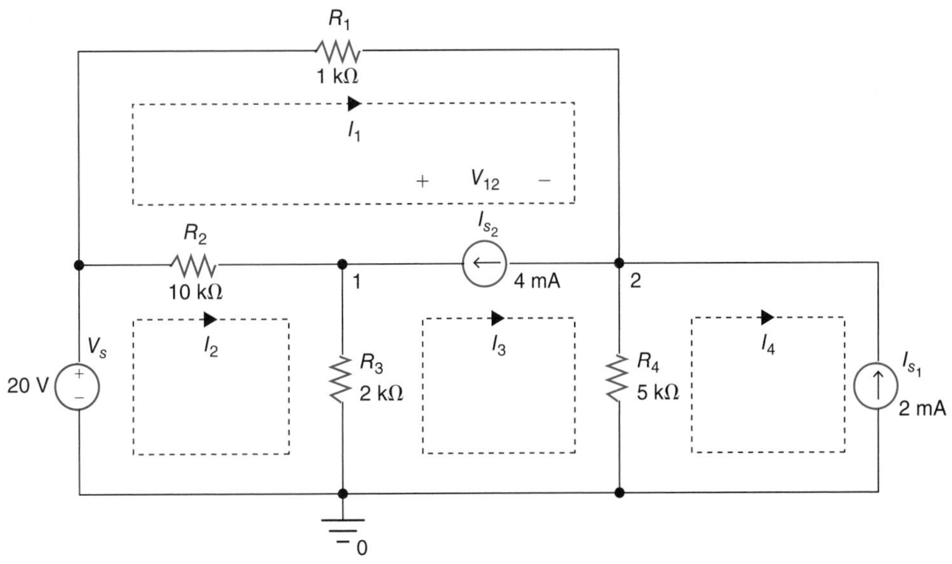

FIGURE 3.40

The circuit for EXAMPLE 3.16.

continued

Example 3.16 continued

Mesh current I_4 is in the opposite direction of current source I_{s_1}. Thus, $I_4 = -I_{s_1} = -2$ mA. Mesh 1 with mesh current I_1 and mesh 3 with mesh current I_3 share a current source I_{s_2}. Since the direction of I_1 coincides with I_{s_2}, we have

$$I_1 - I_3 = I_{s_2} \tag{3.81}$$

Let the voltage across the current source I_{s_2} from node 1 to node 2 be V_{12}, as shown in Figure 3.40. The mesh equation on mesh 1 with mesh current I_1 is given by, starting from the left side of R_1 in the clockwise direction,

$$R_1I_1 - V_{12} + R_2(I_1 - I_2) = 0 \tag{3.82}$$

The mesh equation on mesh 3 with mesh current I_3 is given by, starting from node 1 in the clockwise direction,

$$V_{12} + R_4(I_3 - I_4) + R_3(I_3 - I_2) = 0 \tag{3.83}$$

The unknown voltage V_{12} across the current source is eliminated by adding Equations (3.82) and (3.83):

$$R_1I_1 + R_4(I_3 - I_4) + R_3(I_3 - I_2) + R_2(I_1 - I_2) = 0 \tag{3.84}$$

Equation (3.84) suggests that if a supermesh consisting of meshes 1 and 3 without the current source I_{s_2} is formed as shown in Figure 3.41 (blue dashed line), the sum of voltage drops around this supermesh will be zero. Notice that the original mesh currents I_1 and I_3 are not changed.

For mesh 2 with mesh current I_2, the mesh equation is given by

$$-V_s + R_2(I_2 - I_1) + R_3(I_2 - I_3) = 0 \tag{3.85}$$

Substitution of $V_s = 20$ V, $I_4 = -I_{s_1} = -2$ mA, $I_{s_2} = 4$ mA, $R_1 = 1\ k\Omega$, $R_2 = 10\ k\Omega$, $R_3 = 2\ k\Omega$, $R_4 = 5\ k\Omega$ into Equations (3.81), (3.84), and (3.85) results in

$$I_1 - I_3 = 0.004 \tag{3.86}$$

FIGURE 3.41

The supermesh is outlined in the blue dashed line.

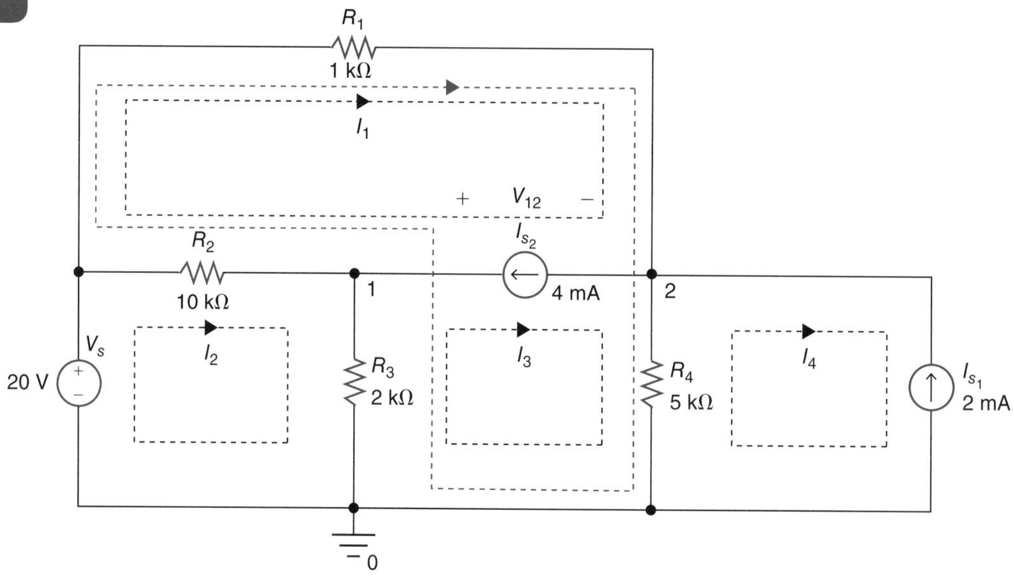

continued

Example 3.16 continued

$$1000I_1 + 5000[I_3 - (-0.002)] + 2000(I_3 - I_2) + 10{,}000(I_1 - I_2) = 0 \quad \textbf{(3.87)}$$

$$-20 + 10{,}000(I_2 - I_1) + 2000(I_2 - I_3) = 0 \quad \textbf{(3.88)}$$

Simplifying Equations (3.86)–(3.88), we obtain

$$I_1 - I_3 = 0.004 \quad \textbf{(3.89)}$$

$$11I_1 - 12I_2 + 7I_3 = -0.01 \quad \textbf{(3.90)}$$

$$-10I_1 + 12I_2 - 2I_3 = 0.02 \quad \textbf{(3.91)}$$

Equations (3.89), (3.90), and (3.91) can be solved using the substitution method:

Solving Equation (3.89) for I_3, we obtain

$$I_3 = I_1 - 0.004$$

Substituting I_3 into Equations (3.90) and (3.91), we obtain

$$18I_1 - 12I_2 = 0.018$$

$$-12I_1 + 12I_2 = 0.012$$

Adding these two equations, we obtain

$$6I_1 = 0.03$$

Thus, we obtain

$$I_1 = \frac{0.03}{6} = 0.005 \text{ A} = 5 \text{ mA}$$

$$I_3 = I_1 - 0.004 = 0.001 \text{ A} = 1 \text{ mA}$$

From Equation (3.90), we get

$$I_2 = \frac{11}{12}I_1 + \frac{7}{12}I_3 + \frac{0.01}{12} = 0.006 \text{ A} = 6 \text{ mA}$$

Alternatively, Equations (3.89), (3.90), and (3.91) can be solved using Cramer's rule:

$$\Delta = \begin{vmatrix} 1 & 0 & -1 \\ 11 & -12 & 7 \\ -10 & 12 & -2 \end{vmatrix} = -(0)\begin{vmatrix} 11 & 7 \\ -10 & -2 \end{vmatrix} + (-12)\begin{vmatrix} 1 & -1 \\ -10 & -2 \end{vmatrix} - 12\begin{vmatrix} 1 & -1 \\ 11 & 7 \end{vmatrix}$$
$$= (-12)[(1)(-2) - (-1)(-10)] - 12[(1)(7) - (-1)(11)]$$
$$= (-12)[-2 - 10] - 12[7 + 11] = 144 - 216 = -72$$

continued

Example 3.16 continued

$$I_1 = \frac{\begin{vmatrix} 0.004 & 0 & -1 \\ -0.01 & -12 & 7 \\ 0.02 & 12 & -2 \end{vmatrix}}{\Delta} = \frac{-(0)\begin{vmatrix} -0.01 & 7 \\ 0.02 & -2 \end{vmatrix} + (-12)\begin{vmatrix} 0.004 & -1 \\ 0.02 & -2 \end{vmatrix} - 12\begin{vmatrix} 0.004 & -1 \\ -0.01 & 7 \end{vmatrix}}{\Delta}$$

$$= \frac{(-12)[(0.004)(-2) - (-1)(0.02)] - 12[(0.004)(7) - (-1)(-0.01)]}{\Delta}$$

$$= \frac{(-12)[-0.008 + 0.02] - 12[0.028 - 0.01]}{\Delta} = \frac{-0.144 - 0.216}{-72} = \frac{0.36}{72} = 0.005 \text{ A} = 5 \text{ mA}$$

$$I_2 = \frac{\begin{vmatrix} 1 & 0.004 & -1 \\ 11 & -0.01 & 7 \\ -10 & 0.02 & -2 \end{vmatrix}}{\Delta} = \frac{(1)\begin{vmatrix} -0.01 & 7 \\ 0.02 & -2 \end{vmatrix} - (0.004)\begin{vmatrix} 11 & 7 \\ -10 & -2 \end{vmatrix} + (-1)\begin{vmatrix} 11 & -0.01 \\ -10 & 0.02 \end{vmatrix}}{\Delta}$$

$$= \frac{(1)[(-0.01)(-2) - (7)(0.02)] - 0.004[(11)(-2) - (7)(-10)] - [(11)(0.02) - (-0.01)(-10)]}{\Delta}$$

$$= \frac{[0.02 - 0.14] - 0.004[-22 + 70] - [0.22 - 0.1]}{\Delta} = \frac{-0.12 - 0.192 - 0.12}{-72}$$

$$= \frac{-0.432}{-72} = 0.006 \text{ A} = 6 \text{ mA}$$

$$I_3 = \frac{\begin{vmatrix} 1 & 0 & 0.004 \\ 11 & -12 & -0.01 \\ -10 & 12 & 0.02 \end{vmatrix}}{\Delta} = \frac{-(0)\begin{vmatrix} 11 & -0.01 \\ -10 & 0.02 \end{vmatrix} + (-12)\begin{vmatrix} 1 & 0.004 \\ -10 & 0.02 \end{vmatrix} - (12)\begin{vmatrix} 1 & 0.004 \\ 11 & -0.01 \end{vmatrix}}{\Delta}$$

$$= \frac{(-12)[(1)(0.02) - (0.004)(-10)] - (12)[(1)(-0.01) - (0.004)(11)]}{\Delta}$$

$$= \frac{(-12)[0.02 + 0.04] - (12)[-0.01 - 0.044]}{\Delta} = \frac{-0.72 + 0.648}{-72} = \frac{-0.072}{-72} = 0.001 \text{ A} = 1 \text{ mA}$$

MATLAB

```
A=[1 0 -1;11 -12 7;-10 12 -2]
b=[0.004;-0.01;0.02]
I=A\b

Answer:
I =
   0.005000000000000
   0.006000000000000
   0.001000000000000
```

Alternative solution:

```
%EXAMPLE 3.16
clear all;
Vs=20;Is1=2e-3;Is2=4e-3;R1=1000;R2=10000;R3=2000;R4=5000;
syms I1 I2 I3 I4
[I1,I2,I3,I4]=solve(R1*I1+R4*(I3-I4)+R3*(I3-I2)+R2*(I1-I2),...
-Vs+R2*(I2-I1)+R3*(I2-I3),...
I1-I3==Is2,...
```

continued

Example 3.16 continued

```
I4==-Is1,I1,I2,I3,I4);
I1=vpa(I1,7)
I2=vpa(I2,7)
I3=vpa(I3,7)
I4=vpa(I4,7)

Answers:
I1 =
0.005
I2 =
0.006
I3 =
0.001
I4 =
-0.002
```

Exercise 3.16

Find mesh currents I_1, I_2, and I_3 and voltages V_1, V_2, and V_3 in the circuit shown in Figure 3.42.

FIGURE 3.42

The circuit for EXERCISE 3.16.

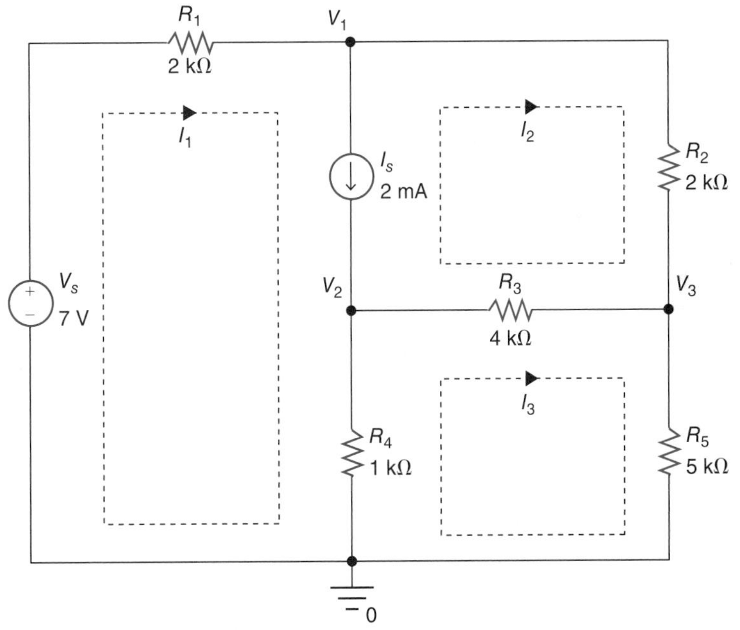

Answer:
$I_1 = 2.3077$ mA, $I_2 = 307.6923$ μA, $I_3 = 353.8462$ μA, $V_1 = 2.3846$ V, $V_2 = 1.9538$ V, $V_3 = 1.7692$ V.

EXAMPLE 3.17

Use mesh analysis to find voltages V_1 and V_2 in the circuit shown in Figure 3.43.

continued

Example 3.17 continued

FIGURE 3.43

The circuit for EXAMPLE 3.17.

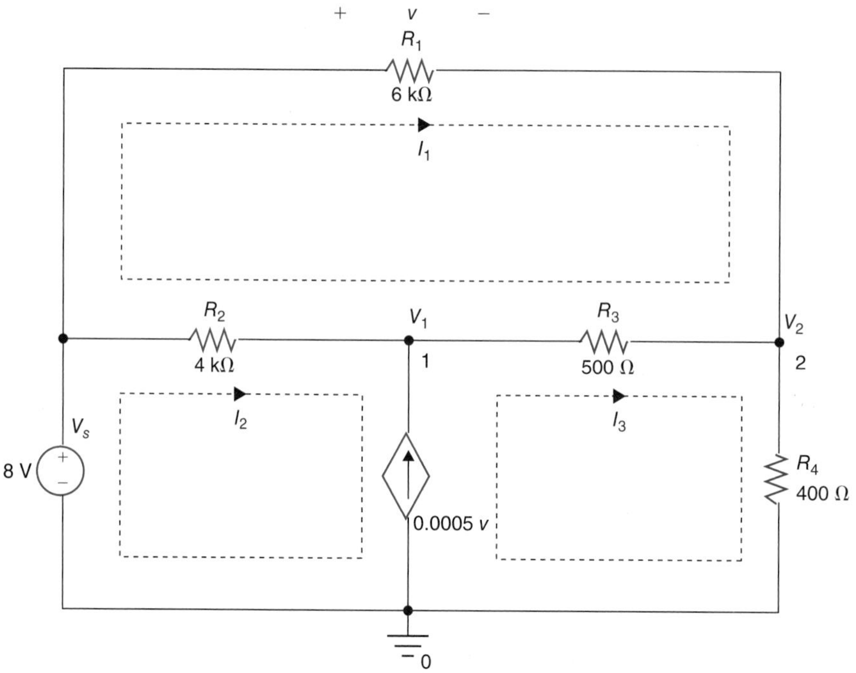

Mesh currents I_1, I_2, and I_3 are shown in Figure 3.43. Due to the presence of a voltage-controlled current source (VCCS) between meshes 2 and 3, meshes 2 and 3 form a supermesh. The current through the VCCS equals $I_3 - I_2$. Thus, we have

$$I_3 - I_2 = 0.0005v = 0.0005 \times 6000I_1$$

or

$$-3I_1 - I_2 + I_3 = 0 \tag{3.92}$$

Summing the voltage drops around mesh 1 in the clockwise direction, we obtain

$$R_1I_1 + R_3(I_1 - I_3) + R_2(I_1 - I_2) = 0$$

Substituting the resistance values, we get

$$6000I_1 + 500(I_1 - I_3) + 4000(I_1 - I_2) = 0$$

or

$$10.5I_1 - 4I_2 - 0.5I_3 = 0 \tag{3.93}$$

Summing the voltage drops around a supermesh consisting of meshes 2 and 3 in the clockwise direction, we obtain

$$-V_s + R_2(I_2 - I_1) + R_3(I_3 - I_1) + R_4I_3 = 0$$

Substituting the resistance and voltage values, we get

$$-8 + 4000(I_2 - I_1) + 500(I_3 - I_1) + 400I_3 = 0$$

continued

Example 3.17 continued

or

$$-4.5I_1 + 4I_2 + 0.9I_3 = 0.008 \quad \textbf{(3.94)}$$

Equations (3.92), (3.93), and (3.94) can be solved using the substitution method: Multiplying Equation (3.92) by 4, we obtain

$$-12I_1 - 4I_2 + 4I_3 = 0$$

Adding this equation and Equation (3.94), we obtain

$$-16.5I_1 + 4.9I_3 = 0.008$$

Adding Equations (3.93) and (3.94), we obtain

$$6I_1 + 0.4I_3 = 0.008$$

Solving this equation for I_3, we get

$$I_3 = 0.02 - 15I_1$$

Substituting I_3 into $-16.5I_1 + 4.9I_3 = 0.008$, we obtain

$$-16.5I_1 + 4.9(0.02 - 15I_1) = 0.008$$

or

$$-90I_1 = -0.09$$

Thus, we get

$$I_1 = \frac{-0.09}{-90} = 0.001 \text{ A} = 1 \text{ mA}$$

$$I_3 = 0.02 - 15I_1 = 5 \text{ mA}$$

From Equation (3.92), we get

$$I_2 = -3I_1 + I_3 = 2 \text{ mA}$$

Alternatively, with the application of Cramer's rule to Equations (3.92)–(3.94), we get

$$\Delta = \begin{vmatrix} -3 & -1 & 1 \\ 10.5 & -4 & -0.5 \\ -4.5 & 4 & 0.9 \end{vmatrix} = (-3)\begin{vmatrix} -4 & -0.5 \\ 4 & 0.9 \end{vmatrix} - (-1)\begin{vmatrix} 10.5 & -0.5 \\ -4.5 & 0.9 \end{vmatrix} + (1)\begin{vmatrix} 10.5 & -4 \\ -4.5 & 4 \end{vmatrix}$$

$$= (-3)[(-4)(0.9) - (-0.5)(4)] + [(10.5)(0.9) - (-0.5)(-4.5)]$$
$$+ [(10.5)(4) - (-4)(-4.5)]$$
$$= (-3)[-3.6 + 2] + [9.45 - 2.25] + [42 - 18] = 4.8 + 7.2 + 24 = 36$$

$$I_1 = \frac{\begin{vmatrix} 0 & -1 & 1 \\ 0 & -4 & -0.5 \\ 0.008 & 4 & 0.9 \end{vmatrix}}{\Delta} = \frac{0.008\begin{vmatrix} -1 & 1 \\ -4 & -0.5 \end{vmatrix}}{36} = \frac{0.008[(-1)(-0.5) - (1)(-4)]}{36}$$

$$= \frac{0.008[0.5 + 4]}{36} = \frac{0.036}{36} = 0.001 \text{ A} = 1 \text{ mA}$$

continued

Example 3.17 continued

$$I_2 = \frac{\begin{vmatrix} -3 & 0 & 1 \\ 10.5 & 0 & -0.5 \\ -4.5 & 0.008 & 0.9 \end{vmatrix}}{\Delta} = \frac{-0.008\begin{vmatrix} -3 & 1 \\ 10.5 & -0.5 \end{vmatrix}}{36}$$

$$= \frac{-0.008[(-3)(-0.5) - (1)(10.5)]}{36}$$

$$= \frac{-0.008[1.5 - 10.5]}{36} = \frac{0.072}{36} = 0.002 \text{ A} = 2 \text{ mA}$$

$$I_3 = \frac{\begin{vmatrix} -3 & -1 & 0 \\ 10.5 & -4 & 0 \\ -4.5 & 4 & 0.008 \end{vmatrix}}{\begin{vmatrix} -3 & -1 & 1 \\ 10.5 & -4 & -0.5 \\ -4.5 & 4 & 0.9 \end{vmatrix}} = \frac{0.008\begin{vmatrix} -3 & 1 \\ 10.5 & -4 \end{vmatrix}}{36} = \frac{0.008[(-3)(-4) - (-1)(10.5)]}{36}$$

$$= \frac{0.008[12 + 10.5]}{36} = \frac{0.18}{36} = 0.005 \text{ A} = 5 \text{ mA}$$

$$V_1 = V_s - R_2(I_2 - I_1) = 8 \text{ V} - 4000\ \Omega \times (0.002 - 0.001) \text{ A} = 4 \text{ V}$$

$$V_2 = R_4 I_3 = 400\ \Omega \times 0.005 \text{ A} = 2 \text{ V}$$

MATLAB

```
A=[-3 -1 1;10.5 -4 -0.5;-4.5 4 0.9]
b=[0;0;0.008]
I=A\b

I =
    0.001000000000000
    0.002000000000000
    0.005000000000000
```

Exercise 3.17

Use mesh analysis to find V_0 in the circuit shown in Figure 3.44.

FIGURE 3.44

The circuit for EXERCISE 3.17.

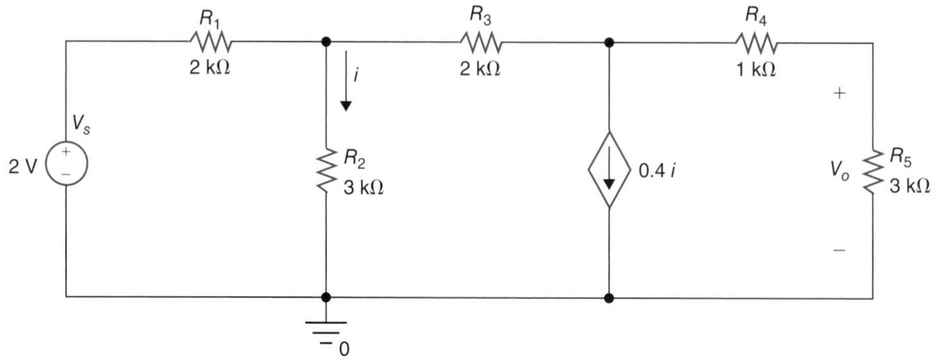

Answer:
$V_0 = 0.3367$ V.

EXAMPLE 3.18

Find the mesh currents I_1, I_2, I_3, and I_4 and voltages V_1, V_2, and V_3 in the circuit shown in Figure 3.45.

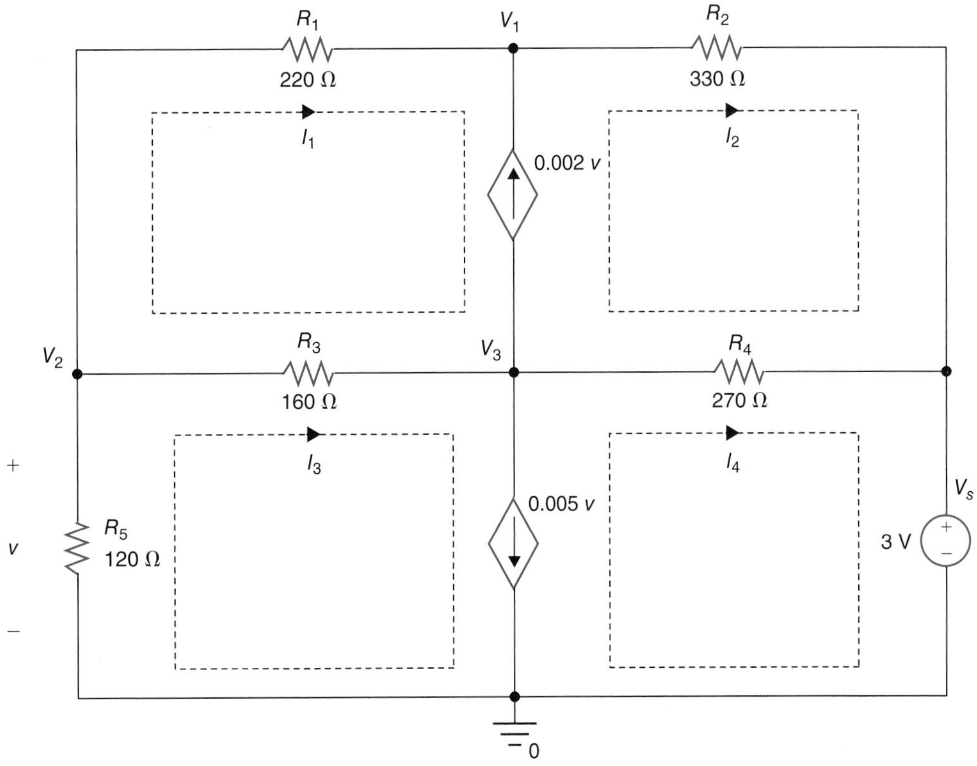

FIGURE 3.45 The circuit for EXAMPLE 3.18.

In this circuit, mesh 1 and mesh 2 form supermesh 1, while mesh 3 and mesh 4 form supermesh 2. From supermesh 1, we have

$$I_2 - I_1 = 0.002v = 0.002 \times 120 \times (-I_3) = -0.24I_3$$

Solving for I_2, we have

$$I_2 = I_1 - 0.24I_3 \tag{3.95}$$

From supermesh 2, we have

$$I_3 - I_4 = 0.005v = 0.005 \times 120 \times (-I_3) = -0.6I_3$$

Thus, we get

$$I_4 = 1.6I_3 \tag{3.96}$$

Summing the voltage drops around supermesh 1, we obtain

$$R_1I_1 + R_2I_2 + R_4(I_2 - I_4) + R_3(I_1 - I_3) = 0$$

Substituting the resistance values, we get

$$220I_1 + 330I_2 + 270(I_2 - I_4) + 160(I_1 - I_3) = 0$$

continued

Example 3.18 continued

which can be simplified to

$$380I_1 + 600I_2 - 160I_3 - 270I_4 = 0 \quad \textbf{(3.97)}$$

Substitution of Equations (3.95) and (3.96) into Equation (3.97) yields

$$980I_1 - 736I_3 = 0 \quad \textbf{(3.98)}$$

Summing the voltage drops around supermesh 2, we obtain

$$R_5I_3 + R_3(I_3 - I_1) + R_4(I_4 - I_2) + V_s = 0$$

Substituting the resistance and voltage values, we get

$$120I_3 + 160(I_3 - I_1) + 270(I_4 - I_2) + 3 = 0$$

which can be simplified to

$$-160I_1 - 270I_2 + 280I_3 + 270I_4 = -3 \quad \textbf{(3.99)}$$

Substitution of Equations (3.95) and (3.96) into Equation (3. 99) yields

$$-430I_1 + 776.8I_3 = -3 \quad \textbf{(3.100)}$$

Equations (3.98) and (3.100) can be solved using the substitution method:

Solving Equation (3.98) for I_3, we obtain

$$I_3 = \frac{980}{736}I_1$$

Substituting I_3 into Equation (3.100), we get

$$-430I_1 + 776.8\left(\frac{980}{736}I_1\right) = -3$$

Thus, we have

$$I_1 = \frac{-3}{-430 + 776.8 \times \dfrac{980}{736}} = -4.9642 \text{ mA}$$

$$I_3 = \frac{980}{736}I_1 = -6.60995 \text{ mA}$$

Alternatively, the application of Cramer's rule to Equations (3.98) and (3.100) yields

$$\Delta = \begin{vmatrix} 980 & -736 \\ -430 & 776.8 \end{vmatrix} = 980 \times 776.8 - (-736) \times (-430) = 4.44784 \times 10^5$$

$$I_1 = \frac{\begin{vmatrix} 0 & -736 \\ -3 & 776.8 \end{vmatrix}}{\Delta} A = \frac{-2208}{4.44784 \times 10^5} A = -4.9642 \text{ mA}$$

continued

Example 3.18 continued

$$I_3 = \frac{\begin{vmatrix} 980 & 0 \\ -430 & -3 \end{vmatrix}}{\Delta} A = \frac{-2940}{4.44784 \times 10^5} A = -6.60995 \text{ mA}$$

$$I_2 = I_1 - 0.24I_3 = -3.3778 \text{ mA}$$

$$I_4 = 1.6I_3 = -10.5759 \text{ mA}$$

The voltages are given by

$$V_1 = V_s + R_2 \times I_2 = 1.8853 \text{ V}$$

$$V_2 = V_1 + R_1 \times I_1 = 0.7932 \text{ V}$$

$$V_3 = V_s - R_4 \times (I_2 - I_4) = 1.0565 \text{ V}$$

MATLAB

```
%EXAMPLE 3.18
clear all;
Vs=3;R1=220;R2=330;R3=160;R4=270;R5=120;
syms I1 I2 I3 I4 v
[I1,I2,I3,I4,v]=solve(v==-R5*I3,I2-I1==0.002*v,I3-I4==0.005*v,...
R1*I1+R2*I2+R4*(I2-I4)+R3*(I1-I3),...
R5*I3+R3*(I3-I1)+R4*(I4-I2)+Vs,I1,I2,I3,I4,v);
V1=R2*I2+Vs;
V2=v;
V3=R4*(I4-I2)+Vs;
I1=vpa(I1,7)
I2=vpa(I2,7)
I3=vpa(I3,7)
I4=vpa(I4,7)
V1=vpa(V1,7)
V2=vpa(V2,7)
V3=vpa(V3,7)

Answers:
I1 =
-0.004964207
I2 =
-0.003377819
I3 =
-0.00660995
I4 =
-0.01057592
V1 =
1.88532
V2 =
0.793194
V3 =
1.056513
```

Exercise 3.18

Find the mesh currents I_1, I_2, and I_3 and voltages V_1, V_2, V_3, and V_4 in the circuit shown in Figure 3.46.

FIGURE 3.46

The circuit for EXERCISE 3.18.

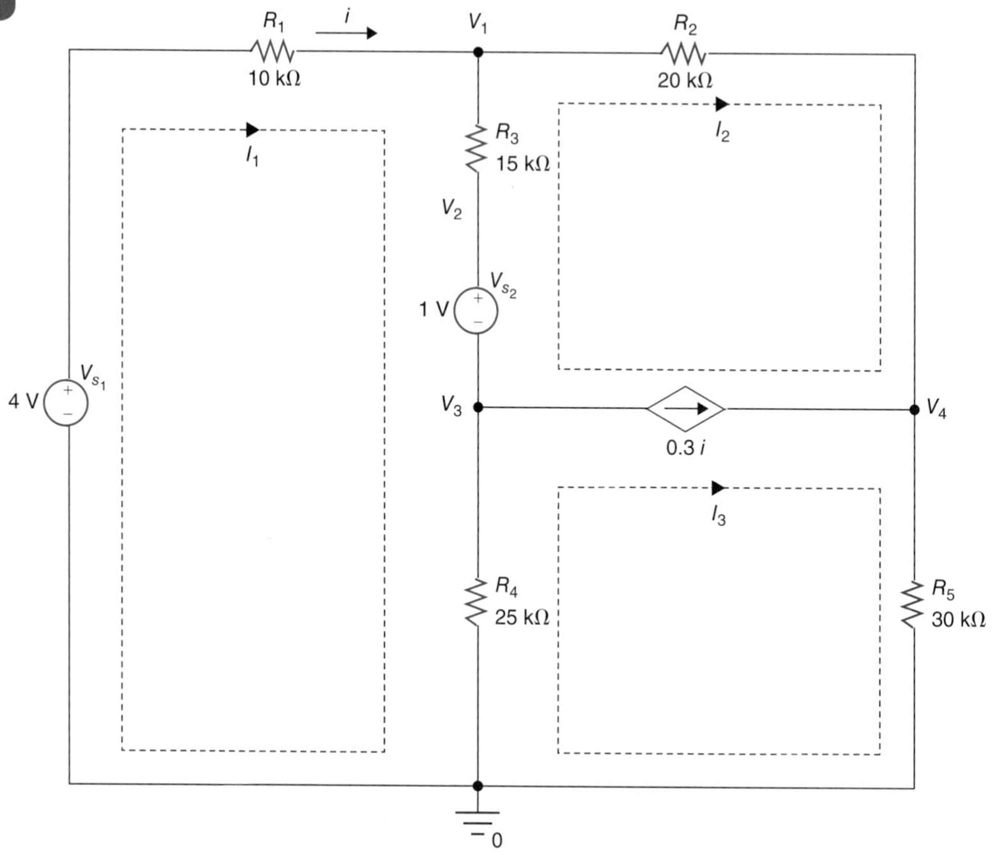

Answer:
$I_1 = 107.4523\ \mu A$, $I_2 = 39.1681\ \mu A$, $I_3 = 71.4038\ \mu A$, $V_1 = 2.9255$ V, $V_2 = 1.9012$ V, $V_3 = 0.9012$ V, $V_4 = 2.1421$ V.

3.6 PSpice and Simulink

3.6.1 PSPICE

We present PSpice simulation for circuits with dependent sources.

3.6.2 VCVS

Consider the circuit shown in Figure 3.47. It can be shown that $V_1 = 1.99938$ V and $V_2 = 5.998017$ V. The PSpice schematic for this circuit is shown in Figure 3.48. The part name of VCVS starts with E. Double-click on E_1 and enter 10,000 (the voltage gain). The controlling voltage v is the voltage across R_1, and the controlled voltage source is placed between node 3 and the ground. From the PSpice simulation, we get $V_1 = 1.999$ V and $V_2 = 5.998$ V, as shown in Figure 3.48.

FIGURE 3.47

A circuit with VCVS.

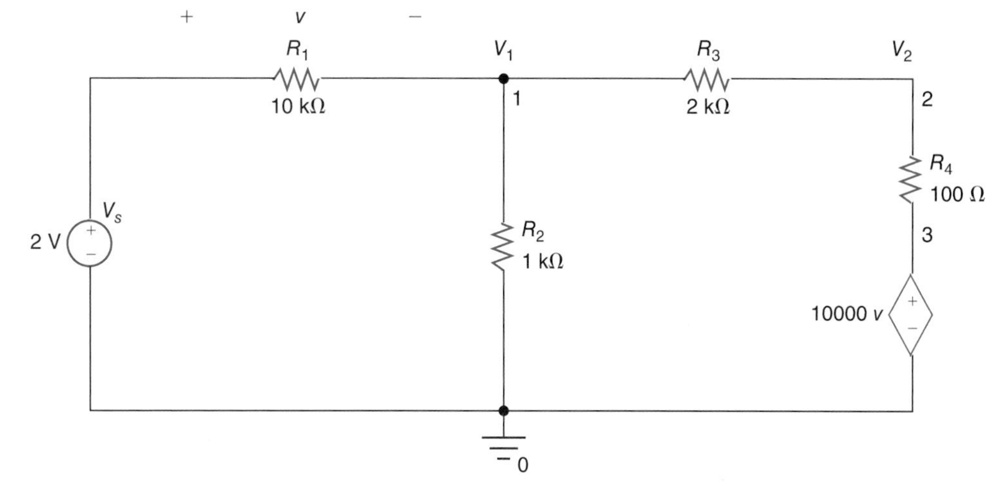

FIGURE 3.48

The PSpice schematic of the circuit shown in Figure 3.47.

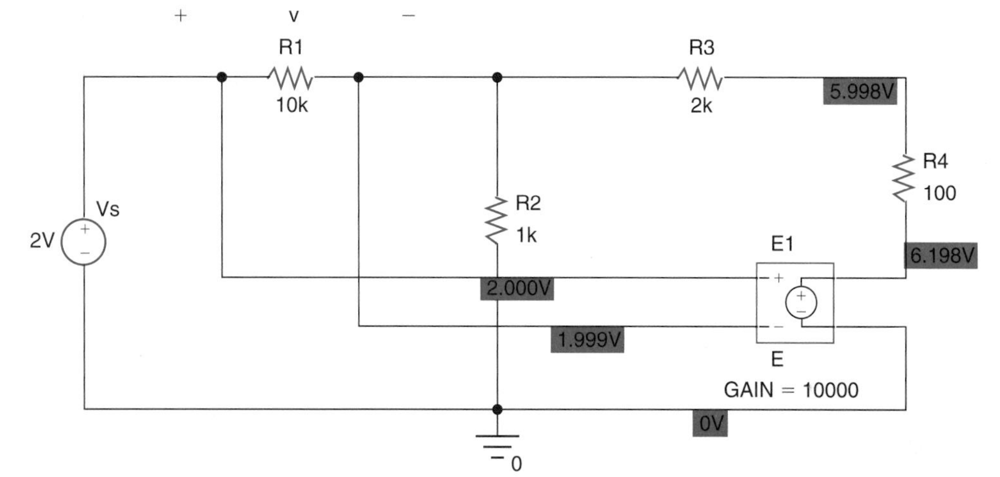

3.6.3 VCCS

Consider the circuit shown in Figure 3.49. It can be shown that $V_1 = 472.0930$ mV and $V_2 = -4.8023$ V. The PSpice schematic for this circuit is shown in Figure 3.50. The part name of VCCS starts with G. Double-click on G_1 and enter 0.005 (the conductance). The controlling voltage v is the voltage across R_2, and the controlled current source is placed between node 2 and ground. From the PSpice simulation, we get $V_1 = 472.1$ mV and $V_2 = -4.802$ V, as shown in Figure 3.50.

FIGURE 3.49

A circuit with VCCS.

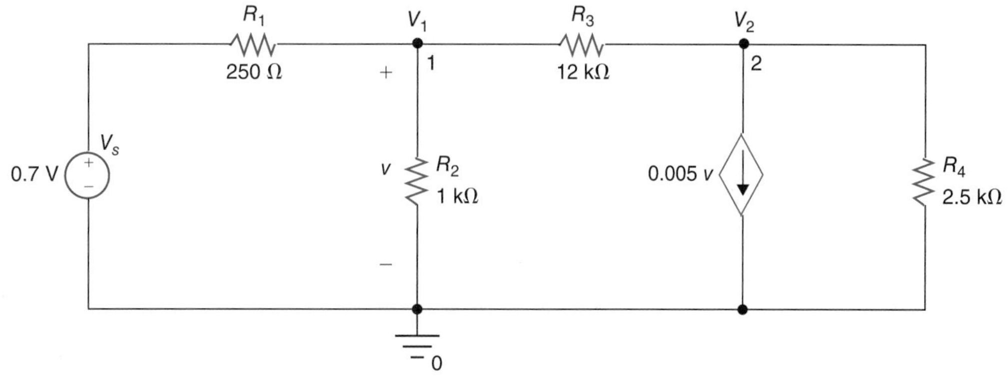

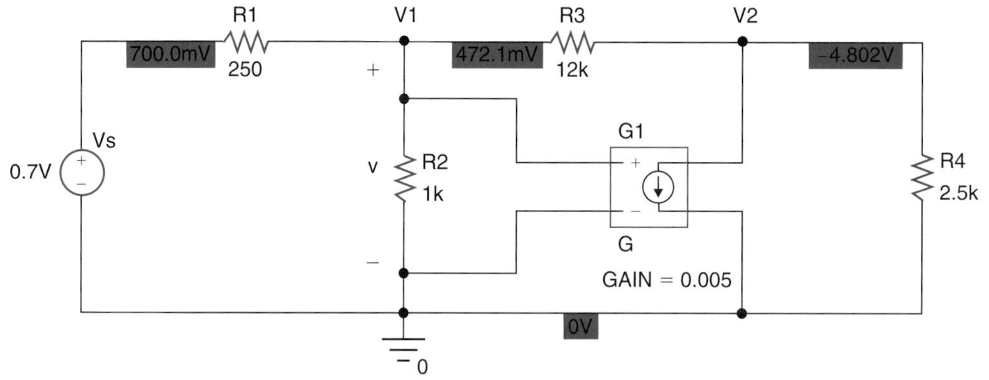

FIGURE 3.50 The PSpice schematic of the circuit shown in Figure 3.49.

3.6.4 CCVS

Consider the circuit shown in Figure 3.51. It can be shown that $V_1 = 856.88943$ mV and $V_2 = 2.1433$ V. The PSpice schematic for this circuit is shown in Figure 3.52. The part name of CCVS starts with H. Double-click on H_1 and enter 3500 (the resistance). The controlling current i is the current through R_2 and the controlled voltage source is placed between node 3 and ground. From the PSpice simulation, we get $V_1 = 856.9$ mV and $V_2 = 2.143$ V, as shown in Figure 3.52.

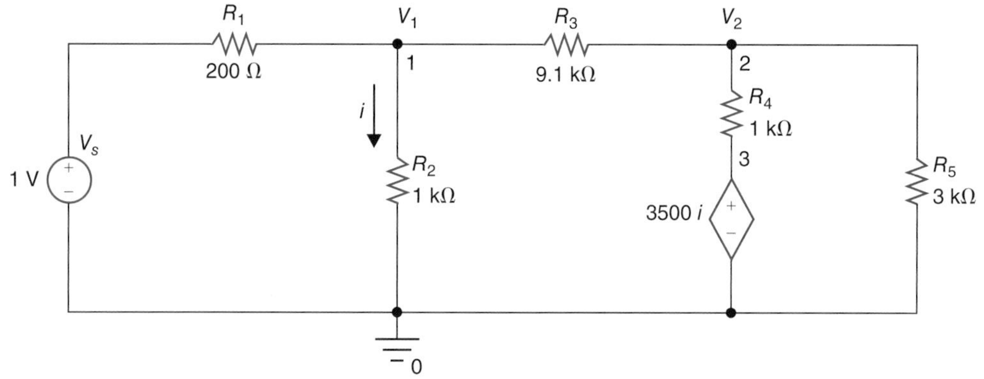

FIGURE 3.51 A circuit with CCVS.

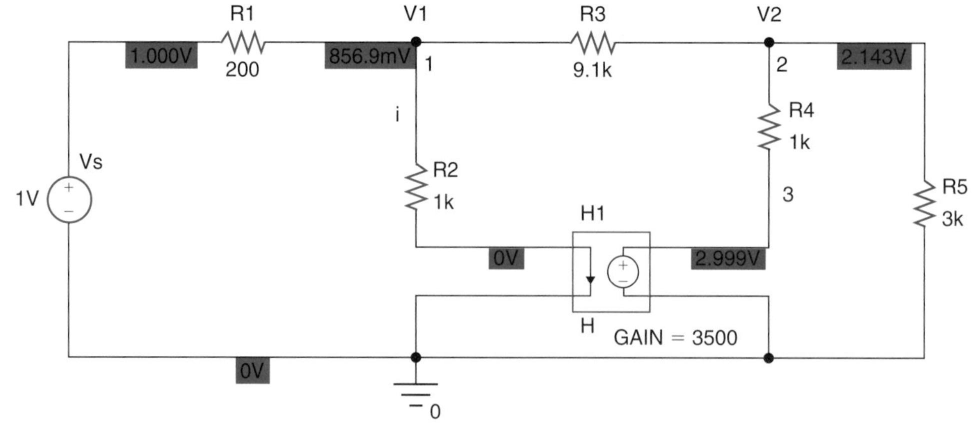

FIGURE 3.52 The PSpice schematic of the circuit shown in Figure 3.51.

3.6.5 CCCS

Consider the circuit shown in Figure 3.53. It can be shown that $V_1 = 2.7068$ V and $V_2 = -6.3158$ V. The PSpice schematic for this circuit is shown in Figure 3.54. The part name of CCCS starts with *F*. Double-click on F_1 and enter 2 (the current gain). The controlling current i is the current through R_2, and the controlled current source is placed between node 2 and the ground. From the PSpice simulation, we get $V_1 = 2.707$ V and $V_2 = -6.316$ V, as shown in Figure 3.54.

FIGURE 3.53

A circuit with CCCS.

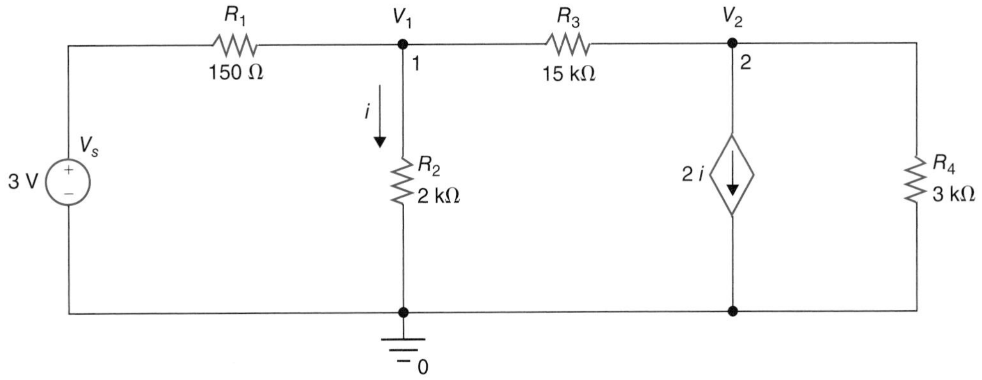

FIGURE 3.54

The PSpice schematic of the circuit shown in Figure 3.53.

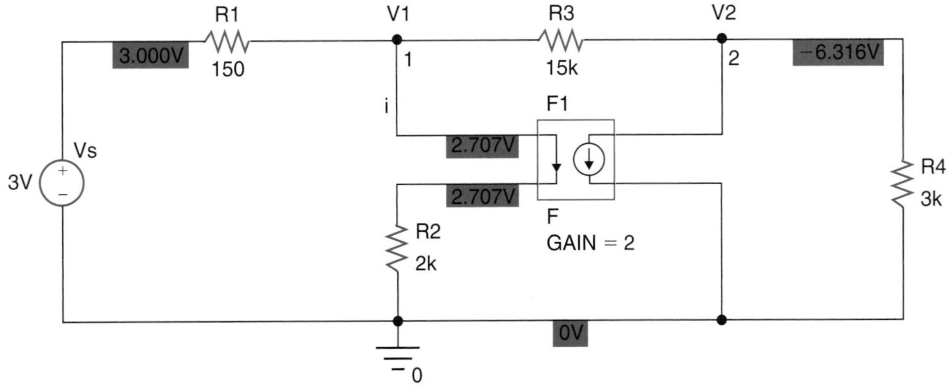

3.6.6 SIMULINK

A simscape of Simulink® makes it possible to simulate a circuit with VCCS, as shown in Figure 3.55.

FIGURE 3.55

A Simulink model for a circuit with VCCS.

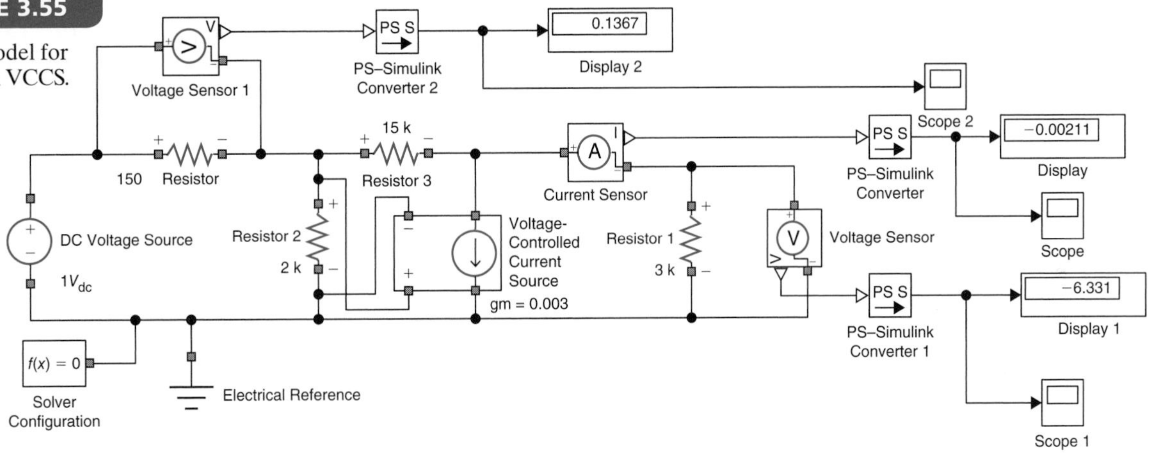

SUMMARY

Two analysis methods, nodal analysis and mesh analysis, are widely used in analyzing electrical and electronic circuits.

The nodal analysis is a circuit analysis method that finds all unknown node voltages. For each node whose voltage is unknown, we sum the currents leaving (or entering) the node. If there are *n* unknown node voltages, we obtain *n* equations in *n* unknowns. The node voltages are found by solving these *n* equations using Cramer's rule or MATLAB. Once all node voltages are found, we can find the branch currents and the powers everywhere.

The mesh analysis is a circuit analysis method that finds all unknown mesh currents. For each mesh whose mesh current is unknown, we sum the voltage drops around the mesh. If there are *n* unknown mesh currents, we obtain *n* equations in *n* unknowns. The mesh currents are found by solving these *n* equations using Cramer's rule or MATLAB. Once all mesh currents are found, we can find the branch currents, the node voltages, and the powers everywhere.

PROBLEMS

Nodal Analysis

3.1 **Find the voltage V_1 in the circuit shown in Figure P3.1.**

FIGURE P3.1

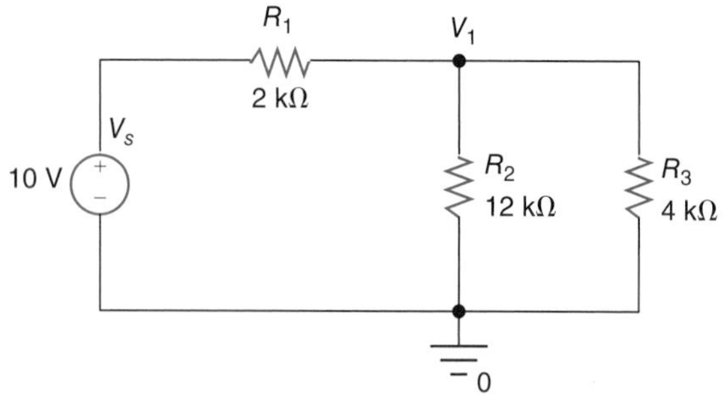

3.2 **Find the voltage V_1 in the circuit shown in Figure P3.2.**

FIGURE P3.2

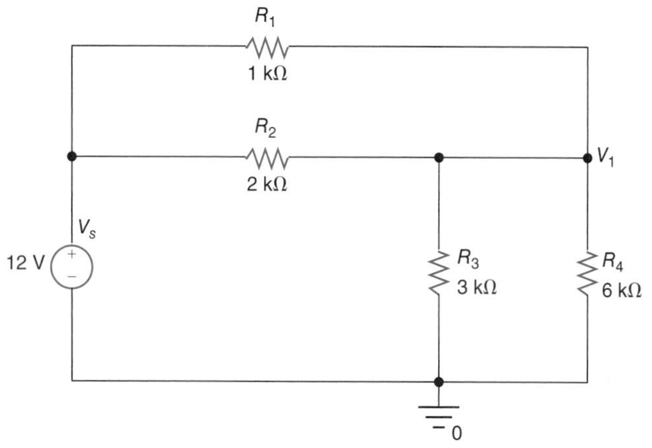

3.3 **Find the voltage V_1 in the circuit shown in Figure P3.3.**

FIGURE P3.3

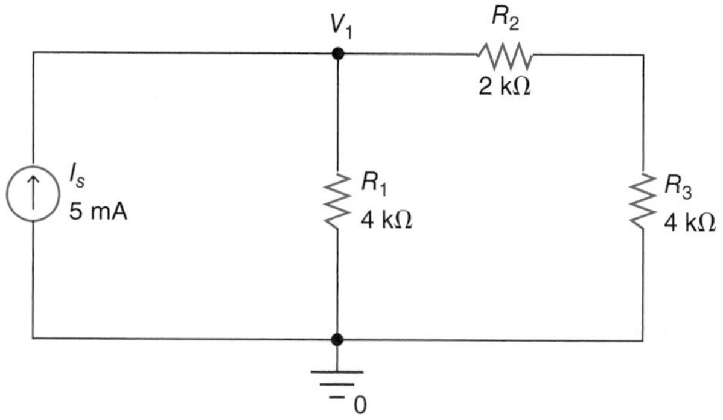

3.4 **Find the voltage V_1 in the circuit shown in Figure P3.4.**

FIGURE P3.4

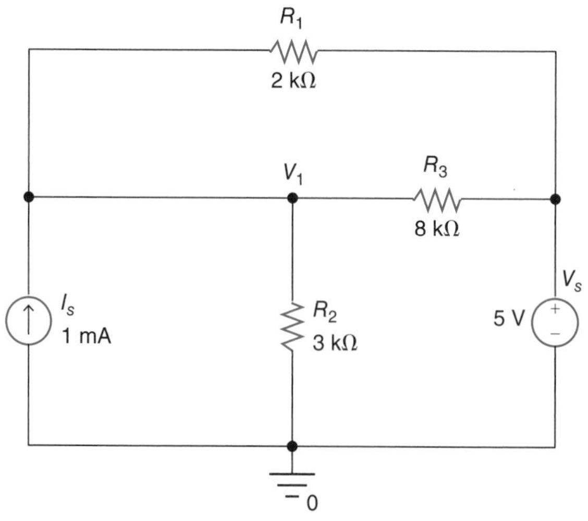

3.5 **Find V_1, I_1, and I_2 in the circuit shown in Figure P3.5.**

FIGURE P3.5

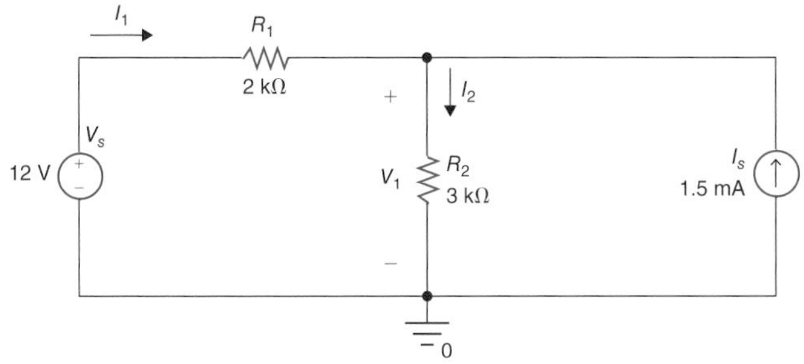

3.6 **Find V_1 and I_1 in the circuit shown in Figure P3.6.**

FIGURE P3.6

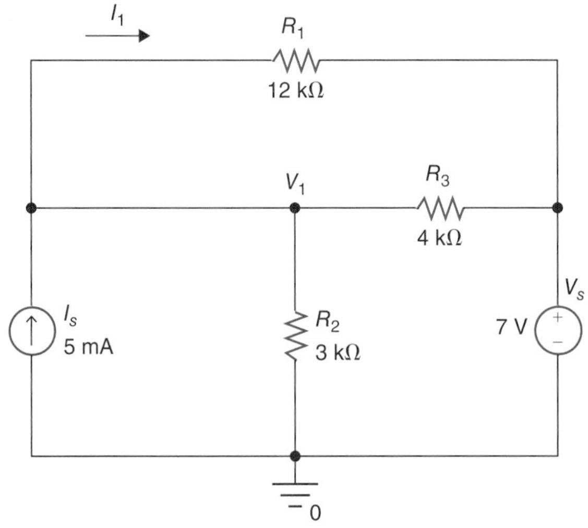

3.7 **Find the voltages V_1 and V_2 in the circuit shown in Figure P3.7.**

FIGURE P3.7

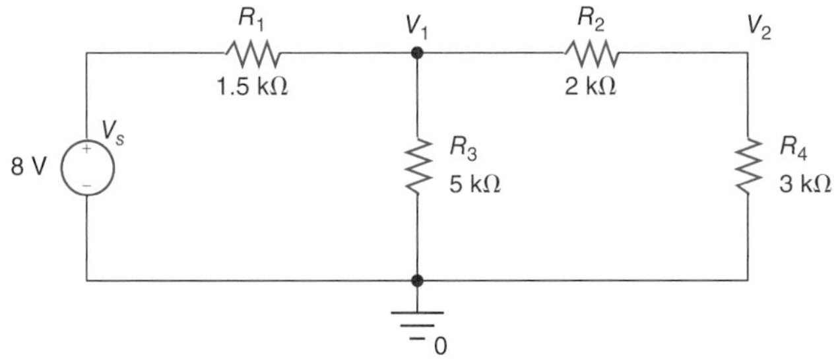

3.8 **Find the voltages V_1 and V_2 in the circuit shown in Figure P3.8.**

FIGURE P3.8

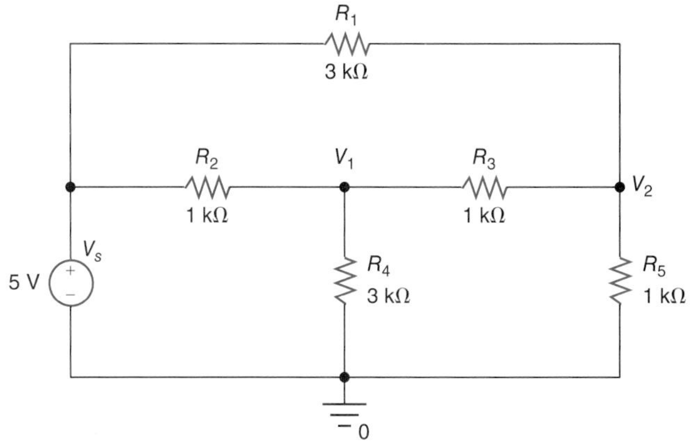

3.9 **Find the voltages V_1 and V_2 in the circuit shown in Figure P3.9.**

FIGURE P3.9

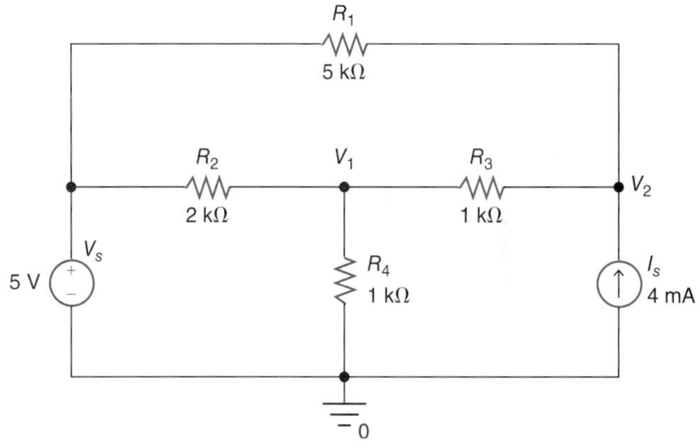

3.10 **Find the voltages V_1 and V_2 in the circuit shown in Figure P3.10.**

FIGURE P3.10

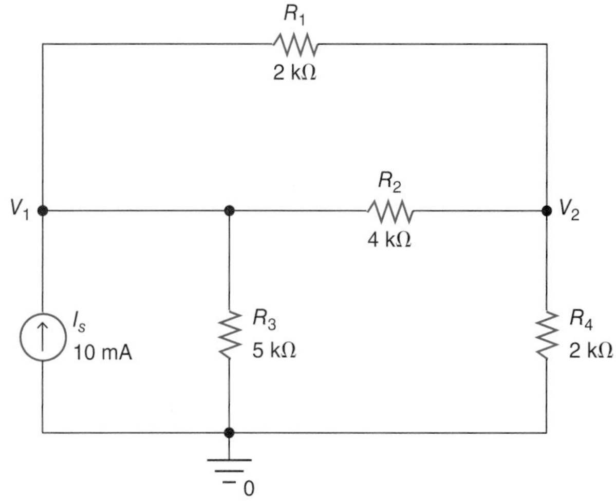

3.11 Find the voltages V_1 and V_2 in the circuit shown in Figure P3.11.

FIGURE P3.11

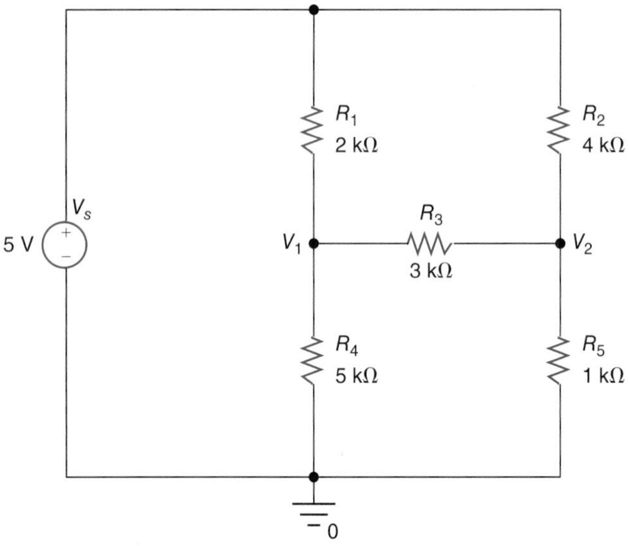

3.12 Find the voltages V_1 and V_2 in the circuit shown in Figure P3.12.

FIGURE P3.12

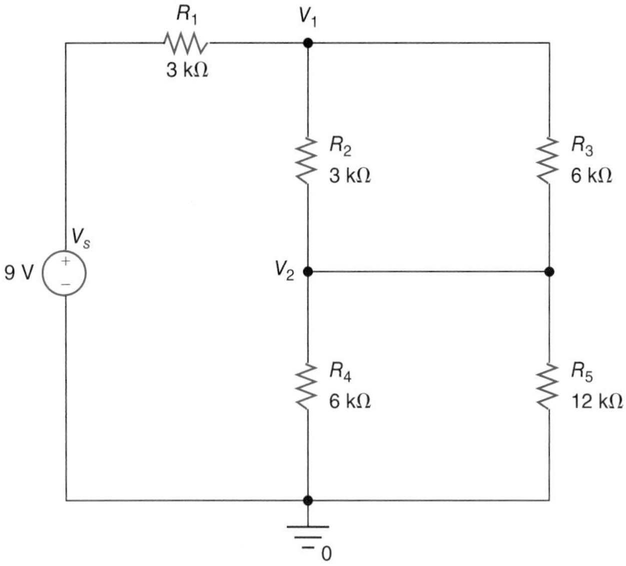

3.13 Find the voltages V_1 and V_2 in the circuit shown in Figure P3.13.

FIGURE P3.13

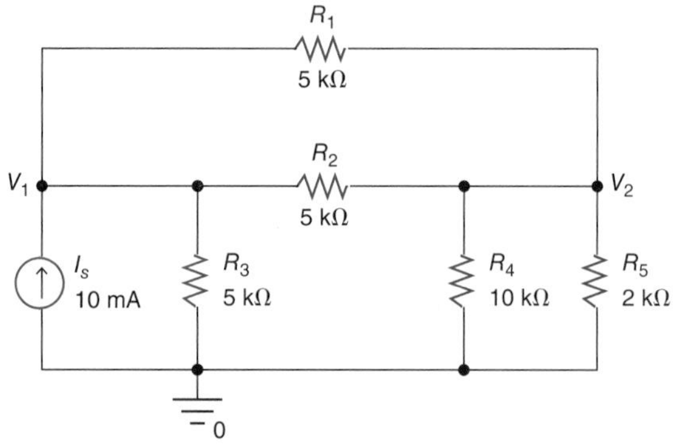

3.14 Determine V_1 and V_2 for the circuit shown in Figure P3.14.

FIGURE P3.14

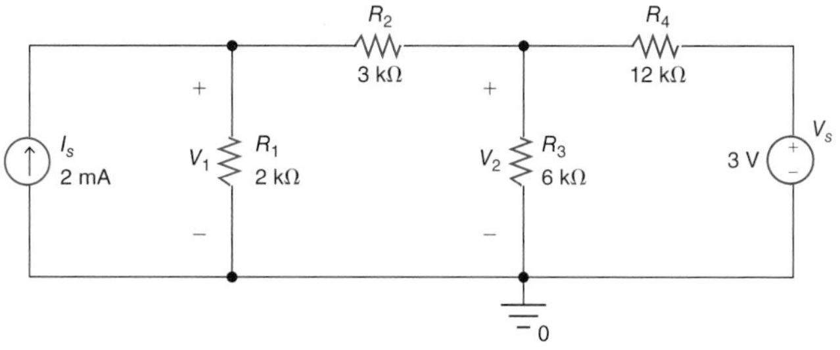

3.15 Find V_1 and V_2 in the circuit shown in Figure P3.15.

FIGURE P3.15

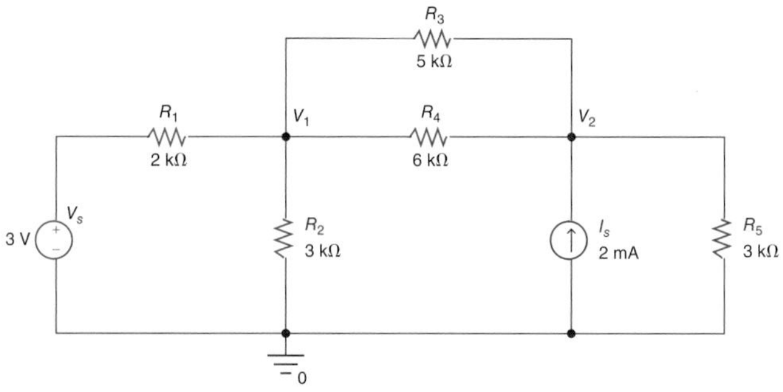

3.16 Find V_1 and V_2 in the circuit shown in Figure P3.16.

FIGURE P3.16

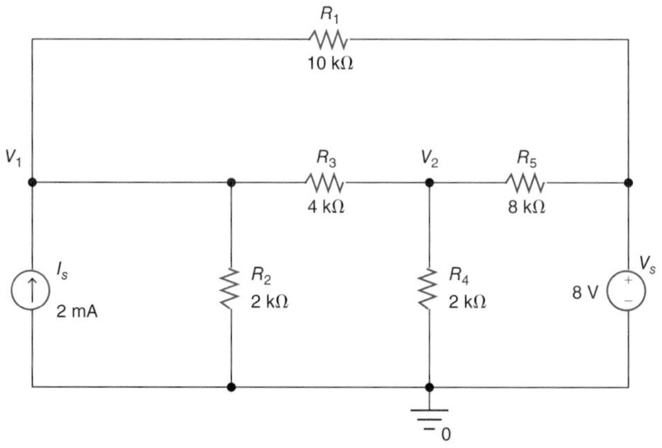

3.17 Find V_1, V_2, I_1, and I_2 in the circuit shown in Figure P3.17.

FIGURE P3.17

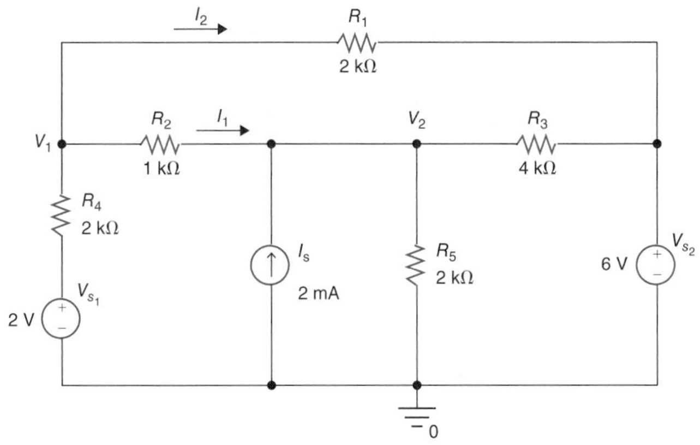

3.18 Find the voltages V_1, V_2, and V_3 in the circuit shown in Figure P3.18.

FIGURE P3.18

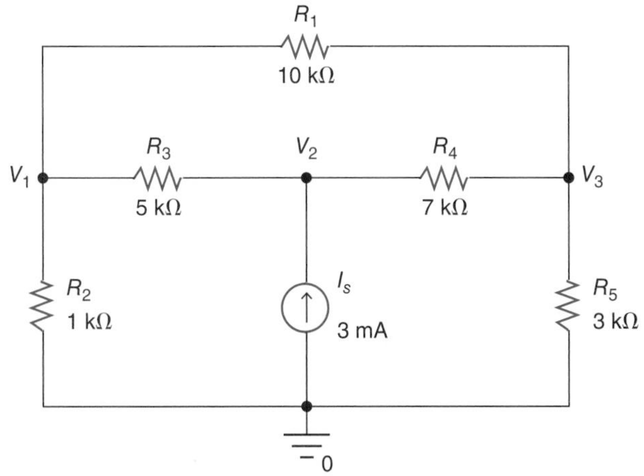

3.19 Find the voltages V_1, V_2, and V_3 in the circuit shown in Figure P3.19.

FIGURE P3.19

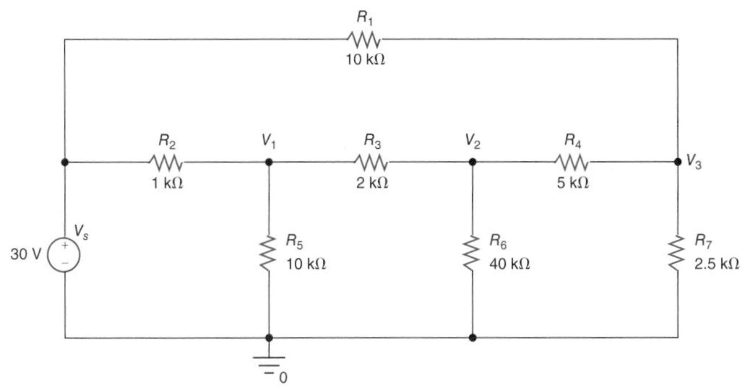

3.20 Find the voltages V_1, V_2, and V_3 in the circuit shown in Figure P3.20.

FIGURE P3.20

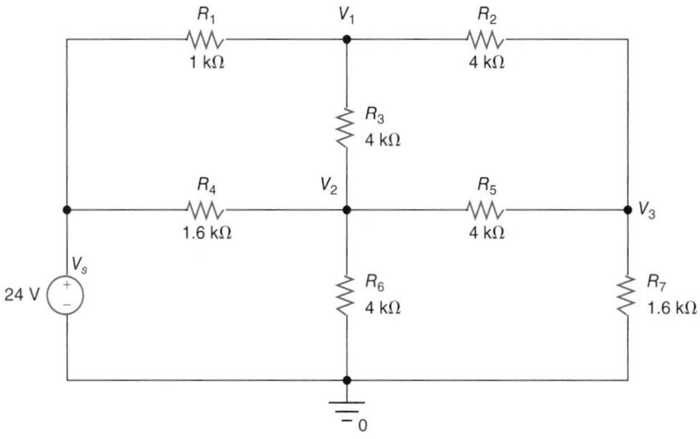

3.21 Find the voltages V_1, V_2, and V_3 in the circuit shown in Figure P3.21.

FIGURE P3.21

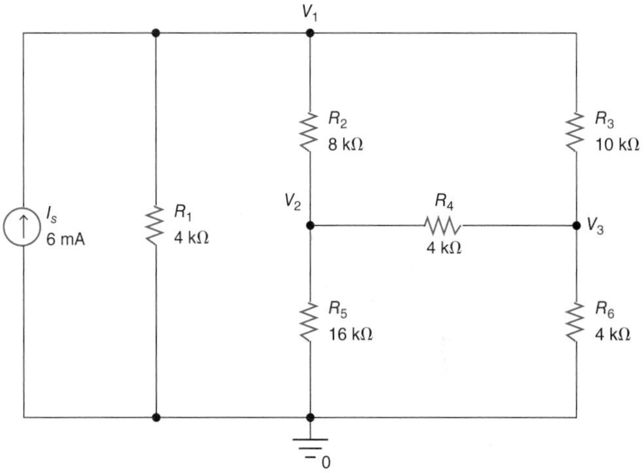

3.22 Find V_1, V_2, V_3, and I in the circuit shown in Figure P3.22.

FIGURE P3.22

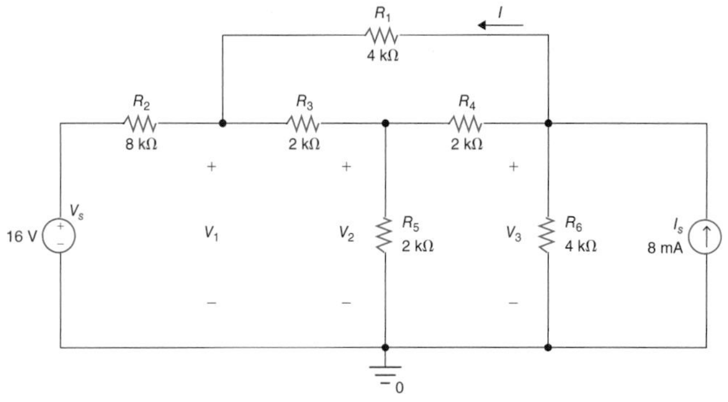

3.23 Find V_1 in the circuit shown in Figure P3.23.

FIGURE P3.23

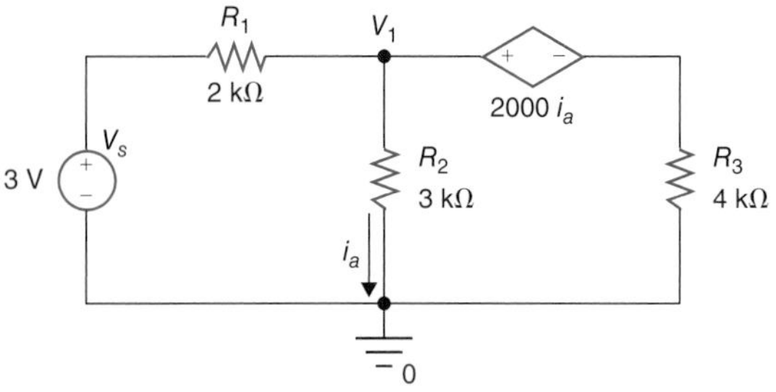

3.24 Find V_1 and V_2 in the circuit shown in Figure P3.24.

FIGURE P3.24

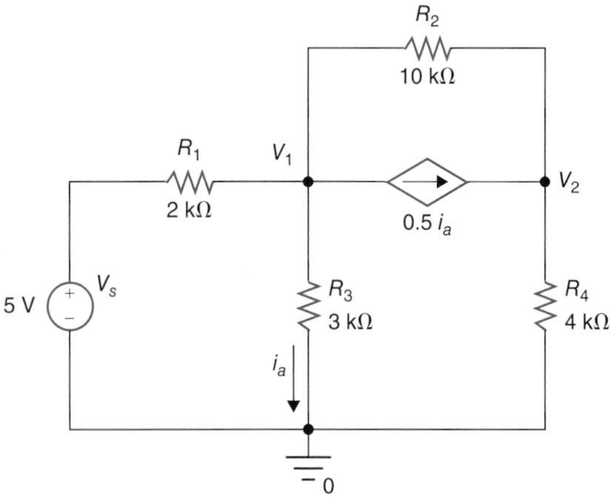

3.25 Find V_1 and V_2 in the circuit shown in Figure P3.25.

FIGURE P3.25

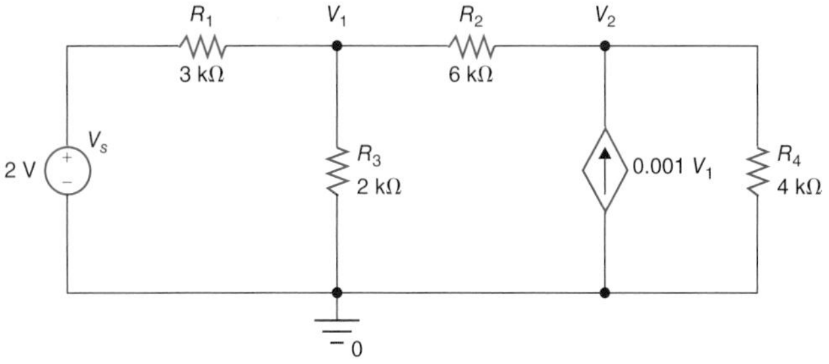

3.26 Find V_1 and V_2 in the circuit shown in Figure P3.26.

FIGURE P3.26

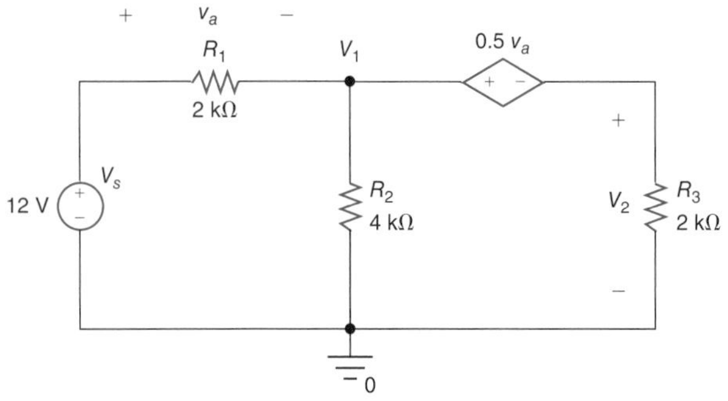

3.27 Find V_1 and V_2 in the circuit shown in Figure P3.27.

FIGURE P3.27

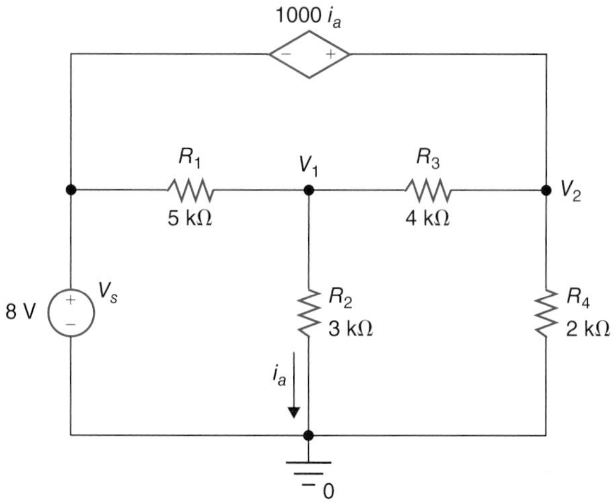

3.28 Find V_1 and V_2 in the circuit shown in Figure P3.28.

FIGURE P3.28

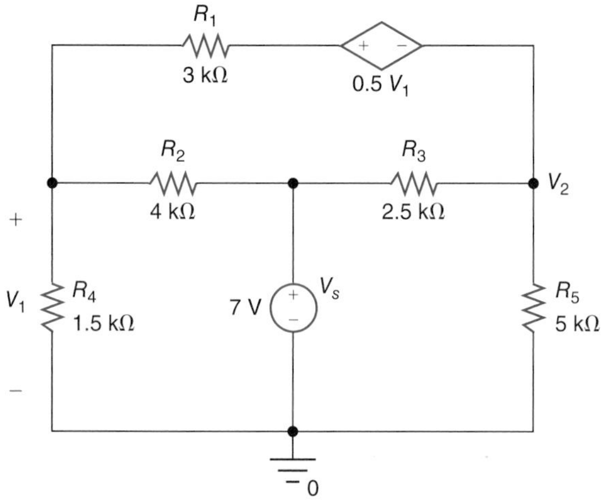

3.29 Find V_1 and V_2 in the circuit shown in Figure P3.29.

FIGURE P3.29

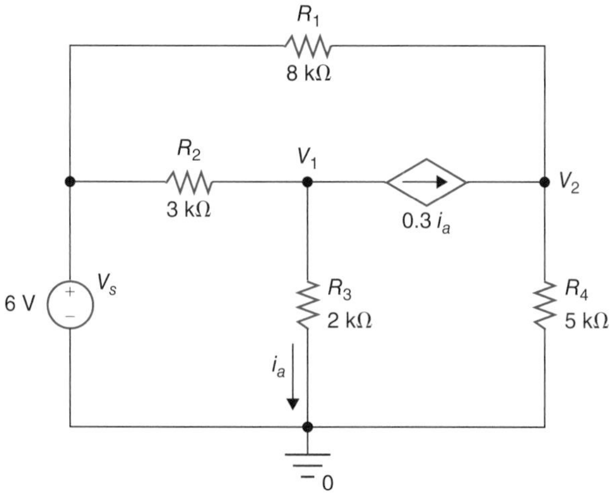

3.30 Find V_1 and V_2 in the circuit shown in Figure P3.30.

FIGURE P3.30

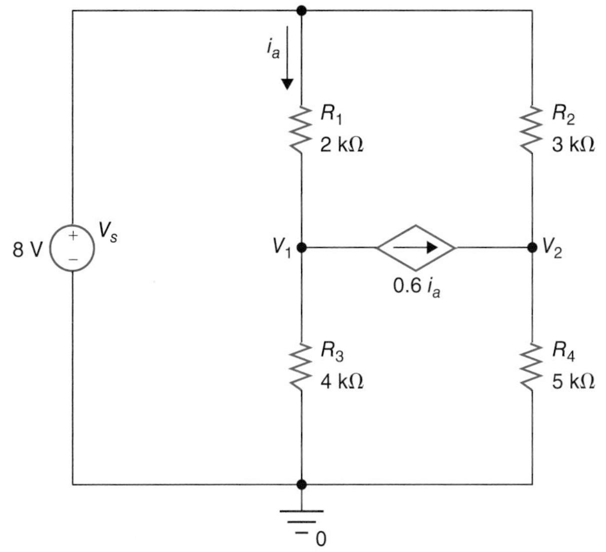

3.31 Find V_1 and V_2 in the circuit shown in Figure P3.31.

FIGURE P3.31

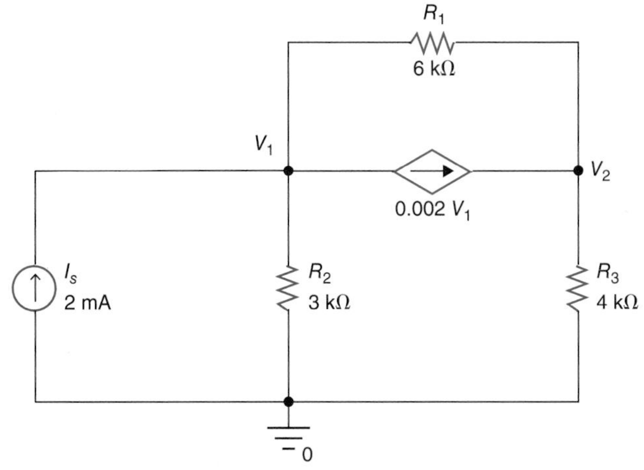

3.32 Find V_1, V_2, and I_1 in the circuit shown in Figure P3.32.

FIGURE P3.32

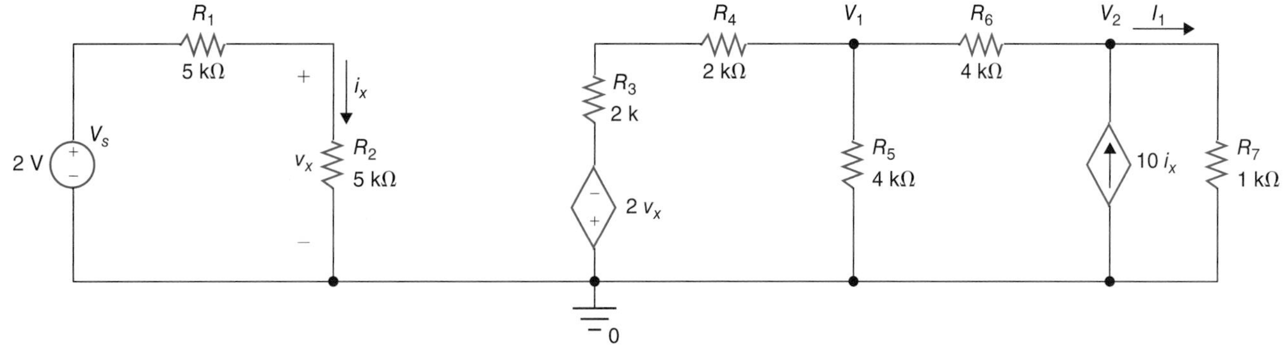

Supernode

3.33 Find V_1 and V_2 in the circuit shown in Figure P3.33.

FIGURE P3.33

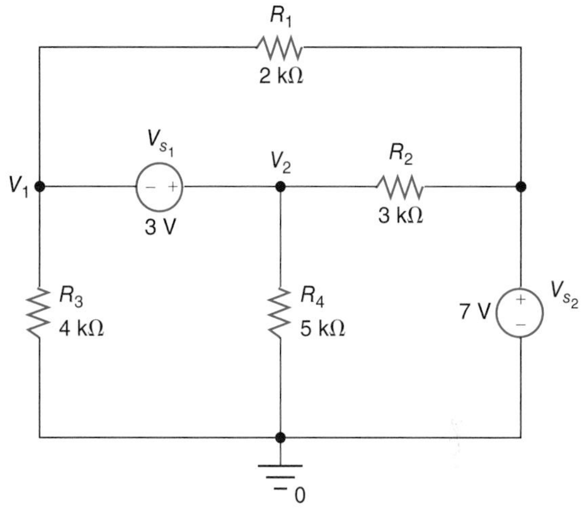

3.34 Find V_1 and V_2 in the circuit shown in Figure P3.34.

FIGURE P3.34

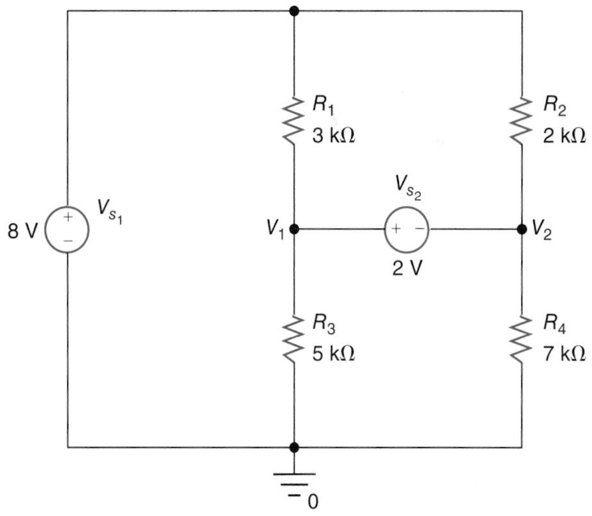

3.35 Find V_1 and V_2 in the circuit shown in Figure P3.35.

FIGURE P3.35

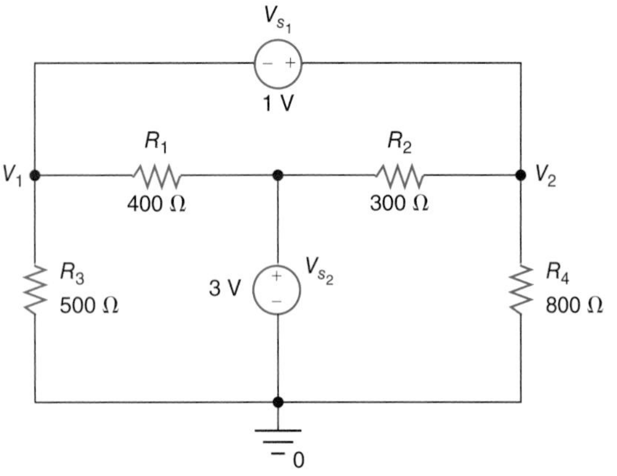

3.36 Find V_1 and V_2 in the circuit shown in Figure P3.36.

FIGURE P3.36

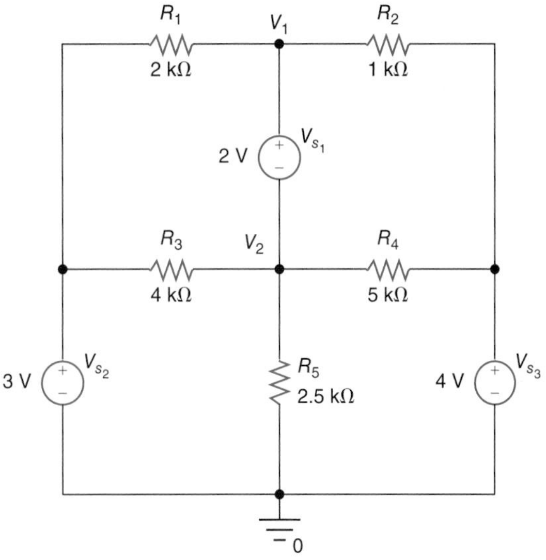

3.37 Find the voltages V_1, V_2, and V_3 in the circuit shown in Figure P3.37.

FIGURE P3.37

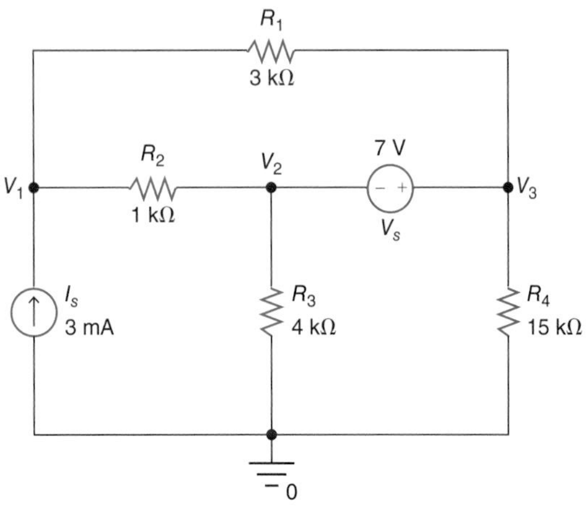

3.38 Find V_1, V_2, and V_3 in the circuit shown in Figure P3.38.

FIGURE P3.38

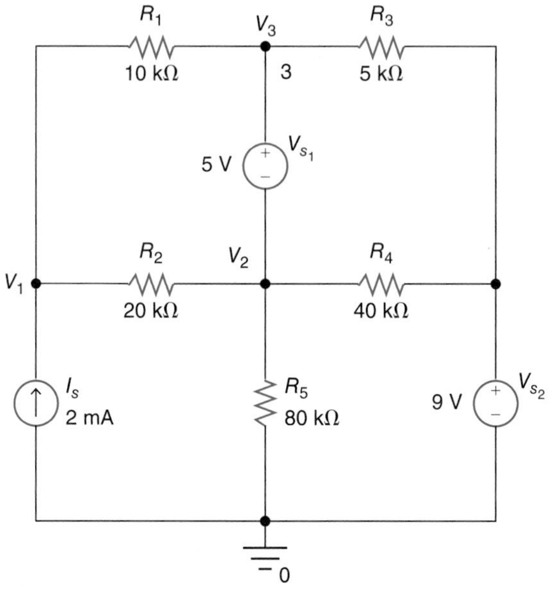

3.39 Find the voltages V_1, V_2, V_3, and V_4 in the circuit shown in Figure P3.39.

FIGURE P3.39

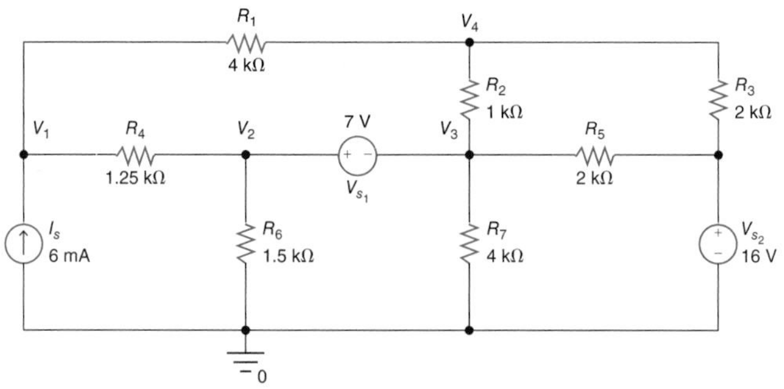

3.40 Find V_1 and V_2 in the circuit shown in Figure P3.40.

FIGURE P3.40

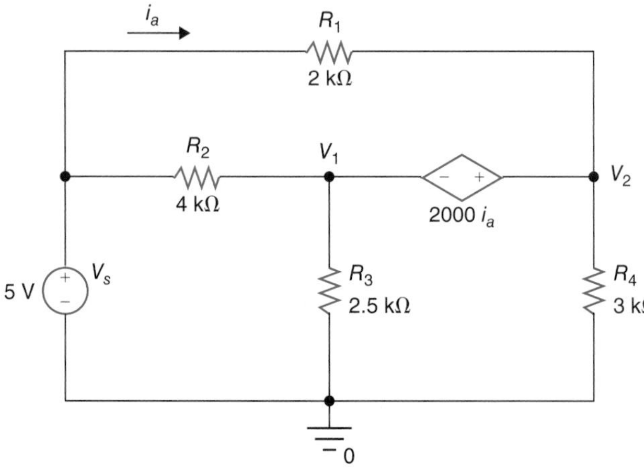

3.41 Find V_1 and V_2 in the circuit shown in Figure P3.41.

FIGURE P3.41

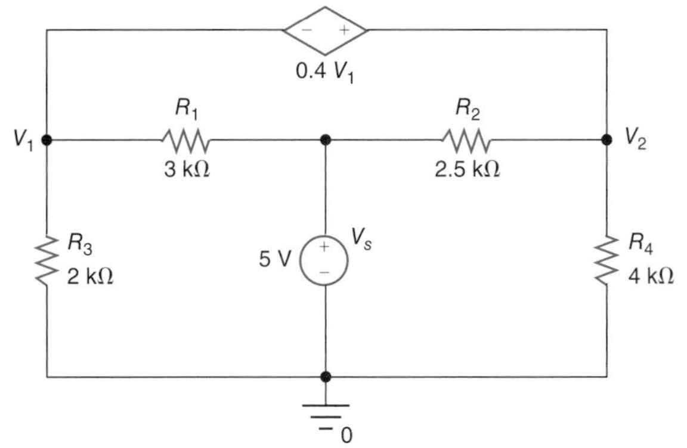

3.42 Find the voltages V_1, V_2, and V_3 in the circuit shown in Figure P3.42.

FIGURE P3.42

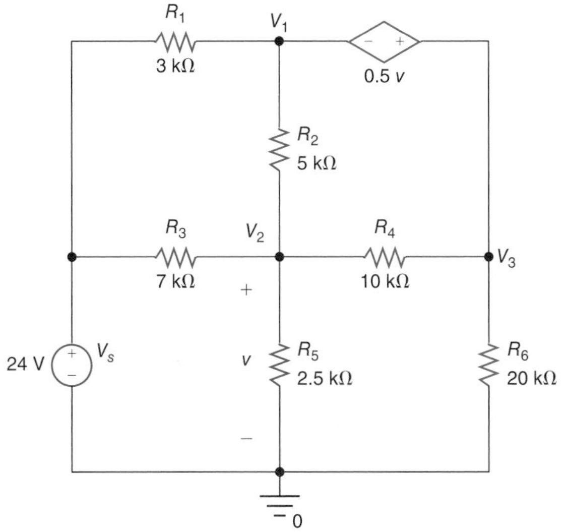

3.43 Find the voltages V_1, V_2, and V_3 in the circuit shown in Figure P3.43.

FIGURE P3.43

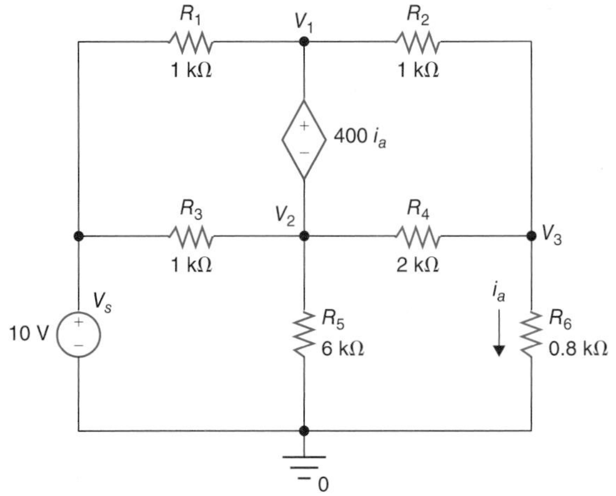

3.44 Find the voltages V_1, V_2, and V_3 in the circuit shown in Figure P3.44.

FIGURE P3.44

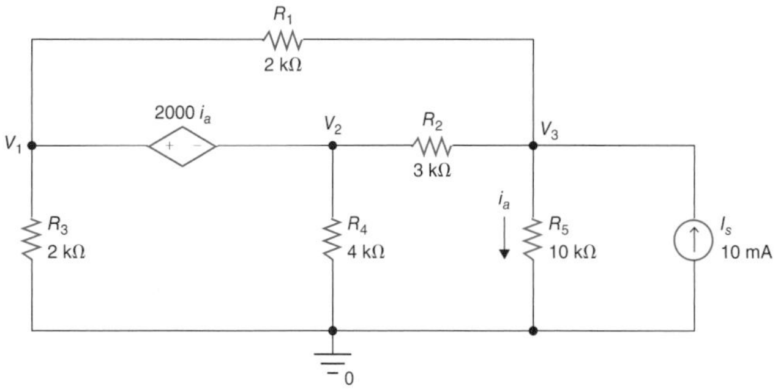

3.45 Find the voltages V_1, V_2, and V_3 in the circuit shown in Figure P3.45.

FIGURE P3.45

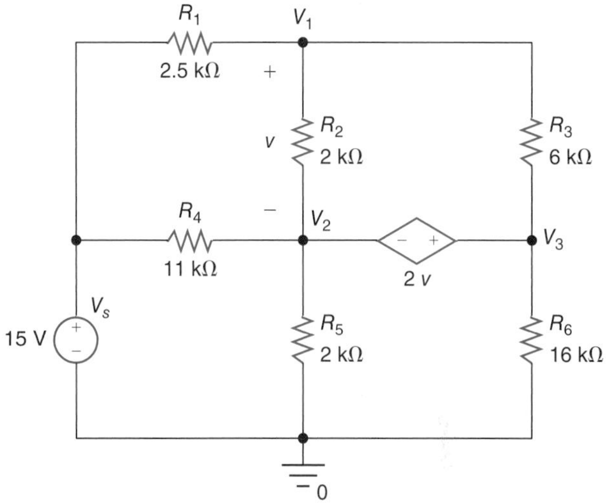

3.46 Find the voltages V_1, V_2, V_3, and V_4 in the circuit shown in Figure P3.46.

FIGURE P3.46

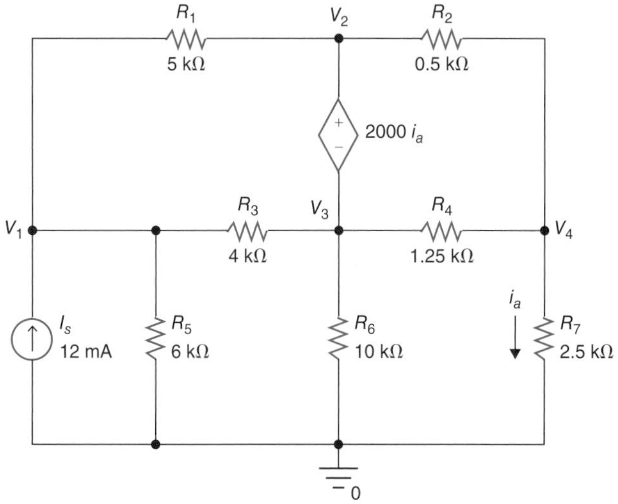

3.47 Find V_1 and V_2 in the circuit shown in Figure P3.47.

FIGURE P3.47

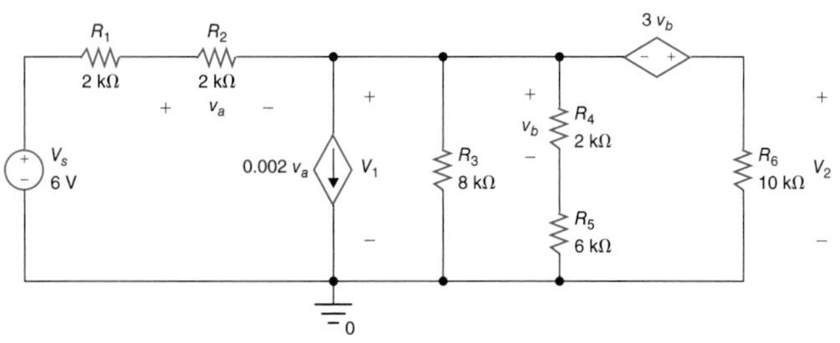

3.48 Find V_1, V_2, and V_3 in the circuit shown in Figure P3.48.

FIGURE P3.48

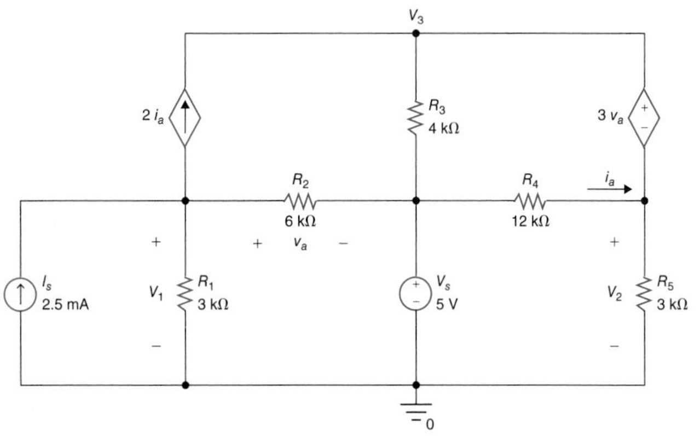

3.49 Find V_1 and V_2 in the circuit shown in Figure P3.49.

FIGURE P3.49

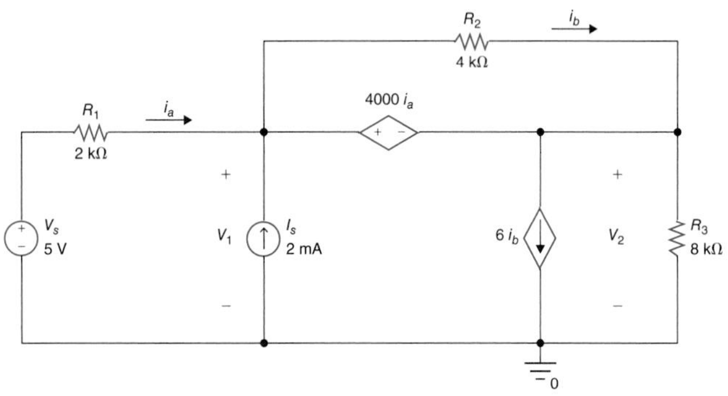

Mesh Analysis

3.50 Find mesh currents I_1 and I_2 and node voltages V_1 and V_2 in the circuit shown in Figure P3.50.

FIGURE P3.50

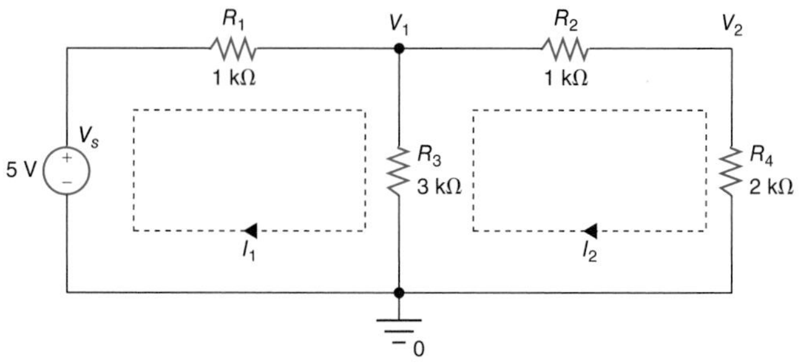

3.51 Find mesh currents I_1 and I_2 and node voltage V_1 in the circuit shown in Figure P3.51.

FIGURE P3.51

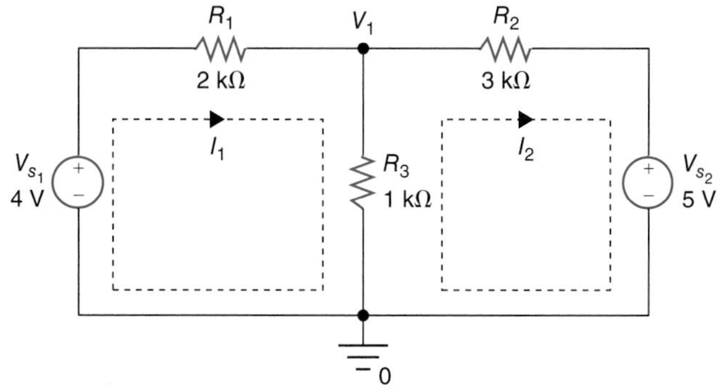

3.52 Find mesh currents I_1, I_2, and I_3 and node voltages V_1, V_2, and V_3 in the circuit shown in Figure P3.52.

FIGURE P3.52

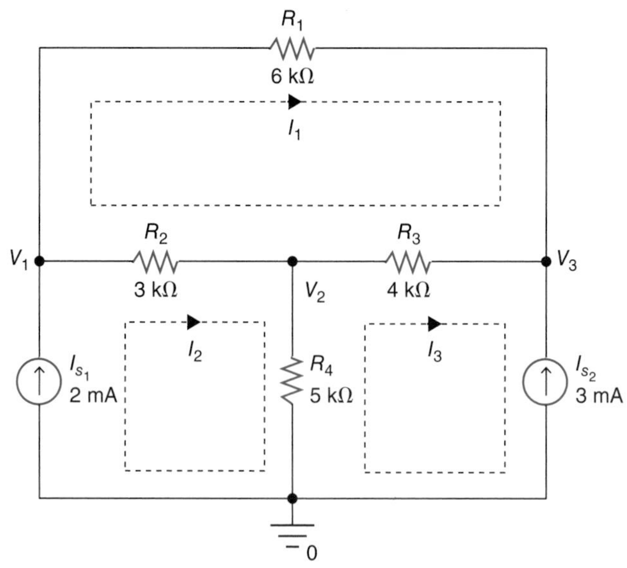

3.53 **Find mesh currents I_1, I_2, and I_3 and node voltages V_1, V_2, and V_3 in the circuit shown in Figure P3.53.**

FIGURE P3.53

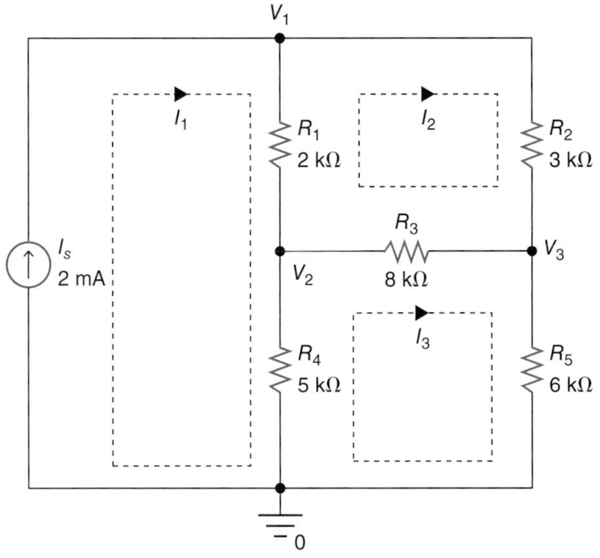

3.54 **Find mesh currents I_1, I_2, and I_3 and node voltages V_1, V_2, and V_3 in the circuit shown in Figure P3.54.**

FIGURE P3.54

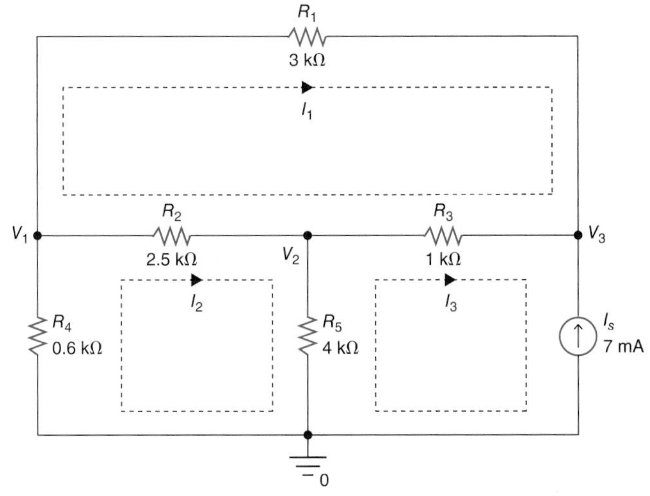

3.55 **Find mesh currents I_1, I_2, and I_3 and node voltages V_1 and V_2 in the circuit shown in Figure P3.55.**

FIGURE P3.55

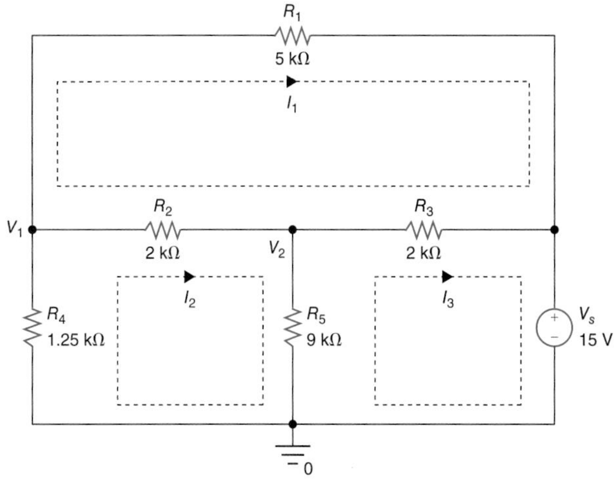

3.56 **Find mesh currents I_1, I_2, I_3, and I_4 and node voltages V_1, V_2, and V_3 in the circuit shown in Figure P3.56.**

FIGURE P3.56

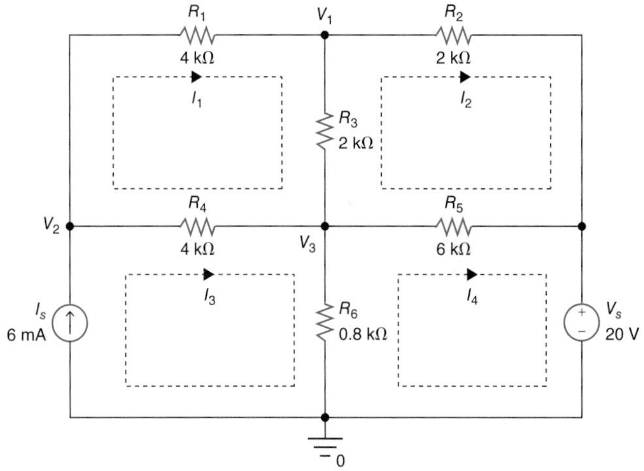

3.57 **Find mesh currents I_1, I_2, and I_3 and node voltages V_1, V_2, and V_3 in the circuit shown in Figure P3.57.**

FIGURE P3.57

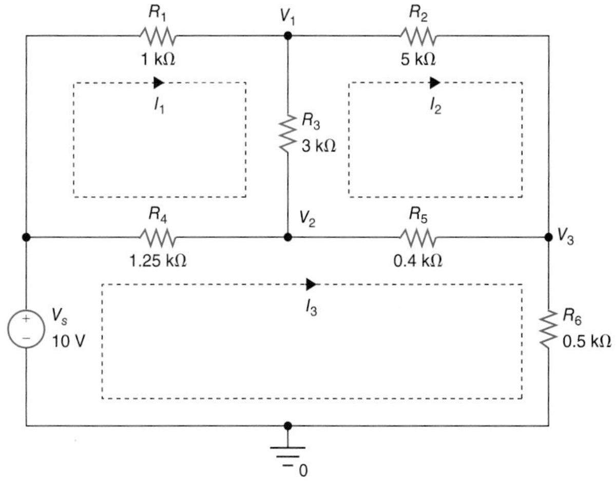

3.58 Use mesh analysis to find the current I in the circuit shown in Figure P3.58.

FIGURE P3.58

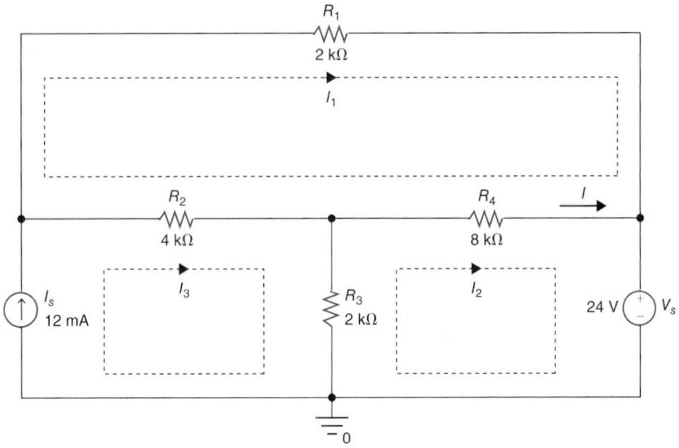

3.59 Find mesh currents I_1 and I_2 and node voltages V_1 and V_2 in the circuit shown in Figure P3.59.

FIGURE P3.59

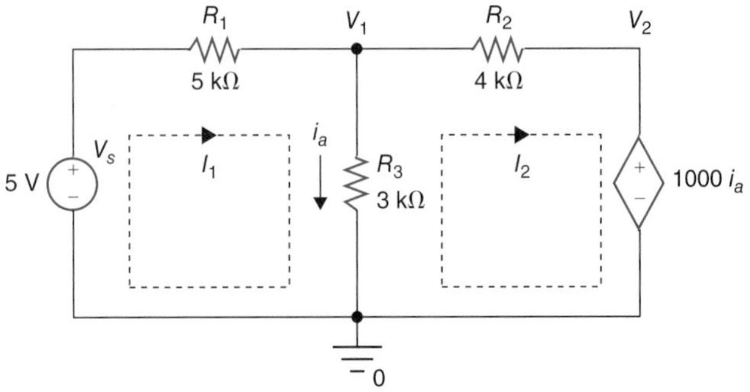

3.60 Find mesh currents I_1 and I_2 and node voltages V_1 and V_2 in the circuit shown in Figure P3.60.

FIGURE P3.60

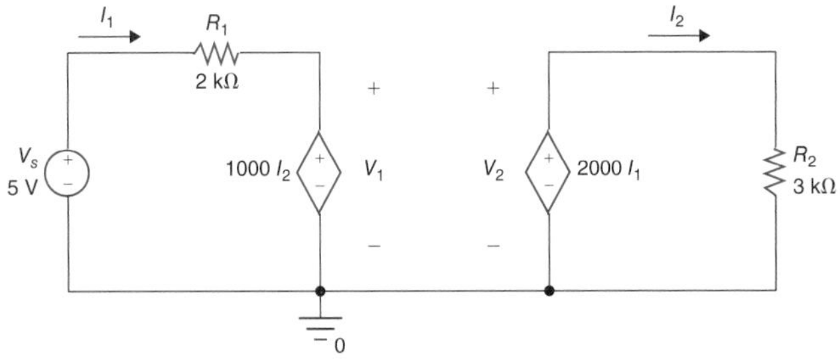

3.61 Find mesh currents I_1 and I_2 and node voltages V_1 and V_2 in the circuit shown in Figure P3.61.

FIGURE P3.61

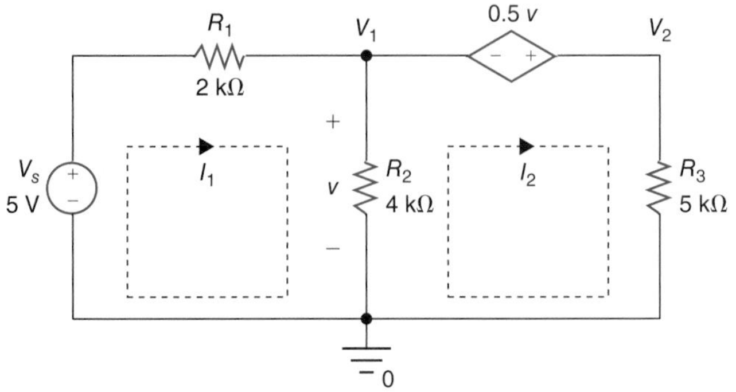

3.62 Use mesh analysis to find V_a in the circuit shown in Figure P3.62.

FIGURE P3.62

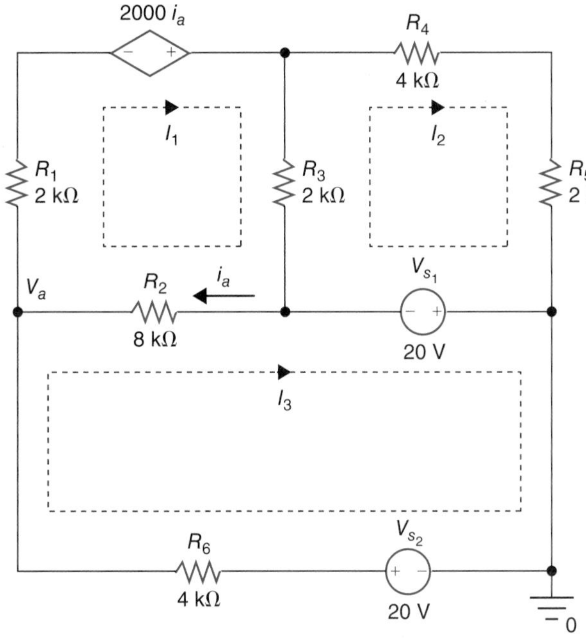

3.63 Use mesh analysis to find v_o in the circuit shown in Figure P3.63.

FIGURE P3.63

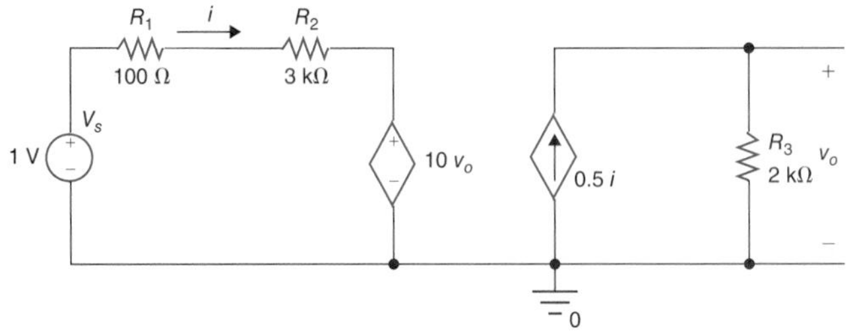

Supermesh

3.64 Find mesh currents I_1, I_2, and I_3, and node voltages V_1 and V_2 in the circuit shown in Figure P3.64.

FIGURE P3.64

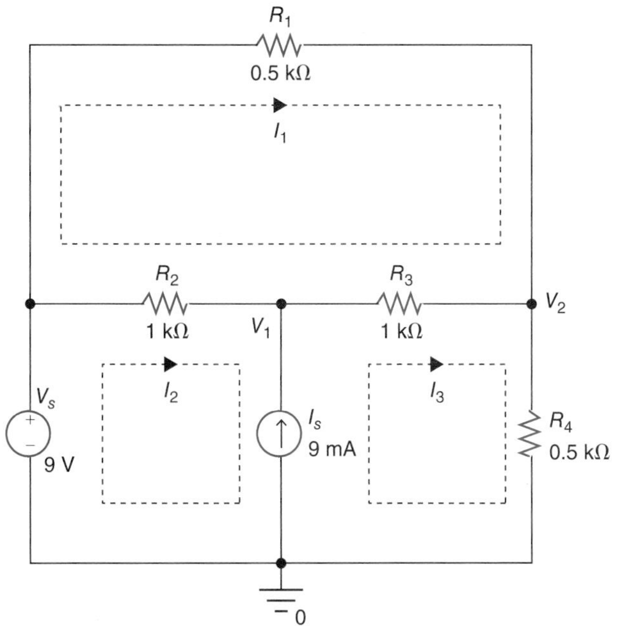

3.65 Find the mesh currents I_1, I_2, I_3, and I_4, and node voltages V_1, V_2, and V_3 in the circuit shown in Figure P3.65.

FIGURE P3.65

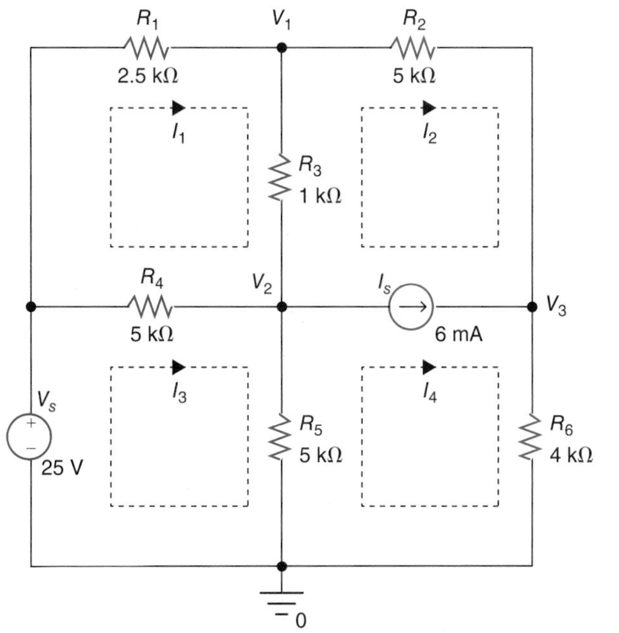

3.66 Use mesh analysis to find the voltage V in the circuit shown in Figure P3.66.

FIGURE P3.66

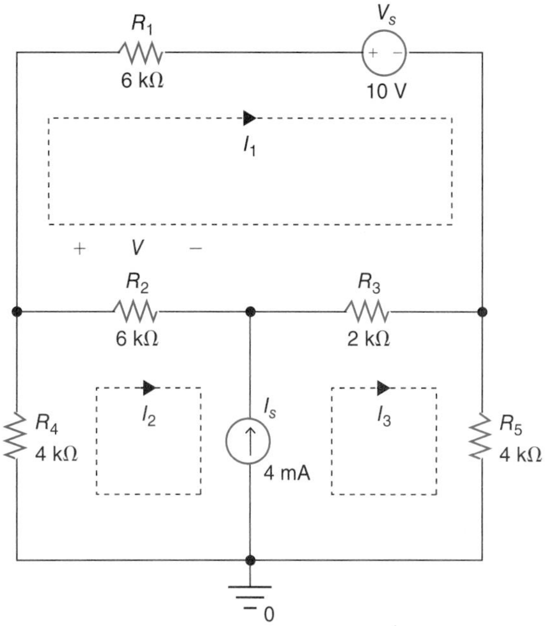

3.67 Find mesh currents I_1, I_2, I_3, and I_4, and node voltages V_1, V_2, V_3, and V_4 in the circuit shown in Figure P3.67.

FIGURE P3.67

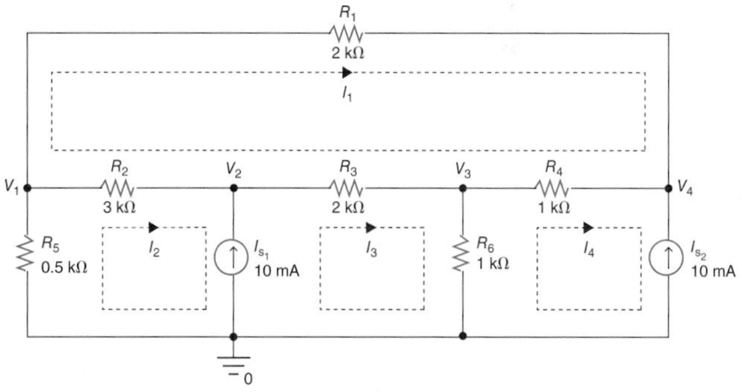

3.68 Find mesh currents I_1, I_2, and I_3, and node voltages V_1 and V_2 in the circuit shown in Figure P3.68.

FIGURE P3.68

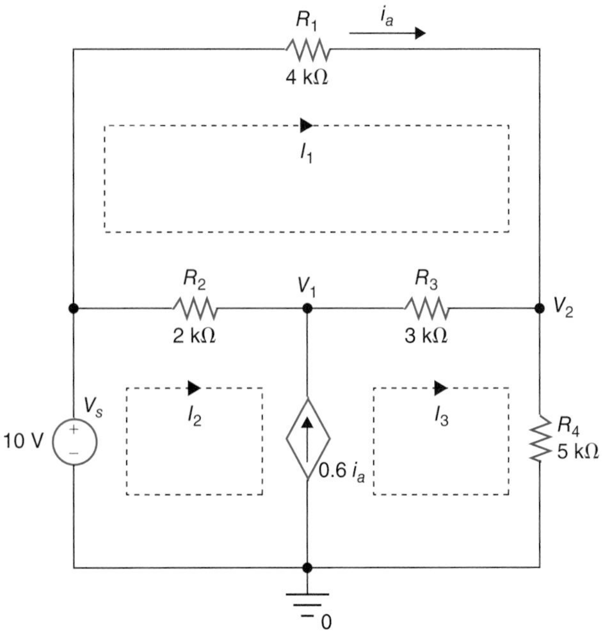

3.69 **Find mesh currents I_1, I_2, and I_3, and node voltages V_1 and V_2 in the circuit shown in Figure P3.69.**

FIGURE P3.69

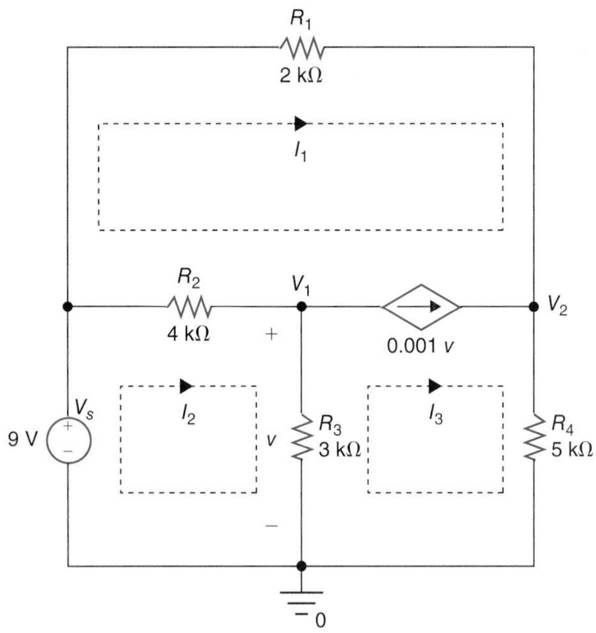

3.70 **Find mesh currents I_1, I_2, and I_3, and node voltages V_1 and V_2 in the circuit shown in Figure P3.70.**

FIGURE P3.70

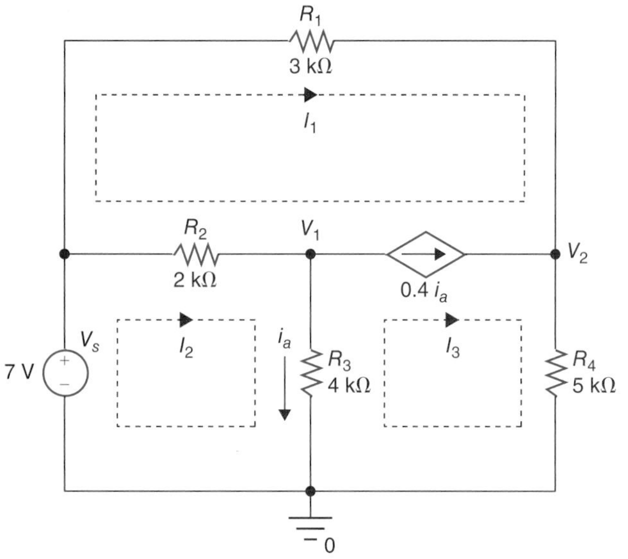

3.71 **Find mesh currents I_1, I_2, I_3, and I_4, and node voltages V_1, V_2, and V_3 in the circuit shown in Figure P3.71.**

FIGURE P3.71

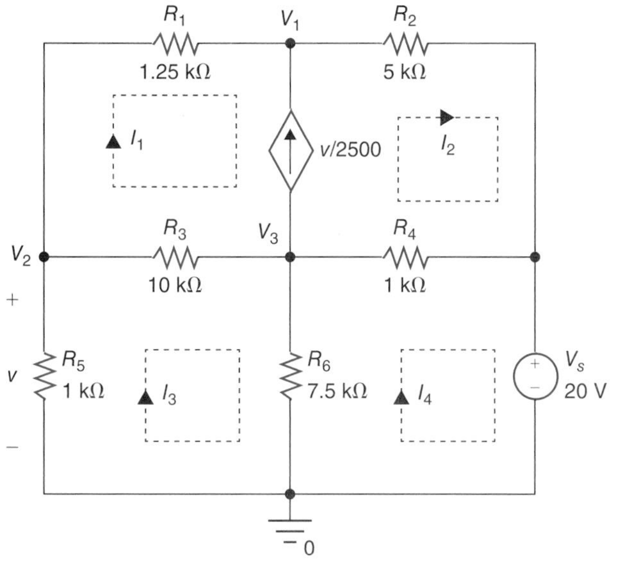

3.72 **Find mesh currents I_1, I_2, I_3, and I_4, and node voltages V_1, V_2, and V_3 in the circuit shown in Figure P3.72.**

FIGURE P3.72

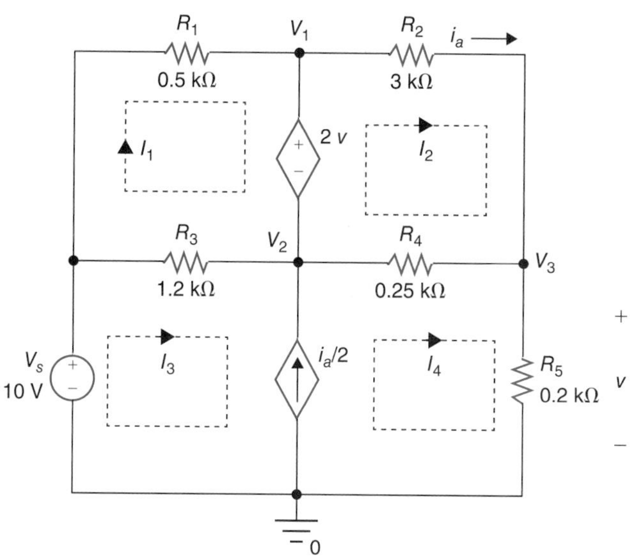

3.73 **Find mesh currents I_1, I_2, I_3, and I_4, and voltages V_1, V_2, and V_3 in the circuit shown in Figure P3.73.**

FIGURE P3.73

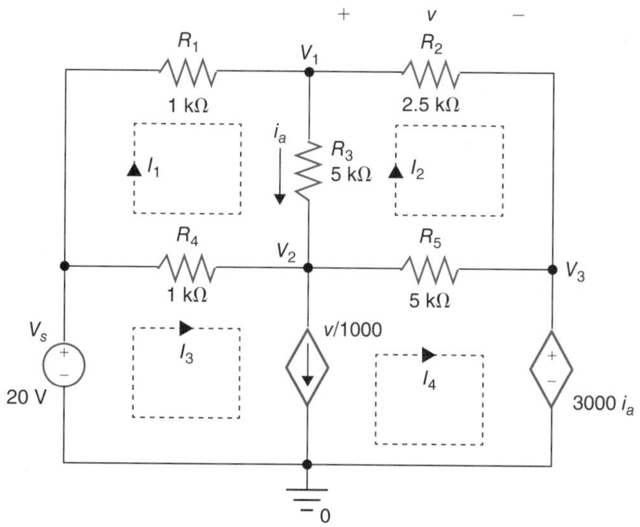

3.74 **Use mesh analysis to find voltage V_3 across R_3 for the circuit shown in Figure P3.74.**

FIGURE P3.74

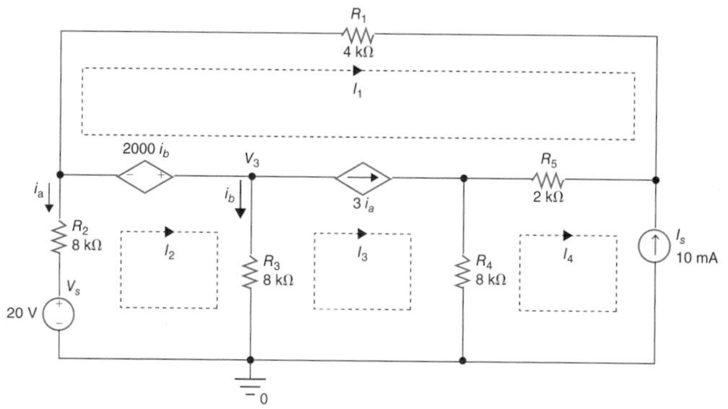

Circuit Theorems

4.1 Introduction

In this chapter, several important theorems related to electric circuits are discussed: superposition principle, source transformation, Thévenin's theorem, Norton's theorem, and maximum power transfer.

If a circuit contains more than one source, the circuit can be analyzed by summing the response from each source with all other sources deactivated. This is called the *superposition principle.* Deactivating a voltage source is equivalent to short-circuiting it, and deactivating a current source is equivalent to open-circuiting it. The superposition principle reveals the contribution of each source to the voltages and currents in the circuit. It makes it easier to interpret the response of the circuit because we can trace the source of the response.

A voltage source with a series resistor is interchangeable with a current source in parallel to a resistor. This is called *source transformation.* The resistance value of both circuits is the same. The source transformation can be used to simplify the given circuit to find the desired voltages and currents.

According to Thévenin's theorem, a given circuit is equivalent to a voltage source V_{th} and a series resistor R_{th} between terminals a and b. The Thévenin equivalent voltage V_{th} can be obtained by finding the open-circuit voltage V_{oc} across a and b without modifying the circuit. The Thévenin equivalent resistance R_{th} can be found in one of the three methods. The first method is to find the equivalent resistance seen from a and b after deactivating the independent sources. The first method can be used only when a circuit does not contain dependent sources. The second method is to find the open-circuit voltage V_{oc} across a and b without modifying the circuit and to find the short-circuit current I_{sc} from a and b after connecting a and b by wire (a short-circuit). The Thévenin equivalent resistance R_{th} is the ratio of V_{oc} to I_{sc}; i.e., $R_{th} = V_{oc}/I_{sc}$. The third method is to apply a test voltage between terminals a and b after deactivating the independent sources and measure the current flowing out of the positive terminal of the test voltage source. The Thévenin equivalent resistance R_{th} is the ratio of the test voltage to the current flowing out of the positive terminal of the test voltage source. A test current source can be used instead of a test voltage source.

According to Norton's theorem, a given circuit is equivalent to a current source I_n and a parallel resistor R_n between terminals a and b. The Norton equivalent current I_n can be obtained by finding the short-circuit current I_{sc} from a to b after connecting a and b by wire. The Norton equivalent resistance R_n can be found in one of the three methods. The first method is to find the equivalent resistance seen from terminals a and b after deactivating the independent sources. The first method can be used only when a circuit does not contain dependent sources. The second method is to measure the open-circuit voltage V_{oc} between a and b without modifying the circuit. The Norton equivalent resistance R_n is the ratio of V_{oc} to I_{sc}; i.e., $R_n = V_{oc}/I_{sc}$. The third method is to apply a test voltage between terminals a and b after deactivating the independent sources and measure the current flowing out of the positive terminal of the test voltage source. The Norton equivalent resistance R_n is the ratio of the test voltage to the current flowing out of the positive terminal of the test voltage source. A test current source can be used instead of a test voltage source.

Suppose that a load with resistance R_L is connected to a circuit between terminals a and b. We can find the Thévenin equivalent circuit with respect to the terminals a and b excluding the load resistance. Let V_{th} be the Thévenin equivalent voltage and R_{th} be the Thévenin equivalent resistance. It can be shown that the load resistance R_L that maximizes the power delivered to the load is given by the Thévenin equivalent resistance R_{th}.

4.2 Superposition Principle

Suppose that a circuit has N independent sources with $N \geq 2$. Create N circuits from the original circuit with only one independent source by deactivating the other $N - 1$ independent sources. Deactivating a current source is to open-circuit it and deactivating a voltage source is to short-circuit it. The unknown voltages and currents of the original circuit can be found by adding the voltages and currents from the N circuits with one independent source. This is the superposition principle. The circuit shown in Figure 4.1 contains one voltage source and one current source. It is desired to find voltage V_1 across R_2, which is also the voltage across the current source, using the superposition principle.

FIGURE 4.1

A circuit with a voltage source and a current source.

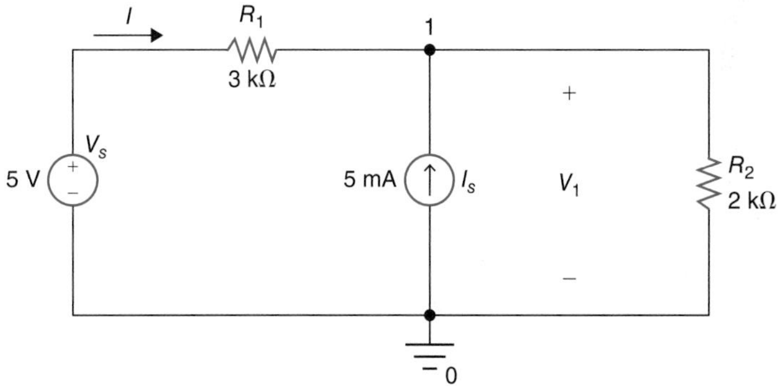

If the current from the current source is reduced to zero (that is, $I_s = 0$), the current source is open-circuited, and the circuit reduces to the one shown in Figure 4.2.

FIGURE 4.2

Circuit shown in Figure 4.1 with the current source deactivated.

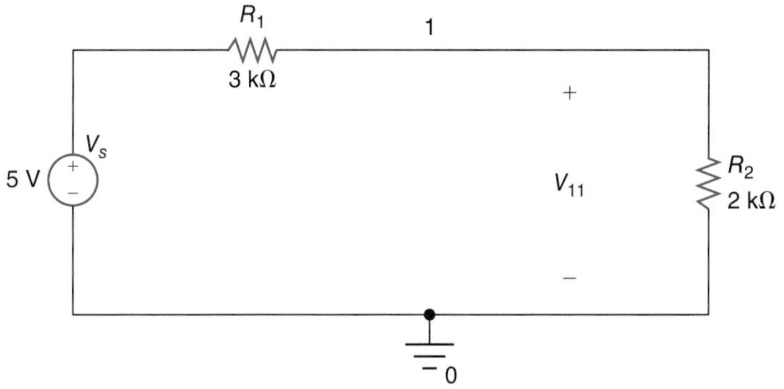

From the voltage divider rule, the voltage at node 1 is given by

$$V_{11} = \frac{R_2}{R_1 + R_2} V_s = \frac{2\,k\Omega}{3\,k\Omega + 2\,k\Omega} \times 5\text{ V} = \frac{2}{5} \times 5\text{ V} = 0.4 \times 5\text{ V} = 2\text{ V}$$

If the voltage from the voltage source is reduced to zero (that is, $V_s = 0$), the voltage source is short-circuited, and the circuit reduces to the one shown in Figure 4.3.

FIGURE 4.3

Circuit shown in Figure 4.1 with the voltage source deactivated.

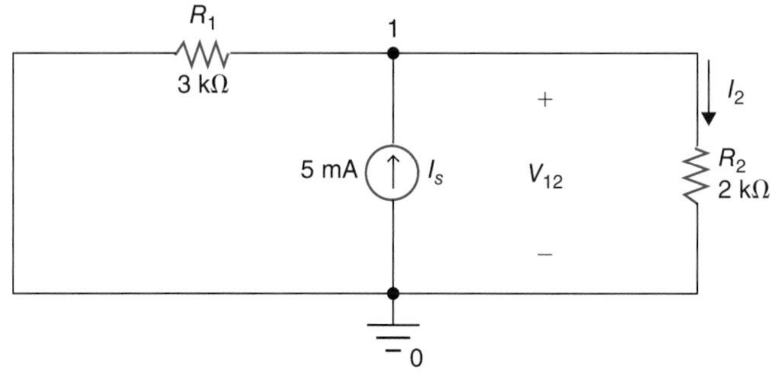

From the current divider rule, the current through R_2 is given by

$$I_2 = \frac{\frac{1}{R_2}}{\frac{1}{R_2} + \frac{1}{R_1}} I_s = \frac{R_1}{R_1 + R_2} I_s = \frac{3000}{3000 + 2000} \times 5\text{ mA} = 3\text{ mA}$$

Thus, the voltage at node 1 is

$$V_{12} = R_2 I_2 = \frac{R_2 R_1}{R_1 + R_2} I_s = \frac{2000 \times 3000}{3000 + 2000} \Omega \times 5 \times 10^{-3}\text{ A} = 1200\,\Omega \times 5 \times 10^{-3}\text{ A} = 6\text{ V}$$

The sum of V_{11} and V_{12} is given by

$$V_1 = V_{11} + V_{12} = \frac{R_2}{R_1 + R_2} V_s + \frac{R_2 R_1}{R_1 + R_2} I_s = 0.4 V_s + 1200 I_s = 2\text{ V} + 6\text{ V} = 8\text{ V} \quad \textbf{(4.1)}$$

As a check, let us find V_1 using nodal analysis. Summing the currents leaving node 1 in the circuit shown in Figure 4.1, we have

$$\frac{V_1 - V_s}{R_1} + \frac{V_1}{R_2} - I_s = 0$$

This equation can be revised as follows:

$$\left(\frac{1}{R_1} + \frac{1}{R_2}\right) V_1 = \frac{V_s}{R_1} + I_s$$

Solving this equation for V_1, we have

$$V_1 = \frac{\frac{V_s}{R_1} + I_s}{\frac{1}{R_1} + \frac{1}{R_2}} = \frac{R_2}{R_1 + R_2} V_s + \frac{R_2 R_1}{R_1 + R_2} I_s = \frac{2000}{3000 + 2000} V_s + \frac{2000 \times 3000}{3000 = 2000} I_s \quad \textbf{(4.2)}$$

$$= 0.4 V_s + 1200 I_s = 0.4 \times 5\text{ V} + 1200 \times 5 \times 10^{-3}\text{ V} = 2\text{ V} + 6\text{ V} = 8\text{ V}$$

Comparison of Equations (4.1) and (4.2) reveals that the answer derived from the superposition principle matches the one from the nodal analysis. This proves that the superposition principle works for the circuit shown in Figure 4.1. In general, if the circuit is linear, the superposition principle holds. Circuits consisting of independent and dependent voltage sources and current sources along with resistors are all linear circuits. Thus, the superposition principle works for these circuits. Equations (4.1) and (4.2) show that the voltage at node 1, V_1, consists of two components. The first component of V_1,

$$V_{11} = \frac{R_2}{R_1 + R_2} V_s$$

is due to the voltage source V_s; and the second component of V_1,

$$V_{12} = \frac{R_2 R_1}{R_1 + R_2} I_s$$

is due to the current source I_s. The voltage at node 1, V_1, is the linear combination of two inputs V_s and I_s. The coefficients a_1 and a_2 in the representation,

$$V_1 = a_1 V_s + a_2 I_s \tag{4.3}$$

are given by

$$a_1 = \frac{R_2}{R_1 + R_2} = 0.4, \quad a_2 = \frac{R_2 R_1}{R_1 + R_2} = 1200\ \Omega \tag{4.4}$$

Equation (4.3) is called linear because the output, V_1, is a linear function of inputs V_s and I_s; that is, V_1 is proportional to V_s and I_s. The proportionality constants are a_1 and a_2. Notice that if I_s is set to zero in Equation (4.3), we obtain V_{11}, and if V_s is set to zero in Equation (4.3), we obtain V_{12}.

Let I be the current from the voltage source, as shown in Figure 4.1. The current I is given by

$$I = \frac{V_s - V_1}{R_1} = \frac{V_s - \dfrac{R_2}{R_1 + R_2} V_s - \dfrac{R_1 R_2}{R_1 + R_2} I_s}{R_1} = \frac{1}{R_1 + R_2} V_s - \frac{R_2}{R_1 + R_2} I_s \tag{4.5}$$

The current I is a linear combination of the two inputs V_s and I_s.

EXAMPLE 4.1

Use the superposition principle to find voltage V_1 in the circuit shown in Figure 4.4.

FIGURE 4.4

Circuit for EXAMPLE 4.1.

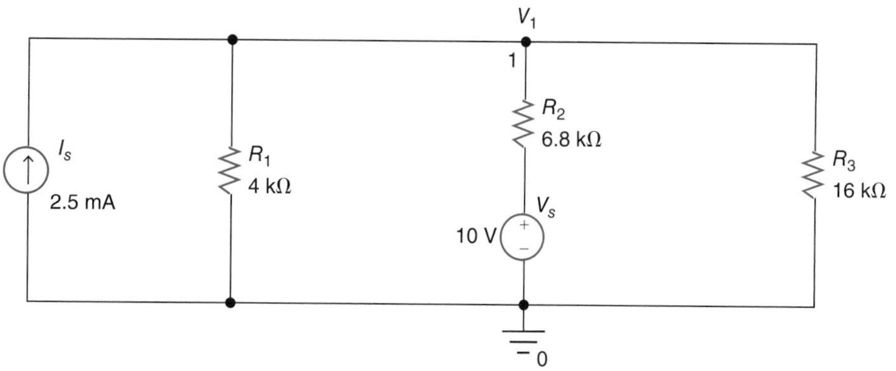

continued

Example 4.1 continued

When the current source is deactivated by open-circuiting it, the circuit shown in Figure 4.4 reduces to the one shown in Figure 4.5.

FIGURE 4.5

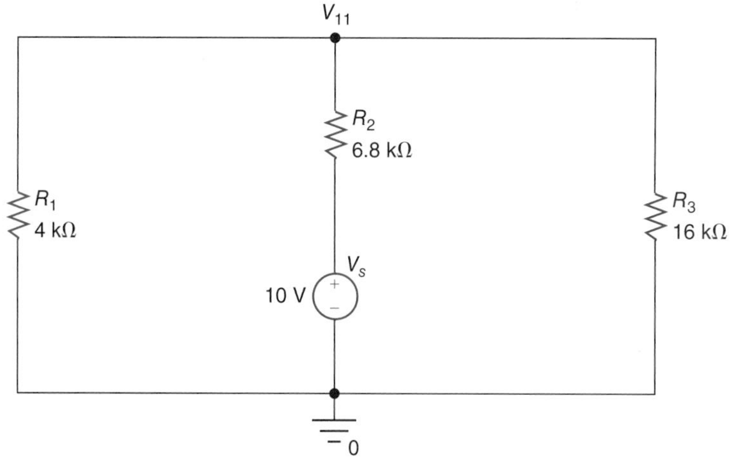

The circuit in Figure 4.4, with the current source deactivated by open-circuiting.

The equivalent resistance of the parallel connection of R_1 and R_3 is given by

$$R_a = R_1 \| R_3 = \frac{R_1 R_3}{R_1 + R_3} = \frac{4\,k\Omega \times 16\,k\Omega}{4\,k\Omega + 16\,k\Omega} = \frac{64}{20}\,k\Omega = 3.2\,k\Omega$$

Application of the voltage divider rule yields

$$V_{11} = \frac{R_a}{R_2 + R_a} V_s = \frac{3.2\,k\Omega}{6.8\,k\Omega + 3.2\,k\Omega} \times 10\text{ V} = \frac{3.2}{10} \times 10\text{ V} = 3.2\text{ V}$$

When the voltage source is deactivated by short-circuiting it, the circuit shown in Figure 4.4 reduces to the one shown in Figure 4.6.

FIGURE 4.6

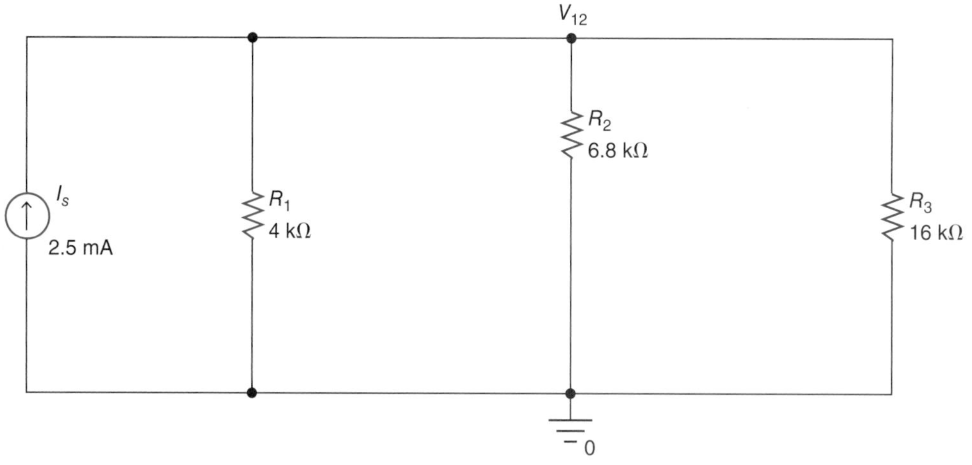

The circuit in Figure 4.4, with the current source deactivated by closed-circuiting.

The equivalent resistance of the parallel connection of R_1, R_3, and R_2 is given by

$$R_b = R_a \| R_2 = \frac{R_a R_2}{R_a + R_2} = \frac{3.2\,k\Omega \times 6.8\,k\Omega}{3.2\,k\Omega + 6.8\,k\Omega} = 2.176\,k\Omega$$

continued

Example 4.1 continued

Voltage V_{12} is the product of R_b and I_s. Thus, we have

$$V_{12} = R_b I_s = 2176\,\Omega \times 2.5 \times 10^{-3}\,\text{A} = 5.44\text{ V}$$

Voltage V_1 is the sum of V_{11} and V_{12}:

$$V_1 = V_{11} + V_{12} = 3.2\text{ V} + 5.44\text{ V} = 8.64\text{ V}$$

As a check, we can find V_1 directly from the circuit shown in Figure 4.4 by applying nodal analysis. Summing the currents leaving node 1, we obtain

$$-2.5 \times 10^{-3} + \frac{V_1}{4000} + \frac{V_1 - 10}{6800} + \frac{V_1}{16{,}000} = 0$$

which can be rearranged as

$$\left(\frac{1}{4000} + \frac{1}{6800} + \frac{1}{16{,}000}\right)V_1 = 2.5 \times 10^{-3} + \frac{10}{6800}$$

Solving for V_1, we obtain

$$V_1 = \frac{2.5 \times 10^{-3} + \dfrac{10}{6800}}{\dfrac{1}{4000} + \dfrac{1}{6800} + \dfrac{1}{16{,}000}} = \frac{3.9705882353 \times 10^{-3}}{4.5955882353 \times 10^{-4}} = 8.64\text{ V}$$

This value is the same as the one obtained from the superposition principle.

Exercise 4.1

Use the superposition principle to find voltage V_1 in the circuit shown in Figure 4.7.

FIGURE 4.7

Circuit for EXERCISE 4.1.

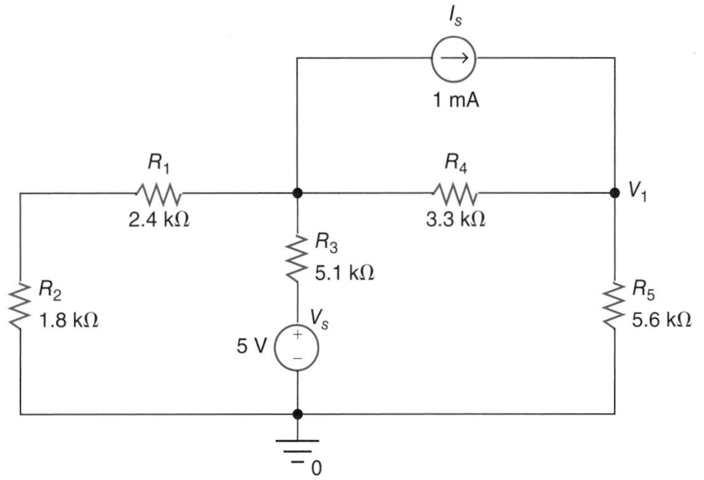

Answer:

$V_1 = 2.7782$ V, 1.1287 V from V_s, 1.6495 V from I_s.

EXAMPLE 4.2

Use the superposition principle to find the voltage across R_2 in the circuit shown in Figure 4.8.

FIGURE 4.8

Circuit for EXAMPLE 4.2.

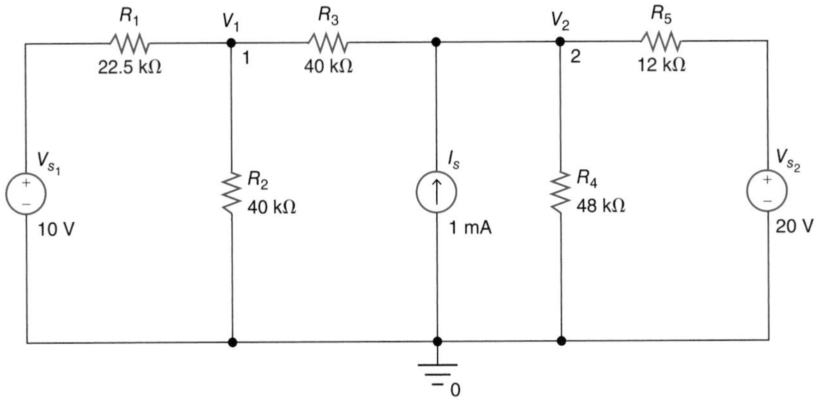

Deactivating the current source from the circuit shown in Figure 4.8 by removing it and also deactivating the voltage source V_{s_2} by short-circuiting it, we obtain the circuit shown in Figure 4.9.

FIGURE 4.9

The circuit shown in Figure 4.8 with I_s and V_{s_2} deactivated.

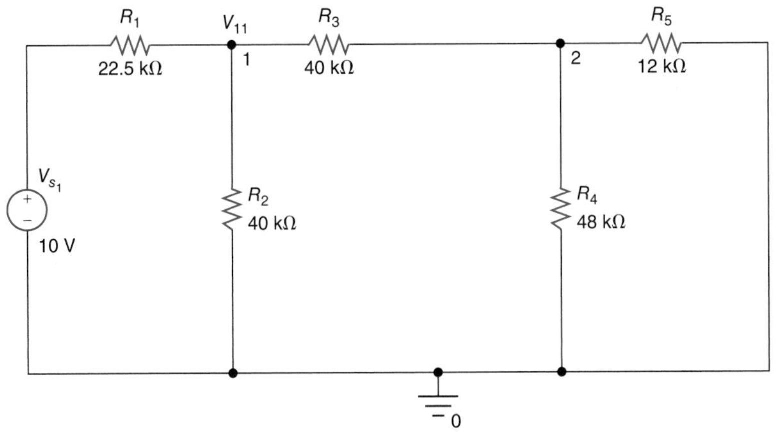

Let R_a be the equivalent resistance of R_2, R_3, R_4, and R_5 seen from R_1. Then, we have

$$R_a = R_2 \| (R_3 + (R_4 \| R_5)) = 40{,}000 \| \left(40{,}000 + \frac{48{,}000 \times 12{,}000}{48{,}000 + 12{,}000}\right)$$

$$= 40{,}000 \| (40{,}000 + 9600) = 40{,}000 \| 49{,}600 = \frac{40{,}000 \times 49{,}600}{40{,}000 + 49{,}600}$$

$$= \frac{1.984 \times 10^9}{89{,}600} = 22.142857\,k\Omega$$

Applying the voltage divider rule, we get voltage V_{11} at node 1:

$$V_{11} = V_{s_1} \times \frac{R_a}{R_1 + R_a} = 10\text{ V} \times \frac{22.142857\ k\Omega}{22.5\ k\Omega + 22.142857\ k\Omega} = 4.96\text{ V} \quad \mathbf{(4.6)}$$

continued

Example 4.2 continued

Deactivating the voltage sources V_{s_1} and V_{s_2} from the circuit shown in Figure 4.8 by short-circuiting both of them, we obtain the circuit shown in Figure 4.10.

FIGURE 4.10

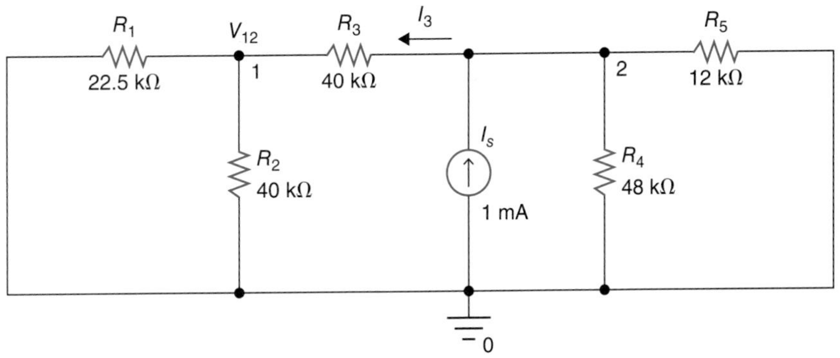

The circuit shown in Figure 4.8 with voltage sources V_{s_1} and V_{s_2} deactivated.

Let R_b be the equivalent resistance of the parallel connection of R_1 and R_2. Then, we have

$$R_b = \frac{R_1 \times R_2}{R_1 + R_2} = \frac{22.5\ k\Omega \times 40\ k\Omega}{22.5\ k\Omega + 40\ k\Omega} = \frac{900}{62.5}\ k\Omega = 14.4\ k\Omega$$

Let R_c be the equivalent resistance of the series connection of R_3 and R_b. Then, we have

$$R_c = R_3 + R_b = 40\ k\Omega + 14.4\ k\Omega = 54.4\ k\Omega$$

Let R_d be the equivalent resistance of the parallel connection of R_4 and R_5. Then, we have

$$R_d = \frac{R_4 \times R_5}{R_4 + R_5} = \frac{48\ k\Omega \times 12\ k\Omega}{48\ k\Omega + 12\ k\Omega} = \frac{576}{60}\ k\Omega = 9.6\ k\Omega$$

From the current divider rule, the current through R_c is given by

$$I_3 = I_s \times \frac{R_d}{R_c + R_d} = 1\ \text{mA} \times \frac{9.6\ k\Omega}{54.4\ k\Omega + 9.6\ k\Omega} = 0.15\ \text{mA}$$

Voltage V_{12} across R_b is

$$V_{12} = I_3 \times R_b = 0.15 \times 10^{-3} \times 14{,}400\ \text{V} = 2.16\ \text{V} \quad \textbf{(4.7)}$$

Deactivating the current source from the circuit shown in Figure 4.8 by removing it and also deactivating the voltage source V_{s_1} by short-circuiting it, we obtain the circuit shown in Figure 4.11.

Let R_e be the equivalent resistance of the parallel connection of R_c and R_4. Then, we have

$$R_e = R_c \| R_4 = 54{,}400 \| 48{,}000 = \frac{54{,}400 \times 48{,}000}{54{,}400 + 48{,}000} = 25.5\ k\Omega$$

Applying the voltage divider rule, we get voltage V_{23} at node 2:

$$V_{23} = V_{s_2} \times \frac{R_e}{R_5 + R_e} = 20\ \text{V} \times \frac{25.5\ k\Omega}{25.5\ k\Omega + 12\ k\Omega} = 13.6\ \text{V}$$

continued

Example 4.2 continued

FIGURE 4.11

The circuit shown in Figure 4.8 with I_s and V_{s_1} deactivated.

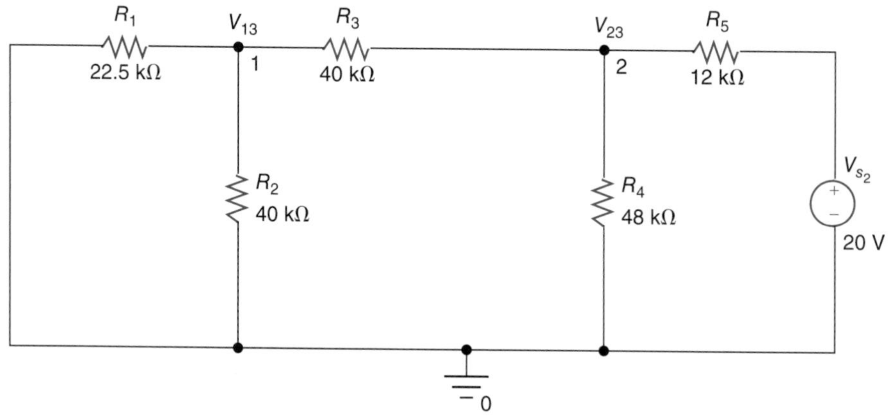

Applying the voltage divider rule, we get voltage V_{13} at node 1:

$$V_{13} = V_{23} \times \frac{R_b}{R_c} = 13.6 \text{ V} \times \frac{14.4\ k\Omega}{54.4\ k\Omega} = 3.6 \text{ V} \quad \textbf{(4.8)}$$

Adding the three voltages, we obtain voltage V_1:

$$V_1 = V_{11} + V_{12} + V_{13} = 4.96 \text{ V} + 2.16 \text{ V} + 3.6 \text{ V} = 10.72 \text{ V}$$

To verify this answer, we can find voltage V_1 across R_2 directly from the original circuit shown in Figure 4.8 using nodal analysis, as shown in the MATLAB script given here:

MATLAB

```
%EXAMPLE 4.2
%Function P.m should be in the same folder as this file.
clear all;format long;
R1=22500;R2=40000;R3=40000;R4=48000;R5=12000;
Vs1=10;Vs2=20;Is=1e-3;
%V11 from Vs1
Ra=P([R2,R3+P([R4,R5])])
V11=Vs1*Ra/(Ra+R1)
%V12 from Is
Rb=P([R1,R2])
Rc=R3+Rb
Rd=P([R4,R5])
I3=Is*Rd/(Rc+Rd)
V12=Rb*I3
%V13 from Vs2
Re=P([Rc,R4])
V23=Vs2*Re/(R5+Re)
V13=V23*Rb/Rc
%Sum of V11, V12, V13
V1b=V11+V12+V13
%Check from nodal analysis
syms V1 V2
[V1,V2]=solve((V1-Vs1)/R1+V1/R2+(V1-V2)/R3,...
(V2-V1)/R3-Is+V2/R4+(V2-Vs2)/R5);
V1=vpa(V1,8)
V2=vpa(V2,8)
```

continued

Example 4.2 continued
MATLAB continued

```
Answers:
Ra =
     2.214285714285715e+04
V11 =
   4.960000000000000
Rb =
         14400
Rc =
         54400
Rd =
          9600
I3 =
     1.500000000000000e-04
V12 =
   2.160000000000000
Re =
         25500
V23 =
  13.600000000000000
V13 =
   3.600000000000000
V1b =
  10.719999999999999
V1 =
10.72
V2 =
22.72
```

Exercise 4.2

Use the superposition principle to find voltage V_1 in the circuit shown in Figure 4.12.

FIGURE 4.12

Circuit for EXERCISE 4.2.

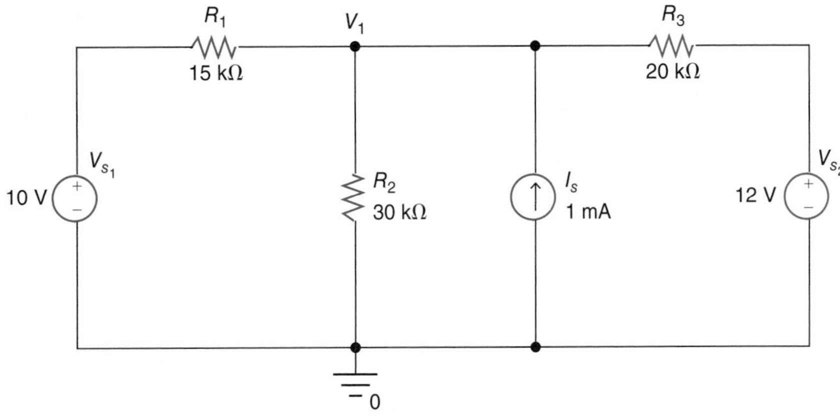

Answer:
V_1 = 15.1111 V. 4.4444 V from V_{s_1}, 4 V from V_{s_2}, 6.6667 V from I_s.

EXAMPLE 4.3

Use the superposition principle to find voltage V_0 across R_4 in the circuit shown in Figure 4.13.

FIGURE 4.13

Circuit for EXAMPLE 4.3.

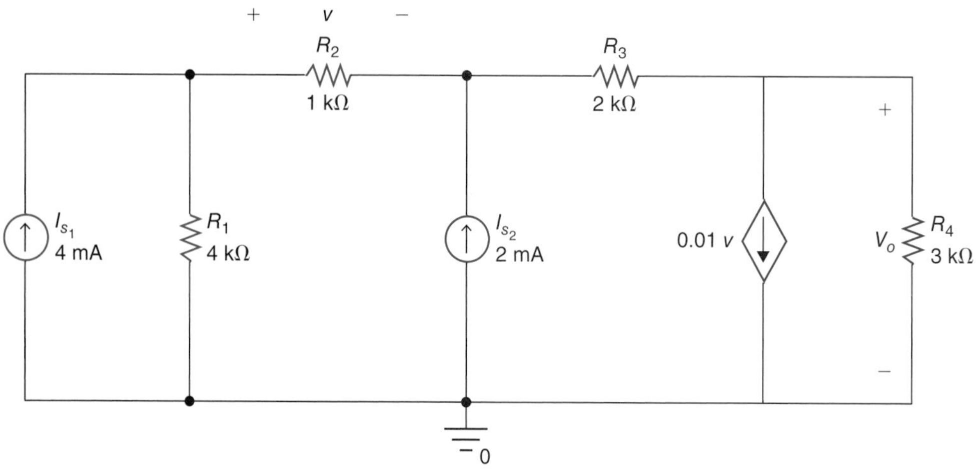

Deactivating the current source I_{s_2} from the circuit shown in Figure 4.13, we obtain the circuit shown in Figure 4.14.

FIGURE 4.14

The circuit shown in Figure 4.13 with the current source I_{s_2} deactivated.

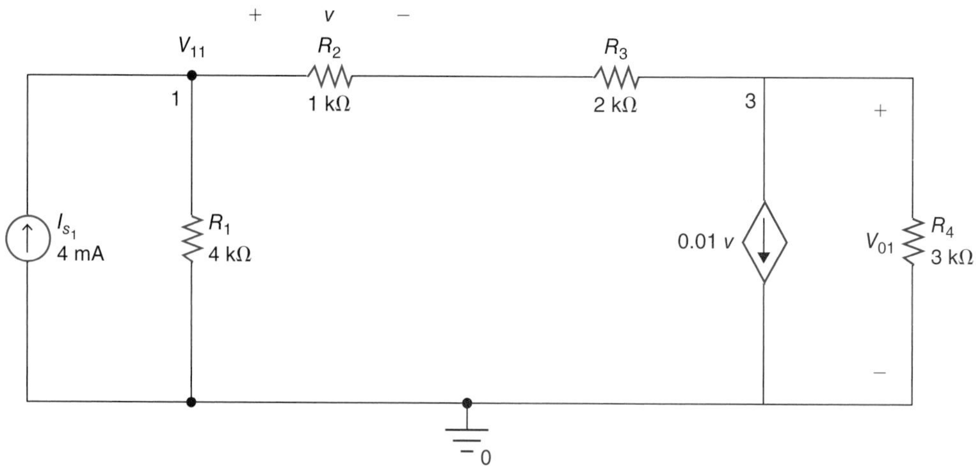

Applying the voltage divider rule, we can find the controlling voltage v to be

$$v = (V_{11} - V_{01})\frac{R_2}{R_2 + R_3} = (V_{11} - V_{01})\frac{1}{3} \tag{4.9}$$

Summing the currents leaving node 1, we obtain

$$-0.004 + \frac{V_{11}}{4000} + \frac{V_{11} - V_{01}}{3000} = 0$$

continued

Example 4.3 continued

Multiplication by 12,000 yields

$$-48 + 3V_{11} + 4(V_{11} - V_{01}) = 0$$

which can be rearranged as

$$7V_{11} - 4V_{01} = 48 \quad \textbf{(4.10)}$$

Summing the currents leaving node 3, we get

$$\frac{V_{01} - V_{11}}{3000} + 0.01\frac{V_{11} - V_{01}}{3} + \frac{V_{01}}{3000} = 0$$

Multiplication by 3000 yields

$$V_{01} - V_{11} + 10(V_{11} - V_{01}) + V_{01} = 0$$

which can be rearranged as

$$9V_{11} - 8V_{01} = 0 \quad \textbf{(4.11)}$$

Solving Equation (4.11) for V_{11}, we obtain

$$V_{11} = \frac{8}{9}V_{01} \quad \textbf{(4.12)}$$

Substitution of Equation (4.12) into Equation (4.10) yields

$$7\frac{8}{9}V_{01} - 4V_{01} = \frac{20}{9}V_{01} = 48$$

Thus, voltage V_{01} across R_4 from I_{s_1} is given by

$$V_{01} = 48\frac{9}{20} = \frac{108}{5} = 21.6 \text{ V} \quad \textbf{(4.13)}$$

Deactivating the current source I_{s_1} from the circuit shown in Figure 4.13, we obtain the circuit shown in Figure 4.15.

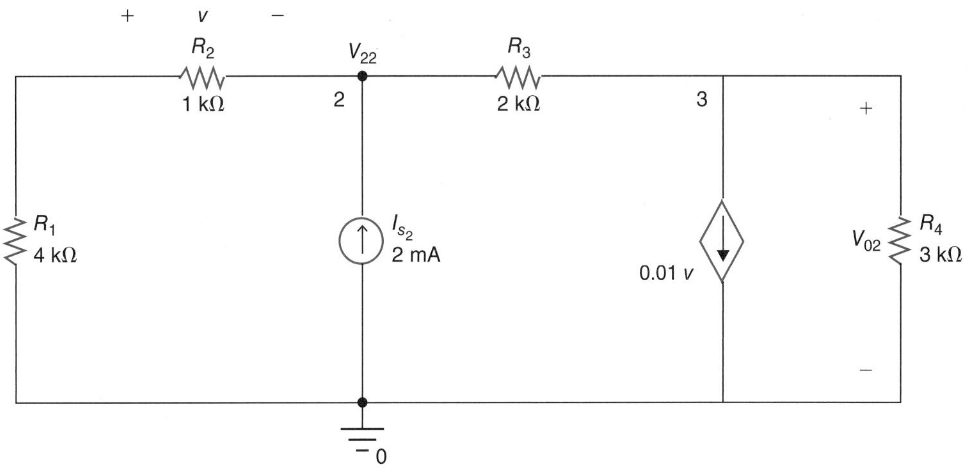

FIGURE 4.15 Circuit shown in Figure 4.13 with the current source I_{s_1} deactivated.

continued

Example 4.3 continued

Applying the voltage divider rule, we can find the controlling voltage v to be

$$v = -V_{22}\frac{R_2}{R_1 + R_2} = \frac{-V_{22}}{5} \quad \textbf{(4.14)}$$

Summing the currents leaving node 2, we obtain

$$\frac{V_{22}}{5000} - 0.002 + \frac{V_{22} - V_{02}}{2000} = 0$$

Multiplication by 10,000 yields

$$2V_{22} - 20 + 5(V_{22} - V_{02}) = 0$$

which can be rearranged as

$$7V_{22} - 5V_{02} = 20 \quad \textbf{(4.15)}$$

Summing the currents leaving node 3, we get

$$\frac{V_{02} - V_{22}}{2000} + 0.01\frac{-V_{22}}{5} + \frac{V_{02}}{3000} = 0$$

Multiplication by 6000 yields

$$3V_{02} - 3V_{22} - 12V_{22} + 2V_{02} = 0$$

which can be rearranged as

$$-15V_{22} + 5V_{02} = 0 \quad \textbf{(4.16)}$$

Solving Equation (4.16) for V_{22}, we obtain

$$V_{22} = \frac{1}{3}V_{02} \quad \textbf{(4.17)}$$

Substitution of Equation (4.17) into Equation (4.15) yields

$$7\frac{1}{3}V_{02} - 5V_{02} = -\frac{8}{3}V_{02} = 20$$

Thus, voltage V_{02} across R_4 from I_{s_2} is given by

$$V_{02} = -20\frac{3}{8} = -\frac{60}{8} = -7.5\text{ V} \quad \textbf{(4.18)}$$

Voltage V_0 across R_4 is the sum of V_{01} and V_{02}. From Equations (4.13) and (4.18), we obtain

$$V_0 = V_{01} + V_{02} = 14.1\text{ V}$$

Exercise 4.3

Use the superposition principle to find voltage V_2 in the circuit shown in Figure 4.16.

FIGURE 4.16

Circuit for EXERCISE 4.3.

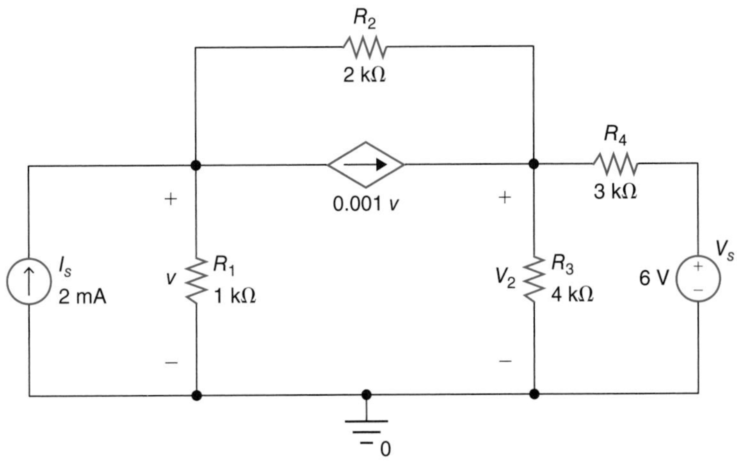

Answer:
$V_2 = 4.0851$ V. 1.5319 V from I_s, 2.5532 V from V_s.

4.3 Source Transformations

A circuit consisting of a voltage source with voltage V_s and a series resistor with resistance R_s, as shown in Figure 4.17(a), is equivalent to a circuit consisting of a current source with current V_s/R_s and a parallel resistor with resistance R_s, as shown in Figure 4.17(b). *Equivalence* means that the circuit shown in Figure 4.17(a) and the circuit shown in Figure 4.17(b) have the same open-circuit voltage across a and b, the same short-circuit current through a and b, and the same resistance looking into the circuit from a and b after deactivating the source. The open-circuit voltage across a and b for the circuit shown in Figure 4.17(a) is V_s, and that for the circuit shown in Figure 4.17(b) is $R_s \times (V_s/R_s) = V_s$. The short-circuit current (with a and b connected by wire) through a and b for the circuit shown in Figure 4.17(a) is V_s/R_s, and that for the circuit shown in Figure 4.17(b) is V_s/R_s based on the current divider rule. After short-circuiting V_s, the resistance across a and b for the circuit shown in Figure 4.17(a) is R_s, and that for the circuit shown in Figure 4.17(b) after open-circuiting the current source is R_s. The circuit shown in Figure 4.17(a) can be replaced by the circuit shown in Figure 4.17(b).

FIGURE 4.17

A voltage source and a series resistor are equivalent to a current source and a parallel resistor.

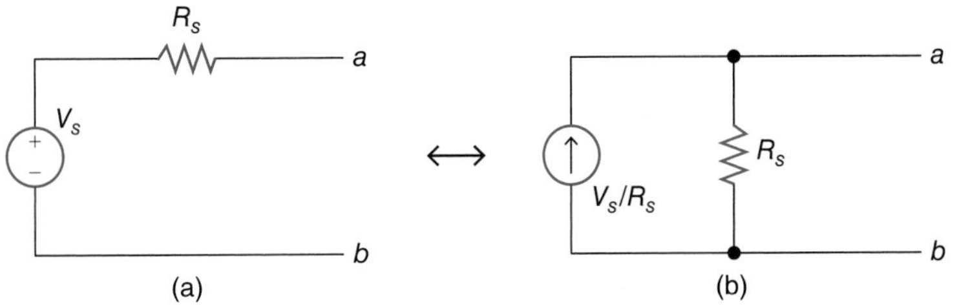

Also, the circuit shown in Figure 4.18(a) is equivalent to the circuit shown in Figure 4.18(b). The circuit shown in Figure 4.18(a) can be replaced by the circuit shown in Figure 4.18(b).

FIGURE 4.18

A current source and a parallel resistor are equivalent to a voltage source and a series resistor.

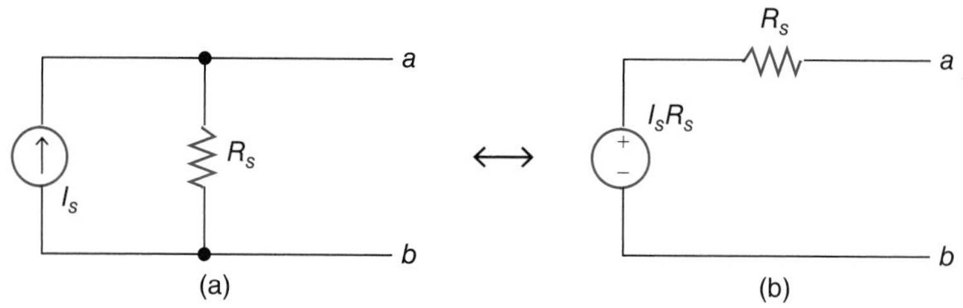

The source transformations apply to dependent sources as well. Figures 4.19 and 4.20 show the equivalence of a voltage source and a series resistor, and a current source and a parallel resistor.

FIGURE 4.19

A dependent voltage source and a series resistor are equivalent to a dependent current source and a parallel resistor.

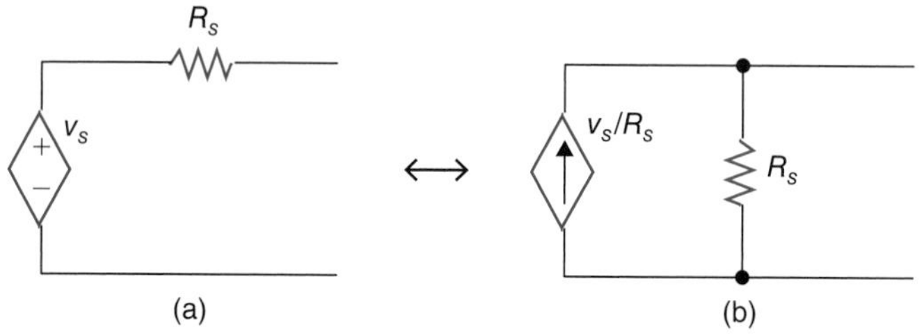

FIGURE 4.20

A dependent current source and a parallel resistor are equivalent to a dependent voltage source and a series resistor.

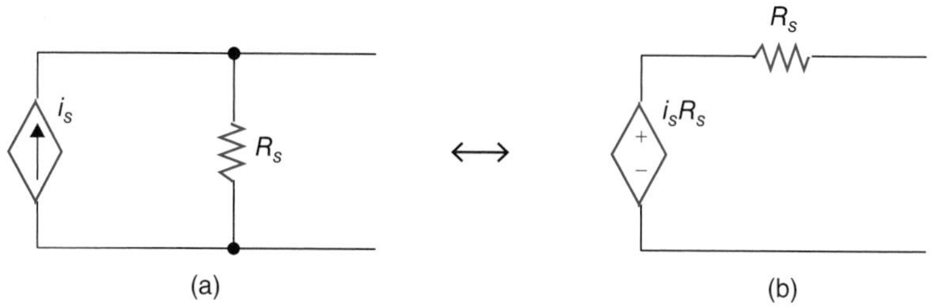

Figure 4.21(a) shows a circuit with a voltage source and a series resistor connected to the rest of the circuit. The voltage v is the voltage at the input of the rest of the circuit, and i is the current into the rest of the circuit. Writing a mesh equation in the clockwise direction, we obtain

$$-v_s + R_s i + v = 0 \tag{4.19}$$

which can be rewritten as

$$v = v_s - R_s i \tag{4.20}$$

Solving Equation (4.20) for i, we get

$$i = \frac{v_s}{R_s} - \frac{v}{R_s} \tag{4.21}$$

The first term, v_s/R_s, on the right side of Equation (4.21) represents a current source with current v_s/R_s, and the second term, v/R_s, on the right side of Equation (4.21) represents a current through a resistor with resistance R_s whose voltage is v. Based on Equation (4.21), we can draw an equivalent circuit, as shown in Figure 4.21(b). Writing a node equation at node 1 of the circuit shown in Figure 4.21(b) as a check, we obtain Equation (4.21). Rearrangement of Equation (4.21) results in Equation (4.20). This shows the equivalence of the circuits shown in Figures 4.21(a–b).

Consider a circuit shown in Figure 4.22. We are interested in finding voltage V_o across R_5 using source transformation.

FIGURE 4.21

Circuit showing proof of equivalence.

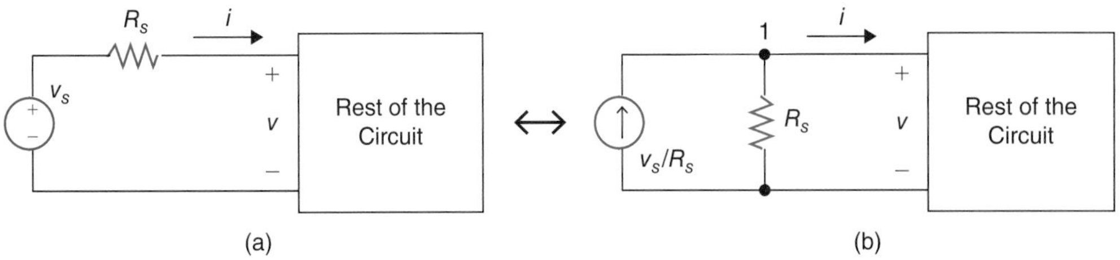

FIGURE 4.22

A circuit used to illustrate source transformation.

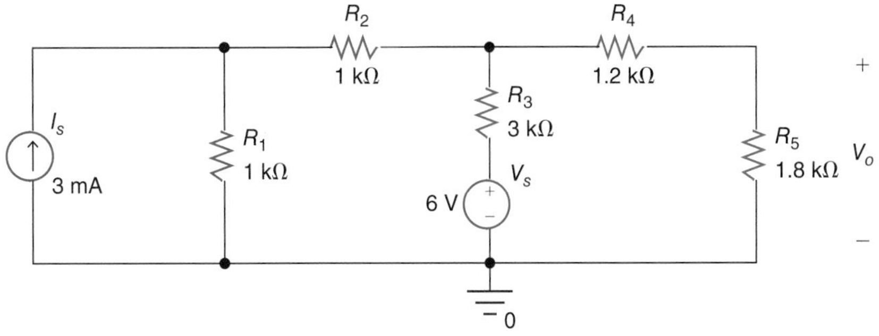

The current source I_s and the parallel resistor R_1 can be transformed into a voltage source with voltage

$$V_{s_1} = R_1 I_s = 1000 \times 3 \times 10^{-3} = 3\ \text{V}$$

and a series resistor R_1, as shown in Figure 4.23.

FIGURE 4.23

I_s and R_1 are transformed into V_{s_1} and R_1.

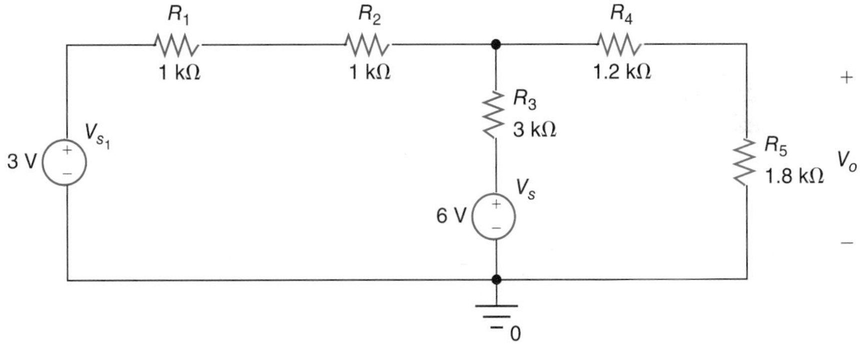

Let R_a be the sum of R_1 and R_2. Then, we have

$$R_a = R_1 + R_2 = 1\ k\Omega + 1\ k\Omega = 2\ k\Omega$$

The voltage source V_{s_1} and the series resistor R_a can be transformed into a current source with current

$$I_{s_1} = \frac{V_{s_1}}{R_a} = \frac{3\ \text{V}}{2\ k\Omega} = 1.5\ \text{mA}$$

and a parallel resistor R_a, as shown in Figure 4.24. Similarly, the voltage source V_s and the series resistor R_3 can be transformed into a current source with current

$$I_{s_2} = \frac{V_s}{R_3} = \frac{6\ \text{V}}{3\ k\Omega} = 2\ \text{mA}$$

and a parallel resistor R_3, as shown in Figure 4.24. Notice that the direction of I_{s_2} is identical to the direction of I_{s_1}.

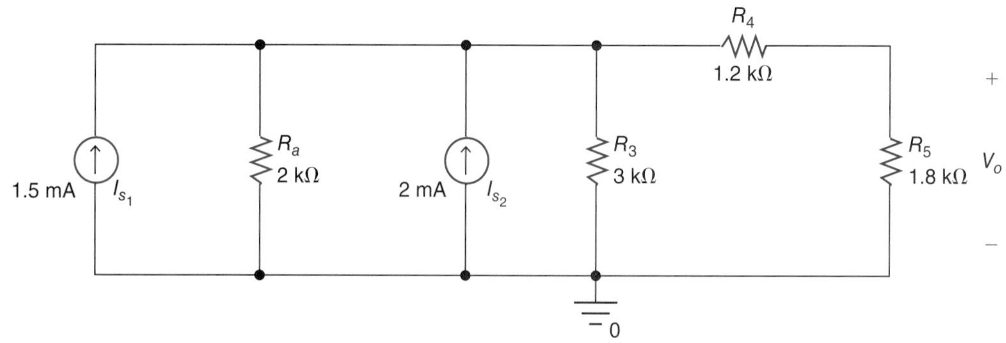

FIGURE 4.24 A circuit after source transformations.

The combined current from the two parallel current sources is

$$I_{s_3} = I_{s_1} + I_{s_2} = 1.5 \text{ mA} + 2 \text{ mA} = 3.5 \text{ mA}$$

The equivalent resistance of the parallel connection of R_a and R_3 is given by

$$R_b = R_a \| R_3 = \frac{R_a \times R_3}{R_a + R_3} = \frac{2000 \times 3000}{2000 + 3000} = 1.2 \; k\Omega$$

Replacing the two current sources I_{s_1} and I_{s_2} and two parallel resistors R_a and R_3 by current source I_{s_3} and parallel resistor R_b, we obtain the circuit shown in Figure 4.25.

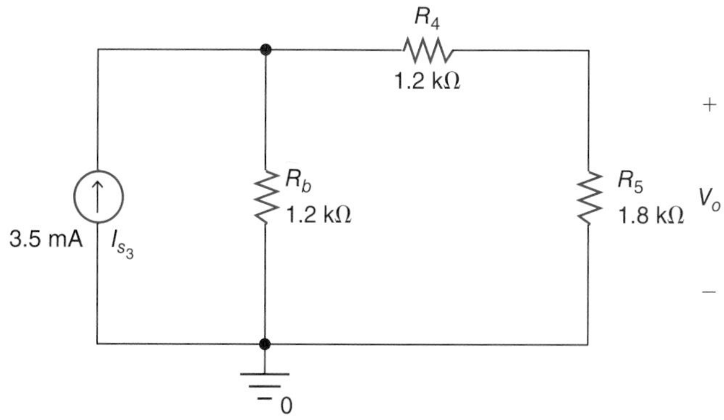

FIGURE 4.25 The circuit in Figure 4.24 with one current source.

The current source I_{s_3} and the parallel resistor R_b can be transformed into a voltage source with voltage

$$V_{s_2} = I_{s_3} \times R_b = 3.5 \times 10^{-3} \times 1200 = 4.2 \text{ V}$$

and a series resistor with resistance $R_b = 1.2 \; k\Omega$. The circuit shown in Figure 4.25 becomes the one shown in Figure 4.26.

Application of the voltage divider rule to the circuit shown in Figure 4.26 yields

$$V_o = V_{s_2} \times \frac{R_5}{R_b + R_4 + R_5} = 4.2 \text{ V} \times \frac{1.8 \; k\Omega}{1.2 \; k\Omega + 1.2 \; k\Omega + 1.8 \; k\Omega}$$

$$= 4.2 \text{ V} \times \frac{1.8}{4.2} = 1.8 \text{ V}$$

FIGURE 4.26

The circuit in Figure 4.25 with one voltage source.

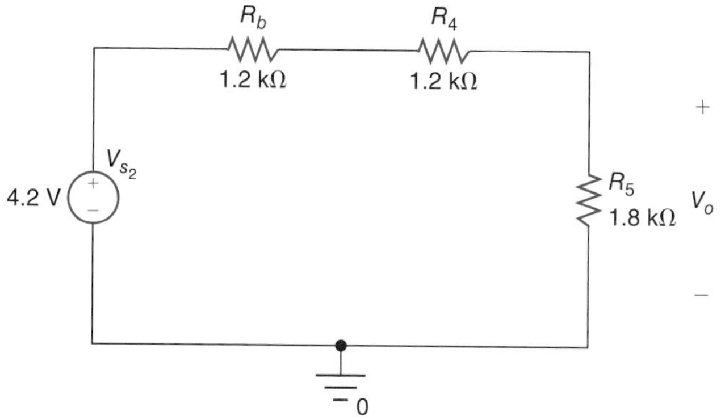

Consider the circuit shown in Figure 4.27. We are interested in finding the voltage V_o using source transformation.

FIGURE 4.27

A circuit with a VCVS.

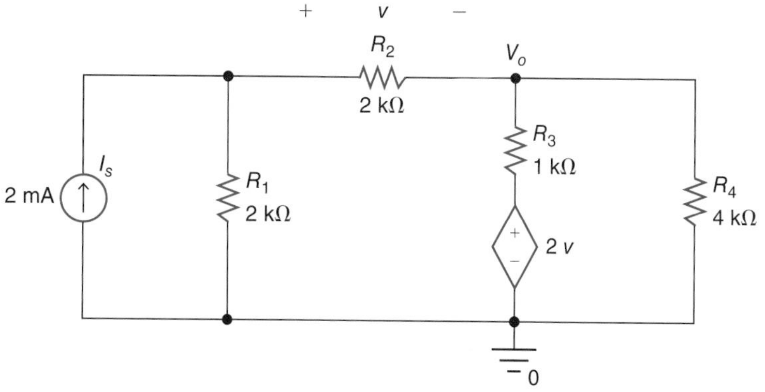

The current source I_s and parallel resistor R_1 can be transformed into a voltage source V_s with voltage

$$V_s = R_1 \times I_s = 2\ k\Omega \times 2\text{ mA} = 4\text{ V}$$

and a series resistor R_1, as shown in Figure 4.28.

FIGURE 4.28

I_s and R_1 are transformed into a voltage source V_s and a series resistor R_1.

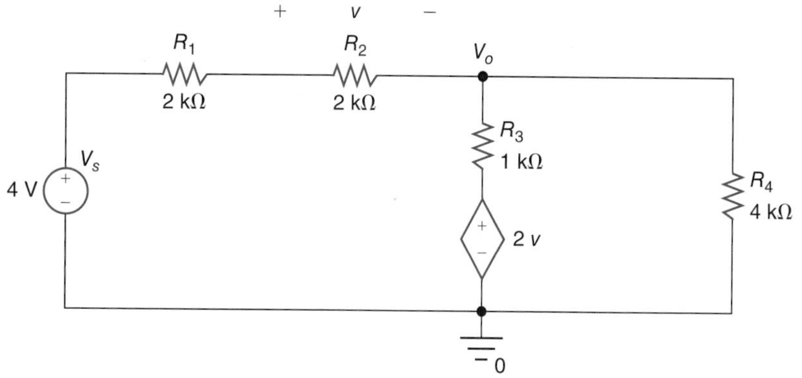

The voltage-controlled voltage source (VCVS) and R_3 can be transformed into a voltage-controlled current source (VCCS) with current $\dfrac{2v}{R_3}$ and a parallel resistor R_3, as shown in Figure 4.29.

FIGURE 4.29

A VCVS and R_3 are transformed into a VCCS and a parallel resistor R_3.

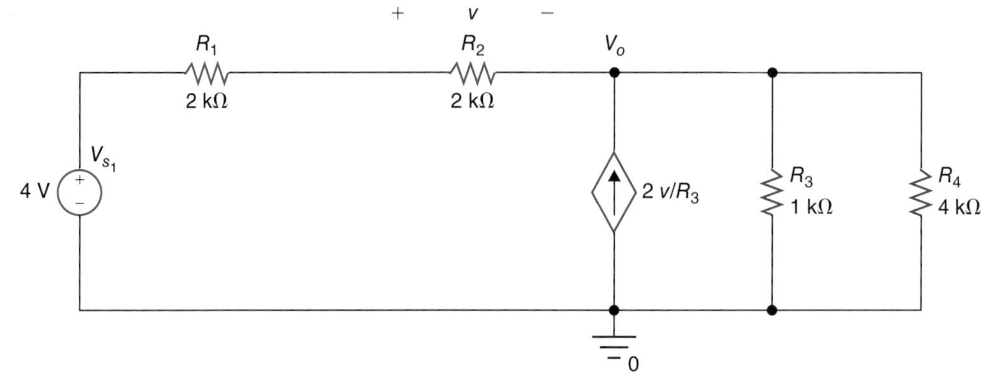

Let R_a be the equivalent resistance of the parallel connection of R_3 and R_4. Then, we have

$$R_a = R_3 \| R_4 = 1\ k\Omega \| 4\ k\Omega = 0.8\ k\Omega$$

FIGURE 4.30

A VCCS and R_a are transformed into a VCVS and a series resistor R_a.

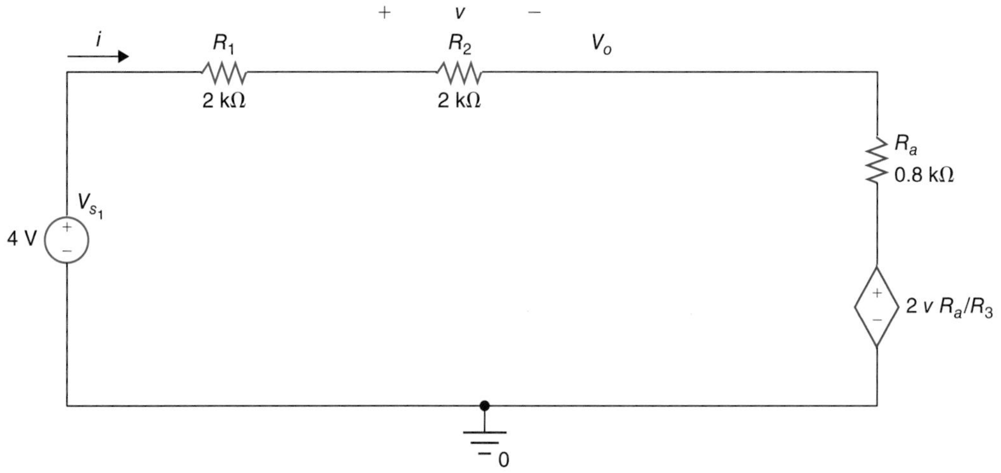

The VCCS and a parallel resistor R_a can be transformed into a VCVS with voltage $\dfrac{2vR_a}{R_3}$ and a series resistor R_a, as shown in Figure 4.30.

The original circuit is transformed into a single mesh. Let the mesh current be i. Then, the controlling voltage v is given by

$$v = R_2 i$$

Collecting the voltage drops around the mesh in the clockwise direction, we obtain

$$-R_1 I_s + R_1 i + R_2 i + R_a i + \frac{2R_2R_a}{R_3} i = 0$$

Solving for i, we obtain

$$i = \frac{R_1 I_s}{R_1 + R_2 + R_a + \dfrac{2R_2R_a}{R_3}} = \frac{4\ \text{V}}{2\ k\Omega + 2\ k\Omega + 0.8\ k\Omega + 3.2\ k\Omega} = 0.5\ \text{mA}$$

Voltage V_o is given by

$$V_o = R_a i + \frac{2R_2R_a}{R_3} i = 2\text{ V}$$

EXAMPLE 4.4

Find voltage V_1 using source transformation for the circuit shown in Figure 4.31.

FIGURE 4.31

Circuit for EXAMPLE 4.4.

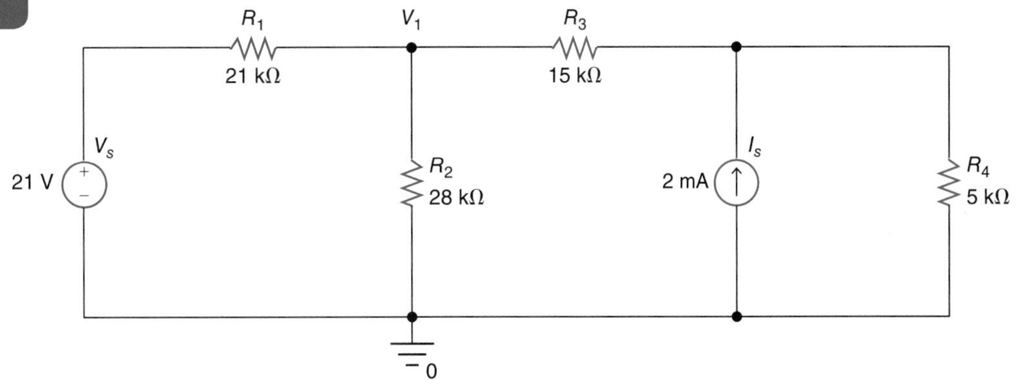

The voltage source V_s and a series resistor R_1 can be transformed into a current source with the current:

$$I_{s_1} = \frac{V_s}{R_1} = \frac{21\text{ V}}{21\ k\Omega} = 1\text{ mA}$$

and a parallel resistor R_1, as shown in Figure 4.32.

FIGURE 4.32

The series connection of V_s and R_1 is transformed into a parallel connection of I_{s_1} and R_1.

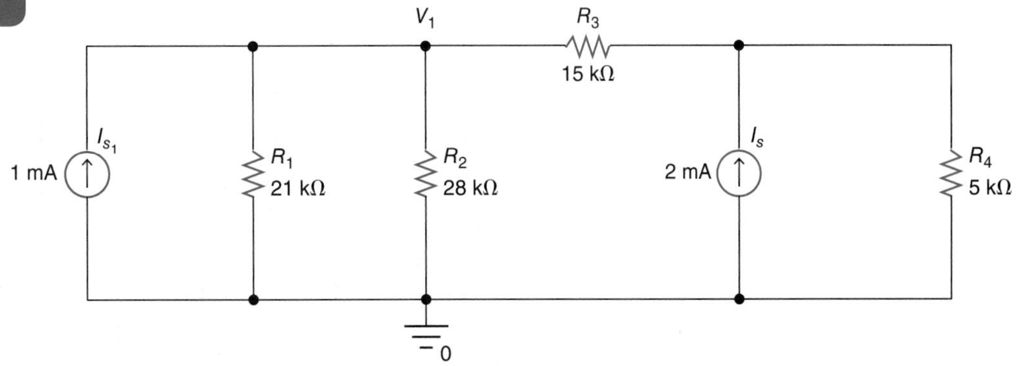

The equivalent resistance of parallel connection of R_1 and R_2 is given by

$$R_a = R_1 \| R_2 = 21\ k\Omega \| 28\ k\Omega = 12\ k\Omega$$

Figure 4.33 shows a circuit with I_{s_1} in parallel with R_a.

continued

Example 4.4 continued

FIGURE 4.33

A circuit with a parallel connection of I_{s_1} and R_a.

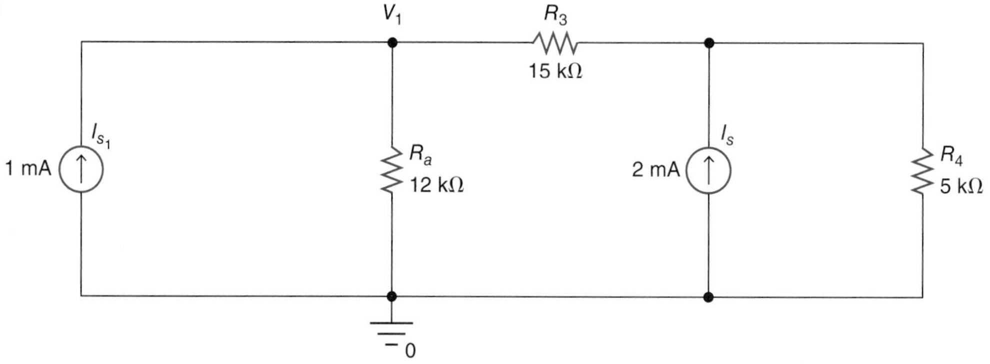

The parallel connection of I_s and R_4 can be transformed into a voltage source with the following voltage:

$$V_{s_1} = R_4 \times I_s = 5\ k\Omega \times 2\ \text{mA} = 10\ \text{V}$$

and a series resistor R_4, as shown in Figure 4.34.

FIGURE 4.34

The circuit in Figure 4.33 with a series connection of V_{s_1} and R_4.

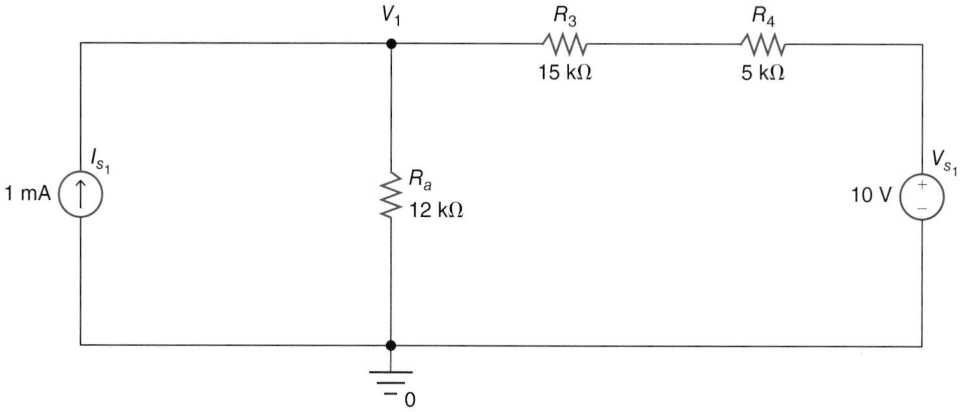

Let R_b be the equivalent resistance of the series connection of R_3 and R_4. Then, we have

$$R_b = R_3 + R_4 = 20\ k\Omega$$

The series connection of V_{s_1} and R_b can be transformed into a parallel connection of a current source with current

$$I_{s_2} = \frac{V_{s_1}}{R_b} = \frac{10\ \text{V}}{20\ k\Omega} = 0.5\ \text{mA}$$

and a parallel resistor R_b, as shown in Figure 4.35.

Let R_c be the equivalent resistance of the parallel connection on R_a and R_b. Then, we have

$$R_c = R_a \| R_b = 12\ k\Omega \| 20\ k\Omega = 7.5\ k\Omega$$

continued

Example 4.4 continued

FIGURE 4.35

The circuit in Figure 4.35 with a parallel connection of I_{s_2} and R_b.

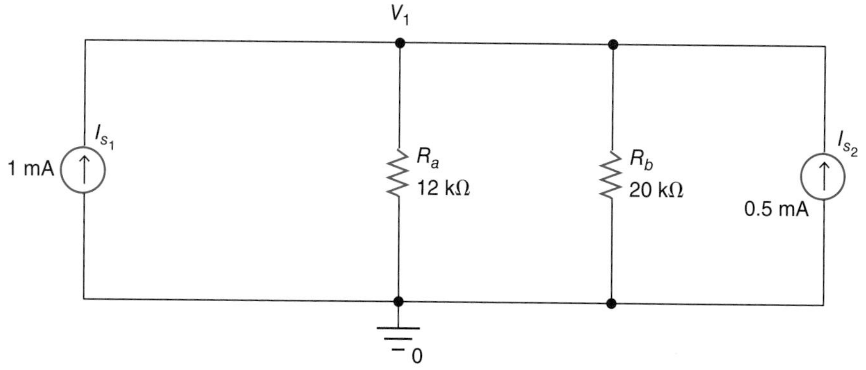

FIGURE 4.36

A circuit with a parallel connection of I_{s_3} and R_c.

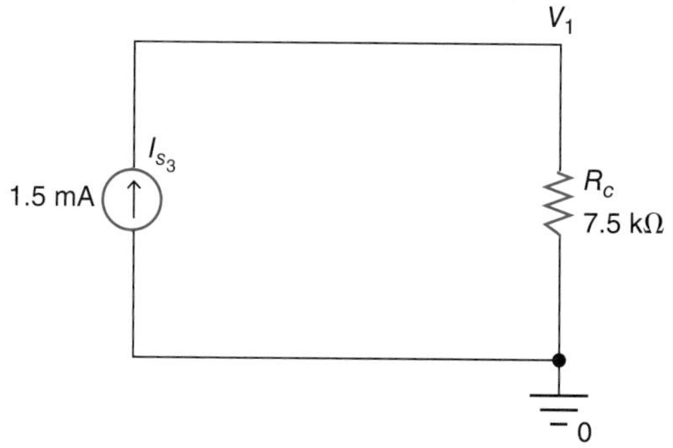

The two current sources can be combined into a single current source with the current given by

$$I_{s_3} = I_{s_1} + I_{s_2} = 1 \text{ mA} + 0.5 \text{ mA} = 1.5 \text{ mA}$$

The circuit with a current source I_{s_3} and a parallel resistor R_c is shown in Figure 4.36.

Voltage V_1 is given by

$$V_1 = R_c \times I_{s_3} = 7.5 \ k\Omega \times 1.5 \text{ mA} = 11.25 \text{ V}$$

Exercise 4.4

Find voltage V_1 using source transformation for the circuit shown in Figure 4.37.

FIGURE 4.37

Circuit for EXERCISE 4.4.

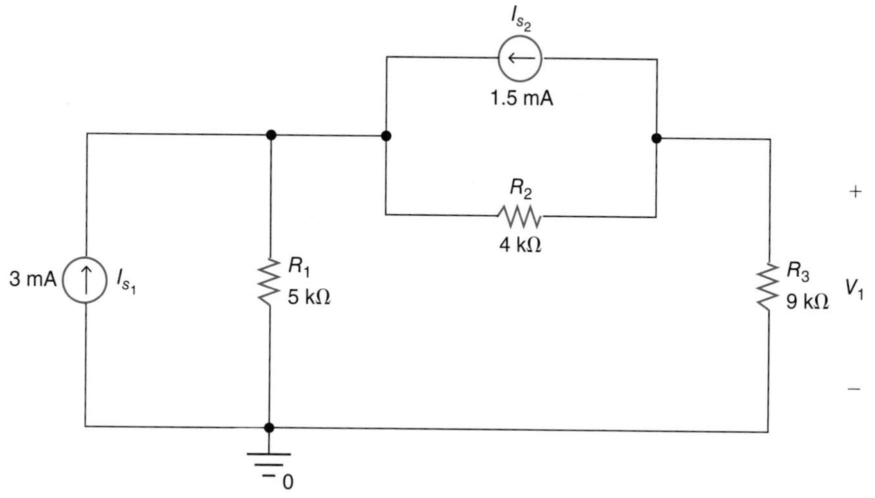

Answer:
$V_1 = 4.5$ V.

EXAMPLE 4.5

Find voltage V and current I using source transformation for the circuit shown in Figure 4.38.

FIGURE 4.38

Circuit for EXAMPLE 4.5.

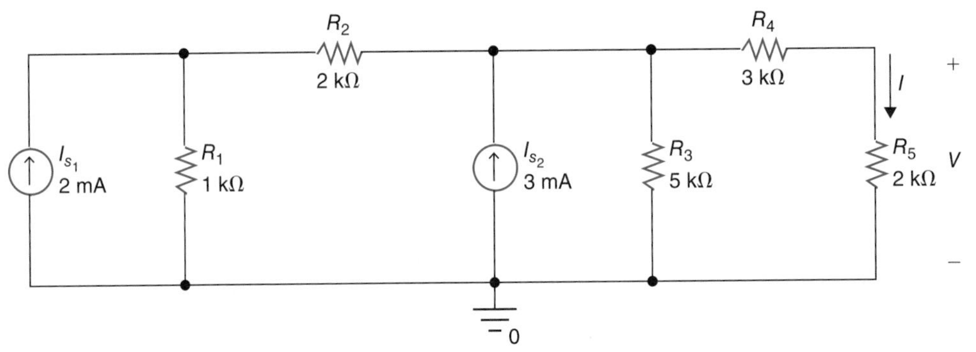

Current source I_{s_1} and resistor R_1 can be transformed into a voltage source with voltage $V_{s_1} = R_1 I_{s_1} = 1\ k\Omega \times 2\ \text{mA} = 2\ \text{V}$ in series with a resistor R_1 with resistance $1\ k\Omega$, as shown in Figure 4.39.

FIGURE 4.39

The circuit in Figure 4.38 with I_{s_1} and R_1 replaced by V_{s_1} and R_1.

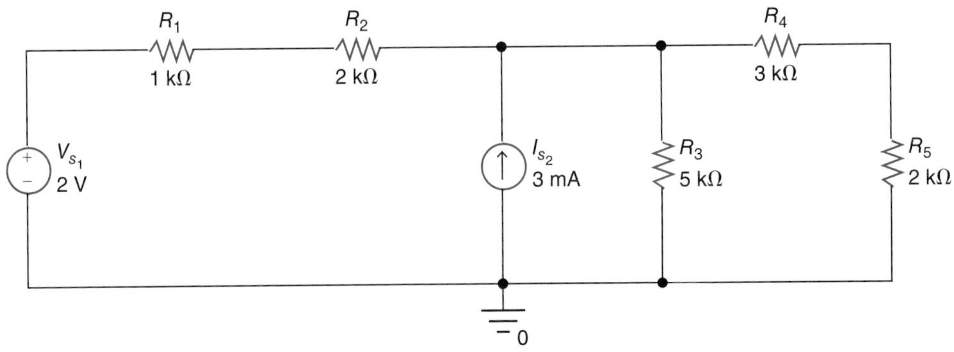

The equivalent resistance of R_1 and R_2 is given by $R_6 = R_1 + R_2 = 3\ k\Omega$. Figure 4.40 shows the circuit with R_6.

FIGURE 4.40

The circuit in Figure 4.39 with $R_6 = R_1 + R_2$.

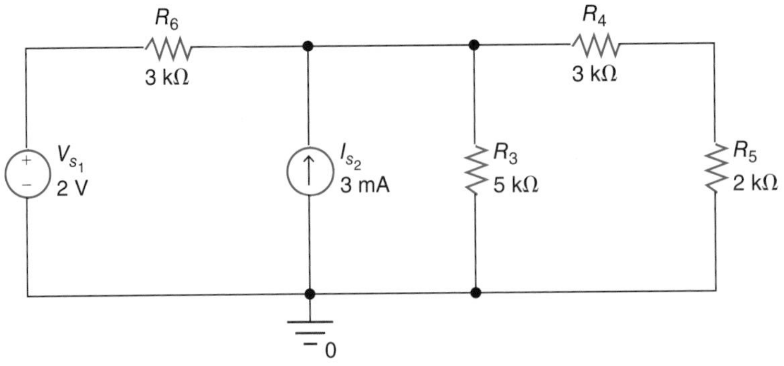

The voltage source V_{s_1} and R_6 can be transformed into a current source with current 2/3 mA and a resistor R_6, as shown in Figure 4.41.

continued

Example 4.5 continued

FIGURE 4.41

The circuit from Figure 4.40 with I_{s_3} in parallel with R_6.

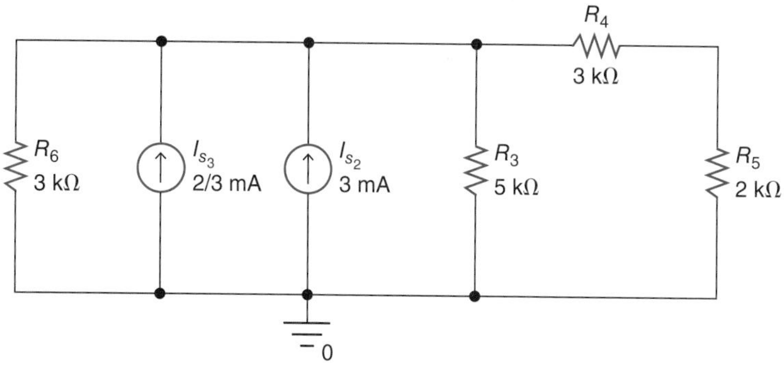

The current sources I_{s_2} and I_{s_3} are in parallel and can be combined into a single current source I_{s_4} with $2/3 + 3 = 11/3$ mA. The two parallel resistors R_3 and R_6 can be combined into a single resistor R_7 with $3\|5 = 15/8\ k\Omega$. The circuit shown in Figure 4.41 simplifies to the one shown in Figure 4.42.

FIGURE 4.42

The circuit from Figure 4.41 with I_{s_4} in parallel with R_7.

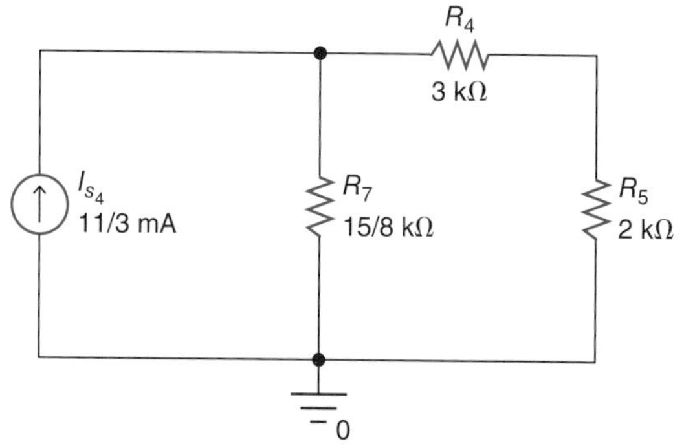

The current source I_{s_4} and the parallel resistor R_7 can be transformed into a voltage source V_{s_2} with voltage (11/3 mA) × (15/8 $k\Omega$) = 55/8 V and the series resistor R_7, as shown in Figure 4.43.

FIGURE 4.43

The circuit from Figure 4.42 with V_{s_2} in series with R_7.

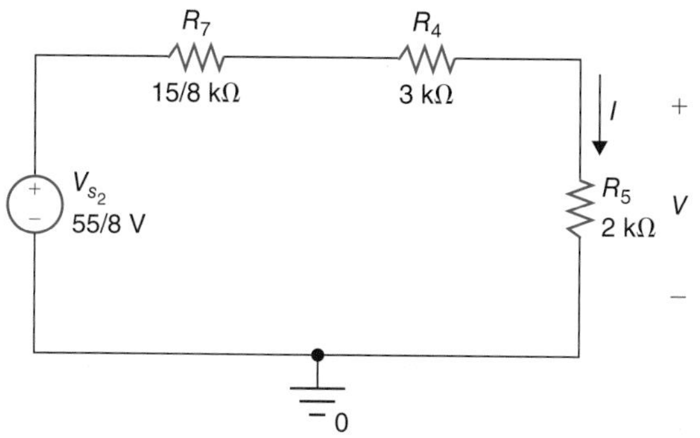

The equivalent resistance of R_7, R_4, and R_5 is

$$R = R_7 + R_4 + R_5 = \frac{55}{8}\ k\Omega$$

Thus, the current I through the circuit shown in Figure 4.43 is given by

$$I = \frac{V_{s_2}}{R} = 1\text{ mA}$$

The voltage V across R_5 is given by

$$V = R_5 I = 2\ k\Omega \times 1\text{ mA} = 2\text{ V}$$

Exercise 4.5

Find voltage V_1 using source transformation for the circuit shown in Figure 4.44.

FIGURE 4.44

Circuit for EXERCISE 4.5.

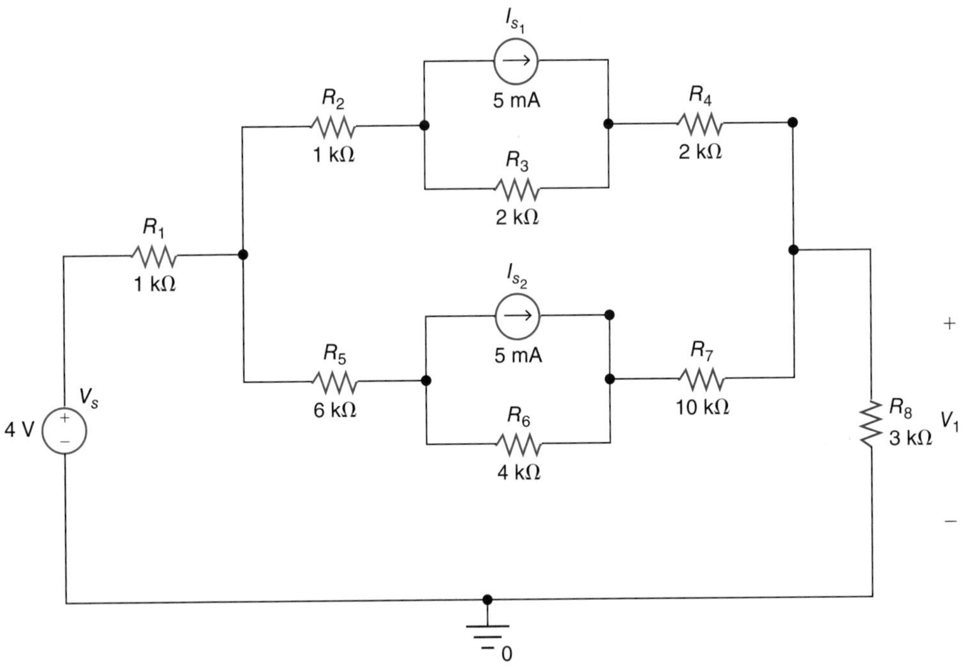

Answer:
$V_1 = 6$ V.

EXAMPLE 4.6

Find voltage V_1 using source transformation for the circuit shown in Figure 4.45.

FIGURE 4.45

Circuit for EXAMPLE 4.6.

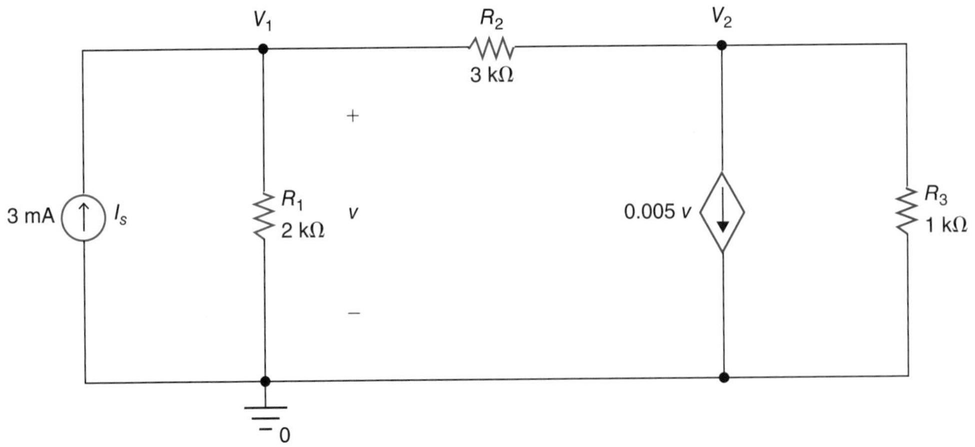

continued

Example 4.6 continued

The VCCS and the parallel resistor R_3 can be transformed into a VCVS with voltage

$$0.005v \times 1000 = 5v$$

and a series resistor R_3, as shown in Figure 4.46.

FIGURE 4.46

The circuit in Figure 4.45 with VCVS.

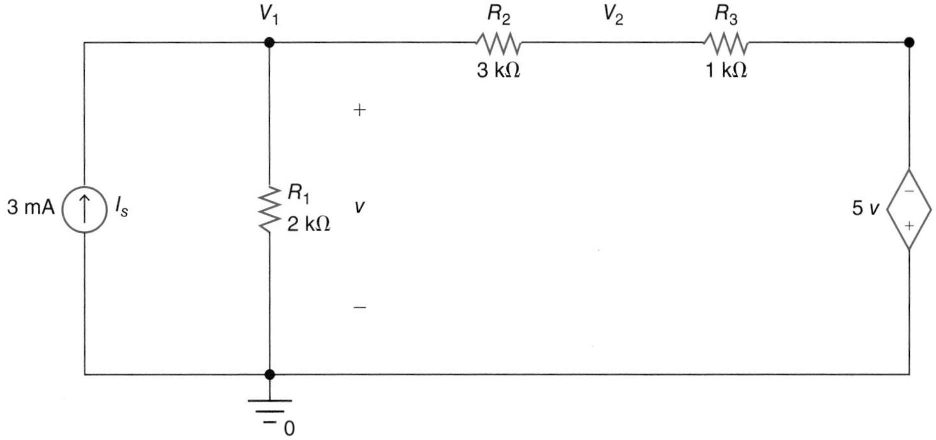

Let R_4 be the sum of R_2 and R_3. Then, we have

$$R_4 = R_2 + R_3 = 4\ k\Omega$$

The VCVS and the series resistor R_4 can be transformed into a VCCS with current

$$\frac{-5v}{4000} = -0.00125v$$

and a parallel resistor R_4, as shown in Figure 4.47.

FIGURE 4.47

The circuit in Figure 4.46 with two current sources in parallel and two resistors in parallel.

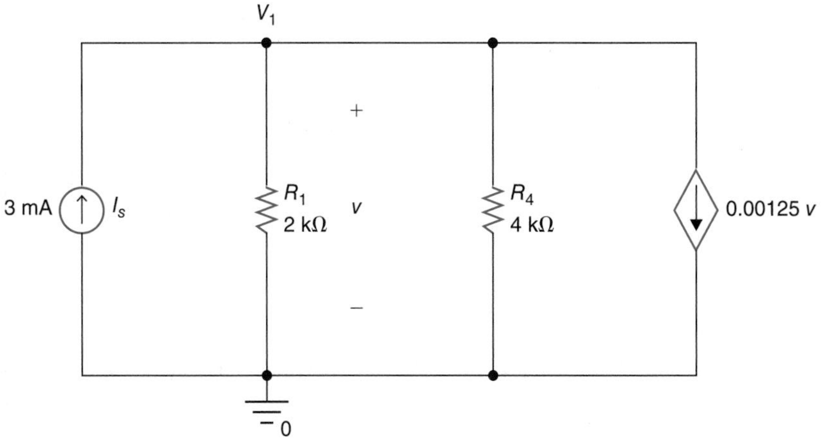

The sum of currents from the two current sources is given by

$$0.003 - 0.00125v$$

continued

Example 4.6 continued

and the equivalent resistance R_{eq} of the parallel connection of R_1 and R_4 is given by

$$R_{eq} = R_1 \| R_4 = \frac{R_1 \times R_4}{R_1 + R_4} = \frac{2000 \times 4000}{2000 + 4000} = \frac{8000}{6} = 1.3333\ k\Omega$$

FIGURE 4.48

The final reduced circuit.

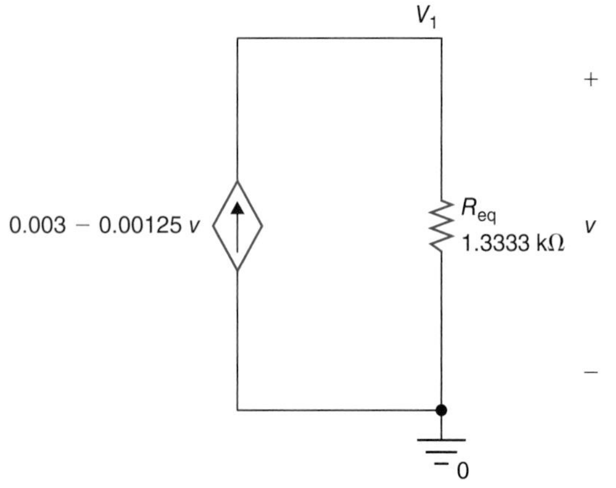

The circuit shown in Figure 4.47 reduces to the one shown in Figure 4.48.

The voltage v, which is V_1, is given by, from Ohm's law,

$$v = (0.003 - 0.00125v) \times 1333.3333 = 4 - 1.6667v$$

Solving for v, we obtain

$$v = V_1 = \frac{4}{2.66667}\ \text{V} = 1.5\ \text{V}.$$

Exercise 4.6

Find voltage v using the source transformation for the circuit shown in Figure 4.49.

FIGURE 4.49

Circuit for EXERCISE 4.6.

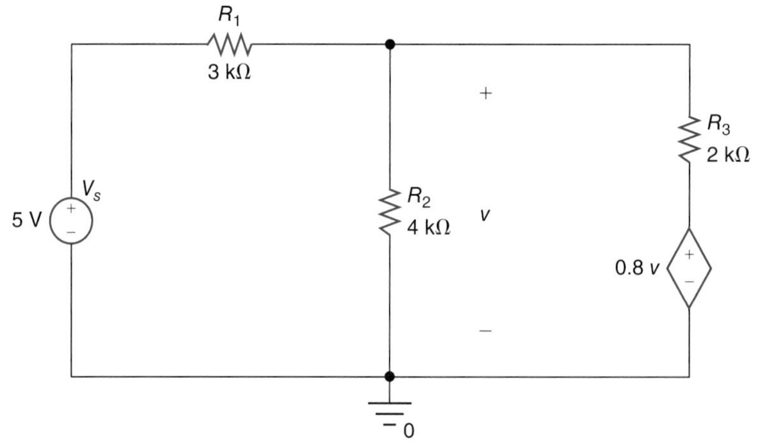

Answer:
$v = 2.4390$ V.

4.4 Thévenin's Theorem

A circuit consisting of a voltage source V_{th} and a series resistor R_{th}, representing the original circuit looking from a pair of terminals, is called a **Thévenin equivalent circuit**. The voltage V_{th} is called **Thévenin equivalent voltage**, and the resistance R_{th} is called **Thévenin equivalent resistance**, as shown in Figure 4.50.

FIGURE 4.50

A Thévenin equivalent circuit.

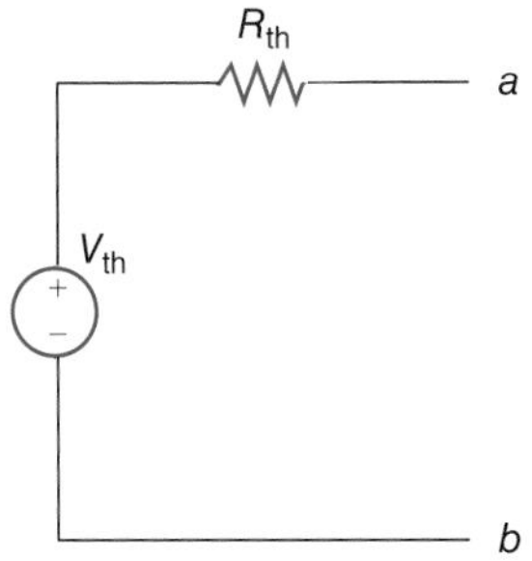

The Thévenin equivalent circuit can be used to simplify the circuit. When a load resistor is connected between terminals a and b, we can find the effects of the circuit on the load from the Thévenin equivalent circuit. We do not need all the details of the original circuit to find the voltage, current, and power on the load.

Let the voltage across terminals a and b of the Thévenin equivalent circuit be V_{oc}. This voltage is called *open-circuit voltage* because terminals a and b are open (with an infinite resistance between a and b). No current flows on the Thévenin equivalent resistor R_{th}. Thus,

$$V_{oc} = V_{th}$$

If the terminals a and b are short-circuited, as shown in Figure 4.51, the current through the short circuit is given by

$$I_{sc} = \frac{V_{th}}{R_{th}} = \frac{V_{oc}}{R_{th}}$$

FIGURE 4.51

Short-circuit current.

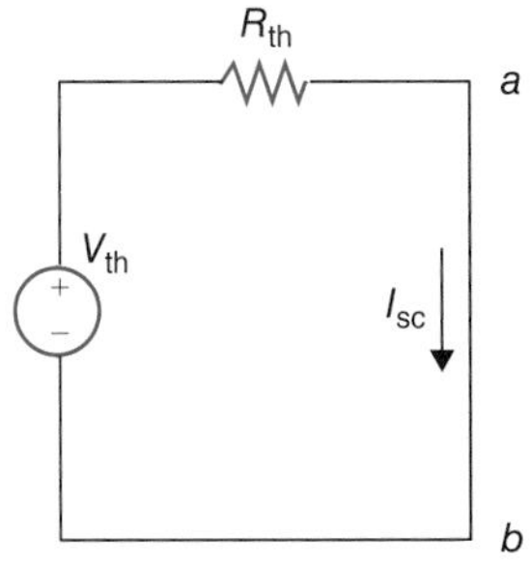

If we solve this equation for R_{th}, we have

$$R_{th} = \frac{V_{oc}}{I_{sc}}$$

This equation can be used to find the Thévenin equivalent resistance R_{th} from the original circuit.

4.4.1 FINDING THE THÉVENIN EQUIVALENT VOLTAGE V_{th}

Given a circuit and terminals a and b, we can find the Thévenin equivalent voltage V_{th} with respect to terminals a and b by finding the open-circuit voltage V_{oc} across terminals a and b. The open-circuit voltage V_{oc} can be found by utilizing circuit analysis methods such as the voltage divider rule, current divider rule, superposition principle, nodal analysis, and mesh analysis. The Thévenin equivalent voltage V_{th} is found from the original circuit without any changes.

4.4.2 FINDING THE THÉVENIN EQUIVALENT RESISTANCE R_{th}

Given a circuit and terminals a and b, we can find the Thévenin equivalent resistance R_{th} with respect to terminals a and b by using one of the three methods listed next.

Method 1

Deactivate all the independent sources by short-circuiting voltage sources and open-circuiting current sources. Find the equivalent resistance looking into the circuit from terminals a and b. This equivalent resistance is the Thévenin equivalent resistance R_{th}. This method can be used if the circuit does not contain dependent sources.

Method 2

Connect terminals a and b by wire (short-circuit). Find the short-circuit current I_{sc} by utilizing circuit analysis methods such as nodal analysis and mesh analysis. The Thévenin equivalent resistance R_{th} is given by

$$R_{th} = \frac{V_{oc}}{I_{sc}}$$

Method 3

Deactivate all the independent sources by open-circuiting current sources and short-circuiting voltage sources. Apply a test voltage of 1 V (or any other value) between terminals a and b with terminal a connected to the positive terminal of the test voltage. Measure the current flowing out of the positive terminal of the test voltage source. The Thévenin equivalent resistance R_{th} is given by the ratio of the test voltage to the current flowing out of the positive terminal of the test voltage source. A test current can be used instead of the test voltage. Apply a test current between terminals a and b after deactivating the independent sources, and measure the voltage across a and b of the test current source. The Thévenin equivalent resistance R_{th} is the ratio of the voltage across a and b to the test current.

Consider a circuit shown in Figure 4.52. We are interested in finding the Thévenin equivalent voltage V_{th} and the Thévenin equivalent resistance R_{th} between terminals a and b. The Thévenin equivalent voltage V_{th} is the open-circuit voltage V_{oc} between terminals a and b. The open-circuit voltage is the voltage across the resistor R_4, which is labeled V_2 in the circuit shown in Figure 4.52.

FIGURE 4.52

A circuit with a pair of terminals.

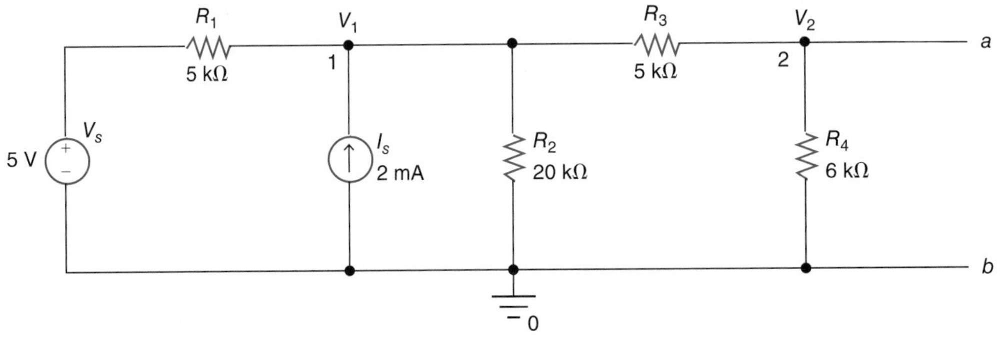

Summing the currents leaving node 1, we obtain

$$\frac{V_1 - 5}{5000} - 0.002 + \frac{V_1}{20{,}000} + \frac{V_1 - V_2}{5000} = 0$$

Multiplication by 20,000 yields

$$4V_1 - 20 - 40 + V_1 + 4V_1 - 4V_2 = 0$$

which can be rearranged as

$$9V_1 - 4V_2 = 60 \tag{4.22}$$

Summing the currents leaving node 2, we obtain

$$\frac{V_2 - V_1}{5000} + \frac{V_2}{6000} = 0$$

Multiplication by 30,000 yields

$$6V_2 - 6V_1 + 5V_2 = 0$$

which can be rearranged as

$$6V_1 = 11V_2$$

or

$$V_1 = \frac{11}{6}V_2 \tag{4.23}$$

Substituting Equation (4.23) into Equation (4.22), we obtain

$$9\frac{11}{6}V_2 - 4V_2 = \frac{75}{6}V_2 = 60$$

Thus, we have

$$V_2 = 4.8\text{ V}$$

Since V_2 is the open-circuit voltage between terminals a and b, it is the Thévenin equivalent voltage V_{th}. Thus, we have

$$V_{\text{th}} = V_{\text{oc}} = 4.8\text{ V}$$

We will try all three methods to find the Thévenin equivalent resistance R_{th}. In method 1, we deactivate all independent sources and find the equivalent resistance looking into the circuit from terminals a and b. Figure 4.53 shows the circuit shown in Figure 4.52 with the voltage source short-circuited and current source open-circuited.

FIGURE 4.53

The circuit from Figure 4.52 with its sources deactivated.

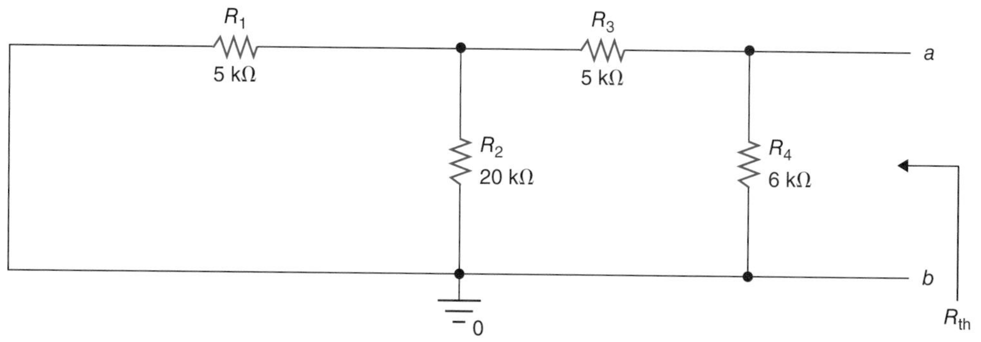

The equivalent resistance of the parallel connection of R_1 and R_2 is

$$R_a = R_1 \| R_2 = 4\ k\Omega$$

The equivalent resistance of the series connection of R_3 and R_a is

$$R_b = R_3 + R_a = R_3 + (R_1 \| R_2) = 5\ k\Omega + 4\ k\Omega = 9\ k\Omega$$

The Thévenin equivalent resistance, R_{th}, is the equivalent resistance of parallel connection of R_b and R_4. Thus, we have

$$R_{\text{th}} = R_b \| R_4 = 9\ k\Omega \| 6\ k\Omega = 3.6\ k\Omega$$

In method 2, we find the open-circuit voltage V_{oc} and the short-circuit current I_{sc}. The Thévenin equivalent resistance between terminals a and b is the ratio of V_{oc} to I_{sc}. The open-circuit voltage is found to be

$$V_{\text{oc}} = V_2 = 4.8\text{ V}$$

To find the short-circuit current, we connect terminals a and b by a wire without changing the rest of the circuit, as shown in Figure 4.54.

FIGURE 4.54

A circuit with a short-circuit between *a* and *b*.

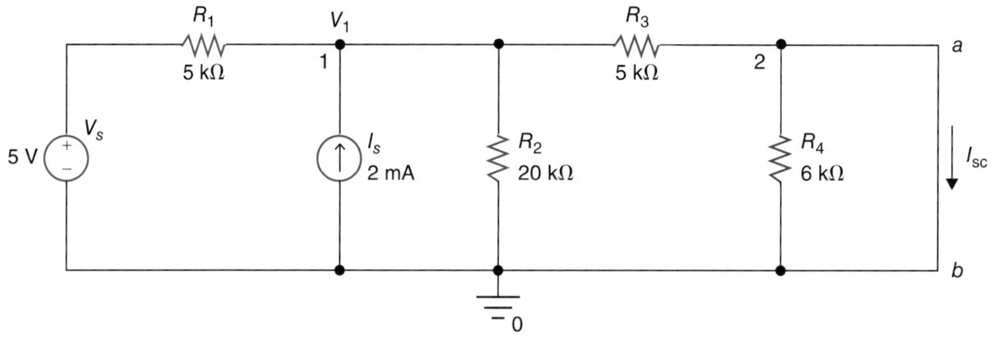

Notice that node 2 is a ground and no current flows through R_4. The short-circuit current I_{sc} is the current through R_3 from left to right (→). Summing the currents leaving node 1, we obtain

$$\frac{V_1 - 5}{5000} - 0.002 + \frac{V_1}{20{,}000} + \frac{V_1}{5000} = 0$$

Multiplication by 20,000 yields

$$4V_1 - 20 - 40 + V_1 + 4V_1 = 0$$

which can be revised as

$$9V_1 = 60$$

The node voltage at node 1 is given by

$$V_1 = \frac{60}{9} = \frac{20}{3}\ \text{V}$$

The current through R_3, which is also I_{sc}, is given by

$$I_{sc} = \frac{V_1}{R_3} = \frac{\frac{20}{3}\ \text{V}}{5\ k\Omega} = \frac{4}{3}\ \text{mA}$$

The Thévenin equivalent resistance is given by

$$R_{th} = \frac{V_{oc}}{I_{sc}} = \frac{4.8\ \text{V}}{\frac{4}{3}\ \text{mA}} = 3.6\ k\Omega$$

In method 3, we deactivate all independent sources and apply a test voltage across terminals *a* and *b*. The Thévenin equivalent resistance is the ratio of the test voltage to the current flowing out of the positive terminal of the test voltage. Figure 4.55 shows a circuit with a test voltage source V_t after deactivating independent sources.

Notice that $R_1 \| R_2 = 4\ k\Omega$ and $R_3 + (R_1 \| R_2) = 9\ k\Omega$. The current through R_3 is

$$I_{R_3} = \frac{V_t}{R_3 + (R_1 \| R_2)} = \frac{1\ \text{V}}{9\ k\Omega} = \frac{1}{9}\ \text{mA}$$

FIGURE 4.55

Circuit with test voltage.

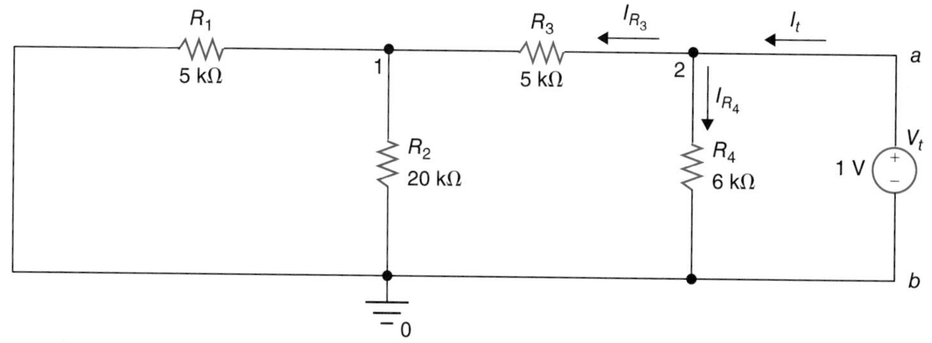

The current through R_4 is

$$I_{R_4} = \frac{V_t}{R_4} = \frac{1\text{ V}}{6\ k\Omega} = \frac{1}{6}\text{mA}$$

The total current flowing out of the positive terminal of the test voltage source is given by

$$I_t = I_{R_3} + I_{R_4} = \frac{1}{9}\text{ mA} + \frac{1}{6}\text{ mA} = \frac{5}{18}\text{ mA}$$

The Thévenin equivalent resistance is given by

$$R_{\text{th}} = \frac{V_t}{I_t} = \frac{1\text{ V}}{\frac{5}{18}\text{mA}} = \frac{18}{5}k\Omega = 3.6\ k\Omega$$

In method 3, instead of a test voltage source, a test current source can be applied after deactivating the independent sources. Figure 4.56 shows a circuit with a test current source I_t after deactivating independent sources.

FIGURE 4.56

Circuit with a test current.

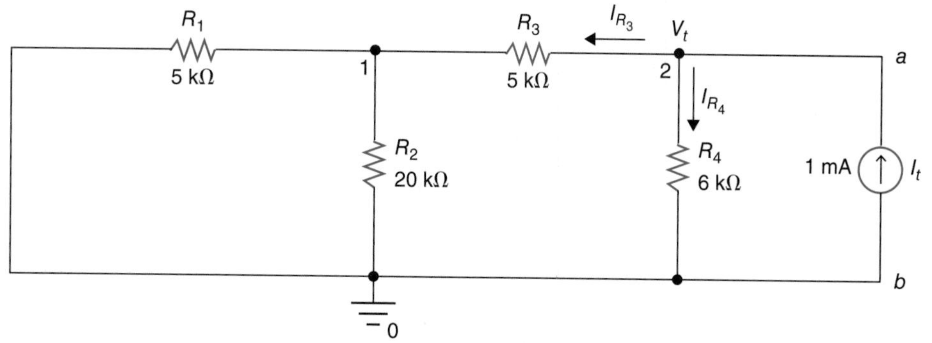

Notice that $R_1 \| R_2 = 4\ k\Omega$ and $R_3 + (R_1 \| R_2) = 9\ k\Omega$. The current through R_4, I_{R_4}, can be obtained by applying the current divider rule:

$$I_{R_4} = I_t \times \frac{9\ k\Omega}{9\ k\Omega + 6\ k\Omega} = 1\text{ mA} \times \frac{9}{15} = \frac{9}{15}\text{ mA}$$

The voltage across R_4, which is also the voltage across the test current source, is given by

$$V_t = R_4 I_{R_4} = 6\ k\Omega \times \frac{9}{15}\text{ mA} = \frac{54}{15}\text{ V} = 3.6\text{ V}$$

The Thévenin equivalent resistance is given by

$$R_{th} = \frac{V_t}{I_t} = \frac{3.6\text{ V}}{1\text{ mA}} = 3.6\ k\Omega$$

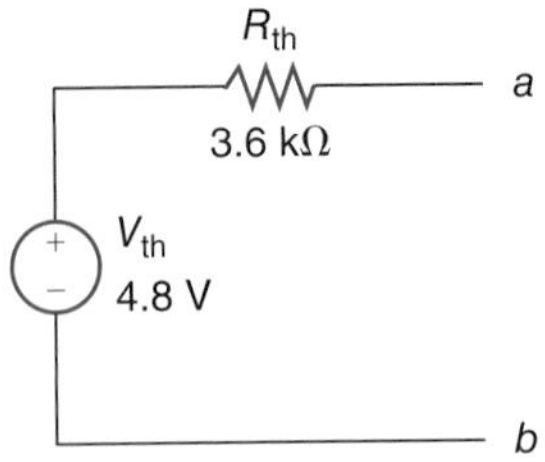

FIGURE 4.57

The Thévenin equivalent circuit.

The Thévenin equivalent circuit is shown in Figure 4.57.

A circuit with VCCS is shown in Figure 4.58. We are interested in finding the Thévenin equivalent voltage V_{th} and the Thévenin equivalent resistance R_{th} between terminals a and b. The Thévenin equivalent voltage V_{th} is the open-circuit voltage V_{oc} between terminals a and b. The open-circuit voltage is the voltage across the resistor R_4, which is labeled V_2 in the circuit shown in Figure 4.58. Notice that the controlling voltage v is equal to V_1.

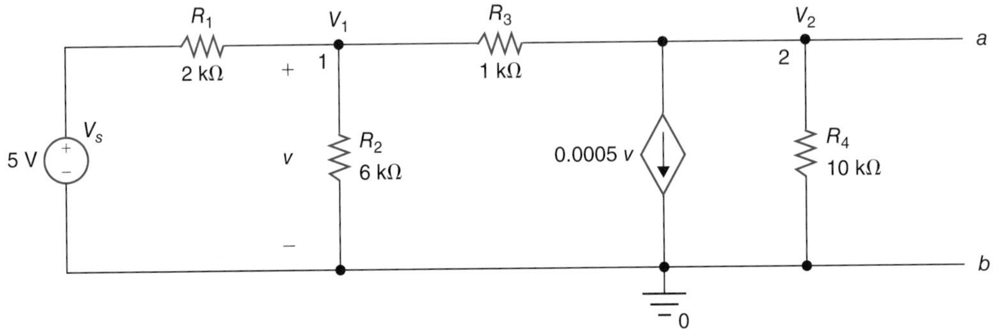

FIGURE 4.58

A circuit with VCCS.

Summing the currents leaving node 1, we obtain

$$\frac{V_1 - 5}{2000} + \frac{V_1}{6000} + \frac{V_1 - V_2}{1000} = 0$$

Multiplication by 6000 yields

$$3V_1 - 15 + V_1 + 6V_1 - 6V_2 = 0,$$

which can be revised as

$$10V_1 - 6V_2 = 15 \tag{4.24}$$

Summing the currents leaving node 2, we obtain

$$\frac{V_2 - V_1}{1000} + 0.0005V_1 + \frac{V_2}{10{,}000} = 0$$

Multiplication by 10,000 yields

$$10V_2 - 10V_1 + 5V_1 + V_2 = 0$$

which can be revised as

$$5V_1 = 11V_2$$

or

$$V_1 = \frac{11}{5}V_2 = 2.2V_2 \tag{4.25}$$

Substituting Equation (4.25) into Equation (4.24), we obtain

$$10 \times 2.2V_2 - 6V_2 = 16V_2 = 15$$

Thus, we have

$$V_2 = \frac{15}{16}\text{ V} = 0.9375\text{ V}$$

Since V_2 is the open-circuit voltage between terminals a and b, it is the Thévenin equivalent voltage V_{th}. Thus, we have

$$V_{th} = 0.9375\text{ V}$$

Since the circuit contains a dependent source, method 1 cannot be used to find the Thévenin equivalent resistance R_{th}. We will try method 2 and method 3. In method 2, we find the open-circuit voltage V_{oc} and the short-circuit current I_{sc}. The Thévenin equivalent resistance between terminals a and b is the ratio of V_{oc} to I_{sc}. The open-circuit voltage is found to be

$$V_{oc} = V_2 = 0.9375\text{ V}$$

To find the short-circuit current, we connect terminals a and b by a wire without changing the rest of the circuit, as shown in Figure 4.59.

FIGURE 4.59

A circuit with a short circuit between a and b.

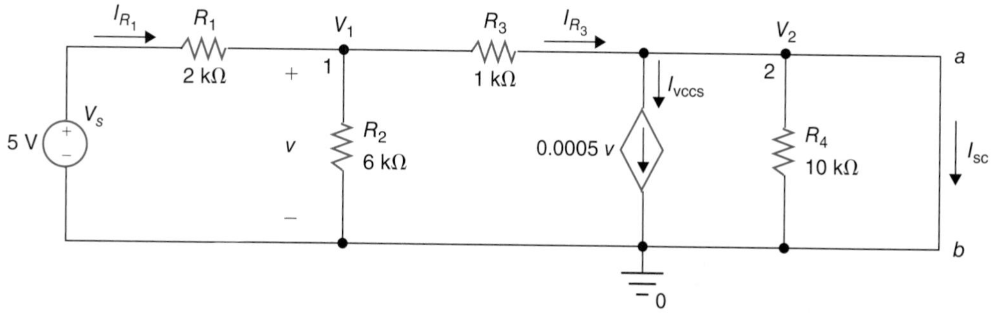

Notice that node 2 is a ground and no current flows through R_4. The short-circuit current I_{sc} is the current through R_3, I_{R_3}, minus the current through VCCS. The equivalent resistance of the parallel connection of R_2 and R_3 is given by

$$R_2 \| R_3 = \frac{6}{7}\, k\Omega = 0.857143\, k\Omega$$

The total resistance seen from the voltage source V_s is given by

$$R_1 + (R_2 \| R_3) = 2\, k\Omega + \frac{6}{7}\, k\Omega = \frac{20}{7}\, k\Omega = 2.857143\, k\Omega$$

The current through R_1 is given by

$$I_{R_1} = \frac{V_s}{R_1 + (R_2 \| R_3)} = \frac{5\text{ V}}{\frac{20}{7}\, k\Omega} = \frac{35}{20}\text{ mA} = 1.75\text{ mA}$$

Voltage V_1 is given by

$$V_1 = V_s - R_1 \times I_{R_1} = 5\text{ V} - 1.75\text{ mA} \times 2\, k\Omega = 1.5\text{ V}$$

Since the controlling voltage v is identical to V_1, we have

$$v = V_1 = 1.5\text{ V}$$

The current through R_3 is given by

$$I_{R_3} = \frac{V_1}{R_3} = \frac{1.5\text{ V}}{1\ k\Omega} = 1.5\text{ mA}$$

The current through VCCS is given by

$$I_{VCCS} = 0.0005 \times V_1 = 0.0005 \times 1.5\text{ A} = 0.75\text{ mA}$$

The short-circuit current is the difference of I_{R_3} and I_{VCCS}. Thus, we have

$$I_{sc} = I_{R_3} - I_{VCCS} = 1.5\text{ mA} - 0.75\text{ mA} = 0.75\text{ mA}$$

The Thévenin equivalent resistance is given by

$$R_{th} = \frac{V_{oc}}{I_{sc}} = \frac{0.9375\text{ V}}{0.75\text{ mA}} = 1.25\ k\Omega$$

In method 3, we deactivate the independent voltage source and apply a test voltage across terminals a and b. The Thévenin equivalent resistance is the ratio of the test voltage to the current flowing out of the positive terminal of the test voltage source. Figure 4.60 shows a circuit with test voltage source V_t after deactivating the independent source.

FIGURE 4.60

A circuit with test voltage.

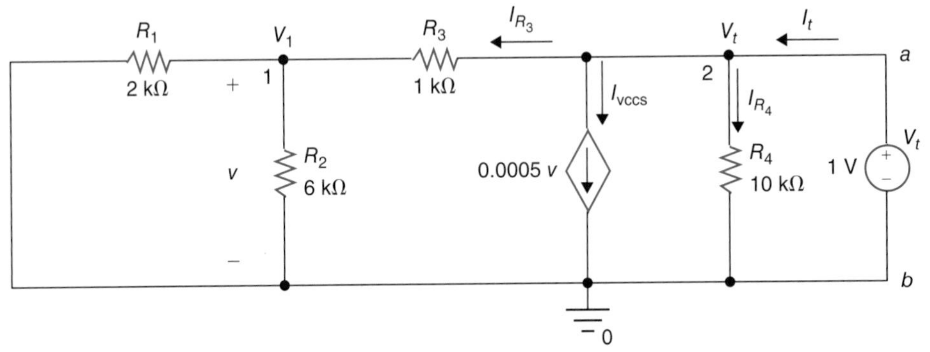

Notice that $R_1 \| R_2 = 1.5\ k\Omega$ and $R_3 + (R_1 \| R_2) = 2.5\ k\Omega$. The current through R_3 is

$$I_{R_3} = \frac{V_t}{R_3 + (R_1 \| R_2)} = \frac{1\text{ V}}{2.5\ k\Omega} = 0.4\text{ mA}$$

The voltage at node 1 is given by

$$V_1 = v = V_t - R_3 \times I_{R_3} = 1\text{ V} - 1\ k\Omega \times 0.4\text{ mA} = 1\text{ V} - 0.4\text{ V} = 0.6\text{ V}$$

The current through the VCCS is given by

$$I_{VCCS} = 0.0005v = 0.0005 \times 0.6\text{ V} = 0.0003\text{ A} = 0.3\text{ mA}$$

The current through R_4 is

$$I_{R_4} = \frac{V_t}{R_4} = \frac{1\text{ V}}{10\ k\Omega} = 0.1\text{ mA}$$

The total current flowing out of the positive terminal of the test voltage source is given by

$$I_t = I_{R_3} + I_{VCCS} + I_{R_4} = 0.4\text{ mA} + 0.3\text{ mA} + 0.1\text{ mA} = 0.8\text{ mA}$$

The Thévenin equivalent resistance is given by

$$R_{\text{th}} = \frac{V_t}{I_t} = \frac{1\text{ V}}{0.8\text{ mA}} = 1.25\ k\Omega$$

In method 3, instead of test voltage, a test current can be applied after deactivating the independent voltage source. Figure 4.61 shows a circuit with test current source I_t after deactivating the independent voltage source.

FIGURE 4.61

A circuit with a test current.

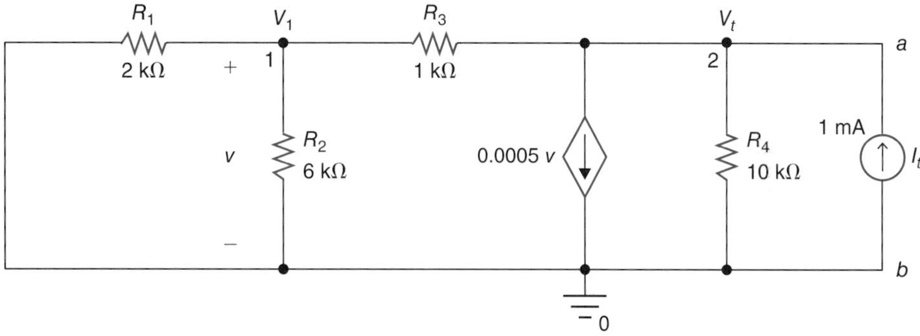

Summing the currents leaving node 1, we obtain

$$\frac{V_1}{2000} + \frac{V_1}{6000} + \frac{V_1 - V_t}{1000} = 0$$

Multiplication by 6000 yields

$$3V_1 + V_1 + 6V_1 - 6V_t = 0$$

which can be revised as

$$10V_1 - 6V_t = 0$$

Solving for V_1, we get

$$V_1 = 0.6V_t \tag{4.26}$$

Summing the currents leaving node 2, we obtain

$$\frac{V_t - V_1}{1000} + 0.0005V_1 + \frac{V_t}{10{,}000} - 0.001 = 0$$

Multiplication by 10,000 yields

$$10V_t - 10V_1 + 5V_1 + V_t - 10 = 0$$

which can be revised as

$$-5V_1 + 11V_t = 10 \tag{4.27}$$

Substitution of Equation (4.26) into Equation (4.27) yields

$$-5(0.6V_t) + 11V_t = 8V_t = 10$$

Thus, we have

$$V_t = \frac{10}{8}\text{ V} = 1.25\text{ V}$$

The Thévenin equivalent resistance is given by

$$R_{\text{th}} = \frac{V_t}{I_t} = \frac{1.25\text{ V}}{1\text{ mA}} = 1.25\ k\Omega$$

A Thévenin equivalent circuit is shown in Figure 4.62.

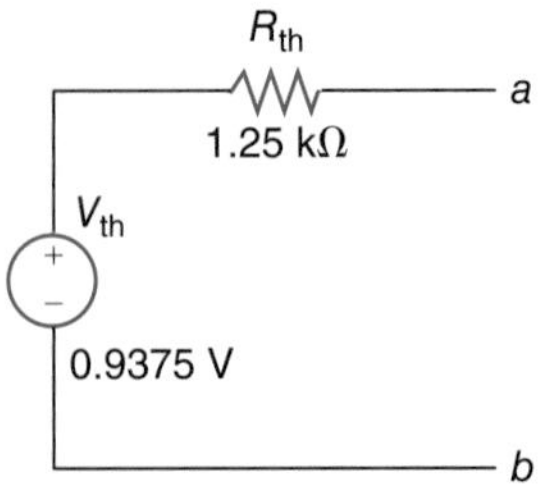

FIGURE 4.62

Thévenin equivalent circuit.

EXAMPLE 4.7

Find the Thévenin equivalent voltage and the Thévenin equivalent resistance between *a* and *b* in the circuit shown in Figure 4.63.

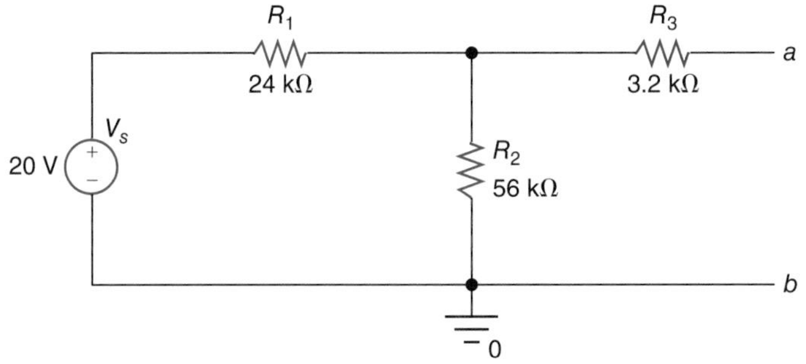

FIGURE 4.63

Circuit for EXAMPLE 4.7.

Since *a* and *b* is an open circuit, resistance between *a* and *b* is infinite. There is no current flowing through R_3. Therefore, there is no voltage drop across R_3. The open-circuit voltage between *a* and *b* is identical to the voltage across R_2. From the voltage divider rule, we have

$$V_{\text{th}} = V_{\text{oc}} = V_s \times \frac{R_2}{R_1 + R_2} = 20\text{ V} \times \frac{56\ k\Omega}{24\ k\Omega + 56\ k\Omega} = 14\text{ V}$$

To find the Thévenin equivalent resistance, we deactivate the voltage source by short-circuiting it, as shown in Figure 4.64.

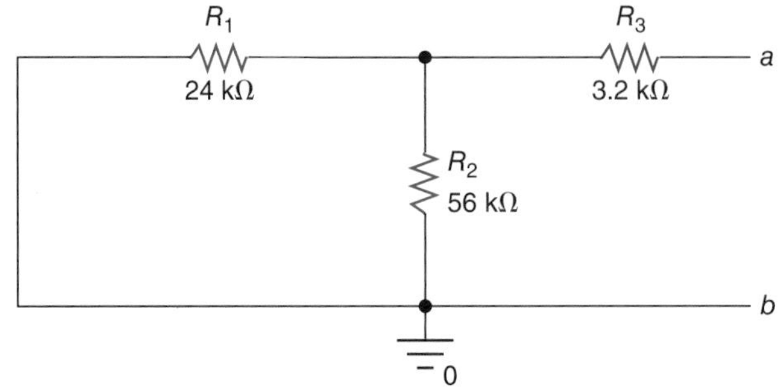

FIGURE 4.64

Circuit in Figure 4.63 after deactivating the voltage source.

continued

Example 4.7 continued

FIGURE 4.65

The Thévenin equivalent circuit.

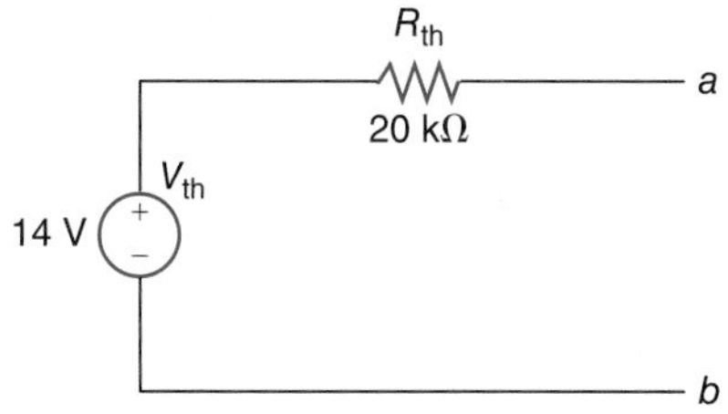

The equivalent resistance to the left of a and b is given by

$$R_{th} = R_3 + (R_1 \| R_2) = R_3 + \frac{R_1 \times R_2}{R_1 + R_2} = 3.2\ k\Omega + \frac{24\ k\Omega \times 56\ k\Omega}{24\ k\Omega + 56\ k\Omega}$$
$$= 3.2\ k\Omega + 16.8\ k\Omega = 20\ k\Omega$$

When the circuit is replaced by a Thévenin equivalent circuit, the original circuit shown in Figure 4.63 becomes the circuit shown in Figure 4.65.

Exercise 4.7

Find the Thévenin equivalent circuit between *a* and *b* in the circuit shown in Figure 4.66.

FIGURE 4.66

Circuit for EXERCISE 4.7.

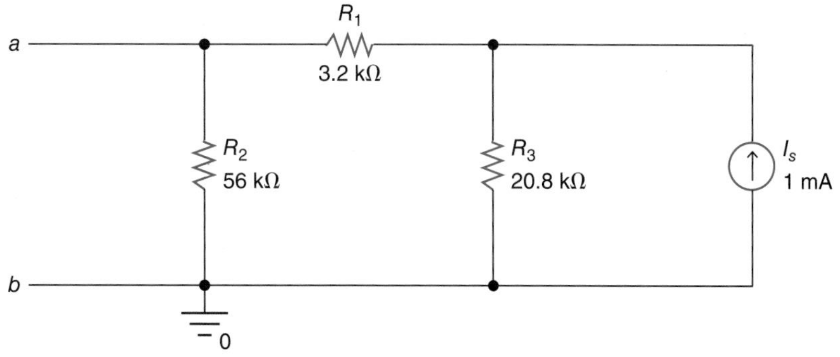

Answer:
$V_{th} = 14.56$ V, $R_{th} = 16.8\ k\Omega$.

EXAMPLE 4.8

Find the Thévenin equivalent voltage and the Thévenin equivalent resistance between *a* and *b* in the circuit shown in Figure 4.67.

FIGURE 4.67

Circuit for EXAMPLE 4.8.

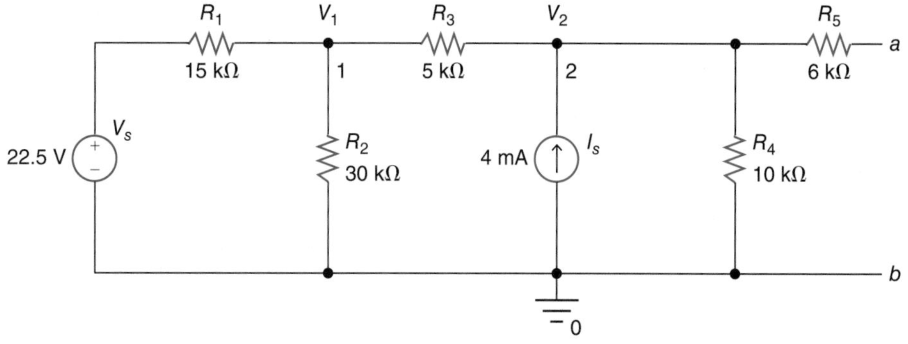

continued

Example 4.8 continued

Since no current flows through R_5, the Thévenin equivalent voltage is the voltage across R_4, which is labeled V_2 in the circuit shown in Figure 4.67. The superposition principle will be applied to find the Thévenin equivalent voltage. When the current source is deactivated, the circuit reduces to the one shown in Figure 4.68.

FIGURE 4.68

The circuit shown in Figure 4.67 with I_s deactivated.

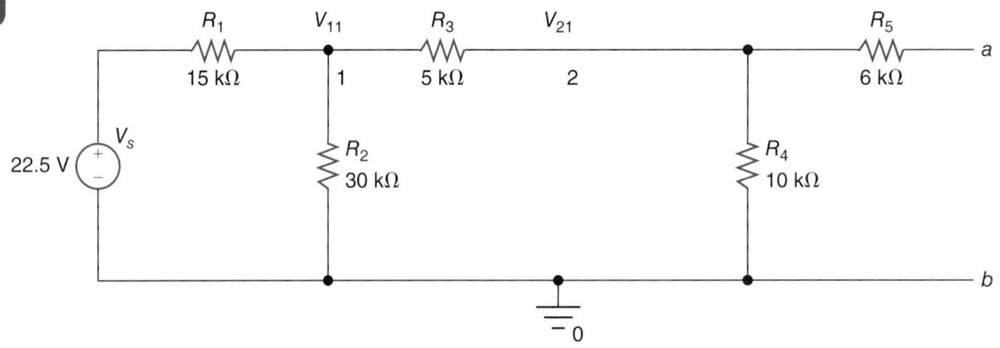

The equivalent resistance of the parallel connection of R_2 and $R_3 + R_4$ is given by

$$R_2 \| (R_3 + R_4) = 30\ k\Omega \| 15\ k\Omega = 10\ k\Omega$$

Application of the voltage divider rule results in

$$V_{11} = V_s \times \frac{R_2 \| (R_3 + R_4)}{R_1 + [R_2 \| (R_3 + R_4)]} = 22.5\ \text{V} \times \frac{10\ k\Omega}{15\ k\Omega + 10\ k\Omega} = 9\ \text{V}$$

Application of the voltage divider rule to R_3 and R_4 results in

$$V_{21} = V_{11} \times \frac{R_4}{R_3 + R_4} = 9\ \text{V} \times \frac{10\ k\Omega}{5\ k\Omega + 10\ k\Omega} = 6\ \text{V}$$

When the voltage source is deactivated, the circuit shown in Figure 4.67 reduces to the one shown in Figure 4.69.

FIGURE 4.69

The circuit shown in Figure 4.67 with V_s deactivated.

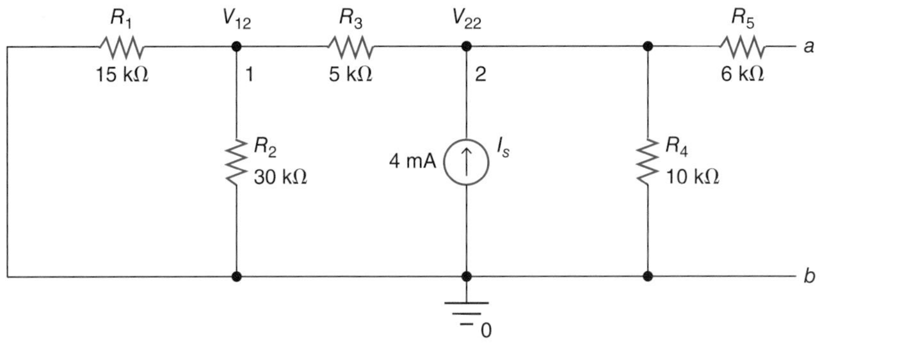

Let R_a be the equivalent resistance of $R_3 + (R_1 \| R_2)$. Then, we have

$$R_a = R_3 + (R_1 \| R_2) = 5\ k\Omega + (15\ k\Omega \| 30\ k\Omega) = 5\ k\Omega + 10\ k\Omega = 15\ k\Omega$$

continued

Example 4.8 continued

From the current divider rule, the current through R_4 is given by

$$I_{R_4} = I_s \times \frac{R_a}{R_a + R_4} = 4\text{ mA} \times \frac{15\ k\Omega}{15\ k\Omega + 10\ k\Omega} = 2.4\text{ mA}$$

The voltage across R_4 is given by

$$V_{22} = R_4 \times I_{R_4} = 10\ k\Omega \times 2.4\text{ mA} = 24\text{ V}$$

The open-circuit voltage, which is the Thévenin equivalent voltage, is the sum of V_{21} and V_{22}:

$$V_{\text{th}} = V_{\text{oc}} = V_2 = V_{21} + V_{22} = 6\text{ V} + 24\text{ V} = 30\text{ V}$$

To find the Thévenin equivalent resistance R_{th}, we deactivate the sources by short-circuiting the voltage source and open-circuiting the current source, as shown in Figure 4.70.

FIGURE 4.70

The circuit shown in Figure 4.67 after deactivating the sources.

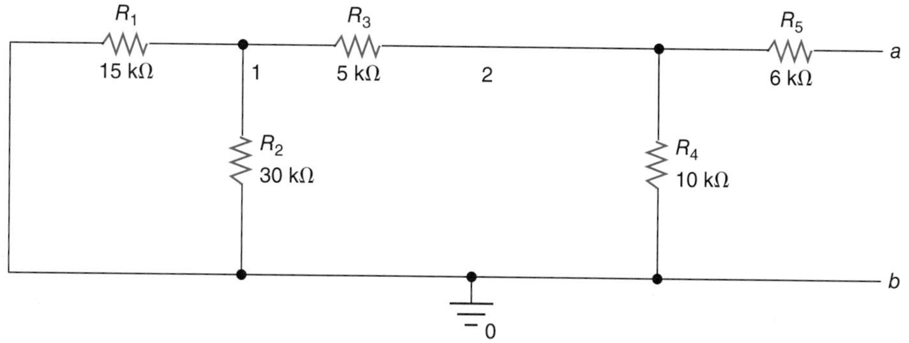

We find the equivalent resistance starting from the left side of the circuit and moving toward terminals a and b. The equivalent resistance of the parallel connection of R_1 and R_2 is given by

$$R_1 \| R_2 = \frac{R_1 \times R_2}{R_1 + R_2} = \frac{15\ k\Omega \times 30\ k\Omega}{15\ k\Omega + 30\ k\Omega} = \frac{30\ k\Omega}{1 + 2} = 10\ k\Omega$$

$R_1 \| R_2$ is in series with R_3. Thus, we have

$$(R_1 \| R_2) + R_3 = 10\ k\Omega + 5\ k\Omega = 15\ k\Omega.$$

$(R_1 \| R_2) + R_3$ is in parallel with R_4. Thus, we have

$$[(R_1 \| R_2) + R_3] \| R_4 = \frac{15\ k\Omega \times 10\ k\Omega}{15\ k\Omega + 10\ k\Omega} = \frac{30\ k\Omega}{5} = 6\ k\Omega$$

The Thévenin equivalent resistance is the sum of 6 $k\Omega$ and R_5; i.e.,

$$R_{\text{th}} = 6\ k\Omega + 6\ k\Omega = 12\ k\Omega$$

The Thévenin equivalent circuit is shown in Figure 4.71.

FIGURE 4.71

The Thévenin equivalent circuit.

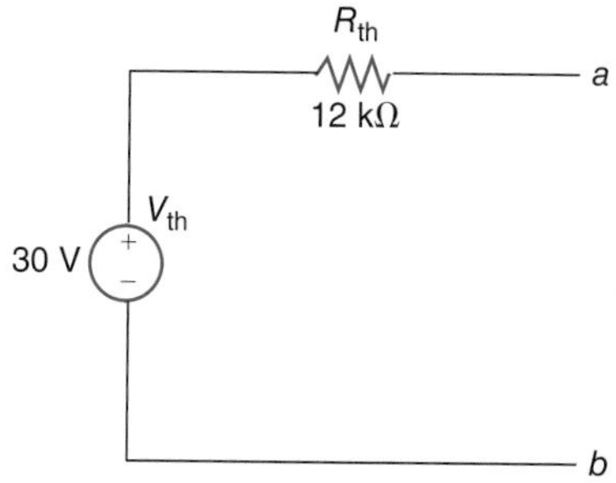

Exercise 4.8

Find the Thévenin equivalent voltage and the Thévenin equivalent resistance between *a* and *b* in the circuit shown in Figure 4.72.

FIGURE 4.72

Circuit for EXERCISE 4.8.

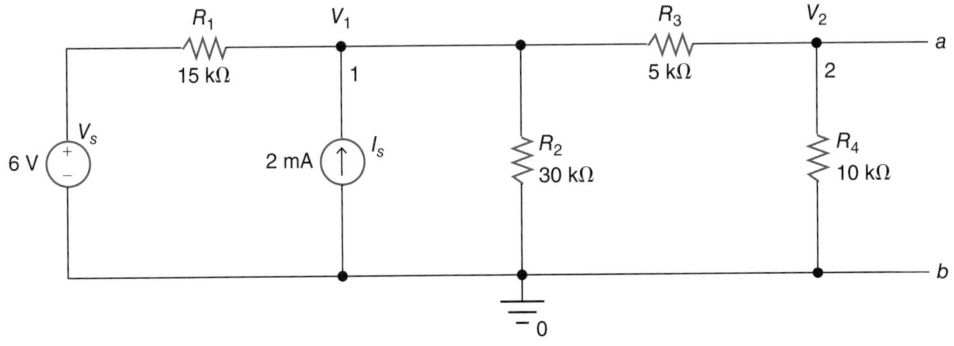

Answer:
$V_{th} = 9.6$ V, $R_{th} = 6\ k\Omega$.

EXAMPLE 4.9

Find the Thévenin equivalent circuit between terminals *a* and *b* for the circuit shown in Figure 4.73.

FIGURE 4.73

Circuit for EXAMPLE 4.9.

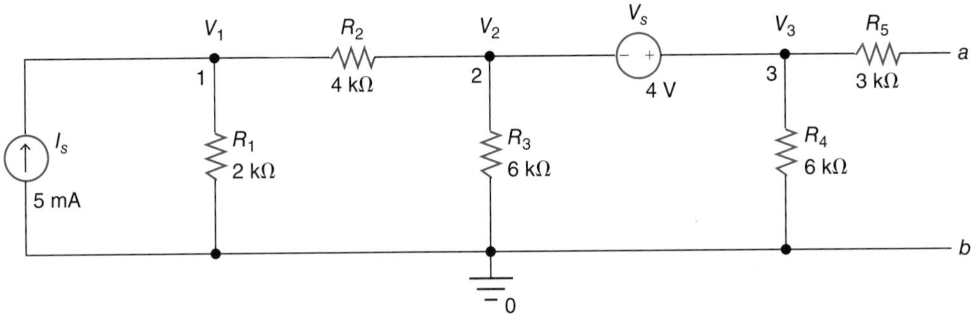

Voltage V_3 at node 3 is 4 V higher than voltage V_2 at node 2. Thus, we have

$$V_3 = V_2 + 4 \quad (4.28)$$

Summing the currents away from node 1, we get

$$-5 \times 10^{-3} + \frac{V_1}{2000} + \frac{V_1 - V_2}{4000} = 0 \quad (4.29)$$

Multiplication of Equation (4.29) by 4000 yields

$$2V_1 + V_1 - V_2 = 20$$

which reduces to

$$3V_1 - V_2 = 20 \quad (4.30)$$

continued

Example 4.9 continued

Summing the currents away from nodes 2 and 3 (i.e., the supernode, as discussed in Chapter 3), we have

$$\frac{V_2 - V_1}{4000} + \frac{V_2}{6000} + \frac{V_3}{6000} = 0 \quad \textbf{(4.31)}$$

Since one end of R_5 is open, there is no current through R_5. Thus, the voltage drop across R_5 is zero. Substitution of Equation (4.28) into Equation (4.31) results in

$$\frac{V_2 - V_1}{4000} + \frac{V_2}{6000} + \frac{V_2 + 4}{6000} = 0 \quad \textbf{(4.32)}$$

Multiplication of Equation (4.32) by 12,000 yields

$$3V_2 - 3V_1 + 2V_2 + 2V_2 + 8 = 0$$

which reduces to

$$-3V_1 + 7V_2 = -8 \quad \textbf{(4.33)}$$

Summing Equations (4.30) and (4.33) results in

$$6V_2 = 12$$

Thus, $V_2 = 2$ V. Substituting this into Equation (4.30), we get $V_1 = 22/3\text{ V} = 7.3333$ V. The Thévenin voltage is V_3, which is the sum of V_2 and 4 V from the voltage source V_s. Thus,

$$V_{\text{th}} = V_3 = V_2 + 4\text{ V} = 2\text{ V} + 4\text{ V} = 6\text{ V}$$

In method 2 of finding the Thévenin equivalent resistance, the terminals a and b are short-circuited, as shown in Figure 4.74.

FIGURE 4.74

Terminals a and b are short-circuited.

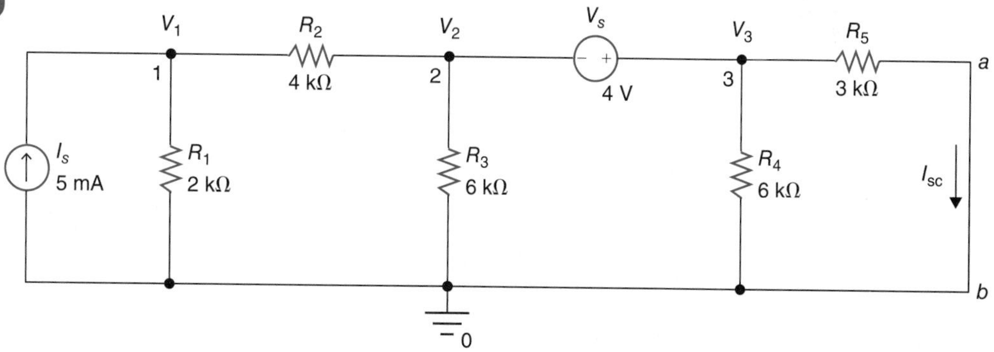

To find the short-circuit current I_{sc} for the circuit shown in Figure 4.46, we modify Equation (4.32) to include the current through the resistor R_5:

$$\frac{V_2 - V_1}{4000} + \frac{V_2}{6000} + \frac{V_3}{6000} + \frac{V_3}{3000} = 0 \quad \textbf{(4.34)}$$

Substitution of Equation (4.28) into Equation (4.34) yields

$$\frac{V_2 - V_1}{4000} + \frac{V_2}{6000} + \frac{V_2 + 4}{6000} + \frac{V_2 + 4}{3000} = 0 \quad \textbf{(4.35)}$$

Multiplying Equation (4.35) by 12,000, we get

$$3V_2 - 3V_1 + 2V_2 + 2V_2 + 8 + 4V_2 + 16 = 0$$

continued

Example 4.9 continued

which can be simplified to

$$-3V_1 + 11V_2 = -24 \qquad (4.36)$$

Summing Equations (4.30) and (4.36) yields

$$10V_2 = -4$$

from which we get $V_2 = -0.4$ V. Thus, $V_3 = V_2 + 4 = 3.6$ V. The short-circuit current is given by

$$I_{sc} = \frac{V_3}{R_3} = \frac{3.6\text{ V}}{3\ k\Omega} = 1.2\text{ mA}$$

Thus, the Thévenin equivalent resistance is given by

$$R_{th} = \frac{V_{th}}{I_{sc}} = \frac{6\text{ V}}{1.2\text{ mA}} = 5\ k\Omega$$

FIGURE 4.75

The Thévenin equivalent circuit.

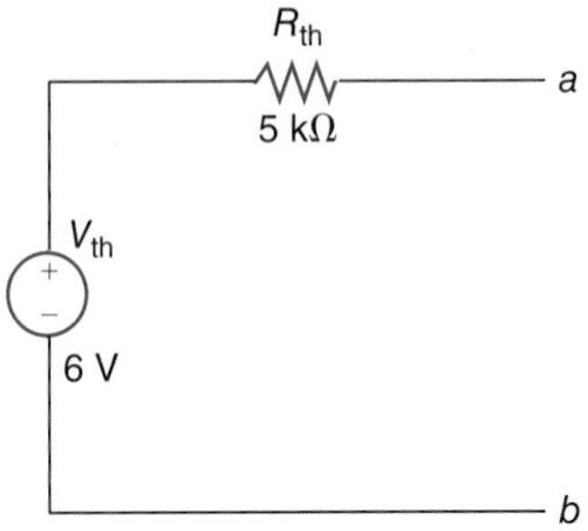

The Thévenin equivalent circuit between terminals a and b is shown in Figure 4.75.

The Thévenin equivalent resistance can also be found using method 1. If current source I_s and voltage source V_s are deactivated (i.e., open-circuit the current source and short-circuit the voltage source) from the circuit shown in Figure 4.73, we obtain the circuit shown in Figure 4.76.

FIGURE 4.76

Circuit shown in Figure 4.73 with the sources deactivated.

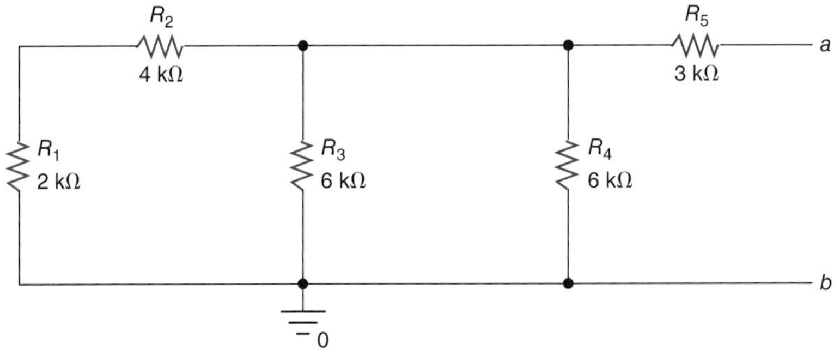

The equivalent resistance of the series connection of R_1 and R_2 is $2\ k\Omega + 4\ k\Omega = 6\ k\Omega$. The equivalent resistance of the parallel connection of $R_1 + R_2$ and R_3 is given by $6\ k\Omega \| 6\ k\Omega = 3\ k\Omega$. The equivalent resistance of this result ($3\ k\Omega$) and R_4 is $3\ k\Omega \| 6\ k\Omega = 18\ k\Omega/9 = 2\ k\Omega$. The series connection of this result ($2\ k\Omega$) and R_5 yields $2\ k\Omega + 3\ k\Omega = 5\ k\Omega$. Thus, the Thévenin equivalent resistance is $5\ k\Omega$.

The Thévenin equivalent resistance can also be found using method 3. If a test voltage source with voltage of 1 V is applied between terminals a and b of the circuit shown in Figure 4.73 after deactivating sources, we obtain the circuit shown in Figure 4.77.

Summing the currents away from node 1 of the circuit shown in Figure 4.77, we obtain

$$\frac{V_1}{2000 + 4000} + \frac{V_1}{6000} + \frac{V_1}{6000} + \frac{V_1 - 1}{3000} = 0 \qquad (4.37)$$

Multiplication of Equation (4.37) by 6000 results in

$$5V_1 = 2 \qquad (4.38)$$

continued

Example 4.9 continued

FIGURE 4.77

Circuit with a test voltage.

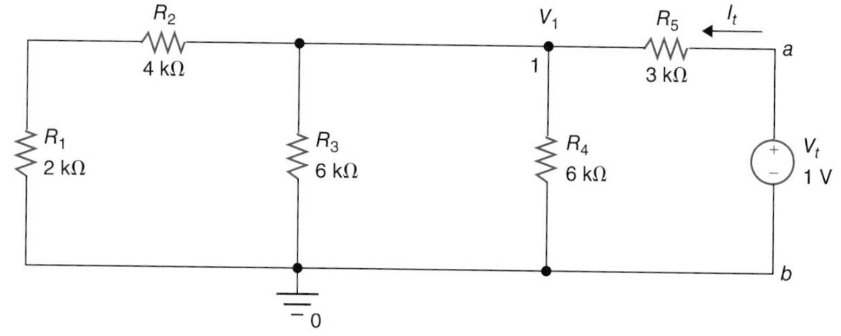

The solution of Equation (4.38) is $V_1 = 2/5\text{ V} = 0.4\text{ V}$. Current I_t is given by

$$I_t = \frac{V_t - V_1}{R_5} = \frac{1\text{ V} - \frac{2}{5}\text{ V}}{3000} = \frac{1}{5}\text{ mA} = 0.2\text{ mA}$$

Thus, the Thévenin equivalent resistance is given by

$$R_{\text{th}} = \frac{V_t}{I_t} = \frac{1\text{ V}}{\frac{1}{5}\text{ mA}} = 5\ k\Omega$$

MATLAB

```
%EXAMPLE 4.9
%Function P.m should be in the same folder as this file.
clear all;
Is=5e-3;Vs=4;R1=2000;R2=4000;R3=6000;R4=6000;R5=3000;Vt=1;
syms V1 V2 V3 Va Vb Vc Vd
%Voc = Vth
[V1,V2,V3]=solve(V3==V2+Vs,...
-Is+V1/R1+(V1-V2)/R2,...
(V2-V1)/R2+V2/R3+V3/R4);
Vth=V3;
%Method 2: Rth2 = Voc/Isc = Vth/Isc
[Va,Vb,Vc]=solve(Vc==Vb+Vs,...
-Is+Va/R1+(Va-Vb)/R2,...
(Vb-Va)/R2+Vb/R3+Vc/R4+Vc/R5);
Isc=Vc/R5;
Rth2=Vth/Isc;
%Method 1: Rth1 = Req
Rth1=R5+P([R4,R3,R2+R1]);
%Method 3: Rth3 = Vt/It (test voltage)
Vd=solve(Vd/(R2+R1)+Vd/R3+Vd/R4+(Vd-Vt)/R5);
It=(Vt-Vd)/R5;
Rth3=Vt/It;
%Display results
V1=vpa(V1,7)
V2=vpa(V2,7)
V3=vpa(V3,7)
Va=vpa(Va,7)
Vb=vpa(Vb,7)
Vc=vpa(Vc,7)
Isc=vpa(Isc,7)
Vd=vpa(Vd,7)
```

continued

Example 4.9 continued

MATLAB continued

```
It=vpa(It,7)
Vth=vpa(Vth,10)
Rth2=vpa(Rth3,10)
Rth1=vpa(Rth1,10)
Rth3=vpa(Rth3,10)

Answers:
V1 =
7.333333
V2 =
2.0
V3 =
6.0
Va =
6.533333
Vb =
-0.4
Vc =
3.6
Isc =
0.0012
Vd =
0.4
It =
0.0002
Vth =
6.0
Rth2 =
5000.0
Rth1 =
5000.0
Rth3 =
5000.0
```

Exercise 4.9

Find the Thévenin equivalent circuit between terminals *a* and *b* for the circuit shown in Figure 4.78.

FIGURE 4.78

Circuit for EXERCISE 4.9.

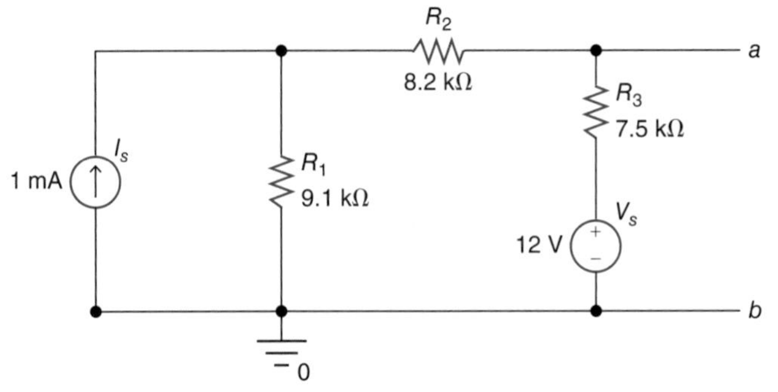

Answer:

$V_{th} = 11.1230$ V, $R_{th} = 5.2319\ k\Omega$.

EXAMPLE 4.10

Find the Thévenin equivalent circuit between terminals *a* and *b* for the circuit shown in Figure 4.79.

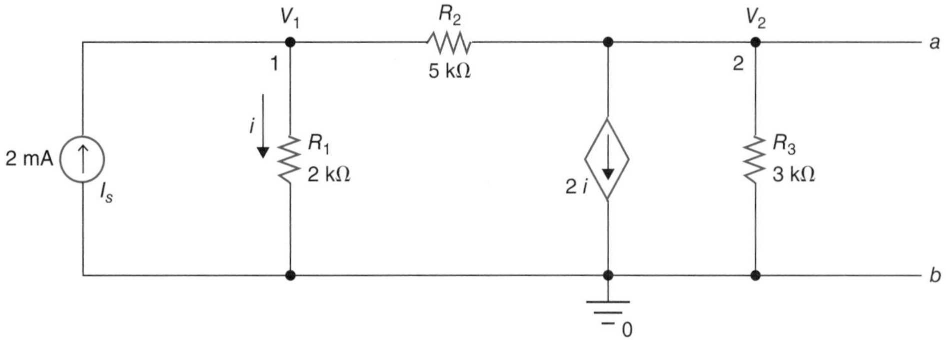

FIGURE 4.79

Circuit with a CCCS.

Notice that the open-circuit voltage between terminals a and b is the voltage across R_3, which is labeled V_2 in the circuit shown in Figure 4.79. Summing the currents away from node 1, we obtain

$$-0.002 + \frac{V_1}{2000} + \frac{V_1 - V_2}{5000} = 0$$

Multiplication of this equation by 10,000 yields

$$-20 + 5V_1 + 2(V_1 - V_2) = 0$$

which can be simplified to

$$7V_1 - 2V_2 = 20 \tag{4.39}$$

Summing the currents away from node 2, we obtain

$$\frac{V_2 - V_1}{5000} + 2\frac{V_1}{2000} + \frac{V_2}{3000} = 0$$

Multiplication by 30,000 yields

$$6V_2 - 6V_1 + 30V_1 + 10V_2 = 0$$

which can be simplified to

$$24V_1 + 16V_2 = 0 \tag{4.40}$$

Solving Equation (4.40) for V_1, we have

$$V_1 = -\frac{2}{3}V_2$$

Substituting this into Equation (4.39), we obtain

$$7V_1 - 2V_2 = 7\left(-\frac{2}{3}V_2\right) - 2V_2 = -\frac{20}{3}V_2 = 20$$

continued

Example 4.10 continued

Thus, $V_2 = V_{oc} = -3$ V. The Thévenin equivalent voltage is the open-circuit voltage between a and b, which is V_2. Therefore,

$$V_{th} = V_{oc} = -3 \text{ V}$$

To find the Thévenin equivalent resistance, we first use method 2. The terminals a and b are short-circuited without changing the rest of the circuit, as shown in Figure 4.80.

FIGURE 4.80

A circuit with terminals a and b short-circuited.

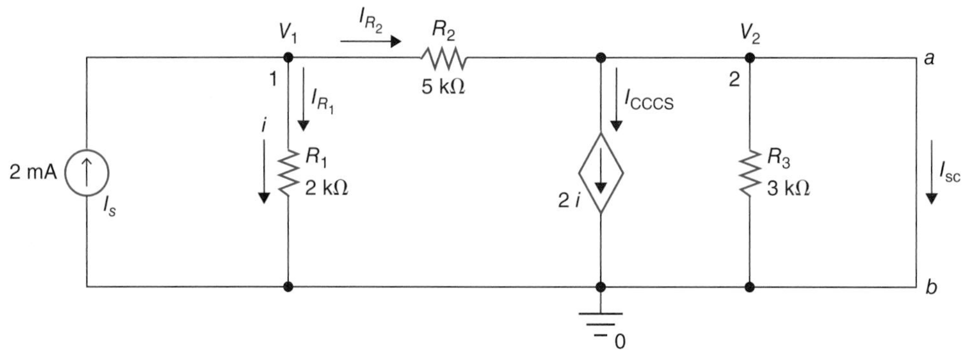

Notice that node 2 is connected to ground and no current flows through R_3. From the current divider rule, the current through R_1 is given by

$$I_{R_1} = i = I_s \times \frac{R_2}{R_1 + R_2} = 2 \text{ mA} \times \frac{5\,k\Omega}{2\,k\Omega + 5\,k\Omega} = \frac{10}{7} \text{ mA}$$

Similarly, the current through R_2 is given by

$$I_{R_2} = I_s \times \frac{R_1}{R_1 + R_2} = 2 \text{ mA} \times \frac{2\,k\Omega}{2\,k\Omega + 5\,k\Omega} = \frac{4}{7} \text{ mA}$$

The current through a current-controlled current source (CCCS) is given by

$$I_{CCCS} = 2i = 2I_{R_1} = \frac{20}{7} \text{ mA}$$

Applying KCL at node 2, we obtain

$$I_{R_2} = I_{CCCS} + I_{sc}$$

Solving for I_{sc}, we have

$$I_{sc} = I_{R_2} - I_{CCCS} = \frac{4}{7} \text{ mA} - \frac{20}{7} \text{ mA} = -\frac{16}{7} \text{ mA} = -2.2857 \text{ mA}$$

The Thévenin equivalent resistance is the ratio of the open-circuit voltage to the short-circuit current. Thus, we have

$$R_{th} = \frac{V_{oc}}{I_{sc}} = \frac{-3 \text{ V}}{-\frac{16}{7} \text{ mA}} = \frac{21}{16} k\Omega = 1.3125\, k\Omega$$

continued

Example 4.10 continued

The Thévenin equivalent resistance can also be found using method 3. After deactivating the current source by removing it from the circuit, a test voltage is applied to the circuit from terminals a and b, as shown in Figure 4.81.

FIGURE 4.81

A circuit with a test voltage source.

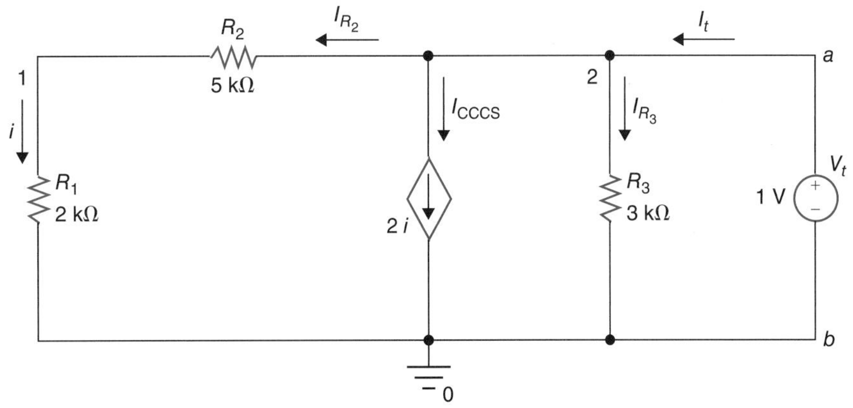

The test voltage V_t is applied across R_3 and $R_2 - R_1$. The current through $R_2 - R_1$, which is also the controlling current, is given by

$$I_{R_2} = i = \frac{V_t}{R_2 + R_1} = \frac{1\text{ V}}{7\,k\Omega} = \frac{1}{7}\text{ mA}$$

The current through R_3 is given by

$$I_{R_3} = \frac{V_t}{R_3} = \frac{1\text{ V}}{3\,k\Omega} = \frac{1}{3}\text{ mA}$$

The current through CCCS is twice the controlling current i. Thus,

$$I_{CCCS} = 2i = 2I_{R_2} = \frac{2}{7}\text{ mA}$$

The total current flowing out of the positive terminal of the test voltage source is given by

$$I_t = I_{R_2} + I_{CCCS} + I_{R_3} = \frac{1}{7}\text{ mA} + \frac{2}{7}\text{ mA} + \frac{1}{3}\text{ mA} = \frac{16}{21}\text{ mA} = 0.7619\text{ mA}$$

The Thévenin equivalent resistance is the ratio of V_t to I_t:

$$R_{\text{th}} = \frac{V_t}{I_t} = \frac{1\text{ V}}{\frac{16}{21}\text{ mA}} = \frac{21}{16}\,k\Omega = 1.3125\,k\Omega$$

After deactivating the current source, instead of a test voltage source, a test current source can be applied to the circuit from terminals a and b, as shown in Figure 4.82. Summing the currents leaving node 2, we obtain

$$\frac{V_t}{7000} + \frac{2V_t}{7000} + \frac{V_t}{3000} - 0.001 = 0$$

Multiplication by 21,000 yields

$$3V_t + 6V_t + 7V_t = 21$$

continued

Example 4.10 continued

FIGURE 4.82

A circuit with a test current source.

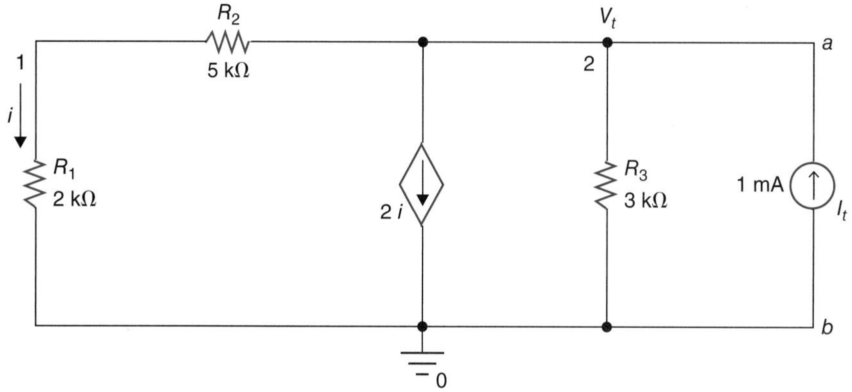

FIGURE 4.83

The Thévenin equivalent circuit.

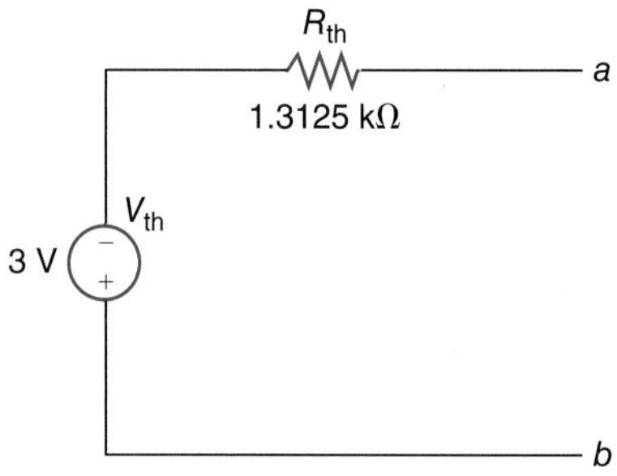

Solving for V_t, we obtain

$$V_t = \frac{21}{16}\text{ V} = 1.3125\text{ V}$$

The Thévenin equivalent resistance is the ratio of V_t to I_t:

$$R_{th} = \frac{V_t}{I_t} = \frac{\frac{21}{16}\text{ V}}{1\text{ mA}} = \frac{21}{16}k\Omega = 1.3125\ k\Omega$$

The Thévenin equivalent circuit is shown in Figure 4.83.

MATLAB

```
%EXAMPLE 4.10
clear all;
Is=2e-3;R1=2000;R2=5000;R3=3000;ki=2;Vta=1;Itb=1e-3;
syms V1 V2 Va Vtb
%Voc = Vth
[V1,V2]=solve(-Is+V1/R1+(V1-V2)/R2,(V2-V1)/R2+ki*V1/R1+V2/R3);
Vth=V2;
%Method 2: Rth2 = Voc/Isc
Va=solve(-Is+Va/R1+Va/R2);
Isc=Va/R2-ki*Va/R1;
Rth2=Vth/Isc;
%Method 3: Rth3a = Vta/Ita (test voltage)
IR2=Vta/(R2+R1);
Icccs=ki*IR2;
IR3=Vta/R3;
Ita=IR2+Icccs+IR3;
Rth3a=Vta/Ita;
Rth3a=vpa(Rth3a,10);
%Method 3: Rth3b = Vtb/Itb (test current)
Vtb=solve(Vtb/(R2+R1)+ki*Vtb/(R2+R1)+Vtb/R3-Itb);
Rth3b=Vtb/Itb;
Rth3b=vpa(Rth3b,10);
%Display results
V1=vpa(V1,7)
V2=vpa(V2,7)
Va=vpa(Va,7)
```

continued

Example 4.10 continued

MATLAB continued

```
Isc=vpa(Isc,7)
Ita=vpa(Ita,7)
Vtb=vpa(Vtb,7)
Vth=vpa(Vth,10)
Rth2=vpa(Rth2,10)
Rth3a=vpa(Rth3a,10)
Rth3b=vpa(Rth3b,10)

Answers:
V1 =
2.0
V2 =
-3.0
Va =
2.857143
Isc =
-0.002285714
Ita =
0.0007619048
Vtb =
1.3125
Vth =
-3.0
Rth2 =
1312.5
Rth3a =
1312.5
Rth3b =
1312.5
```

Exercise 4.10

Find the Thévenin equivalent circuit between terminals *a* and *b* for the circuit shown in Figure 4.84.

FIGURE 4.84

Circuit for EXERCISE 4.10.

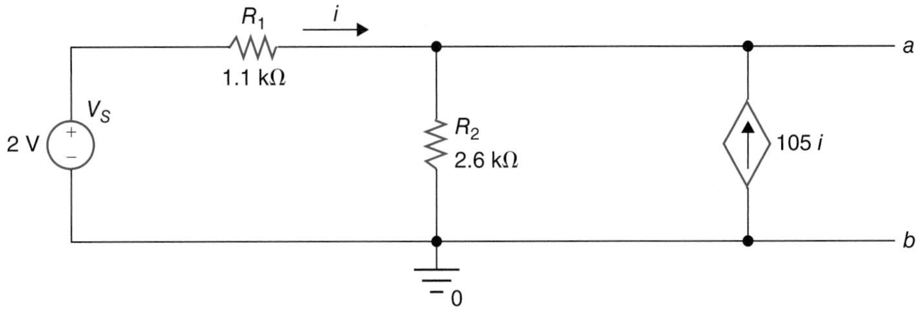

Answer:

$V_{th} = 1.9920$ V, $R_{th} = 10.3361\ \Omega$.

EXAMPLE 4.11

Find the Thévenin equivalent circuit between terminals *a* and *b* for the circuit shown in Figure 4.85.

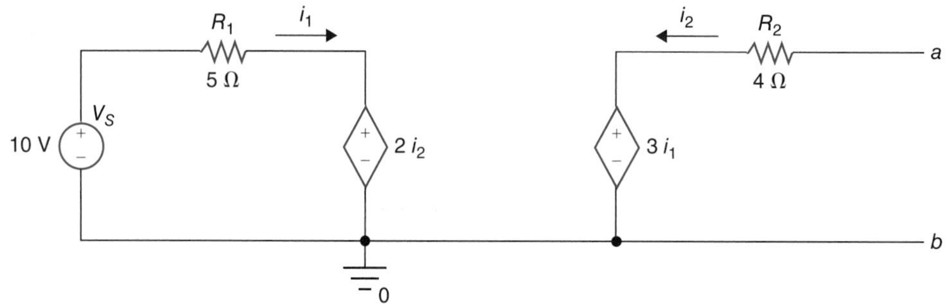

FIGURE 4.85
Circuit for EXAMPLE 4.11.

Since current i_2 equals zero, the voltage across the CCVS on the left side of the circuit is zero. Current i_1 is given by

$$i_1 = \frac{V_s}{R_1} = \frac{10\text{ V}}{5\ \Omega} = 2\text{ A}$$

The Thévenin voltage is the voltage across the CCVS on the right side of the circuit. Thus,

$$V_{\text{th}} = 3i_1 = 3 \times 2\text{ V} = 6\text{ V}$$

To find the Thévenin resistance, we deactivate the voltage source by short-circuiting it, and then apply a test voltage V_t of 1 V across a and b, as shown in Figure 4.86.

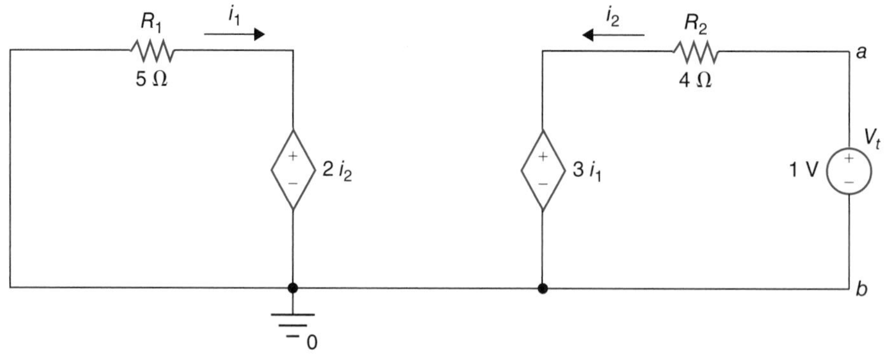

FIGURE 4.86
A circuit with a test voltage.

Collecting the voltage drops around the mesh on the left side in the clockwise direction, we obtain

$$5i_1 + 2i_2 = 0$$

Solving for i_1, we get

$$i_1 = \frac{-2i_2}{5} = -0.4i_2 \tag{4.41}$$

continued

Example 4.11 continued

Collecting the voltage drops around the mesh on the right side in the clockwise direction, we obtain

$$-3i_1 - 4i_2 + 1 = 0 \quad (4.42)$$

Substitution of Equation (4.41) into Equation (4.42) yields

$$1.2i_2 - 4i_2 + 1 = 0$$

Solving for i_2, we get

$$i_2 = \frac{1}{2.8}\ \text{A} = 0.3571\ \text{A}$$

The Thévenin equivalent resistance is given by

$$R_{th} = \frac{V_t}{i_2} = \frac{1\ \text{V}}{\frac{1}{2.8}\ \text{A}} = 2.8\ \Omega$$

The Thévenin equivalent circuit is shown in Figure 4.87.

FIGURE 4.87

The Thévenin equivalent circuit.

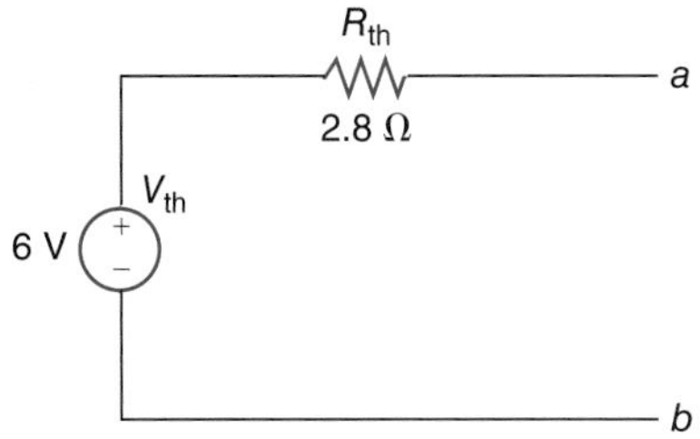

Exercise 4.11

Find the Thévenin equivalent circuit between terminals *a* and *b* for the circuit shown in Figure 4.88.

FIGURE 4.88

Circuit for EXERCISE 4.11.

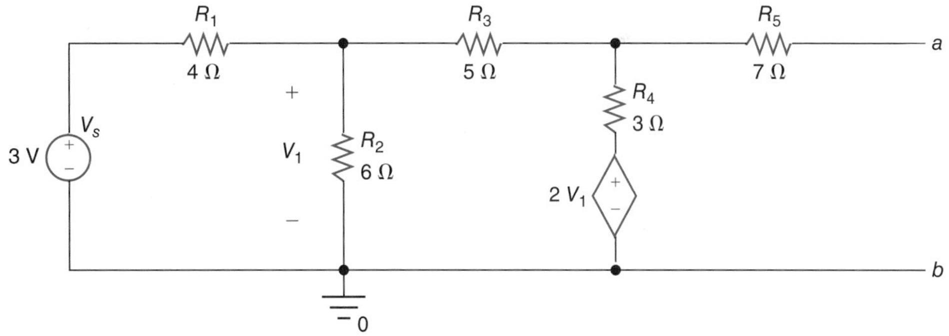

Answer:

$V_{th} = 4.1786$ V, $R_{th} = 10.9643\ \Omega$.

EXAMPLE 4.12

Find the Thévenin equivalent circuit between terminals *a* and *b* for the circuit shown in Figure 4.89.

continued

Example 4.12 continued

FIGURE 4.89

Circuit for EXAMPLE 4.12.

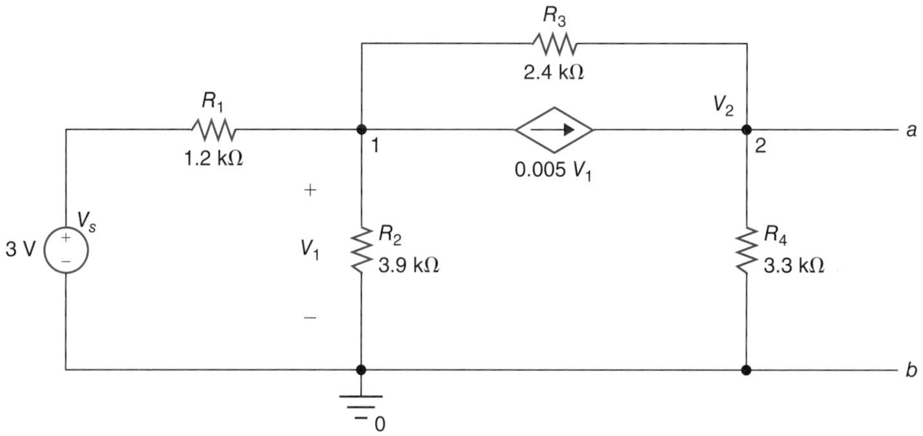

The Thévenin equivalent circuit between terminals a and b can be found by finding the open-circuit voltage V_{oc} and the short-circuit current I_{sc}. The open-circuit voltage V_{oc} is V_2, which is the voltage across R_4 in the circuit shown in Figure 4.89. Summing the currents leaving node 1, we obtain

$$\frac{V_1 - V_s}{R_1} + \frac{V_1}{R_2} + \frac{V_1 - V_2}{R_3} + g_m V_1 = 0 \quad \textbf{(4.43)}$$

which can be rearranged as

$$\left(\frac{1}{R_1} + \frac{1}{R_2} + \frac{1}{R_3} + g_m\right)V_1 - \frac{1}{R_3}V_2 = \frac{V_s}{R_1}$$

Substituting the component values, we get

$$\left(\frac{1}{1200} + \frac{1}{3900} + \frac{1}{2400} + 0.005\right)V_1 - \frac{1}{2400}V_2 = \frac{3}{1200}$$

which can be simplified to

$$6.50641 \times 10^{-3}V_1 - 4.16667 \times 10^{-4}V_2 = 2.5 \times 10^{-3}$$

Multiplication by 1000 yields

$$6.50641V_1 - 0.416667V_2 = 2.5 \quad \textbf{(4.44)}$$

Summing the currents leaving node 2, we obtain

$$\frac{V_2 - V_1}{R_3} - g_m V_1 + \frac{V_2}{R_4} = 0 \quad \textbf{(4.45)}$$

which can be revised as

$$-\left(\frac{1}{R_3} + g_m\right)V_1 + \left(\frac{1}{R_3} + \frac{1}{R_4}\right)V_2 = 0$$

Substituting the component values, we get

$$-\left(\frac{1}{2400} + 0.005\right)V_1 + \left(\frac{1}{2400} + \frac{1}{3300}\right)V_2 = 0$$

continued

Example 4.12 continued

which can be simplified to

$$-5.416667 \times 10^{-3}V_1 + 7.19697 \times 10^{-4}V_2 = 0$$

Multiplication by 1000 yields

$$-5.416667V_1 + 0.719697V_2 = 0 \qquad \textbf{(4.46)}$$

Solving Equation (4.46) for V_2, we obtain

$$V_2 = \frac{5.416667}{0.719697}V_1 = 7.526317V_1$$

Substituting V_2 into Equation (4.44), we get

$$6.50641V_1 - 0.416667(7.526317V_1) = 2.5$$

Thus,

$$V_1 = \frac{2.5}{6.50641 - 0.416667 \times 7.526317} = 0.74174 \text{ V}$$

$$V_2 = 7.526317V_1 = 5.5826 \text{ V}$$

Alternatively, application of Cramer's rule to Equations (4.44) and (4.46) yields

$$V_1 = \frac{\begin{vmatrix} 2.5 & -0.416667 \\ 0 & 0.719697 \end{vmatrix}}{\begin{vmatrix} 6.50641 & -0.416667 \\ -5.416667 & 0.719697 \end{vmatrix}} = \frac{1.7992}{2.4257} = 0.74174 \text{ V}$$

$$V_2 = \frac{\begin{vmatrix} 6.50641 & 2.5 \\ -5.416667 & 0 \end{vmatrix}}{\begin{vmatrix} 6.50641 & -0.416667 \\ -5.416667 & 0.719697 \end{vmatrix}} = \frac{13.541667}{2.4257} = 5.5826 \text{ V}$$

To find the short-circuit current, we short-circuit a and b, as shown in Figure 4.90.

FIGURE 4.90

The circuit from Figure 4.89 with a and b short-circuited.

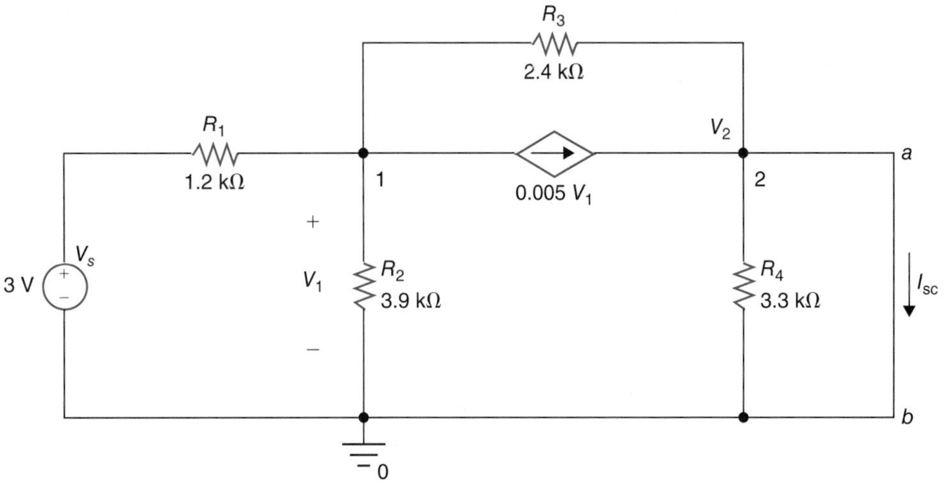

continued

Example 4.12 continued

Notice that node 2 is connected to ground and no current flows through R_4. Summing the currents leaving node 1 of the circuit shown in Figure 4.90, we obtain

$$\frac{V_1 - V_s}{R_1} + \frac{V_1}{R_2} + \frac{V_1}{R_3} + g_m V_1 = 0$$

which can be rearranged as

$$\left(\frac{1}{R_1} + \frac{1}{R_2} + \frac{1}{R_3} + g_m\right)V_1 = \frac{V_s}{R_1}$$

Thus, we obtain

$$V_1 = \frac{\dfrac{V_s}{R_1}}{\dfrac{1}{R_1} + \dfrac{1}{R_2} + \dfrac{1}{R_3} + g_m} = \frac{\dfrac{3}{1200}}{\dfrac{1}{1200} + \dfrac{1}{3900} + \dfrac{1}{2400} + 0.005} = 0.38423645 \text{ V}$$

The short-circuit current is given by

$$I_{sc} = \frac{V_1}{R_3} + g_m V_1 = V_1\left(\frac{1}{R_3} + g_m\right)$$

$$= 0.38423645\left(\frac{1}{2400} + 0.005\right) = 2.0813 \text{ mA}$$

The Thévenin equivalent resistance is given by

$$R_{th} = \frac{V_{oc}}{I_{sc}} = \frac{5.5826 \text{ V}}{2.6823 \text{ mA}} = 2.6823\ k\Omega$$

The Thévenin equivalent circuit is shown in Figure 4.91.

FIGURE 4.91

The Thévenin equivalent circuit.

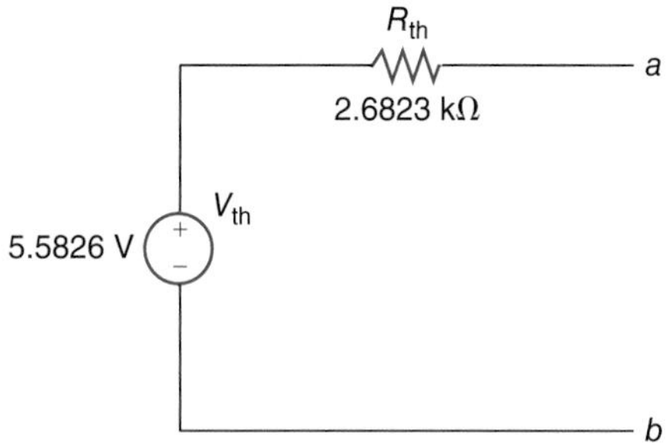

MATLAB

```
%EXAMPLE 4.12
clear all;
Vs=3;R1=1200;R2=3900;R3=2400;R4=3300;gm=0.005;
syms V1 V2 Va Vb
%Voc = Vth
[V1,V2]=solve((V1-Vs)/R1+V1/R2+(V1-V2)/R3+gm*V1,...
(V2-V1)/R3-gm*V1+V2/R4,V1,V2);
Vth=V2;
%Method 2: Rth = Voc/Isc
Va=solve((Va-Vs)/R1+Va/R2+(Va-0)/R3+gm*Va,Va);
Isc=Va/R3+gm*Va;
Rth=Vth/Isc;
%Display results
V1=vpa(V1,8)
V2=vpa(V2,8)
Va=vpa(Va,8)
Isc=vpa(Isc,8)
Vth=vpa(Vth,8)
Rth=vpa(Rth,8)

Answers:
V1 =
0.74174174
```

continued

Example 4.12 continued
MATLAB continued

```
V2 =
5.5825826
Va =
0.38423645
Isc =
0.0020812808
Vth =
5.5825826
Rth =
2682.2823
```

Exercise 4.12

Find the Thévenin equivalent circuit between terminals *a* and *b* for the circuit shown in Figure 4.92.

FIGURE 4.92

Circuit for EXERCISE 4.12.

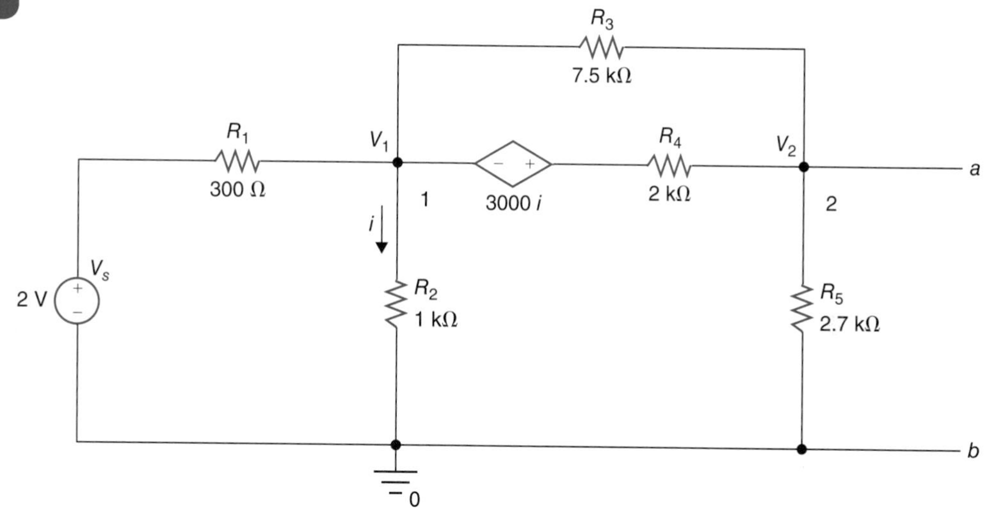

Answer:
$V_{th} = 2.7672$ V, $R_{th} = 1.2582$ $k\Omega$.

4.5 Norton's Theorem

FIGURE 4.93

A Norton equivalent circuit.

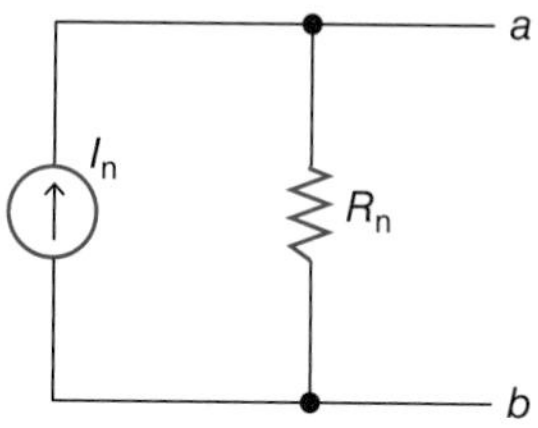

A circuit looking from terminals *a* and *b* can be replaced by a current source with current I_n and a parallel resistor with resistance R_n, as shown in Figure 4.93. This equivalent circuit consisting of a current source and a parallel resistor is called **Norton equivalent circuit**. The current I_n is called **Norton equivalent current** and the resistance R_n is called **Norton equivalent resistance**.

The Norton equivalent circuit can be used to simplify a circuit. When a load resistor is connected between terminals *a* and *b*, we can find the effects of the circuit on the load from the Norton equivalent circuit. We do not need all the details of the original circuit to find the voltage, current, and power on the load.

When the terminals *a* and *b* are short-circuited in the Norton equivalent circuit, as shown in Figure 4.94, the short-circuit current I_{sc} is equal to I_n from the current divider rule. Thus, the Norton equivalent current can be obtained by finding the short-circuit current.

FIGURE 4.94

Short-circuit current.

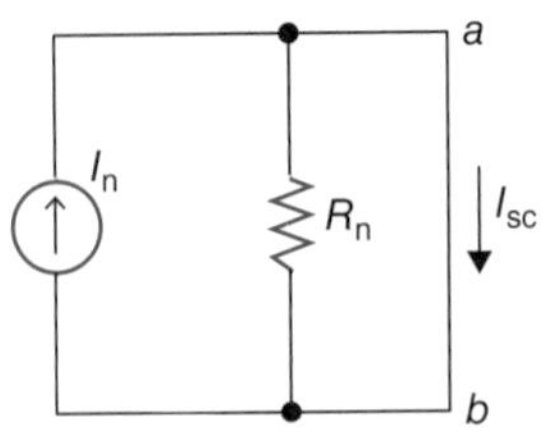

4.5.1 FINDING THE NORTON EQUIVALENT CURRENT I_n

Given a circuit and terminals a and b, we can find the Norton equivalent current I_n with respect to terminals a and b by finding the short-circuit current I_{sc} between terminals a and b. The short-circuit current I_{sc} can be found by utilizing circuit analysis methods such as the voltage divider rule, current divider rule, superposition principle, nodal analysis, and mesh analysis. The Norton equivalent current I_n is found from the original circuit by connecting a wire between a and b without any changes for the rest of the circuit.

4.5.2 FINDING THE NORTON EQUIVALENT RESISTANCE R_n

The Norton equivalent resistance can be obtained by applying the three methods used to find the Thévenin equivalent resistance. These three methods are listed next.

Method 1

Deactivate all the independent sources by short-circuiting the voltage sources and open-circuiting the current sources. Find the equivalent resistance looking into the circuit from terminals a and b. This equivalent resistance is the Norton equivalent resistance R_n. This method can be used if the circuit does not contain dependent sources.

Method 2

Find the open-circuit voltage V_{oc} and the short-circuit current I_{sc} between terminals a and b. The Norton equivalent resistance R_n is given by

$$R_n = \frac{V_{oc}}{I_{sc}}$$

Method 3

Deactivate all the independent sources by open-circuiting current sources and short-circuiting voltage sources. Apply a test voltage of 1 V (or any other value) between terminals a and b with terminal a connected to the positive terminal of the test voltage. Measure the current flowing out of the positive terminal of the test voltage source. The Norton equivalent resistance R_n is given by the ratio of the test voltage to the current flowing out of the positive terminal of the test voltage source. A test current can be used instead of a test voltage. Apply a test current between terminals a and b after deactivating the independent sources and measure the voltage across a and b of the test current source. The Norton equivalent resistance R_n is the ratio of the voltage across a and b to the test current.

4.5.3 RELATION BETWEEN THE THÉVENIN EQUIVALENT CIRCUIT AND THE NORTON EQUIVALENT CIRCUIT

Application of source transformation to the Norton equivalent circuit shown in Figure 4.95(a) yields the Thévenin equivalent circuit shown in Figure 4.95(b). Notice that the Thévenin equivalent voltage is $V_{th} = I_n R_n$ and the Thévenin equivalent resistance is $R_{th} = R_n$. The source transformation does not change the resistance value. Application of source transformation to the Thévenin equivalent circuit shown in Figure 4.96(a) yields the Norton equivalent

FIGURE 4.95

Transformation from Norton equivalent circuit to Thévenin equivalent circuit.

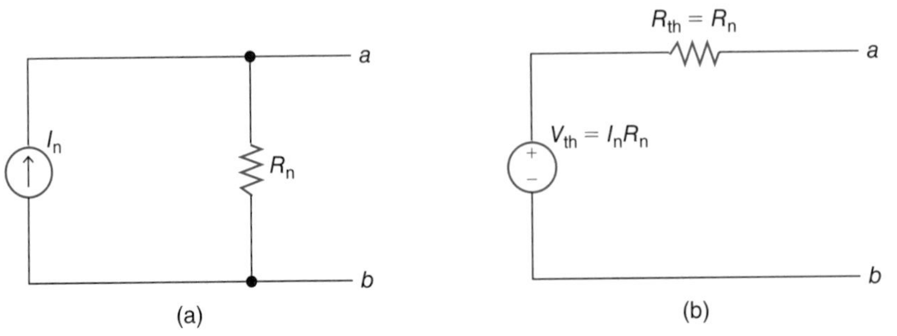

FIGURE 4.96

Transformation from Thévenin equivalent circuit to Norton equivalent circuit.

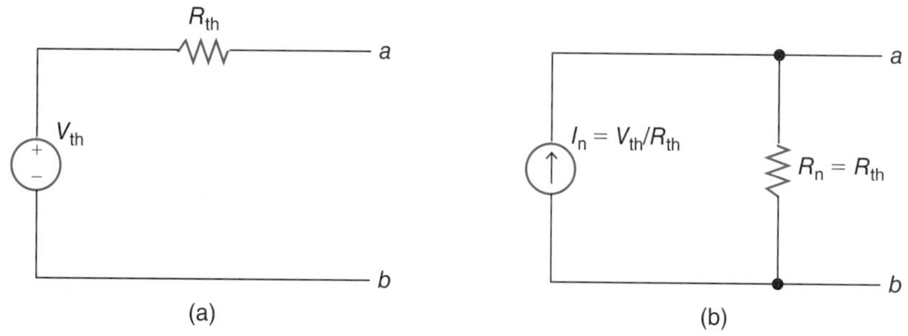

circuit, as shown in Figure 4.96(b). Notice that the Norton equivalent current is $I_n = V_{th}/R_{th}$ and the Norton equivalent resistance is $R_n = R_{th}$. The source transformation does not change the resistance value.

Consider the circuit shown in Figure 4.97. We are interested in finding a Norton equivalent circuit looking into the circuit from terminals a and b.

FIGURE 4.97

A circuit with two sources.

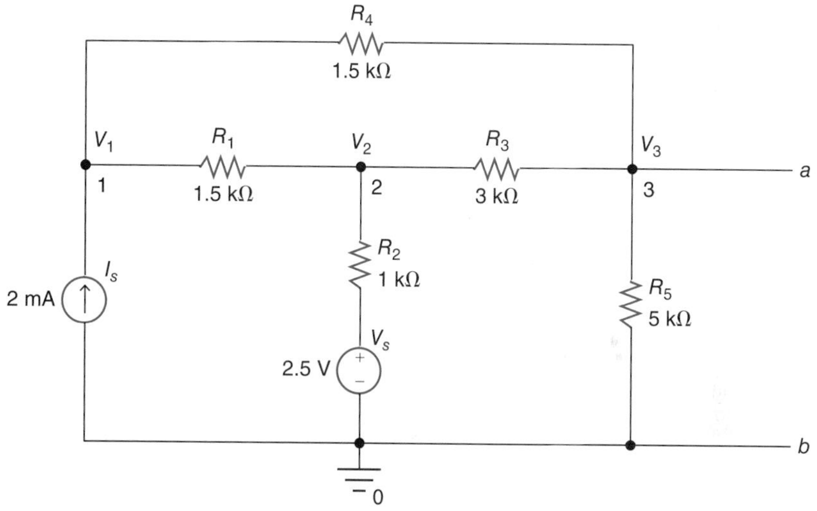

To find the short-circuit current, we short-circuit a and b, as shown in Figure 4.98. Node 3 is a ground, and no current flows through R_5.

FIGURE 4.98

A circuit with a and b shorted.

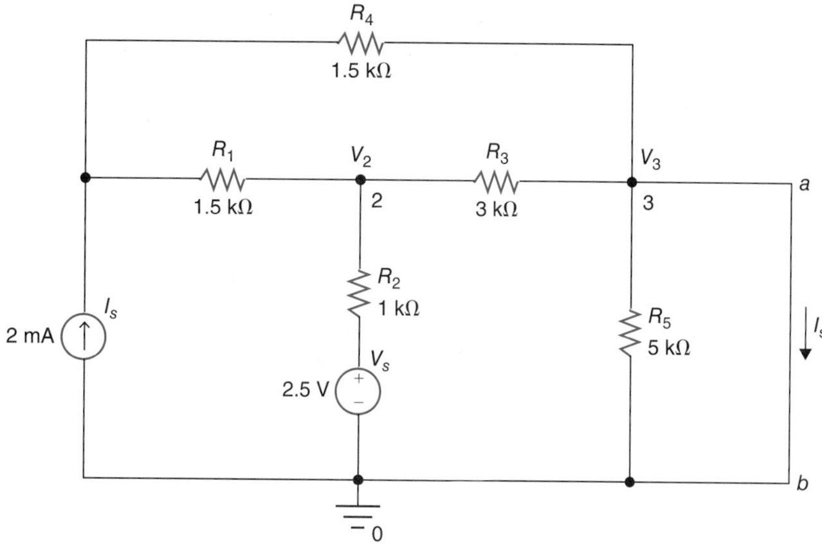

We can find the short-circuit current using the superposition principle. First, we deactivate the voltage source by short-circuiting it, as shown in Figure 4.99.

FIGURE 4.99

A circuit with the voltage source deactivated.

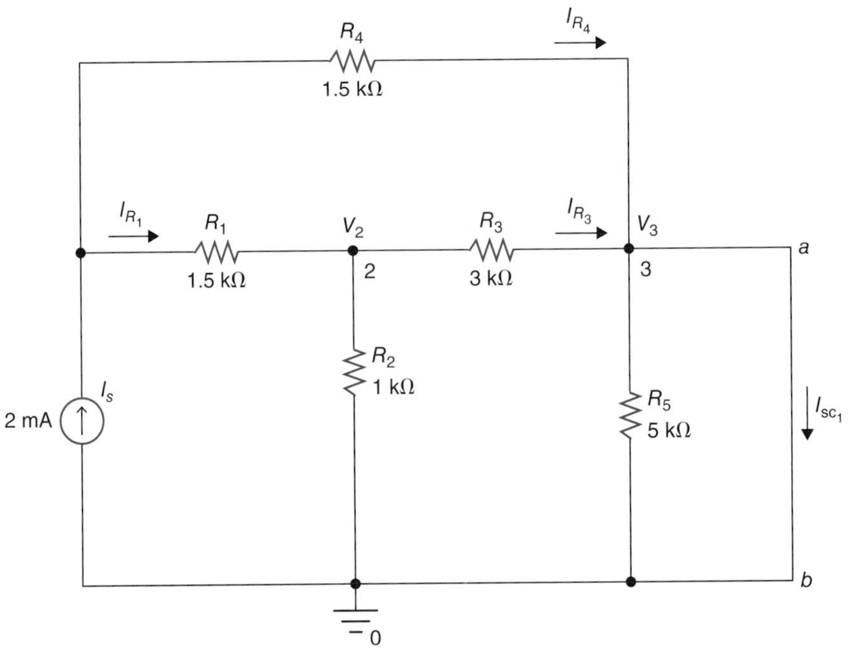

Notice that R_2 and R_3 are connected in parallel. Let R_a be the equivalent resistance of $R_1 + (R_2 \| R_3)$. Then, we have

$$R_a = R_1 + (R_2 \| R_3) = 1.5\ k\Omega + (1\ k\Omega \| 3\ k\Omega) = 1.5\ k\Omega + 0.75\ k\Omega = 2.25\ k\Omega$$

Application of the current divider rule yields

$$I_{R_4} = I_s \times \frac{R_a}{R_a + R_4} = 2\text{ mA} \times \frac{2.25\ k\Omega}{2.25\ k\Omega + 1.5\ k\Omega} = 1.2\text{ mA}$$

The current through R_1 is given by

$$I_{R_1} = I_s - I_{R_4} = 2\text{ mA} - 1.2\text{ mA} = 0.8\text{ mA}$$

Application of the current divider rule to R_2 and R_3 yields

$$I_{R_3} = I_{R_1} \times \frac{R_2}{R_2 + R_3} = 0.8\text{ mA} \times \frac{1\ k\Omega}{1\ k\Omega + 3\ k\Omega} = 0.2\text{ mA}$$

The short-circuit current is given by

$$I_{sc_1} = I_{R_4} + I_{R_3} = 1.2\text{ mA} + 0.2\text{ mA} = 1.4\text{ mA}$$

Now, we deactivate the current source by open-circuiting it, as shown in Figure 4.100. Let R_b be the equivalent resistance of parallel connection of $R_1 + R_4$ and R_3. Then, we have

$$R_b = (R_1 + R_4) \| R_3 = 3\ k\Omega \| 3\ k\Omega = 1.5\ k\Omega$$

FIGURE 4.100

Circuit with the current source deactivated.

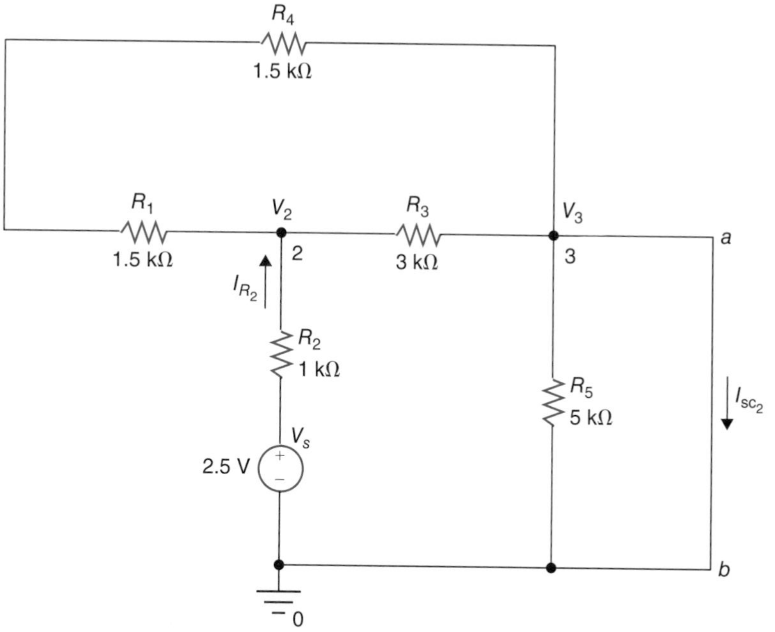

The total resistance seen from the voltage source is $R_2 + R_b$. Thus, the current through R_2, which is also the short-circuit current, is given by

$$I_{R_2} = I_{sc_2} = \frac{V_s}{R_2 + R_b} = \frac{2.5\text{ V}}{1\ k\Omega + 1.5\ k\Omega} = 1\text{ mA}$$

The total short-circuit current from the two sources is

$$I_{sc} = I_{sc_1} + I_{sc_2} = 1.4\text{ mA} + 1\text{ mA} = 2.4\text{ mA}$$

To find the Norton equivalent resistance, we deactivate the two sources in the circuit shown in Figure 4.97 to obtain the circuit shown in Figure 4.101.

FIGURE 4.101

The circuit from Figure 4.97 with sources deactivated.

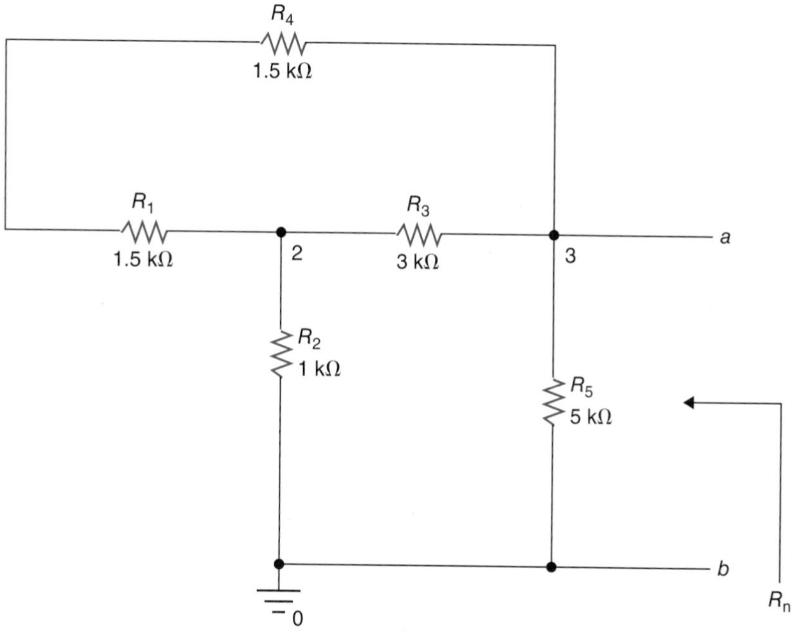

FIGURE 4.102

Norton equivalent circuit.

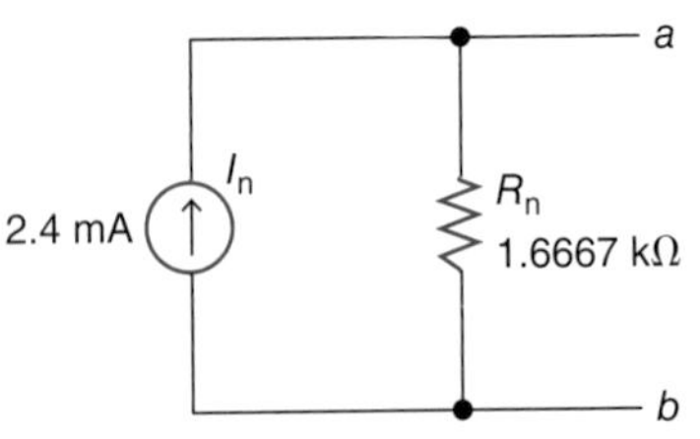

The Norton equivalent resistance is given by the parallel connection of R_5 and $R_b + R_2$, where $R_b = (R_1 + R_4)\|R_3 = 1.5\ k\Omega$. Thus, we have

$$R_n = R_5\|(R_b + R_2) = 5\ k\Omega\|2.5\ k\Omega = 1.6667\ k\Omega.$$

The Norton equivalent circuit is shown in Figure 4.102.

Consider a circuit with a CCVS shown in Figure 4.103. We are interested in finding the Norton equivalent circuit between terminals *a* and *b*.

To find the Norton equivalent current, we short-circuit terminals *a* and *b*, as shown in Figure 4.104. Notice that node 2 is connected to ground, and the current through R_4 is zero.

FIGURE 4.103

A circuit with CCVS.

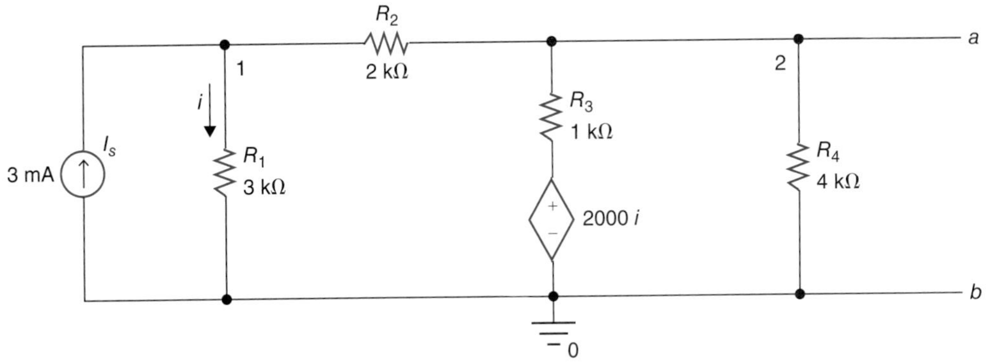

FIGURE 4.104

A circuit with *a* and *b* short-circuited.

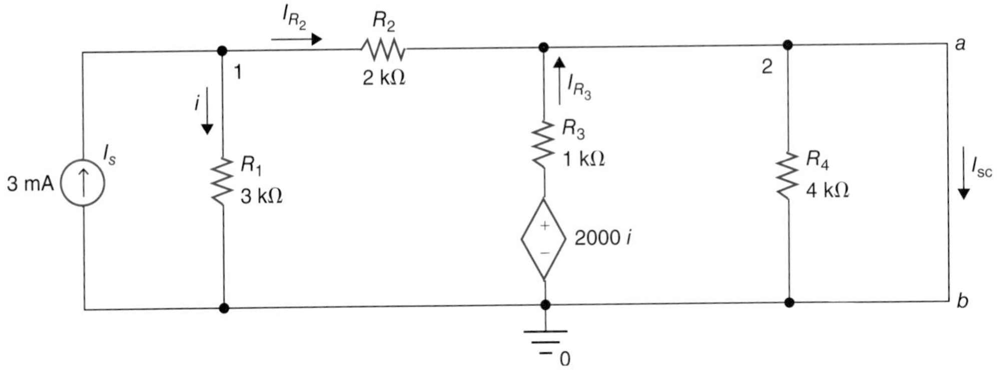

From the current divider rule, the current through R_2 is given by

$$I_{R_2} = I_s \times \frac{R_1}{R_1 + R_2} = 3\text{ mA} \times \frac{3\ k\Omega}{3\ k\Omega + 2\ k\Omega} = 1.8\text{ mA}$$

From KCL, the current through R_1 is given by

$$i = I_s - I_{R_2} = 3\text{ mA} - 1.8\text{ mA} = 1.2\text{ mA}$$

The voltage across the CCVS is

$$V_{CCVS} = 2000i = 2000\,(\text{V/A}) \times 1.2\text{ mA} = 2.4\text{ V}$$

The current through R_3 is

$$I_{R_3} = \frac{V_{CCVS}}{R_3} = \frac{2.4\text{ V}}{1\ k\Omega} = 2.4\text{ mA}$$

The short-circuit current is the sum of I_{R_2} and I_{R_3}:

$$I_{sc} = I_n = I_{R_2} + I_{R_3} = 1.8 \text{ mA} + 2.4 \text{ mA} = 4.2 \text{ mA}$$

To find the Norton equivalent resistance, after deactivating the current source, we apply a test voltage to the circuit at the terminals a and b, as shown in Figure 4.105.

FIGURE 4.105

A circuit with a test voltage source.

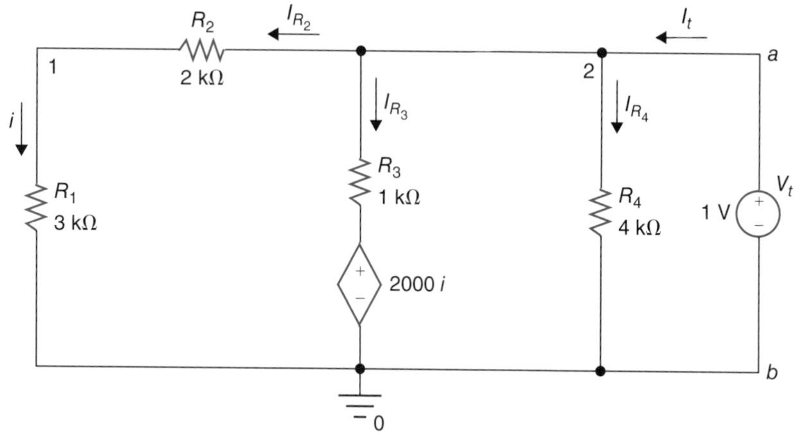

The current through R_2, which is also the controlling current i, is given by

$$I_{R_2} = i = \frac{V_t}{R_2 + R_1} = \frac{1 \text{ V}}{2\ k\Omega + 3\ k\Omega} = 0.2 \text{ mA}$$

The voltage of the CCVS is given by

$$V_{CCVS} = 2000i = 2000\,(\text{V/A}) \times 0.2 \text{ mA} = 0.4 \text{ V}$$

The current through R_3 is given by

$$I_{R_3} = \frac{V_t - V_{CCVS}}{R_3} = \frac{1 \text{ V} - 0.4 \text{ V}}{1\ k\Omega} = 0.6 \text{ mA}$$

The current through R_4 is given by

$$I_{R_4} = \frac{V_t}{R_4} = \frac{1 \text{ V}}{4\ k\Omega} = 0.25 \text{ mA}$$

The total current flowing out of the positive terminal of the test voltage source is given by

$$I_t = I_{R_2} + I_{R_3} + I_{R_4} = 0.2 \text{ mA} + 0.6 \text{ mA} + 0.25 \text{ mA} = 1.05 \text{ mA}$$

The Norton equivalent resistance is the ratio of V_t to I_t:

$$R_n = \frac{V_t}{I_t} = \frac{1 \text{ V}}{1.05 \text{ mA}} = 952.381\ \Omega$$

The Norton equivalent circuit is shown in Figure 4.106.

FIGURE 4.106

Norton equivalent circuit.

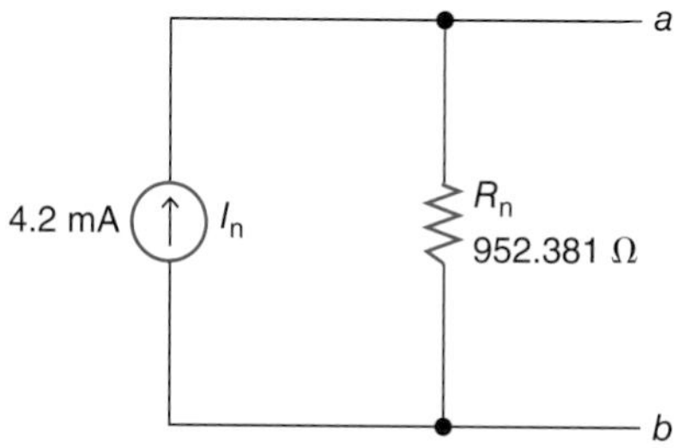

EXAMPLE 4.13

Find the Norton equivalent circuit between terminals *a* and *b* for the circuit shown in Figure 4.107.

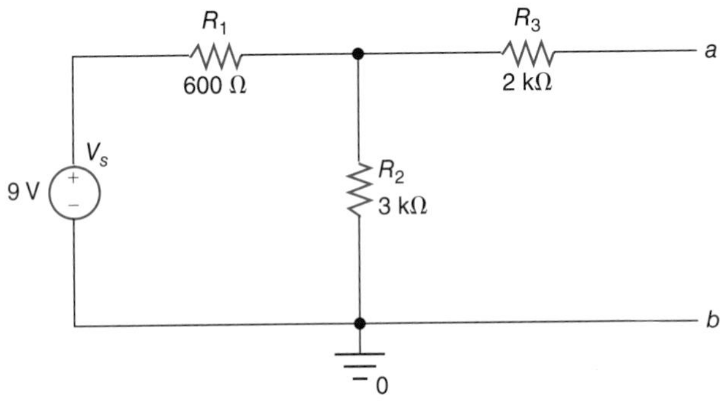

FIGURE 4.107
Circuit for EXAMPLE 4.13.

To find the Norton equivalent current, *a* and *b* are short-circuited, as shown in Figure 4.108.

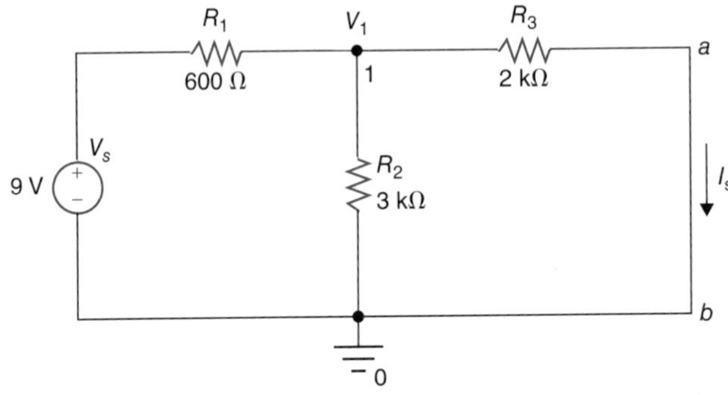

FIGURE 4.108
Circuit shown in Figure 4.107 with *a* and *b* shorted.

Let R_a be the equivalent resistance of the parallel connection of R_2 and R_3. Then, we get

$$R_a = R_2 \| R_3 = \frac{R_2 \times R_3}{R_2 + R_3} = \frac{3\,k\Omega \times 2\,k\Omega}{3\,k\Omega + 2\,k\Omega} = \frac{6}{5}\,k\Omega = 1.2\,k\Omega$$

Application of the voltage divider rule yields

$$V_1 = V_s \times \frac{R_a}{R_1 + R_a} = 9\text{ V} \times \frac{1.2}{0.6 + 1.2} = 9\text{ V} \times \frac{1.2}{1.8} = 6\text{ V}$$

The short-circuit current I_{sc}, which is the Norton equivalent current, is the current through R_3. Thus, we have

$$I_n = I_{sc} = \frac{V_1}{R_3} = \frac{6\text{ V}}{2\,k\Omega} = 3\text{ mA}$$

To find the Norton equivalent resistance, R_n, we deactivate the voltage source by short-circuiting it, as shown in Figure 4.109, and find the equivalent resistance seen from terminals *a* and *b*. The Norton equivalent resistance is given by

continued

Example 4.13 continued

$$R_n = R_3 + (R_1 \| R_2) = 2\,k\Omega + (0.6\,k\Omega \| 3\,k\Omega) = 2\,k\Omega + \frac{0.6 \times 3}{0.6 + 3}\,k\Omega$$

$$= 2\,k\Omega + \frac{1.8}{3.6}\,k\Omega = 2.5\,k\Omega$$

FIGURE 4.109

The circuit shown in Figure 4.107 after deactivating the voltage source.

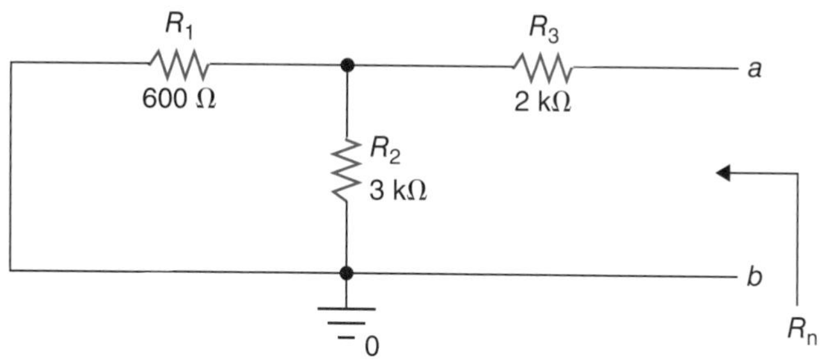

The Norton equivalent circuit is shown in Figure 4.110.

FIGURE 4.110

The Norton equivalent circuit.

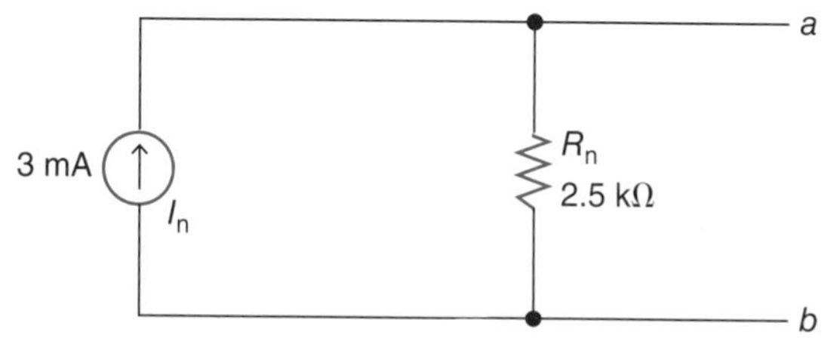

Exercise 4.13

Find the Norton equivalent current I_n and the Norton equivalent resistance R_n between terminals *a* and *b* for the circuit shown in Figure 4.111.

FIGURE 4.111

Circuit for EXERCISE 4.13.

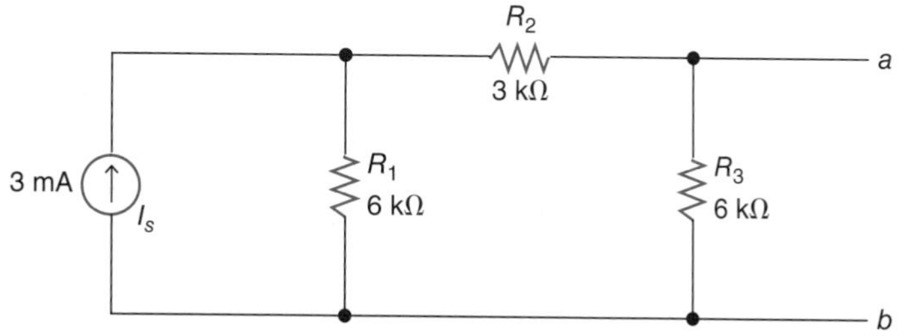

Answer:
$I_n = 2$ mA, $R_n = 3.6\,k\Omega$.

EXAMPLE 4.14

Find the Norton equivalent circuit between terminals *a* and *b* for the circuit shown in Figure 4.112.

continued

Example 4.14 continued

FIGURE 4.112

Circuit for EXAMPLE 4.14.

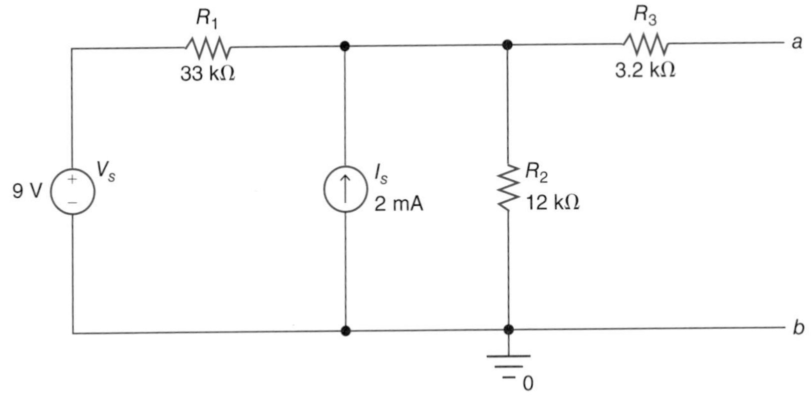

When terminals a and b are short-circuited, we obtain the circuit shown in Figure 4.113. The Norton equivalent current I_n is the short-circuit current I_{sc}. The short-circuit current I_{sc} is the current through R_3. Nodal analysis can be used to find voltage V_1 at node 1.

FIGURE 4.113

The circuit shown in Figure 4.112 with terminals a and b short-circuited.

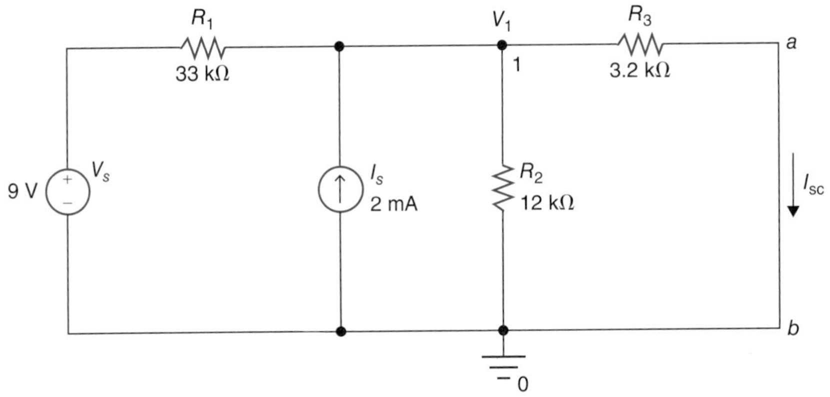

Summing the currents leaving node 1, we obtain

$$\frac{V_1 - 9}{33{,}000} - 0.002 + \frac{V_1}{12{,}000} + \frac{V_1}{3200} = 0$$

Multiplication by 33,000 yields

$$V_1 - 9 - 66 + 2.75V_1 + 10.3125V_1 = 0$$

which can be simplified to

$$14.0625V_1 = 75$$

Thus, we have

$$V_1 = \frac{75}{14.0625} = 5.3333 \text{ V}$$

The short-circuit current I_{sc} is found to be

$$I_n = I_{sc} = \frac{V_1}{R_3} = \frac{5.3333 \text{ V}}{3200\ \Omega} = 1.6667 \text{ mA}$$

continued

Example 4.14 continued

To find the Norton equivalent resistance R_n, we deactivate the voltage source by short-circuiting it and deactivate the current source by open-circuiting it, as shown in Figure 4.114.

FIGURE 4.114

The circuit shown in Figure 4.113 with sources deactivated.

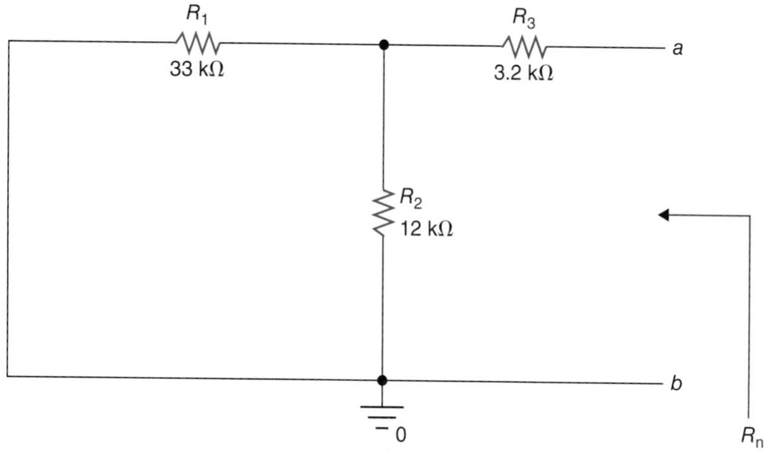

FIGURE 4.115

The Norton equivalent circuit.

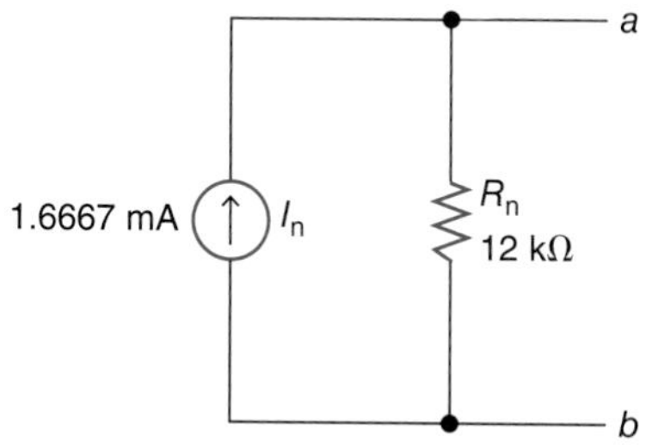

The Norton equivalent resistance R_n is the equivalent resistance of the circuit shown in Figure 4.114 from terminals *a* and *b*. Thus, we have

$$R_n = R_3 + (R_1 \| R_2) = R_3 + \frac{R_1 \times R_2}{R_1 + R_2} = 3.2\,k\Omega + \frac{33\,k\Omega \times 12\,k\Omega}{33\,k\Omega + 12\,k\Omega}$$

$$= 3.2\,k\Omega + \frac{396}{45}\,k\Omega = 12\,k\Omega$$

The Norton equivalent circuit is shown in Figure 4.115.

Exercise 4.14

Find the Norton equivalent circuit between terminals *a* and *b* for the circuit shown in Figure 4.116.

FIGURE 4.116

Circuit for EXERCISE 4.14.

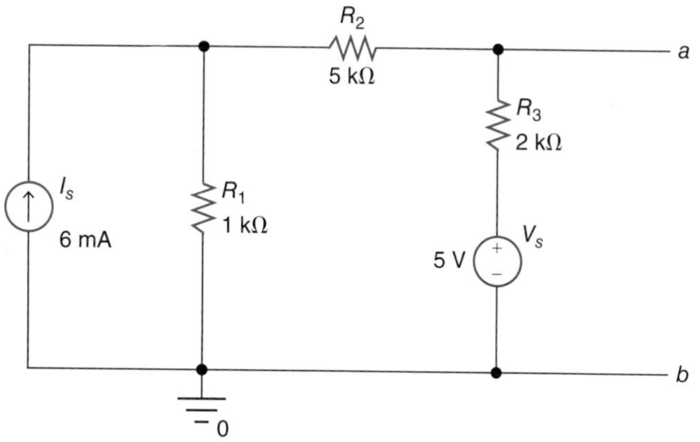

Answer:
$I_n = 3.5$ mA, $R_n = 1.5\,k\Omega$.

EXAMPLE 4.15

Find the Norton equivalent circuit between terminals *a* and *b* for the circuit shown in Figure 4.117.

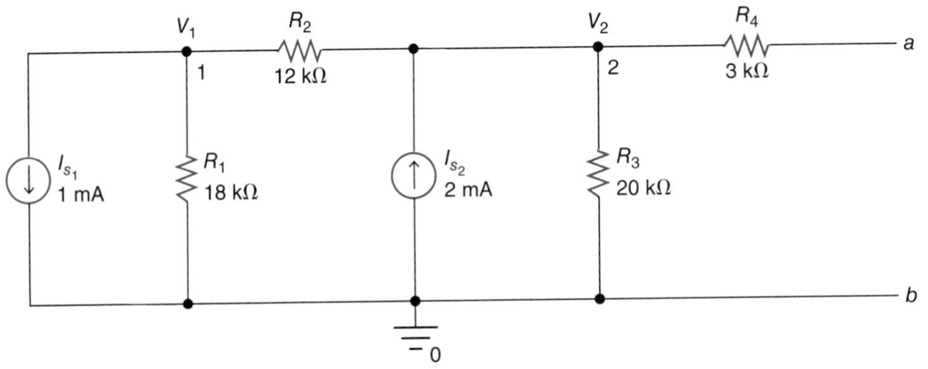

FIGURE 4.117

Circuit for EXAMPLE 4.15.

We will find the open-circuit voltage V_{oc} across *a* and *b*, and the short-circuit current I_{sc} through *a* and *b* to find the Norton equivalent circuit. Since no current flows through R_4, the open-circuit voltage is voltage V_2 at node 2. Summing the currents leaving node 1, we obtain

$$0.001 + \frac{V_1}{18{,}000} + \frac{V_1 - V_2}{12{,}000} = 0$$

Multiplication by 36,000 yields

$$36 + 2V_1 + 3V_1 - 3V_2 = 0$$

which can be simplified to

$$5V_1 - 3V_2 = -36 \tag{4.47}$$

Summing the currents leaving node 2, we obtain

$$\frac{V_2 - V_1}{12{,}000} - 0.002 + \frac{V_2}{20{,}000} = 0$$

Multiplication by 60,000 yields

$$5V_2 - 5V_1 - 120 + 3V_2 = 0$$

which can be simplified to

$$-5V_1 + 8V_2 = 120 \tag{4.48}$$

Adding Equations (4.47) and (4.48), we obtain

$$5V_2 = 84$$

Thus, we have

$$V_2 = 16.8\text{ V}$$

continued

Example 4.15 continued

The open-circuit voltage is V_2. Therefore,

$$V_{oc} = V_2 = 16.8 \text{ V}$$

To find the short-circuit current I_{sc}, a and b are short-circuited, as shown in Figure 4.118.

FIGURE 4.118

Circuit with a and b short-circuited.

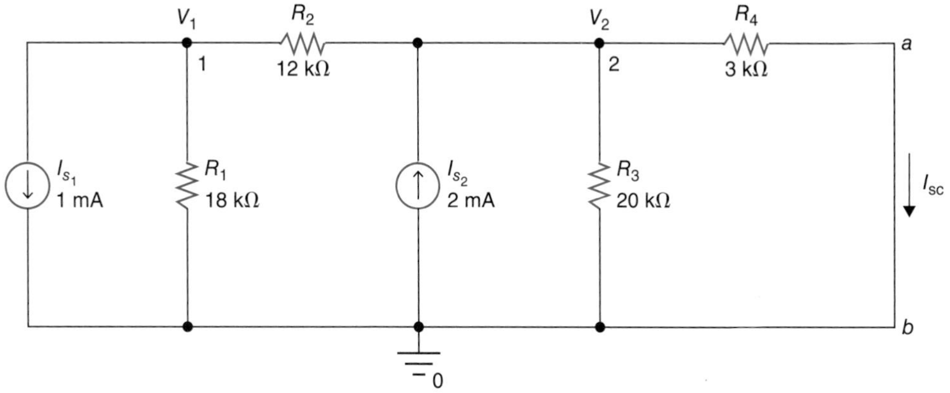

Equation (4.47) is still valid for the circuit shown in Figure 4.118. Summing the currents leaving node 2, we obtain

$$\frac{V_2 - V_1}{12{,}000} - 0.002 + \frac{V_2}{20{,}000} + \frac{V_2}{3000} = 0$$

Multiplication by 60,000 yields

$$5V_2 - 5V_1 - 120 + 3V_2 + 20V_2 = 0,$$

which can be simplified to

$$-5V_1 + 28V_2 = 120 \quad \textbf{(4.49)}$$

Adding Equations (4.47) and (4.49), we obtain

$$25V_2 = 84$$

Thus, we have

$$V_2 = 3.36 \text{ V}$$

The short-circuit current is the current through R_4. Thus, we have

$$I_{sc} = \frac{V_2}{R_4} = \frac{3.36 \text{ V}}{3\ k\Omega} = 1.12 \text{ mA}$$

The Norton equivalent resistance is given by

$$R_n = \frac{V_{oc}}{I_{sc}} = \frac{16.8 \text{ V}}{1.12 \text{ mA}} = 15\ k\Omega$$

The Norton equivalent circuit is shown in Figure 4.119.

FIGURE 4.119

The Norton equivalent circuit.

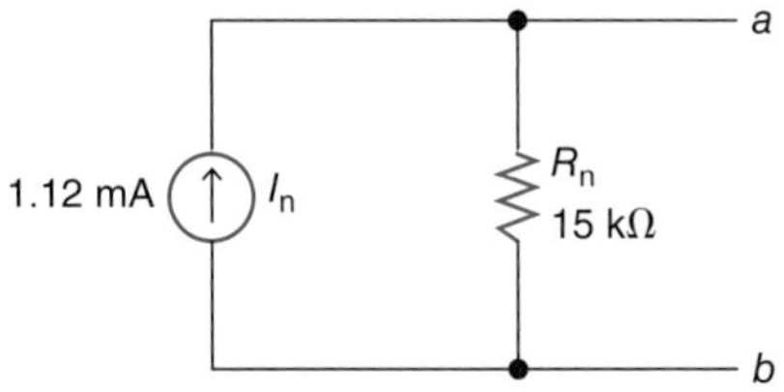

Exercise 4.15

Find the Norton equivalent circuit between terminals *a* and *b* for the circuit shown in Figure 4.120.

FIGURE 4.120

Circuit for EXERCISE 4.15.

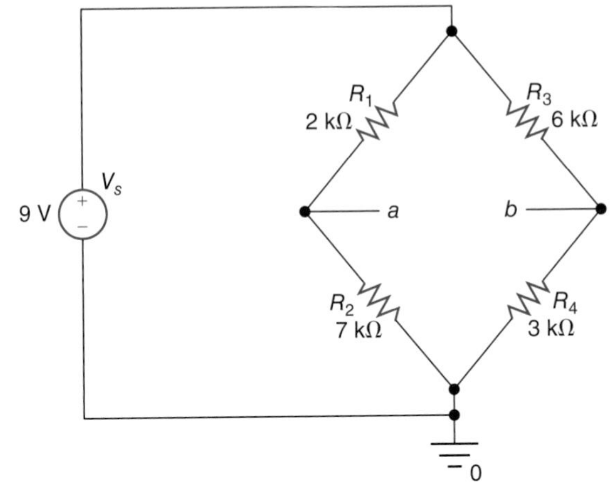

Answer:
$I_n = 1.125$ mA, $R_n = 3.5556\ k\Omega$.

EXAMPLE 4.16

Find the Norton equivalent circuit between terminals *a* and *b* for the circuit shown in Figure 4.121. Assume that $g_m = 3$ (mA/V).

FIGURE 4.121

Circuit for EXAMPLE 4.16.

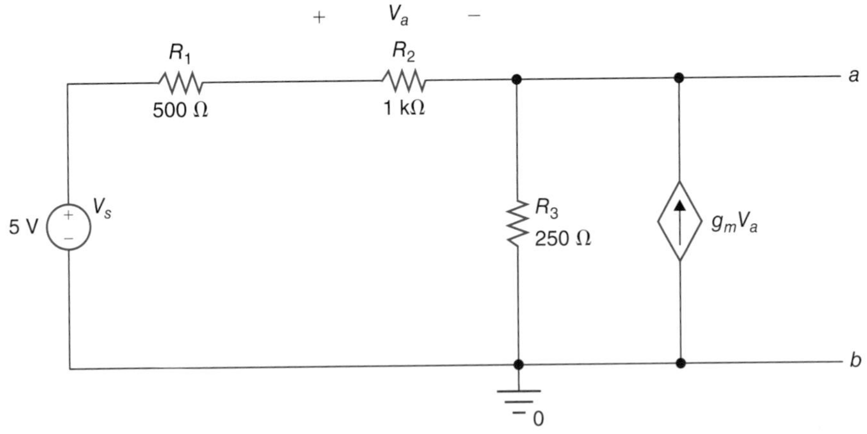

To find the Norton equivalent current, we short-circuit *a* and *b* in the circuit shown in Figure 4.121 to get the circuit shown in Figure 4.122.

The current I through R_1 and R_2 is given by

$$I = \frac{5\text{ V}}{0.5\ k\Omega + 1\ k\Omega} = \frac{10}{3}\text{ mA}$$

continued

Example 4.16 continued

FIGURE 4.122

The circuit from Figure 4.121 with a and b shorted.

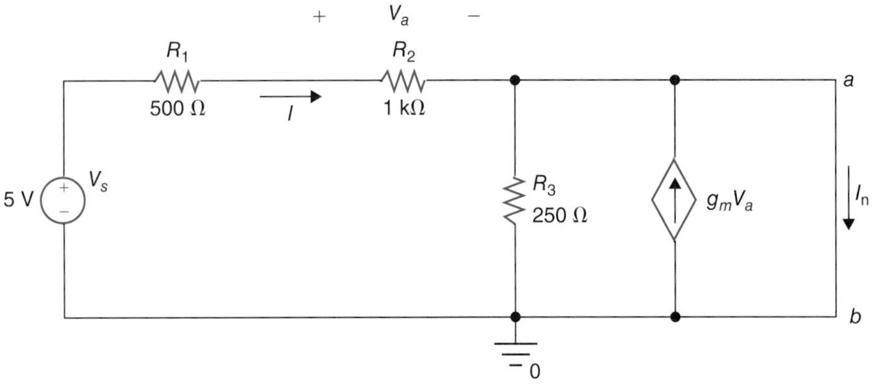

The voltage V_a across R_2 is given by

$$V_a = R_2 I = 1\ k\Omega \times \frac{10}{3}\text{ mA} = \frac{10}{3}\text{ V}$$

The current from the VCCS is

$$g_m V_a = 3\left(\frac{\text{mA}}{\text{V}}\right) \times \frac{10}{3}\text{ (V)} = \frac{30}{3}\text{ mA} = 10\text{ mA}$$

According to the current divider rule, all the currents flowing into node a will flow out through the short circuit. In other words, there is no current through R_3. Thus, we have

$$I_{\text{n}} = I + g_m V_a = \frac{10}{3}\text{ mA} + \frac{30}{3}\text{ mA} = \frac{40}{3}\text{ mA} = 13.3333\text{ mA}$$

To find the Norton equivalent resistance, we short-circuit the independent voltage source V_s and apply a test voltage of 1 V between terminals a and b, as shown in Figure 4.123.

FIGURE 4.123

Circuit with a test voltage.

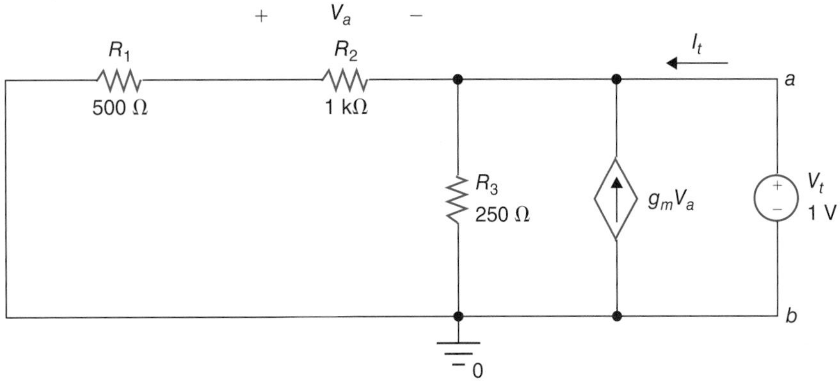

From the voltage divider rule, the voltage V_a of the circuit shown in Figure 4.123 is given by

$$V_a = -\frac{R_2}{R_2 + R_1} \times V_t = -\frac{1\ k\Omega}{1\ k\Omega + 0.5\ k\Omega} \times 1 = -\frac{2}{3}\text{ V}$$

continued

Example 4.16 continued

The current I_t flowing out of the test voltage source V_t is given by

$$I_t = -g_m V_a + \frac{V_t}{R_3} + \frac{V_t}{R_2 + R_1} = (-3\text{ mA/V}) \times \left(-\frac{2}{3}\text{ V}\right) + \frac{1\text{ V}}{0.25\ k\Omega} + \frac{1\text{ V}}{1.5\ k\Omega} = \frac{20}{3}\text{ mA}$$

Thus, the Norton equivalent resistance is given by

$$R_n = \frac{V_t}{I_t} = \frac{1\text{ V}}{\frac{20}{3}\text{ mA}} = \frac{3}{20}\ k\Omega = 150\ \Omega$$

The Norton equivalent circuit is shown in Figure 4.124.

FIGURE 4.124

The Norton equivalent circuit.

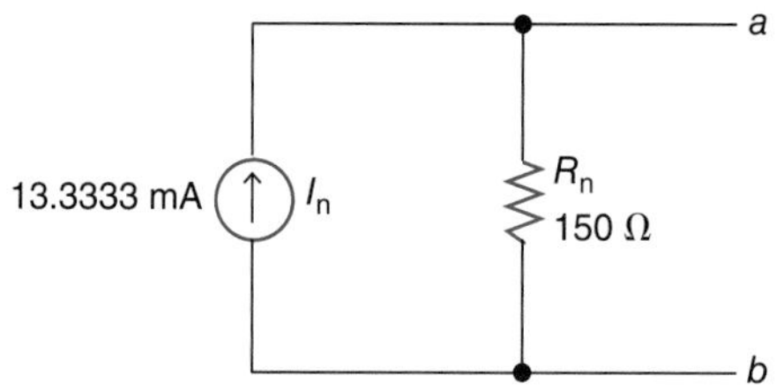

Exercise 4.16

Find the Norton equivalent circuit between terminals *a* and *b* for the circuit shown in Figure 4.125.

FIGURE 4.125

Circuit for EXERCISE 4.16.

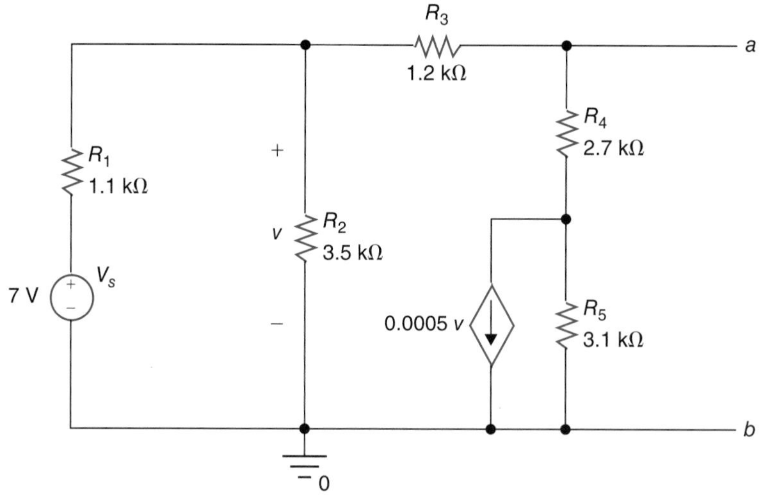

Answer:
$I_n = 1.7762$ mA, $R_n = 1.2934\ k\Omega$

EXAMPLE 4.17

Find the Norton equivalent circuit between terminals *a* and *b* for the circuit shown in Figure 4.126.

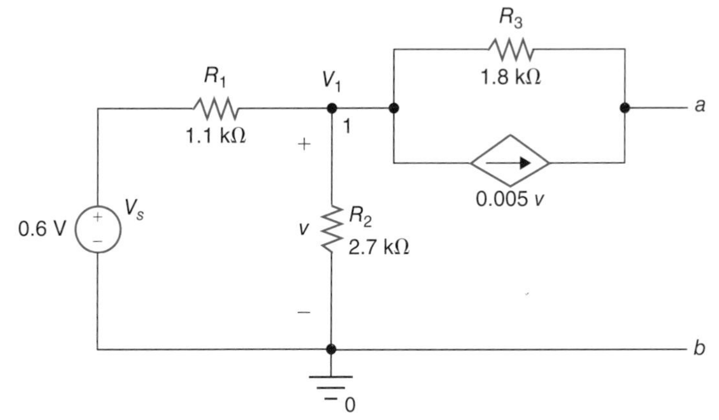

FIGURE 4.126
Circuit for EXAMPLE 4.17.

To find the Norton equivalent current, we short-circuit a and b, as shown in Figure 4.127, and find the current through the short circuit.

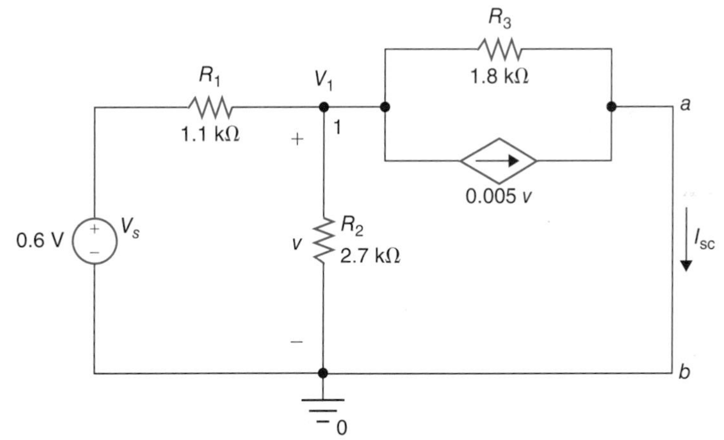

FIGURE 4.127
Circuit shown in Figure 4.126 with a and b short-circuited.

Summing the currents leaving node 1, we obtain

$$\frac{V_1 - 0.6}{1100} + \frac{V_1}{2700} + \frac{V_1}{1800} + 0.005 \times V_1 = 0$$

which can be rearranged as

$$\left(\frac{1}{1100} + \frac{1}{2700} + \frac{1}{1800} + 0.005\right)V_1 = \frac{0.6}{1100}$$

Thus, we have

$$V_1 = \frac{\dfrac{0.6}{1100}}{\dfrac{1}{1100} + \dfrac{1}{2700} + \dfrac{1}{1800} + 0.005} = 0.079803 \text{ V}$$

continued

Example 4.17 continued

The short-circuit current I_{sc}, which is the Norton equivalent current I_n, is given by

$$I_n = I_{sc} = \frac{V_1}{R_3} + 0.005 \times V_1 = \frac{0.079803}{1800} + 0.005 \times 0.079803 = 443.3498\ \mu\text{A}$$

To find the Norton equivalent resistance R_n, we deactivate the voltage source V_s by short-circuiting it and then apply a test voltage V_t of 1 V between terminals a and b, as shown in Figure 4.128.

FIGURE 4.128

Circuit with a test voltage.

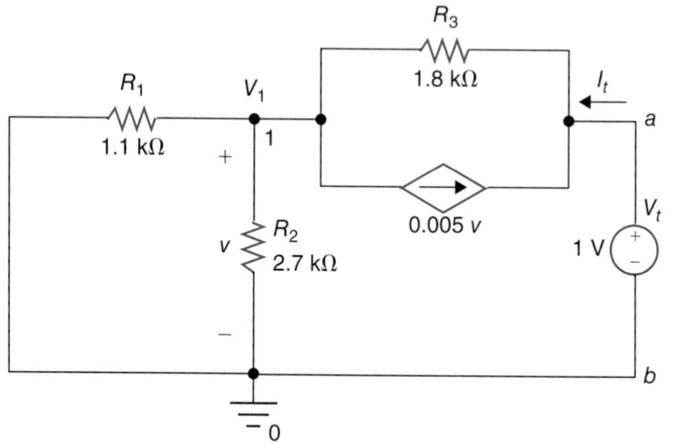

Summing the currents leaving node 1 of the circuit shown in Figure 4.128, we obtain

$$\frac{V_1}{1100} + \frac{V_1}{2700} + \frac{V_1 - 1}{1800} + 0.005 \times V_1 = 0$$

which can be rearranged as

$$\left(\frac{1}{1100} + \frac{1}{2700} + \frac{1}{1800} + 0.005\right)V_1 = \frac{1}{1800}$$

Thus, we have

$$V_1 = \frac{\frac{1}{1800}}{\frac{1}{1100} + \frac{1}{2700} + \frac{1}{1800} + 0.005} = 0.0812808\ \text{V}$$

The current I_t flowing out of the positive terminal of the test voltage is given by

$$I_t = \frac{V_t - V_1}{R_3} - 0.005 \times V_1 = 103.9956\ \mu\text{A}$$

The Norton equivalent resistance is given by

$$R_n = \frac{V_t}{I_t} = \frac{1\ \text{V}}{103.9956\ \mu\text{A}} = 9.6158\ k\Omega$$

The Norton equivalent circuit is shown in Figure 4.129.

FIGURE 4.129

The Norton equivalent circuit.

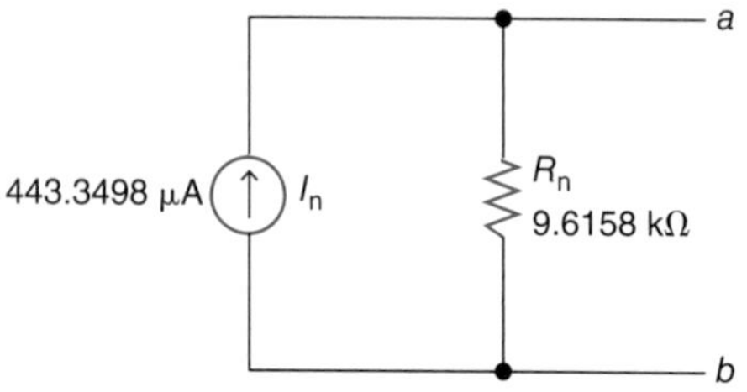

Exercise 4.17

Find the Norton equivalent circuit between terminals *a* and *b* for the circuit shown in Figure 4.130.

FIGURE 4.130

Circuit for EXERCISE 4.17.

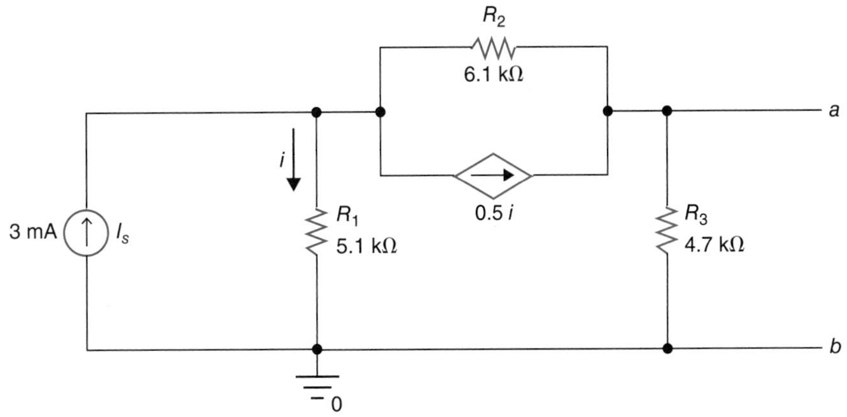

Answer:

$I_n = 1.7158$ mA, $R_n = 3.5343\ k\Omega$

EXAMPLE 4.18

Find the Norton equivalent circuit between terminals *a* and *b* for the circuit shown in Figure 4.131.

FIGURE 4.131

Circuit for EXAMPLE 4.18.

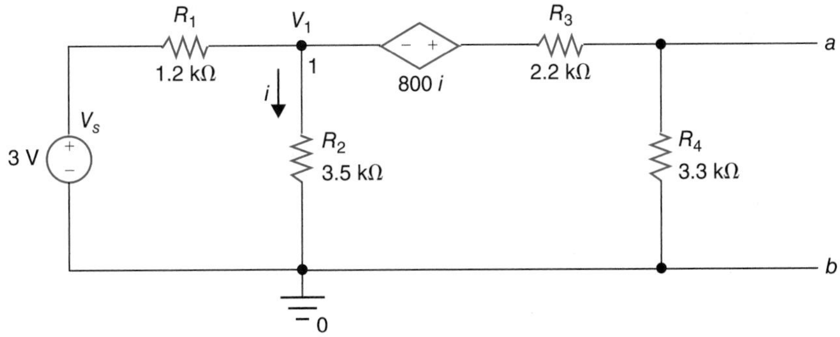

We can find the open-circuit voltage V_{oc} between a and b. The open-circuit voltage is the voltage across R_4. Summing the currents leaving node 1, we obtain

$$\frac{V_1 - 3}{1200} + \frac{V_1}{3500} + \frac{V_1 + 800 \times \dfrac{V_1}{3500}}{2200 + 3300} = 0$$

which can be rearranged as

$$\left(\frac{1}{1200} + \frac{1}{3500} + \frac{1 + \dfrac{800}{3500}}{5500}\right) V_1 = \frac{3}{1200}$$

continued

Example 4.18 continued

Thus, we have

$$V_1 = \frac{\frac{3}{1200}}{\frac{1}{1200} + \frac{1}{3500} + \frac{1 + \frac{800}{3500}}{5500}} = 1.8623 \text{ V}$$

The open-circuit voltage is given by

$$V_{oc} = \left(V_1 + 800 \times \frac{V_1}{3500}\right) \times \frac{R_4}{R_3 + R_4} = 1.8623 \times \left(1 + \frac{800}{3500}\right) \times \frac{3300}{5500} = 1.3728 \text{ V}$$

The Norton equivalent current is the short-circuit current when a and b are short-circuited, as shown in Figure 4.132.

FIGURE 4.132

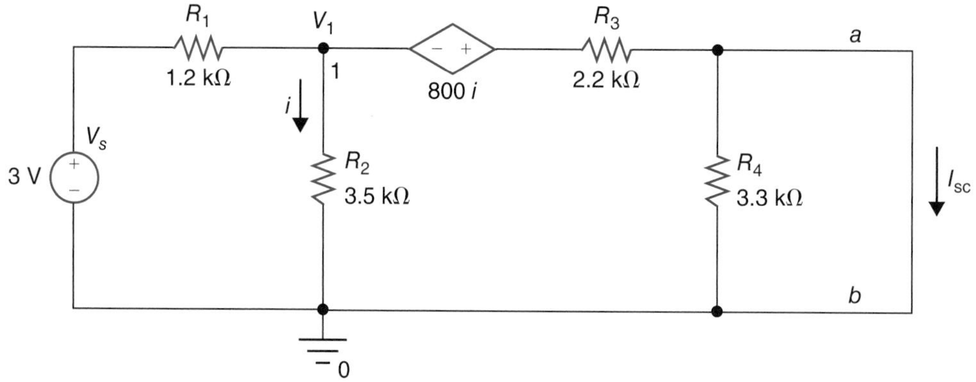

Circuit with a and b short-circuited.

Since the potential difference across R_4 is zero, there is no current through R_4. The short-circuit current is the current through R_3. Summing the currents leaving node 1, we obtain

$$\frac{V_1 - 3}{1200} + \frac{V_1}{3500} + \frac{V_1 + 800 \times \frac{V_1}{3500}}{2200} = 0$$

which can be rearranged as

$$\left(\frac{1}{1200} + \frac{1}{3500} + \frac{1 + \frac{800}{3500}}{2200}\right) V_1 = \frac{3}{1200}$$

Thus, we have

$$V_1 = \frac{\frac{3}{1200}}{\frac{1}{1200} + \frac{1}{3500} + \frac{1 + \frac{800}{3500}}{2200}} = 1.4903 \text{ V}$$

continued

Example 4.18 continued

FIGURE 4.133

The Norton equivalent circuit.

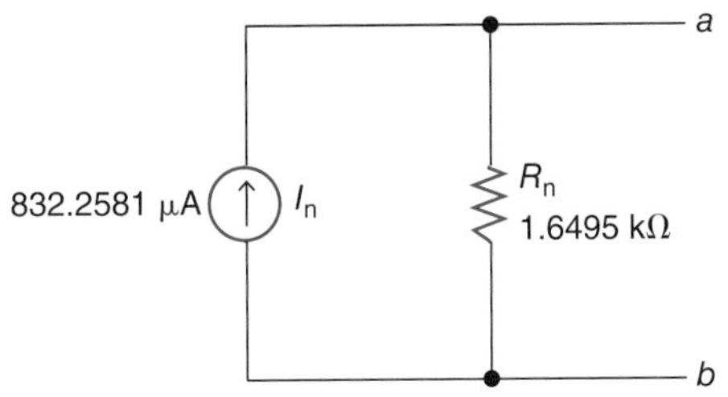

The short-circuit current is given by

$$I_n = I_{sc} = \frac{V_1 + 800 \times \dfrac{V_1}{3500}}{R_3} = \frac{1.8309\ \text{V}}{2200\ \Omega} = 832.2581\ \mu\text{A}$$

The Norton equivalent resistance is given by

$$R_n = \frac{V_{oc}}{I_{sc}} = \frac{1.3728\ \text{V}}{832.2581\ \mu\text{A}} = 1.6495\, k\Omega$$

The Norton equivalent circuit is shown in Figure 4.133.

MATLAB

```
%EXAMPLE 4.18
clear all;format long;
R1=1200;R2=3500;R3=2200;R4=3300;Vs=3;kr=800;
syms V1 Va
%Voc
V1=solve((V1-Vs)/R1+V1/R2+(V1+kr*V1/R2)/(R3+R4));
Voc=(V1+kr*V1/R2)*R4/(R3+R4);
%Isc
Va=solve((Va-Vs)/R1+Va/R2+(Va+kr*Va/R2)/R3);
Isc=(Va+kr*Va/R2)/R3;
Rn=Voc/Isc;
In=Isc;
V1=vpa(V1,7)
Voc=vpa(Voc,7)
Va=vpa(Va,7)
Rn=vpa(Rn,7)
Isc=vpa(Isc,7)
In=vpa(In,7)

Answers:
V1 =
1.862302
Voc =
1.372783
Va =
1.490323
Rn =
1649.468
Isc =
0.0008322581
In =
0.0008322581
```

Exercise 4.18

Find the Norton equivalent circuit between terminals *a* and *b* for the circuit shown in Figure 4.134.

continued

Exercise 4.18 continued

FIGURE 4.134

Circuit for EXERCISE 4.18.

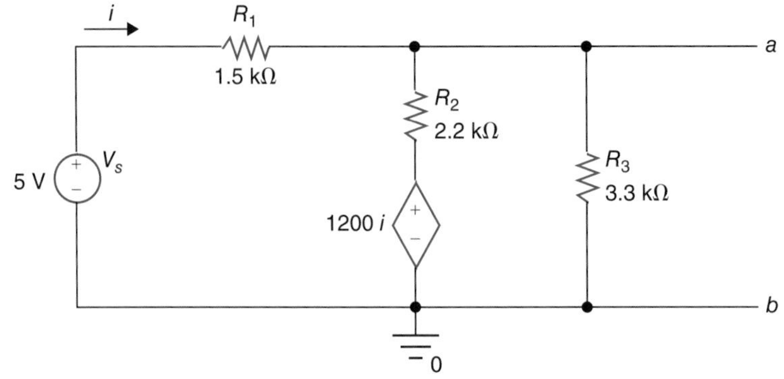

Answer:
$I_n = 5.1515$ mA, $R_n = 559.322\ \Omega$.

4.6 Maximum Power Transfer

FIGURE 4.135

A load connected to the Thévenin equivalent circuit.

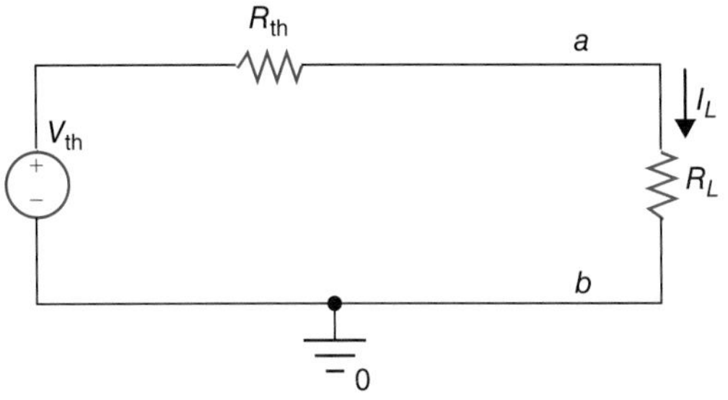

Suppose that a load with resistance R_L is connected to a circuit between terminals a and b. We are interested in finding the power P_L delivered to the load and finding the load resistance R_L that maximizes the power delivered to the load. We first find the Thévenin equivalent circuit with respect to the terminals a and b. Let V_{th} be the Thévenin equivalent voltage and R_{th} be the Thévenin equivalent resistance. With the original circuit replaced by the Thévenin equivalent circuit, we obtain the circuit shown in Figure 4.135.

The current through the load resistor is given by

$$I_L = \frac{V_{th}}{R_{th} + R_L}$$

and the voltage across the load resistor is given by

$$V_L = R_L I_L = \frac{R_L V_{th}}{R_{th} + R_L}$$

Thus, the power delivered to the load is

$$p_L = I_L V_L = \frac{R_L V_{th}^2}{(R_{th} + R_L)^2}$$

When $R_L = 0, p_L = 0$; and when $R_L = \infty, p_L = 0$. The power delivered to the load p_L must peak at a certain value. Figure 4.136 shows p_L as a function of R_L for $0 \le R_L \le 5R_{th}$ ($V_{th} = 1$ V, $R_{th} = 1\ \Omega$).

From this figure, we can see that when p_L is at its maximum, the derivative of p_L with respect to R_L is zero; that is, $dp_L/dR_L = 0$. Using

$$\frac{d}{dt}\left(\frac{u(t)}{v(t)}\right) = \frac{v(t)\dfrac{du(t)}{dt} - u(t)\dfrac{dv(t)}{dt}}{v^2(t)},$$

FIGURE 4.136

Plot of the power on the load as a function of load resistance.

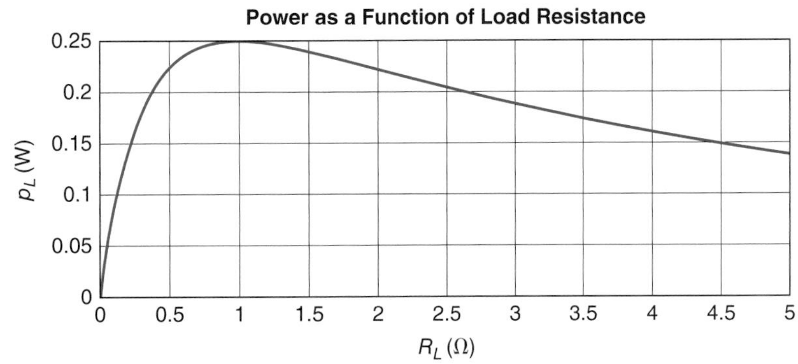

we have

$$\frac{dp_L}{dR_L} = \frac{d}{dR_L}\left(\frac{R_L(V_{th})^2}{(R_{th}+R_L)^2}\right) = \frac{(R_{th}+R_L)^2\frac{dR_L}{dR_L} - R_L\frac{d(R_{th}+R_L)^2}{dR_L}}{(R_{th}+R_L)^4}(V_{th})^2$$

$$= \frac{(R_{th}+R_L)^2 1 - R_L 2(R_{th}+R_L)}{(R_{th}+R_L)^4}(V_{th})^2 = \frac{(R_{th}+R_L)(R_{th}-R_L)}{(R_{th}+R_L)^4}(V_{th})^2$$

Setting $dp_L/dR_L = 0$, we have two solutions: $R_L = -R_{th}$ or $R_L = R_{th}$. We take the positive answer. Thus, the load resistance R_L that maximizes the power delivered to the load is given by the Thévenin equivalent resistance from terminals a and b.

The maximum power delivered to the load when the load resistance is $R_L = R_{th}$ is given by

$$p_{L,\max} = \frac{R_{th}V_{th}^2}{(R_{th}+R_{th})^2} = \frac{V_{th}^2}{4R_{th}} = \frac{V_{th}^2}{4R_L} \quad \textbf{(4.50)}$$

FIGURE 4.137

A Norton equivalent circuit with load R_L.

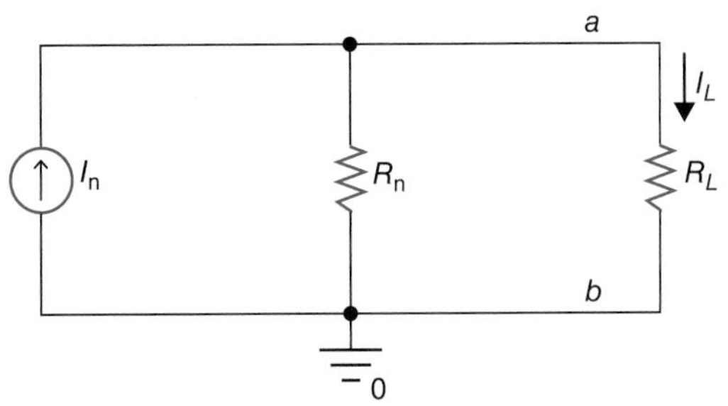

For the Norton equivalent circuit with load R_L shown in Figure 4.137, let the current through the load be I_L. Then, from the current divider rule, we have

$$I_L = \frac{R_n}{R_n + R_L} \times I_n$$

The power on the load is given by

$$p_L = I_L^2 R_L = \frac{R_L}{(R_n+R_L)^2}R_n^2 I_n^2$$

$$\frac{dp_L}{dR_L} = \frac{d}{dR_L}\left(\frac{R_L}{(R_n+R_L)^2}(R_n I_n)^2\right) = \frac{(R_n+R_L)^2\frac{dR_L}{dR_L} - R_L\frac{d(R_n+R_L)^2}{dR_L}}{(R_n+R_L)^4}(R_n I_n)^2$$

$$= \frac{(R_n+R_L)^2 1 - R_L 2(R_n+R_L)}{(R_n+R_L)^4}(R_n I_n)^2 = \frac{(R_n+R_L)(R_n-R_L)}{(R_n+R_L)^4}(R_n I_n)^2$$

Setting $dp_L/dR_L = 0$, we have two solutions: $R_L = -R_n$ and $R_L = R_n$. We take the positive answer. Thus, the load resistance R_L that maximizes the power delivered to the load is given by the Norton equivalent resistance from terminals a and b.

The maximum power delivered to the load when the load resistance is $R_L = R_n$ is given by

$$p_{L,\max} = \frac{R_n}{(R_n + R_n)^2} R_n^2 I_n^2 = \frac{I_n^2 R_n}{4} = \frac{I_n^2 R_L}{4} \tag{4.51}$$

We can get the same result using source transformation.

EXAMPLE 4.19

For the circuit shown in Figure 4.138, find the value of the load resistance R_L that maximizes the power delivered to the load. Also, find the maximum power delivered to R_L.

FIGURE 4.138

Circuit for EXAMPLE 4.19.

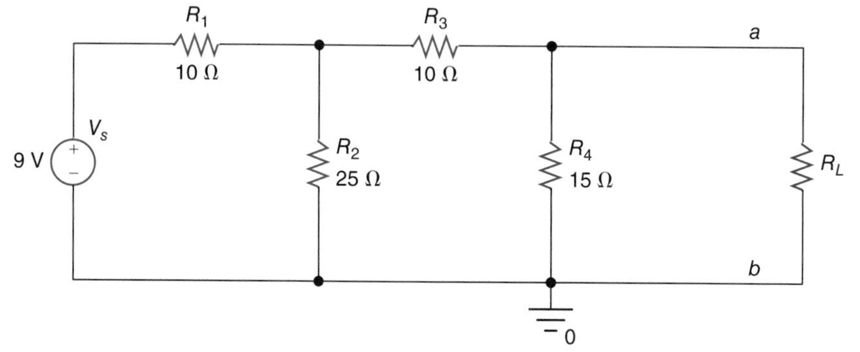

Figure 4.139 shows the circuit in Figure 4.138 without the load resistor.

FIGURE 4.139

Circuit shown in Figure 4.138 without the load resistor.

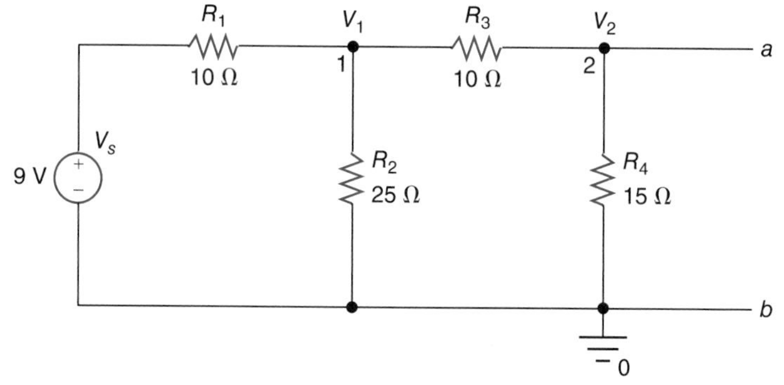

Summing the currents leaving node 1, we obtain

$$\frac{V_1 - 9}{10} + \frac{V_1}{25} + \frac{V_1 - V_2}{10} = 0 \tag{4.52}$$

Multiplication by 50 yields

$$5V_1 - 45 + 2V_1 + 5V_1 - 5V_2 = 0$$

continued

Example 4.19 continued

which can be simplified to

$$12V_1 - 5V_2 = 45 \tag{4.53}$$

Summing the currents leaving node 2, we obtain

$$\frac{V_2 - V_1}{10} + \frac{V_2}{15} = 0 \tag{4.54}$$

Multiplication by 30 results in

$$3V_2 - 3V_1 + 2V_2 = 0$$

which can be simplified to

$$-3V_1 + 5V_2 = 0 \tag{4.55}$$

Adding Equations (4.53) and (4.55), we get

$$9V_1 = 45$$

Thus, $V_1 = 5$ V. Substituting this into Equation (4.55), we obtain

$$5V_2 = 15$$

from which we have $V_2 = 3$ V. The Thévenin equivalent voltage between a and b is voltage V_2. Thus, we have

$$V_{\text{th}} = 3\text{ V}$$

The Thévenin equivalent resistance R_{th} is the equivalent resistance across a and b after deactivating the voltage source by short-circuiting it, as shown in Figure 4.140.

FIGURE 4.140

The circuit shown in Figure 4.139 with the voltage source deactivated.

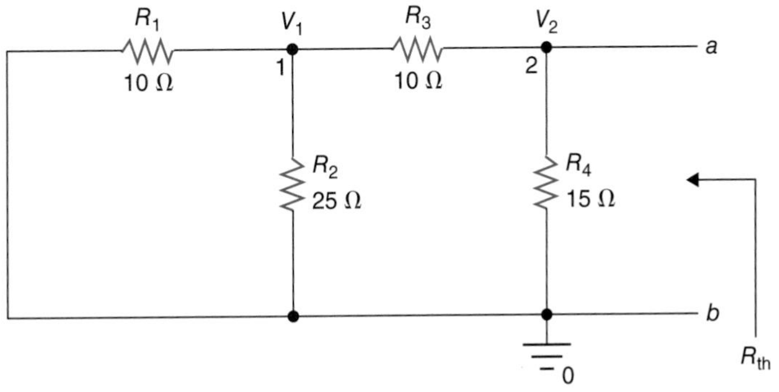

Let the equivalent resistance of the parallel connection of R_1 and R_2 be R_a. Then, R_a is

$$R_a = R_1 \| R_2 = \frac{R_1 \times R_2}{R_1 + R_2} = \frac{10\ \Omega \times 25\ \Omega}{10\ \Omega + 25\ \Omega} = \frac{250}{35}\ \Omega = \frac{50}{7}\ \Omega$$

continued

Example 4.19 continued

FIGURE 4.141

The Thévenin equivalent circuit between *a* and *b*.

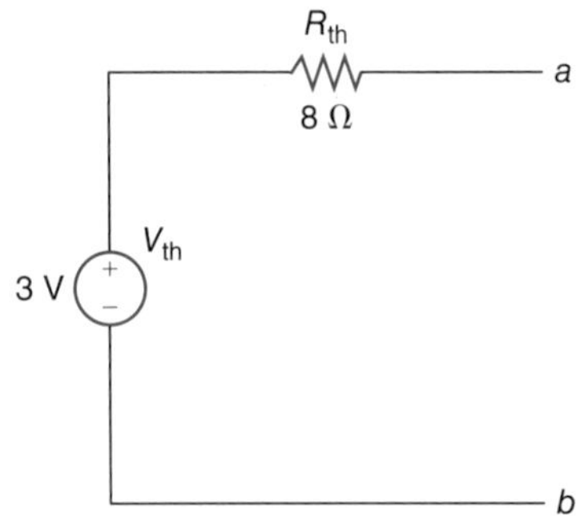

Let R_b be the equivalent resistance of the series connection of R_3 and R_a. Then, we have

$$R_b = R_3 + R_a = 10\ \Omega + 50/7\ \Omega = 120/7\ \Omega$$

The Thévenin equivalent resistance R_{th} is given by the equivalent resistance of the parallel connection of R_4 and R_b. Thus, we obtain

$$R_{th} = R_4 \| R_b = \frac{R_4 \times R_b}{R_4 + R_b} = \frac{15\ \Omega \times \frac{120}{7}\ \Omega}{15\ \Omega + \frac{120}{7}\ \Omega} = \frac{1800}{225}\ \Omega = 8\ \Omega$$

The Thévenin equivalent circuit between *a* and *b* for the circuit shown in Figure 4.139 is shown in Figure 4.141.

The load resistance for maximum power transfer is $R_L = 8\ \Omega$, and the maximum power delivered to the load is given by

$$p_L = \frac{V_{th}^2}{4R_L} = \frac{3^2}{4 \times 8} = \frac{9}{32}\ \mathrm{W} = 0.28125\ \mathrm{W} = 281.25\ \mathrm{mW}$$

Exercise 4.19

For the circuit shown in Figure 4.142, find the value of the load resistance R_L that maximizes the power delivered to the load. Also, find the maximum power delivered to R_L.

FIGURE 4.142

Circuit for EXERCISE 4.19.

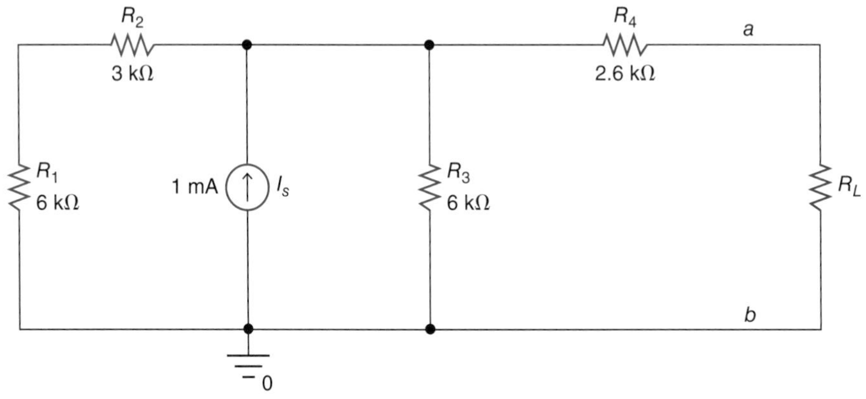

Answer:

$V_{th} = 3.6$ V, $R_{th} = 6.2$ kΩ, $R_L = R_{th} = 6.2$ kΩ, $p_L = 522.5806$ μW.

EXAMPLE 4.20

For the circuit shown in Figure 4.143, find the value of the load resistance R_L that maximizes the power delivered to the load. Also, find the maximum power delivered to R_L.

continued

Example 4.20 continued

FIGURE 4.143

Circuit for EXAMPLE 4.20.

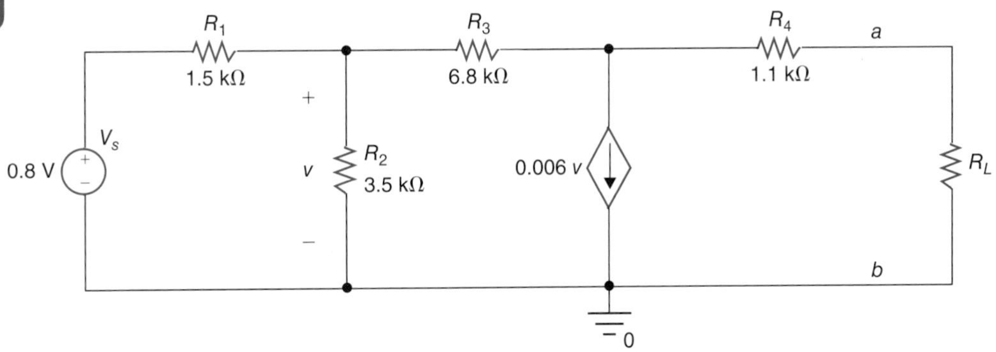

The circuit without the load connected is shown in Figure 4.144.

FIGURE 4.144

The circuit in Figure 4.143 without the load connected.

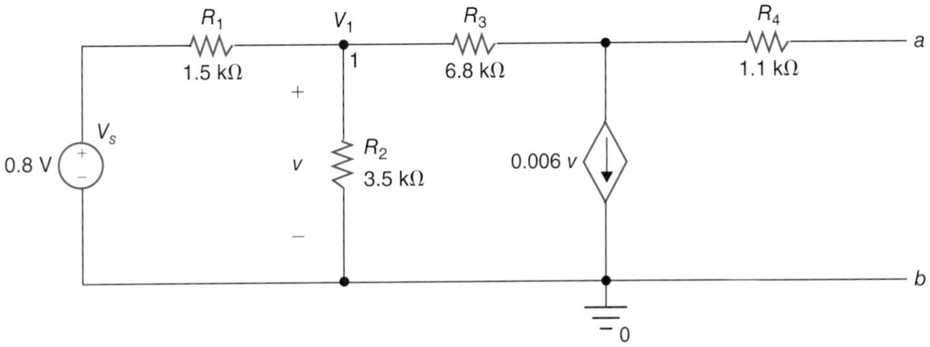

No current flows through R_4. Summing the currents leaving node 1, we obtain

$$\frac{V_1 - 0.8}{1500} + \frac{V_1}{3500} + 0.006 \times V_1 = 0$$

which can be rearranged as

$$\left(\frac{1}{1500} + \frac{1}{3500} + 0.006\right)V_1 = \frac{0.8}{1500}$$

Thus, we have

$$V_1 = \frac{\frac{0.8}{1500}}{\frac{1}{1500} + \frac{1}{3500} + 0.006} \times \frac{10{,}500}{10{,}500} = \frac{7 \times 0.8}{7 + 3 + 63} = \frac{5.6}{73} = 0.07671233 \text{ V}$$

The open-circuit voltage between a and b is given by

$$V_{\text{th}} = V_{\text{oc}} = V_1 - 0.006V_1 \times R_3 = 0.07671233(1 - 0.006 \times 6800) = -3.05315 \text{ V}$$

To find the short-circuit current, a and b are short-circuited, as shown in Figure 4.145. Summing the currents leaving node 1, we obtain

$$\frac{V_1 - 0.8}{1500} + \frac{V_1}{3500} + \frac{V_1 - V_2}{6800} = 0$$

continued

Example 4.20 continued

FIGURE 4.145

The circuit in Figure 4.144 with a and b shorted.

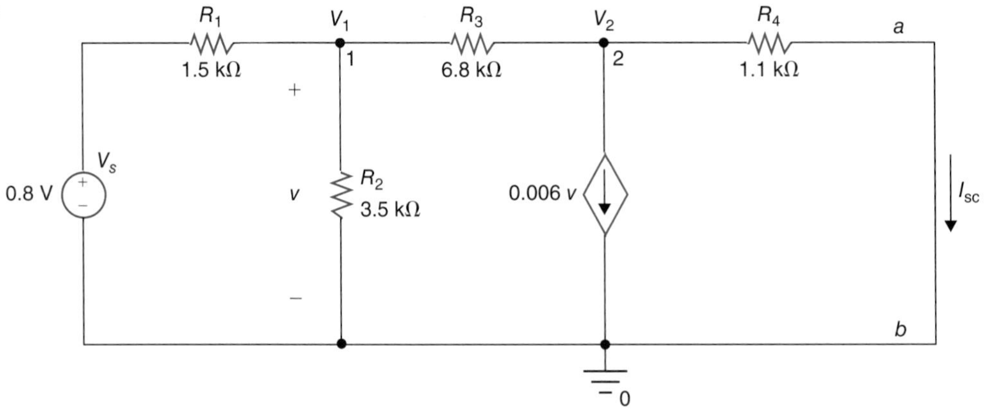

which can be rearranged as

$$\left(\frac{1}{1500}+\frac{1}{3500}+\frac{1}{6800}\right)V_1-\frac{1}{6800}V_2=\frac{0.8}{1500}$$

or

$$0.00109944V_1-1.470588\times10^{-4}V_2=5.33333\times10^{-4} \quad \textbf{(4.56)}$$

Summing the currents leaving node 2, we obtain

$$\frac{V_2-V_1}{6800}+0.006\times V_1+\frac{V_2}{1100}=0$$

which can be rearranged as

$$\left(0.006-\frac{1}{6800}\right)V_1+\left(\frac{1}{6800}+\frac{1}{1100}\right)V_2=0$$

or

$$0.00585294V_1+0.00105615V_2=0 \quad \textbf{(4.57)}$$

Solving Equation (4.57) for V_2, we obtain

$$V_2=-\frac{0.00585294}{0.00105615}V_1=-5.54177V_1$$

Substituting V_2 into Equation (4.56), we get

$$0.00109944V_1-1.470588\times10^{-4}(-5.54177V_1)=5.333333\times10^{-4}$$

Thus,

$$V_1=\frac{5.333333\times10^{-4}}{0.00109944+1.470588\times10^{-4}\times5.54177}=0.27859\text{ V}$$

$$V_2=-5.54177V_1=-1.54388\text{ V}$$

continued

Example 4.20 continued

Alternatively, application of Cramer's rule to Equations (4.56) and (4.57) yields

$$V_1 = \frac{\begin{vmatrix} 5.33333 \times 10^{-4} & -1.470588 \times 10^{-4} \\ 0 & 0.00105615 \end{vmatrix}}{\begin{vmatrix} 0.00109944 & -1.470588 \times 10^{-4} \\ 0.00585294 & 0.00105615 \end{vmatrix}} = 0.27859 \text{ V}$$

$$V_2 = \frac{\begin{vmatrix} 0.00109944 & 5.33333 \times 10^{-4} \\ 0.00585294 & 0 \end{vmatrix}}{\begin{vmatrix} 0.00109944 & -1.470588 \times 10^{-4} \\ 0.00585294 & 0.00105615 \end{vmatrix}} = -1.54388 \text{ V}$$

The short-circuit current is given by

$$I_{sc} = \frac{V_2}{R_4} = -0.00140353 \text{ A}$$

The Thévenin equivalent resistance is given by

$$R_{th} = \frac{V_{oc}}{I_{sc}} = 2.1753 \; k\Omega$$

The load resistance that provides maximum power transfer is $R_L = 2.1753 \; k\Omega$. The maximum power transferred is given by

$$p_L = \frac{V_{th}^2}{4R_L} = 1.0713 \text{ mW}$$

Exercise 4.20

For the circuit shown in Figure 4.146, find the value of the load resistance R_L that maximizes the power delivered to the load. Also, find the maximum power delivered to R_L.

FIGURE 4.146

Circuit for EXERCISE 4.20.

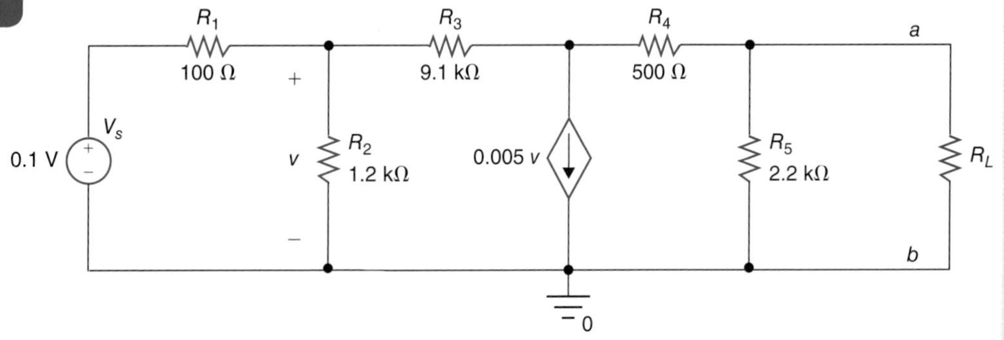

Answer:
$R_L = 1.6616 \; k\Omega, p_L = 71.1816 \; \mu\text{W}$.

EXAMPLE 4.21

For the circuit shown in Figure 4.147, find the value of the load resistance R_L that maximizes the power delivered to the load. Also, find the maximum power delivered to R_L.

FIGURE 4.147
Circuit for EXAMPLE 4.21.

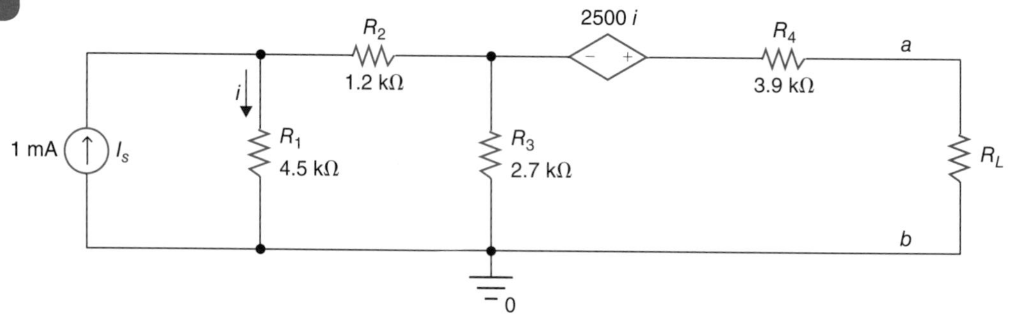

Figure 4.148 shows the circuit without the load resistor. There is no current through R_4.

FIGURE 4.148
The circuit in Figure 4.147 without the load resistor.

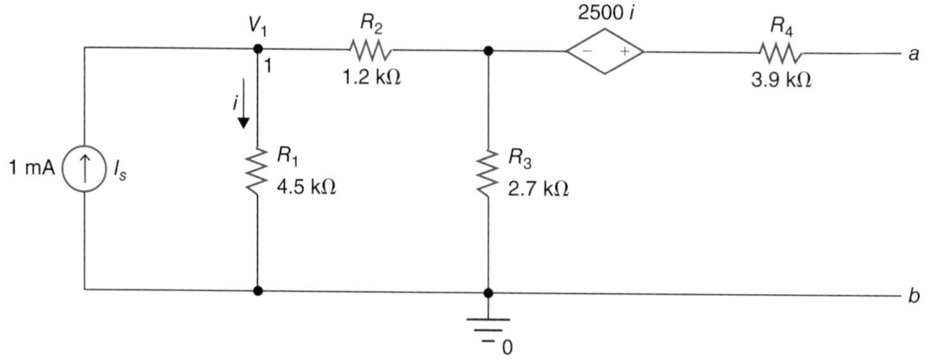

Summing the currents leaving node 1, we obtain

$$-0.001 + \frac{V_1}{4500} + \frac{V_1}{1200 + 2700} = 0$$

which can be rearranged as

$$\left(\frac{1}{4500} + \frac{1}{3900}\right)V_1 = 0.001$$

Thus, we have

$$V_1 = \frac{0.001}{\dfrac{1}{4500} + \dfrac{1}{3900}} = 2.0893 \text{ V}$$

The open-circuit voltage, which is also the Thévenin voltage, is given by

$$V_{th} = V_{oc} = V_1 \times \frac{R_3}{R_2 + R_3} + 2500 \times \frac{V_1}{R_1}$$

$$= 2.0893 \times \frac{2700}{1200 + 2700} + 2500 \times \frac{2.0893}{4500} = 2.6071 \text{ V}$$

continued

Example 4.21 continued

To find the short-circuit current, terminals a and b are short-circuited, as shown in Figure 4.149.

FIGURE 4.149

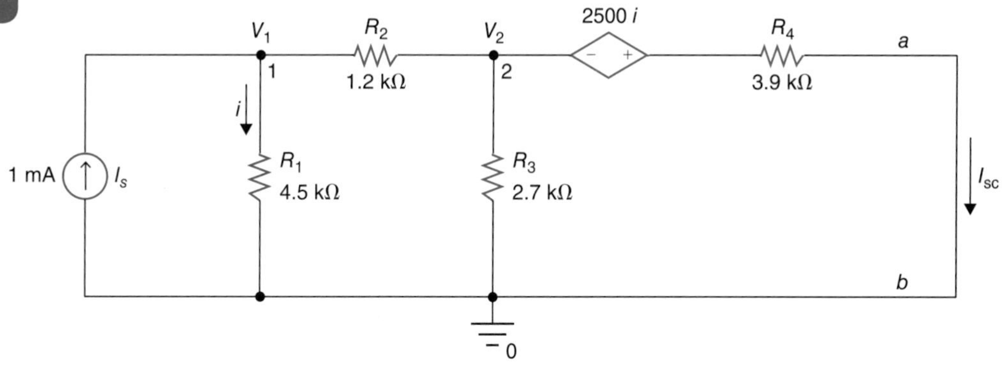

The circuit in Figure 4.148 with a and b shorted.

Summing the currents leaving node 1, we obtain

$$-0.001 + \frac{V_1}{4500} + \frac{V_1 - V_2}{1200} = 0$$

which can be rearranged as

$$\left(\frac{1}{4500} + \frac{1}{1200}\right)V_1 - \frac{1}{1200}V_2 = 0.001$$

or

$$0.00105556V_1 - 8.33333 \times 10^{-4}V_2 = 0.001 \qquad \textbf{(4.58)}$$

Summing the currents leaving node 2, we obtain

$$\frac{V_2 - V_1}{1200} + \frac{V_2}{2700} + \frac{V_2 + 2500 \times \dfrac{V_1}{4500}}{3900} = 0,$$

which can be rearranged as

$$\left(\frac{-1}{1200} + \frac{\frac{5}{9}}{3900}\right)V_1 + \left(\frac{1}{1200} + \frac{1}{2700} + \frac{1}{3900}\right)V_2 = 0$$

or

$$-6.908832 \times 10^{-4}V_1 + 0.001460114V_2 = 0 \qquad \textbf{(4.59)}$$

Solving Equation (4.59) for V_2, we obtain

$$V_2 = \frac{6.908832 \times 10^{-4}}{0.001460114}V_1 = 0.473171V_1$$

Substituting V_2 into Equation (4.58), we get

$$0.00105556V_1 - 8.333333 \times 10^{-4}(0.473171V_1) = 0.001$$

continued

Example 4.21 continued

Thus,

$$V_1 = \frac{0.001}{0.00105556 - 8.333333 \times 10^{-4} \times 0.473171} = 1.5123 \text{ V}$$

$$V_2 = 0.473171 V_1 = 0.7156 \text{ V}$$

Alternatively, application of Cramer's rule to Equations (4.58) and (4.59) yields

$$V_1 = \frac{\begin{vmatrix} 0.001 & -8.33333 \times 10^{-4} \\ 0 & 0.001460114 \end{vmatrix}}{\begin{vmatrix} 0.00105556 & -8.33333 \times 10^{-4} \\ -6.908832 \times 10^{-4} & 0.001460114 \end{vmatrix}} = 1.5123 \text{ V}$$

$$V_2 = \frac{\begin{vmatrix} 0.00105556 & 0.001 \\ -6.908832 \times 10^{-4} & 0 \end{vmatrix}}{\begin{vmatrix} 0.00105556 & -8.33333 \times 10^{-4} \\ -6.908832 \times 10^{-4} & 0.001460114 \end{vmatrix}} = 0.7156 \text{ V}$$

The short-circuit current is given by

$$I_{sc} = \frac{V_2 + 2500 \times \dfrac{V_1}{R_1}}{R_4} = 0.3989 \text{ mA}$$

The Thévenin equivalent resistance is the ratio of V_{oc} to I_{sc}:

$$R_{th} = \frac{V_{oc}}{I_{sc}} = \frac{V_{th}}{I_{sc}} = 6.5357\, k\Omega$$

The load resistance value that provides the maximum power transfer is $R_L = R_{th} = 6.5357\ k\Omega$. The maximum power delivered to the load is given by

$$p_L = \frac{V_{th}^2}{4R_L} = 0.26 \text{ mW}$$

MATLAB

```
%EXAMPLE 4.21
clear all;format long;
Is=1e-3;R1=4500;R2=1200;R3=2700;R4=3900;
syms V1 V2 Va Vb
%Voc = Vth
[V1,V2]=solve(-Is+V1/R1+(V1-V2)/R2,(V2-V1)/R2+V2/R3);
Vth=V2+2500*V1/R1;
%Isc
[Va,Vb]=solve(-Is+Va/R1+(Va-Vb)/R2,...
(Vb-Va)/R2+Vb/R3+(Vb+2500*Va/R1)/R4);
Isc=(Vb+2500*Va/R1)/R4;
Rth=Vth/Isc;
PL=Vth^2/(4*Rth);
V1=vpa(V1,7)
```

continued

Example 4.21 continued
MATLAB continued

```
V2=vpa(V2,7)
Va=vpa(Va,7)
Vb=vpa(Vb,7)
Isc=vpa(Isc,7)
Vth=vpa(Vth,7)
Rth=vpa(Rth,7)
PL=vpa(PL,7)

Answers:
V1 =
2.089286
V2 =
1.446429
Va =
1.512295
Vb =
0.7155738
Isc =
0.0003989071
Vth =
2.607143
Rth =
6535.714
PL =
0.000260002
```

Exercise 4.21

For the circuit shown in Figure 4.150, find the value of the load resistance R_L that maximizes the power delivered to the load. Also, find the maximum power delivered to R_L.

FIGURE 4.150

Circuit for EXERCISE 4.21.

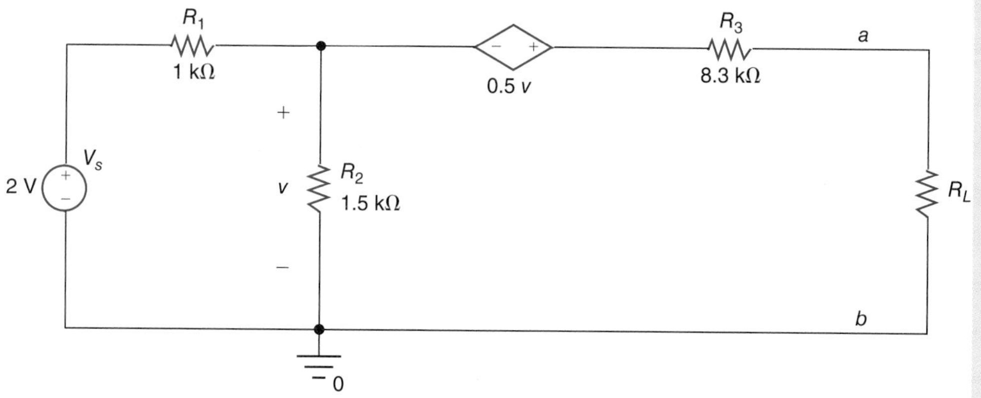

Answer:

$R_L = 9.2\ k\Omega, p_L = 88.0435\ \mu W.$

4.7 PSpice

In PSpice, the **transfer function** can be used to find the Thévenin equivalent resistance. The Thévenin equivalent voltage can be found by finding the open-circuit voltage. In EXAMPLE 4.7, earlier in this chapter, we showed that the Thévenin equivalent voltage is $V_{\text{th}} = 14$ V and the Thévenin equivalent resistance is $R_{\text{th}} = 20\ k\Omega$ between a and b for the circuit shown in Figure 4.151. We can verify these values using PSpice. If we try to run the circuit as shown in Figure 4.151, we get an error message that says that there are fewer than two connections at node a. To avoid this error, we can add a resistor with large resistance ($10^{12}\ \Omega$) between a and b, as shown in Figure 4.152. Because of the large resistance value, it is virtually an open circuit between a and b.

FIGURE 4.151

Circuit for EXAMPLE 4.7.

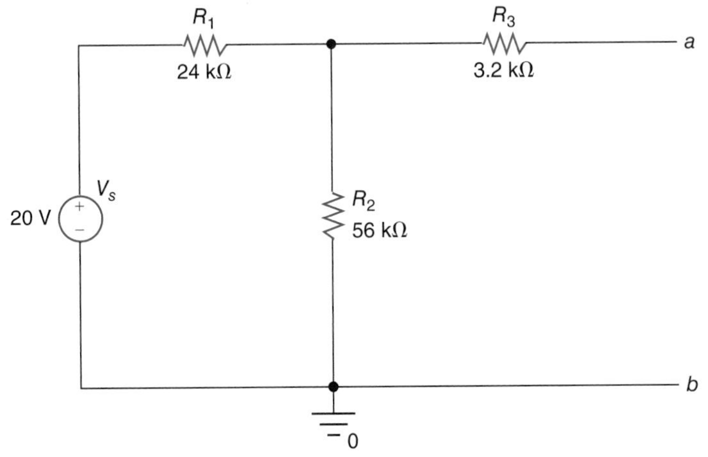

FIGURE 4.152

The circuit shown in Figure 4.151 with a large resistor R_4 between a and b.

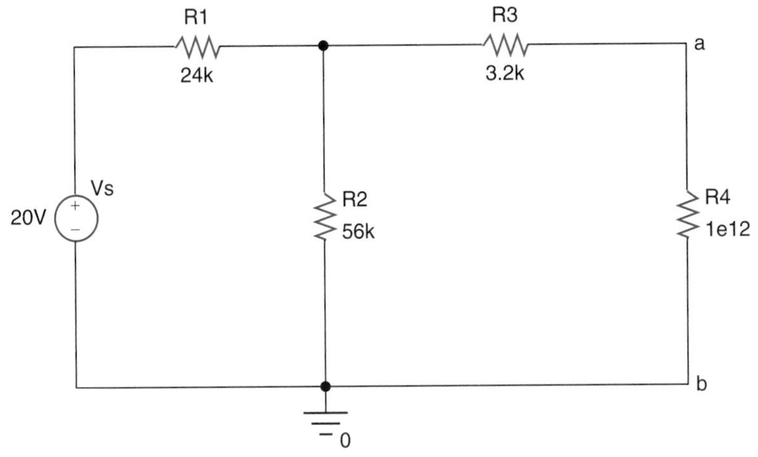

In the Edit Simulation Profile, check Calculate small-signal DC gain (.TF) and enter V_s and $V(R_4)$ respectively for the Input source name and Output variable fields, as shown in Figure 4.153.

After running the simulation, click on the V (Enable Bias Voltage Display) to display the voltages, as shown in Figure 4.154.

The voltage across R_4 is 14 V. This is the open-circuit voltage and the Thévenin equivalent voltage; that is,

$$V_{\text{th}} = V_{\text{oc}} = 14 \text{ V}$$

FIGURE 4.153

Setting the transfer function.
(Source: OrCAD PSpice by Cadence)

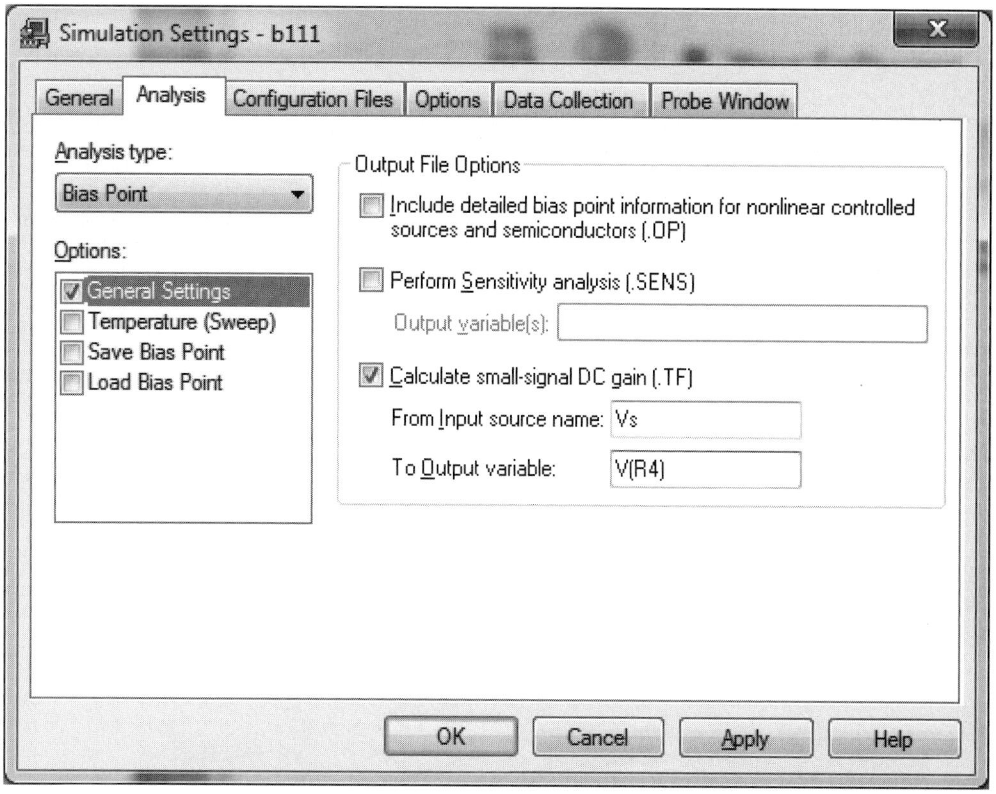

FIGURE 4.154

Node voltages.

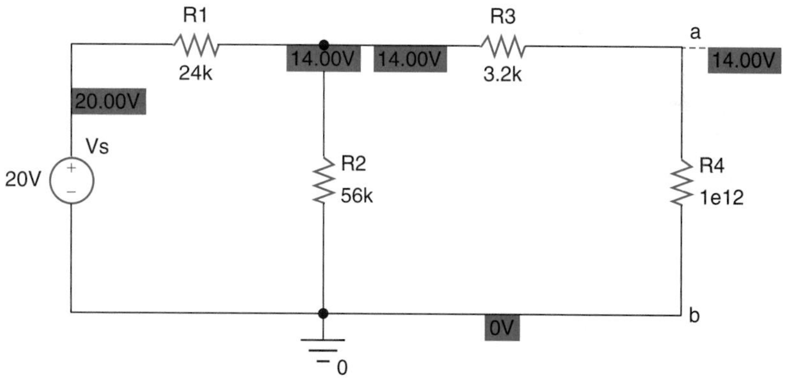

If you click on View Simulation Output File (icon) in the SCHEMATIC1 window, or select View → Output File and scroll down, the following results are shown near the end of the output file:

```
****      SMALL-SIGNAL CHARACTERISTICS

V(R_R4)/V_Vs = 7.000E-01
INPUT RESISTANCE AT V_Vs = 8.000E+04
OUTPUT RESISTANCE AT V(R_R4) = 2.000E+04
```

The first line shows the ratio of the voltage across R_4 to the voltage of the voltage source V_s, called transfer function; that is,

$$\frac{V(R_4)}{V_s} = 0.7$$

Thus, the voltage across R_4 is given by

$$V(R_4) = 0.7 \times V_s = 0.7 \times 20 = 14 \text{ V}$$

which is the Thévenin equivalent voltage. The second line shows the input resistance of 80 $k\Omega$ from the source. Notice that

$$R_{in} = R_1 + [R_2 \| (R_3 + R_4)] = 24{,}000\ \Omega + [56{,}000\ \Omega \| (3200\ \Omega + 10^{12}\ \Omega)]$$
$$= 24{,}000\ \Omega + 56{,}000\ \Omega = 80\ k\Omega$$

The third line shows the output resistance of 20 $k\Omega$ from the output. Notice that

$$R_{out} = R_{\text{th}} = R_3 + (R_1 \| R_2) = 3.2\ k\Omega + (24\ k\Omega \| 56\ k\Omega)$$
$$= 3.2\ k\Omega + 16.8\ k\Omega = 20\ k\Omega$$

If the value of R_4 is changed to $10^{-12}\ \Omega$, as shown in Figure 4.155, we can find the short-circuit current between a and b.

FIGURE 4.155

A short-circuit current between a and b.

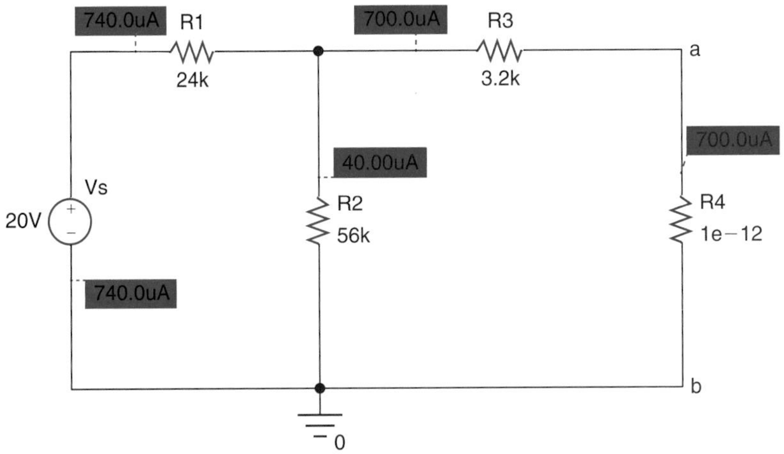

The short-circuit current is $I_{sc} = 700\ \mu\text{A}$. The Thévenin equivalent resistance can also be found by taking the ratio of V_{oc} to I_{sc}:

$$R_{\text{th}} = \frac{V_{oc}}{I_{sc}} = \frac{14 \text{ V}}{0.7 \text{ mA}} = 20\ k\Omega$$

The short-circuit current is the Norton equivalent current I_n.

EXAMPLE 4.22

Use PSpice to find the Thévenin equivalent voltage V_{th} and the Thévenin equivalent resistance R_{th} between a and b for the circuit shown in Figure 4.156.

The Thévenin equivalent voltage is $V_{\text{th}} = -3.053$ V, as shown in Figure 4.157. From the transfer function, we obtain the Thévenin equivalent resistance $R_{\text{th}} = 2.175\ k\Omega$, as shown here.

continued

Example 4.22 continued

FIGURE 4.156

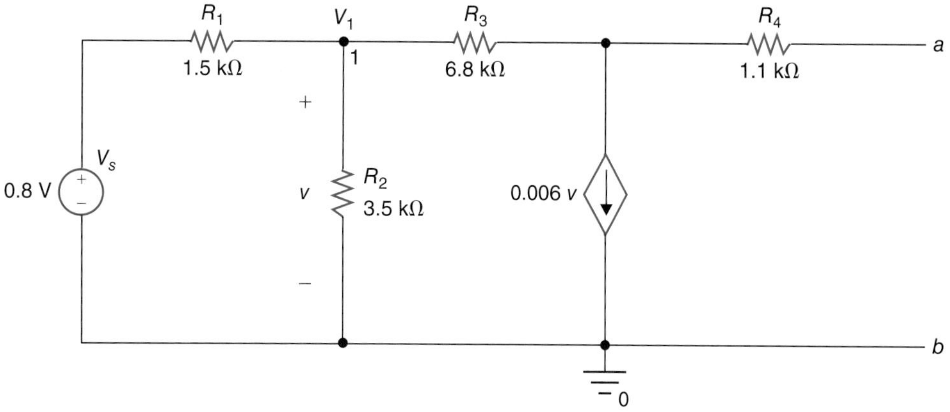

Circuit for EXAMPLE 4.22.

FIGURE 4.157

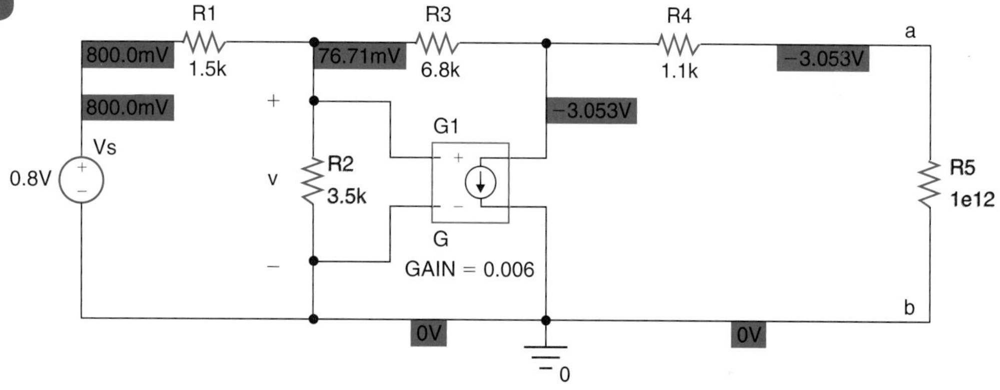

Node voltages.

```
****      SMALL-SIGNAL CHARACTERISTICS

V(R_R5)/V_Vs = -3.816E+00
INPUT RESISTANCE AT V_Vs = 1.659E+03
OUTPUT RESISTANCE AT V(R_R5) = 2.175E+03
```

4.7.1 SIMULINK

Figure 4.158 shows a Simulink model to measure the Thévenin voltage for the circuit shown in Figure 4.151. Figure 4.159 shows a Simulink model to measure the Thévenin resistance for the circuit shown in Figure 4.151. A test voltage of 1 V is applied across the terminals *a* and *b* after deactivating the source. The ratio of the test voltage to the current flowing out of the positive terminal of the test source is the Thévenin resistance. The answer from the model is 20 $k\Omega$.

FIGURE 4.158

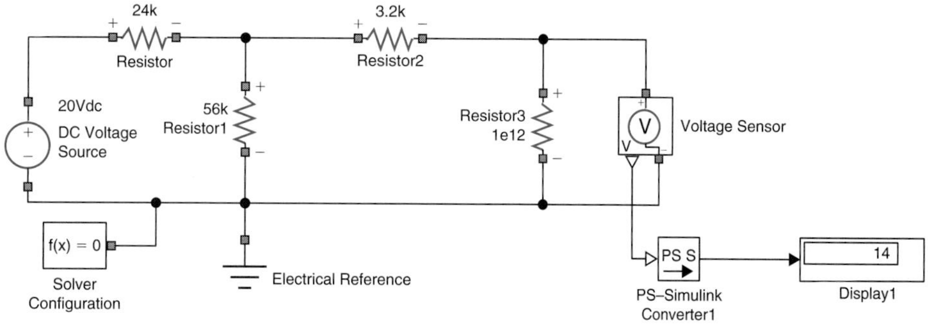

A Simulink model to measure the Thévenin voltage.

FIGURE 4.159

A Simulink model to measure the Thévenin resistance for the circuit shown in Figure 4.151.

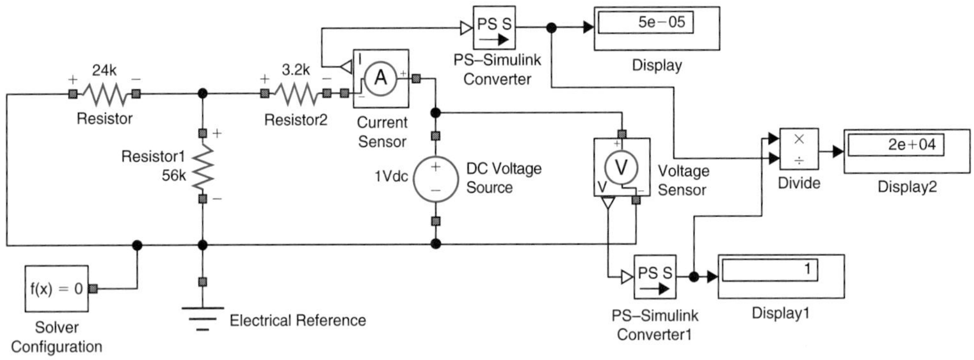

SUMMARY

In this chapter, four important circuit theorems have been presented: the superposition principle, source transformation, Thévenin's theorem, and Norton's theorem. A brief summary is given for each of these theorems.

The *superposition principle* says that if a circuit contains more than one independent source, we can find the response of the circuit from one source at a time while the rest are deactivated and add the responses from individual sources to get the response.

The *source transformation* says that a voltage source with voltage V_s in series with a resistor with resistance R can be replaced by a current source with current V_s/R and a parallel resistor with resistance R. Also, a current source with current I_s in parallel with a resistor with resistance R can be replaced by a voltage source with voltage RI_s and a series resistor with resistance R.

Thévenin's theorem says that given terminals a and b, the circuit looking from the terminals a and b can be represented by a voltage source with voltage V_{th}, called the *Thévenin equivalent voltage*, and a series resistor with resistance R_{th}, called the *Thévenin equivalent resistance.* The Thévenin equivalent voltage V_{th} can be found by calculating the open-circuit voltage between terminals a and b. There are three methods to find the Thévenin equivalent resistance R_{th}. The first method is to find the equivalent resistance looking into the circuit from terminals a and b after deactivating independent sources. The second method is to find the short-circuit current from a to b, and to take the ratio of the open-circuit voltage to the short-circuit current (i.e., $R_{th} = V_{oc}/I_{sc}$). The third method is to apply a test voltage from a to b into the circuit after deactivating the independent sources, measure the current flowing out of the positive terminal of the test voltage source, and then take the ratio of the voltage to current. Alternatively, a test current can be applied to the circuit from terminals a and b after deactivating the independent sources. Then, measure the voltage across the test current source. The Thévenin equivalent resistance is the ratio of the voltage to current.

Norton's theorem says that given terminals a and b, the circuit looking from the terminals a and b can be represented by a current source with current I_n, called the *Norton equivalent current*, and a parallel resistor with resistance R_n, called the *Norton equivalent resistance.* The Norton equivalent current I_n can be found by calculating the short-circuit current from a to b. There are three methods to find the Norton equivalent resistance R_n. The first method is to find the equivalent resistance looking into the circuit from terminals a and b after deactivating independent sources. The second method is to find the open-circuit voltage between terminals a and b, and to take the ratio of the open-circuit voltage to the short-circuit current (i.e., $R_n = V_{oc}/I_{sc}$). The third method is to apply a test voltage from a to b into the circuit after deactivating the independent sources, measure the current flowing out of the positive terminal of the test voltage source, and then take the ratio of the voltage to current. Alternatively, a test current can be applied to the circuit from terminals a and b after deactivating the independent sources. Then, measure the voltage across the test current source. The Norton equivalent resistance is the ratio of the voltage to the current.

Suppose that a load with resistance R_L is connected to a circuit between terminals a and b. We are interested in finding the load resistance R_L that maximizes the power delivered to the load. We first find the Thévenin equivalent circuit with respect to the terminals a and b. Let V_{th} be the Thévenin equivalent voltage and R_{th} be the Thévenin equivalent resistance. Then, the load resistance R_L that maximizes the power delivered to the load is given by the Thévenin equivalent resistance from terminals a and b.

PROBLEMS

Superposition Principle

4.1 Use the superposition principle to find voltage V_1 in the circuit shown in Figure P4.1.

FIGURE P4.1

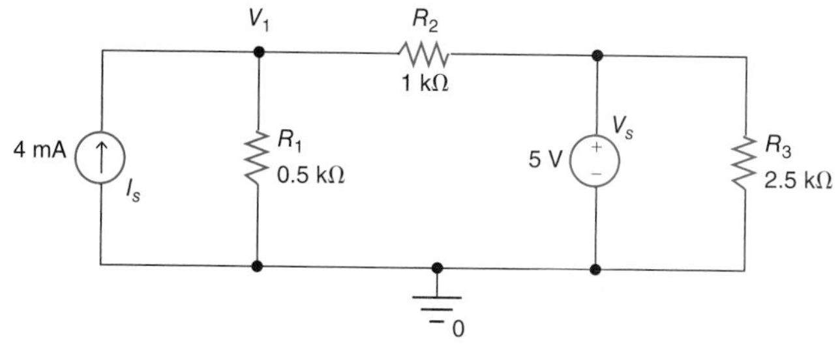

4.2 Use the superposition principle to find voltage V_1 in the circuit shown in Figure P4.2.

FIGURE P4.2

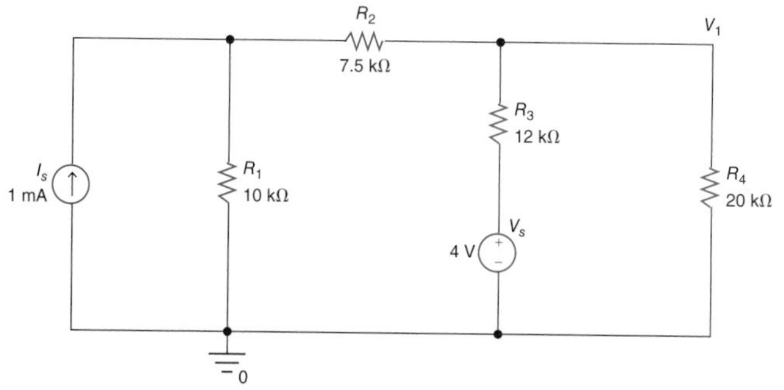

4.3 Use the superposition principle to find voltage V_1 in the circuit shown in Figure P4.3.

FIGURE P4.3

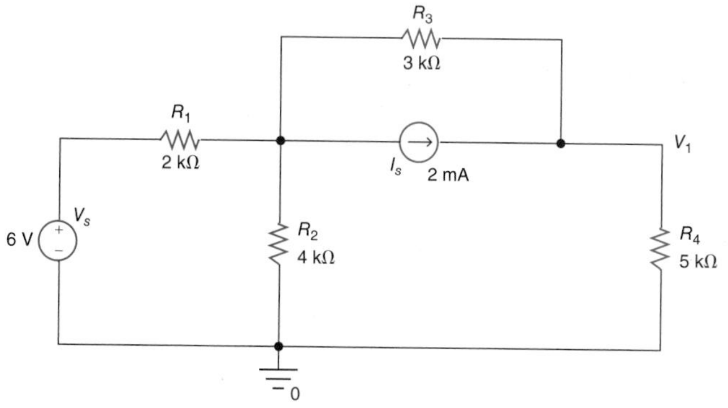

4.4 Use the superposition principle to find voltage V_1 in the circuit shown in Figure P4.4.

FIGURE P4.4

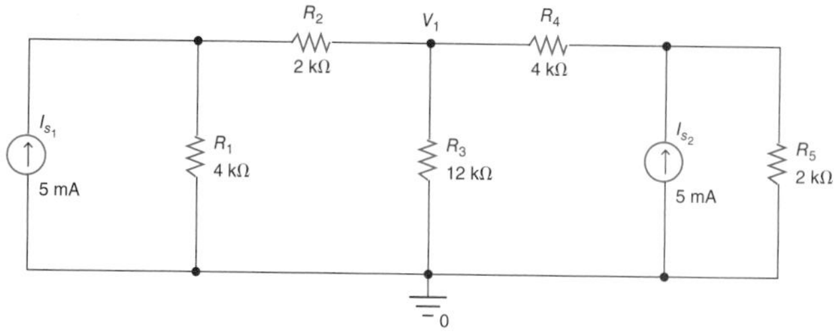

4.5 Use the superposition principle to find voltage V_1 in the circuit shown in Figure P4.5.

FIGURE P4.5

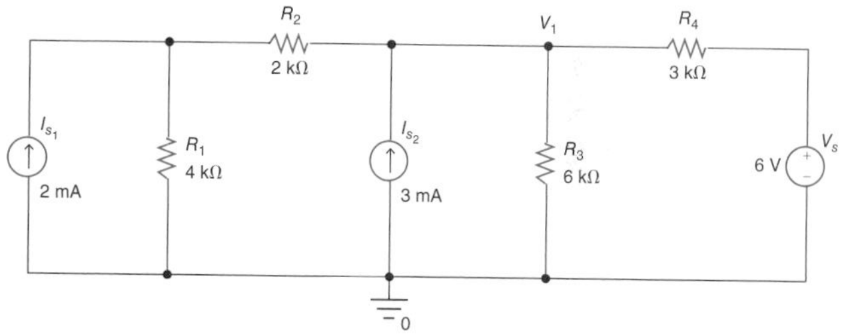

4.6 Use the superposition principle to find voltages V_1 and V_2 in the circuit shown in Figure P4.6.

FIGURE P4.6

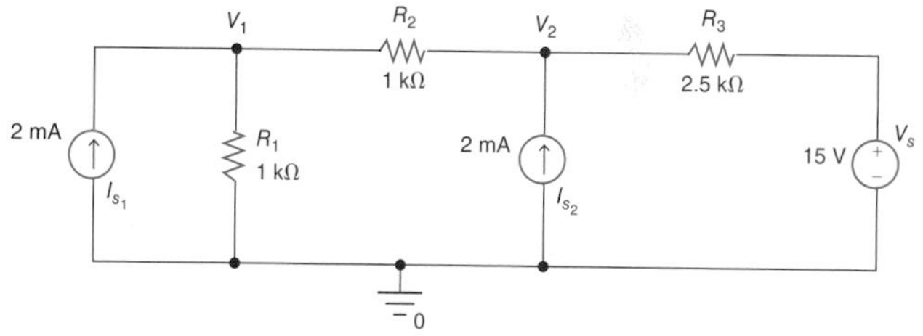

4.7 Use the superposition principle to find voltages V_1 and V_2 in the circuit shown in Figure P4.7.

FIGURE P4.7

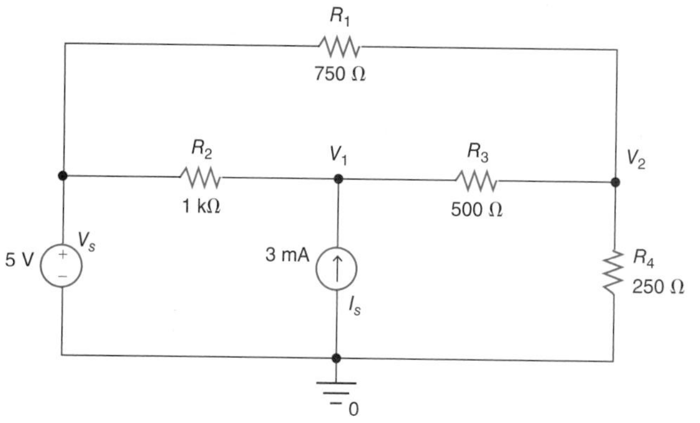

4.8 Use the superposition principle to find voltages V_1, V_2, and V_3 in the circuit shown in Figure P4.8.

FIGURE P4.8

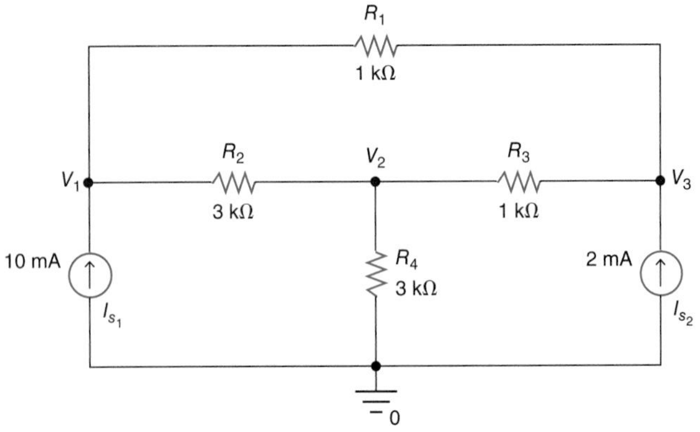

4.9 Use the superposition principle to find voltages V_1 and V_2 in the circuit shown in Figure P4.9.

FIGURE P4.9

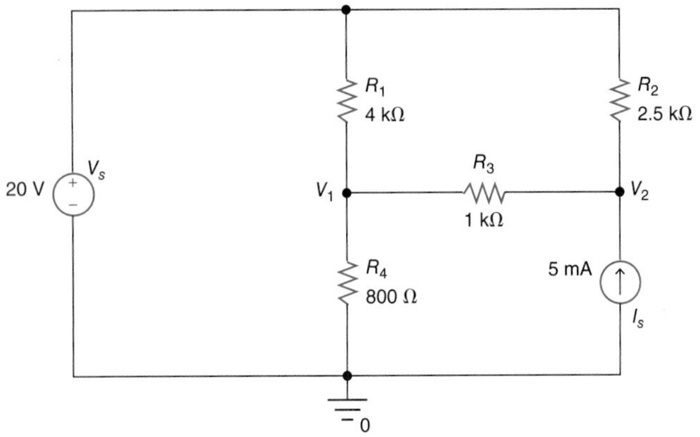

4.10 Use the superposition principle to find voltages V_1, V_2, and V_3 in the circuit shown in Figure P4.10.

FIGURE P4.10

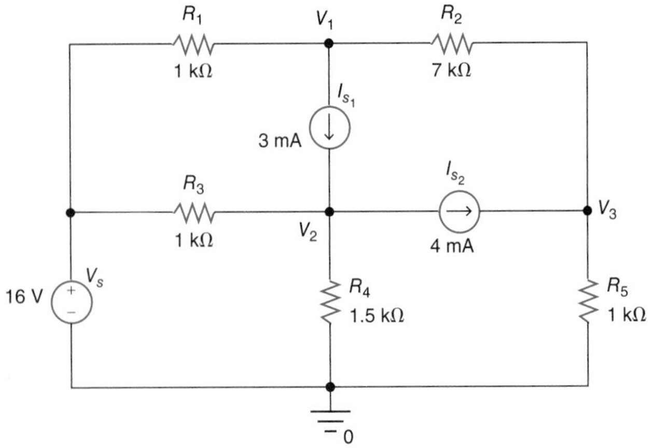

4.11 Use the superposition principle to find voltages V_1 and V_2 in the circuit shown in Figure P4.11.

FIGURE P4.11

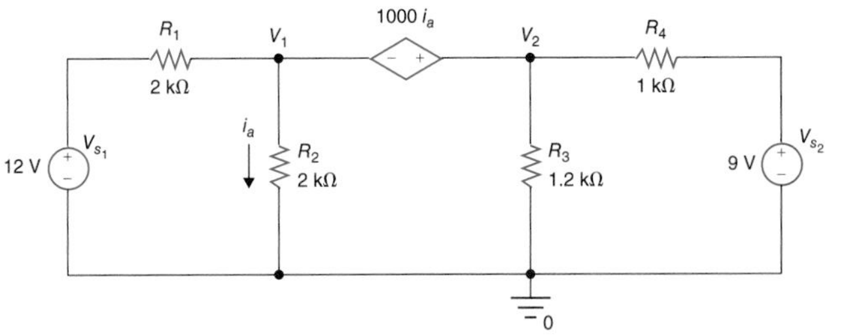

4.12 Use the superposition principle to find voltages V_1 and V_2 in the circuit shown in Figure P4.12.

FIGURE P4.12

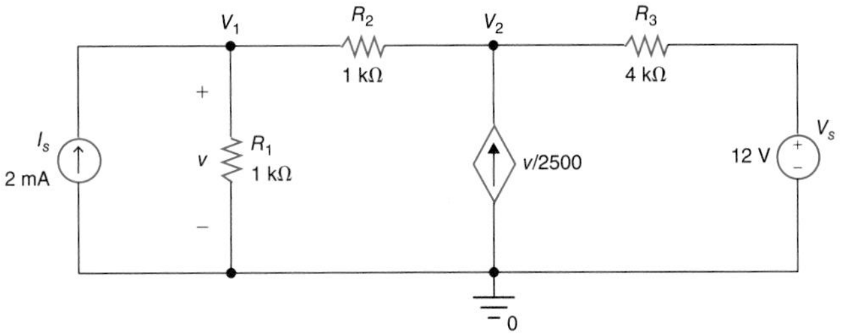

4.13 Use the superposition principle to find voltages V_1 and V_2 in the circuit shown in Figure P4.13.

FIGURE P4.13

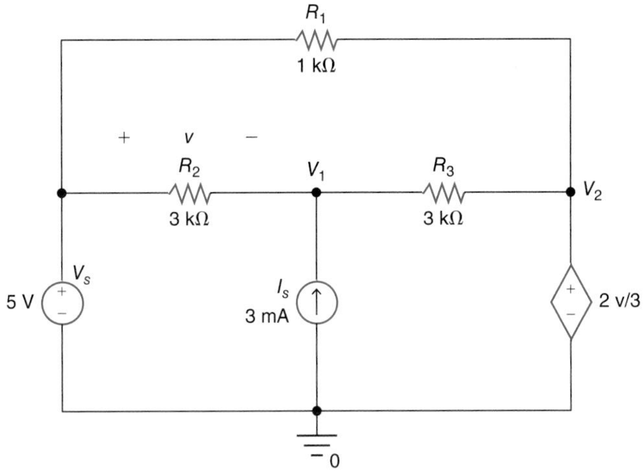

4.14 Use the superposition principle to find voltages V_1, V_2, and V_3 in the circuit shown in Figure P4.14.

FIGURE P4.14

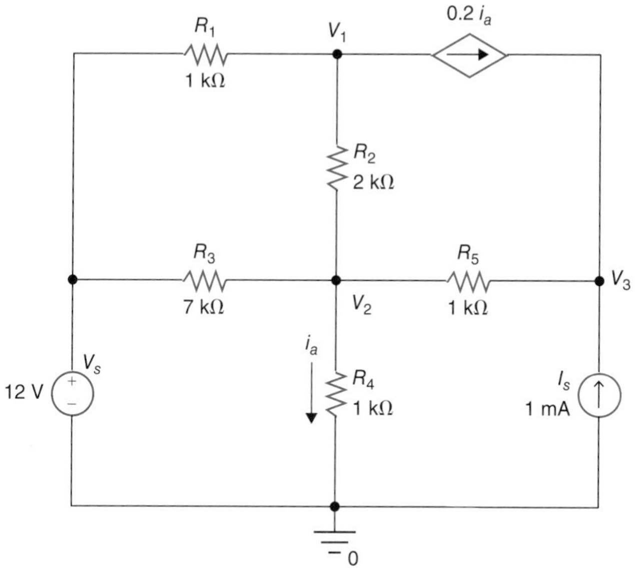

4.15 Use the superposition principle to find voltages V_1, V_2, and V_3 in the circuit shown in Figure P4.15.

FIGURE P4.15

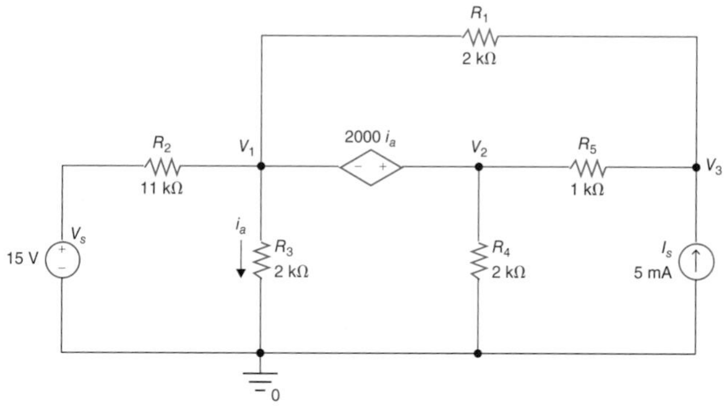

Source Transformation

4.16 Use source transformation to find voltages V_1 and V_2 in the circuit shown in Figure P4.16.

FIGURE P4.16

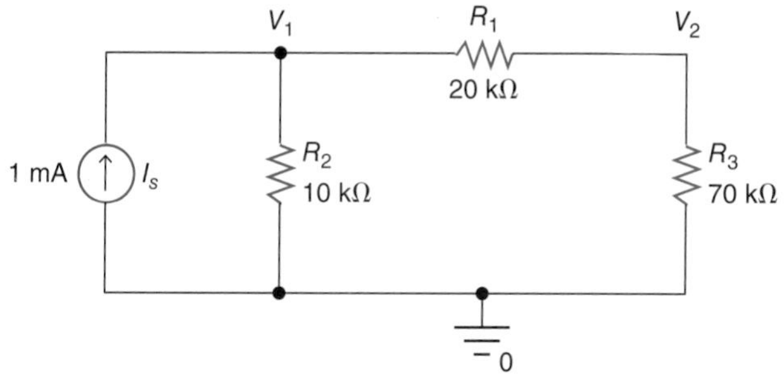

4.17 Use source transformation to find currents I_1 and I_2 in the circuit shown in Figure P4.17.

FIGURE P4.17

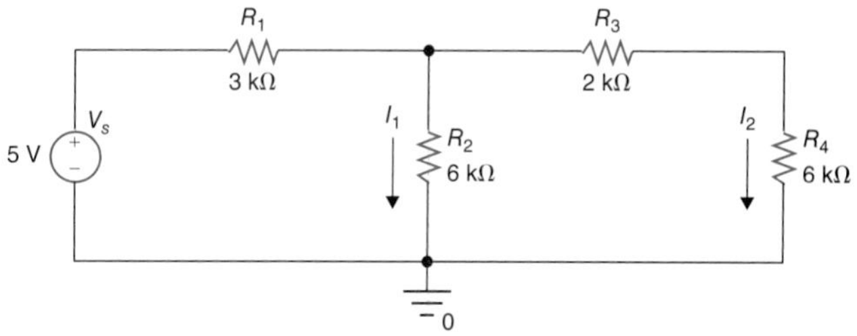

4.18 Use source transformation to find voltage V_o in the circuit shown in Figure P4.18.

FIGURE P4.18

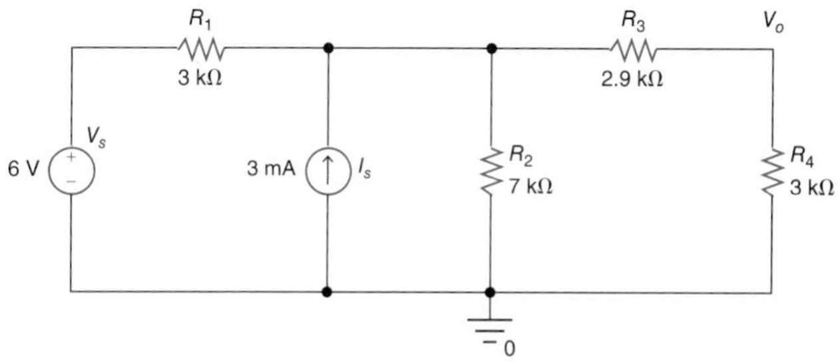

4.19 Use source transformation to find voltage V_o in the circuit shown in Figure P4.19.

FIGURE P4.19

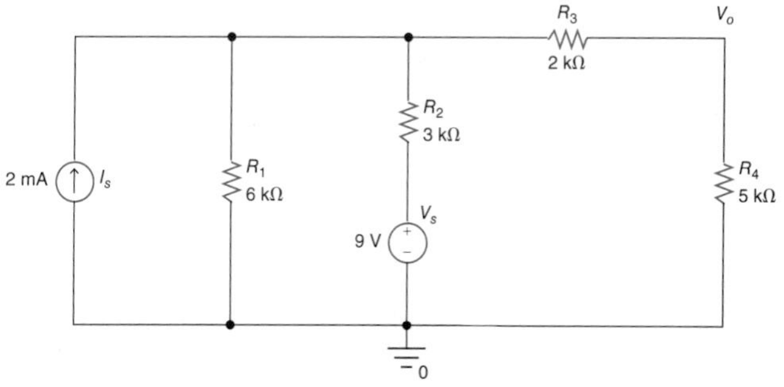

4.20 Use source transformation to find voltage V_o in the circuit shown in Figure P4.20.

FIGURE P4.20

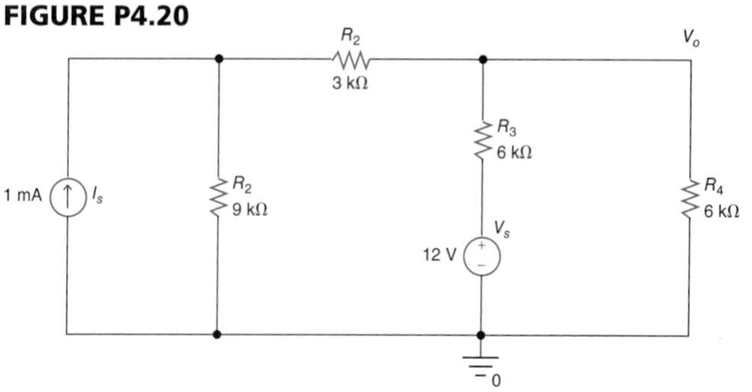

4.21 Use source transformation to find voltage V_o in the circuit shown in Figure P4.21.

FIGURE P4.21

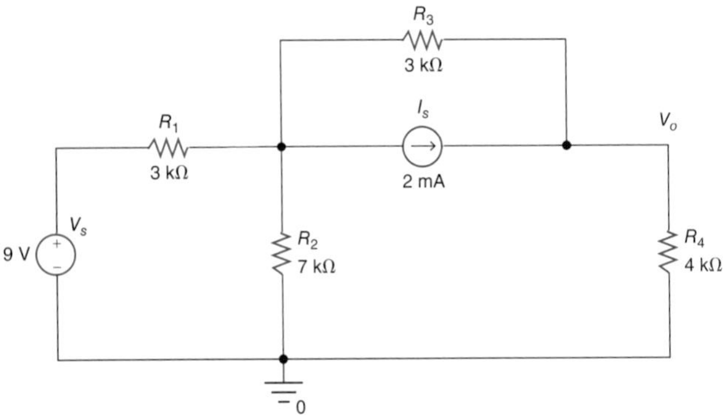

4.22 Use source transformation to find voltage V_o in the circuit shown in Figure P4.22.

FIGURE P4.22

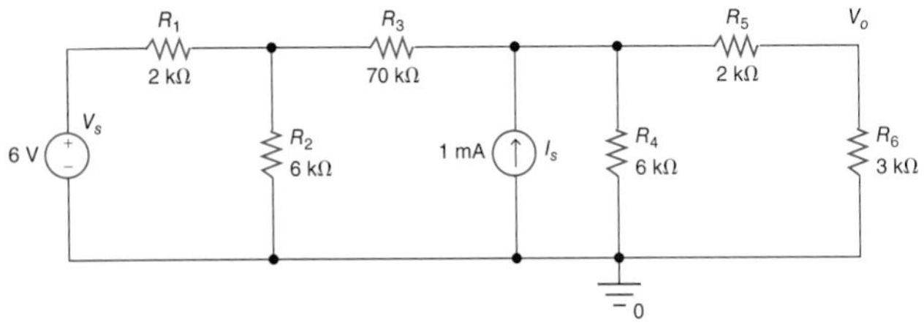

4.23 Use source transformation to find voltage V_o in the circuit shown in Figure P4.23.

FIGURE P4.23

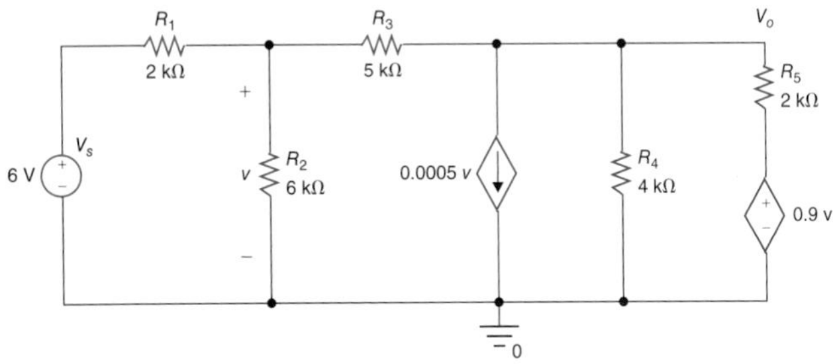

4.24 Use source transformation to find voltages V_1, V_2, and V_3 in the circuit shown in Figure P4.24.

FIGURE P4.24

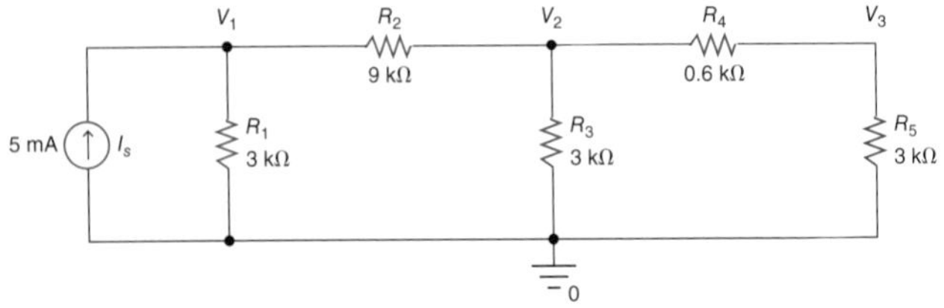

4.25 Use source transformation to find voltages V_1 and V_2 in the circuit shown in Figure P4.25.

FIGURE P4.25

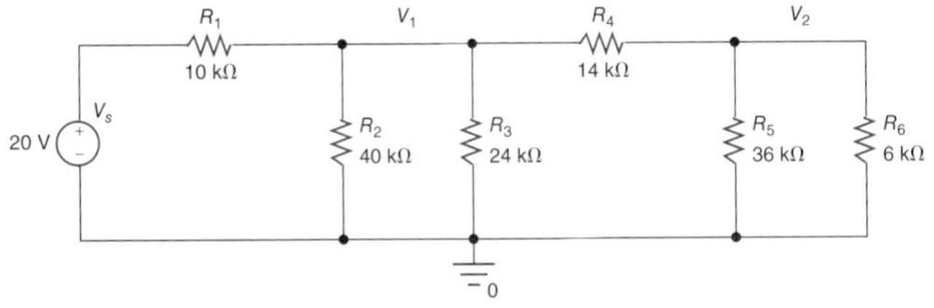

4.26 Use source transformation to find voltages V_1, V_2, and V_3 in the circuit shown in Figure P4.26.

FIGURE P4.26

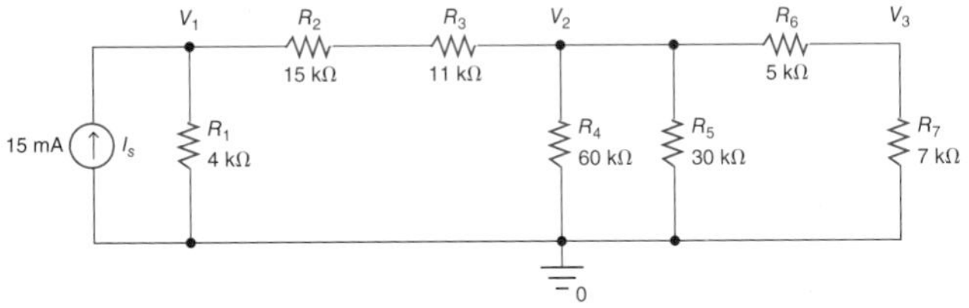

4.27 Use source transformation to find voltages V_1, V_2, and V_3 in the circuit shown in Figure P4.27.

FIGURE P4.27

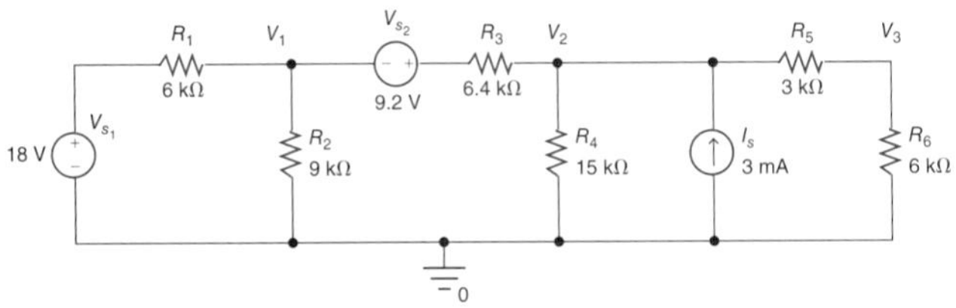

4.28 Use source transformation to find voltages V_1, V_2, V_3, and V_4 in the circuit shown in Figure P4.28.

FIGURE P4.28

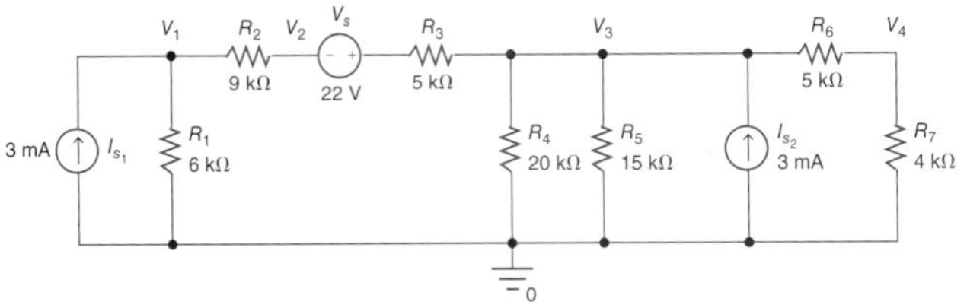

4.29 Use source transformation to find voltages V_1, V_2, and V_3 in the circuit shown in Figure P4.29.

FIGURE P4.29

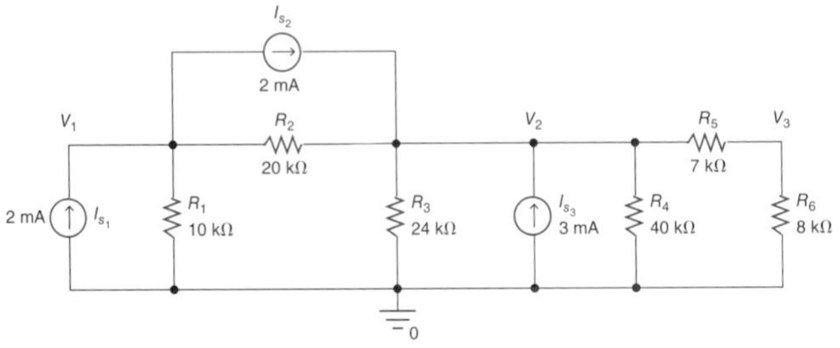

4.30 Use source transformation to find voltages V_1, V_2, and V_3 in the circuit shown in Figure P4.30.

FIGURE P4.30

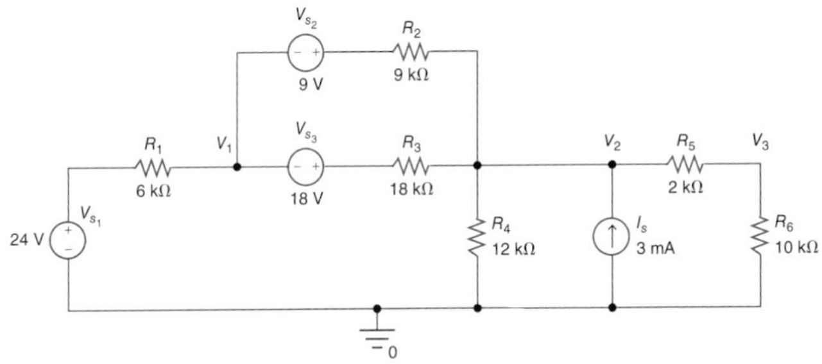

Thévenin Equivalent Circuit

4.31 Find the Thévenin equivalent circuit between *a* and *b* for the circuit shown in Figure P4.31.

FIGURE P4.31

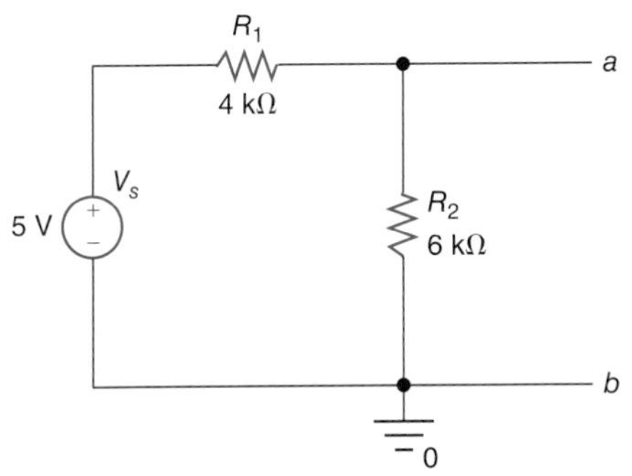

4.32 Find the Thévenin equivalent circuit between *a* and *b* for the circuit shown in Figure P4.32.

FIGURE P4.32

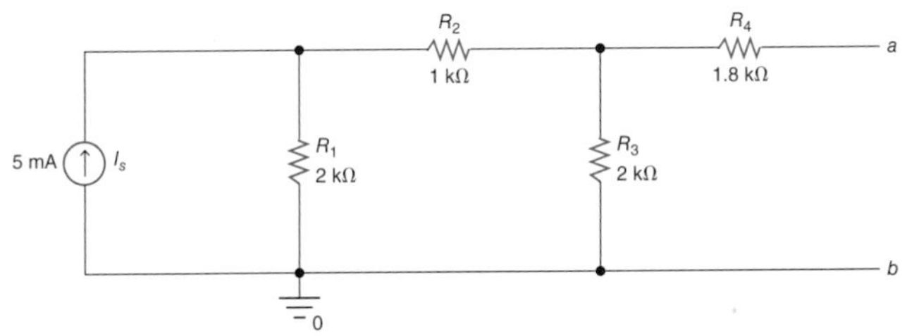

4.33 Find the Thévenin equivalent circuit between *a* and *b* for the circuit shown in Figure P4.33.

FIGURE P4.33

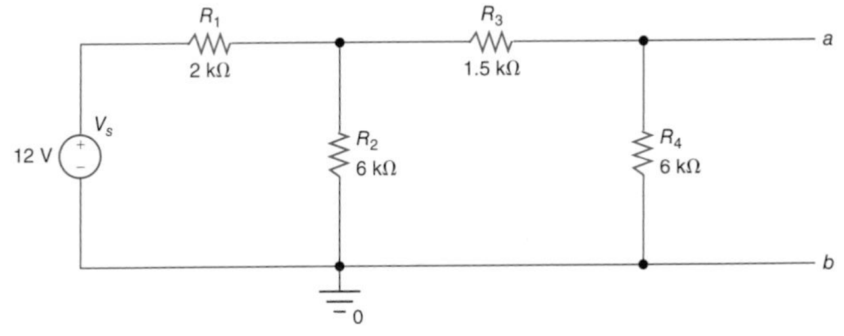

4.34 Find the Thévenin equivalent circuit between *a* and *b* for the circuit shown in Figure P4.34.

FIGURE P4.34

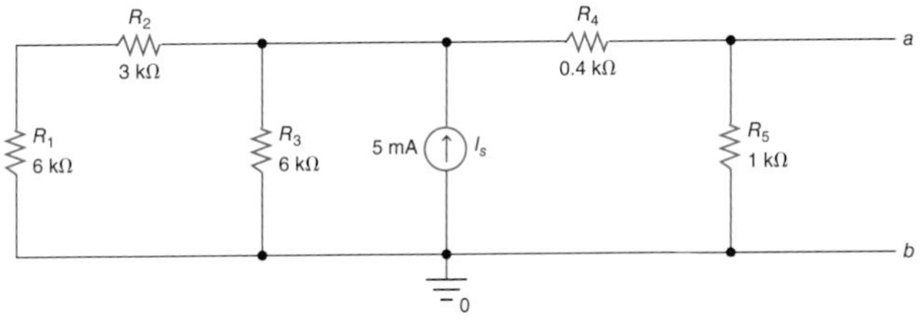

4.35 Find the Thévenin equivalent circuit between *a* and *b* for the circuit shown in Figure P4.35.

FIGURE P4.35

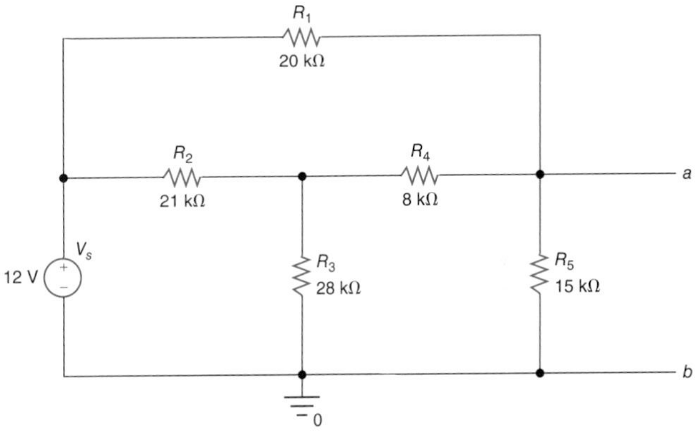

4.36 Find the Thévenin equivalent circuit between *a* and *b* for the circuit shown in Figure P4.36.

FIGURE P4.36

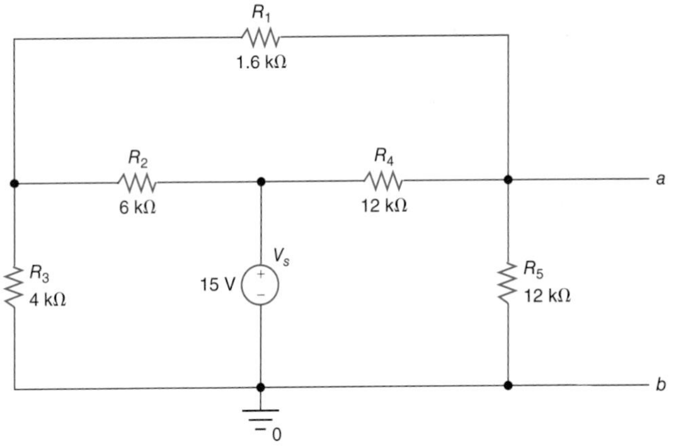

4.37 **Find the Thévenin equivalent circuit between *a* and *b* for the circuit shown in Figure P4.37.**

FIGURE P4.37

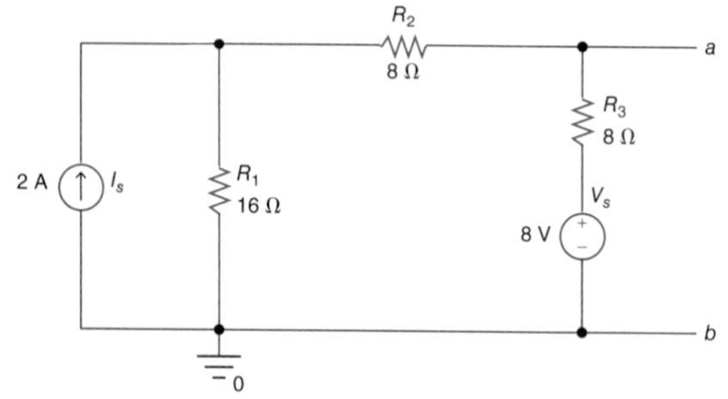

4.38 **Find the Thévenin equivalent circuit between *a* and *b* for the circuit shown in Figure P4.38.**

FIGURE P4.38

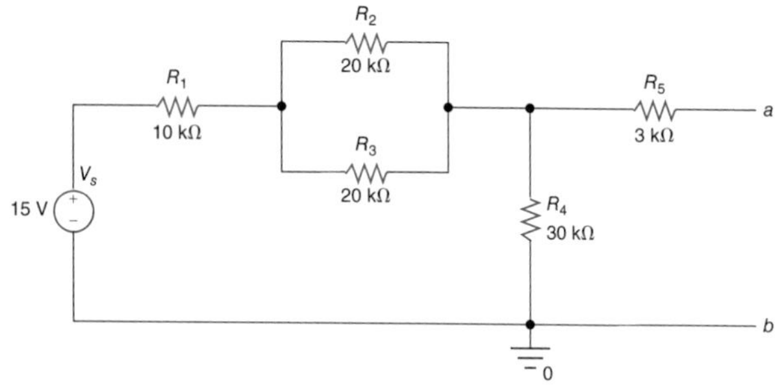

4.39 **Find the Thévenin equivalent circuit between *a* and *b* for the circuit shown in Figure P4.39.**

FIGURE P4.39

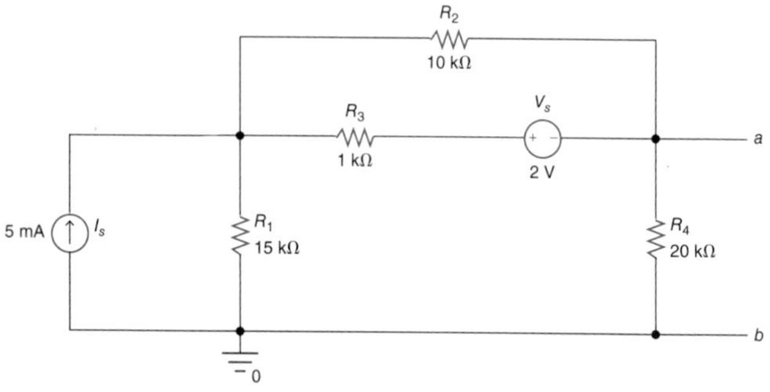

4.40 **Find the Thévenin equivalent circuit between *a* and *b* for the circuit shown in Figure P4.40.**

FIGURE P4.40

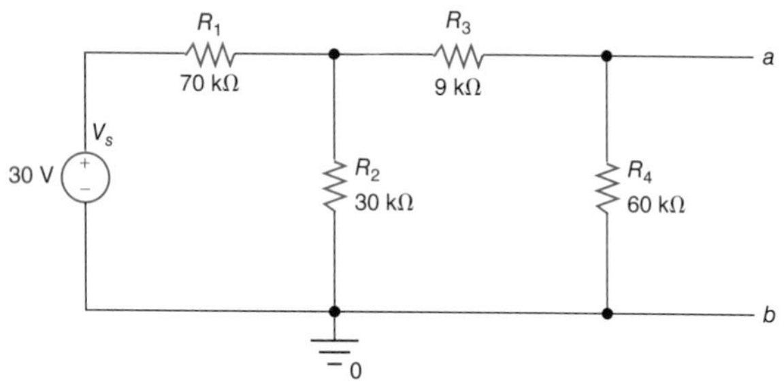

4.41 **Find the Thévenin equivalent circuit between *a* and *b* for the circuit shown in Figure P4.41.**

FIGURE P4.41

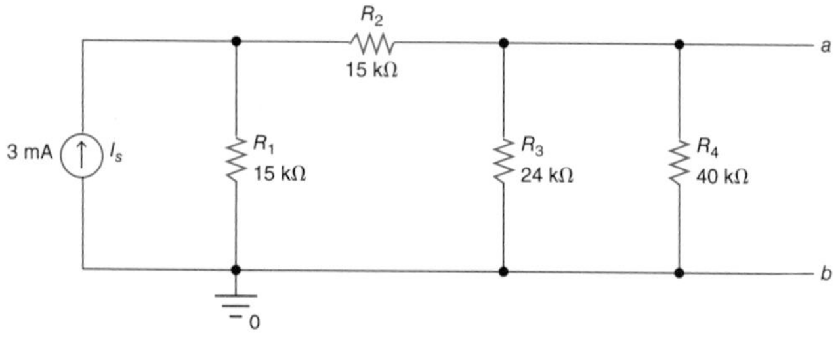

4.42 **Find the Thévenin equivalent circuit between *a* and *b* for the circuit shown in Figure P4.42.**

FIGURE P4.42

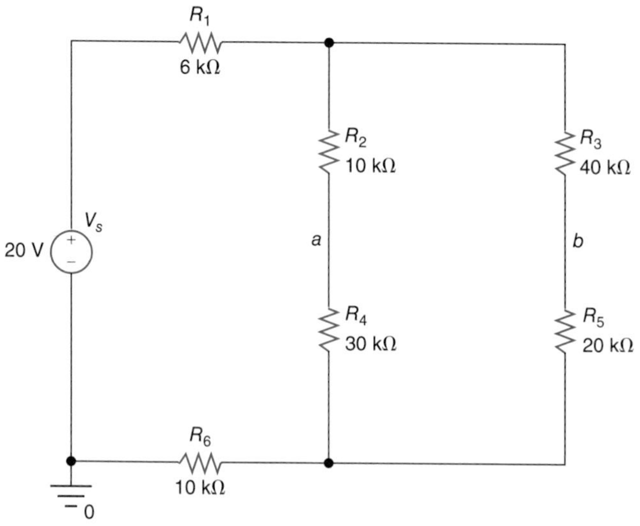

4.43 **Find the Thévenin equivalent circuit between *a* and *b* for the circuit shown in Figure P4.43.**

FIGURE P4.43

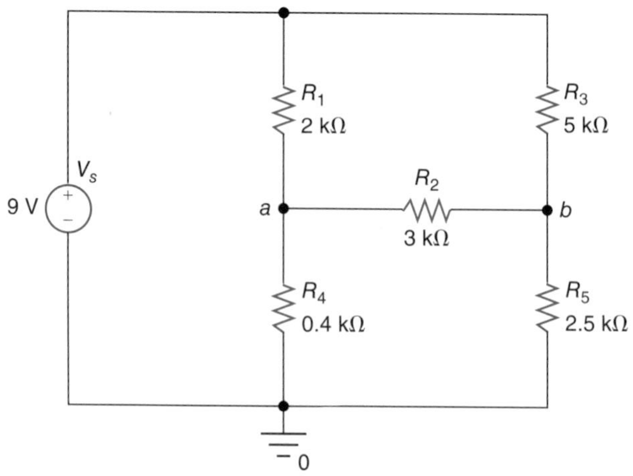

4.44 Find the Thévenin equivalent circuit between *a* and *b* for the circuit shown in Figure P4.44.

FIGURE P4.44

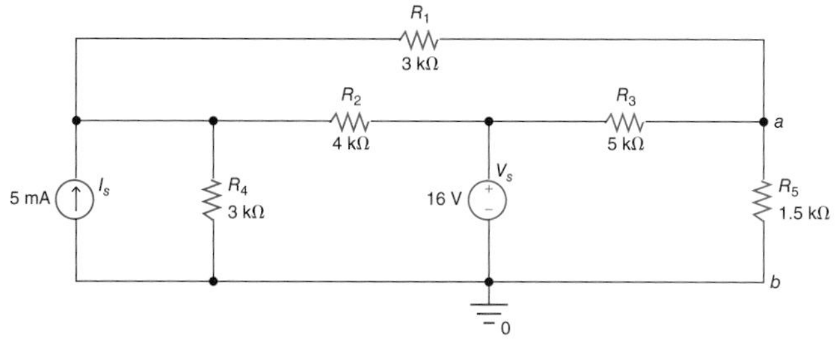

4.45 Find the Thévenin equivalent circuit between *a* and *b* for the circuit shown in Figure P4.45.

FIGURE P4.45

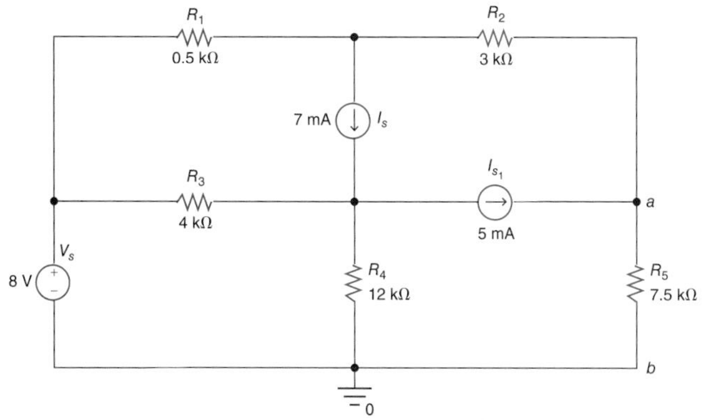

4.46 Find the Thévenin equivalent circuit between *a* and *b* for the circuit shown in Figure P4.46.

FIGURE P4.46

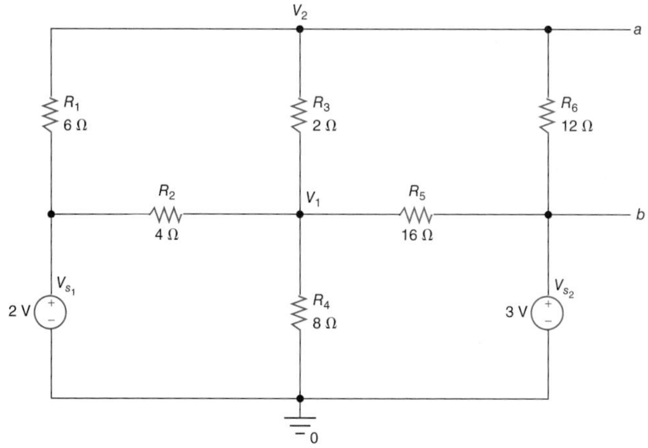

4.47 Find the Thévenin equivalent circuit between *a* and *b* for the circuit shown in Figure P4.47.

FIGURE P4.47

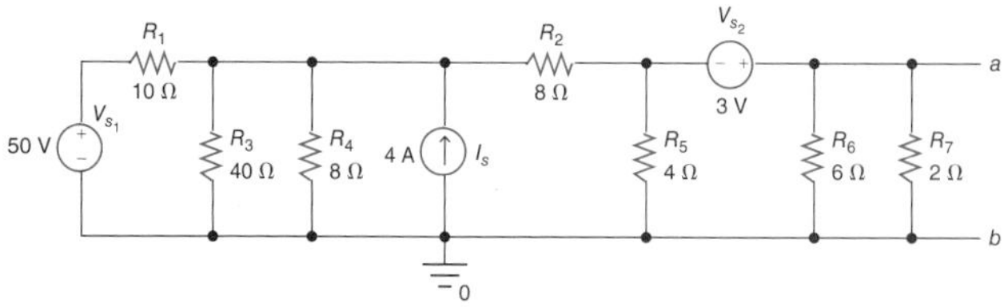

4.48 Find the Thévenin equivalent circuit between *a* and *b* for the circuit shown in Figure P4.48.

FIGURE P4.48

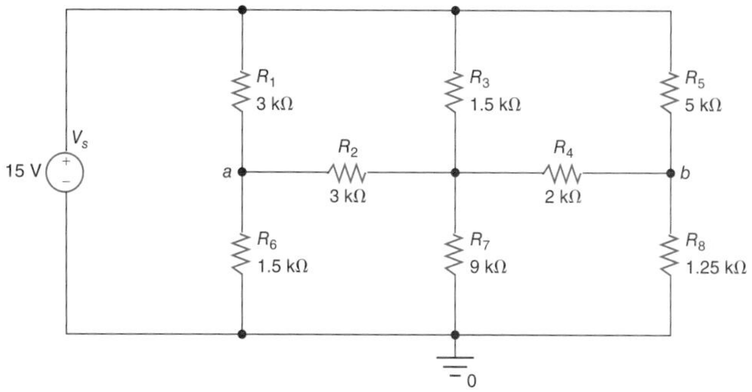

4.49 Find the Thévenin equivalent circuit between *a* and *b* for the circuit shown in Figure P4.49.

FIGURE P4.49

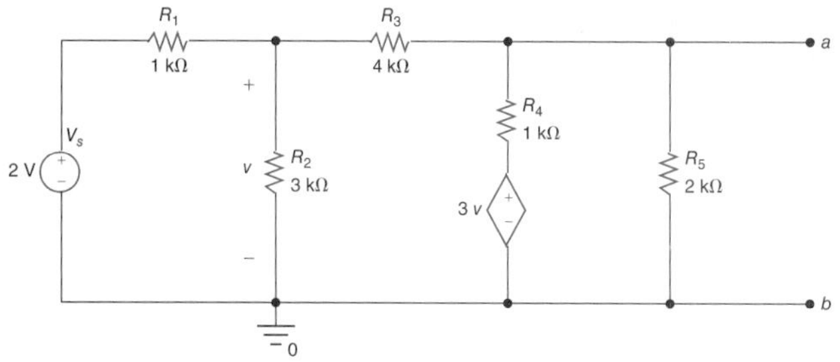

4.50 Find the Thévenin equivalent circuit between *a* and *b* for the circuit shown in Figure P4.50.

FIGURE P4.50

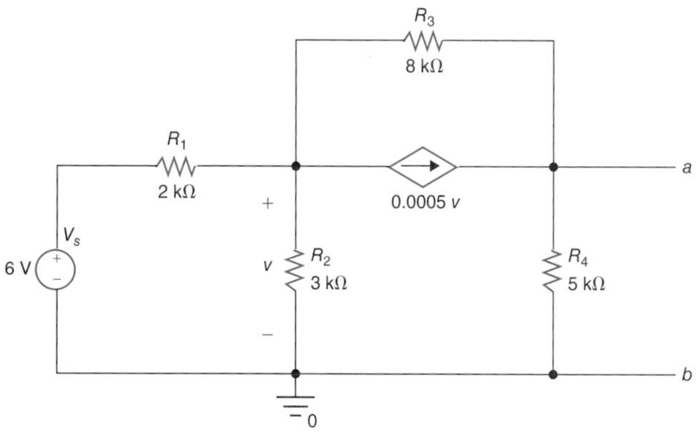

4.51 Find the Thévenin equivalent circuit between *a* and *b* for the circuit shown in Figure P4.51.

FIGURE P4.51

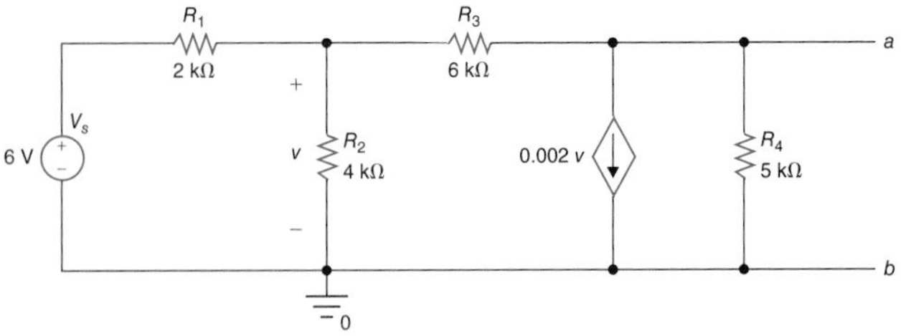

4.52 Find the Thévenin equivalent circuit between *a* and *b* for the circuit shown in Figure P4.52.

FIGURE P4.52

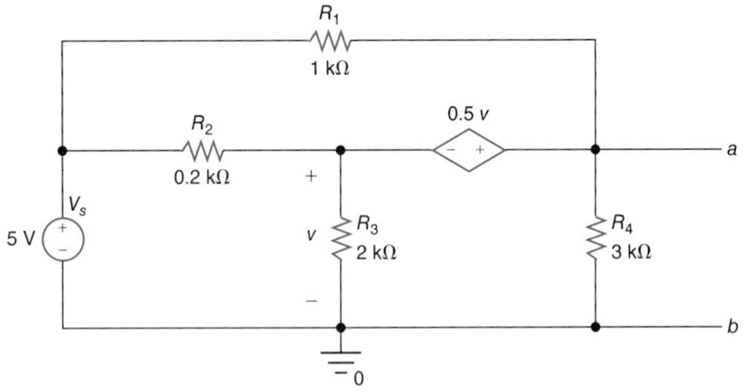

4.53 Find the Thévenin equivalent circuit between *a* and *b* for the circuit shown in Figure P4.53.

FIGURE P4.53

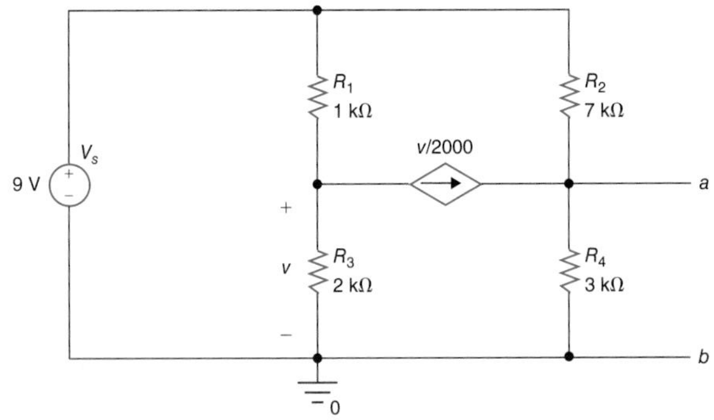

4.54 Find the Thévenin equivalent circuit between *a* and *b* for the circuit shown in Figure P4.54.

FIGURE P4.54

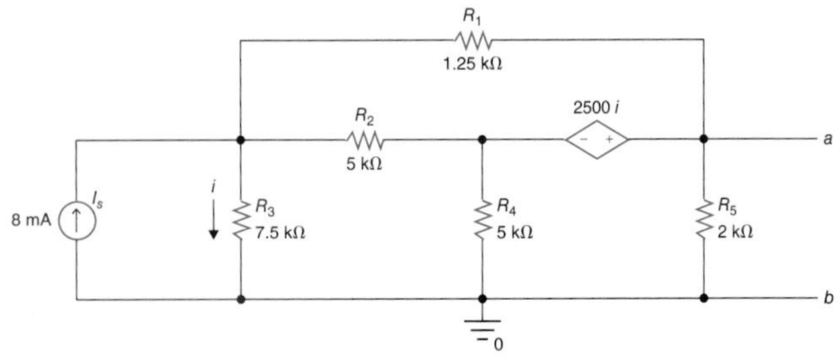

4.55 Find the Thévenin equivalent circuit between *a* and *b* for the circuit shown in Figure P4.55.

FIGURE P4.55

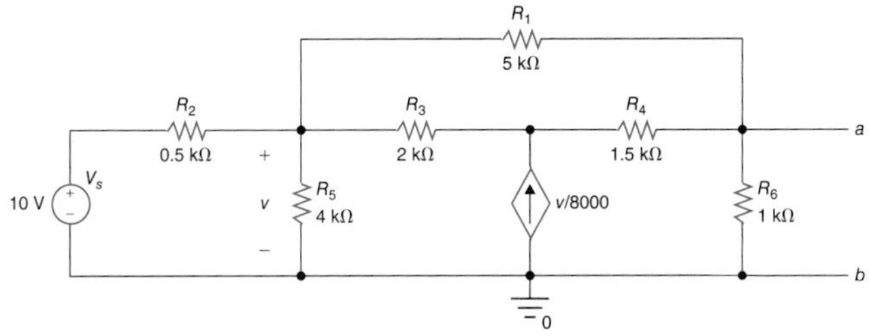

4.56 Find the Thévenin equivalent circuit between *a* and *b* for the circuit shown in Figure P4.56.

FIGURE P4.56

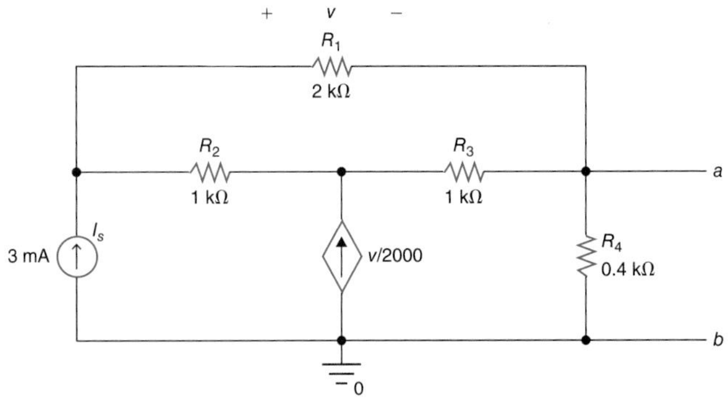

4.57 Find the Thévenin equivalent circuit between *a* and *b* for the circuit shown in Figure P4.57.

FIGURE P4.57

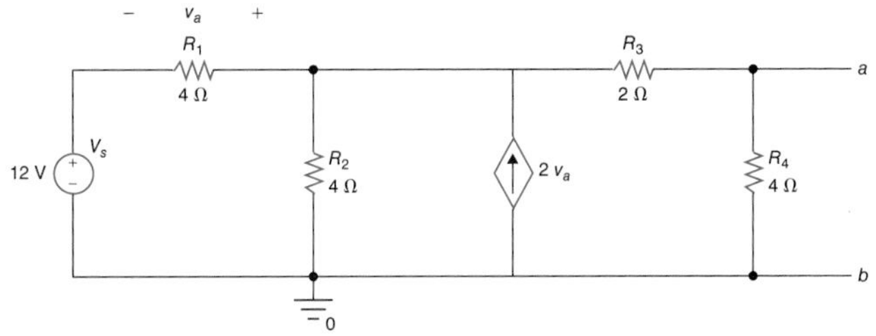

4.58 Find the Thévenin equivalent circuit between *a* nd *b* for the circuit shown in Figure P4.58.

FIGURE P4.58

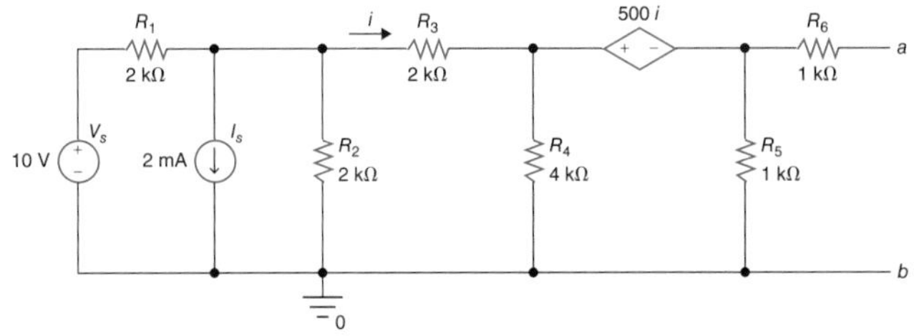

4.59 Find the Thévenin equivalent circuit between *a* and *b* for the circuit shown in Figure P4.59.

FIGURE P4.59

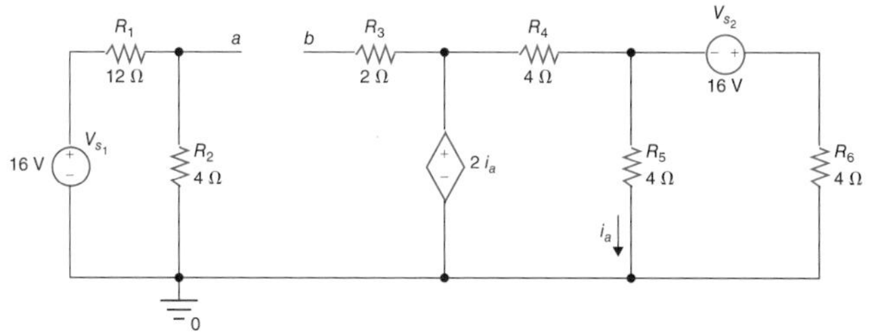

4.60 Find the Thévenin equivalent circuit between *a* and *b* for the circuit shown in Figure P4.60.

FIGURE P4.60

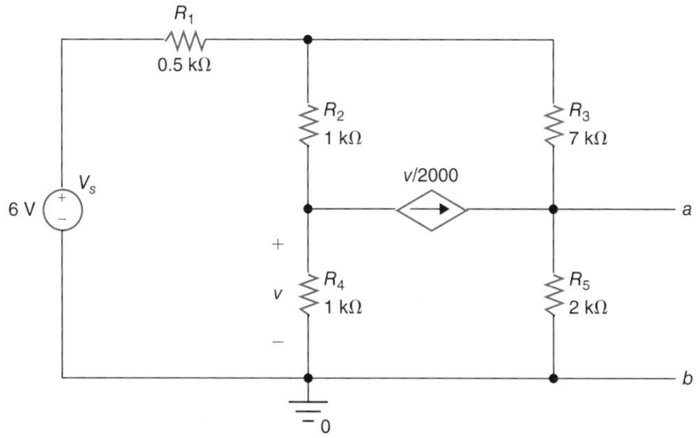

4.61 Find the Thévenin equivalent circuit between *a* and *b* for the circuit shown in Figure P4.61.

FIGURE P4.61

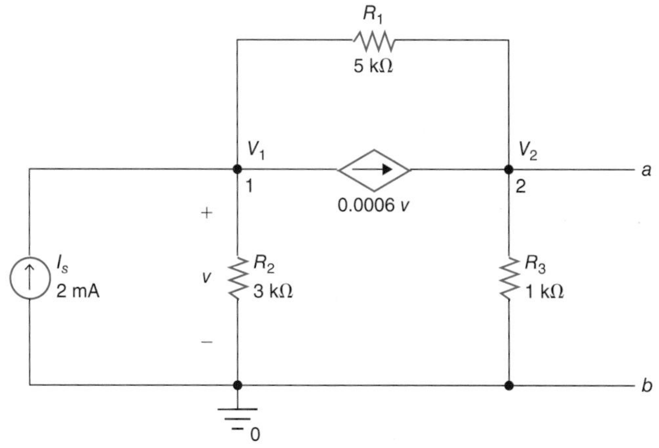

4.62 Find the Thévenin equivalent circuit between *a* and *b* for the circuit shown in Figure P4.62.

FIGURE P4.62

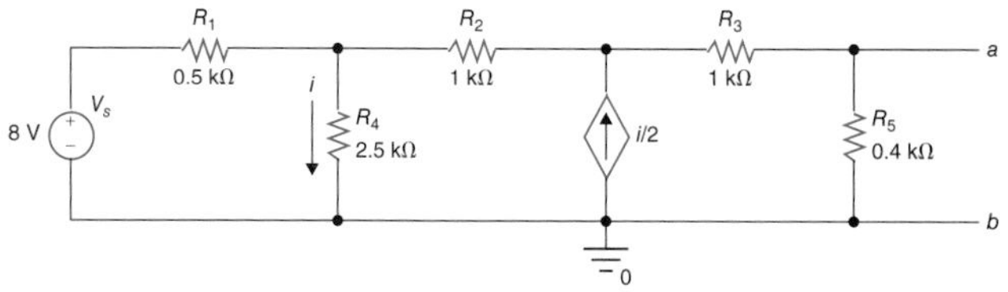

Norton Equivalent Circuit

4.63 Find the Norton equivalent circuit between *a* and *b* for the circuit shown in Figure P4.63.

FIGURE P4.63

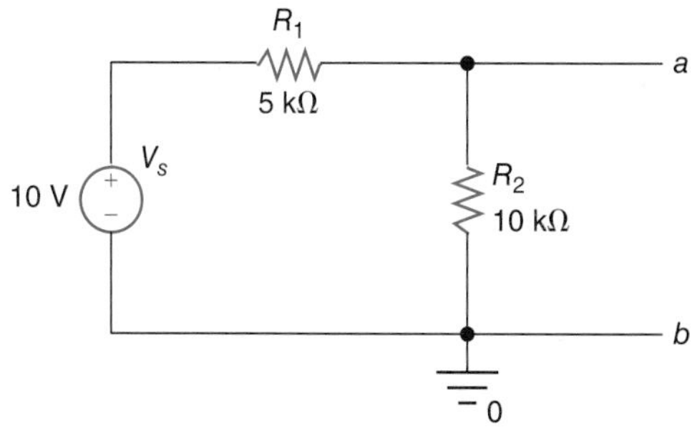

4.64 Find the Norton equivalent circuit between *a* and *b* for the circuit shown in Figure P4.64.

FIGURE P4.64

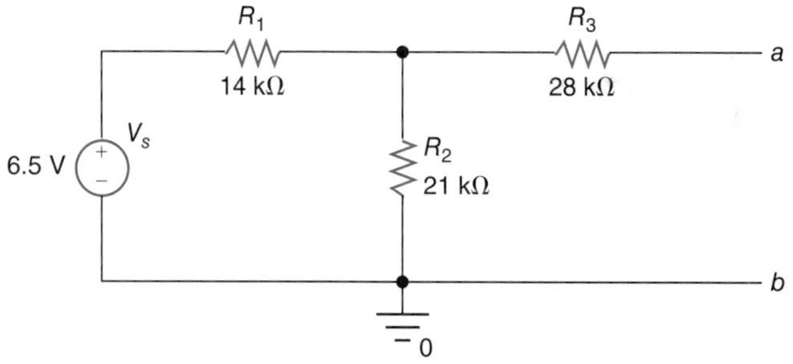

4.65 Find the Norton equivalent circuit between *a* and *b* for the circuit shown in Figure P4.65.

FIGURE P4.65

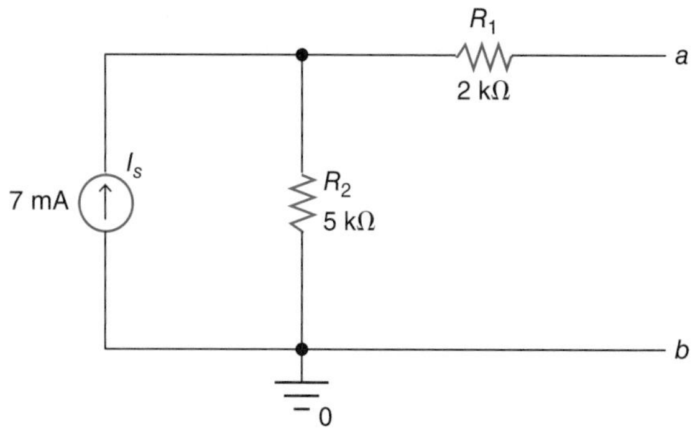

4.66 Find the Norton equivalent circuit between *a* and *b* for the circuit shown in Figure P4.66.

FIGURE P4.66

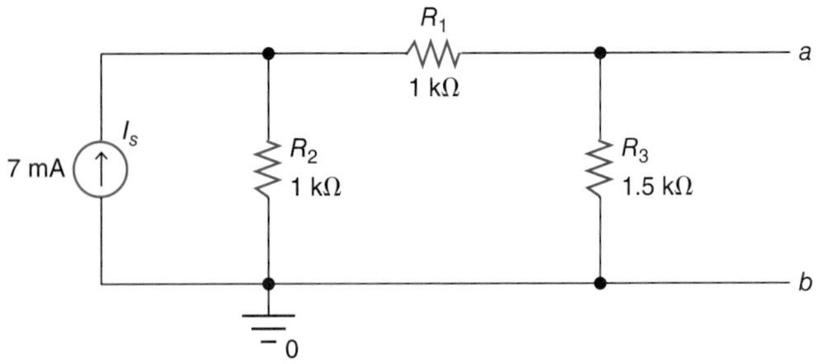

4.67 Find the Norton equivalent circuit between *a* and *b* for the circuit shown in Figure P4.67.

FIGURE P4.67

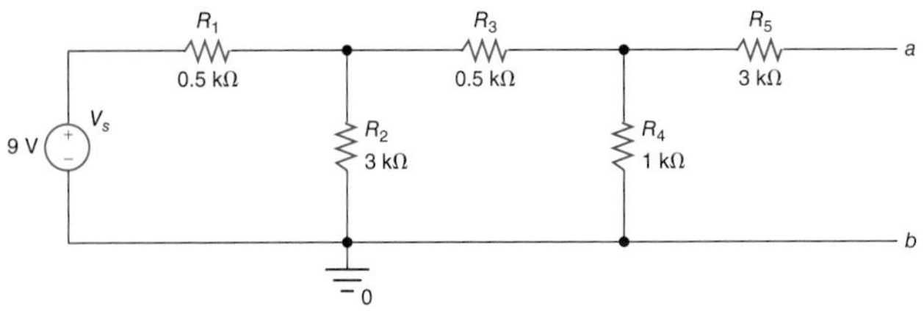

4.68 Find the Norton equivalent circuit between *a* and *b* for the circuit shown in Figure P4.68.

FIGURE P4.68

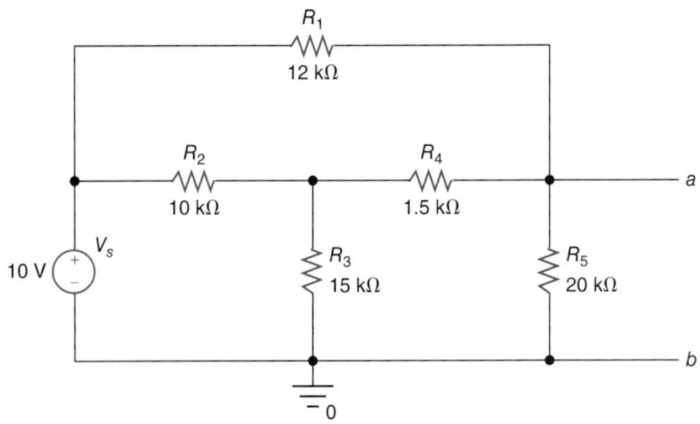

4.69 Find the Norton equivalent circuit between *a* and *b* for the circuit shown in Figure P4.69.

FIGURE P4.69

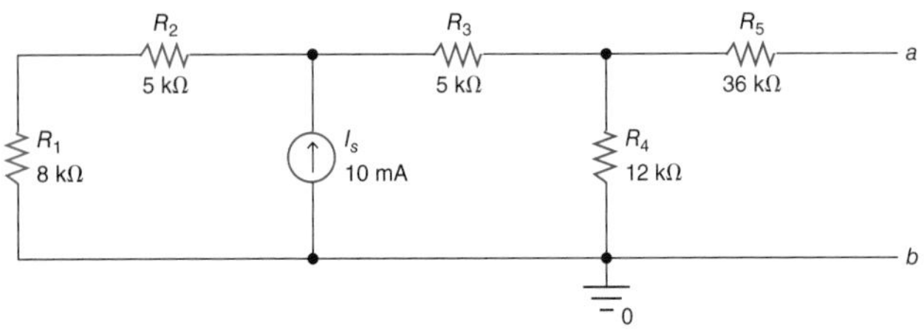

4.70 Find the Norton equivalent circuit between *a* and *b* for the circuit shown in Figure P4.70.

FIGURE P4.70

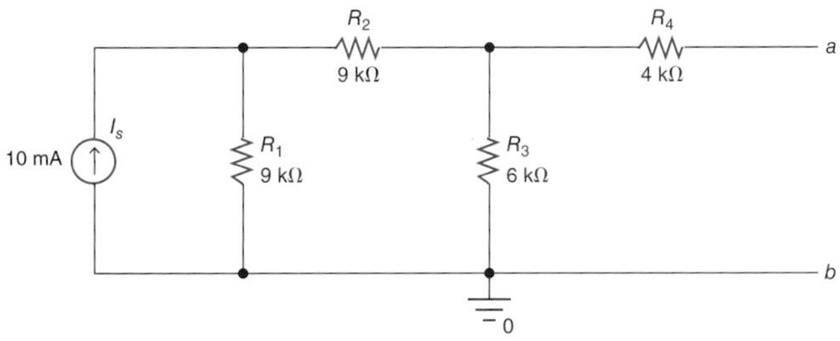

4.71 Find the Norton equivalent circuit between *a* and *b* for the circuit shown in Figure P4.71.

FIGURE P4.71

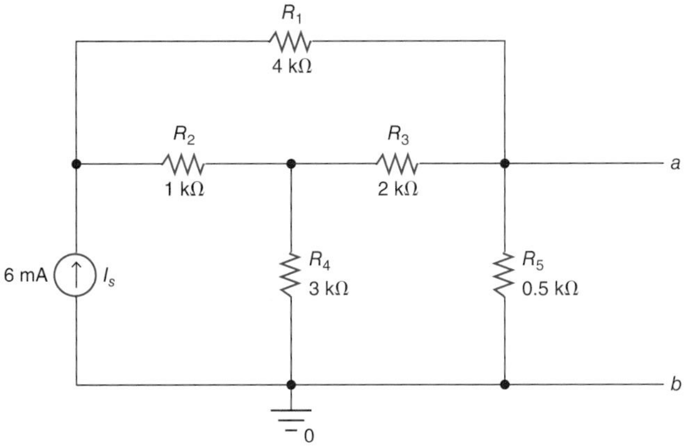

4.72 Find the Norton equivalent circuit between *a* and *b* for the circuit shown in Figure P4.72.

FIGURE P4.72

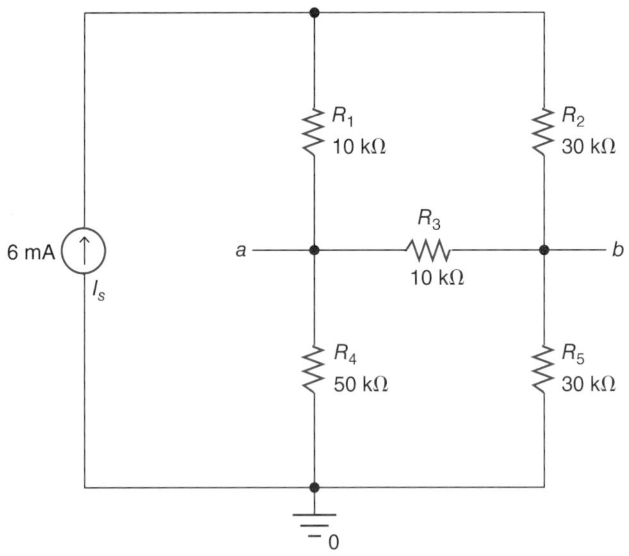

4.73 **Find the Norton equivalent circuit between *a* and *b* for the circuit shown in Figure P4.73.**

FIGURE P4.73

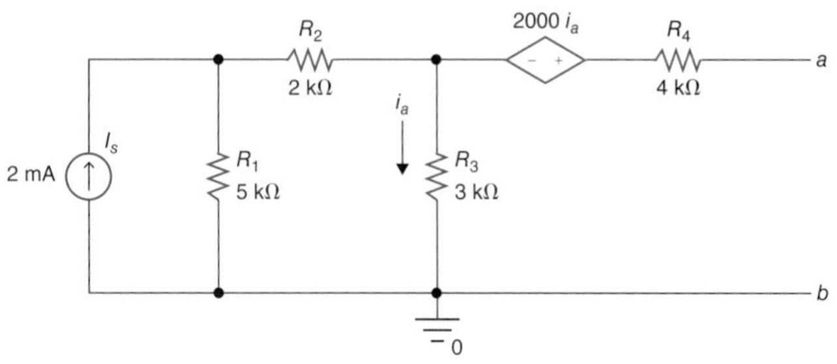

4.74 **Find the Norton equivalent circuit between *a* and *b* for the circuit shown in Figure P4.74.**

FIGURE P4.74

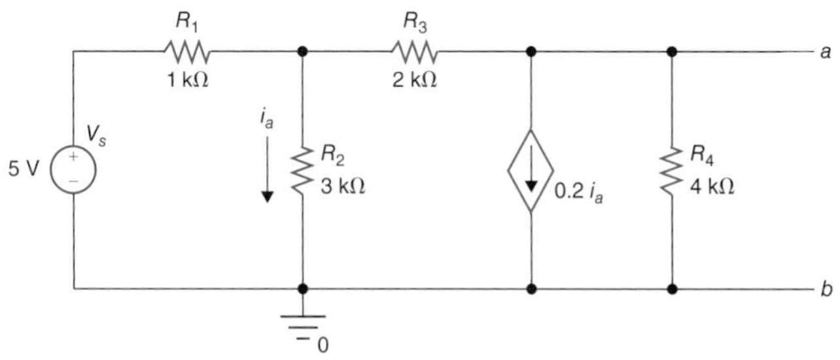

4.75 **Find the Norton equivalent circuit between *a* and *b* for the circuit shown in Figure P4.75.**

FIGURE P4.75

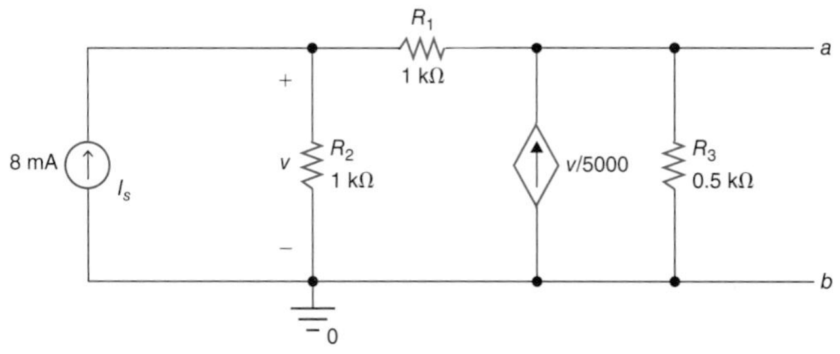

4.76 **Find the Norton equivalent circuit between *a* and *b* for the circuit shown in Figure P4.76.**

FIGURE P4.76

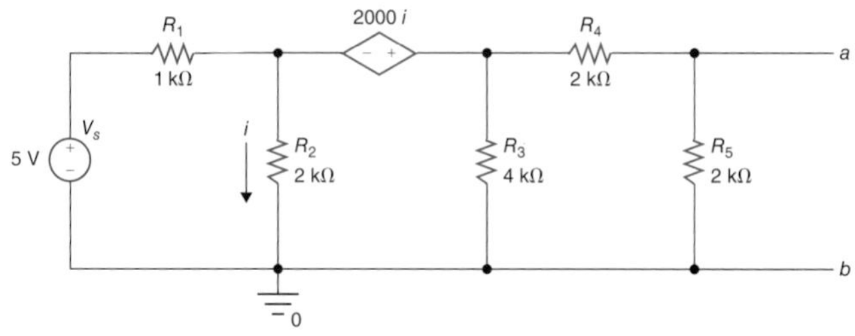

4.77 **Find the Norton equivalent circuit between *a* and *b* for the circuit shown in Figure P4.77.**

FIGURE P4.77

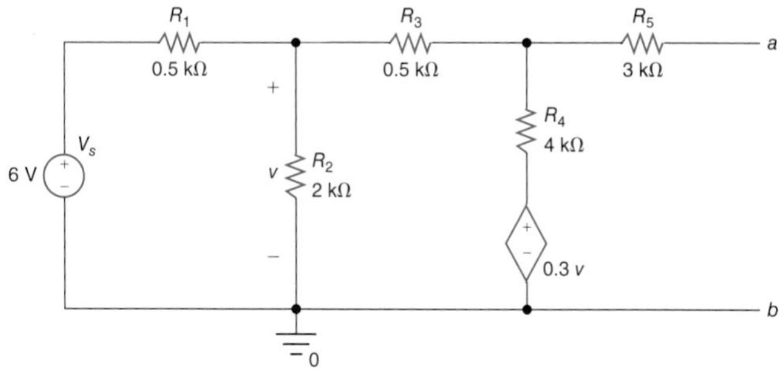

Maximum Power Transfer

4.78 **Find the load resistance value R_L for maximum power transfer, and find the maximum power transferred to the load for the circuit shown in Figure P4.78.**

FIGURE P4.78

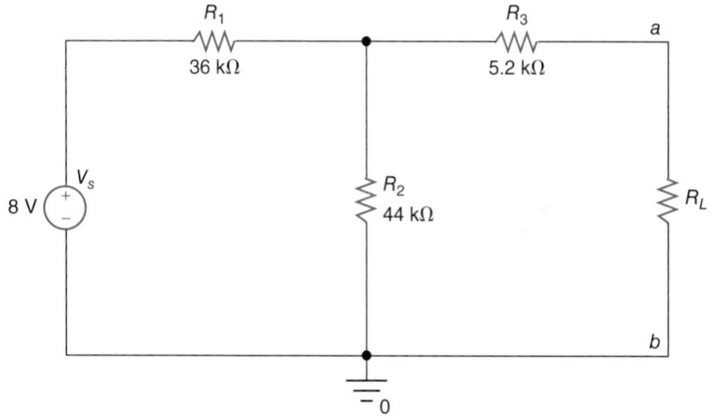

4.79 **Find the load resistance value R_L for maximum power transfer, and find the maximum power transferred to the load for the circuit shown in Figure P4.79.**

FIGURE P4.79

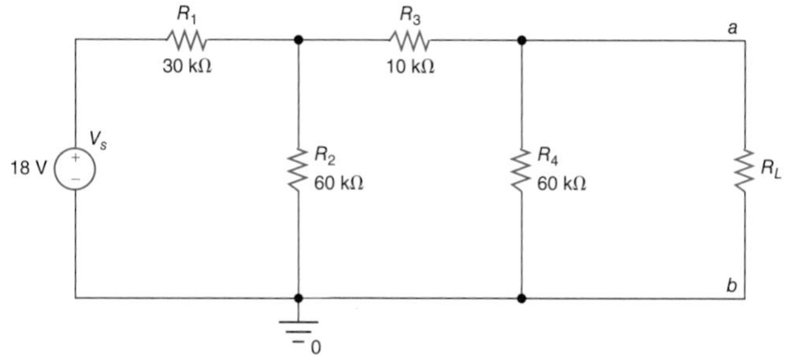

4.80 **Find the load resistance value R_L for maximum power transfer, and find the maximum power transferred to the load for the circuit shown in Figure P4.80.**

FIGURE P4.80

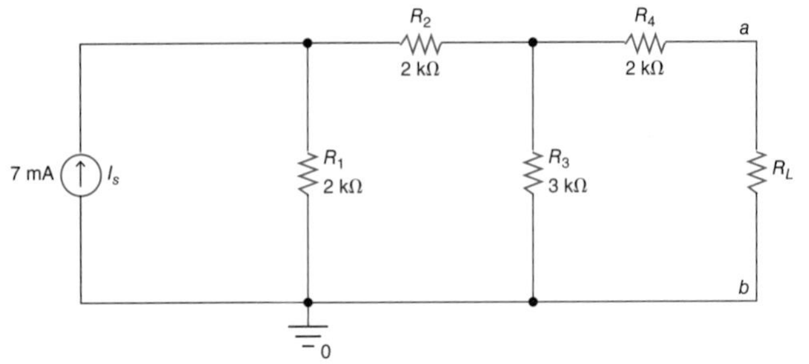

4.81 **Find the load resistance value R_L for maximum power transfer, and find the maximum power transferred to the load for the circuit shown in Figure P4.81.**

FIGURE P4.81

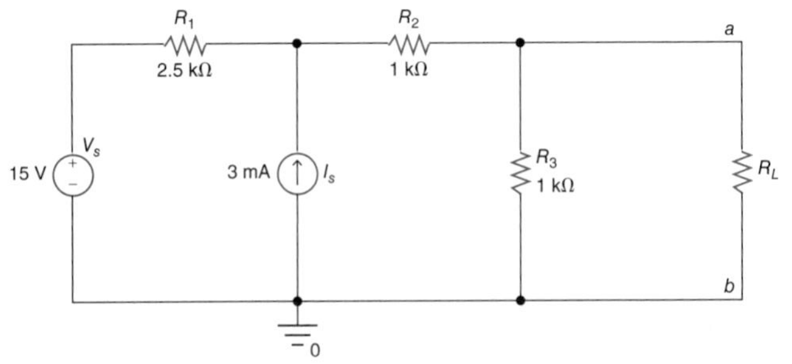

4.82 **Find the load resistance value R_L for maximum power transfer, and find the maximum power transferred to the load for the circuit shown in Figure P4.82.**

FIGURE P4.82

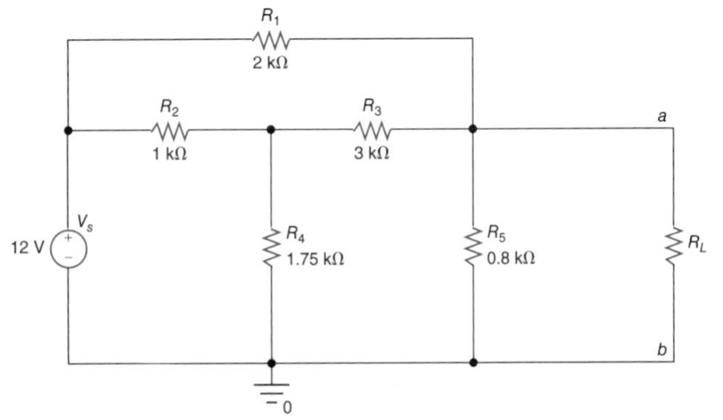

4.83 **Find the load resistance value R_L for maximum power transfer, and find the maximum power transferred to the load for the circuit shown in Figure P4.83.**

FIGURE P4.83

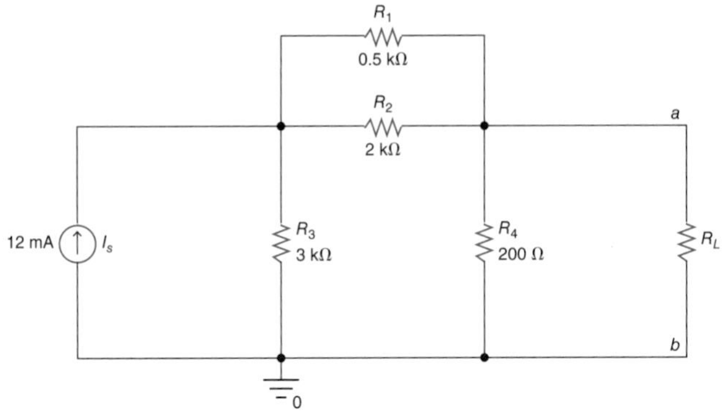

4.84 **Find the load resistance value R_L for maximum power transfer, and find the maximum power transferred to the load for the circuit shown in Figure P4.84.**

FIGURE P4.84

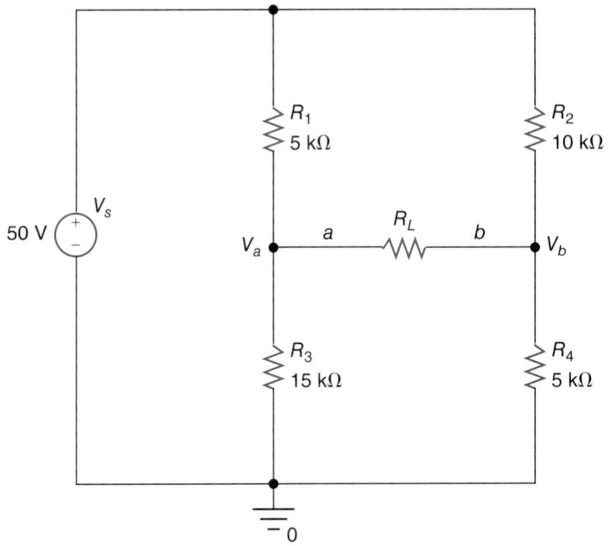

4.85 **Find the load resistance value R_L for maximum power transfer, and find the maximum power transferred to the load for the circuit shown in Figure P4.85.**

FIGURE P4.85

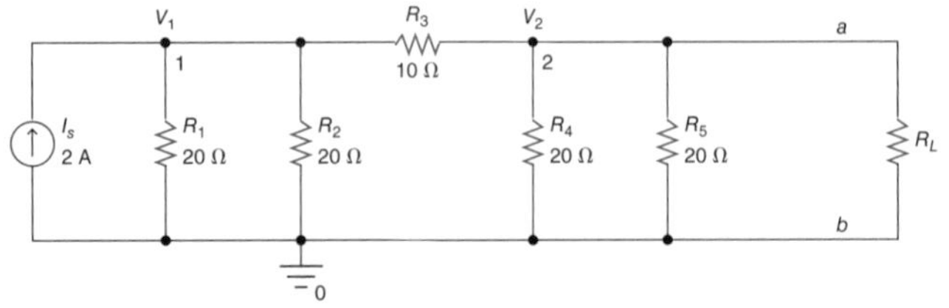

4.86 Find the load resistance value R_L for maximum power transfer, and find the maximum power transferred to the load for the circuit shown in Figure P4.86.

FIGURE P4.86

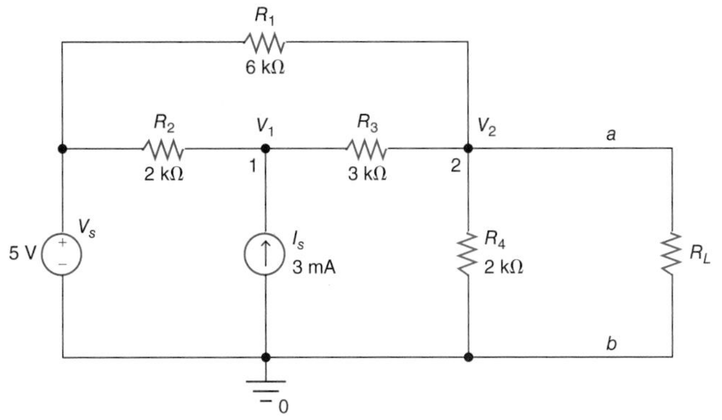

4.87 Find the value of R_L for the maximum power transfer for the circuit shown in Figure P4.87. Also, find the maximum power dissipated at R_L.

FIGURE P4.87

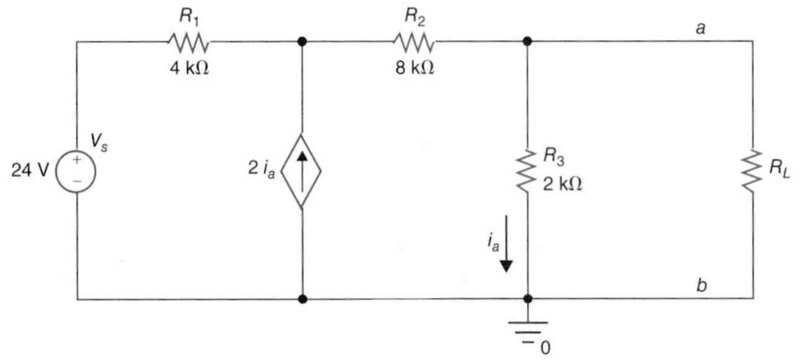

4.88 Find the load resistance value R_L for maximum power transfer, and find the maximum power transferred to the load for the circuit shown in Figure P4.88.

FIGURE P4.88

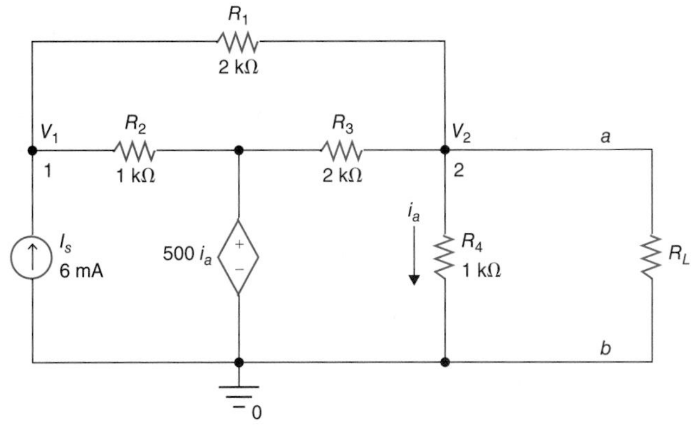

4.89 Find the load resistance value R_L for maximum power transfer, and find the maximum power transferred to the load for the circuit shown in Figure P4.89.

FIGURE P4.89

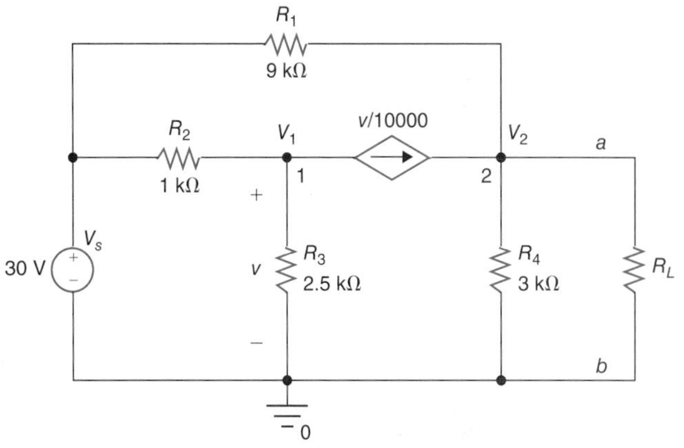

4.90 Find the load resistance value R_L for maximum power transfer, and find the maximum power transferred to the load for the circuit shown in Figure P4.90.

FIGURE P4.90

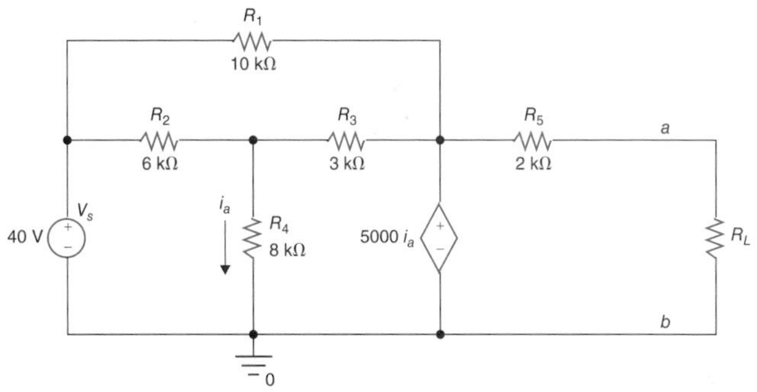

Chapter 5

Operational Amplifier Circuits

5.1 Introduction

An operational amplifier (commonly called *op amp* or *opamp*) is a device that can be used to perform mathematical operations such as addition, subtraction, amplification, attenuation, integration, and differentiation. It is a versatile integrated circuit (IC) chip that is widely used in amplifiers, filters, signal conditioning, and instrumentation circuits. The circuit symbol for an op amp is shown in Figure 5.1.

As can be seen from Figure 5.1, there are two input terminals and one output terminal for an op amp. The voltage at the positive input terminal (i.e., the noninverting input terminal) is v_p, the voltage at the negative input terminal (i.e., the inverting input terminal) is v_n, and the voltage at the output terminal is v_o. Figure 5.2 shows pin configuration for a typical 8-pin package. Pin number 2 is the inverting input terminal, and pin number 3 is the noninverting input terminal. The output signal is available at pin number 6. Pin number 7 is the positive power supply, V_{cc^+}, and pin number 4 is the negative power supply, V_{cc^-}.

FIGURE 5.1

Circuit symbol for an op amp.

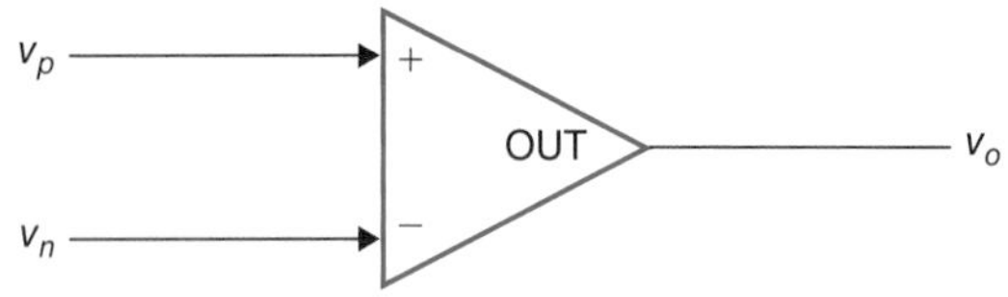

FIGURE 5.2

Pin configuration of typical 8-pin package.

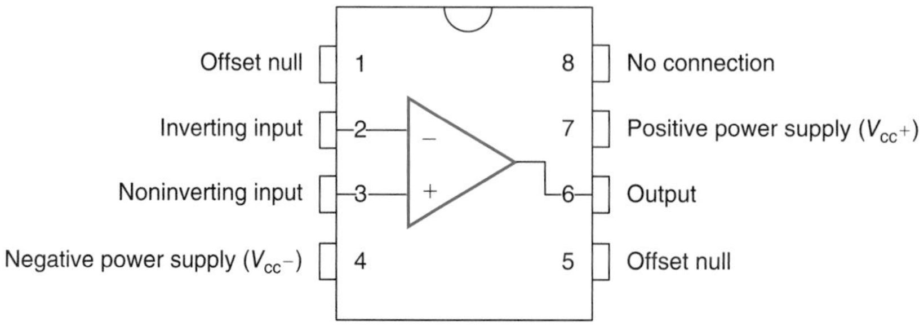

5.2 Ideal Op Amp

Op amps can be modeled as a voltage-controlled voltage source (VCVS), as shown in Figure 5.3.

FIGURE 5.3

A model for op amp.

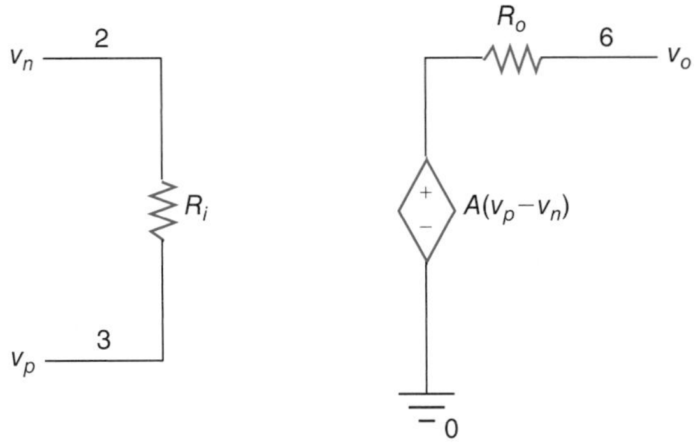

In the model, v_n is the voltage on the inverting input (pin 2) of the op amp, v_p is the voltage on the noninverting input (pin 3) of the op amp, v_o is the voltage on the output (pin 6) of the op amp, R_i is the input resistance, R_o is the output resistance, and A is the unloaded voltage gain. In general, the input resistance R_i is large, the output resistance R_o is small, the gain A is large. For $\mu A741$ (or $uA741$) general-purpose operational amplifier from Texas Instruments, typical value of the input resistance is 2 MΩ, output resistance is 75 Ω, and large-signal differential voltage amplification is 200,000. If $R_i = \infty$, $R_o = 0$, and $A = \infty$, the op amp is called ideal. If an op amp is assumed to be ideal, the analysis of circuits with op amps is greatly simplified. The controlled voltage is proportional to the difference of the two inputs, $v_d = v_p - v_n$. The voltage v_d is called *differential input*.

Figure 5.4 shows the inverting configuration of an op amp.

FIGURE 5.4

Inverting configuration.

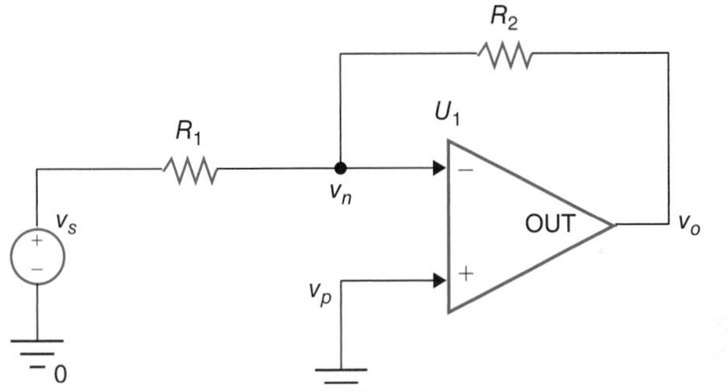

The noninverting input terminal is grounded ($v_p = 0$). The input voltage v_s is applied to the inverting input through a resistor R_1. Resistor R_2 provides a feedback path between the output terminal and the inverting input terminal. When the op amp is replaced by the model shown in Figure 5.3, we obtain the circuit shown in Figure 5.5.

FIGURE 5.5

A model for inverting configuration.

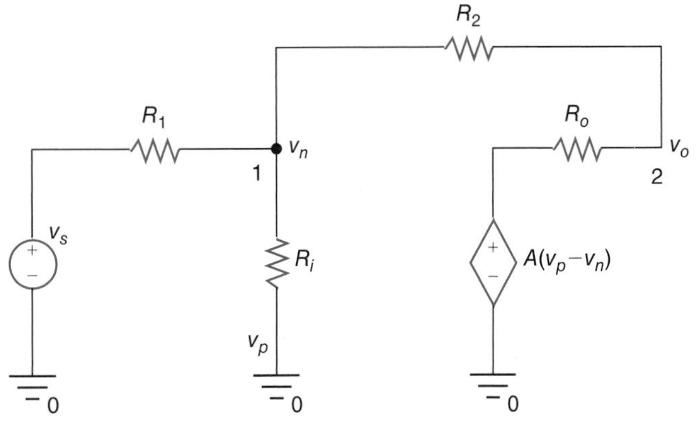

More detailed analysis of the circuit shown in Figure 5.5 is presented in Section 5.6 of this chapter. For now, in this section, we assume $R_o = 0$, $R_i = \infty$ and A is large. Under these conditions, the circuit shown in Figure 5.5 reduces to the one shown in Figure 5.6.

FIGURE 5.6

A simplified model for inverting configuration.

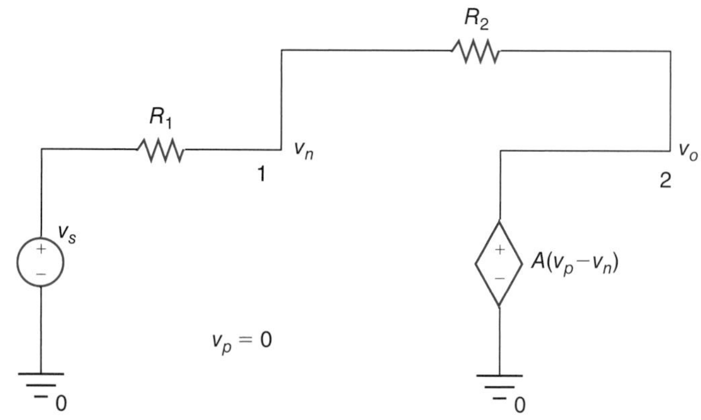

We can apply nodal analysis to the circuit shown in Figure 5.6. Summing the currents leaving node 1, we obtain

$$\frac{v_n - v_s}{R_1} + \frac{v_n - v_o}{R_2} = 0$$

which can be revised as

$$\left(\frac{1}{R_1} + \frac{1}{R_2}\right)v_n = \frac{1}{R_1}v_s + \frac{1}{R_2}v_o$$

Solving for v_n, we obtain

$$v_n = \frac{\dfrac{1}{R_1}}{\dfrac{1}{R_1} + \dfrac{1}{R_2}}v_s + \frac{\dfrac{1}{R_2}}{\dfrac{1}{R_1} + \dfrac{1}{R_2}}v_o = \frac{R_2}{R_1 + R_2}v_s + \frac{R_1}{R_1 + R_2}v_o \tag{5.1}$$

The voltage at the inverting input is the sum of input component, $v_sR_2/(R_1 + R_2)$, and the feedback component from the output, $v_oR_1/(R_1 + R_2)$. Let the voltage difference between noninverting input v_p and the inverting input v_n be v_d. Then, we have

$$v_d = v_p - v_n$$

The output v_o is given by

$$v_o = Av_d = A(v_p - v_n) = A(0 - v_n) = -Av_n \tag{5.2}$$

Substitution of Equation (5.1) into Equation (5.2) yields

$$v_o = -\frac{AR_2}{R_1 + R_2}v_s - \frac{AR_1}{R_1 + R_2}v_o \tag{5.3}$$

Equation (5.3) can be rearranged as

$$v_o = \frac{-\dfrac{AR_2}{R_1 + R_2}}{1 + \dfrac{AR_1}{R_1 + R_2}}v_s = \frac{-\dfrac{R_2}{R_1 + R_2}}{\dfrac{1}{A} + \dfrac{R_1}{R_1 + R_2}}v_s \tag{5.4}$$

Without the negative feedback component from the output, which is the second term of Equation (5.3), the output v_o consists of the first term of Equation (5.3) only, which is due to input v_s. The output v_o will be large due to large gain A. Equation (5.4) shows that the negative feedback component provides comparable gain in the denominator $[AR_1/(R_1 + R_2)]$ to offset the effects of large gain in the numerator $[AR_2/(R_1 + R_2)]$. It is called *negative feedback* because the feedback from the output is subtracted from the positive input v_p.

Equation (5.4) can be written as

$$v_o = -\frac{AR_2}{R_1 + R_2 + AR_1} v_s = -\frac{\dfrac{R_2}{R_1 + R_2}A}{1 + \dfrac{R_1}{R_1 + R_2}A} v_s = -\frac{R_2}{\dfrac{R_1 + R_2}{A} + R_1} v_s \quad (5.5)$$

Substitution of Equation (5.5) into Equation (5.1) results in

$$v_n = \frac{R_2}{R_1 + R_2} v_s - \frac{R_1}{R_1 + R_2}\frac{AR_2}{R_1 + R_2 + AR_1} v_s = \frac{R_2(R_1 + R_2 + AR_1) - R_1R_2A}{(R_1 + R_2)(R_1 + R_2 + AR_1)} v_s$$
$$= \frac{R_2}{R_1 + R_2 + AR_1} v_s = \frac{\dfrac{R_2}{R_1 + R_2}}{1 + \dfrac{R_1}{R_1 + R_2}A} v_s = \frac{\dfrac{R_2}{A}}{\dfrac{R_1 + R_2}{A} + R_1} v_s \quad (5.6)$$

Since $A \gg (R_1 + R_2)$, $(R_1 + R_2)/A \approx 0$ and Equation (5.5) reduces to

$$v_o \cong -\frac{R_2}{R_1} v_s \quad (5.7)$$

Since $A \gg (R_1 + R_2)$, Equation (5.6) reduces to

$$v_n \cong \frac{R_2}{R_1 A} v_s \approx 0 \quad (5.8)$$

The voltage difference is given by

$$v_d = v_p - v_n = 0 - v_n = -\frac{\dfrac{R_2}{A}}{\dfrac{R_1 + R_2}{A} + R_1} v_s \cong -\frac{R_2}{R_1 A} v_s \approx 0 \quad (5.9)$$

The current through R_i can be approximated by

$$i_{R_i} = \frac{v_d}{R_i} = \frac{-v_n}{R_i} \cong -\frac{R_2}{R_1 A R_i} v_s \approx 0 \quad (5.10)$$

Notice that the denominator of Equation (5.10) includes AR_i, which is the product of two large numbers, making the current through R_i very small. From Equations (5.9) and (5.10), we conclude that the current through R_i is close to zero, and the voltage difference, $v_d = v_p - v_n$, is close to zero (i.e., $v_p \approx v_n$). The voltage drop across R_i is close to zero.

Figure 5.7 shows a noninverting configuration of an op amp.

The input voltage v_s is applied to the noninverting input terminal of the op amp ($v_p = v_s$). There is a resistor R_1 between inverting input and the ground, and another resistor R_2 between the inverting input and the output v_o provides feedback from output v_o to v_n. When the op amp is replaced by the model shown in Figure 5.3, we obtain the circuit shown in Figure 5.8.

FIGURE 5.7

Noninverting configuration of an op amp.

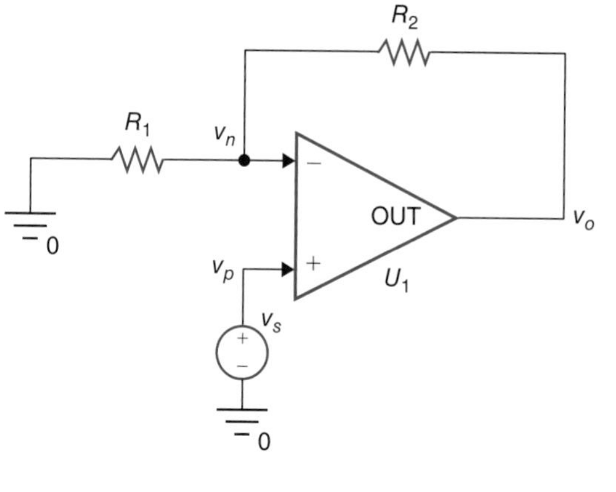

FIGURE 5.8

A model of a noninverting configuration.

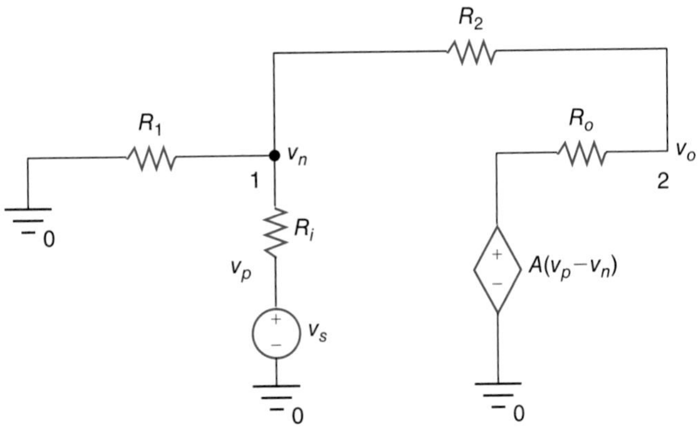

FIGURE 5.9

A simplified model for noninverting configuration.

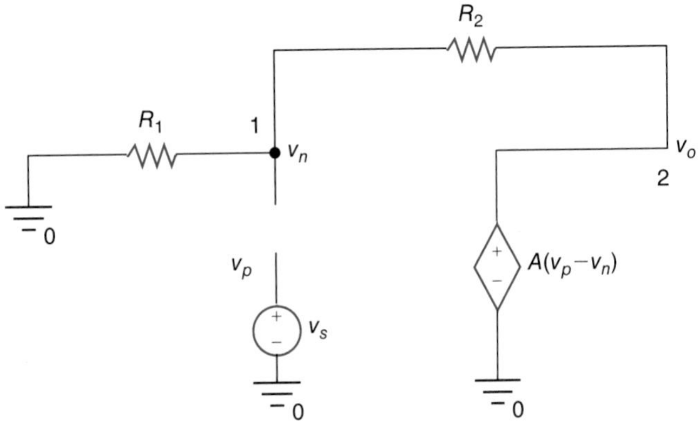

More detailed analysis of the circuit shown in Figure 5.8 is presented in Section 5.7 of this chapter. In this section, we assume that $R_o = 0$, $R_i = \infty$ and A is large. Under these conditions, the circuit shown in Figure 5.8 reduces to the one shown in Figure 5.9.

We can apply nodal analysis to the circuit shown in Figure 5.9. Summing the currents leaving node 1, we obtain

$$\frac{v_n}{R_1} + \frac{v_n - v_o}{R_2} = 0$$

which can be revised as

$$\left(\frac{1}{R_1} + \frac{1}{R_2}\right)v_n = \frac{1}{R_2}v_o$$

Solving for v_n, we obtain

$$v_n = \frac{\frac{1}{R_2}}{\frac{1}{R_1} + \frac{1}{R_2}}v_o = \frac{R_1}{R_1 + R_2}v_o \quad \textbf{(5.11)}$$

The output signal is given by

$$v_o = A(v_p - v_n) = A(v_s - v_n) \quad \textbf{(5.12)}$$

Substitution of Equation (5.11) into Equation (5.12) yields

$$v_o = Av_s - \frac{AR_1}{R_1 + R_2}v_o \quad \textbf{(5.13)}$$

Equation (5.13) can be revised as

$$v_o = \frac{A}{1 + \frac{R_1}{R_1 + R_2}A}v_s = \frac{1}{\frac{1}{A} + \frac{R_1}{R_1 + R_2}}v_s \quad \textbf{(5.14)}$$

Without the negative feedback component, which is the second term of Equation (5.13), the output v_o consists of the first term of Equation (5.13) only, due to input v_s. The output v_o will be large due to large gain A. Equation (5.14) shows that the negative feedback component provides comparable gain $[AR_1/(R_1 + R_2)]$ in the denominator to offset the effects of large gain (A) in the numerator.

Equation (5.14) can be written as

$$v_o = \frac{A(R_1 + R_2)}{R_1 + R_2 + AR_1}v_s = \frac{R_1 + R_2}{\frac{R_1 + R_2}{A} + R_1}v_s \quad \textbf{(5.15)}$$

Substitution of Equation (5.15) into Equation (5.11) results in

$$v_n = \frac{R_1}{R_1 + R_2}\frac{A(R_1 + R_2)}{R_1 + R_2 + AR_1}v_s = \frac{AR_1}{R_1 + R_2 + AR_1}v_s = \frac{R_1}{\frac{R_1 + R_2}{A} + R_1}v_s \quad \textbf{(5.16)}$$

Since $A \gg (R_1 + R_2)$, $(R_1 + R_2)/A \approx 0$ and Equation (5.15) reduces to

$$v_o \cong \frac{R_1 + R_2}{R_1} v_s = \left(1 + \frac{R_2}{R_1}\right) v_s \tag{5.17}$$

Since $A \gg (R_1 + R_2)$, Equation (5.16) reduces to

$$v_n = \frac{R_1}{\dfrac{R_1 + R_2}{A} + R_1} v_s \cong v_s \tag{5.18}$$

The voltage difference v_d is given by

$$\begin{aligned} v_d = v_p - v_n = v_s - \frac{AR_1}{R_1 + R_2 + AR_1} v_s = \frac{R_1 + R_2}{R_1 + R_2 + AR_1} v_s \\ \cong \frac{R_1 + R_2}{AR_1} v_s \approx 0 \end{aligned} \tag{5.19}$$

The current through R_i can be approximated by

$$i_{R_i} = \frac{v_d}{R_i} \cong \frac{R_1 + R_2}{AR_1 R_i} v_s \approx 0 \tag{5.20}$$

Notice that the denominator of Equation (5.20) includes AR_i, which is the product of two large numbers, making the current through R_i very small. From Equations (5.19) and (5.20), we conclude that the current through R_i is close to zero, and the voltage difference $v_d = v_p - v_n$ is close to zero (i.e., $v_p \approx v_n$). The voltage drop across R_i is close to zero.

In summary, in the ideal op amp model for both inverting and noninverting configuration,

a. The current flowing into the op amp from the positive input terminal is zero ($i_p = 0$).
b. The current flowing into the op amp from the negative input terminal is zero ($i_n = 0$).
c. The voltage difference between v_p and v_n is zero; that is, $v_p = v_n$.

The fact that $v_p = v_n$ is called **virtual short**. Figure 5.10 shows an op amp with these three properties for both inverting and noninverting configuration.

FIGURE 5.10 An op amp showing zero input currents and a virtual short.

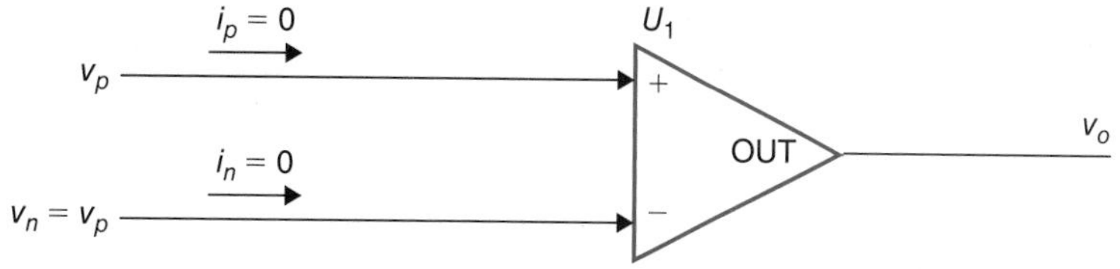

These properties make it simpler to analyze circuits with op amps. Since the error between the exact solution based on practical op amp model and the ideal op amp model is so small, we can use the ideal op amp model for many applications. We can use the ideal op amp model to find the output of the inverting amplifier shown in Figure 5.11.

Since $v_p = v_n$ and $v_p = 0$, we have $v_n = 0$. From Ohm's law, the current i_1 through R_1 is given by

$$i_1 = \frac{v_s - v_n}{R_1} = \frac{v_s - 0}{R_1} = \frac{v_s}{R_1}$$

Since the current i_n flowing into the op amp at the inverting input terminal (node 1) is zero, according to Kirchhoff's current law (KCL), the current through R_2 is the same as i_1;

FIGURE 5.11

Inverting amplifier.

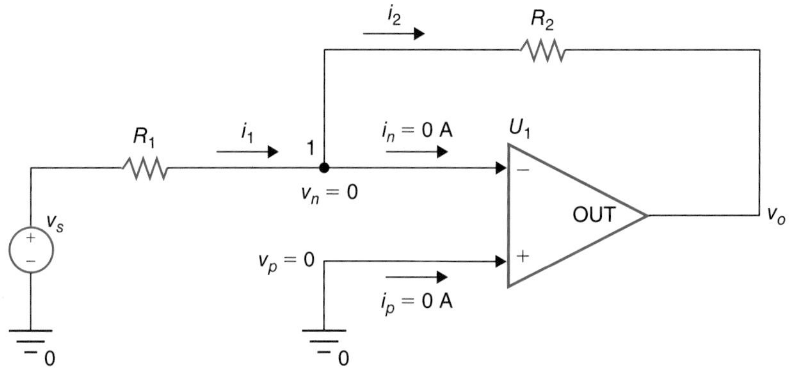

that is, $i_2 = i_1 = v_s/R_1$. The voltage across R_2 is $-i_2R_2$ (positive on the right side); that is, $v_{R_2} = -i_2R_2 = (-R_2/R_1)v_s$. Since $v_o = v_{R_2} + v_n = v_{R_2} + 0$, we have $v_o = (-R_2/R_1)v_s$. Instead of this intuitive approach, we can simply write a node equation at the inverting input of the op amp by summing the currents leaving node 1:

$$\frac{v_n - v_s}{R_1} + \frac{v_n - v_o}{R_2} = 0$$

Setting $v_n = 0$, we have

$$\frac{0 - v_s}{R_1} + \frac{0 - v_o}{R_2} = 0$$

Solving for v_o, we obtain

$$v_o = -\frac{R_2}{R_1}v_s$$

The output voltage is the input voltage multiplied by $-R_2/R_1$. The voltage gain of the amplifier, defined by $G = v_o/v_s$, is

$$G = -\frac{R_2}{R_1} \tag{5.21}$$

with sign inversion.

Figure 5.12 shows two inverting amplifiers that are cascaded. The voltage gain of the first amplifier is $v_a/v_s = -R_2/R_1$ and the voltage gain of the second amplifier is $v_o/v_a = -R_4/R_3$. The overall gain of the two cascaded amplifiers is given by

$$G = \frac{v_o}{v_s} = \frac{v_a}{v_s} \times \frac{v_o}{v_a} = \left(-\frac{R_2}{R_1}\right) \times \left(-\frac{R_4}{R_3}\right) = \frac{R_2R_4}{R_1R_3}$$

FIGURE 5.12

Two inverting amplifiers are cascaded.

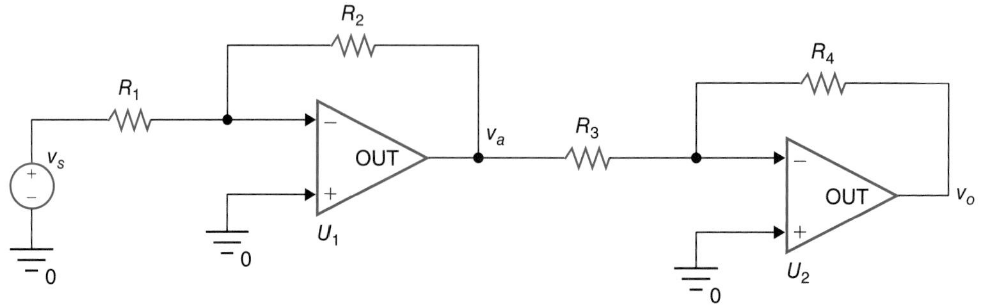

The output signal has the same sign as the input signal, and the voltage gain is the product of the voltage gain of each amplifier.

We can use the ideal op amp model to find the output of the noninverting amplifier shown in Figure 5.13.

FIGURE 5.13

Noninverting amplifier.

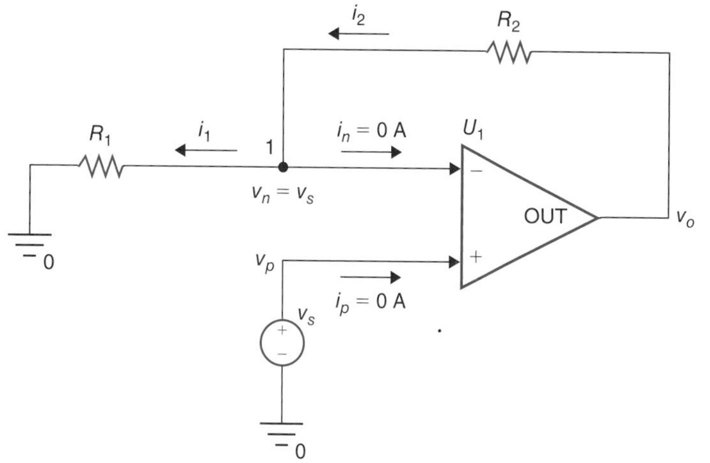

Since $v_p = v_n$ and $v_p = v_s$, we have $v_n = v_s$. The current i_1 through R_1 is given by

$$i_1 = \frac{v_n - 0}{R_1} = \frac{v_s - 0}{R_1} = \frac{v_s}{R_1}$$

Since the current i_n flowing into the op amp at the inverting input terminal (node 1) is zero, according to KCL, the current through R_2 is the same as i_1; that is, $i_2 = i_1 = v_s/R_1$. The voltage across R_2 is i_2R_2 (positive on the right side); that is, $v_{R_2} = i_2R_2 = (R_2/R_1)v_s$. Since $v_o = v_n + v_{R_2}$, we have

$$v_o = v_n + v_{R_2} = v_s + \frac{R_2}{R_1}v_s = \left(1 + \frac{R_2}{R_1}\right)v_s = \frac{R_1 + R_2}{R_1}v_s$$

Instead of this intuitive approach, we can simply write a node equation at the inverting input of the op amp. Summing the currents leaving node 1, we have

$$\frac{v_n - 0}{R_1} + \frac{v_n - v_o}{R_2} = 0$$

Setting $v_n = v_s$, we get

$$\frac{v_s}{R_1} + \frac{v_s - v_o}{R_2} = 0$$

Solving for v_o, we obtain

$$v_o = \left(1 + \frac{R_2}{R_1}\right)v_s = \frac{R_1 + R_2}{R_1}v_s$$

The noninverting configuration can be viewed as a feedback circuit from the output v_o to the input. Since the current into the op amp from the inverting input is zero ($i_n = 0$), we can apply the voltage divider rule: the output voltage v_o is divided between

R_1 and R_2 in proportion to the resistance values. Thus, the voltage at the inverting input is given by

$$v_s = \frac{R_1}{R_1 + R_2} v_o$$

Solving for v_o, we have

$$v_o = \frac{R_1 + R_2}{R_1} v_s = \left(1 + \frac{R_2}{R_1}\right) v_s$$

The output voltage is the input voltage multiplied by $(1 + R_2/R_1)$. The voltage gain of the amplifier, $G = v_o/v_s$, is

$$G = 1 + \frac{R_2}{R_1} \tag{5.22}$$

without sign inversion. The voltage gain is always greater than or equal to 1.

5.2.1 VOLTAGE FOLLOWER

The circuit shown in Figure 5.14 is called a *voltage follower*, or a *buffer amplifier*.

The input signal is applied to the noninverting input terminal of the op amp, and the inverting input terminal is directly connected to the output terminal. Since $v_n = v_p$ and $v_p = v_s$, we have $v_n = v_s$. Since $v_o = v_n$, we have $v_o = v_s$. Thus, the output voltage follows the input voltage exactly. That is why this circuit is called a *voltage follower*. When the op amp in the voltage follower circuit shown in Figure 5.14 is replaced by a VCVS model, we obtain the circuit shown in Figure 5.15.

FIGURE 5.14

A voltage follower circuit.

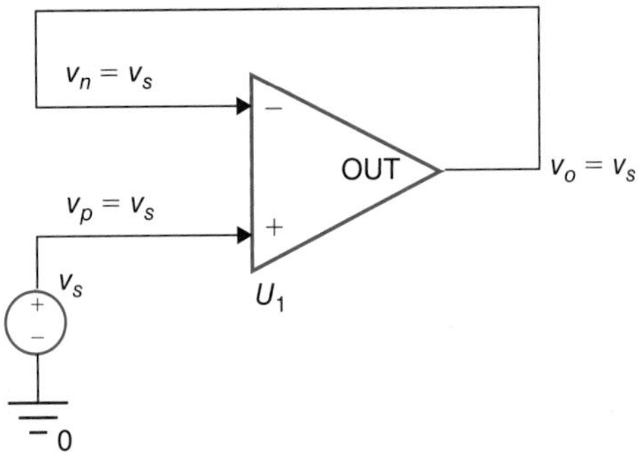

Notice that $v_n = v_o$. Summing the currents leaving node 2, we obtain

$$\frac{v_o - v_s}{R_i} + \frac{v_o - A(v_s - v_o)}{R_o} = 0$$

Solving for v_o, we obtain

$$v_o = v_n = \frac{\dfrac{1}{R_i} + \dfrac{A}{R_o}}{\dfrac{1}{R_i} + \dfrac{1}{R_o} + \dfrac{A}{R_o}} v_s$$

FIGURE 5.15

A voltage follower circuit with the op amp model.

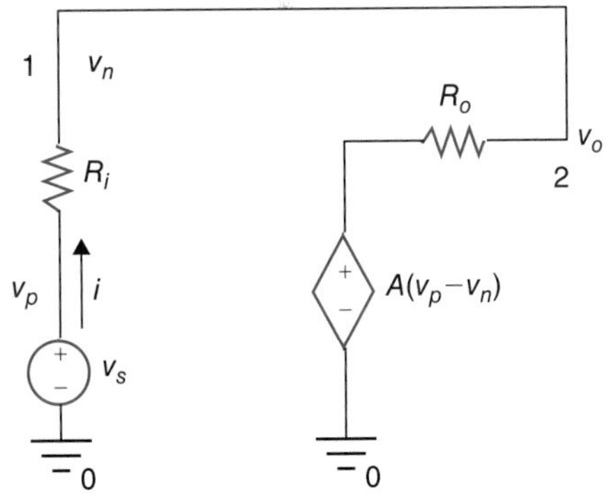

The current flowing out of the positive terminal of v_s is given by

$$i = \frac{v_s - v_n}{R_i} = \frac{1 - \dfrac{\dfrac{1}{R_i} + \dfrac{A}{R_o}}{\dfrac{1}{R_i} + \dfrac{1}{R_o} + \dfrac{A}{R_o}}}{R_i} v_s = \frac{\dfrac{\dfrac{1}{R_o}}{\dfrac{1}{R_i} + \dfrac{1}{R_o} + \dfrac{A}{R_o}}}{R_i} v_s$$

$$= \frac{\dfrac{1}{R_o}}{1 + \dfrac{R_i}{R_o} + \dfrac{AR_i}{R_o}} v_s = \frac{1}{R_o + R_i + AR_i} v_s$$

FIGURE 5.16

A voltage follower circuit with test voltage at the output.

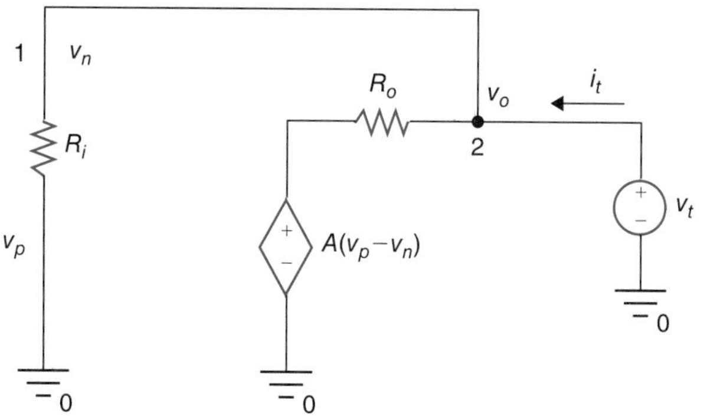

The input resistance R_{in} is defined as the ratio of v_s to i. For the voltage follower model shown in Figure 5.15, the input resistance is given by

$$R_{in} = \frac{v_s}{i} = (A + 1)R_i + R_o \approx AR_i$$

Since both A and R_i are large, the input resistance of the voltage follower is large. To find the output resistance of the voltage follower model shown in Figure 5.15, we deactivate the voltage source and apply a test voltage at the output, as shown in Figure 5.16. Notice that $v_p = 0$ and $v_n = v_o = v_t$.

The current flowing out of the test voltage source is given by

$$i_t = \frac{v_t}{R_i} + \frac{v_t - A(0 - v_t)}{R_o} = \left(\frac{1}{R_i} + \frac{1 + A}{R_o}\right)v_t$$

The output resistance R_{out} is defined as the ratio of v_t to i_t. For the voltage follower model shown in Figure 5.16, the output resistance is given by

$$R_{out} = \frac{v_t}{i_t} = \frac{1}{\dfrac{1}{R_i} + \dfrac{1 + A}{R_o}} \approx \frac{R_o}{A + 1} \approx \frac{R_o}{A}$$

Since R_o is small and A is large, the output resistance of voltage follower is small.

FIGURE 5.17

A source connected directly to the load.

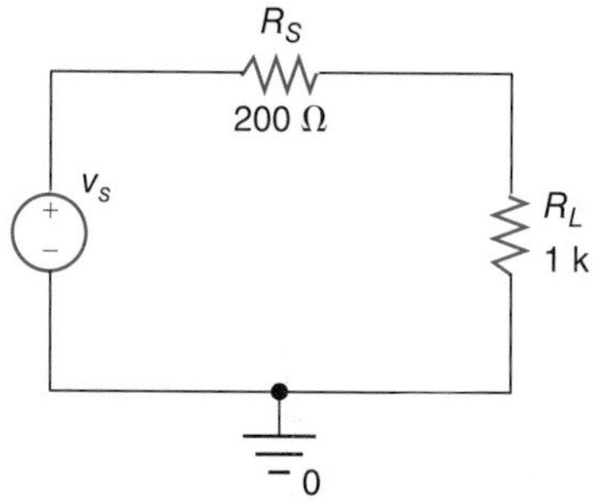

The voltage gain of the voltage follower is 1. What is the purpose of using a unity gain voltage follower? The op amp provides large input resistance and small output resistance, as shown previously. The purpose of the voltage follower is to provide a buffer between the source and load to reduce the loading effect. Consider a voltage source v_s with source resistance $R_S = 200\ \Omega$. If this source is directly applied to a load with resistance $R_L = 1\ k\Omega$, as shown in Figure 5.17, the voltage across the load is $v_s \times 1000/(1000 + 200) = 0.833v_s$. Direct connection of the source and load results in a 16.7% reduction in the load voltage.

Figure 5.18 shows a circuit that connects the voltage source with source resistance R_S to the positive input terminal of the op amp, and the load resistor is connected to the output of the op amp. For the ideal op amp, since the current flowing into the op amp is zero ($i_p = 0$), the voltage drop across R_S is zero. Thus, the voltage at the noninverting input terminal is v_s ($v_p = v_s$). Due to a virtual short, the voltage

FIGURE 5.18

A source connected to the load through voltage follower.

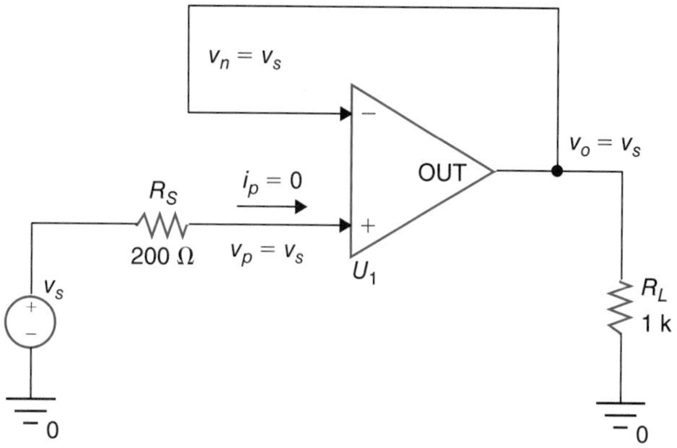

at the negative input terminal is also v_s ($v_n = v_p = v_s$). Since the output terminal is directly connected to negative input terminal, we have $v_o = v_s$. The voltage across the load impedance is v_s, eliminating the effect of R_S. The voltage follower can be used to isolate different sections of circuits.

EXAMPLE 5.1

Express V_o in the circuit shown in Figure 5.19 as a function of V_s, R_1, R_2, R_3, and R_4. Find the numerical value of V_o if $V_s = 0.5$ V, $R_1 = 2\ k\Omega$, $R_2 = 5\ k\Omega$, $R_3 = 1.5\ k\Omega$, $R_4 = 9\ k\Omega$.

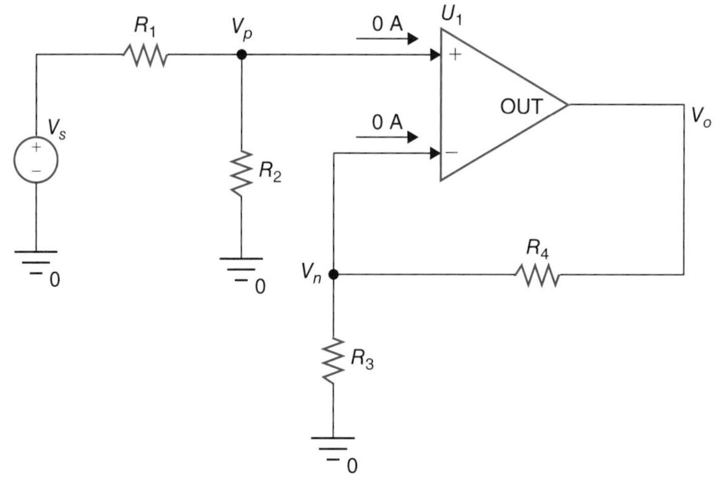

FIGURE 5.19

Circuit for EXAMPLE 5.1.

Since the current entering the positive input of the op amp is zero, according to the voltage divider rule, the voltage at the positive input of the op amp, which is the voltage across R_2, is given by

$$V_p = \frac{R_2}{R_1 + R_2} V_s$$

Similarly, since the current entering the negative input of the op amp is zero, according to the voltage divider rule, the voltage at the negative input of the op amp, which is the voltage across R_3, is given by

$$V_n = \frac{R_3}{R_3 + R_4} V_o$$

Due to a virtual short, we have $V_n = V_p$. Thus, we have

$$\frac{R_3}{R_3 + R_4} V_o = \frac{R_2}{R_1 + R_2} V_s$$

Solving for V_o, we obtain

$$V_o = \left(\frac{R_2}{R_1 + R_2}\right)\left(\frac{R_3 + R_4}{R_3}\right) V_s = \left(\frac{R_2}{R_1 + R_2}\right)\left(1 + \frac{R_4}{R_3}\right) V_s$$

continued

Example 5.1 continued

If $V_s = 0.5$ V, $R_1 = 2\ k\Omega$, $R_2 = 5\ k\Omega$, $R_3 = 1.5\ k\Omega$, $R_4 = 9\ k\Omega$, V_o is given by

$$V_o = \left(\frac{5\ k\Omega}{2\ k\Omega + 5\ k\Omega}\right)\left(1 + \frac{9\ k\Omega}{1.5\ k\Omega}\right) \times 0.5\ \text{V} = 2.5\ \text{V}$$

Exercise 5.1

Express V_o in the circuit shown in Figure 5.20 as a function of V_{s_1}, V_{s_2}, R_1, R_2, R_3, and R_4. Find V_o as a function of V_{s_1} and V_{s_2} if $R_1 = 2\ k\Omega$, $R_2 = 4\ k\Omega$, $R_3 = 1.5\ k\Omega$, $R_4 = 7.5\ k\Omega$.

FIGURE 5.20

Circuit for EXERCISE 5.1.

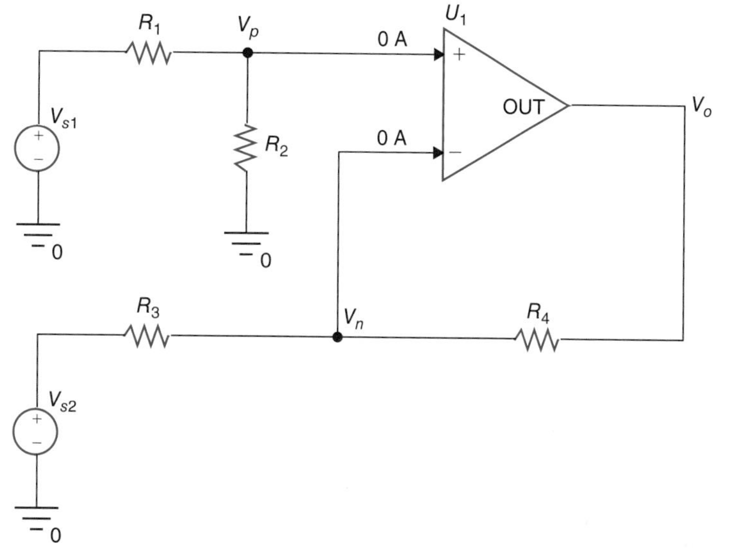

Answer:

From the superposition principle, we have

$$V_o = \left(\frac{R_2}{R_1 + R_2}\right)\left(1 + \frac{R_4}{R_3}\right)V_{s_1} - \frac{R_4}{R_3}V_{s_2}$$

$$V_o = \left(\frac{4}{6}\right)(1 + 5)V_{s_1} - \frac{7.5}{1.5}V_{s_2} = 4V_{s_1} - 5V_{s_2}$$

EXAMPLE 5.2

Find v_o in the circuit shown in Figure 5.21.

The voltage at the inverting input of the op amp is equal to zero due to a virtual short. Thus, we have $v_1 = 0$. Summing the currents leaving node 1 (inverting input of the op amp), we get

$$\frac{0 - v_s}{R_1} + \frac{0 - v_2}{R_2} = 0$$

continued

Example 5.2 continued

FIGURE 5.21

Circuit for EXAMPLE 5.2.

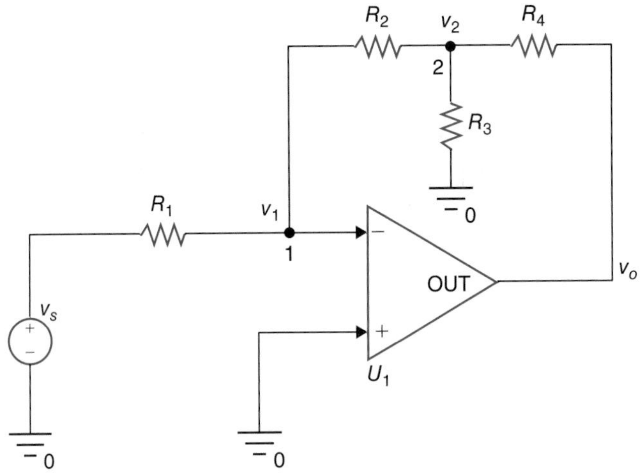

Solving for v_2, we obtain

$$v_2 = -\frac{R_2}{R_1} v_s$$

Summing the currents leaving node 2, we get

$$\frac{v_2 - 0}{R_2} + \frac{v_2 - 0}{R_3} + \frac{v_2 - v_o}{R_4} = 0$$

This equation can be simplified as

$$v_o = R_4\left(\frac{1}{R_2} + \frac{1}{R_3} + \frac{1}{R_4}\right)v_2$$

Substituting $v_2 = -\frac{R_2}{R_1} v_s$ into $v_o = R_4\left(\frac{1}{R_2} + \frac{1}{R_3} + \frac{1}{R_4}\right)v_2$, we get

$$v_o = -\frac{R_2}{R_1} R_4\left(\frac{1}{R_2} + \frac{1}{R_3} + \frac{1}{R_4}\right)v_s = -\frac{1}{R_1}\left(R_4 + R_2 + \frac{R_2 R_4}{R_3}\right)v_s$$

$$= -\frac{1}{R_1}\left[R_4 + R_2\left(1 + \frac{R_4}{R_3}\right)\right]v_s$$

The single feedback resistor is replaced by a T circuit consisting of R_2, R_3, and R_4. The combined resistance of the T circuit is given by $R_4 + R_2\left(1 + \frac{R_4}{R_3}\right)$. If $R_4 > R_3$, the value of the combined resistance will be very large due to the product of R_2 and $1 + \frac{R_4}{R_3}$.

Exercise 5.2

Find v_o in the circuit shown in Figure 5.22 as a function of v_s, R_1, R_2, R_3, k, and R_4. The resistance of the variable resistor is kR_3, where $0 < k < \infty$. Let $v_s = 1$ V, $R_1 = R_2 = R_3 = R_4 = 1$ kΩ. Find v_o for $k = 0.1, 0.5, 1, 10$, and 1000.

FIGURE 5.22

Circuit for EXERCISE 5.2.

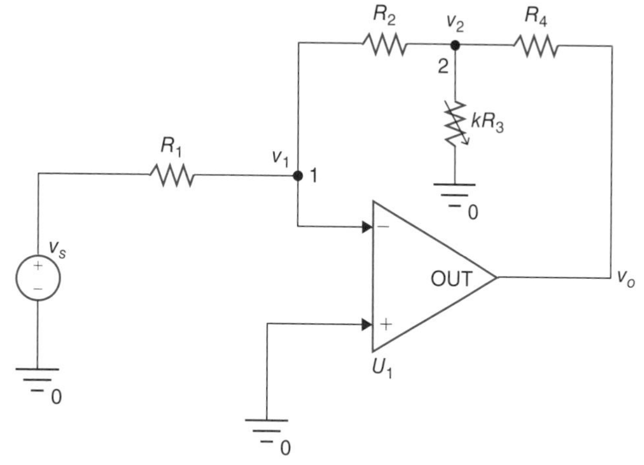

Answer:

$$v_o = -\frac{1}{R_1}\left(R_4 + R_2\left(1 + \frac{R_4}{kR_3}\right)\right)v_s$$

$v_o = -12$ V, -4 V, -3 V, -2.1 V, -2.001 V.

EXAMPLE 5.3

Express V_o in the circuit shown in Figure 5.23 as a function of V_s.

FIGURE 5.23

Circuit for EXAMPLE 5.3.

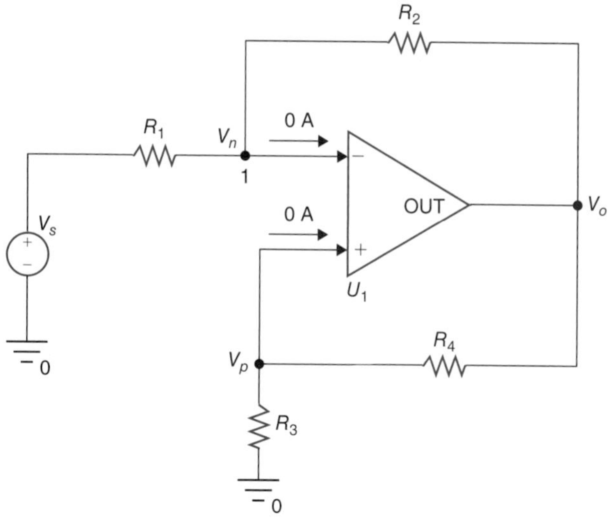

Since the current entering the positive input of the op amp is zero, according to the voltage divider rule, the voltage at the positive input of the op amp, which is the voltage across R_3, is given by

$$V_p = \frac{R_3}{R_3 + R_4}V_o \tag{5.23}$$

Summing the currents leaving node 1, we obtain

$$\frac{V_n - V_s}{R_1} + \frac{V_n - V_o}{R_2} = 0$$

Solving for V_o, we get

$$V_o = \left(1 + \frac{R_2}{R_1}\right)V_n - \frac{R_2}{R_1}V_s \tag{5.24}$$

continued

Example 5.3 continued

Due to a virtual short, we have $V_n = V_p$. Substitution of Equation (5.23) into Equation (5.24) yields

$$V_o = \left(1 + \frac{R_2}{R_1}\right)\frac{R_3}{R_3 + R_4}V_o - \frac{R_2}{R_1}V_s$$

which can be rewritten as

$$V_o = \frac{-\frac{R_2}{R_1}V_s}{1 - \left(\frac{R_1 + R_2}{R_1}\right)\frac{R_3}{R_3 + R_4}} = \frac{-R_2(R_3 + R_4)}{R_1R_4 - R_2R_3}V_s \quad (5.25)$$

Notice that if $R_1R_4 > R_2R_3$, the voltage gain V_o/V_s is negative (i.e., inverting amplifier), and if $R_1R_4 < R_2R_3$, the voltage gain V_o/V_s is positive (i.e., noninverting amplifier).

Exercise 5.3

(a) Let $R_1 = 1\ k\Omega, R_2 = 2\ k\Omega, R_3 = 2\ k\Omega$, and $R_4 = 2.5\ k\Omega$. Find the voltage gain V_o/V_s for the circuit shown in Figure 5.23.

(b) Let $R_1 = 3.25\ k\Omega, R_2 = 2\ k\Omega, R_3 = 2.5\ k\Omega$, and $R_4 = 2\ k\Omega$. Find the voltage gain V_o/V_s for the circuit shown in Figure 5.23.

Answer:

(a) $V_o/V_s = 6$ V/V (a) $V_o/V_s = -6$ V/V

EXAMPLE 5.4

Express V_o in the circuit shown in Figure 5.24 as a function of V_s.

FIGURE 5.24

Circuit for EXAMPLE 5.4.

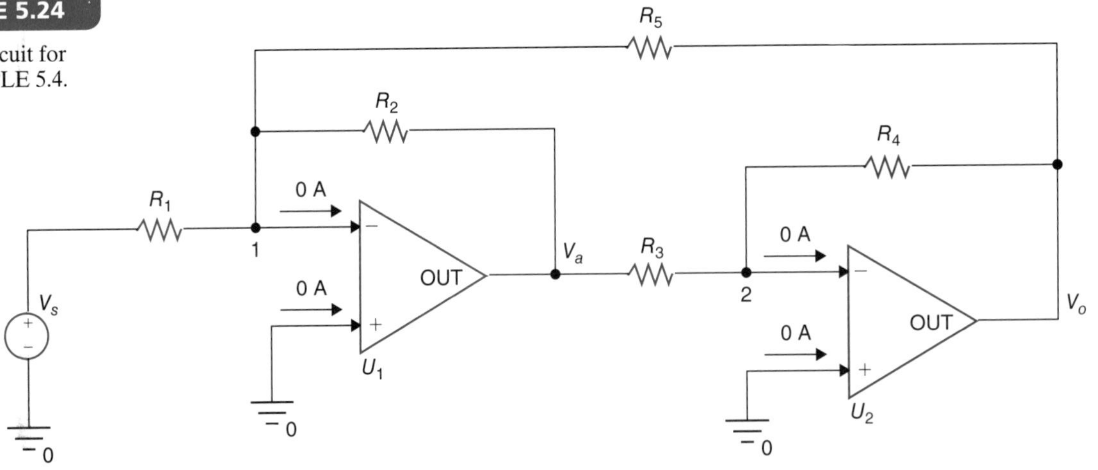

continued

Example 5.4 continued

Summing the currents leaving node 2, we obtain

$$\frac{0 - V_a}{R_3} + \frac{0 - V_o}{R_4} = 0$$

Solving for V_a, we obtain

$$V_a = -\frac{R_3}{R_4}V_o \tag{5.26}$$

Summing the currents leaving node 1, we obtain

$$\frac{0 - V_s}{R_1} + \frac{0 - V_o}{R_5} + \frac{0 - V_a}{R_2} = 0$$

Solving for V_o and substituting Equation (5.26), we obtain

$$V_o = -\frac{R_5}{R_1}V_s - \frac{R_5}{R_2}V_a = -\frac{R_5}{R_1}V_s - \frac{R_5}{R_2}\left(-\frac{R_3}{R_4}V_o\right) = -\frac{R_5}{R_1}V_s + \frac{R_3R_5}{R_2R_4}V_o \tag{5.27}$$

Solving for V_o, we get

$$V_o = \frac{-\dfrac{R_5}{R_1}}{1 - \dfrac{R_3R_5}{R_2R_4}}V_s = -\frac{R_2R_4R_5}{R_1(R_2R_4 - R_3R_5)}V_s \tag{5.28}$$

Notice that $R_3 - R_4$ provides a feedback from V_o to V_a, as shown by Equation (5.26). In turn, this voltage is amplified by $R_2 - R_5$, as shown by the second term of Equation (5.27). The first term of Equation (5.27) is the direct path from V_s to V_o with voltage gain $-R_5/R_1$.

Exercise 5.4

Express V_o in the circuit shown in Figure 5.25 as a function of V_s.

FIGURE 5.25

Circuit for EXERCISE 5.4.

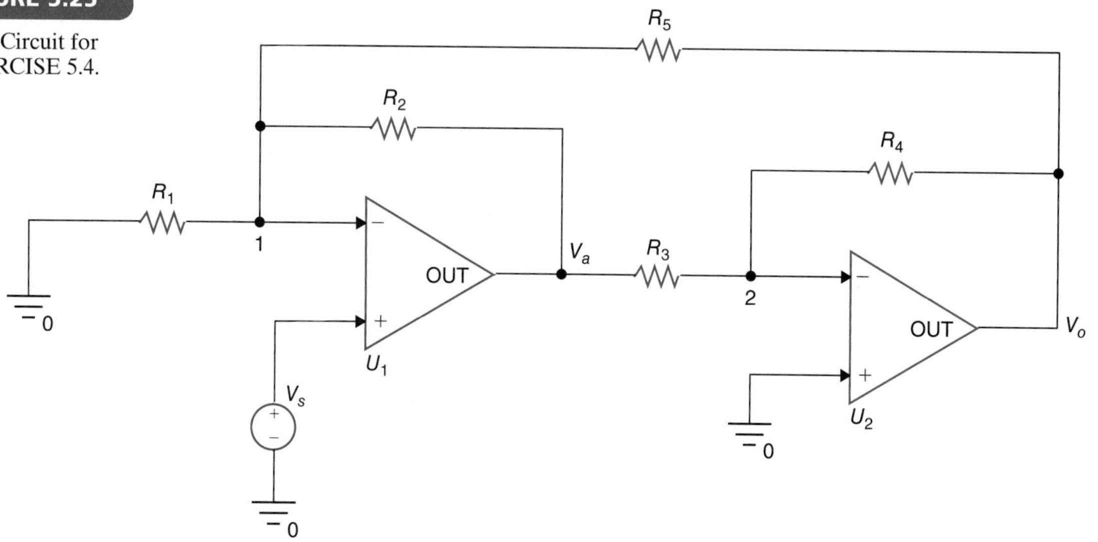

continued

Exercise 5.4 continued

Answer:

$$V_o = \frac{R_4(R_1R_2 + R_1R_5 + R_2R_5)}{R_1(R_2R_4 - R_3R_5)}V_s$$

EXAMPLE 5.5

Express V_o in the circuit shown in Figure 5.26 as a function of V_1 and V_2.

FIGURE 5.26

Circuit for EXAMPLE 5.5.

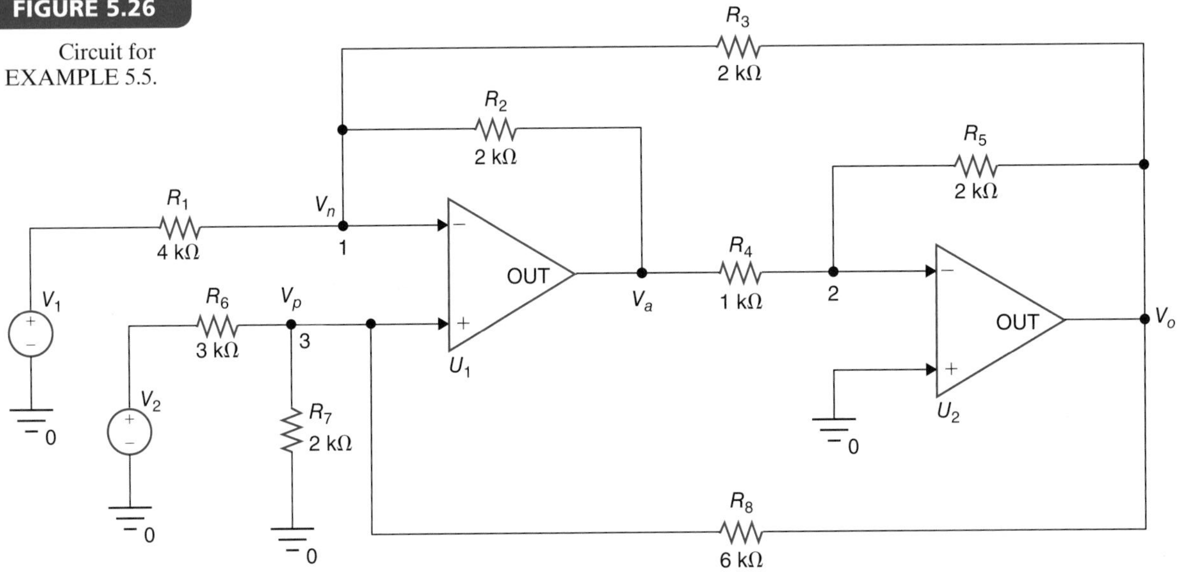

Summing the currents leaving node 2, we obtain

$$\frac{0 - V_a}{1000} + \frac{0 - V_o}{2000} = 0$$

Multiplication by 2000 yields

$$-2V_a - V_o = 0,$$

which can be simplified to

$$V_a = -0.5V_o \qquad (5.29)$$

Summing the currents leaving node 1, we obtain

$$\frac{V_n - V_1}{4000} + \frac{V_n - V_o}{2000} + \frac{V_n - V_a}{2000} = 0$$

continued

Example 5.5 continued

Multiplication by 4000 yields

$$V_n - V_1 + 2(V_n - V_o) + 2(V_n - V_a) = 0$$

which can be simplified to

$$5V_n - V_1 - 2V_o - 2V_a = 0 \tag{5.30}$$

Substitution of Equation (5.29) into Equation (5.30) yields

$$5V_n - V_1 - 2V_o + V_o = 0$$

Solving for V_n, we obtain

$$V_n = 0.2V_1 + 0.2V_o \tag{5.31}$$

Summing the currents leaving node 3, we obtain

$$\frac{V_p - V_2}{3000} + \frac{V_p}{2000} + \frac{V_p - V_o}{6000} = 0$$

Multiplication by 6000 yields

$$2(V_p - V_2) + 3V_p + V_p - V_o = 0$$

which can be simplified to

$$6V_p - 2V_2 - V_o = 0$$

Solving for V_p, we obtain

$$V_p = \frac{1}{3}V_2 + \frac{1}{6}V_o \tag{5.32}$$

Since $V_n = V_p$ due to a virtual short, from Equations (5.31) and (5.32), we obtain

$$\frac{1}{5}V_1 + \frac{1}{5}V_o = \frac{1}{3}V_2 + \frac{1}{6}V_o$$

which can be simplified to

$$\frac{1}{30}V_o = -\frac{1}{5}V_1 + \frac{1}{3}V_2$$

Multiplying by 30, we obtain

$$V_o = -6V_1 + 10V_2$$

Exercise 5.5

Find V_o in the circuit shown in Figure 5.27.

FIGURE 5.27

Circuit for EXERCISE 5.5.

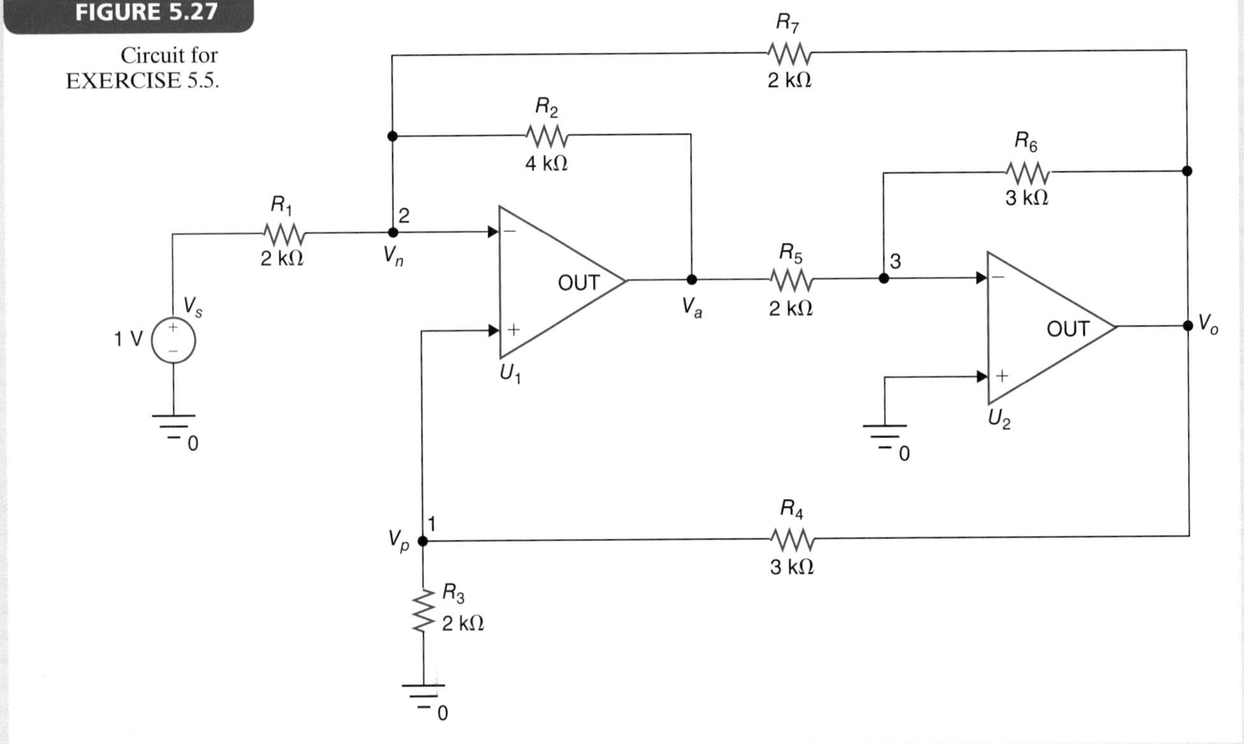

Answer:
$V_o = 3$ V.

EXAMPLE 5.6

Design an op amp circuit that will provide a voltage gain of 0.8.

FIGURE 5.28

Circuit for EXAMPLE 5.6.

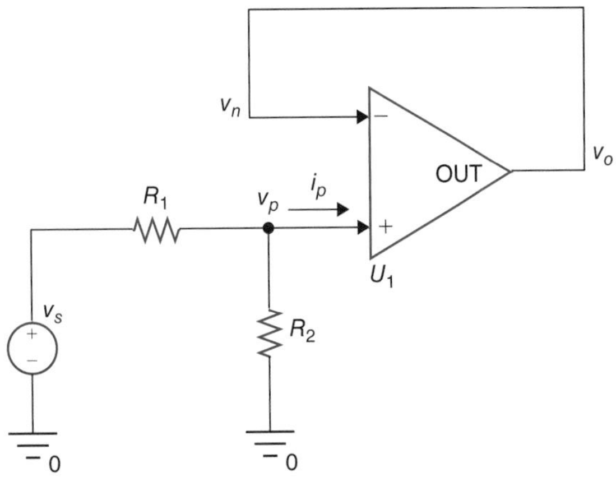

A circuit shown in Figure 5.28 can be used to provide a positive voltage gain of less than 1. Since the current flowing into the positive input terminal of the op amp is zero, according to the voltage divider rule, the voltage v_p is given by

$$v_p = \frac{R_2}{R_1 + R_2} v_s$$

Due to a virtual short, we have $v_n = v_p$. Since the output v_o is connected to v_n, we have

$$v_o = \frac{R_2}{R_1 + R_2} v_s$$

continued

Example 5.6 continued

The voltage gain is

$$G = \frac{R_2}{R_1 + R_2}$$

Taking the inverses on both sides, we obtain

$$\frac{R_1 + R_2}{R_2} = \frac{1}{G}$$

For $G = 0.8, 1/G = 1.25$. Let us choose $R_2 = 10\ k\Omega$. Then, we need $R_1 = 2.5\ k\Omega$.

Exercise 5.6

Design an op amp circuit that will provide a voltage gain of 3.7.

Answer:
Since the gain is positive, we can use the noninverting amplifier shown in Figure 5.13 earlier in this chapter. Since the voltage gain is $G = 1 + \frac{R_2}{R_1} = 3.7$, we can choose $R_1 = 1\ k\Omega$ and $R_2 = 2.7\ k\Omega$. The circuit with these values is shown in Figure 5.29.

FIGURE 5.29

Circuit for a voltage gain of 3.7.

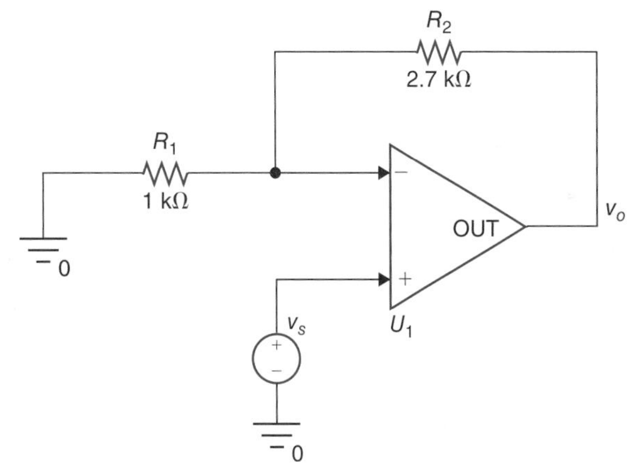

5.3 Sum and Difference

In this section, we discuss op amp circuits that can be used to add input signals or subtract one signal from another. The summing amplifiers can be implemented in the inverting configuration or noninverting configuration.

5.3.1 SUMMING AMPLIFIER (INVERTING CONFIGURATION)

Figure 5.30 shows an inverting amplifier with two inputs, v_1 and v_2.

Due to a virtual short, the voltage at the inverting input of the op amp is zero ($v_n = v_p = 0$). Summing the currents leaving node 1, we obtain

$$\frac{0 - v_1}{R_1} + \frac{0 - v_2}{R_2} + \frac{0 - v_o}{R_f} = 0$$

FIGURE 5.30

A summing amplifier with an inverting configuration.

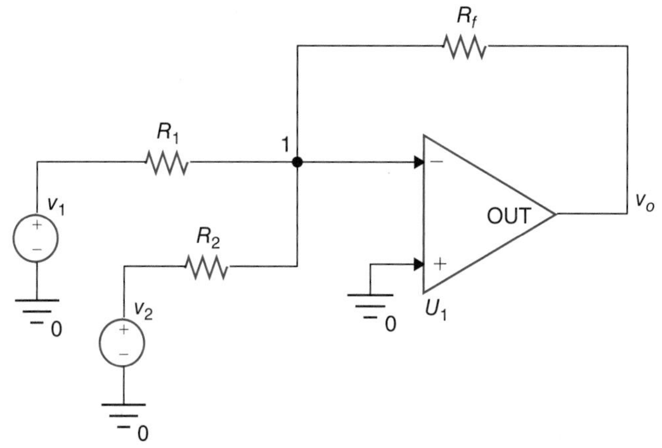

Solving for v_o, we have

$$v_o = -\left(\frac{R_f}{R_1}v_1 + \frac{R_f}{R_2}v_2\right) \quad \textbf{(5.33)}$$

If $R_1 = R_2 = R_f$, we have

$$v_o = -(v_1 + v_2) \quad \textbf{(5.34)}$$

If we want a positive sum instead of a negative sum, we can reverse the sign by applying the output to another inverting amplifier with $R_3 = R_4$, as shown in Figure 5.31.

FIGURE 5.31

Having a sign inverter added to the summing amplifier.

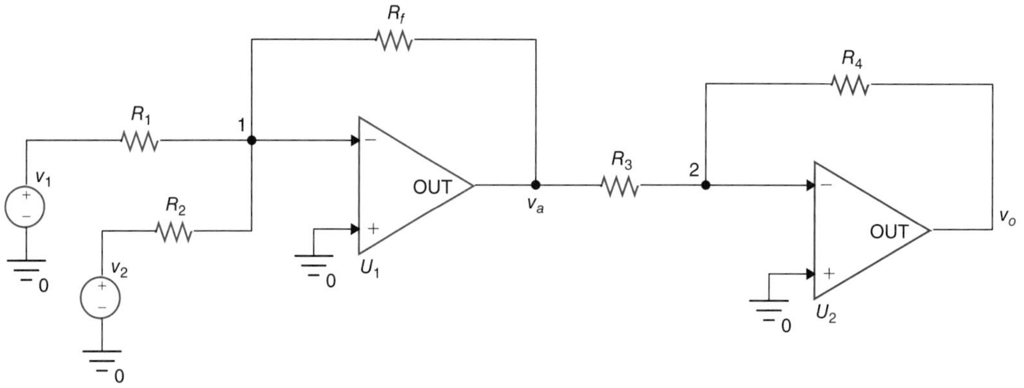

Let v_a be the voltage at the output of the first op amp. Summing the currents leaving node 2, we obtain

$$\frac{0 - v_a}{R_3} + \frac{0 - v_o}{R_4} = 0$$

Solving for v_o, we get

$$v_o = -\frac{R_4}{R_3}v_a$$

If $R_4 = R_3$, we have $v_o = -v_a$. Summing the currents leaving node 1, we obtain

$$\frac{0 - v_1}{R_1} + \frac{0 - v_2}{R_2} + \frac{0 - v_a}{R_f} = 0$$

Solving for v_a, we obtain

$$v_a = -\left(\frac{R_f}{R_1}v_1 + \frac{R_f}{R_2}v_2\right)$$

From $v_o = -\frac{R_4}{R_3}v_a$, the output voltage is given by

$$v_o = \frac{R_4}{R_3}\left(\frac{R_f}{R_1}v_1 + \frac{R_f}{R_2}v_2\right) \tag{5.35}$$

If $R = R_1 = R_2 = R_3 = R_4 = R_f$, the output voltage becomes

$$v_o = v_1 + v_2 \tag{5.36}$$

If $R = R_3 = R_4 = R_f$, $R_1 = R/k_1$, and $R_2 = R/k_2$, the output voltage becomes

$$v_o = k_1v_1 + k_2v_2 \tag{5.37}$$

Figure 5.32 shows an inverting amplifier with N inputs, $v_1, v_2, \ldots, v_N$.

FIGURE 5.32

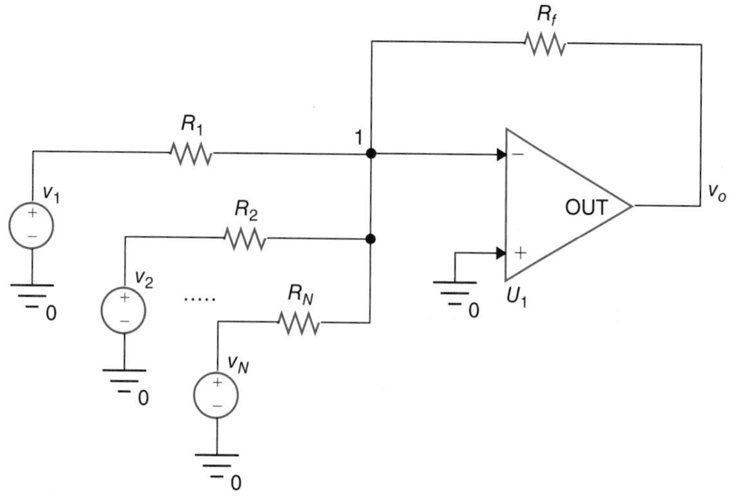

Summing amplifier with n inputs.

Due to a virtual short, the voltage at the inverting input of the op amp is zero ($v_n = v_p = 0$). Summing the currents leaving node 1, we obtain

$$\frac{0 - v_1}{R_1} + \frac{0 - v_2}{R_2} + \frac{0 - v_3}{R_3} + \cdots + \frac{0 - v_N}{R_N} + \frac{0 - v_o}{R_f} = 0$$

Solving for v_o, we have

$$v_o = -\left(\frac{R_f}{R_1}v_1 + \frac{R_f}{R_2}v_2 + \frac{R_f}{R_3}v_3 + \cdots + \frac{R_f}{R_N}v_N\right) \tag{5.38}$$

If $R_1 = R_2 = R_3 = \cdots = R_N = R_f$, we have

$$v_o = -(v_1 + v_2 + v_3 + \cdots + v_N) \tag{5.39}$$

If we want a positive sum instead of a negative sum, we can reverse the sign by applying the output to an inverting amplifier with the same resistance values.

If $R_1 = R/k_1, R_2 = R/k_2, R_3 = R/k_3, \ldots, R_N = R/k_N$, and $R_f = R$, we have

$$v_o = -(k_1 v_1 + k_2 v_2 + k_3 v_3 + \cdots + k_N v_N) \quad \textbf{(5.40)}$$

5.3.2 SUMMING AMPLIFIER (NONINVERTING CONFIGURATION)

The circuit shown in Figure 5.33 can be used to add two signals, v_1 and v_2. There is no sign change in this configuration.

FIGURE 5.33

A summing amplifier with a noninverting configuration.

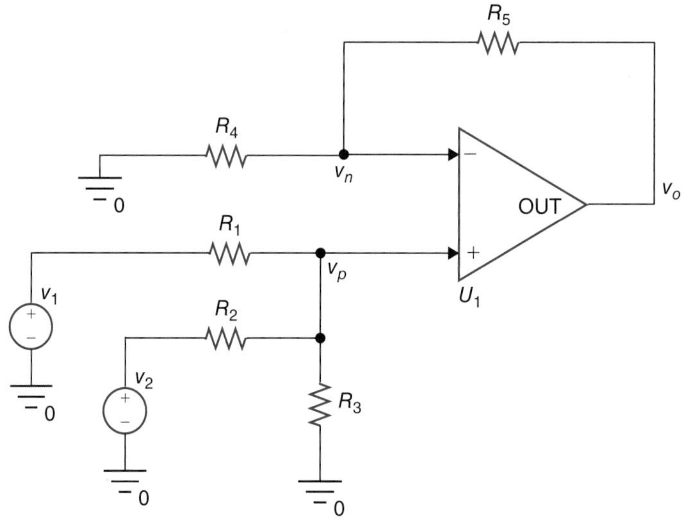

Summing the currents leaving the negative input of the op amp, we have

$$\frac{v_n - 0}{R_4} + \frac{v_n - v_o}{R_5} = 0$$

Solving for v_n, we obtain

$$v_n = \frac{\dfrac{1}{R_5}}{\dfrac{1}{R_4} + \dfrac{1}{R_5}} v_o \quad \textbf{(5.41)}$$

Summing the currents leaving the positive input of the op amp, we have

$$\frac{v_p - v_1}{R_1} + \frac{v_p - v_2}{R_2} + \frac{v_p - 0}{R_3} = 0$$

which can be rewritten as

$$\left(\frac{1}{R_1} + \frac{1}{R_2} + \frac{1}{R_3}\right) v_p = \frac{v_1}{R_1} + \frac{v_2}{R_2} \quad \textbf{(5.42)}$$

Substituting Equation (5.41) into Equation (5.42), we obtain

$$\left(\frac{1}{R_1} + \frac{1}{R_2} + \frac{1}{R_3}\right) \frac{1}{\dfrac{1}{R_4} + \dfrac{1}{R_5}} \frac{v_o}{R_5} = \frac{v_1}{R_1} + \frac{v_2}{R_2}$$

Solving for v_o, we have

$$v_o = \frac{1}{\frac{1}{R_1} + \frac{1}{R_2} + \frac{1}{R_3}}\left(1 + \frac{R_5}{R_4}\right)\left(\frac{v_1}{R_1} + \frac{v_2}{R_2}\right) \tag{5.43}$$

Let $R = R_1 = R_2 = R_3$. Then, Equation (5.43) becomes

$$v_o = \frac{R}{3}\left(\frac{R_4 + R_5}{R_4}\right)\frac{1}{R}(v_1 + v_2) \tag{5.44}$$

Let $R_4 = R$ and $R_5 = 2R$. Then, we have

$$v_o = v_1 + v_2 \tag{5.45}$$

Let $R_3 = R_4 = R$, $R_1 = R/k_1$, $R_2 = R/k_2$, and $R_5 = R(k_1 + k_2)$. Then, Equation (5.43) becomes

$$v_o = \frac{1}{\frac{k_1}{R} + \frac{k_2}{R} + \frac{1}{R}}(1 + k_1 + k_2)\frac{1}{R}(k_1v_1 + k_2v_2) = k_1v_1 + k_2v_2 \tag{5.46}$$

EXAMPLE 5.7

Design an op amp circuit for $v_o = 0.5v_1 + 3v_2$.

Notice that $k_1 = 0.5$ and $k_2 = 3$ in Equation (5.46). Let $R = 3\ k\Omega$. Then, we have

$$R_1 = R/k_1 = 3\ k\Omega/0.5 = 6\ k\Omega, R_2 = R/k_2 = 3\ k\Omega/3 = 1\ k\Omega, R_3 = R_4 = R = 3\ k\Omega,$$
$$R_5 = R(k_1 + k_2) = 3\ k\Omega \times (0.5 + 3) = 10.5\ k\Omega$$

The circuit with these resistance values is shown in Figure 5.34. There are many other choices for the resistance values.

FIGURE 5.34

Circuit for $v_o = 0.5v_1 + 3v_2$.

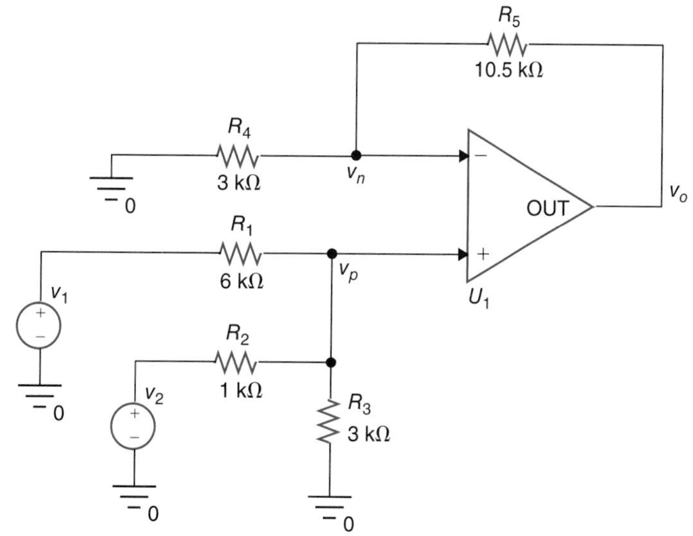

Exercise 5.7

Design an op amp circuit for $v_o = 2v_1 + 5v_2$.

Answer:
$k_1 = 2, k_2 = 5, R = 5\ k\Omega, R_1 = R/k_1 = 2.5\ k\Omega, R_2 = R/k_2 = 1\ k\Omega, R_3 = R_4 = R = 5\ k\Omega,$ $R_5 = R(k_1 + k_2) = 5\ k\Omega \times (2 + 5) = 35\ k\Omega$. The op amp circuit is shown in Figure 5.35.

FIGURE 5.35

Circuit for $v_o = 2v_1 + 5v_2$.

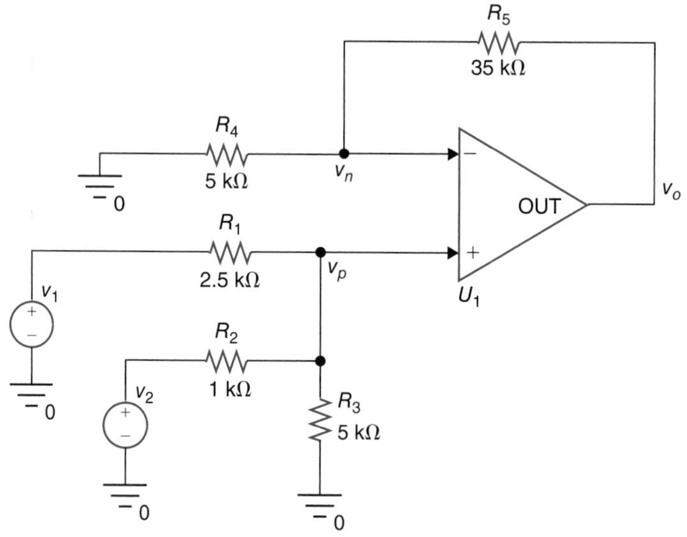

EXAMPLE 5.8

Design a circuit for converting a polar binary signal (1 is represented by a pulse with 5 V, and 0 is represented by a pulse with −5 V) to a unipolar binary signal (1 is represented by a pulse with 5 V, and 0 is represented by a pulse with 0 V).

Let v_1 be the polar signal and v_o be the unipolar signal. Then, adding 5 V to v_1 results in a signal with 10 V (for 1) or 0 V (for 0). We can obtain the unipolar signal by dividing $(v_1 + 5)$ V by 2. Thus, we have

$$v_o = \frac{1}{2}(v_1 + 5)$$

Let v_2 be 5 V. Then choosing $R_4 = 2R$ and $R_5 = R$ in Equation (5.44) results in

$$v_o = \frac{R}{3}\left(\frac{R_4 + R_5}{R_4}\right)\frac{1}{R}(v_1 + v_2) = \frac{R}{3}\left(\frac{2R + R}{2R}\right)\frac{1}{R}(v_1 + v_2) = \frac{1}{2}(v_1 + v_2)$$

In conclusion, if we choose $R_1 = R_2 = R_3 = R_5 = R, R_4 = 2R, V_2 = 5$ V, we can convert a polar signal to a unipolar signal. Figure 5.36 shows the circuit with $R = 10\ k\Omega$, and Figure 5.37 shows sample waveforms for v_1 and v_o. In PSpice®, VPULSE is the voltage source v_1. If the amplitude is other than 5 V, change v_2 to the amplitude of the input signal.

continued

Example 5.8 continued

FIGURE 5.36

Circuit for $v_o = 0.5(v_1 + v_2)$.

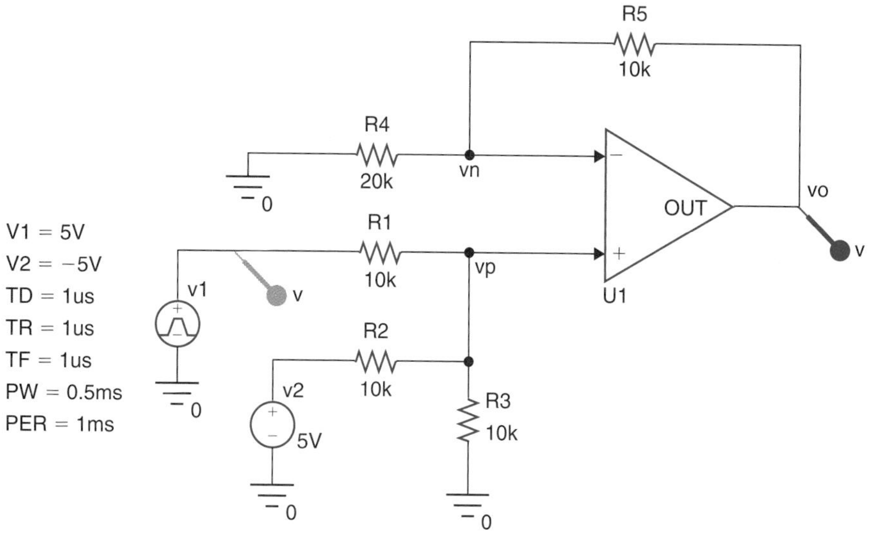

FIGURE 5.37

Sample waveforms.

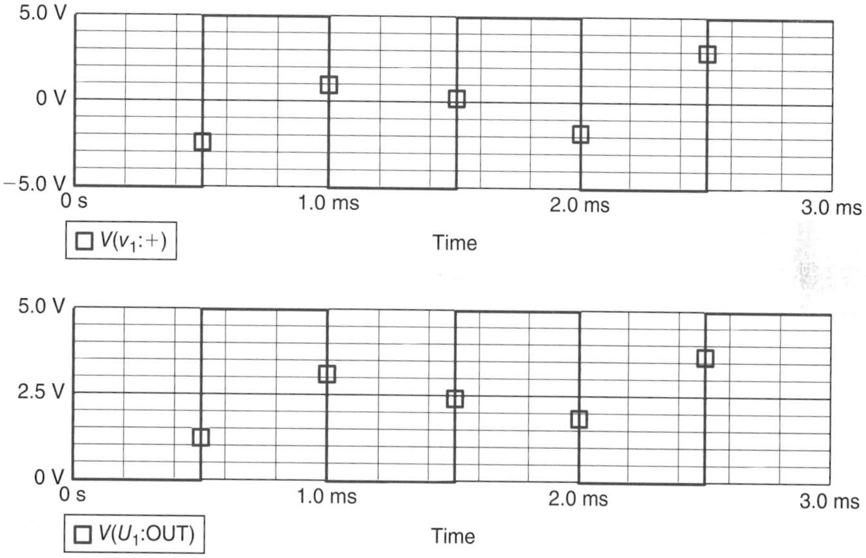

Figure 5.38 shows a noninverting summing amplifier with N inputs. Summing the currents leaving the negative input of the op amp, we have

$$\frac{v_n - 0}{R_{N+2}} + \frac{v_n - v_o}{R_{N+3}} = 0$$

Solving for v_n, we obtain

$$v_n = \frac{\dfrac{1}{R_{N+3}}}{\dfrac{1}{R_{N+2}} + \dfrac{1}{R_{N+3}}} v_o \tag{5.47}$$

continued

Example 5.8 continued

FIGURE 5.38

A summing amplifier with a noninverting configuration with n inputs.

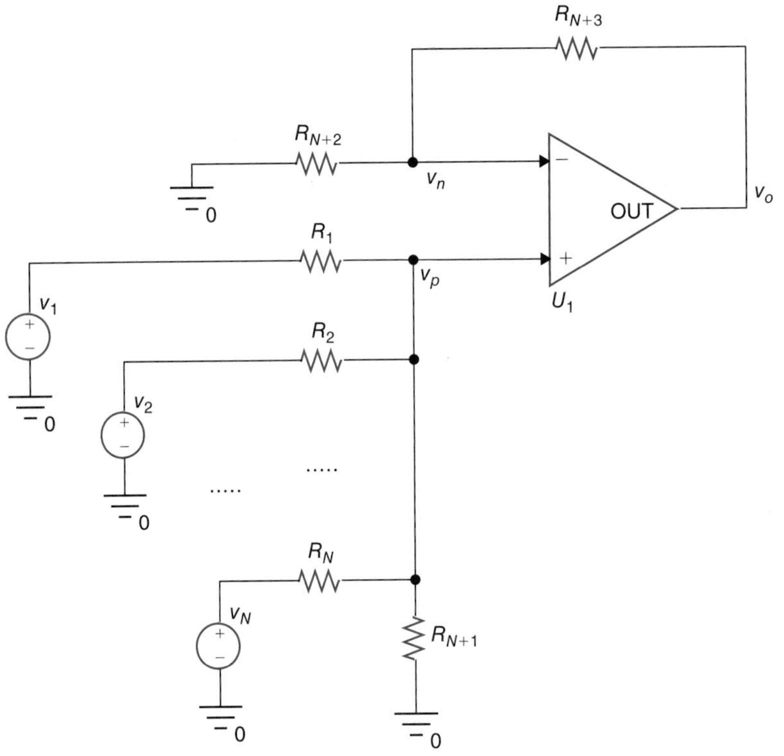

Summing the currents leaving the positive input of the op amp, we have

$$\frac{v_p - v_1}{R_1} + \cdots + \frac{v_p - v_N}{R_N} + \frac{v_p - 0}{R_{N+1}} = 0$$

which can be rewritten as

$$\left(\frac{1}{R_1} + \cdots + \frac{1}{R_N} + \frac{1}{R_{N+1}}\right)v_p = \frac{v_1}{R_1} + \cdots + \frac{v_N}{R_N} \tag{5.48}$$

Substitution of Equation (5.47) into Equation (5.48) yields

$$\left(\frac{1}{R_1} + \cdots + \frac{1}{R_N} + \frac{1}{R_{N+1}}\right)\frac{1}{\dfrac{1}{R_{N+2}} + \dfrac{1}{R_{N+3}}}\frac{v_o}{R_{N+3}} = \frac{v_1}{R_1} + \cdots + \frac{v_N}{R_N}$$

Solving for v_o, we have

$$v_o = \frac{1}{\dfrac{1}{R_1} + \cdots + \dfrac{1}{R_N} + \dfrac{1}{R_{N+1}}}\left(1 + \frac{R_{N+3}}{R_{N+2}}\right)\left(\frac{v_1}{R_1} + \cdots + \frac{v_N}{R_N}\right) \tag{5.49}$$

Let $R = R_1 = R_2 = \cdots = R_N = R_{N+1}$. Then, Equation (5.49) becomes

$$v_o = \frac{R}{N+1}\left(\frac{R_{N+2} + R_{N+3}}{R_{N+2}}\right)\frac{1}{R}(v_1 + \cdots + v_N) \tag{5.50}$$

continued

Example 5.8 continued

Let $R_{N+2} = R$ and $R_{N+3} = NR$. Then, Equation (5.50) becomes

$$v_o = v_1 + \cdots + v_N \tag{5.51}$$

Let $R_{N+1} = R_{N+2} = R, R_1 = R/k_1, R_2 = R/k_2, \ldots, R_N = R/k_N, R_{N+3} = R(k_1 + k_2 + \cdots + k_N)$. Then, Equation (5.49) becomes

$$v_o = \frac{1}{\frac{k_1}{R} + \frac{k_2}{R} + \cdots + \frac{k_N}{R} + \frac{1}{R}}(1 + k_1 + k_2 + \cdots + k_N) \times \frac{1}{R}(k_1v_1 + k_2v_2 + \cdots + k_Nv_N) = k_1v_1 + k_2v_2 + \cdots + k_Nv_N \tag{5.52}$$

Figure 5.39 shows the circuit that produces the output $v_o = v_1 + \cdots + v_N$.

FIGURE 5.39

A summing amplifier with a noninverting configuration with n inputs and gain of 1.

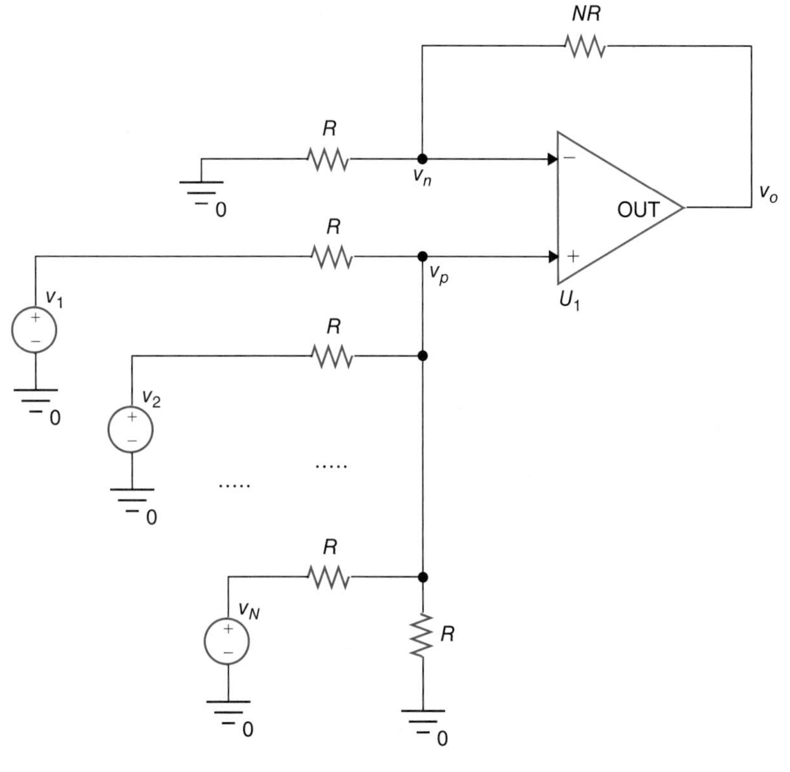

5.3.3 ALTERNATIVE SUMMING AMPLIFIER (NONINVERTING CONFIGURATION)

Another summing amplifier with a noninverting configuration is shown in Figure 5.40.

Application of the voltage divider rule to $R_4 - R_3$ yields

$$v_a = \frac{R_3}{R_3 + R_4}v_o \tag{5.53}$$

Summing the currents leaving the noninverting input of the op amp, we obtain

$$\frac{v_a - v_1}{R_1} + \frac{v_a - v_2}{R_2} = 0$$

FIGURE 5.40

A summing amplifier with a noninverting configuration.

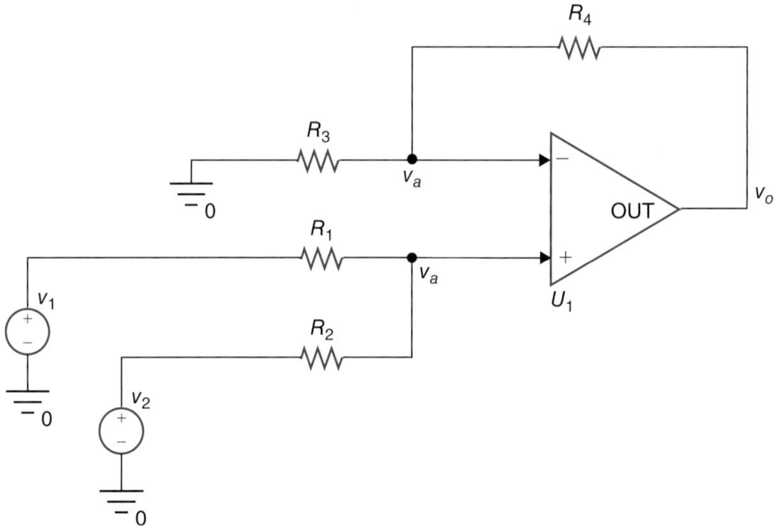

which can be revised as

$$\left(\frac{1}{R_1} + \frac{1}{R_2}\right)v_a = \frac{v_1}{R_1} + \frac{v_2}{R_2} \tag{5.54}$$

Substitution of Equation (5.53) into Equation (5.54) yields

$$\left(\frac{1}{R_1} + \frac{1}{R_2}\right)\frac{R_3}{R_3 + R_4}v_o = \frac{v_1}{R_1} + \frac{v_2}{R_2}$$

Solving for v_o, we obtain

$$v_o = \frac{1}{\dfrac{1}{R_1} + \dfrac{1}{R_2}}\frac{R_3 + R_4}{R_3}\left(\frac{v_1}{R_1} + \frac{v_2}{R_2}\right) \tag{5.55}$$

Let $R = R_1 = R_2$. Then, Equation (5.55) becomes

$$v_o = \frac{R}{2}\frac{R_3 + R_4}{R_3}\frac{1}{R}(v_1 + v_2) \tag{5.56}$$

Let $R_3 = R_4$. Then, we have

$$v_o = \frac{R}{2}\frac{2R_3}{R_3}\frac{1}{R}(v_1 + v_2) = v_1 + v_2 \tag{5.57}$$

If $R_1 = R/k_1, R_2 = R/k_2, R_3 = R, R_4 = (k_1 + k_2 - 1)R$, Equation (5.55) becomes

$$v_o = \frac{1}{\dfrac{k_1}{R} + \dfrac{k_2}{R}}\frac{R + (k_1 + k_2 - 1)R}{R}\left(\frac{k_1v_1}{R} + \frac{k_2v_2}{R}\right) = k_1v_1 + k_2v_2 \tag{5.58}$$

The constants k_1 and k_2 should be selected such that $k_1 + k_2 > 1$.

In general, to design an adder that adds N inputs, $v_1, v_2, v_3, \ldots, v_N$, that is,

$$v_o = v_1 + v_2 + v_3 + \cdots + v_N$$

choose $R = R_1 = R_2 = \cdots = R_N$ and choose $R_{N+2} = (N-1)R_{N+1}$, where R_{N+2} is the feedback resistor connecting the inverting input and output and R_{N+1} is the resistor connecting the inverting input and ground.

5.3.4 DIFFERENCE AMPLIFIER

Figure 5.41 shows a difference amplifier. For an ideal op amp model, the current flowing into the op amp from the noninverting input terminal is zero. Summing the currents leaving the noninverting input, we have

FIGURE 5.41

Difference amplifier.

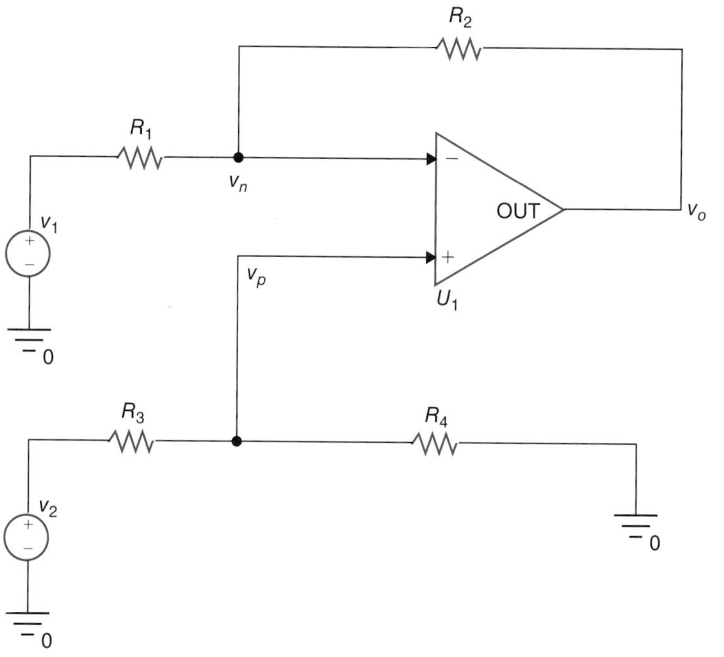

$$\frac{v_p - v_2}{R_3} + \frac{v_p - 0}{R_4} = 0$$

Solving this equation for v_p, we get

$$v_p = \frac{R_4}{R_3 + R_4} v_2$$

Alternatively, we can apply the voltage divider rule to find v_p.

Since $v_n = v_p$ (a virtual short), we have

$$v_n = \frac{R_4}{R_3 + R_4} v_2 \qquad \textbf{(5.59)}$$

Summing the currents leaving the inverting input, we obtain

$$\frac{v_n - v_1}{R_1} + \frac{v_n - v_o}{R_2} = 0$$

Solving for v_o, we get

$$v_o = \left(\frac{R_1 + R_2}{R_1}\right) v_n - \frac{R_2}{R_1} v_1 \qquad \textbf{(5.60)}$$

Substitution of Equation (5.59) into Equation (5.60) yields

$$v_o = \left(\frac{R_1 + R_2}{R_1}\right) \frac{R_4}{R_3 + R_4} v_2 - \frac{R_2}{R_1} v_1 \qquad \textbf{(5.61)}$$

If $R = R_1 = R_2 = R_3 = R_4$, the output voltage becomes

$$v_o = v_2 - v_1 \qquad \textbf{(5.62)}$$

which is the difference between v_2 and v_1.

EXAMPLE 5.9

Design a circuit for converting a unipolar binary signal (1 is represented by a pulse with 5 V, and 0 is represented by a pulse with 0 V) to a polar binary signal (1 is represented by a pulse with 5 V, and 0 is represented by a pulse with −5V).

continued

Example 5.9 continued

Let v_2 be the unipolar signal and v_o be the polar signal. Then, subtracting 2.5 V from v_2 results in a signal with 2.5 V (for 1) or −2.5 V (for 0). We can obtain the polar signal by multiplying (v_2 − 2.5 V) by 2. Thus, we have

$$v_o = 2(v_2 - 2.5)$$

Choosing $R_3 = R_1, R_4 = R_2, R_2 = 2R_1$, and $v_1 = 2.5$ V in Equation (5.61) results in

$$\begin{aligned} v_o &= \left(\frac{R_1 + R_2}{R_1}\right)\frac{R_2}{R_1 + R_2}v_2 - \frac{R_2}{R_1}v_1 = \frac{R_2}{R_1}(v_2 - v_1) \\ &= \frac{2R_1}{R_1}(v_2 - v_1) = 2(v_2 - v_1) = 2(v_2 - 2.5) \end{aligned} \quad \textbf{(5.63)}$$

In conclusion, if we choose $R_3 = R_1, R_4 = R_2, R_2 = 2R_1, v_1 = 2.5$ V, we can convert a unipolar signal to a polar signal. Figure 5.42 shows the circuit with $R_1 = 10\ k\Omega$, and Figure 5.43 shows sample waveforms for v_2 and v_o. If the amplitude is other than 5 V, change v_1 to the amplitude of the input signal divided by 2.

FIGURE 5.42

Circuit for $v_o = 2(v_2 - v_1)$.

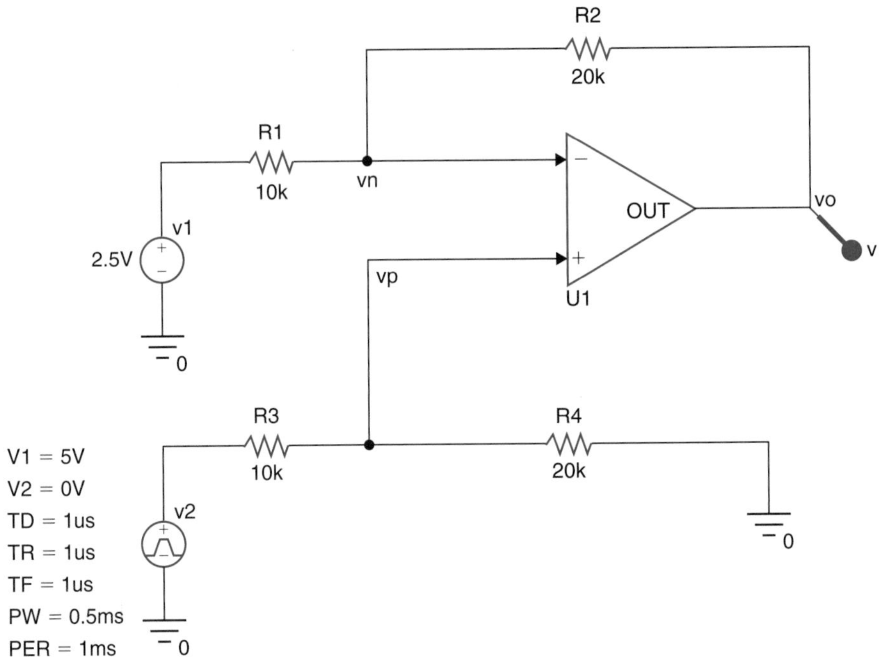

An op amp circuit can be used to generate a signal such as

$$v_o = 7v_1 - 2v_2$$

Signal v_o is the linear combination of the two input signals v_1 and v_2. We can use the superposition principle to show that the circuit shown in Figure 5.44 provides an output voltage given by $7v_1 - 2v_2$. We first deactivate voltage source v_2 by short-circuiting it. When v_2 is short-circuited, the voltage at node 1 is zero. Due to a virtual short, the voltage at the negative input of op amp 1 is zero. The voltage v_a at the output of op amp 1 is zero as well. The circuit reduces to a noninverting amplifier (Figure 5.13). The output due to v_1 is given by

$$v_{o_1} = \left(1 + \frac{R_4}{R_3}\right)v_1 = \left(1 + \frac{6\ k\Omega}{1\ k\Omega}\right)v_1 = 7v_1$$

continued

Example 5.9 continued

FIGURE 5.43

Sample waveforms.

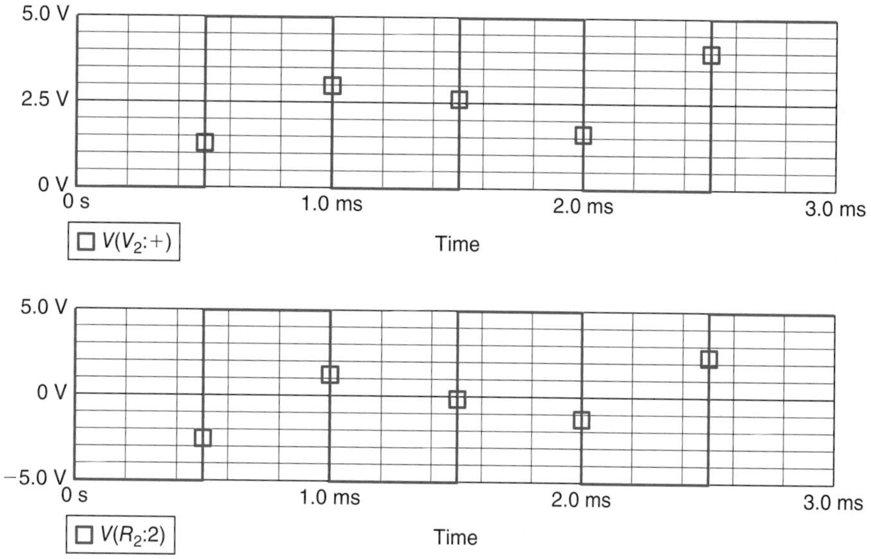

FIGURE 5.44

Circuit to implement $v_o = 7v_1 - 2v_2$.

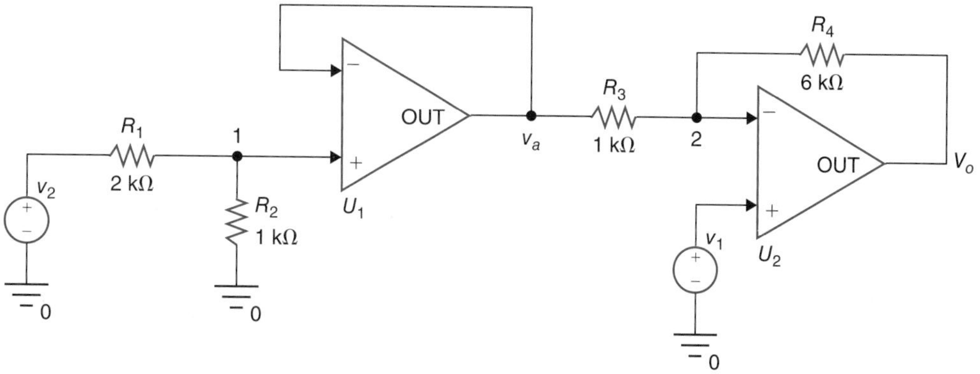

Now, we deactivate voltage source v_1 by short-circuiting it. Op amp 2 becomes an inverting amplifier (Figure 5.11). The input to the inverting amplifier is provided by the output v_a of the voltage follower. Due to a virtual short, v_a is the voltage at node 1. From the voltage divider rule, we obtain

$$v_a = \frac{R_2}{R_1 + R_2} v_2 = \frac{1\ k\Omega}{1\ k\Omega + 2\ k\Omega} v_2 = \frac{1}{3} v_2$$

The output due to v_2 is given by

$$v_{o_2} = -\frac{R_4}{R_3} \times \frac{1}{3} v_2 = -\frac{6\ k\Omega}{1\ k\Omega} \times \frac{1}{3} v_2 = -2v_2$$

Adding v_{o_1} and v_{o_2}, we get

$$v_o = v_{o_1} + v_{o_2} = 7v_1 - 2v_2$$

5.4 Instrumentation Amplifier

Figure 5.45 shows an instrumentation amplifier.

FIGURE 5.45

Instrumentation amplifier.

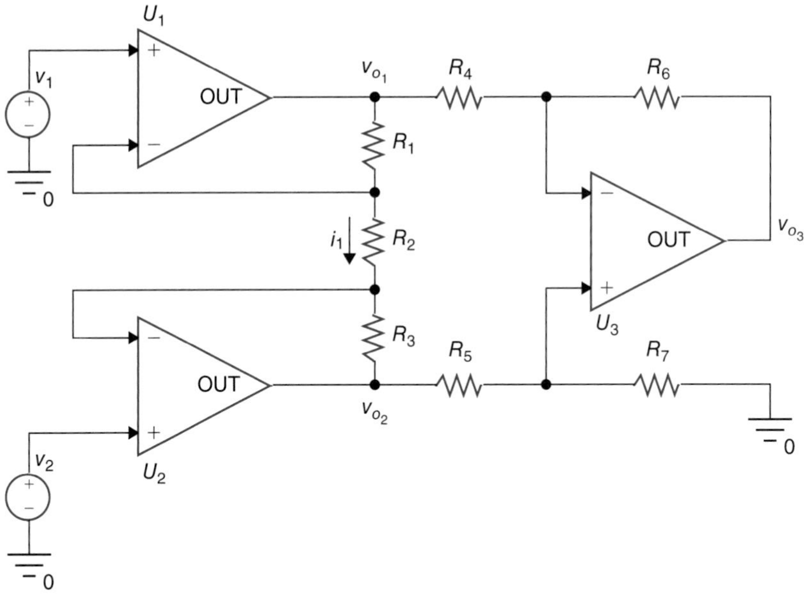

The voltage at the inverting (−) input of the op amp U_1 is v_1 due to a virtual short. Likewise, the voltage at the inverting input of the op amp U_2 is v_2. Thus, the current through R_2 (↓) is given by

$$i_1 = \frac{v_1 - v_2}{R_2} \tag{5.64}$$

Since the current into the inverting (−) input terminal of the op amp U_1 and the current into the inverting input terminal of the op amp U_2 are both zero, the current through R_1 and the current through R_3 are also i_1. The voltage at the output of the op amp U_1 is given by

$$v_{o_1} = i_1 R_1 + v_1 = \frac{R_1}{R_2}(v_1 - v_2) + v_1 = \left(\frac{R_1 + R_2}{R_2}\right)v_1 - \frac{R_1}{R_2}v_2 \tag{5.65}$$

The voltage at the output of the op amp U_2 is given by

$$v_{o_2} = -i_1 R_3 + v_2 = -\frac{R_3}{R_2}(v_1 - v_2) + v_2 = -\frac{R_3}{R_2}v_1 + \left(\frac{R_3 + R_2}{R_2}\right)v_2 \tag{5.66}$$

If $R_3 = R_1$, Equation (5.66) becomes

$$v_{o_2} = -\frac{R_1}{R_2}v_1 + \left(\frac{R_1 + R_2}{R_2}\right)v_2 \tag{5.67}$$

The difference in voltage between the two outputs is given by

$$v_{o_2} - v_{o_1} = \frac{2R_1 + R_2}{R_2}(v_2 - v_1) = \left(1 + \frac{2R_1}{R_2}\right)(v_2 - v_1) \tag{5.68}$$

The voltage drop across $R_1 - R_2 - R_3$ is given by

$$v_{o_1} - v_{o_2} = (R_1 + R_2 + R_3)i_1$$

If $R_3 = R_1$, we have

$$v_{o_1} - v_{o_2} = (2R_1 + R_2)i_1 = \frac{2R_1 + R_2}{R_2}(v_1 - v_2) = \left(1 + \frac{2R_1}{R_2}\right)(v_1 - v_2) \quad \textbf{(5.69)}$$

Since the current into the positive terminal of the op amp U_3 is zero, the voltage at the positive terminal of the op amp U_3 is given by, based on the voltage divider rule,

$$v_{p_3} = \frac{R_7}{R_5 + R_7} v_{o_2} \quad \textbf{(5.70)}$$

Since the current into the negative terminal of the op amp U_3 is zero, the voltage at the negative terminal of the op amp U_3 is given by, based on the voltage divider rule,

$$v_{n_3} = v_{o_1} + \frac{R_4}{R_4 + R_6}(v_{o_3} - v_{o_1}) = \frac{R_6}{R_4 + R_6} v_{o_1} + \frac{R_4}{R_4 + R_6} v_{o_3} \quad \textbf{(5.71)}$$

Since $v_{p_3} = v_{n_3}$, from Equations (5.70) and (5.71), we have

$$\frac{R_7}{R_5 + R_7} v_{o_2} = \frac{R_6}{R_4 + R_6} v_{o_1} + \frac{R_4}{R_4 + R_6} v_{o_3}$$

If $R_4 = R_5$ and $R_6 = R_7$, the output of the op amp U_3 is given by

$$v_{o_3} = \frac{R_6}{R_4}(v_{o_2} - v_{o_1}) = \left(1 + \frac{2R_1}{R_2}\right)\frac{R_6}{R_4}(v_2 - v_1) \quad \textbf{(5.72)}$$

Thus, the output voltage of the op amp U_3 is proportional to the difference between v_2 and v_1. The instrumentation amplifier provides large input resistances for inputs v_1 and v_2. This results in no load down by finite input resistance like the difference circuit shown in Figure 5.41.

5.5 Current Amplifier

FIGURE 5.46

Current amplifier.

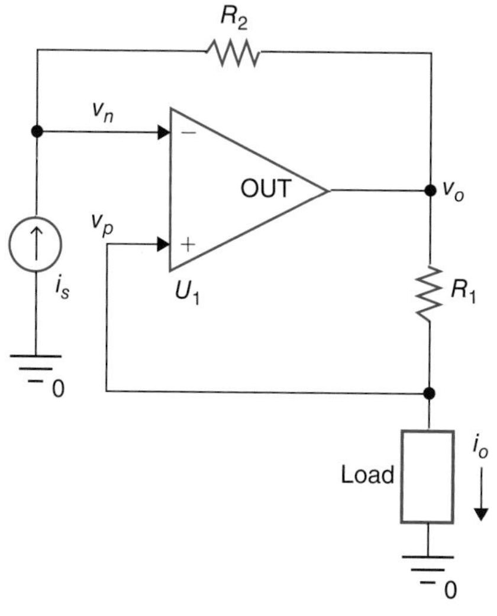

A current amplifier is shown in Figure 5.46.

The current through R_1, which is also the current through the load, is given by

$$i_o = \frac{v_o - v_p}{R_1} \quad \textbf{(5.73)}$$

The voltage at the output of the op amp is given by

$$v_o = v_n - R_2 i_s \quad \textbf{(5.74)}$$

Substitution of Equation (5.74) into Equation (5.73) yields

$$i_o = \frac{v_o - v_p}{R_1} = \frac{v_n - R_2 i_s - v_p}{R_1} \quad \textbf{(5.75)}$$

Since $v_n = v_p$ (i.e., a virtual short), Equation (5.75) reduces to

$$i_o = \frac{-R_2}{R_1} i_s \quad \textbf{(5.76)}$$

FIGURE 5.47

Another current amplifier.

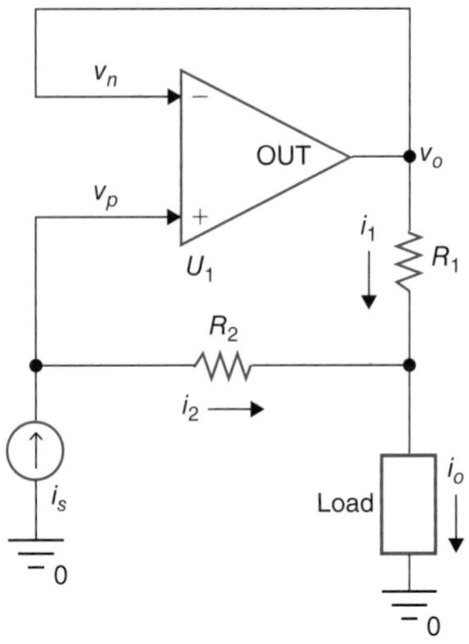

The output current is the input current multiplied by $-R_2/R_1$, independent of the value of the load resistance. The current gain is R_2/R_1 with sign inversion. If $R_2 = R_1$, $i_o = -i_s$. The circuit is called a current reverser or a current mirror.

Figure 5.47 shows another current amplifier.

Since the current flowing into the op amp from the positive input terminal is zero, the current through R_2 is i_s; that is,

$$i_2 = i_s$$

Due to a virtual short, we have $v_n = v_p = v_o$. Thus, the voltage drop across R_1 and R_2 are the same; that is,

$$R_1 i_1 = R_2 i_2$$

The current into the load is the sum of i_1 and i_2. Thus, we have

$$i_o = i_1 + i_2 = \frac{R_2}{R_1} i_2 + i_2 = \left(1 + \frac{R_2}{R_1}\right) i_2 = \left(1 + \frac{R_2}{R_1}\right) i_s \quad \textbf{(5.77)}$$

The output current is the input current multiplied by $(1 + R_2/R_1)$, independent of the value of the load resistance. The current gain is $(1 + R_2/R_1)$ without sign inversion. The current gain is always greater than or equal to 1.

FIGURE 5.48

Current-to-voltage converter.

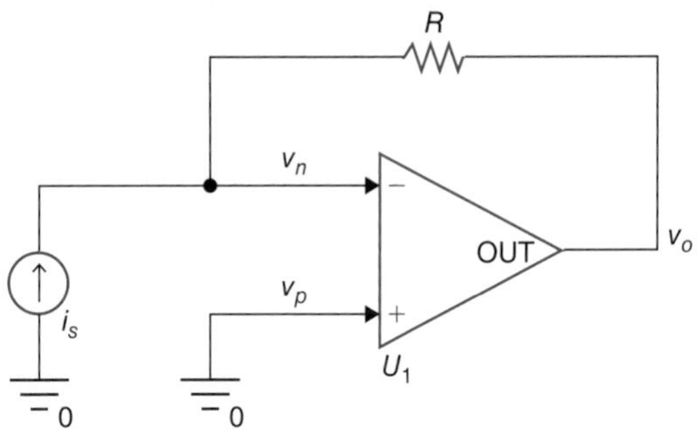

5.5.1 CURRENT TO VOLTAGE CONVERTER (TRANSRESISTANCE AMPLIFIER)

A current-to-voltage converter is shown in Figure 5.48.

Since the current flowing into the op amp from the negative input terminal is zero, the current from the current source flows through R. Thus, we have

$$v_o = -Ri_s \quad \textbf{(5.78)}$$

Another implementation of a current-to-voltage converter is shown in Figure 5.49.

Due to a virtual short, the voltage at the negative input terminal of the op amp is zero. Since the current flowing into the op amp from the negative input terminal is zero, the current from the current source flows through R_1. Thus, we have

$$v_a = -R_1 i_s$$

FIGURE 5.49

Another example of a current-to-voltage converter.

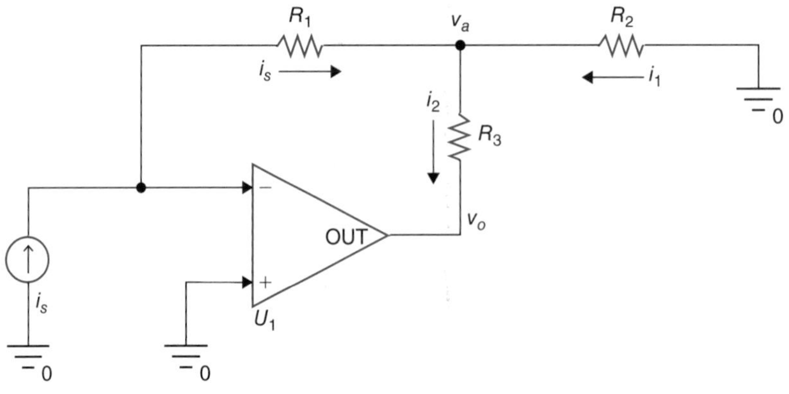

From Ohm's law, the current through R_2 is given by

$$i_1 = \frac{0 - v_a}{R_2} = \frac{R_1}{R_2} i_s$$

The current i_2 is the sum of i_s and i_1. Thus, we get

$$i_2 = i_s + i_1 = \left(1 + \frac{R_1}{R_2}\right) i_s$$

The output voltage of the op amp is given by

$$v_o = -i_2 R_3 + v_a = -R_3\left(1 + \frac{R_1}{R_2}\right)i_s - R_1 i_s = -R_1\left(1 + \frac{R_3}{R_1} + \frac{R_3}{R_2}\right)i_s \quad \textbf{(5.79)}$$

Compared to the circuit shown in Figure 5.48, assuming that $R = R_1$, the gain of the transresistance amplifier is increased by a factor of $\left(1 + \frac{R_3}{R_1} + \frac{R_3}{R_2}\right)$.

FIGURE 5.50

Negative resistance circuit.

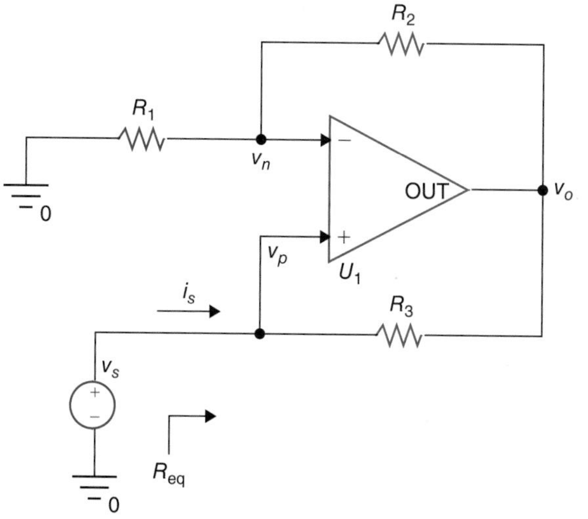

5.5.2 NEGATIVE RESISTANCE CIRCUIT

A negative resistance circuit is shown in Figure 5.50.

Summing the currents leaving the negative input of the op amp, we obtain

$$\frac{v_n - 0}{R_1} + \frac{v_n - v_o}{R_2} = 0$$

Solving for v_o, we have

$$v_o = \left(1 + \frac{R_2}{R_1}\right) v_n$$

Due to a virtual short, the voltage at the inverting input and at the noninverting input is the same as the input voltage v_s; that is,

$$v_n = v_p = v_s$$

Thus, the output voltage is given by

$$v_o = \left(1 + \frac{R_2}{R_1}\right) v_s$$

Notice that this is the output voltage of the noninverting amplifier. The output voltage v_o is greater than the input voltage v_s by $(R_2/R_1)v_s$. The current flows through R_3 from right to left. Thus, the current flowing out of the positive terminal of the voltage source is negative. The current i_s is given by

$$i_s = \frac{v_s - v_o}{R_3} = \frac{v_s - \left(1 + \frac{R_2}{R_1}\right)v_s}{R_3} = \frac{v_s - v_s - \frac{R_2}{R_1}v_s}{R_3} = -\frac{R_2}{R_1 R_3} v_s$$

The equivalent resistance of the circuit from the voltage source v_s is given by

$$R_{eq} = \frac{v_s}{i_s} = -\frac{R_1 R_3}{R_2} \tag{5.80}$$

which proves that the circuit shown in Figure 5.50 provides negative resistance.

5.5.3 VOLTAGE-TO-CURRENT CONVERTER (TRANSCONDUCTANCE AMPLIFIER)

The circuit shown in Figure 5.51 is called a Howland current pump circuit.

Due to a virtual short, we have

$$v_L = v_p = v_n$$

Summing the currents leaving node 1, we have

$$\frac{v_n}{R_1} + \frac{v_n - v_o}{R_2} = 0,$$

which can be rewritten as

$$\left(\frac{1}{R_1} + \frac{1}{R_2}\right) v_n = \frac{v_o}{R_2}$$

Solving for v_n, we obtain

$$v_n = v_p = v_L = \frac{\frac{1}{R_2}}{\frac{1}{R_1} + \frac{1}{R_2}} v_o = \frac{R_1}{R_1 + R_2} v_o \tag{5.81}$$

FIGURE 5.51

Howland current pump circuit.

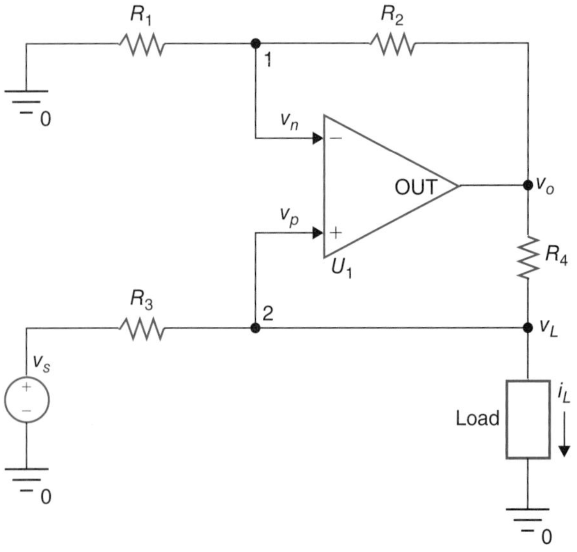

Solving Equation (5.81) for v_o, we obtain

$$v_o = \frac{R_1 + R_2}{R_1} v_n = \left(1 + \frac{R_2}{R_1}\right) v_n \tag{5.82}$$

The current i_L through the load is given by

$$\begin{aligned} i_L &= \frac{v_s - v_L}{R_3} + \frac{v_o - v_L}{R_4} = \frac{v_s - v_L}{R_3} + \frac{\frac{R_1 + R_2}{R_1} v_L - v_L}{R_4} = \frac{v_s}{R_3} - \left(\frac{1}{R_3} - \frac{R_2}{R_1 R_4}\right) v_L \\ &= \frac{v_s}{R_3} - \frac{R_1 R_4 - R_2 R_3}{R_1 R_3 R_4} v_L = \frac{v_s}{R_3} - \frac{1}{\frac{R_1 R_3 R_4}{R_1 R_4 - R_2 R_3}} v_L \\ &= \frac{v_s}{R_3} - \frac{1}{\frac{R_4}{\frac{R_1 R_4}{R_1 R_3} - \frac{R_2 R_3}{R_1 R_3}}} v_L = \frac{v_s}{R_3} - \frac{v_L}{\frac{R_4}{\frac{R_4}{R_3} - \frac{R_2}{R_1}}} \end{aligned} \tag{5.83}$$

Let

$$R_a = \frac{R_4}{\frac{R_4}{R_3} - \frac{R_2}{R_1}} \quad (5.84)$$

Then, Equation (5.83) becomes

$$i_L = \frac{v_s}{R_3} - \frac{v_L}{R_a} \quad (5.85)$$

If

$$\frac{R_4}{R_3} = \frac{R_2}{R_1} \quad (5.86)$$

the denominator of R_a becomes zero, and R_a is infinity ($R_a = \infty$). Thus, the current through the load is given by

$$i_L = \frac{v_s}{R_3} \quad (5.87)$$

The current through the load does not depend on the voltage across the load. The current depends only on the input voltage v_s. For a discussion of an improved Howland current pump circuit, refer to Sergio Franco.[1] One application of constant current source is in the light-emitting diode (LED) driver design. The current through LED is related to voltage across it exponentially. Due to variability of the current-voltage characteristics of LEDs, for the given constant voltage, the current can vary significantly from LED to LED. If a constant voltage driver is used, there are LEDs that will violate the absolute maximum current rating and compromise reliability. If a constant current driver is used, the current through LED will be constant, regardless of the current-voltage characteristics of LED.

5.6 Analysis of Inverting Configuration

Figure 5.4, earlier in this chapter, showed the inverting configuration of an op amp. When the op amp is replaced by the model shown in Figure 5.3, we obtain the circuit shown in Figure 5.5, which is repeated here in Figure 5.52.

The circuit shown in Figure 5.52 is analyzed in Section 5.2, earlier in this chapter, assuming that $R_i = \infty$ and $R_o = 0$. In this section, we analyze the circuit shown in Figure 5.52, including R_i and R_o.

FIGURE 5.52

A model for an inverting configuration.

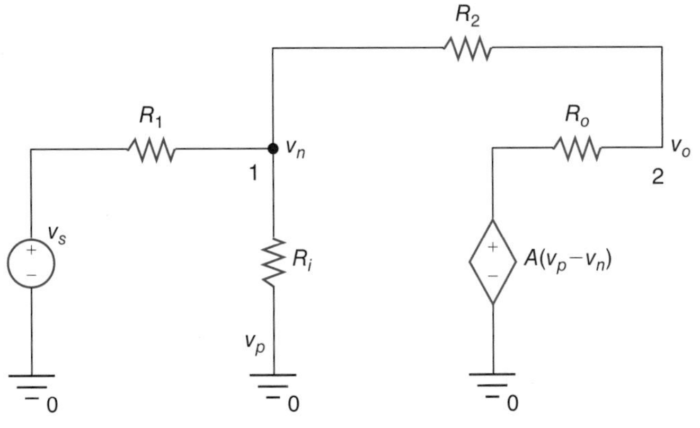

Summing the currents leaving node 2, we obtain

$$\frac{v_o - v_n}{R_2} + \frac{v_o - A(v_p - v_n)}{R_o} = 0 \quad (5.88)$$

Since $v_p = 0$, Equation (5.88) becomes

$$\frac{v_o - v_n}{R_2} + \frac{v_o + Av_n}{R_o} = 0 \quad (5.89)$$

Solving Equation (5.89) for v_n, we obtain

$$v_n = \frac{R_o + R_2}{R_o - R_2 A} v_o \quad (5.90)$$

[1] Sergio Franco, *Design with Operational Amplifiers and Analog Integrated Circuits*, McGraw-Hill, New York, 1988.

Solving Equation (5.90) for v_o, we obtain

$$v_o = \frac{R_o - R_2A}{R_o + R_2} v_n \tag{5.91}$$

Summing the currents leaving node 1, we obtain

$$\frac{v_n - v_s}{R_1} + \frac{v_n}{R_i} + \frac{v_n - v_o}{R_2} = 0 \tag{5.92}$$

which can be revised as

$$\left(\frac{1}{R_1} + \frac{1}{R_i} + \frac{1}{R_2}\right) v_n - \frac{v_o}{R_2} = \frac{v_s}{R_1} \tag{5.93}$$

Substitution of Equation (5.91) into Equation (5.93) yields

$$\left(\frac{R_2R_i + R_1R_2 + R_1R_i}{R_1R_2R_i} - \frac{1}{R_2}\frac{R_o - R_2A}{R_o + R_2}\right) v_n = \frac{v_s}{R_1} \tag{5.94}$$

Solving Equation (5.94) for v_n, we obtain

$$\begin{aligned} v_n &= \frac{\dfrac{v_s}{R_1}}{\dfrac{R_2R_i + R_1R_2 + R_1R_i}{R_1R_2R_i} - \dfrac{1}{R_2}\dfrac{R_o - R_2A}{R_o + R_2}} \\ &= \frac{R_2R_i(R_o + R_2)v_s}{(R_2R_i + R_1R_2 + R_1R_i)(R_o + R_2) - R_1R_i(R_o - R_2A)} \\ &= \frac{R_i(R_o + R_2)v_s}{R_oR_i + R_oR_1 + R_iR_2 + R_1R_2 + R_1R_i + AR_1R_i} \end{aligned} \tag{5.95}$$

Substitution of Equation (5.95) into Equation (5.91) yields

$$v_o = \frac{-R_i(-R_o + R_2A)v_s}{R_oR_i + R_oR_1 + R_iR_2 + R_1R_2 + R_1R_i + AR_1R_i} \tag{5.96}$$

Dividing every term in Equations (5.95) and (5.96) by R_i, we obtain

$$v_n = \frac{(R_o + R_2)v_s}{R_o + \dfrac{R_oR_1}{R_i} + R_2 + \dfrac{R_1R_2}{R_i} + R_1 + AR_1} \tag{5.97}$$

$$v_o = \frac{-(-R_o + R_2A)v_s}{R_o + \dfrac{R_oR_1}{R_i} + R_2 + \dfrac{R_1R_2}{R_i} + R_1 + AR_1} \tag{5.98}$$

If $R_o = 0$ and $R_i = \infty$, Equations (5.97) and (5.98) become, respectively,

$$v_n = \frac{R_2v_s}{R_1 + R_2 + AR_1} = \frac{\dfrac{R_2}{A}v_s}{\dfrac{R_1 + R_2}{A} + R_1} \tag{5.99}$$

$$v_o = \frac{-R_2 A v_s}{R_1 + R_2 + AR_1} = \frac{-R_2 v_s}{\dfrac{R_1 + R_2}{A} + R_1} \tag{5.100}$$

These equations are identical to Equations (5.6) and (5.5), respectively. Since A $\gg$ $(R_1 + R_2)$, Equation (5.99) reduces to

$$v_n \cong \frac{R_2}{R_1 A} v_s \approx 0 \tag{5.101}$$

Equation (5.101) is identical to Equation (5.8). Since A $\gg$ $(R_1 + R_2)$, Equation (5.100) reduces to

$$v_o \cong -\frac{R_2}{R_1} v_s \tag{5.102}$$

Equation (5.102) is identical to Equation (5.7).

Alternatively, if every term in Equations (5.95) and (5.96) is divided by $R_i A$, we obtain

$$v_n = \frac{\dfrac{R_o + R_2}{A} v_s}{\dfrac{R_o}{A} + \dfrac{R_o R_1}{R_i A} + \dfrac{R_2}{A} + \dfrac{R_1 R_2}{R_i A} + \dfrac{R_1}{A} + R_1} \tag{5.103}$$

$$v_o = \frac{-\left(-\dfrac{R_o}{A} + R_2\right) v_s}{\dfrac{R_o}{A} + \dfrac{R_o R_1}{R_i A} + \dfrac{R_2}{A} + \dfrac{R_1 R_2}{R_i A} + \dfrac{R_1}{A} + R_1} \tag{5.104}$$

In typical op amps, the input resistance is large ($R_i \gg 1, R_i \gg R_1, R_i \gg R_2$), the output resistance is small ($R_o \ll R_i$), and the unloaded voltage gain is large ($A \gg 1$, $A \gg R_1$, $A \gg R_2$). If we ignore the terms that are divided by A or $R_i A$, Equations (5.103) and (5.104) become, respectively,

$$v_n \cong \frac{R_2}{R_1 A} v_s \approx 0 \tag{5.105}$$

$$v_o = -\frac{R_2}{R_1} v_s \tag{5.106}$$

The current flowing into the op amp from the negative input terminal (node 1) is obtained by dividing v_n by R_i:

$$i_n = \frac{v_n}{R_i} = \frac{(R_o + R_2) v_s}{R_o R_i + R_o R_1 + R_i R_2 + R_1 R_2 + R_1 R_i + A R_1 R_i} \tag{5.107}$$

If we divide each term of Equation (5.107) by AR_i, we have

$$i_n = \frac{v_n}{R_i} = \frac{\dfrac{R_o + R_2}{AR_i}v_s}{\dfrac{R_oR_i}{AR_i} + \dfrac{R_oR_1}{AR_i} + \dfrac{R_iR_2}{AR_i} + \dfrac{R_1R_2}{AR_i} + \dfrac{R_1R_i}{AR_i} + R_1} = \frac{\dfrac{R_o + R_2}{AR_i}v_s}{\dfrac{R_o}{A} + \dfrac{R_oR_1}{AR_i} + \dfrac{R_2}{A} + \dfrac{R_1R_2}{AR_i} + \dfrac{R_1}{A} + R_1} \tag{5.108}$$

From Equation (5.108), we get

$$i_n = \frac{v_n}{R_i} \cong \frac{(R_o + R_2)v_s}{R_1AR_i} \cong \frac{R_2v_s}{R_1AR_i} \approx 0 \tag{5.109}$$

5.6.1 INPUT RESISTANCE

The current through R_1, from left to right, is given by

$$i_1 = \frac{v_s - v_n}{R_1} = \frac{v_s - \dfrac{R_i(R_o + R_2)v_s}{R_oR_i + R_oR_1 + R_iR_2 + R_1R_2 + R_1R_i + AR_1R_i}}{R_1}$$

$$= \frac{\dfrac{R_oR_i + R_oR_1 + R_iR_2 + R_1R_2 + R_1R_i + AR_1R_i - R_iR_o - R_iR_2}{R_oR_i + R_oR_1 + R_iR_2 + R_1R_2 + R_1R_i + AR_1R_i}v_s}{R_1}$$

$$= \frac{\dfrac{R_oR_1 + R_1R_2 + R_1R_i + AR_1R_i}{R_oR_i + R_oR_1 + R_iR_2 + R_1R_2 + R_1R_i + AR_1R_i}v_s}{R_1}$$

$$= \frac{R_o + R_2 + R_i + AR_i}{R_oR_i + R_oR_1 + R_iR_2 + R_1R_2 + R_1R_i + AR_1R_i}v_s$$

The input resistance R_{in} is given by v_s/i_1:

$$R_{in} = \frac{v_s}{i_1} = \frac{R_oR_i + R_oR_1 + R_iR_2 + R_1R_2 + R_1R_i + AR_1R_i}{R_o + R_2 + R_i + AR_i} \tag{5.110}$$

Dividing every term by R_iA, we get

$$R_{in} = \frac{v_S}{i_1} = \frac{\dfrac{R_o}{A} + \dfrac{R_oR_1}{AR_i} + \dfrac{R_2}{A} + \dfrac{R_1R_2}{AR_i} + \dfrac{R_1}{A} + R_1}{\dfrac{R_o}{AR_i} + \dfrac{R_2}{AR_i} + \dfrac{1}{A} + 1} \approx R_1 \tag{5.111}$$

The input resistance of the inverting configuration is R_1. The noninverting configuration has a higher value of input resistance.

5.6.2 OUTPUT RESISTANCE

Deactivate v_s (i.e., short-circuit it) and apply test voltage v_t to the output of the op amp. The circuit is shown in Figure 5.53.

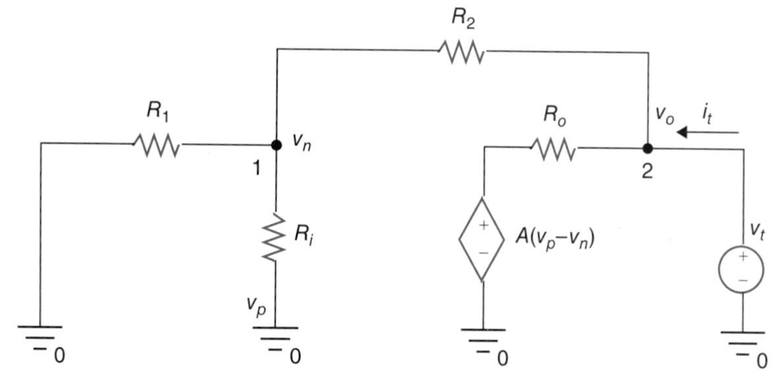

FIGURE 5.53
Circuit for finding the output resistance.

From the voltage divider rule, we obtain

$$v_n = \frac{\frac{R_1R_i}{R_1+R_i}}{R_2 + \frac{R_1R_i}{R_1+R_i}}v_t = \frac{R_1R_i}{R_1R_2+R_2R_i+R_1R_i}v_t \tag{5.112}$$

The current flowing out of the test voltage source is given by

$$i_t = \frac{v_t - v_n}{R_2} + \frac{v_t - A(0-v_n)}{R_o} = \frac{v_t - \frac{R_1R_i}{R_1R_2+R_2R_i+R_1R_i}v_t}{R_2} + \frac{v_t + A\frac{R_1R_i}{R_1R_2+R_2R_i+R_1R_i}v_t}{R_o} \tag{5.113}$$

The output resistance is the ratio of v_t to i_t:

$$R_{out} = \frac{v_t}{i_t} = \frac{1}{\frac{1-\frac{R_1R_i}{R_1R_2+R_2R_i+R_1R_i}}{R_2} + \frac{1+A\frac{R_1R_i}{R_1R_2+R_2R_i+R_1R_i}}{R_o}} \tag{5.114}$$

Equation (5.114) can be simplified to

$$R_{out} = \frac{R_o(R_1R_2+R_2R_i+R_1R_i)}{R_oR_1+R_oR_i+AR_1R_i+R_1R_2+R_2R_i+R_1R_i} \tag{5.115}$$

If we keep only the AR_1R_i term from the denominator and ignore the R_1R_2 term from the numerator of Equation (5.115), we obtain

$$R_{out} \cong \frac{R_o(R_2+R_1)}{AR_1} \approx 0 \tag{5.116}$$

The output resistance is close to zero for the inverting configuration.

EXAMPLE 5.10

Find V_o in the circuit shown in Figure 5.54.

FIGURE 5.54

Circuit for EXAMPLE 5.10.

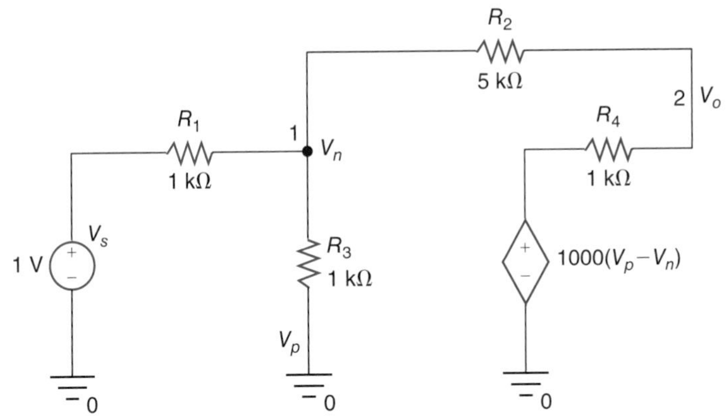

Summing the currents leaving node 1, we obtain

$$\frac{V_n - 1}{1000} + \frac{V_n}{1000} + \frac{V_n - V_o}{5000} = 0$$

Multiplication by 5000 yields

$$5V_n - 5 + 5V_n + V_n - V_o = 0$$

which can be simplified to

$$11V_n - V_o = 5 \quad \textbf{(5.117)}$$

Summing the currents leaving node 2, we obtain

$$\frac{V_o - V_n}{5000} + \frac{V_o - 1000(0 - V_n)}{1000} = 0$$

Multiplication by 5000 yields

$$V_o - V_n + 5V_o + 5000V_n = 0$$

which can be simplified to

$$4999V_n + 6V_o = 0 \quad \textbf{(5.118)}$$

Multiplication of Equation (5.117) by 6 yields

$$66V_n - 6V_o = 30 \quad \textbf{(5.119)}$$

Adding Equations (5.118) and (5.119), we get

$$5065V_n = 30$$

continued

Example 5.10 continued

Thus, we have

$$V_n = 30/5065 = 0.005923 \text{ V}$$

From Equation (5.117), we obtain

$$V_o = 11V_n - 5 = 330/5065 - 5 = -4.934847 \text{ V}$$

For an ideal op amp, $V_o = -5$ V. Since the circuit shown in Figure 5.54 is not ideal (R_i (= R_3) is small, A is small, and R_o (= R_4) is large), the output voltage is not -5 V.

MATLAB

```
% EXAMPLE 5.10
clear all;
R1=1000;R2=5000;R3=1000;R4=1000;Vs=1;A=1000;
syms Vn Vo
[Vn,Vo]=solve((Vn-Vs)/R1+Vn/R3+(Vn-Vo)/R2,...
(Vo-Vn)/R2+(Vo-A*(0-Vn))/R4,Vn,Vo);
Vn=vpa(Vn,11)
Vo=vpa(Vo,11)

Answers:
Vn =
0.0059230009872
Vo =
-4.9348469891
```

Exercise 5.8

Find V_n and I_1 in the circuit shown in Figure 5.55.

FIGURE 5.55

Circuit for EXERCISE 5.8.

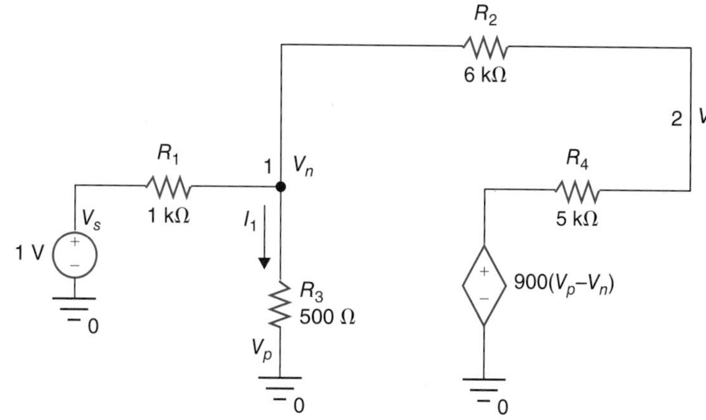

Answer:
$V_n = 11.7773$ mV, $I_1 = 23.5546\ \mu$A.

5.7 Analysis of Noninverting Configuration

Figure 5.7, earlier in this chapter, showed a noninverting configuration of an op amp. When the op amp is replaced by the model shown in Figure 5.3, we obtain the circuit shown in Figure 5.8, which is repeated in Figure 5.56.

FIGURE 5.56

Model of a noninverting configuration.

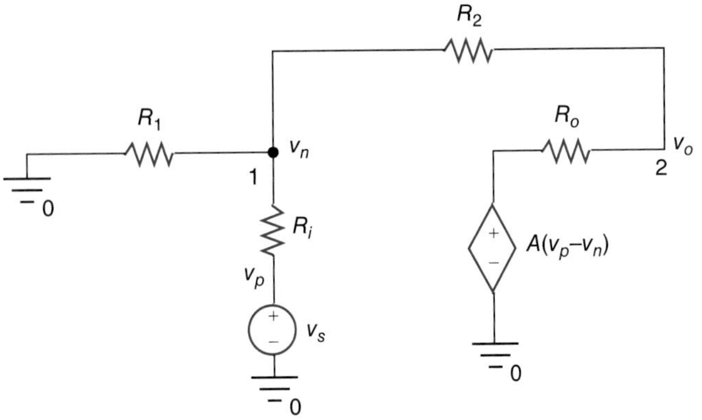

Summing the currents leaving node 2, we obtain

$$\frac{v_o - v_n}{R_2} + \frac{v_o - A(v_p - v_n)}{R_o} = 0 \tag{5.120}$$

Since $v_p = v_s$, Equation (5.120) becomes

$$\frac{v_o - v_n}{R_2} + \frac{v_o - A(v_s - v_n)}{R_o} = 0 \tag{5.121}$$

Summing the currents leaving node 1, we obtain

$$\frac{v_n}{R_1} + \frac{v_n - v_s}{R_i} + \frac{v_n - v_o}{R_2} = 0 \tag{5.122}$$

which can be revised as

$$\left(\frac{1}{R_1} + \frac{1}{R_i} + \frac{1}{R_2}\right)v_n = \frac{v_s}{R_i} + \frac{v_o}{R_2} \tag{5.123}$$

Solving Equations (5.121) and (5.123), we have

$$v_n = \frac{R_1(R_o + R_2 + AR_i)v_s}{R_oR_i + R_oR_1 + R_iR_2 + R_1R_2 + R_1R_i + AR_1R_i} \tag{5.124}$$

$$v_o = \frac{(R_oR_1 + R_1R_iA + R_2R_iA)v_s}{R_oR_i + R_oR_1 + R_iR_2 + R_1R_2 + R_1R_i + AR_1R_i} \tag{5.125}$$

If each term of Equations (5.124) and (5.125) is divided by R_i, we get

$$v_n = \frac{R_1\left(\frac{R_o}{R_i} + \frac{R_2}{R_i} + A\right)v_s}{R_o + \frac{R_oR_1}{R_i} + R_2 + \frac{R_1R_2}{R_i} + R_1 + AR_1} \tag{5.126}$$

$$v_o = \frac{\left(\frac{R_oR_1}{R_i} + R_1A + R_2A\right)v_s}{R_o + \frac{R_oR_1}{R_i} + R_2 + \frac{R_1R_2}{R_i} + R_1 + AR_1} \tag{5.127}$$

If $R_o = 0$ and $R_i = \infty$, Equations (5.126) and (5.127) reduce, respectively, to

$$v_n = \frac{AR_1v_s}{R_1 + R_2 + AR_1} = \frac{R_1v_s}{\frac{R_1 + R_2}{A} + R_1} \tag{5.128}$$

$$v_o = \frac{A(R_1 + R_2)v_s}{R_1 + R_2 + AR_1} = \frac{(R_1 + R_2)v_s}{\frac{R_1 + R_2}{A} + R_1} \tag{5.129}$$

These equations are identical to Equations (5.16) and (5.15), respectively. Since $A \gg (R_1 + R_2)$, Equation (5.129) reduces to

$$v_o \cong \frac{R_1 + R_2}{R_1} v_s = \left(1 + \frac{R_2}{R_1}\right) v_s \tag{5.130}$$

which is identical to Equation (5.17). Since $A \gg (R_1 + R_2)$, Equation (5.128) reduces to

$$v_n = \frac{R_1}{\dfrac{R_1 + R_2}{A} + R_1} v_s \cong v_s \tag{5.131}$$

which is identical to Equation (5.18).

Alternatively, if every term in Equations (5.124) and (5.125) is divided by R_iA, we obtain

$$v_n = \frac{R_1\left(\dfrac{R_o + R_2}{R_iA} + 1\right) v_s}{\dfrac{R_o}{A} + \dfrac{R_oR_1}{R_iA} + \dfrac{R_2}{A} + \dfrac{R_1R_2}{R_iA} + \dfrac{R_1}{A} + R_1} \tag{5.132}$$

$$v_o = \frac{\left(\dfrac{R_oR_1}{R_iA} + R_1 + R_2\right) v_s}{\dfrac{R_o}{A} + \dfrac{R_oR_1}{R_iA} + \dfrac{R_2}{A} + \dfrac{R_1R_2}{R_iA} + \dfrac{R_1}{A} + R_1} \tag{5.133}$$

The current flowing into the op amp from the negative input terminal (node 1) is obtained by dividing $v_n - v_s$ by R_i:

$$i_n = \frac{v_n - v_s}{R_i} = \frac{(R_oR_1 + R_1R_2 + AR_iR_1)v_s}{R_i(R_oR_i + R_oR_1 + R_iR_2 + R_1R_2 + R_1R_i + AR_1R_i)} - \frac{v_s}{R_i} \tag{5.134}$$

If we divide each term by AR_i, we have

$$i_n = \frac{v_n - v_s}{R_i} = \frac{\left(\dfrac{R_oR_1}{AR_i} + \dfrac{R_1R_2}{AR_i} + R_1\right) v_s}{R_i\left(\dfrac{R_o}{A} + \dfrac{R_oR_1}{AR_i} + \dfrac{R_2}{A} + \dfrac{R_1R_2}{AR_i} + \dfrac{R_1}{A} + R_1\right)} - \frac{v_s}{R_i} \tag{5.135}$$

In typical op amps, the input resistance is large ($R_i \gg 1$), the output resistance is small ($R_o \ll R_i$), and the unloaded voltage gain is large ($A \gg 1, A \gg R_1, A \gg R_2$). If we ignore the terms that are divided by A or R_iA, Equations (5.132), (5.133), and (5.135) become, respectively,

$$v_n = v_s \tag{5.136}$$

$$v_o = \frac{R_1 + R_2}{R_1} v_s = \left(1 + \frac{R_2}{R_1}\right) v_s \tag{5.137}$$

$$i_n = \frac{v_s}{R_i} - \frac{v_s}{R_i} = 0 \tag{5.138}$$

5.7.1 INPUT RESISTANCE

The current through R_i, from bottom to top, is given by

$$i_i = \frac{v_s - v_n}{R_i} = \frac{v_s - \dfrac{R_1R_o + R_1R_2 + R_1AR_i}{R_oR_i + R_oR_1 + R_iR_2 + R_1R_2 + R_1R_i + AR_1R_i}v_s}{R_i}$$

$$= \frac{\dfrac{R_oR_i + R_oR_1 + R_iR_2 + R_1R_2 + R_1R_i + AR_1R_i - R_1R_o - R_1R_2 - R_1AR_i}{R_oR_i + R_oR_1 + R_iR_2 + R_1R_2 + R_1R_i + AR_1R_i}}{R_i}v_s$$

$$= \frac{R_o + R_2 + R_1}{R_oR_i + R_oR_1 + R_iR_2 + R_1R_2 + R_1R_i + AR_1R_i}v_s$$

The input resistance R_{in} is given by v_s/i_i:

$$R_{in} = \frac{v_s}{i_i} = \frac{R_oR_i + R_oR_1 + R_iR_2 + R_1R_2 + R_1R_i + AR_1R_i}{R_o + R_2 + R_1} \tag{5.139}$$

Approximate input resistance is given by

$$R_{in} \simeq \frac{AR_iR_1}{R_2 + R_1} \tag{5.140}$$

Due to the multiplication of A and R_i in the numerator, the input resistance is large compared to R_1 for the inverting configuration.

5.7.2 OUTPUT RESISTANCE

Deactivate v_s (i.e., short-circuit it) and apply test voltage v_t to the output of the op amp. The circuit is shown in Figure 5.57.

From the voltage divider rule, we can find voltage v_n at node 1:

$$v_n = \frac{\dfrac{R_1R_i}{R_1 + R_i}}{R_2 + \dfrac{R_1R_i}{R_1 + R_i}}v_t = \frac{R_1R_i}{R_1R_2 + R_2R_i + R_1R_i}v_t$$

The current flowing out of the test voltage source is given by

$$i_t = \frac{v_t - v_n}{R_2} + \frac{v_t - A(0 - v_n)}{R_o} = \frac{v_t - \dfrac{R_1R_i}{R_1R_2 + R_2R_i + R_1R_i}v_t}{R_2} + \frac{v_t + A\dfrac{R_1R_i}{R_1R_2 + R_2R_i + R_1R_i}v_t}{R_o}$$

The output resistance is the ratio of v_t to i_t:

$$R_{out} = \frac{v_t}{i_t} = \frac{1}{\dfrac{1 - \dfrac{R_1R_i}{R_1R_2 + R_2R_i + R_1R_i}}{R_2} + \dfrac{1 + A\dfrac{R_1R_i}{R_1R_2 + R_2R_i + R_1R_i}}{R_o}}$$

which can be simplified to

$$R_{out} = \frac{R_o(R_1R_2 + R_2R_i + R_1R_i)}{R_oR_1 + R_oR_i + AR_1R_i + R_1R_2 + R_2R_i + R_1R_i} \quad (5.141)$$

If the largest terms are kept, Equation (5.141) becomes

$$R_{out} \cong \frac{R_o(R_2 + R_1)R_i}{AR_1R_i} = \frac{R_o(R_2 + R_1)}{AR_1} \approx 0 \quad (5.142)$$

The output resistance is close to zero.

FIGURE 5.57

A circuit with test voltage.

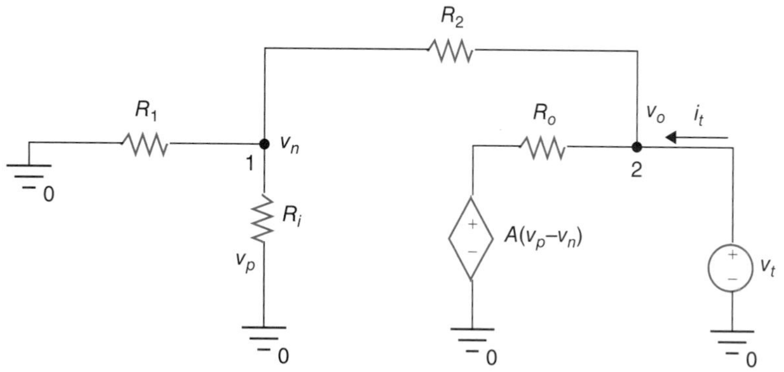

EXAMPLE 5.11

Find V_o in the circuit shown in Figure 5.58.

FIGURE 5.58

Circuit for EXAMPLE 5.11.

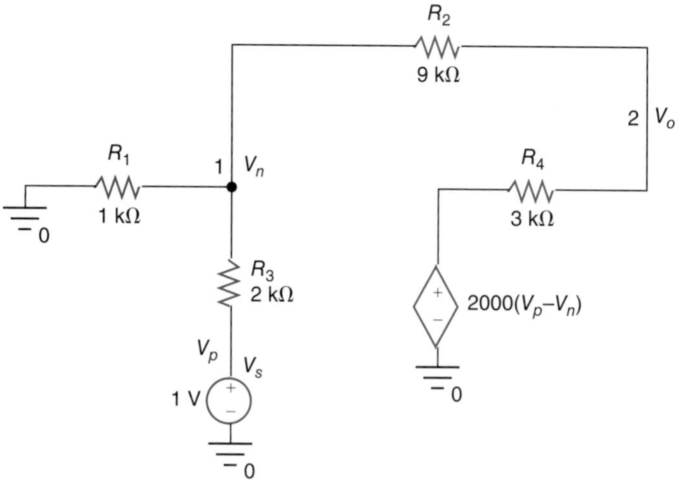

Summing the currents leaving node 1, we obtain

$$\frac{V_n - 1}{2000} + \frac{V_n}{1000} + \frac{V_n - V_o}{9000} = 0$$

Multiplication by 18,000 yields

$$9V_n - 9 + 18V_n + 2V_n - 2V_o = 0$$

which can be simplified to

$$29V_n - 2V_o = 9 \quad (5.143)$$

Summing the currents leaving node 2, we obtain

$$\frac{V_o - V_n}{9000} + \frac{V_o - 2000(1 - V_n)}{3000} = 0$$

Multiplication by 9000 yields

$$V_o - V_n + 3V_o - 6000 + 6000V_n = 0$$

continued

Example 5.11 continued

which can be simplified to

$$5999V_n + 4V_o = 6000 \tag{5.144}$$

Multiplying Equation (5.143) by 2, we obtain

$$58V_n - 4V_o = 18$$

Adding this equation to Equation (5.144), we obtain

$$6057V_n = 6018$$

Thus,

$$V_n = \frac{6018}{6057} = 0.99356117\text{ V}$$

From Equation (5.143), we get

$$V_o = \frac{29}{2}V_n - \frac{9}{2} = 9.90664\text{ V}$$

Alternatively, application of Cramer's rule to Equations (5.143) and (5.144) yields

$$V_o = \frac{\begin{vmatrix} 29 & 9 \\ 5999 & 6000 \end{vmatrix}}{\begin{vmatrix} 29 & -2 \\ 5999 & 4 \end{vmatrix}} = 9.90664\text{ V}$$

For an ideal op amp, $V_o = 10$ V. Since the circuit shown in Figure 5.58 is not ideal (R_i (= R_3) is small, A is small, and R_o (= R_4) is large), the output voltage is not 10 V.

MATLAB

```
% EXAMPLE 5.11
clear all;
R1=1000;R2=9000;R3=2000;R4=3000;Vs=1;A=2000;
Vp=Vs;
syms Vn Vo
[Vn,Vo]=solve(Vn/R1+(Vn-Vs)/R3+(Vn-Vo)/R2,...
(Vo-Vn)/R2+(Vo-A*(Vp-Vn))/R4,Vn,Vo);
Vn=vpa(Vn,11)
Vo=vpa(Vo,11)

Answers:
Vn =
0.9935611689
Vo =
9.906636949
```

Exercise 5.9

Find V_n and I_1 in the circuit shown in Figure 5.59.

FIGURE 5.59

Circuit for EXERCISE 5.9.

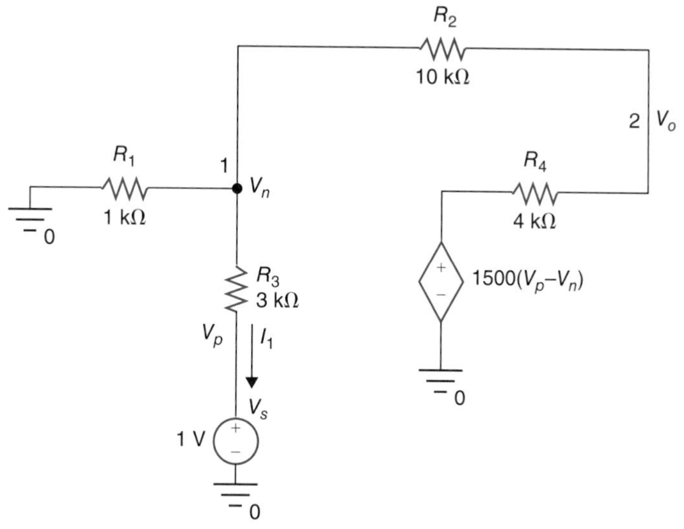

Answer:
$V_n = 0.99013$ V, $I_1 = -3.2902\ \mu$A.

5.8 PSpice and Simulink

The PSpice schematic for the inverting configuration of μA741 op amp is shown in Figure 5.60. Figure 5.61 shows the simulation settings that enable voltage gain, input resistance, and output resistance, as well as the results of the simulation.

FIGURE 5.60

The PSpice schematic for the μA741 op amp in an inverting configuration.

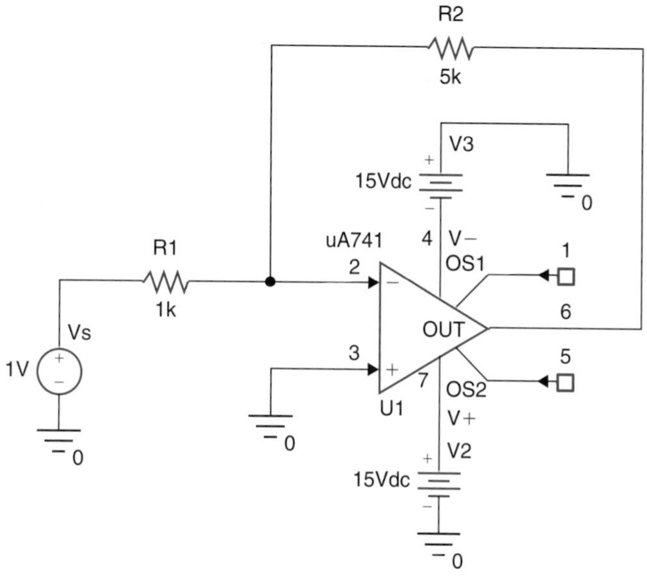

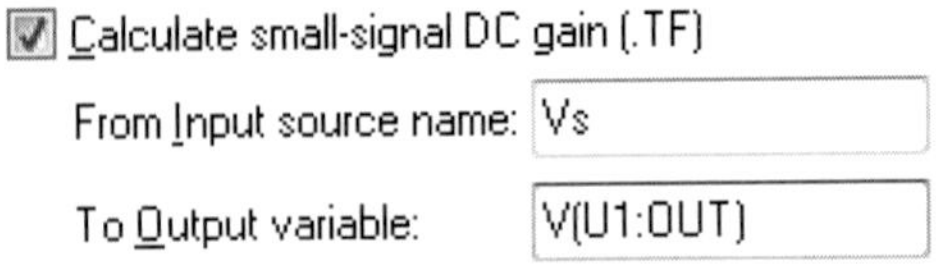

FIGURE 5.61

Simulation settings for the transfer function, and the result of the simulation. (*Source: OrCAD PSpice by Cadence*)

```
****     SMALL-SIGNAL CHARACTERISTICS
      V(N00212)/V_Vs = -5.000E+00
      INPUT RESISTANCE AT V_Vs =  1.000E+03
      OUTPUT RESISTANCE AT V(N00212) = 5.000E+03
```

The PSpice schematic for the inverting configuration of the μA741 op amp is shown in Figure 5.62. Simulation settings for DC sweep analysis are shown in Figure 5.63. The input voltage is swept from -5 V to 5 V at an interval of 0.1 V. Since the gain of the op amp is -5, the output changes from 25 V to -25 V. However, the output cannot exceed the power supply voltages. The output is limited between -15 V and 15 V. Beyond this range, the voltage stays at -15 V or 15 V. The saturation of the output voltage is shown in Figure 5.64.

FIGURE 5.62

Schematic for the μA741 op amp.

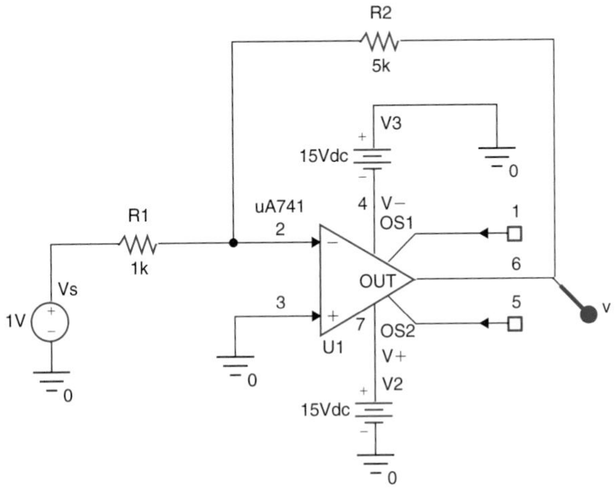

FIGURE 5.63

Simulation settings for DC sweep. (*Source: OrCAD PSpice by Cadence*)

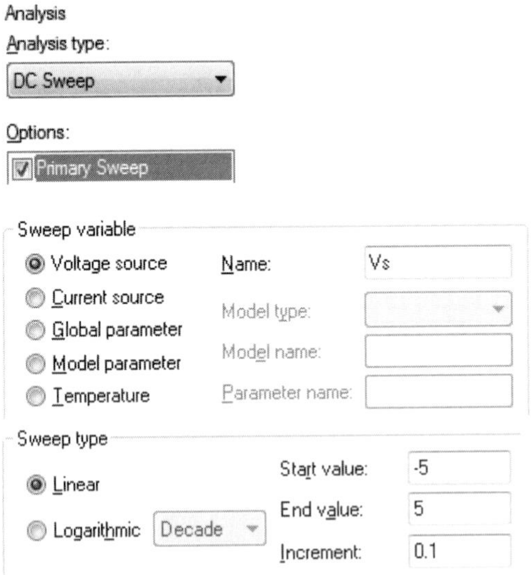

FIGURE 5.64

The voltage at the output of the op amp.

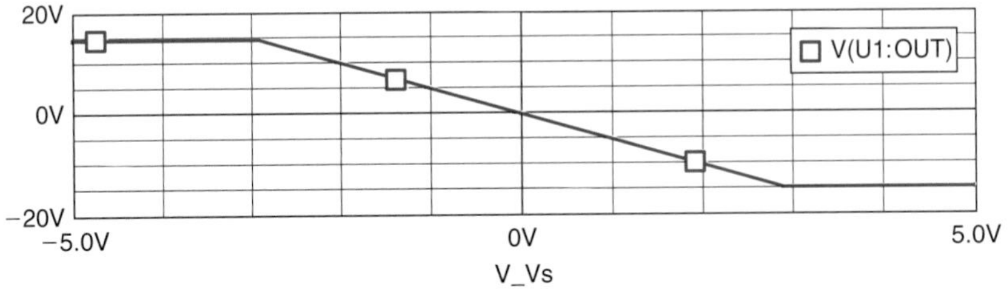

FIGURE 5.65

The PSpice model for a noninverting amplifier using VCVS.

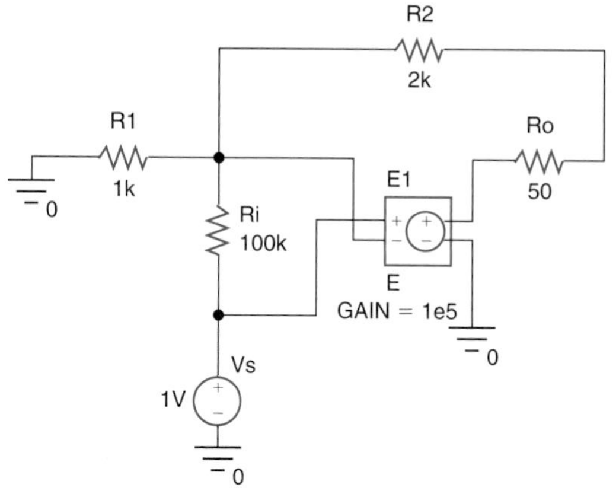

The PSpice model for a noninverting configuration using VCVS is shown in Figure 5.65. The results from the small-signal DC gain are as follows:

```
****      SMALL-SIGNAL CHARACTERISTICS
     V(N00248)/V_Vs = 3.000E+00
     INPUT RESISTANCE AT V_Vs = 3.279E+09
     OUTPUT RESISTANCE AT V(N00248) = 1.510E-03
```

In the Wheatstone bridge circuit shown in Figure 5.66, R_4 represents a sensor. When the bridge is balanced, the resistance values are: $R_1 = 1\ k\Omega$, $R_2 = 1\ k\Omega$, $R_3 = 1\ k\Omega$, $R_4 = 1\ k\Omega$. Under certain physical condition, the resistance value of R_4 increases by $\Delta R = 10\ \Omega$ from $1\ k\Omega$ to $1.010\ k\Omega$. According to Equation (2.54), the voltage difference $v_1 - v_2$ is given by

$$v_o = v_1 - v_2 = \left(\frac{R_2}{R_1 + R_2} - \frac{R_4 + \Delta R}{R_3 + R_4 + \Delta R}\right) \times V_s$$

$$= \left(\frac{1000}{1000 + 1000} - \frac{1010}{1000 + 1010}\right) \times 5\,\text{V} = -0.012437810945274\ \text{V}$$

Thus, the voltage difference $v_2 - v_1$ is given by

$$-v_o = v_2 - v_1 = 0.012437810945274\ \text{V}$$

According to Equation (2.56), the voltage difference $v_2 - v_1$ can be approximated by

$$-v_o = v_2 - v_1 \approx \frac{+\Delta R \times V_s}{4R} = \frac{+10 \times 5}{4000}\,\text{V} = 0.0125\ \text{V}$$

The voltage difference $v_2 - v_1$ is amplified by an instrumentation amplifier as shown in Figure 5.66. According to Equation (5.72), the output of the instrumentation amplifier is given by

$$v_{o_3} = \frac{R_{10}}{R_8}(v_{o_2} - v_{o_1}) = \left(1 + \frac{2R_5}{R_6}\right)\frac{R_{10}}{R_8}(v_2 - v_1)$$

$$= \left(1 + \frac{2 \times 2\ k\Omega}{1\ k\Omega}\right)\frac{4\ k\Omega}{1\ k\Omega}(0.012437811) = 0.24875622\ \text{V}$$

FIGURE 5.66

Wheatstone bridge circuit connected to an instrumentation amplifier.

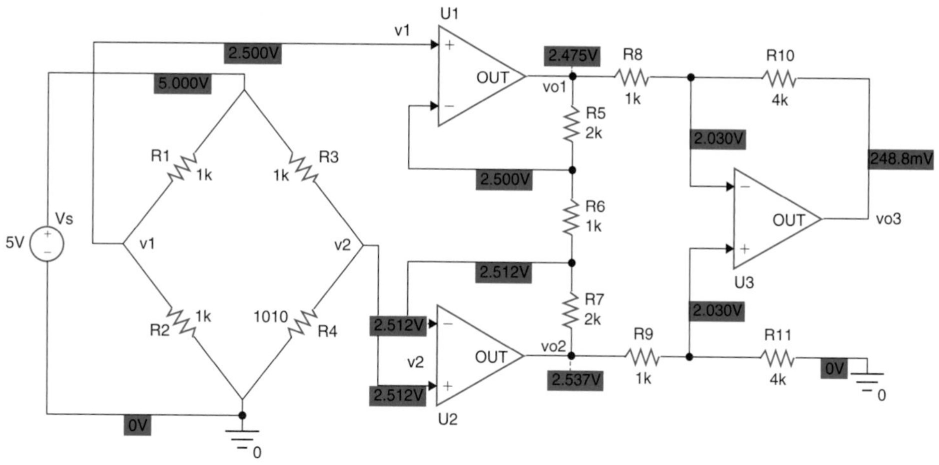

When the approximation 0.0125 V is amplified by the same instrumentation amplifier, we get

$$v_{o_{3b}} = \left(1 + \frac{2 \times 2\,k\Omega}{1\,k\Omega}\right)\frac{4\,k\Omega}{1\,k\Omega}(0.0125) = 0.25\,\text{V}$$

The percent error between approximation and exact output is

$$\%\text{Error} = \frac{v_{o_{3b}} - v_{o_3}}{v_{o_3}} \times 100\% = \frac{0.25 - 0.24875622}{0.24875622} \times 100\% = 0.5\%$$

Figure 5.66 shows the voltage values from PSpice® simulation.

The Simulink model for the noninverting configuration is shown in Figure 5.67.

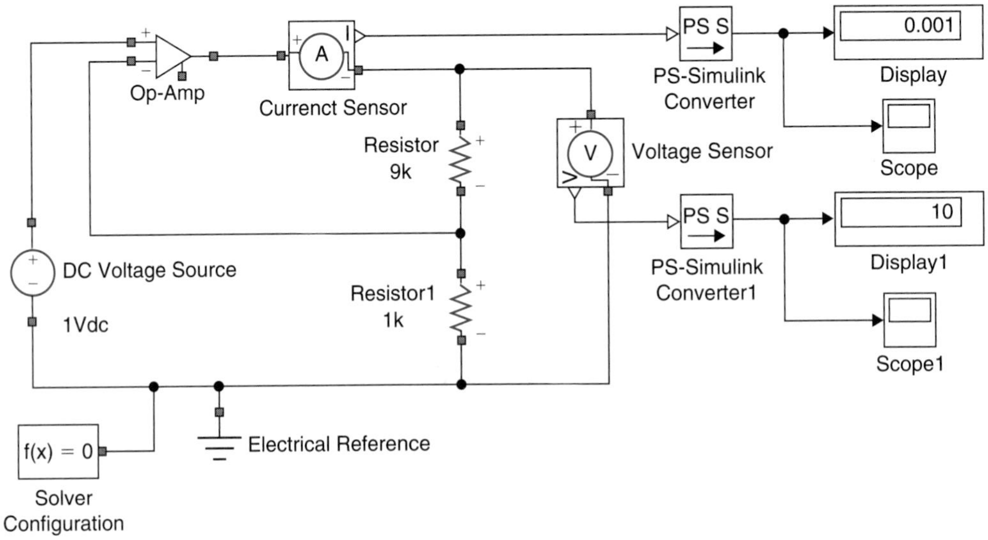

FIGURE 5.67 The Simulink model for a noninverting amplifier.

The model for inverting configuration shown in Figure 5.52 can be simulated in Simulink. From Equation (5.91), we have

$$v_o = \frac{R_2A - R_o}{R_o + R_2}(0 - v_n) = Bv_d \tag{5.145}$$

where

$$B = \frac{R_2A - R_o}{R_o + R_2} \tag{5.146}$$

Since $v_p = 0$, we have $v_d = v_p - v_n = -v_n$, and Equation (5.145) becomes

$$v_o = -Bv_n$$

From Equation (5.93), we get

$$v_n = \frac{\dfrac{1}{R_1}}{\dfrac{1}{R_1} + \dfrac{1}{R_i} + \dfrac{1}{R_2}}v_s + \frac{\dfrac{1}{R_2}}{\dfrac{1}{R_1} + \dfrac{1}{R_i} + \dfrac{1}{R_2}}v_o = \alpha v_s + \beta v_o \tag{5.147}$$

where

$$\alpha = \frac{\frac{1}{R_1}}{\frac{1}{R_1} + \frac{1}{R_i} + \frac{1}{R_2}} \tag{5.148}$$

and

$$\beta = \frac{\frac{1}{R_2}}{\frac{1}{R_1} + \frac{1}{R_i} + \frac{1}{R_2}} \tag{5.149}$$

Since $v_p = 0$, we have $v_o = -Bv_n$, and Equation (5.147) becomes

$$v_o = \frac{-B\alpha}{1 + \beta B} v_s = \frac{-\alpha}{\frac{1}{B} + \beta} v_s \approx \frac{-\alpha}{\beta} v_s$$

The Simulink model for Equations (5.145) and (5.147) is shown in Figure 5.68. The values of coefficients can be calculated in the Command window:

```
>> R1=1e3;R2=20e3;Ri=100e3;Ro=100;A=100e3;
>> B=(R2*A-Ro)/(Ro+R2)
B =
   9.9502e+04
>> alpha=1/R1/(1/R1+1/Ri+1/R2)
alpha =
    0.9434
>> beta=1/R2/(1/R1+1/Ri+1/R2)
beta =
    0.0472
```

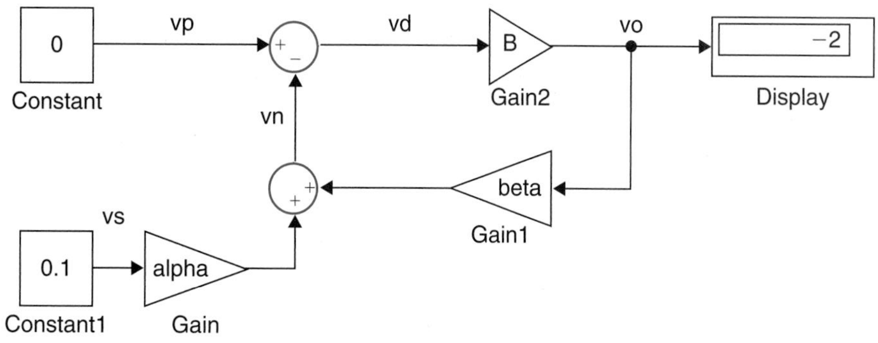

FIGURE 5.68 The Simulink model for the inverting configuration.

If $R_o = 0$ and $R_i = \infty$, we have

$$B = A, \quad \alpha = R_2/(R_1 + R_2), \quad \beta = R_1/(R_1 + R_2), \\ v_o = Av_d, \quad v_n = \alpha v_s + \beta v_o \tag{5.150}$$

The model shown in Figure 5.68 can be redrawn as the one shown in Figure 5.69.

FIGURE 5.69

The Simulink model for the inverting configuration with $R_o = 0$ and $R_i = \infty$.

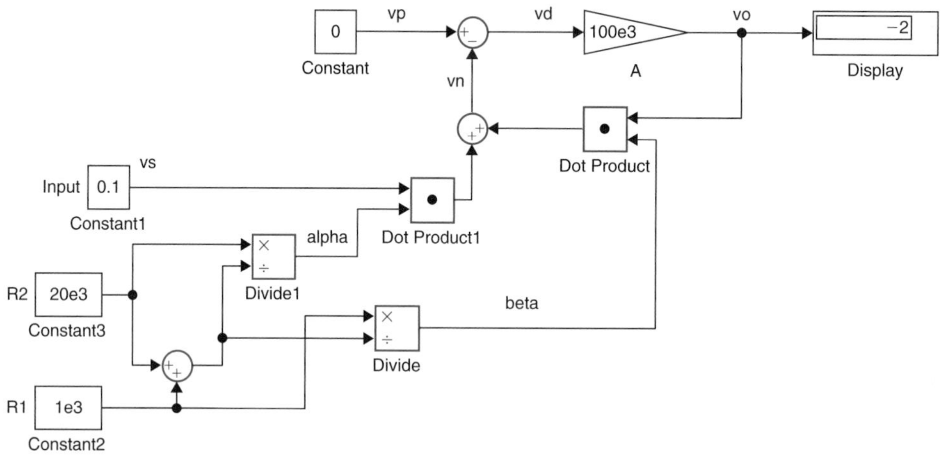

EXAMPLE 5.12

Build a Simulink model similar to the one shown in Figure 5.68 for the noninverting configuration.

From Equation (5.120), we get

$$v_o = \frac{R_o}{R_o + R_2}v_n + \frac{R_2A}{R_o + R_2}(v_p - v_n) = Cv_n + B(v_p - v_n) \quad \textbf{(5.151)}$$

where

$$C = \frac{R_o}{R_o + R_2} \quad \textbf{(5.152)}$$

$$B = \frac{R_2A}{R_o + R_2} \quad \textbf{(5.153)}$$

From Equation (5.123), we get

$$v_n = \frac{\dfrac{1}{R_i}}{\dfrac{1}{R_1} + \dfrac{1}{R_i} + \dfrac{1}{R_2}}v_s + \frac{\dfrac{1}{R_2}}{\dfrac{1}{R_1} + \dfrac{1}{R_i} + \dfrac{1}{R_2}}v_o = \alpha v_s + \beta v_o \quad \textbf{(5.154)}$$

where

$$\alpha = \frac{\dfrac{1}{R_i}}{\dfrac{1}{R_1} + \dfrac{1}{R_i} + \dfrac{1}{R_2}} \quad \textbf{(5.155)}$$

$$\beta = \frac{\dfrac{1}{R_2}}{\dfrac{1}{R_1} + \dfrac{1}{R_i} + \dfrac{1}{R_2}} \quad \textbf{(5.156)}$$

continued

Example 5.12 continued

From Equations (5.151) and (5.154), we obtain

$$v_o = \frac{C\alpha + (1-\alpha)B}{1 - C\beta + \beta B} v_s = \frac{\dfrac{C\alpha}{B} + 1 - \alpha}{\dfrac{1 - C\beta}{B} + \beta} v_s \approx \frac{1-\alpha}{\beta} v_s$$

The Simulink model is shown in Figure 5.70.

```
>> R1=1e3;R2=20e3;Ri=100e3;Ro=100;A=100e3;
>> B=R2*A/(Ro+R2)
B =
   9.9502e+04
>> C=Ro/(Ro+R2)
C =
    0.0050
>> alpha=1/Ri/(1/R1+1/Ri+1/R2)
alpha =
    0.0094
>> beta=1/R2/(1/R1+1/Ri+1/R2)
beta =
    0.0472
```

If $R_o = 0$ and $R_i = \infty$, we have

$$C = 0, \quad B = A, \quad \alpha = 0, \quad \beta = \frac{R_1}{R_1 + R_2}, \quad v_n = \beta v_o, \quad v_o = A(v_p - v_n) \qquad \textbf{(5.157)}$$

The model shown in Figure 5.70 reduces to the one shown in Figure 5.71.

```
>> R1=1e3;R2=20e3;A=100e3;
>> beta=R1/(R1+R2)
beta =
    0.0476
```

FIGURE 5.70

The Simulink model for the noninverting configuration.

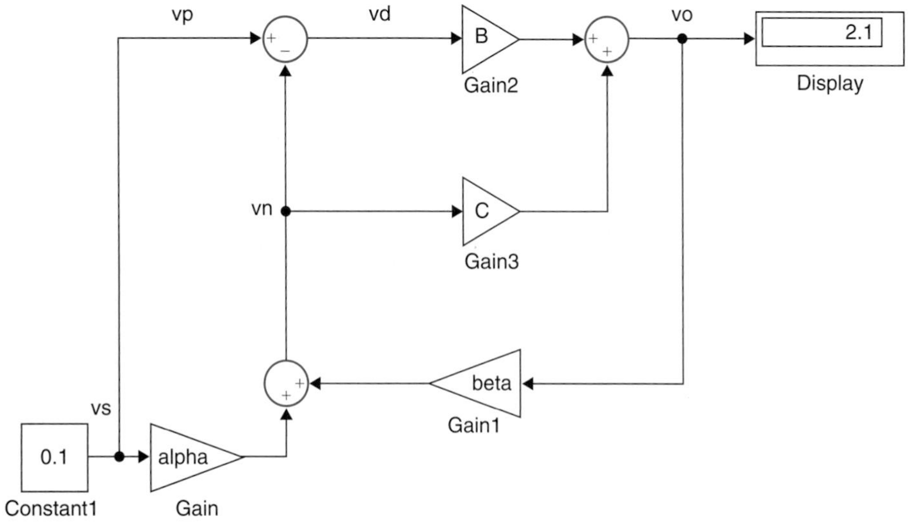

continued

Example 5.12 continued

FIGURE 5.71

The Simulink model for the noninverting configuration for $R_o = 0$ and $R_i = \infty$.

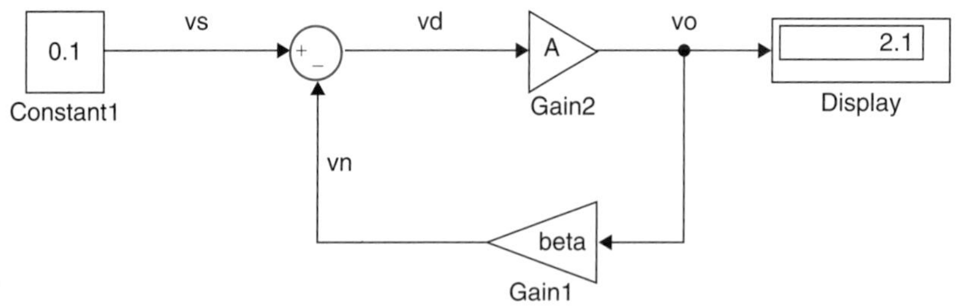

Exercise 5.10

Build a Simulink model for the summing amplifier shown in Figure 5.34 by modifying the model shown in Figure 5.71. Assume that $v_1 = 5$ V and $v_2 = 2$ V.

Answer:

$$v_p = \frac{\frac{1}{R_1}}{\frac{1}{R_1}+\frac{1}{R_2}+\frac{1}{R_3}}v_1 + \frac{\frac{1}{R_2}}{\frac{1}{R_1}+\frac{1}{R_2}+\frac{1}{R_3}}v_2, \beta = \frac{R_4}{R_4+R_5}.$$

The model is shown in Figure 5.72.

```
>> R1=6e3;R2=1e3;R3=3e3;R4=3e3;R5=10.5e3;A=1e6;
```

FIGURE 5.72

The Simulink model for $v_o = 0.5v_1 + 3v_2$.

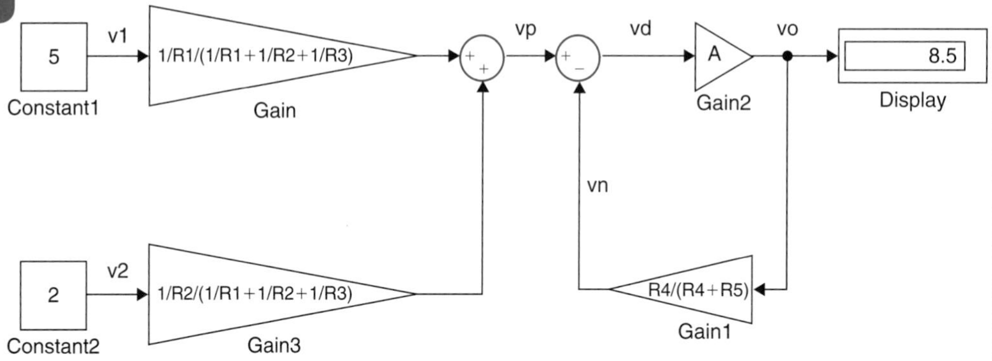

SUMMARY

In this chapter, the analysis and design of circuits that include op amp have been presented. Both the ideal and practical models of op amp are analyzed. In the ideal model, it is found that the values of the current entering the op amp from the positive and the negative input terminals are both zero, and the voltage on the positive input terminal is identical to the voltage on the negative input terminal for both the noninverting configuration and the inverting configuration. Both the noninverting configuration and the inverting configuration provide negative feedback from the output to the input. In the practical model, the effect of output resistance R_o, the input resistance R_i, and the open loop gain A are included.

In the inverting configuration, the output voltage v_o is given by

$$v_o = -\frac{R_2}{R_1}v_s$$

where v_s is the input voltage. In the noninverting configuration, the output voltage v_o is given by

$$v_o = \left(1 + \frac{R_2}{R_1}\right)v_s$$

Applications of the op amp in amplifying signals, adding and subtracting signals, converting current to voltage, and converting voltage to current have been discussed. Circuits with op amps can be modeled in PSpice and Simulink. Several examples of simulating circuits using PSpice and Simulink are presented.

PROBLEMS

Ideal Op Amp

5.1 Find V_o in the circuit shown in Figure P5.1.

FIGURE P5.1

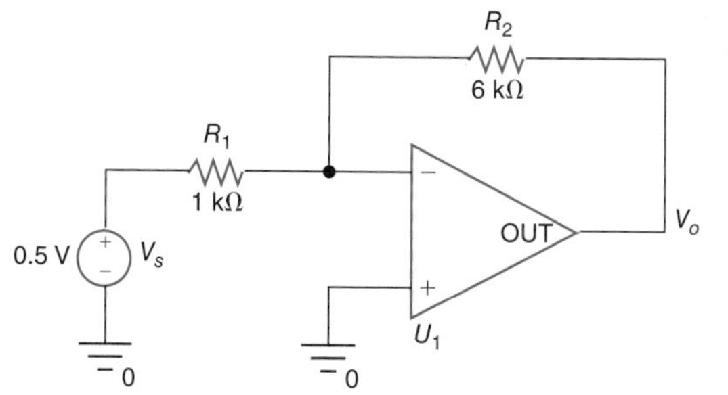

5.2 Find V_o in the circuit shown in Figure P5.2.

FIGURE P5.2

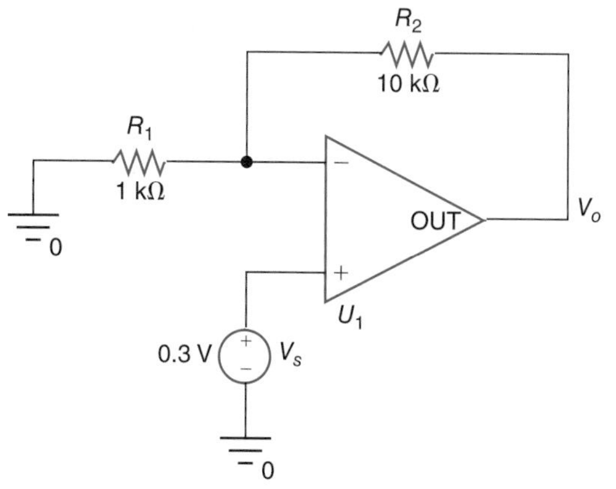

5.3 Find V_o in the circuit shown in Figure P5.3.

FIGURE P5.3

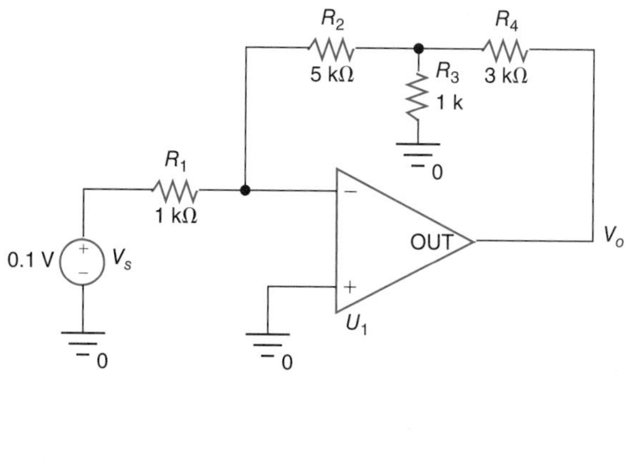

5.4 Find V_o in the circuit shown in Figure P5.4.

FIGURE P5.4

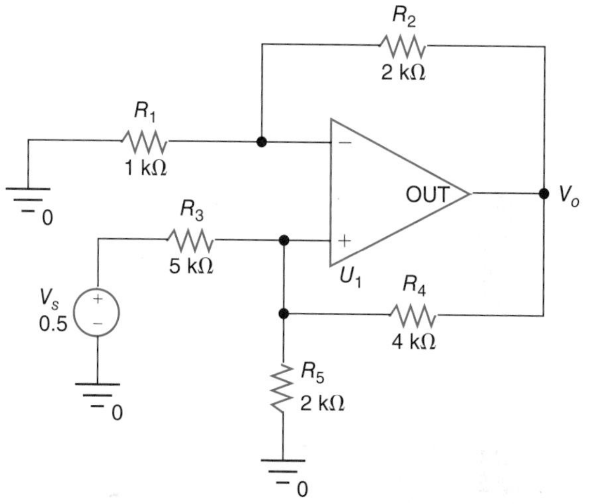

5.5 Find V_o in the circuit shown in Figure P5.5.

FIGURE P5.5

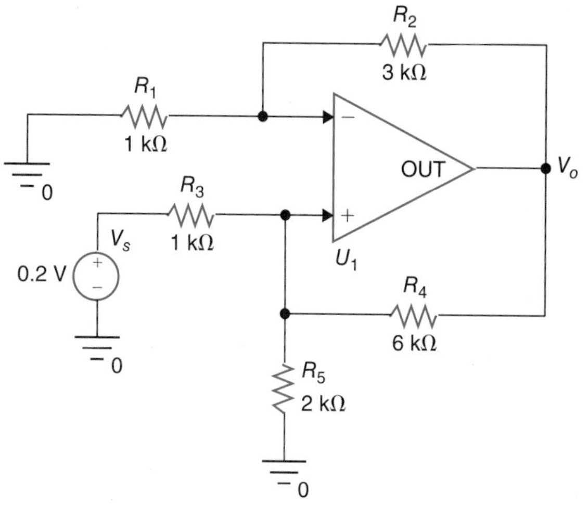

5.6 **Find V_o in the circuit shown in Figure P5.6.**

FIGURE P5.6

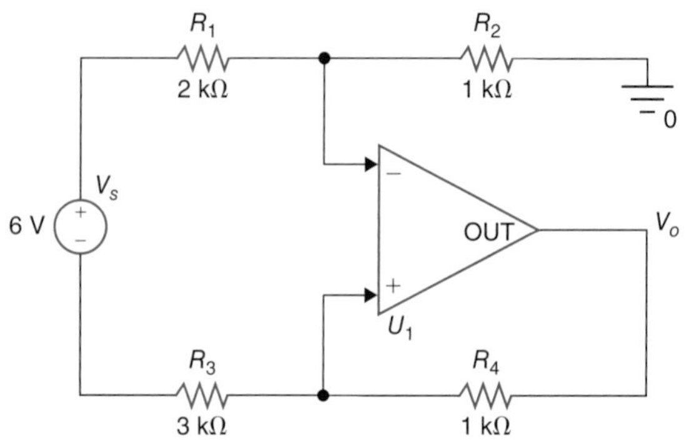

5.7 **Find V_o in the circuit shown in Figure P5.7.**

FIGURE P5.7

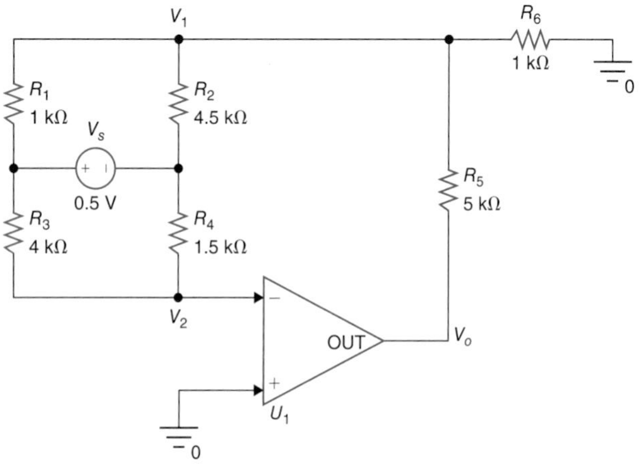

5.8 **An op amp circuit is shown in Figure P5.8.**

FIGURE P5.8

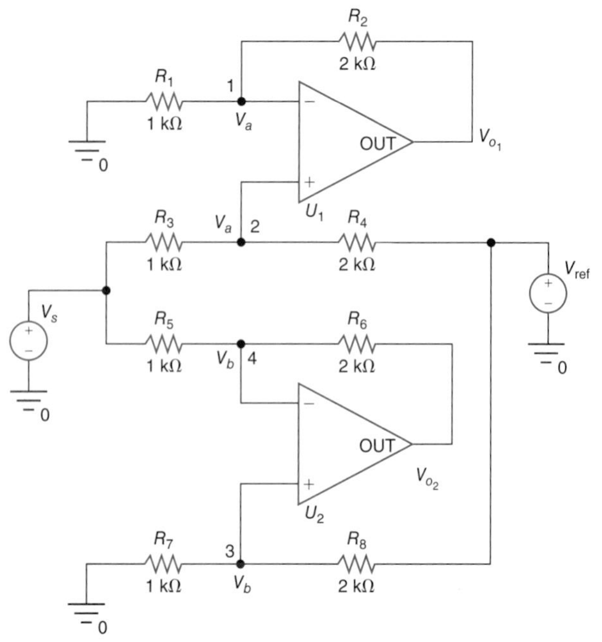

a. Write a node equation at node 1 by summing the currents away from node 1. The voltage at node 1 is V_a.
b. Write a node equation at node 2 by summing the currents away from node 2. The voltage at node 2 is V_a.
c. Solve the two node equations from (a) and (b) and express V_{o_1} as a function of V_s and V_{ref}.
d. Write a node equation at node 3 by summing the currents away from node 3. The voltage at node 3 is V_b.
e. Write a node equation at node 4 by summing the currents away from node 4. The voltage at node 4 is V_b.
f. Solve the two node equations from (d) and (e) and express V_{o_2} as a function of V_s and V_{ref}.
g. If $V_s = 1$ V and $V_{ref} = 0$ V, what are the numerical values of V_{o_1} and V_{o_2}?
h. If $V_s = 1$ V and $V_{ref} = 5$ V, what are the numerical values of V_{o_1} and V_{o_2}?

5.9 **In the circuit shown in Figure P5.9, if $V_s = 0.5$ V, what is the value of V_o?**

FIGURE P5.9

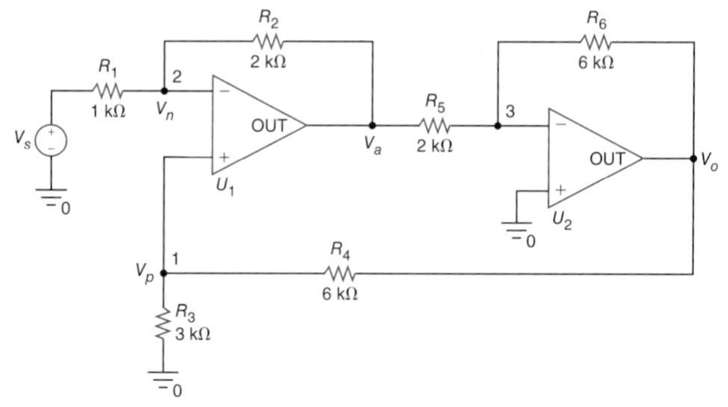

5.10 **In the circuit shown in Figure P5.10,**

FIGURE P5.10

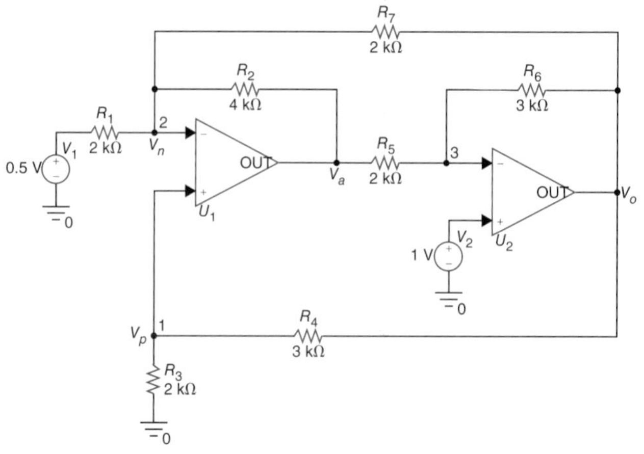

a. write a node equation at node 1 by summing the currents leaving node 1. Express V_p as a function of V_o.
b. write a node equation at node 3 by summing the currents leaving node 3. Express V_a as a function of V_o.
c. write a node equation at node 2 by summing the currents leaving node 2.
d. express V_o as a function of V_1 and V_2
e. find the numerical values of V_o, V_p, and V_a.

5.11 **Design an op amp circuit that amplifies the input signal by −4.5 times (i.e., $v_o = -4.5\ v_{in}$). You can use one op amp and two resistors. The resistance value of the resistor connecting the inverting input and the voltage source is 1 $k\Omega$. Find the value of the other resistor.**

5.12 **Design an op amp circuit that amplifies the input signal by −0.9 times (i.e., $v_o = -0.9\ v_{in}$). You can use one op amp and two resistors. The resistance value of the resistor connecting the inverting input and the voltage source is 10 $k\Omega$. Find the value of the other resistor.**

5.13 **Design an op amp circuit that amplifies the input signal by 0.75 times (i.e., $v_o = 0.75\ v_{in}$). You can use one op amp and two resistors. The resistance value of the resistor connecting the noninverting input and the voltage source is 1 $k\Omega$. Find the value of the other resistor.**

5.14 **Design an op amp circuit that amplifies the input signal by 2.5 times (i.e., $v_o = 2.5\ v_{in}$). You can use one op amp and two resistors. The resistance value of the resistor connecting the inverting input and ground is 1 $k\Omega$. Find the value of the other resistor.**

Sum and Difference

5.15 **Find V_o in the circuit shown in Figure P5.15.**

FIGURE P5.15

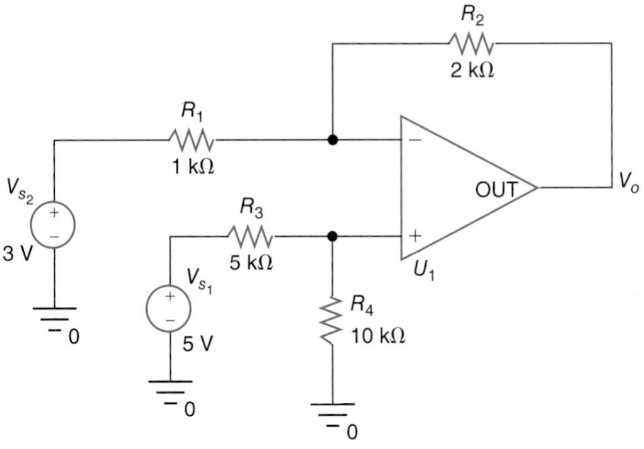

5.16 **Find V_o in the circuit shown in Figure P5.16.**

FIGURE P5.16

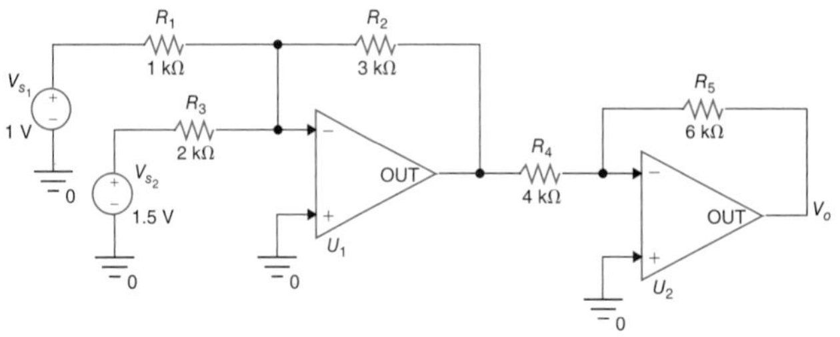

5.17 **Find V_a, V_b, V_c, and V_o in the circuit shown in Figure P5.17.**

FIGURE P5.17

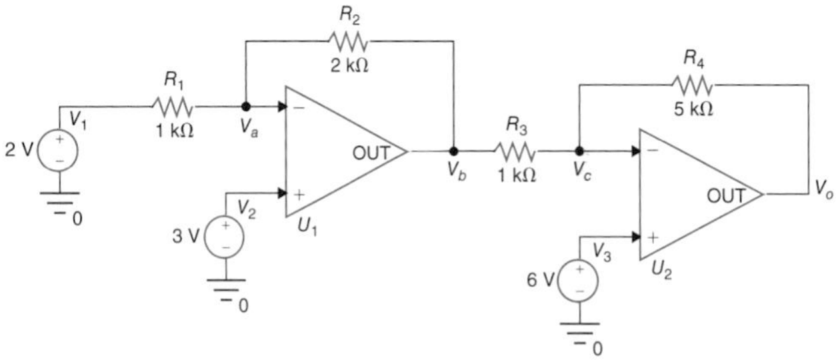

5.18 **In the circuit shown in Figure P5.18,**

FIGURE P5.18

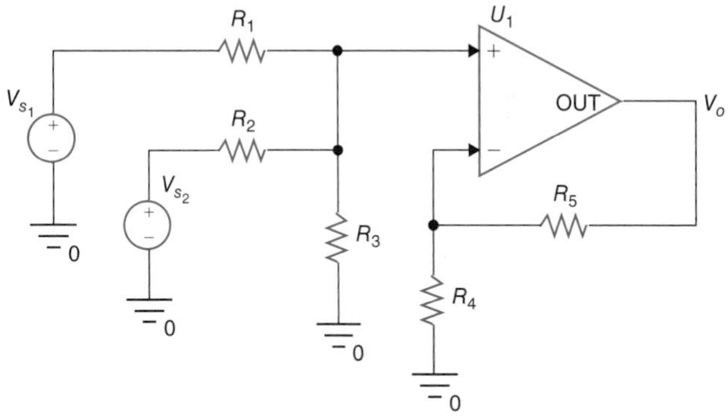

a. find the expression of the output voltage V_o as a function of V_{S_1} and V_{S_2}.
b. find the expression of the output voltage V_o as a function of V_{S_1} and V_{S_2} when $R_3 = R_4 = 12\ k\Omega$, $R_1 = 4\ k\Omega$, $R_2 = 3\ k\Omega$, and $R_5 = 84\ k\Omega$.
c. find the expression of the output voltage V_o as a function of V_{S_1} and V_{S_2} when $R_3 = R_4 = 6\ k\Omega$, $R_1 = 20\ k\Omega$, $R_2 = 1.5\ k\Omega$, and $R_5 = 25.8\ k\Omega$.

5.19 **Find V_o in the circuit shown in Figure P5.19.**

FIGURE P5.19

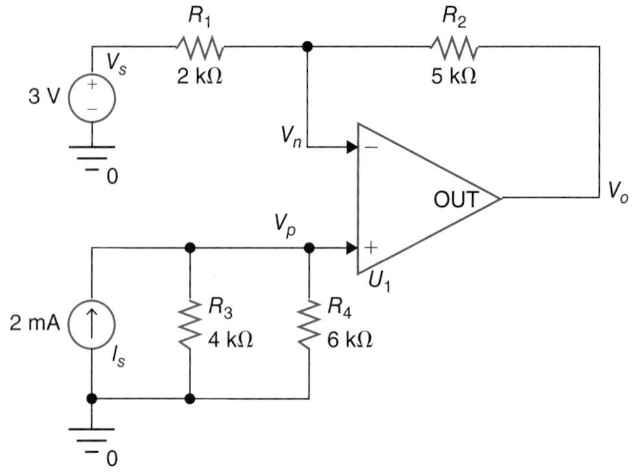

5.20 **The op amp in the circuit shown in Figure P5.20 is ideal.**

FIGURE P5.20

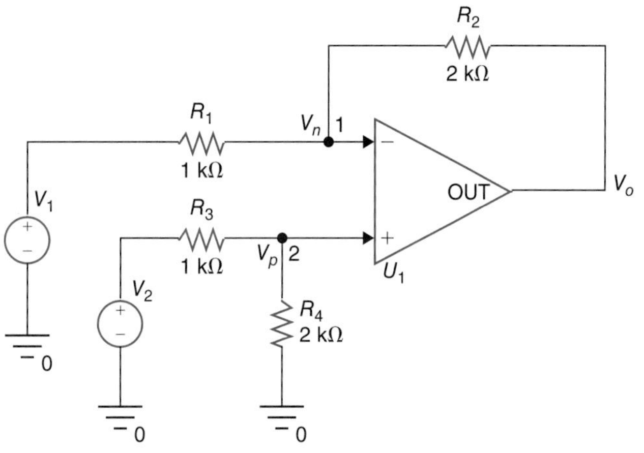

a. Write a node equation at node 1 by summing the currents leaving node 1. Represent V_o as a function of V_1 and V_n.
b. Write a node equation at node 2 by summing the currents leaving node 2. Represent V_p as a function of V_2.
c. Represent V_o as a function of V_1 and V_2.
d. Find the numerical value of V_o when $V_1 = 3.5$ V and $V_2 = 5$ V.
e. Find the numerical value of V_o when $V_1 = 3.5$ V and $V_2 = 0$ V.

5.21 **Design an op amp circuit with two inputs and one output. The output of the op amp is given by $V_o = -2V_{s_1} - 3V_{s_2}$. There is one op amp and three resistors in this circuit. Find the values of the two resistors connected to the input signals when the feedback resistor between the inverting input and the output of the op amp is given by 12 $k\Omega$.**

5.22 **Design an op amp circuit with three inputs and one output. The output of the op amp is given by $V_o = 2V_1 + 3V_2 + 6V_3$. There is one op amp and six resistors in this circuit. Find the values of the resistors.**

5.23 **Design an op amp circuit with three inputs and one output. The output of the op amp is given by $V_o = V_1/2 + V_2/4 + V_3/8$.**

5.24 **Design an op amp circuit with two inputs and one output. The output of the op amp is given by $V_o = 5(V_{s_1} - V_{s_2})$. There is one op amp and four resistors in this circuit. Find the values of the two remaining resistors when the resistors connected to two inputs are 2 $k\Omega$.**

5.25 **Use op amps to design a circuit that provides an output given by**

$$V_o = 0.7V_1 + 1.2V_2 - 2V_3$$

Instrumentation Amplifier

5.26 **In the fully differential amplifier shown in Figure P5.26, there are two outputs, V_{o+} and V_{o-}.**

FIGURE P5.26

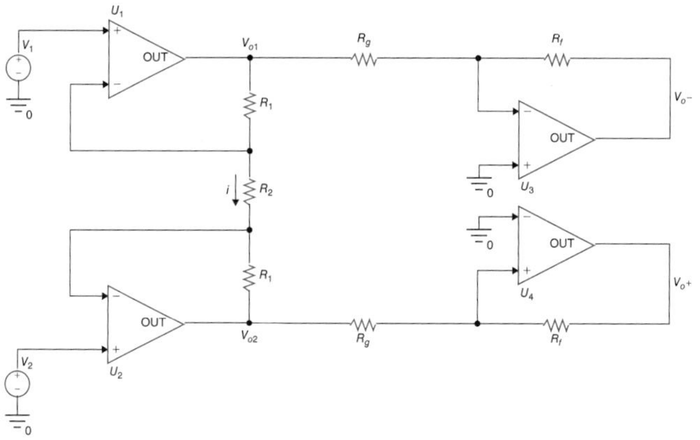

a. Find the current i through R_2 as a function of V_1, V_2, and R_2.
b. Find V_{o_1} as a function of V_1, V_2, R_1, and R_2.
c. Find V_{o_2} as a function of V_1, V_2, R_1, and R_2.
d. Find $V_{o_1} - V_{o_2}$ as a function of V_1, V_2, R_1, and R_2.
e. Find $V_{o+} - V_{o-}$ as a function of V_1, V_2, R_1, R_2, R_g, and R_f.
f. Find V_{o+} and V_{o-} when $V_2 = -V_1$.

5.27 **Consider the circuit shown in Figure P5.27.**

a. Find V_{o_1} as a function of V_s, R_1, and R_2.
b. Find V_{o_2} as a function of V_s, R_2, and R_3.

FIGURE P5.27

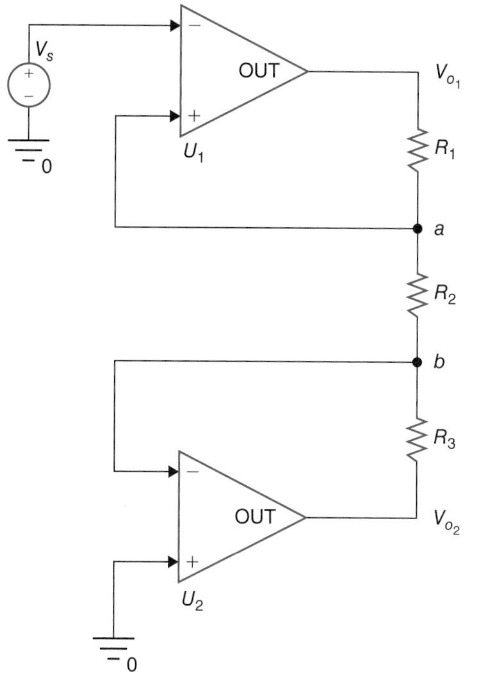

c. Find the numerical values of V_{o_1} and V_{o_2} when $V_s = 1$ V, $R_1 = 4\ k\Omega$, $R_2 = 1\ k\Omega$, and $R_3 = 5\ k\Omega$.

d. It is desired to obtain $V_{o_1} = 10$ V and $V_{o_2} = -5$ V when $V_s = 1$ V. Find one solution for R_1, R_2, and R_3.

5.28 **In the fully differential amplifier shown in Figure P5.28, there are two outputs, V_{o+} and V_{o-}.**

FIGURE P5.28

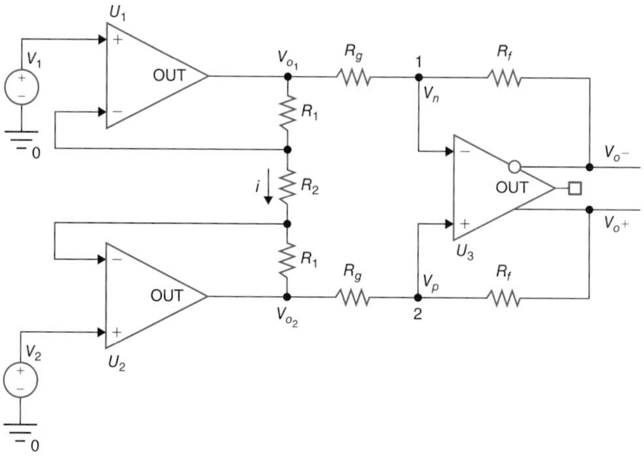

a. Find the current i through R_2 as a function of V_1, V_2, and R_2.

b. Find V_{o_1} as a function of V_1, V_2, R_1, and R_2.

c. Find V_{o_2} as a function of V_1, V_2, R_1, and R_2.

d. Find $V_{o_1} - V_{o_2}$ as a function of V_1, V_2, R_1, and R_2.

e. Write a node equation at node 1 by summing the currents away from node 1.

f. Write a node equation at node 2 by summing the currents away from node 2.

g. Find $V_{o+} - V_{o-}$ as a function of V_1, V_2, R_1, R_2, R_g, and R_f.

h. Find $V_{o+} - V_{o-}$ if $V_1 = 1$ V, $V_2 = 0.5$ V, $R_1 = 1\ k\Omega$, $R_2 = 1\ k\Omega$, $R_g = 1\ k\Omega$, and $R_f = 2\ k\Omega$.

Current Amplifier

5.29 **Find V_o in the circuit shown in Figure P5.29.**

FIGURE P5.29

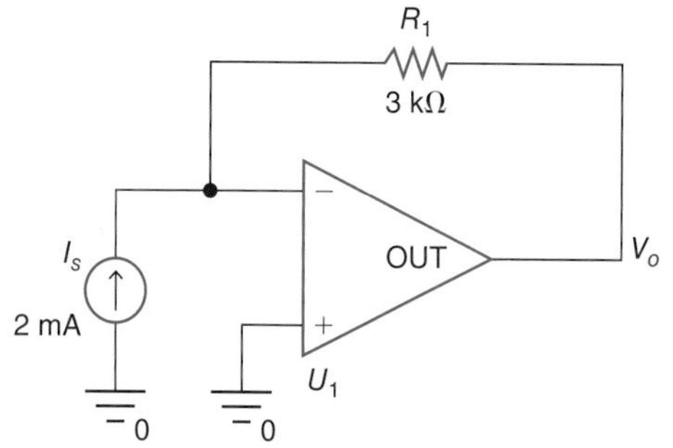

5.30 **Find V_o in the circuit shown in Figure P5.30.**

FIGURE P5.30

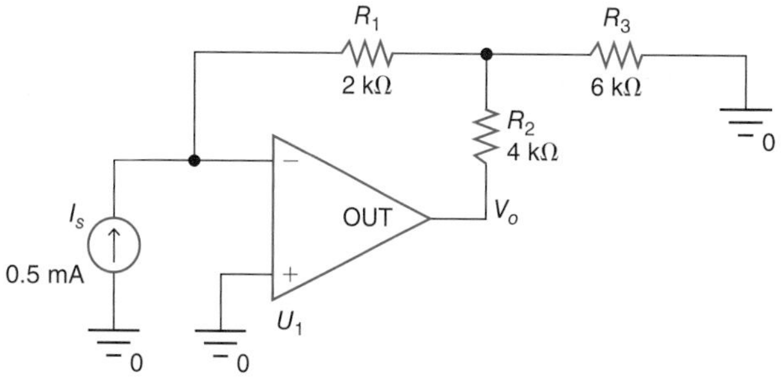

5.31 **Find I_o in the circuit shown in Figure P5.31.**

FIGURE P5.31

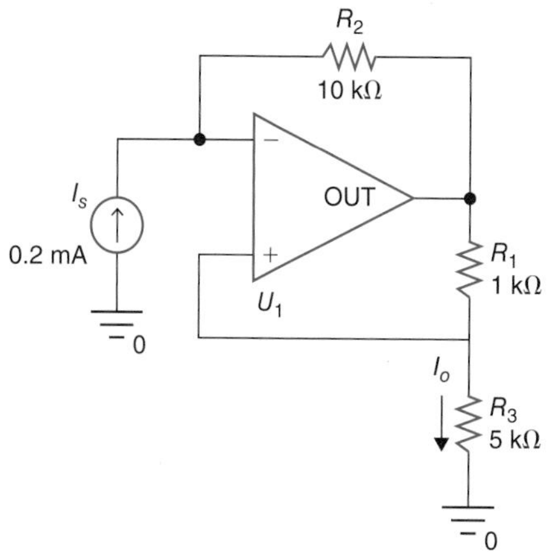

5.32 **Find I_o in the circuit shown in Figure P5.32.**

FIGURE P5.32

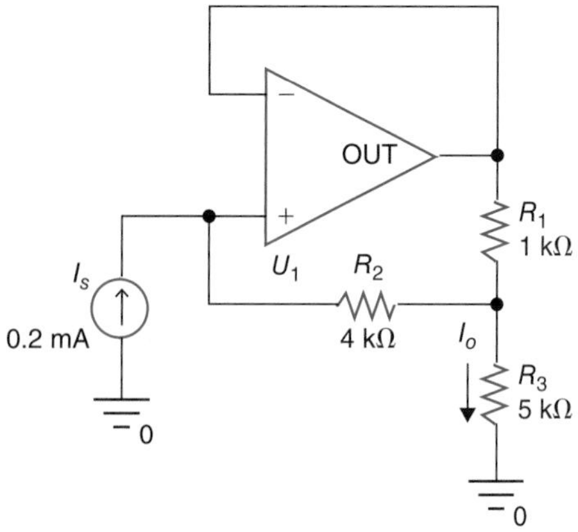

5.33 a. Represent I_o as a function of V_s in the circuit shown in Figure P5.33.
b. If $V_s = 0.5$ V, $R_1 = 1\ k\Omega$, $R_2 = 2\ k\Omega$, $R_3 = 1\ k\Omega$, $R_4 = 2\ k\Omega$, and $R_5 = 5\ k\Omega$, what is I_o?
c. If $V_s = 0.5$ V, $R_1 = 1\ k\Omega$, $R_2 = 2\ k\Omega$, $R_3 = 1\ k\Omega$, $R_4 = 2.1\ k\Omega$, and $R_5 = 5\ k\Omega$, what is I_o?

FIGURE P5.33

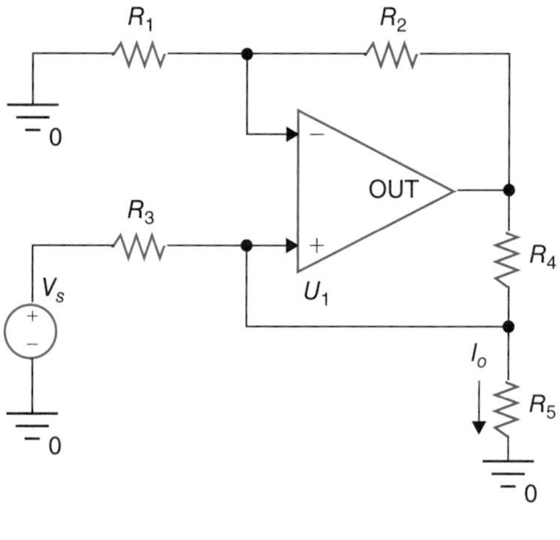

5.34 **Design an op amp current amplifier that amplifies input current by a factor of 3; that is, $i_o = 3i_s$.**

5.35 **Design an op amp current amplifier that amplifies input current by a factor of −2; that is, $i_o = -2i_s$.**

5.36 **Design an op amp circuit that converts input current 1 mA to output voltage of 5 V.**

5.37 **Design an op amp circuit that provides a constant current of 2 mA to a load resistor from a 5-V voltage source.**

Inverting Configuration

5.38 **Find V_o in the circuit shown in Figure P5.38.**

FIGURE P5.38

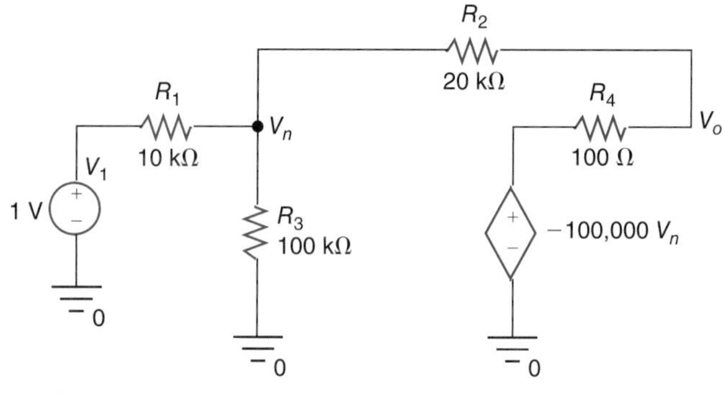

5.39 **Find V_o in the circuit shown in Figure P5.39.**

FIGURE P5.39

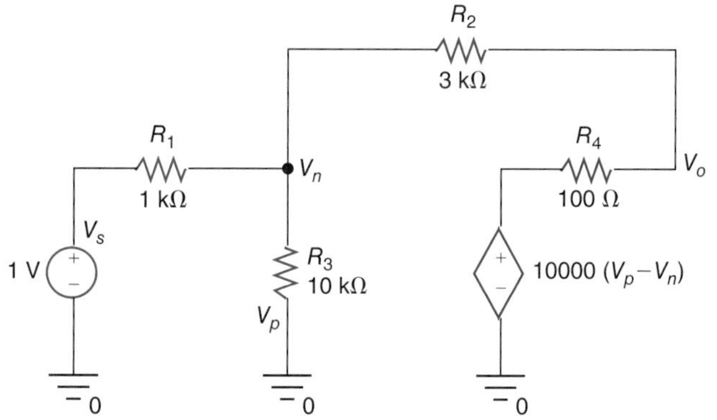

5.40 **In the circuit shown in Figure P5.40, find input resistance R_{in}.**

FIGURE P5.40

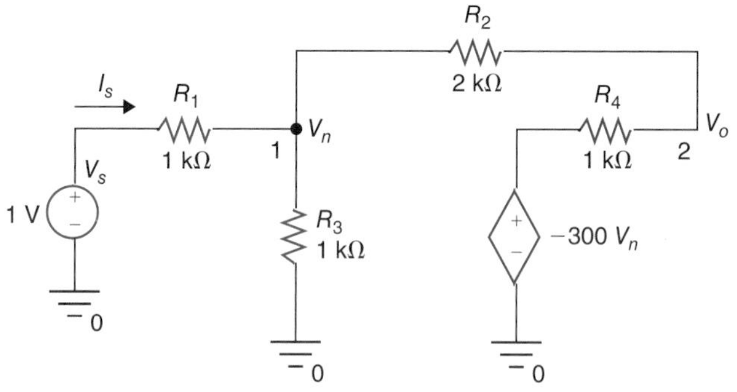

5.41 **In the circuit shown in Figure P5.41, find output resistance R_{out}.**

FIGURE P5.41

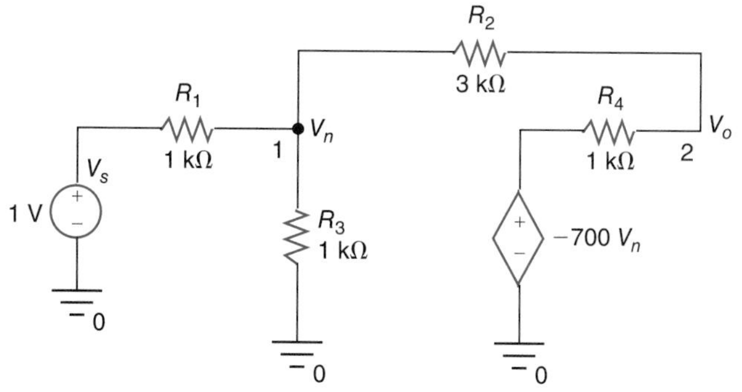

5.42 **In the circuit shown in Figure P5.42,**

FIGURE P5.42

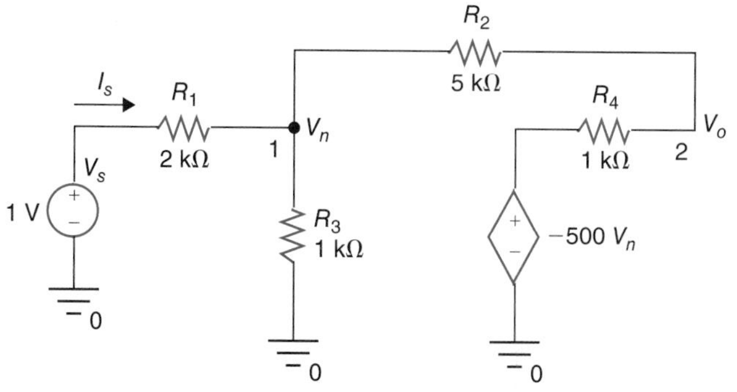

a. write a node equation at node 1 by summing the currents leaving node 1.
b. write a node equation at node 2 by summing the currents leaving node 2.
c. solve the two node equations to find numerical values of V_n and V_o.
d. find the value of I_s through R_1. Find the input resistance $R_{in} = V_s/I_s$.

5.43 **In the circuit shown in Figure P5.43, find input resistance $R_{in} = V_s/I_s$.**

FIGURE P5.43

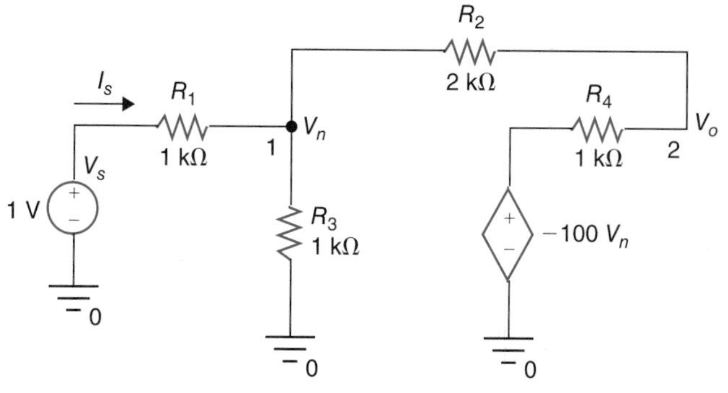

5.44 **In the circuit shown in Figure P5.44,**

FIGURE P5.44

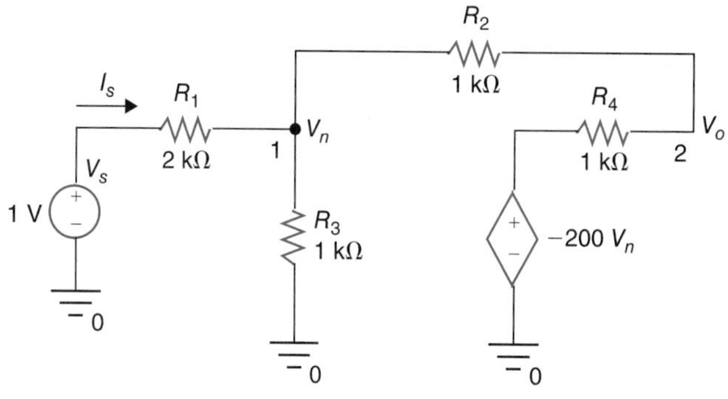

a. write a node equation at node 1 by summing the currents leaving node 1.
b. write a node equation at node 2 by summing the currents leaving node 2.
c. solve the two node equations to find numerical values of V_n and V_o.
d. find the value of I_s through R_1. Find the input resistance $R_{in} = V_s/I_s$.

Noninverting Configuration

5.45 **Find V_o in the circuit shown in Figure P5.45.**

FIGURE P5.45

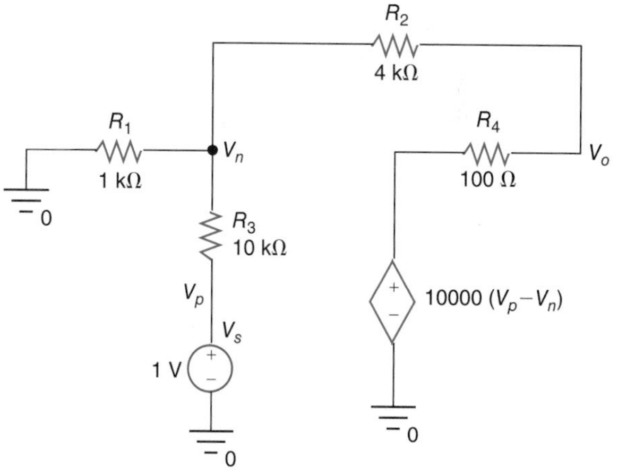

5.46 **In the circuit shown in Figure P5.46, find input resistance R_{in}.**

FIGURE P5.46

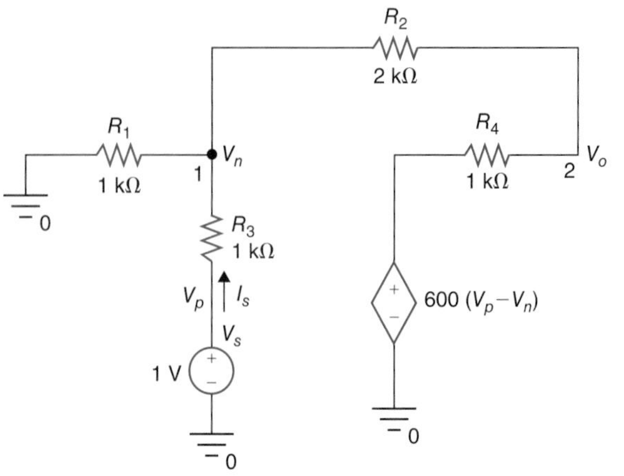

5.47 **In the circuit shown in Figure P5.47, find output resistance R_{out}.**

FIGURE P5.47

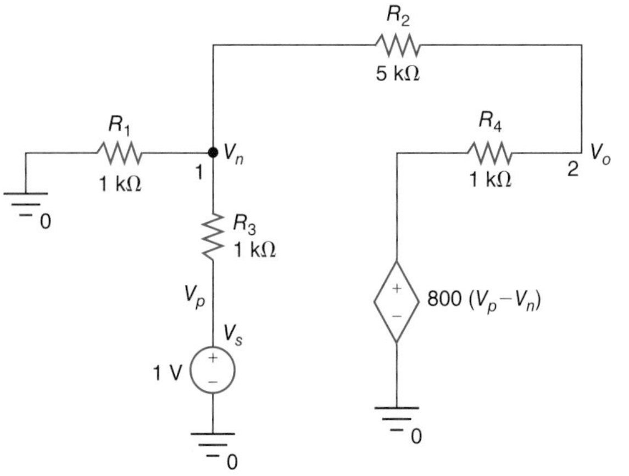

5.48 **In the circuit shown in Figure P5.48,**

FIGURE P5.48

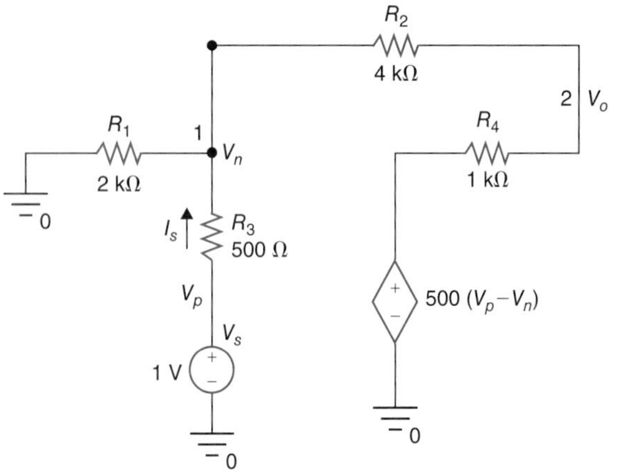

a. write a node equation at node 1 by summing the currents away from node 1.
b. write a node equation at node 2 by summing the currents away from node 2. ($V_p = V_s = 1$ V)
c. solve the two node equations to find numerical values of V_n and V_o.
d. find the value of I_s through R_3. Find the input resistance $R_{in} = V_s/I_s$.

5.49 **In the circuit shown in Figure P5.49, find input resistance $R_{in} = V_s/I_s$.**

FIGURE P5.49

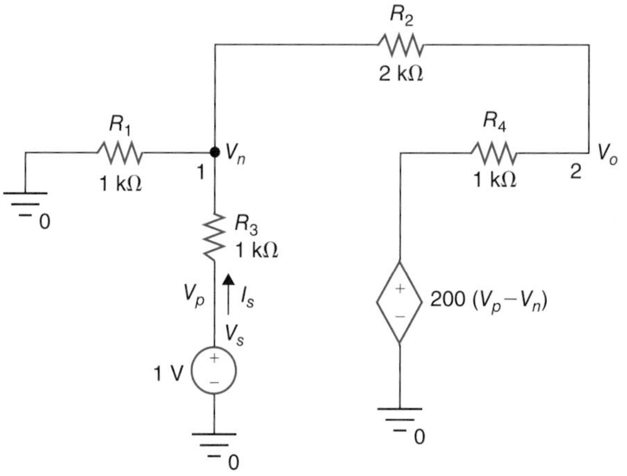

5.50 **In the circuit shown in Figure P5.50,**

FIGURE P5.50

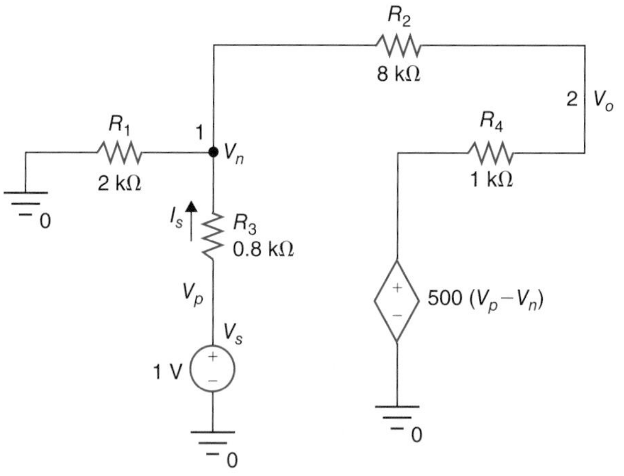

a. write a node equation at node 1 by summing the currents away from node 1.
b. write a node equation at node 2 by summing the currents away from node 2. ($V_p = V_s = 1$ V)
c. solve the two node equations to find numerical values of V_n and V_o.
d. find the value of I_s through R_3. Find the input resistance $R_{in} = V_s/I_s$.

Chapter 6

Capacitors and Inductors

6.1 Introduction

So far, we have analyzed circuits that contain voltage sources, current sources, and resistors. In this chapter, we introduce two new components: capacitors and inductors. Unlike resistors, capacitors and inductors are passive elements that can store and release energy.

A parallel plate capacitor has two plates filled by dielectric material in between. The positive plate holds positive charges, and the negative plate holds negative charges. The capacitance of a capacitor is the amount of charge that the capacitor can hold for the given voltage. The energy is stored in the form of an electric field from the positive plate to the negative plate. The amount of energy stored on the capacitor depends on the capacitance and the voltage across the capacitor plates. The current through the capacitor is given by the product of the capacitance and the time rate of change of the voltage across the capacitor plates. There is no current through the capacitor when the voltage is not changing with time. If a direct current (dc) voltage is applied to a capacitor, the capacitor will charge to the maximum value possible and will stay at that value. When the voltage applied to the capacitor changes with time, the current flowing through the capacitor is proportional to the slope of the voltage waveform.

A simple inductor consists of a solenoid, which is a wire wound in helix. The core can be an air or a ferromagnetic material. When current flows through an inductor, it introduces a magnetic field. According to Faraday's law, a changing magnetic field induces an electromotive force. The inductance of an inductor is the measure of the inductor to generate magnetic flux for the given change of current. The energy is stored in the form of magnetic field. The amount of energy stored in the inductor depends on the inductance and the current through the inductor. The voltage across the inductor is given by the product of the inductance and the time rate of change of the current through the inductor. If a dc current is applied to an inductor, the inductor will increase the flux to the maximum value possible and stay at that value. When the current through the inductor changes with time, the voltage across the inductor is proportional to the slope of the current waveform.

6.2 Capacitors

A *capacitor* is a passive device that can store energy in the form of an electric field. A simple parallel plate capacitor consists of two conductor plates. The space between the conductor plates is filled with insulators called *dielectric materials*. Examples of dielectric materials are ceramic, mica, paraffin-coated paper, and air. The symbol for capacitors is shown in Figure 6.1.

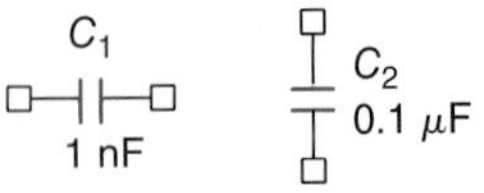

FIGURE 6.1 Symbol for capacitors.

When a capacitor is connected to a voltage source such as a battery, as shown in Figure 6.2, positive charges accumulate on the plate connected to the positive terminal of the battery, and negative charges accumulate on the plate connected to the negative terminal of the battery.

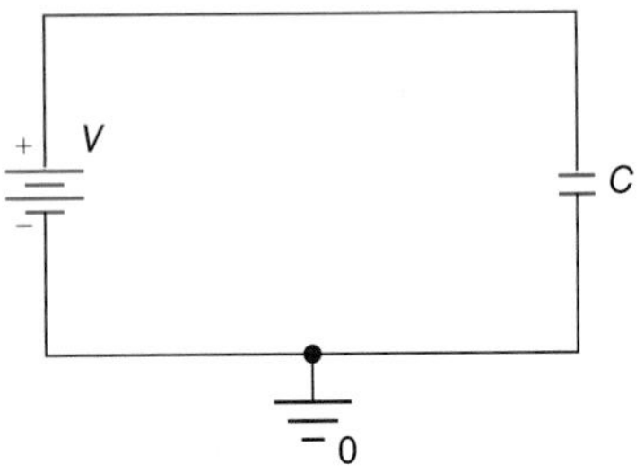

FIGURE 6.2 A capacitor connected to a battery.

The amount of charge Q (in coulombs) on the plates is proportional to the voltage V (in volts) of the voltage source. Let C be the proportionality constant in this linear relation. Then we have

$$Q = CV \tag{6.1}$$

The constant C is called the *capacitance,* measured in farads (F). The unit for measuring farads is coulombs/volt. The capacitance C can be written as

$$C = \frac{Q}{V} \tag{6.2}$$

Capacitance is the capacity of the capacitor to store charges on the plates. According to Equation (6.2), the capacitance is defined as the ratio of the charge stored to the potential difference. The capacitance is the amount of charge that can be stored for 1 V of potential difference (voltage). If $v(t)$ is time varying, Equation (6.1) can be rewritten as

$$Q = C\,v(t)$$

Common capacitors used in electronic circuits have small capacitance values in the microfarad (μF), nanofarad (nF), and picofarad (pF) range. A microfarad is 10^{-6} F, a nanofarad is 10^{-9} F, and a picofarad is 10^{-12} F. The charges on the conductor plates generate an electric field, as shown in Figure 6.3. The electric field originates at the positive charges and sinks on the negative charges.

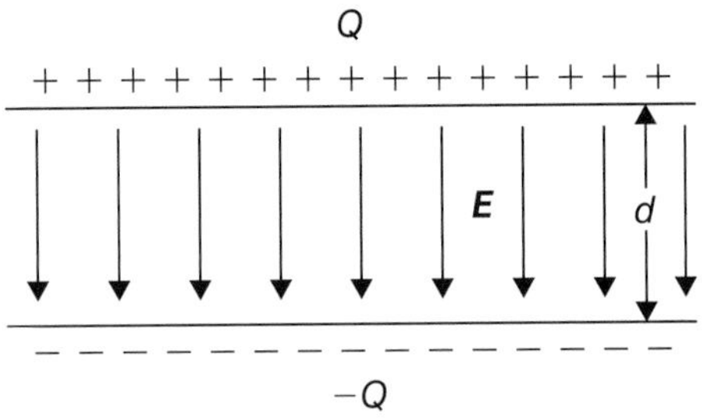

FIGURE 6.3 An electric field between the plates.

Let S be the area of conductor plates, d be the distance between the plates, and ε be the permittivity of the dielectric material between the plates. Typically, the distance d is very small and S is very large. Under this condition, the fringing effects of the electric field at the edges of the parallel conductor plates can be ignored. The electric field intensity $\boldsymbol{E}$ between the conductor plates can be assumed to be uniform. The electric flux density is defined as

$$\boldsymbol{D} = \varepsilon \boldsymbol{E}$$

where ε is the permittivity of the medium. The permittivity of the free space is given by

$$\varepsilon_0 = 8.8541878176 \times 10^{-12}\ \text{F/m}$$

The relative permittivity (also called the *dielectric constant*) ε_r is defined as the ratio of the permittivity of the material to the permittivity of the free space. Then, the permittivity can be written as

$$\varepsilon = \varepsilon_r \varepsilon_0$$

According to Gauss's law, the total electric flux emanating from the positive plate should be equal to the amount of charge enclosed; that is,

$$\oint \boldsymbol{D} \cdot d\boldsymbol{s} = \oint \varepsilon \boldsymbol{E} \cdot d\boldsymbol{s} = Q \quad \textbf{(6.3)}$$

where ε is the permittivity of the dielectric material between the plates and $d\mathbf{s}$ is a vector normal to the surface area where the integration is evaluated. Since the electric field intensity $\boldsymbol{E}$ is constant and $d\boldsymbol{s}$ is in the same direction as $\boldsymbol{E}$ ($\boldsymbol{E} \cdot d\boldsymbol{s} = Eds\cos(0) = Eds$), we have

$$\oint \varepsilon \boldsymbol{E} \cdot d\boldsymbol{s} = \varepsilon E \int ds = \varepsilon ES = Q \quad \textbf{(6.4)}$$

where S is the area of the plate. Thus, the electric field between the parallel plates is given by

$$E = \frac{Q}{\varepsilon S} \quad \textbf{(6.5)}$$

The direction of $\boldsymbol{E}$ is from the plate with positive charges to the plate with negative charges. The voltage is defined as

$$V = -\int \boldsymbol{E} \cdot d\ell \quad \textbf{(6.6)}$$

The negative sign implies that moving against the electric field increases potential. The voltage from the positive plate to the negative plate is given by

$$V = Ed = \frac{Qd}{\varepsilon S} = \frac{Q}{\frac{\varepsilon S}{d}} \quad \textbf{(6.7)}$$

Notice that the voltage V is the work per unit charge required to move a test charge from the negative plate to positive plate against the uniform electric field $\boldsymbol{E}$. Since $V = Q/C$, the capacitance C is given by

$$C = \frac{\varepsilon S}{d} \quad \textbf{(6.8)}$$

Notice that the capacitance is proportional to the area of the plates and the permittivity of the material between the plates and inversely proportional to the distance between the plates. With ε and d held constant, if the area S is doubled, the capacitance value C is also doubled. If two identical capacitors are connected in parallel, the capacitance values will be two times of the capacitance value of each capacitor. If two capacitors with identical ε and d values have areas S_1 and S_2, the capacitances are given by $C_1 = \varepsilon S_1/d$ and $C_2 = \varepsilon S_2/d$, respectively. If these two capacitors are connected in parallel, the total area of the combined capacitor is $S = S_1 + S_2$. Thus, the equivalent capacitance is given by

$$C = \frac{\varepsilon S}{d} = \frac{\varepsilon(S_1 + S_2)}{d} = \frac{\varepsilon S_1}{d} + \frac{\varepsilon S_2}{d} = C_1 + C_2$$

In general, if n capacitors with capacitance values $C_1, C_2, \ldots, C_n$ are connected in parallel, the combined capacitance value is

$$C = C_1 + C_2 + \cdots + C_n.$$

With ε and S held constant, if the distance d is doubled, the capacitance value C becomes half of the original value. If two capacitors with identical ε and S values have lengths d_1 and d_2, the capacitances are given by $C_1 = \varepsilon S/d_1$ and $C_2 = \varepsilon S/d_2$, respectively. If these two capacitors are connected in series, the total length of the combined capacitor is $d = d_1 + d_2$. Thus, the equivalent capacitance is given by

$$C = \frac{\varepsilon S}{d} = \frac{\varepsilon S}{d_1 + d_2} = \frac{1}{\frac{d_1}{\varepsilon S} + \frac{d_2}{\varepsilon S}} = \frac{1}{\frac{1}{\frac{\varepsilon S}{d_1}} + \frac{1}{\frac{\varepsilon S}{d_2}}} = \frac{1}{\frac{1}{C_1} + \frac{1}{C_2}} = \frac{C_1 C_2}{C_1 + C_2}$$

In general, if n capacitors with capacitance values $C_1, C_2, \ldots, C_n$ are connected in series, the combined capacitance value is

$$C = \frac{1}{\frac{1}{C_1} + \frac{1}{C_2} + \cdots + \frac{1}{C_n}}$$

More discussion of the parallel and series connection of capacitors can be found in Section 6.3, later in this chapter.

The cross-sectional view of a cylindrical capacitor with length L is shown in Figure 6.4. An inner conductor with charge Q has a radius a and the outer conductor with charge $-Q$ has a radius b. The space between the conductors are filled with dielectric material with permittivity ε. The direction of the electric field inside the dielectric medium is radial.

FIGURE 6.4

A cylindrical capacitor.

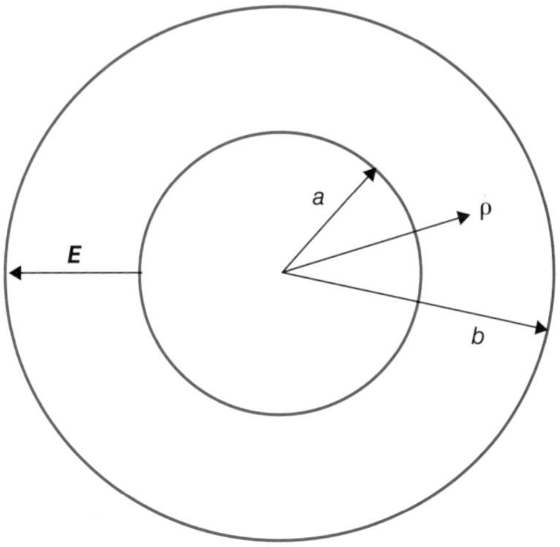

Applications of Gauss's law on the surface of radius ρ results in

$$\oint \varepsilon \mathbf{E} \cdot d\mathbf{s} = \varepsilon E 2\pi\rho L = Q$$

Thus, we have

$$E = \frac{Q}{\varepsilon 2\pi\rho L}$$

The voltage difference between the conductors is

$$V_{ab} = V_a - V_b = -\int_b^a \frac{Q}{\varepsilon 2\pi\rho L}\,d\rho = \frac{Q}{\varepsilon 2\pi L}\left[-\int_b^a \frac{1}{\rho}\,d\rho\right]$$

$$= \frac{Q}{\varepsilon 2\pi L}\left[-\ln\left(\frac{a}{b}\right)\right] = \frac{Q}{\varepsilon 2\pi L}\ln\left(\frac{b}{a}\right)$$

The capacitance of a cylindrical capacitor is given by

$$C = \frac{Q}{V_{ab}} = \frac{2\pi\varepsilon L}{\ln\left(\frac{b}{a}\right)}$$

EXAMPLE 6.1

Find the capacitance of a cylindrical capacitor with length of 15 m, inner radius of 5 mm, and outer radius of 7 mm. The relative permittivity of the medium is 8.87.

The capacitance is given by

$$C = \frac{2\pi\varepsilon_r\varepsilon_0 L}{\ln\left(\frac{b}{a}\right)} = \frac{2 \times \pi \times 8.87 \times 8.8541878176 \times 10^{-12} \times 15}{\ln\left(\frac{0.007}{0.005}\right)}$$

$$= 2.19986 \times 10^{-8}\ \text{F} = 21.9986\ \text{nF}$$

Exercise 6.1

Find the capacitance of a parallel plate capacitor with area of 12 m² and separation of 6 mm. The relative permittivity of the medium is 5.7.

Answer:

$$C = \frac{\varepsilon_r\varepsilon_0 S}{d} = \frac{5.7 \times 8.8541878176 \times 10^{-12} \times 12}{6 \times 10^{-3}} = 1.0094 \times 10^{-7}\ \text{F} = 0.10094\ \mu\text{F}$$

FIGURE 6.5

Voltage across and current through a capacitor.

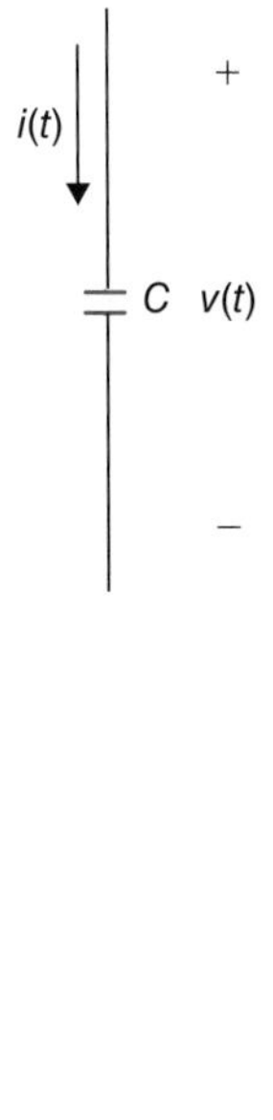

Let $v(t)$ be the voltage across a capacitor and $i(t)$ be the current through the capacitor, as shown in Figure 6.5.

Since the current is defined as the time rate of change of the charge, we have

$$i(t) = \frac{dQ}{dt} \tag{6.9}$$

Substituting $Q = Cv(t)$ into Equation (6.9), we have

$$i(t) = \frac{dCv(t)}{dt} = C\frac{dv(t)}{dt} \tag{6.10}$$

The current through the capacitor is proportional to the time rate of change of the voltage applied to the capacitor. When a dc voltage is applied to a capacitor, the current through the capacitor is zero in the steady state $[dv(t)/dt = 0]$. The capacitor acts as an open circuit to dc input in the steady state.

If Equation (6.10) is integrated, we obtain

$$v(t) = \frac{1}{C}\int_{-\infty}^{t} i(\lambda)d\lambda = v(0) + \frac{1}{C}\int_{0}^{t} i(\lambda)d\lambda \tag{6.11}$$

where $v(0)$ is the voltage across the capacitor at $t = 0$. The voltage across the capacitor is given by the integral of the current through the capacitor.

The instantaneous power on the capacitor is given by

$$p(t) = v(t)i(t) = v(t)C\frac{dv(t)}{dt} = Cv(t)\frac{dv(t)}{dt} \tag{6.12}$$

The energy stored on the capacitor at time t can be evaluated by integrating the instantaneous power:

$$w(t) = \int_{-\infty}^{t} p(\lambda)d\lambda = C\int_{-\infty}^{t} v(\lambda)\frac{dv(\lambda)}{d\lambda}d\lambda = C\int_{-\infty}^{t} v(\lambda)dv(\lambda) = \frac{1}{2}Cv^2(t) \quad \textbf{(6.13)}$$

If the voltage applied across the capacitor is constant [dc input; $v(t) = V$], the energy stored on the capacitor is given by

$$W = \frac{1}{2}CV^2 \quad \textbf{(6.14)}$$

Since $V = Q/C$, the energy stored on the capacitor can also be written as

$$W = \frac{Q^2}{2C} \quad \textbf{(6.15)}$$

For a parallel plate capacitor, substituting $C = \frac{\varepsilon S}{d}$ and $V = Ed$ into Equation (6.14), the energy stored in the capacitor can be written as

$$W = \frac{1}{2}CV^2 = \frac{1}{2}\left(\frac{\varepsilon S}{d}\right)(Ed)^2 = \frac{1}{2}\varepsilon E^2 Sd \quad \textbf{(6.16)}$$

Since Sd is the volume between the plates, the energy stored per volume, called *energy density,* is given by

$$W_d = \frac{1}{2}\varepsilon E^2 \quad \textbf{(6.17)}$$

Equation (6.16) indicates that the energy stored in the capacitor is proportional to the square of the electric field and the volume of the capacitor between the plates.

If $p(t) > 0$, energy is being stored in the capacitor (i.e., energy is being absorbed). The current direction is into the positive terminal of the capacitor. Charges are accumulating on the capacitor plates. If $p(t) < 0$, energy is being released from the capacitor to the rest of the circuit.

EXAMPLE 6.2

The voltage across a capacitor with capacitance of 100 μF is given by

$$v(t) = \begin{cases} 50t, & 0 \le t < 1 \\ -50t + 100, & 1 \le t < 3 \\ 50t - 200, & 3 \le t < 4 \\ 0, & otherwise \end{cases} \text{ V}$$

This waveform is shown in Figure 6.6.

a. Find the current through the capacitor $i(t)$ and plot the current as a function of time.
b. Find and plot the instantaneous power $p(t)$ on the capacitor.
c. Find and plot the instantaneous energy $w(t)$ stored on the capacitor.

continued

Example 6.2 continued

FIGURE 6.6

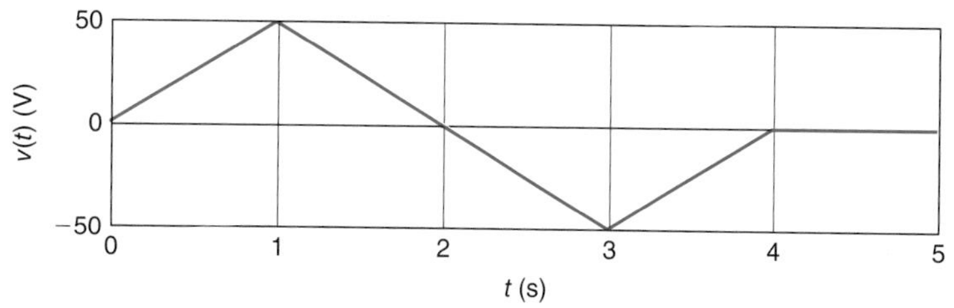

Voltage waveform for EXAMPLE 6.2.

The current through the capacitor is given by

$$i(t) = C\frac{dv(t)}{dt} = \begin{cases} 100 \times 10^{-6} \times 50 \text{ A}, & 0 \le t < 1 \\ 100 \times 10^{-6} \times (-50) \text{ A}, & 1 \le t < 3 \\ 100 \times 10^{-6} \times 50 \text{ A}, & 3 \le t < 4 \\ 0, & otherwise \end{cases} = \begin{cases} 5 \text{ mA}, & 0 \le t < 1 \\ -5 \text{ mA}, & 1 \le t < 3 \\ 5 \text{ mA}, & 3 \le t < 4 \\ 0, & otherwise \end{cases}$$

This waveform for the current is shown in Figure 6.7.

FIGURE 6.7

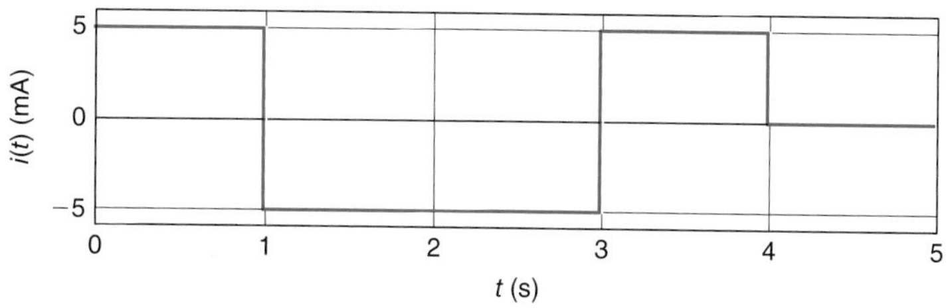

Current through the capacitor as a function of time.

The instantaneous power on the capacitor is given by

$$p(t) = v(t)i(t) = \begin{cases} 50t \times 0.005, & 0 \le t < 1 \\ (-50t + 100) \times (-0.005), & 1 \le t < 3 \\ (50t - 200) \times 0.005, & 3 \le t < 4 \\ 0, & otherwise \end{cases} = \begin{cases} 0.25t, & 0 \le t < 1 \\ 0.25t - 0.5, & 1 \le t < 3 \\ 0.25t - 1, & 3 \le t < 4 \\ 0, & otherwise \end{cases} \text{ W}$$

The waveform for the instantaneous power is shown in Figure 6.8.

FIGURE 6.8

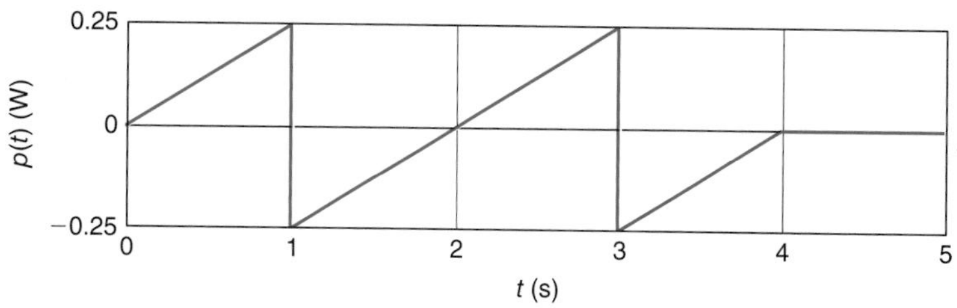

Power as a function of time.

Figure 6.8 reveals that the power is absorbed by the capacitor during time intervals $0 < t < 1$ s and $2 \text{ s} \le t < 3$ s, and the power is released by the capacitor during time intervals $1 \text{ s} \le t < 2$ s and $3 \text{ s} \le t < 4$ s.

The energy stored on the capacitor is given by

$$w(t) = \frac{1}{2}Cv^2(t) = \begin{cases} 0.5 \times 100 \times 10^{-6} \times (50t)^2, & 0 \le t < 1 \\ 0.5 \times 100 \times 10^{-6} \times (-50t + 100)^2, & 1 \le t < 3 \\ 0.5 \times 100 \times 10^{-6} \times (50t - 200)^2, & 3 \le t < 4 \\ 0, & otherwise \end{cases}$$

continued

Example 6.2 continued

$$= \begin{cases} 0.125t^2, & 0 \le t < 1 \\ 0.125(t-2)^2, & 1 \le t < 3 \\ 0.125(t-4)^2, & 3 \le t < 4 \\ 0, & otherwise \end{cases} \text{ J}$$

The waveform for energy is shown in Figure 6.9.

FIGURE 6.9

Energy as a function of time.

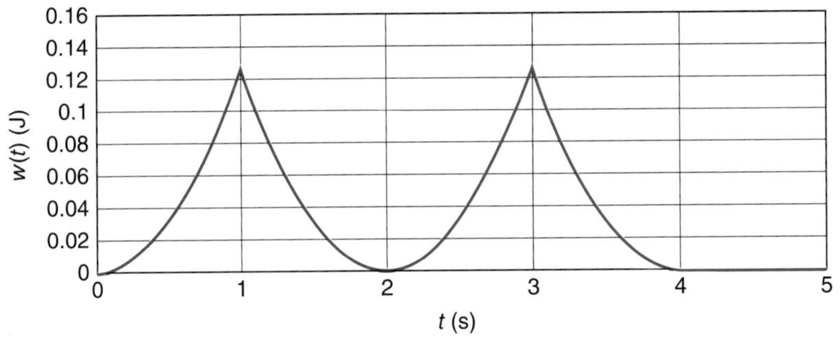

Figure 6.9 reveals that the energy stored in the capacitor increases during time intervals $0 < t < 1$ s and $2 \text{ s} \le t < 3$ s, and the energy stored in the capacitor decreases during time intervals $1 \text{ s} \le t < 2$ s and $3 \text{ s} \le t < 4$ s.

Exercise 6.2

The voltage across a capacitor with capacitance of 47 μF is given by

$$v(t) = \begin{cases} -100t, & 0 \le t < 1 \\ 200t - 300, & 1 \le t < 2 \\ -100t + 300, & 2 \le t < 3 \\ 0, & otherwise \end{cases} \text{ V}$$

This waveform is shown in Figure 6.10.

a. Find the current through the capacitor $i(t)$.
b. Find the instantaneous power $p(t)$ on the capacitor.
c. Find the instantaneous energy $w(t)$ stored on the capacitor.

FIGURE 6.10

Waveform for EXERCISE 6.2.

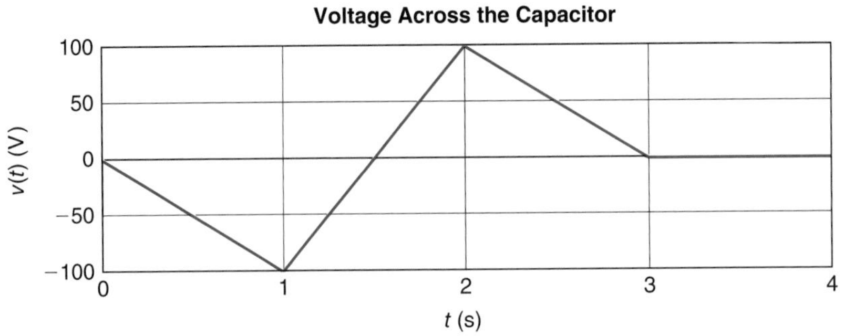

continued

Exercise 6.2 continued

Answer:

$$i(t) = \begin{cases} -4.7, & 0 \le t < 1 \\ 9.4, & 1 \le t < 2 \\ -4.7, & 2 \le t < 3 \\ 0, & otherwise \end{cases} \text{ mA}$$

$$p(t) = \begin{cases} 0.47t, & 0 \le t < 1 \\ 1.88t - 2.82, & 1 \le t < 2 \\ 0.47t - 1.41, & 2 \le t < 3 \\ 0, & otherwise \end{cases} \text{ W}$$

$$w(t) = 0.5 \times C \times v^2(t) = \begin{cases} 0.235t^2, & 0 \le t < 1 \\ 0.94(t-1.5)^2, & 1 \le t < 2 \\ 0.235(t-3)^2, & 2 \le t < 3 \\ 0, & otherwise \end{cases} \text{ J}$$

EXAMPLE 6.3

The current through a capacitor with capacitance 100 μF is shown in Figure 6.11. Find the voltage, power, and energy on the capacitor.

FIGURE 6.11

The current through the capacitor.

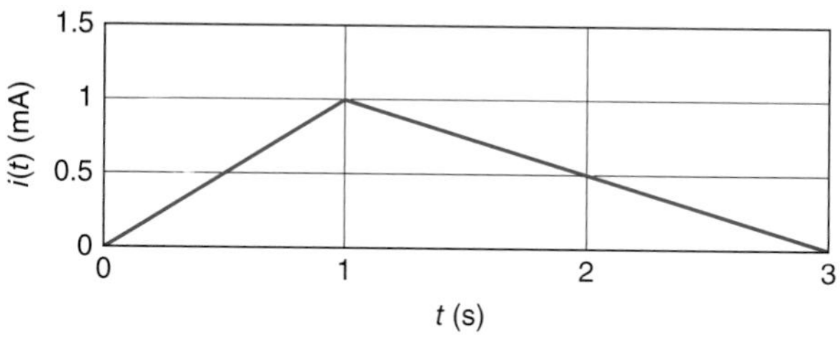

The current through the capacitor is given by

$$i(t) = \begin{cases} t, & 0 \le t < 1 \\ -\frac{t}{2} + \frac{3}{2}, & 1 \le t < 3 \\ 0, & otherwise \end{cases} \text{ mA}$$

The voltage across the capacitor is obtained by

$$v(t) = \frac{1}{C}\int_0^t i(\lambda)d\lambda$$

$$\text{For } 0 \le t < 1, v(t) = \frac{10^{-3}}{100 \times 10^{-6}}\int_0^t \lambda d\lambda = 5t^2 \text{ V}$$

For $1 \le t < 3$,

$$v(t) = 5 + \frac{10^{-3}}{100 \times 10^{-6}}\int_1^t \left(\frac{-\lambda}{2} + \frac{3}{2}\right)d\lambda = 5 + 10\left[\frac{-\lambda^2}{4} + \frac{3}{2}\lambda\right]_1^t$$

$$= 5 + 10\left(\frac{-t^2}{4} + \frac{3}{2}t - \left(\frac{-1}{4} + \frac{3}{2}1\right)\right)$$

$$= -2.5t^2 + 15t - 7.5 = -2.5(t-3)^2 + 15 \text{ V}$$

continued

Example 6.3 continued

Thus, the voltage across the capacitor is given by

$$v(t) = \begin{cases} 5t^2 \text{ V}, & 0 \le t < 1 \\ -2.5(t-3)^2 + 15 \text{ V}, & 1 \le t < 3 \text{ V} \\ 0, & otherwise \end{cases}$$

The waveform for voltage as a function of time is shown in Figure 6.12.

FIGURE 6.12

Voltage as a function of time.

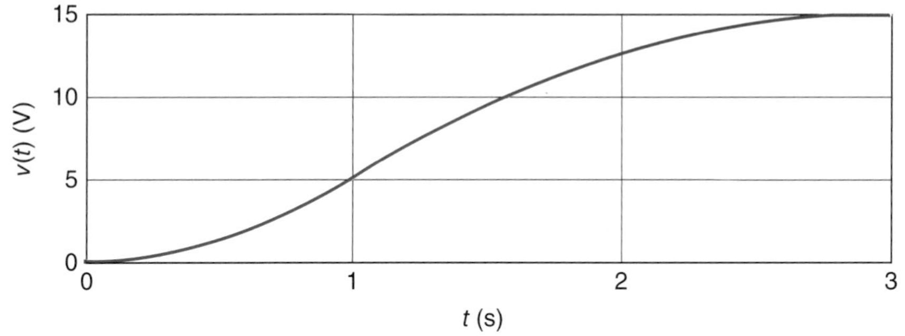

The power on the capacitor is given by

$$p(t) = v(t)i(t) = \begin{cases} \frac{t^3}{200} W, & 0 \le t < 1 \\ [(t-3)(t^2 - 6t + 3)]/800\ W, & 1 \le t < 3 \\ 0, & otherwise \end{cases}$$

The waveform for power as a function of time is shown in Figure 6.13.

FIGURE 6.13

Power as a function of time.

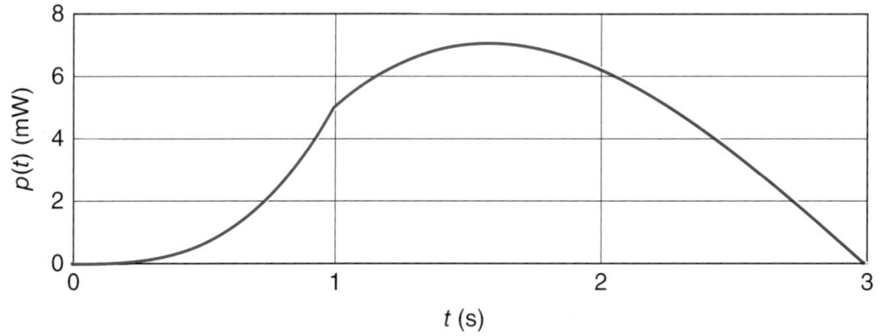

The energy on the capacitor is given by

$$w(t) = \frac{1}{2}Cv^2(t) = \begin{cases} \frac{t^4}{800} J, & 0 \le t < 1 \\ [(t-3)^2 - 6]^2/3200\ J, & 1 \le t < 3 \\ 0, & otherwise \end{cases}$$

The waveform for energy as a function of time is shown in Figure 6.14.

continued

Example 6.3 continued

FIGURE 6.14

Energy as a function of time.

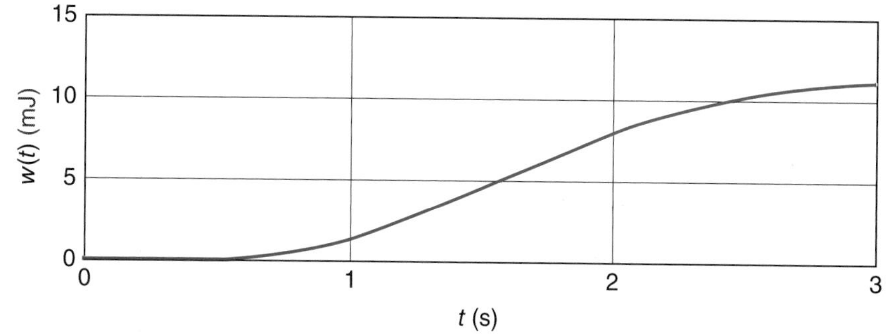

Exercise 6.3

Let the current through a capacitor with capacitance 500 μF be

$$i(t) = \begin{cases} 1\,\text{mA}, & 0 \le t < 1s \\ -1\text{ mA}, & 1 \le t < 2s \\ 0, & \textit{otherwise} \end{cases}$$

Find the voltage across the capacitor.

Answer:

$$v(t) = \begin{cases} 2t, & 0 \le t < 1s \\ -2t + 4, & 1s \le t < 2s \\ 0, & \textit{otherwise} \end{cases} \text{ V}$$

6.2.1 SINUSOIDAL INPUT TO CAPACITOR

A sinusoidal voltage,

$$v(t) = \cos(2\pi 10t)\text{ V}$$

is applied to a capacitor with capacitance $C = 0.01$ F, as shown in Figure 6.15.

The current through the capacitor is given by

$$i(t) = C\frac{dv(t)}{dt} = 0.01 \times (-1) \times (2\pi 10) \times \sin(2\pi 10t)$$
$$= -0.6383 \sin(2\pi 10t)$$
$$= 0.6283 \cos(2\pi 10t + 90°)\text{ A}$$

FIGURE 6.15

Sinusoidal input to a capacitor.

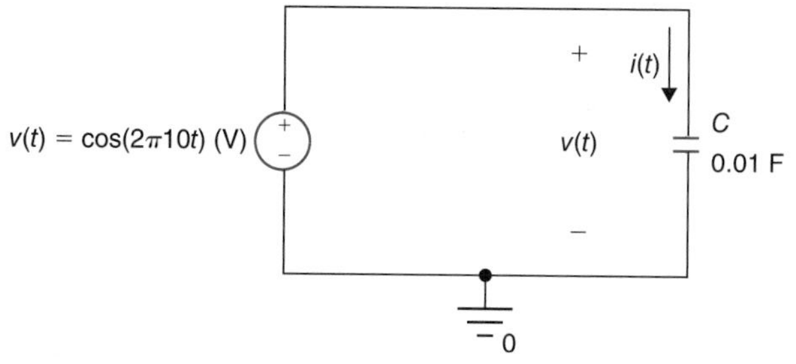

Notice that the phase of current is 90°, compared to 0° for the voltage. The current **leads** the voltage by 90°. Figure 6.16 shows $v(t)$ and $i(t)$ on the same graph. As can be seen from the graph, the current crosses zero $T/4$ s earlier than voltage. Here, T is a period given by $1/10$ s. $T/4$ s is equivalent to $360°/4 = 90°$. Leading means the waveform crosses zero earlier or reaches peak value earlier.

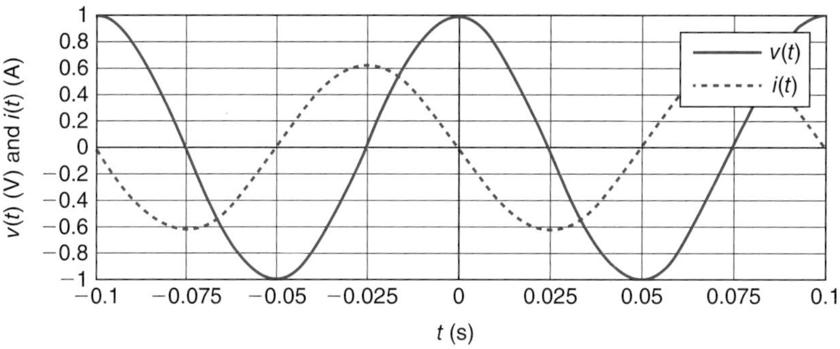

FIGURE 6.16

Voltage across a capacitor and a current through a capacitor.

A sinusoidal voltage $v(t) = V_m\cos(2\pi \mathrm{ft})$ is applied to a capacitor with capacitance C, as shown in Figure 6.17.

The current through the capacitor is given by

$$i(t) = C\frac{dv(t)}{dt} = CV_m \times (-1) \times (2\pi f) \times \sin(2\pi ft)$$
$$= -CV_m 2\pi f \times \sin(2\pi ft)$$

FIGURE 6.17

Sinusoidal input to a capacitor.

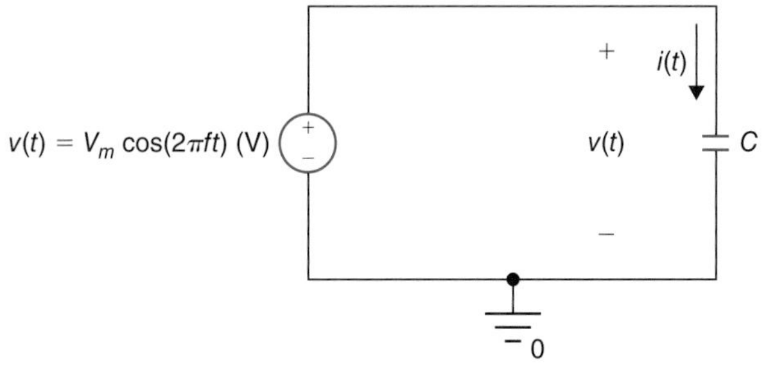

The amplitude of the current is proportional to the frequency of the voltage applied. Figure 6.18 shows $i(t)$ for $f =$ 10 Hz, 5 Hz, and 1 Hz when $V_m = 1$ V, $C = 0.01$ F. As the frequency decreases, the amplitude decreases. As the frequency decreases to zero ($f = 0$ Hz), the input voltage becomes $v(t) = V_m\cos(2\pi 0t) = V_m$, which is a *dc* voltage with amplitude V_m V. The current through the capacitor is $i(t) = -CV_m 2\pi 0\sin(2\pi 0t) = 0$. For *dc* voltage, the current through the capacitor is zero in the steady state. On the other hand, as the frequency increases, the current through the capacitor increases. As the frequency increases to infinity ($f = \infty$), the current increases to infinity as well. The capacitor acts as an open circuit for *dc* voltage and acts as a short circuit for high-frequency voltages.

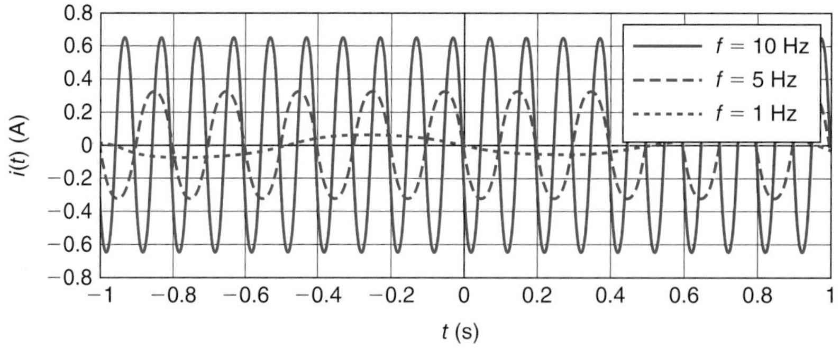

FIGURE 6.18

Current through the capacitor for different frequencies of applied voltage.

6.3 Series and Parallel Connection of Capacitors

6.3.1 SERIES CONNECTION OF CAPACITORS

Figure 6.19(a) shows n capacitors connected in series. The current through the n capacitors is $i(t)$. Let $v_1(t)$ be the voltage across C_1, $v_2(t)$ be the voltage across C_2,..., $v_n(t)$ be the voltage across C_n, and $v(t)$ be the voltage across all n capacitors. Then we have

$$v(t) = v_1(t) + v_2(t) + \cdots + v_n(t) \quad \textbf{(6.18)}$$

Substitution of the capacitor i–v relation

$$v_k(t) = \frac{1}{C_k}\int_{-\infty}^{t} i(\lambda)d\lambda, \quad 1 \le k \le n$$

to Equation (6.18) yields

$$\begin{aligned} v(t) &= \frac{1}{C_1}\int_{-\infty}^{t} i(\lambda)d\lambda + \frac{1}{C_2}\int_{-\infty}^{t} i(\lambda)d\lambda + \cdots + \frac{1}{C_n}\int_{-\infty}^{t} i(\lambda)d\lambda \\ &= \left(\frac{1}{C_1} + \frac{1}{C_2} + \cdots + \frac{1}{C_n}\right)\int_{-\infty}^{t} i(\lambda)d\lambda = \frac{1}{C_{eq}}\int_{-\infty}^{t} i(\lambda)d\lambda \end{aligned} \tag{6.19}$$

where

$$\frac{1}{C_{eq}} = \frac{1}{C_1} + \frac{1}{C_2} + \cdots + \frac{1}{C_n} \tag{6.20}$$

The equivalent capacitance C_{eq} is given by

$$C_{eq} = \frac{1}{\frac{1}{C_1} + \frac{1}{C_2} + \cdots + \frac{1}{C_n}} \tag{6.21}$$

The equivalent capacitor with capacitance C_{eq} is shown in Figure 6.19(b).

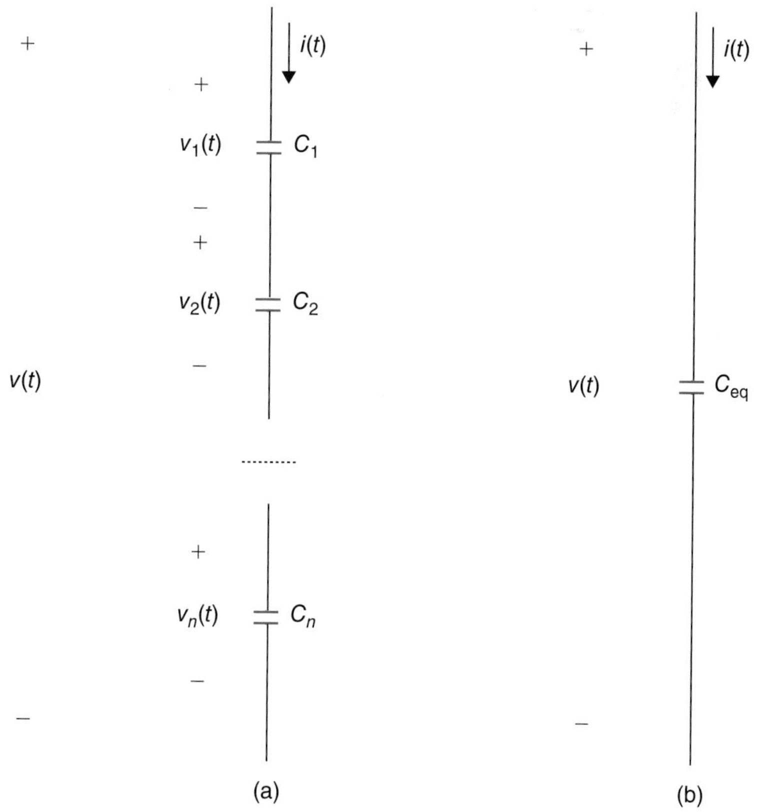

FIGURE 6.19

(a) Series connection of capacitors. (b) Equivalent capacitor.

The equivalent capacitance of two capacitors with capacitances C_1 and C_2 connected in series is given by

$$C_{eq} = \frac{1}{\frac{1}{C_1} + \frac{1}{C_2}} = \frac{C_1 C_2}{C_1 + C_2} \tag{6.22}$$

Notice that the equation for the equivalent capacitance of the series connected capacitors is similar to the equivalent resistance of parallel connected resistors.

6.3.2 PARALLEL CONNECTION OF CAPACITORS

Figure 6.20(a) shows n capacitors connected in parallel. The voltage across the n capacitors is $v(t)$. Let $i_1(t)$ be the current through C_1, $i_2(t)$ be the current through $C_2, \ldots, i_n(t)$ be the current through C_n, and let $i(t)$ be the current through all n capacitors. Then we have

$$i(t) = i_1(t) + i_2(t) + \cdots + i_n(t) \tag{6.23}$$

FIGURE 6.20

(a) Parallel connection of capacitors. (b) Equivalent capacitor.

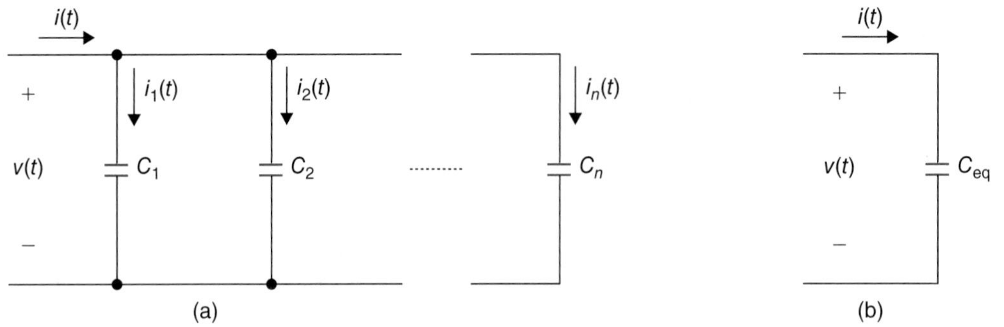

Application of the capacitor i–v relation

$$i_k(t) = C_k \frac{dv(t)}{dt}, \quad 1 \le k \le n$$

to Equation (6.23) yields

$$\begin{aligned} i(t) &= i_1(t) + i_2(t) + \cdots + i_n(t) = C_1 \frac{dv(t)}{dt} + C_2 \frac{dv(t)}{dt} + \cdots + C_n \frac{dv(t)}{dt} \\ &= (C_1 + C_2 + \cdots + C_n) \frac{dv(t)}{dt} = C_{eq} \frac{dv(t)}{dt} \end{aligned} \tag{6.24}$$

where the equivalent capacitance C_{eq} is given by

$$C_{eq} = C_1 + C_2 + \cdots + C_n \tag{6.25}$$

The equivalent capacitor with capacitance C_{eq} is shown in Figure 6.20(b). Notice that the equation for the equivalent capacitance of the parallel connected capacitors is similar to the equivalent resistance of series connected resistors.

EXAMPLE 6.4

Find the equivalent capacitance between *a* and *b* for the circuit shown in Figure 6.21.

FIGURE 6.21

Circuit for EXAMPLE 6.4.

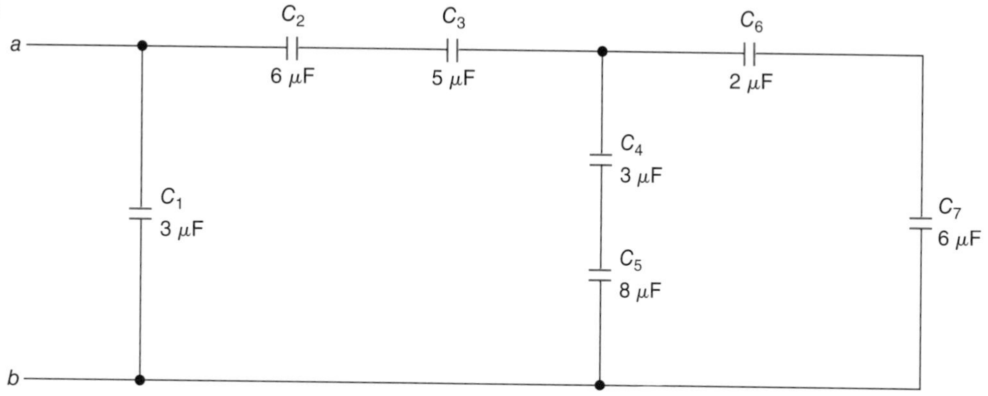

Let C_8 be the equivalent capacitance of the series connection of C_6 and C_7. Then we have

$$C_8 = \frac{C_6C_7}{C_6 + C_7} = \frac{2 \times 6}{2 + 6}\ \mu\text{F} = \frac{12}{8}\ \mu\text{F} = 1.5\ \mu\text{F}$$

Let C_9 be the equivalent capacitance of the series connection of C_4 and C_5. Then we have

$$C_9 = \frac{C_4C_5}{C_4 + C_5} = \frac{3 \times 8}{3 + 8}\ \mu\text{F} = \frac{24}{11}\ \mu\text{F} = 2.1818\ \mu\text{F}$$

Let C_{10} be the equivalent capacitance of the parallel connection of C_8 and C_9. Then we have

$$C_{10} = C_8 + C_{10} = 3.6818\ \mu\text{F}$$

Let C_{11} be the equivalent capacitance of the series connection of C_2, C_3, and C_{10}. Then we have

$$C_{11} = \frac{1}{\frac{1}{C_2} + \frac{1}{C_3} + \frac{1}{C_{10}}} = \frac{1}{\frac{1}{6} + \frac{1}{5} + \frac{1}{3.6818}}\ \mu\text{F} = 1.5667\ \mu\text{F}$$

Since C_1 and C_{11} are connected in parallel, the equivalent capacitance between a and b is given by

$$C_{eq} = C_1 + C_{11} = 3\ \mu\text{F} + 1.5667\ \mu\text{F} = 4.5667\ \mu\text{F}$$

Exercise 6.4

Find the equivalent capacitance between *a* and *b* for the circuit shown in Figure 6.22.

FIGURE 6.22

Circuit for EXERCISE 6.4.

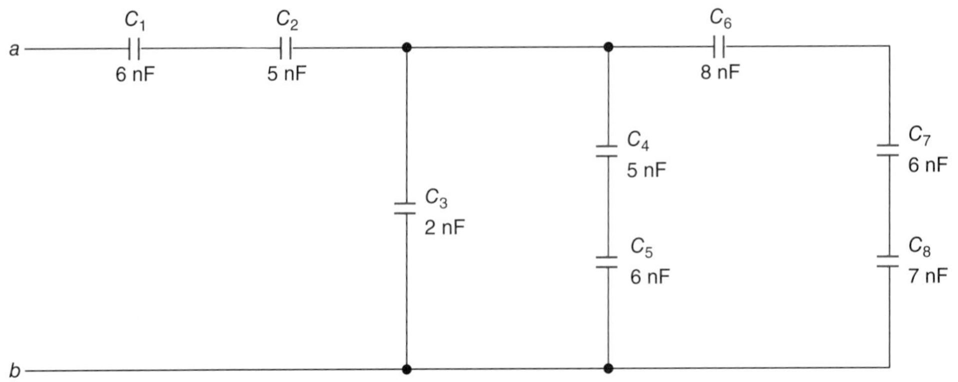

Answer:

$C_{eq} = 1.9348$ nF

EXAMPLE 6.5

You have three capacitors with capacitance values of $C_1 = 0.1\ \mu$F, $C_2 = 0.22\ \mu$F, and $C_3 = 0.47\ \mu$F, respectively. List all the capacitance values that you can get from these three capacitors.

a. Use one capacitor:

There are three values using one capacitor: 0.1 μF, 0.22 μF, 0.47 μF

b. Use two capacitors:

There are six values using two capacitors. For each pair, there is parallel connection and series connection: $C_1 + C_2 = 0.32\ \mu$F, $C_1 + C_3 = 0.57\ \mu$F, $C_2 + C_3 = 0.69\ \mu$F

$$\frac{1}{\frac{1}{C_1}+\frac{1}{C_2}} = 0.06875\ \mu\text{F}, \frac{1}{\frac{1}{C_1}+\frac{1}{C_3}} = 0.0825\ \mu\text{F}, \frac{1}{\frac{1}{C_2}+\frac{1}{C_3}} = 0.1499\ \mu\text{F}$$

c. Use three capacitors:

Parallel connection of all three capacitors: $C_1 + C_2 + C_3 = 0.79\ \mu$F

Series connection of all three capacitors: $\dfrac{1}{\frac{1}{C_1}+\frac{1}{C_2}+\frac{1}{C_3}} = 0.06\ \mu$F

Two capacitors in parallel, series with the third one:

$$\frac{1}{\frac{1}{C_1+C_2}+\frac{1}{C_3}} = 0.1904\ \mu\text{F}, \frac{1}{\frac{1}{C_1+C_3}+\frac{1}{C_2}} = 0.1587\ \mu\text{F}, \frac{1}{\frac{1}{C_2+C_3}+\frac{1}{C_1}} = 0.0873\ \mu\text{F}$$

continued

Example 6.5 continued

Two capacitors in series, parallel with the third one:

$$\frac{1}{\frac{1}{C_1}+\frac{1}{C_2}}+C_3=0.5387\ \mu\text{F},\ \frac{1}{\frac{1}{C_1}+\frac{1}{C_3}}+C_2=0.3025\ \mu\text{F},\ \frac{1}{\frac{1}{C_2}+\frac{1}{C_3}}+C_1=0.2499\ \mu\text{F}$$

There are 17 possible values:

0.06 μF, 0.06875 μF, 0.0825 μF, 0.0873 μF, 0.1 μF, 0.1499 μF, 0.1587 μF, 0.1904 μF, 0.22 μF, 0.2499 μF, 0.3025 μF, 0.32 μF, 0.47 μF, 0.5387 μF, 0.57 μF, 0.69 μF, 0.79 μF.

Exercise 6.5

You have three capacitors with capacitance values of $C_1 = 0.1\ \mu$F, $C_2 = 0.15\ \mu$F, and $C_3 = 0.18\ \mu$F, respectively. Interconnect these three capacitors to create capacitance value of 0.24 μF.

Answer:
Series connection of C_1 and C_2 provides equivalent capacitance of 0.06 μF. Parallel connection of C_3 to the series connection of C_1 and C_2 will create capacitance value of 0.24 μF.

6.4 Op Amp Integrator and Op Amp Differentiator

6.4.1 OP AMP INTEGRATOR

Figure 6.23 shows an op amp integrator.

Summing the currents leaving the inverting input of the op amp, we have

$$\frac{0-v_s(t)}{R}+C\frac{d(0-v_o(t))}{dt}=0$$

Rearrangement of this equation yields

$$\frac{dv_o(t)}{dt}=-\frac{1}{RC}v_s(t)$$

Integrating on both sides, we obtain

$$v_o(t)=-\frac{1}{RC}\int_{-\infty}^{t}v_s(\lambda)d\lambda$$

If the integrator started at $t = 0$, we have

$$v_o(t)=-\frac{1}{RC}\int_{0}^{t}v_s(\lambda)d\lambda+v_o(0)$$

where $v_o(0)$ is the initial voltage across the capacitor at $t = 0$.

If the integrator started at $t = 0$ and $v_0(0) = 0$, we have

$$v_o(t)=-\frac{1}{RC}\int_{0}^{t}v_s(\lambda)d\lambda \tag{6.26}$$

FIGURE 6.23

Op amp integrator.

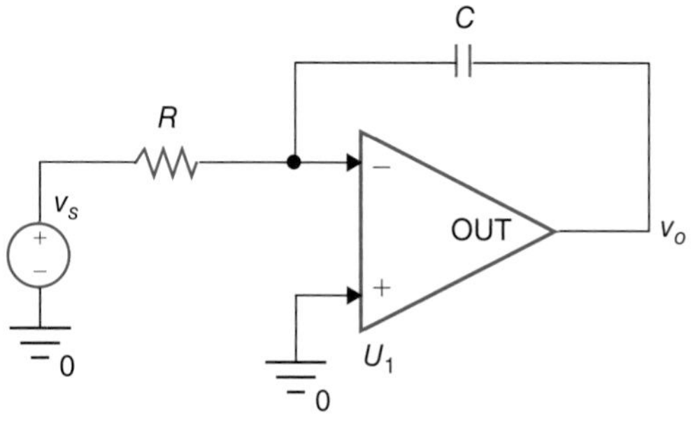

EXAMPLE 6.6

Find the output voltage $v_0(t)$ for the circuit shown in Figure 6.24.

FIGURE 6.24

An op amp integrator with a positive sign.

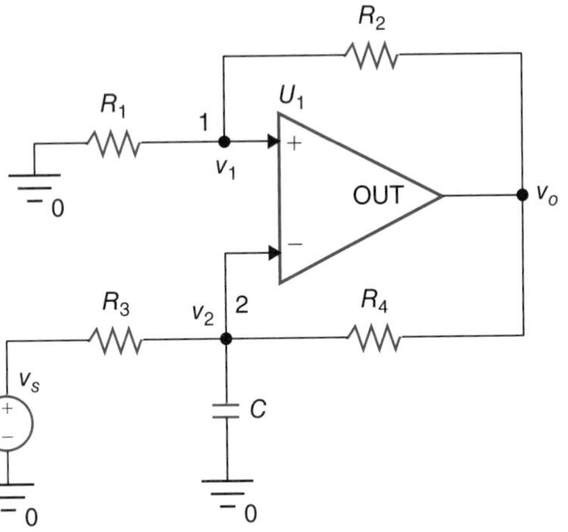

Summing the currents leaving node 1, we obtain

$$\frac{v_1}{R_1} + \frac{v_1 - v_o}{R_2} = 0$$

Solving this equation for v_1, we get

$$v_1 = \frac{R_1}{R_1 + R_2} v_o$$

Summing the currents leaving node 2, we obtain

$$\frac{v_2 - v_s}{R_3} + C\frac{dv_2}{dt} + \frac{v_2 - v_o}{R_4} = 0$$

Since $v_2 = v_1 = \dfrac{R_1}{R_1 + R_2} v_o$, we have

$$\frac{\frac{R_1}{R_1 + R_2} v_o - v_s}{R_3} + C\frac{R_1}{R_1 + R_2}\frac{dv_o}{dt} + \frac{\frac{R_1}{R_1 + R_2} v_o - v_o}{R_4} = 0$$

Assume that $R_1R_4 = R_2R_3$. Then, this equation reduces to

$$\frac{dv_o}{dt} = \frac{1 + \frac{R_2}{R_1}}{R_3C} v_s$$

Integrating on both sides, we obtain

$$v_o(t) = \frac{1 + \frac{R_2}{R_1}}{R_3C} \int_{-\infty}^{t} v_s(\lambda)d\lambda \qquad (6.27)$$

Notice that the output $v_o(t)$ is the integral of the input $v_s(t)$ with a positive gain. Also notice that the circuit shown in Figure 6.24 is the Howland current pump circuit shown in Figure 5.51 with a capacitor as the load. Under the condition $R_1R_4 = R_2R_3$, the current through the capacitor is given by v_s/R_3 as shown in Equation (5.87). Equation (6.27) can also be obtained from the i-v relation of the capacitor.

Exercise 6.6

Let $R = 1\,\text{M}\Omega$ and $C = 1\,\mu\text{F}$ in the circuit shown in Figure 6.23. Find the output signal $v_0(t)$ when the input signal is $v_s(t) = \cos(2t)$ for $t \geq 0$. Assume that $v_0(0) = 0$.

Answer:
$v_o(t) = -0.5\sin(2t)\,u(t)$.

6.4.2 OP AMP DIFFERENTIATOR

Figure 6.25 shows an op amp differentiator.

FIGURE 6.25

Op amp differentiator.

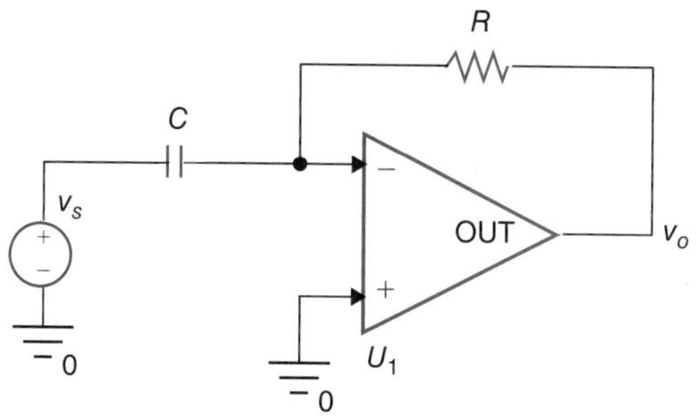

Summing the currents leaving the inverting input of the op amp, we have

$$C\frac{d(0 - v_s(t))}{dt} + \frac{0 - v_o(t)}{R} = 0$$

Solving for $v_o(t)$, we obtain

$$v_o(t) = -RC\frac{dv_s(t)}{dt} \tag{6.28}$$

Notice that the output is the derivative of the input.

6.5 Inductors

FIGURE 6.26

Circuit symbol for inductors.

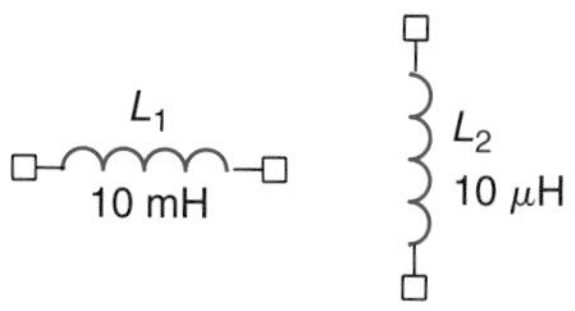

An *inductor* is a passive circuit element that can store energy in the form of a magnetic field. The circuit symbol for inductors is shown in Figure 6.26.

A brief review of magnetic field, Ampere's law, magnetic flux, and Faraday's law are given next. A point charge generates an electric field around it. Moving charges generate a magnetic field. An infinitely long and straight conductor carrying current I generates magnetic flux density $\boldsymbol{B}$ around it, as shown in Figure 6.27. The magnetic flux density $\boldsymbol{B}$ can be written as $\boldsymbol{B} = \mu\boldsymbol{H}$, where μ is permeability of the medium, and $\boldsymbol{H}$ is magnetic field intensity, also called magnetic field strength. The unit for magnetic flux density is tesla (T), the unit for magnetic field intensity is ampere per meter (A/m), and the unit for permeability is henry per meter (H/m). The direction of the magnetic flux density around a wire that carries current I can be found using the right-hand rule, which says that when the wire is wrapped with the right hand with the thumb pointing in the direction of the current, the direction of the magnetic field coincides with the direction of the four fingers. If the wire is formed into a coil of many turns, as shown in Figure 6.28, the magnetic flux density inside the coil will be intensified. If an iron core is provided inside the coil, the magnetic flux density will be even higher.

FIGURE 6.27

A magnetic field around an infinitely long, straight conductor.

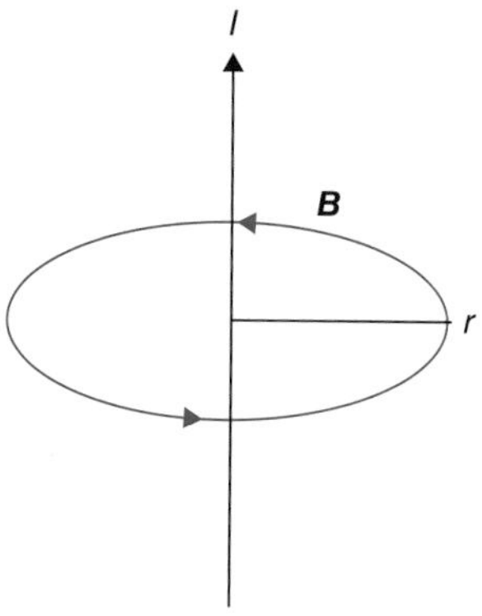

Ampere's law is the magnetic equivalent of Gauss's law. It says that the integral of the magnetic flux density along a closed loop is equal to the product of the permeability of free space μ_0 and the total current I flowing through the loop; that is,

$$\oint \boldsymbol{B} \cdot d\ell = \mu_0 I \tag{6.29}$$

where the permeability of free space μ_0 is given by

$$\mu_0 = 4\pi \times 10^{-7}\text{H/m} \tag{6.30}$$

If the medium is not free space, μ_0 is replaced by the permeability μ of the medium. For a long, straight wire carrying a current I, the magnetic flux density at a distance r on a circle in the tangential direction, as shown in Figure 5.27, is constant. The left side of Equation (6.29) becomes $B2\pi r$. Thus, we have

$$B = \frac{\mu_0 I}{2\pi r} \tag{6.31}$$

The magnetic flux Φ is defined as the integral of magnetic flux density $\boldsymbol{B}$ over an area S; that is,

$$\Phi = \int_S \boldsymbol{B} \cdot d\boldsymbol{s} \quad (6.32)$$

The unit for magnetic flux is Weber (Wb). Notice that

$$1 \text{ Wb} = 1 \text{ Tm}^2$$

The magnetic flux through the surface of area A from a uniform magnetic field B is given by

$$\Phi = \boldsymbol{B} \cdot \boldsymbol{A} = BA\cos(\theta) \quad (6.33)$$

where θ is the angle between $\boldsymbol{B}$ and the flat surface $\boldsymbol{A}$.

The simplest inductor is a solenoid, which consists of an insulated conducting wire made of a substance such as copper, tightly wound as a coil. The core of the coil can be filled with air, nickel-zinc ferrite, manganese-zinc ferrite, iron powder, or silicon steel. A solenoid is shown in Figure 6.28.

FIGURE 6.28

A solenoid.

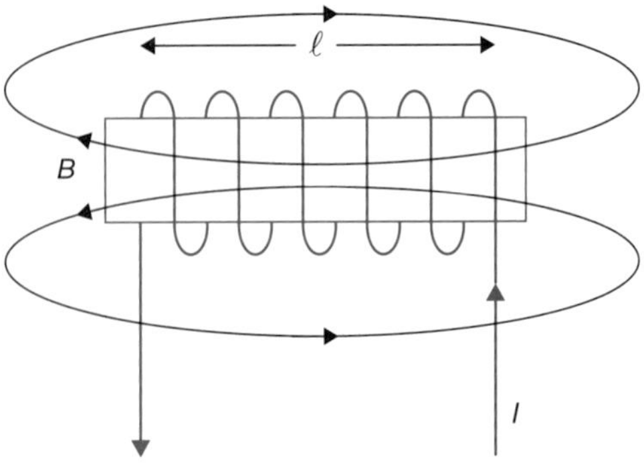

If the wire carries a current I, a magnetic field is generated around the solenoid. The direction of magnetic field follows the right-hand rule. The magnetic field is strongest inside the solenoid. If the diameter of the solenoid is small compared to the length of the solenoid, the magnetic field inside the solenoid can be assumed to be uniform. The strength of the magnetic flux density inside the solenoid can be found by applying Ampere's law. Let N be the number of turns and ℓ be the length of the solenoid, and let the closed path be a rectangle, as shown in Figure 6.29. Then we have

$$B\ell = \mu_0 NI \quad (6.34)$$

Notice that $B\ell$ represents an integral along c from right to left. The magnetic field along path a is negligible and the integral along b and d is zero because the magnetic field is perpendicular to the path of integration.

FIGURE 6.29

A closed path to calculate B inside the solenoid.

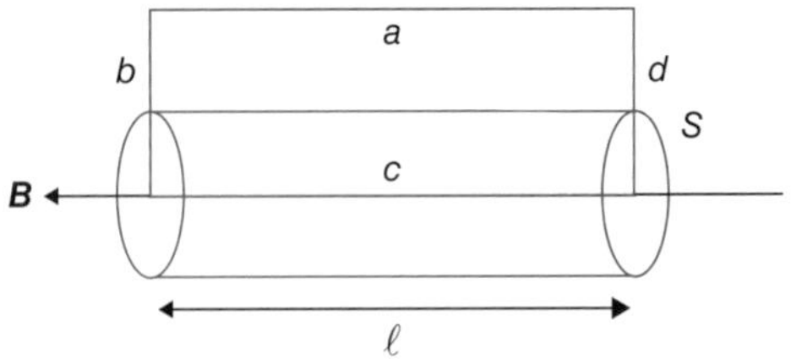

From Equation (6.34), the magnitude of magnetic flux density inside the solenoid is given by

$$B = \frac{\mu_0 NI}{\ell} \quad (6.35)$$

The direction of $\boldsymbol{B}$ is from right to left. Since the magnetic flux density is assumed to be uniform inside the solenoid and the direction of the magnetic field is the normal direction to the cross-sectional area inside the solenoid, the magnetic flux inside the solenoid is given by

$$\Phi = \int_S \boldsymbol{B} \cdot d\boldsymbol{s} = BS = \frac{\mu_0 NI}{\ell} S \quad (6.36)$$

where S is the cross-sectional area inside the solenoid, as shown in Figure 6.29. If the radius of the solenoid is r, the cross-sectional area is $S = \pi r^2$ and the magnetic flux is given by

$$\Phi = \frac{\mu_0 NI\pi r^2}{\ell} \quad (6.37)$$

The magnetic flux density lines shown in Figure 6.28 form closed loops. Each loop links (encircles) all N turns of the solenoid. The magnetic flux passing through the entire core of solenoid links all N turns. The total magnetic flux linkage Λ of the solenoid is defined as the product of the magnetic flux Φ inside the solenoid and the number of turns; that is,

$$\Lambda = N\Phi = \frac{\mu_0 N^2 I \pi r^2}{\ell} \tag{6.38}$$

The inductance, also called *self-inductance*, is defined as the ratio of the total magnetic flux linkage Λ to the current I through the inductor:

$$L = \frac{\Lambda}{I} = \frac{N\Phi}{I} = \frac{\mu_0 N^2 \pi r^2}{\ell} \tag{6.39}$$

where

μ_0 = permeability of free space
N = number of turns
r = radius of the solenoid
ℓ = length of the solenoid

If a ferromagnetic material with permeability μ is put inside the solenoid instead of free space with permeability μ_0, the inductance is given by

$$L = \frac{\mu N^2 \pi r^2}{\ell} \tag{6.40}$$

A toroid with a rectangular cross section, as shown in Figure 6.30, has N turns of wire that carries current I. Let the inner radius be a, the outer radius be b, and the height be h. Let the magnetic flux density inside the toroid be $\boldsymbol{B} = B_\phi \boldsymbol{a}_\phi$, where $\boldsymbol{a}_\phi$ is a unit vector in the circular direction inside the toroid. Then, from Ampere's law, we have

$$\oint \boldsymbol{B} \cdot d\ell = \int_0^{2\pi} (B_\phi \boldsymbol{a}_\phi) \cdot (\rho d\phi \boldsymbol{a}_\phi) = \int_0^{2\pi} B_\phi \rho d\phi = 2\pi B_\phi \rho = \mu_0 NI$$

where ρ is the radius from center of the toroid. The magnetic flux density inside the toroid is

$$B_\phi = \frac{\mu_0 NI}{2\pi\rho}$$

The magnetic flux inside the toroid is

$$\Phi = \int_S \frac{\mu_0 NI}{2\pi\rho} \boldsymbol{a}_\phi \cdot h d\rho \boldsymbol{a}_\phi = \frac{\mu_0 NIh}{2\pi} \int_a^b \frac{1}{\rho} d\rho = \frac{\mu_0 NIh}{2\pi} ln\left(\frac{b}{a}\right) \tag{6.41}$$

The flux linkage Λ is given by

$$\Lambda = N\Phi = \frac{\mu_0 N^2 Ih}{2\pi} \ln\left(\frac{b}{a}\right)$$

Thus, the inductance of a toroid is

$$L = \frac{\Lambda}{I} = \frac{\mu_0 N^2 h}{2\pi} \ln\left(\frac{b}{a}\right) \tag{6.42}$$

FIGURE 6.30

A toroid.

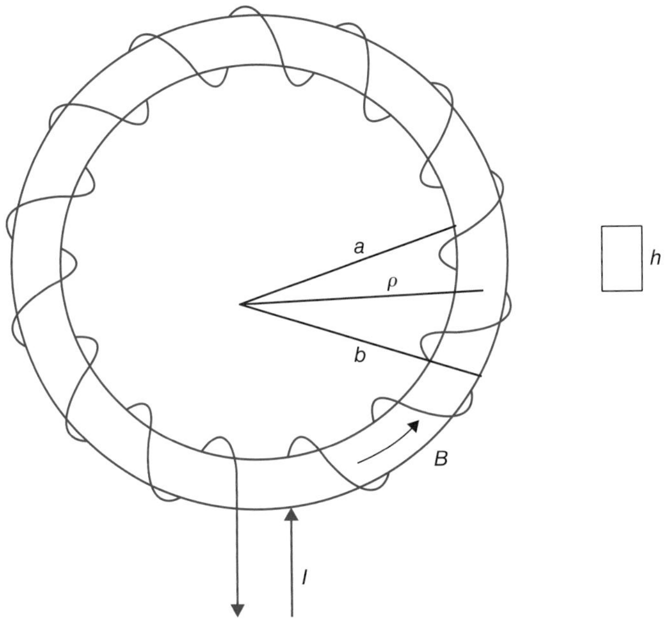

EXAMPLE 6.7

Find the inductance of a toroid with 800 turns, inner radius 50 mm, outer radius 60 mm, and height 15 mm.

From Equation (6.42), we have

$$L = \frac{\mu_0 N^2 h}{2\pi} \ln\left(\frac{b}{a}\right) = \frac{4\pi \times 10^{-7} \times (800)^2 \times 0.015}{2\pi} \ln\left(\frac{0.06}{0.05}\right) = 350.0574\ \mu\text{H}$$

Exercise 6.7

Find the inductance of a solenoid with 2500 turns, radius 30 mm, and length 1 m.

Answer:

$$L = \frac{\mu_0 N^2 \pi r^2}{\ell} = \frac{4\pi \times 10^{-7} \times 2500^2 \times \pi \times (0.03)^2}{1} = 22.2066\ \text{mH}$$

A current carrying conductor generates magnetic field around it. The opposite of this statement is "magnetic field generates current in a closed circuit." This is true if the magnetic field changes with time. The Faraday's law of induction says that changing magnetic flux induces electromotive force (emf) in the closed circuit. The emf is the source of the electric field that enables the current to flow in the closed circuit like loop. The induced electromotive force is proportional to the negative rate of change of magnetic flux.

$$\xi = -\frac{d\Phi}{dt} \tag{6.43}$$

If the Faraday's law is applied to a coil of N turns, an emf appears in every turn and these emfs are to be added. If the coil is so tightly wound that each turn can be said to occupy the same region of space, the flux through each turn will then be the same. The induced emf is given by

$$\xi = -N\frac{d\Phi}{dt} = -\frac{d(N\Phi)}{dt} \tag{6.44}$$

The Lenz's law can be used to find the direction of the induced current. The Lenz's law says that the induced current will appear in such a direction that it opposes the change that produced it. Refer to Figure 6.31.

FIGURE 6.31

Lenz's law.

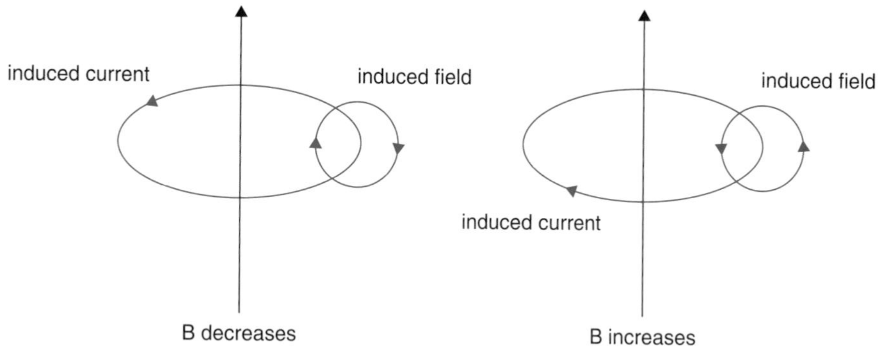

For the solenoid, substituting the magnetic flux from Equation (6.37) into Equation (6.44), we have

$$\xi = -\frac{\mu_0 N^2 \pi r^2}{\ell}\frac{dI}{dt} \tag{6.45}$$

The constant term is inductance L. Using the inductance, Equation (6.45) can be rewritten as

$$\xi = -L\frac{di(t)}{dt} \tag{6.46}$$

where the current I was changed to $i(t)$ to include time variance.

If an inductor is traversed in the direction of current, and the time rate of change of the current is positive, $\frac{di(t)}{dt} > 0$, as shown in Figure 6.32(a), according to Lenz's law, the current $i_{\text{induced}}(t)$ generated from the induced emf will flow in the opposite direction of $i(t)$ to oppose the increase in $i(t)$, as shown in Figure 6.32(b). The inductor can be replaced by an emf, as shown in Figure 6.32(b). To generate the induced current in the direction shown in Figure 6.32(b), the polarity of the induced emf should be the one shown in Figure 6.32(b) since the current originates from the positive terminal of the voltage source into the rest of the circuit and returns to the negative terminal of the voltage source.

FIGURE 6.32

Polarity of the voltage $v(t)$ for $\frac{di(t)}{dt} > 0$.

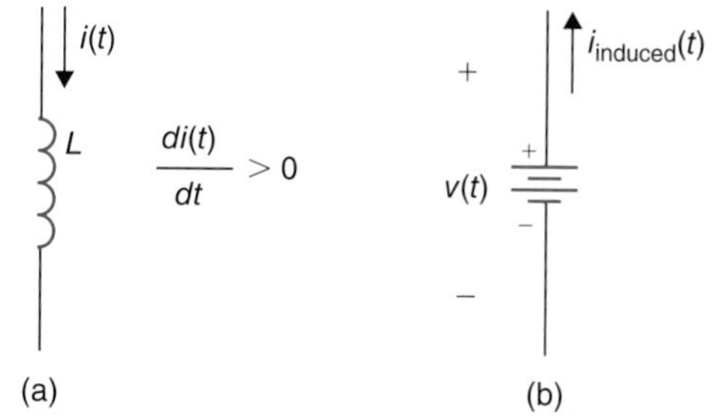

If the positive terminal of the voltage $v(t)$ across the inductor is defined as the terminal where the current is entering the inductor, and if $\frac{di(t)}{dt} > 0$, the voltage across the inductor is given by

$$v(t) = L\frac{di(t)}{dt}$$

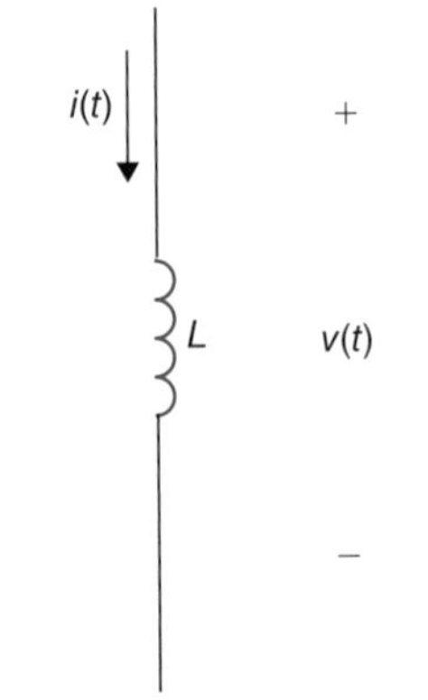

FIGURE 6.33

The direction of $i(t)$ and polarity of $v(t)$.

This is shown in Figure 6.33. The voltage $v(t)$ across the inductor, as defined in Figure 6.33, is positive.

If an inductor is traversed in the direction of current, and the time rate of change of the current is negative, $\frac{di(t)}{dt} < 0$, as shown in Figure 6.34(a), according to Lenz's law, the current $i_{\text{induced}}(t)$ generated from the induced emf will flow in the same direction as $i(t)$ to oppose the decrease in $i(t)$, as shown in Figure 6.34(b). The inductor can be replaced by an emf, as shown in Figure 6.34(b). To generate the induced current in the direction shown in Figure 6.34(b), the polarity of the induced emf should be the one shown in Figure 6.34(b) since the current originates from the positive terminal of the voltage source into the rest of the circuit and returns to the negative terminal of the voltage source.

If the positive terminal of the voltage $v(t)$ across the inductor is defined as the terminal where the current is entering the inductor, and if $\frac{di(t)}{dt} < 0$, the voltage across the inductor is given by

$$v(t) = L\frac{di(t)}{dt}$$

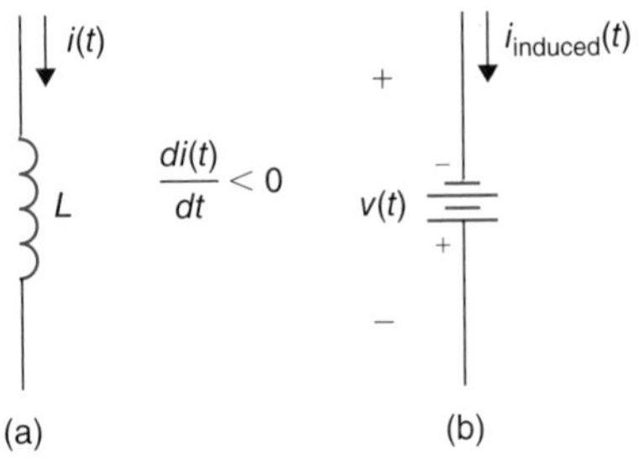

FIGURE 6.34

Polarity of the voltage $v(t)$ for $\frac{di(t)}{dt} < 0$.

This time, the voltage $v(t)$ across the inductor is negative. Thus, the voltage across the inductor can always be written as

$$v(t) = L\frac{di(t)}{dt} \qquad \textbf{(6.47)}$$

where the direction of $i(t)$ and $v(t)$ are shown in Figure 6.33.

According to Equation (6.47), the voltage across the inductor is proportional to the time rate of change of the current through the inductor. When a dc current is applied to an inductor, the voltage across the inductor is zero in the steady state because $di(t)/dt = 0$. The inductor acts as a short circuit to dc current in the steady state.

If Equation (6.47) is integrated, we obtain

$$i(t) = \frac{1}{L}\int_{-\infty}^{t} v(\lambda)d\lambda = i(0) + \frac{1}{L}\int_{0}^{t} v(\lambda)d\lambda \qquad \textbf{(6.48)}$$

where $i(0)$ is the current through the inductor at $t = 0$. The current through the inductor is given by the integral of the voltage across the inductor.

The instantaneous power on the inductor is given by

$$p(t) = i(t)v(t) = i(t)\left[L\frac{di(t)}{dt}\right] = Li(t)\frac{di(t)}{dt} \qquad \textbf{(6.49)}$$

The energy stored on the inductor at time t can be evaluated by integrating the instantaneous power:

$$w(t) = \int_{-\infty}^{t} p(\lambda)d\lambda = L\int_{-\infty}^{t} i(\lambda)\frac{di(\lambda)}{d\lambda}d\lambda = L\int_{-\infty}^{t} i(\lambda)di(\lambda) = \frac{1}{2}Li^2(t) \qquad \textbf{(6.50)}$$

If the current through the inductor is constant (dc input), $i(t) = I$, and the energy stored on the inductor is given by

$$W = \frac{1}{2}LI^2 \qquad \textbf{(6.51)}$$

For a solenoid, by substituting Equation (6.39) and Equation (6.34) into Equation (6.51), the energy stored in the inductor can be written as

$$W = \frac{1}{2}LI^2 = \frac{1}{2}\left(\frac{\mu_0 N^2 \pi r^2}{\ell}\right)\left(\frac{B\ell}{\mu_0 N}\right)^2 = \frac{B^2}{2\mu_0}\pi r^2 \ell \qquad (6.52)$$

Since $\pi r^2 \ell$ is the volume inside the solenoid, the energy density is given by

$$W_d = \frac{1}{2}\frac{B^2}{\mu_0} \qquad (6.53)$$

Equation (6.52) indicates that the energy stored in the inductor is proportional to the square of the magnetic flux density and the volume of the inductor inside the coil.

EXAMPLE 6.8

The current through an inductor with inductance of 100 mH is given by

$$i(t) = \begin{cases} 20t, & 0 \le t < 1 \\ -40t + 60, & 1 \le t < 2 \\ 10t - 40, & 2 \le t < 4 \\ 0, & otherwise \end{cases} \text{ A}$$

This waveform is shown in Figure 6.35.

a. Find the voltage across the inductor $v(t)$ and plot the voltage as a function of time.
b. Find and plot the instantaneous power $p(t)$ on the inductor.
c. Find and plot the instantaneous energy $w(t)$ stored on the inductor.

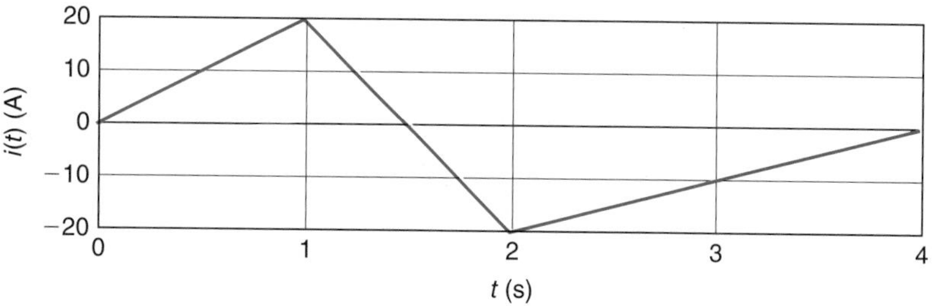

FIGURE 6.35 Current waveform for EXAMPLE 6.8.

$$v(t) = L\frac{di(t)}{dt} = \begin{cases} 0.1 \times 20, & 0 \le t < 1s \\ 0.1 \times (-40), & 1s \le t < 2s \\ 0.1 \times 10, & 2s \le t < 4s \\ 0, & otherwise \end{cases} = \begin{cases} 2, & 0 \le t < 1 \\ -4, & 1 \le t < 2 \\ 1, & 2 \le t < 4 \\ 0, & otherwise \end{cases} \text{ V}$$

The voltage waveform is shown in Figure 6.36.

$$p(t) = i(t)v(t) = \begin{cases} 2 \times 20t, & 0 \le t < 1 \\ -4 \times (-40t + 60), & 1 \le t < 2 \\ 1 \times (10t - 40), & 2 \le t < 4 \\ 0, & otherwise \end{cases} = \begin{cases} 40t, & 0 \le t < 1 \\ 160t - 240, & 1 \le t < 2 \\ 10t - 40, & 2 \le t < 4 \\ 0, & otherwise \end{cases} \text{ W}$$

continued

Example 6.8 continued

FIGURE 6.36

Voltage across the inductor.

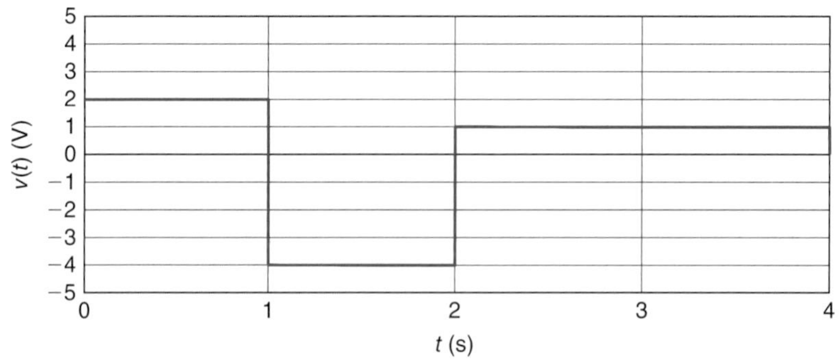

The waveform for the instantaneous power on the inductor is shown in Figure 6.37.

FIGURE 6.37

Power on the inductor.

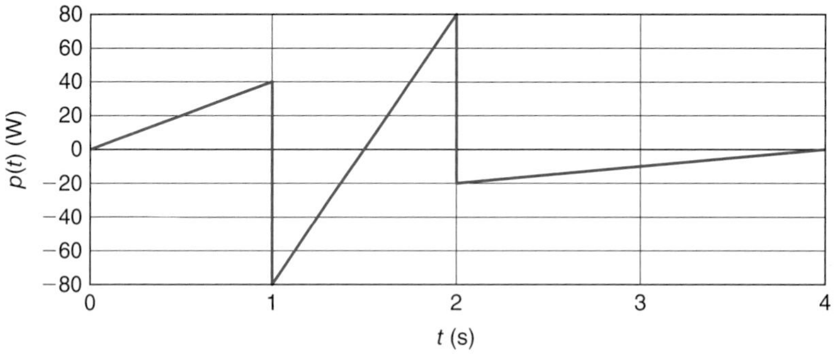

Figure 6.37 reveals that the power is absorbed by the inductor during time intervals $0 < t < 1s$ and $1.5s \leq t < 2s$, and the power is released by the inductor during time intervals $1s \leq t < 1.5s$ and $2s \leq t < 4s$. The instantaneous energy on the inductor is given by

$$w(t) = \frac{1}{2}Li^2(t) = \begin{cases} 0.05 \times (20t)^2, & 0 \leq t < 1 \\ 0.05 \times (-40t + 60)^2, & 1 \leq t < 2 \\ 0.05 \times (10t - 40)^2, & 2 \leq t < 4 \\ 0, & otherwise \end{cases} = \begin{cases} 20t^2, & 0 \leq t < 1 \\ 80(t - 1.5)^2, & 1 \leq t < 2 \\ 5(t - 4)^2, & 2 \leq t < 4 \\ 0, & otherwise \end{cases} \text{ J}$$

The waveform for the instantaneous energy stored on the inductor is shown in Figure 6.38.

FIGURE 6.38

Energy on the inductor.

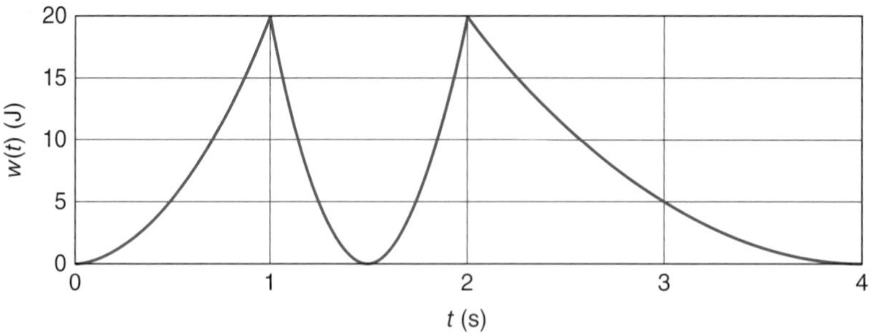

Figure 6.38 reveals that the energy stored in the inductor increases during time intervals $0 < t < 1$ s and 1.5 s $\leq t < 2$ s, and the energy stored in the inductor decreases during time intervals 1 s $\leq t < 1.5$ s and 2 s $\leq t < 4$ s.

Exercise 6.8

The current through an inductor with inductance 0.5 H is given by

$$i(t) = 4\,e^{-2t}\sin(5t), t \geq 0\ (\text{A})$$

Find the voltage across the inductor.

Answer:
$v(t) = e^{-2t}[-4\sin(5t) + 10\cos(5t)]$ V.

EXAMPLE 6.9

A sinusoidal signal

$$v(t) = \cos(2\pi 100\text{t}), t \geq 0, \text{V}$$

is shown in Figure 6.39. This signal is applied to an inductor with inductance of 100 mH. Find and plot the current $i(t)$, power $p(t)$, and energy $w(t)$ on the inductor.

FIGURE 6.39

Voltage across the inductor.

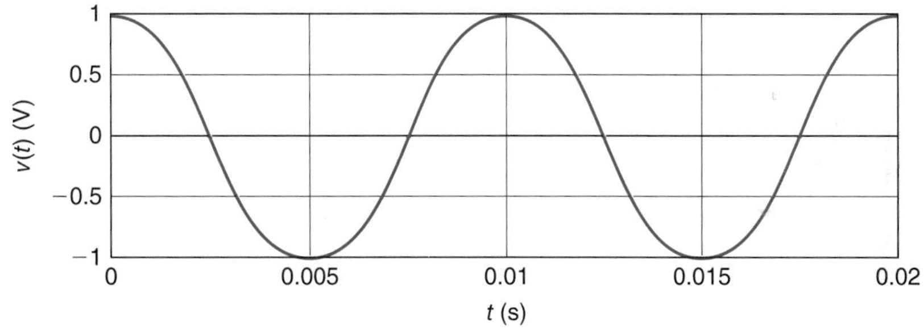

The current through the inductor is given by

$$i(t) = \frac{1}{L}\int_0^t v(\lambda)d\lambda = \frac{1}{0.1}\int_0^t \cos(2\pi 100\lambda)d\lambda = \frac{10}{2\pi 100}\sin(2\pi 100t)$$
$$= 0.01592\sin(2\pi 100t)\ \text{A} = 15.92\sin(2\pi 100t)\ \text{mA}$$

The current through the inductor is shown in Figure 6.40.

The power on the inductor is the product of the voltage and the current. Thus, we have

$$p(t) = v(t)i(t) = \frac{\sin(2\pi 200t)}{400\pi} = 0.0079577472\sin(2\pi 200t)$$
$$= 7.9577472\sin(2\pi 2000t)\ \text{mW}$$

continued

Example 6.9 continued

FIGURE 6.40

Current through the inductor.

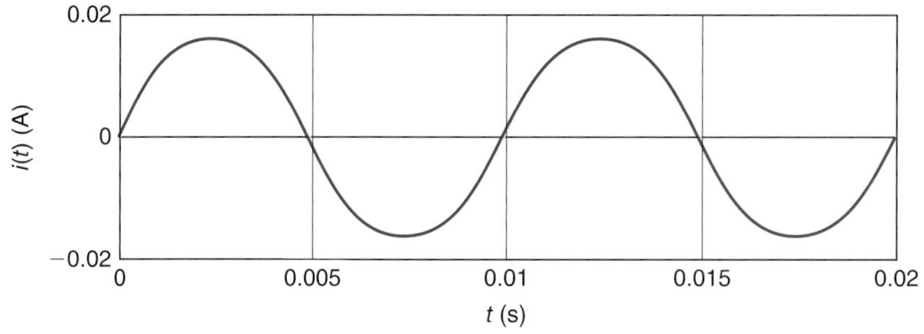

The power on the inductor is shown in Figure 6.41.

FIGURE 6.41

Power on the inductor.

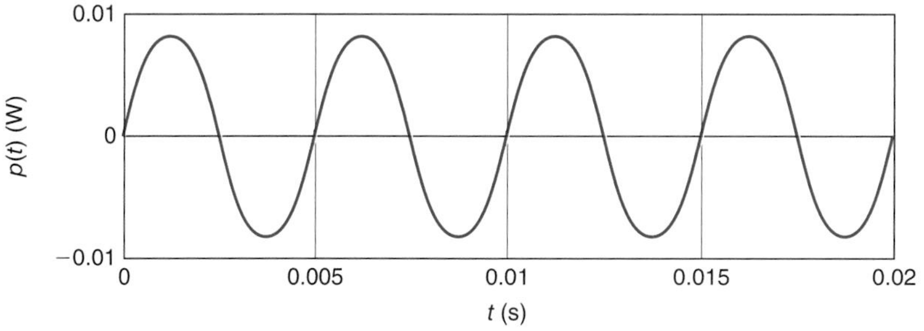

The energy on the inductor is given by

$$w(t) = \frac{1}{2}Li^2(t) = 1.2665 \times 10^{-5}[\sin(2\pi 100t)]^2 = 12.665[\sin(2\pi 100t)]^2 \ \mu\text{J}$$

The energy on the inductor is shown in Figure 6.42.

FIGURE 6.42

Energy on the inductor.

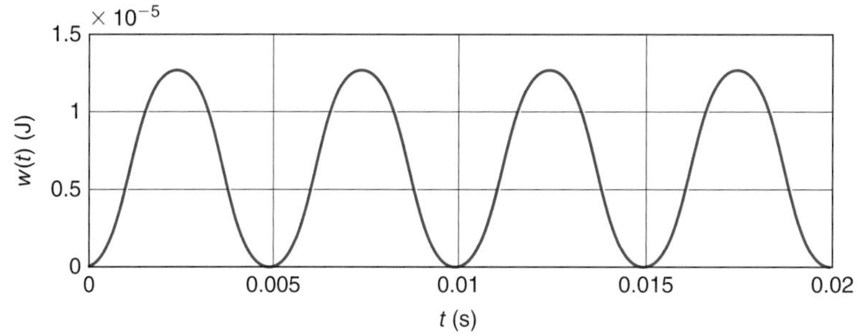

Exercise 6.9

Let the voltage across an inductor with inductance 500 mH be

$$v(t) = \begin{cases} 0.25 \text{ V}, & 0 \le t < 1 \text{ s} \\ -0.25 \text{ V}, & 1 \le t < 2 \text{ s} \\ 0, & \textit{elsewhere} \end{cases}$$

Find the current through the inductor.

continued

Exercise 6.9 continued

Answer:

$$i(t) = \begin{cases} 0.5t \text{ A}, & 0 \le t < 1 \text{ s} \\ -0.5t + 1 \text{ A}, & 1 \le t < 2 \text{ s} \\ 0, & elsewhere \end{cases}$$

6.5.1 SINUSOIDAL INPUT TO INDUCTOR

A sinusoidal current,

$$i(t) = \cos(2\pi 10t) \text{ A}$$

is applied to an inductor with inductance $L = 10$ mH, as shown in Figure 6.43.

The voltage across the inductor is given by

FIGURE 6.43

An inductor with sinusoidal input.

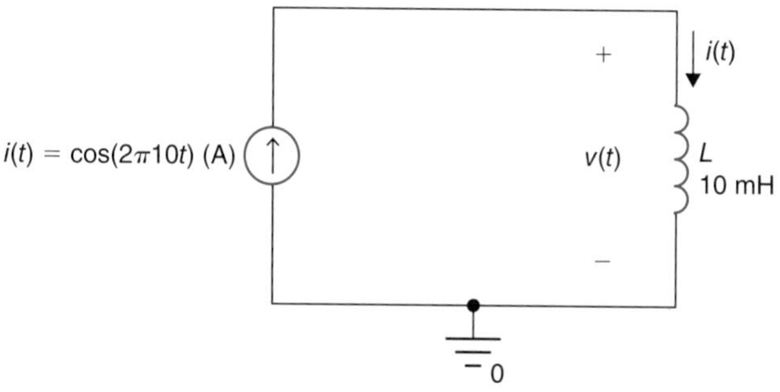

$$v(t) = L\frac{di(t)}{dt} = 0.01 \times (-1) \times (2\pi 10) \times \sin(2\pi 10t)$$
$$= -0.6383 \sin(2\pi 10t)$$
$$= 0.6283 \cos(2\pi 10t + 90°)$$

Notice that the phase of voltage is 90° compared to 0° for the current. The current **lags** the voltage by 90°. Figure 6.44 shows $i(t)$ and $v(t)$ on the same graph. As can be seen from the graph, the current crosses zero $T/4$ s later than the voltage. Here, T is a period given by 1/10 s. $T/4$ s is equivalent to $360°/4 = 90°$. *Lagging* means that the waveform crosses zero later or reaches peak value later.

FIGURE 6.44

The current through and voltage across an inductor.

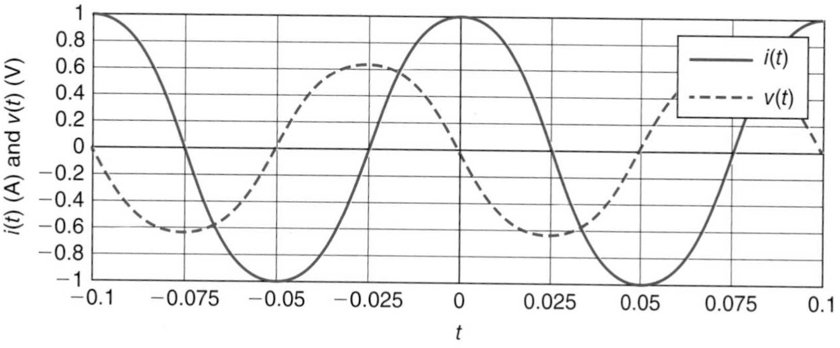

A sinusoidal current,

$$i(t) = I_m \cos(2\pi ft) \text{ A}$$

is applied to an inductor with inductance L, as shown in Figure 6.45.

The voltage across the inductor is given by

$$v(t) = L\frac{di(t)}{dt} = LI_m \times (-1) \times (2\pi f) \times \sin(2\pi ft) = -LI_m 2\pi f \times \sin(2\pi ft)$$

FIGURE 6.45

An inductor with sinusoidal input.

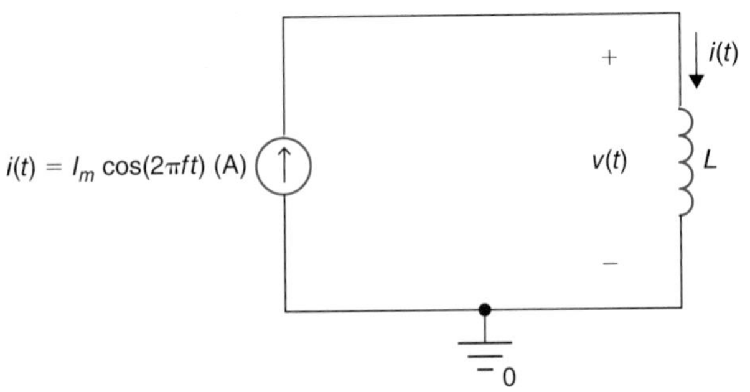

The amplitude of the voltage is proportional to the frequency of the current applied. Figure 6.46 shows $v(t)$ for $f = 10$ Hz, 5 Hz, and 1 Hz when $I_m = 1$ A and $L = 10$ mH. As the frequency decreases, the amplitude decreases. As the frequency decreases to zero ($f = 0$ Hz), the input current becomes $i(t) = I_m \cos(2\pi 0 t) = I_m$, which is a *dc* current with amplitude I_m A. The voltage across the inductor is $v(t) = -LI_m 2\pi 0 \sin(2\pi 0 t) = 0$ V. For *dc* current, the voltage across the inductor is zero in the steady state. On the other hand, as the frequency increases, the voltage across the inductor increases. As the frequency increases to infinity ($f = \infty$), the voltage increases to infinity as well. The inductor acts as a short circuit for dc current and acts as an open circuit for high-frequency current.

FIGURE 6.46

Voltage across the inductor for different frequencies of applied current.

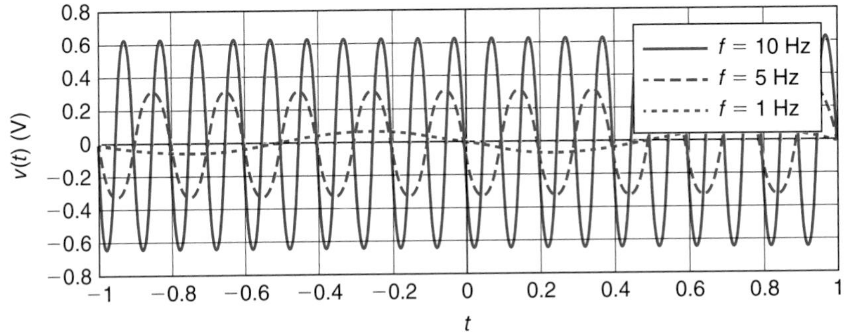

6.6 Series and Parallel Connection of Inductors

6.6.1 SERIES CONNECTION OF INDUCTORS

Figure 6.47(a) shows n inductors connected in series. The current through the n inductors is $i(t)$. Let $v_1(t)$ be the voltage across L_1, $v_2(t)$ be the voltage across $L_2, \ldots, v_n(t)$ be the voltage across L_n, and $v(t)$ be the voltage across all n inductors. Then we have

$$v(t) = v_1(t) + v_2(t) + \cdots + v_n(t) \tag{6.54}$$

Application of the inductor i–v relation

$$v_k(t) = L_k \frac{di(t)}{dt}, \quad 1 \le k \le n$$

to Equation (6.54) yields

$$v(t) = v_1(t) + v_2(t) + \cdots + v_n(t) = L_1 \frac{di(t)}{dt} + L_2 \frac{di(t)}{dt} + \cdots + L_n \frac{di(t)}{dt}$$

$$= (L_1 + L_2 + \cdots + L_n) \frac{di(t)}{dt} = L_{eq} \frac{di(t)}{dt}$$

where

$$L_{eq} = L_1 + L_2 + \cdots + L_n \tag{6.55}$$

is the equivalent inductance.

FIGURE 6.47

(a) Series connection of inductors. (b) Equivalent inductor.

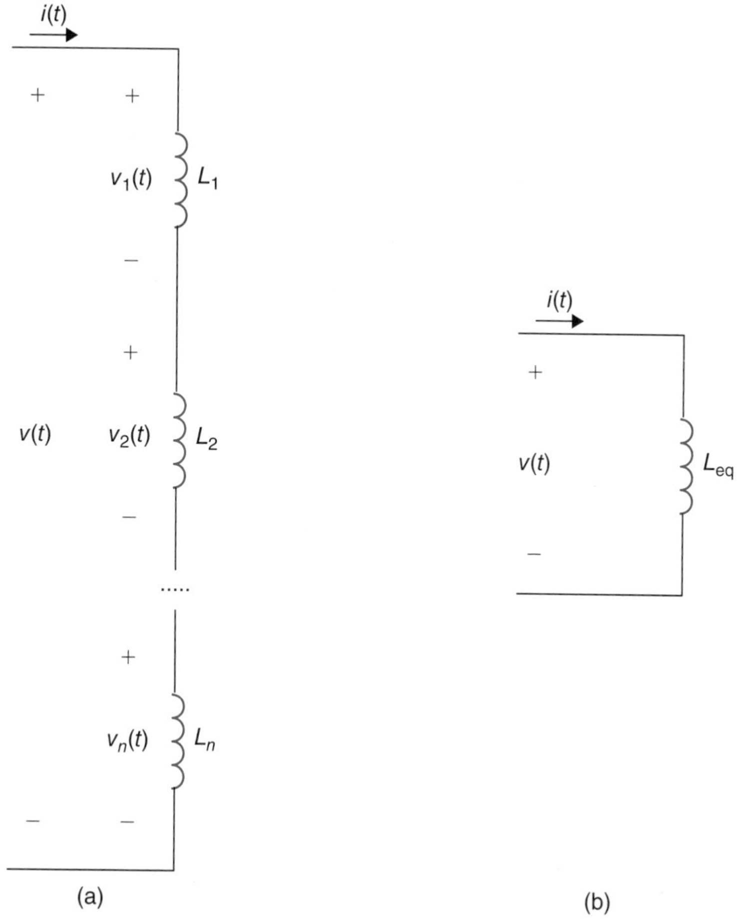

The equivalent inductor with inductance L_{eq} is shown in Figure 6.47(b).

Notice that the equation for the equivalent inductance of the series-connected inductors is similar to equivalent resistance of the series-connected resistors.

6.6.2 PARALLEL CONNECTION OF INDUCTORS

Figure 6.48(a) shows n inductors connected in parallel. The voltage across the n inductors is $v(t)$. Let $i_1(t)$ be the current through L_1, $i_2(t)$ be the current through $L_2, \ldots, i_n(t)$ be the current through L_n, and $i(t)$ be the current through all n inductors. Then, we have

$$i(t) = i_1(t) + i_2(t) + \cdots + i_n(t) \tag{6.56}$$

FIGURE 6.48

(a) Parallel connection of inductors. (b) Equivalent inductor.

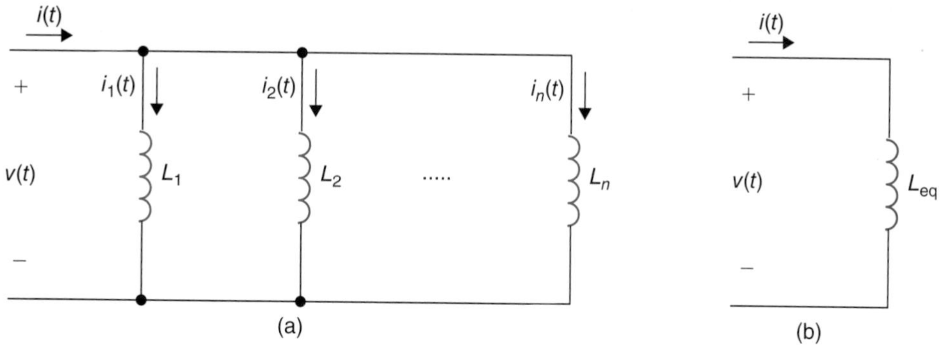

Application of the inductor i–v relation

$$i_k(t) = \frac{1}{L_k}\int_{-\infty}^{t} v(\lambda)d\lambda, \quad 1 \le k \le n$$

to Equation (6.56) yields

$$i(t) = i_1(t) + i_2(t) + \cdots + i_n(t) = \frac{1}{L_1}\int_{-\infty}^{t} v(\lambda)d\lambda + \frac{1}{L_2}\int_{-\infty}^{t} v(\lambda)d\lambda + \cdots + \frac{1}{L_n}\int_{-\infty}^{t} v(\lambda)d\lambda$$

$$= \left(\frac{1}{L_1} + \frac{1}{L_2} + \cdots + \frac{1}{L_n}\right)\int_{-\infty}^{t} v(\lambda)d\lambda = \frac{1}{L_{eq}}\int_{-\infty}^{t} v(\lambda)d\lambda \qquad \textbf{(6.57)}$$

where

$$\frac{1}{L_{eq}} = \frac{1}{L_1} + \frac{1}{L_2} + \cdots + \frac{1}{L_n} \qquad \textbf{(6.58)}$$

Solving Equation (6.58) for the equivalent inductance L_{eq}, we obtain

$$L_{eq} = \frac{1}{\dfrac{1}{L_1} + \dfrac{1}{L_2} + \cdots + \dfrac{1}{L_n}} \qquad \textbf{(6.59)}$$

The equivalent inductor with inductance L_{eq} is shown in Figure 6.48(b).

The equivalent inductance of two inductors with inductances L_1 and L_2 connected in parallel is given by

$$L_{eq} = \frac{1}{\dfrac{1}{L_1} + \dfrac{1}{L_2}} = \frac{L_1 L_2}{L_1 + L_2} \qquad \textbf{(6.60)}$$

Notice that the equation for the equivalent inductance of the parallel-connected inductors is similar to the equivalent resistance of the parallel-connected resistors.

EXAMPLE 6.10

Find the equivalent inductance L_{eq} between *a* and *b* for the circuit shown in Figure 6.49.

FIGURE 6.49

Circuit for EXAMPLE 6.10.

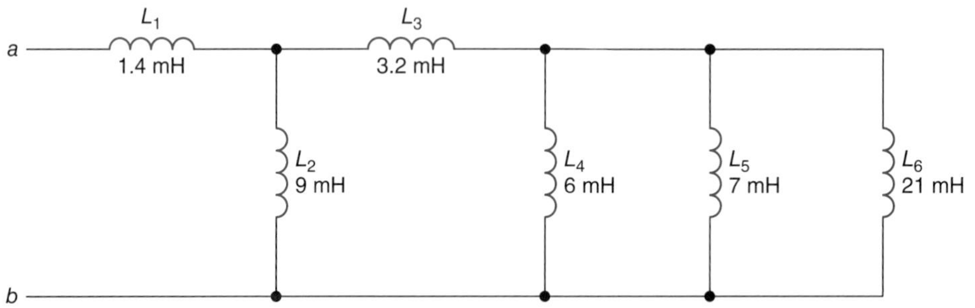

continued

Example 6.10 continued

Let L_a be the equivalent inductance of the parallel connection of L_4, L_5, and L_6. Then we have

$$L_a = \frac{1}{\frac{1}{L_4} + \frac{1}{L_5} + \frac{1}{L_6}} = \frac{1}{\frac{1}{6} + \frac{1}{7} + \frac{1}{21}} \text{ mH} = \frac{42}{\frac{42}{6} + \frac{42}{7} + \frac{42}{21}} \text{ mH}$$

$$= \frac{42}{7 + 6 + 2} \text{ mH} = \frac{42}{15} \text{ mH} = 2.8 \text{ mH}$$

Let L_b be the equivalent inductance of the series connection of L_3 and L_a. Then we have

$$L_b = L_3 + L_a = 3.2 \text{ mH} + 2.8 \text{ mH} = 6 \text{ mH}$$

Let L_c be the equivalent inductance of the parallel connection of L_2 and L_b. Then we have

$$L_c = \frac{L_2 \times L_b}{L_2 + L_b} = \frac{9 \times 6}{9 + 6} \text{ mH} = \frac{54}{15} \text{ mH} = 3.6 \text{ mH}$$

The equivalent inductance between a and b is the sum of L_1 and L_c. Therefore,

$$L_{eq} = L_1 + L_c = 1.4 \text{ mH} + 3.6 \text{ mH} = 5 \text{ mH}$$

Exercise 6.10

Find the equivalent inductance L_{eq} between a and b for the circuit shown in Figure 6.50.

FIGURE 6.50

Circuit for EXERCISE 6.10.

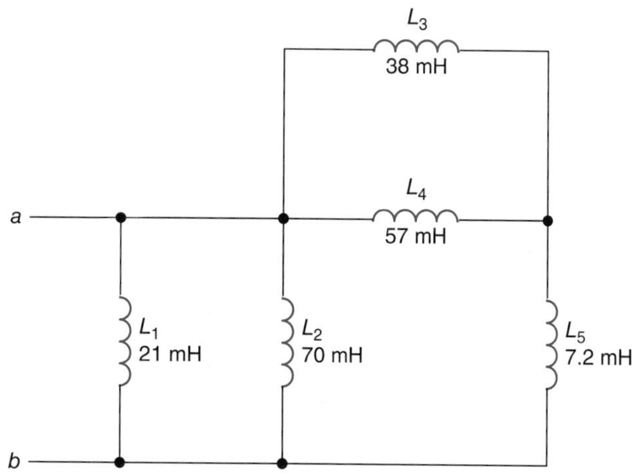

Answer:
$L_{eq} = 10.5$ mH

EXAMPLE 6.11

For the circuit shown in Figure 6.51, find the equivalent circuit consisting of a single inductor and a single capacitor connected in parallel between *a* and *b*.

FIGURE 6.51

Circuit for EXAMPLE 6.11.

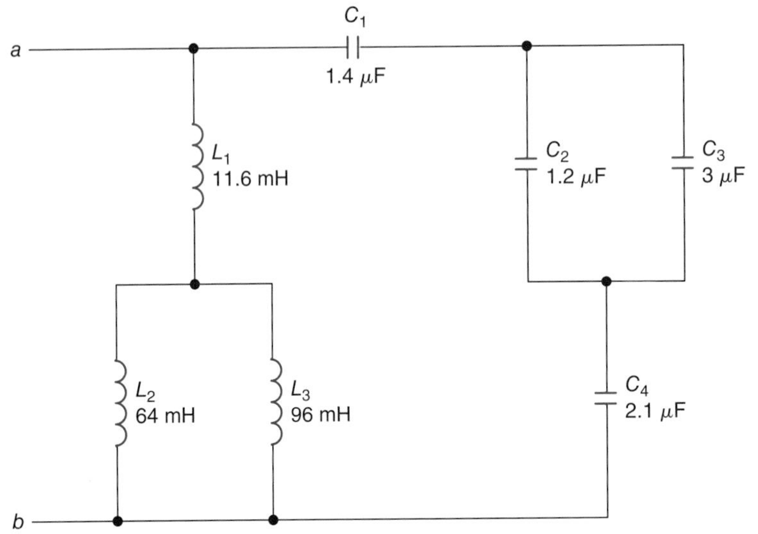

The equivalent inductance is given by

$$L_{eq} = L_1 + (L_2 \| L_3) = L_1 + \frac{L_2 \times L_3}{L_2 + L_3} = 11.6\text{ mH} + \frac{64 \times 96}{64 + 96}\text{ mH}$$
$$= 11.6\text{ mH} + 38.4\text{ mH} = 50\text{ mH}$$

Let C_5 be the equivalent capacitance of the parallel connection of C_2 and C_3. Then we have

$$C_5 = C_2 + C_3 = 1.2\,\mu\text{F} + 3\,\mu\text{F} = 4.2\,\mu\text{F}$$

The equivalent capacitance of the series connection of C_1, C_5, and C_4 is given by

$$C_{eq} = \frac{1}{\frac{1}{C_1} + \frac{1}{C_5} + \frac{1}{C_4}} = \frac{1}{\frac{1}{1.4} + \frac{1}{4.2} + \frac{1}{2.1}}\,\mu\text{F} = 0.7\,\mu\text{F}$$

The equivalent circuit is shown in Figure 6.52.

FIGURE 6.52

The equivalent circuit.

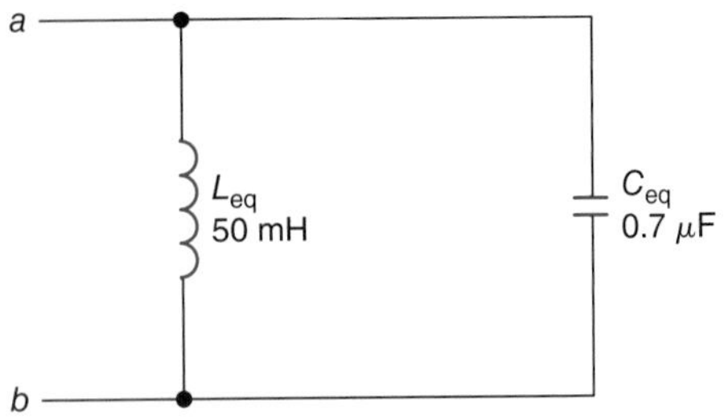

Exercise 6.11

For the circuit shown in Figure 6.53, find the equivalent circuit consisting of a single resistor and a single inductor connected in series between *a* and *b*.

continued

Exercise 6.11 continued

FIGURE 6.53

Circuit for EXERCISE 6.11.

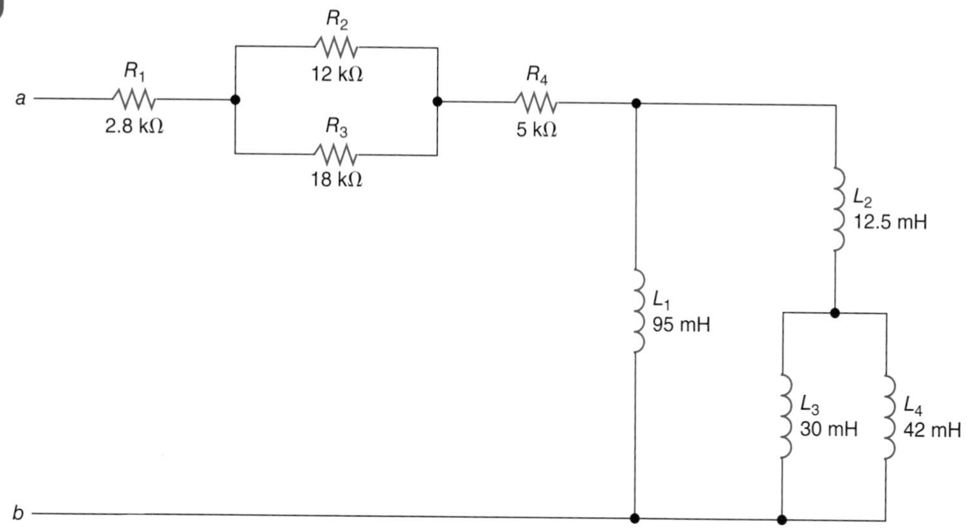

Answer:
$R_{eq} = 15\ k\Omega, L_{eq} = 22.8$ mH.

6.7 PSpice and Simulink

Figure 6.54 shows a circuit with a piecewise linear voltage source with a triangular shape, voltage across the capacitor, and current through the capacitor. Enter VPWL to place the piecewise linear voltage source. Double-click on the voltage source and enter the time and voltage:

FIGURE 6.54

A triangular voltage produces a rectangular current.

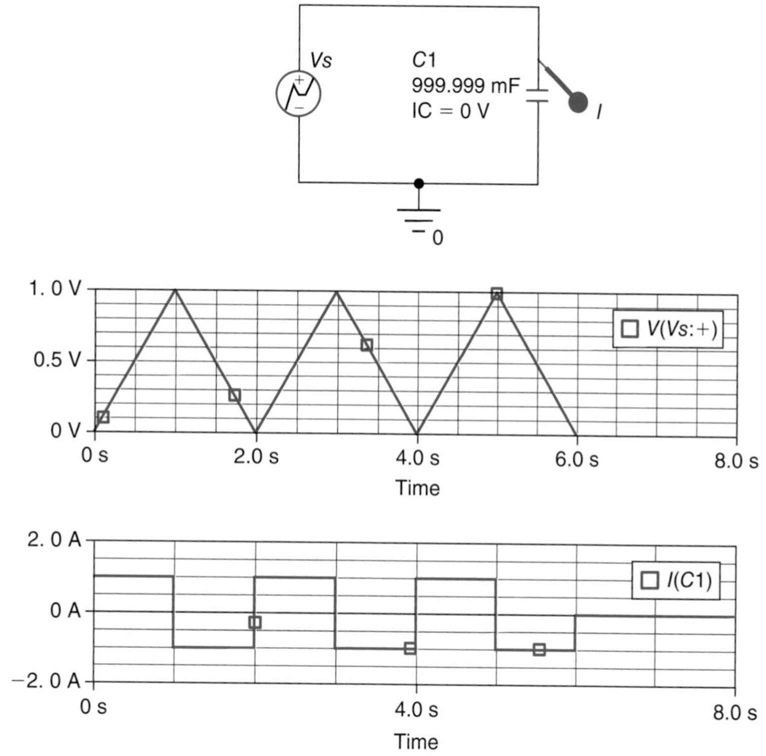

T1	T2	T3	T4	T5	T6	T7	V1	V2	V3	V4	V5	V6	V7
0	1	2	3	4	5	6	0	1	0	1	0	1	0

Double-click on the capacitor and enter 0 under *IC* to set the initial condition to zero. In the Simulation Settings window, enter 6 for the Run to time. The current is a rectangular shape.

If a sinusoidal signal $v(t) = \cos(2\pi 10t)$ is applied to a capacitor with capacitance 0.01 F, the current is given by $i(t) = C\, dv(t)/dt = -0.01 \times 2\pi 10 \sin(2\pi 10t) = -0.6283 \sin(2\pi 10t)$. The circuit, voltage wave, and current wave are shown in Figure 6.55. Enter VSIN to place the sinusoidal voltage source. Double-click on VSIN and enter 90 under PHASE to change sin to cosine.

FIGURE 6.55

The voltage across and the current through the capacitor.

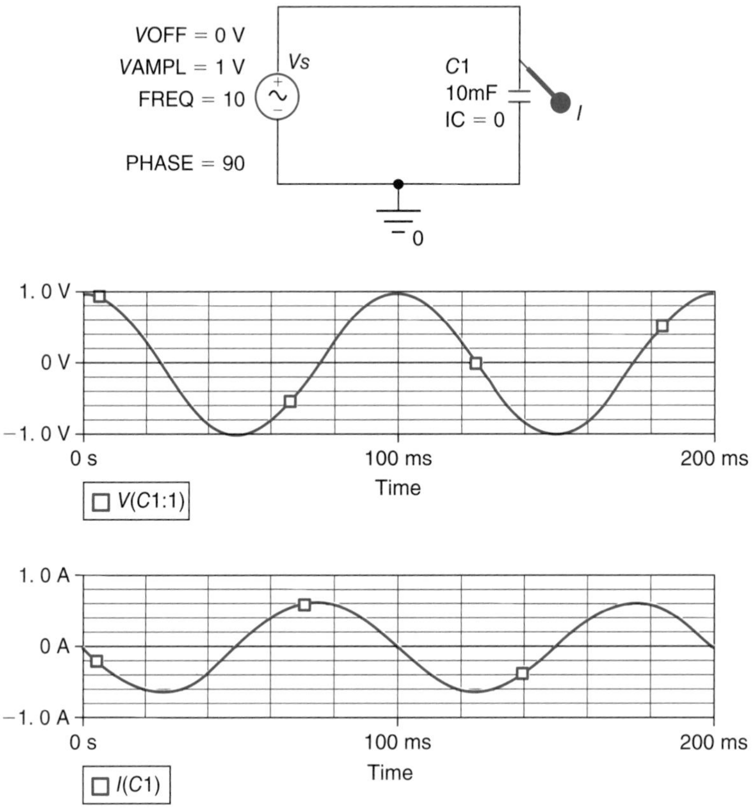

The Simulink model for simulation of the circuit shown in Figure 6.55 is shown in Figure 6.56, along with the plot. Double-click on Sine Wave block and enter Amplitude = 1, Bias = 0, Frequency = 2*pi*10, and Phase = pi/2.

Figure 6.57 shows a Simulink model using blocks from the Simscape library. For the capacitor, enter capacitance of 0.01F, set the initial voltage at 1 V, and enter series resistance of $1e-6$.

FIGURE 6.56

The Simulink model and plot of voltage and current.

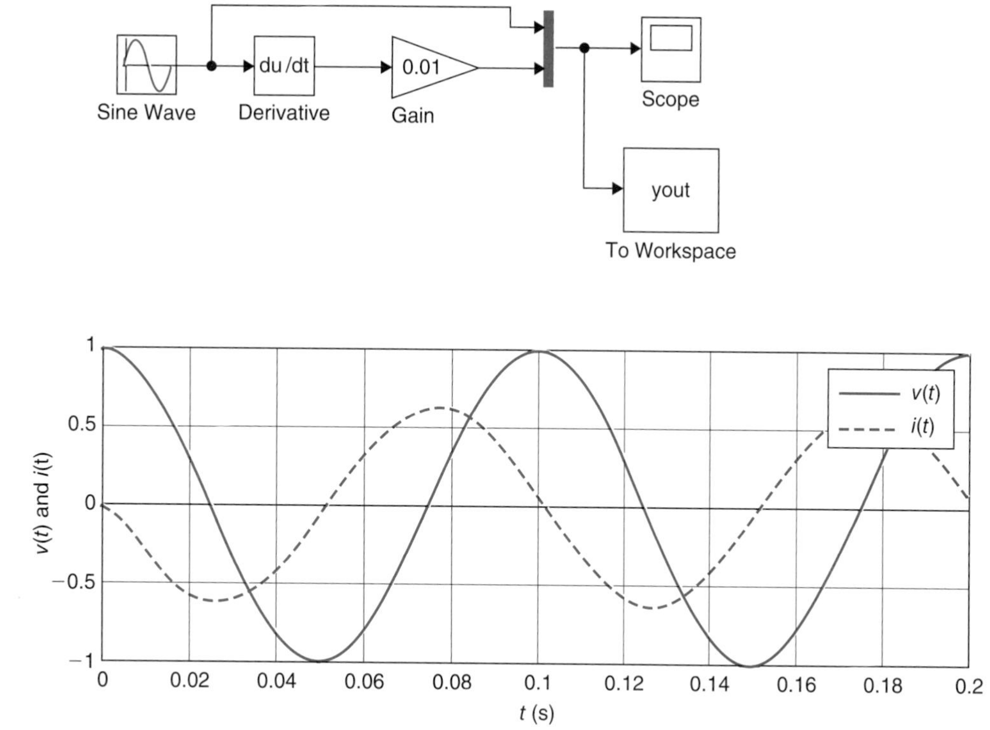

FIGURE 6.57

The Simulink model and plot of voltage and current.

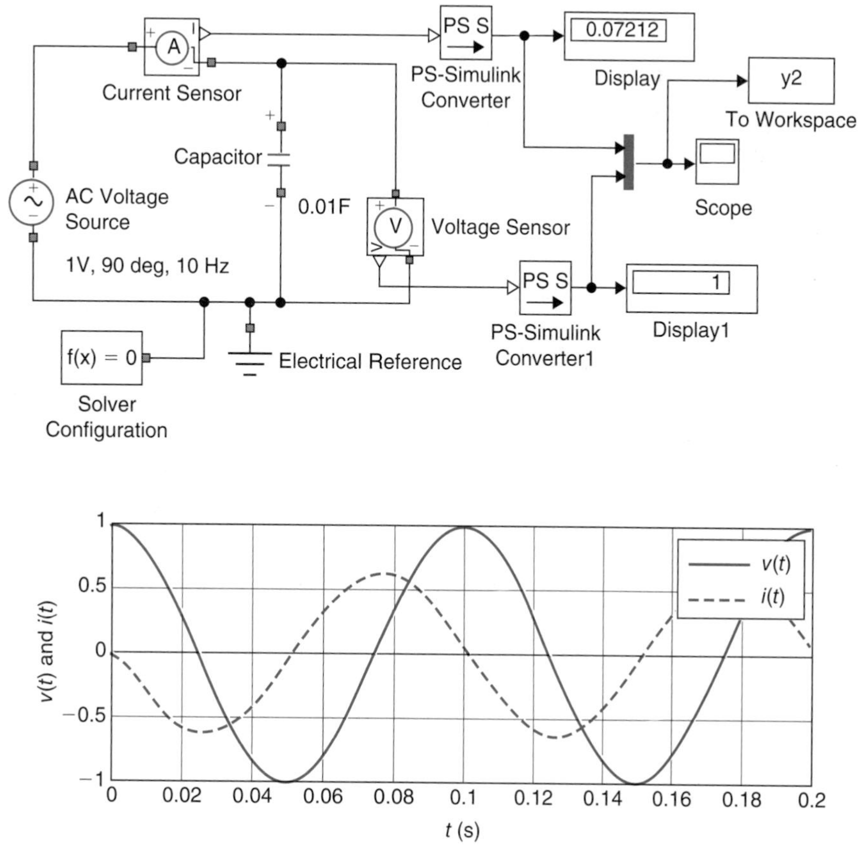

SUMMARY

Two circuit elements, capacitor and inductor, that store and release energy are introduced in this chapter. The capacitor stores energy in the form of an electric field between the plates. *Capacitance* is defined as the ratio of the charge stored to the potential difference, $C = Q/V$. The capacitance depends on the permittivity of the medium and the dimension of the capacitor. The i–v relations of a capacitor are

$$i(t) = C\frac{dv(t)}{dt}$$

and

$$v(t) = \frac{1}{C}\int_{-\infty}^{t} i(\lambda)d\lambda$$

The current through a capacitor is proportional to the rate of change of the voltage across the capacitor.

The inductor stores energy in the form of a magnetic field inside a coil of wire. The i–v relations of an inductor are

$$v(t) = L\frac{di(t)}{dt}$$

and

$$i(t) = \frac{1}{L}\int_{-\infty}^{t} v(\lambda)d\lambda$$

The voltage across an inductor is proportional to the rate of change of the current through the inductor. The proportionality constant is the inductance L, which depends on the number of turns, permeability, and the dimension of the inductor.

PROBLEMS

Capacitors

6.1 **The voltage across a capacitor with capacitance of 10 μF is given by $v(t) = 200t\ u(t)$ V, where $u(t)$ is a unit step function defined as**

$$u(t) = \begin{cases} 1, & t \geq 0 \\ 0, & t < 0 \end{cases}$$

Find the current through the capacitor $i(t)$ and plot the voltage and current as a function of time.

6.2 **The voltage across a capacitor with capacitance of 50 nF is given by $v(t) = e^{-3000t}u(t)$ V. Find the current through the capacitor $i(t)$ and plot the voltage and current as a function of time.**

6.3 **The voltage across a capacitor with capacitance of 0.02 μF is given by $v(t) = \cos(2\pi \times 1000t)\ u(t)$ V. Find the current through the capacitor $i(t)$ and plot the voltage and current as a function of time.**

6.4 **The voltage across a capacitor with capacitance of 0.047 μF is given by $v(t) = \sin(2\pi \times 5000t)\ u(t)$ V. Find the current through the capacitor $i(t)$ and plot the voltage and current as a function of time.**

6.5 **The voltage across a capacitor with capacitance of 2 nF is given by $v(t) = e^{-5000t}\cos(2\pi \times 8000t)\ u(t)$ V. Find the current through the capacitor $i(t)$ and plot the voltage and current as a function of time.**

6.6 **The voltage across a capacitor with capacitance of 7 nF is given by $v(t) = e^{-9000t}\sin(2\pi \times 3000t)\ u(t)$ V. Find the current through the capacitor $i(t)$ and plot the voltage and current as a function of time.**

6.7 **The voltage across a capacitor with capacitance of 0.05 μF is given by $v(t) = \cos(2\pi \times 2000t)\ u(t)$ V.**

a. Find the current through the capacitor $i(t)$ and plot the voltage and current as a function of time.
b. Find and plot the instantaneous power $p(t)$ on the capacitor.
c. Find and plot the instantaneous energy $w(t)$ stored on the capacitor.

6.8 **The voltage across a capacitor with capacitance of 50 μF is given by**

$$v(t) = \begin{cases} -20t, & 0 \leq t < 5 \\ 40t - 300, & 5 \leq t < 10 \\ -10t + 200, & 10 \leq t < 20 \\ 0, & otherwise \end{cases} \text{ V}$$

This waveform is shown in Figure P6.8.

FIGURE P6.8

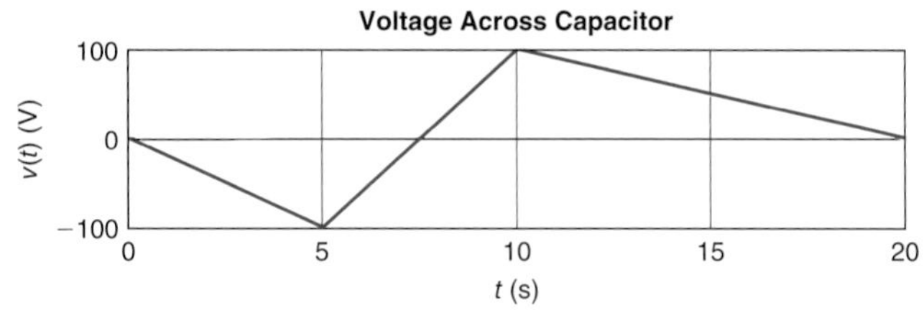

a. Find the current through the capacitor $i(t)$ and plot the voltage and current as a function of time.
b. Find and plot the instantaneous power $p(t)$ on the capacitor. Find the time interval when the power is released to the circuit from the capacitor.
c. Find and plot the instantaneous energy $w(t)$ stored on the capacitor.

6.9 **The voltage across a capacitor with capacitance of 233 μF is given by**

$$v_c(t) = \begin{cases} 100t, & 0 \le t < 1 \\ -200t + 300, & 1 \le t < 2 \\ 100t - 300, & 2 \le t < 4 \\ -50t + 300, & 4 \le t < 6 \\ 0, & otherwise \end{cases} \text{ V}$$

This waveform is shown in Figure P6.9.

FIGURE P6.9

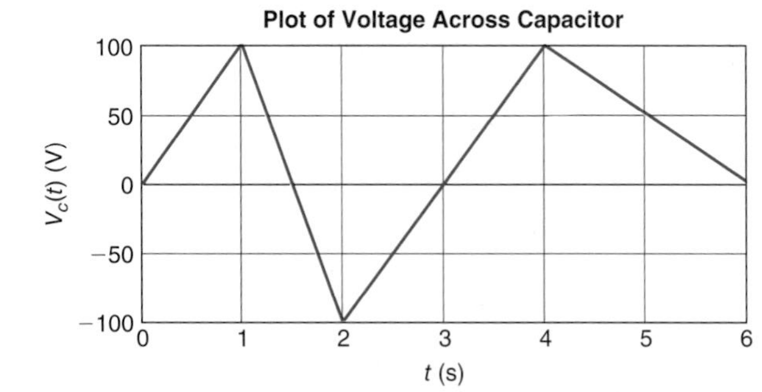

a. Find the current through the capacitor $i_C(t)$ and plot the voltage and current as a function of time.
b. Find and plot the instantaneous power $p(t)$ on the capacitor. Find the time interval when the power is released to the circuit from the capacitor.
c. Find and plot the instantaneous energy $w(t)$ stored on the capacitor.

6.10 **The current through a capacitor with capacitance of 47 μF is given by $i(t) = 0.5\, u(t)$ mA. Find the voltage across the capacitor $v(t)$ and plot the current and voltage as a function of time.**

6.11 **The current through a capacitor with capacitance of 25 nF is given by $i_c(t) = e^{-500t}\, u(t)$ mA.**

a. Find the voltage across the capacitor $v_C(t)$ and plot the current and voltage as a function of time.
b. Find and plot the instantaneous power $p(t)$ on the capacitor.
c. Find and plot the instantaneous energy $w(t)$ stored on the capacitor.

6.12 **The current through a capacitor with capacitance of 354 nF is given by $i_c(t) = \cos(200t)\, u(t)$ mA. Find the voltage across the capacitor $v_C(t)$ and plot the current and voltage as a function of time.**

6.13 **The current through a capacitor with capacitance of 10 μF is given by $i_c(t) = \sin(5000t)\, u(t)$ mA.**

a. Find the voltage across the capacitor $v_C(t)$ and plot the current and voltage as a function of time.
b. Find and plot the instantaneous power $p(t)$ on the capacitor.
c. Find and plot the instantaneous energy $w(t)$ stored on the capacitor.

6.14 **The current through a capacitor with capacitance of 20 μF is given by the triangular pulse shown in Figure P6.14. Find the voltage across the capacitor $v_C(t)$ and plot the current and voltage as a function of time.**

FIGURE P6.14

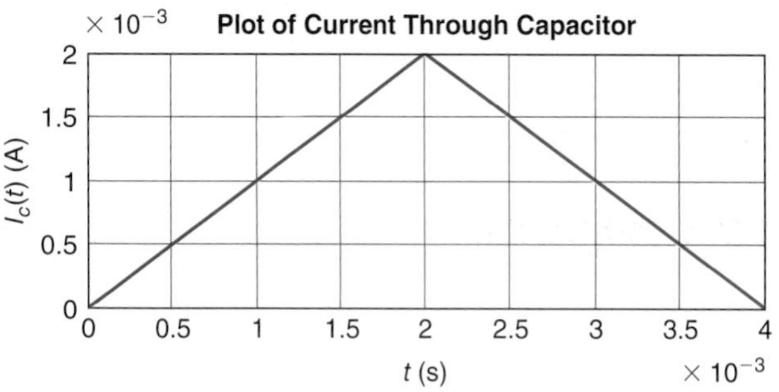

6.15 **The current through a capacitor with capacitance of 150 μF is given by**

$$i_C(t) = \begin{cases} t, & 0 \le t < 1 \\ 1, & 1 \le t < 3 \\ -t + 4, & 3 \le t < 4 \\ 0, & otherwise \end{cases} \text{ mA}$$

This waveform is shown in Figure P6.15. Find the voltage across the capacitor $v_C(t)$ and plot the voltage as a function of time.

FIGURE P6.15

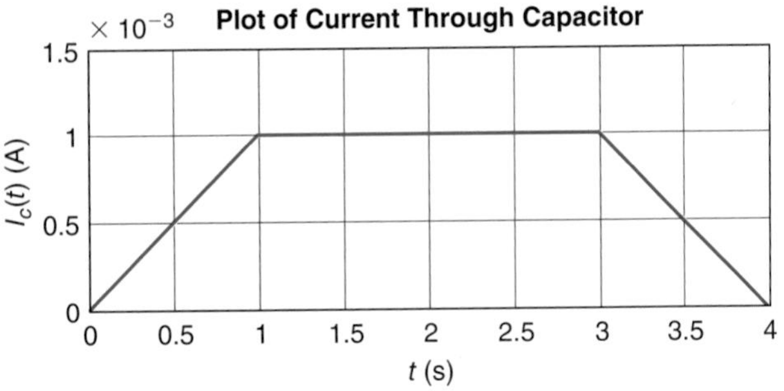

6.16 **The current through a capacitor with capacitance of 20 μF is given by**

$$i_C(t) = \begin{cases} \frac{3t}{2}, & 0 \le t < 2 \\ 9 - 3t, & 2 \le t < 3 \quad \text{mA} \\ 0, & otherwise \end{cases}$$

This waveform is shown in Figure P6.16.

FIGURE P6.16

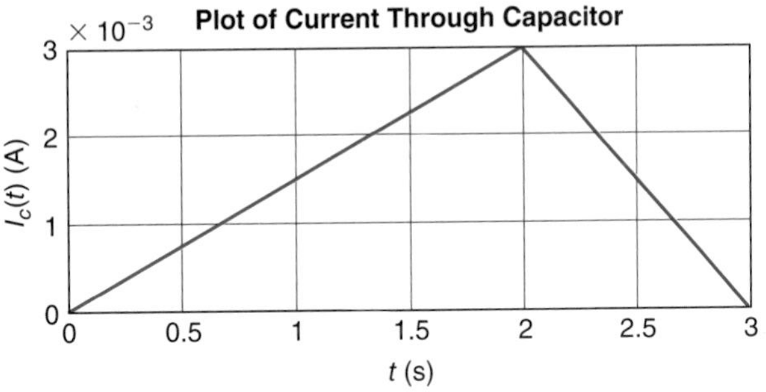

a. Find the voltage across the capacitor $v_C(t)$ and plot the current and voltage as a function of time.
b. Find and plot the instantaneous power $p(t)$ on the capacitor.
c. Find and plot the instantaneous energy $w(t)$ stored on the capacitor.

6.17 **The voltage across a capacitor with capacitance $C = 270\ \mu$F is given by**

$$v_C(t) = \begin{cases} 4t, & 0 \le t < 1 \\ -8t + 12, & 1 \le t < 2 \\ 8t - 20, & 2 \le t < 3 \quad \text{V} \\ -2t + 10, & 3 \le t < 5 \\ 0, & otherwise \end{cases}$$

This voltage is shown in Figure P6.17.

FIGURE P6.17

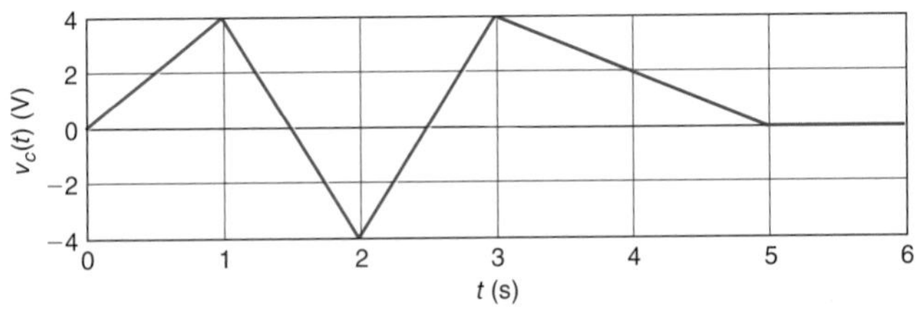

a. Find the current $i_C(t)$ through the capacitor and plot $i_C(t)$ for $0 \le t \le 5$ s.
b. Find the instantaneous power $p_C(t)$ on the capacitor and plot $p_C(t)$ for $0 \le t \le 5$ s.
c. Find the instantaneous energy $w_C(t)$ on the capacitor and plot $w_C(t)$ for $0 \le t \le 5$ s.

6.18 **The voltage across a capacitor with capacitance $C = 650\ \mu$F is given by**

$$v(t) = \begin{cases} 20t, & 0 \le t < 1 \\ -40t + 60, & 1 \le t < 2 \\ 10t - 40, & 2 \le t < 4 \\ 0, & otherwise \end{cases} \quad \text{V}$$

This voltage is shown in Figure P6.18.

FIGURE P6.18

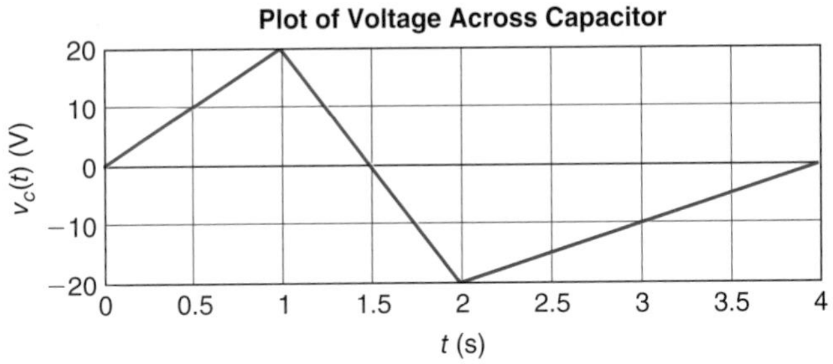

a. Find the current $i(t)$ through the capacitor and plot $i(t)$ for $0 \le t \le 4$ s.
b. Find the instantaneous power $p(t)$ on the capacitor and plot $p(t)$ for $0 \le t \le 4$ s.
c. Find the instantaneous energy $w(t)$ on the capacitor and plot $w(t)$ for $0 \le t \le 4$ s.

6.19 **The voltage across a capacitor with capacitance $C = 350\ \mu$F is given by**

$$v_C(t) = \begin{cases} 40t, & 0 \le t < 2 \\ -80t + 240, & 2 \le t < 4 \\ 40t - 240, & 4 \le t < 6 \\ 0, & otherwise \end{cases} \quad \text{V}$$

This voltage is shown in Figure P6.19.

FIGURE P6.19

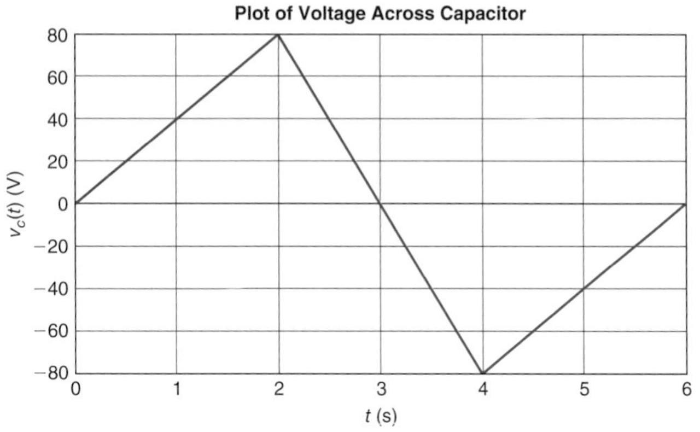

a. Find the current $i_C(t)$ through the capacitor and plot $i_C(t)$ for $0 \le t \le 6$ s.
b. Find the instantaneous power $p_C(t)$ on the capacitor and plot $p_C(t)$ for $0 \le t \le 6$ s.
c. Find the instantaneous energy $w_C(t)$ on the capacitor and plot $w_C(t)$ for $0 \le t \le 6$ s.

6.20 **The voltage across a capacitor with capacitance $C = 550\ \mu F$ is given by**

$$v_C(t) = \begin{cases} 20t, & 0 \le t < 2 \\ -40t + 120, & 2 \le t < 4 \\ 20t - 120, & 4 \le t < 6 \\ 0, & otherwise \end{cases} \text{ V}$$

This voltage is shown in Figure P6.20.

FIGURE P6.20

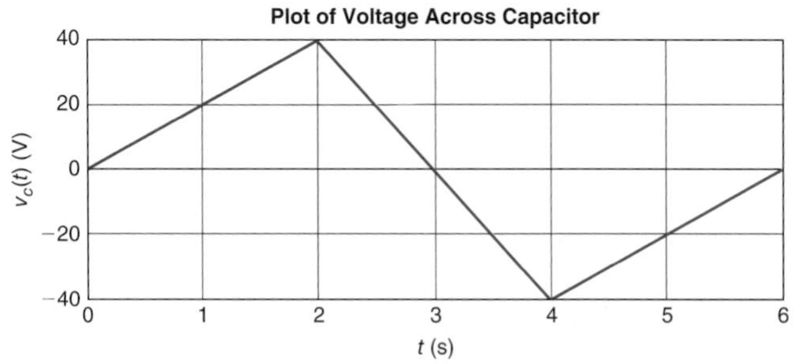

a. Find the current $i_C(t)$ through the capacitor and plot $i_C(t)$ for $0 \le t \le 6$ s.
b. Find the instantaneous power $p_C(t)$ on the capacitor and plot $p_C(t)$ for $0 \le t \le 6$ s.
c. Find the instantaneous energy $w_C(t)$ on the capacitor and plot $w_C(t)$ for $0 \le t \le 6$ s.
d. Specify the time interval when the energy is being stored on the capacitor and the time interval when the energy is being released from the capacitor to the circuit.

6.21 **The voltage across a capacitor with capacitance $C = 330\ \mu F$ is given by**

$$v(t) = \begin{cases} 30t, & 0 \le t < 2 \\ -60t + 180, & 2 \le t < 4 \\ 30t - 180, & 4 \le t < 6 \\ 0, & otherwise \end{cases} \text{ V}$$

This voltage is shown in Figure P6.21.

FIGURE P6.21

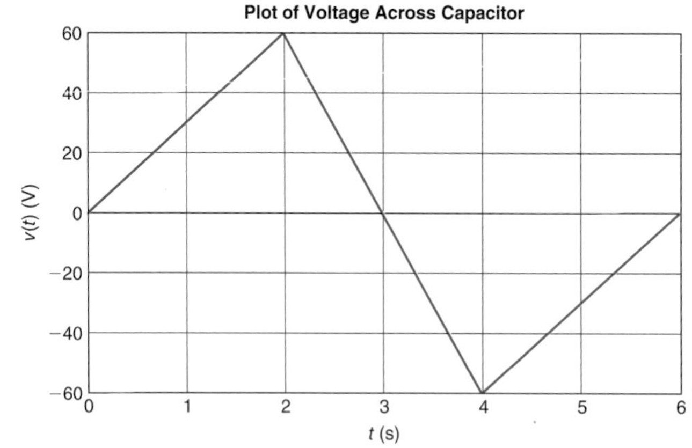

a. Find the current $i(t)$ through the capacitor and plot $i(t)$ for $0 \le t \le 6$ s.
b. Find the instantaneous power $p(t)$ on the capacitor and plot $p(t)$ for $0 \le t \le 6$ s.
c. Find the instantaneous energy $w(t)$ on the capacitor and plot $w(t)$ for $0 \le t \le 6$ s.

Series and Parallel Connection of Capacitors

6.22 **Find the equivalent capacitance between *a* and *b* for the circuit shown in Figure P6.22.**

FIGURE P6.22

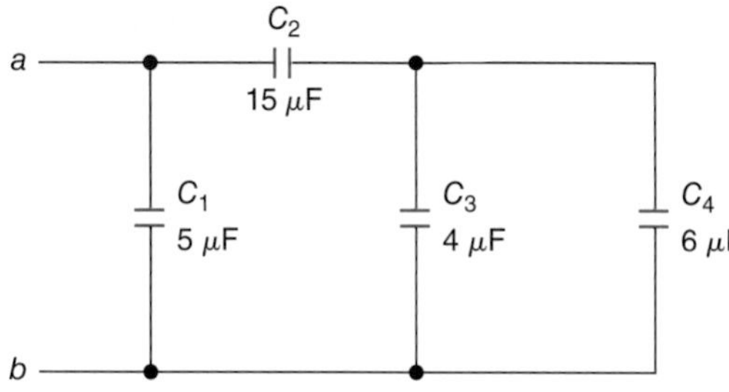

6.23 Find the equivalent capacitance between a and b for the circuit shown in Figure P6.23.

FIGURE P6.23

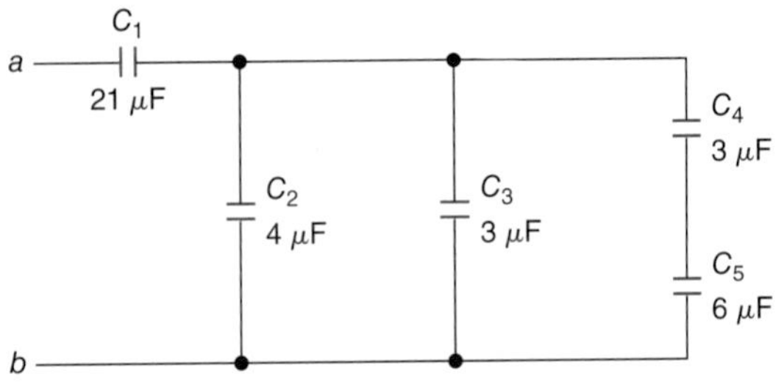

6.24 Find the equivalent capacitance between a and b for the circuit shown in Figure P6.24.

FIGURE P6.24

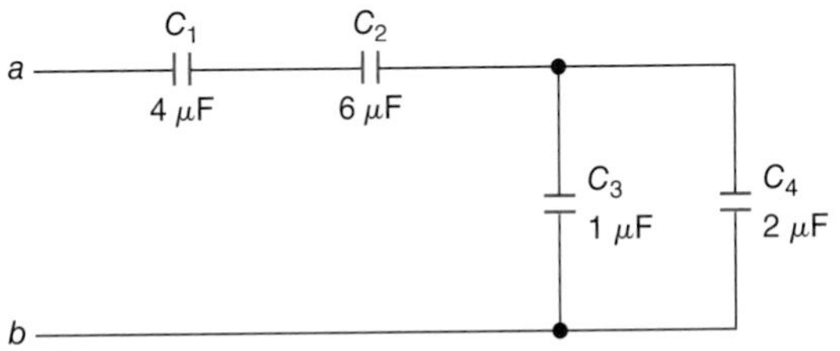

6.25 Find the equivalent capacitance between a and b for the circuit shown in Figure P6.25.

FIGURE P6.25

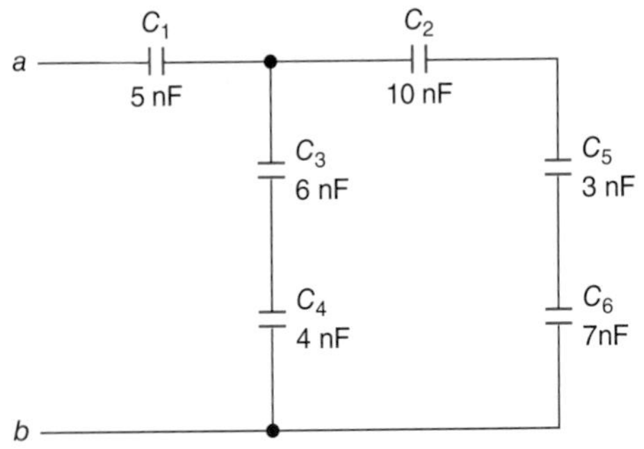

6.26 Find the equivalent capacitance between a and b for the circuit shown in Figure P6.26.

FIGURE P6.26

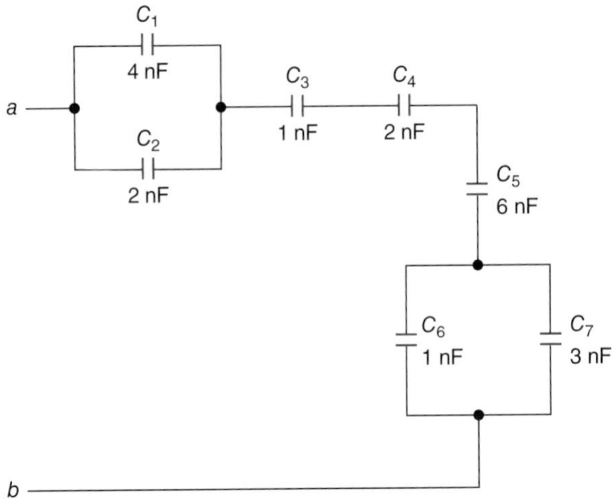

6.27 Find the equivalent capacitance between a and b for the circuit shown in Figure P6.27.

FIGURE P6.27

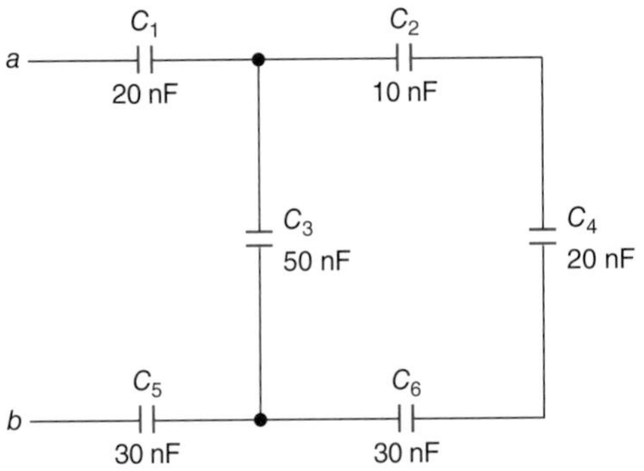

6.28 Find the equivalent capacitance between a and b for the circuit shown in Figure P6.28.

FIGURE P6.28

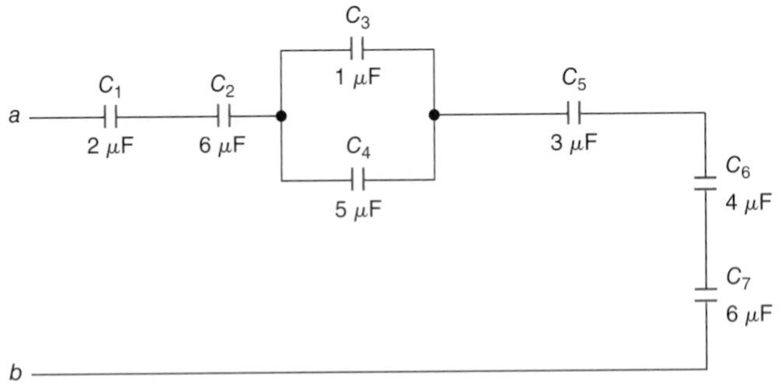

6.29 Find the equivalent capacitance between a and b for the circuit shown in Figure P6.29.

FIGURE P6.29

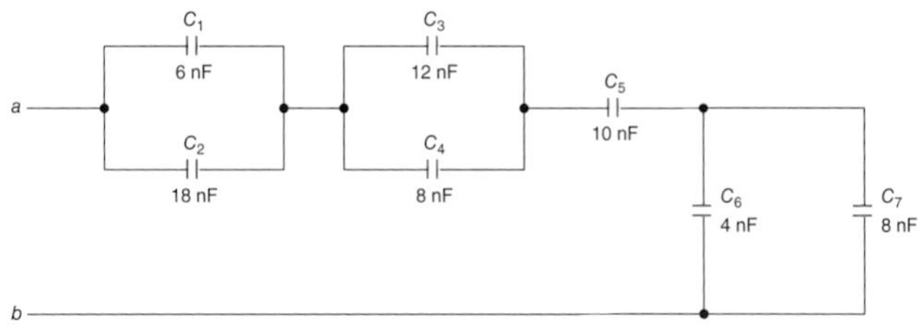

Inductors

6.30 The current through an inductor with inductance of 20 mH is given by $i_L(t) = 50t\, u(t)$ mA. Find the voltage across the inductor $v_L(t)$ and plot the current and voltage as a function of time.

6.31 The current through an inductor with inductance of 1.2 mH is given by $i_L(t) = e^{-2000t}\, u(t)$ A.

a. Find the voltage across the inductor $v_L(t)$ and plot the voltage and current as a function of time.
b. Find and plot the instantaneous power $p(t)$ on the inductor.
c. Find and plot the instantaneous energy $w(t)$ stored on the inductor.

6.32 The current through an inductor with inductance of 0.11 mH is given by $i_L(t) = \cos(2\pi \times 1000t)\, u(t)$ A.

a. Find the voltage across the inductor $v_L(t)$ and plot the voltage and current as a function of time.
b. Find and plot the instantaneous power $p(t)$ on the inductor.
c. Find and plot the instantaneous energy $w(t)$ stored on the inductor.

6.33 The current through an inductor with inductance of 0.27 mH is given by $i_L(t) = \sin(2\pi \times 5000t)\, u(t)$ A. Find the voltage across the inductor $v_L(t)$ and plot the voltage and current as a function of time.

6.34 The current through an inductor with inductance of 15 mH is given by $i_L(t) = e^{-4000t} \cos(2\pi \times 3000t)\, u(t)$ mA. Find the voltage across the inductor $v_L(t)$ and plot the voltage and current as a function of time.

6.35 The current through an inductor with inductance of 22 mH is given by $i_L(t) = e^{-2000t} \sin(2\pi \times 1000t)\, u(t)$ mA. Find the voltage across the inductor $v_L(t)$ and plot the voltage and current as a function of time.

6.36 The current through an inductor with inductance of 27 mH is given by

$$i_L(t) = \begin{cases} 5t, & 0 \le t < 1 \\ -5t + 10, & 1 \le t < 3 \\ 5t - 20, & 3 \le t < 4 \\ 0, & otherwise \end{cases} \text{ A}$$

This waveform is shown in Figure P6.36.

FIGURE P6.36

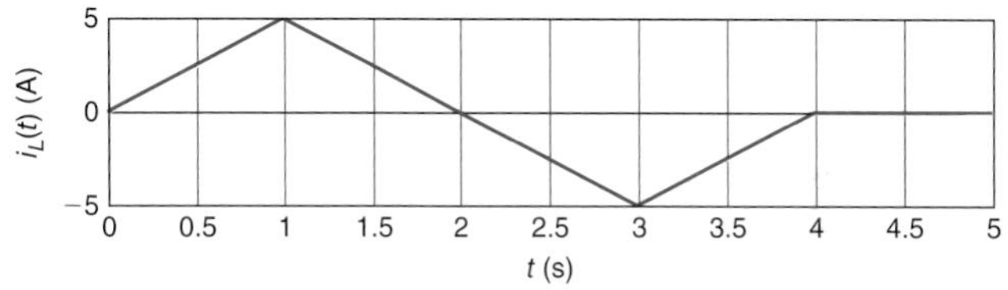

a. Find the voltage across the inductor $v_L(t)$ and plot the voltage and current as a function of time.
b. Find and plot the instantaneous power $p(t)$ on the inductor.
c. Find and plot the instantaneous energy $w(t)$ stored on the inductor.

6.37 The current through an inductor with inductance of 56 mH is given by

$$i_L(t) = \begin{cases} -10t, & 0 \le t < 1 \\ 20t - 30, & 1 \le t < 2 \\ -10t + 30, & 2 \le t < 3 \\ 0, & otherwise \end{cases} \text{ A}$$

This waveform is shown in Figure P6.37.

FIGURE P6.37

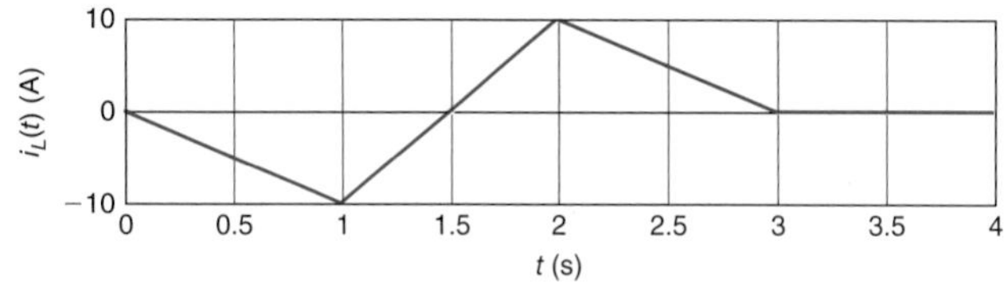

a. Find the voltage across the inductor $v_L(t)$ and plot the voltage and current as a function of time.
b. Find and plot the instantaneous power $p(t)$ on the inductor.
c. Find and plot the instantaneous energy $w(t)$ stored on the inductor.

6.38 **The voltage across an inductor with inductance of 160 mH is given by $v_L(t) = 0.05\,u(t)$ V. Find the current through the inductor $i_L(t)$ and plot the current and voltage as a function of time.**

6.39 **The voltage across an inductor with inductance of 18 mH is given by $v_L(t) = e^{-5t}\,u(t)$ V. Find the current through the inductor $i_L(t)$ and plot the current and voltage as a function of time.**

6.40 **The voltage across an inductor with inductance of 240 mH is given by $v_L(t) = \cos(200t)\,u(t)$ V. Find the current through the inductor $i_L(t)$ and plot the current and voltage as a function of time.**

6.41 **The voltage across an inductor with inductance of 16 mH is given by $v_L(t) = \sin(50t)\,u(t)$ V.**

a. Find the current through the inductor $i_L(t)$ and plot the current and voltage as a function of time.
b. Find and plot the instantaneous power $p(t)$ on the inductor.
c. Find and plot the instantaneous energy $w(t)$ stored on the inductor.

6.42 **The voltage across an inductor with inductance of 510 mH is given by**

$$v_L(t) = \begin{cases} t, & 0 \le t < 1 \\ 1, & 1 \le t < 3 \\ -t+4, & 3 \le t < 4 \\ 0, & otherwise \end{cases} \text{ V}$$

This waveform is shown in Figure P6.42.

FIGURE P6.42

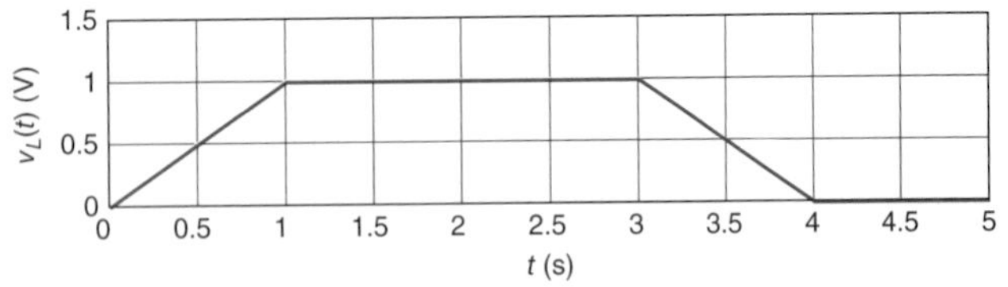

a. Find the current through the inductor $i_L(t)$ and plot the current and voltage as a function of time.
b. Find and plot the instantaneous power $p(t)$ on the inductor.
c. Find and plot the instantaneous energy $w(t)$ stored on the inductor.

6.43 **The voltage across an inductor with inductance of 330 mH is given by**

$$v_L(t) = \begin{cases} \frac{3t}{2}, & 0 \le t < 2 \\ 9 - 3t, & 2 \le t < 3 \text{ V} \\ 0, & otherwise \end{cases}$$

This waveform is shown in Figure P6.43.

FIGURE P6.43

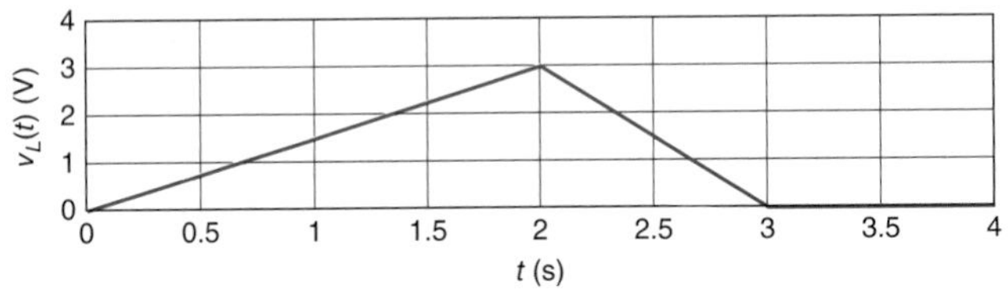

a. Find the current through the inductor $i_L(t)$ and plot the current and voltage as a function of time.
b. Find and plot the instantaneous power $p(t)$ on the inductor.
c. Find and plot the instantaneous energy $w(t)$ stored on the inductor.

Series and Parallel Connection of Inductors

6.44 **Find the equivalent inductance between *a* and *b* for the circuit shown in Figure P6.44.**

FIGURE P6.44

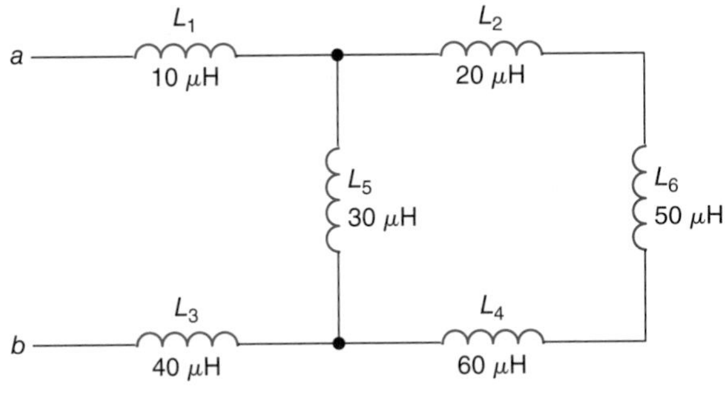

6.45 **Find the equivalent inductance between *a* and *b* for the circuit shown in Figure P6.45.**

FIGURE P6.45

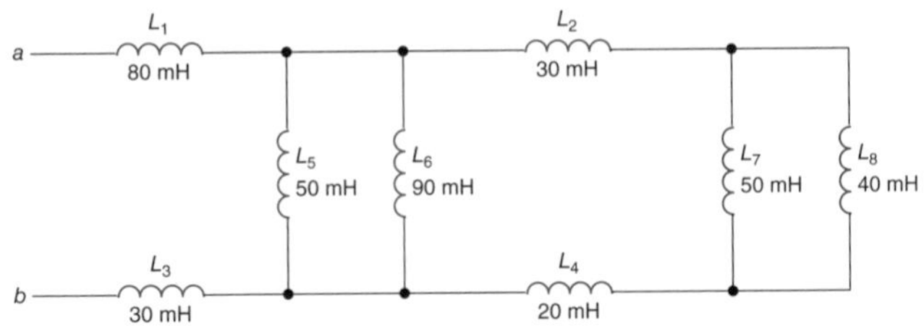

6.46 Find the equivalent inductance between a and b for the circuit shown in Figure P6.46.

FIGURE P6.46

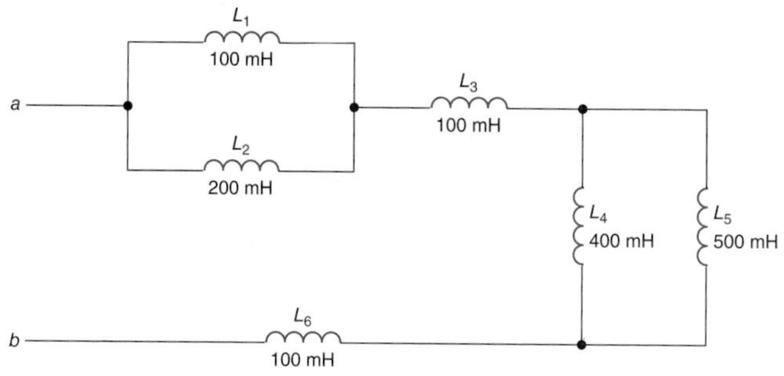

6.47 Find the equivalent inductance between a and b for the circuit shown in Figure P6.47.

FIGURE P6.47

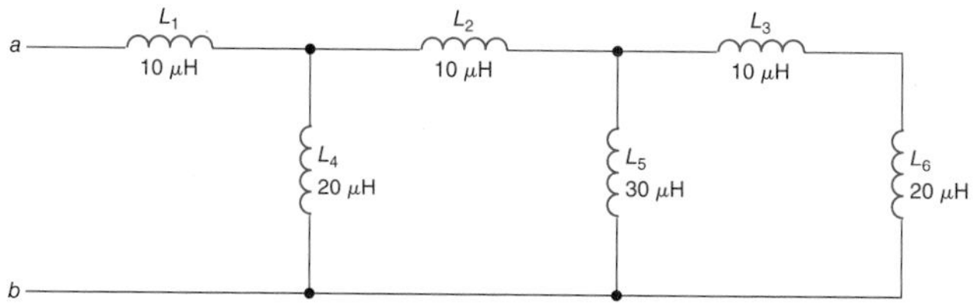

6.48 Find the equivalent inductance between a and b for the circuit shown in Figure P6.48.

FIGURE P6.48

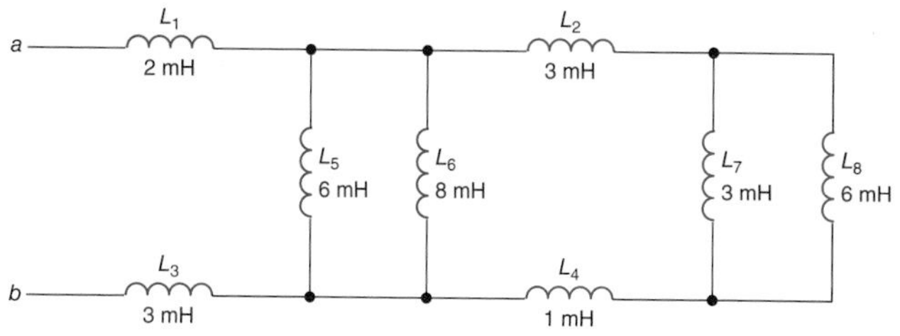

Chapter 7

RC and *RL* Circuits

7.1 Introduction

In Chapter 6, voltage-current relation of capacitor and inductor, along with instantaneous power and energy stored, were discussed. In this chapter, we find the voltages, currents, instantaneous power, and energy stored on a capacitor or inductor when the capacitor or inductor is connected with resistors, sources, and switches. When the capacitor or inductor possesses initial energy, the circuit responds to the initial energy until all the energy is spent, even when there is no input source. The response of a circuit due to initial energy only is called a *natural response* (also *transient response*, *zero input response*, and *source-free response*). The response of a circuit to a dc input signal (step input) is called a *step response.* The step response includes the response due to the initial energy stored in the capacitor or inductor. In this chapter, the natural response and step response of *RC* circuit and *RL* circuit are discussed, as well as responses of general first-order circuits to exponential function, ramp function, and sinusoidal function. Simulation of first-order circuits using PSpice and Simulink are presented as well.

7.2 Natural Response of *RC* Circuit

The switch in Figure 7.1 has been in position a for a long time before it is moved to position b at $t = 0$.

At $t = 0$, the voltage across the capacitor is equal to the voltage of the source V_S; that is, $v(0) = V_0 = V_S$. For $t \geq 0$, the circuit shown in Figure 7.1 becomes the circuit shown in Figure 7.2, with initial voltage of $v(0) = V_0 = V_S$.

For $t \geq 0$, the initial voltage across the capacitor $v(0) = V_0$ will introduce current around the circuit. The current on the circuit implies that the charge on the capacitor plates will discharge through the resistor. As time progresses, the amount of charge on the plates of the capacitor will decrease. As the amount of charge decreases, the voltage ($V = Q/C$) across the capacitor will decrease. As shown next, the voltage $v(t)$ decays exponentially from

FIGURE 7.1

RC circuit.

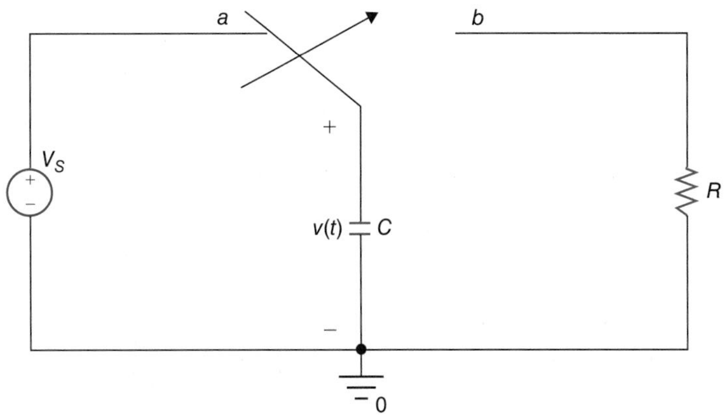

FIGURE 7.2

An *RC* circuit with initial voltage of $v(0) = V_0 = V_S$.

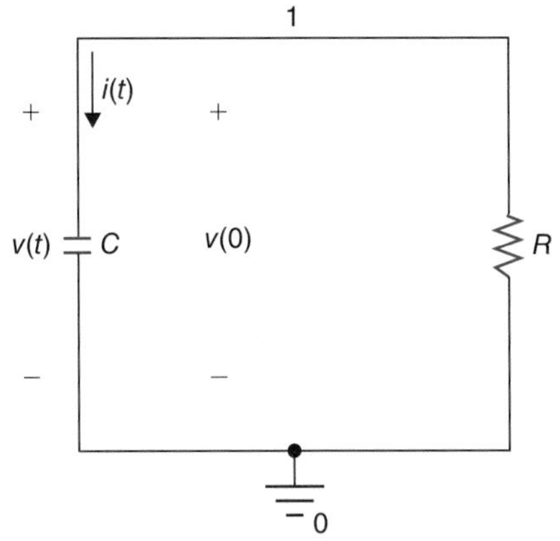

the initial value of V_0. The current $i(t)$ through the capacitor is proportional to the time rate of change of the voltage, $i(t) = Cdv(t)/dt$. Since the voltage decays exponentially, the current does as well. The magnitude of the current is at its maximum at $t = 0$. When the capacitance value C is large, the capacitor holds more charges for the given voltage ($Q = CV$), and it takes a longer time to discharge. When the resistance value R is increased, the amount of current flowing through the resistor decreases, and it takes a longer time to discharge, $q(t) = \int i(t)dt$. The component values C and R control the speed of the discharge. The product of R and C is called a *time constant* ($\tau = RC$), as discussed later in this chapter. As time progresses, the initial energy stored in the capacitor, $\frac{Cv^2(0)}{2}$, will dissipate as heat in the resistor. Eventually, after a long period of time, all the energy stored in the capacitor will be dissipated.

Summing the currents leaving node 1, we have

$$i(t) + \frac{v(t)}{R} = 0 \tag{7.1}$$

From the voltage-current relation on the capacitor, we have

$$i(t) = C\frac{dv(t)}{dt} \tag{7.2}$$

Substitution of Equation (7.2) into Equation (7.1) yields

$$C\frac{dv(t)}{dt} + \frac{v(t)}{R} = 0 \tag{7.3}$$

Dividing by C everywhere, we get

$$\frac{dv(t)}{dt} + \frac{1}{RC}v(t) = 0 \tag{7.4}$$

Dividing by $v(t)$, we obtain

$$\frac{\frac{dv(t)}{dt}}{v(t)} + \frac{1}{RC} = 0 \tag{7.5}$$

Equation (7.5) can be rewritten as

$$\frac{\frac{dv(t)}{dt}}{v(t)} = -\frac{1}{RC} \tag{7.6}$$

The left side of the equation is the derivative of $\ln[v(t)]$. Thus, we have

$$\frac{d}{dt}\ln[v(t)] = -\frac{1}{RC} \tag{7.7}$$

Integrating on both sides of Equation (7.7), we obtain

$$\ln[v(t)] = -\int_0^t \frac{1}{RC}dt + K = \frac{-t}{RC} + K \tag{7.8}$$

where K is a constant. Exponentiation on both sides of Equation (7.8) yields

$$e^{\ln[v(t)]} = v(t) = e^{\left(-\frac{t}{RC} + K\right)} = e^K e^{-\frac{t}{RC}} \tag{7.9}$$

Let $e^K = A$. Then, Equation (7.9) becomes

$$v(t) = Ae^{-\frac{t}{RC}} \tag{7.10}$$

Evaluation of Equation (7.10) at $t = 0$ yields

$$v(0) = Ae^{-\frac{0}{RC}} = A$$

Therefore, $A = v(0) = V_0$, which is the initial voltage across the capacitor. Overall, the voltage across the capacitor (and resistor) for $t \geq 0$ is given by

$$v(t) = v(0)e^{-\frac{t}{RC}} = V_0 e^{-\frac{t}{RC}} \tag{7.11}$$

where $v(0)$ is the voltage across the capacitor at $t = 0$. Equation (7.11) is valid for $t \geq 0$. A unit step function defined by

$$u(t) = \begin{cases} 1, & t \geq 0 \\ 0, & elsewhere \end{cases} \tag{7.12}$$

can be used to indicate that the solution given by Equation (7.11) is valid for $t \geq 0$ by multiplying Equation (7.11) by Equation (7.12); that is,

$$v(t) = v(0)e^{-\frac{t}{RC}}u(t) = V_0 e^{-\frac{t}{RC}}u(t) \tag{7.13}$$

Refer to Chapter 1 for the definition of the unit step function. The current $i(t)$ through the capacitor for $t \geq 0$ is given by

$$i(t) = C\frac{dv(t)}{dt} = C\left(-\frac{1}{RC}\right)v(0)\, e^{-\frac{t}{RC}} = -\frac{v(0)}{R}e^{-\frac{t}{RC}} = -\frac{V_0}{R}e^{-\frac{t}{RC}}u(t) \tag{7.14}$$

The negative sign in the current indicates that the actual current flows from the positive terminal (top) of the capacitor to node 1, then to the resistor, and enters the capacitor from the negative terminal (bottom).

The instantaneous power on the capacitor is given by

$$p_C(t) = v(t)i(t) = v(0)e^{-\frac{t}{RC}}\left[\frac{-v(0)}{R}e^{-\frac{t}{RC}}\right] = -\frac{v^2(0)}{R}e^{-\frac{2t}{RC}}u(t) \quad \textbf{(7.15)}$$

The negative sign indicates that the power is being released from the capacitor to the resistor. The instantaneous power on the resistor for $t \geq 0$ is given by

$$p_R(t) = v(t)[-i(t)] = v(0)e^{-\frac{t}{RC}}\left[-\frac{-v(0)}{R}e^{-\frac{t}{RC}}\right] = \frac{v^2(0)}{R}e^{-\frac{2t}{RC}}u(t) \quad \textbf{(7.16)}$$

The positive sign indicates that the power is being absorbed by the resistor. Equations (7.15) and (7.16) are identical except for the sign. The instantaneous power released from the capacitor is absorbed at the resistor. The instantaneous energy on the resistor is given by

$$w_R(t) = \int_0^t p_R(\lambda)d\lambda = \frac{v^2(0)}{R}\frac{RC}{2}\left(1 - e^{-\frac{2t}{RC}}\right) = \frac{Cv^2(0)}{2}\left(1 - e^{-\frac{2t}{RC}}\right)u(t) \quad \textbf{(7.17)}$$

At $t = \infty$, the total energy absorbed by the resistor is given by

$$w(\infty) = \frac{Cv^2(0)}{2} = \frac{1}{2}CV_0^2 \quad \textbf{(7.18)}$$

Equation (7.18) shows that the total energy supplied by the capacitor, given by $\frac{Cv^2(0)}{2}$, has been absorbed by the resistor.

EXAMPLE 7.1

Let $V_S = 5$ V, $R = 2$ kΩ, $C = 1$ μF in the circuit shown in Figure 7.2 earlier in the chapter. Find the expression of $v(t)$, $i(t)$, $p_C(t)$ and plot $v(t)$, $i(t)$, and $p_C(t)$ for $t \geq 0$.

Since $v(0) = V_S = 5$ V and $RC = 2000 \times 10^{-6} = 2 \times 10^{-3}$ s $= 2$ ms, we have

$$v(t) = v(0)e^{-\frac{t}{RC}} = 5e^{-\frac{t}{2\times10^{-3}}}\text{ V}$$

$$i(t) = -\frac{v(0)}{R}e^{-\frac{t}{RC}} = -2.5e^{-\frac{t}{2\times10^{-3}}}\text{ mA}$$

$$p_C(t) = -\frac{v^2(0)}{R}e^{-\frac{2t}{RC}}u(t) = -12.5e^{-\frac{2t}{2\times10^{-3}}}u(t)\text{ mW}$$

Figure 7.3 shows the plot of $v(t)$, $i(t)$, and $p_C(t)$ for $t \geq 0$.

continued

Example 7.1 continued

FIGURE 7.3

Plot of $v(t)$, $i(t)$, and $p_C(t)$ for EXAMPLE 7.1.

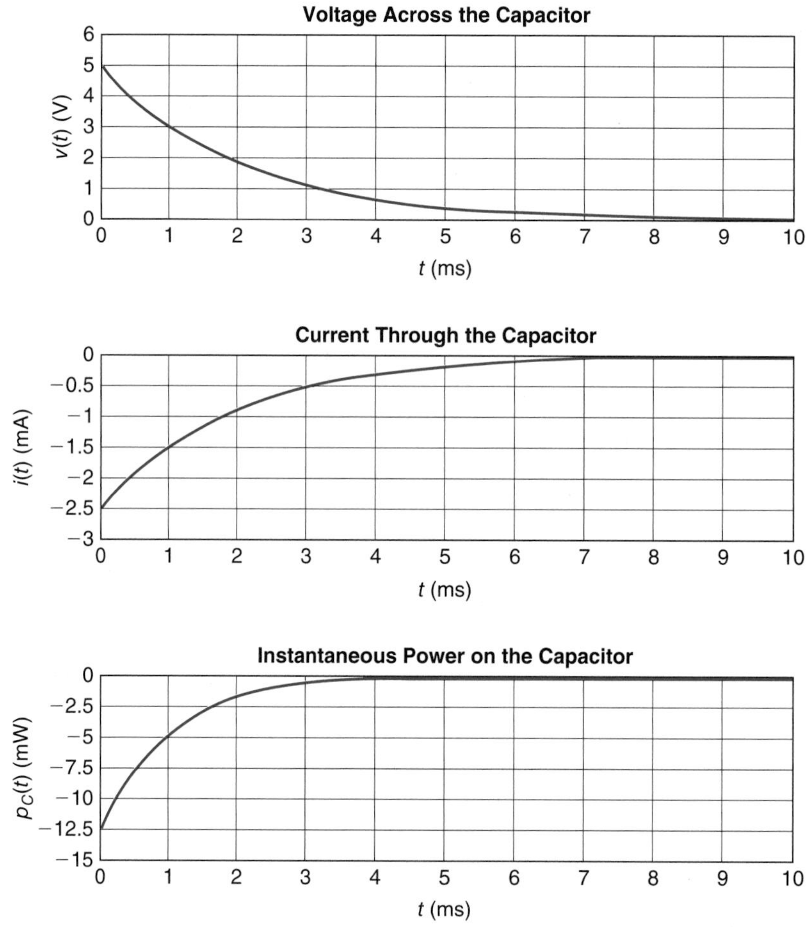

7.2.1 TIME CONSTANT

The voltage $v(t)$ across the capacitor resulting from initial voltage $v(0)$ decays exponentially. The voltage $v(t)$ is called **natural response**, **source-free response**, **zero input response (ZIR)**, or **transient response**, since it lasts for a brief period of time. The product of R and C, RC, is measured in seconds and is called a **time constant**, denoted by τ. Thus,

$$\tau = RC \tag{7.19}$$

In terms of the time constant τ, Equations (7.13)–(7.15) become, respectively,

$$v(t) = v(0)e^{-\frac{t}{\tau}} \tag{7.20}$$

$$i(t) = -\frac{v(0)}{R}e^{-\frac{t}{\tau}} \tag{7.21}$$

$$p_C(t) = -\frac{v^2(0)}{R}e^{-\frac{2t}{\tau}} \tag{7.22}$$

Figure 7.4 shows $v(t)$ for $\tau = 1, 2, 3, 4,$ and 5 [$(v(0) = 1$ V].

FIGURE 7.4

Voltage across the capacitor as time constant is varied.

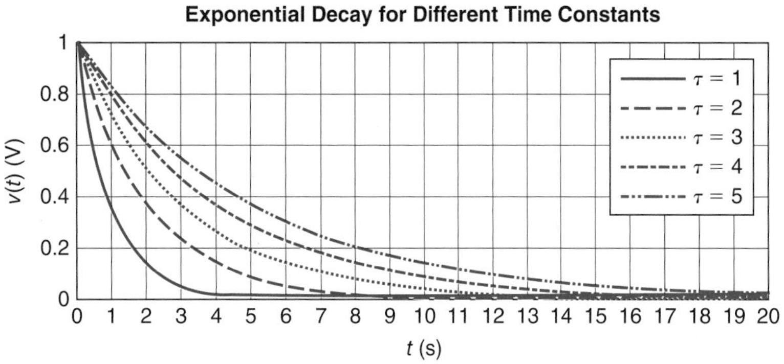

As can be seen from Figure 7.4, as the time constant τ increases, it takes longer time for the voltage across the capacitor to decay. The speed at which the charges stored on the capacitor plates are discharged is controlled by the time constant τ. At time zero ($t = 0$), the voltage is at its peak value $v(0)$. At time $t = \tau$, the voltage is

$$v(\tau) = v(0)e^{-\frac{\tau}{\tau}} = v(0)e^{-1} = 0.3678794412\, v(0)$$

At time $t = \tau$, the voltage across the capacitor drops to 36.788% of the initial value at $t = 0$. For $t = 2\tau, 3\tau, 4\tau, 5\tau, \ldots, 10\tau$, we have the values shown in Table 7.1.

TABLE 7.1

Voltage Across the Capacitor Normalized to V_0 at $t = n\tau$.

n	exp(−n)
0	1.000000000000000
1.000000000000000	0.367879441171442
2.000000000000000	0.135335283236613
3.000000000000000	0.049787068367864
4.000000000000000	0.018315638888734
5.000000000000000	0.006737946999085
6.000000000000000	0.002478752176666
7.000000000000000	0.000911881965555
8.000000000000000	0.000335462627903
9.000000000000000	0.000123409804087
10.000000000000000	0.000045399929762

At five times the time constant, the voltage across the capacitor due to initial energy on the capacitor is less than 1% of the initial voltage (0.6738%). For all practical purposes, the transient response can be ignored after about five times the time constant. The time derivative of the voltage across the capacitor is given by

$$\frac{dv(t)}{dt} = v(0)\frac{d}{dt}e^{-\frac{t}{\tau}} = -\frac{v(0)}{\tau}e^{-\frac{t}{\tau}} \qquad \textbf{(7.23)}$$

which is plotted in Figure 7.5 for $v(0) = 1$ V. Since the current through the capacitor is given by $Cdv(t)/dt$, the plot of current $i(t)$ has the same shape as the one shown in Figure 7.5 with different scale.

The rate of decay of the voltage across the capacitor is at its maximum at $t = 0$ and slows down as time progresses. The rate of decay at $t = 0$ is $-v(0)/\tau$. If the voltage decreases at this rate, $v(t)$ is given by

$$v_1(t) - v(0) = -\frac{v(0)}{\tau}(t - 0)$$

FIGURE 7.5

The rate of decay of the voltage across the capacitor.

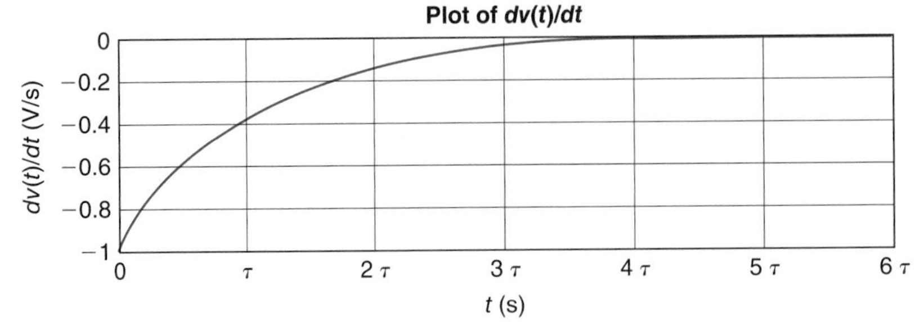

or

$$v_1(t) = v(0) - \frac{v(0)}{\tau}t = v(0)\left(1 - \frac{t}{\tau}\right)$$

At $t = \tau$, $v_1(t)$ is zero. Thus, if the rate of decay of $v(t)$ is kept at the rate of decay at $t = 0$, the voltage will be decreased to zero at one time constant. This equation is plotted in Figure 7.6 for $V_0 = 1$ V.

FIGURE 7.6

The plot of linear decay at the maximum rate, as well as the original decay.

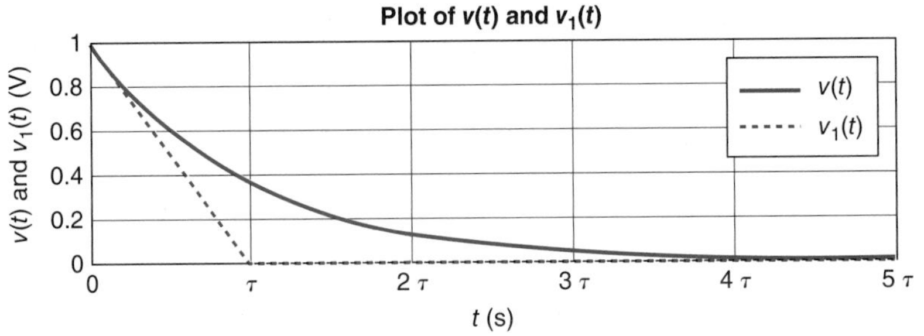

If there is one resistor with resistance R and one capacitor with capacitance C, as in the circuit shown in Figure 7.2, the time constant is the product of the R and C; that is, $\tau = RC$. If there is one capacitor with capacitance C and more than one resistor in the circuit, find the equivalent resistance R_{eq} of all the resistors in the circuit that connects in parallel to the capacitor. Then, the circuit reduces to one capacitor with capacitance C and one resistor with resistance R_{eq} connected in parallel. The time constant is given by $\tau = R_{eq}C$. The voltage across the capacitor is given by

$$v(t) = v(0)e^{-\frac{t}{\tau}}u(t) \text{ V}$$

where $v(0)$ is the initial voltage across the capacitor. The switches in the circuit provide changes in the circuit. For example, they may change a circuit with sources to one without the sources. When a switch is opened or closed, it may provide initial conditions on the capacitor, the inductor, or both.

Finding the voltage $v(t)$ across the capacitor for the given circuit involves finding the initial voltage $v(0)$ across the capacitor, finding the equivalent resistance R_{eq}, and finding the time constant $\tau = R_{eq}C$, as illustrated by the following examples.

EXAMPLE 7.2

In the circuit shown in Figure 7.7, the switch has been closed for a long time before it is opened at $t = 0$. Find the voltage $v(t)$ across the capacitor and the current $i(t)$ through the capacitor for $t \geq 0$ and plot $v(t)$ and $i(t)$ for $t \geq 0$.

continued

Example 7.2 continued

FIGURE 7.7

Circuit for EXAMPLE 7.2.

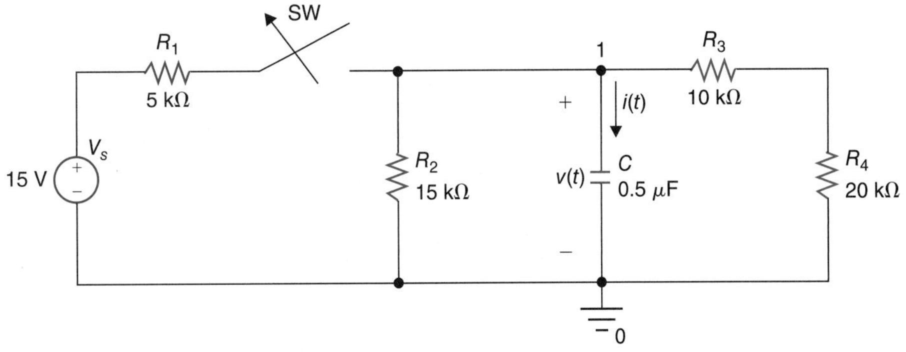

Since the switch has been closed for a long time before it is opened at $t = 0$, the circuit already reached steady state just before the switch is opened. The voltage $v(t)$ across the capacitor is a constant. If it is a constant, the current $i(t)$ through the capacitor is zero, since $i(t) = Cdv(t)/dt = C \times 0 = 0$. The capacitor can be treated as an open circuit when the sources in the circuit are dc and the circuit is in steady state. When the capacitor is treated as an open circuit, the capacitor can be deleted from the circuit temporarily. Resistor R_2 and series combination of R_3 and R_4 are connected in parallel to the capacitor. The equivalent resistance of the parallel combination of R_2 and $R_3 + R_4$ is

$$R_{eq} = \frac{R_2(R_3 + R_4)}{R_2 + R_3 + R_4} = \frac{15(10 + 20)}{15 + 10 + 20} k\Omega = 10\ k\Omega$$

FIGURE 7.8

The Thévenin equivalent circuit for $t < 0$.

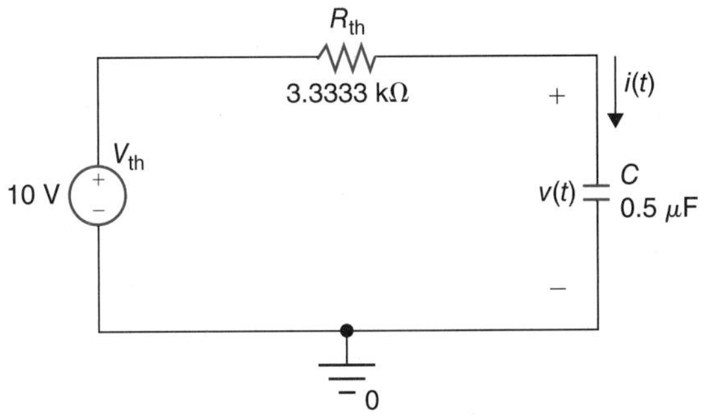

The voltage across the capacitor is the voltage across the equivalent resistor R_{eq}. Applying the voltage divider rule, we obtain the initial voltage across the capacitor, given by

$$v(0) = \frac{R_{eq}}{R_1 + R_{eq}} V_s = \frac{10}{5 + 10} \times 15\ \text{V} = 10\ \text{V}$$

FIGURE 7.9

The equivalent circuit for $t \geq 0$.

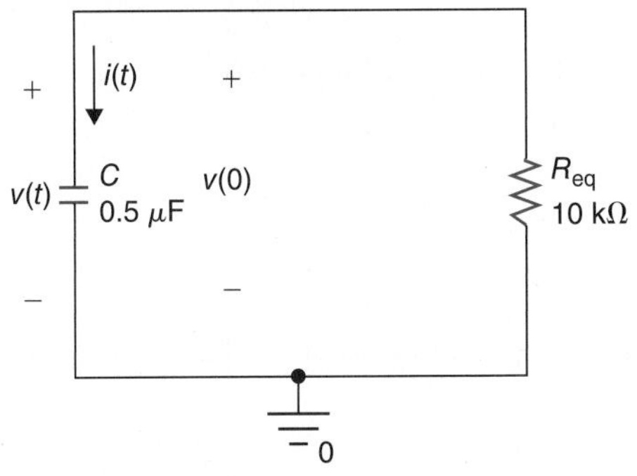

Notice that the Thévenin equivalent circuit for $t < 0$ consists of a voltage source with voltage of $V_{th} = 10$ V, Thévenin resistance $R_{th} = R_1 \| R_{eq} = R_1 \times R_{eq}/(R_1 + R_{eq}) = 50/15\ k\Omega = 10/3\ k\Omega$, and the capacitor, as shown in Figure 7.8. After capacitor charges to 10 V, the current through the resistor R_{th} will be zero; that is, $i(t) = (10 - 10)/R_{th} = 0$ A. Since the current through the capacitor is zero, it can be treated as an open circuit.

The equivalent circuit for $t \geq 0$ consists of the capacitor with initial voltage $v(0) = 10$ V and a resistor R_{eq} with resistance $10\ k\Omega$, as shown in Figure 7.9.

The time constant of the circuit for $t \geq 0$ is given by $\tau = R_{eq}C = 10{,}000 \times 5 \times 10^{-7} = 5 \times 10^{-3} = 5$ ms. Notice that $1/\tau = 200$. Thus, the voltage across the capacitor for $t \geq 0$ is given by

$$v(t) = v(0)e^{-\frac{t}{\tau}}u(t) = 10e^{-\frac{t}{0.005}}u(t) = 10e^{-200t}u(t)\ \text{V}$$

continued

Example 7.2 continued

The current through the capacitor is given by

$$i(t) = C\frac{dv(t)}{dt} = 5 \times 10^{-7} \times 10 \times (-200)e^{-200t}u(t) = -e^{-200t}u(t) \text{ mA}$$

Figure 7.10 shows $v(t)$ and $i(t)$.

FIGURE 7.10

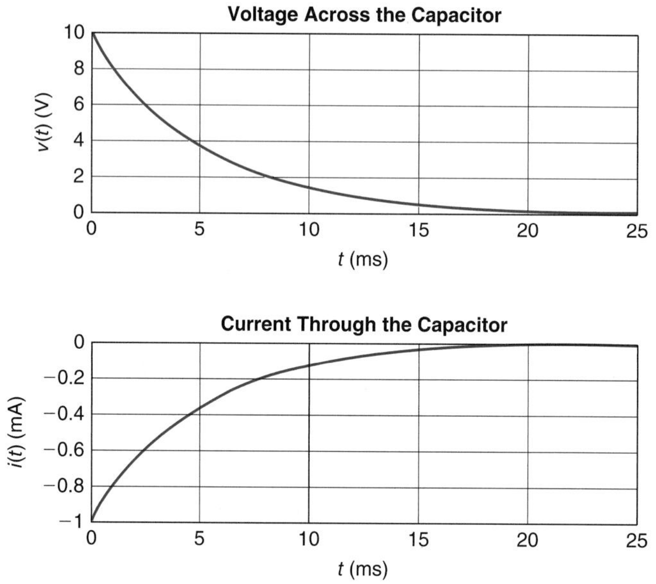

Plot of $v(t)$ and $i(t)$.

For $t \geq 0$, summing the currents leaving node 1 in the circuit shown in Figure 7.7, we obtain

$$\frac{v(t)}{R_2} + C\frac{dv(t)}{dt} + \frac{v(t)}{R_3 + R_4} = 0$$

The MATLAB function **dsolve** can be used to solve this differential equation as shown here.

MATLAB

```
%EXAMPLE 7.2
clear all;
C=0.5e-6;R2=15000;R3=10000;R4=20000;V0=10;
syms v(t)
Dv=diff(v);
v=dsolve(v/R2+C*Dv+v/(R3+R4),v(0)==V0);
i=C*diff(v);
taui=-log(subs(v,t,1)/subs(v,t,0));tau=1/taui;Req=tau/C;
v=vpa(v,10)
i=vpa(i,10)
ta=0:1e-5:25e-3;
v=subs(v,t,ta);
i=subs(i,t,ta);
figure(1)
plot(ta,v,'LineWidth',2);grid;xlabel('t (s)');ylabel('v(t) (V)');
figure(2)
plot(ta,i,'LineWidth',2);grid;xlabel('t (s)');ylabel('i(t) (A)');
taui=vpa(taui,10)
tau=vpa(tau,10)
Req=vpa(Req,10)
```

continued

Example 7.2 continued

MATLAB continued

```
Answers:
v =
10.0*exp(-200.0*t)
i =
-0.001*exp(-200.0*t)
taui =
200.0
tau =
0.005
Req =
10000.0
```

The plot of $v(t)$ and the plot of $i(t)$ are identical to those shown in Figure 7.10.

Exercise 7.1

The initial voltage across the capacitor is $v(0) = 5$ V in the circuit shown in Figure 7.11. Find the voltage $v(t)$ across the capacitor for $t \geq 0$. Also, find the voltage $v_1(t)$ across the resistor R_2 for $t \geq 0$.

FIGURE 7.11

Circuit for EXERCISE 7.1.

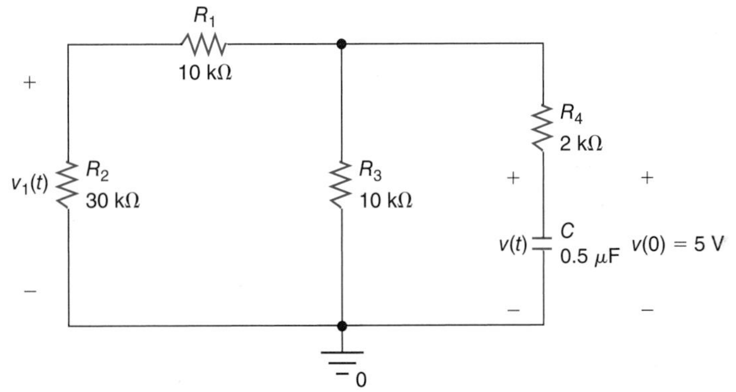

Answer:

$$v(t) = 5e^{-\frac{t}{0.005}}u(t) = 5e^{-200t}u(t)\text{ V}, v_1(t) = 3e^{-\frac{t}{0.005}}u(t) = 3e^{-200t}u(t)\text{ V}.$$

EXAMPLE 7.3

The switch in the circuit shown in Figure 7.12 has been closed for a long time before it is opened at $t = 0$. Find the voltage $v(t)$ across the capacitor for $t \geq 0$.

For $t < 0$, the capacitor can be treated as an open circuit for the dc source. The current from the current source is split between R_1 and $R_2 + R_3$. The current through R_3 is given by

$$I_S \times \frac{R_1}{R_1 + R_2 + R_3} = 5\text{ mA} \times \frac{3}{5} = 3\text{ mA}$$

continued

Example 7.3 continued

FIGURE 7.12

Circuit for EXAMPLE 7.3.

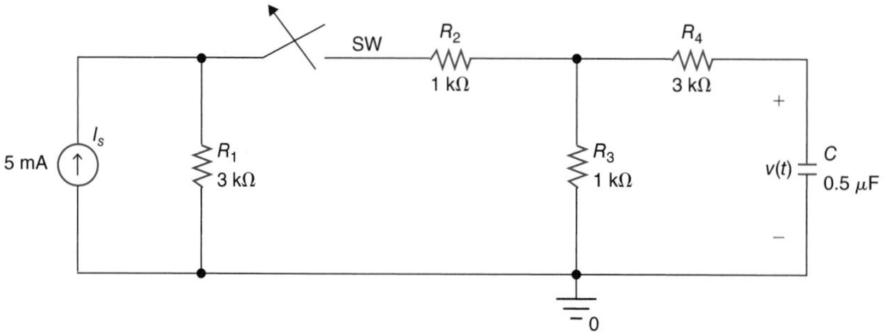

FIGURE 7.13

The RC circuit for $t \geq 0$.

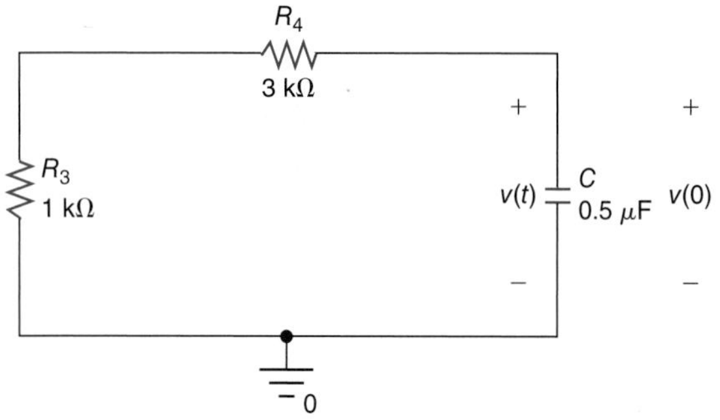

The voltage across R_3, which is the voltage across the capacitor, is

$$v(0) = 3\,\text{mA} \times 1\,k\Omega = 3\,\text{V}$$

For $t \geq 0$, the circuit consists of R_3, R_4, and C, as shown in Figure 7.13.

The time constant is given by

$$\tau = (R_3 + R_4)C = 4000\,\Omega \times 5 \times 10^{-7}\,\text{F} = 2 \times 10^{-3}\,\text{s}$$

The voltage across the capacitor is given by

$$v(t) = 3e^{-\frac{t}{0.002}}u(t) = 3e^{-500t}u(t)\,\text{V}$$

Exercise 7.2

The initial voltage across the capacitor is $v(0) = 12$ V in the circuit shown in Figure 7.14. Find the voltage $v(t)$ across the capacitor for $t \geq 0$.

FIGURE 7.14

Circuit for EXERCISE 7.2.

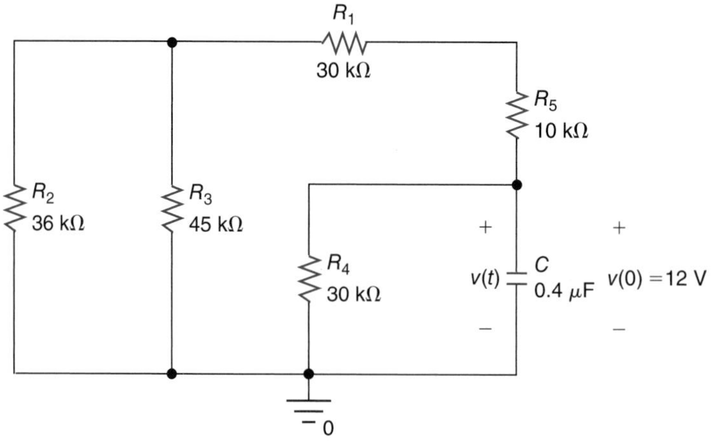

Answer:

$$v(t) = 12e^{-\frac{t}{0.008}}u(t) = 12e^{-125t}u(t)\text{ V.}$$

7.3 Step Response of *RC* Circuit

The switch in the circuit shown in Figure 7.15 is closed at $t = 0$. At $t = 0$, the voltage across the capacitor is $v(0) = V_0$.

Assume that $v(0) = 0$ (no charge on the capacitor plates at $t = 0$) for now. Without the resistor ($R = 0$), the time constant is zero ($\tau = RC = 0$). The voltage across the capacitor reaches its final value in no time [$v(t) = V_S$ for $t \geq 0$]. With the resistor present, when the switch is closed at $t = 0$, there is no charge on the capacitor plates, and the voltage across the capacitor is zero [$v(t) = q(t)/C = 0$]. Thus, the current through the resistor at $t = 0$ is V_S/R. As the current starts flowing in the circuit, the capacitor begins charging $\left[q(t) = \int i(t)dt\right]$. The presence of charges on the capacitor plates implies nonzero voltage across the capacitor [$v(t) = q(t)/C$]. The voltage across the resistor is reduced to $V_S - v(t)$, making the current through the circuit [$(V_S - v(t))/R$] smaller. The charging of the capacitor slows down more as time progresses until the capacitor is fully charged. In the steady state (after about five times the time constant is elapsed), the voltage across the capacitor is close to the applied voltage [$v(t) \approx V_S$] and the current through the circuit is zero [$i(t) = 0$]. When the sources (voltage source and current source) are dc, at $t = \infty$, the current through the capacitor is zero because there is no variation in the capacitor voltage [$i(t) = C\,dv(t)/dt$], and the capacitor can be treated as an open circuit.

FIGURE 7.15

RC circuit with step input.

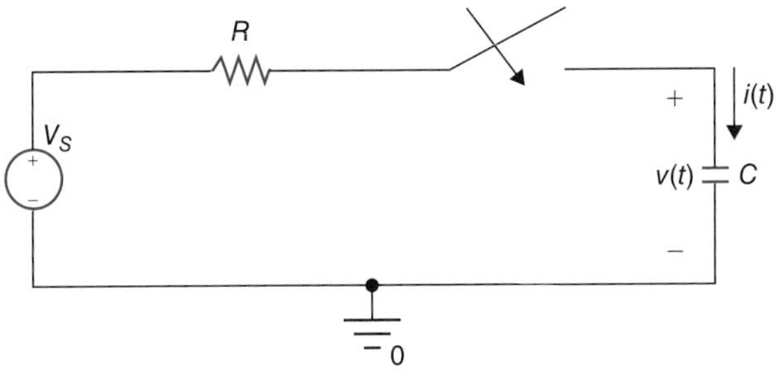

When the capacitance value C is fixed, as the resistance value is increased (i.e., the time constant is increased), the current through the resistor $[V_S - v(t)]/R$ is decreased, reducing the amount of charge delivered to the capacitor. It will take a longer time to charge the capacitor to full extent. When the resistance value R is fixed, as the capacitance value is increased (time constant is increased), the voltage across the capacitor is reduced for the same amount of charge delivered to the capacitor [$v(t) = q(t)/C$]. It will take a longer time to charge the capacitor to the full extent. The time constant indicates the speed at which the voltage across the capacitor reaches its final value. If the time constant is reduced, the voltage across the capacitor reaches its final value faster than before.

Collecting the voltage drops around the mesh in the clockwise direction for $t \geq 0$, we have

$$-V_S + Ri(t) + v(t) = 0 \quad (7.24)$$

The current $i(t)$ through the capacitor is given by

$$i(t) = C\frac{dv(t)}{dt} \quad (7.25)$$

Substitution of Equation (7.25) into Equation (7.24) yields

$$-V_S + RC\frac{dv(t)}{dt} + v(t) = 0 \quad (7.26)$$

Dividing by RC everywhere and rearranging Equation (7.26), we obtain

$$\frac{dv(t)}{dt} + \frac{1}{RC}v(t) = \frac{1}{RC}V_S \quad (7.27)$$

which can be rearranged as

$$\frac{dv(t)}{dt} = -\frac{1}{RC}[v(t) - V_S]$$

Dividing by $v(t) - V_S$, we have

$$\frac{\frac{dv(t)}{dt}}{v(t) - V_S} = \frac{-1}{RC} \quad \textbf{(7.28)}$$

The left side is the derivative of $\ln[|v(t) - V_S|]$. Notice that if $v(t) - V_S > 0$, then we have

$$\frac{d}{dt}[\ln|v(t) - V_S|] = \frac{d}{dt}[\ln(v(t) - V_S)] = \frac{\frac{dv(t)}{dt}}{v(t) - V_S}$$

On the other hand, if $v(t) - V_S < 0$, we get

$$\frac{d}{dt}[\ln|v(t) - V_S|] = \frac{d}{dt}[\ln(-v(t) + V_S)] = \frac{-\frac{dv(t)}{dt}}{-v(t) + V_S} = \frac{\frac{dv(t)}{dt}}{v(t) - V_S}$$

Thus,

$$\frac{d}{dt}[\ln|v(t) - V_S|] = -\frac{1}{RC}$$

Integrating on both sides, we have

$$\ln|v(t) - V_S| = \int_0^t \left(-\frac{1}{RC}\right)dt + K = -\frac{t}{RC} + K \quad \textbf{(7.29)}$$

Exponentiation on both sides of Equation (7.29) yields

$$e^{\ln|v(t)-V_S|} = |v(t) - V_S| = e^K e^{-\frac{t}{RC}}$$

If $v(t) - V_S > 0$, we have $|v(t) - V_S| = v(t) - V_S = e^K e^{-\frac{t}{RC}} = Ae^{-\frac{t}{RC}}$, where $A = e^K$. On the other hand, if $v(t) - V_S < 0$, we have $|v(t) - V_S| = -v(t) + V_S = e^K e^{-\frac{t}{RC}}$ or $v(t) - V_S = -e^K e^{-\frac{t}{RC}} = Ae^{-\frac{t}{RC}}$ where $A = -e^K$. Therefore, we have

$$v(t) - V_S = \pm e^K e^{-\frac{t}{RC}} = Ae^{-\frac{t}{RC}} \quad \textbf{(7.30)}$$

The solution for $v(t)$ is given by

$$v(t) = V_S + Ae^{-\frac{t}{RC}} \quad \textbf{(7.31)}$$

The constant term A is determined from the fact that the voltage across the capacitor is V_0 at $t = 0$. Setting $t = 0$ in Equation (7.31), we have

$$v(0) = V_S + Ae^{-\frac{0}{RC}} = V_S + A = V_0$$

Thus, $A = V_0 - V_S$. Substituting this value of A into Equation (7.31), we get the final solution:

$$v(t) = V_S + (V_0 - V_S)e^{-\frac{t}{RC}} \quad \textbf{(7.32)}$$

This solution is valid for $t \geq 0$. At $t = 0$, the voltage is $v(0) = V_0$, and at $t = \infty$, the voltage is $v(\infty) = V_S$. The voltage across the capacitor changes from the initial value of $v(0) = V_0$ at $t = 0$ to the final value of $v(\infty) = V_S$ at $t = \infty$.

Let the time constant of the *RC* circuit shown in Figure 7.15 be

$$\tau = RC \tag{7.33}$$

Then the first-order differential equation given by Equation (7.27) can be written as

$$\frac{dv(t)}{dt} + \frac{1}{\tau}v(t) = \frac{1}{\tau}V_S \tag{7.34}$$

In general, the solution to the first-order differential equation with a constant coefficient and dc input such as Equation (7.34) can be written as

$$v(t) = (\textit{Final Value}) + (\textit{Initial Value} - \textit{Final Value})e^{-(t-\textit{Delay})/(\textit{Time Constant})} \tag{7.35}$$

Here, the *Final Value* is the value of $v(t)$ at $t = \infty$, the *Initial Value* is the value of $v(t)$ at $t = \textit{Delay}$, the *Delay* is the time of change of conditions in the circuit, and *Time Constant* represents the speed at which the circuit adjusts itself after the change at $t = \textit{Delay}$. The change of circuit at $t = \textit{Delay}$ is an event such as turning a switch on or off. For the circuit shown in Figure 7.15, since there is no current in the circuit for dc input, the voltage across the capacitor at $t = \infty$ is V_S. Thus, Final Value $= V_S$. The initial voltage at $t = 0$ is V_0. Thus, Initial Value $= V_0$. Since the switch is changed at $t = 0$, there is no delay. Thus, $\textit{Delay} = 0$. The Time Constant is RC. Thus, we have

$$\begin{aligned} v(t) &= (\textit{Final Value}) + (\textit{Initial Value} - \textit{Final Value})e^{-(t-\textit{Delay})/(\textit{Time Constant})} \\ &= V_S + (V_0 - V_S)e^{-\frac{t-0}{RC}} \end{aligned}$$

The current $i(t)$ through the capacitor is given by

$$i(t) = C\frac{dv(t)}{dt} = C\frac{V_S - V_0}{RC}e^{-\frac{t}{\tau}} = \frac{V_S - V_0}{R}e^{-\frac{t}{RC}}u(t)$$

The current decays exponentially. At $t = \infty$, the current through the capacitor is zero. Thus, the capacitor can be treated as an open circuit for $t = \infty$.

The instantaneous power on the capacitor is

$$\begin{aligned} p(t) &= \left[V_S + (V_0 - V_S)e^{-\frac{t}{RC}}\right]\left(\frac{V_S - V_0}{R}e^{-\frac{t}{RC}}\right) \\ &= \left[V_S\frac{V_S - V_0}{R}e^{-\frac{t}{RC}} - \frac{(V_S - V_0)^2}{R}e^{-\frac{2t}{RC}}\right]u(t) \end{aligned}$$

The energy on the capacitor is

$$\begin{aligned} w(t) &= \frac{1}{2}Cv^2(t) = \frac{1}{2}C\left[V_S + (V_0 - V_S)e^{-\frac{t}{RC}}\right]^2 \\ &= \frac{1}{2}CV_S^2 + (CV_SV_0 - CV_S^2)e^{-\frac{t}{RC}} + \frac{1}{2}C(V_0 - V_S)^2e^{-\frac{2t}{RC}} \end{aligned}$$

At $t = 0$, the energy is $w(0) = \frac{1}{2}CV_S^2 + (CV_SV_0 - CV_S^2) + \frac{1}{2}C(V_0 - V_S)^2 = \frac{1}{2}CV_0^2$, which is the energy from the initial voltage V_0 on the capacitor. At $t = \infty$, the energy is $w(\infty) = \frac{1}{2}CV_S^2$, which is the energy from the final voltage of V_S.

If the initial voltage across the capacitor is zero at $t = 0$ (i.e., $V_0 = 0$), the voltage for $t \geq 0$ is given by

$$v(t) = V_S\left(1 - e^{-\frac{t}{RC}}\right)u(t) \tag{7.36}$$

At $t = \infty$, the current through the capacitor is zero; that is, $Cdv(t)/dt = 0$. The differential equation $\frac{dv(t)}{dt} + \frac{1}{RC}v(t) = \frac{1}{RC}V_S$ reduces to $\frac{1}{RC}v(t) = \frac{1}{RC}V_S$. The final value of the voltage across the capacitor is found by solving $\frac{1}{RC}v(t) = \frac{1}{RC}V_S$ for $v(t)$. Thus, we have $v(\infty) = V_S$. The current $i(t)$ through the capacitor for $V_0 = 0$ is given by

$$i(t) = C\frac{dv(t)}{dt} = C\frac{V_S}{\tau}e^{-\frac{t}{\tau}} = \frac{V_S}{R}e^{-\frac{t}{RC}}u(t) \tag{7.37}$$

7.3.1 INITIAL VALUE

If the initial voltage across the capacitor is not given, it should be calculated from the circuit. Usually, there is a change in the circuit, such as switch setting, that requires the calculation of the initial voltage. For simplicity, let us assume that the switch is closed or opened at $t = 0$. At that point, the circuit has already reached steady state. The current through the capacitor is zero, and the capacitor can be treated as an open circuit. Also, resistors connected in series with the capacitor are treated as an open circuit. If necessary, draw the circuit for $t < 0$ to evaluate the initial voltage across the capacitor. The initial voltage $v(0)$ across the capacitor is the voltage across the path that includes the capacitor. If necessary, use one or more of the following circuit laws and theorems to find the initial value: the voltage divider rule, the current divider rule, Thévenin's theorem, Norton's theorem, nodal analysis, mesh analysis, source transformation, and the superposition principle.

7.3.2 FINAL VALUE

The final value of the voltage across the capacitor is the voltage at $t = \infty$. Since the circuit is at steady state at $t = \infty$, the current through the capacitor is zero, and the capacitor can be treated as an open circuit. Also, resistors connected in series with the capacitor are treated as an open circuit. If necessary, draw the circuit for $t = \infty$ to evaluate the final voltage across the capacitor. The final voltage $v(\infty)$ across the capacitor is the voltage across the path that includes the capacitor. If necessary, use one or more of the following circuit laws and theorems to find the final value: the voltage divider rule, the current divider rule, Thévenin's theorem, Norton's theorem, nodal analysis, mesh analysis, source transformation, and the superposition principle.

7.3.3 TIME CONSTANT

For *RC* circuits with one capacitor in the circuit, the time constant is given by

$$\tau = R_{eq}C$$

where R_{eq} is the equivalent resistance seen from the capacitor. The equivalent resistance R_{eq} is the Thévenin equivalent resistance when the rest of the circuit (excluding the capacitor) is converted to the Thévenin equivalent circuit. In general, R_{eq} can be found by deactivating independent sources (short-circuit voltage sources and open-circuit current sources) and finding the equivalent resistance seen from the capacitor. Other methods, such as test voltage and test current, can also be used.

Figure 7.16 shows $v(t) = V_S + (V_0 - V_S)e^{-\frac{t}{RC}}$ for different values of time constants for $V_S = 1$ V and $V_0 = 0$ V.

FIGURE 7.16

Step response of an *RC* circuit for different time constants.

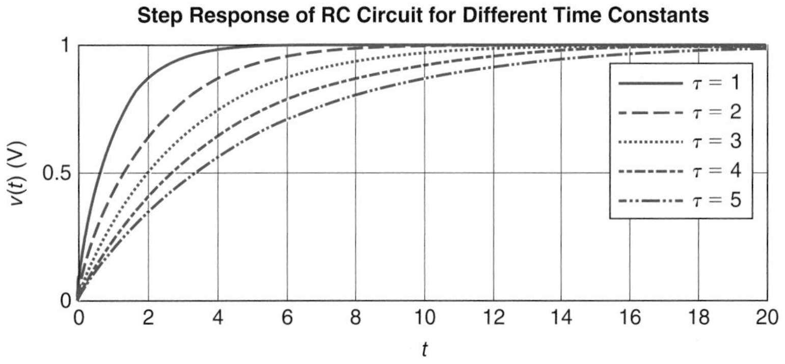

At $t = \tau$, the voltage across the capacitor is given by

$$v(\tau) = V_S\left(1 - e^{-\frac{\tau}{\tau}}\right) = V_S(1 - e^{-1}) = 0.63212\ V_S$$

At $t = \tau$ (one time constant), the voltage reaches 63.212% of the final value. Table 7.2 shows $v(t)/V_S$ for $t = 0, \tau, 2\tau, \ldots, 10\tau$.

TABLE 7.2

Normalized Voltage at $t = n\tau$.

n	$1 - \exp(-n)$
0	0
1.000000000000000	0.632120558828558
2.000000000000000	0.864664716763387
3.000000000000000	0.950212931632136
4.000000000000000	0.981684361111266
5.000000000000000	0.993262053000915
6.000000000000000	0.997521247823334
7.000000000000000	0.999088118034446
8.000000000000000	0.999664537372097
9.000000000000000	0.999876590195913
10.000000000000000	0.999954600070238

At five times the time constant, the voltage reaches 99.3262% of the final value. The time that it takes for the voltage $v(t)$ to reach 10% of the final value can be evaluated as follows:

$$0.1\ V_S = V_S\left(1 - e^{-\frac{t}{\tau}}\right) \quad \Rightarrow \quad t = -\tau \ln(0.9) = 0.10536\tau$$

Similarly, the time that it takes for the voltage $v(t)$ to reach 90% of the final value is found to be

$$0.9\ V_S = V_S\left(1 - e^{-\frac{t}{\tau}}\right) \quad \Rightarrow \quad t = -\tau \ln(0.1) = 2.30259\tau$$

It takes about 2.19722τ for the voltage $v(t)$ to rise from 10% of the final value to 90% of the final value. The time for the voltage $v(t)$ to rise from 10% of the final value to 90% of the final value is called *rise time* (T_r). For the first-order *RC* circuit with zero initial condition, the rise time is given by

$$T_r = 2.19722\tau$$

where the time constant τ is given by $\tau = RC$.

The time derivative of $v(t)$, $dv(t)/dt$, provides the rate of increase of $v(t)$ as a function of time. From Equation (7.36), for $V_0 = 0$, we have

$$\frac{dv(t)}{dt} = \frac{V_S}{\tau}e^{-\frac{t}{\tau}}$$

The rate of increase of $v(t)$ is at its maximum at $t = 0$. Then, the rate of increase decreases exponentially, as shown in Figure 7.17 for $V_S = 1$ V, $V_0 = 0$ V, $\tau = 1$ s.

FIGURE 7.17

The rate of increase of the voltage across the capacitor.

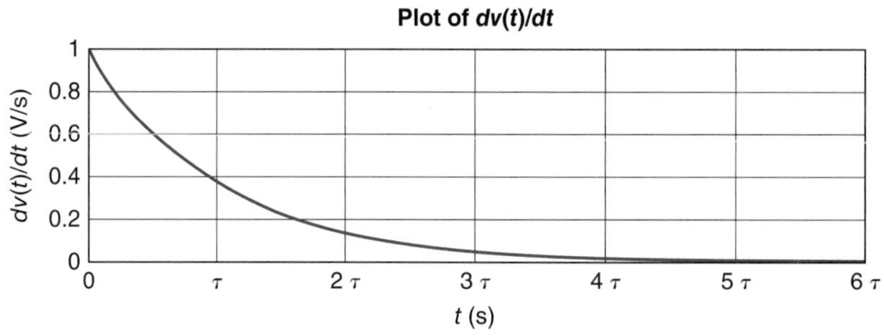

If the rate of increase is kept at $t = 0$, V_S/τ, the voltage increases linearly as

$$v_1(t) = \frac{V_S}{\tau}t \qquad \textbf{(7.38)}$$

At $t = \tau$, the voltage reaches the final value V_S. Equations (7.36) and (7.38) are plotted in Figure 7.18 for $V_S = 1$ V, $V_0 = 0$ V.

FIGURE 7.18

The step response $v(t)$ of *RC* circuit under normal condition and under maximum rate of increase.

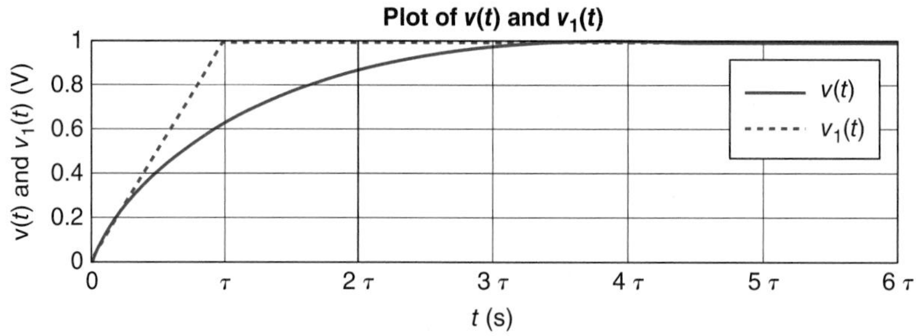

7.3.4 SOLUTION TO GENERAL FIRST-ORDER DIFFERENTIAL EQUATION WITH CONSTANT COEFFICIENT AND CONSTANT INPUT

A general first-order differential equation with a constant coefficient and constant input can be written as

$$\frac{dv(t)}{dt} + av(t) = b \qquad \textbf{(7.39)}$$

In Equation (7.39), a and b are constants. In the steady state, there is no change in the voltage due to the dc source. Thus, we have $dv(t)/dt = 0$. The differential equation is given by $a\,v(t) = b$ for $t = \infty$. Thus, the final value is given by $v(\infty) = b/a$. A comparison of Equations (7.34) and (7.39) reveals that the coefficient of $v(t)$ is the inverse of the time constant (i.e., $a = 1/\tau$). Thus, $\tau = 1/a$. The solution of the differential equation given by Equation (7.39) is

$$v(t) = \left[\frac{b}{a} + \left(v(0) - \frac{b}{a}\right)e^{-at}\right]u(t) \qquad \textbf{(7.40)}$$

This solution is derived in Section 7.6, later in this chapter, as well. Equations (7.39) and (7.40) suggest an alternative method of finding the step response of the *RC* circuit. Rather than finding $v(\infty)$, R_{eq}, τ, and Equation (7.34), find the differential Equation (7.39) of the circuit for $t \geq 0$ using nodal analysis or mesh analysis or any other method, and solve the differential equation using Equation (7.40).

EXAMPLE 7.4

Let $V_S = 1$ V, $R = 4\ k\Omega$, $C = 1\ \mu$F, $v(0) = 0$ V for the circuit shown in Figure 7.15. Find the voltage $v(t)$ across the capacitor and the current $i(t)$ through the capacitor for $t \geq 0$. Plot $v(t)$ and $i(t)$ for $t \geq 0$.

The time constant is $\tau = RC = 4000 \times 1 \times 10^{-6}$ s $= 4 \times 10^{-3}$ s $= 4$ ms. Thus, the voltage across the capacitor is given by

$$v(t) = \left[V_S + (V_0 - V_S)e^{-\frac{t-0}{RC}}\right]u(t) = \left[1 + (0 - 1)e^{-\frac{t}{0.004}}\right]u(t)\text{ V} = \left(1 - e^{-250t}\right)u(t)\text{ V}$$

and the current through the capacitor is given by

$$i(t) = C\frac{dv(t)}{dt} = C\frac{V_S}{\tau}e^{-\frac{t}{\tau}} = \frac{V_S}{R}e^{-\frac{t}{RC}}u(t) = \frac{1}{4000}e^{-\frac{t}{0.004}}u(t)\text{ A} = 250e^{-250t}u(t)\ \mu\text{A}$$

Figure 7.19 shows $v(t)$ and $i(t)$ for $t \geq 0$.

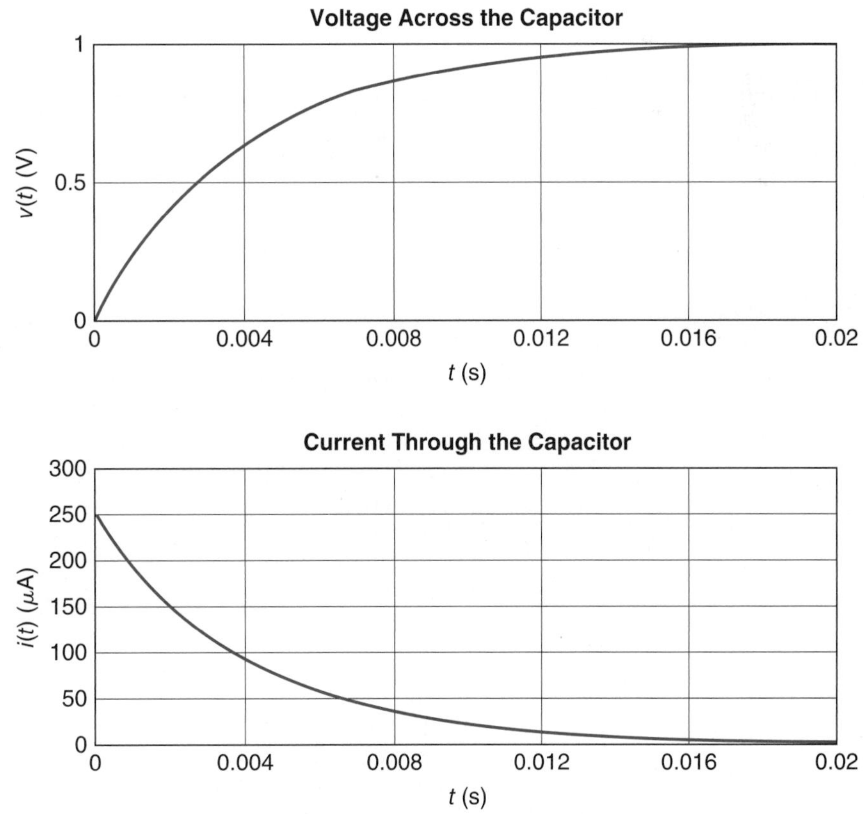

FIGURE 7.19

Plot of the voltage $v(t)$ and current $i(t)$.

continued

Example 7.4 continued

The MATLAB function **dsolve** can be used to solve the differential equation given by Equation (7.27), as shown here:

MATLAB

```
%EXAMPLE 7.4
clear all;
R=4000;C=1e-6;V0=0;Vs=1;
syms v(t)
Dv=diff(v);
v=dsolve(Dv+v/(R*C)==Vs/(R*C),v(0)==V0);
i=C*diff(v);
v=vpa(v,10)
i=vpa(i,10)
ta=0:1e-5:20e-3;
v=subs(v,t,ta);
i=subs(i,t,ta);
figure(1)
plot(ta,v,'LineWidth',2);grid;xlabel('t (s)');ylabel('v(t) (V)');
figure(2)
plot(ta,i,'LineWidth',2);grid;xlabel('t (s)');ylabel('i(t) (A)');

Answers:
v =
1.0 - 1.0*exp(-250.0*t)
i =
0.00025*exp(-250.0*t)
```

The plot of $v(t)$ and the plot of $i(t)$ are identical to those shown in Figure 7.19.

EXAMPLE 7.5

The switch in the circuit shown in Figure 7.20 is closed at $t = 0$. At $t = 0$, the voltage across the capacitor is $v(0) = V_0$. Find the voltage $v(t)$ across the capacitor and the current $i(t)$ through the capacitor.

FIGURE 7.20

An *RC* circuit.

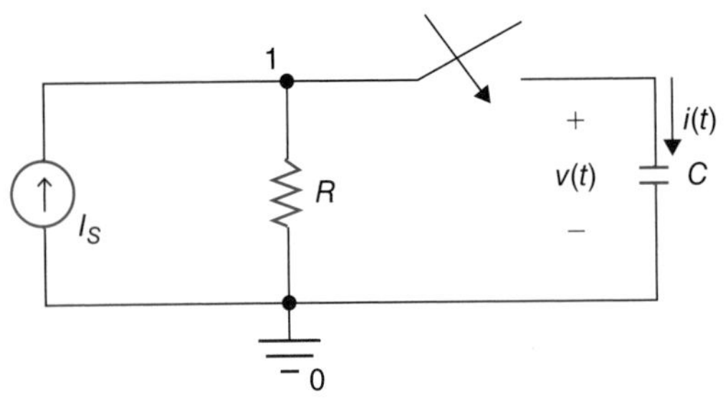

The voltage $v(t)$ across the capacitor for $t \geq 0$ can be found using Equation (7.35). Since the switch is closed at $t = 0$, *Delay* is 0. The voltage $v(t)$ at $t = 0$ is given by *Initial Value* $= v(0) = V_0$. The voltage $v(t)$ at $t = \infty$ is the voltage across the resistor R given by *Final Value* $= v(\infty) = I_S R$. The time constant τ is $\tau = RC$. Therefore, the voltage $v(t)$ for $t \geq 0$ is given by

$$\begin{aligned} v(t) &= (\textit{Final Value}) + [(\textit{Initial Value}) \\ &\quad - (\textit{Final Value})]e^{-(t-\textit{Delay})/(\textit{Time Constant})} \\ &= \left[I_S R + (V_0 - I_S R)e^{-\frac{t}{RC}}\right]u(t) \end{aligned} \tag{7.41}$$

Alternatively, the circuit shown in Figure 7.20 can be transformed to the circuit shown in Figure 7.21 by source transformation. The Thévenin equivalent of I_S and R is given by a voltage source with voltage $I_S R$ and a series resistor with resistance R.

continued

Example 7.5 continued

FIGURE 7.21

Thévenin equivalent circuit.

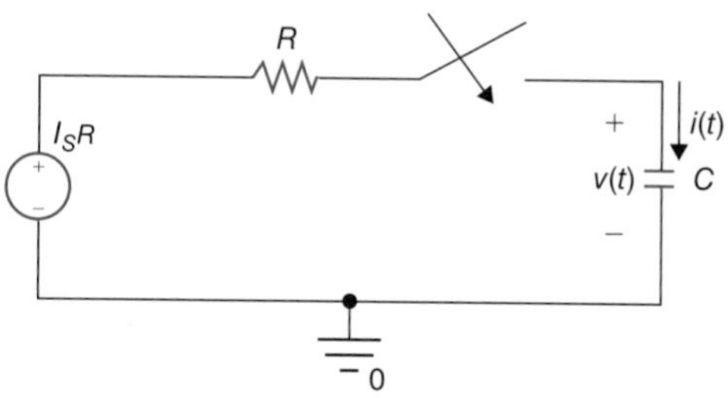

The initial value of the voltage across the capacitor is given by $v(0) = V_0$, and the final value of the voltage across the capacitor is $v(\infty) = I_SR$. The time constant is $\tau = RC$. From Equation (7.35), the voltage across the capacitor is given by

$$v(t) = (Final\ Value) + [(Initial\ Value - Final\ Value)]e^{-(t-Delay)/(Time\ Constant)}$$
$$= \left[I_SR + (V_0 - I_SR)e^{-\frac{t}{RC}}\right]u(t)$$

EXAMPLE 7.6

Switch 1 is opened at $t = 0$ and switch 2 is closed at $t = 0.5$ ms in the circuit shown in Figure 7.22. Find the voltage $v(t)$ across the capacitor for $t \geq 0$.

FIGURE 7.22

Circuit for EXAMPLE 7.6.

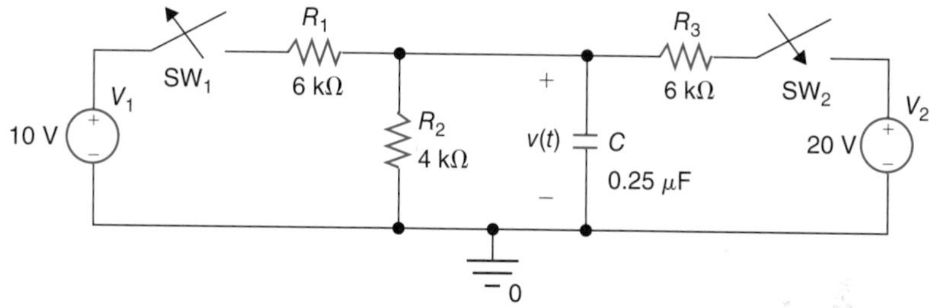

Since the circuit is in the steady state, there is no current through the capacitor and the capacitor can be treated as an open circuit for $t < 0$. Based on the voltage divider rule, the initial voltage across the capacitor is

$$v(0) = V_1 \times \frac{R_2}{R_1 + R_2} = 10\text{ V} \times \frac{4\ k\Omega}{6\ k\Omega + 4\ k\Omega} = 4\text{ V}$$

Thus, $v(0) = 4$ V. For $0 \leq t < 0.5$ ms, the circuit consists of R_2 and C. The time constant is given by

$$\tau = R_2C = 4000 \times 0.25 \times 10^{-6} = 10^{-3} = 1\text{ ms. } 1/\tau = 1000$$

Therefore, for $0 \leq t \leq 0.5$ ms, the voltage across the capacitor, $v(t)$, is given by

$$v(t) = V_0e^{-\frac{t}{\tau}} = 4e^{-1000t}[u(t) - u(t - 0.5 \times 10^{-3})]$$

At $t = 0.5$ ms, the voltage across the capacitor is

$$v(0.5\text{ ms}) = 4e^{-1000\times0.5\times0.001} = 4e^{-0.5} = 2.4261\text{ V}$$

For $t \geq 0.5$ ms, the initial voltage across the capacitor is $V_0 = 2.4261$ V. For $t \geq 0.5$ ms, the circuit consists of R_2, R_3, C, and V_2. At $t = \infty$, the circuit is in steady state. The current through the capacitor is zero and the capacitor can be treated as an open circuit. Based on the voltage divider rule, the final value of $v(t)$ is given by

continued

Example 7.6 continued

$$v(\infty) = V_2 \times \frac{R_2}{R_2 + R_3} = 20\text{ V} \times \frac{4\,k\Omega}{4\,k\Omega + 6\,k\Omega} = 8\text{ V}$$

To find the equivalent resistance, we deactivate the voltage source by short-circuiting it and find the equivalent resistance of the parallel connection of R_2 and R_3. Thus, we have

$$R_{eq} = R_2 \| R_3 = \frac{R_2 \times R_3}{R_2 + R_3} = \frac{4\,k\Omega \times 6\,k\Omega}{4\,k\Omega + 6\,k\Omega} = 2.4\,k\Omega$$

The time constant is given by

$$\tau = R_{eq}C = 2400 \times 0.25 \times 10^{-6}\text{ s} = 6 \times 10^{-4}\text{ s} = 600\,\mu\text{s}$$

For 0.5 ms $\leq t$, the voltage across the capacitor is given by

$$\begin{aligned} v(t) &= (\textit{Final Value}) + [(\textit{Initial Value}) - (\textit{Final Value})]e^{-(t-\textit{Delay})/(\textit{Time Constant})} \\ &= \left[8 + (2.4261 - 8)e^{-\frac{t-0.5\times10^{-3}}{6\times10^{-4}}}\right]u(t - 0.5 \times 10^{-3}) \\ &= \left[8 - 5.5739e^{-1666.6667\,(t-0.5\times10^{-3})}\right]u(t - 0.5 \times 10^{-3}) \end{aligned}$$

The plot of $v(t)$ for $t \geq 0$ is shown in Figure 7.23.

FIGURE 7.23

Voltage $v(t)$ across the capacitor.

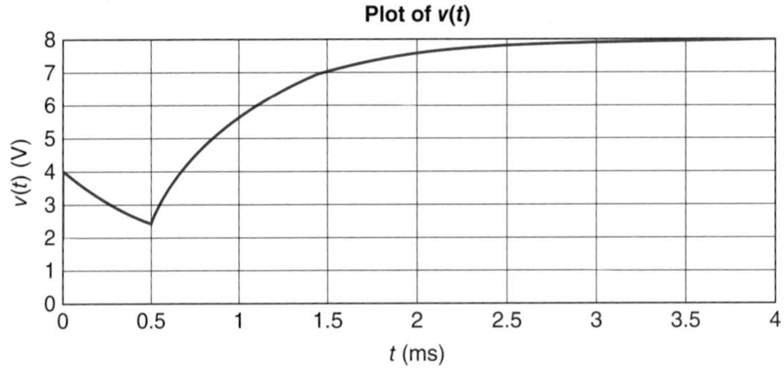

Exercise 7.3

The switch in the circuit shown in Figure 7.24 has been open for a long time before it is closed at $t = 0$. Find the voltage $v(t)$ across the capacitor for $t \geq 0$.

FIGURE 7.24

Circuit for EXERCISE 7.3.

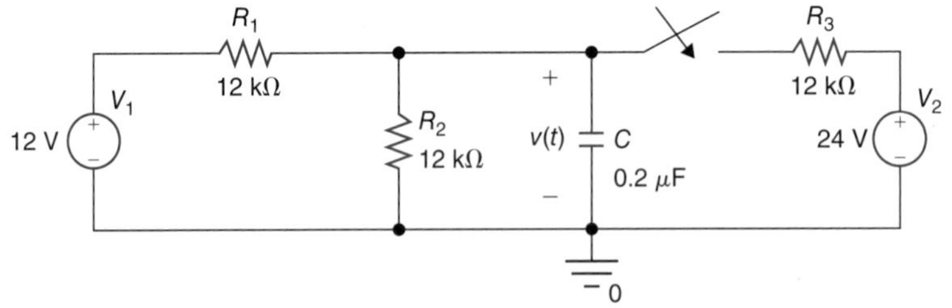

Answer:
$v(0) = 6$ V, $R_{eq} = R_1 \| R_2 \| R_3 = 4\,k\Omega$, $\tau = R_{eq}C = 8 \times 10^{-4}$. Applying the superposition principle, we have $v(\infty) = 12\text{ V} \times 6\,k\Omega/18\,k\Omega + 24\text{ V} \times 6\,k\Omega/18\,k\Omega = 12$ V.

$$v(t) = \left[12 + (6 - 12)e^{-\frac{t}{0.0008}}\right]u(t)\text{ V} = \left[12 - 6e^{-1250t}\right]u(t)\text{ V}$$

EXAMPLE 7.7

Switch 1 is opened at $t = 0$ and switch 2 is closed at $t = 0$ in the circuit shown in Figure 7.25. Find the voltage $v(t)$ across the capacitor and voltage $v_a(t)$ at node *a* for $t \geq 0$.

FIGURE 7.25

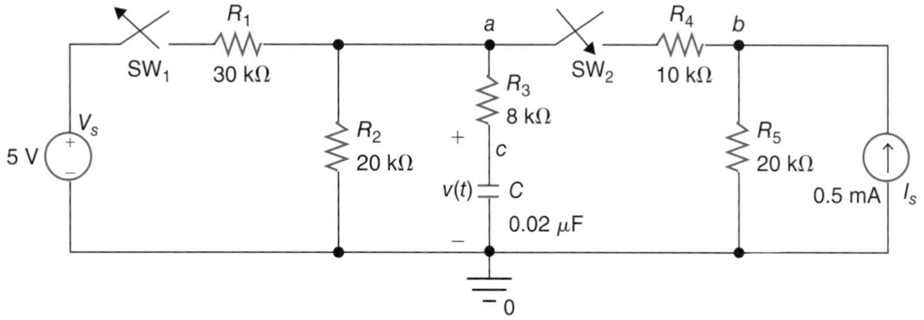

Circuit for EXAMPLE 7.7.

For $t < 0$, switch 1 is closed and switch 2 is open. Since the current through the capacitor is zero, the capacitor can be treated as an open circuit. According to the voltage divider rule, the input voltage $V_s = 5$ V is divided between R_1 and R_2 in proportion to the resistance values. Thus, we have

$$v(0) = V_s \times \frac{R_2}{R_1 + R_2} = 5\text{ V} \times \frac{20\,k\Omega}{30\,k\Omega + 20\,k\Omega} = 2\text{ V}$$

For $t \geq 0$, switch 1 is open and switch 2 is closed. At $t = \infty$, the current through the capacitor is zero, and the capacitor acts as an open circuit. According to the current divider rule, the input current $I_s = 0.5$ mA is divided between R_5 and $R_4 + R_2$. The current through $R_4 + R_2$ is given by

$$I_s \times \frac{R_5}{R_4 + R_2 + R_5} = 0.5\text{ mA} \times \frac{20\,k\Omega}{10\,k\Omega + 20\,k\Omega + 20\,k\Omega} = 0.2\text{ mA}$$

The voltage across R_2, which is also the voltage across the capacitor, is given by

$$v(\infty) = 0.2\text{ mA} \times 20\,k\Omega = 4\text{ V}$$

The equivalent resistance connected to the capacitor is obtained by deactivating the current source (open-circuit it, or remove it) and find the equivalent resistance.

$$R_{eq} = R_3 + \left[R_2 \| (R_4 + R_5)\right] = R_3 + \frac{R_2(R_4 + R_5)}{R_2 + R_4 + R_5}$$

$$= 8\,k\Omega + \frac{20\,k\Omega \times (10\,k\Omega + 20\,k\Omega)}{20\,k\Omega + 10\,k\Omega + 20\,k\Omega} = 20\,k\Omega$$

The time constant is

$$\tau = R_{eq}C = 20000 \times 2 \times 10^{-8} = 4 \times 10^{-4} = 0.4 \times 10^{-3}\text{s}$$

Notice that $1/\tau = 2500$. The voltage $v(t)$ for $t \geq 0$ is given by

$$v(t) = (\textit{Final Value}) + [(\textit{Initial Value}) - (\textit{Final Value})]e^{-(t-\textit{Delay})/(\textit{Time Constant})}$$

$$= \left[4 + (2 - 4)e^{-\frac{t}{4\times10^{-4}}}\right]u(t)\text{ V}$$

continued

Example 7.7 continued

or

$$v(t) = (4 - 2e^{-2500t})u(t) \text{ V}$$

The current $i(t)$ through the capacitor (from top to bottom) is given by

$$i(t) = C\frac{dv(t)}{dt} = 0.02 \times 10^{-6}(-2)(-2500)e^{-2500t}u(t) = 0.1e^{-2500t}u(t) \text{ mA}$$

The voltage at node a is given by

$$v_a(t) = R_3 i(t) + v(t) = 0.8e^{-2500t}u(t) + (4 - 2e^{-2500t})u(t) = (4 - 1.2e^{-2500t})u(t) \text{ V}$$

Summing the currents leaving nodes a, b, and c, we obtain respectively

$$\frac{v_a(t)}{R_2} + \frac{v_a(t) - v(t)}{R_3} + \frac{v_a(t) - v_b(t)}{R_4} = 0$$

$$\frac{v_b(t) - v_a(t)}{R_4} + \frac{v_b(t)}{R_5} - I_s = 0$$

$$\frac{v(t) - v_a(t)}{R_3} + C\frac{dv(t)}{dt} = 0$$

MATLAB can be used to solve these equations as shown here:

MATLAB

```
%EXAMPLE 7.7
clear all;
C=0.02e-6;R1=30000;R2=20000;R3=8000;R4=10000;R5=20000;
Is=0.5e-3;Vs=5;V0=Vs*R2/(R1+R2)
syms v(t) va vb
Dv=diff(v);
[va,vb]=solve(va/R2+(va-v)/R3+(va-vb)/R4,...
(vb-va)/R4+vb/R5-Is,va,vb);
v=dsolve((v-va)/R3+C*Dv,v(0)==V0);
i=C*diff(v);
va=R3*i+v;
taui=-log(subs(i,t,1)/subs(i,t,0));tau=1/taui;Req=tau/C;
v=vpa(v,10)
i=vpa(i,10)
va=vpa(va,10)
taui=vpa(taui,10)
tau=vpa(tau,10)
Req=vpa(Req,10)

Answers:
V0 =
     2
v =
4.0 - 2.0*exp(-2500.0*t)
i =
0.0001*exp(-2500.0*t)
va =
4.0 - 1.2*exp(-2500.0*t)
taui =
2500.0
tau =
0.0004
Req =
20000.0
```

Exercise 7.4

Find the voltage $v(t)$ across the capacitor in the circuit shown in Figure 7.26. The initial voltage across the capacitor is $v(0) = 5$ V.

FIGURE 7.26

Circuit for EXERCISE 7.4.

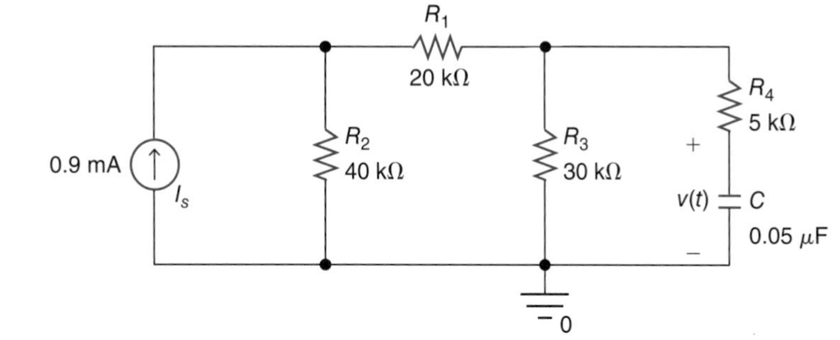

Answer:

$$v(t) = \left[12 + (5 - 12)e^{-\frac{t}{0.00125}}\right]u(t) = (12 - 7e^{-800t})u(t) \text{ V}.$$

EXAMPLE 7.8

Find the voltage $v(t)$ across the capacitor in the circuit shown in Figure 7.27. The initial voltage across the capacitor is $v(0) = 0$ V.

FIGURE 7.27

Circuit for EXAMPLE 7.8.

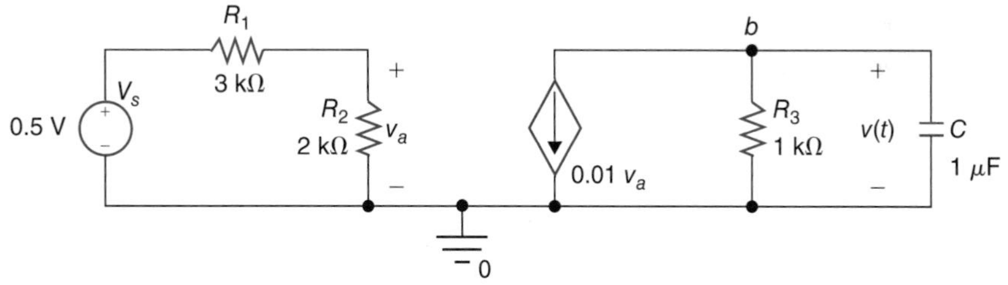

Based on the voltage divider rule, the voltage v_a across R_2 is given by

$$v_a = V_s \times \frac{R_2}{R_1 + R_2} = 0.5 \text{ V} \times \frac{2\,k\Omega}{3\,k\Omega + 2\,k\Omega} = 0.2 \text{ V}$$

The current of the voltage-controlled current source is $0.01v_a = 2$ mA. The final value of the voltage across the capacitor is given by

$$v(\infty) = -0.01v_aR_3 = -0.002 \times 1000 = -2 \text{ V}$$

The time constant is

$$\tau = R_3C = 1000 \times 10^{-6} = 10^{-3} = 1 \text{ ms}$$

The voltage $v(t)$ across the capacitor is given by

$$v(t) = \left[-2 + (0 + 2)e^{-\frac{t}{0.001}}\right]u(t) \text{ V} = -2(1 - e^{-1000t})u(t) \text{ V}$$

Exercise 7.5

Find the voltage $v(t)$ across the capacitor in the circuit shown in Figure 7.28. The initial voltage across the capacitor is $v(0) = 0$ V.

FIGURE 7.28

Circuit for EXERCISE 7.5.

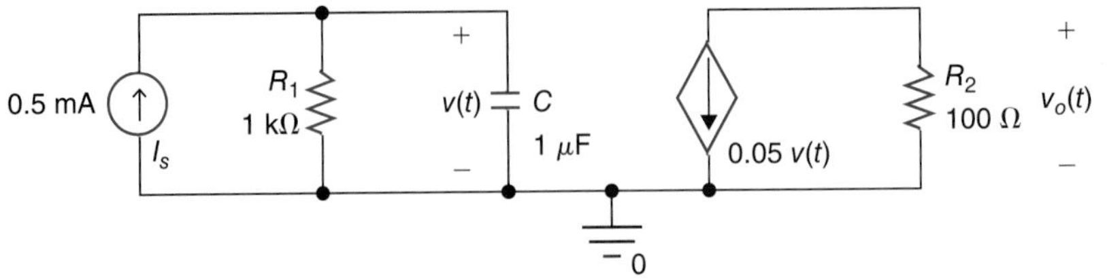

Answer:

$v_o(t) = -2.5(1 - e^{-1000t})u(t)$ V.

7.4 Natural Response of *RL* Circuit

The switch in Figure 7.29 has been closed for a long time before it is opened at $t = 0$. For $t \geq 0$, the circuit shown in Figure 7.29 becomes the circuit shown in Figure 7.30.

FIGURE 7.29

An *RL* circuit.

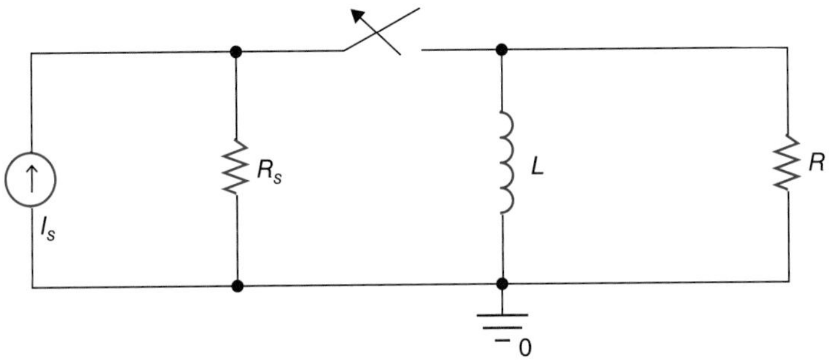

FIGURE 7.30

An *RL* circuit for $t \geq 0$.

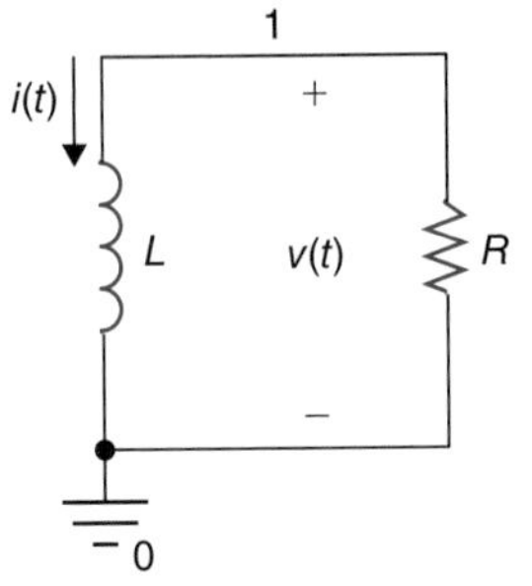

Before the switch is opened at $t = 0$, the current source I_s is connected to resistors R_s and R and an inductor L in parallel. Since the inductor can be treated as a short circuit for the dc current source I_s, according to the current divider rule, all the current from the current source will flow through the inductor, and not through the resistors R_s and R. Thus, the current through the inductor at $t = 0$ is given by

$$i(0) = I_0 = I_s$$

For $t \geq 0$, the initial current through the inductor $i(0)$ starts flowing through the resistor R. At $t = 0$, when the switch is opened, the initial current I_0 flows through the resistor R from bottom to top. The initial voltage $v(0)$ across the resistor (and inductor) is $-I_0R$. As shown here, the current through the inductor decreases exponentially as time progresses [i.e., $i(t) = I_0e^{-\frac{t}{\tau}}u(t) = I_0e^{-\frac{Rt}{L}}u(t)$], where the time constant is given by $\tau = L/R$. The time rate of change of the current is given by $\dfrac{di(t)}{dt} = -\dfrac{I_0}{\tau}e^{-\frac{t}{\tau}}u(t) = -\dfrac{RI_0}{L}e^{-\frac{t}{\tau}}u(t)$. The negative sign indicates that the current decreases. The rate of decrease is smaller as time progresses.

Summing the currents away from node 1 in the circuit shown in Figure 7.30, we have

$$\frac{v(t)}{R} + i(t) = 0 \qquad \textbf{(7.42)}$$

From the voltage-current relation on the inductor, we have

$$v(t) = L\frac{di(t)}{dt} \tag{7.43}$$

Substituting Equation (7.43) into Equation (7.42), we get

$$\frac{L}{R}\frac{di(t)}{dt} + i(t) = 0 \tag{7.44}$$

Rearrangement of Equation (7.44) yields

$$\frac{di(t)}{dt} + \frac{R}{L}i(t) = 0 \tag{7.45}$$

Dividing by $i(t)$, we have

$$\frac{\frac{di(t)}{dt}}{i(t)} + \frac{R}{L} = 0 \tag{7.46}$$

Equation (7.46) can be rewritten as

$$\frac{\frac{di(t)}{dt}}{i(t)} = -\frac{R}{L} \tag{7.47}$$

The left side is the derivative of $\ln[i(t)]$. Thus, we have

$$\frac{d}{dt}\ln[i(t)] = -\frac{R}{L} \tag{7.48}$$

Integrating on both sides of (7.48), we have

$$\ln[i(t)] = -\int_0^t \frac{R}{L}dt + K = -\frac{R}{L}t + K \tag{7.49}$$

where K is a constant. Taking the exponential on both sides, we get

$$i(t) = e^{\left(-\frac{Rt}{L}+K\right)} = e^K e^{-\frac{Rt}{L}} \tag{7.50}$$

Let $e^K = A$. Then, Equation (7.50) becomes

$$i(t) = Ae^{-\frac{Rt}{L}} \tag{7.51}$$

Evaluation of Equation (7.51) at $t = 0$ yields

$$i(0) = Ae^{-\frac{R0}{L}} = A$$

Therefore, $A = i(0) = I_0$. Overall, the current through the inductor for $t \geq 0$ is given by

$$i(t) = i(0)e^{-\frac{Rt}{L}} = I_0e^{-\frac{Rt}{L}} = I_0e^{-\frac{t}{L/R}} \tag{7.52}$$

where $i(0)$ is the current through the inductor at $t = 0$. The voltage $v(t)$ across the inductor for $t \geq 0$ is give by

$$v(t) = L\frac{di(t)}{dt} = L\left(-\frac{R}{L}\right)i(0)e^{-\frac{Rt}{L}} = -Ri(0)e^{-\frac{Rt}{L}} = -RI_0e^{-\frac{t}{L/R}} \tag{7.53}$$

The power on the inductor for $t \geq 0$ is given by

$$p(t) = v(t)i(t) = -Ri(0)e^{-\frac{Rt}{L}}i(0)e^{-\frac{Rt}{L}} = -i^2(0)Re^{-\frac{2Rt}{L}} = -I_0^2Re^{-\frac{2t}{L/R}}$$

Since $p(t) < 0$, the inductor is releasing power to the resistor.

The power absorbed by the resistor for $t \geq 0$ is given by

$$p(t) = v(t)[-i(t)] = Ri(0)e^{-\frac{Rt}{L}}i(0)e^{-\frac{Rt}{L}} = i^2(0)Re^{-\frac{2Rt}{L}} = I_0^2Re^{-\frac{2t}{L/R}}\ \text{W} \tag{7.54}$$

The total energy spent on the resistor as a function of time is given by

$$w(t) = \int_0^t p(\lambda)d\lambda = \int_0^t I_0^2Re^{-\frac{2t}{L/R}}d\lambda = -I_0^2R\frac{L}{2R}\left(e^{-\frac{2t}{L/R}} - 1\right) = \frac{1}{2}LI_0^2\left(1 - e^{-\frac{2t}{L/R}}\right) \tag{7.55}$$

At $t = \infty$, the total energy dissipated by the resistor is $\frac{1}{2}LI_0^2$ J, which is the energy stored on the inductor at $t = 0$. The current through the resistor from node 1 to 0 (top to bottom) is $-i(t)$. The current $i(t)$ through the inductor resulting from initial current $i(0)$ decays exponentially. The current $i(t)$ is called **transient response**, since it lasts for a brief period of time.

The ratio of L over R, L/R, has a unit of seconds and is called a time constant of the RL circuit. The time constant is denoted by τ. Thus,

$$\tau = \frac{L}{R} \tag{7.56}$$

In terms of the time constant τ, Equations (7.52)–(7.54) become

$$i(t) = I_0e^{-\frac{t}{\tau}} \tag{7.57}$$

$$v(t) = -RI_0e^{-\frac{t}{\tau}} \tag{7.58}$$

$$p(t) = I_0^2Re^{-\frac{2t}{\tau}} \tag{7.59}$$

7.4.1 TIME CONSTANT

In terms of time constant τ, the current through the inductor for $t \geq 0$ is given by Equation (7.57):

$$i(t) = I_0e^{-\frac{t}{\tau}}$$

Figure 7.31 shows $i(t)$ for $\tau = 1, 2, 3, 4,$ and 5 s for $I_0 = 1$ A.

As can be seen from Figure 7.31, as the time constant τ increases, it takes a longer time for the current through the inductor to decay. The speed at which the current decays depends on the time constant τ. At time zero ($t = 0$), the current is at its peak value, I_0. At time $t = \tau$, the current is

$$i(\tau) = I_0e^{-\frac{\tau}{\tau}} = I_0e^{-1} = 0.3678794412I_0$$

FIGURE 7.31

The current through the inductor of an *RL* circuit for different time constants.

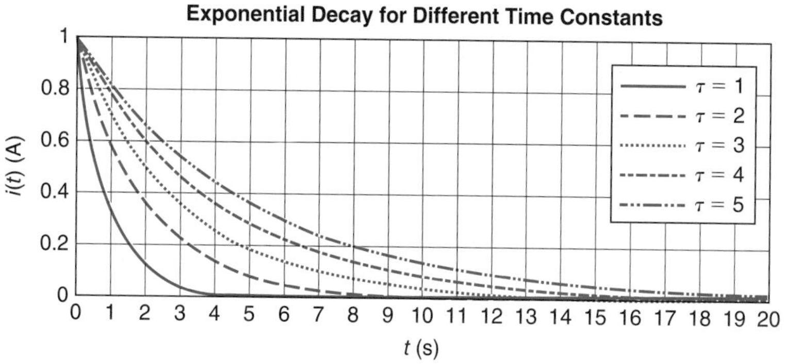

The current through the inductor drops to 36.788% of the initial value at $t = \tau$. For $t = 2\tau, 3\tau, 4\tau, 5\tau, \ldots, 10\tau$, we have the values shown in Table 7.3.

TABLE 7.3

Current Through the Inductor Normalized to I_0 at $t = n\tau$.

n	$\exp(-n)$
0	1.000000000000000
1.000000000000000	0.367879441171442
2.000000000000000	0.135335283236613
3.000000000000000	0.049787068367864
4.000000000000000	0.018315638888734
5.000000000000000	0.006737946999085
6.000000000000000	0.002478752176666
7.000000000000000	0.000911881965555
8.000000000000000	0.000335462627903
9.000000000000000	0.000123409804087
10.000000000000000	0.000045399929762

At five times the time constant, the current through the inductor due to initial energy on the inductor is less than 1% of the initial current (0.6738%). For all practical purposes, the transient response can be ignored after about five times the time constant. The time derivative of the current through the inductor is given by

$$\frac{di(t)}{dt} = I_0 \frac{d}{dt} e^{-\frac{t}{\tau}} = -\frac{I_0}{\tau} e^{-\frac{t}{\tau}} = -\frac{RI_0}{L} e^{-\frac{t}{\tau}} \tag{7.60}$$

which is plotted in Figure 7.32 for $I_0 = 1$ A and $\tau = 1$ s.

FIGURE 7.32

Plot of $di(t)/dt$.

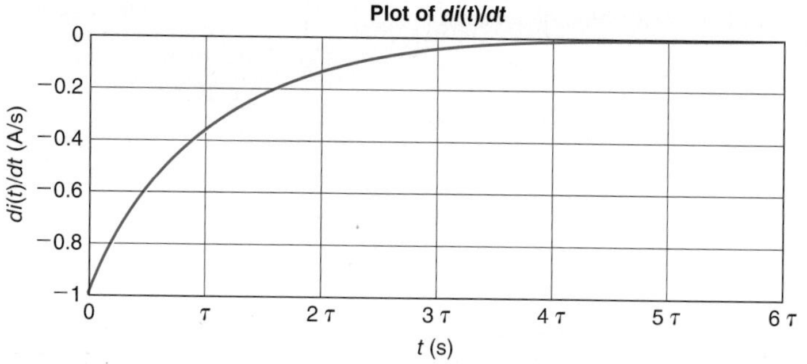

The rate of decay of the current through the inductor is at its maximum at $t = 0$ and slows down as time progresses. The rate of decay at $t = 0$ is $-I_0/\tau$. If the current decreases at this rate, $i(t)$ is given by

$$i_1(t) - I_0 = -\frac{I_0}{\tau}(t - 0)$$

or

$$i_1(t) = I_0 - \frac{I_0}{\tau}t = I_0\left(1 - \frac{t}{\tau}\right)$$

At $t = \tau$, $i_1(t)$ is zero. Thus, if the rate of decay of $i(t)$ is kept at the rate of decay at $t = 0$, the current will decrease to zero in one time constant. This equation is plotted in Figure 7.33, along with the original $i(t)$ given by Equation (7.57) for $I_0 = 1$ A.

FIGURE 7.33

Plot of $i(t)$ and $i_1(t)$.

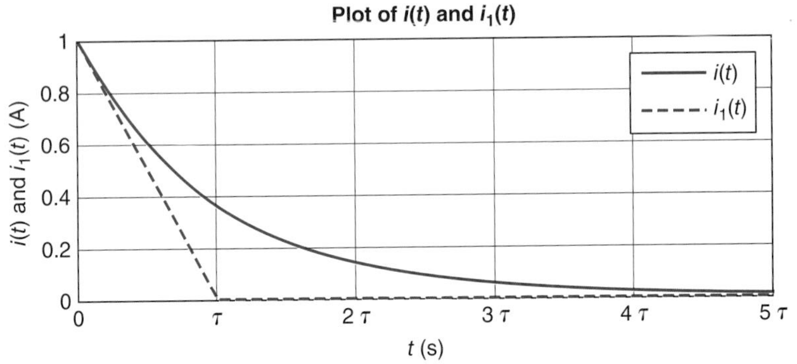

If there is one resistor with resistance R and one inductor with inductance L, as in the circuit shown in Figure 7.30, the time constant is the ratio of L over R (i.e., $\tau = L/R$). If there is one inductor with inductance L and more than one resistor in the circuit, find the equivalent resistance R_{eq} of all the resistors in the circuit that connects in parallel to the inductor. Then, the circuit reduces to one inductor with inductance L and one resistor with resistance R_{eq} connected in parallel. The time constant is given by $\tau = L/R_{eq}$. The current $i(t)$ through the inductor is given by

$$i(t) = i(0)e^{-\frac{t}{\tau}}u(t) \text{ A}$$

where $i(0)$ is the initial current through the inductor. The switches in the circuit provide changes in the circuit. For example, they may change a circuit with sources to one without the sources. When a switch is opened or closed, it may provide initial conditions on the capacitor, the inductor, or both.

Finding the current $i(t)$ through the inductor for the given circuit involves finding the initial current $i(0)$ through the inductor, finding the equivalent resistance R_{eq}, and finding the time constant $\tau = L/R_{eq}$, as illustrated by the following examples.

EXAMPLE 7.9

The switch in the circuit shown in Figure 7.34 has been closed for a long time before it is opened at $t = 0$. Find the current $i(t)$ through the inductor and voltage $v(t)$ across the inductor for $t \geq 0$. Also, plot $i(t)$ and $v(t)$ for $t \geq 0$.

In the steady state, the inductor can be treated as a short circuit for the dc source. For $t < 0$, current from the source flows through R_1 and L bypassing R_2. Thus, the initial current $i(0)$ through the inductor is given by

$$i(0) = \frac{V_s}{R_1} = \frac{12 \text{ V}}{4 \text{ k}\Omega} = 3 \text{ mA}$$

continued

Example 7.9 continued

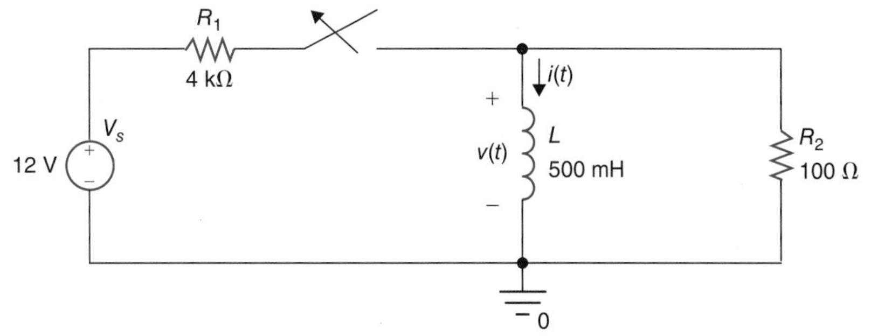

FIGURE 7.34

Circuit for EXAMPLE 7.9.

The time constant is given by

$$\tau = \frac{L}{R_2} = \frac{0.5}{100}\,\text{s} = 0.005\,\text{s} = 5\ \text{ms}$$

Thus, the current through the inductor for $t \geq 0$ is given by

$$i(t) = 3e^{-\frac{t}{0.005}}u(t)\ \text{mA}$$

The voltage across the inductor is given by

$$v(t) = L\frac{di(t)}{dt} = 0.5 \times 3 \times 10^{-3} \times \left(-\frac{1}{5 \times 10^{-3}}\right) \times e^{-\frac{t}{0.005}}u(t)\ \text{V} = -0.3e^{-\frac{t}{0.005}}u(t)\ \text{V}$$

Figure 7.35 shows $i(t)$ and $v(t)$ for $t \geq 0$.

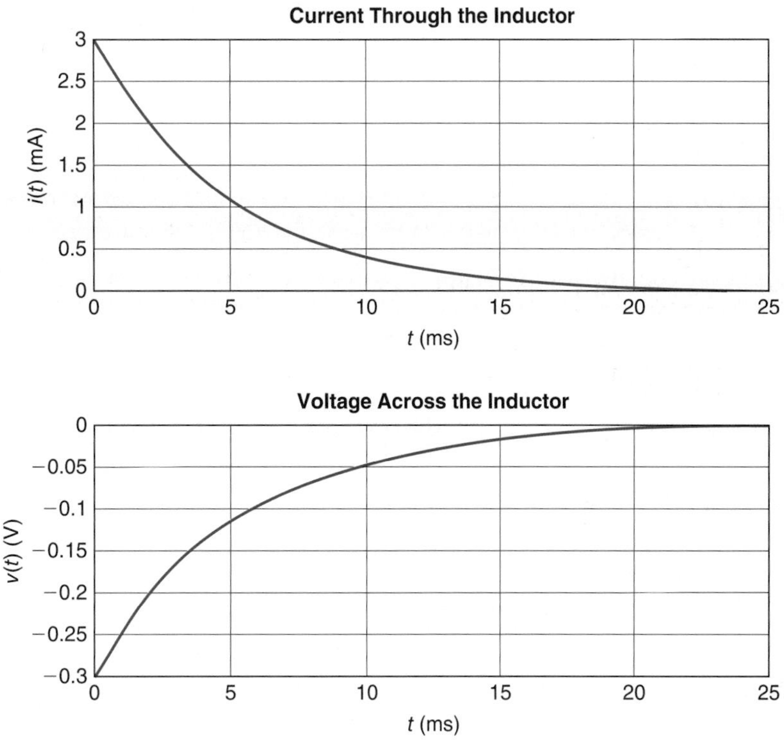

FIGURE 7.35

Plot of $i(t)$ and $v(t)$ for the circuit shown in Figure 7.34.

EXAMPLE 7.10

The switch in the circuit shown in Figure 7.36 has been closed for a long time before it is opened at $t = 0$. Find the current $i(t)$ through the inductor for $t \geq 0$.

FIGURE 7.36

Circuit for EXAMPLE 7.10.

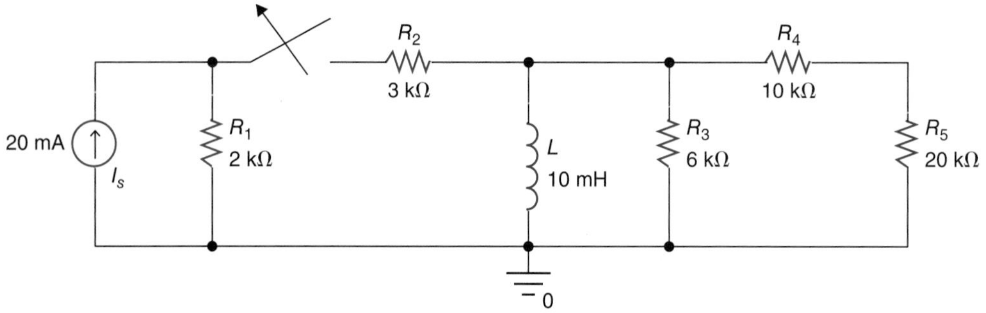

FIGURE 7.37

Circuit for $t \geq 0$.

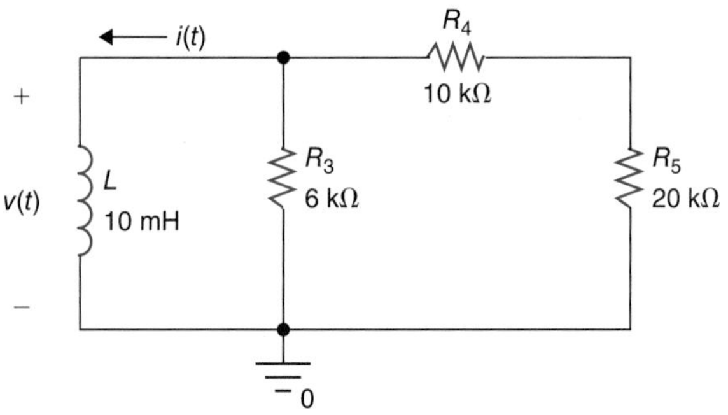

Before the switch is opened at $t = 0$, the circuit is at steady state. For dc current, the voltage across the inductor is zero, since $LdI/dt = 0$. The inductor can be treated as a short circuit for $t < 0$. Based on the current divider rule, the initial current through the inductor is $i(0) = I_0 = 20 \text{ mA} \times 2/(2+3) = 8 \text{ mA}$.

After the switch is opened at $t = 0$, the circuit becomes the one shown in Figure 7.37.

The equivalent resistance of R_3, R_4, and R_5 that is connected to the inductor is given by $R_{eq} = 6\,k\Omega \parallel (10\,k\Omega + 20\,k\Omega) = 6\,k\Omega \parallel 30\,k\Omega = 5\,k\Omega$. The time constant is $\tau = L/R_{eq} = 10 \times 10^{-3}/5 \times 10^{3} = 2 \times 10^{-6} = 2\,\mu\text{s}$. $1/\tau = 1/(2 \times 10^{-6}) = 500{,}000\,(1/\text{s})$. The current through the inductor for $t \geq 0$ is given by

$$i(t) = I_0 e^{-\frac{t}{\tau}} = 8e^{-\frac{t}{2\times10^{-6}}} \text{ mA} = 8e^{-500{,}000t} \text{ mA}$$

Exercise 7.6

The switch in the circuit shown in Figure 7.38 has been closed for a long time before it is opened at $t = 0$. Find the current $i(t)$ through the inductor for $t \geq 0$.

FIGURE 7.38

Circuit for EXERCISE 7.6.

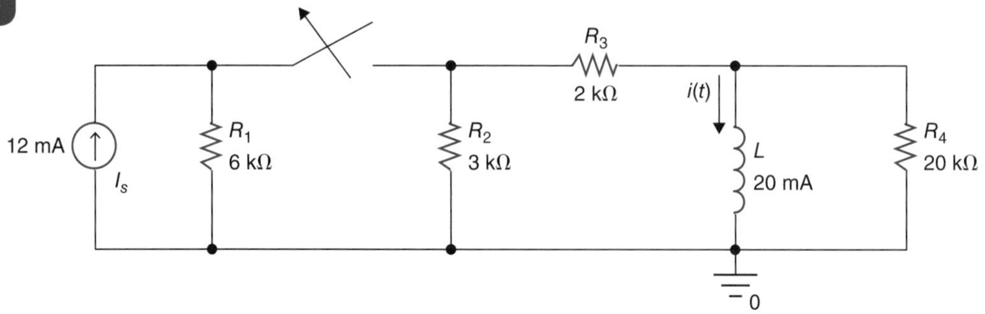

continued

Exercise 7.6 continued

Answer:

$$i(t) = I_0 e^{-\frac{t}{\tau}} = 6e^{-\frac{t}{5\times10^{-6}}} \text{ mA} = 6e^{-200{,}000t} \text{ mA}.$$

EXAMPLE 7.11

The switch in the circuit shown in Figure 7.39 has been closed for a long time before it is opened at $t = 0$. Find the current $i(t)$ through the inductor for $t \geq 0$.

FIGURE 7.39

Circuit for EXAMPLE 7.11.

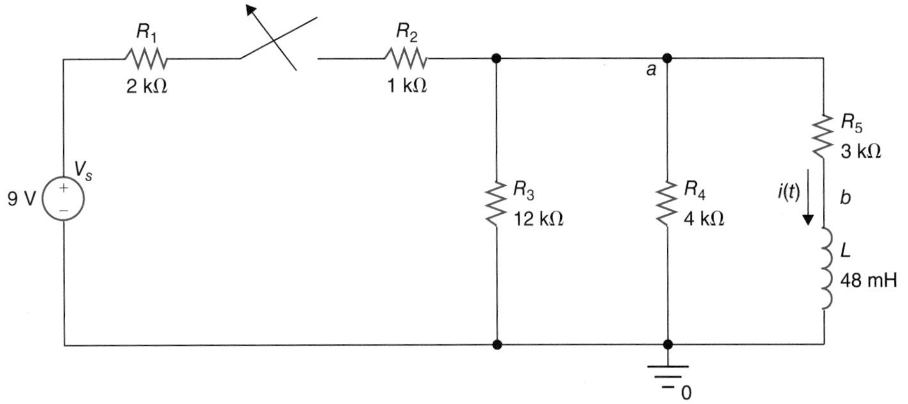

Before the switch is opened at $t = 0$, the circuit reached steady state. For dc current, the voltage across the inductor is zero, since $L dI/dt = 0$. The inductor can be treated as a short circuit for $t < 0$. Let R_a be the equivalent resistance of the parallel connection of R_3, R_4, and R_5. Then, we have

$$R_a = \frac{1}{\frac{1}{R_3} + \frac{1}{R_4} + \frac{1}{R_5}} = \frac{1}{\frac{1}{12{,}000} + \frac{1}{4000} + \frac{1}{3000}} = \frac{12{,}000}{8} = 1500\ \Omega = 1.5\ k\Omega$$

From the voltage divider rule, the voltage at node a is given by

$$V_a = \frac{R_a}{R_1 + R_2 + R_a} \times V_S = \frac{1500}{2000 + 1000 + 1500} \times 9 \text{ V} = 3 \text{ V}$$

The current through the inductor at $t = 0$ is given by

$$i(0) = I_0 = \frac{V_a}{R_5} = \frac{3 \text{ V}}{3\ k\Omega} = 1 \text{ mA}$$

The equivalent resistance of R_5, R_3, and R_4 that is connected to the inductor is given by

$$R_{\text{eq}} = R_5 + (R_3 \parallel R_4) = 3\ k\Omega + (12\ k\Omega \parallel 4\ k\Omega) = 3\ k\Omega + \frac{12\ k\Omega \times 4\ k\Omega}{12\ k\Omega + 4\ k\Omega}$$
$$= 3\ k\Omega + 3\ k\Omega = 6\ k\Omega$$

The time constant is given by $\tau = \dfrac{L}{R_{\text{eq}}} = \dfrac{48 \text{ mH}}{6\ k\Omega} = \dfrac{48 \times 10^{-3}}{6000} = 8 \times 10^{-6} \text{s} = 8\ \mu\text{s}.$

continued

Example 7.11 continued

The current through the inductor for $t \geq 0$ is

$$i(t) = I_0 e^{-\frac{t}{\tau}} = 1 \times e^{-\frac{t}{8\times10^{-6}}} \text{ mA} = 1 \times e^{-125{,}000t} \text{ mA}$$

For $t \geq 0$, the voltage at node b is given by

$$v_b(t) = L\frac{di(t)}{dt}$$

Summing the currents leaving nodes a and b, we obtain respectively

$$\frac{v_a(t)}{R_3} + \frac{v_a(t)}{R_4} + \frac{v_a(t) - v_b(t)}{R_5} = 0$$

$$\frac{v_b(t) - v_a(t)}{R_5} + i(t) = 0$$

MATLAB can be used to solve these equations as shown here.

MATLAB

```
%EXAMPLE 7.11
clear all;
L=48e-3;R1=2000;R2=1000;R3=12000;R4=4000;R5=3000;Vs=9;
Ra=P([R3,R4,R5]) %P.m should be in the current folder.
IR1=Vs/(R1+R2+Ra)
I0=IR1*(1/R5)/(1/R3+1/R4+1/R5)
syms i(t) va vb
Di=diff(i);
[va vb]=solve(va/R3+va/R4+(va-vb)/R5,...
(vb-va)/R5+i,va,vb);
i=dsolve(vb==L*Di,i(0)==I0);
v=L*diff(i);
va=R5*i+v;
taui=-log(subs(i,t,1)/subs(i,t,0));tau=1/taui;Req=L/tau;
i=vpa(i,12)
v=vpa(v,12)
va=vpa(va,12)
taui=vpa(taui,12)
tau=vpa(tau,12)
Req=vpa(Req,12)

Answers:
Ra =
        1500
IR1 =
    0.0020
I0 =
   1.0000e-03
i =
0.001*exp(-125000.0*t)
v =
-6.0*exp(-125000.0*t)
va =
-3.0*exp(-125000.0*t)
taui =
125000.0
tau =
0.000008
Req =
6000.0
```

Exercise 7.7

The initial current through the inductor shown in Figure 7.40 is $i(0) = 5$ mA. Find the current $i(t)$ through the inductor for $t \geq 0$.

FIGURE 7.40

Circuit for EXERCISE 7.7.

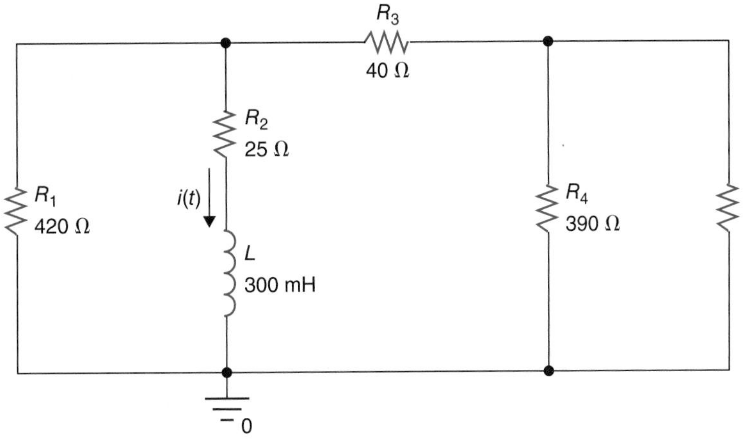

Answer:

$i(t) = 5e^{-\frac{t}{1.5\times10^{-3}}}u(t)$ mA.

EXAMPLE 7.12

The switch in the circuit shown in Figure 7.41 has been closed for a long time before it is opened at $t = 0$. Find the current $i(t)$ through the inductor and the voltage $v(t)$ across the inductor for $t \geq 0$.

FIGURE 7.41

Circuit for EXAMPLE 7.12.

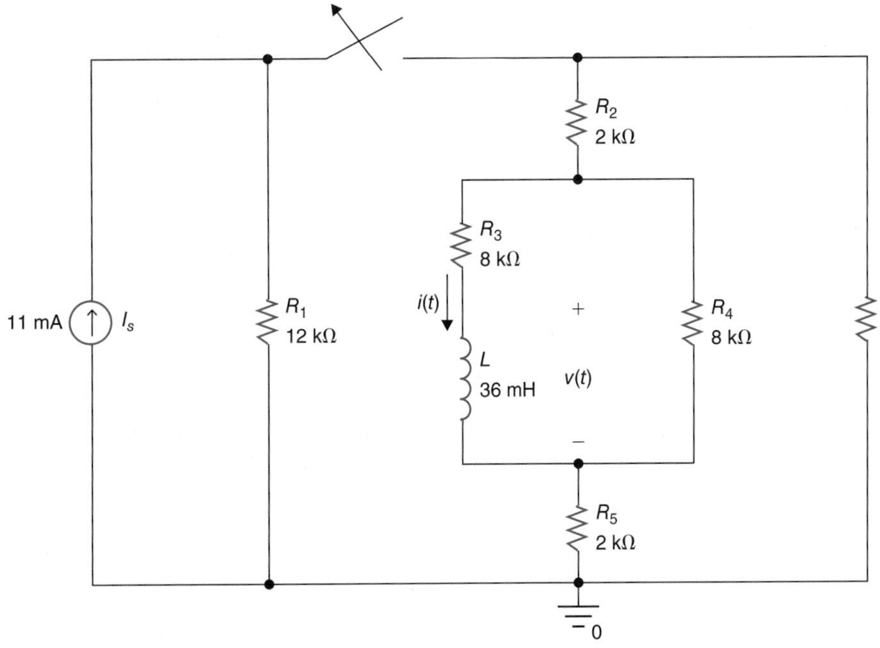

continued

Example 7.12 continued

For $t < 0$, since the circuit is in the steady state, the inductor can be treated as a short circuit. The equivalent resistance of R_2, R_3, R_4, and R_5 is given by

$$R_7 = R_2 + (R_3 \| R_4) + R_5 = 2\,k\Omega + (8\,k\Omega \| 8\,k\Omega) + 2\,k\Omega$$
$$= 2\,k\Omega + 4\,k\Omega + 2\,k\Omega = 8\,k\Omega.$$

The current from the current source is divided among three branches based on the current divider rule. The current through the middle branch with equivalent resistance R_7 is given by

$$I_{R_7} = \frac{\frac{1}{R_7}}{\frac{1}{R_1} + \frac{1}{R_7} + \frac{1}{R_6}} \times I_S = \frac{\frac{1}{8}}{\frac{1}{12} + \frac{1}{8} + \frac{1}{4}} \times 11\text{ mA} = \frac{3}{2+3+6} \times 11\text{ mA} = 3\text{ mA}$$

Half of this current will go through R_3 and L. Thus, the initial current through the inductor is given by

$$i(0) = 1.5\text{ mA}$$

For $t \geq 0$, notice that R_4 is parallel to $R_2 + R_6 + R_5$. Let the equivalent resistance be R_8. Then, we have

$$R_8 = R_4 \| (R_2 + R_6 + R_5) = 8\,k\Omega \| 8\,k\Omega = 4\,k\Omega$$

The equivalent resistance R_{eq} seen from the inductor is the sum of R_3 and R_8; that is,

$$R_{eq} = R_3 + R_8 = 12\,k\Omega$$

The time constant is given by

$$\tau = \frac{L}{R_{eq}} = \frac{36\text{ mH}}{12\,k\Omega} = 3 \times 10^{-6}\text{ s} = 3\,\mu\text{s}$$

The current through the inductor for $t \geq 0$ is

$$i(t) = i(0)e^{-\frac{t}{\tau}}u(t)\text{ A} = 1.5e^{-\frac{t}{3\times10^{-6}}}u(t)\text{ mA}$$

The voltage across the inductor is

$$v(t) = L\frac{di(t)}{dt} = 36 \times 10^{-3} \times 1.5 \times 10^{-3}\left(-\frac{1}{3 \times 10^{-6}}\right)e^{-\frac{t}{3\times10^{-6}}}u(t)\text{ V}$$
$$= -18e^{-\frac{t}{3\times10^{-6}}}u(t)\text{ V}$$

Exercise 7.8

The initial current through the inductor shown in Figure 7.42 is $i(0) = 5$ mA. Find the current $i(t)$ through the inductor for $t \geq 0$.

continued

Exercise 7.8 continued

FIGURE 7.42

Circuit for EXERCISE 7.8.

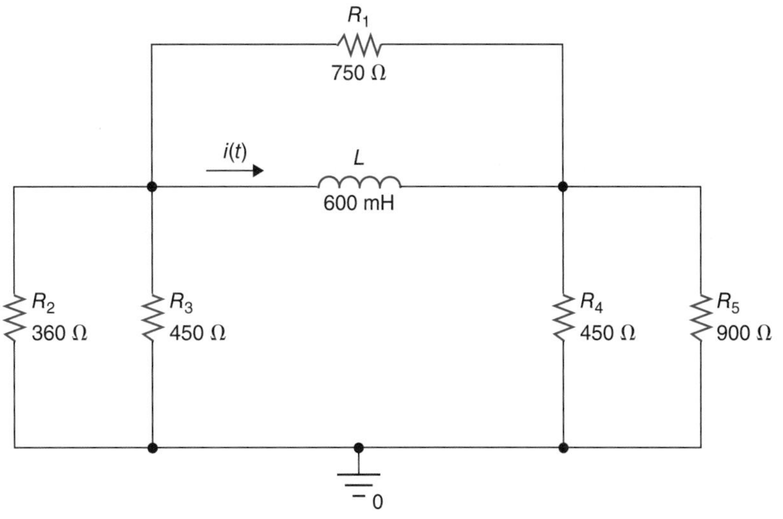

Answer:
$i(t) = 5e^{-\frac{t}{0.002}}u(t)$ mA.

7.5 Step Response of *RL* Circuit

The switch in the circuit shown in Figure 7.43 is closed at $t = 0$. The initial current through the inductor at $t = 0$ is $i(0) = I_0\,A$.

Before the switch is closed at $t = 0$, current I_S from the current source flows through the resistor R $[i_R(t) = I_S]$, and the voltage across the resistor is $I_S R$. Assume that the initial current through the inductor is zero $[i(0) = I_0 = 0]$. When the switch is closed at $t = 0$, current will start flowing through the inductor. If the inductor is ideal, the resistance value of the inductor is zero. According to the current divider rule, all the current I_S from the current source will flow through the inductor $[I_S \times R/(R+0) = I_S]$ eventually. But the current through the inductor $i(t)$ does not change instantly from zero to I_S. The current through the inductor starts to increase and the current through the resistor starts to decrease while maintaining $i(t) + i_R(t) = I_S$. The current $i_R(t)$ through the resistor decreases exponentially from I_S to zero $[i_R(t) = I_S e^{-\frac{t}{\tau}}u(t)]$, where the time constant τ is given by $\tau = L/R$. On the other hand, the current $i(t)$ through the inductor increases from zero to I_S according to $i(t) = I_S - i_R(t) = \left(I_S - I_S e^{-\frac{t}{\tau}}\right)u(t)$.

FIGURE 7.43

RL circuit with dc input.

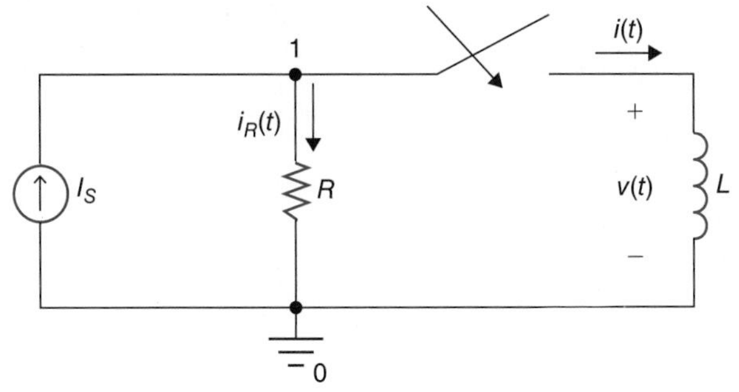

According to Faraday's law, the change in current induces voltage on the inductor. The induced voltage is proportional to the time rate of change in the current $[v(t) = L\,di(t)/dt]$. The time rate of change in current through the inductor is given by

$$\frac{di(t)}{dt} = \frac{RI_S}{L}e^{-\frac{t}{\tau}}u(t) = \frac{Ri_R(t)}{L} = \frac{v(t)}{L}$$

Notice that the time rate of change in current through the inductor decreases exponentially from RI_S/L to zero. Since the change is largest at $t = 0$, the induced voltage is maximum at $t = 0$. As time progresses, the rate of change in current, $di(t)/dt$, decreases, making both the induced voltage $v(t)$ and the current $i_R(t)$ through the resistor $(v(t)/R)$ smaller. After about five times the time constant has elapsed, the induced voltage on the inductor is close to zero $[v(t) \approx 0]$. At this point, the current $i_R(t)$ through the resistor is $v(t)/R \approx 0$, and the current through the inductor is $i(t) \approx I_S$.

When L is large, the rate of change of current $i(t)$ is small, and it takes a longer time to reach steady-state value. Also, when R is small, the rate of change is small, and it takes a longer time to reach steady-state value.

For $t \geq 0$, summing the currents away from node 1 in Figure 7.43, we have

$$-I_S + \frac{v(t)}{R} + i(t) = 0 \tag{7.61}$$

The voltage $v(t)$ across the inductor is given by

$$v(t) = L\frac{di(t)}{dt} \tag{7.62}$$

Substituting Equation (7.62) into Equation (7.61), we get

$$-I_S + \frac{L}{R}\frac{di(t)}{dt} + i(t) = 0 \tag{7.63}$$

Rearrangement of Equation (7.63) results in

$$\frac{di(t)}{dt} + \frac{R}{L}i(t) = \frac{R}{L}I_S \tag{7.64}$$

Equation (7.64) can be rewritten as

$$\frac{di(t)}{dt} = -\frac{R}{L}[i(t) - I_S] \tag{7.65}$$

If Equation (7.65) is divided by $i(t) - I_S$, we have

$$\frac{\frac{di(t)}{dt}}{i(t) - I_S} = \frac{-R}{L}$$

The left side of this equation is the derivative of $\ln[|i(t) - I_S|]$. Thus,

$$\frac{d}{dt}[\ln|i(t) - I_S|] = -\frac{R}{L} \tag{7.66}$$

Integrating on both sides, we have

$$\ln|i(t) - I_S| = \int_0^t \left(-\frac{R}{L}\right)dt = -\frac{Rt}{L} + K \tag{7.67}$$

where K is a constant. Taking the exponential on both sides, we have

$$i(t) - I_S = \pm e^K e^{-\frac{Rt}{L}} = Ae^{-\frac{Rt}{L}} \tag{7.68}$$

Notice that if $i(t) > I_S, A = e^K$; and if $i(t) < I_S, A = -e^K$. The solution for $i(t)$ is given by

$$i(t) = I_S + Ae^{-\frac{Rt}{L}} \tag{7.69}$$

The constant A is found from the fact that the current through the inductor is I_0 at $t = 0$. Setting $t = 0$ in Equation (7.69), we have

$$i(0) = I_S + Ae^{-\frac{R0}{L}} = I_S + A = I_0$$

Thus, $A = I_0 - I_S$. Substituting this value of A into Equation (7.69), we get the final solution:

$$i(t) = I_S + (I_0 - I_S)e^{-\frac{Rt}{L}} = I_S + (I_0 - I_S)e^{-\frac{t}{L/R}} \quad \textbf{(7.70)}$$

This solution is valid for $t \geq 0$. After all the initial energy stored in the inductor is dissipated as heat in the resistor, the voltage across the inductor will be zero because the current through the inductor will be invariant to time. With $v(t) = 0$, the inductor acts as a short circuit. All the current from the current source flows through the inductor. Thus, the final value of $i(t)$ at $t = \infty$ is I_S [i.e., $i(\infty) = I_S$]. This is verified from Equation (7.70). At $t = \infty$, the exponential term is zero, and we have $i(\infty) = I_S$.

Let the time constant of the *RL* circuit shown in Figure 7.43 be

$$\tau = \frac{L}{R} \quad \textbf{(7.71)}$$

Then, Equation (7.64) can be written as

$$\frac{di(t)}{dt} + \frac{1}{\tau}i(t) = \frac{1}{\tau}I_S \quad \textbf{(7.72)}$$

and Equation (7.70) can be written as

$$i(t) = I_S + (I_0 - I_S)e^{-\frac{t}{\tau}} \quad \textbf{(7.73)}$$

In general, the solution to a differential equation such as Equation (7.72) is given by

$$i(t) = (\textit{Final Value}) + (\textit{Initial Value} - \textit{Final Value})e^{-(t-\textit{Delay})/(\textit{Time Constant})} \quad \textbf{(7.74)}$$

Here, the *Final Value* is the value of $i(t)$ at $t = \infty$, the *Initial Value* is the value of $i(t)$ at $t = \textit{Delay}$, the *Delay* is the time of change of conditions in the circuit, and *Time Constant* represents the speed at which the circuit adjusts itself after the change at $t = \textit{Delay}$. The change of circuit at $t = \textit{Delay}$ is an event such as turning a swtich on or off. For the circuit shown in Figure 7.43, the inductor acts as a short circuit for dc. Based on the current divider rule, all the current from the current source flows through the inductor. Therefore, the current through the inductor at $t = \infty$ is I_S. Thus, Final Value $= I_S$. The initial current through the inductor at $t = 0$ is given by I_0. Thus, Initial Value $= I_0$. Since the switch is changed at $t = 0$, there is no delay. Thus, Delay $= 0$. The Time Constant is L/R. Thus, we have

$$\begin{aligned} i(t) &= (\textit{Final Value}) + (\textit{Initial Value} - \textit{Final Value})e^{-(t-\textit{Delay})/(\textit{Time Constant})} \\ &= I_S + (I_0 - I_S)e^{-\frac{t-0}{L/R}} = I_S + (I_0 - I_S)e^{-\frac{t}{L/R}} = I_S + (I_0 - I_S)e^{-\frac{t}{\tau}} \end{aligned} \quad \textbf{(7.75)}$$

From Equation (7.62), the voltage across the inductor, which is also the voltage across the resistor, is given by

$$v(t) = L\frac{di(t)}{dt} = L(I_0 - I_S)\left(-\frac{R}{L}\right)e^{-\frac{Rt}{L}} = -(I_0 - I_S)Re^{-\frac{Rt}{L}} = (I_S - I_0)Re^{-\frac{Rt}{L}} \quad \textbf{(7.76)}$$

The voltage decays exponentially. At $t = \infty$, the voltage across the inductor is zero. Thus, the inductor can be treated as a short circuit for $t = \infty$.

The instantaneous power on the inductor is

$$\begin{aligned} p(t) &= \left[I_S + (I_0 - I_S)e^{-\frac{Rt}{L}}\right]\left[(I_S - I_0)Re^{-\frac{Rt}{L}}\right] \\ &= \left[I_S(I_S - I_0)Re^{-\frac{Rt}{L}} - (I_S - I_0)^2Re^{-\frac{2Rt}{L}}\right]u(t) \end{aligned} \tag{7.77}$$

The energy on the inductor is

$$\begin{aligned} w(t) &= \frac{1}{2}Li^2(t) = \frac{1}{2}L\left[I_S + (I_0 - I_S)e^{-\frac{Rt}{L}}\right]^2 \\ &= \frac{1}{2}LI_S^2 + (LI_SI_0 - LI_S^2)e^{-\frac{Rt}{L}} + \frac{1}{2}L(I_0 - I_S)^2e^{-\frac{2Rt}{L}} \end{aligned} \tag{7.78}$$

At $t = 0$, the energy is $w(0) = \frac{1}{2}LI_S^2 + (LI_SI_0 - LI_S^2) + \frac{1}{2}L(I_0 - I_S)^2 = \frac{1}{2}LI_0^2$, which is the energy from the initial current I_0 through the inductor. At $t = \infty$, the energy is $w(\infty) = \frac{1}{2}LI_S^2$, which is the energy from the current I_S.

The current $i_R(t)$ through the resistor R is given by

$$i_R(t) = \frac{v(t)}{R} = (I_S - I_0)e^{-\frac{Rt}{L}} = (I_S - I_0)e^{-\frac{t}{\tau}} \tag{7.79}$$

7.5.1 INITIAL VALUE

If the initial current through the inductor is not given, it should be calculated from the circuit. Usually, there is a change in the circuit, such as switch setting, that requires the calculation of the initial voltage. For simplicity, let us assume that the switch is closed or opened at $t = 0$. At that point, the circuit is already in the steady state. The voltage across the inductor is zero, and the inductor can be treated as a short circuit. If necessary, draw the circuit for $t < 0$ to evaluate the initial current through the inductor. The initial current, $i(0)$, through the inductor is the current through the path that includes the inductor. If necessary, use one or more of the following circuit laws and theorems to find the final value: the voltage divider rule, the current divider rule, Thévenin's theorem, Norton's theorem, nodal analysis, mesh analysis, source transformation, and the superposition principle.

7.5.2 FINAL VALUE

The final value of the current through the inductor is the current at $t = \infty$. Since the circuit is at steady state at $t = \infty$, the voltage across the inductor is zero, and the inductor can be treated as a short circuit. If necessary, draw the circuit for $t = \infty$ to evaluate the final current through the inductor. The final current $i(\infty)$ through the inductor is the current through the path that includes the inductor. If necessary, use one or more of the following circuit laws and theorems to find the final value: the voltage divider rule, the current divider rule, Thévenin's theorem, Norton's theorem, nodal analysis, mesh analysis, source transformation, and the superposition principle.

7.5.3 TIME CONSTANT

For RL circuits with one inductor in the circuit, the time constant is given by

$$\tau = \frac{L}{R_{eq}} \tag{7.80}$$

where R_{eq} is the equivalent resistance seen from the inductor. The equivalent resistance R_{eq} is the Thévenin equivalent resistance when the rest of the circuit (excluding the inductor) is converted to the Thévenin equivalent circuit. In general, R_{eq} can be found by deactivating independent sources (open-circuit voltage sources and short-circuit current sources) and finding the equivalent resistance seen from the inductor. Other methods, such as test voltage and test current, can also be used to find the equivalent resistance.

The time constant indicates the speed at which the current through the inductor reaches its final value. If the time constant is reduced, the current through the inductor reaches its final value faster than before. Figure 7.44 shows $i(t)$ for different values of time constant for $I_S = 1$ A and $I_0 = 0$ A.

FIGURE 7.44

The response of the *RL* circuit for different time constants for dc input.

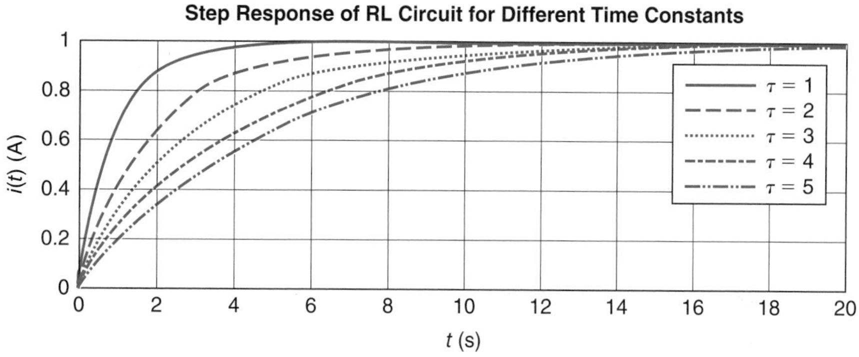

At $t = \tau$, the current through the inductor is given by

$$i(\tau) = I_S\left(1 - e^{-\frac{\tau}{\tau}}\right) = I_S(1 - e^{-1}) = 0.63212I_S$$

Table 7.4 shows $i(t)/I_S$ for $t = 0, \tau, 2\tau, \ldots, 10\tau$.

TABLE 7.4

Normalized Current at $t = n\tau$.

n	$1-\exp(-n)$
0	0
1.000000000000000	0.632120558828558
2.000000000000000	0.864664716763387
3.000000000000000	0.950212931632136
4.000000000000000	0.981684361111266
5.000000000000000	0.993262053000915
6.000000000000000	0.997521247823334
7.000000000000000	0.999088118034446
8.000000000000000	0.999664537372097
9.000000000000000	0.999876590195913
10.000000000000000	0.999954600070238

At five times the time constant, the current reaches 99.3262% of the final value.

The time that it takes for the current $i(t)$ to reach 10% of the final value can be evaluated as follows:

$$0.1I_S = I_S\left(1 - e^{-\frac{t}{\tau}}\right) \quad \Rightarrow \quad t = -\tau \ln(0.9) = 0.10536\tau$$

Similarly, the time that it takes for the current $i(t)$ to reach 90% of the final value is found to be

$$0.9I_S = I_S\left(1 - e^{-\frac{t}{\tau}}\right) \quad \Rightarrow \quad t = -\tau \ln(0.1) = 2.30259\tau$$

It takes about 2.19722τ for the current $i(t)$ to rise from 10% of the final value to 90% of the final value. The time for the current $i(t)$ to rise from 10% of the final value to 90% of the final value is called *rise time* (T_r). For the first-order *RL* circuit with zero initial condition, the rise time is given by

$$T_r = 2.19722\tau$$

where the time constant τ is given by $\tau = L/R$.

The time derivative of $i(t)$, $di(t)/dt$, provides the rate of increase of $i(t)$ as a function of time. From Equation (7.73), we have, for $I_0 = 0$,

$$\frac{di(t)}{dt} = \frac{I_S}{\tau}e^{-\frac{t}{\tau}}$$

The rate of increase of $i(t)$ is at its maximum at $t = 0$. Then, the rate of increase decreases exponentially. If the rate of increase is kept at the value at $t = 0$, I_S/τ, the current increases linearly as

$$i(t) = \frac{I_S}{\tau}t$$

At $t = \tau$, the current reaches the final value I_S.

7.5.4 SOLUTION TO GENERAL FIRST-ORDER DIFFERENTIAL EQUATION WITH CONSTANT COEFFICIENT AND CONSTANT INPUT

A general first-order differential equation with constant coefficient and constant input can be written as

$$\frac{di(t)}{dt} + a\,i(t) = b \tag{7.81}$$

In Equation (7.81), a and b are constants. In the steady state, there is no change in the current due to the dc source. Thus, we have $di(t)/dt = 0$. The differential equation is given by $a\,i(t) = b$. Thus, the final value is given by $i(\infty) = b/a$. Comparison of Equations (7.72) and (7.81) reveals that the coefficient of $i(t)$ is the inverse of the time constant (i.e., $a = 1/\tau$). Thus, $\tau = 1/a$. The solution of the differential equation given by Equation (7.81) is

$$i(t) = \left[\frac{b}{a} + \left(i(0) - \frac{b}{a}\right)e^{-at}\right]u(t) \tag{7.82}$$

This solution is derived in Section 7.6, later in this chapter, as well.

EXAMPLE 7.13

Let $L = 700$ mH, $R = 200\ \Omega$, $I_S = 5$ mA, and $I_0 = 0$ A in the circuit shown in Figure 7.43. Find the current $i(t)$ through the inductor and voltage $v(t)$ across the inductor for $t \geq 0$, and plot $i(t)$ and $v(t)$.

The time constant is $\tau = L/R = 0.7/200 = 0.0035$ s $= 3.5$ ms.

From Equation (7.73), the current through the inductor is given by

$$i(t) = I_S + (I_0 - I_S)e^{-\frac{t}{\tau}} = \left(5 + (0-5)e^{-\frac{t}{0.0035}}\right)u(t)\text{ mA} = 5\left(1 - e^{-\frac{t}{0.0035}}\right)u(t)\text{ mA}$$

continued

Example 7.13 continued

The voltage across the inductor is given by

$$v(t) = L\frac{di(t)}{dt} = 0.7 \times 5 \times 10^{-3} \times (-1) \times \left(-\frac{1}{0.0035}\right) \times e^{-\frac{t}{0.0035}}u(t)\ \text{V} = e^{-\frac{t}{0.0035}}u(t)\ \text{V}$$

Figure 7.45 shows $i(t)$ and $v(t)$ for $t \geq 0$.

FIGURE 7.45

Plot of $i(t)$ and $v(t)$.

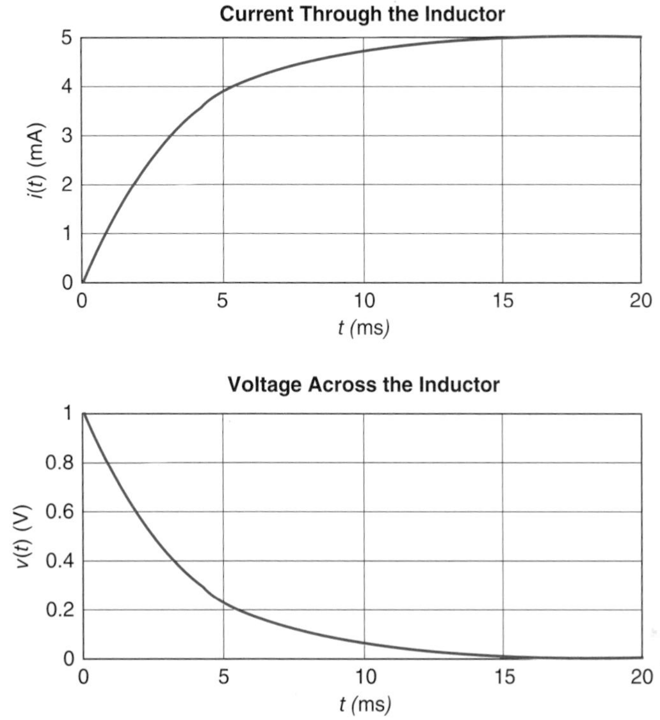

EXAMPLE 7.14

The switch in the circuit shown in Figure 7.46 is closed at $t = 0$. The initial current through the inductor is $i(0) = I_0$. Find the current $i(t)$ through the inductor and the voltage $v(t)$ across the inductor.

FIGURE 7.46

An *RL* circuit for EXAMPLE 7.14.

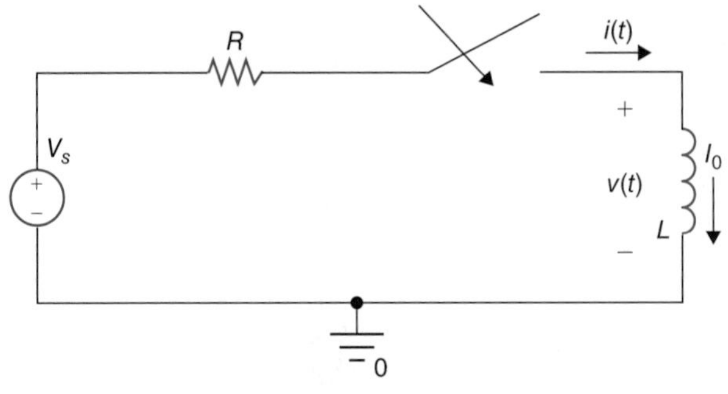

Summing the voltage drops around the mesh, we obtain

$$-V_S + Ri(t) + v(t) = 0$$

Substitution of $v(t) = L\frac{di(t)}{dt}$ yields

$$-V_S + Ri(t) + L\frac{di(t)}{dt} = 0$$

which can be rearranged as

$$\frac{di(t)}{dt} + \frac{R}{L}i(t) = \frac{V_S}{L}$$

continued

Example 7.14 continued

FIGURE 7.47

Circuit shown in Figure 7.46 after source transformation.

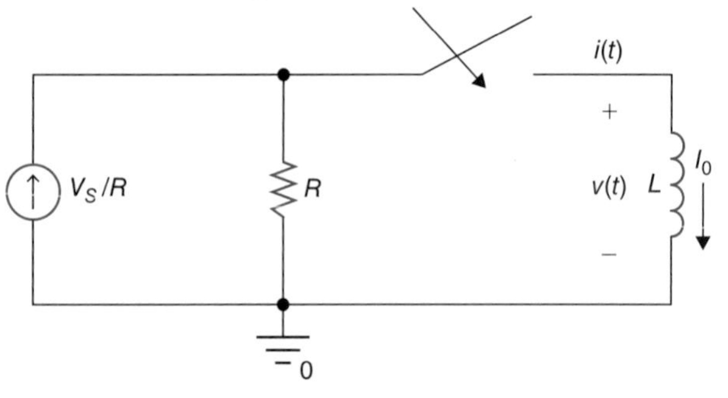

Comparing this equation with Equation (7.81) reveals that

$$a = \frac{R}{L},\ b = \frac{V_S}{L},\ \tau = \frac{1}{a} = \frac{L}{R}$$

The final value of the current is given by

$$i(\infty) = \frac{b}{a} = \frac{V_S}{R}$$

The current through the inductor is given by

$$i(t) = \left[\frac{V_S}{R} + \left(I_0 - \frac{V_S}{R}\right)e^{-\frac{t}{L/R}}\right]u(t)\ \text{A}$$

Alternatively, we can apply the source transformation to find $i(t)$. Applying source transformation to V_S and R, we obtain the circuit shown in Figure 7.47.

The final value of the current through the inductor is V_S/R, and the time constant is $\tau = L/R$. Thus, the current through the inductor for $t \geq 0$ is given by

$$i(t) = (Final\,Value) + (Initial\,Value - Final\,Value)e^{-(t-Delay)/(Time\ Constant)}$$
$$= \frac{V_S}{R} + \left(I_0 - \frac{V_S}{R}\right)e^{-\frac{t-0}{L/R}} = \left[\frac{V_S}{R} + \left(I_0 - \frac{V_S}{R}\right)e^{-\frac{t}{L/R}}\right]u(t)\ \text{A}$$

The voltage across the inductor for $t \geq 0$ is given by

$$v(t) = L\frac{di(t)}{dt} = L\left(\frac{-R}{L}\right)\left(I_0 - \frac{V_S}{R}\right)e^{-\frac{t}{L/R}} = (V_S - RI_0)e^{-\frac{t}{L/R}}u(t)$$

EXAMPLE 7.15

In the circuit shown in Figure 7.48, switch 1 has been closed for a long time before it is opened at $t = 0$. Switch 2 is closed at $t = 12\ \mu$s. Find the current $i(t)$ through the inductor for $t \geq 0$.

FIGURE 7.48

Circuit for EXAMPLE 7.15.

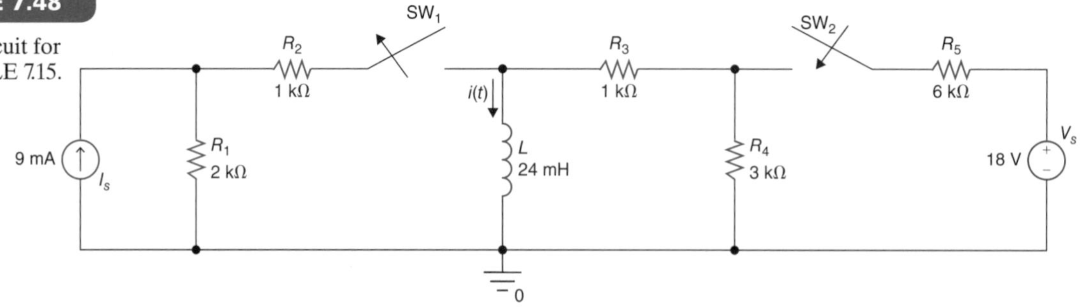

From the current divider rule, the initial current through the inductor, which is treated as a short circuit for $t < 0$, is $i(0) = I_0 = I_S \times R_1/(R_1 + R_2) = 9\ \text{mA} \times 2\ k\Omega/(2\ k\Omega + 1\ k\Omega) = 6\ \text{mA}$.

continued

Example 7.15 continued

For $0 \le t < 12\ \mu s$, the circuit consists of L, R_3, and R_4. Since there is no input source, the current through the inductor decreases exponentially from the initial value of $i(0) = 6$ mA. The equivalent resistance of the series connection of R_3 and R_4 is $R_{eq} = R_3 + R_4 = 4\ k\Omega$. The time constant is $\tau = L/R_{eq} = 24\ \text{mH}/4\ k\Omega = 0.024/4000 = 6 \times 10^{-6}$ s $= 6\ \mu s$. Thus, the current through the inductor is given by

$$i(t) = I_0 e^{-\frac{t}{\tau}} = 6e^{-\frac{t}{6\times10^{-6}}}\ \text{mA}$$

At $t = 12\ \mu s$, the value of the current through the inductor is given by

$$i(12 \times 10^{-6}) = 6e^{-\frac{12\times10^{-6}}{6\times10^{-6}}} = 6e^{-2} = 0.8120\ \text{mA}$$

FIGURE 7.49

Circuit for $t \ge 12\ \mu s$.

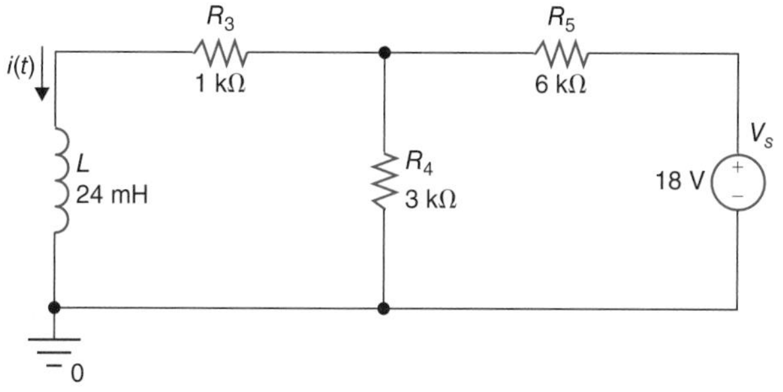

For $t \ge 12\ \mu s$, the circuit reduces to the one shown in Figure 7.49.

At $t = \infty$, the inductor can be treated as a short circuit. The equivalent resistance of the parallel connection of R_3 and R_4 is given by $R_3 \| R_4 = R_3 \times R_4/(R_3 + R_4) = 1\ k\Omega \times 3\ k\Omega/(1\ k\Omega + 3\ k\Omega) = 0.75\ k\Omega$. According to the voltage divider rule, the voltage from the voltage source V_s is split between R_5 and $R_3 \| R_4$ in proportion to the resistance. Thus, the voltage across R_4, which is also the voltage across R_3, is given by 18 V $\times$ $0.75\ k\Omega/(6\ k\Omega + 0.75\ k\Omega) = 2$ V. The current through the inductor, which is also the current through R_3, is

$$i(\infty) = 2\ \text{V}/1\ k\Omega = 2\ \text{mA}$$

To find the equivalent resistance R_{eq} connected to the inductor, the voltage source is deactivated by short-circuiting it. The equivalent resistance is given by

$$R_{eq} = R_3 + (R_4 \| R_5) = 1\ k\Omega + 3\ k\Omega \times 6\ k\Omega/(3\ k\Omega + 6\ k\Omega)$$
$$= 1\ k\Omega + 2\ k\Omega = 3\ k\Omega$$

The time constant τ is given by $\tau = L/R_{eq} = 24\ \text{mH}/3\ k\Omega = 24 \times 10^{-3}/3 \times 10^3 = 8 \times 10^{-6}$ s $= 8\ \mu s$. Notice that $1/\tau = 125{,}000$ (1/s). The current through the inductor for $12\ \mu s \le t$ is given by

$$i(t) = i(\infty) + (i(12\mu s) - i(\infty))e^{-\frac{t-12\times10^{-6}}{\tau}} = 2 + (0.8120 - 2)e^{-125000(t-12\times10^{-6})}$$
$$= \left[2 - 1.1880e^{-125000(t-12\times10^{-6})}\right]u(t - 12 \times 10^{-6})\ \text{mA}$$

Figure 7.50 shows the current through the inductor for $t \ge 0$.

FIGURE 7.50

Plot of current $i(t)$ through the inductor.

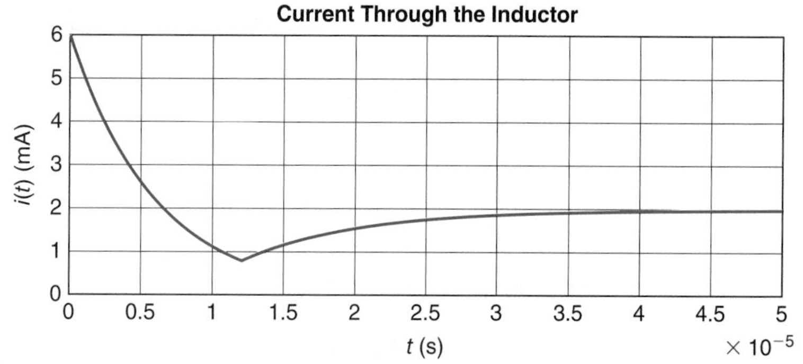

Exercise 7.9

Find the current $i(t)$ through the inductor for $t \geq 0$ for the circuit shown in Figure 7.48 when switch 2 is closed at $t = 0$.

Answer:

$$i(t) = \left[2 + (6-2)e^{-\frac{t}{8\times10^{-6}}}\right]u(t)\text{ mA} = \left(2 + 4e^{-\frac{t}{8\times10^{-6}}}\right)u(t)\text{ mA}.$$

EXAMPLE 7.16

In the circuit shown in Figure 7.51, switch 1 has been closed for a long time before it is opened at $t = 0$. Switch 2 is closed at $t = 2\ \mu s$.

FIGURE 7.51

Circuit for EXAMPLE 7.16.

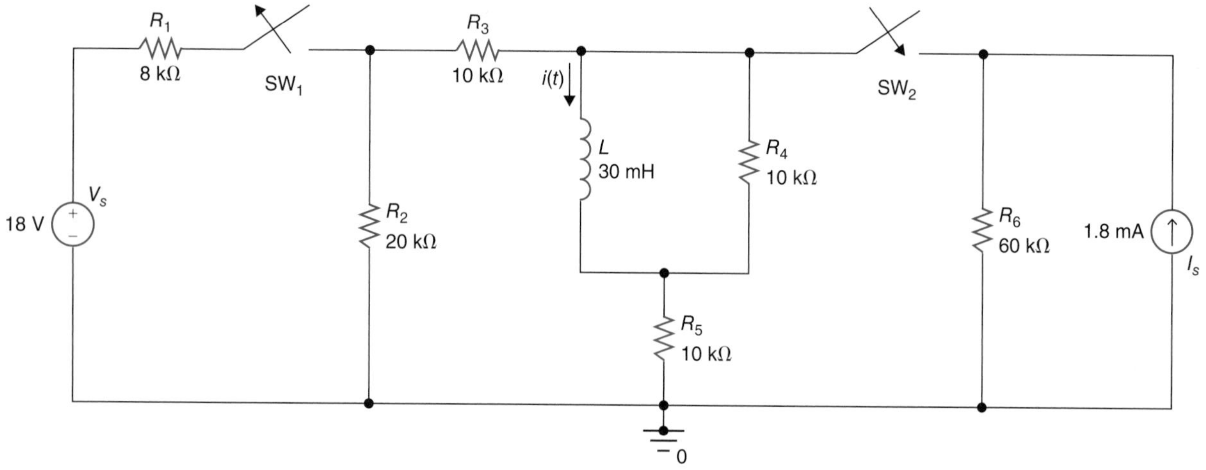

For $t < 0$, switch 1 is closed and switch 2 is opened. Since the circuit is in a steady state, the inductor can be treated as a short circuit. Since R_4 is in parallel with the short circuit, all currents will flow through the inductor. R_4 is irrelevant in finding $i(0)$. Figure 7.52 shows the circuit for $t < 0$. The equivalent resistance of the parallel connection of R_2 and $R_3 + R_5$ is

$$R_7 = R_2 \parallel (R_3 + R_5) = 20\ k\Omega \parallel (10\ k\Omega + 10\ k\Omega) = 20\ k\Omega \parallel 20\ k\Omega = 10\ k\Omega$$

Applying the voltage divider rule, we find the voltage across R_2, which is also the voltage across $R_3 + R_5$:

$$V_{R_2} = V_S \frac{R_7}{R_1 + R_7} = 18\text{ V} \times \frac{10\ k\Omega}{8\ k\Omega + 10\ k\Omega} = 10\text{ V}$$

The initial current $i(0)$ through the inductor is

$$i(0) = \frac{V_{R_2}}{R_3 + R_5} = \frac{10\text{ V}}{20\ k\Omega} = 0.5\text{ mA}$$

continued

Example 7.16 continued

FIGURE 7.52

Circuit for $t < 0$.

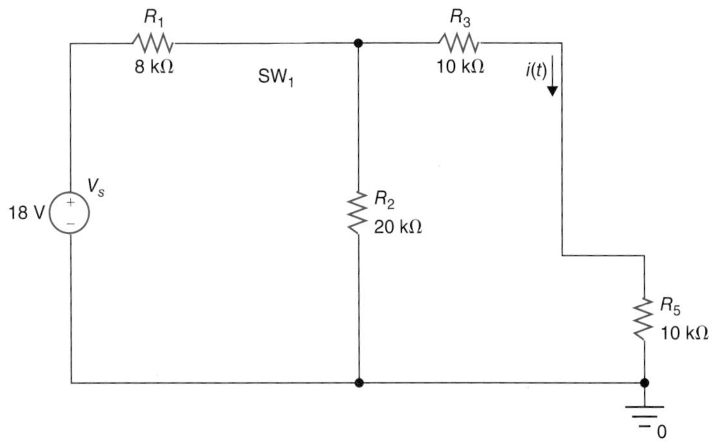

For $0 \le t < 2\ \mu s$, the circuit consists of L, R_2, R_3, R_4, and R_5. Resistors R_3, R_2, and R_5 are connected in series. R_4 is connected in parallel to $R_3 + R_2 + R_5$. Thus, the equivalent resistance R_{eq1} is given by

$$R_{eq1} = R_4 \| (R_3 + R_2 + R_5) = 10\ k\Omega \| 40\ k\Omega = \frac{10\ k\Omega \times 40\ k\Omega}{10\ k\Omega + 40\ k\Omega} = 8\ k\Omega$$

The time constant is given by

$$\tau_1 = \frac{L}{R_{eq1}} = \frac{30\text{ mH}}{8\ k\Omega} = 3.75 \times 10^{-6}\text{s} = 3.75\ \mu\text{s}$$

Thus, the current through the inductor for $0 \le t < 2\ \mu s$ is

$$i(t) = i(0)e^{-\frac{t}{\tau_1}} = 0.5e^{-\frac{t}{3.75\times10^{-6}}}[u(t) - u(t - 2 \times 10^{-6})]\text{ mA}$$

At $t = 2\ \mu s$, the current is

$$i(2 \times 10^{-6}) = 0.5e^{-\frac{2\times10^{-6}}{3.75\times10^{-6}}}\text{ mA} = 0.2933\text{ mA}$$

For $t > 2\ \mu s$, the circuit shown in Figure 7.51 reduces to the one shown in Figure 7.53. In the steady state, the inductor can be treated as a short circuit. The parallel connection of L and R_4 can be treated as a short circuit. The current from the

FIGURE 7.53

Circuit for $t > 2\ \mu s$.

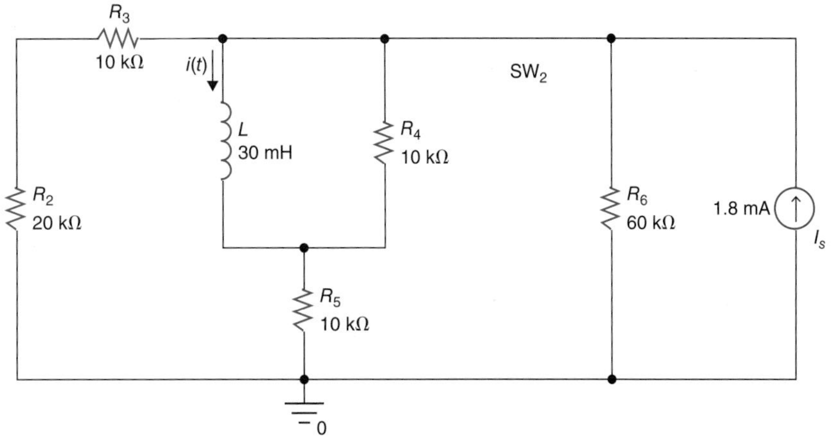

continued

Example 7.16 continued

current source I_S is split into three paths (R_6, $L - R_5$, $R_3 - R_2$) based on the current divider rule. The current through R_5 is

$$i(\infty) = I_S \times \frac{\frac{1}{R_5}}{\frac{1}{R_6} + \frac{1}{R_5} + \frac{1}{R_3 + R_2}} = 1.8\text{ mA} \times \frac{\frac{1}{10}}{\frac{1}{60} + \frac{1}{10} + \frac{1}{30}}$$

$$= 1.8\text{ mA} \times \frac{6}{9} = 1.2\text{ mA}$$

The equivalent resistance R_{eq2} parallel to the inductor can be found by deactivating the current source by open-circuiting it and finding the resistance. Let R_8 be the equivalent resistance of the parallel connection of $R_3 + R_2$ and R_6. Then, we have

$$R_8 = \frac{(R_3 + R_2) \times R_6}{R_3 + R_2 + R_6} = \frac{(10\,k\Omega + 20\,k\Omega) \times 60\,k\Omega}{10\,k\Omega + 20\,k\Omega + 60\,k\Omega} = \frac{1800}{90}\,k\Omega = 20\,k\Omega$$

Let R_9 be the sum of R_8 and R_5. Then, $R_9 = R_8 + R_5 = 20\,k\Omega + 10\,k\Omega = 30\,k\Omega$. The equivalent resistance R_{eq2} is given by the parallel connection of R_4 and R_9. Thus, we get

$$R_8 = \frac{R_4 \times R_9}{R_4 + R_9} = \frac{10\,k\Omega \times 30\,k\Omega}{10\,k\Omega + 30\,k\Omega} = \frac{300}{40}\,k\Omega = 7.5\,k\Omega$$

The time constant τ_2 is given by

$$\tau_2 = \frac{L}{R_{eq2}} = \frac{30\text{ mH}}{7.5\,k\Omega} = 4 \times 10^{-6}\text{s} = 4\,\mu\text{s}$$

The current through the inductor for $t \geq 2\,\mu$s is given by

$$i(t) = i(\infty) + (i(2\mu s) - i(\infty))e^{-\frac{t-2\times10^{-6}}{\tau^2}} = 1.2 + (0.2933 - 1.2)e^{-\frac{t-2\times10^{-6}}{4\times10^{-6}}}$$

$$= \left[1.2 - 0.9067e^{-\frac{t-2\times10^{-6}}{4\times10^{-6}}}\right]u(t - 2 \times 10^{-6})\text{ mA}$$

Figure 7.54 shows the plot of $i(t)$.

FIGURE 7.54

Plot of $i(t)$.

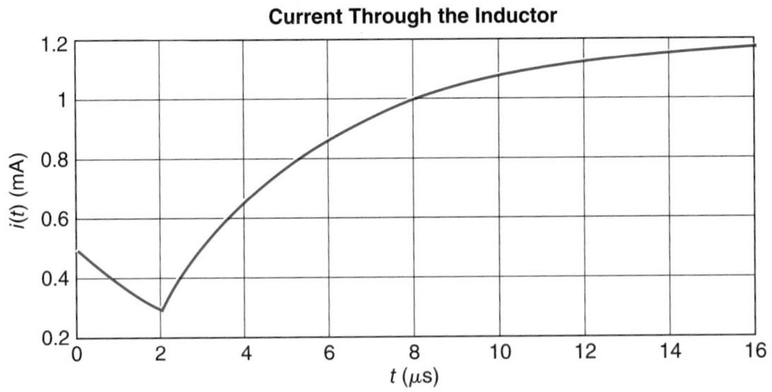

Exercise 7.10

Find the current $i(t)$ through the inductor for $t \geq 0$ for the circuit shown in Figure 7.51 when switch 2 is closed at $t = 0$.

continued

Exercise 7.10 continued

Answer:

$$i(t) = \left[1.2 + (0.5 - 1.2)e^{-\frac{t}{4\times10^{-6}}}\right]u(t)\text{ mA} = \left(1.2 - 0.7e^{-\frac{t}{4\times10^{-6}}}\right)u(t)\text{ mA}.$$

EXAMPLE 7.17

In the circuit shown in Figure 7.55, the switch has been closed for a long time before it is opened at $t = 0$. Find the current $i(t)$ through the inductor for $t \geq 0$.

FIGURE 7.55

Circuit for EXAMPLE 7.17.

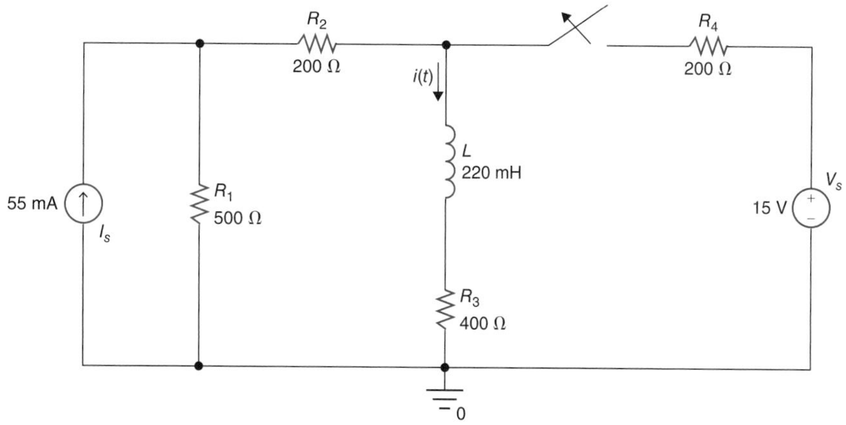

Because the sources are dc, the inductor acts as a short circuit for $t < 0$. Figure 7.56 shows the circuit for $t < 0$.

FIGURE 7.56

Circuit for $t < 0$.

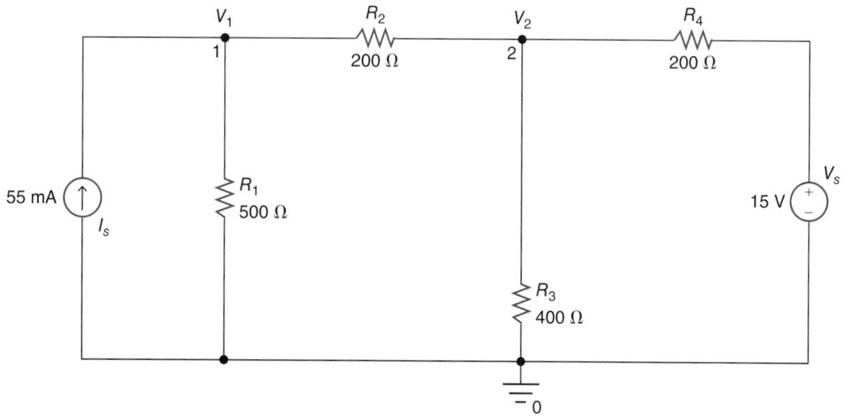

Summing the currents leaving node 1, we obtain

$$-55 \times 10^{-3} + \frac{V_1}{500} + \frac{V_1 - V_2}{200} = 0 \tag{7.83}$$

Multiplying by 1000 on every term of Equation (7.83), we get

$$-55 + 2V_1 + 5V_1 - 5V_2 = 0 \tag{7.84}$$

continued

Example 7.17 continued

Summing the currents leaving node 2, we obtain

$$\frac{V_2 - V_1}{200} + \frac{V_2}{400} + \frac{V_2 - 15}{200} = 0 \quad \textbf{(7.85)}$$

Multiplication of 400 on every term of Equation (7.85) yields

$$2V_2 - 2V_1 + V_2 + 2V_2 - 30 = 0 \quad \textbf{(7.86)}$$

Equations (7.84) and (7.86) can be rewritten, respectively, as

$$7V_1 - 5V_2 = 55 \quad \textbf{(7.87)}$$

$$-2V_1 + 5V_2 = 30 \quad \textbf{(7.88)}$$

FIGURE 7.57

Circuit for $t \geq 0$.

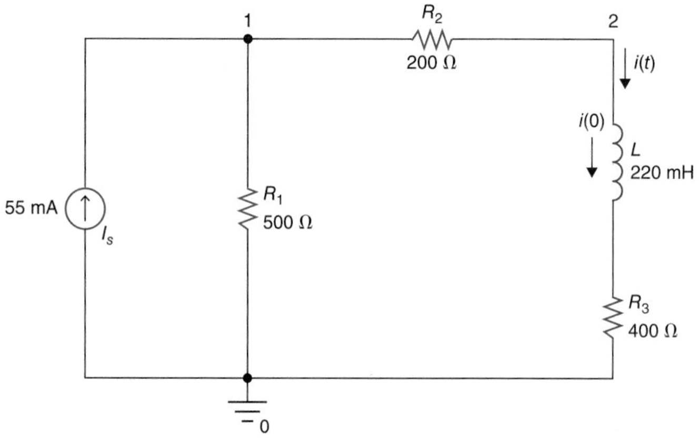

Summing Equations (7.87) and (7.88) provides $5V_1 = 85$ or $V_1 = 17$ V. Substitution of this in Equation (7.88) yields $5V_2 = 64$ or $V_2 = 12.8$ V. The current through R_3, which is also the current through the inductor, is given by

$$i(0) = \frac{V_2}{R_3} = \frac{12.8\text{ V}}{400\,\Omega} = 32\text{ mA}$$

Figure 7.57 shows the circuit for $t \geq 0$ after the switch is opened. Notice that the initial current through the inductor is 32 mA.

In the steady state (say, at $t = \infty$), the current through the inductor is a constant and the voltage across the inductor is zero ($Ldi/dt = 0$). Thus, the inductor can be treated as a short circuit. The final value of the current through the inductor is given by

$$\begin{aligned} i(\infty) &= \frac{R_1}{R_1 + R_2 + R_3} \times I_S = \frac{500}{500 + 200 + 400} \times 55\text{ mA} \\ &= \frac{500}{1100} \times 55\text{ mA} = 25\text{ mA} \end{aligned}$$

To find the time constant, we deactivate the current source by open-circuiting it and find the equivalent resistance connected to the inductor. The equivalent resistance seen from the inductor is

$$R_{eq} = R_2 + R_1 + R_3 = 200 + 500 + 400 = 1100\,\Omega$$

The time constant is given by $\tau = L/R_{eq} = 0.22/1100 = 2 \times 10^{-4}$ s. The current through the inductor for $t \geq 0$ is given by

$$\begin{aligned} i(t) &= \{i(\infty) + [i(0) - i(\infty)]e^{-\frac{t}{\tau}}\}u(t) = \left[25 + (32 - 25)e^{-\frac{t}{2\times10^{-4}}}\right]u(t)\text{ mA} \\ &= \left[25 + 7e^{-\frac{t}{2\times10^{-4}}}\right]u(t)\text{ mA} \end{aligned}$$

For $t \geq 0$, the voltage at node 2 of the circuit shown in Figure 7.57 is given by

$$v_2(t) = L\frac{di(t)}{dt} + R_3 i(t)$$

continued

Example 7.17 continued

The sum of currents leaving node 1 and node 2 are given respectively by

$$-I_s + \frac{v_1(t)}{R_1} + \frac{v_1(t) - v_2(t)}{R_2} = 0$$

$$\frac{v_2(t) - v_1(t)}{R_2} + i(t) = 0$$

The MATLAB script given here can be used to solve these equations:

MATLAB

```
%EXAMPLE 7.17
clear all;
L=220e-3;R1=500;R2=200;R3=400;R4=200;Vs=15;Is=55e-3;
Ra=P([R3,R4]) %P.m should be in the current folder.
Rb=R2+Ra
IR2=Is*R1/(R1+Rb)
I01=IR2*R4/(R3+R4)
Rc=P([R1+R2,R3])
IR4=Vs/(R4+Rc)
I02=IR4*(R1+R2)/(R1+R2+R3)
I0=I01+I02
syms i(t) v1 v2
Di=diff(i);
[v1,v2]=solve(-Is+v1/R1+(v1-v2)/R2,(v2-v1)/R2+i,v1,v2);
i=dsolve(v2==L*Di+R3*i,i(0)==I0);
v=L*diff(i);
v2=v+R3*i;
v1=R2*i+v2;
taui=-log(subs(v,t,1)/subs(v,t,0));tau=1/taui;Req=L/tau;
i=vpa(i,12)
v=vpa(v,12)
v2=vpa(v2,12)
v1=vpa(v1,12)
taui=vpa(taui,12)
tau=vpa(tau,12)
Req=vpa(Req,12)
```

```
Answers:
Ra =
  133.3333
Rb =
  333.3333
IR2 =
    0.0330
I01 =
    0.0110
Rc =
  254.5455
IR4 =
    0.0330
I02 =
    0.0210
I0 =
    0.0320
i =
0.007*exp(-5000.0*t) + 0.025
v =
-7.7*exp(-5000.0*t)
v2 =
10.0 - 4.9*exp(-5000.0*t)
```

continued

Example 7.17 continued

MATLAB continued

```
v1 =
15.0 - 3.5*exp(-5000.0*t)
taui =
5000.0
tau =
0.0002
Req =
1100.0
```

Exercise 7.11

Switch 1 in the circuit shown in Figure 7.58 has been closed for a long time before it is opened at $t = 0$. Switch 2 is closed at $t = 0$ right after switch 1 is opened. Find the current $i(t)$ for $t \geq 0$.

FIGURE 7.58

Circuit for EXERCISE 7.11.

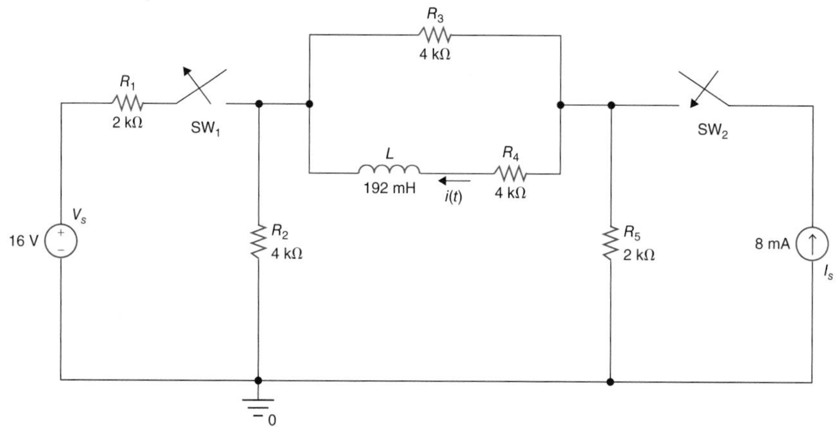

Answer:

$$i(t) = \left[1 + (-1 - 1)e^{-\frac{t}{3\times10^{-5}}}\right]u(t) = \left(1 - 2e^{-\frac{t}{3\times10^{-5}}}\right)u(t) \text{ mA.}$$

EXAMPLE 7.18

The initial current through the inductor is $i(0) = 1$ A in the circuit shown in Figure 7.59. Find the current $i_o(t)$ through R_2.

FIGURE 7.59

Circuit for EXAMPLE 7.18.

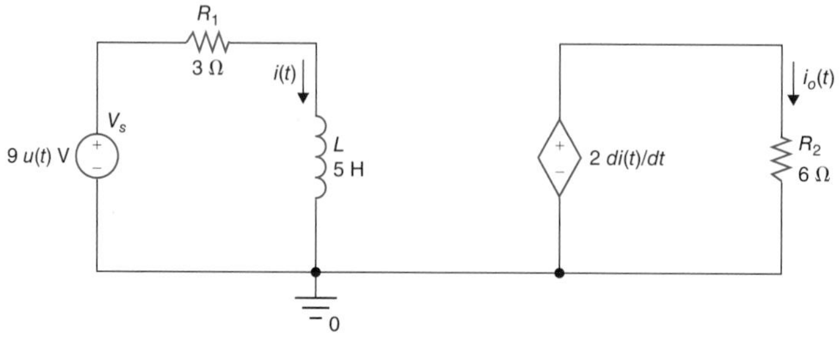

continued

Example 7.18 continued

Summing the voltage drops around the mesh on the left side of the circuit, we obtain

$$-9 + 3i(t) + 5\frac{di(t)}{dt} = 0$$

which can be rearranged as

$$\frac{di(t)}{dt} + 0.6i(t) = 1.8 \qquad \textbf{(7.89)}$$

In the steady state, the inductor can be treated as a short circuit, and the current through the inductor is $i(\infty) = V_S/R_1 = 9/3 = 3$ A. The final value can also be found from Equation (7.89). Since $di(t)/dt = 0$ in the steady state, the current at $t = \infty$ is obtained from $0.6\, i(\infty) = 1.8$. Thus, we have $i(\infty) = 1.8/0.6 = 3$ A. The time constant for the RL circuit is given by $\tau = L/R_1 = 5/3$ s. The time constant also can be obtained from Equation (7.89). The coefficient of $i(t)$ is $1/\tau$. Thus, $\tau = 1/0.6 = 5/3$ s. The solution to Equation (7.89) is

$$i(t) = \left\{i(\infty) + [i(0) - i(\infty)]e^{-\frac{t}{\tau}}\right\}u(t) = \left\{3 + [1 - 3]e^{-0.6t}\right\}u(t) = (3 - 2e^{-0.6t})u(t) \text{ A}$$

The controlled voltage on the right side of the circuit is given by

$$2\frac{di(t)}{dt} = 2(-2)(-0.6)e^{-0.6t}u(t) = 2.4e^{-0.6t}u(t) \text{ A}$$

The current through R_2 is given by

$$i_o(t) = \frac{2.4e^{-0.6t}u(t)}{6} = 0.4e^{-0.6t}u(t) \text{ A}$$

Notice that the controlled voltage is strongest when the input voltage changes from 0 to 9 V at $t = 0$.

Exercise 7.12

The initial current through the inductor is $i(0) = 1$ A in the circuit shown in Figure 7.60. Find the voltage $v_o(t)$.

FIGURE 7.60

Circuit for EXERCISE 7.12.

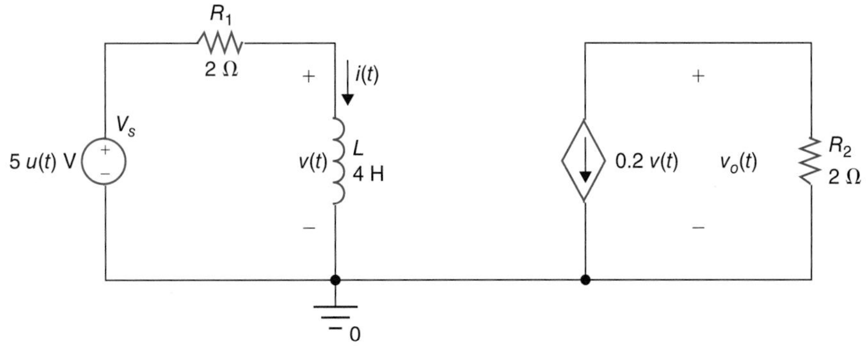

Answer:

$i(t) = (2.5 - 1.5e^{-0.5t})u(t)$ A, $v(t) = 3e^{-0.5t}u(t)$ V, $v_o(t) = -1.2e^{-0.5t}u(t)$ V.

7.6 Solving General First-Order Differential Equations

In this section, we are interested in solving general first-order differential equations that can be applied to circuit analysis. A general first-order differential equation can be written as

$$\frac{dv(t)}{dt} + P(t)v(t) = Q(t) \tag{7.90}$$

where $P(t)$ and $Q(t)$ are functions of t. Multiplying both sides of Equation (7.90) by a positive function $\mu(t)$, we obtain

$$\mu(t)\frac{dv(t)}{dt} + P(t)\mu(t)v(t) = \mu(t)Q(t) \tag{7.91}$$

The positive function $\mu(t)$ is chosen to satisfy

$$\frac{d}{dt}[\mu(t)v(t)] = \mu(t)Q(t) \tag{7.92}$$

Notice that $\frac{d}{dt}[\mu(t)v(t)]$ can be written as

$$\frac{d}{dt}[\mu(t)v(t)] = \mu(t)\frac{dv(t)}{dt} + v(t)\frac{d\mu(t)}{dt} \tag{7.93}$$

Substitution of Equation (7.93) into Equation (7.92) yields

$$\mu(t)\frac{dv(t)}{dt} + v(t)\frac{d\mu(t)}{dt} = \mu(t)Q(t) \tag{7.94}$$

Since the right sides of Equations (7.91) and (7.94) are identical, the left sides must be the same. Thus, we have

$$\mu(t)\frac{dv(t)}{dt} + v(t)\frac{d\mu(t)}{dt} = \mu(t)\frac{dv(t)}{dt} + P(t)\mu(t)v(t)$$

Deleting the common term $\mu(t)\frac{dv(t)}{dt}$ from both sides and dividing by $v(t)$ on both sides, we obtain

$$\frac{d\mu(t)}{dt} = P(t)\mu(t)$$

Dividing by $\mu(t)$on both sides, we obtain

$$\frac{\frac{d\mu(t)}{dt}}{\mu(t)} = P(t)$$

This equation can be expressed as

$$\frac{d}{dt}[\ln(\mu(t))] = P(t)$$

Integrating on both sides, we obtain

$$\ln[\mu(t)] = k_1 + \int P(t)dt$$

Exponentiation on both sides yields

$$\exp\{\ln[\mu(t)]\} = \exp\left[k_1 + \int P(t)dt\right] = \exp(k_1)\exp\left[\int P(t)dt\right]$$

Therefore, $\mu(t)$ is

$$\mu(t) = \exp(k_1)\exp\left[\int P(t)dt\right] \tag{7.95a}$$

Integral of Equation (7.92) results in

$$\mu(t)v(t) = \int \mu(t)Q(t)dt + k_2$$

Dividing by $\mu(t)$, we obtain

$$v(t) = \frac{\int \mu(t)Q(t)dt + k_2}{\mu(t)}$$

Substitution of Equation (7.95a) for $\mu(t)$ yields

$$\begin{aligned} v(t) &= \frac{\int \mu(t)Q(t)dt + k_2}{\mu(t)} = \frac{\int \exp(k_1)\exp\left[\int P(t)dt\right]Q(t)dt + k_2}{\exp(k_1)\exp\left[\int P(t)dt\right]} \\ &= \frac{\int \exp\left[\int P(t)dt\right]Q(t)dt + \dfrac{k_2}{\exp(k_1)}}{\exp\left[\int P(t)dt\right]} = \frac{\int \exp\left[\int P(t)dt\right]Q(t)dt + K}{\exp\left[\int P(t)dt\right]} \end{aligned} \tag{7.96}$$

At $t = 0$, we have $\int_0^0 P(t)dt = 0$ and $\int_0^0 \exp\left[\int P(t)dt\right]Q(t)dt = 0$. Thus, $K = v(0)$. We redefine $\mu(t)$ as

$$\mu(t) = \exp\left[\int P(t)dt\right] \tag{7.95b}$$

Let $P(t) = a$ (constant). If the integral starts at $t = 0$, we have

$$\exp\left[\int P(t)dt\right] = \exp\left[\int_0^t a\,dt\right] = \exp(at) = e^{at}$$

and Equation (7.96) becomes

$$v(t) = \frac{\int e^{at}Q(t)dt + K}{e^{at}} \tag{7.97}$$

If $Q(t) = b$, in addition to $P(t) = a$, then the differential equation reduces to

$$\frac{dv(t)}{dt} + av(t) = b \tag{7.98}$$

The solution to the differential equation [Equation (7.98)] is, from Equation (7.97),

$$v(t) = \frac{\int e^{at} b\,dt + K}{e^{at}} = \frac{b\int e^{at} dt + K}{e^{at}} = \frac{b\frac{e^{at}-1}{a} + K}{e^{at}} \tag{7.99}$$

At $t = 0$, we have

$$v(0) = \frac{b\frac{e^{a0}-1}{a} + K}{e^{a0}} = K$$

With $K = v(0)$, Equation (7.99) becomes

$$v(t) = \frac{b\frac{e^{at}-1}{a} + v(0)}{e^{at}} = \frac{b}{a} - \frac{b}{a}e^{-at} + v(0)e^{-at} = \frac{b}{a} + \left[v(0) - \frac{b}{a}\right]e^{-at} \tag{7.100}$$

Notice that b/a is the final value of $v(t)$, $v(0)$ is the initial value of $v(t)$, and $1/a$ is the time constant. The solution given by Equation (7.100) matches the solution from

$$v(t) = (Final\,Value) + [(Initial\,Value) - (Final\,Value)]e^{-\frac{t}{\tau}} \tag{7.101}$$

EXAMPLE 7.19

In the circuit shown in Figure 7.61, find the voltage $v(t)$ across the capacitor for $t \geq 0$. The initial voltage across the capacitor is $v(0) = 1$ V.

FIGURE 7.61

An *RC* circuit.

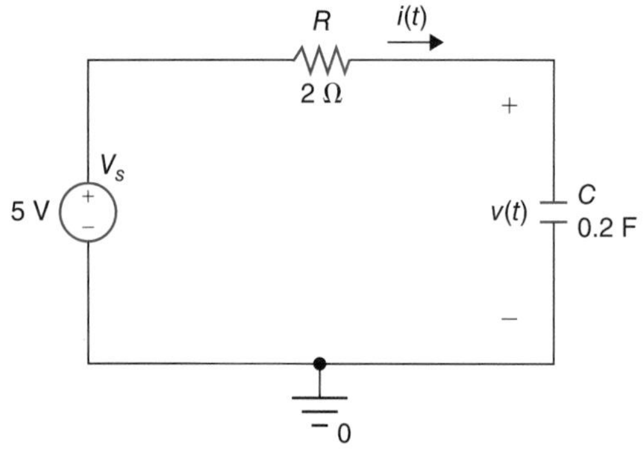

Collecting the voltage drops around the circuit in the clockwise direction, we obtain

$$-5 + 2i(t) + v(t) = 0 \tag{7.102}$$

The current $i(t)$ through the capacitor is given by

$$i(t) = C\frac{dv(t)}{dt} = 0.2\frac{dv(t)}{dt} \tag{7.103}$$

Substitution of Equation (7.103) into Equation (7.102) yields

$$-5 + 2 \times 0.2\frac{dv(t)}{dt} + v(t) = 0$$

which can be rearranged as

$$\frac{dv(t)}{dt} + 2.5v(t) = 12.5 \tag{7.104}$$

continued

Example 7.19 continued

Comparison of Equations (7.104) and (7.90) reveals that $P(t) = 2.5$ and $Q(t) = 12.5$. The function $\mu(t)$ can be found from Equation (7.95b):

$$\mu(t) = \exp\left[\int P(t)dt\right] = \exp\left[\int_0^t 2.5dt\right] = \exp(2.5t) = e^{2.5t}$$

The voltage across the capacitor can be found from Equation (7.96):

$$v(t) = \frac{\int \mu(t)Q(t)dt + K}{\mu(t)} = \frac{\int_0^t e^{2.5\lambda}12.5d\lambda + K}{e^{2.5t}} = \frac{12.5\frac{e^{2.5t}-1}{2.5} + K}{e^{2.5t}}$$
$$= 5 - 5e^{-2.5t} + Ke^{-2.5t} \quad \textbf{(7.105)}$$

The constant K can be found from the initial condition $v(0) = 1$ V. Setting $t = 0$ in Equation (7.105), we obtain

$$5 - 5e^{-2.5\times0} + Ke^{-2.5\times0} = 5 - 5 + K = 1$$

Thus, $K = 1$. The final answer for the voltage across the capacitor is given by

$$v(t) = (5 - 4e^{-2.5t})u(t) \quad \textbf{(7.106)}$$

The time constant of the RC circuit is $\tau = RC = 2 \times 0.2 = 0.4$. The solution based on Equation (7.35) is

$$v_o(t) = (Final\ Value) + [(Initial\ Value) - (Final\ Value)]e^{-\frac{t}{\tau}}$$
$$= 5 + (1-5)e^{-2.5t} = 5 - 4e^{-2.5t} \text{ V}$$

for $t \geq 0$. This answer is identical to the one given in Equation (7.106).

Exercise 7.13

Find the current $i(t)$ through the inductor for the circuit shown in Figure 7.62 for $t \geq 0$. The initial current through the inductor is $i(0) = 1$ mA.

FIGURE 7.62

Circuit for EXERCISE 7.13.

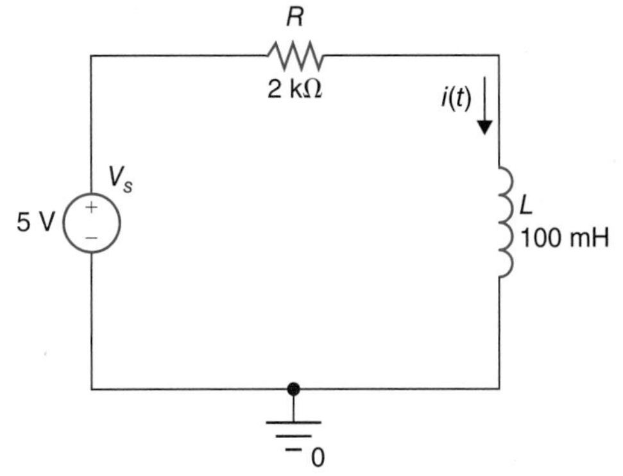

Answer:

$i(t) = \left(2.5 - 1.5e^{-\frac{t}{5\times10^{-5}}}\right)u(t)$ mA.

EXAMPLE 7.20

Find the voltage $v(t)$ across the capacitor in the circuit shown in Figure 7.63. The input voltage is given by $v_s(t) = 6e^{-2t}u(t)$.

FIGURE 7.63

Circuit for EXAMPLE 7.20.

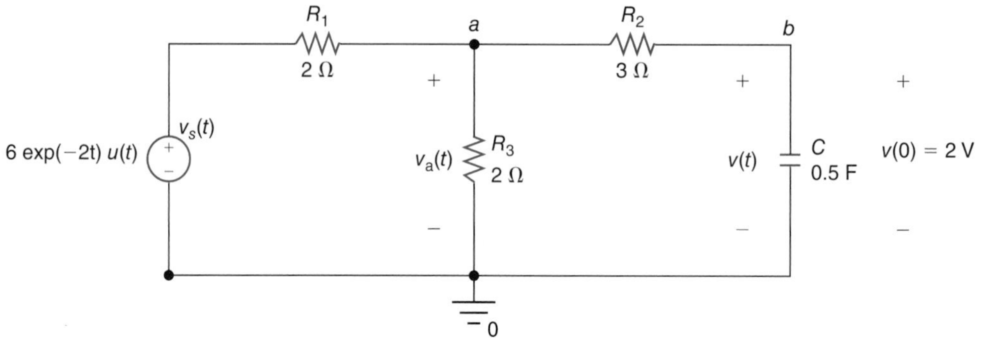

Summing the currents leaving node a, we obtain

$$\frac{v_a - v_s}{2} + \frac{v_a}{2} + \frac{v_a - v}{3} = 0 \tag{7.107}$$

where v_a refers to $v_a(t)$, v_s refers to $v_s(t)$, and v refers to $v(t)$. Summing the currents leaving node b, we obtain

$$\frac{v - v_a}{3} + \frac{1}{2}\frac{dv}{dt} = 0 \tag{7.108}$$

Multiplication by 6 on both sides of Equation (7.107) yields

$$3v_a - 3v_s + 3v_a + 2v_a - 2v = 0 \tag{7.109}$$

Solving Equation (7.109) for v_a, we get

$$v_a = \frac{3}{8}v_s + \frac{1}{4}v \tag{7.110}$$

Multiplication by 2 on both sides of Equation (7.108) yields

$$\frac{dv}{dt} + \frac{2}{3}v - \frac{2}{3}v_a = 0 \tag{7.111}$$

Substitution of Equation (7.110) into Equation (7.111) yields

$$\frac{dv}{dt} + \frac{2}{3}v - \frac{2}{3}\left(\frac{3}{8}v_s + \frac{1}{4}v\right) = 0$$

which reduces to

$$\frac{dv}{dt} + \frac{1}{2}v = \frac{1}{4}v_s$$

continued

Example 7.20 continued

Substituting $v_s = 6e^{-2t}$ in this expression, we obtain the first-order differential equation

$$\frac{dv}{dt} + \frac{1}{2}v = \frac{3}{2}e^{-2t} \tag{7.112}$$

The solution of Equation (7.112) is given by Equation (7.97) with $a = 0.5$ and $Q(t) = (3/2)e^{-2t}$. Equation (7.97) becomes

$$v(t) = \frac{\int e^{at}Q(t)dt + K}{e^{at}} = \frac{\int_0^t e^{0.5\lambda}\frac{3}{2}e^{-2\lambda}d\lambda + K}{e^{0.5t}} = \frac{\frac{3}{2}\int_0^t e^{-1.5\lambda}d\lambda + K}{e^{0.5t}}$$

$$= \frac{\frac{3}{2}\frac{e^{-1.5\lambda}\big|_0^t}{-1.5} + K}{e^{0.5t}} = \frac{-(e^{-1.5t} - 1) + K}{e^{0.5t}} = \frac{(1 - e^{-1.5t}) + K}{e^{0.5t}}$$

At $t = 0$, we have

$$v(0) = \frac{(1 - e^{-1.5\times 0}) + K}{e^{0.5\times 0}} = \frac{(1 - 1) + K}{1} = K = 2$$

Thus, the solution for the first-order differential Equation (7.112) is given by

$$v(t) = \frac{(1 - e^{-1.5t}) + 2}{e^{0.5t}} = e^{-0.5t} - e^{-2t} + 2e^{-0.5t} = 3e^{-0.5t} - e^{-2t}$$

for $t \geq 0$, or, using the unit step function, we have

$$v(t) = (3e^{-0.5t} - e^{-2t})u(t) \text{ V} \tag{7.113}$$

The input and output signals are plotted in Figure 7.64.

FIGURE 7.64

Input voltage and output voltage for the circuit shown in Figure 7.63.

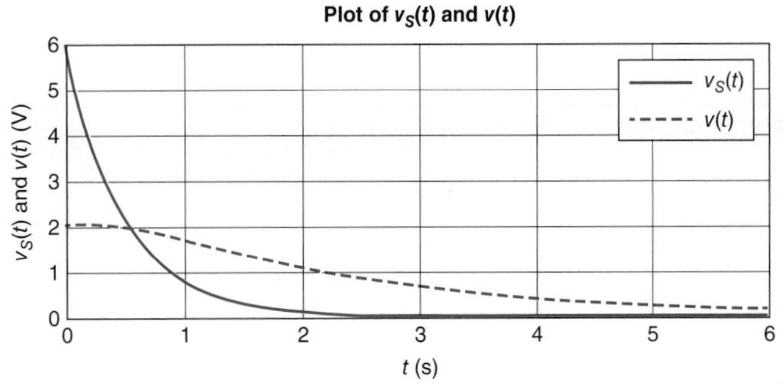

MATLAB

```
%EXAMPLE 7.20
clear all;
C=0.5;R1=2;R2=3;R3=2;V0=2;
syms v(t) va
vs=6*exp(-2*t);
Dv=diff(v);
va=solve((va-vs)/R1+va/R3+(va-v)/R2,va); %Equation (7.107)
v=dsolve((v-va)/R2+C*Dv,v(0)==V0); %Equation (7.108)
i=C*diff(v);
va=R2*i+v;
v=simplify(vpa(v,10))
i=simplify(vpa(i,10))
va=simplify(vpa(va,10))
```

continued

Example 7.20 continued

MATLAB continued

```
Answers:
v =
3.0*exp(-0.5*t) - 1.0*exp(-2.0*t)
i =
exp(-2.0*t) - 0.75*exp(-0.5*t)
va =
0.75*exp(-0.5*t) + 2.0*exp(-2.0*t)
```

Exercise 7.14

Find the current $i(t)$ through the inductor for $t \geq 0$ for the circuit shown in Figure 7.65. The initial value of the current is $i(0) = 3$ A.

FIGURE 7.65

Circuit for EXERCISE 7.14.

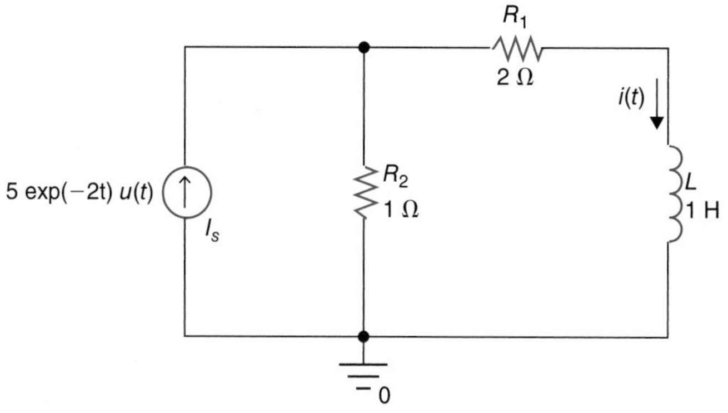

Answer:

$i(t) = (5e^{-2t} - 2e^{-3t})u(t)$ A.

EXAMPLE 7.21

Find the voltage $v(t)$ across the capacitor for the circuit shown in Figure 7.66. The input signal is given by $v_s(t) = 6\cos(5t)u(t)$.

FIGURE 7.66

Circuit for EXAMPLE 7.21.

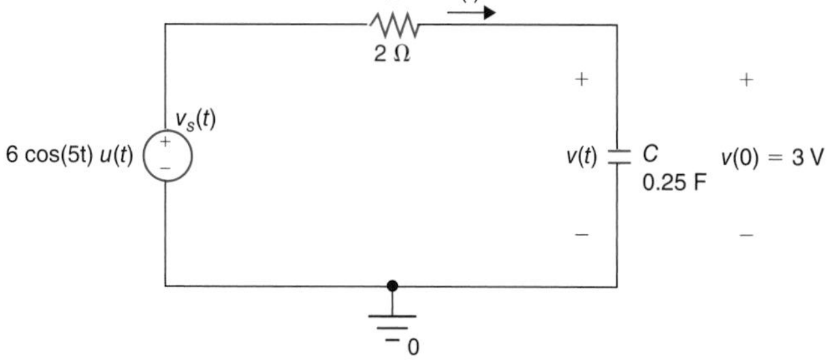

Summing the voltage drops around the mesh in the clockwise direction, we obtain

$$-v_s + Ri + v = 0 \tag{7.114}$$

The current-voltage relation on the capacitor is given by

$$i = C\frac{dv}{dt} \tag{7.115}$$

continued

Example 7.21 continued

Substitution of Equation (7.115) into Equation (7.114) results in

$$-v_s + RC\frac{dv}{dt} + v = 0 \tag{7.116}$$

Equation (7.116) can be rearranged as

$$\frac{dv}{dt} + \frac{1}{RC}v = \frac{1}{RC}v_s \tag{7.117}$$

Substituting $R = 2\ \Omega$, $C = 0.25$ F, and $v_s(t) = 6\cos(5t)$ into Equation (7.117), we obtain

$$\frac{dv}{dt} + 2v = 12\cos(5t) \tag{7.118}$$

Notice that $P(t) = a = 2$ and $Q(t) = 12\cos(5t)$. From Equation (7.97), the solution to the differential equation [Equation (7.118)] is given by

$$v(t) = \frac{\int e^{at}Q(t)dt + K}{e^{at}} = \frac{12\int_0^t e^{2\lambda}\cos(5\lambda)d\lambda + K}{e^{2t}} \tag{7.119}$$

Application of the integral formula

$$\int e^{\alpha x}\cos(\beta x)dx = \frac{1}{\alpha^2 + \beta^2}e^{\alpha x}[\alpha\cos(\beta x) + \beta\sin(\beta x)]$$

to $12\int_0^t e^{2\lambda}\cos(5\lambda)d\lambda$ results in

$$12\int_0^t e^{2\lambda}\cos(5\lambda)d\lambda = \frac{12}{2^2 + 5^2}e^{2\lambda}[2\cos(5x) + 5\sin(5x)]\Big|_0^t$$

$$= \frac{12e^{2t}[2\cos(5t) + 5\sin(5t)] - 24}{29}$$

Equation (7.119) becomes

$$v(t) = \frac{\frac{12e^{2t}[2\cos(5t) + 5\sin(5t)] - 24}{29} + K}{e^{2t}} \tag{7.120}$$

At $t = 0$, Equation (7.120) becomes

$$v(0) = 3 = \frac{\frac{12e^{2\times 0^-}\left[2\cos(5\times 0^-) + 5\sin(5\times 0^-)\right] - 24}{29} + K}{e^{2\times 0^-}}$$

$$= \frac{\frac{12[2 + 0] - 24}{29} + K}{1} = K$$

continued

Example 7.21 continued

Thus, $K = 3$. The solution to the differential equation is given by

$$v(t) = \frac{\dfrac{12e^{2t}[2\cos(5t) + 5\sin(5t)] - 24}{29} + 3}{e^{2t}}$$
$$= \left\{\frac{63}{29}e^{-2t} + \frac{12}{29}[2\cos(5t) + 5\sin(5t)]\right\}u(t) \tag{7.121}$$
$$= [2.1724e^{-2t} + 0.8276\cos(5t) + 2.069\sin(5t)]u(t)$$

Notice that the cosine and sine terms can be combined into a single cosine term, as shown here:

$$\frac{12}{29}[2\cos(5t) + 5\sin(5t)] = \frac{12\sqrt{29}}{29}\left[\frac{2}{\sqrt{29}}\cos(5t) + \frac{5}{\sqrt{29}}\sin(5t)\right]$$
$$= \frac{12}{\sqrt{29}}\cos\left[5t - \tan^{-1}\left(\frac{5}{2}\right)\right]$$

The solution can also be written as

$$v(t) = \left[\frac{63}{29}e^{-2t} + \frac{12}{\sqrt{29}}\cos(5t - 68.1986°)\right]u(t) \tag{7.122}$$
$$= [2.1724e^{-2t} + 2.2283\cos(5t - 68.1986°)]u(t)$$

The voltage across the capacitor consists of two terms, the exponential term and the sinusoidal term. The exponential term, $2.1724e^{-2t}u(t)$, is due to the initial condition $v(0) = 3$V and disappears in about five times the time constant. This term represents transient response. On the other hand, the cosine term, $2.2283\cos(5t - 68.1986°)u(t)$, is due to the input signal $6\cos(5t)\,u(t)$ and lasts continuously as long as input signal is present. This term represents the steady-state response. Notice that the circuit changed the amplitude and phase of the input cosine signal. The voltage $v(t)$ across the capacitor is plotted in Figure 7.67. Notice that the period of the input and output sinusoids is $T = 2\pi/5 = 1.25664$ s. Due to the initial conditions, $v(t)$ starts at 3 V.

FIGURE 7.67

Input voltage and output voltage for the circuit shown in Figure 7.66.

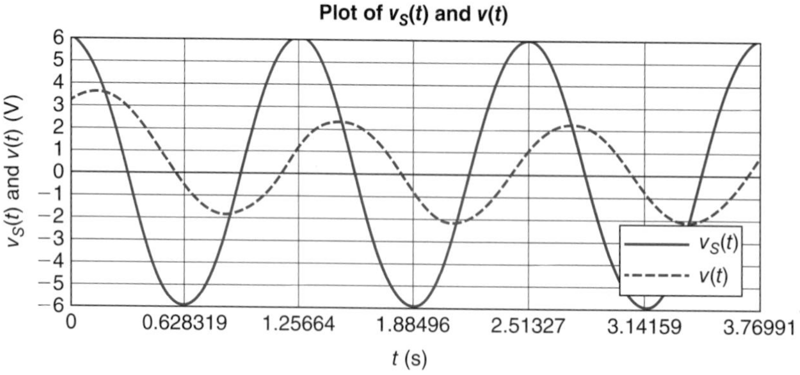

MATLAB

```
%EXAMPLE 7.21
clear all;
C=0.25;R=2;V0=3;
syms v(t)
vs=6*cos(5*t);
Dv=diff(v);
```

continued

Example 7.21 continued

MATLAB continued

```
v=dsolve(-vs+R*C*Dv+v,v(0)==V0);
i=C*diff(v);
v=vpa(v,10);
i=vpa(i,10);
v=expand(v);
i=expand(i);
v=vpa(v,7)
i=vpa(i,7)

Answers:
v =
0.8275862*cos(5.0*t) + 2.068966*sin(5.0*t) + 2.172414*exp(-2.0*t)
i =
2.586207*cos(5.0*t) - 1.034483*sin(5.0*t) - 1.086207*exp(-2.0*t)
```

Exercise 7.15

Find the current $i(t)$ through the inductor for $t \geq 0$ for the circuit shown in Figure 7.68. The initial value of the current is $i(0) = 2$ A.

FIGURE 7.68

Circuit for EXERCISE 7.15.

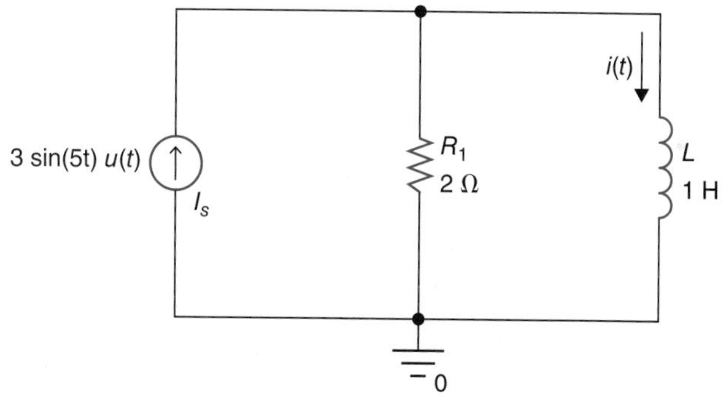

Answer:

$$i(t) = \left\{\frac{88}{29}e^{-2t} - \frac{6}{29}[5\cos(5t) - 2\sin(5t)]\right\}u(t) \text{ A}$$

$$i(t) = \left\{\frac{88}{29}e^{-2t} - \frac{6}{\sqrt{29}}\cos\left[5t + \tan^{-1}\left(\frac{2}{5}\right)\right]\right\}u(t) \text{ A.}$$

EXAMPLE 7.22

Find the current $i(t)$, $t \geq 0$, through the inductor for the circuit shown in Figure 7.69. The input signal is $i_S(t) = 8t\, u(t)$ A and $i(0) = 2$ A.

continued

Example 7.22 continued

FIGURE 7.69

Circuit for EXAMPLE 7.22.

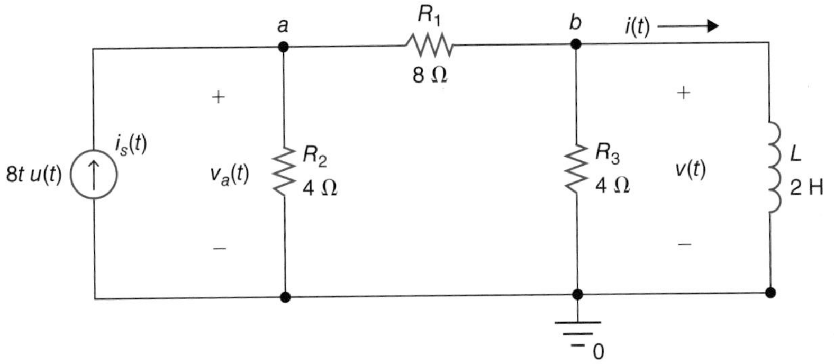

Summing the currents leaving node a, we obtain

$$-i_s + \frac{v_a}{4} + \frac{v_a - v}{8} = 0 \tag{7.123}$$

where v_a refers to $v_a(t)$, i_s refers to $i_s(t)$, and v refers to $v(t)$. Summing the currents leaving node b, we obtain

$$\frac{v - v_a}{8} + \frac{v}{4} + i = 0 \tag{7.124}$$

Multiplication by 8 on both sides of Equation (7.123) yields

$$-8i_s + 2v_a + v_a - v = 0 \tag{7.125}$$

Solving Equation (7.125) for v_a, we get

$$v_a = \frac{8}{3}i_s + \frac{1}{3}v \tag{7.126}$$

Multiplication by 8 on both sides of Equation (7.124) yields

$$v - v_a + 2v + 8i = 0 \tag{7.127}$$

Substitution of Equation (7.126) into Equation (7.127) yields

$$v + 3i = i_s$$

The voltage $v(t)$ across the inductor is given by

$$v = L\frac{di}{dt} = 2\frac{di}{dt}$$

Substitution of this into $v + 3i = i_s$ yields

$$\frac{di}{dt} + 1.5i = 0.5i_s$$

Substituting $i_s = 8t$, we obtain the first-order differential equation:

$$\frac{di}{dt} + 1.5i = 4t \tag{7.128}$$

continued

Example 7.22 continued

The solution of Equation (7.128) is given by Equation (7.97) with $a = 1.5$ and $Q(t) = 4t$. Equation (7.97) becomes

$$i(t) = \frac{\int e^{at}Q(t)dt + K}{e^{at}} = \frac{\int_0^t e^{1.5\lambda}4\lambda d\lambda + K}{e^{1.5t}} = \frac{4\int_0^t \lambda e^{1.5\lambda}d\lambda + K}{e^{1.5t}}$$

$$= \frac{4\frac{e^{1.5\lambda}(1.5\lambda - 1)|_0^t}{1.5^2} + K}{e^{1.5t}} = \frac{\frac{4e^{1.5t}(1.5t - 1) + 4}{2.25} + K}{e^{1.5t}}$$

Since $K = i(0) = 2\text{A}$, the solution for the first-order differential equation [Equation (7.128)] is given by

$$i(t) = \frac{\frac{4e^{1.5t}(1.5t - 1) + 4}{2.25} + 2}{e^{1.5t}} = \frac{4}{2.25}1.5t - \frac{4}{2.25} + \frac{4}{2.25}e^{-1.5t} + 2e^{-1.5t}$$

Thus, we have

$$i(t) = \left(\frac{8}{3}t - \frac{16}{9} + \frac{34}{9}e^{-1.5t}\right)u(t) = (2.6667t - 1.7778 + 3.7778e^{-1.5t})u(t) \quad \textbf{(7.129)}$$

The input and output signals are plotted in Figure 7.70.

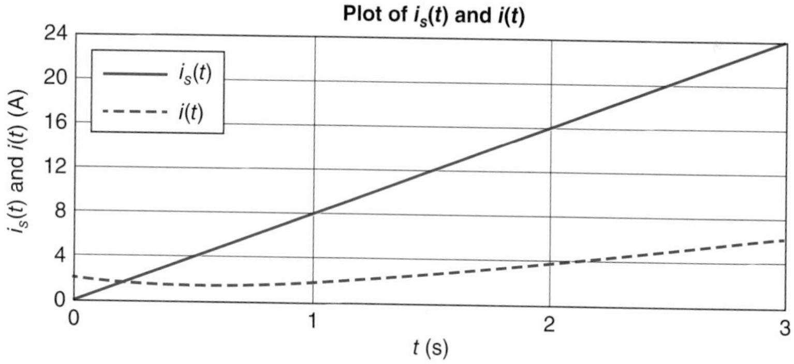

FIGURE 7.70

Plot of input $i_s(t)$ and output $i(t)$.

MATLAB

```
%EXAMPLE 7.22
clear all;
b741 = [0 0.447 0.741];
L=2;R1=8;R2=4;R3=4;I0=2;
syms i(t) va v
is=8*t;
Di=diff(i);
[va,v]=solve(-is+va/R2+(va-v)/R1,...
(v-va)/R1+v/R3+i,va,v); %Equations (7.123) and (7.124)
i=dsolve(v==L*Di,i(0)==I0);
v=L*diff(i);
i=vpa(i,10)
v=vpa(v,10)
ta=0:1e-3:3;
i=subs(i,t,ta);
figure(1)
plot(ta,i,'--','LineWidth',2,'Color',b741);grid;
```

continued

Example 7.22 continued

MATLAB continued

```
xlabel('t (s)');ylabel('i_s(t) and i(t) (A)');hold on;
plot(ta,8*ta,'LineWidth',2,'Color',b741);hold off;
legend('i_s(t)','i(t)','Location','NorthWest');

Answers:
i =
2.666666667*t + 3.777777778*exp(-1.5*t) - 1.777777778
v =
5.333333333 - 11.33333333*exp(-1.5*t)
```

Exercise 7.16

Find the voltage $v(t)$ across the capacitor for $t \geq 0$ for the circuit shown in Figure 7.71. The initial value of the voltage is $v(0) = 2$ V.

FIGURE 7.71

Circuit for EXERCISE 7.16.

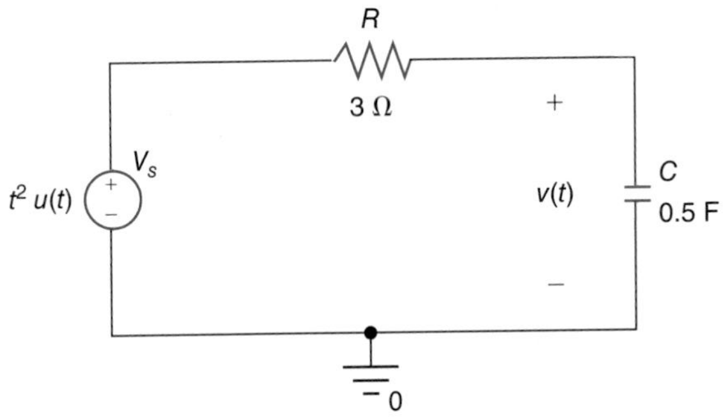

Answer:

$v(t) = (t^2 - 2.5e^{-\frac{2}{3}t} - 3t + 4.5)u(t)$ V.

7.7 PSpice and Simulink

PSpice can be used to simulate the first-order circuits with or without initial conditions. As a start, let us consider the circuit shown in Figure 7.72. The initial voltage across the capacitor is 10 V. Since the time constant is $\tau = RC = 2$ ms, the voltage across the capacitor is given by

FIGURE 7.72

An *RC* circuit with initial condition.

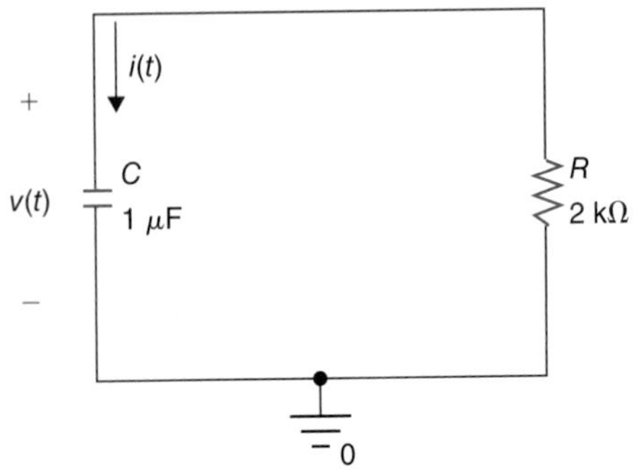

$$v(t) = 10e^{-\frac{t}{0.002}}u(t) = 10e^{-500t}u(t) \text{ V}$$

The current through the capacitor in the direction shown in Figure 7.72 is given by

$$i(t) = C\frac{dv(t)}{dt} = 10^{-6} \times 10 \times (-500)e^{-500t}u(t) = -5e^{-500t}u(t) \text{ mA}$$

When a part such as a resistor, capacitor, or inductor is placed horizontally, the left side is labeled 1 and the right side is labeled 2. If you want to place the part vertically, rotate the part before (Control+R) or after the part is placed. The rotation rotates the part in the counterclockwise direction. If you rotate once, 1 is at the bottom and 2 is at the top. You have to

FIGURE 7.73

Parts from the ANALOG_P library.

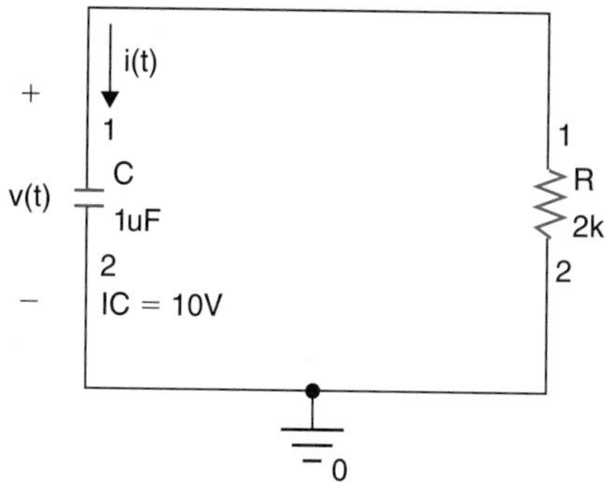

rotate three times to place 1 on the top and 2 at the bottom. When the initial condition is specified, 1 is the positive side. If the parts are selected from the ANALOG_P library, numbers 1 and 2 show up, as shown in Figure 7.73.

To set the initial condition on the capacitor, double-click in the middle of the capacitor symbol. Property Editor pops up. Enter 10V under (or to the right of) IC, highlight the IC column and click Apply, Display → Name and Value → OK, as shown in Figure 7.74.

Click on the Edit Simulation Profile, enter 8ms (four times the time constant) in the Run to time box, as shown in Figure 7.75, and press the Run button.

To add a second y-axis for current, click Plot → Add Y Axis → Add Trace → Select I(C). The plots are shown in Figure 7.76.

The circuit shown in Figure 7.72 can be simulated in Simulink. Summing the currents leaving the top node, we obtain

$$i(t) + \frac{v(t)}{R} = 0 \tag{7.130}$$

FIGURE 7.74

Entering the initial condition. (*Source: OrCAD PSpice by Cadence*)

Color	Designator	Graphic	IC
Default		c.Normal	10V

FIGURE 7.75

Simulation profile for a zero-input RC circuit. (*Source: OrCAD PSpice by Cadence*)

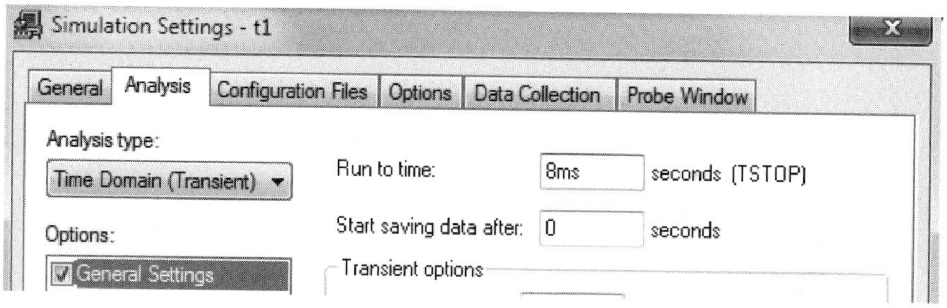

FIGURE 7.76

Voltage $v(t)$ across the capacitor and current $i(t)$ through the capacitor.

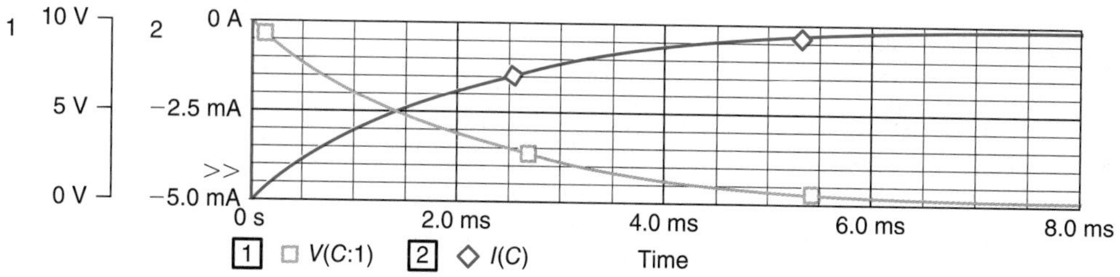

Since the current through the capacitor is $i(t) = C\frac{dv(t)}{dt}$, Equation (7.130) becomes

$$C\frac{dv(t)}{dt} + \frac{v(t)}{R} = 0 \tag{7.131}$$

Dividing by C everywhere in Equation (7.131), we obtain

$$\frac{dv(t)}{dt} + \frac{1}{RC}v(t) = 0 \tag{7.132}$$

Equation (7.132) can be rearranged as

$$\frac{dv(t)}{dt} = -\frac{1}{RC}v(t) \tag{7.133}$$

The voltage $v(t)$ across the capacitor is obtained by integrating Equation (7.133):

$$v(t) = v(0) - \frac{1}{RC}\int_0^t v(\lambda)d\lambda$$

where $v(0)$ is the initial condition. The input to the integrator is given by Equation (7.133). In the Command Window, type

```
>> R=2000;C=1e-6;
```

The variables R and C are in the Workspace and can be used in the Simulink model. Build the model shown in Figure 7.77. To enter the initial condition, double-click on the integrator and enter 10 in the Initial condition box. Set the Simulation stop time to 8*e*-3. The current $i(t)$ is given by $Cdv(t)/dt$. Figure 7.78 shows $v(t)$ and $i(t)$.

FIGURE 7.77

The Simulink model for solving the first-order differential equation without a source.

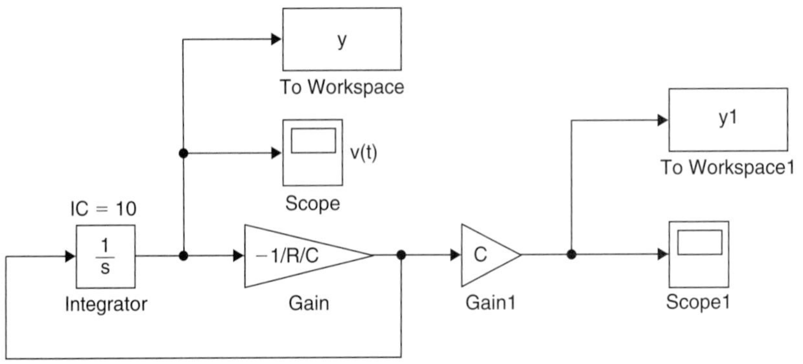

FIGURE 7.78

Voltage $v(t)$ is shown by Scope and current $i(t)$ is shown by Scope 1.

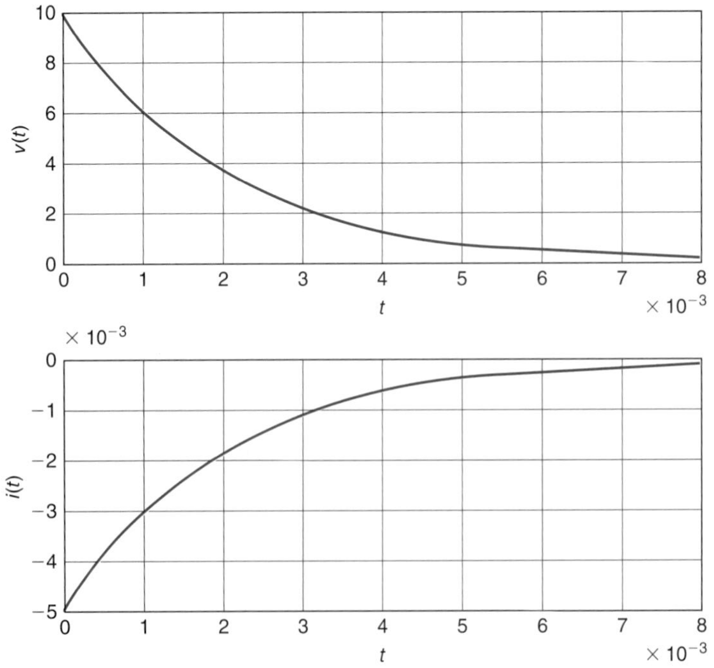

The circuit shown in Figure 7.79 can be used to plot the step response of an *RC* circuit for five different values of *C* (1 μF, 3 μF, 5 μF, 7 μF, 9 μF) on the same graph. After connecting the voltage source, resistor, capacitor, and ground, enter {cval} for the value of the capacitor. Double-click on the middle of the capacitor and enter 0 for the IC (initial condition). Display the initial condition (IC = 0). In the Place Part box, enter PARAM in the SPECIAL library and place it in the circuit, as shown in Figure 7.79. Double-click on PARAMETERS:; click on New Property; enter Name = cval, Value = 1 uF; and click Apply and OK, as shown in Figure 7.80. Double-click on the PARAMETERS: and highlight cval column, right-click and choose Display… to display cval.

FIGURE 7.79

An *RC* circuit with step input.

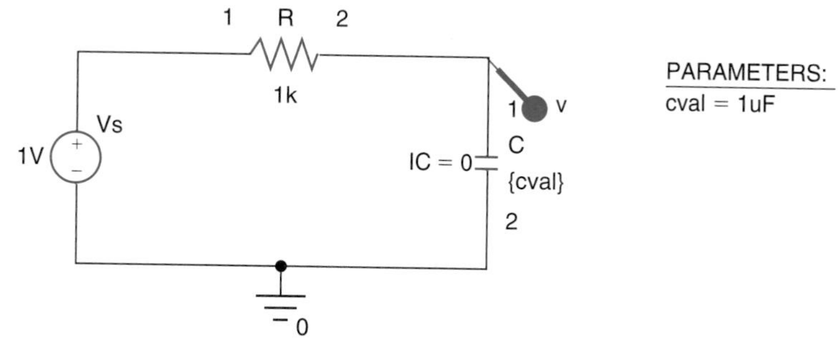

FIGURE 7.80

Adding the Name and Value of the parameters. (*Source: OrCAD PSpice by Cadence*)

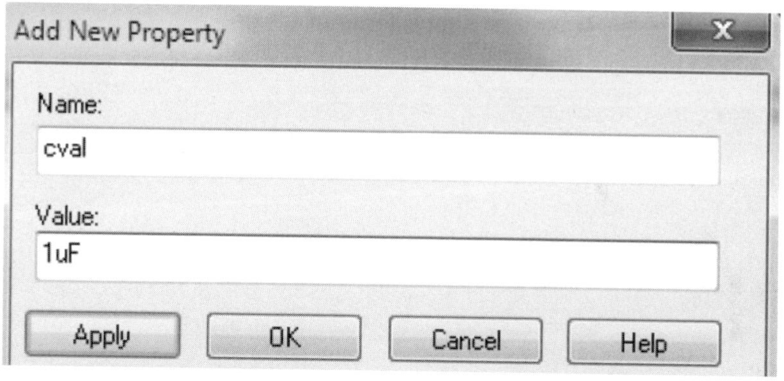

Click on the New Simulation Profile and enter a name such as tran1. In the General Settings box, enter 20ms for Run to time and select Parameter Sweep, and then select Global parameter and enter cval for the Parameter name. Select Linear for the Sweep type and enter Start value = 1uF, End value = 9uF, and Increment = 2uF, as shown in Figure 7.81.

FIGURE 7.81

Setting the parameter sweep. (*Source: OrCAD PSpice by Cadence*)

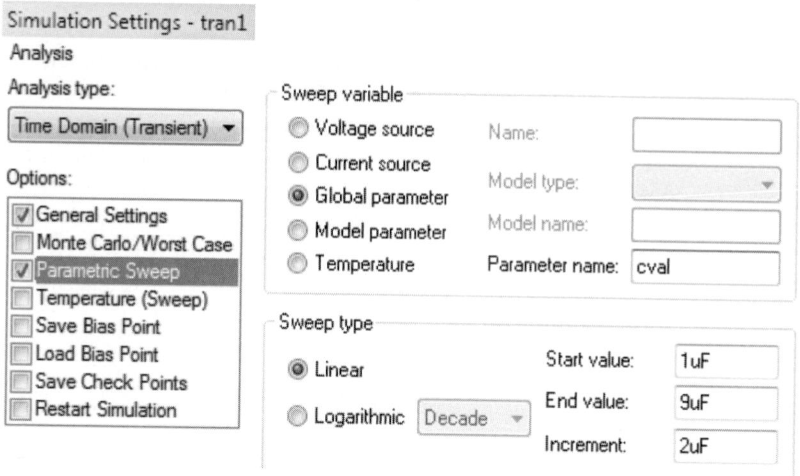

The plot shown in Figure 7.82 is obtained. As can be seen from Figure 7.82, as the C value is increased, it takes a longer time for the voltage to reach the steady-state value of 1 V.

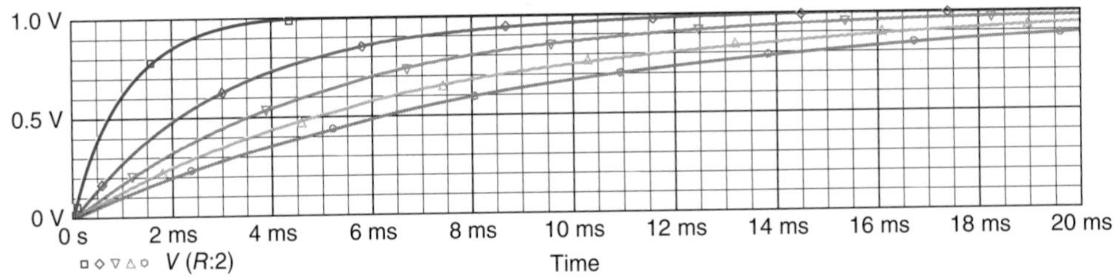

FIGURE 7.82 The step response of an RC circuit with a zero initial condition.

Simulink can be used to obtain the step response of an RC circuit shown in Figure 7.79. Assume that $R = 1\ \text{k}\Omega$ and $C = 1\ \mu\text{F}$. The differential equation is given by

$$\frac{dv(t)}{dt} + \frac{1}{RC}v(t) = \frac{1}{RC}V_S$$

Let $V_S = 1$ V and the initial condition be zero. Then the model shown in Figure 7.83 produces $v(t)$ and $i(t)$, as shown in Figure 7.84. Set the Simulation stop time to 6e-3. In the Command Window, enter

```
>> R=1000;C=1e-6;
```

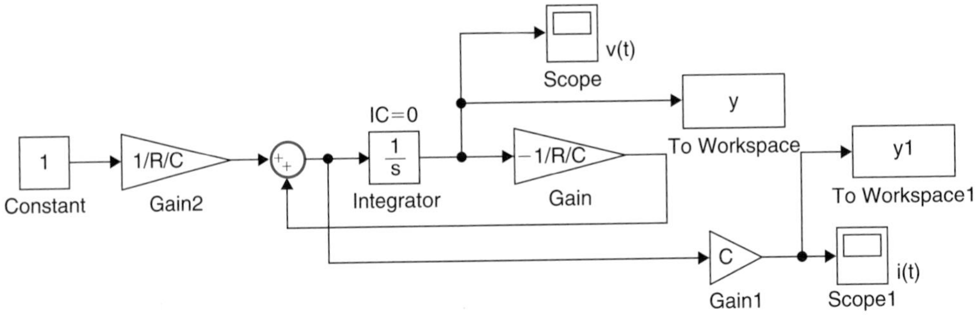

FIGURE 7.83 The Simulink model for an RC circuit.

so that Workspace contains the variables R and C. R and C can be used in the Simulink models.

Simscape makes it possible to simulate physical systems, such as electrical systems, in Simulink. It is similar to PSpice except that in Simscape, you can interface Simulink blocks to physical systems. Figure 7.85 shows a model for the RC circuit shown in Figure 7.79. The voltage across the capacitor (Scope) and the current through the capacitor (Scope1) show the same graphs shown in Figure 7.84.

Figure 7.86 shows a PSpice schematic of a circuit with switches. Switch U2 is opened at $t = 50\ \mu$s and switch U1 is closed at $t = 50\ \mu$s. The initial current through the inductor from top to bottom at $t = 0$ is 1 mA. The switches Sw_tClose and Sw_tOpen are in the EVAL library.

For $0 \le t \le 50\ \mu$s, the time constant is $\tau = L/R_1 = 10\ \mu$s. The final value of the current through the inductor is $5\ \text{V}/10{,}000\ \Omega = 0.5$ mA. Thus, the equation for the current through the inductor is

$$i(t) = \left(0.5 + 0.5e^{-\frac{t}{10\times10^{-6}}}\right)u(t)\ \text{mA}$$

FIGURE 7.84

Voltage across the capacitor (Scope) and current through the capacitor (Scope 1).

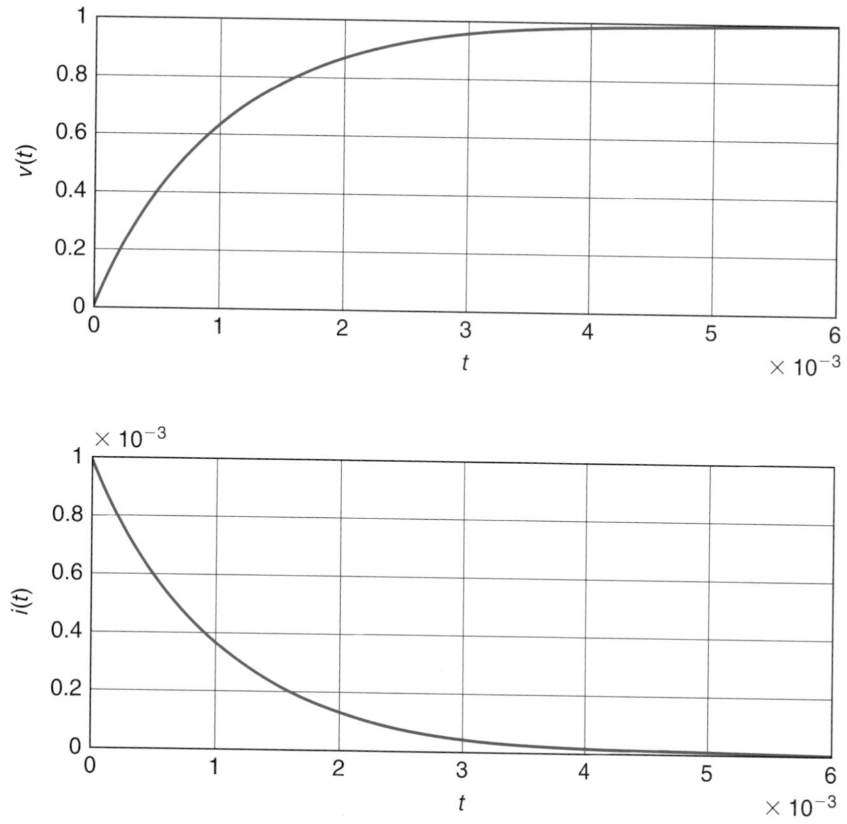

FIGURE 7.85

The model for an *RC* circuit using Simscape.

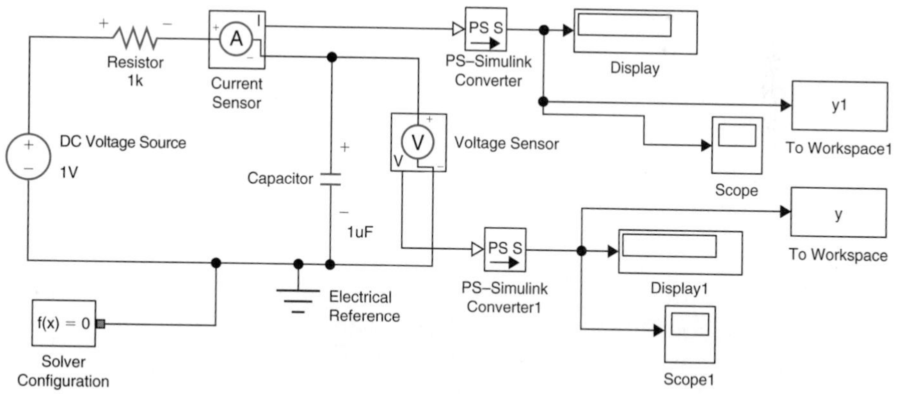

FIGURE 7.86

A circuit with switches

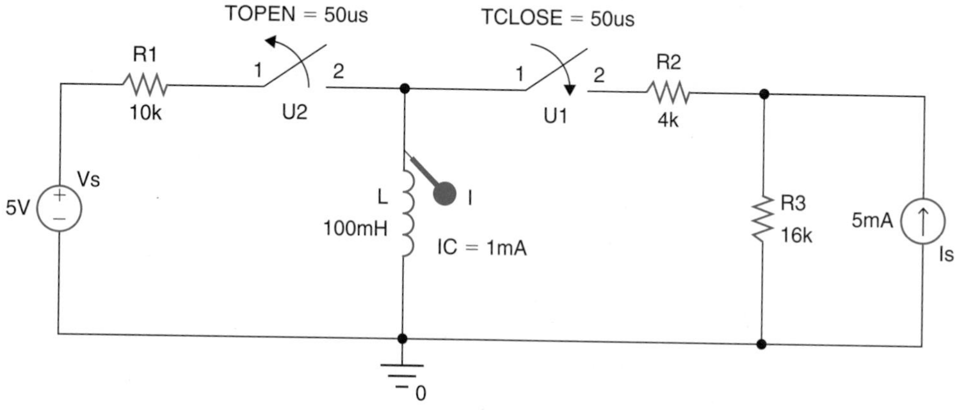

At $t = 50\ \mu s$, $i(50 \times 10^{-6}) = 0.5034$ mA. For $t \geq 50\ \mu s$, the time constant is $\tau = L/(R_2 + R_3) = 5\ \mu s$. The final value of the current through the inductor is 5 mA × 16/(4 + 16) = 4 mA. Thus, the current through the inductor is given by

$$i(t) = \left[4 + (0.5034 - 4)e^{-\frac{t-50\times10^{-6}}{5\times10^{-6}}}\right]u(t - 50 \times 10^{-6})$$

$$= \left(4 - 3.4966e^{-\frac{t-50\times10^{-6}}{5\times10^{-6}}}\right)u(t - 50 \times 10^{-6})\ \text{mA}$$

The current through the inductor is shown in Figure 7.87.

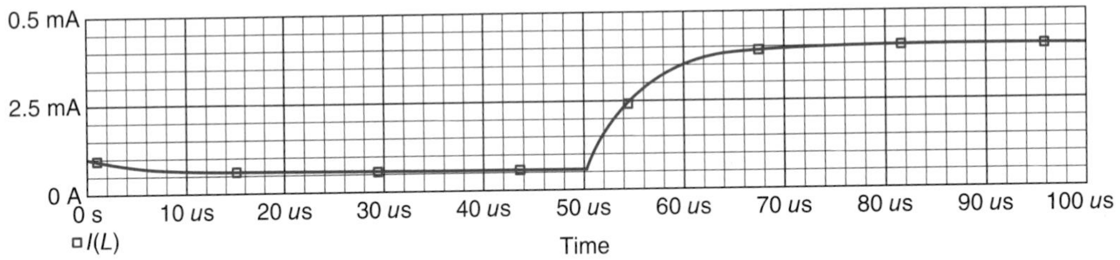

FIGURE 7.87
Current through the inductor.

SUMMARY

Figure 7.2 shows a circuit with one capacitor with capacitance C and one resistor with resistance R connected in parallel. Let the initial voltage across the capacitor at $t = 0$ be $v(0)$. Then, the voltage $v(t)$ across the capacitor for $t \geq 0$ is given by

$$v(t) = v(0)e^{-\frac{t}{\tau}}u(t)\ \text{V}$$

where the time constant τ is given by

$$\tau = RC$$

Figure 7.30 shows a circuit with one inductor with inductance L and one resistor with resistance R connected in parallel. Let the initial current through the inductor at $t = 0$ be $i(0)$. Then, the current $i(t)$ through the inductor for $t \geq 0$ is given by

$$i(t) = i(0)e^{-\frac{t}{\tau}}u(t)\ \text{A}$$

where the time constant τ is given by

$$\tau = \frac{L}{R}$$

Figure 7.15 shows a circuit with a voltage source with voltage V_S, a resistor with resistance R, and a capacitor with capacitance C connected in series. Let the initial voltage across the capacitor at $t = 0$ be $v(0) = V_0$. Then, the voltage $v(t)$ across the capacitor for $t \geq 0$ is given by

$$v(t) = V_S + (V_0 - V_S)e^{-\frac{t}{\tau}}$$

where τ is the time constant given by

$$\tau = RC$$

Figure 7.43 shows a circuit with a current source with current I_S, a resistor with resistance R, and an inductor with inductance L connected in parallel. Let the initial current through the inductor at $t = 0$ be $i(0) = I_0$. Then, the current $i(t)$ through the inductor for $t \geq 0$ is given by

$$i(t) = I_S + (I_0 - I_S)e^{-\frac{t}{\tau}}$$

where τ is the time constant given by

$$\tau = \frac{L}{R}$$

PROBLEMS

Natural Response of *RC* Circuit

7.1 **The initial voltage across the capacitor at $t = 0$ in the circuit shown in Figure P7.1 is 5 V.**

FIGURE P7.1

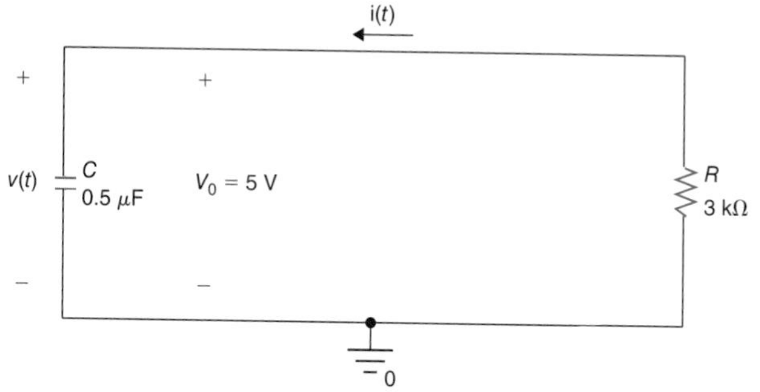

a. Find voltage $v(t), t \geq 0$, across the capacitor and plot $v(t)$.
b. Find current $i(t), t \geq 0$, through the capacitor and plot $i(t)$.

7.2 **The initial voltage across the capacitor at $t = 0$ in the circuit shown in Figure P7.2 is $V_0 = 3$ V.**

FIGURE P7.2

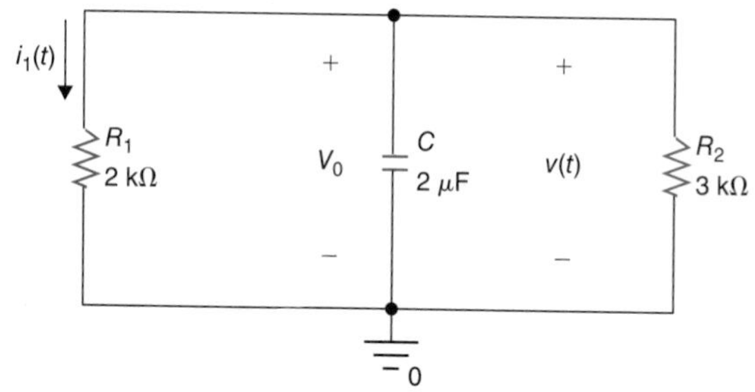

a. Find voltage $v(t), t \geq 0$, across capacitor C.
b. Find current $i_1(t), t \geq 0$, through resistor R_1.

7.3 **The initial voltage across the capacitor at $t = 0$ in the circuit shown in Figure P7.3 is $v(0) = V_0 = 5$ V.**

a. Find voltage $v(t), t \geq 0$, across capacitor C.
b. Find voltage $v_o(t), t \geq 0$, across resistor R_3.

FIGURE P7.3

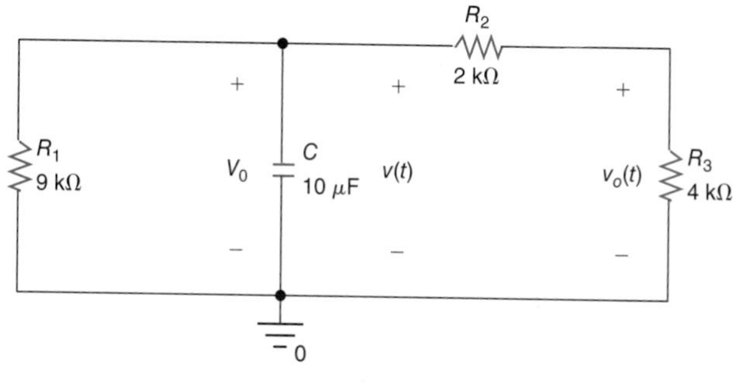

7.4 **The initial voltage across the capacitor at $t = 0$ in the circuit shown in Figure P7.4 is 10 V.**

FIGURE P7.4

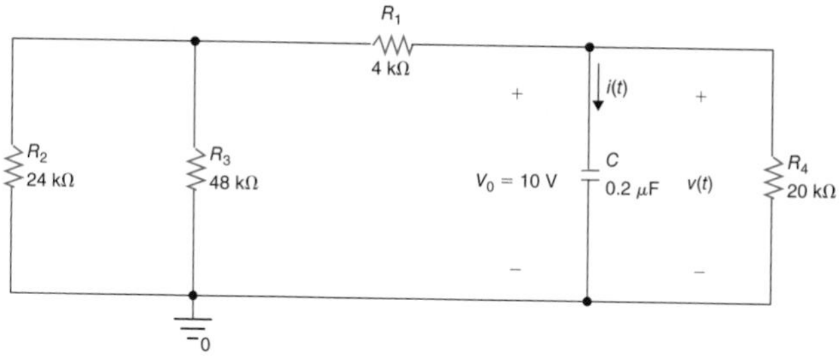

a. Find voltage $v(t), t \geq 0$, across the capacitor and plot $v(t)$.
b. Find current $i(t), t \geq 0$, through the capacitor and plot $i(t)$.

7.5 **The initial voltage across the capacitor at $t = 0$ in the circuit shown in Figure P7.5 is $v(0) = V_0 = 5$ V.**

FIGURE P7.5

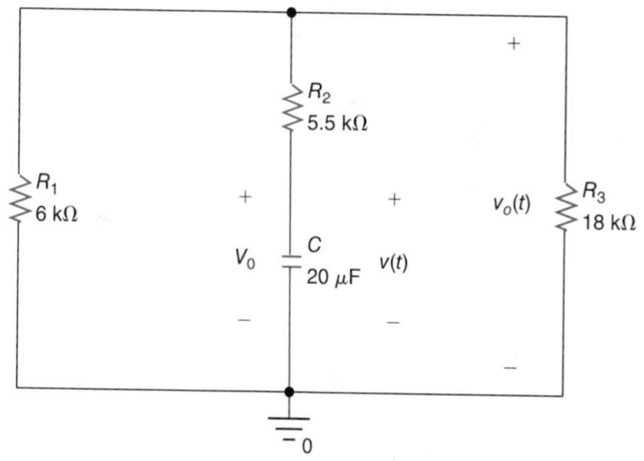

a. Find voltage $v(t), t \geq 0$, across capacitor C.
b. Find voltage $v_o(t), t \geq 0$, across resistor R_3.

7.6 **The initial voltage across the capacitor at $t = 0$ in the circuit shown in Figure P7.6 is 10 V.**

FIGURE P7.6

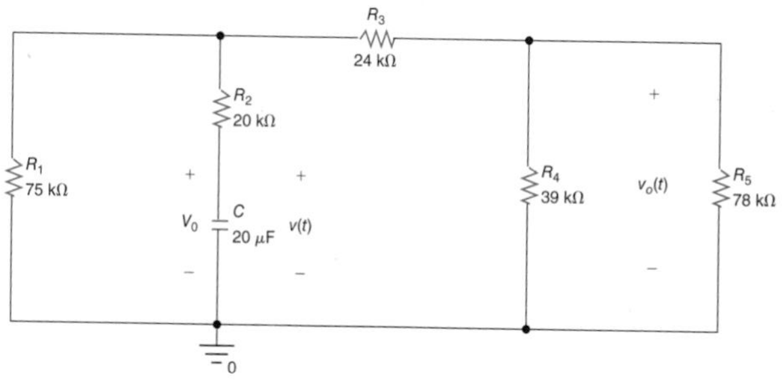

a. Find voltage $v(t)$, $t \geq 0$, across capacitor C.
b. Find voltage $v_o(t)$, $t \geq 0$, across resistor R_5.

7.7 **The initial voltage across the capacitor at $t = 0$ in the circuit shown in Figure P7.7 is 12 V.**

FIGURE P7.7

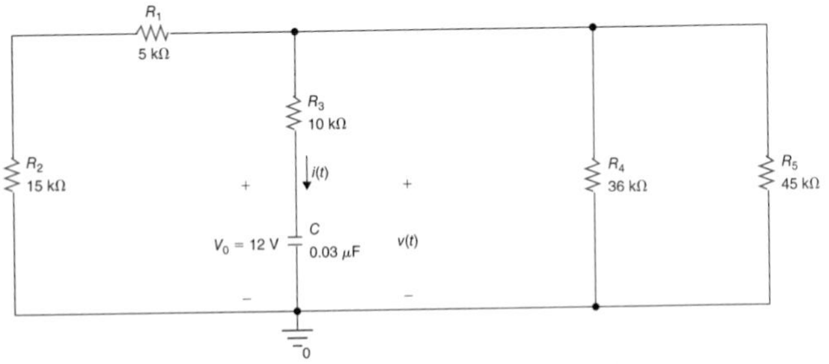

a. Find voltage $v(t)$, $t \geq 0$, across the capacitor and plot $v(t)$.
b. Find current $i(t)$, $t \geq 0$, through the capacitor and plot $i(t)$.

7.8 **The initial voltage across the capacitor at $t = 0$ in the circuit shown in Figure P7.8 is 8 V.**

FIGURE P7.8

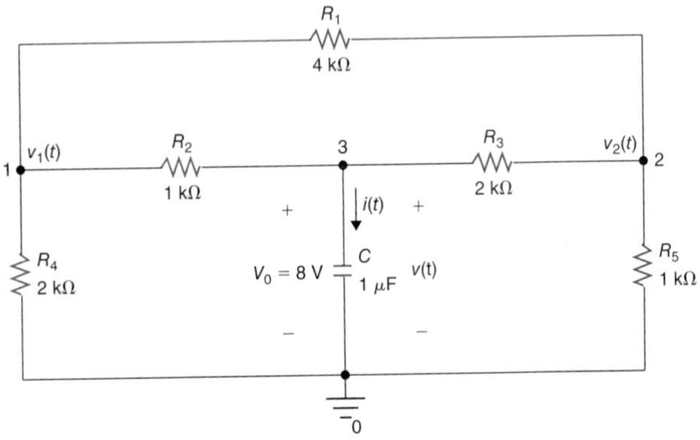

a. Write a node equation at node 1 by summing the currents away from node 1. Notice that the voltage at node 3 is given by $v_3(t) = v(t)$.
b. Write a node equation at node 2 by summing the currents away from node 2. Notice that the voltage at node 3 is given by $v_3(t) = v(t)$.
c. Solve the two node equations from (a) and (b) to express $v_1(t)$ as a function of $v(t)$, and $v_2(t)$ as a function of $v(t)$.
d. Write a node equation at node 3 by summing the currents away from node 3. Use the results from (c) to simplify the equation as a first-order differential equation of $v(t)$.
e. Solve the differential equation to find the voltage $v(t)$, $t \geq 0$, across the capacitor and plot $v(t)$.
f. Find current $i(t)$, $t \geq 0$, through the capacitor and plot $i(t)$.
g. Find voltage $v_1(t)$, $t \geq 0$.
h. Find voltage $v_2(t)$, $t \geq 0$.

7.9 **Find $v(t)$ for $t \geq 0$ for the circuit shown in Figure P7.8 by finding Thévenin equivalent resistance using the test voltage method.**

7.10 **The switch in the circuit shown in Figure P7.10 has been closed for a long time before it is opened at $t = 0$.**

FIGURE P7.10

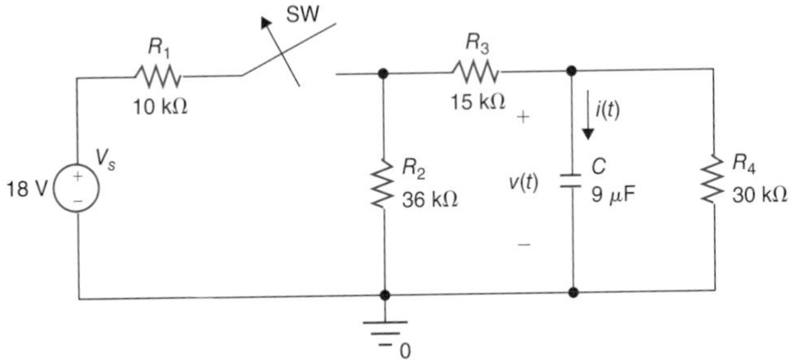

a. Find the initial voltage V_0 across the capacitor at $t = 0$.
b. After switch is opened, find the resistance seen from the capacitor.
c. Find the time constant τ.
d. Find voltage $v(t)$ across the capacitor for $t \geq 0$ and plot $v(t)$ for $t \geq 0$.

Step Response of *RC* Circuit

7.11 **The initial voltage across the capacitor at $t = 0$ in the circuit shown in Figure P7.11 is $v(0) = V_0 = 3$ V. Voltage V_s is applied at $t = 0$; that is, $V_s = 10\,u(t)$ V.**

FIGURE P7.11

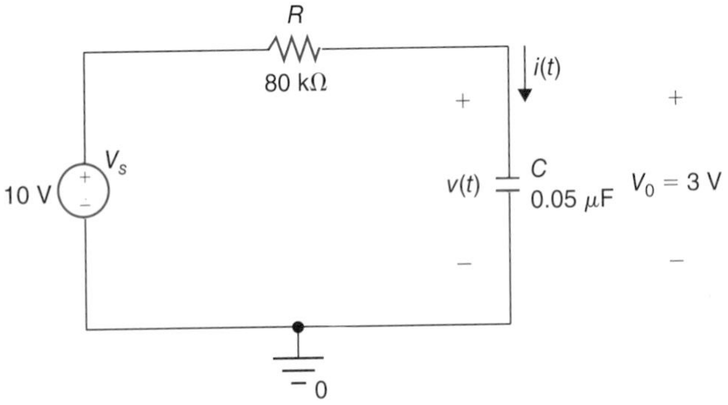

a. Find voltage $v(t)$, $t \geq 0$, across the capacitor and plot $v(t)$.
b. Find current $i(t)$, $t \geq 0$, through the capacitor and plot $i(t)$.

7.12 **The initial voltage across the capacitor at $t = 0$ in the circuit shown in Figure P7.12 is 1 V. Voltage V_s is applied at $t = 0$; that is, $V_s = 8\,u(t)$ V. Find voltage $v(t)$, $t \geq 0$, across the capacitor.**

FIGURE P7.12

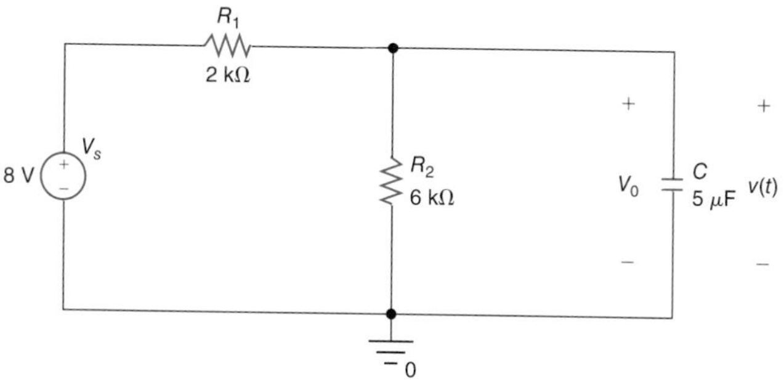

7.13 **The initial voltage across the capacitor at $t = 0$ in the circuit shown in Figure P7.13 is 2 V. Voltage V_s is applied at $t = 0$; that is, $V_s = 10\,u(t)$ V. Find voltage $v(t)$, $t \geq 0$, across the capacitor.**

FIGURE P7.13

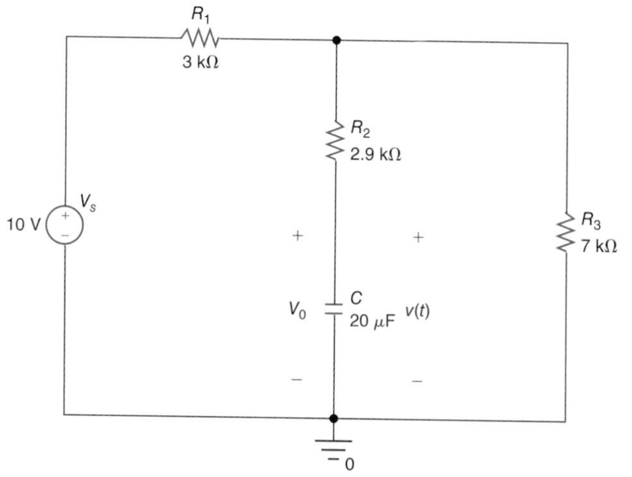

7.14 **The initial voltage across the capacitor at $t = 0$ in the circuit shown in Figure P7.14 is 16 V. Voltage V_s is applied at $t = 0$; that is, $V_s = 10\,u(t)$ V.**

FIGURE P7.14

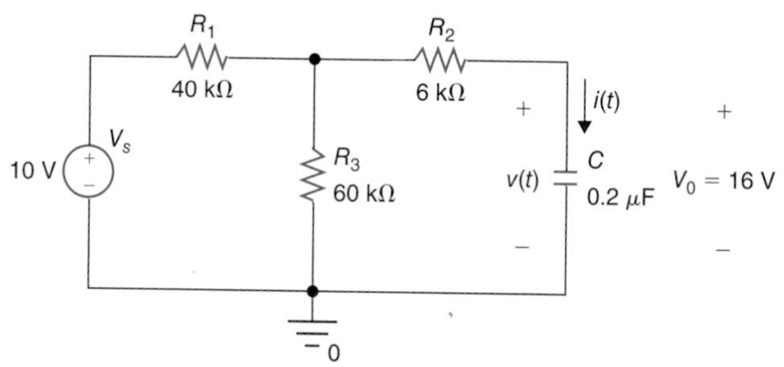

a. Find voltage $v(t)$, $t \geq 0$, across the capacitor and plot $v(t)$.
b. Find current $i(t)$, $t \geq 0$, through the capacitor and plot $i(t)$.

7.15 **The initial voltage across the capacitor at $t = 0$ in the circuit shown in Figure P7.15 is $v(0) = V_0 = 3$ V. Voltage V_s is applied at $t = 0$; that is, $V_s = 15\,u(t)$ V. Find voltage $v(t)$, $t \geq 0$, across the capacitor.**

FIGURE P7.15

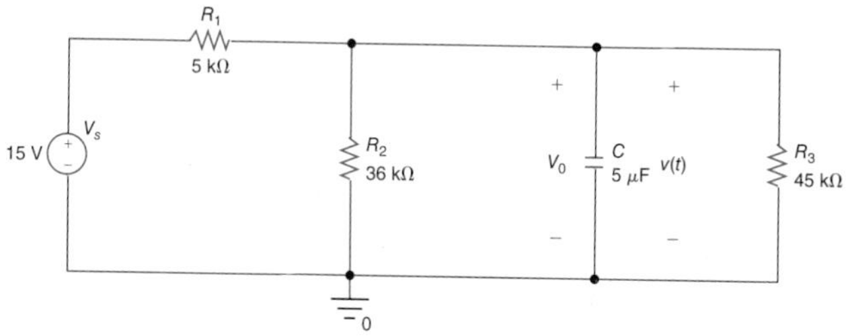

7.16 **The initial voltage across the capacitor at $t = 0$ in the circuit shown in Figure P7.16 is $v(0) = V_0 = 1$ V. Voltage V_s is applied at $t = 0$; that is, $V_s = 9\,u(t)$ V. Find voltage $v(t)$, $t \geq 0$, across the capacitor.**

FIGURE P7.16

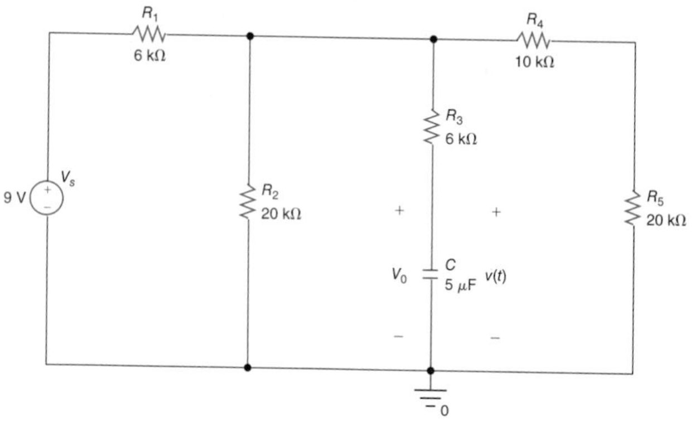

7.17 **The initial voltage across the capacitor at $t = 0$ in the circuit shown in Figure P7.17 is 4 V. The current I_s is applied at $t = 0$; that is, $I_s = 5\,u(t)$ mA.**

FIGURE P7.17

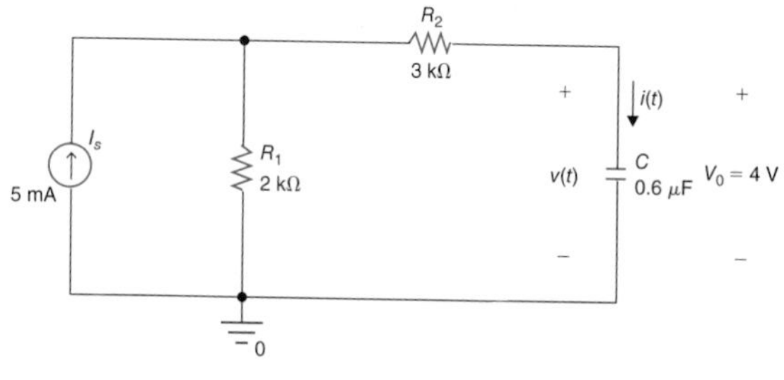

a. Find voltage $v(t)$, $t \geq 0$, across the capacitor and plot $v(t)$.
b. Find current $i(t)$, $t \geq 0$, through the capacitor and plot $i(t)$.

7.18 **The initial voltage across the capacitor at $t = 0$ in the circuit shown in Figure P7.18 is $v(0) = V_0 = 1$ V. Current I_s is applied at $t = 0$; that is, $I_s = 5\,u(t)$ mA. Find voltage $v(t)$, $t \geq 0$, across the capacitor.**

FIGURE P7.18

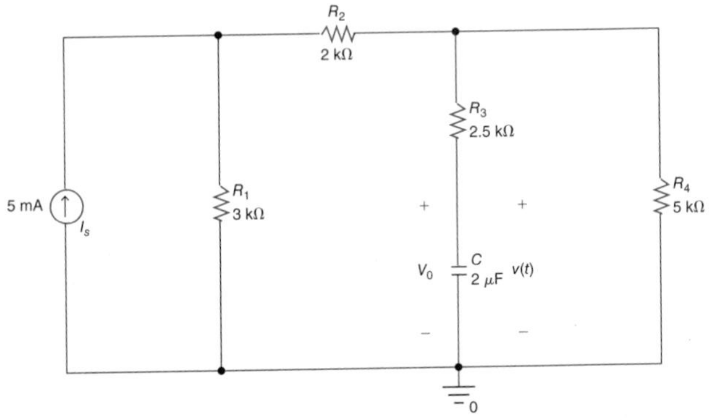

7.19 **The initial voltage across the capacitor at $t = 0$ in the circuit shown in Figure P7.19 is 3 V. Voltage V_s is applied at $t = 0$; that is, $V_s = 10\,u(t)$. Find voltage $v(t)$, $t \geq 0$, across the capacitor.**

FIGURE P7.19

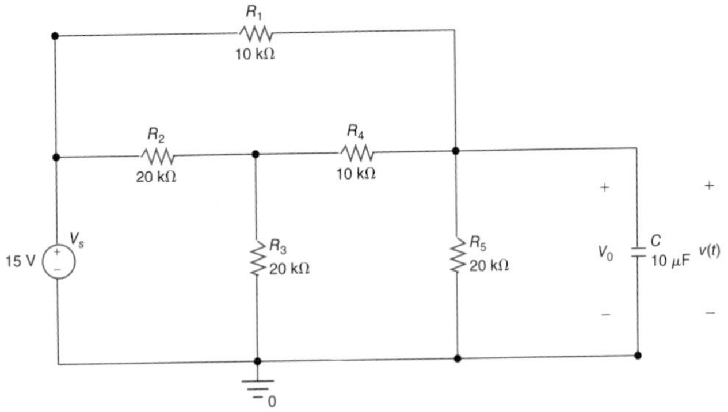

7.20 **The initial voltage across the capacitor at $t = 0$ in the circuit shown in Figure P7.20 is 4 V.**

a. Write a node equation at node 1 by summing the currents away from node 1. Notice that the voltage at node 3 is given by $v_3(t) = v(t)$.
b. Write a node equation at node 2 by summing the currents away from node 2. Notice that the voltage at node 3 is given by $v_3(t) = v(t)$.
c. Solve the two node equations from (a) and (b) to express $v_1(t)$ as a function of $v(t)$, and $v_2(t)$ as a function of $v(t)$.
d. Write a node equation at node 3 by summing the currents away from node 3. Use the results from (c) to simplify the equation as a first-order differential equation of $v(t)$.
e. Solve the differential equation to find voltage $v(t)$, $t \geq 0$, across the capacitor and plot $v(t)$.
f. Find current $i(t)$, $t \geq 0$, through the capacitor and plot $i(t)$.
g. Find voltage $v_1(t)$, $t \geq 0$.
h. Find voltage $v_2(t)$, $t \geq 0$.

FIGURE P7.20

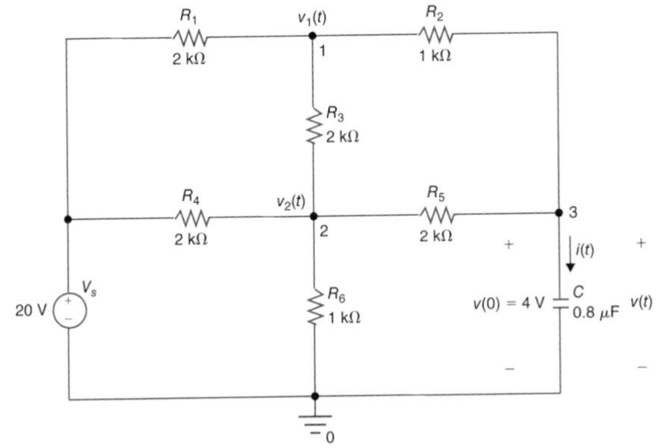

7.21 **Switch 1 in the circuit shown in Figure P7.21 has been closed for a long time before it is opened at $t = 0$. Switch 2 is closed at $t = 4$ ms.**

FIGURE P7.21

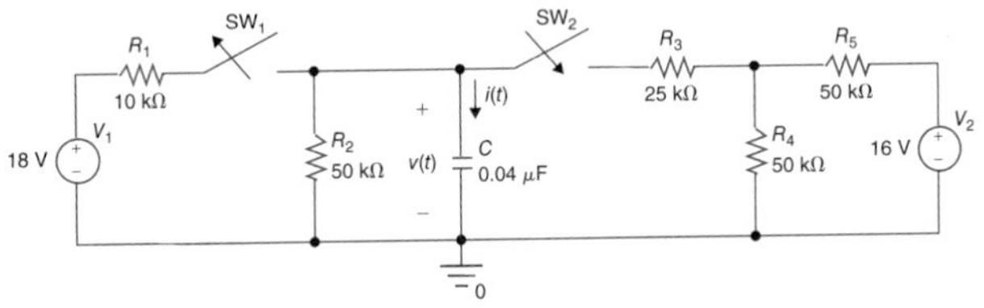

a. Find the initial voltage V_0 across the capacitor at $t = 0$.
b. Find voltage $v(t)$ across the capacitor for $0 \leq t < 4$ ms.
c. Find voltage $v(t)$ across the capacitor for $4 \text{ ms} \leq t$.
d. Plot $v(t)$ for $0 \leq t < 10$ ms.

7.22 **Switch 1 in the circuit shown in Figure P7.22 has been closed for a long time before it is opened at $t = 0$. Switch 2 is closed at $t = 0.2$ s.**

FIGURE P7.22

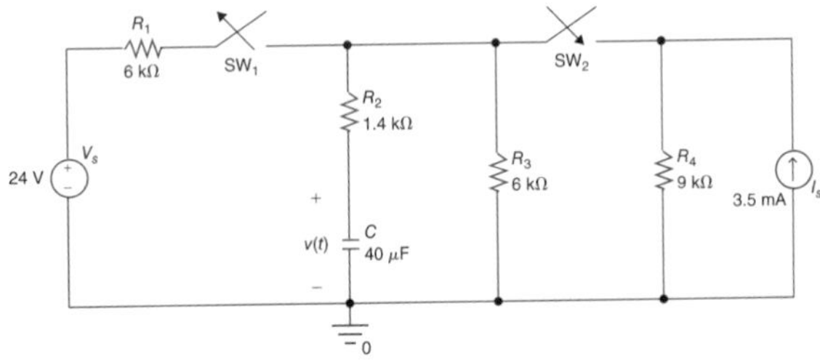

a. Find the initial voltage across the capacitor at $t = 0$; that is, $v(0) = V_0$.
b. Find the voltage $v(t)$ across the capacitor for $0 \le t < 0.2$ s.
c. Find the voltage $v(t)$ across the capacitor at $t = 0.2$ s.
d. Find the voltage $v(t)$ across the capacitor for $0.2 \le t$.

7.23 **Switch 1 in the circuit shown in Figure P7.23 has been closed for a long time before it is opened at $t = 0$. Switch 2 is closed at $t = 0$ right after switch 1 is opened.**

FIGURE P7.23

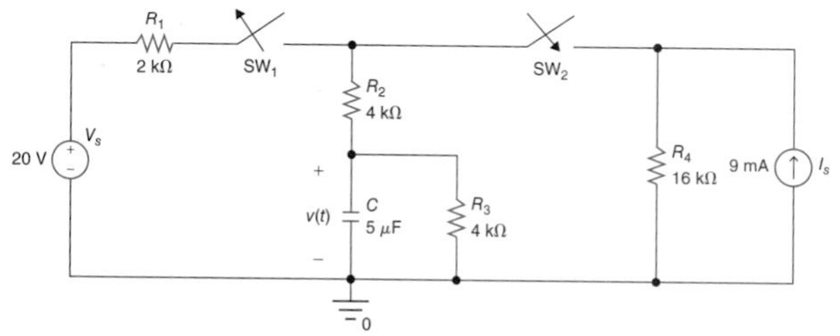

a. Find the initial voltage $v(0) = V_0$ across the capacitor at $t = 0$.
b. Find the final voltage $v(\infty) = V_f$ across the capacitor at $t = \infty$.
c. Find the resistance R_{eq} seen from the capacitor after deactivating the current source for $t \ge 0$.
d. Find the voltage $v(t)$, $t \ge 0$, across the capacitor.

7.24 **Switch 1 in the circuit shown in Figure P7.24 has been closed for a long time before it is opened at $t = 0^-$. Switch 2 is closed at $t = 0$.**

FIGURE P7.24

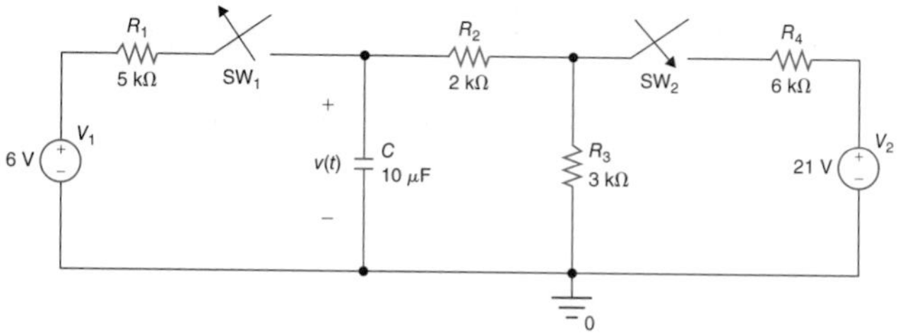

a. Find the initial voltage $v(0^-) = V_0$ across the capacitor at $t = 0$.
b. Find the final voltage $v(\infty) = V_f$ across the capacitor at $t = \infty$.
c. Find resistance R_{eq} seen from the capacitor after deactivating voltage source V_2 for $t \ge 0$.
d. Find voltage $v(t)$, $t \ge 0$, across the capacitor.

7.25 **Switch 1 in the circuit shown in Figure P7.25 has been closed for a long time before it is opened at $t = 0$. Switch 2 is closed at $t = 0.12$ s.**

FIGURE P7.25

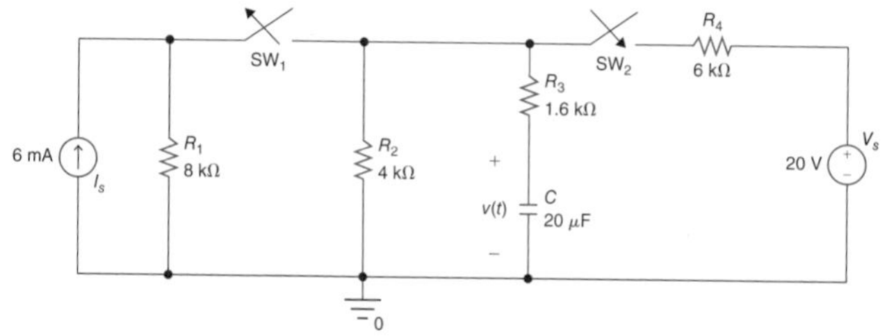

a. Find the initial voltage across the capacitor at $t = 0$; that is, $v(0) = V_0$.
b. Find voltage $v(t)$ across the capacitor for $0 \le t < 0.12$ s.
c. Find voltage $v(t)$ across the capacitor at $t = 0.12$ s.
d. Find voltage $v(t)$ across the capacitor for $0.12 \le t$.

Natural Response of *RL* Circuit

7.26 **The initial current through the inductor at $t = 0$ in the circuit shown in Figure P7.26 is 5 mA.**

FIGURE P7.26

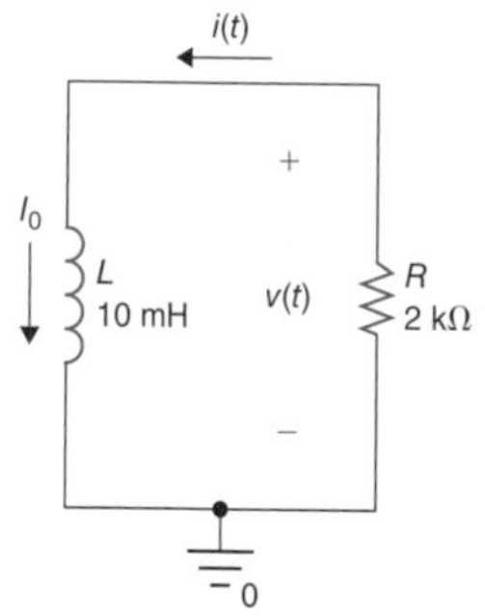

a. Find the current $i(t)$, $t \ge 0$, through the inductor and plot $i(t)$.
b. Find voltage $v(t)$, $t \ge 0$, across the inductor and plot $v(t)$.

7.27 **The initial current through the inductor at $t = 0$ in the circuit shown in Figure P7.27 is $i(0) = I_0 = 5$ mA.**

FIGURE P7.27

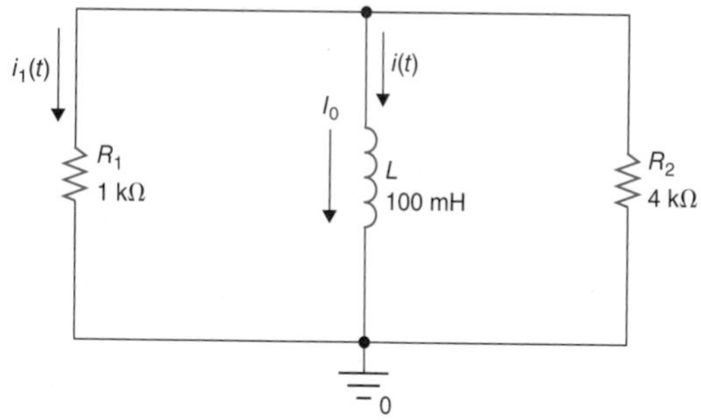

a. Find current $i(t)$, $t \geq 0$, through the inductor.
b. Find current $i_1(t)$, $t \geq 0$, through resistor R_1.

7.28 **The initial current through the inductor at $t = 0$ in the circuit shown in Figure P7.28 is $i(0) = I_0 =$ 8 mA.**

FIGURE P7.28

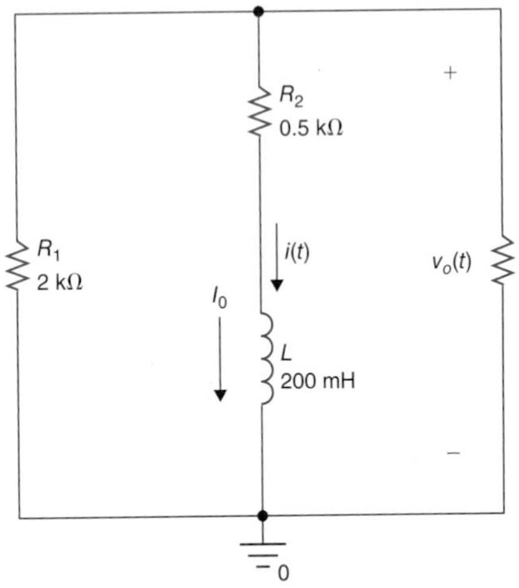

a. Find current $i(t)$, $t \geq 0$, through the inductor.
b. Find voltage $v_o(t)$, $t \geq 0$, across resistor R_3.

7.29 **The initial current through the inductor at $t = 0$ in the circuit shown in Figure P7.29 is 1 mA.**

FIGURE P7.29

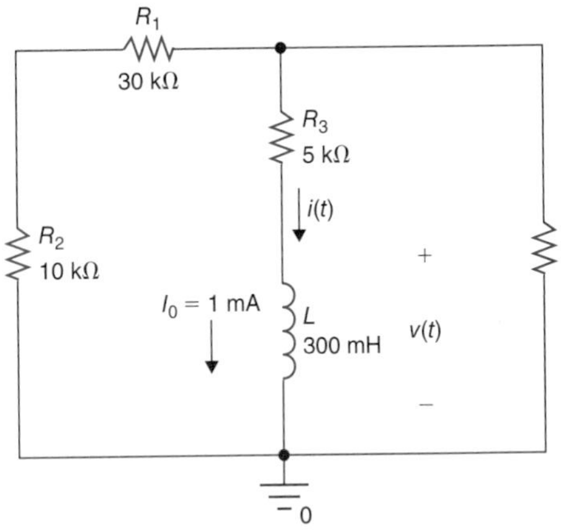

a. Find current $i(t)$, $t \geq 0$, through the inductor and plot $i(t)$.
b. Find voltage $v(t)$, $t \geq 0$, across the inductor and plot $v(t)$.

7.30 **The initial current through the inductor at $t = 0$ in the circuit shown in Figure P7.30 is 0.2 mA.**

a. Find current $i(t)$, $t \geq 0$, through the inductor and plot $i(t)$.
b. Find voltage $v(t)$, $t \geq 0$, across the inductor and plot $v(t)$.

FIGURE P7.30

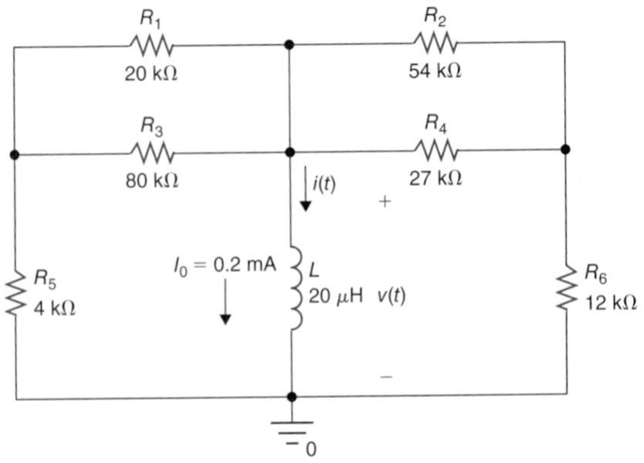

7.31 **The switch in the circuit shown in Figure P7.31 has been closed for a long time before it is opened at $t = 0$.**

FIGURE P7.31

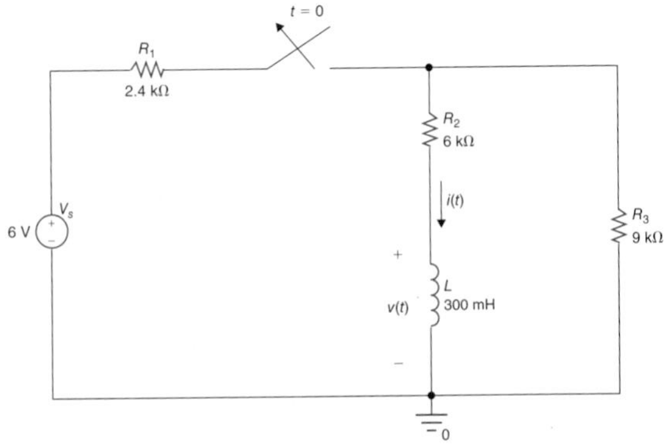

a. Find the initial current $i(0) = I_0$ through the inductor at $t = 0$.
b. After switch is opened, find the resistance seen from the inductor.
c. Find the time constant τ.
d. Find current $i(t)$ through the inductor for $t \geq 0$.

7.32 **The initial current through the inductor at $t = 0$ in the circuit shown in Figure P7.32 is 1 mA.**

FIGURE P7.32

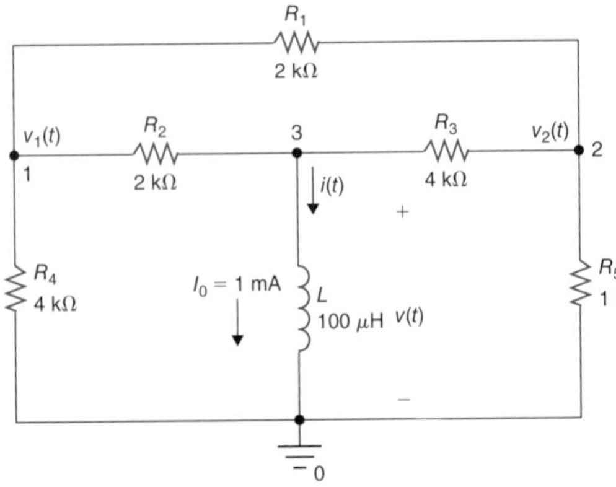

a. Write a node equation at node 1 by summing the currents away from node 1. Notice that the voltage at node 3 is given by $v_3(t) = v(t)$.
b. Write a node equation at node 2 by summing the currents away from node 2. Notice that the voltage at node 3 is given by $v_3(t) = v(t)$.
c. Solve the two node equations from (a) and (b) to express $v_1(t)$ as a function of $v(t)$, and $v_2(t)$ as a function of $v(t)$.
d. Write a node equation at node 3 by summing the currents away from node 3. Use the results from (c) to simplify the equation as a first-order differential equation of $i(t)$.
e. Solve the differential equation to find the current $i(t)$, $t \geq 0$, through the inductor and plot $i(t)$.
f. Find voltage $v(t)$, $t \geq 0$, across the inductor and plot $v(t)$.
g. Find voltage $v_1(t)$, $t \geq 0$.
h. Find voltage $v_2(t)$, $t \geq 0$.

7.33 **Find $i(t)$ for the circuit shown in Figure P7.32 by finding the Thévenin resistance using the test voltage method.**

7.34 **The switch in the circuit shown in Figure P7.34 has been closed for a long time before it is opened at $t = 0$.**

FIGURE P7.34

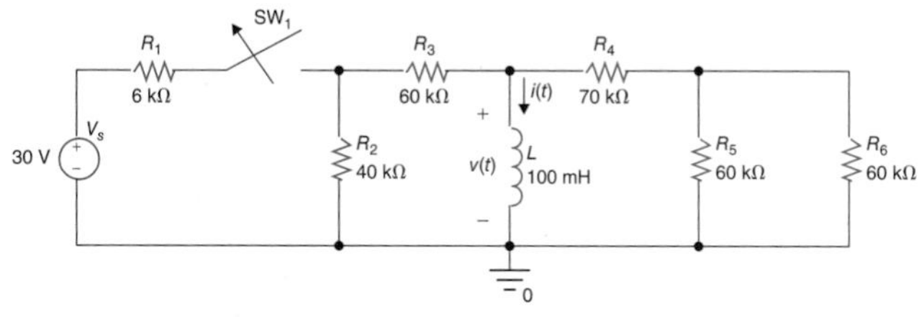

a. Find the initial current I_0 through the inductor at $t = 0$.
b. After switch is opened, find the resistance seen from the inductor.
c. Find the time constant τ.
d. Find the current $i(t)$ through the inductor for $t \geq 0$ and plot $i(t)$ for $t \geq 0$.

7.35 **The switch in the circuit shown in Figure P7.35 has been closed for a long time before it is opened at $t = 0$.**

a. Find the initial current $i(0) = I_0$ through the inductor at $t = 0$.
b. After switch is opened, find the resistance seen from the inductor.
c. Find time constant τ.
d. Find the current $i(t)$ through the inductor for $t \geq 0$.

FIGURE P7.35

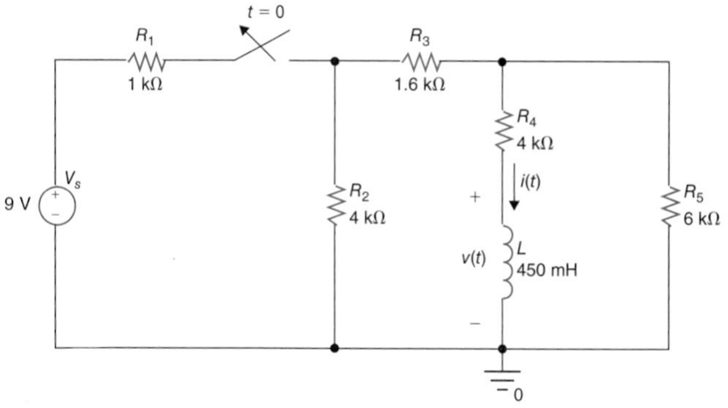

Step Response of *RL* Circuit

7.36 **The initial current through the inductor at $t = 0$ in the circuit shown in Figure P7.36 is 1 mA. The voltage V_s is applied at $t = 0$; that is, $V_s = 10\,u(t)$ V.**

FIGURE P7.36

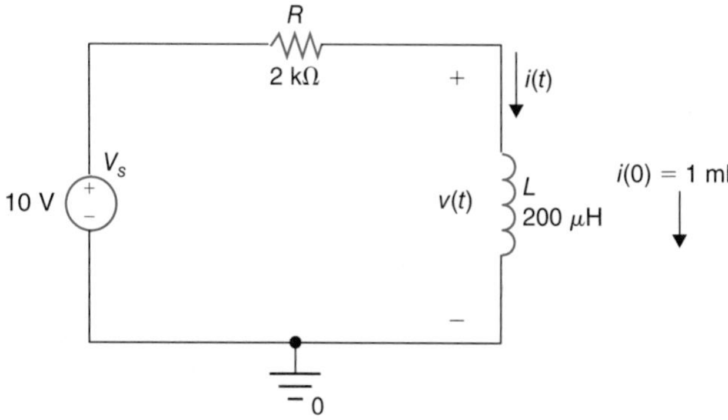

a. Find current $i(t)$, $t \geq 0$, through the inductor and plot $i(t)$.
b. Find voltage $v(t)$, $t \geq 0$, across the inductor and plot $v(t)$.

7.37 **The initial current through the inductor at $t = 0$ in the circuit shown in Figure P7.37 is $i(0) = I_0 =$ 1 mA. The current I_s is applied at $t = 0$; that is, $I_s = 10\,u(t)$ mA. Find the current $i(t)$, $t \geq 0$, through the inductor.**

FIGURE P7.37

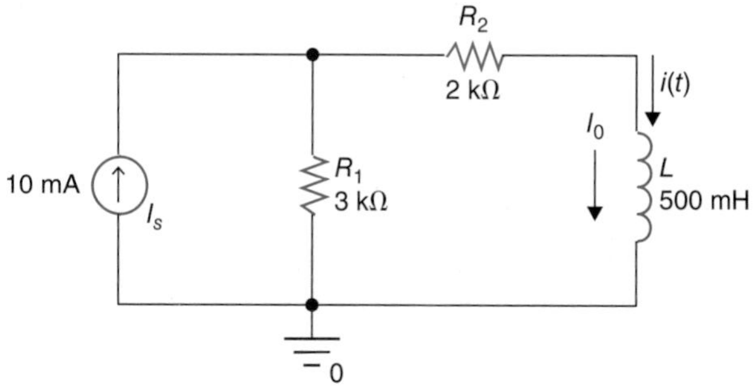

7.38 **The initial current through the inductor at $t = 0$ in the circuit shown in Figure P7.38 is $i(0) = I_0 = 1$ mA. Voltage V_s is applied at $t = 0$; that is, $V_s = 8\,u(t)$ V. Find the current $i(t)$, $t \geq 0$, through the inductor.**

FIGURE P7.38

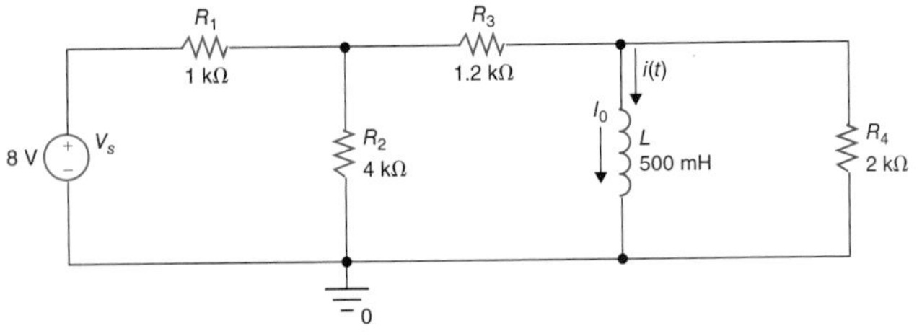

7.39 **The initial current through the inductor at $t = 0$ in the circuit shown in Figure P7.39 is $i(0) = I_0 = 0.5$ mA. Current I_s is applied at $t = 0$; that is, $I_s = 6\,u(t)$ mA. Find current $i(t)$, $t \geq 0$, through the inductor.**

FIGURE P7.39

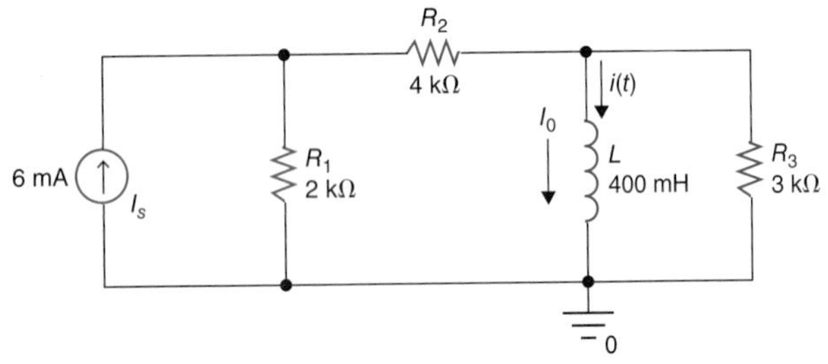

7.40 **The initial current through the inductor at $t = 0$ in the circuit shown in Figure P7.40 is $i(0) = I_0 = 0.7$ mA. Voltage V_s is applied at $t = 0$; that is, $V_s = 10\,u(t)$ V. Find current $i(t)$, $t \geq 0$, through the inductor.**

FIGURE P7.40

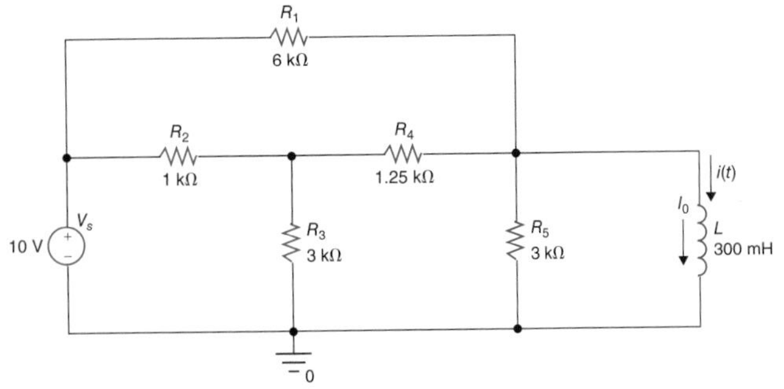

7.41 **The initial current through the inductor at $t = 0$ in the circuit shown in Figure P7.41 is 2 mA. Current I_s is applied at $t = 0$; that is, $I_s = 2.5\,u(t)$ mA.**

FIGURE P7.41

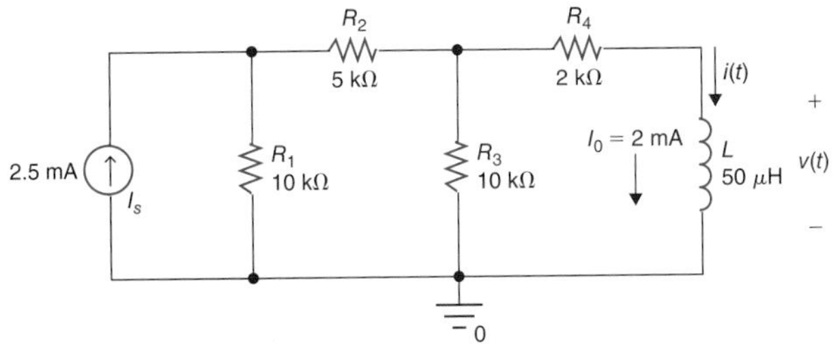

a. Find current $i(t)$, $t \geq 0$, through the inductor and plot $i(t)$.
b. Find voltage $v(t)$, $t \geq 0$, across the inductor and plot $v(t)$.

7.42 **The initial current through the inductor at $t = 0$ in the circuit shown in Figure P7.42 is 0.2 mA. Voltage V_s is applied at $t = 0$; that is, $V_s = 30\,u(t)$ V.**

FIGURE P7.42

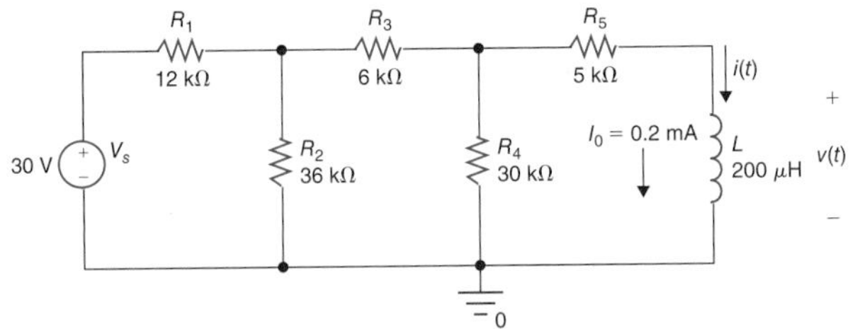

a. Find current $i(t)$, $t \geq 0$, through the inductor and plot $i(t)$.
b. Find voltage $v(t)$, $t \geq 0$, across the inductor and plot $v(t)$.

7.43 **The initial current through the inductor at $t = 0$ in the circuit shown in Figure P7.43 is $I_0 = 4$ mA.**

FIGURE P7.43

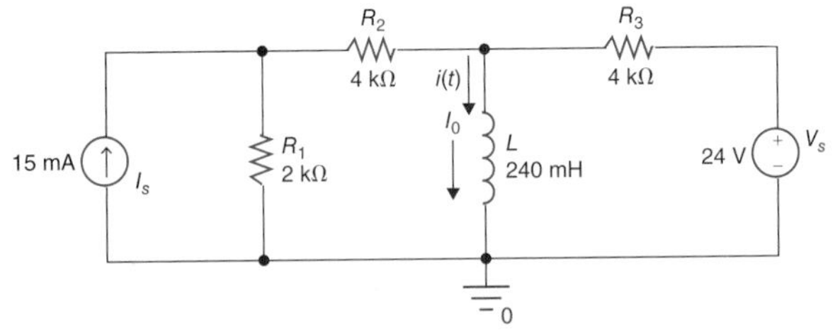

a. Find the final value of current $i(\infty)$ through the inductor at $t = \infty$.
b. Find resistance R_{eq} seen from the inductor after deactivating the sources.
c. Find time constant τ.
d. Find the current $i(t)$ through the inductor for $t \geq 0$.

7.44 **Switch 1 in the circuit shown in Figure P7.44 has been closed for a long time before it is opened at $t = 0$. Switch 2 is closed at $t = 20\ \mu s$.**

FIGURE P7.44

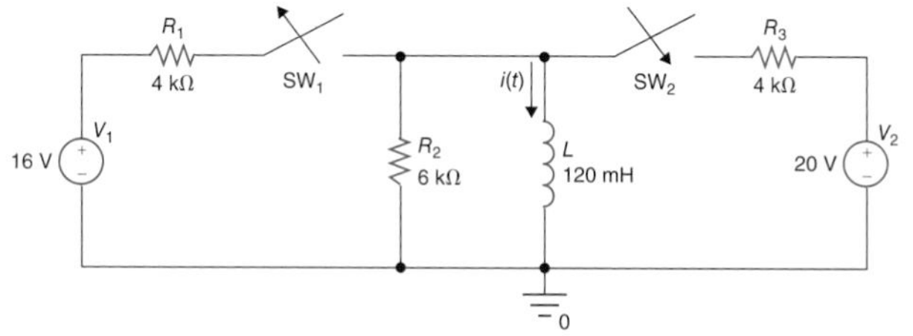

a. Find the initial current through the inductor at $t = 0$; that is, $i(0) = I_0$.
b. Find current $i(t)$ through the inductor for $0 \le t < 20\ \mu s$.
c. Find current $i(t)$ through the inductor at $t = 20\ \mu s$.
d. Find current $i(t)$ through the inductor for $20\ \mu s \le t$.

7.45 **In the circuit shown in Figure P7.45, switch 1 is opened at $t = 0$, and switch 2 is closed at $t = 10\ \mu s$.**

FIGURE P7.45

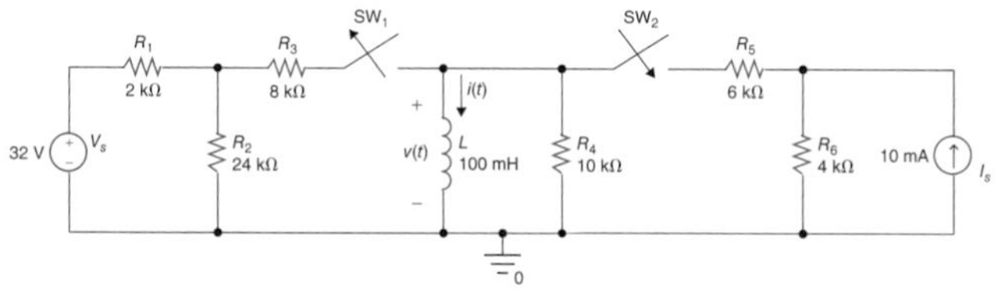

a. Find the initial current I_0 through the inductor at time $t = 0$.
b. Find current $i(t)$ through the inductor for $0 \le t < 10\ \mu s$.
c. Find current $i(t)$ through the inductor for $10\ \mu s \le t$.
d. Plot current $i(t)$ through the inductor for $0 \le t < 80\ \mu s$.

7.46 **Switch 1 in the circuit shown in Figure P7.46 has been closed for a long time before it is opened at $t = 0$. Switch 2 is closed at $t = 40\ \mu s$.**

FIGURE P7.46

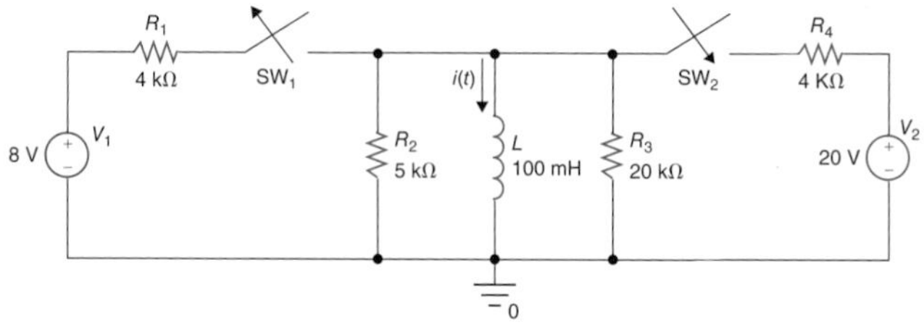

a. Find the initial current through the inductor at $t = 0$; that is, $i(0) = I_0$.
b. Find current $i(t)$ through the inductor for $0 \le t < 40\ \mu s$.
c. Find current $i(t)$ through the inductor at $t = 40\ \mu s$.
d. Find current $i(t)$ through the inductor for $40\ \mu s \le t$.

7.47 **The initial current through the inductor at $t = 0$ in the circuit shown in Figure P7.47 is $I_0 = 1$ mA.**

FIGURE P7.47

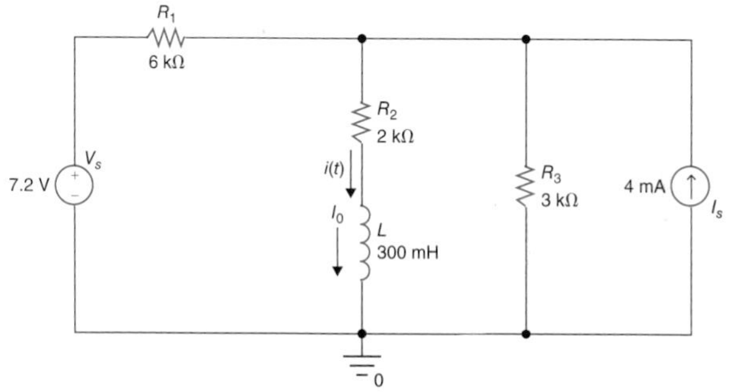

a. Find the final value of the current $i(\infty)$ through the inductor at $t = \infty$.
b. Find resistance R_{eq} seen from the inductor after deactivating the sources.
c. Find time constant τ.
d. Find the current $i(t)$ through the inductor for $t \ge 0$.

General First Order

7.48 **The initial voltage across the capacitor in the circuit shown in Figure P7.48 is given by $v(0) = 2$ V. Find the voltage $v(t)$ across the capacitor for $t \ge 0$. The input signal is $v_s(t) = 12t\ u(t)$.**

FIGURE P7.48

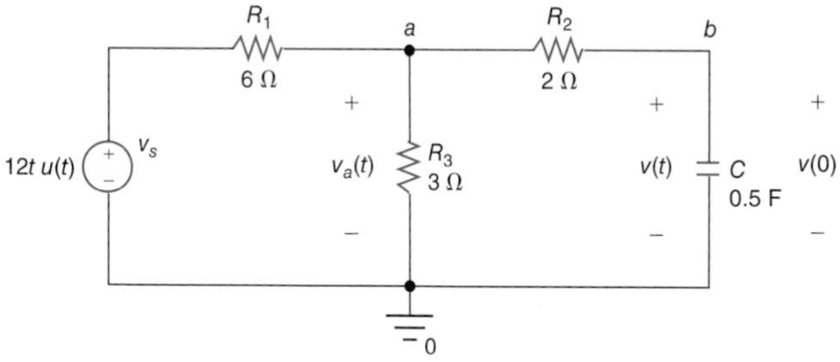

a. Write a node equation at node a by summing the currents leaving node a.
b. Write a node equation at node b by summing the currents leaving node b.
c. Find the first-order differential equation as a function of $v(t)$.
d. Solve the differential equation for $v(t)$, $t \ge 0$.

7.49 **For the circuit shown in Figure P7.49,**

FIGURE P7.49

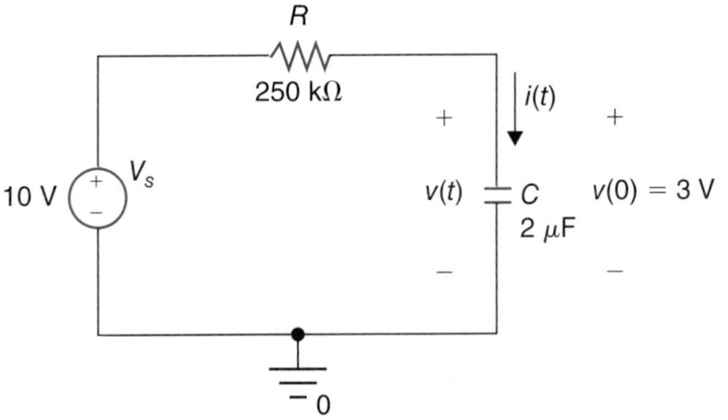

a. find the differential equation of the voltage $v(t)$ across the capacitor.
b. find $v(t)$ and plot $v(t)$.
c. design a circuit using an op amp integrator to implement the differential equation.
d. verify your design via a PSpice simulation.

7.50 **For the circuit shown in Figure P7.50,**

FIGURE P7.50

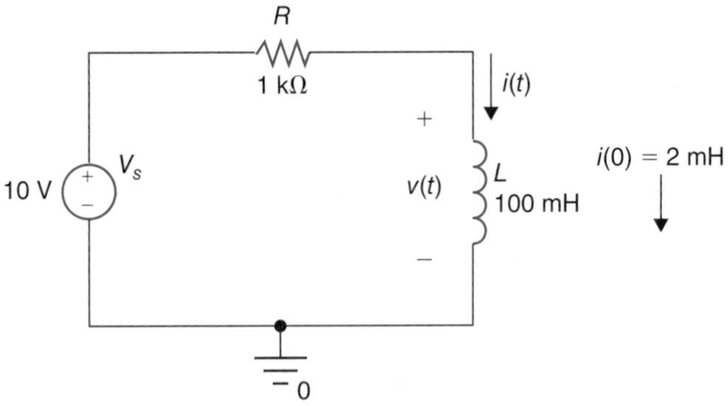

a. find the differential equation of the current $i(t)$ through the inductor.
b. find $i(t)$ and plot $i(t)$.
c. design a circuit using an op amp integrator to implement the differential equation.
d. verify your design by a PSpice simulation.

Chapter 8

RLC Circuits

8.1 Introduction

A circuit with one energy storage element (either a capacitor or an inductor) can be analyzed by solving first-order differential equations, as shown in Chapter 7. In this chapter, we analyze circuits with two energy storage elements, an inductor and a capacitor, or two capacitors, or two inductors, by solving second-order differential equations without and with the input signals.

If there is no input signal, the response of the circuit is called the *zero input response.* The zero input response is also called *source free response, natural response,* and *transient response.* The initial conditions on the energy storage elements cause the zero input response. Once the initial energy is exhausted, the response becomes zero. Depending on the coefficients of the differential equation called the *homogeneous equation,* we have three responses: overdamped, critically damped, and underdamped. If the response is underdamped, the response displays oscillation.

If there are inputs, the solution of the differential equation is the sum of the complementary solution and the particular solution. The complementary solution is the solution to the homogeneous equation, which is the differential equation without the input. Depending on the coefficients, the complementary solution can be overdamped, critically damped, or underdamped. The particular solution is the solution to the entire differential equation including the input. The form of the particular solution is similar to the input.

8.2 Zero Input Response of Second-Order Differential Equations

The second-order differential equation with constant coefficients and zero input can be written as

$$c_2 \frac{d^2 v(t)}{dt^2} + c_1 \frac{dv(t)}{dt} + c_0 v(t) = 0 \tag{8.1}$$

Assume that the second-order differential equation has initial conditions $v(0) = V_0$ and $\frac{dv(0)}{dt} = Dv_0$. Finding the solution of a differential equation that satisfies the initial conditions is called an **initial value problem**. If Equation (8.1) is divided by c_2, we obtain

$$\frac{d^2v(t)}{dt^2} + \frac{c_1}{c_2}\frac{dv(t)}{dt} + \frac{c_0}{c_2}v(t) = 0 \tag{8.2}$$

Let

$$a_2 = 1,\ a_1 = \frac{c_1}{c_2},\ a_0 = \frac{c_0}{c_2}$$

Then, Equation (8.2) becomes

$$\frac{d^2v(t)}{dt^2} + a_1\frac{dv(t)}{dt} + a_0v(t) = 0 \tag{8.3}$$

A differential equation like Equation (8.3) that does not have an input signal is called a **homogeneous differential equation**. The solution to the homogeneous differential equation is called **complementary solution**. Let the solution to the differential equation given by Equation (8.3) be

$$Ae^{st}$$

Substitution of this proposed solution to Equation (8.3) yields

$$s^2Ae^{st} + a_1sAe^{st} + a_0Ae^{st} = (s^2 + a_1s + a_0)Ae^{st} = 0 \tag{8.4}$$

Since Ae^{st} is not zero for all t, $s^2 + a_1s + a_0$ must be zero. Thus, we have

$$s^2 + a_1s + a_0 = 0 \tag{8.5}$$

Equation (8.5) is called a **characteristic equation** (also called an **auxiliary equation**). From the quadratic formula, the two roots of the characteristic equation are given by

$$\begin{aligned} s_1 &= \frac{-a_1 + \sqrt{a_1^2 - 4a_0}}{2} = -\frac{a_1}{2} + \sqrt{\left(\frac{a_1}{2}\right)^2 - a_0} \\ s_2 &= \frac{-a_1 - \sqrt{a_1^2 - 4a_0}}{2} = -\frac{a_1}{2} - \sqrt{\left(\frac{a_1}{2}\right)^2 - a_0} \end{aligned} \tag{8.6}$$

Let

$$\alpha = \frac{a_1}{2},\ \omega_0 = \sqrt{a_0} \tag{8.7}$$

Then, Equation (8.6) can be rewritten as

$$s_1 = -\alpha + \sqrt{\alpha^2 - \omega_0^2},\ s_2 = -\alpha - \sqrt{\alpha^2 - \omega_0^2} \tag{8.8}$$

From Equation (8.7), we have

$$a_1 = 2\alpha,\ a_0 = \omega_0^2 \tag{8.9}$$

FIGURE 8.1

Roots of the characteristic equation: (a) overdamped; (b) critically damped; (c) underdamped.

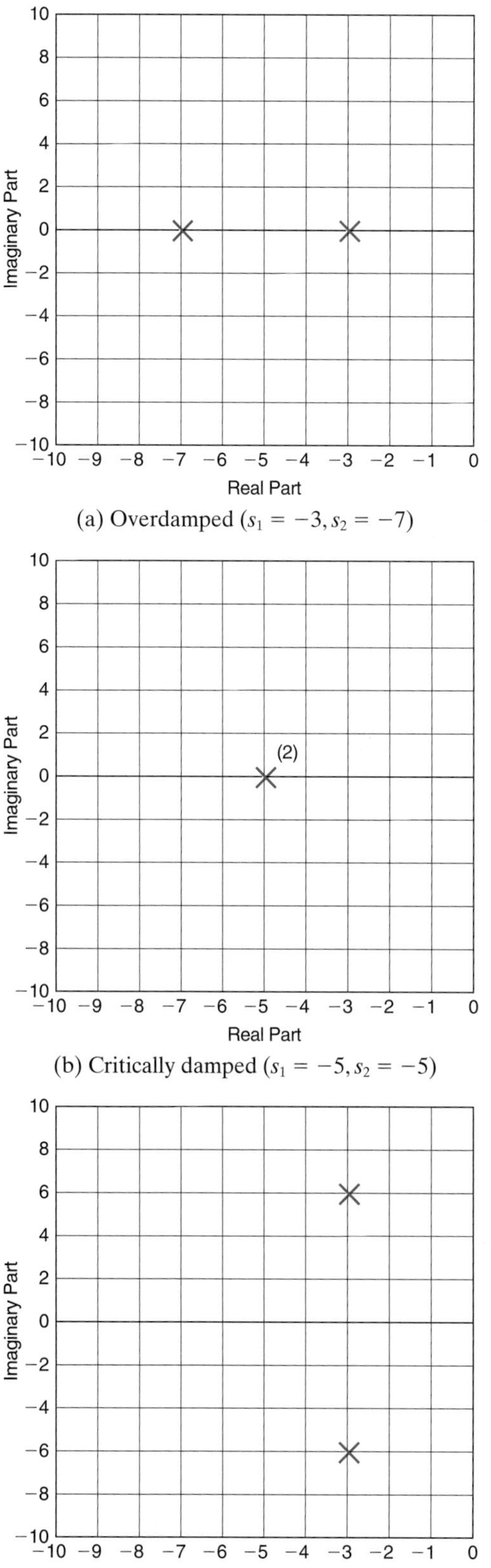

(a) Overdamped ($s_1 = -3, s_2 = -7$)

(b) Critically damped ($s_1 = -5, s_2 = -5$)

(c) Underdamped ($s_1 = -3 + 6j, s_2 = -3 - 6j$)

The parameter α is called **neper frequency** or **damping coefficient** and is measured in nepers per second (Np/s). The parameter ω_0 is called **resonant frequency** or **undamped natural frequency** and is measured in radians per second (rad/s). A **damping factor**, also called **damping ratio**, is defined as

$$\zeta = \frac{\alpha}{\omega_0}$$

The $-\alpha$ term in the roots of the characteristic equation introduces exponential decay $e^{-\alpha t}$ in the response $v(t)$. Since $\alpha > 0$, as time progresses, the value of $e^{-\alpha t}$ decreases exponentially, attenuating the other part of $v(t)$ resulting from $\pm\sqrt{\alpha^2 - \omega_0^2}$. The term $-\alpha$ damps the signal $v(t)$ as time progresses. This is why α is called the *damping coefficient*. The larger the value of α, the faster the attenuation of the response $v(t)$. Notice that *neper* is a logarithmic unit of measurement. Let V_1 be the measurement of voltage and V_0 be the reference voltage. Then, $\ln(V_1/V_0)$ is the measurement of voltage V_1 in nepers. Let $v(t) = Ae^{-\alpha t}$. Then, $-\alpha t = \ln[v(t)/A]$ and $\alpha = -\ln[v(t)/A]/t$. The parameter α has a unit of nepers per second. Since the neper is unitless, α has the same unit as ω_0, which is rad/s. The damping factor is unitless.

Depending on the relative magnitude of α and ω_0, we have three cases for the solution to the differential equation given by Equation (8.3).

8.2.1 CASE 1: OVERDAMPED ($\alpha > \omega_0$ or $a_1 > 2\sqrt{a_0}$ or $\zeta > 1$)

If $\alpha > \omega_0$, the two roots of the characteristic equation given by

$$s_1 = -\alpha + \sqrt{\alpha^2 - \omega_0^2} = -\alpha + \omega_0\sqrt{\zeta^2 - 1}$$
$$s_2 = -\alpha - \sqrt{\alpha^2 - \omega_0^2} = -\alpha - \omega_0\sqrt{\zeta^2 - 1}$$

are real and distinct. The response is called **overdamped**. Since $0 < \sqrt{\alpha^2 - \omega_0^2} < \alpha$, s_1 is negative. Also, s_2 is negative. Figure 8.1(a) shows the two roots of the characteristic equation for an overdamped case in the complex s-plane. The roots are called **poles**.

Since $\alpha > \omega_0$, the two real roots of the characteristic equation are given by Equation (8.8). The two solutions for the homogeneous differential equation [Equation (8.3)] are $e^{s_1 t}$ and $e^{s_2 t}$. From the superposition principle, the linear combination of $e^{s_1 t}$ and $e^{s_2 t}$ is also a solution to the homogeneous differential equation [Equation (8.3)]. Thus, the solution $v(t)$ can be written as

$$v(t) = A_1 e^{s_1 t} + A_2 e^{s_2 t} \tag{8.10}$$

where the coefficients A_1 and A_2 of the linear combination are to be determined by initial conditions. One of the initial conditions is $v(0) = V_0$. Setting $t = 0$ in Equation (8.10), we have

$$v(0) = A_1 e^{s_1 0} + A_2 e^{s_2 0} = A_1 + A_2 = V_0 \tag{8.11}$$

Taking the derivative of Equation (8.10), we obtain

$$\frac{dv(t)}{dt} = A_1 s_1 e^{s_1 t} + A_2 s_2 e^{s_2 t} \tag{8.12}$$

Setting $t = 0$ in Equation (8.12), we have

$$\frac{dv(0)}{dt} = A_1 s_1 e^{s_1 0} + A_2 s_2 e^{s_2 0} = A_1 s_1 + A_2 s_2 = Dv_0 \tag{8.13}$$

Solving Equation (8.11) for A_2, we obtain

$$A_2 = V_0 - A_1$$

Substituting A_2 into Equation (8.13), we get

$$A_1 s_1 + (V_0 - A_1) s_2 = Dv_0$$

Solving for A_1, we obtain

$$A_1 = \frac{Dv_0 - V_0 s_2}{s_1 - s_2} = \frac{V_0 s_2 - Dv_0}{s_2 - s_1}$$

From $A_2 = V_0 - A_1$, we get

$$A_2 = \frac{Dv_0 - V_0 s_1}{s_2 - s_1}$$

Alternatively, from Equations (8.11) and (8.13), the constants A_1 and A_2 are found using Cramer's rule:

$$A_1 = \frac{\begin{vmatrix} V_0 & 1 \\ Dv_0 & s_2 \end{vmatrix}}{\begin{vmatrix} 1 & 1 \\ s_1 & s_2 \end{vmatrix}} = \frac{V_0 s_2 - Dv_0}{s_2 - s_1} \tag{8.14}$$

$$A_2 = \frac{\begin{vmatrix} 1 & V_0 \\ s_1 & Dv_0 \end{vmatrix}}{\begin{vmatrix} 1 & 1 \\ s_1 & s_2 \end{vmatrix}} = \frac{Dv_0 - V_0 s_1}{s_2 - s_1} \tag{8.15}$$

The solution can also be written as

$$v(t) = A_1 e^{s_1 t} + A_2 e^{s_2 t} = \frac{V_0 s_2 - Dv_0}{s_2 - s_1} e^{s_1 t} + \frac{Dv_0 - V_0 s_1}{s_2 - s_1} e^{s_2 t} \tag{8.16}$$

Substituting s_1 and s_2 from Equation (8.8), we obtain

$$\begin{aligned}
v(t) &= \frac{V_0 s_2 - Dv_0}{-\alpha - \sqrt{\alpha^2 - \omega_0^2} + \alpha - \sqrt{\alpha^2 - \omega_0^2}} e^{(-\alpha + \sqrt{\alpha^2 - \omega_0^2})t} \\
&\quad + \frac{Dv_0 - V_0 s_1}{-\alpha - \sqrt{\alpha^2 - \omega_0^2} + \alpha - \sqrt{\alpha^2 - \omega_0^2}} e^{(-\alpha - \sqrt{\alpha^2 - \omega_0^2})t} \\
&= \frac{V_0 s_2 - Dv_0}{-2\sqrt{\alpha^2 - \omega_0^2}} e^{(-\alpha + \sqrt{\alpha^2 - \omega_0^2})t} + \frac{Dv_0 - V_0 s_1}{-2\sqrt{\alpha^2 - \omega_0^2}} e^{(-\alpha - \sqrt{\alpha^2 - \omega_0^2})t} \\
&= \frac{Dv_0}{2\sqrt{\alpha^2 - \omega_0^2}} e^{-\alpha t}\left(e^{\sqrt{\alpha^2 - \omega_0^2}t} - e^{-\sqrt{\alpha^2 - \omega_0^2}t}\right) + \frac{\alpha V_0}{2\sqrt{\alpha^2 - \omega_0^2}} e^{-\alpha t}\left(e^{\sqrt{\alpha^2 - \omega_0^2}t} - e^{-\sqrt{\alpha^2 - \omega_0^2}t}\right) \\
&\quad + \frac{V_0}{2} e^{-\alpha t}\left(e^{\sqrt{\alpha^2 - \omega_0^2}t} + e^{-\sqrt{\alpha^2 - \omega_0^2}t}\right)
\end{aligned}$$

$$= \frac{\alpha V_0 + Dv_0}{\sqrt{\alpha^2 - \omega_0^2}} e^{-\alpha t} \sinh(\sqrt{\alpha^2 - \omega_0^2} t) + V_0 e^{-\alpha t} \cosh(\sqrt{\alpha^2 - \omega_0^2} t)$$

$$= e^{-\alpha t} \left[\frac{\alpha V_0 + Dv_0}{\sqrt{\alpha^2 - \omega_0^2}} \sinh(\sqrt{\alpha^2 - \omega_0^2} t) + V_0 \cosh(\sqrt{\alpha^2 - \omega_0^2} t) \right] \tag{8.17}$$

Due to $e^{-\alpha t}$, the solution $v(t)$ decays exponentially toward zero without oscillation.

8.2.2 CASE 2: CRITICALLY DAMPED ($\alpha = \omega_0$ or $a_1 = 2\sqrt{a_0}$ or $\zeta = 1$)

If $\alpha = \omega_0$, from Equation (8.8), the two roots are given by

$$s_1 = -\alpha = -\frac{a_1}{2}, \quad s_2 = -\alpha = -\frac{a_1}{2} \tag{8.18}$$

The two roots of the characteristic equation are real and equal. The response is called **critically damped**. Figure 8.1(b) shows the two roots of the characteristic equation for a critically damped case in the complex s-plane.

The solutions for the differential equation from these roots are $e^{-\alpha t}$ and $e^{-\alpha t}$. Since the solutions from the two roots are identical, we need to find another independent solution for the differential equation. The second solution is given by $te^{-\alpha t}$. To show that this solution satisfies the differential equation, we substitute $te^{-\alpha t}$ into the differential equation [Equation (8.3)]:

$$\frac{d^2v(t)}{dt^2} + a_1\frac{dv(t)}{dt} + a_0 v(t) = \frac{d^2v(t)}{dt^2} + 2\alpha\frac{dv(t)}{dt} + \alpha^2 v(t)$$
$$= (-\alpha e^{-\alpha t} - \alpha e^{-\alpha t} + \alpha^2 te^{-\alpha t}) + 2\alpha(e^{-\alpha t} - \alpha te^{-\alpha t}) + \alpha^2(te^{-\alpha t}) = 0$$

This proves that $te^{-\alpha t}$ is a solution to the differential equation [Equation (8.3)]. Thus, the solution $v(t)$ for the critically damped case can be written as a linear combination of $e^{-\alpha t}$ and $te^{-\alpha t}$:

$$v(t) = A_1 e^{s_1 t} + A_2 te^{s_2 t} \tag{8.19}$$

Here, the coefficients A_1 and A_2 are to be determined by initial conditions. One of the initial conditions is $v(0) = V_0$. Setting $t = 0$ in Equation (8.19), we have

$$v(0) = A_1 e^{s_1 \times 0} + A_2 0 e^{s_2 \times 0} = A_1 = V_0 \tag{8.20}$$

Taking the derivative of Equation (8.19), we obtain

$$\frac{dv(t)}{dt} = A_1 s_1 e^{s_1 t} + A_2 e^{s_2 t} + A_2 s_2 te^{s_2 t} \tag{8.21}$$

Setting $t = 0$ in Equation (8.21), we have

$$\frac{dv(0)}{dt} = A_1 s_1 e^{s_1 \times 0} + A_2 e^{s_2 \times 0} + A_2 s_2 0 e^{s_2 \times 0} = A_1 s_1 + A_2 = Dv_0 \tag{8.22}$$

From Equations (8.20) and (8.22), constants A_1 and A_2 are found to be

$$A_1 = V_0 \tag{8.23}$$

$$A_2 = Dv_0 - A_1 s_1 = Dv_0 - V_0 s_1 = Dv_0 + V_0 \alpha = Dv_0 + \frac{V_0 a_1}{2} \tag{8.24}$$

The critically damped case is the boundary between the overdamped case without oscillation and underdamped case with oscillation.

8.2.3 CASE 3: UNDERDAMPED ($\alpha < \omega_0$ or $a_1 < 2\sqrt{a_0}$ or $\zeta < 1$)

If $\alpha < \omega_0$, s_1 and s_2 are complex conjugates. The response is called **underdamped.** Let **damped resonant frequency** be

$$\beta = \sqrt{\omega_0^2 - \alpha^2} = \omega_0\sqrt{1 - \zeta^2} \tag{8.25}$$

Then, from Equation (8.8), the two roots of the characteristic equation for the underdamped case are given by

$$s_1 = -\alpha + j\beta, \quad s_2 = -\alpha - j\beta \tag{8.26}$$

If the roots of the characteristic equation are complex conjugates, the response is underdamped. The damping coefficient α is the negative of the real part of the roots and β is the magnitude of the imaginary part of the roots. Figure 8.1(c) shows the two roots of the characteristic equation for an underdamped case in the complex s-plane. The solution $v(t)$ includes $\cos(\beta t)$ and $\sin(\beta t)$ that oscillates with radian frequency β rad/s. The sinusoidal components are multiplied by a term $e^{-\alpha t}$ that forces the sinusoidal components decay exponentially toward zero.

The solution $v(t)$ can be written as

$$v(t) = A_1 e^{s_1 t} + A_2 e^{s_2 t} \tag{8.27}$$

Substitution of Equations (8.26) into Equation (8.27) yields

$$v(t) = A_1 e^{(-\alpha + j\beta)t} + A_2 e^{(-\alpha - j\beta)t} = e^{-\alpha t}(A_1 e^{j\beta t} + A_2 e^{-j\beta t}) \tag{8.28}$$

From Euler's rule, we have

$$e^{j\beta t} = \cos(\beta t) + j\sin(\beta t), \quad e^{-j\beta t} = \cos(\beta t) - j\sin(\beta t) \tag{8.29}$$

Substitution of Equation (8.29) into Equation (8.28) yields

$$\begin{aligned} v(t) &= e^{-\alpha t}[A_1\cos(\beta t) + jA_1\sin(\beta t) + A_2\cos(\beta t) - jA_2\sin(\beta t)] \\ &= e^{-\alpha t}[(A_1 + A_2)\cos(\beta t) + (jA_1 - jA_2)\sin(\beta t)] \end{aligned} \tag{8.30}$$

Let

$$B_1 = A_1 + A_2, \quad B_2 = jA_1 - jA_2 \tag{8.31}$$

Then, Equation (8.30) becomes

$$v(t) = e^{-\alpha t}\,[B_1\cos(\beta t) + B_2\sin(\beta t)] \tag{8.32}$$

where B_1 and B_2 are constants to be determined by initial conditions. One of the initial conditions is $v(0) = V_0$. Setting $t = 0$ in Equation (8.32), we have

$$v(0) = e^{-\alpha \times 0}[B_1\cos(\beta \times 0) + B_2\sin(\beta \times 0)] = B_1 = V_0 \tag{8.33}$$

Taking the derivative of Equation (8.32), we obtain

$$\begin{aligned} \frac{dv(t)}{dt} &= -\alpha e^{-\alpha t}B_1\cos(\beta t) - \beta e^{-\alpha t}B_1\sin(\beta t) \\ &\quad - \alpha e^{-\alpha t}B_2\sin(\beta t) + \beta e^{-\alpha t}B_2\cos(\beta t) \\ &= e^{-\alpha t}[(-\alpha B_1 + \beta B_2)\cos(\beta t) + (-\beta B_1 - \alpha B_2)\sin(\beta t)] \end{aligned} \tag{8.34}$$

Setting $t = 0$ in Equation (8.34), we have

$$\frac{dv(0)}{dt} = -\alpha B_1 + \beta B_2 = Dv_0 \tag{8.35}$$

Substitution of $B_1 = V_0$ into Equation (8.35) results in

$$B_2 = \frac{Dv_0 + \alpha B_1}{\beta} = \frac{Dv_0 + \alpha V_0}{\beta} \tag{8.36}$$

Notice that substitution of Equations (8.14) and (8.15) into Equation (8.31) results in B_1 and B_2 given by Equations (8.33) and (8.36):

$$B_1 = A_1 + A_2 = \frac{Dv_0 - V_0 s_2}{s_1 - s_2} + \frac{V_0 s_1 - Dv_0}{s_1 - s_2} = \frac{V_0(s_1 - s_2)}{s_1 - s_2} = V_0$$

$$B_2 = jA_1 - jA_2 = j\frac{Dv_0 - V_0 s_2}{s_1 - s_2} - j\frac{V_0 s_1 - Dv_0}{s_1 - s_2} = j\frac{Dv_0 - V_0 s_2}{j2\beta} - j\frac{V_0 s_1 - Dv_0}{j2\beta}$$

$$= \frac{2Dv_0 - V_0(s_1 + s_2)}{2\beta} = \frac{2Dv_0 - V_0(-2\alpha)}{2\beta} = \frac{Dv_0 + \alpha V_0}{\beta}$$

If needed, the solution for $v(t)$, given by Equation (8.32), can be rewritten as a single cosine, as shown here:

$$\begin{aligned} v(t) &= e^{-\alpha t}\sqrt{B_1^2 + B_2^2}\left[\frac{B_1}{\sqrt{B_1^2 + B_2^2}}\cos(\beta t) + \frac{B_2}{\sqrt{B_1^2 + B_2^2}}\sin(\beta t)\right] \\ &= e^{-\alpha t}\sqrt{B_1^2 + B_2^2}\,[\cos(\theta)\cos(\beta t) + \sin(\theta)\sin(\beta t)] \\ &= e^{-\alpha t}\sqrt{B_1^2 + B_2^2}\cos(\beta t - \theta) \end{aligned} \tag{8.37}$$

where

$$\theta = \begin{cases} 0, & B_1 \geq 0 \text{ and } B_2 = 0 \\ 180°, & B_1 < 0 \text{ and } B_2 = 0 \\ 90°, & B_1 = 0 \text{ and } B_2 > 0 \\ -90°, & B_1 = 0 \text{ and } B_2 < 0 \\ \tan^{-1}\left(\frac{B_2}{B_1}\right), & B_1 > 0 \text{ and } B_2 \neq 0 \\ 180° + \tan^{-1}\left(\frac{B_2}{B_1}\right), & B_1 < 0 \text{ and } B_2 > 0 \\ -180° + \tan^{-1}\left(\frac{B_2}{B_1}\right), & B_1 < 0 \text{ and } B_2 < 0 \end{cases} \tag{8.38}$$

8.3 Zero Input Response of Series RLC Circuit

FIGURE 8.2

A series RLC circuit.

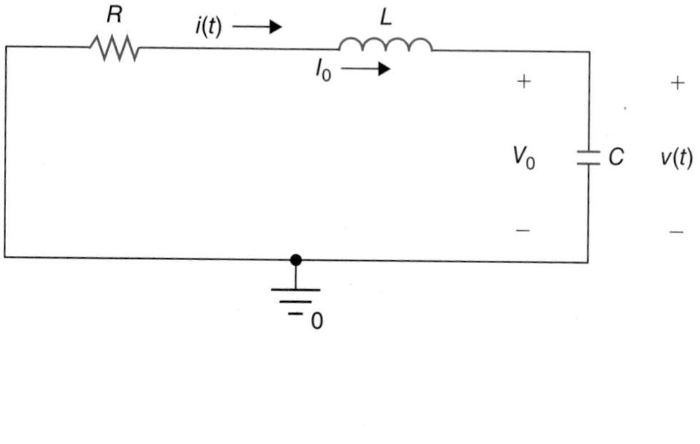

Figure 8.2 shows a series RLC circuit without any independent source.

If there are no energies stored in the inductor and capacitor at $t = 0$, the voltage and current on each element will be zero for $t > 0$. If the voltage across the capacitor at $t = 0$ is not zero, or the current through the inductor at $t = 0$ is not zero, or both are not zero, there will be transient response in the circuit. Let the initial voltage across the capacitor at $t = 0$ be V_0, and the initial current through the inductor at $t = 0$ be I_0. Let $i(t)$ be the mesh current of the series RLC circuit shown in Figure 8.2 and $v(t)$ be the voltage across the capacitor. Then, adding the voltage drops around the mesh in the clockwise direction starting from the resistor, we have

$$R\,i(t) + L\frac{di(t)}{dt} + v(t) = 0 \tag{8.39}$$

The current through the capacitor $i(t)$ is related to the voltage across the capacitor $v(t)$ by

$$i(t) = C\frac{dv(t)}{dt} \tag{8.40}$$

Substituting Equation (8.40) into Equation (8.39), we obtain

$$RC\frac{dv(t)}{dt} + LC\frac{d^2v(t)}{dt^2} + v(t) = 0 \tag{8.41}$$

Rearrangement of Equation (8.41) yields

$$\frac{d^2v(t)}{dt^2} + \frac{R}{L}\frac{dv(t)}{dt} + \frac{1}{LC}v(t) = 0 \tag{8.42}$$

Comparison of Equations (8.3) and (8.42) reveals

$$a_2 = 1,\ a_1 = \frac{R}{L},\ a_0 = \frac{1}{LC} \tag{8.43}$$

Since $\alpha = a_1/2$ and $\omega_0 = \sqrt{a_0}$ from Equation (8.7), we have

$$\alpha = \frac{R}{2L},\ \omega_0 = \frac{1}{\sqrt{LC}} \tag{8.44}$$

The damping factor is given by

$$\zeta = \frac{\alpha}{\omega_0} = \frac{R}{2}\sqrt{\frac{C}{L}}$$

The characteristic equation of the differential equation given by Equation (8.42) is

$$s^2 + a_1 s + a_0 = s^2 + \frac{R}{L}s + \frac{1}{LC} = 0 \tag{8.45}$$

The initial voltage across the capacitor is given by

$$v(0) = V_0 \tag{8.46}$$

From Equation (8.40), we obtain

$$\frac{dv(t)}{dt} = \frac{i(t)}{C} \tag{8.47}$$

Setting $t = 0$, we obtain the second initial condition:

$$Dv_0 = \frac{dv(0)}{dt} = \frac{i(0)}{C} = \frac{I_0}{C} \tag{8.48}$$

Depending on the relative magnitude of α and ω_0, we have three cases, described in the next sections.

8.3.1 CASE 1: OVERDAMPED ($\alpha > \omega_0$ or $a_1 > 2\sqrt{a_0}$ or $\zeta > 1$)

From Equations (8.8), (8.10), (8.14), and (8.15), we have

$$s_1 = -\alpha + \sqrt{\alpha^2 - \omega_0^2},\ s_2 = -\alpha - \sqrt{\alpha^2 - \omega_0^2} \tag{8.49}$$

$$v(t) = A_1 e^{s_1 t} + A_2 e^{s_2 t} \tag{8.50}$$

$$A_1 = \frac{V_0 s_2 - Dv_0}{s_2 - s_1} \tag{8.51}$$

$$A_2 = \frac{Dv_0 - V_0 s_1}{s_2 - s_1} \tag{8.52}$$

where $V_0 = v(0)$, $Dv_0 = \dfrac{dv(0)}{dt}$.

8.3.2 CASE 2: CRITICALLY DAMPED ($\alpha = \omega_0$ or $a_1 = 2\sqrt{a_0}$ or $\zeta = 1$)

From Equations (8.18), (8.19), (8.20), and (8.24), we have

$$s_1 = -\alpha,\ \ s_2 = -\alpha \tag{8.53}$$

$$v(t) = A_1 e^{s_1 t} + A_2 t e^{s_2 t} \tag{8.54}$$

$$A_1 = V_0 \tag{8.55}$$

$$A_2 = Dv_0 + V_0\alpha \tag{8.56}$$

where $V_0 = v(0)$, $Dv_0 = \dfrac{dv(0)}{dt}$.

8.3.3 CASE 3: UNDERDAMPED ($\alpha < \omega_0$ or $a_1 < 2\sqrt{a_0}$ or $\zeta < 1$)

From Equations (8.25), (8.26), (8.32), (8.33), and (8.36), we have

$$\beta = \sqrt{\omega_o^2 - \alpha^2} = \omega_0\sqrt{1 - \zeta^2} \tag{8.57}$$

$$s_1 = -\alpha + j\beta, s_2 = -\alpha - j\beta \tag{8.58}$$

$$v(t) = e^{-\alpha t}[B_1 \cos(\beta t) + B_2 \sin(\beta t)] \tag{8.59}$$

$$B_1 = V_0 \tag{8.60}$$

$$B_2 = \frac{Dv_0 + \alpha V_0}{\beta} \tag{8.61}$$

where $V_0 = v(0)$, $Dv_0 = \dfrac{dv(0)}{dt}$.

EXAMPLE 8.1

Let $R = 900\ \Omega$, $L = 100$ mH, $C = 0.5\ \mu$F, $V_0 = 10$ V, and $I_0 = 20$ mA in the circuit shown in Figure 8.2. Find voltage $v(t)$ across the capacitor, current $i(t)$ through the capacitor, voltage $v_R(t)$ across the resistor, and the voltage $v_L(t)$ across the inductor. Plot $v(t)$, $i(t)$, v_R(t), and $v_L(t)$.

The coefficients are given by

$$a_2 = 1,\ a_1 = \frac{R}{L} = 9000,\ a_0 = \frac{1}{LC} = 2 \times 10^7$$

The neper frequency is given by

$$\alpha = \frac{a_1}{2} = \frac{R}{2L} = 4500 \text{ Np/s}$$

The resonant frequency is

$$\omega_0 = \sqrt{a_0} = \frac{1}{\sqrt{LC}} = 4472.135955 \text{ rad/s}$$

Since $\alpha > \omega_0$, this is case 1, and $v(t)$ is overdamped.

The characteristic equation is

$$s^2 + a_1 s + a_0 = s^2 + \frac{R}{L}s + \frac{1}{LC} = s^2 + 9000s + 2 \times 10^7 = (s + 4000)(s + 5000) = 0$$

The roots of the characteristic equation are given by

$$s_1 = -4000, s_2 = -5000$$

Notice that the roots can also be found from

$$s_1 = -\alpha + \sqrt{\alpha^2 - \omega_0^2} = -4000$$

$$s_2 = -\alpha - \sqrt{\alpha^2 - \omega_0^2} = -5000$$

The differential equation given by Equation (8.42) becomes

$$\frac{d^2v(t)}{dt^2} + 9000\frac{dv(t)}{dt} + 2 \times 10^7 v(t) = 0 \qquad \textbf{(8.62)}$$

From Equation (8.50), the solution to the differential equation given by Equation (8.62) can be written as

$$v(t) = A_1 e^{s_1 t} + A_2 e^{s_2 t} = A_1 e^{-4000t} + A_2 e^{-5000t} \qquad \textbf{(8.63)}$$

Coefficients A_1 and A_2 are found from the initial conditions. Since $v(0) = V_0 = 10$ V, we have

$$v(0) = A_1 e^{-4000\times 0} + A_2 e^{-5000\times 0} = A_1 + A_2 = V_0 = 10 \qquad \textbf{(8.64)}$$

continued

Example 8.1 continued

From Equation (8.40), we have

$$\frac{dv(t)}{dt} = \frac{i(t)}{C} \tag{8.65}$$

At $t = 0$, Equation (8.65) becomes

$$\frac{dv(0)}{dt} = \frac{i(0)}{C} = \frac{I_0}{C} = \frac{20 \times 10^{-3}}{5 \times 10^{-7}} = 40{,}000 \tag{8.66}$$

Taking the derivative of Equation (8.63), we obtain

$$\frac{dv(t)}{dt} = -4000A_1e^{-4000t} - 5000A_2e^{-5000t} \tag{8.67}$$

At $t = 0$, Equation (8.67) becomes

$$\frac{dv(0)}{dt} = -4000A_1 - 5000A_2 = 40{,}000 \tag{8.68}$$

Dividing Equation (8.68) by 1000, we get

$$-4A_1 - 5A_2 = 40 \tag{8.69}$$

Solving Equation (8.64) for A_2, we obtain

$$A_2 = 10 - A_1$$

Substituting A_2 into Equation (8.69), we get

$$-4A_1 - 5(10 - A_1) = 40$$

Thus,

$$A_1 = 90$$

and

$$A_2 = -80$$

Alternatively, application of Cramer's rule to Equations (8.64) and (8.69) yields

$$A_1 = \frac{\begin{vmatrix} 10 & 1 \\ 40 & -5 \end{vmatrix}}{\begin{vmatrix} 1 & 1 \\ -4 & -5 \end{vmatrix}} = \frac{-50 - 40}{-5 + 4} = \frac{-90}{-1} = 90$$

$$A_2 = \frac{\begin{vmatrix} 1 & 10 \\ -4 & 40 \end{vmatrix}}{\begin{vmatrix} 1 & 1 \\ -4 & -5 \end{vmatrix}} = \frac{40 + 40}{-5 + 4} = \frac{80}{-1} = -80$$

continued

Example 8.1 continued

Therefore, the voltage across the capacitor is given by

$$v(t) = (90e^{-4000t} - 80e^{-5000t})u(t)\ \text{V}$$

The current through the capacitor is

$$i(t) = C\frac{dv(t)}{dt} = 5 \times 10^{-7} \times (-4000 \times 90e^{-4000t} + 5000 \times 80e^{-5000t})u(t)\ \text{A}$$
$$= (-180e^{-4000t} + 200e^{-5000t})u(t)\ \text{mA}$$

where $u(t)$ is defined as

$$u(t) = \begin{cases} 1, & t \geq 0 \\ 0, & t < 0 \end{cases}$$

As discussed in Chapter 1, $u(t)$ is called a *unit step function.*

The voltage across the resistor is given by

$$v_R(t) = Ri(t) = 900 \times i(t) = (-162e^{-4000t} + 180e^{-5000t})u(t)\ \text{V}$$

The voltage across the inductor is

$$v_L(t) = L\frac{di(t)}{dt} = 0.1(0.18 \times 4000e^{-4000t} - 0.2 \times 5000e^{-5000t})\,u(t)\ \text{V}$$
$$= (72e^{-4000t} - 100e^{-5000t})u(t)\ \text{V}$$

Notice that the sum of voltages across the resistor, inductor, and capacitor is equal to zero; that is,

$$v_R(t) + v_L(t) + v(t) = 0$$

Also, notice that at $t = 0$, the voltage across the capacitor is $v(0) = 10$ V (which matches the initial condition $V_0 = 10$ V); the current through the inductor is 20 mA (which matches the initial condition of $I_0 = 20$ mA); the voltage across the resistor is $v_R(0) = 900 \times I_0 = 18$ V; and the voltage across the inductor is $v_L(0) = L \times di(0)/dt = -18$ V. At $t = \infty$, all the voltages and current are zero. At $t = 2$ ms, all voltages and currents are less than 1% of the value at $t = 0$.

Figure 8.3 shows the plots of $v(t)$, $i(t)$, $v_R(t)$, and $v_L(t)$.

FIGURE 8.3

Plots for EXAMPLE 8.1.

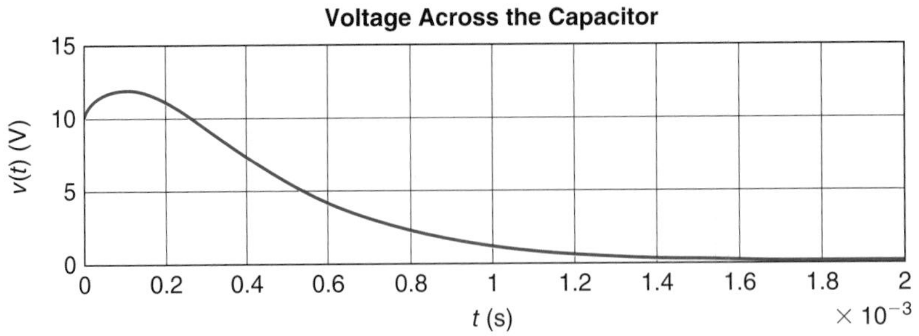

continued

Example 8.1 continued

FIGURE 8.3

continued

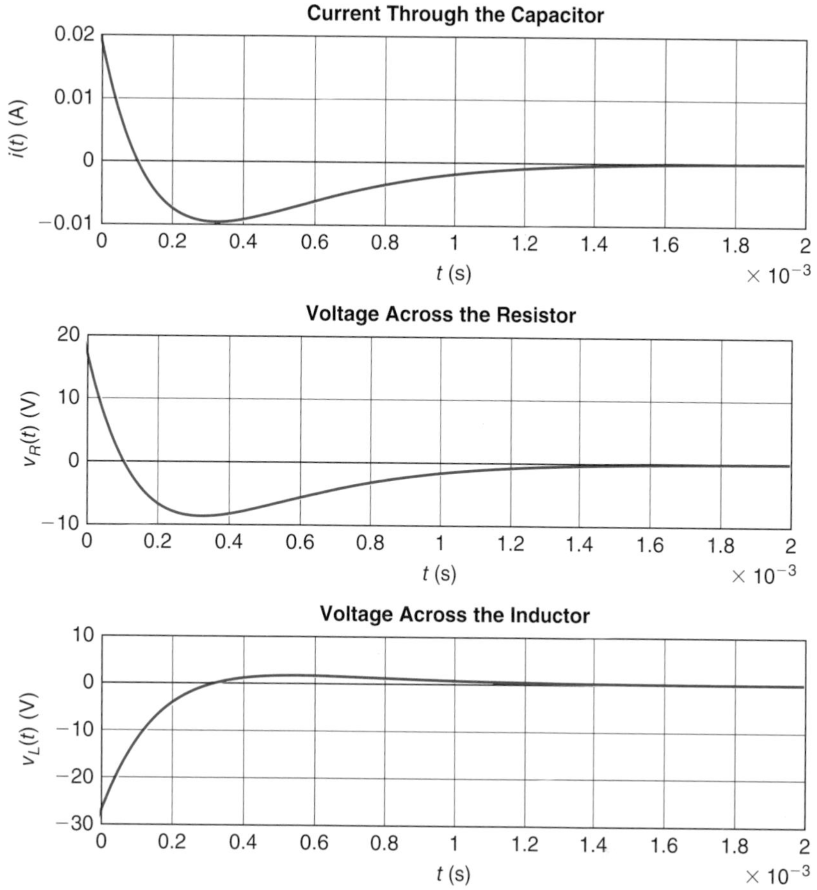

The answers can be checked with the MATLAB script given here:

MATLAB

```
% EXAMPLE 8.1, EXAMPLE 8.2, EXAMPLE 8.3
clear all;format long;
R=900;L=100e-3;C=0.5e-6;V0=10;I0=20e-3;Dv0=I0/C %EXAMPLE 8.1
% R=1000;L=100e-3;C=0.4e-6;V0=10;I0=5e-3;Dv0=I0/C %EXAMPLE 8.2
% R=1000;L=100e-3;C=0.2e-6;V0=10;I0=5e-3;Dv0=I0/C %EXAMPLE 8.3
a2=1
a1=R/L
a0=1/(L*C)
alpha=a1/2
w0=sqrt(a0)
t=0:1/(100*alpha):8/alpha;
Th=1e-10;
syms ta
if (alpha - w0) >= Th
    disp('Overdamped (Case 1)');
    s=roots([a2 a1 a0])
    s1=-alpha+sqrt(alpha^2-w0^2)
    s2=-alpha-sqrt(alpha^2-w0^2)
    syms A1 A2
    [A1,A2]=solve(A1+A2==V0,s1*A1+s2*A2==Dv0,A1,A2)
    A1=eval(A1)
```

continued

Example 8.1 continued
MATLAB continued

```
        A2=eval(A2)
        if A2 < 0 sn = '- '; else sn = '+ '; end
        disp(['v(t) = ',num2str(A1),' exp(',num2str(s1),'t) u(t) ',...
        sn,num2str(abs(A2)),' exp(',num2str(s2),'t) u(t)']);
        v=A1*exp(s1*t)+A2*exp(s2*t);
        va=A1*exp(s1*ta)+A2*exp(s2*ta);
    elseif abs(alpha - w0) < Th
        disp('Critically Damped (Case 2)');
        s=roots([a2 a1 a0])
        s1=-alpha
        s2=-alpha
        syms A1 A2
        [A1,A2]=solve(A1==V0,s1*A1+A2==Dv0,A1,A2)
        A1=eval(A1)
        A2=eval(A2)
        if A2 < 0 sn = '- '; else sn = '+ '; end
        disp(['v(t) = ',num2str(A1),' exp(',num2str(s1),'t) u(t) ',...
        sn,num2str(abs(A2)),' t exp(',num2str(s2),'t) u(t)']);
        v=A1*exp(s1*t)+A2*t.*exp(s2*t);
        va=A1*exp(s1*ta)+A2*ta*exp(s2*ta);
    else
        disp('Underdamped (Case 3)');
        beta=sqrt(w0^2-alpha^2)
        s=roots([1 a1 a0])
        s1=-alpha+j*beta
        s2=-alpha-j*beta
        s=roots([1 a1 a0])
        B1=V0
        B2=(Dv0+alpha*B1)/beta
        if B2 < 0 sn = '- '; else sn = '+ '; end
        disp(['v(t) = ',num2str(B1),' exp(',num2str(real(s1)),...
        't) cos(',num2str(beta),'t) u(t) ',sn,num2str(abs(B2)),...
        ' exp(',num2str(real(s1)),'t) sin(',num2str(beta),'t) u(t)']);
        v=B1*exp(-alpha*t).*(cos(beta*t))+B2*exp(-alpha*t).*(sin(beta*t));
        va=exp(-alpha*ta)*B1*cos(beta*ta)+exp(-alpha*ta)*B2*sin(beta*ta);
    end
    figure(1)
    plot(t,v,'LineWidth',2);grid;xlabel('t (s)');ylabel('v(t) (V)');
    title('Voltage Across the Capacitor');
    i=C*diff(va,ta);
    vR=i*R;
    vL=L*diff(i,ta);
    i=vpa(i,10)
    vR=vpa(vR,10)
    vL=vpa(vL,10)
    ia=subs(i,ta,t);
    vRa=subs(vR,ta,t);
    vLa=subs(vL,ta,t);
    va=vpa(va,10)
    figure(2)
    plot(t,ia,'LineWidth',2);grid;xlabel('t (s)');ylabel('i(t) (A)');
    title('Current Through the Capacitor')
    figure(3)
    plot(t,vRa,'LineWidth',2);grid;xlabel('t (s)');ylabel('v_R(t) (V)');
```

continued

Example 8.1 continued

MATLAB continued

```
title('Voltage Across the Resistor')
figure(4)
plot(t,vLa,'LineWidth',2);grid;xlabel('t (s)');ylabel('v_L(t) (V)');
title('Voltage Across the Inductor')
```

EXAMPLE 8.2

Let $R = 1\ k\Omega$, $L = 100$ mH, $C = 0.4\ \mu$F, $V_0 = 10$ V, and $I_0 = 5$ mA in the circuit shown in Figure 8.2. Find voltage $v(t)$ across the capacitor, current $i(t)$ through the capacitor, voltage $v_R(t)$ across the resistor, and the voltage $v_L(t)$ across the inductor. Plot $v(t)$ and $i(t)$.

The coefficients are given by

$$a_2 = 1,\ \ a_1 = \frac{R}{L} = 10{,}000,\ \ a_0 = \frac{1}{LC} = 25 \times 10^6$$

The neper frequency is given by

$$\alpha = \frac{a_1}{2} = \frac{R}{2L} = 5000 \text{ Np/s}$$

The resonant frequency is

$$\omega_o = \sqrt{a_0} = \frac{1}{\sqrt{LC}} = 5000 \text{ rad/s}$$

Since $\alpha = \omega_0$, this is case 2, and $v(t)$ is critically damped. The characteristic equation is

$$s^2 + a_1 s + a_0 = s^2 + \frac{R}{L}s + \frac{1}{LC} = s^2 + 10{,}000s + 25 \times 10^6 = (s + 5000)^2 = 0$$

The roots of the characteristic equation are given by

$$s_1 = -5000, s_2 = -5000$$

Notice that the roots can also be found from

$$s_1 = -\alpha = -5000$$

$$s_2 = -\alpha = -5000$$

The differential equation given by Equation (8.42) becomes

$$\frac{d^2v(t)}{dt^2} + 10{,}000\frac{dv(t)}{dt} + 25 \times 10^6 v(t) = 0 \tag{8.70}$$

continued

Example 8.2 continued

The solution to the differential equation given by Equation (8.70) can be written as

$$v(t) = A_1 e^{s_1 t} + A_2 t e^{s_1 t} = A_1 e^{-5000t} + A_2 t e^{-5000t} \quad \textbf{(8.71)}$$

Coefficients A_1 and A_2 are found from the initial conditions. Since $v(0) = V_0 = 10$ V, we have

$$v(0) = A_1 e^{-5000\times 0} + A_2 \times 0 \times e^{-5000\times 0} = A_1 = V_0 = 10 \text{ V} \quad \textbf{(8.72)}$$

From Equation (8.40), we have

$$\frac{dv(t)}{dt} = \frac{i(t)}{C} \quad \textbf{(8.73)}$$

At $t = 0$, Equation (8.73) becomes

$$\frac{dv(0)}{dt} = \frac{i(0)}{C} = \frac{I_0}{C} = \frac{5 \times 10^{-3}}{4 \times 10^{-7}} = 12{,}500 \quad \textbf{(8.74)}$$

Taking the derivative of Equation (8.71), we obtain

$$\frac{dv(t)}{dt} = -5000A_1 e^{-5000t} + A_2 e^{-5000t} - 5000A_2 t e^{-5000t} \quad \textbf{(8.75)}$$

At $t = 0$, Equation (8.75) becomes

$$\frac{dv(0)}{dt} = -5000A_1 + A_2 = 12{,}500 \quad \textbf{(8.76)}$$

Solving Equation (8.76) for A_2, we obtain

$$A_2 = 12{,}500 + 5000A_1 = 62{,}500 \quad \textbf{(8.77)}$$

Therefore, the voltage across the capacitor is given by

$$v(t) = (10e^{-5000t} + 62{,}500te^{-5000t})u(t) \text{ V}$$

The current through the capacitor is

$$\begin{aligned} i(t) = C\frac{dv(t)}{dt} &= 4 \times 10^{-7} \times (-5000 \times 10e^{-5000t} \\ &\quad + 62{,}500e^{-5000t} - 62{,}500 \times 5000te^{-5000t})u(t) \text{ A} \\ &= (0.005e^{-5000t} - 125te^{-5000t})u(t) \text{ A} \end{aligned}$$

The voltage across the resistor is given by

$$v_R(t) = Ri(t) = 1000 \times i(t) = (5e^{-5000t} - 125{,}000te^{-5000t})u(t) \text{ V}$$

The voltage across the inductor is

$$\begin{aligned} v_L(t) = L\frac{di(t)}{dt} &= 0.1(-0.005 \times 5000e^{-5000t} - 125e^{-5000t} + 125 \times 5000te^{-5000t})u(t) \text{ V} \\ &= (-15e^{-5000t} + 62{,}500te^{-5000t})u(t) \text{ V} \end{aligned}$$

continued

Example 8.2 continued

Notice that the sum of voltages across the resistor, inductor, and capacitor equals zero; that is,

$$v_R(t) + v_L(t) + v(t) = 0$$

Also, notice that at $t = 0$, the voltage across the capacitor is $v(0) = 10$ V (which matches the initial condition $V_0 = 10$ V); the current through the inductor is 5 mA (which matches the initial condition of $I_0 = 5$ mA); the voltage across the resistor is $v_R(0) = 1000 \times I_0 = 5$ V; and the voltage across the inductor is $v_L(0) = L \times di(0)/dt = -15$ V. At $t = \infty$, all the voltages and current are zero. At $t = 2$ ms, all voltages and currents are less than 1% of the value at $t = 0$.

Figure 8.4 shows the plots of $v(t)$, $i(t)$, $v_R(\text{t})$, and $v_L(\text{t})$.

FIGURE 8.4

Plots for EXAMPLE 8.2.

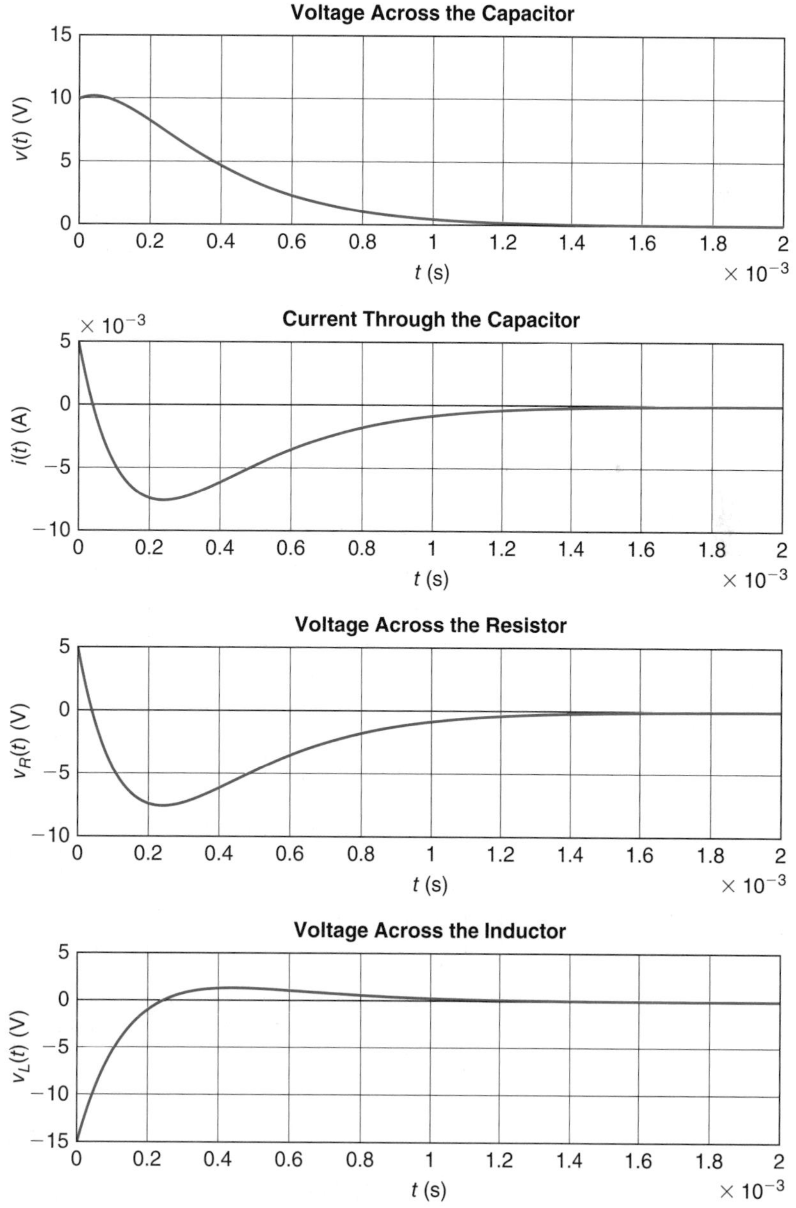

EXAMPLE 8.3

Let $R = 1\ k\Omega$, $L = 100$ mH, $C = 0.2\ \mu$F, $V_0 = 10$ V, and $I_0 = 5$ mA in the circuit shown in Figure 8.2. Find voltage $v(t)$ across the capacitor, current $i(t)$ through the capacitor, voltage $v_R(t)$ across the resistor, and the voltage $v_L(t)$ across the inductor. Plot $v(t)$ and $i(t)$.

The coefficients are given by

$$a_2 = 1,\ a_1 = \frac{R}{L} = 10{,}000,\ a_0 = \frac{1}{LC} = 5 \times 10^7$$

The neper frequency is given by

$$\alpha = \frac{a_1}{2} = \frac{R}{2L} = 5000 \text{ Np/s}$$

The resonant frequency is

$$\omega_o = \sqrt{a_0} = \frac{1}{\sqrt{LC}} = 7071.067812 \text{ rad/s}$$

Since $\alpha < \omega_0$, this is case 3, and $v(t)$ is underdamped. Parameter β is given by

$$\beta = \sqrt{\omega_0^2 - \alpha^2} = 5000 \text{ rad/s}$$

The characteristic equation is

$$s^2 + \frac{R}{L}s + \frac{1}{LC} = s^2 + 10{,}000s + 5 \times 10^7 = (s + 5000 - j5000)(s + 5000 + j5000) = 0$$

The roots of the characteristic equation are given by

$$s_1 = -5000 + j5000, s_2 = -5000 - j5000$$

Notice that the roots can also be found from

$$s_1 = -\alpha + j\beta = -5000 + j5000$$

$$s_2 = -\alpha - j\beta = -5000 - j5000$$

The differential equation given by Equation (8.42) becomes

$$\frac{d^2v(t)}{dt^2} + 10{,}000\frac{dv(t)}{dt} + 5 \times 10^7 v(t) = 0 \quad \textbf{(8.78)}$$

The solution to the differential equation [Equation (8.78)] can be written as

$$\begin{aligned} v(t) &= e^{-\alpha t}[B_1 \cos(\beta t) + B_2 \sin(\beta t)] \\ &= e^{-5000t}[B_1 \cos(5000t) + B_2 \sin(5000t)] \end{aligned} \quad \textbf{(8.79)}$$

continued

Example 8.3 continued

Coefficients B_1 and B_2 are found from the initial conditions. Since $v(0) = V_0 = 10$ V, we have

$$v(0) = e^{-5000\times 0}[B_1 \cos(5000 \times 0) + B_2 \sin(5000 \times 0)] = B_1 = V_0 = 10 \text{ V} \quad \textbf{(8.80)}$$

From Equation (8.40), we have

$$\frac{dv(t)}{dt} = \frac{i(t)}{C} \quad \textbf{(8.81)}$$

At $t = 0$, Equation (8.78) becomes

$$\frac{dv(0)}{dt} = \frac{i(0)}{C} = \frac{I_0}{C} = \frac{5 \times 10^{-3}}{2 \times 10^{-7}} = 25{,}000 \quad \textbf{(8.82)}$$

Taking the derivative of Equation (8.79), we obtain

$$\frac{dv(t)}{dt} = e^{-\alpha t}[(-\alpha B_1 + \beta B_2)\cos(\beta t) + (-\beta B_1 - \alpha B_2)\sin(\beta t)] \quad \textbf{(8.83)}$$

At $t = 0$, Equation (8.83) becomes

$$\frac{dv(0)}{dt} = -\alpha B_1 + \beta B_2 = -5000B_1 + 5000B_2 = 25{,}000 \quad \textbf{(8.84)}$$

Solving Equation (8.84) for B_2, we obtain

$$B_2 = \frac{25{,}000 + 5000B_1}{5000} = \frac{75{,}000}{5000} = 15 \text{ V} \quad \textbf{(8.85)}$$

Thus, the voltage across the capacitor is given by

$$v(t) = e^{-5000t}[10\cos(5000t) + 15\sin(5000t)]u(t) \text{ V}$$

The current through the capacitor is

$$\begin{aligned} i(t) &= C\frac{dv(t)}{dt} = Ce^{-\alpha t}[(-\alpha B_1 + \beta B_2)\cos(\beta t) - (\beta B_1 + \alpha B_2)\sin(\beta t)] \\ &= e^{-5000t}[5\cos(5000t) - 25\sin(5000t)]u(t) \text{ mA} \end{aligned}$$

The voltage across the resistor is given by

$$v_R(t) = Ri(t) = 1000 \times i(t) = e^{-5000t}[5\cos(5000t) - 25\sin(5000t)]u(t) \text{ V}$$

The voltage across the inductor is

$$\begin{aligned} v_L(t) &= L\frac{di(t)}{dt} \\ &= LCe^{-\alpha t}\{[(\alpha^2 - \beta^2)B_1 - 2\alpha\beta B_2]\cos(\beta t) + [2\alpha\beta B_1 + (\alpha^2 - \beta^2)B_2]\sin(\beta t)\} \\ &= e^{-5000t}[-15\cos(5000t) + 10\sin(5000t)]\,u(t) \text{ V} \end{aligned}$$

Notice that the sum of voltages across the resistor, inductor, and capacitor is equal to zero; that is,

$$v_R(t) + v_L(t) + v(t) = 0$$

continued

Example 8.3 continued

Also, notice that at $t = 0$, the voltage across the capacitor is $v(0) = 10$ V (which matches the initial condition $V_0 = 10$ V); the current through the inductor is 5 mA (which matches the initial condition of $I_0 = 20$ mA); the voltage across the resistor is $v_R(0) = 1000 \times I_0 = 5$ V; and the voltage across the inductor is $v_L(0) = L \times di(0)/dt = -15$ V. At $t = \infty$, all the voltages and current are zero. At $t = 2$ ms, all voltages and currents are less than 1% of the value at $t = 0$.

Figure 8.5 shows the plots of $v(t)$, $i(t)$, $v_R(t)$, and $v_L(t)$.

FIGURE 8.5

Plots for EXAMPLE 8.3.

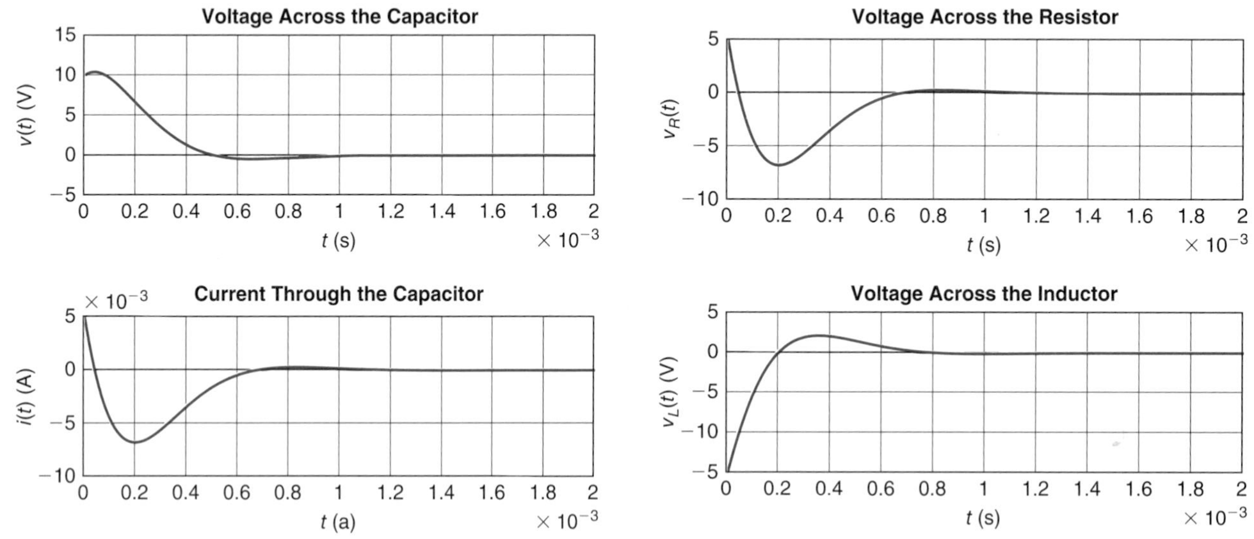

A MATLAB script that can be used to get $v(t)$, $i(t)$, $v_R(t)$, and $v_L(t)$ using **dsolve** is shown here:

MATLAB

```
% EXAMPLE 8.3
clear all;
R=1000;L=100e-3;C=0.2e-6;V0=10;I0=5e-3;
syms v(t)
Dv=diff(v)
Dv2=diff(v,2)
v=dsolve(Dv2+R/L*Dv+1/(L*C)*v,v(0)==V0,Dv(0)==I0/C);
i=C*diff(v);
vR=R*i;
vL=L*diff(i);
v=vpa(v,10)
i=vpa(i,10)
vR=vpa(vR,10)
vL=vpa(vL,10)

Answers:
Dv(t) =
diff(v(t), t)
Dv2(t) =
diff(v(t), t, t)
v =
10.0*exp(-5000.0*t)*cos(5000.0*t) + 15.0*exp(-5000.0*t)*sin(5000.0*t)
```

continued

Example 8.3 continued

MATLAB continued

```
i =
0.005*exp(-5000.0*t)*cos(5000.0*t) - 0.025*exp(-5000.0*t)*sin(5000.0*t)
vR =
5.0*exp(-5000.0*t)*cos(5000.0*t) - 25.0*exp(-5000.0*t)*sin(5000.0*t)
vL =
10.0*exp(-5000.0*t)*sin(5000.0*t) - 15.0*exp(-5000.0*t)*cos(5000.0*t)
```

EXAMPLE 8.4

Derive a differential equation for the circuit shown in Figure 8.2 in terms of $i(t)$, and solve the differential equation for $i(t)$ when $R = 1\ k\Omega$, $L = 100$ mH, $C = 0.2\ \mu$F, $V_0 = 10$ V, $I_0 = 5$ mA. Compare the answer with the one found in EXAMPLE 8.3.

Writing a mesh equation around the mesh in Figure 8.2, we obtain

$$Ri(t) + L\frac{di(t)}{dt} + v(t) = 0 \tag{8.86}$$

Since the voltage across the capacitor is given by $v(t) = \frac{1}{C}\int_{-\infty}^{t} i(\lambda)d\lambda$, Equation (8.86) can be written as

$$Ri(t) + L\frac{di(t)}{dt} + \frac{1}{C}\int_{-\infty}^{t} i(\lambda)d\lambda = 0 \tag{8.87}$$

Differentiating Equation (8.87) with respect to t, we get

$$R\frac{di(t)}{dt} + L\frac{d^2i(t)}{dt^2} + \frac{1}{C}i(t) = 0 \tag{8.88}$$

Rearrangement of Equation (8.88) results in

$$\frac{d^2i(t)}{dt^2} + \frac{R}{L}\frac{di(t)}{dt} + \frac{1}{LC}i(t) = 0 \tag{8.89}$$

The characteristic equation of Equation (8.89) is given by

$$s^2 + \frac{R}{L}s + \frac{1}{LC} = 0 \tag{8.90}$$

The differential equation given by Equation (8.89) and the characteristic equation given by Equation (8.90) are the same as those given by Equations (8.42) and (8.45), except that the variable is $i(t)$ instead of $v(t)$. The initial conditions are different as well. The initial current through the inductor is given by

$$i(0) = I_0 \tag{8.91}$$

continued

Example 8.4 continued

If t is set to zero in Equation (8.86), we obtain

$$Ri(0) + L\frac{di(0)}{dt} + v(0) = 0 \quad \textbf{(8.92)}$$

Solving for the initial condition $\frac{di(0)}{dt}$, we obtain

$$\frac{di(0)}{dt} = \frac{-v(0) - Ri(0)}{L} = \frac{-V_0 - RI_0}{L} \quad \textbf{(8.93)}$$

The neper frequency is given by

$$\alpha = \frac{R}{2L} = 5000 \text{ Np/s}$$

The resonant frequency is

$$\omega_0 = \frac{1}{\sqrt{LC}} = 7071.067812 \text{ rad/s}$$

Since $\alpha < \omega_0$, this is case 3, and $i(t)$ is underdamped. Parameter β is given by

$$\beta = \sqrt{\omega_0^2 - \alpha^2} = 5000 \text{ rad/s}$$

The characteristic equation is

$$s^2 + \frac{R}{L}s + \frac{1}{LC} = s^2 + 10{,}000s + 5 \times 10^7 = (s + 5000 - j5000)(s + 5000 + j5000) = 0$$

The roots of the characteristic equation are given by

$$s_1 = -5000 + j5000, s_2 = -5000 - j5000$$

Notice that the roots can also be found from

$$s_1 = -\alpha + j\beta = -5000 + j5000$$

$$s_2 = -\alpha - j\beta = -5000 - j5000$$

The solution to the differential equation [Equation (8.89)] can be written as

$$\begin{aligned} i(t) &= e^{-\alpha t}[B_1 \cos(\beta t) + B_2 \sin(\beta t)] \\ &= e^{-5000t}[B_1 \cos(5000t) + B_2 \sin(5000t)] \end{aligned} \quad \textbf{(8.94)}$$

Coefficients B_1 and B_2 are found from the initial conditions. Since $i(0) = I_0 = 5$ mA, we have

$$i(0) = e^{-5000\times 0}[B_1 \cos(5000 \times 0) + B_2 \sin(5000 \times 0)] = B_1 = I_0 = 0.005 \text{ A}$$

continued

Example 8.4 continued

From Equation (8.92), we have

$$\frac{di(0)}{dt} = \frac{-v(0) - Ri(0)}{L} = \frac{-V_0 - RI_0}{L} = \frac{-10 - 1000 \times 0.005}{0.1} = -150 \quad \textbf{(8.95)}$$

Taking the derivative of Equation (8.94), we obtain

$$\frac{di(t)}{dt} = e^{-\alpha t}[(-\alpha B_1 + \beta B_2)\cos(\beta t) + (-\beta B_1 - \alpha B_2)\sin(\beta t)] \quad \textbf{(8.96)}$$

At $t = 0$, Equation (8.96) becomes

$$\frac{di(0)}{dt} = -\alpha B_1 + \beta B_2 = -5000B_1 + 5000B_2 = -150 \quad \textbf{(8.97)}$$

Solving Equation (8.97) for B_2, we obtain

$$B_2 = \frac{-150 + 25}{5000} = \frac{-125}{5000} = -0.025 \text{ A} \quad \textbf{(8.98)}$$

Thus, the current through the mesh is given by

$$i(t) = e^{-5000t}[5\cos(5000t) - 25\sin(5000t)]u(t) \text{ mA}$$

This answer is identical to the one obtained in EXAMPLE 8.3.

EXAMPLE 8.5

The switch in the circuit shown in Figure 8.6 has been closed for a long time before it is opened at $t = 0$. Find voltage $v(t)$ across the capacitor and the current $i(t)$ through the capacitor for $t \geq 0$.

FIGURE 8.6

Circuit for EXAMPLE 8.5.

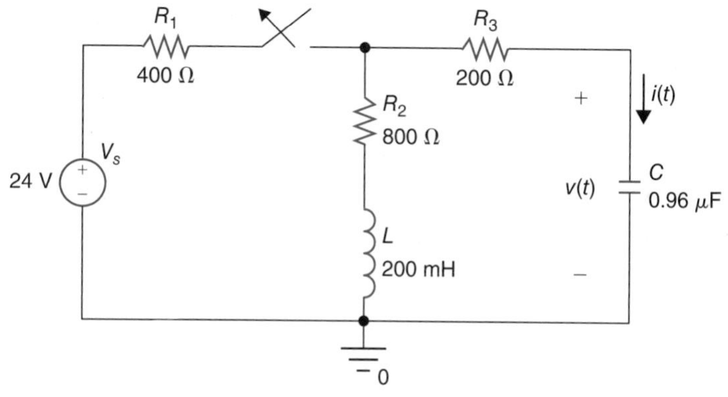

For $t < 0$, the circuit is in steady state. The inductor can be treated as a short circuit, and the capacitor can be treated as an open circuit. Since the capacitor can be treated as an open circuit, there is no current through R_3. The circuit consists of V_s, R_1, R_2, and L, which can be treated as a short circuit. The current through the inductor is $I_0 = 24 \text{ V}/(400 + 800)\ \Omega = 20$ mA. The voltage across the capacitor equals the voltage across R_2. Thus, we have $V_0 = 0.02 \text{ A} \times 800\ \Omega = 16$ V.

For $t \geq 0$, the circuit consists of L, R_2, R_3, and C. Let $R = R_2 + R_3 = 1\ k\Omega$. Then, the circuit reduces to a series RLC circuit.

The neper frequency is given by

$$\alpha = \frac{R}{2L} = 2500 \text{ Np/s}$$

continued

Example 8.5 continued

The resonant frequency is

$$\omega_0 = \frac{1}{\sqrt{LC}} = 2282.1773 \text{ rad/s}$$

Since $\alpha > \omega_0$, this is case 1, and $v(t)$ is overdamped. The characteristic equation is

$$s^2 + \frac{R}{L}s + \frac{1}{LC} = s^2 + 5{,}000s + 5.208333 \times 10^6 = (s + 1479.3793)(s + 3520.6207) = 0$$

The roots of the characteristic equation are given by

$$s_1 = -1479.3793, \quad s_2 = -3520.6207$$

Notice that the roots can also be found from

$$s_1 = -\alpha + \sqrt{\alpha^2 - \omega_0^2} = -1479.3793$$

$$s_2 = -\alpha - \sqrt{\alpha^2 - \omega_0^2} = -3520.6207$$

The differential equation on $v(t)$ is given by

$$\frac{d^2v(t)}{dt^2} + 5000\frac{dv(t)}{dt} + 5.208333 \times 10^6 v(t) = 0 \tag{8.99}$$

The solution to the differential equation [Equation (8.99)] can be written as

$$v(t) = A_1e^{s_1t} + A_2e^{s_2t} = A_1e^{-1479.3793t} + A_2e^{-3520.6207t} \tag{8.100}$$

Coefficients A_1 and A_2 are found from the initial conditions. Since $v(0) = V_0 = 16$ V, we have

$$v(0) = A_1e^{-1479.3793\times0} + A_2e^{-3520.6207\times0} = A_1 + A_2 = V_0 = 16 \tag{8.101}$$

From Equation (8.40), we have

$$\frac{dv(t)}{dt} = \frac{i(t)}{C} \tag{8.102}$$

At $t = 0$, Equation (8.102) becomes

$$\frac{dv(0)}{dt} = \frac{i(0)}{C} = \frac{-I_0}{C} = \frac{-20 \times 10^{-3}}{9.6 \times 10^{-7}} = -20{,}833.3333 \tag{8.103}$$

Taking the derivative of Equation (8.100), we obtain

$$\frac{dv(t)}{dt} = -1479.3793A_1e^{-1479.3793t} - 3520.6207A_2e^{-3520.6207t} \tag{8.104}$$

At $t = 0$, Equation (8.104) becomes

$$\frac{dv(0)}{dt} = -1479.3793A_1 - 3520.6207A_2 = -20{,}833.3333 \tag{8.105}$$

continued

Example 8.5 continued

Solving Equation (8.101) for A_2, we obtain

$$A_2 = 16 - A$$

Substituting A_2 into Equation (8.105), we get

$$-1479.3793A_1 - 3520.6207(16 - A_1) = -20{,}833.3333$$

Solving for A_1, we obtain

$$A_1 = 17.3897$$

Substituting A_1 into $A_2 = 16 - A_1$, we get

$$A_2 = -1.3897.$$

Alternatively, application of Cramer's rule to Equations (8.101) and (8.105) yields

$$A_1 = \frac{\begin{vmatrix} 16 & 1 \\ -20{,}833.3333 & -3520.6207 \end{vmatrix}}{\begin{vmatrix} 1 & 1 \\ -1479.3793 & -3520.6207 \end{vmatrix}} = 17.3897$$

$$A_2 = \frac{\begin{vmatrix} 1 & 16 \\ -1479.3793 & -20{,}833.3333 \end{vmatrix}}{\begin{vmatrix} 1 & 1 \\ -1479.3793 & -3520.6207 \end{vmatrix}} = -1.3897$$

Therefore, the voltage across the capacitor is given by

$$v(t) = (17.3897e^{-1479.3793t} - 1.3897e^{-3520.6207t})u(t)\text{ V}$$

The current through the capacitor is

$$\begin{aligned} i(t) = C\frac{dv(t)}{dt} &= 9.6 \times 10^{-7} \times (-1479.3793 \times 17.8021e^{-4000t} \\ &\quad + 3520.6207 \times 1.3897e^{-5000t})u(t)\text{ A} \\ &= (-24.697e^{-1479.3793t} + 4.697e^{-3520.6207t})u(t)\text{ mA} \end{aligned}$$

MATLAB

```
% EXAMPLE 8.5
clear all;
R1=400;R2=800;R3=200;L=200e-3;C=0.96e-6;Vs=24;
I0=Vs/(R1+R2)
V0=R2*I0
R=R2+R3
syms v(t)
Dv=diff(v);
Dv2=diff(v,2);
v=dsolve(Dv2+R/L*Dv+1/(L*C)*v,v(0)==V0,Dv(0)==-I0/C);
i=C*diff(v);
```

continued

Example 8.5 continued
MATLAB continued

```
v=vpa(v,10)
i=vpa(i,10)

Answers:
I0 =
         0.0200
V0 =
         16
R =
                1000
v =
17.38971068*exp(-1479.379274*t) - 1.389710681*exp(-3520.620726*t)
i =
0.004696938457*exp(-3520.620726*t) - 0.02469693846*exp(-1479.379274*t)
```

Exercise 8.1

The switch in the circuit shown in Figure 8.7 has been closed for a long time before it is opened at $t = 0$. Find the voltage $v(t)$ across the capacitor for $t \geq 0$.

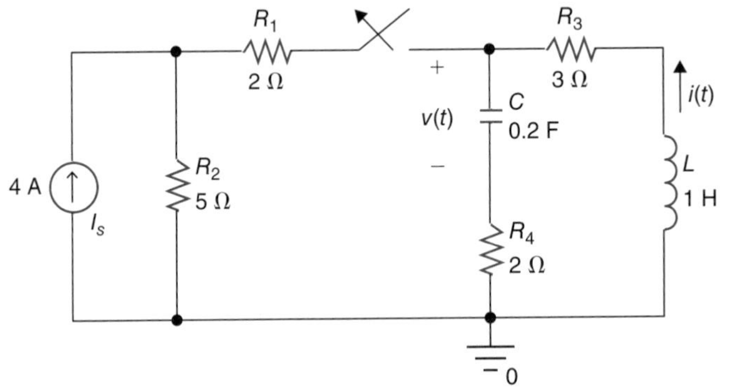

FIGURE 8.7

Circuit for EXERCISE 8.1

Answer:
$v(0) = 6$ V, $i(0) = -2$ A, and $v(t) = (0.7639e^{-3.618t} + 5.2360e^{-1.382t})\,u(t)$ V.

FIGURE 8.8

A parallel RLC circuit.

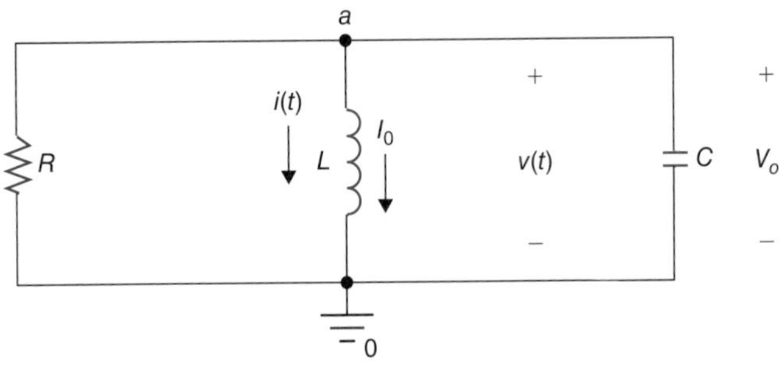

8.4 Zero Input Response of Parallel RLC Circuit

Figure 8.8 shows a parallel RLC circuit without any independent source.

If there are no energies stored in the inductor and capacitor at $t = 0$, the voltage and current on each element will be zero for $t > 0$. If the voltage across the capacitor at $t = 0$ is not zero, or current through the inductor at $t = 0$ is not zero, or both, there will be a transient response in the circuit. Let the initial voltage across the capacitor at $t = 0$ be V_0 and the initial current through the inductor at $t = 0$ be I_0. Let $v(t)$ be the voltage across the resistor, inductor, and capacitor, and $i(t)$ be the current through the inductor in the circuit shown in Figure 8.8.

Summing the currents away from node a (top of the circuit), we have

$$\frac{v(t)}{R} + i(t) + C\frac{dv(t)}{dt} = 0 \tag{8.106}$$

On the inductor, the voltage-current relation is given by

$$v(t) = L\frac{di(t)}{dt} \tag{8.107}$$

Substitution of Equation (8.107) into Equation (8.106) yields

$$\frac{L}{R}\frac{di(t)}{dt} + i(t) + LC\frac{d^2i(t)}{dt^2} = 0 \tag{8.108}$$

Rearrangement of Equation (8.108) results in

$$\frac{d^2i(t)}{dt^2} + \frac{1}{RC}\frac{di(t)}{dt} + \frac{1}{LC}i(t) = 0 \tag{8.109}$$

Comparison of Equations (8.3) and (8.109) reveals that

$$a_2 = 1,\ a_1 = \frac{1}{RC},\ a_0 = \frac{1}{LC} \tag{8.110}$$

Since $\alpha = a_1/2$ and $\omega_0 = \sqrt{a_0}$, we have

$$\alpha = \frac{1}{2RC},\ \omega_0 = \frac{1}{\sqrt{LC}} \tag{8.111}$$

The damping factor is given by

$$\zeta = \frac{\alpha}{\omega_0} = \frac{1}{2R}\sqrt{\frac{L}{C}}$$

The characteristic equation of the differential equation given by Equation (8.109) is

$$s^2 + a_1s + a_0 = s^2 + \frac{1}{RC}s + \frac{1}{LC} = 0 \tag{8.112}$$

The initial current through the inductor is given by

$$i(0) = I_0 \tag{8.113}$$

From Equation (8.107), we obtain

$$\frac{di(t)}{dt} = \frac{v(t)}{L} \tag{8.114}$$

Setting $t = 0$, we obtain the second initial condition:

$$\frac{di(0)}{dt} = \frac{v(0)}{L} = \frac{V_0}{L} \tag{8.115}$$

Depending on the relative magnitude of α and ω_0, we have three cases, described in the next sections.

8.4.1 CASE 1: OVERDAMPED ($\alpha > \omega_0$ or $a_1 > 2\sqrt{a_0}$ or $\zeta > 1$)

From Equations (8.8), (8.10), (8.14), and (8.15), we have

$$s_1 = -\alpha + \sqrt{\alpha^2 - \omega_0^2},\ s_2 = -\alpha - \sqrt{\alpha^2 - \omega_0^2} \tag{8.116}$$

$$i(t) = A_1 e^{s_1 t} + A_2 e^{s_2 t} \tag{8.117}$$

$$A_1 = \frac{I_0 s_2 - Di_0}{s_2 - s_1} \tag{8.118}$$

$$A_2 = \frac{Di_0 - I_0 s_1}{s_2 - s_1} \tag{8.119}$$

where $I_0 = i(0), Di_0 = \dfrac{di(0)}{dt}$.

8.4.2 CASE 2: CRITICALLY DAMPED ($\alpha = \omega_0$ or $a_1 = 2\sqrt{a_0}$ or $\zeta = 1$)

From Equations (8.18), (8.19), (8.20), and (8.24), we have

$$s_1 = -\alpha,\ \ s_2 = -\alpha \tag{8.120}$$

$$i(t) = A_1 e^{s_1 t} + A_2 t e^{s_2 t} \tag{8.121}$$

$$A_1 = I_0 \tag{8.122}$$

$$A_2 = Di_0 - A_1 s_1 \tag{8.123}$$

where $I_0 = i(0), Di_0 = \dfrac{di(0)}{dt}$.

8.4.3 CASE 3: UNDERDAMPED ($\alpha < \omega_0$ or $a_1 < 2\sqrt{a_0}$ or $\zeta < 1$)

From Equations (8.25), (8.26), (8.32), (8.33), and (8.36), we have

$$\beta = \sqrt{\omega_0^2 - \alpha^2} = \omega_0\sqrt{1 - \zeta^2} \tag{8.124}$$

$$s_1 = -\alpha + j\beta,\ \ s_2 = -\alpha - j\beta \tag{8.125}$$

$$i(t) = e^{-\alpha t}[B_1 \cos(\beta t) + B_2 \sin(\beta t)] \tag{8.126}$$

$$B_1 = I_0 \tag{8.127}$$

$$B_2 = \frac{Di_0 + \alpha B_1}{\beta} \tag{8.128}$$

where $I_0 = i(0), Di_0 = \dfrac{di(0)}{dt}$.

EXAMPLE 8.6

Let $R = 1\ k\Omega$, $L = 25$ mH, $C = 5$ nF, $V_0 = 5$ V, and $I_0 = 1$ mA in the circuit shown in Figure 8.8. Find current $i(t)$ through the inductor and voltage $v(t)$ across the inductor.

The coefficients are given by

$$a_2 = 1,\ \ a_1 = \frac{1}{RC} = 2 \times 10^5,\ \ a_0 = \frac{1}{LC} = 8 \times 10^9$$

The neper frequency is given by

$$\alpha = \frac{a_1}{2} = \frac{1}{2RC} = 10^5 \text{ Np/s}$$

The resonant frequency is

$$\omega_0 = \sqrt{a_0} = \frac{1}{\sqrt{LC}} = 89{,}442.7191 \text{ rad/s}$$

Since $\alpha > \omega_o$, this is case 1, and $i(t)$ is overdamped. The characteristic equation is

$$s^2 + \frac{1}{RC}s + \frac{1}{LC} = s^2 + 2\times 10^5 s + 8\times 10^9 = (s + 55{,}278.6405)(s + 144{,}721.3595) = 0$$

The roots of the characteristic equation are given by

$$s_1 = -55{,}278.6405, s_2 = -144{,}721.3595$$

Notice that the roots can also be found from

$$s_1 = -\alpha + \sqrt{\alpha^2 - \omega_0^2} = -55{,}278.6405$$

$$s_2 = -\alpha - \sqrt{\alpha^2 - \omega_0^2} = -144{,}721.3595$$

The differential equation given by Equation (8.109) becomes

$$\frac{d^2i(t)}{dt^2} + 2 \times 10^5 \frac{di(t)}{dt} + 8 \times 10^9 i(t) = 0 \quad \textbf{(8.129)}$$

From Equation (8.117), the solution to the differential equation [Equation (8.129)] can be written as

$$i(t) = A_1e^{s_1t} + A_2e^{s_2t} = A_1e^{-55{,}278.6405t} + A_2e^{-144{,}721.3595t} \quad \textbf{(8.130)}$$

Coefficients A_1 and A_2 are found from the initial conditions. Since $i(0) = I_0 = 1 \times 10^{-3}$ A, we have

$$i(0) = A_1e^{-55{,}279\times 0} + A_2e^{-144{,}721\times 0} = A_1 + A_2 = I_0 = 0.001 \quad \textbf{(8.131)}$$

continued

Example 8.6 continued

From Equation (8.107), we have

$$\frac{di(t)}{dt} = \frac{v(t)}{L} \tag{8.132}$$

At $t = 0$, Equation (8.132) becomes

$$\frac{di(0)}{dt} = \frac{v(0)}{L} = \frac{V_0}{L} = \frac{5}{0.025} = 200 \tag{8.133}$$

Taking the derivative of Equation (8.130), we obtain

$$\frac{di(t)}{dt} = -55{,}278.6405A_1e^{-55{,}278.6405t} - 144{,}721.3595A_2e^{-144{,}721.3595t} \tag{8.134}$$

At $t = 0$, Equation (8.134) becomes

$$\frac{di(0)}{dt} = -55{,}278.6405A_1 - 144{,}721.3595A_2 = 200 \tag{8.135}$$

Solving Equation (8.131) for A_2, we obtain

$$A_2 = 0.001 - A_1$$

Substituting A_2 into Equation (8.135), we get

$$-55{,}278.6405A_1 - 144{,}721.3595(0.001 - A_1) = 200$$

Solving for A_1, we obtain

$$A_1 = 0.0038541 \text{ A}$$

Substituting A_1 into $A_2 = 0.001 - A_1$, we get

$$A_2 = -0.0028541 \text{ A}$$

Alternatively, application of Cramer's rule to Equations (8.131) and (8.135) yields

$$A_1 = \frac{\begin{vmatrix} 0.001 & 1 \\ 200 & -144{,}721.3595 \end{vmatrix}}{\begin{vmatrix} 1 & 1 \\ -55{,}278.6405 & -144{,}721.3595 \end{vmatrix}} = 0.0038541 \text{ A}$$

$$A_2 = \frac{\begin{vmatrix} 1 & 0.001 \\ -55{,}278.6405 & 200 \end{vmatrix}}{\begin{vmatrix} 1 & 1 \\ -55{,}278.6405 & -144{,}721.3595 \end{vmatrix}} = -0.0028541 \text{ A}$$

Therefore, the current through the inductor is given by

$$i(t) = A_1e^{s_1t} + A_2e^{s_2t} = (3.8541e^{-55{,}278.6405t} - 2.8541e^{-144{,}721.3595t})\,u(t) \text{ mA} \tag{8.136}$$

continued

Example 8.6 continued

The voltage across the inductor is

$$v(t) = L\frac{di(t)}{dt} = (-5.3263e^{-55{,}278.6405t} + 10.3262e^{-144{,}721.3595t})u(t) \text{ V} \tag{8.137}$$

The answers can be checked with the MATLAB script given here:

MATLAB

```
% EXAMPLE 8.6, EXAMPLE 8.7, EXAMPLE 8.8
clear all;format long;
R=1000;L=25e-3;C=5e-9;V0=5;I0=1e-3;Di0=V0/L      %EXAMPLE 8.6
% R=1000;L=10e-3;C=2.5e-9;V0=15;I0=12e-3;Di0=V0/L %EXAMPLE 8.7
% R=1000;L=50e-3;C=25e-9;V0=10;I0=8e-3;Di0=V0/L   %EXAMPLE 8.8
a2=1
a1=1/(R*C)
a0=1/(L*C)
alpha=a1/2
w0=sqrt(a0)
t=0:1/(100*alpha):8/alpha;
Th=1e-10;
syms ta
if (alpha - w0) >= Th
    disp('Overdamped (Case 1)');
    s=roots([a2 a1 a0])
    s1=-alpha+sqrt(alpha^2-w0^2)
    s2=-alpha-sqrt(alpha^2-w0^2)
    syms A1 A2
    [A1,A2]=solve(A1+A2==I0,s1*A1+s2*A2==Di0,A1,A2)
    A1=eval(A1)
    A2=eval(A2)
    if A2 < 0 sn = '- '; else sn = '+ '; end
    disp(['i(t) = ',num2str(A1),' exp(',num2str(s1),'t) u(t) ',...
    sn,num2str(abs(A2)),' exp(',num2str(s2),'t) u(t)']);
    i=A1*exp(s1*t)+A2*exp(s2*t);
    ia=A1*exp(s1*ta)+A2*exp(s2*ta);
elseif abs(alpha - w0) < Th
    disp('Critically Damped (Case 2)');
    s=roots([a2 a1 a0])
    s1=-alpha
    s2=-alpha
    syms A1 A2
    [A1,A2]=solve(A1==I0,s1*A1+A2==Di0,A1,A2)
    A1=eval(A1)
    A2=eval(A2)
    if A2 < 0 sn = '- '; else sn = '+ '; end
    disp(['i(t) = ',num2str(A1),' exp(',num2str(s1),'t) u(t) ',...
    sn,num2str(abs(A2)),' t exp(',num2str(s2),'t) u(t)']);
    i=A1*exp(s1*t)+A2*t.*exp(s2*t);
    ia=A1*exp(s1*ta)+A2*ta*exp(s2*ta);
else
    disp('Underdamped (Case 3)');
    beta=sqrt(w0^2-alpha^2)
    s=roots([1 a1 a0])
    s1=-alpha+j*beta
    s2=-alpha-j*beta
    B1=I0
```

continued

Example 8.6 continued
MATLAB continued

```
        B2=(Di0+alpha*B1)/beta
        if B2 < 0 sn = '- '; else sn = '+ '; end
        disp(['i(t) = ',num2str(B1),' exp(',num2str(real(s1)),...
        't) cos(',num2str(beta),'t) u(t) ',sn,num2str(abs(B2)),...
        ' exp(',num2str(real(s1)),'t) sin(',num2str(beta),'t) u(t)']);
        i=B1*exp(-alpha*t).*(cos(beta*t))+B2*exp(-alpha*t).*(sin(beta*t));
        ia=exp(-alpha*ta)*B1*cos(beta*ta)+exp(-alpha*ta)*B2*sin(beta*ta);
    end
    figure(1)
    plot(t,i,'LineWidth',2);grid;xlabel('t (s)');ylabel('i(t) (A)');
    title('Current Through the Inductor');
    v=L*diff(ia,ta);
    v=vpa(v,10)
    va=subs(v,ta,t);
    ia=vpa(ia,10)
    figure(2)
    plot(t,va,'LineWidth',2);grid;xlabel('t (s)');ylabel('v(t) (V)');
    title('Voltage Across the Inductor')
```

EXAMPLE 8.7

Let $R = 1\ k\Omega$, $L = 10$ mH, $C = 2.5$ nF, $V_0 = 15$ V, and $I_0 = 12$ mA in the circuit shown in Figure 8.8. Find current $i(t)$ through the inductor and voltage $v(t)$ across the inductor.

The coefficients are given by

$$a_2 = 1,\ \ a_1 = \frac{1}{RC} = 4 \times 10^5,\ \ a_0 = \frac{1}{LC} = 4 \times 10^{10}$$

The neper frequency is given by

$$\alpha = \frac{a_1}{2} = \frac{1}{2RC} = 2 \times 10^5 \text{ Np/s}$$

The resonant frequency is

$$\omega_0 = \sqrt{a_0} = \frac{1}{\sqrt{LC}} = 2 \times 10^5 \text{ rad/s}$$

Since $\alpha = \omega_0$, this is case 2, and $i(t)$ is critically damped. The characteristic equation is

$$s^2 + \frac{1}{RC}s + \frac{1}{LC} = s^2 + 4 \times 10^5 s + 4 \times 10^{10} = (s + 2 \times 10^5)^2 = 0$$

The roots of the characteristic equation are given by

$$s_1 = -2 \times 10^5,\ \ s_2 = -2 \times 10^5$$

continued

Example 8.7 continued

Notice that the roots also can be found from

$$s_1 = -\alpha = -2 \times 10^5$$

$$s_2 = -\alpha = -2 \times 10^5$$

The differential equation given by Equation (8.109) becomes

$$\frac{d^2i(t)}{dt^2} + 4 \times 10^5 \frac{di(t)}{dt} + 4 \times 10^{10} i(t) = 0 \tag{8.138}$$

From Equation (8.121), the solution to the differential equation [Equation (8.138)] can be written as

$$i(t) = A_1 e^{s_1 t} + A_2 t e^{s_2 t} = A_1 e^{-200{,}000t} + A_2 t e^{-200{,}000t} \tag{8.139}$$

Coefficients A_1 and A_2 are found from the initial conditions. Since $i(0) = I_0 = 12 \times 10^{-3}$ A, we have

$$i(0) = A_1 e^{-200{,}000 \times 0} + A_2 0 e^{-200{,}000 \times 0} = A_1 = I_0 = 0.012 \text{ A} \tag{8.140}$$

From Equation (8.107), we have

$$\frac{di(t)}{dt} = \frac{v(t)}{L} \tag{8.141}$$

At $t = 0$, Equation (8.141) becomes

$$\frac{di(0)}{dt} = \frac{v(0)}{L} = \frac{V_0}{L} = \frac{15}{0.01} = 1500 \tag{8.142}$$

Taking the derivative of Equation (8.139), we obtain

$$\frac{di(t)}{dt} = -200{,}000 A_1 e^{-200{,}000t} + A_2 e^{-200{,}000t} - 200{,}000 A_2 t e^{-200{,}000t} \tag{8.143}$$

At $t = 0$, Equation (8.143) becomes

$$\begin{aligned}\frac{di(0)}{dt} &= -200{,}000 A_1 e^{-200{,}000 \times 0} + A_2 e^{-200{,}000 \times 0} - 200{,}000 A_2 0 e^{-200{,}000 \times 0} \\ &= -200{,}000 A_1 + A_2 = 1500\end{aligned} \tag{8.144}$$

Solving for A_2, we obtain

$$A_2 = 1500 + 200{,}000 \times 0.012 = 3900 \text{ A} \tag{8.145}$$

Therefore, the current through the inductor is given by

$$i(t) = A_1 e^{s_1 t} + A_2 t e^{s_2 t} = (0.012 e^{-200{,}000t} + 3900 t e^{-200{,}000t}) u(t) \text{ A} \tag{8.146}$$

The voltage across the inductor is

$$v(t) = L \frac{di(t)}{dt} = (15 e^{-200{,}000t} - 7.8 \times 10^6 t e^{-200{,}000t}) u(t) \text{ V} \tag{8.147}$$

EXAMPLE 8.8

Let $R = 1\ k\Omega$, $L = 50$ mH, $C = 25$ nF, $V_0 = 10$ V, and $I_0 = 8$ mA in the circuit shown in Figure 8.8. Find current $i(t)$ through the inductor and voltage $v(t)$ across the inductor.

The coefficients are given by

$$a_2 = 1,\ a_1 = \frac{1}{RC} = 4 \times 10^4,\ a_0 = \frac{1}{LC} = 8 \times 10^8$$

The neper frequency is given by

$$\alpha = \frac{a_1}{2} = \frac{1}{2RC} = 20{,}000 \text{ Np/s}$$

The resonant frequency is

$$\omega_0 = \sqrt{a_0} = \frac{1}{\sqrt{LC}} = 28{,}284.2712 \text{ rad/s}$$

Since $\alpha < \omega_0$, this is case 3, and $i(t)$ is underdamped. Parameter β is given by

$$\beta = \sqrt{\omega_0^2 - \alpha^2} = 20{,}000 \text{ rad/s}$$

The characteristic equation is

$$s^2 + \frac{1}{RC}s + \frac{1}{LC} = s^2 + 40{,}000s + 8 \times 10^8 = (s + 20{,}000 - j20{,}000) \times (s + 20{,}000 + j20{,}000) = 0$$

The roots of the characteristic equation are given by

$$s_1 = -20{,}000 + j20{,}000,\quad s_2 = -20{,}000 - j20{,}000$$

Notice that the roots can also be found from

$$s_1 = -\alpha + j\beta = -20{,}000 + j20{,}000$$

$$s_2 = -\alpha - j\beta = -20{,}000 - j20{,}000$$

The differential equation given by Equation (8.109) becomes

$$\frac{d^2i(t)}{dt^2} + 40{,}000\frac{di(t)}{dt} + 8 \times 10^8 i(t) = 0 \tag{8.148}$$

The solution to the differential equation [Equation (8.148)] can be written as

$$i(t) = e^{-\alpha t}[B_1 \cos(\beta t) + B_2 \sin(\beta t)] = e^{-20{,}000t}[B_1 \cos(20{,}000t) + B_2 \sin(20{,}000t)] \tag{8.149}$$

continued

Example 8.8 continued

Coefficients B_1 and B_2 are found from the initial conditions. Since $i(0) = I_0 = 0.008$ A, we have

$$i(0) = e^{-20,000\times 0}[B_1 \cos(20,000 \times 0) + B_2 \sin(20,000 \times 0)] \\ = B_1 = I_0 = 0.008 \text{ A} \tag{8.150}$$

From Equation (8.107), we have

$$\frac{di(t)}{dt} = \frac{v(t)}{L} \tag{8.151}$$

At $t = 0$, Equation (8.151) becomes

$$\frac{di(0)}{dt} = \frac{v(0)}{L} = \frac{V_0}{L} = \frac{10}{0.05} = 200 \tag{8.152}$$

Taking the derivative of Equation (8.149), we obtain

$$\frac{di(t)}{dt} = e^{-\alpha t}[(-\alpha B_1 + \beta B_2)\cos(\beta t) + (-\beta B_1 - \alpha B_2)\sin(\beta t)] \tag{8.153}$$

At $t = 0$, Equation (8.153) becomes

$$\frac{di(0)}{dt} = -\alpha B_1 + \beta B_2 = -20,000B_1 + 20,000B_2 = 200 \tag{8.154}$$

Solving Equation (8.154) for B_2, we obtain

$$B_2 = \frac{200 + 20,000 \times 0.008}{20,000} = 0.018 \text{ A} \tag{8.155}$$

Thus, the current through the inductor is given by

$$i(t) = e^{-20,000t}[8\cos(20,000t) + 18\sin(20,000t)]u(t) \text{ mA} \tag{8.156}$$

The voltage across the inductor is

$$v(t) = L\frac{di(t)}{dt} = e^{-20,000t}[10\cos(20,000t) - 26\sin(20,000t)]u(t) \text{ V} \tag{8.157}$$

EXAMPLE 8.9

Derive a differential equation to the circuit shown in Figure 8.8 in $v(t)$, and solve the differential equation for $v(t)$ and $i(t)$ when $R = 1\ k\Omega$, $L = 50$ mH, $C = 25$ nF, $V_0 = 10$ V, and $I_0 = 8$ mA. Compare the answer with the one found in EXAMPLE 8.8.

Summing the currents away from node a, we have

$$\frac{v(t)}{R} + i(t) + C\frac{dv(t)}{dt} = 0 \tag{8.158}$$

continued

Example 8.9 continued

Since $i(t) = \frac{1}{L}\int_{-\infty}^{t} v(\lambda)d\lambda$, Equation (8.158) becomes

$$\frac{v(t)}{R} + \frac{1}{L}\int_{-\infty}^{t} v(\lambda)d\lambda + C\frac{dv(t)}{dt} = 0 \tag{8.159}$$

Taking the derivative of Equation (8.159), we get

$$\frac{1}{R}\frac{dv(t)}{dt} + \frac{1}{L}v(t) + C\frac{d^2v(t)}{dt^2} = 0 \tag{8.160}$$

Rearrangement of Equation (8.160) yields

$$\frac{d^2v(t)}{dt^2} + \frac{1}{RC}\frac{dv(t)}{dt} + \frac{1}{LC}v(t) = 0 \tag{8.161}$$

From Equation (8.158), we have

$$\frac{dv(t)}{dt} = \frac{-\frac{v(t)}{R} - i(t)}{C} \tag{8.162}$$

Setting $t = 0$ in Equation (8.162), we get

$$\frac{dv(0)}{dt} = \frac{-\frac{v(0)}{R} - i(0)}{C} = \frac{-\frac{V_0}{R} - I_0}{C} \tag{8.163}$$

The neper frequency is given by

$$\alpha = \frac{1}{2RC} = 20{,}000 \text{ Np/s}$$

The resonant frequency is

$$\omega_0 = \frac{1}{\sqrt{LC}} = 28{,}284.2712 \text{ rad/s}$$

Since $\alpha < \omega_0$, this is case 3, and $v(t)$ is underdamped. Parameter β is given by

$$\beta = \sqrt{\omega_0^2 - \alpha^2} = 20{,}000 \text{ rad/s}$$

The characteristic equation is

$$s^2 + \frac{1}{RC}s + \frac{1}{LC} = s^2 + 40{,}000s + 8 \times 10^8 = (s + 20{,}000 - j20{,}000) \times (s + 20{,}000 + j20{,}000) = 0$$

The roots of the characteristic equation are given by

$$s_1 = -20{,}000 + j20{,}000, s_2 = -20{,}000 - j20{,}000$$

continued

Example 8.9 continued

Notice that the roots can also be found from

$$s_1 = -\alpha + j\beta = -20{,}000 + j20{,}000$$

$$s_2 = -\alpha - j\beta = -20{,}000 - j20{,}000$$

The differential equation given by Equation (8.161) becomes

$$\frac{d^2v(t)}{dt^2} + 40{,}000\frac{dv(t)}{dt} + 8 \times 10^8 v(t) = 0 \tag{8.164}$$

The solution to the differential equation [Equation (8.164)] can be written as

$$v(t) = e^{-\alpha t}[B_1\cos(\beta t) + B_2\sin(\beta t)] = e^{-20{,}000t}[B_1\cos(20{,}000t) + B_2\sin(20{,}000t)] \tag{8.165}$$

Coefficients B_1 and B_2 are found from the initial conditions. Since $v(0) = V_0 = 10$ V, we have

$$v(0) = e^{-20{,}000\times 0}[B_1\cos(20{,}000 \times 0) + B_2\sin(20{,}000 \times 0)] = B_1 = V_0 = 10\text{ V} \tag{8.166}$$

From Equation (8.163), we obtain

$$\frac{dv(0)}{dt} = \frac{-\frac{v(0)}{R} - i(0)}{C} = \frac{-\frac{V_0}{R} - I_0}{C} = \frac{-\frac{10}{1000} - 0.008}{2.5 \times 10^{-8}} = \frac{-0.018}{2.5 \times 10^{-8}} = -720{,}000 \tag{8.167}$$

Taking the derivative of Equation (8.165), we obtain

$$\frac{dv(t)}{dt} = e^{-\alpha t}[(-\alpha B_1 + \beta B_2)\cos(\beta t) + (-\beta B_1 - \alpha B_2)\sin(\beta t)] \tag{8.168}$$

At $t = 0$, Equation (8.168) becomes

$$\frac{dv(0)}{dt} = -\alpha B_1 + \beta B_2 = -20{,}000B_1 + 20{,}000B_2 = -720{,}000 \tag{8.169}$$

Solving Equation (8.169) for B_2, we obtain

$$B_2 = \frac{-720{,}000 + 20{,}000 \times 10}{20{,}000} = -26\text{ V} \tag{8.170}$$

Thus, the voltage across the inductor is given by

$$v(t) = e^{-20{,}000t}[10\cos(20{,}000t) - 26\sin(20{,}000t)]u(t)\text{ V} \tag{8.171}$$

The current through the inductor is given by

$$i(t) = \frac{1}{L}\int_{-\infty}^{t} v(\lambda)d\lambda \tag{8.172}$$

continued

Example 8.9 continued

Application of the integral formulas

$$\int e^{ax}\cos(bx)dx = \frac{e^{ax}[a\cos(bx) + b\sin(bx)]}{a^2 + b^2}$$

$$\int e^{ax}\sin(bx)dx = \frac{e^{ax}[a\sin(bx) - b\cos(bx)]}{a^2 + b^2}$$

results in

$$\begin{aligned} i(t) &= 20\frac{10}{8\times 10^8}[-20{,}000e^{-20{,}000t}\cos(20{,}000t) + 20{,}000e^{-20{,}000t}\sin(20{,}000t)] \\ &\quad + 20\frac{-26}{8\times 10^8}[-20{,}000e^{-20{,}000t}\sin(20{,}000t) - 20{,}000e^{-20{,}000t}\cos(20{,}000t)] \\ &= e^{-20{,}000t}\frac{-200\times 20{,}000 + 520\times 20{,}000}{8\times 10^8}\cos(20{,}000t) \\ &\quad + e^{-20{,}000t}\frac{200\times 20{,}000 + 520\times 20{,}000}{8\times 10^8}\sin(20{,}000t) \\ &= e^{-20{,}000t}[8\cos(20{,}000t) + 18\sin(20{,}000t)]u(t)\ \text{mA} \end{aligned} \tag{8.173}$$

EXAMPLE 8.10

The switch in the circuit shown in Figure 8.9 has been closed for a long time before it is opened at $t = 0$. Find $i(t)$ for $t \geq 0$.

FIGURE 8.9

Circuit for EXAMPLE 8.10.

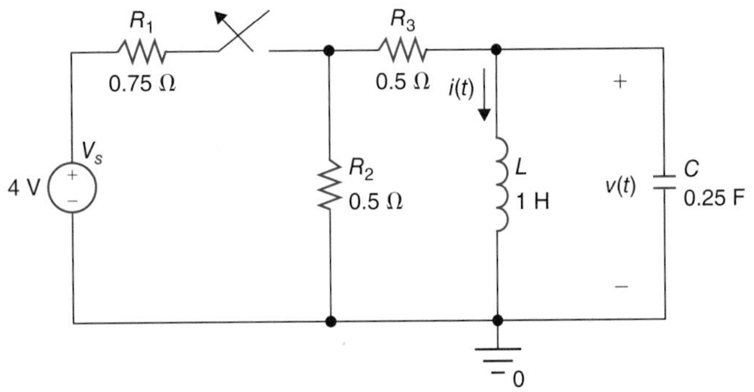

For $t < 0$, since the circuit is in the steady state, the inductor can be treated as a short circuit and the capacitor can be treated as an open circuit. The circuit consists of the voltage source, R_1, R_2, and R_3 with the inductor short-circuited. The equivalent resistance is

$$\begin{aligned} R_{eq} &= R_1 + (R_2 \| R_3) \\ &= 0.75 + (0.5\|0.5) = 1\ \Omega \end{aligned}$$

The current through R_1 is

$$I_{R_1} = \frac{V_s}{R_{eq}} = \frac{4\ \text{V}}{1\Omega} = 4\ \text{A}$$

This current splits between R_2 and R_3 equally. Thus, the current through the inductor, which is the current through R_3, is

$$i(0) = 2\ \text{A}$$

The initial voltage across the capacitor, which is the voltage across the inductor, is zero. Thus, we have

$$v(0) = 0\ \text{V}$$

continued

Example 8.10 continued

Let

$$R = R_3 + R_2 = 1\ \Omega$$

The circuit for $t \geq 0$ is a parallel RLC circuit, as shown in Figure 8.8. From Equation (8.109), the differential equation is given by

$$\frac{d^2i(t)}{dt^2} + \frac{1}{RC}\frac{di(t)}{dt} + \frac{1}{LC}i(t) = \frac{d^2i(t)}{dt^2} + 4\frac{di(t)}{dt} + 4i(t) = 0$$

The characteristic equation is given by

$$s^2 + 4s + 4 = (s + 2)^2 = 0$$

The roots of the characteristic equation are

$$s_1 = -2, \quad s_2 = -2$$

This is case 2. The solution is given by

$$i(t) = A_1e^{-2t} + A_2te^{-2t}$$

for $t \geq 0$. At $t = 0$, we have

$$i(0) = A_1 = 2\ \text{A}$$

The voltage across the inductor, which is the voltage across the capacitor, is given by

$$v(t) = L\frac{di(t)}{dt}$$

At $t = 0$, we have

$$\frac{di(0)}{dt} = \frac{v(0)}{L} = \frac{0}{1} = 0$$

Taking the derivative of $i(t) = A_1e^{-2t} + A_2te^{-2t}$ with respect to t, we obtain

$$\frac{di(t)}{dt} = -2A_1e^{-2t} + A_2e^{-2t} - 2A_2te^{-2t}$$

Setting $t = 0$, we obtain

$$\frac{di(0)}{dt} = -2A_1 + A_2 = 0$$

Thus, we have

$$A_2 = 2A_1 = 4\ \text{A}$$

The current through the inductor for $t \geq 0$ is given by

$$i(t) = (2e^{-2t} + 4te^{-2t})u(t)\ \text{A}$$

continued

Example 8.10 continued

MATLAB

```
%EXAMPLE 8.10
clear all;
R1=0.75;R2=0.5;R3=0.5;L=1;C=0.25;Vs=4;
Req=R1+R2*R3/(R2+R3)
I0=Vs/Req*R2/(R2+R3)
V0=0
R=R2+R3
syms i(t)
Di=diff(i);
Di2=diff(i,2);
i=dsolve(Di2+1/(R*C)*Di+1/(L*C)*i,i(0)==I0,Di(0)==V0/L);
v=L*diff(i);
i=vpa(i,10)
v=vpa(v,10)

Answers:
Req =
     1
I0 =
     2
V0 =
     0
R =
     1
i =
2.0*exp(-2.0*t) + 4.0*t*exp(-2.0*t)
v =
-8.0*t*exp(-2.0*t)
```

Exercise 8.2

The switch in the circuit shown in Figure 8.10 has been closed for a long time before it is opened at $t = 0$. Find $i(t)$ for $t \geq 0$.

FIGURE 8.10

Circuit for EXERCISE 8.2.

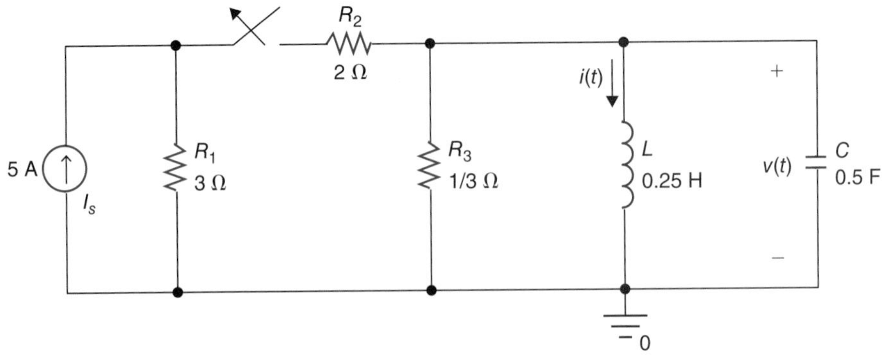

Answer:

$i(0) = 3\text{ A}, \quad v(0) = 0\text{ V}, \quad i(t) = (6e^{-2t} - 3e^{-4t})\,u(t).$

8.5 Solution of the Second-Order Differential Equations to Constant Input

The second-order differential equations with constant coefficients with constant input can be written as

$$c_2\frac{d^2v(t)}{dt^2} + c_1\frac{dv(t)}{dt} + c_0v(t) = d_0 \quad \textbf{(8.174)}$$

The initial conditions $v(0) = V_0$ and $\frac{dv(0)}{dt} = Dv_0$ are known. If Equation (8.174) is divided by c_2, we obtain

$$\frac{d^2v(t)}{dt^2} + \frac{c_1}{c_2}\frac{dv(t)}{dt} + \frac{c_0}{c_2}v(t) = \frac{d_0}{c_2} \quad \textbf{(8.175)}$$

Let

$$a_2 = 1,\ a_1 = \frac{c_1}{c_2},\ a_0 = \frac{c_0}{c_2},\ b_0 = \frac{d_0}{c_2}$$

Then Equation (8.175) becomes

$$\frac{d^2v(t)}{dt^2} + a_1\frac{dv(t)}{dt} + a_0v(t) = b_0 \quad \textbf{(8.176)}$$

The solution to the differential equation [Equation (8.176)] is the sum of the complementary solution $v_c(t)$ and the particular solution $v_p(t)$. The complementary solution, $v_c(t)$, is the solution to the homogeneous differential equation given by

$$\frac{d^2v(t)}{dt^2} + a_1\frac{dv(t)}{dt} + a_0v(t) = 0 \quad \textbf{(8.177)}$$

In Section 8.2, earlier in this chapter, we found the complementary solutions for case 1, case 2, and case 3. The complementary solutions are given by Equations (8.10), (8.19), and (8.32). The coefficients are found by applying the initial conditions to the sum of the particular solution and the complementary solution, as shown next.

8.5.1 PARTICULAR SOLUTION

The particular solution is the solution to the original differential equation given by Equation (8.176), including the input. The form of the particular solution is similar to the input signal. For the constant input, the particular solution will be a constant. Let the particular solution be

$$v_p(t) = K$$

Substitution of this proposed solution to Equation (8.176) results in

$$\frac{d^2K}{dt^2} + a_1\frac{dK}{dt} + a_0K = b_0$$

Since $\frac{dK}{dt} = 0$ and $\frac{d^2K}{dt^2} = 0$ for constant K, we have $a_0K = b_0$. Thus, the particular solution is given by

$$v_p(t) = K = \frac{b_0}{a_0} \quad \textbf{(8.178)}$$

The complete solution is given by the sum of the particular solution and the complementary solution:

$$v(t) = v_p(t) + v_c(t) = \frac{b_0}{a_0} + v_c(t)$$

The complementary solution will be one of the three solutions given in Equations (8.10), (8.19), and (8.32).

The coefficients can be found using the initial conditions $v(0) = V_0$ and $dv(0)/dt = Dv_0$ to the sum of the particular solution and the complementary solution.

8.5.2 CASE 1: OVERDAMPED ($\alpha > \omega_0$ or $a_1 > 2\sqrt{a_0}$ or $\zeta > 1$)

Characteristic equation: $s^2 + a_1 s + a_0 = 0$

Roots of the characteristic equation:

$$s_1 = -\frac{a_1}{2} + \sqrt{\left(\frac{a_1}{2}\right)^2 - a_0} = -\alpha + \sqrt{\alpha^2 - \omega_0^2},\ s_2 = -\frac{a_1}{2} - \sqrt{\left(\frac{a_1}{2}\right)^2 - a_0} = -\alpha - \sqrt{\alpha^2 - \omega_0^2} \quad (8.179)$$

where

$$\alpha = \frac{a_1}{2} \text{ and } \omega_0 = \sqrt{a_0}$$

Particular solution: $v_p(t) = \dfrac{b_0}{a_0}$

$$\text{Complementary solution: } v_c(t) = A_1 e^{s_1 t} + A_2 e^{s_2 t} \quad (8.180)$$

$$\text{Complete solution: } v(t) = v_p(t) + v_c(t) = \frac{b_0}{a_0} + A_1 e^{s_1 t} + A_2 e^{s_2 t} \quad (8.181)$$

The coefficients A_1 and A_2 are found from initial conditions $v(0) = V_0$ and $dv(0)/dt = Dv_0$. Setting $t = 0$ in Equation (8.181), we get

$$v(0) = \frac{b_0}{a_0} + A_1 + A_2 = V_0$$

which can be rewritten as

$$A_1 + A_2 = V_0 - \frac{b_0}{a_0} \quad (8.182)$$

Differentiating Equation (8.181) with respect to t, we get

$$\frac{dv(t)}{dt} = A_1 s_1 e^{s_1 t} + A_2 s_2 e^{s_2 t} \quad (8.183)$$

Setting $t = 0$ in Equation (8.183), we get

$$\frac{dv(0)}{dt} = A_1 s_1 + A_2 s_2 = Dv_0 \quad (8.184)$$

Coefficients A_1 and A_2 are found by solving Equations (8.182) and (8.184) for A_1 and A_2. Solving Equation (8.182) for A_2, we obtain

$$A_2 = -A_1 + V_0 - \frac{b_0}{a_0}$$

Substituting A_2 into Equation (8.184), we get

$$A_1 s_1 + \left(-A_1 + V_0 - \frac{b_0}{a_0}\right) s_2 = Dv_0$$

Solving for A_1, we get

$$A_1 = \frac{Dv_0 - s_2\left(V_0 - \frac{b_0}{a_0}\right)}{s_1 - s_2} = \frac{s_2\left(V_0 - \frac{b_0}{a_0}\right) - Dv_0}{s_2 - s_1}$$

Substituting A_1 into $A_2 = -A_1 + V_0 - \frac{b_0}{a_0}$, we obtain

$$A_2 = \frac{Dv_0 - s_1\left(V_0 - \frac{b_0}{a_0}\right)}{s_2 - s_1}$$

Alternatively, Cramer's rule can be used, as shown here:

$$A_1 = \frac{\begin{vmatrix} V_0 - \frac{b_0}{a_0} & 1 \\ Dv_0 & s_2 \end{vmatrix}}{\begin{vmatrix} 1 & 1 \\ s_1 & s_2 \end{vmatrix}} = \frac{s_2\left(V_0 - \frac{b_0}{a_0}\right) - Dv_0}{s_2 - s_1} \tag{8.185}$$

$$A_2 = \frac{\begin{vmatrix} 1 & V_0 - \frac{b_0}{a_0} \\ s_1 & Dv_0 \end{vmatrix}}{\begin{vmatrix} 1 & 1 \\ s_1 & s_2 \end{vmatrix}} = \frac{Dv_0 - s_1\left(V_0 - \frac{b_0}{a_0}\right)}{s_2 - s_1} \tag{8.186}$$

8.5.3 CASE 2: CRITICALLY DAMPED ($\alpha = \omega_0$ or $a_1 = 2\sqrt{a_0}$ or $\zeta = 1$)

Characteristic equation: $s^2 + a_1 s + a_0 = 0$

Roots of the characteristic equation:

$$s_1 = s_2 = -\frac{a_1}{2} = -\alpha$$

Particular solution: $v_p(t) = \frac{b_0}{a_0}$

$$\text{Complementary solution: } v_c(t) = A_1 e^{s_1 t} + A_2 t e^{s_2 t} \tag{8.187}$$

$$\text{Complete solution: } v(t) = v_p(t) + v_c(t) = \frac{b_0}{a_0} + A_1 e^{s_1 t} + A_2 t e^{s_2 t} \tag{8.188}$$

The coefficients A_1 and A_2 are found from initial conditions $v(0) = V_0$ and $dv(0)/dt = Dv_0$. Setting $t = 0$ in Equation (8.188), we get

$$v(0) = \frac{b_0}{a_0} + A_1 = V_0 \tag{8.189}$$

Solving Equation (8.189) for A_1, we get

$$A_1 = V_0 - \frac{b_0}{a_0} \tag{8.190}$$

Differentiating Equation (8.188) with respect to t, we get

$$\frac{dv(t)}{dt} = A_1 s_1 e^{s_1 t} + A_2 e^{s_2 t} + A_2 s_2 t e^{s_2 t} \tag{8.191}$$

Setting $t = 0$ in Equation (8.191), we get

$$\frac{dv(0)}{dt} = A_1 s_1 + A_2 = Dv_0 \tag{8.192}$$

Solving Equation (8.192) for A_2, we obtain

$$A_2 = Dv_0 - A_1 s_1 \tag{8.193}$$

8.5.4 CASE 3: UNDERDAMPED ($\alpha < \omega_0$ or $a_1 < 2\sqrt{a_0}$ or $\zeta < 1$)

Characteristic equation: $s^2 + a_1 s + a_0 = 0$

Roots of the characteristic equation:

$$s_1 = -\alpha + j\beta, \quad s_2 = -\alpha - j\beta$$

where

$$\alpha = \frac{a_1}{2}$$

$$\beta = \sqrt{a_0 - \left(\frac{a_1}{2}\right)^2} = \sqrt{\omega_0^2 - \alpha^2}$$

Particular solution: $v_p(t) = \dfrac{b_0}{a_0}$

Complementary solution: $v_c(t) = e^{-\alpha t}[B_1 \cos(\beta t) + B_2 \sin(\beta t)]$ **(8.194)**

Complete solution: $v(t) = v_p(t) + v_c(t) = \dfrac{b_0}{a_0} + e^{-\alpha t}[B_1 \cos(\beta t) + B_2 \sin(\beta t)]$ **(8.195)**

Coefficients B_1 and B_2 are found from initial conditions $v(0) = V_0$ and $dv(0)/dt = Dv_0$. Setting $t = 0$ in Equation (8.195), we get

$$v(0) = \frac{b_0}{a_0} + B_1 = V_0 \tag{8.196}$$

Solving Equation (8.196) for B_1, we get

$$B_1 = V_0 - \frac{b_0}{a_0} \tag{8.197}$$

Differentiating Equation (8.195) with respect to t, we get

$$\frac{dv(t)}{dt} = e^{-\alpha t}[(-\alpha B_1 + \beta B_2)\cos(\beta t) + (-\beta B_1 - \alpha B_2)\sin(\beta t)] \quad \textbf{(8.198)}$$

Setting $t = 0$ in Equation (8.198), we get

$$\frac{dv(0)}{dt} = -\alpha B_1 + \beta B_2 = Dv_0 \quad \textbf{(8.199)}$$

Solving Equation (8.199) for B_2, we obtain

$$B_2 = \frac{Dv_0 + \alpha B_1}{\beta} \quad \textbf{(8.200)}$$

8.6 Step Response of a Series RLC Circuit

The switch in the circuit shown in Figure 8.11 is closed at time $t = 0$.

FIGURE 8.11

A series RLC circuit with step input.

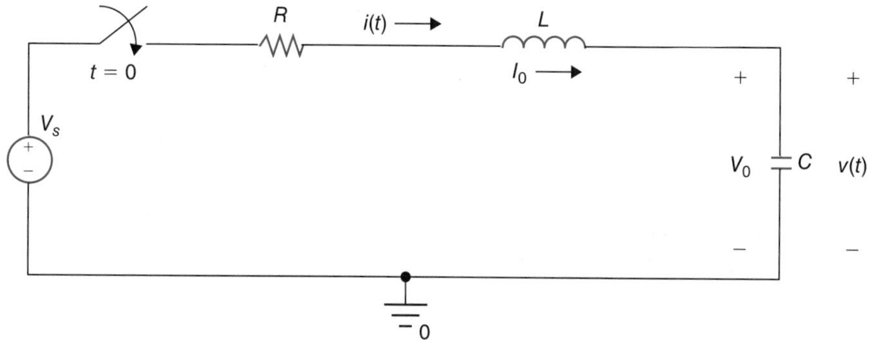

Let the initial voltage across the capacitor at $t = 0$ be V_0, and the initial current through the inductor at $t = 0$ be I_0. Let $i(t)$ be the mesh current of the series RLC circuit shown in Figure 8.11 and $v(t)$ be the voltage across the capacitor. Then, adding the voltage drops around the mesh in the clockwise direction starting from the voltage source, we have

$$-V_s + Ri(t) + L\frac{di(t)}{dt} + v(t) = 0 \quad \textbf{(8.201)}$$

The current through the capacitor $i(t)$ is related to the voltage across the capacitor $v(t)$ by

$$i(t) = C\frac{dv(t)}{dt} \quad \textbf{(8.202)}$$

Substituting Equation (8.202) into Equation (8.201), we have

$$RC\frac{dv(t)}{dt} + LC\frac{d^2v(t)}{dt^2} + v(t) = V_s \quad \textbf{(8.203)}$$

Rearrangement of Equation (8.203) yields

$$\frac{d^2v(t)}{dt^2} + \frac{R}{L}\frac{dv(t)}{dt} + \frac{1}{LC}v(t) = \frac{V_s}{LC} \quad \textbf{(8.204)}$$

Comparison of Equation (8.204) with Equation (8.176) reveals that

$$a_1 = \frac{R}{L},\ a_0 = \frac{1}{LC},\ b_0 = \frac{V_s}{LC} \tag{8.205}$$

The neper frequency is given by

$$\alpha = \frac{a_1}{2} = \frac{R}{2L}\ \text{Np/s} \tag{8.206}$$

and the resonant frequency is given by

$$\omega_0 = \sqrt{a_0} = \frac{1}{\sqrt{LC}}\ \text{rad/s} \tag{8.207}$$

The particular solution is, from Equations (8.178) and (8.205),

$$v_p(t) = \frac{b_0}{a_0} = \frac{\dfrac{V_s}{LC}}{\dfrac{1}{LC}} = V_s \tag{8.208}$$

The complementary solution contains the exponential decay term $\exp(-\alpha t)$ with $\alpha > 0$. As time t increases, the complementary solution vanishes, and the general solution consists only of the particular solution. In the circuit shown in Figure 8.11, since the voltage provided by the source is constant, the current through the capacitor will be zero after the complementary solution disappears. Without any current in the circuit, the voltage drop across the resistor and inductor is zero. Therefore, the voltage across the capacitor will be the same as the voltage from the source, V_s. Thus, the particular solution must be V_s. The particular solution is the voltage across the capacitor at time $t = \infty$. Therefore, we have

$$v_p(t) = v(\infty) = V_s$$

From Equation (8.202), we have

$$\frac{dv(t)}{dt} = \frac{i(t)}{C} \tag{8.209}$$

Setting $t = 0$, we get

$$Dv_0 = \frac{dv(0)}{dt} = \frac{i(0)}{C} = \frac{I_0}{C} \tag{8.210}$$

The complete solutions with the coefficients are listed below for the three cases for the series RLC circuit shown in Figure 8.11.

8.6.1 CASE 1: OVERDAMPED ($\alpha > \omega_0$ or $a_1/2 > \sqrt{a_0}$ or $\zeta > 1$)

Characteristic equation: $s^2 + a_1 s + a_0 = s^2 + \frac{R}{L}s + \frac{1}{LC} = 0$

Neper frequency: $\alpha = \frac{a_1}{2} = \frac{R}{2L}$ Np/s

Resonant frequency: $\omega_0 = \sqrt{a_0} = \frac{1}{\sqrt{LC}}$ rad/s

Roots of the characteristic equation:

$$s_1 = -\frac{a_1}{2} + \sqrt{\left(\frac{a_1}{2}\right)^2 - a_0} = -\alpha + \sqrt{\alpha^2 - \omega_0^2} = -\frac{R}{2L} + \sqrt{\left(\frac{R}{2L}\right)^2 - \frac{1}{LC}}$$
$$s_2 = -\frac{a_1}{2} - \sqrt{\left(\frac{a_1}{2}\right)^2 - a_0} = -\alpha - \sqrt{\alpha^2 - \omega_0^2} = -\frac{R}{2L} - \sqrt{\left(\frac{R}{2L}\right)^2 - \frac{1}{LC}} \qquad \textbf{(8.211)}$$

Particular solution: $v_p(t) = \frac{b_0}{a_0} = \frac{\frac{V_s}{LC}}{\frac{1}{LC}} = V_s$

Complementary solution: $v_c(t) = A_1 e^{s_1 t} + A_2 e^{s_2 t}$

$$\text{Complete solution: } v(t) = v_p(t) + v_c(t) = V_s + A_1 e^{s_1 t} + A_2 e^{s_2 t} \qquad \textbf{(8.212)}$$

Initial conditions: $v(0) = V_0, \quad Dv_0 = \frac{dv(0)}{dt} = \frac{i(0)}{C} = \frac{I_0}{C}$

The coefficients A_1 and A_2 are found from initial conditions. Setting $t = 0$ in Equation (8.212), we get

$$v(0) = V_s + A_1 + A_2 = V_0$$

which can be rewritten as

$$A_1 + A_2 = V_0 - V_s \qquad \textbf{(8.213)}$$

Differentiating Equation (8.212) with respect to t, we get

$$\frac{dv(t)}{dt} = A_1 s_1 e^{s_1 t} + A_2 s_2 e^{s_2 t} \qquad \textbf{(8.214)}$$

Setting $t = 0$ in Equation (8.214), we get

$$\frac{dv(0)}{dt} = A_1 s_1 + A_2 s_2 = Dv_0 \qquad \textbf{(8.215)}$$

Coefficients A_1 and A_2 are found by solving Equations (8.213) and (8.215) for A_1 and A_2. Multiplying Equation (8.213) by s_2, we obtain

$$s_2 A_1 + s_2 A_2 = s_2(V_0 - V_s)$$

Subtracting Equation (8.215) from this equation, we get

$$(s_2 - s_1)A_1 = s_2(V_0 - V_s) - Dv_0$$

Thus,

$$A_1 = \frac{s_2(V_0 - V_s) - Dv_0}{s_2 - s_1}$$

$$A_2 = V_0 - V_s - A_1 = V_0 - V_s - \frac{s_2(V_0 - V_s) - Dv_0}{s_2 - s_1} = \frac{Dv_0 - s_1(V_0 - V_s)}{s_2 - s_1}$$

Alternatively, Cramer's rule can be used, as shown here:

$$A_1 = \frac{\begin{vmatrix} V_0 - V_s & 1 \\ Dv_0 & s_2 \end{vmatrix}}{\begin{vmatrix} 1 & 1 \\ s_1 & s_2 \end{vmatrix}} = \frac{s_2(V_0 - V_s) - Dv_0}{s_2 - s_1} \quad (8.216)$$

$$A_2 = \frac{\begin{vmatrix} 1 & V_0 - V_s \\ s_1 & Dv_0 \end{vmatrix}}{\begin{vmatrix} 1 & 1 \\ s_1 & s_2 \end{vmatrix}} = \frac{Dv_0 - s_1(V_0 - V_s)}{s_2 - s_1} \quad (8.217)$$

The current through the mesh is given by

$$i(t) = C\frac{dv(t)}{dt} = C(A_1 s_1 e^{s_1 t} + A_2 s_2 e^{s_2 t})u(t) \quad (8.218)$$

8.6.2 CASE 2: CRITICALLY DAMPED ($\alpha = \omega_0$ or $a_1 = 2\sqrt{a_0}$ or $\zeta = 1$)

Characteristic equation: $s^2 + a_1 s + a_0 = s^2 + \frac{R}{L}s + \frac{1}{LC} = 0$

Neper frequency: $\alpha = \frac{a_1}{2} = \frac{R}{2L}$ Np/s

Resonant frequency: $\omega_0 = \sqrt{a_0} = \frac{1}{\sqrt{LC}}$ rad/s

Roots of the characteristic equation:

$$s_1 = s_2 = -\frac{a_1}{2} = -\alpha = -\frac{R}{2L}$$

Particular solution: $v_p(t) = \frac{b_0}{a_0} = \frac{\frac{V_s}{LC}}{\frac{1}{LC}} = V_s$

Complementary solution: $v_c(t) = A_1 e^{s_1 t} + A_2 t e^{s_2 t}$

$$\text{Complete solution: } v(t) = v_p(t) + v_c(t) = V_s + A_1 e^{s_1 t} + A_2 t e^{s_2 t} \quad (8.219)$$

Initial conditions: $v(0) = V_0, Dv_0 = \frac{dv(0)}{dt} = \frac{i(0)}{C} = \frac{I_0}{C}$

Coefficients A_1 and A_2 are found from initial conditions. Setting $t = 0$ in Equation (8.219), we get

$$v(0) = V_s + A_1 = V_0 \quad (8.220)$$

Solving Equation (8.220) for A_1, we get

$$A_1 = V_0 - V_s \quad (8.221)$$

Differentiating Equation (8.219) with respect to t, we get

$$\frac{dv(t)}{dt} = A_1 s_1 e^{s_1 t} + A_2 e^{s_2 t} + A_2 s_2 t e^{s_2 t} \tag{8.222}$$

Setting $t = 0$ in Equation (8.222), we get

$$\frac{dv(0)}{dt} = A_1 s_1 + A_2 = Dv_0 \tag{8.223}$$

Solving Equation (8.223) for A_2, we obtain

$$A_2 = Dv_0 - A_1 s_1 = Dv_0 - (V_0 - V_s)s_1 \tag{8.224}$$

The current through the mesh is given by

$$i(t) = C\frac{dv(t)}{dt} = C[(A_1 s_1 + A_2)e^{s_1 t} + A_2 s_2 t e^{s_2 t}]u(t) \tag{8.225}$$

8.6.3 CASE 3: UNDERDAMPED ($\alpha < \omega_0$ or $a_1 < 2\sqrt{a_0}$ or $\zeta < 1$)

Characteristic equation: $s^2 + a_1 s + a_0 = s^2 + \frac{R}{L}s + \frac{1}{LC} = 0$

Neper frequency: $\alpha = \frac{a_1}{2} = \frac{R}{2L}$ Np/s

Resonant frequency: $\omega_0 = \sqrt{a_0} = \frac{1}{\sqrt{LC}}$ rad/s

Damped resonant frequency: $\beta = \sqrt{a_0 - \left(\frac{a_1}{2}\right)^2} = \sqrt{\omega_0^2 - \alpha^2} = \sqrt{\frac{1}{LC} - \left(\frac{R}{2L}\right)^2}$

Roots of the characteristic equation:

$$s_1 = -\alpha + j\beta, \quad s_2 = -\alpha - j\beta$$

Particular solution: $v_p(t) = \frac{b_0}{a_0} = \frac{\frac{V_s}{LC}}{\frac{1}{LC}} = V_s$

Complementary solution: $v_c(t) = e^{-\alpha t}[B_1 \cos(\beta t) + B_2 \sin(\beta t)]$

$$\text{Complete solution: } v(t) = v_p(t) + v_c(t) = V_s + e^{-\alpha t}[B_1 \cos(\beta t) + B_2 \sin(\beta t)] \tag{8.226}$$

Initial conditions: $v(0) = V_0$, $Dv_0 = \frac{dv(0)}{dt} = \frac{i(0)}{C} = \frac{I_0}{C}$

Coefficients B_1 and B_2 are found from initial conditions. Setting $t = 0$ in Equation (8.226), we get

$$v(0) = V_s + B_1 = V_0 \tag{8.227}$$

Solving Equation (8.227) for B_1, we get

$$B_1 = V_0 - V_s \tag{8.228}$$

Differentiating Equation (8.226) with respect to t, we get

$$\frac{dv(t)}{dt} = e^{-\alpha t}[(-\alpha B_1 + \beta B_2)\cos(\beta t) + (-\beta B_1 - \alpha B_2)\sin(\beta t)] \tag{8.229}$$

Setting $t = 0$ in Equation (8.229), we get

$$\frac{dv(0)}{dt} = -\alpha B_1 + \beta B_2 = Dv_0 \tag{8.230}$$

Solving Equation (8.230) for B_2, we obtain

$$B_2 = \frac{Dv_0 + \alpha B_1}{\beta} = \frac{Dv_0 + \alpha(V_0 - V_s)}{\beta} \tag{8.231}$$

The current through the mesh is given by

$$i(t) = C\frac{dv(t)}{dt} = Ce^{-\alpha t}[(-\alpha B_1 + \beta B_2)\cos(\beta t) + (-\beta B_1 - \alpha B_2)\sin(\beta t)]u(t) \tag{8.232}$$

EXAMPLE 8.11

Let $R = 1.5\ k\Omega$, $L = 50$ mH, $C = 0.1\ \mu$F, $V_0 = 5$ V, $I_0 = 6$ mA, and $V_s = 10$ V in the circuit shown in Figure 8.11. Find the voltage $v(t)$ across the capacitor, current $i(t)$ through the capacitor, voltage $v_R(t)$ across the resistor, and the voltage $v_L(t)$ across the inductor. Plot $v(t)$, $i(t)$, $v_R(t)$, and $v_L(t)$.

The coefficients are given by

$$a_2 = 1,\ a_1 = \frac{R}{L} = 30{,}000,\ a_0 = \frac{1}{LC} = 2 \times 10^8,\ b_0 = \frac{V_s}{LC} = 2 \times 10^9$$

The neper frequency is given by

$$\alpha = \frac{a_1}{2} = \frac{R}{2L} = 15{,}000 \text{ Np/s}$$

The resonant frequency is

$$\omega_0 = \sqrt{a_0} = \frac{1}{\sqrt{LC}} = 14{,}142.1356 \text{ rad/s}$$

Since $\alpha > \omega_0$, this is case 1, and $v(t)$ is overdamped. The characteristic equation is

$$s^2 + a_1 s + a_0 = s^2 + \frac{R}{L}s + \frac{1}{LC} = s^2 + 30{,}000s + 2 \times 10^8$$
$$= (s + 10{,}000)(s + 20{,}000) = 0$$

The roots of the characteristic equation are

$$s_1 = -10{,}000,\quad s_2 = -20{,}000$$

continued

Example 8.11 continued

The roots of the characteristic equation can also be found from

$$s_1 = -\alpha + \sqrt{\alpha^2 - \omega_0^2} = -10{,}000$$

$$s_2 = -\alpha - \sqrt{\alpha^2 - \omega_0^2} = -20{,}000$$

The complementary solution can be written as

$$y_c(t) = A_1 e^{-10{,}000t} + A_2 e^{-20{,}000t}$$

The particular solution is given by

$$v_p(t) = V_s = 10\text{ V}$$

The solution to the differential equation is the sum of the particular solution and the complementary solution:

$$v(t) = v_p(t) + v_c(t) = 10 + A_1 e^{-10{,}000t} + A_2 e^{-20{,}000t} \tag{8.233}$$

At $t = 0$, Equation (8.233) becomes

$$v(0) = 10 + A_1 e^{-10{,}000\times 0} + A_2 e^{-20{,}000\times 0} = 10 + A_1 + A_2 = V_0 = 5$$

which can be rearranged as

$$A_1 + A_2 = -5 \tag{8.234}$$

Taking the derivative of Equation (8.233), we obtain

$$\frac{dv(t)}{dt} = -10{,}000 A_1 e^{-10{,}000t} - 20{,}000 A_2 e^{-20{,}000t} \tag{8.235}$$

Setting $t = 0$ in Equation (8.235) and from Equation (8.210), we get

$$\frac{dv(0)}{dt} = -10{,}000 A_1 - 20{,}000 A_2 = \frac{I_0}{C} = \frac{6 \times 10^{-3}}{10^{-7}} = 60{,}000$$

which simplifies to

$$-A_1 - 2A_2 = 6 \tag{8.236}$$

Sum of Equations (8.234) and (8.236) results in $-A_2 = 1$ or $A_2 = -1$. Substitution of this result into Equation (8.236) yields $A_1 = -4$. Thus, the voltage across the capacitor, Equation (8.233), becomes

$$v(t) = (10 - 4e^{-10{,}000t} - e^{-20{,}000t})\,u(t)\text{ V} \tag{8.237}$$

The current through the capacitor is

$$\begin{aligned} i(t) &= C\frac{dv(t)}{dt} = 10^{-7}(-10{,}000(-4)e^{-10{,}000t} - 20{,}000(-1)e^{-20{,}000t}) \\ &= (4e^{-10{,}000t} + 2e^{-20{,}000t})u(t)\text{ mA} \end{aligned} \tag{8.238}$$

continued

Example 8.11 continued

The voltage across the resistor is given by

$$v_R(t) = Ri(t) = 1500(4 \times 10^{-3}e^{-10{,}000t} + 2 \times 10^{-3}e^{-20{,}000t})u(t)$$
$$= (6e^{-10{,}000t} + 3e^{-20{,}000t})u(t) \text{ V} \quad \textbf{(8.239)}$$

The voltage across the inductor is given by

$$v_L(t) = L\frac{di(t)}{dt} = 0.05(-10{,}000 \times 4 \times 10^{-3}e^{-10{,}000t}$$
$$- 20{,}000 \times 2 \times 10^{-3}e^{-20{,}000t})u(t) \quad \textbf{(8.240)}$$
$$= (-2e^{-10{,}000t} - 2e^{-20{,}000t})u(t) \text{ V}$$

Notice that the sum of $v(t)$, $v_R(t)$, and $v_L(t)$ equals the source voltage V_s. Figure 8.12 shows the plots of $v(t)$, $i(t)$, $v_R(t)$, and $v_L(t)$.

FIGURE 8.12

Plot for EXAMPLE 8.11.

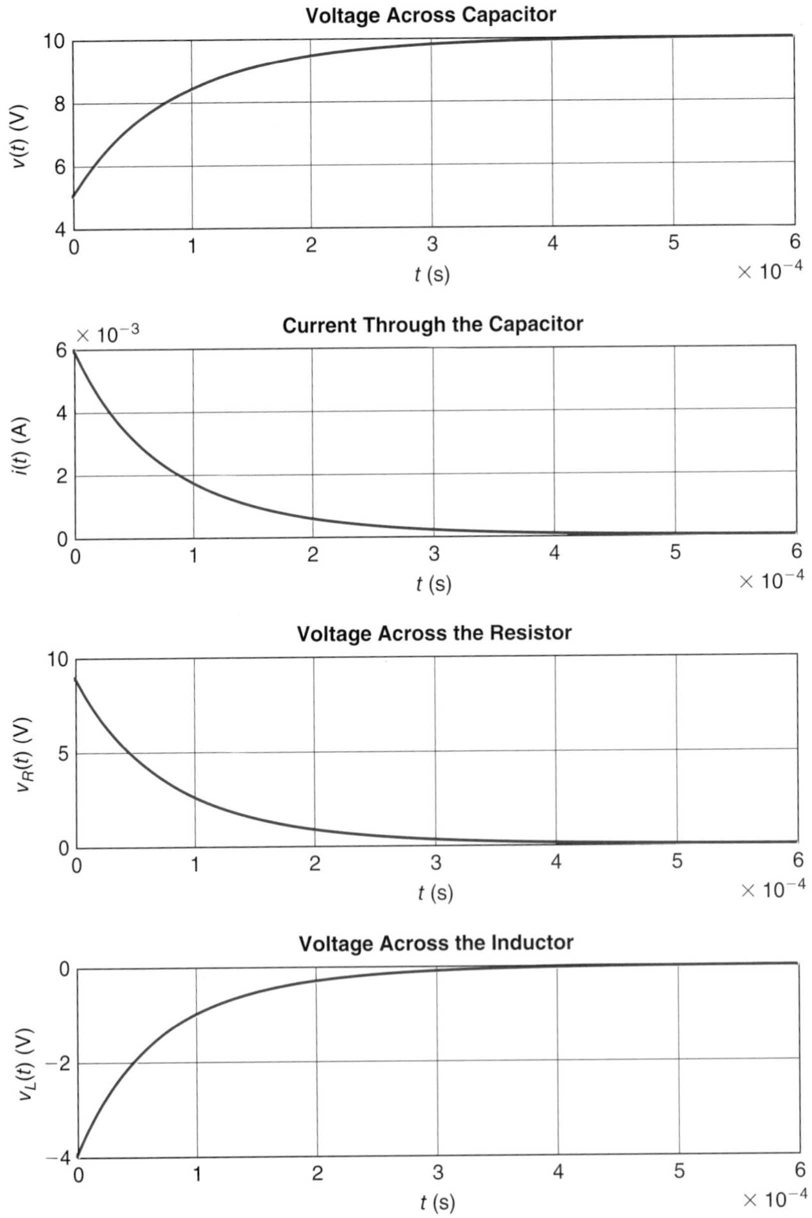

EXAMPLE 8.12

Let $R = 400\ \Omega$, $L = 100$ mH, $C = 2.5\ \mu$F, $V_0 = 6$ V, $I_0 = 5$ mA, and $V_s = 16$ V in the circuit shown in Figure 8.11. Find voltage $v(t)$ across the capacitor, current $i(t)$ through the capacitor, voltage $v_R(t)$ across the resistor, and voltage $v_L(t)$ across the inductor. Plot $v(t)$, $i(t)$, $v_R(t)$, and $v_L(t)$.

The coefficients are given by

$$a_2 = 1,\ a_1 = \frac{R}{L} = 4000,\ a_0 = \frac{1}{LC} = 4 \times 10^6,\ b_0 = \frac{V_s}{LC} = 64 \times 10^6$$

The neper frequency is given by

$$\alpha = \frac{a_1}{2} = \frac{R}{2L} = 2000 \text{ Np/s}$$

The resonant frequency is

$$\omega_0 = \sqrt{a_0} = \frac{1}{\sqrt{LC}} = 2000 \text{ rad/s}$$

Since $\alpha = \omega_0$, this is case 2, and $v(t)$ is critically damped. The characteristic equation is

$$s^2 + a_1 s + a_0 = s^2 + \frac{R}{L}s + \frac{1}{LC} = s^2 + 4000s + 4 \times 10^6 = (s + 2000)(s + 2000) = 0$$

The roots of the characteristic equation are

$$s_1 = -2000,\quad s_2 = -2000$$

The roots of the characteristic equation can also be found from

$$s_1 = -a_1/2 = -2000,\quad s_2 = -a_1/2 = -2000$$

The complementary solution can be written as

$$y_c(t) = A_1 e^{-2000t} + A_2 t e^{-2000t}$$

The particular solution is given by

$$v_p(t) = V_s = 16 \text{ V}$$

The solution to the differential equation is the sum of the particular solution and the complementary solution:

$$v(t) = v_p(t) + v_c(t) = 16 + A_1 e^{-2000t} + A_2 t e^{-2000t} \qquad \textbf{(8.241)}$$

At $t = 0$, Equation (8.241) becomes

$$v(0) = 16 + A_1 e^{-2000\times 0} + A_2 \times 0 \times e^{-2000\times 0} = 16 + A_1 = V_0 = 6$$

Thus, $A_1 = -10$ V.

continued

Example 8.12 continued

Taking the derivative of Equation (8.241), we obtain

$$\frac{dv(t)}{dt} = -2000A_1e^{-2000t} + A_2e^{-2000t} - 2000A_2te^{-2000t} \quad \textbf{(8.242)}$$

Setting $t = 0$ in Equation (8.242) and from Equation (8.210), we get

$$\frac{dv(0)}{dt} = -2000A_1 + A_2 = \frac{I_0}{C} = \frac{5 \times 10^{-3}}{2.5 \times 10^{-6}} = 2000$$

from which we obtain

$$A_2 = 2000A_1 + 2000 = -18{,}000$$

Thus, the voltage across the capacitor, Equation (8.241), becomes

$$v(t) = 16 - 10e^{-2000t} - 18{,}000te^{-2000t} \quad \textbf{(8.243)}$$

The current through the capacitor is

$$\begin{aligned} i(t) = C\frac{dv(t)}{dt} &= 2.5 \times 10^{-6}\big(-2000(-10)e^{-2000t} - 18{,}000e^{-2000t} \\ &\quad - 18{,}000(-2000)te^{-2000t}\big) \\ &= (0.005e^{-2000t} + 90te^{-2000t})u(t)\text{ A} \end{aligned} \quad \textbf{(8.244)}$$

The voltage across the resistor is given by

$$\begin{aligned} v_R(t) = Ri(t) &= 400(0.005e^{-2000t} + 90te^{-2000t})u(t) \\ &= (2e^{-2000t} + 36{,}000te^{-2000t})u(t)\text{ V} \end{aligned} \quad \textbf{(8.245)}$$

The voltage across the inductor is given by

$$\begin{aligned} v_L(t) = L\frac{di(t)}{dt} &= 0.1\big(-2000 \times 5 \times 10^{-3}e^{-10{,}000t} + 90e^{-10{,}000t} \\ &\quad - 2000 \times 90te^{-20{,}000t}\big)u(t) \\ &= \big(8e^{-2000t} - 18{,}000te^{-2000t}\big)u(t)\text{ V} \end{aligned} \quad \textbf{(8.246)}$$

Notice that the sum of $v(t)$, $v_R(t)$, and $v_L(t)$ equals the source voltage V_S. Figure 8.13 shows the plots of $v(t)$, $i(t)$, $v_R(t)$, and $v_L(t)$.

FIGURE 8.13

Plots for EXAMPLE 8.12.

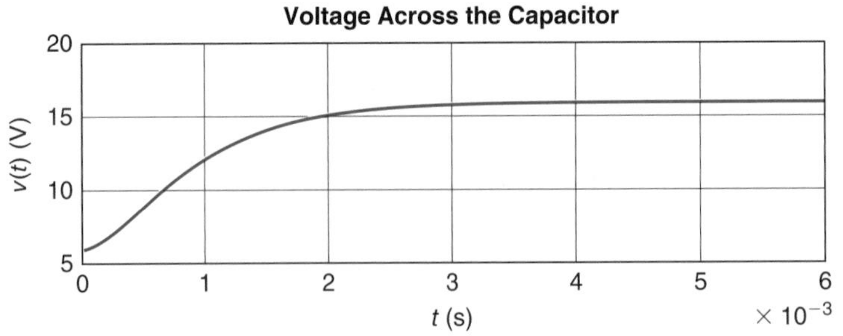

continued

Example 8.12 continued

FIGURE 8.13

continued

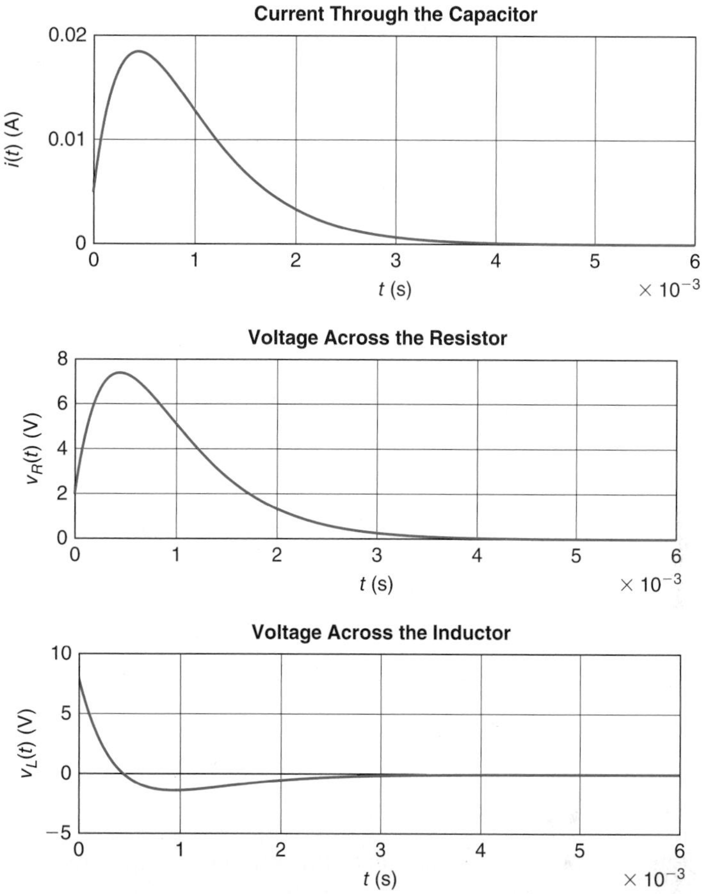

EXAMPLE 8.13

Let $R = 400\ \Omega$, $L = 100$ mH, $C = 0.5\ \mu$F, $V_0 = 5$ V, $I_0 = 4$ mA, and $V_s = 15$ V in the circuit shown in Figure 8.11. Find voltage $v(t)$ across the capacitor, current $i(t)$ through the capacitor, voltage $v_R(t)$ across the resistor, and voltage $v_L(t)$ across the inductor. Plot $v(t)$, $i(t)$, $v_R(t)$, and $v_L(t)$.

The coefficients are given by

$$a_2 = 1,\ \ a_1 = \frac{R}{L} = 4000,\ \ a_0 = \frac{1}{LC} = 2 \times 10^7,\ \ b_0 = \frac{V_s}{LC} = 30 \times 10^7$$

The neper frequency is given by

$$\alpha = \frac{a_1}{2} = \frac{R}{2L} = 2000 \text{ Np/s}$$

The resonant frequency is

$$\omega_0 = \sqrt{a_0} = \frac{1}{\sqrt{LC}} = 4472.135955 \text{ rad/s}$$

continued

Example 8.13 continued

Since $\alpha < \omega_0$, this is case 3, and $v(t)$ is underdamped. The damped resonant frequency is

$$\beta = \sqrt{a_0 - \left(\frac{a_1}{2}\right)^2} = \sqrt{\omega_0^2 - \alpha^2} = \sqrt{\frac{1}{LC} - \left(\frac{R}{2L}\right)^2} = 4000 \text{ rad/s}$$

The characteristic equation is

$$s^2 + a_1 s + a_0 = s^2 + \frac{R}{L}s + \frac{1}{LC} = s^2 + 4000s + 2 \times 10^7$$
$$= (s + 2000 - j4000)(s + 2000 + j4000) = 0$$

The roots of the characteristic equation are

$$s_1 = -2000 + j4000, \quad s_2 = -2000 - j4000$$

The roots of the characteristic equation can also be found from

$$s_1 = -\alpha + j\beta = -2000 + j4000, \quad s_2 = -\alpha - j\beta = -2000 - j4000$$

The complementary solution can be written as

$$y_c(t) = e^{-2000t}[B_1 \cos(4000t) + B_2 \sin(4000t)]$$

The particular solution is given by

$$v_p(t) = V_s = 15 \text{ V}$$

The solution to the differential equation is the sum of the particular solution and the complementary solution:

$$v(t) = v_p(t) + v_c(t) = 15 + e^{-2000t}[B_1 \cos(4000t) + B_2 \sin(4000t)] \quad \textbf{(8.247)}$$

At $t = 0$, Equation (8.247) becomes

$$v(0) = 15 + e^{-2000 \times 0}[B_1 \cos(4000 \times 0) + B_2 \sin(4000 \times 0)] = 15 + B_1 = V_0 = 5$$

Thus, $B_1 = -10$ V.
Taking the derivative of Equation (8.247), we obtain

$$\begin{aligned}\frac{dv(t)}{dt} &= e^{-\alpha t}[(-\alpha B_1 + \beta B_2)\cos(\beta t) + (-\beta B_1 - \alpha B_2)\sin(\beta t)] \\ &= e^{-2000t}[(-2000B_1 + 4000B_2)\cos(4000t) \\ &\quad + (-4000B_1 - 2000B_2)\sin(4000t)]\end{aligned} \quad \textbf{(8.248)}$$

Setting $t = 0$ in Equation (8.248) and from Equation (8.210), we get

$$\frac{dv(0)}{dt} = -2000B_1 + 4000B_2 = \frac{I_0}{C} = \frac{4 \times 10^{-3}}{0.5 \times 10^{-6}} = 8000$$

from which we obtain

$$B_2 = \frac{B_1 + 4}{2} = \frac{-10 + 4}{2} = -3$$

continued

Example 8.13 continued

Thus, the voltage across the capacitor, Equation (8.247), becomes

$$v(t) = v_p(t) + v_c(t) = 15 + e^{-2000t}[-10\cos(4000t) - 3\sin(4000t)]\text{ V} \quad \textbf{(8.249)}$$

The current through the capacitor is

$$\begin{aligned} i(t) &= C\frac{dv(t)}{dt} = Ce^{-\alpha t}[(-\alpha B_1 + \beta B_2)\cos(\beta t) + (-\beta B_1 - \alpha B_2)\sin(\beta t)] \\ &= 5\times 10^{-7}e^{-2000t}[(20{,}000 - 12{,}000)\cos(4000t) + (40{,}000 + 6000)\sin(4000t)] \\ &= e^{-2000t}[4\cos(4000t) + 23\sin(4000t)]u(t)\text{ mA} \end{aligned} \quad \textbf{(8.250)}$$

The voltage across the resistor is given by

$$\begin{aligned} v_R(t) &= Ri(t) = 400e^{-2000t}[4\times 10^{-3}\cos(4000t) + 23\times 10^{-3}\sin(4000t)]u(t)\text{ V} \\ &= e^{-2000t}[1.6\cos(4000t) + 9.2\sin(4000t)]u(t)\text{ V} \end{aligned} \quad \textbf{(8.251)}$$

The voltage across the inductor is given by

$$\begin{aligned} v_L(t) &= L\frac{di(t)}{dt} \\ &= LCe^{-\alpha t}\{[(\alpha^2 - \beta^2)B_1 - 2\alpha\beta B_2]\cos(\beta t) + [2\alpha\beta B_1 + (\alpha^2 - \beta^2)B_2]\sin(\beta t)\} \\ &= e^{-2000t}[8.4\cos(4000t) - 6.2\sin(4000t)]u(t)\text{ V} \end{aligned} \quad \textbf{(8.252)}$$

Notice that the sum of $v(t)$, $v_R(t)$, and $v_L(t)$ equals the source voltage V_S. Figure 8.14 shows the plots of $v(t)$, $i(t)$, $v_R(t)$, and $v_L(t)$.

FIGURE 8.14

Plots for EXAMPLE 8.13.

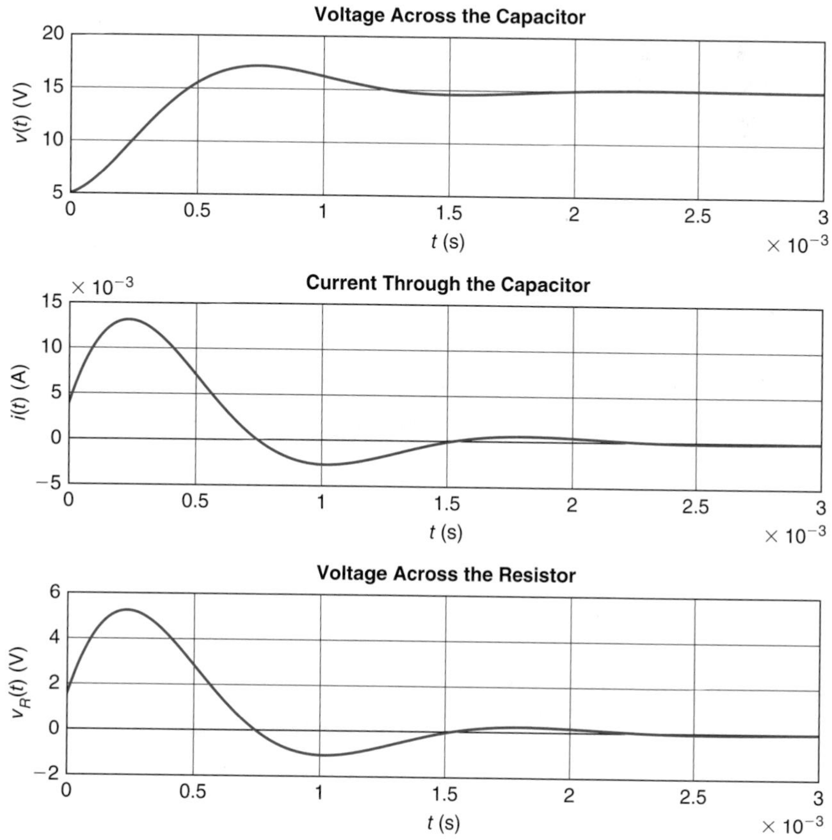

continued

Example 8.13 continued

FIGURE 8.14

continued

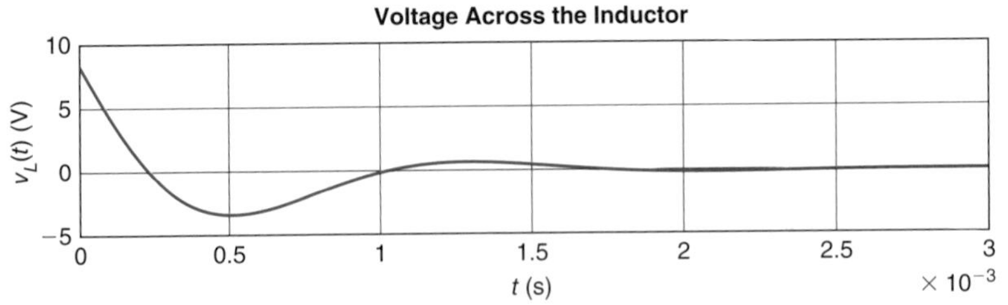

MATLAB

```
% EXAMPLE 8.13
clear all;
R=400;L=100e-3;C=0.5e-6;V0=5;I0=4e-3;Vs=15;
syms v(t)
Dv=diff(v);
Dv2=diff(v,2);
v=dsolve(Dv2+R/L*Dv+1/(L*C)*v==Vs/(L*C),...
    v(0)==V0,Dv(0)==I0/C);
i=C*diff(v);
vR=R*i;
vL=L*diff(i);
v=vpa(v,10)
i=vpa(i,10)
vR=vpa(vR,10)
vL=vpa(vL,10)
ta=0:1e-5:3e-3;
v=subs(v,t,ta);
i=subs(i,t,ta);
vR=subs(vR,t,ta);
vL=subs(vL,t,ta);
figure(1)
plot(ta,v,'LineWidth',2);grid;xlabel('t (s)');ylabel('v(t) (V)');
figure(2)
plot(ta,i,'LineWidth',2);grid;xlabel('t (s)');ylabel('i(t) (A)');
figure(3)
plot(ta,vR,'LineWidth',2);grid;xlabel('t (s)');ylabel('v_R(t) (V)');
figure(4)
plot(ta,vL,'LineWidth',2);grid;xlabel('t (s)');ylabel('v_L(t) (V)');

Answers:
v =
15.0 - 3.0*exp(-2000.0*t)*sin(4000.0*t) - 10.0*exp(-2000.0*t)*cos(4000.0*t)
i =
0.004*exp(-2000.0*t)*cos(4000.0*t) + 0.023*exp(-2000.0*t)*sin(4000.0*t)
vR =
1.6*exp(-2000.0*t)*cos(4000.0*t) + 9.2*exp(-2000.0*t)*sin(4000.0*t)
vL =
8.4*exp(-2000.0*t)*cos(4000.0*t) - 6.2*exp(-2000.0*t)*sin(4000.0*t)
```

EXAMPLE 8.14

Find voltage $v(t)$ across the capacitor for $t \geq 0$ in the circuit shown in Figure 8.15. Assume that $v(0) = V_0 = 3$ V and $i(0) = I_0 = 2$ A. The voltage source is applied to the circuit at $t = 0$.

continued

Example 8.14 continued

The circuit to the left of the capacitor consisting of the voltage source and resistors R_1 and R_2 can be replaced by its Thévenin equivalent circuit. The Thévenin equivalent voltage is given by $V_{th} = V_S R_2/(R_1 + R_2) = 18 \times 3/(6 + 3) = 6$ V. The Thévenin equivalent resistance is $R_{eq} = R_1 \| R_2 = R_1 R_2/(R_1 + R_2) = 6 \times 3/(6 + 3) = 2\ \Omega$. When the circuit consisting of V_s, R_1, and R_2 are replaced by its Thévenin equivalent circuit, we obtain the circuit shown in Figure 8.16.

FIGURE 8.15

Circuit for EXAMPLE 8.14

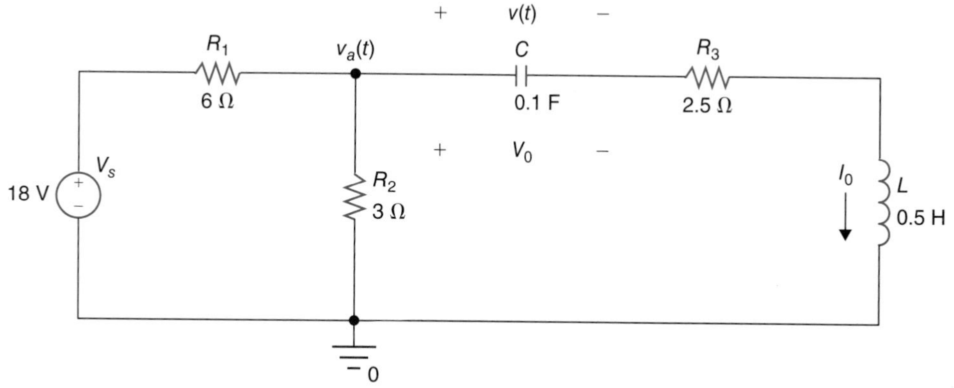

FIGURE 8.16

The simplified circuit.

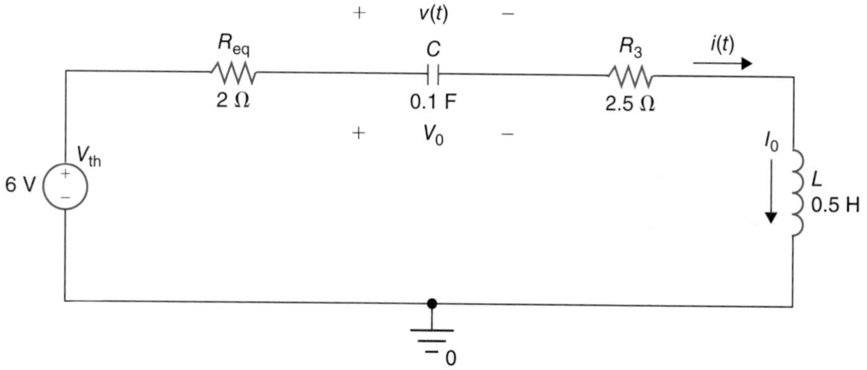

Let $i(t)$ be the current through the mesh, as shown in Figure 8.16. Collecting the voltage drops around the mesh in the clockwise direction, we obtain

$$-V_{th} + R_{eq}i(t) + v(t) + R_3 i(t) + L\frac{di(t)}{dt} = 0 \tag{8.253}$$

The current through the capacitor is given by

$$i(t) = C\frac{dv(t)}{dt} \tag{8.254}$$

Substitution of Equation (8.254) into Equation (8.253) yields

$$-V_{th} + (R_{eq} + R_3)C\frac{dv(t)}{dt} + v(t) + LC\frac{d^2v(t)}{dt^2} = 0 \tag{8.255}$$

Dividing every term of Equation (8.255) by LC, we obtain

$$\frac{d^2v(t)}{dt^2} + \frac{R_{eq} + R_3}{L}\frac{dv(t)}{dt} + \frac{1}{LC}v(t) = \frac{1}{LC}V_{th} \tag{8.256}$$

continued

Example 8.14 continued

or

$$\frac{d^2v(t)}{dt^2} + 9\frac{dv(t)}{dt} + 20v(t) = 120 \tag{8.257}$$

The characteristic equation is given by

$$s^2 + 9s + 20 = (s + 4)(s + 5) = 0$$

The roots of the characteristic equation are $s = -4$ and $s = -5$. This is an overdamped case. The complementary solution may be written as

$$v_c(t) = A_1e^{-4t} + A_2e^{-5t}$$

Coefficients A_1 and A_2 will be found later in this EXAMPLE. Let the particular solution be $v_p(t) = K$. Substitution of this solution in Equation (8.257) results in

$$\frac{d^2K}{dt^2} + 9\frac{dK}{dt} + 20K = 120$$

from which we get $K = 6$. Thus, the particular solution is

$$v_p(t) = K = 6\text{ V}$$

The particular solution is the voltage across the capacitor in the steady state ($t = \infty$). The particular solution can be found from Figure 8.16 by short-circuiting the inductor and open-circuiting the capacitor. The current around the mesh is zero, resulting in no voltage drops on the resistors. Thus, the voltage across the capacitor in the steady state is $V_{th} = 6$ V. The solution to the differential equation [Equation (8.257)] is the sum of the particular solution and the complementary solution:

$$v(t) = v_p(t) + v_c(t) = 6 + A_1e^{-4t} + A_2e^{-5t} \tag{8.258}$$

Since $v(0) = V_0 = 3$ V, we obtain

$$v(0) = 6 + A_1e^{-4\times0} + A_2e^{-5\times0} = 6 + A_1 + A_2 = 3 \tag{8.259}$$

From Equation (8.254), we have $\frac{dv(t)}{dt} = \frac{i(t)}{C}$. Setting $t = 0$, we get $\frac{dv(0)}{dt} = \frac{i(0)}{C} = \frac{2}{0.1} = 20$. Taking derivatives and setting $t = 0$ in Equation (8.258), we obtain

$$\frac{dv(0)}{dt} = -4A_1e^{-4\times0} - 5A_2e^{-5\times0} = -4A_1 - 5A_2 = 20 \tag{8.260}$$

Solving Equation (8.259) for A_2, we obtain

$$A_2 = -A_1 - 3$$

Substituting A_2 into Equation (8.260), we get

$$-4A_1 - 5(-A_1 - 3) = 20$$

continued

Example 8.14 continued

Thus,

$$A_1 = 5$$

and

$$A_2 = -8.$$

The final solution for $v(t)$ is

$$v(t) = (6 + 5e^{-4t} - 8e^{-5t})u(t)\ \text{V} \tag{8.261}$$

MATLAB

```
% EXAMPLE 8.14
clear all;
R1=6;R2=3;R3=2.5;L=0.5;C=0.1;V0=3;I0=2;Vs=18;
syms  va i(t) v(t)
va=solve((va-Vs)/R1+va/R2+i,va)
Di=diff(i);
Dv=diff(v);
[i,v]=dsolve(va==v+R3*i+L*Di,...
    i==C*Dv,i(0)==I0,v(0)==V0);
i=vpa(i,10);
v=vpa(v,10);
i=expand(i);
v=expand(v);
i=vpa(i,10)
v=vpa(v,10)

Answers:
va =
6 - 2*i(t)
i =
4.0*exp(-5.0*t) - 2.0*exp(-4.0*t) + 1.040834086e-17
v =
5.0*exp(-4.0*t) - 8.0*exp(-5.0*t) + 6.0
```

Exercise 8.3

Find the voltage $v(t)$ across the capacitor for $t \geq 0$ in the circuit shown in Figure 8.17. Assume that $v(0) = V_0 = 1$ V and $i(0) = I_0 = 1$ A. The current source is applied to the circuit at $t = 0$.

FIGURE 8.17

Circuit for EXERCISE 8.3.

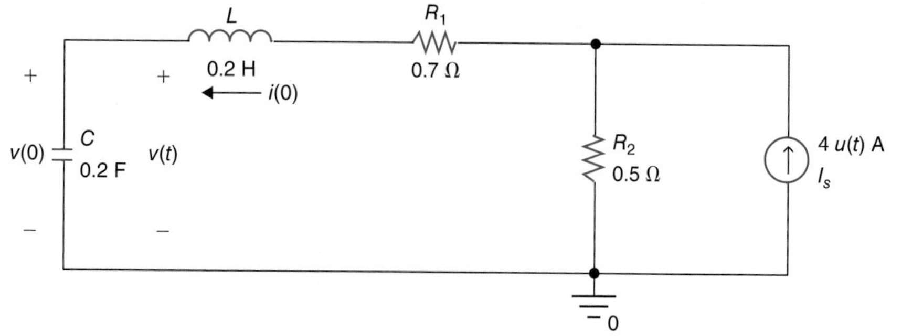

Answer:

$v(t) = \{2 + e^{-3t}[-\cos(4t) + 0.5\sin(4t)]\}\ u(t)$ V.

8.7 Step Response of a Parallel RLC Circuit

Figure 8.18 shows a parallel *RLC* circuit with a current source I_s. The current source is applied to the circuit at time $t = 0$. At $t = 0$, the current through the inductor is $i(0) = I_0$ and the voltage across the capacitor is $v(0) = V_0$.

FIGURE 8.18

Parallel RLC circuit with a dc current source.

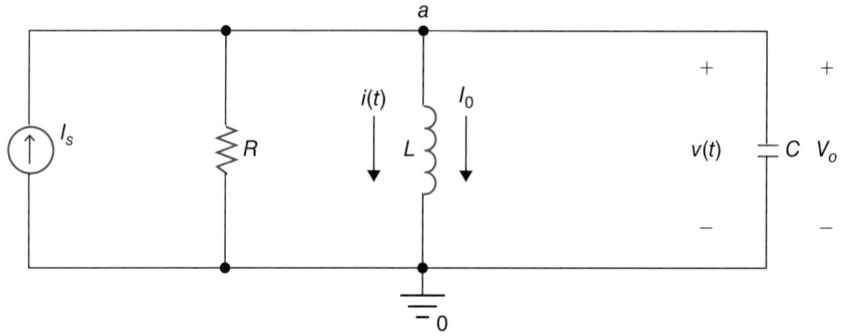

Summing the currents away from node a, we have

$$-I_s + \frac{v(t)}{R} + i(t) + C\frac{dv(t)}{dt} = 0 \quad (8.262)$$

On the inductor, the voltage-current relation is given by

$$v(t) = L\frac{di(t)}{dt} \quad (8.263)$$

Substitution of Equation (8.263) into Equation (8.262) yields

$$-I_s + \frac{L}{R}\frac{di(t)}{dt} + i(t) + LC\frac{d^2i(t)}{dt^2} = 0 \quad (8.264)$$

Rearrangement of Equation (8.264) results in

$$\frac{d^2i(t)}{dt^2} + \frac{1}{RC}\frac{di(t)}{dt} + \frac{1}{LC}i(t) = \frac{I_s}{LC} \quad (8.265)$$

Comparison of Equation (8.265) with Equation (8.176) reveals that

$$a_1 = \frac{1}{RC},\ a_0 = \frac{1}{LC},\ b_0 = \frac{I_s}{LC} \quad (8.266)$$

The neper frequency is given by

$$\alpha = \frac{a_1}{2} = \frac{1}{2RC}\ \text{Np/s} \quad (8.267)$$

and the resonant frequency is given by

$$\omega_0 = \sqrt{a_0} = \frac{1}{\sqrt{LC}}\ \text{rad/s} \quad (8.268)$$

The particular solution is, from Equations (8.178) and (8.266),

$$i_p(t) = \frac{b_0}{a_0} = \frac{\frac{I_s}{LC}}{\frac{1}{LC}} = I_s \qquad \textbf{(8.269)}$$

The complementary solution contains exponential decay term $\exp(-\alpha t)$ with $\alpha > 0$. As time t increases, the complementary solution vanishes, and the general solution consists only of the particular solution. In the circuit shown in Figure 8.18, since the current provided by the source is constant, the voltage across the inductor will be zero after the complementary solution disappears. The inductor behaves like a short circuit. All the currents supplied by the current source will flow through the inductor. Thus, the particular solution must be I_s. The particular solution is the current through the inductor at time $t = \infty$. Therefore, we have

$$i_p(t) = i(\infty) = I_s$$

From Equation (8.263), we have

$$\frac{di(t)}{dt} = \frac{v(t)}{L} \qquad \textbf{(8.270)}$$

Setting $t = 0$, we get

$$Di_0 = \frac{di(0)}{dt} = \frac{v(0)}{L} = \frac{V_0}{L} \qquad \textbf{(8.271)}$$

The complete solutions with the coefficients are listed next for the three cases for the parallel *RLC* circuit shown in Figure 8.18.

8.7.1 CASE 1: OVERDAMPED ($\alpha > \omega_0$ or $a_1 > 2\sqrt{a_0}$ or $\zeta > 1$)

$$\text{Characteristic equation: } s^2 + a_1 s + a_0 = s^2 + \frac{1}{RC}s + \frac{1}{LC} = 0 \qquad \textbf{(8.272)}$$

$$\text{Neper frequency: } \alpha = \frac{a_1}{2} = \frac{1}{2RC}\ \text{Np/s} \qquad \textbf{(8.273)}$$

$$\text{Resonant frequency: } \omega_0 = \sqrt{a_0} = \frac{1}{\sqrt{LC}}\ \text{rad/s} \qquad \textbf{(8.274)}$$

Roots of the characteristic equation:

$$s_1 = -\frac{a_1}{2} + \sqrt{\left(\frac{a_1}{2}\right)^2 - a_0} = -\alpha + \sqrt{\alpha^2 - \omega_0^2} = -\frac{1}{2RC} + \sqrt{\left(\frac{1}{2RC}\right)^2 - \frac{1}{LC}}$$

$$s_2 = -\frac{a_1}{2} - \sqrt{\left(\frac{a_1}{2}\right)^2 - a_0} = -\alpha - \sqrt{\alpha^2 - \omega_0^2} = -\frac{1}{2RC} - \sqrt{\left(\frac{1}{2RC}\right)^2 - \frac{1}{LC}}$$

(8.275)

$$\text{Particular solution: } i_p(t) = \frac{b_0}{a_0} = \frac{\dfrac{I_s}{LC}}{\dfrac{1}{LC}} = I_s \tag{8.276}$$

Complementary solution: $i_c(t) = A_1 e^{s_1 t} + A_2 e^{s_2 t}$

$$\text{Complete solution: } i(t) = i_p(t) + i_c(t) = I_s + A_1 e^{s_1 t} + A_2 e^{s_2 t} \text{ A} \tag{8.277}$$

$$\text{Initial conditions: } i(0) = I_0, \quad Di_0 = \frac{di(0)}{dt} = \frac{v(0)}{L} = \frac{V_0}{L} \tag{8.278}$$

Coefficients A_1 and A_2 are found from initial conditions given by Equation (8.278). Setting $t = 0$ in Equation (8.277), we get

$$i(0) = I_s + A_1 + A_2 = I_0$$

which can be rewritten as

$$A_1 + A_2 = I_0 - I_s \tag{8.279}$$

Differentiating Equation (8.277) with respect to t, we get

$$\frac{di(t)}{dt} = A_1 s_1 e^{s_1 t} + A_2 s_2 e^{s_2 t} \tag{8.280}$$

Setting $t = 0$ in Equation (8.280), we get

$$\frac{di(0)}{dt} = A_1 s_1 + A_2 s_2 = Di_0 \tag{8.281}$$

Coefficients A_1 and A_2 are found by solving Equations (8.279) and (8.281) for A_1 and A_2. Solving Equation (8.279) for A_2, we obtain

$$A_2 = -A_1 + I_0 - I_s$$

Substituting A_2 into Equation (8.281), we get

$$A_1 s_1 + (-A_1 + I_0 - I_s)s_2 = Di_0$$

Solving for A_1, we get

$$A_1 = \frac{Di_0 - s_2(I_0 - I_s)}{s_1 - s_2} = \frac{s_2(I_0 - I_s) - Di_0}{s_2 - s_1}$$

Substituting A_1 into $A_2 = -A_1 + I_0 - I_s$, we obtain

$$A_2 = \frac{Di_0 - s_1(I_0 - I_s)}{s_2 - s_1}$$

Alternatively, Cramer's rule can be used, as shown here:

$$A_1 = \frac{\begin{vmatrix} I_0 - I_s & 1 \\ Di_0 & s_2 \end{vmatrix}}{\begin{vmatrix} 1 & 1 \\ s_1 & s_2 \end{vmatrix}} = \frac{s_2(I_0 - I_s) - Di_0}{s_2 - s_1} \tag{8.282}$$

$$A_2 = \frac{\begin{vmatrix} 1 & I_0 - I_s \\ s_1 & Di_0 \end{vmatrix}}{\begin{vmatrix} 1 & 1 \\ s_1 & s_2 \end{vmatrix}} = \frac{Di_0 - s_1(I_0 - I_s)}{s_2 - s_1} \tag{8.283}$$

The voltage across the inductor is given by

$$v(t) = L\frac{di(t)}{dt} = L(A_1 s_1 e^{s_1 t} + A_2 s_2 e^{s_2 t})u(t) \text{ V} \tag{8.284}$$

The voltage given by Equation (8.284) is also the voltage across the capacitor and the resistor.

8.7.2 CASE 2: CRITICALLY DAMPED ($\alpha = \omega_0$ or $a_1 = 2\sqrt{a_0}$ or $\zeta = 1$)

Characteristic equation: $s^2 + a_1 s + a_0 = s^2 + \frac{1}{RC}s + \frac{1}{LC} = 0$

$$\text{Neper frequency: } \alpha = \frac{a_1}{2} = \frac{1}{2RC} \text{ Np/s} \tag{8.285}$$

$$\text{Resonant frequency: } \omega_0 = \sqrt{a_0} = \frac{1}{\sqrt{LC}} \text{ rad/s} \tag{8.286}$$

Roots of the characteristic equation:

$$s_1 = s_2 = -\frac{a_1}{2} = -\alpha = -\frac{1}{2RC}$$

$$\text{Particular solution: } i_p(t) = \frac{b_0}{a_0} = \frac{\frac{I_s}{LC}}{\frac{1}{LC}} = I_s \tag{8.287}$$

Complementary solution: $i_c(t) = A_1 e^{s_1 t} + A_2 t e^{s_2 t}$

$$\text{Complete solution: } i(t) = i_p(t) + i_c(t) = I_s + A_1 e^{s_1 t} + A_2 t e^{s_2 t} \text{ A} \tag{8.288}$$

$$\text{Initial conditions: } i(0) = I_0,\ Di_0 = \frac{di(0)}{dt} = \frac{v(0)}{L} = \frac{V_0}{L} \tag{8.289}$$

Coefficients A_1 and A_2 are found from initial conditions given by Equation (8.289). Setting $t = 0$ in Equation (8.288), we get

$$i(0) = I_s + A_1 = I_0 \tag{8.290}$$

Solving Equation (8.290) for A_1, we get

$$A_1 = I_0 - I_s \tag{8.291}$$

Differentiating Equation (8.288) with respect to t, we get

$$\frac{di(t)}{dt} = A_1 s_1 e^{s_1 t} + A_2 e^{s_2 t} + A_2 s_2 t e^{s_2 t} \tag{8.292}$$

Setting $t = 0$ in Equation (8.292), we get

$$\frac{di(0)}{dt} = A_1 s_1 + A_2 = Di_0 \quad (8.293)$$

Solving Equation (8.293) for A_2, we obtain

$$A_2 = Di_0 - A_1 s_1 = Di_0 - (I_0 - I_s)s_1 \quad (8.294)$$

The voltage across the inductor is given by

$$v(t) = L\frac{di(t)}{dt} = L[(A_1 s_1 + A_2)e^{s_1 t} + A_2 s_2 t e^{s_2 t}]u(t) \text{ V} \quad (8.295)$$

The voltage given by Equation (8.295) is also the voltage across the capacitor and the resistor.

8.7.3 CASE 3: UNDERDAMPED ($\alpha < \omega_0$ or $a_1 < 2\sqrt{a_0}$ or $\zeta < 1$)

Characteristic equation: $s^2 + a_1 s + a_0 = s^2 + \frac{1}{RC}s + \frac{1}{LC} = 0$

$$\text{Neper frequency: } \alpha = \frac{a_1}{2} = \frac{1}{2RC} \text{ Np/s} \quad (8.296)$$

$$\text{Resonant frequency: } \omega_0 = \sqrt{a_0} = \frac{1}{\sqrt{LC}} \text{ rad/s} \quad (8.297)$$

Damped resonant frequency:

$$\beta = \sqrt{a_0 - \left(\frac{a_1}{2}\right)^2} = \sqrt{\omega_0^2 - \alpha^2} = \sqrt{\frac{1}{LC} - \left(\frac{1}{2RC}\right)^2} \text{ rad/s} \quad (8.298)$$

Roots of the characteristic equation:

$$s_1 = -\alpha + j\beta, \quad s_2 = -\alpha - j\beta$$

$$\text{Particular solution: } i_p(t) = \frac{b_0}{a_0} = \frac{\frac{I_s}{LC}}{\frac{1}{LC}} = I_s \quad (8.299)$$

Complementary solution: $i_c(t) = e^{-\alpha t}[B_1 \cos(\beta t) + B_2 \sin(\beta t)]$

$$\text{Complete solution: } i(t) = i_p(t) + i_c(t) = I_s + e^{-\alpha t}[B_1 \cos(\beta t) + B_2 \sin(\beta t)] \text{ A} \quad (8.300)$$

$$\text{Initial conditions: } i(0) = I_0,\ Di_0 = \frac{di(0)}{dt} = \frac{v(0)}{L} = \frac{V_0}{L} \quad (8.301)$$

Coefficients B_1 and B_2 are found from initial conditions given by Equation (8.301). Setting $t = 0$ in Equation (8.300), we get

$$i(0) = I_s + B_1 = I_0 \quad (8.302)$$

Solving Equation (8.302) for B_1, we get

$$B_1 = I_0 - I_s \tag{8.303}$$

Differentiating Equation (8.300) with respect to t, we get

$$\frac{di(t)}{dt} = e^{-\alpha t}[(-\alpha B_1 + \beta B_2)\cos(\beta t) + (-\beta B_1 - \alpha B_2)\sin(\beta t)] \tag{8.304}$$

Setting $t = 0$ in Equation (8.304), we get

$$\frac{di(0)}{dt} = -\alpha B_1 + \beta B_2 = Di_0 \tag{8.305}$$

Solving Equation (8.305) for B_2, we obtain

$$B_2 = \frac{Di_0 + \alpha B_1}{\beta} = \frac{Di_0 + \alpha(I_0 - I_s)}{\beta} \tag{8.306}$$

The voltage across the inductor is given by

$$v(t) = L\frac{di(t)}{dt} = Le^{-\alpha t}[(-\alpha B_1 + \beta B_2)\cos(\beta t) + (-\beta B_1 - \alpha B_2)\sin(\beta t)]\text{ V} \tag{8.307}$$

The voltage given by Equation (8.307) is also the voltage across the capacitor and the resistor. The current through the capacitor can be found by taking the derivative of $v(t)$ given by Equation (8.307):

$$\begin{aligned} i_C(t) &= C\frac{dv(t)}{dt} \\ &= LCe^{-\alpha t}\{[(\alpha^2 - \beta^2)B_1 - 2\alpha\beta B_2]\cos(\beta t) \\ &\quad + [2\alpha\beta B_1 + (\alpha^2 - \beta^2)B_2]\sin(\beta t)\}\text{ A} \end{aligned} \tag{8.308}$$

The current through the resistor is given by

$$i_R(t) = \frac{v(t)}{R} = \frac{L}{R}e^{-\alpha t}[(-\alpha B_1 + \beta B_2)\cos(\beta t) + (-\beta B_1 - \alpha B_2)\sin(\beta t)]\text{ A} \tag{8.309}$$

EXAMPLE 8.15

Let $R = 200\ \Omega$, $L = 20$ mH, $C = 0.1\ \mu$F, $I_0 = 4$ mA, $V_0 = 5$ V, $I_s = 12$ mA in the circuit shown in Figure 8.18. Find the current $i(t)$ through the inductor and voltage $v(t)$ across the inductor and plot $i(t)$ and $v(t)$.

The coefficients are given by

$$a_2 = 1,\ a_1 = \frac{1}{RC} = 50{,}000,\ a_0 = \frac{1}{LC} = 5 \times 10^8,\ b_0 = \frac{I_s}{LC} = 48 \times 10^5$$

The neper frequency is given by

$$\alpha = \frac{a_1}{2} = \frac{1}{2RC} = 25{,}000\text{ NP/s}$$

continued

Example 8.15 continued

The resonant frequency is

$$\omega_0 = \sqrt{a_0} = \frac{1}{\sqrt{LC}} = 22{,}360.6798 \text{ rad/s}$$

Since $\alpha > \omega_0$, this is case 1, and $i(t)$ is overdamped. The characteristic equation is

$$s^2 + a_1 s + a_0 = s^2 + \frac{1}{RC}s + \frac{1}{LC} = s^2 + 50{,}000s + 4 \times 10^8$$
$$= (s + 13{,}819.6601)(s + 36{,}180.3399) = 0$$

The roots of the characteristic equation are $s_1 = -13{,}819.6601$ and $s_2 = -36{,}180.3399$. The roots of the characteristic equation can also be found from

$$s_1 = -\alpha + \sqrt{\alpha^2 - \omega_0^2} = -13{,}819.6601$$

$$s_2 = -\alpha - \sqrt{\alpha^2 - \omega_0^2} = -36{,}180.3399$$

The complementary solution can be written as

$$y_c(t) = A_1 e^{-13{,}819.6601t} + A_2 e^{-36{,}180.3399t}$$

The particular solution is given by

$$i_p(t) = I_s = 12 \text{ mA}$$

The solution to the differential equation is the sum of the particular solution and the complementary solution:

$$i(t) = i_p(t) + i_c(t) = 0.012 + A_1 e^{-13{,}819.6601t} + A_2 e^{-36{,}180.3399t} \quad \textbf{(8.310)}$$

At $t = 0$, Equation (8.310) becomes

$$i(0) = 0.012 + A_1 + A_2 = 0.004$$

which can be rearranged as

$$A_1 + A_2 = -0.008 \quad \textbf{(8.311)}$$

Taking the derivative of Equation (8.310), we obtain

$$\frac{di(t)}{dt} = -13{,}819.6601 A_1 e^{-13{,}819.6601t} - 36{,}180.3399 A_2 e^{-36{,}180.3399t} \quad \textbf{(8.312)}$$

Setting $t = 0$ in Equation (8.312) and from Equation (8.278), we get

$$\frac{di(0)}{dt} = A_1 s_1 e^{s_1 0} + A_2 s_2 e^{s_2 0} = -13{,}819.6601 A_1 - 36{,}180.3399 A_2 = \frac{V_0}{L} = 250 \quad \textbf{(8.313)}$$

Solving Equation (8.311) for A_2, we obtain

$$A_2 = -A_1 - 0.008$$

continued

Example 8.15 continued

Substituting A_2 into Equation (8.313), we get

$$-13{,}819.6601A_1 - 36{,}180.3399(-A_1 - 0.008) = 250$$

Thus,

$$A_1 = -0.0017639\text{ A} = -1.7639\text{ mA}$$

and

$$A_2 = -0.0062361\text{ A} = -6.2361\text{ mA}$$

The current through the inductor is given by

$$i(t) = I_S + A_1 e^{s_1 t} + A_2 e^{s_2 t} = (12 - 1.7639e^{-13{,}820t} - 6.2361e^{-36{,}180t})u(t)\text{ mA} \quad \textbf{(8.314)}$$

The voltage across the inductor is

$$v(t) = L\, di(t)/dt = [0.48754\exp(-13{,}820t) + 4.5125\exp(-36{,}180t)]u(t)\text{ V} \quad \textbf{(8.315)}$$

The plot of $i(t)$ and $v(t)$ is shown in Figure 8.19.

FIGURE 8.19

Plot of $i(t)$ and $v(t)$ for EXAMPLE 8.15.

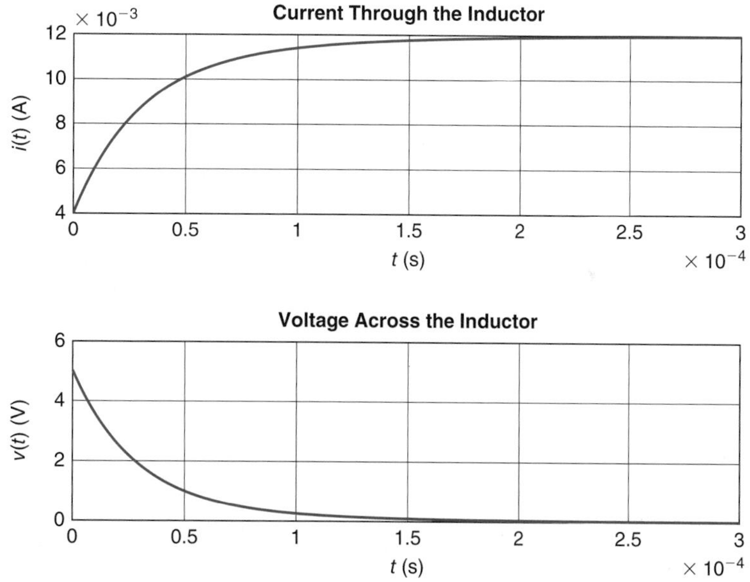

EXAMPLE 8.16

Let $R = 2.5\ k\Omega$, $L = 40$ mH, $C = 1.6$ nF, $I_0 = 2$ mA, $V_0 = 5$ V, $I_s = 8$ mA in the circuit shown in Figure 8.18. Find the current $i(t)$ through the inductor and voltage $v(t)$ across the inductor, and plot $i(t)$ and $v(t)$.

The coefficients are given by

$$a_2 = 1,\ a_1 = \frac{1}{RC} = 250{,}000,\ a_0 = \frac{1}{LC} = 1.5625 \times 10^{10},\ b_0 = \frac{I_s}{LC} = 12.5 \times 10^7$$

continued

Example 8.16 continued

The neper frequency is given by

$$\alpha = \frac{a_1}{2} = \frac{1}{2RC} = 125{,}000 \text{ Np/s}$$

The resonant frequency is

$$\omega_0 = \sqrt{a_0} = \frac{1}{\sqrt{LC}} = 125{,}000 \text{ rad/s}$$

Since $\alpha = \omega_0$, this is case 2, and $i(t)$ is critically damped. The characteristic equation is

$$s^2 + a_1 s + a_0 = s^2 + \frac{1}{RC}s + \frac{1}{LC} = s^2 + 2.5 \times 10^5 s + 1.5625 \times 10^{10}$$
$$= (s + 125{,}000)(s + 125{,}000) = 0$$

The roots of the characteristic equation are

$$s_1 = -\alpha = -125{,}000$$

$$s_2 = -\alpha = -125{,}000$$

The complementary solution can be written as

$$y_c(t) = A_1 e^{-125{,}000t} + A_2 t e^{-125{,}000t}$$

The particular solution is given by

$$i_p(t) = I_s = 8 \text{ mA}$$

The solution to the differential equation is the sum of the particular solution and the complementary solution:

$$i(t) = i_p(t) + i_c(t) = 0.008 + A_1 e^{-125{,}000t} + A_2 t e^{-125{,}000t} \quad \textbf{(8.316)}$$

At $t = 0$, Equation (8.316) becomes

$$i(0) = 0.008 + A_1 = 0.002$$

Solving for A_1, we obtain

$$A_1 = -0.006 \text{ A} = -6 \text{ mA} \quad \textbf{(8.317)}$$

Taking the derivative of Equation (8.316), we obtain

$$\frac{di(t)}{dt} = -125{,}000 A_1 e^{-125{,}000t} + A_2 e^{-125{,}000t} - 125{,}000 A_2 t e^{-125{,}000t} \quad \textbf{(8.318)}$$

Setting $t = 0$ in Equation (8.318), and from Equation (8.278), we get

$$\frac{di(0)}{dt} = -125{,}000 A_1 + A_2 = \frac{V_0}{L} = 125 \quad \textbf{(8.319)}$$

continued

Example 8.16 continued

Solving Equations (8.319) for A_2, we obtain

$$A_2 = -625 \text{ A}$$

The current through the inductor is given by

$$i(t) = 0.008 - 0.006e^{-125{,}000t} - 625te^{-125{,}000t} \text{ A} \quad \textbf{(8.320)}$$

The voltage across the inductor is

$$v(t) = L\, di(t)/dt = [5e^{-125{,}000t} + 3{,}125{,}000te^{-125{,}000t}]u(t) \text{ V} \quad \textbf{(8.321)}$$

The plot of $i(t)$ and $v(t)$ is shown in Figure 8.20.

FIGURE 8.20

Plot of $i(t)$ and $v(t)$ for EXAMPLE 8.16.

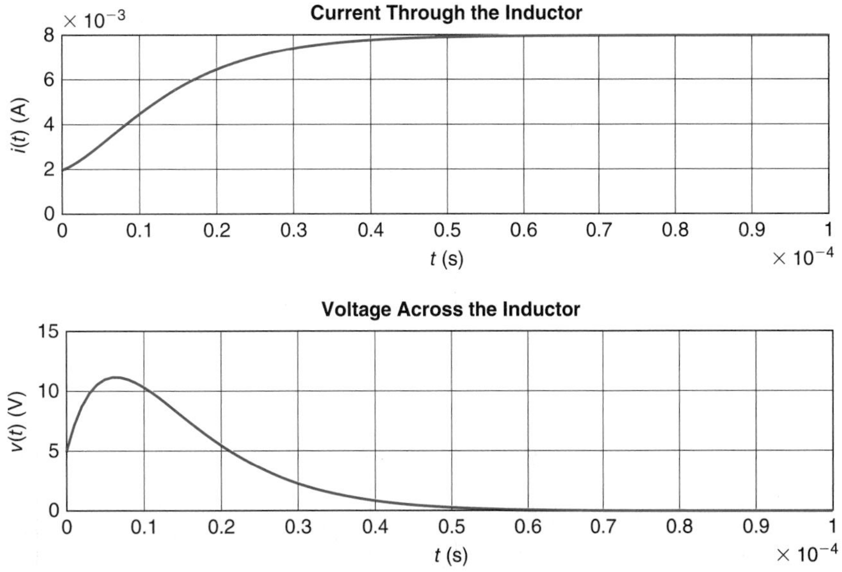

EXAMPLE 8.17

Let $R = 5\ k\Omega$, $L = 20$ mH, $C = 5$ nF, $I_0 = 1$ mA, $V_0 = 12$ V, $I_s = 10$ mA in the circuit shown in Figure 8.18. Find the current $i(t)$ through the inductor and voltage $v(t)$ across the inductor, and plot $i(t)$ and $v(t)$.

The coefficients are given by

$$a_2 = 1,\ a_1 = \frac{1}{RC} = 40{,}000,\ a_0 = \frac{1}{LC} = 1 \times 10^{10},\ b_0 = \frac{I_s}{LC} = 1 \times 10^8$$

The neper frequency is given by

$$\alpha = \frac{a_1}{2} = \frac{1}{2RC} = 20{,}000 \text{ Np/s}$$

The resonant frequency is

$$\omega_0 = \sqrt{a_0} = \frac{1}{\sqrt{LC}} = 100{,}000 \text{ rad/s}$$

continued

Example 8.17 continued

Since $\alpha < \omega_0$, this is case 3, and $i(t)$ is underdamped. The damped resonant frequency is

$$\beta = \sqrt{a_0 - \left(\frac{a_1}{2}\right)^2} = \sqrt{\omega_0^2 - \alpha^2} = \sqrt{\frac{1}{LC} - \left(\frac{1}{2RC}\right)^2} = 97{,}979.59 \text{ rad/s}$$

The characteristic equation is

$$s^2 + a_1 s + a_0 = s^2 + \frac{1}{RC}s + \frac{1}{LC} = s^2 + 40{,}000s + 1 \times 10^{10}$$
$$= (s + 20{,}000 - j97{,}979.59)(s + 20{,}000 + j97{,}979.59) = 0$$

The roots of the characteristic equation are

$$s_1 = -\alpha + j\beta = -20{,}000 + j97{,}979.59$$

$$s_2 = -\alpha - j\beta = -20{,}000 - j97{,}979.59$$

The complementary solution can be written as

$$y_c(t) = e^{-20{,}000t}[B_1 \cos(97{,}979.59t) + B_2 \sin(97{,}979.59t)]u(t)$$

The particular solution is given by

$$i_p(t) = I_s = 10 \text{ mA}$$

The solution to the differential equation is the sum of the particular solution and the complementary solution:

$$i(t) = i_p(t) + i_c(t) = 0.01 + e^{-20{,}000t}[B_1 \cos(97{,}979.59t) + B_2 \sin(97{,}979.59t)] \quad \textbf{(8.322)}$$

At $t = 0$, Equation (8.322) becomes

$$i(0) = 0.01 + B_1 = 0.001$$

Solving for B_1, we obtain

$$A_1 = -0.009 = -9 \text{ mA} \quad \textbf{(8.323)}$$

Taking the derivative of Equation (8.322), we obtain

$$\frac{di(t)}{dt} = e^{-\alpha t}[(-\alpha B_1 + \beta B_2)\cos(\beta t) + (-\beta B_1 - \alpha B_2)\sin(\beta t)] \quad \textbf{(8.324)}$$

Setting $t = 0$ in Equation (8.324) and from Equation (8.278), we get

$$\frac{di(0)}{dt} = -\alpha B_1 + \beta B_2 = \frac{V_0}{L} = 600 \quad \textbf{(8.325)}$$

Solving Equation (8.325) for B_2, we obtain

$$B_2 = 4.2866 \text{ mA}$$

continued

Example 8.17 continued

The current through the inductor is given by

$$i(t) = 10 + e^{-20,000t}[-9\cos(97,979.59t) + 4.2866\sin(97,979.59t)]\ \text{mA} \quad \textbf{(8.326)}$$

The voltage across the inductor is

$$v(t) = L\,di(t)/dt = e^{-20,000t}[12\cos(97,979.59t) + 15.9217\sin(97,979.59t)]u(t)\ \text{V} \quad \textbf{(8.327)}$$

The plot of $i(t)$ and $v(t)$ is shown in Figure 8.21.

FIGURE 8.21

Plot of $i(t)$ and $v(t)$ for EXAMPLE 8.17.

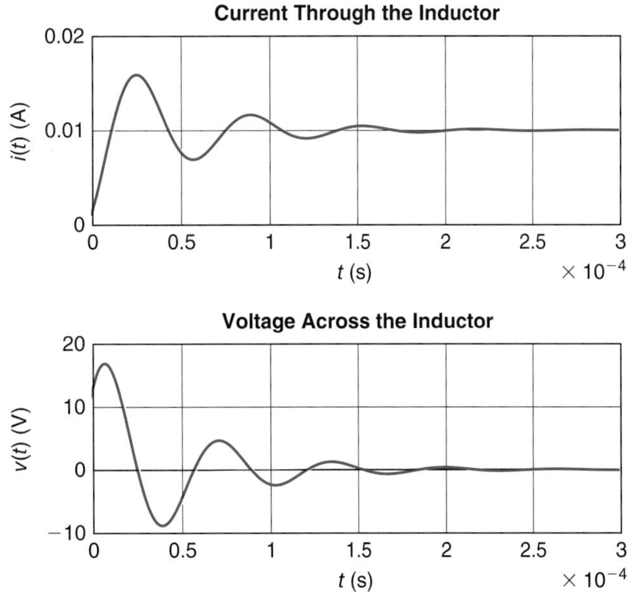

EXAMPLE 8.18

The initial current through the inductor is $i(0) = 1$ A and the initial voltage across the capacitor is $v(0) = 2$ V in the circuit shown in Figure 8.22. Find the current $i(t)$ through the inductor for $t \geq 0$.

FIGURE 8.22

Circuit for EXAMPLE 8.18.

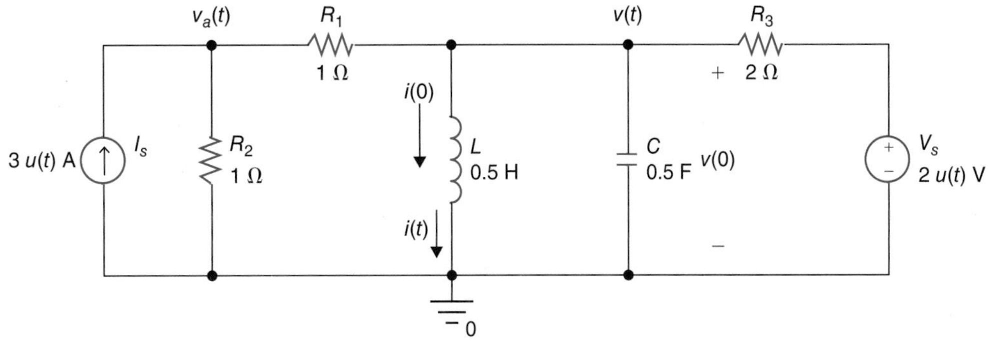

continued

Example 8.18 continued

The Norton equivalent circuit to the left of the inductor consists of a current source with current 1.5 A and a parallel resistor with resistance 2 Ω, as shown in Figure 8.23. Application of the source transformation to V_s and R_3 results in a current source with current 1 A and a parallel resistor with resistance of 2 Ω, as shown in Figure 8.23.

FIGURE 8.23

Circuit after transformation.

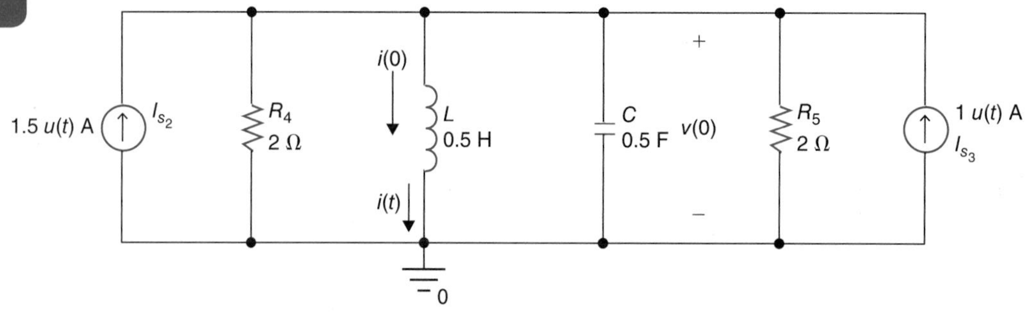

The two current sources can be combined into a single current source with current 2.5 A, and the two parallel resistors R_4 and R_5 can be combined into a single resistor with resistance 1 Ω, as shown in Figure 8.24.

FIGURE 8.24

Parallel *RLC* circuit.

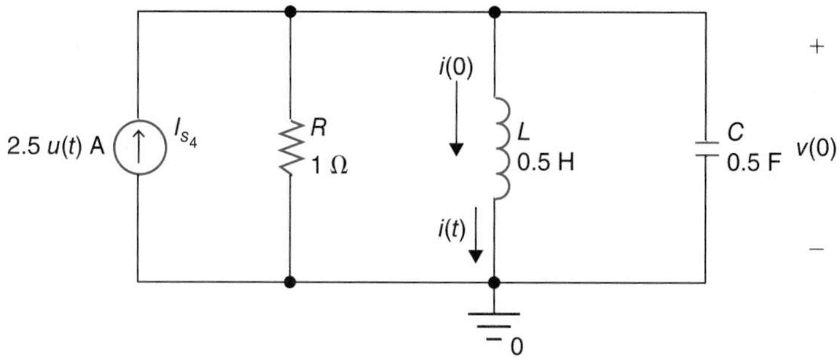

The circuit shown in Figure 8.24 is identical to the circuit shown in Figure 8.18. Thus the differential equation in $i(t)$ is given by Equation (8.265):

$$\frac{d^2i(t)}{dt^2} + \frac{1}{RC}\frac{di(t)}{dt} + \frac{1}{LC}i(t) = \frac{I_s}{LC}$$

Notice that

$$a_1 = \frac{1}{RC} = \frac{1}{1 \times \frac{1}{2}} = 2,\ a_0 = \frac{1}{LC} = \frac{1}{\frac{1}{2} \times \frac{1}{2}} = 4,\ \alpha = a_1/2 = 1,\ \omega_0 = \sqrt{a_0} = 2$$

Since $\alpha < \omega_0$, this is case 3, and $i(t)$ is underdamped. The damped resonant frequency is

$$\beta = \sqrt{a_0 - \left(\frac{a_1}{2}\right)^2} = \sqrt{4 - \left(\frac{2}{2}\right)^2} = \sqrt{3} \text{ rad/s}$$

The characteristic equation is given by

$$s^2 + a_1 s + a_0 = s^2 + 2s + 4 = (s + 1 - j\sqrt{3})(s + 1 + j\sqrt{3}) = 0$$

continued

Example 8.18 continued

Thus, the roots of the characteristic equation are $s = -1 + j\sqrt{3}, s = -1 - j\sqrt{3}$. The roots can also be found from

$$s = -\alpha + j\beta = -1 + j\sqrt{3}, s = -\alpha - j\beta = -1 - j\sqrt{3}$$

The solution can be written as

$$i(t) = \{2.5 + e^{-t}[B_1 \cos(\sqrt{3}t) + B_2 \sin(\sqrt{3}t)]\}\, u(t) \text{ A}$$

The current at $t = 0$, we have

$$i(0) = 1 = 2.5 + B_1$$

Thus, $B_1 = -1.5$A. From Equation (8.325), we obtain

$$\frac{di(0)}{dt} = -\alpha B_1 + \beta B_2 = \frac{V_0}{L} = 4$$

or

$$1.5 + \sqrt{3}B_2 = 4$$

Thus, $B_2 = 2.5/\sqrt{3} = 1.4434$. The current through the inductor is given by

$$i(t) = \{2.5 + e^{-t}[-1.5 \cos(\sqrt{3}t) + 1.4434 \sin(\sqrt{3}t)]\}\, u(t) \text{ A}$$

MATLAB

```
% EXAMPLE 8.18
clear all;
R1=1;R2=1;R3=2;L=0.5;C=0.5;V0=2;I0=1;Is=3;Vs=2;
syms  va i(t) v(t)
va=solve(-Is+va/R2+(va-v)/R1,va)
Di=diff(i);
Dv=diff(v);
[i,v]=dsolve((v-va)/R1+i+C*Dv+(v-Vs)/R3,...
    v==L*Di,i(0)==I0,v(0)==V0);
i=vpa(i,7)
v=vpa(v,7)

Answers:
va =
v(t)/2 + 3/2
i =
1.443376*exp(-1.0*t)*sin(1.732051*t) - 1.5*exp(-1.0*t)*cos(1.732051*t) + 2.5
v =
0.5773503*exp(-1.0*t)*sin(1.732051*t) + 2.0*exp(-1.0*t)*cos(1.732051*t)
```

Exercise 8.4

The initial current through the inductor is $i(0) = 1$ A and the initial voltage across the capacitor is 2 V in the circuit shown in Figure 8.25. Find the current $i(t)$ through the inductor for $t \geq 0$.

FIGURE 8.25

Circuit for EXERCISE 8.4.

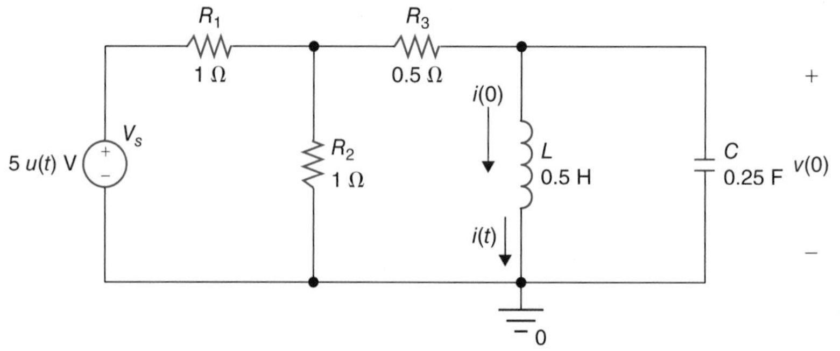

Answer:

$a_1 = 4, \quad a_0 = 8, \quad \alpha = 2, \quad \omega_0 = 2\sqrt{2}, \quad \beta = 2.$

$i(t) = \{2.5 + e^{-2t}[-1.5\cos(2t) + 0.5\sin(2t)]\}\,u(t)$ A.

8.8 General Second-Order Circuits

In this section, we analyze circuits that cannot be reduced to a series *RLC* circuit or parallel *RLC* circuit, and circuits that may contain two capacitors or two inductors, rather than one inductor and one capacitor. Also, we analyze circuits that may contain inputs other than constant (step) signal. The inputs can be exponential, sinusoidal, or ramp, etc. If the input signal is exponential $A\exp(-at)$, the form of the particular solution is also exponential $K\exp(-at)$. If the input signal is $A\cos(\omega t)$ or $A\sin(\omega t)$ or $A_1\cos(\omega t) + A_2\sin(\omega t)$, then the form of the particular solution is $K_1\cos(\omega t) + K_2\sin(\omega t)$. Table 8.1 lists the particular solution for commonly used inputs. In analyzing general second-order circuits, it may be simpler to choose the variables to be voltage across the capacitor and current through the inductor. Several examples, given here, illustrate solving general second-order circuit problems.

TABLE 8.1

Particular Solution

Input	Particular Solution
A	K
Ae^{-at}	Ke^{-at} if a is not a root of a characteristic equation Kte^{-at} if a is a simple root of a characteristic equation Kt^2e^{-at} if a is a double root of a characteristic equation Kt^re^{-at} if there are r roots of a characteristic equation at a
$A\cos(\omega t)$	$K_1\cos(\omega t) + K_2\sin(\omega t)$
$A\sin(\omega t)$	$K_1\cos(\omega t) + K_2\sin(\omega t)$
$A_1\cos(\omega t) + A_2\sin(\omega t)$	$K_1\cos(\omega t) + K_2\sin(\omega t)$
$A_2t^2 + A_1t + A_0$	$K_2t^2 + K_1t + K_0$ if 0 is not a root of a characteristic equation
$A_2t^2 + A_1t + A_0$	$K_2t^3 + K_1t^2 + K_0t$ if 0 is a simple root of a characteristic equation
$A_2t^2 + A_1t + A_0$	$K_2t^4 + K_1t^3 + K_0t^2$ if 0 is a double root of a characteristic equation
$A_nt^n + A_{n-1}t^{n-1} + \cdots + A_1t + A_0$	$K_nt^n + K_{n-1}t^{n-1} + \cdots + K_1t + K_0$ if 0 is not a root of a characteristic equation

EXAMPLE 8.19

Find $v_2(t)$ for $t \geq 0$ for the circuit shown in Figure 8.26. The initial conditions are $v_1(0) = V_{10} = 3$ V and $v_2(0) = V_{20} = 1$ V.

FIGURE 8.26

Circuit for EXAMPLE 8.19.

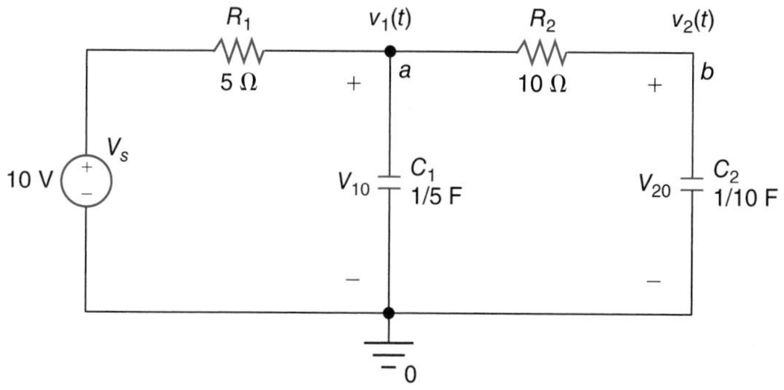

Summing the currents leaving node a, we get

$$\frac{v_1(t) - V_s}{R_1} + C_1\frac{dv_1(t)}{dt} + \frac{v_1(t) - v_2(t)}{R_2} = 0 \tag{8.328}$$

Writing node equation at node b, we obtain

$$\frac{v_2(t) - v_1(t)}{R_2} + C_2\frac{dv_2(t)}{dt} = 0 \tag{8.329}$$

Rearrangement of Equation (8.329) yields

$$v_1(t) = R_2C_2\frac{dv_2(t)}{dt} + v_2(t) \tag{8.330}$$

At $t = 0$, Equation (8.330) becomes

$$\frac{dv_2(0)}{dt} = \frac{v_1(0) - v_2(0)}{R_2C_2} = \frac{V_{10} - V_{20}}{R_2C_2} \tag{8.331}$$

Substitution of Equation (8.330) into Equation (8.328) and simplification results in

$$\frac{d^2v_2(t)}{dt^2} + \left(\frac{1}{R_1C_1} + \frac{1}{R_2C_1} + \frac{1}{R_2C_2}\right)\frac{dv_2(t)}{dt} + \frac{1}{R_1R_2C_1C_2}v_2(t) = \frac{1}{R_1R_2C_1C_2}V_S \tag{8.332}$$

Substitution of component values to Equation (8.332) yields

$$\frac{d^2v_2(t)}{dt^2} + 2.5\frac{dv_2(t)}{dt} + v_2(t) = 10 \tag{8.333}$$

continued

Example 8.19 continued

The characteristic equation is given by

$$s^2 + 2.5s + 1 = (s + 0.5)(s + 2) = 0 \tag{8.334}$$

The roots of the characteristic equation are $s = -0.5$ and $s = -2$. This is an overdamped case. Let the particular solution to Equation (8.333) be $v_{2p}(t) = K$. Substituting this proposed solution to Equation (8.333), we obtain

$$\frac{d^2K}{dt^2} + 2.5\frac{dK}{dt} + K = 10$$

Since $\frac{d^2K}{dt^2} = 0$ and $\frac{dK}{dt} = 0$, $K = 10$. Thus, the particular solution is $v_{2p}(t) = 10$ V. The complete solution to the differential equation given by Equation (8.333) is

$$v_2(t) = 10 + A_1e^{-0.5t} + A_2e^{-2t}$$

for $t \geq 0$.

Since $v_2(0) = 1$ V and, from Equation (8.331), $dv_2(0)/dt = [v_1(0) - v_2(0)]/(R_2C_2) = 2$ V/s, we have

$$v_2(0) = 10 + A_1e^{-0.5\times 0} + A_2e^{-2\times 0} = 10 + A_1 + A_2 = 1$$

$$\frac{dv_2(0)}{dt} = -0.5A_1e^{-0.5\times 0} - 2A_2e^{-2\times 0} = -0.5A_1 - 2A_2 = 2$$

Solving these two equations, we get $A_1 = -32/3 = -10.6667$ and $A_2 = 5/3 = 1.6667$. Therefore, $v_2(t)$ becomes

$$v_2(t) = \left(10 - \frac{32}{3}e^{-0.5t} + \frac{5}{3}e^{-2t}\right)u(t)\text{ V}$$

MATLAB

```
% EXAMPLE 8.19
clear all;
R1=5;R2=10;R3=2;C1=0.2;C2=0.1;V10=3;V20=1;Vs=10;
syms v1(t) v2(t)
Dv1=diff(v1);
Dv2=diff(v2);
[v1,v2]=dsolve((v1-Vs)/R1+C1*Dv1+(v1-v2)/R2,...
    (v2-v1)/R2+C2*Dv2,v1(0)==V10,v2(0)==V20,...
    Dv1(0)==-(V10-Vs)/(R1*C1)-(V10-V20)/(R2*C1),...
    Dv2(0)==(V10-V20)/(R2*C2));
v1=vpa(v1,7);
v2=vpa(v2,7);
v1=expand(v1);
v2=expand(v2);
v1=vpa(v1,7)
v2=vpa(v2,7)

Answers:
v1 =
10.0 - 1.666667*exp(-2.0*t) - 5.333333*exp(-0.5*t)
v2 =
1.666667*exp(-2.0*t) - 10.66667*exp(-0.5*t) + 10.0
```

EXAMPLE 8.20

Find $i(t)$ for $t \geq 0$ for the circuit shown in Figure 8.27, assuming that

$$i(0) = I_0 = 2\text{A},\ v(0) = V_0 = 4\text{ V}$$

FIGURE 8.27

Circuit for EXAMPLE 8.20.

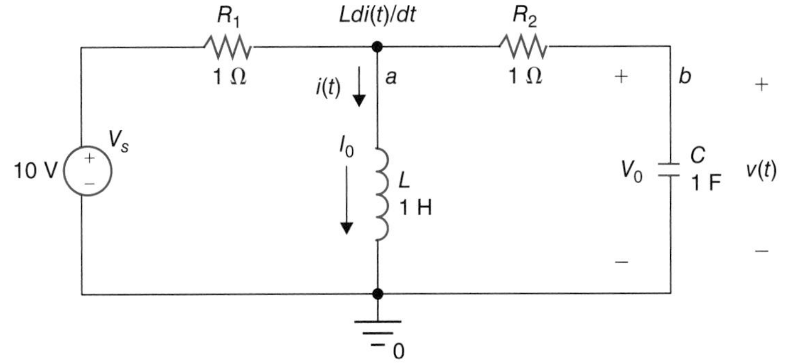

Let $i(t)$ be the current through the inductor and $v(t)$ be the voltage across the capacitor. The voltage across the inductor is given by $Ldi(t)/dt$, and the current through the capacitor is $Cdv(t)/dt$. Summing the currents leaving node a, we have

$$\frac{L\dfrac{di(t)}{dt} - V_s}{R_1} + i(t) + \frac{L\dfrac{di(t)}{dt} - v(t)}{R_2} = 0 \tag{8.335}$$

Summing the currents leaving node b, we have

$$\frac{v(t) - L\dfrac{di(t)}{dt}}{R_2} + C\frac{dv(t)}{dt} = 0 \tag{8.336}$$

Equation (8.335) can be rewritten as

$$\left(\frac{L}{R_1} + \frac{L}{R_2}\right)\frac{di(t)}{dt} + i(t) - \frac{v(t)}{R_2} = \frac{V_s}{R_1} \tag{8.337}$$

Factoring out $i(t)$, we obtain

$$\left[\left(\frac{L}{R_1} + \frac{L}{R_2}\right)\frac{d}{dt} + 1\right]i(t) - \frac{v(t)}{R_2} = \frac{V_s}{R_1} \tag{8.338}$$

or

$$\left(2\frac{d}{dt} + 1\right)i(t) - v(t) = 10 \tag{8.339}$$

Equation (8.336) can be rewritten as

$$-\frac{L}{R_2}\frac{di(t)}{dt} + C\frac{dv(t)}{dt} + \frac{v(t)}{R_2} = 0 \tag{8.340}$$

continued

Example 8.20 continued

Factoring out $v(t)$, we obtain

$$\left(-\frac{L}{R_2}\frac{d}{dt}\right)i(t) + \left(C\frac{d}{dt} + \frac{1}{R_2}\right)v(t) = 0 \qquad \textbf{(8.341)}$$

or

$$\left(-\frac{d}{dt}\right)i(t) + \left(\frac{d}{dt} + 1\right)v(t) = 10 \qquad \textbf{(8.342)}$$

We can apply Cramer's rule to Equations (8.339) and (8.342) to find $i(t)$.

$$i(t) = \frac{\begin{vmatrix} 10 & -1 \\ 0 & \dfrac{d}{dt} + 1 \end{vmatrix}}{\begin{vmatrix} 2\dfrac{d}{dt} + 1 & -1 \\ -\dfrac{d}{dt} & \dfrac{d}{dt} + 1 \end{vmatrix}} = \frac{\dfrac{d10}{dt} + 10}{2\dfrac{d^2}{dt^2} + 2\dfrac{d}{dt} + 1} \qquad \textbf{(8.343)}$$

Multiplying $2\frac{d^2}{dt^2} + 2\frac{d}{dt} + 1$ on both sides, we obtain the second-order differential equation given by

$$2\frac{d^2i(t)}{dt^2} + 2\frac{di(t)}{dt} + i(t) = 10$$

Dividing by 2 on both sides, we have

$$\frac{d^2i(t)}{dt^2} + \frac{di(t)}{dt} + 0.5i(t) = 5 \qquad \textbf{(8.344)}$$

The characteristic equation is given by

$$s^2 + s + 0.5 = (s + 0.5)^2 + 0.5^2 = (s + 0.5 - j0.5)(s + 0.5 + j0.5) = 0$$

Notice that the characteristic equation can be found by replacing d/dt by s and setting the determinant of the denominator of Equation (8.343) equal to zero as shown here:

$$\begin{vmatrix} 2\dfrac{d}{dt} + 1 & -1 \\ -\dfrac{d}{dt} & \dfrac{d}{dt} + 1 \end{vmatrix} = \begin{vmatrix} 2s + 1 & -1 \\ -s & s + 1 \end{vmatrix} = (2s + 1)(s + 1) - s = 2s^2 + 2s + 1 = 0$$

The roots of the characteristic equation are $s = -0.5 + j0.5$ and $s = -0.5 - j0.5$. This is underdamped case with $\alpha = 0.5$ and $\beta = 0.5$. Solving $0.5i_p(t) = 5$, we find the particular solution: $i_p(t) = 10$ A. From the circuit, since the inductor is a short circuit and the capacitor is an open circuit in the steady state, the current through the inductor in the steady state is $i(\infty) = V_s/R_1 = 10$ A. The solution to the differential equation is written as

$$i(t) = 10 + e^{-0.5t}[B_1 \cos(0.5t) + B_2 \sin(0.5t)] \qquad \textbf{(8.345)}$$

continued

Example 8.20 continued

Since $i(0) = 2$ A, we have $2 = 10 + B_1$ or $B_1 = -8$ A. From Equation (8.337), we have

$$\frac{di(t)}{dt} = \frac{-i(t) + \dfrac{v(t)}{R_2} + \dfrac{V_s}{R_1}}{\dfrac{L}{R_1} + \dfrac{L}{R_2}}$$

Setting $t = 0$, we have

$$\frac{di(0)}{dt} = \frac{-i(0) + \dfrac{v(0)}{R_2} + \dfrac{V_s}{R_1}}{\dfrac{L}{R_1} + \dfrac{L}{R_2}} = \frac{-2 + \dfrac{4}{1} + \dfrac{10}{1}}{2} = 6$$

Differentiating Equation (8.345), we have

$$\frac{di(t)}{dt} = -0.5e^{-0.5t}[B_1 \cos(0.5t) + B_2 \sin(0.5t)] + e^{-0.5t}[-0.5B_1 \sin(0.5t) + 0.5B_2 \cos(0.5t)]$$

At $t = 0$, we have

$$\frac{di(0)}{dt} = -0.5B_1 + 0.5B_2 = 6$$

Since $B_1 = -8$ A, $B_2 = 4$ A. The solution for $i(t)$ becomes

$$i(t) = \{10 + e^{-0.5t}[-8 \cos(0.5t) + 4 \sin(0.5t)]\}u(t) \text{ A} \quad \textbf{(8.346)}$$

for $t \geq 0$. The voltage across the inductor is given by

$$v_L(t) = L\frac{di(t)}{dt} = -0.5e^{-0.5t}[-8 \cos(0.5t) + 4 \sin(0.5t)] + e^{-0.5t}[4 \sin(0.5t) + 2 \cos(0.5t)]$$

or

$$v_L(t) = e^{-0.5t}[6 \cos(0.5t) + 2 \sin(0.5t)] \text{ V}$$

for $t \geq 0$.

MATLAB

```
% EXAMPLE 8.20
clear all;
R1=1;R2=1;L=1;C=1;V0=4;I0=2;Vs=10;
syms i(t) v(t)
Di=diff(i);
Dv=diff(v);
[i,v]=dsolve((L*Di-Vs)/R1+i+(L*Di-v)/R2,...
    (v-L*Di)/R2+C*Dv,v(0)==V0,i(0)==I0,...
    Di(0)==(Vs/R1-I0+V0/R2)/(L/R1+L/R2),...
    Dv(0)==(-V0/R2+(L/R2)*Di(0))/(C));
```

continued

Example 8.20 continued
MATLAB continued

```
vL=L*diff(i);
vL=combine(vL);
i=vpa(i,7)
v=vpa(v,7)
vL=vpa(vL,7)

Answers:
i =
(4.0*sin(0.5*t))/exp(t)^(1/2) - (8.0*cos(0.5*t))/exp(t)^(1/2) + 10.0
v =
(4.0*cos(0.5*t))/exp(t)^(1/2) + (8.0*sin(0.5*t))/exp(t)^(1/2)
vL =
(6.0*cos(0.5*t))/exp(t)^(1/2) + (2.0*sin(0.5*t))/exp(t)^(1/2)
```

EXAMPLE 8.21

Switch 1 in the circuit shown in Figure 8.28 has been closed for a long time before it is opened at $t = 0$. Switch 2 has been open for a long time before it is closed at $t = 0.1$ s.

a. Find the initial current $i(0)$ through the inductor and the initial voltage $v(0)$ across the capacitor.
b. Find the voltage $v(t)$ across the capacitor and the current $i(t)$ through the inductor for $0 \le t \le 0.1$ s.
c. Find the voltage $v(0.1)$ across the capacitor and the current $i(0.1)$ through the inductor for $t = 0.1$ s.
d. Find the voltage $v(t)$ across the capacitor and the current $i(t)$ through the inductor for $t \ge 0.1$ s.
e. Plot $v(t)$ and $i(t)$ for $t \ge 0$.

FIGURE 8.28

Circuit for EXAMPLE 8.21.

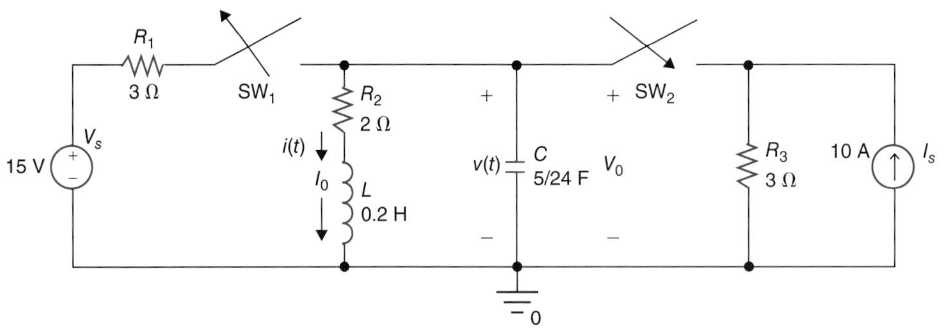

a. For $t \le 0$, switch 1 is closed and switch 2 is open. The inductor is a short circuit, and the capacitor is an open circuit. The circuit reduces to the one shown in Figure 8.29. In Figure 8.29, the inductor is represented as a wire connecting R_2 and the ground. The current around the mesh in Figure 8.29 is $15\text{ V}/(3 + 2)\Omega = 3$ A. Thus, the initial current through the inductor is $i(0) = I_0 = 3$ A. The voltage across R_2 is $3\text{ A} \times 2\ \Omega = 6$ V. Since the capacitor is connected parallel to R_2, the initial voltage across the capacitor is $v(0) = V_0 = 6$ V.

continued

Example 8.21 continued

b. For $0 \le t \le 0.1$ s, the circuit consists of L, R_2, and C, as shown in Figure 8.30. Summing the voltage drops in the clockwise direction in the circuit shown in Figure 8.30, we obtain

FIGURE 8.29

Circuit for $t \le 0$.

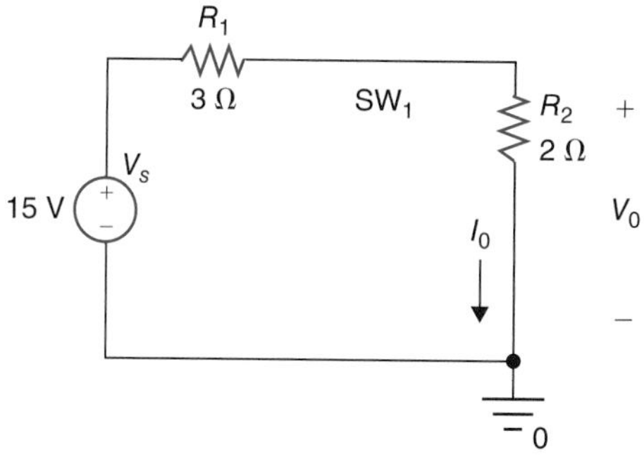

$$-L\frac{di(t)}{dt} - R_2 i(t) + v(t) = 0 \quad \textbf{(8.347)}$$

The current through the capacitor is given by

$$-i(t) = C\frac{dv(t)}{dt} \quad \textbf{(8.348)}$$

Substitution of Equation (8.348) into Equation (8.347) yields

$$LC\frac{d^2v(t)}{dt^2} + R_2C\frac{dv(t)}{dt} + v(t) = 0 \quad \textbf{(8.349)}$$

Dividing by LC everywhere, we get

FIGURE 8.30

Circuit For $0 \le t \le 0.1$ s.

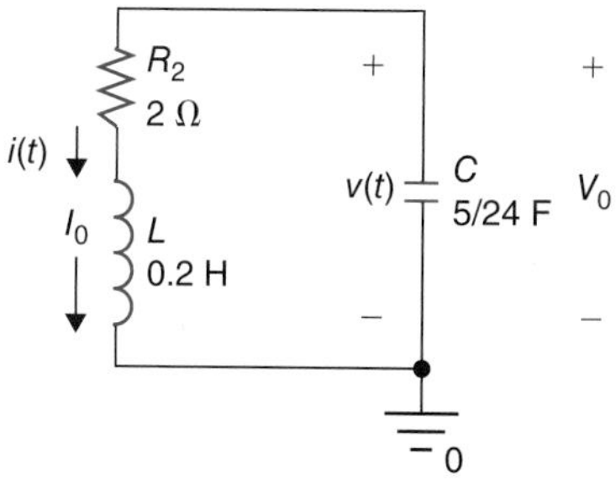

$$\frac{d^2v(t)}{dt^2} + \frac{R_2}{L}\frac{dv(t)}{dt} + \frac{1}{LC}v(t) = 0 \quad \textbf{(8.350)}$$

or

$$\frac{d^2v(t)}{dt^2} + 10\frac{dv(t)}{dt} + 24v(t) = 0 \quad \textbf{(8.351)}$$

The characteristic equation is

$$s^2 + 10s + 24 = (s+4)(s+6) = 0$$

The roots of the characteristic equation are $s = -4$ and $s = -6$. This is an overdamped case. The solution to the differential equation given by Equation (8.351) is

$$v(t) = A_1e^{-4t} + A_2e^{-6t} \quad \textbf{(8.352)}$$

Since $v(0) = V_0 = 6$ V, we have

$$v(0) = A_1e^{-4\times 0} + A_2e^{-6\times 0} = A_1 + A_2 = 6 \quad \textbf{(8.353)}$$

From Equation (8.348), we get $\frac{dv(0)}{dt} = \frac{-i(0)}{C} = \frac{-3}{\frac{5}{24}} = -\frac{72}{5} = -14.4$. Taking the derivative of Equation (8.352) and setting $t = 0$, we obtain

$$\frac{dv(0)}{dt} = -4A_1e^{-4\times 0} - 6A_2e^{-6\times 0} = -4A_1 - 6A_2 = -14.4 \quad \textbf{(8.354)}$$

Coefficients A_1 and A_2 are found by solving Equations (8.353) and (8.354). Multiplying Equation (8.353) by 6, we obtain

$$6A_1 + 6A_2 = 36$$

Adding this equation and Equation (8.354), we get

$$2A_1 = 21.6$$

continued

Example 8.21 continued

Thus,

$$A_1 = 10.8$$

and

$$A_2 = 6 - A_1 = -4.8$$

The solution to the differential equation [Equation (8.351)] is given by

$$v(t) = 10.8e^{-4t} - 4.8e^{-6t}\ \text{V} \tag{8.355}$$

The current through the capacitor from top to bottom (↓) is given by

$$i_C(t) = C\frac{dv(t)}{dt} = \frac{5}{24}[10.8(-4)e^{-4t} - 4.8(-6)e^{-6t}] = -9e^{-4t} + 6e^{-6t}\ \text{A}$$

The current $i(t)$ through the inductor from top to bottom (↓) is $-i_C(t)$. Thus, we have

$$i(t) = -i_C(t) = 9e^{-4t} - 6e^{-6t}\ \text{A}$$

c. At $t = 0.1$ s, the voltage across the capacitor is

$$v(0.1) = V_1 = 10.8e^{-4\times0.1} - 4.8e^{-6\times0.1} = 4.6052\ \text{V}$$

The current through the capacitor from top to bottom (↓) is given by

$$i_C(t) = C\frac{dv(t)}{dt} = \frac{5}{24}[10.8(-4)e^{-4t} - 4.8(-6)e^{-6t}]$$

At $t = 0.1$ s, the current through the capacitor is

$$i_C(0.1) = I_1 = \frac{5}{24}[10.8(-4)e^{-4\times0.1} - 4.8(-6)e^{-6\times0.1}] = -2.74\ \text{A}$$

The current $i(t)$ through the inductor (↓) is $-i_C(t)$. At $t = 0.1$ s, the current through the inductor is

$$i(0.1) = 2.74\ \text{A}$$

d. At $t = 0.1$ s, switch 2 is closed. Figure 8.31 shows the circuit for $t \geq 0.1$ s.

FIGURE 8.31

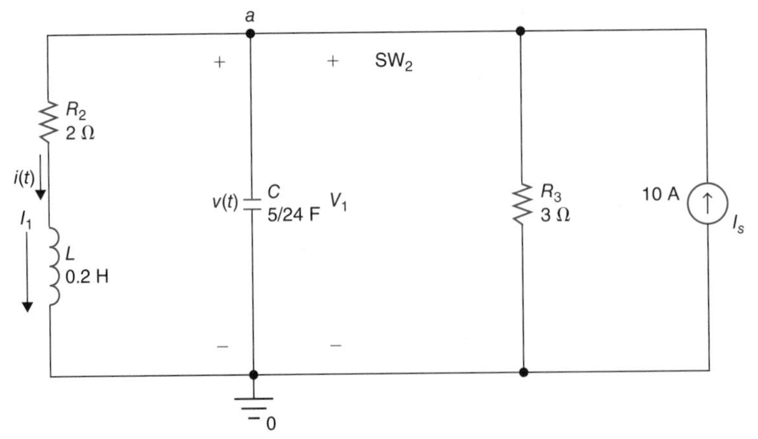

Circuit for $t \geq 0.1$ s.

continued

Example 8.21 continued

Summing the currents leaving node a, we obtain

$$-I_s + \frac{v(t)}{R_3} + C\frac{dv(t)}{dt} + i(t) = 0 \tag{8.356}$$

The voltage across $R_2 - L$ is given by

$$v(t) = R_2 i(t) + L\frac{di(t)}{dt} \tag{8.357}$$

Substitution of Equation (8.357) into Equation (8.356) yields

$$-I_s + \frac{R_2}{R_3}i(t) + \frac{L}{R_3}\frac{di(t)}{dt} + R_2C\frac{di(t)}{dt} + LC\frac{d^2i(t)}{dt^2} + i(t) = 0$$

which can be simplified to

$$\frac{d^2i(t)}{dt^2} + \left(\frac{1}{R_3C} + \frac{R_2}{L}\right)\frac{di(t)}{dt} + \left(\frac{1}{LC} + \frac{R_2}{R_3LC}\right)i(t) = \frac{I_s}{LC}$$

or

$$\frac{d^2i(t)}{dt^2} + 11.6\frac{di(t)}{dt} + 40i(t) = 240 \tag{8.358}$$

The characteristic equation is

$$s^2 + 11.6\,s + 40 = 0$$

The roots of the characteristic equation are $s = -5.8 + j2.5219$ and $s = -5.8 - j2.5219$. This is an underdamped case with $\alpha = 5.8$ and $\beta = 2.5219$. Let the particular solution be $i_p(t) = K$. Substituting this proposed solution to Equation (8.358), we obtain

$$\frac{d^2K}{dt^2} + 11.6\frac{dK}{dt} + 40K = 240$$

Thus, $K = 6$. In the steady state (at $t = \infty$), the capacitor is an open circuit and the inductor is a short circuit in the circuit shown in Figure 8.31. The current from the current source is split between R_2 and R_3. Application of the current divider rule reveals that the current through the inductor is $i(\infty) = I_sR_3/(R_2 + R_3) = 10 \times 3/(2 + 3) = 6$ A in the steady state. The solution to the differential equation [Equation (8.358)] may be expressed as

$$i(t) = 6 + e^{-5.8(t-0.1)}\{B_1\cos[2.5219(t-0.1)] + B_2\sin[2.5219(t-0.1)]\} \tag{8.359}$$

At $t = 0.1$ s, $i(t)$ becomes

$$i(0.1) = 6 + e^{-5.8(0.1-0.1)}\{B_1\cos[2.5219(0.1-0.1)]$$
$$+ B_2\sin[2.5219(0.1-0.1)]\} = 6 + B_1 = 2.74\text{ A}$$

Thus, $B_1 = -3.26$. The voltage across the inductor is given by

$$L\frac{di(t)}{dt} = \frac{1}{5}(-5.8)e^{-5.8(t-0.1)}\{B_1\cos[2.5219(t-0.1)] + B_2\sin[2.5219(t-0.1)]\}$$
$$+\frac{1}{5}e^{-5.8(t-0.1)}\{-2.5219B_1\sin[2.5219(t-0.1)] + 2.5219B_2\cos[2.5219(t-0.1)]\}$$

continued

Example 8.21 continued

At $t = 0.1$ s, the voltage across the inductor is

$$L\frac{di(0.1)}{dt} = \frac{1}{5}(-5.8)B_1 + \frac{1}{5}2.5219B_2$$

The voltage $v(t)$ across the capacitor is $R_2 i(t) + L\frac{di(t)}{dt}$. At $t = 0.1$ s, $v(t)$ becomes

$$R_2 i(0) + L\frac{di(0)}{dt} = 2 \times 2.74 + \frac{1}{5}(-5.8)(-3.26) + \frac{1}{5}2.5219B_2 = 4.6052$$

From this equation, we find $B_2 = -9.2319$. The final solution for $i(t)$ for $t \geq 0.1$ s is given by

$$i(t) = 6 + e^{-5.8(t-0.1)}\{-3.26\cos[2.5219(t-0.1)] - 9.2319\sin[2.5219(t-0.1)]\} \quad \textbf{(8.360)}$$

The voltage across the capacitor is given by

$$\begin{aligned} v(t) = R_2 i(t) + L\frac{di(t)}{dt} &= 12 + 2e^{-5.8(t-0.1)}\{-3.26\cos[2.5219(t-0.1)] \\ &- 9.2319\sin[2.5219(t-0.1)]\} \\ &+ \frac{1}{5}(-5.8)e^{-5.8(t-0.1)}\{-3.26\cos[2.5219(t-0.1)] - 9.2319\sin[2.5219(t-0.1)]\} \\ &+ \frac{1}{5}e^{-5.8(t-0.1)}\{8.2214\sin[2.5219(t-0.1)] - 23.2819\cos[2.5219(t-0.1)]\} \end{aligned}$$

or

$$v(t) = 12 + e^{-5.8(t-0.1)}\{-7.3948\cos[2.5219(t-0.1)] - 6.1105\sin[2.5219(t-0.1)]\}$$

for $t \geq 0.1$ s.

e. Plots of $v(t)$ and $i(t)$ are shown in Figure 8.32.

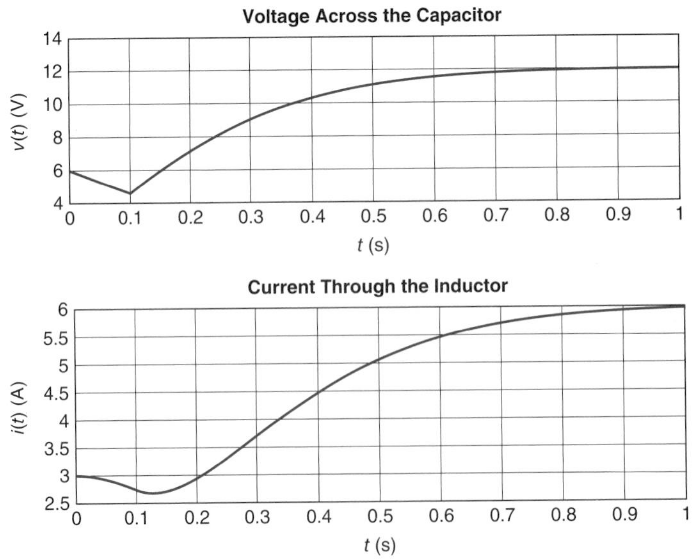

FIGURE 8.32

Plot of $v(t)$ and $i(t)$.

continued

EXAMPLE 8.22

Find $i_1(t)$ and $v_1(t)$ for $t \geq 0$ for the circuit shown in Figure 8.33, assuming that $i_1(0) = I_1 = 2A$, $i_2(0) = I_2 = 6A$.

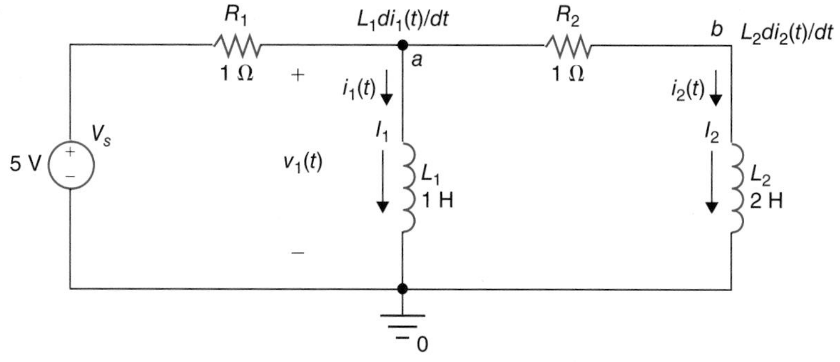

FIGURE 8.33 Circuit for EXAMPLE 8.22.

Let $i_1(t)$ be the current through the inductor L_1, $i_2(t)$ be the current through the inductor L_2, $v(t)$ be the voltage across the inductor L_1, I_1 be the initial current through L_1, I_2 be the initial current through L_2. The voltage across the inductor L_1 is given by $L_1 di_1(t)/dt$, and the voltage across the inductor L_2 is given by $L_2 di_2(t)/dt$. Summing the currents leaving node a, we have

$$\frac{L_1\frac{di_1(t)}{dt} - V_s}{R_1} + i_1(t) + i_2(t) = 0 \tag{8.361}$$

Summing the currents leaving node b, we have

$$\frac{L_2\frac{di_2(t)}{dt} - L_1\frac{di_1(t)}{dt}}{R_2} + i_2(t) = 0 \tag{8.362}$$

Equation (8.361) can be rewritten as

$$\frac{L_1}{R_1}\frac{di_1(t)}{dt} + i_1(t) + i_2(t) = \frac{V_s}{R_1} \tag{8.363}$$

Factoring out $i_1(t)$, we obtain

$$\left(\frac{L_1}{R_1}\frac{d}{dt} + 1\right)i_1(t) + i_2(t) = \frac{V_s}{R_1} \tag{8.364}$$

or

$$\left(\frac{d}{dt} + 1\right)i_1(t) + i_2(t) = 5 \tag{8.365}$$

Equation (8.362) can be rewritten as

$$-\frac{L_1}{R_2}\frac{di_1(t)}{dt} + \frac{L_2}{R_2}\frac{di_2(t)}{dt} + i_2(t) = 0 \tag{8.366}$$

continued

Example 8.22 continued

Factoring out $i_2(t)$, we obtain

$$-\frac{L_1}{R_2}\frac{d}{dt}i_1(t) + \left(\frac{L_2}{R_2}\frac{d}{dt} + 1\right)i_2(t) = 0 \quad \textbf{(8.367)}$$

or

$$\left(-\frac{d}{dt}\right)i_1(t) + \left(2\frac{d}{dt} + 1\right)i_2(t) = 0 \quad \textbf{(8.368)}$$

We can apply Cramer's rule to Equations (8.365) and (8.368) to find $i_1(t)$.

$$i_1(t) = \frac{\begin{vmatrix} 5 & 1 \\ 0 & 2\frac{d}{dt} + 1 \end{vmatrix}}{\begin{vmatrix} \frac{d}{dt} + 1 & 1 \\ -\frac{d}{dt} & 2\frac{d}{dt} + 1 \end{vmatrix}} = \frac{2\frac{d5}{dt} + 5}{2\frac{d^2}{dt^2} + 4\frac{d}{dt} + 1} = \frac{2.5}{\frac{d^2}{dt^2} + 2\frac{d}{dt} + 0.5} \quad \textbf{(8.369)}$$

Multiplying $\frac{d^2}{dt^2} + 2\frac{d}{dt} + 0.5$ on both sides, we obtain the second-order differential equation given by

$$\frac{d^2i_1(t)}{dt^2} + 2\frac{di_1(t)}{dt} + 0.5i_1(t) = 2.5 \quad \textbf{(8.370)}$$

The characteristic equation is given by

$$s^2 + 2s + 0.5 = (s + 1)^2 - (\sqrt{0.5})^2 = (s + 1 - \sqrt{0.5})(s + 1 + \sqrt{0.5})$$
$$= (s + 0.2929)(s + 1.7071)$$

Notice that the characteristic equation can be found by replacing d/dt by s and setting the determinant of the denominator of Equation (8.369) equal to zero.

$$\begin{vmatrix} \frac{d}{dt} + 1 & 1 \\ -\frac{d}{dt} & 2\frac{d}{dt} + 1 \end{vmatrix} = \begin{vmatrix} s + 1 & 1 \\ -s & 2s + 1 \end{vmatrix} = (s + 1)(2s + 1) + s = 2s^2 + 4s + 1 = 0$$

The roots of the characteristic equation are $s = -0.2929$ and $s = -1.7071$. This is an overdamped case. Solving $0.5i_p(t) = 2.5$, we find the particular solution: $i_p(t) = 5$ A. From the circuit, since both inductors are short circuit in the steady state, the current through the inductor L_1 in the steady state is $i_1(\infty) = V_s/R_1 = 5$ A. The solution to the differential equation is written as

$$i_1(t) = 5 + A_1e^{-0.2929t} + A_2e^{-1.7071t} \quad \textbf{(8.371)}$$

Since $i_1(0) = 2$ A, we have

$$i_1(0) = 5 + A_1e^{-0.2929\times 0} + A_2e^{-1.7071\times 0} = 5 + A_1 + A_2 = 2$$

continued

Example 8.22 continued

From Equation (8.363), we have

$$\frac{L_1\frac{di_1(t)}{dt} - V_s}{R_1} + i_1(t) + i_2(t) = 0$$

Solving for $\frac{di_1(t)}{dt}$, we obtain

$$\frac{di_1(t)}{dt} = \frac{-R_1 i_1(t) - R_1 i_2(t) + V_s}{L_1}$$

Setting $t = 0$, we have

$$\frac{di_1(0)}{dt} = \frac{-R_1 i_1(0) - R_1 i_2(0) + V_s}{L_1} = \frac{-1 \times 2 - 1 \times 6 + 5}{1} = -3$$

Differentiating Equation (8.371), we have

$$\frac{di_1(t)}{dt} = -0.2929A_1 e^{-0.2929t} - 1.7071A_2 e^{-1.7071t}$$

At $t = 0$, we have

$$\frac{di_1(0)}{dt} = -0.2929A_1 - 1.7071A_2 = -3$$

The coefficients A_1 and A_2 are found to be $A_1 = -5.7427$, $A_2 = 2.7427$. The solution for $i_1(t)$ becomes

$$i_1(t) = 5 - 5.7427e^{-0.2929t} + 2.7427e^{-1.7071t}\ \text{A} \qquad \textbf{(8.372)}$$

for $t \geq 0$. The voltage across the inductor L_1 is given by

$$v(t) = L_1\frac{di_1(t)}{dt} = (-5.7427)(-0.2929)e^{-0.2929t} + (2.7427)(-1.7071)e^{-1.7071t}$$

or

$$v(t) = L_1\frac{di_1(t)}{dt} = 1.6820e^{-0.2929t} - 4.6820e^{-1.7071t}$$

As a check, at $t = 0$, the current through R_1 from left to right is $(5 - (-3))/1 = 8$ A. This current is identical to the sum of initial currents through L_1 and L_2; that is, $8 = 2 + 6$, satisfying the Kirchhoff's current law.

EXAMPLE 8.23

Find the current $i(t)$ through the inductor and voltage $v_a(t)$ across the inductor in the circuit shown in Figure 8.34. Assume $i(0) = I_0 = 4$ A and $v(0) = V_0 = 5$ V.

continued

Example 8.23 continued

FIGURE 8.34

Circuit for EXAMPLE 8.23.

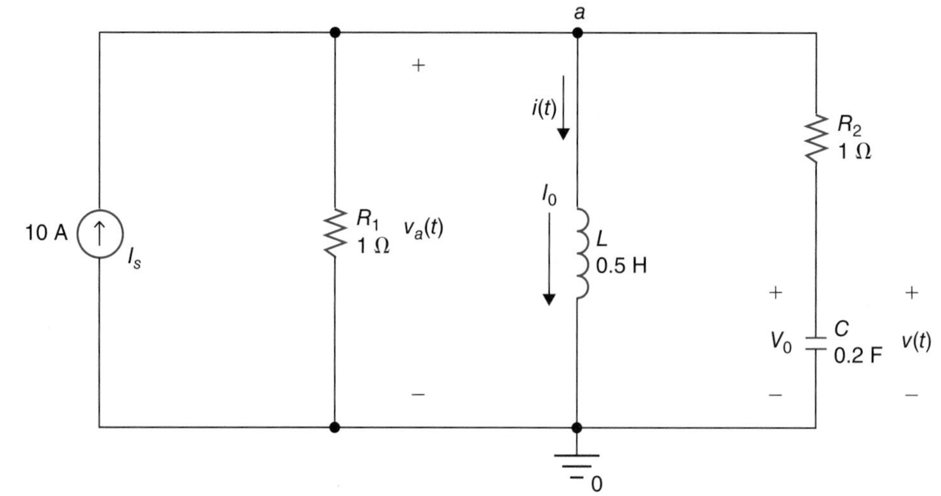

Let $i(t)$ be the current through the inductor, $v_a(t)$ be the voltage across the inductor, and $v(t)$ be the voltage across the capacitor. Notice that $v_a(t) = L\frac{di(t)}{dt}$. Summing the currents leaving node a, we get

$$-I_s + \frac{L\frac{di(t)}{dt}}{R_1} + i(t) + \frac{L\frac{di(t)}{dt} - v(t)}{R_2} = 0 \quad \textbf{(8.373)}$$

The voltage across $R_2 - C$ path must be identical to the voltage across L. This provides another equation, given by

$$L\frac{di(t)}{dt} = R_2C\frac{dv(t)}{dt} + v(t) \quad \textbf{(8.374)}$$

Solving Equation (8.373) for $v(t)$, we obtain

$$v(t) = -R_2I_s + R_2i(t) + L\left(1 + \frac{R_2}{R_1}\right)\frac{di(t)}{dt} \quad \textbf{(8.375)}$$

Taking the derivative of Equation (8.375), we get

$$\frac{dv(t)}{dt} = R_2\frac{di(t)}{dt} + L\left(1 + \frac{R_2}{R_1}\right)\frac{d^2i(t)}{dt^2} \quad \textbf{(8.376)}$$

Substitution of Equations (8.375) and (8.376) into Equation (8.374) yields

$$L\frac{di(t)}{dt} = R_2^2C\frac{di(t)}{dt} + R_2CL\left(1 + \frac{R_2}{R_1}\right)\frac{d^2i(t)}{dt^2} - R_2I_S + R_2i(t) + L\left(1 + \frac{R_2}{R_1}\right)\frac{di(t)}{dt}$$

which can be rewritten as

$$LC\left(1 + \frac{R_2}{R_1}\right)\frac{d^2i(t)}{dt^2} + \left(R_2C + \frac{L}{R_1}\right)\frac{di(t)}{dt} + i(t) = I_s$$

continued

Example 8.23 continued

Dividing by $LC\left(1+\frac{R_2}{R_1}\right)$ on both sides, we obtain

$$\frac{d^2i(t)}{dt^2}+\frac{R_1R_2C+L}{LC(R_1+R_2)}\frac{di(t)}{dt}+\frac{R_1}{LC(R_1+R_2)}i(t)=\frac{R_1I_S}{LC(R_1+R_2)} \quad \textbf{(8.377)}$$

or

$$\frac{d^2i(t)}{dt^2}+3.5\frac{di(t)}{dt}+5i(t)=50 \quad \textbf{(8.378)}$$

The characteristic equation is

$$s^2+3.5s+5=0$$

The roots of the characteristic equation are $s=-1.75+j1.3919$ and $s=-1.75-j1.3919$. This is an underdamped case, with $\alpha=1.75$ and $\beta=1.3919$. Let the particular solution be $i_p(t)=K$. Substituting this proposed solution to Equation (8.378), we obtain

$$\frac{d^2K}{dt^2}+3.5\frac{dK}{dt}+5K=50$$

Thus, $K=10$. In the steady state (at $t=\infty$), the capacitor is an open circuit and the inductor is a short circuit in the circuit shown in Figure 8.34. All the current from the current source flows through the short circuit (inductor). Thus, $i(\infty)=I_s=10$ A. The solution to the differential equation [Equation (8.378)] may be expressed as

$$i(t)=10+e^{-1.75t}[B_1\cos(1.3919t)+B_2\sin(1.3919t)] \quad \textbf{(8.379)}$$

At $t=0$, $i(t)$ becomes

$$i(0)=10+e^{-1.75\times 0}[B_1\cos(1.3919\times 0)+B_2\sin(1.3919\times 0)]=10+B_1=4$$

Thus, $B_1=-6$. Solving Equation (8.373) for $\frac{di(t)}{dt}$, we obtain

$$\frac{di(t)}{dt}=\frac{I_s-i(t)+\frac{v(t)}{R_2}}{\frac{L}{R_1}+\frac{L}{R_2}}$$

Setting $t=0$, we obtain

$$\frac{di(0)}{dt}=\frac{I_s-i(0)+\frac{v(0)}{R_2}}{\frac{L}{R_1}+\frac{L}{R_2}}=\frac{10-4+\frac{5}{1}}{\frac{0.5}{1}+\frac{0.5}{1}}=11$$

Taking the derivative of Equation (8.379), we get

$$\frac{di(t)}{dt}=-1.75e^{-1.75t}[B_1\cos(1.3919t)+B_2\sin(1.3919t)]$$
$$+e^{-1.75t}[-1.3919B_1\sin(1.3919t)+1.3919B_2\cos(1.3919t)]$$

continued

Example 8.23 continued

Setting $t = 0$, we obtain

$$\frac{di(0)}{dt} = -1.75B_1 + 1.3919B_2 = 11$$

from which we find $B_2 = 0.3582$. The solution to the differential equation [Equation (8.378)] becomes

$$i(t) = 10 + e^{-1.75t}[-6\cos(1.3919t) + 0.3582\sin(1.3919t)] \quad \textbf{(8.380)}$$

The voltage across the inductor is given by

$$v_a(t) = L\frac{di(t)}{dt} = (0.5)(-1.75)e^{-1.75t}[-6\cos(1.3919t) + 0.3592\sin(1.3919t)]$$
$$+ (0.5)e^{-1.75t}[-1.3919 \times (-6)\sin(1.3919t) + 1.3919 \times 0.3592\cos(1.3919t)]$$

which can be simplified to

$$v_a(t) = L\frac{di(t)}{dt} = e^{-1.75t}[5.5\cos(1.3919t) + 3.8614\sin(1.3919t)]$$

As a check, at $t = 0$, the voltage at node a is 5.5 V. The current through R_1 is 5.5 V/1 Ω = 5.5 A (↓), the current through the inductor is 4 A (↓), and the current through R_2 is (5.5 V − 5 V)/1 Ω = 0.5 A (↓). The sum of these three currents equals the current from the current source; that is, 5.5 A + 4 A + 0.5 A = 10 A.

EXAMPLE 8.24

Find the current $i(t)$ through the inductor and voltage $v_a(t)$ at node a in the circuit shown in Figure 8.35. Assume that $i(0) = I_0 = 1$ A and $v(0) = V_0 = 3$ V.

FIGURE 8.35

Circuit for EXAMPLE 8.24.

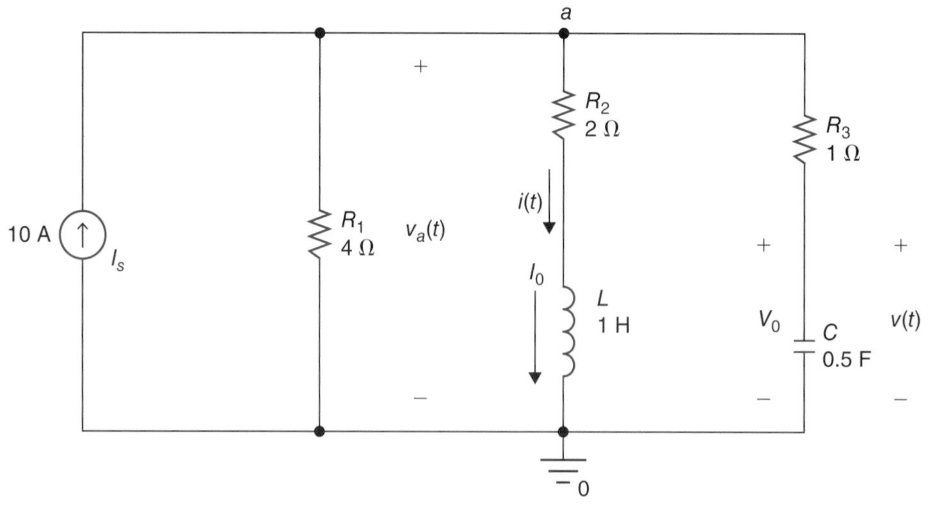

continued

Example 8.24 continued

Let $i(t)$ be the current through the inductor, $v_a(t)$ be the voltage at node a, and $v(t)$ be the voltage across the capacitor. Summing the currents leaving node a, we get

$$-I_s + \frac{v_a(t)}{R_1} + i(t) + \frac{v_a(t) - v(t)}{R_3} = 0 \tag{8.381}$$

The current through R_3 can be written as

$$C\frac{dv(t)}{dt} = \frac{v_a(t) - v(t)}{R_3} \tag{8.382}$$

The voltage across $R_2 - L$ is

$$v_a(t) = R_2 i(t) + L\frac{di(t)}{dt} \tag{8.383}$$

Substitution of Equation (8.383) into Equation (8.381) yields

$$-I_s + \frac{R_2 i(t) + L\frac{di(t)}{dt}}{R_1} + i(t) + \frac{R_2 i(t) + L\frac{di(t)}{dt} - v(t)}{R_3} = 0$$

which can be rewritten as

$$\left(\frac{L}{R_1} + \frac{L}{R_3}\right)\frac{di(t)}{dt} + \left(\frac{R_2}{R_1} + \frac{R_2}{R_3} + 1\right)i(t) - \frac{v(t)}{R_3} = I_s \tag{8.384}$$

Substitution of Equation (8.383) into Equation (8.382) yields

$$C\frac{dv(t)}{dt} = \frac{R_2 i(t) + L\frac{di(t)}{dt} - v(t)}{R_3}$$

which can be simplified to

$$-\frac{L}{R_3}\frac{di(t)}{dt} - \frac{R_2}{R_3}i(t) + C\frac{dv(t)}{dt} + \frac{v(t)}{R_3} = 0 \tag{8.385}$$

Equations (8.384) and (8.385) can be rewritten, respectively, as

$$\left[\left(\frac{L}{R_1} + \frac{L}{R_3}\right)\frac{d}{dt} + \left(\frac{R_2}{R_1} + \frac{R_2}{R_3} + 1\right)\right]i(t) - \frac{v(t)}{R_3} = I_S \tag{8.386}$$

$$\left(-\frac{L}{R_3}\frac{d}{dt} - \frac{R_2}{R_3}\right)i(t) + \left(C\frac{d}{dt} + \frac{1}{R_3}\right)v(t) = 0 \tag{8.387}$$

When component values are substituted, Equations (8.386) and (8.387) reduce, respectively, as

$$\left(1.25\frac{d}{dt} + 3.5\right)i(t) - v(t) = 10 \tag{8.388}$$

$$\left(-\frac{d}{dt} - 2\right)i(t) + \left(0.5\frac{d}{dt} + 1\right)v(t) = 0 \tag{8.389}$$

continued

Example 8.24 continued

The current $i(t)$ can be found by applying Cramer's rule. Solving for $i(t)$, we obtain

$$i(t) = \frac{\begin{vmatrix} 10 & -1 \\ 0 & 0.5\frac{d}{dt} + 1 \end{vmatrix}}{\begin{vmatrix} 1.25\frac{d}{dt} + 3.5 & -1 \\ -\frac{d}{dt} - 2 & 0.5\frac{d}{dt} + 1 \end{vmatrix}} = \frac{0.5\frac{d10}{dt} + 10}{\frac{5}{8}\frac{d^2}{dt^2} + 2\frac{d}{dt} + \frac{3}{2}} = \frac{16}{\frac{d^2}{dt^2} + 3.2\frac{d}{dt} + 2.4}$$

Multiplying the denominator on both sides, we obtain the second-order differential equation, given by

$$\frac{d^2i(t)}{dt^2} + 3.2\frac{di(t)}{dt} + 2.4i(t) = 16 \quad \textbf{(8.390)}$$

The characteristic equation is

$$s^2 + 3.2s + 2.4 = 0$$

The roots of the characteristic equation are $s = -1.2$ and $s = -2$. This is an overdamped case. Let the particular solution be $i_p(t) = K$. Substituting this proposed solution into Equation (8.390), we obtain

$$\frac{d^2K}{dt^2} + 3.2\frac{dK}{dt} + 2.4K = 16$$

Thus, $K = 20/3 = 6.6667$. In the steady state (at $t = \infty$), the capacitor is an open circuit and the inductor is a short circuit in the circuit shown in Figure 8.35. The current from the current source is split between R_1 and R_2. Application of the current divider rule reveals that the current through the inductor is $i(\infty) = I_S R_1/(R_1 + R_2) = 10 \times 4/(4 + 2) = 20/3$ A in the steady state. The solution to the differential equation [Equation (8.390)] may be expressed as

$$i(t) = \frac{20}{3} + A_1e^{-1.2t} + A_2e^{-2t} \quad \textbf{(8.391)}$$

Setting $t = 0$ in Equation (8.391), we have

$$i(0) = \frac{20}{3} + A_1 + A_2 = 1$$

Setting $t = 0$ in Equation (8.388), we obtain

$$\frac{di(0)}{dt} = \frac{10 + v(0) - 3.5i(0)}{1.25} = \frac{10 + 3 - 3.5 \times 1}{1.25} = 7.6$$

Taking the derivative of Equation (8.391) and setting $t = 0$, we get

$$\frac{di(0)}{dt} = -1.2A_1e^{-1.2\times0} - 2A_2e^{-2\times0} = -1.2A_1 - 2A_2 = 7.6$$

continued

Example 8.24 continued

Coefficients A_1 and A_2 are found to be $A_1 = -14/3 = -4.6667$ and $A_2 = -1$. The solution to the differential equation [Equation (8.390)] becomes

$$i(t) = \left(\frac{20}{3} - \frac{14}{3}e^{-1.2t} - e^{-2t}\right)u(t) \tag{8.392}$$

The voltage across the inductor is given by

$$v_L(t) = L\frac{di(t)}{dt} = 1 \times \left(1.2 \times \frac{14}{3}e^{-1.2t} + 2e^{-2t}\right)u(t) = (5.6e^{-1.2t} + 2e^{-2t})u(t)$$

The voltage at node a is given by

$$v_a(t) = R_2 i(t) + L\frac{di(t)}{dt} = 2\left(\frac{20}{3} - \frac{14}{3}e^{-1.2t} - e^{-2t}\right)u(t) + (5.6e^{-1.2t} + 2e^{-2t})u(t)$$
$$= \left(\frac{40}{3} - 3.7333e^{-1.2t}\right)u(t) \tag{8.393}$$

As a check, at $t = 0$, the voltage at node a is 9.6 V. The current through R_1 is $9.6\ \text{V}/4\ \Omega = 2.4\ \text{A}\ (\downarrow)$, the current through the inductor is $1\ \text{A}\ (\downarrow)$, and the current through R_3 is $(9.6\ \text{V} - 3\ \text{V})/1\ \Omega = 6.6\ \text{A}\ (\downarrow)$. The sum of these three currents equals the current from the current source; that is, $2.4\ \text{A} + 1\ \text{A} + 6.6\ \text{A} = 10\ \text{A}$.

MATLAB

```
% EXAMPLE 8.24
clear all;
R1=4;R2=2;R3=1;L=1;C=0.5;V0=3;I0=1;Is=10;
syms i(t) v(t)
Di=diff(i);
Dv=diff(v);
[i,v]=dsolve(-Is+(R2*i+L*Di)/R1+i+(R2*i+L*Di-v)/R3,...
    C*Dv==(R2*i+L*Di-v)/R3,v(0)==V0,i(0)==I0,...
    Di(0)==(Is+V0/R3-(R2/R1+R2/R3+1)*I0)/(L/R1+L/R3),...
    Dv(0)==(R2*I0+L*Di(0)-V0)/(R3*C));
i=vpa(i,7);
v=vpa(v,7);
i=expand(i);
v=expand(v);
i=vpa(i,7)
v=vpa(v,7)
va=R2*i+L*diff(i);
va=combine(va);
va=vpa(va,7)

Answers:
i =
6.666667 - 1.0*exp(-2.0*t) - 4.666667*exp(-1.2*t)
v =
13.33333 - 1.0*exp(-2.0*t) - 9.333333*exp(-1.2*t)
va =
13.33333 - 3.733333*exp(-1.2*t)
```

8.9 PSpice and Simulink

8.9.1 SOLVING DIFFERENTIAL EQUATIONS USING SIMULINK

Simulink can be used to solve differential equations such as

$$\frac{d^2v(t)}{dt^2} + 5\frac{dv(t)}{dt} + 6v(t) = 6, \quad v(0) = 0.2, \frac{dv(0)}{dt} = 0.5$$

Rearrangement of the differential equation results in

$$\frac{d^2v(t)}{dt^2} = 6 - 5\frac{dv(t)}{dt} - 6v(t)$$

In Simulink, integrator block 1/s integrates the input. If $\frac{d^2v(t)}{dt^2}$ is applied to the input of an integrator block, the output is given by $\frac{dv(t)}{dt}$. When $\frac{dv(t)}{dt}$ is applied to the input of the second integrator block, the output is given by $v(t)$. The Simulink model shown in Figure 8.36 plots $v(t)$. $\frac{d^2v(t)}{dt^2}$ is obtained by adding 6, $-5\frac{dv(t)}{dt}$, and $-6v(t)$. Since the output of the first integrator is $\frac{dv(t)}{dt}$, $-5\frac{dv(t)}{dt}$ is obtained by applying $\frac{dv(t)}{dt}$ to a gain of -5. Similarly, $-6v(t)$ is obtained by applying $v(t)$ to a gain of -6. The initial condition $v(0) = 0.2$ and the initial condition $\frac{dv(0)}{dt} = 0.5$ are entered into integrator blocks by double-clicking on the blocks and entering the numbers. Comments can be added to a Simulink model by double-clicking any space and entering the text. Waveform $v(t)$ is shown in Figure 8.37.

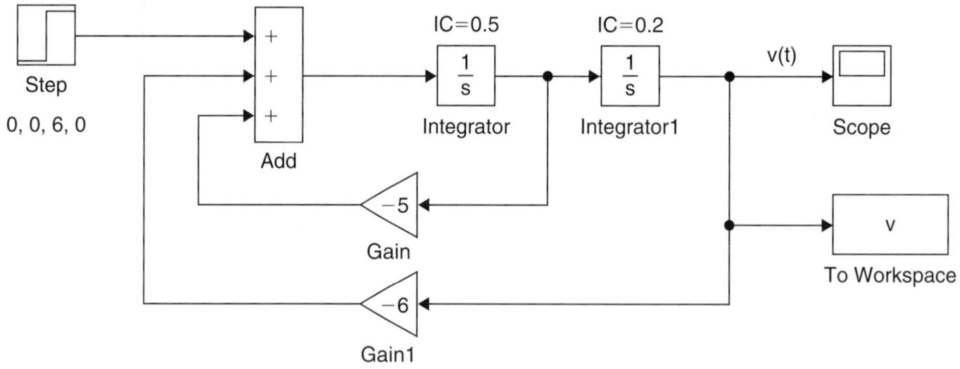

FIGURE 8.36
A Simulink model to solve a differential equation.

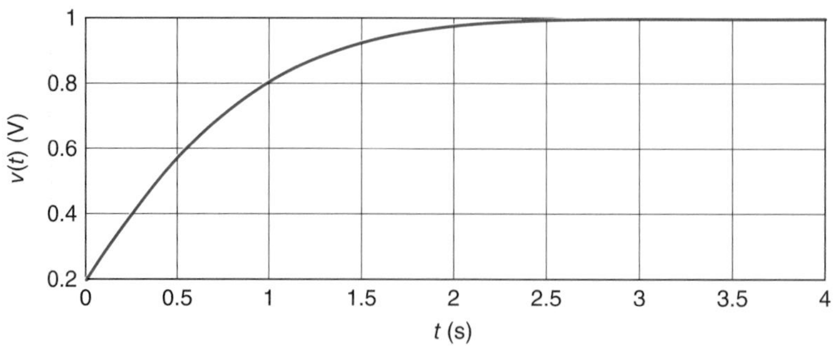

FIGURE 8.37
Waveform $v(t)$ from Scope.

The MATLAB function **dsolve** can be used to find the solution of the same differential equation, as shown here:

```
clear all;
syms v(t)
Dv=diff(v);
Dv2=diff(v,2);
v=dsolve(Dv2+5*Dv+6*v==6,v(0)==0.2,Dv(0)==0.5);
v=vpa(v,10)
ta=0:1e-2:4;
v=subs(v,t,ta);
figure(1)
plot(ta,v,'LineWidth',2);grid;xlabel('t (s)');ylabel('v(t) (V)');
```

The solution of the differential equation is

```
v =
1.1*exp(-3.0*t) - 1.9*exp(-2.0*t) +1.0
```

and the plot is shown in Figure 8.38.

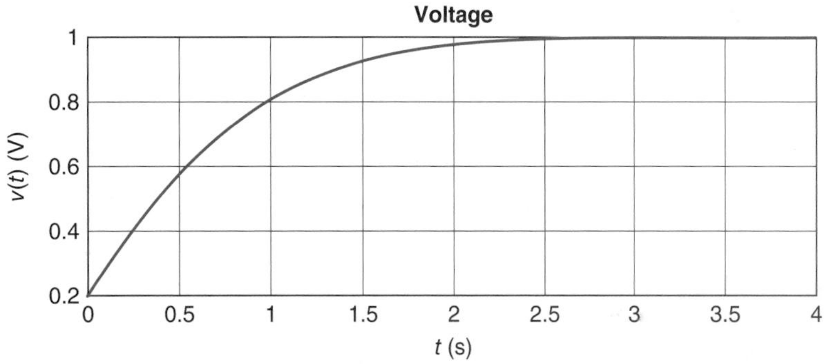

FIGURE 8.38 The plot of $v(t)$ from MATLAB.

The plot obtained from the Simulink model exactly matches the one obtained from MATLAB.

8.9.2 SOLVING DIFFERENTIAL EQUATIONS USING PSpice

The differential equations can also be solved using an op amp circuit. Rewriting the differential equation, we get

$$\frac{d^2v(t)}{dt^2} = 6 - 5\frac{dv(t)}{dt} - 6v(t), \quad v(0) = 0.2, \frac{dv(0)}{dt} = 0.5$$

When $\frac{d^2v(t)}{dt^2}$ is applied to the input of an op amp RC integrator, the output is $-\frac{dv(t)}{dt}$. When $-\frac{dv(t)}{dt}$ is applied to the input of another op amp RC integrator, the output is $v(t)$. The circuit is similar to the Simulink model, but there is a sign change for the integrator. The input to the first integrator is $6 - 5\frac{dv(t)}{dt} - 6v(t)$. The initial conditions are set in the capacitors. The PSpice schematic and the output waveform are shown in Figure 8.39.

PSpice can be used to simulate the step response of a series RLC circuit. Figure 8.40 shows the circuit, voltage across the capacitor, and current through the capacitor. The initial conditions for the inductor and capacitor are set to zero.

FIGURE 8.39

The PSpice schematic and the output waveform.

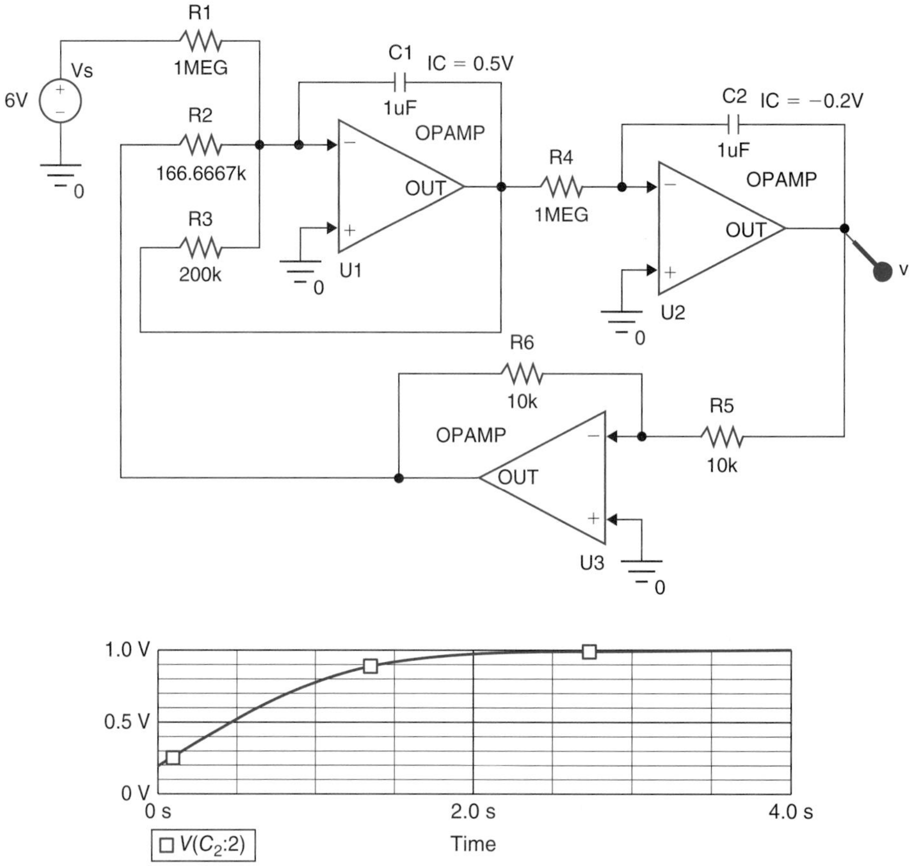

FIGURE 8.40

A series RLC circuit, voltage across the capacitor, and current through the capacitor.

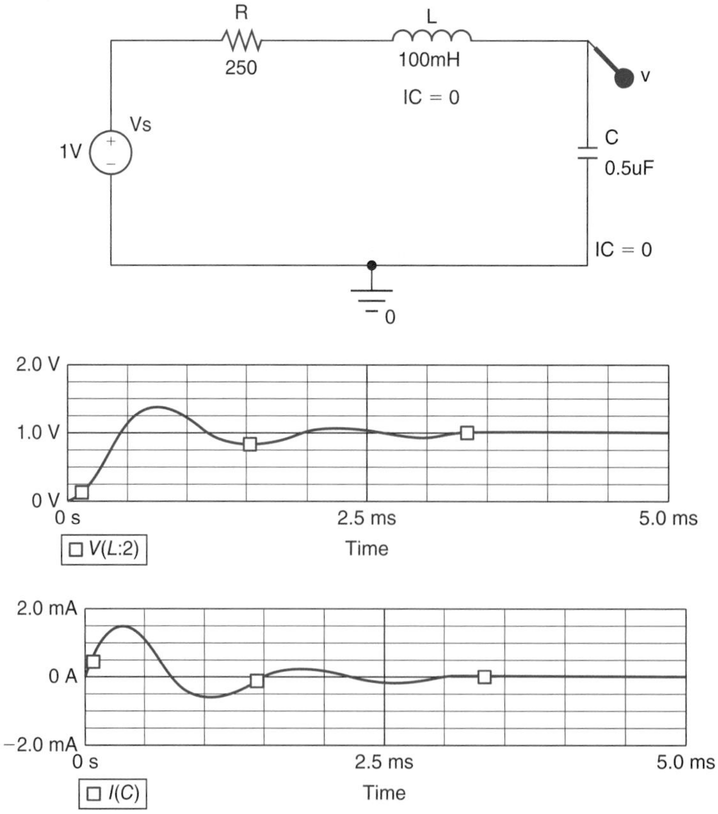

Figure 8.41 shows the Simulink model for the step response of the series RLC circuit shown in Figure 8.40.

FIGURE 8.41

The series RLC circuit from Figure 8.40, with step input and voltage and current on the capacitor.

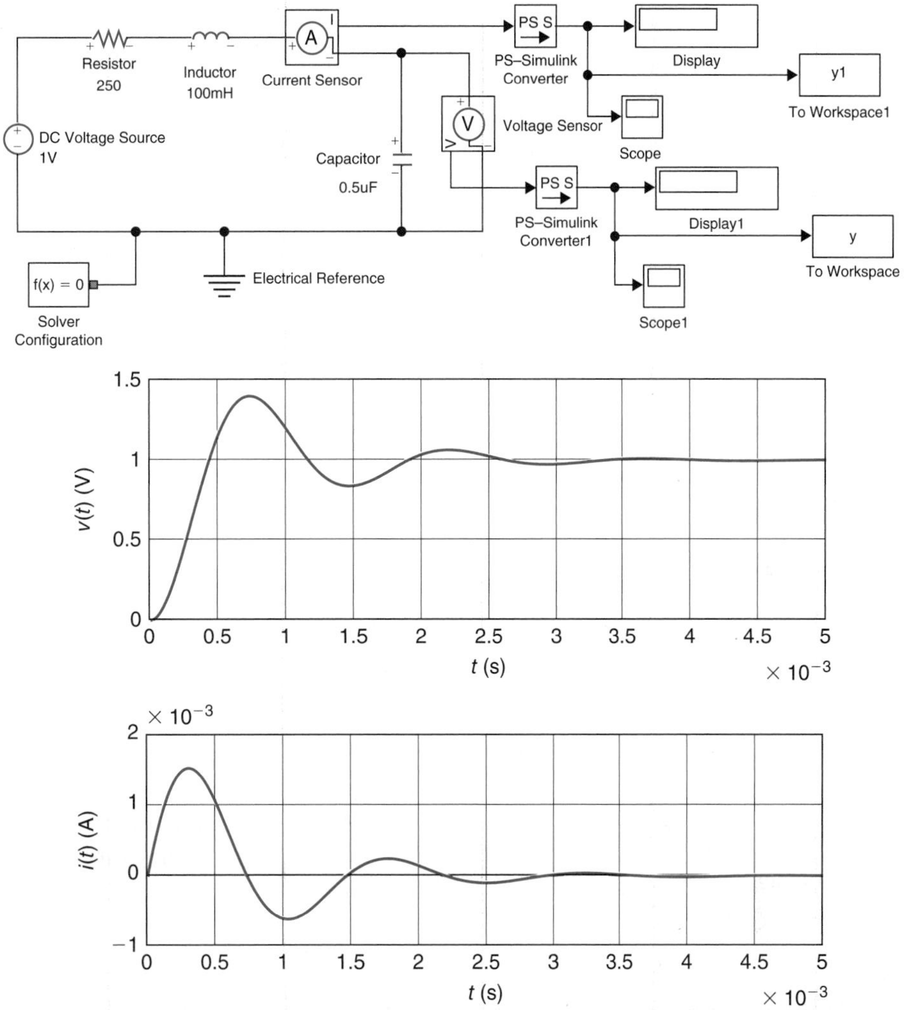

SUMMARY

Finding the voltages and currents of circuits containing two energy storage elements that result in second-order differential equations is discussed in this chapter. If there is no input to the circuit, the differential equation becomes a homogeneous differential equation:

$$\frac{d^2v(t)}{dt^2} + a_1\frac{dv(t)}{dt} + a_0v(t) = 0$$

The characteristic equation of the homogeneous differential equation is given by

$$s^2 + a_1s + a_0 = 0$$

The roots of the characteristic equation are given by

$$s_1 = -\alpha + \sqrt{\alpha^2 - \omega_0^2},\ s_2 = -\alpha - \sqrt{\alpha^2 - \omega_0^2}$$

where

$$\alpha = \frac{a_1}{2},\quad \omega_0 = \sqrt{a_0}$$

The parameter α is called *neper frequency* and is measured in nepers per second (Np/s). The parameter ω_0 is called *resonant frequency* or *undamped natural frequency* and is measured in radians per second (rad/s). A *damping factor* is defined as

$$\zeta = \frac{\alpha}{\omega_0}$$

and the *damped resonant frequency* is defined as

$$\beta = \sqrt{\omega_0^2 - \alpha^2} = \omega_0\sqrt{1-\zeta^2}$$

For the series RLC circuit, the parameters are given by

$$\alpha = \frac{R}{2L}, \quad \omega_0 = \frac{1}{\sqrt{LC}}$$

and for the parallel RLC circuit, the parameters are given by

$$\alpha = \frac{1}{2RC}, \quad \omega_0 = \frac{1}{\sqrt{LC}}$$

Notice that the neper frequency α is different.

The solution of the homogeneous differential equation is called the *complementary solution.* Depending on the coefficients of the homogeneous equation, the complementary solution can be overdamped ($\alpha > \omega_0$), critically damped ($\alpha = \omega_0$), or underdamped ($\alpha < \omega_0$). For each case, the solution takes a different form:

Overdamped: $v(t) = A_1e^{s_1t} + A_2e^{s_2t}$

Critically damped: $v(t) = A_1e^{s_1t} + A_2te^{s_2t}$

Underdamped: $v(t) = e^{-\alpha t}[B_1\cos(\beta t) + B_2\sin(\beta t)]$

The coefficients are found to satisfy the initial conditions $v(0) = V_0$ and $dv(0)/dt = Dv_0$.

If there is input into the circuit, the solution to the differential equation has a particular solution in addition to the complementary solution. The particular solution is the solution to the original differential equation, including the input. The form of the particular solution is similar to the input signal. For the constant input, the particular solution will be a constant. The particular solution for the differential equation

$$\frac{d^2v(t)}{dt^2} + a_1\frac{dv(t)}{dt} + a_0v(t) = b_0$$

is b_0/a_0. If the input is a constant, the solution to the differential equation takes one of the following forms:

Overdamped: $v(t) = \frac{b_0}{a_0} + A_1e^{s_1t} + A_2e^{s_2t}$

Critically damped: $v(t) = \frac{b_0}{a_0} + A_1e^{s_1t} + A_2te^{s_2t}$

Underdamped: $v(t) = \frac{b_0}{a_0} + e^{-\alpha t}[B_1\cos(\beta t) + B_2\sin(\beta t)]$

The coefficients are found to satisfy the initial conditions $v(0) = V_0$ and $dv(0)/dt = Dv_0$. If the input is not a constant, the form of the particular solution is similar to the input as shown in Table 8.1.

PROBLEMS

Zero Input Response of Series RLC Circuit

8.1 **In the circuit shown in Figure P8.1,**

FIGURE P8.1

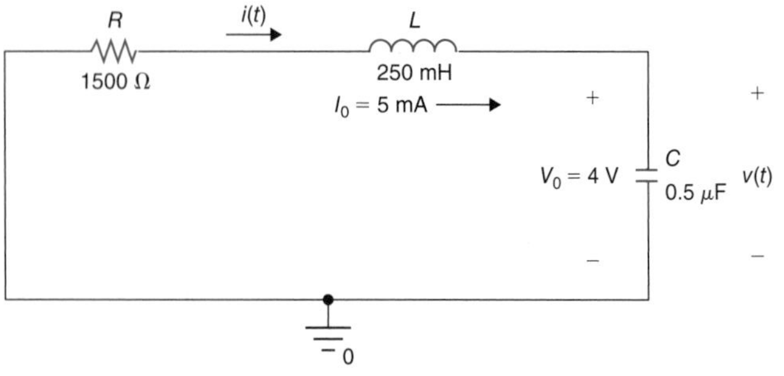

a. write a differential equation in $v(t)$.
b. find α and ω_0.
c. find the characteristic equation.
d. find the roots of the characteristic equation.
e. find voltage $v(t)$ across the capacitor for $t \geq 0$ and plot $v(t)$.
f. find current $i(t)$ for $t \geq 0$ and plot $i(t)$.

8.2 **In the circuit shown in Figure P8.2,**

FIGURE P8.2

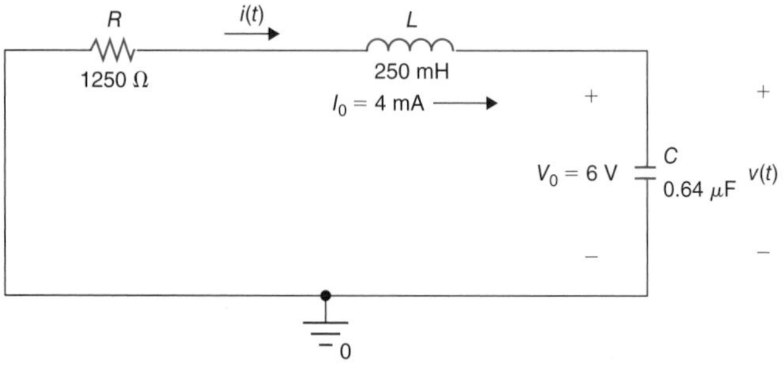

a. write a differential equation in $v(t)$.
b. find α and ω_0.
c. find the characteristic equation.
d. find the roots of the characteristic equation.
e. find voltage $v(t)$ across the capacitor for $t \geq 0$ and plot $v(t)$.
f. find current $i(t)$ for $t \geq 0$ and plot $i(t)$.

8.3 **In the circuit shown in Figure P8.3,**

FIGURE P8.3

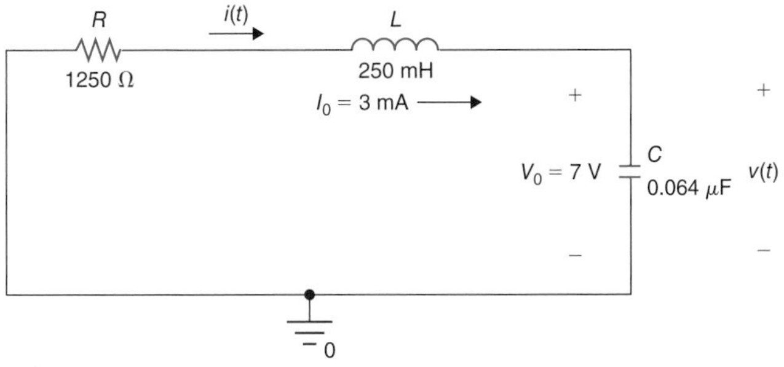

a. write a differential equation in $v(t)$.
b. find α and ω_0.
c. find the characteristic equation.
d. find the roots of the characteristic equation.
e. find voltage $v(t)$ across the capacitor for $t \geq 0$ and plot $v(t)$.
f. find current $i(t)$ for $t \geq 0$ and plot $i(t)$.

8.4 **In the circuit shown in Figure P8.4,**

FIGURE P8.4

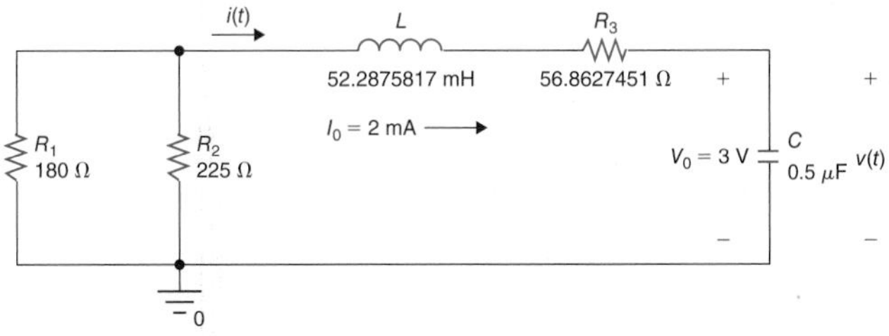

a. write a differential equation in $v(t)$.
b. find α and ω_0.
c. find the characteristic equation.
d. find the roots of the characteristic equation.
e. find voltage $v(t)$ across the capacitor for $t \geq 0$ and plot $v(t)$.
f. find current $i(t)$ for $t \geq 0$ and plot $i(t)$.

8.5 **In the circuit shown in Figure P8.5,**

FIGURE P8.5

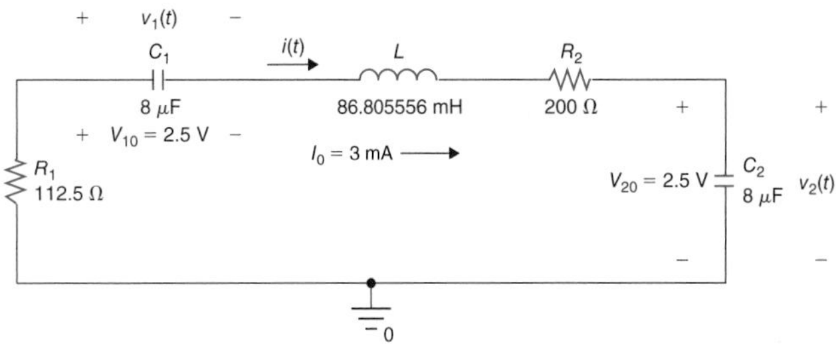

a. write a differential equation in $i(t)$.
b. find α and ω_0.
c. find the characteristic equation.
d. find the roots of the characteristic equation.
e. find current $i(t)$ for $t \geq 0$ and plot $i(t)$.
f. find voltage $v_2(t)$ across the capacitor C_2 for $t \geq 0$ and plot $v_2(t)$.

8.6 **In the circuit shown in Figure P8.6,**

FIGURE P8.6

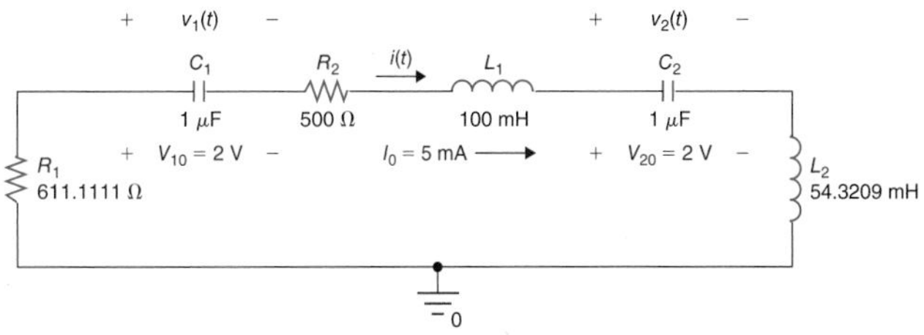

a. write a differential equation in $i(t)$.
b. find α and ω_0.
c. find the characteristic equation.
d. find the roots of the characteristic equation.
e. find current $i(t)$ for $t \geq 0$ and plot $i(t)$.

8.7 **In the circuit shown in Figure P8.7,**

FIGURE P8.7

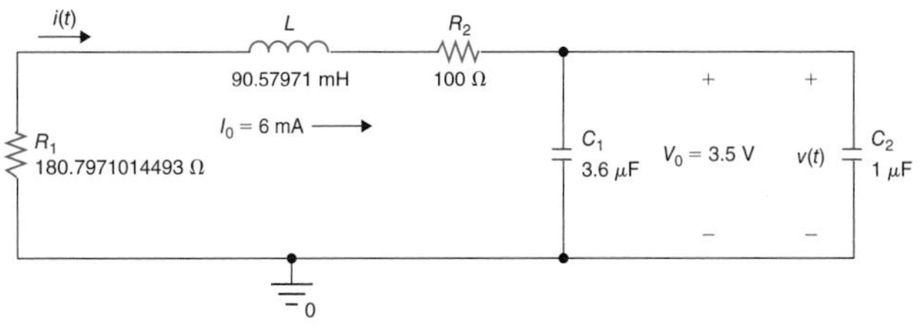

a. write a differential equation in $v(t)$.
b. find α and ω_0.
c. find the characteristic equation.
d. find the roots of the characteristic equation.

e. find voltage $v(t)$ across the capacitor for $t \geq 0$ and plot $v(t)$.
f. find current $i(t)$ for $t \geq 0$ and plot $i(t)$.

8.8 **In the circuit shown in Figure P8.8,**

FIGURE P8.8

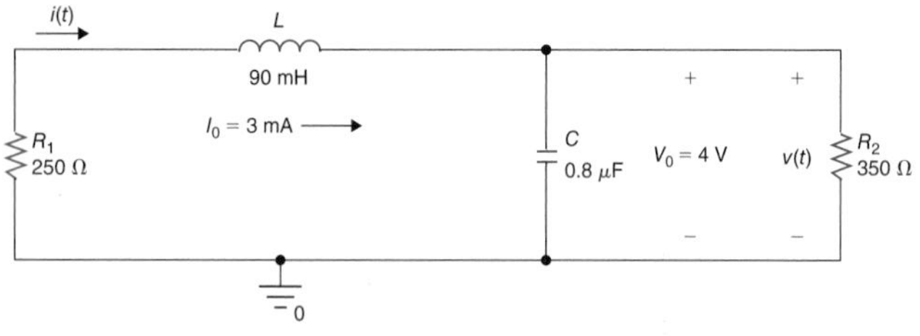

a. write a differential equation in $v(t)$.
b. find α and ω_0.
c. find the characteristic equation.
d. find the roots of the characteristic equation.
e. find the voltage $v(t)$ across the capacitor for $t \geq 0$ and plot $v(t)$.
f. find the current $i(t)$ for $t \geq 0$ and plot $i(t)$.

8.9 **In the circuit shown in Figure P8.9,**

FIGURE P8.9

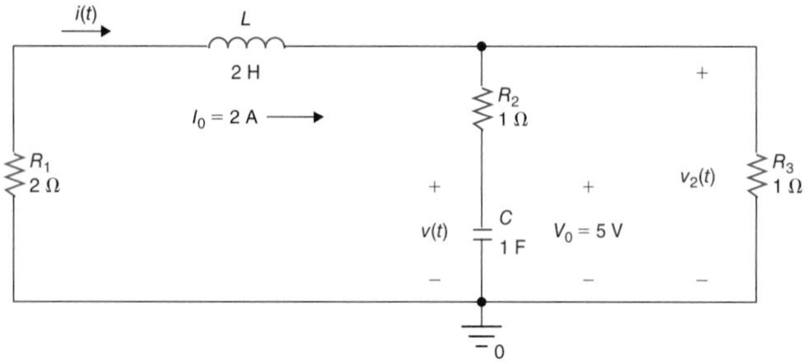

a. write a differential equation in $v(t)$.
b. find α and ω_0.
c. find the characteristic equation.
d. find the roots of the characteristic equation.
e. find the voltage $v(t)$ across the capacitor for $t \geq 0$ and plot $v(t)$.
f. find the current $i(t)$ for $t \geq 0$ and plot $i(t)$.

8.10 **The switch in the circuit shown in Figure P8.10 has been closed for a long time before it is opened at $t = 0$. Find voltage $v(t)$ across the capacitor for $t \geq 0$.**

FIGURE P8.10

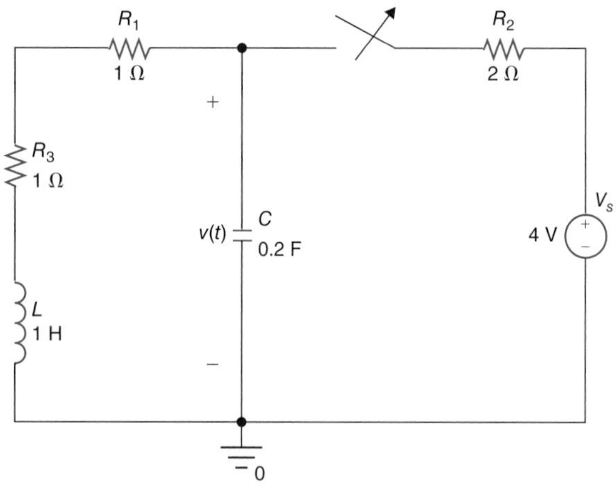

8.11 **The switch in the circuit shown in Figure P8.11 has been closed for a long time before it is opened at $t = 0$. Find voltage $v(t)$ across the capacitor for $t \geq 0$.**

FIGURE P8.11

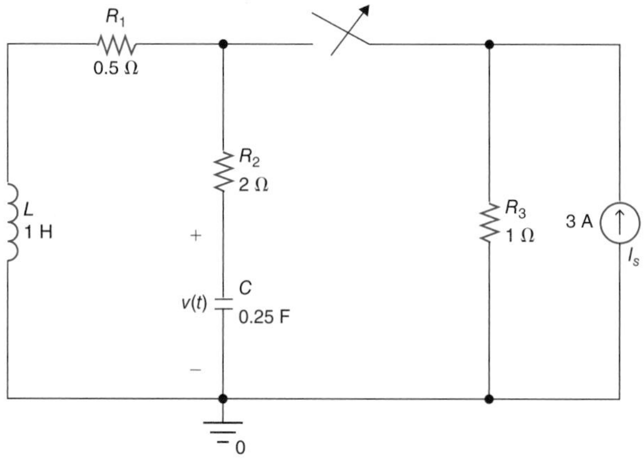

8.12 **The switch in the circuit shown in Figure P8.12 has been closed for a long time before it is opened at $t = 0$. Find voltage $v(t)$ across the capacitor for $t \geq 0$.**

FIGURE P8.12

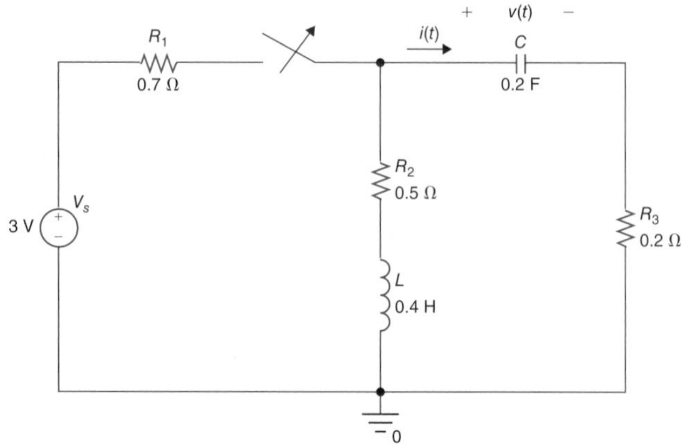

8.13 **The switch in the circuit shown in Figure P8.13 has been closed for a long time before it is opened at $t = 0$. Find voltage $v(t)$ across the capacitor for $t \geq 0$.**

FIGURE P8.13

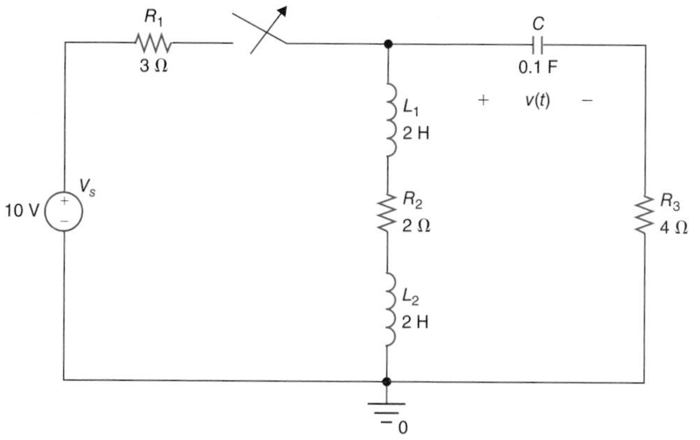

Zero Input Response of Parallel RLC Circuit

8.14 **In the circuit shown in Figure P8.14,**

FIGURE P8.14

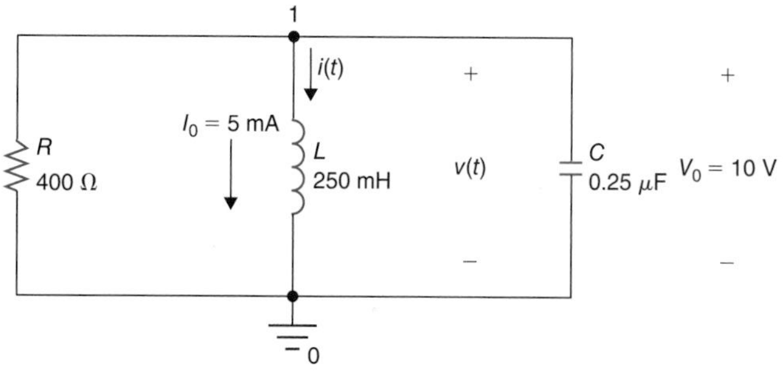

a. find α.
b. find ω_0.
c. write a differential equation in $i(t)$.
d. find the characteristic equation.
e. find the roots of the characteristic equation.
f. find current $i(t)$ through the inductor for $t \geq 0$ and plot $i(t)$.
g. find voltage $v(t)$ for $t \geq 0$ and plot $v(t)$.

8.15 **In the circuit shown in Figure P8.15,**

FIGURE P8.15

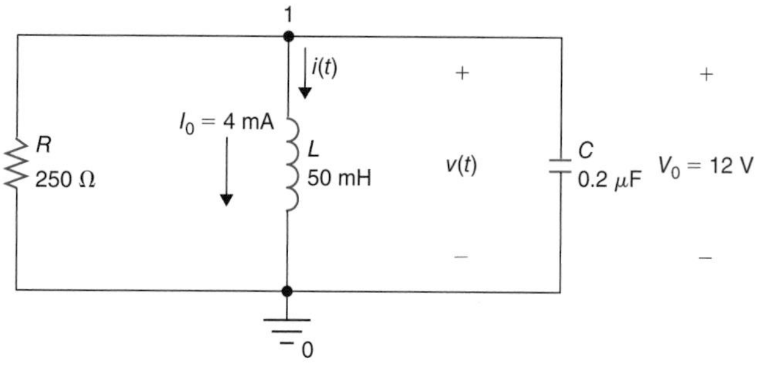

a. find α.
b. find ω_0.
c. write a differential equation in $i(t)$.
d. find the characteristic equation.
e. find the roots of the characteristic equation.
f. find current $i(t)$ through the inductor for $t \geq 0$ and plot $i(t)$.
g. find voltage $v(t)$ for $t \geq 0$ and plot $v(t)$

8.16 **In the circuit shown in Figure P8.16,**

FIGURE P8.16

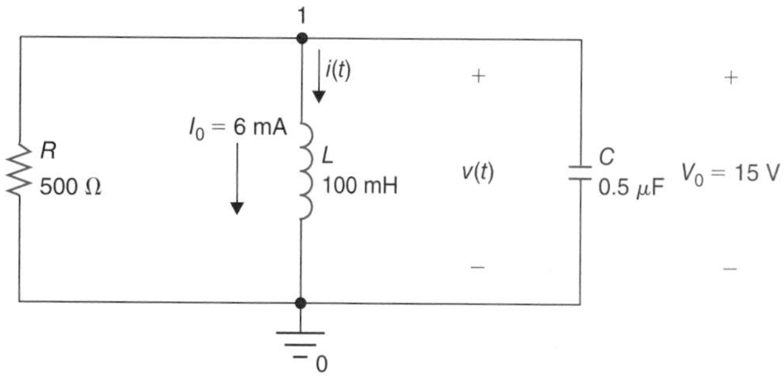

a. find α.
b. find ω_0.
c. write a differential equation in $i(t)$.
d. find the characteristic equation.
e. find the roots of the characteristic equation.
f. find current $i(t)$ through the inductor for $t \geq 0$ and plot $i(t)$.
g. find voltage $v(t)$ for $t \geq 0$ and plot $v(t)$.

8.17 **The switch in the circuit shown in Figure P8.17 has been closed for a long time before it is opened at $t = 0$. Find current $i(t)$ through the inductor for $t \geq 0$.**

FIGURE P8.17

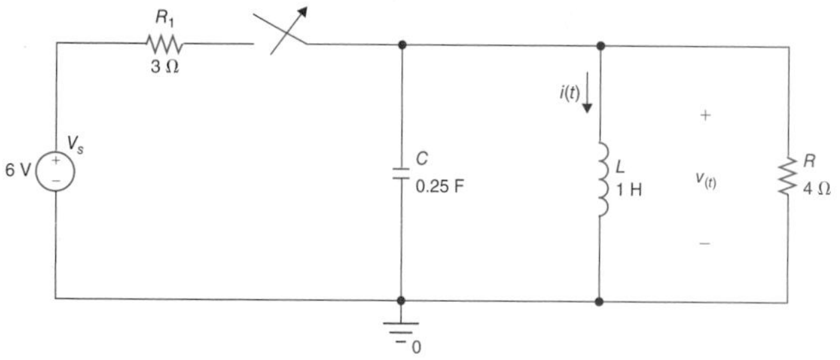

8.18 **The switch in the circuit shown in Figure P8.18 has been closed for a long time before it is opened at $t = 0$. Find current $i(t)$ through the inductor for $t \geq 0$.**

FIGURE P8.18

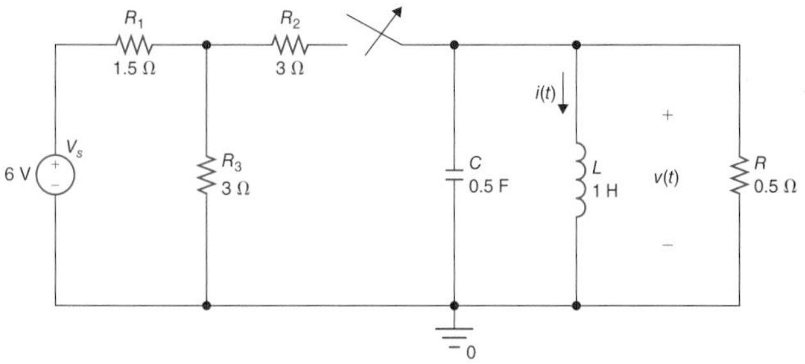

8.19 **The switch in the circuit shown in Figure P8.19 has been closed for a long time before it is opened at $t = 0$. Find current $i(t)$ through the inductor for $t \geq 0$.**

FIGURE P8.19

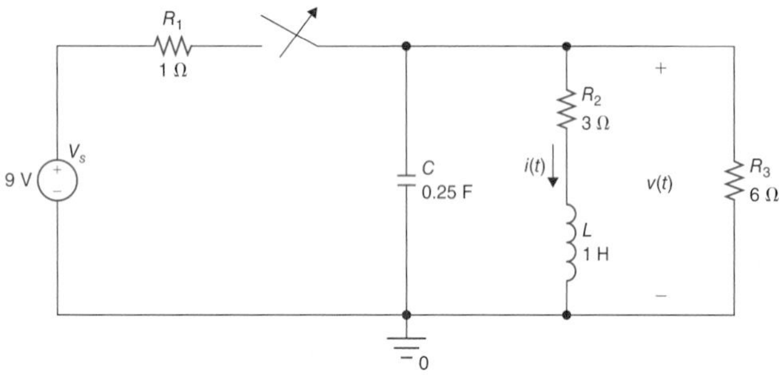

8.20 **The switch in the circuit shown in Figure P8.20 has been closed for a long time before it is opened at $t = 0$. Find current $i(t)$ through the inductor for $t \geq 0$.**

FIGURE P8.20

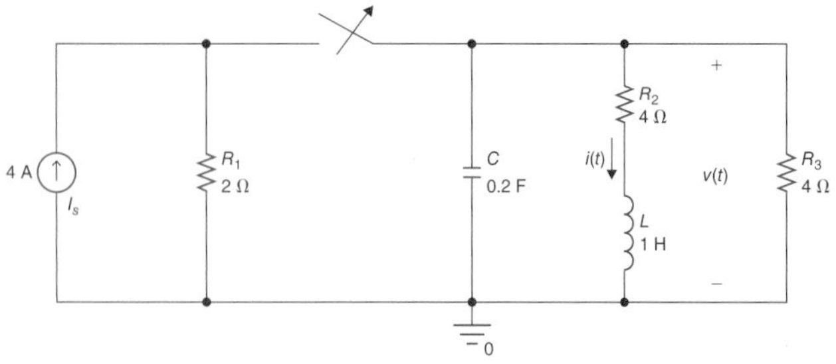

8.21 **The switch in the circuit shown in Figure P8.21 has been closed for a long time before it is opened at $t = 0$. Find current $i(t)$ through the inductor for $t \geq 0$.**

FIGURE P8.21

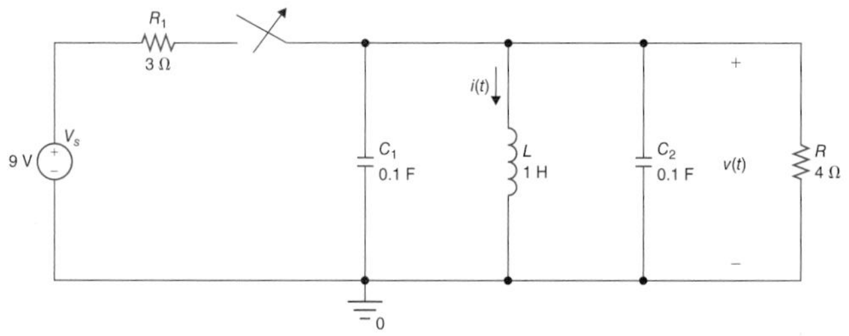

8.22 **The switch in the circuit shown in Figure P8.22 has been closed for a long time before it is opened at $t = 0$. Find current $i(t)$ through the inductor for $t \geq 0$.**

FIGURE P8.22

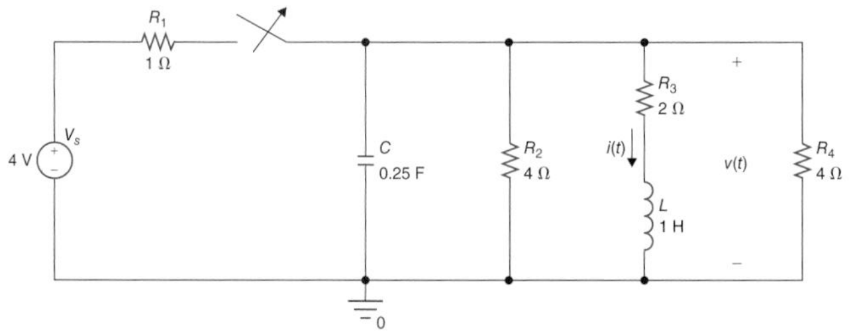

Step Response of a Series RLC Circuit

8.23 **In the circuit shown in Figure P8.23,**

FIGURE P8.23

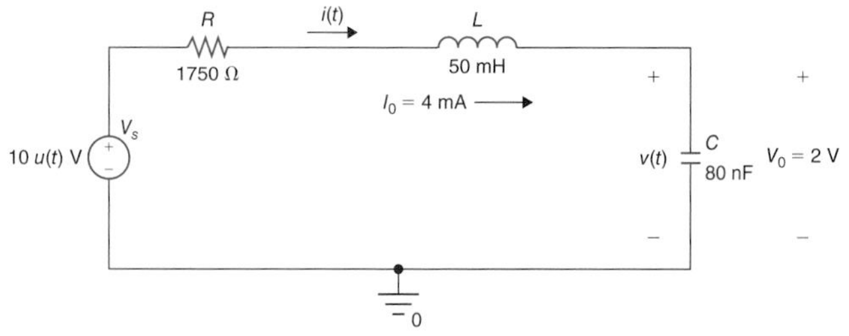

a. find α.
b. find ω_0.
c. write a differential equation in $v(t)$.
d. find the characteristic equation.
e. find the roots of the characteristic equation.
f. find the final value of voltage $v(t)$ across the capacitor.
g. find voltage $v(t)$ across the capacitor for $t \geq 0$ and plot $v(t)$.
h. find current $i(t)$ for $t \geq 0$ and plot $i(t)$.

8.24 **In the circuit shown in Figure P8.24,**

FIGURE P8.24

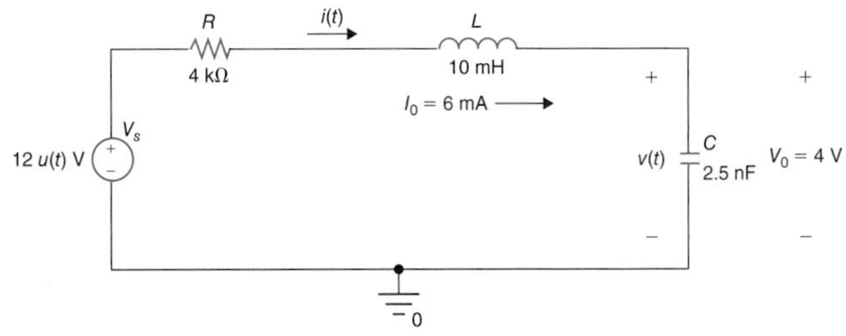

a. find α.
b. find ω_0.
c. write a differential equation in $v(t)$.
d. find the characteristic equation.
e. find the roots of the characteristic equation.
f. find the final value of voltage $v(t)$ across the capacitor.
g. find voltage $v(t)$ across the capacitor for $t \geq 0$ and plot $v(t)$.
h. find current $i(t)$ for $t \geq 0$ and plot $i(t)$.

8.25 **In the circuit shown in Figure P8.25,**

FIGURE P8.25

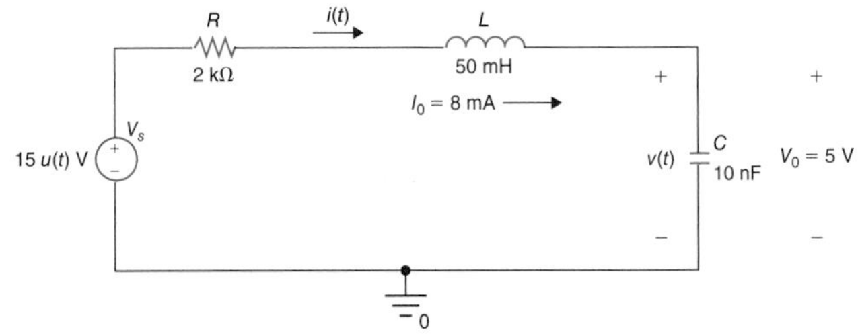

a. find α.
b. find ω_0.
c. write a differential equation in $v(t)$.
d. find the characteristic equation.
e. find the roots of the characteristic equation.
f. find the final value of voltage $v(t)$ across the capacitor.
g. find voltage $v(t)$ across the capacitor for $t \geq 0$ and plot $v(t)$.
h. find current $i(t)$ for $t \geq 0$ and plot $i(t)$.

8.26 **In the circuit shown in Figure P8.26,**

FIGURE P8.26

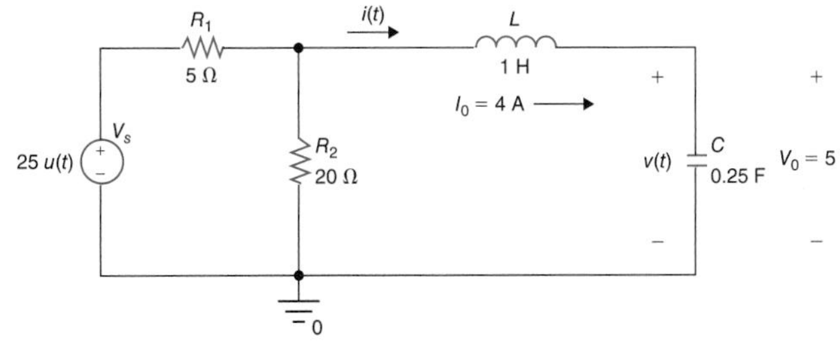

a. find voltage $v(t)$ for $t \geq 0$.
b. find current $i(t)$ for $t \geq 0$.

8.27 **In the circuit shown in Figure P8.27, the initial current through the inductor is $I_0 = 2$ A and the initial voltage across the capacitor is $V_0 = 4$ V when the input signal is applied at $t = 0$.**

FIGURE P8.27

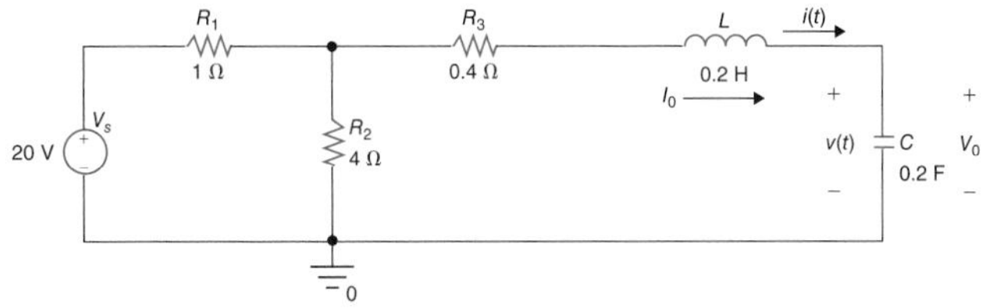

a. Find voltage $v(t)$, $t \geq 0$.
b. Find current $i(t)$, $t \geq 0$.

8.28 **Switch 1 in the circuit shown in Figure P8.28 has been closed for a long time before it is opened at $t = 0$. Switch 2 is closed at $t = 0$. Find voltage $v(t)$ across the capacitor for $t \geq 0$.**

FIGURE P8.28

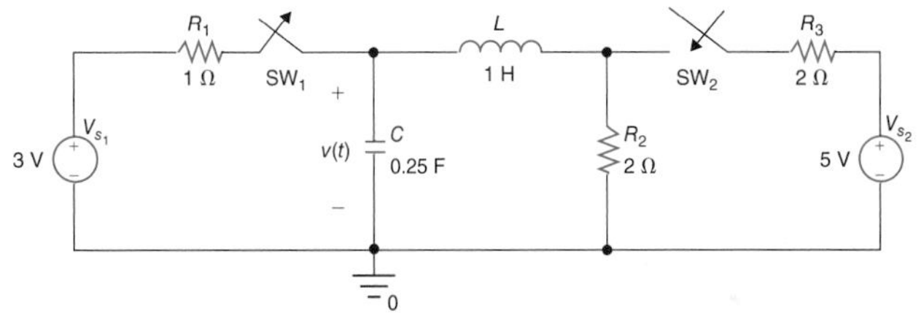

8.29 **Switch 1 in the circuit shown in Figure P8.29 has been closed for a long time before it is opened at $t = 0$. Switch 2 is closed at $t = 0$. Find voltage $v(t)$ across the capacitor for $t \geq 0$.**

FIGURE P8.29

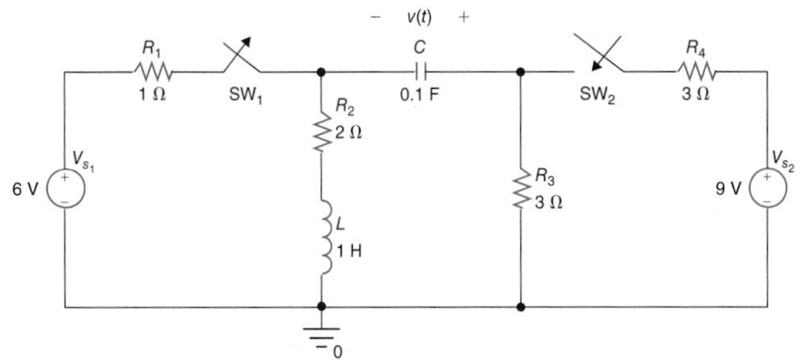

Step Response of a Parallel RLC Circuit

8.30 **In the circuit shown in Figure P8.30,**

a. find α.
b. find ω_0.
c. write a differential equation in $i(t)$.

FIGURE P8.30

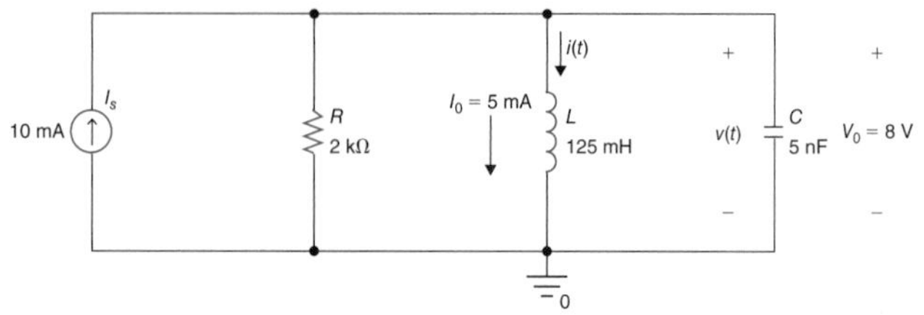

d. find the characteristic equation.
e. find the roots of the characteristic equation.
f. find the final value of current $i(t)$ through the inductor.
g. find current $i(t)$ through the inductor for $t \geq 0$ and plot $i(t)$.
h. find voltage $v(t)$ for $t \geq 0$ and plot $v(t)$.

8.31 In the circuit shown in Figure P8.31,

FIGURE P8.31

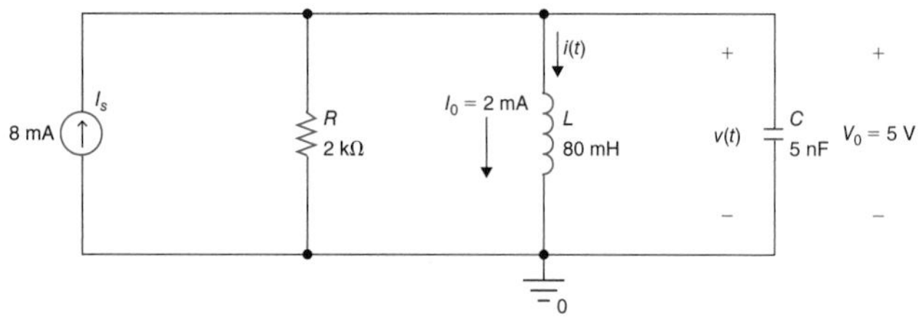

a. find α.
b. find ω_0.
c. write a differential equation in $i(t)$.
d. find the characteristic equation.
e. find the roots of the characteristic equation.
f. find the final value of current $i(t)$ through the inductor.
g. find current $i(t)$ through the inductor for $t \geq 0$ and plot $i(t)$.
h. find voltage $v(t)$ for $t \geq 0$ and plot $v(t)$.

8.32 In the circuit shown in Figure P8.32,

FIGURE P8.32

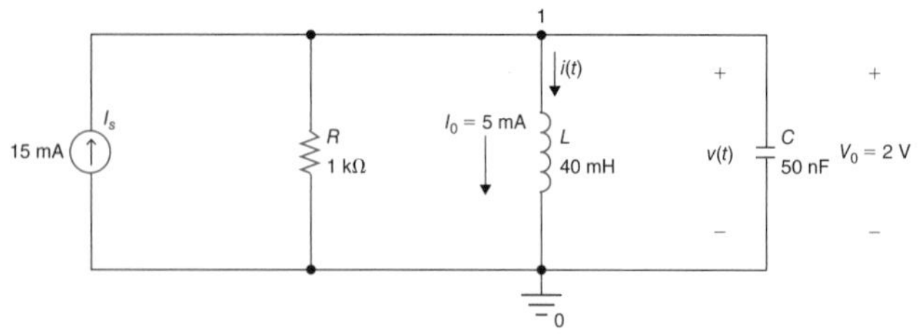

a. find α.
b. find ω_0.
c. write a differential equation in $i(t)$.
d. find the characteristic equation.
e. find the roots of the characteristic equation.
f. find the final value of current $i(t)$ through the inductor.
g. find current $i(t)$ through the inductor for $t \geq 0$ and plot $i(t)$.
h. find voltage $v(t)$ for $t \geq 0$ and plot $v(t)$.

8.33 In the circuit shown in Figure P8.33,

FIGURE P8.33

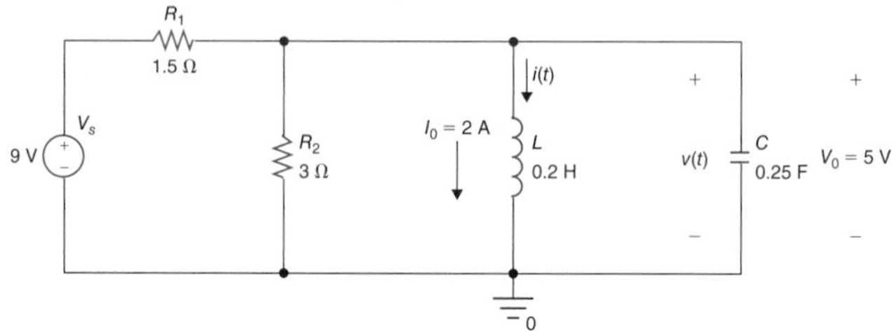

a. find current $i(t)$ for $t \geq 0$.
b. find voltage $v(t)$ for $t \geq 0$.

8.34 Switch 1 in the circuit shown in Figure P8.34 has been closed for a long time before it is opened at $t = 0$. Switch 2 is closed at $t = 0$. Find current $i(t)$ through the inductor for $t \geq 0$.

FIGURE P8.34

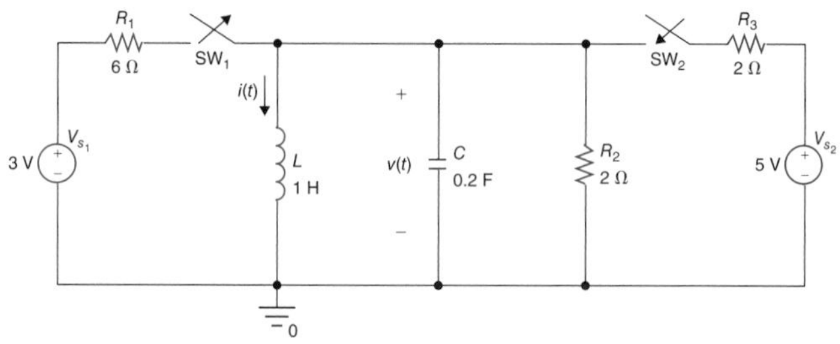

8.35 Switch 1 in the circuit shown in Figure P8.35 has been closed for a long time before it is opened at $t = 0$. Switch 2 is closed at $t = 0$. Find current $i(t)$ through the inductor for $t \geq 0$.

FIGURE P8.35

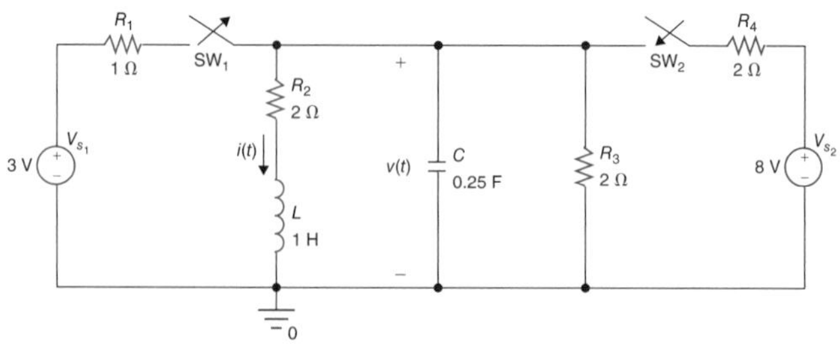

General Second-Order Circuits

8.36 In the circuit shown in Figure P8.36,

FIGURE P8.36

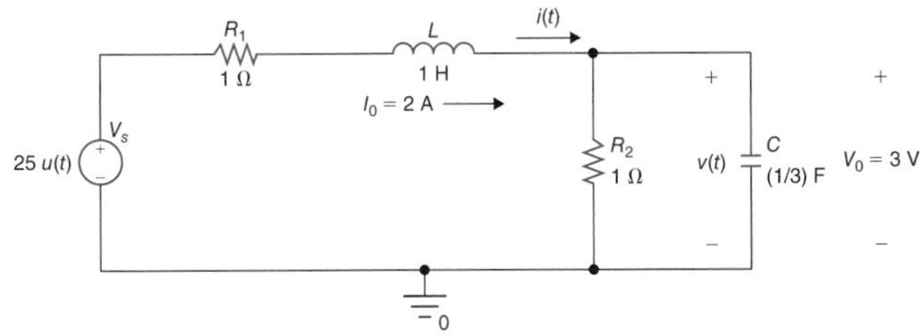

a. find voltage $v(t)$ for $t \geq 0$.
b. find current $i(t)$ for $t \geq 0$.

8.37 The switch in the circuit shown in Figure P8.37 has been in position *a* for a long time before it is moved to position *b* at time $t = 0$.

FIGURE P8.37

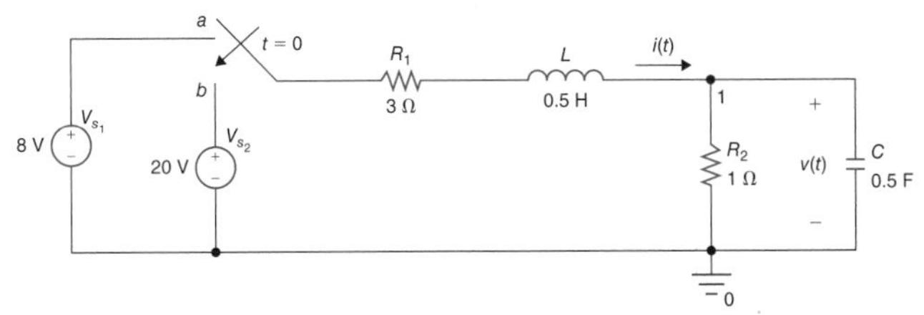

a. Find the current through the inductor I_0 (from left to right) and the voltage across the capacitor V_0 (from top to bottom) at $t = 0$.
b. Write a mesh equation on the left side of the circuit for $t \geq 0$.
c. Write a node equation on top of the capacitor for $t \geq 0$.
d. Find the second-order differential equation in $v(t)$ for $t \geq 0$.
e. Find the voltage across the capacitor, $v(t)$, for $t \geq 0$.
f. Find the current through the inductor, $i(t)$, for $t \geq 0$.

8.38 In the circuit shown in Figure P8.38, the initial current through the inductor is $i(0) = I_0 = 2$ A, and the initial voltage across the capacitor is $v(0) = V_0 = 3$ V.

FIGURE P8.38

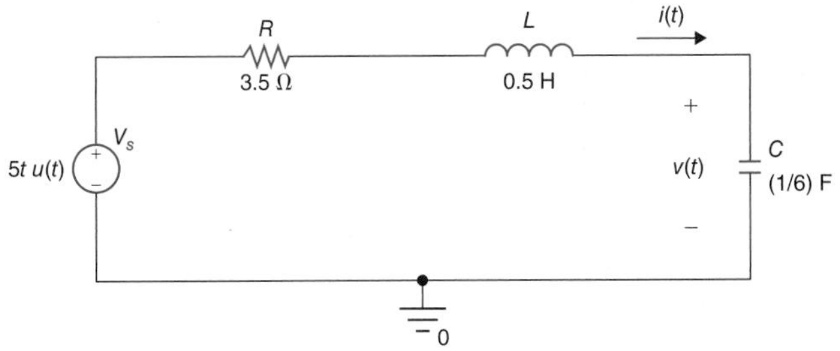

a. Write a mesh equation by collecting the voltage drops in the clockwise direction.
b. Find the second-order differential equation in $v(t)$.
c. Find the characteristic equation and roots of the characteristic equation.
d. Let the particular solution be $v_p(t) = K_1 t + K_2$. Substitute this proposed particular solution for $v(t)$ in the differential equation and find K_1 and K_2.
e. Find the voltage across the capacitor, $v(t)$, for $t \geq 0$.

8.39 In the circuit shown in Figure P8.39,

FIGURE P8.39

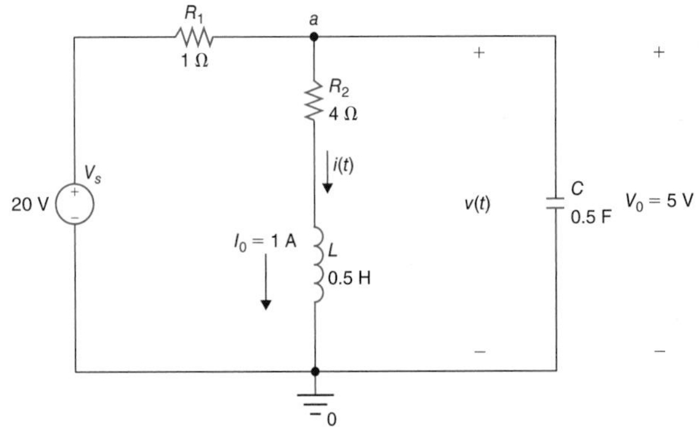

a. Find the node equation at node *a* by summing the currents away from node *a*.
b. On the series connection of R_2 and L, represent $v(t)$ as a function of $i(t)$ and $\dfrac{di(t)}{dt}$.
c. Find a second-order differential equation in $i(t)$.
d. Solve the differential equation to obtain $i(t)$ for $t \geq 0$.

8.40 In the circuit shown in Figure P8.40,

FIGURE P8.40

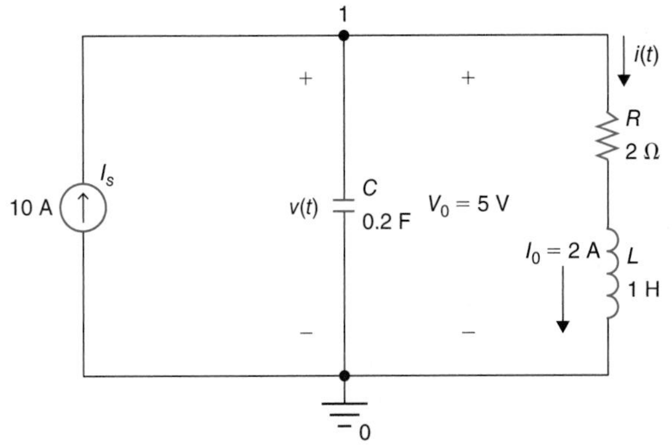

a. find voltage $v(t)$ for $t \geq 0$.
b. find current $i(t)$ for $t \geq 0$.

8.41 **In the circuit shown in Figure P8.41, let**

FIGURE P8.41

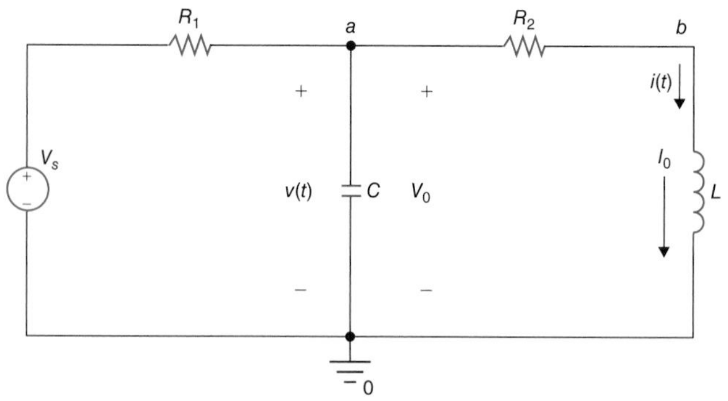

$$V_s = 12\text{ V},\quad R_1 = 1\ \Omega,\quad R_2 = 2\ \Omega,\quad L = 1\text{ H}$$
$$C = 0.2\text{ F},\quad v(0) = V_0 = 3\text{ V},\quad i(0) = I_0 = 1\text{ A}$$

a. Write a node equation at node a by summing the currents away from node a.
b. Represent voltage $v(t)$ as a function of L and $i(t)$, and find the initial condition $di(0)/dt$.
c. Find the second-order differential equation in $i(t)$.
d. Find the final value $i(\infty)$ of $i(t)$
e. Solve the differential equation to obtain $i(t)$ for $t \geq 0$.

8.42 **In the circuit shown in Figure P8.42, the initial voltage across the capacitor C_1 is $v_1(0) = V_{10} = 2$ V, and the initial voltage across the capacitor C_2 is $v_2(0) = V_{20} = 1$ V.**

FIGURE P8.42

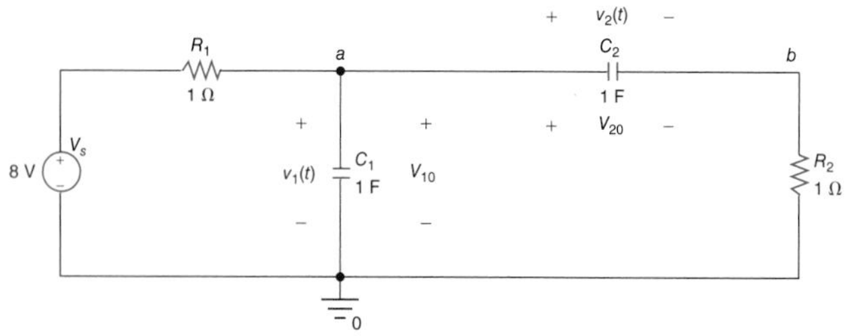

a. Write a node equation at node a by summing the currents leaving node a.
b. On the series connection of C_2 and R_2, represent the voltage at node a as a function of $v_2(t)$ and $dv_2(t)/dt$.
c. Find the second-order differential equation in $v_2(t)$ and find $dv_2(0)/dt$.
d. Solve the differential equation to obtain $v_2(t)$ for $t \geq 0$.

8.43 **The initial conditions for the circuit shown in Figure P8.43 are $i(0) = I_0 = 1$ A, $v(0) = V_0 = 2$ V.**

FIGURE P8.43

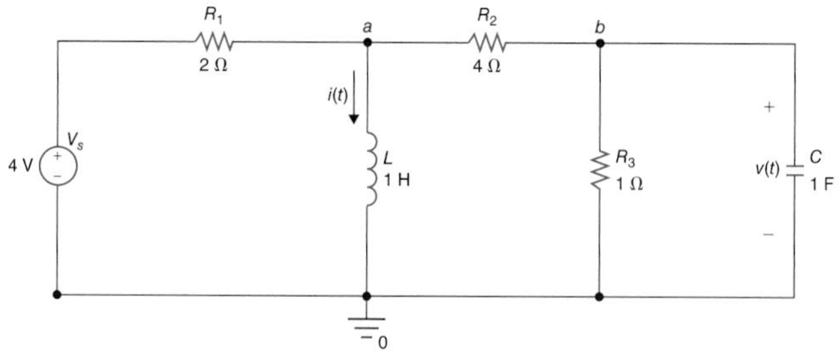

a. Write a node equation at node a by summing the currents leaving node a for $t \geq 0$. Find $di(0)/dt$.
b. Write a node equation at node b by summing the currents leaving node b for $t \geq 0$.
c. Find the differential equation in $i(t)$ and find the roots of the characteristic equation.
d. Find $i(t)$ for $t \geq 0$.

8.44 **In the circuit shown in Figure P8.44, let**

FIGURE P8.44

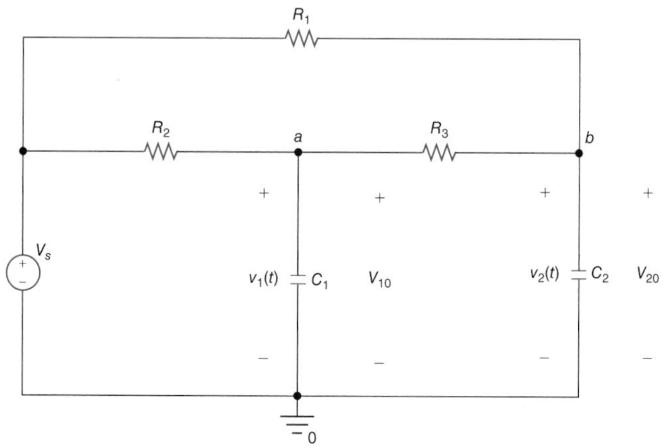

$$V_s = 10\text{ V},\quad R_1 = 1\ \Omega,\quad R_2 = 1\ \Omega,\quad R_3 = 2\ \Omega$$
$$C_1 = 1\text{ F},\quad C_2 = 0.2\text{ F},\quad V_{10} = 1\text{ V},\quad V_{20} = 2\text{ V}$$

a. Write a node equation at node a by summing the currents away from node a.
b. Write a node equation at node b by summing the currents away from node b. Find $dv_2(0)/dt$.
c. Find the second-order differential equation in $v_2(t)$.
d. Find the roots of the characteristic equation, and find the final value $v_2(\infty)$ of $v_2(t)$.
e. Solve the differential equation to obtain $v_2(t)$ for $t \geq 0$.

8.45 **In the circuit shown in Figure P8.45, the initial current through the inductor L_1 is $i_1(0) = I_{10} = 2$ A, and the initial current through the inductor L_2 is $i_2(0) = I_{20} = 1$ A.**

FIGURE P8.45

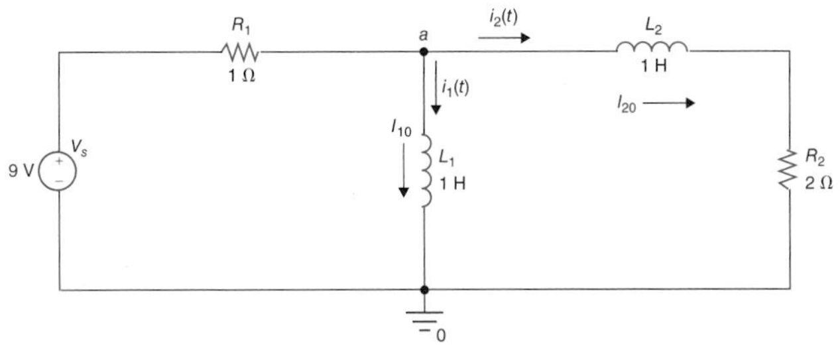

a. Write a node equation at node a by summing the currents leaving node a. Find $di_1(0)/dt$.
b. On the series connection of L_2 and R_2, represent the voltage at node a as a function of $i_2(t)$ and $di_2(t)/dt$.
c. Find the second-order differential equation in $i_1(t)$.
d. Solve the differential equation to obtain $i_1(t)$ for $t \geq 0$.

8.46 **In the circuit shown in Figure P8.46, the initial current through the inductor L is $i(0) = I_0 = 1$ A, and the initial voltage across the capacitor is $v(0) = V_0 = 3$ V.**

FIGURE P8.46

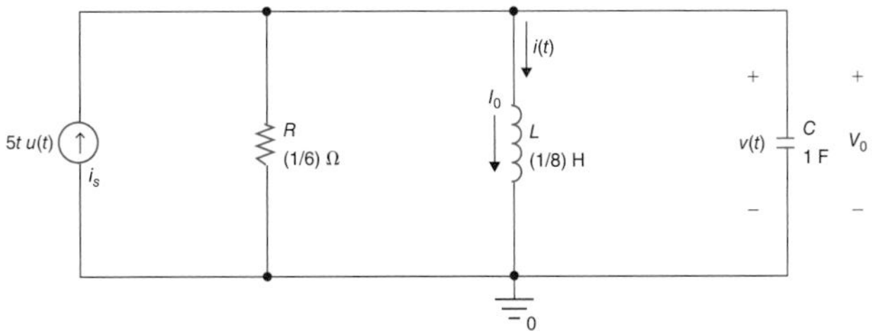

a. Find the second-order differential equation in $i(t)$.
b. Find the characteristic equation and roots of the characteristic equation.
c. Let the particular solution be $i_p(t) = K_1 t + K_2$. Substitute this proposed particular solution for $i(t)$ in the differential equation and find K_1 and K_2.
d. Find the current through the inductor, $i(t)$, for $t \geq 0$.

8.47 **The switch in the circuit shown in Figure P8.47 has been opened for a long time before it is closed at $t = 0$.**

a. Find the initial current $i(0) = I_0$ through the inductor at $t = 0$, and the initial voltage $v(0) = V_0$ across the capacitor at $t = 0$.

FIGURE P8.47

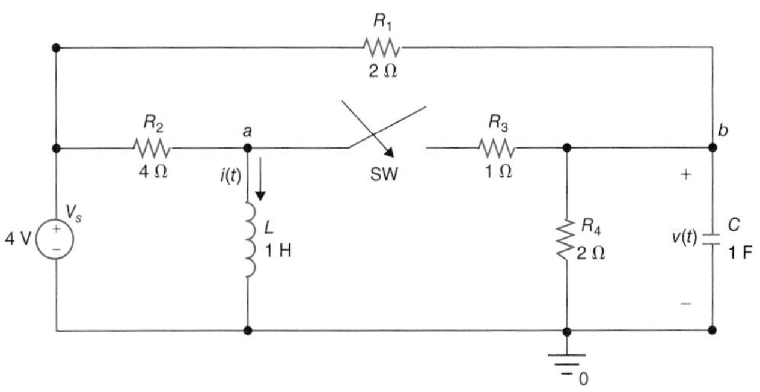

b. Write a node equation at node a by summing the currents leaving node a for $t \geq 0$. Find $di(0)/dt$.
c. Write a node equation at node b by summing the currents leaving node b for $t \geq 0$.
d. Find the differential equation in $i(t)$, and find the roots of the characteristic equation.
e. Find $i(t)$ for $t \geq 0$.

8.48 **In the circuit shown in Figure P8.48, switch 1 has been closed for a long time before it is opened at $t = 0$, and switch 2 has been opened for a long time before it is closed at $t = 0$.**

FIGURE P8.48

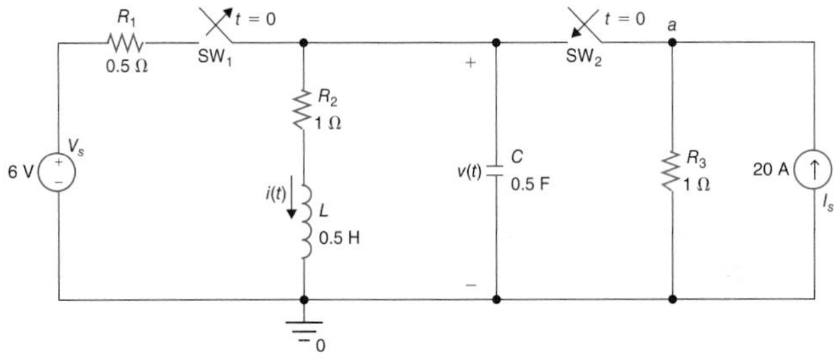

a. Find the initial current $i(0)$ through the inductor and the initial voltage $v(0)$ across the capacitor at $t = 0$.
b. Write a node equation at node a for $t \geq 0$.
c. Represent $v(t)$ as a function of $i(t)$ on the series connection of R_2 and L.
d. Find $dv(t)/dt$.
e. Derive a second-order differential equation in $i(t)$.
f. Find the particular solution to the differential equation; or find the final value of $i(t)$, $i(\infty)$, from the circuit.
g. Find the characteristic equation.
h. Find the roots of the characteristic equation.
i. Find the solution of the differential equation $i(t)$ and plot $i(t)$ for $t \geq 0$.
j. Find $v(t)$ and plot $v(t)$ for $t \geq 0$.

8.49 **In the circuit shown in Figure P8.49, switch 1 has been closed for a long time before it is opened at $t = 0$, and switch 2 has been opened for a long time before it is closed at $t = 0$.**

FIGURE P8.49

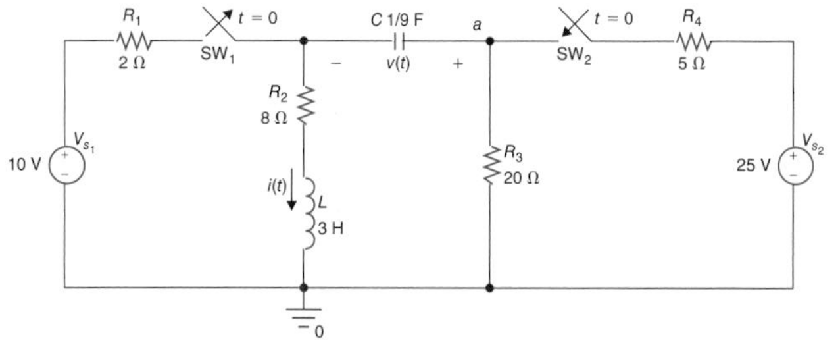

a. Find the initial current $i(0)$ through the inductor and the initial voltage $v(0)$ across the capacitor at $t = 0$.
b. Find the Thévenin equivalent resistance and Thévenin equivalent voltage on the right side of a (V_{S_2}, R_3, R_4).
c. Use mesh analysis to find a differential equation in $v(t)$ for $t \geq 0$.
d. Find the particular solution to the differential equation; or find the final value of $v(t)$, $v(\infty)$, from the circuit.
e. Find the characteristic equation.
f. Find the roots of the characteristic equation.
g. Find the solution of the differential equation, $v(t)$, for $t \geq 0$.

8.50 **In the circuit shown in Figure P8.50, switch 1 has been closed for a long time before it is opened at $t = 0$, and switch 2 has been opened for a long time before it is closed at $t = 0$.**

FIGURE P8.50

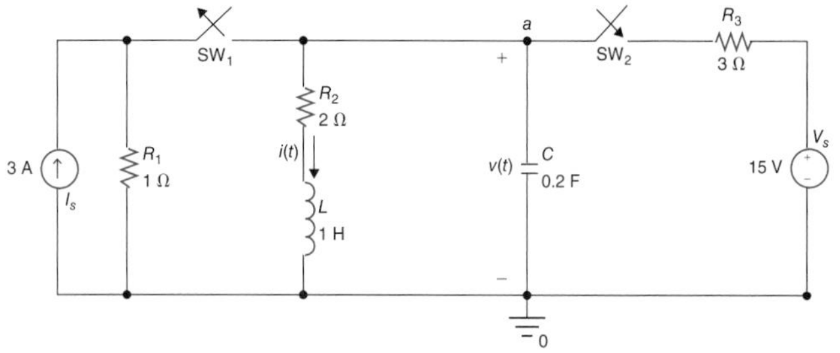

a. Find the initial current through the inductor $i(0) = I_0$ at $t = 0$ and initial voltage $v(0) = V_0$ across the capacitor.
b. For $t \geq 0$, write a node equation by summing the currents away from node a.
c. On the series connection of R_2 and L, represent $v(t)$ as a function of $i(t)$ and $di(t)/dt$.
d. Find the second-order differential equation in $i(t)$ and find the roots of the characteristic equation.
e. Find $i(t)$ for $t \geq 0$.

8.51 **In the circuit shown in Figure P8.51, switch 1 has been closed for a long time before it is opened at $t = 0$, and switch 2 has been opened for a long time before it is closed at $t = 0$.**

FIGURE P8.51

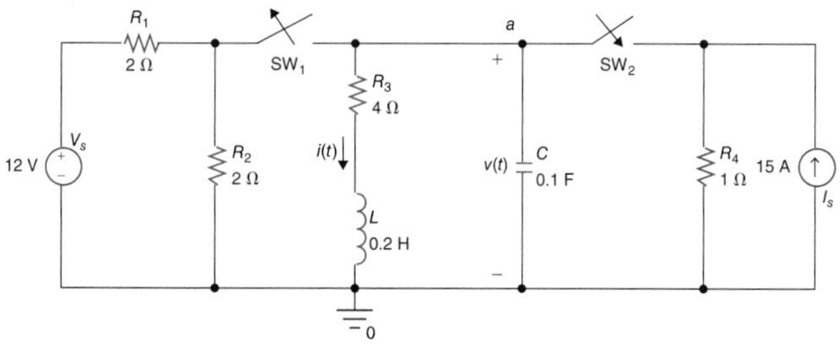

a. Find the initial voltage $v(0) = V_0$ across the capacitor and initial current through the inductor $i(0) = I_0$ at $t = 0$.
b. For $t \geq 0$, write a node equation by summing the currents away from node a.
c. On the series connection of R_3 and L, represent $v(t)$ as a function of $i(t)$ and $di(t)/dt$.
d. Find the second-order differential equation in $i(t)$ and find the roots of the characteristic equation.
e. Find $i(t)$ for $t \geq 0$.

Chapter 9

Phasors and Impedances

9.1 Introduction

In this chapter, we analyze circuits in the steady state when the input signal is a sinusoid. The three parameters (amplitude, frequency, and phase) of a sinusoidal signal are presented, followed by the definition of phasors. For a given frequency, the magnitude and the phase describe the sinusoidal signal completely. The phasor in a polar coordinate system is the magnitude and phase. The phasor can also be represented as a complex number in the Cartesian coordinate system. If the sources of a circuit are sinusoids, the circuit is called an alternating current (ac) circuit. The phasor is a useful tool in analyzing ac circuits.

Impedance is the equivalent of resistance, but depends on frequency. Admittance is the equivalent of conductance and also depends on frequency. Impedance and admittance are defined for the resistor, inductor, and capacitor, and are used to analyze ac circuits. For a given frequency, the impedances for inductors and capacitors are complex numbers. If a circuit is driven by a sinusoidal signal, the steady-state response of the circuit can be found by transforming the circuit to the phasor domain first, and then applying the circuit laws and theorems. The circuit laws and theorems for resistive circuits apply to the phasor-transformed circuits also. The differences are that impedances are used instead of resistances, and the values are complex instead of real numbers.

9.2 Sinusoidal Signals

9.2.1 COSINE WAVE

A cosine wave

$$v(t) = V_m \cos\left(\frac{2\pi}{T}t + \phi\right) \quad (9.1)$$

has three parameters: the amplitude V_m, the period T, and the phase ϕ. As the independent variable t (time) is increased, the angle of the cosine wave, $\frac{2\pi}{T}t + \phi$, increases. At $t = 0$, the angle of the cosine wave is ϕ. As t is increased from $t = 0$ to $t = T$, the angle increases linearly from ϕ to $2\pi + \phi$. Since

$$\begin{aligned}\cos(2\pi + \phi) &= \cos(2\pi)\cos(\phi) - \sin(2\pi)\sin(\phi) \\ &= 1 \times \cos(\phi) - 0 \times \sin(\phi) = \cos(\phi)\end{aligned}$$

the value of the cosine wave at $t = T$ is identical to the value at $t = 0$. As t is increased from $t = T$ to $t = 2T$, the value of the cosine wave repeats that from $t = 0$ to $t = T$. The cosine wave repeats itself every T seconds [as shown in Figure 9.1]. It is a periodic wave with a period of T seconds. In one second, there are

$$f = \frac{1}{T} \tag{9.2}$$

periods (cycles, waves) of the cosine wave. The parameter f is called the **frequency** of the cosine wave and has a unit of one over second (1/s), called hertz (Hz). In terms of frequency, the cosine wave becomes

$$v(t) = V_m \cos(2\pi f t + \phi) \tag{9.3}$$

Because the angle changes by 2π radians in one period, and there are f periods in one second, the change in angle in one second is given by

$$\omega = 2\pi f = \frac{2\pi}{T} \tag{9.4}$$

The parameter ω is called the *angular velocity* of the cosine wave and has a unit of radians per second (rad/s). In terms of radian frequency ω, the cosine wave becomes

$$v(t) = V_m \cos(\omega t + \phi) \tag{9.5}$$

FIGURE 9.1

Cosine wave with $V_m = 5$ V, $T = 1$ ms ($f = 1$ kHz), $\phi = 0$ rad.

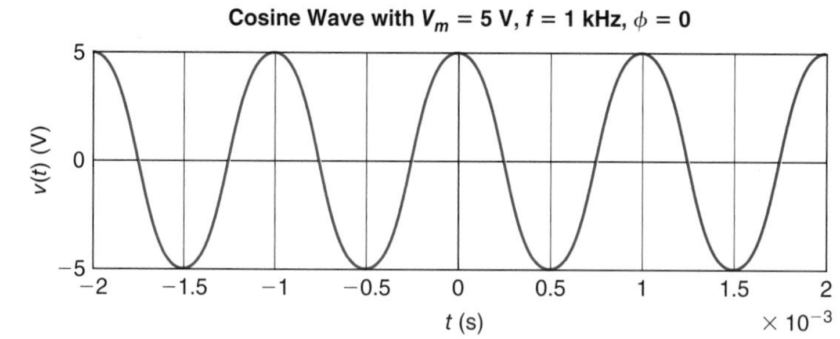

Regardless of the value of period and phase, the value of cosine is bounded between −1 and 1; that is,

$$-1 \le \cos\left(\frac{2\pi}{T}t + \phi\right) \le 1$$

Thus, the cosine wave given by Equation (9.1) is bounded between $-V_m$ and V_m, assuming $V_m > 0$; that is,

$$-V_m \le V_m \cos\left(\frac{2\pi}{T}t + \phi\right) \le V_m$$

The maximum value of the cosine wave is V_m, and the minimum value of the cosine wave is $-V_m$. The amplitude V_m is called the **peak value** of the cosine wave. The distance between the maximum value and the minimum value is $2V_m$. This is called the **peak-to-peak amplitude**. Since the cosine wave is periodic with period T, the average value V_{av} of the cosine wave can be obtained by integrating it for one period and then dividing the result by the period T. Thus,

$$V_{av} = \frac{1}{T}\int_0^T V_m \cos\left(\frac{2\pi}{T}t + \phi\right)dt = \frac{V_m}{T}\frac{\sin\left(\frac{2\pi}{T}t + \phi\right)\Big|_0^T}{\frac{2\pi}{T}}$$

$$= \frac{V_m}{2\pi}[\sin(2\pi + \phi) - \sin(\phi)] = 0$$

The average value of the cosine wave is zero. The average value is called the **dc component** or **dc offset**. If a dc offset V_{dc} is added to the cosine wave given by Equation (9.1), we have

$$v(t) = V_{dc} + V_m \cos\left(\frac{2\pi}{T}t + \phi\right) \tag{9.6}$$

Figure 9.2 shows a cosine wave with five different dc offsets (V_{dc} = −2 V, −1 V, 0 V, 1 V, 2 V) for V_m = 1 V, T = 1 ms (f = 1000 Hz), and $\phi = 0$.

FIGURE 9.2

Cosine waves with different dc offsets.

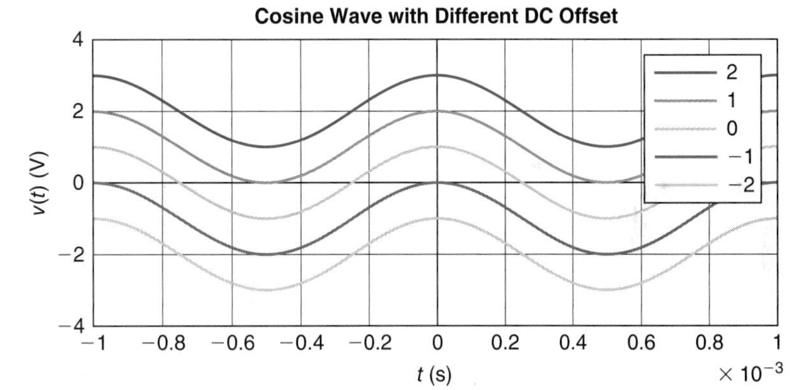

Figure 9.3 shows a cosine wave as the frequency is varied from f = 1000 Hz to f = 3000 Hz at an interval of 1000 Hz with an amplitude of V_m = 1 V and a phase of $\phi = 0$. As the frequency is increased, the cosine wave changes faster, and there are more cycles per a given amount of time.

The phase ϕ of a cosine wave determines the starting value of the cosine wave at $t = 0$. At $t = 0$, the cosine wave has a value of $V_m \cos(\phi)$. If $\phi = 0$, the value at $t = 0$ is $V_m \cos(0) = V_m$; if $\phi = \pi/2$, the value at $t = 0$ is $V_m \cos(\pi/2) = 0$; and if $\phi = \pi$, the value at $t = 0$ is $V_m \cos(\pi) = -V_m$. Rewriting the angle of Equation (9.1), we have

$$v(t) = V_m \cos\left(\frac{2\pi}{T}t + \frac{2\pi}{T}\phi\frac{T}{2\pi}\right) = V_m \cos\left[\frac{2\pi}{T}\left(t + \frac{\phi}{2\pi}T\right)\right] \tag{9.7}$$

The cosine wave $V_m \cos\left(\frac{2\pi}{T}t + \phi\right)$ is a time shift of $V_m \cos\left(\frac{2\pi}{T}t\right)$ by $\phi T/(2\pi)$ seconds. If ϕ is positive, the shift is to the left; if ϕ is negative, the shift is to the right. The shift to the left ($\phi > 0$) is called **lead** and the shift to the right ($\phi < 0$) is called **lag**. When ϕ is a leading phase ($\phi > 0$), the zero crossing point (or the peak) leads the original ($\phi = 0$) cosine wave.

FIGURE 9.3

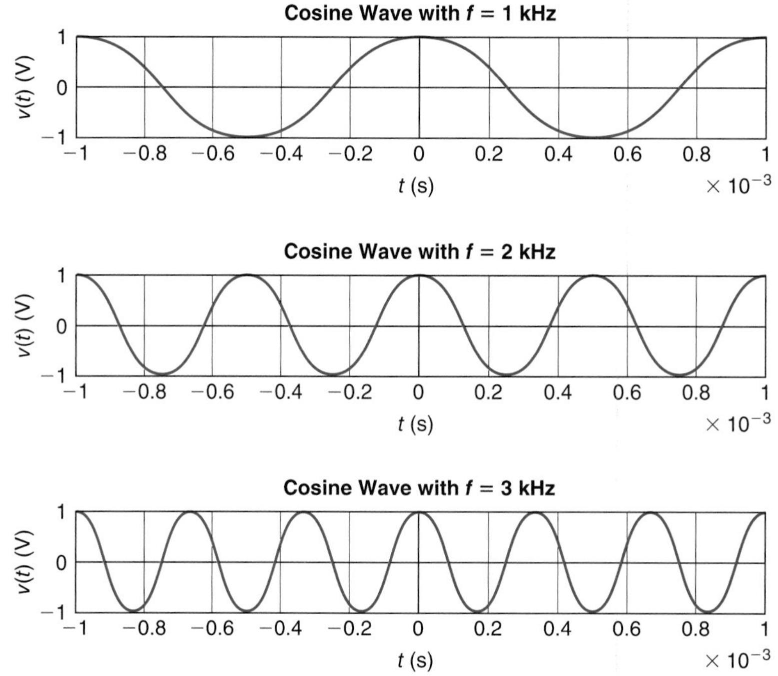

Cosine waves with frequencies 1kHz, 2kHz, and 3kHz.

When ϕ is a lagging phase ($\phi < 0$), the zero crossing point (or the peak) lags the original ($\phi = 0$) cosine wave. Figure 9.4 shows a cosine wave with three different phases ($\phi = -45° = -\pi/4, 0° = 0, 45° = \pi/4$) for $V_m = 1$ V and $f = 1$ Hz ($T = 1$ s).

FIGURE 9.4

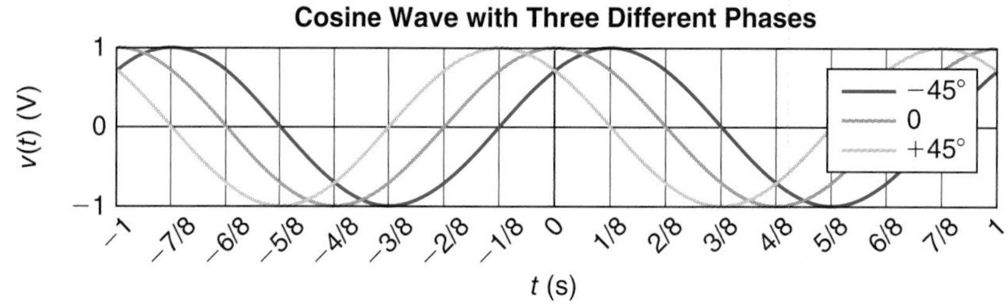

Cosine waves with different phase values ($\phi = 0, \phi = -45°, \phi = 45°$).

When $\phi = 0$, the cosine wave crosses zero at $t = -0.25$ s. When $\phi = \pi/4$, the cosine wave crosses zero at $t = -0.375$ s. This is earlier than $\phi = 0$ by 0.125 s or $T/8$. It is leading. When $\phi = -\pi/4$, the cosine wave crosses zero at $t = -0.125$ s. This is later than $\phi = 0$ by 0.125 s or $T/8$. It is lagging. If we look at the peaks of the three cosine waves around $t = 0$, we arrive at the same conclusion. When $\phi = 0$, the peak of the cosine wave occurs at $t = 0$. When $\phi = \pi/4$, the peak of the cosine wave occurs at $t = -0.125$ s. This is earlier than $\phi = 0$ by 0.125 s or $T/8$. It is leading. When $\phi = -\pi/4$, the peak of the cosine wave occurs at $t = 0.125$ s. This is later than $\phi = 0$ by 0.125 s or $T/8$. It is lagging.

9.2.2 SINE WAVE

A sine wave

$$v(t) = V_m \sin\left(\frac{2\pi}{T}t + \phi\right) \tag{9.8}$$

can be viewed as a time shift of a cosine wave $V_m \cos\left(\frac{2\pi}{T}t + \phi\right)$ by 90° ($= \pi/2$) to the right:

$$V_m \cos\left(\frac{2\pi}{T}t + \phi - \frac{\pi}{2}\right) = V_m \cos\left(\frac{2\pi}{T}t + \phi\right)\cos\left(\frac{\pi}{2}\right) + V_m \sin\left(\frac{2\pi}{T}t + \phi\right)\sin\left(\frac{\pi}{2}\right)$$

$$= V_m \cos\left(\frac{2\pi}{T}t + \phi\right) \times 0 + V_m \sin\left(\frac{2\pi}{T}t + \phi\right) \times 1$$

$$= V_m \sin\left(\frac{2\pi}{T}t + \phi\right)$$

As discussed earlier, the amount of shift is $(\pi/2)T/(2\pi) = T/4$ seconds and the direction of shift is to the right. The amount of shift is equivalent to $(\pi/2)/(2\pi) = 1/4$ of one period ($T/4$ seconds). Figure 9.5 shows the cosine wave and the sine wave with the same amplitude ($V_m = 1$ V) and frequency ($f = 1$ Hz), and $\phi = 0$.

FIGURE 9.5

Cosine and sine as a function of time.

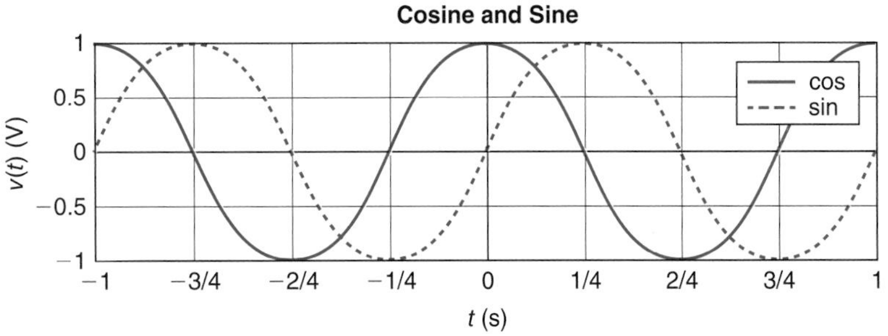

Since

$$V_m \sin\left(\frac{2\pi}{T}t + \phi + \frac{\pi}{2}\right) = V_m \sin\left(\frac{2\pi}{T}t + \phi\right)\cos\left(\frac{\pi}{2}\right) + V_m \cos\left(\frac{2\pi}{T}t + \phi\right)\sin\left(\frac{\pi}{2}\right)$$

$$= V_m \sin\left(\frac{2\pi}{T}t + \phi\right) \times 0 + V_m \cos\left(\frac{2\pi}{T}t + \phi\right) \times 1$$

$$= V_m \cos\left(\frac{2\pi}{T}t + \phi\right)$$

a sine wave can be converted to a cosine wave by shifting it by $T/4$ (90 degrees) to the left.

EXAMPLE 9.1

A sinusoidal signal is shown in Figure 9.6. Find the equation of this signal.

The dc offset is

$$V_{dc} = \frac{3\text{ V} + (-1\text{ V})}{2} = 1\text{ V}$$

continued

Example 9.1 continued

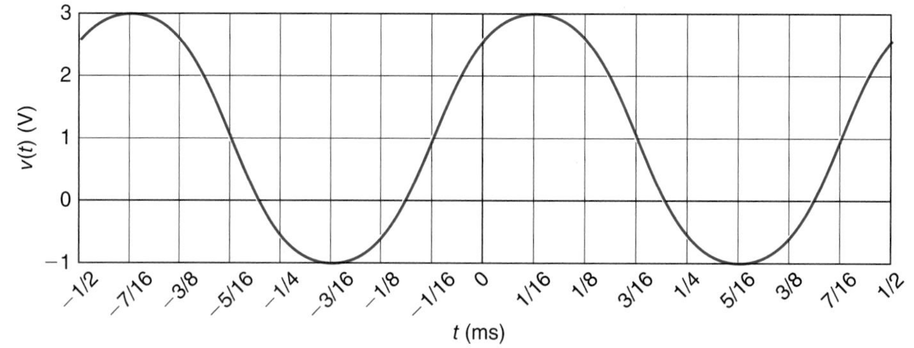

FIGURE 9.6

A sinusoidal signal for Example 9.1.

The peak amplitude is given by

$$V_m = \frac{3\text{ V} - (-1\text{ V})}{2} = 2\text{ V}$$

The period is given by

$$T = 1/16\text{ ms} - (-7/16\text{ ms}) = 0.5\text{ ms}$$

Thus, the frequency is $f = 1/T = 2000$ Hz. The peak value of $v(t)$ occurs at $t = 1/16$ ms. This is a shift to the right by $T(1/16\text{ ms})/(1/2\text{ ms}) = T/8$ corresponding to $\pi/4$. Therefore, the equation of the sinusoid is

$$v(t) = 1 + 2\cos(2\pi \times 2000t - \pi/4) = 1 + 2\cos(2\pi \times 2000t - 45°)$$

Exercise 9.1

Find the equation of the sinusoidal waveform shown in Figure 9.7.

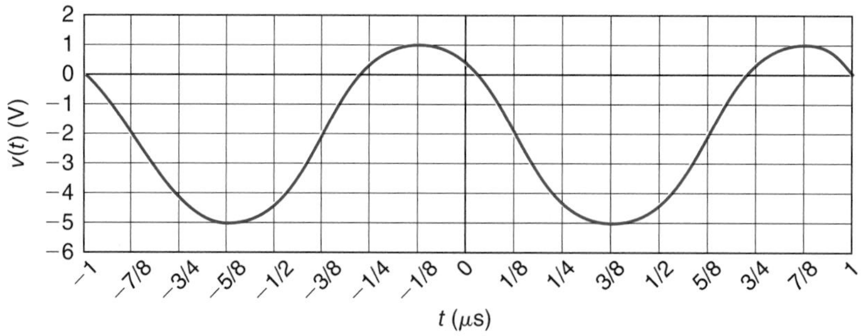

FIGURE 9.7

Signal for EXERCISE 9.1

Answer:
$v(t) = -2 + 3\cos(2\pi \times 10^6 t + 45°)$ V.

9.3 *RMS* Value

Let the sinusoidal voltage across a circuit element be

$$v(t) = V_m \cos(\omega t + \theta_v) \tag{9.9}$$

The peak amplitude of $v(t)$ is

$$V_p = V_m \tag{9.10}$$

The peak-to-peak amplitude of $v(t)$ is

$$V_{p-p} = 2V_m \tag{9.11}$$

If the voltage $v(t)$ is squared, we obtain

$$v^2(t) = V_m^2 \cos^2(\omega t + \theta_v) = \frac{V_m^2}{2} + \frac{V_m^2}{2}\cos(2\omega t + 2\theta_v) \tag{9.12}$$

The **mean square** value of $v(t)$ is defined as the average value of $v^2(t)$:

$$V_{ms} = \lim_{T_0\to\infty} \frac{1}{T_o}\int_{-\frac{T_0}{2}}^{\frac{T_0}{2}} v^2(t)dt \tag{9.13}$$

If $v(t)$ is periodic with period T, the mean square value simplifies to

$$V_{ms} = \frac{1}{T}\int_{-\frac{T}{2}}^{\frac{T}{2}} v^2(t)dt \tag{9.14}$$

The second term of Equation (9.12), $\frac{V_m^2}{2}\cos(2\omega t + 2\theta_v)$, is a periodic cosine wave with period $T_2 = \pi/\omega = T/2$. The average value of this term is zero. The average value of $v^2(t)$ is equal to the first term of Equation (9.12), $\frac{V_m^2}{2}$. Thus, the mean square value of $v(t)$ is given by

$$V_{ms} = \frac{V_m^2}{2} \tag{9.15}$$

Alternatively, we can substitute $v^2(t)$ from Equation (9.12) into Equation (9.14) to get

$$V_{ms} = \frac{1}{T}\int_{-\frac{T}{2}}^{\frac{T}{2}} v^2(t)dt = \frac{1}{T}\int_{-\frac{T}{2}}^{\frac{T}{2}} \frac{V_m^2}{2}dt + \frac{1}{T}\int_{-\frac{T}{2}}^{\frac{T}{2}} \frac{V_m^2}{2}\cos(2\omega t + 2\theta_v)dt$$

$$= \frac{V_m^2}{2} + \frac{V_m^2}{2T}\frac{\sin(2\omega t + 2\theta_v)\Big|_{-\frac{T}{2}}^{\frac{T}{2}}}{2\omega}$$

$$= \frac{V_m^2}{2} + \frac{V_m^2}{2T}\frac{\sin\left(2\frac{2\pi}{T}\frac{T}{2} + 2\theta_v\right) - \sin\left(2\frac{2\pi}{T}\frac{-T}{2} + 2\theta_v\right)}{2\omega}$$

$$= \frac{V_m^2}{2} + \frac{V_m^2}{2T}\frac{\sin(2\theta_v) - \sin(2\theta_v)}{2\omega} = \frac{V_m^2}{2} + 0 = \frac{V_m^2}{2}$$

The **root mean square (rms)** value of $v(t)$ is defined as the square root of mean square value of $v(t)$:

$$V_{rms} = \sqrt{\lim_{T_0 \to \infty} \frac{1}{T_0} \int_{-T_0/2}^{T_0/2} v^2(t)dt} \tag{9.16}$$

If $v(t)$ is periodic with period T, the *rms* value simplifies to

$$V_{rms} = \sqrt{\frac{1}{T} \int_{-T/2}^{T/2} v^2(t)dt} \tag{9.17}$$

The limits of integration can be 0 and T instead of $-T/2$ and $T/2$. For the voltage $v(t)$ given by Equation (9.9), the *rms* value is given by

$$V_{rms} = \sqrt{\frac{V_m^2}{2}} = \frac{V_m}{\sqrt{2}} = 0.7071V_m \tag{9.18}$$

Equation (9.18) states that the *rms* value of a sinusoid with amplitude V_m (peak value) is given by $V_m/\sqrt{2}$ regardless of the frequency and phase of the sinusoid. To get the peak amplitude from the *rms* amplitude, simply multiply $\sqrt{2}$ by the *rms* amplitude. Thus,

$$V_m = \sqrt{2}V_{rms} \tag{9.19}$$

In terms of *rms* amplitude, Equation (9.9) can be written as

$$v(t) = V_{rms} \cos(\omega t + \theta_v) \tag{9.20}$$

where V_{rms} is given by $V_m/\sqrt{2}$. Similarly, for current $i(t)$ defined by

$$i(t) = I_m \cos(\omega t + \theta_i) \tag{9.21}$$

the *rms* value is given by

$$I_{rms} = \frac{I_m}{\sqrt{2}} \tag{9.22}$$

The peak amplitude can be obtained from the *rms* amplitude by

$$I_m = \sqrt{2}I_{rms} \tag{9.23}$$

In terms of *rms* amplitude, Equation (9.21) can be written as

$$i(t) = I_{rms} \cos(\omega t + \theta_i) \tag{9.24}$$

where I_{rms} is given by $I_m/\sqrt{2}$.

EXAMPLE 9.2

Find the *rms* value of the triangular waveform shown in Figure 9.8.

continued

Example 9.2 continued

FIGURE 9.8

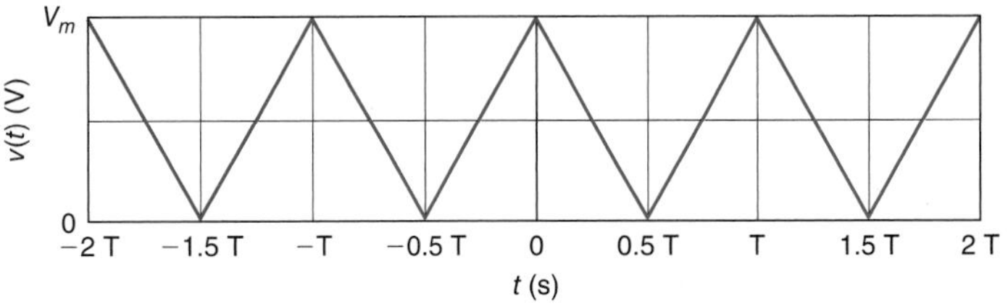

A triangular waveform.

The equation of $v(t)$ for $0 \le t < T/2$ is $v(t) = -2V_m t/T + V_m$. Since $v(t)$ is an even function of t, we can integrate $v^2(t)$ from 0 to $T/2$ and double it. From Equation (9.17), the *rms* value is found to be

$$V_{rms} = \sqrt{\frac{1}{T}\int_0^T v^2(t)dt} = \sqrt{\frac{2}{T}\int_0^{T/2}\left(\frac{-2V_m t}{T} + V_m\right)^2 dt} = \sqrt{\frac{2V_m^2}{T}\int_0^{T/2}\left(\frac{4t^2}{T^2} - \frac{4t}{T} + 1\right)dt}$$

$$= \sqrt{\frac{2V_m^2}{T}\left(\frac{4\left(\frac{T}{2}\right)^3}{3T^2} - \frac{4\left(\frac{T}{2}\right)^2}{2T} + \frac{T}{2}\right)} = \frac{V_m}{\sqrt{3}}$$

Exercise 9.2

Find the *rms* value of the sawtooth waveform shown in Figure 9.9.

FIGURE 9.9

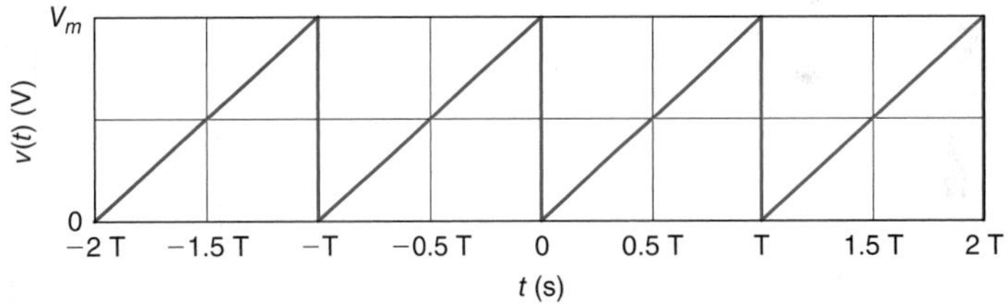

A sawtooth waveform.

Answer:
$V_m/\sqrt{3}$.

EXAMPLE 9.3

Find the *rms* value of the half-wave rectified cosine waveform shown in Figure 9.10.

FIGURE 9.10

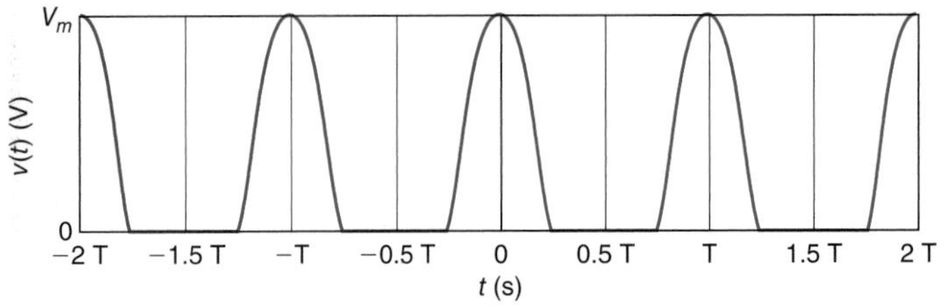

A half-wave rectified cosine waveform.

continued

Example 9.3 continued

The equation of $v(t)$ for $-T/4 \le t < T/4$ is $v(t) = V_m \cos(2\pi t/T)$. Since $v(t)$ is an even function of t, we can integrate $v^2(t)$ from 0 to $T/2$ and double it. From Equation (9.17), the *rms* value is found to be

$$V_{rms} = \sqrt{\frac{1}{T}\int_0^T v^2(t)dt} = \sqrt{\frac{2}{T}\int_0^{T/4}\left(V_m \cos\left(\frac{2\pi}{T}t\right)\right)^2 dt} = \sqrt{\frac{2V_m^2}{T}\int_0^{T/4}\left(\frac{1}{2}+\frac{1}{2}\cos\left(\frac{4\pi}{T}t\right)\right)dt}$$

$$= \sqrt{\frac{2V_m^2}{T}\int_0^{T/4}\left(\frac{1}{2}+\frac{1}{2}\cos\left(\frac{4\pi}{T}t\right)\right)dt} = \sqrt{\frac{V_m^2}{4}} = \frac{V_m}{2}$$

Exercise 9.3

Find the *rms* value of the full-wave rectified cosine waveform shown in Figure 9.11.

FIGURE 9.11

A full-wave rectified cosine waveform.

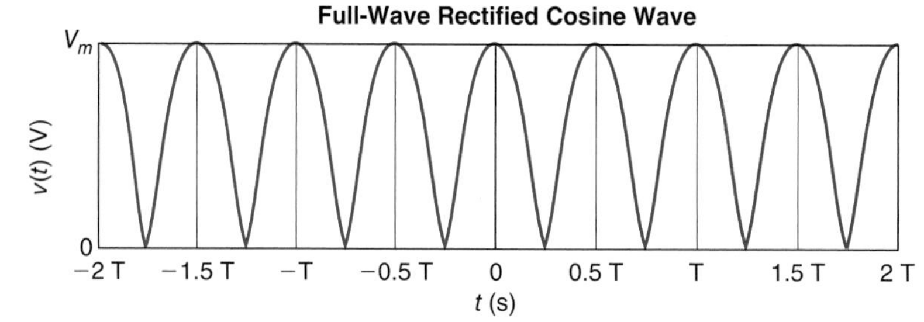

Answer:
$V_m/\sqrt{2}$.

9.4 Phasors

Let $\theta_i(t)$ be the angle of the cosine wave given by Equation (9.1). Then, $\theta_i(t)$ can be expressed as

$$\theta_i(t) = \frac{2\pi}{T}t + \phi = 2\pi ft + \phi = \omega t + \phi \qquad \textbf{(9.25)}$$

The angle $\theta_i(t)$ is called the **instantaneous phase** of the cosine wave. The instantaneous phase of the cosine wave increases linearly as a function of time [as shown in Figure 9.12].

The time rate of increase of the instantaneous phase is called the **instantaneous frequency** (rad/s) and is denoted by $\omega_i(t)$; that is,

$$\omega_i(t) = \frac{d\theta_i(t)}{dt} \qquad \textbf{(9.26)}$$

For the cosine wave, the instantaneous frequency is given by

$$\omega_i(t) = \frac{2\pi}{T} = 2\pi f = \omega \qquad \textbf{(9.27)}$$

FIGURE 9.12

Instantaneous phase $\theta_i(t)$ for $\phi = 0$ and $\phi = \pi/4$.

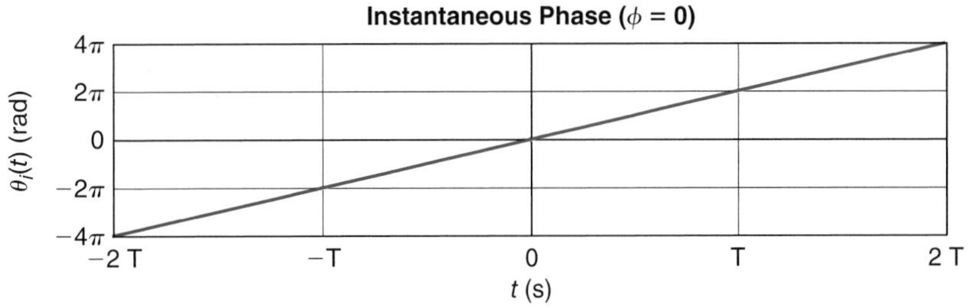

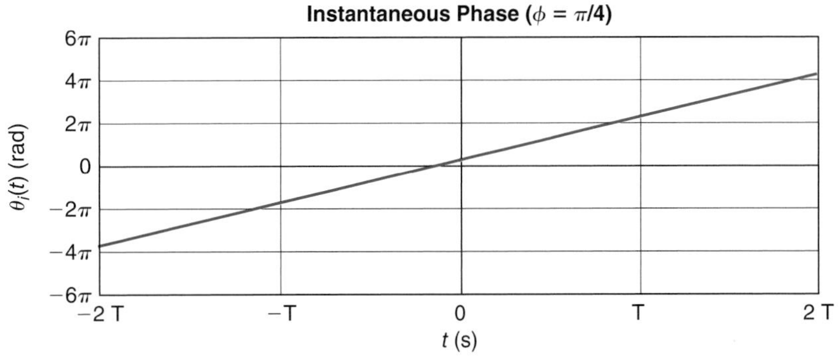

For the cosine wave, the instantaneous frequency is constant regardless of the value of the phase ϕ. The instantaneous frequency $\omega_i(t)$ is shown in Figure 9.13.

FIGURE 9.13

Instantaneous frequency as a function of time.

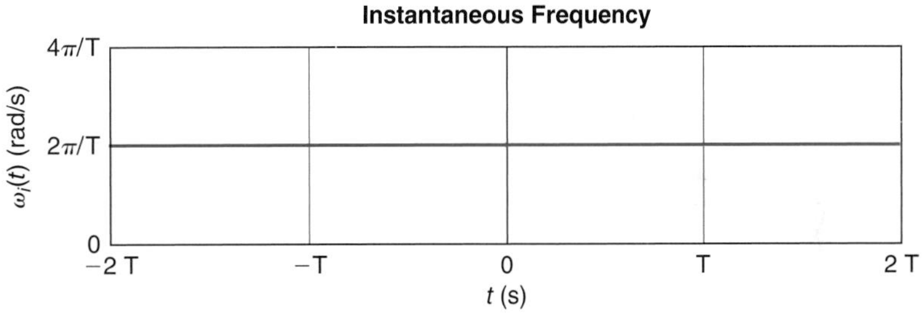

FIGURE 9.14

Rotation of a point P around a circle of radius V_m.

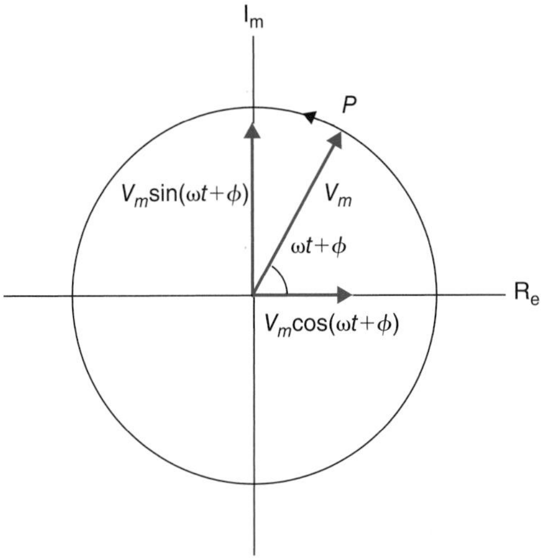

The instantaneous phase $\theta_i(t)$ can be obtained by integrating the instantaneous frequency $\omega_i(t)$; that is,

$$\theta_i(t) = \int_{-\infty}^{t} \omega_i(\lambda)d\lambda = \int_{0}^{t} \omega_i(\lambda)d\lambda + \phi \quad \textbf{(9.28)}$$

where ϕ is the instantaneous phase at $t = 0$. For the cosine wave, we have

$$\theta_i(t) = \int_{0}^{t} \omega d\lambda + \phi = \omega t + \phi \quad \textbf{(9.29)}$$

Physically, the cosine wave given by $v(t) = V_m \cos(\omega t + \phi)$ can be viewed as a projection of a point on a circle of radius V_m to the horizontal axis as the point P rotates at a constant angular velocity of ω rad/s in the counterclockwise direction [as shown in Figure 9.14].

In $T\ (= 2\pi/\omega)$ seconds, the point P makes one complete revolution and returns to the same point where it started T seconds earlier. It makes 2π radians in T seconds. In one second, the point makes $2\pi/T = \omega$ radians. Notice that the projection of P onto the horizontal axis (real axis) is $V_m \cos(\omega t + \phi)$, and the projection of P onto the vertical axis (imaginary axis) is $V_m \sin(\omega t + \phi)$. In T seconds, both $V_m \cos(\omega t + \phi)$ and $V_m \sin(\omega t + \phi)$ return to the same position.

Euler's rule says that

$$e^{j\theta} = \cos(\theta) + j\sin(\theta) \tag{9.30}$$

Euler's rule can be proved using the Maclaurin series. The Maclaurin series of e^x, $\cos(x)$, and $\sin(x)$ are given by

$$e^x = 1 + x + \frac{1}{2!}x^2 + \frac{1}{3!}x^3 + \frac{1}{4!}x^4 + \frac{1}{5!}x^5 + \frac{1}{6!}x^6 + \cdots$$

$$\cos(x) = 1 - \frac{1}{2!}x^2 + \frac{1}{4!}x^4 - \frac{1}{6!}x^6 + \cdots$$

$$\sin(x) = x - \frac{1}{3!}x^3 + \frac{1}{5!}x^5 - \frac{1}{7!}x^7 + \cdots$$

Since $j^2 = -1, j^3 = -j, j^4 = 1$, we have

$$\begin{aligned} e^{j\theta} &= 1 + j\theta - \frac{1}{2!}\theta^2 - j\frac{1}{3!}\theta^3 + \frac{1}{4!}\theta^4 + j\frac{1}{5!}\theta^5 - \frac{1}{6!}\theta^6 + \cdots \\ &= \left(1 - \frac{1}{2!}\theta^2 + \frac{1}{4!}\theta^4 - \frac{1}{6!}\theta^6 + \cdots\right) + j\left(\theta - \frac{1}{3!}\theta^3 + \frac{1}{5!}\theta^5 - \cdots\right) = \cos(\theta) + j\sin(\theta) \end{aligned}$$

In complex notation, point P can be written as

$$V_m e^{j(\omega t+\phi)} = V_m e^{j\phi} e^{j\omega t} \tag{9.31}$$

Applying Euler's rule to Equation (9.31), we obtain

$$V_m e^{j(\omega t+\phi)} = V_m \cos(\omega t + \phi) + jV_m \sin(\omega t + \phi) \tag{9.32}$$

The real part of the complex exponential signal $V_m e^{j(\omega t+\phi)}$ is $V_m \cos(\omega t + \phi)$, and the imaginary part of $V_m e^{j(\omega t+\phi)}$ is $V_m \sin(\omega t + \phi)$; that is,

$$\text{Re}[V_m e^{j(\omega t+\phi)}] = V_m \cos(\omega t + \phi) \tag{9.33}$$

$$\text{Im}[V_m e^{j(\omega t+\phi)}] = V_m \sin(\omega t + \phi) \tag{9.34}$$

At $t = 0$, the complex exponential signal given by Equation (9.31) becomes

$$V_m e^{j(\omega 0+\phi)} = V_m e^{j\phi} = V_m \cos(\phi) + jV_m \sin(\phi) \tag{9.35}$$

This is a single point in two-dimensional space [as shown in Figure 9.15]. The starting point of revolution around the circle given by Equation (9.35) is called the **phasor** of $v(t) = V_m \cos(\omega t + \phi)$.

The phasor is written as

$$\mathbf{V} = V_m e^{j\phi} = V_m \angle \phi \tag{9.36}$$

FIGURE 9.15

Phasor representation of $V_m \cos(\omega t + \phi)$.

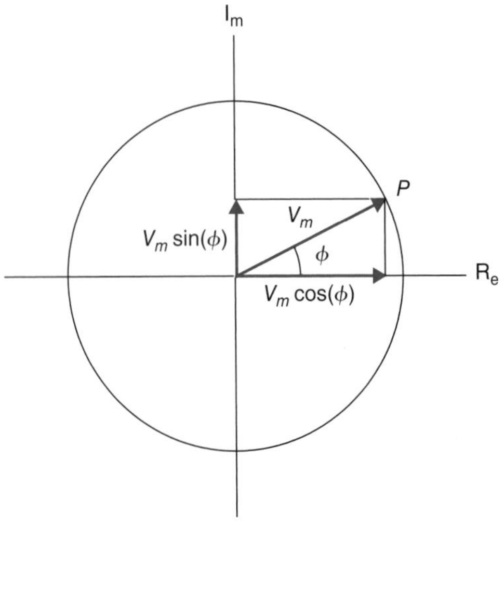

Two of the three parameters of the cosine signal $V_m\cos(\omega t + \phi)$ are specified by the phasor. The other parameter, angular velocity ω rad/s, determines the speed at which point P revolves around the circle of radius V_m, as shown in Figure 9.14. The phasor specifies the starting point. The initial angle of point P at $t = 0$ is ϕ, as shown in Figure 9.15. Given the angular velocity ω rad/s, the shape of the cosine signal is determined by the magnitude V_m (radius of a circle) and the phase angle ϕ. Since V_m is the radius, it cannot be negative; that is, $V_m \geq 0$. The phasor is represented as a vector with magnitude V_m and angle ϕ, as shown in Figure 9.15. The reference for the phase is the positive real axis.

9.4.1 REPRESENTING SINUSOIDS IN PHASOR

The magnitude is nonnegative. If the amplitude of a sinusoid is given by a negative number, the negative number is converted to a positive magnitude, and a phase of 180° (π radians) or −180° ($-\pi$ radians) is added.

$$-\cos(\omega t) = \cos(\omega t + 180°) = \cos(\omega t - 180°)$$
$$-\cos(\omega t + \phi) = \cos(\omega t + \phi + 180°) = \cos(\omega t + \phi - 180°)$$
$$-\sin(\omega t) = \sin(\omega t + 180°) = \sin(\omega t - 180°)$$
$$-\sin(\omega t + \phi) = \sin(\omega t + \phi + 180°) = \sin(\omega t + \phi - 180°)$$

The reference for sinusoids is cosine. If the sinusoid is given as a sine, it should be changed to cosine by subtracting 90 degrees.

$$\sin(\omega t) = \cos(\omega t - 90°)$$
$$-\sin(\omega t) = \cos(\omega t - 90° + 180°) = \cos(\omega t + 90°)$$
$$\sin(\omega t + \phi) = \cos(\omega t + \phi - 90°)$$
$$-\sin(\omega t + \phi) = \cos(\omega t + \phi - 90° + 180°) = \cos(\omega t + \phi + 90°)$$

EXAMPLE 9.4

Find the phasors of the following signals, and draw phasors for (a), (b), and (f).

a. $v(t) = -110\cos(2\pi 60t + 210°)$ V
b. $v(t) = -110\cos(2\pi 60t - 60°)$ V
c. $v(t) = 220\sin(2\pi 50t + 30°)$ V
d. $v(t) = -220\sin(2\pi 50t - 120°)$ V
e. $i(t) = 15\sin(2\pi 60t - 60°)$ A
f. $i(t) = -20\sin(2\pi 60t + 120°)$ A

a. $v(t) = 110\cos(2\pi 60t + 210° - 180°) = 110\cos(2\pi 60t + 30°)$
$\mathbf{V} = 110\angle 30°$ V
b. $v(t) = 110\cos(2\pi 60t - 60° + 180°) = 110\cos(2\pi 60t + 120°)$
$\mathbf{V} = 110\angle 120°$ V
c. $v(t) = 220\cos(2\pi 50t + 30° - 90°) = 220\cos(2\pi 50t - 60°)$
$\mathbf{V} = 220\angle -60°$ V
d. $v(t) = 220\cos(2\pi 50t - 120° - 90° + 180°) = 220\cos(2\pi 50t - 30°)$
$\mathbf{V} = 220\angle -30°$ V
e. $i(t) = 15\cos(2\pi 60t - 60° - 90°) = 15\cos(2\pi 60t - 150°)$
$\mathbf{I} = 15\angle -150°$ A
f. $i(t) = 20\cos(2\pi 60t + 120° - 90° - 180°) = 20\cos(2\pi 60t - 150°)$
$\mathbf{I} = 20\angle -150°$ A

The phasors for (a), (b), and (c) are shown in Figure 9.16.

continued

Example 9.4 continued

FIGURE 9.16

Phasor diagram for (a) $\mathbf{V} = 110\angle 30°$ (b) $\mathbf{V} = 110\angle 120°)$ (c) $\mathbf{I} = 20\angle -150°$.

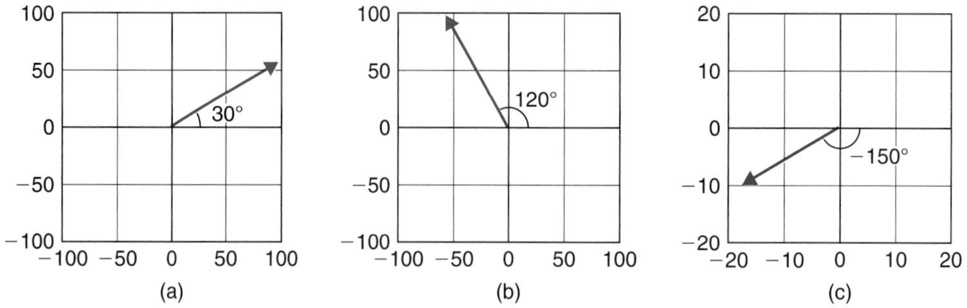

Exercise 9.4

Find the phasor representation of

a. $v(t) = -170\sin(2\pi 60t - 60°)$ V
b. $v(t) = 170\sin(2\pi 60t + 150°)$ V

Answer:

a. $v(t) = -170\sin(2\pi 60t - 60°) = -170\cos(2\pi 60t - 60° - 90°) = 170\cos(2\pi 60t - 60° - 90° + 180°) = 170\cos(2\pi 60t + 30°)$, $\mathbf{V} = 170\angle 30°$ V.
b. $v(t) = 170\sin(2\pi 60t + 150°) = 170\cos(2\pi 60t + 150° - 90°) = 170\cos(2\pi 60t + 60°)$, $\mathbf{V} = 170\angle 60°$ V.

EXAMPLE 9.5

Find the waveform for the following phasors when the frequency is 60 Hz.

a. $\mathbf{V} = 110\angle 120°$ V
b. $\mathbf{V} = 120\angle -30°$ V

a. $v(t) = 110\cos(2\pi 60t + 120°)$ V
b. $v(t) = 120\cos(2\pi 60t - 30°)$ V

Exercise 9.5

Find the waveform for the following phasors when the frequency is 60 Hz.

a. $\mathbf{I} = 15\angle -120°$ A
b. $\mathbf{I} = 25\angle 210°$ A

Answer:

a. $i(t) = 15\cos(2\pi 60t - 120°)$ A,
b. $i(t) = 25\cos(2\pi 60t + 210°)$ A,

9.4.2 CONVERSION BETWEEN CARTESIAN COORDINATE SYSTEM (RECTANGULAR COORDINATE SYSTEM) AND POLAR COORDINATE SYSTEM

Cartesian to Polar

Let $a > 0$ and $b > 0$. Then, a complex number $z = a + jb$ represented in Cartesian coordinate system can be written as

$$z = a + jb = \sqrt{a^2 + b^2}\angle\tan^{-1}\left(\frac{b}{a}\right) = re^{j\phi} = r\angle\phi$$

where the magnitude (radius) is given by

$$r = \sqrt{a^2 + b^2}$$

and the phase angle is given by

$$\phi = \tan^{-1}\left(\frac{b}{a}\right)$$

FIGURE 9.17

A complex number $z = a + jb$ represented in Cartesian coordinate system and in polar coordinate system $r\angle\phi$.

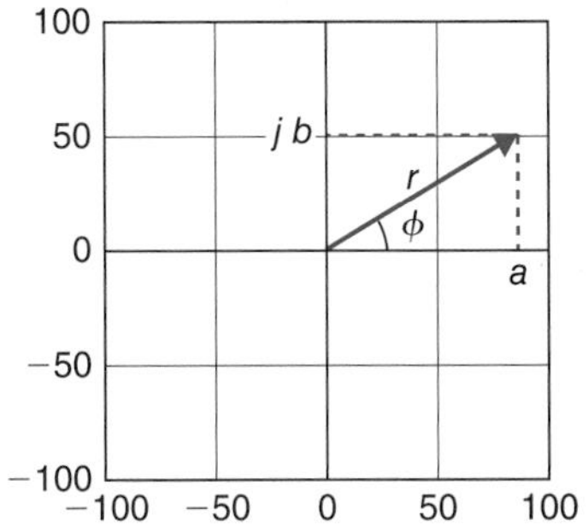

Notice that the phase angle is $0° < \phi < 90°$. The phasor for the complex number $z = a + jb = r\angle\phi$ for $a > 0$ and $b > 0$ is shown in Figure 9.17.

The complex number $z = a + jb$ can also be written as

$$z = a + ib$$
$$z = a + bj$$
$$z = a + bi$$

In MATLAB, the complex numbers can be entered as

```
z=3+j*4
z=3+i*4
z=3+4j
z=3+4i
```

Let $a > 0$ and $b > 0$. If a complex number is in the second quadrant, it can be represented as

$$z = -a + jb = \sqrt{a^2 + b^2}\angle 180° - \tan^{-1}\left(\frac{b}{a}\right) = r\angle\phi$$

where $r = \sqrt{a^2 + b^2}$ and $\phi = 180° - \tan^{-1}\left(\frac{b}{a}\right)$. Notice that the phase angle is $90° < \phi < 180°$.

Let $a > 0$ and $b > 0$. If a complex number is in the third quadrant, it can be represented as

$$z = -a - jb = \sqrt{a^2 + b^2}\angle -180° + \tan^{-1}\left(\frac{b}{a}\right) = r\angle\phi$$

where $r = \sqrt{a^2 + b^2}$ and $\phi = -180° + \tan^{-1}\left(\frac{b}{a}\right)$. Notice that the phase angle is $180° < \phi < 270°$ $(-180° < \phi < -90°)$.

Let $a > 0$ and $b > 0$. If a complex number is in the fourth quadrant, it can be represented as

$$z = a - jb = \sqrt{a^2 + b^2}\angle -\tan^{-1}\left(\frac{b}{a}\right) = r\angle\phi$$

where $r = \sqrt{a^2 + b^2}$ and $\phi = -\tan^{-1}\left(\frac{b}{a}\right)$. Notice that the phase angle is $270° < \phi < 360°$ $(-90° < \phi < 0°)$.

Let $a > 0$ and $b > 0$. If z is real and positive $(z = a)$, it is represented as

$$z = a = a\angle 0°$$

and if z is real and negative $(z = -a)$, it is represented as

$$z = -a = a\angle 180°$$

If z is an imaginary number $(z = jb)$, it is represented as

$$z = jb = b\angle 90°$$

and if z is an imaginary number $(z = -jb)$, it is represented as

$$z = -jb = b\angle -90°$$

If a phase is given in degrees, it can be converted to radians through

$$\phi_r = \phi_d \times \pi/180$$

and if a phase is given in radians, it can be converted to degrees through

$$\phi_d = \phi_r \times 180/\pi$$

In MATLAB, `atan(b/a)` provides angles between $-90°$ and $90°$. If a complex number is in the second quadrant, add 180° to get the correct angle. If a complex number is in the third quadrant, subtract 180° to get the correct angle. `atan2(b,a)` provides the correct angle. `[φ,A]=cart2pol(a,b)` provides direct conversion from Cartesian coordinates to polar coordinates. `abs(a+bj)` provides magnitude and `angle(a+bj)` provides the phase. The MATLAB function R2P converts a complex number in Cartesian coordinates to polar coordinates. The input is a complex number, and the output consists of magnitude, phase in degrees, and phase in radians.

```
function y= R2P( x )
%Rectangular to polar conversion, y=[mag,phase(deg),phase(rad)]
%EX: R2P(3+4j), ans = 5.0000    53.1301    0.9273
y=[abs(x),angle(x)*180/pi,angle(x)];
end
```

The file R2P.m should be in the current folder.

Figure 9.18 shows four complex numbers in four different quadrants. The complex number $1 + j\sqrt{3} = 2\angle 60°$ in the first quadrant is shown in Figure 9.18(a), the complex number $-1 + j\sqrt{3} = 2\angle 120°$ in the second quadrant is shown in Figure 9.18(b), the complex number $-1 - j\sqrt{3} = 2\angle -120°$ in the third quadrant is shown in Figure 9.18(c), and the complex number $1 - j\sqrt{3} = 2\angle -60°$ in the fourth quadrant is shown in Figure 9.18(d).

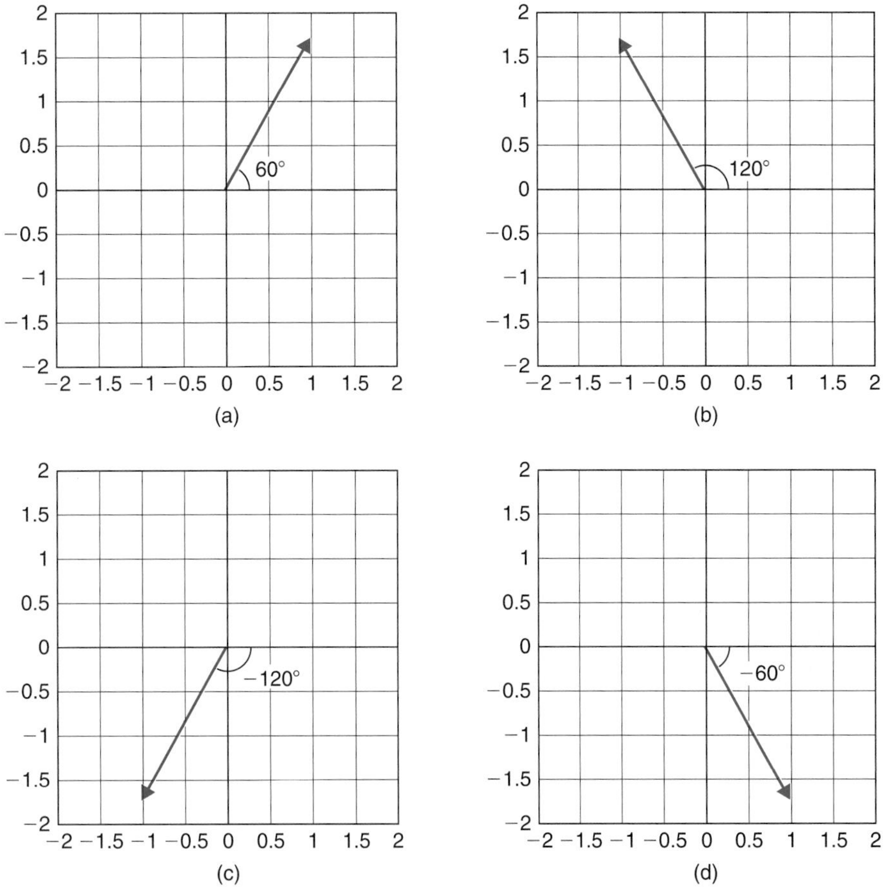

FIGURE 9.18

(a) $1 + j\sqrt{3} = 2\angle 60°$ (b) $-1 + j\sqrt{3} = 2\angle 120°$ (c) $-1 - j\sqrt{3} = 2\angle -120°$ (d) $1 - j\sqrt{3} = 2\angle -60°$.

EXAMPLE 9.6

Convert the following numbers in Cartesian coordinates to polar coordinates.

a. $z = 3 + j4$ **b.** $z = -4 + j3$ **c.** $z = -3 - j4$ **d.** $z = 4 - j3$

a. $z = 3 + j4 = \sqrt{3^2 + 4^2}\angle \tan^{-1}\left(\frac{4}{3}\right) = \sqrt{25}\angle \tan^{-1}\left(\frac{4}{3}\right) = 5\angle 53.1301°$

```
 >>abs(3+4j)
ans =
     5
>>atan(4/3)*180/pi
ans =
  53.130102354155980
```

Instead of `atan`, `angle` can be used:

```
>>angle(3+4j)*180/pi
ans =
  53.130102354155980
```

continued

Example 9.6 continued

Using `cart2pol`, we have

```
>> [phi,A]=cart2pol(3,4)
phi =
   0.927295218001612
A =
     5
>>phid=phi*180/pi
phid =
  53.130102354155980
```

Function **R2P** can be used to find the polar coordinates:

```
>>R2P(3+4j)
ans =
5.000000000000000  53.130102354155980   0.927295218001612
```

The first number is the magnitude, the second is the angle in degrees, and the third is the angle in radians.

b. $z = -4 + j3 = \sqrt{(-4)^2 + 3^2}\angle\tan^{-1}\left(\frac{3}{-4}\right) + 180° = \sqrt{25}\angle -\tan^{-1}\left(\frac{3}{4}\right) + 180°$

$= 5\angle -36.8699° + 180° = 5\angle 143.1301°$

```
>>abs(-4+3j)
ans =
     5
>>angle(-4+3j)*180/pi
ans =
    1.431301023541560e+002
>>R2P(-4+3j)
ans =
  1.0e+002 *
   0.050000000000000   1.431301023541560   0.024980915447965
```

c. $z = -3 - j4 = \sqrt{(-3)^2 + (-4)^2}\angle\tan^{-1}\left(\frac{-4}{-3}\right) - 180° = \sqrt{25}\angle\tan^{-1}\left(\frac{4}{3}\right) - 180°$

$= 5\angle 53.1301° - 180° = 5\angle -126.8699°$

```
>>abs(-3-4j)
ans =
     5
>>angle(-3-4j)*180/pi
ans =
   -1.268698976458440e+002
>>R2P(-3-4j)
ans =
  1.0e+002 *
0.050000000000000  -1.268698976458440  -0.022142974355882
```

d. $z = 4 - j3 = \sqrt{4^2 + (-3)^2}\angle\tan^{-1}\left(\frac{-3}{4}\right) = \sqrt{25}\angle\tan^{-1}\left(\frac{-3}{4}\right) = 5\angle -36.8699°$

continued

Example 9.6 continued

```
>>abs(4-3j)
ans =
     5
>>angle(4-3j)*180/pi
ans =
 -36.869897645844020
>>R2P(4-3j)
ans =
    5.000000000000000 -36.869897645844020  -0.643501108793284
```

Exercise 9.6

Convert the following numbers in Cartesian coordinates to polar coordinates.

a. $z = 5 + j9$ **b.** $z = -6 + j8$ **c.** $z = -7 - j3$ **d.** $z = 8 - j5$

Answer:
a. $z = 10.2956\angle 60.9454°$ **b.** $z = 10\angle 126.8699°$ **c.** $z = 7.6158\angle -156.8014°$
d. $z = 9.4340\angle -32.0054°$

Polar to Cartesian

A phasor given by $z = A\angle\phi$ represents a complex number $Ae^{j\phi}$. According to Euler's rule, we have

$$e^{j\phi} = \cos(\phi) + j\sin(\phi)$$

Thus, z becomes

$$z = A\angle\phi = Ae^{j\phi} = A[\cos(\phi) + j\sin(\phi)] = A\cos(\phi) + jA\sin(\phi)$$

The real part of z is $A\cos(\phi)$, and the imaginary part is $A\sin(\phi)$. The MATLAB function `[a,b]=pol2cart(φ,r)` provides direct conversion from polar coordinates to Cartesian coordinates. The phase angle ϕ is in radians.

The MATLAB function **P2Rd** converts a complex number in polar coordinates to Cartesian coordinates. The phase entered is in degrees. The output consists of a complex number.

```
function y = P2Rd( Mag,Pha )
%Polar to rectangular conversion, phase in degrees
%EX: P2Rd(1,45), ans=0.7071 + 0.7071i
y=Mag*cos(Pha*pi/180)+j*Mag*sin(Pha*pi/180);
end
```

The MATLAB function **P2Rr** converts a complex number in polar coordinates to Cartesian coordinates. The phase entered is in radians. The output consists of a complex number.

```
function y = P2Rr( Mag,Pha )
%Polar to rectangular conversion, phase in radians
%EX: P2Rr(100,pi/3), ans=50.0000 +86.6025i
y=Mag*cos(Pha)+j*Mag*sin(Pha);
end
```

The files P2Rd.m and P2Rr.m should be in the current folder.

EXAMPLE 9.7

Convert the following phasors to Cartesian coordinates.

a. $\mathbf{V} = 110\angle 120°$ **b.** $\mathbf{V} = 240\angle -120°$ **c.** $\mathbf{V} = 480\angle 150°$ **d.** $\mathbf{V} = 880\angle -60°$

a. $\mathbf{V} = 110\cos(120°) + j110\sin(120°) = 110 \times (-1/2) + j110 \times (\sqrt{3}/2) = -55 + j95.2628$

```
>> [a,b]=pol2cart(120*pi/180,110)
a =
  -55.0000
b =
    95.2628
>>P2Rd(110,120)
ans =
 -55.0000 +95.2628i
```

b. $\mathbf{V} = 240\cos(-120°) + j240\sin(-120°) = 240 \times (-1/2) - j240 \times (\sqrt{3}/2) = -120 - j207.8461$

```
>> [a,b]=pol2cart(-120*pi/180,240)
a =
   -1.199999999999999e+002
b =
   -2.078460969082653e+002
>>P2Rd(240,-120)
ans =
    -1.199999999999999e+002 -2.078460969082653e+002i
```

c. $\mathbf{V} = 480\cos(150°) + j480\sin(150°) = 480 \times (-\sqrt{3}/2) + j480 \times (1/2) = -415.6922 + j240$

```
>> [a,b]=pol2cart(150*pi/180,480)
a =
   -4.156921938165306e+002
b =
    2.400000000000000e+002
>>P2Rd(480,150)
ans =
    -4.156921938165306e+002 +2.400000000000000e+002i
```

d. $\mathbf{V} = 880\cos(-60°) + j880\sin(-60°) = 880 \times (1/2) + j880 \times (-\sqrt{3}/2) = 440 - j762.1024$

```
>> [a,b]=pol2cart(-60*pi/180,880)
a =
    4.400000000000001e+002
b =
   -7.621023553303060e+002
>>P2Rd(880,-60)
ans =
     4.400000000000001e+002 -7.621023553303060e+002i
```

Exercise 9.7

Convert the following numbers in polar coordinates to Cartesian coordinates.

a. $\mathbf{V} = 220\angle 60°$ **b.** $\mathbf{V} = 330\angle -150°$ **c.** $\mathbf{V} = 450\angle 120°$
d. $\mathbf{V} = 880\angle -30°$

Answer:
a. $\mathbf{V} = 110 + j190.5256$ **b.** $\mathbf{V} = -285.7884 - j165$ **c.** $\mathbf{V} = -225 + j389.7114$
d. $\mathbf{V} = 762.1024 - j440$

9.4.3 PHASOR ARITHMETIC

Addition

The phasors in polar coordinates are converted to Cartesian coordinates before being added. The sum can be converted back to polar coordinates. Let

$$\begin{aligned} A &= 5\angle 60° = 5e^{j60°} = 5\cos(60°) + j5\sin(60°) \\ &= 5/2 + j5\sqrt{3}/2 = 2.5 + j4.3301 \end{aligned} \tag{9.37}$$

and

$$\begin{aligned} B &= 10\angle -45° = 10e^{-j45°} = 10\cos(45°) - j10\sin(45°) \\ &= 10/\sqrt{2} - j10/\sqrt{2} = 7.0711 - j7.0711 \end{aligned} \tag{9.38}$$

Then, we have

$$C = A + B = 9.5711 - j2.7409 = 9.9558\angle -15.9804°$$

```
>> A=P2Rd(5,60)
A =
  2.500000000000000 + 4.330127018922193i
>> B=P2Rd(10,-45)
B =
  7.071067811865476 - 7.071067811865475i
>> C=A+B
C =
  9.571067811865476 - 2.740940792943282i
>>Cp=R2P(C)
Cp =
   9.955807123972820 -15.980534084003448  -0.278912935993036
```

Subtraction

The phasors in polar coordinates are converted to Cartesian coordinates before being subtracted. The result can be converted back to polar coordinates. For A and B given by Equations (9.37) and (9.38), respectively, we have

$$D = A - B = -4.5711 + j11.4012 = 12.2834\angle 111.8473°$$

```
>> D=A-B
D =
 -4.571067811865476 +11.401194830787666i
>>Dp=R2P(D)
```

```
Dp =
  1.0e+002 *
   0.122833995502162    1.118472728735474    0.019521031710199
```

Multiplication

The magnitude of the product of two phasors in polar coordinates is the product of two magnitudes, and the phase of the product is the sum of the phases. For A and B given by Equations (9.37) and (9.38), respectively, we have

$$E = AB = (5\angle 60°)(10\angle -45°) = 5e^{j60°} \times 10e^{-j45°} = 50e^{j15°} = 50\angle 15°$$
$$= 48.2063 + j12.9410$$

```
>> E=A*B
E =
 48.296291314453413 +12.940952255126035i
>> Ep=R2P(E)
Ep =
50.000000000000000  14.999999999999996   0.261799387799149
```

Division

The magnitude of the division of two phasors in polar coordinates is the division of two magnitudes, and the phase of the division is the difference of the phases. For A and B given by Equations (9.37) and (9.38), respectively, we have

$$F = \frac{A}{B} = \frac{5\angle 60°}{10\angle -45°} = \frac{5e^{j60°}}{10e^{-j45°}} = 0.5e^{j105°} = 0.5\angle 105° = -0.1294 + j0.4830.$$

```
>> F=A/B
F =
 -0.129409522551260 + 0.482962913144534i
>>Fp=R2P(F)
Fp =
  1.0e+002 *
   0.005000000000000    1.050000000000000    0.018325957145940
```

Sum of Sinusoids

Two or more sinusoids with same frequency can be added using phasors. As an example, let

$$v_1(t) = 10\cos(2\pi 100t + 30°)\text{ V and } v_2(t) = -5\sin(2\pi 100t - 45°)\text{ V}$$

Then, the sum of $v_1(t)$ and $v_2(t)$ is given by

$$v(t) = v_1(t) + v_2(t) = 10\cos(2\pi 100t + 30°) - 5\sin(2\pi 100t - 45°)\text{ V}$$

The phasor representation of $v_1(t)$ and $v_2(t)$ are given, respectively,

$$\mathbf{V}_1 = 10\angle 30°$$

$$\mathbf{V}_2 = 5\angle -45° - 90° + 180° = 5\angle 45°$$

The sum of $\mathbf{V}_1$ and $\mathbf{V}_2$ is given by

$$\mathbf{V} = \mathbf{V}_1 + \mathbf{V}_2 = 10\angle 30° + 5\angle 45° = 10\cos(30°) + j10\sin(30°)$$
$$+ 5\cos(45°) + j5\sin(45°)$$
$$= 8.6603 + j5 + 3.5355 + j3.5355 = 12.1958 + j8.5355 = 14.8860\angle 34.9872°\text{ V}$$

Converting back to the time domain, we obtain

$$v(t) = 14.8860\ \cos(2\pi 100t + 34.9872°)$$

In MATLAB, we have

```
clear all;format long;
A1=10;phi1=30;A2=5;phi2=45;
V1=P2Rd(A1,phi1)
V2=P2Rd(A2,phi2)
V=V1+V2
Vp=R2P(V)
```

The answers are given by

```
V1 =
  8.660254037844387 + 4.999999999999999i
V2 =
  3.535533905932738 + 3.535533905932737i
V =
 12.195787943777125 + 8.535533905932736i
Vp =
  14.885986115434436  34.987234725632469   0.610642442130382
```

The same answer can be obtained by applying trigonometric identities:

$$\cos(\alpha + \beta) = \cos(\alpha)\cos(\beta) - \sin(\alpha)\sin(\beta)$$
$$\sin(\alpha - \beta) = \sin(\alpha)\cos(\beta) - \cos(\alpha)\sin(\beta)$$

The sum of $v_1(t)$ and $v_2(t)$ is given by

$$\begin{aligned}
v(t) = v_1(t) + v_2(t) &= 10\cos(30°)\cos(2\pi 100t) - 10\sin(30°)\sin(2\pi 100t) \\
&\quad - 5\cos(45°)\sin(2\pi 100t) + 5\sin(45°)\cos(2\pi 100t) \\
&= [10\cos(30°) + 5\sin(45°)]\cos(2\pi 100t) - [10\sin(30°) + 5\cos(45°)]\sin(2\pi 100t) \\
&= 12.1957\cos(2\pi 100t) - 8.5355\sin(2\pi 100t) \\
&= \sqrt{12.1957^2 + 8.5355^2}\left[\frac{12.1957}{\sqrt{12.1957^2 + 8.5355^2}}\cos(2\pi 100t)\right. \\
&\quad \left. - \frac{8.5355}{\sqrt{12.1957^2 + 8.5355^2}}\sin(2\pi 100t)\right] \\
&= 14.8860\left[\frac{12.1957}{14.8860}\cos(2\pi 100t) - \frac{8.5355}{14.8860}\sin(2\pi 100t)\right]
\end{aligned}$$

The angle is

$$\phi = \tan^{-1}\left(\frac{8.5355}{12.1957}\right) = 34.9872°$$

The sum can be written as

$$v(t) = 14.8860[\cos(34.9872°)\cos(2\pi 100t) - \sin(34.9872°)\sin(2\pi 100t)]$$

or

$$v(t) = 14.8860\ \cos(2\pi 100t + 34.9872°)$$

EXAMPLE 9.8

Represent

$$v(t) = 150\cos(2\pi 60t - 60°) + 100\sin(2\pi 60 + 120°)\text{ V}$$

by a single sinusoid.

The phasor representation of $v(t)$ is

$$\mathbf{V} = 150\angle -60° + 100\angle 30° = 180.2776\angle -26.3099°\text{ V}$$

Thus, $v(t)$ is given by

$$v(t) = 180.2776\cos(2\pi 60t - 26.3099°)\text{ V}$$

Exercise 9.8

Represent

$$v(t) = 120\cos(2\pi 60t + 60°) - 110\cos(2\pi 60 - 135°) - 90\sin(2\pi 60 + 120°)\text{ V}$$

with a single sinusoid.

Answer:
$v(t) = 149.2279\cos(2\pi 60t + 66.3597°)$ V.

9.5 Impedance and Admittance

If the input voltage to a resistor, a capacitor, an inductor, or combination of these is a sinusoid, the current through it is also a sinusoid of same frequency as the input voltage. But the magnitude and phase of the current may be different. The voltage can be transformed into voltage phasor **V**, and the current can be transformed into current phasor **I**. The ratio of **V** to **I** is defined as impedance **Z** of the component; that is,

$$\mathbf{Z} = \frac{\mathbf{V}}{\mathbf{I}} \tag{9.39}$$

The SI unit for impedance is ohm (Ω). The impedance is similar to resistance, but the impedance can be a function of frequency, and in general, is a complex quantity representing both the magnitude and phase. Solving Equation (9.39) for **V**, we obtain

$$\mathbf{V} = \mathbf{ZI} \tag{9.40}$$

Solving Equation (9.39) for **I**, we get

$$\mathbf{I} = \frac{\mathbf{V}}{\mathbf{Z}} \tag{9.41}$$

Equations (9.39)–(9.41) are **Ohm's law** for circuits with phasors and impedances. Since **Z** is complex in general, it can be represented as

$$\mathbf{Z} = R + jX \tag{9.42}$$

The real part of the impedance is defined as resistance, denoted by R, and the imaginary part of the impedance is defined as **reactance**, denoted by X. The SI unit for both R and X is **ohm**.

The ratio of current **I** to voltage **V** is defined as the admittance and is denoted as **Y**. The admittance is the inverse of the impedance:

$$\mathbf{Y} = \frac{\mathbf{I}}{\mathbf{V}} = \frac{1}{\mathbf{Z}} \tag{9.43}$$

The SI unit for admittance is the siemens (S). Solving Equation (9.43) for **I**, we obtain

$$\mathbf{I} = \mathbf{YV} \tag{9.44}$$

Solving Equation (9.43) for **V**, we get

$$\mathbf{V} = \frac{\mathbf{I}}{\mathbf{Y}} \tag{9.45}$$

In general, admittance is a complex quantity. The real part of the admittance is defined as the conductance, denoted by G, and the imaginary part of admittance is defined as the susceptance, denoted by B. Thus, we have

$$\mathbf{Y} = G + jB \tag{9.46}$$

The unit for both G and B is **siemens**.

9.5.1 RESISTOR

FIGURE 9.19

Voltage across and current through a resistor.

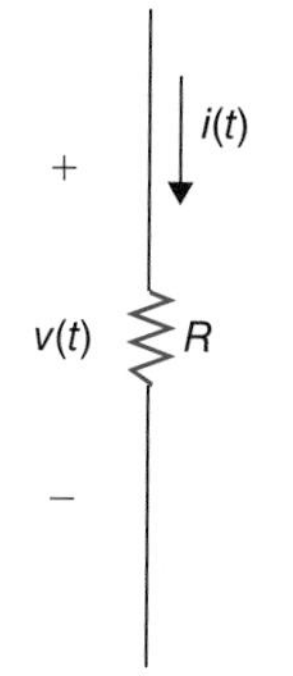

Let $v(t)$ be the voltage across the resistor and $i(t)$ be the current through the resistor [as shown in Figure 9.19].

According to Ohm's law, we have

$$v(t) = R\,i(t)$$

Let the voltage across the resistor be

$$v(t) = V_m \cos(\omega t + \phi) = \mathrm{Re}[V_m e^{j\phi} e^{j\omega t}] = \mathrm{Re}[\mathbf{V} e^{j\omega t}]$$

where **V** is the phasor of the voltage given by

$$\mathbf{V} = V_m e^{j\phi}$$

Then, the current through the resistor in the steady state is given by

$$i(t) = \frac{1}{R} v(t) = \frac{V_m}{R} \cos(\omega t + \phi) = \mathrm{Re}\left[\frac{V_m}{R} e^{j\phi} e^{j\omega t}\right] = \mathrm{Re}[\mathbf{I} e^{j\omega t}]$$

where **I** is the phasor of the current given by

$$\mathbf{I} = \frac{V_m}{R} e^{j\phi} = \frac{V_m e^{j\phi}}{R} = \frac{\mathbf{V}}{R}$$

The impedance is given by

$$\mathbf{Z} = \frac{\mathbf{V}}{\mathbf{I}} = \frac{V_m e^{j\phi}}{\frac{V_m}{R} e^{j\phi}} = R \tag{9.47}$$

FIGURE 9.20

The impedance of a resistor.

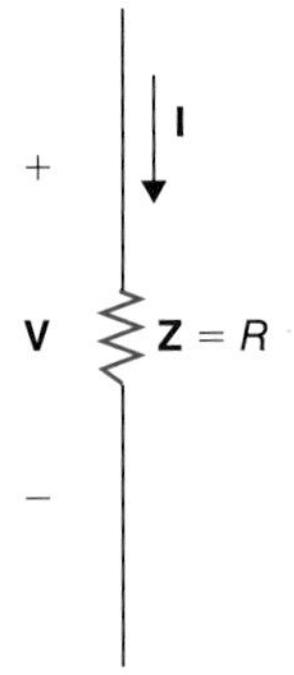

The impedance $\mathbf{Z}$ for a resistor is a real number and is the same as the resistance R. Figure 9.20 shows the resistor in the phasor domain.

The admittance of a resistor is

$$\mathbf{Y} = \frac{\mathbf{I}}{\mathbf{V}} = \frac{1}{\mathbf{Z}} = \frac{1}{R} \tag{9.48}$$

Since $\mathbf{V} = R\mathbf{I}$ and $\mathbf{I} = \mathbf{V}/R$ and R is a scalar quantity (not a vector), the current through the resistor and the voltage across the resistor have the same phase (called in-phase) [as shown in Figure 9.21]. The typical waveform for the voltage across the resistor and the current through the resistor are shown in Figure 9.22.

For a resistor with resistance R, the impedance is given by

$$\mathbf{Z} = \frac{\mathbf{V}}{\mathbf{I}} = R = R + jX \text{ ohms}$$

FIGURE 9.21

Phasors $\mathbf{V}$ and $\mathbf{I}$ for the resistor.

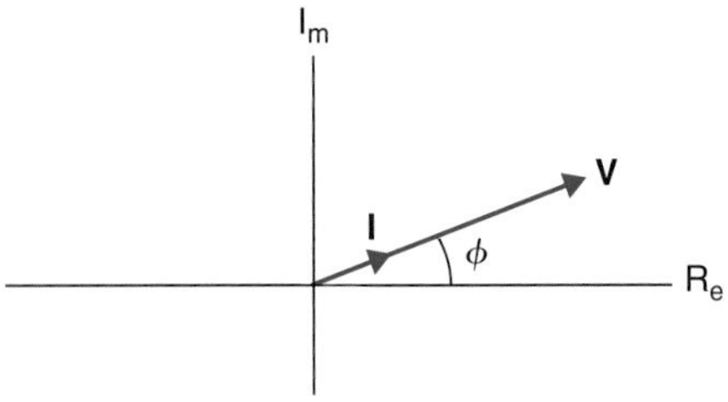

and the admittance is given by

$$\mathbf{Y} = \frac{\mathbf{I}}{\mathbf{V}} = \frac{1}{\mathbf{Z}} = \frac{1}{R} = G + jB \text{ siemens}$$

Thus, we have

$$R = R, \quad X = 0, \quad G = 1/R, \quad B = 0 \tag{9.49}$$

FIGURE 9.22

Waveform for voltage across the resistor and current through the resistor.

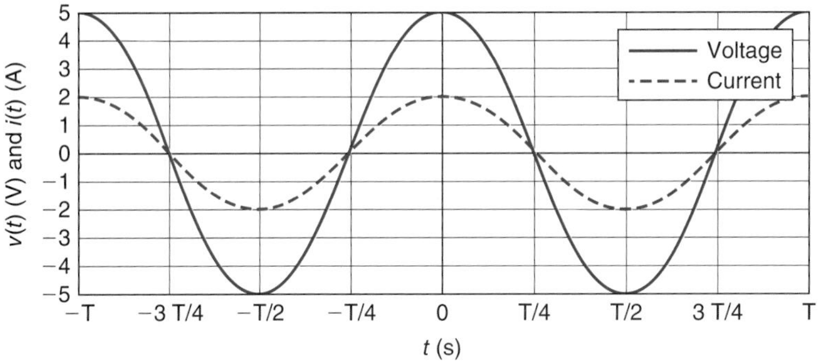

9.5.2 CAPACITOR

Let $v(t)$ be voltage across the capacitor and $i(t)$ be current through the capacitor as shown in Figure 9.23.

FIGURE 9.23

Voltage across and current through a capacitor.

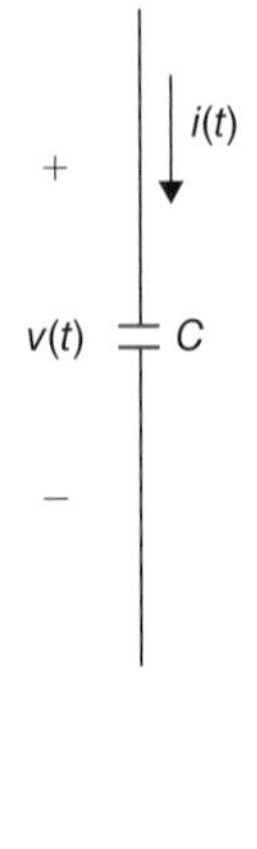

As shown in Chapter 6, the current $i(t)$ through the capacitor is related to the voltage $v(t)$ across the capacitor by

$$i(t) = C\frac{dv(t)}{dt}$$

Let the voltage across the capacitor with capacitance C be

$$v(t) = V_m \cos(\omega t + \phi) = \text{Re}[V_m e^{j\phi} e^{j\omega t}] = \text{Re}[\mathbf{V} e^{j\omega t}]$$

where $\mathbf{V}$ is the phasor of the voltage given by

$$\mathbf{V} = V_m e^{j\phi} = V_m \angle \phi$$

Then, the current through the capacitor in steady state is given by

$$i(t) = C\frac{dv(t)}{dt} = \mathrm{Re}\left[V_m e^{j\phi} C \frac{d}{dt} e^{j\omega t}\right] = \mathrm{Re}[V_m e^{j\phi} j\omega C e^{j\omega t}] = \mathrm{Re}\left[V_m \omega C e^{j\left(\phi + \frac{\pi}{2}\right)} e^{j\omega t}\right]$$

$$= \mathrm{Re}[\mathbf{I}e^{j\omega t}] = V_m \omega C \cos\left(\omega t + \phi + \frac{\pi}{2}\right) = -V_m \omega C \sin(\omega t + \phi)$$

where the phasor of the current $\mathbf{I}$ is given by

$$\mathbf{I} = j\omega C V_m e^{j\phi} = j\omega C\mathbf{V}$$

If $i(t)$ is replaced by phasor $\mathbf{I}$, $v(t)$ is replaced by phasor $\mathbf{V}$, and d/dt is replaced by $j\omega$, the differential equation

$$i(t) = C\frac{dv(t)}{dt}$$

becomes

$$\mathbf{I} = j\omega C\mathbf{V}$$

The impedance of the capacitor is given by

$$\mathbf{Z} = \frac{\mathbf{V}}{\mathbf{I}} = \frac{1}{j\omega C} = -j\left(\frac{1}{\omega C}\right) = \frac{1}{\omega C}\angle -90^\circ \text{ ohms} \quad (9.50)$$

Notice that when $\omega = 0$ (dc), the impedance of the capacitor is infinity ($Z = \infty$), and when $\omega = \infty$, the impedance of the capacitor is zero ($Z = 0$). The capacitor acts as an open circuit (current through the capacitor is zero) for a dc signal, and acts as a short circuit (wire) for a high-frequency ac signal.

The admittance is given as

$$\mathbf{Y} = \frac{\mathbf{I}}{\mathbf{V}} = \frac{1}{\mathbf{Z}} = j\omega C = \omega C\angle 90^\circ \text{ siemens} \quad (9.51)$$

FIGURE 9.24

The impedance of a capacitor.

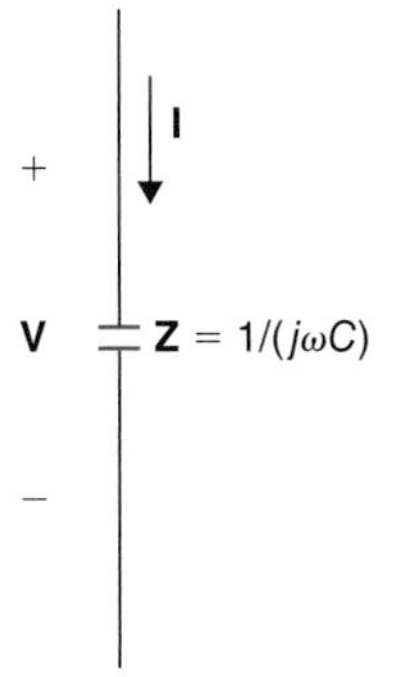

Figure 9.24 shows the capacitor in the phasor domain.

For the sinusoidal inputs, the voltage-current relation for the capacitor is given by

$$\mathbf{V} = \mathbf{ZI}$$

where $\mathbf{Z} = \frac{1}{j\omega C}$. The voltage-current relation for the capacitor ($\mathbf{V} = \mathbf{ZI}$) is identical to the voltage-current relation of the resistor ($\mathbf{V} = \mathbf{ZI}$). The only difference is the impedance. The impedance of the resistor is $\mathbf{Z} = R$, and the impedance of the capacitor is $\mathbf{Z} = \frac{1}{j\omega C}$. From this observation, we conclude that if we are interested in the sinusoidal steady-state response of a circuit, the capacitor can be treated as a resistor with impedance $\mathbf{Z} = \frac{1}{j\omega C}$. The magnitude of the impedance of a capacitor is $1/(\omega C)$, and the phase of the impedance is $1/j = -j = e^{-j\pi/2} = -90^\circ$. The voltage-current relation $\mathbf{V} = \mathbf{ZI}$ can be written as

$$|\mathbf{V}|e^{j\angle\mathbf{V}} = |\mathbf{Z}|e^{j\angle\mathbf{Z}}|\mathbf{I}|e^{j\angle\mathbf{I}} = |\mathbf{Z}|\,|\mathbf{I}|\,e^{j(\angle\mathbf{Z}+\angle\mathbf{I})}$$

FIGURE 9.25

Phasors **V** and **I** for the capacitor.

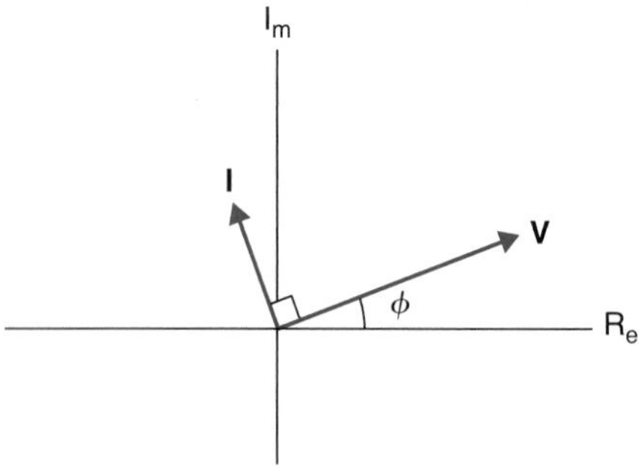

Thus, $|\mathbf{V}| = |\mathbf{Z}|\,|\mathbf{I}|$ and $\angle\mathbf{V} = \angle\mathbf{Z} + \angle\mathbf{I}$. Since $\angle\mathbf{Z} = -90°$, we have $\angle\mathbf{V} = -90° + \angle\mathbf{I}$ or $\angle\mathbf{I} = \angle\mathbf{V} + 90°$. The phase of the current through the capacitor is 90 degrees greater than the phase of the voltage across the capacitor, or the current through the capacitor is **leading** the voltage across the capacitor by 90 degrees, as shown in Figure 9.25. The typical waveform for the voltage across the capacitor and the current through the capacitor are shown in Figure 9.26.

For a capacitor with capacitance C, the impedance is given by

$$\mathbf{Z} = \frac{\mathbf{V}}{\mathbf{I}} = \frac{1}{j\omega C} = j\left(\frac{-1}{\omega C}\right) = R + jX \text{ ohms}$$

and the admittance is given by

$$\mathbf{Y} = \frac{\mathbf{I}}{\mathbf{V}} = \frac{1}{\mathbf{Z}} = j\omega C = G + jB \text{ siemens}$$

Thus, we have

$$R = 0, \quad X = -1/(\omega C), \quad G = 0, \quad B = \omega C \tag{9.52}$$

FIGURE 9.26

Waveform for voltage across the capacitor and current through the capacitor.

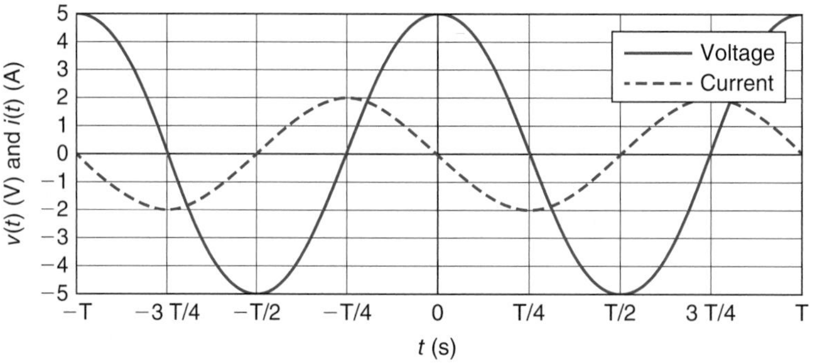

9.5.3 INDUCTOR

Let $v(t)$ be the voltage across the inductor and $i(t)$ be the current through the inductor as shown in Figure 9.27.

FIGURE 9.27

Voltage across and current through an inductor.

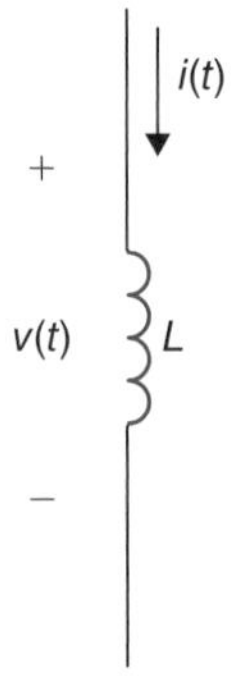

As discussed in Chapter 6, the current $i(t)$ through the inductor is related to the voltage $v(t)$ across the inductor as

$$i(t) = \frac{1}{L}\int_{-\infty}^{t} v(\lambda)\,d\lambda$$

Let the voltage across the inductor with inductance L be

$$v(t) = V_m\cos(\omega t + \phi) = \mathrm{Re}[V_m e^{j\phi} e^{j\omega t}] = \mathrm{Re}[\mathbf{V}e^{j\omega t}]$$

where **V** is the phasor of the voltage given by

$$\mathbf{V} = V_m e^{j\phi}$$

Then, the current through the inductor is given by

$$i(t) = \frac{1}{L}\int_{-\infty}^{t} v(\lambda)d\lambda = \frac{1}{L}\int_{-\infty}^{t} \mathrm{Re}[V_m e^{j\phi} e^{j\omega\lambda}]d\lambda = \mathrm{Re}\left[V_m e^{j\phi}\frac{1}{L}\int_{-\infty}^{t} e^{j\omega\lambda}d\lambda\right] = \mathrm{Re}\left[V_m e^{j\phi}\frac{1}{L}\frac{e^{j\omega t}}{j\omega}\right]$$

$$= \mathrm{Re}[\mathbf{I}e^{j\omega t}] = \frac{V_m}{\omega L}\cos(\omega t + \phi - 90°) = \frac{V_m}{\omega L}\sin(\omega t + \phi)$$

where the phasor of the current is

$$\mathbf{I} = \frac{V_m e^{j\phi}}{j\omega L} = \frac{\mathbf{V}}{j\omega L}$$

If $i(t)$ is replaced by phasor $\mathbf{I}$, $v(t)$ is replaced by phasor $\mathbf{V}$, and $\int_{-\infty}^{t} d\lambda$ is replaced by $\frac{1}{j\omega}$, the equation

$$i(t) = \frac{1}{L}\int_{-\infty}^{t} v(\lambda)d\lambda$$

becomes

$$\mathbf{I} = \frac{\mathbf{V}}{j\omega L}$$

The impedance of the inductor is given by

$$\mathbf{Z} = \frac{\mathbf{V}}{\mathbf{I}} = j\omega L = \omega L\angle 90° \text{ ohms} \tag{9.53}$$

Notice that when $\omega = 0$ (dc), the impedance of the inductor is zero ($\mathbf{Z} = 0$), and when $\omega = \infty$, the impedance of the inductor is infinity ($\mathbf{Z} = \infty$). The inductor acts as a short circuit (wire) for a dc signal, and acts as an open circuit (current through the inductor is zero) for a high-frequency ac signal.

The admittance of the inductor is given by

$$\mathbf{Y} = \frac{\mathbf{I}}{\mathbf{V}} = \frac{1}{\mathbf{Z}} = \frac{1}{j\omega L} = \frac{1}{\omega L}\angle -90° \text{ siemens} \tag{9.54}$$

FIGURE 9.28

The impedance of an inductor.

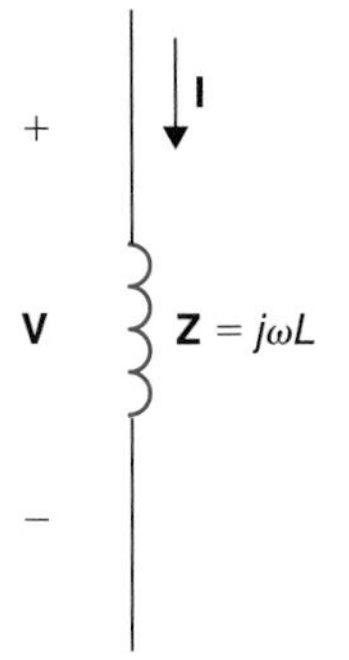

Figure 9.28 shows the inductor in the phasor domain.

For the sinusoidal inputs, the voltage-current relation for the inductor is given by

$$\mathbf{V} = \mathbf{ZI}$$

where $\mathbf{Z} = j\omega L$. The voltage-current relation for the inductor ($\mathbf{V} = \mathbf{ZI}$) is identical to the voltage-current relation of the resistor ($\mathbf{V} = \mathbf{ZI}$). The only difference is the impedance. The impedance of a resistor is $\mathbf{Z} = R$, and the impedance of an inductor is $\mathbf{Z} = j\omega L$. From this observation, we conclude that if we are interested in the sinusoidal steady-state response of a circuit, the inductor can be treated as a resistor with impedance $\mathbf{Z} = j\omega L$.

The magnitude of the impedance of an inductor is ωL, and the phase of the impedance is $j = e^{j\pi/2} = 90°$. The voltage-current relation $\mathbf{V} = \mathbf{ZI}$ can be written as

$$|\mathbf{V}|e^{j\angle\mathbf{V}} = |\mathbf{Z}|e^{j\angle\mathbf{Z}}|\mathbf{I}|e^{j\angle\mathbf{I}} = |\mathbf{Z}|\,|\mathbf{I}|\,e^{j(\angle\mathbf{Z}+\angle\mathbf{I})}$$

FIGURE 9.29

Phasors **V** and **I** for inductor.

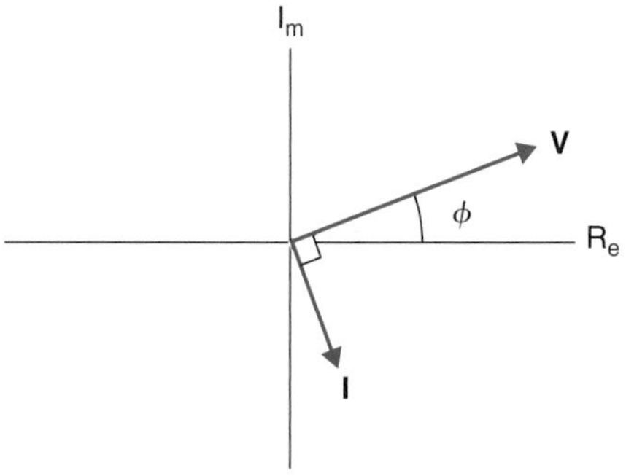

Thus, $|\mathbf{V}| = |\mathbf{Z}|\,|\mathbf{I}|$ and $\angle\mathbf{V} = \angle\mathbf{Z} + \angle\mathbf{I}$. Since $\angle\mathbf{Z} = 90°$, we have $\angle\mathbf{V} = 90° + \angle\mathbf{I}$ or $\angle\mathbf{I} = \angle\mathbf{V} - 90°$. Thus, the phase of the current through the inductor is **lagging** the voltage across the inductor by 90 degrees as shown in Figure 9.29. The typical waveform for voltage across the inductor and current through the inductor are shown in Figure 9.30.

For an inductor with inductance L, the impedance is given by

$$\mathbf{Z} = \frac{\mathbf{V}}{\mathbf{I}} = j\omega L = R + jX \text{ ohms}$$

and the admittance is given by

$$\mathbf{Y} = \frac{\mathbf{I}}{\mathbf{V}} = \frac{1}{\mathbf{Z}} = \frac{1}{j\omega L} = \frac{-j}{\omega L} = G + jB \text{ siemens}$$

Thus, we have

$$R = 0, \quad X = \omega L, \quad G = 0, \quad B = -1/(\omega L) \qquad \textbf{(9.55)}$$

FIGURE 9.30

Waveform for voltage across the inductor and current through the inductor.

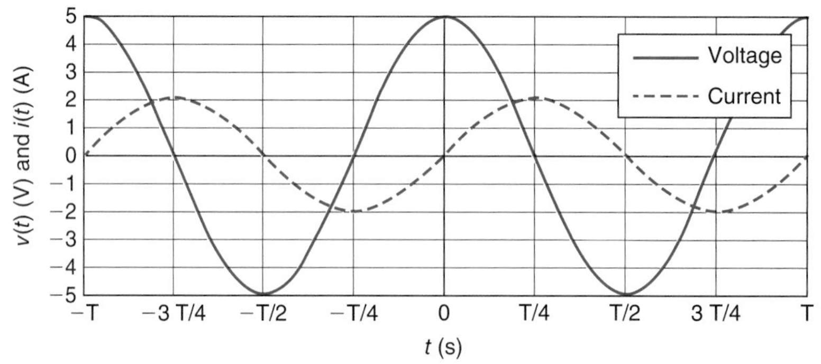

9.6 Phasor-Transformed Circuit

If the voltage sources and current sources are sinusoids, these sources can be transformed into voltage phasors and current phasors, respectively. The circuit elements can be transformed into the impedances. Circuits consisting of voltage phasors, current phasors, and impedances are called **phasor-transformed circuits**. Circuit laws and theorems for resistive circuits can be applied to phasor-transformed circuits, including Kirchhoff's current law (KCL), Kirchhoff's voltage law (KVL), the voltage divider rule, the current divider rule, nodal analysis, mesh analysis, source transformation, Thévenin's theorem, and Norton's theorem. The unknown voltages and currents in the phasor-transformed circuit can be found by applying these circuit laws and theorems.

EXAMPLE 9.9

Draw the phasor-transformed circuit for the circuit shown in Figure 9.31. The AC voltage source is given by $v_s(t) = 150\cos(2\pi 60t + 60°)$ V.

continued

Example 9.9 continued

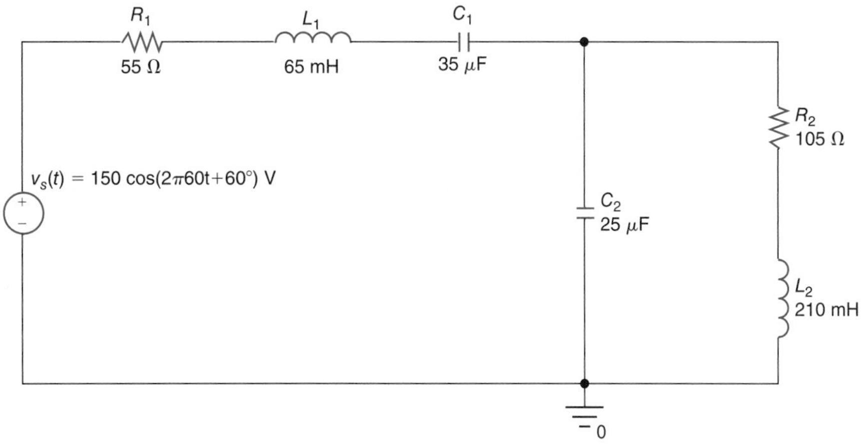

FIGURE 9.31 Circuit for EXAMPLE 9.9.

The phasor of the voltage source is given by

$$\mathbf{V_s} = 150\angle 60°$$

The impedances are given by

$$\mathbf{Z}_{R_1} = R_1 = 55\ \Omega$$

$$\mathbf{Z}_{R_2} = R_2 = 105\ \Omega$$

$$\mathbf{Z}_{L_1} = j\omega L_1 = j2\pi \times 60 \times 65 \times 10^{-3} = j24.5044\ \Omega$$

$$\mathbf{Z}_{L_2} = j\omega L_2 = j2\pi \times 60 \times 210 \times 10^{-3} = j79.1681\ \Omega$$

$$\mathbf{Z}_{C_1} = \frac{1}{j\omega C_1} = \frac{1}{j2\pi \times 60 \times 35 \times 10^{-6}} = -j75.7881\ \Omega$$

$$\mathbf{Z}_{C_2} = \frac{1}{j\omega C_2} = \frac{1}{j2\pi \times 60 \times 25 \times 10^{-6}} = -j106.1033\ \Omega$$

The phasor-transformed circuit is shown in Figure 9.32.

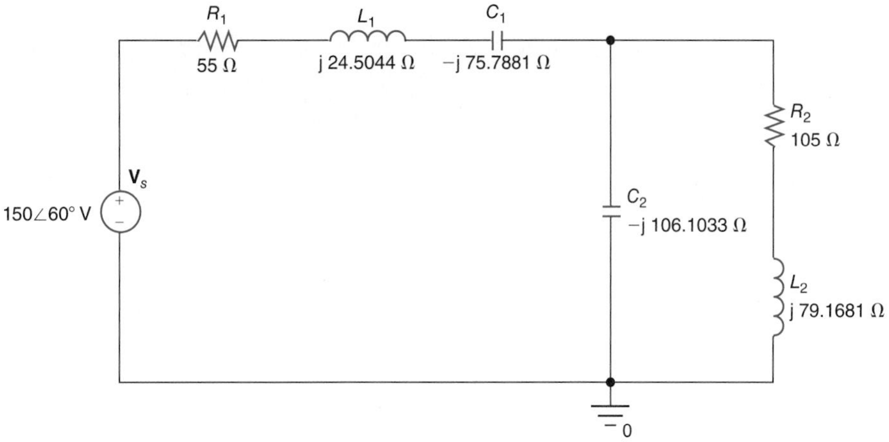

FIGURE 9.32 Phasor-transformed circuit.

Exercise 9.9

Draw the phasor-transformed circuit for the circuit shown in Figure 9.33.

FIGURE 9.33

Circuit for EXERCISE 9.9.

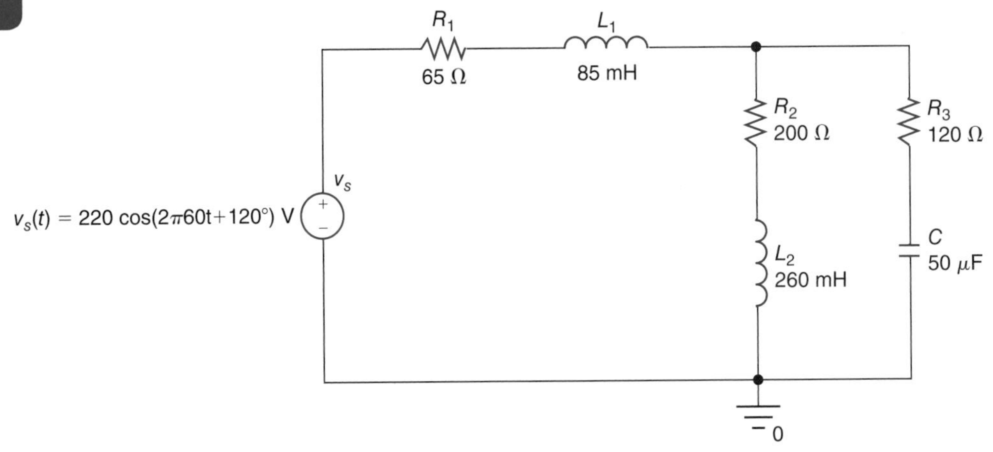

Answer:
The phasor-transformed circuit is shown in Figure 9.34.

FIGURE 9.34

Phasor-transformed circuit for EXERCISE 9.8.

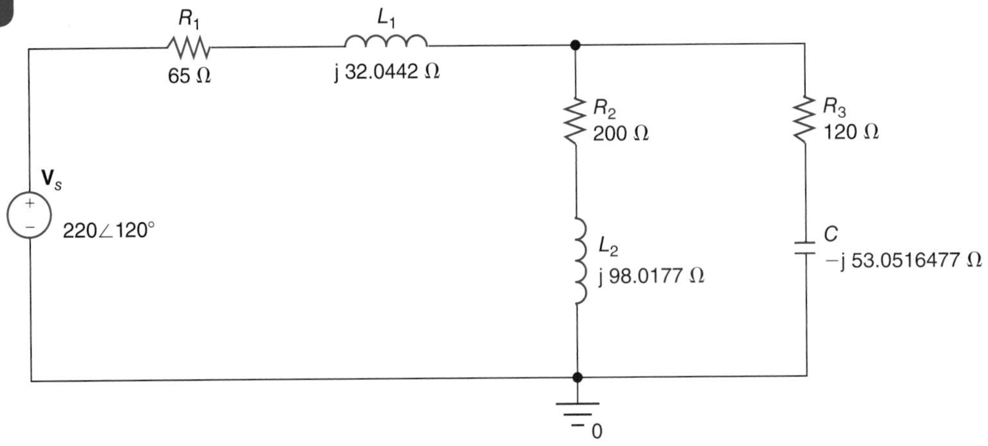

EXAMPLE 9.10

For the circuit shown in Figure 9.35, draw the phasor-transformed circuit, and find the phasors $\mathbf{I}$, $\mathbf{V}_w$, $\mathbf{V}_o$; current $i(t)$; and voltages $v_w(t)$ and $v_o(t)$. $v_s(t) = 200\cos(2\pi 60t)$ V.

The phasor of the voltage source is given by

$$\mathbf{V_s} = 200\angle 0° = 200 \text{ V}$$

The impedances of R_1, L_1, R_2, L_2 are given, respectively, by

$$\mathbf{Z_{R_1}} = R_1 = 50\ \Omega$$

continued

Example 9.10 continued

FIGURE 9.35

Circuit for EXAMPLE 9.10.

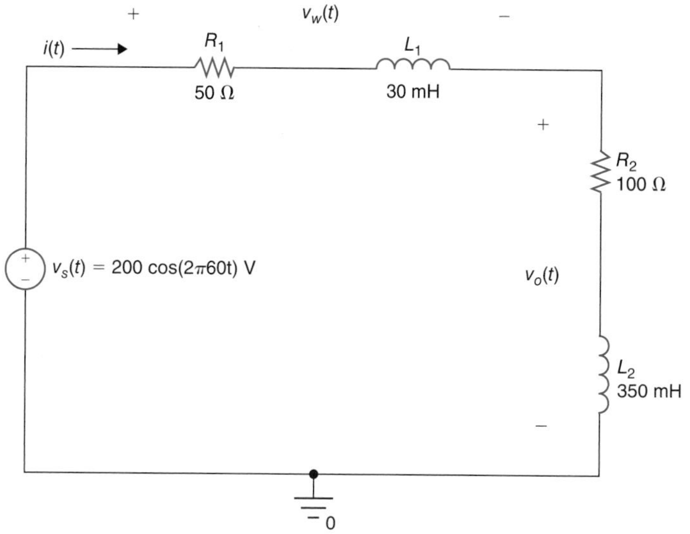

$$\mathbf{Z}_{\mathbf{L}_1} = j\omega L_1 = j2\pi 60 \times 30 \times 10^{-3} = j11.3097\ \Omega$$

$$\mathbf{Z}_{\mathbf{R}_2} = R_2 = 100\ \Omega$$

$$\mathbf{Z}_{\mathbf{L}_2} = j\omega L_2 = j2\pi 60 \times 350 \times 10^{-3} = j131.9469\ \Omega$$

The phasor-transformed circuit is shown in Figure 9.36.

FIGURE 9.36

Phasor-transformed circuit.

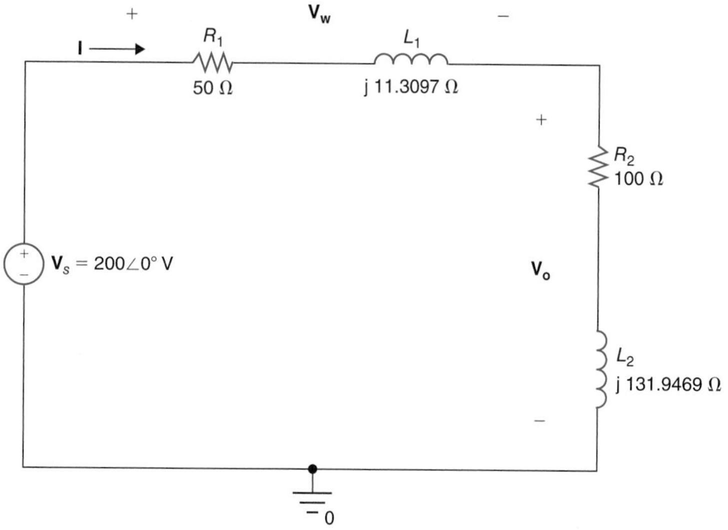

The total impedance seen from the voltage source is

$$\mathbf{Z} = \mathbf{Z}_{\mathbf{R}_1} + \mathbf{Z}_{\mathbf{L}_1} + \mathbf{Z}_{\mathbf{R}_2} + \mathbf{Z}_{\mathbf{L}_2} = 50 + j11.3097 + 100 + j131.9469$$
$$= 150 + j143.2566 = 207.4186\angle 43.6827^\circ\ \Omega$$

The current through the phasor-transformed circuit is given by

$$\mathbf{I} = \frac{\mathbf{V_s}}{\mathbf{Z}} = \frac{200\angle 0^\circ}{207.4186\angle 43.6827^\circ} = \frac{200}{207.4186}\angle 0^\circ - 43.6827^\circ$$
$$= 0.9642\angle -43.6827^\circ = 0.6973 - j0.6660\ \text{A}$$

continued

Example 9.10 continued

The impedance of the series connection of $R_1 - L_1$ path is given by

$$\mathbf{Z_w} = \mathbf{Z_{R_1}} + \mathbf{Z_{L_1}} = 50 + j11.3097 = 51.2631\angle 12.7455°\ \Omega$$

The impedance of the series connection of $R_2 - L_2$ path is given by

$$\mathbf{Z_o} = \mathbf{Z_{R_2}} + \mathbf{Z_{L_2}} = 100 + j131.9469 = 165.5596\angle 52.8422°\ \Omega$$

The voltage across the series connection of $R_1 - L_1$ path is given by

$$\mathbf{V_w} = \mathbf{I} \times \mathbf{Z_w} = (0.9642\angle -43.6827°) \times (51.2631\angle 12.7455°)$$
$$= 49.4297\angle -30.9372°\text{ V} = 42.3974 - j25.4117\text{ V}$$

The voltage across the series connection of $R_2 - L_2$ path is given by

$$\mathbf{V_o} = \mathbf{I} \times \mathbf{Z_o} = (0.9642\angle -43.6827°) \times (165.5596\angle 52.8422°)$$
$$= 159.6382\angle 9.1595°\text{ V} = 157.6026 + j25.4117\text{ V}$$

Thus, the current $i(t)$ and the voltages $v_w(t)$, $v_o(t)$ are given, respectively, as

$$i(t) = 0.9642\cos(2\pi 60t - 43.6827°)\text{ A}$$
$$v_w(t) = 49.4297\cos(2\pi 60t - 30.9372°)\text{ V}$$
$$v_o(t) = 159.6382\cos(2\pi 60t + 9.1595°)\text{ V}$$

Exercise 9.10

For the circuit shown in Figure 9.37, draw the phasor-transformed circuit, and find the phasors I and $\mathbf{V}_o$, the current $i(t)$, and the voltage $v_o(t)$.

FIGURE 9.37

Circuit for EXERCISE 9.10.

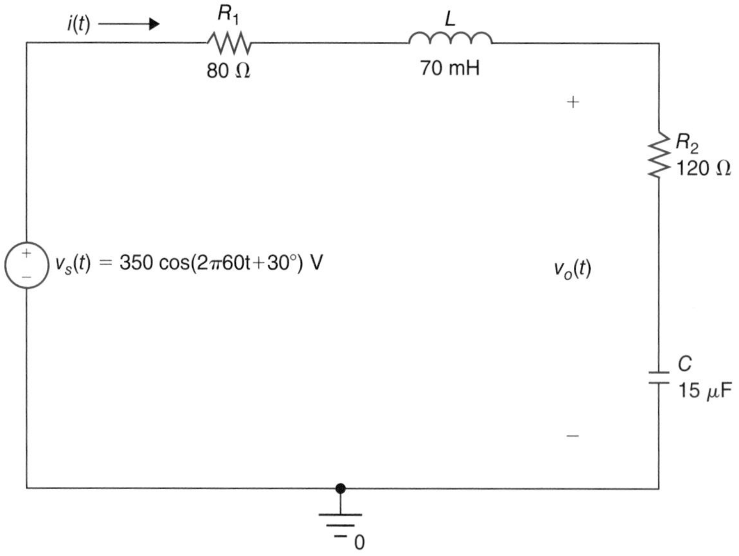

continued

Exercise 9.10 continued

Answer:

The phasor-transformed circuit is shown in Figure 9.38.

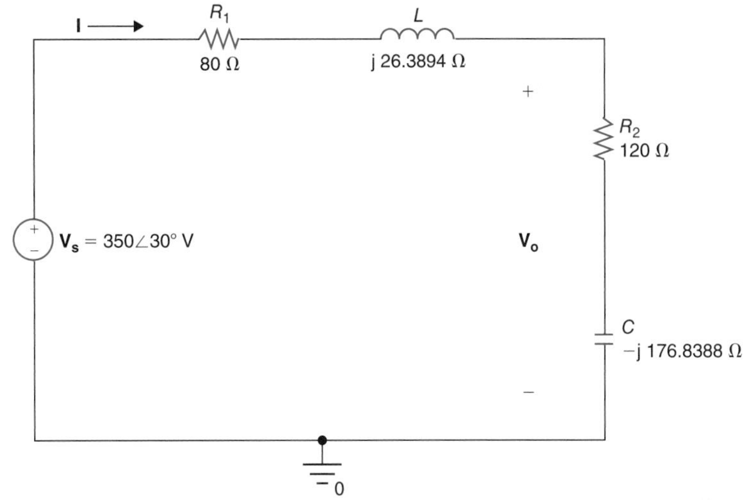

FIGURE 9.38

Phasor-transformed circuit.

$$\mathbf{I} = 1.3985\angle 66.9522°\text{ A},\quad \mathbf{V_o} = 298.8713\angle 11.1124°\text{ V},$$
$$i(t) = 1.3985\cos(2\pi 60t + 66.9522°)\text{ A},\quad v_o(t) = 298.8713\cos(2\pi 60t + 11.1124°)\text{ V}$$

9.7 Kirchhoff's Current Law and Kirchhoff's Voltage Law for Phasors

Kirchhoff's current law (KCL) and Kirchhoff's voltage law (KVL) apply to phasor-transformed circuits just as they do to dc circuits. Restating Kirchhoff's current law (KCL) from Chapter 2, we have the following:

The sum of currents entering a node equals the sum of currents leaving the same node.

Another way to describe Kirchhoff's current law (KCL) is as follows:

The algebraic sum of currents leaving a node equals zero.

Still another way to describe Kirchhoff's current law (KCL) is as follows:

The algebraic sum of currents entering a node equals zero.

EXAMPLE 9.11

Find the currents $\mathbf{I}_1$, $\mathbf{I}_2$, and $\mathbf{I}$ in the circuit shown in Figure 9.39.

Let $\mathbf{Z}_1$ be

$$\mathbf{Z}_1 = \mathbf{Z}_{R_1} + \mathbf{Z}_L = 125 + j180 = 219.1461\angle 55.2222°\ \Omega$$

and $\mathbf{Z}_2$ be

$$\mathbf{Z}_2 = \mathbf{Z}_{R_2} + \mathbf{Z}_C = 200 - j260 = 328.0244\angle -52.4314°\ \Omega$$

continued

Example 9.11 continued

FIGURE 9.39

Circuit for EXAMPLE 9.11.

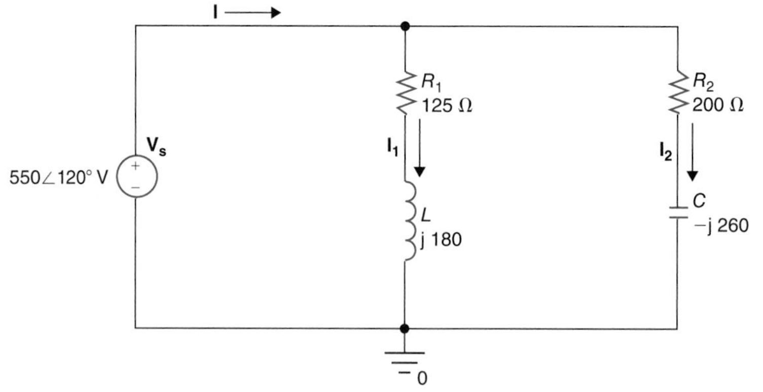

The current through $\mathbf{Z}_1$ is given by

$$\mathbf{I}_1 = \frac{\mathbf{V}_s}{\mathbf{Z}_1} = \frac{550\angle 120^\circ}{219.1461\angle 55.2222^\circ} = 2.5097\angle 64.7778^\circ = 1.0695 + j2.2705 \text{ A}$$

The current through $\mathbf{Z}_2$ is given by

$$\mathbf{I}_2 = \frac{\mathbf{V}_s}{\mathbf{Z}_2} = \frac{550\angle 120^\circ}{328.0244\angle -52.4314^\circ} = 1.6767\angle 172.4314^\circ = -1.6621 + j0.2208 \text{ A}$$

From KCL, the current $\mathbf{I}$ is the sum of $\mathbf{I}_1$ and $\mathbf{I}_2$:

$$\mathbf{I} = \mathbf{I}_1 + \mathbf{I}_2 = -0.5926 + j2.4913 = 2.5608\angle 103.3806^\circ \text{ A}$$

Exercise 9.11

Let $\mathbf{I}_2 = 4.6762\angle 8.7155^\circ$ in the circuit shown in Figure 9.40. Find $\mathbf{V}_a$, $\mathbf{I}_1$, $\mathbf{I}_3$, $\mathbf{V}_b$, $\mathbf{I}_4$, and $\mathbf{I}_5$.

FIGURE 9.40

Circuit for EXERCISE 9.11.

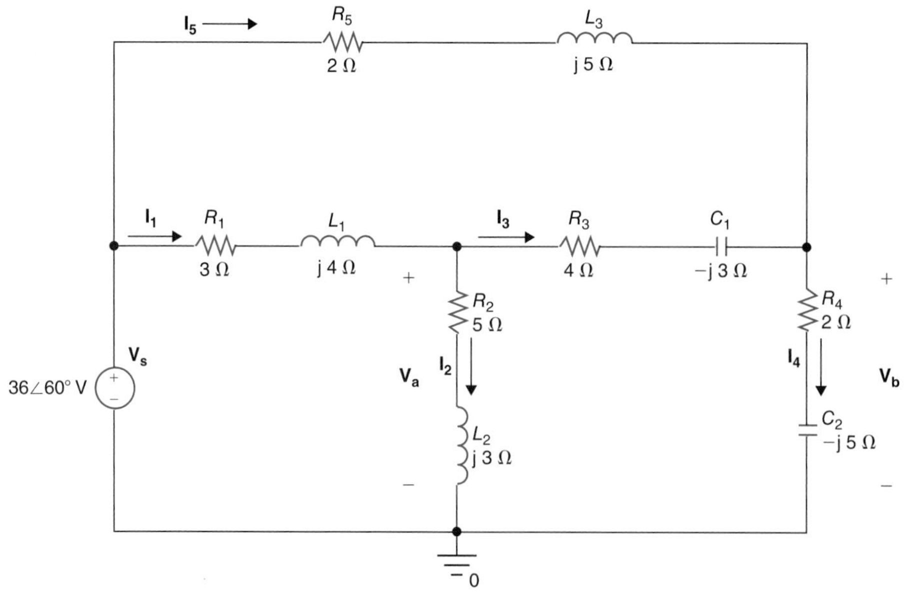

continued

Exercise 9.11 continued

Answer:
$\mathbf{V}_a = 20.9854 + j17.4096 = 27.2669\angle 39.6793°$,
$\mathbf{I}_1 = 1.8445 + j2.1297 = 2.8175\angle 49.1049°$,
$\mathbf{I}_3 = -2.7777 + j1.4212 = 3.1202\angle 152.9044°$,
$\mathbf{V}_b = 27.8328 + j3.3919 = 28.0387\angle 6.9481°$,
$\mathbf{I}_4 = 1.3347 + j5.0327 = 5.2067\angle 75.1467°$,
$\mathbf{I}_5 = 4.1124 + j3.6115 = 5.4731\angle 41.2896°$.

Restating Kirchhoff's voltage law (KVL) from Chapter 2, we have the following:

The sum of voltage drops around a loop equals the sum of voltage rises of the same loop.

Another way to describe Kirchhoff's voltage law (KVL) is as follows:

The algebraic sum of voltage drops around a loop equals zero.

Still another way to describe Kirchhoff's voltage law (KVL) is as follows:

The algebraic sum of voltage rises around the loop equals zero.

EXAMPLE 9.12

Find I, $\mathbf{V}_1$, $\mathbf{V}_2$, $\mathbf{V}_3$, $\mathbf{V}_4$, and $\mathbf{V}_5$ in the circuit shown in Figure 9.41.

FIGURE 9.41

Circuit for EXAMPLE 9.12.

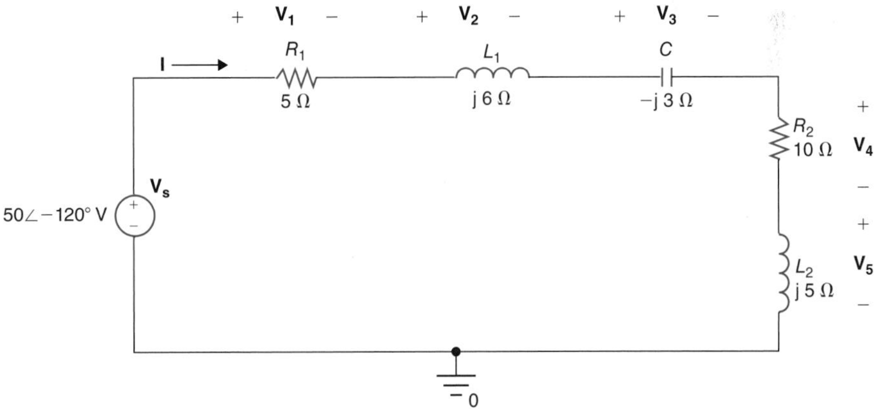

From KVL, the algebraic sum of voltage drops around a mesh is equal to zero. Summing the voltage drops around the mesh in the clockwise direction starting from the negative terminal of the voltage source, we obtain

$$-50\angle -120° + 5\mathbf{I} + j6\mathbf{I} + (-j3)\,\mathbf{I} + 10\,\mathbf{I} + j5\mathbf{I} = 0$$

Solving for I, we get

$$\mathbf{I} = \frac{50\angle -120°}{5 + j6 - j3 + 10 + j5} = \frac{50\angle -120°}{15 + j8} = \frac{50\angle -120°}{17\angle 28.0725°} = 2.9412\angle -148.0725°$$
$$= -2.4962 - j1.5554 \text{ A}$$

continued

Example 9.12 continued

According to Ohm's law ($\mathbf{V} = \mathbf{ZI}$), the voltages are given by

$$\mathbf{V}_1 = 5 \times \mathbf{I} = 14.7059\angle -148.0725° \text{ V}$$

$$\mathbf{V}_2 = j6 \times \mathbf{I} = 17.6471\angle -58.0725° \text{ V}$$

$$\mathbf{V}_3 = (-j3) \times \mathbf{I} = 8.8235\angle 121.9275° \text{ V}$$

$$\mathbf{V}_4 = 10 \times \mathbf{I} = 29.4118\angle -148.0725° \text{ V}$$

$$\mathbf{V}_5 = (j5) \times \mathbf{I} = 14.7059\angle -58.0725° \text{ V}$$

Exercise 9.12

Let $\mathbf{V}_a = 26.8476\angle 19.6392°$ in the circuit shown in Figure 9.42. Find $\mathbf{V}_b$.

FIGURE 9.42

Circuit for EXERCISE 9.12.

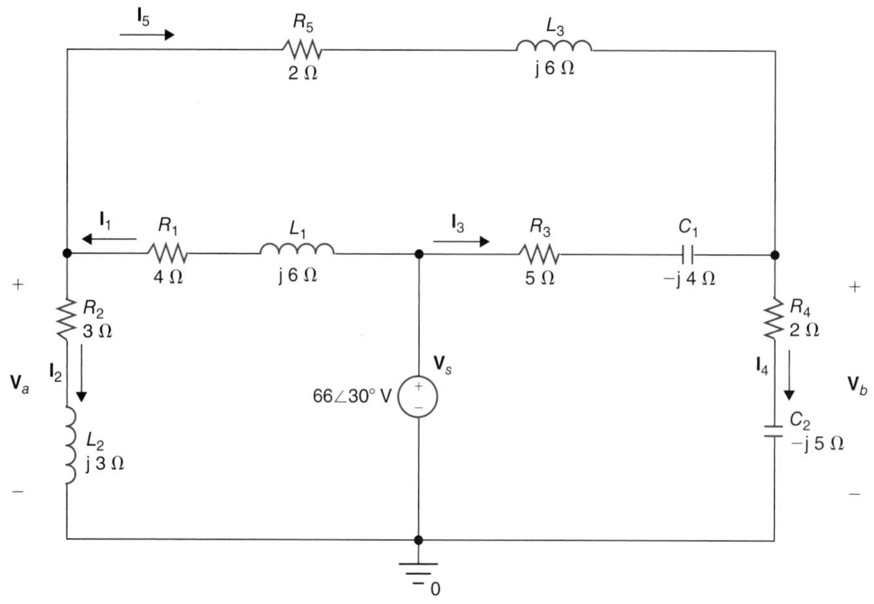

Answer:
$\mathbf{V}_b = 31.5492 + j10.2688 = 33.1783\angle 18.0294°$.

9.8 Series and Parallel Connection of Impedances

If n impedances are connected in series as shown in Figure 9.43(a), the current through all the impedances is $\mathbf{I}$. Thus, according to KVL, the voltage across all the impedances is given by

$$\mathbf{V} = \mathbf{V}_1 + \mathbf{V}_2 + \cdots + \mathbf{V}_n = \mathbf{Z}_1\mathbf{I} + \mathbf{Z}_2\mathbf{I} + \cdots + \mathbf{Z}_n\mathbf{I} = (\mathbf{Z}_1 + \mathbf{Z}_2 + \cdots + \mathbf{Z}_n)\mathbf{I} = \mathbf{Z}_{eq}\mathbf{I}$$

where $\mathbf{Z}_{eq}$ is given by

$$\mathbf{Z}_{eq} = \mathbf{Z}_1 + \mathbf{Z}_2 + \cdots + \mathbf{Z}_n \tag{9.56}$$

The impedance $\mathbf{Z}_{eq}$ is the equivalent impedance of the n impedances connected in series. Figure 9.43(b) shows the equivalent circuit with single impedance $\mathbf{Z}_{eq}$.

FIGURE 9.43

(a) Series connection of impedances. (b) Equivalent impedance.

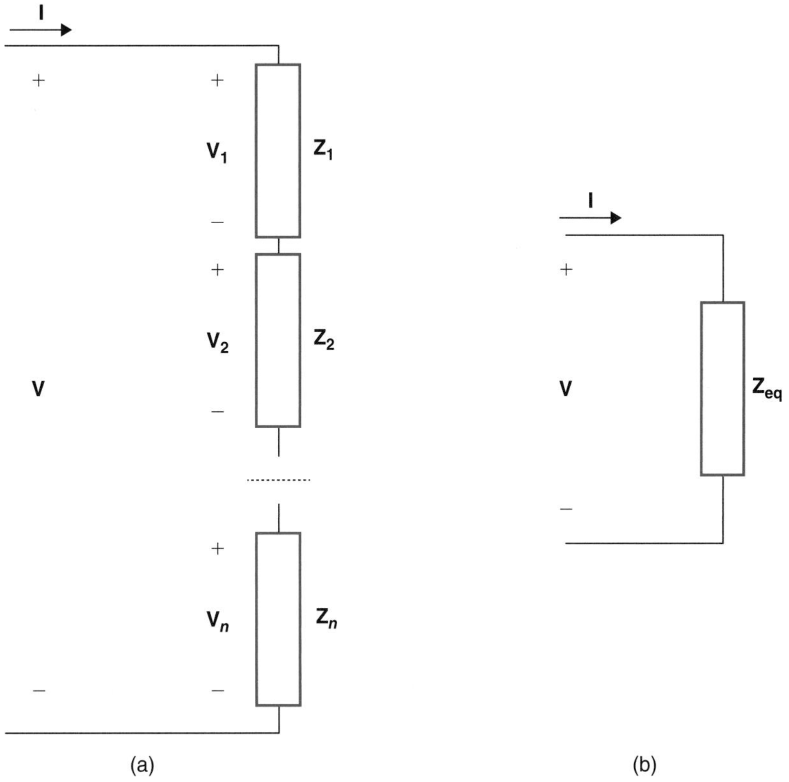

If n impedances are connected in parallel as shown in Figure 9.44(a), the voltage across all the impedances is $\mathbf{V}$. Thus, according to KCL, the current through all the impedances is given by

$$\mathbf{I} = \mathbf{I}_1 + \mathbf{I}_2 + \cdots + \mathbf{I}_n = \frac{\mathbf{V}}{\mathbf{Z}_1} + \frac{\mathbf{V}}{\mathbf{Z}_2} + \cdots + \frac{\mathbf{V}}{\mathbf{Z}_n} = \frac{\mathbf{V}}{\dfrac{1}{\dfrac{1}{\mathbf{Z}_1} + \dfrac{1}{\mathbf{Z}_2} + \cdots + \dfrac{1}{\mathbf{Z}_n}}} = \frac{\mathbf{V}}{\mathbf{Z}_{eq}}$$

where $\mathbf{Z}_{eq}$ is given by

$$\mathbf{Z}_{eq} = \frac{1}{\dfrac{1}{\mathbf{Z}_1} + \dfrac{1}{\mathbf{Z}_2} + \cdots + \dfrac{1}{\mathbf{Z}_n}} \qquad (9.57)$$

FIGURE 9.44

(a) Parallel connection of impedances. (b) Equivalent impedance.

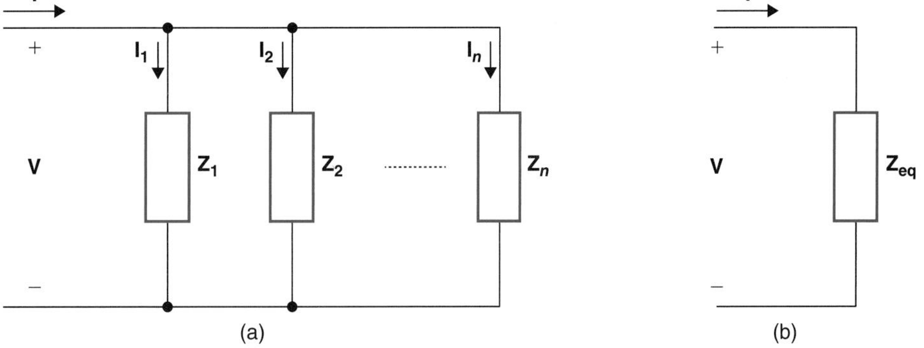

The impedance $\mathbf{Z}_{eq}$ is the equivalent impedance of n impedances connected in parallel. Figure 9.44(b) shows the equivalent circuit with single impedance $\mathbf{Z}_{eq}$.

In terms of the admittances, the equivalent admittance $\mathbf{Y}_{eq}$ is given by

$$\mathbf{Y}_{eq} = \mathbf{Y}_1 + \mathbf{Y}_2 + \cdots + \mathbf{Y}_n \tag{9.58}$$

where

$$\mathbf{Y}_{eq} = \frac{1}{\mathbf{Z}_{eq}}, \mathbf{Y}_1 = \frac{1}{\mathbf{Z}_1}, \mathbf{Y}_2 = \frac{1}{\mathbf{Z}_2}, \ldots, \mathbf{Y}_n = \frac{1}{\mathbf{Z}_n}$$

EXAMPLE 9.13

Find $v_a(t)$ and $v_b(t)$ in the circuit shown in Figure 9.45.

FIGURE 9.45

Circuit for EXAMPLE 9.13.

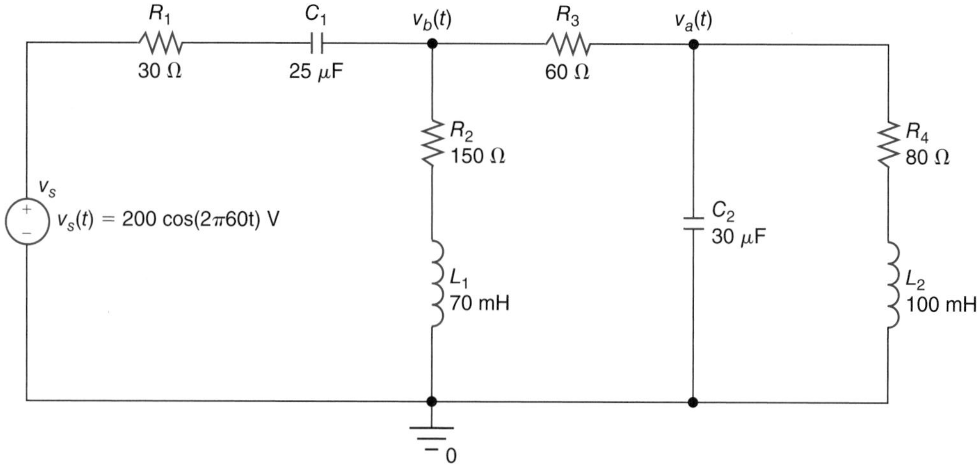

The phasor of the voltage source is

$$\mathbf{V}_s = 200\angle 0° = 200\text{ V}$$

The impedances are given by

$$\mathbf{Z}_{R_1} = 30\ \Omega, \mathbf{Z}_{R_2} = 150\ \Omega, \mathbf{Z}_{R_3} = 60\ \Omega, \mathbf{Z}_{R_4} = 80\ \Omega, \mathbf{Z}_{C_1} = -j106.1033\ \Omega,$$
$$\mathbf{Z}_{C_2} = -j88.4194\ \Omega, \mathbf{Z}_{L_1} = j26.3894\ \Omega, \mathbf{Z}_{L_2} = j37.7\ \Omega$$

The phasor-transformed circuit is shown in Figure 9.46.

Let $\mathbf{Z}_a$ be the equivalent impedance of the parallel connection of $\mathbf{Z}_{C_2}$ and $\mathbf{Z}_{R_4} + \mathbf{Z}_{L_2}$. Then, we have

$$\mathbf{Z}_a = \frac{\mathbf{Z}_{C_2} \times (\mathbf{Z}_{R_4} + \mathbf{Z}_{L_2})}{\mathbf{Z}_{C_2} + \mathbf{Z}_{R_4} + \mathbf{Z}_{L_2}} = \frac{-j88.4194 \times (80 + j37.7)}{-j88.4194 + 80 + j37.7} = 69.7059 - j44.2256\ \Omega$$

Let $\mathbf{Z}_b$ be the equivalent impedance of the parallel connection of $\mathbf{Z}_{R_2} + \mathbf{Z}_{L_1}$ and $\mathbf{Z}_{R_3} + \mathbf{Z}_a$. Then we have

$$\mathbf{Z}_b = \frac{(\mathbf{Z}_{R_2} + \mathbf{Z}_{L_1}) \times (\mathbf{Z}_{R_3} + \mathbf{Z}_a)}{\mathbf{Z}_{R_2} + \mathbf{Z}_{L_1} + \mathbf{Z}_{R_3} + \mathbf{Z}_a} = 74.1614 - j6.7508\ \Omega$$

continued

Example 9.13 continued

FIGURE 9.46

Phasor-transformed circuit.

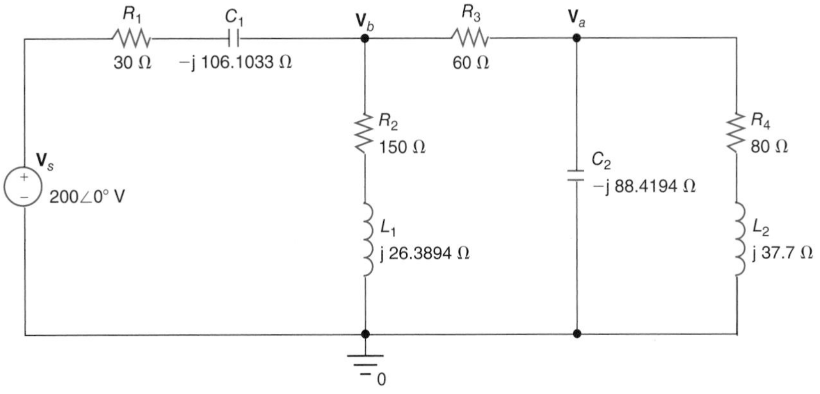

The total impedance seen from the voltage source is given by

$$\mathbf{Z}_t = \mathbf{Z}_{R_1} + \mathbf{Z}_{C_1} + \mathbf{Z}_b = 104.1614 - j112.8540\ \Omega$$

The current through $R_1 - C_1$ path is given by

$$\mathbf{I} = \frac{\mathbf{V}_s}{\mathbf{Z}_t} = 0.8833 + j0.9579 = 1.3023\angle 47.2938°\ \text{A}$$

The voltage $\mathbf{V}_b$ is given by

$$\mathbf{V}_b = \mathbf{V}_s - \mathbf{I} \times (\mathbf{Z}_{R_1} + \mathbf{Z}_{C_1}) = 71.9642 + j65.0078 = 96.9786\angle 42.0926°\ \text{V}$$

The current through R_3 is given by

$$\mathbf{I}_{R_3} = \frac{\mathbf{V}_b}{\mathbf{Z}_{R_3} + \mathbf{Z}_a} = 0.3439 + j0.6185 = 0.7077\angle 60.9204°\ \text{A}$$

The voltage $\mathbf{V}_a$ is given by

$$\mathbf{V}_a = \mathbf{V}_b - \mathbf{I}_{R_3} \times \mathbf{Z}_{R_3} = 51.3273 + j27.8996 = 58.4199\angle 28.5269°\ \text{V}$$

The voltages $v_a(t)$ and $v_b(t)$ are given, respectively, by

$$v_a(t) = 58.4199\cos(2\pi \times 60t + 28.5269°)\ \text{V}$$

$$v_b(t) = 96.9786\cos(2\pi \times 60t + 42.0926°)\ \text{V}$$

The MATLAB script shown below can be used to evaluate the impedances, currents, and voltages efficiently.

MATLAB

```
% EXAMPLE 9.13
% P.m, P2Rd.m, R2P.m should be in the current folder.
clear all;format long;
f=60;Am=200;phi=0;R1=30;R2=150;R3=60;R4=80;
L1=70e-3;L2=100e-3;C1=25e-6;C2=30e-6;w=2*pi*f;
ZR1=R1;ZR2=R2;ZR3=R3;ZR4=R4;
ZL1=j*w*L1
ZL2=j*w*L2
ZC1=1/(j*w*C1)
ZC2=1/(j*w*C2)
Vs=P2Rd(Am,phi)
```

continued

Example 9.13 continued

MATLAB continued

```
Za=P([ZC2,ZR4+ZL2])
Zb=P([ZR2+ZL1,ZR3+Za])
Zt=ZR1+ZC1+Zb
I=Vs/Zt
Ip=R2P(I)
Vb=Zb*I
Vbp=R2P(Vb)
IR3=Vb/(ZR3+Za)
IR3p=R2P(IR3)
Va=Za*IR3
Vap=R2P(Va)
```

The function P.m is given in Chapter 2, and functions P2Rd.m and R2P.m are given in Section 9.4.

Exercise 9.13

Find $v_a(t)$ and $v_b(t)$ in the circuit shown in Figure 9.47.

FIGURE 9.47

Circuit for EXERCISE 9.13.

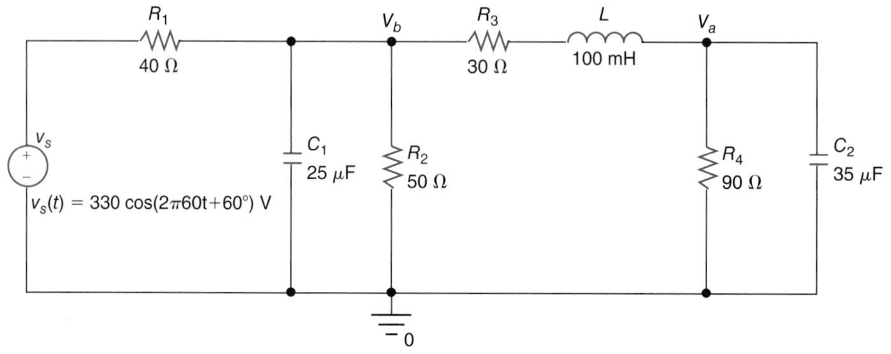

Answer:

$v_a(t) = 116.4594\cos(2\pi 60t + 5.4118°)$ V, $v_b(t) = 135.9389\cos(2\pi 60t + 49.6765°)$ V.

9.9 Delta-Wye (Δ-Y) and Wye-Delta (Y-Δ) Transformation

The impedances $\mathbf{Z}_1$, $\mathbf{Z}_2$, and $\mathbf{Z}_3$ in Figure 9.48(a) are configured in a triangular shape. This configuration is called a delta (Δ). If a delta configuration appears in a circuit, it may be difficult to reduce the circuit directly using a series or parallel combination of impedances. It may be easier to work with the wye (**Y**) configuration shown in Figure 9.48(b), where the three impedances $\mathbf{Z}_a$, $\mathbf{Z}_b$, and $\mathbf{Z}_c$ are arranged in **Y** shape. In other cases, the delta configuration may be easier to work with than the **Y** configuration.

The transformations from delta to wye and wye to delta can be found by equating the impedances seen from the two terminals of wye and delta. Let us find the impedance for terminals a and b, $\mathbf{Z}_{ab}$, with terminal c open. From the wye circuit, we have $\mathbf{Z}_{ab} = \mathbf{Z}_a + \mathbf{Z}_b$. From the delta circuit, we have $\mathbf{Z}_{ab} = \mathbf{Z}_1 \| (\mathbf{Z}_2 + \mathbf{Z}_3)$. These two are equal. Thus, we have

$$\mathbf{Z}_{ab} = \mathbf{Z}_a + \mathbf{Z}_b = \frac{\mathbf{Z}_1(\mathbf{Z}_2 + \mathbf{Z}_3)}{\mathbf{Z}_1 + \mathbf{Z}_2 + \mathbf{Z}_3} \tag{9.59}$$

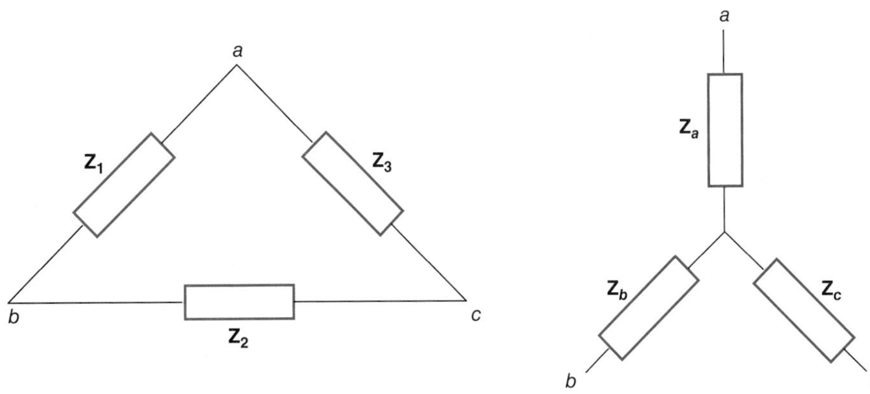

FIGURE 9.48

(a) Δ configuration.
(b) **Y** configuration.

Similarly, $\mathbf{Z}_{bc}$ and $\mathbf{Z}_{ca}$ are given, respectively, by

$$\mathbf{Z}_{bc} = \mathbf{Z}_b + \mathbf{Z}_c = \frac{\mathbf{Z}_2(\mathbf{Z}_1 + \mathbf{Z}_3)}{\mathbf{Z}_2 + \mathbf{Z}_1 + \mathbf{Z}_3} \tag{9.60}$$

$$\mathbf{Z}_{ca} = \mathbf{Z}_c + \mathbf{Z}_a = \frac{\mathbf{Z}_3(\mathbf{Z}_1 + \mathbf{Z}_2)}{\mathbf{Z}_3 + \mathbf{Z}_1 + \mathbf{Z}_2} \tag{9.61}$$

When Equation (9.60) is subtracted from the sum of Equation (9.59) and Equation (9.61), we have

$$\mathbf{Z}_a = \frac{\mathbf{Z}_1\mathbf{Z}_3}{\mathbf{Z}_1 + \mathbf{Z}_2 + \mathbf{Z}_3} \tag{9.62}$$

When Equation (9.61) is subtracted from the sum of Equation (9.59) and Equation (9.60), we have

$$\mathbf{Z}_b = \frac{\mathbf{Z}_1\mathbf{Z}_2}{\mathbf{Z}_1 + \mathbf{Z}_2 + \mathbf{Z}_3} \tag{9.63}$$

When Equation (9.59) is subtracted from the sum of Equation (9.60) and Equation (9.61), we have

$$\mathbf{Z}_c = \frac{\mathbf{Z}_2\mathbf{Z}_3}{\mathbf{Z}_1 + \mathbf{Z}_2 + \mathbf{Z}_3} \tag{9.64}$$

If $\mathbf{Z}_1 = \mathbf{Z}_2 = \mathbf{Z}_3 = \mathbf{Z}_\Delta$, then we have

$$\mathbf{Z}_Y = \mathbf{Z}_a = \mathbf{Z}_b = \mathbf{Z}_c = \mathbf{Z}_\Delta/3$$

The sum of the products of Equations (9.62) and (9.63), (9.63) and (9.64), and (9.62) and (9.64) [(9.62)(9.63) + (9.63)(9.64) + (9.62)(9.64)] results in

$$\mathbf{Z}_a\mathbf{Z}_b + \mathbf{Z}_b\mathbf{Z}_c + \mathbf{Z}_a\mathbf{Z}_c = \frac{\mathbf{Z}_1\mathbf{Z}_2\mathbf{Z}_3}{\mathbf{Z}_1 + \mathbf{Z}_2 + \mathbf{Z}_3} \tag{9.65}$$

When Equation (9.65) is divided by Equation (9.64), we get

$$\mathbf{Z}_1 = \frac{\mathbf{Z}_a\mathbf{Z}_b + \mathbf{Z}_b\mathbf{Z}_c + \mathbf{Z}_a\mathbf{Z}_c}{\mathbf{Z}_c} \tag{9.66}$$

When Equation (9.65) is divided by Equation (9.62), we get

$$\mathbf{Z}_2 = \frac{\mathbf{Z}_a\mathbf{Z}_b + \mathbf{Z}_b\mathbf{Z}_c + \mathbf{Z}_a\mathbf{Z}_c}{\mathbf{Z}_a} \tag{9.67}$$

When Equation (9.65) is divided by Equation (9.63), we get

$$\mathbf{Z}_3 = \frac{\mathbf{Z}_a\mathbf{Z}_b + \mathbf{Z}_b\mathbf{Z}_c + \mathbf{Z}_a\mathbf{Z}_c}{\mathbf{Z}_b} \tag{9.68}$$

If $\mathbf{Z}_a = \mathbf{Z}_b = \mathbf{Z}_c = \mathbf{Z}_Y$, then we have

$$\mathbf{Z}_\Delta = \mathbf{Z}_1 = \mathbf{Z}_2 = \mathbf{Z}_3 = 3\mathbf{Z}_Y$$

In summary, the Δ to **Y** conversion formulas are given by

$$\mathbf{Z}_a = \frac{\mathbf{Z}_1\mathbf{Z}_3}{\mathbf{Z}_1 + \mathbf{Z}_2 + \mathbf{Z}_3}$$

$$\mathbf{Z}_b = \frac{\mathbf{Z}_1\mathbf{Z}_2}{\mathbf{Z}_1 + \mathbf{Z}_2 + \mathbf{Z}_3}$$

$$\mathbf{Z}_c = \frac{\mathbf{Z}_2\mathbf{Z}_3}{\mathbf{Z}_1 + \mathbf{Z}_2 + \mathbf{Z}_3}$$

and the **Y** to Δ conversion formulas are given by

$$\mathbf{Z}_1 = \frac{\mathbf{Z}_a\mathbf{Z}_b + \mathbf{Z}_b\mathbf{Z}_c + \mathbf{Z}_a\mathbf{Z}_c}{\mathbf{Z}_c}$$

$$\mathbf{Z}_2 = \frac{\mathbf{Z}_a\mathbf{Z}_b + \mathbf{Z}_b\mathbf{Z}_c + \mathbf{Z}_a\mathbf{Z}_c}{\mathbf{Z}_a}$$

$$\mathbf{Z}_3 = \frac{\mathbf{Z}_a\mathbf{Z}_b + \mathbf{Z}_b\mathbf{Z}_c + \mathbf{Z}_a\mathbf{Z}_c}{\mathbf{Z}_b}$$

EXAMPLE 9.14

Use Y to Δ conversion to find $\mathbf{V}_o$ in the circuit shown in Figure 9.49.

The **Y** connected impedances $\mathbf{Z}_a$, $\mathbf{Z}_b$, and $\mathbf{Z}_c$ can be converted to Δ-connected impedances $\mathbf{Z}_1$, $\mathbf{Z}_2$, and $\mathbf{Z}_3$, as shown in Figure 9.50.

The Δ-connected impedances are given by

$$\mathbf{Z}_1 = \frac{\mathbf{Z}_a\mathbf{Z}_b + \mathbf{Z}_b\mathbf{Z}_c + \mathbf{Z}_a\mathbf{Z}_c}{\mathbf{Z}_c} = \frac{(4)\times(-j3) + (-j3)\times(j5) + (4)\times(j5)}{j5}$$
$$= 1.6 - j3\ \Omega$$

$$\mathbf{Z}_2 = \frac{\mathbf{Z}_a\mathbf{Z}_b + \mathbf{Z}_b\mathbf{Z}_c + \mathbf{Z}_a\mathbf{Z}_c}{\mathbf{Z}_a} = \frac{(4)\times(-j3) + (-j3)\times(j5) + (4)\times(j5)}{4}$$
$$= 3.75 + j2\ \Omega$$

continued

Example 9.14 continued

FIGURE 9.49

Circuit for EXAMPLE 9.14.

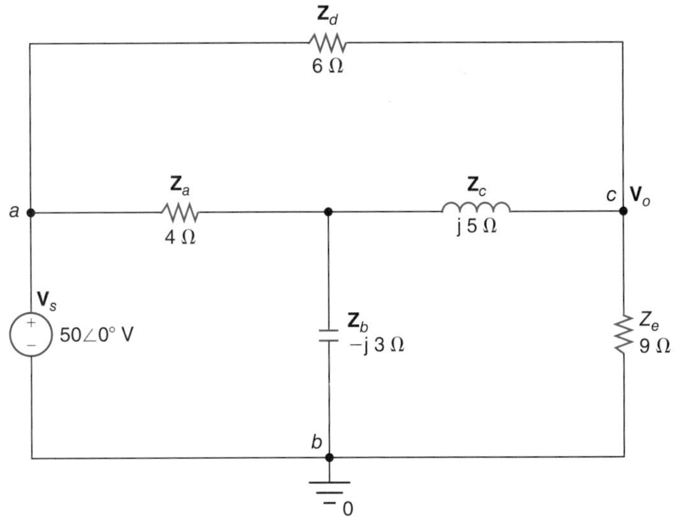

FIGURE 9.50

After **Y** to Δ conversion.

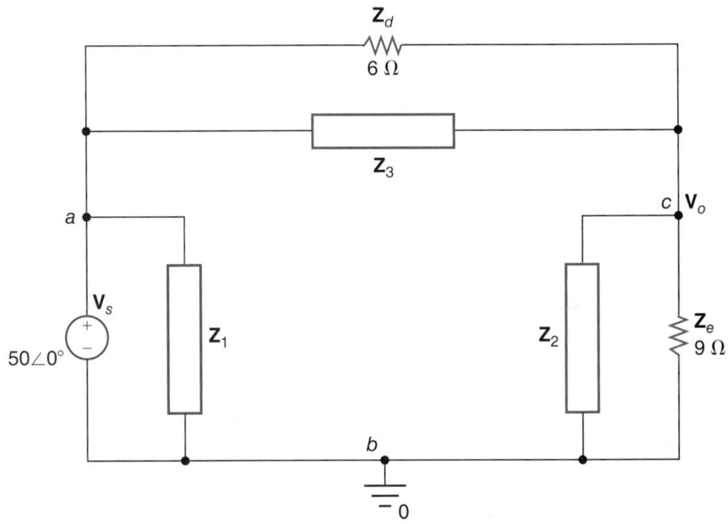

$$\mathbf{Z_3} = \frac{\mathbf{Z_aZ_b} + \mathbf{Z_bZ_c} + \mathbf{Z_aZ_c}}{\mathbf{Z_b}} = \frac{(4) \times (-j3) + (-j3) \times (j5) + (4) \times (j5)}{-j3}$$
$$= -2.6667 + j5\ \Omega$$

Let $\mathbf{Z}_4$ be the equivalent impedance of the parallel connection of $\mathbf{Z}_2$ and $\mathbf{Z}_e$. Then, we have

$$\mathbf{Z}_4 = \mathbf{Z}_2 \| \mathbf{Z}_e = \frac{\mathbf{Z}_2 \times \mathbf{Z}_e}{\mathbf{Z}_2 + \mathbf{Z}_e} = \frac{(3.75 + j2) \times (9)}{3.75 + j2 + 9} = 2.7996 + j0.9726\ \Omega$$

Let $\mathbf{Z}_5$ be the equivalent impedance of the parallel connection of $\mathbf{Z}_3$ and $\mathbf{Z}_d$. Then, we have

$$\mathbf{Z}_5 = \mathbf{Z}_3 \| \mathbf{Z}_d = \frac{\mathbf{Z}_3 \times \mathbf{Z}_d}{\mathbf{Z}_3 + \mathbf{Z}_d} = \frac{(-2.6667 + j5) \times (6)}{-2.6667 + j5 + 6} = 2.6769 + j4.9846\ \Omega$$

continued

Example 9.14 continued

The total impedance seen from the voltage source is

$$\mathbf{Z}_t = \mathbf{Z}_4 + \mathbf{Z}_5 = 5.4765 + j5.9572\ \Omega$$

The current through $\mathbf{Z}_4$ and $\mathbf{Z}_5$ is given by

$$\mathbf{I} = \frac{\mathbf{V}_s}{\mathbf{Z}_t} = 4.1818 - j4.5488\ \text{A}$$

Thus, the voltage phasor $\mathbf{V}_o$ is given by

$$\mathbf{V}_o = \mathbf{I} \times \mathbf{Z}_4 = 16.1316 - j8.6677 = 18.3128\angle{-28.2497°}\ \text{V}$$

MATLAB

```
% EXAMPLE 9.14
% P.m, P2Rd.m, Y2D.m, R2P.m should be in the current folder.
clear all;format long;
Vm=50;phi=0;Za=4;Zb=-3j;Zc=5j;Zd=6;Ze=9;
Vs=P2Rd(Vm,phi)
[Z1 Z2 Z3]=Y2D([Za Zb Zc])
Z4=P([Z2,Ze])
Z5=P([Z3,Zd])
Zt=Z4+Z5
I=Vs/Zt
Vo=I*Z4
Vop=R2P(Vo)
```

Exercise 9.14

Use Δ to Y conversion to find $\mathbf{V}_o$ in the circuit shown in Figure 9.51.

FIGURE 9.51

Circuit for EXERCISE 9.14.

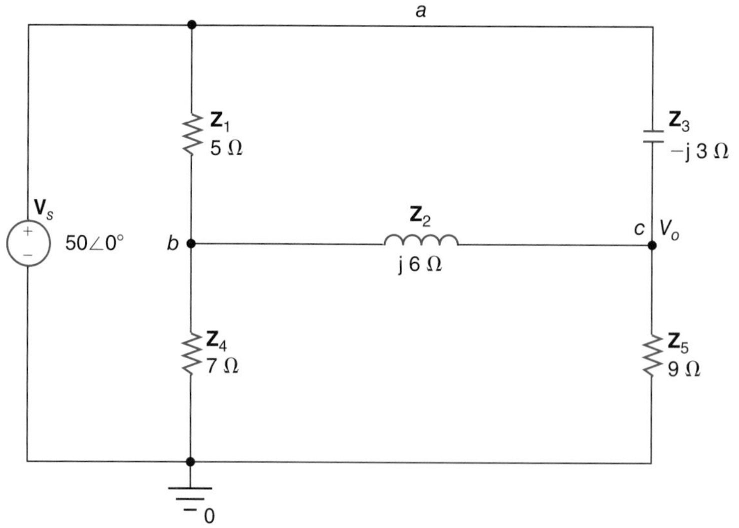

Answer:
$\mathbf{V}_o = 48.4595\angle 33.1265°$ V.

9.10 PSpice and Simulink

The voltages and currents calculated from the phasor-transformed circuit can be obtained from PSpice by adding IPRINT, VPRINT1, and VPRINT2 on the schematic of the circuit being simulated. IPRINT provides the values related to a current in the output file. VPRINT1 provides node voltages, and VPRINT2 provides voltages between two nodes. After adding the parts IPRINT, VPRINT1, and VPRINT2 as shown in Figure 9.52, double-click on IPRINT (or right-click and select Edit Properties). Type Y inside the box under AC, DB, REAL, IMAG, MAG, PHASE and click Apply. Repeat for VPRINT1 and VPRINT2. Double-click on the VAC, and enter 0 for ACPHASE.

FIGURE 9.52

AC analysis.

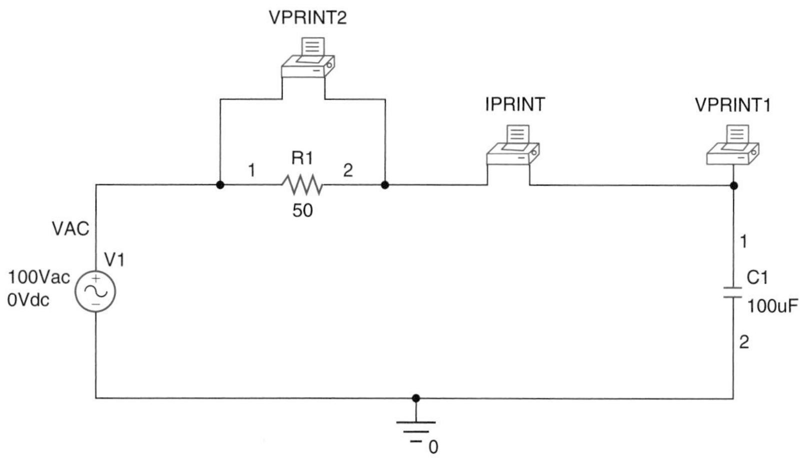

Select AC Sweep/Noise as Analysis type and select
Linear, Start Frequency: 60, End Frequency: 60, Total Points: 1 and Run the simulation.
The results of the simulation are available in the output file as shown here:

```
FREQ          IM(V_PRINT1)        IP(V_PRINT1)
6.000E+01     1.767E+00           2.795E+01

FREQ          IR(V_PRINT1)        II(V_PRINT1)        IDB(V_PRINT1)
6.000E+01     1.561E+00           8.280E-01           4.944E+00

FREQ          VM(N02738)          VP(N02738)
6.000E+01     4.687E+01           -6.205E+01

FREQ          VR(N02738)          VI(N02738)          VDB(N02738)
6.000E+01     2.196E+01           -4.140E+01          3.342E+01

FREQ          VM(N02590,N02718)   VP(N02590,N02718)
6.000E+01     8.834E+01           2.795E+01

FREQ          VR(N02590,N02718)   VI(N02590,N02718)
6.000E+01     7.804E+01           4.140E+01

FREQ          VDB(N02590,N02718)
6.000E+01     3.892E+01
```

The impedance of the capacitor C_1 is

$$\mathbf{Z}_{C_1} = \frac{1}{j\omega C} = \frac{1}{j2\pi 60 \times 100 \times 10^{-6}} = -j26.5258\ \Omega$$

The total impedance from the source is

$$\mathbf{Z} = 50 - j26.5158 = 56.6005\angle -27.9467^\circ\ \Omega$$

The current phasor is

$$\mathbf{I} = \frac{\mathbf{V}_{AC}}{\mathbf{Z}} = \frac{100\angle 0°}{56.6005\angle -27.9467°} = 1.5607 + j0.828 \text{ A} = 1.7668\angle 27.9467° \text{ A}$$

Notice that
magnitude of **I** = IM = 1.7667
phase of **I** = IP = 27.9467°
real part of **I** = IR = 1.5607
imaginary part of **I** = II = 0.828
The magnitude of **I** in dB = IDB = 20 $\log_{10}(1.7667)$ = 4.9436 dB
These values match those from PSpice simulation.
The voltage phasor across C_1 is

$$\mathbf{V}_{C_1} = \mathbf{I} \times \mathbf{Z}_{C_1} = 21.9633 - j41.3998 = 46.8650\angle -62.0533 \text{ V}$$

Thus, we have
magnitude of $\mathbf{V}_{C_1}$ = VM = 46.8650
phase of $\mathbf{V}_{C_1}$ = VP = −62.0533°
real part of $\mathbf{V}_{C_1}$ = VR = 21.9633
imaginary part of $\mathbf{V}_{C_1}$ = VI = −41.3998
The magnitude of $\mathbf{V}_{C_1}$ in dB = VDB = 20 $\log_{10}(46.8650)$ = 33.4170 dB
The voltage phasor across R_1 is

$$\mathbf{V}_{R_1} = \mathbf{I} \times \mathbf{Z}_{R_1} = 78.0367 + j41.3998 = 88.3384\angle 27.9467° \text{ V}$$

Thus, we have
magnitude of $\mathbf{V}_{C_1}$ = VM = 88.3384
phase of $\mathbf{V}_{C_1}$ = VP = 27.9467°
real part of $\mathbf{V}_{C_1}$ = VR = 78.0367
imaginary part of $\mathbf{V}_{C_1}$ = VI = 41.3998
The magnitude of $\mathbf{V}_{C_1}$ in dB = VDB = 20 $\log_{10}(88.3384)$ = 38.923 dB
These values match those obtained from PSpice.

Exercise 9.15

Find the magnitude and phase of the current through L_1 and the voltage across C_1 in the circuit shown in Figure 9.53. The frequency of the voltage source is 60 Hz.

FIGURE 9.53

Circuit for EXERCISE 9.15.

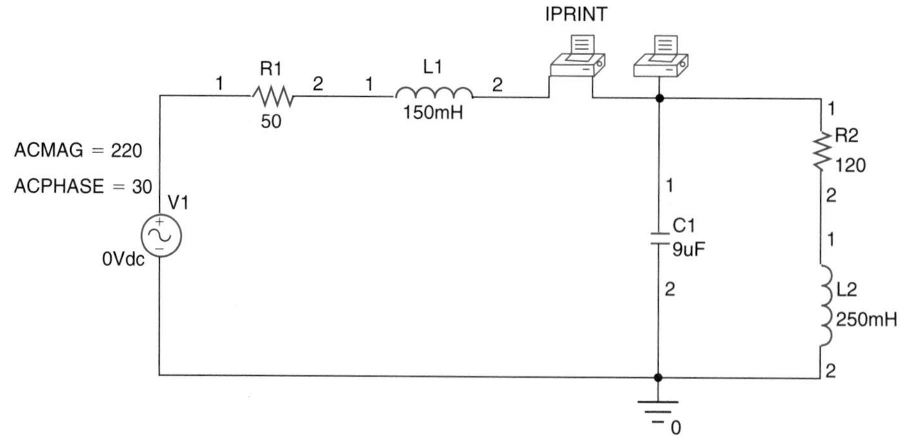

continued

Exercise 9.15 continued

Answer:

```
FREQ         IM(V_PRINT2)    IP(V_PRINT2)
 6.000E+01    8.657E-01       1.146E+01

FREQ         VM(N02487)      VP(N02487)
 6.000E+01    1.666E+02       1.870E+01
```

The voltages and currents of phasor-transformed circuit can be found using Simulink. The Simulink model of the circuit for EXAMPLE 9.10 (Figure 9.35) is shown in Figure 9.54.

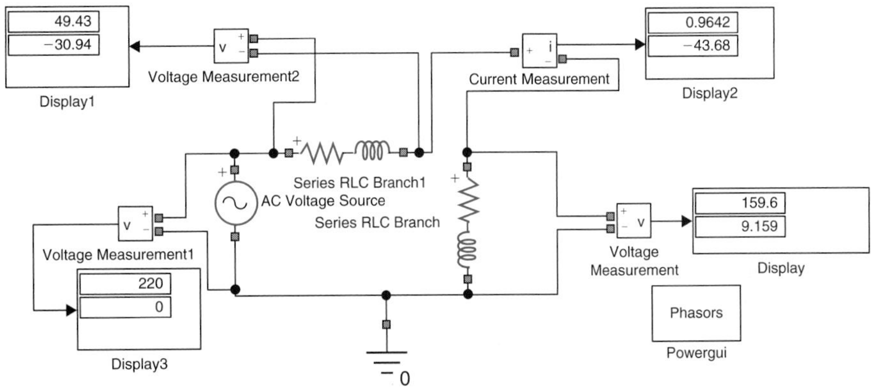

FIGURE 9.54 Simulink model of the circuit for EXAMPLE 9.10.

The AC Voltage Source, Series RLC Branch, Voltage Measurement, Current Measurement, and powergui are available in the SimPowerSystems blockset. Double-click on the **powergui** and click on Configure parameters. Select **Phasor** for Simulation type. Double-click on the Voltage Measurement and Current Measurement and select Magnitude-Angle (or Real-Imag or Complex) for Output signal. In Figure 9.54, Magnitude-Angle is selected. The magnitude and angle shown in the Displays match those calculated in EXAMPLE 9.10. The steady-state voltages and currents are available in powergui → Steady-State Voltages and Currents, or powergui → Generate Report.

Exercise 9.16

Use Simulink to find the voltage $v_o(t)$ and current $i(t)$ for the circuit shown in Figure 9.55.

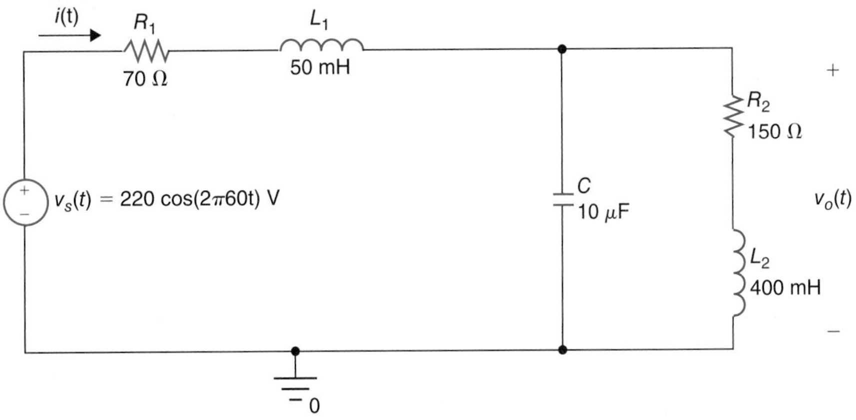

FIGURE 9.55 Circuit for EXERCISE 9.16.

continued

Exercise 9.16 continued

Answer:
$v_o(t) = 179.2\cos(2\pi \times 60t - 4.348°)$ V, $i(t) = 0.5994\cos(2\pi 60t + 3.153°)$ A. The Simulink model is shown in Figure 9.56.

FIGURE 9.56

Simulink model for EXERCISE 9.16.

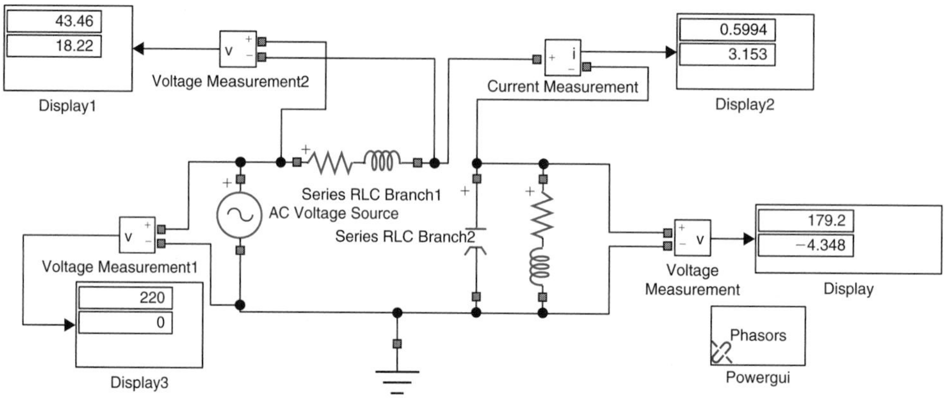

SUMMARY

A phasor represents the magnitude and phase of a sinusoid. Phasors are useful in finding voltages and currents in the steady state when the input signals are sinusoids. The ratio of the voltage phasor **V** to the current phasor **I** is defined as the impedance **Z** of circuit elements. The impedance is similar to resistance, but the impedance is a function of frequency and is a complex quantity representing both the magnitude and phase of the sinusoid.

If voltage sources and current sources are sinusoids, these sources can be transformed into voltage phasors and current phasors, respectively. Circuit elements can be transformed into impedances. Circuits consisting of voltage phasors, current phasors, and impedances are called phasor-transformed circuits. Circuit laws and theorems for resistive circuits can be applied to phasor-transformed circuits, including Kirchhoff's current law (KCL), Kirchhoff's voltage law (KVL), the voltage divider rule, the current divider rule, nodal analysis, mesh analysis, source transformation, Thévenin's theorem, and Norton's theorem. The unknown voltages and currents in the phasor-transformed circuit can be found by applying these circuit laws and theorems. Voltage phasors and current phasors can be transformed into voltages and currents in the time domain.

PROBLEMS

9.1 **Given a sinusoidal signal $v(t) = 7\cos(2\pi 1000t - 120°)$ V,**

a. find the peak amplitude.
b. find the peak-to-peak amplitude.
c. find the average amplitude.
d. find the *rms* amplitude.

9.2 **Given a sinusoidal signal $v(t) = 5\cos(2\pi 500t + 120°)$ V,**

a. find the frequency in hertz.
b. find the period in seconds.
c. find the frequency in radians/s.
d. plot the instantaneous frequency as a function of time.

9.3 **Given a sinusoidal signal $v(t) = 10\cos(2\pi 2000t - 135°)$ V,**

a. find the phase angle at $t = 0$.
b. find the amount of shift in seconds from $10\cos(2\pi 2000t)$.
c. plot the instantaneous phase as a function of time.
d. plot the sinusoidal signal.

9.4 **Plot $v(t) = 2\cos(2\pi 5000t + 60°)$ V.**

9.5 **Plot $v(t) = 110\cos(2\pi 60t)$ V.**

9.6 **Plot $v(t) = 10\sin(2\pi 50t - 135°)$ V.**

9.7 Plot $v(t) = 5\cos(2\pi 1{,}000{,}000t + 72°)$ V.

9.8 A sinusoidal signal $v(t)$ is shown in Figure P9.8. From this figure,

FIGURE P9.8

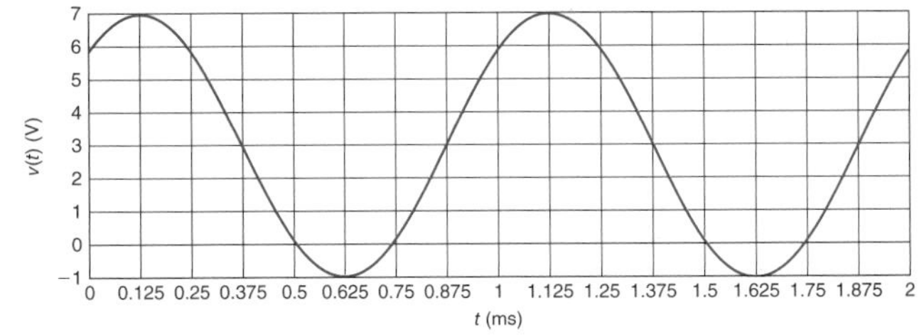

a. find the peak amplitude.
b. find the peak-to-peak amplitude.
c. find the average amplitude.
d. find the *rms* amplitude.
e. find the frequency in hertz.
f. find the frequency in radians/s.
g. find the phase angle at $t = 0$.
h. write the equation of the sinusoid.

9.9 Plot $v(t) = 2\cos(2\pi 10{,}000t) + 5\cos(2\pi 1000t)$ V.

9.10 Plot $v(t) = 10\cos(2\pi 1000t)\cos(2\pi 10{,}000t)$ V.

9.11 Plot $v(t) = 10(1 + 0.5\cos(2\pi 1000t)) \times \cos(2\pi 10{,}000t)$ V.

9.12 Plot $v(t) = 5\cos(2\pi 10{,}000t + 5\sin(2\pi 1000t))$ V.

9.13 Find the *rms* amplitude of the periodic signal with period T shown in Figure P9.13.

FIGURE P9.13

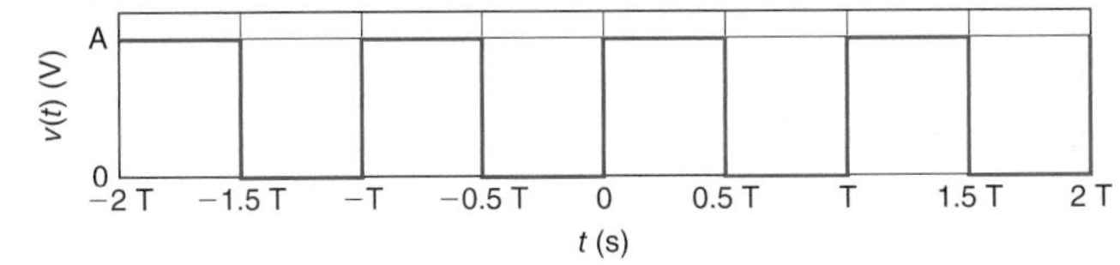

9.14 Find the *rms* amplitude of the periodic signal with period T shown in Figure P9.14.

FIGURE P9.14

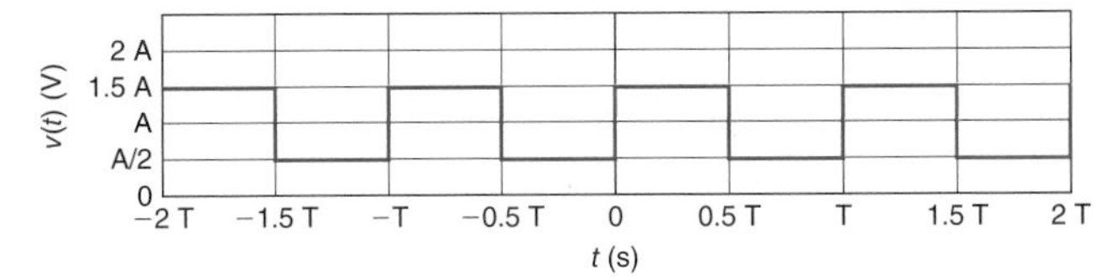

9.15 Find the *rms* amplitude of the periodic signal with period T shown in Figure P9.15.

FIGURE P9.15

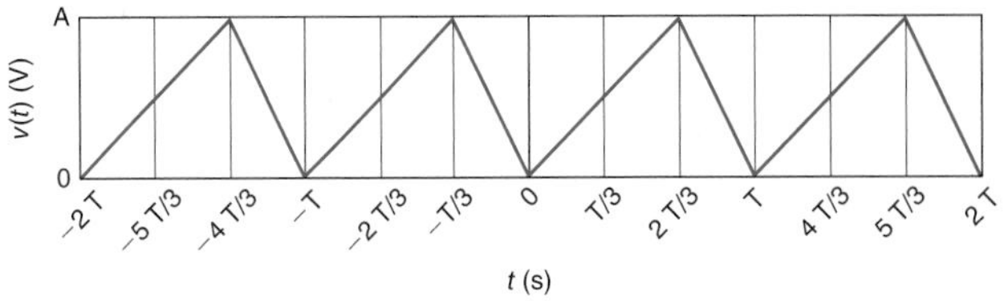

9.16 Find the *rms* amplitude of the periodic signal with period T shown in Figure P9.16. The equation of $v(t)$ for $0 \leq t < T$ is given by

$$v(t) = Ae^{-at}, \quad a > 0$$

FIGURE P9.16

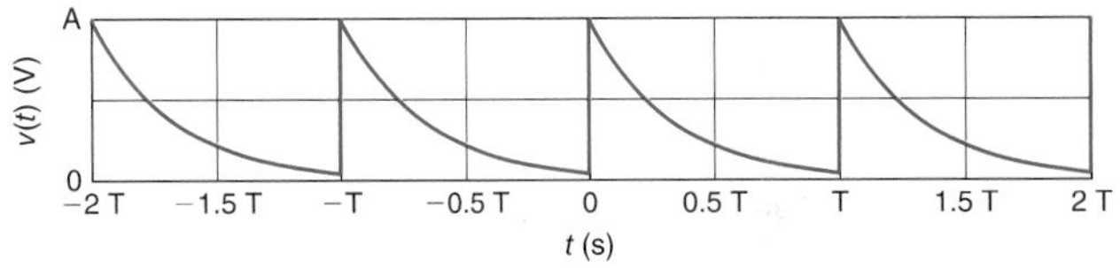

9.17 Represent $v(t) = 110\cos(2\pi 60t + 120°)$ V by its phasor and plot the phasor.

9.18 Represent $v(t) = 110\sin(2\pi 60t - 120°)$ V by its phasor and plot the phasor.

9.19 Represent $i(t) = -10\sin(2\pi 60t + 150°)$ A by its phasor and plot the phasor.

9.20 Represent $i(t) = -2\cos(2\pi 50t - 30°)$ A by its phasor and plot the phasor.

9.21 Find the time domain expression $v(t)$ for the following voltage phasors when the frequency of the sinusoid is $f = 60$ Hz.

a. $\mathbf{V} = 120\angle 120°$ V
b. $\mathbf{V} = 220\angle -150°$ V

9.22 Find the time domain expression $i(t)$ for the following current phasors when the frequency of the sinusoid is $f = 60$ Hz.

a. $\mathbf{I} = 5\angle -120°$ V
b. $\mathbf{I} = 7\angle 150°$ V

9.23 Convert the following complex numbers in Cartesian coordinates (rectangular coordinates) to polar coordinates.

a. $\mathbf{V} = 10 + j20$ V
b. $\mathbf{V} = -30 + j50$ V
c. $\mathbf{V} = -200 - j100$ V
d. $\mathbf{V} = 500 - j300$ V

9.24 Convert the following complex numbers in Cartesian coordinates (rectangular coordinates) to polar coordinates.

a. $\mathbf{I} = 5 + j9$ A
b. $\mathbf{I} = -8 + j5$ A
c. $\mathbf{I} = -10 - j30$ A
d. $\mathbf{I} = 20 - j60$ A

9.25 Convert the following complex numbers in polar coordinates to Cartesian coordinates (rectangular coordinates).

a. $\mathbf{V} = 110\angle 60°$ V
b. $\mathbf{V} = 220\angle -150°$ V
c. $\mathbf{V} = 480\angle 210°$ V
d. $\mathbf{V} = 880\angle -30°$ V

9.26 Convert the following complex numbers in polar coordinates to Cartesian coordinates (rectangular coordinates).

a. $\mathbf{I} = 5\angle 30°$ A
b. $\mathbf{I} = 6\angle 120°$ A
c. $\mathbf{I} = 10\angle -120°$ A
d. $\mathbf{I} = 20\angle -60°$ A

9.27 Evaluate the following operations in phasors and find the time domain expression $i(t)$ for the current when the frequency is $f = 60$ Hz.

a. $\mathbf{I} = 2\angle 60° + 3\angle 30°$ A
b. $\mathbf{I} = 4\angle 120° - 5\angle -120°$ A
c. $\mathbf{I} = (6\angle 60°)\,(7\angle -150°)$ A
d. $\mathbf{I} = (20\angle -120°)/(12\angle -60°)$ A

9.28 Represent **V** in polar notation as $\mathbf{V} = V_m\angle\phi$, and find the time domain expression $v(t)$ for the voltage when the frequency is $f = 60$ Hz.

$$\mathbf{V} = \frac{50\angle 120° - 40\angle -60°}{20\angle 30°}$$

9.29 Use phasors to express the following equations as single sinusoids.

a. $v(t) = -200\cos(2\pi 60t - 30°) - 300\sin(2\pi 60t + 120°)$ V
b. $v(t) = -50\sin(2\pi 60t - 60°) + 40\sin(2\pi 60t - 120°)$ V

9.30 Use phasors to express the following equations as single sinusoids.

a. $v(t) = -600\cos(2\pi 50t - 150°) - 500\sin(2\pi 50t + 30°) + 400\cos(2\pi 50t - 75°)$ V
b. $v(t) = 5\cos(2\pi 20{,}000t - 5\pi/6) + 7\cos(2\pi 20{,}000t + 4\pi/7) + 8\sin(2\pi 20{,}000t + 7\pi/5)$ V

9.31 For the circuit shown in Figure P9.31,

FIGURE P9.31

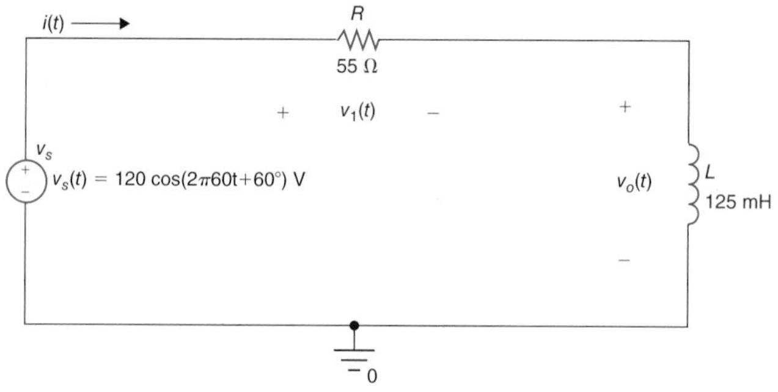

a. draw the phasor-transformed circuit.
b. find the phasors for $i(t)$, $v_1(t)$, and $v_o(t)$.
c. find the time domain waveforms $i(t)$, $v_1(t)$, and $v_o(t)$.

9.32 For the circuit shown in Figure P9.32,

FIGURE P9.32

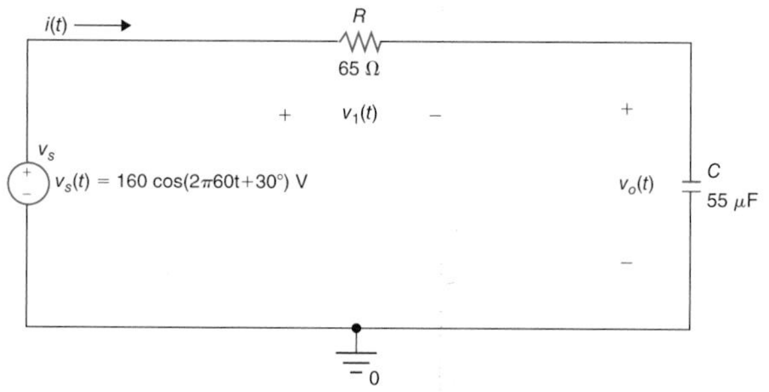

a. draw the phasor-transformed circuit.
b. find the phasors for $i(t)$, $v_1(t)$, and $v_o(t)$.
c. find the time domain waveforms $i(t)$, $v_1(t)$, and $v_o(t)$.

9.33 For the circuit shown in Figure P9.33,

FIGURE P9.33

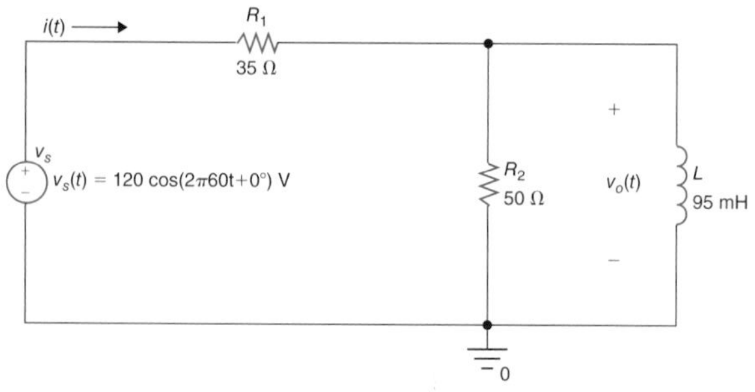

a. draw the phasor-transformed circuit.
b. find the phasors for $i(t)$ and $v_o(t)$.
c. find the time domain waveforms $i(t)$ and $v_o(t)$.

9.34 For the circuit shown in Figure P9.34,

FIGURE P9.34

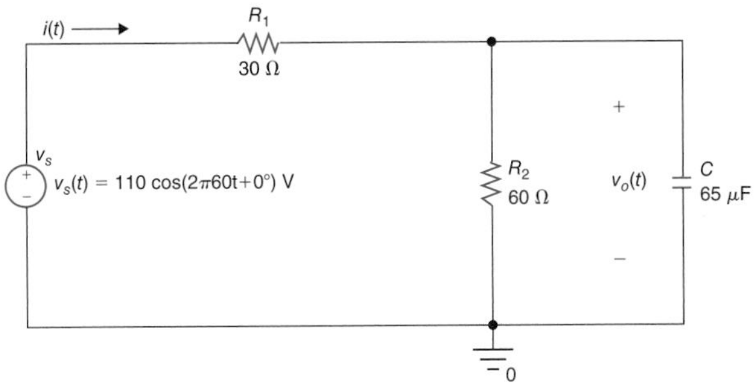

a. draw the phasor-transformed circuit.
b. find the phasors for $i(t)$ and $v_o(t)$.
c. find the time domain waveforms $i(t)$ and $v_o(t)$.

9.35 For the circuit shown in Figure P9.35,

FIGURE P9.35

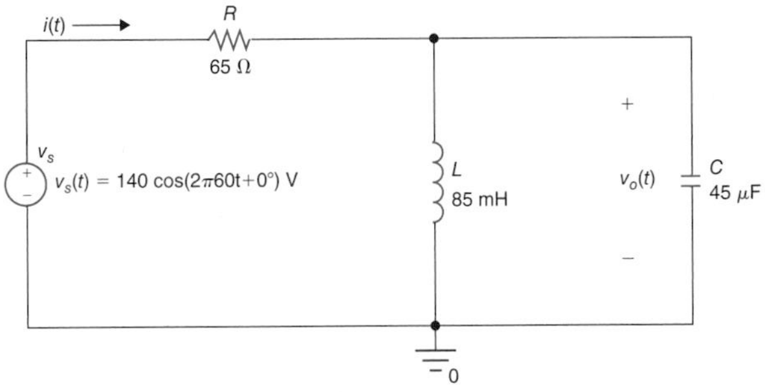

a. draw the phasor-transformed circuit.
b. find the phasors for $i(t)$ and $v_o(t)$.
c. find the time domain waveforms $i(t)$ and $v_o(t)$.

9.36 For the circuit shown in Figure P9.36,

FIGURE P9.36

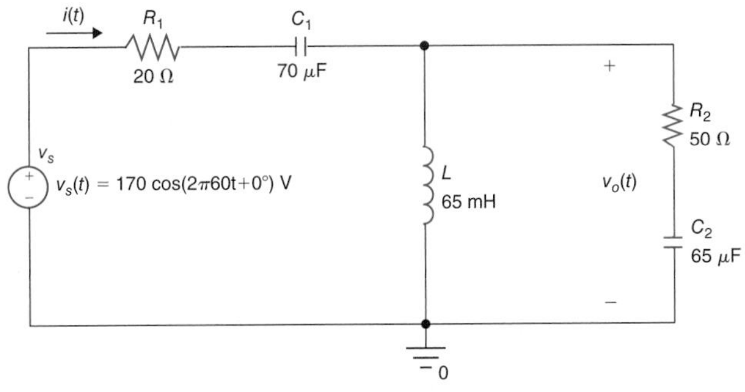

a. draw the phasor-transformed circuit.
b. find the phasors for $i(t)$ and $v_o(t)$.
c. find the time domain waveforms $i(t)$ and $v_o(t)$.

9.37 For the circuit shown in Figure P9.37, find the voltage phasor $\mathbf{V}_o$.

FIGURE P9.37

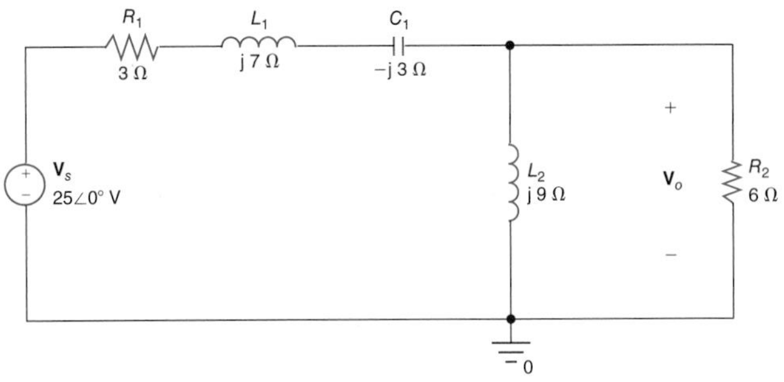

9.38 For the circuit shown in Figure P9.38, find the voltage phasor V_o.

FIGURE P9.38

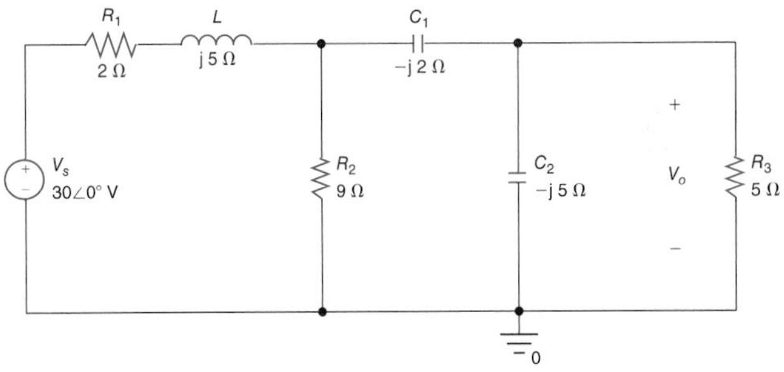

9.39 For the circuit shown in Figure P9.39, find the voltage phasor V_o.

FIGURE P9.39

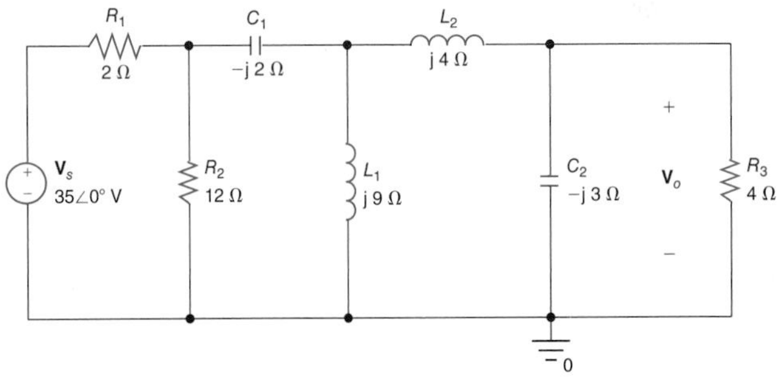

Chapter 10

Analysis of Phasor-Transformed Circuits

10.1 Introduction

The analysis of a circuit with sinusoidal input in the steady state is simplified by transforming all the components of the circuit to the phasor domain. Then, one can apply basic circuit laws and theorems to find the voltages and currents in phasors. The voltages and currents in phasors can be transformed into the sinusoidal signals in the time domain.

The basic circuit laws and theorems include Ohm's law, Kirchhoff's current law (KCL), Kirchhoff's voltage law (KVL), the voltage divider rule, the current divider rule, source transformation, the superposition principle, Thévenin's theorem, and Norton's theorem. Also, a nodal analysis and a mesh analysis can be applied to find the desired voltages and currents.

The difference between the analysis of the resistive circuits and the phasor-transformed circuits is that the component values and the voltages and currents are in general complex quantities in the phasor-transformed circuits. The complex voltages and currents can be transformed into sinusoids in the time domain from the magnitude and phase. The frequencies of the voltages and currents are the same as the frequency of the input signal.

The transfer function is the ratio of the output to input in the frequency domain. The output can be a voltage or a current, and the input can be a voltage or a current. The absolute value of the transfer function is called the *magnitude response,* and the angle of the transfer function is called the *phase response.* The magnitude response is the gain of the circuit as a function of the frequency. The transfer function provides the characteristics of the circuit in the frequency domain.

10.2 Phasor-Transformed Circuits

As discussed in Chapter 9, a sinusoidal signal

$$v(t) = V_m \cos(\omega t + \phi) \tag{10.1}$$

can be transformed into a phasor:

$$\mathbf{V} = V_m e^{j\phi} = V_m \angle \phi \tag{10.2}$$

The impedance of a resistor with resistance R is given by

$$\mathbf{Z}_R = R \tag{10.3}$$

The impedance of a resistor is not a function of frequency ω. For an inductor with inductance L, the impedance is given by

$$\mathbf{Z}_L = j\omega L \tag{10.4}$$

The impedance of a capacitor with capacitance C is given by

$$\mathbf{Z}_C = \frac{1}{j\omega C} \tag{10.5}$$

If the inputs to a circuit are sinusoidal, transform the sinusoidal signals to phasors. Also find impedances for all the rest of the elements of the circuit. A circuit consisting of phasors for inputs and impedances for the rest of the components is called a **phasor-transformed circuit**. The unknown voltages and currents of the circuit in the steady state can be found by applying circuit laws and theorems to the phasor-transformed circuit, including Ohm's law, KCL, KVL, the voltage divider rule, the current divider rule, source transformation, the superposition principle, nodal analysis, and mesh analysis. Unlike resistive circuits, the voltages and currents are generally complex. The complex voltages and currents can be transformed into sinusoidal signals in the time domain.

10.3 Voltage Divider Rule

Suppose that two impedances $\mathbf{Z}_1$ and $\mathbf{Z}_2$ are connected in series to a voltage source with voltage $\mathbf{V}_s$ volts as shown in Figure 10.1. The equivalent impedance is $\mathbf{Z}_1 + \mathbf{Z}_2$. The current through the circuit is $\mathbf{I} = \mathbf{V}_s/(\mathbf{Z}_1 + \mathbf{Z}_2)$. Thus, the voltage $\mathbf{V}_1$ across the first impedance $\mathbf{Z}_1$ is

$$\mathbf{V}_1 = \mathbf{I}\mathbf{Z}_1 = \frac{\mathbf{V}_s}{\mathbf{Z}_1 + \mathbf{Z}_2}\mathbf{Z}_1 = \mathbf{V}_s \times \frac{\mathbf{Z}_1}{\mathbf{Z}_1 + \mathbf{Z}_2} \tag{10.6}$$

and the voltage $\mathbf{V}_2$ across the second impedance $\mathbf{Z}_2$ is

$$\mathbf{V}_2 = \mathbf{I}\mathbf{Z}_2 = \frac{\mathbf{V}_s}{\mathbf{Z}_1 + \mathbf{Z}_2}\mathbf{Z}_2 = \mathbf{V}_s \times \frac{\mathbf{Z}_2}{\mathbf{Z}_1 + \mathbf{Z}_2} \tag{10.7}$$

This result shows that when two impedances are connected in series to a voltage source with voltage $\mathbf{V}_s$ volts, the total voltage $\mathbf{V}_s$ is divided between $\mathbf{V}_1$ and $\mathbf{V}_2$ in proportion to impedance values $\mathbf{Z}_1$ and $\mathbf{Z}_2$. This is referred to as the **voltage divider rule**.

The voltage divider rule can be generalized to include more than two impedances. If a voltage source with voltage $\mathbf{V}_s$ is connected to a series connection of n impedances $\mathbf{Z}_1$, $\mathbf{Z}_2, \ldots, \mathbf{Z}_n$, the voltage across each impedance $\mathbf{Z}_i$, $1 \le i \le n$, is given by

$$\mathbf{V}_i = \mathbf{V}_s \times \frac{\mathbf{Z}_i}{\mathbf{Z}_1 + \mathbf{Z}_2 + \cdots + \mathbf{Z}_n} \tag{10.8}$$

FIGURE 10.1

Circuit with two resistors in series.

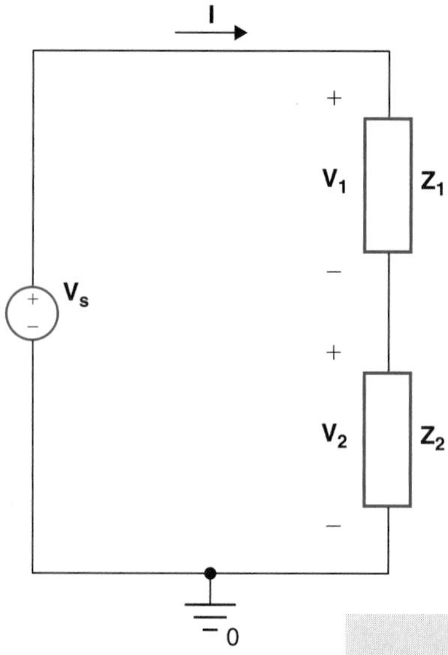

The voltage divider rule can be described by using admittance rather than impedance. Suppose that two impedances $\mathbf{Z}_1$ and $\mathbf{Z}_2$ are connected in series to a voltage source with voltage $\mathbf{V}_s$ volts. The voltage across $\mathbf{Z}_1$ is given by

$$\mathbf{V}_1 = \mathbf{V}_s \times \frac{\mathbf{Z}_1}{\mathbf{Z}_1 + \mathbf{Z}_2} = \mathbf{V}_s \times \frac{\frac{1}{\mathbf{Y}_1}}{\frac{1}{\mathbf{Y}_1} + \frac{1}{\mathbf{Y}_2}} = \mathbf{V}_s \times \frac{\mathbf{Y}_2}{\mathbf{Y}_1 + \mathbf{Y}_2} \quad (10.9)$$

Similarly, the voltage across $\mathbf{Z}_2$ is given by

$$\mathbf{V}_2 = \mathbf{V}_s \times \frac{\mathbf{Z}_2}{\mathbf{Z}_1 + \mathbf{Z}_2} = \mathbf{V}_s \times \frac{\frac{1}{\mathbf{Y}_2}}{\frac{1}{\mathbf{Y}_1} + \frac{1}{\mathbf{Y}_2}} = \mathbf{V}_s \times \frac{\mathbf{Y}_1}{\mathbf{Y}_1 + \mathbf{Y}_2} \quad (10.10)$$

If a voltage source with voltage $\mathbf{V}_s$ is connected to a series connection of n impedances $\mathbf{Z}_1, \mathbf{Z}_2, \ldots, \mathbf{Z}_n$, the voltage across each impedance $\mathbf{Z}_i$, $1 \le i \le n$, is given by

$$\mathbf{V}_i = \mathbf{V}_s \times \frac{\mathbf{Z}_i}{\mathbf{Z}_1 + \mathbf{Z}_2 + \cdots + \mathbf{Z}_n} = \mathbf{V}_s \times \frac{\frac{1}{\mathbf{Z}_1 + \mathbf{Z}_2 + \cdots + \mathbf{Z}_n}}{\frac{1}{\mathbf{Z}_i}} = \mathbf{V}_s \times \frac{\mathbf{Y}}{\mathbf{Y}_i} \quad (10.11)$$

where

$$\mathbf{Y} = \frac{1}{\frac{1}{\mathbf{Y}_1} + \frac{1}{\mathbf{Y}_2} + \cdots + \frac{1}{\mathbf{Y}_n}} = \frac{1}{\mathbf{Z}_1 + \mathbf{Z}_2 + \cdots + \mathbf{Z}_n} \quad (10.12)$$

EXAMPLE 10.1

Find $v_o(t)$ in the circuit shown in Figure 10.2.

FIGURE 10.2

Circuit for EXAMPLE 10.1.

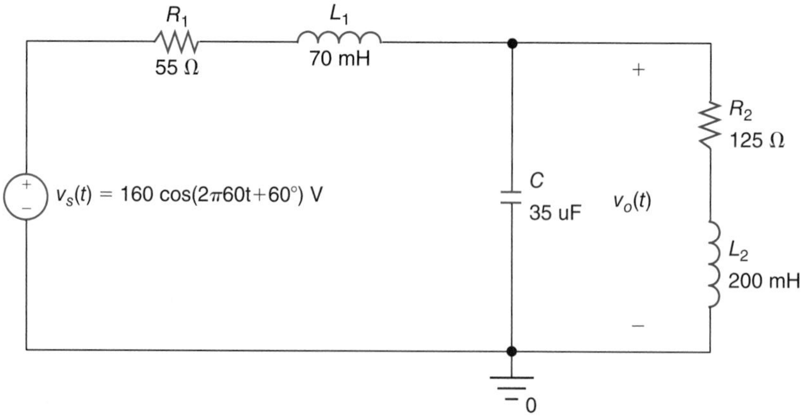

continued

Example 10.1 continued

The phasor of the input signal is given by

$$\mathbf{V}_s = 160\angle 60° = 80 + j138.5641 \text{ V}$$

The radian frequency is $\omega = 2\pi f = 2\pi \times 60 = 376.9911$ rad/s. The impedances are given by

$$\mathbf{Z}_{R_1} = R_1 = 55\ \Omega, \mathbf{Z}_{R_2} = R_2 = 125\ \Omega, \mathbf{Z}_{L_1} = j\omega L_1 = j26.3894\ \Omega, \mathbf{Z}_{L_2} = j\omega L_2 = j75.3982\ \Omega$$

$$\mathbf{Z}_C = 1/(j\omega C) = -j75.7881\ \Omega$$

The phasor-transformed circuit is shown in Figure 10.3.

FIGURE 10.3

Phasor-transformed circuit.

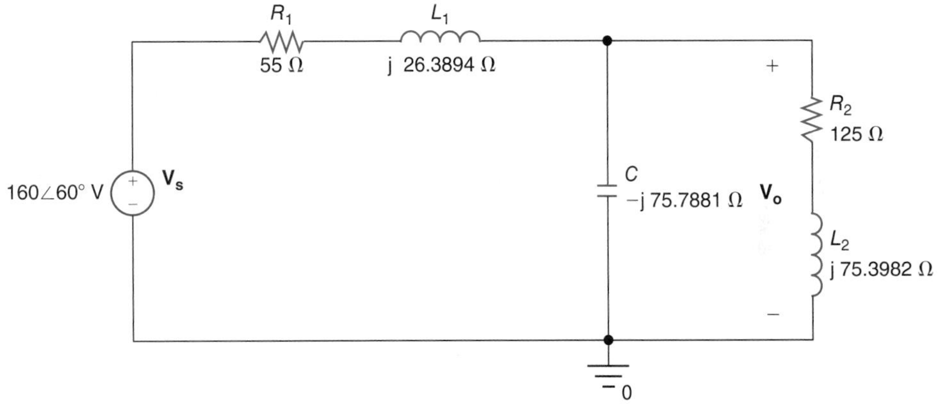

The equivalent impedance of the parallel connection of $\mathbf{Z}_C$ and $\mathbf{Z}_{R_2} + \mathbf{Z}_{L_2}$ is given by

$$\mathbf{Z}_o = \frac{\mathbf{Z}_C \times (\mathbf{Z}_{R_2} + \mathbf{Z}_{L_2})}{\mathbf{Z}_C + \mathbf{Z}_{R_2} + \mathbf{Z}_{L_2}} = 45.9502 - j75.6448\ \Omega$$

The total impedance seen from the voltage source is given by

$$\mathbf{Z}_t = \mathbf{Z}_{R_1} + \mathbf{Z}_{L_1} + \mathbf{Z}_o = 100.9502 - j49.2554\ \Omega$$

The output voltage $\mathbf{V}_o$ can be obtained by applying the voltage divider rule:

$$\mathbf{V}_o = \mathbf{V}_s \times \frac{\mathbf{Z}_o}{\mathbf{Z}_t} = 112.0454 + j57.7939 = 126.0726\angle 27.285° \text{ V}$$

The output voltage in the time domain is given by

$$v_o(t) = 126.0726 \cos(2\pi 60t + 27.285°) \text{ V}$$

MATLAB

```
% EXAMPLE 10.1
clear all;format long;
Vm=160;phi=60;R1=55;R2=125;L1=70e-3;L2=200e-3;C=35e-6;f=60;
w=2*pi*f
ZL1=j*w*L1
ZL2=j*w*L2
ZC=1/(j*w*C)
Vs=P2Rd(Vm,phi)
```

continued

Example 10.1 continued

MATLAB continued

```
Zo=P([ZC,R2+ZL2])
Zt=R1+ZL1+Zo
Vo=Vs*Zo/Zt
Vop=R2P(Vo)
```

Exercise 10.1

Find the voltage $\mathbf{V}_o$ in the circuit shown in Figure 10.4.

FIGURE 10.4

Circuit for EXERCISE 10.1.

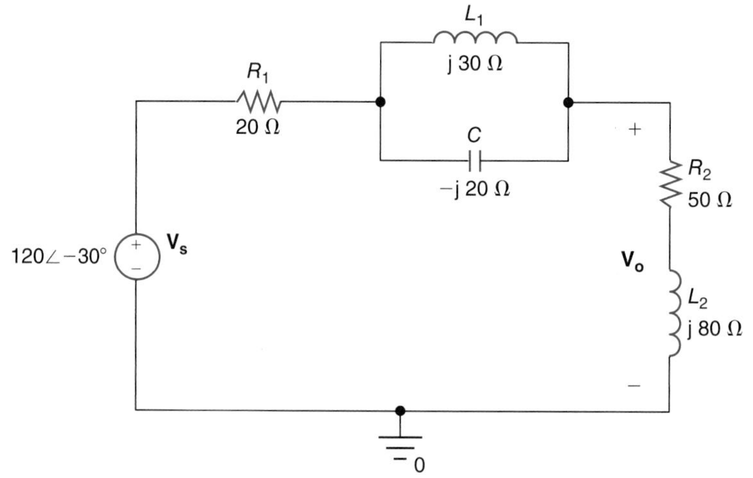

Answer:

$\mathbf{V}_o = 155.5028\angle 12.0492°$ V.

10.4 Current Divider Rule

Suppose that a current source with $\mathbf{I}_s$ amperes of current is connected in parallel to a pair of impedances $\mathbf{Z}_1$ and $\mathbf{Z}_2$, as shown in Figure 10.5.

The equivalent impedance value is

$$\mathbf{Z} = \mathbf{Z}_1 \| \mathbf{Z}_2 = \frac{1}{\dfrac{1}{\mathbf{Z}_1} + \dfrac{1}{\mathbf{Z}_2}} = \frac{\mathbf{Z}_1\mathbf{Z}_2}{\mathbf{Z}_1 + \mathbf{Z}_2}$$

FIGURE 10.5

Circuit with two impedances in parallel.

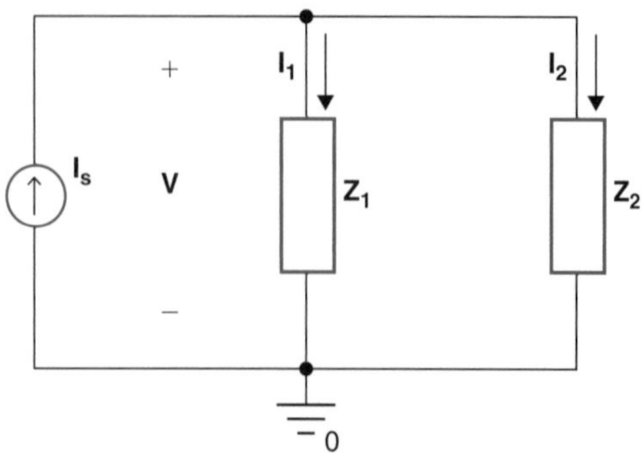

Thus, the voltage across the impedances is, according to Ohm's law,

$$\mathbf{V} = \mathbf{Z}\mathbf{I}_s = \frac{\mathbf{I}_s}{\dfrac{1}{\mathbf{Z}_1} + \dfrac{1}{\mathbf{Z}_2}} = \mathbf{I}_s \times \frac{\mathbf{Z}_1\mathbf{Z}_2}{\mathbf{Z}_1 + \mathbf{Z}_2}$$

The current through $\mathbf{Z}_1$ is

$$\mathbf{I}_1 = \frac{\mathbf{V}}{\mathbf{Z}_1} = \mathbf{I}_s \times \frac{\dfrac{1}{\dfrac{1}{\mathbf{Z}_1} + \dfrac{1}{\mathbf{Z}_2}}}{\mathbf{Z}_1} = \mathbf{I}_s \times \frac{\dfrac{1}{\mathbf{Z}_1}}{\dfrac{1}{\mathbf{Z}_1} + \dfrac{1}{\mathbf{Z}_2}} = \mathbf{I}_s \times \frac{\mathbf{Z}_2}{\mathbf{Z}_1 + \mathbf{Z}_2} \quad \text{(10.13)}$$

and the current through $\mathbf{Z}_2$ is

$$\mathbf{I}_2 = \frac{\mathbf{V}}{\mathbf{Z}_2} = \mathbf{I}_s \times \frac{\dfrac{1}{\dfrac{1}{\mathbf{Z}_1} + \dfrac{1}{\mathbf{Z}_2}}}{\mathbf{Z}_2} = \mathbf{I}_s \times \frac{\dfrac{1}{\mathbf{Z}_2}}{\dfrac{1}{\mathbf{Z}_1} + \dfrac{1}{\mathbf{Z}_2}} = \mathbf{I}_s \times \frac{\mathbf{Z}_1}{\mathbf{Z}_1 + \mathbf{Z}_2} \quad \text{(10.14)}$$

From Kirchhoff's current law, the sum of $\mathbf{I}_1$ and $\mathbf{I}_2$ is $\mathbf{I}_s$; that is,

$$\mathbf{I}_s = \mathbf{I}_1 + \mathbf{I}_2$$

This result suggests that the current through $\mathbf{Z}_1$ is given by $\mathbf{I}_s$ times the ratio of $\mathbf{Z}_2$ to $\mathbf{Z}_1 + \mathbf{Z}_2$, and the current through $\mathbf{Z}_2$ is given by $\mathbf{I}_s$ times the ratio of $\mathbf{Z}_1$ to $\mathbf{Z}_1 + \mathbf{Z}_2$. This is referred to as the **current divider rule**.

The current divider rule can be generalized to include more than two impedances. If a current source with current $\mathbf{I}_s$ is connected to a parallel connection of n impedances $\mathbf{Z}_1$, $\mathbf{Z}_2, \ldots, \mathbf{Z}_n$, then the voltage across the impedances is given by

$$\mathbf{V} = \mathbf{Z}_{\text{eq}}\mathbf{I}_s = \frac{1}{\dfrac{1}{\mathbf{Z}_1} + \dfrac{1}{\mathbf{Z}_2} + \cdots + \dfrac{1}{\mathbf{Z}_n}}\mathbf{I}_s$$

The current through impedance $\mathbf{Z}_i$, $i = 1, 2, \ldots, n$, is given by

$$\mathbf{I}_i = \frac{\mathbf{V}}{\mathbf{Z}_i} = \mathbf{I}_s \times \frac{\dfrac{1}{\mathbf{Z}_i}}{\dfrac{1}{\mathbf{Z}_1} + \dfrac{1}{\mathbf{Z}_2} + \cdots + \dfrac{1}{\mathbf{Z}_n}} \quad \text{(10.15)}$$

The current divider rule can be described by using admittance rather than impedance. Suppose that two impedances $\mathbf{Z}_1$ and $\mathbf{Z}_2$, respectively, are connected in parallel to a current source with a current of $\mathbf{I}_s$ amperes. The current through $\mathbf{Z}_1$ is given by

$$\mathbf{I}_1 = \mathbf{I}_s \times \frac{\dfrac{1}{\mathbf{Z}_1}}{\dfrac{1}{\mathbf{Z}_1} + \dfrac{1}{\mathbf{Z}_2}} = \mathbf{I}_s \times \frac{\mathbf{Y}_1}{\mathbf{Y}_1 + \mathbf{Y}_2} \quad \text{(10.16)}$$

Similarly, the current through $\mathbf{Z}_2$ is given by

$$\mathbf{I}_2 = \mathbf{I}_s \times \frac{\dfrac{1}{\mathbf{Z}_2}}{\dfrac{1}{\mathbf{Z}_1} + \dfrac{1}{\mathbf{Z}_2}} = \mathbf{I}_s \times \frac{\mathbf{Y}_2}{\mathbf{Y}_1 + \mathbf{Y}_2} \quad \text{(10.17)}$$

This result shows that when two admittances are connected in parallel to a current source with a current of $\mathbf{I}_s$ amperes, the total current $\mathbf{I}_s$ is divided between $\mathbf{I}_1$ and $\mathbf{I}_2$ in proportion to the admittance values $\mathbf{Y}_1$ and $\mathbf{Y}_2$.

If a current source with current $\mathbf{I}_s$ is connected to a parallel connection of n impedances $\mathbf{Z}_1, \mathbf{Z}_2, \ldots, \mathbf{Z}_n$, the current through each impedance $\mathbf{Z}_i$, $1 \le i \le n$, is given by

$$\mathbf{I}_i = \mathbf{I}_s \times \frac{\frac{1}{\mathbf{Z}_i}}{\frac{1}{\mathbf{Z}_1} + \frac{1}{\mathbf{Z}_2} + \cdots + \frac{1}{\mathbf{Z}_n}} = \mathbf{I}_s \times \frac{\mathbf{Y}_i}{\mathbf{Y}_1 + \mathbf{Y}_2 + \cdots + \mathbf{Y}_n} \quad \textbf{(10.18)}$$

where $\mathbf{Y}_i = 1/\mathbf{Z}_i$.

EXAMPLE 10.2

Use the current divider rule to find the currents $\mathbf{I}_1$, $\mathbf{I}_2$, $\mathbf{I}_3$, and $\mathbf{I}_4$ in the circuit shown in Figure 10.6.

FIGURE 10.6

Circuit for EXAMPLE 10.2.

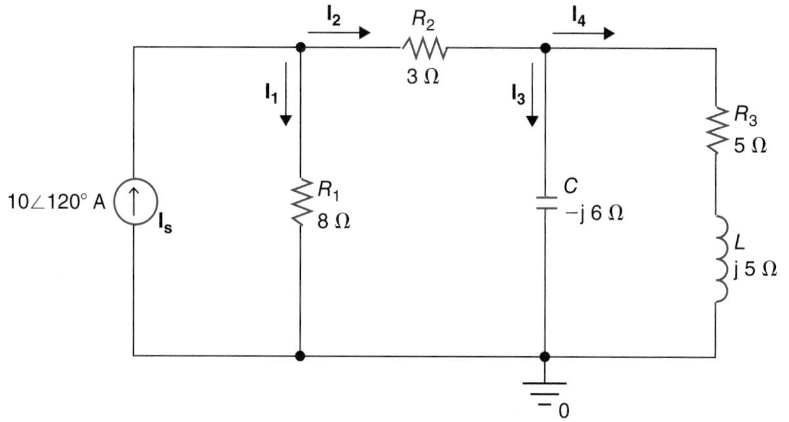

Let $\mathbf{Z}_1$ be the equivalent impedance of the series connection of R_3 and $\mathbf{Z}_L$. Then we have

$$\mathbf{Z}_1 = 5 + j5\ \Omega$$

Let $\mathbf{Z}_2$ be the equivalent impedance of the parallel connection of $\mathbf{Z}_1$ and $\mathbf{Z}_C$. Then we have

$$\mathbf{Z}_2 = (-j6)\|(5 + j5) = \frac{(-j6) \times (5 + j5)}{(-j6) + (5 + j5)} = 6.9231 - j4.6154\ \Omega$$

Let $\mathbf{Z}_3$ be the equivalent impedance of the series connection of R_2 and $\mathbf{Z}_2$. Then we have

$$\mathbf{Z}_3 = 3 + 6.9231 - j4.6154 = 9.9231 - j4.6154\ \Omega$$

According to the current divider rule, the current through R_1 is given by

$$\mathbf{I}_1 = \mathbf{I}_s \times \frac{\mathbf{Z}_3}{\mathbf{R}_1 + \mathbf{Z}_3} = -1.9735 + j5.5741 = 5.9131\angle 109.4966°\ \text{A}$$

continued

Example 10.2 continued

Similarly, the current through $\mathbf{Z}_3$ is given by

$$\mathbf{I}_2 = \mathbf{I}_s \times \frac{R_1}{R_1 + \mathbf{Z}_3} = -3.0265 + j3.0862 = 4.3225\angle 134.4405^\circ \text{ A}$$

The currents $\mathbf{I}_3$ and $\mathbf{I}_4$ can also be found by applying the current divider rule.

$$\mathbf{I}_3 = \mathbf{I}_2 \times \frac{\mathbf{Z}_1}{\mathbf{Z}_C + \mathbf{Z}_1} = -5.8890 - j1.1181 = 5.9942\angle -169.2495^\circ \text{ A}$$

$$\mathbf{I}_4 = \mathbf{I}_2 \times \frac{\mathbf{Z}_C}{\mathbf{Z}_C + \mathbf{Z}_1} = 2.8625 + j4.2043 = 5.0863\angle 55.7505^\circ \text{ A}$$

MATLAB

```
% EXAMPLE 10.2
clear all;format long;
Im=10;phi=120;R1=8;R2=3;R3=5;ZL=5j;ZC=-6j;
Is=P2Rd(Im,phi)
Z1=R3+ZL
Z2=P([Z1,ZC])
Z3=R2+Z2
I1=Is*Z3/(R1+Z3)
I1p=R2P(I1)
I2=Is*R1/(R1+Z3)
I2p=R2P(I2)
I3=I2*Z1/(Z1+ZC)
I3p=R2P(I3)
I4=I2*ZC/(Z1+ZC)
I4p=R2P(I4)
```

Exercise 10.2

Use the current divider rule to find the currents $\mathbf{I}_1$, $\mathbf{I}_2$, and $\mathbf{I}_3$ in the circuit shown in Figure 10.7.

FIGURE 10.7

Circuit for EXERCISE 10.2.

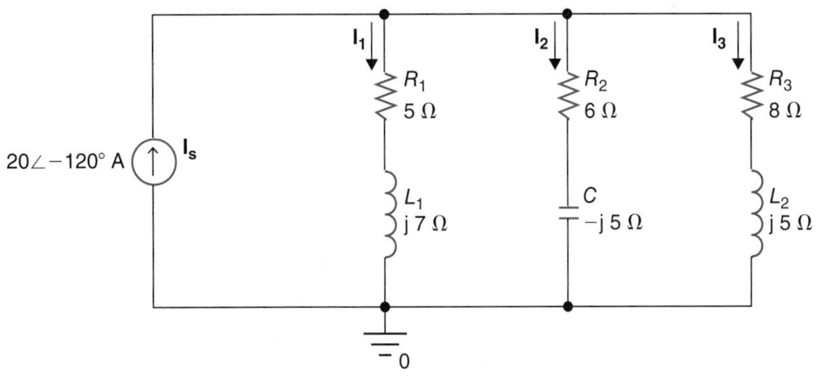

Answer:

$\mathbf{I}_1 = -8.2157 - j3.0868 = 8.7765\angle -159.4077^\circ$ A,
$\mathbf{I}_2 = 4.0639 - j8.7708 = 9.6665\angle -65.1398^\circ$ A,
$\mathbf{I}_3 = -5.8482 - j5.4629 = 8.0028\angle -136.9508^\circ$ A.

10.5 Nodal Analysis

The nodal analysis provides us voltages at all nodes of a given circuit. If a voltage source is connected between a node and the ground, the node voltage is already known and we do not need to find the node voltage on this node. The nodal analysis applies to phasor-transformed circuit.

Excluding the nodes whose voltages are known from the voltage sources, we assign variables such as $\mathbf{V}_1, \mathbf{V}_2, \ldots, \mathbf{V}_n$ to each unknown node voltage. For each node with an assigned variable, write a node equation by applying KCL. Any of the three interpretations of KCL given in Chapters 2 and 9 can be used. For example, the sum of the currents leaving a node must be zero. If there are n unknown node voltages, we get n equations in n unknowns. Thus, we can solve this n system of linear equations with constant coefficients to find the unique solution for the unknown node voltages $\mathbf{V}_1, \mathbf{V}_2, \ldots, \mathbf{V}_n$. An algorithm to find the solution of the n system of linear equations with constant coefficients is Cramer's rule. MATLAB is useful in finding the solution of an n system of linear equations with constant coefficients.

The presence of dependent sources in the circuit requires extra equations describing the controlling voltages or currents, and controlled voltages or currents in terms of node voltages.

EXAMPLE 10.3

Use nodal analysis to find the voltages $\mathbf{V}_1$ and $\mathbf{V}_2$ in the circuit shown in Figure 10.8. Also, find the time domain representation of the voltages assuming that $f = 1$ Hz.

FIGURE 10.8

Circuit for EXAMPLE 10.3.

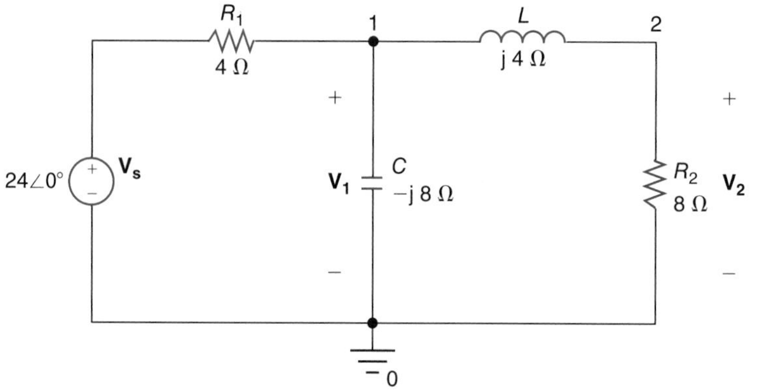

Summing the currents leaving node 1, we obtain

$$\frac{\mathbf{V}_1 - 24\angle 0^\circ}{4} + \frac{\mathbf{V}_1}{-j8} + \frac{\mathbf{V}_1 - \mathbf{V}_2}{j4} = 0 \tag{10.19}$$

Multiplication by 8 yields

$$2\mathbf{V}_1 - 48 + j\mathbf{V}_1 - j2\mathbf{V}_1 + j2\mathbf{V}_2 = 0$$

which reduces to

$$(2 - j)\mathbf{V}_1 + j2\mathbf{V}_2 = 48 \tag{10.20}$$

continued

Example 10.3 continued

Summing the currents leaving node 2, we obtain

$$\frac{\mathbf{V}_2 - \mathbf{V}_1}{j4} + \frac{\mathbf{V}_2}{8} = 0 \tag{10.21}$$

Multiplication by $j8$ results in

$$2\mathbf{V}_2 - 2\mathbf{V}_1 + j\mathbf{V}_2 = 0$$

which reduces to

$$-2\mathbf{V}_1 + (2 + j)\mathbf{V}_2 = 0 \tag{10.22}$$

Solving Equation (10.22) for $\mathbf{V}_1$, we obtain

$$\mathbf{V}_1 = (1 + j0.5)\mathbf{V}_2$$

Substituting $\mathbf{V}_1$ into Equation (10.20), we get

$$(2 - j)(1 + j0.5)\mathbf{V}_2 + j2\mathbf{V}_2 = 48$$

Solving for $\mathbf{V}_2$, we obtain

$$\mathbf{V}_2 = \frac{48}{(2 - j)(1 + j0.5) + j2} = \frac{48}{2 + j1 - j1 + 0.5 + j2} = \frac{48}{2.5 + j2}$$
$$= 11.7073 - j9.3659 = 14.9927\angle -38.6598^\circ \text{ V}$$

Substituting $\mathbf{V}_2$ into $\mathbf{V}_1 = (1 + j0.5)\mathbf{V}_2$, we get

$$\mathbf{V}_1 = 16.3902 - j3.5122 = 16.7623\angle -12.0948^\circ \text{ V}$$

Alternatively, applying Cramer's rule on Equations (10.20) and (10.22), we obtain

$$\mathbf{V}_1 = \frac{\begin{vmatrix} 48 & j2 \\ 0 & 2 + j \end{vmatrix}}{\begin{vmatrix} 2 - j & j2 \\ -2 & 2 + j \end{vmatrix}} = \frac{96 + j48}{(2 - j)(2 + j) + j4} = \frac{96 + j48}{5 + j4}$$
$$= 16.3902 - j3.5122 = 16.7623\angle -12.0948^\circ$$

$$\mathbf{V}_2 = \frac{\begin{vmatrix} 2 - j & 48 \\ -2 & 0 \end{vmatrix}}{\begin{vmatrix} 2 - j & j2 \\ -2 & 2 + j \end{vmatrix}} = \frac{96}{(2 - j)(2 + j) + j4} = \frac{96}{5 + j4}$$
$$= 11.7073 - j9.3659 = 14.9927\angle -38.6598^\circ$$

For $f = 1$ Hz, $v_1(t)$ and $v_2(t)$ are given, respectively, as

$$v_1(t) = 16.7623 \cos(2\pi t - 12.0948^\circ) \text{ V}$$
$$v_2(t) = 14.9927 \cos(2\pi t - 38.6598^\circ) \text{ V}$$

continued

Example 10.3 continued

MATLAB

```
% EXAMPLE 10.3
clear all;format long;
Vm=24;phi=0;R1=4;R2=8;ZC=-8j;ZL=4j;
Vs=P2Rd(Vm,phi);
syms V1 V2
[V1,V2]=solve((V1-Vs)/R1+V1/ZC+(V1-V2)/ZL,...
    (V2-V1)/ZL+V2/R2,V1,V2);
V1p=R2P(V1);
V2p=R2P(V2);
V1=vpa(V1,7)
V2=vpa(V2,7)
V1p=vpa(V1p,7)
V2p=vpa(V2p,7)
```

Exercise 10.3

Find the voltage $v_o(t)$ in the circuit shown in Figure 10.9 in the steady state.

FIGURE 10.9

Circuit for EXERCISE 10.3.

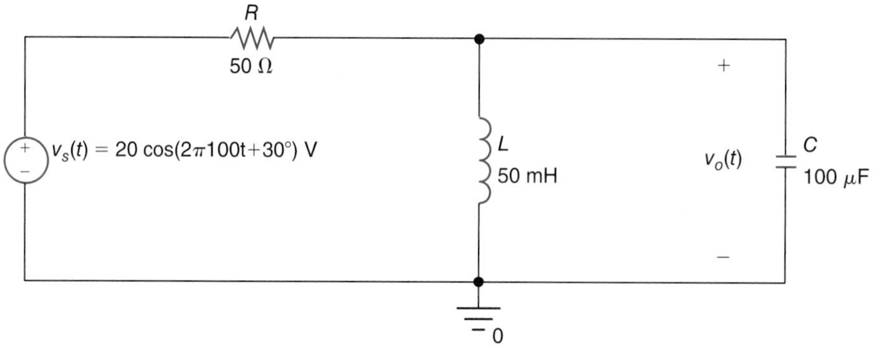

Answer:
$v_o(t) = 10.8423\cos(2\pi 100t - 27.1722°)$ V.

10.6 Mesh Analysis

The mesh analysis provides mesh currents at all meshes of the given circuit. From the mesh currents, we can find currents on every branch of the circuit. The mesh analysis applies to phasor-transformed circuits.

The mesh analysis is based on KVL. We assign mesh current variables such as $\mathbf{I}_1$, $\mathbf{I}_2$, $\mathbf{I}_3, \ldots, \mathbf{I}_n$ on the meshes whose current is unknown. Then, for each mesh with an unknown mesh current, we apply KVL. Specifically, we sum the voltage drops around the mesh and let that equal zero. If a mesh contains a current source that is not shared by other meshes, the mesh current is the same as the current from the current source if they point the same direction. If the direction is opposite, the mesh current is the negative of the current from the current source.

If there are n unknown mesh currents, we get n equations in n unknowns. Thus, we can solve this n system of linear equations with constant coefficients to find the unique solution for the unknown mesh currents $\mathbf{I}_1, \mathbf{I}_2, \mathbf{I}_3, \ldots, \mathbf{I}_n$. A simple algorithm to find the solution of the n system of linear equations with constant coefficients is Cramer's rule. MATLAB is useful in finding the solution of an n system of linear equations with constant coefficients.

The presence of dependent sources in the circuit requires extra equations describing the controlling voltages or currents, and controlled voltages or currents in terms of mesh currents.

EXAMPLE 10.4

In the circuit shown in Figure 10.10, find the mesh currents $\mathbf{I}_1$ and $\mathbf{I}_2$. Also find the voltages across L and R_2 and the time domain representation of the voltages assuming $f = 1$ Hz.

FIGURE 10.10

Circuit for EXAMPLE 10.4.

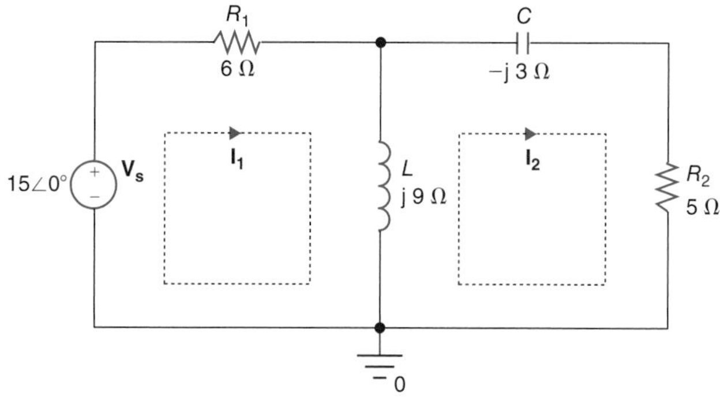

Collecting the voltage drops around mesh 1 in the clockwise direction, we obtain

$$-15 + 6\mathbf{I}_1 + j9\,(\mathbf{I}_1 - \mathbf{I}_2) = 0 \qquad \textbf{(10.23)}$$

which can be reduced to

$$(6 + j9)\mathbf{I}_1 - j9\mathbf{I}_2 = 15 \qquad \textbf{(10.24)}$$

Collecting the voltage drops around mesh 2 in the clockwise direction, we obtain

$$j9(\mathbf{I}_2 - \mathbf{I}_1) - j3\mathbf{I}_2 + 5\mathbf{I}_2 = 0 \qquad \textbf{(10.25)}$$

which can be reduced to

$$-j9\mathbf{I}_1 + (5 + j6)\mathbf{I}_2 = 0 \qquad \textbf{(10.26)}$$

Solving Equation (10.26) for $\mathbf{I}_1$, we obtain

$$\mathbf{I}_1 = \frac{5 + j6}{j9}\mathbf{I}_2 = \frac{6 - j5}{9}\mathbf{I}_2$$

Substituting $\mathbf{I}_1$ into Equation (10.24), we get

$$\frac{(6 - j5)(6 + j9)}{9}\mathbf{I}_2 - j9\mathbf{I}_2 = 15$$

Solving for $\mathbf{I}_2$, we obtain

$$\mathbf{I}_2 = \frac{15}{\dfrac{(6 - j5)(6 + j9)}{9} - j9} = \frac{135}{36 + j54 - j30 + 45 - j81} = \frac{135}{81 - j57}$$
$$= 1.1147 + j0.7844 = 1.3630\angle 35.1342^\circ \text{ A}$$

Substituting $\mathbf{I}_2$ into

$$\mathbf{I}_1 = \frac{5 + j6}{j9}\mathbf{I}_2 = \frac{6 - j5}{9}\mathbf{I}_2,$$

we get

$$\mathbf{I}_1 = 1.1789 - j0.09633 = 1.1828\angle -4.6714^\circ \text{ A}$$

continued

Example 10.4 continued

Alternatively, applying Cramer's rule to Equations (10.24) and (10.26), we obtain

$$\mathbf{I}_1 = \frac{\begin{vmatrix} 15 & -j9 \\ 0 & 5+j6 \end{vmatrix}}{\begin{vmatrix} 6+j9 & -j9 \\ -j9 & 5+j6 \end{vmatrix}} = \frac{75+j90}{(6+j9)(5+j6)+81} = \frac{75+j90}{57+j81}$$
$$= 1.1789 - j0.09633 = 1.1828\angle -4.6714° \text{ A}$$

$$\mathbf{I}_2 = \frac{\begin{vmatrix} 6+j9 & 15 \\ -j9 & 0 \end{vmatrix}}{\begin{vmatrix} 6+j9 & -j9 \\ -j9 & 5+j9 \end{vmatrix}} = \frac{j135}{(6+j9)(5+j6)+81} = \frac{j135}{57+j81}$$
$$= 1.1147 + j0.7844 = 1.3630\angle 35.1342° \text{ A}$$

The voltage across L is given by

$$\mathbf{V}_L = j9(\mathbf{I}_1 - \mathbf{I}_2) = 7.9266 + j0.5780 = 7.9476\angle 4.1704° \text{ V}$$

The voltage across R_2 is given by

$$\mathbf{V}_{R_2} = 5\,\mathbf{I}_2 = 5.5734 + j3.9220 = 6.8151\angle 35.1342° \text{ V}$$

For $f = 1$ Hz, $i_1(t)$, $i_2(t)$, $v_L(t)$, and $v_{R_2}(t)$ are given, respectively, as

$$i_1(t) = 1.1828\cos(2\pi t - 4.6714°) \text{ A}$$
$$i_2(t) = 1.3630\cos(2\pi t + 35.1342°) \text{ A}$$
$$v_L(t) = 7.9476\cos(2\pi t + 4.1704°) \text{ V}$$
$$v_{R_2}(t) = 6.8151\cos(2\pi t + 35.1342°) \text{ V}$$

MATLAB

```
% EXAMPLE 10.4
Vm=15;phi=0;R1=6;R2=5;ZC=-3j;ZL=9j;
Vs=P2Rd(Vm,phi);
syms I1 I2
[I1,I2]=solve(-Vs+R1*I1+ZL*(I1-I2),...
    ZL*(I2-I1)+ZC*I2+R2*I2,I1,I2);
VL=ZL*(I1-I2);
VR2=R2*I2;
I1p=R2P(I1);
I2p=R2P(I2);
VLp=R2P(VL);
VR2p=R2P(VR2);
I1=vpa(I1,7)
I2=vpa(I2,7)
VL=vpa(VL,7)
VR2=vpa(VR2,8)
I1p=vpa(I1p,8)
I2p=vpa(I2p,8)
VLp=vpa(VLp,8)
VR2p=vpa(VR2p,8)
```

Exercise 10.4

Find the mesh currents $\mathbf{I}_1$, $\mathbf{I}_2$, and $\mathbf{I}_3$ in the circuit shown in Figure 10.11.

FIGURE 10.11

Circuit for EXERCISE 10.4.

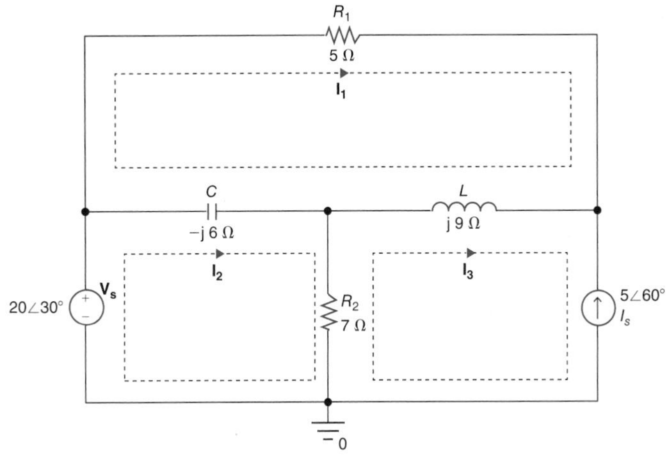

Answer:
$\mathbf{I}_1 = 0.6161 - j4.3225 = 4.3662\angle{-81.8878°}$ A, $\mathbf{I}_2 = -0.4560 - j3.8205 = 3.8476\angle{-96.8058°}$ A, $\mathbf{I}_3 = -2.5 - j4.3301 = 5\angle{-120°}$ A.

10.7 Superposition Principle

Suppose that a circuit has N independent sources with $N \geq 2$. Create N circuits from the original circuit with only one independent source by deactivating the other N–1 independent sources. Deactivating a current source is to open-circuit it, and deactivating a voltage source is to short-circuit it. The unknown voltages and currents of the original circuit can be found by adding the voltages and currents from the N circuits with one independent source. This is called the **superposition principle**.

EXAMPLE 10.5

Use the superposition principle to find V in the circuit shown in Figure 10.12.

FIGURE 10.12

Circuit for EXAMPLE 10.5.

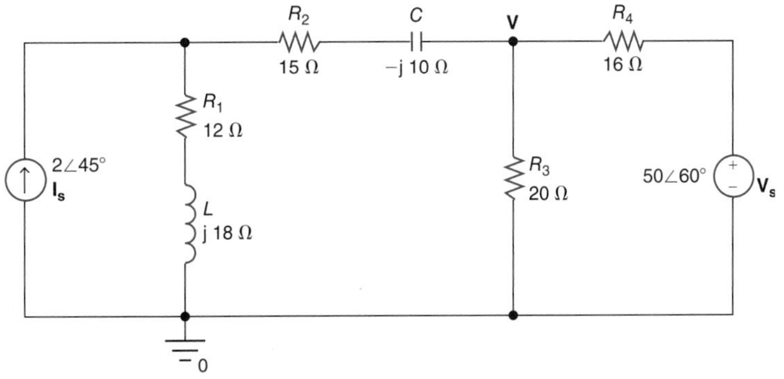

continued

Example 10.5 continued

First, we deactivate the voltage source by short-circuiting it, as shown in Figure 10.13.

FIGURE 10.13

Circuit with voltage source deactivated.

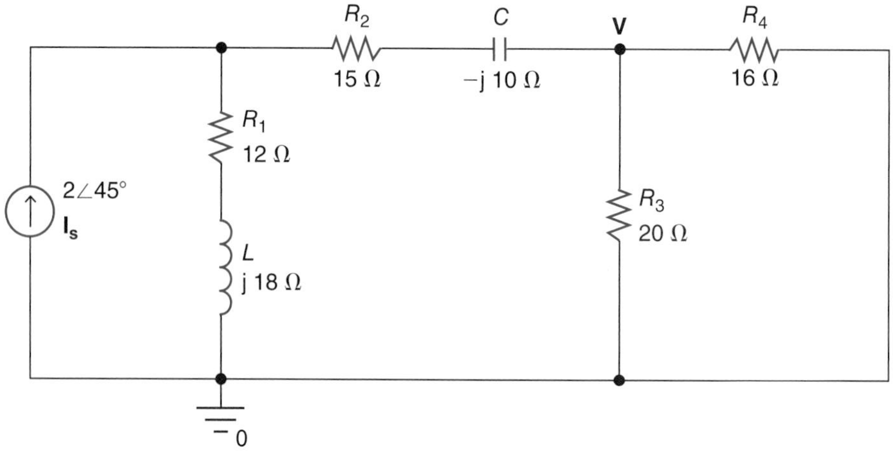

The equivalent impedance of R_3 and R_4 is given by

$$R_{eq} = \frac{R_3 \times R_4}{R_3 + R_4} = \frac{20 \times 16}{20 + 16}\ \Omega = 8.8889\ \Omega$$

Let $\mathbf{Z_a}$ be the equivalent impedance of the series connection of R_1 and $\mathbf{Z_L}$. Then, we have

$$\mathbf{Z_a} = 12 + j18\ \Omega$$

Let $\mathbf{Z_b}$ be the equivalent impedance of the series connection of R_2, $\mathbf{Z_C}$, and R_{eq}. Then, we have

$$\mathbf{Z_b} = 15 - j10 + 8.8889 = 23.8889 - j10\ \Omega$$

The current through the series connection of R_2, $\mathbf{Z_C}$, and R_{eq} can be obtained by applying the current divider rule:

$$\mathbf{I}_b = \mathbf{I}_s \times \frac{\mathbf{Z_a}}{\mathbf{Z_a} + \mathbf{Z_b}} = (1.4142 + j1.4142) \times \frac{12 + j18}{35.8889 + j8}$$

$$= 0.0258 + j1.1764 = 1.1767\angle 88.7436°\ \text{A}$$

The voltage across R_{eq} is given by

$$\mathbf{V_A} = \mathbf{I_b} \times R_{eq} = 0.2293 + j10.457$$

$$= 10.4595\angle 88.7436°\ \text{V}$$

FIGURE 10.14

Circuit with current source deactivated.

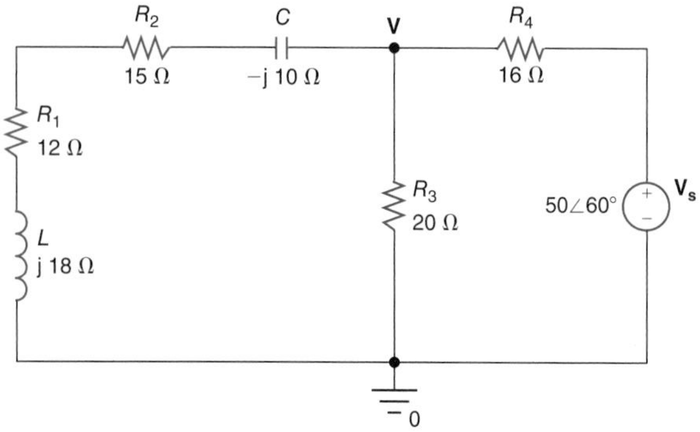

Next, we deactivate the current source by open-circuiting it, as shown in Figure 10.14.

Let $\mathbf{Z_c}$ be the equivalent impedance of the series connection of $\mathbf{Z_C}$, R_2, R_1, and $\mathbf{Z_L}$. Then, $\mathbf{Z_c}$ is given by

$$\mathbf{Z_c} = -j10 + 15 + 12 + j18 = 27 + j8\ \Omega$$

continued

Example 10.5 continued

Let $\mathbf{Z_d}$ be the equivalent impedance of the parallel connection of $\mathbf{Z_c}$ and R_3. Then, we have

$$\mathbf{Z_d} = \frac{\mathbf{Z_c} \times R_3}{\mathbf{Z_c} + R_3} = 11.729 + j1.4078\ \Omega$$

The voltage across $\mathbf{Z_d}$ is obtained by applying the voltage divider rule. Thus, we have

$$\mathbf{V_B} = \mathbf{V_s}\frac{\mathbf{Z_d}}{\mathbf{Z_d} + R_4} = (25 + j43.3013) \times \frac{11.729 + j1.4078}{11.729 + j1.4078 + 16}$$
$$= 9.3465 + j19.1106\ \text{V}$$

According to the superposition principle, the voltage $\mathbf{V}$ is the sum of $\mathbf{V_A}$ and $\mathbf{V_B}$; that is,

$$\mathbf{V} = \mathbf{V_A} + \mathbf{V_B} = 9.5758 + j29.5676 = 31.0795\angle 72.0548°\ \text{V}$$

This answer can be verified by finding $\mathbf{V}$ directly from the original circuit shown in Figure 10.12 by applying nodal analysis.

Exercise 10.5

Use the superposition principle to find V in the circuit shown in Figure 10.15.

FIGURE 10.15

Circuit for EXERCISE 10.5.

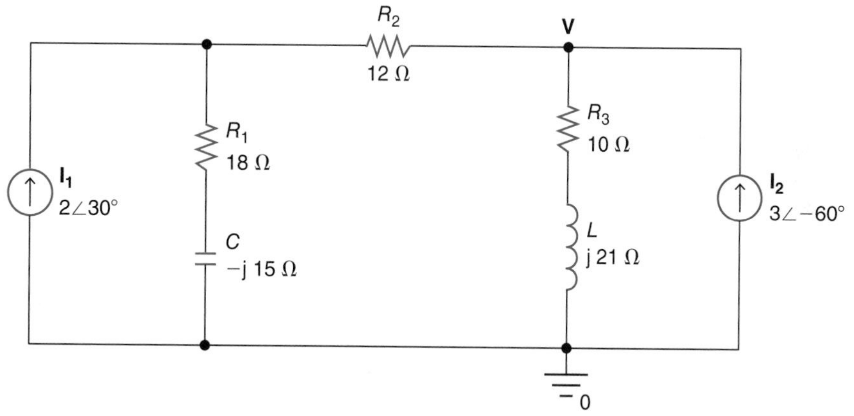

Answer:
$\mathbf{V}$ from $\mathbf{I}_1 = \mathbf{V}_A = 18.6516 + j19.4499 = 26.9478\angle 46.2°$ V,
$\mathbf{V}$ from $\mathbf{I}_2 = \mathbf{V}_B = 49.8265 - j29.4194 = 57.8635\angle -30.5592°$ V,
$\mathbf{V} = \mathbf{V}_A + \mathbf{V}_B = 68.4781 - j9.9695 = 69.2\angle -8.2833°$ V.

10.8 Source Transformation

A circuit consisting of a voltage source with voltage $\mathbf{V_s}$ and a series impedance $\mathbf{Z_s}$ [as shown in Figure 10.16(a)] is equivalent to a circuit consisting of a current source with a current $\mathbf{V_s}/\mathbf{Z_s}$ and a parallel impedance $\mathbf{Z_s}$ [as shown in Figure 10.16(b)]. Equivalence means the circuit shown in Figure 10.16(a) and the circuit shown in Figure 10.16(b) have the same open-circuit voltage across a and b, the same short-circuit current through a

and b, and the same impedance looking into the circuit from a and b after deactivating the source.

FIGURE 10.16

A voltage source and a series impedance is equivalent to a current source and a parallel impedance.

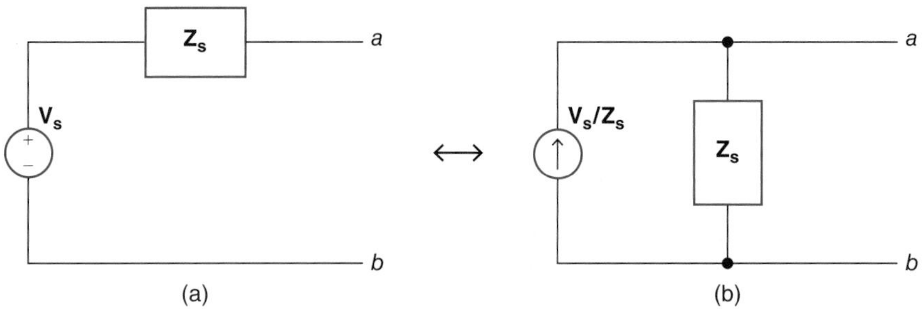

Also, the circuit shown in Figure 10.17(a) is equivalent to the circuit shown in Figure 10.17(b). The circuit shown in Figure 10.17(a) can be replaced by the circuit shown in Figure 10.17(b).

FIGURE 10.17

A current source and a parallel impedance is equivalent to a voltage source and a series impedance.

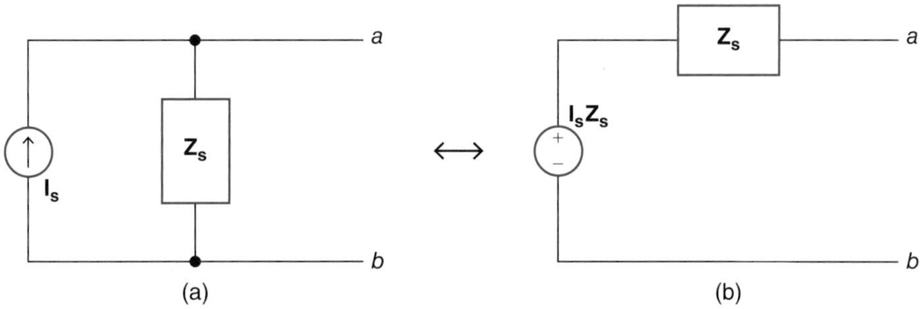

The source transformations apply to dependent sources also. Figures 10.18 and 10.19 show the equivalence of a voltage source and a series impedance, and a current source and a parallel impedance.

FIGURE 10.18

A dependent voltage source and a series impedance is equivalent to a dependent current source and a parallel impedance.

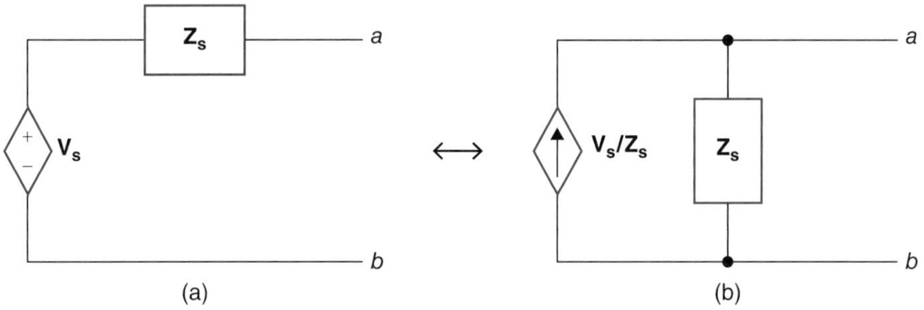

FIGURE 10.19

A dependent current source and a parallel impedance is equivalent to a dependent voltage source and a series impedance.

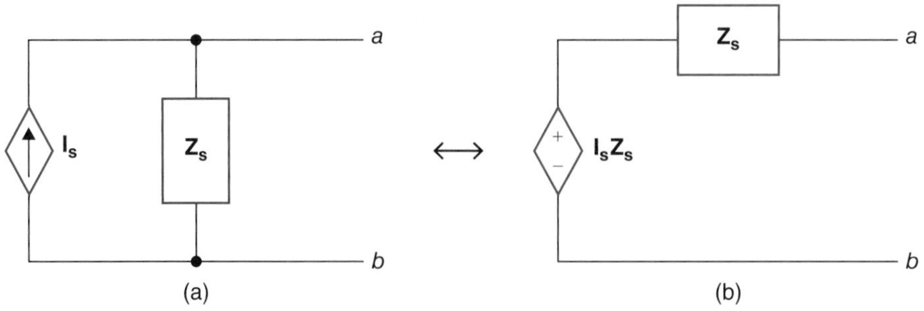

EXAMPLE 10.6

Use source transformation to find the voltage $\mathbf{V_o}$ across R_2 in the circuit shown in Figure 10.20.

FIGURE 10.20

Circuit for EXAMPLE 10.6.

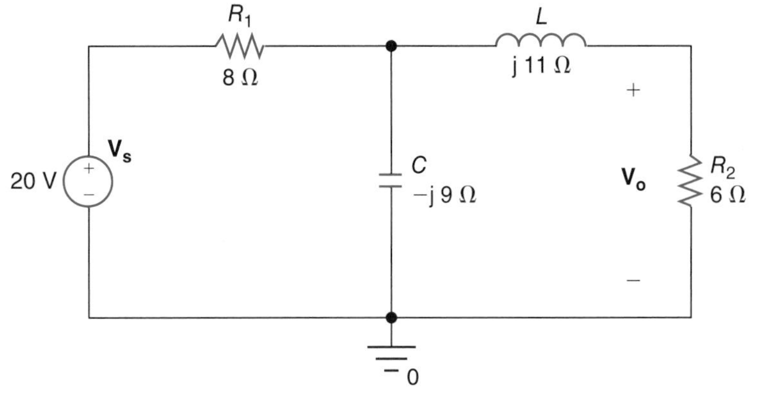

The voltage source $\mathbf{V}_s$ and the series resistor R_1 can be transformed into a current source with current 20 $\mathbf{V}$/8 Ω = 2.5 A and a parallel resistor with resistance 8 Ω, as shown in Figure 10.21.

FIGURE 10.21

Circuit with a current source and a parallel resistor.

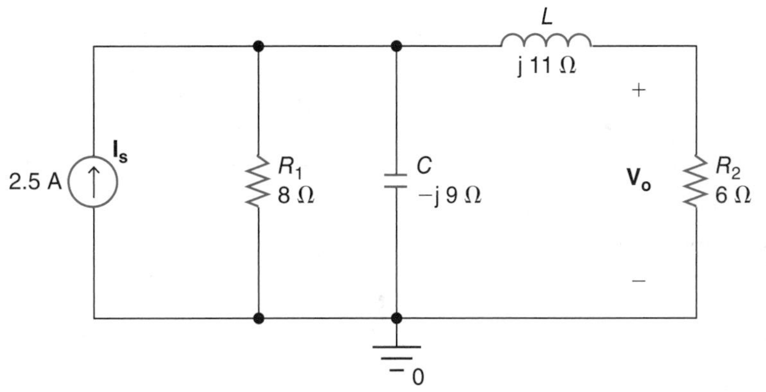

Let $\mathbf{Z_1}$ be the impedance of the parallel combination of R_1 and $\mathbf{Z_C}$. Then we have

$$\mathbf{Z_1} = \frac{-j72}{8 - j9} = 4.469 - j3.972\ \Omega$$

The current source $\mathbf{I_s}$ and $\mathbf{Z_1}$ can be transformed into a voltage source with voltage

$$\mathbf{V}_{\mathbf{s}_2} = \mathbf{I_s} \times \mathbf{Z_1} = 11.1724 - j9.9310 = 14.9482\angle{-41.6335°}\ \text{V}$$

and a series impedance $\mathbf{Z_1}$, as shown in Figure 10.22.

continued

Example 10.6 continued

FIGURE 10.22

After transforming to a voltage source and a series impedance.

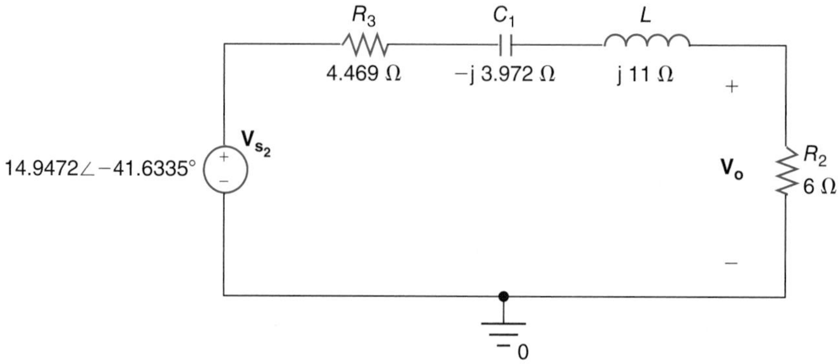

The voltage across R_2 can be obtained by applying the voltage divider rule:

$$\mathbf{V}_o = (11.1724 - j9.9310) \times \frac{6}{4.469 - j3.972 + j11 + 6} = 1.7802 - j6.8867$$

$$= 7.1131\angle{-75.5061°}\text{ V}$$

Exercise 10.6

Use source transformation to find $\mathbf{V}_o$ in the circuit shown in Figure 10.23.

FIGURE 10.23

Circuit for EXERCISE 10.6.

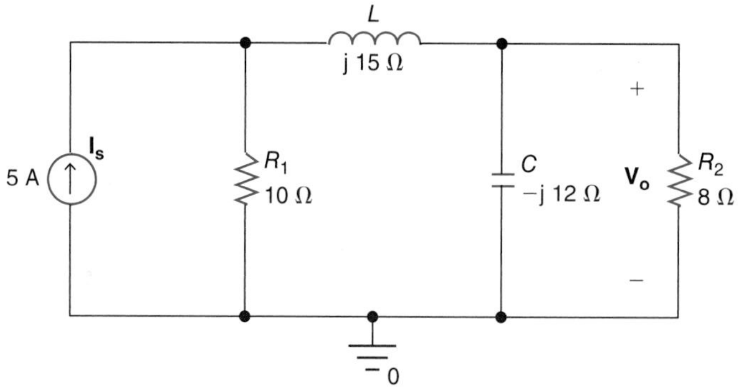

Answer:
$\mathbf{V}_o = 5.9988 - j16.2466 = 17.3187\angle{-69.7343°}$ V.

10.9 Thévenin Equivalent Circuit

A circuit consisting of a voltage source $\mathbf{V}_{th}$ and a series impedance $\mathbf{Z}_{th}$ representing the original circuit looking from a pair of terminals is called a **Thévenin equivalent circuit**. The voltage $\mathbf{V}_{th}$ is called the **Thévenin equivalent voltage** and the impedance $\mathbf{Z}_{th}$ is called the **Thévenin equivalent impedance** [as shown in Figure 10.24].

The Thévenin equivalent circuit can be used to simplify a circuit. When a load impedance is connected between terminals *a* and *b*, we can find the effects of the circuit on the load from the Thévenin equivalent circuit. We do not need all the details of the original circuit to find the voltage, current, and power on the load.

FIGURE 10.24

Thévenin equivalent circuit.

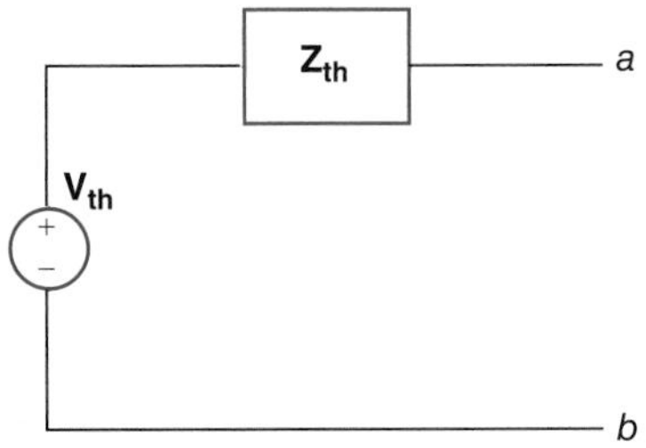

FIGURE 10.25

Short-circuit current.

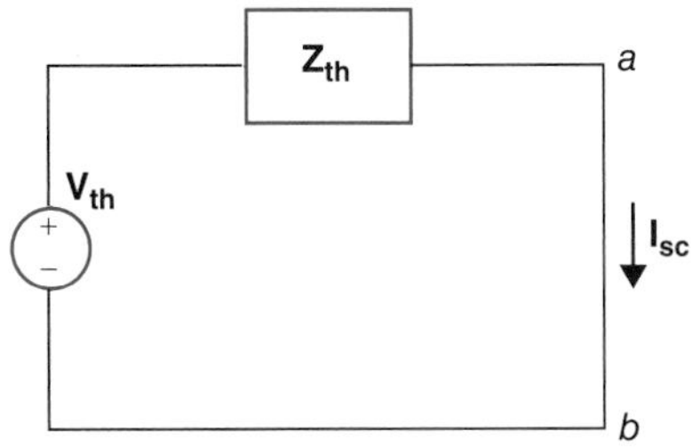

Let the voltage across terminals *a* and *b* of the Thévenin equivalent circuit be $\mathbf{V}_{oc}$. This voltage is called the open-circuit voltage because terminals *a* and *b* are open (there is infinite resistance between *a* and *b*). No current flows on the Thévenin equivalent impedance $\mathbf{Z}_{th}$. Thus,

$$\mathbf{V}_{oc} = \mathbf{V}_{th}$$

If the terminals *a* and *b* are short-circuited as shown in Figure 10.25, the current through the short-circuit is given by

$$\mathbf{I}_{sc} = \frac{\mathbf{V}_{th}}{\mathbf{Z}_{th}} = \frac{\mathbf{V}_{oc}}{\mathbf{Z}_{th}}$$

If we solve this equation for $\mathbf{Z}_{th}$, we have

$$\mathbf{Z}_{th} = \frac{\mathbf{V}_{oc}}{\mathbf{I}_{sc}}$$

This equation can be used in finding the Thévenin equivalent impedance $\mathbf{Z}_{th}$ from the original circuit.

10.9.1 FINDING THE THÉVENIN EQUIVALENT VOLTAGE V_{th}

Given a circuit and terminals *a* and *b*, we can find the Thévenin equivalent voltage $\mathbf{V}_{th}$ with respect to terminals *a* and *b* by finding the open-circuit voltage $\mathbf{V}_{oc}$ across terminals *a* and *b*. The open-circuit voltage $\mathbf{V}_{oc}$ can be found by utilizing circuit analysis methods such as the voltage divider rule, the current divider rule, the superposition principle, nodal analysis, and mesh analysis.

10.9.2 FINDING THE THÉVENIN EQUIVALENT IMPEDANCE Z_{th}

Given a circuit and terminals *a* and *b*, we can find the Thévenin equivalent impedance $\mathbf{Z}_{th}$ with respect to terminals *a* and *b* by using one of the three methods listed below.

Method 1

Deactivate all the independent sources by short-circuiting voltage sources and open-circuiting current sources. Find the equivalent impedance looking into the circuit from terminals *a* and *b*. This equivalent impedance is the Thévenin equivalent impedance $\mathbf{Z}_{th}$.

Method 2

Connect terminals *a* and *b* by wire (short-circuit). Find the short-circuit current $\mathbf{I}_{sc}$ by utilizing circuit analysis methods such as nodal analysis and mesh analysis. The Thévenin equivalent impedance $\mathbf{Z}_{th}$ is given by

$$\mathbf{Z}_{th} = \frac{\mathbf{V}_{oc}}{\mathbf{I}_{sc}}$$

Method 3

Deactivate all the independent sources by open-circuiting current sources and short-circuiting voltage sources. Apply a test voltage of 1 *V* (or any other value) between terminals *a* and *b* with terminal *a* connected to the positive terminal of the test voltage. Measure the current out of the positive terminal of the test voltage source. The Thévenin equivalent impedance $\mathbf{Z}_{th}$ is given by the ratio of the test voltage to the test current. Instead of a test voltage, a test current (1A or 1mA or any other value) can be applied between terminals *a* and *b*. Measure the voltage across the terminals *a* and *b*. The ratio of the voltage to the test current is the Thévenin equivalent impedance $\mathbf{Z}_{th}$.

EXAMPLE 10.7

Find the Thévenin equivalent voltage $\mathbf{V}_{th}$ and Thévenin equivalent impedance $\mathbf{Z}_{th}$ between *a* and *b* for the circuit shown in Figure 10.26. If the frequency of the source is 10 Hz, what are the component values of the Thévenin impedance?

FIGURE 10.26

Circuit for EXAMPLE 10.7.

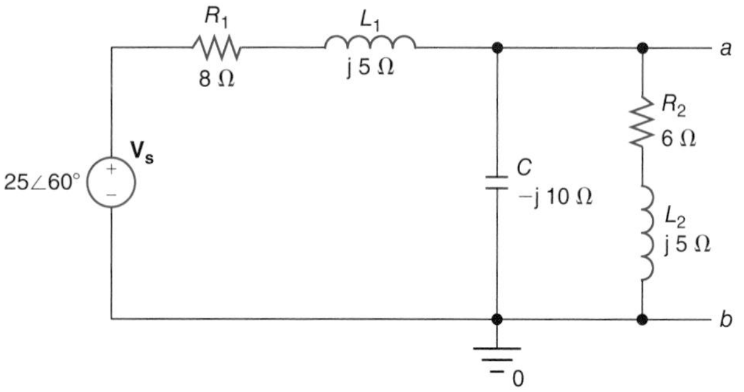

The equivalent impedance $\mathbf{Z_P}$ of the parallel connection of $\mathbf{Z_C}$ and $\mathbf{Z_{R_2}} + \mathbf{Z_{L_2}}$ is given by

$$\mathbf{Z_p} = \mathbf{Z_c} \| (\mathbf{Z_{R_2}} + \mathbf{Z_{L_2}}) = \frac{(-j10)(6 + j5)}{-j10 + 6 + j5}$$

$$= \frac{50 - j60}{6 - j5} = 9.8361 - j1.8033$$

$$= 10\angle -10.3889° \ \Omega$$

The total impedance seen from the voltage source is

$$\mathbf{Z} = 8 + j5 + 9.8361 - j1.8033 = 17.8361$$

$$+ j3.1967 = 18.1203\angle 10.1611° \ \Omega$$

The Thévenin voltage across *a* and *b* can be found by applying the voltage divider rule.

$$\mathbf{V_{th}} = \mathbf{V_s} \times \frac{\mathbf{Z_P}}{\mathbf{Z}} = 25\angle 60° \times \frac{10\angle -10.3889°}{18.1203\angle 10.1611°} = 13.7967\angle 39.4500° \ \mathrm{V}$$

The Thévenin impedance is given by the equivalent impedance of parallel connection of $\mathbf{Z_P}$ and $\mathbf{Z_{R_1}} + \mathbf{Z_{L_1}}$ after short-circuiting the voltage source. Thus,

$$\mathbf{Z_{th}} = (\mathbf{Z_{R_1}} + \mathbf{Z_{L_1}}) \| \mathbf{Z_P} = \frac{(\mathbf{Z_{R_1}} + \mathbf{Z_{L_1}}) \times \mathbf{Z_P}}{\mathbf{Z_{R_1}} + \mathbf{Z_{L_1}} + \mathbf{Z_P}} = \frac{(9.4340\angle 32.0054°) \times (10\angle -10.3889°)}{18.1203\angle 10.1611°}$$

$$= 5.1026 + j1.034 = 5.2063\angle 11.4554° \ \Omega$$

FIGURE 10.27

The Thévenin equivalent circuit.

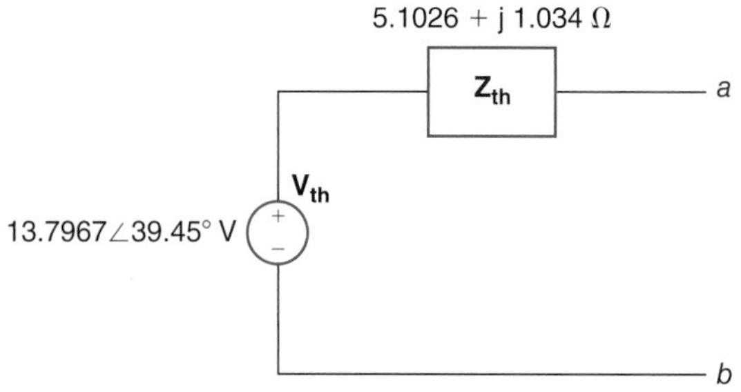

The Thévenin equivalent circuit is shown in Figure 10.27. The resistance value of the Thévenin equivalent impedance is the real part of $\mathbf{Z_{th}}$:

$$R_{th} = Re(\mathbf{Z_{th}}) = 5.1026 \ \Omega$$

If the frequency of the source is 10 Hz, the inductance value of the Thévenin equivalent impedance is

$$L_{th} = \frac{1.034}{2\pi 10} = 16.4566 \ \mathrm{mH}$$

Exercise 10.7

Find the Thévenin equivalent voltage $\mathbf{V}_{th}$ and Thévenin equivalent impedance $\mathbf{Z}_{th}$ between *a* and *b* for the circuit shown in Figure 10.28.

FIGURE 10.28

Circuit for EXERCISE 10.7.

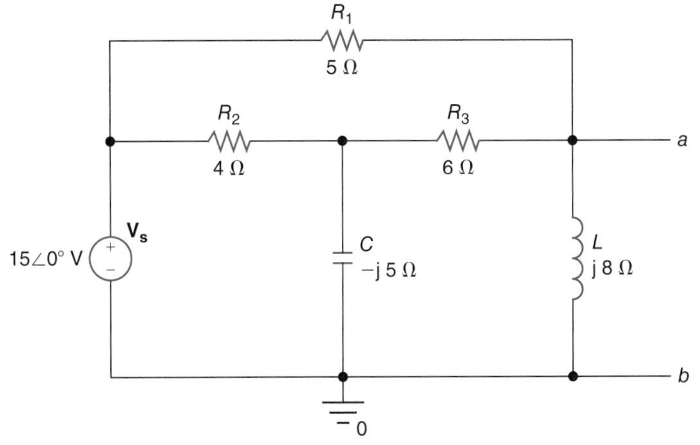

Answer:

$\mathbf{V}_{th} = 12.8096 + j2.1853 = 12.9947\angle 9.6814°$ V, $\mathbf{Z}_{th} = 2.9083 + j0.9213 = 3.0507\angle 17.5778°\ \Omega$.

10.10 Norton Equivalent Circuit

A circuit looking from terminals *a* and *b* can be replaced by a current source with current $\mathbf{I}_n$ and a parallel impedance $\mathbf{Z}_n$ [as shown in Figure 10.29(a)]. This equivalent circuit consisting of a current source and a parallel impedance is called a Norton equivalent circuit. When the terminals *a* and *b* are short-circuited in the Norton equivalent circuit, the short-circuit current $\mathbf{I}_{sc}$ is equal to $\mathbf{I}_n$, according to the current divider rule. Thus, the Norton equivalent current can be obtained by finding the short-circuit current. Application of source transformation to the Norton equivalent circuit yields the Thévenin equivalent circuit as shown in Figure 10.29(b). Notice that the Thévenin equivalent voltage is $\mathbf{V}_{th} = \mathbf{I}_n\mathbf{Z}_n$, and the Thévenin equivalent impedance is $\mathbf{Z}_{th} = \mathbf{Z}_n$. The source transformation does not change the impedance value. The Norton equivalent impedance can be obtained by applying the three methods used to find the Thévenin equivalent impedance. These three methods are listed next.

Method 1

Deactivate all the independent sources by short-circuiting voltage sources and open-circuiting current sources. Find the equivalent impedance looking into the circuit from terminals *a* and *b*. This equivalent impedance is the Norton equivalent impedance $\mathbf{Z}_n$.

Method 2

Find the open-circuit voltage $\mathbf{V}_{oc}$ and the short-circuit current $\mathbf{I}_{sc}$ between terminals *a* and *b*. The Norton equivalent impedance $\mathbf{Z}_n$ is given by

$$\mathbf{Z}_n = \frac{\mathbf{V}_{oc}}{\mathbf{I}_{sc}}$$

Method 3

Deactivate all the independent sources by open-circuiting current sources and short-circuiting voltage sources. Apply a test voltage of 1 *V* (or any other value) between terminals *a* and *b* with terminal *a* connected to the positive terminal of the test voltage. Measure the current out of the positive terminal of the test voltage source. The Norton equivalent impedance $\mathbf{Z_n}$ is given by the ratio of the test voltage to the test current. Instead of a test voltage, a test current (1 A or 1 mA or any other value) can be applied between terminals *a* and *b*. Measure the voltage across the terminals *a* and *b*. The ratio of the voltage to the test current is the Norton equivalent impedance $\mathbf{Z_n}$.

Application of source transformation to the Thévenin equivalent circuit yields the Norton equivalent circuit as shown in Figure 10.30. Notice that the Norton equivalent current is $\mathbf{I_n} = \mathbf{V_{th}}/\mathbf{Z_{th}}$ and the Norton equivalent impedance is $\mathbf{Z_n} = \mathbf{Z_{th}}$.

FIGURE 10.29

Norton equivalent circuit (a) to Thévenin equivalent circuit (b).

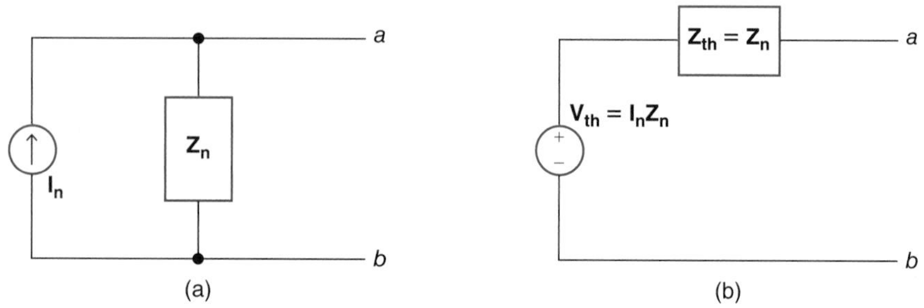

FIGURE 10.30

A Thévenin equivalent circuit transformed into a Norton equivalent circuit.

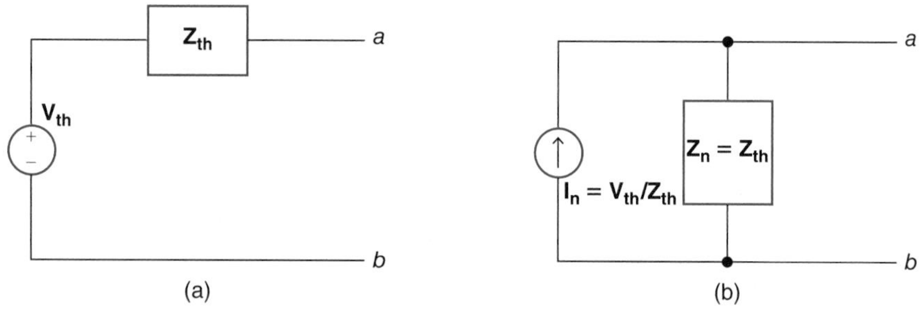

EXAMPLE 10.8

Find the Norton equivalent current $\mathbf{I_n}$ and Norton equivalent impedance $\mathbf{Z_n}$ between *a* and *b* for the circuit shown in Figure 10.31.

FIGURE 10.31

Circuit for EXAMPLE 10.8.

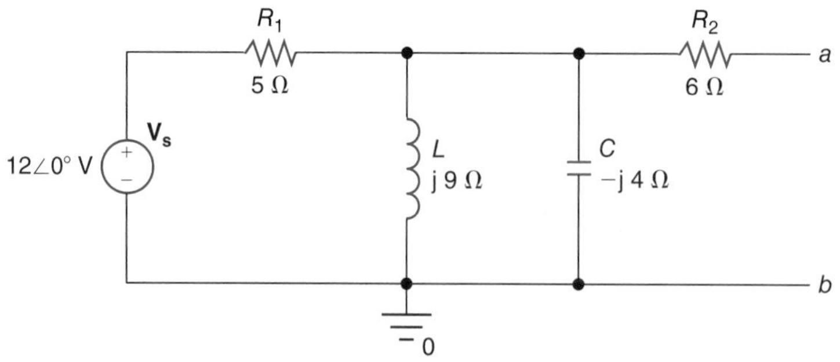

The Norton equivalent current is the short-circuit current between *a* and *b*. When *a* and *b* are short-circuited, the circuit shown in Figure 10.31 becomes the one shown in Figure 10.32.

continued

Example 10.8 continued

Summing the currents leaving node 1, we obtain

$$\frac{\mathbf{V}_1 - 12}{5} + \frac{\mathbf{V}_1}{j9} + \frac{\mathbf{V}_1}{-j4} + \frac{\mathbf{V}_1}{6} = 0$$

FIGURE 10.32

Circuit shown in Figure 10.31 with a and b shorted.

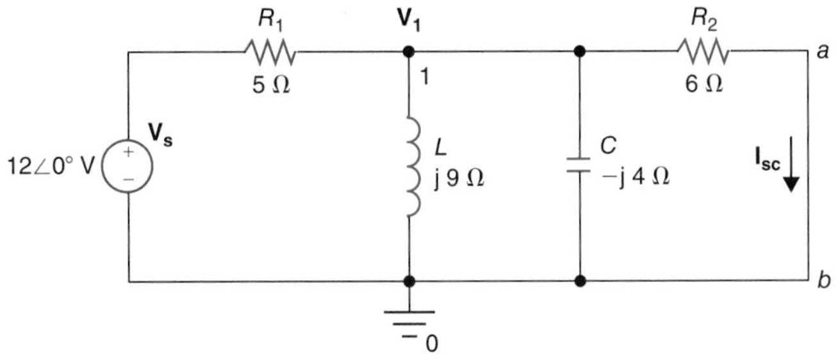

Multiplication by $j180$ results in

$$j36\mathbf{V}_1 - j432 + 20\mathbf{V}_1 - 45\mathbf{V}_1 + j30\mathbf{V}_1 = 0$$

which can be simplified to

$$(-25 + j66)\mathbf{V}_1 = j432$$

Thus, we have

$$\mathbf{V}_1 = \frac{j432}{-25 + j66} = 5.7242 - j2.1682 \text{ V}$$

The short-circuit current $\mathbf{I}_{sc}$, which is the Norton equivalent current, is given by

$$\mathbf{I}_n = \mathbf{I}_{sc} = \frac{\mathbf{V}_1}{R_2} = \frac{j72}{-25 + j66} = 0.9540 - j0.3614$$
$$= 1.0202\angle -20.746° \text{ A}$$

FIGURE 10.33

Circuit shown in Figure 10.31 with the voltage source short-circuited.

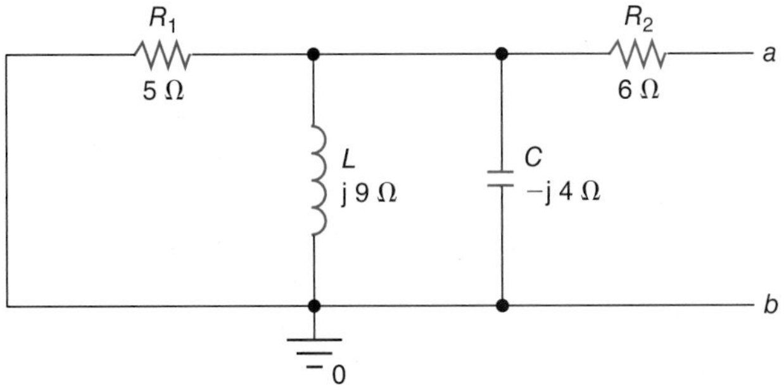

To find the Norton equivalent impedance, the voltage source is short-circuited, as shown in Figure 10.33.

The Norton equivalent impedance is the equivalent impedance looking into the circuit from a and b after the voltage source is short-circuited. The Norton impedance is the sum of R_2 and the parallel connection of R_1, $\mathbf{Z}_L$, and $\mathbf{Z}_C$. Thus, we have

$$\mathbf{Z}_n = R_2 + (R_1 \| \mathbf{Z}_L \| \mathbf{Z}_C) = R_2 + \frac{1}{\frac{1}{R_1} + \frac{1}{\mathbf{Z}_L} + \frac{1}{\mathbf{Z}_C}}$$

$$= 6 + \frac{1}{\frac{1}{5} + \frac{1}{j9} + \frac{1}{-j4}}$$

$$= 9.3732 - j2.3425 = 9.6615\angle -14.0318° \ \Omega$$

FIGURE 10.34

The Norton equivalent circuit.

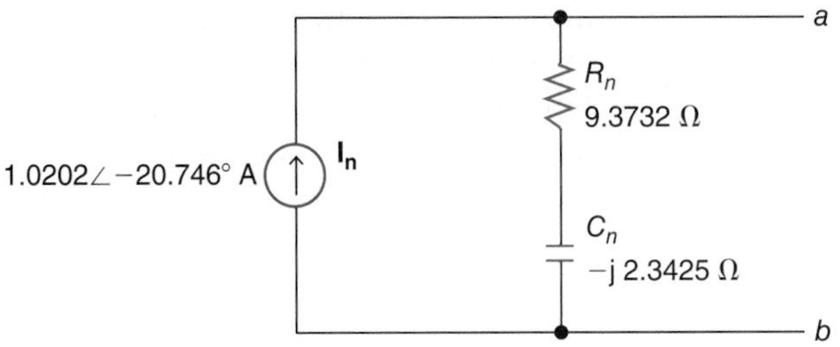

The Norton equivalent circuit is shown in Figure 10.34.

Exercise 10.8

Find the Norton equivalent current $\mathbf{I}_n$ and Norton equivalent impedance $\mathbf{Z}_n$ between *a* and *b* for the circuit shown in Figure 10.35.

FIGURE 10.35

Circuit for EXERCISE 10.8.

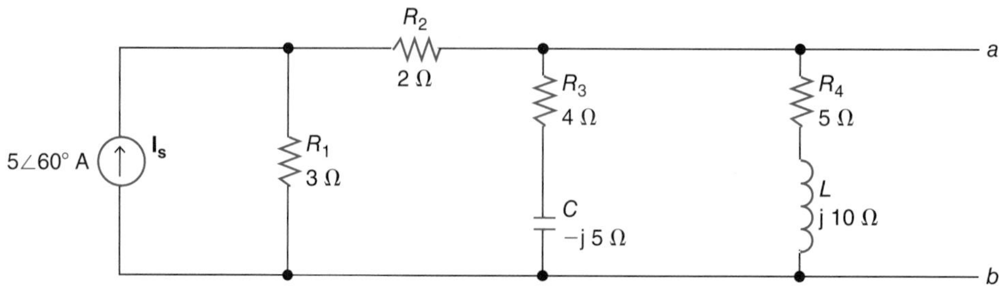

Answer:
$\mathbf{I}_n = 3\angle 60°$ A, $\mathbf{Z}_n = 2.9174 - j0.3626\ \Omega = 2.9398\angle -7.0843°\ \Omega$.

10.11 Transfer Function

Phasor-transformed circuits discussed so far assume a fixed frequency for sources. The frequency of the sources is assumed to be 60 Hz or 50 Hz for ac power systems. But, the impedance of the capacitor, $1/(j\omega C)$, and the impedance of the inductor, $j\omega L$, are frequency-dependent. It will be interesting to see how the circuits respond to sinusoids as the frequency ω is scanned from zero to infinity. This can be achieved by transforming the circuit components to impedances, $1/(j\omega C)$ and $j\omega L$, without specifying values of ω, and designating the input and output of the circuit as a function of ω. Then, one finds the ratio of the output to input. The input signal can be voltage or current, and the output signal can be voltage or current. This ratio is called the transfer function $H(\omega)$. The transfer function is a function of radian frequency ω that characterizes the circuit or system. By plotting the magnitude of the transfer function $|H(\omega)|$, called the *magnitude response*, and the phase of the transfer function $\angle H(\omega)$, called the *phase response*, we can gain more insight into the behavior of the circuit as the frequency is varied. The magnitude is called the *gain*.

A filter is a device that passes certain frequencies and blocks other frequencies. Common types of filters are lowpass filter (LPF), highpass filter (HPF), bandpass filter (BPF), and bandstop filter (BSF). The filters that cannot be physically realizable are called ideal filters. The magnitude responses of the ideal filters are shown in Figure 10.36. The ideal LPF [shown in Figure 10.36(a)] passes input signals with frequencies below ω_c with gain of one and blocks input signals with frequencies above ω_c with gain of zero. The frequency ω_c is called the cutoff frequency. If the filter is non-ideal, which is called a practical filter, the gain in the passband $(0 \le \omega \le \omega_c)$ is not one for the entire band, and the gain in the stopband $(\omega_c \le \omega < \infty)$ is not zero for the entire band. Also, the transition from the passband to stopband is more gradual. The ideal HPF [shown in Figure 10.36(b)] passes input signals with frequencies above the cutoff frequency ω_c with a gain of one and blocks input signals with frequencies below ω_c with a gain of zero. The ideal bandpass filter [shown in Figure 10.36(c)] passes input signals with frequencies between the low cutoff frequency ω_{c_1} and the high cutoff frequency ω_{c_2} with a gain of one and blocks input signals with frequencies outside the passband $(\omega_{c_1} \le \omega \le \omega_{c_2})$. The ideal bandstop filter [shown in Figure 10.36(d)] blocks input signals with frequencies between the low cutoff frequency ω_{c_1} and the high cutoff frequency ω_{c_2} with a gain of zero and passes input signals with frequencies outside the stopband $(\omega_{c_1} < \omega < \omega_{c_2})$ with gain of 1.

FIGURE 10.36

Magnitude responses of ideal (a) LPF, (b) HPF, (c) BPF, and (d) BSF.

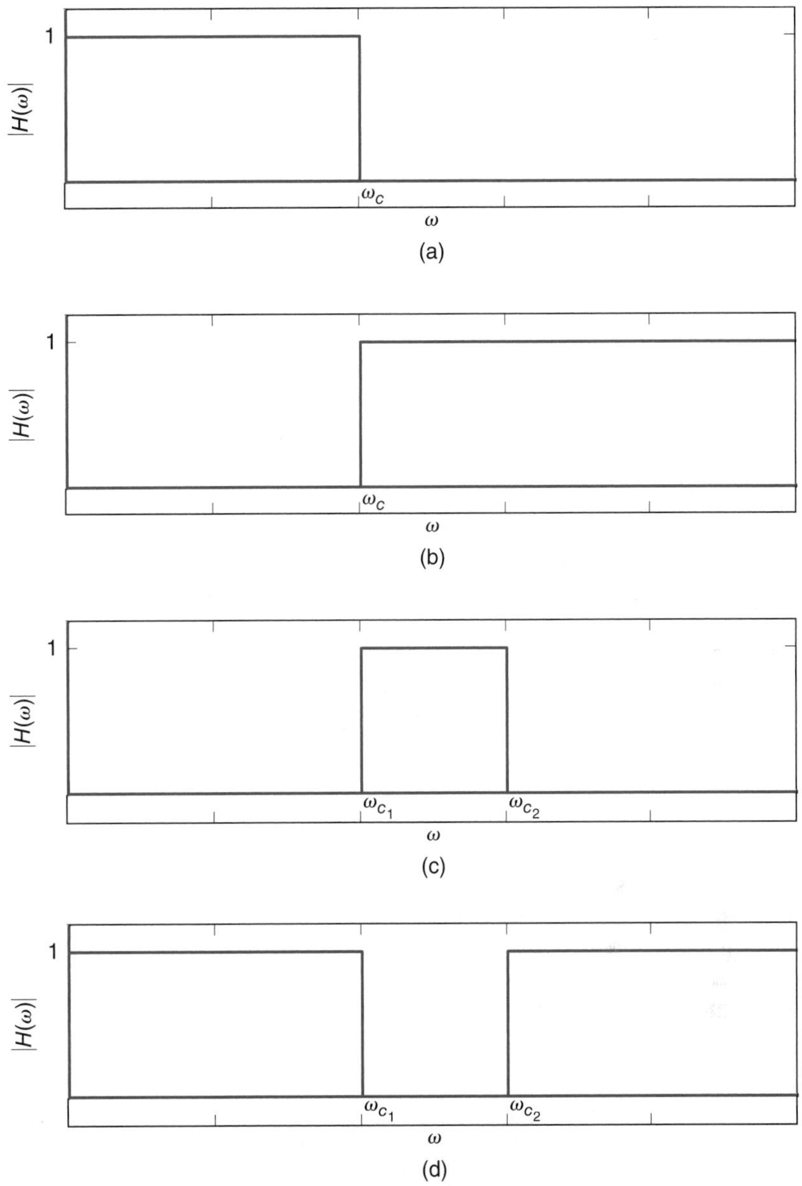

The ideal filters shown in Figure 10.36 have a gain of one in the passband and a gain of zero in the stopband. Also, the transitions from passband to stopband, and stopband to passband, are abrupt. These characteristics of ideal filters cannot be realized in practical filters. In practical filters, the gain in the passband cannot be one for all frequencies, and the gain in the stopband cannot be zero for all frequencies. Also, the transitions from passband to stopband and stopband to passband are gradual. Figure 10.37 shows the magnitude responses of practical LPF, HPF, BPF, and BSF. As can be seen, the magnitude response changes continuously as a function of radian frequency ω, and the transition from passband to stopband and stopband to passband happens over a finite width of ω.

The transfer function in the s-domain is presented in Chapter 15, and the first-order and the second-order analog filters are presented in more detail in Chapter 16.

FIGURE 10.37

Magnitude responses of practical (a) LPF, (b) HPF, (c) BPF, (d) BSF.

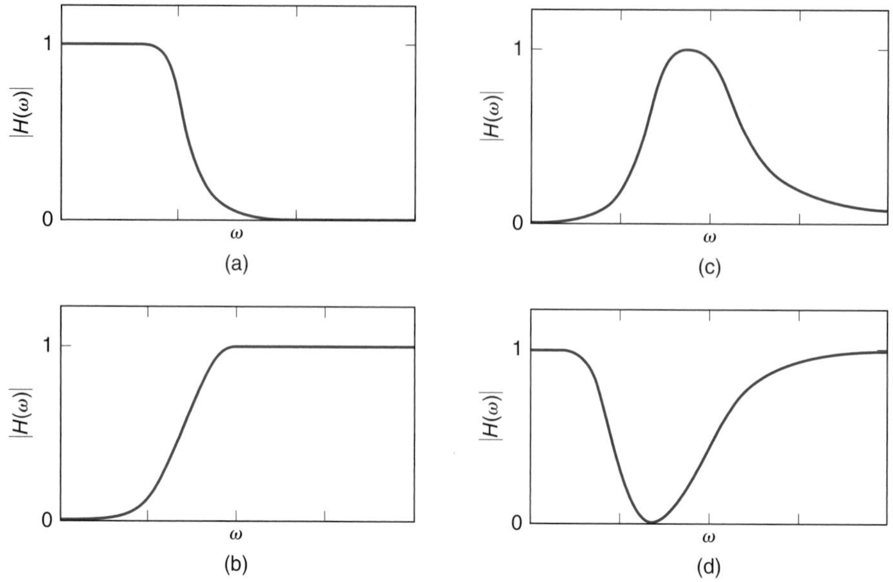

FIGURE 10.38

RC circuit.

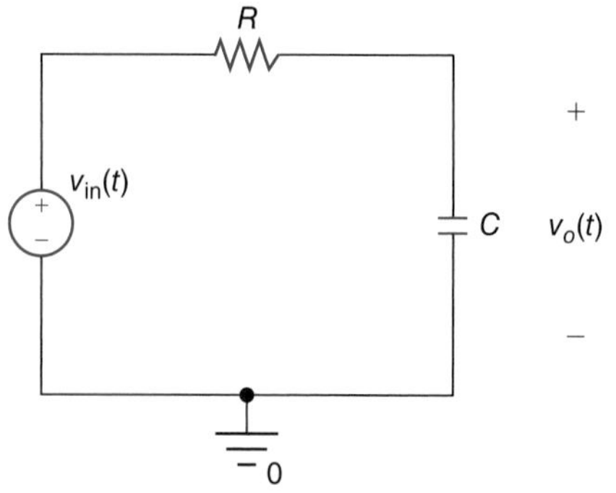

Consider an *RC* circuit shown in Figure 10.38.

In the phasor-transformed circuit shown in Figure 10.39, the input voltage source is transformed into $V_{\text{in}}(\omega)$, the output voltage across the capacitor is transformed into $V_o(\omega)$, the capacitor is transformed into impedance $1/(j\omega C)$. Since the impedance of the resistor is R, there is no change for the resistor.

Application of the voltage divider rule yields

$$V_o(\omega) = \frac{\dfrac{1}{j\omega C}}{R + \dfrac{1}{j\omega C}} V_{\text{in}}(\omega) = \frac{1}{RCj\omega + 1} V_{\text{in}}(\omega) = \frac{\dfrac{1}{RC}}{j\omega + \dfrac{1}{RC}} V_{\text{in}}(\omega)$$

Dividing by $V_{\text{in}}(\omega)$ on both sides, we obtain the transfer function $H(\omega)$:

$$H(\omega) = \frac{V_o(\omega)}{V_{\text{in}}(\omega)} = \frac{\dfrac{1}{RC}}{j\omega + \dfrac{1}{RC}} \tag{10.27}$$

FIGURE 10.39

RC circuit transformed into the ω domain.

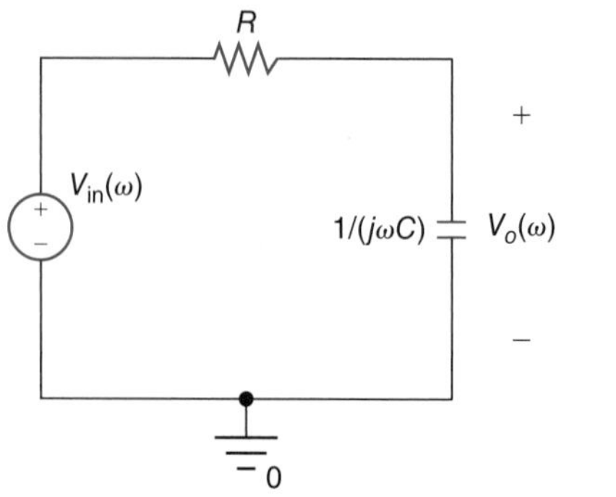

Changing the numerator and denominator to polar coordinates, we obtain

$$\begin{aligned} H(\omega) = \frac{V_o(\omega)}{V_{\text{in}}(\omega)} &= \frac{\dfrac{1}{RC} e^{j0}}{\sqrt{\omega^2 + \left(\dfrac{1}{RC}\right)^2}\, e^{j\tan^{-1}\left(\frac{\omega}{\frac{1}{RC}}\right)}} \\ &= \frac{\dfrac{1}{RC}}{\sqrt{\omega^2 + \left(\dfrac{1}{RC}\right)^2}} e^{-j\tan^{-1}(\omega RC)} \end{aligned} \tag{10.28}$$

The magnitude response is given by

$$|H(\omega)| = \frac{\frac{1}{RC}}{\sqrt{\omega^2 + \left(\frac{1}{RC}\right)^2}} \tag{10.29}$$

and the phase response is given by

$$\angle H(\omega) = -\tan^{-1}\left(\frac{\omega}{\frac{1}{RC}}\right) = -\tan^{-1}(\omega RC) \tag{10.30}$$

If $\omega = 0$ (dc input signal), $|H(\omega)| = 1$ and $\angle H(\omega) = 0$. The output signal is identical to the input signal. When $\omega = 0$, the impedance of the capacitor is $1/(j0C) = \infty$, and the capacitor can be treated as an open circuit. There is no current through the resistor, and the voltage drop across the resistor is zero. Thus, the output voltage is the same as the input voltage. If $\omega = \infty$, $|H(\omega)| = 0$ and $\angle H(\omega) = -90°$. When $\omega = \infty$, the impedance of the capacitor is $1/(j\infty C) = 0$, and the capacitor can be treated as a short circuit. The voltage drop across the capacitor is zero. Thus, the output voltage is zero.

If $\omega = 1/(RC)$, $|H(\omega)| = 1/\sqrt{2} = 0.7071$ and $\angle H(\omega) = -\tan^{-1}(1) = -45°$. At $\omega = 1/(RC)$, the magnitude is decreased to $1/\sqrt{2}$ from 1, and the phase is decreased from 0° to $-45°$. The magnitude of $1/\sqrt{2}$ can be represented in logarithm as

$$20\log_{10}\left(\frac{1}{\sqrt{2}}\right) = 20\log_{10}\left(2^{-\frac{1}{2}}\right) = -10\log_{10}(3) = -3.01dB$$

where dB is decibel. Since $20 \log_{10}(1) = 0$ dB, the gain dropped by 3.01 dB at $\omega = 1/(RC)$. This frequency is called the 3-dB cutoff frequency and is denoted by ω_0. Thus, we have

$$\omega_0 = \frac{1}{RC} \text{ rad/s} \tag{10.31}$$

In terms of the 3-dB cutoff frequency, the transfer function can be rewritten as

$$H(\omega) = \frac{\omega_0}{j\omega + \omega_0} \tag{10.32}$$

Let $R = 2\ k\Omega$ and $C = 0.08\ \mu$F. Then, the 3-dB cutoff frequency is given by

$$\omega_0 = \frac{1}{RC} \text{ rad/s} = 6250 \text{ rad/s} \tag{10.33}$$

Notice that the 3-dB cutoff frequency in Hz is

$$f_0 = \frac{1}{2\pi RC} \text{ Hz} = 994.7184 \text{ Hz} \tag{10.34}$$

The magnitude response and the phase response are plotted in Figure 10.40 for $R = 2\ k\Omega$ and $C = 0.08\ \mu$F. The frequency axis is in log scale for ω.

For frequencies below the 3-dB cutoff frequency ω_0, the magnitude is greater than 0.7071. But for frequencies above ω_0, the magnitude is smaller than 0.7071. The circuit shown in Figure 10.39 passes signals with frequencies below ω_0, and attenuates signals with frequencies above ω_0. The circuit is an LPF. It passes low frequencies and suppresses high frequencies.

FIGURE 10.40

Magnitude response and phase response for the circuit shown in Figure 10.39 with $R = 2\ k\Omega$ and $C = 0.08\ \mu F$.

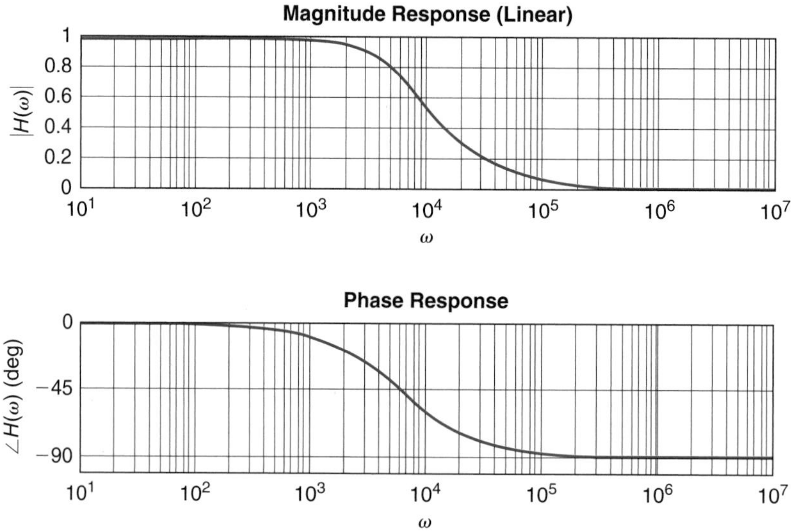

FIGURE 10.41

CR circuit.

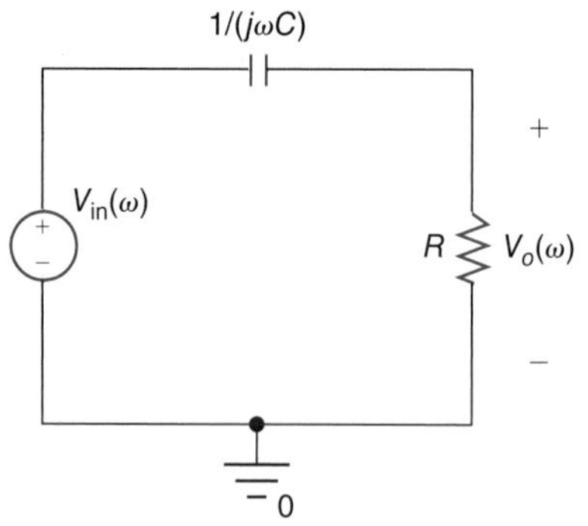

If the capacitor and the resistor are swapped in the circuit shown in Figure 10.39, we obtain the circuit shown in Figure 10.41.

Application of the voltage divider rule yields

$$V_o(\omega) = \frac{R}{R + \frac{1}{j\omega C}} V_{\text{in}}(\omega) = \frac{RCj\omega}{RCj\omega + 1} V_{\text{in}}(\omega) = \frac{j\omega}{j\omega + \frac{1}{RC}} V_{\text{in}}(\omega)$$

Dividing by $V_{\text{in}}(\omega)$ on both sides, we obtain the transfer function $H(\omega)$:

$$H(\omega) = \frac{V_o(\omega)}{V_{\text{in}}(\omega)} = \frac{j\omega}{j\omega + \frac{1}{RC}} \tag{10.35}$$

Changing the numerator and denominator to polar coordinates, we obtain

$$H(\omega) = \frac{V_o(\omega)}{V_{\text{in}}(\omega)} = \frac{\omega e^{j\frac{\pi}{2}}}{\sqrt{\omega^2 + \left(\frac{1}{RC}\right)^2}\, e^{j\tan^{-1}\left(\frac{\omega}{\frac{1}{RC}}\right)}} = \frac{\omega}{\sqrt{\omega^2 + \left(\frac{1}{RC}\right)^2}} e^{j\left[\frac{\pi}{2} - \tan^{-1}(\omega RC)\right]} \tag{10.36}$$

The magnitude response is given by

$$|H(\omega)| = \frac{\omega}{\sqrt{\omega^2 + \left(\frac{1}{RC}\right)^2}} \tag{10.37}$$

and the phase response is given by

$$\angle H(\omega) = \frac{\pi}{2} - \tan^{-1}\left(\frac{\omega}{\frac{1}{RC}}\right) = \frac{\pi}{2} - \tan^{-1}(\omega RC) \tag{10.38}$$

If $\omega = 0$ (dc input signal), $|H(\omega)| = 0$ and $\angle H(\omega) = 0$. The output signal is zero. When $\omega = 0$, the impedance of the capacitor is $1/(j0C) = \infty$, and the capacitor can be treated as an open circuit. There is no current through the resistor, and the voltage drop across the resistor is zero. Thus, the output voltage is zero. If $\omega = \infty$,

$$|H(\omega)| = \lim_{\omega \to \infty} \frac{\omega}{\sqrt{\omega^2 + \left(\frac{1}{RC}\right)^2}} = \lim_{\omega \to \infty} \frac{\omega}{\omega\sqrt{1 + \left(\frac{1}{\omega RC}\right)^2}}$$

$$= \lim_{\omega \to \infty} \frac{1}{\sqrt{1 + \left(\frac{1}{\omega RC}\right)^2}} = \frac{1}{\sqrt{1 + \left(\frac{1}{\infty RC}\right)^2}} = 1$$

and $\angle H(\omega) = 0°$. When $\omega = \infty$, the impedance of the capacitor is $1/(j\infty C) = 0$, and the capacitor can be treated as a short circuit. The voltage drop across the capacitor is zero. Thus, the output voltage is identical to the input voltage, and the transfer function is one.

If $\omega = 1/(RC), |H(\omega)| = 1/\sqrt{2} = 0.7071$ and $\angle H(\omega) = 90° - \tan^{-1}(1) = 45°$. At $\omega = 1/(RC)$, the magnitude is increased to $1/\sqrt{2}$ from zero, and the phase is decreased from 90° (at $\omega = 0^+$) to 45°. The 3-dB cutoff frequency is given by

$$\omega_0 = \frac{1}{RC} \text{rad/s} \tag{10.39}$$

In terms of the 3-dB cutoff frequency, the transfer function can be rewritten as

$$H(\omega) = \frac{\omega}{j\omega + \omega_0} \tag{10.40}$$

Let $R = 2\ k\Omega$ and $C = 0.08\ \mu F$. Then, the 3-dB cutoff frequency is $\omega_0 = 6250$ rad/s and the 3-dB cutoff frequency in Hz is $f_0 = \omega_0/(2\pi) = 994.7184$ Hz. The magnitude response and the phase response are plotted in Figure 10.42 for $R = 2\ k\Omega$ and $C = 0.08\ \mu F$. The frequency axis is in log scale for ω.

FIGURE 10.42

Magnitude response and phase response for the circuit shown in Figure 10.41 with $R = 2\ k\Omega$ and $C = 0.08\ \mu F$.

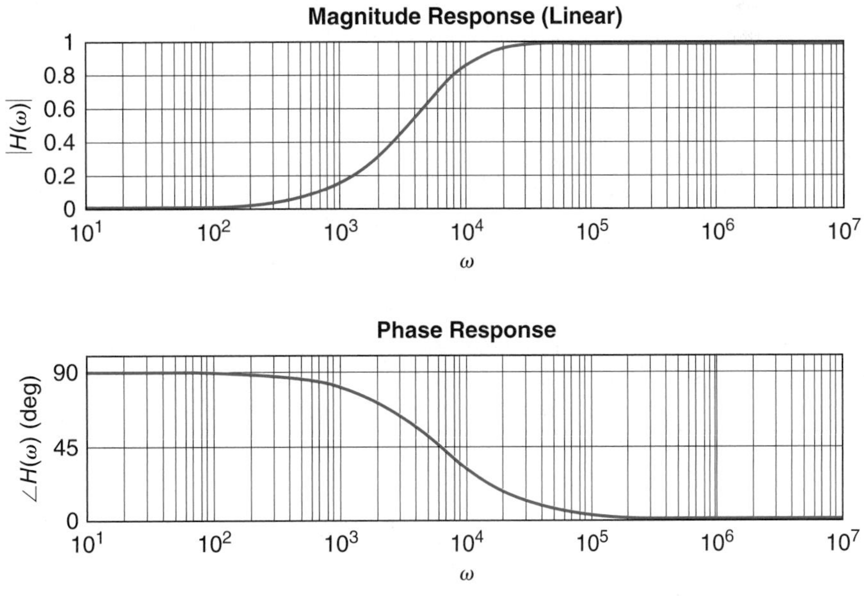

For frequencies below the 3-dB cutoff frequency ω_0, the magnitude is less than 0.7071. But for frequencies above ω_0, the magnitude is greater than 0.7071. The circuit shown in Figure 10.41 passes signals with frequencies above ω_0, and attenuates signals with frequencies below ω_0. The circuit is an HPF. It passes high frequencies and suppresses low frequencies.

EXAMPLE 10.9

Find the transfer function $H(\omega) = V_o(\omega)/V_{in}(\omega)$ of the LR circuit shown in Figure 10.43 and state whether the circuit is an LPF or an HPF.

FIGURE 10.43

LR circuit.

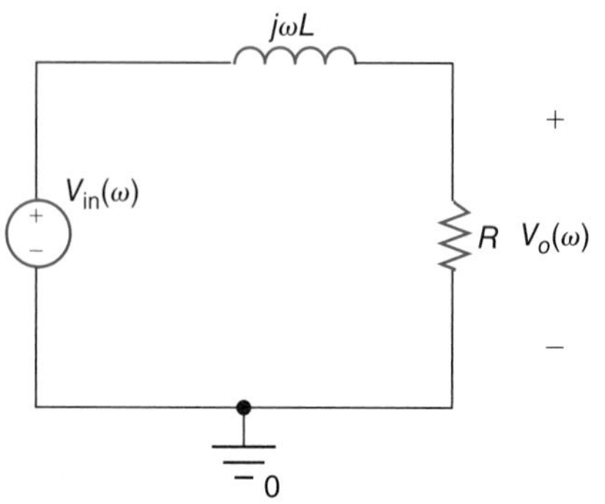

Application of the voltage divider rule yields

$$V_o(\omega) = \frac{R}{j\omega L + R} V_{in}(\omega) = \frac{\frac{R}{L}}{j\omega + \frac{R}{L}} V_{in}(\omega)$$

Dividing by $V_{in}(\omega)$ on both sides, we obtain the transfer function $H(\omega)$:

$$H(\omega) = \frac{V_o(\omega)}{V_{in}(\omega)} = \frac{\frac{R}{L}}{j\omega + \frac{R}{L}} \quad \textbf{(10.41)}$$

Changing the numerator and denominator to polar coordinates, we obtain

$$H(\omega) = \frac{V_o(\omega)}{V_{in}(\omega)} = \frac{\frac{R}{L} e^{j0}}{\sqrt{\omega^2 + \left(\frac{R}{L}\right)^2}\, e^{j\tan^{-1}\left(\frac{\omega}{R/L}\right)}} = \frac{\frac{R}{L}}{\sqrt{\omega^2 + \left(\frac{R}{L}\right)^2}} e^{-j\tan^{-1}\left(\frac{L}{R}\omega\right)} \quad \textbf{(10.42)}$$

The magnitude response is given by

$$|H(\omega)| = \frac{\frac{R}{L}}{\sqrt{\omega^2 + \left(\frac{R}{L}\right)^2}} \quad \textbf{(10.43)}$$

and the phase response is given by

$$\angle H(\omega) = -\tan^{-1}\left(\frac{L}{R}\omega\right) \quad \textbf{(10.44)}$$

If $\omega = 0$ (dc input signal), $|H(\omega)| = 1$ and $\angle H(\omega) = 0$. The output signal is identical to the input signal. When $\omega = 0$, the impedance of the inductor is $j0L = 0$, and the inductor can be treated as a short-circuit. The voltage drop across the inductor is zero. Thus, the output voltage is the same as the input voltage. If $\omega = \infty$, $|H(\omega)| = 0$ and $\angle H(\omega) = -90°$. When $\omega = \infty$, the impedance of the inductor is $j\infty L = \infty$, and the inductor can be treated as an open circuit. There is no current through the circuit, and the voltage drop across the resistor is zero. Thus, the output voltage is zero.

If $\omega = R/L$, $|H(\omega)| = 1/\sqrt{2} = 0.7071$ and $\angle H(\omega) = -\tan^{-1}(1) = -45°$. At $\omega = R/L$, the magnitude is decreased to $1/\sqrt{2}$ from 1, and the phase is decreased from 0° to −45°. The magnitude of $1/\sqrt{2}$ can be represented in logarithm as

$$20\log_{10}\left(\frac{1}{\sqrt{2}}\right) = 20\log_{10}\left(2^{-\frac{1}{2}}\right) = -10\log_{10}(3) = -3.01\text{ dB}$$

continued

Example 10.9 continued

where dB is decibel. Since $20 \log_{10}(1) = 0$ dB, the gain dropped by 3.01 dB at $\omega = R/L$. This frequency is called the *3-dB cutoff frequency* and denoted by ω_0. Thus, we have

$$\omega_0 = \frac{R}{L} \text{rad/s} \qquad \textbf{(10.45)}$$

In terms of the 3-dB cutoff frequency, the transfer function can be rewritten as

$$H(\omega) = \frac{\omega_0}{j\omega + \omega_0} \qquad \textbf{(10.46)}$$

The circuit shown in Figure 10.43 is an LPF.

Exercise 10.9

Find the transfer function $H(\omega) = V_o(\omega)/V_{in}(\omega)$ of the *RL* circuit shown in Figure 10.44 and state whether the circuit is an LPF or an HPF.

FIGURE 10.44

RL circuit.

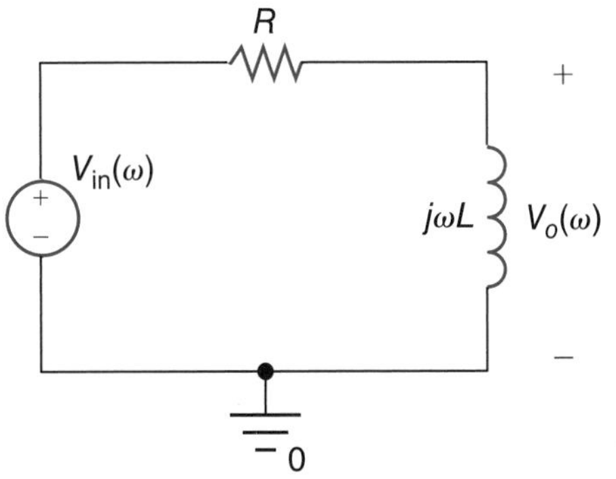

Answer:

$$H(\omega) = \frac{j\omega}{j\omega + \frac{R}{L}}, \text{ HPF.}$$

EXAMPLE 10.10

Find the transfer function $H(\omega) = V_o(\omega)/V_{in}(\omega)$ of the circuit shown in Figure 10.45, and state whether the circuit is an LPF or an HPF.

Summing the currents leaving node 1, we obtain

$$\frac{0 - V_{in}(\omega)}{R_1} + \frac{0 - V_o(\omega)}{R_2} + \frac{0 - V_o(\omega)}{\frac{1}{j\omega C}} = 0$$

continued

Example 10.10 continued

FIGURE 10.45

Circuit for EXAMPLE 10.10.

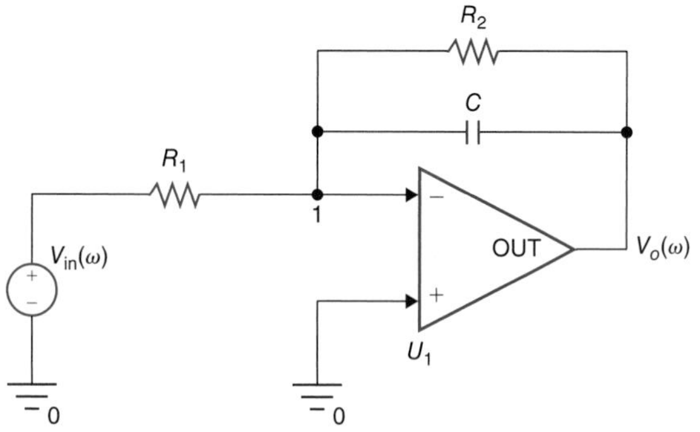

The circuit is an LPF.

which can be revised as

$$\left(j\omega C + \frac{1}{R_2}\right)V_o(\omega) = \frac{-V_{in}(\omega)}{R_1}$$

Taking the ratio of $V_o(\omega)$ to $V_{in}(\omega)$, we obtain

$$\frac{V_o(\omega)}{V_{in}(\omega)} = \frac{-1}{R_1}\frac{1}{j\omega C + \frac{1}{R_2}} = -\frac{\frac{1}{R_1 C}}{j\omega + \frac{1}{R_2 C}} = -\frac{R_2}{R_1}\frac{\frac{1}{R_2 C}}{j\omega + \frac{1}{R_2 C}} \quad \textbf{(10.47)}$$

Exercise 10.10

Find the transfer function $H(\omega) = V_o(\omega)/V_{in}(\omega)$ of the circuit shown in Figure 10.46, and state whether the circuit is an LPF or an HPF.

FIGURE 10.46

Circuit for EXERCISE 10.10.

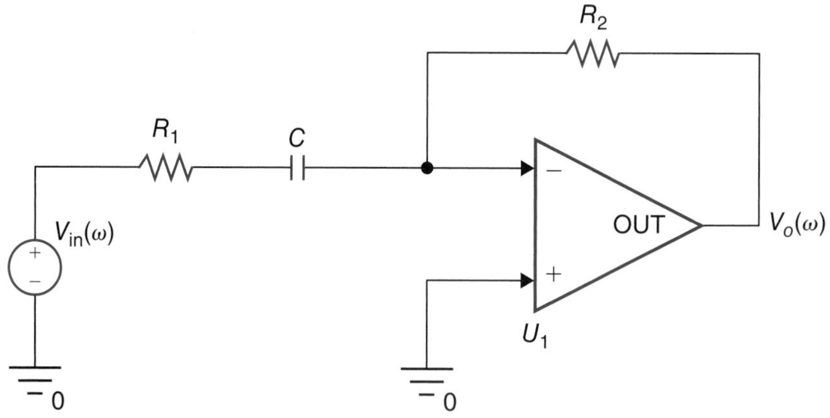

Answer:

$$H(\omega) = -\frac{R_2}{R_1}\frac{j\omega}{j\omega + \frac{1}{R_1 C}}. \quad \textbf{(10.48)}$$

The circuit is an HPF.

10.11.1 SERIES RLC CIRCUITS

A series RLC circuit with capacitor at the output is shown in Figure 10.47. Application of the voltage divider rule yields

FIGURE 10.47

A series RLC circuit.

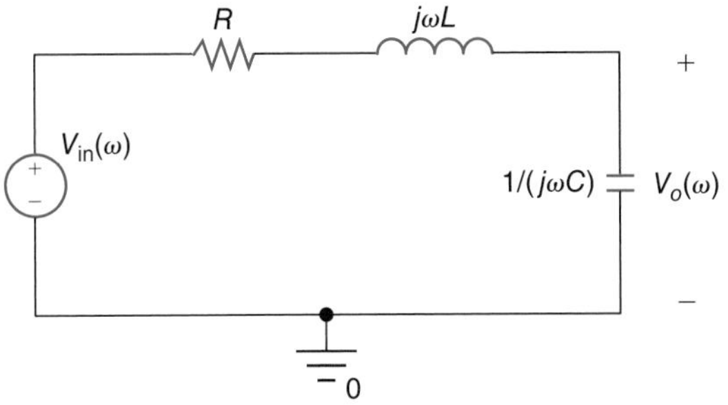

$$H(\omega) = \frac{\dfrac{1}{j\omega C}}{j\omega L + R + \dfrac{1}{j\omega C}} = \frac{1}{(j\omega)^2 LC + RCj\omega + 1}$$
$$= \frac{\dfrac{1}{LC}}{-\omega^2 + \dfrac{R}{L}j\omega + \dfrac{1}{LC}} \tag{10.49}$$

Notice that at $\omega = 0$, $H(\omega) = 1$, and at $\omega = \infty$, $H(\omega) = 0$. This result can also be obtained directly from the circuit. At $\omega = 0$, impedance of the inductor is $j\omega L = j0L = 0$ and impedance of the capacitor is $1/(j\omega C) = 1/(j0C) = \infty$. The inductor can be treated as a short-circuit, and the capacitor can be treated as an open circuit. No current flows through the circuit. Thus, the output voltage is equal to the input voltage, making the transfer function equal to 1. At $\omega = \infty$, the impedance of the inductor is $j\omega L = j\infty L = \infty$, and the impedance of the capacitor is $1/(j\omega C) = 1/(j\infty C) = 0$. The inductor can be treated as an open circuit, and the capacitor can be treated as a short-circuit. No current flows through the circuit. The output voltage across the capacitor is zero, making the transfer function equal to zero. This is an LPF.

The magnitude response is given by

$$|H(\omega)| = \frac{\dfrac{1}{LC}}{\sqrt{\left(\dfrac{1}{LC} - \omega^2\right)^2 + \left(\dfrac{R}{L}\omega\right)^2}} \tag{10.50}$$

and the phase response is given by

$$\angle H(\omega) = -\tan^{-1}\left(\frac{\dfrac{R}{L}\omega}{\dfrac{1}{LC} - \omega^2}\right) \text{ (rad)} \tag{10.51}$$

The magnitude response and the phase response are shown in Figure 10.48 for $R = 2\,k\Omega$, $L = 50$ mH, and $C = 5$ nF.

The sum of the impedances of the inductor and the capacitor is given by

$$j\omega L + \frac{1}{j\omega C} = j\omega L - j\frac{1}{\omega C} = j\left(\omega L - \frac{1}{\omega C}\right) = j\frac{\omega^2 LC - 1}{\omega C}$$

The sum of impedances is zero when $\omega^2 LC - 1 = 0$ or $\omega = \dfrac{1}{\sqrt{LC}}$. This frequency is called the **corner frequency** or **resonant frequency** and is denoted by

$$\omega_0 = \frac{1}{\sqrt{LC}} \text{rad/s} \tag{10.52}$$

FIGURE 10.48

Magnitude and phase response of an LPF.

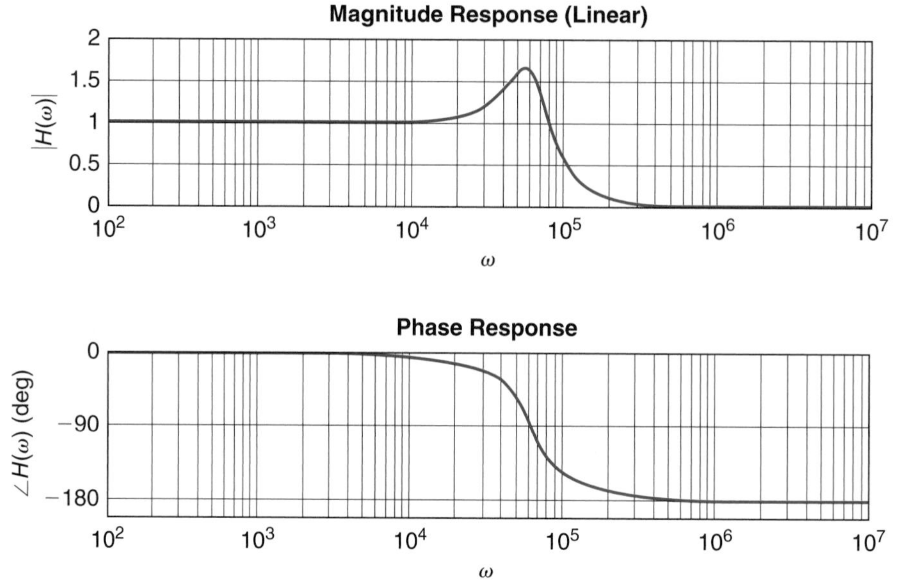

At the resonant frequency, the impedance of the inductor is

$$j\omega_0 L = j\frac{1}{\sqrt{LC}}L = j\sqrt{\frac{L}{C}}$$

and the impedance of the capacitor is

$$-j\frac{1}{\omega_0 C} = -j\frac{1}{\dfrac{1}{\sqrt{LC}}C} = -j\sqrt{\frac{L}{C}}$$

The impedances have the same magnitude but opposite phases (+90° and −90°). The sum of the impedances is equal to zero. At the resonant frequency, the current through the circuit is given by $I = V_{in}(\omega)/R$. Thus, the output voltage is given by

$$V_o(\omega) = \frac{V_{in}(\omega)}{R}\left(-j\sqrt{\frac{L}{C}}\right)$$

Thus, the transfer function is given by

$$H(\omega_0) = -j\frac{1}{R}\sqrt{\frac{L}{C}}$$

This result can also be obtained by substituting $\omega = \omega_0 = 1/\sqrt{LC}$ into Equation (10.49).

$$H(\omega_0) = \frac{\dfrac{1}{LC}}{-\omega_0^2 + \dfrac{R}{L}j\omega_0 + \dfrac{1}{LC}} = \frac{\dfrac{1}{LC}}{-\dfrac{1}{LC} + \dfrac{R}{L}j\dfrac{1}{\sqrt{LC}} + \dfrac{1}{LC}} = \frac{\dfrac{1}{LC}}{\dfrac{R}{L}j\dfrac{1}{\sqrt{LC}}} = -j\frac{1}{R}\sqrt{\frac{L}{C}}$$

A series RLC circuit with an inductor at the output is shown in Figure 10.49. Application of the voltage divider rule yields

FIGURE 10.49

A series RCL circuit.

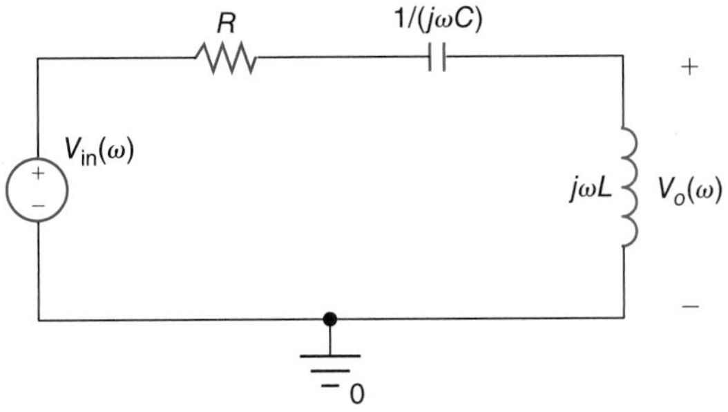

$$H(\omega) = \frac{j\omega L}{j\omega L + R + \dfrac{1}{j\omega C}} = \frac{(j\omega)^2 LC}{(j\omega)^2 LC + RCj\omega + 1} = \frac{-\omega^2}{-\omega^2 + \dfrac{R}{L} j\omega + \dfrac{1}{LC}} \quad \textbf{(10.53)}$$

At $\omega = 0$, $H(\omega) = 0$. At $\omega = \infty$, we have

$$H(\omega) = \lim_{\omega \to \infty} \frac{j\omega L}{j\omega L + R + \dfrac{1}{j\omega C}} = \lim_{\omega \to \infty} \frac{1}{1 + \dfrac{R}{j\omega L} + \dfrac{1}{-\omega^2 LC}} = 1$$

This result can also be obtained directly from the circuit. At $\omega = 0$, the impedance of the inductor is $j\omega L = j0L = 0$, and the impedance of the capacitor is $1/(j\omega C) = 1/(j0C) = \infty$. The inductor can be treated as a short-circuit, and the capacitor can be treated as an open circuit. The voltage across the inductor is equal to zero, making the transfer function equal to zero. At $\omega = \infty$, the impedance of the inductor is $j\omega L = j\infty L = \infty$ and the impedance of the capacitor is $1/(j\omega C) = 1/(j\infty C) = 0$. The inductor can be treated as an open circuit, and the capacitor can be treated as a short-circuit. No current flows through the circuit. The output voltage across the inductor is equal to the input voltage, making the transfer function equal to 1. This is an HPF.

The magnitude response is given by

$$|H(\omega)| = \frac{\omega^2}{\sqrt{\left(\dfrac{1}{LC} - \omega^2\right)^2 + \left(\dfrac{R}{L}\omega\right)^2}} \quad \textbf{(10.54)}$$

and the phase response is given by

$$\angle H(\omega) = \pi - \tan^{-1}\left(\frac{\dfrac{R}{L}\omega}{\dfrac{1}{LC} - \omega^2}\right) \text{ (rad)} \quad \textbf{(10.55)}$$

The magnitude response and the phase response are shown in Figure 10.50 for $R = 2\ k\Omega$, $L = 50$ mH, and $C = 5$ nF.

$$\text{At } \omega = 1/\sqrt{LC}, \quad H(\omega) = j\frac{1}{R}\sqrt{\frac{L}{C}}.$$

A series RLC circuit with a resistor at the output is shown in Figure 10.51. Application of the voltage divider rule yields

$$H(\omega) = \frac{R}{j\omega L + R + \dfrac{1}{j\omega C}} = \frac{RCj\omega}{(j\omega)^2 LC + RCj\omega + 1} = \frac{\dfrac{R}{L} j\omega}{-\omega^2 + \dfrac{R}{L} j\omega + \dfrac{1}{LC}} \quad \textbf{(10.56)}$$

FIGURE 10.50

Magnitude and phase response of an HPF.

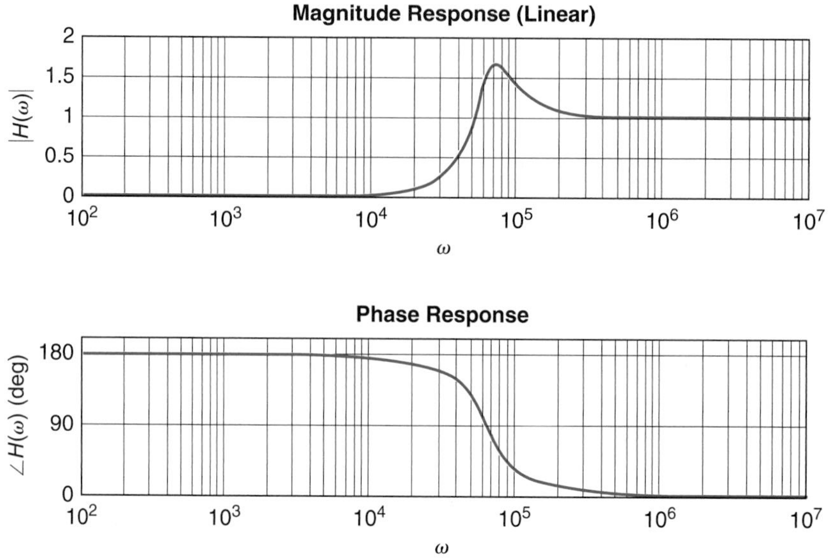

FIGURE 10.51

A series LCR circuit.

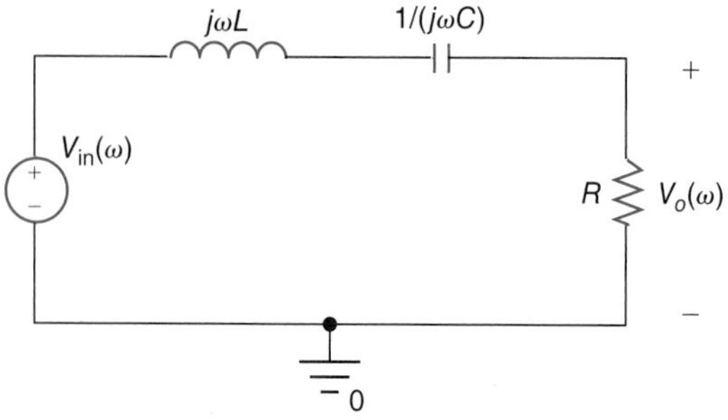

At $\omega = 0$, $H(\omega) = 0$. At $\omega = \infty$, $H(\omega) = 0$. At $\omega = 1/\sqrt{LC}$, $H(\omega) = 1$. This result can also be obtained directly from the circuit. At $\omega = 0$, the impedance of the inductor is $j\omega L = j0L = 0$, and the impedance of the capacitor is $1/(j\omega C) = 1/(j0C) = \infty$. The inductor can be treated as a short-circuit, and the capacitor can be treated as an open circuit. Since the current through the circuit is zero, the voltage across the resistor is equal to zero, making the transfer function equal to zero. At $\omega = \infty$, the impedance of the inductor is $j\omega L = j\infty L = \infty$, and the impedance of the capacitor is $1/(j\omega C) = 1/(j\infty C) = 0$. The inductor can be treated as an open circuit, and the capacitor can be treated as a short-circuit. No current flows through the circuit. The output voltage across the resistor is equal to zero, making the transfer function equal to zero. At the resonant frequency $\omega = \omega_0 = 1/\sqrt{LC}$, the impedance of the inductor is

$$j\omega L = j\frac{1}{\sqrt{LC}}L = j\sqrt{\frac{L}{C}}$$

and the impedance of the capacitor is

$$\frac{1}{j\omega C} = \frac{-j}{\dfrac{1}{\sqrt{LC}}C} = -j\sqrt{\frac{L}{C}}$$

The combined impedance of the inductor and the capacitor is equal to zero. Thus, the voltage across the resistor is equal to the input voltage. Thus, the transfer function is one at $\omega = \omega_0 = 1/\sqrt{LC}$. The magnitude at the resonant frequency is one, and the magnitude decreases below and above the resonant frequency. This is a bandpass filter (BPF).

The magnitude response is given by

$$|H(\omega)| = \frac{\dfrac{R}{L}\omega}{\sqrt{\left(\dfrac{1}{LC} - \omega^2\right)^2 + \left(\dfrac{R}{L}\omega\right)^2}} \tag{10.57}$$

The phase response is given by

$$\angle H(\omega) = \frac{\pi}{2} - \tan^{-1}\left(\frac{\frac{R}{L}\omega}{\frac{1}{LC} - \omega^2}\right) \text{ (rad)} \tag{10.58}$$

The magnitude response and the phase response are shown in Figure 10.52 for $R = 2\ k\Omega$, $L = 50$ mH, and $C = 5$ nF.

FIGURE 10.52

Magnitude and phase response of a bandpass filter.

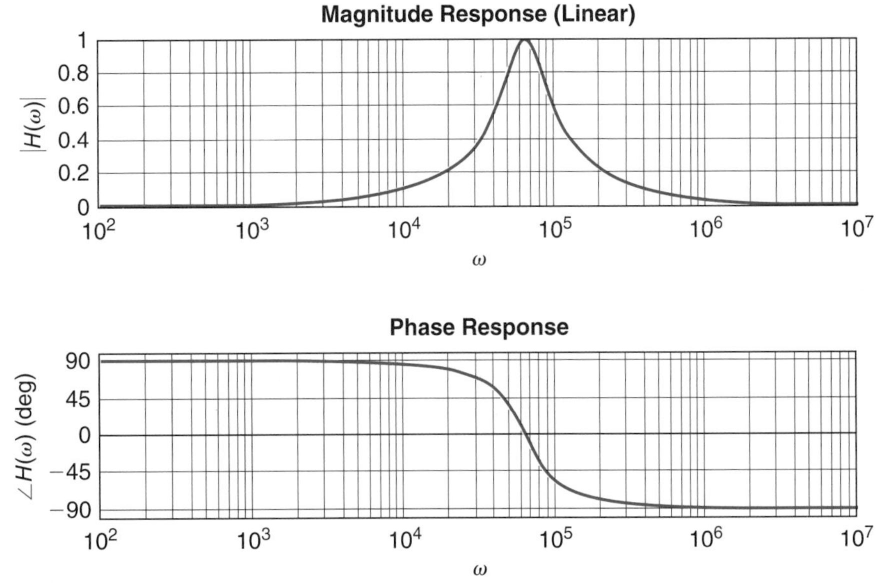

It will be discussed in Chapter 16 that the 3-dB cutoff frequencies are

$$\omega_1 = -\frac{R}{2L} + \frac{R}{2L}\sqrt{\frac{4L}{R^2C} + 1} \tag{10.59}$$

$$\omega_2 = \frac{R}{2L} + \frac{R}{2L}\sqrt{\frac{4L}{R^2C} + 1} \tag{10.60}$$

The difference between ω_2 and ω_1 is defined as the 3-dB bandwidth. Thus, the 3-dB bandwidth of the series RLC bandpass filter is given by

$$\omega_{3dB} = \omega_2 - \omega_1 = \frac{R}{L} \tag{10.61}$$

Notice that R/L is the coefficient of the $j\omega$ term in the denominator and numerator. For $R = 2\ k\Omega$, $L = 50$ mH, and $C = 5$ nF, we have

$$\omega_1 = 46{,}332.4958 \text{ rad/s}, \omega_2 = 86{,}332.4958 \text{ rad/s}, \omega_{3dB} = 40{,}000 \text{ rad/s}$$

A series RLC circuit with capacitor and inductor at the output is shown in Figure 10.53.

FIGURE 10.53

RCL circuit.

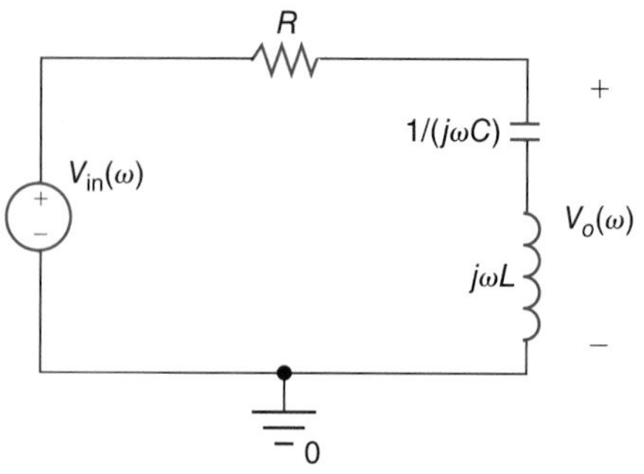

Application of the voltage divider rule yields

$$H(\omega) = \frac{j\omega L + \dfrac{1}{j\omega C}}{j\omega L + R + \dfrac{1}{j\omega C}} = \frac{(j\omega)^2 LC + 1}{(j\omega)^2 LC + RCj\omega + 1}$$
$$= \frac{-\omega^2 + \dfrac{1}{LC}}{-\omega^2 + \dfrac{R}{L}j\omega + \dfrac{1}{LC}} \tag{10.62}$$

At $\omega = 0$, $H(\omega) = 1$. At $\omega = \infty$, $H(\omega) = 1$. At $\omega = 1/\sqrt{LC}$, $H(\omega) = 0$. This result can also be obtained directly from the circuit. At $\omega = 0$, the impedance of the inductor is $j\omega L = j0L = 0$, and the impedance of the capacitor is $1/(j\omega C) = 1/(j0C) = \infty$. The inductor can be treated as a short-circuit, and the capacitor can be treated as an open circuit. Since the current through the circuit is zero, the voltage across the resistor is equal to zero, and the output voltage is equal to the source voltage, making the transfer function equal to 1. At $\omega = \infty$, the impedance of the inductor is $j\omega L = j\infty L = \infty$ and the impedance of the capacitor is $1/(j\omega C) = 1/(j\infty C) = 0$. The inductor can be treated as an open circuit, and the capacitor can be treated as a short-circuit. No current flows through the circuit. The voltage across the resistor is equal to zero, and the output voltage is equal to the source voltage, making the transfer function equal to 1. At the resonant frequency $\omega = \omega_0 = 1/\sqrt{LC}$, the impedance of the series connection of the capacitor and inductor is

$$\frac{1}{j\omega C} + j\omega L = \frac{1 - \omega^2 LC}{j\omega C} = \frac{1 - \dfrac{1}{LC}LC}{j\omega C} = \frac{1-1}{j\omega C} = 0$$

The combined impedance of the capacitor and the inductor equals zero (short-circuit). Thus, the output voltage is equal to zero. The gain of the filter at the resonant frequency is zero. This is a bandstop filter (BSF).

The magnitude response is given by

$$|H(\omega)| = \frac{\left| -\omega^2 + \dfrac{1}{LC} \right|}{\sqrt{\left(\dfrac{1}{LC} - \omega^2\right)^2 + \left(\dfrac{R}{L}\omega\right)^2}} \tag{10.63}$$

and the phase response is given by

$$\angle H(\omega) = -\tan^{-1}\left(\frac{\dfrac{R}{L}\omega}{\dfrac{1}{LC} - \omega^2}\right) \text{ (rad)} \tag{10.64}$$

The magnitude response and the phase response are shown in Figure 10.54 for $R = 2\ k\Omega$, $L = 50$ mH, and $C = 5$ nF.

FIGURE 10.54

Magnitude and phase response of a bandstop filter.

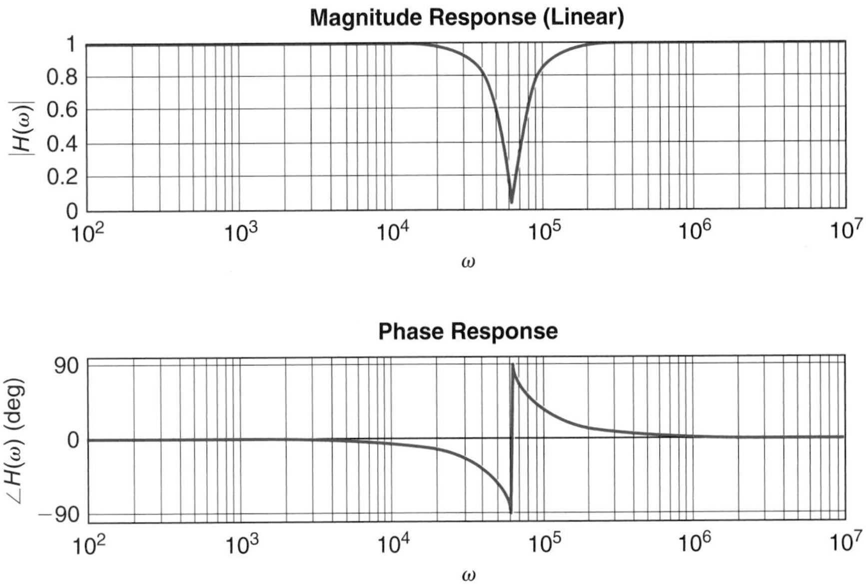

FIGURE 10.55

Parallel LRC circuit.

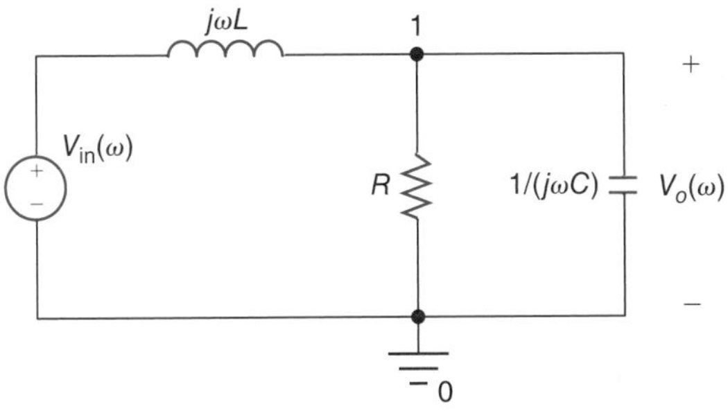

10.11.2 PARALLEL RLC CIRCUITS

A parallel RLC circuit with a capacitor and a resistor in parallel at the output is shown in Figure 10.55.

Summing the currents leaving node 1, we obtain

$$\frac{V_o(\omega) - V_{\text{in}}(\omega)}{j\omega L} + \frac{V_o(\omega)}{R} + \frac{V_o(\omega)}{\dfrac{1}{j\omega C}} = 0$$

which can be revised as

$$\left(j\omega C + \frac{1}{j\omega L} + \frac{1}{R}\right)V_o(\omega) = \frac{1}{j\omega L}V_{\text{in}}(\omega)$$

The transfer function is given by

$$H(\omega) = \frac{V_o(\omega)}{V_{\text{in}}(\omega)} = \frac{\dfrac{1}{j\omega L}}{j\omega C + \dfrac{1}{j\omega L} + \dfrac{1}{R}} = \frac{1}{(j\omega)^2 LC + \dfrac{L}{R}j\omega + 1}$$
$$= \frac{\dfrac{1}{LC}}{-\omega^2 + \dfrac{1}{RC}j\omega + \dfrac{1}{LC}} \quad \textbf{(10.65)}$$

At $\omega = 0, H(\omega) = 1$. At $\omega = \infty, H(\omega) = 0$. This result can also be obtained directly from the circuit. At $\omega = 0$, the impedance of the inductor is $j\omega L = j0L = 0$, and the impedance of the capacitor is $1/(j\omega C) = 1/(j0C) = \infty$. The inductor can be treated as a short-circuit, and the capacitor can be treated as an open circuit. The circuit reduces to a voltage source and a parallel resistor. Thus, the output voltage is equal to the input voltage, making the transfer function equal to 1. At $\omega = \infty$, the impedance of the inductor is $j\omega L = j\infty L = \infty$, and the impedance of the capacitor is $1/(j\omega C) = 1/(j\infty C) = 0$. The inductor can be treated as an open circuit, and the capacitor can be treated as a short-circuit. The output voltage across the capacitor is zero, making the transfer function equal to zero. This is an LPF. At the resonant frequency ($\omega = \omega_0 = 1/\sqrt{LC}$), $H(\omega) = -jR\sqrt{\dfrac{C}{L}}$. The magnitude response is given by

$$|H(\omega)| = \frac{\dfrac{1}{LC}}{\sqrt{\left(\omega^2 - \dfrac{1}{LC}\right)^2 + \left(\dfrac{\omega}{RC}\right)^2}} \tag{10.66}$$

The phase response is given by

$$\angle H(\omega) = -\tan^{-1}\left(\frac{\dfrac{1}{RC}\omega}{\dfrac{1}{LC} - \omega^2}\right) \text{ (rad)} \tag{10.67}$$

The magnitude response and the phase response are shown in Figure 10.56 for $R = 2\ k\Omega$, $L = 50$ mH, and $C = 5$ nF.

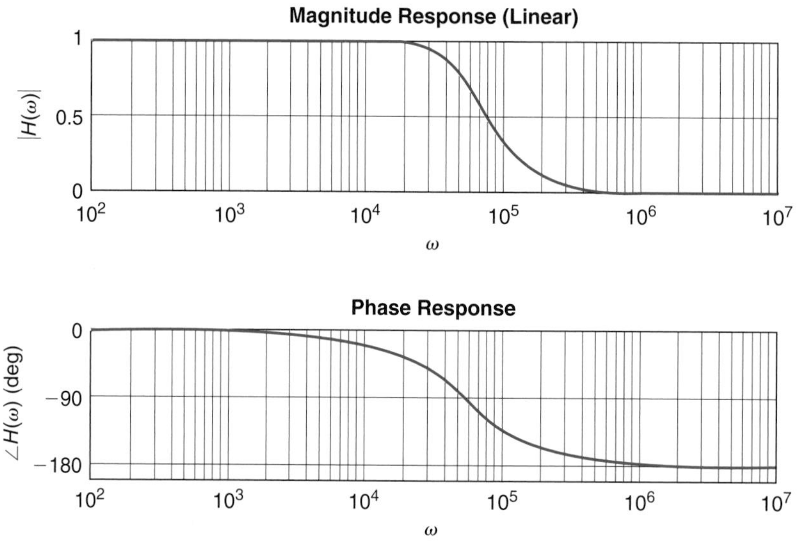

FIGURE 10.56
Magnitude and phase response of an LPF.

A parallel RLC circuit with an inductor and a resistor in parallel at the output is shown in Figure 10.57.

Summing the currents leaving node 1, we obtain

$$\frac{V_o(\omega) - V_{\text{in}}(\omega)}{\dfrac{1}{j\omega C}} + \frac{V_o(\omega)}{R} + \frac{V_o(\omega)}{j\omega L} = 0$$

which can be revised as

$$\left(j\omega C + \frac{1}{R} + \frac{1}{j\omega L}\right)V_o(\omega) = j\omega C V_{\text{in}}(\omega)$$

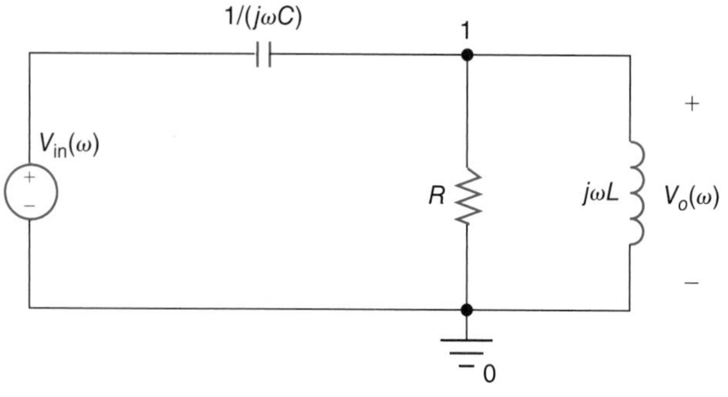

FIGURE 10.57
Parallel CRL circuit.

The transfer function is given by

$$H(\omega) = \frac{V_o(\omega)}{V_{in}(\omega)} = \frac{j\omega C}{j\omega C + \frac{1}{R} + \frac{1}{j\omega L}} = \frac{(j\omega)^2 LC}{(j\omega)^2 LC + \frac{L}{R} j\omega + 1}$$
$$= \frac{-\omega^2}{-\omega^2 + \frac{1}{RC} j\omega + \frac{1}{LC}} \quad \textbf{(10.68)}$$

At $\omega = 0$, $H(\omega) = 0$. At $\omega = \infty$, $H(\omega) = 1$. This result can also be obtained directly from the circuit. At $\omega = 0$, the impedance of the inductor is $j\omega L = j0L = 0$ and the impedance of the capacitor is $1/(j\omega C) = 1/(j0C) = \infty$. The inductor can be treated as a short-circuit, and the capacitor can be treated as an open circuit. The voltage across the inductor is equal to zero, making the transfer function equal to zero. At $\omega = \infty$, the impedance of the inductor is $j\omega L = j\infty L = \infty$, and the impedance of the capacitor is $1/(j\omega C) = 1/(j\infty C) = 0$. The inductor can be treated as an open circuit, and the capacitor can be treated as a short-circuit. The output voltage across the resistor is equal to the input voltage, making the transfer function equal to 1. This is an HPF.

At the resonant frequency $\omega = \omega_0 = 1/\sqrt{LC}$, $H(\omega) = jR\sqrt{\frac{C}{L}}$. The magnitude response is given by

$$|H(\omega)| = \frac{\omega^2}{\sqrt{\left(\omega^2 - \frac{1}{LC}\right)^2 + \left(\frac{\omega}{RC}\right)^2}} \quad \textbf{(10.69)}$$

and the phase response is given by

$$\angle H(\omega) = \pi - \tan^{-1}\left(\frac{\frac{\omega}{RC}}{\frac{1}{LC} - \omega^2}\right) \text{(rad)} \quad \textbf{(10.70)}$$

The magnitude response and the phase response are shown in Figure 10.58 for $R = 2\ k\Omega$, $L = 50$ mH, and $C = 5$ nF.

FIGURE 10.58

Magnitude and phase response of an HPF.

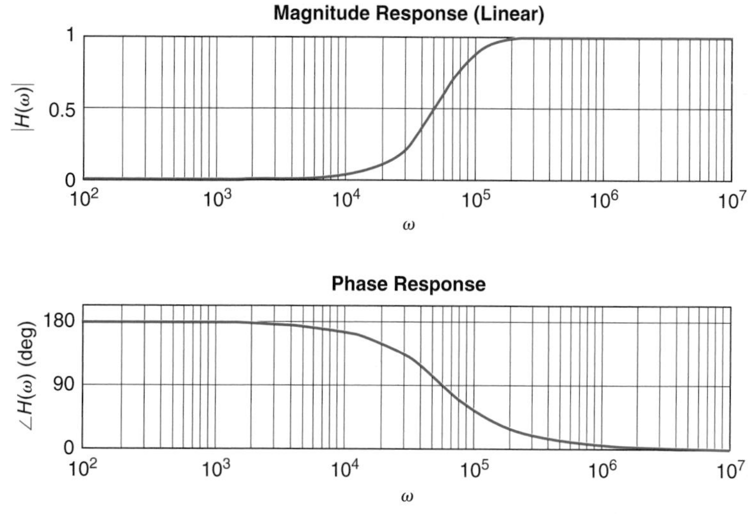

A parallel RLC circuit with a capacitor and an inductor in parallel at the output is shown in Figure 10.59.

FIGURE 10.59

Parallel RCL circuit.

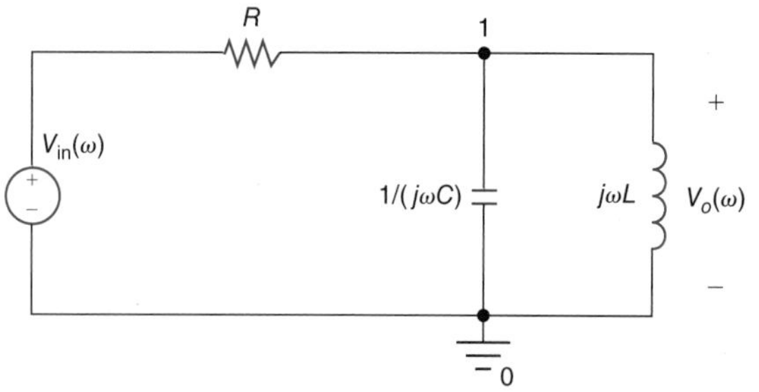

Summing the currents leaving node 1, we obtain

$$\frac{V_o(\omega) - V_{in}(\omega)}{R} + \frac{V_o(\omega)}{\frac{1}{j\omega C}} + \frac{V_o(\omega)}{j\omega L} = 0$$

which can be revised as

$$\left(j\omega C + \frac{1}{j\omega L} + \frac{1}{R}\right)V_o(\omega) = \frac{1}{R}V_{in}(\omega)$$

The transfer function is given by

$$H(\omega) = \frac{V_o(\omega)}{V_{in}(\omega)} = \frac{\frac{1}{R}}{j\omega C + \frac{1}{R} + \frac{1}{j\omega L}} = \frac{\frac{L}{R}j\omega}{(j\omega)^2 LC + \frac{L}{R}j\omega + 1}$$
$$= \frac{\frac{1}{RC}j\omega}{-\omega^2 + \frac{1}{RC}j\omega + \frac{1}{LC}} \quad \textbf{(10.71)}$$

At $\omega = 0$, $H(\omega) = 0$. At $\omega = \infty$, $H(\omega) = 0$. At $\omega = 1/\sqrt{LC}$, $H(\omega) = 1$. This result can also be obtained directly from the circuit. At $\omega = 0$, the impedance of the inductor is $j\omega L = j0L = 0$, and the impedance of the capacitor is $1/(j\omega C) = 1/(j0C) = \infty$. The inductor can be treated as a short-circuit, and the capacitor can be treated as an open circuit. The output voltage across the inductor is equal to zero, making the transfer function equal to zero. At $\omega = \infty$, impedance of the inductor is $j\omega L = j\infty L = \infty$, and the impedance of the capacitor is $1/(j\omega C) = 1/(j\infty C) = 0$. The inductor can be treated as an open circuit, and the capacitor can be treated as a short-circuit. The output voltage across the capacitor is equal to zero, making the transfer function equal to zero. At the resonant frequency, $\omega = \omega_0 = 1/\sqrt{LC}$, the equivalent impedance of the parallel connection of the capacitor and inductor is given by

$$\frac{j\omega L \times \frac{1}{j\omega C}}{j\omega L + \frac{1}{j\omega C}} = \frac{j\omega L}{-\omega^2 LC + 1} = \frac{j\omega L}{-\frac{1}{LC}LC + 1} = \frac{j\omega L}{-1 + 1} = \frac{j\omega L}{0} = \infty$$

No current flows through the circuit. The output voltage across the parallel connection of the capacitor and inductor is equal to the source voltage, making the transfer function equal to 1. This is a bandpass filter.

The magnitude response is given by

$$|H(\omega)| = \frac{\frac{1}{RC}\omega}{\sqrt{\left(\omega^2 - \frac{1}{LC}\right)^2 + \left(\frac{1}{RC}\omega\right)^2}} \quad \textbf{(10.72)}$$

The phase response is given by

$$\angle H(\omega) = \frac{\pi}{2} - \tan^{-1}\left(\frac{\frac{1}{RC}\omega}{\frac{1}{LC} - \omega^2}\right) \text{ (rad)} \quad \textbf{(10.73)}$$

The magnitude response and the phase response are shown in Figure 10.60 for $R = 2\,k\Omega$, $L = 50$ mH, and $C = 5$ nF.

FIGURE 10.60

Magnitude and phase response of a bandpass filter.

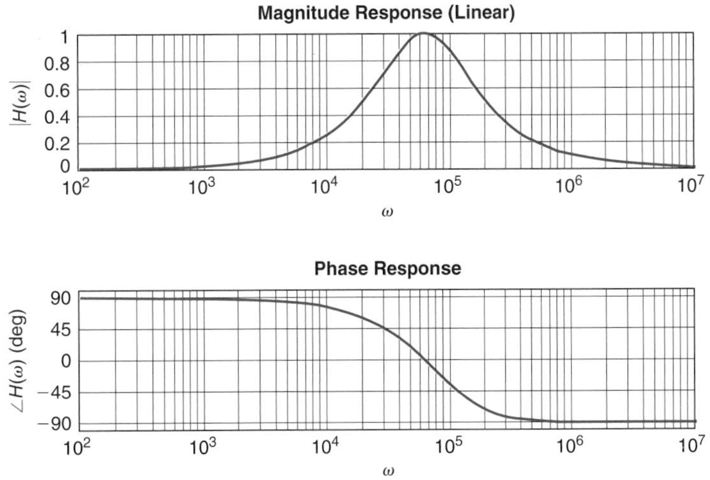

The magnitude at the resonant frequency is one, and the magnitude decreases below and above the resonant frequency. It will be discussed in Chapter 16 in more detail that the 3-dB cutoff frequencies are

$$\omega_1 = -\frac{1}{2RC} + \frac{1}{2RC}\sqrt{\frac{4R^2C}{L} + 1} \quad \textbf{(10.74)}$$

$$\omega_2 = \frac{1}{2RC} + \frac{1}{2RC}\sqrt{\frac{4R^2C}{L} + 1} \quad \textbf{(10.75)}$$

The difference between ω_2 and ω_1 is defined as the 3-dB bandwidth. Thus, the 3-dB bandwidth of the second-order bandpass filter is given by

$$\omega_{3dB} = \omega_2 - \omega_1 = \frac{1}{RC} \quad \textbf{(10.76)}$$

The bandwidth measures the width of the passband. Notice that $1/(RC)$ is the coefficient of the $j\omega$ term in the denominator and numerator. For the circuit shown in Figure 10.59 with $R = 2\,k\Omega$, $L = 50$ mH, and $C = 5$ nF, we have

$$\omega_1 = 46{,}332.4958 \text{ rad/s}, \omega_2 = 86{,}332.4958 \text{ rad/s}, \omega_{3dB} = 40{,}000 \text{ rad/s}$$

FIGURE 10.61

Parallel LCR circuit.

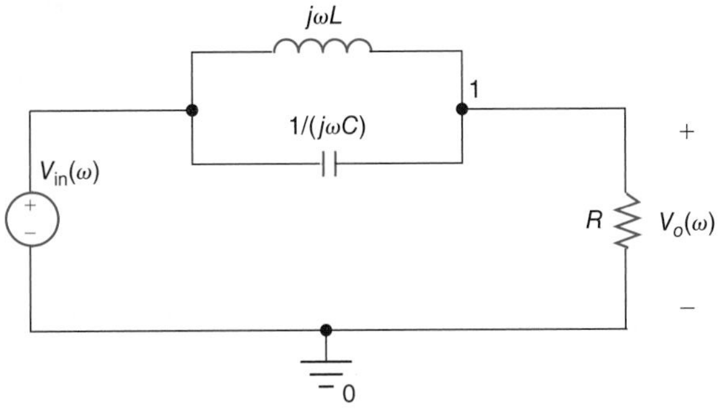

A parallel RLC circuit with a resistor at the output is shown in Figure 10.61.

Summing the currents leaving node 1, we obtain

$$\frac{V_o(\omega) - V_{in}(\omega)}{j\omega L} + \frac{V_o(\omega) - V_{in}(\omega)}{\dfrac{1}{j\omega C}} + \frac{V_o(\omega)}{R} = 0$$

which can be revised as

$$\left(j\omega C + \frac{1}{j\omega L} + \frac{1}{R}\right)V_o(\omega) = \left(j\omega C + \frac{1}{j\omega L}\right)V_{in}(\omega)$$

The transfer function is given by

$$H(\omega) = \frac{V_o(\omega)}{V_{\text{in}}(\omega)} = \frac{j\omega C + \dfrac{1}{j\omega L}}{j\omega C + \dfrac{1}{R} + \dfrac{1}{j\omega L}} = \frac{(j\omega)^2 LC + 1}{(j\omega)^2 LC + \dfrac{L}{R} j\omega + 1}$$
$$= \frac{-\omega^2 + \dfrac{1}{LC}}{-\omega^2 + \dfrac{1}{RC} j\omega + \dfrac{1}{LC}} \tag{10.77}$$

At $\omega = 0$, $H(\omega) = 1$. At $\omega = \infty$, $H(\omega) = 1$. At $\omega = 1/\sqrt{LC}$, $H(\omega) = 0$. This result can also be obtained directly from the circuit. At $\omega = 0$, the impedance of the inductor is $j\omega L = j0L = 0$, and the impedance of the capacitor is $1/(j\omega C) = 1/(j0C) = \infty$. The inductor can be treated as a short-circuit, and the capacitor can be treated as an open circuit. The circuit reduces to a voltage source and a parallel resistor. Thus, the output voltage across the resistor is equal to the source voltage, making the transfer function equal to 1. At $\omega = \infty$, the impedance of the inductor is $j\omega L = j\infty L = \infty$, and the impedance of the capacitor is $1/(j\omega C) = 1/(j\infty C) = 0$. The inductor can be treated as an open circuit, and the capacitor can be treated as a short-circuit. The circuit reduces to a voltage source and a parallel resistor. Thus, the output voltage across the resistor is equal to the source voltage, making the transfer function equal to 1. At the resonant frequency, $\omega = \omega_0 = 1/\sqrt{LC}$, the equivalent impedance of the parallel connection of the inductor and capacitor is given by

$$\frac{j\omega L \times \dfrac{1}{j\omega C}}{j\omega L + \dfrac{1}{j\omega C}} = \frac{j\omega L}{-\omega^2 LC + 1} = \frac{j\omega L}{-\dfrac{1}{LC} LC + 1} = \frac{j\omega L}{-1 + 1} = \frac{j\omega L}{0} = \infty$$

No current flows through the circuit. The output voltage across the resistor is equal to zero, making the transfer function equal to zero. This is a bandstop filter.

The magnitude response is given by

$$|H(\omega)| = \frac{\left| -\omega^2 + \dfrac{1}{LC} \right|}{\sqrt{\left(\omega^2 - \dfrac{1}{LC}\right)^2 + \left(\dfrac{1}{RC}\omega\right)^2}} \tag{10.78}$$

The phase response is given by

$$\angle H(\omega) = -\tan^{-1}\left(\frac{\dfrac{1}{RC}\omega}{\dfrac{1}{LC} - \omega^2}\right) \text{(rad)} \tag{10.79}$$

The magnitude response and the phase response are shown in Figure 10.62 for $R = 2\ k\Omega$, $L = 50$ mH, and $C = 5$ nF.

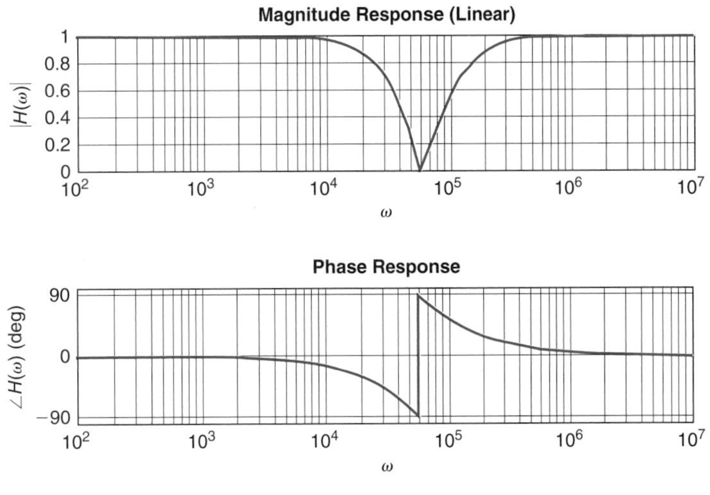

FIGURE 10.62

Magnitude response and phase response of a bandstop filter.

The magnitude at the resonant frequency is zero, and the magnitude increases below and above the resonant frequency. It will be discussed in Chapter 16 in more detail that the 3-dB cutoff frequencies are

$$\omega_1 = -\frac{1}{2RC} + \frac{1}{2RC}\sqrt{\frac{4R^2C}{L} + 1} \tag{10.80}$$

$$\omega_2 = \frac{1}{2RC} + \frac{1}{2RC}\sqrt{\frac{4R^2C}{L} + 1} \tag{10.81}$$

The difference between ω_2 and ω_1 is defined as the 3-dB bandwidth. Thus, the 3-dB bandwidth of the second-order bandstop filter is given by

$$\omega_{3dB} = \omega_2 - \omega_1 = \frac{1}{RC} \tag{10.82}$$

The bandwidth measures the width of the null. Notice that $1/(RC)$ is the coefficient of the $j\omega$ term in the denominator. For the circuit shown in Figure 10.61 with $R = 2\ k\Omega$, $L = 50$ mH, and $C = 5$ nF, we have

$$\omega_1 = 46{,}332.4958 \text{ rad/s}, \omega_2 = 86{,}332.4958 \text{ rad/s}, \omega_{3dB} = 40{,}000 \text{ rad/s}$$

EXAMPLE 10.11

Find the transfer function for the circuit shown in Figure 10.63, and state the type of filter (LPF, HPF, BPF, BSF).

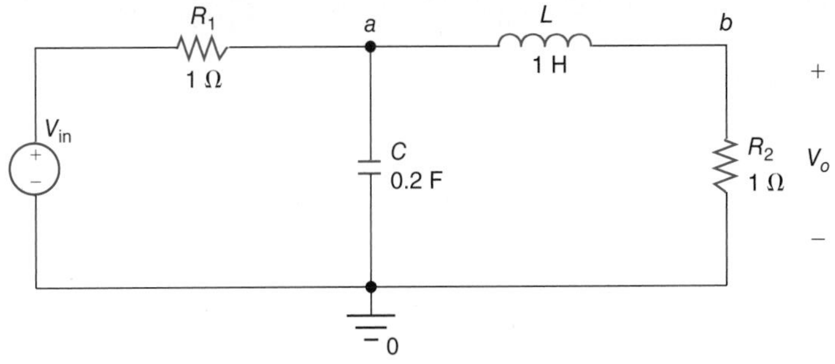

FIGURE 10.63

Circuit for EXAMPLE 10.11.

continued

Example 10.11 continued

Summing the currents leaving node b, we obtain

$$\frac{V_o - V_a}{j\omega \times 1} + \frac{V_o}{1} = 0$$

Multiplication by $j\omega$ yields

$$V_o - V_a + j\omega V_o = 0$$

Solving for V_a, we obtain

$$V_a = (j\,\omega + 1)V_o \tag{10.83}$$

Summing the currents leaving node a, we obtain

$$\frac{V_a - V_{\text{in}}}{1} + \frac{V_a}{\dfrac{1}{j\omega 0.2}} + \frac{V_a - V_o}{j\omega} = 0$$

which can be revised as

$$\left(j\omega 0.2 + 1 + \frac{1}{j\omega}\right)V_a - \frac{1}{j\omega}V_o = V_{\text{in}} \tag{10.84}$$

Substitution of Equation (10.83) into (10.84) yields

$$\left(j\omega 0.2 + 1 + \frac{1}{j\omega}\right)(j\omega + 1)V_o - \frac{1}{j\omega}V_o = V_{\text{in}} \tag{10.85}$$

Taking the ratio of V_o to V_{in} in Equation (10.85), we obtain

$$\begin{aligned} H(\omega) = \frac{V_o}{V_{\text{in}}} &= \frac{1}{\left(j\omega 0.2 + 1 + \dfrac{1}{j\omega}\right)(j\omega + 1) - \dfrac{1}{j\omega}} \\ &= \frac{1}{(j\omega)^2 0.2 + j\omega + 1 + j\omega 0.2 + 1 + \dfrac{1}{j\omega} - \dfrac{1}{j\omega}} \\ &= \frac{5}{(j\omega)^2 + 6j\omega + 10} = \frac{5}{-\omega^2 + 6j\omega + 10} \end{aligned} \tag{10.86}$$

At $\omega = 0, H(\omega) = 0.5$. At $\omega = \infty, H(\omega) = 0$. This is an LPF. If the numerator is a constant, it is an LPF.

Exercise 10.11

Find the transfer function for the circuit shown in Figure 10.64 and state the type of filter (LPF, HPF, BPF, or BSF).

continued

Exercise 10.11 continued

FIGURE 10.64

Circuit for EXERCISE 10.11.

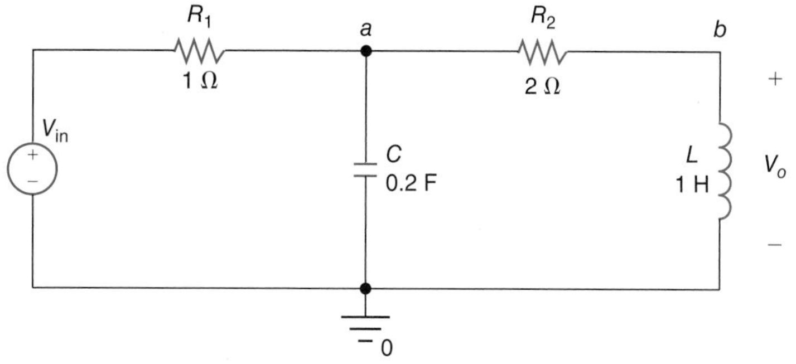

Answer:

$$H(\omega) = \frac{5j\omega}{(j\omega)^2 + 7j\omega + 15}.$$ Bandpass filter.

EXAMPLE 10.12

Find the transfer function for the circuit shown in Figure 10.65.

FIGURE 10.65

Circuit for EXAMPLE 10.12.

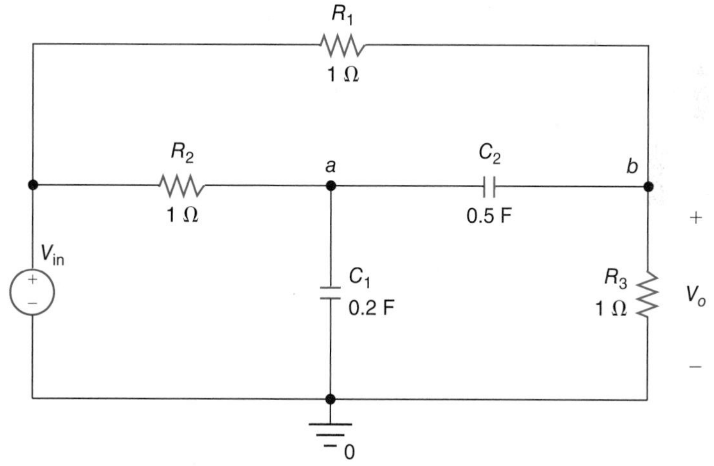

Summing the currents leaving node a, we obtain

$$\frac{V_a - V_{\text{in}}}{1} + \frac{V_a}{\frac{1}{j\omega 0.2}} + \frac{V_a - V_o}{\frac{1}{j\omega 0.5}} = 0$$

which can be revised as

$$(0.7\, j\omega + 1)\, V_a - 0.5\, j\omega V_o = V_{\text{in}} \quad \textbf{(10.87)}$$

continued

Example 10.12 continued

Summing the currents leaving node b, we obtain

$$\frac{V_o - V_{\text{in}}}{1} + \frac{V_o - V_a}{\frac{1}{j\omega 0.5}} + \frac{V_o}{1} = 0$$

which can be revised as

$$-0.5j\omega V_a + (j\omega 0.5 + 2)V_o = V_{\text{in}} \tag{10.88}$$

Application of Cramer's rule on Equations (10.87) and (10.88) yields

$$V_o = \frac{\begin{vmatrix} 0.7j\omega + 1 & 1 \\ -0.5j\omega & 1 \end{vmatrix}}{\begin{vmatrix} 0.7j\omega + 1 & -0.5j\omega \\ -0.5j\omega & 0.5j\omega + 2 \end{vmatrix}} V_{\text{in}} = \frac{1.2j\omega + 1}{0.35(j\omega)^2 + 1.9j\omega + 2 - 0.25(j\omega)^2} V_{\text{in}} \tag{10.89}$$

Taking the ratio of V_o to V_{in} in Equation (10.89), we obtain

$$H(\omega) = \frac{V_o}{V_{\text{in}}} = \frac{1.2j\omega + 1}{0.1(j\omega)^2 + 1.9j\omega + 2} = \frac{12j\omega + 10}{(j\omega)^2 + 19j\omega + 20} \tag{10.90}$$

At $\omega = 0$, $H(\omega) = 0.5$. At $\omega = \infty$, $H(\omega) = 0$. This is an LPF.

Exercise 10.12

Find the transfer function for the circuit shown in Figure 10.66.

FIGURE 10.66

Circuit for EXERCISE 10.12.

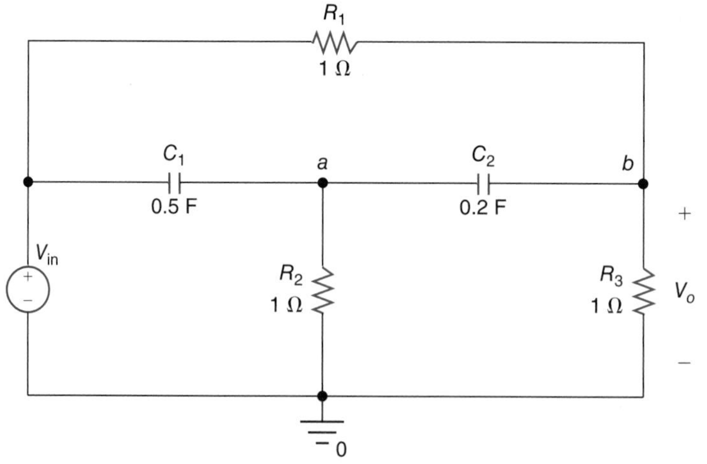

Answer:

$$H(\omega) = \frac{(j\omega)^2 + 7j\omega + 10}{(j\omega)^2 + 16j\omega + 20}. \text{ HPF.}$$

EXAMPLE 10.13

Find the transfer function for the circuit shown in Figure 10.67.

FIGURE 10.67

Circuit for EXAMPLE 10.13.

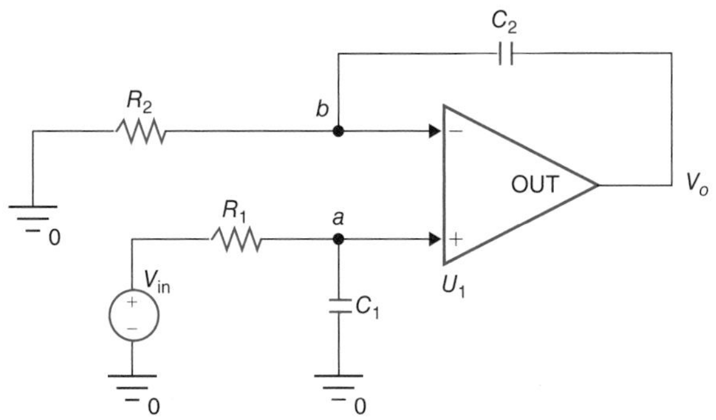

We can find the voltage at node a by applying the voltage divider rule.

$$V_a = \frac{\frac{1}{j\omega C_1}}{R_1 + \frac{1}{j\omega C_1}} V_{\text{in}} = \frac{\frac{1}{R_1 C_1}}{j\omega + \frac{1}{R_1 C_1}} V_{\text{in}} \tag{10.91}$$

Summing the currents leaving node b, we obtain

$$\frac{V_b}{R_2} + \frac{V_b - V_o}{\frac{1}{j\omega C_2}} = 0$$

which can be revised as

$$\left(j\omega C_2 + \frac{1}{R_2}\right) V_b = j\omega C_2 V_o$$

Solving for V_b, we obtain

$$V_b = \frac{j\omega}{j\omega + \frac{1}{R_2 C_2}} V_o \tag{10.92}$$

Since $V_a = V_b$, from Equations (10.91) and (10.92), we obtain

$$\frac{\frac{1}{R_1 C_1}}{j\omega + \frac{1}{R_1 C_1}} V_{\text{in}} = \frac{j\omega}{j\omega + \frac{1}{R_2 C_2}} V_o \tag{10.93}$$

Taking the ratio of V_o to V_{in}, we obtain

$$H(\omega) = \frac{V_o}{V_{\text{in}}} = \frac{1}{R_1 C_1} \frac{j\omega + \frac{1}{R_2 C_2}}{j\omega\left(j\omega + \frac{1}{R_1 C_1}\right)} \tag{10.94}$$

Exercise 10.13

Find the transfer function for the circuit shown in Figure 10.68.

FIGURE 10.68

Circuit for EXERCISE 10.13.

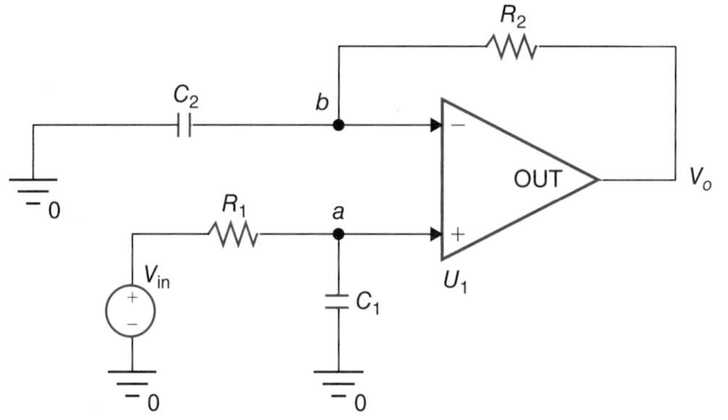

Answer:

$$H(s) = \frac{R_2C_2}{R_1C_1}\frac{j\omega + \dfrac{1}{R_2C_2}}{j\omega + \dfrac{1}{R_1C_1}}. \quad \textbf{(10.95)}$$

10.12 PSpice and Simulink

A series RLC circuit is shown in Figure 10.69. The magnitude response and the phase response of this circuit can be obtained by selecting AC Sweep/Noise as the analysis type and choosing the simulation settings as follows:

Logarithmic, Start frequency: 1e3, End Frequency: 1e5, Points/Decade: 100

To get the magnitude response, put the voltage level marker at pin 1 of R_1 as shown in Figure 10.69.

To get the phase response, select PSpice → Markers → Advanced → Phase of Voltage and place the voltage level marker at pin 1 of R_1. The plots are shown in Figure 10.70.

FIGURE 10.69

Series RLC circuit.

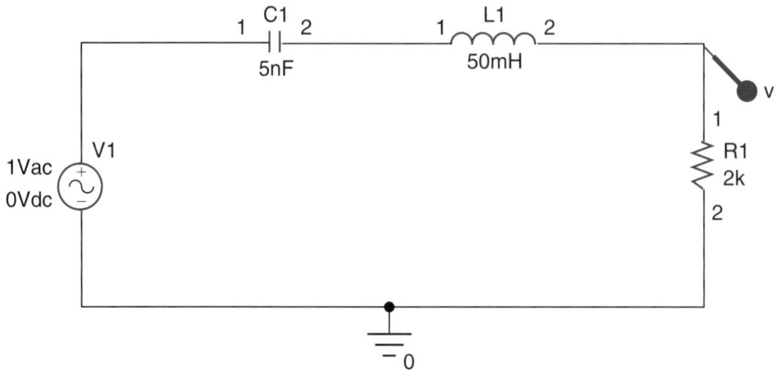

FIGURE 10.70

The magnitude response and the phase response of a BPF.

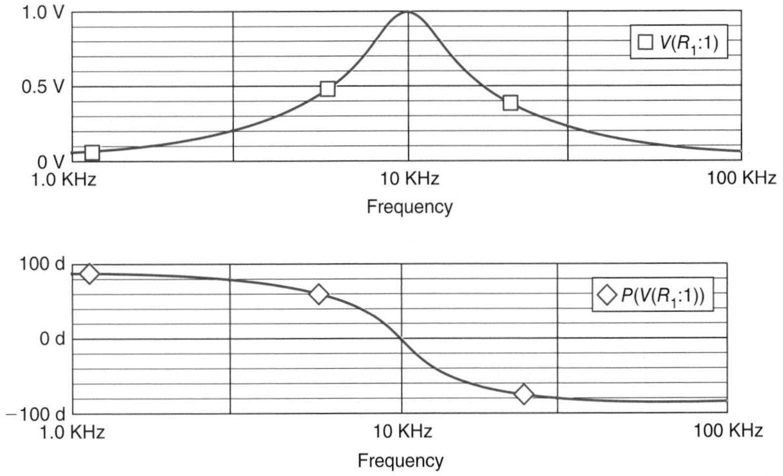

Exercise 10.14

Plot the magnitude response and the phase response of the circuit shown in Figure 10.71.

FIGURE 10.71

Circuit for EXERCISE 10.14.

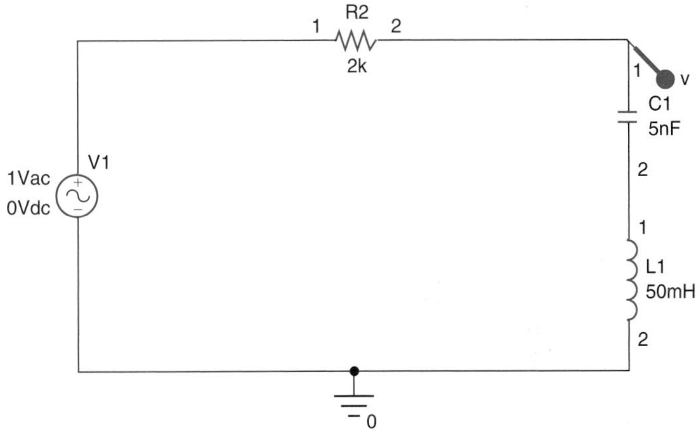

Answer:

The magnitude response and the phase response are shown in Figure 10.72.

FIGURE 10.72

Magnitude response and the phase response.

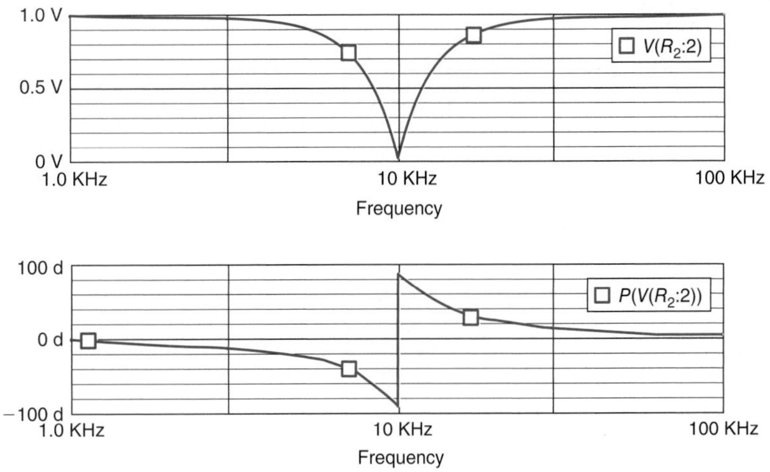

A Simulink model to plot the magnitude response of a series RLC bandpass filter is shown in Figure 10.73.

FIGURE 10.73

A Simulink model to plot the magnitude response.

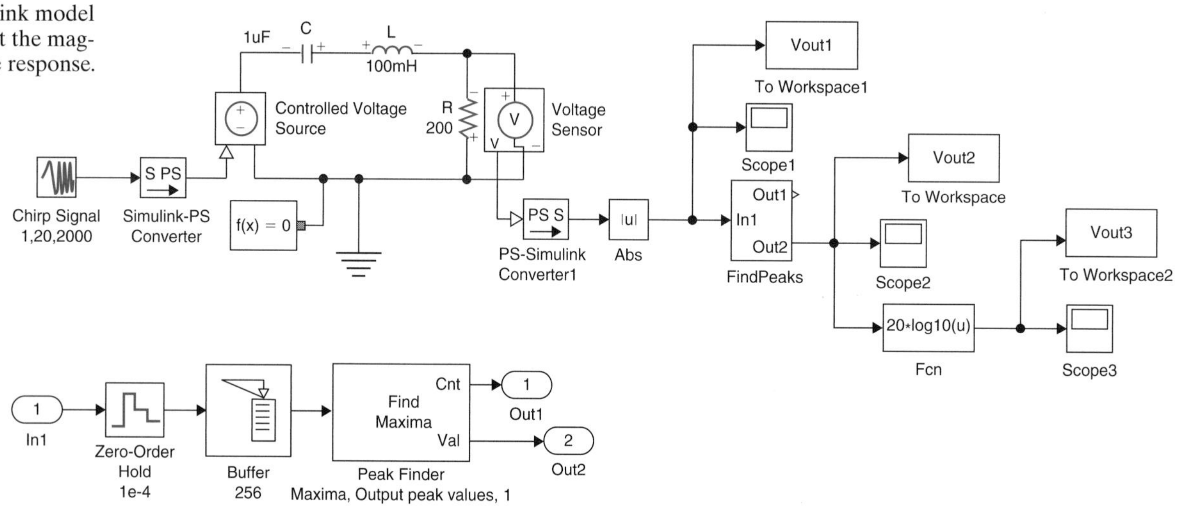

The details of the subsystem titled FindPeaks is the model shown under the RLC circuit. The subsystem consists of Zero-Order Hold, Buffer, and Peak Finder. The settings are shown under each block. The circuit is driven by a Chirp Signal whose frequency is scanned to 2000 Hz. The target time of 20s (Simulation stop time) matches the frequency of 2000 Hz. The data from the model are exported to Workspace. From this data, the magnitude response can be plotted by typing

```
Vo=Vout1.Data;
k=0:length(Vo)-1;
f=2000*k/length(Vo);
plot(f,Vo);grid on;
xlabel('f (Hz)');ylabel('V_o_u_t_1(f)');
set(gca,'YTick',0:0.2:1.2)
set(gca,'YTickLabel',0:0.2:1.2)
axis([0,2000,0,1.2])
```

The Save format is set for Timeseries. Change Vout1 to Vout2, and then again to Vout3 to get all three plots.

FIGURE 10.74

Magnitude response of a series RLC bandpass filter.

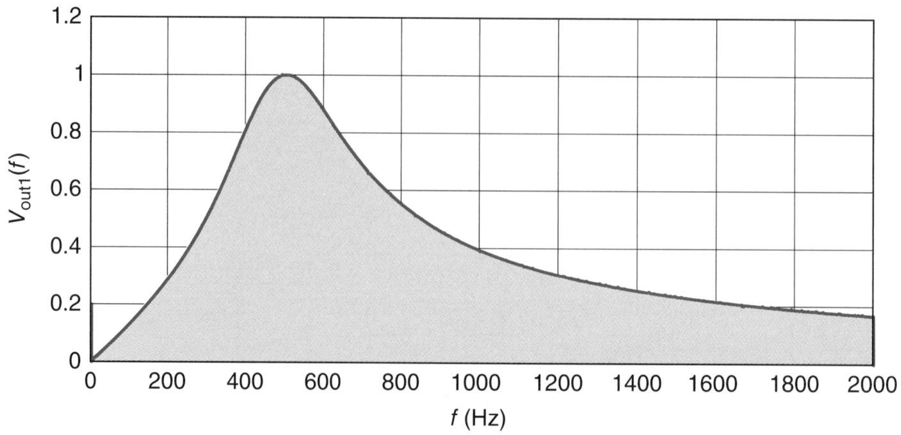

FIGURE 10.74

continued

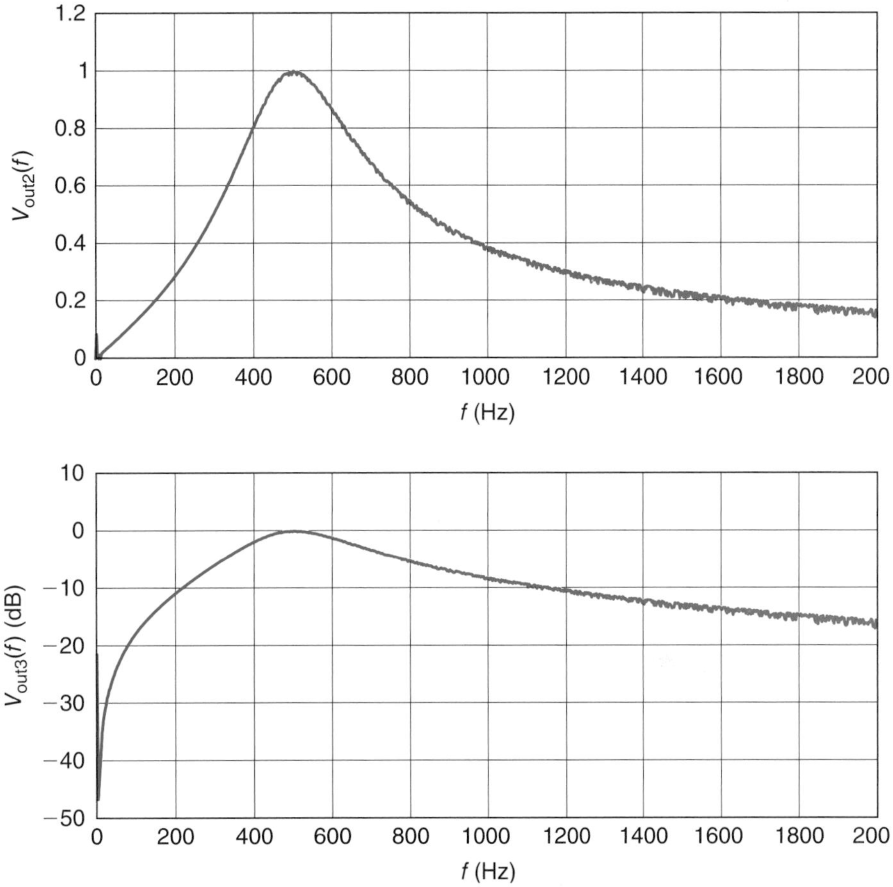

SUMMARY

If the sources of a circuit are transformed into phasors and all the components are transformed into impedances, the circuit becomes a phasor-transformed circuit. The phasor-transformed circuit can be analyzed by applying the same laws and theorems derived for resistive circuits. The difference is that the impedances, voltages, and currents are complex quantities in general. The complex quantities translate into the magnitude and phase of sinusoids.

The circuit laws and theorems that apply to phasor-transformed circuits include Ohm's law, Kirchhoff's voltage law (KVL), Kirchhoff's current law (KCL), the voltage divider rule, the current divider rule, Thévenin's theorem, and Norton's theorem. The nodal analysis and mesh analysis provide a systematic approach for analyzing phasor-transformed circuits.

The *transfer function* is defined as the ratio of the output to input in the frequency domain. The magnitude of the transfer function is called the *magnitude response,* and the phase of the transfer function is the *phase response.* The shape of the magnitude response and the phase response as a function of frequency reveal the characteristics of the circuit in the frequency domain. If the circuit is applied in filtering out unwanted signals, the circuit designers have options for the arrangement of components and the choice of values to achieve the goal.

PROBLEMS

Voltage Divider Rule

10.1 For the circuit shown in Figure P10.1, draw the phasor-transformed circuit and use the voltage divider rule to find the phasor $\mathbf{V}_o$ and time domain signal $v_o(t)$.

FIGURE P10.1

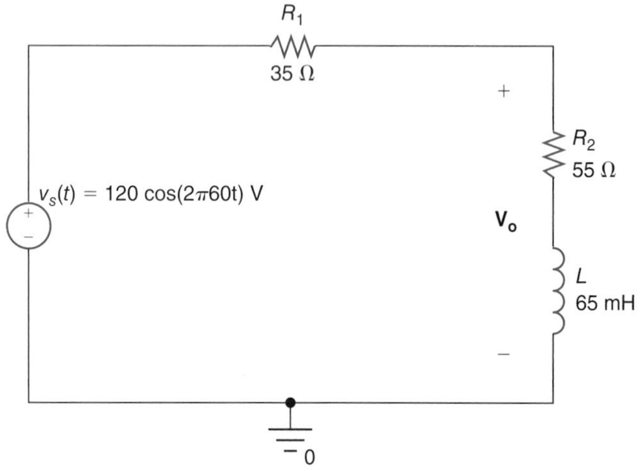

10.2 For the circuit shown in Figure P10.2, draw the phasor-transformed circuit and use the voltage divider rule to find the phasor $\mathbf{V}_o$ and time domain signal $v_o(t)$.

FIGURE P10.2

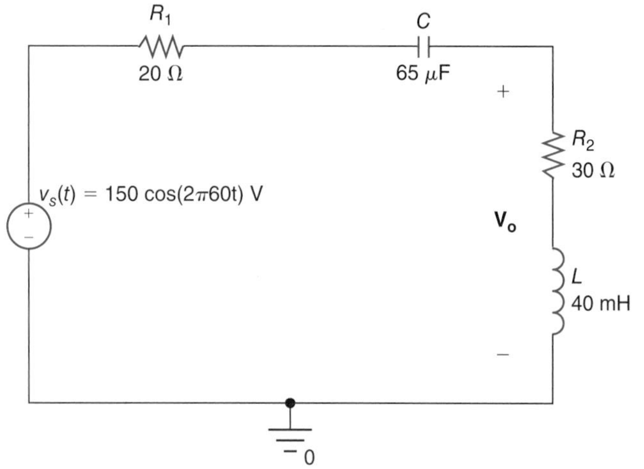

10.3 For the circuit shown in Figure P10.3, draw the phasor-transformed circuit and use the voltage divider rule to find the phasor $\mathbf{V}_o$ and time domain signal $v_o(t)$.

FIGURE P10.3

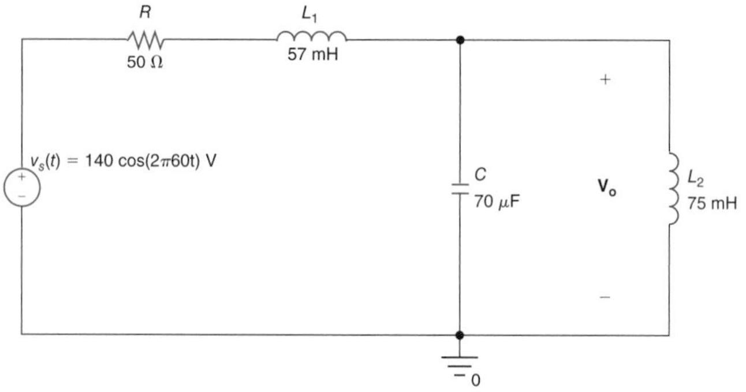

10.4 For the circuit shown in Figure P10.4, draw the phasor-transformed circuit and use the voltage divider rule to find the phasor $\mathbf{V}_o$ and time domain signal $v_o(t)$.

FIGURE P10.4

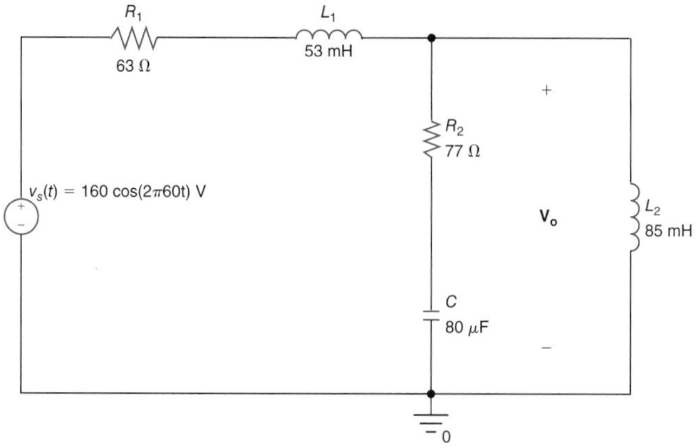

10.5 For the circuit shown in Figure P10.5, draw the phasor-transformed circuit and use the voltage divider rule to find the phasor $\mathbf{V}_o$ and time domain signal $v_o(t)$.

FIGURE P10.5

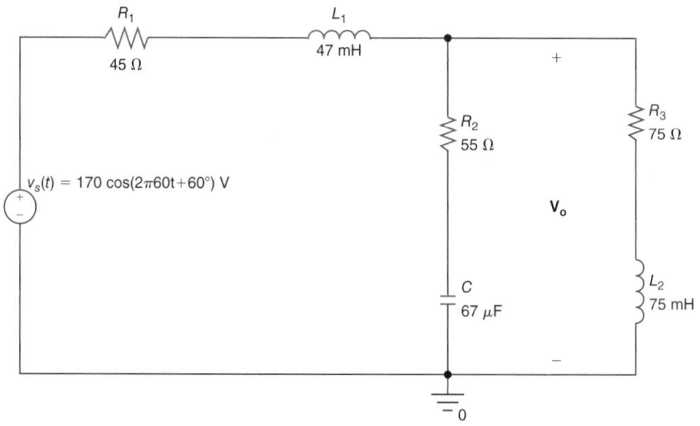

10.6 For the circuit shown in Figure P10.6, use the voltage divider rule to find the phasor $\mathbf{V}_o$.

FIGURE P10.6

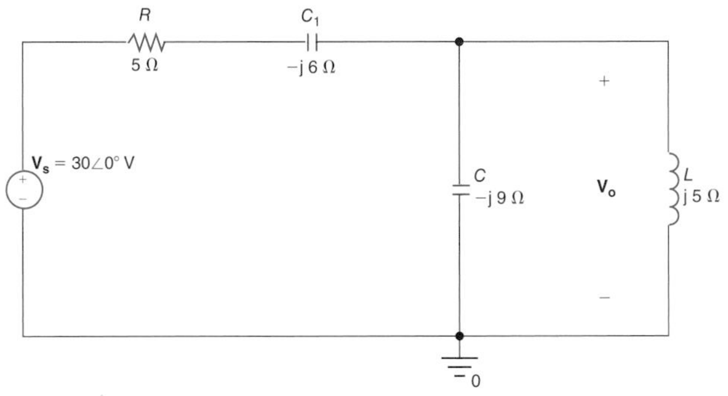

10.7 For the circuit shown in Figure P10.7, use the voltage divider rule to find the phasor $\mathbf{V}_o$.

FIGURE P10.7

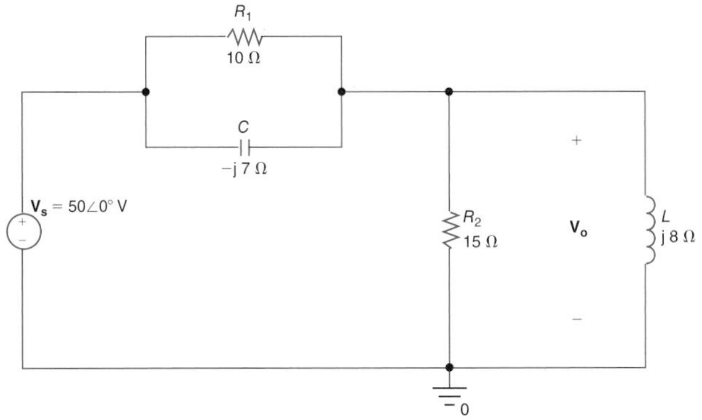

10.8 For the circuit shown in Figure P10.8, use the voltage divider rule to find the phasor $\mathbf{V}_o$.

FIGURE P10.8

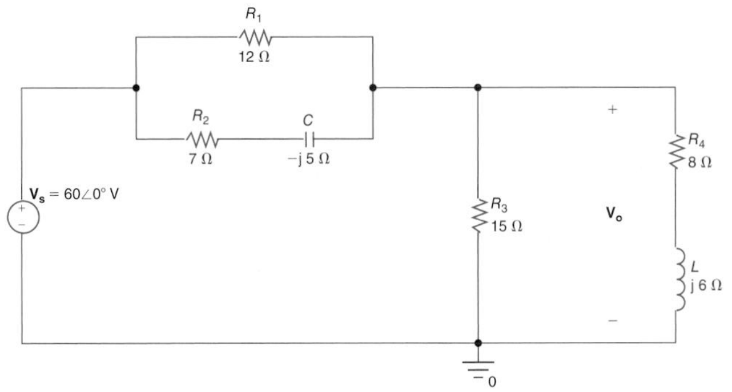

Current Divider Rule

10.9 For the circuit shown in Figure P10.9, draw the phasor-transformed circuit and use the current divider rule to find the phasors $\mathbf{I}_1$ and $\mathbf{V}_o$ and time domain signal $v_o(t)$.

FIGURE P10.9

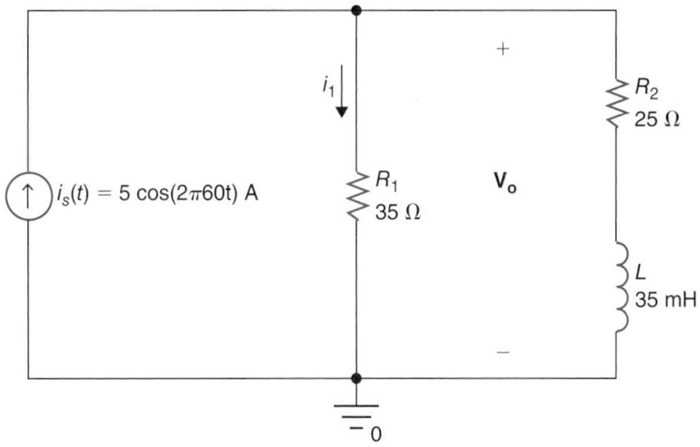

10.10 For the circuit shown in Figure P10.10, draw the phasor-transformed circuit and use the current divider rule to find the phasors $\mathbf{I}_1$ and $\mathbf{V}_o$ and time domain signal $v_o(t)$.

FIGURE P10.10

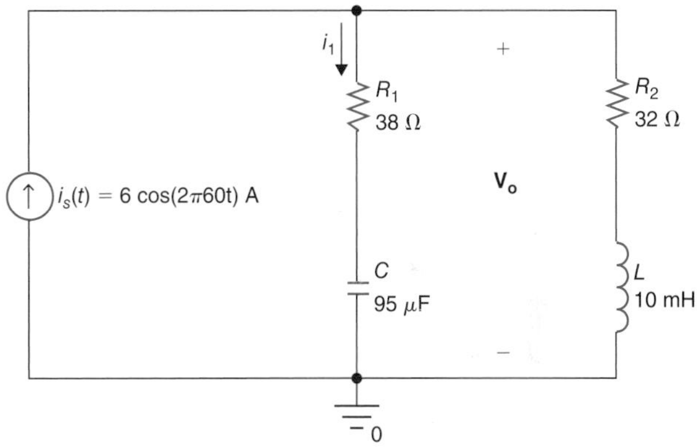

10.11 For the circuit shown in Figure P10.11, draw the phasor-transformed circuit and use the current divider rule to find the phasors $\mathbf{I}_2$ and $\mathbf{V}_o$ and time domain signal $v_o(t)$.

FIGURE P10.11

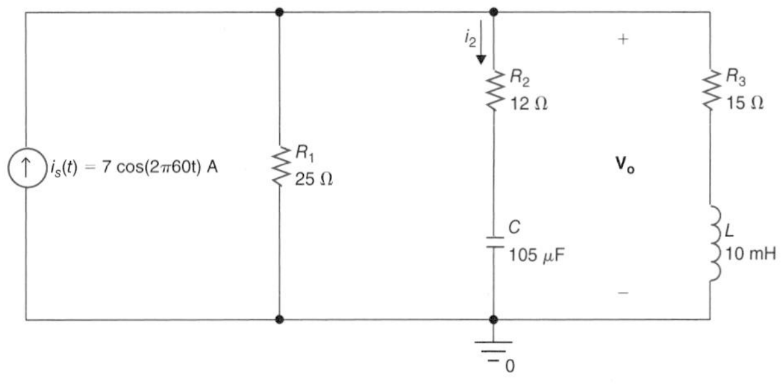

10.12 For the circuit shown in Figure P10.12, use the current divider rule to find the phasors $\mathbf{I}_3$ and $\mathbf{V}_o$.

FIGURE P10.12

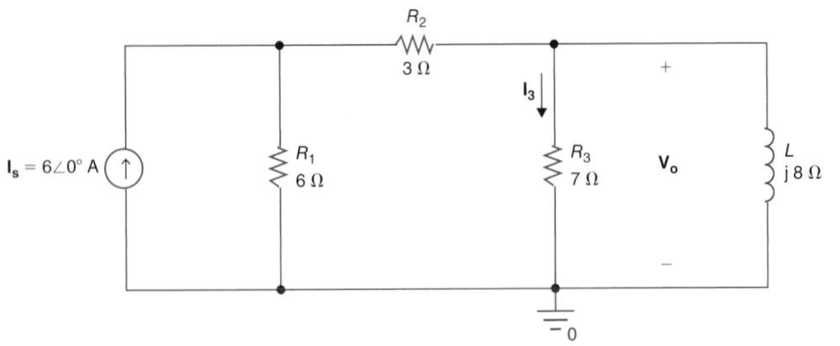

10.13 For the circuit shown in Figure P10.13, use the current divider rule to find the phasors $\mathbf{I}_3$ and $\mathbf{V}_o$.

FIGURE P10.13

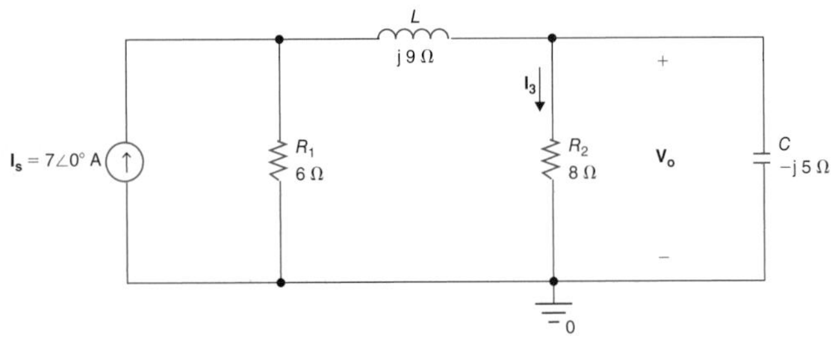

10.14 For the circuit shown in Figure P10.14, use the current divider rule to find the phasors $\mathbf{I}_3$ and $\mathbf{V}_o$.

FIGURE P10.14

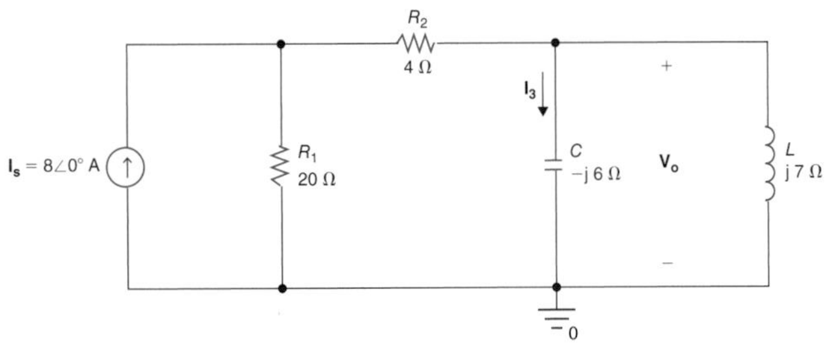

10.15 For the circuit shown in Figure P10.15, use the current divider rule to find the phasors $\mathbf{I}_3$ and $\mathbf{V}_o$.

FIGURE P10.15

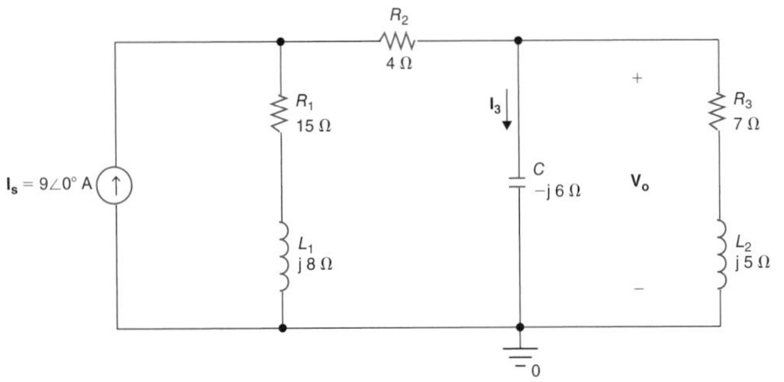

10.16 For the circuit shown in Figure P10.16, use the current divider rule to find the phasors $\mathbf{I}_3$ and $\mathbf{V}_o$.

FIGURE P10.16

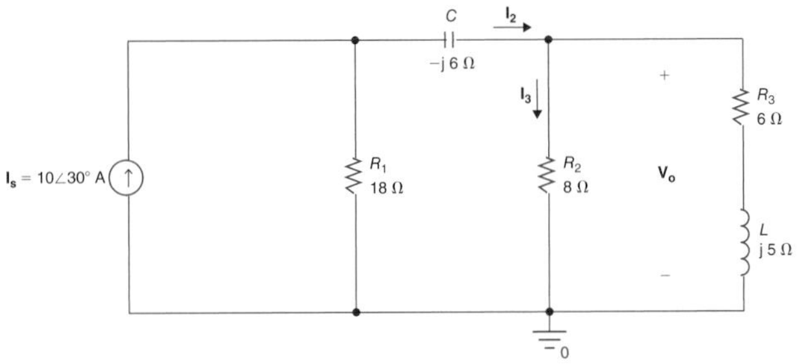

Nodal Analysis

10.17 For the circuit shown in Figure P10.17, use nodal analysis to find the output phasor $\mathbf{V}_o$.

FIGURE P10.17

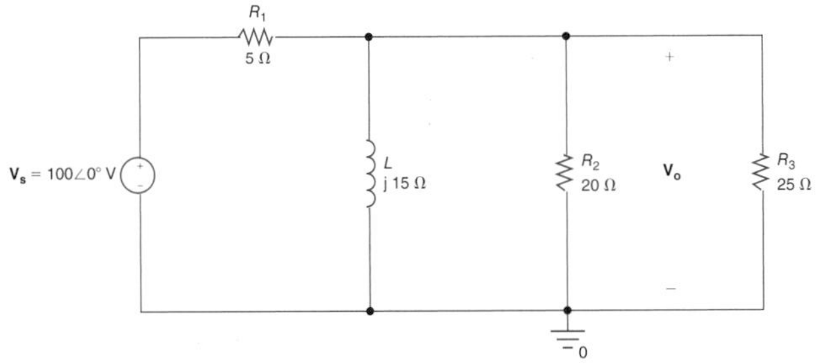

10.18 For the circuit shown in Figure P10.18, use nodal analysis to find the output phasor $\mathbf{V}_o$.

FIGURE P10.18

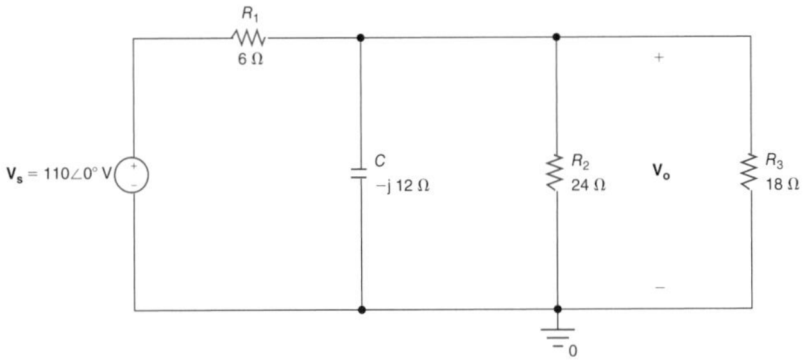

10.19 For the circuit shown in Figure P10.19, use nodal analysis to find the output phasor $\mathbf{V}_o$.

FIGURE P10.19

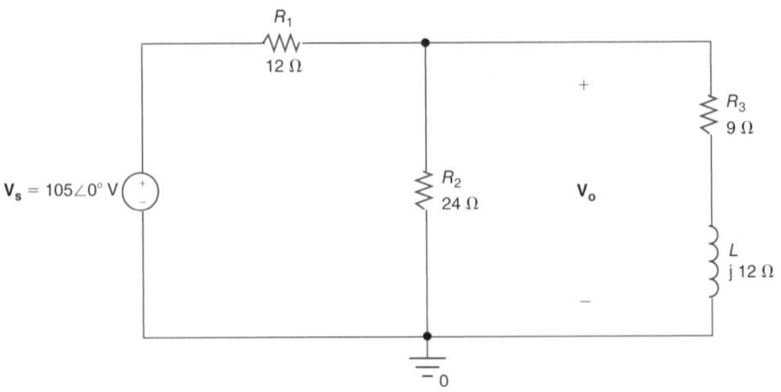

10.20 For the circuit shown in Figure P10.20, use nodal analysis to find the output phasor $\mathbf{V}_o$.

FIGURE P10.20

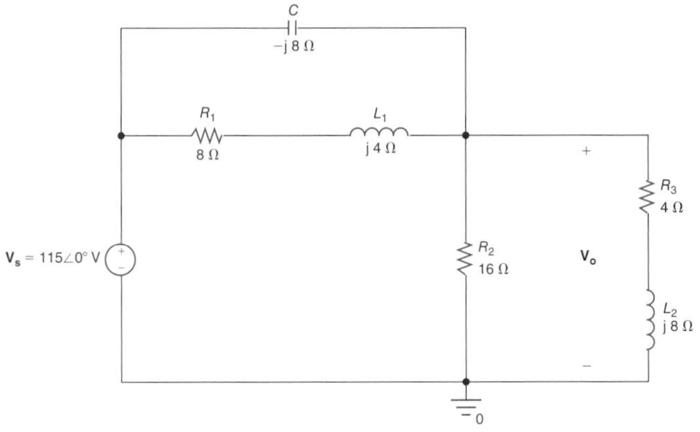

10.21 For the circuit shown in Figure P10.21, use nodal analysis to find the phasors $\mathbf{V}_1$ and $\mathbf{V}_2$.

FIGURE P10.21

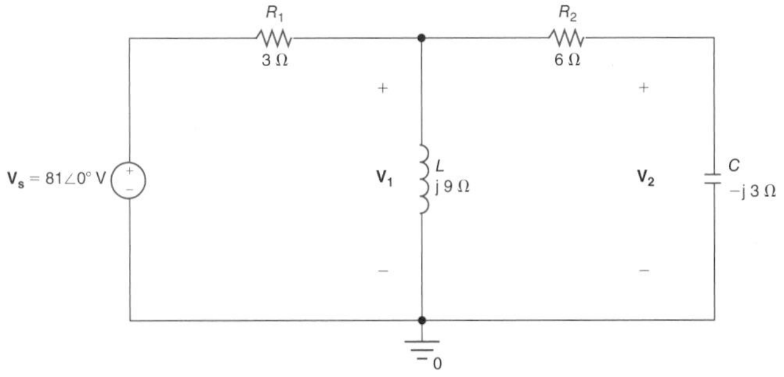

10.22 For the circuit shown in Figure P10.22, use nodal analysis to find the phasors $\mathbf{V}_1$ and $\mathbf{V}_2$.

FIGURE P10.22

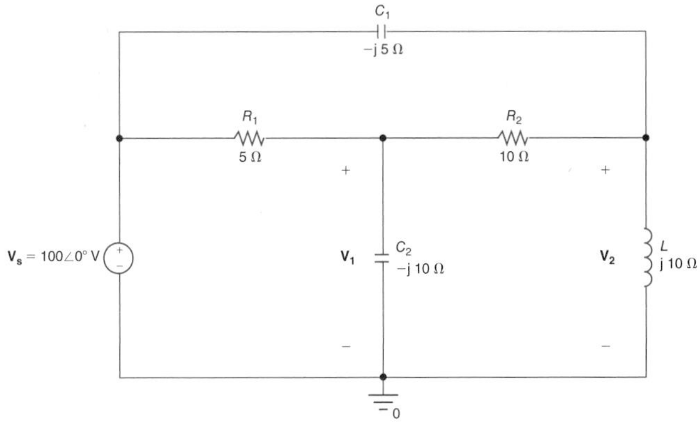

10.23 For the circuit shown in Figure P10.23, use nodal analysis to find the phasors $\mathbf{V}_1$ and $\mathbf{V}_2$.

FIGURE P10.23

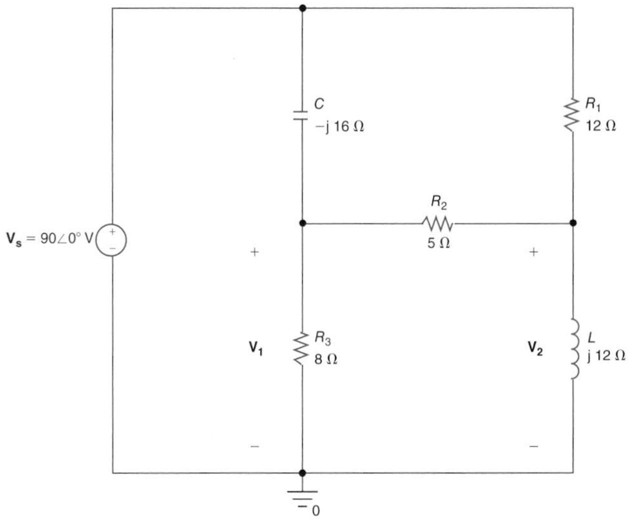

10.24 For the circuit shown in Figure P10.24, use nodal analysis to find the phasors $\mathbf{V}_1$ and $\mathbf{V}_2$.

FIGURE P10.24

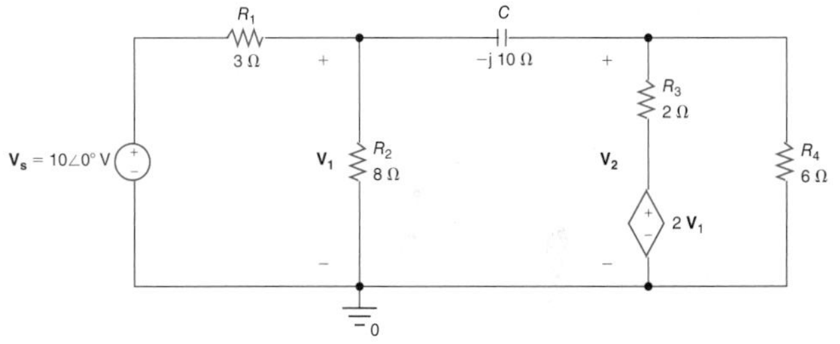

10.25 For the circuit shown in Figure P10.25, use nodal analysis to find the phasors $\mathbf{V}_1$ and $\mathbf{V}_2$.

FIGURE P10.25

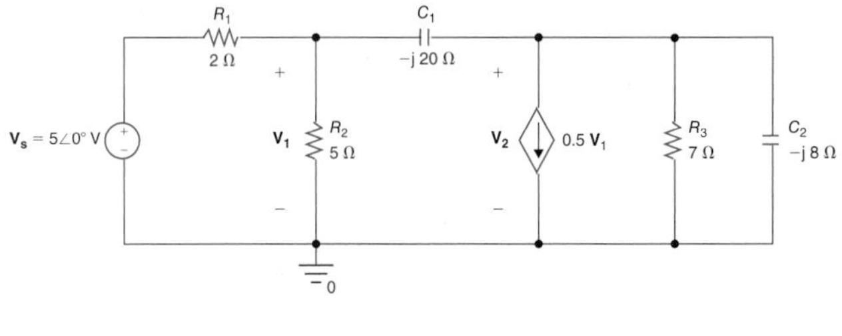

10.26 For the circuit shown in Figure P10.26, use nodal analysis to find the phasors $\mathbf{V}_1$ and $\mathbf{V}_2$.

FIGURE P10.26

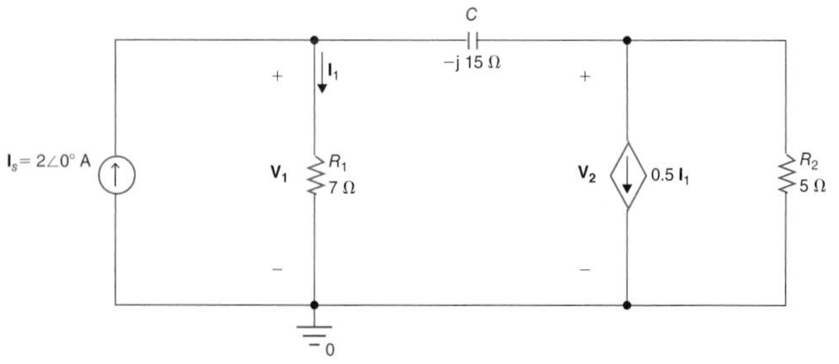

10.27 For the circuit shown in Figure P10.27, use nodal analysis to find the phasors $\mathbf{V}_1$, $\mathbf{V}_2$, and $\mathbf{V}_3$.

FIGURE P10.27

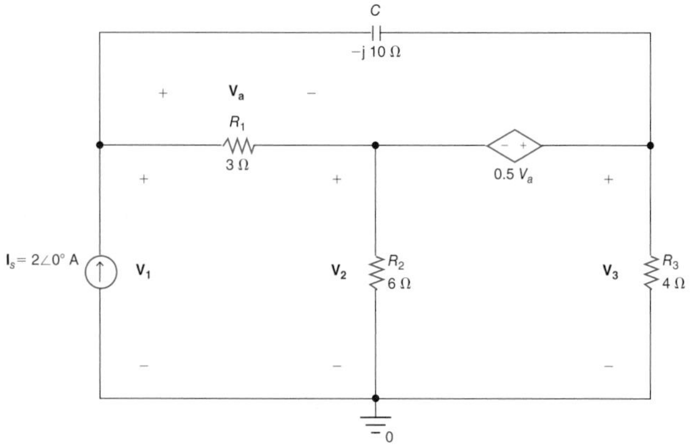

Mesh Analysis

10.28 For the circuit shown in Figure P10.28, find the phasors for mesh currents $\mathbf{I}_1$ and $\mathbf{I}_2$. Also find phasors for voltages $\mathbf{V}_1$ and $\mathbf{V}_2$.

FIGURE P10.28

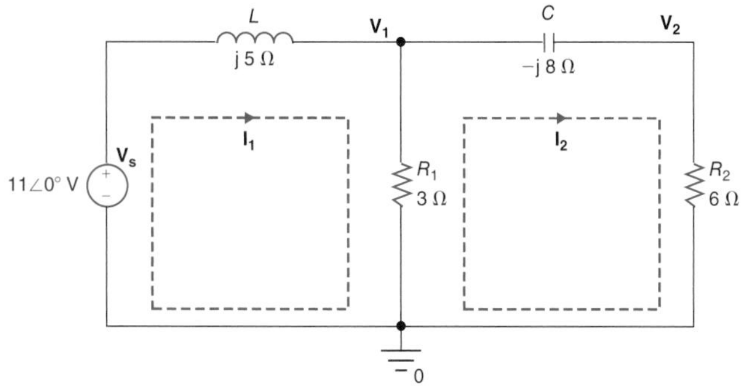

10.29 For the circuit shown in Figure P10.29, find the phasors for mesh currents $\mathbf{I}_1$, $\mathbf{I}_2$, and $\mathbf{I}_3$. Also find phasors for voltages $\mathbf{V}_1$ and $\mathbf{V}_2$.

FIGURE P10.29

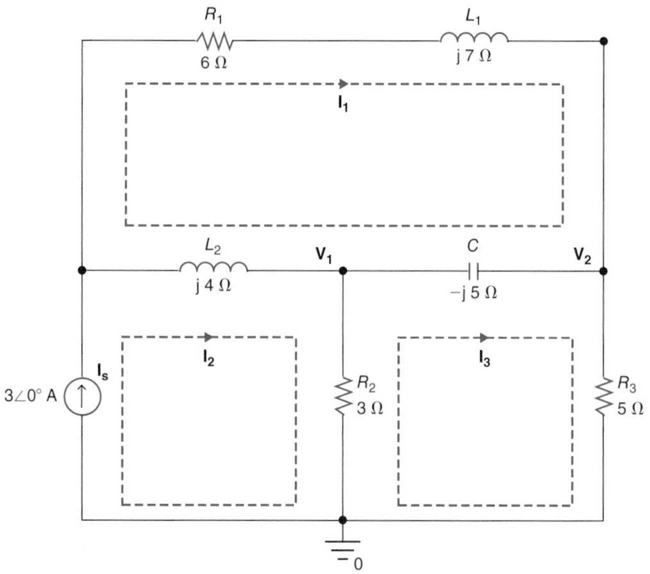

10.30 For the circuit shown in Figure P10.30, find the phasors for mesh currents $\mathbf{I}_1$, $\mathbf{I}_2$, and $\mathbf{I}_3$. Also find phasors for voltages $\mathbf{V}_1$ and $\mathbf{V}_2$.

FIGURE P10.30

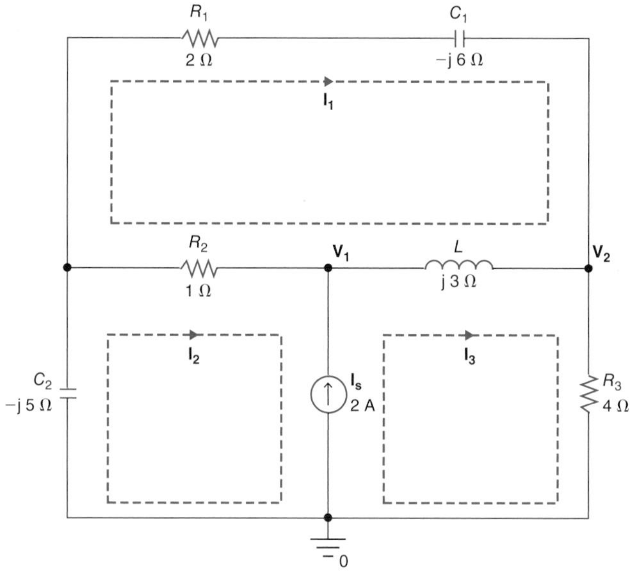

10.31 For the circuit shown in Figure P10.31, find the phasors for mesh currents $\mathbf{I}_1$ and $\mathbf{I}_2$. Also find phasors for voltages $\mathbf{V}_1$ and $\mathbf{V}_2$.

FIGURE P10.31

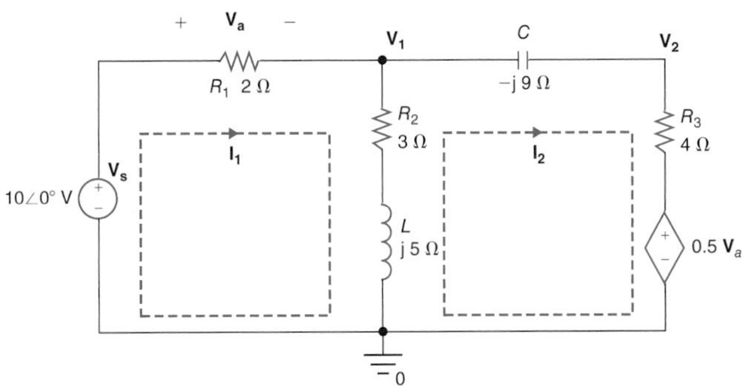

10.32 For the circuit shown in Figure P10.32, find the phasors for mesh currents I_1, I_2, and I_3. Also find phasors for voltages V_1 and V_2.

FIGURE P10.32

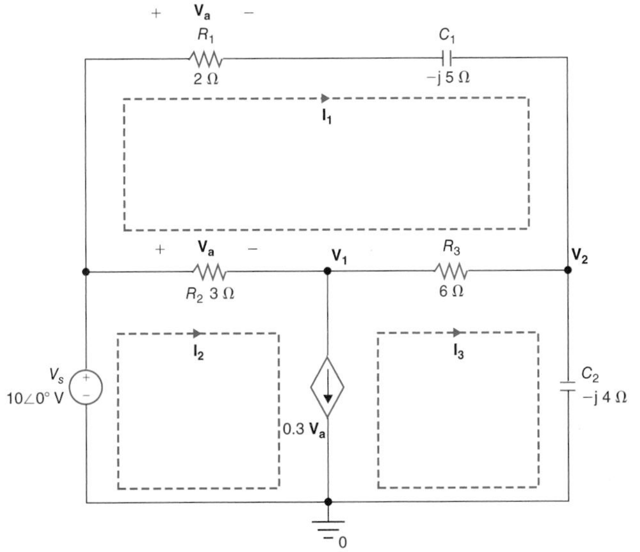

Superposition Principle

10.33 For the circuit shown in Figure P10.33, use the superposition principle to find the phasors for voltages V_1 and V_2.

FIGURE P10.33

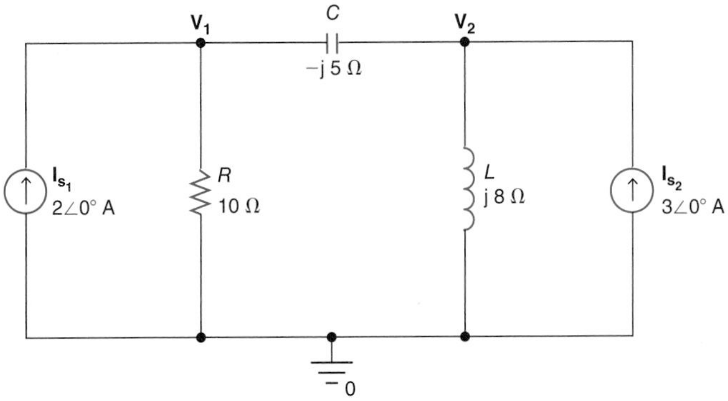

10.34 For the circuit shown in Figure P10.34, use the superposition principle to find the phasor for voltage V_1.

FIGURE P10.34

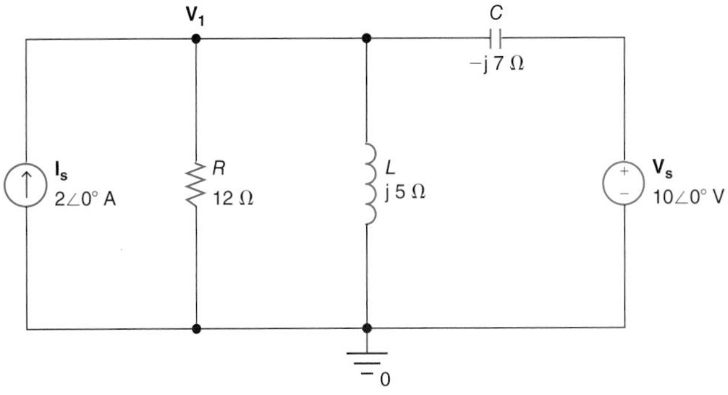

10.35 For the circuit shown in Figure P10.35, use the superposition principle to find the phasors for voltages V_1 and V_2.

FIGURE P10.35

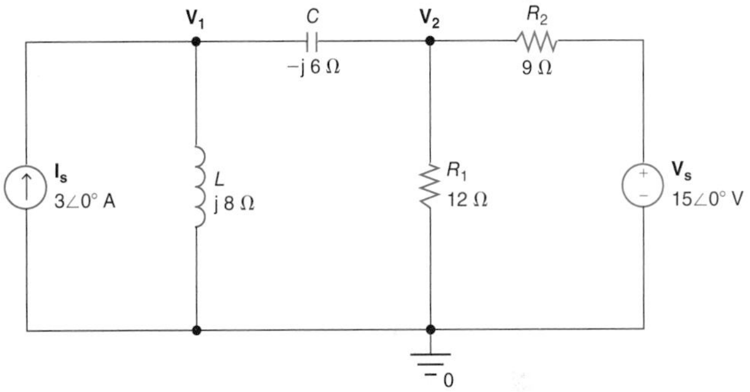

10.36 For the circuit shown in Figure P10.36, use the superposition principle to find the phasors for voltages V_1 and V_2.

FIGURE P10.36

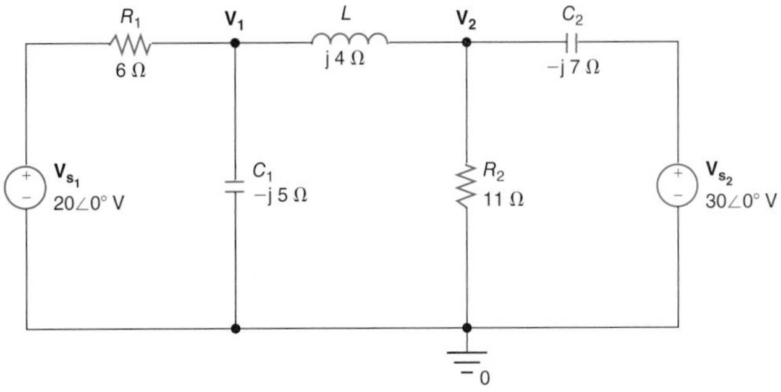

10.37 For the circuit shown in Figure P10.37, use the superposition principle to find the phasors for voltages V_1 and V_2.

FIGURE P10.37

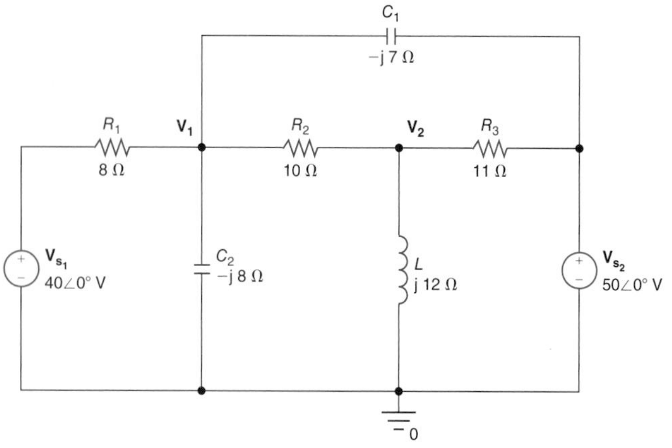

Source Transformation

10.38 For the circuit shown in Figure P10.38, use the source transformation to find the phasor for voltage $\mathbf{V}_o$.

FIGURE P10.38

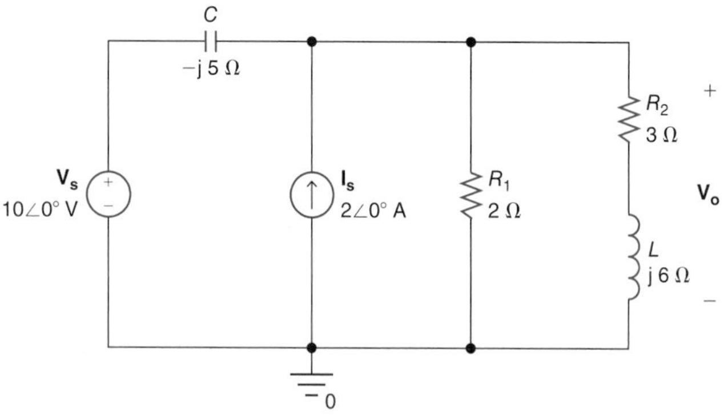

10.39 For the circuit shown in Figure P10.39, use the source transformation to find the phasor for voltage $\mathbf{V}_o$.

FIGURE P10.39

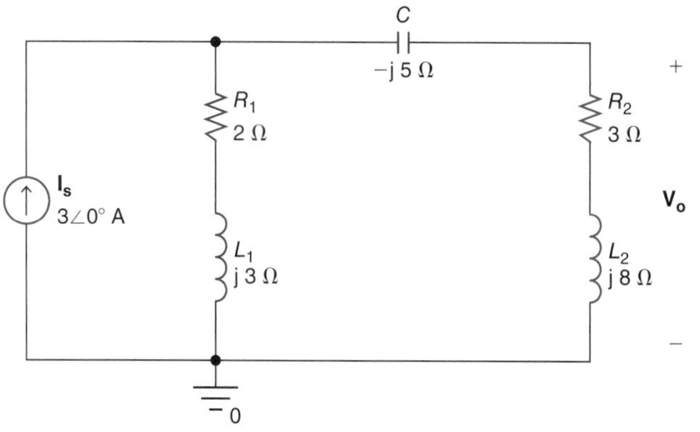

10.40 For the circuit shown in Figure P10.40, use the source transformation to find the phasor for voltage $\mathbf{V}_o$.

FIGURE P10.40

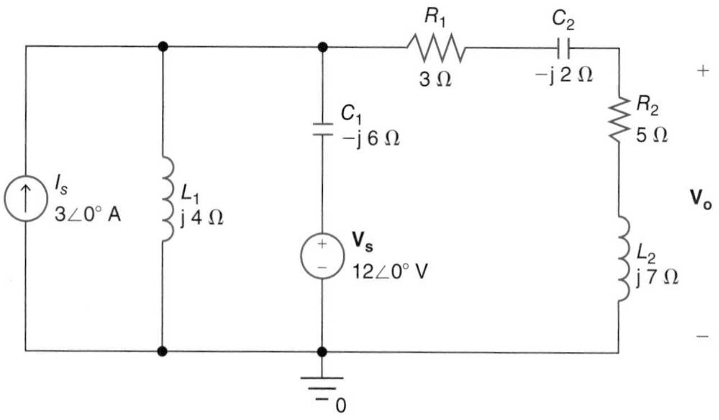

10.41 For the circuit shown in Figure P10.41, use the source transformation to find the phasor for voltage $\mathbf{V}_o$.

FIGURE P10.41

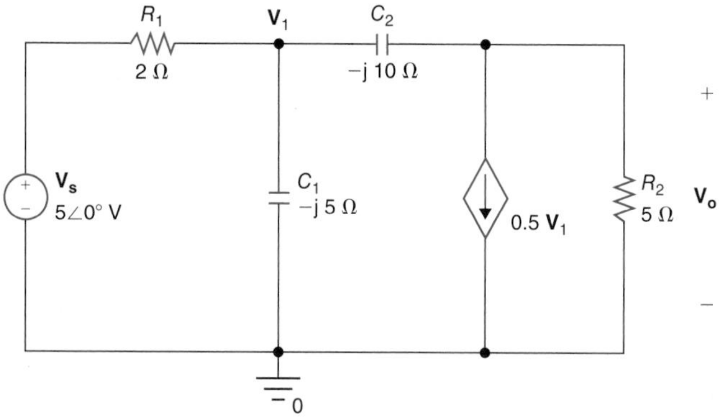

Thévenin Equivalent Circuit

10.42 For the circuit shown in Figure P10.42, find the Thévenin equivalent circuit between *a* and *b*.

FIGURE P10.42

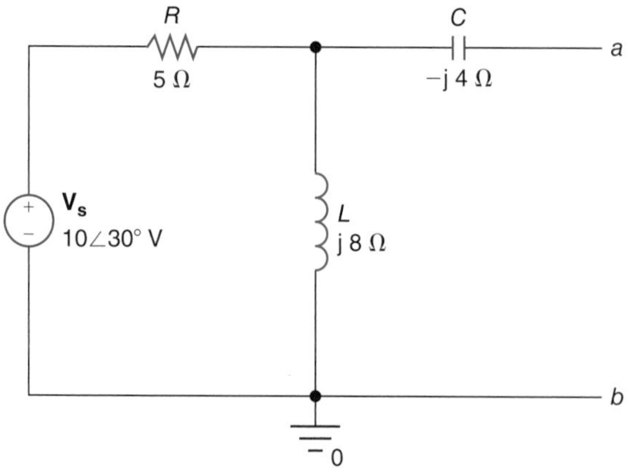

10.43 For the circuit shown in Figure P10.43, find the Thévenin equivalent circuit between *a* and *b*.

FIGURE P10.43

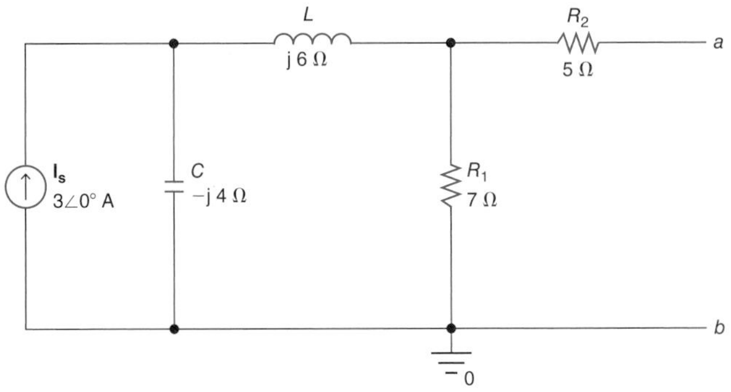

10.44 For the circuit shown in Figure P10.44, find the Thévenin equivalent circuit between *a* and *b*.

FIGURE P10.44

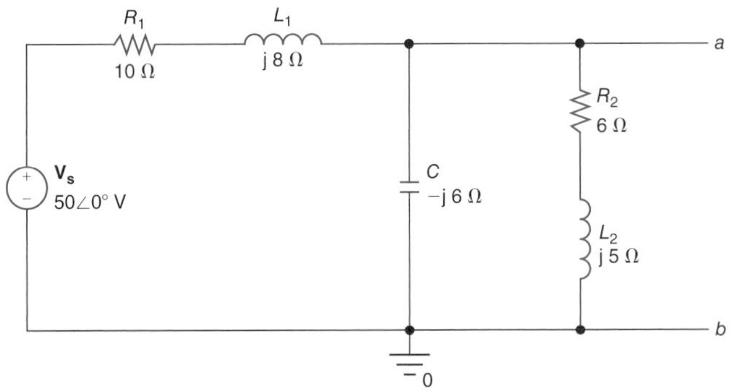

10.45 For the circuit shown in Figure P10.45, find the Thévenin equivalent circuit between *a* and *b*.

FIGURE P10.45

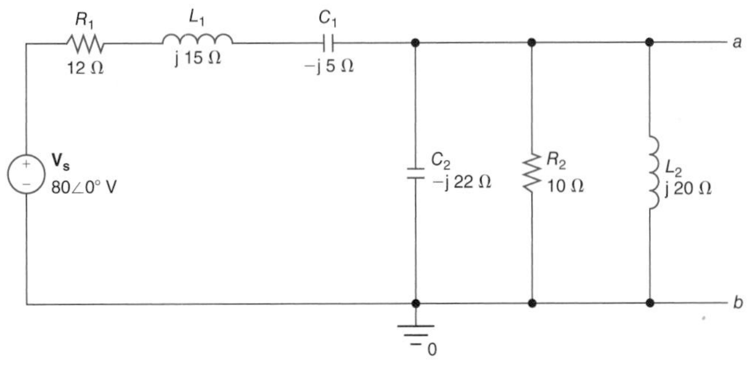

10.46 For the circuit shown in Figure P10.46, find the Thévenin equivalent circuit between *a* and *b*.

FIGURE P10.46

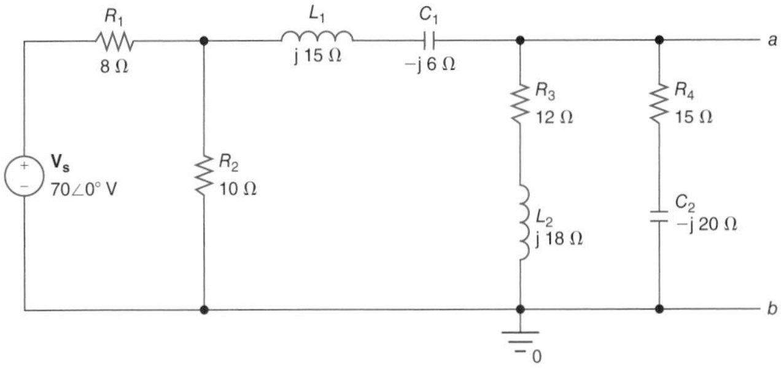

10.47 For the circuit shown in Figure P10.47, find the Thévenin equivalent circuit between *a* and *b*.

FIGURE P10.47

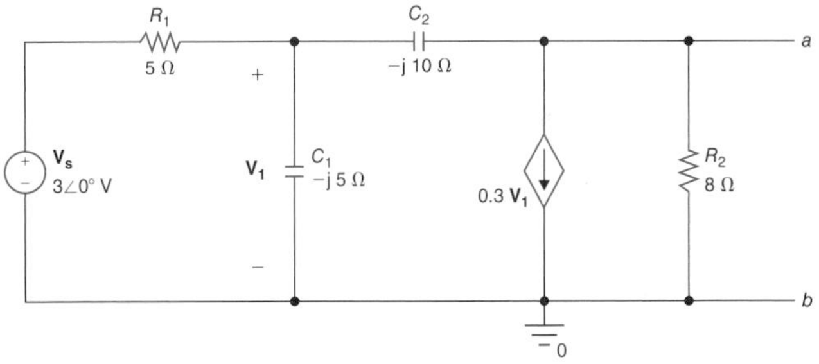

Norton Equivalent Circuit

10.48 For the circuit shown in Figure P10.48, find the Norton equivalent circuit between *a* and *b*.

FIGURE P10.48

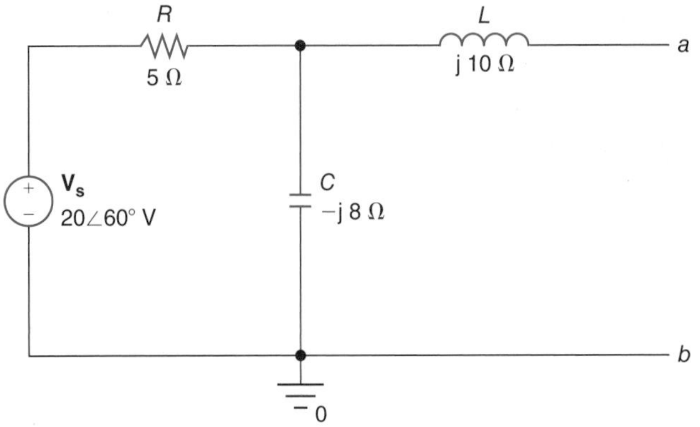

10.49 For the circuit shown in Figure P10.49, find the Norton equivalent circuit between *a* and *b*.

FIGURE P10.49

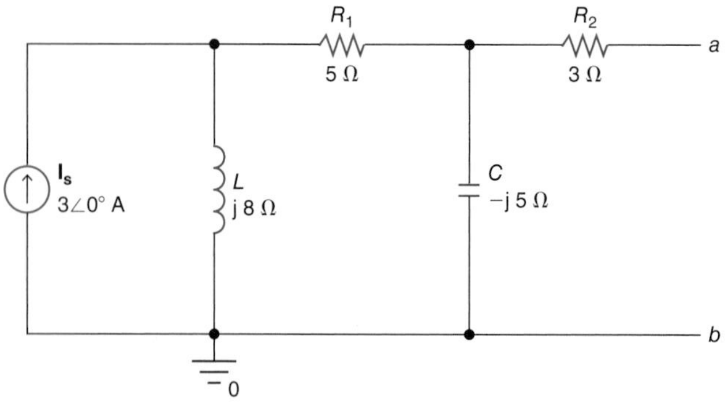

10.50 For the circuit shown in Figure P10.50, find the Norton equivalent circuit between *a* and *b*.

FIGURE P10.50

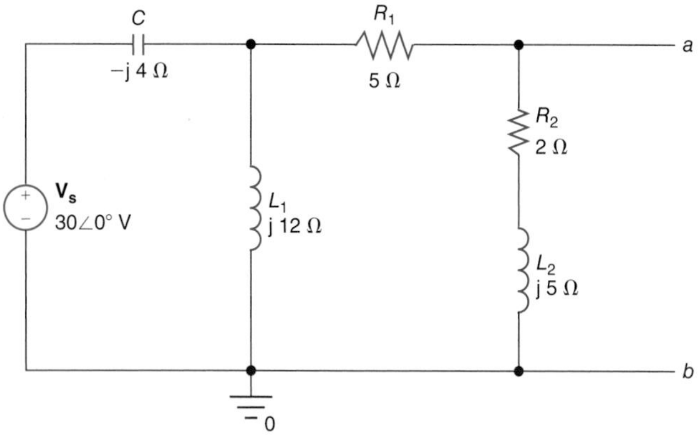

10.51 For the circuit shown in Figure P10.51, find the Norton equivalent circuit between *a* and *b*.

FIGURE P10.51

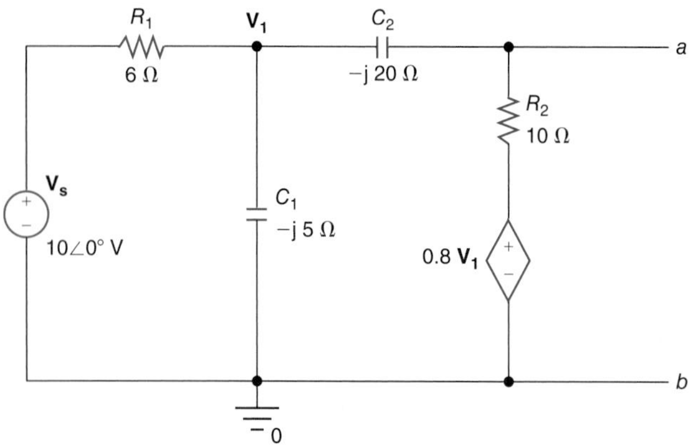

Transfer Function

10.52 Find the transfer function $H(\omega) = V_o(\omega)/V_{in}(\omega)$ for the circuit shown in Figure P10.52 and state the type of the filter (LPF, HPF, BPF, or BSF).

FIGURE P10.52

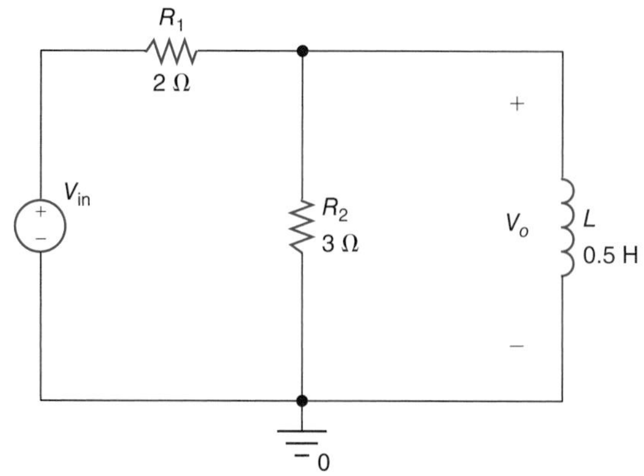

10.53 Find the transfer function $H(\omega) = V_o(\omega)/V_{in}(\omega)$ for the circuit shown in Figure P10.53 and state the type of the filter (LPF, HPF, BPF, or BSF).

FIGURE P10.53

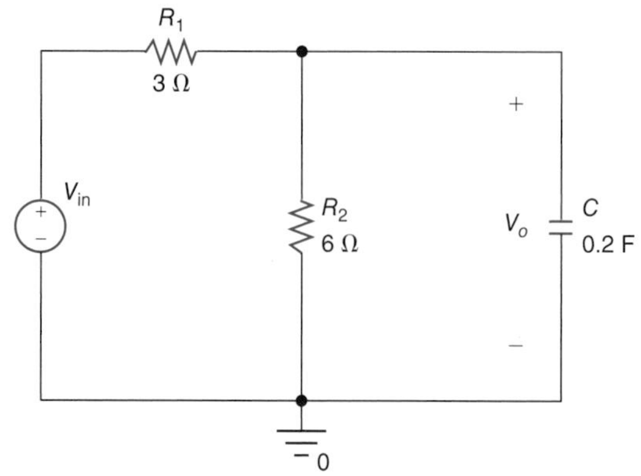

10.54 For the RLC series bandpass filter circuit shown in Figure P10.54, find the corner frequency ω_o, 3 dB cutoff frequencies ω_1 and ω_2, and the 3 dB bandwidth ω_{3dB}.

FIGURE P10.54

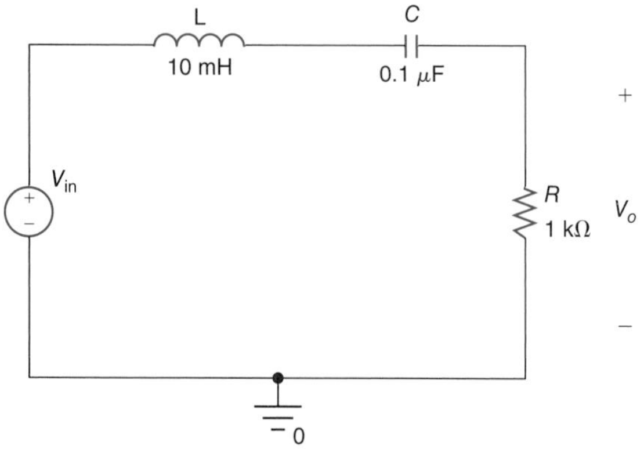

10.55 Find the transfer function $H(\omega) = V_o(\omega)/V_{in}(\omega)$ for the circuit shown in Figure P10.55 and state the type of the filter (LPF, HPF, BPF, or BSF).

FIGURE P10.55

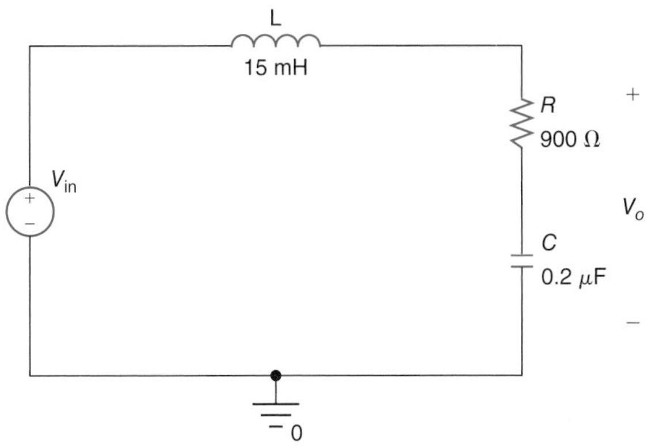

10.56 Find the transfer function $H(\omega) = V_o(\omega)/V_{in}(\omega)$ for the circuit shown in Figure P10.56 and state the type of the filter (LPF, HPF, BPF, or BSF).

FIGURE P10.56

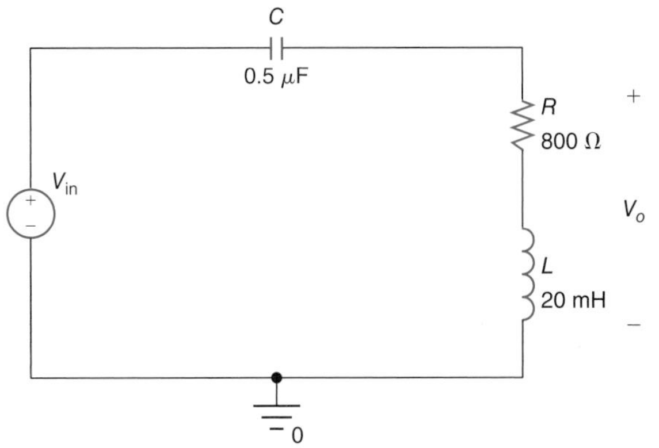

10.57 Find the transfer function $H(\omega) = V_o(\omega)/V_{in}(\omega)$ for the circuit shown in Figure P10.57 and state the type of the filter (LPF, HPF, BPF, or BSF).

FIGURE P10.57

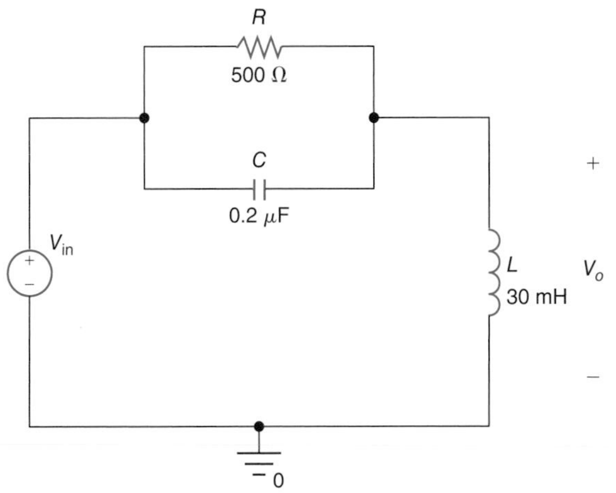

10.58 Find the transfer function $H(\omega) = V_o(\omega)/V_{in}(\omega)$ for the circuit shown in Figure P10.58 and state the type of the filter (LPF, HPF, BPF, or BSF).

FIGURE P10.58

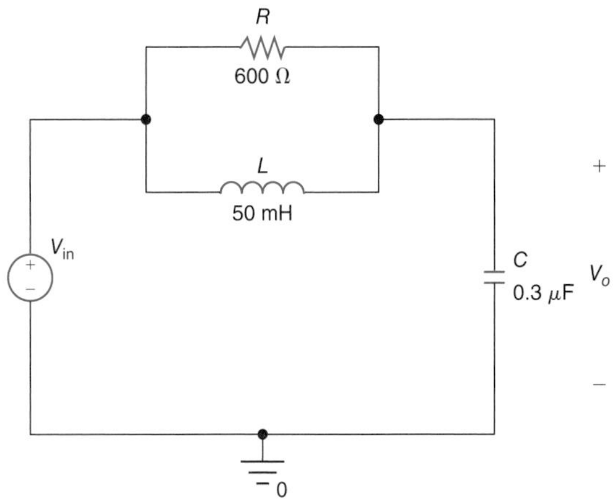

10.59 Find the transfer function $H(\omega) = V_o(\omega)/V_{in}(\omega)$ for the circuit shown in Figure P10.59 and state the type of the filter (LPF, HPF, BPF, or BSF).

FIGURE P10.59

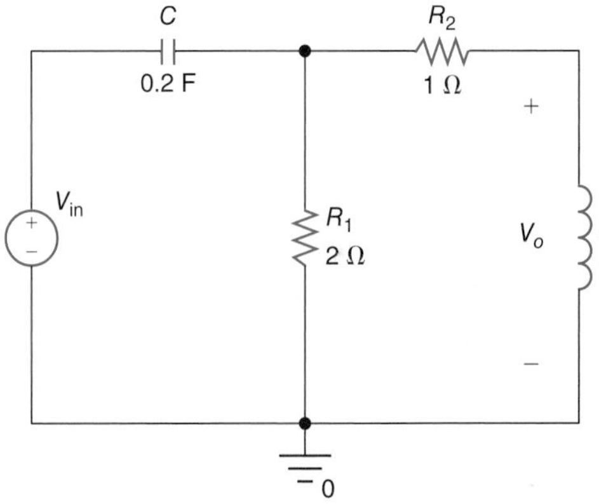

10.60 Find the transfer function $H(\omega) = V_o(\omega)/V_{in}(\omega)$ for the circuit shown in Figure P10.60 and state the type of the filter (LPF, HPF, BPF, or BSF).

FIGURE P10.60

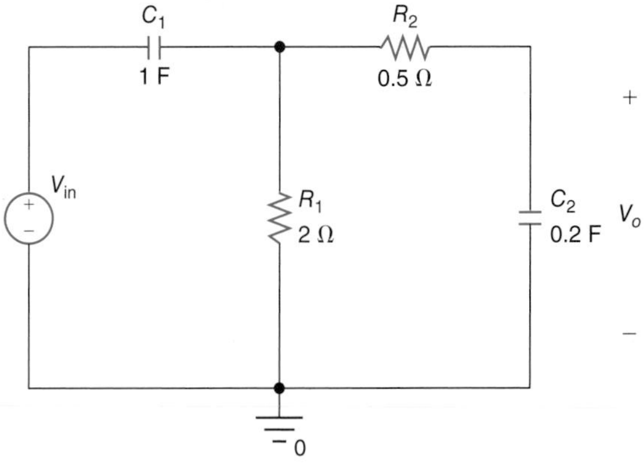

10.61 Find the transfer function $H(\omega) = V_o(\omega)/V_{in}(\omega)$ for the circuit shown in Figure P10.61 and state the type of the filter (LPF, HPF, BPF, or BSF).

FIGURE P10.61

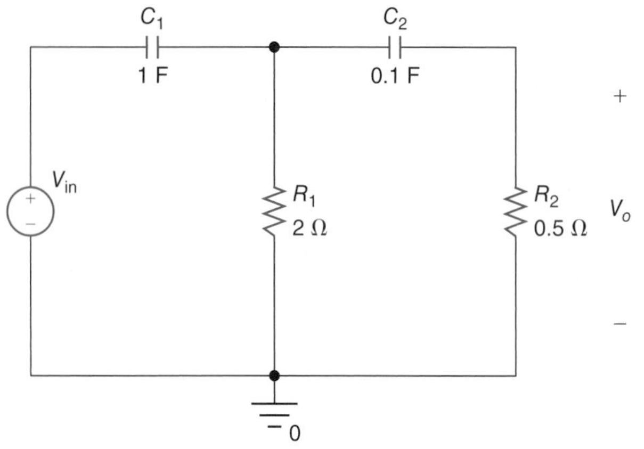

10.62 Find the transfer function $H(\omega) = V_o(\omega)/V_{in}(\omega)$ for the circuit shown in Figure P10.62 and state the type of the filter (LPF, HPF, BPF, or BSF).

FIGURE P10.62

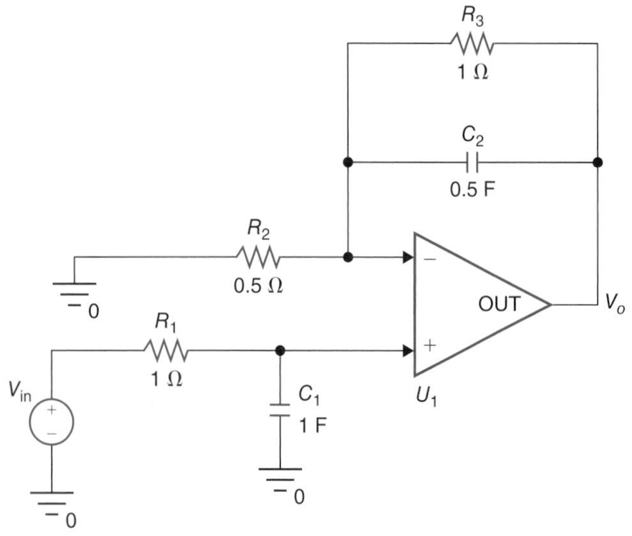

10.63 Find the transfer function $H(\omega) = V_o(\omega)/V_{in}(\omega)$ for the circuit shown in Figure P10.63.

FIGURE P10.63

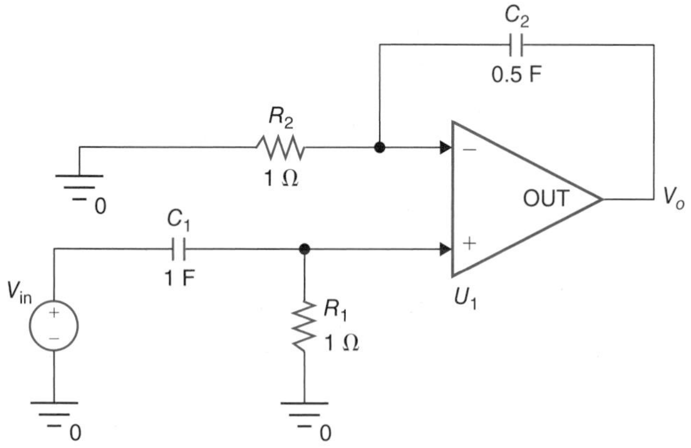

Chapter 11

AC Power

11.1 Introduction

In this chapter, alternating current (ac) power is discussed. The definitions of instantaneous power $p(t)$, average power P, reactive power Q, complex power $\mathbf{S}$, apparent power $|\mathbf{S}|$, and power factor pf are given. The instantaneous power $p(t)$ is the power as a function of time. It is the product of the voltage $v(t)$ and the current $i(t)$. The average value of the instantaneous power is P. The instantaneous power is the sum of the average power P and the time-varying component. The reactive power is the power exchanged between the source and the reactive components (i.e., inductors and capacitors) of the load. The average value of the reactive power is zero.

The complex power is defined as $\mathbf{S} = \mathbf{V}_{\text{rms}} \mathbf{I}^*_{\text{rms}}$. It can be shown that the real part of the complex power is the average power P, and the imaginary part of the complex power is the reactive power Q. The magnitude of the complex power $|\mathbf{S}| = |\mathbf{V}_{\text{rms}}|\,|\mathbf{I}_{\text{rms}}|$ is called the apparent power. A right triangle formed by two sides P (horizontal) and Q (vertical) and hypotenuse $|\mathbf{S}|$ is called a *power triangle*. The phase θ of $\mathbf{S}$ is the difference of phase θ_v of the voltage $v(t)$ and the phase θ_i of current $i(t)$. The angle θ is the angle made by P and $\mathbf{S}$. The power factor is $\cos(\theta)$; that is, $pf = \cos(\theta)$. If the reactive power Q is increased, the angle θ is increased, and the power factor pf is decreased. The reactive power does not contribute to the real power delivered to the load. It contributes to the power loss on the transmission lines. The power factor correction refers to the reduction in the reactive power by connecting a parallel capacitor to the inductive load to offset the Q value of the load.

11.2 Instantaneous Power, Average Power, Reactive Power, Apparent Power

Let $v(t)$ be the voltage across a circuit and $i(t)$ be the current through the circuit from the positive terminal to the negative terminal [as shown in Figure 11.1]. The circuit can be an element, an interconnection of elements, or a load.

FIGURE 11.1

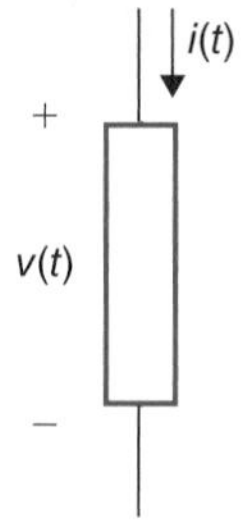

Voltage across and current through a circuit.

The instantaneous power of a circuit is defined as the product of the voltage and current; that is,

$$p(t) = v(t)i(t) \quad (11.1)$$

If the excitation of a circuit is a sinusoid of radian frequency ω, the voltage and current are also sinusoids of radian frequency $\omega = 2\pi/T$ (T = period). Let $v(t)$ and $i(t)$ be

$$v(t) = V_m \cos(\omega t + \theta_v) \quad (11.2)$$

and

$$i(t) = I_m \cos(\omega t + \theta_i) \quad (11.3)$$

respectively. Note that the peak amplitude of $v(t)$ is V_m and the phase of $v(t)$ is θ_v and that the peak amplitude of $i(t)$ is I_m and the phase is θ_i. The instantaneous power can be written as

$$p(t) = v(t)i(t) = V_m I_m \cos(\omega t + \theta_v)\cos(\omega t + \theta_i) \quad (11.4)$$

Using the trigonometric identity

$$\cos(\alpha)\cos(\beta) = \frac{1}{2}\cos(\alpha - \beta) + \frac{1}{2}\cos(\alpha + \beta)$$

we can rewrite Equation (11.4) as

$$p(t) = \frac{V_m I_m}{2}\cos(\theta_v - \theta_i) + \frac{V_m I_m}{2}\cos(2\omega t + \theta_v + \theta_i) \quad (11.5)$$

Let θ be the difference of phases θ_v and θ_i; that is,

$$\theta = \theta_v - \theta_i \quad (11.6)$$

Then, Equation (11.5) can be rewritten as

$$p(t) = \frac{V_m I_m}{2}\cos(\theta) + \frac{V_m I_m}{2}\cos(2\omega t + \theta_v + \theta_i) \quad (11.7)$$

Note that $\cos(\theta)$ is an even function of θ, and $\sin(\theta)$ is an odd function of θ:

$$\cos(\theta) = \cos(\theta_v - \theta_i) = \cos(\theta_i - \theta_v) = \cos(-\theta)$$

$$\sin(\theta) = \sin(\theta_v - \theta_i) = -\sin(\theta_i - \theta_v) = -\sin(-\theta)$$

In general, the **average power** is defined as the average of the instantaneous power:

$$P = \lim_{T_o \to \infty} \frac{1}{T_o}\int_{-\frac{T_o}{2}}^{\frac{T_o}{2}} p(t)dt \quad (11.8)$$

If $p(t)$ is periodic with period T, the average power is calculated by finding the average power of one period:

$$P = \frac{1}{T}\int_{-\frac{T}{2}}^{\frac{T}{2}} p(t)dt \quad (11.9)$$

Since the second term of Equation (11.7) is a periodic cosine wave with period $T_2 = \pi/\omega = T/2$, the average value is zero. The first term of Equation (11.7) is not a function of t. Thus, the average power is given by

$$P = \frac{V_m I_m}{2}\cos(\theta) \tag{11.10}$$

Alternatively, substitution of Equation (11.7) into Equation (11.9) results in

$$P = \frac{1}{T}\int_{-\frac{T}{2}}^{\frac{T}{2}} p(t)dt = \frac{1}{T}\int_{-\frac{T}{2}}^{\frac{T}{2}} \frac{V_m I_m}{2}\cos(\theta)dt + \frac{1}{T}\int_{-\frac{T}{2}}^{\frac{T}{2}} \frac{V_m I_m}{2}\cos(2\omega t + \theta_v + \theta_i)dt$$

$$= \frac{V_m I_m}{2}\cos(\theta) + \frac{V_m I_m}{2T}\frac{\sin(2\omega t + \theta_v + \theta_i)\Big|_{-\frac{T}{2}}^{\frac{T}{2}}}{2\omega}$$

$$= \frac{V_m I_m}{2}\cos(\theta) + \frac{V_m I_m}{2T}\frac{\sin\left(2\frac{2\pi}{T}\frac{T}{2} + \theta_v + \theta_i\right) - \sin\left(2\frac{2\pi}{T}\frac{-T}{2} + \theta_v + \theta_i\right)}{2\omega}$$

$$= \frac{V_m I_m}{2}\cos(\theta) + \frac{V_m I_m}{2T}\frac{\sin(\theta_v + \theta_i) - \sin(\theta_v + \theta_i)}{2\omega} = \frac{V_m I_m}{2}\cos(\theta)$$

In terms of rms values, the average power can be expressed as

$$P = \frac{V_m I_m}{2}\cos(\theta) = \frac{V_m}{\sqrt{2}}\frac{I_m}{\sqrt{2}}\cos(\theta) = V_{rms} I_{rms}\cos(\theta) \tag{11.11}$$

where

$$V_{rms} = \frac{V_m}{\sqrt{2}}, \quad I_{rms} = \frac{I_m}{\sqrt{2}}$$

If the voltage and current are in phase, the phase difference is zero [$\theta = 0$]. Then $\cos(\theta) = 1$, and the average power becomes

$$\frac{V_m I_m}{2} = V_{rms} I_{rms}$$

If the average power is positive ($P > 0$), the element is **absorbing** power on the average. On the other hand, if the average power is negative ($P < 0$), the element is **releasing** power on the average. If the phase difference θ ($\theta = \theta_v - \theta_i$) is between $-90°$ and $90°$ ($-90° < \theta < 90°$), $\cos(\theta)$ is positive ($\cos(\theta) > 0$), and the element is absorbing power. If the phase difference θ ($\theta = \theta_v - \theta_i$) is between $90°$ and $+270°$ ($90° < \theta < 270°$), $\cos(\theta)$ is negative ($\cos(\theta) < 0$), and the element is releasing power. If $\theta = 90°$ or $270°$, $\cos(\theta) = 0$, and the average power is zero. The **power factor (pf)** is defined as

$$pf = \cos(\theta) \tag{11.12}$$

In terms of power factor, the average power is given by

$$P = \frac{V_m I_m}{2}\cos(\theta) = \frac{V_m I_m}{2}pf = V_{rms} I_{rms} pf \tag{11.13}$$

The instantaneous power $p(t)$ given by Equation (11.7) can be rewritten as

$$\begin{aligned} p(t) &= \frac{V_m I_m}{2}\cos(\theta_v - \theta_i) + \frac{V_m I_m}{2}\cos(2\omega t + 2\theta_i + \theta_v - \theta_i) \\ &= \frac{V_m I_m}{2}\cos(\theta) + \frac{V_m I_m}{2}\cos(2\omega t + 2\theta_i + \theta) \\ &= \frac{V_m I_m}{2}\cos(\theta) + \frac{V_m I_m}{2}\cos(\theta)\cos(2\omega t + 2\theta_i) - \frac{V_m I_m}{2}\sin(\theta)\sin(2\omega t + 2\theta_i) \end{aligned} \quad (11.14)$$

Let Q be

$$Q = \frac{V_m I_m}{2}\sin(\theta) = \frac{V_m}{\sqrt{2}}\frac{I_m}{\sqrt{2}}\sin(\theta) = V_{rms} I_{rms}\sin(\theta) \quad (11.15)$$

Then, Equation (11.14) can be written as

$$\begin{aligned} p(t) &= P + P\cos(2\omega t + 2\theta_i) - Q\sin(2\omega t + 2\theta_i) \\ &= P[1 + \cos(2\omega t + 2\theta_i)] - Q\sin(2\omega t + 2\theta_i) \end{aligned} \quad (11.16)$$

The first term, $P[1 + \cos(2\omega t + 2\theta_i)]$, of Equation (11.16) is nonnegative for all t [as shown in Figure 11.2]. This term represents the power absorbed. The average value of the first term is P. This term represents the power on the resistive component (R) of the load. The second term, $-Q\sin(2\omega t + 2\theta_i)$, of Equation (11.16) oscillates between positive (power is absorbed) and negative (power is released). The same amount of power is absorbed and released back and forth. The net power absorbed (or released) is zero. This term represents the power on the reactive components (C and L) of the load. The quantity Q is called the **reactive power**. The unit for reactive power is volt-ampere reactive (VAR). Figure 11.2 shows the average power P (horizontal line), $P[1 + \cos(2\omega t + 2\theta_i)]$ (dash), $-Q\sin(2\omega t + 2\theta_i)$ (dotted line), and the instantaneous power $p(t)$ (solid line) for (a) $\theta_i = 0$, $\theta_v = 30°$, $V_m = 2$ V, $I_m = 1$ A, and (b) $\theta_i = 0$, $\theta_v = 60°$, $V_m = 2$ V, $I_m = 1$ A.

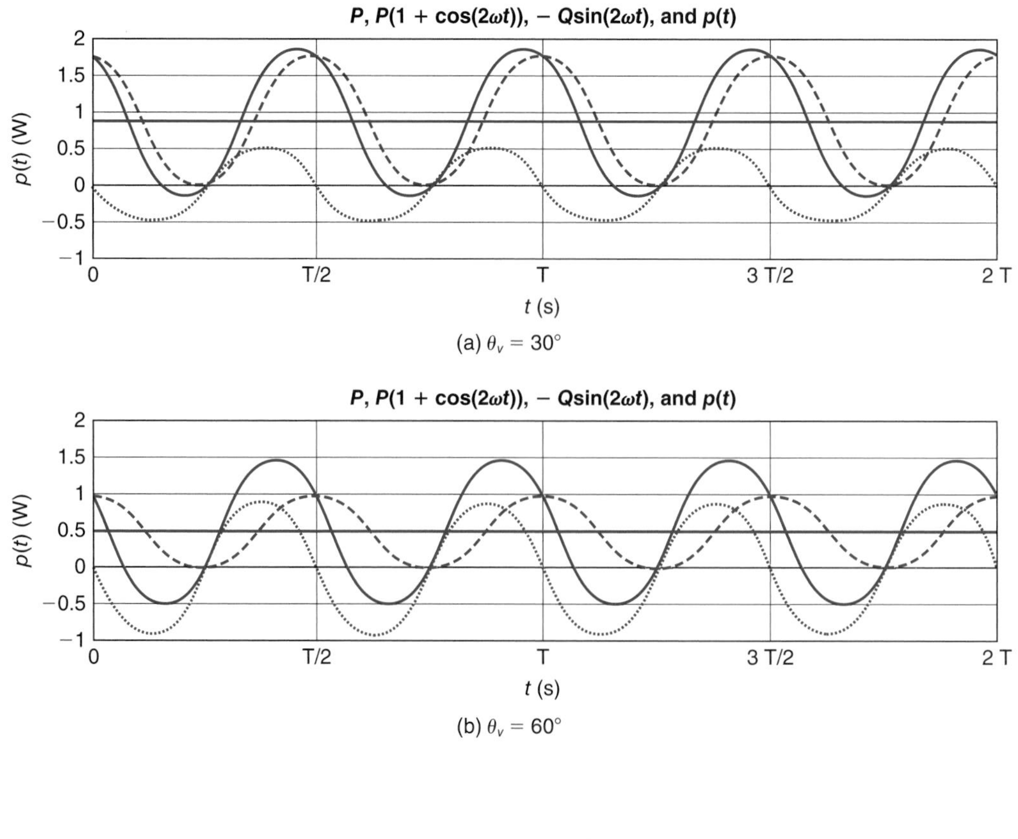

FIGURE 11.2

Plot of P, $P(1 + \cos(2\omega t + \theta_i))$, $-Q\sin(2\omega t + \theta_i)$, $p(t)$ for $\theta_i = 0$ and (a) $\theta_v = 30°$, (b) $\theta_v = 60°$.

According to Equation (11.14), the time-varying part of the instantaneous power is given by

$$\begin{aligned} p_{tv}(t) &= \frac{V_m I_m}{2}\cos(\theta)\cos(2\omega t + 2\theta_i) - \frac{V_m I_m}{2}\sin(\theta)\sin(2\omega t + 2\theta_i) \\ &= P\cos(2\omega t + 2\theta_i) - Q\sin(2\omega t + 2\theta_i) \\ &= \sqrt{P^2 + Q^2}\left[\frac{P}{\sqrt{P^2+Q^2}}\cos(2\omega t + 2\theta_i) - \frac{Q}{\sqrt{P^2+Q^2}}\sin(2\omega t + 2\theta_i)\right] \end{aligned}$$

Notice that

$$\sqrt{P^2 + Q^2} = \frac{V_m I_m}{2}, \quad \frac{P}{\sqrt{P^2+Q^2}} = \cos(\theta), \quad \frac{Q}{\sqrt{P^2+Q^2}} = \sin(\theta),$$

$$\tan(\theta) = \frac{Q}{P}, \quad \theta = \tan^{-1}\left(\frac{Q}{P}\right)$$

Thus, the time-varying part of the instantaneous power is given by

$$p_{tv}(t) = \sqrt{P^2 + Q^2}\cos(2\omega t + 2\theta_i + \theta)$$

If $\theta_i = 0$, the phasor representation of the time-varying part $v_{tv}(t)$ is given by

$$\sqrt{P^2 + Q^2}\angle\theta$$

The time-varying part of the instantaneous power can also be written as

$$\begin{aligned} p_{tv}(t) &= \frac{V_m I_m}{2}\cos(\theta)\cos(2\omega t + 2\theta_i) + \frac{V_m I_m}{2}\sin(\theta)\cos(2\omega t + 2\theta_i + 90^\circ) \\ &= P\cos(2\omega t + 2\theta_i) + Q\cos(2\omega t + 2\theta_i + 90^\circ) \end{aligned}$$

FIGURE 11.3

Phasor diagram for P, Q, and $\mathbf{S}$.

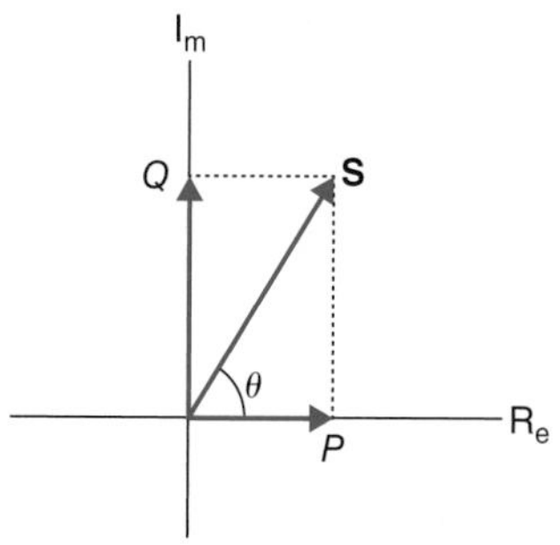

The phasor for the first term is $P\angle 0^\circ$, and the phasor for the second term is $Q\angle 90^\circ$ when $\theta_i = 0$. The reactive power Q is ahead by 90°, as shown in Figure 11.3, where $\mathbf{S}$ is the vector sum of phasors $P\angle 0^\circ$ and $Q\angle 90^\circ$; that is, $\mathbf{S} = P + jQ$. In polar notation, $\mathbf{S}$ is given by

$$\mathbf{S} = P + jQ = \sqrt{P^2 + Q^2}\angle\tan^{-1}\left(\frac{Q}{P}\right) = \sqrt{P^2 + Q^2}\angle\theta \tag{11.17}$$

Notice that $\mathbf{S}$ is the phasor representation of the time-varying part of the instantaneous power.

The phasor $\mathbf{S}$ is called the **complex power**. The unit for the complex power is volt-ampere (VA). Notice that the magnitude $|\mathbf{S}|$ of the complex power $\mathbf{S}$ is given by

$$\begin{aligned} |\mathbf{S}| &= \sqrt{P^2 + Q^2} = \sqrt{V_{rms}^2 I_{rms}^2\cos^2(\theta) + V_{rms}^2 I_{rms}^2\sin^2(\theta)} \\ &= V_{rms} I_{rms} = \frac{V_m I_m}{2} \end{aligned} \tag{11.18}$$

The magnitude $|\mathbf{S}|$ of the complex power is called the **apparent power**. The unit for the apparent power is volt-ampere (VA). The apparent power can be interpreted

FIGURE 11.4

Phasor representation of the instantaneous power.

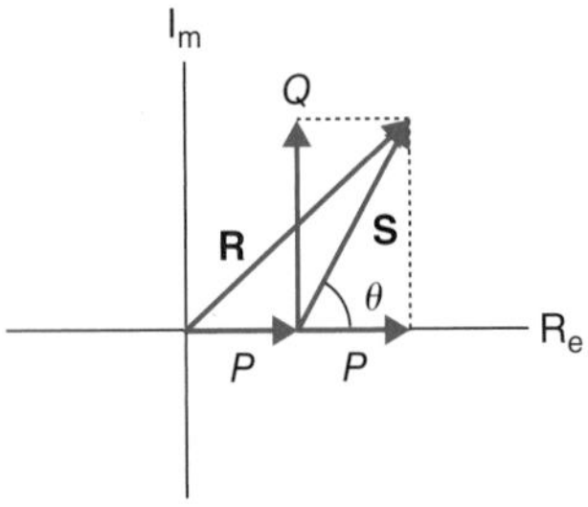

FIGURE 11.5

Stationary P and rotating P.

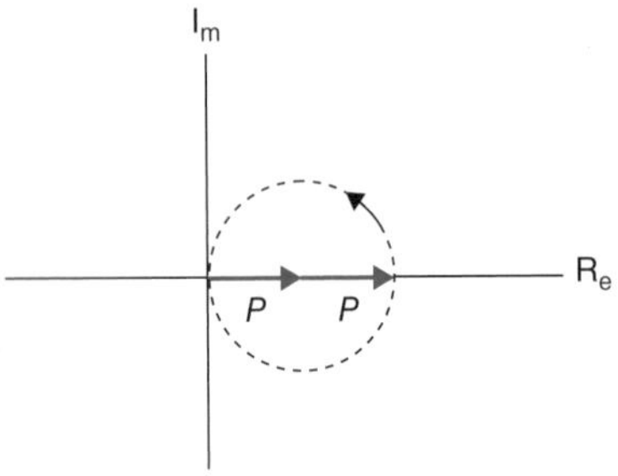

FIGURE 11.6

Rotating Q.

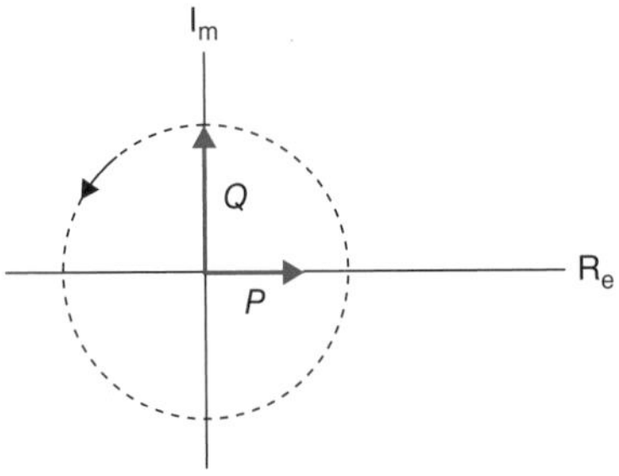

as the power from a dc voltage V_{rms} and a dc current I_{rms}. In terms of the apparent power, we can rewrite the average power and the reactive power given by Equations (11.13) and (11.15) as

$$P = \frac{V_m I_m}{2}\cos(\theta) = \frac{V_m I_m}{2} pf = V_{rms} I_{rms} pf = |\mathbf{S}|\cos(\theta) = |\mathbf{S}|\, pf \tag{11.19}$$

$$\begin{aligned} Q &= \frac{V_m I_m}{2}\sin(\theta) = \frac{V_m}{\sqrt{2}}\frac{I_m}{\sqrt{2}}\sin(\theta) = V_{rms} I_{rms}\sin(\theta) \\ &= |\mathbf{S}|\sin(\theta) = \pm\,|\mathbf{S}|\sqrt{1 - pf^2} \end{aligned} \tag{11.20}$$

respectively.

The angle of the complex power is given by

$$\angle \mathbf{S} = \theta = \tan^{-1}\left(\frac{Q}{P}\right) \tag{11.21}$$

If the average power P is added to **S**, we get the phasor **R** for the instantaneous power as shown in Figure 11.4. The first P vector from the origin to $(P, 0)$ represents the average power. The second P vector from $(P, 0)$ to $(2P, 0)$ represents $P\cos(2\omega t + 2\theta_i)$. The vector Q represents $Q\cos(2\omega t + 2\theta_i + 90°)$.

Figure 11.5 shows the first P and the second P. The first P is stationary. The second P rotates on a circle of radius P counterclockwise at the speed of 2ω radians/s. The projection of the point on the circle to the real axis represents the power $P[1 + \cos(2\omega t + 2\theta_i)]$. The current phase θ_i is assumed to be zero. As time progresses from zero, the projection to the real axis decreases from $2P$ to zero and back to $2P$, and repeats. This matches $P[1 + \cos(2\omega t + 2\theta_i)]$.

Figure 11.6 shows the phasor Q. The phasor Q rotates on a circle of radius Q counterclockwise at the speed of 2ω radians/s. The projection of the point on the circle to the real axis represents $Q\cos(2\omega t + 2\theta_i + 90°)$. The current phase θ_i is assumed to be zero. As time progresses from zero, the projection to the real axis decreases from zero to $-Q$, then increases from $-Q$ to Q, Q to zero, and repeats. This matches $Q\cos(2\omega t + 2\theta_i + 90°)$. The instantaneous power is the sum of the two projections shown in Figure 11.5 and Figure 11.6.

EXAMPLE 11.1

Let $v(t) = 100\cos(2\pi 60t + 120°)$ V and $i(t) = 8\cos(2\pi 60t + 60°)$ A. Find the difference in the phase angle, power factor, average power, reactive power, complex power, and apparent power, and plot the phasor of the complex power.

The difference in phase angle is

$$\theta = \theta_v - \theta_i = 120° - 60° = 60°$$

The power factor is given by

$$pf = \cos(\theta) = \cos(60°) = 0.5$$

continued

Example 11.1 continued

FIGURE 11.7

Phasor of complex power **S**.

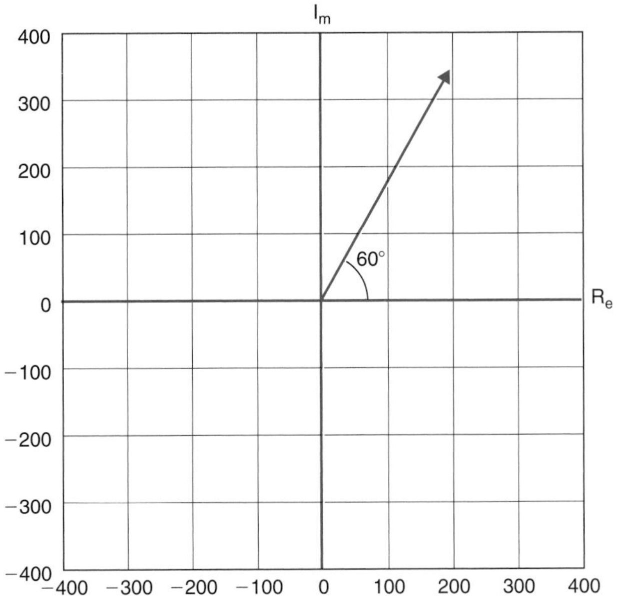

The average power is

$$P = \frac{V_m I_m}{2} pf = \frac{100 \times 8}{2} \times 0.5 = 200 \text{ W}$$

The reactive power is

$$Q = \frac{V_m I_m}{2} \sin(\theta) = \frac{100 \times 8}{2} \sin(60°)$$
$$= \frac{100 \times 8}{2} \times 0.866 = 346.4102 \text{ VAR}$$

The complex power is given by

$$\mathbf{S} = P + jQ = 200 + j346.4102$$
$$= 400\angle 60° \text{ VA}$$

The apparent power is given by

$$|\mathbf{S}| = 400 \text{ VA}$$

The complex power **S** is shown in Figure 11.7.

Exercise 11.1

Let $v(t) = 200 \cos(2\pi 60t + 30°)$ V and $i(t) = 12 \cos(2\pi 60t + 60°)$ A. Find the difference in phase angle, power factor, average power, reactive power, complex power, and apparent power.

Answer:
$\theta = -30°$, $pf = 0.866$, $P = 1039.2305$ W, $Q = -600$ VAR,
$\mathbf{S} = 1039.2305 - j600$ VA, $|\mathbf{S}| = 1200$ VA.

11.3 Complex Power

Consider a circuit with a load as shown in Figure 11.8.

The phasor representation of $v(t) = V_m \cos(\omega t + \theta_v)$ is given by

$$\mathbf{V} = V_m e^{j\theta_v} \quad (11.22)$$

and the phasor representation of $i(t) = I_m \cos(\omega t + \theta_i)$ is given by

$$\mathbf{I} = I_m e^{j\theta_i} \quad (11.23)$$

FIGURE 11.8

A circuit with a load.

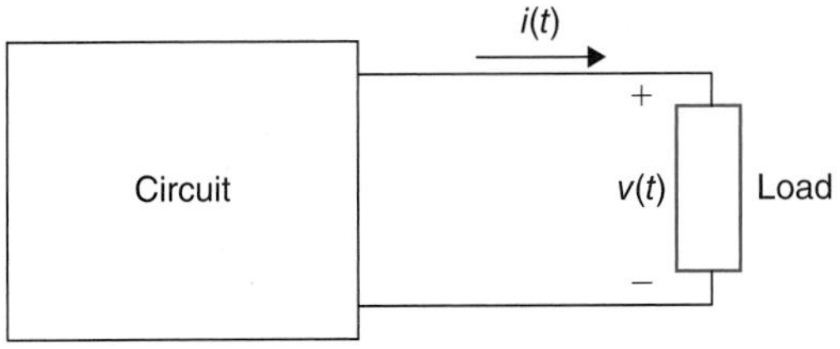

Let **Z** be the impedance of the load. Then, in terms of **V**, **I**, and **Z**, the circuit shown in Figure 11.8 becomes the one shown in Figure 11.9. According to Ohm's law, we have

$$\mathbf{V} = \mathbf{ZI}, \quad \mathbf{I} = \frac{\mathbf{V}}{\mathbf{Z}}, \quad \mathbf{Z} = \frac{\mathbf{V}}{\mathbf{I}} \quad (11.24)$$

FIGURE 11.9

The circuit shown in Figure 11.8 with **V**, **I**, and **Z**.

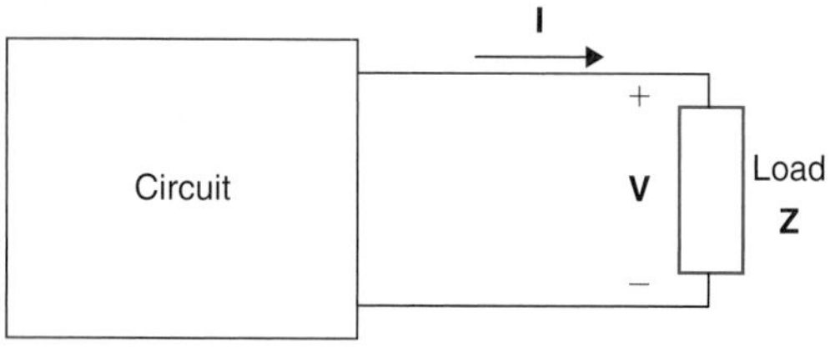

The complex power on the load is defined as

$$\mathbf{S} = \frac{1}{2}\mathbf{V}\mathbf{I}^* \tag{11.25}$$

where $1/2$ is the result of multiplying two cosines as shown in Equation (11.14).

Substituting Equations (11.22) and (11.23) into Equation (11.25), we have

$$\begin{aligned}\mathbf{S} &= \frac{1}{2}V_m e^{j\theta_v} I_m e^{-j\theta_i} = \frac{1}{2}V_m I_m e^{j(\theta_v - \theta_i)} = \frac{1}{2}V_m I_m e^{j(\theta_v - \theta_i)} \\ &= \frac{1}{2}V_m I_m \cos(\theta) + j\frac{1}{2}V_m I_m \sin(\theta) = P + jQ\end{aligned} \tag{11.26}$$

where

$$P = \frac{1}{2}V_m I_m \cos(\theta) = \frac{V_m}{\sqrt{2}}\frac{I_m}{\sqrt{2}}\cos(\theta) = V_{rms}I_{rms}\cos(\theta) = |\mathbf{S}|\,pf \tag{11.27}$$

$$Q = \frac{1}{2}V_m I_m \sin(\theta) = \frac{V_m}{\sqrt{2}}\frac{I_m}{\sqrt{2}}\sin(\theta) = V_{rms}I_{rms}\sin(\theta) = |\mathbf{S}|\sin(\theta) \tag{11.28}$$

and $\theta = \theta_v - \theta_i$. According to Equation (11.26), we can conclude that the real part of the complex power is the **average power** P,

$$\text{Re}(\mathbf{S}) = P$$

and the imaginary part of the complex power is the **reactive power** Q,

$$\text{Im}(\mathbf{S}) = Q$$

The average power is the power delivered from the source to the resistive load and absorbed by the resistive load. The reactive power is the power on the reactance. This is the power exchanged between the source and the reactive part of the load.

Let us define **complex rms voltage** as

$$\mathbf{V}_{\text{rms}} = \frac{V_m}{\sqrt{2}}e^{j\theta_v} = |\mathbf{V}_{\text{rms}}|e^{j\theta_v} \tag{11.29}$$

and **complex rms current** as

$$\mathbf{I}_{\text{rms}} = \frac{I_m}{\sqrt{2}}e^{j\theta_i} = |\mathbf{I}_{\text{rms}}|e^{j\theta_i} \tag{11.30}$$

Notice that if we need to convert rms voltage and rms current to peak voltage and peak current, we multiply by $\sqrt{2}$; that is,

$$\mathbf{V} = \sqrt{2}\mathbf{V}_{\text{rms}} = \sqrt{2}|\mathbf{V}_{\text{rms}}|e^{j\theta_v} = V_m e^{j\theta_v} \tag{11.31}$$

$$\mathbf{I} = \sqrt{2}\mathbf{I}_{\text{rms}} = \sqrt{2}|\mathbf{I}_{\text{rms}}|e^{j\theta_i} = I_m e^{j\theta_i} \tag{11.32}$$

In terms of complex rms voltage and complex rms current, the complex power given by Equation (11.25) becomes

$$\mathbf{S} = \frac{1}{2}V_m e^{j\theta_v} I_m e^{-j\theta_i} = \frac{V_m}{\sqrt{2}} e^{j\theta_v} \frac{I_m}{\sqrt{2}} e^{-j\theta_i} = |\mathbf{V}_{\mathrm{rms}}| e^{j\theta_v} |\mathbf{I}_{\mathrm{rms}}| e^{-j\theta_i} = \mathbf{V}_{\mathrm{rms}} \mathbf{I}_{\mathrm{rms}}^* \quad (11.33)$$

$$\mathbf{S} = |\mathbf{V}_{\mathrm{rms}}||\mathbf{I}_{\mathrm{rms}}| e^{j\theta} = |\mathbf{V}_{\mathrm{rms}}||\mathbf{I}_{\mathrm{rms}}| \cos(\theta) + j|\mathbf{V}_{\mathrm{rms}}||\mathbf{I}_{\mathrm{rms}}| \sin(\theta) \quad (11.34)$$

$$\mathbf{S} = |\mathbf{S}| \cos(\theta) + j|\mathbf{S}| \sin(\theta) = |\mathbf{S}| pf \pm j|\mathbf{S}|\sqrt{1 - pf^2} = P + jQ \quad (11.35)$$

Notice that if $Q > 0$, the imaginary part of $\mathbf{S}$ is positive $(+j|\mathbf{S}|\sqrt{1 - pf^2})$, and if $Q < 0$, the imaginary part of $\mathbf{S}$ is negative $(-j|\mathbf{S}|\sqrt{1 - pf^2})$.

Taking the absolute value of the complex power $\mathbf{S}$, we obtain the apparent power $|\mathbf{S}|$, given by

$$|\mathbf{S}| = \frac{V_m I_m}{2} = |\mathbf{V}_{\mathrm{rms}}||\mathbf{I}_{\mathrm{rms}}| = \frac{P}{pf} = \frac{Q}{\sqrt{1 - pf^2}} \quad (11.36)$$

FIGURE 11.10

The circuit shown in Figure 11.9 with $\mathbf{V}_{\mathrm{rms}}$, $\mathbf{I}_{\mathrm{rms}}$, and $\mathbf{Z}$.

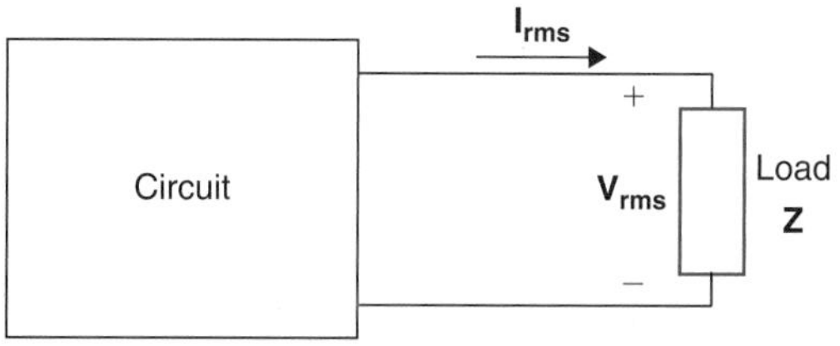

In terms of $\mathbf{I}_{\mathrm{rms}}$, $\mathbf{V}_{\mathrm{rms}}$, and $\mathbf{Z}$, the circuit shown in Figure 11.9 becomes the one shown in Figure 11.10. The load impedance $\mathbf{Z}$ can be represented as

$$\mathbf{Z} = R + jX$$

where R is **resistance** and X is **reactance**

Since $\mathbf{V}_{\mathrm{rms}} = \mathbf{Z}\mathbf{I}_{\mathrm{rms}}$ according to Ohm's law, the complex power $\mathbf{S} = \mathbf{V}_{\mathrm{rms}} \mathbf{I}_{\mathrm{rms}}^*$ can be written as

$$\begin{aligned}\mathbf{S} &= \mathbf{V}_{\mathrm{rms}}\mathbf{I}_{\mathrm{rms}}^* = \mathbf{Z}\mathbf{I}_{\mathrm{rms}}\mathbf{I}_{\mathrm{rms}}^* = \mathbf{Z}|\mathbf{I}_{\mathrm{rms}}|^2 = (R + jX)|\mathbf{I}_{\mathrm{rms}}|^2 \\ &= R|\mathbf{I}_{\mathrm{rms}}|^2 + jX|\mathbf{I}_{\mathrm{rms}}|^2\end{aligned} \quad (11.37)$$

$$P = R|\mathbf{I}_{\mathrm{rms}}|^2, \quad Q = X|\mathbf{I}_{\mathrm{rms}}|^2 \quad (11.38)$$

Since $\mathbf{I}_{\mathrm{rms}} = \mathbf{V}_{\mathrm{rms}}/\mathbf{Z}$, the complex power $\mathbf{S} = \mathbf{V}_{\mathrm{rms}} \mathbf{I}_{\mathrm{rms}}^*$ can be written as

$$\begin{aligned}\mathbf{S} &= \mathbf{V}_{\mathrm{rms}}\left(\frac{\mathbf{V}_{\mathrm{rms}}}{\mathbf{Z}}\right)^* = \frac{\mathbf{V}_{\mathrm{rms}}\mathbf{V}_{\mathrm{rms}}^*}{\mathbf{Z}^*} = \frac{|\mathbf{V}_{\mathrm{rms}}|^2}{\mathbf{Z}^*} = \frac{|\mathbf{V}_{\mathrm{rms}}|^2\mathbf{Z}}{\mathbf{Z}^*\mathbf{Z}} = \frac{|\mathbf{V}_{\mathrm{rms}}|^2\mathbf{Z}}{|\mathbf{Z}|^2} \\ &= \frac{|\mathbf{V}_{\mathrm{rms}}|^2(R + jX)}{|R + jX|^2} = \frac{|\mathbf{V}_{\mathrm{rms}}|^2(R + jX)}{R^2 + X^2}\end{aligned} \quad (11.39)$$

$$P = \frac{|\mathbf{V}_{\mathrm{rms}}|^2 R}{R^2 + X^2}, \quad Q = \frac{|\mathbf{V}_{\mathrm{rms}}|^2 X}{R^2 + X^2} \quad (11.40)$$

According to Equation (11.37), we get

$$\mathbf{Z} = \frac{\mathbf{S}}{|\mathbf{I}_{\mathrm{rms}}|^2} \quad (11.41)$$

According to Equation (11.39), we get

$$\mathbf{Z} = \frac{|\mathbf{V}_{\mathrm{rms}}|^2}{\mathbf{S}^*} = \frac{|\mathbf{V}_{\mathrm{rms}}|^2\mathbf{S}}{\mathbf{S}^*\mathbf{S}} = \frac{|\mathbf{V}_{\mathrm{rms}}|^2\mathbf{S}}{|\mathbf{S}|^2} \quad (11.42)$$

According to Equation (11.37), we have

$$|\mathbf{I}_{rms}| = \sqrt{\frac{\mathbf{S}}{\mathbf{Z}}} \tag{11.43}$$

According to Equation (11.39), we have

$$|\mathbf{V}_{rms}| = \sqrt{\mathbf{Z}^*\mathbf{S}} = \sqrt{\mathbf{Z}\mathbf{S}^*} \tag{11.44}$$

According to Equation (11.33), $\mathbf{S} = \mathbf{V}_{rms}\mathbf{I}^*_{rms}$, we get

$$\mathbf{I}_{rms} = \frac{\mathbf{S}^*}{\mathbf{V}^*_{rms}} \tag{11.45}$$

According to Equation (11.33), $\mathbf{S} = \mathbf{V}_{rms}\mathbf{I}^*_{rms}$, we get

$$\mathbf{V}_{rms} = \frac{\mathbf{S}}{\mathbf{I}^*_{rms}} \tag{11.46}$$

FIGURE 11.11

A power triangle.

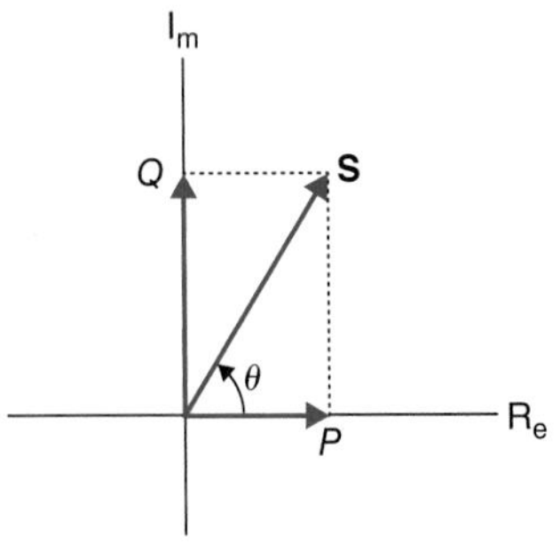

The complex power **S** is a vector in the complex plane [as shown in Figure 11.11]. This diagram is called a **power triangle**.

From the power triangle shown in Figure 11.11, we conclude that the projection of the complex power **S** onto the real axis is the average power P and the projection of the complex power **S** onto the imaginary axis is the reactive power Q; that is,

$$P = \mathrm{Re}(\mathbf{S}) = |\mathbf{S}|\cos(\theta) = |\mathbf{S}|\,pf \tag{11.47}$$

$$Q = \mathrm{Im}(\mathbf{S}) = |\mathbf{S}|\sin(\theta) = \pm|\mathbf{S}|\sqrt{1 - pf^2} \tag{11.48}$$

The power factor is said to be **lagging** if the power factor angle $\theta = \theta_v - \theta_i$ is positive. It is called *lagging* because the phase of the current θ_i is smaller than the phase of the voltage θ_v, making $\theta = \theta_v - \theta_i > 0$. If the power factor angle is negative, it is called **leading**. If the reactance is positive ($X > 0$), the reactive power is positive ($Q > 0$), the load is inductive ($R + j\omega L$), $\theta = \theta_v - \theta_i > 0$, and the power factor is lagging (the current phase θ_i is smaller than the voltage phase θ_v). Since $\mathbf{I} = \mathbf{V}/\mathbf{Z} = \mathbf{V}/(R + j\omega L)$, the phase of the current θ_i is $\theta_v - \tan^{-1}(\omega L/R)$, which is smaller than θ_v. On the other hand, if the reactance is negative ($X < 0$), the reactive power is negative ($Q < 0$), the load is capacitive ($R - j/(\omega C)$), $\theta = \theta_v - \theta_i < 0$, and the power factor is leading (the current phase θ_i is greater than the voltage phase θ_v). Since $\mathbf{I} = \mathbf{V}/\mathbf{Z} = \mathbf{V}/(R - j/(\omega C))$, the phase of the current θ_i is $\theta_v + \tan^{-1}(1/(\omega CR))$, which is greater than θ_v. Refer to Figure 11.12.

FIGURE 11.12

Lagging power factor and the leading power factor.

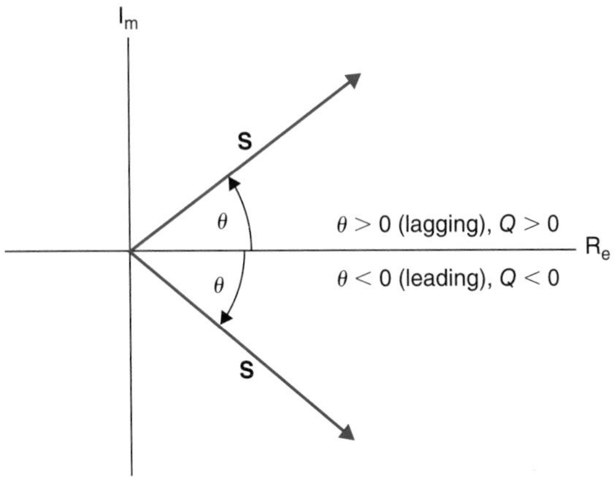

Applying the Pythagorean theorem to the power triangle, we obtain

$$|\mathbf{S}|^2 = P^2 + Q^2 \tag{11.49}$$

$$|\mathbf{S}| = \sqrt{P^2 + Q^2} \tag{11.50}$$

$$P = \sqrt{|\mathbf{S}|^2 - Q^2} \tag{11.51}$$

$$Q = \sqrt{|\mathbf{S}|^2 - P^2} \tag{11.52}$$

According to Equation (11.47), we obtain

$$pf = \cos(\theta) = \frac{P}{|\mathbf{S}|}, \quad |\mathbf{S}| = \frac{P}{pf}, \quad P = |\mathbf{S}|pf \tag{11.53}$$

According to Equation (11.48), we obtain

$$\sin(\theta) = \frac{Q}{|\mathbf{S}|}, \quad |\mathbf{S}| = \frac{|Q|}{\sqrt{1-pf^2}}, \quad pf = \sqrt{1-\frac{|Q|^2}{|\mathbf{S}|^2}}, \quad Q = |\mathbf{S}|\sqrt{1-pf^2}$$
$$= \frac{P}{pf}\sqrt{1-pf^2}, \quad P = \frac{Q \times pf}{\sqrt{1-pf^2}} = \frac{Q}{\sqrt{\frac{1}{pf^2}-1}}, \quad pf \neq 1 \tag{11.54}$$

EXAMPLE 11.2

The average power of a load is 1.63 *k*W, with a lagging power factor of 0.862. Find the apparent power, reactive power, and complex power of the load.

$$|\mathbf{S}| = \frac{P}{pf} = 1890.9513 \text{ VA}, \quad Q = |\mathbf{S}|\sqrt{1-pf^2} = 485.8913 \text{ VAR}$$

$$\mathbf{S} = P + jQ = 1630 + j485.8913 \text{ VA}$$

Exercise 11.2

The reactive power of a load is 1.64 *k*VAR with a lagging power factor of 0.928. Find the apparent power, average power, and complex power of the load.

Answer:

$$|\mathbf{S}| = \frac{|Q|}{\sqrt{1-pf^2}} = 4401.7369 \text{ VA}, \quad P = |\mathbf{S}| \times pf = 4084.8119 \text{ W},$$

$$\mathbf{S} = P + jQ = 4084.8119 + j1640 \text{ VA}.$$

EXAMPLE 11.3

The apparent power of a load is 23.7 *k*VA, with a lagging power factor of 0.876. Find the average power, reactive power, and complex power of the load.

$$P = |\mathbf{S}| \times pf = 20{,}761.2 \text{ W}, \quad Q = |\mathbf{S}|\sqrt{1-pf^2} = 5513.1888 \text{ VAR},$$
$$\mathbf{S} = P + jQ = 20{,}761.2 + j5513.1888 \text{ VA}$$

Exercise 11.3

The apparent power of a load is 22.7 *k*VA and the reactive power is 4680 VAR. Find the power factor, average power, and complex power of the load.

Answer:

$$pf = \sqrt{1 - \frac{|Q|^2}{|\mathbf{S}|^2}} = 0.97852, \quad P = |\mathbf{S}| \times pf = 22{,}212.3299\ \text{W},$$

$$\mathbf{S} = P + jQ = 22{,}212.3299 + j4680\ \text{VA}.$$

According to Equations (11.53) and (11.54), or according to the power triangle, we have

$$\tan(\theta) = \frac{\sin(\theta)}{\cos(\theta)} = \frac{Q}{P} \tag{11.55}$$

According to Equations (11.53) and (11.39), the power factor can be written as

$$pf = \frac{P}{|\mathbf{S}|} = \frac{\text{Re}(\mathbf{S})}{|\mathbf{S}|} = \frac{\dfrac{|\mathbf{V}_{\text{rms}}|^2\,\text{Re}(\mathbf{Z})}{|\mathbf{Z}|^2}}{\dfrac{|\mathbf{V}_{\text{rms}}|^2|\mathbf{Z}|}{|\mathbf{Z}|^2}} = \frac{\text{Re}(\mathbf{Z})}{|\mathbf{Z}|} = \frac{R}{\sqrt{R^2 + X^2}} \tag{11.56}$$

Since the imaginary part of **S** is the reactive power, we have, using $\sin^2(\theta) + \cos^2(\theta) = 1$,

$$Q = \text{Im}(\mathbf{S}) = |\mathbf{S}|\sin(\theta) = \pm|\mathbf{S}|\sqrt{1 - (pf)^2} \tag{11.57}$$

Since $|\mathbf{S}| = \dfrac{P}{pf}$ according to Equation (11.53), Q becomes

$$Q = \pm\frac{P}{pf}\sqrt{1 - (pf)^2} = \pm P\sqrt{\frac{1}{(pf)^2} - 1} \tag{11.58}$$

Solving for pf, we obtain

$$pf = \frac{1}{\sqrt{\left(\dfrac{Q}{P}\right)^2 + 1}} \tag{11.59}$$

From Equations (11.53) and (11.54), the complex power **S** can be written as

$$\begin{aligned} \mathbf{S} &= P + jQ = |\mathbf{S}|\left(pf \pm j\sqrt{1 - pf^2}\right) = \frac{P}{pf}\left(pf \pm j\sqrt{1 - pf^2}\right) \\ &= P\left(1 \pm j\sqrt{\frac{1 - pf^2}{pf^2}}\right) = P\left(1 \pm j\sqrt{\frac{1}{pf^2} - 1}\right) \end{aligned} \tag{11.60}$$

EXAMPLE 11.4

Find the complex power, average power, and reactive power of an inductor with inductance L when the magnitude of the voltage across the inductor is $|\mathbf{V}_{\text{rms}}|$ and the radian frequency is ω.

continued

Example 11.4 continued

According to Equation (11.39), we have

$$\mathbf{S} = \frac{|\mathbf{V}_{\text{rms}}|^2 \mathbf{Z}}{|\mathbf{Z}|^2} = \frac{|\mathbf{V}_{\text{rms}}|^2 j\omega L}{|j\omega L|^2} = j\frac{|\mathbf{V}_{\text{rms}}|^2 \omega L}{\omega^2 L^2} = j\frac{|\mathbf{V}_{\text{rms}}|^2}{\omega L} \tag{11.61}$$

$$P = \text{Re}(\mathbf{S}) = 0$$

$$Q = \text{Im}(\mathbf{S}) = \frac{|\mathbf{V}_{\text{rms}}|^2}{\omega L} \tag{11.62}$$

Exercise 11.4

Find the complex power, average power, and reactive power of an inductor with inductance L when the magnitude of the current through the inductor is $|\mathbf{I}_{\text{rms}}|$.

Answer:

$$\mathbf{S} = \mathbf{Z}|\mathbf{I}_{\text{rms}}|^2 = j\omega L|\mathbf{I}_{\text{rms}}|^2, \quad P = \text{Re}(\mathbf{S}) = 0, \quad Q = \text{Im}(\mathbf{S}) = \omega L|\mathbf{I}_{\text{rms}}|^2 \tag{11.63}$$

EXAMPLE 11.5

Find the complex power, average power, and reactive power of a capacitor with capacitance C when the magnitude of the voltage across the capacitor is $|\mathbf{V}_{\text{rms}}|$ and the radian frequency is ω.

According to Equation (11.39), we have

$$\mathbf{S} = \frac{|\mathbf{V}_{\text{rms}}|^2 \mathbf{Z}}{|\mathbf{Z}|^2} = \frac{|\mathbf{V}_{\text{rms}}|^2 \dfrac{1}{j\omega C}}{\left|\dfrac{1}{j\omega C}\right|^2} = \frac{|\mathbf{V}_{\text{rms}}|^2 \dfrac{1}{j\omega C}}{\dfrac{1}{\omega^2 C^2}} = -j|\mathbf{V}_{\text{rms}}|^2\omega C \tag{11.64}$$

$$P = \text{Re}(\mathbf{S}) = 0$$

$$Q = \text{Im}(\mathbf{S}) = -|\mathbf{V}_{\text{rms}}|^2\omega C \tag{11.65}$$

Exercise 11.5

Find the complex power, average power, and reactive power of a capacitor with capacitance C when the magnitude of the current through the capacitor is $|\mathbf{I}_{\text{rms}}|$.

Answer:

$$\mathbf{S} = \mathbf{Z}|\mathbf{I}_{\text{rms}}|^2 = \frac{1}{j\omega C}|\mathbf{I}_{\text{rms}}|^2 = -j\frac{|\mathbf{I}_{\text{rms}}|^2}{\omega C}, \quad P = \text{Re}(\mathbf{S}) = 0, \quad Q = \text{Im}(\mathbf{S}) = -\frac{|\mathbf{I}_{\text{rms}}|^2}{\omega C} \tag{11.66}$$

EXAMPLE 11.6

Find the complex power, average power, reactive power, and power factor of the load shown in Figure 11.13 when the magnitude of the voltage across the capacitor is $|\mathbf{V}_{rms}| = 220$ V and the radian frequency is $\omega = 2\pi 50$ rad/s.

FIGURE 11.13

Load for EXAMPLE 11.6.

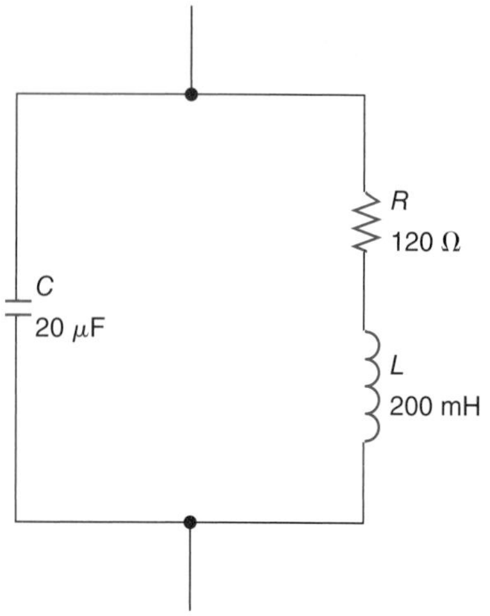

The impedances of R, L, and C are found to be

$$\mathbf{Z_R} = 120\ \Omega$$

$$\mathbf{Z_L} = j\omega L = j2\pi 50 \times 0.2 = j62.8319\ \Omega$$

$$\mathbf{Z_C} = \frac{1}{j\omega C} = \frac{1}{j2\pi 50 \times 20 \times 10^{-6}} = -j159.1549\ \Omega$$

The total impedance of the load is

$$\mathbf{Z} = \mathbf{Z_C} \| (\mathbf{Z_R} + \mathbf{Z_L}) = \frac{\mathbf{Z_C} \times (\mathbf{Z_R} + \mathbf{Z_L})}{\mathbf{Z_C} + \mathbf{Z_R} + \mathbf{Z_L}} = 128.3731 - j56.1108$$
$$= 140.1002\angle -23.6098°\ \Omega$$

The complex power of the load is given by

$$\mathbf{S} = \frac{|\mathbf{V}_{rms}|^2 \mathbf{Z}}{|\mathbf{Z}|^2} = \frac{220^2(128.3731 - j56.1108)}{140.1002^2} = 316.5495 - j138.3612$$
$$= 345.4670\angle -23.6098°\ \text{VA}$$

The average power is the real part of the complex power, and the reactive power is the imaginary part of the complex power. Thus, we have

$$P = \text{Re}(\mathbf{S}) = 316.5495\ \text{W}$$

$$Q = \text{Im}(\mathbf{S}) = -138.3612\ \text{VAR}$$

$$|\mathbf{S}| = 345.4670\ \text{VA}$$

$$pf = \text{Re}(\mathbf{S})/|\mathbf{S}| = 0.9163$$

Exercise 11.6

Find the complex power, average power, reactive power, apparent power, and power factor of the load shown in Figure 11.14 when the magnitude of the voltage across the inductor is $|\mathbf{V}_{rms}| = 200$ V and the radian frequency is $\omega = 2\pi 60$ rad/s.

continued

Exercise 11.6 continued

FIGURE 11.14

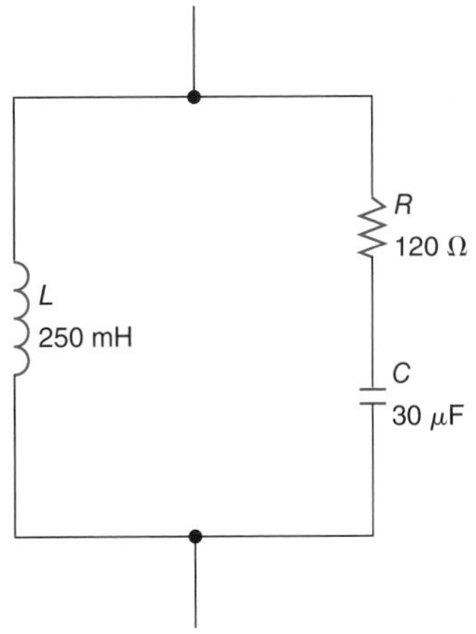

Load for EXERCISE 11.6.

Answer:
$\mathbf{S} = 216.0411 + j265.2279$ VA, $P = 216.0411$ W, $Q = 265.2279$ VAR,
$|\mathbf{S}| = 342.0813$ VA, $pf = P/|\mathbf{S}| = 0.6315$.

EXAMPLE 11.7

The average power of a load is 2.76 *k*W with a lagging power factor of 0.914. The magnitude of the rms voltage across the load is given by $|\mathbf{V}_{rms}| = 355$ V. Find the apparent power, reactive power, and complex power of the load. If the load is replaced by an equivalent load consisting of a series connection of a resistor and an inductor, what are the values of the resistance and the inductance? Assume that $f = 60$ Hz.

According to Equations (11.53), (11.57), and (11.35), the apparent power, reactive power, and complex power are given, respectively, by

$$|\mathbf{S}| = P/pf = 3019.6937 \text{ VA}$$

$$Q = |\mathbf{S}|\sqrt{1 - pf^2} = 1225.1326 \text{ VAR}$$

$$\mathbf{S} = P + jQ = 2760 + j1225.1326 \text{ VA}$$

According to Equation (11.42), we have

$$\mathbf{Z} = \frac{|\mathbf{V}_{rms}|^2}{\mathbf{S}^*} = 38.1452 + j16.9322 \ \Omega$$

Thus, we have

$$R = \text{Re}(\mathbf{Z}) = 38.1452 \ \Omega, \quad X = \text{Im}(\mathbf{Z}) = 16.9322 \ \Omega$$

The inductance value is given by

$$L = \frac{X}{\omega} = 44.9141 \text{ mH}$$

Exercise 11.7

The reactive power of a load is 1.67 *k*VAR with a lagging power factor of 0.904. The magnitude of the rms current through the load is given by $|\mathbf{I}_{rms}| = 10.6$ A. Find the apparent power, average power, and complex power of the load. If the load is replaced by an equivalent load consisting of a series connection of a resistor and an inductor, what are the values of the resistance and the inductance? Assume that $f = 60$ Hz.

Answer:

$$|\mathbf{S}| = \frac{|Q|}{\sqrt{1 - pf^2}} = 3906.1362 \text{ VA}, \quad P = |\mathbf{S}| \times pf = 3531.1471 \text{ W}, \quad \mathbf{S} = 3531.1471 + j1670 \text{ VA}$$

$$|\mathbf{V}_{rms}| = |\mathbf{S}|/|\mathbf{I}_{rms}| = 368.5034 \text{ V}, \quad \mathbf{Z} = \frac{|\mathbf{V}_{rms}|^2}{\mathbf{S}^*} = 31.4271 + j14.8629\ \Omega,$$

$$R = 31.4271\ \Omega, \quad L = 39.4252 \text{ mH}.$$

EXAMPLE 11.8

The average power of a load is 3.37 *k*W with a leading power factor of 0.835. The magnitude of the rms voltage across the load is given by $|\mathbf{V}_{rms}| = 450$ V. Find the apparent power, reactive power, and complex power of the load. If the load is replaced by an equivalent load consisting of a series connection of a resistor and a capacitor, what are the values of the resistance and the capacitance? Assume that $f = 60$ Hz.

According to Equations (11.53), (11.57), and (11.35), the apparent power, reactive power, and complex power are given, respectively, by

$$|\mathbf{S}| = P/pf = 4035.9281 \text{ VA}$$

$$Q = -|\mathbf{S}|\sqrt{1 - pf^2} = -2220.7692 \text{ VAR}$$

$$\mathbf{S} = P + jQ = 3370 - j2220.7692 \text{ VA}$$

According to Equation (11.42), we have

$$\mathbf{Z} = \frac{|\mathbf{V}_{rms}|^2}{\mathbf{S}^*} = 41.8956 - j27.6084\ \Omega$$

Thus, we have

$$R = \text{Re}(Z) = 41.8956\ \Omega, \quad X = -27.6084\ \Omega$$

The inductance value is given by

$$C = \frac{-1}{X\omega} = 96.0787\ \mu\text{F}$$

Exercise 11.8

The reactive power of a load is -3.36 kVAR with a leading power factor of 0.857. The magnitude of the rms current through the load is given by $|\mathbf{I}_{rms}| = 16$ A. Find the apparent power, average power, complex power, and magnitude of the rms voltage of the load. If the load is replaced by an equivalent load consisting of a series connection of a resistor and a capacitor, what are the values of the resistance and the capacitance? Assume that $f = 60$ Hz.

Answer:

$$|\mathbf{S}| = \frac{|Q|}{\sqrt{1 - pf^2}} = 6520.2659 \text{ VA}, \quad P = |\mathbf{S}| \times pf = 5587.8679 \text{ W},$$

$$\mathbf{S} = 5587.8679 - j3360 \text{ VA}, \quad |\mathbf{V}_{rms}| = |\mathbf{S}|/|\mathbf{I}_{rms}| = 407.5166 \text{ V},$$

$$\mathbf{Z} = \frac{\mathbf{S}}{|\mathbf{I}_{rms}|^2} = 21.8276 - j13.125\ \Omega, \quad R = 21.8276\ \Omega, \quad C = 202.1015\ \mu\text{F}.$$

11.4 Conservation of AC Power

FIGURE 11.15

Circuit with a voltage source and equivalent impedance.

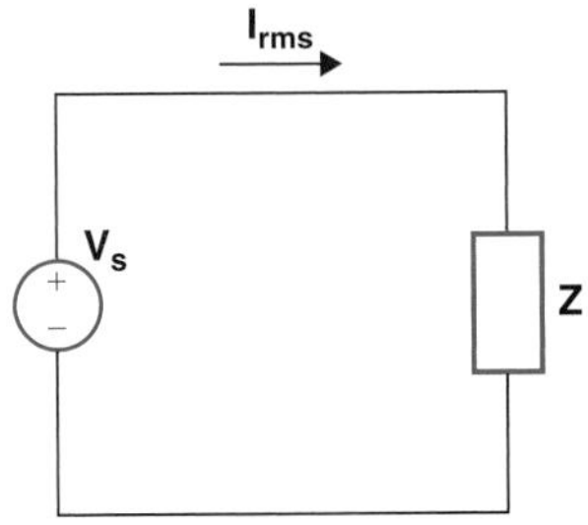

Consider a circuit with a voltage source with voltage phasor $\mathbf{V}_s$(rms) and interconnection of impedances. Let $\mathbf{Z}$ be the equivalent impedance of the interconnection of impedances. Then the circuit reduces to one voltage source and one impedance as shown in Figure 11.15.

The current through the voltage source from the positive terminal to the negative terminal is $-\mathbf{I}_{rms}$. Thus, the complex power on the source is

$$\mathbf{S}_{V_s} = -\mathbf{V}_s\mathbf{I}_{rms}^*$$

The negative sign indicates that the source releases complex power to the rest of the circuit. The complex power on the equivalent impedance is

$$\mathbf{S}_Z = \mathbf{V}_s\mathbf{I}_{rms}^*$$

The positive sign indicates that the impedance absorbs complex power. The complex power released by the source equals the complex power absorbed by the impedance. Since the complex power consists of average power and the reactive power, the average power released by the source equals the average power absorbed by the impedance, and the reactive power released by the source equals the reactive power absorbed by the impedance. This is called conservation of power. If there are more than one source, the superposition principle can be applied to show the conservation of ac power for each source.

EXAMPLE 11.9

Find the complex powers of all the elements in the circuit shown in Figure 11.16 and verify the conservation of complex powers. $v_s(t) = 300\cos(2\pi 60t)$ V.

The phasor of the voltage source is given by

$$V_s = 300\angle 0°\ V$$

continued

Example 11.9 continued

FIGURE 11.16

Circuit for EXAMPLE 11.9.

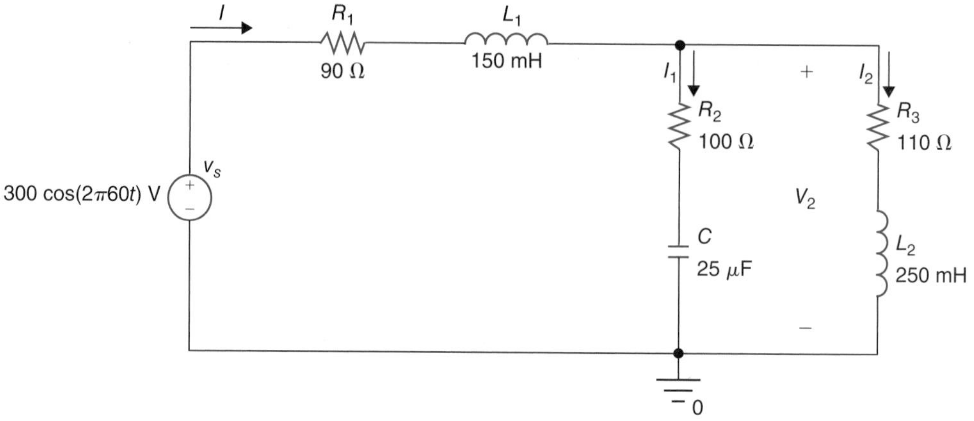

The impedances are given by

$$\mathbf{Z}_{R_1} = 90\ \Omega, \quad \mathbf{Z}_{R_2} = 100\ \Omega, \quad \mathbf{Z}_{R_3} = 110\ \Omega$$

$$\mathbf{Z}_{L_1} = j2\pi 60 \times 0.15 = j56.5487\ \Omega$$

$$\mathbf{Z}_{L_2} = j2\pi 60 \times 0.25 = j94.2478\ \Omega$$

$$\mathbf{Z}_C = \frac{1}{j2\pi \times 60 \times 25 \times 10^{-6}} = -j106.1033\ \Omega$$

Let $\mathbf{Z}_{\text{eq}}$ be the impedance of the parallel connection of the R_2–C and R_3–L_2 paths. Then, we have

$$\mathbf{Z}_{\text{eq}} = (\mathbf{Z}_{R_2} + \mathbf{Z}_C)\|(\mathbf{Z}_{R_3} + \mathbf{Z}_{L_2}) = \frac{(\mathbf{Z}_{R_2} + \mathbf{Z}_C) \times (\mathbf{Z}_{R_3} + \mathbf{Z}_{L_2})}{\mathbf{Z}_{R_2} + \mathbf{Z}_C + \mathbf{Z}_{R_3} + \mathbf{Z}_{L_2}}$$
$$= 100.2843 - j5.0365\ \Omega = 100.4107\angle -2.8751°\ \Omega$$

The total impedance seen from the source is

$$\mathbf{Z} = \mathbf{Z}_{R_1} + \mathbf{Z}_{L_1} + \mathbf{Z}_{\text{eq}} = 190.2843 + j51.5122 = 197.1335\angle 15.1476°\ \Omega$$

The phasor of the current flowing out of the voltage source is given by

$$\mathbf{I} = \frac{\mathbf{V}_s}{\mathbf{Z}} = 1.4689 - j0.3977 = 1.5218\angle -15.1476°\ \text{A}$$

The voltage across the parallel connection of R_2–C and R_3–L_2 paths is given by

$$\mathbf{V}_2 = \mathbf{I} \times \mathbf{Z}_{\text{eq}} = 145.3086 - j47.2772\ \text{V} = 152.8062\angle -18.0227°\ \text{V}$$

The current phasors I_1 and I_2 are given by

$$\mathbf{I}_1 = 0.9195 + j0.5029 = 1.0480\angle 28.6735°\ \text{A}$$

$$\mathbf{I}_2 = 0.5494 - j0.9005 = 1.0549\angle -58.6126°\ \text{A}$$

continued

Example 11.9 continued

The voltage phasors are given by

$$\mathbf{V}_{R_1} = \mathbf{Z}_{R_1} \times \mathbf{I} = 132.2044 - j35.7893 = 136.963\angle -15.1476° \text{ V}$$

$$\mathbf{V}_{L_1} = \mathbf{Z}_{L_1} \times \mathbf{I} = 22.4871 + j83.0664 = 86.0564\angle 74.8524° \text{ V}$$

$$\mathbf{V}_{R_2} = \mathbf{Z}_{R_2} \times \mathbf{I}_1 = 91.9522 + j50.2872 = 104.8047\angle 28.6735° \text{ V}$$

$$\mathbf{V}_{C} = \mathbf{Z}_{C} \times \mathbf{I}_1 = 53.3563 - j97.5644 = 111.2012\angle -61.3265° \text{ V}$$

$$\mathbf{V}_{R_3} = \mathbf{Z}_{R_3} \times \mathbf{I}_2 = 60.4356 - j99.0583 = 116.0389\angle -58.6126° \text{ V}$$

$$\mathbf{V}_{L_2} = \mathbf{Z}_{L_2} \times \mathbf{I}_2 = 84.873 + j51.7811 = 99.4219\angle 31.3874° \text{ V}$$

The complex powers on all the elements of the circuit are

$$\mathbf{S}_{V_s} = \frac{1}{2}\mathbf{V}_s \times (-\mathbf{I}^*) = -220.3406 - j59.6488 = 228.2717\angle -164.8524° \text{ VA}$$

$$\mathbf{S}_{R_1} = \frac{1}{2}\mathbf{V}_{R_1} \times \mathbf{I}^* = 104.2159 = 104.2159\angle 0° \text{ VA}$$

$$\mathbf{S}_{L_1} = \frac{1}{2}\mathbf{V}_{L_1} \times \mathbf{I}^* = j65.4808 = 65.4808\angle 90° \text{ VA}$$

$$\mathbf{S}_{R_2} = \frac{1}{2}\mathbf{V}_{R_2} \times \mathbf{I}_1^* = 54.9201 = 54.9201\angle 0° \text{ VA}$$

$$\mathbf{S}_{C} = \frac{1}{2}\mathbf{V}_{C} \times \mathbf{I}_1^* = -j58.2720 = 58.2720\angle -90° \text{ VA}$$

$$\mathbf{S}_{R_3} = \frac{1}{2}\mathbf{V}_{R_3} \times \mathbf{I}_2^* = 61.2046 = 61.2046\angle 0° \text{ VA}$$

$$\mathbf{S}_{L_2} = \frac{1}{2}\mathbf{V}_{L_2} \times \mathbf{I}_2^* = j52.44 = 52.44\angle 90° \text{ VA}$$

The sum of all the complex powers is equal to zero.

$$\mathbf{S}_{Vs} + \mathbf{S}_{R_1} + \mathbf{S}_{L_1} + \mathbf{S}_{R_2} + \mathbf{S}_{C} + \mathbf{S}_{R_3} + \mathbf{S}_{L_2} = 0$$

The real part of the complex power of the voltage source, $P_{V_s} = -220.3406$ W, is the average power released from the voltage source to the rest of the circuit. The sum of the average powers from the three resistors is

$$P_{R_1} + P_{R_2} + P_{R_3} = 220.3406 \text{ W}$$

which is the power absorbed by the three resistors. The imaginary part of the complex power of the voltage source, $Q_{V_s} = -59.6488$ VAR, is the reactive power released from the voltage source to the rest of the circuit. The sum of the reactive powers from L_1, L_2, and C is

$$Q_{L_1} + Q_{L_2} + Q_C = 59.6488 \text{ VAR}$$

which is the reactive power absorbed by L_1, L_2, and C. This proves the conservation of complex power.

Exercise 11.9

Find the complex powers of all the elements in the circuit shown in Figure 11.17 and verify the conservation of complex powers.

FIGURE 11.17

Circuit for EXERCISE 11.9.

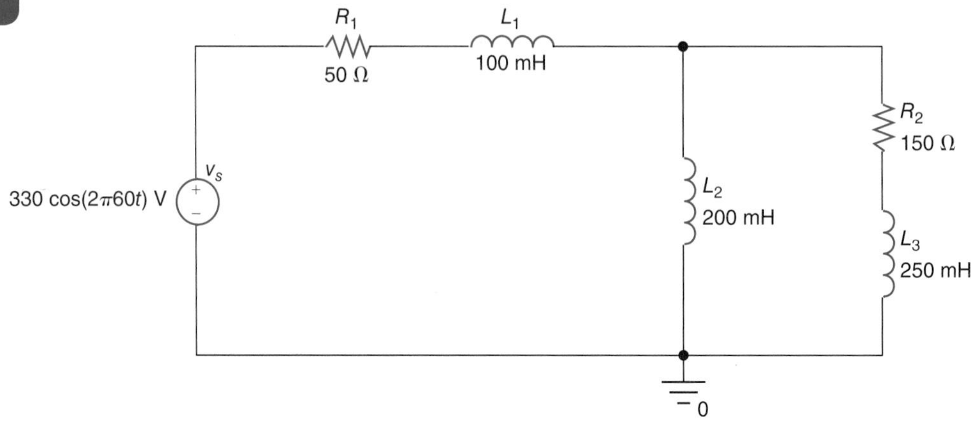

Answer:
$\mathbf{S}_{V_s} = -272.1625 - j385.1516$ VA, $\mathbf{S}_{R_1} = 204.2371$ VA, $\mathbf{S}_{L_1} = j153.9911$ VA, $\mathbf{S}_{L_2} = j188.4817$ VA, $\mathbf{S}_{R_2} = 67.9254$ VA, $\mathbf{S}_{L_3} = j42.6788$ VA, $\mathbf{S}_{V_s} + \mathbf{S}_{R_1} + \mathbf{S}_{L_1} + \mathbf{S}_{L_2} + \mathbf{S}_{R_2} + \mathbf{S}_{L_3} = 0.$

11.5 Maximum Power Transfer

Let the Thévenin equivalent impedance seen from the load toward the source be

$$\mathbf{Z}_{\text{TH}} = R_{TH} + jX_{TH}$$

and the Thévenin equivalent voltage be $\mathbf{V}_{\text{TH}}$. Let the load impedance be

$$\mathbf{Z}_{\text{L}} = R_L + jX_L$$

The Thévenin equivalent voltage and impedance, along with the load, are shown in Figure 11.18.

FIGURE 11.18

Thévenin equivalent circuit with a load impedance.

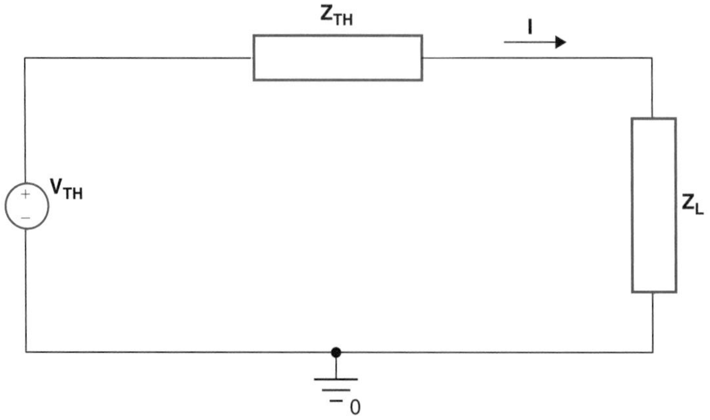

The current **I** through the circuit is given by

$$\mathbf{I} = \frac{\mathbf{V}_{\mathrm{TH}}}{\mathbf{Z}_{\mathrm{TH}} + \mathbf{Z}_{\mathrm{L}}} = \frac{\mathbf{V}_{\mathrm{TH}}}{R_{TH} + jX_{TH} + R_L + jX_L}$$

According to Equation (11.38), the average power on the load is given by

$$P = \frac{1}{2}|\mathbf{I}|^2 R_L = \frac{|\mathbf{V}_{\mathrm{TH}}|^2 \frac{R_L}{2}}{(R_{TH} + R_L)^2 + (X_{TH} + X_L)^2} \quad \textbf{(11.67)}$$

The voltage and current are peak values (not rms). To find the load reactance X_L that maximizes the average power, we take the partial differential of P with respect to X_L and set that equal to zero:

$$\frac{\partial P}{\partial X_L} = \frac{-|\mathbf{V}_{\mathrm{TH}}|^2 R_L(X_{TH} + X_L)}{[(R_{TH} + R_L)^2 + (X_{TH} + X_L)^2]^2} = 0$$

Notice that $\frac{d}{dx}\left(\frac{u}{v}\right) = \frac{v\frac{du}{dx} - u\frac{dv}{dx}}{v^2}$.

From this, we conclude that

$$X_L = -X_{TH} \quad \textbf{(11.68)}$$

This result says that for the maximum average power for the load, the reactance of the load must be the negative of the Thévenin reactance. If the Thévenin reactance is inductive, the load reactance is capacitive, and vice versa. Similarly, to find the load resistance R_L that maximizes the average power, we take the partial differential of P with respect to R_L and set that equal to zero:

$$\frac{\partial P}{\partial R_L} = \frac{|\mathbf{V}_{\mathrm{TH}}|^2[(R_{TH} + R_L)^2 + (X_{TH} + X_L)^2 - 2R_L(R_{TH} + R_L)]}{2[(R_{TH} + R_L)^2 + (X_{TH} + X_L)^2]^2} = 0$$

For the numerator to be zero, the load resistance should be identical to the Thévenin resistance as can be seen from

$$(R_{TH} + R_L)^2 + (X_{TH} - X_{TH})^2 - 2R_L(R_{TH} + R_L) = (R_{TH} + R_L)^2 - 2R_L(R_{TH} + R_L) = 0$$

$$(R_{TH} + R_L)(R_{TH} + R_L - 2R_L) = 0$$

$$(R_{TH} + R_L)(R_{TH} - R_L) = 0$$

Thus, we have

$$R_L = R_{TH} \; or \; R_L = -R_{TH}$$

Since $R_{TH} > 0$, we have

$$R_L = R_{TH} \quad \textbf{(11.69)}$$

Therefore, the load impedance that maximizes the average power of the load is given by the complex conjugate of the Thévenin impedance; that is,

$$\mathbf{Z}_{\mathrm{L}} = R_L + jX_L = R_{TH} - jX_{TH} = \mathbf{Z}_{\mathrm{TH}}^*$$

When Equations (11.68) and (11.69) are substituted into Equation (11.67), the maximum average power of the load becomes

$$P_{\max} = \frac{1}{2}|I|^2 R_L = \frac{|\mathbf{V}_{\mathrm{TH}}|^2 \frac{R_L}{2}}{(R_{TH} + R_L)^2 + (X_{TH} + X_L)^2}$$

$$= \frac{|\mathbf{V}_{\mathrm{TH}}|^2 \frac{R_{TH}}{2}}{(R_{TH} + R_{TH})^2 + (X_{TH} - X_{TH})^2} = \frac{|\mathbf{V}_{\mathrm{TH}}|^2}{8R_{TH}}$$

or

$$P_{\max} = \frac{|\mathbf{V}_{\mathrm{TH}}|^2}{8R_{TH}} \qquad (11.70)$$

This result indicates that the maximum average power delivered to the load is given by the magnitude squared of the Thévenin voltage divided by eight times the Thévenin resistance when the load impedance is set to be the complex conjugate of the Thévenin impedance. The voltage $\mathbf{V}_{\mathrm{TH}}$ is a peak value. For rms value, since $\mathbf{V}_{\mathrm{TH,\,rms}} = \mathbf{V}_{\mathrm{TH}}/\sqrt{2}$, we have

$$P_{\max} = \frac{|\mathbf{V}_{\mathrm{TH,\,rms}}|^2}{4R_{TH}} \qquad (11.71)$$

EXAMPLE 11.10

For the circuit shown in Figure 11.19, find the load impedance Z_{Load} that will maximize the power transfer to the load. Also, find the maximum power absorbed by the load.

FIGURE 11.19

Circuit for EXAMPLE 11.10.

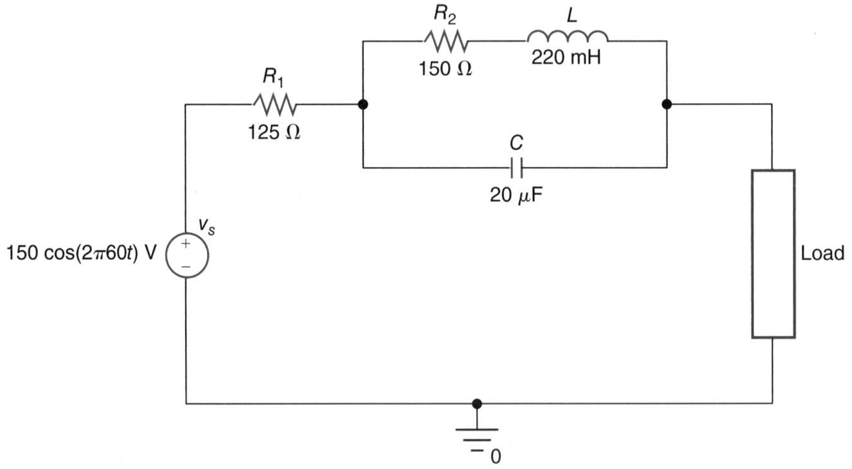

The phasor representation of the voltage source is given by

$$\mathbf{V}_s = 150\angle 0°$$

The impedances are given by

$$\mathbf{Z}_{R_1} = 125\ \Omega, \quad \mathbf{Z}_{R_2} = 150\ \Omega, \quad \mathbf{Z}_L = j\omega L = j2\pi \times 60 \times 0.22 = j82.9380\ \Omega,$$

continued

Example 11.10 continued

$$\mathbf{Z}_C = \frac{1}{j\omega C} = \frac{1}{j2\pi \times 60 \times 20 \times 10^{-6}} = -j132.6291\ \Omega$$

Without the load, the Thévenin equivalent voltage is the voltage of the voltage source. Thus,

$$\mathbf{V}_{TH} = \mathbf{V}_s = 150\angle 0° = 150 + j0\ \mathbf{V}$$

The Thévenin equivalent impedance is the equivalent impedance seen from the load with the voltage source short-circuited. Therefore, we have

$$\mathbf{Z}_{TH} = \mathbf{Z}_{R_1} + [(\mathbf{Z}_{R_2} + \mathbf{Z}_L)||\mathbf{Z}_C] = \mathbf{Z}_{R_1} + \frac{(\mathbf{Z}_{R_2} + \mathbf{Z}_L) \times \mathbf{Z}_C}{\mathbf{Z}_{R_2} + \mathbf{Z}_L + \mathbf{Z}_C} = 230.6731 - j97.6224\ \Omega$$

The load impedance Z_{Load} that will maximize the power transfer to the load is the complex conjugate of the Thévenin equivalent impedance. Thus,

$$\mathbf{Z}_{\text{Load}} = \mathbf{Z}_{TH}^* = 230.6731 + j97.6224$$

The maximum power absorbed by the load is given by

$$P_{\max} = \frac{|V_{TH}|^2}{8R_{TH}} = \frac{150^2}{8 \times 230.6731} = 12.1926\ \text{W}$$

Exercise 11.10

For the circuit shown in Figure 11.20, find the load impedance Z_{Load} that will maximize the power transfer to the load. Also, find the maximum power absorbed by the load.

FIGURE 11.20

Circuit for EXERCISE 11.10.

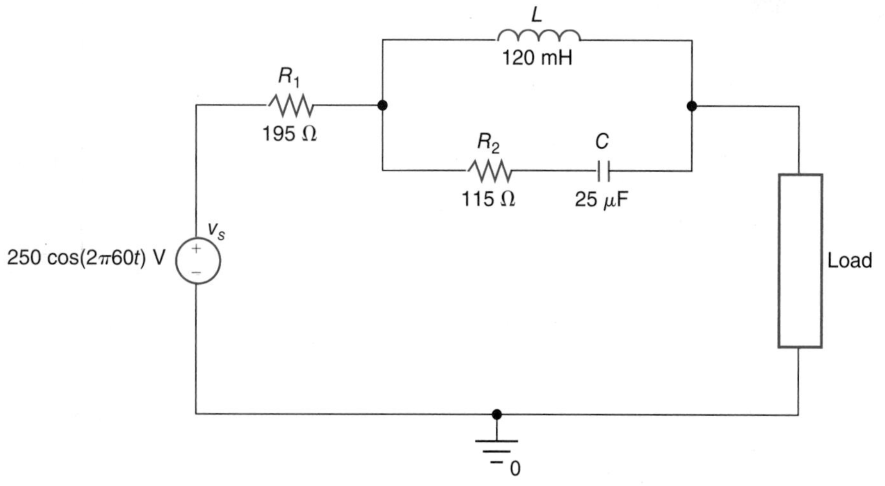

Answer:

$\mathbf{Z}_{\text{Load}} = 208.9021 - j52.5967$, $P_{\max} = 37.3979$ W.

FIGURE 11.21

Norton equivalent circuit with load $\mathbf{Z}_L$.

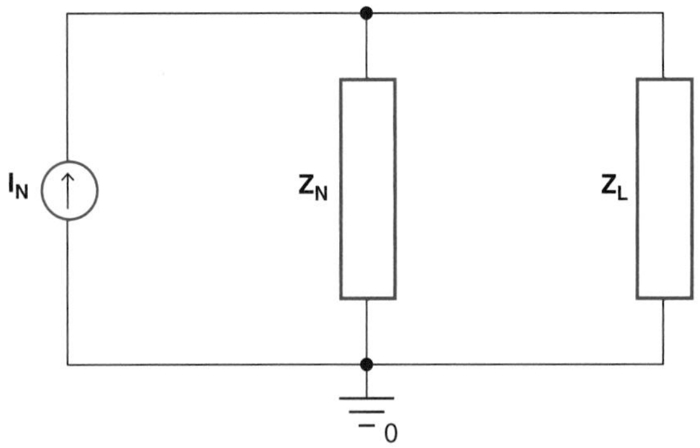

11.5.1 MAXIMUM POWER TRANSFER FOR NORTON EQUIVALENT CIRCUIT

Figure 11.21 shows a Norton equivalent circuit with the Norton equivalent current $\mathbf{I}_N$, the Norton equivalent impedance $\mathbf{Z}_N$, and a load with impedance $\mathbf{Z}_L$. A derivation similar to the one given above for the maximum power transfer for a Thévenin equivalent circuit shows that, for the maximum power transfer, the load impedance is selected to be

$$\mathbf{Z}_L = \mathbf{Z}_N^* \tag{11.72}$$

The maximum power of the load is given by

$$P_{max} = \frac{|\mathbf{I}_N|^2 R_L}{8} \tag{11.73}$$

where $\mathbf{I}_N$ is a peak value. If the rms value is used, Equation (11.73) becomes

$$P_{max} = \frac{|\mathbf{I}_{N,\,rms}|^2 R_L}{4} \tag{11.74}$$

11.6 Power Factor Correction (PFC)

The reactive power Q is the power exchanged between the reactive load and the source. As the power is exchanged between the source and the load, current flows on the transmission lines connecting the source and the load. The current introduces loss on the transmission lines. Let P be the average power on the load, $|\mathbf{V}_{rms}|$ be the magnitude of the rms voltage across the load, $|\mathbf{I}_{rms}|$ be the magnitude of the rms current through the load, and pf be the power factor. Then, according to Equation (11.36), we have

$$|\mathbf{I}_{rms}| = \frac{|\mathbf{S}|}{|\mathbf{V}_{rms}|} = \frac{\dfrac{P}{pf}}{|\mathbf{V}_{rms}|} = \frac{P}{|\mathbf{V}_{rms}| \cdot pf} \tag{11.75}$$

With P and $|\mathbf{V}_{rms}|$ fixed, when the power factor pf is increased, the magnitude of the current $|\mathbf{I}_{rms}|$ is decreased, and the loss on the transmission lines is decreased. If R_w is the resistance of the transmission lines, the average power on the transmission lines is given by, according to Equation (11.38),

$$P_w = |\mathbf{I}_{rms}|^2 R_w = \left(\frac{P}{|\mathbf{V}_{rms}| \cdot pf}\right)^2 R_w \tag{11.76}$$

The power loss is a function of $1/(pf)^2$. As the power factor is increased, the loss on the transmission lines is decreased. Since the maximum value of the power factor is one, the loss on the transmission lines will be minimized when the power factor is increased to 1. The reactive power of the load is related to the average power P and the power factor through

$$pf = \frac{1}{\sqrt{\left(\dfrac{Q}{P}\right)^2 + 1}}$$

according to Equation (11.59). With P and $|\mathbf{V}_{rms}|$ fixed, when the magnitude of Q is decreased, pf is increased, $|\mathbf{I}_{rms}|$ is decreased, and the loss is decreased. When the magnitude of Q is decreased to zero, $pf = 1$ and $|\mathbf{I}_{\text{rms}}|$ will be at its minimum, resulting in minimum loss on the transmission lines.

If the load is inductive (lagging power factor), Q is positive. The Q value can be reduced by connecting a capacitor in parallel to the load. Since the Q value of the capacitor is negative, the final Q value will be smaller than the original Q value. Let Q_i be the original Q value of the load, Q_f be the final Q value, and Q_C be the Q value of the capacitor. Then, we have

$$Q_C = Q_f - Q_i \tag{11.77}$$

The process of increasing the power factor by adding a parallel capacitor is called power factor correction. Let pf_f be the final power factor with the power factor correction. Then the final Q value is given by

$$Q_f = P \times \sqrt{\frac{1}{(pf_f)^2} - 1} \tag{11.78}$$

according to Equation (11.58). According to Equation (11.65), we can find the necessary capacitance value C for the power factor correction:

$$C = \frac{-Q_C}{|\mathbf{V}_{\text{rms}}|^2 \omega} \tag{11.79}$$

FIGURE 11.22

(a) Inductive load. (b) Inductive load with parallel capacitor.

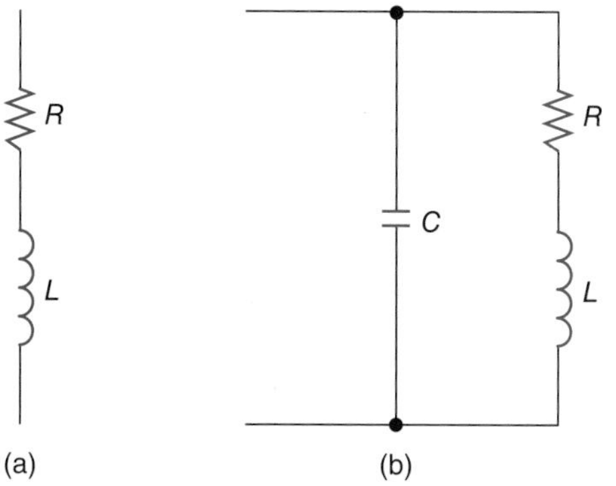

If the load consists of a series connection of a resistor with resistance R and an inductor with inductance L [as shown in Figure 11.22(a)], adding a parallel capacitor with capacitance value C [as shown in Figure 11.22(b)] can increase the power factor to the desired value. Equation (11.79) can be used to find the capacitance value. Alternatively, the capacitance value can be found from Equation (11.56). The equivalent impedance of the circuit shown in Figure 11.22(b) is given by

$$Z_{\text{eq}} = \frac{\dfrac{1}{j\omega C}(R + j\omega L)}{\dfrac{1}{j\omega C} + R + j\omega L} = \frac{R + j\omega L}{1 - \omega^2 LC + j\omega RC}$$

$$= \frac{R(1 - \omega^2 LC) + \omega^2 RLC + j[\omega L(1 - \omega^2 LC) - \omega R^2 C]}{(1 - \omega^2 LC)^2 + (\omega RC)^2}$$

The power factor of the circuit shown in Figure 11.22(b) is given by

$$pf_f = \frac{\text{Re}(Z_{\text{eq}})}{|Z_{\text{eq}}|} = \frac{R}{\sqrt{R^2 + \omega^2(L - \omega^2 L^2 C - R^2 C)^2}}$$

Solving for C, we obtain

$$C = \frac{L - \sqrt{\dfrac{\left(\dfrac{R}{pf_f}\right)^2 - R^2}{\omega^2}}}{R^2 + \omega^2 L^2} = \frac{L - \dfrac{R}{\omega}\sqrt{\dfrac{1}{(pf_f)^2} - 1}}{R^2 + \omega^2 L^2} \tag{11.80}$$

If the load is capacitive (leading power factor), Q is negative. The magnitude of Q value can be reduced by connecting an inductor in parallel to the load. Since the Q value of the inductor is positive, the magnitude of the final Q value will be smaller than the magnitude of the original Q value. Let Q_i be the original Q value of the load, Q_f be the final Q value, and Q_L be the Q value of the inductor. Then, we have

$$Q_L = Q_f - Q_i \tag{11.81}$$

The inductance value L is obtained from Equation (11.62):

$$L = \frac{|\mathbf{V}_{\text{rms}}|^2}{\omega Q_L} \tag{11.82}$$

EXAMPLE 11.11

In the circuit shown in Figure 11.23, find the current phasor I; the voltage phasor across R_1, R_2, and L; the load consisting of R_2 and L; and the complex power, average power, reactive power, and power factor of the load consisting of R_2 and L. Find the capacitance value C in the circuit shown in Figure 11.24 to increase the power factor to 0.98. Find the average power on R_1 before and after the addition of the capacitor. The input voltage is $v_s(t) = 330\cos(2\pi 60t)$ V (rms). All voltages and currents are in rms.

FIGURE 11.23

Circuit for EXAMPLE 11.11.

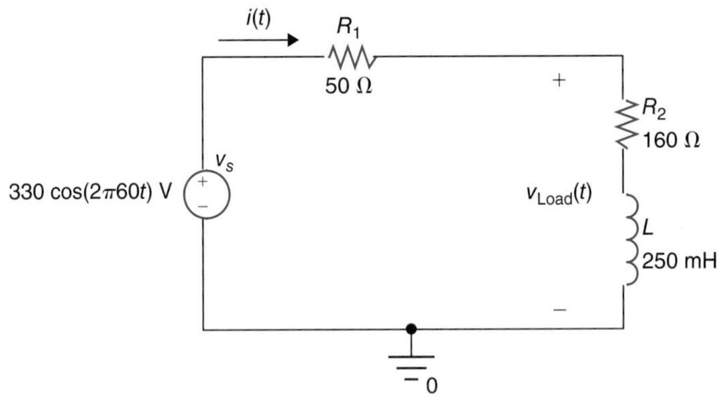

FIGURE 11.24

The circuit shown in Figure 11.23 with a parallel capacitor C.

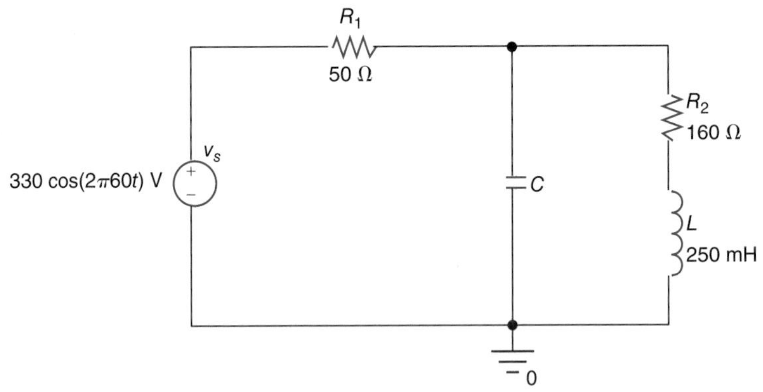

The phasor of the voltage source is

$$\mathbf{V}_s = 330\angle 0° = 330 + j0 \text{ V}$$

The impedances are given by

$$\mathbf{Z}_{R_1} = 50\ \Omega, \quad \mathbf{Z}_{R_2} = 160\ \Omega, \quad \mathbf{Z}_L = j2\pi \times 60 \times 0.25 = j94.2478\ \Omega$$

continued

Example 11.11 continued

The total impedance seen from the voltage source is

$$\mathbf{Z} = \mathbf{Z}_{R_1} + \mathbf{Z}_{R_2} + \mathbf{Z}_L = 210 + j94.2478\ \Omega = 230.1796\angle 24.1705°\ \Omega$$

The impedance of the load is

$$\mathbf{Z}_{\text{Load}} = \mathbf{Z}_{R_2} + \mathbf{Z}_L = 160 + j94.2478\ \Omega = 185.6950\angle 30.5002°\ \Omega$$

The phasor of the current $\mathbf{I}$ is given by

$$\mathbf{I} = \frac{\mathbf{V}_s}{\mathbf{Z}} = \frac{330\angle 0°}{230.1796\angle 24.1705°} = 1.4337\angle -24.1705° = 1.3080 - j0.587\text{ A (rms)}$$

The voltages across R_1, R_2, and L, are given, respectively, as

$$\mathbf{V}_{R_1} = \mathbf{I} \times \mathbf{Z}_{R_1} = 71.6832\angle -24.1705° = 65.3988 - j29.3509\text{ V (rms)}$$

$$\mathbf{V}_{R_2} = \mathbf{I} \times \mathbf{Z}_{R_2} = 229.3861\angle -24.1705° = 209.2761 - j93.9229\text{ V (rms)}$$

$$\mathbf{V}_L = \mathbf{I} \times \mathbf{Z}_L = 135.1196\angle 65.8295° = 55.3251 + j123.2738\text{ V (rms)}$$

$$\mathbf{V}_{\text{Load}} = \mathbf{I} \times \mathbf{Z}_{\text{Load}} = 266.2241\angle 6.3297° = 264.6012 + j29.3509\text{ V (rms)}$$

The complex powers of all elements are:

$$\mathbf{S}_{V_s} = -\mathbf{V}_s \times \mathbf{I}^* = -431.6319 - j193.7159 = 473.1088\angle -155.8295°\text{ VA}$$

$$\mathbf{S}_{R_1} = \mathbf{V}_{R_1} \times \mathbf{I}^* = 102.7695 = 102.7695\angle 0°\text{ VA}$$

$$\mathbf{S}_{R_2} = \mathbf{V}_{R_2} \times \mathbf{I}^* = 328.8624 = 328.8624\angle 0°\text{ VA}$$

$$\mathbf{S}_L = \mathbf{V}_L \times \mathbf{I}^* = j193.7159 = 193.7159\angle 90°\text{ VA}$$

$$\mathbf{S}_{\text{Load}} = \mathbf{V}_{\text{Load}} \times \mathbf{I}^* = 328.8624 + j193.7159 = 381.6757\angle 30.5002°\text{ VA}$$

The average power of the load is $P_{\text{Load}} = 328.8624$ W, the reactive power of the load is $Q_{\text{Load}} = 193.7159$ VAR, and the apparent power of the load is $|\mathbf{S}_{\text{Load}}| = 381.6757$ VA. The power factor of the load is

$$pf_1 = \frac{P_{\text{Load}}}{|\mathbf{S}_{\text{Load}}|} = 0.8616$$

For the final power factor of $pf_f = 0.98$, the final reactive power Q_f will be

$$Q_f = P_{\text{Load}} \times \sqrt{\frac{1}{(pf_f)^2} - 1} = 328.8624 \times \sqrt{\frac{1}{0.98^2} - 1} = 66.7784\text{ VAR}$$

The reactive power required from the parallel capacitor is

$$Q_C = Q_f - Q_{\text{Load}} = 66.7784 - 193.7159 = -126.9376\text{ VAR}$$

continued

Example 11.11 continued

The capacitance value is

$$C = \frac{-Q_C}{|\mathbf{V}_{\text{Load}}|^2\omega} = \frac{126.9376}{266.2241^2 \times 2\pi \times 60} = 4.7508\,\mu\text{F}$$

If Equation (11.80) is used, we obtain the same value:

$$C = \frac{L - \dfrac{R_2}{\omega}\sqrt{\dfrac{1}{(pf_f)^2} - 1}}{R_2^2 + \omega^2 L^2} = 4.7508\ \mu\text{F}$$

Let $\mathbf{Z}_{\text{eq}}$ be the equivalent impedance of the parallel connection of the load and the capacitor C in the circuit shown in Figure 11.24. Then, we have

$$\mathbf{Z}_{\text{eq}} = (\mathbf{Z}_{R_2} + \mathbf{Z}_L)\|\mathbf{Z}_C = \frac{(\mathbf{Z}_{R_2} + \mathbf{Z}_L) \times \mathbf{Z}_C}{\mathbf{Z}_{R_2} + \mathbf{Z}_L + \mathbf{Z}_C} = 206.9821 + j42.0295$$

$$= 211.2062\angle 11.4783^\circ\ \Omega$$

According to Equation (11.56), we have

$$pf = \frac{\text{Re}(\mathbf{Z}_{\text{eq}})}{|\mathbf{Z}_{\text{eq}}|} = \frac{206.9821}{211.2062} = 0.98$$

This verifies that the power factor is raised to 0.98. The total impedance seen from the voltage source is

$$\mathbf{Z}_t = \mathbf{Z}_{R_1} + \mathbf{Z}_{\text{eq}} = 256.9821 + j42.0295 = 260.3964\angle 9.2885^\circ\ \Omega$$

The current through the resistor R_1 is

$$\mathbf{I}_{\text{R}_1} = \frac{\mathbf{V}_s}{\mathbf{Z}_t} = 1.2673\angle -9.2885^\circ = 1.2507 - j0.2045\ \text{A}$$

After the addition of the capacitor, the magnitude of the current through R_1 decreased from 1.4337 A (rms) to 1.2673 A (rms). The voltage across R_1 is

$$\mathbf{V}_{\text{R}_1} = \mathbf{I}_{\text{R}_1} \times \mathbf{Z}_{\text{R}_1} = 63.3649\angle -9.2885^\circ\ \text{V}$$

The complex power on R_1 is given by

$$\mathbf{S}_{\text{R}_1} = \mathbf{V}_{\text{R}_1}\mathbf{I}^*_{\text{R}_1} = 80.3023\ \text{VA}$$

After the addition of the capacitor, the average power absorbed by R_1 decreased from 102.7695 W to 80.3023 W. The power lost in the transmission line decreased by increasing the power factor to 0.98.

Exercise 11.11

Find the value of the capacitance C in the circuit shown in Figure 11.24 to increase the power factor to 1.

Answer:

$C = 7.25\ \mu$F.

EXAMPLE 11.12

In the circuit shown in Figure 11.25, find the inductance value L to increase the power factor to 0.99. The input voltage is $v_s(t) = 220\cos(2\pi 60t)$ V (rms). All voltages and currents are in rms.

FIGURE 11.25

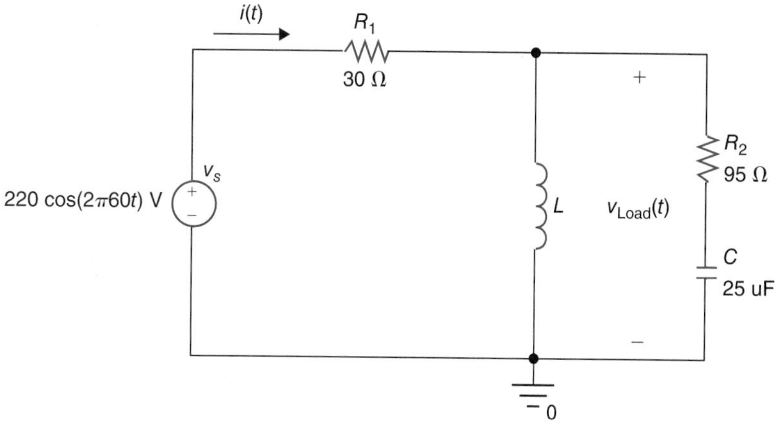

Circuit for EXAMPLE 11.12.

The phasor of the voltage source is

$$\mathbf{V}_s = 220\angle 0° = 220 + j0 \text{ V}$$

The impedances are given by

$$\mathbf{Z}_{R_1} = 30\ \Omega,\quad \mathbf{Z}_{R_2} = 95\ \Omega,\quad \mathbf{Z}_C = \frac{1}{j\omega C} = \frac{1}{j2\pi 60 \times 25 \times 10^{-6}} = -j106.1033\ \Omega$$

The total impedance seen from the voltage source is

$$\mathbf{Z} = \mathbf{Z}_{R_1} + \mathbf{Z}_{R_2} + \mathbf{Z}_C = 125 - j106.1033\ \Omega = 163.9601\angle -40.3255°\ \Omega$$

The impedance of the load is

$$\mathbf{Z}_{\text{Load}} = \mathbf{Z}_{R_2} + \mathbf{Z}_C = 95 - j106.1033\ \Omega = 142.4181\angle -48.1602°\ \Omega$$

The phasor of the current $\mathbf{I}$ is given by

$$\mathbf{I} = \frac{\mathbf{V}_s}{\mathbf{Z}} = 1.3418\angle 40.3255° = 1.0230 + j0.8683 \text{ A (rms)}$$

The voltage across the load is given by

$$\mathbf{V}_{\text{Load}} = \mathbf{I} \times \mathbf{Z}_{\text{Load}} = 191.0952\angle -7.8347° = 189.3114 - j26.0493 \text{ V (rms)}$$

The complex power of the load is

$$\mathbf{S}_{\text{Load}} = \mathbf{V}_{\text{Load}} \times \mathbf{I}^* = 171.0380 - j191.0284 = 256.4096\angle -48.1602° \text{ VA}$$

continued

Example 11.12 continued

The average power of the load is $P_{\text{Load}} = 171.0380$ W, the reactive power of the load is $Q_{\text{Load}} = -191.0284$, and the apparent power of the load is $|S_{\text{Load}}| = 256.4096$. The power factor of the load is

$$pf_1 = \frac{P_{\text{Load}}}{|\mathbf{S}_{\text{Load}}|} = 0.6671$$

Since $Q_{\text{Load}} < 0$, the pf_1 is a leading power factor. For the final power factor of $pf_f = 0.99$, the final reactive power Q_f will be

$$Q_f = -P_{\text{Load}} \times \sqrt{\frac{1}{(pf_f)^2} - 1} = -171.0380 \times \sqrt{\frac{1}{0.99^2} - 1} = -24.3716\,\text{VAR}$$

The reactive power required from the parallel inductor is

$$Q_L = Q_f - Q_{\text{Load}} = -24.3716 + 191.0284 = 166.6568\ \text{VAR}$$

The inductance value is

$$L = \frac{|\mathbf{V}_{\text{Load}}|^2}{Q_L \omega} = \frac{191.0952^2}{166.6568 \times 2\pi \times 60} = 581.2262\ \text{mH}$$

Exercise 11.12

Find the value of the inductance L in the circuit shown in Figure 11.25 to increase the power factor to 1.

Answer:

$L = 507.0727$ mH.

EXAMPLE 11.13

The average power on the load is 2.75 kW with a lagging power factor of 0.763. The voltage across the load is $\mathbf{V}_{\text{rms}} = 320\angle 0°$ V (rms). Thus, $|\mathbf{V}_{\text{rms}}| = 320$ V (rms). Find the equivalent resistance and inductance of the load. A capacitor is added in parallel to the load to raise the power factor to 0.995. Find the capacitance value, given that $f = 60$ Hz.

The apparent power of the load without the parallel capacitor is given by

$$|\mathbf{S}| = P/pf_i = 2750/0.763 = 3604.194\ \text{VA}$$

The initial reactive power of the load without the parallel capacitor is given by

$$Q_i = |\mathbf{S}|\sqrt{1 - pf_i^2} = 2329.7455\ \text{VAR}$$

The complex power is given by

$$\mathbf{S} = P + jQ_i = 2750 + j2329.7455\ \text{VA}$$

continued

Example 11.13 continued

The impedance of the load is given by

$$\mathbf{Z} = \frac{|\mathbf{V}_{\text{rms}}|^2}{\mathbf{S}^*} = 21.6779 + j18.3651\ \Omega$$

The real part of the impedance is resistance, and the imaginary part of the impedance is reactance. Since the reactance is positive, it is from the inductor. Thus, we have

$$R = \text{Re}(\mathbf{Z}) = 21.6779\ \Omega, \quad X = \omega L = 18.3651\ \Omega, \quad L = X/\omega = 48.7148\ \text{mH}$$

The final reactive power of the load with the parallel capacitor is given by

$$Q_f = P \times \sqrt{\frac{1}{(pf_f)^2} - 1} = 276.0362\ \text{VAR}$$

The reactive power required from the parallel capacitor is

$$Q_C = Q_f - Q_i = -2053.7093\ \text{VAR}$$

The required capacitance value is

$$C = \frac{-Q_C}{|\mathbf{V}_{\text{rms}}|^2\omega} = 53.1995\ \mu\text{F}$$

MATLAB

```
% EXAMPLE 11.13
% P.m (Ch 2), P2Rd.m, R2P.m (Ch 9) should be in the current folder
clear all;format long;
f=60                        %f = frequency in Hz
w=2*pi*f                    %w = frequency in rad/s
Pa=2750                     %Pa = average power
VrmsAbs=320                 %|Vrms|
phi=0                       %phi = phase of Vrms
Vrms=P2Rd(VrmsAbs,phi)      %Vrms in Cartesian coordinates
pfi=0.763                   %pfi = initial power factor
pff=0.995                   %pff = final power factor
AP=Pa/pfi                   %AP = apparent power
Qi=AP*sqrt(1-pfi^2)         %Qi = initial reactive power
S=Pa+j*Qi                   %S = complex power before correction
Z=VrmsAbs^2/conj(S)         %Z = equivalent impedance of load
Irms=Vrms/Z                 %Irms
Irmsp=R2P(Irms)             %Irmsp = Irms in polar coordinates
R=real(Z)                   %R = resistance
X=imag(Z)                   %X = reactance
L=X/w                       %L = inductance
Qf=Pa*sqrt(1/pff^2-1)       %Qf = final reactive power
QC=Qf-Qi                    %QC = reactive power from capacitor
C=-QC/(w*VrmsAbs^2)         %C = capacitance value
Ca=(L-(R/w)*sqrt(1/pff^2-1))/(R^2+w^2*L^2) %Alternate method for C
ZL=j*w*L                    %ZL = impedance of L
ZC=1/(j*w*C)                %ZC = impedance of C
Zeqa=P([R+ZL,ZC])           %Zeqa = impedance of load
pf2a=real(Zeqa)/abs(Zeqa)   %pf2a = power factor of load from impedance
```

Exercise 11.13

The apparent power on the load is 5.66 kVA with a lagging power factor of 0.785. The voltage across the load is $\mathbf{V}_{rms} = 530\angle 0°$ V (rms). Thus, $|\mathbf{V}_{rms}| = 530$ V (rms). Find the equivalent resistance and inductance of the load. A capacitor is added in parallel to the load to raise the power factor to 1. Find the capacitance value, given $f = 60$ Hz.

Answer:

$R = 38.9587\ \Omega,\quad L = 81.5535\text{ mH},\quad C = 33.111\ \mu\text{F}.$

EXAMPLE 11.14

The apparent power on the load is 2.43 kVA with a lagging power factor of 0.683. The current through the load is $\mathbf{I}_{rms} = 11.3\angle 0°$ A (rms). Thus, $|\mathbf{I}_{rms}| = 11.3$ A (rms). Find the equivalent resistance and inductance of the load. A capacitor is added in parallel to the load to raise the power factor to 0.99. Find the capacitance value. $f = 60$ Hz.

The average power of the load without the parallel capacitor is given by

$$P = |\mathbf{S}| \times pf_i = 2430 \times 0.683 = 1659.69\text{ W}$$

The initial reactive power of the load without the parallel capacitor is given by

$$Q_i = |\mathbf{S}|\sqrt{1 - pf_i^2} = 1774.9166\text{ VAR}$$

The complex power is given by

$$\mathbf{S} = P + jQ_i = 1659.69 + j1774.9166\text{ VA}$$

The impedance of the load is given by

$$\mathbf{Z} = \frac{\mathbf{S}}{|\mathbf{I}_{rms}|^2} = 12.9978 + j13.9002\ \Omega$$

The real part of the impedance is resistance, and the imaginary part of the impedance is reactance. Since the reactance is positive, it is from the inductor. Thus, we have

$$R = \text{Re}(\mathbf{Z}) = 12.9978\ \Omega,\quad X = \omega L = 13.9002\ \Omega,\quad L = X/\omega = 36.8714\text{ mH}$$

The magnitude of the voltage across the load is given by

$$|\mathbf{V}_{rms}| = \frac{|\mathbf{S}|}{|\mathbf{I}_{rms}|} = \frac{2430\text{ VA}}{11.3\text{ A}} = 215.0442\text{ V (rms)}$$

The final reactive power of the load with the parallel capacitor is given by

$$Q_f = P \times \sqrt{\frac{1}{(pf_f)^2} - 1} = 236.4930\text{ VAR}$$

continued

Example 11.14 continued

The reactive power required from the parallel capacitor is

$$Q_C = Q_f - Q_i = -1538.4236 \text{ VAR}$$

The required capacitance value is

$$C = \frac{-Q_C}{|\mathbf{V}_{\text{rms}}|^2\omega} = 88.2448 \ \mu\text{F}$$

Exercise 11.14

The average power on the load is 2.85 kW with a lagging power factor of 0.825. The current through the load is $\mathbf{I}_{\text{rms}} = 15\angle 0°$ A (rms). Thus, $|\mathbf{I}_{\text{rms}}| = 15$ A (rms). Find the equivalent resistance and inductance of the load. A capacitor is added in parallel to the load to raise the power factor to 0.99. Find the capacitance value, given $f = 60$ Hz.

Answer:

$R = 12.6667\ \Omega,\quad L = 23.0159 \text{ mH},\quad C = 77.3264\ \mu\text{F}.$

EXAMPLE 11.15

The average power on the load is 3.15 kW, with a leading power factor of 0.775. The voltage across the load is $\mathbf{V}_{\text{rms}} = 470\angle 0°$ V (rms). Thus, $|\mathbf{V}_{\text{rms}}| = 470$ V (rms). Find the equivalent resistance and capacitance of the load. An inductor is added in parallel to the load to raise the power factor to 0.99. Find the inductance value, given $f = 60$ Hz.

The apparent power of the load without the parallel inductor is given by

$$|\mathbf{S}| = P/pf_i = 3050/0.775 = 4064.5161 \text{ VA}$$

The initial reactive power of the load without the parallel inductor is given by

$$Q_i = -|\mathbf{S}|\sqrt{1 - pf_i^2} = -2568.6166 \text{ VAR}$$

The complex power is given by

$$\mathbf{S} = P + jQ_i = 3150 - j2568.6166 \text{ VA}$$

The impedance of the load is given by

$$\mathbf{Z} = \frac{|\mathbf{V}_{\text{rms}}|^2}{\mathbf{S}^*} = 42.12 - j34.3461\ \Omega$$

The real part of the impedance is resistance, and the imaginary part of the impedance is reactance. Since the reactance is negative, it is from capacitor. Thus, we have

$$R = \text{Re}(\mathbf{Z}) = 42.12\ \Omega, \quad -X = 1/(\omega C) = 34.3461\ \Omega,$$
$$C = -1/(\omega X) = 77.231\ \mu\text{F}$$

continued

Example 11.15 continued

The final reactive power of the load with the parallel inductor is given by

$$Q_f = -P \times \sqrt{\frac{1}{(pf_f)^2} - 1} = -448.8507 \text{ VAR}$$

The reactive power required from the parallel inductor is

$$Q_L = Q_f - Q_i = 2119.7659 \text{ VAR}$$

The required inductance value is

$$L = \frac{|\mathbf{V}_{\text{rms}}|^2}{\omega Q_L} = 276.4246 \text{ mH}$$

Exercise 11.15

The apparent power on the load is 4.75 *k*VA with a leading power factor of 0.785. The current through the load is $\mathbf{I}_{\text{rms}} = 17\angle 0°$ A (rms). Thus, $|\mathbf{I}_{\text{rms}}| = 17$ A (rms). Find the equivalent resistance and capacitance of the load. An inductor is added in parallel to the load to raise the power factor to 1. Find the inductance value, given $f = 60$ Hz.

Answer:

$R = 12.9022\ \Omega$, $C = 260.5162\ \mu\text{F}$, $L = 70.3763$ mH.

MATLAB

```
% Exercise 11.15
% P.m (Ch 2), P2Rd.m (Ch 9) should be in the current folder
clear all;format long;
f=60
w=2*pi*f
IrmsAbs=17
phi=0
Irms=P2Rd(IrmsAbs,phi)
AP=4750
pfi=0.785
pff=1
Pa=AP*pfi
Qi=-AP*sqrt(1-pfi^2)
S=Pa+j*Qi
Z=S/IrmsAbs^2
R=real(Z)
X=imag(Z)
C=-1/(w*X)
Qf=-Pa*sqrt(1/pff^2-1)
QL=Qf-Qi
Vrms=Irms*Z
VrmsAbs=AP/IrmsAbs
L=VrmsAbs^2/(w*QL)
ZC=1/(j*w*C)
ZL=j*w*L
Zeqa=P([R+ZC,ZL])
pf2a=real(Zeqa)/abs(Zeqa)
```

11.7 PSpice and Simulink

A simple circuit with two resistors and two inductors is shown in Figure 11.26. The AC voltage source with frequency 60 Hz has a voltage of 120 V and phase of 120°.

FIGURE 11.26

A circuit with AC source.

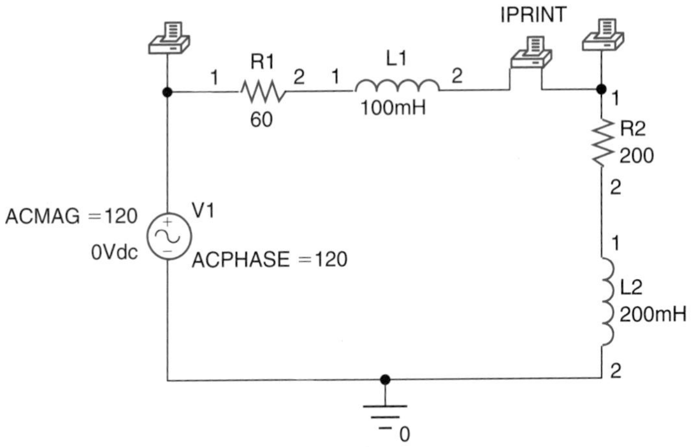

The impedances are given by

$$\mathbf{Z}_{R_1} = 60\ \Omega,\quad \mathbf{Z}_{R_2} = 200\ \Omega,\quad \mathbf{Z}_{L_1} = j37.6991\ \Omega,\quad \mathbf{Z}_{L_2} = j75.3982\ \Omega$$

The impedance across the R_1–L_1 path is

$$\mathbf{Z}_a = 60 + j37.6991\ \Omega = 70.8606\angle 32.1419°\ \Omega$$

and the impedance across the R_2–L_2 path is

$$\mathbf{Z}_b = 200 + j75.3982\ \Omega = 213.7402\angle 20.656°\ \Omega$$

The total impedance from the source is

$$\mathbf{Z} = 260 + j113.0973 = 283.5331\angle 23.5085°\ \Omega$$

The current through the circuit is given by

$$\mathbf{I} = \mathbf{V}_1/\mathbf{Z} = -0.04785 + j0.4205\ \text{A} = 0.4232\angle 96.4915°\ \text{A}$$

The voltage across the R_1–L_1 path is given by

$$\mathbf{V}_a = \mathbf{I} \times \mathbf{Z}_a = -18.7240 + j23.4272 = 29.9904\angle 128.6334°\ \text{V}$$

The voltage across the R_2–L_2 path is given by

$$\mathbf{V}_b = \mathbf{I} \times \mathbf{Z}_b = -41.276 + j80.4958 = 90.4615\angle 117.1475°\ \text{V}$$

The power factor on the R_1–L_1 path can be obtained by taking the cosine of the phase difference between V_a and I; that is,

$$pf_1 = \cos(\angle\mathbf{V}_a - \angle\mathbf{I}) = \cos(128.6334° - 96.4915°) = \cos(32.1419°) = 0.8467$$

Notice that the power factor is the cosine of the angle of Z_a. Similarly, the power factor on the R_2–L_3 path can be obtained by taking the cosine of the phase difference between V_2 and I; that is,

$$pf_2 = \cos(\angle \mathbf{V}_b - \angle \mathbf{I}) = \cos(117.1475° - 96.4915°) = \cos(20.656°) = 0.9357$$

The power factor can be obtained in PSpice by calculating the cosine of the phase difference between the voltage and the current. Set the analysis type to

AC Sweep/Noise, Linear, Start Frequency = 59, End Frequency = 61,
Total Points = 201

After running the simulation, click on Add Trace and type inside Trace Expression:

cos((*P*(V(R1:1)-V(L1:2))-*P*(I(R1:1)))*pi/180)

The power factor pf_1 is plotted as a function of frequency, as shown in Figure 11.27. As the frequency is increased, the power factor is decreased because the impedance of the inductor is increased, resulting in higher reactive power.

FIGURE 11.27

Plot of power factor as a function of frequency.

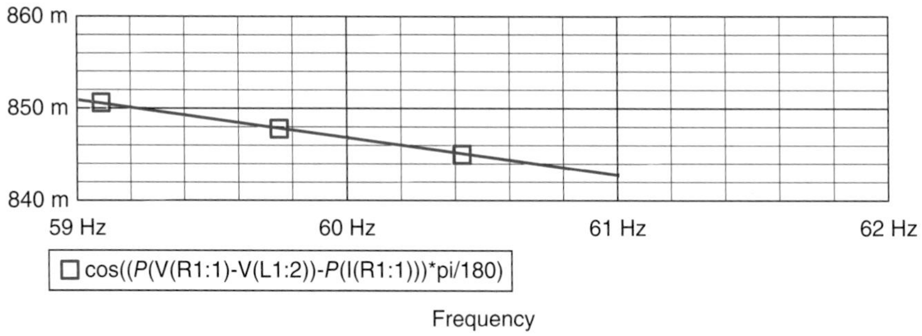

FIGURE 11.28

Power factor at 60 Hz.
(Source: OrCAD PSpice by Cadence)

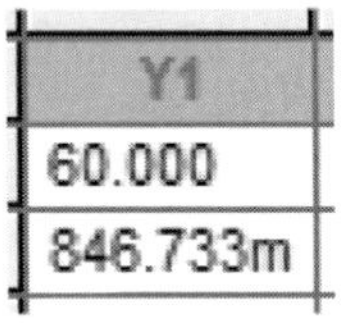

Select Trace → Cursor → Display and click the curve at 60 Hz. The value of the power factor is 0.846733, as shown in Figure 11.28.

We can view the data by selecting Edit → Select All → Copy, and paste in Word. Alternatively, click on the equation (it turns red), then select Edit → Copy and paste. We get the following result at 60 Hz:

```
Frequency            cos((P(V(R1:1)-V(L1:2))-P( I(R1:1)))*pi/180)
59.9999999999998     0.846733016201826
```

Repeating the procedure for pf_2, we obtain the results shown in Figure 11.29.

```
Frequency                 cos((P(V(R2:1))-P(I(R2:1)))*pi/180)
59.9999999999998          0.935715228034157
```

FIGURE 11.29

Power factor for Z_b.
(Source: OrCAD PSpice by Cadence)

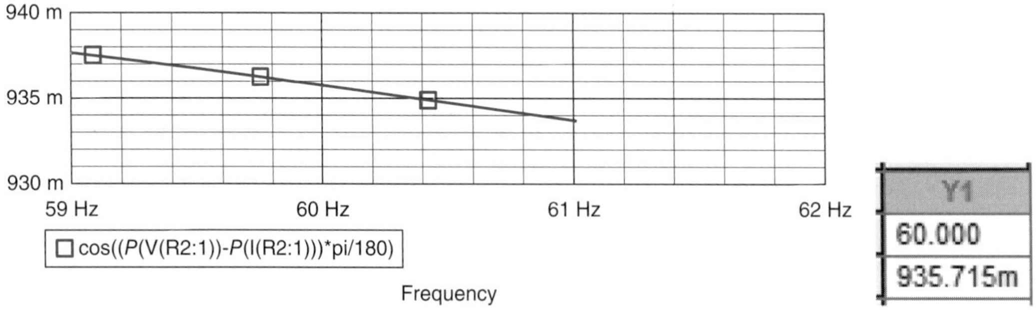

Exercise 11.16

Find the power factor on the load consisting of R_2, L_2, and C_1 in the circuit shown in Figure 11.30. The frequency of the voltage source is 60 Hz.

FIGURE 11.30

Circuit for EXERCISE 11.16.

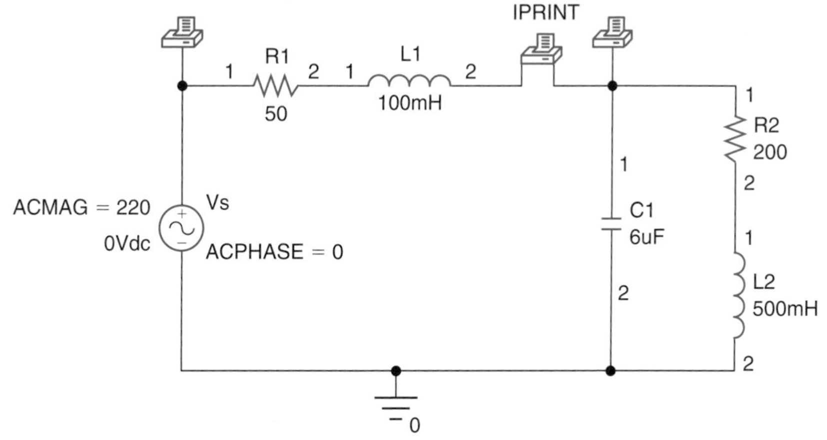

Answer:
$pf = 0.9961$.

A Simulink model for the circuit shown in Figure 11.26 is shown in Figure 11.31. The Voltage Measurement and Current Measurement are set to **Magnitude-Angle** in the model. The power factors are measured by calculating the cosine of the phase difference between voltage and current. All the measurements match with those obtained by direct calculation.

FIGURE 11.31

Simulink model for the circuit shown in Figure 11.26.

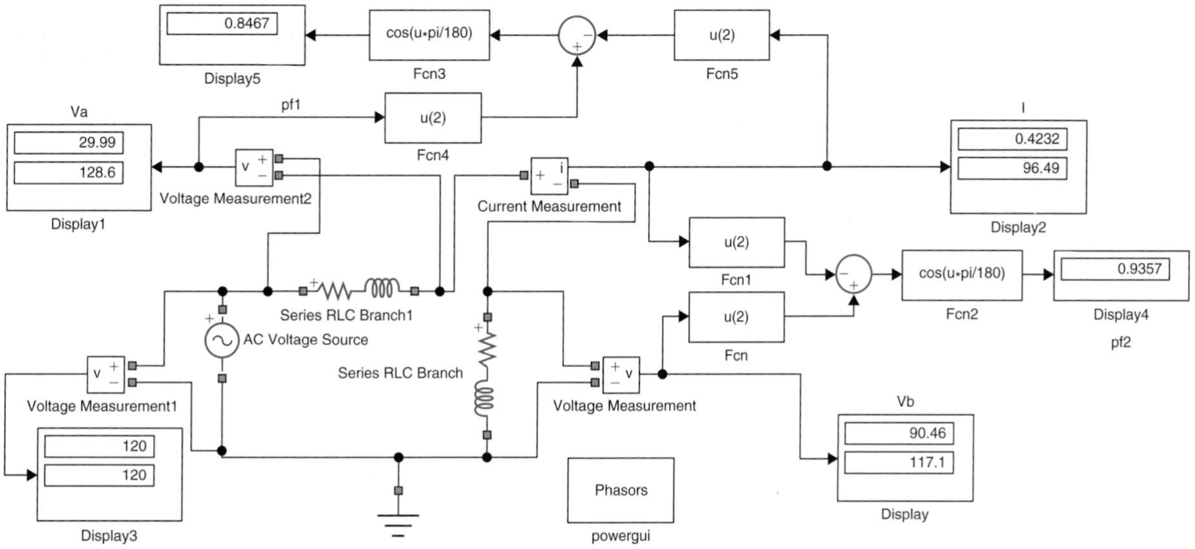

Figure 11.32 shows the Simulink model for the circuit shown in Figure 11.30 with complex powers and power factors. The Voltage Measurement and Current Measurement are set to **Complex** in the model.

FIGURE 11.32

Simulink model for the circuit shown in Figure 11.30.

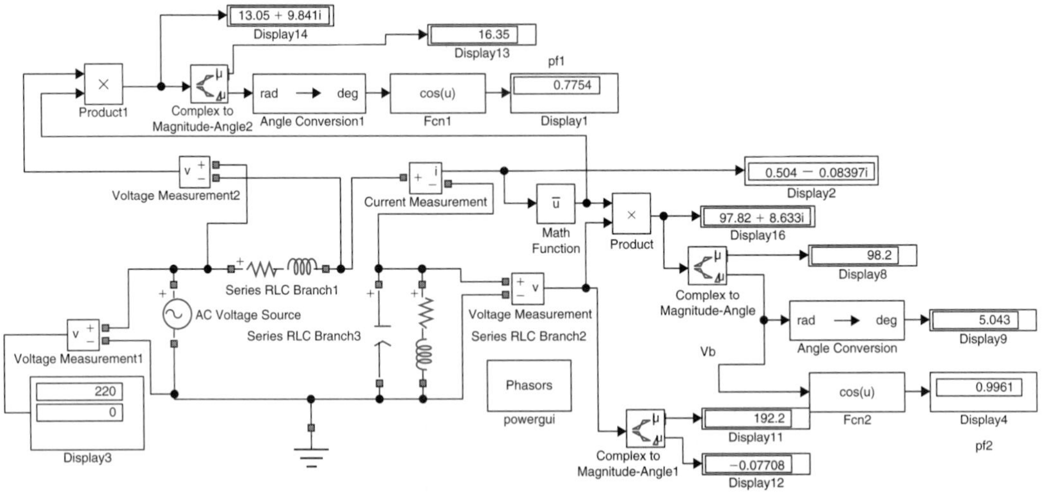

SUMMARY

The definitions of instantaneous power, average power, reactive power, apparent power, and complex power are presented in this chapter. The real part of the complex power is the average power, and the imaginary part of the complex power is the reactive power:

$$\mathbf{S} = P + jQ$$

The magnitude of the complex power, $|\mathbf{S}|$, is the apparent power, and the angle of the complex power, $\angle\mathbf{S}$, is the phase difference between the voltage and the current:

$$\angle S = \theta = \theta_v - \theta_i$$

The cosine of θ is defined as the power factor. The reactive power is the power exchanged between the generator and the reactive components of the load. It increases the currents on the transmission lines and the loss on the transmission lines. To minimize the power loss on the transmission lines, a parallel capacitor is added to the inductive load to decrease the reactive component of the complex power, thereby increasing the power factor close to one. This process is called the power factor correction.

When the load impedance is the complex conjugate of the Thévenin impedance, the power transferred to the load is maximized.

PROBLEMS

AC Power

11.1 **Given the following voltage and current on the load, find the complex power, apparent power, average power, reactive power, and power factor. Draw the phasors for P, Q, and S. The voltage and current are peak values.**

$$v(t) = 120\cos(2\pi 60t + 120°)\text{ V},$$

$$i(t) = 5\cos(2\pi 60t + 60°)\text{ A}$$

11.2 **Given the following voltage and current on the load, find the complex power, apparent power, average power, reactive power, and power factor. Draw the phasors for P, Q, and S. The voltage and current are peak values.**

$$v(t) = 120\cos(2\pi 60t + 60°)\text{ V},$$

$$i(t) = 3\cos(2\pi 60t + 35°)\text{ A}$$

11.3 **Given the following voltage and current on the load, find the complex power, apparent power, average power, reactive power, and power factor. Draw the phasors for P, Q, and S. The voltage and current are peak values.**

$$v(t) = 5\cos(2\pi 50t + 60°)\text{ V},$$

$$i(t) = 2\cos(2\pi 50t + 30°)\text{ A}$$

11.4 **Given the following voltage and current on the load, find the complex power, apparent power, average power, reactive power, and power factor.**

Draw the phasors for *P*, *Q*, and *S*. The voltage and current are peak values.

$$v(t) = 220\cos(2\pi 50t + 120°)\ \text{V},$$
$$i(t) = 5\cos(2\pi 50t + 60°)\ \text{A}$$

11.5 **Given the following voltage and current on the load, find the complex power, apparent power, average power, reactive power, and power factor. Draw the phasors for *P*, *Q*, and *S*. The voltage and current are peak values.**

$$v(t) = 80\cos(2\pi 60t + 135°)\ \text{V},$$
$$i(t) = 2\cos(2\pi 60t + 120°)\ \text{A}$$

Complex Power

11.6 **Find the impedance, complex power, apparent power, average power, reactive power, and power factor of the load shown in Figure P11.6 when the magnitude of the voltage across the load is $|\mathbf{V}_{rms}| = 240$ V and the radian frequency is $\omega = 2\pi 60$ rad/s.**

FIGURE P11.6

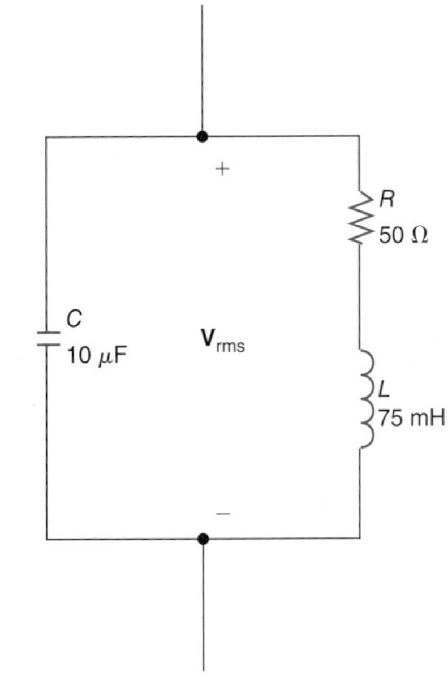

11.7 **Find the impedance, complex power, apparent power, average power, reactive power, and power factor of the load shown in Figure P11.7 when the magnitude of the voltage across the load is $|\mathbf{V}_{rms}| = 250$ V and the radian frequency is $\omega = 2\pi 60$ rad/s.**

FIGURE P11.7

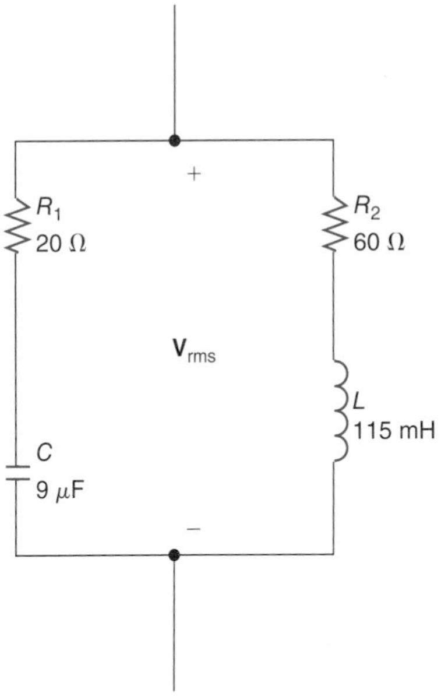

11.8 **Find the impedance, complex power, apparent power, average power, reactive power, and power factor of the load shown in Figure P11.8 when the magnitude of the voltage across the load is $|\mathbf{V}_{rms}| = 150$ V and the radian frequency is $\omega = 2\pi 60$ rad/s.**

FIGURE P11.8

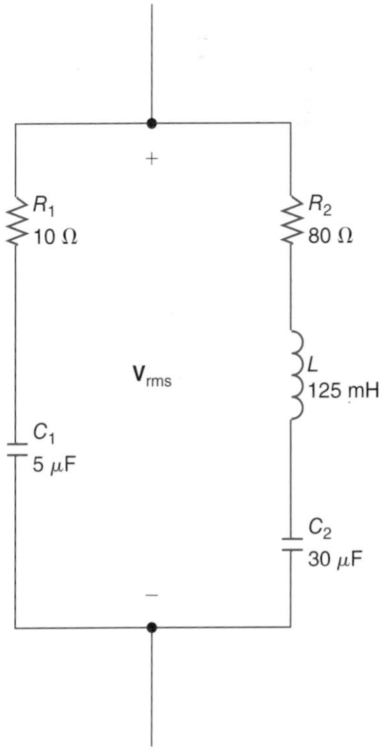

11.9 Find the impedance, complex power, apparent power, average power, reactive power, and power factor of the load shown in Figure P11.9 when the magnitude of the current through the load is $|\mathbf{I}_{\text{rms}}| = 5$ A and the radian frequency is $\omega = 2\pi 60$ rad/s.

FIGURE P11.9

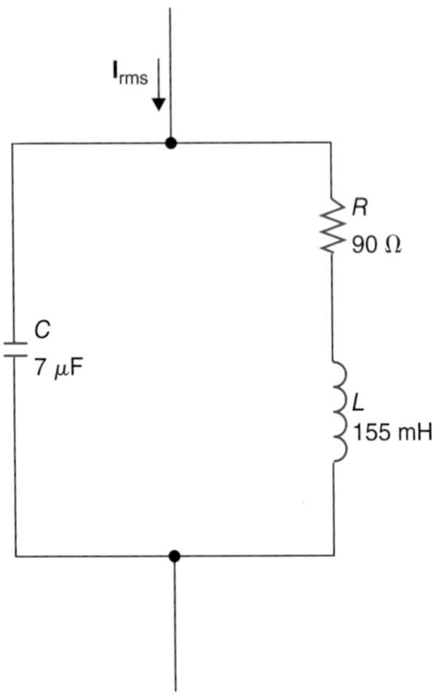

11.10 Find the impedance, complex power, apparent power, average power, reactive power, and power factor of the load shown in Figure P11.10 when the magnitude of the current through the load is $|\mathbf{I}_{\text{rms}}| = 6$ A and the radian frequency is $\omega = 2\pi 50$ rad/s.

FIGURE P11.10

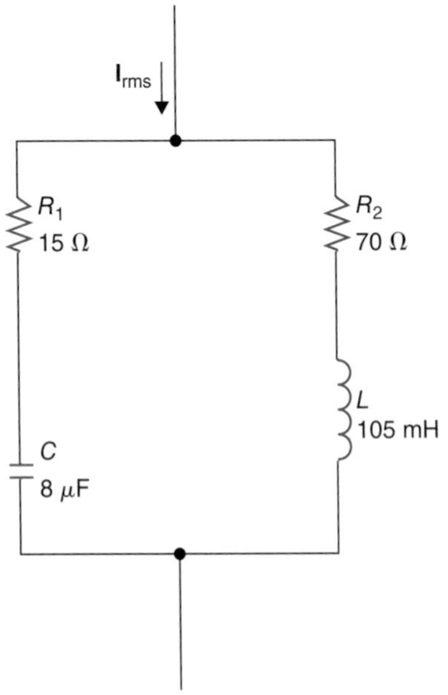

11.11 Find the impedance, complex power, apparent power, average power, reactive power, and power factor of the load shown in Figure P11.11 when the magnitude of the current through the load is $|\mathbf{I}_{\text{rms}}| = 7$ A and the radian frequency is $\omega = 2\pi 50$ rad/s.

FIGURE P11.11

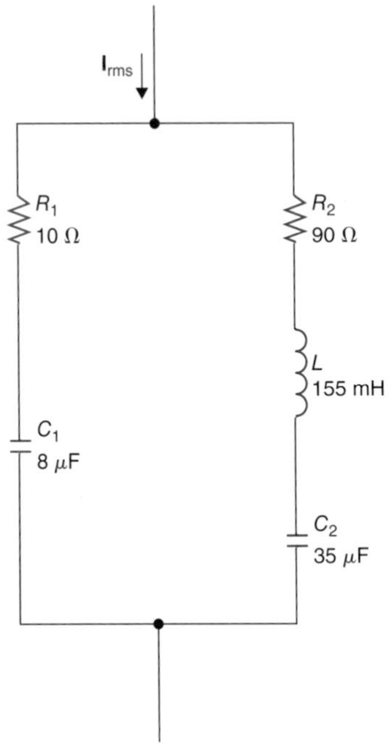

11.12 Find the power factor of the load shown in Figure P11.12 when the radian frequency is $\omega = 2\pi 60$ rad/s. If the load is replaced by an equivalent load consisting of a series connection of a resistor and an inductor, what are the values of the resistance and the inductance?

FIGURE P11.12

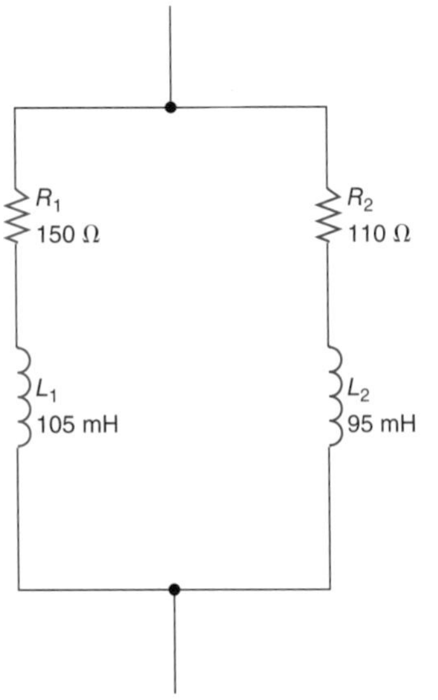

11.13 Find the power factor of the load shown in Figure P11.13 when the radian frequency is $\omega = 2\pi 60$ rad/s. If the load is replaced by an equivalent load consisting of a series connection of a resistor and an inductor, what are the values of the resistance and the inductance?

FIGURE P11.13

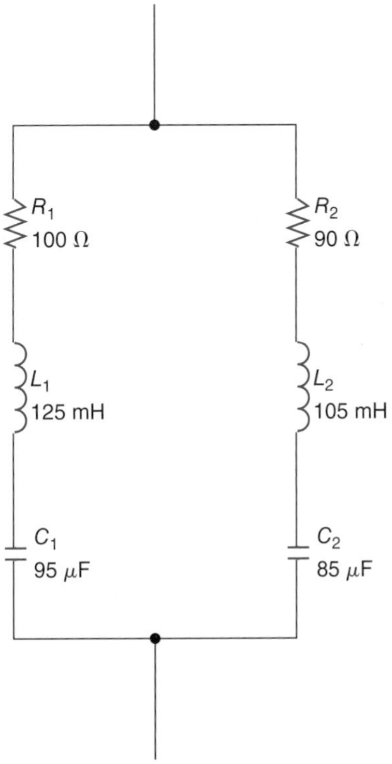

11.14 The average power of a load is 29.5 kW with a lagging power factor of 0.915. Find the reactive power and the complex power of the load.

11.15 The reactive power of a load is 4.53 $kVAR$ with a lagging power factor of 0.874. Find the average power and the complex power of the load.

11.16 The apparent power of a load is 43.7 kVA with a lagging power factor of 0.784. Find the complex power, average power, and reactive power of the load.

11.17 The apparent power of a load is 26.8 kVA with a leading power factor of 0.952. Find the complex power, average power, and reactive power of the load.

11.18 The average power of a load is 3.74 kW with a lagging power factor of 0.926. The magnitude of the rms voltage across the load is given by $|\mathbf{V}_{\text{rms}}| = 530$ V. Find the complex power and the reactive power of the load. If the load is replaced by an equivalent load consisting of a series connection of a resistor and an inductor, what are the values of the resistance and the inductance? Assume that $f = 60$ Hz.

11.19 The reactive power of a load is 1.42 $kVAR$ with a lagging power factor of 0.947. The magnitude of the rms voltage across the load is given by $|\mathbf{V}_{\text{rms}}| = 470$ V. Find the apparent power, average power, and complex power of the load. If the load is replaced by an equivalent load consisting of a series connection of a resistor and an inductor, what are the values of the resistance and the inductance? Assume that $f = 60$ Hz.

11.20 The apparent power of a load is 17.3 kVA with a lagging power factor of 0.837. The magnitude of the rms current through the load is given by $|\mathbf{I}_{\text{rms}}| = 14.3$ A. Find the average power, reactive power, and complex power of the load. If the load is replaced by an equivalent load consisting of a series connection of a resistor and an inductor, what are the values of the resistance and the inductance? Assume that $f = 60$ Hz.

11.21 The apparent power of a load is 22.3 kVA with a leading power factor of 0.941. The magnitude of the rms current through the load is given by $|\mathbf{I}_{\text{rms}}| = 16.6$ A. Find the average power, reactive power, and complex power of the load. If the load is replaced by an equivalent load consisting of a series connection of a resistor and a capacitor, what are the values of the resistance and the capacitance? Assume that $f = 60$ Hz.

11.22 The input signal applied to the circuit shown in Figure P11.22 is given by

$$v_s(t) = 120\cos(2\pi 60t + 20°)\ \text{V (rms)}$$

a. Find the phasors $\mathbf{Z}_{R_1}$, $\mathbf{Z}_{L_1}$, $\mathbf{Z}_{R_2}$, $\mathbf{Z}_{L_2}$, $\mathbf{I}$, $\mathbf{V}_{R_1}$, $\mathbf{V}_{L_1}$, $\mathbf{V}_{R_2}$, and $\mathbf{V}_{L_2}$.
b. Find the average power absorbed by R_1 and R_2.
c. Find the complex power on L_1 and L_2.
d. Find the complex power and power factor of R_1 and L_1 combined and the complex power and the power factor of R_2 and L_2 combined, and state whether the power is leading or lagging.

FIGURE P11.22

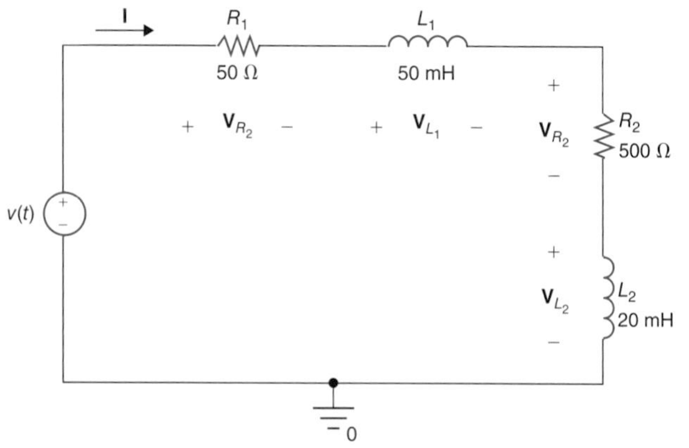

11.23 The input signal applied to the circuit shown in Figure P11.23 is given by

FIGURE P11.23

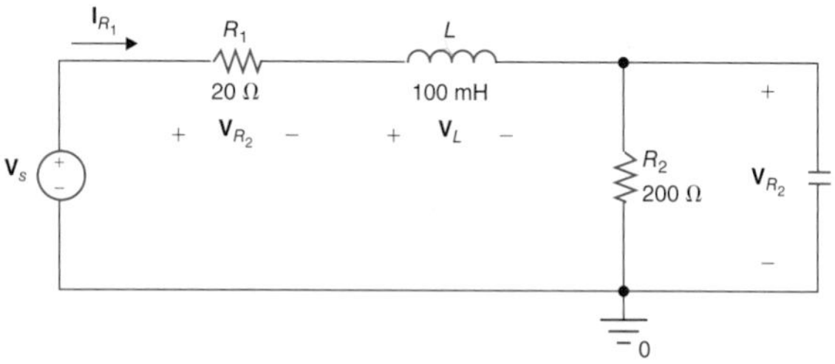

$$v_s(t) = 120\cos(2\pi 60t + 30°)\text{ V (rms)}$$

a. Find the phasors $\mathbf{Z}_{R_1}$, $\mathbf{Z}_L$, $\mathbf{Z}_{R_2}$, $\mathbf{Z}_C$, $\mathbf{I}_{R_1}$, $\mathbf{V}_{R_1}$, $\mathbf{V}_L$, and $\mathbf{V}_{R_2}$.
b. Find the average power absorbed by R_1 and R_2.
c. Find the complex power and power factor of R_1 and L combined and the complex power and the power factor of R_2 and C combined.

11.24 The input signal applied to the circuit shown in Figure P11.24 is given by

FIGURE P11.24

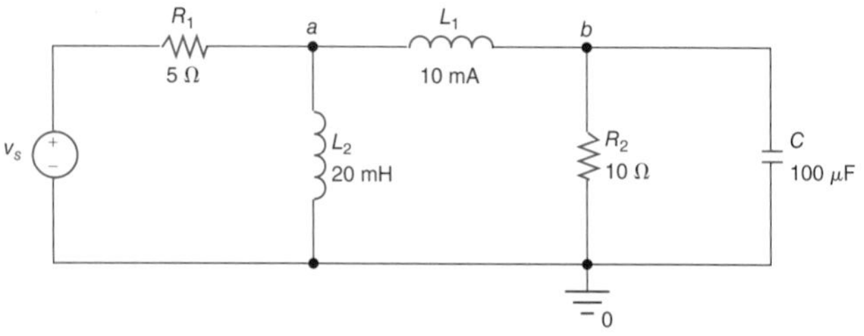

$$v_s(t) = 120\cos(2\pi 60t + 60°)\text{ V (rms)}$$

a. Find the impedances of all elements.
b. Find the voltages at nodes a and b.
c. Find the average power absorbed by R_2.
d. Find the complex power and power factor of the parallel combination of R_2 and C.

11.25 In Figure P11.25, the load voltage is given by $\mathbf{V}_L = 660\angle 30°$ V (rms). All voltages and currents are in rms.

a. Find the current $\mathbf{I}_1$.
b. Find the current $\mathbf{I}_2$.
c. Find the current $\mathbf{I}_3$.
d. Find the voltage phasor $\mathbf{V}_s$ at the source.
e. Find the complex power on $\mathbf{Z}_{R_1} + \mathbf{Z}_{C_1}$, $\mathbf{Z}_L$, and $\mathbf{Z}_{R_2} + \mathbf{Z}_{C_2}$.
f. Find the complex power of the source $\mathbf{V}_s$.

FIGURE P11.25

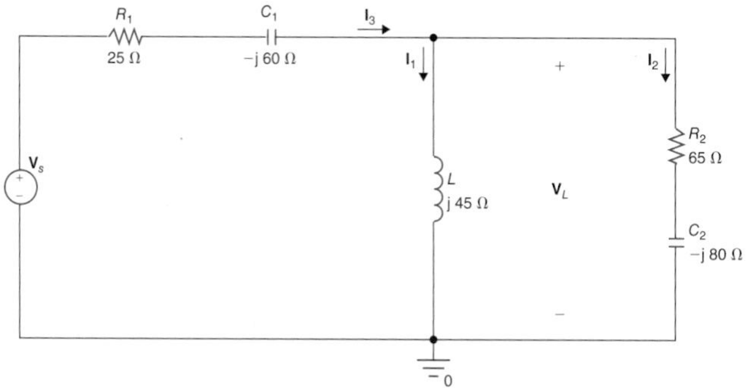

11.26 The input signal $v_s(t)$ in the circuit shown in Figure P11.26 is given by

FIGURE P11.26

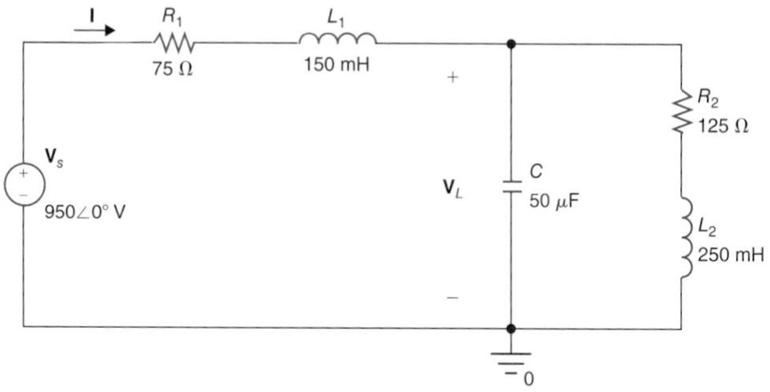

$$v_s(t) = 950\cos(2\pi 60t + 0°)\text{ V (rms)}$$

a. Find the impedances $\mathbf{Z}_{R_1}$, $\mathbf{Z}_{R_2}$, $\mathbf{Z}_{L_1}$, $\mathbf{Z}_{L_2}$, and $\mathbf{Z}_C$.
b. Find the wire impedance $\mathbf{Z}_W = \mathbf{Z}_{R_1} + \mathbf{Z}_{L_1}$ and load impedance $\mathbf{Z}_L = (\mathbf{Z}_{R_2} + \mathbf{Z}_{L_2}) \,||\, \mathbf{Z}_C$.
c. Find the total impedance $\mathbf{Z}_T = \mathbf{Z}_W + \mathbf{Z}_L$.
d. Find the current $\mathbf{I}$, wire voltage $\mathbf{V}_W$, and load voltage $\mathbf{V}_L$.
e. Find the complex power $\mathbf{S}_W$ of the wire $\mathbf{Z}_W$.
f. Find the complex power $\mathbf{S}_L$ of the load $\mathbf{Z}_L$.
g. Find the complex power $\mathbf{S}_{V_s}$ of the source $\mathbf{V}_S$.

Conservation of AC Power

11.27 Find the complex powers on v_s, R_1, R_2, L_1, and L_2 and prove conservation of complex power for the circuit shown in Figure P11.27. Assume $v_s(t) = 170\cos(2\pi 60 + 0°)$ V (rms).

FIGURE P11.27

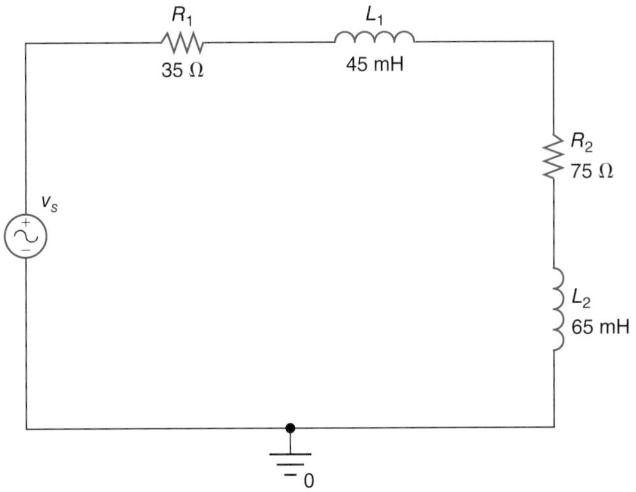

11.28 Find the complex power on v_s, R_1, R_2, L_1, L_2, and C, and prove conservation of complex power for the circuit shown in Figure P11.28. Assume $v_s(t) = 180\cos(2\pi 60 + 0°)$ V (rms).

FIGURE P11.28

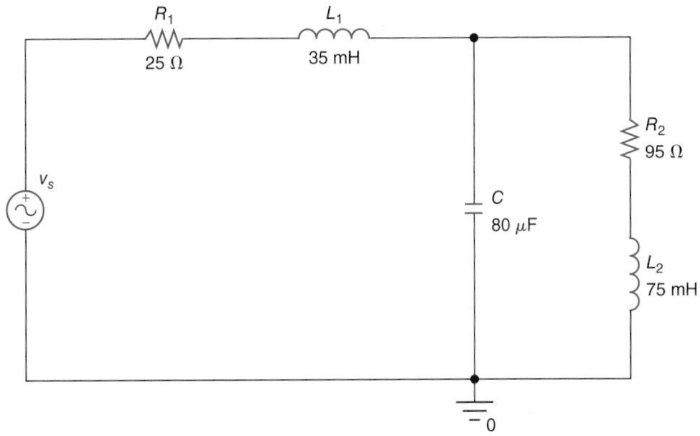

11.29 Find the complex power on v_s, R_1, R_2, L_1, L_2, and C, and prove conservation of complex power for the circuit shown in Figure P11.29. Assume $v_s(t) = 160\cos(2\pi 60 + 0°)$ V (rms).

FIGURE P11.29

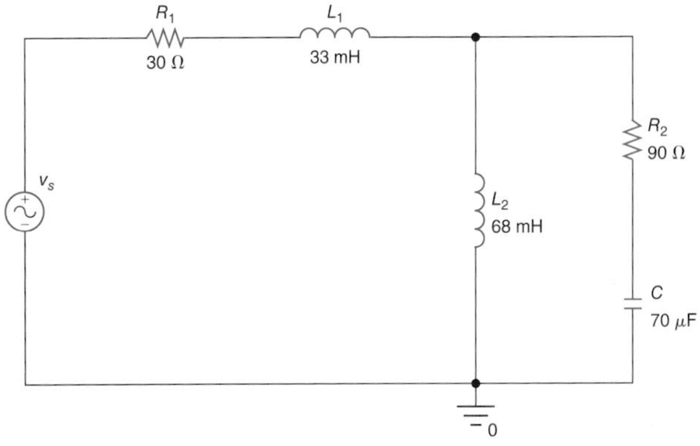

11.30 Find the complex power on v_s, R_1, R_2, R_3, L_1, L_2, and L_3 and prove conservation of complex power for the circuit shown in Figure P11.30. Assume $v_s(t) = 190\cos(2\pi 60 + 0°)$ V (rms).

FIGURE P11.30

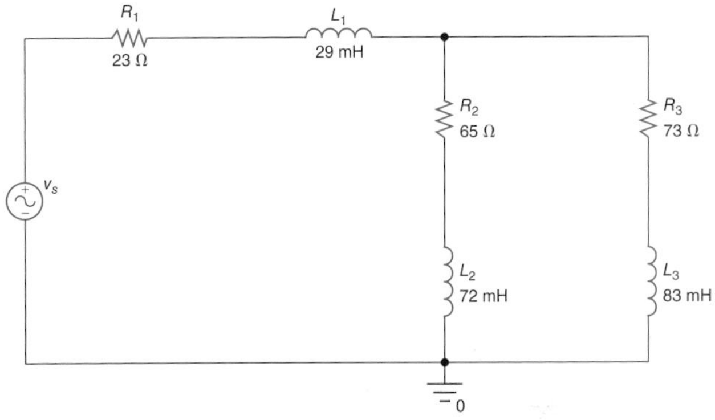

11.31 Find the complex power on v_s, R_1, R_2, L_1, L_2, C_1, and C_2, and prove conservation of complex power for the circuit shown in Figure P11.31. Assume that $v_s(t) = 200\cos(2\pi 60 + 0°)$ V (rms).

FIGURE P11.31

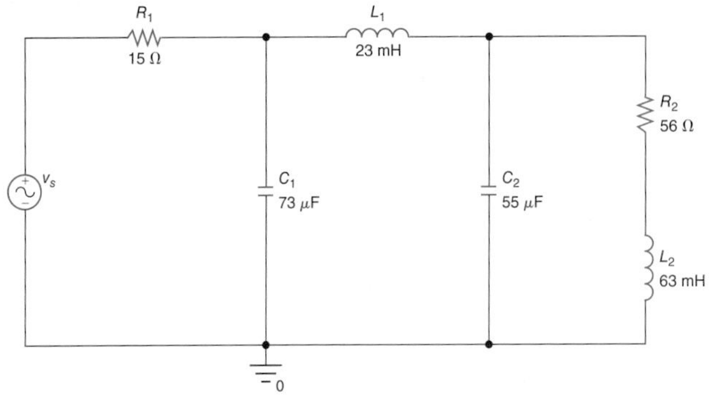

Maximum Power Transfer

11.32 Find the load impedance in the circuit shown in Figure P11.32 for maximum power transfer. Also, find the maximum power delivered to the load. Assume $v_s(t) = 50\cos(2\pi 60t + 0°)$ V (rms).

FIGURE P11.32

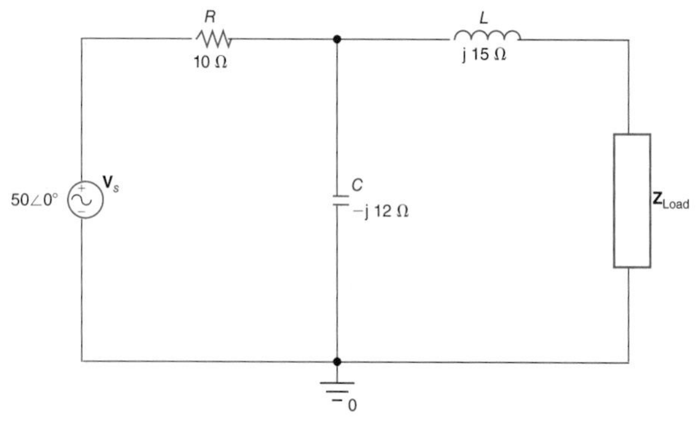

11.33 Find the load impedance in the circuit shown in Figure P11.33 for maximum power transfer. Also, find the maximum power delivered to the load. Assume that $v_s(t) = 60\cos(2\pi 60t + 0°)$ V (rms).

FIGURE P11.33

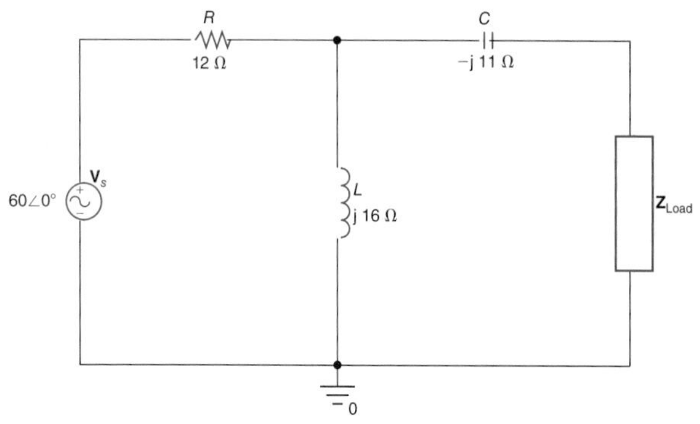

11.34 Find the load impedance in the circuit shown in Figure P11.34 for maximum power transfer. Also, find the maximum power delivered to the load. Assume that $v_s(t) = 70\cos(2\pi 60t + 45°)$ V (rms).

FIGURE P11.34

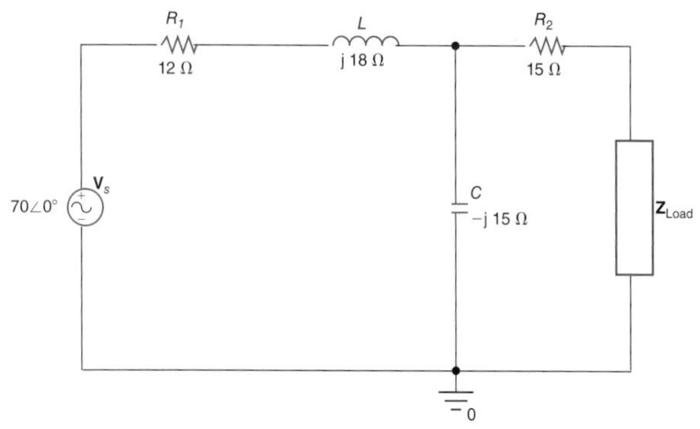

11.35 Find the load impedance in the circuit shown in Figure P11.35 for maximum power transfer. Also, find the maximum power delivered to the load. Assume that $v_s(t) = 80\cos(2\pi 60t + 0°)$ V (rms).

FIGURE P11.35

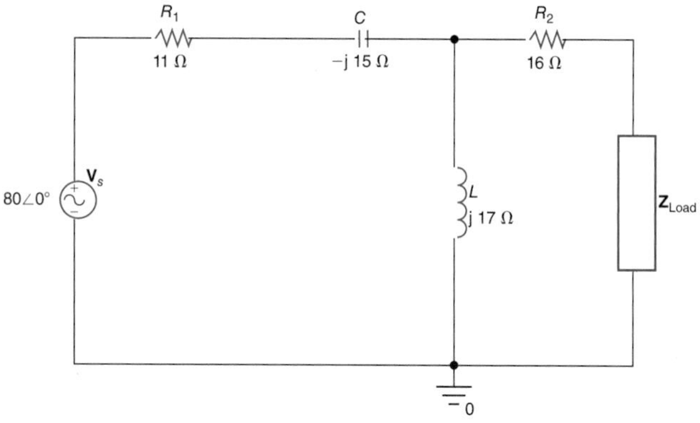

11.36 Find the load impedance in the circuit shown in Figure P11.36 for maximum power transfer. Also, find the maximum power delivered to the load. Assume that $i_s(t) = 10\cos(2\pi 60t + 0°)$ A (rms).

FIGURE P11.36

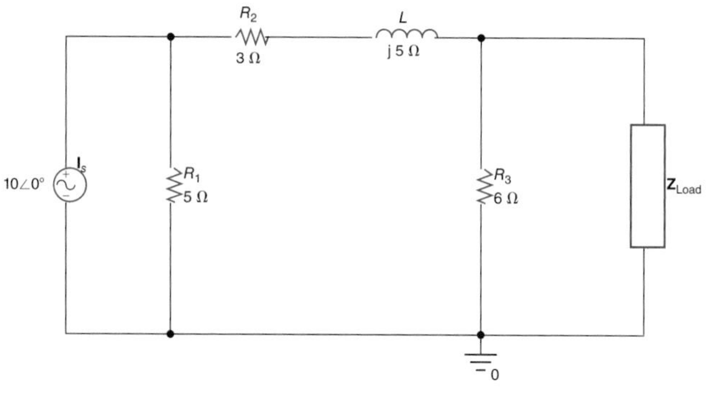

Power Factor Correction

11.37 A load consists of a resistor with resistance $R = 73\ \Omega$ and an inductor with inductance $L = 127$ mH as shown in Figure P11.37(a). Find the power factor of the load shown in Figure P11.37(a). A capacitor is connected in parallel to the load to raise the power factor to 0.98 as shown in Figure P11.37(b). Find the capacitance value C.

FIGURE P11.37

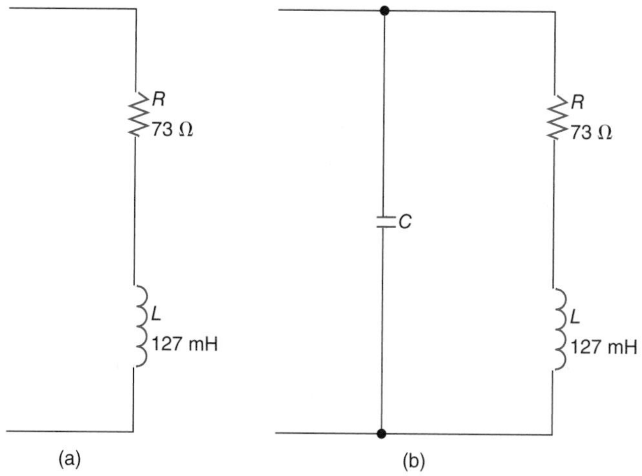

11.38 A load consists of a resistor with resistance $R = 56\ \Omega$ and an inductor with inductance $L = 135$ mH as shown in Figure P11.38(a). Find the power factor of the load shown in Figure P11.38(a). A capacitor is connected in parallel to the load to raise the power factor to 0.99 as shown in Figure P11.38(b). Find the capacitance value C.

FIGURE P11.38

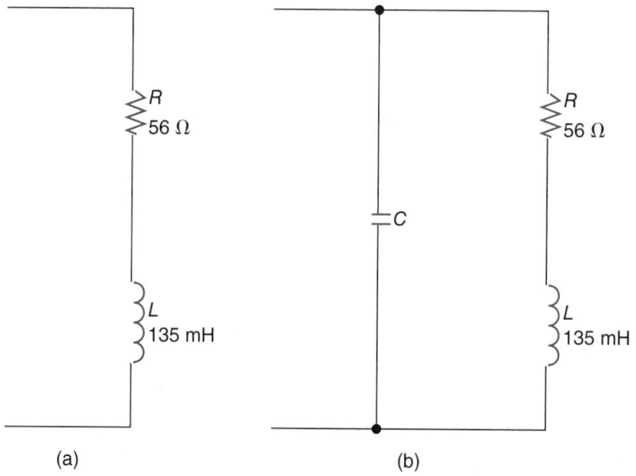

11.39 The average power on the load is 2.45 kW with a lagging power factor of 0.755. The voltage across the load is $\mathbf{V}_{rms} = 580\angle 0°$ V (rms). Thus, $|\mathbf{V}_{rms}| = 580$ V (rms). A capacitor is added in parallel to the load to raise the power factor to 0.99. Find the capacitance value, given $f = 60$ Hz.

11.40 The apparent power on the load is 2.65 kVA with a lagging power factor of 0.835. The current through the load is $\mathbf{I}_{rms} = 10.75\angle 0°$ A (rms). Thus, $|\mathbf{I}_{rms}| = 10.75$ A (rms). A capacitor is added in parallel to the load to raise the power factor to 0.98. Find the capacitance value, given $f = 60$ Hz.

11.41 The average power on the load is 1.73 kW with a lagging power factor of 0.684. The current through the load is $\mathbf{I}_{rms} = 11.2\angle 0°$ A (rms). Thus, $|\mathbf{I}_{rms}| = 11.2$ A (rms). A capacitor is added in parallel to the load to raise the power factor to 0.97. Find the capacitance value, given $f = 60$ Hz.

11.42 The apparent power on the load is 2.47 kVA with a lagging power factor of 0.863. The voltage across the load is $\mathbf{V}_{rms} = 420\angle 0°$ V (rms). Thus, $|\mathbf{V}_{rms}| = 420$ V (rms). A capacitor is added in parallel to the load to raise the power factor to 1. Find the capacitance value, given $f = 60$ Hz.

Chapter 12

Three-Phase Systems

12.1 Introduction

In this chapter, circuits connecting balanced three-phase sources and balanced three-phase loads are analyzed. The balanced three-phase sources are three voltages with the same amplitude and frequency, but three different phases separated by 120°. The sources can be arranged in a wye (Y) shape or a delta (Δ) shape. The balanced loads refer to three loads with identical impedances and also can be arranged in a Y or Δ shape.

There are four combinations in connecting three-phase sources and three-phase loads. These are Y-Y connection, Y-Δ connection, Δ-Δ connection, and Δ-Y connection. The delta connected sources have equivalent wye connected sources, and vice versa. The delta connected loads have equivalent wye connected loads, and vice versa. The Y-Δ connection can be converted to a Y-Y connection or a Δ-Δ connection before analysis. Similarly, a Δ-Y connection can be converted to a Y-Y connection or a Δ-Δ connection before analysis.

12.2 Three-Phase Sources

The three-phase generator can be modeled as three independent sources with the same amplitude and frequency, but three different phases separated by 120°:

$$v_{an}(t) = V_p \cos(\omega t + 0°) \tag{12.1}$$

$$v_{bn}(t) = V_p \cos(\omega t - 120°) \tag{12.2}$$

$$v_{cn}(t) = V_p \cos(\omega t - 240°) \tag{12.3}$$

Notice that V_p (rms) is the **magnitude of the phase voltage**. If 360°, which is equal to 0°, is added to the phase of Equation (12.3), we obtain

$$v_{cn}(t) = V_p \cos(\omega t + 120°) \tag{12.4}$$

Figure 12.1 shows the three waveforms for the frequency of 60 Hz and the amplitude of 120 V (rms). For the frequency of 60 Hz, the period is $1/60$ s = 16.6667 ms. For the frequency of 50 Hz, the period is $1/50$ s = 20 ms.

FIGURE 12.1

Three-phase sources in the time domain.

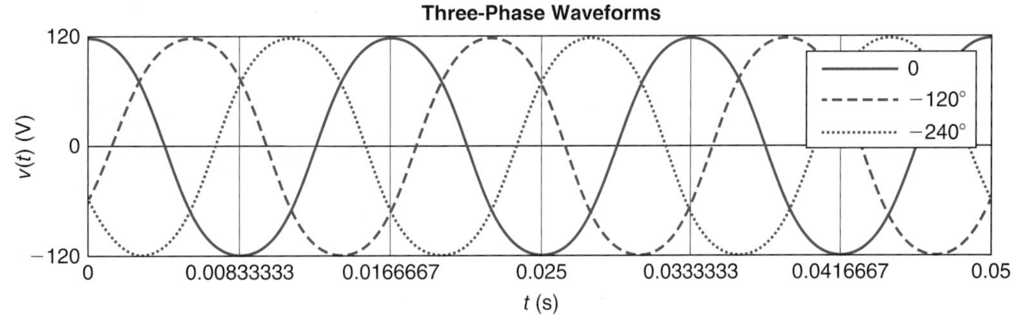

When the amplitudes and frequencies of the three voltage sources are the same and the phases are separated by 120°, the three-phase generator is called **balanced**. The three voltage sources can be connected in Y shape (wye shape) or Δ shape (delta shape), as shown in Figure 12.2.

FIGURE 12.2

(a) Y-shaped source.
(b) Δ-shaped source.

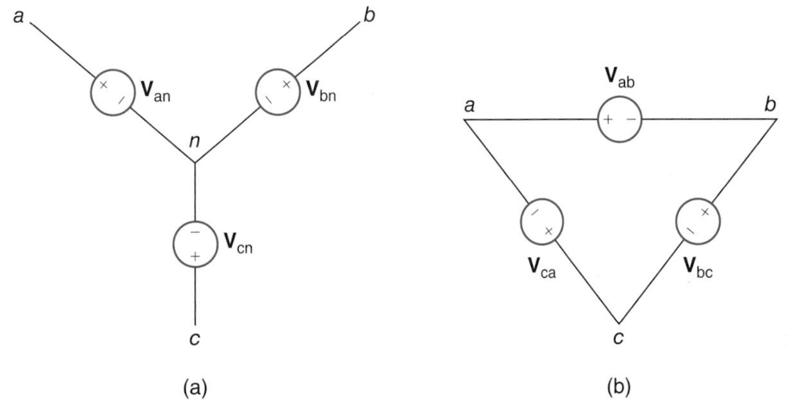

The three voltages $\mathbf{V}_{an}$, $\mathbf{V}_{bn}$, and $\mathbf{V}_{cn}$ in the Y-connected source shown in Figure 12.2(a) are called **phase voltages**. The magnitude of phase voltage in rms is written as V_P. The phase voltages in rms are written in phasor notation as

$$\mathbf{V}_{an} = V_P \angle 0° \tag{12.5}$$

$$\mathbf{V}_{bn} = V_P \angle -120° \tag{12.6}$$

$$\mathbf{V}_{cn} = V_P \angle -240° = V_P \angle 120° \tag{12.7}$$

FIGURE 12.3

Three phasors for the three-phase sources.

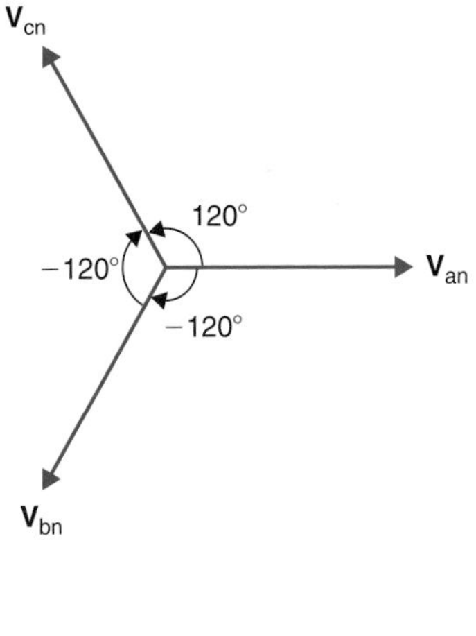

The three phasors given by Equations (12.5)–(12.7) are shown in Figure 12.3. The phasor sum of these three voltages is

$$\mathbf{V}_{an} + \mathbf{V}_{bn} + \mathbf{V}_{cn} = V_p e^{j0°} + V_p e^{-j120°} + V_p e^{j120°}$$
$$= V_p + V_p\left(-\frac{1}{2} - j\frac{\sqrt{3}}{2}\right) + V_p\left(-\frac{1}{2} + j\frac{\sqrt{3}}{2}\right) = 0 + j0$$

The phases of the three voltages are separated by 120°. The phase of $\mathbf{V}_{bn}$ lags the phase of $\mathbf{V}_{an}$ by 120°, and the phase of $\mathbf{V}_{cn}$ lags the phase of $\mathbf{V}_{an}$ by 240°.

The voltages between the lines, $\mathbf{V}_{ab}$, $\mathbf{V}_{bc}$, and $\mathbf{V}_{ca}$, of the Y-connected source shown in Figure 12.2(a) are given by

$$\begin{aligned}\mathbf{V}_{ab} &= \mathbf{V}_{an} + \mathbf{V}_{nb} = \mathbf{V}_{an} - \mathbf{V}_{bn} = V_p\angle 0° - V_p\angle -120° \\ &= V_p\angle 0° - V_p[\cos(-120°) + j\sin(-120°)] = V_p - V_p\left(-\frac{1}{2} - j\frac{\sqrt{3}}{2}\right) \\ &= V_p + \frac{1}{2}V_p + j\frac{\sqrt{3}}{2}V_p = V_p\left(\frac{3}{2} + j\frac{\sqrt{3}}{2}\right) = \sqrt{3}V_p\left(\frac{\sqrt{3}}{2} + j\frac{1}{2}\right) \\ &= \sqrt{3}V_p\angle 30°\end{aligned} \quad (12.8)$$

The **magnitude of the line voltage** is defined as

$$V_L = \sqrt{3}V_p \quad (12.9)$$

The magnitude of the phase voltage is expressed as a function of the magnitude of the line voltage as

$$V_p = \frac{V_L}{\sqrt{3}} \quad (12.10)$$

The line voltage $\mathbf{V}_{ab}$ is given by

$$\mathbf{V}_{ab} = \sqrt{3}V_p\angle 30° = V_L\angle 30° \quad (12.11)$$

The line voltage $\mathbf{V}_{bc}$ is given by

$$\begin{aligned}\mathbf{V}_{bc} &= \mathbf{V}_{bn} + \mathbf{V}_{nc} = \mathbf{V}_{bn} - \mathbf{V}_{cn} = V_p\angle -120° - V_p\angle 120° \\ &= V_p\left(-\frac{1}{2} - j\frac{\sqrt{3}}{2} + \frac{1}{2} - j\frac{\sqrt{3}}{2}\right) = V_p(-j\sqrt{3}) \\ &= \sqrt{3}V_p(-j) = \sqrt{3}V_p\angle -90° = V_L\angle -90°\end{aligned} \quad (12.12)$$

The line voltage $\mathbf{V}_{ca}$ is given by

$$\begin{aligned}\mathbf{V}_{ca} &= \mathbf{V}_{cn} + \mathbf{V}_{na} = \mathbf{V}_{cn} - \mathbf{V}_{an} = V_p\angle 120° - V_p\angle 0° \\ &= V_p\left(-\frac{1}{2} + j\frac{\sqrt{3}}{2}\right) - V_p = -\frac{1}{2}V_p + j\frac{\sqrt{3}}{2}V_p - V_p \\ &= V_p\left(-\frac{3}{2} + j\frac{\sqrt{3}}{2}\right) = \sqrt{3}V_p\left(-\frac{\sqrt{3}}{2} + j\frac{1}{2}\right) \\ &= \sqrt{3}V_p\angle 150° = \sqrt{3}V_p\angle -210° = V_L\angle -210° = V_L\angle 150°\end{aligned} \quad (12.13)$$

The voltages between the lines, $\mathbf{V}_{ab}$, $\mathbf{V}_{bc}$, and $\mathbf{V}_{ca}$, are called **line voltages**. The voltages $\mathbf{V}_{ab}$, $\mathbf{V}_{bc}$, and $\mathbf{V}_{ca}$ are the voltages on the Δ-connected source shown in Figure 12.2(b). The three voltages $\mathbf{V}_{ab}$, $\mathbf{V}_{bc}$, and $\mathbf{V}_{ca}$ have magnitude of $\sqrt{3}V_p$ and phases are separated by 120°, as shown in Figure 12.4. When the Y-connected source shown in Figure 12.2(a) is converted to the Δ-connected source in Figure 12.2(b), the magnitudes are multiplied by $\sqrt{3}$. Thus, we have

$$|\mathbf{V}_{ab}| = \sqrt{3}|\mathbf{V}_{an}| \quad \text{or} \quad V_L = \sqrt{3}\,V_p \quad (12.14)$$

FIGURE 12.4

Phasors for line voltages.

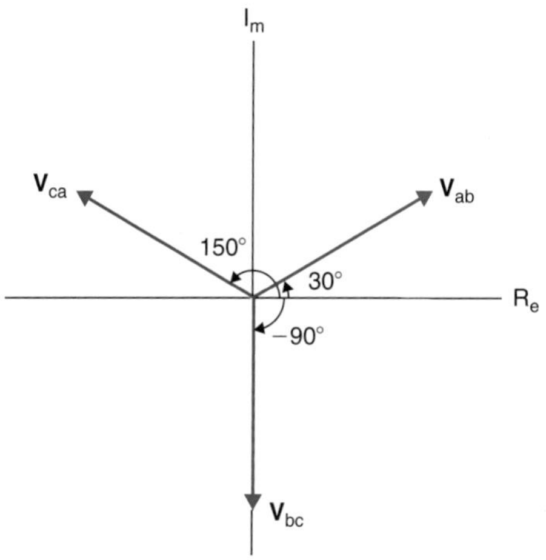

$$|\mathbf{V}_{bc}| = \sqrt{3}|\mathbf{V}_{bn}| \quad \text{or} \quad V_L = \sqrt{3}V_p \tag{12.15}$$

$$|\mathbf{V}_{ca}| = \sqrt{3}|\mathbf{V}_{cn}| \quad \text{or} \quad V_L = \sqrt{3}V_p \tag{12.16}$$

The phases of the Δ-connected sources are related to the phases of the Y-connected sources by

$$\angle \mathbf{V}_{ab} = \angle \mathbf{V}_{an} + 30° \tag{12.17}$$

$$\angle \mathbf{V}_{bc} = \angle \mathbf{V}_{bn} + 30° \tag{12.18}$$

$$\angle \mathbf{V}_{ca} = \angle \mathbf{V}_{cn} + 30° \tag{12.19}$$

EXAMPLE 12.1

If $\mathbf{V}_{cn} = 230\angle -150°$ V, what is $\mathbf{V}_{an}$ and $\mathbf{V}_{ab}$?

$$\mathbf{V}_{bn} = 230\angle -150° + 120° = 230\angle -30°\text{ V}$$

$$\mathbf{V}_{an} = 230\angle -30° + 120° = 230\angle 90°\text{ V}$$

$$\mathbf{V}_{ab} = \sqrt{3} \times 230\angle 90° + 30° = 398.3717\angle 120°\text{ V}$$

Exercise 12.1

If $\mathbf{V}_{bc} = 280\angle -100°$ V, what is $\mathbf{V}_{an}$?

Answer:
$\mathbf{V}_{ab} = 280\angle 20°$ V, $\mathbf{V}_{an} = 280/\sqrt{3}\angle 20° - 30° = 161.6581\angle -10°$ V.

12.2.1 NEGATIVE PHASE SEQUENCE

The Y-connected source voltages given by Equations (12.5)–(12.7) are called **positive phase sequence** or **abc phase sequence.** In the **negative phase sequence** or **acb phase sequence**, the source voltages are defined by

$$\mathbf{V}_{an} = V_p e^{j0°} = V_p\angle 0° \tag{12.20}$$

$$\mathbf{V}_{bn} = V_p e^{j120°} = V_p\angle 120° = V_p\angle -240° \tag{12.21}$$

$$\mathbf{V}_{cn} = V_p e^{j240°} = V_p e^{-j120°} = V_p\angle -120° \tag{12.22}$$

The voltage $\mathbf{V}_{cn}$ of the positive phase sequence is $\mathbf{V}_{bn}$ in the negative phase sequence and the voltage $\mathbf{V}_{bn}$ of the positive phase sequence is $\mathbf{V}_{cn}$ in the negative phase sequence. In the negative phase sequence, the phase of $\mathbf{V}_{bn}$ lags the phase of $\mathbf{V}_{an}$ by 240° and the phase of $\mathbf{V}_{cn}$ lags the phase of $\mathbf{V}_{an}$ by 120°. The line voltages of the negative phase sequence are given by

$$\mathbf{V}_{ab} = \mathbf{V}_{an} + \mathbf{V}_{nb} = \mathbf{V}_{an} - \mathbf{V}_{bn} = V_p\angle 0° - V_p\angle 120° = \sqrt{3}V_p\angle -30° \quad \textbf{(12.23)}$$

$$\mathbf{V}_{bc} = \mathbf{V}_{bn} + \mathbf{V}_{nc} = \mathbf{V}_{bn} - \mathbf{V}_{cn} = V_p\angle 120° - V_p\angle -120° = \sqrt{3}V_p\angle 90° \quad \textbf{(12.24)}$$

$$\mathbf{V}_{ca} = \mathbf{V}_{cn} + \mathbf{V}_{na} = \mathbf{V}_{cn} - \mathbf{V}_{an} = V_p\angle -120° - V_p\angle 0° = \sqrt{3}V_p\angle -150° \quad \textbf{(12.25)}$$

When the Y-connected three-phase source with negative phase sequence is converted to the Δ-connected three-phase source, the magnitudes are multiplied by $\sqrt{3}$. Unless mentioned otherwise, we will discuss only the positive phase sequence in the rest of this chapter.

12.3 Balanced Y-Y Circuit

The three loads can be connected in a Y shape or a Δ shape, as shown in Figure 12.5. The loads are balanced if the three impedances are identical; that is, $\mathbf{Z}_Y = \mathbf{Z}_{AN} = \mathbf{Z}_{BN} = \mathbf{Z}_{CN}$ and $\mathbf{Z}_\Delta = \mathbf{Z}_{AB} = \mathbf{Z}_{BC} = \mathbf{Z}_{CA}$.

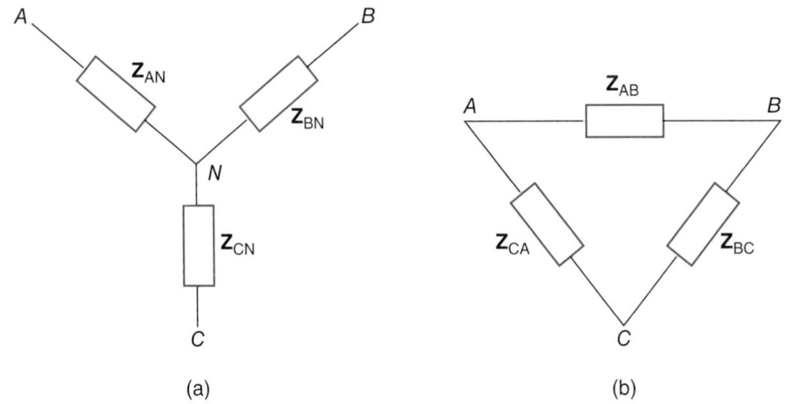

FIGURE 12.5

(a) Y-connected load.
(b) Δ-connected load.

Since there are two methods to connect the sources and two methods to connect the loads, there are four different methods to connect the source to the load. These are Y-connected source and Y-connected load, Y-connected source and Δ-connected load, Δ-connected source and Y-connected load, and Δ-connected source and Δ-connected load.

A balanced Y-connected source connected to the balanced Y-connected load is shown in Figure 12.6. The impedances on the wires and internal impedances on the generators are ignored in this model. Notice that there are three independent circuits in the circuit shown in Figure 12.6. Each of the three circuits consists of a phase voltage source and a load with impedance $\mathbf{Z}_Y$, as shown in Figure 12.7.

FIGURE 12.6

Y-connected source connected to Y-connected load.

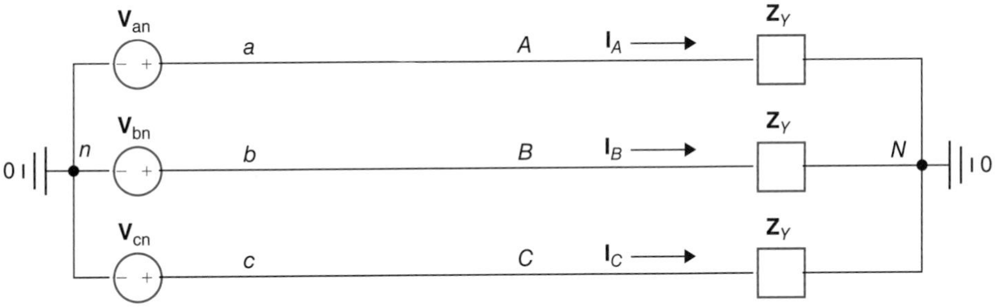

FIGURE 12.7

Y-Y connection separated for each phase.

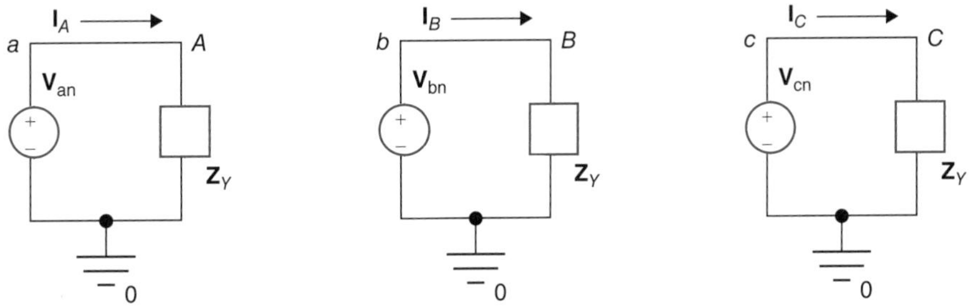

Let the phase voltages be

$$\mathbf{V}_{an} = V_P\angle 0^\circ \text{ V}$$
$$\mathbf{V}_{bn} = V_P\angle -120^\circ \text{ V}$$
$$\mathbf{V}_{cn} = V_P\angle -240^\circ \text{ V}$$

The impedance of the load, for each phase, is represented as

$$\mathbf{Z}_Y = |\mathbf{Z}_Y|\angle\theta_Y$$

The current $\mathbf{I}_A$ is given by

$$\mathbf{I}_A = \frac{\mathbf{V}_{an}}{\mathbf{Z}_Y} = \frac{V_P\angle 0^\circ}{|\mathbf{Z}_Y|\angle\theta_Y} = \frac{V_P}{|\mathbf{Z}_Y|}\angle -\theta_Y = I_L\angle -\theta_Y \text{ A} \quad (12.26)$$

The magnitude of $\mathbf{I}_A$ is given by

$$I_L = \frac{V_P}{|\mathbf{Z}_Y|} \text{A} \quad (12.27)$$

and is called **magnitude of line current**. Similarly, the other two currents are given by

$$\mathbf{I}_B = \frac{V_P\angle -120^\circ}{|\mathbf{Z}_Y|\angle\theta_Y} = \frac{V_P}{|\mathbf{Z}_Y|}\angle -\theta_Y - 120^\circ = I_L\angle -\theta_Y - 120^\circ \text{ A} \quad (12.28)$$

$$\mathbf{I}_C = \frac{V_P\angle -240^\circ}{|\mathbf{Z}_Y|\angle\theta_Y} = \frac{V_P}{|\mathbf{Z}_Y|}\angle -\theta_Y - 240^\circ = I_L\angle -\theta_Y - 240^\circ \text{ A} \quad (12.29)$$

Notice that $\mathbf{I}_B$ and $\mathbf{I}_C$ have the same magnitude as $\mathbf{I}_A$, but the phase is decreased from the phase of $\mathbf{I}_A$ by 120° and 240°, respectively. The magnitude of the line voltages is, from Equation (12.9),

$$V_L = \sqrt{3}V_P \text{ V} \quad (12.30)$$

From Equations (12.11)–(12.13), the line voltages are given by

$$\mathbf{V}_{AB} = V_L \angle 30^\circ \text{ V} \tag{12.31}$$

$$\mathbf{V}_{BC} = V_L \angle -90^\circ \text{ V} \tag{12.32}$$

$$\mathbf{V}_{CA} = V_L \angle -210^\circ \text{ V} \tag{12.33}$$

The complex power of the balanced load is given by

$$\mathbf{S}_L = 3I_L^2\mathbf{Z}_Y = 3I_LI_L\mathbf{Z}_Y = 3\frac{V_P}{|\mathbf{Z}_Y|}I_L\mathbf{Z}_Y = 3\frac{\frac{V_L}{\sqrt{3}}}{|\mathbf{Z}_Y|}I_L|\mathbf{Z}_Y|\angle\theta_Y = \sqrt{3}V_LI_L\angle\theta_Y \text{ VA} \tag{12.34}$$

The complex power of the load can be rewritten as

$$\mathbf{S}_L = \sqrt{3}V_LI_L\cos(\theta_Y) + j\sqrt{3}V_LI_L\sin(\theta_Y) \text{ VA} \tag{12.35}$$

The absolute value of complex power of the load $\mathbf{S}_L$ is given by

$$|\mathbf{S}_L| = \sqrt{3}V_LI_L \text{ VA} \tag{12.36}$$

and is the apparent power of the load.

The average power of the load is

$$P_L = \sqrt{3}V_LI_L\cos(\theta_Y) = |\mathbf{S}_L|\cos(\theta_Y) \text{ W} \tag{12.37}$$

and the reactive power of the load is

$$Q_L = \sqrt{3}V_LI_L\sin(\theta_Y) = |\mathbf{S}_L|\sin(\theta_Y) \text{ VAR} \tag{12.38}$$

EXAMPLE 12.2

In the balanced Y-Y connection shown in Figure 12.6, let the magnitude of the line voltages be $V_L = 880$ V (rms), the load impedance per phase be $\mathbf{Z}_Y = 50 + j25\ \Omega$, and the phase angle of $\mathbf{V}_{an}$ be 0°. Find the phase voltages, the line currents, and line voltages. Also, find the complex power, apparent power, average power, reactive power, and power factor of the load.

The load impedance can be written in polar form as

$$\mathbf{Z}_Y = 50 + j25 = 55.9017\angle 26.5651^\circ$$

The magnitude of the phase voltage is

$$V_P = \frac{V_L}{\sqrt{3}} = 508.0682 \text{ V (rms)}$$

The phase voltages are

$$\mathbf{V}_{an} = 508.0682\angle 0^\circ, \quad \mathbf{V}_{bn} = 508.0682\angle -120^\circ, \quad \mathbf{V}_{cn} = 508.0682\angle -240^\circ$$

continued

Example 12.2 continued

The magnitude of line currents is

$$I_L = \frac{V_P}{|\mathbf{Z}_Y|} = \frac{508.0682}{55.9017} = 9.0886$$

The line currents are given by

$$\mathbf{I}_A = \frac{V_P\angle 0^\circ}{|\mathbf{Z}_Y|\angle\theta_Y} = \frac{508.0682\angle 0^\circ}{55.9017\angle 26.5651^\circ} = 9.0886\angle -26.5651^\circ \text{ A (rms)}$$

$$\mathbf{I}_B = 9.0886\angle -146.5651^\circ \text{ A (rms)}$$

$$\mathbf{I}_C = 9.0886\angle -266.5651^\circ \text{ A (rms)} = 9.0886\angle 93.4349^\circ \text{ A (rms)}$$

The phase of the line voltage $\mathbf{V}_{ab}$ leads the phase of the phase voltage $\mathbf{V}_{an}$ by 30°. Since the magnitude of the line voltages are 880 V (rms), the line voltages are given by

$$\mathbf{V}_{ab} = 880\angle 30^\circ \text{ V (rms)}$$

$$\mathbf{V}_{bc} = 880\angle -90^\circ \text{ V (rms)}$$

$$\mathbf{V}_{ca} = 880\angle -210^\circ \text{ V (rms)}$$

The apparent power of the load is given by

$$|\mathbf{S}_L| = \sqrt{3}V_L I_L = 13{,}852.8883 \text{ VA}$$

The average power of the load is given by

$$P_L = |\mathbf{S}_L|\cos(\theta_Y) = 12{,}390.4 \text{ W}$$

The reactive power of the load is given by

$$Q_L = |\mathbf{S}_L|\sin(\theta_Y) = 6195.2 \text{ VAR}$$

The power factor of the load is given by

$$pf = \cos(\theta_Y) = 0.894427190999916$$

Exercise 12.2

In the balanced Y-Y connection shown in Figure 12.6, let $\mathbf{V}_{an} = 990\angle 0^\circ$ V (rms), and the load impedance per phase be $\mathbf{Z}_Y = 60 + j35\ \Omega$. Find the complex power, apparent power, average power, reactive power of the load.

Answer:
$\mathbf{S}_L = 36{,}563.3161 + j21{,}328.6010$ VA, $|\mathbf{S}_L| = 42{,}329.4850$ VA, $P_L = 36{,}563.3161$ W, $Q_L = 21{,}328.6010$ VAR.

12.3.1 BALANCED Y-Y CIRCUIT WITH WIRE IMPEDANCE

A Y-Y connection that includes impedances on the wires $\mathbf{Z}_w$ is shown in Figure 12.8. The wire impedances include internal impedances on the generators. The Y-Y connection separated for each phase with wire impedances is shown in Figure 12.9.

FIGURE 12.8

Y-Y connection that includes impedances on the wires.

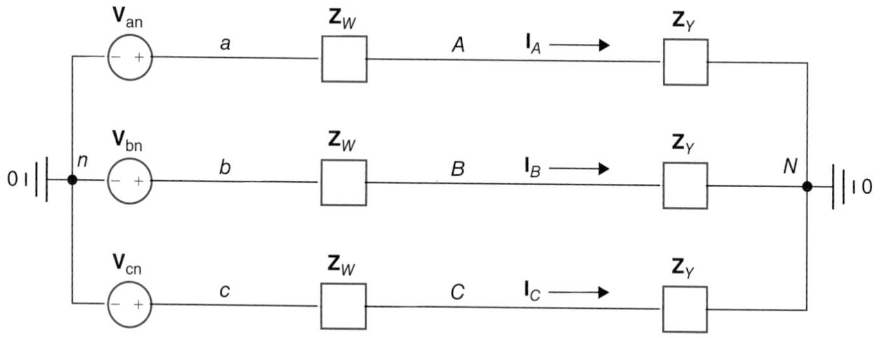

FIGURE 12.9

Y-Y connection separated for each phase with wire impedances.

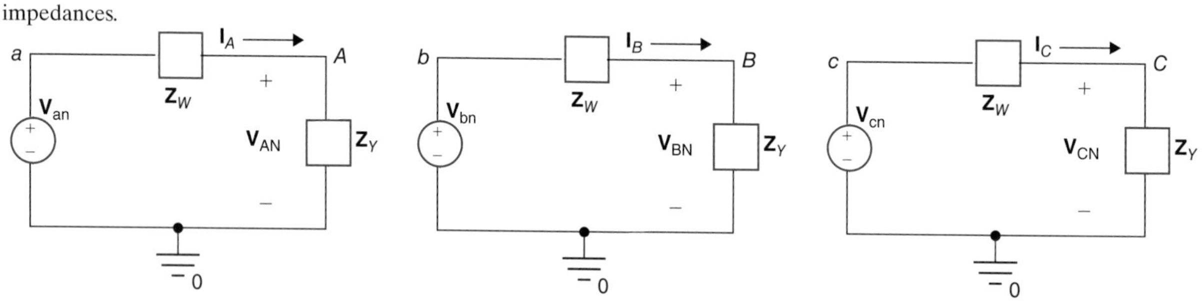

Let the phase voltages be

$$\mathbf{V}_{an} = V_P\angle 0°$$
$$\mathbf{V}_{bn} = V_P\angle -120°$$
$$\mathbf{V}_{cn} = V_P\angle -240°$$

The impedance of the load, for each phase, is represented as

$$\mathbf{Z}_Y = |\mathbf{Z}_Y|\angle\theta_Y$$

The total impedance for each voltage phase is given by

$$\mathbf{Z}_T = \mathbf{Z}_W + \mathbf{Z}_Y = |\mathbf{Z}_T|\angle\theta_T$$

The magnitude of line currents is

$$I_L = \frac{V_P}{|\mathbf{Z}_T|} \tag{12.39}$$

The line current I_A is given by

$$\mathbf{I}_A = \frac{V_P\angle 0°}{|\mathbf{Z}_T|\angle\theta_T} = \frac{V_P}{|\mathbf{Z}_T|}\angle -\theta_T = I_L\angle -\theta_T \tag{12.40}$$

Similarly, the other two line currents are given by

$$\mathbf{I}_B = I_L\angle -\theta_T - 120° \tag{12.41}$$

$$\mathbf{I}_C = I_L\angle -\theta_T - 240° \tag{12.42}$$

The voltage between A and N at the load, $\mathbf{V}_{AN}$, across $\mathbf{Z}_Y$ is given by

$$\mathbf{V}_{AN} = \mathbf{I}_A\mathbf{Z}_Y = (I_L\angle -\theta_T)(|\mathbf{Z}_Y|\angle\theta_Y) = I_L|\mathbf{Z}_Y|\angle -\theta_T + \theta_Y \tag{12.43}$$

Similarly, we have

$$\mathbf{V}_{BN} = \mathbf{I}_B\mathbf{Z}_Y = I_L|\mathbf{Z}_Y|\angle -\theta_T + \theta_Y - 120° \tag{12.44}$$

$$\mathbf{V}_{CN} = \mathbf{I}_C\mathbf{Z}_Y = I_L|\mathbf{Z}_Y|\angle -\theta_T + \theta_Y - 240° \tag{12.45}$$

The magnitude of line voltages is given by

$$V_L = \sqrt{3}|\mathbf{V}_{AN}| = \sqrt{3}I_L|\mathbf{Z}_Y| \tag{12.46}$$

The line voltages are given by

$$\mathbf{V}_{AB} = V_L\angle -\theta_T + \theta_Y + 30° \tag{12.47}$$

$$\mathbf{V}_{BC} = V_L\angle -\theta_T + \theta_Y - 90° \tag{12.48}$$

$$\mathbf{V}_{CA} = V_L\angle -\theta_T + \theta_Y - 210° \tag{12.49}$$

Notice that the line voltages can also be found using

$$\mathbf{V}_{AB} = \mathbf{V}_{AN} - \mathbf{V}_{BN}$$
$$\mathbf{V}_{BC} = \mathbf{V}_{BN} - \mathbf{V}_{CN}$$
$$\mathbf{V}_{CA} = \mathbf{V}_{CN} - \mathbf{V}_{AN}$$

The magnitude of $\mathbf{V}_{AB}$ is V_L; that is, $|\mathbf{V}_{AB}| = V_L$.
The complex power of the balanced load is given by

$$\mathbf{S}_L = 3I_L^2\mathbf{Z}_Y \tag{12.50}$$

The average power is

$$P_L = \text{Re}(\mathbf{S}_L) \tag{12.51}$$

and the reactive power is

$$Q_L = \text{I}_\text{m}(\mathbf{S}_L) \tag{12.52}$$

Notice that the complex power $\mathbf{S}_L$ can also be written as

$$\mathbf{S}_L = \mathbf{V}_{AN}\mathbf{I}_A^* + \mathbf{V}_{BN}\mathbf{I}_B^* + \mathbf{V}_{CN}\mathbf{I}_C^*$$

Also, notice that the apparent power of the three-phase load is

$$|\mathbf{S}_L| = \sqrt{3}V_LI_L \tag{12.53}$$

The complex power on the three wires is

$$\mathbf{S}_W = 3I_L^2\mathbf{Z}_W \tag{12.54}$$

The total complex power of the three-phase load and the three wires is given by

$$\mathbf{S}_T = \mathbf{S}_W + \mathbf{S}_L \tag{12.55}$$

The total average power is

$$P_T = \text{Re}(\mathbf{S}_T) \tag{12.56}$$

and the total reactive power is

$$Q_T = \text{I}_\text{m}(\mathbf{S}_T) \tag{12.57}$$

The complex power of the three-phase generator is

$$\mathbf{S}_S = -\mathbf{S}_T \tag{12.58}$$

In this context, *efficiency* can be defined as the ratio of the average power absorbed by the three-phase load to the total average power. The efficiency in percent is given by

$$Eff = \frac{P_L}{P_T} \times 100\% \tag{12.59}$$

EXAMPLE 12.3

In the balanced Y-Y connection shown in Figure 12.8, let the magnitude of the phase voltages be $V_P = 510$ V (rms), the load impedance per phase be $\mathbf{Z}_Y = 70 + j55\ \Omega$, the wire impedance per phase be $\mathbf{Z}_W = 9 + j7\ \Omega$, and the phase angle of $\mathbf{V}_{an}$ be 0°. Find the phase voltages, the line currents, and line voltages. Find the complex power, apparent power, average power, reactive power, and power factor of the load. Find the complex power of the wire, the total complex power, and efficiency.

The load impedance per phase in polar form is given by

$$\mathbf{Z}_Y = 70 + j55 = 89.0225\angle 38.1572°$$

The total impedance per phase is given by

$$\mathbf{Z}_T = \mathbf{Z}_W + \mathbf{Z}_Y = 79 + j62 = 100.4241\angle 38.1252°$$

$$|\mathbf{Z}_T| = 100.4241, \theta_T = 38.1252°$$

The phase voltages are

$$\mathbf{V}_{an} = V_P\angle 0° = 510\angle 0° = 510 + j0$$

$$\mathbf{V}_{bn} = V_P\angle -120° = 510\angle -120° = -255 - j441.6730$$

$$\mathbf{V}_{cn} = V_P\angle -240° = 510\angle -240° = -255 + j441.6730$$

continued

Example 12.3 continued

The magnitude of line currents is

$$I_L = \frac{V_P}{|\mathbf{Z}_T|} = 5.0785$$

The line currents are given by

$$\mathbf{I}_A = I_L\angle -\theta_T = 5.0785\angle -38.1252° = 3.9950 - j3.13535$$

Notice that $\mathbf{I}_A = \mathbf{V}_{an}/\mathbf{Z}_T$ provides the same answer. The other two line currents are given by

$$\mathbf{I}_B = I_L\angle -\theta_T - 120° = 5.0785\angle -158.1252° = -4.7128 - j1.8921$$

$$\mathbf{I}_C = I_L\angle -\theta_T - 240° = 5.0785\angle 81.8748° = 0.7178 + j5.0275$$

The voltage between A and N, $\mathbf{V}_{AN}$, across $\mathbf{Z}_Y$ is given by

$$\mathbf{V}_{AN} = \mathbf{I}_A\mathbf{Z}_Y = 452.0972 + j0.2529 = 452.0972\angle 0.03204°$$

Similarly, we have

$$\mathbf{V}_{BN} = \mathbf{I}_B\mathbf{Z}_Y = -225.8296 - j391.6541 = 452.0972\angle -119.9680°$$

$$\mathbf{V}_{CN} = \mathbf{I}_C\mathbf{Z}_Y = -226.2676 + j391.4012 = 452.0972\angle 120.0320°$$

The magnitude of line voltages is given by

$$V_L = \sqrt{3}|\mathbf{V}_{AN}| = 783.0554$$

The line voltages are given by

$$\mathbf{V}_{AB} = V_L\angle \mathbf{V}_{AN} + 30° = 677.9268 + j391.9069 = 783.0554\angle 30.0320°$$

$$\mathbf{V}_{BC} = V_L\angle \mathbf{V}_{AN} - 90° = 0.4380 - j783.0553 = 783.0554\angle -89.9680°$$

$$\mathbf{V}_{CA} = V_L\angle \mathbf{V}_{AN} - 210° = -678.3647 + j391.1484 = 783.0554\angle 150.0320°$$

The complex power of the balanced load is given by

$$\mathbf{S}_L = 3I_L^2\mathbf{Z}_Y = 5416.0635 + j4255.4784 = 6887.8763\angle 38.1572°$$

The average power is

$$P_L = \mathrm{Re}(\mathbf{S}_L) = 5416.0635 \text{ W}$$

and the reactive power is

$$Q_L = \mathrm{I_m}(\mathbf{S}_L) = 4255.4784 \text{ VAR}$$

The apparent power is

$$|\mathbf{S}_L| = 6887.8763 \text{ VA}$$

continued

Example 12.3 continued

Notice that the complex power $\mathbf{S}_L$ can also be found from

$$\mathbf{S}_L = \mathbf{V}_{AN}\mathbf{I}_A^* + \mathbf{V}_{BN}\mathbf{I}_B^* + \mathbf{V}_{CN}\mathbf{I}_C^* = 5416.0635 + j4255.4784 = 6887.8763\angle 38.1572°$$

The power factor of the load is

$$pf = \cos(38.1572°) = 0.786318338822423$$

Let $\theta_1 = \angle \mathbf{Z}_Y$. Then, we have

$$|\mathbf{S}_L| = \sqrt{3}V_L I_L = 6887.8763 \text{ VA}$$

$$P_L = |\mathbf{S}_L|\cos(\theta_1) = 5416.0635 \text{ W}$$

$$Q_L = |\mathbf{S}_L|\sin(\theta_1) = 4255.4784 \text{ VAR}$$

$$pf = \cos(\theta_1) = 0.786318338822423$$

The complex power of the wires is

$$\mathbf{S}_W = 3I_L^2\mathbf{Z}_W = 696.3510 + j541.6063 = 882.1804\angle 37.8750°$$

The total complex power from the source to load is

$$\mathbf{S}_T = \mathbf{S}_W + \mathbf{S}_L = 6112.4145 + j4797.0848 = 7770.0472\angle 38.1252°$$

The total average power is given by

$$P_T = \text{Re}(\mathbf{S}_T) = 6112.4145 \text{ W}$$

The total average power is given by

$$Q_T = \text{I}_\text{m}(\mathbf{S}_T) = 4797.0848 \text{ VAR}$$

The efficiency is given by

$$Eff = \frac{P_L}{P_T} \times 100\% = 100 \times 5416.0635/6112.4145 = 88.6076\%$$

MATLAB

```
% EXAMPLE 12.3
% P2Rd.m, R2P.m should be in the current folder.
f=60;w=2*pi*f;VP=510;ZW=9+7j;ZY=70+55j;
Van=P2Rd(VP,0);Vbn=P2Rd(VP,-120);Vcn=P2Rd(VP,-240);
ZYp=R2P(ZY)
Zt=ZW+ZY
Ztp=R2P(Zt)
theta=Ztp(3)
thetad=Ztp(2)
IL=VP/Ztp(1)
IA=P2Rd(IL,-thetad)
IAp=R2P(IA)
IB=P2Rd(IL,-thetad-120)
IBp=R2P(IB)
IC=P2Rd(IL,-thetad-240)
ICp=R2P(IC)
VAN=IA*ZY
VANp=R2P(VAN)
```

continued

Example 12.3 continued

MATLAB continued

```
VBN=IB*ZY
VBNp=R2P(VBN)
VCN=IC*ZY
VCNp=R2P(VCN)
VAB=VAN-VBN
VABp=R2P(VAB)
VBC=VBN-VCN
VBCp=R2P(VBC)
VCA=VCN-VAN
VCAp=R2P(VCA)
VL=VABp(1)
SL=3*IL^2*ZY
absSL=abs(SL)
SLp=R2P(SL)
PL=real(SL)
QL=imag(SL)
pf=cos(SLp(3))
SW=3*IL^2*ZW
absSW=abs(SW)
SWp=R2P(SW)
PW=real(SW)
QW=imag(SW)
pfw=cos(SWp(3))
ST=SW+SL
absST=abs(ST)
STp=R2P(ST)
PT=real(ST)
QT=imag(ST)
pft=cos(STp(3))
Eff=PL/PT
Effpc=Eff*100
```

Exercise 12.3

In the balanced Y-Y connection shown in Figure 12.8, let the magnitude of the phase voltages be $V_P = 750$ V (rms), the load impedance per phase be $\mathbf{Z}_Y = 60 + j45\ \Omega$, the wire impedance per phase be $\mathbf{Z}_W = 20 + j25\ \Omega$, and the phase angle of $\mathbf{V}_{an}$ be 0°. Find the complex power of the load and the wire.

Answer:

$\mathbf{S}_L = 8960.1770 + j6720.1327 = 11{,}200.2212\angle 36.8699°$ VA,

$\mathbf{S}_W = 2986.7257 + j3733.4071 = 4781.0939\angle 51.3402°$ VA.

A Y-Y connection that includes impedances on the wires $\mathbf{Z}_w$ and the impedances on the generators $\mathbf{Z}_s$ is shown in Figure 12.10. The Y-Y connection separated for each phase with wire impedances and generator impedances is shown in Figure 12.11.

The total impedance for each voltage phase is given by

$$\mathbf{Z}_T = \mathbf{Z}_S + \mathbf{Z}_W + \mathbf{Z}_Y = |\mathbf{Z}_T|\angle\theta_T$$

Equations (12.39)–(12.59) apply to the Y-Y connection with wire impedances and generator impedances with the total impedance given by $\mathbf{Z}_T = \mathbf{Z}_S + \mathbf{Z}_W + \mathbf{Z}_Y = |\mathbf{Z}_T|\angle\theta_T$.

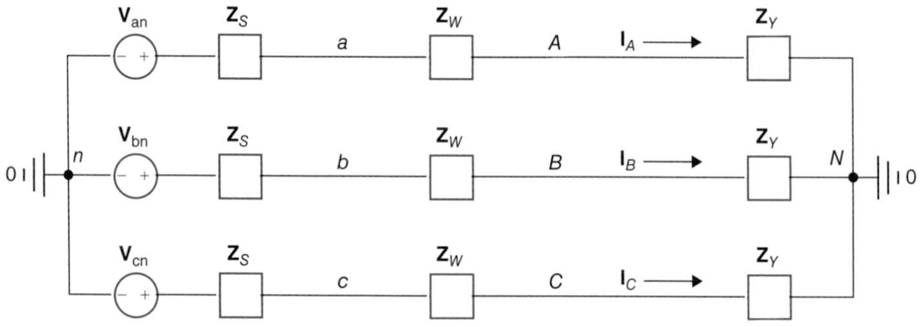

FIGURE 12.10

A Y-Y connection that includes impedances on the wires and generators.

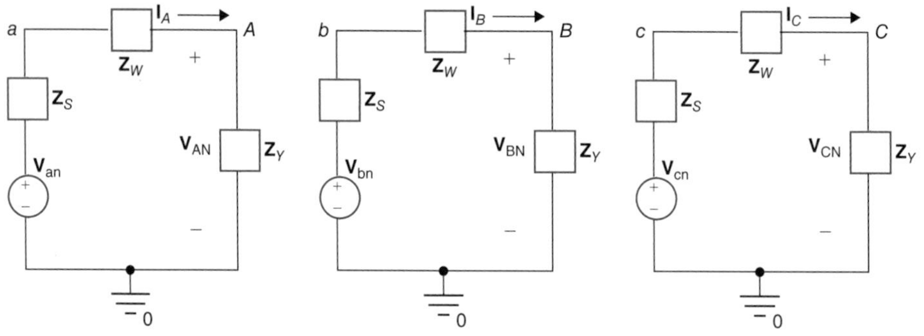

FIGURE 12.11

A Y-Y connection separated for each phase with wire and generator impedances.

12.4 Balanced Y-Δ Circuit

A balanced Y-connected source connected to the balanced Δ-connected load is shown in Figure 12.12. The impedances on the wires and internal impedances on the generators are ignored in this model.

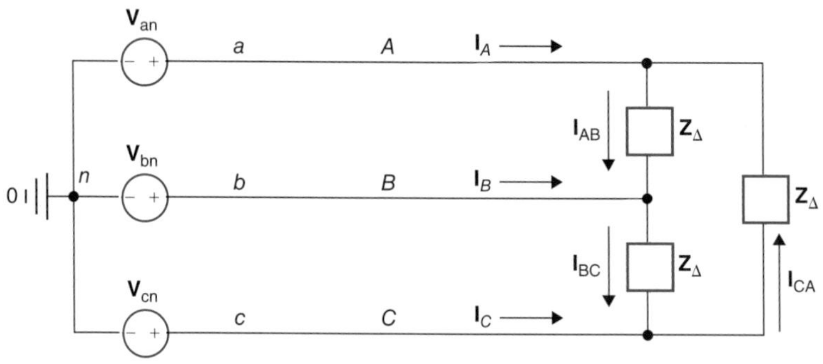

FIGURE 12.12

A Y-Δ connection.

The impedance of the load, for each phase, is represented as

$$\mathbf{Z}_\Delta = |\mathbf{Z}_\Delta|\angle\theta_\Delta$$

Let the phase voltages of the sources be

$$\mathbf{V}_{an} = V_p\angle 0^\circ = \frac{V_L}{\sqrt{3}}\angle 0^\circ$$

$$\mathbf{V}_{bn} = V_p\angle -120^\circ = \frac{V_L}{\sqrt{3}}\angle -120^\circ$$

$$\mathbf{V}_{cn} = V_p\angle -240^\circ = \frac{V_L}{\sqrt{3}}\angle -240^\circ$$

The magnitude of the line voltages is

$$V_L = \sqrt{3}\,V_P$$

The line voltages are given by

$$\mathbf{V}_{AB} = V_L\angle 30°$$

$$\mathbf{V}_{BC} = V_L\angle -90°$$

$$\mathbf{V}_{CA} = V_L\angle -210°$$

The phase current $\mathbf{I}_{AB}$ is given by

$$\mathbf{I}_{AB} = \frac{\mathbf{V}_{AB}}{\mathbf{Z}_\Delta} = \frac{V_L\angle 30°}{|\mathbf{Z}_\Delta|\angle\theta_\Delta} = \frac{V_L}{|\mathbf{Z}_\Delta|}\angle -\theta_\Delta + 30° \tag{12.60}$$

The current $\mathbf{I}_{AB}$ is called the *phase current* because it represents current through one of the phases of the three-phase load. The magnitude of the phase current is given by

$$I_P = \frac{V_L}{|\mathbf{Z}_\Delta|} \tag{12.61}$$

The phase current $\mathbf{I}_{AB}$ is given by

$$\mathbf{I}_{AB} = I_P\angle -\theta_\Delta + 30° \tag{12.62}$$

The phase currents lead line currents by 30°.

$$\angle\mathbf{I}_{AB} = \angle\mathbf{I}_A + 30° \tag{12.63}$$

Similarly, the other two phase currents are given by

$$\mathbf{I}_{BC} = I_P\angle -\theta_\Delta - 90° \tag{12.64}$$

$$\mathbf{I}_{CA} = I_P\angle -\theta_\Delta - 210° \tag{12.65}$$

When the load is converted from Δ to Y, the balanced load for each phase in the Y-connected load is given by

$$\mathbf{Z}_Y = \frac{\mathbf{Z}_\Delta}{3} = \frac{|\mathbf{Z}_\Delta|\angle\theta_\Delta}{3} = \frac{|\mathbf{Z}_\Delta|}{3}\angle\theta_\Delta = |\mathbf{Z}_Y|\angle\theta_\Delta \tag{12.66}$$

The line current $\mathbf{I}_A$ is given by

$$\mathbf{I}_A = \frac{V_P\angle 0°}{\mathbf{Z}_Y} = \frac{\dfrac{V_L}{\sqrt{3}}\angle 0°}{\dfrac{|\mathbf{Z}_\Delta|}{3}\angle\theta_\Delta} = \sqrt{3}\frac{V_L}{|\mathbf{Z}_\Delta|}\angle -\theta_\Delta = \sqrt{3}I_P\angle -\theta_\Delta = I_L\angle -\theta_\Delta \tag{12.67}$$

Notice that the magnitude of line currents, I_L, is $\sqrt{3}$ times the magnitude of the phase currents I_P. The magnitude of line current is given by

$$I_L = \sqrt{3}I_P = \sqrt{3}\frac{V_L}{|\mathbf{Z}_\Delta|} = \sqrt{3}\frac{V_L}{3|\mathbf{Z}_Y|} = \frac{\dfrac{V_L}{\sqrt{3}}}{|\mathbf{Z}_Y|} = \frac{V_P}{|\mathbf{Z}_Y|} \tag{12.68}$$

The other two line currents are given by

$$\mathbf{I}_B = I_L\angle -\theta_\Delta - 120° \tag{12.69}$$

$$\mathbf{I}_C = I_L\angle -\theta_\Delta - 240° \tag{12.70}$$

The complex power of the balanced load is given by

$$\mathbf{S}_L = 3I_P^2\mathbf{Z}_\Delta = 3\frac{V_L}{|\mathbf{Z}_\Delta|}\frac{I_L}{\sqrt{3}}\mathbf{Z}_\Delta = \sqrt{3}V_LI_L\frac{|\mathbf{Z}_\Delta|\angle\theta_\Delta}{|\mathbf{Z}_\Delta|} = \sqrt{3}V_LI_L\angle\theta_\Delta \tag{12.71}$$

The complex power can also be written as

$$\mathbf{S}_L = \sqrt{3}V_LI_L\cos(\theta_\Delta) + j\sqrt{3}V_LI_L\sin(\theta_\Delta) \tag{12.72}$$

The absolute value of $\mathbf{S}_L$, which is the apparent power of the load, is given by

$$|\mathbf{S}_L| = \sqrt{3}V_LI_L \tag{12.73}$$

The average power of the load is

$$P_L = \sqrt{3}V_LI_L\cos(\theta_\Delta) \tag{12.74}$$

and the reactive power of the load is

$$Q_L = \sqrt{3}V_LI_L\sin(\theta_\Delta) \tag{12.75}$$

EXAMPLE 12.4

In the balanced Y-Δ connection shown in Figure 12.12, let the magnitude of the line voltages be $V_L = 1500$ V (rms), the load impedance per phase be $\mathbf{Z}_\Delta = 75 + j33\ \Omega$, and the phase angle of $\mathbf{V}_{an}$ be 0°. Find the phase voltages, line voltages, phase currents, and line currents. Also, find the complex power, apparent power, average power, reactive power, and power factor of the load.

The delta load impedance per phase in polar form is given by

$$\mathbf{Z}_\Delta = 75 + j33 = 81.9390\angle 23.7495°\ \Omega$$

$$\mathbf{Z}_\Delta = |\mathbf{Z}_\Delta|\angle\theta_\Delta = 81.9390\angle 23.7495°, \quad |\mathbf{Z}_\Delta| = 81.9390, \quad \theta_\Delta = 23.7495°$$

The wye load impedance per phase is given by

$$\mathbf{Z}_Y = \frac{\mathbf{Z}_\Delta}{3} = 25 + j11 = 27.3130\angle 23.7495°\ \Omega$$

The magnitude of the phase voltage is

$$V_P = \frac{V_L}{\sqrt{3}} = 866.0254 \text{ V (rms)}$$

continued

Example 12.4 continued

The phase voltages are

$$\mathbf{V}_{an} = V_P\angle 0° = 866.0254\angle 0°\text{ V}$$
$$\mathbf{V}_{bn} = V_P\angle -120° = 866.0254\angle -120°\text{ V}$$
$$\mathbf{V}_{cn} = V_P\angle -140° = 866.0254\angle -240°\text{ V}$$

The line voltages are

$$\mathbf{V}_{AB} = V_L\angle 30° = 1500\angle 30°\text{ V}$$
$$\mathbf{V}_{BC} = V_L\angle -90° = 1500\angle -90°\text{ V}$$
$$\mathbf{V}_{CA} = V_L\angle -210° = 1500\angle -210° = 1500\angle 150°\text{ V}$$

The magnitude of phase currents is

$$I_P = \frac{V_L}{|Z_\Delta|} = 18.3063\text{ A}$$

The phase currents are given by

$$\mathbf{I}_{AB} = I_P\angle -\theta_\Delta + 30° = 18.3063\angle 6.2505°\text{ A}$$
$$\mathbf{I}_{BC} = I_P\angle -\theta_\Delta - 90° = 18.3063\angle -113.7494°\text{ A}$$
$$\mathbf{I}_{CA} = I_P\angle -\theta_\Delta - 210° = 18.3063\angle 126.2505°\text{ A}$$

The magnitude of line currents is

$$I_L = \sqrt{3}I_P = 31.7074\text{ A}$$

The line currents are given by

$$\mathbf{I}_A = I_L\angle -\theta_\Delta = 31.7074\angle -23.7495°\text{ A}$$
$$\mathbf{I}_B = I_L\angle -\theta_\Delta - 120° = 31.7074\angle -143.7495°\text{ A}$$
$$\mathbf{I}_C = I_L\angle -\theta_\Delta - 240° = 31.7074\angle -263.7495° = 31.7074\angle 96.2505°\text{ A}$$

The apparent power of the load is given by

$$|\mathbf{S}_L| = \sqrt{3}V_L I_L = 82{,}378.3529\text{ VA}$$

The average power of the load is given by

$$P_L = |\mathbf{S}_L|\cos(\theta_\Delta) = 75{,}402.1448\text{ W}$$

The reactive power of the load is given by

$$Q_L = |\mathbf{S}_L|\sin(\theta_\Delta) = 33{,}176.9437\text{ VAR}$$

The power factor of the load is given by

$$pf = \cos(\theta_\Delta) = 0.915315032422766$$

Exercise 12.4

In the balanced Y-Δ connection shown in Figure 12.12, let the magnitude of the line voltages be $V_L = 1200$ V (rms), the load impedance per phase be $\mathbf{Z}_\Delta = 66 + j36\ \Omega$, and the phase angle of $\mathbf{V}_{an}$ be 0°. Find the complex power of the load.

Answer:
$\mathbf{S}_L = 50{,}445.8599 + j27{,}515.9236 = 57{,}462.2557\angle 28.6105°$ VA.

12.4.1 BALANCED Y-Δ CIRCUIT WITH WIRE IMPEDANCE

A Y-Δ connection that includes impedances on the wires $\mathbf{Z}_w$ is shown in Figure 12.13. The wire impedances include internal impedances on the generators. The three-phase delta load shown in Figure 12.13 can be converted to the three-phase wye load, as shown in Figure 12.14.

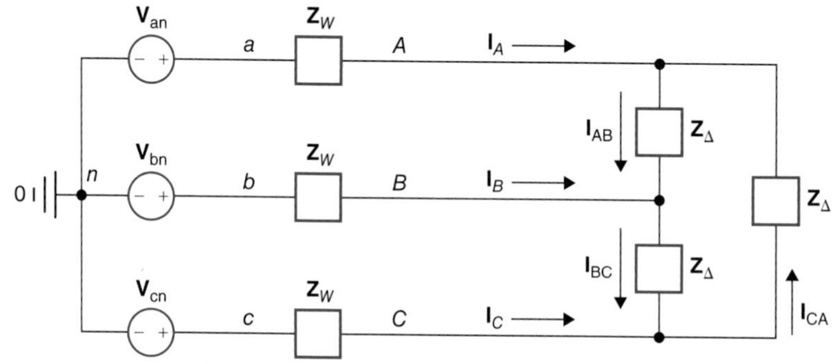

FIGURE 12.13
A Y-Δ connection with wire impedances.

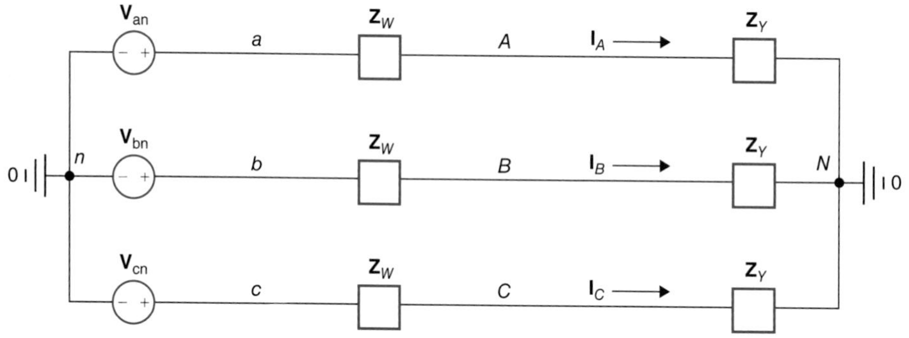

FIGURE 12.14
A Y-Δ connection converted to a Y-Y connection.

The impedance of the three-phase wye load per phase is given by

$$\mathbf{Z}_Y = \frac{\mathbf{Z}_\Delta}{3} \tag{12.76}$$

Let the phase voltages be

$$\mathbf{V}_{an} = V_P\angle 0°$$
$$\mathbf{V}_{bn} = V_P\angle -120°$$
$$\mathbf{V}_{cn} = V_P\angle -240°$$

The total impedance for each voltage phase is given by

$$\mathbf{Z}_T = \mathbf{Z}_W + \mathbf{Z}_Y = |\mathbf{Z}_T|\angle\theta_T \tag{12.77}$$

The magnitude of line currents is

$$I_L = \frac{V_P}{|\mathbf{Z}_T|} \quad (12.78)$$

The line current $\mathbf{I}_A$ is given by

$$\mathbf{I}_A = \frac{V_P\angle 0°}{|\mathbf{Z}_T|\angle\theta_T} = \frac{V_P}{|\mathbf{Z}_T|}\angle -\theta_T = I_L\angle -\theta_T \quad (12.79)$$

Similarly, the other two line currents are given by

$$\mathbf{I}_B = I_L\angle -\theta_T - 120° \quad (12.80)$$

$$\mathbf{I}_C = I_L\angle -\theta_T - 240° \quad (12.81)$$

The voltage between A and N, $\mathbf{V}_{AN}$, across $\mathbf{Z}_Y$ is given by

$$\mathbf{V}_{AN} = \mathbf{I}_A\mathbf{Z}_Y \quad (12.82)$$

Similarly, we have

$$\mathbf{V}_{BN} = \mathbf{I}_B\mathbf{Z}_Y \quad (12.83)$$

$$\mathbf{V}_{CN} = \mathbf{I}_C\mathbf{Z}_Y \quad (12.84)$$

The magnitude of line voltages is given by

$$V_L = \sqrt{3}|\mathbf{V}_{AN}| \quad (12.85)$$

The line voltages are given by

$$\mathbf{V}_{AB} = V_L\angle\mathbf{V}_{AN} + 30° \quad (12.86)$$

$$\mathbf{V}_{BC} = V_L\angle\mathbf{V}_{AN} - 90° \quad (12.87)$$

$$\mathbf{V}_{CA} = V_L\angle\mathbf{V}_{AN} - 210° \quad (12.88)$$

Notice that the line voltages can also be found using

$$\mathbf{V}_{AB} = \mathbf{V}_{AN} - \mathbf{V}_{BN}$$

$$\mathbf{V}_{BC} = \mathbf{V}_{BN} - \mathbf{V}_{CN}$$

$$\mathbf{V}_{CA} = \mathbf{V}_{CN} - \mathbf{V}_{AN}$$

Notice that the magnitude of $\mathbf{V}_{AB}$ is V_L; that is, $|\mathbf{V}_{AB}| = V_L$.
The complex power of the balanced load is given by

$$\mathbf{S}_L = 3I_L^2\mathbf{Z}_Y \quad (12.89)$$

The average power of the load is

$$P_L = \text{Re}(\mathbf{S}_L) \quad (12.90)$$

and the reactive power of the load is

$$Q_L = \text{I}_\text{m}(\mathbf{S}_L) \quad (12.91)$$

Notice that the complex power $\mathbf{S}_L$ can also be obtained from

$$\mathbf{S}_L = \mathbf{V}_{AN}\mathbf{I}_A^* + \mathbf{V}_{BN}\mathbf{I}_B^* + \mathbf{V}_{CN}\mathbf{I}_C^* \tag{12.92}$$

Also, notice that the apparent power of the load is given by

$$|\mathbf{S}_L| = \sqrt{3}V_L I_L \tag{12.93}$$

The complex power of the wires is given by

$$\mathbf{S}_W = 3I_L^2\mathbf{Z}_W \tag{12.94}$$

The total complex power is given by

$$\mathbf{S}_T = \mathbf{S}_W + \mathbf{S}_L = P_T + jQ_T \tag{12.95}$$

The efficiency is given by

$$Eff = \frac{P_L}{P_T} \times 100\% \tag{12.96}$$

EXAMPLE 12.5

In the balanced Y-Δ connection shown in Figure 12.13, let the magnitude of the phase voltages be $V_P = 550$ V (rms), the load impedance per phase be $\mathbf{Z}_\Delta = 75 + j54\ \Omega$, the wire impedance per phase be $\mathbf{Z}_W = 5 + j3\ \Omega$, and the phase angle of $\mathbf{V}_{an}$ be 0°. Find the phase voltages, line currents, and line voltages. Find the complex power, apparent power, average power, reactive power, and power factor of the load. Find the complex power of the wire, the total complex power, and efficiency. Also, find $\mathbf{I}_{AB}$, $\mathbf{I}_{BC}$, and $\mathbf{I}_{CA}$.

The Δ load is converted to a Y load, as shown in Figure 12.14. The Y load impedance per phase is given by

$$\mathbf{Z}_Y = \mathbf{Z}_\Delta/3 = 25 + j18 = 30.8058\angle 35.7539^\circ\ \Omega$$

The total impedance is given by,

$$\mathbf{Z}_T = \mathbf{Z}_W + \mathbf{Z}_Y = 30 + j21 = 36.6197\angle 34.9920^\circ\ \Omega$$

The magnitude and the phase of the total impedance are given, respectively, by

$$|\mathbf{Z}_T| = 36.6197\ \Omega,\ \theta_T = 34.9920^\circ$$

The phase voltages are given by

$$\mathbf{V}_{an} = V_P\angle 0^\circ = 550\angle 0^\circ = 550 + j0\ \text{V}$$

$$\mathbf{V}_{bn} = V_P\angle -120^\circ = 550\angle -120^\circ = -275 - j476.3140\ \text{V}$$

$$\mathbf{V}_{cn} = V_P\angle -240^\circ = 550\angle -240^\circ = -275 + j476.3140\ \text{V}$$

The magnitude of the line currents is

$$I_L = \frac{V_P}{|\mathbf{Z}_T|} = 15.0193\ \text{A}$$

continued

Example 12.5 continued

The line current $\mathbf{I}_A$ is given by

$$\mathbf{I}_A = I_L\angle -\theta_T = 15.0193\angle -34.9920° = 12.3043 - j8.6130 \text{ A}$$

Notice that $\mathbf{I}_A = \mathbf{V}_{an}/\mathbf{Z}_T$ provides the same answer. The other two line currents are given by

$$\mathbf{I}_B = I_L\angle -\theta_T - 120° = 15.0193\angle -154.9920° = -13.6112 - j6.3493 \text{ A}$$

$$\mathbf{I}_C = I_L\angle -\theta_T - 240° = 15.0193\angle 85.0080° = 1.3069 + j14.9623 \text{ A}$$

The voltage between A and N, $\mathbf{V}_{AN}$, across $\mathbf{Z}_Y$ is given by

$$\mathbf{V}_{AN} = \mathbf{I}_A\mathbf{Z}_Y = 462.6398 + j6.1521 = 462.6807\angle 0.7619° \text{ V}$$

Similarly, we have

$$\mathbf{V}_{BN} = \mathbf{I}_B\mathbf{Z}_Y = -225.8296 - j391.6541 = 462.6807\angle -119.2381° \text{ V}$$

$$\mathbf{V}_{CN} = \mathbf{I}_C\mathbf{Z}_Y = -226.2676 + j391.4012 = 462.6807\angle 120.7619° \text{ V}$$

The magnitude of line voltages is given by

$$V_L = \sqrt{3}|\mathbf{V}_{AN}| = 801.3865 \text{ V}$$

The line voltages are given by

$$\mathbf{V}_{AB} = V_L\angle \mathbf{V}_{AN} + 30° = 688.6318 + j409.8860 = 801.3865\angle 30.7619°$$

$$\mathbf{V}_{BC} = V_L\angle \mathbf{V}_{AN} - 90° = 10.6558 - j801.3157 = 801.3865\angle -89.2381°$$

$$\mathbf{V}_{CA} = V_L\angle \mathbf{V}_{AN} - 210° = -699.2876 + j391.4296 = 801.3865\angle 150.7619°$$

The complex power of the balanced load is given by

$$\mathbf{S}_L = 3I_L^2\mathbf{Z}_Y = 16{,}918.3445 + j12{,}181.2081 = 20{,}847.3550\angle 35.7539° \text{ VA}$$

The average power is

$$P_L = \text{Re}(\mathbf{S}_L) = 16{,}918.3445 \text{ W}$$

and the reactive power is

$$Q_L = \text{I}_\text{m}(\mathbf{S}_L) = 12{,}181.2081 \text{ VAR}$$

The apparent power is

$$|\mathbf{S}_L| = 20{,}847.3550 \text{ VA}$$

Notice that the complex power $\mathbf{S}_L$ can also be found from

$$\mathbf{S}_L = \mathbf{V}_{AN}\mathbf{I}_A^* + \mathbf{V}_{BN}\mathbf{I}_B^* + \mathbf{V}_{CN}\mathbf{I}_C^* = 16{,}918.3445 + j12{,}181.2081$$
$$= 20{,}847.3550\angle 35.7539° \text{ VA}$$

continued

Example 12.5 continued

The power factor of the load is given by

$$pf = \cos(35.7539°) = 0.811534341451494$$

Let $\theta_1 = \angle \mathbf{Z}_Y$. Then, we have

$$|\mathbf{S}_L| = \sqrt{3}V_L I_L = 20{,}847.3550 \text{ VA}$$

$$P_L = |\mathbf{S}_L|\cos(\theta_1) = 16{,}918.3445 \text{ W}$$

$$Q_L = |\mathbf{S}_L|\sin(\theta_1) = 12{,}181.2081 \text{ VAR}$$

$$pf = \cos(\theta_1) = 0.811534341451494$$

The complex power of the wires is

$$\mathbf{S}_W = 3I_L^2\mathbf{Z}_W = 3383.6689 + j2030.2013 = 3946.0021\angle 30.9638° \text{ VA}$$

The total complex power from the source to load is

$$\mathbf{S}_T = \mathbf{S}_W + \mathbf{S}_L = 20{,}302.0134 + j14{,}211.4094 = 24{,}781.7656\angle 34.9920° \text{ VA}$$

The total average power from the source to load is

$$P_T = \text{Re}(\mathbf{S}_T) = 20{,}302.0134 \text{ W}$$

The total reactive power from the source to load is

$$Q_T = \text{I}_\text{m}(\mathbf{S}_T) = 14{,}211.4094 \text{ VAR}$$

The efficiency is given by

$$Eff = \frac{P_L}{P_T} \times 100\% = 100 \times 16{,}918.3445/20{,}302.0134 = 83.3333\%$$

The currents $\mathbf{I}_{AB}$, $\mathbf{I}_{BC}$, and $\mathbf{I}_{CA}$ are given, respectively, as

$$\mathbf{I}_{AB} = \frac{\mathbf{V}_{AB}}{\mathbf{Z}_\Delta} = 8.6385 - j0.7546 = 8.6714\angle -4.9920° \text{ A}$$

$$\mathbf{I}_{BC} = \frac{\mathbf{V}_{BC}}{\mathbf{Z}_\Delta} = -4.9727 - j7.1039 = 8.6714\angle -124.9920° \text{ A}$$

$$\mathbf{I}_{CA} = \frac{\mathbf{V}_{CA}}{\mathbf{Z}_\Delta} = -3.6658 + j7.8584 = 8.6714\angle 115.008° \text{ A}$$

Exercise 12.5

In the balanced Y-Δ connection shown in Figure 12.13, let the magnitude of the phase voltages be $V_P = 750$ V (rms), the load impedance per phase be $\mathbf{Z}_\Delta = 72 + j57\ \Omega$, the wire impedance per phase be $\mathbf{Z}_W = 9 + j6\ \Omega$, and the phase angle of $\mathbf{V}_{an}$ be 0°. Find the complex power of the load.

Answer:
$\mathbf{S}_L = 23{,}628.9382 + j18{,}706.2427 = 30{,}137.1902\angle 38.3675°$ VA.

A Y-Δ connection that includes impedances on the wires ($\mathbf{Z}_W$) and the impedances on the generators ($\mathbf{Z}_S$) is shown in Figure 12.15.

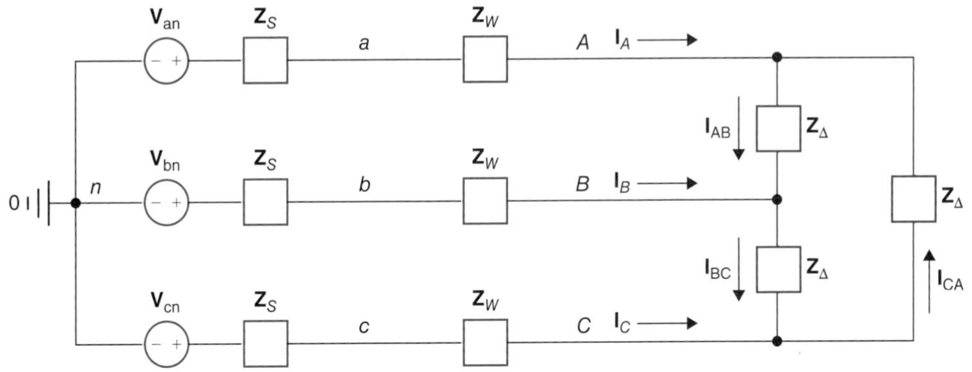

FIGURE 12.15 The Y-Δ connection that includes impedances on the wires and generators.

When the Δ load is converted to a Y load with $\mathbf{Z}_Y = \mathbf{Z}_\Delta/3$, the circuit shown in Figure 12.15 becomes the one shown in Figure 12.16.

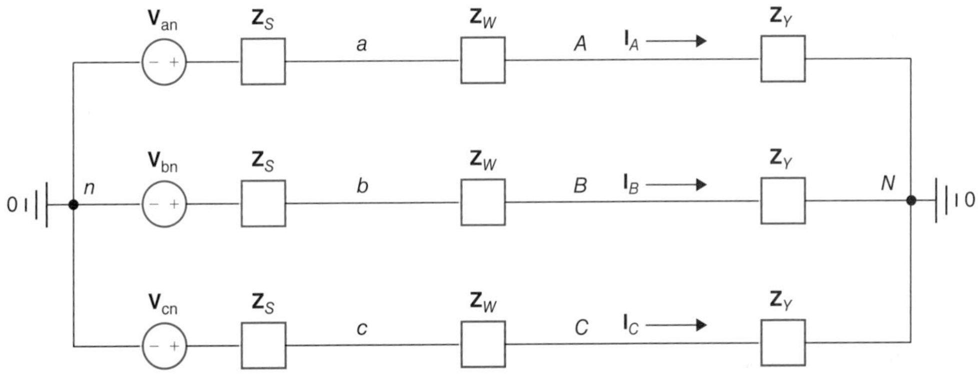

FIGURE 12.16 A Y-Δ connection that includes impedances on the wires and generators with the Δ load converted to a Y load.

12.5 Balanced Δ-Δ Circuit

A balanced Δ-Δ connection is shown in Figure 12.17.

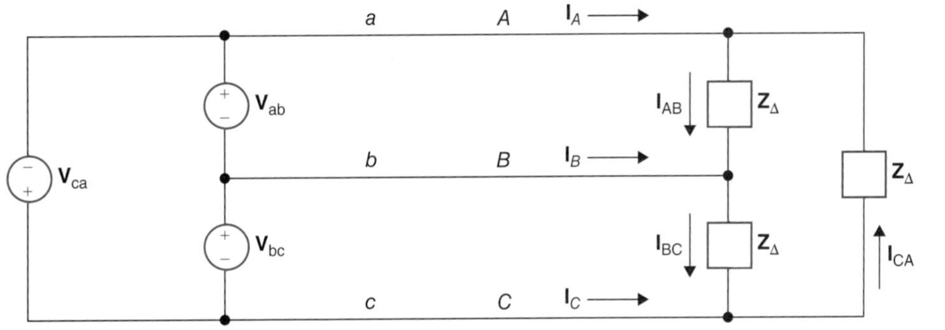

FIGURE 12.17 A Δ-Δ connection.

The impedance of the load, for each phase, is represented as

$$\mathbf{Z}_\Delta = |\mathbf{Z}_\Delta| \angle \theta_\Delta$$

Let V_P be the magnitude of the phase voltages, and let the phase voltages of the three-phase sources be

$$\mathbf{V}_{ab} = V_P\angle 0°$$
$$\mathbf{V}_{bc} = V_P\angle -120°$$
$$\mathbf{V}_{ca} = V_P\angle -240°$$

The voltages across the three balanced loads are the same as the three-phase voltage sources; that is,

$$\mathbf{V}_{AB} = \mathbf{V}_{ab}$$
$$\mathbf{V}_{BC} = \mathbf{V}_{bc}$$
$$\mathbf{V}_{CA} = \mathbf{V}_{ca}$$

The phase currents are given by

$$\mathbf{I}_{AB} = \frac{\mathbf{V}_{AB}}{\mathbf{Z}_\Delta} = \frac{\mathbf{V}_{ab}}{\mathbf{Z}_\Delta} = \frac{V_P\angle 0°}{|\mathbf{Z}_\Delta|\angle\theta_\Delta} = \frac{V_P}{|\mathbf{Z}_\Delta|}\angle 0° - \theta_\Delta = \frac{V_P}{|\mathbf{Z}_\Delta|}\angle -\theta_\Delta \quad \textbf{(12.97)}$$

The magnitude of the phase current is

$$I_P = \frac{V_P}{|\mathbf{Z}_\Delta|} \quad \textbf{(12.98)}$$

The phase current $\mathbf{I}_{AB}$ can be written as

$$\mathbf{I}_{AB} = I_P\angle 0° - \theta_\Delta = I_P\angle -\theta_\Delta \quad \textbf{(12.99)}$$

The other two phase currents are given by

$$\mathbf{I}_{BC} = \frac{\mathbf{V}_{BC}}{\mathbf{Z}_\Delta} = \frac{\mathbf{V}_{bc}}{\mathbf{Z}_\Delta} = \frac{V_P\angle -120°}{|\mathbf{Z}_\Delta|\angle\theta_\Delta} = \frac{V_P}{|\mathbf{Z}_\Delta|}\angle -120° - \theta_\Delta = I_P\angle -\theta_\Delta - 120° \quad \textbf{(12.100)}$$

$$\mathbf{I}_{CA} = \frac{\mathbf{V}_{CA}}{\mathbf{Z}_\Delta} = \frac{\mathbf{V}_{ca}}{\mathbf{Z}_\Delta} = \frac{V_P\angle 120°}{|\mathbf{Z}_\Delta|\angle\theta_\Delta} = \frac{V_P}{|\mathbf{Z}_\Delta|}\angle 120° - \theta_\Delta = I_P\angle -\theta_\Delta + 120° \quad \textbf{(12.101)}$$

The line current $\mathbf{I}_A$ is given by

$$\begin{aligned}
\mathbf{I}_A = \mathbf{I}_{AB} - \mathbf{I}_{CA} &= \frac{V_P}{|\mathbf{Z}_\Delta|}\angle 0° - \theta_\Delta - \frac{V_P}{|\mathbf{Z}_\Delta|}\angle 120° - \theta_\Delta = \frac{V_P}{|\mathbf{Z}_\Delta|}(e^{-j\theta_\Delta} - e^{j120° - j\theta_\Delta}) \\
&= \frac{V_P}{|\mathbf{Z}_\Delta|}e^{-j\theta_\Delta}(1 - e^{j120°}) = \frac{V_P}{|\mathbf{Z}_\Delta|}e^{-j\theta_\Delta}\left[1 - \left(-\frac{1}{2} + j\frac{\sqrt{3}}{2}\right)\right] \\
&= \frac{V_P}{|\mathbf{Z}_\Delta|}e^{-j\theta_\Delta}\left(\frac{3}{2} - j\frac{\sqrt{3}}{2}\right) = \frac{V_P}{|\mathbf{Z}_\Delta|}\sqrt{3}e^{-j\theta_\Delta}\left(\frac{\sqrt{3}}{2} - j\frac{1}{2}\right) \\
&= \frac{V_P}{|\mathbf{Z}_\Delta|}\sqrt{3}e^{-j\theta_\Delta}e^{-j30°} = \frac{V_P}{|\mathbf{Z}_\Delta|}\sqrt{3}e^{-j\theta_\Delta - j30°} = \frac{V_P}{|\mathbf{Z}_\Delta|}\sqrt{3}\angle -\theta_\Delta - 30° \\
&= \sqrt{3}I_P\angle -\theta_\Delta - 30° = \sqrt{3}I_P\angle \mathbf{I}_{AB} - 30°
\end{aligned} \quad \textbf{(12.102)}$$

The magnitude of the line currents is

$$I_L = \sqrt{3} I_P \tag{12.103}$$

The phase of line current $\mathbf{I}_A$ lags the phase of the phase current $\mathbf{I}_{AB}$ by 30°. Similarly, we have

$$\mathbf{I}_B = \sqrt{3} I_P \angle \mathbf{I}_{AB} - 150° \tag{12.104}$$

$$\mathbf{I}_C = \sqrt{3} I_P \angle \mathbf{I}_{AB} + 90° \tag{12.105}$$

The complex power of the load is

$$\mathbf{S}_L = 3 I_P^2 \mathbf{Z}_\Delta \tag{12.106}$$

The average power of the load is

$$P_L = \mathrm{Re}(\mathbf{S}_L) \tag{12.107}$$

The reactive power of the load is

$$Q_L = \mathrm{I_m}(\mathbf{S}_L) \tag{12.108}$$

Notice that $\mathbf{S}_L$ can also be found using

$$\mathbf{S}_L = \mathbf{V}_{ab}\mathbf{I}_{AB}^* + \mathbf{V}_{bc}\mathbf{I}_{BC}^* + \mathbf{V}_{ca}\mathbf{I}_{CA}^* \tag{12.109}$$

EXAMPLE 12.6

In the balanced Δ-Δ connection shown in Figure 12.17, let the magnitude of the phase voltages be $V_P = 550$ V (rms), the load impedance per phase be $\mathbf{Z}_\Delta = 35 + j20\ \Omega$, and the phase angle of $\mathbf{V}_{ab}$ be 0°. Find the phase voltages, phase currents, and line currents. Also, find the complex power, apparent power, average power, reactive power, and power factor of the load.

The impedance of the load, for each phase, is represented as

$$\mathbf{Z}_\Delta = |\mathbf{Z}_\Delta| \angle \theta_\Delta = 40.3113\angle 29.7449°\ \Omega, \quad |\mathbf{Z}_\Delta| = 40.3113, \quad \theta_\Delta = 29.7449°$$

The phase voltages are given by

$$\mathbf{V}_{ab} = V_P \angle 0° = 550\angle 0°\ \mathrm{V}$$
$$\mathbf{V}_{bc} = V_P \angle -120° = 550\angle -120°\ \mathrm{V}$$
$$\mathbf{V}_{ca} = V_P \angle -240° = 550\angle -240° = 550\angle 120°\ \mathrm{V}$$

The voltages across the three balanced loads are the same as the three-phase voltage sources; that is,

$$\mathbf{V}_{AB} = \mathbf{V}_{ab} = 550\angle 0°\ \mathrm{V}$$
$$\mathbf{V}_{BC} = \mathbf{V}_{bc} = 550\angle -120°\ \mathrm{V}$$
$$\mathbf{V}_{CA} = \mathbf{V}_{ca} = 550\angle 120°\ \mathrm{V}$$

continued

Example 12.6 continued

The phase currents are given by

$$\mathbf{I}_{AB} = \frac{\mathbf{V}_{AB}}{\mathbf{Z}_\Delta} = \frac{\mathbf{V}_{ab}}{\mathbf{Z}_\Delta} = 13.6438\angle -29.7449° \text{ A}$$

$$\mathbf{I}_{BC} = \frac{\mathbf{V}_{BC}}{\mathbf{Z}_\Delta} = \frac{\mathbf{V}_{bc}}{\mathbf{Z}_\Delta} = 13.6438\angle -149.7449° \text{ A}$$

$$\mathbf{I}_{CA} = \frac{\mathbf{V}_{CA}}{\mathbf{Z}_\Delta} = \frac{\mathbf{V}_{ca}}{\mathbf{Z}_\Delta} = 13.6438\angle 90.2551° \text{ A}$$

The magnitude of the phase currents at the load is

$$I_P = \frac{V_P}{|\mathbf{Z}_\Delta|} = 13.6438 \text{ A}$$

The phase currents can also be found from

$$\mathbf{I}_{AB} = I_P\angle 0° - \theta_\Delta = 13.6438\angle -29.7449° \text{ A}$$

$$\mathbf{I}_{BC} = I_P\angle -120° - \theta_\Delta = 13.6438\angle -149.7449° \text{ A}$$

$$\mathbf{I}_{CA} = I_P\angle 120° - \theta_\Delta = 13.6438\angle 90.2551° \text{ A}$$

The magnitude of the line currents is

$$I_L = \sqrt{3}I_P = 23.6318 \text{ A}$$

The line currents are given by

$$\mathbf{I}_A = \sqrt{3}I_P\angle \mathbf{I}_{AB} - 30° = 23.6318\angle -59.7449° \text{ A}$$

$$\mathbf{I}_B = \sqrt{3}I_P\angle \mathbf{I}_{AB} - 150° = 23.6318\angle -179.7449° \text{ A}$$

$$\mathbf{I}_C = \sqrt{3}I_P\angle \mathbf{I}_{AB} + 90° = 23.6318\angle 60.2551° \text{ A}$$

The complex power of the load is

$$\mathbf{S}_L = 3I_P^2\mathbf{Z}_\Delta = 19{,}546.1538 + j11{,}169.2308 \text{ VA}$$

The average power of the load is

$$P_L = \text{Re}(\mathbf{S}_L) = 19{,}546.1538 \text{ W}$$

The reactive power of the load is

$$Q_L = \text{I}_\text{m}(\mathbf{S}_L) = 11{,}169.2308 \text{ VAR}$$

Notice that complex power $\mathbf{S}_L$ can also be found using

$$\mathbf{S}_L = \mathbf{V}_{ab}\mathbf{I}_{AB}^* + \mathbf{V}_{bc}\mathbf{I}_{BC}^* + \mathbf{V}_{ca}\mathbf{I}_{CA}^*$$

Exercise 12.6

In the balanced Δ-Δ connection shown in Figure 12.17, let the magnitude of the phase voltages be $V_P = 760$ V (rms), the load impedance per phase be $\mathbf{Z}_\Delta = 55 + j35\ \Omega$, and the phase angle of $\mathbf{V}_{ab}$ be 0°. Find the complex power of the load.

Answer:
$\mathbf{S}_L = 22{,}424.4706 + j14{,}270.1176 = 26{,}579.9387\angle 32.4712^{u}$ VA.

12.5.1 BALANCED Δ-Δ CIRCUIT WITH WIRE IMPEDANCE

A Δ-Δ connection that includes impedances on the wires $\mathbf{Z}_W$ is shown in Figure 12.18. The three-phase delta load and source shown in Figure 12.18 can be converted to the three-phase wye load and source, as shown in Figure 12.19.

FIGURE 12.18

A Δ-Δ connection with wire impedances.

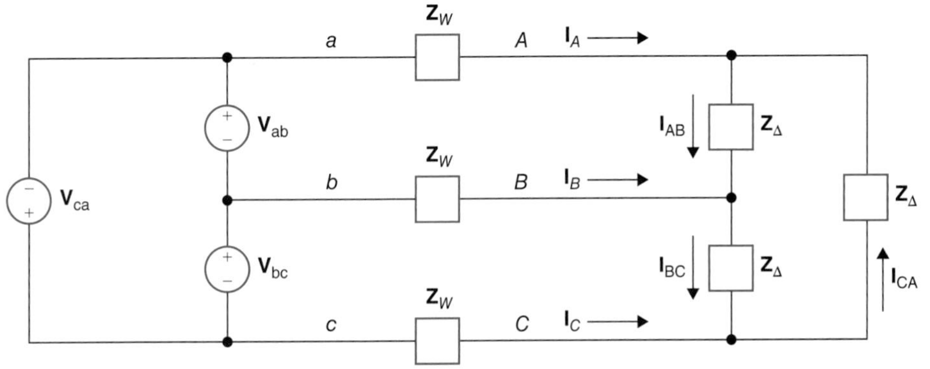

FIGURE 12.19

A Δ-Δ connection with wire impedances converted to Y-Y connection.

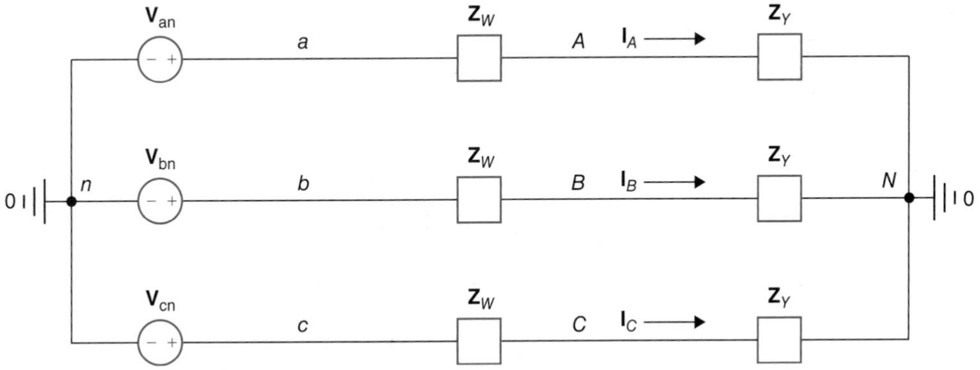

The impedance of the Y load per phase is given by

$$\mathbf{Z}_Y = \frac{\mathbf{Z}_\Delta}{3} \tag{12.110}$$

The phase voltages of the delta sources are given by

$$\mathbf{V}_{ab} = V_P\angle 0^\circ$$
$$\mathbf{V}_{bc} = V_P\angle -120^\circ$$
$$\mathbf{V}_{ca} = V_P\angle -240^\circ$$

The delta sources can be converted to wye sources with voltages

$$\mathbf{V}_{an} = \frac{V_P}{\sqrt{3}}\angle -30° \tag{12.111}$$

$$\mathbf{V}_{bn} = \frac{V_P}{\sqrt{3}}\angle -150° \tag{12.112}$$

$$\mathbf{V}_{cn} = \frac{V_P}{\sqrt{3}}\angle -270° = \frac{V_P}{\sqrt{3}}\angle 90° \tag{12.113}$$

The total impedance for each voltage phase is given by

$$\mathbf{Z}_T = \mathbf{Z}_W + \mathbf{Z}_Y = |\mathbf{Z}_T|\angle \theta_T \tag{12.114}$$

The magnitude of line currents is

$$I_L = \frac{\frac{V_P}{\sqrt{3}}}{|\mathbf{Z}_T|} \tag{12.115}$$

The line current $\mathbf{I}_A$ is given by

$$\mathbf{I}_A = \frac{\frac{V_P}{\sqrt{3}}\angle -30°}{|\mathbf{Z}_T|\angle \theta_T} = \frac{\frac{V_P}{\sqrt{3}}}{|\mathbf{Z}_T|}\angle -30° - \theta_T = I_L\angle -\theta_T - 30° \tag{12.116}$$

Similarly, the other two line currents are given by

$$\mathbf{I}_B = I_L\angle -\theta_T - 150° \tag{12.117}$$

$$\mathbf{I}_C = I_L\angle -\theta_T + 90° \tag{12.118}$$

The voltage between A and N, $\mathbf{V}_{AN}$, across $\mathbf{Z}_Y$ is given by

$$\mathbf{V}_{AN} = \mathbf{I}_A\mathbf{Z}_Y \tag{12.119}$$

Similarly, we have

$$\mathbf{V}_{BN} = \mathbf{I}_B\mathbf{Z}_Y \tag{12.120}$$

$$\mathbf{V}_{CN} = \mathbf{I}_C\mathbf{Z}_Y \tag{12.121}$$

The magnitude of line voltages is given by

$$V_L = \sqrt{3}|\mathbf{V}_{AN}| \tag{12.122}$$

The line voltages are given by

$$\mathbf{V}_{AB} = V_L\angle \mathbf{V}_{AN} + 30° \tag{12.123}$$

$$\mathbf{V}_{BC} = V_L\angle \mathbf{V}_{AN} - 90° \tag{12.124}$$

$$\mathbf{V}_{CA} = V_L\angle \mathbf{V}_{AN} - 210° \tag{12.125}$$

Notice that the line voltages can also be found using

$$\mathbf{V}_{AB} = \mathbf{V}_{AN} - \mathbf{V}_{BN}$$
$$\mathbf{V}_{BC} = \mathbf{V}_{BN} - \mathbf{V}_{CN}$$
$$\mathbf{V}_{CA} = \mathbf{V}_{CN} - \mathbf{V}_{AN}$$

The magnitude of $\mathbf{V}_{AB}$ is V_L; that is, $|\mathbf{V}_{AB}| = V_L$.
The complex power of the balanced load is given by

$$\mathbf{S}_L = 3I_L^2\mathbf{Z}_Y \quad (12.126)$$

The average power is

$$P_L = \text{Re}(\mathbf{S}_L) \quad (12.127)$$

and the reactive power is

$$Q_L = \text{I}_\text{m}(\mathbf{S}_L) \quad (12.128)$$

Notice that the complex power $\mathbf{S}_L$ can also be written as

$$\mathbf{S}_L = \mathbf{V}_{AN}\mathbf{I}_A^* + \mathbf{V}_{BN}\mathbf{I}_B^* + \mathbf{V}_{CN}\mathbf{I}_C^* \quad (12.129)$$

Also, notice that the apparent power is given by

$$|\mathbf{S}_L| = \sqrt{3}V_LI_L \quad (12.130)$$

The complex power of wires is given by

$$\mathbf{S}_W = 3I_L^2\mathbf{Z}_W \quad (12.131)$$

The total complex power is given by

$$\mathbf{S}_T = \mathbf{S}_W + \mathbf{S}_L = P_T + jQ_T \quad (12.132)$$

The efficiency is given by

$$Eff = \frac{P_L}{P_T} \times 100\% \quad (12.133)$$

EXAMPLE 12.7

In the balanced Δ-Δ connection shown in Figure 12.18, let the magnitude of the phase voltages be $V_P = 650$ V (rms), the load impedance per phase be $\mathbf{Z}_\Delta = 63 + j36\ \Omega$, the wire impedance per phase be $\mathbf{Z}_W = 5 + j4\ \Omega$, and the phase angle of $\mathbf{V}_{ab}$ be 0°. Find the voltages of Δ and Y at the source, the line currents, and line voltages at the load. Find the complex power, apparent power, average power, reactive power, and power factor of the load. Find the complex power of the wire, the total complex power, and efficiency.

The impedance of the Y load per phase is given by

$$\mathbf{Z}_Y = \mathbf{Z}_\Delta/3 = 21 + j12\ \Omega = 24.1868\angle 29.7449°\ \Omega$$
$$|\mathbf{Z}_Y| = 24.1868, \quad \theta_Y = 29.7449°$$

continued

Example 12.7 continued

The phase voltages of the sources are given by

$$\mathbf{V}_{ab} = 650\angle 0° \text{ V}$$
$$\mathbf{V}_{bc} = 650\angle -120° \text{ V}$$
$$\mathbf{V}_{ca} = 650\angle -240° \text{ V}$$

Converting to Y sources, we obtain

$$\mathbf{V}_{an} = \frac{V_P}{\sqrt{3}}\angle -30° = 375.2777\angle -30° \text{ V}$$

$$\mathbf{V}_{bn} = \frac{V_P}{\sqrt{3}}\angle -150° = 375.2777\angle -150° \text{ V}$$

$$\mathbf{V}_{cn} = \frac{V_P}{\sqrt{3}}\angle -270° = \frac{V_P}{\sqrt{3}}\angle 90° = 375.2777\angle 90° \text{ V}$$

The total impedance for each voltage phase is given by

$$\mathbf{Z}_T = \mathbf{Z}_W + \mathbf{Z}_Y = |\mathbf{Z}_T|\angle\theta_T = 26 + j16 = 30.5287\angle 31.6075° \ \Omega$$

The magnitude of line currents is

$$I_L = \frac{\frac{V_P}{\sqrt{3}}}{|\mathbf{Z}_T|} = 12.2926 \text{ A}$$

The line currents are given by

$$\mathbf{I}_A = I_L\angle -\theta_T - 30° = 12.2926\angle -61.6075° \text{ A}$$
$$\mathbf{I}_B = I_L\angle -\theta_T - 150° = 12.2926\angle 178.3925° \text{ A}$$
$$\mathbf{I}_C = I_L\angle -\theta_T + 90° = 12.2926\angle 58.3925° \text{ A}$$

The voltage between A and N, $\mathbf{V}_{AN}$, across $\mathbf{Z}_Y$ is given by

$$\mathbf{V}_{AN} = \mathbf{I}_A\mathbf{Z}_Y = 252.5179 - j156.9501 = 297.3190\angle -31.8626° \text{ V}$$

Similarly, we have

$$\mathbf{V}_{BN} = \mathbf{I}_B\mathbf{Z}_Y = -262.1817 - j140.2119 = 297.3190\angle -151.8626° \text{ V}$$
$$\mathbf{V}_{CN} = \mathbf{I}_C\mathbf{Z}_Y = 9.6638 + j297.1619 = 297.3190\angle 88.1374° \text{ V}$$

The magnitude of line voltages is given by

$$V_L = \sqrt{3}|\mathbf{V}_{AN}| = 514.9717 \text{ V}$$

continued

Example 12.7 continued

The line voltages are given by

$$\mathbf{V}_{AB} = V_L \angle \mathbf{V}_{AN} + 30° = 514.9717\angle -1.8626° \text{ V}$$

$$\mathbf{V}_{BC} = V_L \angle \mathbf{V}_{AN} - 90° = 514.9717\angle -121.8626° \text{ V}$$

$$\mathbf{V}_{CA} = V_L \angle \mathbf{V}_{AN} - 210° = 514.9717\angle 118.1374° \text{ V}$$

Notice that the line voltages can also be found using

$$\mathbf{V}_{AB} = \mathbf{V}_{AN} - \mathbf{V}_{BN} = 514.9717\angle -1.8626° \text{ V}$$

$$\mathbf{V}_{BC} = \mathbf{V}_{BN} - \mathbf{V}_{CN} = 514.9717\angle -121.8626° \text{ V}$$

$$\mathbf{V}_{CA} = \mathbf{V}_{CN} - \mathbf{V}_{AN} = 514.9717\angle 118.1374° \text{ V}$$

The magnitude of $\mathbf{V}_{AB}$ is V_L; that is, $|\mathbf{V}_{AB}| = V_L$.
The complex power of the balanced load is given by

$$\mathbf{S}_L = 3I_L^2 \mathbf{Z}_Y = 9519.8498 + j5439.9142 = 10{,}964.4975\angle 29.7449° \text{ VA}$$

The average power is

$$P_L = \text{Re}(\mathbf{S}_L) = 9519.8498 \text{ W}$$

and the reactive power is

$$Q_L = \text{I}_\text{m}(\mathbf{S}_L) = 5439.9142 \text{ VAR}$$

Notice that the complex power $\mathbf{S}_L$ can also be obtained from

$$\mathbf{S}_L = \mathbf{V}_{AN}\mathbf{I}_A^* + \mathbf{V}_{BN}\mathbf{I}_B^* + \mathbf{V}_{CN}\mathbf{I}_C^* = 9519.8498 + j5439.9142$$
$$= 10{,}964.4975\angle 29.7449° \text{ VA}$$

Also, notice that the apparent power is given by

$$|\mathbf{S}_L| = \sqrt{3} V_L I_L = 10{,}964.4975 \text{ VA}$$

The complex power on the wires is given by

$$\mathbf{S}_W = 3I_L^2 \mathbf{Z}_W = 2266.6309 + j1813.3047 \text{ VA}$$

The total complex power is given by

$$\mathbf{S}_T = \mathbf{S}_W + \mathbf{S}_L = P_T + jQ_T = 11{,}786.4807 + j7253.2189 \text{ VA}$$

The efficiency is given by

$$Eff = \frac{P_L}{P_T} \times 100\% = 80.7692\%$$

continued

Example 12.7 continued

Alternatively, we can find mesh currents $\mathbf{I}_1$, $\mathbf{I}_2$, and $\mathbf{I}_3$ in the circuit shown in Figure 12.20. The three mesh equations are given, respectively, by

$$-\mathbf{V}_{ab} + \mathbf{Z}_w\mathbf{I}_1 + \mathbf{Z}_\Delta(\mathbf{I}_1 - \mathbf{I}_3) + \mathbf{Z}_w(\mathbf{I}_1 - \mathbf{I}_2) = 0$$

$$-\mathbf{V}_{bc} + \mathbf{Z}_w(\mathbf{I}_2 - \mathbf{I}_1) + \mathbf{Z}_\Delta(\mathbf{I}_2 - \mathbf{I}_3) + \mathbf{Z}_w\mathbf{I}_2 = 0$$

$$\mathbf{Z}_\Delta(\mathbf{I}_3 - \mathbf{I}_2) + \mathbf{Z}_\Delta(\mathbf{I}_3 - \mathbf{I}_1) + \mathbf{Z}_\Delta\mathbf{I}_3 = 0$$

These three equations can be rearranged as

$$(\mathbf{Z}_w + \mathbf{Z}_\Delta + \mathbf{Z}_w)\mathbf{I}_1 - \mathbf{Z}_w\mathbf{I}_2 - \mathbf{Z}_\Delta\mathbf{I}_3 = \mathbf{V}_{ab}$$

$$-\mathbf{Z}_w\mathbf{I}_1 + (\mathbf{Z}_w + \mathbf{Z}_\Delta + \mathbf{Z}_w)\mathbf{I}_2 - \mathbf{Z}_\Delta\mathbf{I}_3 = \mathbf{V}_{bc}$$

$$-\mathbf{Z}_\Delta\mathbf{I}_1 - \mathbf{Z}_\Delta\mathbf{I}_2 + 3\mathbf{Z}_\Delta\mathbf{I}_3 = 0$$

FIGURE 12.20

A Δ-Δ circuit with three meshes.

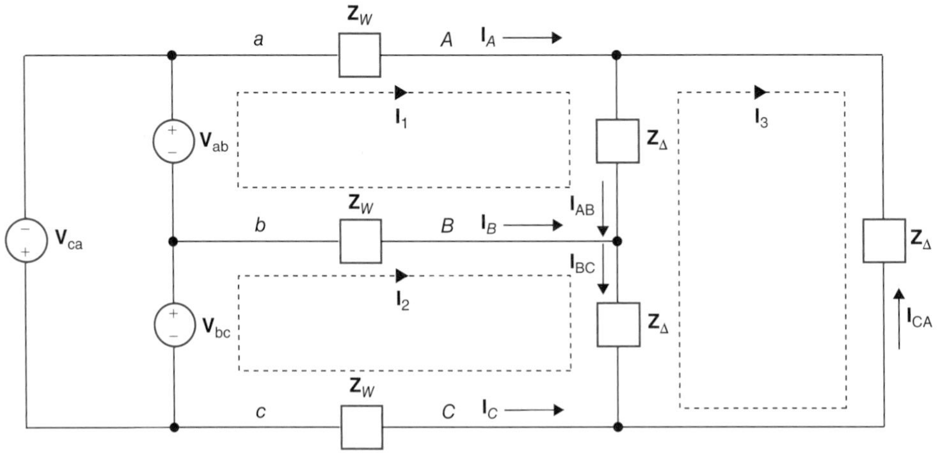

In MATLAB, we have

MATLAB

```
% EXAMPLE 12.7
clear all;
VP=650;ZD=63+36j;ZW=5+4j;Vab=P2Rd(VP,0);
Vbc=P2Rd(VP,-120);Vca=P2Rd(VP,-240);
syms I1 I2 I3
[I1,I2,I3]=solve(-Vab+ZW*I1+ZD*(I1-I3)+ZW*(I1-I2),...
    -Vbc+ZW*(I2-I1)+ZD*(I2-I3)+ZW*I2,...
    ZD*(I3-I2)+ZD*(I3-I1)+ZD*I3,I1,I2,I3);
IA=I1;IB=I2-I1;IC=-I2;
IAp=R2P(IA);IBp=R2P(IB);ICp=R2P(IC);
IAB=I1-I3;IBC=I2-I3;ICA=-I3;
IABp=R2P(IAB);IBCp=R2P(IBC);ICAp=R2P(ICA);
VAB=ZD*IAB;VBC=ZD*IBC;VCA=ZD*ICA;
VABp=R2P(VAB);VBCp=R2P(VBC);VCAp=R2P(VCA);
SL=VAB*conj(IAB)+VBC*conj(IBC)+VCA*conj(ICA);
SLp=R2P(SL);
SW=3*IAp(1)^2*ZW;SWp=R2P(SW);
ST=SW+SL;STp=R2P(ST);
IA=vpa(IA,10)
IB=vpa(IB,10)
IC=vpa(IC,10)
IAp=vpa(IAp,10)
```

continued

Example 12.7 continued

MATLAB continued

```
IBp=vpa(IBp,10)
ICp=vpa(ICp,10)
IAB=vpa(IAB,10)
IBC=vpa(IBC,10)
ICA=vpa(ICA,10)
IABp=vpa(IABp,10)
IBCp=vpa(IBCp,10)
ICAp=vpa(ICAp,10)
VAB=vpa(VAB,10)
VBC=vpa(VBC,10)
VCA=vpa(VCA,10)
VABp=vpa(VABp,10)
VBCp=vpa(VBCp,10)
VCAp=vpa(VCAp,10)
SL=vpa(SL,10)
SLp=vpa(SLp,10)
SW=vpa(SW,10)
SWp=vpa(SWp,10)
ST=vpa(ST,10)
STp=vpa(STp,10)
```

The answers match those obtained earlier.

Exercise 12.7

In the balanced Δ-Δ connection shown in Figure 12.18, let the magnitude of the phase voltages be $V_P = 520$ V (rms), the load impedance per phase be $\mathbf{Z}_\Delta = 75 + j45\ \Omega$, the wire impedance per phase be $\mathbf{Z}_W = 7 + j5\ \Omega$, and the phase angle of $\mathbf{V}_{ab}$ be 0°. Find the complex power of the load.

Answer:

$\mathbf{S}_L = 4747.1910 + j2848.3146 = 5536.1285\angle 30.9638°$ VA.

EXAMPLE 12.8

In a Δ-Δ circuit with balanced sources and unbalanced loads shown in Figure 12.21, the source voltage is $\mathbf{V}_{ab} = 500\angle 0°$ V (rms). The wire impedance is $\mathbf{Z}_w = 8 + j5$, and the load impedances are $\mathbf{Z}_{\Delta 1} = 25\angle 0°\ \Omega$, $\mathbf{Z}_{\Delta 2} = 30\angle 0°\ \Omega$, $\mathbf{Z}_{\Delta 3} = 35\angle 0°\ \Omega$. Find the line currents and phase currents of the load.

The three mesh equations are given, respectively, by

$$-\mathbf{V}_{ab} + \mathbf{Z}_w\mathbf{I}_1 + \mathbf{Z}_{\Delta 1}(\mathbf{I}_1 - \mathbf{I}_3) + \mathbf{Z}_w(\mathbf{I}_1 - \mathbf{I}_2) = 0$$

$$-\mathbf{V}_{bc} + \mathbf{Z}_w(\mathbf{I}_2 - \mathbf{I}_1) + \mathbf{Z}_{\Delta 2}(\mathbf{I}_2 - \mathbf{I}_3) + \mathbf{Z}_w\mathbf{I}_2 = 0$$

$$\mathbf{Z}_{\Delta 2}(\mathbf{I}_3 - \mathbf{I}_2) + \mathbf{Z}_{\Delta 1}(\mathbf{I}_3 - \mathbf{I}_1) + \mathbf{Z}_{\Delta 3}\mathbf{I}_3 = 0$$

continued

Example 12.8 continued

FIGURE 12.21

Δ-Δ circuit with unbalanced load impedances.

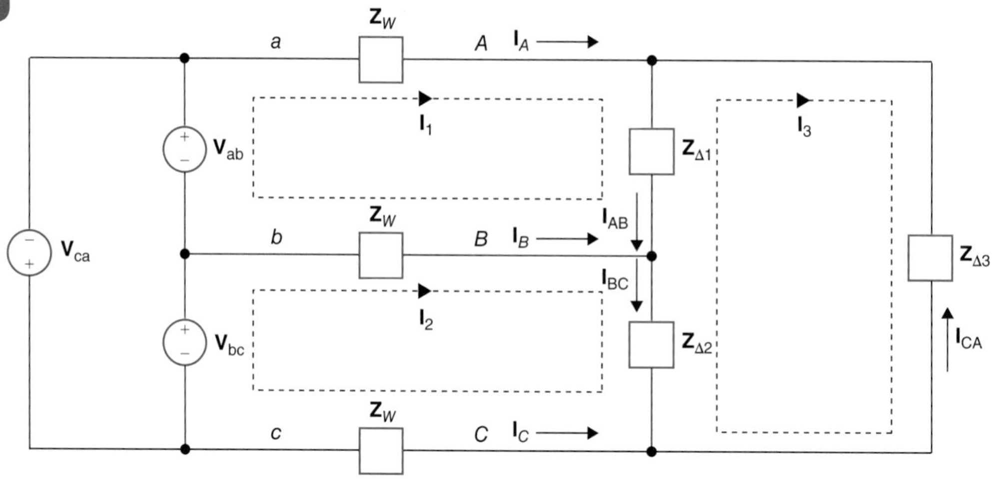

A MATLAB script can be used to find the currents:

MATLAB

```
% EXAMPLE 12.8
clear all;
VP=500;ZW=8+5j;Vab=P2Rd(VP,0);ZD1=P2Rd(25,0);
ZD2=P2Rd(30,0);ZD3=P2Rd(35,0);
Vbc=P2Rd(VP,-120);Vca=P2Rd(VP,-240);
syms I1 I2 I3
[I1,I2,I3]=solve(-Vab+ZW*I1+ZD1*(I1-I3)+ZW*(I1-I2),...
     -Vbc+ZW*(I2-I1)+ZD2*(I2-I3)+ZW*I2,...
     ZD2*(I3-I2)+ZD1*(I3-I1)+ZD3*I3);
IA=I1;
IB=I2-I1;
IC=-I2;
IAB=I1-I3;
IBC=I2-I3;
ICA=-I3;
IAp=R2P(IA);
IBp=R2P(IB);
ICp=R2P(IC);
IABp=R2P(IAB);
IBCp=R2P(IBC);
ICAp=R2P(ICA);
IAp=vpa(IAp,7)
IBp=vpa(IBp,7)
ICp=vpa(ICp,7)
IABp=vpa(IABp,7)
IBCp=vpa(IBCp,7)
ICAp=vpa(ICAp,7)

Answers:
IAp =
[ 15.87175, -42.95465, -0.7497001]
IBp =
[ 16.08302, -167.9248, -2.93084]
```

continued

Example 12.8 continued

MATLAB continued

```
ICp =
[ 14.76365, 73.83309, 1.28863]
IABp =
[ 10.23569, -17.53831, -0.3061012]
IBCp =
[ 8.786207, -132.7784, -2.31742]
ICAp =
[ 7.950656, 103.5037, 1.80648]
```

The answers are given by

$$\mathbf{I}_A = 15.8717\angle -42.9547^\circ\ \text{A},\quad \mathbf{I}_B = 16.0830\angle -167.9248^\circ\ \text{A},$$
$$\mathbf{I}_C = 14.7636\angle 73.8331^\circ\ \text{A}$$

$$\mathbf{I}_{AB} = 10.2357\angle -17.5383^\circ\ \text{A},\quad \mathbf{I}_{BC} = 8.7862\angle -132.7784^\circ\ \text{A},$$
$$\mathbf{I}_{CA} = 7.9507\angle 103.5037^\circ\ \text{A}$$

Exercise 12.8

In a Δ-Δ circuit with balanced sources and unbalanced loads shown in Figure 12.21, the source voltage is $\mathbf{V}_{ab} = 530\angle 0^\circ$ V (rms). The wire impedance is $\mathbf{Z}_w = 9 + j7$, and the load impedances are $\mathbf{Z}_{\Delta 1} = 30\angle 0^\circ\ \Omega$, $\mathbf{Z}_{\Delta 2} = 30\angle 5^\circ\ \Omega$, $\mathbf{Z}_{\Delta 3} = 30\angle 10^\circ\ \Omega$. Find the line currents of the load.

Answer:
$\mathbf{I}_{AB} = 8.7883\angle -18.6779^\circ$ A, $\mathbf{I}_{BC} = 8.3729\angle -142.9935^\circ$ A, $\mathbf{I}_{CA} = 8.6766\angle 94.0324^\circ$ A.

12.6 Balanced Δ-Y Circuit

A balanced Δ-Y circuit is shown in Figure 12.22. When the delta sources are transformed to Y sources, we obtain the circuit shown in Figure 12.23.

FIGURE 12.22

A Δ-Y connection.

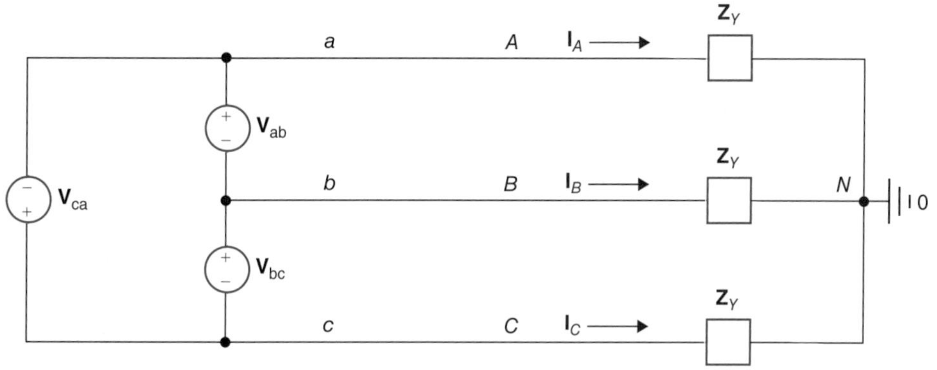

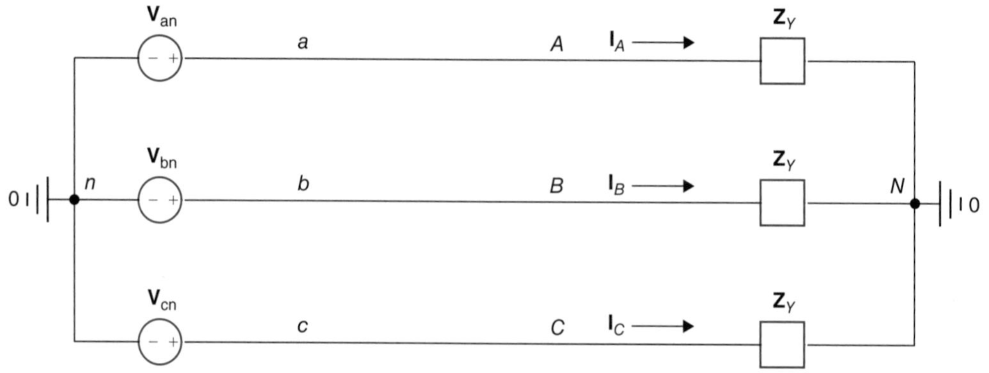

FIGURE 12.23 The Δ-Y circuit after conversion to Y-connected sources.

The impedance of the load, for each phase, is represented as

$$\mathbf{Z}_Y = |\mathbf{Z}_Y|\angle\theta_Y$$

Let the magnitude of the phase voltages at the balanced three-phase delta source be V_p. Then, the phase voltages are given by

$$\mathbf{V}_{ab} = V_P\angle 0°$$

$$\mathbf{V}_{bc} = V_P\angle -120°$$

$$\mathbf{V}_{ca} = V_P\angle -240°$$

When the balanced three-phase delta sources are converted to balanced three-phase wye sources, the phase voltages are given by

$$\mathbf{V}_{an} = \frac{V_P}{\sqrt{3}}\angle -30° \quad (12.134)$$

$$\mathbf{V}_{bn} = \frac{V_P}{\sqrt{3}}\angle -150° \quad (12.135)$$

$$\mathbf{V}_{cn} = \frac{V_P}{\sqrt{3}}\angle -270° = \frac{V_P}{\sqrt{3}}\angle 90° \quad (12.136)$$

The magnitude of the line current is given by

$$I_L = \frac{\frac{V_P}{\sqrt{3}}}{|\mathbf{Z}_Y|} \quad (12.137)$$

The line currents are given by

$$\mathbf{I}_A = \frac{\mathbf{V}_{an}}{\mathbf{Z}_Y} \quad (12.138)$$

$$\mathbf{I}_B = \frac{\mathbf{V}_{bn}}{\mathbf{Z}_Y} \quad (12.139)$$

$$\mathbf{I}_C = \frac{\mathbf{V}_{cn}}{\mathbf{Z}_Y} \quad (12.140)$$

Alternatively, the line currents can be found from

$$\mathbf{I}_A = I_L\angle -30^\circ - \theta_Y \quad \textbf{(12.141)}$$

$$\mathbf{I}_B = I_L\angle -150^\circ - \theta_Y \quad \textbf{(12.142)}$$

$$\mathbf{I}_C = I_L\angle 90^\circ - \theta_Y \quad \textbf{(12.143)}$$

The complex power of the load is given by

$$\mathbf{S}_L = 3I_L^2\mathbf{Z}_Y \quad \textbf{(12.144)}$$

The average power of the load is given by

$$P_L = \mathrm{Re}(\mathbf{S}_L) \quad \textbf{(12.145)}$$

The complex power of the load is given by

$$Q_L = \mathrm{I_m}(\mathbf{S}_L) \quad \textbf{(12.146)}$$

EXAMPLE 12.9

In the balanced Δ-Y connection shown in Figure 12.22, let the magnitude of the phase voltages be $V_P =$ 660 V (rms), the load impedance per phase be $\mathbf{Z}_Y = 60 + j30\ \Omega$, and the phase angle of $\mathbf{V}_{ab}$ be 0°. Find the phase voltages in Δ and Y and line currents. Also, find the complex power, apparent power, average power, reactive power, and power factor of the load.

The delta phase voltages are given by

$$\mathbf{V}_{ab} = V_p\angle 0^\circ = 660\angle 0^\circ\ \mathrm{V}$$

$$\mathbf{V}_{bc} = V_p\angle -120^\circ = 660\angle -120^\circ\ \mathrm{V}$$

$$\mathbf{V}_{ca} = V_p\angle -240^\circ = 660\angle 120^\circ\ \mathrm{V}$$

The Y-connected three-phase sources have the following voltages:

$$\mathbf{V}_{an} = \frac{V_P}{\sqrt{3}}\angle -30^\circ = 381.0512\angle -30^\circ\ \mathrm{V}$$

$$\mathbf{V}_{bn} = \frac{V_P}{\sqrt{3}}\angle -150^\circ = 381.0512\angle -150^\circ\ \mathrm{V}$$

$$\mathbf{V}_{cn} = \frac{V_P}{\sqrt{3}}\angle -270^\circ = \frac{V_P}{\sqrt{3}}\angle 90^\circ = 381.0512\angle 90^\circ\ \mathrm{V}$$

The magnitude of the line voltage is

$$V_L = V_P = 660\ \mathrm{V}$$

The magnitude of the line current is

$$I_L = \frac{\frac{V_P}{\sqrt{3}}}{|\mathbf{Z}_Y|} = 5.6804\ \mathrm{A}$$

continued

Example 12.9 continued

The line currents are given by

$$\mathbf{I}_A = \frac{\mathbf{V}_{an}}{\mathbf{Z}_Y} = 5.6804\angle -56.5651^\circ \text{ A}$$

$$\mathbf{I}_B = \frac{\mathbf{V}_{bn}}{\mathbf{Z}_Y} = 5.6804\angle -176.5651^\circ \text{ A}$$

$$\mathbf{I}_C = \frac{\mathbf{V}_{cn}}{\mathbf{Z}_Y} = 5.6804\angle 63.4349^\circ \text{ A}$$

Alternatively, the line currents can be found from

$$\mathbf{I}_A = I_L\angle -30^\circ - \theta_Y = 5.6804\angle -56.5651^\circ \text{ A}$$

$$\mathbf{I}_B = I_L\angle -150^\circ - \theta_Y = 5.6804\angle -176.5651^\circ \text{ A}$$

$$\mathbf{I}_C = I_L\angle 90^\circ - \theta_Y = 5.6804\angle 63.4349^\circ \text{ A}$$

The complex power of the load is given by

$$\mathbf{S}_L = 3I_L^2\mathbf{Z}_Y = 5808 + j2904 = 6493.5414\angle 26.5651^\circ \text{ VA}$$

The average power of the load is given by

$$P_L = \text{Re}(\mathbf{S}_L) = 5808 \text{ W}$$

The reactive power of the load is given by

$$Q_L = \text{I}_{\text{m}}(\mathbf{S}_L) = 2904 \text{ VAR}$$

The power factor of the load is given by

$$pf = \cos(26.5651^\circ) = 0.8944$$

Exercise 12.9

In the balanced Δ-Y connection shown in Figure 12.22, let the magnitude of the phase voltages be $V_P = 470$ V (rms), the load impedance per phase be $\mathbf{Z}_Y = 57 + j33\ \Omega$, and the phase angle of $\mathbf{V}_{ab}$ be 0°. Find the complex power of the load.

Answer:
$\mathbf{S}_L = 2902.5588 + j1680.4288 = 3353.9064\angle 30.0686^\circ$ VA.

12.6.1 BALANCED Δ-Y CIRCUIT WITH WIRE IMPEDANCE

A Δ-Y connection that includes impedances on the wires $\mathbf{Z}_w$ is shown in Figure 12.24. The three-phase delta source shown in Figure 12.24 can be converted to the three-phase wye source, as shown in Figure 12.25.

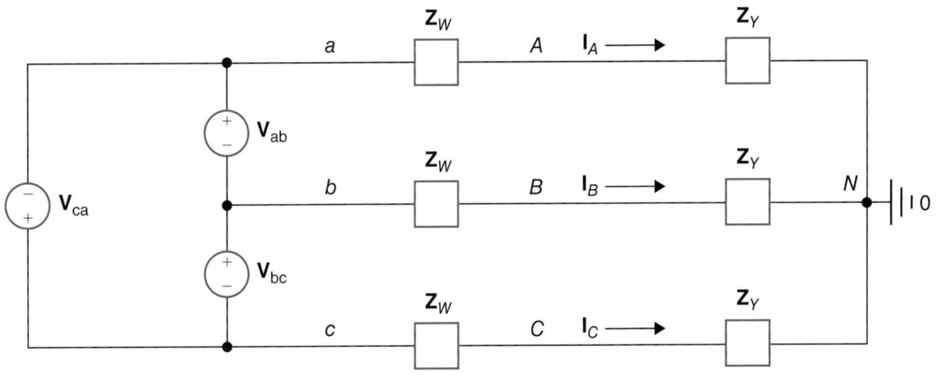

FIGURE 12.24

A Δ-Y circuit with wire impedances.

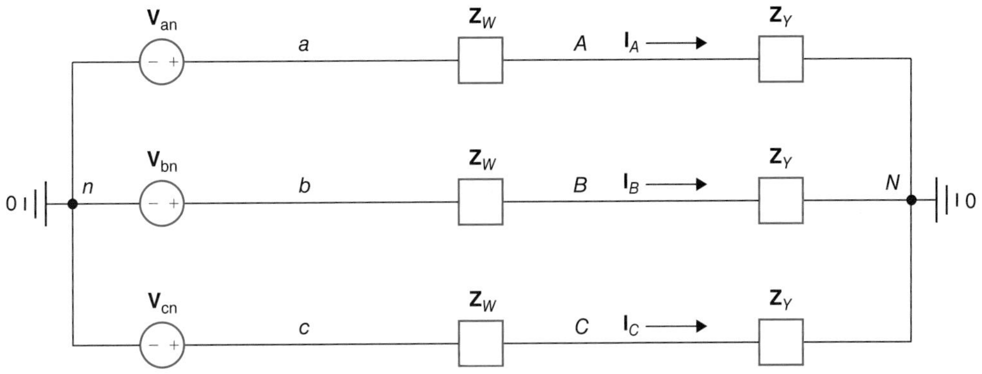

FIGURE 12.25

The Δ-Y circuit after conversion to Y-connected sources.

The phase voltages are given by

$$\mathbf{V}_{ab} = V_P\angle 0°$$

$$\mathbf{V}_{bc} = V_P\angle -120°$$

$$\mathbf{V}_{ca} = V_P\angle -240°$$

The Y-connected three-phase sources have the following voltages:

$$\mathbf{V}_{an} = \frac{V_P}{\sqrt{3}}\angle -30° \tag{12.147}$$

$$\mathbf{V}_{bn} = \frac{V_P}{\sqrt{3}}\angle -150° \tag{12.148}$$

$$\mathbf{V}_{cn} = \frac{V_P}{\sqrt{3}}\angle -270° = \frac{V_P}{\sqrt{3}}\angle 90° \tag{12.149}$$

The total impedance for each voltage phase is given by

$$\mathbf{Z}_T = \mathbf{Z}_W + \mathbf{Z}_Y = |\mathbf{Z}_T|\angle \theta_T \tag{12.150}$$

The magnitude of line currents is

$$I_L = \frac{\frac{V_P}{\sqrt{3}}}{|\mathbf{Z}_T|} \tag{12.151}$$

The line current $\mathbf{I}_A$ is given by

$$\mathbf{I}_A = \frac{\frac{V_P}{\sqrt{3}}\angle -30^\circ}{|\mathbf{Z}_T|\angle\theta_T} = \frac{\frac{V_P}{\sqrt{3}}}{|\mathbf{Z}_T|}\angle -30^\circ - \theta_T = I_L\angle -30^\circ - \theta_T \quad (12.152)$$

Similarly, the other two line currents are given by

$$\mathbf{I}_B = I_L\angle -\theta_T - 150^\circ \quad (12.153)$$

$$\mathbf{I}_C = I_L\angle -\theta_T + 90^\circ \quad (12.154)$$

The voltage between A and N, $\mathbf{V}_{AN}$, across $\mathbf{Z}_Y$ is given by

$$\mathbf{V}_{AN} = \mathbf{I}_A\mathbf{Z}_Y \quad (12.155)$$

Similarly, we have

$$\mathbf{V}_{BN} = \mathbf{I}_B\mathbf{Z}_Y \quad (12.156)$$

$$\mathbf{V}_{CN} = \mathbf{I}_C\mathbf{Z}_Y \quad (12.157)$$

The magnitude of line voltages is given by

$$V_L = \sqrt{3}|\mathbf{V}_{AN}| \quad (12.158)$$

The line voltages are given by

$$\mathbf{V}_{AB} = V_L\angle\mathbf{V}_{AN} + 30^\circ \quad (12.159)$$

$$\mathbf{V}_{BC} = V_L\angle\mathbf{V}_{AN} - 90^\circ \quad (12.160)$$

$$\mathbf{V}_{CA} = V_L\angle\mathbf{V}_{AN} - 210^\circ \quad (12.161)$$

Notice that the line voltages can also be found using

$$\mathbf{V}_{AB} = \mathbf{V}_{AN} - \mathbf{V}_{BN}$$
$$\mathbf{V}_{BC} = \mathbf{V}_{BN} - \mathbf{V}_{CN}$$
$$\mathbf{V}_{CA} = \mathbf{V}_{CN} - \mathbf{V}_{AN}$$

The complex power of the balanced load is given by

$$\mathbf{S}_L = 3I_L^2\mathbf{Z}_Y \quad (12.162)$$

The average power is

$$P_L = \mathrm{Re}(\mathbf{S}_L) \quad (12.163)$$

and the reactive power is

$$Q_L = \mathrm{I_m}(\mathbf{S}_L) \quad (12.164)$$

Notice that the complex power $\mathbf{S}_L$ can also be written as

$$\mathbf{S}_L = \mathbf{V}_{AN}\mathbf{I}_A^* + \mathbf{V}_{BN}\mathbf{I}_B^* + \mathbf{V}_{CN}\mathbf{I}_C^* \quad (12.165)$$

The apparent power of the load is given by

$$|\mathbf{S}_L| = \sqrt{3}V_L I_L \tag{12.166}$$

The complex power of the wires is given by

$$\mathbf{S}_W = 3I_L^2\mathbf{Z}_W \tag{12.167}$$

The total complex power is given by

$$\mathbf{S}_T = \mathbf{S}_W + \mathbf{S}_L = P_T + jQ_T \tag{12.168}$$

The efficiency is given by

$$Eff = \frac{P_L}{P_T} \times 100\% \tag{12.169}$$

EXAMPLE 12.10

In the balanced Δ-Y connection shown in Figure 12.24, let the magnitude of the phase voltages be $V_P = 750$ V (rms), the load impedance per phase be $\mathbf{Z}_Y = 65 + j35\ \Omega$, the wire impedance per phase be $\mathbf{Z}_W = 9 + j8\ \Omega$, and the phase angle of $\mathbf{V}_{ab}$ be 0°. Find the voltages of Δ and Y at the source, the line currents, and line voltages at the load. Find the complex power, apparent power, average power, reactive power, and power factor of the load. Find the complex power of the wire and the total complex power.

The phase voltages are given by

$$\mathbf{V}_{ab} = V_P\angle 0° = 750\angle 0°\ \text{V}$$

$$\mathbf{V}_{bc} = V_P\angle -120° = 750\angle -120°\ \text{V}$$

$$\mathbf{V}_{ca} = V_P\angle -240° = 750\angle -240°\ \text{V}$$

The Y-connected three-phase sources have the following voltages:

$$\mathbf{V}_{an} = \frac{V_P}{\sqrt{3}}\angle -30° = 433.0127\angle -30°\ \text{V}$$

$$\mathbf{V}_{bn} = \frac{V_P}{\sqrt{3}}\angle -150° = 433.0127\angle -150°\ \text{V}$$

$$\mathbf{V}_{cn} = \frac{V_P}{\sqrt{3}}\angle -270° = \frac{V_P}{\sqrt{3}}\angle 90° = 433.0127\angle 90°\ \text{V}$$

The total impedance for each voltage phase is given by

$$\mathbf{Z}_T = \mathbf{Z}_W + \mathbf{Z}_Y = |\mathbf{Z}_T|\angle\theta_T = 74 + j43 = 85.5862\angle 30.1601°\ \Omega$$

The magnitude of line currents is

$$I_L = \frac{\frac{V_P}{\sqrt{3}}}{|\mathbf{Z}_T|} = 5.0594\ \text{A}$$

continued

Example 12.10 continued

The line current $\mathbf{I}_A$ is given by

$$\mathbf{I}_A = \frac{\frac{V_P}{\sqrt{3}}\angle -30°}{|\mathbf{Z}_T|\angle\theta_T} = \frac{\frac{V_P}{\sqrt{3}}}{|\mathbf{Z}_T|}\angle -30° - \theta_T = I_L\angle -30° - \theta_T = 5.0594\angle -60.1601°\text{ A}$$

Similarly, the other two line currents are given by

$$\mathbf{I}_B = I_L\angle -\theta_T - 150° = 5.0594\angle 179.8399°\text{ A}$$

$$\mathbf{I}_C = I_L\angle -\theta_T + 90° = 5.0594\angle 59.8399°\text{ A}$$

The voltage between A and N, $\mathbf{V}_{AN}$, across $\mathbf{Z}_Y$ is given by

$$\mathbf{V}_{AN} = I_A\mathbf{Z}_Y = 373.5038\angle -31.8593°\text{ V}$$

Similarly, we have

$$\mathbf{V}_{BN} = \mathbf{I}_B\mathbf{Z}_Y = 373.5038\angle -151.8593°\text{ V}$$

$$\mathbf{V}_{CN} = \mathbf{I}_C\mathbf{Z}_Y = 373.5038\angle 88.1407°\text{ V}$$

The magnitude of line voltages is given by

$$V_L = \sqrt{3}|\mathbf{V}_{AN}| = 646.9276\text{ V}$$

The line voltages are given by

$$\mathbf{V}_{AB} = V_L\angle\mathbf{V}_{AN} + 30° = 646.9276\angle -1.8593°\text{ V}$$

$$\mathbf{V}_{BC} = V_L\angle\mathbf{V}_{AN} - 90° = 646.9276\angle -121.8593°\text{ V}$$

$$\mathbf{V}_{CA} = V_L\angle\mathbf{V}_{AN} - 210° = 646.9276\angle 118.1407°\text{ V}$$

The complex power of the balanced load is given by

$$\mathbf{S}_L = 3I_L^2\mathbf{Z}_Y = 4991.4676 + j2687.7133 = 5669.0874\angle 28.3008°\text{ VA}$$

The average power is

$$P_L = \text{Re}(\mathbf{S}_L) = 4991.4676\text{ W}$$

and the reactive power is

$$Q_L = \text{I}_\text{m}(\mathbf{S}_L) = 2687.7133\text{ VAR}$$

The apparent power of the load is given by

$$|\mathbf{S}_L| = \sqrt{3}V_L I_L = 5669.0874\text{ VA}$$

The power factor of the load is

$$pf = P_L/|\mathbf{S}_L| = 0.8805$$

continued

Example 12.10 continued

The complex power of the wires is given by

$$\mathbf{S}_W = 3I_L^2\mathbf{Z}_W = 691.1263 + j614.3345 \text{ VA}$$

The total complex power is given by

$$\mathbf{S}_T = \mathbf{S}_W + \mathbf{S}_L = P_T + jQ_T = 5682.5939 + j3302.0478 = 6572.3202\angle 30.1601 \text{ VA}$$

Exercise 12.10

In the balanced Δ-Y connection shown in Figure 12.24, let the magnitude of the phase voltages be $V_P = 970$ V (rms), the load impedance per phase be $\mathbf{Z}_Y = 86 + j41\ \Omega$, the wire impedance per phase be $\mathbf{Z}_W = 12 + j9\ \Omega$, and the phase angle of $\mathbf{V}_{ab}$ be 0°. Find the power factor of the load.

Answer:
$pf = 0.9027.$

12.7 PSpice and Simulink

The PSpice schematic for the simulation of the balanced Y-Y connection discussed in EXAMPLE 12.3 is shown in Figure 12.26. Since $\mathbf{Z}_w = 9 + j7\ \Omega$ and $\mathbf{Z}_Y = 70 + j55\ \Omega$, we have

$$R_1 = R_2 = R_3 = 9\ \Omega, \quad R_4 = R_5 = R_6 = 70\ \Omega$$

$$L_1 = L_2 = L_3 = 7/(2\pi \times 60) = 18.5680767 \text{ mH}$$

$$L_4 = L_5 = L_6 = 55/(2\pi \times 60) = 145.8920312 \text{ mH}$$

FIGURE 12.26

A PSpice schematic to simulate a Y-Y circuit.

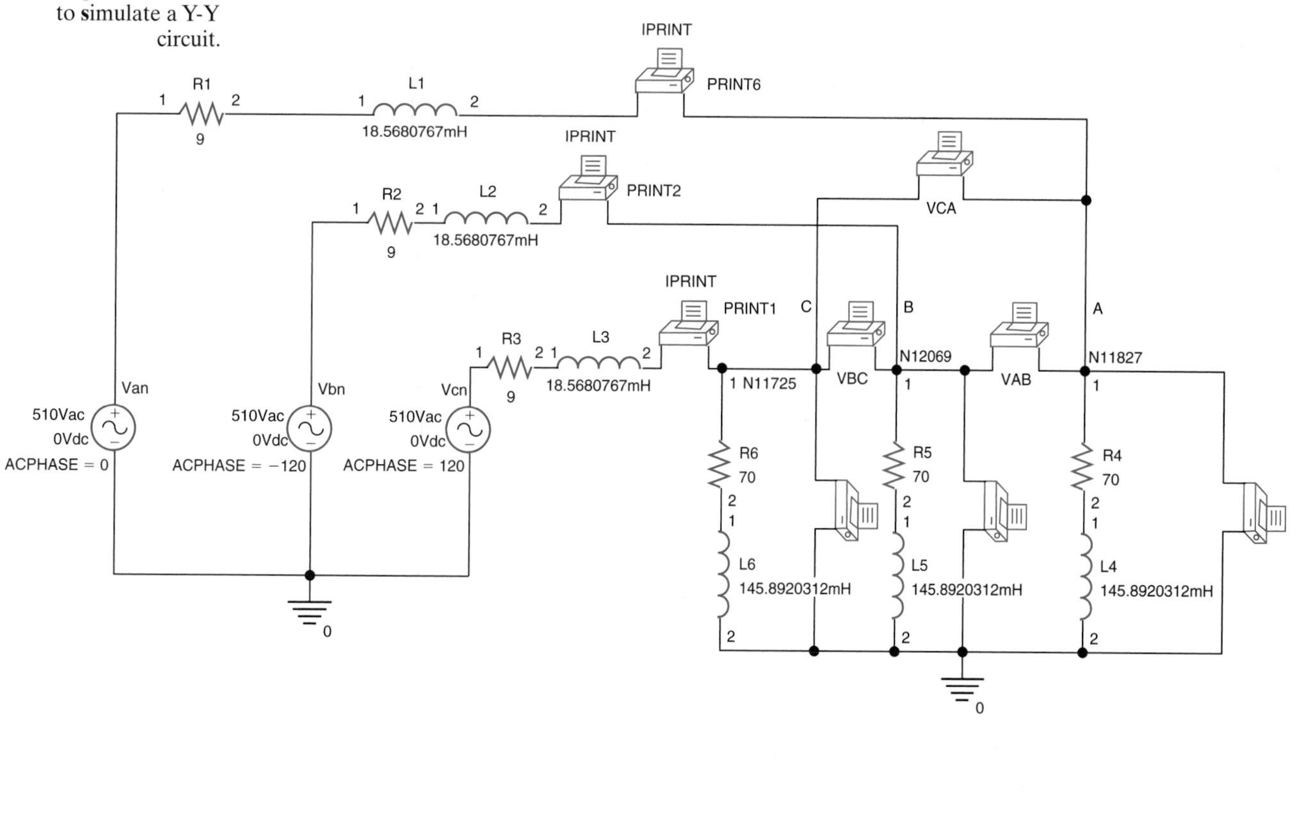

The results of the simulation are available in the output file. The values of $\mathbf{I}_A$, $\mathbf{I}_B$, $\mathbf{I}_C$, $\mathbf{V}_{AN}$, $\mathbf{V}_{BN}$, $\mathbf{V}_{CN}$, $\mathbf{V}_{AB}$, $\mathbf{V}_{BC}$, and $\mathbf{V}_{CA}$ are shown here. The values match those given in EXAMPLE 12.3. The notations are as follows:

IM = magnitude of current, IP = phase of current, IR = real part of current, and II = imaginary part of current

VM = magnitude of voltage, VP = phase of voltage, VR = real part of voltage, and VI = imaginary part of voltage

```
IA
FREQ         IM(V_PRINT6)   IP(V_PRINT6)  IR(V_PRINT6)  II(V_PRINT6)
6.000E+01    5.078E+00      -3.813E+01    3.995E+00     -3.135E+00

IB
FREQ         IM(V_PRINT2)   IP(V_PRINT2)  IR(V_PRINT2)  II(V_PRINT2)
6.000E+01    5.078E+00      -1.581E+02    -4.713E+00    -1.892E+00

IC
FREQ         IM(V_PRINT1)   IP(V_PRINT1)  IR(V_PRINT1)  II(V_PRINT1)
6.000E+01    5.078E+00      8.187E+01     7.178E-01     5.027E+00

VAN
FREQ         VM(N11827,0)   VP(N11827,0)  VR(N11827,0)  VI(N11827,0)
6.000E+01    4.521E+02      3.204E-02     4.521E+02     2.529E-01

VBN
FREQ         VM(N12069,0)   VP(N12069,0)  VR(N12069,0)  VI(N12069,0)
6.000E+01    4.521E+02      -1.200E+02    -2.258E+02    -3.917E+02

VCN
FREQ         VM(N11725,0)   VP(N11725,0)  VR(N11725,0)  VI(N11725,0)
6.000E+01    4.521E+02      1.200E+02     -2.263E+02     3.914E+02

VAB
FREQ       VM(N11827,N12069)  VP(N11827,N12069)  VR(N11827,N12069)  VI(N11827,N12069)
6.000E+01  7.831E+02          3.003E+01          6.779E+02          3.919E+02

VBC
FREQ       VM(N12069,N11725)  VP(N12069,N11725)  VR(N12069,N11725)  VI(N12069,N11725)
6.000E+01  7.831E+02          -8.997E+01         4.380E-01          -7.831E+02

VCA
FREQ       VM(N11725,N11827)  VP(N11725,N11827)  VR(N11725,N11827)  VI(N11725,N11827)
6.000E+01  7.831E+02          1.500E+02          -6.784E+02         3.911E+02
```

The PSpice schematic for the simulation of the Y-Δ connection discussed in EXAMPLE 12.5 is shown in Figure 12.27. Since $\mathbf{Z}_w = 5 + j3\ \Omega$ and $\mathbf{Z}_\Delta = 75 + j54\ \Omega$, we have

$$R_1 = R_2 = R_3 = 5\ \Omega, \quad R_4 = R_5 = R_6 = 75\ \Omega$$

$$L_1 = L_2 = L_3 = 3/(2\pi \times 60) = 7.9577471546 \text{ mH}$$

$$L_4 = L_5 = L_6 = 54/(2\pi \times 60) = 143.2394488 \text{ mH}$$

FIGURE 12.27

A PSpice schematic to simulate a Y-Δ circuit.

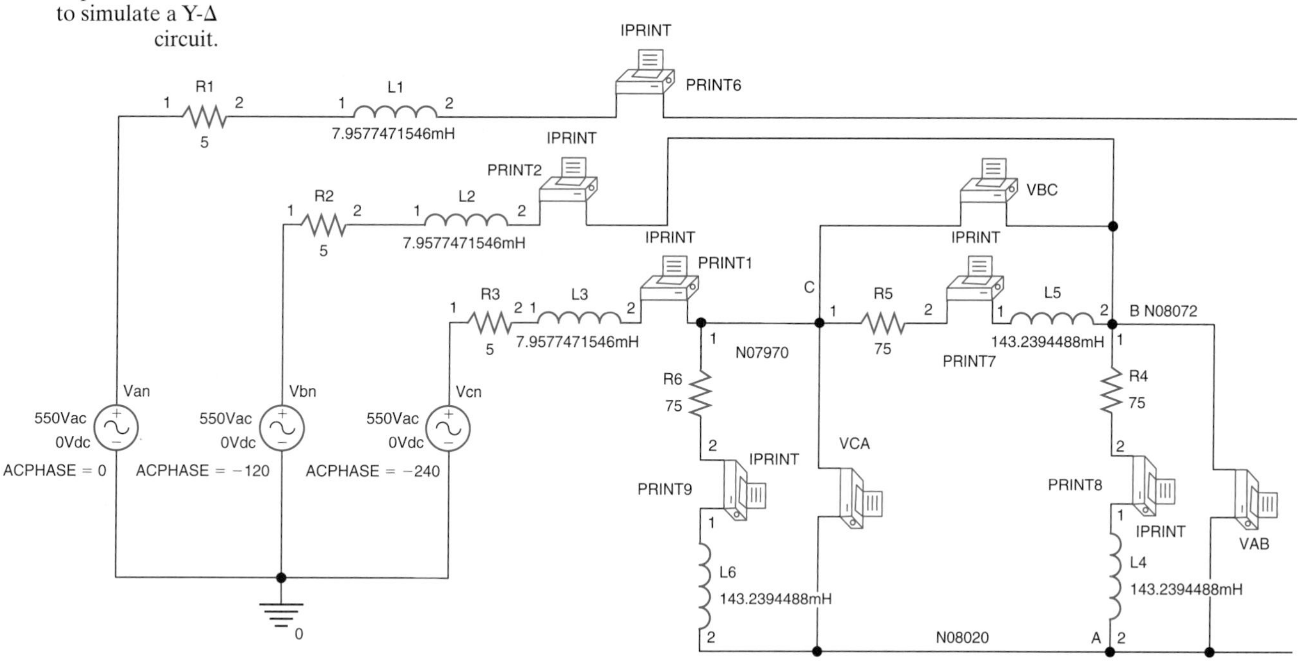

The results of the simulation are available in the output file. The values of $\mathbf{I}_A$, $\mathbf{I}_B$, $\mathbf{I}_C$, $\mathbf{V}_{AB}$, $\mathbf{V}_{BC}$, $\mathbf{V}_{CA}$, $\mathbf{I}_{AB}$, $\mathbf{I}_{BC}$, and $\mathbf{I}_{CA}$ are shown here. The values match those given in EXAMPLE 12.5. The notations are as follows:

IM = magnitude of current, IP = phase of current, IR = real part of current, and II = imaginary part of current

VM = magnitude of voltage, VP = phase of voltage, VR = real part of voltage, and VI = imaginary part of voltage

I_A

FREQ	IM(V_PRINT6)	IP(V_PRINT6)	IR(V_PRINT6)	II(V_PRINT6)
6.000E+01	1.502E+01	-3.499E+01	1.230E+01	-8.613E+00

I_B

FREQ	IM(V_PRINT2)	IP(V_PRINT2)	IR(V_PRINT2)	II(V_PRINT2)
6.000E+01	1.502E+01	-1.550E+02	-1.361E+01	-6.349E+00

I_C

FREQ	IM(V_PRINT1)	IP(V_PRINT1)	IR(V_PRINT1)	II(V_PRINT1)
6.000E+01	1.502E+01	8.501E+01	1.307E+00	1.496E+01

V_{AB}

FREQ	VM(N08020,N08072)	VP(N08020,N08072)	VR(N08020,N08072)	VI(N08020,N08072)
6.000E+01	8.014E+02	3.076E+01	6.886E+02	4.099E+02

V_{BC}

FREQ	VM(N08072,N07970)	VP(N08072,N07970)	VR(N08072,N07970)	VI(N08072,N07970)
6.000E+01	8.014E+02	-8.924E+01	1.066E+01	-8.013E+02

V_{CA}

FREQ	VM(N07970,N08020)	VP(N07970,N08020)	VR(N07970,N08020)	VI(N07970,N08020)
6.000E+01	8.014E+02	1.508E+02	-6.993E+02	3.914E+02

I_{AB}

FREQ	IM(V_PRINT8)	IP(V_PRINT8)	IR(V_PRINT8)	II(V_PRINT8)
6.000E+01	8.671E+00	-4.992E+00	8.638E+00	-7.546E-01

I_{BC}

FREQ	IM(V_PRINT7)	IP(V_PRINT7)	IR(V_PRINT7)	II(V_PRINT7)
6.000E+01	8.671E+00	-1.250E+02	-4.973E+00	-7.104E+00

I_{CA}

FREQ	IM(V_PRINT9)	IP(V_PRINT9)	IR(V_PRINT9)	II(V_PRINT9)
6.000E+01	8.671E+00	1.150E+02	-3.666E+00	7.858E+00

A Simulink model to simulate the Y-Y circuit in EXAMPLE 12.3 is shown in Figure 12.28. The source resistance and the source inductance are set to zero in this model. Since the three-phase source requires phase-to-phase rms voltage, the phase voltage 510 V (rms) is multiplied by $\sqrt{3}$. All the voltages and currents match those derived from direct calculation.

FIGURE 12.28

A Simulink model for the Y-Y circuit.

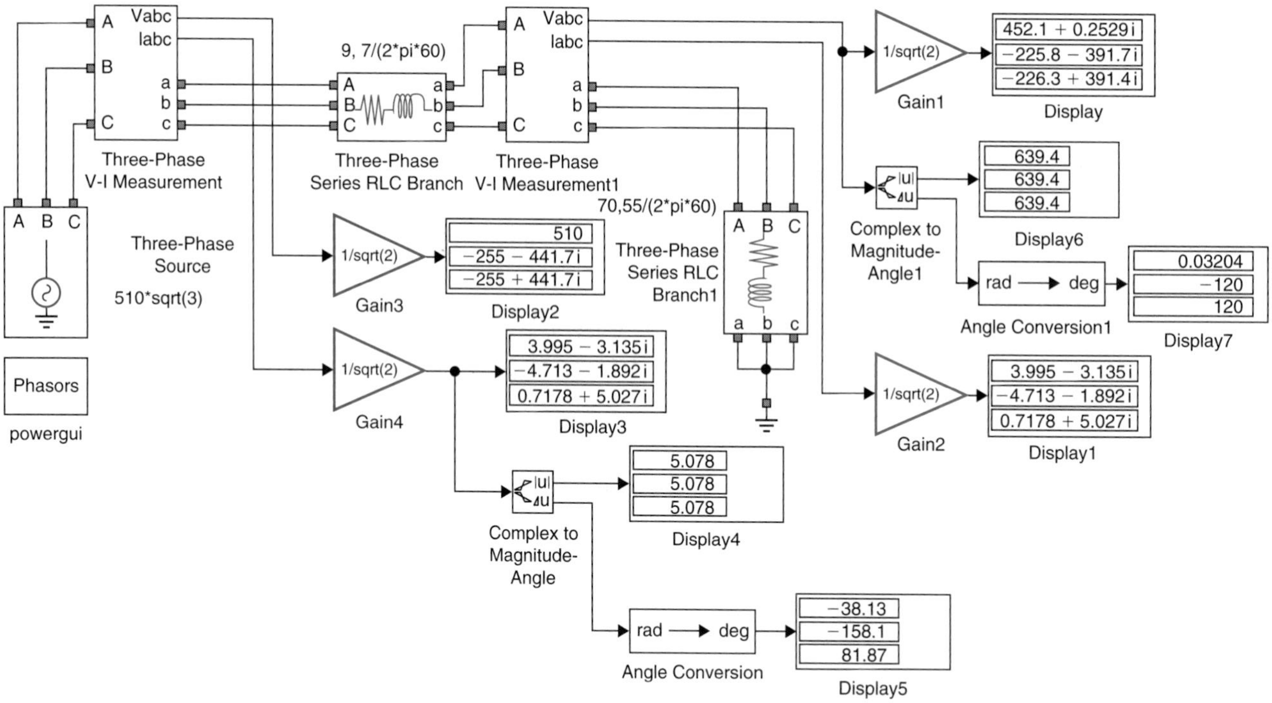

SUMMARY

In this chapter, circuits that connect the balanced three-phase sources and the balanced three-phase loads are treated. The source can be a Y shape or a Δ shape, and the load can be a Y shape or a Δ shape. There are four combinations to connect the source to the load: Y-Y, Y-Δ, Δ-Y, and Δ-Δ. There are conversions between Y and Δ at the source and load. There are wire impedances and the internal impedances at the sources.

In the Y-Y circuit, there are three independent circuits connecting three sources to the three loads. If the circuit is balanced, the magnitude of voltages and currents are the same at the same point for all three phases, but the phase angles are separated by 120°. For the Y-Δ circuit, the Δ-shaped load can be converted to a Y-shaped load, and the same method as the Y-Y circuit can be applied. The Δ-Δ circuit and the Δ-Y circuit also can be converted to a Y-Y circuit.

PROBLEMS

Three-Phase Sources

12.1 **Find the phase voltages $\mathbf{V}_{bn}$ and $\mathbf{V}_{cn}$ and the line voltages $\mathbf{V}_{ab}$, $\mathbf{V}_{bc}$, and $\mathbf{V}_{ca}$ if $\mathbf{V}_{an} = 460\angle 0°$ V (rms) for the balanced three-phase sources.**

12.2 **Find the phase voltages $\mathbf{V}_{an}$, $\mathbf{V}_{bn}$, and $\mathbf{V}_{cn}$ and the line voltages $\mathbf{V}_{bc}$ and $\mathbf{V}_{ca}$ if $\mathbf{V}_{ab} = 360\angle 0°$ V (rms) for the balanced three-phase sources.**

12.3 **If $\mathbf{V}_{bn} = 250\angle -150°$ V, what are $\mathbf{V}_{an}$ and $\mathbf{V}_{ab}$ for the balanced three-phase sources?**

12.4 **If $\mathbf{V}_{ca} = 270\angle -110°$ V, what is $\mathbf{V}_{an}$ for the balanced three-phase sources?**

12.5 **The voltages of Y sources are given by $\mathbf{V}_{an} = 100\angle 0°$ V(rms), $\mathbf{V}_{bn} = 100\angle -115°$ V (rms), $\mathbf{V}_{cn} = 100\angle -230°$ V (rms). Find the voltages $\mathbf{V}_{ab}$, $\mathbf{V}_{bc}$, and $\mathbf{V}_{ca}$ of the Δ sources.**

12.6 **The voltages of Y sources are given by $\mathbf{V}_{an} = 100\angle 0°$ V (rms), $\mathbf{V}_{bn} = 105\angle -120°$ V (rms), $\mathbf{V}_{cn} = 110\angle -240°$ V (rms). Find the voltages $\mathbf{V}_{ab}$, $\mathbf{V}_{bc}$, and $\mathbf{V}_{ca}$ of the Δ sources.**

Y-Y Circuit

12.7 **In a balanced Y-Y circuit shown in Figure P12.7, the magnitude of the line voltage is $V_L = 410$ V (rms). The phase impedance is $\mathbf{Z}_Y = 72 + j42\ \Omega$ per phase. Find $\mathbf{I}_A$, $\mathbf{V}_{AN}$, and $\mathbf{V}_{AB}$, assuming that $\angle\mathbf{V}_{an} = 0°$.**

FIGURE P12.7

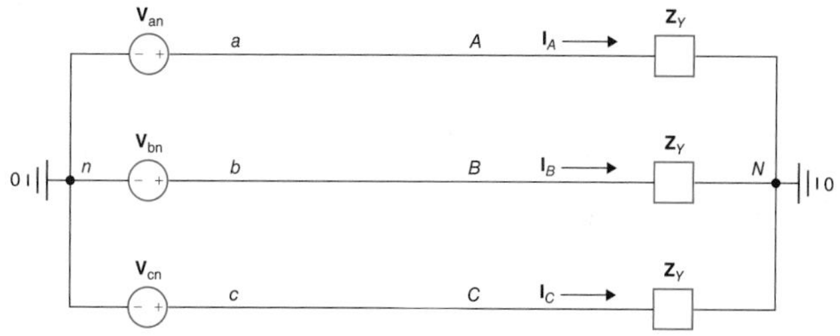

12.8 **In the balanced Y-Y circuit shown in Figure P12.8, the magnitude of the line voltage is $V_L = 420$ V (rms). The phase impedance is $\mathbf{Z}_Y = 69 + j36\ \Omega$ per phase, and the wire impedance is $9 + j6$ per phase. Find $\mathbf{I}_A$, $\mathbf{V}_{AN}$ and $\mathbf{V}_{AB}$, assuming that $\angle\mathbf{V}_{an} = 0°$.**

FIGURE P12.8

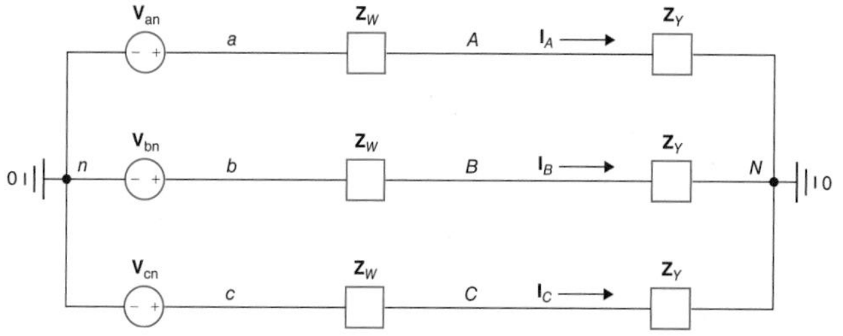

12.9 **In a balanced Y-Y circuit shown in Figure P12.9, the magnitude of the line voltage is $V_L = 430$ V (rms). The phase impedance is $\mathbf{Z}_Y = 69 + j36\ \Omega$ per phase, the wire impedance is $9 + j6$ per phase, and the source impedance is $7 + j8$ per phase. Find $\mathbf{I}_A$, $\mathbf{V}_{AN}$, and $\mathbf{V}_{AB}$, assuming that $\angle\mathbf{V}_{an} = 0°$.**

FIGURE P12.9

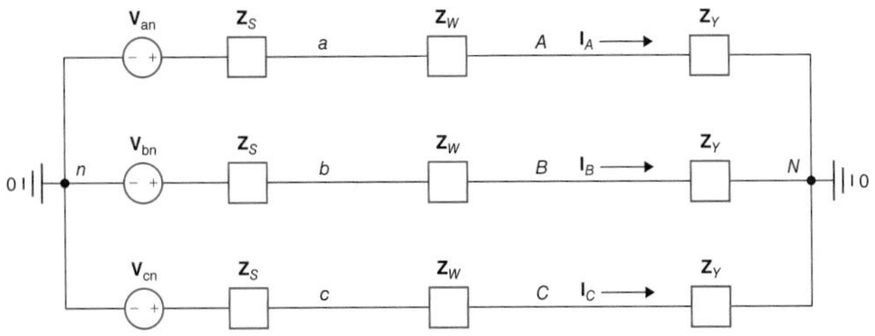

12.10 In a balanced Y-Y circuit shown in Figure P12.10, the magnitude of the phase voltage is $V_P = 550$ V (rms). The phase impedance is $\mathbf{Z}_Y = 72 + j48\ \Omega$ per phase, and the wire impedance is $\mathbf{Z}_W = 12 + j7$ per wire. Find $\mathbf{I}_A$ and $\mathbf{V}_{AB}$, assuming that $\angle \mathbf{V}_{an} = 0°$.

FIGURE P12.10

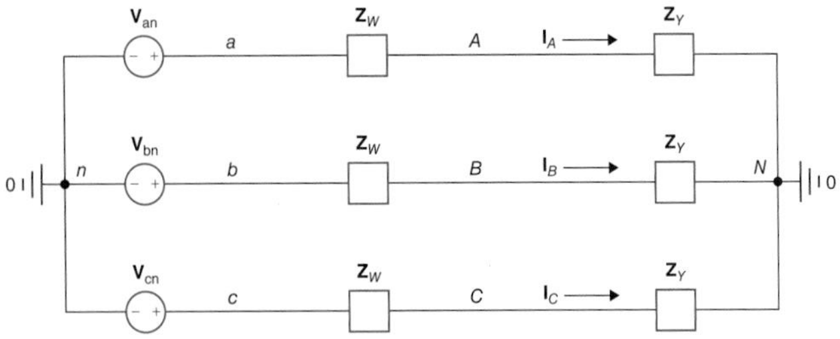

12.11 In the Y-Y circuit shown in Figure P12.11, the phase voltages are given by $\mathbf{V}_{an} = 200\angle 0°$ V (rms), $\mathbf{V}_{bn} = 200\angle -120°$ V (rms), and $\mathbf{V}_{cn} = 200\angle 120°$ V (rms). The source impedance is $5 + j7$ per phase, and the wire impedance is $8 + j10$ per phase. Let the load impedances be $\mathbf{Z}_{Y1} = 50\angle 0°\ \Omega$, $\mathbf{Z}_{Y2} = 50\angle 10°\ \Omega$, and $\mathbf{Z}_{Y3} = 50\angle 20°\ \Omega$. Find $\mathbf{I}_A$, $\mathbf{I}_B$, $\mathbf{I}_C$, $\mathbf{V}_{AN}$, $\mathbf{V}_{BN}$, and $\mathbf{V}_{CN}$, and $\mathbf{V}_{AB}$, $\mathbf{V}_{BC}$, and $\mathbf{V}_{CA}$.

FIGURE P12.11

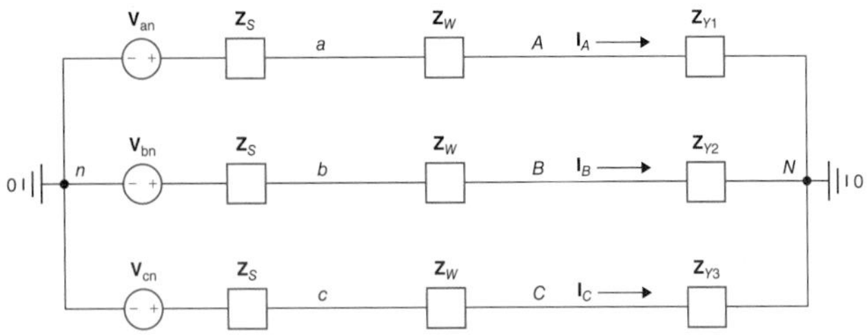

12.12 In the Y-Y circuit shown in Figure P12.12, the phase voltages are given by $\mathbf{V}_{an} = 200\angle 0°$ V (rms), $\mathbf{V}_{bn} = 200\angle -120°$ V (rms), and $\mathbf{V}_{cn} = 200\angle 120°$ V (rms). The source impedance is $5 + j7$ per phase, and the wire impedance is $8 + j10$ per phase. Let the load impedances be $\mathbf{Z}_{Y1} = 50\angle 0°\ \Omega$, $\mathbf{Z}_{Y2} = 60\angle 0°\ \Omega$, $\mathbf{Z}_{Y3} = 70\angle 0°\ \Omega$. Find $\mathbf{I}_A$, $\mathbf{I}_B$, $\mathbf{I}_C$, $\mathbf{V}_{AN}$, $\mathbf{V}_{BN}$, and $\mathbf{V}_{CN}$, and $\mathbf{V}_{AB}$, $\mathbf{V}_{BC}$, and $\mathbf{V}_{CA}$.

FIGURE P12.12

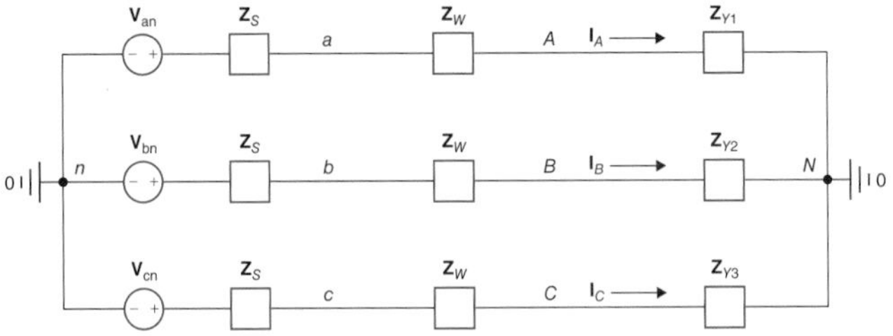

12.13 In the Y-Y circuit shown in Figure P12.13, the phase voltages are given by $\mathbf{V}_{an} = 200\angle 0°$ V (rms), $\mathbf{V}_{bn} = 200\angle -130°$ V (rms), and $\mathbf{V}_{cn} = 200\angle 100°$ V (rms). The source impedance is $5 + j7$ per phase, and the wire impedance is $8 + j10$ per phase. Let the load impedances be $\mathbf{Z}_{Y1} = 50\angle 0°\ \Omega$, $\mathbf{Z}_{Y2} = 50\angle 0°\ \Omega$, $\mathbf{Z}_{Y3} = 50\angle 0°\ \Omega$. Find $\mathbf{I}_A$, $\mathbf{I}_B$, $\mathbf{I}_C$, $\mathbf{V}_{AN}$, $\mathbf{V}_{BN}$, and $\mathbf{V}_{CN}$, and $\mathbf{V}_{AB}$, $\mathbf{V}_{BC}$, and $\mathbf{V}_{CA}$.

FIGURE P12.13

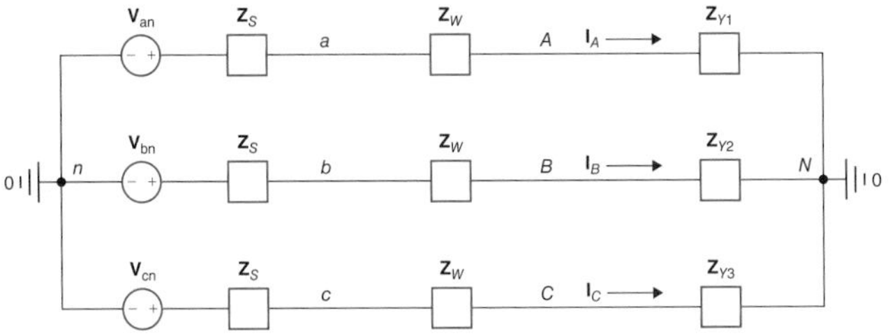

Y-Δ Circuit

12.14 In the balanced Y-Δ circuit shown in Figure P12.14, the magnitude of the line voltage is $V_L = 520$ V (rms). The phase impedance of the load is $\mathbf{Z}_\Delta = 48 + j24\ \Omega$ per phase. Find the line current and phase current, assuming that $\angle \mathbf{V}_{an} = 0°$.

FIGURE P12.14

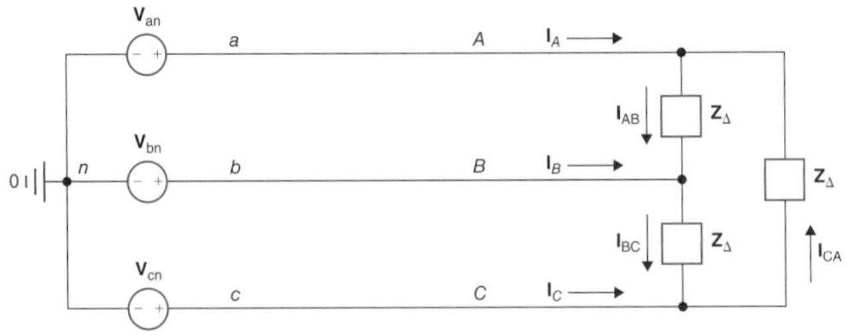

12.15 In the balanced Y-Δ circuit shown in Figure P12.15, the magnitude of the phase voltage is $V_P = 680$ V (rms). The phase impedance of the load is $\mathbf{Z}_\Delta = 69 + j36\ \Omega$ per phase. The wire impedance is $\mathbf{Z}_W = 8 + j6$ per wire. Find the line current and phase current, assuming that $\angle \mathbf{V}_{an} = 0°$.

FIGURE P12.15

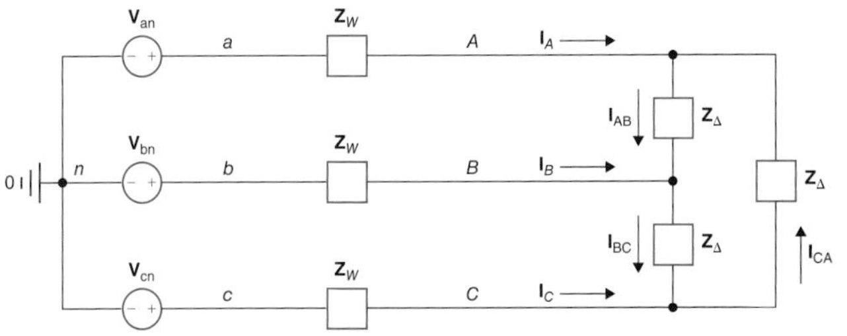

12.16 In a Y-Δ circuit shown in Figure P12.16, the magnitude of the phase voltage is $V_P = 640$ V (rms). The phase impedances of the load are $\mathbf{Z}_{\Delta 1} = 50\angle 0°\ \Omega$, $\mathbf{Z}_{\Delta 2} = 50\angle 5°\ \Omega$, $\mathbf{Z}_{\Delta 3} = 50\angle 10°\ \Omega$. The wire impedance is $\mathbf{Z}_W = 8 + j7$ per wire. Find the line current and phase current, assuming that $\angle \mathbf{V}_{an} = 0°$.

FIGURE P12.16

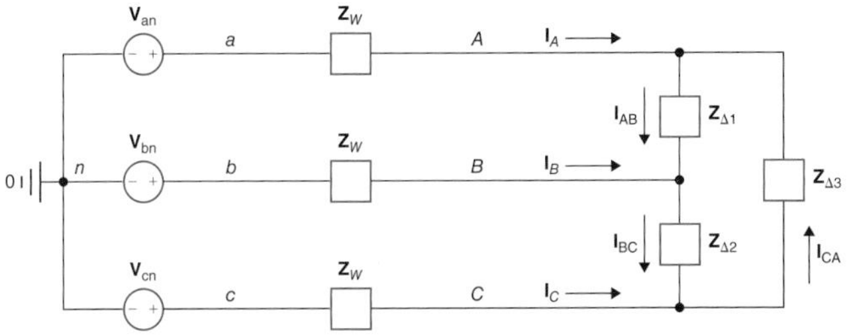

12.17 In a Y-Δ circuit shown in Figure P12.17, the magnitude of the phase voltage is $V_P = 620$ V (rms). The load impedances of the load are $\mathbf{Z}_{\Delta 1} = 40\angle 0°\ \Omega$, $\mathbf{Z}_{\Delta 2} = 45\angle 0°\ \Omega$, $\mathbf{Z}_{\Delta 3} = 50\angle 0°\ \Omega$. The wire impedance is $\mathbf{Z}_W = 8 + j7$ per wire. Find the line current and phase current, assuming that $\angle \mathbf{V}_{an} = 0°$.

FIGURE P12.17

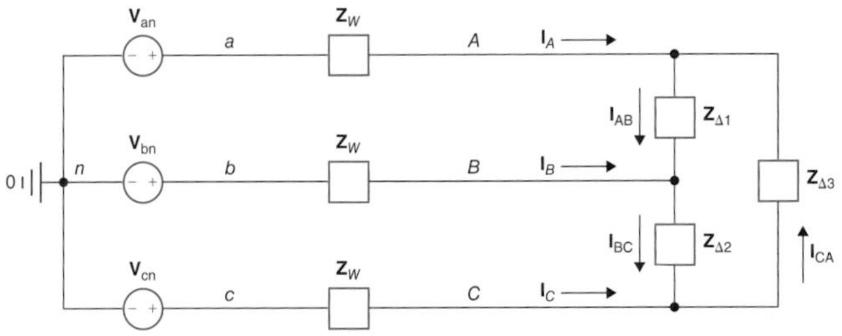

12.18 In a Y-Δ circuit shown in Figure P12.18, the phase voltages are given by $\mathbf{V}_{an} = 630\angle 0°$ V (rms), $\mathbf{V}_{bn} = 600\angle -125°$ V (rms), and $\mathbf{V}_{cn} = 600\angle 110°$ V (rms). The phase impedances of the load are $\mathbf{Z}_{\Delta 1} = 40\angle 0°\ \Omega$, $\mathbf{Z}_{\Delta 2} = 40\angle 0°\ \Omega$, $\mathbf{Z}_{\Delta 3} = 40\angle 0°\ \Omega$. The wire impedance is $\mathbf{Z}_W = 8 + j6$ per wire. Find the line current and phase current, assuming that $\angle \mathbf{V}_{an} = 0°$.

FIGURE P12.18

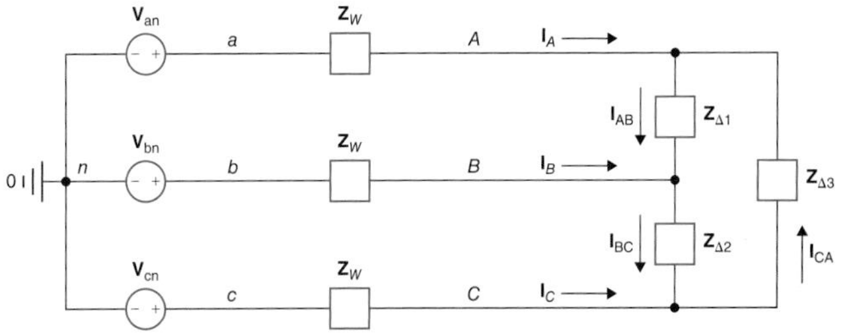

Δ-Δ Circuit

12.19 In the balanced Δ-Δ circuit shown in Figure P12.19, the source voltage is $\mathbf{V}_{ab} = 260\angle 0°$ V (rms). The phase impedance of the load is $\mathbf{Z}_\Delta = 24 + j12\ \Omega$ per phase. Find the phase currents, line currents, and complex power of the load.

FIGURE P12.19

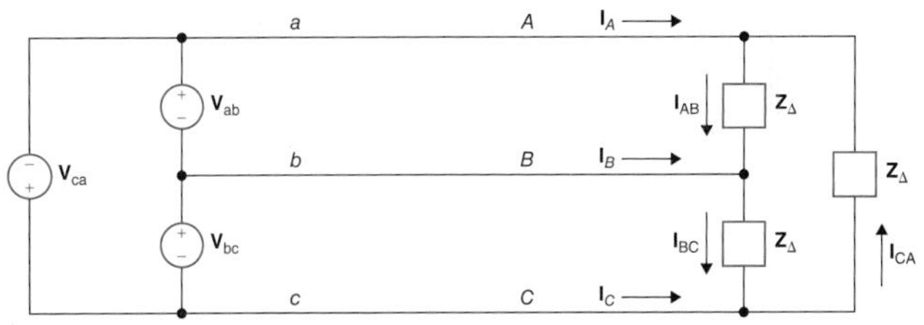

12.20 In the balanced Δ-Δ circuit shown in Figure P12.20, the source voltage is $\mathbf{V}_{ab} = 500\angle 0°$ V (rms). The wire impedance is $\mathbf{Z}_w = 7 + j5$, and the phase impedance is $\mathbf{Z}_\Delta = 36 + j24\ \Omega$ per phase. Find the phase currents, line currents, and complex power of the load.

FIGURE P12.20

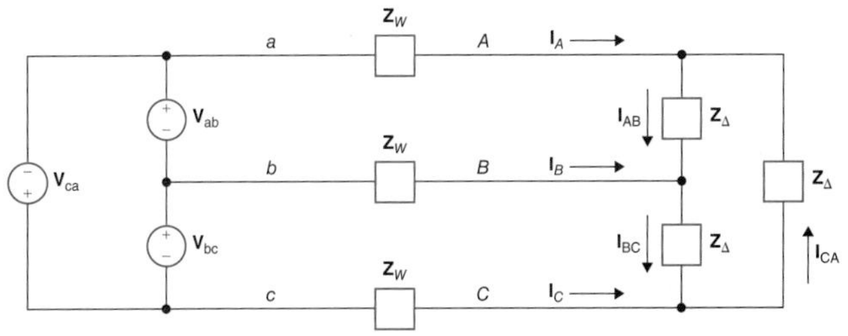

12.21 In the Δ-Δ circuit shown in Figure P12.21, the source voltage is $\mathbf{V}_{ab} = 420\angle 0°$ V (rms). The wire impedance is $\mathbf{Z}_w = 7 + j5$, and the load impedances are $\mathbf{Z}_{\Delta 1} = 50\angle 0°\ \Omega$, $\mathbf{Z}_{\Delta 2} = 50\angle 15°\ \Omega$, $\mathbf{Z}_{\Delta 3} = 50\angle 30°\ \Omega$. Find the line currents, phase currents, and complex power of the load.

FIGURE P12.21

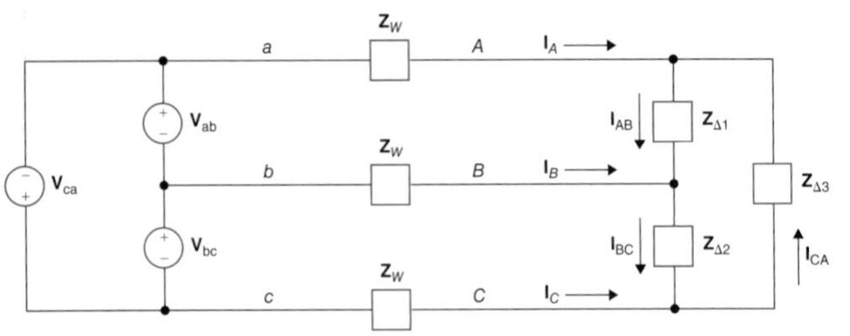

12.22 In the Δ-Δ circuit shown in Figure P12.22, the source voltage is $\mathbf{V}_{ab} = 360\angle 0°$ V (rms). The wire impedance is $\mathbf{Z}_w = 7 + j6$, and the load impedances are $\mathbf{Z}_{\Delta 1} = 30\angle 0°\ \Omega$, $\mathbf{Z}_{\Delta 2} = 35\angle 0°\ \Omega$, $\mathbf{Z}_{\Delta 3} = 40\angle 0°\ \Omega$. Find the phase currents, line currents, and complex power of the load.

FIGURE P12.22

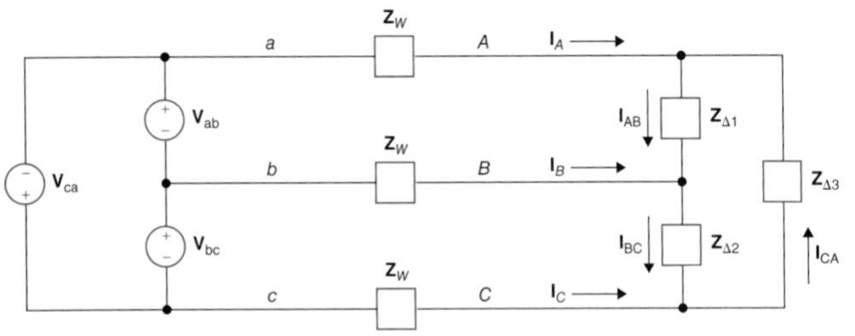

Δ-Y Circuit

12.23 In the balanced Δ-Y circuit shown in Figure P12.23, the magnitude of the phase voltage is $V_P = 450$ V (rms). The phase impedance is $\mathbf{Z}_Y = 27 + j16\ \Omega$ per phase. Find the line currents and voltages of the equivalent Y-connected source, assuming that $\angle\mathbf{V}_{ab} = 0°$. Also, find the complex power of the load.

FIGURE P12.23

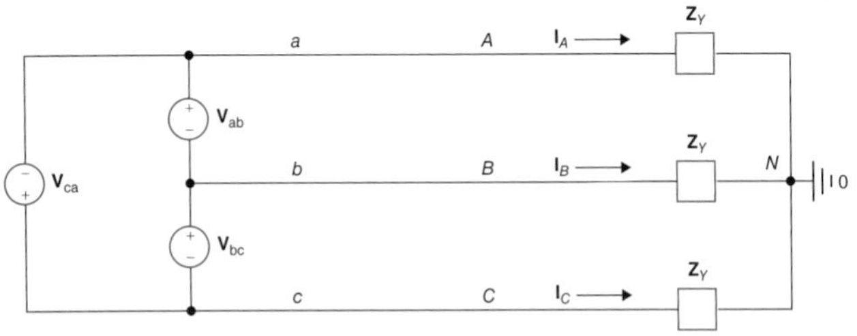

12.24 In the balanced Δ-Y circuit shown in Figure P12.24, the magnitude of the phase voltage is $V_P = 460$ V (rms). The wire impedance is $\mathbf{Z}_W = 9 + j5$, and the phase impedance is $\mathbf{Z}_Y = 25 + j12\ \Omega$ per phase. Find the line currents and voltages of the equivalent Y-connected source, assuming that $\angle\mathbf{V}_{ab} = 0°$. Also, find the complex power of the load.

FIGURE P12.24

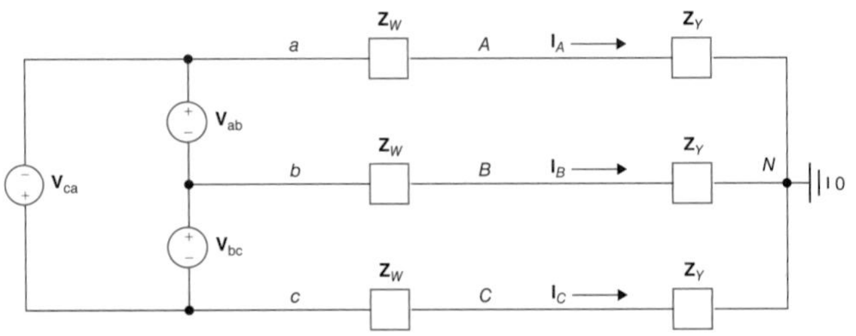

12.25 In the Δ-Y circuit shown in Figure P12.25, the magnitude of the phase voltage is 470 V (rms) for the balanced sources with $\angle\mathbf{V}_{ab} = 0°$. The wire impedance is $\mathbf{Z}_w = 12 + j8$ and the load impedances are $\mathbf{Z}_{Y1} = 30\angle 0°$, $\mathbf{Z}_{Y2} = 30\angle 10°$, $\mathbf{Z}_{Y3} = 30\angle 20°$. Find the line currents $\mathbf{I}_A$, $\mathbf{I}_B$, and $\mathbf{I}_C$ and the voltages across the load impedances. Also, find the complex power of each load.

FIGURE P12.25

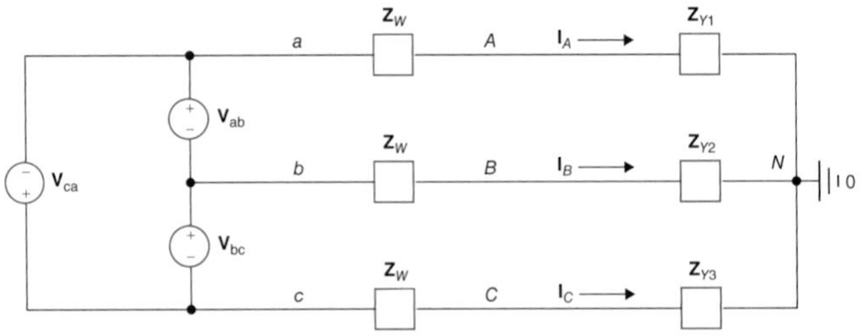

12.26 In the Δ-Y circuit shown in Figure P12.26, the magnitude of the phase voltage is 480 V (rms) for the balanced sources with $\angle\mathbf{V}_{ab} = 0°$. The wire impedance is $\mathbf{Z}_w = 11 + j7$ and the load impedances are $\mathbf{Z}_{Y1} = 30\angle 0°$, $\mathbf{Z}_{Y2} = 35\angle 0°$, $\mathbf{Z}_{Y3} = 40\angle 0°$. Find the line currents $\mathbf{I}_A$, $\mathbf{I}_B$, and $\mathbf{I}_C$ and the voltages across the load impedances. Also, find the complex power of each load.

FIGURE P12.26

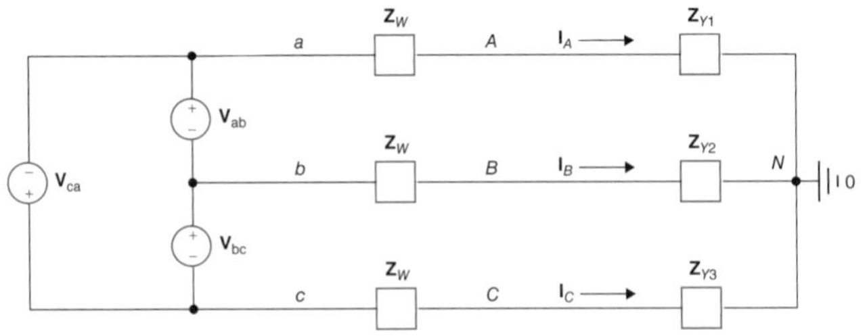

Magnetically Coupled Circuits

13.1 Introduction

When two coils, coil 1 and coil 2, are placed in proximity, the magnetic flux from coil 1 links with coil 1, but some of the magnetic flux also links with coil 2. When the magnetic flux changes with time, in addition to the voltage induced in coil 1, a voltage is also induced in coil 2. Unless all the magnetic flux that links coil 1 also links coil 2, the induced voltage at coil 2 will be smaller than the voltage induced at coil 1. The induced voltage at coil 1 is proportional to the self-inductance L_1 of coil 1, and the induced voltage at coil 2 is proportional to the mutual inductance M. Similarly, the magnetic flux originated from coil 2 will induce voltages at coil 2 and coil 1. The induced voltage at coil 2 is proportional to the self-inductance L_2 of coil 2, and the induced voltage at coil 1 is proportional to the mutual inductance M. The voltage current relations of coupled coils are derived in this chapter. The voltage current relations can be transformed to the phasor domain for the sinusoidal steady-state analysis.

A transformer is a device that changes the voltage from one level to another level. When the voltage is transformed from a high level to a low level, it is called a *step-down transformer*, and when the voltage is transformed from a low level to a high level, it is called a *step-up transformer*. The ideal transformer is a transformer without losses on the windings or core and has perfect coupling between the coils. If the coupling coefficient is one, the flux is confined within magnetic core without leakage. All the flux passes through the core and links the primary winding and the secondary winding. The windings are perfect conductors without power loss.

13.2 Mutual Inductance

Imposing an electric field on a conductor gives rise to a current, which in turn generates a magnetic field around it. This is the magnetic equivalent of Coulomb's law. The direction of the magnetic field around a wire that carries current $i(t)$ can be found using the right-hand rule, which says that when the wire is wrapped with the right hand with the thumb pointing in the direction of the current, the direction of the magnetic field coincides with the direction of the four fingers. Ampere's law says that the integral of the magnetic flux density around a closed loop is equal to the product of the permeability μ_0, and the total current i flowing through the loop; that is,

FIGURE 13.1

Magnetic flux density around an infinitely long straight conductor.

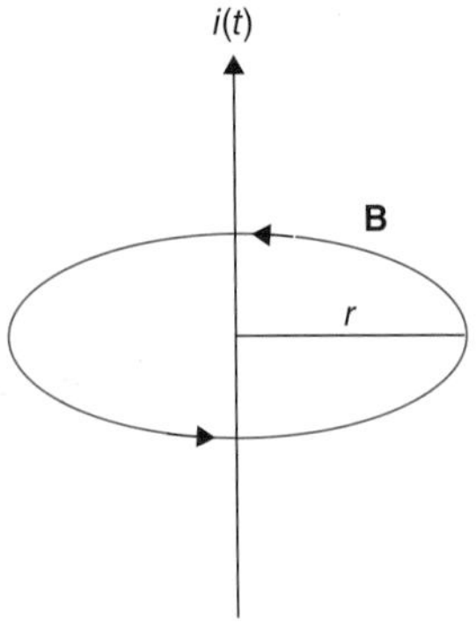

$$\oint \boldsymbol{B} \cdot dl = \mu_0 i \tag{13.1}$$

Ampere's law is the magnetic equivalent of Gauss's law. For a long straight wire carrying a current $i(t)$, as shown in Figure 13.1, the magnetic flux density at a distance r on a circle perpendicular to the wire is constant. The left side of Equation (13.1) becomes $B2\pi r$. Thus, we have

$$B = \frac{\mu_0 i}{2\pi r} \tag{13.2}$$

The unit for magnetic flux density $\boldsymbol{B}$ is tesla (T).

The magnetic flux Φ is defined as the integral of magnetic flux density $\boldsymbol{B}$ over an area $\boldsymbol{S}$

$$\Phi = \int_S \boldsymbol{B} \cdot d\boldsymbol{s} \tag{13.3}$$

The unit for magnetic flux is weber (Wb). The magnetic flux through the surface of area S from a uniform magnetic field $\boldsymbol{B}$ is given by

$$\Phi = \boldsymbol{B} \cdot \boldsymbol{S} = BS\cos(\theta)$$

where θ is the angle between $\boldsymbol{B}$ and the flat surface $\boldsymbol{S}$. If the vectors $\boldsymbol{B}$ and $\boldsymbol{S}$ point to the same direction, $\theta = 0$ and

$$\Phi = \boldsymbol{B} \cdot \boldsymbol{S} = BS$$

The magnetic effect of a current can be intensified by forming the wire into a coil of many turns and by providing an iron core. The total magnetic flux linkage Λ is defined as the product of the magnetic flux Φ and the number of turns

$$\Lambda = N\Phi \tag{13.4}$$

The inductance, also called *self-inductance,* is defined as the ratio of the total magnetic flux linkage Λ to the current I through the inductor:

$$L = \frac{\Lambda}{I} = \frac{N\Phi}{I} \tag{13.5}$$

13.2.1 FARADAY'S LAW

A current-carrying conductor generates a magnetic field around it. The opposite of this statement is "a magnetic field generates a current in a closed circuit." According to Faraday's law of induction, this is true if the magnetic field changes with time. Faraday's law of induction says that changing the magnetic flux induces an electromotive force (emf) in the closed circuit. The emf is the source of the electric field that enables the current to flow in the closed circuit-like loop. The induced electromotive force ξ is proportional to the negative rate of change of magnetic flux.

$$\xi = -\frac{d\Phi}{dt} \tag{13.6}$$

If Faraday's law is applied to a coil of N turns, an emf appears in every turn, and these emfs are to be added. If the coil is so tightly wound that each turn can be said to occupy the same region of space, the flux through each turn will then be the same. The flux through each turn is also the same for toroids and solenoids. The induced emf in these devices is given by

$$\xi = -N\frac{d\Phi}{dt} = -\frac{d(N\Phi)}{dt} = -\frac{d\Lambda}{dt} \quad (13.7)$$

Lenz's law can be used to find the direction of the induced current. Lenz's law says that the induced current will appear in such a direction that it opposes the change that produced it.

13.2.2 MUTUAL INDUCTANCE

Two coils are placed in proximity, as shown in Figure 13.2, in two different ways. Coil 1 has N_1 turns, and coil 2 has N_2 turns. If coil 1 carries a current i_1, a magnetic field will be produced in coil 1. Some of the magnetic field lines produced in coil 1 will pass through coil 2. Let Φ_{12} be the magnetic flux that passes through and links N_2 turns in coil 2. Then, the total flux linkage in coil 2 due to current in coil 1 is given by

$$\Lambda_{12} = N_2\Phi_{12} \qquad \text{(Wb)}$$

From Faraday's law, the induced voltage on coil 2 is given by

$$v_{12}(t) = \frac{d\Lambda_{12}}{dt} = N_2\frac{d\Phi_{12}}{dt} \quad (13.8)$$

FIGURE 13.2

Two coils in proximity.

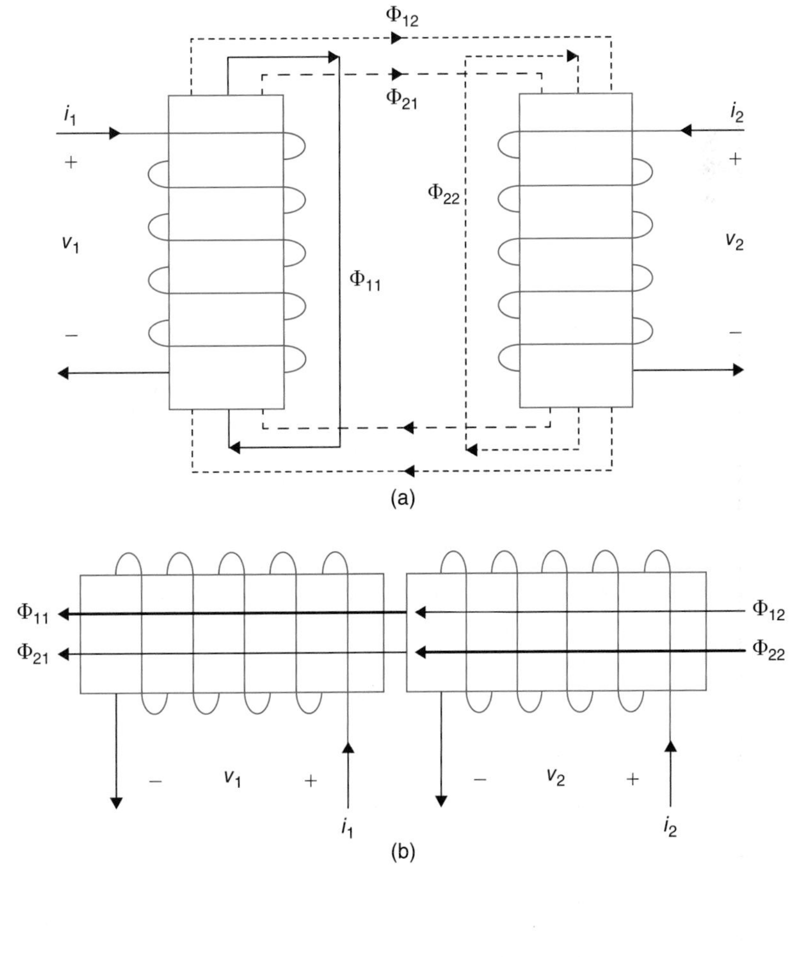

The magnetic flux density $\boldsymbol{B}_1$ is directly proportional to i_1. Since the magnetic flux Φ_{12} is the surface integral of $\boldsymbol{B}_1$, the magnetic flux Φ_{12} is proportional to the current i_1. Equation (13.8) can be rewritten as

$$v_{12}(t) = N_2 \frac{d\Phi_{12}}{di_1}\frac{di_1}{dt} = M_{12}\frac{di_1}{dt} \tag{13.9}$$

where M_{12} is defined as mutual inductance from coil 1 to coil 2. Thus, we have

$$M_{12} = N_2 \frac{d\Phi_{12}}{di_1} \qquad \text{(H)} \tag{13.10}$$

Let Φ_{11} be the magnetic flux that passes through and links N_1 turns in coil 1. Then, the total flux linkage in coil 1 due to current in coil 1 is given by

$$\Lambda_{11} = N_1\Phi_{11} \qquad \text{(Wb)}$$

From Faraday's law, the induced voltage on coil 1 is given by

$$v_{11}(t) = \frac{d\Lambda_{11}}{dt} = N_1\frac{d\Phi_{11}}{dt} \tag{13.11}$$

The magnetic flux density $\boldsymbol{B}_1$ is directly proportional to i_1. Since the magnetic flux Φ_{11} is the surface integral of $\boldsymbol{B}_1$, the magnetic flux Φ_{11} is proportional to the current i_1. Equation (13.11) can be rewritten as

$$v_{11}(t) = N_1 \frac{d\Phi_{11}}{di_1}\frac{di_1}{dt} = L_1\frac{di_1}{dt} \tag{13.12}$$

where L_1 is defined as inductance (self-inductance) of coil 1. Thus, we have

$$L_1 = N_1 \frac{d\Phi_{11}}{di_1} \qquad \text{(H)} \tag{13.13}$$

On the other hand, if coil 2 carries a current i_2, a magnetic field will be produced in coil 2. Some of the magnetic field lines produced in coil 2 will pass through coil 1. Let Φ_{21} be the magnetic flux that passes through and links N_1 turns in coil 1. Then, the total flux linkage in coil 1 due to current in coil 2 is given by

$$\Lambda_{21} = N_1\Phi_{21} \qquad \text{(Wb)}$$

From Faraday's law, the induced voltage on coil 1 is given by

$$v_{21}(t) = \frac{d\Lambda_{21}}{dt} = N_1\frac{d\Phi_{21}}{dt} \tag{13.14}$$

The magnetic flux density $\boldsymbol{B}_2$ is directly proportional to i_2. Since the magnetic flux Φ_{21} is the surface integral of $\boldsymbol{B}_2$, the magnetic flux Φ_{21} is proportional to the current i_2. Equation (13.14) can be rewritten as

$$v_{21}(t) = N_1 \frac{d\Phi_{21}}{di_2}\frac{di_2}{dt} = M_{21}\frac{di_2}{dt} \tag{13.15}$$

where M_{21} is defined as mutual inductance from coil 2 to coil 1. Thus, we have

$$M_{21} = N_1 \frac{d\Phi_{21}}{di_2} \qquad \text{(H)} \tag{13.16}$$

Let Φ_{22} be the magnetic flux that passes through and links N_2 turns in coil 2 due to i_2. Then, the total flux linkage in coil 2 due to current in coil 2 is given by

$$\Lambda_{22} = N_2\Phi_{22} \qquad \text{(Wb)}$$

From Faraday's law, the induced voltage on coil 2 is given by

$$v_{22}(t) = \frac{d\Lambda_{22}}{dt} = N_2\frac{d\Phi_{22}}{dt} \tag{13.17}$$

The magnetic flux density $\boldsymbol{B}_2$ is directly proportional to i_2. Since the magnetic flux Φ_{22} is the surface integral of $\boldsymbol{B}_2$, the magnetic flux Φ_{22} is proportional to the current i_2. Equation (13.17) can be rewritten as

$$v_{22}(t) = N_2\frac{d\Phi_{22}}{di_2}\frac{di_2}{dt} = L_2\frac{di_2}{dt} \tag{13.18}$$

where L_2 is defined as inductance (self-inductance) of coil 2. Thus, we have

$$L_2 = N_2\frac{d\Phi_{22}}{di_2} \qquad \text{(H)} \tag{13.19}$$

From the reciprocity theorem, we have

$$M_{12} = M_{21} = M \tag{13.20}$$

The voltage $v_1(t)$ across coil 1 and the voltage $v_2(t)$ across coil 2 are given respectively as

$$v_1(t) = v_{11}(t) + v_{21}(t) = L_1\frac{di_1(t)}{dt} + M_{21}\frac{di_2(t)}{dt} \tag{13.21}$$

$$v_2(t) = v_{12}(t) + v_{22}(t) = M_{12}\frac{di_1(t)}{dt} + L_2\frac{di_2(t)}{dt} \tag{13.22}$$

Since $M_{12} = M_{21} = M$, Equations (13.21) and (13.22) become, respectively,

$$v_1(t) = v_{11}(t) + v_{21}(t) = L_1\frac{di_1(t)}{dt} + M\frac{di_2(t)}{dt} \tag{13.23}$$

$$v_2(t) = v_{12}(t) + v_{22}(t) = M\frac{di_1(t)}{dt} + L_2\frac{di_2(t)}{dt} \tag{13.24}$$

13.2.3 MUTUAL INDUCTANCE OF A SECOND COIL WRAPPED AROUND A SOLENOID

A solenoid of length ℓ, cross-sectional area S, N_1 turns carries a current $i_1(t)$. A second coil with N_2 turns is wrapped around it, as shown in Figure 13.3.

The fluxes inside the solenoid are given from Equation (6.36)

$$\Phi_{11} = \Phi_{12} = B_1S = \frac{\mu_0 N_1 i_1}{\ell}S \tag{13.25}$$

$$\Phi_{21} = \Phi_{22} = B_2S = \frac{\mu_0 N_2 i_2}{\ell}S \tag{13.26}$$

FIGURE 13.3

A solenoid with second coil wrapped around it.

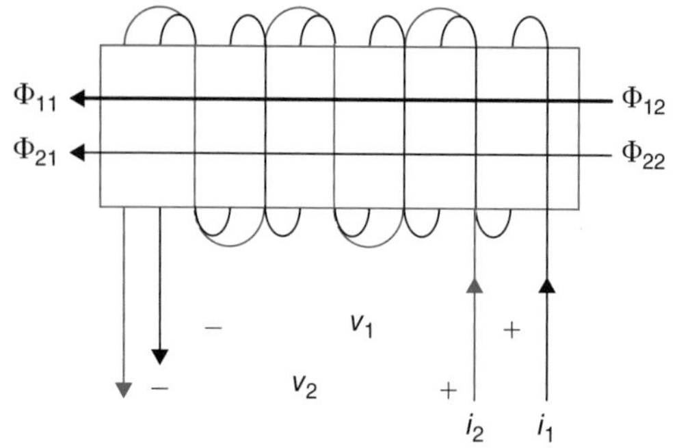

The mutual inductances are given by

$$M_{12} = N_2\frac{d\Phi_{12}}{di_1} = N_2\frac{d}{di_1}\left(\frac{\mu_0 N_1 i_1}{\ell}S\right) = \frac{N_2\mu_0 N_1 S}{\ell} \quad \textbf{(13.27)}$$

$$M_{21} = N_1\frac{d\Phi_{21}}{di_2} = N_1\frac{d}{di_2}\left(\frac{\mu_0 N_2 i_2}{\ell}S\right) = \frac{N_1\mu_0 N_2 S}{\ell} \quad \textbf{(13.28)}$$

Notice that

$$M = M_{12} = M_{21} = \frac{\mu_0 N_1 N_2 S}{\ell} \quad \textbf{(13.29)}$$

The self-inductance of the solenoid is given by

$$L_1 = \frac{N_1\Phi_{11}}{i_1} = \frac{\mu_0 N_1^2 S}{\ell} \quad \textbf{(13.30)}$$

The self-inductance of the second coil is given by

$$L_2 = \frac{N_2\Phi_{22}}{i_2} = \frac{\mu_0 N_2^2 S}{\ell} \quad \textbf{(13.31)}$$

Thus, from Equations (13.29) through (13.31), we have

$$M = \sqrt{L_1L_2} \quad \textbf{(13.32)}$$

In general, the mutual inductance M is given by

$$M \le \sqrt{L_1L_2} \quad \textbf{(13.33)}$$

The mutual inductance can also be written as

$$M = k\sqrt{L_1L_2} \quad \textbf{(13.34)}$$

where k is called the *coupling coefficient*. The value of the coupling coefficient k is bounded by

$$0 \le k \le 1 \quad \textbf{(13.35)}$$

FIGURE 13.4

The direction of $i_2(t)$ is reversed.

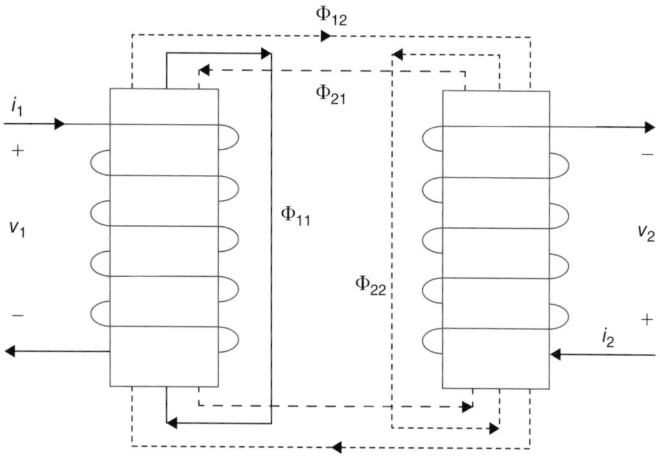

In the solenoid example, $k = 1$.

In Figure 13.2(a), the fluxes Φ_{11} and Φ_{21} are in the same direction (↑) inside the left coil, and the fluxes are additive. Similarly, the fluxes Φ_{22} and Φ_{12} are in the same direction (↓) inside the right coil, and the fluxes are additive. The flux linkage is increased in each coil. When the fluxes are additive, the mutual inductance term in Equations (13.23) and (13.24) is positive. If the direction of $i_2(t)$ is reversed in Figure 13.2(a), according to the right-hand rule, the directions of Φ_{22} and Φ_{21} are reversed, as shown in Figure 13.4. In this case, the fluxes Φ_{11} and Φ_{21} are in the opposite direction inside the left coil, and the fluxes are subtractive. Similarly, the fluxes Φ_{22} and Φ_{12} are in the opposite direction inside the right coil, so the fluxes are subtractive.

When the fluxes are subtractive, the mutual inductance term in Equations (13.23) and (13.24) is

FIGURE 13.5

The direction of $i_1(t)$ is reversed.

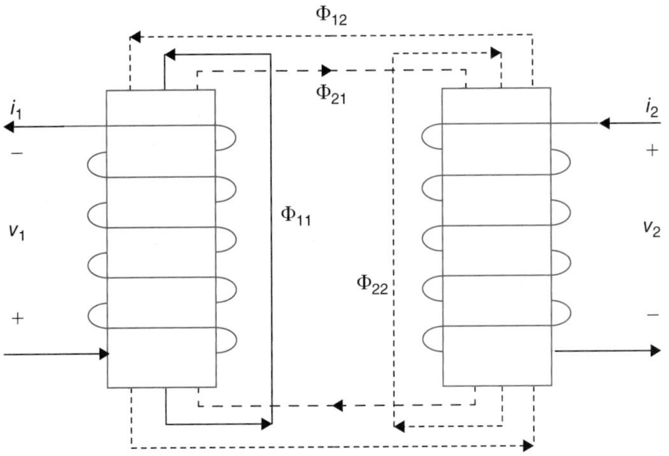

FIGURE 13.6

The direction of $i_1(t)$ and $i_2(t)$ are reversed.

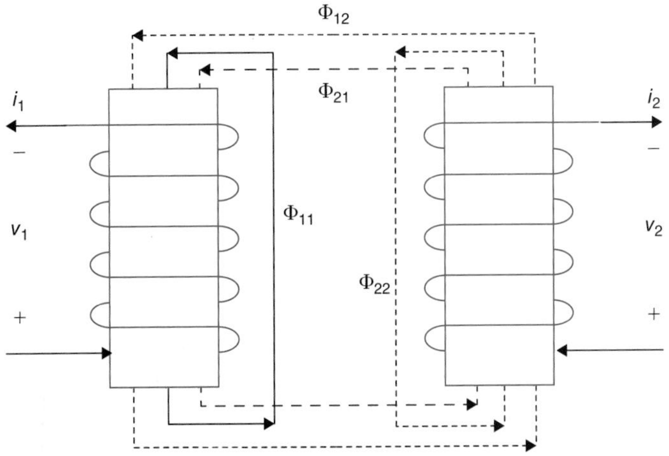

negative. Thus, Equations (13.23) and (13.24) are modified to Equations (13.36) and (13.37), respectively.

$$v_1(t) = L_1\frac{di_1(t)}{dt} - M\frac{di_2(t)}{dt} \tag{13.36}$$

$$v_2(t) = -M\frac{di_1(t)}{dt} + L_2\frac{di_2(t)}{dt} \tag{13.37}$$

The direction of $v_2(t)$ is reversed to make the self-inductance term, which is the second term of Equation (13.37), positive.

If the direction of $i_1(t)$ is reversed in Figure 13.2(a), according to right-hand rule, the directions of Φ_{11} and Φ_{12} are reversed, as shown in Figure 13.5. In this case, the fluxes Φ_{11} and Φ_{21} are in the opposite direction inside the left coil, and the fluxes are subtractive. Similarly, the fluxes Φ_{22} and Φ_{12} are in the opposite direction inside the right coil, and the fluxes are subtractive.

When the fluxes are subtractive, the mutual inductance term in Equations (13.23) and (13.24) is negative. Thus, Equations (13.23) and (13.24) are modified to Equations (13.36) and (13.37), respectively. The direction of $v_1(t)$ is reversed to make the self-inductance term, which is the first term of Equation (13.36), positive.

If the direction of $i_1(t)$ and $i_2(t)$ are reversed in Figure 13.2(a), according to the right-hand rule, the directions of all four fluxes are reversed, as shown in Figure 13.6. In this case, the fluxes Φ_{11} and Φ_{21} are in the same direction inside the left coil, and the fluxes are additive. Similarly, the fluxes Φ_{22} and Φ_{12} are in the same direction inside the right coil, and the fluxes are additive. When the fluxes are additive, the mutual inductance term in Equations (13.23) and (13.24) is positive. The direction of $v_1(t)$ and $v_2(t)$ are reversed to make the self-inductance terms in Equations (13.23) and (13.24) positive.

FIGURE 13.7

The circuit symbol for the coupled coils shown in Figure 13.2(a).

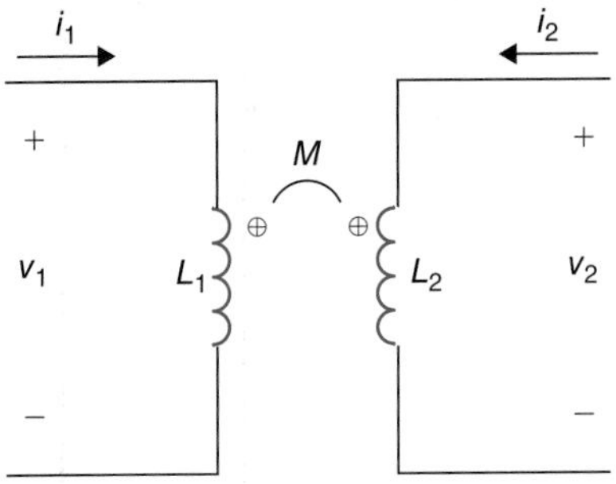

The circuit symbol for the coupled coils, shown in Figure 13.2(a), is shown in Figure 13.7. Notice that L_1 is the self-inductance of coil 1, L_2 is the self-inductance of coil 2, and M is the mutual inductance. The dots placed on the coils are explained in the next section.

13.3 Dot Convention and Induced Voltage

In a circuit for coupled coils, a dot convention is used to clarify the correct polarity of induced voltage. If a current enters into one coil through the dotted terminal, the positive terminal of the induced voltage at the second coil is at the dotted terminal. If a current enters into one coil through the undotted terminal, the positive terminal of the induced voltage at the second coil is at the undotted terminal. Figure 13.8 shows four different cases of induced voltages with $i_2(t) = 0$. In the circuit shown in Figure 13.8(a), the current $i_1(t)$ enters the first

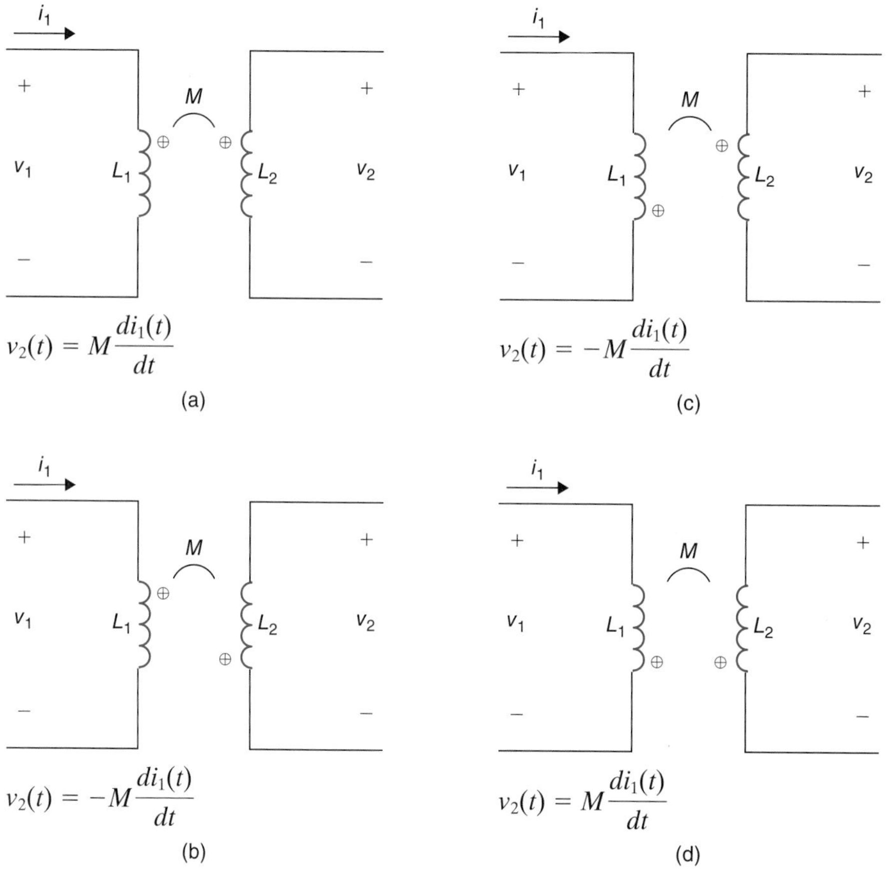

FIGURE 13.8

Four different combinations of dot placement.

coil through the dotted terminal. This means that the induced voltage on the second coil is positive on the dotted side. Since $v_2(t)$ is defined to be positive at the dotted terminal, the induced voltage is positive. Thus, we have

$$v_2(t) = M\frac{di_1(t)}{dt} \tag{13.38}$$

In the circuit shown in Figure 13.8(b), the current $i_1(t)$ enters the first coil through the dotted terminal. This means that the induced voltage on the second coil is positive on the dotted side. Since $v_2(t)$ is defined to be positive at the undotted terminal, the induced voltage is negative. Thus, we have

$$v_2(t) = -M\frac{di_1(t)}{dt} \tag{13.39}$$

In the circuit shown in Figure 13.8(c), the current $i_1(t)$ enters the first coil through the undotted terminal. This means that the induced voltage on the second coil is positive on the undotted side. Since $v_2(t)$ is defined to be positive at the dotted terminal, the induced voltage is negative. Thus, we have

$$v_2(t) = -M\frac{di_1(t)}{dt} \tag{13.40}$$

In the circuit shown in Figure 13.8(d), the current $i_1(t)$ enters the first coil through the undotted terminal. This means the induced voltage on the second coil is positive on the undotted side. Since $v_2(t)$ is defined to be positive at the undotted terminal, the induced voltage is positive. Thus, we have

$$v_2(t) = M\frac{di_1(t)}{dt} \tag{13.41}$$

If $i_1(t)$ is set to zero, and $i_2(t)$ is applied, we can find the polarity of the induced voltage by applying the same rule. If both $i_1(t)$ and $i_2(t)$ are present, the same dot convention applies to $i_1(t)$ and $i_2(t)$.

EXAMPLE 13.1

A current $i_1(t)$ is shown in Figure 13.9. This current is applied to the circuit shown in Figure 13.8(a). Plot the induced voltage $v_2(t)$ when $M = 500$ mH.

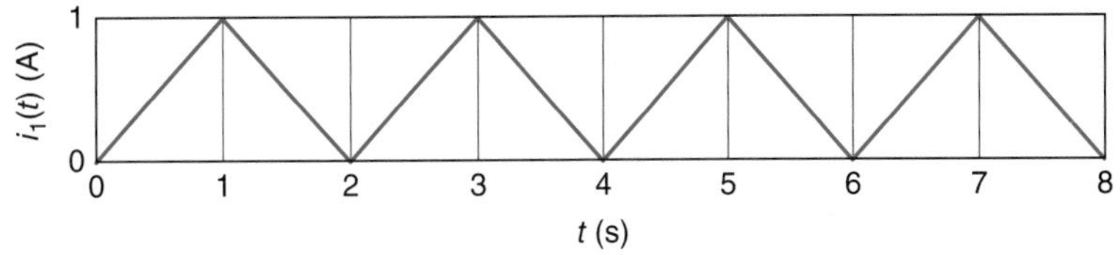

FIGURE 13.9

Waveform for $i_1(t)$.

Since $v_2(t) = M\frac{di_1(t)}{dt}$, $v_2(t)$ is either $0.5 \times 1 = 0.5$ V or $0.5 \times (-1) = -0.5$ V, as shown in Figure 13.10.

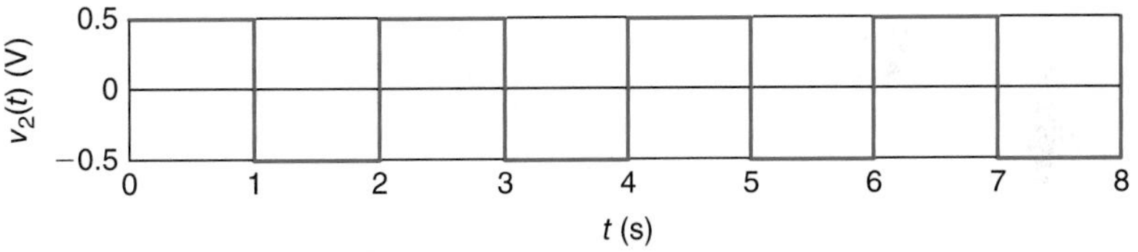

FIGURE 13.10

Waveform for $v_2(t)$.

Exercise 13.1

A current $i_1(t)$ is shown in Figure 13.11. This current is applied to the circuit shown in Figure 13.8(b). Plot the induced voltage $v_2(t)$ when $M = 200$ mH.

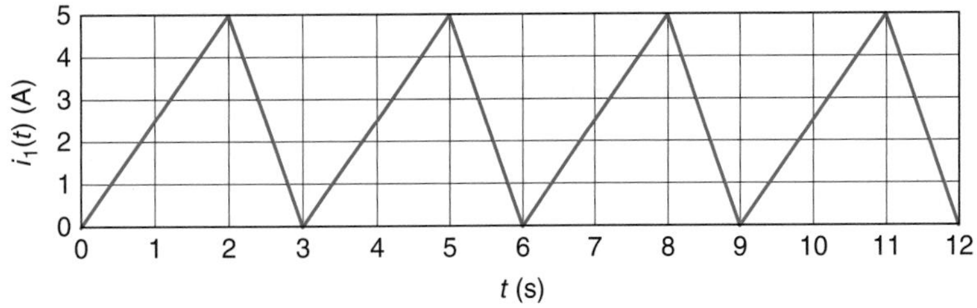

FIGURE 13.11

Waveform for $i_1(t)$.

continued

Exercise 13.1 continued

Answer:
The induced voltage is shown in Figure 13.12.

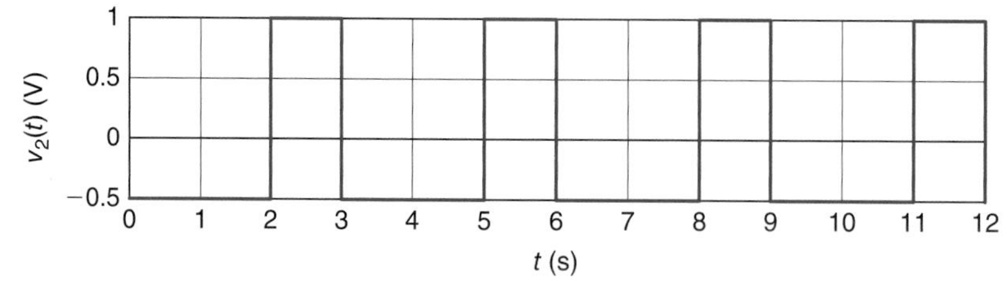

FIGURE 13.12

Waveform for $v_2(t)$.

13.3.1 COMBINED MUTUAL AND SELF-INDUCTION VOLTAGE

As shown in Chapter 6, if the positive terminal of the voltage $v(t)$ across the inductor is defined as the terminal where the current is entering the inductor, the voltage across the inductor is given by

$$v(t) = L\frac{di(t)}{dt}$$

This is the voltage due to self-induction. It follows passive sign convention. If the voltage due to mutual induction is added to the voltage of self-induction, we obtain the combined voltage for each terminal. For the mutual inductance voltage, the dot convention applies in both directions.

In the circuit shown in Figure 13.13, since the positive terminal of $v_1(t)$ is defined as the terminal where the current $i_1(t)$ enters coil 1, the voltage from the self-induction is positive; that is, $+L_1\frac{di_1(t)}{dt}$. Since $i_2(t)$ enters coil 2 through the dotted terminal, the induced voltage on coil 1 is positive on the dotted terminal. Since the positive terminal of $v_1(t)$ is at the dotted terminal, the voltage induced from $i_2(t)$ is also positive; that is, $+M\frac{di_2(t)}{dt}$.Thus, we have

FIGURE 13.13

Coupled coils in time domain.

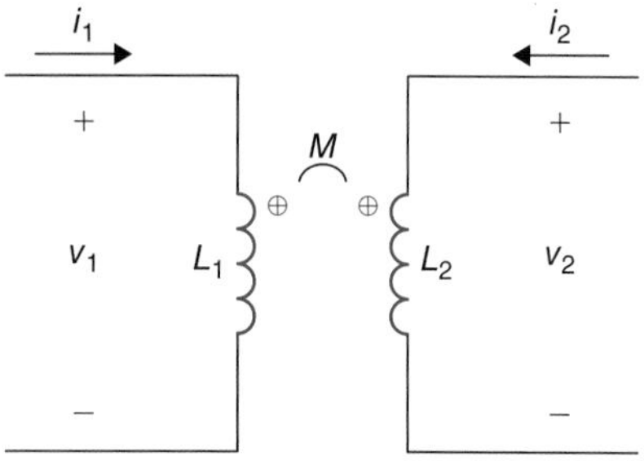

$$v_1(t) = L_1\frac{di_1(t)}{dt} + M\frac{di_2(t)}{dt} \tag{13.42}$$

Similarly, since the positive terminal of $v_2(t)$ is defined as the terminal where the current $i_2(t)$ enters coil 2, the voltage from the self-induction is positive; that is,$+L_2\frac{di_2(t)}{dt}$. Since $i_1(t)$ enters coil 1 through the dotted terminal, the induced voltage on coil 2 is positive on the dotted terminal. Since the positive terminal of $v_2(t)$ is at the dotted terminal, the voltage induced from $i_1(t)$ is also positive; that is, $+M\frac{di_1(t)}{dt}$. Thus, we have

$$v_2(t) = M\frac{di_1(t)}{dt} + L_2\frac{di_2(t)}{dt} \tag{13.43}$$

Let

$$v_1(t) = \text{Re}[\mathbf{V}_1e^{j\omega t}], \quad v_2(t) = \text{Re}[\mathbf{V}_2e^{j\omega t}], \quad i_1(t) = \text{Re}[\mathbf{I}_1e^{j\omega t}], \quad i_2(t) = \text{Re}[\mathbf{I}_2e^{j\omega t}]$$

where $\mathbf{V}_1, \mathbf{V}_2, \mathbf{I}_1, \mathbf{I}_2$ are phasors for $v_1(t)$, $v_2(t)$, $i_1(t)$, and $i_2(t)$, respectively. Since

$$\frac{di_1(t)}{dt} = \text{Re}[j\omega \mathbf{I}_1 e^{j\omega t}], \quad \frac{di_2(t)}{dt} = \text{Re}[j\omega \mathbf{I}_2 e^{j\omega t}],$$

Equation (13.42) can be written as

$$\text{Re}[\mathbf{V}_1 e^{j\omega t}] = \text{Re}[j\omega L_1 \mathbf{I}_1 e^{j\omega t} + j\omega M \mathbf{I}_2 e^{j\omega t}]$$

FIGURE 13.14

Coupled coils in phasor domain.

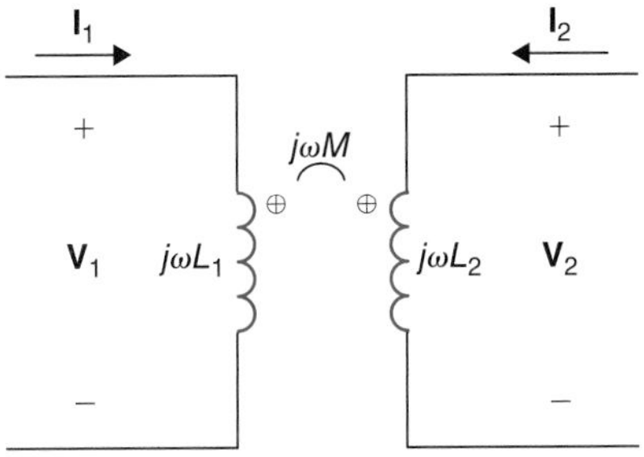

Thus, we have

$$\mathbf{V}_1 = j\omega L_1 \mathbf{I}_1 + j\omega M \mathbf{I}_2 \tag{13.44}$$

Similarly, in phasor domain, Equation (13.43) becomes

$$\mathbf{V}_2 = j\omega M \mathbf{I}_1 + j\omega L_2 \mathbf{I}_2 \tag{13.45}$$

In the phasor domain, the circuit shown in Figure 13.13 can be redrawn as the one shown in Figure 13.14.

The circuit shown in Figure 13.13 with Equations (13.42) and (13.43) can be modeled using dependent voltage sources, as shown in Figure 13.15. In the circuit shown in Figure 13.15, the two inductors are not coupled.

FIGURE 13.15

A model for the circuit shown in Figure 13.13.

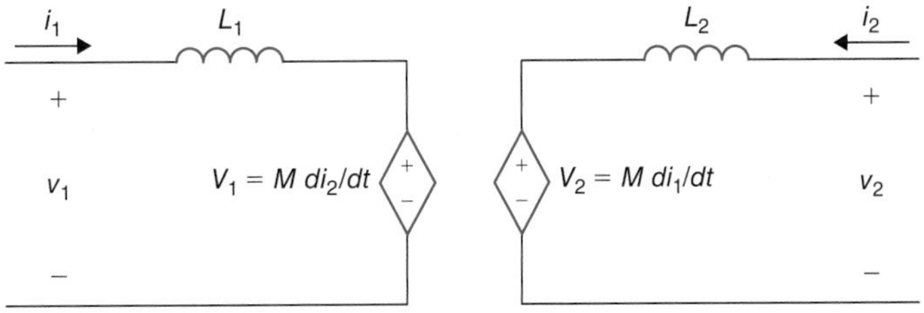

In the phasor domain, the circuit shown in Figure 13.15 can be redrawn as shown in Figure 13.16.

FIGURE 13.16

A model for the circuit shown in Figure 13.13 in the phasor domain.

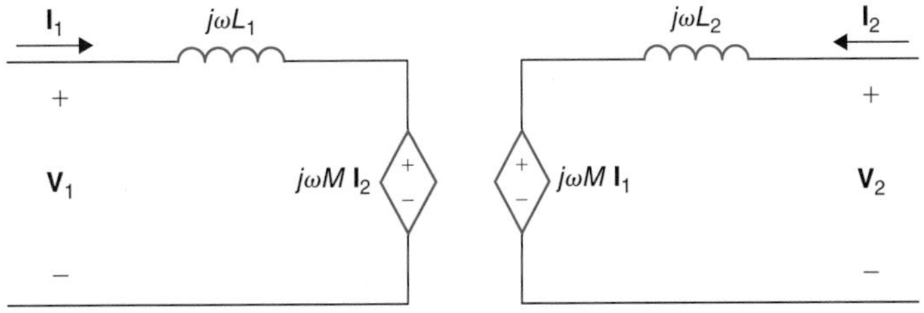

FIGURE 13.17

Circuit with $i_2(t)$ in the reverse direction.

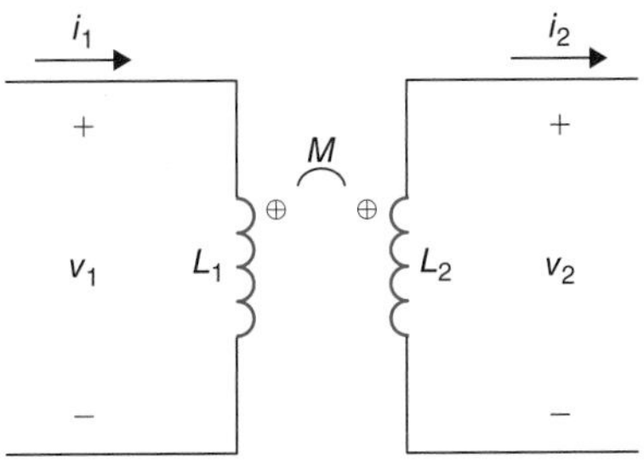

If the direction of $i_2(t)$ in the circuit shown in Figure 13.13 is reversed, we obtain the circuit shown in Figure 13.17. The only difference between the circuit shown in Figure 13.13 and the circuit shown in Figure 13.17 is the direction of $i_2(t)$. If $i_2(t)$ is substituted by $-i_2(t)$ in Equations (13.42) and (13.43), we obtain Equations (13.46) and (13.47).

$$v_1(t) = L_1 \frac{di_1(t)}{dt} - M \frac{di_2(t)}{dt} \tag{13.46}$$

$$v_2(t) = M\frac{di_1(t)}{dt} - L_2\frac{di_2(t)}{dt} \tag{13.47}$$

In the phasor domain, Equations (13.46) and (13.47) become, respectively,

$$\mathbf{V}_1 = j\omega L_1\mathbf{I}_1 - j\omega M\mathbf{I}_2 \tag{13.48}$$

$$\mathbf{V}_2 = j\omega M\mathbf{I}_1 - j\omega L_2\mathbf{I}_2 \tag{13.49}$$

FIGURE 13.18

Phasor representation of the circuit shown in Figure 13.17.

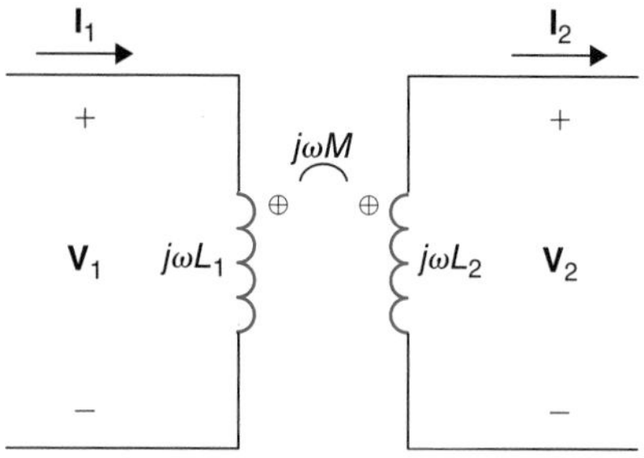

In the phasor domain, the circuit shown in Figure 13.17 can be redrawn as the one shown in Figure 13.18.

In the circuit shown in Figure 13.19, since the positive terminal of $v_1(t)$ is defined as the terminal where the current $i_1(t)$ leaves coil 1, the voltage from the self-induction is negative. Since $i_2(t)$ leaves coil 2 through the dotted terminal, the induced voltage on coil 1 is negative on the dotted terminal. Since the positive terminal of $v_1(t)$ is at the undotted terminal, the voltage induced from $i_2(t)$ is positive. Thus, we have

FIGURE 13.19

Mutual inductance is additive.

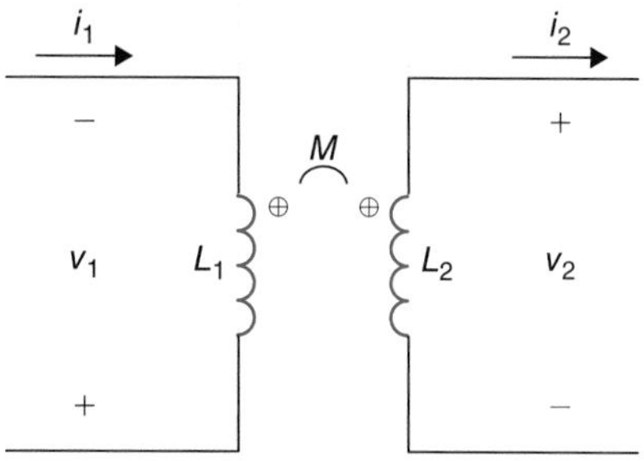

$$v_1(t) = -L_1\frac{di_1(t)}{dt} + M\frac{di_2(t)}{dt} \tag{13.50}$$

Since the positive terminal of $v_2(t)$ is defined as the terminal where the current $i_2(t)$ leaves coil 2, the voltage from the self-induction is negative. Since $i_1(t)$ enters coil 1 through the dotted terminal, the induced voltage on coil 2 is positive on the dotted terminal. Since the positive terminal of $v_2(t)$ is at the dotted terminal, the voltage induced from $i_1(t)$ is positive. Thus, we have

$$v_2(t) = M\frac{di_1(t)}{dt} - L_2\frac{di_2(t)}{dt} \tag{13.51}$$

FIGURE 13.20

Phasor representation of the circuit shown in Figure 13.19.

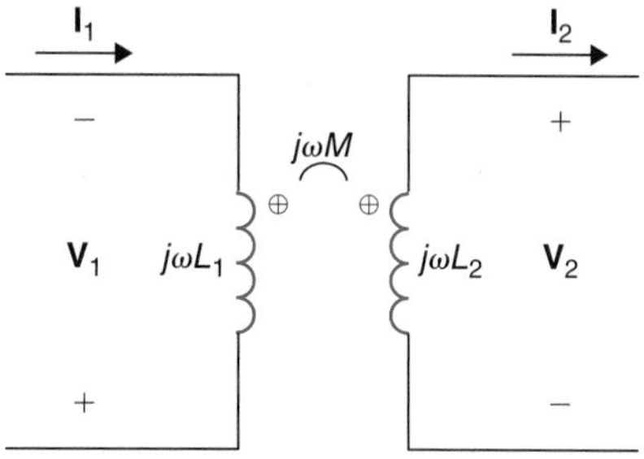

In the phasor domain, Equations (13.50) and (13.51) become, respectively,

$$\mathbf{V}_1 = -j\omega L_1\mathbf{I}_1 + j\omega M\mathbf{I}_2 \tag{13.52}$$

$$\mathbf{V}_2 = j\omega M\mathbf{I}_1 - j\omega L_2\mathbf{I}_2 \tag{13.53}$$

In the phasor domain, the circuit shown in Figure 13.19 can be redrawn as shown in Figure 13.20.

In the circuit shown in Figure 13.21, since the positive terminal of $v_1(t)$ is defined as the terminal where the current $i_1(t)$ enters coil 1, the voltage from the self-induction is positive. Since $i_2(t)$ leaves coil 2 through the dotted terminal, the induced voltage on coil 1 is negative on the dotted terminal. Since the positive terminal of $v_1(t)$ is at the dotted terminal, the voltage induced from $i_2(t)$ is negative. Thus, we have

FIGURE 13.21

Mutual inductance is subtractive.

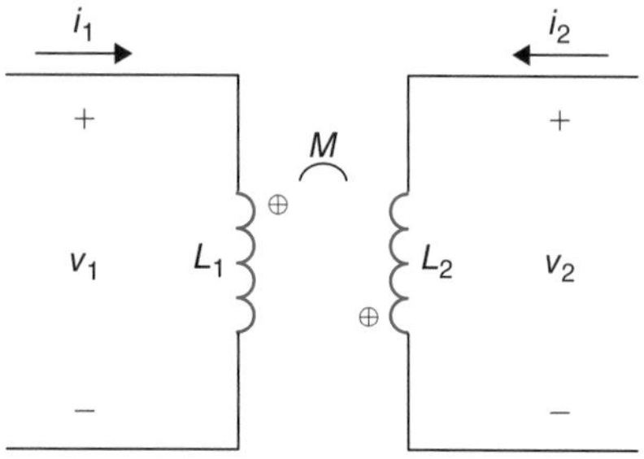

$$v_1(t) = L_1\frac{di_1(t)}{dt} - M\frac{di_2(t)}{dt} \tag{13.54}$$

Since the positive terminal of $v_2(t)$ is defined as the terminal where the current $i_2(t)$ enters coil 2, the voltage from the self-induction is positive. Since $i_1(t)$ enters coil 1 through the dotted terminal, the induced voltage on coil 2 is positive on the dotted terminal. Since the positive terminal of $v_2(t)$ is at the undotted terminal, the voltage induced from $i_1(t)$ is negative. Thus, we have

$$v_2(t) = -M\frac{di_1(t)}{dt} + L_2\frac{di_2(t)}{dt} \tag{13.55}$$

In the phasor domain, Equations (13.54) and (13.55) become, respectively,

$$\mathbf{V}_1 = j\omega L_1\mathbf{I}_1 - j\omega M\mathbf{I}_2 \tag{13.56}$$

$$\mathbf{V}_2 = -j\omega M\mathbf{I}_1 + j\omega L_2\mathbf{I}_2 \tag{13.57}$$

In the phasor domain, the circuit shown in Figure 13.21 can be redrawn as shown in Figure 13.22.

FIGURE 13.22

Phasor representation of the circuit shown in Figure 13.21.

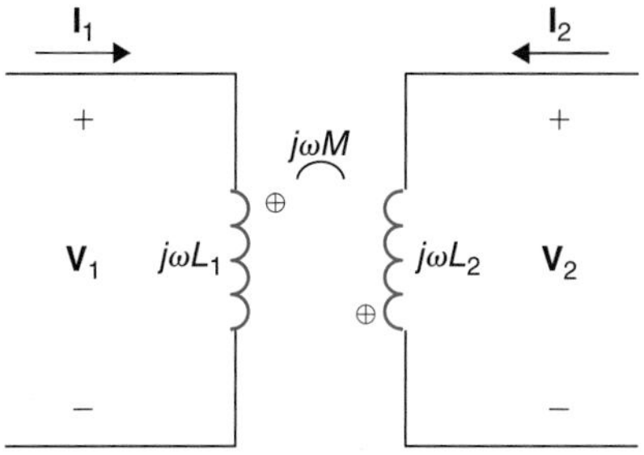

EXAMPLE 13.2

Draw the circuit symbol for the coupled coils shown in Figure 13.4.

Since the fluxes are subtractive, we obtain the circuit shown in Figure 13.23. The circuit shown in Figure 13.23 can be redrawn as the one shown in Figure 13.24. This circuit is identical to the one shown in Figure 13.21. Equations (13.54) and (13.55) describe the circuit shown in Figure 13.24.

FIGURE 13.23

Circuit symbol for the coupled coils shown in Figure 13.4.

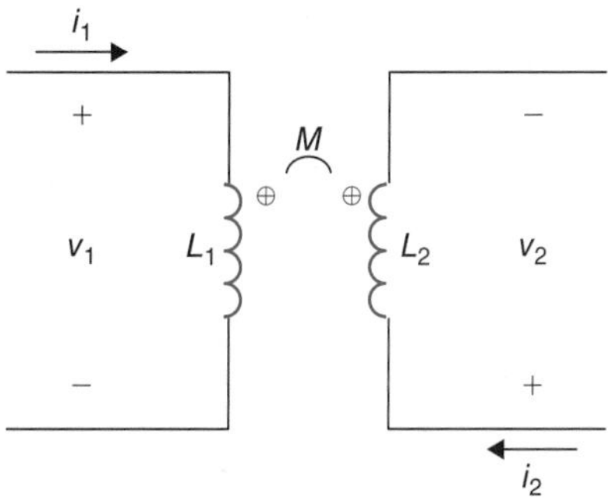

FIGURE 13.24

The right side of the circuit is flipped.

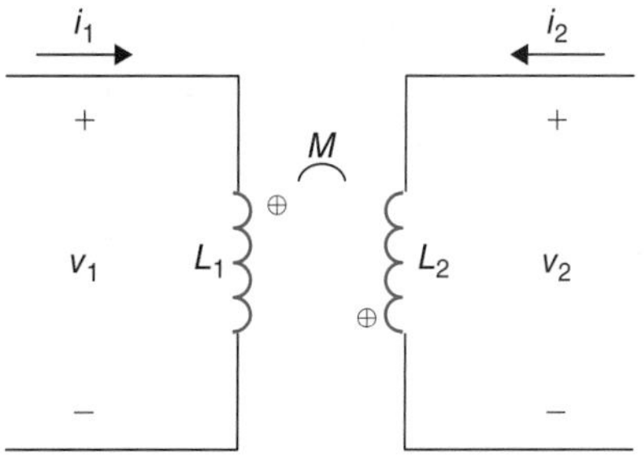

Exercise 13.2

Draw the circuit symbol for the coupled coils shown in Figure 13.5.

Answer:
The circuit is shown in Figure 13.25(a) or (b). Equations (13.54) and (13.55) describe the circuit shown in Figure 13.25(b).

FIGURE 13.25

Circuit symbol for the coupled coils shown in Figure 13.5.

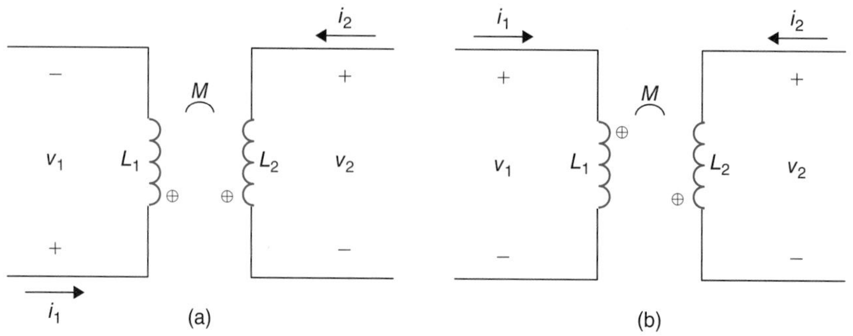

EXAMPLE 13.3

Find $\mathbf{I}_1$, $\mathbf{I}_2$, $\mathbf{V}_1$, and $\mathbf{V}_2$ in the circuit shown in Figure 13.26.

FIGURE 13.26

Circuit for EXAMPLE 13.3.

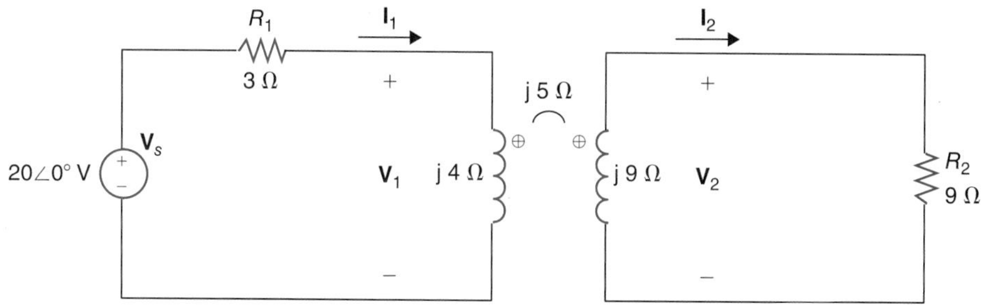

Summing the voltage drops around mesh 1 (left side), we obtain

$$-\mathbf{V}_s + R_1\mathbf{I}_1 + \mathbf{V}_1 = 0$$

which can be rearranged as

$$\mathbf{V}_1 = \mathbf{V}_s - R_1\mathbf{I}_1 = 20 - 3\mathbf{I}_1 \quad (13.58)$$

The voltage across R_2 is given by

$$\mathbf{V}_2 = R_2\mathbf{I}_2 = 9\mathbf{I}_2 \quad (13.59)$$

From Equation (13.48), we have

$$\mathbf{V}_1 = j4\,\mathbf{I}_1 - j5\,\mathbf{I}_2 \quad (13.60)$$

continued

Example 13.3 continued

Substitution of Equation (13.58) into Equation (13.60) yields

$$20 - 3\mathbf{I}_1 = j4\,\mathbf{I}_1 - j5\,\mathbf{I}_2$$

which can be rewritten as

$$(3 + j4)\,\mathbf{I}_1 - j5\,\mathbf{I}_2 = 20 \tag{13.61}$$

Similarly, from Equation (13.49), we have

$$\mathbf{V}_2 = j5\,\mathbf{I}_1 - j9\,\mathbf{I}_2 \tag{13.62}$$

Substitution of Equation (13.59) into Equation (13.62) yields

$$9\,\mathbf{I}_2 = j5\,\mathbf{I}_1 - j9\,\mathbf{I}_2$$

which can be rewritten as

$$j5\,\mathbf{I}_1 - (9 + j9)\,\mathbf{I}_2 = 0 \tag{13.63}$$

Solving Equation (13.61) for $\mathbf{I}_2$, we obtain

$$\mathbf{I}_2 = (0.8 - j0.6)\mathbf{I}_1 + j4$$

Substituting $\mathbf{I}_2$ into Equation (13.63), we get

$$j5\,\mathbf{I}_1 - (9 + j9)(0.8 - j0.6)\mathbf{I}_1 - (9 + j9)(j4) = 0$$

Solving for $\mathbf{I}_1$, we obtain

$$\mathbf{I}_1 = \frac{-36 + j36}{j5 - 7.2 + j5.4 - j7.2 - 5.4} = \frac{-36 + j36}{-12.6 + j3.2} = 3.3657 - j2.0024$$
$$= 3.9163\angle -30.75^\circ \text{ A}$$

Substituting $\mathbf{I}_1$ into $\mathbf{I}_2 = (0.8 - j\,0.6)\mathbf{I}_1 + j4$, we get

$$\mathbf{I}_2 = 1.4911 + j0.3787 = 1.5385\angle 14.25^\circ \text{ A}$$

Alternatively, Cramer's rule can be used to find $\mathbf{I}_1$ and $\mathbf{I}_2$:

$$\mathbf{I}_1 = \frac{\begin{vmatrix} 20 & -j5 \\ 0 & -9 - j9 \end{vmatrix}}{\begin{vmatrix} 3 + j4 & -j5 \\ j5 & -9 - j9 \end{vmatrix}} = \frac{-180 - j180}{-16 - j63} = 3.3657 - j2.0024 = 3.9163\angle -30.75^\circ \text{ A}$$

$$\mathbf{I}_2 = \frac{\begin{vmatrix} 3 + j4 & 20 \\ j5 & 0 \end{vmatrix}}{\begin{vmatrix} 3 + j4 & -j5 \\ j5 & -9 - j9 \end{vmatrix}} = \frac{-j100}{-16 - j63} = 1.4911 + j0.3787 = 1.5385\angle 14.25^\circ \text{ A}$$

continued

Example 13.3 continued

Voltages $\mathbf{V}_1$ and $\mathbf{V}_2$ are found from Equations (13.58) and (13.59):

$$\mathbf{V}_1 = \mathbf{V}_s - R_1\mathbf{I}_1 = 20 - 3\mathbf{I}_1 = 9.9030 + j6.0071 = 11.5825\angle 31.2409° \text{ V}$$

$$\mathbf{V}_2 = R_2\mathbf{I}_2 = 9\mathbf{I}_2 = 13.4201 + j3.4083 = 13.8462\angle 14.25° \text{ V}$$

MATLAB

```
% EXAMPLE 13.3
clear all;
ZL1=4j;ZL2=9j;ZM=5j;R1=3;R2=9;Vm=20;phi=0;
Vs=P2Rd(Vm,phi)
syms I1 I2 V1 V2
[I1,I2,V1,V2]=solve(-Vs+R1*I1+V1,-V2+R2*I2,...
V1==ZL1*I1-ZM*I2,V2==ZM*I1-ZL2*I2);
I1p=R2P(I1);
I2p=R2P(I2);
V1p=R2P(V1);
V2p=R2P(V2);
I1=vpa(I1,6)
I1p=vpa(I1p,7)
I2=vpa(I2,6)
I2p=vpa(I2p,7)
V1=vpa(V1,6)
V1p=vpa(V1p,7)
V2=vpa(V2,6)
V2p=vpa(V2p,7)
```

Exercise 13.3

Find $\mathbf{I}_1$, $\mathbf{I}_2$, $\mathbf{V}_1$, $\mathbf{V}_2$ in the circuit shown in Figure 13.27.

FIGURE 13.27

Circuit for EXERCISE 13.3.

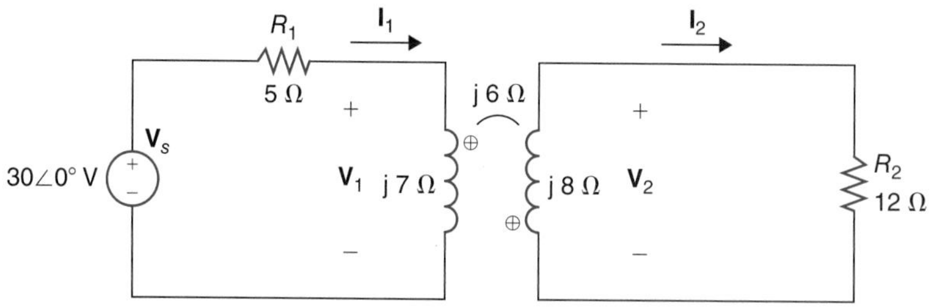

Answer:

$\mathbf{I}_1 = 2.6013 - j2.0641 = 3.3207\angle -38.4312°$ A,

$\mathbf{I}_2 = -1.3148 - j0.4241 = 1.3815\angle -162.1213°$ A,

$\mathbf{V}_1 = 16.9934 + j10.3205 = 19.8818\angle 31.2712°$ V,

$\mathbf{V}_2 = -15.7776 - j5.08954 = 16.5782\angle -162.1213°$ V.

Figure 13.28 shows three coils coupled to each other. The self-inductances of the three coils are L_1, L_2, and L_3, respectively. The mutual inductance between coil 1 and coil 2 is denoted by $M_{12} = M_{21}$. Similarly, the mutual inductance between coil 1 and coil 3 is denoted by $M_{13} = M_{31}$, and the mutual inductance between coil 2 and coil 3 is denoted by $M_{23} = M_{32}$. The dot convention applies to each pair of the coils.

FIGURE 13.28

Three coupled coils.

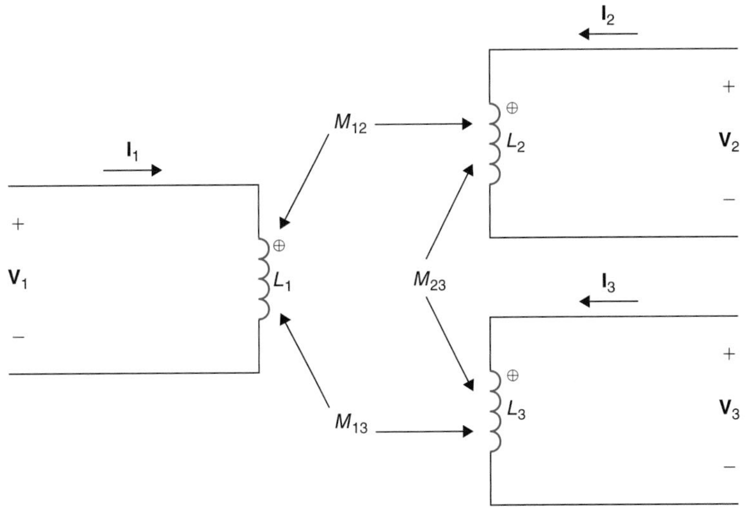

The voltage v_1 across coil 1 is the sum of the self-induced voltage, the induced voltage from coil 2, and the induced voltage from coil 3. Therefore, we have

$$v_1(t) = L_1\frac{di_1(t)}{dt} + M_{12}\frac{di_2(t)}{dt} + M_{13}\frac{di_3(t)}{dt} \tag{13.64}$$

Similarly, the voltage across coil 2 and coil 3 are given, respectively, by

$$v_2(t) = M_{12}\frac{di_1(t)}{dt} + L_2\frac{di_2(t)}{dt} + M_{23}\frac{di_3(t)}{dt} \tag{13.65}$$

$$v_3(t) = M_{13}\frac{di_1(t)}{dt} + M_{23}\frac{di_2(t)}{dt} + L_3\frac{di_3(t)}{dt} \tag{13.66}$$

In the phasor domain, Equations (13.64) through (13.66) become, respectively,

$$\mathbf{V}_1 = j\omega L_1\mathbf{I}_1 + j\omega M_{12}\mathbf{I}_2 + j\omega M_{13}\mathbf{I}_3 \tag{13.67}$$

$$\mathbf{V}_2 = j\omega M_{12}\mathbf{I}_1 + j\omega L_2\mathbf{I}_2 + j\omega M_{23}\mathbf{I}_3 \tag{13.68}$$

$$\mathbf{V}_3 = j\omega M_{13}\mathbf{I}_1 + j\omega M_{23}\mathbf{I}_2 + j\omega L_3\mathbf{I}_3 \tag{13.69}$$

EXAMPLE 13.4

Find $\mathbf{I}_1$, $\mathbf{I}_2$, and $\mathbf{I}_3$ in the circuit shown in Figure 13.29.

Summing the voltage drops around mesh 1 (left side), we obtain

$$-\mathbf{V}_s + R_1\mathbf{I}_1 + \mathbf{V}_1 = 0$$

which can be rearranged as

$$\mathbf{V}_1 = \mathbf{V}_s - R_1\mathbf{I}_1 = 60 - 5\mathbf{I}_1 \tag{13.70}$$

continued

Example 13.4 continued

FIGURE 13.29

Circuit for EXAMPLE 13.4.

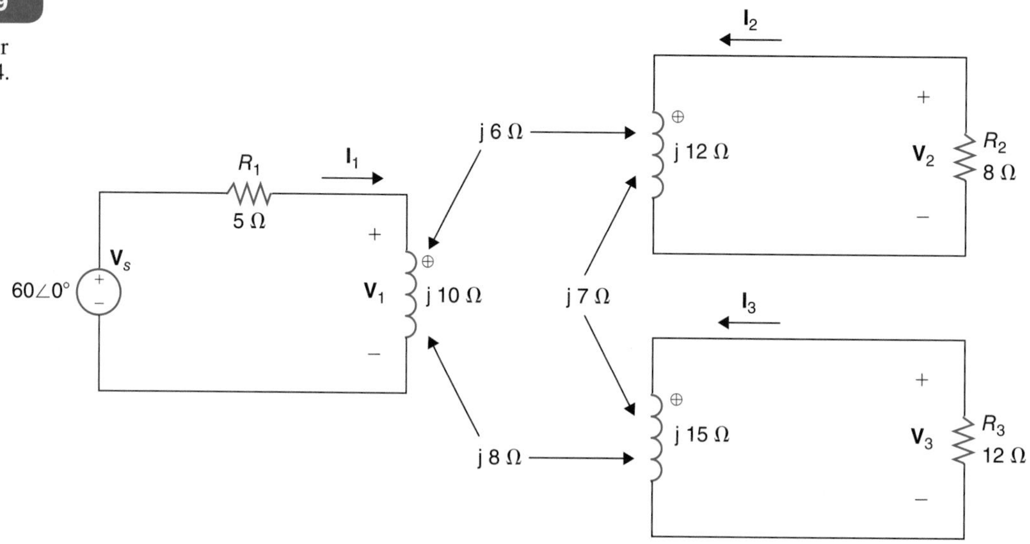

The voltage across R_2 is given by

$$\mathbf{V}_2 = -R_2\mathbf{I}_2 = -8\mathbf{I}_2 \quad (13.71)$$

The voltage across R_3 is given by

$$\mathbf{V}_3 = -R_3\mathbf{I}_3 = -12\mathbf{I}_3 \quad (13.72)$$

From Equation (13.67), we have

$$\mathbf{V}_1 = j10\,\mathbf{I}_1 + j6\,\mathbf{I}_2 + j8\,\mathbf{I}_3 \quad (13.73)$$

Substitution of Equation (13.70) into Equation (13.73) yields

$$60 - 5\mathbf{I}_1 = j10\,\mathbf{I}_1 + j6\,\mathbf{I}_2 + j8\,\mathbf{I}_3$$

which can be rewritten as

$$(5 + j10)\,\mathbf{I}_1 + j6\,\mathbf{I}_2 + j8\,\mathbf{I}_3 = 60 \quad (13.74)$$

From Equation (13.68), we have

$$\mathbf{V}_2 = j6\,\mathbf{I}_1 + j12\,\mathbf{I}_2 + j7\,\mathbf{I}_3 \quad (13.75)$$

Substitution of Equation (13.71) into Equation (13.75) yields

$$-8\mathbf{I}_2 = j6\,\mathbf{I}_1 + j12\,\mathbf{I}_2 + j7\,\mathbf{I}_3$$

which can be rewritten as

$$j6\,\mathbf{I}_1 + (8 + j12)\,\mathbf{I}_2 + j7\,\mathbf{I}_3 = 0 \quad (13.76)$$

From Equation (13.69), we have

$$\mathbf{V}_3 = j8\,\mathbf{I}_1 + j7\,\mathbf{I}_2 + j15\,\mathbf{I}_3 \quad (13.77)$$

continued

Example 13.4 continued

Substitution of Equation (13.72) into Equation (13.77) yields

$$-12\mathbf{I}_3 = j8\,\mathbf{I}_1 + j7\,\mathbf{I}_2 + j15\,\mathbf{I}_3$$

which can be rewritten as

$$j8\,\mathbf{I}_1 + j7\,\mathbf{I}_2 + (12 + j15)\,\mathbf{I}_3 = 0 \tag{13.78}$$

Solving Equations (13.74), (13.76), and (13.78) for $\mathbf{I}_1$, $\mathbf{I}_2$, and $\mathbf{I}_3$, we obtain

$$\mathbf{I}_1 = \frac{\begin{vmatrix} 60 & j6 & j8 \\ 0 & 8 + j12 & j7 \\ 0 & j7 & 12 + j15 \end{vmatrix}}{\begin{vmatrix} 5 + j10 & j6 & j8 \\ j6 & 8 + j12 & j7 \\ j8 & j7 & 12 + j15 \end{vmatrix}} = 4.8304 - j4.3198 = 6.4802\angle -41.8064°\text{ A}$$

$$\mathbf{I}_2 = \frac{\begin{vmatrix} 5 + j10 & 60 & j8 \\ j6 & 0 & j7 \\ j8 & 0 & 12 + j15 \end{vmatrix}}{\begin{vmatrix} 5 + j10 & j6 & j8 \\ j6 & 8 + j12 & j7 \\ j8 & j7 & 12 + j15 \end{vmatrix}} = -1.7689 + 0.7906 = 1.9375\angle 155.9193°\text{ A}$$

$$\mathbf{I}_3 = \frac{\begin{vmatrix} 5 + j10 & j6 & 60 \\ j6 & 8 + j12 & 0 \\ j8 & j7 & 0 \end{vmatrix}}{\begin{vmatrix} 5 + j10 & j6 & j8 \\ j6 & 8 + j12 & j7 \\ j8 & j7 & 12 + j15 \end{vmatrix}} = -2.0114 + j0.3259 = 2.0376\angle 170.7976°\text{ A}$$

Voltages $\mathbf{V}_1$, $\mathbf{V}_2$, and $\mathbf{V}_3$ are found from Equations (13.70) through (13.72):

$$\mathbf{V}_1 = \mathbf{V}_s - R_1\mathbf{I}_1 = 35.8481 + j21.5992 = 41.8522\angle 31.0698°\text{ V}$$

$$\mathbf{V}_2 = -R_2\mathbf{I}_2 = 14.1513 - j6.3244 = 15.5002\angle -24.0807°\text{ V}$$

$$\mathbf{V}_3 = -R_3\mathbf{I}_3 = 24.1368 - j3.9103 = 24.4515\angle -9.2024°\text{ V}$$

MATLAB

```
% EXAMPLE 13.4
clear all;
ZL1=10j;ZL2=12j;ZL3=15j;ZM12=6j;ZM13=8j;ZM23=7j;
Vm=60;phi=0;R1=5;R2=8;R3=12;
Vs=P2Rd(Vm,phi)
syms I1 I2 I3 V1 V2 V3
[I1,I2,I3,V1,V2,V3]=solve(-Vs+R1*I1+V1,...
V2==-R2*I2,V3==-R3*I3,...
V1==ZL1*I1+ZM12*I2+ZM13*I3,V2==ZM12*I1+ZL2*I2+ZM23*I3,...
V3==ZM13*I1+ZM23*I2+ZL3*I3);
I1p=R2P(I1);I2p=R2P(I2);I3p=R2P(I3);V1p=R2P(V1);
V2p=R2P(V2);V3p=R2P(V3);
I1=vpa(I1,6)
```

continued

Example 13.4 continued

MATLAB continued

```
I1p=vpa(I1p,7)
I2=vpa(I2,6)
I2p=vpa(I2p,7)
I3=vpa(I3,6)
I3p=vpa(I3p,7)
V1=vpa(V1,6)
V1p=vpa(V1p,7)
V2=vpa(V2,6)
V2p=vpa(V2p,7)
V3=vpa(V3,6)
V3p=vpa(V3p,7)
```

Exercise 13.4

Find $\mathbf{I}_1$, $\mathbf{I}_2$, and $\mathbf{I}_3$ in the circuit shown in Figure 13.30.

FIGURE 13.30

Circuit for EXERCISE 13.4.

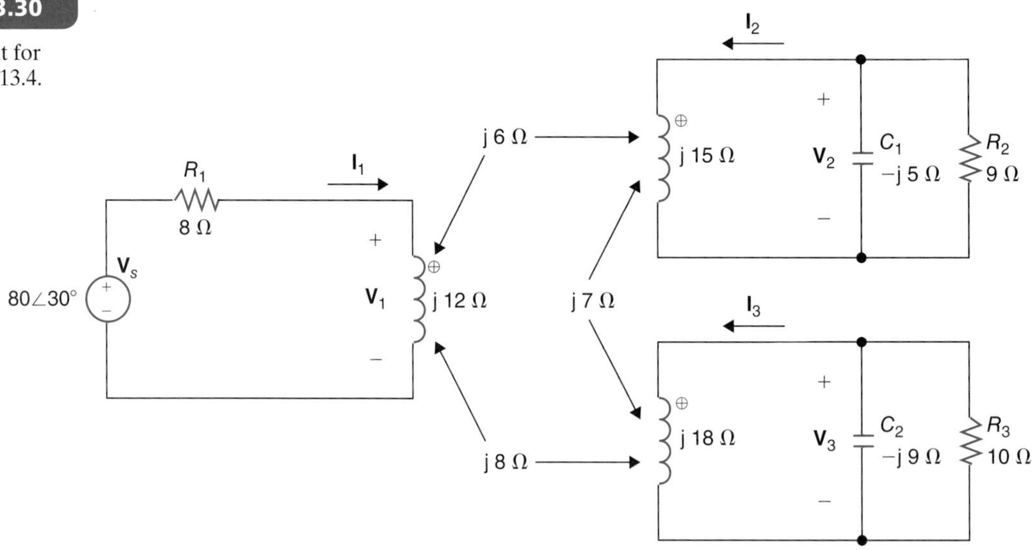

Answer:
$\mathbf{I}_1 = 6.9274 - j0.9645$ A, $\mathbf{I}_2 = -1.9984 + j0.5984$ A, $\mathbf{I}_3 = -2.9277 - j0.735$ A.

13.4 Equivalent Circuits

Figure 13.31 shows the T-equivalent circuit for the circuit shown in Figure 13.13. Writing the mesh equation on the left-side mesh, we have

$$v_1(t) = L_a\frac{di_1(t)}{dt} + L_b\frac{di_1(t)}{dt} + L_b\frac{di_2(t)}{dt} = (L_a + L_b)\frac{di_1(t)}{dt} + L_b\frac{di_2(t)}{dt}$$

Comparing this equation with Equation (13.42), we conclude that

$$L_b = M \tag{13.79}$$

FIGURE 13.31

T-equivalent circuit.

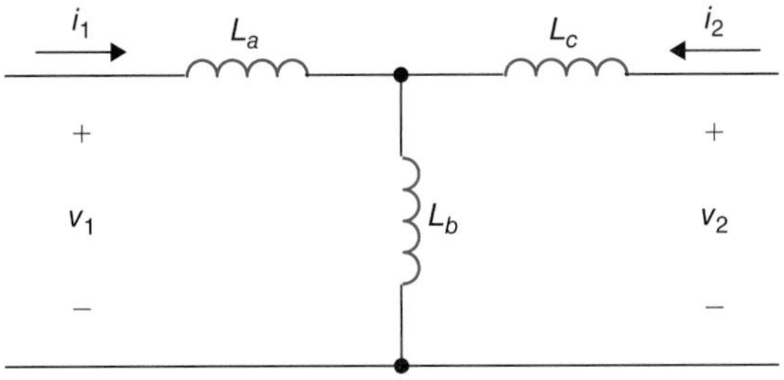

$$L_a = L_1 - M \tag{13.80}$$

Writing the mesh equation on the right-side mesh, we get

$$\begin{aligned} v_2(t) &= L_c\frac{di_2(t)}{dt} + L_b\frac{di_2(t)}{dt} + L_b\frac{di_1(t)}{dt} \\ &= L_b\frac{di_1(t)}{dt} + (L_c + L_b)\frac{di_2(t)}{dt} \end{aligned}$$

Comparing this equation with Equation (13.43), we conclude that

$$L_b = M$$

FIGURE 13.32

T-equivalent circuit with inductance values.

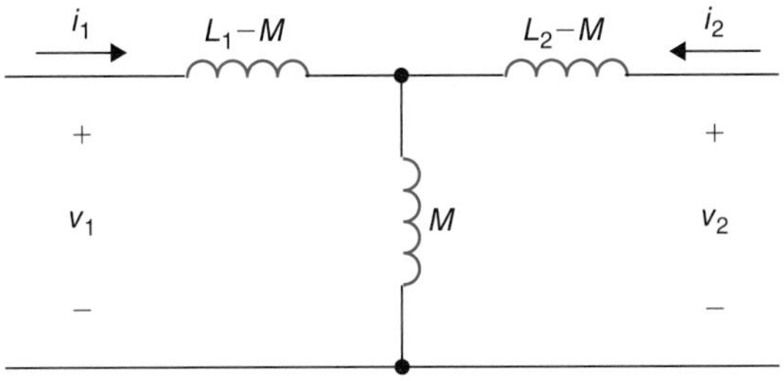

$$L_c = L_2 - M \tag{13.81}$$

Using the values $L_a = L_1 - M$, $L_b = M$, $L_c = L_2 - M$, the circuit shown in Figure 13.31 can be redrawn as that shown in Figure 13.32. Notice that the three inductors in the circuit shown in Figure 13.32 are not coupled. One of the inductance values $L_1 - M$ or $L_2 - M$ in the equivalent circuit can be negative.

Figure 13.33 shows a Π-equivalent (pi-equivalent) circuit for the circuit shown in Figure 13.13.

FIGURE 13.33

Π-equivalent circuit.

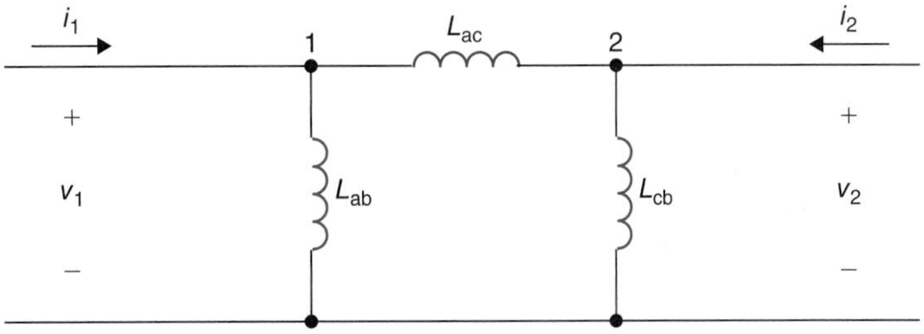

The Π-equivalent circuit shown in Figure 13.33 can be viewed as a Δ-network. Applying the *Y* (*T*-equivalent circuit shown in Figure 13.31) to Δ (Π-equivalent circuit shown in Figure 13.33) transformation, we have

$$\begin{aligned} L_{ab} &= \frac{L_aL_b + L_bL_c + L_cL_a}{L_c} = \frac{(L_1 - M)M + M(L_2 - M) + (L_2 - M)(L_1 - M)}{L_2 - M} \\ &= \frac{L_1L_2 - M^2}{L_2 - M} \end{aligned} \tag{13.82}$$

$$\begin{aligned} L_{cb} &= \frac{L_aL_b + L_bL_c + L_cL_a}{L_a} = \frac{(L_1 - M)M + M(L_2 - M) + (L_2 - M)(L_1 - M)}{L_1 - M} \\ &= \frac{L_1L_2 - M^2}{L_1 - M} \end{aligned} \tag{13.83}$$

$$\begin{aligned} L_{ac} &= \frac{L_aL_b + L_bL_c + L_cL_a}{L_b} = \frac{(L_1 - M)M + M(L_2 - M) + (L_2 - M)(L_1 - M)}{M} \\ &= \frac{L_1L_2 - M^2}{M} \end{aligned} \tag{13.84}$$

EXAMPLE 13.5

Find the equivalent inductance L between a and b for the coupled coils connected in series (additive), as shown in Figure 13.34(a).

Notice that $i_2 = i_1 = i$ and $v = v_1 + v_2$. The circuit shown in Figure 13.34(a) can also be drawn as that shown in Figure 13.34(b). The voltage between a and b is given by

$$v(t) = L\frac{di(t)}{dt} = L_1\frac{di(t)}{dt} + M\frac{di(t)}{dt} + L_2\frac{di(t)}{dt} + M\frac{di(t)}{dt} = (L_1 + L_2 + 2M)\frac{di(t)}{dt}$$

Thus, the equivalent inductance L is given by

$$L = L_1 + L_2 + 2M$$

FIGURE 13.34

Circuit for EXAMPLE 13.5.

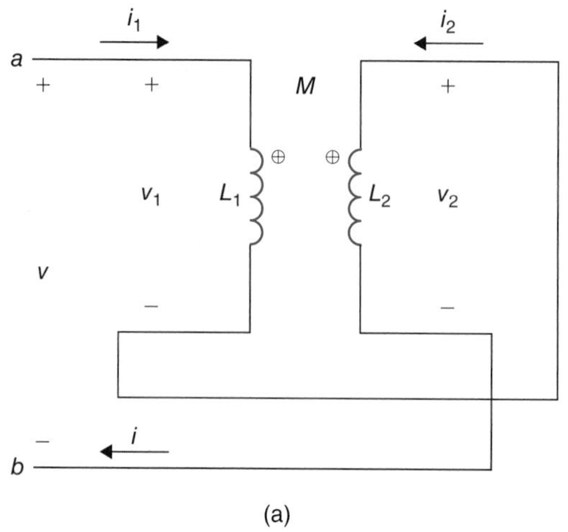

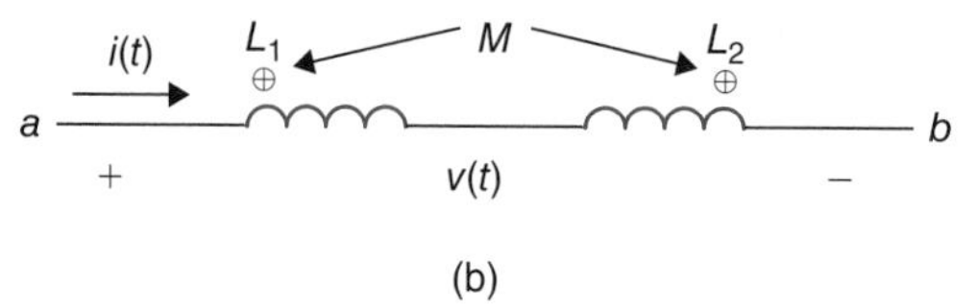

Exercise 13.5

Find the equivalent inductance L between a and b for the coupled coils connected in series (subtractive) shown in Figure 13.35.

FIGURE 13.35

Circuit for EXERCISE 13.5.

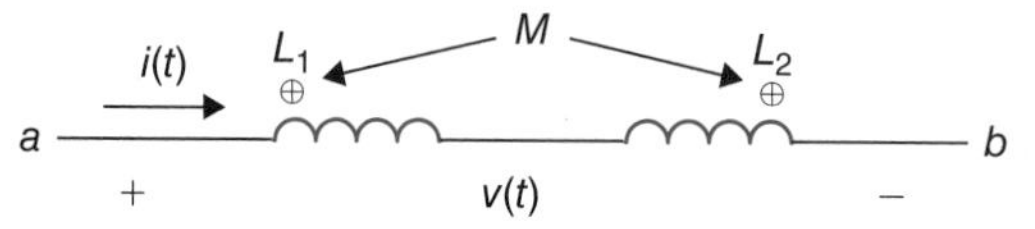

Answer:
$L = L_1 + L_2 - 2M$

EXAMPLE 13.6

Find $\mathbf{I}_1$ and $\mathbf{I}_2$ in the circuit shown in Figure 13.36.

FIGURE 13.36

Circuit for EXAMPLE 13.6.

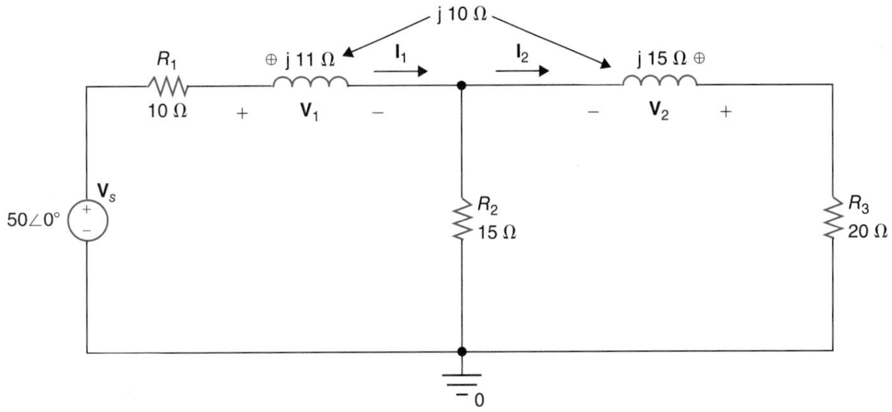

The circuit shown in Figure 13.36 can be redrawn as shown in Figure 13.37.

FIGURE 13.37

Circuit for EXAMPLE 13.6.

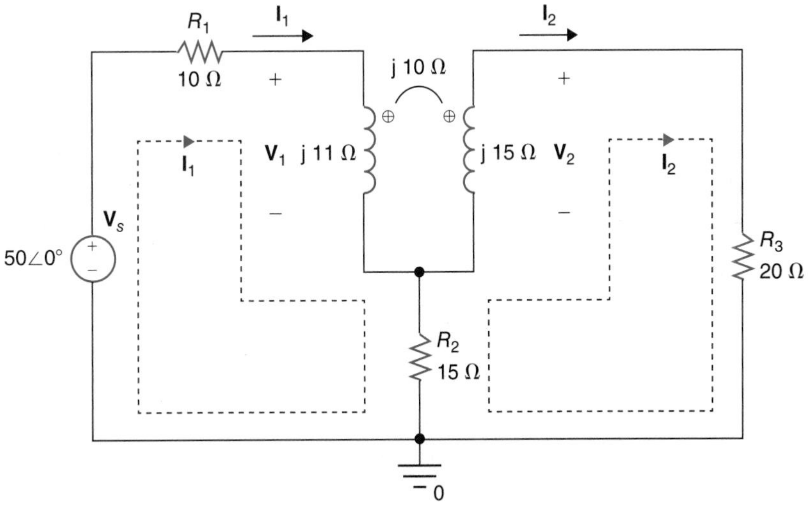

Summing the voltage drops around mesh 1 (left side), we obtain

$$-50 + 10\mathbf{I}_1 + \mathbf{V}_1 + 15(\mathbf{I}_1 - \mathbf{I}_2) = 0 \quad \textbf{(13.85)}$$

Summing the voltage drops around mesh 2 (right side), we obtain

$$15(\mathbf{I}_2 - \mathbf{I}_1) - \mathbf{V}_2 + 20\mathbf{I}_2 = 0 \quad \textbf{(13.86)}$$

On the coupled coils, we have

$$\mathbf{V}_1 = j11\,\mathbf{I}_1 - j10\,\mathbf{I}_2 \quad \textbf{(13.87)}$$

$$\mathbf{V}_2 = j10\,\mathbf{I}_1 - j15\,\mathbf{I}_2 \quad \textbf{(13.88)}$$

continued

Example 13.6 continued

Substitution of Equations (13.87) and (13.88) into Equations (13.85) and (13.86), respectively, yields

$$-50 + 10\mathbf{I}_1 + j11\,\mathbf{I}_1 - j10\,\mathbf{I}_2 + 15(\mathbf{I}_1 - \mathbf{I}_2) = 0$$

$$15(\mathbf{I}_2 - \mathbf{I}_1) - j10\,\mathbf{I}_1 + j15\,\mathbf{I}_2 + 20\mathbf{I}_2 = 0$$

which can be rearranged as

$$(25 + j11)\mathbf{I}_1 + (-15 - j10)\mathbf{I}_2 = 50 \quad (13.89)$$

$$(-15 - j10)\,\mathbf{I}_1 + (35 + j15)\,\mathbf{I}_2 = 0 \quad (13.90)$$

Application of Cramer's rule yields

$$\mathbf{I}_1 = \frac{\begin{vmatrix} 50 & -15 - j10 \\ 0 & 35 + j15 \end{vmatrix}}{\begin{vmatrix} 25 + j11 & -15 - j10 \\ -15 - j10 & 35 + j15 \end{vmatrix}} = \frac{1750 + j750}{585 + j450} = 2.4714 - j0.6613$$
$$= 2.5584\angle -14.9803^\circ \text{ A}$$

$$\mathbf{I}_2 = \frac{\begin{vmatrix} 25 + j11 & 50 \\ -15 - j10 & 0 \end{vmatrix}}{\begin{vmatrix} 25 + j11 & -15 - j10 \\ -15 - j10 & 35 + j15 \end{vmatrix}} = \frac{750 + j500}{585 + j450} = 1.2075 - j0.0948$$
$$= 1.2112\angle -4.4888^\circ \text{ A}$$

MATLAB

```
clear all;
ZL1=11j;ZL2=15j;ZM=10j;R1=10;R2=15;R3=20;Vm=50;phi=0;
Vs=P2Rd(Vm,phi)
syms I1 I2 V1 V2
[I1,I2,V1,V2]=solve(-Vs+R1*I1+V1+R2*(I1-I2),...
R2*(I2-I1)-V2+R3*I2,...
V1==ZL1*I1-ZM*I2,V2==ZM*I1-ZL2*I2);
I1p=R2P(I1);
I2p=R2P(I2);
V1p=R2P(V1);
V2p=R2P(V2);
I1=vpa(I1,6)
I1p=vpa(I1p,7)
I2=vpa(I2,6)
I2p=vpa(I2p,7)
V1=vpa(V1,6)
V1p=vpa(V1p,7)
V2=vpa(V2,6)
V2p=vpa(V2p,7)
```

Exercise 13.6

Find $\mathbf{I}_1$ and $\mathbf{I}_2$ in the circuit shown in Figure 13.38.

FIGURE 13.38

Circuit for EXERCISE 13.6.

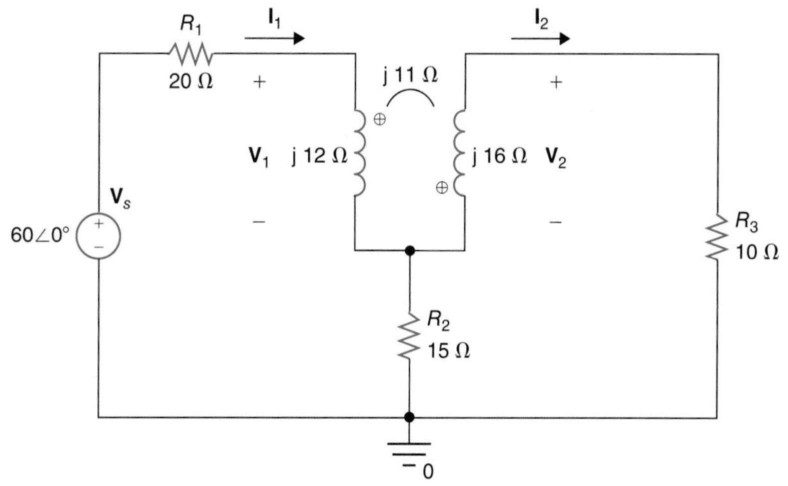

Answer:
$\mathbf{I}_1 = 1.3457\angle -31.4353^\circ$ A, $\mathbf{I}_2 = 0.8433\angle -100.3084^\circ$ A.

13.5 Energy of Coupled Coils

In the circuit shown in Figure 13.13, the instantaneous power absorbed by the coupled coils is given by

$$\begin{aligned} p(t) &= i_1(t)v_1(t) + i_2(t)v_2(t) = i_1(t)\left[L_1\frac{di_1(t)}{dt} + M\frac{di_2(t)}{dt}\right] + i_2(t)\left[M\frac{di_1(t)}{dt} + L_2\frac{di_2(t)}{dt}\right] \\ &= L_1\left[i_1(t)\frac{di_1(t)}{dt}\right] + M\left[i_1(t)\frac{di_2(t)}{dt} + i_2(t)\frac{di_1(t)}{dt}\right] + L_2\left[i_2(t)\frac{di_2(t)}{dt}\right] \\ &= L_1\left[i_1(t)\frac{di_1(t)}{dt}\right] + M\frac{d}{dt}[i_1(t)i_2(t)] + L_2\left[i_2(t)\frac{di_2(t)}{dt}\right] \end{aligned} \quad \textbf{(13.91)}$$

The total energy stored in the coupled coils is obtained by integrating Equation (13.91)

$$\begin{aligned} w(t) &= L_1\int i_1(t)\frac{di_1(t)}{dt}dt + M\int \frac{d}{dt}[i_1(t)i_2(t)]\,dt + L_2\int i_2(t)\frac{di_2(t)}{dt}dt \\ &= L_1\int i_1(t)di_1(t) + Mi_1(t)i_2(t) + L_2\int i_2(t)di_2(t) \\ w(t) &= \frac{1}{2}L_1i_1^2(t) + Mi_1(t)i_2(t) + \frac{1}{2}L_2i_2^2(t) \end{aligned} \quad \textbf{(13.92)}$$

For the circuit shown in Figure 13.21, the total energy in the coupled coils is given by

$$w(t) = \frac{1}{2}L_1i_1^2(t) - Mi_1(t)i_2(t) + \frac{1}{2}L_2i_2^2(t) \quad \textbf{(13.93)}$$

For other configurations, the total energy in the coupled coils is given either by Equation (13.92) or Equation (13.93). Combining Equations (13.92) and (13.93), the total energy in the coupled coils can be written as

$$w(t) = \frac{1}{2}L_1 i_1^2(t) \pm M i_1(t) i_2(t) + \frac{1}{2}L_2 i_2^2(t) \tag{13.94}$$

If both currents enter the dotted terminals, or both currents leave the dotted terminals, the sign is positive. Otherwise, the sign is negative. Equation (13.94) can be written as

$$w(t) = \frac{1}{2}i_1^2(t)\left[L_1 \pm 2M\frac{i_2(t)}{i_1(t)} + L_2\frac{i_2^2(t)}{i_1^2(t)}\right] = \frac{1}{2}L_2 i_1^2(t)\left[\frac{L_1}{L_2} \pm 2\frac{M}{L_2}\frac{i_2(t)}{i_1(t)} + \frac{i_2^2(t)}{i_1^2(t)}\right]$$

$$= \frac{1}{2}L_2 i_1^2(t)\left[\left(\frac{i_2(t)}{i_1(t)} \pm \frac{M}{L_2}\right)^2 - \left(\frac{M}{L_2}\right)^2 + \frac{L_1}{L_2}\right] = \frac{1}{2}L_2 i_1^2(t)\left[\left(\frac{i_2(t)}{i_1(t)} \pm \frac{M}{L_2}\right)^2 + \frac{-M^2 + L_1 L_2}{L_2^2}\right]$$

$$w(t) = \frac{1}{2}L_2 i_1^2(t)\left(\frac{i_2(t)}{i_1(t)} \pm \frac{M}{L_2}\right)^2 + \frac{i_1^2(t)}{2L_2}(-M^2 + L_1 L_2) \tag{13.95}$$

For the given value of $i_1(t)$, the energy is a parabola as a function of $i_2(t)$. The energy is minimized when

$$i_2(t) = \mp\frac{M}{L_2}i_1(t) \tag{13.96}$$

The minimum value of the energy is given by

$$w(t)_{\min} = \frac{i_1^2(t)}{2L_2}(-M^2 + L_1 L_2) \tag{13.97}$$

Since the energy cannot be negative, we have

$$-M^2 + L_1 L_2 \geq 0$$

which can be rewritten as

$$M \leq \sqrt{L_1 L_2} \tag{13.98}$$

The coupling coefficient is defined as

$$k = \frac{M}{\sqrt{L_1 L_2}} \tag{13.99}$$

From Equation (13.98), we have

$$0 \leq k \leq 1 \tag{13.100}$$

The coupling coefficient k is between zero (no coupling) and one (perfect coupling). If k is greater than 0.5, the coils are called *tightly coupled*, and if k is less than 0.5, the coils are called *loosely coupled*.

If $k = 1$, the coupling between the coils is perfect. In this case, the mutual inductance is the square root of the product of L_1 and L_2

$$M = \sqrt{L_1 L_2} \tag{13.101}$$

For the perfect coupling, the minimum energy given by Equation (13.97) is zero.

$$w(t)_{\min} = \frac{i_1^2(t)}{2L_2}(-M^2 + L_1L_2) = 0 \tag{13.102}$$

If Equation (13.96) is satisfied

$$i_2(t) = \mp\frac{M}{L_2}i_1(t)$$

The energy stored on the coupled coils given by Equation (13.95) is zero for $k = 1$. Equation (13.96) can be rewritten as

$$\frac{i_2(t)}{i_1(t)} = \mp\frac{M}{L_2} = \mp\frac{\sqrt{L_1L_2}}{L_2} = \mp\sqrt{\frac{L_1}{L_2}} \tag{13.103}$$

For the coil wrapped around solenoid, the coupling is perfect ($k = 1$), and we have, from Equations (13.29) through (13.31),

$$L_1 = \frac{\mu_0 S}{\ell}N_1^2, \quad L_2 = \frac{\mu_0 S}{\ell}N_2^2, \quad M = \frac{\mu_0 S}{\ell}N_1N_2 \tag{13.104}$$

Thus, Equation (13.103) becomes

$$\frac{i_2(t)}{i_1(t)} = \mp\sqrt{\frac{L_1}{L_2}} = \mp\sqrt{\frac{\frac{\mu_0 S}{\ell}N_1^2}{\frac{\mu_0 S}{\ell}N_2^2}} = \mp\frac{N_1}{N_2} = \mp\frac{1}{\frac{N_2}{N_1}} = \mp\frac{1}{n} \tag{13.105}$$

where

$$n = \frac{N_2}{N_1} \tag{13.106}$$

is called the *turns ratio*. Equation (13.105) says that for perfect coupling, the ratio of currents $i_2(t)/i_1(t)$ is given by the ratio $\mp N_1/N_2 = \mp 1/n$. Equation (13.105) can be written as

$$i_1(t)N_1 = -i_2(t)N_2 \tag{13.107}$$

Since $p(t) = dw(t)/dt$, if energy is zero, the power is also zero. If the coupled coils have perfect coupling, the power loss on the coils is zero. According to Equation (13.105), we have

$$n = \sqrt{\frac{L_2}{L_1}}$$

13.6 Linear Transformer

A *transformer* is a device that changes voltage from one level to another level. When the voltage is transformed from a high level to a low level, it is called a *step-down transformer*, and when the voltage is transformed from a low level to a high level, it is called a *step-up transformer*. Most of the transformers are step-down transformers. Two windings around a ferromagnetic core are used as transformers. The input ac signal is applied to the primary winding, and the output is taken from the secondary winding. The transformer is said to be linear if the permeability of the material used for the core is constant. A typical linear transformer with source impedance Z_s and load impedance Z_L is shown in Figure 13.39.

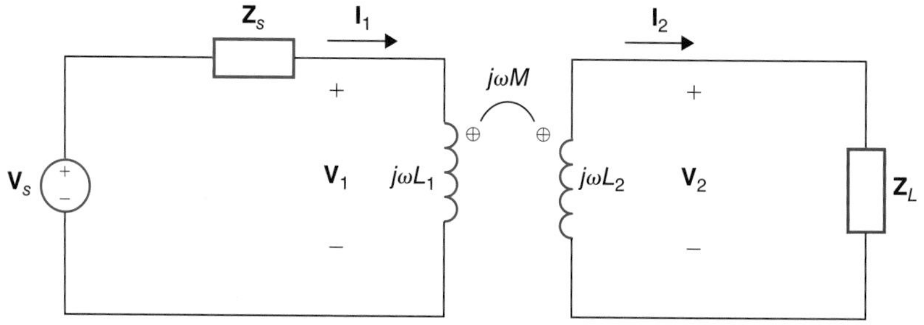

FIGURE 13.39
A typical linear transformer.

From Equations (13.48) and (13.49), we have

$$\mathbf{V}_1 = j\omega L_1\mathbf{I}_1 - j\omega M\mathbf{I}_2 \quad \textbf{(13.108)}$$

$$\mathbf{V}_2 = j\omega M\mathbf{I}_1 - j\omega L_2\mathbf{I}_2 \quad \textbf{(13.109)}$$

Summing the voltage drops around mesh 1 (left side), we obtain

$$-\mathbf{V}_s + \mathbf{Z}_s\mathbf{I}_1 + \mathbf{V}_1 = 0 \quad \textbf{(13.110)}$$

Substituting Equation (13.108) into Equation (13.110), we obtain

$$-\mathbf{V}_s + \mathbf{Z}_s\mathbf{I}_1 + j\omega L_1\mathbf{I}_1 - j\omega M\mathbf{I}_2 = 0$$

which can be rearranged as

$$\mathbf{V}_s = (\mathbf{Z}_s + j\omega L_1)\mathbf{I}_1 - j\omega M\mathbf{I}_2 \quad \textbf{(13.111)}$$

Summing the voltage drops around mesh 2 on the right side, we obtain

$$-\mathbf{V}_2 + \mathbf{Z}_L\mathbf{I}_2 = 0 \quad \textbf{(13.112)}$$

Substituting Equation (13.109) into Equation (13.112), we obtain

$$-j\omega M\mathbf{I}_1 + j\omega L_2\mathbf{I}_2 + \mathbf{Z}_L\mathbf{I}_2 = 0$$

which can be rearranged as

$$-j\omega M\mathbf{I}_1 + (j\omega L_2 + \mathbf{Z}_L)\mathbf{I}_2 = 0 \quad \textbf{(13.113)}$$

Let

$$\mathbf{Z}_{11} = \mathbf{Z}_s + j\omega L_1 \quad \textbf{(13.114)}$$

and

$$\mathbf{Z}_{22} = j\omega L_2 + \mathbf{Z}_L \quad \textbf{(13.115)}$$

Then, Equations (13.111) and (13.113) become, respectively,

$$\mathbf{V}_s = \mathbf{Z}_{11}\mathbf{I}_1 - j\omega M\mathbf{I}_2 \quad \textbf{(13.116)}$$

$$-j\omega M\mathbf{I}_1 + \mathbf{Z}_{22}\mathbf{I}_2 = 0 \quad \textbf{(13.117)}$$

Solving Equation (13.117) for $\mathbf{I}_2$, we obtain

$$\mathbf{I}_2 = \frac{j\omega M\mathbf{I}_1}{\mathbf{Z}_{22}} \quad \textbf{(13.118)}$$

Substitution of Equation (13.118) into Equation (13.116) yields

$$\mathbf{V}_s = \mathbf{Z}_{11}\mathbf{I}_1 - j\omega M \frac{j\omega M\mathbf{I}_1}{\mathbf{Z}_{22}} = \left(\mathbf{Z}_{11} + \frac{(\omega M)^2}{\mathbf{Z}_{22}}\right)\mathbf{I}_1 \tag{13.119}$$

The input impedance from the voltage source is given by

$$\mathbf{Z}_{\text{in}} = \frac{\mathbf{V}_s}{\mathbf{I}_1} = \mathbf{Z}_{11} + \frac{(\omega M)^2}{\mathbf{Z}_{22}} \tag{13.120}$$

The second term is the reflected impedance from the output side to the input side. Equation (13.120) can also be written in terms of resistance and reactance:

$$\begin{aligned}\mathbf{Z}_{\text{in}} &= \frac{\mathbf{V}_s}{\mathbf{I}_1} = R_{11} + jX_{11} + \frac{(\omega M)^2}{R_{22} + jX_{22}} = R_{11} + jX_{11} + \frac{(\omega M)^2 R_{22}}{R_{22}^2 + X_{22}^2} - j\frac{(\omega M)^2 X_{22}}{R_{22}^2 + X_{22}^2}\\ &= R_{11} + \frac{(\omega M)^2 R_{22}}{R_{22}^2 + X_{22}^2} + j\left(X_{11} - \frac{(\omega M)^2 X_{22}}{R_{22}^2 + X_{22}^2}\right)\end{aligned} \tag{13.121}$$

From Equation (13.119), we have

$$\mathbf{I}_1 = \frac{\mathbf{V}_s}{\mathbf{Z}_{11} + \dfrac{(\omega M)^2}{\mathbf{Z}_{22}}} \tag{13.122}$$

From Equation (13.118), we have

$$\mathbf{I}_2 = \frac{j\omega M}{\mathbf{Z}_{22}}\mathbf{I}_1 = \frac{j\omega M}{\mathbf{Z}_{22}} \frac{\mathbf{V}_s}{\mathbf{Z}_{11} + \dfrac{(\omega M)^2}{\mathbf{Z}_{22}}} = \frac{j\omega M}{\mathbf{Z}_{11}\mathbf{Z}_{22} + (\omega M)^2}\mathbf{V}_s$$

The voltage $\mathbf{V}_1$ is given by

$$\mathbf{V}_1 = \mathbf{V}_s - \mathbf{Z}_s\mathbf{I}_1 = \left(\mathbf{Z}_{11} + \frac{(\omega M)^2}{\mathbf{Z}_{22}}\right)\mathbf{I}_1 - \mathbf{Z}_s\mathbf{I}_1 = \left(j\omega L_1 + \frac{(\omega M)^2}{\mathbf{Z}_{22}}\right)\mathbf{I}_1$$

The impedance to the right of $\mathbf{Z}_s$ is given by

$$\frac{\mathbf{V}_1}{\mathbf{I}_1} = j\omega L_1 + \frac{(\omega M)^2}{\mathbf{Z}_{22}} \tag{13.123}$$

The voltage $\mathbf{V}_2$ is given by

$$\mathbf{V}_2 = \mathbf{Z}_L\mathbf{I}_2 = \frac{j\omega M\mathbf{Z}_L}{\mathbf{Z}_{11}\mathbf{Z}_{22} + (\omega M)^2}\mathbf{V}_s \tag{13.124}$$

We are interested in finding the Thévenin equivalent circuit to the left of terminals a and b in the circuit shown in Figure 13.40.

Notice that $\mathbf{I}_2 = 0$. From Equations (13.108) and (13.109), we obtain

$$\mathbf{V}_1 = j\omega L_1\mathbf{I}_1 \tag{13.125}$$

$$\mathbf{V}_2 = j\omega M\mathbf{I}_1 \tag{13.126}$$

Summing the voltage drops around mesh 1 (left side), we obtain

$$-\mathbf{V}_s + \mathbf{Z}_s\mathbf{I}_1 + \mathbf{V}_1 = 0$$

FIGURE 13.40

A linear transformer.

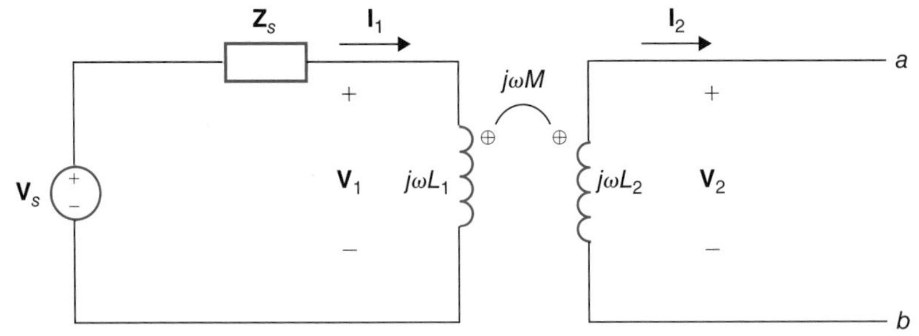

which can be rearranged as

$$\mathbf{V}_1 = \mathbf{V}_s - \mathbf{Z}_s\mathbf{I}_1$$

Substituting this equation into Equation (13.125), we get

$$\mathbf{V}_s - \mathbf{Z}_s\mathbf{I}_1 = j\omega L_1\mathbf{I}_1$$

Solving for $\mathbf{I}_1$, we obtain

$$\mathbf{I}_1 = \frac{\mathbf{V}_s}{\mathbf{Z}_s + j\omega L_1}$$

Substituting this equation into Equation (13.126), we obtain the Thévenin-equivalent voltage:

$$\mathbf{V}_{\text{th}} = \mathbf{V}_2 = \frac{j\omega M}{\mathbf{Z}_s + j\omega L_1}\mathbf{V}_s \qquad \textbf{(13.127)}$$

To find the Thévenin equivalent impedance, we deactivate the voltage source by short-circuiting it and apply a test voltage across a and b, as shown in Figure 13.41.

FIGURE 13.41

Circuit with test voltage with source deactivated.

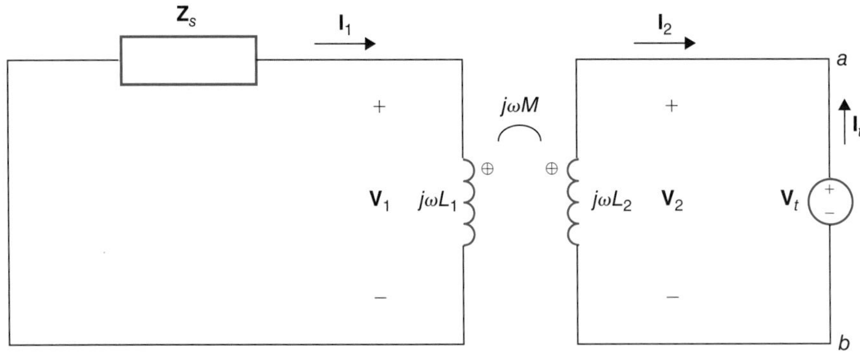

From Equations (13.108) and (13.109), we obtain

$$\mathbf{V}_1 = j\omega L_1\mathbf{I}_1 - j\omega M\mathbf{I}_2 \qquad \textbf{(13.128)}$$

$$\mathbf{V}_t = j\omega M\mathbf{I}_1 - j\omega L_2\mathbf{I}_2 \qquad \textbf{(13.129)}$$

Summing the voltage drops around mesh 1 (left side), we obtain

$$\mathbf{Z}_s\mathbf{I}_1 + \mathbf{V}_1 = 0$$

which can be rearranged as

$$\mathbf{V}_1 = -\mathbf{Z}_s\mathbf{I}_1$$

Substituting this equation into Equation (13.128), we get

$$-\mathbf{Z}_s\mathbf{I}_1 = j\omega L_1\mathbf{I}_1 - j\omega M\mathbf{I}_2$$

Solving for $\mathbf{I}_1$, we obtain

$$\mathbf{I}_1 = \frac{j\omega M}{\mathbf{Z}_s + j\omega L_1}\mathbf{I}_2$$

Substituting this equation into Equation (13.129), we obtain

$$\mathbf{V}_t = j\omega M\frac{j\omega M}{\mathbf{Z}_s + j\omega L_1}\mathbf{I}_2 - j\omega L_2\mathbf{I}_2 = -\mathbf{I}_2\left(j\omega L_2 + \frac{(\omega M)^2}{\mathbf{Z}_s + j\omega L_1}\right)$$

The Thévenin equivalent impedance is given by

$$\mathbf{Z}_{\text{th}} = \frac{\mathbf{V}_t}{\mathbf{I}_t} = \frac{\mathbf{V}_t}{-\mathbf{I}_2} = j\omega L_2 + \frac{(\omega M)^2}{\mathbf{Z}_s + j\omega L_1} \tag{13.130}$$

Notice that the second term of Equation (13.130) is the reflected impedance of $\mathbf{Z}_s + j\omega L_1$ from the input side to the output side. The Thévenin equivalent circuit is shown in Figure 13.42. Let $\mathbf{Z}_a$ be the sum of the impedances around the input side, and let $\mathbf{Z}_b$ be the sum of the impedances around the output side, excluding the load impedance. In general, if $\mathbf{I}_2 = 0$, we have

FIGURE 13.42

Thévenin-equivalent circuit.

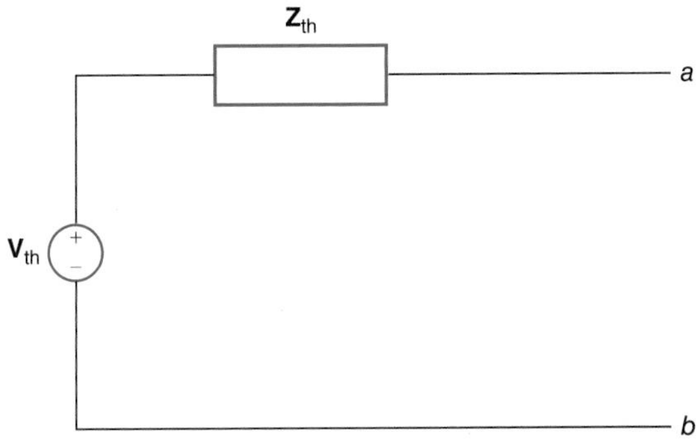

$$\mathbf{V}_{\text{th}} = \frac{j\omega M}{\mathbf{Z}_a}\mathbf{V}_s \tag{13.131}$$

and

$$\mathbf{Z}_{\text{th}} = \mathbf{Z}_b + \frac{(\omega M)^2}{\mathbf{Z}_a} \tag{13.132}$$

The first term represents the impedance of the output side looking from terminals a and b, and the second term represents the reflected impedance from the input side to the output side.

When a load with impedance $\mathbf{Z}_L$ is connected between a and b, we obtain the circuit shown in Figure 13.43.

The current through the load is given by

$$\begin{aligned}\mathbf{I}_2 = \frac{\mathbf{V}_{\text{th}}}{\mathbf{Z}_{\text{th}} + \mathbf{Z}_L} &= \frac{\dfrac{j\omega M}{\mathbf{Z}_s + j\omega L_1}\mathbf{V}_s}{\mathbf{Z}_L + j\omega L_2 + \dfrac{(\omega M)^2}{\mathbf{Z}_s + j\omega L_1}} \\ &= \frac{j\omega M}{(\mathbf{Z}_s + j\omega L_1)(\mathbf{Z}_L + j\omega L_2) + (\omega M)^2}\mathbf{V}_s \\ &= \frac{j\omega M}{\mathbf{Z}_{11}\mathbf{Z}_{22} + (\omega M)^2}\mathbf{V}_s\end{aligned} \tag{13.133}$$

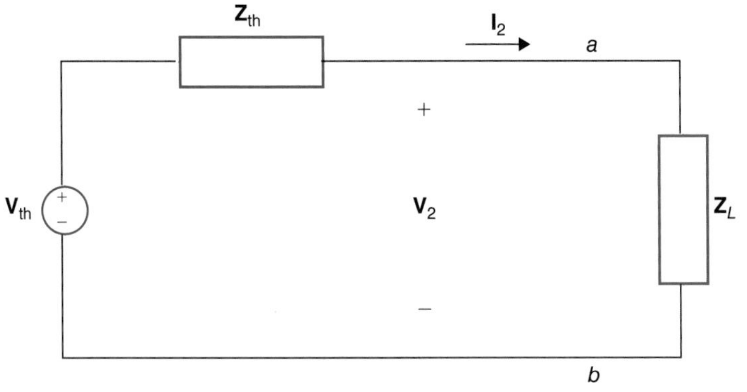

FIGURE 13.43
Thévenin equivalent circuit with load impedance $\mathbf{Z}_L$.

where

$$\mathbf{Z}_{11} = \mathbf{Z}_s + j\omega L_1, \quad \mathbf{Z}_{22} = \mathbf{Z}_L + j\omega L_2 \tag{13.134}$$

The voltage across the load is given by

$$\begin{aligned}\mathbf{V}_2 = \mathbf{Z}_L\mathbf{I}_2 &= \frac{\mathbf{V}_{th}\mathbf{Z}_L}{\mathbf{Z}_{th} + \mathbf{Z}_L} = \frac{j\omega M\mathbf{Z}_L}{(\mathbf{Z}_s + j\omega L_1)(\mathbf{Z}_L + j\omega L_2) + (\omega M)^2}\mathbf{V}_s \\ &= \frac{j\omega M\mathbf{Z}_L}{\mathbf{Z}_{11}\mathbf{Z}_{22} + (\omega M)^2}\mathbf{V}_s\end{aligned} \tag{13.135}$$

EXAMPLE 13.7

Find $\mathbf{I}_1$, $\mathbf{I}_2$, $\mathbf{V}_1$, and $\mathbf{V}_2$ in the circuit shown in Figure 13.44.

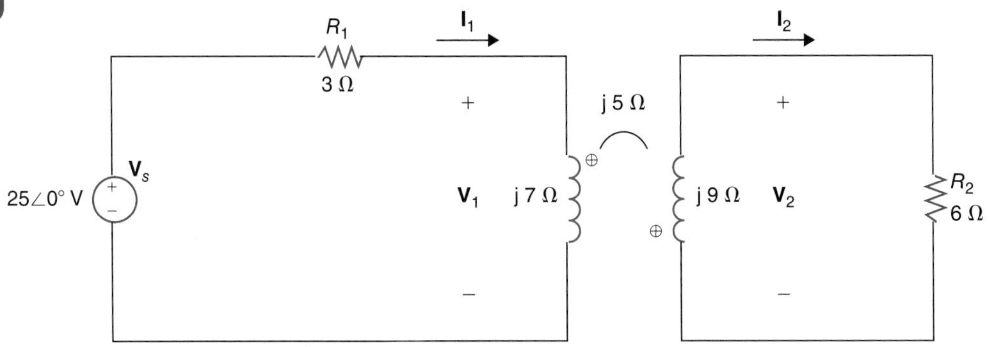

FIGURE 13.44
Circuit for EXAMPLE 13.7.

When $\mathbf{I}_2$ is replaced by $-\mathbf{I}_2$ in Equations (13.56) and (13.57), we obtain

$$\mathbf{V}_1 = j\omega L_1\mathbf{I}_1 + j\omega M\mathbf{I}_2 = j7\,\mathbf{I}_1 + j5\,\mathbf{I}_2 \tag{13.136}$$

$$\mathbf{V}_2 = -j\omega M\mathbf{I}_1 - j\omega L_2\mathbf{I}_2 = -j5\,\mathbf{I}_1 - j9\,\mathbf{I}_2 \tag{13.137}$$

Summing the voltage drops around mesh 1 (left side), we obtain

$$-25 + 3\,\mathbf{I}_1 + \mathbf{V}_1 = 0 \tag{13.138}$$

continued

Example 13.7 continued

Substituting Equation (13.136) into Equation (13.138), we obtain

$$-25 + 3\,\mathbf{I}_1 + j7\,\mathbf{I}_1 + j5\,\mathbf{I}_2 = 0$$

which can be rearranged as

$$(3 + j7)\,\mathbf{I}_1 + j5\,\mathbf{I}_2 = 25 \quad \textbf{(13.139)}$$

Summing the voltage drops around mesh 2 on the right side, we obtain

$$-\mathbf{V}_2 + 6\,\mathbf{I}_2 = 0 \quad \textbf{(13.140)}$$

Substituting Equation (13.137) into Equation (13.140), we obtain

$$j5\,\mathbf{I}_1 + j9\,\mathbf{I}_2 + 6\,\mathbf{I}_2 = 0$$

which can be rearranged as

$$j5\,\mathbf{I}_1 + (6 + j9)\,\mathbf{I}_2 = 0 \quad \textbf{(13.141)}$$

Solving Equation (13.141) for $\mathbf{I}_1$, we obtain

$$\mathbf{I}_1 = (-1.8 + j1.2)\mathbf{I}_2$$

Substituting $\mathbf{I}_1$ into Equation (13.139), we get

$$(3 + j7)\,(-1.8 + j1.2)\mathbf{I}_2 + j5\,\mathbf{I}_2 = 25$$

Solving for $\mathbf{I}_2$, we get

$$\mathbf{I}_2 = -1.6712 + j0.4844 = 1.74\angle 163.8355^\circ \text{ A}$$

Substituting $\mathbf{I}_2$ into $\mathbf{I}_1 = (-1.8 + j1.2)\mathbf{I}_2$, we get

$$\mathbf{I}_1 = 2.4269 - j2.8773 = 3.7641\angle -49.8546^\circ \text{ A}$$

Alternatively, Cramer's rule can be used to solve Equations (13.139) and (13.141):

$$\mathbf{I}_1 = \frac{\begin{vmatrix} 25 & j5 \\ 0 & 6 + j9 \end{vmatrix}}{\begin{vmatrix} 4 + j7 & j5 \\ j5 & 6 + j9 \end{vmatrix}} = 2.4269 - j2.8773 = 3.7641\angle -49.8546^\circ \text{ A}$$

$$\mathbf{I}_2 = \frac{\begin{vmatrix} 4 + j7 & 25 \\ j5 & 0 \end{vmatrix}}{\begin{vmatrix} 4 + j7 & j5 \\ j5 & 6 + j9 \end{vmatrix}} = -1.6712 + j0.4844 = 1.74\angle 163.8355^\circ \text{ A}$$

Voltages $\mathbf{V}_1$ and $\mathbf{V}_2$ are given, respectively, by

$$\mathbf{V}_1 = 25 - 3\,\mathbf{I}_1 = 17.7194 + j8.6320 = 19.7102\angle 25.9731^\circ \text{ V}$$

$$\mathbf{V}_2 = 6\,\mathbf{I}_2 = -10.0271 + j2.9064 = 10.4399\angle 163.8355^\circ \text{ V}$$

Exercise 13.7

Let $\mathbf{V}_s = 50$ V, $\mathbf{Z}_s = 6 + j10$, $\mathbf{Z}_L = 3 + j6$, $j\omega L_1 = j9$, $j\omega L_2 = j10$, $j\omega M = j7$ in the circuit shown in Figure 13.45. Find $\mathbf{V}_2$.

FIGURE 13.45

Circuit for EXERCISE 13.7.

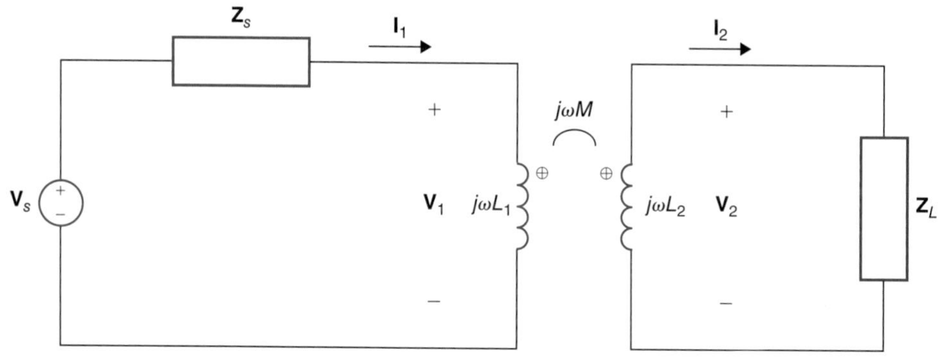

Answer:
$\mathbf{V}_2 = 8.273 + j0.9104\ \text{V} = 8.323\angle 6.28°\ \text{V}.$

Consider a circuit shown in Figure 13.46. The resistors R_1 and R_2 represent the winding resistances of a transformer.

FIGURE 13.46

A typical linear transformer.

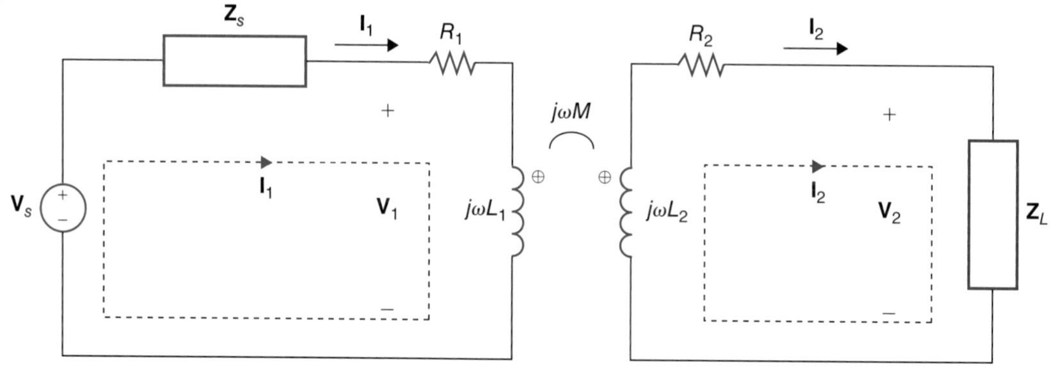

Summing the voltage drops around mesh 1 on the left side, we obtain

$$-\mathbf{V}_s + \mathbf{Z}_s\mathbf{I}_1 + R_1\mathbf{I}_1 + j\omega L_1\mathbf{I}_1 - j\omega M\mathbf{I}_2 = 0 \tag{13.142}$$

Summing the voltage drops around mesh 2 on the right side, we obtain

$$j\omega L_2\mathbf{I}_2 - j\omega M\mathbf{I}_1 + R_2\mathbf{I}_2 + \mathbf{Z}_L\mathbf{I}_2 = 0 \tag{13.143}$$

Let

$$\mathbf{Z}_{11} = \mathbf{Z}_s + R_1 + j\omega L_1 \tag{13.144}$$

and

$$\mathbf{Z}_{22} = j\omega L_2 + R_2 + \mathbf{Z}_L \tag{13.145}$$

Then, Equations (13.142) and (13.143) become, respectively,

$$\mathbf{V}_s = \mathbf{Z}_{11}\mathbf{I}_1 - j\omega M\mathbf{I}_2 \tag{13.146}$$

$$0 = -j\omega M\mathbf{I}_1 + \mathbf{Z}_{22}\mathbf{I}_2 \tag{13.147}$$

Solving Equation (13.147) for $\mathbf{I}_2$, we obtain

$$\mathbf{I}_2 = \frac{j\omega M\mathbf{I}_1}{\mathbf{Z}_{22}} \tag{13.148}$$

Substitution of Equation (13.148) into Equation (13.146) yields

$$\mathbf{V}_s = \mathbf{Z}_{11}\mathbf{I}_1 - j\omega M\frac{j\omega M\mathbf{I}_1}{\mathbf{Z}_{22}} = \left(\mathbf{Z}_{11} + \frac{(\omega M)^2}{\mathbf{Z}_{22}}\right)\mathbf{I}_1 \tag{13.149}$$

The input impedance from the voltage source is given by

$$\mathbf{Z}_{\text{in}} = \frac{\mathbf{V}_s}{\mathbf{I}_1} = \mathbf{Z}_{11} + \frac{(\omega M)^2}{\mathbf{Z}_{22}} \tag{13.150}$$

The second term is the reflected impedance. Equation (13.150) can also be written in terms of resistance and reactance

$$\begin{aligned}\mathbf{Z}_{\text{in}} = \frac{\mathbf{V}_s}{\mathbf{I}_1} &= R_{11} + jX_{11} + \frac{(\omega M)^2}{R_{22} + jX_{22}} = R_{11} + jX_{11} + \frac{(\omega M)^2 R_{22}}{R_{22}^2 + X_{22}^2} - j\frac{(\omega M)^2 X_{22}}{R_{22}^2 + X_{22}^2} \\ &= R_{11} + \frac{(\omega M)^2 R_{22}}{R_{22}^2 + X_{22}^2} + j\left(X_{11} - \frac{(\omega M)^2 X_{22}}{R_{22}^2 + X_{22}^2}\right)\end{aligned} \tag{13.151}$$

From Equation (13.149), we have

$$\mathbf{I}_1 = \frac{\mathbf{V}_s}{\mathbf{Z}_{11} + \dfrac{(\omega M)^2}{\mathbf{Z}_{22}}} \tag{13.152}$$

From Equation (13.148), we have

$$\mathbf{I}_2 = \frac{j\omega M}{\mathbf{Z}_{22}}\mathbf{I}_1 \tag{13.153}$$

Voltage $\mathbf{V}_1$ is given by

$$\mathbf{V}_1 = \mathbf{V}_s - \mathbf{Z}_s\mathbf{I}_1 = \left(\mathbf{Z}_{11} + \frac{(\omega M)^2}{\mathbf{Z}_{22}}\right)\mathbf{I}_1 - \mathbf{Z}_s\mathbf{I}_1 = \left(R_1 + j\omega L_1 + \frac{(\omega M)^2}{\mathbf{Z}_{22}}\right)\mathbf{I}_1 \tag{13.154}$$

The impedance to the right of $\mathbf{Z}_s$ is given by

$$\frac{\mathbf{V}_1}{\mathbf{I}_1} = R_1 + j\omega L_1 + \frac{(\omega M)^2}{\mathbf{Z}_{22}} \tag{13.155}$$

EXAMPLE 13.8

Let

$$\mathbf{V}_s = 20\angle 0^\circ\text{ V},\quad \mathbf{Z}_s = 5 + j6,\quad R_1 = 5,\quad L_1 = 4\text{ mH},\quad L_2 = 3\text{ mH},$$
$$M = 3\text{ mH},\quad R_2 = 3,\quad \mathbf{Z}_L = 6 + j5,\quad f = 100\text{ Hz}.$$

in the circuit shown in Figure 13.46. Find $\mathbf{I}_1, \mathbf{I}_2, \mathbf{V}_1, \mathbf{V}_2$, and the input impedance $\mathbf{Z}_{in}$.

The radian frequency is $\omega = 2\pi f = 2\pi 100 = 628.3185$ rad/s. The impedances are given by

$$\mathbf{Z}_{L_1} = j\omega L_1 = j2.5133\ \Omega,\quad \mathbf{Z}_{L_2} = j\omega L_2 = j1.8850\ \Omega,\quad \mathbf{Z}_M = j\omega M = j1.8850\ \Omega,$$

Notice that $R_s = 5\ \Omega$, $L_s = 6/(2\pi 100) = 9.5493$ mH, $R_L = 6\ \Omega$, $L_L = 5/(2\pi 100) = 7.9577$ mH.

The impedance $\mathbf{Z}_{11}$ is given by

$$\mathbf{Z}_{11} = \mathbf{Z}_s + R_1 + j\omega L_1 = 10 + j8.5133\ \Omega$$

The impedance $\mathbf{Z}_{22}$ is given by

$$\mathbf{Z}_{22} = j\omega L_2 + R_2 + \mathbf{Z}_L = 9 + j6.8850\ \Omega$$

Application of Cramer's rule on Equations (13.146) and (13.147) yields

$$\mathbf{I}_1 = \frac{\begin{vmatrix} \mathbf{V}_s & -j\omega M \\ 0 & \mathbf{Z}_{22} \end{vmatrix}}{\begin{vmatrix} \mathbf{Z}_{11} & -j\omega M \\ -j\omega M & \mathbf{Z}_{22} \end{vmatrix}} = \frac{\mathbf{V}_s\mathbf{Z}_{22}}{\mathbf{Z}_{11}\mathbf{Z}_{22} + (\omega M)^2} = 1.1759 - j0.9549 = 1.5148\angle -39.0784^\circ\text{ A}$$

$$\mathbf{I}_2 = \frac{\begin{vmatrix} \mathbf{Z}_{11} & \mathbf{V}_s \\ -j\omega M & 0 \end{vmatrix}}{\begin{vmatrix} \mathbf{Z}_{11} & -j\omega M \\ -j\omega M & \mathbf{Z}_{22} \end{vmatrix}} = \frac{j\omega M\mathbf{V}_s}{\mathbf{Z}_{11}\mathbf{Z}_{22} + (\omega M)^2} = 0.2450 + j0.0589 = 0.2520\angle 13.5058^\circ\text{ A}$$

Voltage $\mathbf{V}_1$ is given by

$$\mathbf{V}_1 = \mathbf{V}_s - \mathbf{Z}_s\mathbf{I}_1 = 8.3907 - j2.2810 = 8.6952\angle -15.2086^\circ\text{ V}$$

Voltage $\mathbf{V}_2$ is given by

$$\mathbf{V}_2 = \mathbf{Z}_L\mathbf{I}_2 = 1.1759 + j1.5782 = 1.9681\angle 53.3114^\circ\text{ V}$$

The input impedance is given by

$$\mathbf{Z}_{in} = \frac{\mathbf{V}_s}{\mathbf{I}_1} = \mathbf{Z}_{11} + \frac{\omega^2 M^2}{\mathbf{Z}_{22}} = 10.2490 + j8.3228\ \Omega$$

Exercise 13.8

Let

$$\mathbf{V}_s = 15\angle 0°\ \text{V},\ \ \mathbf{Z}_s = 2 + j3,\ \ R_1 = 3,\ \ L_1 = 3\ \text{mH},\ \ L_2 = 2\ \text{mH},\ \ M = 2\ \text{mH},$$
$$R_2 = 1,\ \ \mathbf{Z}_L = 3 + j2,\ \ f = 200\ \text{Hz}$$

in the circuit shown in Figure 13.46. Find the input impedance $\mathbf{Z}_{\text{in}}$, $\mathbf{I}_1$, and $\mathbf{I}_2$.

Answer:

$\mathbf{Z}_{\text{in}} = 5.6947 + j5.9861\ \Omega$, $\mathbf{I}_1 = 1.2514 - j1.3154 = 1.8155\angle -46.4288°\ \text{A}$,
$\mathbf{I}_2 = 0.7539 - j0.06435 = 0.7566\angle -4.8791°\ \text{A}$.

13.7 Ideal Transformer

An *ideal transformer* is a transformer without losses on the windings or the core and has perfect coupling between the coils. If the coupling coefficient is one ($k = 1$), the flux is confined within the magnetic core without leakage. All the flux passes through the core and links the primary winding and the secondary winding. The windings are perfect conductors without power loss (R_1 and R_2 in Figure 13.46 are zero). Also, the permeability of the core is infinity. An ideal transformer is shown in Figure 13.47. The number of turns of the primary winding is N_1, and the number of turns of the secondary winding is N_2. The turns ratio n is defined as the ratio of N_2 to N_1:

FIGURE 13.47

An ideal transformer.

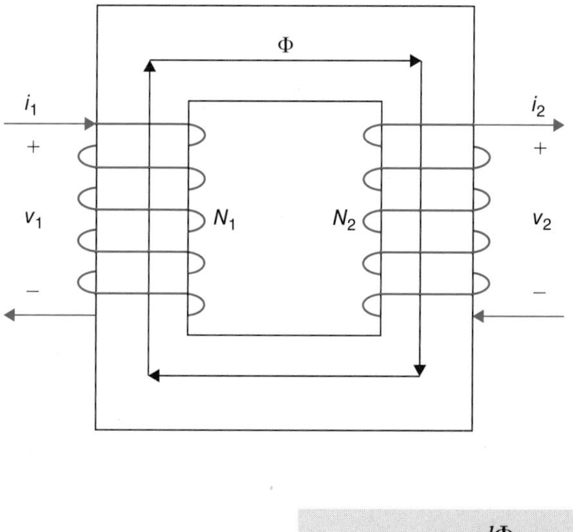

$$n = \frac{N_2}{N_1} \tag{13.156}$$

The direction of the magnetic field inside the core is based on the application of the right-hand rule to the direction of current i_1. The direction of the induced current i_2 is based on Lenz's law. Assuming $di_1/dt > 0$, the magnetic field from the induced current i_2 will be in the opposite direction to the magnetic field generated by i_1. The induced voltage v_1 in the primary winding is positive on the side where current i_1 enters. The induced magnetic field will be in the opposite direction to the magnetic field generated by i_1. The voltage v_2 on the secondary winding is positive on the side where current i_2 leaves. Since the same flux Φ links the primary winding and the secondary winding, the voltages v_1 and v_2 are given, respectively, by

$$v_1 = N_1 \frac{d\Phi}{dt} \tag{13.157}$$

and

$$v_2 = N_2 \frac{d\Phi}{dt} \tag{13.158}$$

Dividing Equation (13.158) by Equation (13.157), we obtain

$$\frac{v_2}{v_1} = \frac{N_2}{N_1} = n \tag{13.159}$$

Equation (13.159) can also be written as

$$v_2 = \frac{N_2}{N_1} v_1 = n v_1 \tag{13.160}$$

and

$$v_1 = \frac{N_1}{N_2} v_2 = \frac{1}{n} v_2 \tag{13.161}$$

Applying Ampere's law around the center of the core, we obtain

$$\oint B \cdot dl = \mu i_{enc} \tag{13.162}$$

where μ is the permeability of the core, and i_{enc} is the current enclosed. Equation (13.162) can be rewritten as

$$\frac{B\ell}{\mu} = N_1 i_1 - N_2 i_2 \tag{13.163}$$

where ℓ is the length of the path inside the core. For an ideal transformer, the permeability (μ) is infinity. Thus, we have

$$N_1 i_1 - N_2 i_2 = 0 \tag{13.164}$$

Equation (13.164) can be rewritten as

$$\frac{i_2}{i_1} = \frac{N_1}{N_2} = \frac{1}{n}, \quad i_1 = n i_2, \quad i_2 = \frac{1}{n} i_1 \tag{13.165}$$

The power on the ideal transformer is given by

$$p = i_1 v_1 - i_2 v_2 = i_1 v_1 - \frac{1}{n} i_1 n v_1 = i_1 v_1 - i_1 v_1 = 0 \tag{13.166}$$

The power loss on the ideal transformer is zero. In the phasor domain, Equation (13.159) becomes

$$\frac{\mathbf{V}_2}{\mathbf{V}_1} = \frac{N_2}{N_1} = n \tag{13.167}$$

Similarly, in the phasor domain, Equation (13.165) becomes

$$\frac{\mathbf{I}_2}{\mathbf{I}_1} = \frac{N_1}{N_2} = \frac{1}{n} \tag{13.168}$$

If $n > 1$ ($N_2 > N_1$), the transformer is called a *step-up transformer.* The voltage on the secondary winding $v_2(t)$ is larger than the voltage on the primary winding $v_1(t)$. If $n < 1$ ($N_2 < N_1$), the transformer is called a *step-down transformer.* The voltage on the secondary winding $v_2(t)$ is smaller than the voltage on the primary winding $v_1(t)$. The voltage from the power generator is increased using the step-up transformer. The high voltage is transmitted on the transmission lines. The high voltage results in smaller currents on the transmission lines. This reduces losses (I^2R) on the transmission lines. At the destination, the voltage is reduced using a step-down transformer.

The voltage ratio v_2/v_1 is either n or $-n$. The current ratio i_2/i_1 is either $1/n$ or $-1/n$. Dot convention is used to indicate the polarity of voltages and currents in the circuit symbols for ideal transformers.

If both $v_1(t)$ and $v_2(t)$ are positive or negative at the dotted terminals, then v_2/v_1 is n. If one of the voltages [$v_1(t)$ or $v_2(t)$] is positive and the other is negative at the dotted terminals, then v_2/v_1 is $-n$.

If both $i_1(t)$ and $i_2(t)$ enter into the dotted terminals or both $i_1(t)$ and $i_2(t)$ leave the dotted terminals, i_2/i_1 is $-1/n$. If one of the currents [$i_1(t)$ or $i_2(t)$] enters the dotted terminal and the other current leaves the dotted terminal, i_2/i_1 is $1/n$.

The circuit symbol for the ideal transformer is shown in Figure 13.48(a) in phasors. Since both $\mathbf{V}_1$ and $\mathbf{V}_2$ are positive at the dotted terminals, we have $\mathbf{V}_2 = n\mathbf{V}_1 = (N_2/N_1)\mathbf{V}_1$. Since $\mathbf{I}_1$ enters into the dotted terminal and $\mathbf{I}_2$ leaves the dotted terminal, we have $\mathbf{I}_2 = \mathbf{I}_1/n = (N_1/N_2)\mathbf{I}_1$. The complex power of the ideal transformer shown in Figure 13.48(a) is given by

$$\mathbf{S} = \mathbf{V}_1\mathbf{I}_1^* - \mathbf{V}_2\mathbf{I}_2^* = \mathbf{V}_1\mathbf{I}_1^* - (n\mathbf{V}_1)(\mathbf{I}_1/n)^* = \mathbf{V}_1\mathbf{I}_1^* - \mathbf{V}_1\mathbf{I}_1^* = 0$$

The second circuit symbol for the ideal transformer is shown in Figure 13.48(b). Since both $\mathbf{V}_1$ and $\mathbf{V}_2$ are positive at the dotted terminals, we have $\mathbf{V}_2 = n\mathbf{V}_1 = (N_2/N_1)\mathbf{V}_1$. Since both $\mathbf{I}_1$ and $\mathbf{I}_2$ enter into the dotted terminals, we have $\mathbf{I}_2 = -\mathbf{I}_1/n = -(N_1/N_2)\mathbf{I}_1$. The third circuit symbol for the ideal transformer is shown in Figure 13.48(c). Since $\mathbf{V}_1$ is positive at the dotted terminal and $\mathbf{V}_2$ is positive at the undotted terminal, we have $\mathbf{V}_2 = -n\mathbf{V}_1 = -(N_2/N_1)\mathbf{V}_1$. Since both $\mathbf{I}_1$ and $\mathbf{I}_2$ enter into the dotted terminals, we have $\mathbf{I}_2 = -\mathbf{I}_1/n = -(N_1/N_2)\mathbf{I}_1$. The voltage current relation of ideal transformers is shown in Figure 13.49. The complex power of the ideal transformers shown in Figure 13.49 is zero.

FIGURE 13.48

Circuit symbols for ideal transformer.

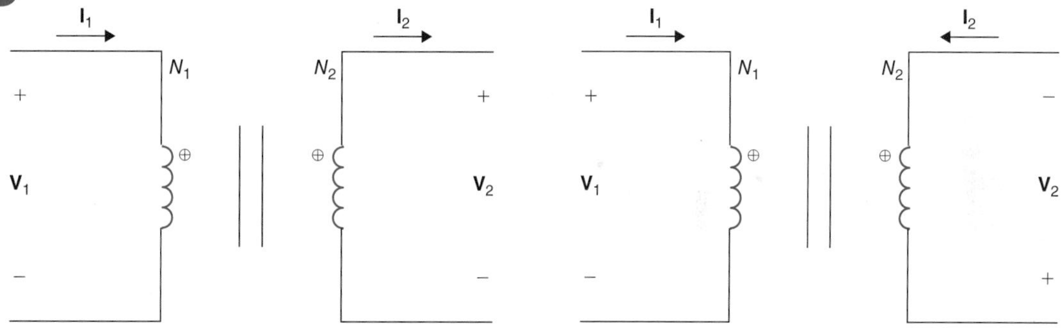

$$\frac{\mathbf{V}_2}{\mathbf{V}_1} = \frac{N_2}{N_1} = n, \quad \frac{\mathbf{I}_2}{\mathbf{I}_1} = \frac{N_1}{N_2} = \frac{1}{n}$$

(a)

$$\frac{\mathbf{V}_2}{\mathbf{V}_1} = -\frac{N_2}{N_1} = -n, \quad \frac{\mathbf{I}_2}{\mathbf{I}_1} = -\frac{N_1}{N_2} = -\frac{1}{n}$$

(c)

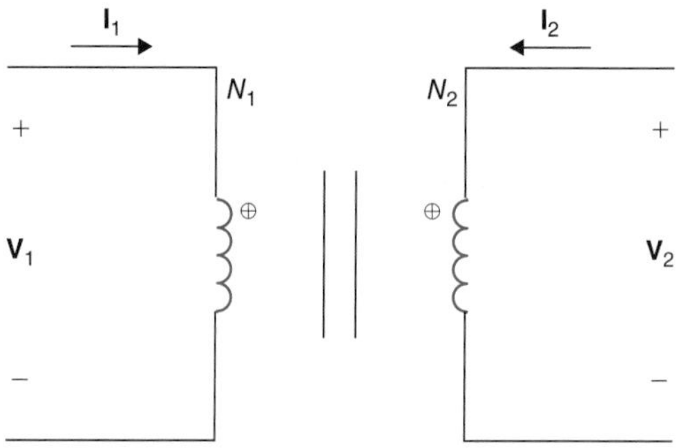

$$\frac{\mathbf{V}_2}{\mathbf{V}_1} = \frac{N_2}{N_1} = n, \quad \frac{\mathbf{I}_2}{\mathbf{I}_1} = -\frac{N_1}{N_2} = -\frac{1}{n}$$

(b)

FIGURE 13.49

Voltage and current relation for ideal transformer.

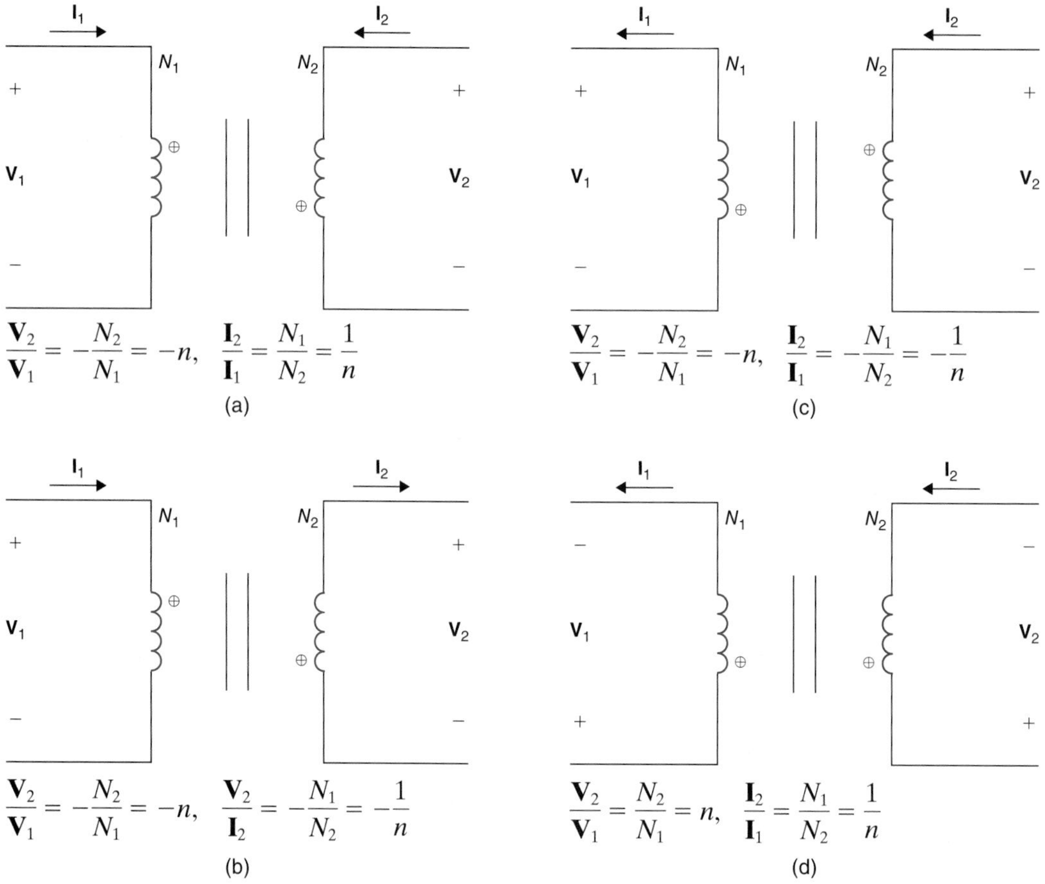

FIGURE 13.50

An ideal transformer with load impedance.

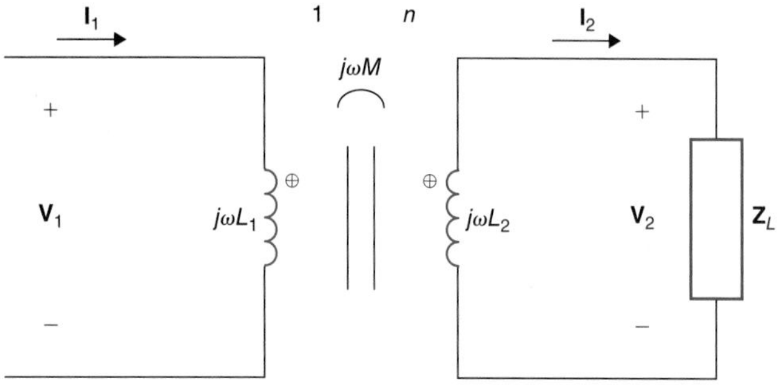

Figure 13.50 shows an ideal transformer in the phasor domain with load impedance $\mathbf{Z}_L$.

Summing the voltage drops around mesh 1, we obtain

$$-\mathbf{V}_1 + j\omega L_1\mathbf{I}_1 - j\omega M\mathbf{I}_2 = 0 \tag{13.169}$$

Summing the voltage drops around mesh 2, we obtain

$$j\omega L_2\mathbf{I}_2 - j\omega M\mathbf{I}_1 + \mathbf{Z}_L\mathbf{I}_2 = 0 \tag{13.170}$$

Solving Equation (13.170) for $\mathbf{I}_2$, we obtain

$$\mathbf{I}_2 = \frac{j\omega M}{j\omega L_2 + \mathbf{Z}_L}\mathbf{I}_1 \tag{13.171}$$

Substitution of Equation (13.171) into Equation (13.169) yields

$$\mathbf{V}_1 = j\omega L_1\mathbf{I}_1 - j\omega M\frac{j\omega M}{j\omega L_2 + \mathbf{Z}_L}\mathbf{I}_1 = \left(j\omega L_1 + \frac{\omega^2 M^2}{j\omega L_2 + \mathbf{Z}_L}\right)\mathbf{I}_1$$

The voltage $\mathbf{V}_2$ is given by

$$\mathbf{V}_2 = \mathbf{Z}_L\mathbf{I}_2 = \frac{j\omega M \times \mathbf{Z}_L}{j\omega L_2 + \mathbf{Z}_L}\mathbf{I}_1$$

The voltage ratio $\mathbf{V}_2/\mathbf{V}_1$ is given by

$$\frac{\mathbf{V}_2}{\mathbf{V}_1} = \frac{\dfrac{j\omega M \times \mathbf{Z}_L}{j\omega L_2 + \mathbf{Z}_L}\mathbf{I}_1}{\left(j\omega L_1 + \dfrac{\omega^2 M^2}{j\omega L_2 + \mathbf{Z}_L}\right)\mathbf{I}_1} = \frac{j\omega M \times \mathbf{Z}_L}{-(L_1L_2 - M^2)\omega^2 + j\omega L_1 \times \mathbf{Z}_L} \tag{13.172}$$

If the coupling coefficient is one ($k = 1$), we have $M^2 = L_1L_2$ and $\mathbf{V}_2/\mathbf{V}_1 = M/L_1 = n$. If the load impedance is small ($\mathbf{Z}_L = 0$), the voltage ratio is $\mathbf{V}_2/\mathbf{V}_1 = 0$. If the load impedance is large ($\mathbf{Z}_L = \infty$), the voltage ratio is $\mathbf{V}_2/\mathbf{V}_1 = M/L_1 = k\sqrt{L_1L_2}/L_1 = kn$.

The input impedance is given by

$$\mathbf{Z}_{in} = \frac{\mathbf{V}_1}{\mathbf{I}_1} = j\omega L_1 + \frac{\omega^2 M^2}{j\omega L_2 + \mathbf{Z}_L} = \frac{j\omega L_1\mathbf{Z}_L - \omega^2 L_1L_2 + \omega^2M^2}{j\omega L_2 + \mathbf{Z}_L}$$
$$= \frac{j\omega L_1\mathbf{Z}_L - \omega^2(L_1L_2 - M^2)}{j\omega L_2 + \mathbf{Z}_L} \tag{13.173}$$

If the coupling coefficient is one ($k = 1$), we have $M^2 = L_1L_2$ and $\mathbf{Z}_{in}$ becomes

$$\mathbf{Z}_{in} = \frac{j\omega L_1\mathbf{Z}_L}{j\omega L_2 + \mathbf{Z}_L} = \frac{\mathbf{Z}_L}{n^2\left(1 + \dfrac{\mathbf{Z}_L}{j\omega L_2}\right)}$$

If the load impedance is small ($\mathbf{Z}_L = 0$), the input impedance becomes $\mathbf{Z}_{in} = j\omega L_1(1 - k^2)$. If the load impedance is large ($\mathbf{Z}_L = \infty$), the input impedance becomes $\mathbf{Z}_{in} = j\omega L_1$.

If $\omega L_2 >> |\mathbf{Z}_L|$, the input impedance becomes

$$\mathbf{Z}_{in} = \frac{j\omega L_1\mathbf{Z}_L}{j\omega L_2} = \frac{\mathbf{Z}_L}{\dfrac{L_2}{L_1}} = \frac{\mathbf{Z}_L}{n^2}$$

The voltage $\mathbf{V}_1$ becomes $\mathbf{V}_1 = \mathbf{Z}_{in}\mathbf{I}_1 = \mathbf{Z}_L\mathbf{I}_1/n^2$ and the voltage $\mathbf{V}_2$ becomes $\mathbf{V}_2 = \mathbf{Z}_L\mathbf{I}_1/n$. Thus, the voltage ratio becomes $\mathbf{V}_2/\mathbf{V}_1 = n$. Therefore, if $\omega L_2 >> |\mathbf{Z}_L|$, the transformer becomes ideal.

Figure 13.51 shows an ideal transformer connected to a source with impedance $\mathbf{Z}_s$ and a load with impedance $\mathbf{Z}_L$.

FIGURE 13.51

An ideal transformer connected to a load and a source with impedance $\mathbf{Z}_s$.

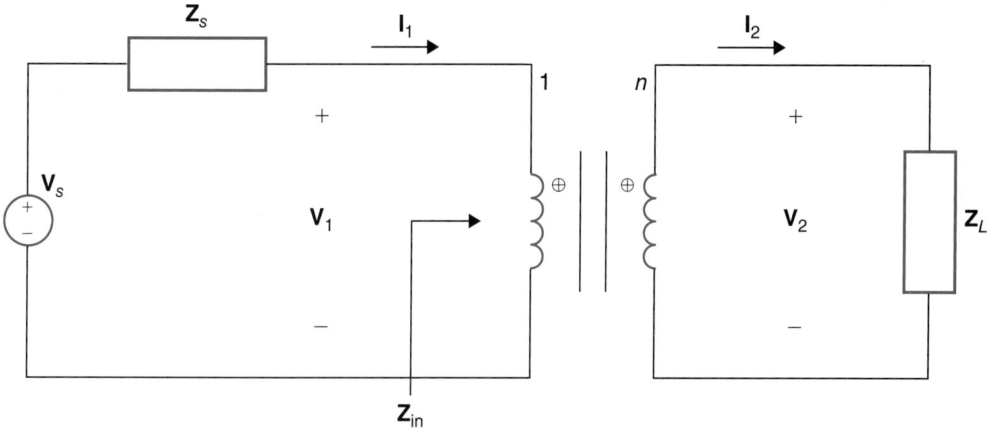

From the second mesh on the right side, we have

$$\mathbf{V}_2 = \mathbf{I}_2\mathbf{Z}_L$$

which can be rearranged as

$$\frac{\mathbf{V}_2}{\mathbf{I}_2} = \mathbf{Z}_L$$

The input impedance of the transformer is given by

$$\mathbf{Z}_{\text{in}} = \frac{\mathbf{V}_1}{\mathbf{I}_1} = \frac{\frac{\mathbf{V}_2}{n}}{n\mathbf{I}_2} = \frac{\frac{\mathbf{V}_2}{\mathbf{I}_2}}{n^2} = \frac{\mathbf{Z}_L}{n^2} \quad (13.174)$$

The input impedance $\mathbf{Z}_{\text{in}}$ is the load impedance $\mathbf{Z}_L$ reflected to the input side of the transformer. The mesh equation around the mesh on the left side is given by

$$-\mathbf{V}_s + \mathbf{Z}_s\mathbf{I}_1 + \mathbf{Z}_{\text{in}}\mathbf{I}_1 = 0$$

Solving for $\mathbf{I}_1$, we obtain

$$\mathbf{I}_1 = \frac{\mathbf{V}_s}{\mathbf{Z}_s + \mathbf{Z}_{\text{in}}} \quad (13.175)$$

The voltage $\mathbf{V}_1$ is given by

$$\mathbf{V}_1 = \mathbf{I}_1\mathbf{Z}_{\text{in}} = \frac{\mathbf{Z}_{\text{in}}}{\mathbf{Z}_s + \mathbf{Z}_{\text{in}}}\mathbf{V}_s \quad (13.176)$$

Voltage $\mathbf{V}_1$ can be obtained by applying the voltage divider rule. The ideal transformer and the load can be replaced by $\mathbf{Z}_{\text{in}}$ to calculate $\mathbf{V}_1$ and $\mathbf{I}_1$.

EXAMPLE 13.9

Let

$$\mathbf{V}_s = 70\angle 0° \text{ V}, \quad \mathbf{Z}_L = 5 + j3\ \Omega, \quad n = 1/5, \quad \mathbf{Z}_s = 25 + j10\ \Omega$$

in the circuit shown in Figure 13.51. Find $\mathbf{Z}_{\text{in}}$, $\mathbf{I}_1$, $\mathbf{V}_1$, $\mathbf{V}_2$, and $\mathbf{I}_2$.

The input impedance looking into the transformer is given by

$$\mathbf{Z}_{\text{in}} = \frac{\mathbf{Z}_L}{n^2} = \frac{5 + j3}{\frac{1}{25}} = 125 + 75\ \Omega$$

Current $\mathbf{I}_1$ is given by

$$\mathbf{I}_1 = \frac{\mathbf{V}_s}{\mathbf{Z}_s + \mathbf{Z}_{\text{in}}} = 0.3331 - j0.2109 = 0.3942\angle -32.3474° \text{ A}$$

Voltage $\mathbf{V}_1$ is given by

$$\mathbf{V}_1 = \mathbf{Z}_{\text{in}} \times \mathbf{I}_1 = 57.4544 - j1.3878 = 57.4712\angle -1.3837° \text{ V}$$

continued

Example 13.9 continued

Voltage $\mathbf{V}_2$ is found from

$$\mathbf{V}_2 = n\mathbf{V}_1 = \frac{\mathbf{V}_1}{5} = 11.4909 - j0.2776 = 11.4942\angle -1.3837° \text{ V}$$

Current $\mathbf{I}_2$ is given by

$$\mathbf{I}_2 = \frac{\mathbf{I}_1}{n} = 5\mathbf{I}_1 = 1.6653 - j1.05472 = 1.9712\angle - 32.3474° \text{ A}$$

Exercise 13.9

Let $\mathbf{V}_s = 150\angle 0°$ V, $\mathbf{Z}_L = 3 + j2\ \Omega$, $n = 1/6$, $\mathbf{Z}_s = 50 + j30\ \Omega$ in the circuit shown in Figure 13.52. Find $\mathbf{Z}_{in}$, $\mathbf{I}_1$, $\mathbf{V}_1$, $\mathbf{V}_2$, $\mathbf{I}_2$.

FIGURE 13.52
Circuit for EXERCISE 13.9.

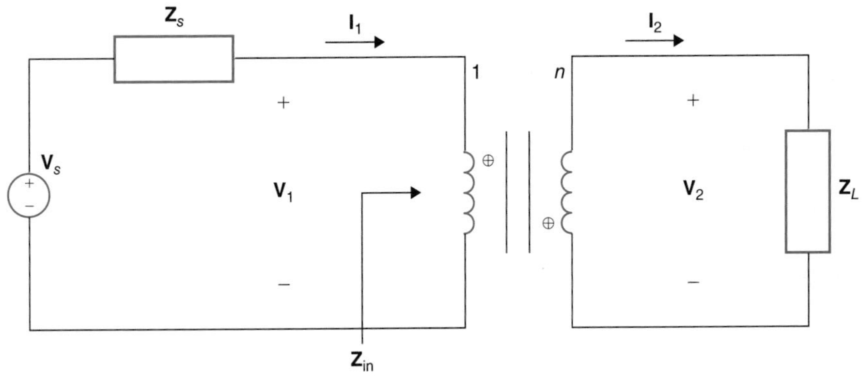

Answer:

$\mathbf{Z}_{in} = 108 + j72$, $\mathbf{I}_1 = 0.6701 - j0.4326 = 0.7976\angle -32.8451°$ A,

$\mathbf{V}_1 = 103.5173 - j1.5268 = 103.5173\angle 0.8450°$ V,

$\mathbf{V}_2 = -17.2529 - j0.2545 = 17.2548\angle -179.1550°$ V,

$\mathbf{I}_2 = -4.0206 + j2.5956 = 4.7856\angle 147.1549°$ A.

As shown in Chapter 11, for maximum power transfer, the impedance of the load should be the complex conjugate of the Thévenin equivalent impedance. When the load is connected to the secondary winding of an ideal transformer, as shown in Figure 13.53(a), the load impedance can be reflected to the input side of the transformer. The input impedance $\mathbf{Z}_{in}$ should be the complex conjugate of the Thévenin equivalent impedance $\mathbf{Z}_{th}$, as shown in Figure 13.53(b), for maximum power transfer. From equation (13.174), we have

$$\mathbf{Z}_{in} = \frac{\mathbf{Z}_L}{n^2}$$

For maximum power transfer, we have

$$\mathbf{Z}_{in} = \mathbf{Z}_{th}^* \tag{13.177}$$

For the given $\mathbf{Z}_L$, we can find the turns ratio n:

$$n = \sqrt{\frac{\mathbf{Z}_L}{\mathbf{Z}_{th}^*}} \tag{13.178}$$

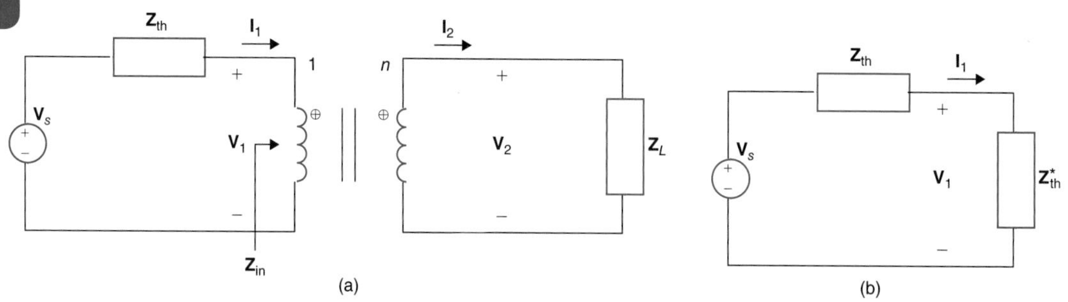

FIGURE 13.53

(a) Thévenin-equivalent circuit connected to the load through an ideal transformer. (b) Z_{in} for maximum power transfer.

If the load impedance is small and the source impedance is large, n is less than $1(N_1 > N_2)$. The matching of the load impedance to the source impedance through a transformer is called *impedance matching.*

EXAMPLE 13.10

A speaker with a resistance of 7.5 Ω is to be matched to an audio amplifier with an output resistance of 1080 Ω through an ideal transformer, as shown in Figure 13.54. Determine the turns ratio of the transformer to obtain the maximum power transfer to the speaker. If $v_s(t) = 10 \sin(2\pi 1000t)$ V, what is the voltage $v_2(t)$?

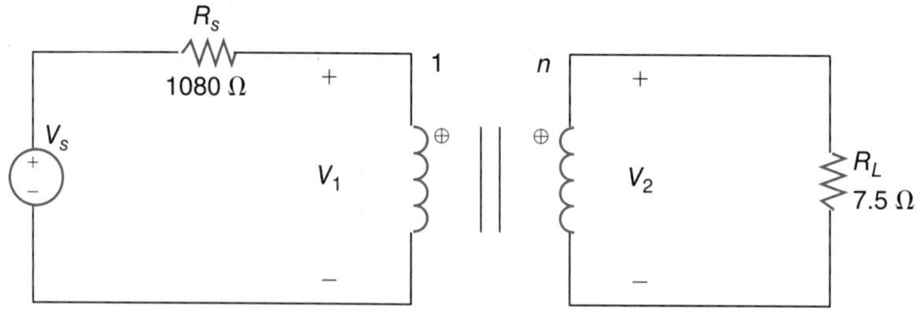

FIGURE 13.54

Impedance matching of an audio system.

From Equation (13.178), the turns ratio is given by

$$n = \sqrt{\frac{7.5}{1080}} = 0.083333$$

Thus, we have

$$\frac{1}{n} = \frac{N_1}{N_2} = 12$$

The ratio of N_1 to N_2 is 12. The input resistance is 1080 Ω. Thus, $v_1(t) = 5 \sin(2\pi 1000t)$ V, and $v_2(t) = (5/12) \sin(2\pi 1000t)$ V $= 0.4167 \sin(2\pi 1000t)$ V.

Exercise 13.10

Let $R_L = 9.5$ Ω and $R_s = 1149.5$ Ω in the circuit shown in Figure 13.54. Find the turns ratio n for the maximum power transfer.

Answer:

$n = 0.09091, 1/n = 11.$

EXAMPLE 13.11

Let $\mathbf{V}_s = 110\angle 0°$ V rms, $\mathbf{Z}_s = 10 + j20\ \Omega$, $\mathbf{Z}_a = 15 - j10\ \Omega$, and $\mathbf{Z}_L = 40 + j25\ \Omega$, $N_1 = 150$, $N_2 = 100$ in the circuit shown in Figure 13.55. Find $\mathbf{I}_2$, $\mathbf{V}_L$, and the complex power on the load.

FIGURE 13.55

Circuit for EXAMPLE 13.11.

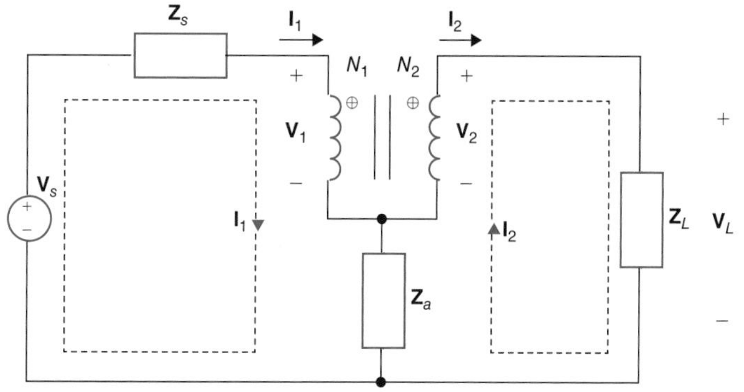

From the dot convention of ideal transformers [refer to Figure 13.48(a)], we have

$$n = N_2/N_1, \quad \mathbf{V}_1 = \mathbf{V}_2/n, \quad \mathbf{I}_1 = n\mathbf{I}_2 \tag{13.179}$$

Summing the voltage drops around mesh 1 (left side), we obtain

$$-\mathbf{V}_s + \mathbf{Z}_s\,\mathbf{I}_1 + \mathbf{V}_1 + \mathbf{Z}_a(\mathbf{I}_1 - \mathbf{I}_2) = 0 \tag{13.180}$$

Notice that the mesh current $\mathbf{I}_1$ for mesh 1 is identical to the current through the primary winding. Similarly, the mesh current $\mathbf{I}_2$ for mesh 2 is identical to the current through the secondary winding.

Substitution of equations $\mathbf{V}_1 = \mathbf{V}_2/n, \mathbf{I}_1 = n\mathbf{I}_2$ into Equation (13.180) yields

$$-\mathbf{V}_s + \mathbf{Z}_s\,n\mathbf{I}_2 + \mathbf{V}_2/n + \mathbf{Z}_a(n-1)\mathbf{I}_2 = 0 \tag{13.181}$$

Summing the voltage drops around mesh 2 (right side), we obtain

$$\mathbf{Z}_a(\mathbf{I}_2 - \mathbf{I}_1) - \mathbf{V}_2 + \mathbf{Z}_L\,\mathbf{I}_2 = 0 \tag{13.182}$$

which can be rearranged as

$$\mathbf{V}_2 = \mathbf{Z}_a(\mathbf{I}_2 - \mathbf{I}_1) + \mathbf{Z}_L\,\mathbf{I}_2 = \mathbf{Z}_a(1-n)\,\mathbf{I}_2 + \mathbf{Z}_L\,\mathbf{I}_2 = [(1-n)\mathbf{Z}_a + \mathbf{Z}_L]\mathbf{I}_2 \tag{13.183}$$

Substitution of Equation (13.183) into Equation (13.181) yields

$$\mathbf{I}_2 = \frac{\mathbf{V}_s}{n\mathbf{Z}_s + \dfrac{1}{n}[(1-n)\mathbf{Z}_a + \mathbf{Z}_L] + (n-1)\mathbf{Z}_a} = \frac{\mathbf{V}_s}{n\mathbf{Z}_s + \dfrac{n^2 - 2n + 1}{n}\mathbf{Z}_a + \dfrac{1}{n}\mathbf{Z}_L}$$

Substituting the values, we obtain

$$\mathbf{I}_2 = 1.0565 - j0.7510 = 1.2962\angle -35.4069°\text{ A}$$

The voltage across the load is

$$\mathbf{V}_L = \mathbf{Z}_L\mathbf{I}_2 = 61.0357 - j3.6278 = 61.1434\angle -3.4015°\text{ V}$$

continued

Example 13.11 continued

The complex power on the load is given by

$$\mathbf{S}_L = |\mathbf{I}_2|^2 \mathbf{Z}_L = 67.2093 + j42.0058 = 79.2564\angle 32.0054° \text{ VA}$$

Exercise 13.11

Let $\mathbf{V}_s = 100\angle 0°$ V rms, $\mathbf{Z}_s = 15 + j20\ \Omega$, $\mathbf{Z}_a = 15 - j20\ \Omega$, and $\mathbf{Z}_L = 50 + j35\ \Omega$, $N_1 = 150$, $N_2 = 250$ in the circuit shown in Figure 13.56. Find the complex power on the load.

FIGURE 13.56

Circuit for EXERCISE 13.11.

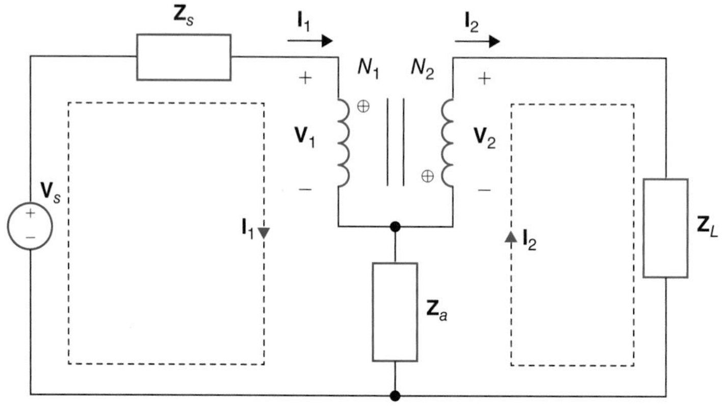

Answers:
$\mathbf{S}_L = 33.0644 + j23.1451 = 40.3603\angle 34.9920°$ VA.

FIGURE 13.57

Voltage and current relation for autotransformer.

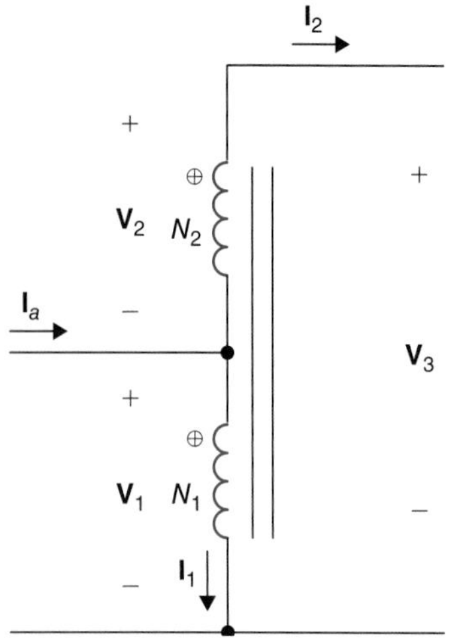

$$\frac{\mathbf{V}_2}{\mathbf{V}_1} = \frac{N_2}{N_1} = n, \quad \frac{\mathbf{I}_2}{\mathbf{I}_1} = \frac{N_1}{N_2} = \frac{1}{n},$$
$$\mathbf{I}_a = (n+1)\mathbf{I}_2, \quad \mathbf{V}_3 = \mathbf{V}_1 + \mathbf{V}_2 = (1+n)\mathbf{V}_1.$$

13.7.1 AUTOTRANSFORMER

In a regular transformer, there are two separate windings called the *primary winding* and the *secondary winding.* This provides electrical isolation between the primary winding and the secondary winding. In the autotransformer, there is a single winding with a tap between the terminals. Depending on where the input is applied and the output is taken, the autotransformer can be either step-up or step-down. In the circuit shown in Figure 13.57, voltage $\mathbf{V}_2$ is related to $\mathbf{V}_1$ through

$$\mathbf{V}_2 = \frac{N_2}{N_1}\mathbf{V}_1 = n\mathbf{V}_1$$

and current $\mathbf{I}_2$ is related to $\mathbf{I}_1$ through

$$\mathbf{I}_2 = \frac{N_1}{N_2}\mathbf{I}_1 = \frac{1}{n}\mathbf{I}_1$$

Voltage $\mathbf{V}_3$ is the sum of $\mathbf{V}_1$ and $\mathbf{V}_2$. Thus, we have

$$\mathbf{V}_3 = \mathbf{V}_1 + \mathbf{V}_2 = (1+n)\mathbf{V}_1 \tag{13.184}$$

Since $\mathbf{V}_3 > \mathbf{V}_1$, this is a step-up transformer. Current $\mathbf{I}_a$ is the sum of $\mathbf{I}_1$ and $\mathbf{I}_2$. Thus, we obtain

$$\mathbf{I}_a = \mathbf{I}_1 + \mathbf{I}_2 = (1 + n)\mathbf{I}_2 = \left(1 + \frac{1}{n}\right)\mathbf{I}_1 \tag{13.185}$$

EXAMPLE 13.12

Let $\mathbf{V}_s = 55\angle 60°$ V, $\mathbf{Z}_s = 6 + 9j\ \Omega$, and $\mathbf{Z}_L = 15 + 25j\ \Omega$, $N_1 = 100$, $N_2 = 150$ in the circuit shown in Figure 13.58. Find $\mathbf{V}_1$, $\mathbf{V}_2$, $\mathbf{V}_3$, $\mathbf{I}_1$, $\mathbf{I}_2$, $\mathbf{I}_a$.

FIGURE 13.58

Autotransformer for EXAMPLE 13.12 and EXERCISE 13.12.

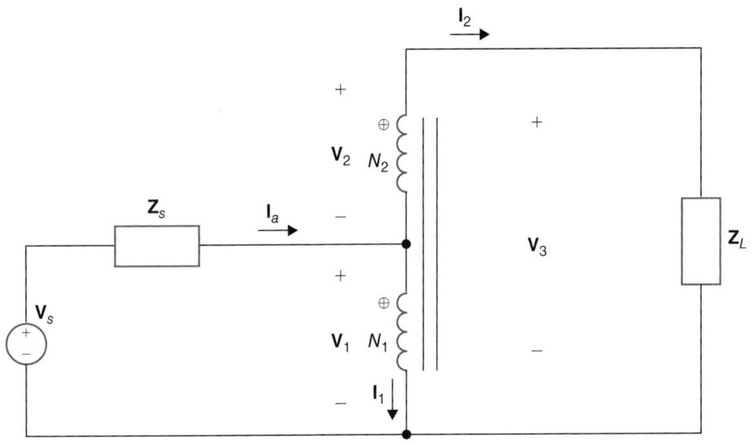

The turns ration is

$$n = \frac{N_2}{N_1} = \frac{150}{100} = 1.5$$

From Equation (13.184), we have

$$\mathbf{V}_3 = \mathbf{V}_1 + \mathbf{V}_2 = (1 + n)\mathbf{V}_1 = 2.5\mathbf{V}_1$$

The current $\mathbf{I}_2$ through load $\mathbf{Z}_L$ is given by

$$\mathbf{I}_2 = \frac{\mathbf{V}_3}{\mathbf{Z}_L} = \frac{2.5\mathbf{V}_1}{15 + j25}$$

Solving for $\mathbf{V}_1$, we obtain

$$\mathbf{V}_1 = \frac{\mathbf{Z}_L}{n + 1}\mathbf{I}_2 = \frac{15 + j25}{2.5}\mathbf{I}_2 = (6 + j10)\mathbf{I}_2 \tag{13.186}$$

From Equation (13.185), we have

$$\mathbf{I}_a = \mathbf{I}_1 + \mathbf{I}_2 = (1 + n)\mathbf{I}_2 = 2.5\mathbf{I}_2$$

Summing the voltage drops around the mesh on the left side, we obtain

$$-\mathbf{V}_s + \mathbf{I}_a\mathbf{Z}_s + \mathbf{V}_1 = 0$$

continued

Example 13.12 continued

which can be rewritten as

$$-27.5 + j47.6314 + 2.5\,(6 + j9)\mathbf{I}_2 + \mathbf{V}_1 = 0 \qquad \textbf{(13.187)}$$

Substitution of Equation (13.186) into Equation (13.187) yields

$$-27.5 + j47.6314 + 2.5\,(6 + j9)\mathbf{I}_2 + (6 + j10)\mathbf{I}_2 = 0$$

Solving for $\mathbf{I}_2$, we obtain

$$\mathbf{I}_2 = 1.4196 + j0.0711 = 1.4214\angle 2.8687^\circ \text{ A}$$

Since $\mathbf{I}_1 = n\mathbf{I}_2$, we have

$$\mathbf{I}_1 = 2.1294 + j0.1067 = 2.1321\angle 2.8687^\circ \text{ A}$$

Substitution of $\mathbf{I}_2$ into Equation (13.186) results in

$$\mathbf{V}_1 = 7.8063 + j14.6230 = 16.5762\angle 61.9049^\circ \text{ V}$$

Since $\mathbf{V}_2 = n\mathbf{V}_1$, we have

$$\mathbf{V}_2 = 11.7095 + j21.9345 = 24.8643\angle 61.9049^\circ \text{ V}$$

The complex power of the load is given by

$$\mathbf{S} = \frac{1}{2}\mathbf{V}_3\mathbf{I}_2^* = 15.1528 + j25.2546 = 29.4517\angle 59.0362^\circ \text{ VA}$$

Exercise 13.12

Find $\mathbf{V}_1$, $\mathbf{V}_2$, $\mathbf{I}_1$, and $\mathbf{I}_2$ for the circuit shown in Figure 13.58 when

$$\mathbf{V}_s = 75\angle 30^\circ \text{ V}, \quad \mathbf{Z}_s = 12 + 16j\,\Omega, \quad \mathbf{Z}_L = 25 + 35j\,\Omega, \quad N_1 = 100, \quad N_2 = 180$$

Answers:

$\mathbf{V}_1 = 13.8326 + j8.3264 = 16.1453\angle 31.0455^\circ$ V

$\mathbf{V}_2 = 24.8987 + j14.9876 = 29.0619\angle 31.0455^\circ$ V

$\mathbf{I}_1 = 1.7360 - j0.7519 = 1.8919\angle -23.4169^\circ$ A

$\mathbf{I}_2 = 0.9645 - j0.4177 = 1.0510\angle -23.4169^\circ$ A.

In the circuit shown in Figure 13.59, we have

$$N_1 = N_2 + N_3$$

The entire coil with N_1 turns can be treated as the primary winding, and the bottom portion of the coil with N_2 turns where the load is connected can be treated as the secondary winding. Let

$$n = \frac{N_2}{N_1} = \frac{N_2}{N_2 + N_3} \qquad \textbf{(13.188)}$$

FIGURE 13.59

Voltage and current relation for autotransformer.

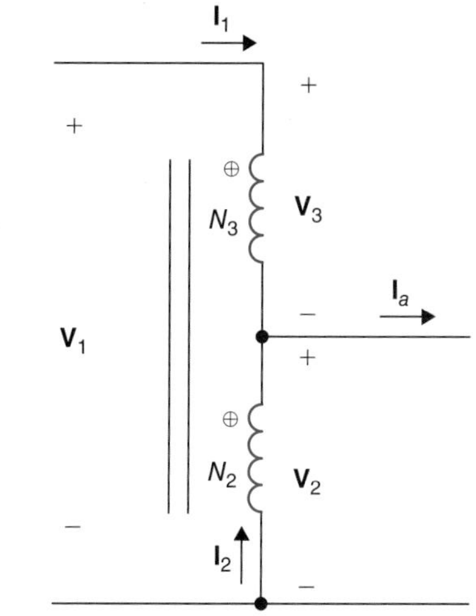

$$N_1 = N_2 + N_3, \quad \frac{\mathbf{V}_2}{\mathbf{V}_1} = \frac{N_2}{N_1} = n,$$

$$\frac{\mathbf{I}_2}{\mathbf{I}_1} = \frac{N_3}{N_2}, \quad \mathbf{I}_a = \mathbf{I}_1 + \mathbf{I}_2 = \frac{N_1}{N_2}\mathbf{I}_1 = \frac{N_1}{N_3}\mathbf{I}_2$$

Then, the voltage $\mathbf{V}_2$ is related to $\mathbf{V}_1$ through

$$\mathbf{V}_2 = \frac{N_2}{N_1}\mathbf{V}_1 = n\mathbf{V}_1, \quad \mathbf{V}_1 = \frac{\mathbf{V}_2}{n} \tag{13.189}$$

Notice that $n < 1$. Since $\mathbf{V}_2 < \mathbf{V}_1$, this is a step-down transformer. Voltage $\mathbf{V}_3$ is the difference between $\mathbf{V}_1$ and $\mathbf{V}_2$. Thus, we have

$$\mathbf{V}_3 = \mathbf{V}_1 - \mathbf{V}_2 = (1-n)\mathbf{V}_1 = \left(\frac{1}{n} - 1\right)\mathbf{V}_2$$

or

$$\mathbf{V}_1 = \frac{1}{1-n}\mathbf{V}_3 \quad \text{and} \quad \mathbf{V}_2 = \frac{n}{1-n}\mathbf{V}_3$$

Current $\mathbf{I}_2$ is related to $\mathbf{I}_1$ through

$$\frac{\mathbf{I}_2}{\mathbf{I}_1} = \frac{N_3}{N_2}, \quad \mathbf{I}_1 = \frac{N_2}{N_3}\mathbf{I}_2, \quad \mathbf{I}_2 = \frac{N_3}{N_2}\mathbf{I}_1 \tag{13.190}$$

Current $\mathbf{I}_a$ is the sum of $\mathbf{I}_1$ and $\mathbf{I}_2$. Thus, we obtain

$$\mathbf{I}_a = \mathbf{I}_1 + \mathbf{I}_2 = \left(\frac{N_2}{N_3} + 1\right)\mathbf{I}_2 = \left(1 + \frac{N_3}{N_2}\right)\mathbf{I}_1 = \frac{N_2 + N_3}{N_3}\mathbf{I}_2 = \frac{N_2 + N_3}{N_2}\mathbf{I}_1$$

or

$$\mathbf{I}_a = \frac{N_1}{N_3}\mathbf{I}_2 = \frac{N_1}{N_2}\mathbf{I}_1 = \frac{1}{n}\mathbf{I}_1 \tag{13.191}$$

Notice that

$$\mathbf{I}_1 = n\mathbf{I}_a, \quad \mathbf{I}_2 = \frac{N_3}{N_1}\mathbf{I}_a$$

EXAMPLE 13.13

Let $\mathbf{V}_s = 75\angle 0°$ V rms, $\mathbf{Z}_s = 9 + j15\ \Omega$, and $\mathbf{Z}_L = 25 + j20\ \Omega$, $N_2 = 250$, $N_3 = 1000$ in the circuit shown in Figure 13.60. Find $\mathbf{V}_1$, $\mathbf{V}_2$, $\mathbf{V}_3$, $\mathbf{I}_1$, $\mathbf{I}_2$, $\mathbf{I}_a$, and the complex power on the load.

The n value is given by, from Equation (13.188),

$$n = \frac{N_2}{N_1} = \frac{N_2}{N_2 + N_3} = 0.2$$

From Equations (13.189) through (13.191), we have

$$\mathbf{V}_1 = \frac{\mathbf{V}_2}{n}, \quad \mathbf{I}_1 = \frac{N_2}{N_3}\mathbf{I}_2, \quad \mathbf{I}_a = \frac{N_1}{N_3}\mathbf{I}_2$$

Summing the voltage drops around the mesh on the left side, we obtain

$$-\mathbf{V}_s + \mathbf{Z}_s\mathbf{I}_1 + \mathbf{V}_1 = 0$$

continued

Example 13.13 continued

FIGURE 13.60

Autotransformer for EXAMPLE 13.13 and EXERCISE 13.13.

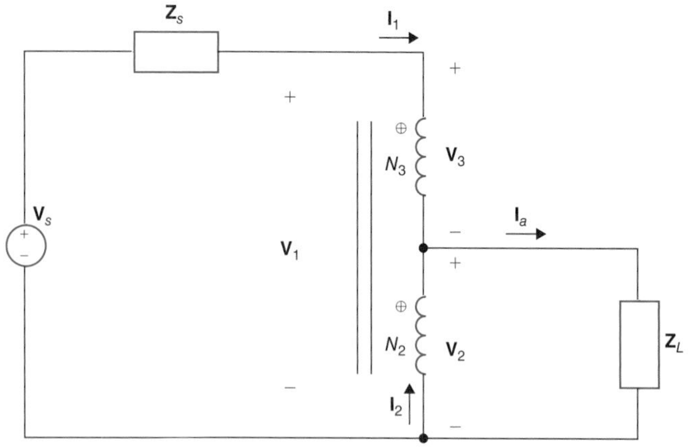

which can be rewritten as

$$-\mathbf{V}_s + \mathbf{Z}_s\frac{N_2}{N_3}\mathbf{I}_2 + \frac{\mathbf{V}_2}{n} = 0 \qquad \textbf{(13.192)}$$

Summing the voltage drops around the mesh on the right side, we obtain

$$-\mathbf{V}_2 + \mathbf{Z}_L\mathbf{I}_a = 0$$

which can be rewritten as

$$-\mathbf{V}_2 + \mathbf{Z}_L\frac{N_1}{N_3}\mathbf{I}_2 = 0 \qquad \textbf{(13.193)}$$

Solving Equation (13.193), we obtain

$$\mathbf{V}_2 = \mathbf{Z}_L\frac{N_1}{N_3}\mathbf{I}_2 \qquad \textbf{(13.194)}$$

Substitution of Equation (13.194) into Equation (13.192) yields

$$-\mathbf{V}_s + \mathbf{Z}_s\frac{N_2}{N_3}\mathbf{I}_2 + \frac{1}{n}\mathbf{Z}_L\frac{N_1}{N_3}\mathbf{I}_2 = 0$$

Solving for $\mathbf{I}_2$, we get

$$\mathbf{I}_2 = \frac{\mathbf{V}_s}{\mathbf{Z}_s\frac{N_2}{N_3} + \frac{1}{n}\mathbf{Z}_L\frac{N_1}{N_3}} = 0.2851 - j0.2316 = 0.3673\angle -39.0871°\text{ A}$$

From Equation (13.194), we have

$$\mathbf{V}_2 = \mathbf{Z}_L\frac{N_1}{N_3}\mathbf{I}_2 = 14.6980 - j0.1096 = 16.6984\angle -0.4274°$$

The rest of voltages and currents are given by

$$\mathbf{V}_1 = \frac{\mathbf{V}_2}{n} = 73.4902 - j0.5480 = 73.4922\angle -0.4272°\text{ V}$$

$$\mathbf{V}_3 = \mathbf{V}_1 - \mathbf{V}_2 = (1-n)\mathbf{V}_1 = \left(\frac{1}{n} - 1\right)\mathbf{V}_2 = 58.7921 - j0.4384 = 58.7938\angle -0.4272°\text{ V}$$

$$\mathbf{I}_1 = \frac{N_2}{N_3}\mathbf{I}_2 = 0.0713 - j0.0579 = 0.0918\angle -39.0871°\text{ A}$$

$$\mathbf{I}_a = \frac{N_1}{N_3}\mathbf{I}_2 = 0.3564 - j0.2895 = 0.4591\angle -39.0871°\text{ A}$$

The complex power on the load is given by

$$\mathbf{S}_L = \mathbf{V}_2\mathbf{I}_a^* = 5.2694 + j4.2155 = 6.7481\angle 38.6598°\text{ VA}$$

Exercise 13.13

Find the complex power on the load for the circuit shown in Figure 13.60 when

$$\mathbf{V}_s = 100\angle 30°\text{ V rms},\quad \mathbf{Z}_s = 12 + j15\ \Omega,\quad \text{and } \mathbf{Z}_L = 35 + j25\ \Omega,$$
$$N_2 = 150,\quad N_3 = 450$$

Answers:

$\mathbf{S}_L = 13.2127 + j9.4376 = 16.2371\angle 35.5377°\text{ VA}$

13.8 PSpice and Simulink

The coupled coils can be simulated in PSpice using XFRM_LINEAR/ANALOG. The PSpice schematic for the circuit shown in Figure 13.46 with the component values

$$\mathbf{V}_s = 250\angle 0°\text{ V},\quad \mathbf{Z}_s = 5 + j6,\quad R_1 = 30\ \Omega,\quad L_1 = 100\text{ mH},\quad L_2 = 70\text{ mH},$$
$$M = 60\text{ mH},\quad R_2 = 35\ \Omega,\quad \mathbf{Z}_L = 20 + j10,\quad f = 60\text{ Hz}$$

is shown in Figure 13.61. Notice that $L_s = 6/(2\pi 60) = 15.9155$ mH and $L_L = 10/(2\pi 60) = 26.5258$ mH. Double-click on the XFRM_LINEAR, and enter $k = 0.717137$, $L_1 = 100$ mH, and $L_2 = 70$ mH. The k value is obtained from $k = M/\sqrt{L_1L_2} = 0.717137$. The results of simulation are available in the simulation output file.

FIGURE 13.61

PSpice schematic for the circuit shown in Figure 13.46.

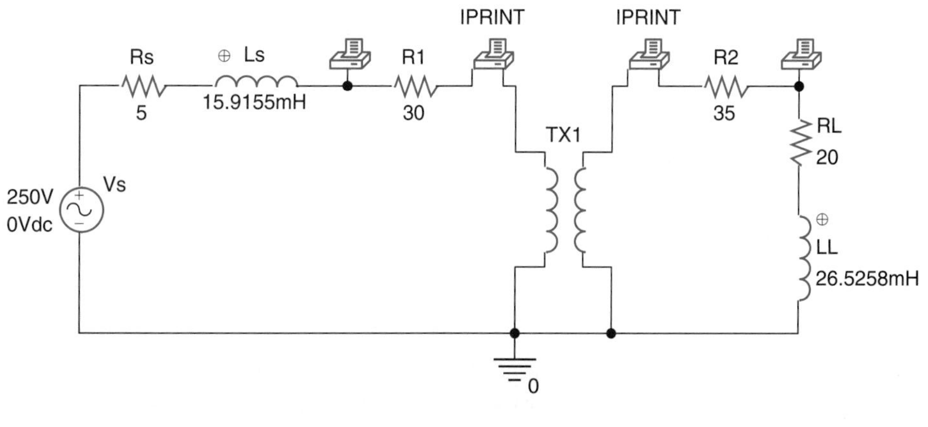

The coupled coils can be simulated in Simulink using a mutual inductance block. The Simulink model for the circuit shown in Figure 13.46 with the component values

$$\mathbf{V}_s = 250\angle 0°\text{ V},\quad \mathbf{Z}_s = 5 + j6,\quad R_1 = 30\ \Omega,\quad L_1 = 100\text{ mH},\quad L_2 = 70\text{ mH},$$
$$M = 60\text{ mH},\quad R_2 = 35\ \Omega,\quad \mathbf{Z}_L = 20 + j10,\quad f = 60\text{ Hz}$$

is shown in Figure 13.62. The winding resistances R_1 and R_2 can be entered in the mutual inductance block along with L_1, L_2, and M. The values obtained from the Simulink match those from PSpice.

The PSpice schematic for the circuit shown in Figure 13.54 is shown in Figure 13.63.

The part name for the transformer is XFRM_LINEAR/ANALOG. To get $n = 1/12$, the ratio of L_1 to L_2 should be $L_1/L_2 = 12^2 = 144$. Choose large values for L_1 and L_2, and set $k = 1$ to make the transformer model ideal. VSIN/SOURCE is used for the sinusoidal input.

The maximum step size is set at 1us = 1μs.

FIGURE 13.62

Simulink model for the circuit shown in Figure 13.46.

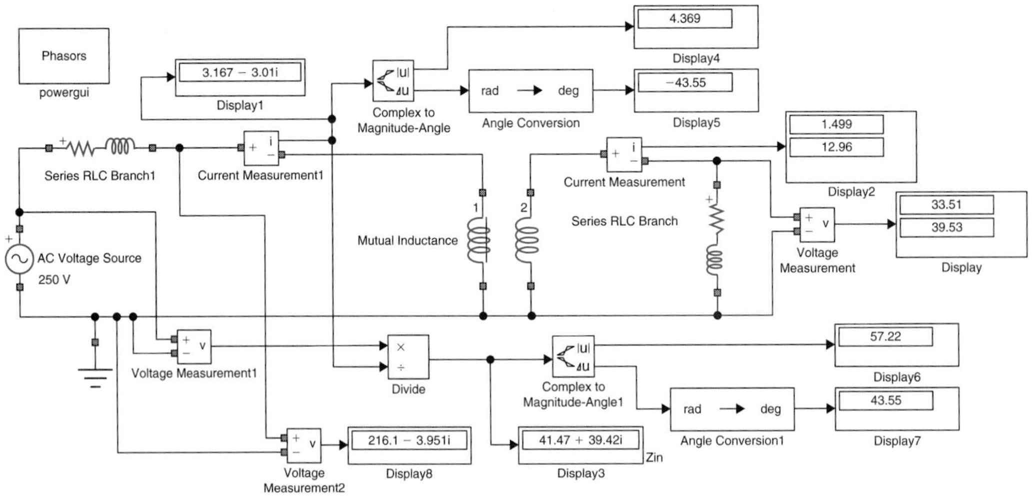

FIGURE 13.63

PSpice schematic for the circuit shown in Figure 13.54.

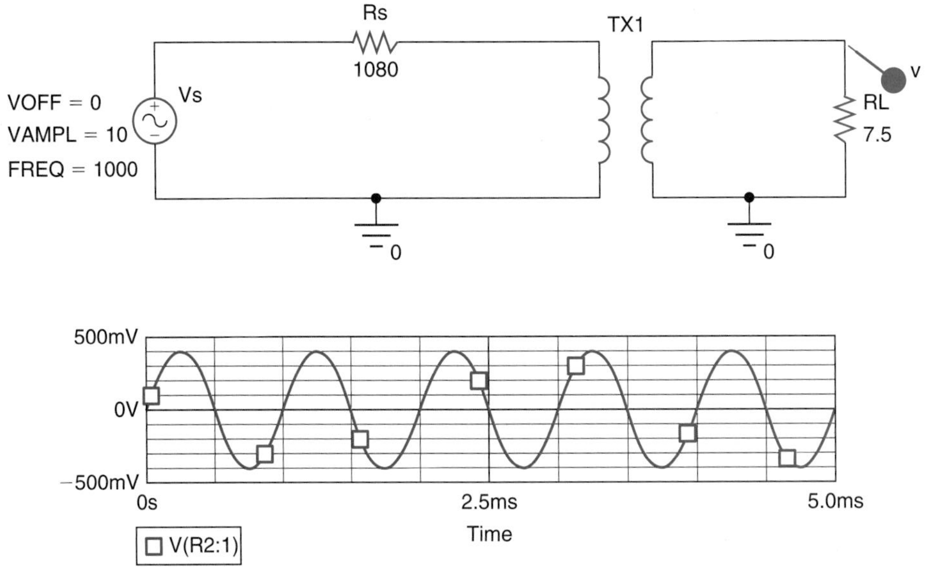

The Simulink model for the circuit shown in Figure 13.54 is shown in Figure 13.64. For the winding ratio, enter 12 for $1/n$.

FIGURE 13.64

A Simulink model for the circuit shown in Figure 13.54.

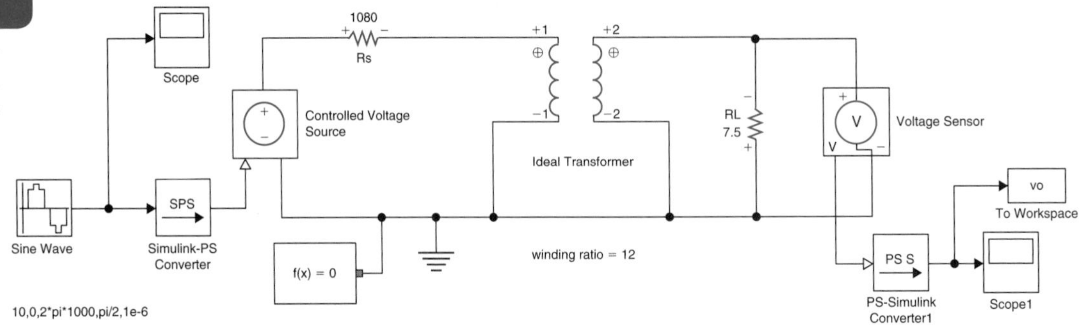

continued

FIGURE 13.64

continued

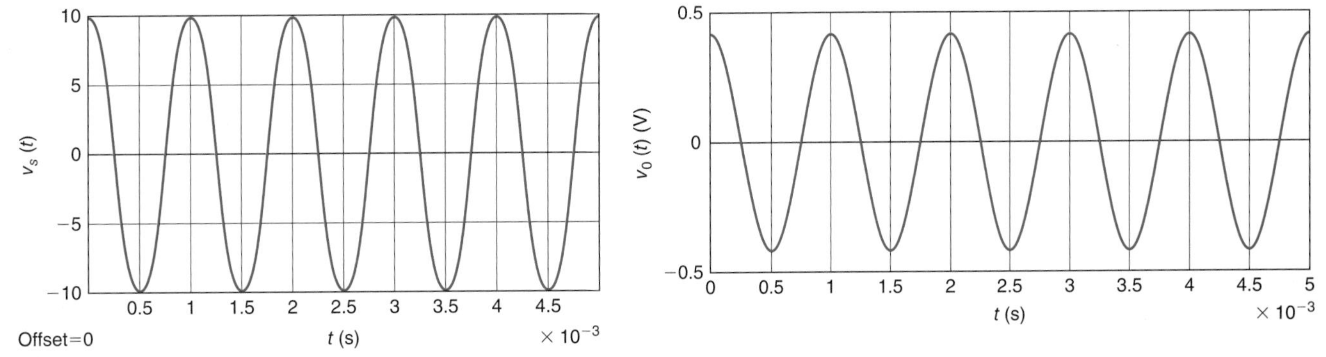

SUMMARY

When two coils are in proximity, part of the magnetic flux from one coil links to the other coil, producing an induced voltage on the second coil. The flux also links its own coil, producing an induced voltage on the first coil. When the fluxes are additive, the voltages and currents are related by

$$v_1(t) = L_1\frac{di_1(t)}{dt} + M\frac{di_2(t)}{dt} \quad v_2(t) = M\frac{di_1(t)}{dt} + L_2\frac{di_2(t)}{dt}$$

When the fluxes are subtractive, the induced term is negative. In the phasor notation, the voltages and the currents are related by, for the additive case,

$$V_1 = j\omega L_1 I_1 + j\omega M I_2 \quad V_2 = j\omega M I_1 + j\omega L_2 I_2$$

For an ideal transformer, the voltage v_1 across the primary winding and the voltage v_2 across the secondary winding are related by

$$v_2 = \frac{N_2}{N_1}v_1 = \pm n v_1$$

where n is called the *turns ratio*. For an ideal transformer, the current i_1 through the primary winding and the current i_2 through the secondary winding are related by

$$i_2 = \pm\frac{1}{n}i_1$$

The input impedance of an ideal transformer is given by

$$Z_{in} = \frac{Z_L}{n^2}$$

where Z_L is the load impedance.

PROBLEMS

Mutual Inductance

13.1 **A second coil with 300 turns is wrapped around a solenoid with a length of 0.5 m, cross-sectional area of 0.05 m^2, and 500 turns. Find the mutual inductance.**

13.2 **The magnetic flux density from a circular loop of radius b that carries current I at a distance d from the center in the normal direction ($\mathbf{a}_x$) is given by**

$$\boldsymbol{B} = \frac{\mu_0 I b^2}{2(d^2 + b^2)^{3/2}}\boldsymbol{a}_x$$

Coil 1 with N_1 turns and radius b is located at $x = 0$ and coil 2 with N_2 turns and radius b is located at $x = d$, as shown in Figure P13.2. Assume that $b < d$, and the magnetic flux density inside the coils is equal to that at the center.

FIGURE P13.2

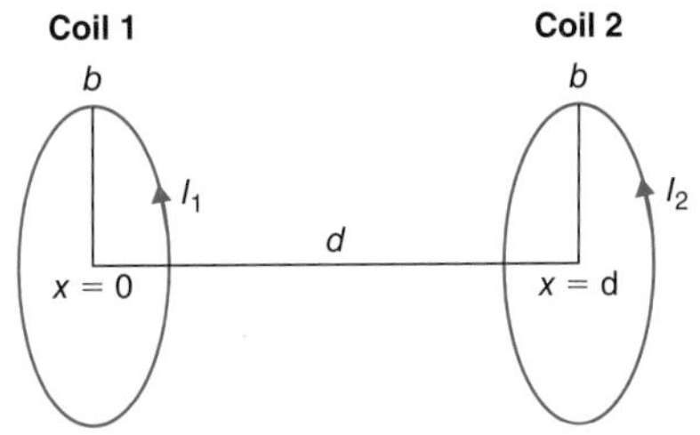

a. Find the fluxes Φ_{11}, Φ_{12}, Φ_{21}, and Φ_{22}.

b. Find the inductances L_1, M_{12}, M_{21}, and L_2.

13.3 A second wire with 300 turns is wrapped around a toroid with a rectangular cross section. The inner radius of the toroid is 0.3 m, the outer radius is 0.5 m, and the height is 0.2 m. The number of turns of the primary wire is 500. Find the mutual inductance.

Dot Convention and Induced Voltage

13.4 Write equations for v_1 and v_2 as a function of i_1, i_2, L_1, L_2, and M for the coupled coils shown in Figure P13.4.

FIGURE P13.4

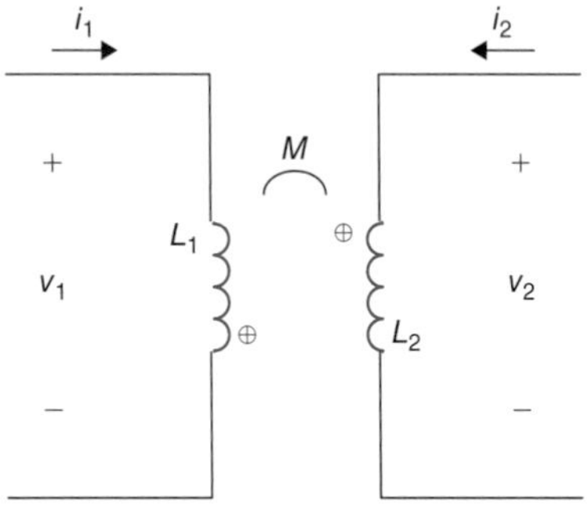

13.5 Write equations for v_1 and v_2 as a function of i_1, i_2, L_1, L_2, and M for the coupled coils shown in Figure P13.5.

FIGURE P13.5

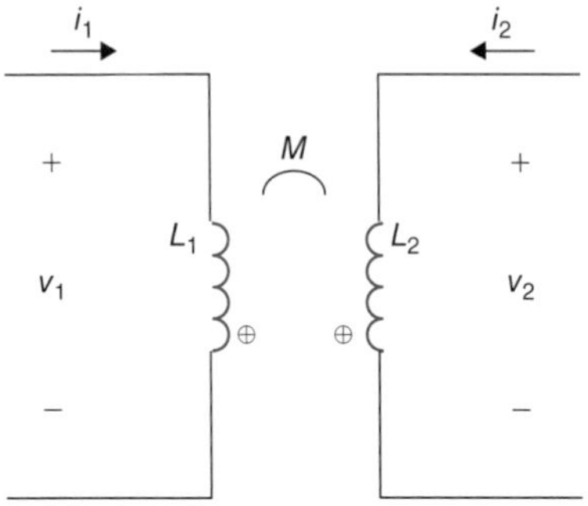

13.6 Write equations for v_1 and v_2 as a function of i_1, i_2, L_1, L_2, and M for the coupled coils shown in Figure P13.6.

FIGURE P13.6

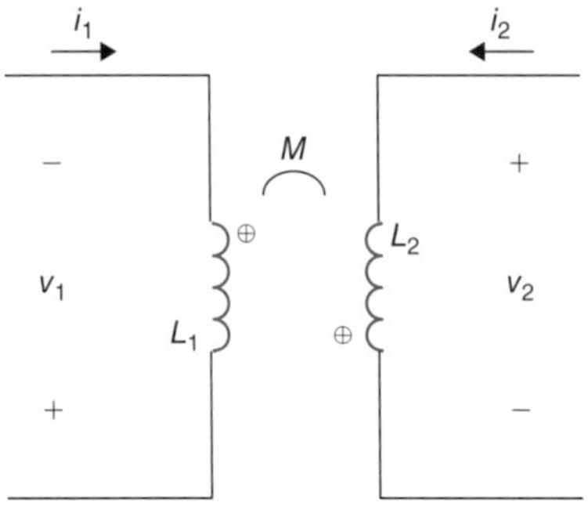

13.7 Write equations for v_1 and v_2 as a function of i_1, i_2, L_1, L_2, and M for the coupled coils shown in Figure P13.7.

FIGURE P13.7

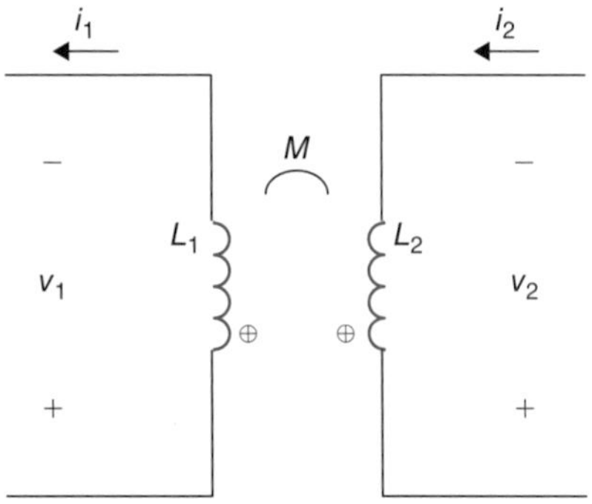

13.8 Find the equivalent inductance L between a and b for the coupled coils connected in parallel (additive), as shown in Figure P13.8.

FIGURE P13.8

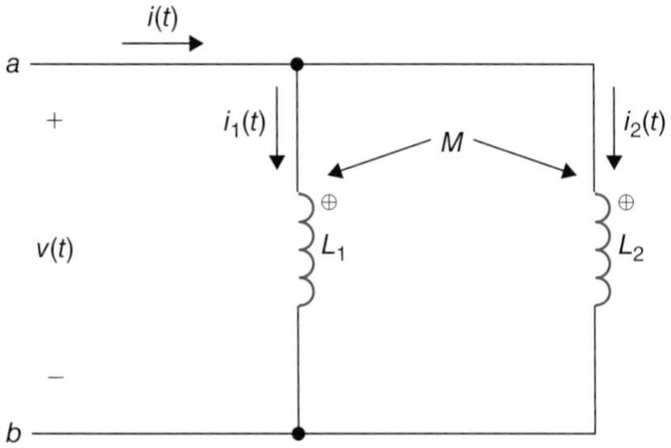

13.9 Find the equivalent inductance L between a and b for the coupled coils connected in parallel (subtractive), as shown in Figure P13.9.

FIGURE P13.9

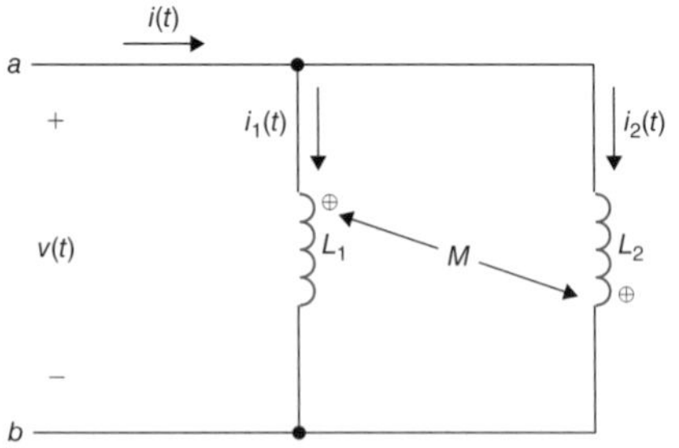

13.10 Find the equivalent inductance L for three coupled coils connected in parallel. The mutual inductance between any two is M (additive).

13.11 Find the equivalent inductance L for three coupled coils connected in parallel. The mutual inductance between any two is M (subtractive).

13.12 **Find $\mathbf{I}_1$, $\mathbf{I}_2$, $\mathbf{V}_1$, and $\mathbf{V}_2$ for the circuit shown in Figure P13.12.**

FIGURE P13.12

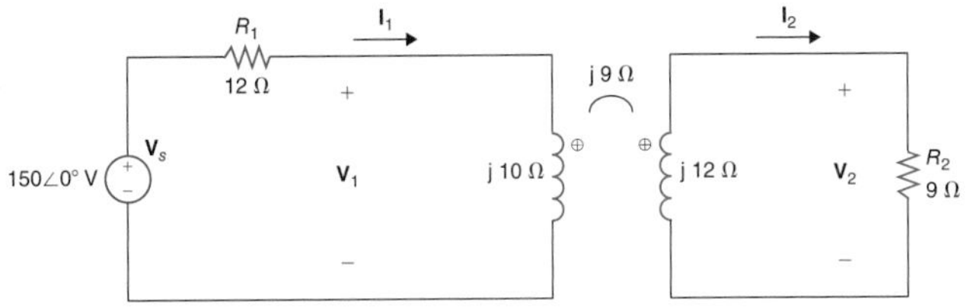

13.13 **Find the phasors $\mathbf{I}_1$, $\mathbf{I}_2$, $\mathbf{V}_1$, and $\mathbf{V}_2$ in the circuit shown in Figure P13.13.**

FIGURE P13.13

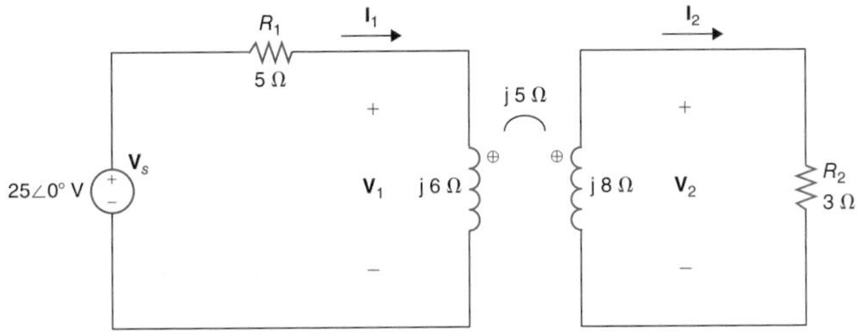

13.14 **Find the phasors $\mathbf{I}_1$, $\mathbf{I}_2$, $\mathbf{V}_1$, and $\mathbf{V}_2$ in the circuit shown in Figure P13.14.**

FIGURE P13.14

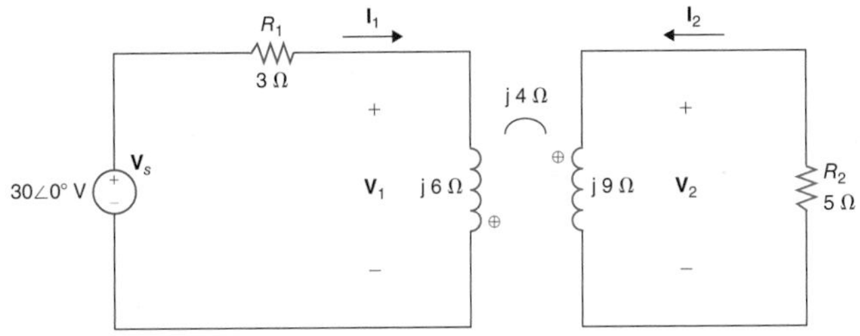

13.15 **Find $\mathbf{V}_1$, $\mathbf{V}_2$, $\mathbf{I}_1$, and $\mathbf{I}_2$ for the circuit shown in Figure P13.15.**

FIGURE P13.15

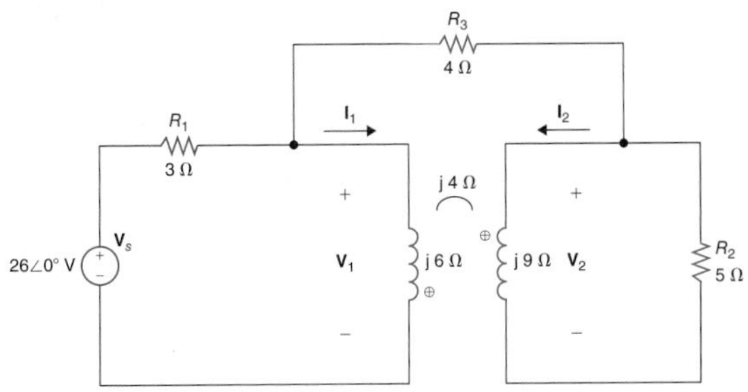

13.16 **Find $\mathbf{V}_1$, $\mathbf{V}_2$, $\mathbf{I}_1$, and $\mathbf{I}_2$ for the circuit shown in Figure P13.16.**

FIGURE P13.16

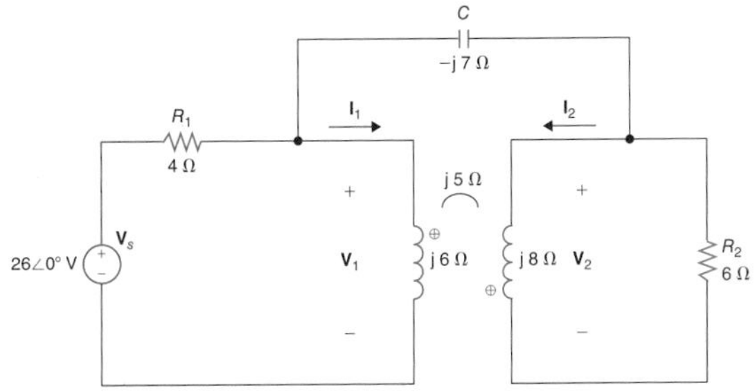

13.17 **Find $\mathbf{I}_1$, $\mathbf{I}_2$, and $\mathbf{I}_3$ in the circuit shown in Figure P13.17.**

FIGURE P13.17

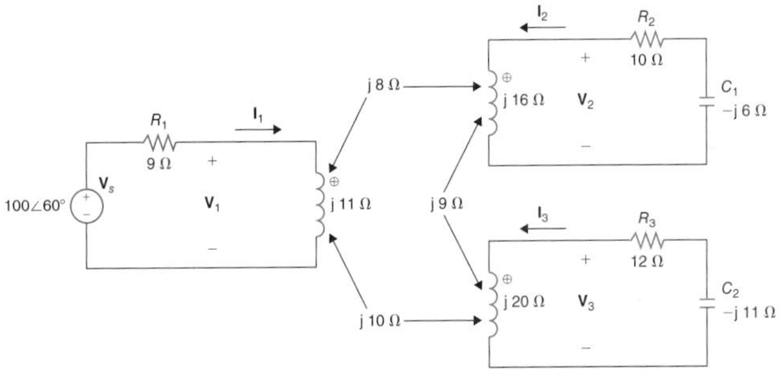

13.18 **Find $\mathbf{I}_1$, $\mathbf{I}_2$, and $\mathbf{I}_3$ in the circuit shown in Figure P13.18.**

FIGURE P13.18

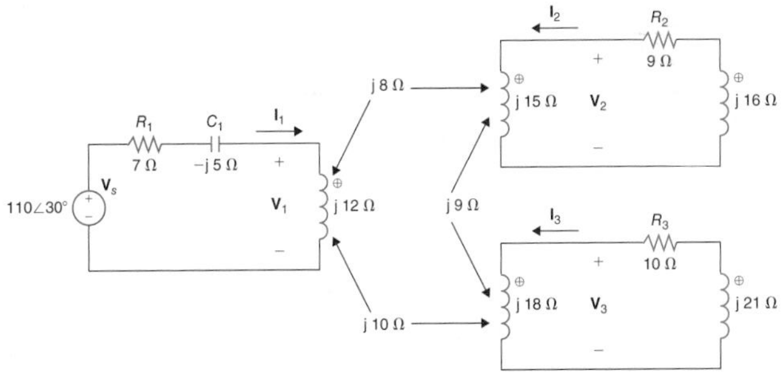

Linear Transformer

13.19 **Find the Thévenin equivalent circuit to the left of terminals *a* and *b* in the circuit shown in Figure P13.19.**

FIGURE P13.19

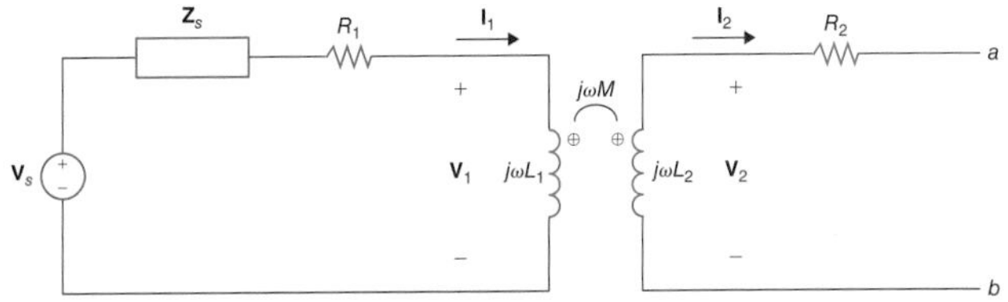

13.20 Find the Thévenin equivalent circuit to the left of terminals a and b in the circuit shown in Figure P13.20. Use the Thévenin equivalent circuit to find the voltage $\mathbf{V}_o$ across the load R_L.

FIGURE P13.20

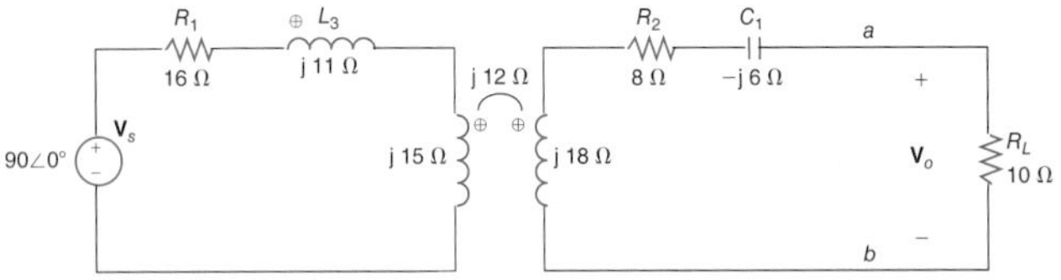

13.21 Find the Thévenin equivalent circuit to the left of terminals a and b in the circuit shown in Figure P13.21. Use the Thévenin equivalent circuit to find the voltage $\mathbf{V}_o$ across the load R_L.

FIGURE P13.21

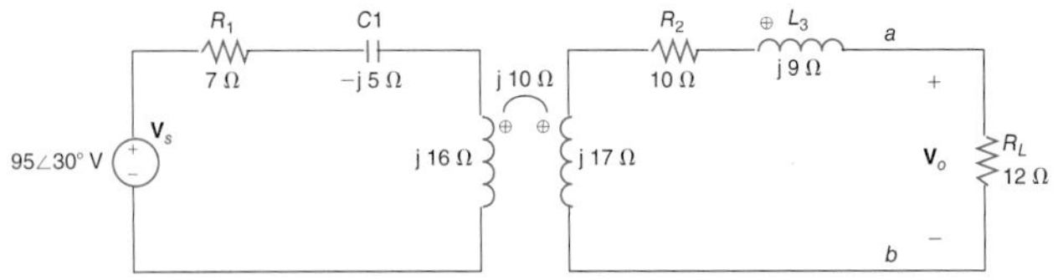

13.22 Find the Thévenin equivalent circuit to the left of terminals a and b in the circuit shown in Figure P13.22. Use the Thévenin equivalent circuit to find the voltage $\mathbf{V}_o$ across the load R_L.

FIGURE P13.22

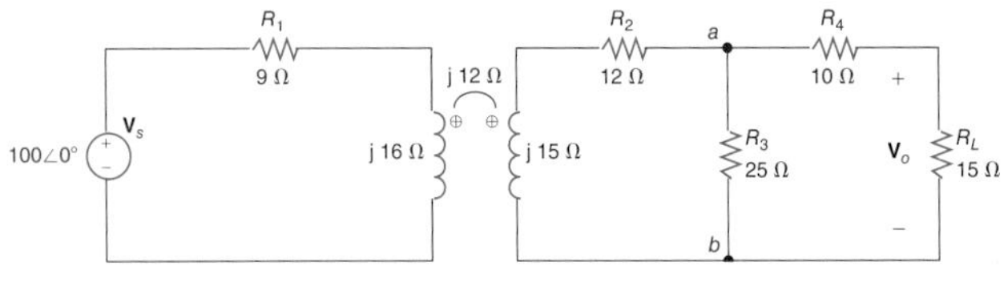

13.23 Find $\mathbf{V}_1$, $\mathbf{V}_2$, $\mathbf{I}_1$, and $\mathbf{I}_2$ in the circuit shown in Figure P13.23.

FIGURE P13.23

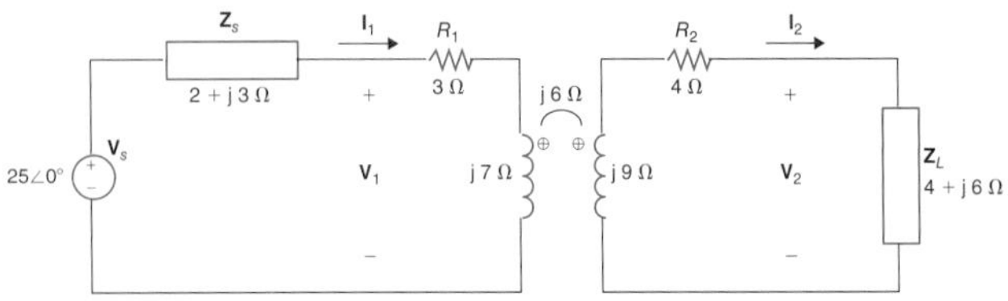

13.24 Find $\mathbf{V}_1$, $\mathbf{V}_2$, $\mathbf{I}_1$, and $\mathbf{I}_2$ in the circuit shown in Figure P13.24.

FIGURE P13.24

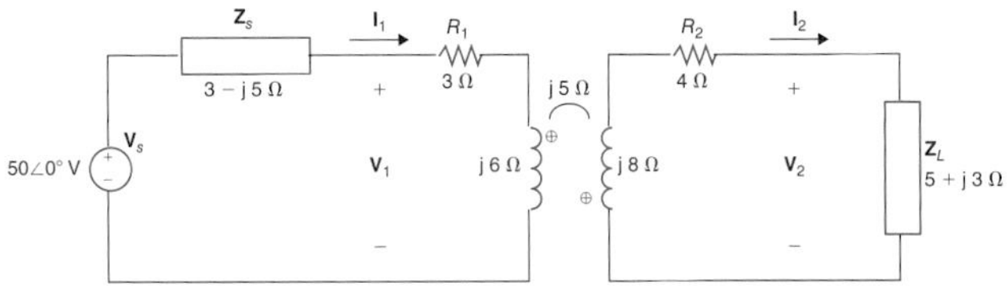

Ideal Transformer

13.25 A speaker with a resistance of 8.5 Ω is to be matched to an audio amplifier with an output resistance of 688.5 Ω through an ideal transformer, as shown in Figure P13.25. Determine the turns ratio of the transformer to obtain the maximum power transfer to the speaker.

FIGURE P13.25

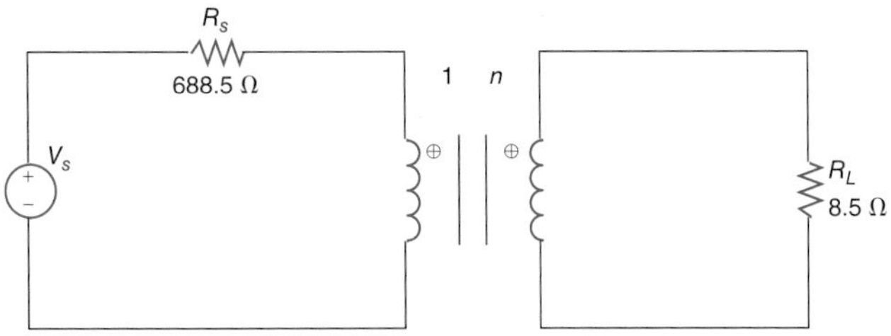

13.26 A speaker with a resistance of 10.5 Ω is to be matched to an audio amplifier with an output resistance of 2362.5 Ω through an ideal transformer, as shown in Figure P13.26. Determine the turns ratio of the transformer to obtain the maximum power transfer to the speaker.

FIGURE P13.26

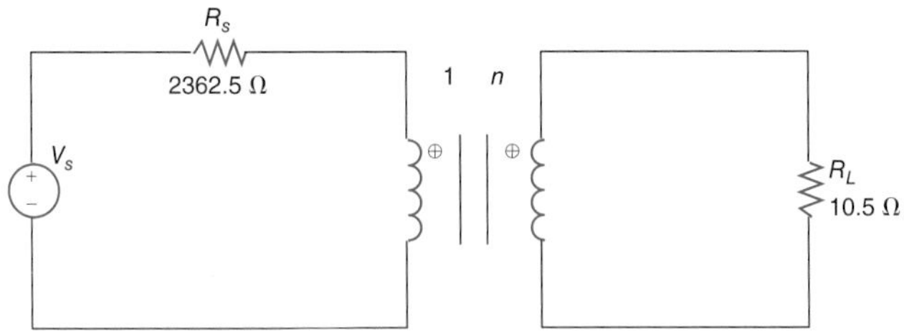

13.27 Find R_{in} in the circuit shown in Figure P13.27.

FIGURE P13.27

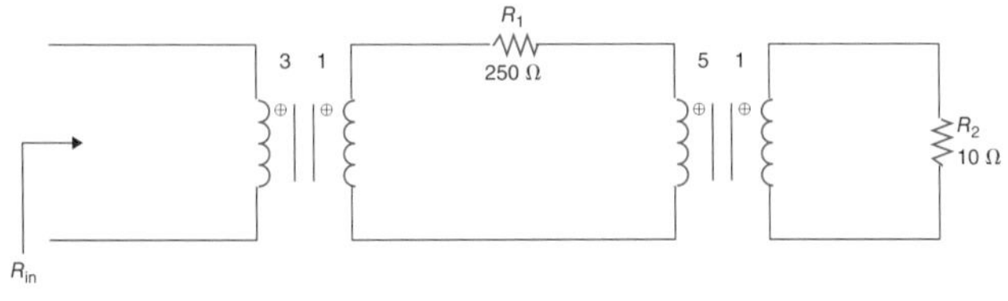

13.28 **Find R_{in} in the circuit shown in Figure P13.28.**

FIGURE P13.28

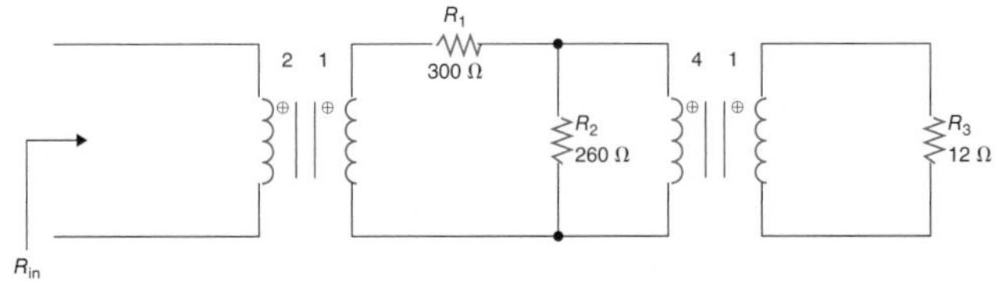

13.29 **Find R_{in} in the circuit shown in Figure P13.29.**

FIGURE P13.29

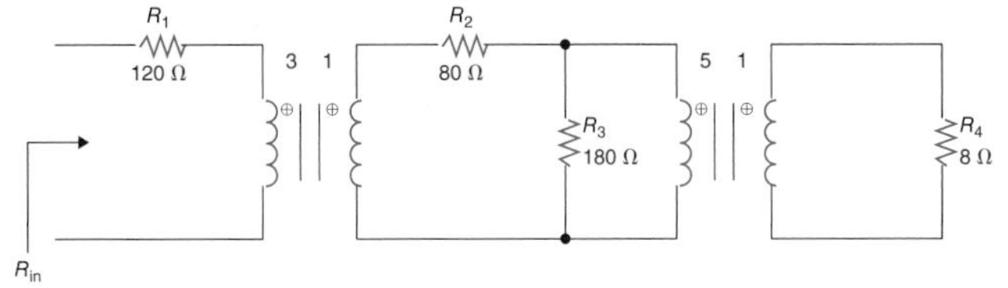

13.30 **Find V_1, V_2, V_3, I_1, I_2, and I_a in the circuit shown in Figure P13.30 when $n = 4$.**

FIGURE P13.30

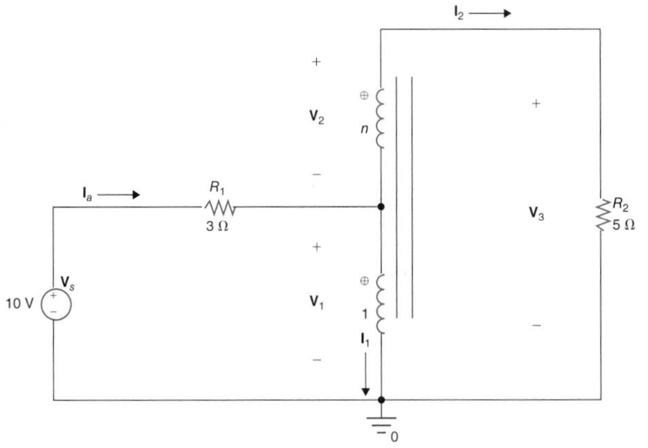

13.31 **Find V_1, V_2, V_3, I_1, I_2, and I_a in the circuit shown in Figure P13.31 when $n = 5$.**

FIGURE P13.31

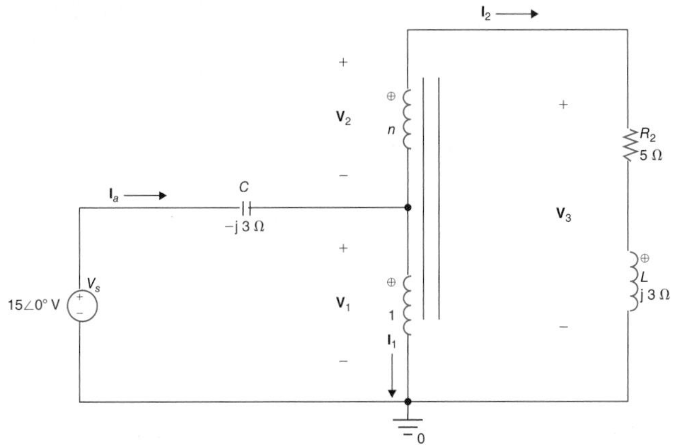

13.32 **Find V_1, V_2, V_3, I_1, I_2, and I_a in the circuit shown in Figure P13.32 when $n = 3$.**

FIGURE P13.32

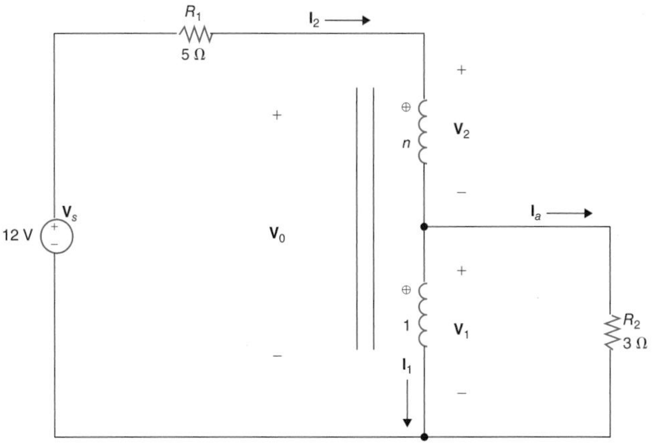

13.33 **Find V_1, V_2, V_3, I_1, I_2, and I_a in the circuit shown in Figure P13.33 when $n = 6$.**

FIGURE P13.33

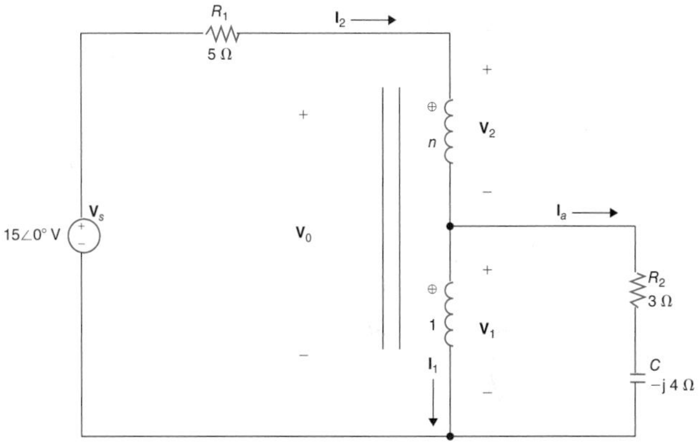

13.34 **Let $\mathbf{V}_s = 220\angle 0°$ V rms, $\mathbf{Z}_s = 25 + j20\ \Omega$, $\mathbf{Z}_a = 20 - j25\ \Omega$, and $\mathbf{Z}_L = 75 + j55\ \Omega$, $N_1 = 150$, $N_2 = 350$ in the circuit shown in Figure P13.34. Find the complex power on the load.**

FIGURE P13.34

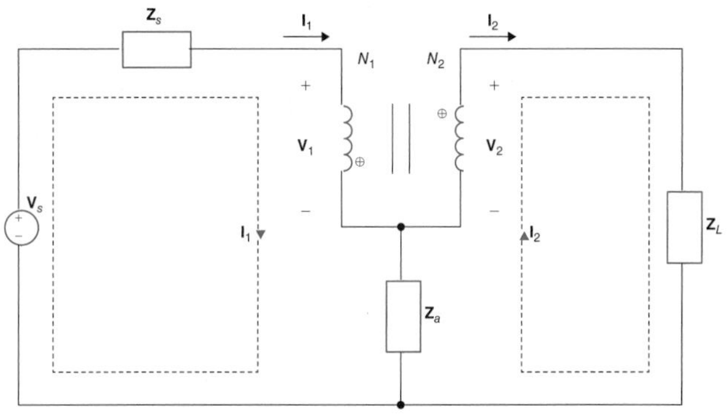

Chapter 14

The Laplace Transform

14.1 Introduction

The phasors and the phasor transformed networks discussed in Chapters 9–11 are useful for circuits or systems with sinusoidal input. The concept of the phasor can be generalized by adding a real part σ to the imaginary part $j\omega$ to define a complex variable $s = \sigma + j\omega$. The transform based on the complex variable s is called Laplace transform. The Laplace transform $F(s)$ of a signal $f(t)$ reveals characteristics of the signal in the s-domain.

The Laplace transform is invertible [that is, if $F(s)$ is the Laplace transform of a time domain signal $f(t)$], we can recover $f(t)$ from $F(s)$ by taking the inverse Laplace transform of $F(s)$. Residue method, partial fraction expansion, and table lookup are used to recover $f(t)$ from $F(s)$.

One of the applications of the Laplace transform is solving differential equations. The Laplace transform converts differential equations to algebraic equations in s. The unknown variable such as voltage $V(s)$ is represented as a rational function of s. The time domain expression $v(t)$ is found by taking the inverse Laplace transform of $V(s)$. Another application of the Laplace transform is evaluating convolution. The convolution property of the Laplace transform says that the convolution in the time domain results in the multiplication in the s-domain. The convolution in the time domain can be obtained by taking the inverse Laplace transform of the product of the Laplace transforms of the two functions being convolved as shown in Chapter 15. The Laplace transform makes it possible to define the transfer function of a linear, time-invariant system in the s-domain. The system can be analyzed in the s-domain and can be designed in the s-domain. The Laplace transform is applied in many different disciplines, including circuit theory, control theory, and filter design.

14.2 Definition of the Laplace Transform

The **one-sided (unilateral) Laplace transform** of a continuous time signal $f(t)$ is defined as

$$F(s) = \int_{0^-}^{\infty} f(t)e^{-st}dt \tag{14.1}$$

The lower limit of the integration is $t = 0^-$, which is the time just before $t = 0$. The integration does include $t = 0$. The value of the signal at $t = 0^-$ is written as $f(0^-)$ and is called the initial condition of the signal. The **two-sided (bilateral) Laplace transform** of a continuous time signal $f(t)$ is defined as

$$F(s) = \int_{-\infty}^{\infty} f(t)e^{-st}dt \tag{14.2}$$

The lower limit of the integration is $t = -\infty$ rather than $t = 0^-$. If we are interested in the analysis of systems or circuits for $t \geq 0$ with the initial condition $f(0^-)$, the one-sided Laplace transform is enough.

The complex variable s can be written as

$$s = \sigma + j\omega \tag{14.3}$$

where σ is the real part, and ω is the complex part. Substituting $s = \sigma + j\omega$ into Equation (14.1), we obtain

$$F(s) = F(\sigma + j\omega) = \int_{0^-}^{\infty} f(t)e^{-st}\,dt = \int_{0^-}^{\infty} f(t)e^{-(\sigma+j\omega)t}\,dt = \int_{0^-}^{\infty} f(t)e^{-\sigma t}e^{-j\omega t}\,dt \tag{14.4}$$

Since the one-sided Fourier transform of $f(t)$ is defined by

$$F(\omega) = \int_{0^-}^{\infty} f(t)e^{-j\omega t}\,dt \tag{14.5}$$

Equation (14.4) indicates that the Laplace transform of $f(t)$ is the Fourier transform of $f(t)e^{-\sigma t}$. For signals $f(t)$ whose Fourier integral $\int_{0^-}^{\infty} f(t)e^{-j\omega t}dt$ does not converge, depending on the choice of σ, multiplication of $f(t)$ by $e^{-\sigma t}$ may attenuate $f(t)$ enough so that the integral $\int_{0^-}^{\infty} f(t)e^{-\sigma t}e^{-j\omega t}dt$ may converge. This implies that the Laplace transform may exist for the signals whose Fourier integral does not converge. If the Fourier transform of a signal exists, it implies the Laplace transform of the signal exists, but the converse is not necessarily true.

The Laplace transform is denoted by L, and the inverse Laplace transform is denoted by L^{-1}. Thus

$$L[f(t)] = F(s)$$

$$L^{-1}[F(s)] = f(t)$$

Also, the Laplace transform pair is denoted by double arrows as

$$f(t) \leftrightarrow F(s)$$

EXAMPLE 14.1

Find the Laplace transform of unit impulse function (called the *Dirac delta function*) $f(t) = \delta(t)$.

Applying the sifting property of Dirac delta function (discussed in Chapter 1), we have

$$F(s) = \int_{0^-}^{\infty} \delta(t)e^{-st}\,dt = e^{-s0} = e^0 = 1$$

Exercise 14.1

Find the Laplace transform of

a. $f(t) = 2\delta(t - 7)$ **b.** $f(t) = -3\delta(t - 2)$

Answer:
a. $F(s) = 2e^{-7s}$ **b.** $F(s) = -3e^{-2s}$

EXAMPLE 14.2

Find the Laplace transform of a unit step function $f(t) = u(t)$.

$$F(s) = \int_{0^-}^{\infty} u(t)e^{-st}\,dt = \int_{0^-}^{\infty} e^{-st}\,dt = \frac{e^{-st}\big|_{0^-}^{\infty}}{-s} = \frac{e^{-(\sigma+j\omega)\infty} - e^{-s0^-}}{-s}$$
$$= \frac{\dfrac{e^{-j\omega\infty}}{e^{\sigma\infty}} - 1}{-s} = \frac{\dfrac{e^{-j\omega\infty}}{\infty} - 1}{-s} = \frac{0-1}{-s} = \frac{1}{s} \qquad \textbf{(14.6)}$$

FIGURE 14.1

Region of convergence for $L[u(t)] = 1/s$.

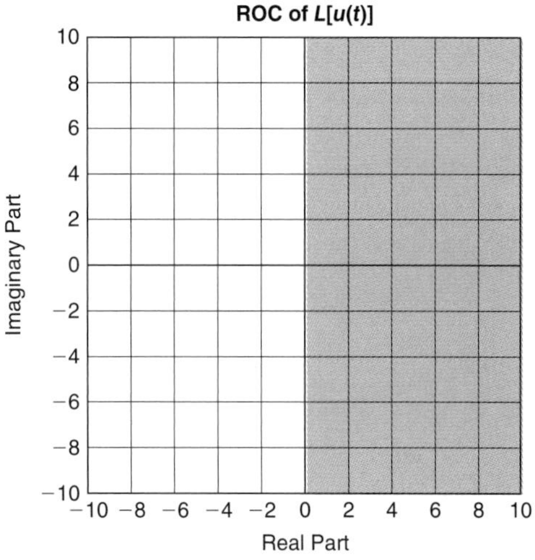

Notice that $e^{\sigma\infty} = \infty$ if $\sigma > 0$. Thus, $L[u(t)] = 1/s$ if $\sigma > 0$. The region in the complex s-plane where the Laplace integral converges is called **region of convergence (ROC).** Thus, the region of convergence for $L[u(t)] = 1/s$ is $\sigma > 0$, which represents the right half of the s-plane, as shown in Figure 14.1. The line defining the boundary, $\sigma = 0$, is called the **abscissa of absolute convergence**. The line defining the boundary, $\sigma = 0$, is not included in the region of convergence. Notice that if $\sigma \leq 0$, the Laplace integral does not converge. In this case, since we cannot represent the signal in the s-domain, there is no use for the Laplace transform.

In general, the Laplace transform for real signals can be written as

$$F(s) = \frac{N(s)}{D(s)}$$

where $N(s)$ is the numerator polynomial with real coefficients, and $D(s)$ is the denominator polynomial with real coefficients.

continued

Example 14.2 continued

FIGURE 14.2

Pole-zero diagram of $1/s$.

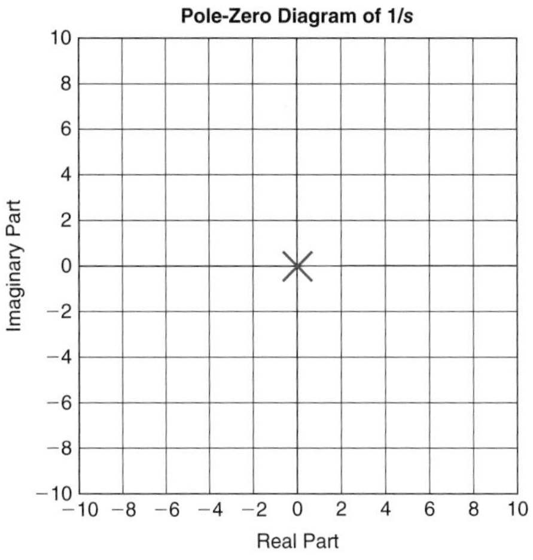

The roots of $D(s) = 0$ are called *poles*, and the roots of $N(s) = 0$ are called *zeros*. At the location of poles, $F(s)$ becomes infinity, and at the location of zeros, $F(s)$ becomes zero. For the unit step function, $D(s) = s$ and $N(s) = 1$. Thus, the location of a pole is $s = 0$. Since $1 \neq 0$, we cannot find a zero from $N(s) = 0$, but, when $s = \infty$, $1/s = 1/\infty = 0$. Thus, the location of a zero for $F(s) = 1/s$ is $s = \infty$. Including poles and zeros at zero and at infinity, there are equal numbers of poles and zeros for the given $F(s)$. The poles are denoted by $\times$, and the zeros are denoted by $\bigcirc$. Figure 14.2 shows the pole at $s = 0$. The graph that shows the locations of poles and zeros is called **pole-zero diagram**. The poles and zeros at infinity are not shown in the pole-zero diagram. The boundary of the region of convergence must have at least one pole. If there is no pole in the boundary, the region of convergence can be extended.

Exercise 14.2

Find the Laplace transform of

a. $f(t) = 3u(t - 5)$ **b.** $f(t) = -6u(t - 9)$.

Answer:

a. $F(s) = \dfrac{3e^{-5s}}{s}$ **b.** $F(s) = \dfrac{-6e^{-9s}}{s}$

EXAMPLE 14.3

Find the Laplace transform of $f(t) = e^{-at}u(t)$.

$$F(s) = \int_{0^-}^{\infty} e^{-at}u(t)e^{-st}dt = \int_{0^-}^{\infty} e^{-(s+a)t}dt = \frac{e^{-(s+a)t}\big|_{0^-}^{\infty}}{-(s+a)} = \frac{e^{-(\sigma+j\omega+a)\infty} - e^{-(s+a)0^-}}{-(s+a)}$$
$$= \frac{\dfrac{e^{-j\omega\infty}}{e^{(\sigma+a)\infty}} - 1}{-(s+a)} = \frac{\dfrac{e^{-j\omega\infty}}{\infty} - 1}{-(s+a)} = \frac{0-1}{-(s+a)} = \frac{1}{s+a} \qquad (14.7)$$

if $\sigma + a > 0$ ($e^{(\sigma+a)\infty} = \infty$). Thus, the region of convergence for $e^{-at}u(t)$ is $\sigma > -a$. Notice that if a is positive ($a > 0$), the region of convergence contains the imaginary axis. If the imaginary axis is inside the region of convergence, the Fourier integral defined by Equation (14.5) converges also. On the other hand, if a is negative ($a < 0$), the ROC, $\sigma > -a$, does not include the imaginary axis. In this case, the Fourier integral defined by Equation (14.5) does not converge. This example illustrates the difference between the Fourier transform and the Laplace transform. If $a < 0$, as $t \to \infty$,

continued

Example 14.3 continued

FIGURE 14.3

Region of convergence and pole location for $F(s) = L[e^{-6t}u(t)] = 1/(s+6)$.

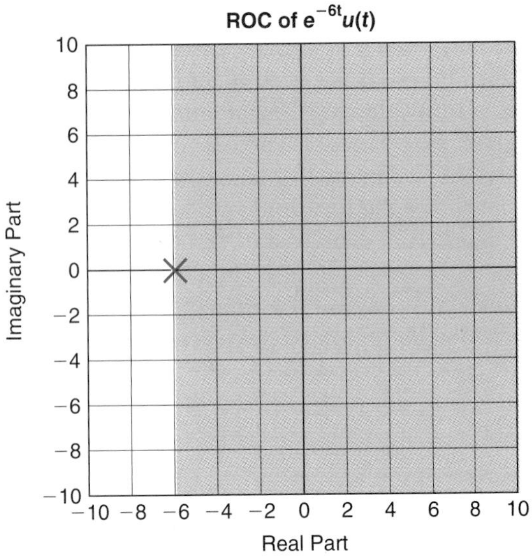

$e^{-at}u(t) \to \infty$. Still, the Laplace transform exists for $\sigma > -a$. $F(s)$ has a pole at $s = -a$ and a zero at $s = \infty$. Figure 14.3 shows the region of convergence and location of pole for $a = 6$. The pole is located at $s = -6$.

In MATLAB, symbolic math can be used to find the Laplace transform:

```
clear all;
syms t a
f=exp(-a*t)
F=laplace(f)
pretty(F)

Answer:
f =
exp(-a*t)
F =
1/(a + s)

  1
-----
a + s
```

Exercise 14.3

Find the Laplace transform of the following signals, and specify the region of convergences.

a. $f(t) = 2e^{-5t}u(t)$ **b.** $f(t) = e^{5t}u(t)$

Answer:

a. $F(s) = \dfrac{2}{s+5}$, ROC: $\sigma > -5$ **b.** $F(s) = \dfrac{1}{s-5}$, ROC: $\sigma > 5$

In general, the Laplace integral defined by Equation (14.1) converges if

$$\int_{0^-}^{\infty} |f(t)e^{-\sigma t}|dt < \infty \tag{14.8}$$

Equation (14.8) is the condition for the existence of the Laplace transform. This condition is less strict than the condition for the existence of the Fourier transform, which is

$$\int_{0^-}^{\infty} |f(t)|dt < \infty \tag{14.9}$$

The signal that satisfies Equation (14.9) is absolutely integrable.

14.3 Properties of the Laplace Transform

Various properties of the Laplace transform are presented in this section. These properties enable us to gain some insights on the Laplace transform. Also, these properties make it easier to calculate the Laplace transform and inverse Laplace transform. Assume

$$L[f(t)] = F(s), \quad L[f_1(t)] = F_1(s), \quad L[f_2(t)] = F_2(s)$$

Table 14.1 lists various properties of the Laplace transform, and Table 14.2 shows the Laplace transform pairs for commonly used signals.

TABLE 14.1 Properties of the Laplace Transform

Property	$f(t)$	$F(s)$
Linearity	$a_1f_1(t) + a_2f_2(t)$	$a_1F_1(s) + a_2F_2(s)$
Time shifting	$f(t-a)u(t-a), a>0$	$e^{-as}F(s)$
Time scaling	$f(at)$	$\frac{1}{\lvert a\rvert}F\left(\frac{s}{a}\right)$
Time integration	$\int_{0^-}^{t} f(\lambda)d\lambda$	$\frac{1}{s}F(s)$
	$\int_{-\infty}^{t} f(\lambda)d\lambda$	$\frac{1}{s}F(s) + \frac{1}{s}\int_{-\infty}^{0^-} f(t)dt$
Time differentiation	$\frac{df(t)}{dt}$	$sF(s) - f(0^-)$
	$\frac{d^2f(t)}{dt^2}$	$s^2F(s) - sf(0^-) - f'(0^-)$
	$\frac{d^3f(t)}{dt^3}$	$s^3F(s) - s^2f(0^-) - sf'(0^-) - f''(0^-)$
Time convolution	$f_1(t)*f_2(t)$	$F_1(s)F_2(s)$
Frequency translation	$e^{-at}f(t)$	$F(s+a)$
Multiplication by $\cos(at)$	$f(t)\cos(at)$	$\frac{1}{2}F(s-ja) + \frac{1}{2}F(s+ja)$
Multiplication by $\sin(at)$	$f(t)\sin(at)$	$\frac{1}{2j}F(s-ja) - \frac{1}{2j}F(s+ja)$
Frequency differentiation	$tf(t)$	$-\frac{dF(s)}{ds}$
Frequency integration	$\frac{f(t)}{t}$	$\int_{s}^{\infty} F(s)ds$
Frequency scaling	$\frac{1}{\lvert a\rvert}f\left(\frac{t}{a}\right)$	$F(as)$
Initial value	$f(0^+)$	$\lim_{s\to\infty} sF(s)$
Final value	$f(\infty)$	$\lim_{s\to 0} sF(s)$
Periodic signal ($t>0$)	$f(t) = f(t+nT)$	$\frac{1}{1-e^{-Ts}}F_1(s)$ where $F_1(s) = \int_{0^-}^{T} f(t)e^{-st}dt$

TABLE 14.2 Laplace Transform Pairs

$f(t)$	$F(s)$	ROC
$\delta(t)$	1	Everywhere
$u(t)$	$\frac{1}{s}$	$\sigma > 0$
$e^{-at}u(t)$	$\frac{1}{s+a}$	$\sigma > -a$
$tu(t)$	$\frac{1}{s^2}$	$\sigma > 0$
$t^n u(t)$	$\frac{n!}{s^{n+1}}$	$\sigma > 0$
$te^{-at}u(t)$	$\frac{1}{(s+a)^2}$	$\sigma > -a$
$t^n e^{-at}u(t)$	$\frac{n!}{(s+a)^{n+1}}$	$\sigma > -a$
$\frac{t^{n-1}}{(n-1)!}e^{-at}u(t)$	$\frac{1}{(s+a)^n}$	$\sigma > -a$
$\sin(\omega t)u(t)$	$\frac{\omega}{s^2+\omega^2}$	$\sigma > 0$
$\cos(\omega t)u(t)$	$\frac{s}{s^2+\omega^2}$	$\sigma > 0$
$e^{-at}\sin(\omega t)u(t)$	$\frac{\omega}{(s+a)^2+\omega^2}$	$\sigma > -a$
$e^{-at}\cos(\omega t)u(t)$	$\frac{s+a}{(s+a)^2+\omega^2}$	$\sigma > -a$
$\sin(\omega t+\theta)u(t)$	$\frac{s\sin(\theta)+\omega\cos(\theta)}{s^2+\omega^2}$	$\sigma > 0$
$\cos(\omega t+\theta)u(t)$	$\frac{s\cos(\theta)-\omega\sin(\theta)}{s^2+\omega^2}$	$\sigma > 0$
$e^{-at}\sin(\omega t+\theta)u(t)$	$\frac{(s+a)\sin(\theta)+\omega\cos(\theta)}{(s+a)^2+\omega^2}$	$\sigma > -a$
$e^{-at}\cos(\omega t+\theta)u(t)$	$\frac{(s+a)\cos(\theta)-\omega\sin(\theta)}{(s+a)^2+\omega^2}$	$\sigma > -a$
$t\sin(\omega t)u(t)$	$\frac{2\omega s}{(s^2+\omega^2)^2}$	$\sigma > 0$
$t\cos(\omega t)u(t)$	$\frac{s^2-\omega^2}{(s^2+\omega^2)^2}$	$\sigma > 0$
$te^{-at}\sin(\omega t)u(t)$	$\frac{2\omega(s+a)}{[(s+a)^2+\omega^2]^2}$	$\sigma > -a$
$te^{-at}\cos(\omega t)u(t)$	$\frac{(s+a)^2-\omega^2}{[(s+a)^2+\omega^2]^2}$	$\sigma > -a$
$te^{-at}\sin(\omega t+\theta)u(t)$	$\frac{[(s+a)^2-\omega^2]\sin(\theta)+2\omega(s+a)\cos(\theta)}{[(s+a)^2+\omega^2]^2}$	$\sigma > -a$
$te^{-at}\cos(\omega t+\theta)u(t)$	$\frac{[(s+a)^2-\omega^2]\cos(\theta)-2\omega(s+a)\sin(\theta)}{[(s+a)^2+\omega^2]^2}$	$\sigma > -a$

14.3.1 LINEARITY PROPERTY (SUPERPOSITION PRINCIPLE)

The Laplace transform of a linear combination of two signals, $f_1(t)$ and $f_2(t)$, is a linear combination of the Laplace transforms $F_1(s)$ and $F_2(s)$:

$$L[a_1 f_1(t) + a_2 f_2(t)] = a_1 F_1(s) + a_2 F_2(s) \tag{14.10}$$

Proof

From the definition of Laplace transform, Equation (14.1), we have

$$L[a_1 f_1(t) + a_2 f_2(t)] = \int_{0^-}^{\infty} [a_1 f_1(t) + a_2 f_2(t)] e^{-st} dt = a_1 \int_{0^-}^{\infty} f_1(t) e^{-st} dt + a_2 \int_{0^-}^{\infty} f_2(t) e^{-st} dt$$

$$= a_1 F_1(s) + a_2 F_2(s)$$

The region of convergence for $a_1 f_1(t) + a_2 f_2(t)$ is the intersection of the region of convergence of $F_1(s)$ and the region of convergence of $F_2(s)$.

EXAMPLE 14.4

Find the Laplace transform of $f(t) = 2e^{-3t}u(t) + 5e^{-4t}u(t)$.

Since

$$L[2e^{-3t}u(t)] = \frac{2}{s+3}, \quad \text{ROC: } \sigma > -3$$

and

$$L[5e^{-4t}u(t)] = \frac{5}{s+4}, \quad \text{ROC: } \sigma > -4$$

we have

$$L[2e^{-3t}u(t) + 5e^{-4t}u(t)] = \frac{2}{s+3} + \frac{5}{s+4} = \frac{7s+23}{(s+3)(s+4)}, \quad \text{ROC: } \sigma > -3$$

Exercise 14.4

Find the Laplace transform of $f(t) = 6e^{-2t}u(t) + 2e^{4t}u(t)$.

Answer:

$$F(s) = \frac{6}{s+2} + \frac{2}{s-4} = \frac{8s-20}{(s+2)(s-4)} = \frac{8s-20}{s^2-2s-8}, \quad \text{ROC: } \sigma > 4.$$

14.3.2 TIME-SHIFTING PROPERTY

Let the one-sided Laplace transform of a signal $f(t)$ be $F(s)$. Then, when $f(t)u(t)$ is shifted to the right by t_d, the Laplace transform of $f(t - t_d)u(t - t_d)$, $t_d > 0$, is given by

$$L[f(t - t_d)u(t - t_d)] = e^{-t_d s}F(s) \tag{14.11}$$

Proof

Since $u(t - t_d) = 0$ for $t < t_d$, $f(t - t_d)u(t - t_d) = 0$ for $t < t_d$. Let $t - t_d = \tau$. Then, $t = t_d + \tau$, $dt = d\tau$, and when $t = t_d^-$, $\tau = 0^-$. Thus, the Laplace transform of $f(t - t_d)$ becomes

$$L[f(t - t_d)] = \int_{0^-}^{\infty} f(t - t_d)u(t - t_d)e^{-st}dt = \int_{t_d^-}^{\infty} f(t - t_d)e^{-st}dt = \int_{0^-}^{\infty} f(\tau)e^{-s(t_d+\tau)}d\tau$$

$$= e^{-t_d s}\int_{0^-}^{\infty} f(\tau)e^{-s\tau}d\tau = e^{-t_d s}F(s)$$

Let the one-sided Laplace transform of a signal $f(t)$ be $F(s)$. Then, when $f(t)u(t)$ is shifted to the left by t_d, the one-sided Laplace transform of $f(t + t_d)u(t + t_d)$, $t_d > 0$, is given by

$$L[f(t + t_d)u(t + t_d)] = \int_{0^-}^{\infty} f(t + t_d)u(t + t_d)e^{-st}dt = \int_{0^-}^{\infty} f(t + t_d)e^{-st}dt$$

Notice that the lower limit of the integral stays at $t = 0^-$ for one-sided Laplace transform.

EXAMPLE 14.5

Find the Laplace transform of

a. $f(t) = \delta(t - 12)$ **b.** $f(t) = u(t - 3)$ **c.** $f(t) = e^{-5(t-2)}u(t - 2)$

Applying the time-shifting property, we obtain

a. $F(s) = L[\delta(t - 12)] = e^{-12s}L[\delta(t)] = e^{-12s} \times 1 = e^{-12s}$

b. $F(s) = L[u(t - 3)] = e^{-3s}L[u(t)] = e^{-3s} \times \dfrac{1}{s} = \dfrac{e^{-3s}}{s}$

c. $F(s) = L[e^{-5(t-2)}u(t - 2)] = e^{-2s}L[e^{-5t}u(t)] = e^{-2s} \times \dfrac{1}{s + 5} = \dfrac{e^{-2s}}{s + 5}$

Exercise 14.5

Find the Laplace transform of

a. $f(t) = \delta(t - 7)$ **b.** $f(t) = u(t - 12)$ **c.** $f(t) = e^{-3t+7}u(t - 2)$

Answer:

a. $F(s) = e^{-7s}$ **b.** $F(s) = \dfrac{e^{-12s}}{s}$ **c.** $f(t) = e \times e^{-3(t-2)}u(t - 2)$, $F(s) = \dfrac{e \times e^{-2s}}{s + 3}$

EXAMPLE 14.6

Find the Laplace transform of $f(t) = e^{-2(t+1)}u(t + 1)$.

Figure 14.4 shows $f(t)$. The one-sided Laplace transform of $f(t)$ is obtained by integrating $f(t)$ from 0^- to infinity. The signal from $t = -1$ to $t = 0^-$ is ignored. Thus, we have

$$F(s) = \int_{0^-}^{\infty} e^{-2(t+1)}u(t+1)e^{-st}\,dt = \int_{0^-}^{\infty} e^{-2(t+1)}e^{-st}\,dt = e^{-2}\int_{0^-}^{\infty} e^{-2t}e^{-st}\,dt = \frac{e^{-2}}{s+2}$$

FIGURE 14.4

Plot of $f(t) = e^{-2(t+1)}u(t + 1)$.

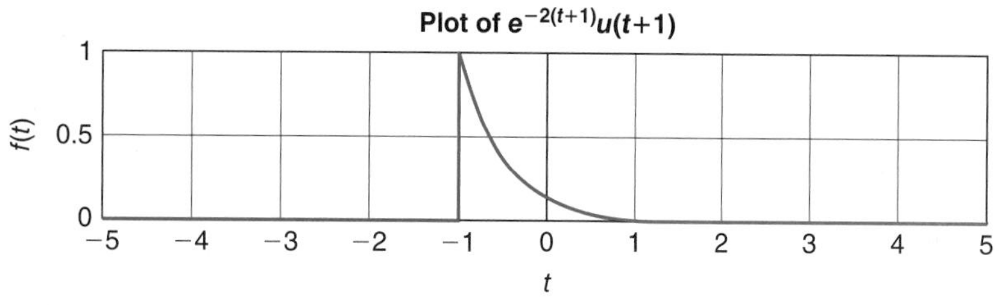

Exercise 14.6

Find the Laplace transform of $f(t) = 6e^{-(t+2)}u(t + 2)$.

Answer:

$$F(s) = \int_{0^-}^{\infty} 6e^{-(t+2)}e^{-st}\,dt = 6e^{-2}\int_{0^-}^{\infty} e^{-t}e^{-st}\,dt = 6e^{-2}\int_{0^-}^{\infty} e^{-(s+1)t}\,dt = \frac{6e^{-2}}{s+1} = \frac{0.812}{s+1}$$

14.3.3 FREQUENCY TRANSLATION PROPERTY

Let the one-sided Laplace transform of a signal $f(t)$ be $F(s)$. Then, when $f(t)$ is multiplied by e^{-at}, the Laplace transform of $e^{-at}f(t)$ is given by

$$L[e^{-at}f(t)] = F(s + a) \quad \textbf{(14.12)}$$

Proof

From the definition of Laplace transform given by Equation (14.1), we have

$$L[e^{-at}f(t)] = \int_{0^-}^{\infty} e^{-at}f(t)e^{-st}\,dt = \int_{0^-}^{\infty} f(t)e^{-at}e^{-st}\,dt = \int_{0^-}^{\infty} f(t)e^{-(s+a)t}\,dt = F(s + a)$$

Multiplication of $f(t)$ by e^{-at} in the time domain results in translation in the s-domain from $F(s)$ to $F(s + a)$. If a is a real number, $F(s + a)$ is a horizontal translation of $F(s)$ in the s-plane. If a is a purely imaginary number, $F(s + a)$ is a vertical translation of $F(s)$ in the s-plane. If a is a complex number, $F(s + a)$ is a translation of $F(s)$ in both horizontal and vertical directions.

As an example, the Laplace transform of $f(t) = e^{-at}u(t)$ can be obtained from frequency translation property. Since $L[u(t)] = \frac{1}{s}$, application of the frequency translation property to this Laplace transform pair results in

$$L[e^{-at}u(t)] = \frac{1}{s + a}$$

The frequency translation property is handy in finding the Laplace transform of $\cos(\omega_0 t)u(t)$ and $\sin(\omega_0 t)u(t)$. Writing $\cos(\omega_0 t)$ as

$$\cos(\omega_0 t) = \frac{e^{j\omega_0 t} + e^{-j\omega_0 t}}{2}$$

and applying the frequency translation property to $L[u(t)] = \frac{1}{s}$, we have

$$\begin{aligned} L[\cos(\omega_0 t)u(t)] &= \frac{1}{2}L[e^{j\omega_0 t}u(t) + e^{-j\omega_0 t}u(t)] \\ &= \frac{1}{2}\left(\frac{1}{s - j\omega_0} + \frac{1}{s + j\omega_0}\right) = \frac{s}{s^2 + \omega_0^2} \end{aligned} \tag{14.13}$$

The region of convergence (ROC) is the same as the ROC for $L[u(t)] = \frac{1}{s}$; that is, $\sigma > 0$.

Similarly, writing $\sin(\omega_0 t)$ as

$$\sin(\omega_0 t) = \frac{e^{j\omega_0 t} - e^{-j\omega_0 t}}{2j}$$

and applying the frequency translation property to $L[u(t)] = \frac{1}{s}$, we have

$$\begin{aligned} L[\sin(\omega_0 t)u(t)] &= \frac{1}{2j}L[e^{j\omega_0 t}u(t) - e^{-j\omega_0 t}u(t)] \\ &= \frac{1}{2j}\left(\frac{1}{s - j\omega_0} - \frac{1}{s + j\omega_0}\right) = \frac{\omega_0}{s^2 + \omega_0^2} \end{aligned} \tag{14.14}$$

The ROC is the same as the ROC for $L[u(t)] = \frac{1}{s}$; that is, $\sigma > 0$.

Application of the frequency translation property to $L[\cos(\omega_0 t)u(t)] = \frac{s}{s^2 + \omega_0^2}$ yields

$$L[e^{-at}\cos(\omega_0 t)u(t)] = \frac{s + a}{(s + a)^2 + \omega_0^2}, \quad \text{ROC: } \sigma > 0$$

Similarly, application of the frequency translation property to $L[\sin(\omega_0 t)u(t)] = \frac{\omega_0}{s^2 + \omega_0^2}$ yields

$$L[e^{-at}\sin(\omega_0 t)u(t)] = \frac{\omega_0}{(s + a)^2 + \omega_0^2}, \quad \text{ROC: } \sigma > 0$$

The Laplace transforms of $\cos(\omega_0 t)u(t)$ and $\sin(\omega_0 t)u(t)$ can be used to find the Laplace transforms of $\cos(\omega_0 t + \theta)u(t)$ and $\sin(\omega_0 t + \theta)u(t)$. Using trigonometric identity,

$$\cos(\alpha + \beta) = \cos(\alpha)\cos(\beta) - \sin(\alpha)\sin(\beta)$$

we can expand $\cos(\omega_0 t + \theta)u(t)$ as

$$\cos(\omega_0 t + \theta)u(t) = \cos(\omega_0 t)\cos(\theta)u(t) - \sin(\omega_0 t)\sin(\theta)u(t)$$
$$= \cos(\theta)\cos(\omega_0 t)u(t) - \sin(\theta)\sin(\omega_0 t)u(t)$$

Application of the Laplace transforms of $\cos(\omega_0 t)u(t)$ and $\sin(\omega_0 t)u(t)$ yields

$$L[\cos(\omega_0 t + \theta)u(t)] = \cos(\theta)\frac{s}{s^2 + \omega_0^2} - \sin(\theta)\frac{\omega_0}{s^2 + \omega_0^2}$$
$$= \frac{s\cos(\theta) - \omega_0 \sin(\theta)}{s^2 + \omega_0^2} \quad \text{ROC: } \sigma > 0 \tag{14.15}$$

Similarly, using trigonometric identity,

$$\sin(\alpha + \beta) = \sin(\alpha)\cos(\beta) + \cos(\alpha)\sin(\beta)$$

we can expand $\sin(\omega_0 t + \theta)u(t)$ as

$$\sin(\omega_0 t + \theta)u(t) = \sin(\omega_0 t)\cos(\theta)u(t) + \cos(\omega_0 t)\sin(\theta)u(t)$$
$$= \cos(\theta)\sin(\omega_0 t)u(t) + \sin(\theta)\cos(\omega_0 t)u(t)$$

Application of the Laplace transforms of $\cos(\omega_0 t)u(t)$ and $\sin(\omega_0 t)u(t)$ yields

$$L[\sin(\omega_0 t + \theta)u(t)] = \cos(\theta)\frac{\omega_0}{s^2 + \omega_0^2} + \sin(\theta)\frac{s}{s^2 + \omega_0^2}$$
$$= \frac{\omega_0 \cos(\theta) + s\sin(\theta)}{s^2 + \omega_0^2} \quad \text{ROC: } \sigma > 0 \tag{14.16}$$

Applying the frequency translation property, we obtain

$$L[e^{-at}\cos(\omega_0 t + \theta)u(t)] = \frac{(s + a)\cos(\theta) - \omega_0 \sin(\theta)}{(s + a)^2 + \omega_0^2}, \quad \text{ROC: } \sigma > 0 \tag{14.17}$$

$$L[e^{-at}\sin(\omega_0 t + \theta)u(t)] = \frac{\omega_0 \cos(\theta) + (s + a)\sin(\theta)}{(s + a)^2 + \omega_0^2}, \quad \text{ROC: } \sigma > 0 \tag{14.18}$$

EXAMPLE 14.7

Find the Laplace transform of the following signals.

a. $f(t) = 2\cos(5t)u(t)$

b. $f(t) = 2\cos\left(5t + \frac{\pi}{3}\right)u(t)$

c. $f(t) = 2e^{-3t}\cos(5t)u(t)$

d. $f(t) = 2e^{-3t}\cos\left(5t + \frac{\pi}{3}\right)u(t)$

continued

Example 14.7 continued

Answer:

a. $F(s) = \dfrac{2s}{s^2 + 5^2} = \dfrac{2s}{s^2 + 25}$

b. $F(s) = \dfrac{2s\cos\left(\dfrac{\pi}{3}\right) - 2 \times 5\sin\left(\dfrac{\pi}{3}\right)}{s^2 + 5^2} = \dfrac{2s\dfrac{1}{2} - 2 \times 5 \times \dfrac{\sqrt{3}}{2}}{s^2 + 25} = \dfrac{s - 5\sqrt{3}}{s^2 + 25} = \dfrac{s - 8.66}{s^2 + 25}$

c. $F(s) = \dfrac{2(s + 3)}{(s + 3)^2 + 25} = \dfrac{2s + 6}{s^2 + 6s + 34}$

d. $F(s) = \dfrac{(s + 3) - 5\sqrt{3}}{(s + 3)^2 + 25} = \dfrac{s - 5.66}{s^2 + 6s + 34}$

Exercise 14.7

Find the Laplace transform of the following signals.

a. $f(t) = 2\sin(5t)u(t)$

b. $f(t) = 2\sin\left(5t + \dfrac{\pi}{3}\right)u(t)$

c. $f(t) = 2e^{-3t}\sin(5t)u(t)$

d. $f(t) = 2e^{-3t}\sin\left(5t + \dfrac{\pi}{3}\right)u(t)$

Answer:

a. $F(s) = \dfrac{2 \times 5}{s^2 + 5^2} = \dfrac{10}{s^2 + 25}$

b. $F(s) = \dfrac{2 \times 5\cos\left(\dfrac{\pi}{3}\right) + 2 \times s \times \sin\left(\dfrac{\pi}{3}\right)}{s^2 + 5^2} = \dfrac{10\dfrac{1}{2} + 2s\dfrac{\sqrt{3}}{2}}{s^2 + 25} = \dfrac{5 + \sqrt{3}s}{s^2 + 25} = \dfrac{\sqrt{3}s + 5}{s^2 + 25}$

c. $F(s) = \dfrac{10}{(s + 3)^2 + 25} = \dfrac{10}{s^2 + 6s + 34}$

d. $F(s) = \dfrac{\sqrt{3}(s + 3) + 5}{(s + 3)^2 + 25} = \dfrac{\sqrt{3}s + 3\sqrt{3} + 5}{s^2 + 6s + 34} = \dfrac{\sqrt{3}s + 10.196}{s^2 + 6s + 34}$

14.3.4 MULTIPLICATION BY $\cos(\omega_0 t)$

Let the one-sided Laplace transform of a signal $f(t)$ be $F(s)$. Then, when $f(t)$ is multiplied by $\cos(\omega_0 t)$, the Laplace transform of $f(t)\cos(\omega_0 t)$ is given by

$$L[f(t)\cos(\omega_0 t)] = \frac{1}{2}F(s - j\omega_0) + \frac{1}{2}F(s + j\omega_0) \quad \textbf{(14.19)}$$

Proof

Using the frequency translation property, we have

$$L[f(t)\cos(\omega_0 t)] = L\left[f(t)\frac{e^{j\omega_0 t} + e^{-j\omega_0 t}}{2}\right] = L\left[\frac{e^{j\omega_0 t}}{2}f(t)\right] + L\left[\frac{e^{-j\omega_0 t}}{2}f(t)\right]$$

$$= \frac{1}{2}F(s - j\omega_0) + \frac{1}{2}F(s + j\omega_0)$$

EXAMPLE 14.8

Find the Laplace transform of $g(t) = e^{-at}\cos(\omega_0 t)u(t)$.

Since the Laplace transform of $f(t) = e^{-at}u(t)$ is given by

$$F(s) = \frac{1}{s+a} \quad \text{ROC:}\sigma > -a$$

we have

$$G(s) = \frac{1}{2}F(s - j\omega_0) + \frac{1}{2}F(s + j\omega_0) = \frac{1}{2}\frac{1}{s - j\omega_0 + a} + \frac{1}{2}\frac{1}{s + j\omega_0 + a}$$

$$= \frac{s+a}{(s+a)^2 + \omega_0^2} \quad \text{ROC:}\sigma > -a$$

Exercise 14.8

Find the Laplace transform of $g(t) = \cos(\omega_0 t)[u(t) - u(t - T)]$.

Answer:

$$L[u(t) - u(t - T)] = \frac{1 - e^{-sT}}{s},$$

$$G(s) = \frac{1}{2}\frac{1 - e^{-(s - j\omega_0)T}}{s - j\omega_0} + \frac{1}{2}\frac{1 - e^{-(s + j\omega_0)T}}{s + j\omega_0} = \frac{s - e^{-sT}[s\cos(\omega_0 T) - \omega_0\sin(\omega_0 T)]}{s^2 + \omega_0^2}.$$

14.3.5 MULTIPLICATION BY $\sin(\omega_0 t)$

Let the one-sided Laplace transform of a signal $f(t)$ be $F(s)$. Then, when $f(t)$ is multiplied by $\sin(\omega_0 t)$, the Laplace transform of $f(t)\sin(\omega_0 t)$ is given by

$$L[f(t)\sin(\omega_0 t)] = \frac{1}{2j}F(s - j\omega_0) - \frac{1}{2j}F(s + j\omega_0) \qquad \textbf{(14.20)}$$

Proof

Using the frequency translation property, we have

$$L[f(t)\sin(\omega_0 t)] = L\left[f(t)\frac{e^{j\omega_0 t} - e^{-j\omega_0 t}}{2j}\right] = L\left[\frac{e^{j\omega_0 t}}{2j}f(t)\right] - L\left[\frac{e^{-j\omega_0 t}}{2j}f(t)\right]$$

$$= \frac{1}{2j}F(s - j\omega_0) - \frac{1}{2j}F(s + j\omega_0)$$

EXAMPLE 14.9

Find the Laplace transform of $g(t) = e^{-at}\sin(\omega_0 t)u(t)$.

continued

Example 14.9 continued

Since the Laplace transform of $f(t) = e^{-at}u(t)$ is given by

$$F(s) = \frac{1}{s+a} \qquad \text{ROC: } \sigma > -a$$

we have

$$G(s) = \frac{1}{2j}F(s - j\omega_0) - \frac{1}{2j}F(s + j\omega_0) = \frac{1}{2j}\frac{1}{s - j\omega_0 + a} - \frac{1}{2j}\frac{1}{s + j\omega_0 + a}$$

$$= \frac{\omega_0}{(s+a)^2 + \omega_0^2} \qquad \text{ROC: } \sigma > -a$$

Exercise 14.9

Find the Laplace transform of $g(t) = \sin(\omega_0 t)[u(t) - u(t - T)]$.

Answer:

$$G(s) = \frac{1}{2j}\left[\frac{1 - e^{-(s-j\omega_0)T}}{s - j\omega_0} - \frac{1 - e^{-(s+j\omega_0)T}}{s + j\omega_0}\right] = \frac{\omega_0 - e^{-sT_0}[s\sin(\omega_0 T) + \omega_0\cos(\omega_0 T)]}{s^2 + \omega_0^2}$$

14.3.6 TIME DIFFERENTIATION PROPERTY

As shown in Section 14.5, later in this chapter, the time differentiation property is used to transform differential equations into algebraic equations. The time differentiation property is the foundation for the circuit analysis in the s-domain.

The Laplace transform of the derivative of $f(t)$, $df(t)/dt$, is given by

$$L\left[\frac{df(t)}{dt}\right] = sF(s) - f(0^-) \qquad \textbf{(14.21)}$$

Proof

$$L\left[\frac{df(t)}{dt}\right] = \int_{0^-}^{\infty}\frac{df(t)}{dt}e^{-st}dt$$

Let $u(t) = e^{-st}$, $dv(t) = \frac{df(t)}{dt}dt$. Then, $du(t) = -se^{-st}dt, v(t) = f(t)$. Using the integration by parts, $\int u(t)dv(t) = u(t)v(t) - \int v(t)du(t)$, we have

$$\int_{0^-}^{\infty}\frac{df(t)}{dt}e^{-st}dt = e^{-st}f(t)\Big|_{0^-}^{\infty} - \int_{0^-}^{\infty} f(t)[-se^{-st}]dt$$

$$= e^{-(\sigma+j\omega)\infty}f(\infty) - e^{-s0^-}f(0^-) + s\int_{0^-}^{\infty} f(t)e^{-st}dt$$

$$= 0 - f(0^-) + sF(s) = sF(s) - f(0^-)$$

Notice that $e^{-(\sigma+j\omega)\infty}f(\infty) = 0$ for $\sigma > 0$ and $f(\infty) < \infty$.

The Laplace transform for second derivative of $f(t)$ is given by

$$L\left[\frac{d^2f(t)}{dt^2}\right] = s^2F(s) - sf(0^-) - \frac{df(0^-)}{dt} \tag{14.22}$$

Proof

$$L\left[\frac{d^2f(t)}{dt^2}\right] = \int_{0^-}^{\infty} \frac{d^2f(t)}{dt^2} e^{-st}dt$$

Let $u(t) = e^{-st}$, $dv(t) = \frac{d^2f(t)}{dt^2}dt$. Then, $du(t) = -se^{-st}dt, v(t) = \frac{df(t)}{dt}$.

$$\int_{0^-}^{\infty} \frac{d^2f(t)}{dt^2} e^{-st}dt = e^{-st}\frac{df(t)}{dt}\bigg|_{0^-}^{\infty} - \int_{0^-}^{\infty} \frac{df(t)}{dt}[-se^{-st}]dt = e^{-st}\frac{df(t)}{dt}\bigg|_{0^-}^{\infty} + s\int_{0^-}^{\infty} \frac{df(t)}{dt} e^{-st}dt$$

Substituting

$$L\left[\frac{df(t)}{dt}\right] = \int_{0^-}^{\infty} \frac{df(t)}{dt} e^{-st}dt = sF(s) - f(0^-)$$

we have

$$\int_{0^-}^{\infty} \frac{d^2f(t)}{dt^2} e^{-st}dt = -\frac{df(0^-)}{dt} + s[sF(s) - f(0^-)] = s^2F(s) - sf(0^-) - \frac{df(0^-)}{dt}$$

For $\frac{d^3f(t)}{dt^3}$, we have

$$L\left[\frac{d^3f(t)}{dt^3}\right] = s^3F(s) - s^2f(0^-) - s\frac{df(0^-)}{dt} - \frac{d^2f(0^-)}{dt^2} \tag{14.23}$$

In general, we have

$$L\left[\frac{d^nf(t)}{dt^n}\right] = s^nF(s) - s^{n-1}f(0^-) - s^{n-2}\frac{df(0^-)}{dt} - s^{n-3}\frac{d^2f(0^-)}{dt^2} - \cdots - \frac{d^{n-1}f(0^-)}{dt^{n-1}} \tag{14.24}$$

EXAMPLE 14.10

Let $f(t) = \sin(\omega_0 t)u(t)$ with $f(0^-) = 0$. Find $\frac{df(t)}{dt}$ and also find the Laplace transform of $\frac{df(t)}{dt}$.

continued

Example 14.10 continued

The derivative of $f(t) = \sin(\omega_0 t)u(t)$ is given by

$$\frac{df(t)}{dt} = \omega_0 \cos(\omega_0 t)u(t)$$

The Laplace transform of $\frac{df(t)}{dt}$ is given by

$$L\left[\frac{df(t)}{dt}\right] = \omega_0 L[\cos(\omega_0 t)u(t)] = \omega_0 \frac{s}{s^2+\omega_0^2} = s\frac{\omega_0}{s^2+\omega_0^2} = s\,L[\sin(\omega_0 t)u(t)] = sF(s)$$

Exercise 14.10

Find the Laplace transform of $f(t) = t\sin(\omega_0 t)u(t)$.

Answer:

$$F(s) = \frac{2\omega_0 s}{(s^2+\omega_0^2)^2}$$

EXAMPLE 14.11

Find the Laplace transform of $\frac{d\delta(t)}{dt}$.

Since $L[\delta(t)] = 1$, we have

$$\frac{d\delta(t)}{dt} = s \times 1 = s$$

Exercise 14.11

Find the Laplace transform of $\frac{d^n\delta(t)}{dt^n}$.

Answer:

$$\frac{d^n\delta(t)}{dt^n} = s^n \quad (14.25)$$

14.3.7 INTEGRAL PROPERTY

The Laplace transform of an integral of $f(t)$ is $F(s)/s$:

$$L\left[\int_{0^-}^{t} f(\lambda)d\lambda\right] = \frac{F(s)}{s} \quad (14.26)$$

Proof

$$L\left[\int_{0^-}^{t} f(\lambda)d\lambda\right] = \int_{0^-}^{\infty}\left[\int_{0^-}^{t} f(\lambda)d\lambda\right]e^{-st}dt$$

Let $u(t) = \int_{0^-}^{t} f(\lambda)d\lambda$, $dv(t) = e^{-st}dt$. Then, $du(t) = f(t)dt, v(t) = \frac{e^{-st}}{-s}$. Using the integration by parts, $\int u(t)dv(t) = u(t)v(t) - \int v(t)du(t)$, we have

$$L\left[\int_{0^-}^{t} f(\lambda)d\lambda\right] = \int_{0^-}^{\infty}\left[\int_{0^-}^{t} f(\lambda)d\lambda\right]e^{-st}dt = \left[\int_{0^-}^{t} f(\lambda)d\lambda\right]\frac{e^{-st}}{-s}\Bigg|_{0^-}^{\infty} - \int_{0^-}^{\infty} f(t)\frac{e^{-st}}{-s}dt$$

$$= \left[\int_{0^-}^{\infty} f(\lambda)d\lambda\right]\frac{e^{-(\sigma+j\omega)\infty}}{-s} - \left[\int_{0^-}^{0^-} f(\lambda)d\lambda\right]\frac{e^{-s0^-}}{-s} + \frac{1}{s}\int_{0^-}^{\infty} f(t)e^{-st}dt = \frac{F(s)}{s}$$

Notice that $e^{-(\sigma+j\omega)\infty} = 0$ for $\sigma > 0$ and $\int_{0^-}^{0^-} f(\lambda)d\lambda = 0$ because the interval of integration is zero.

EXAMPLE 14.12

Let $f(t) = \cos(\omega_0 t)u(t)$. Find $\int_{0^-}^{t} f(\lambda)d\lambda$ and also find the Laplace transform of $\int_{0^-}^{t} f(\lambda)d\lambda$.

The integral of $f(t) = \cos(\omega_0 t)u(t)$ is given by

$$\int_{0^-}^{t} f(\lambda)d\lambda = \frac{1}{\omega_0}\sin(\omega_0 t)u(t)$$

The Laplace transform of $\int_{0^-}^{t} f(\lambda)d\lambda$ is given by

$$L\left[\int_{0^-}^{t} f(\lambda)d\lambda\right] = \frac{1}{\omega_0}L[\sin(\omega_0 t)u(t)] = \frac{1}{\omega_0}\frac{\omega_0}{s^2+\omega_0^2} = \frac{1}{s^2+\omega_0^2} = \frac{1}{s}\frac{s}{s^2+\omega_0^2}$$

$$= \frac{1}{s}L[\cos(\omega_0 t)u(t)]$$

Exercise 14.12

Let $f(t) = u(t) - u(t-2)$. Find the Laplace transform of the integral of $f(t)$.

Answer:

$$L\left[\int_{0^-}^{t} f(\lambda)d\lambda\right] = L[tu(t) - (t-2)u(t-2)] = \frac{1 - e^{-2s}}{s^2}$$

14.3.8 FREQUENCY DIFFERENTIATION PROPERTY

$$L[tf(t)] = -\frac{dF(s)}{ds} \tag{14.27}$$

Proof

Taking the derivative with respect to s on both sides of the definition of the Laplace transform

$$F(s) = \int_{0^-}^{\infty} f(t)e^{-st}dt$$

we obtain

$$\frac{dF(s)}{ds} = \int_{0^-}^{\infty} f(t)\frac{d(e^{-st})}{ds}dt = -\int_{0^-}^{\infty} tf(t)e^{-st}dt = -L[tf(t)]$$

Thus, we have

$$L[tf(t)] = -\frac{dF(s)}{ds}$$

If this process continues, we have

$$L[t^n f(t)] = (-1)^n \frac{d^n F(s)}{ds^n} \tag{14.28}$$

EXAMPLE 14.13

Find the Laplace transform of the following signals.

a. $f(t) = tu(t)$
b. $f(t) = t^2u(t)$
c. $f(t) = t^nu(t)$
d. $f(t) = te^{-at}u(t)$
e. $f(t) = t^ne^{-at}u(t)$

Answers:

a. Since $L[u(t)] = \frac{1}{s}$ with ROC: $\sigma > 0$, the Laplace transform of $tu(t)$ is given by

$$L[tu(t)] = -\frac{d}{ds}\left[\frac{1}{s}\right] = \frac{1}{s^2}, \quad \text{ROC: } \sigma > 0$$

continued

Example 14.13 continued

b. Since $L[u(t)] = \frac{1}{s}$ with ROC: $\sigma > 0$, the Laplace transform of $t^2u(t)$ is given by

$$L[t^2u(t)] = \frac{d^2}{ds^2}\left[\frac{1}{s}\right] = -\frac{d}{ds}\left[\frac{1}{s^2}\right] = \frac{2}{s^3}, \quad \text{ROC: } \sigma > 0$$

c. Continuing this process, we have

$$L[t^3u(t)] = -\frac{d}{ds}\left[\frac{2}{s^3}\right] = \frac{3 \times 2}{s^4} = \frac{3!}{s^4}, \quad \text{ROC: } \sigma > 0$$

$$L[t^4u(t)] = -\frac{d}{ds}\left[\frac{3!}{s^4}\right] = \frac{4!}{s^5}, \quad \text{ROC: } \sigma > 0$$

Therefore, we have

$$L[t^nu(t)] = \frac{n!}{s^{n+1}}, \quad \text{ROC: } \sigma > 0$$

d. Since $L[e^{-at}u(t)] = \frac{1}{s+a}$ with ROC: $\sigma > -a$, the Laplace transform of $te^{-at}u(t)$ is given by

$$L[te^{-at}u(t)] = -\frac{d}{ds}\left[\frac{1}{s+a}\right] = \frac{1}{(s+a)^2}, \quad \text{ROC: } \sigma > -a \qquad \textbf{(14.29)}$$

e. Repeating the procedure shown in Equation (14.29) $n - 1$ times more, we have

$$L[t^ne^{-at}u(t)] = \frac{n!}{(s+a)^{n+1}}, \quad \text{ROC: } \sigma > -a$$

Exercise 14.13

Find the Laplace transform of

a. $f(t) = t\cos(\omega_0 t)u(t)$

b. $f(t) = t\sin(\omega_0 t)u(t)$

Answer:

a. Applying the frequency differentiation property to the Laplace transform pair

$$L[\cos(\omega_0 t)u(t)] = \frac{s}{s^2 + \omega_0^2}$$

we obtain

$$L[t\cos(\omega_0 t)u(t)] = -\frac{d}{ds}\left[\frac{s}{s^2 + \omega_0^2}\right] = \frac{s^2 - \omega_0^2}{(s^2 + \omega_0^2)^2}$$

where use has been made of

$$\frac{d}{ds}\left[\frac{b(s)}{a(s)}\right] = \frac{b'(s)a(s) - b(s)a'(s)}{a^2(s)}$$

continued

Exercise 14.13 continued

b. Applying the frequency differentiation property to the Laplace transform pair

$$L[\sin(\omega_0 t)u(t)] = \frac{\omega_0}{s^2 + \omega_0^2}$$

we obtain

$$L[t\sin(\omega_0 t)u(t)] = -\frac{d}{ds}\left[\frac{\omega_0}{s^2 + \omega_0^2}\right] = \frac{2\omega_0 s}{(s^2 + \omega_0^2)^2}$$

EXAMPLE 14.14

Find the Laplace transform of $f(t)$ shown in Figure 14.5.

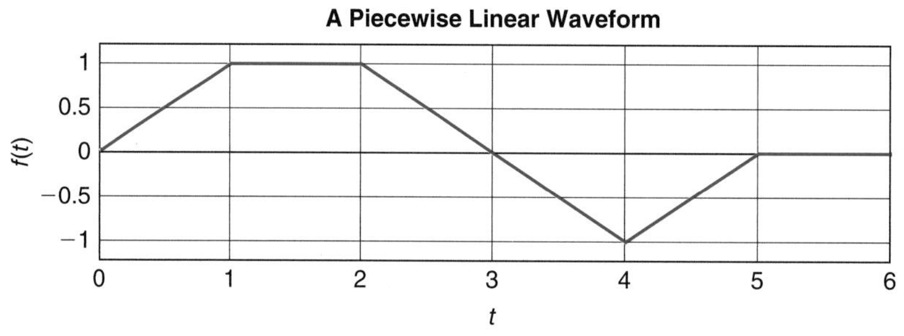

FIGURE 14.5

Waveform for EXAMPLE 14.14.

The piecewise linear waveform shown in Figure 14.5 can be written as

$$f(t) = tu(t) - (t-1)u(t-1) - (t-2)u(t-2) + 2(t-4)u(t-4) - (t-5)u(t-5)$$

Each of the terms is plotted in Figure 14.6.

Using $L[tu(t)] = 1/s^2$ and the time-shifting property $L[f(t - t_0)] = F(s)e^{-t_0 s}$, we have

$$F(s) = (1 - e^{-s} - e^{-2s} + 2e^{-4s} - e^{-5s})/s^2$$

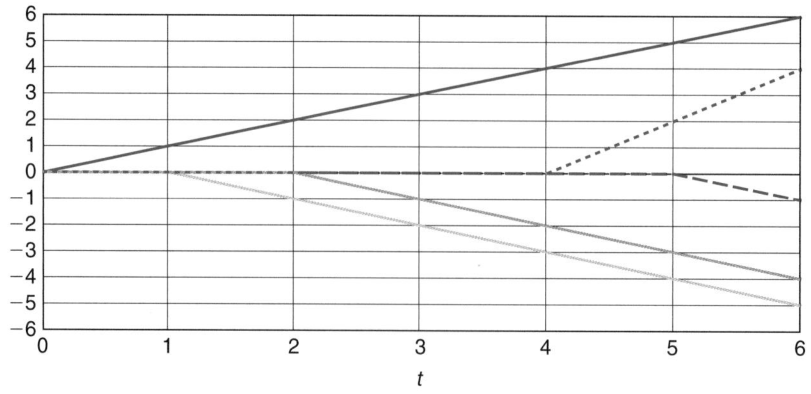

FIGURE 14.6

Plot of each term.

Exercise 14.14

Find the Laplace transform of $f(t)$ shown in Figure 14.7.

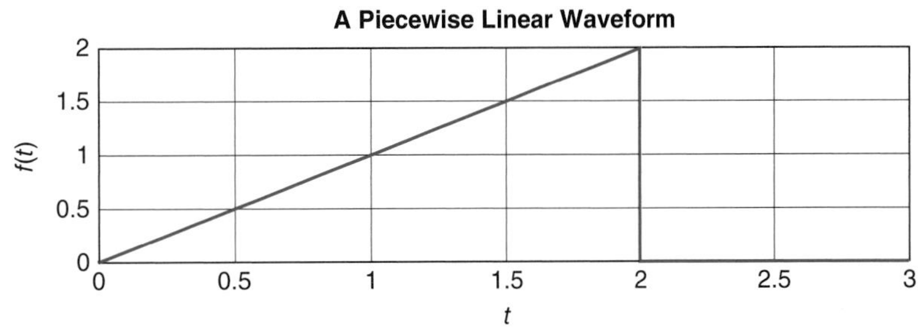

FIGURE 14.7

Waveform for EXERCISE 14.14.

The piecewise linear waveform shown in Figure 14.7 can be written as

$$f(t) = tu(t) - (t - 2)u(t - 2) - 2u(t - 2)$$

Each of the terms is plotted in Figure 14.8.

Using $L[u(t)] = 1/s$, $L[tu(t)] = 1/s^2$ and the time-shifting property $L[f(t - t_0)] = F(s)e^{-t_0 s}$, we have

$$F(s) = (1 - e^{-2s} - 2se^{-2s})/s^2$$

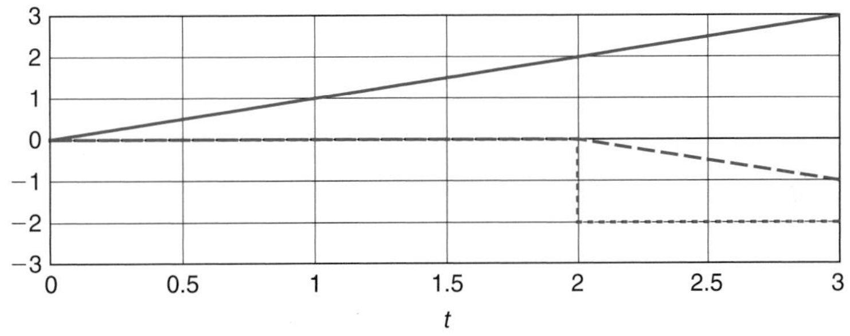

FIGURE 14.8

Plot of each term.

14.3.9 FREQUENCY INTEGRATION PROPERTY

Let the one-sided Laplace transform of a signal $f(t)$ be $F(s)$; that is,

$$F(s) = \int_{0^-}^{\infty} f(t)e^{-st}dt \tag{14.30}$$

Then, we have

$$L\left[\frac{f(t)}{t}\right] = \int_{s}^{\infty} F(s)ds \tag{14.31}$$

This property is called the *frequency integration property.* It is the complement of the frequency differentiation property. Integration in the s-domain results in division in the time domain.

Proof

Let us integrate both sides of Equation (14.30) with respect to s from s to ∞. Then, we have

$$\int_s^\infty F(s)ds = \int_{0^-}^\infty f(t)\left[\int_s^\infty e^{-st}ds\right]dt = \int_{0^-}^\infty f(t)\left[\frac{e^{-st}\big|_s^\infty}{-t}\right]dt = \int_{0^-}^\infty f(t)\left[\frac{e^{-st}}{t}\right]dt = \int_{0^-}^\infty\left[\frac{f(t)}{t}\right]e^{-st}dt$$

This is the Laplace transform of $\frac{f(t)}{t}$. Thus, we conclude that

$$L\left[\frac{f(t)}{t}\right] = \int_s^\infty F(s)ds$$

EXAMPLE 14.15

Find the Laplace transform of $f(t) = \frac{tu(t)}{t} = u(t)$.

We already know that the answer is $1/s$, but let us use the frequency integration property to see whether we get the same answer. Since $L[tu(t)] = \frac{1}{s^2}$, we have

$$L\left[\frac{tu(t)}{t}\right] = \int_s^\infty \frac{1}{s^2}ds = \frac{\frac{1}{s}\Big|_s^\infty}{-1} = \frac{1}{s}$$

Exercise 14.15

Find the Laplace transform of $f(t) = \frac{te^{-at}u(t)}{t}$ using the frequency integration property.

Answer:

$$L\left[\frac{te^{-at}u(t)}{t}\right] = \int_s^\infty \frac{1}{(s+a)^2}ds = \frac{\frac{1}{s+a}\Big|_s^\infty}{-1} = \frac{1}{s+a}$$

14.3.10 TIME-SCALING PROPERTY

Let the one-sided Laplace transform of a signal $f(t)$ be $F(s)$; that is,

$$F(s) = \int_{0^-}^\infty f(t)e^{-st}dt$$

Then, we have

$$L[f(at)] = \frac{1}{a}F\left(\frac{s}{a}\right) \tag{14.32}$$

This is called the *time-scaling property.*

Proof

Let $\tau = at$. Then, $t = \tau/a$, $dt = d\tau/a$.

$$L[f(at)] = \int_{0^-}^{\infty} f(at)e^{-st}dt = \int_{0^-}^{\infty} f(\tau)e^{-s\frac{\tau}{a}}\frac{d\tau}{a} = \frac{1}{a}\int_{0^-}^{\infty} f(\tau)e^{-\frac{s}{a}\tau}d\tau = \frac{1}{a}F\left(\frac{s}{a}\right)$$

EXAMPLE 14.16

Let $f(t)$ be a rectangular pulse with height 1, width 5, and centered at $t = 2.5$. The rectangular pulse $f(t)$ can be written as

$$f(t) = \text{rect}\left(\frac{t - 2.5}{5}\right) = u(t) - t(t - 5)$$

Find the Laplace transform of $f(2t)u(t)$.

Application of the time-shifting property to $L[u(t)] = 1/s$ yields

$$F(s) = \frac{1 - e^{-5s}}{s}$$

Application of the time-scaling property results in

$$L[f(2t)] = \frac{1}{2}F\left(\frac{s}{2}\right) = \frac{1 - e^{-5\frac{s}{2}}}{2\frac{s}{2}} = \frac{1 - e^{-2.5s}}{s}$$

Notice that

$$f(2t) = \text{rect}\left(\frac{2t - 2.5}{5}\right) = \text{rect}\left(\frac{2(t - 1.25)}{5}\right) = \text{rect}\left(\frac{t - 1.25}{2.5}\right) = u(t) - t(t - 2.5)$$

Exercise 14.16

Let $f(t) = u(t) - u(t - 5)$. Find the Laplace transform of $f\left(\frac{t}{2}\right)u(t)$.

Answer:

$$L\left[f\left(\frac{t}{2}\right)\right] = \frac{1 - e^{-10s}}{s}.$$

14.3.11 INITIAL VALUE THEOREM AND FINAL VALUE THEOREM

If the Laplace transform $F(s)$ of a signal $f(t)$ is known, we can find the value of $f(t)$ at $t = 0^+$, denoted by $f(0^+)$, and the value of $f(t)$ at $t = \infty$, denoted by $f(\infty)$ without finding the inverse Laplace transform. The initial value theorem makes it possible to evaluate $f(t)$ at $t = 0^+$ from $F(s)$, and the final value theorem makes it possible to evaluate $f(t)$ at $t = \infty$ from $F(s)$. If we find the inverse Laplace transform, the value of $f(t)$ is available for all t from 0 to ∞. Thus, the inverse Laplace transform provides complete information about $f(t)$. If we are only interested in checking the value of $f(t)$ for $t = 0^+$, we can use the initial value theorem. Likewise, if we are only interested in finding the value of $f(t)$ at $t = \infty$, we can use the final value theorem.

14.3.12 INITIAL VALUE THEOREM

If the Laplace transform of $f(t)$ is $F(s)$, the initial value of $f(t)$, $f(0^+)$, can be evaluated using

$$f(0^+) = \lim_{s\to\infty} sF(s) \tag{14.33}$$

Proof

From the time differentiation property of the Laplace transform, we have

$$L\left[\frac{df(t)}{dt}\right] = \int_{0^-}^{\infty} \frac{df(t)}{dt} e^{-st}\,dt = sF(s) - f(0^-)$$

The limits of integral $(0^-, \infty)$ can be broken into $(0^-, 0^+)$ and $(0^+, \infty)$. Thus,

$$\int_{0^-}^{0^+} \frac{df(t)}{dt} e^{-st}\,dt + \int_{0^+}^{\infty} \frac{df(t)}{dt} e^{-st}\,dt = sF(s) - f(0^-) \tag{14.34}$$

In the first integral, since the interval $(0^-, 0^+)$ is essentially zero, we can set $t = 0$ in e^{-st}. Thus,

$$\int_{0^-}^{0^+} \frac{df(t)}{dt} e^{-st}\,dt = \int_{0^-}^{0^+} \frac{df(t)}{dt} e^{-s0}\,dt = \int_{0^-}^{0^+} \frac{df(t)}{dt} 1\,dt = \int_{0^-}^{0^+} \frac{df(t)}{dt}\,dt = f(t)\big|_{0^-}^{0^+} = f(0^+) - f(0^-)$$

Substitution of this result into Equation (14.34) results in

$$f(0^+) + \int_{0^+}^{\infty} \frac{df(t)}{dt} e^{-st}\,dt = sF(s) \tag{14.35}$$

In the limit as $s \to \infty$, Equation (14.35) becomes

$$\lim_{s\to\infty} f(0^+) + \lim_{s\to\infty} \int_{0^+}^{\infty} \frac{df(t)}{dt} e^{-st}\,dt = \lim_{s\to\infty} sF(s) \tag{14.36}$$

Since $f(0^+)$ is constant, the limit $(s \to \infty)$ has no effect on this term, as $s \to \infty$, $e^{-st} \to 0$. Thus, the term $\lim_{s\to\infty} \int_{0^+}^{\infty} \frac{df(t)}{dt} e^{-st}\,dt$ is zero. Therefore, Equation (14.36) becomes

$$f(0^+) = \lim_{s\to\infty} sF(s)$$

This concludes the proof.

If the degree of the numerator polynomial is equal to or greater than the degree of the denominator polynomial, the initial value is infinity.

EXAMPLE 14.17

Find the initial value $f(0^+)$ for the signal $f(t)$ whose Laplace transform is given by

a. $F(s) = \dfrac{1}{s}$

b. $F(s) = \dfrac{1}{s^2}$

c. $F(s) = \dfrac{2s+3}{s(s+1)(s+2)}$

d. $F(s) = \dfrac{2s+7}{(s+5)(s+6)}$

Answer:

a. $f(0^+) = \lim_{s\to\infty} sF(s) = \lim_{s\to\infty}\left[s\dfrac{1}{s}\right] = \lim_{s\to\infty}[1] = 1$

Notice that the inverse Laplace transform of $F(s) = \dfrac{1}{s}$ is $f(t) = u(t)$. Thus, $f(0^+) = u(0^+) = 1$. The initial value theorem provides only one value of $f(t)$, $f(0^+)$, while the inverse Laplace transform provides $f(t)$ for $0 \le t < \infty$.

b. $f(0^+) = \lim_{s\to\infty} sF(s) = \lim_{s\to\infty}\left[s\dfrac{1}{s^2}\right] = \lim_{s\to\infty}\left[\dfrac{1}{s}\right] = 0$

Notice that the inverse Laplace transform of $F(s) = \dfrac{1}{s^2}$ is the ramp function given by $f(t) = tu(t)$. Thus, $f(0^+) = 0^+u(0^+) = 0$.

c. $f(0^+) = \lim_{s\to\infty} sF(s) = \lim_{s\to\infty}\left[s\dfrac{2s+3}{s(s+1)(s+2)}\right] = \lim_{s\to\infty}\dfrac{2s+3}{(s+1)(s+2)}$

$$= \lim_{s\to\infty}\frac{2+\dfrac{3}{s}}{s\left(1+\dfrac{1}{s}\right)\left(1+\dfrac{2}{s}\right)} = 0$$

The inverse Laplace transform of $F(s)$ is given by

$$f(t) = \frac{3}{2} - e^{-t}u(t) - \frac{1}{2}e^{-2t}u(t).$$

The value of $f(t)$ at $t = 0^+$ is given by

$$f(0^+) = \frac{3}{2} - e^{-0^+}u(0^+) - \frac{1}{2}e^{-2\times 0^+}u(0^+) = 0$$

continued

Example 14.17 continued

d. $$f(0^+) = \lim_{s\to\infty} sF(s) = \lim_{s\to\infty}\left[s\frac{2s+7}{(s+5)(s+6)}\right] = \lim_{s\to\infty}\frac{(1)\left(2+\frac{7}{s}\right)}{\left(1+\frac{5}{s}\right)\left(1+\frac{6}{s}\right)} = 2$$

The inverse Laplace transform of $F(s)$ is given by

$$f(t) = -3e^{-5t}u(t) + 5e^{-6t}u(t).$$

The value of $f(t)$ at $t = 0^+$ is given by

$$f(0^+) = -3e^{-50^+}u(0^+) + 5e^{-6\times0^+}u(0^+) = 2$$

Exercise 14.17

Find the initial value $f(0^+)$ for the signal $f(t)$, whose Laplace transform is given by

$$F(s) = \frac{s(s+3)}{(s+2)(s+6)}$$

Answer:

$$f(0^+) = \lim_{s\to\infty} sF(s) = \lim_{s\to\infty}\left[s\frac{s(s+3)}{(s+2)(s+6)}\right] = \lim_{s\to\infty}\left[s\frac{(1)\left(1+\frac{3}{s}\right)}{\left(1+\frac{2}{s}\right)\left(1+\frac{6}{s}\right)}\right] = \infty$$

14.3.13 FINAL VALUE THEOREM

If the Laplace transform of $f(t)$ is $F(s)$, the final value of $f(t)$, $f(\infty)$, can be evaluated using

$$f(\infty) = \lim_{s\to0} sF(s) \tag{14.37}$$

Proof

From the time differentiation property of the Laplace transform, we have

$$L\left[\frac{df(t)}{dt}\right] = \int_{0^-}^{\infty}\frac{df(t)}{dt}e^{-st}dt = sF(s) - f(0^-) \tag{14.38}$$

Taking the limit $s \to 0$ on both sides of Equation (14.38), we have

$$\lim_{s\to0}\int_{0^-}^{\infty}\frac{df(t)}{dt}e^{-st}dt = \lim_{s\to0} sF(s) - \lim_{s\to0} f(0^-) \tag{14.39}$$

As $s \to 0, e^{-st} \to 1$. Thus, the term on the left side of Equation (14.39) becomes

$$\lim_{s\to 0}\int_{0^-}^{\infty}\frac{df(t)}{dt}e^{-st}dt = \lim_{s\to 0}\int_{0^-}^{\infty}\frac{df(t)}{dt}e^{-0t}dt = \int_{0^-}^{\infty}\frac{df(t)}{dt}dt = f(t)\Big|_{0^-}^{\infty} = f(\infty) - f(0^-) \quad \textbf{(14.40)}$$

Since $f(0^-)$ is a constant, taking the limit $s \to 0$ has no effect. Thus, Equation (14.39) becomes

$$f(\infty) - f(0^-) = \lim_{s\to 0} sF(s) - f(0^-) \quad \textbf{(14.41)}$$

Canceling $f(0^-)$ from both sides of Equation (14.41), we have

$$f(\infty) = \lim_{s\to 0} sF(s)$$

EXAMPLE 14.18

Find the final value $f(\infty)$ for the signal $f(t)$, whose Laplace transform is given by

a. $F(s) = \dfrac{1}{s}$

b. $F(s) = \dfrac{1}{s^2}$

c. $F(s) = \dfrac{2s+3}{s(s+1)(s+2)}$

d. $F(s) = \dfrac{2s+7}{(s+5)(s+6)}$

Answer:

a. $f(\infty) = \lim_{s\to 0} sF(s) = \lim_{s\to 0}\left[s\dfrac{1}{s}\right] = \lim_{s\to 0}[1] = 1$

Notice that the inverse Laplace transform of $F(s) = \dfrac{1}{s}$ is $f(t) = u(t)$. Thus, $f(\infty) = u(\infty) = 1$. The final value theorem provides only one value of $f(t)$, $f(\infty)$, while the inverse Laplace transform provides $f(t)$ for $0 \le t < \infty$.

b. $f(\infty) = \lim_{s\to 0} sF(s) = \lim_{s\to 0}\left[s\dfrac{1}{s^2}\right] = \lim_{s\to 0}\left[\dfrac{1}{s}\right] = \infty$

Notice that the inverse Laplace transform of $F(s) = \dfrac{1}{s^2}$ is the ramp function given by $f(t) = tu(t)$. Thus, $f(\infty) = \infty u(\infty) = \infty$.

c. $f(\infty) = \lim_{s\to 0} sF(s) = \lim_{s\to 0}\left[s\dfrac{2s+3}{s(s+1)(s+2)}\right] = \lim_{s\to 0}\dfrac{2s+3}{(s+1)(s+2)} = \dfrac{3}{2}$

The inverse Laplace transform of $F(s)$ is given by

$$f(t) = \frac{3}{2} - e^{-t}u(t) - \frac{1}{2}e^{-2t}u(t).$$

The value of $f(t)$ at $t = \infty$ is given by

$$f(\infty) = \frac{3}{2} - e^{-\infty}u(\infty) - \frac{1}{2}e^{-2\infty}u(\infty) = \frac{3}{2}$$

continued

Example 14.18 continued

d. $f(\infty) = \lim_{s\to 0} sF(s) = \lim_{s\to 0}\left[s\frac{2s+7}{(s+5)(s+6)}\right] = 0 \times \frac{2\times 0+7}{(0+5)(0+6)} = 0$

The inverse Laplace transform of $F(s)$ is given by

$$f(t) = -3e^{-5t}u(t) + 5e^{-6t}u(t).$$

The value of $f(t)$ at $t = \infty$ is given by

$$f(\infty) = -3e^{-5\infty}u(\infty) + 5e^{-6\infty}u(\infty) = 0$$

Exercise 14.18

Find the final value $f(\infty)$ for the signal $f(t)$, whose Laplace transform is given by

$$F(s) = \frac{s(s+3)}{(s+2)(s+6)}$$

Answer:

$$f(\infty) = \lim_{s\to 0} sF(s) = \lim_{s\to 0}\left[s\frac{s(s+3)}{(s+2)(s+6)}\right] = 0 \times \frac{0\times(0+3)}{(0+2)(0+6)} = 0.$$

14.4 Inverse Laplace Transform

Recovering the time domain signal $f(t)$ from the s-domain transform $F(s)$ is the inverse Laplace transform. As shown in Equation (14.4), the Laplace transform is the Fourier transform of $f(t)e^{-\sigma t}$:

$$F(s) = F(\sigma + j\omega) = \int_{0^-}^{\infty} f(t)e^{-\sigma t}e^{-j\omega t}\,dt \tag{14.42}$$

The inverse Fourier transform of $F(\sigma + j\omega)$ is given by

$$f(t)e^{-\sigma t} = \frac{1}{2\pi}\int_{-\infty}^{\infty} F(\sigma + j\omega)e^{j\omega t}\,d\omega \tag{14.43}$$

Multiplication of $e^{\sigma t}$ on both sides of Equation (14.43) yields

$$f(t) = \frac{1}{2\pi}\int_{-\infty}^{\infty} F(\sigma + j\omega)e^{\sigma t}e^{j\omega t}d\omega = \frac{1}{2\pi}\int_{-\infty}^{\infty} F(\sigma + j\omega)e^{(\sigma + j\omega)t}\,d\omega \tag{14.44}$$

Since the complex variable s is defined as $s = \sigma + j\omega$, we have

$$d\omega = \frac{ds}{j}$$

and when $\omega = -\infty, s = \sigma - j\infty$, and when $\omega = \infty, s = \sigma + j\infty$, Equation (14.44) becomes

$$f(t) = \frac{1}{2\pi j}\int_{\sigma-j\infty}^{\sigma+j\infty} F(s)e^{st}ds \quad \textbf{(14.45)}$$

The path of integral is a vertical line $s = \sigma + j\omega$ $(-\infty < \omega < \infty)$. The value of σ must be selected so that the path of integral is inside the region of convergence (ROC). According to the **residue theorem**, the inverse Laplace transform given by Equation (14.45) can be evaluated by summing the residues of $F(s)e^{st}$ at all the poles of $F(s)$. The transform $F(s)$ can be represented as a ratio of numerator polynomial $N(s)$ to denominator polynomial $D(s)$

$$F(s) = \frac{N(s)}{D(s)}$$

If $F(s)$ has a simple pole at $s = -a$, or the denominator of $F(s)$ contain a factor $(s + a)$ so that

$$F(s) = \frac{N(s)}{D(s)} = \frac{N(s)}{(s+a)D_1(s)}$$

Then, $F(s)e^{st}$ can be written as

$$F(s)e^{st} = \frac{N(s)e^{st}}{(s+a)D_1(s)}$$

The residue at $s = -a$ is given by

$$R = (s+a)\frac{N(s)e^{st}}{(s+a)D_1(s)}\bigg|_{s=-a} = \frac{N(s)e^{st}}{D_1(s)}\bigg|_{s=-a} = \frac{N(-a)e^{-at}}{D_1(-a)} \quad \textbf{(14.46)}$$

If $F(s)$ has a multiple pole at $s = -a$ with multiplicity r, or the denominator of $F(s)$ contains a factor $(s + a)^r$ so that

$$F(s)e^{st} = \frac{N(s)e^{st}}{(s+a)^r D_1(s)}$$

The residue at $s = -a$ is given by

$$R = \frac{1}{(r-1)!}\frac{d^{r-1}}{ds^{r-1}}\left[(s+a)^r\frac{N(s)e^{st}}{(s+a)^r D_1(s)}\right]\bigg|_{s=-a} = \frac{1}{(r-1)!}\frac{d^{r-1}}{ds^{r-1}}\left[\frac{N(s)e^{st}}{D_1(s)}\right]\bigg|_{s=-a} \quad \textbf{(14.47)}$$

The residue theorem states that the inverse Laplace transform given by Equation (14.45) is the sum of the residues

$$f(t) = \frac{1}{2\pi j}\int_{\sigma-j\infty}^{\sigma+j\infty} F(s)e^{st}ds = \sum_{i=1}^{P} R_i \quad \textbf{(14.48)}$$

where R_i is a residue for the ith pole of $F(s)$.

In summary, the inverse Laplace transform of $F(s) = N(s)/D(s)$ can be found as follows:

a. Multiply $F(s)$ by e^{st} to obtain $F(s)e^{st}$.

b. Factor the denominator polynomial $D(s)$ and find the roots of $D(s) = 0$. The roots of $D(s) = 0$ are called *poles*.

c. Find the residues using Equations (14.46) and (14.47).
d. The inverse Laplace transform $f(t)$ is the sum of the residues.

EXAMPLE 14.19

Find the inverse Laplace transform of

$$F(s) = \frac{1}{s+a}, \qquad \sigma > -a$$

a. Multiplying $F(s)$ by e^{st}, we have

$$F(s)e^{st} = \frac{e^{st}}{s+a}$$

b. The denominator polynomial is already factored. The only factor is $(s + a)$. The pole is at $s = -a$.
c. The residue of the pole at $s = -a$ is

$$R_1 = (s+a)\frac{e^{st}}{s+a}\bigg|_{s=-a} = e^{st}\bigg|_{s=-a} = e^{-at}, \qquad t \geq 0$$

Therefore,

$$f(t) = e^{-at}u(t)$$

A function **residue** can be used to find the inverse Laplace transform. The residue function

$$[\mathrm{R}, \mathrm{P}, \mathrm{Q}] = \mathrm{residue}(b, a)$$

returns residue vector R, pole locations vector P, and quotient Q, given the numerator coefficients vector b and the denominator coefficients vector a in descending order (from highest order of s to the lowest order of s). The quotient Q is a null set unless the degree of the numerator is greater than or equal to the degree of the denominator. If the degree of the numerator is greater than or equal to the degree of the denominator, Q provides the quotient when the numerator polynomial $N(s)$ is divided by the denominator polynomial $D(s)$. Let

$$F(s) = \frac{1}{s+2}, \qquad \sigma > -2$$

Then, $b = 1, a = [1 \quad 2]$, and the following MATLAB script returns the residue and location of pole.

MATLAB

```
clear all;
b=1;a=[1 2];
[R,P,Q]=residue(b,a)
R =
     1
P =
    -2
Q =
     []
```

continued

Example 14.19 continued

The residue is 1 and the location of the pole is -2. Thus, the inverse z-transform is $e^{-2t}u(t)$. Q is a null set because the degree of the numerator polynomial is less than the degree of the denominator polynomial. (Refer to EXAMPLE 14.22, later in this chapter, for the case where Q is not a null set.)

Instead of using **residue**, symbolic math can be used to find the inverse Laplace transform in MATLAB. For

$$F(s) = \frac{1}{s+a}, \qquad \sigma > -a$$

the inverse transform is obtained as follows:

```
clear all;
syms s a
F=1/(s+a)
f=ilaplace(F)

F =
1/(a + s)
f =
exp(-a*t)
```

Exercise 14.19

Find the inverse Laplace transform of

$$F(s) = \frac{e^{-3s}}{s+2}$$

Answer:
$f(t) = e^{-2(t-3)}u(t-3)$.

EXAMPLE 14.20

Find the inverse Laplace transform of

$$F(s) = \frac{1}{s^2+5s+6}, \qquad \sigma > -3$$

Multiplying $F(s)$ by e^{st} and factoring the denominator polynomial, we have

$$F(s)e^{st} = \frac{e^{st}}{(s+2)(s+3)}$$

The residue at $s = -2$ is

$$R_1 = (s+2)\frac{e^{st}}{(s+2)(s+3)}\bigg|_{s=-2} = \frac{e^{st}}{(s+3)}\bigg|_{s=-2} = \frac{e^{-2t}}{(-2+3)} = e^{-2t}$$

continued

Example 14.20 continued

and the residue at $s = -3$ is

$$R_2 = (s+3)\frac{e^{st}}{(s+2)(s+3)}\bigg|_{s=-3} = \frac{e^{st}}{(s+2)}\bigg|_{s=-3} = \frac{e^{-3t}}{(-3+2)} = -e^{-3t}$$

Thus

$$f(t) = (e^{-2t} - e^{-3t})u(t)$$

MATLAB

```
clear all;
b=1;a=[1 5 6];
[R,P,Q]=residue(b,a)

R =
   -1.0000
    1.0000
P =
   -3.0000
   -2.0000
Q =
     []
```

The inverse Laplace transform is $(-e^{-3t} + e^{-2t})u(t)$. Alternatively, using symbolic math, we have

```
clear all;
syms s
F=1/(s^2+5*s+6)
f=ilaplace(F)

F =
1/(s^2 + 5*s + 6)
f =
exp(-2*t) - exp(-3*t)
```

Exercise 14.20

Find the inverse Laplace transform of

$$F(s) = \frac{3}{s^2 + 3.5s + 3}$$

Answer:
$f(t) = (6e^{-1.5t} - 6e^{-2t})u(t)$.

EXAMPLE 14.21

Find the inverse Laplace transform of

$$F(s) = \frac{1}{(s+2)^2(s+3)}, \quad \sigma > -3$$

continued

Example 14.21 continued

Since

$$F(s)e^{st} = \frac{e^{st}}{(s+2)^2(s+3)}$$

the residue at $s = -3$ is

$$R_1 = (s+3)\frac{e^{st}}{(s+2)^2(s+3)}\bigg|_{s=-3} = \frac{e^{st}}{(s+2)^2}\bigg|_{s=-3} = \frac{e^{-3t}}{(-1)^2} = e^{-3t}$$

and the residue at $s = -2$ is

$$R_2 = \frac{d}{ds}\left[(s+2)^2\frac{e^{st}}{(s+2)^2(s+3)}\right]\bigg|_{s=-2} = \frac{d}{ds}[e^{st}(s+3)^{-1}]\bigg|_{s=-2}$$
$$= [te^{st}(s+3)^{-1} + e^{st}(-1)(s+3)^{-2}]|_{s=-2} = te^{-2t}(-2+3)^{-1} + e^{-2t}(-1)(-2+3)^{-2}$$
$$= te^{-2t} - e^{-2t}$$

Thus, the inverse Laplace transform is

$$f(t) = (e^{-3t} - e^{-2t} + te^{-2t})u(t)$$

If V is a vector, poly(V) is a vector whose elements are the coefficients of the polynomial whose roots are the elements of V. poly(V) can be used to change the factored form to a polynomial form. poly([−2, −2, −3]) expands $(s+2)^2(s+3)$ into $s^3 + 7s^2 + 16s + 12$.

MATLAB

```
clear all;
b=1;a=poly([-2 -2 -3])
[R,P,Q]=residue(b,a)
a =
     1     7    16    12
R =
    1.0000
   -1.0000
    1.0000
P =
   -3.0000
   -2.0000
   -2.0000
Q =
     []
```

According to this result, $F(s)$ can be expanded as

$$F(s) = \frac{1}{s+3} - \frac{1}{s+2} + \frac{1}{(s+2)^2}$$

Thus, the inverse z-transform is given by

$$f(t) = (e^{-3t} - e^{-2t} + te^{-2t})u(t)$$

Notice that the Laplace transform pair

$$L[t^n e^{-at}u(t)] = \frac{n!}{(s+a)^{n+1}}$$

continued

Example 14.21 continued

is used in finding $f(t)$. Using symbolic math, we have

```
clear all;
syms s
F=1/((s+2)^2*(s+3))
f=ilaplace(F)

F =
1/((s + 2)^2*(s + 3))
f =
exp(-3*t) - exp(-2*t) + t*exp(-2*t)
```

Exercise 14.21

Find the inverse Laplace transform of

$$F(s) = \frac{5s+8}{(s+2)^3(s+6)}$$

Answer:

$$f(t) = \left(\frac{11}{32}e^{-6t} - \frac{11}{32}e^{-2t} + \frac{11}{8}te^{-2t} - \frac{1}{4}t^2e^{-2t}\right)u(t).$$

EXAMPLE 14.22

Find the inverse Laplace transform of

$$F(s) = \frac{s^2+5s+11}{s+3}, \qquad \sigma > -3$$

When the numerator polynomial is divided by the denominator polynomial using long division, we obtain

$$\begin{array}{r|l} & s+2 \\ \hline s+3 & s^2+5s+11 \\ & s^2+3s \\ \hline & \quad 2s+11 \\ & \quad 2s+6 \\ \hline & \qquad 5 \end{array}$$

$$F(s) = \frac{s^2+5s+11}{s+3} = s+2+\frac{5}{s+3}$$

Since $L[\delta'(t)] = s$ and $L[\delta(t)] = 1$, we have

$$f(t) = \delta'(t) + 2\delta(t) + 5e^{-3t}u(t)$$

continued

Example 14.22 continued

MATLAB

```
clear all;
b=[1 5 11];a=[1 3]
[R,P,Q]=residue(b,a)
R =
     5
P =
    -3
Q =
     1     2
```

Based on this result, we can expand $F(s)$ as

$$F(s) = s + 2 + \frac{5}{s+3}$$

Thus, the inverse Laplace transform of $F(s)$ is given by

$$f(t) = \delta'(t) + 2\delta(t) + 5e^{-3t}u(t)$$

Exercise 14.22

Find the inverse Laplace transform of

$$F(s) = \frac{s^3 + 6.5s^2 + 8s + 11}{s^2 + 6.5s + 10}$$

Answer:

$$f(t) = \left(\delta'(t) + \frac{32}{3}e^{-2.5t} - \frac{38}{3}e^{-4t}\right)u(t) = (\delta'(t) + 10.6667e^{-2.5t} - 12.6667e^{-4t})u(t).$$

EXAMPLE 14.23

Find the inverse Laplace transform of

$$F(s) = \frac{5}{s^2 + 2s + 5}, \qquad \sigma > -1$$

The residue of $\dfrac{5e^{st}}{(s+1-j2)(s+1+j2)}$ at $s = -1 + j2$ is given by

$$R_1 = (s+1-j2)\frac{5e^{st}}{(s+1+j2)(s+1-j2)}\bigg|_{s=-1+j2} = \frac{5e^{st}}{(s+1+j2)}\bigg|_{s=-1+j2} = 5\frac{e^{(-1+j2)t}}{j4}$$

continued

Example 14.23 continued

The residue of $\dfrac{5e^{st}}{(s+1-j2)(s+1+j2)}$ at $s=-1-j2$ is given by

$$R_2 = (s+1+j2)\frac{5e^{st}}{(s+1+j2)(s+1-j2)}\bigg|_{s=-1-j2}$$

$$= \frac{5e^{st}}{(s+1-j2)}\bigg|_{s=-1-j2} = 5\frac{e^{(-1-j2)t}}{-j4}$$

Thus, the inverse Laplace transform of $F(s)$ is given by

$$f(t) = 5\frac{e^{(-1+j2)t}}{j4} - 5\frac{e^{(-1-j2)t}}{j4} = \frac{5}{2}e^{-t}\left(\frac{e^{j2t}-e^{-j2t}}{2j}\right) = 2.5e^{-t}\sin(2t)u(t)$$

MATLAB

```
clear all;
b=5;a=[1 2 5];
[R,P,Q]=residue(b,a)

R =
   0.0000 - 1.2500i
   0.0000 + 1.2500i
P =
  -1.0000 + 2.0000i
  -1.0000 - 2.0000i
Q =
     []

clear all;
syms s
F=5/(s^2+2*s+5)
f=ilaplace(F)
pretty(f)

F =
5/(s^2 + 2*s + 5)
f =
(5*sin(2*t)*exp(-t))/2

  5 sin(2 t) exp(-t)
  ------------------
          2
```

Exercise 14.23

Find the inverse Laplace transform of

$$F(s) = \frac{20}{s^2 + 5s + 31.25}$$

Answer:
$f(t) = 4e^{-2.5t}\sin(5t)u(t)$.

We can find the inverse Laplace transform using partial fraction expansion and Table 14.2.

14.4.1 PARTIAL FRACTION EXPANSION

In general, the Laplace transform $F(s)$ can be written as

$$F(s) = \frac{N(s)}{D(s)} = \frac{b_m s^m + b_{m-1}s^{m-1} + b_{m-2}s^{m-2} + \cdots + b_1 s + b_0}{s^n + a_{n-1}s^{n-1} + a_{n-2}s^{n-2} + \cdots + a_1 s + a_0} \tag{14.49}$$

The degree of the numerator polynomial $N(s)$ is m, and the degree of the denominator polynomial $D(s)$ is n. The coefficients $a_{n-1}, \ldots, a_1, a_0$ of the denominator polynomial and the coefficients $b_m, b_{m-1}, \ldots, b_1, b_0$ of the numerator polynomial are real numbers. The Laplace transform $F(s)$ given by Equation (14.49) is called **rational polynomial** because it is the ratio of the numerator polynomial $N(s)$ to the denominator polynomial $D(s)$. The rational polynomial $F(s)$ is referred to as **proper rational polynomial** if the degree of the numerator polynomial is smaller than the degree of the denominator polynomial. The rational polynomial $F(s)$ is called **improper rational polynomial** if the degree of the numerator polynomial is greater than or equal to the degree of the denominator polynomial. If $F(s)$ is an improper rational polynomial, we divide the numerator polynomial $N(s)$ by the denominator polynomial $D(s)$ to get the quotient polynomial $Q(s)$ and the remainder polynomial $R(s)$. The maximum degree of the remainder polynomial is one less than the degree of the denominator polynomial, that is, $n - 1$. Thus, $R(s)/D(s)$ is a proper rational polynomial. The degree of the quotient polynomial $Q(s)$ is $n - m$.

Partial fraction expansion (PFE) is the representation of a proper rational polynomial as a linear combination of proper rational polynomials of smaller degrees. After we represent the given proper rational polynomial as a sum of proper rational polynomials of smaller degrees, we use Table 14.2 to find the inverse Laplace transform of each proper rational polynomial. The sum of the inverse transforms is the inverse Laplace transform of $F(s)$. To represent a proper rational polynomial in partial fraction expansion, we have to factor the denominator polynomial. The factors of the denominator polynomial $D(s)$ can be found by finding the roots of $D(s) = 0$. The roots of $D(s) = 0$ are called the poles of $F(s)$. The MATLAB function **roots** can be used to find the roots of $D(s) = 0$, as shown in EXAMPLE 14.24. The MATLAB function **poly** can be used to find the polynomial from the roots, as shown in EXAMPLE 14.25.

EXAMPLE 14.24

Use MATLAB to find the roots of

$$s^3 + 5s + 11s + 15 = 0$$

```
>> roots([1 5 11 15])
ans =
  -3.0000
  -1.0000 + 2.0000i
  -1.0000 - 2.0000i
```

The polynomial can be factored as $s^3 + 5s + 11s + 15 = (s + 3)(s + 1 - j2)(s + 1 + j2)$.

Exercise 14.24

Use MATLAB to find the roots of

$$s^6 + 12s^5 + 87s^4 + 352s^3 + 991s^2 + 1540s + 1625 = 0$$

Answer:

```
>> roots([1 12 87 352 991 1540 1625])
ans =
  -3.0000 + 4.0000i
  -3.0000 - 4.0000i
  -2.0000 + 3.0000i
  -2.0000 - 3.0000i
  -1.0000 + 2.0000i
  -1.0000 - 2.0000i
```

EXAMPLE 14.25

Use MATLAB to find the polynomial with roots

$$-2, -1.5 \pm j2$$

```
>> poly([-2,-1.5+2j,-1.5-2j])
ans =
    1.0000    5.0000    12.2500    12.5000
```

The polynomial is $s^3 + 5s^2 + 12.25s + 12.5$.

Exercise 14.25

Use MATLAB to expand the equation

$$(s + 1.5 - j2.5)(s + 1.5 + j2.5)(s + 2 - j3)(s + 2 + j3)(s + 2.5 - j4)(s + 2.5 + j4)$$

Answer:

```
>> poly([-1.5+2.5j,-1.5-2.5j,-2+3j,-2-3j,-2.5+4j,-2.5-4j])
ans =
   1.0e+03 *
  Columns 1 through 3
   0.001000000000000   0.012000000000000   0.090750000000000
  Columns 4 through 6
   0.396250000000000   1.220875000000000   2.176750000000000
  Column 7
   2.458625000000000
```

The equation is written as $s^6 + 12s^5 + 90.75s^4 + 396.25s^3 + 1220.875s^2 + 2176.75s + 2458.625$.

Let the n roots of $D(s) = 0$ be

$$-p_1, -p_2, -p_3, \ldots, -p_n$$

These are the poles of $F(s)$. Then, the proper rational polynomial $F(s) = N(s)/D(s)$ can be written as

$$F(s) = \frac{N(s)}{D(s)} = \frac{N(s)}{(s + p_1)(s + p_2)\dots(s + p_n)}$$

14.4.2 SIMPLE REAL POLES

Suppose all the poles $-p_1, -p_2, \dots, -p_n$ of a proper rational polynomial $F(s)$ are simple or distinct (no repeated poles). Then, $F(s)$ can be written as

$$\begin{aligned} H(s) &= \frac{N(s)}{D(s)} = \frac{N(s)}{(s + p_1)(s + p_2)\dots(s + p_n)} \\ &= \frac{R_1}{s + p_1} + \frac{R_2}{s + p_2} + \dots + \frac{R_n}{s + p_n} \end{aligned} \tag{14.50}$$

To find the coefficient R_1, we multiply $(s + p_1)$ on both sides of Equation (14.50):

$$\frac{N(s)(s+p_1)}{D(s)} = \frac{N(s)(s + p_1)}{(s+p_1)(s+p_2)\dots(s+p_n)} = \frac{R_1(s+p_1)}{s + p_1} + \frac{R_2(s+p_1)}{s + p_2} + \dots + \frac{R_n(s+p_1)}{s + p_n}$$

Since $(s + p_1)$ is canceled in two places, this equation reduces to

$$\frac{N(s)}{(s + p_2)\dots(s + p_n)} = R_1 + \frac{R_2(s + p_1)}{s + p_2} + \dots + \frac{R_n(s + p_1)}{s + p_n}$$

Setting $s = -p_1$ everywhere, we have

$$\frac{N(-p_1)}{(-p_1 + p_2)\dots(-p_1 + p_n)} = R_1 + \frac{R_2(-p_1 + p_1)}{-p_1 + p_2} + \dots + \frac{R_n(-p_1 + p_1)}{-p_1 + p_n}$$

Since all the terms on the right side (except the first term) are zero, we have

$$R_1 = \left.\frac{N(s)(s + p_1)}{D(s)}\right|_{s=-p_1} = \frac{N(-p_1)}{(-p_1 + p_2)\dots(-p_1 + p_n)}$$

In general, the coefficients R_i, $1 \le i \le n$, can be found by

$$R_i = \left.\frac{N(s)(s + p_i)}{D(s)}\right|_{s=-p_i} \tag{14.51}$$

The coefficients R_i are called *residues*.

EXAMPLE 14.26

Find the inverse Laplace transform of

$$F(s) = \frac{10s + 38}{s^2 + 7s + 10}$$

continued

Example 14.26 continued

Since the degree of the numerator polynomial ($m = 1$) is smaller than the degree of the denominator polynomial ($n = 2$), $F(s)$ is a proper rational polynomial. Factoring the denominator polynomial $D(s) = s^2 + 7s + 10$, we have

$$D(s) = s^2 + 7s + 10 = (s + 2)(s + 5)$$

Notice that the two numbers that multiply to 10 and add to 7 are 2 and 5. The roots of $D(s) = 0$ are $s = -2$ and $s = -5$. Thus, the locations of poles for $F(s)$ are $-p_1 = -2$ and $-p_2 = -5$. The partial fraction expansion of $F(s)$ can be written as

$$F(s) = \frac{10s + 38}{s^2 + 7s + 10} = \frac{10s + 38}{(s + 2)(s + 5)} = \frac{A}{s + 2} + \frac{B}{s + 5}$$

Also, notice that both terms in the partial fraction expansion are proper rational polynomials with a degree of one for the denominator polynomials. To find the coefficient A, we multiply $(s + 2)$ on both sides:

$$\frac{(10s + 38)(s + 2)}{(s + 2)(s + 5)} = \frac{A(s + 2)}{s + 2} + \frac{B(s + 2)}{s + 5}$$

Simplifying this equation, we get

$$\frac{10s + 38}{s + 5} = A + \frac{B(s + 2)}{s + 5}$$

When we set $s = -2$, the last term on the right side is zero. Thus,

$$A = \left.\frac{10s + 38}{s + 5}\right|_{s=-2} = \frac{10(-2) + 38}{-2 + 5} = \frac{18}{3} = 6$$

To find the coefficient B, we multiply $(s + 5)$ on both sides:

$$\frac{(10s + 38)(s + 5)}{(s + 2)(s + 5)} = \frac{A(s + 5)}{s + 2} + \frac{B(s + 5)}{s + 5}$$

Simplifying this equation, we get

$$\frac{10s + 38}{s + 2} = \frac{A(s + 5)}{s + 2} + B$$

When we set $s = -5$, the first term on the right side is zero. Thus,

$$B = \left.\frac{10s + 38}{s + 2}\right|_{s=-5} = \frac{10(-5) + 38}{-5 + 2} = \frac{-12}{-3} = 4$$

Thus, the partial fraction expansion of $F(s)$ is given by

$$F(s) = \frac{10s + 38}{s^2 + 7s + 10} = \frac{6}{s + 2} + \frac{4}{s + 5}$$

continued

Example 14.26 continued

Using the Laplace transform pair $L[e^{-at}u(t)] = \frac{1}{s+a}$, we have

$$f(t) = 6e^{-2t}u(t) + 4e^{-5t}u(t)$$

Alternatively, the right side of the partial fraction expansion

$$F(s) = \frac{10s + 38}{s^2 + 7s + 10} = \frac{10s + 38}{(s+2)(s+5)} = \frac{A}{s+2} + \frac{B}{s+5}$$

can be combined into a single term as

$$F(s) = \frac{10s + 38}{s^2 + 7s + 10} = \frac{10s + 38}{(s+2)(s+5)} = \frac{A}{s+2} + \frac{B}{s+5} = \frac{A(s+5) + B(s+2)}{(s+2)(s+5)}$$
$$= \frac{(A+B)s + 5A + 2B}{(s+2)(s+5)}$$

Equating the coefficients of the numerator polynomial on the right side of this equation to the coefficients of the numerator polynomial on the left side, we have

$$A + B = 10$$
$$5A + 2B = 38$$

Solving these equations, we obtain $A = 6$, $B = 4$. These answers match those found earlier.

MATLAB

```
>> b=[10 38];a=[1 7 10];
>> [R,P,Q]=residue(b,a)
R =
     4
     6
P =
    -5
    -2
Q =
     []
```

Thus, the partial fraction expansion is given by

$$F(s) = \frac{10s + 38}{s^2 + 7s + 10} = \frac{4}{s+5} + \frac{6}{s+2}$$

$[b, a] = \text{residue}(R, P, Q)$

When there are three variables inside the residue function, the numerator coefficients vector b and the denominator coefficients vector a are obtained from the vector of residues R, vector of poles P, and vector of direct terms Q.

```
>> [b1,a1]=residue(R,P,Q)
b1 =
    10    38
a1 =
     1     7    10
```

continued

Example 14.26 continued

Alternatively, a MATLAB function **ilaplace** can be used to find the inverse Laplace transform.

```
>> syms s
>> F=(10*s+38)/(s^2+7*s+10);
>> f=ilaplace(F)
f =
6/exp(2*t) + 4/exp(5*t)
```

Exercise 14.26

Find the inverse Laplace transform of

$$F(s) = \frac{s+6}{s^2+5s+5.25}$$

Answer:
$f(t) = (2.25e^{-1.5t} - 1.25e^{-3.5t})u(t)$.

14.4.3 COMPLEX POLES

Since the coefficients of the denominator polynomial $D(s)$ are real, if a pole $-p_1$ is complex, its complex conjugate $-p_1^*$ is also a pole. Otherwise, the coefficients of $D(s)$ are not all real. Suppose that a pole $-p_2$ is a complex conjugate of a pole $-p_1 = -a + jb$, that is, $-p_2 = -p_1^* = -a - jb$. Then, a proper rational polynomial $F(s) = N(s)/D(s)$ can be written as

$$F(s) = \frac{N(s)}{D(s)} = \frac{N(s)}{(s+p_1)(s+p_2)D_1(s)} = \frac{N(s)}{(s+a-jb)(s+a+jb)D_1(s)}$$
$$= \frac{A}{s+a-jb} + \frac{B}{s+a+jb} + \frac{N_1(s)}{D_1(s)} \tag{14.52}$$

To find A, we multiply $(s + p_1) = (s + a - jb)$ on both sides and set $s = -p_1 = -a + jb$:

$$A = \left.\frac{N(s)}{(s+a+jb)D_1(s)}\right|_{s=-a+jb} = \frac{N(-a+jb)}{(-a+jb+a+jb)D_1(-a+jb)}$$
$$= \frac{N(-a+jb)}{(j2b)D_1(-a+jb)}$$

To find B, we multiply $(s + p_2) = (s + a + jb)$ on both sides and set $s = -p_2 = -a - jb$:

$$B = \left.\frac{N(s)}{(s+a-jb)D_1(s)}\right|_{s=-a-jb} = \frac{N(-a-jb)}{(-a-jb+a-jb)D_1(-a-jb)}$$
$$= \frac{N(-a-jb)}{(-j2b)D_1(-a-jb)}$$

We can show that the coefficients A and B are complex conjugates. The inverse Laplace transform of the two complex conjugate terms $\frac{A}{s+a-jb}+\frac{B}{s+a+jb}$ in Equation (14.52) is given by

$$L^{-1}\left[\frac{A}{s+a-jb}+\frac{B}{s+a+jb}\right]=[Ae^{(-a+jb)t}+Be^{(-a-jb)t}]u(t)$$

In general, A is a complex number. Let $A=|A|e^{j\angle A}$. Then, $B=A^*=|A|e^{-j\angle A}$. Then, the inverse Laplace transform becomes

$$\begin{aligned}&[|A|e^{j\angle A}e^{(-a+jb)t}+|A|e^{-j\angle A}e^{(-a-jb)t}]u(t)=|A|e^{-at}(e^{j(bt+\angle A)}+e^{-j(bt+\angle A)})u(t)\\&=2|A|e^{-at}\cos(bt+\angle A)u(t)\end{aligned} \tag{14.53}$$

Alternatively, let $A=R+jI$ where R is the real part of A, and I is the imaginary part of A. Then, $B=A^*=R-jI$. The inverse Laplace transform becomes

$$\begin{aligned}&[(R+jI)e^{(-a+jb)t}+(R-jI)e^{(-a-jb)t}]u(t)\\&=e^{-at}[R(e^{jbt}+e^{-jbt})+jI(e^{jbt}-e^{-jbt})]u(t)=2e^{-at}[R\cos(bt)-I\sin(bt)]u(t)\end{aligned} \tag{14.54}$$

EXAMPLE 14.27

Find the inverse Laplace transform of

$$F(s)=\frac{8s+4}{s^2+6s+34}$$

The denominator polynomial $D(s)=s^2+6s+34$ can be factored by completing the square:

$$\begin{aligned}D(s)&=s^2+6s+34=(s^2+6s+9)+25=(s+3)^2+5^2\\&=(s+3-j5)(s+3+j5)\end{aligned}$$

The location of poles are $-p_1=-3+j5$ and $-p_2=-p_1^*=-3-j5$. The partial fraction expansion can be written as

$$F(s)=\frac{8s+4}{s^2+6s+34}=\frac{8s+4}{(s+3-j5)(s+3+j5)}=\frac{A}{s+3-j5}+\frac{A^*}{s+3+j5}$$

Multiplying $(s+3-j5)$ on both sides, we have

$$\frac{(8s+4)(s+3-j5)}{(s+3-j5)(s+3+j5)}=\frac{A(s+3-j5)}{s+3-j5}+\frac{A^*(s+3-j5)}{s+3+j5}$$

Canceling $(s+3-j5)$, we get

$$\frac{(8s+4)}{(s+3+j5)}=A+\frac{A^*(s+3-j5)}{s+3+j5}$$

continued

Example 14.27 continued

Setting $s = -3 + j5$, we obtain

$$A = \left.\frac{8s+4}{s+3+j5}\right|_{s=-3+j5} = \frac{8(-3+j5)+4}{-3+j5+3+j5} = \frac{-20+j40}{j10} = 4 + j2$$
$$= \sqrt{20}e^{j\tan^{-1}(1/2)} = 4.4721e^{j0.4636}$$
$$= 4.4721e^{j26.5651^\circ}$$

Since $A = R + jI = 4 + j2$, we have $R = 4$ and $I = 2$. The inverse Laplace transform is given by

$$f(t) = 2e^{-at}[R\cos(bt) - I\sin(bt)]u(t) = 2e^{-3t}[4\cos(5t) - 2\sin(5t)]u(t)$$

Also, since $A = |A|e^{j\angle A} = 4.4721e^{j0.4636} = 4.4721e^{j26.5651^\circ}$, we get

$$f(t) = 2|A|e^{-at}\cos(bt + \angle A)u(t) = 2 \times 4.4721e^{-3t}\cos(5t + 0.4636)u(t)$$
$$= 8.9443e^{-3t}\cos(5t + 0.4636)u(t) = 8.9443e^{-3t}\cos(5t + 26.5651^\circ)u(t)$$

Alternatively, $F(s)$ can be rearranged as

$$F(s) = \frac{8s+4}{s^2+3s+34} = \frac{8(s+3)-20}{(s+3)^2+5^2} = 8\frac{(s+3)}{(s+3)^2+5^2} - 4\frac{5}{(s+3)^2+5^2}.$$

Using the Laplace transform pairs

$$L\left[\frac{s}{s^2+\omega^2}\right] = \cos(\omega t) \quad \text{and} \quad L\left[\frac{\omega}{s^2+\omega^2}\right] = \sin(\omega t)$$

and the frequency translation property, we have

$$f(t) = 8e^{-3t}\cos(5t)u(t) - 4e^{-3t}\sin(5t)u(t)$$

Transforming this into a single cosine, we obtain

$$f(t) = 4e^{-3t}[2\cos(5t) - \sin(5t)]u(t) = 4\sqrt{5}e^{-3t}\left[\frac{2}{\sqrt{5}}\cos(5t) - \frac{1}{\sqrt{5}}\sin(5t)\right]u(t)$$
$$= 4\sqrt{5}e^{-3t}\cos\left[5t + \tan^{-1}\left(\frac{1}{2}\right)\right]u(t) = 8.9443e^{-3t}\cos(5t + 26.5651^\circ)u(t)$$

This is the same answer obtained earlier in this example using partial fraction expansion.

MATLAB

```
clear all;
b=[8 4];a=[1 6 34];
[R,P,Q]=residue(b,a)
alpha=real(P(1))
beta=abs(imag(P(1)))
Mag=2*abs(R(1))
Phase=angle(R(1))*180/pi
if imag(R(1)) >= 0 sn1 = '- '; else sn1 = '+ '; end;
```

continued

Example 14.27 continued

MATLAB continued

```
if Phase >= 0 sn2 = '+ '; else sn2 = '- '; end;
disp(['f(t) = exp(',num2str(alpha),'t) [',num2str(real(2*R(1))),...
    ' cos(',num2str(beta),'t) ',sn1,num2str(abs(imag(2*R(1)))),...
    ' sin(',num2str(beta),'t)] u(t)']);
disp(['f(t) = ',num2str(Mag),' exp(',num2str(alpha),...
    't) cos(',num2str(beta),'t ',sn2,num2str(abs(Phase)),') u(t)']);

R =
   4.0000 + 2.0000i
   4.0000 - 2.0000i
P =
  -3.0000 + 5.0000i
  -3.0000 - 5.0000i
Q =
     []
alpha =
    -3
beta =
    5.0000
Mag =
    8.9443
Phase =
   26.5651
f(t) = exp(-3t) [8 cos(5t) - 4 sin(5t)] u(t)
f(t) = 8.9443 exp(-3t) cos(5t + 26.5651) u(t)
```

Notice that $R = 4, I = 2$. Thus, we have

$$f(t) = 2e^{-3t}[4\cos(5t) - 2\sin(5t)]u(t)$$

Since $|4 + j2| = \sqrt{20} = 4.4721$ and $\angle(4 + j2) = 0.4636$ rad $= 26.5651°$, the inverse Laplace transform can also be written as

$$f(t) = 8.9443e^{-3t}\cos(5t + 26.5651°)u(t)$$

```
clear all;
syms s
F=(8*s+4)/(s^2+6*s+34)
f=ilaplace(F)
pretty(f)

F =
(8*s + 4)/(s^2 + 6*s + 34)
f =
8*exp(-3*t)*(cos(5*t) - sin(5*t)/2)

                 /             sin(5 t) \
  8 exp(-3 t) | cos(5 t) - -------- |
                 \                 2    /
```

Exercise 14.27

Find the inverse Laplace transform of

$$F(s) = \frac{s + 5}{s^2 + 7s + 48.25}$$

Answer:

$f(t) = e^{-3.5t}[\cos(6t) + 0.25\sin(6t)]u(t)$.

EXAMPLE 14.28

Find the inverse Laplace transform of

$$F(s) = \frac{9s + 5}{(s + 5)(s^2 + 4s + 20)}$$

$F(s)$ can be expanded as

$$F(s) = \frac{9s + 5}{(s + 5)(s^2 + 4s + 20)} = \frac{A}{s + 5} + \frac{Bs + C}{s^2 + 4s + 20} \tag{14.55}$$

Coefficient A is given by

$$A = (s + 5)\frac{9s + 5}{(s + 5)(s^2 + 4s + 20)}\bigg|_{s=-5} = \frac{9s + 5}{(s^2 + 4s + 20)}\bigg|_{s=-5}$$

$$= \frac{9(-5) + 5}{(-5)^2 + 4(-5) + 20} = \frac{-40}{25} = -1.6$$

Combining the two terms on the right side of Equation (14.55), we obtain

$$\begin{aligned} F(s) &= \frac{9s + 5}{(s + 5)(s^2 + 4s + 20)} = \frac{-1.6}{s + 5} + \frac{Bs + C}{s^2 + 4s + 20} \\ &= \frac{-1.6(s^2 + 4s + 20) + (Bs + C)(s + 5)}{(s + 5)(s^2 + 4s + 20)} \\ &= \frac{(-1.6 + B)s^2 + (-6.4 + 5B + C)s + (-32 + 5C)}{(s + 5)(s^2 + 4s + 20)} \end{aligned} \tag{14.56}$$

Comparing the coefficients of the numerator terms of Equation (14.56), we obtain

$$-1.6 + B = 0 \tag{14.57}$$

$$-6.4 + 5B + C = 9 \tag{14.58}$$

$$-32 + 5C = 5 \tag{14.59}$$

continued

Example 14.28 continued

From Equation (14.57), we obtain $B = 1.6$. From Equation (14.59), we get $C = 37/5 = 7.4$. As a check, substituting these values into Equation (14.58), we have $-6.4 + 8 + 7.4 = 9$. Equation (14.55) becomes

$$F(s) = \frac{-1.6}{s+5} + \frac{1.6s + 7.4}{s^2 + 4s + 20} \quad \textbf{(14.60)}$$

The inverse Laplace transform of the first term of Equation (14.60) is $-1.6e^{-5t}u(t)$. To find the inverse Laplace transform of the second term of Equation (14.60), we expand the second term as

$$\frac{1.6s + 7.4}{s^2 + 4s + 20} = \frac{1.6s + 7.4}{(s+2)^2 + 4^2} = \frac{1.6s + 7.4}{(s+2-j4)(s+2+j4)} = \frac{A}{s+2-j4} + \frac{A^*}{s+2+j4}$$

Coefficient A is given by

$$A = (s+2-j4)\frac{1.6s + 7.4}{(s+2-j4)(s+2+j4)}\bigg|_{s=-2+j4} = \frac{1.6s + 7.4}{(s+2+j4)}\bigg|_{s=-2+j4}$$

$$= \frac{1.6(-2+j4) + 7.4}{j8} = 0.8 - j0.525 = 0.9569\angle -33.2749°$$

The real part of A is $R = 0.8$ and the imaginary part of A is $I = -0.525$. The inverse transform of $F(s)$ is given by

$$\begin{aligned} f(t) &= -1.6e^{-5t}u(t) + 2e^{-2t}[0.8\cos(4t) + 0.525\sin(4t)]u(t) \\ &= -1.6e^{-5t}u(t) + e^{-2t}[1.6\cos(4t) + 1.05\sin(4t)]u(t) \\ &= -1.6e^{-5t}u(t) + 1.9138e^{-2t}\cos(4t - 33.2749°)u(t) \end{aligned}$$

Alternatively, the second term of Equation (14.60) can be revised as

$$\begin{aligned} \frac{1.6s + 7.4}{s^2 + 4s + 20} &= \frac{1.6(s+2) + 4.2}{(s+2)^2 + 4^2} = \frac{1.6(s+2) + 1.05 \times 4}{(s+2)^2 + 4^2} \\ &= 1.6\frac{(s+2)}{(s+2)^2 + 4^2} + 1.05\frac{4}{(s+2)^2 + 4^2} \end{aligned}$$

Using the Laplace transform pairs

$$L\left[\frac{s}{s^2 + \omega^2}\right] = \cos(\omega t) \quad \text{and} \quad L\left[\frac{\omega}{s^2 + \omega^2}\right] = \sin(\omega t)$$

and the frequency translation property, we have

$$L^{-1}\left[\frac{1.6s + 7.4}{s^2 + 4s + 20}\right] = 1.6e^{-2t}\cos(4t)u(t) + 1.05e^{-2t}\sin(4t)u(t)$$

Thus, the inverse Laplace transform is given by

$$f(t) = -1.6e^{-5t}u(t) + 1.9138e^{-2t}\cos(4t - 33.2749°)u(t)$$

continued

Example 14.28 continued

MATLAB

```
clear all;
syms s
F=(9*s+5)/((s+5)*(s^2+4*s+20));
f=ilaplace(F);
f=vpa(f,7)
f=expand(f);
f=vpa(f,7)

Answer:
f =
1.6*exp(-2.0*t)*(cos(4.0*t) + 0.65625*sin(4.0*t)) - 1.6*exp(-5.0*t)
f =
1.6*exp(-2.0*t)*cos(4.0*t) - 1.6*exp(-5.0*t) + 1.05*exp(-2.0*t)*sin(4.0*t)
```

Exercise 14.28

Find the inverse Laplace transform of

$$F(s) = \frac{20s + 90}{(s + 2)(s^2 + 6s + 45)}$$

Answer:

$$f(t) = \frac{50}{37}e^{-2t}u(t) - \frac{50}{37}e^{-3t}[\cos(6t) - 2.3\sin(6t)]u(t)$$
$$= 1.3514e^{-2t}u(t) - 1.3514e^{-3t}[\cos(6t) - 2.3\sin(6t)]u(t)$$
$$= 1.3514e^{-2t}u(t) - e^{-3t}[1.3514\cos(6t) - 3.1081\sin(6t)]u(t).$$

14.4.4 REPEATED POLES

If the denominator polynomial $D(s)$ of a proper rational polynomial $F(s) = N(s)/D(s)$ includes a factor $(s + p)^r$, $r > 1$, then $F(s)$ has r poles at $s = -p$. The partial fraction expansion of $F(s)$ can be written as

$$F(s) = \frac{N(s)}{(s + p)^r D_1(s)} = \frac{R_1}{(s + p)^r} + \frac{R_2}{(s + p)^{r-1}} + \frac{R_3}{(s + p)^{r-2}} + \cdots + \frac{R_r}{s + p} + \frac{N_1(s)}{D_1(s)}$$

The last term in the expansion, $N_1(s)/D_1(s)$, represents the rest of the terms. To find coefficient R_1, we multiply $(s + p)^r$ on both sides:

$$\frac{N(s)}{D_1(s)} = R_1 + R_2(s+p) + R_3(s+p)^2 + \cdots + R_r(s+p)^{r-1} + \frac{N_1(s)}{D_1(s)}(s+p)^r \quad \textbf{(14.61)}$$

Setting $s = -p$, we have

$$R_1 = \left.\frac{N(s)}{D_1(s)}\right|_{s=-p}$$

Differentiating Equation (14.61) with respect to s on both sides, we obtain

$$\frac{d}{ds}\left[\frac{N(s)}{D_1(s)}\right] = R_2 + R_3 2(s+p) + \cdots + R_r(r-1)(s+p)^{r-2} + (s+p)^r \frac{d}{ds}\left[\frac{N_1(s)}{D_1(s)}\right] + \frac{N_1(s)}{D_1(s)} r(s+p)^{r-1} \tag{14.62}$$

Setting $s = -p$ on both sides of Equation (14.62), we get

$$R_2 = \frac{d}{ds}\left[\frac{N(s)}{D_1(s)}\right]\Bigg|_{s=-p}$$

Differentiating Equation (14.62) with respect to s on both sides, we have

$$\frac{d^2}{ds^2}\left[\frac{N(s)}{D_1(s)}\right] = R_3 2 + \cdots + R_r(r-1)(r-2)(s+p)^{r-3} + \cdots \tag{14.63}$$

Setting $s = -p$ on both sides of Equation (14.63), we get

$$R_3 = \frac{1}{2}\frac{d^2}{ds^2}\left[\frac{N(s)}{D_1(s)}\right]\Bigg|_{s=-p}$$

When this procedure is repeated, we can show that, for $2 \le m \le r$,

$$R_m = \frac{1}{(m-1)!}\frac{d^{m-1}}{ds^{m-1}}\left[\frac{N(s)}{D_1(s)}\right]\Bigg|_{s=-p} \tag{14.64}$$

EXAMPLE 14.29

Find the inverse Laplace transform of

$$F(s) = \frac{s+6}{(s+2)^2(s+3)}$$

The partial fraction expansion can be expressed as

$$F(s) = \frac{s+6}{(s+2)^2(s+3)} = \frac{R_1}{(s+2)^2} + \frac{R_2}{s+2} + \frac{C}{s+3} \tag{14.65}$$

Multiplying both sides of Equation (14.65) by $(s+3)$ and setting $s = -3$, we get

$$C = \frac{s+6}{(s+2)^2}\Bigg|_{s=-3} = \frac{-3+6}{(-3+2)^2} = 3$$

Multiplying both sides of Equation (14.65) by $(s+2)^2$ and setting $s = -2$, we get

$$R_1 = \frac{s+6}{s+3}\Bigg|_{s=-2} = \frac{-2+6}{-2+3} = 4$$

continued

Example 14.29 continued

Equation (14.64) can be applied to find the coefficient R_2. The derivative of rational polynomial can be evaluated using

$$\frac{d}{ds}\left[\frac{u(s)}{v(s)}\right] = \frac{v(s)\frac{du(s)}{ds} - u(s)\frac{dv(s)}{ds}}{v^2(s)} = \frac{1}{v(s)}\frac{du(s)}{ds} - \frac{u(s)}{v^2(s)}\frac{dv(s)}{ds}$$

or

$$\frac{d}{ds}\left[\frac{u(s)}{v(s)}\right] = \frac{d}{ds}[u(s)v(s)^{-1}] = v(s)^{-1}\frac{du(s)}{ds} + u(s)\frac{dv(s)^{-1}}{ds}$$

After multiplying $(s+2)^2$ on the left side of Equation (14.65), we have $(s+6)/(s+3)$. Coefficient R_2 is found by differentiating $(s+6)/(s+3)$ with respect to s and then setting $s=-2$:

$$R_2 = \frac{d}{ds}\left[\frac{s+6}{s+3}\right]\Bigg|_{s=-2} = \left[\frac{1}{s+3}\frac{d}{ds}(s+6) - \frac{s+6}{(s+3)^2}\frac{d}{ds}(s+3)\right]\Bigg|_{s=-2}$$

$$= \left[\frac{1}{s+3}\times 1 - \frac{s+6}{(s+3)^2}\times 1\right]\Bigg|_{s=-2} = \frac{1}{-2+3} - \frac{-2+6}{(-2+3)^2} = -3$$

Alternatively, R_2 can be calculated from

$$R_2 = \frac{d}{ds}\left[\frac{s+6}{s+3}\right]\Bigg|_{s=-2} = \frac{d}{ds}[(s+6)(s+3)^{-1}]\Big|_{s=-2}$$

$$= \left[\frac{d}{ds}(s+6)\right](s+3)^{-1}\Big|_{s=-2} + \left[\frac{d}{ds}(s+3)^{-1}\right](s+6)\Big|_{s=-2}$$

$$= 1\times(s+3)^{-1}\big|_{s=-2} + [(-1)(s+3)^{-2}](s+6)\big|_{s=-2}$$

$$= \frac{1}{-2+3} + (-1)(-2+3)^{-2}(-2+6)$$

$$= 1 - (1)^{-2}(4) = 1 - \frac{4}{(1)^2} = 1 - 4 = -3$$

The partial fraction expansion is given by

$$F(s) = \frac{s+6}{(s+2)^2(s+3)} = \frac{4}{(s+2)^2} - \frac{3}{s+2} + \frac{3}{s+3} \quad \textbf{(14.66)}$$

When n is replaced by $n-1$ in the Laplace transform pair

$$L[t^n e^{-at}u(t)] = \frac{n!}{(s+a)^{n+1}}$$

we obtain

$$L[t^{n-1}e^{-at}u(t)] = \frac{(n-1)!}{(s+a)^n} \quad \textbf{(14.67)}$$

continued

Example 14.29 continued

Thus,

$$L^{-1}\left[\frac{1}{(s+a)^n}\right] = \frac{1}{(n-1)!}t^{n-1}e^{-at}u(t)$$

The inverse Laplace transform of the first term is given by

$$L^{-1}\left[\frac{4}{(s+2)^2}\right] = \frac{4}{(2-1)!}t^{2-1}e^{-2t}u(t) = 4te^{-2t}u(t)$$

The inverse Laplace transform of $F(s)$ is given by

$$f(t) = 4te^{-2t}u(t) - 3e^{-2t}u(t) + 3e^{-3t}u(t)$$

Alternatively, the coefficients of the partial fraction expansion

$$F(s) = \frac{s+6}{(s+2)^2(s+3)} = \frac{A}{(s+2)^2} + \frac{B}{s+2} + \frac{C}{s+3}$$

can be found by combining the right side as

$$\begin{aligned}\frac{s+6}{(s+2)^2(s+3)} &= \frac{A}{(s+2)^2} + \frac{B}{s+2} + \frac{C}{s+3}\\ &= \frac{A(s+3) + B(s+2)(s+3) + C(s+2)^2}{(s+2)^2(s+3)}\\ &= \frac{(B+C)s^2 + (A+5B+4C)s + 3A + 6B + 4C}{(s+2)^2(s+3)}\end{aligned}$$

Equating the coefficients of the numerator polynomial on the right side of this equation to the coefficients of the numerator polynomial on the left side, we have

$$B + C = 0$$

$$A + 5B + 4C = 1$$

$$3A + 6B + 4C = 6$$

Solving these equations, we obtain $A = 4$, $B = -3$, and $C = 3$. These answers match those found earlier.

MATLAB

Using **ilaplace**:

```
clear all;
syms s
F=(s+6)/((s+2)^2*(s+3))
f=ilaplace(F)
F =
(s + 6)/((s + 2)^2*(s + 3))
f =
3*exp(-3*t) - 3*exp(-2*t) + 4*t*exp(-2*t)
```

continued

Example 14.29 continued

MATLAB continued

Using **residue:**

```
clear all;
b=[1 6];a=poly([-2 -2 -3]);
[R,P,Q]=residue(b,a)
R =
    3.0000
   -3.0000
    4.0000
P =
   -3.0000
   -2.0000
   -2.0000
Q =
     []
```

This result shows that the partial fraction expansion is given by

$$H(s) = \frac{3}{s+3} - \frac{3}{s+2} + \frac{4}{(s+2)^2}$$

which is identical to the one given by Equation (14.66).

Exercise 14.29

Find the inverse Laplace transform of

$$F(s) = \frac{2s+5}{(s+3)^3(s+4)}$$

Answer:

$$f(t) = \left(3e^{-4t} - 3e^{-3t} + 3te^{-3t} - \frac{1}{2}t^2e^{-3t}\right)u(t).$$

EXAMPLE 14.30

Find the inverse Laplace transform of

a. $F(s) = \frac{2}{s+2}$ **b.** $F(s) = \frac{2}{s(s+2)}$ **c.** $F(s) = \frac{2}{s^2(s+2)}$

a. $f(t) = 2e^{-2t}u(t)$

b. $F(s) = \frac{2}{s(s+2)} = \frac{A}{s} + \frac{B}{s+2}$ $\quad A = s\frac{2}{s(s+2)}\Big|_{s=0} = \frac{2}{(s+2)}\Big|_{s=0} = \frac{2}{(0+2)} = \frac{2}{2} = 1$

$B = (s+2)\frac{2}{s(s+2)}\Big|_{s=-2} = \frac{2}{s}\Big|_{s=-2} = \frac{2}{-2} = -1$ $\quad F(s) = \frac{2}{s(s+2)} = \frac{1}{s} - \frac{1}{s+2}$

$f(t) = u(t) - e^{-2t}u(t) = (1 - e^{-2t})u(t)$

continued

Example 14.30 continued

c. $F(s) = \dfrac{2}{s^2(s+2)} = \dfrac{A}{s^2} + \dfrac{B}{s} + \dfrac{C}{s+2}$

$$C = (s+2)\frac{2}{s^2(s+2)}\bigg|_{s=-2} = \frac{2}{s^2}\bigg|_{s=-2} = \frac{2}{4} = 0.5$$

$$A = s^2\frac{2}{s^2(s+2)}\bigg|_{s=0} = \frac{2}{(s+2)}\bigg|_{s=0} = \frac{2}{(0+2)} = \frac{2}{2} = 1$$

$$B = \frac{d}{ds}s^2\frac{2}{s^2(s+2)}\bigg|_{s=0} = \frac{d}{ds}\frac{2}{(s+2)}\bigg|_{s=0} = 2\frac{d}{ds}(s+2)^{-1}\bigg|_{s=0}$$

$$= 2(-1)(s+2)^{-2}|_{s=0} = \frac{-2}{(2)^2} = -0.5$$

$$F(s) = \frac{1}{s^2(s+2)} = \frac{1}{s^2} - \frac{0.5}{s} + \frac{0.5}{s+2}$$

When n is replaced by $n-1$ in the Laplace transform pair

$$L[t^n u(t)] = \frac{n!}{s^{n+1}}$$

we obtain

$$L[t^{n-1}u(t)] = \frac{(n-1)!}{s^n}$$

Thus,

$$L^{-1}\left[\frac{1}{s^n}\right] = \frac{1}{(n-1)!}t^{n-1}u(t)$$

The inverse Laplace transform of the first term is given by

$$L^{-1}\left[\frac{1}{s^2}\right] = \frac{1}{(2-1)!}t^{2-1}u(t) = tu(t)$$

The inverse Laplace transform is given by

$$f(t) = tu(t) - 0.5u(t) + 0.5e^{-2t}u(t) = (t - 0.5 + 0.5e^{-2t})u(t)$$

MATLAB

```
clear all;
syms s
F=2/(s^2*(s+2))
f=ilaplace(F)
f=vpa(f,4)
Answer:
f =
t + 0.5*exp(-2.0*t) - 0.5
```

Exercise 14.30

Find the inverse Laplace transform of

$$F(s) = \frac{2}{s^3(s+2)}$$

Answer:
$f(t) = (0.5t^2 - 0.5t + 0.25 - 0.25e^{-2t})u(t).$

EXAMPLE 14.31

Find the inverse Laplace transform of

a. $F(s) = \dfrac{29}{s^2 + 4s + 29}$ **b.** $F(s) = \dfrac{29}{s(s^2 + 4s + 29)}$ **c.** $F(s) = \dfrac{29}{s^2(s^2 + 4s + 29)}$

a. $F(s) = \dfrac{29}{s^2 + 4s + 29} = \dfrac{29}{s^2 + 4s + 4 + 25} = 5.8\dfrac{5}{(s+2)^2 + 5^2}$

$f(t) = 5.8e^{-2t}\sin(5t)u(t)$

b. $F(s) = \dfrac{29}{s(s^2 + 4s + 29)} = \dfrac{A}{s} + \dfrac{Bs + C}{s^2 + 4s + 29}$

$$A = s\frac{29}{s(s^2 + 4s + 29)}\bigg|_{s=0} = \frac{29}{s^2 + 4s + 29}\bigg|_{s=0} = \frac{29}{29} = 1$$

$F(s)$ can be rewritten as

$$F(s) = \frac{29}{s(s^2 + 4s + 29)} = \frac{1}{s} + \frac{Bs + C}{s^2 + 4s + 29} = \frac{(1 + B)s^2 + (4 + C)s + 29}{s(s^2 + 4s + 29)}$$

Comparing the coefficients of the numerator polynomials, we obtain $1 + B = 0$ and $4 + C = 0$. Thus, we have $B = -1$ and $C = -4$. $F(s)$ becomes

$$F(s) = \frac{1}{s} + \frac{-s - 4}{s^2 + 4s + 29} = \frac{1}{s} - \frac{(s + 2) + 0.4 \times 5}{(s + 2)^2 + 5^2}$$
$$= \frac{1}{s} - \frac{(s + 2)}{(s + 2)^2 + 5^2} - 0.4\frac{5}{(s + 2)^2 + 5^2}$$

The inverse Laplace transform is given by

$$f(t) = [1 - e^{-2t}\cos(5t) - 0.4e^{-2t}\sin(5t)]u(t)$$

c. $F(s) = \dfrac{29}{s^2(s^2 + 4s + 29)} = \dfrac{A}{s^2} + \dfrac{B}{s} + \dfrac{Cs + D}{s^2 + 4s + 29}$

$$A = s^2\frac{29}{s^2(s^2 + 4s + 29)}\bigg|_{s=0} = \frac{29}{s^2 + 4s + 29}\bigg|_{s=0} = \frac{29}{29} = 1$$

continued

Example 14.31 continued

$$B = \frac{d}{ds}s^2\frac{29}{s^2(s^2+4s+29)}\bigg|_{s=0} = \frac{d}{ds}\frac{29}{s^2+4s+29}\bigg|_{s=0} = 29\frac{d}{ds}(s^2+4s+29)^{-1}\bigg|_{s=0}$$

$$= 29(-1)(s^2+4s+29)^{-2}(2s+4)|_{s=0} = -\frac{29\times 4}{29^2} = -\frac{4}{29} = -0.1379$$

$F(s)$ can be rewritten as

$$F(s) = \frac{29}{s^2(s^2+4s+29)} = \frac{A}{s^2} + \frac{B}{s} + \frac{Cs+D}{s^2+4s+29}$$

$$= \frac{\left(-\frac{4}{29}+C\right)s^3 + \left(1-\frac{16}{29}+D\right)s^2 + 29}{s^2(s^2+4s+29)}$$

Comparing the coefficients of the numerator polynomials, we obtain

$$-\frac{4}{29} + C = 0 \quad \text{and} \quad 1 - \frac{16}{29} + D = 0$$

Thus, we have

$$C = \frac{4}{29} = 0.1379 \quad \text{and} \quad D = -\frac{13}{29} = -0.4483$$

$F(s)$ becomes

$$F(s) = \frac{29}{s^2(s^2+4s+29)} = \frac{1}{s^2} - \frac{\frac{4}{29}}{s} + \frac{4}{29}\frac{s-\frac{13}{4}}{s^2+4s+29} = \frac{1}{s^2} - \frac{\frac{4}{29}}{s} + \frac{4}{29}\frac{s+2-\frac{21}{4}}{(s+2)^2+5^2}$$

$$= \frac{1}{s^2} - \frac{\frac{4}{29}}{s} + \frac{4}{29}\frac{s+2}{(s+2)^2+5^2} - \frac{21}{145}\frac{5}{(s+2)^2+5^2}$$

The inverse Laplace transform is given by

$$f(t) = [t - 0.1379 + 0.1379e^{-2t}\cos(5t) - 0.1448e^{-2t}\cos(5t)]u(t)$$

MATLAB

```
clear all;
syms s
F=29/(s^2*(s^2+4*s+29))
f=ilaplace(F)
f=vpa(f,4)
Answer:
f =
t + 0.1379*exp(-2.0*t)*(cos(5.0*t) - 1.05*sin(5.0*t)) - 0.1379
```

Exercise 14.31

Find the inverse Laplace transform of

$$F(s) = \frac{29}{s^3(s^2 + 4s + 29)}$$

Answer:
$f(t) = [0.5t^2 - 0.1379t - 0.01546 + 0.01546e^{-2t}\cos(5t) + 0.03377e^{-2t}\sin(5t)]u(t)$.

EXAMPLE 14.32

Use MATLAB to find the inverse Laplace transform of

$$F(s) = \frac{2s + 11}{s^3(s + 3)^2}$$

```
clear all;
syms s
F=(2*s+11)/(s^3*(s+3)^2)
f=ilaplace(F)
f=vpa(f,4)
```

```
Answer:
f =
0.6111*t^2 - 0.2593*exp(-3.0*t) - 0.1852*t*exp(-3.0*t) - 0.5926*t + 0.2593
```

$$f(t) = (0.6111t^2 - 0.5926t + 0.2593 - 0.1852te^{-3t} - 0.2593e^{-3t})u(t)$$

Exercise 14.32

Use MATLAB to find the inverse Laplace transform of

$$F(s) = \frac{3s + 175}{(s + 3)^3(s^2 + 4s + 29)}$$

Answer:
$f(t) = (3.1923t^2e^{-3t} - 0.1989e^{-3t} + 0.6065te^{-3t} + 0.1989e^{-2t}\cos(5t) - 0.1611e^{-2t}\sin(5t))u(t)$.

14.5 Solving Differential Equations Using the Laplace Transform

Solving first- and second-order differential equations is discussed in Chapters 7 and 8. In this section, we will discuss an alternative approach to solving differential equations. The differential equations can be transformed to the s-domain by applying the time differentiation property discussed in Section 14.3, earlier in this chapter. The initial conditions are included in this transformation. For convenience, the time differentiation properties are repeated here.

$$L\left[\frac{df(t)}{dt}\right] = sF(s) - f(0^-) \tag{14.68}$$

$$L\left[\frac{d^2f(t)}{dt^2}\right] = s^2F(s) - sf(0^-) - \frac{df(0^-)}{dt} \tag{14.69}$$

$$L\left[\frac{d^3f(t)}{dt^3}\right] = s^3F(s) - s^2f(0^-) - s\frac{df(0^-)}{dt} - \frac{d^2f(0^-)}{dt^2} \tag{14.70}$$

$$L\left[\frac{d^nf(t)}{dt^n}\right] = s^nF(s) - s^{n-1}f(0^-) - s^{n-2}\frac{df(0^-)}{dt} - s^{n-3}\frac{d^2f(0^-)}{dt^2} - \cdots - \frac{d^{n-1}f(0^-)}{dt^{n-1}} \tag{14.71}$$

The transformed equations are algebraic equations of s and transformed variables such as $F(s)$. The algebraic equations can be solved for the variables such as $F(s)$. The time domain expression can be obtained by taking the inverse Laplace transform, $f(t) = L^{-1}(s)$. Consider an RC circuit, as shown in Figure 14.9.

FIGURE 14.9

An RC circuit.

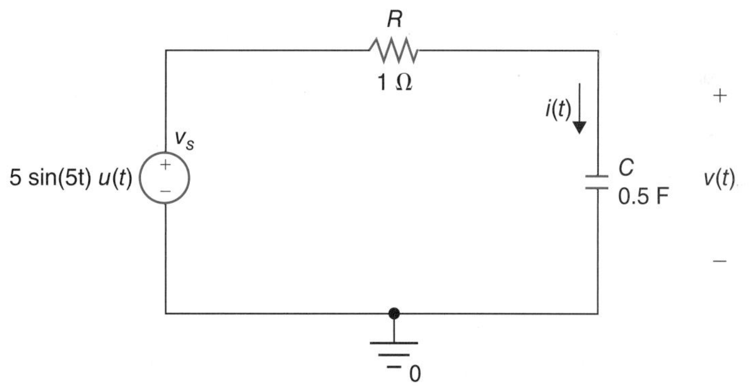

Let $v(t)$ be the voltage across the capacitor and $i(t)$ be the current through the capacitor. The initial voltage across the capacitor is $v(0^-) = 1$ V, and the input voltage is $v_s(t) = 5\sin(5t)\,u(t)$. Collecting the voltage drops around the mesh in the clockwise direction, we obtain

$$-v_s(t) + Ri(t) + v(t) = 0 \tag{14.72}$$

The current-voltage relation of the capacitor is

$$i(t) = C\frac{dv(t)}{dt} \tag{14.73}$$

Substitution of Equation (14.73) into Equation (14.72) and simplification yields

$$\frac{dv(t)}{dt} + 2v(t) = 10\sin(5t)u(t) \tag{14.74}$$

Taking the one-sided Laplace transform on both sides of Equation (14.74), we obtain

$$sV(s) - v(0^-) + 2V(s) = 10\frac{5}{s^2 + 5^2}$$

Since the initial voltage across the capacitor is $v(0^-) = 1$ V, we have

$$(s + 2)V(s) = 10\frac{5}{s^2 + 5^2} + 1 = \frac{s^2 + 75}{s^2 + 5^2}$$

Solving for $V(s)$, we obtain

$$V(s) = \frac{s^2 + 75}{(s + 2)(s^2 + 5^2)}$$

Expanding $V(s)$ by partial fraction, we obtain

$$V(s) = \frac{s^2 + 75}{(s + 2)(s^2 + 5^2)} = \frac{A}{s + 2} + \frac{Bs + C}{s^2 + 5^2}$$

Coefficient A is given by

$$A = (s + 2)\frac{s^2 + 75}{(s + 2)(s^2 + 5^2)}\bigg|_{s=-2} = \frac{s^2 + 75}{(s^2 + 5^2)}\bigg|_{s=-2} = \frac{(-2)^2 + 75}{(-2)^2 + 5^2} = \frac{79}{29} = 2.7241$$

Combining the two terms on the right side of $V(s)$, we obtain

$$V(s) = \frac{s^2 + 75}{(s + 2)(s^2 + 5^2)} = \frac{\frac{79}{29}}{s + 2} + \frac{Bs + C}{s^2 + 5^2}$$
$$= \frac{\left(\frac{79}{29} + B\right)s^2 + (2B + C)s + \left(2C + \frac{79 \times 25}{29}\right)}{(s + 2)(s^2 + 5^2)}$$

Comparing the coefficients of the numerator polynomials, we obtain

$$\frac{79}{29} + B = 1 \tag{14.75}$$

$$2C + \frac{79 \times 25}{29} = 75 \tag{14.76}$$

$$2B + C = 0 \tag{14.77}$$

From Equation (14.75), we get

$$B = 1 - \frac{79}{29} = -\frac{50}{29} = -1.7241$$

From Equation (14.76), we get

$$C = \frac{\frac{75 \times 29 - 79 \times 25}{29}}{2} = \frac{200}{58} = 3.4483$$

Substituting B and C into Equation (14.77), we have $2B + C = 0$, verifying B and C. The partial fraction expansion of $V(s)$ is given by

$$V(s) = \frac{\frac{79}{29}}{s+2} + \frac{-\frac{50}{29}s + \frac{40}{58} \times 5}{s^2 + 5^2}$$

Taking the inverse Laplace transform of $V(s)$, we obtain

$$v(t) = \left[\frac{79}{29}e^{-2t} - \frac{50}{29}\cos(5t) + \frac{40}{58}\sin(5t)\right]u(t)$$
$$= [2.7241e^{-2t} - 1.7241\cos(5t) + 0.6897\sin(5t)]u(t)$$

As shown in Chapter 15, we can transform components and sources of circuits to the s-domain directly and find the equations in the s-domain without finding the differential equations.

EXAMPLE 14.33

Use the Laplace transform to find $i(t)$ through the inductor in the circuit shown in Figure 14.10. The initial voltage across the capacitor is $v(0^-) = 2$ V, and the initial current through the inductor is $i(0^-) = 1$ A.

FIGURE 14.10

Circuit for EXAMPLE 14.33.

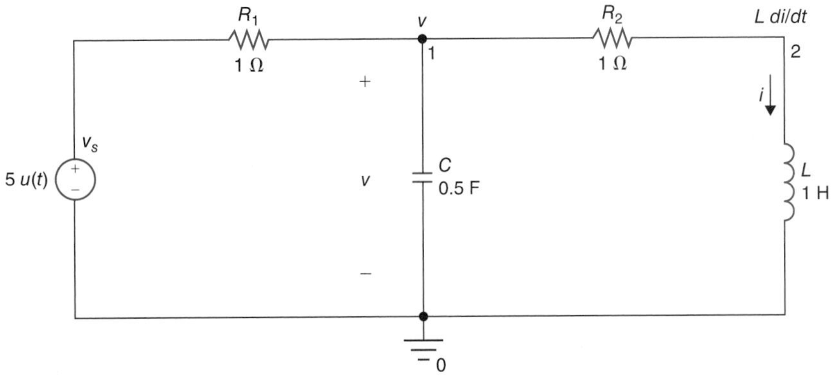

Summing the currents leaving node 1, we obtain

$$\frac{v(t) - v_s(t)}{R_1} + C\frac{dv(t)}{dt} + i(t) = 0 \quad \textbf{(14.78)}$$

Summing the currents leaving node 2, we obtain

$$\frac{L\frac{di(t)}{dt} - v(t)}{R_2} + i(t) = 0 \quad \textbf{(14.79)}$$

continued

Example 14.33 continued

After substitution of the component values, Equations (14.78) and (14.79) become, respectively,

$$v(t) + 0.5\frac{dv(t)}{dt} + i(t) = v_s(t) \quad \textbf{(14.80)}$$

$$\frac{di(t)}{dt} - v(t) + i(t) = 0 \quad \textbf{(14.81)}$$

From Equation (14.81), we obtain

$$\frac{di(t)}{dt} = v(t) - i(t)$$

Setting $t = 0^-$, we get

$$\frac{di(0^-)}{dt} = v(0^-) - i(0^-) = 2 - 1 = 1 \text{ A/}s$$

Solving Equation (14.81) for $v(t)$, we obtain

$$v(t) = \frac{di(t)}{dt} + i(t) \quad \textbf{(14.82)}$$

Substitution of Equation (14.82) into Equation (14.80) yields

$$\frac{di(t)}{dt} + i(t) + 0.5\frac{d^2i(t)}{dt^2} + 0.5\frac{di(t)}{dt} + i(t) = v_s(t)$$

which can be revised as

$$\frac{d^2i(t)}{dt^2} + 3\frac{di(t)}{dt} + 4i(t) = 2v_s(t) \quad \textbf{(14.83)}$$

Taking the Laplace transform on both sides of Equation (14.83), we obtain

$$s^2I(s) - si(0^-) - \frac{di(0^-)}{dt} + 3[sI(s) - i(0^-)] + 4I(s) = 2\frac{5}{s} \quad \textbf{(14.84)}$$

Since $i(0^-) = 1$ A and $\frac{di(0^-)}{dt} = 1$ A/s, we can rewrite Equation (14.84) as

$$(s^2 + 3s + 4)I(s) = \frac{s^2 + 4s + 10}{s}$$

Solving for $I(s)$, we obtain

$$I(s) = \frac{s^2 + 4s + 10}{s(s^2 + 3s + 4)}$$

Taking the inverse Laplace transform using MATLAB, we obtain

$$i(t) = [2.5 - 1.5e^{-1.5t}\cos(1.3229t) - 0.9449e^{-1.5t}\sin(1.3229t)]u(t)$$

Exercise 14.33

Use the Laplace transform to find $v(t)$ across the capacitor in the circuit shown in Figure 14.11. The initial voltage across the capacitor is $v(0^-) = 2$ V, and the initial current through the inductor is $i(0^-) = 1$ A.

FIGURE 14.11

Circuit for EXERCISE 14.33.

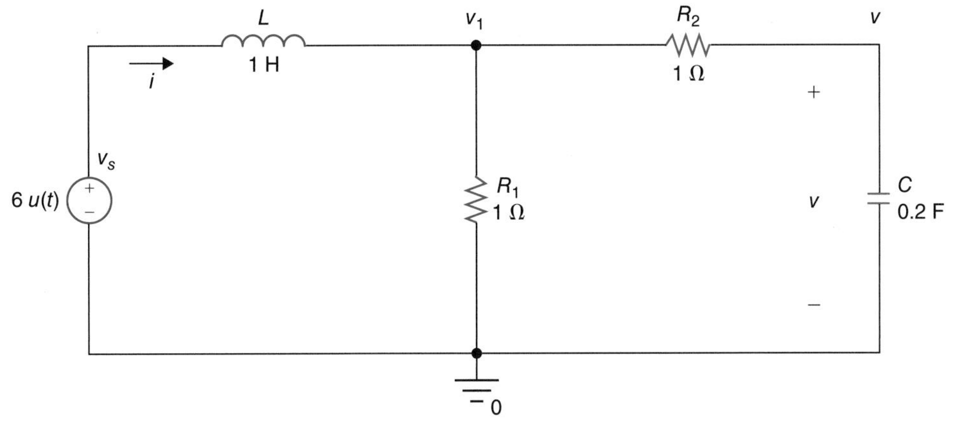

Answer:
$v(t) = [6 - 4e^{-1.5t}\cos(0.5t) - 17e^{-1.5t}\sin(0.5t)]u(t)$.

14.6 PSpice and Simulink

As will be shown in Chapter 15, the impulse response is the inverse Laplace transform of the transfer function. The inverse Laplace transform of

$$H(s) = \frac{1}{s+1} \tag{14.85}$$

can be displayed from the PSpice model shown in Figure 14.12.

Dividing each term on the right side of Equation (14.85) by s everywhere, we obtain

$$H(s) = \frac{V_o(s)}{V_{in}(s)} = \frac{\frac{1}{s}}{1 + \frac{1}{s}}$$

FIGURE 14.12

RC circuit.

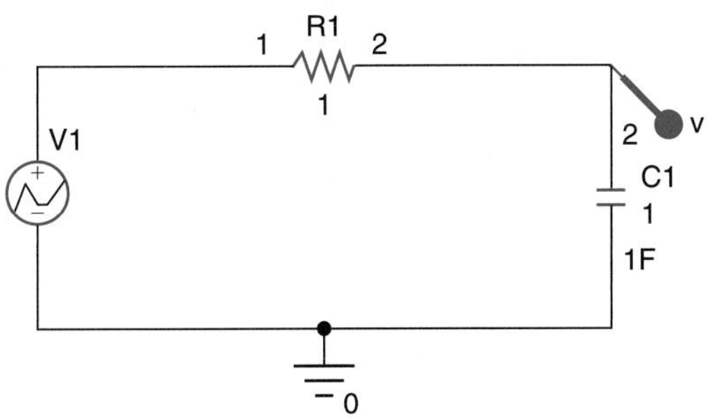

This circuit can be designed by a series connection of a 1 Ω resistor [(1 in $H(s)$] and a 1-F capacitor [$1/s$ in $H(s)$], as shown in Figure 14.12.

The inverse Laplace transform of $1/(s + 1)$ is equivalent to the impulse response of $H(s)$. When $v_{in}(t) = \delta(t)$, $V_{in}(s) = L[\delta(t)] = 1$, and $V_o(s) = [1/(s + 1)]V_{in}(s) = [1/(s + 1)] \times 1 = 1/(s + 1)$. Thus, the inverse Laplace transform of $1/(s + 1)$ is obtained by applying an impulse to the circuit shown in Figure 14.12. The impulse is generated by a piecewise linear voltage source (VPWL). Double-click on VPWL (V1). Figure 14.13 shows the time and voltage to create an impulse.

FIGURE 14.13

The time and voltage to simulate an impulse. (*Source: OrCAD PSpice by Cadence*)

T1	T2	T3	T4
0	1u	10u	11u

V1	V2	V3	V4
0	100000	100000	0

Notice that a rectangular pulse with width 10 μs and amplitude 100,000 V has an area of 1. This rectangle simulates $\delta(t)$. The voltage across the capacitor is given by the plot shown in Figure 14.14.

FIGURE 14.14

The inverse Laplace transform of $1/(s + 1)$.

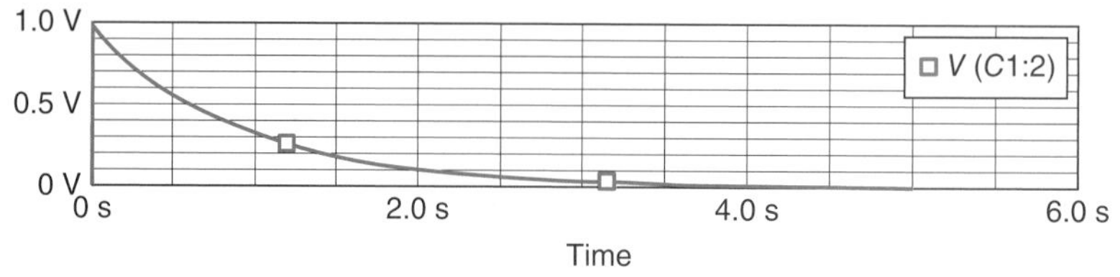

Notice that the inverse Laplace transform is $e^{-t}u(t)$. The MATLAB plot of $e^{-t}u(t)$ shown in Figure 14.15 confirms that the output is correct.

FIGURE 14.15

The inverse Laplace transform from MATLAB.

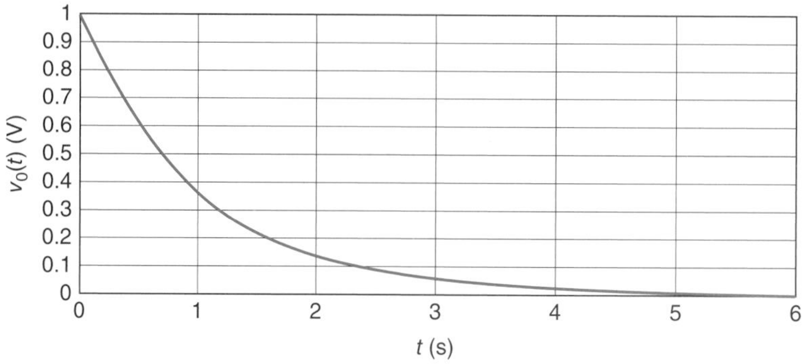

The inverse Laplace transform can be displayed from the Simulink model shown in Figure 14.16. The impulse is simulated by a pulse with width 10^{-5}, which is the sample time, and height 10^5 to make the area equal to 1.

$$\delta(t) \approx 10^5 u(t) - 10^5 u(t - 10^{-5})$$

Double-click on the step block and set
Step time = 0, initial value = 0, final value = 1e5, sample time = 1e−5

Double-click on the step 1 block and set
Step time = 1e−5, initial value = 0, final value = 1e5, sample time = 1e−5

FIGURE 14.16

A Simulink model to display the inverse Laplace transform.

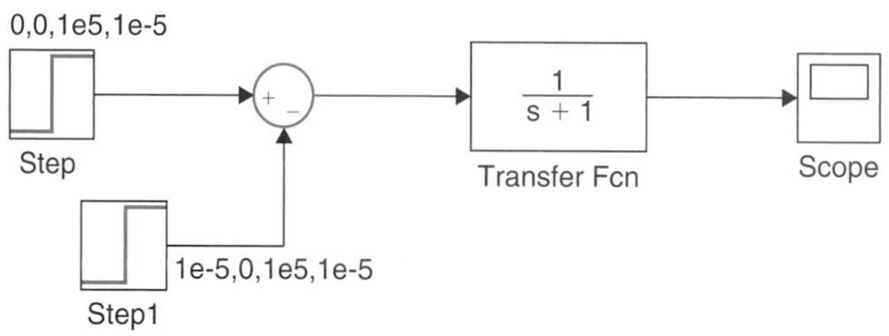

The inverse Laplace transform can also be obtained by taking the derivative of the step response, as shown in Figure 14.17. The inverse Laplace transform is shown in Figure 14.18.

FIGURE 14.17

A Simulink model to plot the inverse Laplace transform.

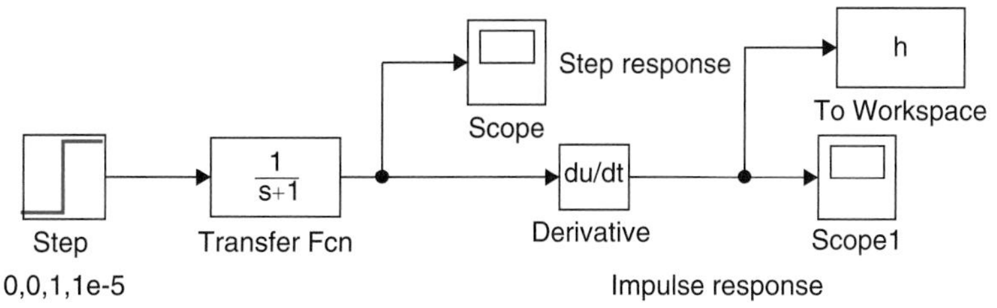

FIGURE 14.18

The inverse Laplace transform of $1/(s+1)$, which is $e^{-t}u(t)$.

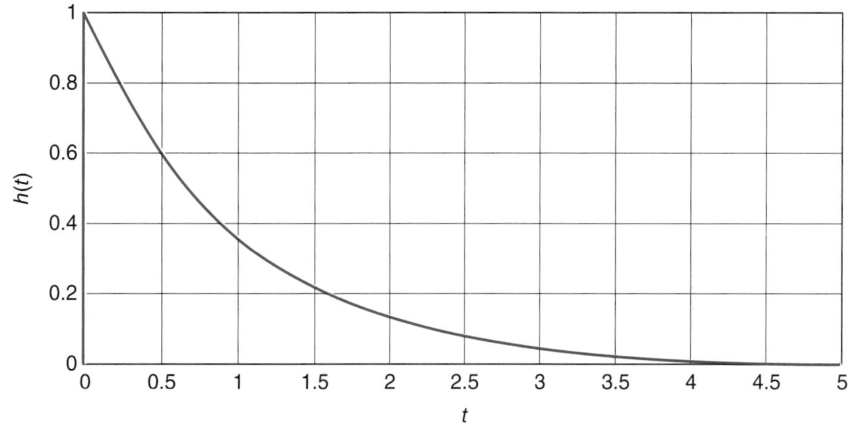

EXAMPLE 14.34

Plot the inverse Laplace transform of

$$H(s) = \frac{1}{s^2 + 0.2s + 1} = \frac{\frac{1}{s}}{s + 0.2 + \frac{1}{s}} \tag{14.86}$$

using PSpice.

Figure 14.19 shows the PSpice simulation to get the inverse Laplace transform of $H(s)$, given by Equation (14.86). The inverse Laplace transform is shown in Figure 14.20.

FIGURE 14.19

The *RLC* circuit.

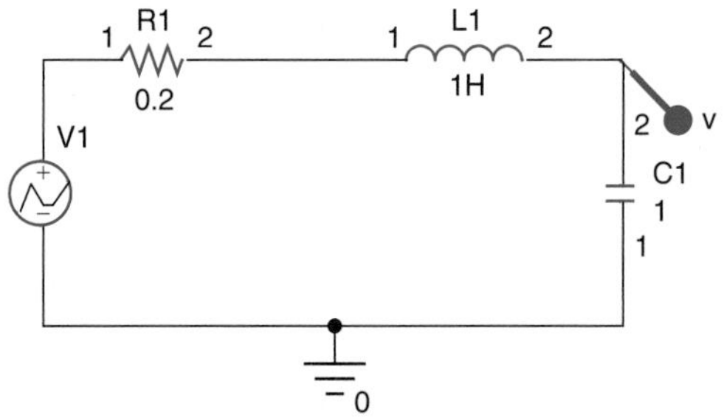

FIGURE 14.20

The plot of the inverse Laplace transform.

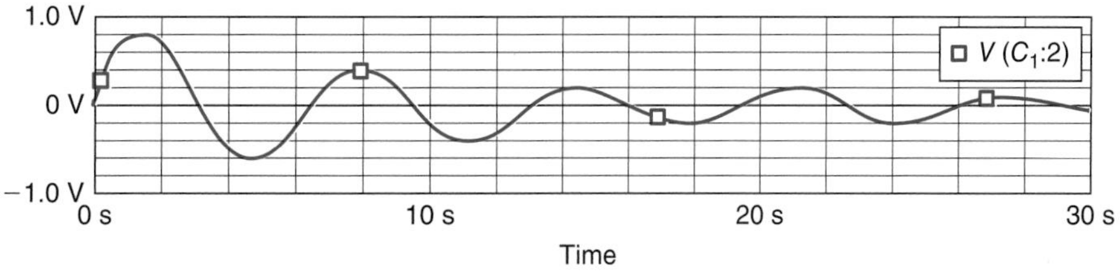

Exercise 14.34

Plot the inverse Laplace transform of

$$H(s) = \frac{1}{s^2 + 0.2s + 1} = \frac{\frac{1}{s}}{s + 0.2 + \frac{1}{s}}$$

using Simulink.

Answer:

Figure 14.21 shows the Simulink model, and Figure 14.22 shows the output waveform.

FIGURE 14.21

The Simulink model.

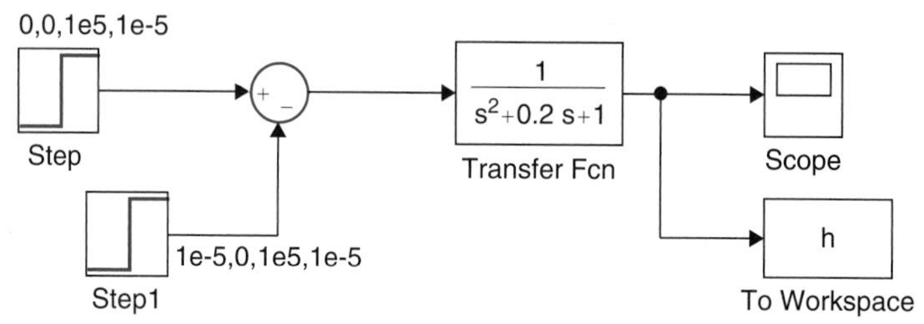

FIGURE 14.22

The inverse Laplace transform.

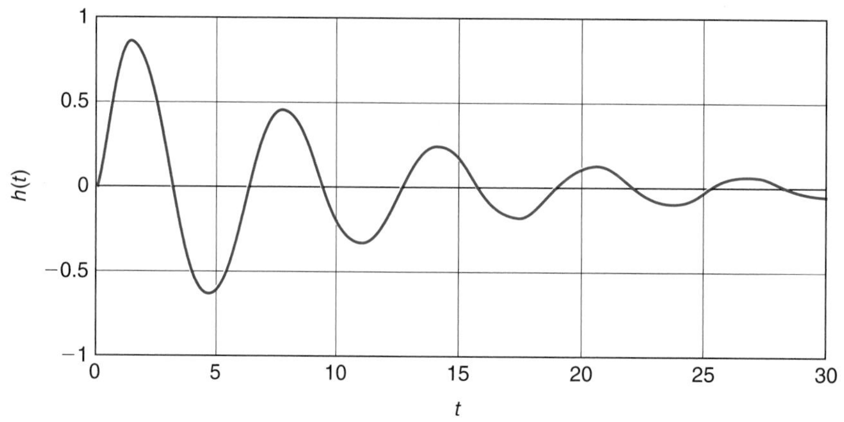

SUMMARY

In this chapter, the definition of the Laplace transform, the properties of the Laplace transform, and the inverse Laplace transform are presented. The one-sided Laplace transform is defined as

$$F(s) = \int_{0^-}^{\infty} f(t)e^{-st}dt$$

Application of the properties of the Laplace transform makes it easier to find the Laplace transform. Also, the properties provide a mathematical foundation for the applications of the Laplace transform. As an example, the time differentiation property for a one-sided Laplace transform with initial conditions is the basis for the circuit analysis in the s-domain.

The process of recovering the time domain signal $f(t)$ from the Laplace transform $F(s)$ is called the *inverse Laplace transform.* The residue method and the partial fraction expansion can be used to find $f(t)$ given $F(s)$. Also, MATLAB is a useful tool in finding the Laplace transform and the inverse Laplace transform.

PROBLEMS

Laplace Transform

14.1 Find the one-sided Laplace transform of

$$f(t) = 7\delta(t - 5)$$

14.2 Find the one-sided Laplace transform of

$$f(t) = 3u(t - 9)$$

14.3 Find the one-sided Laplace transform of

$$f(t) = 5u(t + 3)$$

14.4 Find the one-sided Laplace transform of

$$f(t) = u(-t + 5)$$

14.5 Find the one-sided Laplace transform of

$$f(t) = (t + 2)\,u(t - 3)$$

14.6 Find the one-sided Laplace transform of

$$f(t) = 4e^{-3t}u(t)$$

14.7 Find the one-sided Laplace transform of

$$f(t) = 4e^{-2t}u(t) + 3e^{-5t}u(t)$$

14.8 Find the one-sided Laplace transform of

$$f(t) = \frac{1}{2}e^{2t}u(t) + \frac{1}{2}e^{-2t}u(t)$$

14.9 Find the one-sided Laplace transform of

$$f(t) = 6e^{-2t+11}u(t - 5)$$

14.10 Find the one-sided Laplace transform of

$$f(t) = e^{-1.5(t+1)}u(t + 1)$$

14.11 Find the one-sided Laplace transform of

$$f(t) = e^{-2(t+2)}u(t + 2)$$

14.12 Find the one-sided Laplace transform of

$$f(t) = 5te^{-3t+7}u(t - 2)$$

14.13 Find the one-sided Laplace transform of

$$f(t) = 5^t$$

14.14 Find the one-sided Laplace transform of

$$f(t) = e^{-5t}\cosh(3t)u(t)$$

14.15 Find the one-sided Laplace transform of

$$f(t) = e^{-3t}\sinh(5t)u(t)$$

14.16 Find the one-sided Laplace transform of

$$f(t) = 5\cos\left(10t + \frac{\pi}{3}\right)u(t)$$

14.17 Find the one-sided Laplace transform of

$$f(t) = 7e^{-3t}\sin\left(5t - \frac{\pi}{6}\right)u(t)$$

14.18 Find the one-sided Laplace transform of

$$f(t) = 5te^{-3t}\cos\left(5t - \frac{\pi}{4}\right)u(t)$$

14.19 Find the one-sided Laplace transform of

$$f(t) = 6t^2e^{-3t}\sin\left(5t - \frac{2\pi}{3}\right)u(t)$$

14.20 Find the one-sided Laplace transform of

$$f(t) = u(t) - 2u(t - 1) + u(t - 2)$$

14.21 Find the one-sided Laplace transform of

$$f(t) = tu(t) - (t - 1)u(t - 1) - (t - 4)u(t - 4) + (t - 5)u(t - 5)$$

14.22 Find the one-sided Laplace transform of

$$f(t) = tu(t) - 2(t - 1)u(t - 1) + 3(t - 3)u(t - 3) - 2(t - 4)u(t - 4)$$

14.23 Find the one-sided Laplace transform of

$$f(t) = -tu(t) + 3(t - 1)u(t - 1) - 3(t - 2)u(t - 2) + (t - 3)u(t - 3)$$

14.24 Find the one-sided Laplace transform of

$$f(t) = t\,u(t) - 1.5(t-1)u(t-1) + 1.5(t-5) \times u(t-5) - (t-6)u(t-6)$$

14.25 Let $f(t) = 5\sin(3t)\,u(t)$. Find the one-sided Laplace transform of $f(t)$ and $g(t) = \dfrac{d^2f(t)}{dt^2}$. The initial conditions are zero.

14.26 Let $f(t) = tu(t) - 2(t-1)u(t-1) + (t-2) \times u(t-2)$. Find the one-sided Laplace transform of $f(t)$, $g(t) = f(2t)$, $h(t) = f(t/2)$.

Inverse Laplace Transform

14.27 Find the inverse Laplace transform of

$$F(s) = \frac{2s+9}{s^2+7s+12}$$

14.28 Find the inverse Laplace transform of

$$F(s) = \frac{s^2+4s+12}{s(s+2)(s+6)}$$

14.29 Find the inverse Laplace transform of

$$F(s) = \frac{s+5}{(s+2)(s+3)(s+4)}$$

14.30 Find the inverse Laplace transform of

$$F(s) = \frac{2s+11}{(s+3)^2(s+6)}$$

14.31 Find the inverse Laplace transform of

$$F(s) = \frac{s+6}{s^2(s+3)}$$

14.32 Find the inverse Laplace transform of

$$F(s) = \frac{2s+5}{s^2+2s+5}$$

14.33 Find the inverse Laplace transform of

$$F(s) = \frac{2s+9}{s^2+4s+29}$$

14.34 Find the inverse Laplace transform of

$$F(s) = \frac{3s+17}{s^2+6s+25}$$

14.35 Find the inverse Laplace transform of

$$F(s) = \frac{2s+24}{(s+1)(s^2+6s+34)}$$

14.36 Find the inverse Laplace transform of

$$F(s) = \frac{7s^2+52s+161}{(s+2)(s^2+6s+25)}$$

14.37 Find the inverse Laplace transform of

$$F(s) = \frac{3s^2+10s+120}{(s+3)(s^2+8s+41)}$$

14.38 Find the inverse Laplace transform of

$$F(s) = \frac{8}{(s+2)^3}$$

14.39 Find the inverse Laplace transform of

$$F(s) = \frac{s^4+18s^3+10s^2+15s+20}{s^2+12s+35}$$

14.40 Find the inverse Laplace transform of

$$F(s) = \frac{s^3+7s^2+12s+12}{s^2+6s+25}$$

14.41 Find the inverse Laplace transform of

$$F(s) = \frac{5s^2+6s+12}{(s+2)^3(s+3)}$$

14.42 Find the inverse Laplace transform of

$$F(s) = \frac{(5s+12)e^{-2s}}{s^2+11s+30}$$

14.43 Find the inverse Laplace transform of

$$F(s) = \frac{(s+6)e^{-3s}}{s^2+8s+41}$$

14.44 Find the inverse Laplace transform of

$$F(s) = \frac{12(s+1)}{(s^2+2s+10)^2}$$

14.45 Find the inverse Laplace transform of

$$F(s) = \frac{12s}{(s+1)^2(s+3)^2}$$

14.46 Find the inverse Laplace transform of

$$F(s) = \frac{10s + 300}{s^3 + 135s^2 + 3800s + 30{,}000}$$

14.47 Find the initial value and the final value of $f(t)$ if

$$F(s) = \frac{3s^2 + 19s + 36}{s^3 + 5s^2 + 6s}$$

14.48 Find the initial value and the final value of $f(t)$ if

$$F(s) = \frac{2s^2 + 45}{s^3 + 9s}$$

14.49 Find the initial value and the final value of $f(t)$ if

$$F(s) = \frac{2s^3 + 6.6s^2 + 5.4s + 1.4}{s^4 + 1.7s^3 + 0.95s^2 + 0.175s}$$

14.50 Find the initial value and the final value of $f(t)$ if

$$F(s) = \frac{3s + 15}{s^2 + 10s + 89}$$

Chapter 15

Circuit Analysis in the *s*-Domain

15.1 Introduction

The time differentiation property of the Laplace transform discussed in Chapter 14 makes it possible to transform differential equations describing circuits into algebraic equations. The algebraic equations are functions of complex variable s. The time differentiation property transforms voltages and currents from the time domain to the s-domain. This transformation even includes initial conditions on voltages and currents. Thus, the solution of the algebraic equations contains both the transient response and the steady-state response. We do not have to find the complementary solution and the particular solution separately and then combine them.

Rather than writing differential equations and then transforming to algebraic equations, a circuit in the time domain can be transformed to the s-domain by transforming each element to the s-domain. This process is similar to finding the phasor-transformed circuit. Once the s-domain circuit is obtained, familiar circuit analysis techniques such as Ohm's law, voltage divider rule, current divider rule, nodal analysis, and mesh analysis can be applied to obtain the s-domain expression of the unknown circuit variables such as voltages and currents. The inverse Laplace transforms of these unknown circuit variables provide the time domain expressions. The solution includes both the transient response and the steady-state response. The input signals are not restricted to be step functions or sinusoids. The input signal can be arbitrary as long as its Laplace integral converges.

The Thévenin and Norton equivalent circuits in the s-domain are presented in this chapter, along with transfer functions in the s-domain. Also, convolution, linear and time invariant systems, and Bode diagrams are introduced.

15.2 Laplace-Transformed Circuit Elements

Consider a circuit element (or a combination of elements) shown as a rectangle in Figure 15.1(a). Let $i(t)$ be the current through the element, and let $I(s)$ be the Laplace transform of $i(t)$. Also, let $v(t)$ be the voltage across the element and $V(s)$ be the Laplace transform of $v(t)$. Then, we define the impedance $Z(s)$ of the element as the ratio of $V(s)$ to $I(s)$. Thus, we have

$$Z(s) = \frac{V(s)}{I(s)} \tag{15.1}$$

Equation (15.1) can be written as

$$V(s) = Z(s)I(s) \tag{15.2}$$

Equation (15.2) is Ohm's law in the s-domain. Equation (15.1) can also be written as

$$I(s) = \frac{V(s)}{Z(s)} \tag{15.3}$$

The ratio of $I(s)$ to $V(s)$ is defined as admittance $Y(s)$. Thus, we have

$$Y(s) = \frac{I(s)}{V(s)} = \frac{1}{Z(s)} \tag{15.4}$$

The admittance is the inverse of impedance. Equation (15.4) can also be written as

$$V(s) = \frac{I(s)}{Y(s)} \tag{15.5}$$

and

$$I(s) = Y(s)V(s) \tag{15.6}$$

Figure 15.1(b) shows the circuit element represented in the s-domain.

FIGURE 15.1

(a) Circuit element in the time domain. (b) Circuit element in the s-domain.

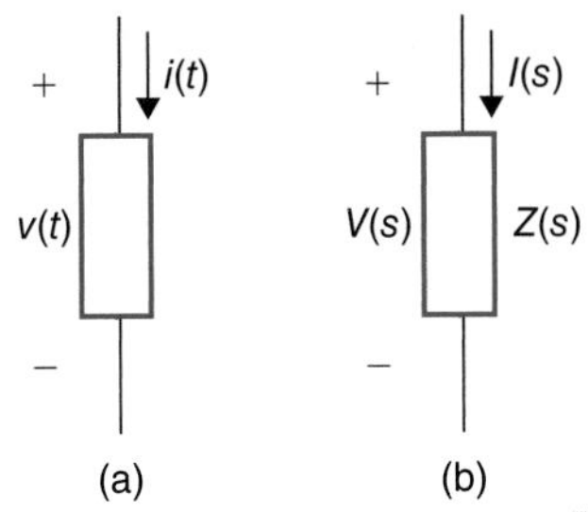

15.2.1 RESISTOR

The voltage-current relation of a resistor is given by

$$v(t) = Ri(t) \tag{15.7}$$

where R is the resistance value in ohms. Taking the Laplace transform on both sides, we obtain

$$V(s) = RI(s) \tag{15.8}$$

The impedance of a resistor is given by

$$Z(s) = \frac{V(s)}{I(s)} = R \tag{15.9}$$

Thus, the impedance of a resistor is R. The impedance of a resistor is a constant and is not a function of s. The admittance of a resistor is

$$Y(s) = \frac{I(s)}{V(s)} = \frac{1}{R} \tag{15.10}$$

The resistor representation in the time domain and in the s-domain is shown in Figure 15.2.

FIGURE 15.2

(a) Resistor in the time domain. (b) Resistor in the s-domain.

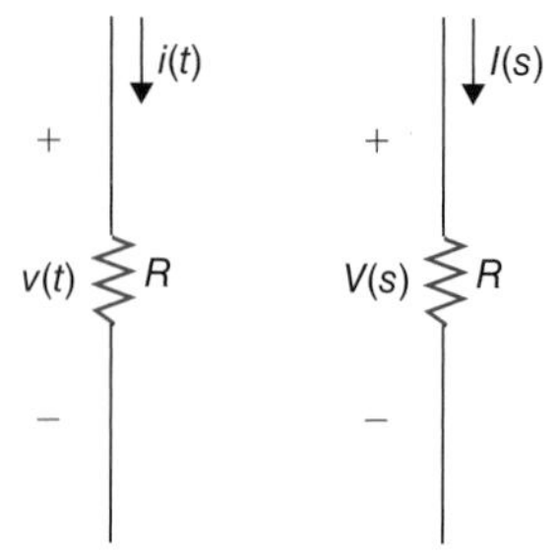

15.2.2 CAPACITOR

The current-voltage relation of a capacitor is given by

$$i(t) = C\frac{dv(t)}{dt} \tag{15.11}$$

where C is the capacitance value in farads. From the time differentiation property with initial conditions, the Laplace transform of Equation (15.11) is given by

$$I(s) = C[sV(s) - v(0^-)] = CsV(s) - Cv(0^-) = sCV(s) - Cv(0^-) \tag{15.12}$$

where $v(0^-)$ is the voltage across the capacitor at $t = 0^-$. Solving Equation (15.12) for $V(s)$, we get

$$V(s) = \frac{1}{sC}I(s) + \frac{v(0^-)}{s} \tag{15.13}$$

If the initial voltage across the capacitor is zero ($v(0^-) = 0$), Equation (15.13) reduces to

$$V(s) = \frac{1}{sC}I(s) \tag{15.14}$$

The impedance of a capacitor is defined as

$$Z(s) = \frac{V(s)}{I(s)} = \frac{1}{sC} \tag{15.15}$$

If s is replaced by $j\omega$, the impedance of a capacitor becomes $Z = 1/(j\omega C)$ given by Equation (9.50). Equation (15.12) can be rewritten as

$$I(s) = \frac{V(s)}{\dfrac{1}{sC}} - Cv(0^-) \tag{15.16}$$

According to Equation (15.13), in the s-domain, a capacitor can be represented by an impedance of $1/(sC)$ in series with a voltage source with voltage $v(0^-)/s$. According to Equation (15.16), in the s-domain, a capacitor can be represented by an impedance of $1/(sC)$ in parallel with a current source with current $-Cv(0^-)$. The capacitor representation in the time domain and in the s-domain is shown in Figure 15.3.

FIGURE 15.3

(a) Capacitor in the time domain. (b) Capacitor in the s-domain.

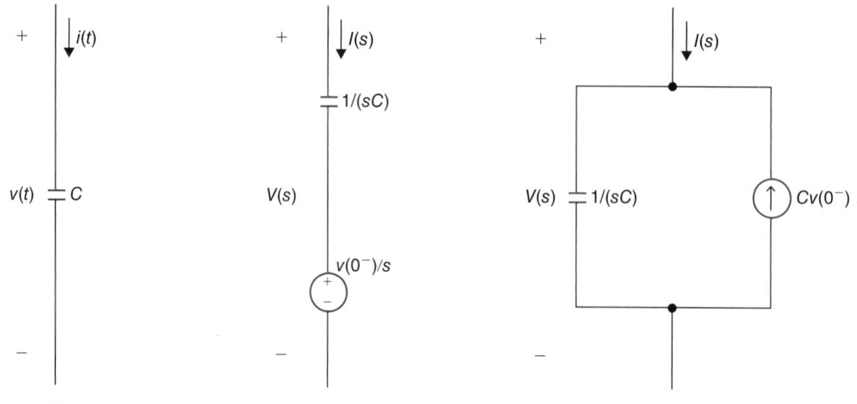

15.2.3 INDUCTOR

The voltage-current relation of an inductor is given by

$$v(t) = L\frac{di(t)}{dt} \tag{15.17}$$

where L is the inductance value in henrys. From the time differentiation property with initial conditions, the Laplace transform of Equation (15.17) is given by

$$V(s) = L[sI(s) - i(0^-)] = LsI(s) - Li(0^-) = sLI(s) - Li(0^-) \tag{15.18}$$

where $i(0^-)$ is the current through the inductor at $t = 0^-$. Solving Equation (15.18) for $I(s)$, we get

$$I(s) = \frac{V(s)}{sL} + \frac{i(0^-)}{s} \tag{15.19}$$

If the initial current through the inductor is zero ($i(0^-) = 0$), Equation (15.19) reduces to

$$I(s) = \frac{V(s)}{sL} \tag{15.20}$$

The impedance of the inductor is defined by

$$Z(s) = \frac{V(s)}{I(s)} = sL \tag{15.21}$$

Since the admittance is the inverse of the impedance, the admittance of an inductor is given by

$$Y(s) = \frac{I(s)}{V(s)} = \frac{1}{sL} \tag{15.22}$$

According to Equation (15.18), in the s-domain, an inductor can be represented by an impedance of sL in series with a voltage source with voltage $-Li(0^-)$. Alternatively, according to Equation (15.19), an inductor can be represented by an impedance of sL in parallel with a current source with current $i(0^-)/s$. The inductor representation in the time domain and in the s-domain is shown in Figure 15.4.

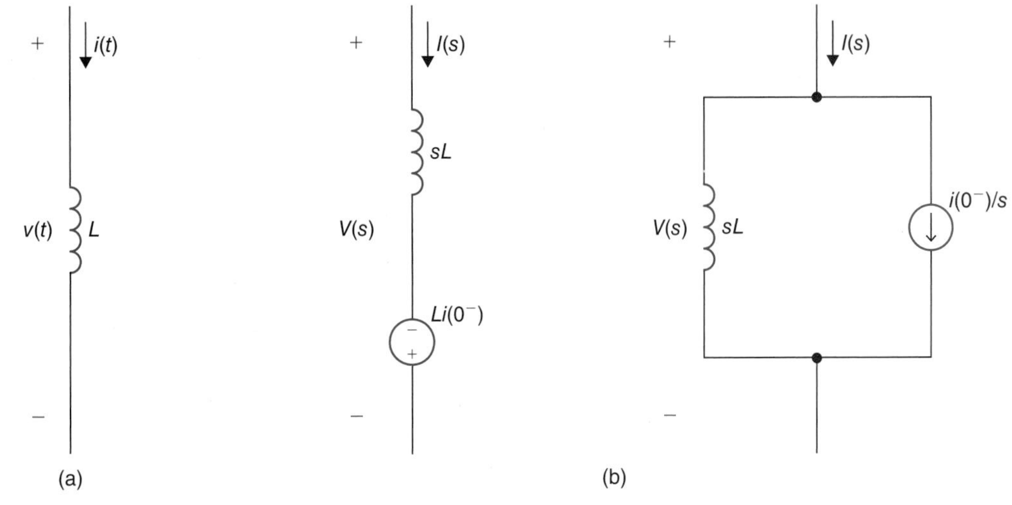

FIGURE 15.4

(a) Inductor in the time domain.
(b) Inductor in the s-domain.

15.3 Laplace-Transformed Circuit

A circuit can be transformed to the s-domain by replacing capacitors and inductors to the s-domain-equivalent circuits, as shown in Figure 15.3 and Figure 15.4. The voltage sources and current sources are replaced by their Laplace transforms. The s-domain circuits are called *Laplace-transformed circuits*, and they can be analyzed by applying circuit laws and circuit theorems.

Once the given circuit is transformed into the s-domain, any of the circuit analysis methods such as Ohm's law, Kirchhoff's current law (KCL), Kirchhoff's voltage law (KVL), voltage divider rule, current divider rule, nodal analysis, mesh analysis, Thévenin's theorem, or Norton's theorem discussed in Chapters 2−4 for resistive circuits can be applied to find the unknown voltages and currents in the s-domain. The only differences are the use of impedances instead of resistances and Laplace transforms for the sources. The time domain expressions for voltages and currents are obtained by taking the inverse Laplace transforms of s-domain voltages and currents.

15.3.1 VOLTAGE DIVIDER RULE

A voltage source, $V(s)$, is applied to a series connection of impedances $Z_1(s)$ and $Z_2(s)$, as shown in Figure 15.5.

Let $I(s)$ be the mesh current. Then, applying KVL around the mesh, we have

$$-V(s) + V_1(s) + V_2(s) = -V(s) + I(s)Z_1(s) + I(s)Z_2(s) = 0 \quad \textbf{(15.23)}$$

Solving Equation (15.23) for $I(s)$, we have

$$I(s) = \frac{V(s)}{Z_1(s) + Z_2(s)} \quad \textbf{(15.24)}$$

The voltage across $Z_1(s)$ is

$$V_1(s) = Z_1(s)I(s) = \frac{Z_1(s)}{Z_1(s) + Z_2(s)}V(s) \quad \textbf{(15.25)}$$

The voltage across $Z_2(s)$ is

$$V_2(s) = Z_2(s)I(s) = \frac{Z_2(s)}{Z_1(s) + Z_2(s)}V(s) \quad \textbf{(15.26)}$$

FIGURE 15.5

A voltage source connected in series with two impedances.

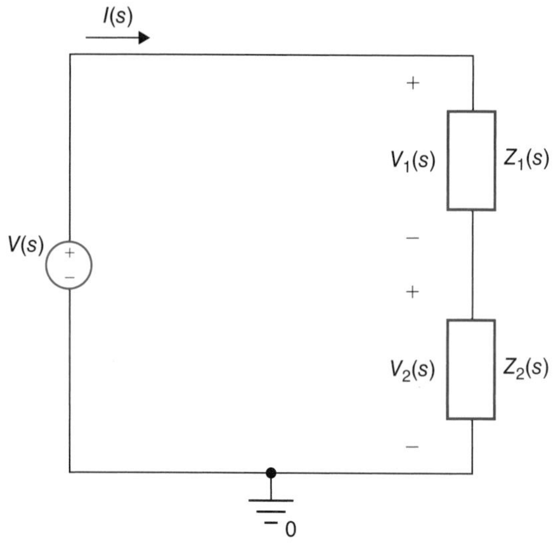

The voltage across $Z_1(s)$ is $V(s)$ times the ratio of $Z_1(s)$ to the total impedance $Z_1(s) + Z_2(s)$ seen from the voltage source $V(s)$. Likewise, the voltage across $Z_2(s)$ is $V(s)$ times the ratio of $Z_2(s)$ to the total impedance $Z_1(s) + Z_2(s)$ seen from the source $V(s)$. This result indicates that the voltage $V(s)$ across the total impedance $Z_1(s) + Z_2(s)$ is split between $V_1(s)$ and $V_2(s)$ in proportion to the impedances $Z_1(s)$ and $Z_2(s)$. In general, if N impedances $Z_1(s), Z_2(s), \ldots, Z_N(s)$ are connected in series to a voltage source $V(s)$, the voltage across $Z_n(s)$ is given by

$$V_n(s) = \frac{Z_n(s)}{Z_1(s) + Z_2(s) + \cdots + Z_N(s)}V(s) \quad \textbf{(15.27)}$$

EXAMPLE 15.1

The initial current through the inductor shown in Figure 15.6 is $i(0^-) = 3$ A. Find the voltage across the resistor $v_R(t)$ for $v_s(t) = u(t)$.

When the circuit shown in Figure 15.6 is transformed to an s-domain, we obtain the circuit shown in Figure 15.7.

From the voltage divider rule, the voltage $V_R(s)$ across R is given by

$$V_R(s) = \frac{R}{R + sL}\{V(s) - [-Li(0^-)]\} = \frac{R}{R + sL}[V(s) + Li(0^-)]$$

$$= \frac{2}{2 + s0.5}\left[\frac{1}{s} + 1.5\right]$$

$$= \frac{4}{s+4}\,\frac{1.5s + 1}{s} = \frac{6s + 4}{s(s+4)} = \frac{A}{s} + \frac{B}{s+4}$$

$$A = \left.\frac{6s+4}{(s+4)}\right|_{s=0} = 1$$

$$B = \left.\frac{6s+4}{s}\right|_{s=-4} = \frac{6(-4)+4}{-4} = \frac{-6+1}{-1} = 5$$

Thus,

$$V_R(s) = \frac{1}{s} + \frac{5}{s+4}$$

Taking the inverse Laplace transform, we obtain

$$v_R(t) = (1 + 5e^{-4t})u(t)$$

FIGURE 15.6

Circuit for EXAMPLE 15.1.

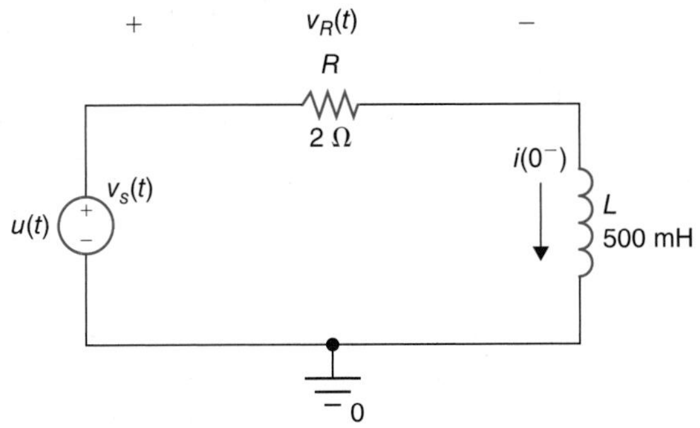

FIGURE 15.7

Circuit in the s-domain.

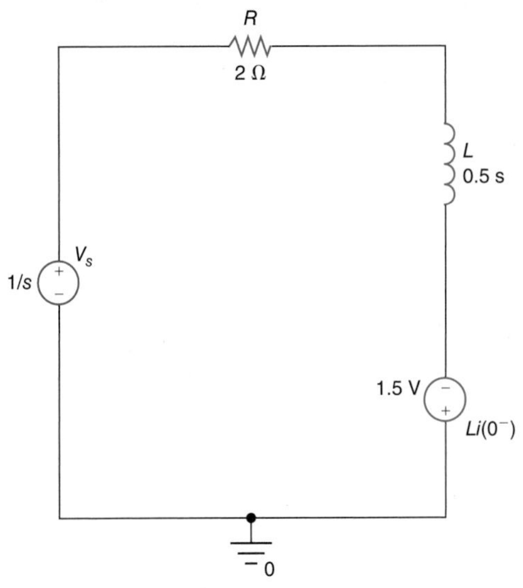

Exercise 15.1

Find $v(t)$ for $t \geq 0$ for the circuit shown in Figure 15.8.

FIGURE 15.8

Circuit for EXERCISE 15.1.

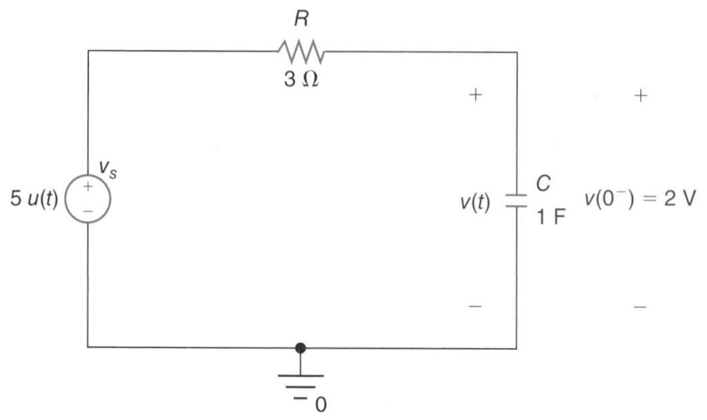

Answer:
$v(t) = (5 - 3e^{-t/3})\,u(t)$ V.

EXAMPLE 15.2

The switch in the circuit shown in Figure 15.9 has been closed for a long time before it is opened at $t = 0^-$. Find the voltage $v(t)$ across the capacitor for $t \geq 0$.

FIGURE 15.9

Circuit for EXAMPLE 15.2

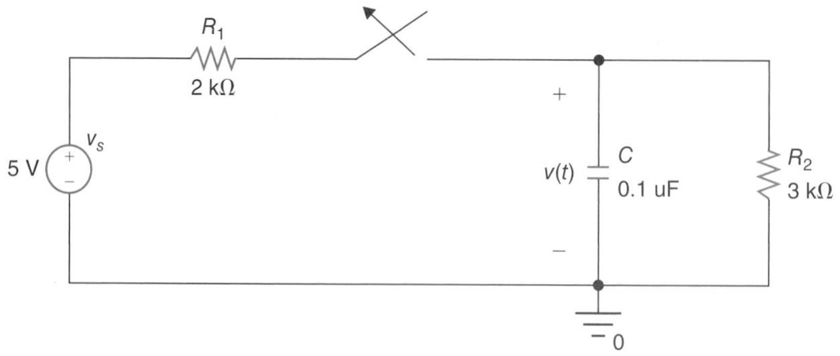

The capacitor can be treated as an open circuit for a dc input in the steady state. The voltage across the capacitor is the voltage across the resistor R_2. Thus, at $t = 0^-$, from the voltage divider rule, the voltage across the capacitor is given by

$$v(0^-) = v_s \times \frac{R_2}{R_1 + R_2} = 5\text{ V} \times \frac{3k\Omega}{2k\Omega + 3k\Omega} = 3\text{ V}$$

continued

Example 15.2 continued

FIGURE 15.10

The RC circuit in the s-domain.

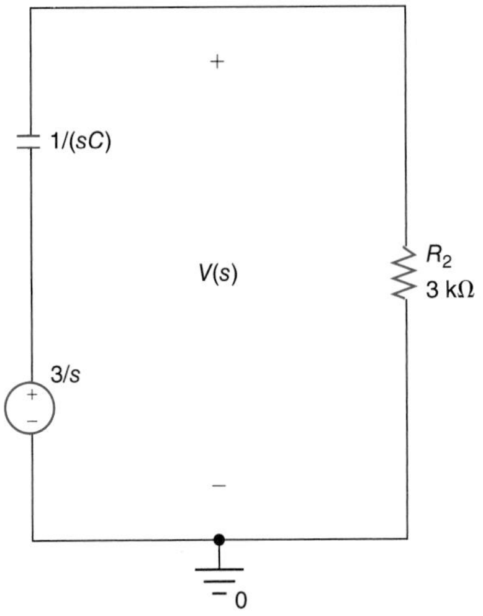

When the switch is opened, the initial voltage across the capacitor is given by $v(0^-) = 3$ V. Transforming the RC circuit to the s-domain, we obtain the circuit shown in Figure 15.10.

Application of the voltage divider rule to the circuit shown in Figure 15.10 yields

$$V(s) = \frac{3}{s} \times \frac{R_2}{R_2 + \dfrac{1}{sC}} = \frac{3}{s} \times \frac{sR_2}{sR_2 + \dfrac{1}{C}} = \frac{3}{s} \times \frac{s}{s + \dfrac{1}{R_2 C}}$$

$$= \frac{3}{s + 3333.3333}$$

Taking the inverse Laplace transform, we obtain

$$v(t) = 3e^{-3333.3333t}u(t)$$

Exercise 15.2

The switch in the circuit shown in Figure 15.11 has been closed for a long time before it is opened at $t = 0^-$. Find the voltage $v(t)$ across the resistor for $t \geq 0$.

FIGURE 15.11

Circuit for EXERCISE 15.2.

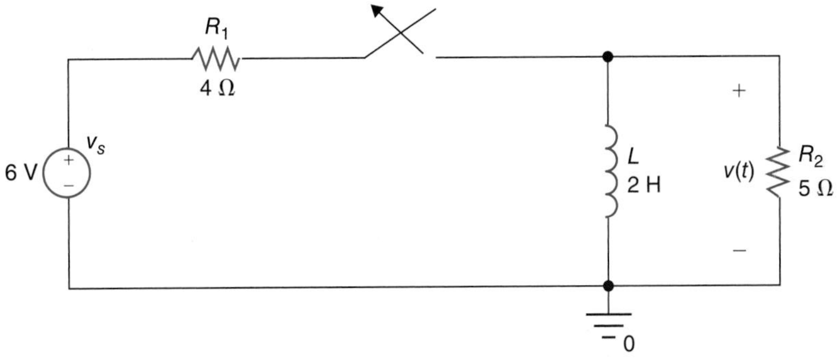

Answer:

$v(t) = -7.5e^{-2.5t}\,u(t)$ V.

15.3.2 CURRENT DIVIDER RULE

A current source $I(s)$ is applied to a parallel connection of impedances $Z_1(s)$ and $Z_2(s)$, as shown in Figure 15.12.

Let $V(s)$ be the voltage across $Z_1(s)$ and $Z_2(s)$. Then, summing the currents leaving node 1, we have

$$-I(s) + I_1(s) + I_1(s) = -I(s) + \frac{V(s)}{Z_1(s)} + \frac{V(s)}{Z_2(s)} = 0 \qquad (15.28)$$

FIGURE 15.12

A current source connected in parallel with two impedances.

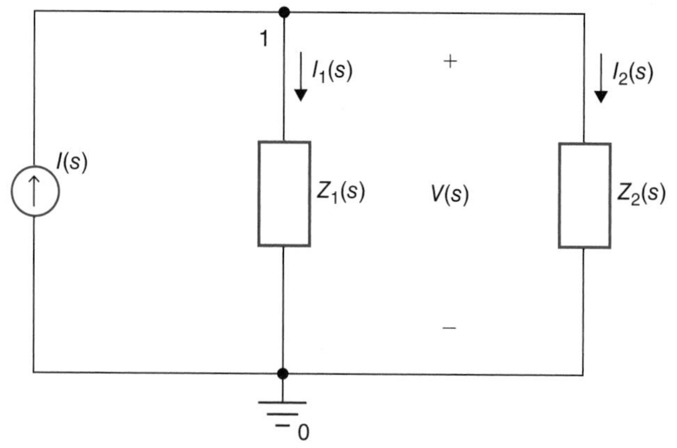

Solving Equation (15.28) for $V(s)$, we have

$$V(s) = \frac{1}{\dfrac{1}{Z_1(s)} + \dfrac{1}{Z_2(s)}} I(s) = \frac{Z_1(s)Z_2(s)}{Z_1(s) + Z_2(s)} I(s) \tag{15.29}$$

The current through $Z_1(s)$ is

$$\begin{aligned} I_1(s) &= \frac{V(s)}{Z_1(s)} = \frac{\dfrac{1}{Z_1(s)}}{\dfrac{1}{Z_1(s)} + \dfrac{1}{Z_2(s)}} I(s) = \frac{Z_2(s)}{Z_1(s) + Z_2(s)} I(s) \\ &= \frac{Y_1(s)}{Y_1(s) + Y_2(s)} I(s) \end{aligned} \tag{15.30}$$

The current through $Z_2(s)$ is

$$\begin{aligned} I_2(s) &= \frac{V(s)}{Z_2(s)} = \frac{\dfrac{1}{Z_2(s)}}{\dfrac{1}{Z_1(s)} + \dfrac{1}{Z_2(s)}} I(s) = \frac{Z_1(s)}{Z_1(s) + Z_2(s)} I(s) \\ &= \frac{Y_2(s)}{Y_1(s) + Y_2(s)} I(s) \end{aligned} \tag{15.31}$$

The current through $Z_1(s)$ is $I(s)$ times the ratio of $Z_2(s)$ to the sum of impedances $Z_1(s) + Z_2(s)$. Likewise, the current through $Z_2(s)$ is $I(s)$ times the ratio of $Z_1(s)$ to the sum of impedances $Z_1(s) + Z_2(s)$. In terms of admittances, the current through $Z_1(s)$ is $I(s)$ times the ratio of $Y_1(s)$ to the sum of admittances $Y_1(s) + Y_2(s)$, and the current through $Z_2(s)$ is $I(s)$ times the ratio of $Y_2(s)$ to the sum of admittances $Y_1(s) + Y_2(s)$. In general, if N impedances $Z_1(s), Z_2(s), \ldots, Z_N(s)$ are connected in parallel to a current source $I(s)$, the current through $Z_n(s)$ is given by

$$\begin{aligned} I_n(s) &= \frac{\dfrac{1}{Z_n(s)}}{\dfrac{1}{Z_1(s)} + \dfrac{1}{Z_2(s)} + \cdots + \dfrac{1}{Z_N(s)}} I(s) \\ &= \frac{Y_n(s)}{Y_1(s) + Y_2(s) + \cdots + Y_N(s)} I(s) \end{aligned} \tag{15.32}$$

EXAMPLE 15.3

Let $i(0^-) = 2$ A and $v(0^-) = 6$ V in the circuit shown in Figure 15.13. Find the current $i_R(t)$ through the resistor R for $t \geq 0$.

FIGURE 15.13

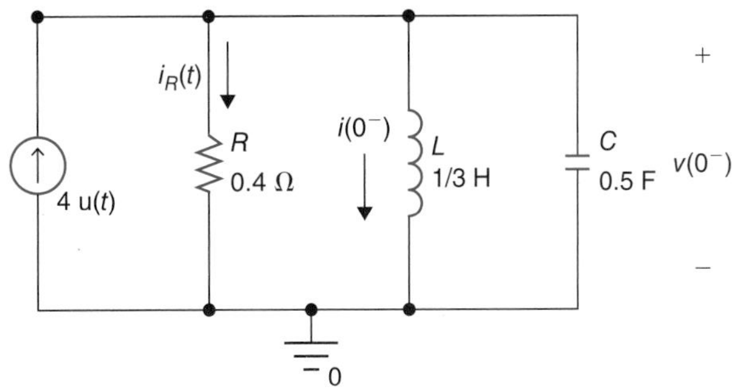

Circuit for EXAMPLE 15.3.

The s-domain equivalent circuit is shown in Figure 15.14. The sum of currents from the three current sources is

$$\frac{4}{s} - \frac{2}{s} + 3 = \frac{2}{s} + 3 = \frac{3s + 2}{s} \text{ A}$$

FIGURE 15.14

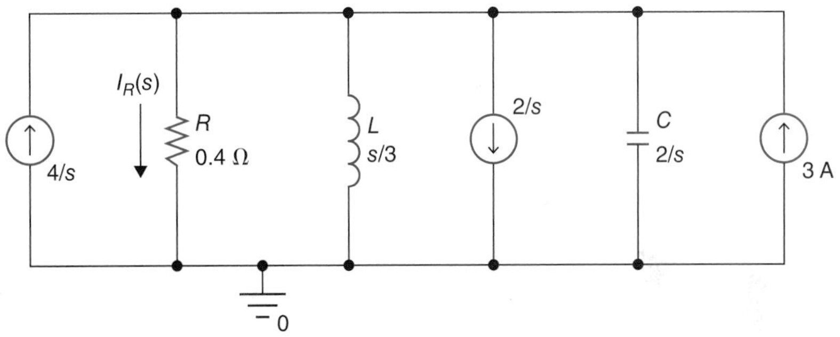

The circuit in the s-domain.

The direction of the current is up (↑). Since the three impedances are in parallel, the current through R is given by, from the current divider rule,

$$I_R(s) = \frac{\dfrac{1}{0.4}}{\dfrac{1}{0.4} + \dfrac{3}{s} + \dfrac{s}{2}} \times \frac{3s + 2}{s} = \frac{7.5s + 5}{0.5s^2 + 2.5s + 3} = \frac{15s + 10}{s^2 + 5s + 6} = \frac{15s + 10}{(s + 2)(s + 3)}$$

$$= \frac{A}{s + 2} + \frac{B}{s + 3}$$

Coefficients A and B are given by

$$A = \left.\frac{15s + 10}{(s + 3)}\right|_{s=-2} = -20$$

$$B = \left.\frac{15s + 10}{(s + 2)}\right|_{s=-3} = 35$$

continued

Example 15.3 continued

The current through the resistor is

$$i_R(t) = (-20e^{-2t} + 35e^{-3t})\, u(t)$$

Notice that the current through the resistor is 15 A at $t = 0$, and it decays exponentially to zero. The 15 A is the result of the initial voltage, $v(0^-) = 6$ V across the capacitor (also across the resistor), for example, $6\text{ V}/0.4\Omega = 15$ A.

Exercise 15.3

The switch in the circuit shown in Figure 15.15 has been closed for a long time before it is opened at $t = 0^-$. Find the current $i(t)$ through the inductor for $t \geq 0$.

FIGURE 15.15

Circuit for EXERCISE 15.3.

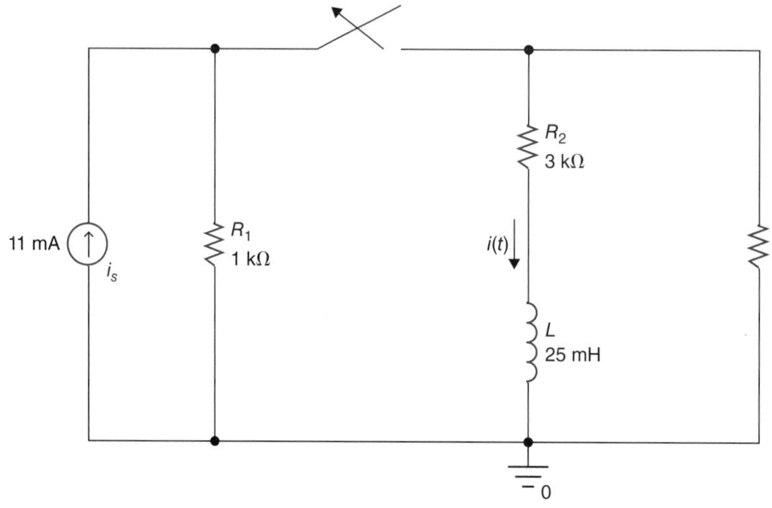

Answer:
$i(t) = 2\, e^{-200{,}000t}\, u(t)$ mA.

15.4 Nodal Analysis

Nodal analysis provides voltages at all nodes of a given circuit. If a voltage source is connected between a node and the ground, the node voltage is already known, and we do not need to find the node voltage on this node. The nodal analysis applies to the Laplace-transformed circuit. All the elements in the circuit are represented in the s-domain. Inputs and outputs are represented by Laplace transforms in the s-domain. The time domain expressions are obtained by taking the inverse Laplace transforms of the s-domain expressions. This method can be used to find the response of a circuit for arbitrary input as long as the Laplace integral of the input signal converges.

Excluding the nodes whose voltages are known from the voltage sources, we assign variables such as $V_1, V_2, \ldots, V_n$ to each unknown node voltage. For each node with assigned variable, write a node equation by applying Kirchhoff's current law (KCL). Any of the three interpretations of KCL given in Chapter 2 can be used. For example, the sum of currents leaving a node must be zero. If there are n unknown node voltages, we get n equations in n unknowns. Thus, we can solve these n system of linear equations to find the unique solution for the unknown node voltages $V_1, V_2, \ldots, V_n$ in the s-domain. An algorithm to find the solution of the n system of linear equations is Cramer's rule. MATLAB is useful in finding the solution of the n system of linear equations.

EXAMPLE 15.4

Let $v(0^-) = 1$ V and $i(0^-) = 1$ A in the circuit shown in Figure 15.16. Find the voltage $v_o(t)$ across the capacitor for $t \geq 0$ using nodal analysis.

FIGURE 15.16

Circuit for EXAMPLE 15.4.

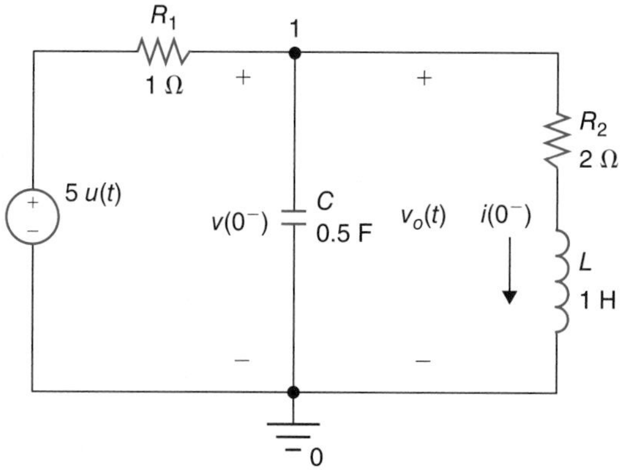

The circuit in the s-domain is shown in Figure 15.17. Summing the currents leaving node 1, we obtain

$$\frac{V_o - \frac{5}{s}}{1} + \frac{V_o - \frac{1}{s}}{\frac{2}{s}} + \frac{V_o - (-1)}{s + 2} = 0$$

Rearrangement of this equation yields

$$\left(1 + \frac{s}{2} + \frac{1}{s+2}\right)V_o = \frac{5}{s} + \frac{1}{2} - \frac{1}{s+2}$$

$$= \frac{10(s+2) + s^2 + 2s - 2s}{2s(s+2)} = \frac{s^2 + 10s + 20}{2s(s+2)}$$

Solving for V_o, we have

$$V_o = \frac{s^2 + 10s + 20}{2s(s+2)\left(1 + \frac{s}{2} + \frac{1}{s+2}\right)}$$

$$= \frac{s^2 + 10s + 20}{2s(s+2)\left(\frac{2s + 4 + s^2 + 2s + 2}{2(s+2)}\right)}$$

$$= \frac{s^2 + 10s + 20}{s(s^2 + 4s + 6)}$$

FIGURE 15.17

Circuit in the s-domain.

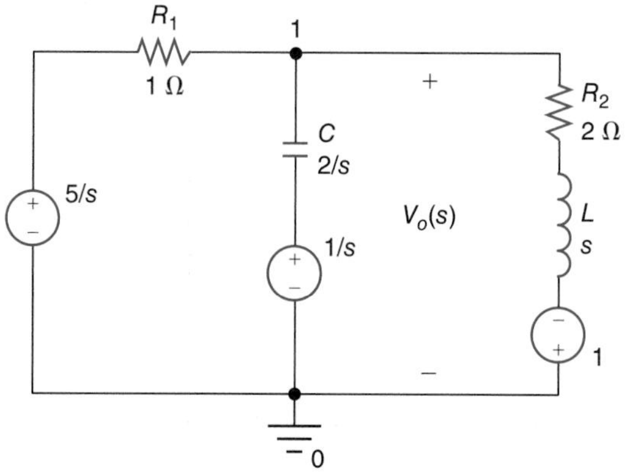

V_o can be expanded in a partial fraction expansion as

$$V_o = \frac{s^2 + 10s + 20}{s(s^2 + 4s + 6)} = \frac{A}{s} + \frac{Bs + C}{s^2 + 4s + 6}$$

$$= \frac{(A+B)s^2 + (4A + C)s + 6A}{s(s^2 + 4s + 6)}$$

where

$$A = \left.\frac{s^2 + 10s + 20}{(s^2 + 4s + 6)}\right|_{s=0} = \frac{20}{6} = \frac{10}{3} = 3.3333$$

$$A + B = 1, \quad B = 1 - A = -7/3 = -2.3333$$

$$4A + C = 10, \quad C = 10 - 4A = -10/3 = -3.3333$$

continued

Example 15.4 continued

Thus, we have

$$V_o = \frac{\frac{10}{3}}{s} + \frac{-\frac{7}{3}s - \frac{10}{3}}{s^2 + 4s + 6} = \frac{\frac{10}{3}}{s} + \frac{-\frac{7}{3}(s+2) + \frac{14}{3} - \frac{10}{3}}{(s+2)^2 + (\sqrt{2})^2}$$

$$= \frac{\frac{10}{3}}{s} + \frac{-\frac{7}{3}(s+2) + \frac{4}{3\sqrt{2}}\sqrt{2}}{(s+2)^2 + (\sqrt{2})^2}$$

Taking the inverse Laplace transform, we obtain

$$v_o(t) = \left[\frac{10}{3} - \frac{7}{3}e^{-2t}\cos(\sqrt{2}t) + \frac{4}{3\sqrt{2}}e^{-2t}\sin(\sqrt{2}t)\right]u(t)$$

or

$$v_o(t) = [3.3333 - 2.3333e^{-2t}\cos(1.4142t) + 0.9428e^{-2t}\sin(1.4142t)]u(t)$$

Notice that in the steady state ($t = \infty$), the capacitor can be treated as an open circuit, and the inductor can be treated as a short circuit. The current through R_1 and R_2 is 5/3 A, and the voltage across R_2, which is also the voltage across C, is $2 \times 5/3 = 10/3$ V = 3.3333 V.

MATLAB

```
% EXAMPLE 15.4
clear all;
R1=1;R2=2;C=0.5;L=1;v0=1;i0=1;
syms s Vo
Vo=solve((Vo-5/s)/(R1)+(Vo-v0/s)*(s*C)+(Vo+1)/(R2+s*L),Vo)
[Num,Den]=numden(Vo)
N1=sym2poly(Num)
D1=sym2poly(Den)
N=N1/D1(1)
D=D1/D1(1)
vo=ilaplace(Vo)
vo=vpa(vo,5)
vo=vpa(expand(vo),7)
[R,P,Q]=residue(N,D)
coeff=2*R(2)
```

Exercise 15.4

Let $v(0^-) = 2$ V and $i(0^-) = 1$ A in the circuit shown in Figure 15.18. Find the voltage $v_o(t)$ across the capacitor for $t \geq 0$ using nodal analysis.

FIGURE 15.18

Circuit for EXERCISE 15.4.

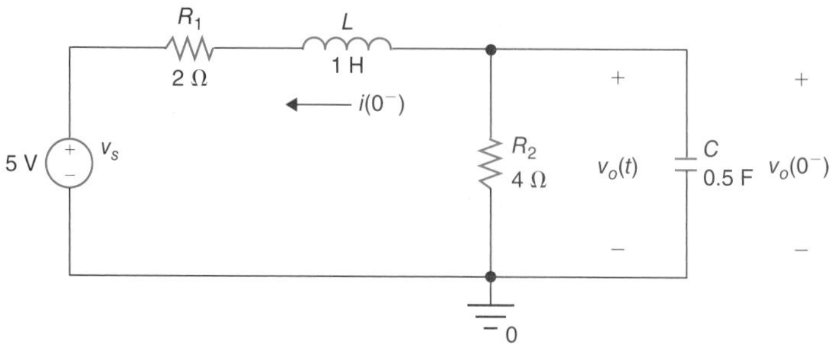

continued

Exercise 15.4 continued

Answer:

$v_o(t) = [3.3333 - 1.3333e^{-1.25t}\cos(1.199t) - 3.8923e^{-1.25t}\sin(1.199t)]\,u(t).$

EXAMPLE 15.5

The switch in the circuit shown in Figure 15.19 has been opened for a long time before it is closed at $t = 0$. Find the voltage $v_o(t)$ across R_3 for $t \geq 0$.

FIGURE 15.19

Circuit for EXAMPLE 15.5.

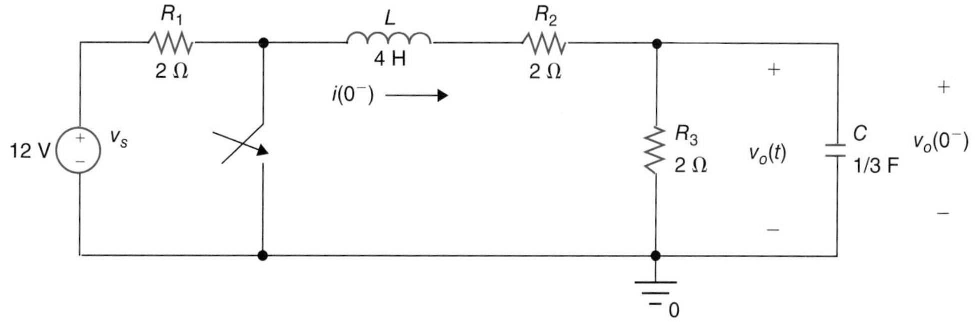

Before the switch is closed, the circuit is in the steady state. The inductor can be treated as a short circuit, and the capacitor can be treated as an open circuit. For $t < 0$, the circuit consists of the voltage source and a series connection of R_1, R_2, and R_3. The current through the mesh is $i(0^-) = 12\text{ V}/6\ \Omega = 2\text{ A}$. The initial voltage across the capacitor, which is the voltage across R_3, is given by $v(0^-) = 2\text{ A} \times 2\ \Omega = 4\text{ V}$. The circuit in the s-domain is shown in Figure 15.20.

FIGURE 15.20

Circuit in the s-domain.

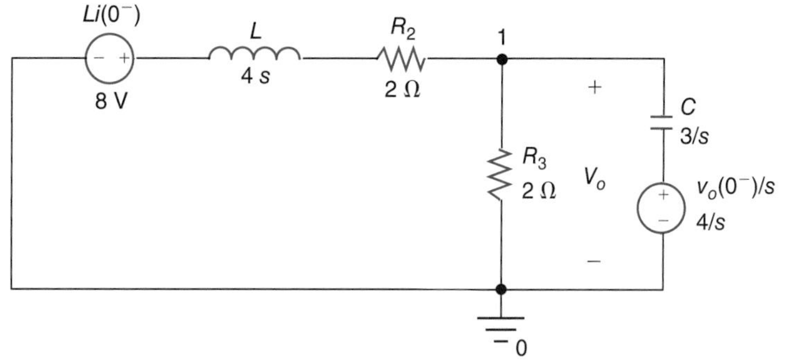

Summing the currents leaving node 1, we obtain

$$\frac{V_o - 8}{4s + 2} + \frac{V_o}{2} + \frac{V_o - \frac{4}{s}}{\frac{3}{s}} = 0$$

which can be rearranged as

$$V_o\left(\frac{1}{4s + 2} + \frac{1}{2} + \frac{s}{3}\right) = \frac{8}{4s + 2} + \frac{4}{3} = \frac{8 \times 3 + 4(4s + 2)}{3(4s + 2)} = \frac{16s + 32}{3(4s + 2)}$$

continued

Example 15.5 continued

Solving for V_o, we obtain

$$V_o = \frac{1}{\frac{1}{4s+2} + \frac{1}{2} + \frac{s}{3}} \times \frac{16s+32}{3(4s+2)} = \frac{1}{\frac{6+3(4s+2)+2s(4s+2)}{6(4s+2)}} \times \frac{16s+32}{3(4s+2)}$$

$$= \frac{2(16s+32)}{8s^2+16s+12} = \frac{4(s+2)}{s^2+2s+1.5} = \frac{4(s+1)+4\sqrt{2}\frac{1}{\sqrt{2}}}{(s+1)^2+\left(\frac{1}{\sqrt{2}}\right)^2}$$

$$= 4\frac{(s+1)}{(s+1)^2+\left(\frac{1}{\sqrt{2}}\right)^2} + 4\sqrt{2}\frac{\frac{1}{\sqrt{2}}}{(s+1)^2+\left(\frac{1}{\sqrt{2}}\right)^2}$$

Taking the inverse Laplace transform, we obtain

$$v_o(t) = 4\left[e^{-t}\cos\left(\frac{1}{\sqrt{2}}t\right) + \sqrt{2}e^{-t}\sin\left(\frac{1}{\sqrt{2}}t\right)\right]u(t)$$

Exercise 15.5

The switch in the circuit shown in Figure 15.21 has been opened for a long time before it is closed at $t = 0$. Find the voltage $v_o(t)$ across R_2 for $t \geq 0$.

FIGURE 15.21

Circuit for EXERCISE 15.5.

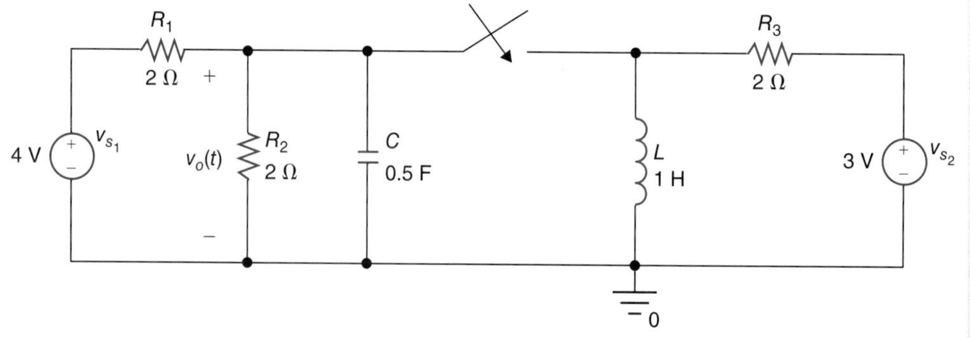

Answer:

$v_o(t) = 2e^{-t}u(t)$.

EXAMPLE 15.6

The initial voltage across the capacitor in the circuit shown in Figure 15.22 is $v(0^-) = 1$ V. The input signal is $v_s(t) = tu(t)$. Find the voltage $v(t)$ across the capacitor for $t \geq 0$.

The circuit in the s-domain is shown in Figure 15.23.

continued

Example 15.6 continued

FIGURE 15.22

RC circuit with ramp input.

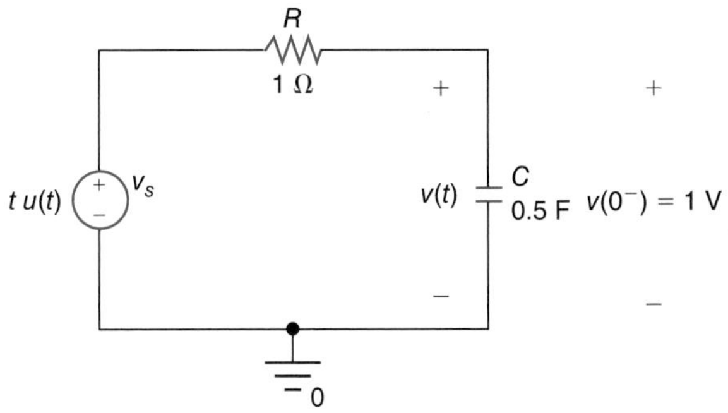

FIGURE 15.23

Circuit in the *s*-domain.

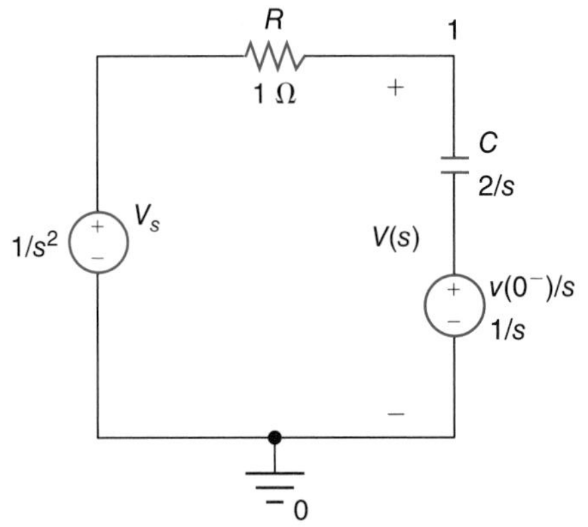

Summing the currents leaving node 1, we obtain

$$\frac{V-\frac{1}{s^2}}{1}+\frac{V-\frac{1}{s}}{\frac{2}{s}}=0$$

which can be rearranged as

$$\left(1+\frac{s}{2}\right)V=\frac{1}{s^2}+\frac{1}{2}=\frac{s^2+2}{2s^2}$$

Solving for V, we obtain

$$V=\frac{\frac{s^2+2}{2s^2}}{\frac{s+2}{2}}=\frac{s^2+2}{s^2(s+2)}=\frac{A}{s^2}+\frac{B}{s}+\frac{C}{s+2}$$

where

$$A=\left.\frac{s^2+2}{s+2}\right|_{s=0}=1,\quad C=\left.\frac{s^2+2}{s^2}\right|_{s=-2}=1.5,$$

$$B=\left.\frac{d}{ds}\frac{s^2+2}{s+2}\right|_{s=0}=\left.\frac{d}{ds}(s^2+2)(s+2)^{-1}\right|_{s=0}$$

$$=\left.(2s)(s+2)^{-1}-(s^2+2)(s+2)^{-2}\right|_{s=0}=-0.5$$

$$v(t)=(t-0.5+1.5e^{-2t})\,u(t)$$

At $t=\infty$, after the transient response $1.5e^{-2t}u(t)$ vanishes, there is a steady-state error of 0.5 V between the input and the output voltage.

Exercise 15.6

The initial current through the inductor in the circuit shown in Figure 15.24 is $i(0^-)=1$ A. The input signal is $v_s(t)=5e^{-3t}\,u(t)$. Find the voltage $v(t)$ across R for $t\geq 0$.

FIGURE 15.24

RL circuit with exponential input.

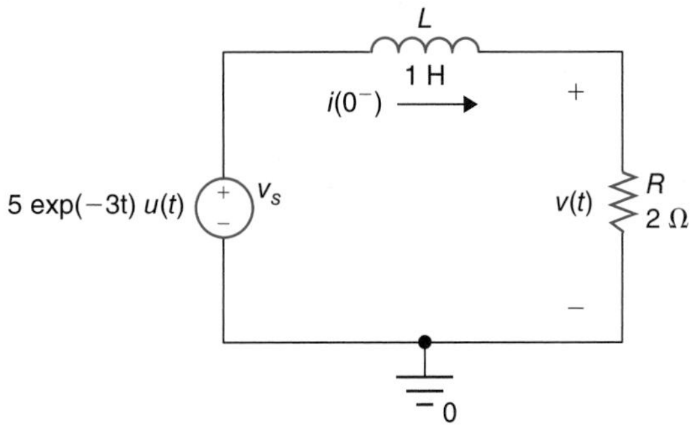

Answer:
$v(t)=(12e^{-2t}-10e^{-3t})u(t)$.

EXAMPLE 15.7

The initial voltage across the capacitor in the circuit shown in Figure 15.25 is $v_o(0^-) = 1$ V. Find the voltage $v_o(t)$ across the capacitor for $t \geq 0$.

FIGURE 15.25

Circuit for EXAMPLE 15.7.

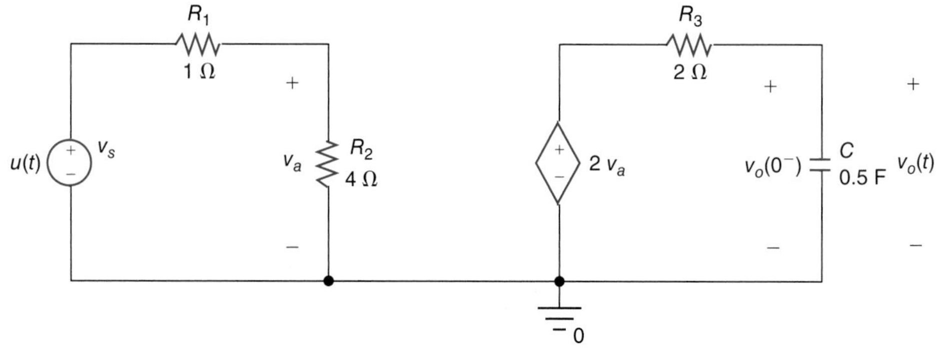

The circuit in the s-domain is shown in Figure 15.26.

FIGURE 15.26

Circuit in the s-domain.

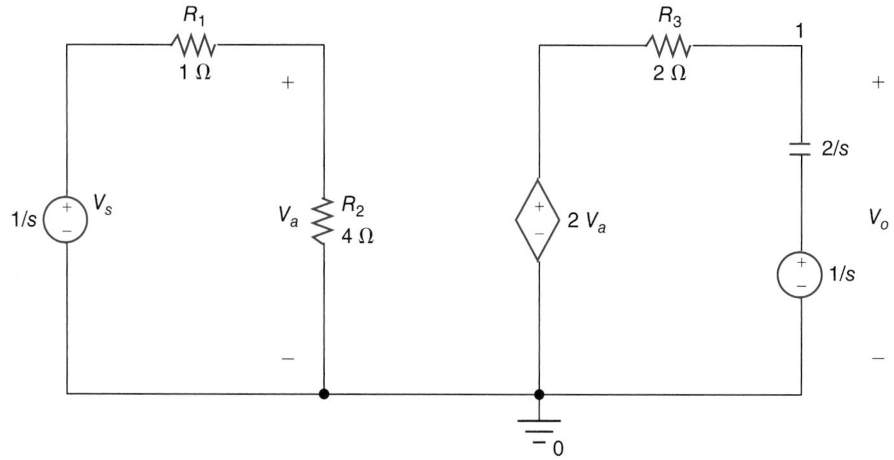

From the voltage divider rule, the voltage v_a across R_2 is given by

$$V_a = \frac{R_2}{R_1 + R_2} \times V_s = \frac{4}{1+4} \times \frac{1}{s} = \frac{4}{5s}$$

Summing the currents leaving node 1, we obtain

$$\frac{V_o - 2 \times \dfrac{4}{5s}}{2} + \frac{V_o - \dfrac{1}{s}}{\dfrac{2}{s}} = 0$$

which can be rearranged as

$$\left(\frac{1}{2} + \frac{s}{2}\right)V_o = \frac{4}{5s} + \frac{1}{2} = \frac{8 + 5s}{10s}$$

continued

Example 15.7 continued

Solving for V_o, we get

$$V_o = \frac{\dfrac{8+5s}{10s}}{\dfrac{s+1}{2}} = \frac{s+1.6}{s(s+1)} = \frac{A}{s} + \frac{B}{s+1}$$

where

$$A = \left.\frac{s+1.6}{s+1}\right|_{s=0} = 1.6, \quad B = \left.\frac{s+1.6}{s}\right|_{s=-1} = -0.6$$

Taking the inverse Laplace transform, we obtain

$$v_o(t) = (1.6 - 0.6e^{-t})u(t)$$

Exercise 15.7

The initial voltage across the capacitor in the circuit shown in Figure 15.27 is $v_o(0^-) = 1$ V. Find the voltage $v_o(t)$ across the capacitor for $t \geq 0$.

FIGURE 15.27

Circuit for EXERCISE 15.7.

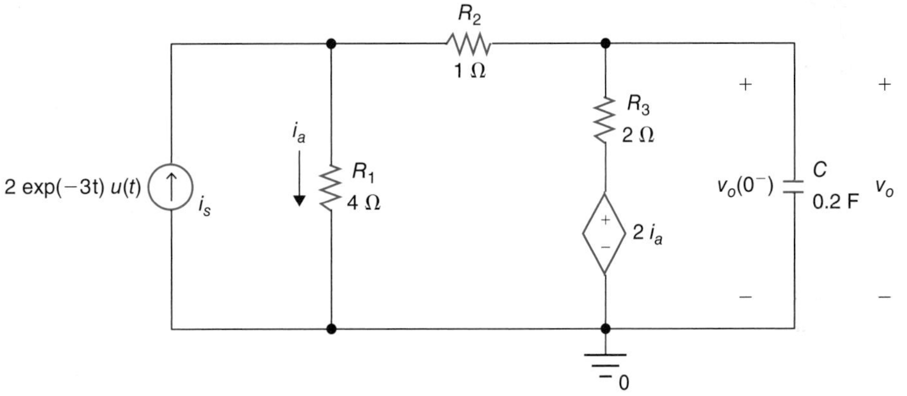

Answer:
$v_o(t) = (21e^{-2.5t} - 20e^{-3t})u(t)$.

15.5 Mesh Analysis

Mesh analysis provides mesh currents at all meshes of the given circuit. From the mesh currents, we can find the currents on every branch of the circuit. If a mesh contains a current source, the mesh current is the same as the current from the current source if they point the same direction. If the direction is opposite, the mesh current is the negative of the current from the current source. The mesh analysis applies to the Laplace-transformed circuit. In the Laplace-transformed circuit, all the elements in the circuit are represented in the s-domain. Inputs and outputs are represented by Laplace transforms in the s-domain. The time domain expressions are obtained by taking the inverse Laplace transforms of the s-domain expressions. This method can be used to find the response of a circuit for arbitrary input as long as the Laplace integral of the input signal converges.

The mesh analysis is based on Kirchhoff's voltage law (KVL). We assign mesh current variables such as $I_1, I_2, I_3, \ldots, I_n$ on the meshes whose current is unknown. Then, for each mesh with unknown mesh current, we apply Kirchhoff's voltage law (KVL). Specifically, we sum the voltage drops around the mesh and let that equal to zero.

If there are n unknown mesh currents, we get n equations in n unknowns. Thus, we can solve these n system of linear equations to find the unique solution for the unknown mesh currents $I_1, I_2, I_3, \ldots, I_n$. A simple algorithm to find the solution of the n system of linear equations is Cramer's rule. MATLAB is useful in finding the solution of n system of linear equations.

EXAMPLE 15.8

Let $v(0^-) = 1$ V in the circuit shown in Figure 15.28. Find the voltage $v(t)$ across the capacitor for $t \geq 0$ using mesh analysis.

FIGURE 15.28

Circuit for EXAMPLE 15.8.

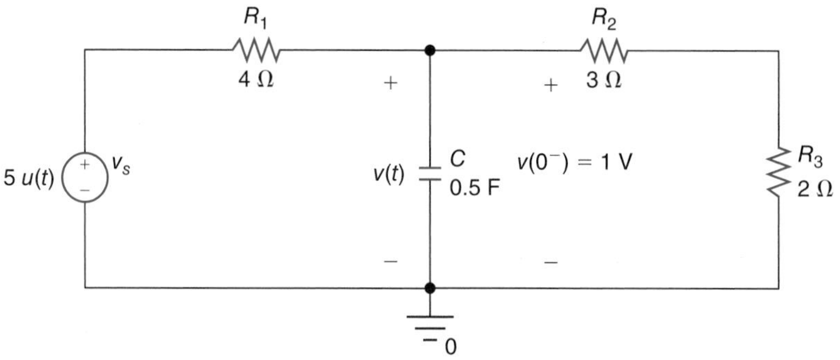

The Laplace-transformed circuit is shown in Figure 15.29.

FIGURE 15.29

Circuit in the s-domain.

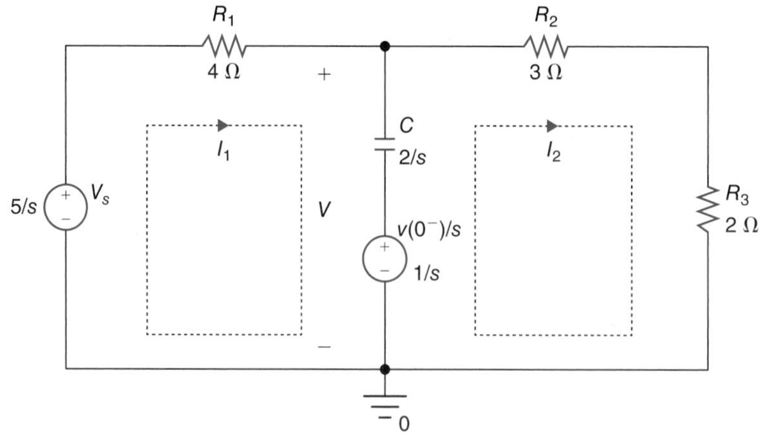

Collecting the voltage drops around mesh 1 in the clockwise direction, we obtain

$$-\frac{5}{s} + 4I_1 + \frac{2}{s}(I_1 - I_2) + \frac{1}{s} = 0$$

continued

Example 15.8 continued

Multiplication by s yields

$$-5 + 4sI_1 + 2I_1 - 2I_2 + 1 = 0$$

which can be rewritten as

$$(4s + 2)I_1 - 2I_2 = 4 \tag{15.33}$$

Collecting the voltage drops around mesh 2 in the clockwise direction, we obtain

$$-\frac{1}{s} + \frac{2}{s}(I_2 - I_1) + 3I_2 + 2I_2 = 0$$

Multiplication by s yields

$$-1 + 2I_2 - 2I_1 + 3sI_2 + 2sI_2 = 0$$

which can be rewritten as

$$-2I_1 + (5s + 2)I_2 = 1 \tag{15.34}$$

Application of Cramer's rule to Equations (15.33) and (15.34) yields

$$I_1 = \frac{\begin{vmatrix} 4 & -2 \\ 1 & 5s+2 \end{vmatrix}}{\begin{vmatrix} 4s+2 & -2 \\ -2 & 5s+2 \end{vmatrix}} = \frac{20s + 10}{20s^2 + 18s} = \frac{s + 0.5}{s(s + 0.9)}$$

$$I_2 = \frac{\begin{vmatrix} 4s+2 & 4 \\ -2 & 1 \end{vmatrix}}{\begin{vmatrix} 4s+2 & -2 \\ -2 & 5s+2 \end{vmatrix}} = \frac{4s + 10}{20s^2 + 18s} = \frac{0.2s + 0.5}{s(s + 0.9)}$$

The Laplace transform of the voltage $v(t)$ is given by

$$V = \frac{1}{s} + \frac{2}{s}(I_1 - I_2) = \frac{s + 2.5}{s(s + 0.9)} = \frac{A}{s} + \frac{B}{s + 0.9}$$

where the coefficients are given by

$$A = \left.\frac{s + 2.5}{(s + 0.9)}\right|_{s=0} = \frac{2.5}{0.9} = 2.7778, \quad B = \left.\frac{s + 2.5}{s}\right|_{s=-0.9} = \frac{1.6}{-0.9} = -1.7778$$

Thus, the voltage $v(t)$ is

$$v(t) = (2.7778 - 1.7778e^{-0.9t})\, u(t) \text{ V}$$

Exercise 15.8

Let $i(0^-) = 1$ A in the circuit shown in Figure 15.30. Find the voltage $v(t)$ across the inductor for $t \geq 0$.

FIGURE 15.30

Circuit for EXERCISE 15.8.

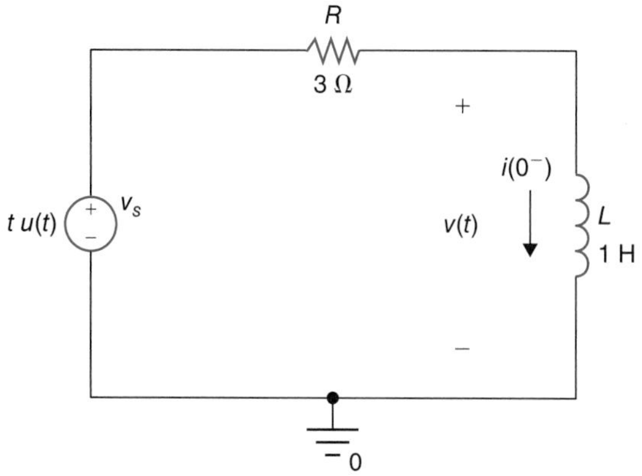

Answer:

$$v(t) = \left(\frac{1}{3} - \frac{10}{3}e^{-3t}\right)u(t)\,\text{V} = (0.3333 - 3.3333e^{-3t})u(t)\,\text{V}.$$

EXAMPLE 15.9

The switch in the circuit shown in Figure 15.31 has been closed for a long time before it is opened at $t = 0$. Find the voltage $v_o(t)$ across the capacitor for $t \geq 0$.

FIGURE 15.31

Circuit for EXAMPLE 15.9.

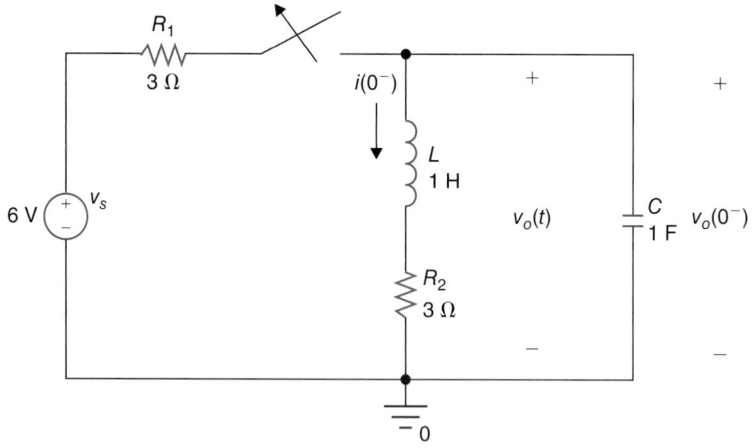

Since the circuit is in the steady state at $t = 0^-$, the inductor can be treated as a short circuit, and the capacitor can be treated as an open circuit. The circuit consists of the voltage source, R_1, and R_2. The current through R_2 is $i(0^-) = 6\,\text{V}/6\,\Omega = 1$ A. The voltage across the capacitor is the voltage across R_2. Thus, we have $v_o(0^-) = 1\,\text{A} \times 3\,\Omega = 3$ V. The Laplace-transformed circuit is shown in Figure 15.32.

continued

Example 15.9 continued

FIGURE 15.32

Circuit in the *s*-domain.

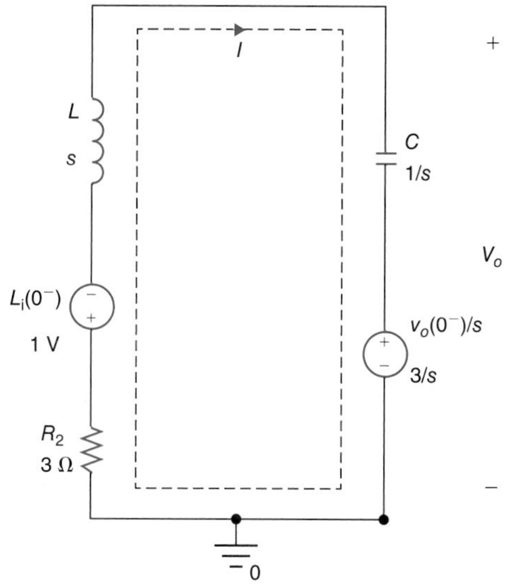

Collecting the voltage drops around mesh in the clockwise direction, we obtain

$$3I + 1 + sI + \frac{1}{s}I + \frac{3}{s} = 0$$

Multiplication by s yields

$$3sI + s + s^2I + I + 3 = 0$$

which can be rewritten as

$$(s^2 + 3s + 1)I = -s - 3$$

Solving for I, we obtain

$$I = \frac{-s - 3}{s^2 + 3s + 1}$$

The voltage across the capacitor is given by

$$V_o = \frac{1}{s}I + \frac{3}{s} = \frac{1}{s} \times \frac{-s-3}{s^2+3s+1} + \frac{3}{s} = \frac{-s - 3 + 3s^2 + 9s + 3}{s(s^2 + 3s + 1)} = \frac{3s + 8}{s^2 + 3s + 1}$$

$$= \frac{3.0652}{s + 0.382} - \frac{0.0652}{s + 2.618}$$

Thus, the voltage $v_o(t)$ is

$$v_o(t) = (3.0652e^{-0.382t} - 0.0652e^{-2.618t})\, u(t)\text{ V}$$

Exercise 15.9

The switch in the circuit shown in Figure 15.33 has been closed for a long time before it is opened at $t = 0$. Find the initial conditions $i(0^-)$ and $v(0^-)$ and the voltage $v_o(t)$ across the capacitor for $t \geq 0$.

FIGURE 15.33

Circuit for EXERCISE 15.9.

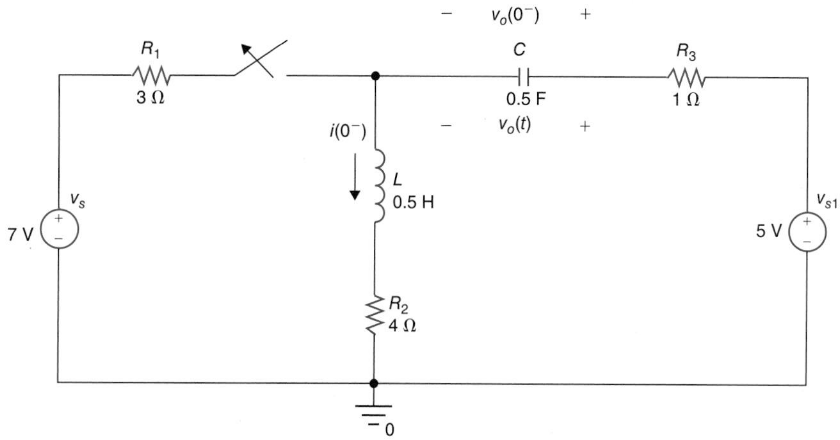

Answer:
$i(0^-) = 1$ A, $v(0^-) = 1$ V, $v_o(t) = (5 - 0.036e^{-9.5826t} - 3.964e^{-0.4174t})u(t)$ V.

EXAMPLE 15.10

Let $v(0^-) = 2$ V and $i(0^-) = 1$ A in the circuit shown in Figure 15.34. Find the voltage $v_o(t)$ across the inductor for $t \geq 0$ using mesh analysis.

FIGURE 15.34

Circuit for EXAMPLE 15.10.

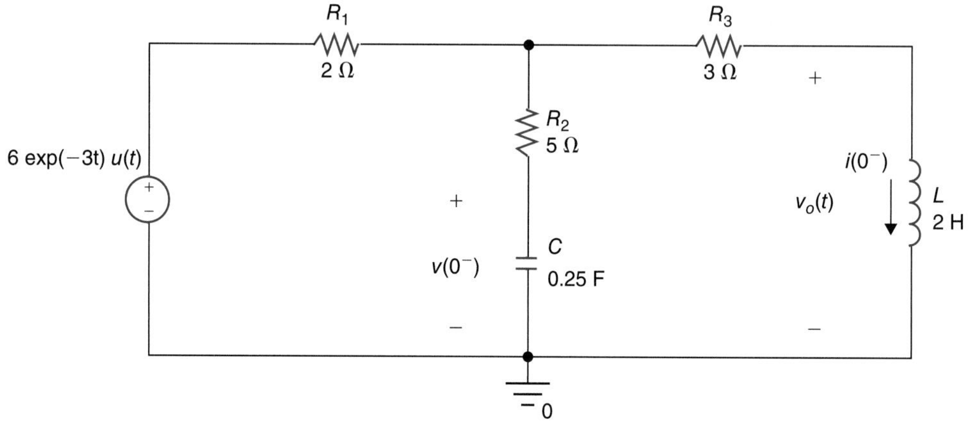

The circuit in the s-domain is shown in Figure 15.35.

FIGURE 15.35

Circuit in the s-domain.

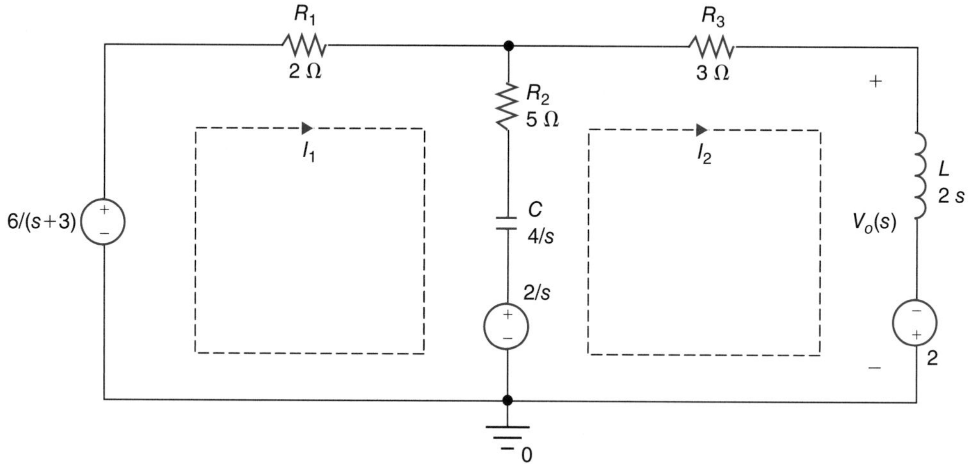

Summing the voltages drops around mesh 1 in the clockwise direction, we obtain

$$-\frac{6}{s+3} + 2I_1 + 5(I_1 - I_2) + \frac{4}{s}(I_1 - I_2) + \frac{2}{s} = 0$$

Multiplication by $s(s + 3)$ yields

$$-6s + (2s^2 + 6s)I_1 + (5s^2 + 15s)(I_1 - I_2) + (4s + 12)(I_1 - I_2) + 2s + 6 = 0$$

which can be rearranged as

$$(7s^2 + 25s + 12)I_1 + (-5s^2 - 19s - 12)I_2 = 4s - 6 \tag{15.35}$$

continued

Example 15.10 continued

Summing the voltages drops around mesh 2 in the clockwise direction, we obtain

$$-\frac{2}{s} + \frac{4}{s}(I_2 - I_1) + 5(I_2 - I_1) + 3I_2 + 2sI_2 - 2 = 0$$

Multiplication by s yields

$$-2 + 4(I_2 - I_1) + 5s(I_2 - I_1) + 3sI_2 + 2s^2I_2 - 2s = 0$$

which can be rearranged as

$$(-5s - 4)I_1 + (2s^2 + 8s + 4)I_2 = 2s + 2 \qquad \textbf{(15.36)}$$

Applying Cramer's rule to Equations (15.35) and (15.36), we can find I_2.

$$I_2 = \frac{\begin{vmatrix} 7s^2 + 25s + 12 & 4s - 6 \\ -5s - 4 & 2s + 2 \end{vmatrix}}{\begin{vmatrix} 7s^2 + 25s + 12 & -5s^2 - 19s - 12 \\ -5s - 4 & 2s^2 + 8s + 4 \end{vmatrix}}$$

$$= \frac{14s^3 + 64s^2 + 74s + 24 + 20s^2 - 14s - 24}{14s^4 + 106s^3 + 252s^2 + 196s + 48 - 25s^3 - 115s^2 - 136s - 48}$$

$$= \frac{14s^3 + 84s^2 + 60s}{14s^4 + 81s^3 + 137s^2 + 60s} = \frac{14s^2 + 84s + 60}{14s^3 + 81s^2 + 137s + 60}$$

The voltage across L is given by

$$V_o = 2sI_2 - 2 = 2s\frac{14s^2 + 84s + 60}{14s^3 + 81s^2 + 137s + 60} - 2\frac{14s^3 + 81s^2 + 137s + 60}{14s^3 + 81s^2 + 137s + 60}$$

$$= \frac{6s^2 - 154s - 120}{14s^3 + 81s^2 + 137s + 60}$$

```
>>b=[6 -154 -120];a=[14 81 137 60];
>> [R,P,Q]=residue(b,a)
R =
   13.6552
  -12.9496
   -0.2770
P =
   -3.0000
   -2.1080
   -0.6777
Q =
     []
```

Therefore, $v_o(t)$ is given by

$$v_o(t) = (13.6552e^{-3t} - 12.9496e^{-2.1080t} - 0.2770e^{-0.6777t})u(t)$$

continued

Example 15.10 continued

MATLAB

```
% EXAMPLE 15.10
clear all;
R1=2;R2=5;R3=3;C=0.25;L=2;v0=2;i0=1;
syms s I1 I2 Vo
[I1,I2]=solve(-6/(s+3)+R1*I1+R2*(I1-I2)+1/(s*C)*(I1-I2)+v0/s,...
-v0/s+1/(s*C)*(I2-I1)+R2*(I2-I1)+R3*I2+s*L*I2-L*i0,I1,I2)
Vo=s*L*I2-L*i0
[Num,Den]=numden(Vo)
N1=sym2poly(Num)
D1=sym2poly(Den)
N=N1/D1(1)
D=D1/D1(1)
syms t
vo=ilaplace(Vo)
vo=vpa(vo,7)
[R,P,Q]=residue(N,D)
```

Exercise 15.10

Let $v(0^-) = 1$ V in the circuit shown in Figure 15.36. Find the voltage $v(t)$ across the capacitor for $t \geq 0$.

FIGURE 15.36

Circuit for EXERCISE 15.10.

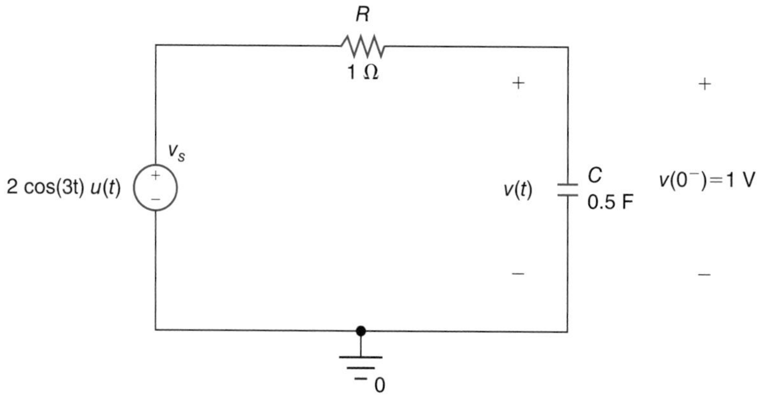

Answer:
$v(t) = [0.38462\, e^{-2t} + 0.61538 \cos(3t) + 0.92308 \sin(3t)]\, u(t)$ V.

EXAMPLE 15.11

The initial voltage across the capacitor in the circuit shown in Figure 15.37 is given by $v(0^-) = 2$ V. Find the voltage $v_o(t)$ across the capacitor for $t \geq 0$.

continued

Example 15.11 continued

FIGURE 15.37

Circuit for EXAMPLE 15.11.

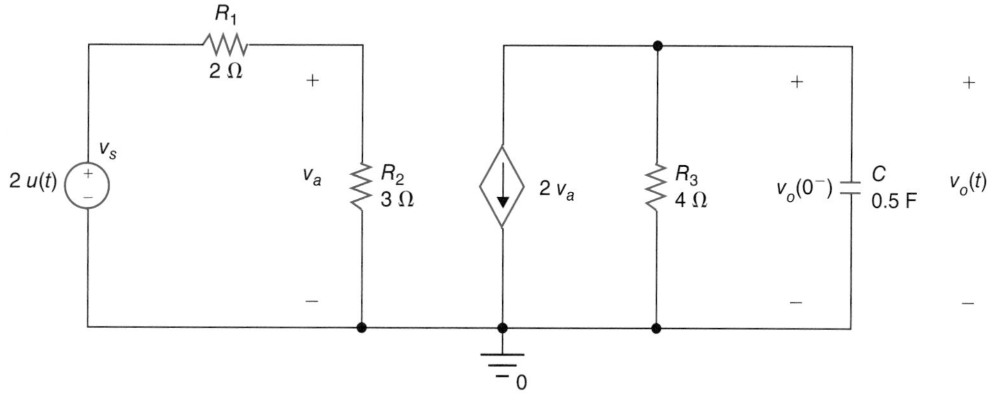

From the source transformation, the VCCS and a parallel resistor R_3 can be transformed to a VCVS and a series resistor R_3. The circuit in the s-domain is shown in Figure 15.38.

FIGURE 15.38

Circuit in the s-domain.

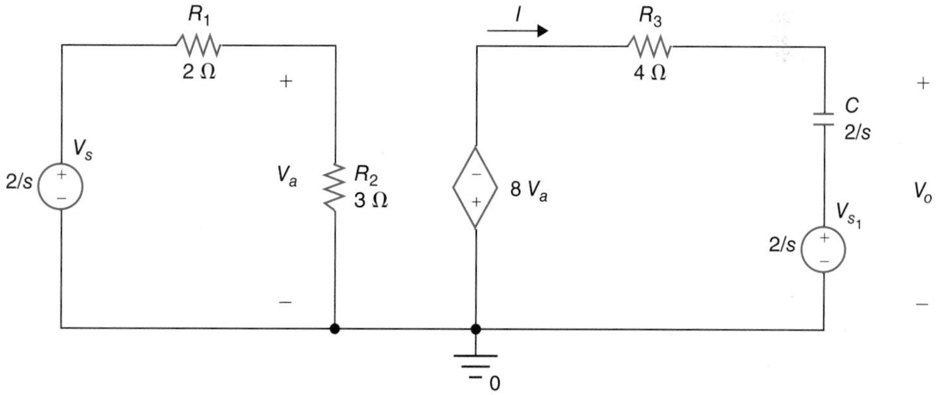

From the voltage divider rule, the voltage across R_2 is given by

$$V_a = \frac{R_2}{R_1 + R_2} \times \frac{2}{s} = \frac{3}{2+3} \times \frac{2}{s} = \frac{6}{5s} = \frac{1.2}{s}$$

Summing the voltage drops around the mesh on the right side, we obtain

$$8 \times \frac{1.2}{s} + 4I + \frac{2}{s}I + \frac{2}{s} = 0$$

Solving for I, we obtain

$$I = \frac{-\frac{9.6}{s} - \frac{2}{s}}{4 + \frac{2}{s}} = \frac{-11.6}{4s+2} = -\frac{2.9}{s+0.5}$$

The voltage V_o is given by

$$V_o = \frac{2}{s}I + \frac{2}{s} = \frac{2}{s}\left(-\frac{2.9}{s+0.5}\right) + \frac{2}{s} = \frac{2s - 4.8}{s(s+0.5)}$$

continued

Example 15.11 continued

Expanding V_o by a partial fraction, we get

$$V_o = \frac{2s - 4.8}{s(s + 0.5)} = \frac{A}{s} + \frac{B}{s + 0.5}$$

where

$$A = \left.\frac{2s - 4.8}{s + 0.5}\right|_{s=0} = -9.6, \quad B = \left.\frac{2s - 4.8}{s}\right|_{s=-0.5} = 11.6$$

Thus, the voltage across the capacitor is given by

$$v_o(t) = (-9.6 + 11.6e^{-0.5t})u(t)\ V$$

Exercise 15.11

The initial voltage across the capacitor in the circuit shown in Figure 15.39 is given by $v(0^-) = 1$ V. Find the voltage $v_o(t)$ for $t \geq 0$ using mesh analysis.

FIGURE 15.39

Circuit for EXERCISE 15.11.

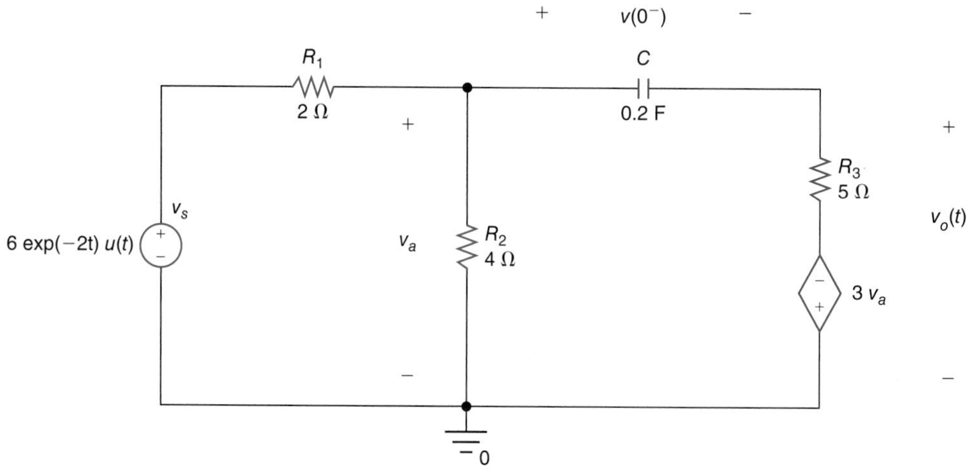

Answer:
$v_o(t) = (6.383e^{-2t} - 5.3185e^{-0.4839t})u(t)$ V.

FIGURE 15.40

A circuit with terminals a and b.

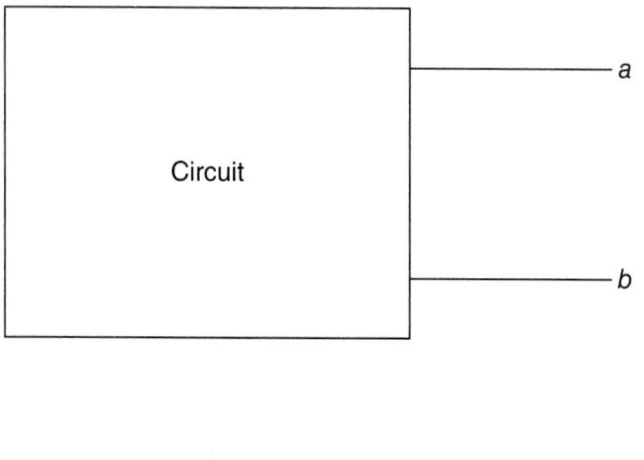

15.6 Thévenin Equivalent Circuit in the s-Domain

Figure 15.40 shows a circuit with a pair of terminals a and b. If the circuit consists of resistors and sources, as shown in Chapter 4, the circuit can be represented by a voltage source in series with a resistor. If the circuit is a Laplace-transformed circuit consisting of impedances and sources in the s-domain, the circuit can be represented by a source in series with an impedance, as shown in Figure 15.41.

FIGURE 15.41

Thévenin equivalent circuit.

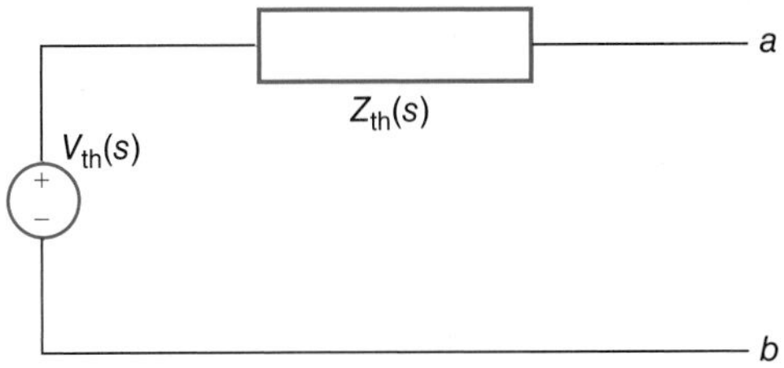

In Figure 15.41, $V_{th}(s)$ is the Thévenin equivalent voltage, and $Z_{th}(s)$ is the Thévenin equivalent impedance. The Thévenin equivalent voltage $V_{th}(s)$ is obtained by finding the open-circuit voltage between a and b from the original circuit, as shown in Figure 15.42.

The Thévenin equivalent impedance $Z_{th}(s)$ is found by applying one of the following three methods.

(a) Find the short-circuit current $I_{sc}(s)$ between a and b, as shown in Figure 15.43. The Thévenin equivalent impedance is given by

$$Z_{th}(s) = \frac{V_{th}(s)}{I_{sc}(s)}$$

(b) Deactivate the independent sources in the circuit by short-circuiting the voltage sources and open-circuiting the current sources. Then, find the equivalent impedance seen from terminals a and b.

(c) Deactivate the independent sources in the circuit by short-circuiting the voltage sources and open-circuiting the current sources. Apply a test voltage $V(s)$ between the terminals a and b, as shown in Figure 15.44. Find the current $I(s)$ flowing into the circuit from the positive terminal of the test voltage source $V(s)$. Then, the Thévenin equivalent impedance is given by

$$Z_{th}(s) = \frac{V(s)}{I(s)}$$

The test voltage can be $V(s) = 1$ $(v(t) = \delta(t))$, $V(s) = 1/s$ $(v(t) = u(t))$ or any other signal. A test current can be used instead of the text voltage.

FIGURE 15.42

Open-circuit voltage.

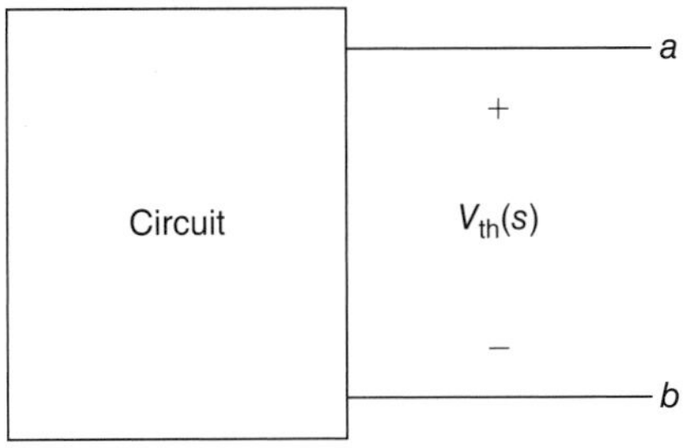

FIGURE 15.43

Short-circuit current.

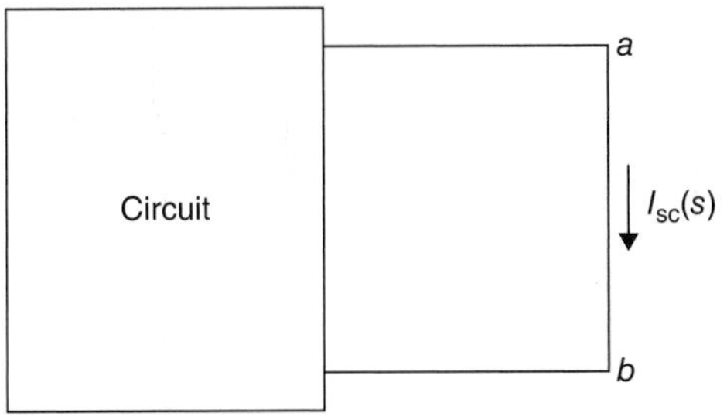

FIGURE 15.44

A test voltage.

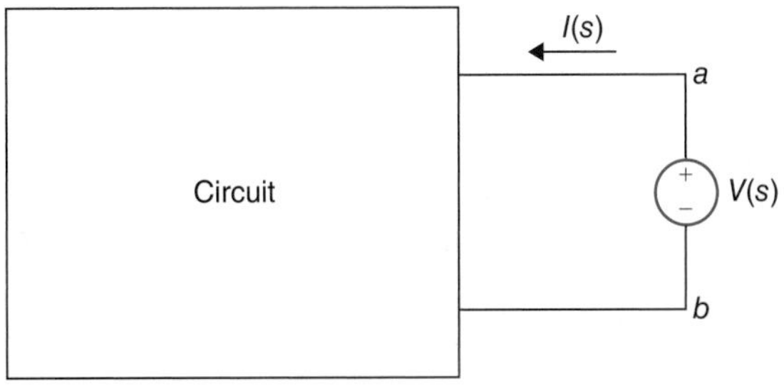

EXAMPLE 15.12

The initial current through the inductor is $i(0^-) = 2$ A in the circuit shown in Figure 15.45. Find the Thévenin equivalent voltage and the Thévenin equivalent impedance for the circuit shown in Figure 15.45 between terminals a and b.

continued

Example 15.12 continued

FIGURE 15.45

Circuit for EXAMPLE 15.12.

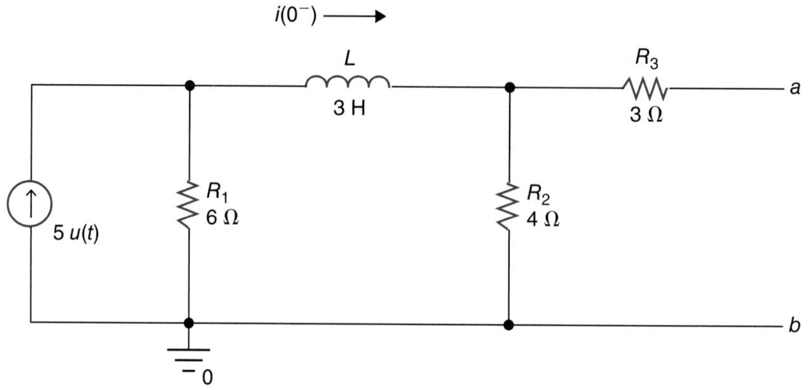

The s-domain equivalent circuit is shown in Figure 15.46.

FIGURE 15.46

Circuit in the s-domain.

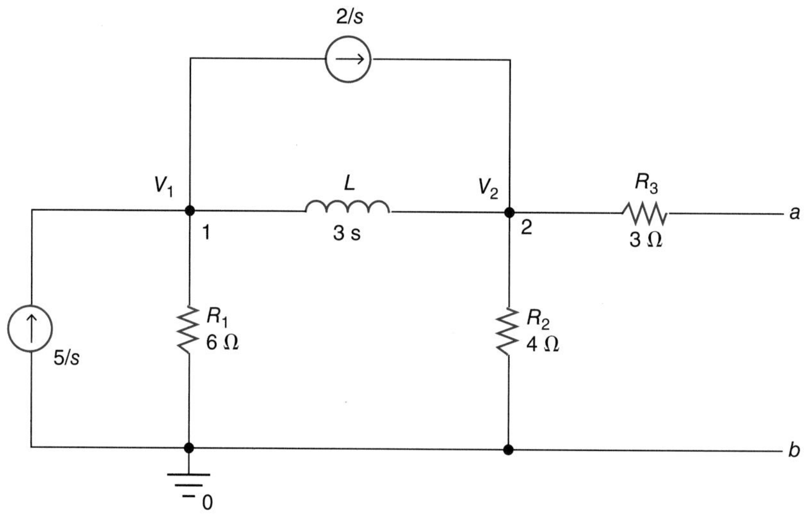

Summing the currents leaving node 1, we obtain

$$-\frac{5}{s} + \frac{2}{s} + \frac{V_1}{6} + \frac{V_1 - V_2}{3s} = 0$$

Multiplication by $6s$ yields

$$-30 + 12 + sV_1 + 2V_1 - 2V_2 = 0$$

which can be rearranged as

$$(s + 2)V_1 - 2V_2 = 18 \tag{15.37}$$

Summing the currents leaving node 2, we obtain

$$-\frac{2}{s} + \frac{V_2 - V_1}{3s} + \frac{V_2}{4} = 0$$

continued

Example 15.12 continued

Multiplication by 12s yields

$$-24 + 4V_2 - 4V_1 + 3sV_2 = 0$$

which can be rearranged as

$$-4V_1 + (3s + 4)V_2 = 24 \quad \textbf{(15.38)}$$

The voltage V_2 can be found using Cramer's rule:

$$V_2 = \frac{\begin{vmatrix} s+2 & 18 \\ -4 & 24 \end{vmatrix}}{\begin{vmatrix} s+2 & -2 \\ -4 & 3s+4 \end{vmatrix}} = \frac{24s + 48 + 72}{3s^2 + 10s + 8 - 8} = \frac{24s + 120}{3s^2 + 10s} = \frac{8s + 40}{s^2 + \frac{10}{3}s} = \frac{8s + 40}{s\left(s + \frac{10}{3}\right)}$$

$$= \frac{8s + 40}{s(s + 3.3333)}$$

V_2 is the Thévenin voltage V_{th}. Thus, we have

$$V_{th} = \frac{24s + 120}{3s^2 + 10s} = \frac{8s + 40}{s^2 + \frac{10}{3}s} = \frac{8s + 40}{s\left(s + \frac{10}{3}\right)} = \frac{8s + 40}{s(s + 3.3333)}$$

The short-circuit current I between a and b is found by writing node equations at nodes 1 and 2 in the circuit shown in Figure 15.47. Since the difference between the circuit shown in Figures 15.46 and 15.47 is the inclusion of R_3, the node equation at node 2 is modified as

$$-\frac{2}{s} + \frac{V_2 - V_1}{3s} + \frac{V_2}{4} + \frac{V_2}{3} = 0$$

Multiplication by 12s yields

$$-24 + 4V_2 - 4V_1 + 3sV_2 + 4sV_2 = 0$$

which can be rearranged as

$$-4V_1 + (7s + 4)V_2 = 24 \quad \textbf{(15.39)}$$

FIGURE 15.47

Finding a short-circuit current.

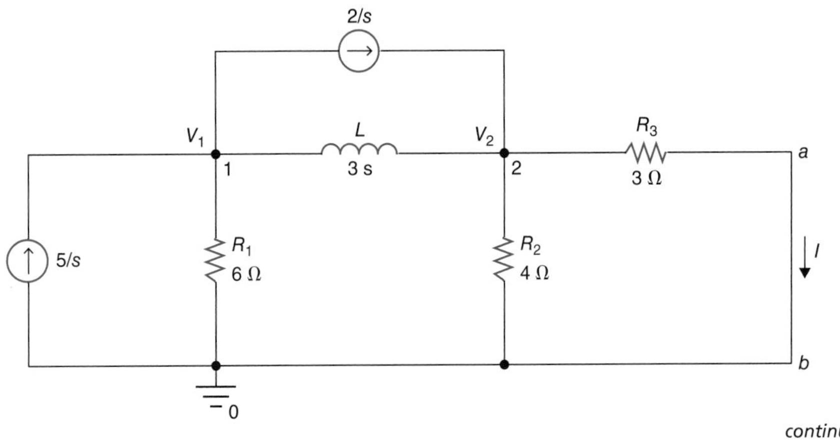

continued

Example 15.12 continued

The voltage V_2 can be found by applying Cramer's rule on Equations (15.37) and (15.39).

$$V_2 = \frac{\begin{vmatrix} s+2 & 18 \\ -4 & 24 \end{vmatrix}}{\begin{vmatrix} s+2 & -2 \\ -4 & 7s+4 \end{vmatrix}} = \frac{24s + 48 + 72}{7s^2 + 18s + 8 - 8} = \frac{24s + 120}{7s^2 + 18s} = \frac{\frac{24}{7}s + \frac{120}{7}}{s^2 + \frac{18}{7}s}$$

$$= \frac{3.4286s + 17.1429}{s(s + 2.5714)}$$

The short-circuit current I is obtained by dividing V_2 by R_3. Thus, we have

$$I = \frac{V_2}{R_3} = \frac{\dfrac{\frac{24}{7}s + \frac{120}{7}}{s^2 + \frac{18}{7}s}}{3} = \frac{8s + 40}{7s^2 + 18s} = \frac{\frac{8}{7}s + \frac{40}{7}}{s^2 + \frac{18}{7}s} = \frac{1.1429s + 5.7143}{s(s + 2.5714)}$$

The Thévenin impedance is given by the ratio of the Thévenin voltage to the short-circuit current. Thus, we have

$$Z_{th} = \frac{V_{th}}{I} = \frac{\dfrac{24s + 120}{3s^2 + 10s}}{\dfrac{8s + 40}{7s^2 + 18s}} = 3\frac{7s^2 + 18s}{3s^2 + 10s} = \frac{21s + 54}{3s + 10} = \frac{7s + 18}{s + \frac{10}{3}} \tag{15.40}$$

$$= \frac{7s + 18}{s + 3.3333}$$

The Thévenin impedance can also be found by finding the equivalent impedance between a and b after deactivating independent sources. For current sources, deactivating implies removing them from the circuit, as shown in Figure 15.48.

FIGURE 15.48

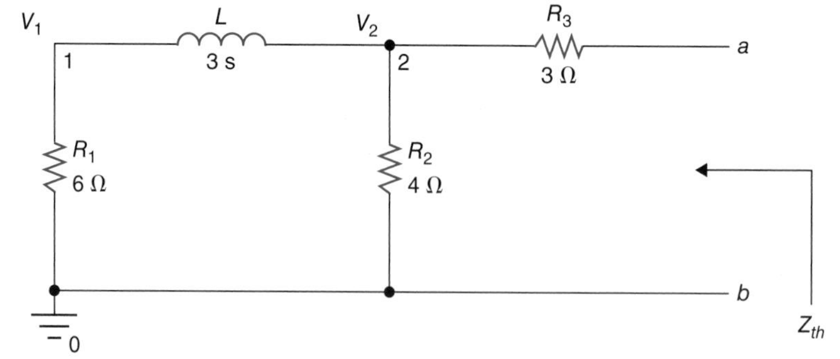

Circuit for finding Z_{th}.

The Thévenin impedance is given by

$$Z_{th} = 3 + [4 \| (3s + 6)] = 3 + \frac{4 \times (3s + 6)}{4 + 3s + 6} = 3 + \frac{12s + 24}{3s + 10} = \frac{21s + 54}{3s + 10} = \frac{7s + 18}{s + \frac{10}{3}s}$$

continued

Example 15.12 continued

which is identical to the one given by Equation (15.40). The Thévenin equivalent circuit is shown in Figure 15.49.

FIGURE 15.49

Thévenin equivalent circuit.

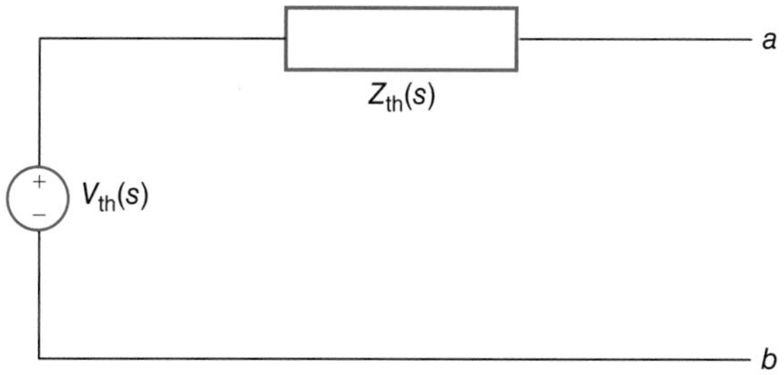

Exercise 15.12

The initial voltage across the capacitor in the circuit shown in Figure 15.50 is $v(0^-) = 2$ V. Find the Thévenin equivalent circuit between terminals *a* and *b* for the circuit shown in Figure 15.50.

FIGURE 15.50

Circuit for EXERCISE 15.12.

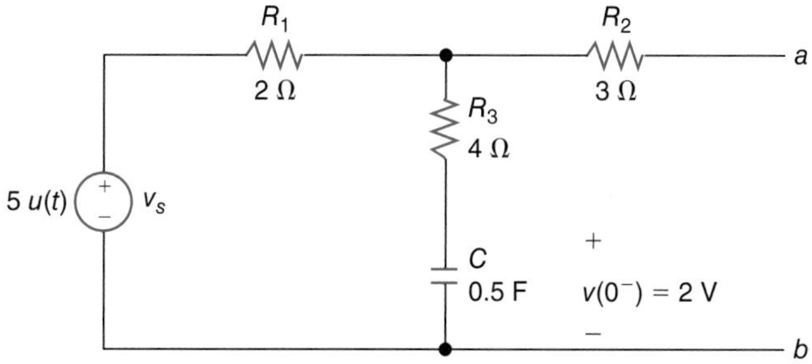

Answer:

$$V_{th} = \frac{12s + 5}{s(3s + 1)}, \quad Z_{th} = \frac{13s + 5}{3s + 1}.$$

EXAMPLE 15.13

The initial voltages across the capacitors are zero in the circuit shown in Figure 15.51. Find the Thévenin equivalent voltage and the Thévenin equivalent impedance between terminals *a* and *b*.

FIGURE 15.51

Circuit for EXAMPLE 15.13.

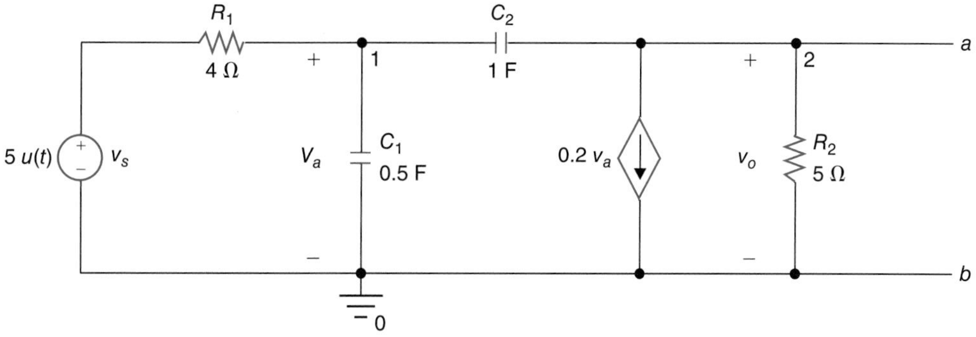

continued

Example 15.13 continued

The s-domain equivalent circuit is shown in Figure 15.52.

FIGURE 15.52

Circuit for EXAMPLE 15.13.

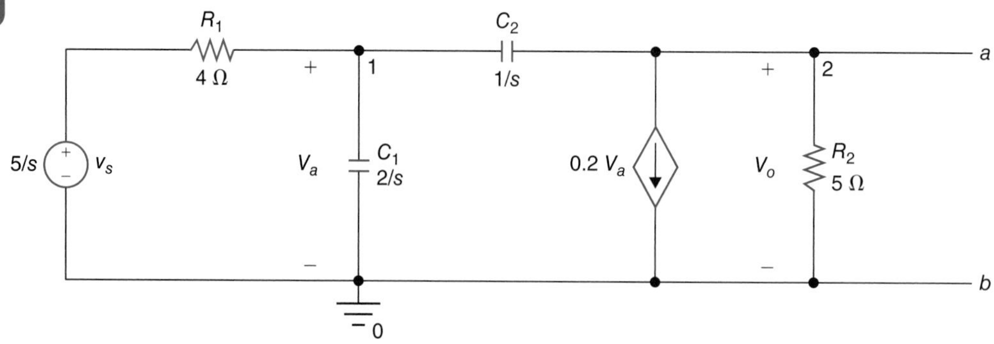

Summing the currents leaving node 1, we obtain

$$\frac{V_a - \frac{5}{s}}{4} + \frac{V_a}{\frac{2}{s}} + \frac{V_a - V_o}{\frac{1}{s}} = 0$$

which can be rearranged as

$$(1.5s + 0.25)V_a = sV_o + \frac{1.25}{s}$$

Solving for V_a, we obtain

$$V_a = \frac{s}{1.5s + 0.25}V_o + \frac{1.25}{s(1.5s + 0.25)} \quad \textbf{(15.41)}$$

Summing the currents leaving node 2, we obtain

$$\frac{V_o - V_a}{\frac{1}{s}} + 0.2V_a + \frac{V_o}{5} = 0 \quad \textbf{(15.42)}$$

Substituting Equation (15.41) into Equation (15.42) and simplifying, we get

$$(s + 0.2)V_o + \frac{s(-s + 0.2)}{1.5s + 0.25}V_o + \frac{1.25(-s + 0.2)}{s(1.5s + 0.25)} = 0$$

Solving for V_o, we obtain

$$V_o = \frac{\frac{1.25(s - 0.2)}{s(1.5s + 0.25)}}{\frac{(s + 0.2)(1.5s + 0.25) + s(-s + 0.2)}{1.5s + 0.25}}$$

$$= \frac{1.25(s - 0.2)}{s[(s + 0.2)(1.5s + 0.25) + s(-s + 0.2)]} = \frac{2.5s - 0.5}{s(s^2 + 1.5s + 0.1)}$$

continued

Example 15.13 continued

Since the open circuit voltage is V_o, the Thévenin equivalent voltage is V_o. Thus, we have

$$V_{th} = \frac{2.5s - 0.5}{s(s^2 + 1.5s + 0.1)}$$

To find the short-circuit current, we short-circuit *a* and *b*, as shown in Figure 15.53. Notice that node 2 is a ground, and no current flows through R_2.

FIGURE 15.53

Circuit with *a* and *b* shorted.

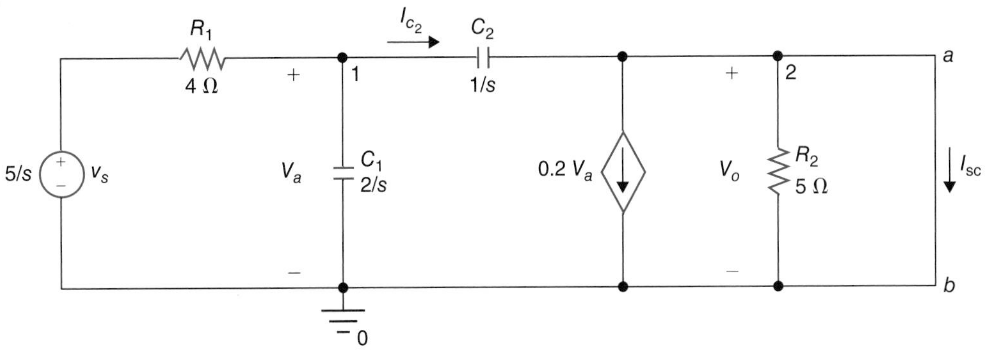

Summing the currents leaving node 1, we obtain

$$\frac{V_a - \frac{5}{s}}{4} + \frac{V_a}{\frac{2}{s}} + \frac{V_a}{\frac{1}{s}} = 0$$

which can be rearranged as

$$(1.5s + 0.25)V_a = \frac{1.25}{s}$$

Solving for V_a, we obtain

$$V_a = \frac{1.25}{s(1.5s + 0.25)} = \frac{\frac{5}{6}}{s\left(s + \frac{1}{6}\right)}$$

The current through C_2 is given by

$$I_{C_2} = \frac{V_a}{\frac{1}{s}} = \frac{\frac{5}{6}}{s + \frac{1}{6}}$$

The current through VCCS is given by

$$I_{VCCS} = 0.2V_a = \frac{\frac{1}{6}}{s\left(s + \frac{1}{6}\right)}$$

continued

Example 15.13 continued

The short-circuit current is the difference of I_{C_2} and I_{VCCS}. Thus, we have

$$I_{SC} = I_{C_2} - I_{VCCS} = \frac{\frac{5}{6}}{s + \frac{1}{6}} - \frac{\frac{1}{6}}{s\left(s + \frac{1}{6}\right)} = \frac{\frac{5}{6}(s - 0.2)}{s\left(s + \frac{1}{6}\right)}$$

The Thévenin equivalent impedance is the ratio of $V_{oc} = V_o$ to I_{sc}:

$$Z_{th} = \frac{V_o}{I_{sc}} = \frac{\frac{2.5s - 0.5}{s(s^2 + 1.5s + 0.1)}}{\frac{\frac{5}{6}(s - 0.2)}{s\left(s + \frac{1}{6}\right)}} = \frac{\frac{6}{5} \times 2.5(s - 0.2)\left(s + \frac{1}{6}\right)}{(s^2 + 1.5s + 0.1)(s - 0.2)} = \frac{3s + 0.5}{s^2 + 1.5s + 0.1}$$

Alternatively, a test voltage $u(t)$ can be applied across a and b after deactivating the voltage source, as shown in Figure 15.54.

FIGURE 15.54

Circuit with a and b shorted.

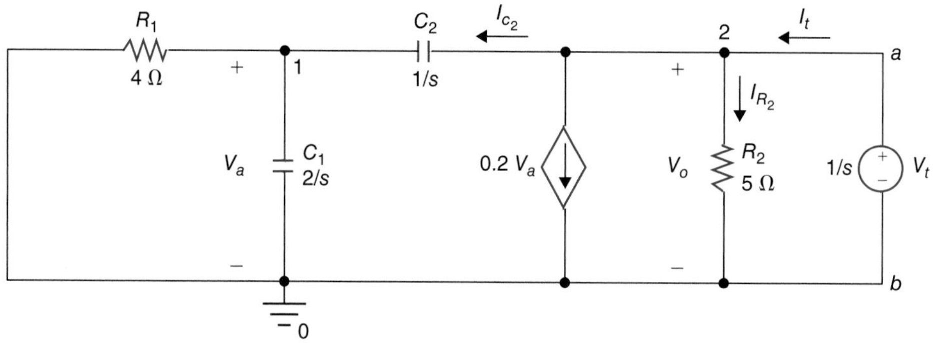

Summing the currents leaving node 1, we obtain

$$\frac{V_a}{4} + \frac{V_a}{\frac{2}{s}} + \frac{V_a - \frac{1}{s}}{\frac{1}{s}} = 0$$

which can be rearranged as

$$(1.5s + 0.25)V_a = 1$$

Solving for V_a, we obtain

$$V_a = \frac{1}{1.5s + 0.25} = \frac{\frac{4}{6}}{s + \frac{1}{6}}$$

continued

Example 15.13 continued

The current flowing out of the positive terminal of the test voltage source is given by

$$I_t = I_{C_2} + I_{VCCS} + I_{R_2} = \frac{\dfrac{1}{s} - \dfrac{\frac{4}{6}}{s+\frac{1}{6}}}{\dfrac{1}{s}} + 0.2\frac{\frac{4}{6}}{s+\frac{1}{6}} + \frac{\frac{1}{s}}{5} = \frac{\frac{1}{3}(s^2+1.5s+0.1)}{s\left(s+\frac{1}{6}\right)}$$

Taking the ratio of V_t to I_t, we obtain the Thévenin equivalent impedance:

$$Z_{th} = \frac{\dfrac{1}{s}}{\dfrac{\frac{1}{3}(s^2+1.5s+0.1)}{s\left(s+\frac{1}{6}\right)}} = \frac{3\left(s+\frac{1}{6}\right)}{(s^2+1.5s+0.1)}$$

$$= \frac{3s+0.5}{s^2+1.5s+0.1}$$

FIGURE 15.55

Thévenin equivalent circuit.

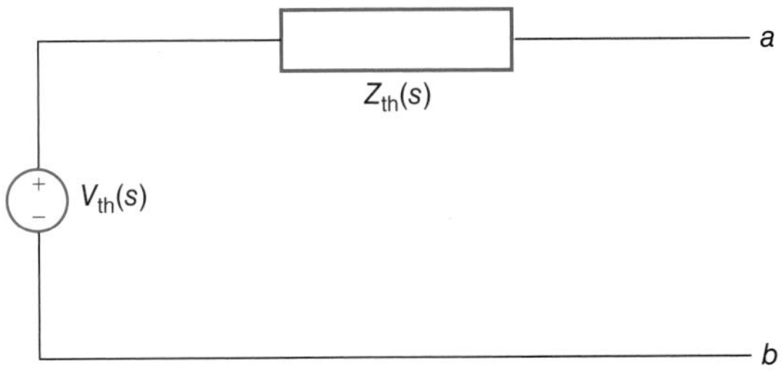

This is the same result obtained previously. The Thévenin equivalent circuit is shown in Figure 15.55.

$$V_{th}(s) = \frac{2.5s - 0.5}{s(s^2+1.5s+0.1)}, \quad Z_{th}(s) = \frac{3s+0.5}{s^2+1.5s+0.1}$$

Exercise 15.13

The initial voltages across the capacitors are zero in the circuit shown in Figure 15.56. Find the Thévenin equivalent voltage and the Thévenin equivalent impedance between terminals *a* and *b*.

FIGURE 15.56

Circuit for EXERCISE 15.13.

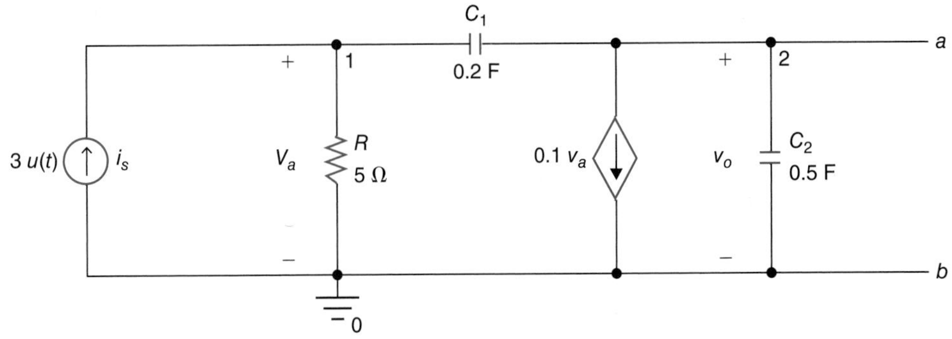

Answer:

$$V_{th} = \frac{6(s-0.5)}{s^2(s+1.6)}, \quad Z_{th} = \frac{2(s+1)}{s(s+1.6)}.$$

15.7 Norton Equivalent Circuit in the s-Domain

FIGURE 15.57

Circuit with a pair of terminals a and b.

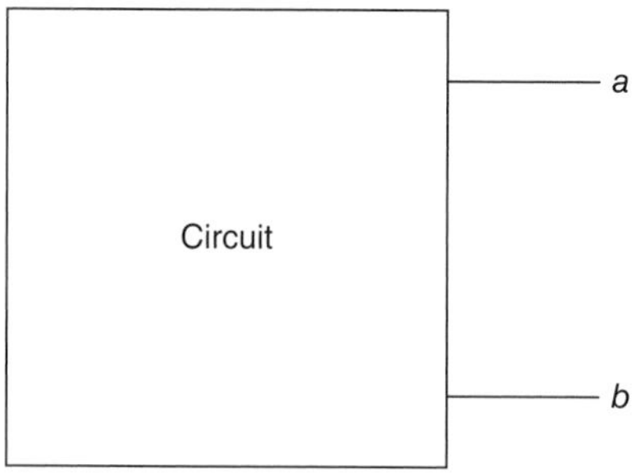

Figure 15.57 shows a circuit with a pair of terminals a and b. If the circuit consists of resistors and sources, as shown in Chapter 4, the circuit can be represented by a current source and a parallel resistor. If the circuit is a Laplace-transformed circuit consisting of impedances and sources in the s-domain, the circuit can be represented by a source and a parallel impedance, as shown in Figure 15.58. This representation is called a *Norton equivalent circuit.*

In Figure 15.58, $I_n(s)$ is the Norton equivalent current, and $Z_n(s)$ is the Norton equivalent impedance. The Norton equivalent current $I_n(s)$ is obtained by finding the short-circuit current between a and b from the original circuit, as shown in Figure 15.59.

The Norton equivalent impedance is found by using the same methods used in finding the Thévenin equivalent impedance. When the terminals a and b are short-circuited in the Thévenin equivalent circuit and the Norton equivalent circuit, the current through the short circuit (from a to b) is

$$I_n(s) = \frac{V_{th}(s)}{Z_{th}(s)}$$

The Norton-equivalent current can be found by dividing the Thévenin equivalent voltage by the Thévenin equivalent impedance. The open-circuit voltage across the terminals in the Thévenin equivalent circuit and the Norton equivalent circuit is

$$V_{th}(s) = I_n(s)Z_n(s)$$

FIGURE 15.58

Norton-equivalent circuit.

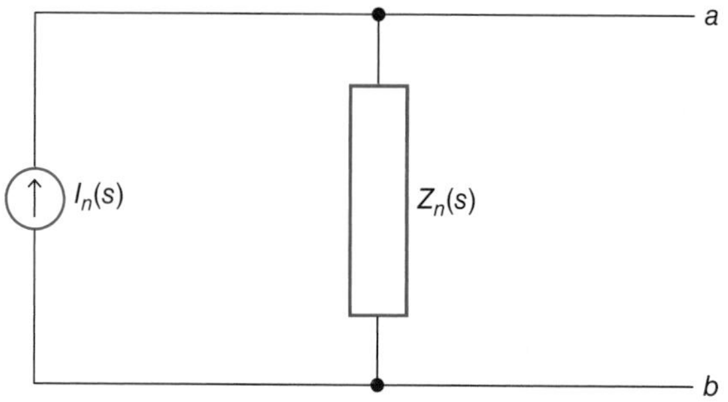

FIGURE 15.59

Short-circuit current.

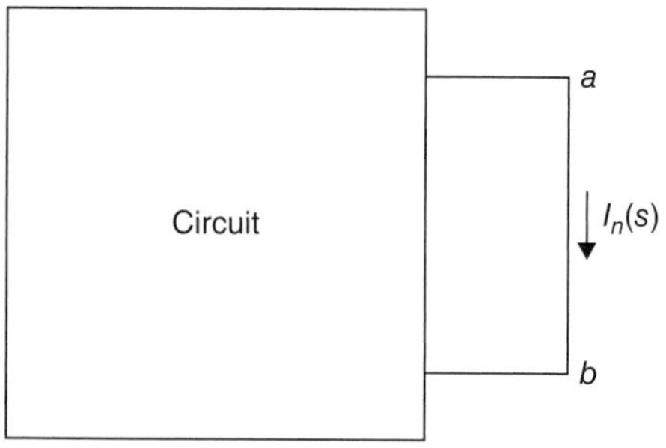

EXAMPLE 15.14

The initial voltage across the capacitor is $v(0^-) = 1$ V in the circuit shown in Figure 15.60. Find the Norton equivalent current and the Norton equivalent impedance for the circuit shown in Figure 15.60.

The s-domain equivalent circuit is shown in Figure 15.61.

Summing the currents leaving node 1, we obtain

$$\frac{V_1 - \frac{6}{s}}{4} + \frac{V_1 - \frac{1}{s}}{\frac{5}{s}} + \frac{V_1}{6} = 0$$

continued

Example 15.14 continued

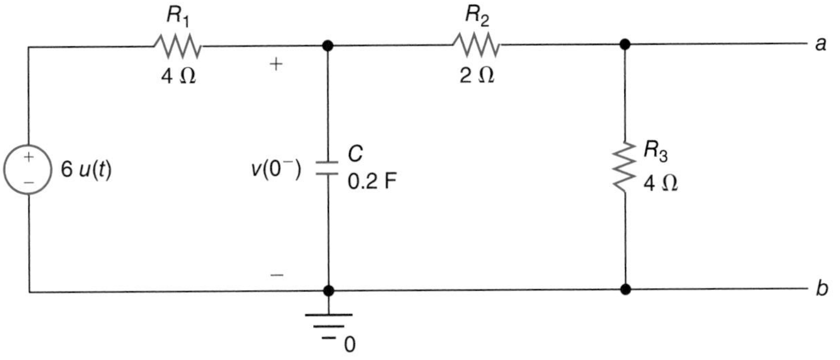

FIGURE 15.60

Circuit for EXAMPLE 15.14.

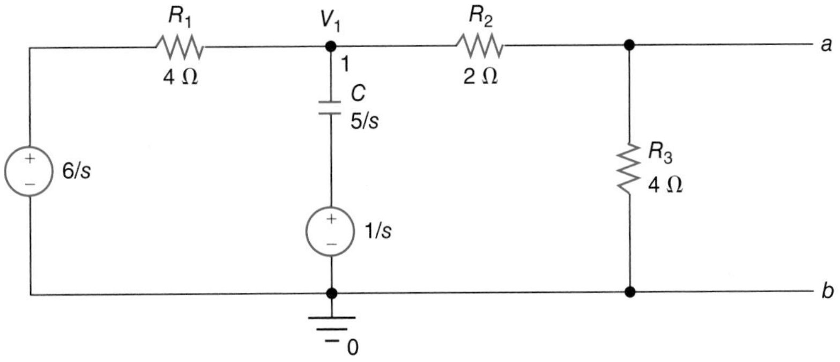

FIGURE 15.61

S-domain-equivalent circuit.

Multiplication by 60 yields

$$15V_1 - \frac{90}{s} + 12sV_1 - 12 + 10V_1 = 0$$

which can be rearranged as

$$(12s + 25)V_1 = \frac{90}{s} + 12 = \frac{12s + 90}{s}$$

Solving for V_1, we get

$$V_1 = \frac{12s + 90}{s(12s + 25)} = \frac{s + 7.5}{s\left(s + \frac{25}{12}\right)}$$

The open-circuit voltage between a and b can be obtained by applying the voltage divider rule across R_2 and R_3:

$$V_{oc} = V_1 \times \frac{R_4}{R_3 + R_4} = \frac{s + 7.5}{s\left(s + \frac{25}{12}\right)} \times \frac{4}{6} = \frac{2}{3}\frac{s + 7.5}{s\left(s + \frac{25}{12}\right)}$$

continued

Example 15.14 continued

Figure 15.62 shows the circuit in the *s*-domain with *a* and *b* short-circuited.

FIGURE 15.62

Circuit in the *s*-domain with *a* and *b* short-circuited.

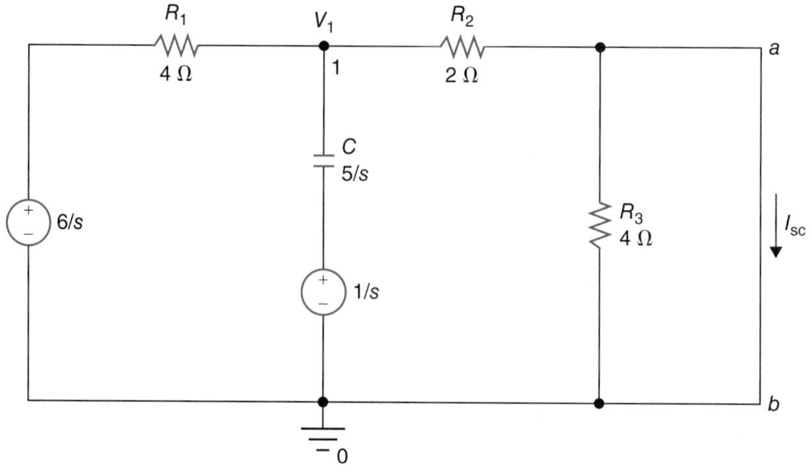

Notice that there is no current through R_3. Summing the currents leaving node 1, we obtain

$$\frac{V_1 - \frac{6}{s}}{4} + \frac{V_1 - \frac{1}{s}}{\frac{5}{s}} + \frac{V_1}{2} = 0$$

Multiplication by 20 yields

$$5V_1 - \frac{30}{s} + 4sV_1 - 4 + 10V_1 = 0$$

which can be rearranged as

$$(4s + 15)V_1 = \frac{30}{s} + 4 = \frac{4s + 30}{s}$$

Solving for V_1, we get

$$V_1 = \frac{4s + 30}{s(4s + 15)} = \frac{s + 7.5}{s(s + 3.75)}$$

The short-circuit current is the current through R_2. Thus, we have

$$I_{sc} = \frac{V_1}{2} = \frac{1}{2}\frac{s + 7.5}{s(s + 3.75)}$$

The short-circuit current is the Norton equivalent current. The Norton impedance, which is also the Thévenin impedance, is the ratio of the open-circuit voltage to short-circuit current. Thus, we have

continued

Example 15.14 continued

$$Z_N = \frac{V_{oc}}{I_{sc}} = \frac{\dfrac{2}{3}\dfrac{s+7.5}{s\left(s+\dfrac{25}{12}\right)}}{\dfrac{1}{2}\dfrac{s+7.5}{s(s+3.75)}} = \frac{4}{3}\frac{s+3.75}{s+\dfrac{25}{12}}$$

The Norton equivalent impedance can also be found from the circuit shown in Figure 15.48 after deactivating the voltage sources, as shown in Figure 15.63.

FIGURE 15.63

Circuit with sources deactivated.

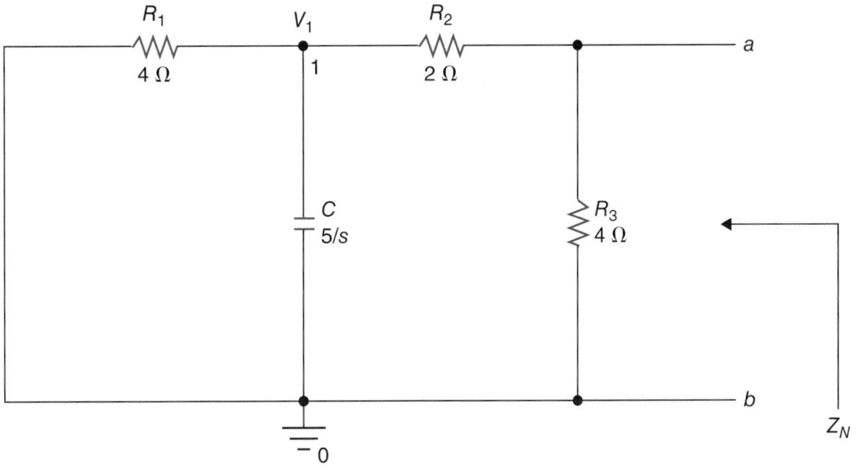

The Norton equivalent impedance is given by

$$Z_N = \left[\left(4\|\frac{5}{s}\right)+2\right]\|4 = \left[\frac{4\times\dfrac{5}{s}}{4+\dfrac{5}{s}}+2\right]\|4 = \left[\frac{20}{4s+5}+2\right]\|4$$

$$= \left[\frac{8s+30}{4s+5}\right]\|4 = \frac{4\dfrac{8s+30}{4s+5}}{\dfrac{8s+30}{4s+5}+4}$$

$$= \frac{32s+120}{24s+50} = \frac{4}{3}\frac{s+3.75}{s+\dfrac{25}{12}}$$

FIGURE 15.64

Norton-equivalent circuit.

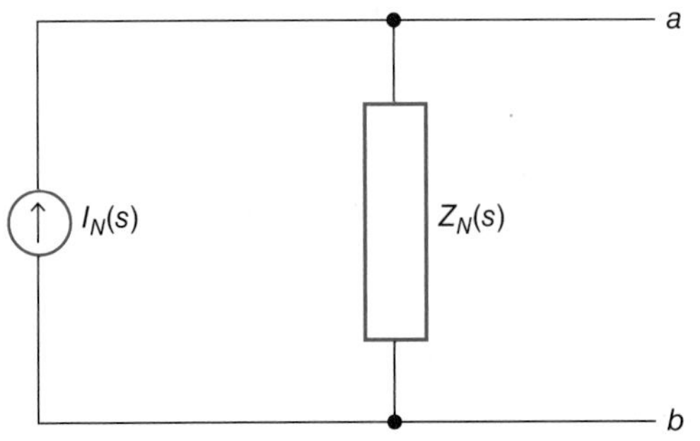

The Norton equivalent circuit is shown in Figure 15.64.

Exercise 15.14

The initial voltage across the capacitor in the circuit shown in Figure 15.65 is $v(0^-) = 2$ V, and the initial current through the inductor is $i(0^-) = 1$ A. Find the Norton equivalent circuit between terminals a and b for the circuit shown in Figure 15.65.

continued

Exercise 15.14 continued

FIGURE 15.65

Circuit for EXERCISE 15.14.

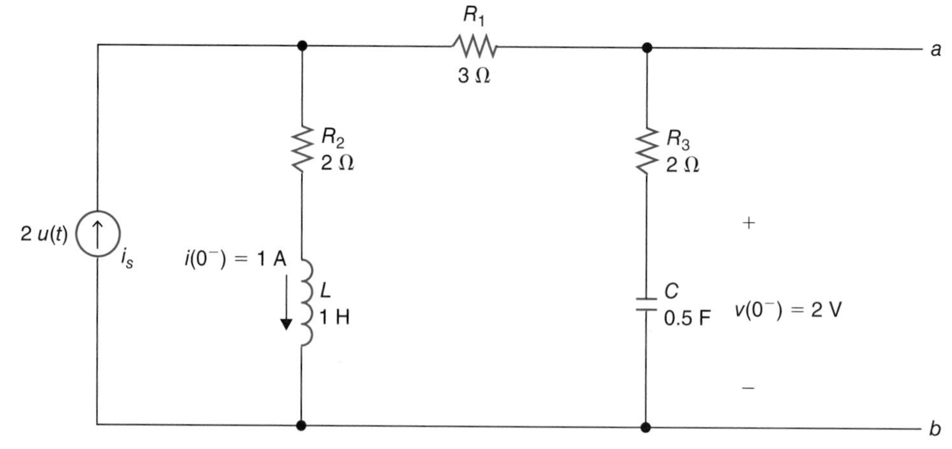

Answer:

$$I_n = \frac{s+4}{s(s+5)}, \quad Z_n = \frac{2(s^2+6s+5)}{s^2+7s+2}.$$

EXAMPLE 15.15

The initial voltages across the capacitors are zero in the circuit shown in Figure 15.66. Find the Norton equivalent current and the Norton equivalent impedance between terminals *a* and *b*.

FIGURE 15.66

Circuit for EXAMPLE 15.15.

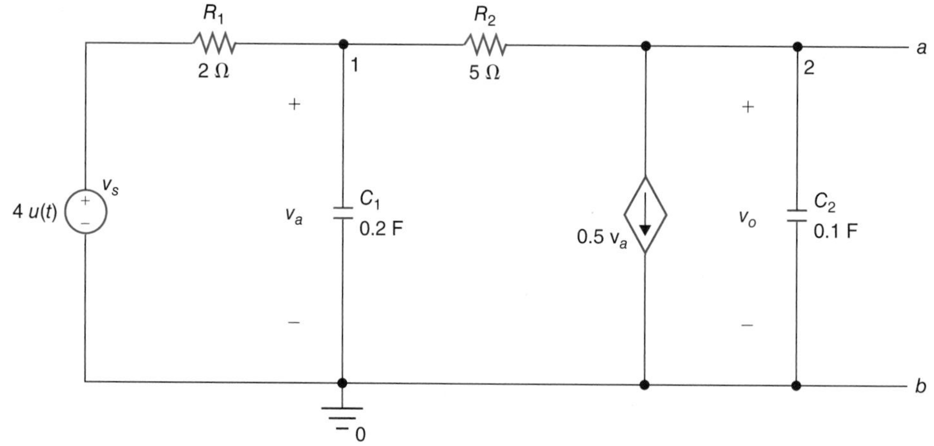

The circuit in the s-domain is shown in Figure 15.67.

We will find the open-circuit voltage V_{oc} and the short-circuit current I_{sc} between terminals a and b to find the Norton equivalent current and the Norton equivalent impedance. Summing the currents leaving node 1, we obtain

$$\frac{V_a - \frac{4}{s}}{2} + \frac{V_a}{\frac{5}{s}} + \frac{V_a - V_o}{5} = 0$$

continued

Example 15.15 continued

FIGURE 15.67

Circuit in the s-domain.

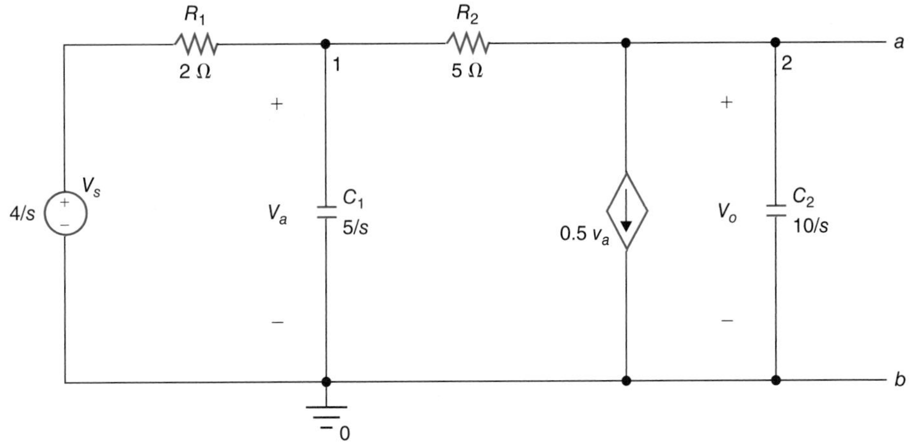

which can be rearranged as

$$(0.2s + 0.7)V_a = 0.2V_o + \frac{2}{s}$$

Solving for V_a, we obtain

$$V_a = \frac{1}{s + 3.5}V_o + \frac{2}{s(0.2s + 0.7)} = \frac{1}{s + 3.5}V_o + \frac{10}{s(s + 3.5)} \quad \textbf{(15.43)}$$

Summing the currents leaving node 2, we obtain

$$\frac{V_o - V_a}{5} + 0.5V_a + \frac{sV_o}{10} = 0 \quad \textbf{(15.44)}$$

Substituting Equation (15.43) into Equation (15.44) and simplifying, we get

$$(0.1s + 0.2)V_o + \frac{0.3}{s + 3.5}V_o + \frac{3}{s(s + 3.5)} = 0$$

Solving for V_o, we obtain

$$V_o = V_{oc} = \frac{-\dfrac{3}{s(s + 3.5)}}{\dfrac{(0.1s + 0.2)(s + 3.5) + 0.3}{s + 3.5}} = \frac{-3}{s[(0.1s + 0.2)(s + 3.5) + 0.3]}$$

$$= \frac{-3}{s[0.1s^2 + 0.55s + 1]} = \frac{-30}{s(s^2 + 5.5s + 10)}$$

To find the short-circuit current, we short-circuit a and b, as shown in Figure 15.68. Notice that node 2 is a ground, and no current flows through C_2.

Summing the currents leaving node 1, we obtain

$$\frac{V_a - \dfrac{4}{s}}{2} + \frac{V_a}{\dfrac{5}{s}} + \frac{V_a}{5} = 0$$

continued

Example 15.15 continued

FIGURE 15.68

Circuit with a and b shorted.

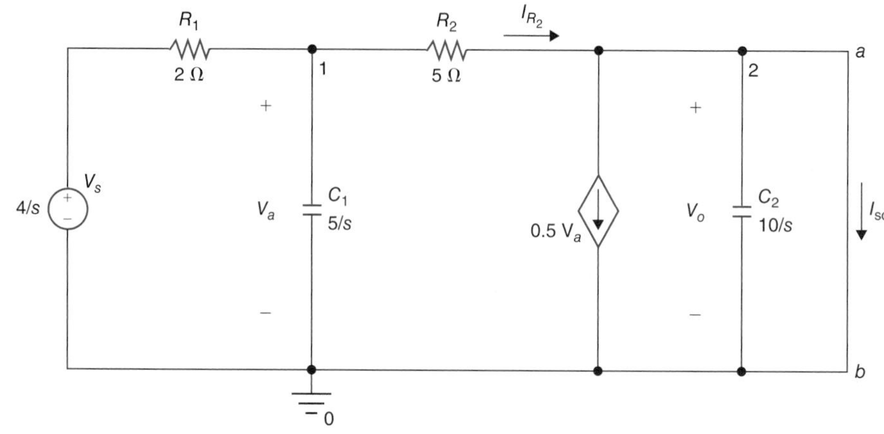

which can be rearranged as

$$(0.2s + 0.7)V_a = \frac{2}{s}$$

Solving for V_a, we obtain

$$V_a = \frac{2}{s(0.2s + 0.7)} = \frac{10}{s(s + 3.5)}$$

The current through R_2 is given by

$$I_{R_2} = \frac{V_a}{5} = \frac{2}{s(s + 3.5)}$$

The current through VCCS is given by

$$I_{VCCS} = 0.5V_a = \frac{5}{s(s + 3.5)}$$

The short-circuit current is the difference of I_{R_2} and I_{VCCS}. Thus, we have

$$I_{sc} = I_{R_2} - I_{VCCS} = \frac{2}{s(s + 3.5)} - \frac{5}{s(s + 3.5)} = \frac{-3}{s(s + 3.5)}$$

The short-circuit current is the Norton equivalent current. Thus, we have

$$I_n = I_{sc} = \frac{-3}{s(s + 3.5)}$$

FIGURE 15.69

Norton equivalent circuit.

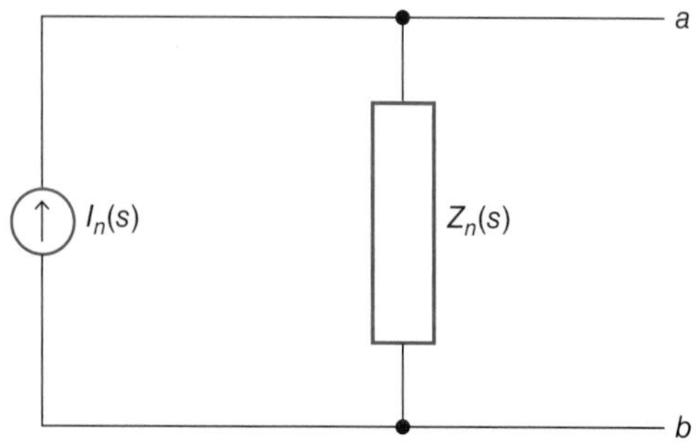

The Norton equivalent impedance is the ratio of $V_{oc} = V_o$ to I_{sc}:

$$Z_n = \frac{V_o}{I_{sc}} = \frac{\dfrac{-30}{s(s^2 + 5.5s + 10)}}{\dfrac{-3}{s(s + 3.5)}} = \frac{10(s + 3.5)}{s^2 + 5.5s + 10}$$

The Norton equivalent circuit is shown in Figure 15.69.

$$I_n = \frac{-3}{s(s + 3.5)}, \quad Z_n = \frac{10(s + 3.5)}{s^2 + 5.5s + 10}$$

Exercise 15.15

The initial voltages across the capacitors are zero in the circuit shown in Figure 15.70. Find the Norton equivalent current and the Norton equivalent impedance between terminals *a* and *b*.

FIGURE 15.70

Circuit for EXERCISE 15.15.

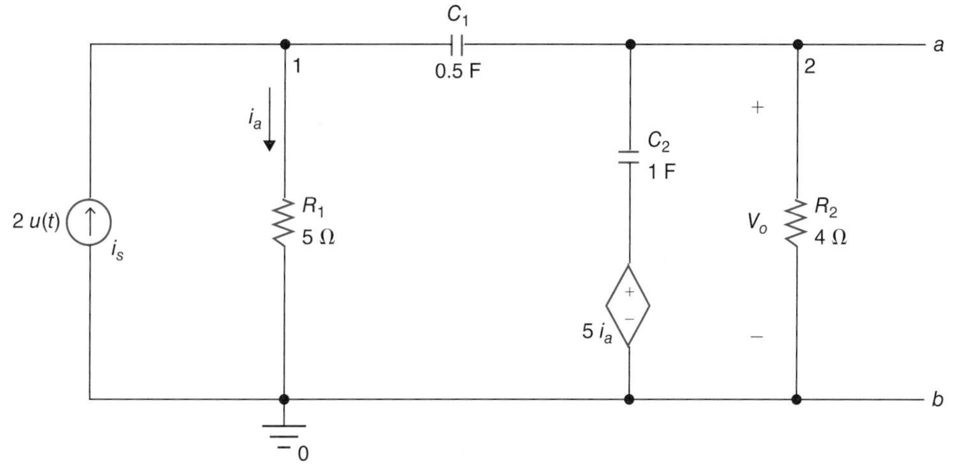

Answer:

$$I_n = \frac{6}{s + 0.4}, \quad Z_n = \frac{1.1765s + 0.4706}{s + 0.1176}.$$

15.8 Transfer Function

In the Laplace-transformed circuit with zero initial conditions, the ratio of the output in the s-domain to the input in the s-domain is defined as the transfer function. Let $X(s)$ be the s-domain representation of the input, and let $Y(s)$ be the s-domain representation of the output. Then, the transfer function is given by

$$H(s) = \frac{Y(s)}{X(s)}$$

The input can be a voltage or a current, and the output can be a voltage or a current. If both the input and the output are voltages, the transfer function can be written as

$$H(s) = \frac{V_o(s)}{V_{in}(s)}$$

If both the input and the output are currents, the transfer function can be written as

$$H(s) = \frac{I_o(s)}{I_{in}(s)}$$

Other combinations are possible also. The output $Y(s)$ can be written as

$$Y(s) = H(s)X(s)$$

The transfer function $H(s)$ transfers input $X(s)$ into output $Y(s)$. In the process of the transfer, the input signal is changed. The nature of the change is specified in the transfer function. If the input signal is a sinusoid, the transfer function may change the amplitude and the phase of the input signal. If the input signal is a sum of many sinusoids, the transfer function changes the amplitude and phase of each of the sinusoids in a certain way. The transfer function can be selected to change the input signal to the desired signal at the output.

FIGURE 15.71

Black box representation of a circuit.

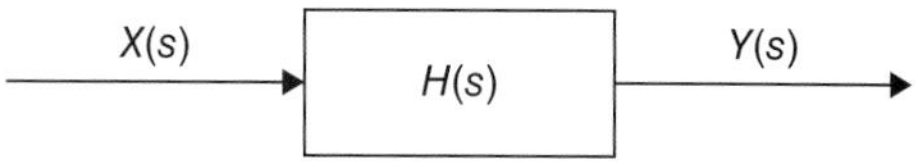

The transfer function $H(s)$ represents the effects of the circuit (or system in general) to the input $X(s)$. It transforms input $X(s)$ to output $Y(s)$. It hides the details of the circuit and mathematically describes the effects of the circuit to the input signal. The given circuit can now be represented as a black box with $H(s)$, as shown in Figure 15.71.

Finding the transfer function $H(s)$ of a given circuit is called analysis, and finding the circuit implementation using available parts for a given $H(s)$ is called synthesis.

If a time domain response is desired, we first find $X(s)$ by taking the Laplace transform of $x(t)$ and then multipling the transfer function $H(s)$ and $X(s)$ to get $Y(s) = H(s)X(s)$. The output $y(t)$ is obtained by taking the inverse Laplace transform of $Y(s)$

$$y(t) = L^{-1}[Y(s)] = L^{-1}[H(s)X(s)].$$

For the given circuit, the transfer function $H(s)$ is fixed. For different input, only $X(s)$ is changed.

If an impulse function $\delta(t)$ is applied at the input to a system with transfer function $H(s)$, $X(s) = L[\delta(t)] = 1$. The output $Y(s)$ is $Y(s) = H(s)\ X(s) = H(s) \times 1 = H(s)$. The output $y(t)$ in the time domain is

$$y(t) = L^{-1}[Y(s)] = L^{-1}[H(s)] = h(t) \qquad \textbf{(15.45)}$$

The signal $h(t)$ is called the **impulse response**. The impulse response is the inverse Laplace transform of the transfer function.

If a unit step function $u(t)$ is applied at the input to a system with transfer function $H(s)$, $X(s) = L[u(t)] = 1/s$. The output $Y(s)$ is $Y(s) = H(s)\ X(s) = H(s)/s$. The output $y(t)$ in the time domain is

$$h_s(t) = L^{-1}\left[\frac{H(s)}{s}\right] \qquad \textbf{(15.46)}$$

The signal $h_s(t)$ is called the **step response**. The step response is the inverse Laplace transform of $H(s)/s$.

If a unit ramp function $tu(t)$ is applied at the input to a system with transfer function $H(s)$, $X(s) = L[tu(t)] = 1/s^2$. The output $Y(s)$ is $Y(s) = H(s)\ X(s) = H(s)/s^2$. The output $y(t)$ in the time domain is

$$h_r(t) = L^{-1}\left[\frac{H(s)}{s^2}\right] \qquad \textbf{(15.47)}$$

The signal $h_r(t)$ is called the **ramp response**. The ramp response is the inverse Laplace transform of $H(s)/s^2$.

15.8.1 SINUSOIDAL INPUT

If the input signal to a circuit is a sinusoid, the voltages and currents everywhere in the circuit are also sinusoids of the same frequency as the input frequency. A phasor-transformed circuit can be used to analyze the circuit. Remember that the capacitor is replaced by $1/(j\omega C)$, and the inductor is replaced by $j\omega L$ in the phasor-transformed circuit. In the Laplace-transformed circuit, the capacitor is replaced by $1/(sC)$, and the inductor is replaced by sL. Thus, if s is replaced by $j\omega$, a Laplace-transformed circuit becomes a phasor-transformed circuit. Notice that $s = j\omega$ represents the imaginary axis. $H(\omega)$ is the transfer function for sinusoidal input, and $H(s)$ is the transfer function for any input including sinusoidal input. $H(\omega)$ is also called the *frequency response.* The absolute value of $H(\omega)$, $|H(\omega)|$, is called the *magnitude response,* and the angle of $H(\omega)$, $\angle H(\omega)$, is called the *phase response.* The transfer function $H(s)$ can be transformed to $H(\omega)$ by replacing s by $j\omega$:

$$H(\omega) = H(s)\Big|_{s=j\omega} \tag{15.48}$$

In calculating the transfer function, the initial conditions are ignored. To find the transfer function, we first transform the circuit to a Laplace-transformed circuit. The input can be represented by $X(s)$ or X, and the output can be represented by $Y(s)$ or Y. Then, apply any of the circuit laws and theorems to find the ratio of the output to input.

15.8.2 POLES AND ZEROS

The transfer function $H(s)$ is given by a ratio of the numerator polynomial $N(s)$ to the denominator polynomial $D(s)$. Thus, we have

$$H(s) = \frac{N(s)}{D(s)}$$

The roots of $D(s) = 0$ are called poles, and the roots of $N(s) = 0$ are called zeros. At the location of poles, $H(s)$ becomes infinity, and at the location of zeros, $H(s)$ becomes zero. A plot that shows poles and zeros is called a **pole zero diagram.** The location of poles and zeros correlates with the frequency response and the impulse response of the system. The transfer function can be written in factored form as

$$H(s) = K\frac{(s + z_1)(s + z_2)\cdots(s + z_m)}{(s + p_1)(s + p_2)\cdots(s + p_n)} \tag{15.49}$$

where $-p_1, -p_2, \ldots, -p_n$ are poles, $-z_1, -z_2, \ldots, -z_m$ are zeros, and K is a constant. The frequency response $H(\omega)$ is obtained by evaluating the transfer function at $s = j\omega$.

$$H(\omega) = H(s)\Big|_{s=j\omega} = K\frac{(j\omega + z_1)(j\omega + z_2)\cdots(j\omega + z_m)}{(j\omega + p_1)(j\omega + p_2)\cdots(j\omega + p_n)} \tag{15.50}$$

The magnitude response is obtained by taking the absolute value of $H(\omega)$:

$$|H(\omega)| = |K|\frac{|j\omega + z_1||j\omega + z_2|\cdots|j\omega + z_m|}{|j\omega + p_1||j\omega + p_2|\cdots|j\omega + p_n|} \tag{15.51}$$

Each term in the denominator is the distance from $j\omega$ to the location of a pole, and each term in the numerator is the distance from $j\omega$ to the location of a zero. The magnitude response can be obtained by multiplying the distances from $j\omega$ to all zeros and gain $|K|$, and then dividing it by the product of distances from $j\omega$ to all poles. Equation (15.51) can be rewritten as

$$|H(\omega)| = |K|\frac{\sum_{k=1}^{m} DZ_k}{\sum_{k=1}^{n} DP_k} \tag{15.52}$$

where DZ_k is the distance from $j\omega$ to kth zero, and DP_k is the distance from $j\omega$ to kth pole. As ω is scanned from zero to infinity in the $j\omega$ axis, we can evaluate Equation (15.52) to get the magnitude at frequency ω. If there is a pole close to the $j\omega$ axis, the magnitude peaks at a frequency ω closest to the pole. If there is a zero close to the $j\omega$ axis, the magnitude dips at a frequency ω closest to the zero.

The phase response is obtained by taking the angle of $H(\omega)$:

$$\angle H(\omega) = \angle K \frac{e^{j\angle j\omega + z_1} e^{j\angle j\omega + z_2} \ldots e^{j\angle j\omega + z_m}}{e^{j\angle j\omega + p_1} e^{j\angle j\omega + p_2} \ldots e^{j\angle j\omega + p_n}} \tag{15.53}$$

Each term in the denominator is the angle from $j\omega$ to the location of a pole, and each term in the numerator is the angle from $j\omega$ to the location of a zero. The phase response can be obtained by adding the angles from $j\omega$ to all zeros and angle of K, $\angle K$, and then subtracting it by the sum of angles from $j\omega$ to all poles. Equation (15.53) can be rewritten as

$$\angle H(\omega) = \angle K + \angle j\omega + z_1 + \angle j\omega + z_2 + \cdots + \angle j\omega + z_m$$
$$- \angle j\omega + p_1 - \angle j\omega + p_2 - \cdots - \angle j\omega + p_n$$

or

$$\angle H(\omega) = \angle K + \sum_{k=1}^{m} AZ_k - \sum_{k=1}^{n} AP_k \tag{15.54}$$

where AZ_k is the angle from $j\omega$ to the kth zero, and AP_k is the angle from $j\omega$ to kth pole. As ω is scanned from zero to infinity in the $j\omega$ axis, we can evaluate Equation (15.54) to get the angle at frequency ω. Each zero contributes 90° to $\angle H(\omega)$ at $\omega = \infty$, and each pole contributes −90° to $\angle H(\omega)$ at $\omega = \infty$.

EXAMPLE 15.16

Find the transfer function $H(s) = V_o(s)/V_{in}(s)$ of the circuit shown in Figure 15.72.

FIGURE 15.72

RC circuit.

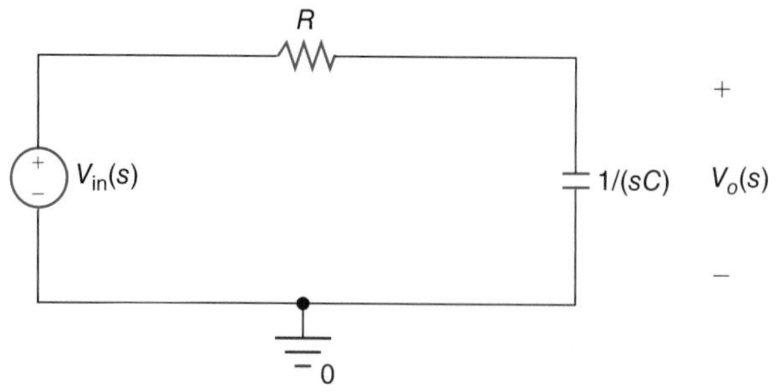

Application of the voltage divider rule yields

$$H(s) = \frac{V_o(s)}{V_{in}(s)} = \frac{\frac{1}{sC}}{R + \frac{1}{sC}} = \frac{1}{sRC + 1} = \frac{\frac{1}{RC}}{s + \frac{1}{RC}} \tag{15.55}$$

Exercise 15.16

Find the transfer function $H(s) = V_o(s)/V_{in}(s)$ of the circuit shown in Figure 15.73.

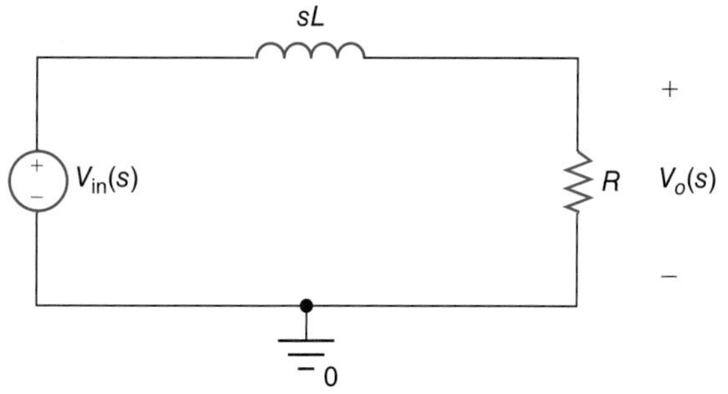

FIGURE 15.73

LR circuit.

Answer:

$$H(s) = \frac{\frac{R}{L}}{s + \frac{R}{L}} \tag{15.56}$$

EXAMPLE 15.17

Find the transfer function $H(s) = V_o(s)/V_{in}(s)$ of the circuit shown in Figure 15.74.

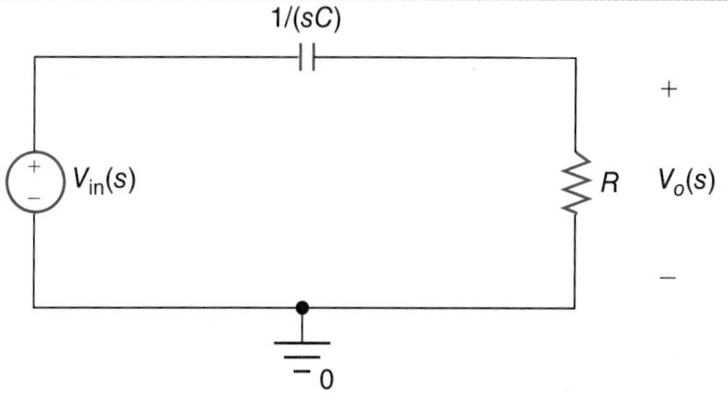

FIGURE 15.74

CR circuit.

Application of the voltage divider rule yields

$$H(s) = \frac{V_o(s)}{V_{in}(s)} = \frac{R}{R + \frac{1}{sC}} = \frac{sRC}{sRC + 1} = \frac{s}{s + \frac{1}{RC}} \tag{15.57}$$

Exercise 15.17

Find the transfer function $H(s) = V_o(s)/V_{in}(s)$ of the circuit shown in Figure 15.75.

FIGURE 15.75

RL circuit.

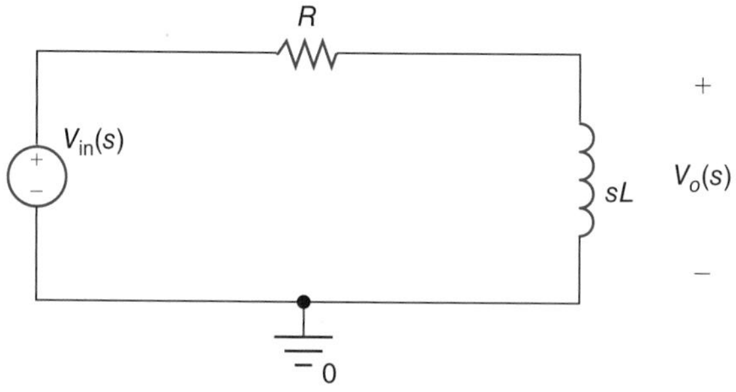

Answer:

$$H(s) = \frac{s}{s + \frac{R}{L}} \tag{15.58}$$

EXAMPLE 15.18

Find the transfer function $H(s) = V_o(s)/V_{in}(s)$ of the circuit shown in Figure 15.76.

FIGURE 15.76

RLC circuit.

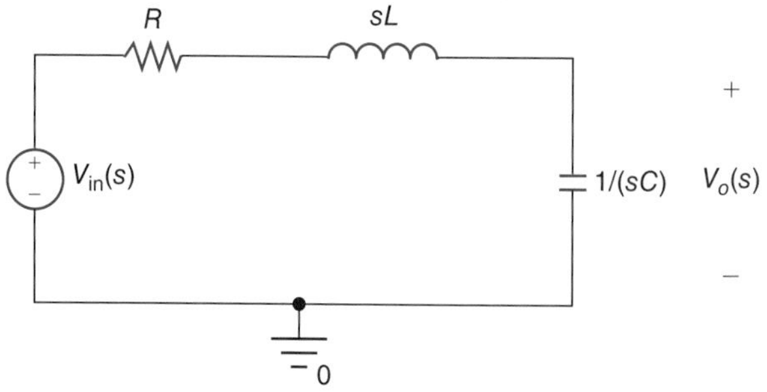

Application of the voltage divider rule yields

$$H(s) = \frac{V_o(s)}{V_{in}(s)} = \frac{\frac{1}{sC}}{sL + R + \frac{1}{sC}} = \frac{1}{s^2LC + sRC + 1}$$
$$= \frac{\frac{1}{LC}}{s^2 + \frac{R}{L}s + \frac{1}{LC}} \tag{15.59}$$

Exercise 15.18

Find the transfer function $H(s) = V_o(s)/V_{in}(s)$ of the circuit shown in Figure 15.77.

FIGURE 15.77

RCL circuit.

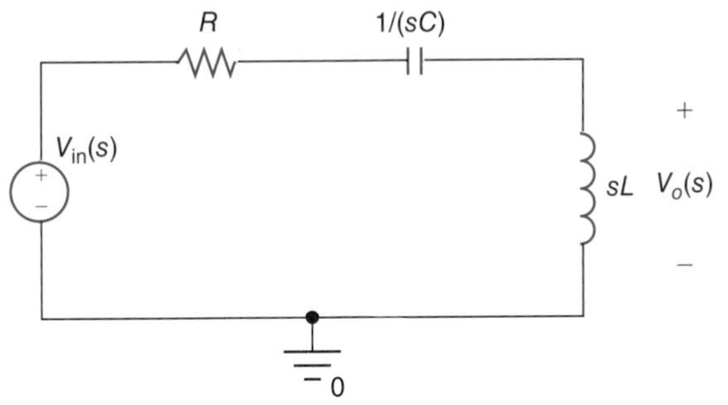

Answer:

$$H(s) = \frac{sL}{sL + R + \frac{1}{sC}} = \frac{s^2LC}{s^2LC + sRC + 1} = \frac{s^2}{s^2 + \frac{R}{L}s + \frac{1}{LC}} \quad \textbf{(15.60)}$$

EXAMPLE 15.19

Find the transfer function $H(s) = V_o(s)/V_{in}(s)$ of the circuit shown in Figure 15.78.

FIGURE 15.78

LCR circuit.

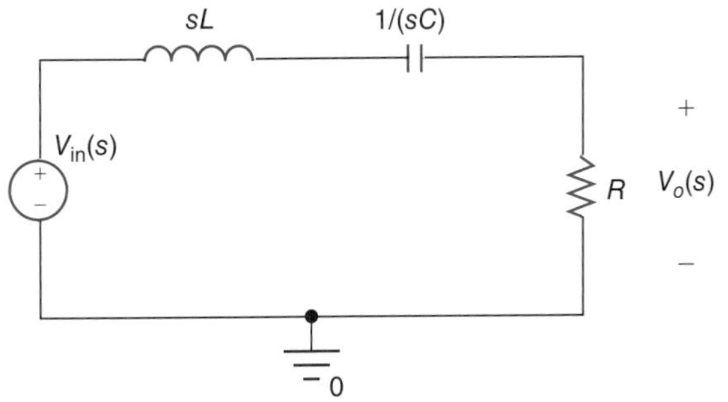

Application of the voltage divider rule yields

$$H(s) = \frac{V_o(s)}{V_{in}(s)} = \frac{R}{sL + R + \frac{1}{sC}} = \frac{sRC}{s^2LC + sRC + 1} = \frac{\frac{R}{L}s}{s^2 + \frac{R}{L}s + \frac{1}{LC}} \quad \textbf{(15.61)}$$

Exercise 15.19

Find the transfer function $H(s) = V_o(s)/V_{in}(s)$ of the circuit shown in Figure 15.79.

FIGURE 15.79

RLC circuit.

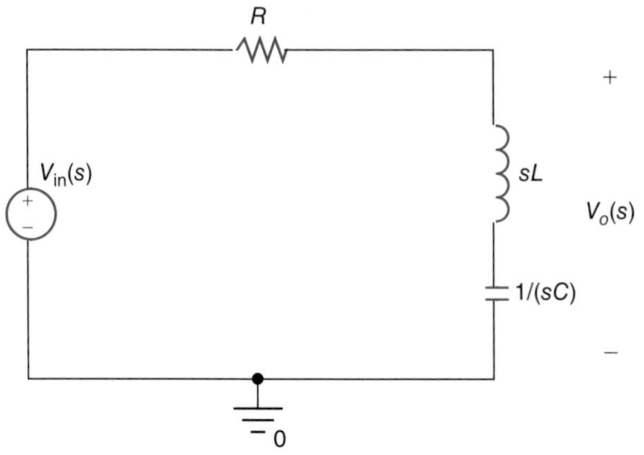

Answer:

$$H(s) = \frac{sL + \dfrac{1}{sC}}{sL + R + \dfrac{1}{sC}} = \frac{s^2LC + 1}{s^2LC + sRC + 1} = \frac{s^2 + \dfrac{1}{LC}}{s^2 + \dfrac{R}{L}s + \dfrac{1}{LC}} \tag{15.62}$$

EXAMPLE 15.20

Find the transfer function $H(s) = V_o(s)/V_{in}(s)$ of the circuit shown in Figure 15.80.

FIGURE 15.80

LRC circuit.

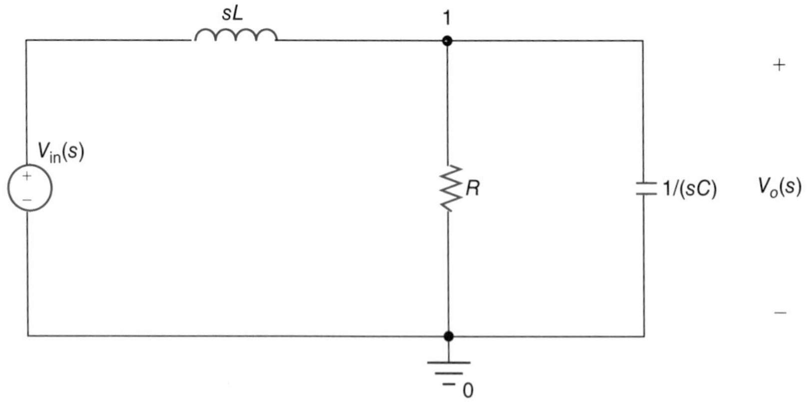

Summing the currents leaving node 1, we obtain

$$\frac{V_o(s) - V_{in}(s)}{sL} + \frac{V_o(s)}{R} + \frac{V_o(s)}{\dfrac{1}{sC}} = 0$$

continued

Example 15.20 continued

which can be rearranged as

$$V_o(s)\left(\frac{1}{sL} + \frac{1}{R} + sC\right) = \frac{V_{in}(s)}{sL}$$

Taking the ratio of $V_o(s)$ to $V_{in}(s)$, we obtain

$$H(s) = \frac{V_o(s)}{V_{in}(s)} = \frac{\frac{1}{sL}}{\frac{1}{sL} + \frac{1}{R} + sC} = \frac{1}{s^2LC + \frac{L}{R}s + 1}$$
$$= \frac{\frac{1}{LC}}{s^2 + \frac{1}{RC}s + \frac{1}{LC}} \qquad \textbf{(15.63)}$$

Exercise 15.20

Find the transfer function $H(s) = V_o(s)/V_{in}(s)$ of the circuit shown in Figure 15.81.

FIGURE 15.81

CRL circuit.

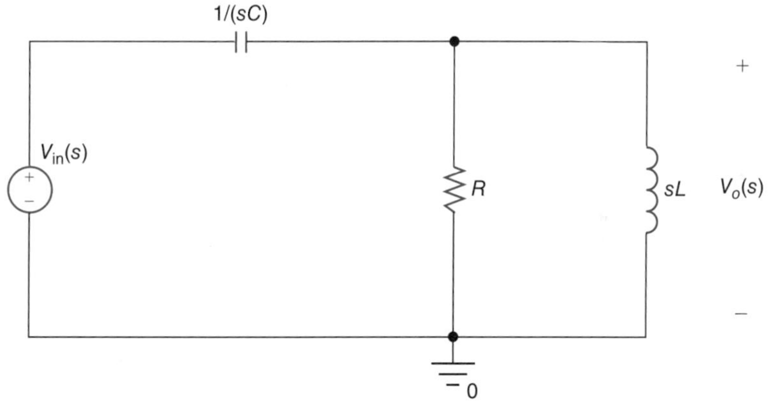

Answer:

$$H(s) = \frac{V_o(s)}{V_{in}(s)} = \frac{s^2}{s^2 + \frac{1}{RC}s + \frac{1}{LC}} \qquad \textbf{(15.64)}$$

EXAMPLE 15.21

Find the transfer function $H(s) = V_o(s)/V_{in}(s)$ of the circuit shown in Figure 15.82.

continued

Example 15.21 continued

FIGURE 15.82

RLC circuit.

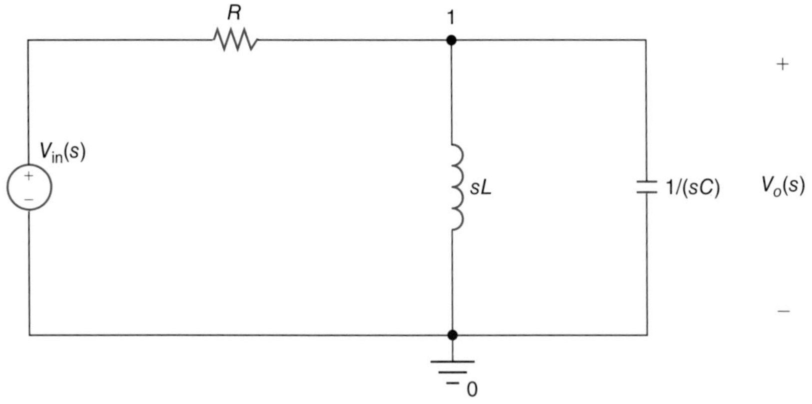

Summing the currents leaving node 1, we obtain

$$\frac{V_o(s) - V_{in}(s)}{R} + \frac{V_o(s)}{sL} + \frac{V_o(s)}{\frac{1}{sC}} = 0$$

which can be rearranged as

$$V_o(s)\left(\frac{1}{R} + \frac{1}{sL} + sC\right) = \frac{V_{in}(s)}{R}$$

Taking the ratio of $V_o(s)$ to $V_{in}(s)$, we obtain

$$H(s) = \frac{V_o(s)}{V_{in}(s)} = \frac{\frac{1}{R}}{\frac{1}{sL} + \frac{1}{R} + sC} = \frac{\frac{1}{R}sL}{s^2LC + \frac{L}{R}s + 1} = \frac{\frac{1}{RC}s}{s^2 + \frac{1}{RC}s + \frac{1}{LC}} \tag{15.65}$$

Exercise 15.21

Find the transfer function $H(s) = V_o(s)/V_{in}(s)$ of the circuit shown in Figure 15.83.

FIGURE 15.83

LCR circuit.

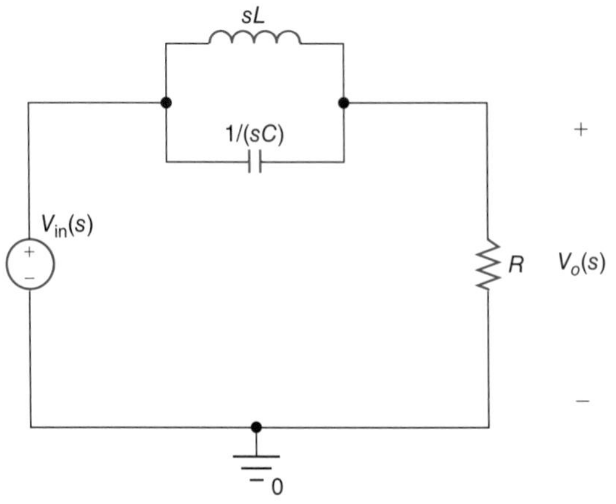

continued

Exercise 15.21 continued

Answer:

$$H(s) = \frac{V_o(s)}{V_{in}(s)} = \frac{s^2 + \dfrac{1}{LC}}{s^2 + \dfrac{1}{RC}s + \dfrac{1}{LC}} \qquad (15.66)$$

EXAMPLE 15.22

Find the output $v_o(t)$ and plot it for the circuit shown in Figure 15.84.

FIGURE 15.84

RLC circuit.

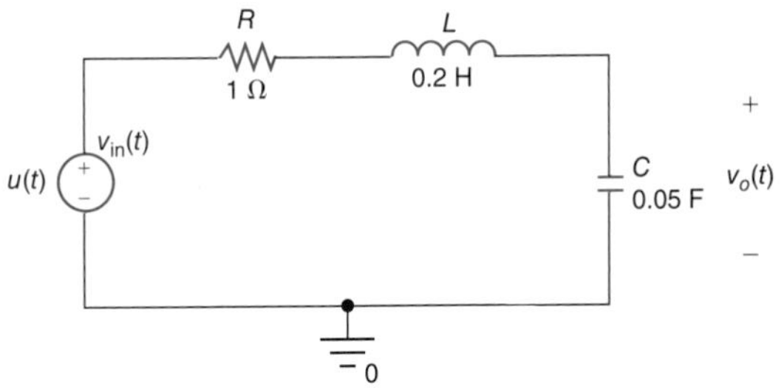

From Equation (15.59), the transfer function of the circuit shown in Figure 15.84 is given by

$$H(s) = \frac{V_o(s)}{V_{in}(s)} = \frac{\dfrac{1}{LC}}{s^2 + \dfrac{R}{L}s + \dfrac{1}{LC}} = \frac{100}{s^2 + 5s + 100}$$

The Laplace transform of the input signal is

$$V_{in}(s) = \frac{1}{s}$$

Thus, the Laplace transform of the output is

$$V_o(s) = H(s)V_{in}(s) = \frac{100}{s(s^2 + 5s + 100)} = \frac{A}{s} + \frac{Bs + C}{s^2 + 5s + 100}$$

$$= \frac{(A + B)s^2 + (5A + C)s + 100A}{s(s^2 + 5s + 100)}$$

Since $100A = 100$, we have $A = 1$. Since $A + B = 0$, we get $B = -1$. Since $5A + C = 0$, $C = -5$. Thus, $V_o(s)$ can be written as

$$V_o(s) = H(s)V_{in}(s) = \frac{1}{s} + \frac{-s - 5}{s^2 + 5s + 100} = \frac{1}{s} + \frac{-(s + 2.5) - 0.2582 \times 9.6825}{(s + 2.5)^2 + (9.6825)}$$

$$= \frac{1}{s} - \frac{(s + 2.5)}{(s + 2.5)^2 + (9.6825)} - \frac{0.2582 \times 9.6825}{(s + 2.5)^2 + (9.6825)}$$

continued

Example 15.22 continued

Taking the inverse Laplace transform, we obtain

$$v_o(t) = [1 - e^{-2.5t}\cos(9.6825t) - 0.2582e^{-2.5t}\sin(9.6825t)]\, u(t) \text{ V}$$

The output signal $v_o(t)$, along with the input signal $v_{in}(t)$, is shown in Figure 15.85.

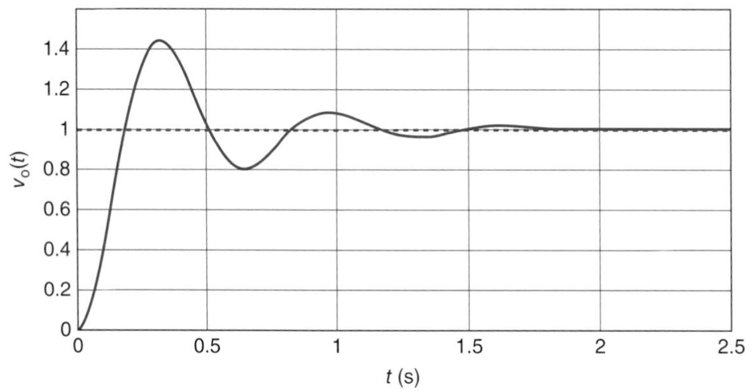

FIGURE 15.85

Plot of $v_o(t)$ (solid line) and $v_{in}(t)$ (dotted line).

Exercise 15.22

Find the output $v_o(t)$, and plot it for the circuit shown in Figure 15.86.

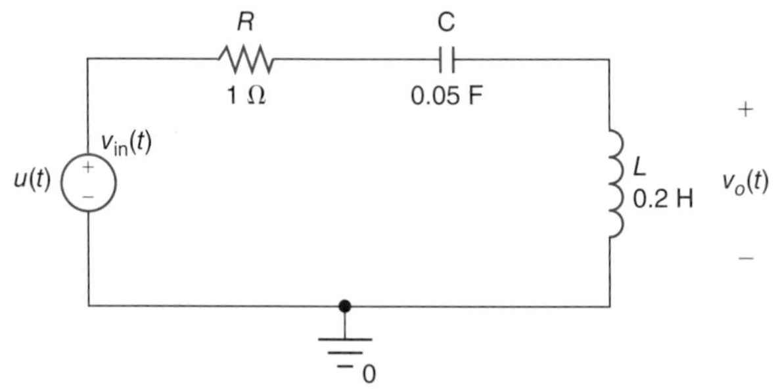

FIGURE 15.86

RCL circuit.

Answer:

$$v_o(t) = [e^{-2.5t}\cos(9.6825t) - 0.2582e^{-2.5t}\sin(9.6825t)]\, u(t) \text{ V}$$

The output signal $v_o(t)$, along with the input signal $v_{in}(t)$, is shown in Figure 15.87.

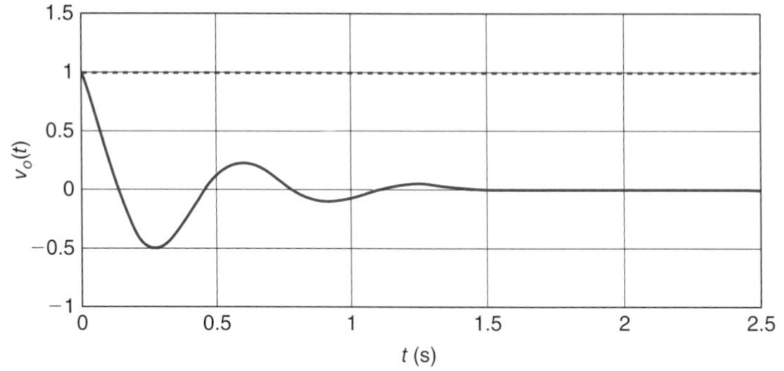

FIGURE 15.87

Plot of $v_o(t)$ and $v_{in}(t)$.

EXAMPLE 15.23

Find the transfer function $H(s) = V_o(s)/V_{in}(s)$ of the circuit shown in Figure 15.88.

FIGURE 15.88

Circuit for EXAMPLE 15.23.

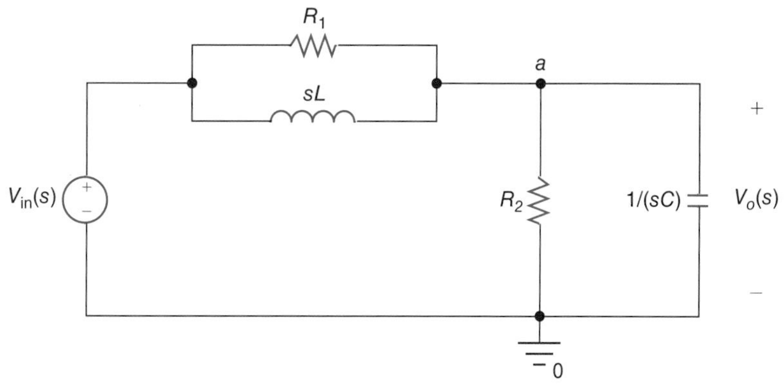

Summing the currents leaving node a, we have

$$\frac{V_o(s) - V_{in}(s)}{R_1} + \frac{V_o(s) - V_{in}(s)}{sL} + \frac{V_o(s)}{R_2} + \frac{V_o(s)}{\frac{1}{sC}} = 0$$

which can be rearranged as

$$V_o(s)\left(\frac{1}{R_1} + \frac{1}{sL} + \frac{1}{R_2} + sC\right) = \left(\frac{1}{R_1} + \frac{1}{sL}\right)V_{in}(s)$$

The transfer function is given by

$$\frac{V_o(s)}{V_{in}(s)} = \frac{\frac{1}{R_1} + \frac{1}{sL}}{\frac{1}{R_1} + \frac{1}{sL} + \frac{1}{R_2} + sC}$$

Multiplication by R_1R_2Ls of every term in the numerator and denominator yields

$$H(s) = \frac{V_o(s)}{V_{in}(s)} = \frac{R_2Ls + R_1R_2}{R_2Ls + R_1R_2 + R_1Ls + R_1R_2LCs^2}$$

Division by R_1R_2LC of every term in the numerator and denominator yields

$$H(s) = \frac{V_o(s)}{V_{in}(s)} = \frac{\frac{R_2L}{R_1R_2LC}s + \frac{R_1R_2}{R_1R_2LC}}{\frac{R_2L}{R_1R_2LC}s + \frac{R_1R_2}{R_1R_2LC} + \frac{R_1L}{R_1R_2LC}s + s^2}$$

$$= \frac{\frac{1}{R_1C}s + \frac{1}{LC}}{\frac{1}{R_1C}s + \frac{1}{LC} + \frac{1}{R_2C}s + s^2}$$

continued

Example 15.23 continued

which can be rearranged as

$$H(s) = \frac{V_o(s)}{V_{in}(s)} = \frac{\frac{1}{R_1C}s + \frac{1}{LC}}{s^2 + \left(\frac{1}{R_1C} + \frac{1}{R_2C}\right)s + \frac{1}{LC}}$$

Exercise 15.23

Find the transfer function $H(s) = V_o(s)/V_{in}(s)$ of the circuit shown in Figure 15.89.

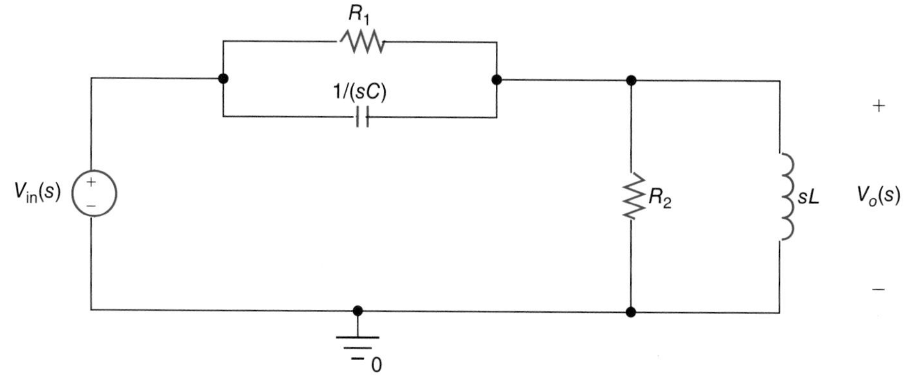

FIGURE 15.89
Circuit for EXERCISE 15.23.

Answer:

$$H(s) = \frac{V_o(s)}{V_{in}(s)} = \frac{s^2 + \frac{1}{R_1C}s}{s^2 + \left(\frac{1}{R_1C} + \frac{1}{R_2C}\right)s + \frac{1}{LC}}.$$

EXAMPLE 15.24

Find the transfer function $H(s) = V_o(s)/V_{in}(s)$ of the circuit shown in Figure 15.90.

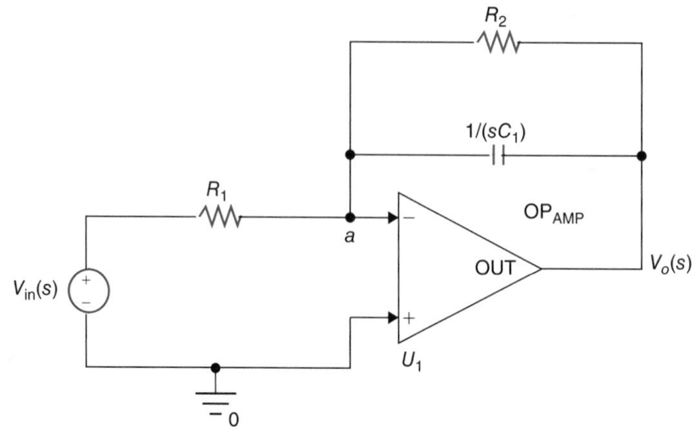

FIGURE 15.90
Circuit for EXAMPLE 15.24.

continued

Example 15.24 continued

Notice that the voltage at node a is zero. Summing the currents leaving node a, we get

$$\frac{0 - V_{in}(s)}{R_1} + \frac{0 - V_o(s)}{R_2} + \frac{0 - V_o(s)}{\frac{1}{sC_1}} = 0$$

$$\left(sC_1 + \frac{1}{R_2}\right)V_o(s) = -\frac{1}{R_1}V_{in}(s)$$

$$H(s) = \frac{V_o(s)}{V_{in}(s)} = -\frac{\frac{1}{R_1}}{s + \frac{1}{R_2}} = -\frac{\frac{1}{R_1C_1}}{s + \frac{1}{R_2C_1}} \tag{15.67}$$

There is one pole at $s = -1/R_2C_1$ and one zero at $s = \infty$. At $s = 0$, $H(s) = -R_2/R_1$, and at $s = \infty$, $H(s) = 0$. This is a low-pass filter.

Exercise 15.24

Find the transfer function $H(s) = V_o(s)/V_{in}(s)$ of the circuit shown in Figure 15.91.

FIGURE 15.91

Circuit for EXERCISE 15.24.

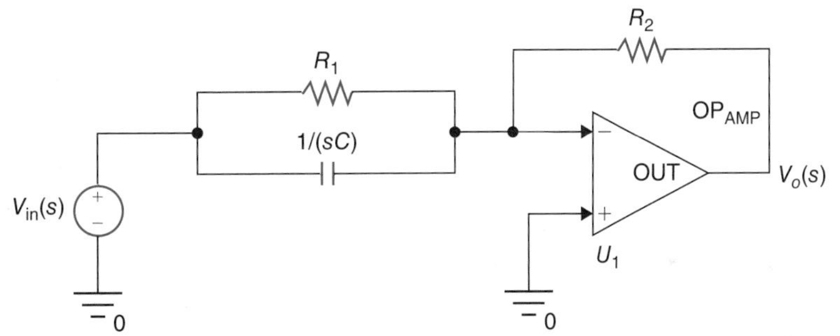

Answer:

$$H(s) = \frac{V_o(s)}{V_{in}(s)} = -R_2C\left(s + \frac{1}{R_1C}\right).$$

EXAMPLE 15.25

Find the transfer function $H(s) = V_o(s)/V_{in}(s)$ of the circuit shown in Figure 15.92.

Notice that the voltage at node a is zero. Summing the currents leaving node a, we get

$$\frac{0 - V_{in}(s)}{R_1 + \frac{1}{sC_1}} + \frac{0 - V_o(s)}{R_2} = 0$$

continued

Example 15.25 continued

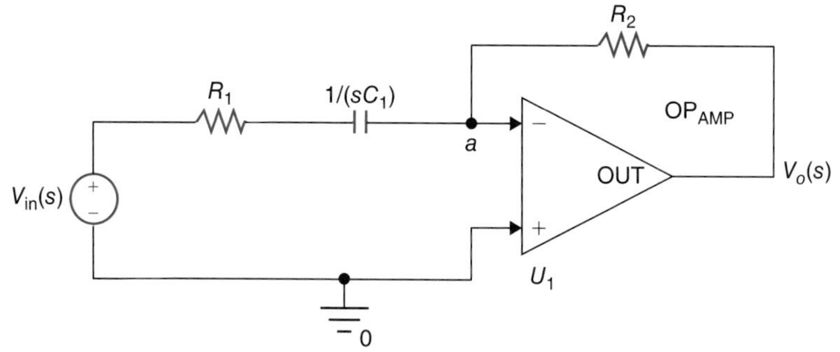

FIGURE 15.92

Circuit for EXAMPLE 15.25.

which can be rearranged as

$$\frac{1}{R_2}V_o(s) = -\frac{1}{R_1 + \dfrac{1}{sC_1}}V_{in}(s)$$

The transfer function is given by

$$H(s) = \frac{V_o(s)}{V_{in}(s)} = -\frac{\dfrac{1}{R_1 + \dfrac{1}{sC_1}}}{\dfrac{1}{R_2}} = -\frac{sR_2C_1}{sR_1C_1 + 1} = -\frac{R_2}{R_1}\frac{s}{s + \dfrac{1}{R_1C_1}} \tag{15.68}$$

There is one pole at $s = -1/R_1C_1$ and one zero at $s = 0$. At $s = 0$, $H(s) = 0$, and at $s = \infty$, $H(s) = -R_2/R_1$. This is a high-pass filter.

Exercise 15.25

Find the transfer function $H(s) = V_o(s)/V_{in}(s)$ of the circuit shown in Figure 15.93.

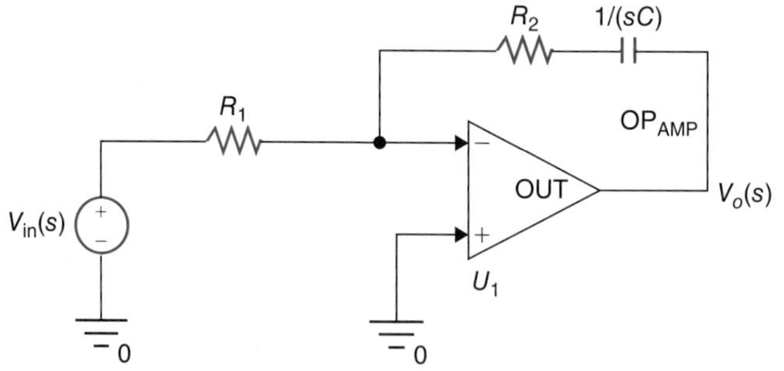

FIGURE 15.93

Circuit for EXERCISE 15.25.

Answer:

$$H(s) = \frac{V_o(s)}{V_{in}(s)} = -\frac{R_2}{R_1}\frac{s + \dfrac{1}{R_2C}}{s}.$$

EXAMPLE 15.26

Find the transfer function $H(s) = V_o(s)/V_{in}(s)$ of the circuit shown in Figure 15.94.

FIGURE 15.94

Circuit for EXAMPLE 15.26.

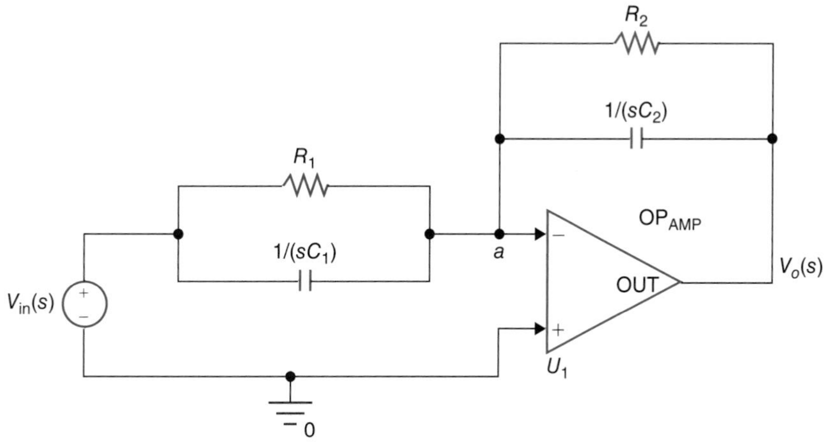

Notice that the voltage at node a is zero. Summing the currents leaving node a, we get

$$\frac{0 - V_{in}(s)}{R_1} + \frac{0 - V_{in}(s)}{\frac{1}{sC_1}} + \frac{0 - V_o(s)}{R_2} + \frac{0 - V_o(s)}{\frac{1}{sC_2}} = 0$$

which can be rearranged as

$$\left(sC_2 + \frac{1}{R_2}\right)V_o(s) = -\left(sC_1 + \frac{1}{R_1}\right)V_{in}(s)$$

The transfer function is given by

$$H(s) = \frac{V_o(s)}{V_{in}(s)} = -\frac{sC_1 + \frac{1}{R_1}}{sC_2 + \frac{1}{R_2}} = -\frac{C_1}{C_2}\frac{s + \frac{1}{R_1C_1}}{s + \frac{1}{R_2C_2}} \quad \textbf{(15.69)}$$

There is one pole at $s = -1/R_2C_2$ and one zero at $s = -1/R_1C_1$. At $s = 0$, $H(s) = -R_2/R_1$, and at $s = \infty$, $H(s) = -C_1/C_2$.

Exercise 15.26

Find the transfer function $H(s) = V_o(s)/V_{in}(s)$ of the circuit shown in Figure 15.95.

continued

Exercise 15.26 continued

FIGURE 15.95

Circuit for EXERCISE 15.26.

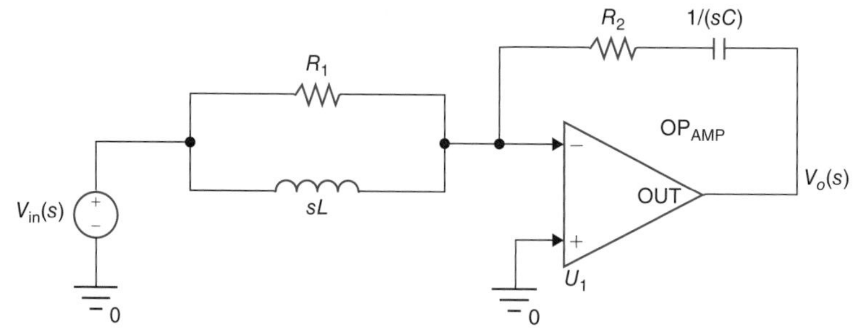

Answer:

$$H(s) = \frac{V_o(s)}{V_{in}(s)} = -\frac{R_2}{R_1}\left(\frac{s + \dfrac{1}{R_2C}}{s}\right)\left(\frac{s + \dfrac{R_1}{L}}{s}\right).$$

EXAMPLE 15.27

Find the transfer function $H(s) = V_o(s)/V_{in}(s)$ for the circuit shown in Figure 15.96.

FIGURE 15.96

Circuit for EXAMPLE 15.27.

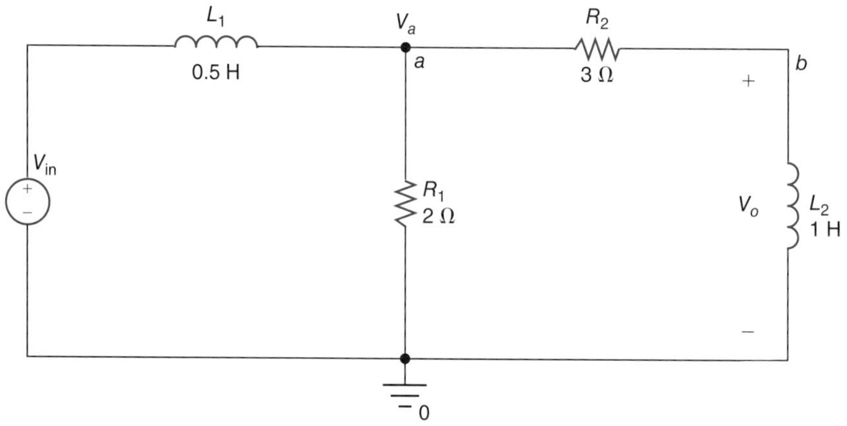

Summing the currents leaving node a, we obtain

$$\frac{V_a - V_{in}}{sL_1} + \frac{V_a}{R_1} + \frac{V_a - V_o}{R_2} = 0 \quad \textbf{(15.70)}$$

Summing the currents leaving node b, we obtain

$$\frac{V_o - V_a}{R_2} + \frac{V_o}{sL_2} = 0 \quad \textbf{(15.71)}$$

Solving Equation (15.71) for V_a, we obtain

$$V_a = \left(1 + \frac{R_2}{sL_2}\right)V_o \quad \textbf{(15.72)}$$

continued

Example 15.27 continued

Equation (15.70) can be rearranged as

$$\left(\frac{1}{sL_1}+\frac{1}{R_1}+\frac{1}{R_2}\right)V_a-\frac{1}{R_2}V_o=\frac{V_{in}}{sL_1} \tag{15.73}$$

Substitution of Equation (15.72) into Equation (15.73) yields

$$\left(\frac{1}{sL_1}+\frac{1}{R_1}+\frac{1}{R_2}\right)\left(1+\frac{R_2}{sL_2}\right)V_o-\frac{1}{R_2}V_o=\frac{V_{in}}{sL_1} \tag{15.74}$$

Taking the ratio of V_o to V_{in}, we obtain

$$H(s)=\frac{V_o(s)}{V_{in}(s)}=\frac{\frac{1}{sL_1}}{\left(\frac{1}{sL_1}+\frac{1}{R_1}+\frac{1}{R_2}\right)\left(1+\frac{R_2}{sL_2}\right)-\frac{1}{R_2}} \tag{15.75}$$

Multiplication by $sL_1sL_2R_1R_2$ yields

$$H(s)=\frac{V_o(s)}{V_{in}(s)}=\frac{sL_2R_1R_2}{(sR_1R_2+sL_1R_2+sL_1R_1)(sL_2+R_2)-sL_1sL_2R_1}$$

$$=\frac{sL_2R_1R_2}{s^2R_2L_1L_2+sR_1R_2L_2+sL_1R_2^2+sL_1R_1R_2+R_1R_2^2}$$

Dividing by $L_1L_2R_2$, we obtain

$$H(s)=\frac{V_o(s)}{V_{in}(s)}=\frac{\frac{R_1}{L_1}s}{s^2+\left(\frac{R_1}{L_1}+\frac{R_2}{L_2}+\frac{R_1}{L_2}\right)s+\frac{R_1R_2}{L_1L_2}} \tag{15.76}$$

For $R_1=2\ \Omega$, $R_2=3\ \Omega$, $L_1=0.5$ H, $L_2=1$ H, Equation (15.76) becomes

$$H(s)=\frac{V_o(s)}{V_{in}(s)}=\frac{4s}{s^2+9s+12} \tag{15.77}$$

MATLAB

```
% EXAMPLE 15.27
clear all;
L1=0.5;L2=1;R1=2;R2=3;Vin=1;
syms Va Vo s
[Va,Vo]=solve((Va-Vin)/(s*L1)+Va/R1+(Va-Vo)/R2,...
(Vo-Va)/R2+Vo/(s*L2),Va,Vo)
[Num,Den]=numden(Vo)
N1=sym2poly(Num)
D1=sym2poly(Den)
N=N1/D1(1)
D=D1/D1(1)
H=tf(N,D)
```

Exercise 15.27

Find the transfer function $H(s)=V_o(s)/V_{in}(s)$ of the circuit shown in Figure 15.97.

continued

Exercise 15.27 continued

FIGURE 15.97

Circuit for EXERCISE 15.27.

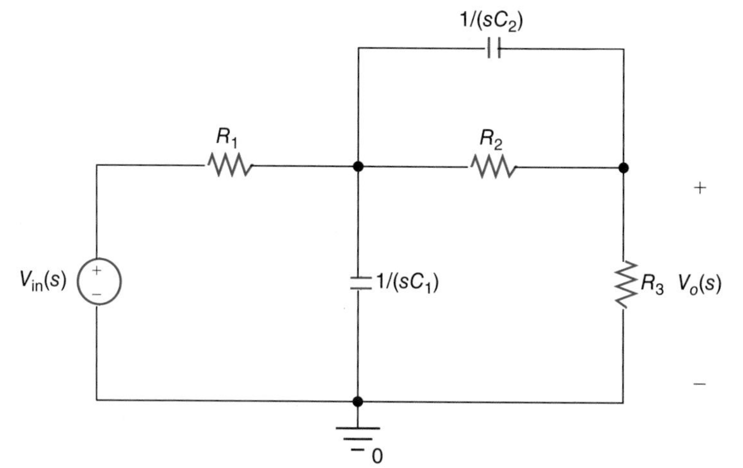

Answer:

$$H(s) = \frac{V_o(s)}{V_{in}(s)} = \frac{1}{R_1C_1} \frac{s + \dfrac{1}{R_2C_2}}{s^2 + \left(\dfrac{1}{R_1C_1} + \dfrac{1}{R_2C_2} + \dfrac{1}{R_3C_1} + \dfrac{1}{R_3C_2}\right)s + \dfrac{R_1 + R_2 + R_3}{R_1R_2R_3C_1C_2}}.$$

EXAMPLE 15.28

Find the transfer function $H(s) = V_o(s)/V_{in}(s)$ for the circuit shown in Figure 15.98.

FIGURE 15.98

Circuit for EXAMPLE 15.28.

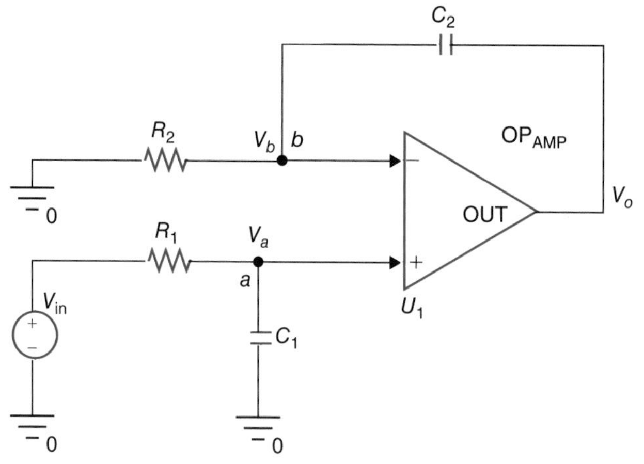

Summing the currents leaving node a, we obtain

$$\frac{V_a - V_{in}}{R_1} + \frac{V_a}{\dfrac{1}{sC_1}} = 0 \tag{15.78}$$

continued

Example 15.28 continued

Solving Equation (15.78) for V_a, we obtain

$$V_a = \frac{1}{sR_1C_1 + 1}V_{in} \quad (15.79)$$

Summing the currents leaving node b, we obtain

$$\frac{V_b}{R_2} + \frac{V_b - V_o}{\frac{1}{sC_2}} = 0 \quad (15.80)$$

Solving Equation (15.80) for V_b, we obtain

$$V_b = \frac{sR_2C_2}{sR_2C_2 + 1}V_o \quad (15.81)$$

Since $V_a = V_b$ due to a virtual shirt, from Equations (15.79) and (15.81), we obtain

$$\frac{sR_2C_2}{sR_2C_2 + 1}V_o = \frac{1}{sR_1C_1 + 1}V_{in} \quad (15.82)$$

The transfer function is given by

$$H(s) = \frac{V_o(s)}{V_{in}(s)} = \frac{sR_2C_2 + 1}{sR_2C_2(sR_1C_1 + 1)} = \frac{1}{R_1C_1}\frac{s + \frac{1}{R_2C_2}}{s\left(s + \frac{1}{R_1C_1}\right)} \quad (15.83)$$

Exercise 15.28

Find the transfer function $H(s) = V_o(s)/V_{in}(s)$ of the circuit shown in Figure 15.99.

FIGURE 15.99

Circuit for EXERCISE 15.28.

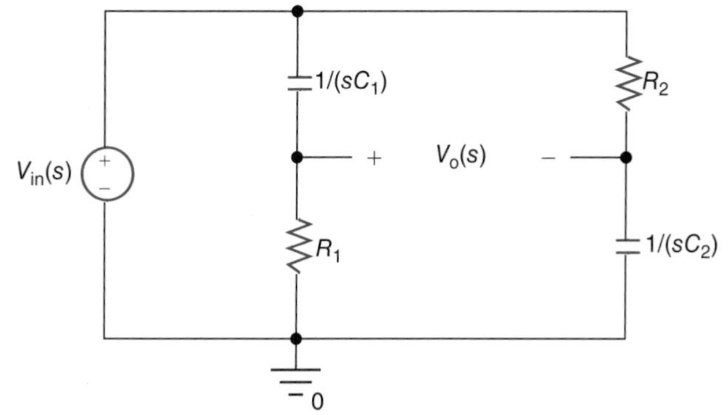

Answer:

$$H(s) = \frac{V_o(s)}{V_{in}(s)} = \frac{s^2 - \frac{1}{R_1RC_1C_2}}{s^2 + \left(\frac{1}{R_1C_1} + \frac{1}{R_2C_2}\right)s + \frac{1}{R_1R_2C_1C_2}}.$$

EXAMPLE 15.29

Find the transfer function $H(s) = V_o(s)/V_{in}(s)$ for the circuit shown in Figure 15.100.

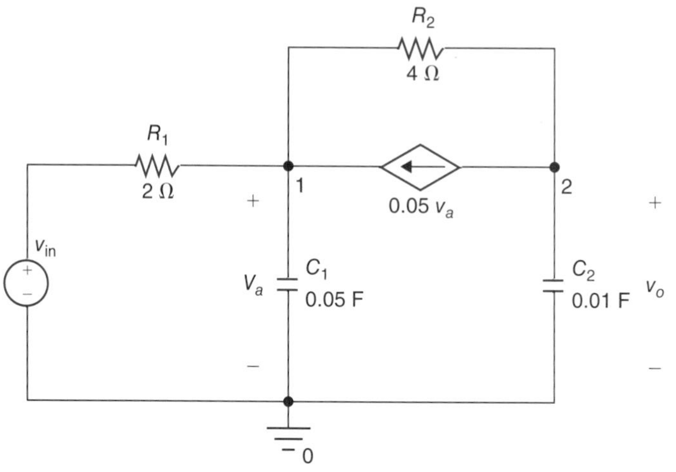

FIGURE 15.100 Circuit for EXAMPLE 15.29.

Summing the currents leaving node 1, we obtain

$$\frac{V_a - V_{in}}{R_1} + \frac{V_a}{\frac{1}{sC_1}} + \frac{V_a - V_o}{R_2} - g_m V_a = 0$$

which can be rearranged as

$$\left(sC_1 + \frac{1}{R_1} + \frac{1}{R_2} - g_m\right)V_a = \frac{V_{in}}{R_1} + \frac{V_o}{R_2}$$

Solving for V_a, we obtain

$$V_a = \frac{\frac{V_{in}}{R_1} + \frac{V_o}{R_2}}{sC_1 + \frac{1}{R_1} + \frac{1}{R_2} - g_m} \quad \textbf{(15.84)}$$

Summing the currents leaving node 2, we obtain

$$\frac{V_o - V_a}{R_2} + g_m V_a + \frac{V_o}{\frac{1}{sC_2}} = 0$$

which can be rearranged as

$$\left(sC_2 + \frac{1}{R_2}\right)V_o + \left(-\frac{1}{R_2} + g_m\right)V_a = 0 \quad \textbf{(15.85)}$$

continued

Example 15.29 continued

Substituting Equation (15.84) into Equation (15.85), we get

$$\left[\left(sC_2+\frac{1}{R_2}\right)+\left(-\frac{1}{R_2}+g_m\right)\left(\frac{\frac{1}{R_2}}{sC_1+\frac{1}{R_1}+\frac{1}{R_2}-g_m}\right)\right]V_o$$

$$=\left(\frac{1}{R_2}-g_m\right)\left(\frac{\frac{V_{in}}{R_1}}{sC_1+\frac{1}{R_1}+\frac{1}{R_2}-g_m}\right)$$

Solving for V_o/V_{in}, we obtain

$$H(s)=\frac{V_o}{V_{in}}=\frac{\left(\frac{1}{R_2}-g_m\right)\left(\frac{\frac{1}{R_1}}{sC_1+\frac{1}{R_1}+\frac{1}{R_2}-g_m}\right)}{\left[\left(sC_2+\frac{1}{R_2}\right)+\left(-\frac{1}{R_2}+g_m\right)\left(\frac{\frac{1}{R_2}}{sC_1+\frac{1}{R_1}+\frac{1}{R_2}-g_m}\right)\right]}$$

$$=\frac{\frac{1}{R_1}\left(\frac{1}{R_2}-g_m\right)}{\left(sC_2+\frac{1}{R_2}\right)\left(sC_1+\frac{1}{R_1}+\frac{1}{R_2}-g_m\right)+\frac{1}{R_2}\left(-\frac{1}{R_2}+g_m\right)}$$

$$=\frac{\frac{1}{R_1C_1C_2}\left(\frac{1}{R_2}-g_m\right)}{\left(s+\frac{1}{R_2C_2}\right)\left(s+\frac{1}{R_1C_1}+\frac{1}{R_2C_1}-\frac{g_m}{C_1}\right)+\frac{1}{R_2C_1C_2}\left(-\frac{1}{R_2}+g_m\right)}$$

$$=\frac{\frac{1}{R_1C_1C_2}\left(\frac{1}{R_2}-g_m\right)}{s^2+\left(\frac{1}{R_2C_2}+\frac{1}{R_1C_1}+\frac{1}{R_2C_1}-\frac{g_m}{C_1}\right)s+\frac{1}{R_2R_2C_1C_2}}$$

Notice that the constant term in the numerator is smaller than the constant term of the denominator due to the VCCS that flows from the output to the input. The magnitude of $H(s)$ at $s = 0$ is less than 1. For the values given, the transfer function becomes

$$H(s)=\frac{200}{s^2+39s+250}$$

Exercise 15.29

Find the transfer function $H(s) = V_o(s)/V_{in}(s)$ for the circuit shown in Figure 15.101 for $g_m = 0$ and $g_m = 0.05$.

continued

Exercise 15.29 continued

FIGURE 15.101

Circuit for EXERCISE 15.29.

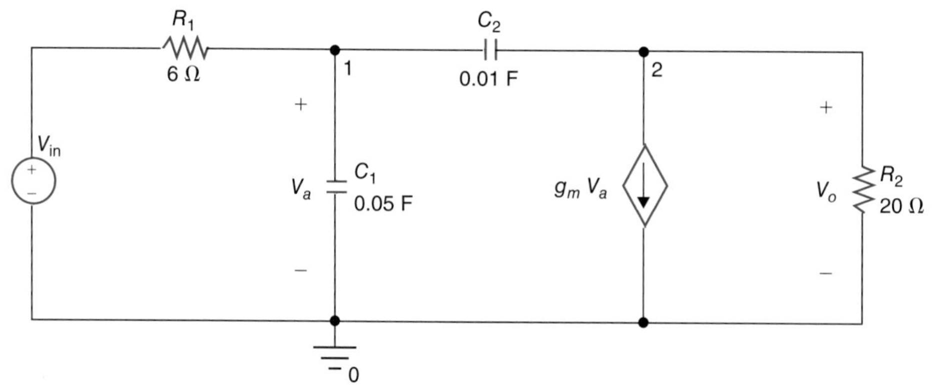

Answer:

For $g_m = 0$,

$$H(s) = \frac{3.3333s}{s^2 + 9.3333s + 16.6667}$$

For $g_m = 0.05$,

$$H(s) = \frac{3.3333s - 16.6667}{s^2 + 10.3333s + 16.6667}.$$

15.9 Convolution

The convolution of two continuous time signals $f_1(t)$ and $f_2(t)$ is defined as

$$f(t) = \int_{-\infty}^{\infty} f_1(\lambda) f_2(t - \lambda) d\lambda \tag{15.86}$$

and is denoted by $f(t) = f_1(t) * f_2(t)$.

From Equation (15.86), it is apparent that the computation of convolution involves the following steps:

1. Change the independent variable from t to λ (or any other letter) for the two signals.
2. Fold (flip) one of the two signals, say $f_2(\lambda)$, with respect to the ordinate to get $f_2(-\lambda)$. The folded signal $f_2(-\lambda)$ is the mirror image of $f_2(\lambda)$ with respect to the ordinate. Choose the simpler of the two signals to fold.
3. Shift the folded signal $f_2(-\lambda)$ by t to get $f_2[-(\lambda - t)] = f_2(t - \lambda)$. If t is positive, the mirror image $f_2(-\lambda)$ will be moved to the right by t, and if t is negative, the mirror image $f_2(-\lambda)$ will be moved to the left by $|t|$. Remember that the amount of movement from the mirror image $f_2(-\lambda)$ is t.
4. Multiply $f_1(\lambda)$ and $f_2(t - \lambda)$.
5. Integrate the product $f_1(\lambda)f_2(t - \lambda)$ from $-\infty$ to $+\infty$ with respect to λ. The limits of integration can be restricted to the values where the product $f_1(\lambda)f_2(t - \lambda)$ is nonzero. Notice that $f_1(\lambda)$ is stationary throughout the computation.
6. Repeat steps 3 through 5 for different values of the shift t.

The Laplace transform of $f(t)$ is given by

$$F(s) = L[f(t)] = \int_{-\infty}^{\infty}\left[\int_{-\infty}^{\infty} f_1(\lambda)f_2(t-\lambda)d\lambda\right]e^{-st}dt$$

Let $\tau = t - \lambda$. Then, $t = \tau + \lambda$ and $dt = d\tau$. Thus, we have

$$\begin{aligned} F(s) &= \int_{-\infty}^{\infty}\left[\int_{-\infty}^{\infty} f_1(\lambda)f_2(\tau)d\lambda\right]e^{-s(\tau+\lambda)}d\tau = \int_{-\infty}^{\infty} f_1(\lambda)e^{-s\lambda}d\lambda \int_{-\infty}^{\infty} f_2(\tau)e^{-s\tau}d\tau \\ &= F_1(s)F_2(s) \end{aligned} \tag{15.87}$$

This result indicates that the convolution in the time domain results in the multiplication of Laplace transforms in the s-domain. Several properties of convolution are listed in the next sections.

15.9.1 COMMUTATIVE PROPERTY

$$f_1(t)*f_2(t) = f_2(t)*f_1(t)$$

Proof

Let $t - \lambda = \beta$ in Equation (15.86). Then, $\lambda = t - \beta$ and $d\lambda = -d\beta$. When $\lambda = -\infty, \beta = \infty$, and when $\lambda = \infty, \beta = -\infty$. The convolution $f(t)$ given by Equation (15.86) becomes

$$f(t) = \int_{\infty}^{-\infty} f_1(t-\beta)f_2(\beta)(-d\beta) = \int_{-\infty}^{\infty} f_1(t-\beta)f_2(\beta)d\beta = f_2(t)*f_1(t) \tag{15.88}$$

15.9.2 ASSOCIATIVE PROPERTY

$$[f_1(t)*f_2(t)]*f_3(t) = f_1(t)*[f_2(t)*f_3(t)]$$

15.9.3 DISTRIBUTIVE PROPERTY

$$f_1(t)*[f_2(t) + f_3(t)] = f_1(t)*f_2(t) + f_1(t)*f_3(t)$$

15.9.4 TIME-SHIFTING PROPERTY

If $f(t) = f_1(t)*f_2(t)$, then $f_1(t - t_d)*f_2(t) = f(t - t_d)$

EXAMPLE 15.30

Find the convoluton $f(t)$ of $f_1(t) = u(t)$ and $f_2(t) = u(t)$.

continued

Example 15.30 continued

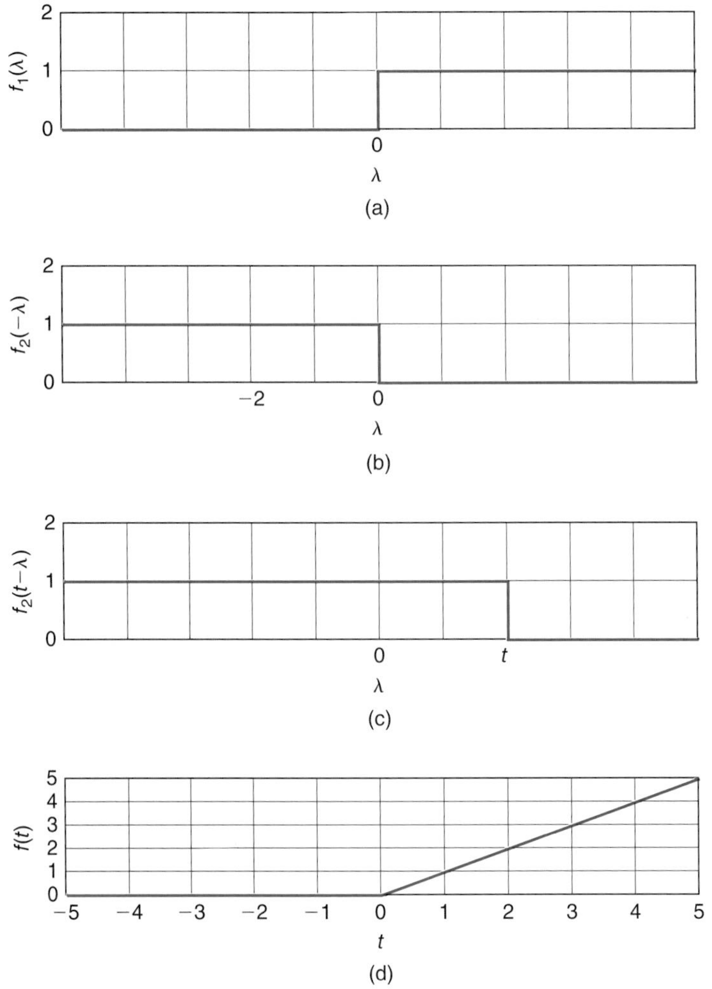

FIGURE 15.102

(a) $f_1(\lambda) = u(\lambda)$
(b) $f_2(-\lambda) = u(-\lambda)$
(c) $f_2(t - \lambda) = u(t - \lambda)$
(d) $f(t)$

Figure 15.102 shows $f_1(\lambda)$ and $f_2(-\lambda)$. If $f_2(-\lambda)$ is shifted to the left ($t < 0$), $f_1(\lambda)\,f_2(t - \lambda) = 0$ everywhere, and the convolution is zero. Thus

$$\text{for } t \leq 0,\ f(t) = 0$$

If $t > 0$, $f_2(-\lambda)$ is shifted to the right, as shown in Figure 15.102(c), $f_1(\lambda)\,f_2(t - \lambda) = 1$ for $0 \leq \lambda < t$ and zero elsewhere. Thus, we have

$$\text{for } t \geq 0,\ f(t) = \int_0^t 1 \times 1\, d\lambda = \lambda \Big|_0^t = t$$

Therefore, $f(t) = t\,u(t)$.

In the s-domain, we have $F_1(s) = 1/s$ and $F_2(s) = 1/s$. Thus, $F(s) = F_1(s)F_2(s) = 1/s^2$. The convolution is given by

$$f(t) = L^{-1}[F(s)] = L^{-1}\left[\frac{1}{s^2}\right] = t\,u(t)$$

This answer matches the one obtained in the time domain.

Exercise 15.30

Let $f_1(t) = tu(t)$ and $f_2(t) = u(t)$. Find the convolution of $f_1(t)$ and $f_2(t)$, for example, find $f(t) = f_1(t) * f_2(t)$.

Answer:
$f(t) = 0.5t^2u(t)$.

EXAMPLE 15.31

Find the convoluton $f(t)$ of $f_1(t) = 5e^{-3t}u(t)$ and $f_2(t) = u(t)$.

Figure 15.103 shows $f_1(\lambda), f_2(-\lambda), f_2(t-\lambda)$, and $f(t)$.

FIGURE 15.103

Waveforms for $f_1(\lambda)$, $f_2(-\lambda), f_2(t-\lambda)$, and $f(t)$.

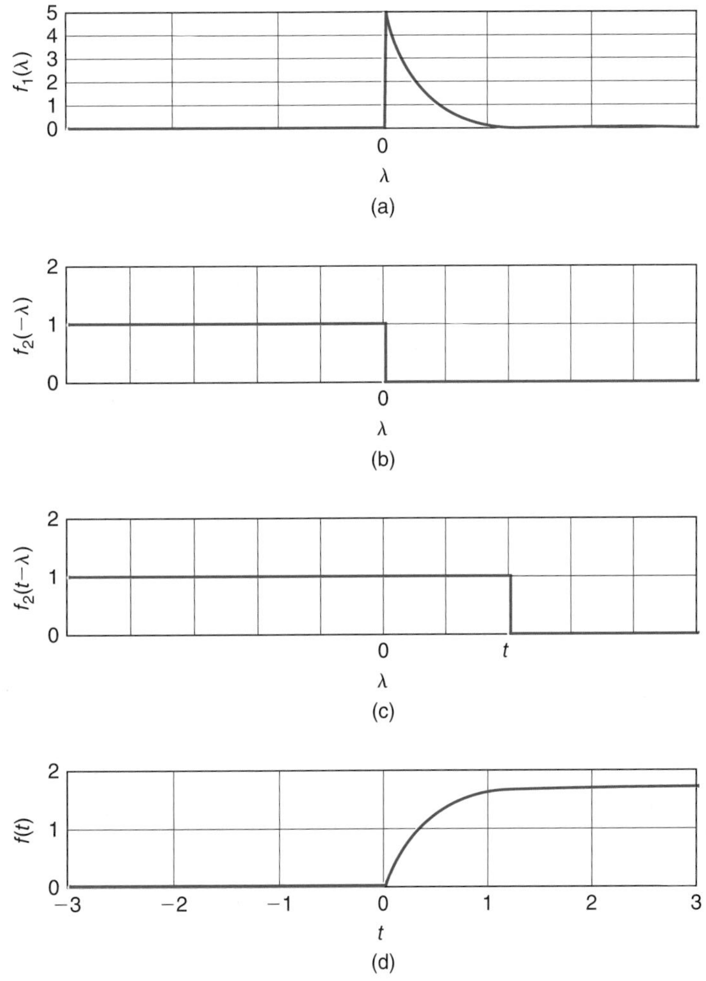

For $t < 0, f(t) = 0$.

For $t \geq 0$, $f(t) = \int_0^t 5e^{-3\lambda} \times 1\,d\lambda = 5\left.\frac{e^{-3\lambda}}{-3}\right|_0^t = \frac{5}{-3}(e^{-3t} - 1) = \frac{5}{3}(1 - e^{-3t})$.

continued

Example 15.31 continued

Thus, $f(t) = \frac{5}{3}(1 - e^{-3t})u(t)$.

Notice that $f(t)$ is the integral of $f_1(t)$ when $f_2(t)$ is a unit step function. In general, the convolution of signal $f(t)$ with unit step function $u(t)$ results in the integral of $f(t)$.

In the s-domain, we have $F_1(s) = 5/(s + 3)$ and $F_2(s) = 1/s$. Thus, $F(s) = F_1(s)F_2(s) = 5/[s(s + 3)]$. The convolution is given by

$$f(t) = L^{-1}[F(s)] = L^{-1}\left[\frac{5}{s(s+3)}\right] = L^{-1}\left[\frac{\frac{5}{3}}{s} - \frac{\frac{5}{3}}{s+3}\right] = \frac{5}{3}(1 - e^{-3t})u(t)$$

This answer matches the one obtained in the time domain.

Exercise 15.31

Use the Laplace transform to find the convoluton $f(t)$ of $f_1(t) = 3e^{-2t}u(t)$ and $f_2(t) = tu(t)$.

Answer:

$f(t) = (-0.75 + 1.5t + 0.75e^{-2t})u(t)$.

EXAMPLE 15.32

Find the convoluton $f(t)$ of $f_1(t) = 3e^{-t}u(t)$ and $f_2(t) = u(t - 2)$.

Figure 15.104 shows $f_1(\lambda), f_2(-\lambda), f_2(t - \lambda)$, and $f(t)$.

For $t < 2$, $f(t) = 0$.

$$\text{For } t \geq 2,\ f(t) = \int_0^{t-2} 3e^{-\lambda} \times 1\, d\lambda = 3\frac{e^{-\lambda}}{-1}\Bigg|_0^{t-2} = 3[1 - e^{-(t-2)}]$$

Thus, $f(t) = 3[1 - e^{-(t-2)}]u(t - 2)$.

FIGURE 15.104

Waveforms for $f_1(\lambda)$, $f_2(-\lambda), f_2(t - \lambda)$, and $f(t)$.

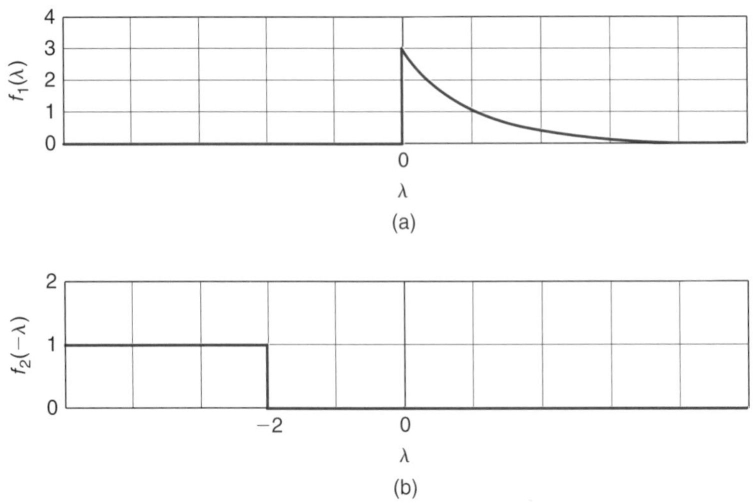

continued

Example 15.32 continued

FIGURE 15.104

continued

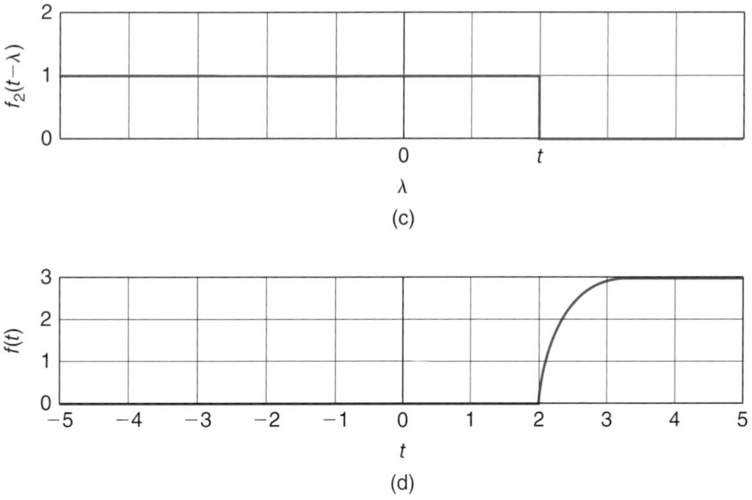

The upper limit of the integral is $t - 2$ because the amount of movement from the folded signal $f_2(-\lambda)$ is t and 2 out of t is the interval where $f_1(\lambda) = 0$. Refer to Figure 15.104.

In the s-domain, we have $F_1(s) = 3/(s + 1)$ and $F_2(s) = e^{-2s}/s$. Thus, $F(s) = F_1(s)F_2(s) = 3e^{-2s}/[s(s + 1)]$. Since

$$L^{-1}\left[\frac{3}{s(s+1)}\right] = L^{-1}\left[\frac{3}{s} - \frac{3}{s+1}\right] = 3(1 - e^{-t})u(t)$$

we have, from the time-shifting property,

$$f(t) = L^{-1}\left[\frac{3e^{-2s}}{s(s+1)}\right] = 3[1 - e^{-(t-2)}]u(t-2)$$

This answer matches the one obtained in the time domain.

Exercise 15.32

Find the convoluton $f(t)$ of $f_1(t) = u(t - 1)$ and $f_2(t) = u(t - 2)$.

Answer:

$f(t) = (t - 3)u(t - 3)$.

EXAMPLE 15.33

Find the convoluton $f(t)$ of $f_1(t) = 3e^{-2t}u(t)$ and $f_2(t) = 2e^{-5t}u(t)$.

Figure 15.105 shows $f_1(\lambda)$, $f_2(-\lambda)$, $f_2(t - \lambda)$, and $f(t)$.

For $t < 0$, $f(t) = 0$

continued

Example 15.33 continued

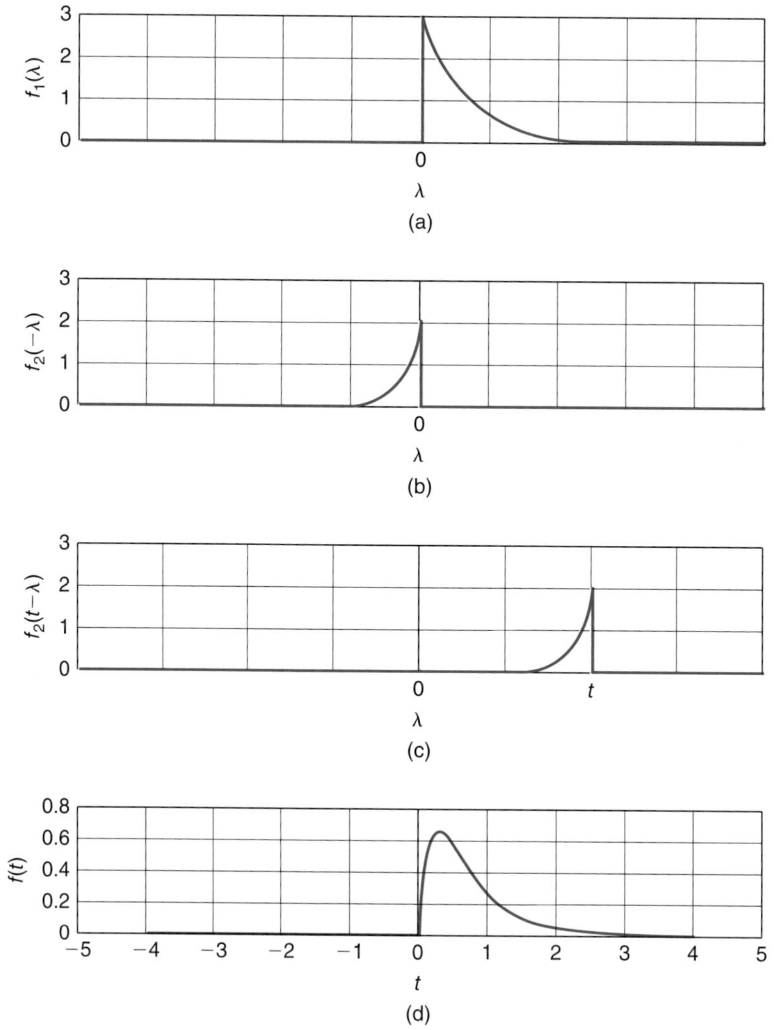

FIGURE 15.105

Waveforms for $f_1(\lambda)$, $f_2(-\lambda), f_2(t-\lambda)$, and $f(t)$.

$$\text{For } t \geq 0,\ f(t) = \int_0^t 3e^{-2\lambda} \times 2e^{5(\lambda-t)}\, d\lambda = 6e^{-5t}\int_0^t e^{3\lambda}\, d\lambda = 6e^{-5t}\left.\frac{e^{3\lambda}}{3}\right|_0^t = 2e^{-5t}(e^{3t}-1)$$

$$= 2(e^{-2t} - e^{-5t})$$

Thus, $f(t) = 2(e^{-2t} - e^{-5t})u(t)$.

In the s-domain, we have $F_1(s) = 3/(s+2)$ and $F_2(s) = 2/(s+5)$. Thus, $F(s) = F_1(s)F_2(s) = 6/[(s+2)(s+5)]$. The convolution is given by

$$f(t) = L^{-1}[F(s)] = L^{-1}\left[\frac{6}{(s+2)(s+5)}\right] = L^{-1}\left[\frac{2}{s+2} - \frac{2}{s+5}\right] = 2(e^{-2t} - e^{-5t})u(t)$$

This answer matches the one obtained in the time domain.

Exercise 15.33

Find the convoluton $f(t)$ of $f_1(t) = 3e^{-2(t-3)}u(t-3)$ and $f_2(t) = 2e^{-5(t-4)}u(t-4)$.

Answer:

$f(t) = 2[e^{-2(t-7)} - e^{-5(t-7)}]u(t-7).$

EXAMPLE 15.34

Find the convoluton $f(t)$ of $f_1(t) = 5e^{-2t}u(t)$ and $f_2(t) = u(t) - u(t-3)$.

Figure 15.106 shows $f_1(\lambda), f_2(-\lambda), f_2(t-\lambda)$, and $f(t)$.

FIGURE 15.106

Waveforms for $f_1(\lambda)$, $f_2(-\lambda), f_2(t-\lambda)$, and $f(t)$.

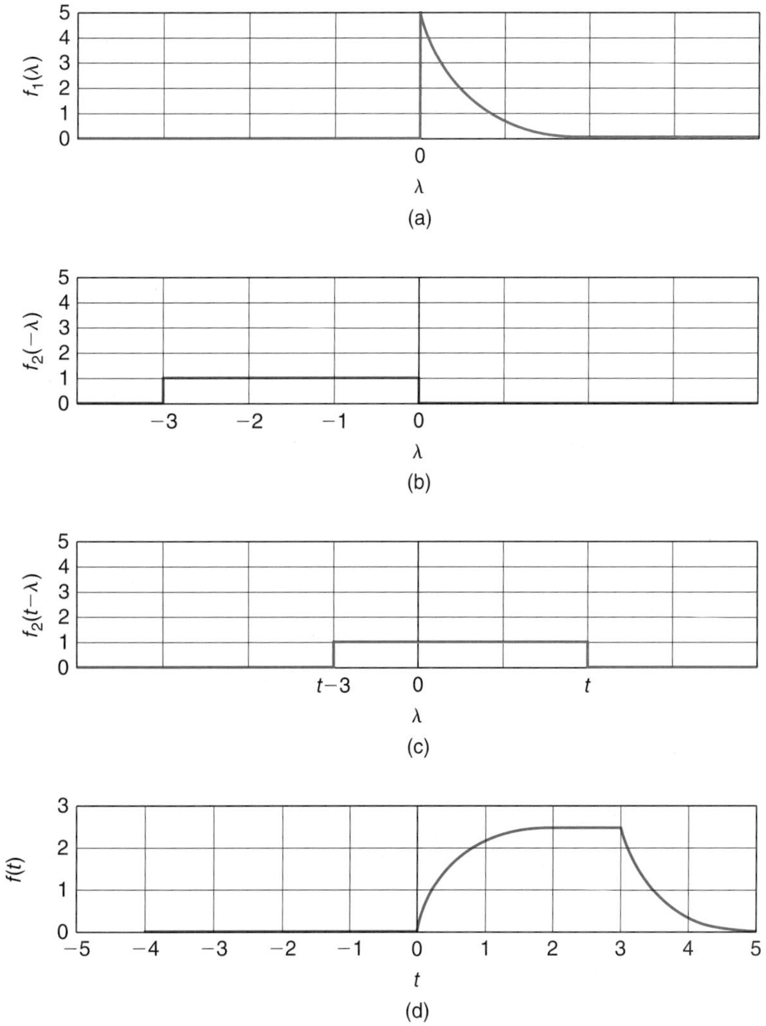

For $t < 0, f(t) = 0$.

$$\text{For } 0 \le t < 3,\ f(t) = \int_0^t 5e^{-2\lambda} \times 1\, d\lambda = 5\frac{e^{-2\lambda}}{-2}\Bigg|_0^t = \frac{5}{2}(1 - e^{-2t}).$$

continued

Example 15.34 continued

$$\text{For } 3 \le t,\ f(t) = \int_{t-3}^{t} 5e^{-2\lambda} \times 1\, d\lambda = 5\left.\frac{e^{-2\lambda}}{-2}\right|_{t-3}^{t} = \frac{5}{-2}[e^{-2t} - e^{-2(t-3)}]$$

$$= 2.5(e^{6} - 1)e^{-2t}.$$

In the s-domain, we have $F_1(s) = 5/(s+2)$ and $F_2(s) = (1 - e^{-3s})/s$. Thus, $F(s) = F_1(s)F_2(s) = 5(1 - e^{-3s})/[s(s+2)]$. From

$$L^{-1}\left[\frac{5}{s(s+2)}\right] = L^{-1}\left[\frac{\frac{5}{2}}{s} - \frac{\frac{5}{2}}{s+2}\right] = \frac{5}{2}(1 - e^{-2t})u(t)$$

we have

$$L^{-1}\left[\frac{5(1 - e^{-3s})}{s(s+2)}\right] = \frac{5}{2}(1 - e^{-2t})u(t) - \frac{5}{2}(1 - e^{-2(t-3)})u(t-3)$$

For $0 \le t < 3$, $f(t) = \frac{5}{2}(1 - e^{-2t})u(t)$.

For $3 \le t$,

$$f(t) = \left[\frac{5}{2}(1 - e^{-2t}) - \frac{5}{2}(1 - e^{-2(t-3)})\right]u(t-3) = \left[-\frac{5}{2}e^{-2t} + \frac{5}{2}e^{-2(t-3)}\right]u(t-3)$$

$$= \frac{5}{2}(e^{6} - 1)e^{-2t}u(t-3)$$

This answer matches the one obtained in the time domain.

Exercise 15.34

Find the convoluton $f(t)$ of $f_1(t) = 5e^{-2t}u(t)$ and $f_2(t) = u(t-1) - u(t-4)$.

Answer:

$$f(t) = \begin{cases} 0, & t < 1 \\ \frac{5}{2}(1 - e^{-2(t-1)}), & 1 \le t < 4 \\ 2.5(e^{6} - 1)e^{-2(t-4)}u(t-4), & 4 \le t \end{cases}.$$

EXAMPLE 15.35

Determine the convolution of $f_1(t) = e^{-2t}u(t)$ and $f_2(t) = e^{5t}u(-t)$ shown in Figures 15.107(a) and 15.107(b).

Note that $f_2(-\lambda) = e^{-5\lambda}u(\lambda)$ and $f_2(t - \lambda) = e^{5(t-\lambda)}u(\lambda - t)$

continued

Example 15.35 continued

FIGURE 15.107

Waveforms for $f_1(\lambda)$, $f_2(\lambda), f_2(-\lambda), f_2(t-\lambda)$, and $f(t)$.

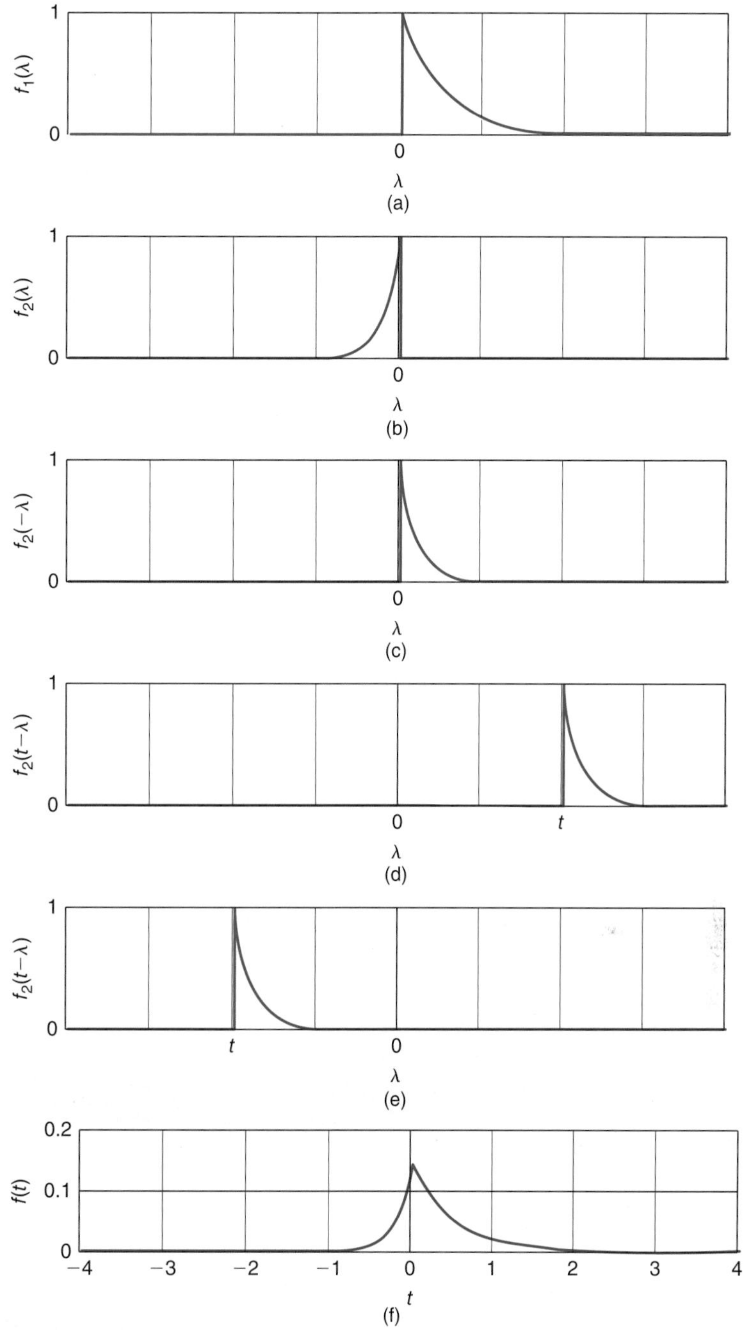

Thus, for $t < 0$,

$$f(t) = \int_0^{\infty} e^{-2\lambda} e^{5(t-\lambda)} d\lambda = e^{5t} \int_0^{\infty} e^{-7\lambda} d\lambda = \frac{1}{7} e^{5t}$$

For $t \geq 0$,

$$f(t) = \int_t^{\infty} e^{-2\lambda} e^{5(t-\lambda)} d\lambda = e^{5t} \int_t^{\infty} e^{-7\lambda} d\lambda = e^{5t} \frac{0 - e^{-7t}}{-7} = \frac{1}{7} e^{-2t}$$

continued

Example 15.35 continued

Therefore,

$$f(t) = \frac{1}{7}[e^{5t}u(-t) + e^{-2t}u(t)]$$

This $f(t)$ is plotted in Figure 15.107(f).

The convolution $f(t)$ can be obtained using the Laplace transform by taking the inverse Laplace transform of the product of the Laplace tansforms $F_1(s)$ and $F_2(s)$. $F_1(s)$, $F_2(s)$, and $F(s)$ are given, respectively, as

$$F_1(s) = \frac{1}{s+2}, \quad F_2(s) = \frac{1}{-s+5}, \quad F(s) = \frac{1}{(s+2)(-s+5)}$$

Using partial fraction expansion, we have

$$\frac{1}{(s+2)(-s+5)} = \frac{A}{s+2} + \frac{B}{-s+5}, \quad A = \left.\frac{1}{-s+5}\right|_{s=-2} = \frac{1}{7}, \quad B = \left.\frac{1}{s+2}\right|_{s=5} = \frac{1}{7}$$

Thus, the convolution $f(t)$ is given by

$$f(t) = L^{-1}\left[\frac{1}{(s+2)(-s+5)}\right] = L^{-1}\left[\frac{\frac{1}{7}}{s+2} + \frac{\frac{1}{7}}{-s+5}\right] = \frac{1}{7}e^{-2t}u(t) + \frac{1}{7}e^{5t}u(-t)$$

Exercise 15.35

Determine the convolution of $f_1(t) = e^{-2t}u(t)$ and $f_2(t) = te^{-5t}u(t)$.

Answer:

$$f(t) = \left(-\frac{1}{3}te^{-5t} + \frac{1}{9}e^{-2t} - \frac{1}{9}e^{-5t}\right)u(t).$$

EXAMPLE 15.36

Find the convolution of $f_1(t) = 5\sin(10t)u(t)$ and $f_2(t) = u(t)$.

Figure 15.108 shows $f_1(\lambda), f_2(-\lambda), f_2(t-\lambda)$, and $f(t)$.

When $f_2(\lambda)$ is flipped, we get $f_2(-\lambda) = u(-\lambda)$. Shifting $f_2(-\lambda)$ by t, we get $f_2(t-\lambda) = u(t-\lambda)$. For $t < 0, x(t) = 0$. For $t \geq 0$, we have

$$f(t) = \int_0^t 5\sin(10\lambda)d\lambda = 5\left[\frac{-\cos(10\lambda)\Big|_0^t}{10}\right] = \frac{1}{2}[1 - \cos(10t)]$$

or

$$f(t) = \frac{1}{2}[1 - \cos(10t)]u(t)$$

continued

Example 15.36 continued

FIGURE 15.108

Waveforms for $f_1(\lambda)$, $f_2(-\lambda)$, $f_2(t-\lambda)$, and $f(t)$.

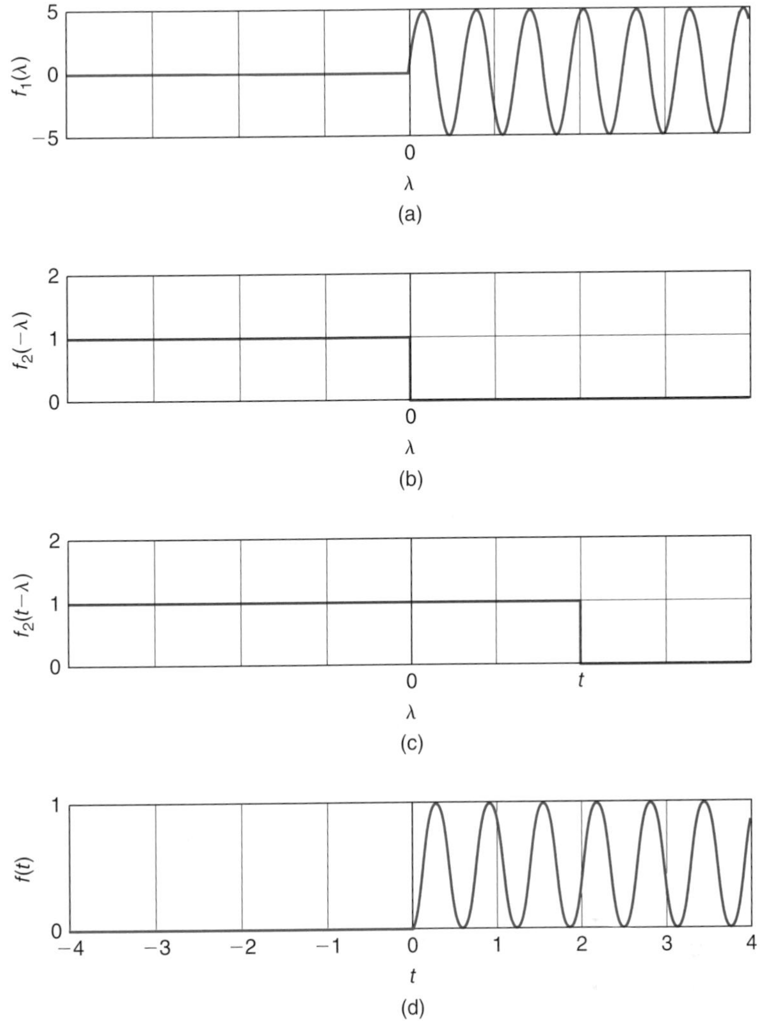

Alternatively, the convolution can be found in the s-domain:

$$F(s) = F_1(s)F_2(s) = \frac{50}{s(s^2 + 10^2)}$$

Using partial fraction expansion, we have

$$F(s) = \frac{50}{s(s^2 + 10^2)} = \frac{A}{s} + \frac{Bs + C}{s^2 + 10^2} = \frac{(A + B)s^2 + Cs + 100A}{s(s^2 + 10^2)}$$

$$A = \left.\frac{50}{s^2 + 100}\right|_{s=0} = \frac{1}{2}, \quad C = 0, \quad B = \frac{-1}{2}$$

$$F(s) = \frac{50}{s(s^2 + 10^2)} = \frac{\frac{1}{2}}{s} + \frac{\frac{-1}{2}s}{s^2 + 10^2}$$

$$f(t) = \frac{1}{2}u(t) - \frac{1}{2}\cos(10t)\,u(t) = \frac{1}{2}[1 - \cos(10t)]u(t)$$

Exercise 15.36

Determine the convolution $f(t)$ of $f_1(t) = 5\sin(5t)u(t)$ and $f_2(t) = 3\cos(12t)u(t)$.

Answer:

$$f(t) = \left[\frac{75}{119}\cos(5t) - \frac{75}{119}\cos(12t)\right]u(t).$$

EXAMPLE 15.37

Determine the convolution $f(t)$ of $f_1(t) = u(t) - u(t-2)$ and $f_2(t) = 2u(t) - 2u(t-2)$.

Figure 15.109 shows $f_1(\lambda), f_2(\lambda), f_2(-\lambda)$, and $f_2(t-\lambda)$. If $f_2(-\lambda)$ is shifted to the left, the product of $f_1(\lambda)$ and $f_2(t-\lambda)$ will be zero. Therefore, the convolution $f(t)$ is zero for $t < 0$. Now, we shift $f_2(-\lambda)$ to the right. If the amount of shift t is less than or equal to 2, $0 \le t < 2$, the convolution value $f(t)$ increases linearly as t increases since

$$f(t) = \int_0^t 1 \times 2d\lambda = 2t \quad \text{for} \quad 0 \le t < 2$$

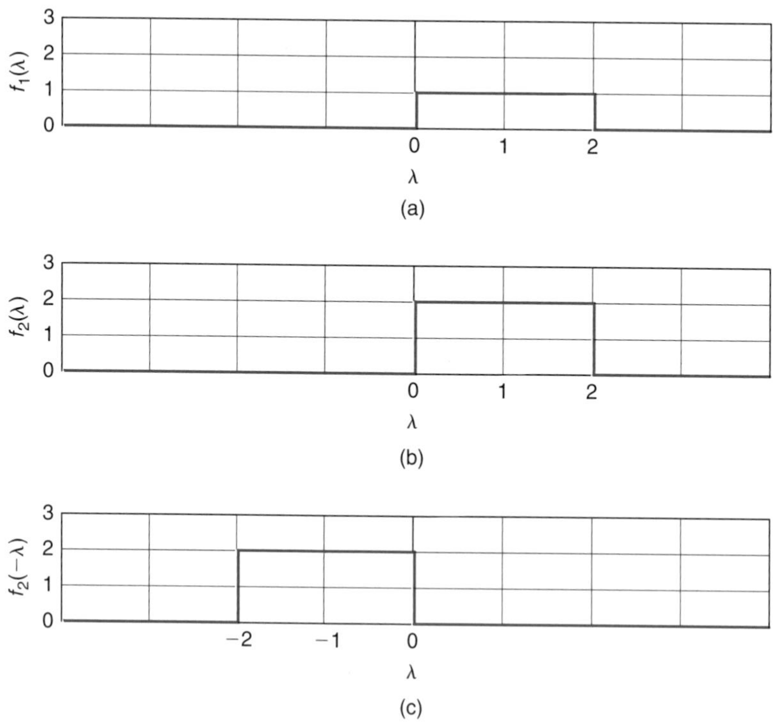

FIGURE 15.109

Waveforms for $f_1(\lambda)$, $f_2(\lambda), f_2(-\lambda), f_2(t-\lambda)$, and $f(t)$.

continued

Example 15.37 continued

FIGURE 15.109 *continued*

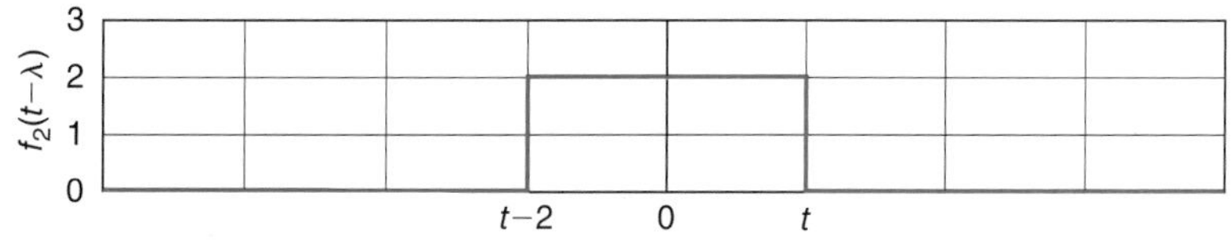

(d)

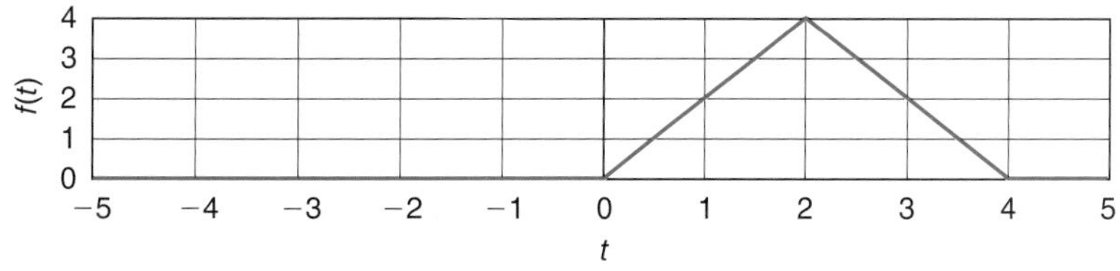

(e)

If the amount of shift to the right is greater than 2 and less than or equal to 4, the convolution value $f(t)$ will decrease linearly as t increases because the overlapping interval of the two signals decreases.

$$f(t) = \int_{t-2}^{2} 1 \times 2d\lambda = -2t + 8 \quad \text{for} \quad 2 \le t < 4$$

We integrate from $(t - 2)$ to 2 because the two signals overlap in the interval $(t - 2, 2)$.

If the amount of shift to the right is greater than 4 ($t \ge 4$), the two signals $f_1(\lambda)$ and $f_2(t - \lambda)$ do not overlap, making the convolution $f(t)$ to be zero. Therefore,

$$f(t) = \begin{cases} 0, & t < 0 \\ 2t, & 0 \le t < 2 \\ -2t + 8, & 2 \le t < 4 \\ 0, & 4 < t \end{cases}$$

This result is plotted in Figure 15.109(e).

The convolution $f(t)$ can also be obtained using the Laplace transform by taking the inverse Laplace transform of the product of the Laplace tansforms $F_1(s)$ and $F_2(s)$.

$$f_1(t) = u(t) - u(t - 2), F_1(s) = (1 - e^{-2s})/s$$

$$f_2(t) = 2u(t) - 2u(t - 2), F_2(s) = 2(1 - e^{-2s})/s$$

$$F(s) = F_1(s)F_2(s) = 2(1 - 2e^{-2s} + e^{-4s})/s^2$$

$$f(t) = L^{-1}[F(s)] = 2tu(t) - 4(t - 2)u(t - 2) + 2(t - 4)u(t - 4)$$

Exercise 15.37

Determine the convolution $f(t)$ of $f_1(t) = u(t - 2) - u(t - 6)$ and $f_2(t) = 2u(t - 3) - 2u(t - 5)$.

continued

Exercise 15.37 continued

Answer:

$$f(t) = \begin{cases} 2t - 10, & 5 \le t < 7 \\ 4, & 7 \le t < 9 \\ -2t + 22, & 9 \le t < 11 \\ 0, & \text{elsewhere} \end{cases}.$$

EXAMPLE 15.38

Determine the convolution $f(t)$ of the two rectangular pulses shown in Figure 15.110(a) and 15.110(b).

FIGURE 15.110

Waveforms for $f_1(\lambda)$, $f_2(\lambda), f_2(-\lambda), f_2(t-\lambda)$, and $f(t)$.

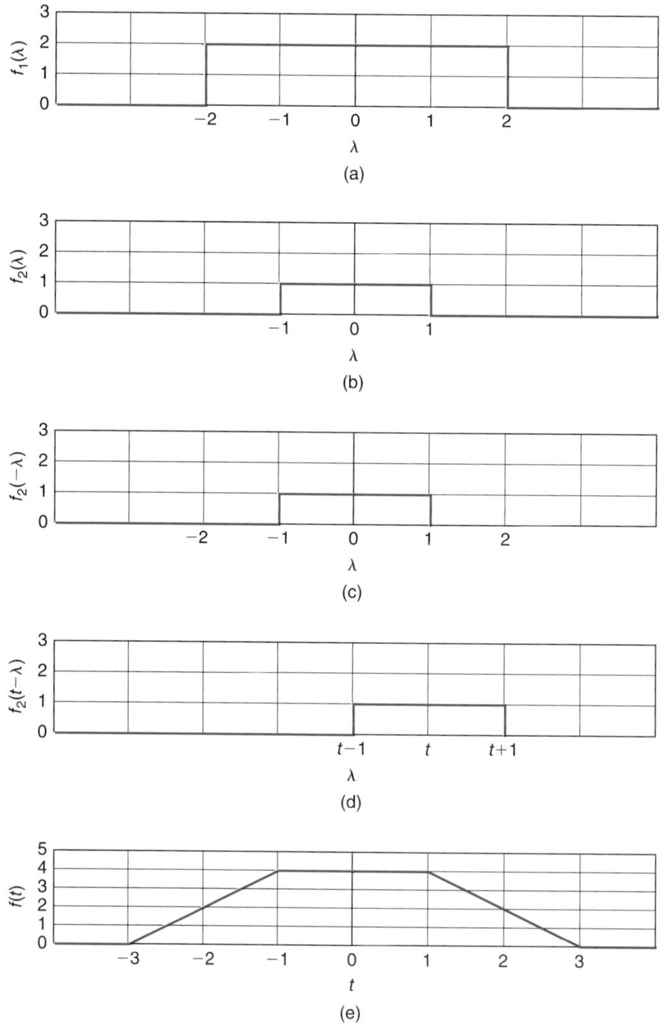

When $f_2(-\lambda)$ is shifted to the left by 3 or more, $f_2(t-\lambda)$, where $t < -3$, does not overlap with $f_1(\lambda)$. Thus, $f(t) = 0$, for $t < -3$.

For $-3 \le t < -1$,

$$f(t) = \int_{-2}^{t+1} 1 \times 2 d\lambda = 2t + 6$$

continued

Example 15.38 continued

For $-1 \le t < 1$,

$$f(t) = \int_{t-1}^{t+1} 1 \times 2d\lambda = 4$$

For $1 \le t < 3$,

$$f(t) = \int_{t-1}^{2} 1 \times 2d\lambda = -2t + 6$$

For $t \ge 3$, $f(t) = 0$. $f(t)$ is plotted in Figure 15.110(e).

The convolution $f(t)$ can be obtained using Laplace transform by taking the inverse Laplace transform of the product of the Laplace tansforms $F_1(s)$ and $F_2(s)$.

$$f_1(t) = u(t+1) - u(t-1), F_1(s) = (e^s - e^{-s})/s$$

$$f_2(t) = 2u(t+2) - 2u(t-2), F_2(s) = 2(e^{2s} - e^{-2s})/s$$

$$F(s) = F_1(s)F_2(s) = 2(e^{3s} - e^s - e^{-s} + e^{-3s})/s^2$$

$$f(t) = L^{-1}[F(s)] = 2(t+3)u(t+3) - 2(t+1)u(t+1) - 2(t-1)u(t-1) + 2(t-3)u(t-3)$$

Exercise 15.38

Determine the convolution $f(t)$ of $f_1(t) = u(t+3) - u(t-5)$ and $f_2(t) = 2u(t+1) - 2u(t-2)$.

Answer:

$$f(t) = \begin{cases} 2t + 8, & -4 \le t < -1 \\ 6, & -1 \le t < 4 \\ -2t + 14, & 4 \le t < 7 \\ 0, & \text{elsewhere} \end{cases}.$$

The Meaning of Convolution

If we treat one of the signals $f_2(t)$ [or $f_1(t)$] as a weighting function, then the convolution $f(t)$ gives us a weighted running mean of $f_1(t)$ [or $f_2(t)$]. Thus, if the weighting function $f_2(t)$ is a smooth function of time, or a low-pass signal, then the convolution $f(t)$ is a smoothed or blurred version of $f_1(t)$. If $f_2(t)$ is a high-pass signal, then the convolution $f(t)$ is a high-frequency emphasized version of $f_1(t)$. We are extracting the information we want from the signal $f_1(t)$ based on the shape of the weighting function $f_2(t)$.

Convolutions Involving Dirac Delta Function

From the sifting property of the Dirac delta function, we obtain

$$\delta(t)*f(t) = \int_{-\infty}^{\infty} \delta(\lambda)f(t-\lambda)d\lambda = f(t) \qquad \textbf{(15.89)}$$

This equation states that the convolution of an arbitrary signal $f(t)$ with a Dirac delta function results in the signal $f(t)$ itself. Again, applying the sifting property, we get

$$f(t)*\delta(t-a) = \int_{-\infty}^{\infty} f(\lambda)\delta(t-\lambda-a)d\lambda = f(t-a) \tag{15.90}$$

This equation says that the convolution of an arbitrary signal $f(t)$ with a time-shifted Dirac delta function $\delta(t-a)$ gives us $f(t-a)$, which is the time-shifted version of $f(t)$. The signal $f(t)$ is moved to the location of the Dirac delta function.

EXAMPLE 15.39

Determine the convolution of $f_1(t) = u(t)$ and $f_2(t) = \delta(t+1) - 2\delta(t-1) + \delta(t-2)$.

Let $f(t)$ be the convolution of $f_1(t)$ and $f_2(t)$. Then, we have

$$\begin{aligned} f(t) &= u(t)*\delta(t+1) - 2u(t)*\delta(t-1) + u(t)*\delta(t-2) \\ &= u(t+1) - 2u(t-1) + u(t-2) \end{aligned}$$

Figure 15.111 shows $f_1(t)$, $f_2(t)$, and $f(t)$.

FIGURE 15.111

Waveform for EXAMPLE 15.39.

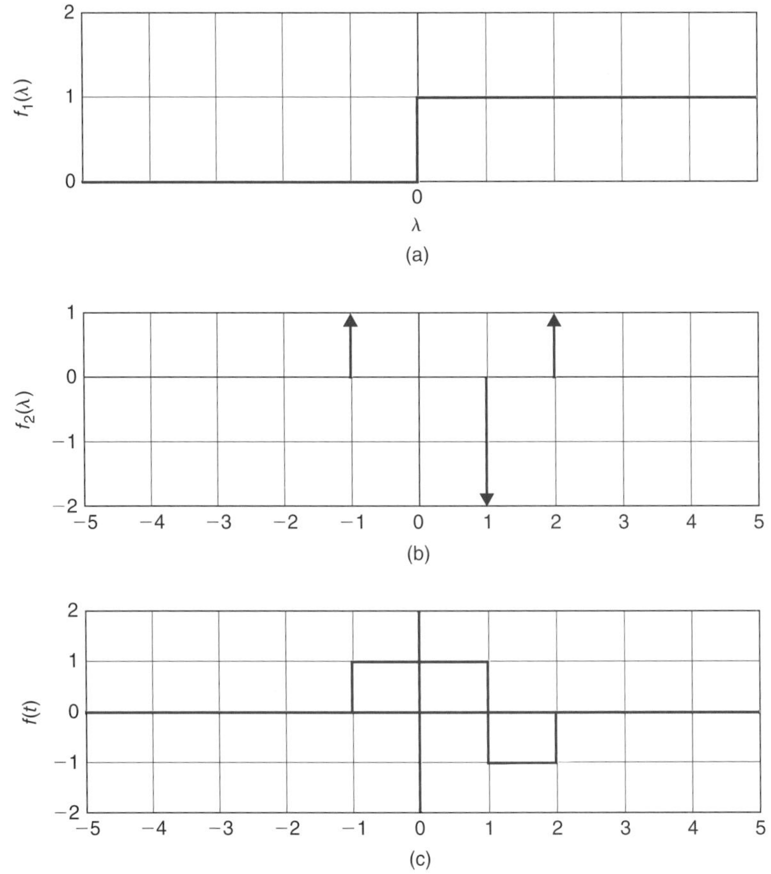

Exercise 15.39

Determine the convolution of $f_1(t) = 2tri\left(\frac{t}{2}\right)$ and $f_2(t) = 2\delta(t-4)$.

Answer:

$$f(t) = 2\,tri\left(\frac{t}{2}\right) * 2\delta(t-4) = 4\,tri\left(\frac{t-4}{2}\right).$$

Figure 15.112 shows $f_1(t), f_2(t)$, and $f(t)$.

FIGURE 15.112

Waveforms for EXERCISE 15.39.

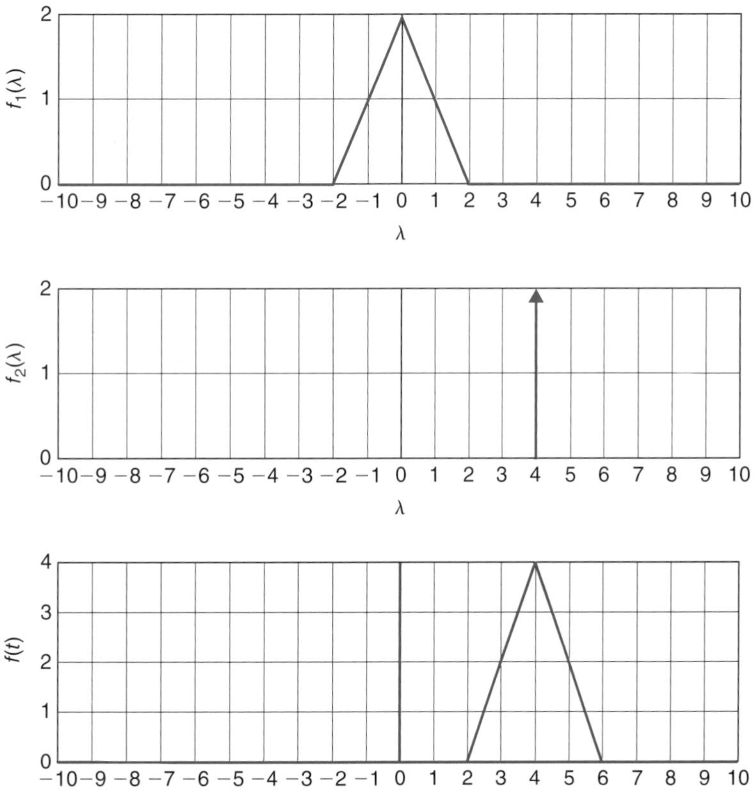

15.10 Linear, Time-Invariant (LTI) System

A system can be thought of as an interconnection of components such as voltage sources, current sources, resistors, capacitors, and inductors. The circuits consisting of voltage sources, current sources, resistors, inductors, and capacitors are systems. A circuit and a system can be used interchangeably. A system can be represented as a black box, as shown in Figure 15.113.

FIGURE 15.113

A system represented as a black box.

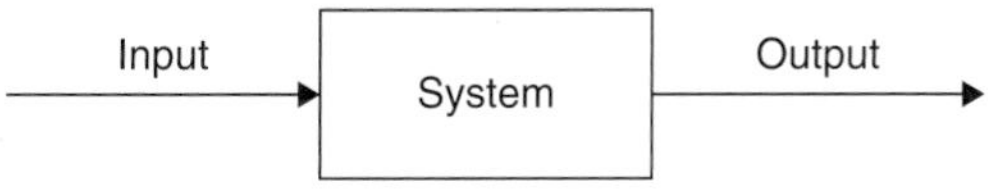

Let $y_1(t)$ be the output of a system when $x_1(t)$ is applied at the input. Let $y_2(t)$ be the output of the same system when $x_2(t)$ is applied at the input. The system is said to be **linear** if the response (output) of the system to excitation (input) $a_1x_1(t) + a_2x_2(t)$ is given by $a_1y_1(t) + a_2y_2(t)$. If the superposition principle holds in the system, the system is linear. Since the superposition principle holds in all the circuits we discussed so far, these circuits are examples of a linear system.

Let $y(t)$ be the output of a system when $x(t)$ is applied at the input. The system is said to be **time-invariant** if the response of the system to the excitation $x(t - t_0)$ is given by $y(t - t_0)$. If a system is time-invariant, the response of the system to a given input is the same no matter when the input signal is applied. The system is time-invariant if the underlying physical mechanism of the system is stationary.

If a system is both linear and time-invariant, the system is called a **linear, time-invariant (LTI)** system.

15.10.1 IMPULSE RESPONSE

The impulse response of a system is defined as the response of the system to an impulse. Since it is difficult to generate an impulse, direct measurement of an impulse response is not an easy task. The best approximation to an impulse is a pulse with an area of one and a width as small as possible. Alternatively, the impulse response can be measured by differentiating the step response of the system. The impulse response of the system is written as $h(t)$. If the system is time invariant, the response of the system to $\delta(t - t_0)$ is $h(t - t_0)$. If the system is causal, $h(t) = 0$ for $t < 0$, since $\delta(t)$ is applied at $t = 0$.

15.10.2 OUTPUT OF LINEAR TIME-INVARIANT SYSTEM

The response (output) of a system to an input of impulse $\delta(t)$ is called an **impulse response** and is denoted by $h(t)$. Let $x(t)$ be an input to an LTI with an impulse response $h(t)$. Then, the output of the system $y(t)$ is given by the convolution of the input and the impulse response of the system; that is,

$$y(t) = h(t) * x(t) = \int_{-\infty}^{\infty} x(\lambda)h(t - \lambda)d\lambda$$

Since the convolution operation is commutative, the output signal $y(t)$ can also be written as

$$y(t) = x(t) * h(t) = \int_{-\infty}^{\infty} h(\lambda)x(t - \lambda)d\lambda$$

To prove this result, we start by approximating the input signal as a sum of rectangles as shown in Figure 15.114:

$$x(t) = \lim_{T \to 0} \sum_{n=-\infty}^{\infty} x(nT)\,\text{rect}\left(\frac{t - nT}{T}\right) \tag{15.91}$$

FIGURE 15.114

Approximation of $x(t)$ as a sum of rectangles.

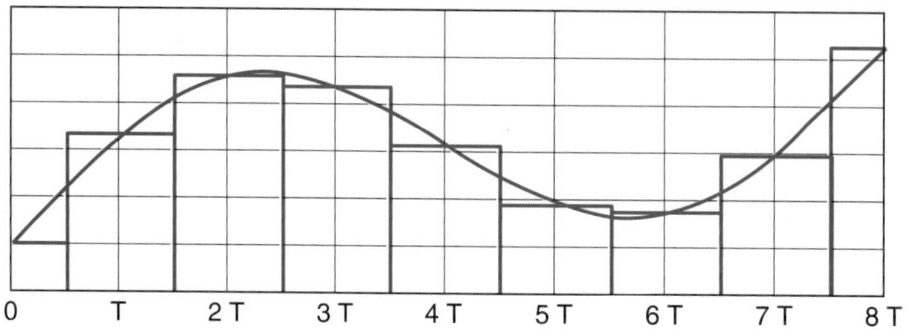

where $x(nT)$ is the sample of $x(t)$ at $t = nT$, and the rectangular pulse is defined as

$$\text{rect}\left(\frac{t-nT}{T}\right) = \begin{cases} 1, & nT - \dfrac{1}{2T} \leq t < nT + \dfrac{1}{2T} \\ 0, & \text{otherwise} \end{cases}$$

Notice that the height of rectangle at $t = nT$ is $x(nT)$.

As $T \rightarrow 0$, the rectangular pulse with height one and width T becomes an impulse with area T, that is

$$\lim_{T\rightarrow 0} \text{rect}\left(\frac{t-nT}{T}\right) \rightarrow T\,\delta(t - nT) \tag{15.92}$$

Substitution of Equation (15.92) into Equation (15.91) yields

$$x(t) = \lim_{T\rightarrow 0} \sum_{n=-\infty}^{\infty} x(nT)\,\delta(t - nT)T \tag{15.93}$$

Since the impulse response of the system is $h(t)$, the response of the system to $\delta(t)$ is $h(t)$. Since the system is time-invariant, the response of the system to $\delta(t - nT)$ is $h(t - nT)$. Since the system is linear, the output of the system is the sum of the responses of the system to $\delta(t - nT)$; that is, the sum of $h(t - nT)$:

$$y(t) = \lim_{T\rightarrow 0} \sum_{n=-\infty}^{\infty} x(nT)h(t - nT)T \tag{15.94}$$

As $T \rightarrow 0$, the discrete samples $x(nT)$ and $h(t - nT)$ merge to form continuous signals $x(\lambda)$ and $h(t - \lambda)$, T can be written as dt, the summation (Σ) becomes integral ($\int$), and Equation (15.94) can be rewritten as

$$y(t) = \int x(\lambda)h(t - \lambda)d\lambda \tag{15.95}$$

As shown in Section 15.9, earlier in this chapter, convolution in the time domain results in multiplication in the s-domain. Taking the Laplace transform of Equation (15.95), we obtain

$$Y(s) = H(s)X(s) \tag{15.96}$$

where $H(s)$ is the Laplace transform of the impulse response. $H(s)$ is called the *transfer function*, as discussed in Section 15.8 earlier in this chapter.

15.10.3 STEP RESPONSE OF LTI SYSTEM

If a unit step function $u(t)$ is applied to a linear, time-invariant (LTI) system with an impulse response $h(t)$, the response of the system is given by

$$y(t) = h(t)*u(t) = \int_{-\infty}^{\infty} h(\lambda)u(t - \lambda)d\lambda = \int_{-\infty}^{t} h(\lambda)d\lambda$$

If the LTI system is also causal, then $h(t) = 0$ for $t < 0$. Thus, the output of the system becomes

$$y(t) = \int_{0}^{t} h(\lambda)d\lambda$$

If we denote the unit step response as $y_s(t)$, then we have

$$y_s(t) = \int_0^t h(\lambda)d\lambda \tag{15.97}$$

This result shows that the unit step response of a causal, linear, time-invariant system is the integral of the impulse response. If the unit step response $h_s(t)$ is differentiated with respect to t, we obtain the impulse response $h(t)$.

Since the integral in the time domain is the division by s in the s-domain, we have

$$H_s(s) = \frac{H(s)}{s} \tag{15.98}$$

The step response is the inverse Laplace transform of the transfer function $H(s)$ divided by s.

15.11 Bode Diagram

The Bode diagram is an asymptotic plot of the magnitude response in decibels (dB) and the phase response of a system described by the transfer function $H(s)$. The Bode diagram is a straight-line approximation of the true magnitude response in decibels and the true phase response. It is a fast way of plotting the frequency response. MATLAB can be used to plot an exact Bode diagram.

15.11.1 LINEAR SCALE

Consider a transfer function given by

$$H(s) = \frac{100}{s + 100} \tag{15.99}$$

The frequency response of $H(s)$ is obtained by replacing s by $j\omega$ [evaluating $H(s)$ on the imaginary axis]; that is,

$$H(\omega) = H(s)|_{s=j\omega} = \frac{100}{j\omega + 100} \tag{15.100}$$

The magnitude response of Equation (15.100) in linear scale is given by

$$|H(\omega)| = \frac{100}{|j\omega + 100|} = \frac{100}{\sqrt{\omega^2 + 100^2}} \tag{15.101}$$

At $\omega = 0$, the magnitude is 1. At $\omega = \infty$, the magnitude is zero. The magnitude is decreased from one to zero as the frequency is increased from zero to infinity. This is a low-pass filter. At $\omega = 100$ rad/s, the magnitude becomes

$$|H(100)| = \frac{100}{|j100 + 100|} = \frac{1}{\sqrt{2}} = 0.7071 \tag{15.102}$$

At $\omega = 100$ rad/s, the magnitude is $1/\sqrt{2}$ times the magnitude at $\omega = 0$. This frequency is called cutoff frequency (corner frequency) of the low-pass filter and is denoted by ω_0. Thus, we have $\omega_0 = 100$ rad/s. In terms of ω_0, Equation (15.100) can be written as

$$H(\omega) = \frac{\omega_0}{j\omega + \omega_0} \tag{15.103}$$

The phase response of Equation (15.100) is given by

$$\angle H(\omega) = -\tan^{-1}\left(\frac{\omega}{\omega_0}\right) = -\tan^{-1}\left(\frac{\omega}{100}\right) \quad \textbf{(15.104)}$$

15.11.2 dB SCALE

The frequency response given by Equation (15.100) can be rewritten as

$$H(\omega) = \frac{100}{j\omega + 100} = \frac{100}{100\left(1 + \frac{j\omega}{100}\right)} = \frac{1}{1 + \frac{j\omega}{100}} \quad \textbf{(15.105)}$$

The magnitude response in decibels is defined as

$$M_{dB}(\omega) = 20\log_{10}(|H(\omega)|) = 20\log_{10}\left(\left|\frac{1}{1 + j\frac{\omega}{100}}\right|\right) = 20\log_{10}\left(\frac{1}{\left|1 + j\frac{\omega}{100}\right|}\right)$$
$$= -20\log_{10}\left(\left|1 + j\frac{\omega}{100}\right|\right) \quad \textbf{(15.106)}$$

At $\omega = 0$, $\mathrm{H}(\omega) = 1$, and $M_{dB}(0) = 0$ dB.

At $\omega = \infty$, $\mathrm{H}(\omega) = 0$, and $M_{dB}(\infty) = -\infty$ dB.

At $\omega = \omega_0 = 100$ rad/s, the magnitude in decibels is given by

$$M_{dB}(\omega_0) = 20\log_{10}\left(\frac{1}{\sqrt{2}}\right) = -20\log_{10}2^{\frac{1}{2}} = -10\log_{10}2 = -3.01\,\text{dB} \quad \textbf{(15.107)}$$

At the cutoff frequency ω_0, the magnitude is 3.01 dB lower than the magnitude at $\omega = 0$.

When $\omega < 100$ rad/s, $\omega/100 < 1$, and we have

$$M_{dB}(\omega) = -20\log_{10}\left(\left|1 + j\frac{\omega}{100}\right|\right) \approx -20\log_{10}1 = 0\,\text{dB}$$

On the other hand, when $\omega > 100$, $\omega/100 > 1$, and we have

$$M_{dB}(\omega) = -20\log_{10}\left(\left|1 + j\frac{\omega}{100}\right|\right) \approx -20\log_{10}\left(\frac{\omega}{100}\right) \quad \textbf{(15.108)}$$

Approximate (straight-line) values are as follows

At $\omega = 100$ rad/s, $M_{dB}(100) \approx -20\log_{10}\left(\frac{100}{100}\right) = 0$ dB

At $\omega = 1000$ rad/s, $M_{dB}(1000) \approx -20\log_{10}\left(\frac{1000}{100}\right) = -20$ dB

At $\omega = 10{,}000$ rad/s, $M_{dB}(10{,}000) \approx -20\log_{10}\left(\frac{10{,}000}{100}\right) = -40$ dB

$$\text{At } \omega = 100{,}000 \text{ rad/s, } M_{dB}(100{,}000) \approx -20\log_{10}\left(\frac{100{,}000}{100}\right) = -60 \text{ dB}$$

$$\text{At } \omega = 200 \text{ rad/s, } M_{dB}(200) \approx -20\log_{10}\left(\frac{200}{100}\right) = -6.0206 \text{ dB}$$

$$\text{At } \omega = 400 \text{ rad/s, } M_{dB}(400) \approx -20\log_{10}\left(\frac{400}{100}\right) = -12.0412 \text{ dB}$$

As the frequency is increased by tenfold, the magnitude is decreased by 20 dB. The rate of decrease in magnitude is 20 dB/dec (where *dec* means "decade"). As the frequency is doubled, the magnitude is decreased by 6 dB. The rate of decrease in magnitude is 6 dB/oct, where oct is for octave. The straight-line approximation for the magnitude response is obtained by drawing a horizontal line at 0 dB for $\omega \le \omega_0$ and a straight line starting from $(\omega_0, 0)$ at a rate of -20 dB/dec for $\omega > \omega_0$. The straight-line approximation can be written as

$$M_{dB}(\omega) \approx \begin{cases} 0 \text{ dB}, & \omega \le \omega_0 \\ -20\log_{10}(\omega/\omega_0) \text{ dB}, & \omega > \omega_0 \end{cases}$$

The exact values of the magnitude response are:

$$\text{At } \omega = 10 \text{ rad/s, } M_{dB}(10) = -20\log_{10}\left(\left|1 + j\frac{10}{100}\right|\right) = -0.0432 \text{ dB}$$

$$\text{At } \omega = 50 \text{ rad/s, } M_{dB}(50) = -20\log_{10}\left(\left|1 + j\frac{50}{100}\right|\right) = -0.9691 \text{ dB}$$

$$\text{At } \omega = 100 \text{ rad/s, } M_{dB}(100) = -20\log_{10}\left(\left|1 + j\frac{100}{100}\right|\right) = -3.0103 \text{ dB}$$

$$\text{At } \omega = 200 \text{ rad/s, } M_{dB}(200) = -20\log_{10}\left(\left|1 + j\frac{200}{100}\right|\right) = -6.9897 \text{ dB}$$

$$\text{At } \omega = 1000 \text{ rad/s, } M_{dB}(1000) = -20\log_{10}\left(\left|1 + j\frac{1000}{100}\right|\right) = -20.0432 \text{ dB}$$

The maximum error in the straight-line approximation is -3.0103 dB at $\omega = \omega_0 = 100$ rad/s. The error at $\omega = \omega_0/2 = 50$ and at $\omega = 2\omega_0 = 200$ is 0.9691 dB. The straight-line approximation (dashed line) and the exact magnitude response (solid line) are shown in Figure 15.115.

FIGURE 15.115

Straight-line approximation and exact Bode magnitude plot.

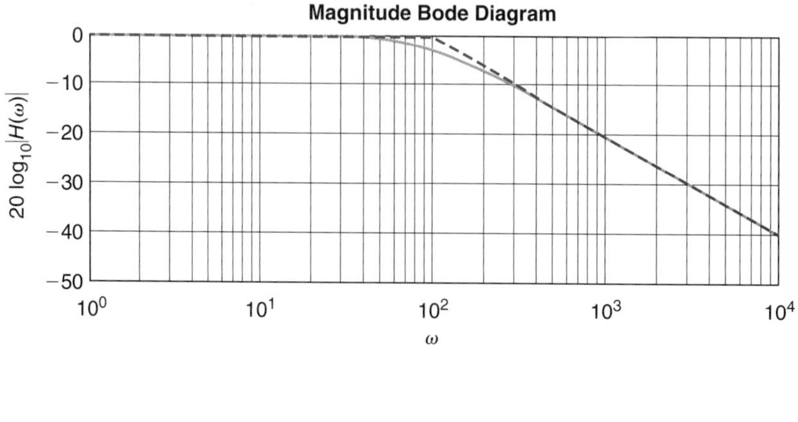

The phase response is

$$\angle H(\omega) = -\tan^{-1}\left(\frac{\omega}{100}\right) \tag{15.109}$$

At $\omega = 0$, the phase is $\angle H(\omega) = -\tan^{-1}(0) = 0$ rad $= 0°$. At $\omega = 100$, the phase is $\angle H(\omega) = -\tan^{-1}(1) = -\frac{\pi}{4}$rad $= -45°$. At $\omega = \infty$, the phase is $\angle H(\omega) = -\tan^{-1}(\infty) = -\frac{\pi}{2}$ rad $= -90°$. As the radian frequency ω is increased from zero to infinity, the phase response decreases from zero to –90°.

The straight-line approximation for the phase response is obtained by drawing a straight line from zero degrees at one-tenth of the cutoff frequency ($0.1\omega_0$) to –90° at 10 times the cutoff frequency ($10\omega_0$); that is, from a point $(0.1\omega_0, 0)$ to a point $(10\omega_0, -90°)$. The difference between the exact phase angle in degrees to the straight-line approximation for several frequencies is given in Table 15.1.

TABLE 15.1 Exact Phase Value and Approximation (in Degrees)

Exact Value	Approximation	Difference
$\theta(0.1\omega_0) = -5.7106$	$\theta_s(0.1\omega_0) = 0$	$\theta(0.1\omega_0) - \theta_s(0.1\omega_0) = -5.7106$
$\theta(0.5\omega_0) = -26.5651$	$\theta_s(0.5\omega_0) = -31.4537$	$\theta(0.5\omega_0) - \theta_s(0.5\omega_0) = 4.8886$
$\theta(\omega_0) = -45$	$\theta_s(\omega_0) = -45$	$\theta(\omega_0) - \theta_s(\omega_0) = 0$
$\theta(2\omega_0) = -63.4349$	$\theta_s(2\omega_0) = -58.5463$	$\theta(2\omega_0) - \theta_s(2\omega_0) = -4.8886$
$\theta(10\omega_0) = -84.2894$	$\theta_s(10\omega_0) = -90$	$\theta(10\omega_0) - \theta_s(10\omega_0) = 5.7106$

The maximum error occurs at $\omega = 0.1\omega_0$ and $\omega = 10\omega_0$, and the magnitude of the maximum error is 0.1 radians, which is 5.7106°. The phase response is shown in Figure 15.116. The dashed line is the straight-line approximation, and the solid line is the exact phase plot.

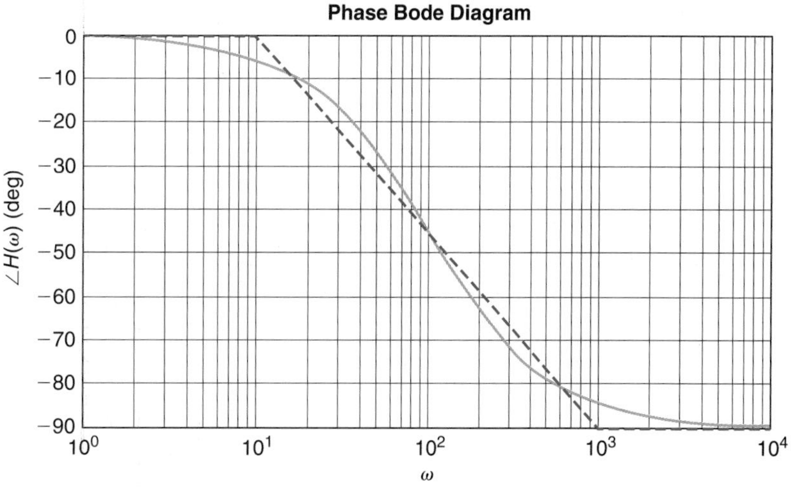

FIGURE 15.116 The phase Bode diagram.

In MATLAB, we can get the exact Bode diagram using function **bode**, as shown here in several different ways:

```
H=tf(100,[1 100])
bode(H);grid;
```

or

```
s = tf('s');
H = 100/(s+100)
bode(H);grid;
```

or

```
H=zpk([],-100,100)
bode(H);grid;
```

The frequency range can be specified using the **logspace** function:

```
H=tf(100,[1 100])
w=logspace(-1,5);
bode(H,w);grid;
```

or

```
s = tf('s');
H = 100/(s+100)
w=logspace(-1,5);
bode(H,w);grid;
```

or

```
H=zpk([],-100,100)
w=logspace(-1,5);
bode(H,w);grid;
```

The function **bodemag** plots the magnitude response only in decibels without the phase response:

```
clear all;
H=tf(100,[1 100])
bodemag(H,{0.1,1e5});grid;
```

15.11.3 BODE DIAGRAM OF CONSTANT TERM

The magnitude response of a constant term K is

$$M_{dB}(\omega) = 20 \log_{10}(|K|) \text{ dB}$$

for all frequencies. If the constant term K is positive, the contribution to phase angle is zero. If K is negative, the constant term adds (or subtracts) 180° (= π radians) to the phase response.

15.11.4 BODE DIAGRAM OF $H(s) = s + 1000$

The transfer function of the first-order zero term is given by

$$H(s) = s + 1000 \tag{15.110}$$

The frequency response of the transfer function given by Equation (15.110) is given by

$$H(\omega) = H(s)\Big|_{s=j\omega} = j\omega + 1000 \tag{15.111}$$

This transfer function can be rewritten as

$$H(\omega) = j\omega + 1000 = 1000\left(1 + \frac{j\omega}{1000}\right) \tag{15.112}$$

The magnitude response in decibels is given by

$$\begin{aligned} M_{dB}(\omega) &= 20\log_{10}(|H(\omega)|) = 20\log_{10}\left(\left|1000\left(1 + \frac{j\omega}{1000}\right)\right|\right) \\ &= 20\log_{10}(1000) + 20\log_{10}\left(\left|1 + j\frac{\omega}{1000}\right|\right) \\ &= 60 + 20\log_{10}\left(\left|1 + j\frac{\omega}{1000}\right|\right) \end{aligned} \tag{15.113}$$

When $\omega < 1000$ rad/s, $\omega/1000 < 1$, and we have

$$M_{dB}(\omega) = 60 + 20\log_{10}\left(\left|1 + j\frac{\omega}{1000}\right|\right) \approx 60 + 20\log_{10}1 = 60 \text{ dB}$$

On the other hand, when $\omega > 1000$, $\omega/1000 > 1$, and we have

$$M_{dB}(\omega) = 60 + 20\log_{10}\left(\left|1 + j\frac{\omega}{1000}\right|\right) \approx 60 + 20\log_{10}\left(\frac{\omega}{1000}\right)$$

Thus, the straight-line approximation is given by

$$Ms_{dB}(\omega) = \begin{cases} 60, & \omega \le 1000 \\ 60 + 20\log_{10}\left(\dfrac{\omega}{1000}\right), & \omega > 1000 \end{cases} \tag{15.114}$$

At $\omega = 1000$ rad/s, $M_{dB} = 60$ dB.

At $\omega = 10{,}000$ rad/s, $M_{dB} = 80$ dB.

At $\omega = 100{,}000$ rad/s, $M_{dB} = 100$ dB.

For $\omega > 1000$ rad/s, as the frequency is increased by tenfold, the magnitude is increased by 20 dB. The rate of increase is 20 dB/dec. Figure 15.117 shows a Bode diagram, including a straight-line approximation.

15.11.5 BODE DIAGRAM OF $H(s) = 100/s$

Let the transfer function be

$$H(s) = \frac{100}{s}$$

The frequency response is given by

$$H(\omega) = H(s)\Big|_{s=j\omega} = \frac{100}{j\omega}$$

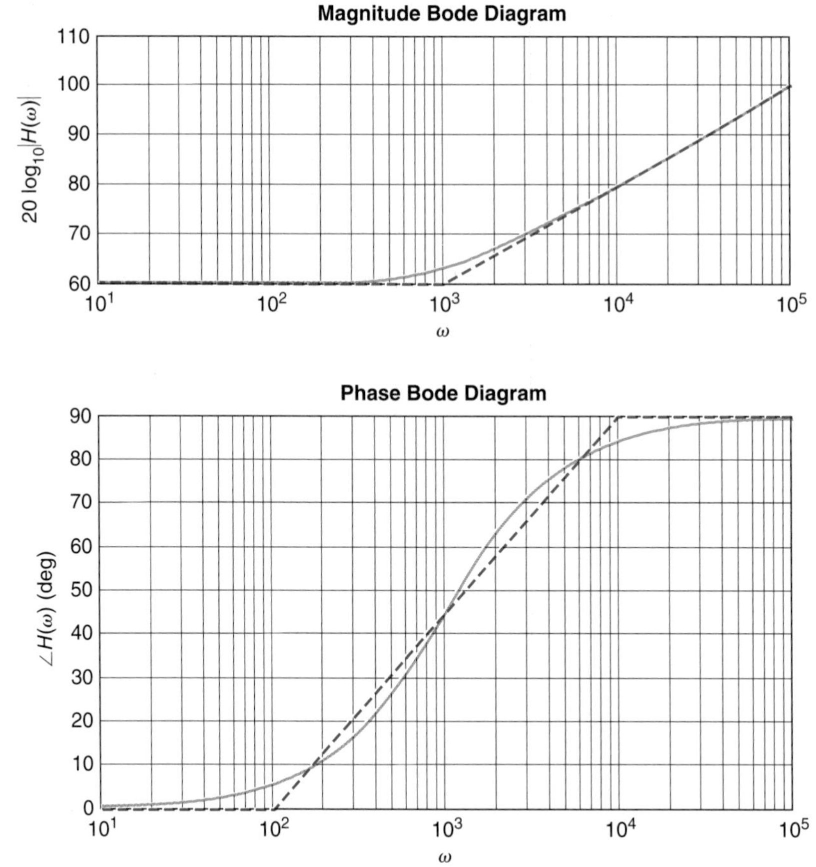

FIGURE 15.117

Bode diagram for $s + 1000$.

The magnitude response is given by

$$M_{dB}(\omega) = 20\log_{10}\left|H(\omega)\right| = 20\log_{10}\left|\frac{100}{j\omega}\right| = 40 - 20\log_{10}\omega$$

At $\omega = 100$ rad/s, $H(\omega) = -j$, and $M_{dB}(\omega) = 0$ dB. The magnitude response decreases at a rate of 20 dB/dec. The phase response is given by

$$\angle H(\omega) = -\frac{\pi}{2}\text{rad} = -90°$$

The phase response is $-90°$ for all ω. The Bode diagram is shown in Figure 15.118. There is no difference between the exact Bode plot and the straight-line approximation.

15.11.6 BODE DIAGRAM OF $H(s) = s/1000$

Let the transfer function be

$$H(s) = \frac{s}{1000}$$

The frequency response is given by

$$H(\omega) = H(s)\Big|_{s=j\omega} = \frac{j\omega}{1000}$$

FIGURE 15.118

Bode diagram of $H(s) = 100/s$.

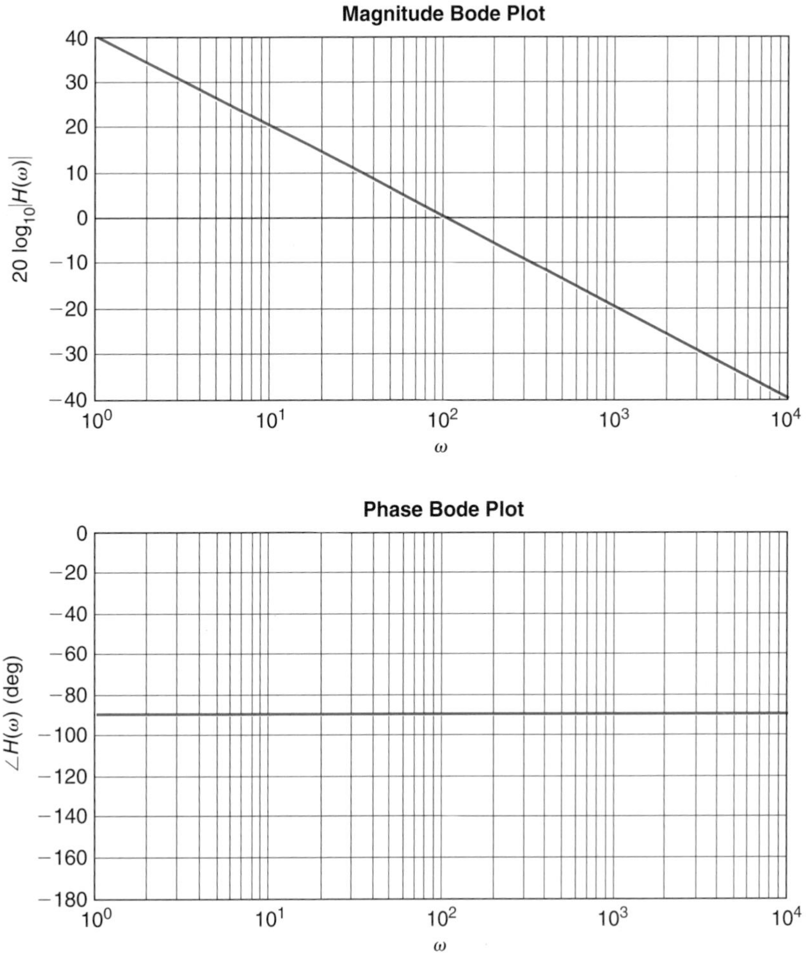

The magnitude response in decibels is given by

$$M_{dB}(\omega) = 20\log_{10}\left|H(\omega)\right| = 20\log_{10}\left|\frac{j\omega}{1000}\right| = 20\log_{10}\omega - 60$$

At $\omega = 1000$ rad/s, $M_{dB}(\omega) = 0$ dB. The magnitude response increases at a rate of 20 dB/dec. The phase response is given by

$$\angle H(\omega) = \frac{\pi}{2}\text{rad} = 90°$$

The phase response is 90° for all ω. The Bode diagram is shown in Figure 15.119. There is no difference between the exact Bode plot and the straight-line approximation.

15.11.7 BODE DIAGRAM OF $H(s) = 10^4/(s + 100)^2$

Let the transfer function be

$$H(s) = \frac{10{,}000}{(s+100)^2} = \frac{1}{\left(1 + \frac{s}{100}\right)^2}$$

FIGURE 15.119

Bode diagram of $H(s) = s/1000$.

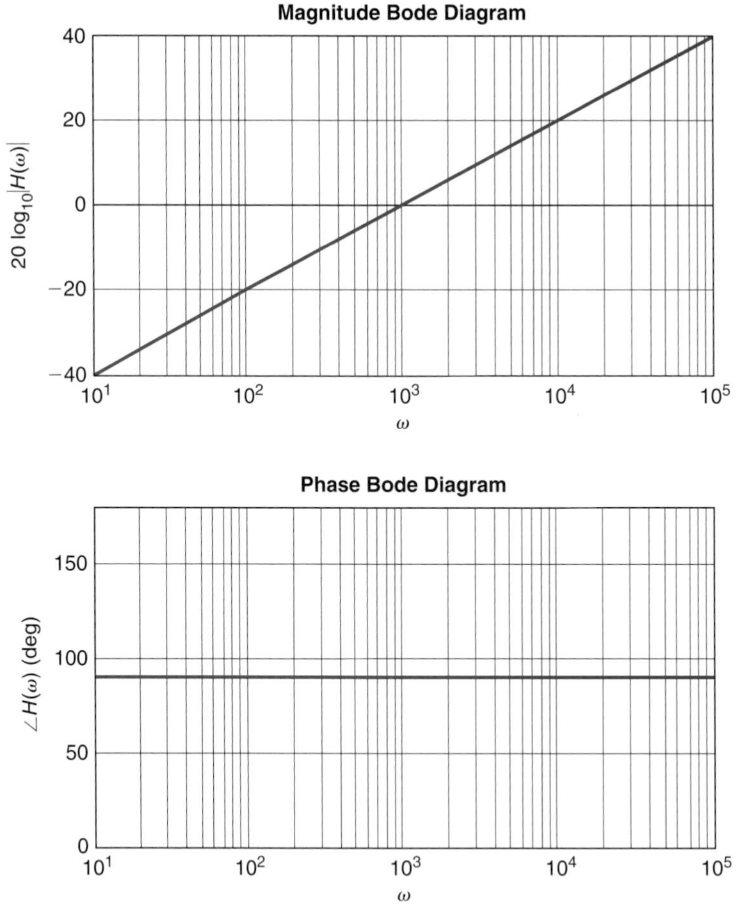

The frequency response is given by

$$H(\omega) = H(s)\Big|_{s=j\omega} = \frac{1}{\left(1 + \dfrac{j\omega}{100}\right)^2}$$

The magnitude response is given by

$$M_{dB}(\omega) = 20\log_{10}|H(\omega)| = -40\log_{10}\left|1 + \frac{j\omega}{100}\right| \approx -40\log_{10}\left(\frac{\omega}{100}\right)$$

At $\omega = 100$ rad/s, $M_{dB}(\omega) = 0$ dB. For $\omega > 100$ rad/s, the magnitude response decreases at a rate of 40 dB/dec. The phase response is given by

$$\angle H(\omega) = -2\tan^{-1}\left(\frac{\omega}{100}\right)$$

The phase response is $-90°$ at $\omega = 100$ rad/s, and $-180°$ at $\omega = \infty$. The Bode diagram is shown in Figure 15.120.

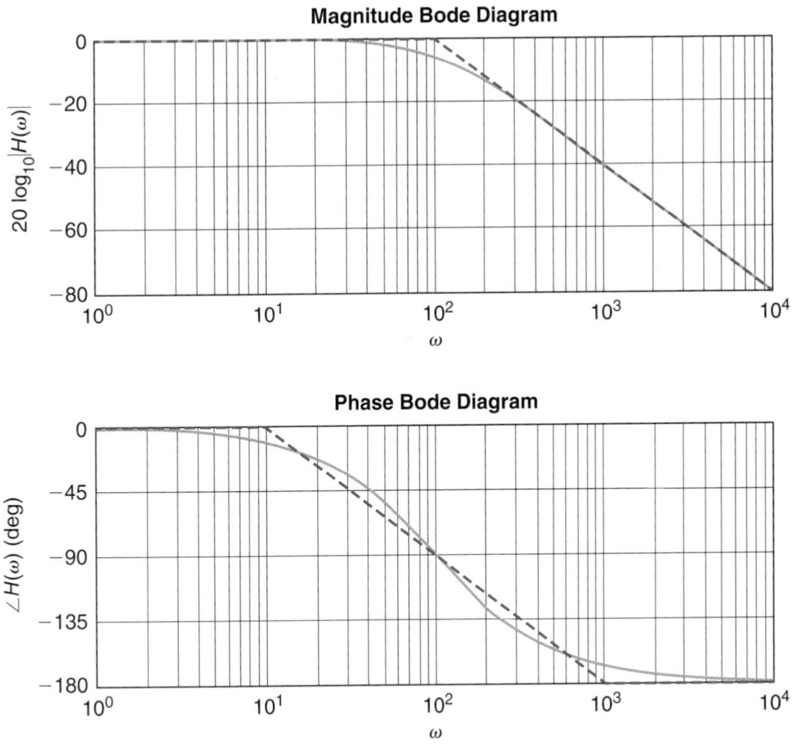

FIGURE 15.120

Bode diagram of $H(s) = 10^4/(s + 100)^2$.

EXAMPLE 15.40

Plot the magnitude and the phase Bode diagram for the transfer function given by

$$H(s) = \frac{10{,}000(s + 1000)}{(s + 100)(s + 10{,}000)}$$

Dividing by a constant for each term, we obtain

$$H(s) = \frac{10{,}000 \times 1000\left(1 + \dfrac{s}{1000}\right)}{100\left(1 + \dfrac{s}{100}\right)10{,}000\left(1 + \dfrac{s}{10{,}000}\right)} = \frac{10^7\left(1 + \dfrac{s}{1000}\right)}{10^6\left(1 + \dfrac{s}{100}\right)\left(1 + \dfrac{s}{10{,}000}\right)}$$

$$= 10\frac{1 + \dfrac{s}{1000}}{\left(1 + \dfrac{s}{100}\right)\left(1 + \dfrac{s}{10{,}000}\right)}$$

Setting $s = j\omega$, we have

$$H(\omega) = H(s)\Big|_{s=j\omega} = 10\frac{1 + \dfrac{j\omega}{1000}}{\left(1 + \dfrac{j\omega}{100}\right)\left(1 + \dfrac{j\omega}{10{,}000}\right)}$$

continued

Example 15.40 continued

Taking the logarithm of $|H(\omega)|$, we obtain

$$M_{dB}(\omega) = 20\log_{10}(10) - 20\log_{10}\left|1 + j\frac{\omega}{100}\right| + 20\log_{10}\left|1 + j\frac{\omega}{1000}\right| - 20\log_{10}\left|1 + j\frac{\omega}{10^4}\right|$$

The approximate magnitude in decibels for the four terms are given, respectively, by

$$M0_{dB}(\omega) = 20\log_{10}(10) = 20 \text{ dB}$$

$$M1_{dB}(\omega) = -20\log_{10}\left|1 + j\frac{\omega}{100}\right| \approx -20\log_{10}\left(\frac{\omega}{100}\right)$$

$$M2_{dB}(\omega) = 20\log_{10}\left|1 + j\frac{\omega}{1000}\right| \approx 20\log_{10}\left(\frac{\omega}{1000}\right)$$

$$M3_{dB}(\omega) = -20\log_{10}\left|1 + j\frac{\omega}{10^4}\right| \approx -20\log_{10}\left(\frac{\omega}{10^4}\right)$$

The straight-line magnitude Bode diagrams for these terms are shown in Figure 15.121.

FIGURE 15.121

Straight-line Bode diagram for four terms.

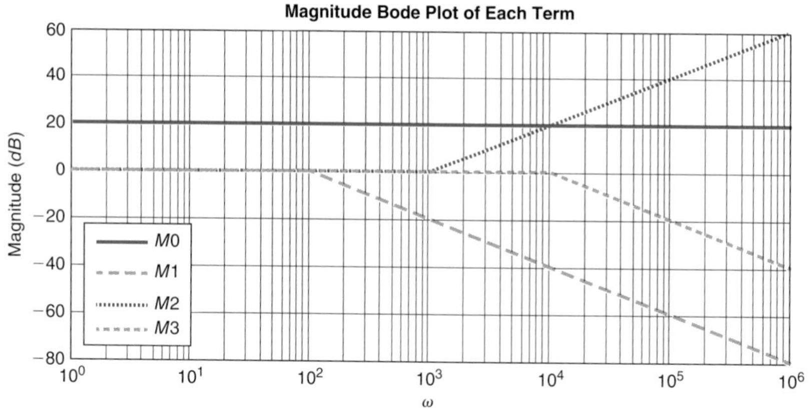

The straight-line magnitude Bode diagram is the sum of the four separate Bode diagrams, as shown in Figure 15.122.

FIGURE 15.122

Straight-line Bode diagram for four terms and the sum.

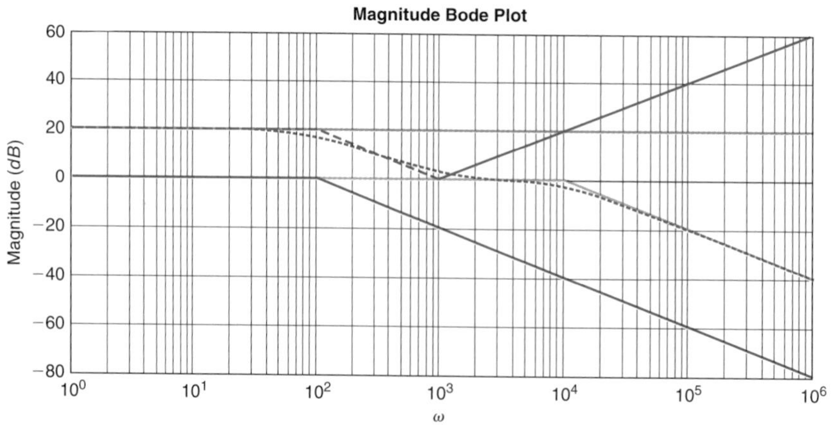

continued

Example 15.40 continued

FIGURE 15.123

Bode diagram for EXAMPLE 15.40.

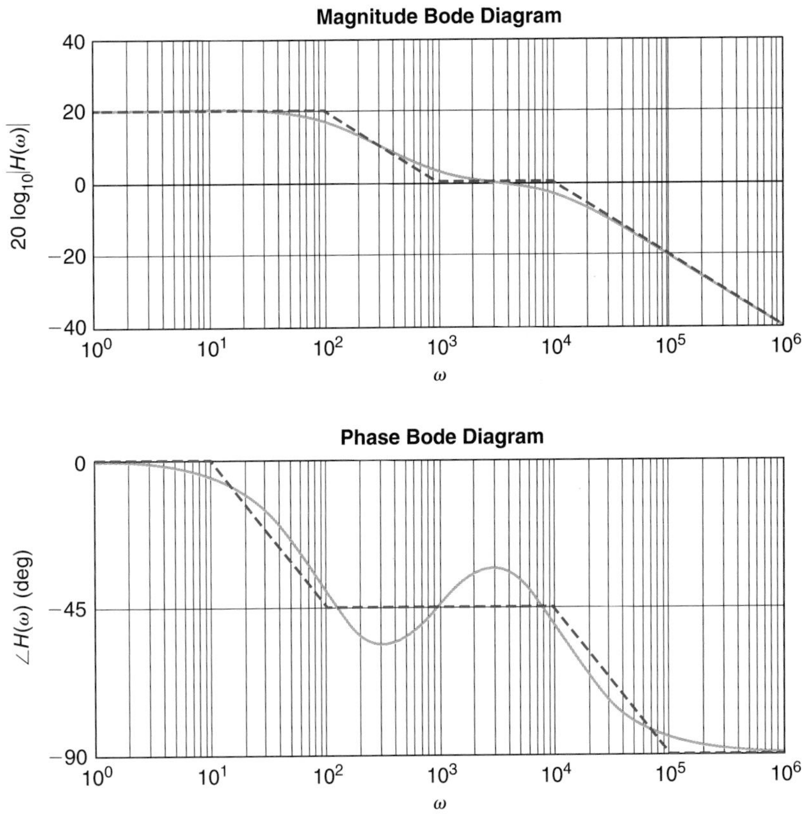

The exact Bode diagram as well as the straight-line approximation are shown in Figure 15.123.

In MATLAB,

```
clear all;
s = tf('s');
H = 10000*(s+1000)/(s+100)/(s+10000)
bode(H);grid;
```

or

```
clear all;
H=zpk(-1000,[-100 -10000],10000)
bodeplot(H);grid;
ltiview('bode',H)
ltiview({'bode','impulse','step'},H)
```

Exercise 15.40

Plot the magnitude and phase Bode diagram for the transfer function given by

$$H(s) = \frac{10^3(s + 10^2)}{s(s + 10^3)}$$

continued

Exercise 15.40 continued

Answer:

FIGURE 15.124

Bode diagram for EXERCISE 15.40.

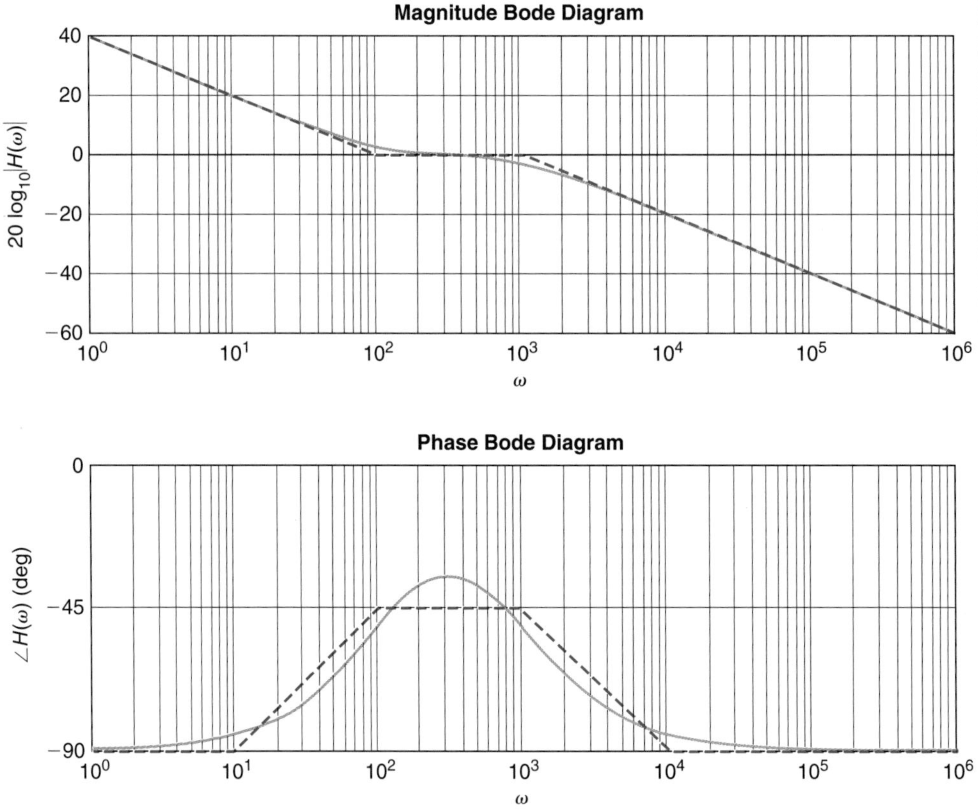

EXAMPLE 15.41

Plot the magnitude and phase Bode diagram for the transfer function given by

$$H(s) = \frac{10{,}000s}{(s + 100)(s + 10{,}000)}$$

The transfer function can be rewritten as

$$H(s) = \frac{10{,}000s}{(s + 100)(s + 10{,}000)} = \frac{\dfrac{s}{100}}{\left(1 + \dfrac{s}{100}\right)\left(1 + \dfrac{s}{10{,}000}\right)}$$

The magnitude increases at a rate of 20 dB/dec due to term $s/100$ in the numerator, reaching 0 dB at $\omega = 100$ rad/s. The term $(1 + s/100)$ in the denominator provides -20 dB/dec for $\omega \geq 100$ rad/s. This cancels 20 dB/dec from $s/100$, making the magnitude equal to 0 dB for $\omega \geq 0$ until $\omega = 10^4$ rad/s. The term $(1 + s/10^4)$ in the denominator provides -20 dB/dec for $\omega \geq 10^4$ rad/s. The Bode diagram is shown in

continued

Example 15.41 continued

FIGURE 15.125

Bode diagram for EXAMPLE 15.41.

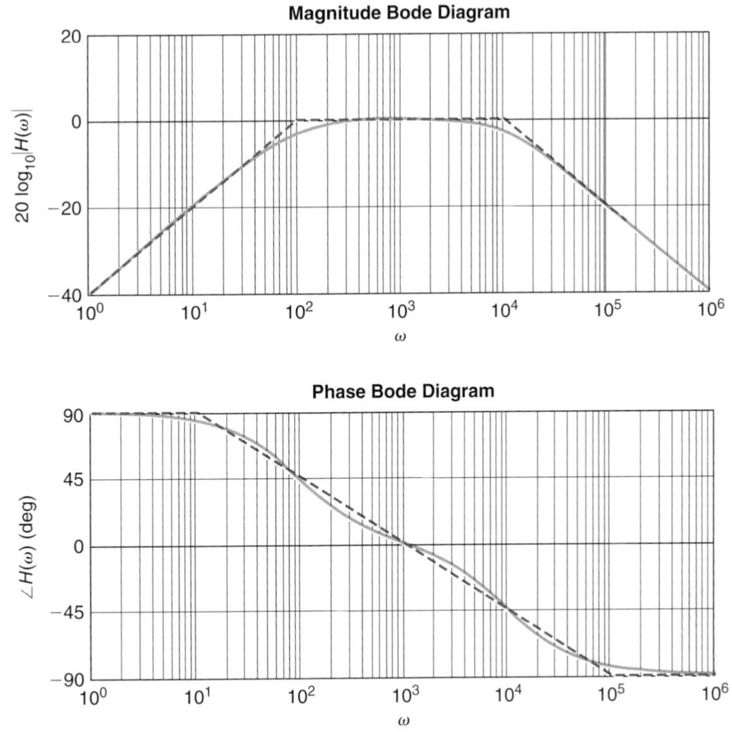

Figure 15.125. Since the gain is 0 dB for $100 \le \omega < 10^4$ and lower outside this interval, the transfer function represents a bandpass filter.

Exercise 15.41

Plot the magnitude and phase Bode diagram for the transfer function given by

$$H(s) = \frac{10^6(s + 10^2)}{s(s + 10^3)^2}$$

Answer:

FIGURE 15.126

Bode diagram for EXERCISE 15.41.

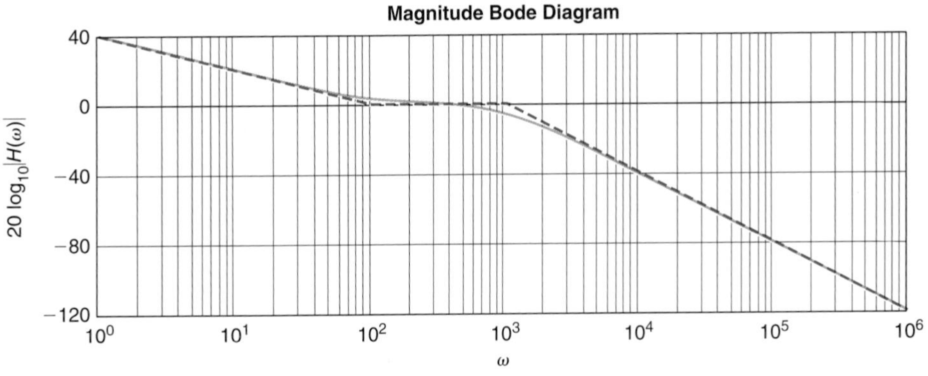

continued

Exercise 15.41 continued

FIGURE 15.126

continued

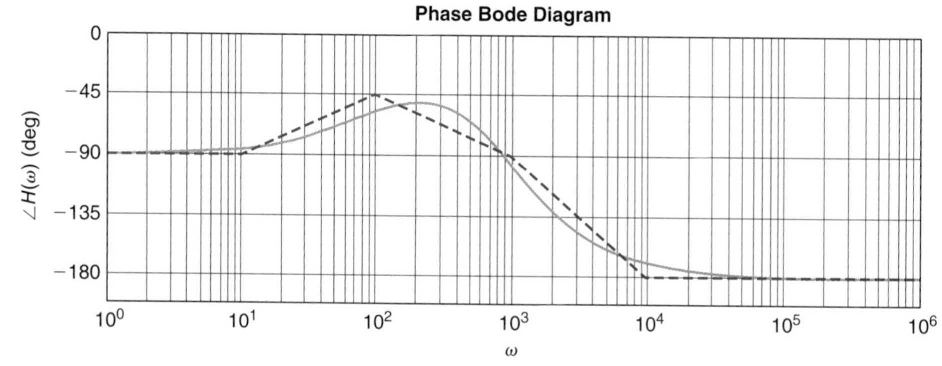

EXAMPLE 15.42

Plot the magnitude and phase Bode diagram for the transfer function given by

$$H(s) = \frac{(s + 10^4)^2}{(s + 1000)(S + 10^5)}$$

The transfer function can be rewritten as

$$H(s) = \frac{(s + 10^4)^2}{(s + 1000)(S + 10^5)} = \frac{\left(1 + \dfrac{s}{10^4}\right)^2}{\left(1 + \dfrac{s}{1000}\right)\left(1 + \dfrac{s}{10^5}\right)}$$

For $\omega < 1000$ rad/s, the magnitude is 0 dB. At $\omega = 1000$ rad/s, due to $(1 + s/1000)$ in the denominator, the magnitude starts decreasing by 20 dB/dec. At $\omega = 10^4$ rad/s, the magnitude starts increasing by 20 dB/s until $\omega = 10^5$ rad/s. For $\omega \geq 10^5$, the magnitude stays at 0 dB, as shown in Figure 15.127.

FIGURE 15.127

Bode diagram for EXAMPLE 15.42.

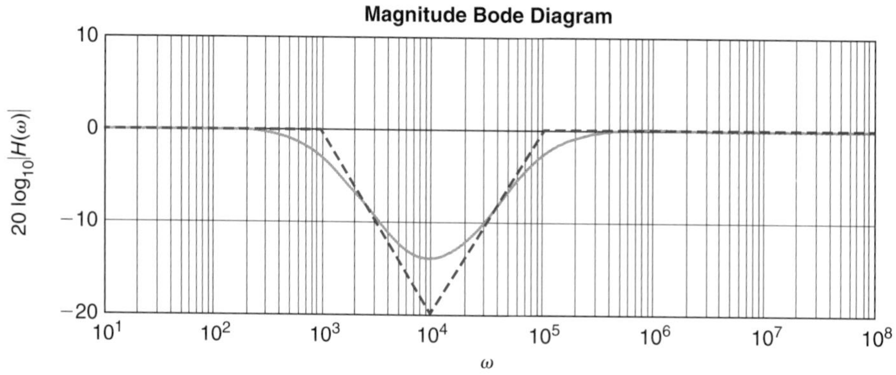

continued

Example 15.42 continued

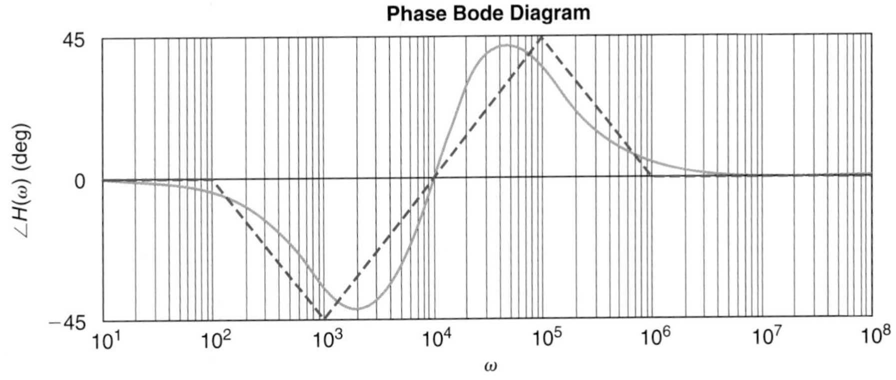

FIGURE 15.127

continued

Exercise 15.42

Plot the magnitude and phase Bode diagram for the transfer function given by

$$H(s) = \frac{s^2}{(s + 10^3)^2}$$

Answer:

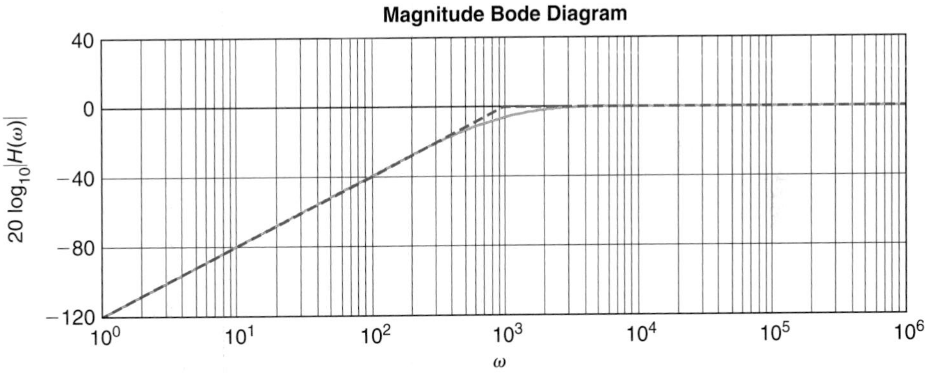

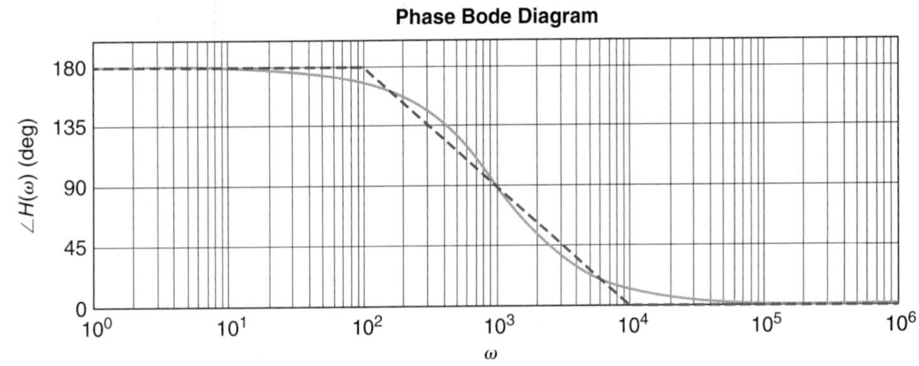

FIGURE 15.128

Bode diagram for EXERCISE 15.42.

EXAMPLE 15.43

Plot the magnitude and phase Bode diagram for the transfer function given by

$$H(s) = \frac{(s + 10^5)(s + 10^6)}{(s + 10^3)(s + 10^8)}$$

The transfer function can be rewritten as

$$H(s) = \frac{\left(1 + \dfrac{s}{10^5}\right)\left(1 + \dfrac{s}{10^6}\right)}{\left(1 + \dfrac{s}{10^3}\right)\left(1 + \dfrac{s}{10^8}\right)}$$

For $\omega < 1000$ rad/s, the magnitude is 0 dB. At $\omega = 1000$ rad/s, due to $(1 + s/1000)$ in the denominator, the magnitude starts decreasing by 20 dB/dec. For $10^5 < \omega < 10^6$ rad/s, the magnitude is -40 dB. At $\omega = 10^6$ rad/s, the magnitude starts increasing by 20 dB/s until $\omega = 10^8$ rad/s. For $\omega \geq 10^8$, the magnitude stays at 0 dB, as shown in Figure 10.129. This is a band-stop filter.

FIGURE 15.129

Bode diagram for EXAMPLE 15.43.

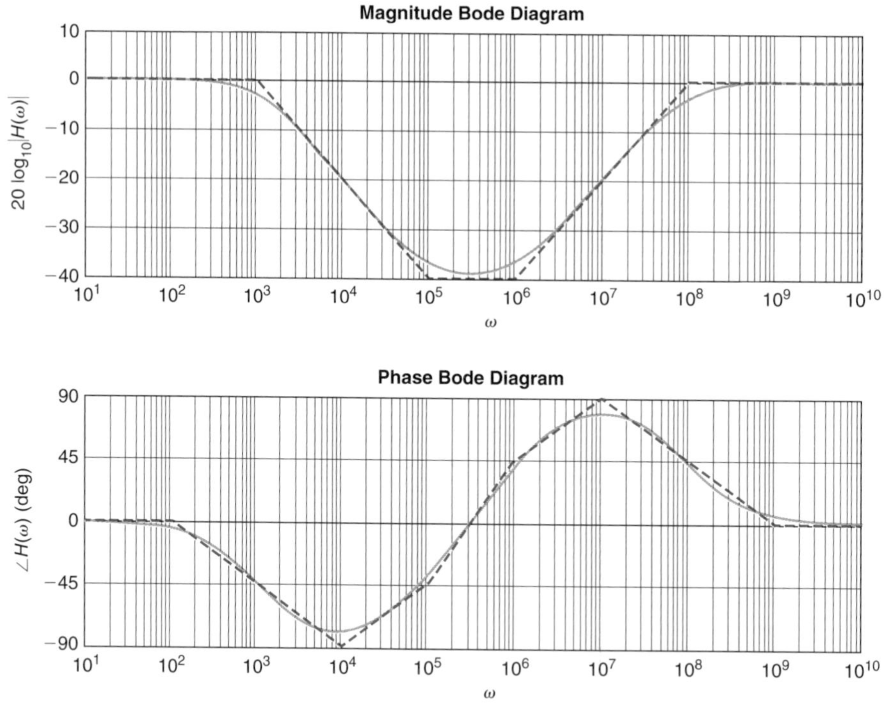

Exercise 15.43

Plot the magnitude and phase Bode diagram for the transfer function given by

$$H(s) = \frac{s^2}{(s + 10^3)^2}$$

continued

Exercise 15.43 continued

Answer:

FIGURE 15.130

Bode diagram for EXERCISE 15.43.

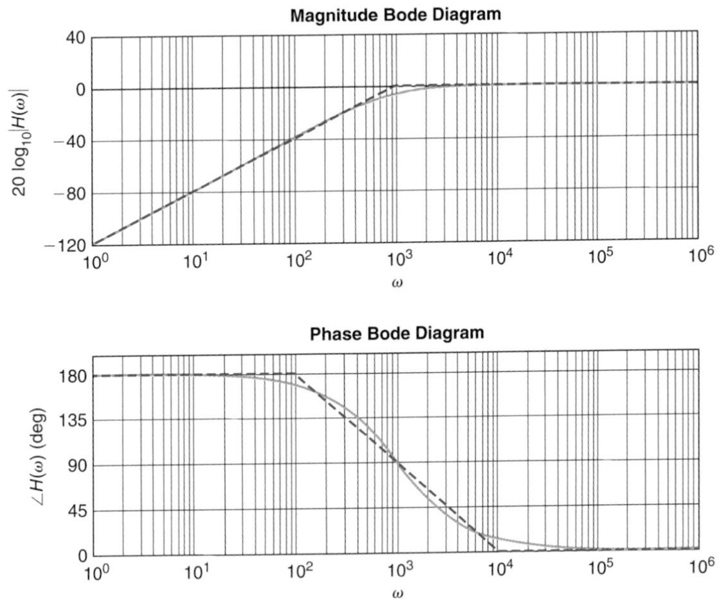

EXAMPLE 15.44

Plot the magnitude and phase Bode diagram for the transfer function given by

$$H(s) = \frac{10^{12}(s + 10^5)}{(s + 10^3)(s + 10^6)^2}$$

The transfer function can be rewritten as

$$H(s) = \frac{10^{12}(s + 10^5)}{(s + 10^3)(s + 10^6)^2} = \frac{10^2\left(1 + \dfrac{s}{10^5}\right)}{\left(1 + \dfrac{s}{10^3}\right)\left(1 + \dfrac{s}{10^6}\right)^2}$$

The Bode diagram is shown in Figure 15.131.

FIGURE 15.131

Bode diagram for EXAMPLE 15.44.

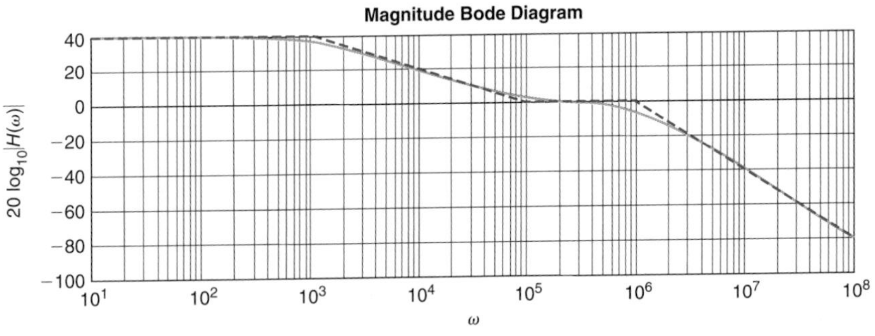

continued

Example 15.44 continued

FIGURE 15.131

continued

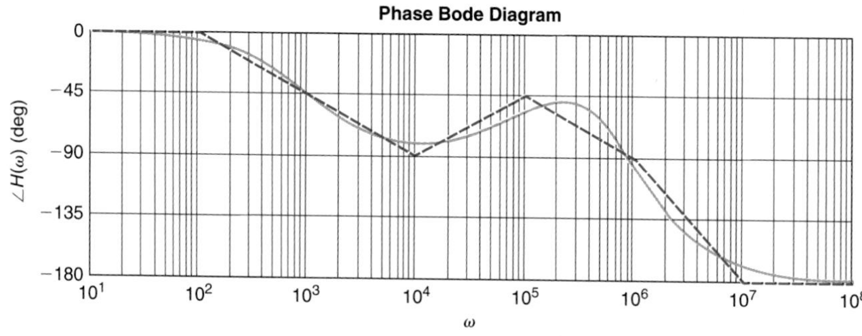

Exercise 15.44

Plot the magnitude and phase Bode diagram for the transfer function given by

$$H(s) = \frac{10^8 s^2}{(s + 10^3)^2(s + 10^4)^2}$$

Answer:

FIGURE 15.132

Bode diagram for EXERCISE 15.44.

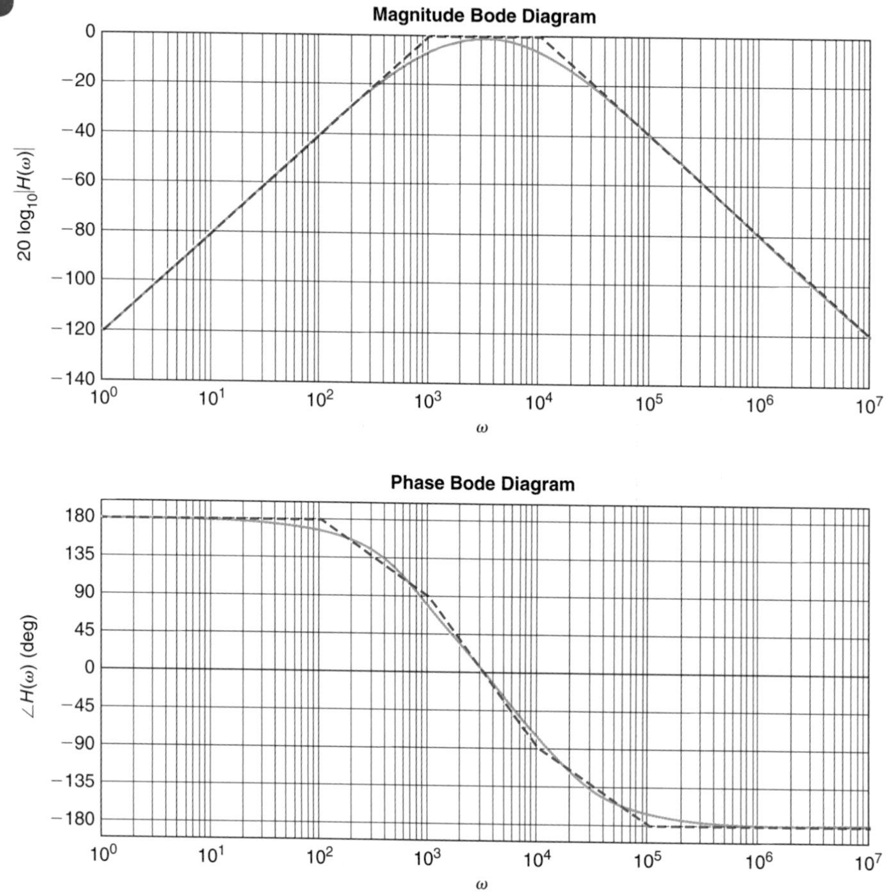

15.11.8 COMPLEX POLES AND ZEROS

The transfer function of the second-order system (low-pass filter) can be written as

$$H(s) = \frac{\omega_0^2}{s^2 + 2\zeta\omega_0 s + \omega_0^2}$$

where ζ is the damping ratio, and ω_0 is the corner frequency. The frequency response is given by

$$H(\omega) = H(s)\Big|_{s=j\omega} = \frac{\omega_0^2}{-\omega^2 + 2\zeta\omega_0 j\omega + \omega_0^2}$$

Dividing by ω_0^2, we have

$$H(\omega) = \frac{1}{1 - \dfrac{\omega^2}{\omega_0^2} + j2\zeta\dfrac{\omega}{\omega_0}}$$

Taking the log, we obtain

$$M_{dB}(\omega) = 20\log_{10}(|H(\omega)|) = -20\log_{10}\left(\left|1 - \frac{\omega^2}{\omega_0^2} + j2\zeta\frac{\omega}{\omega_0}\right|\right)$$

$$= -10\log_{10}\left[\left(1 - \frac{\omega^2}{\omega_0^2}\right)^2 + \left(2\zeta\frac{\omega}{\omega_0}\right)^2\right].$$

When $\omega < \omega_0$, we have $\omega/\omega_0 < 1$, and the magnitude response in decibels can be approximated by

$$M_{dB}(\omega) \approx -20\log(1) = 0 \text{ dB}.$$

When $\omega > \omega_0$, 1 and ω/ω_0 is negligible compared to $\dfrac{\omega^2}{\omega_0^2}$. Thus, the magnitude response in decibels can be approximated by

$$M_{dB}(\omega) = 20\log(|H(\omega)|) \approx -20\log\left(\frac{\omega^2}{\omega_0^2}\right) = -40\log\left(\frac{\omega}{\omega_0}\right)$$

From this approximation, we obtain:

At $\omega = \omega_0$, $M_{dB} = 0$ dB

At $\omega = 10\omega_0$, $M_{dB} = -40$ dB

At $\omega = 100\omega_0$, $M_{dB} = -80$ dB

As the radian frequency is increased by ten times, the magnitude is decreased by 40 dB. The slope of the straight-line approximation is -40 dB/dec. The difference between the exact magnitude in decibels and the straight-line approximation depends on the value of the damping ratio ζ. When the damping ratio is small, the magnitude response shows a peak around the corner frequency ω_0. Figure 15.133 shows the exact magnitude response in decibels along with the straight-line approximation for several values of ζ for $\omega_0 = 1000$ rad/s. When $\zeta \geq 1/\sqrt{2}$, there is no peak.

FIGURE 15.133

The Bode diagram for the second-order transfer function.

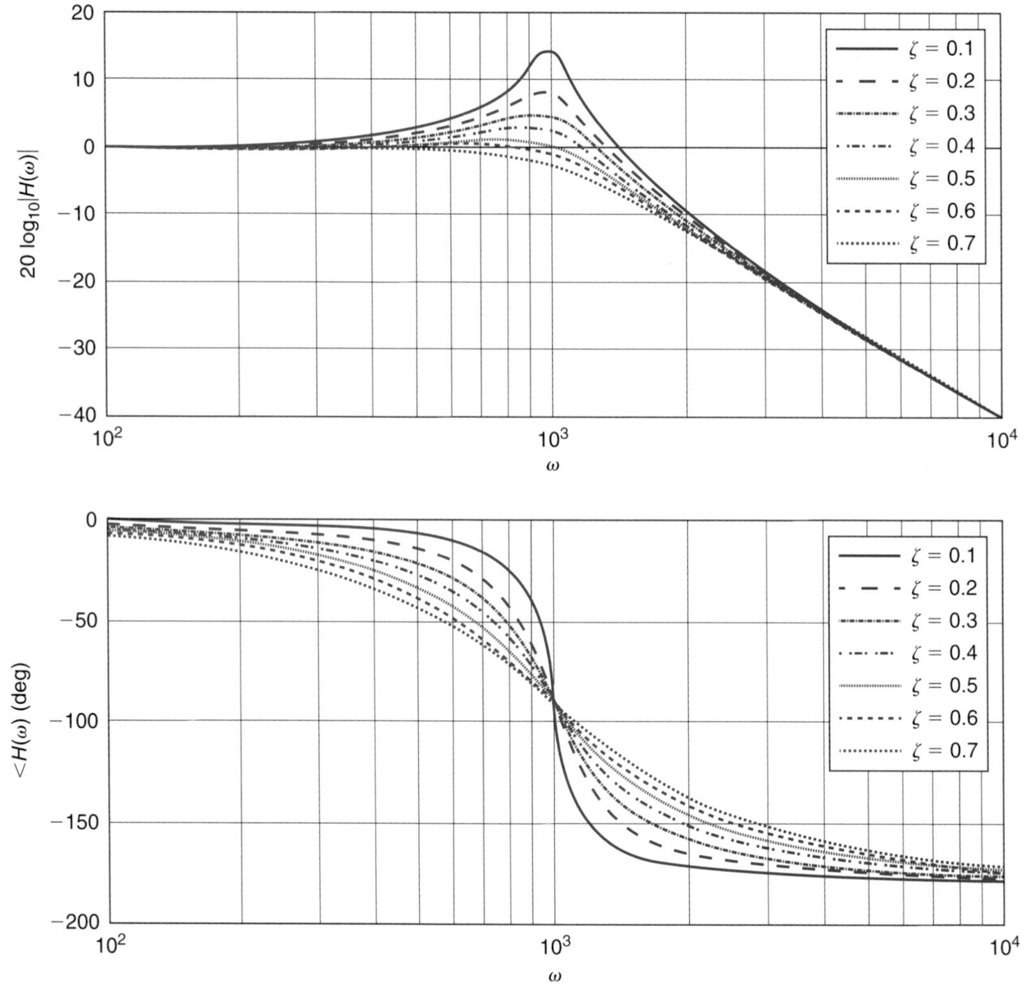

Let ω_{peak} be the frequency where the magnitude response is at its maximum. Solving

$$\frac{d}{d\omega}\left[\left(1-\frac{\omega^2}{\omega_0^2}\right)^2+\left(2\zeta\frac{\omega}{\omega_0}\right)^2\right]=4\omega\frac{\omega^2-\omega_0^2+2\zeta^2\omega_0^2}{\omega_0^4}=0$$

we have

$$\omega_{peak}=\omega_0\sqrt{1-2\zeta^2} \tag{15.115}$$

The frequency ω_z where M_{dB} crosses zero is obtained by solving

$$\left(1-\frac{\omega^2}{\omega_0^2}\right)^2+\left(2\zeta\frac{\omega}{\omega_0}\right)^2=1$$

The answer we get is

$$\omega_z=\omega_0\sqrt{2-4\zeta^2}=\sqrt{2}\omega_{peak} \tag{15.116}$$

The magnitude in decibels at $\omega = \omega_{peak}$ is given by

$$M_{dB}(\omega_{peak}) = -10\log\left[\left(1-\frac{\omega_{peak}^2}{\omega_0^2}\right)^2 + \left(2\zeta\frac{\omega_{peak}}{\omega_0}\right)^2\right]$$
$$= -10\log\left[\left(1-\frac{\omega_0^2(1-2\zeta^2)}{\omega_0^2}\right)^2 + \left(2\zeta\frac{\omega_0\sqrt{1-2\zeta^2}}{\omega_0}\right)^2\right] \quad \textbf{(15.117)}$$
$$= -10\log(4\zeta^2 - 4\zeta^4)$$

The phase response is given by

$$\angle H(\omega) = -\tan^{-1}\left(\frac{2\zeta\frac{\omega}{\omega_0}}{1-\frac{\omega^2}{\omega_0^2}}\right). \quad \textbf{(15.118)}$$

At $\frac{\omega}{\omega_0} = 1$ or $\omega = \omega_0$, $\angle H(\omega) = -\tan^{-1}(\infty) = -\frac{\pi}{2} = -90°$.

For $\omega < \omega_0$, $\angle H(j\omega) \approx -\tan^{-1}(0) = 0$ rad. For $\omega > \omega_0$, $\frac{\omega^2}{\omega_0^2} > 2\zeta\frac{\omega}{\omega_0}$. Thus, we have

$$H(\omega) = \frac{1}{1-\frac{\omega^2}{\omega_0^2} + j2\zeta\frac{\omega}{\omega_0}} \approx \frac{1}{-\frac{\omega^2}{\omega_0^2} + j2\zeta\frac{\omega}{\omega_0}} \approx \frac{1}{-\frac{\omega^2}{\omega_0^2} + j0} = -\frac{\omega_0^2}{\omega^2}$$

Therefore, $\angle H(j\omega) \approx -\pi$ rad $= -180°$. The phase decreases from 0° to −180°.

EXAMPLE 15.45

Plot the Bode diagram of the transfer function given by

$$H(s) = \frac{10^{10}}{(s^2 + 200s + 10^6)(s + 10^4)}$$

The Bode diagram is shown in Figure 15.134.

FIGURE 15.134

Magnitude Bode diagram and phase Bode diagram.

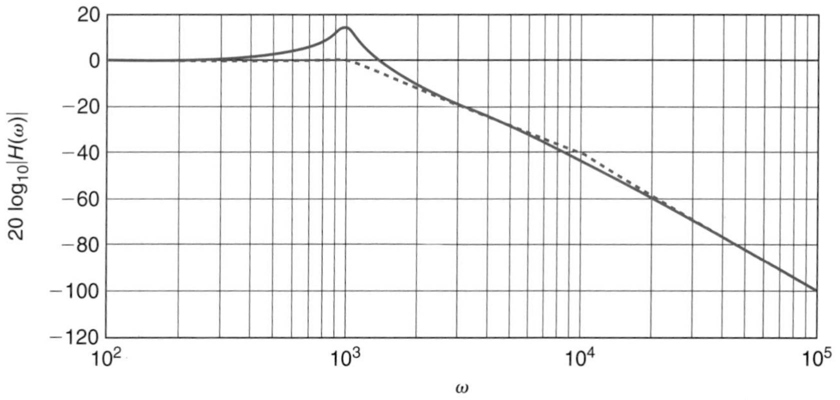

continued

Example 15.45 continued

FIGURE 15.134

continued

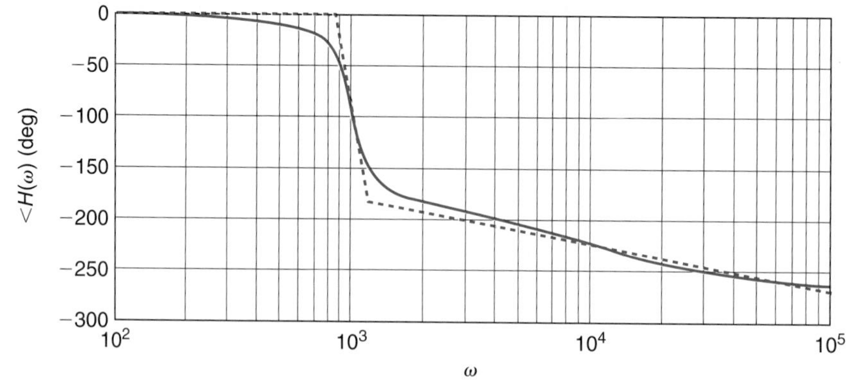

Exercise 15.45

Plot the magnitude and phase Bode diagram for the transfer function given by

$$H(s) = \frac{s^2}{s^2 + 200s + 10^6}$$

Answer:

FIGURE 15.135

Bode diagram for EXERCISE 15.45.

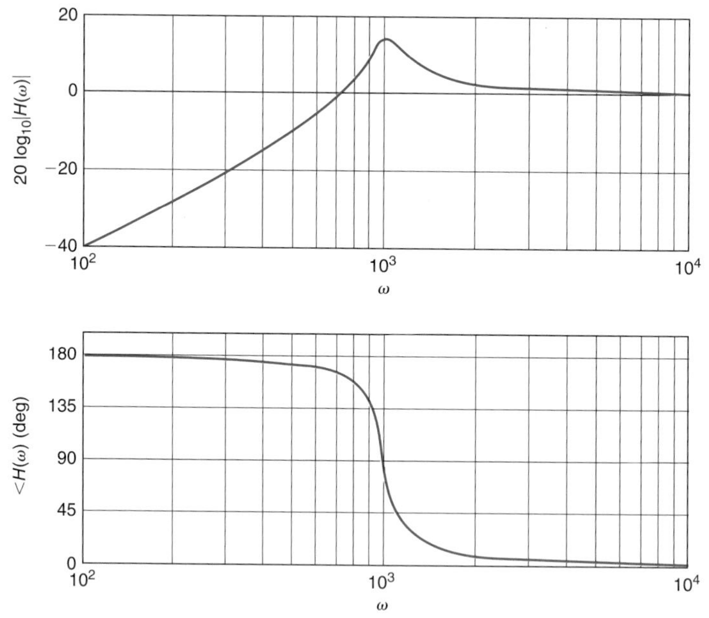

15.12 Simulink

The linear analysis tool of Simulink can be used to plot impulse response, step response, Bode diagram, and pole zero diagram of a system. As an example, consider a system with transfer function

$$H(s) = \frac{10}{s^2 + 2s + 10}$$

FIGURE 15.136

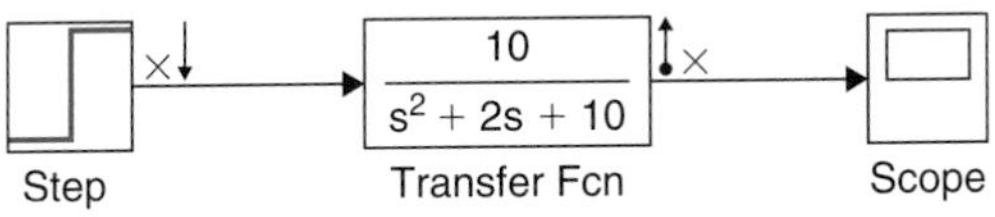

A Simulink model.

Build a Simulink model shown in Figure 15.136.

Click on the wire connecting the Step block to the Transfer Fcn block. Select the open-loop input point. An arrow shows up, as shown in Figure 15.136. Similarly, select the open-loop output point on the wire connecting Transfer Fcn and Scope. The linear analysis is on the system between the input and the output points. Select Analysis → Control Design → Linear Analysis to open the Linear Analysis Tool window. Select New Step, and click on Linearize to get the step response shown in Figure 15.137.

Select New Impulse, and click on Linearize. We obtain the impulse response shown in Figure 15.138. The Bode diagram is discussed in Chapter 16.

FIGURE 15.137

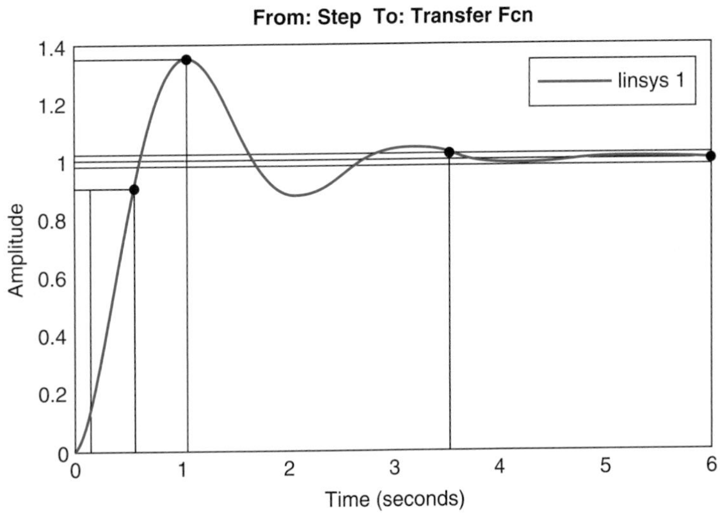

Step response.

FIGURE 15.138

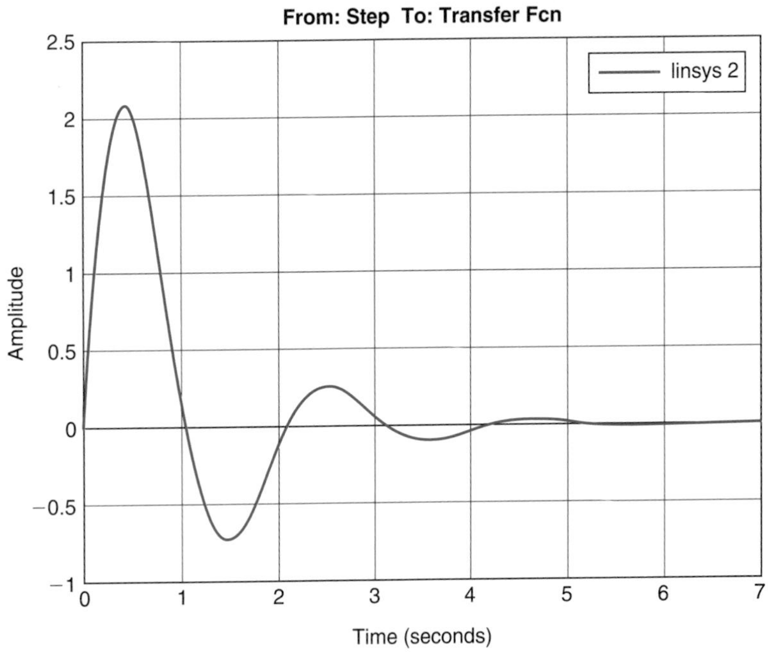

Impulse response.

SUMMARY

One of the applications of the Laplace transform is in the circuit analysis. The circuit components and sources are transformed to the s-domain to convert the original circuit to a Laplace-transformed circuit. The initial conditions are incorporated in the Laplace-transformed circuit. The desired signal in the s-domain is obtained by applying circuit laws and theorems to the Laplace-transformed circuit. Instead of solving differential equations, we are solving algebraic equations in s. The desired signal in the time domain can be found by taking the inverse Laplace transform of the desired signal in the s-domain.

The transfer function of a circuit or a system is defined as the ratio of the output signal in the s-domain to the input signal in the s-domain. The transfer function provides the characteristics of the circuit in the s-domain. It provides poles and zeros of the system and the frequency response of the system. The output, called response, of the system in the s-domain is the product of the transfer function and the input, called excitation, in the s-domain. The output in the time domain is found by taking the inverse Laplace transform of the output in the s-domain. The response of a system to Dirac delta function is called the impulse response. The impulse response is the inverse Laplace transform of the transfer function. The response of a system to unit step function is called the *step response.* The step response is the inverse Laplace transform of the transfer function divided by s. The output of an LTI system is the convolution of the input and the impulse response. The Bode diagram is a useful tool in system analysis and design. It shows the magnitude response and the phase response as a function of frequency ω. It correlates the location of poles and zeros to the frequency response.

PROBLEMS

Circuit Analysis in the s-Domain

15.1 **The initial current through the inductor in the circuit shown in Figure P15.1 is $i(0^-) = 2$ A. Find the voltage $v(t)$ across the resistor for $t \geq 0$.**

FIGURE P15.1

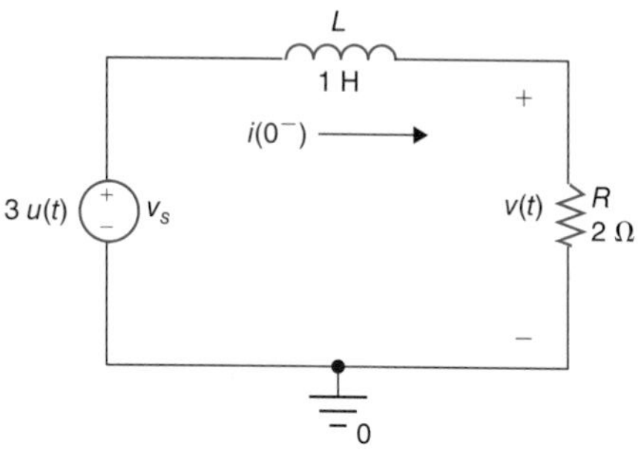

15.2 **The initial voltage across the capacitor in the circuit shown in Figure P15.2 is $v(0^-) = 2$ V. Find the voltage $v(t)$ across the capacitor for $t \geq 0$.**

FIGURE P15.2

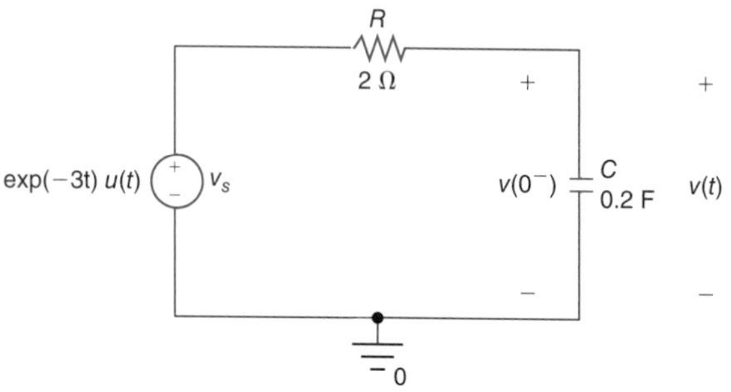

15.3 **The initial voltage across the capacitor in the circuit shown in Figure P15.3 is $v(0^-) = 1$ V, and the initial current through the inductor is $i(0^-) = 0$ mA. Find the voltage $v_o(t)$ across the capacitor for $t \geq 0$.**

FIGURE P15.3

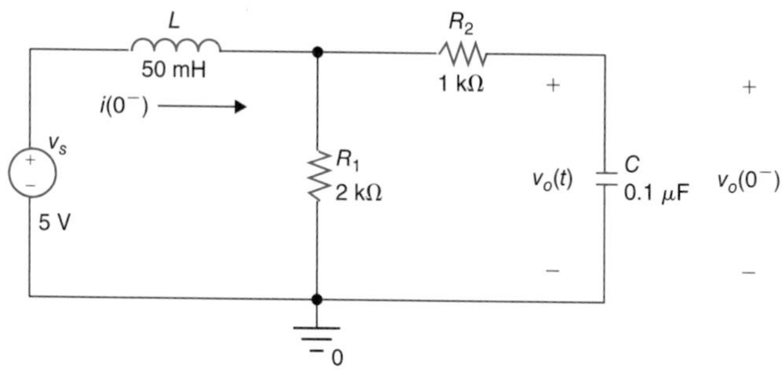

15.4 **Find the $v_o(t)$ for the circuit shown in Figure P15.4, assuming that all initial conditions are zero.**

FIGURE P15.4

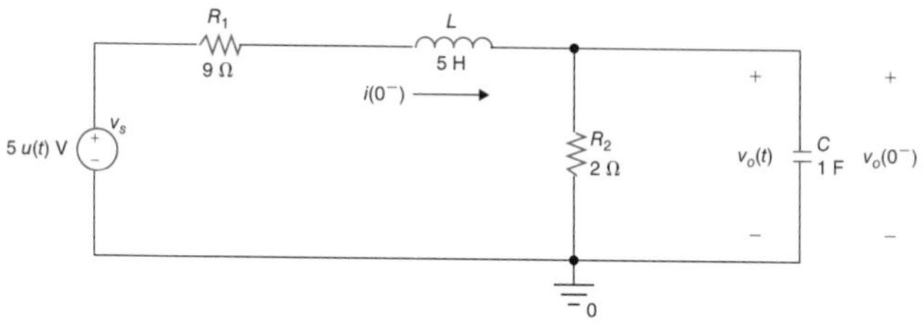

15.5 **Repeat Problem 15.4 when the initial current through the inductor is $i(0^-) = 1$ A. The initial voltage across the capacitor is zero.**

15.6 **The switch in the circuit shown in Figure P15.6 has been closed for a long time before it is opened at $t = 0$. Find the voltage $v_o(t)$ for $t \geq 0$.**

FIGURE P15.6

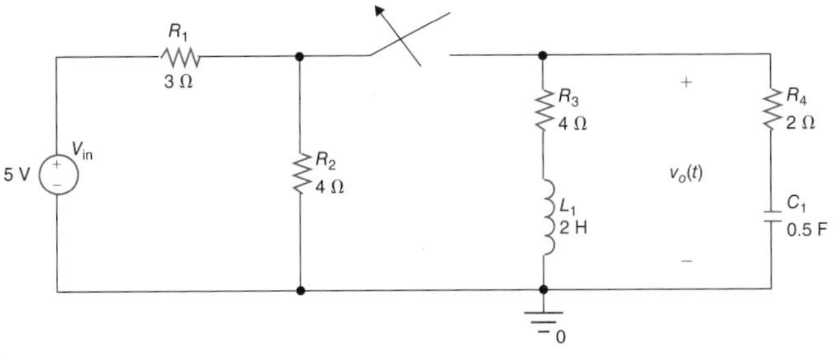

15.7 **The switch in the circuit shown in Figure P15.7 has been closed for a long time before it is opened at $t = 0$. Find the voltage $v_o(t)$ for $t \geq 0$.**

FIGURE P15.7

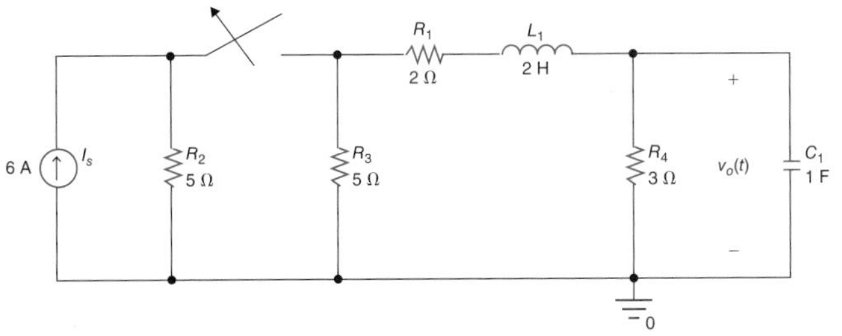

15.8 **In the circuit shown in Figure P15.8, let** $v_s(t) = 5\,u(t)$, $R_1 = 1\ \Omega$, $R_2 = 2\ \Omega$, $R_3 = 3\ \Omega$, $L = 1$ H, $C = (1/2)$F, $i(0^-) = 1$ A, and $v(0^-) = 2$ V.

FIGURE P15.8

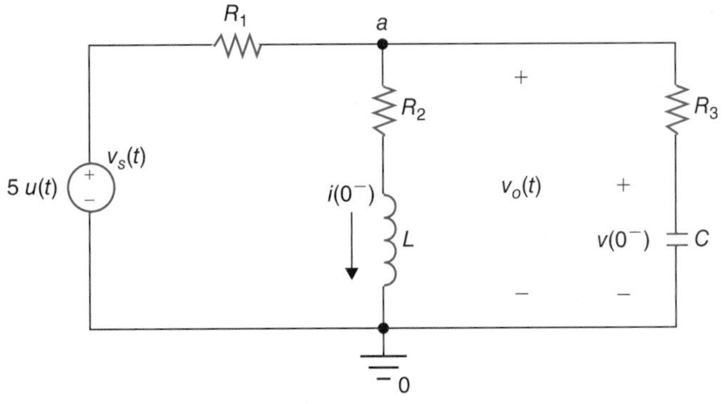

a. Draw the circuit in the s-domain for $t \geq 0$.
b. Write a node equation at node a by summing the currents leaving node a.
c. Find $V_o(s)$ in the s-domain.
d. Find $v_o(t)$ in the time domain by taking the inverse Laplace transform of $V_o(s)$.

15.9 **In the circuit shown in Figure P15.9, let** $v_s(t) = 6\,u(t)$, $R_1 = 2\ \Omega$, $R_2 = 3\ \Omega$, $L = 5$ H, $C = 0.5$ F, $i(0^-) = 1$ A, and $v(0^-) = 2$ V.

FIGURE P15.9

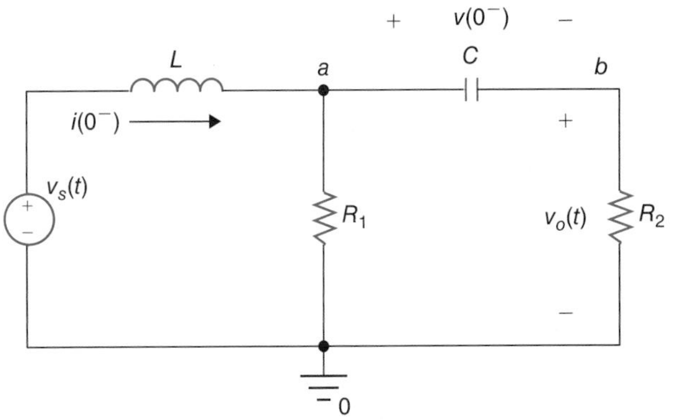

a. Draw the circuit in the s-domain for $t \geq 0$.
b. Write a node equation at node a by summing the currents leaving node a.
c. Write a node equation at node b by summing the currents leaving node b.
d. Find $V_o(s)$ in the s-domain.
e. Find $v_o(t)$ in the time domain by taking the inverse Laplace transform of $V_o(s)$.

15.10 **The switch in the circuit shown in Figure P15.10 has been open for a long time before it is closed at $t = 0$.**

Let $V_s = 12$ V dc, $R_1 = 2\ \Omega$, $R_2 = 2\ \Omega$, $R_3 = 2\ \Omega$, $L = 4$ H, and $C = (1/3)$F.

FIGURE P15.10

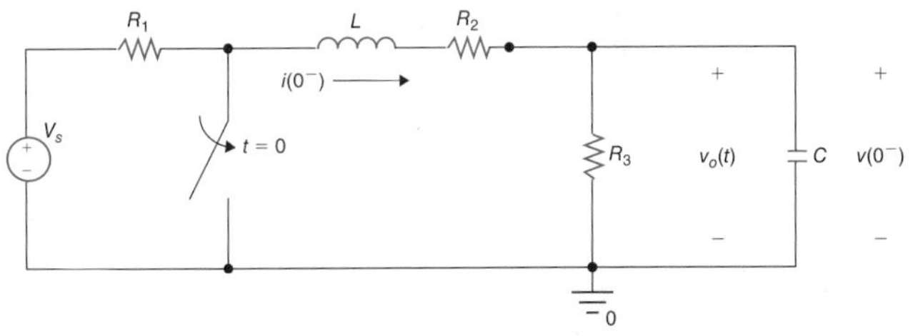

a. Find the initial current $i(0^-)$ through the inductor and the initial voltage $v(0^-)$ across the capacitor.
b. Draw the circuit in the s-domain for $t \geq 0$.
c. Find $V_o(s)$ in the s-domain.
d. Find $v_o(t)$ in the time domain.

15.11 **In the circuit shown in Figure P15.11, let** $v_s(t) = 5\,u(t)$, $R_1 = 2\ \Omega$, $R_2 = 2\ \Omega$, $L_1 = 1$ H, $L_2 = 2$ H, $C = (1/2)$F, $i_1(0^-) = 2$ A, $i_2(0^-) = 1$ A, and $v(0^-) = 2$ V.

Find $v_o(t)$ for $t \geq 0$.

FIGURE P15.11

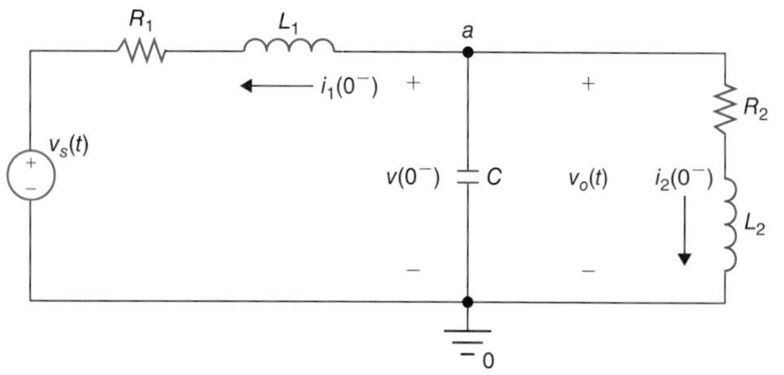

15.12 The switch in the circuit shown in Figure P15.12 has been closed for a long time before it is opened at $t = 0$. Let

$V_s = 9$ V dc, $R_1 = 5\ \Omega$, $R_2 = 3\ \Omega$, $L = 2$ H, $C = (1/3)$F.

FIGURE P15.12

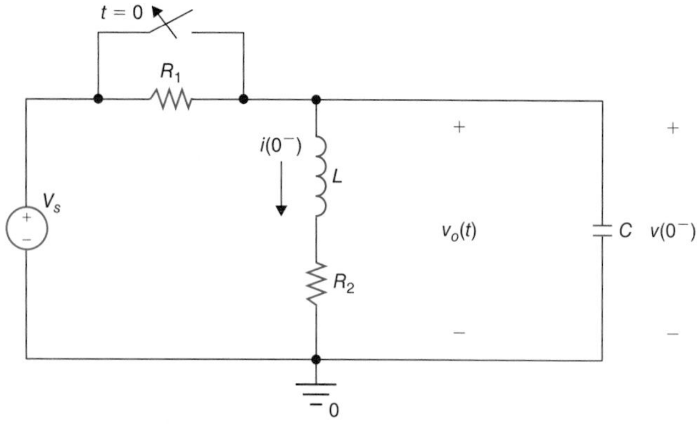

a. Find the initial current $i(0^-)$ through the inductor and the initial voltage $v(0^-)$ across the capacitor.
b. Draw the circuit in the s-domain for $t \geq 0$.
c. Find $V_o(s)$ in the s-domain.
d. Find $v_o(t)$ in the time domain.

15.13 Let $C = 0.1$ F, $R_1 = 1\ \Omega$, $R_2 = 2\ \Omega$, $L = 2$ H, initial voltage across the capacitor = $v(0^-)$ = 5 V, and initial current through the inductor = $i(0^-) = 1$ A, in the circuit shown in Figure P15.13. Find $v_o(t)$ for $t \geq 0$.

FIGURE P15.13

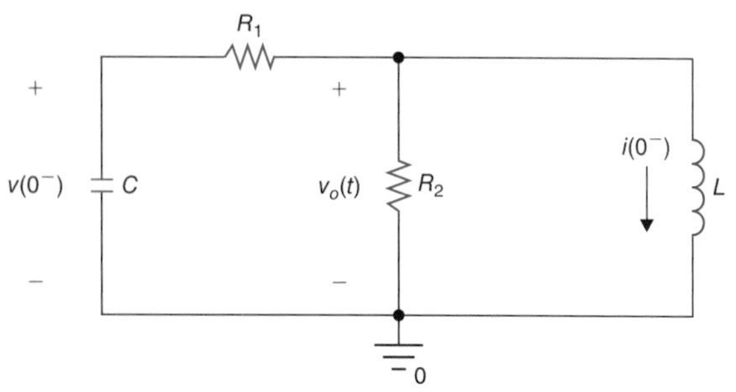

15.14 In the circuit shown in Figure P15.14, let

$R_1 = 1\ \Omega$, $R_2 = 2\ \Omega$, $C = 5$ F, $L = 1$ H, $v(0^-) = 3$ V, $i(0^-) = 1$ A, and $i(t) = 6\,u(t)$, $V_b = V_o$

FIGURE P15.14

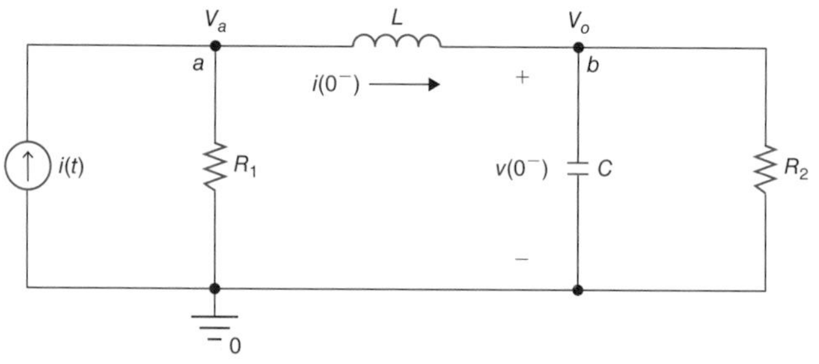

a. Draw the circuit in the s-domain for $t \geq 0$.
b. Write a node equation at node a by summing the currents leaving node a.
c. Write a node equation at node b by summing the currents leaving node b.
d. Find $V_o(s)$ in the s-domain.
e. Find $v_o(t)$ in the time domain.

15.15 In the circuit shown in Figure P15.15, let

$R_1 = 1\ \Omega$, $R_2 = 3\ \Omega$, $C = (1/2)$F, $L = 1$ H, $v(0^-) = 3$ V, $i(0^-) = 1$ A, $i(t) = 2\,u(t)$, and $V_b = V_o$.
Find $v_o(t)$ for $t \geq 0$.

FIGURE P15.15

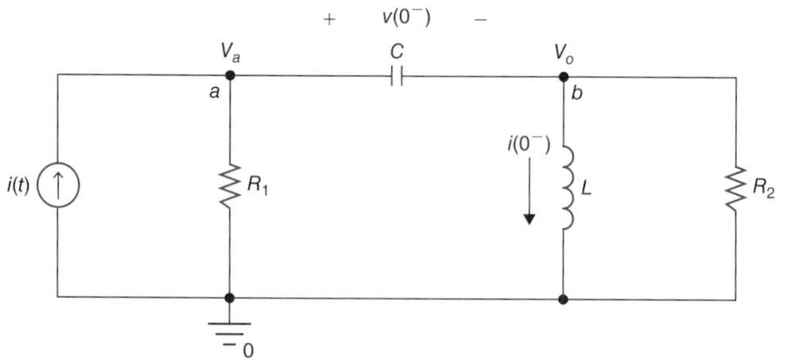

15.16 The switch in the circuit shown in Figure P15.16 has been in position A for a long time before it is moved to position B at time $t = 0$. Let

$I_1 = 4$ A dc, $V_1 = 10$ V dc, $R_1 = 3\ \Omega$, $R_2 = 3\ \Omega$, $L = 1$ H, and $C = (1/2)$F.

FIGURE P15.16

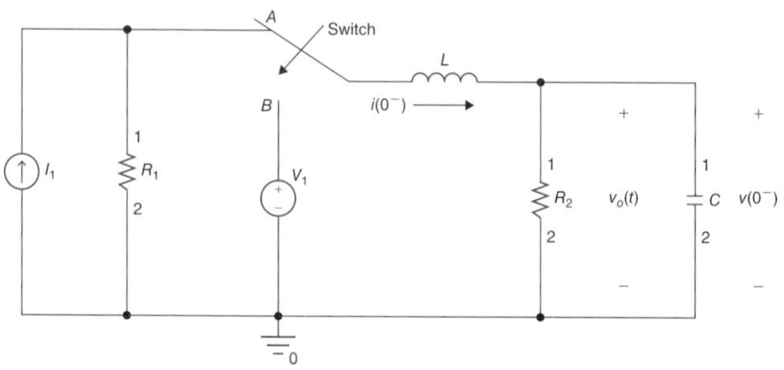

a. Find the initial current $i(0^-)$ through the inductor and the initial voltage $v(0^-)$ across the capacitor.
b. Draw the circuit in the s-domain for $t \geq 0$.
c. Find $V_o(s)$ in the s-domain.
d. Find $v_o(t)$ in the time domain.

15.17 **In the circuit shown in Figure P15.17, let** $R = 3\ \Omega$, $L = 1$ H, $C = (1/2)$F, $i(0^-) = 1$ A, $v(0^-) = 4$ V, and $v_s(t) = 2\ u(t)$.

Find $v_o(t)$ for $t \geq 0$.

FIGURE P15.17

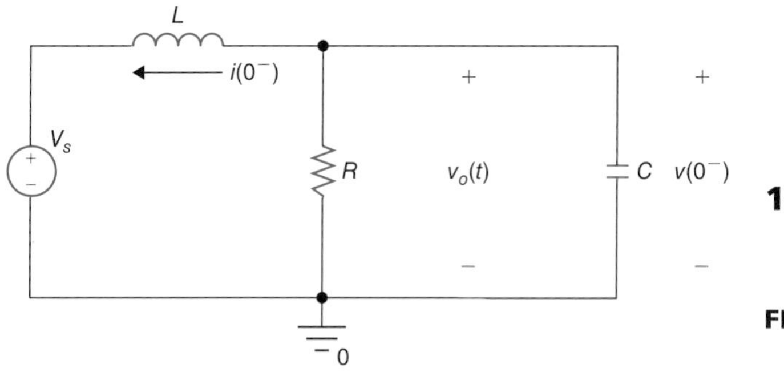

15.18 **The switch in the circuit shown in Figure P15.18 has been in position *A* for a long time before it is moved to position *B* at $t = 0$. Let**

$V_s = 9$ V dc, $R_1 = 1\ \Omega$, $R_2 = 1\ \Omega$, $R_3 = 2\ \Omega$, $L = 1$ H, $C = (1/2)$F

FIGURE P15.18

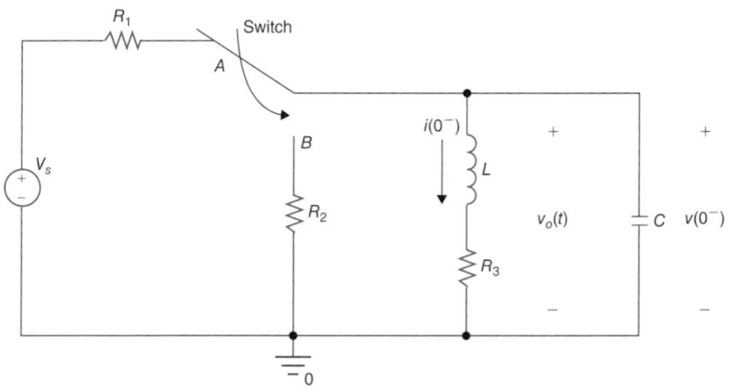

a. Find the initial current $i(0^-)$ through the inductor and the initial voltage $v(0^-)$ across the capacitor.
b. Draw the circuit in the s-domain for $t \geq 0$.
c. Find $V_o(s)$ in the s-domain.
d. Find $v_o(t)$ in the time domain.

15.19 **In the circuit shown in Figure P15.19, the initial voltage across the capacitor is $v(0^-) = 2$ V and the input signal is given by $i(t) = e^{-3t}u(t)$. $R_1 = 1\ \Omega$, $R_2 = 1\ \Omega$, $C = 0.5$ F.**

FIGURE P15.19

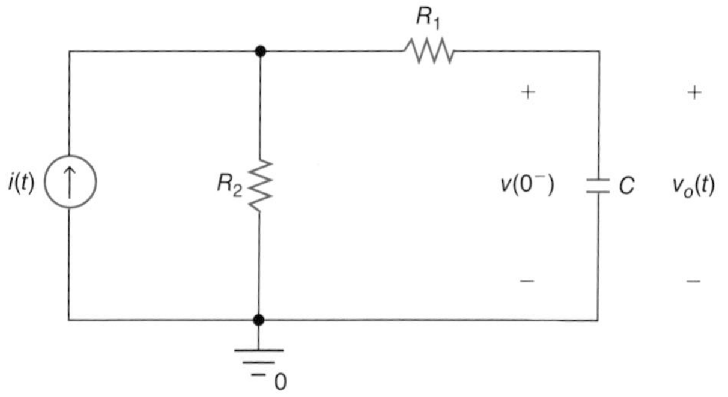

a. Draw the circuit in the s-domain.
b. Find the Laplace transform $V_o(s)$ of the output signal.
c. Find $v_o(t)$ in the time domain by taking the inverse Laplace transform of $V_o(s)$.

15.20 **Let the initial voltage across the capacitor be $g(0^-) = 2$ V and $f(t) = e^{-20{,}000t}\ u(t)$ for the circuit shown in Figure P15.20. Find $g(t)$.**

FIGURE P15.20

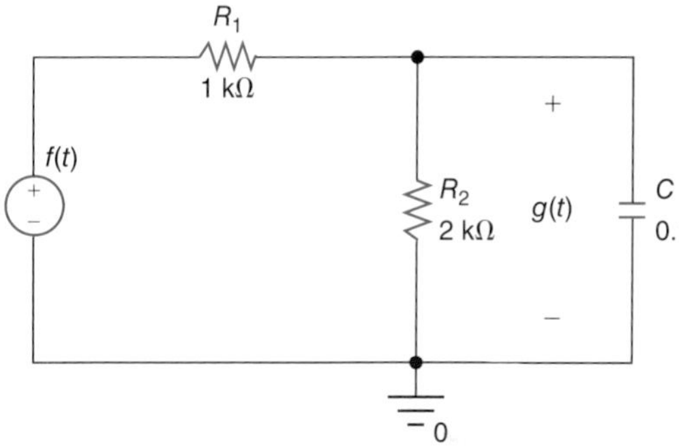

15.21 **Let $f(t) = e^{-20{,}000t}\ u(t)$ in the circuit shown in Figure P15.21. Find $g(t)$ for $t \geq 0$, assuming that initial voltages across the capacitors are zero.**

FIGURE P15.21

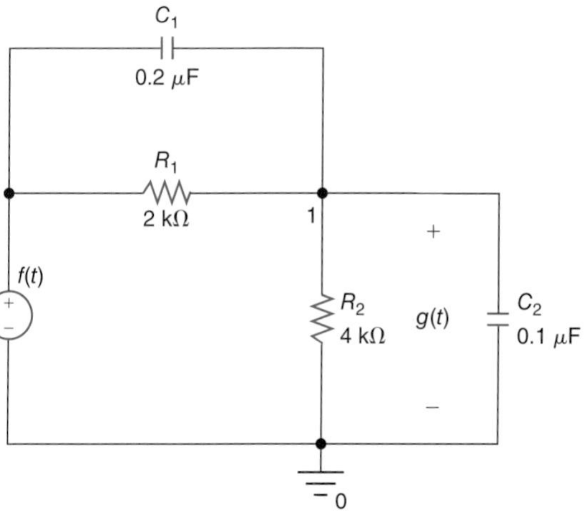

Thévenin Equivalent Circuit

15.22 Find the Thévenin equivalent voltage and the Thévenin equivalent impedance for the circuit shown in Figure P15.22. The initial voltage across the capacitors is $v(0^-) = 1$ V.

FIGURE P15.22

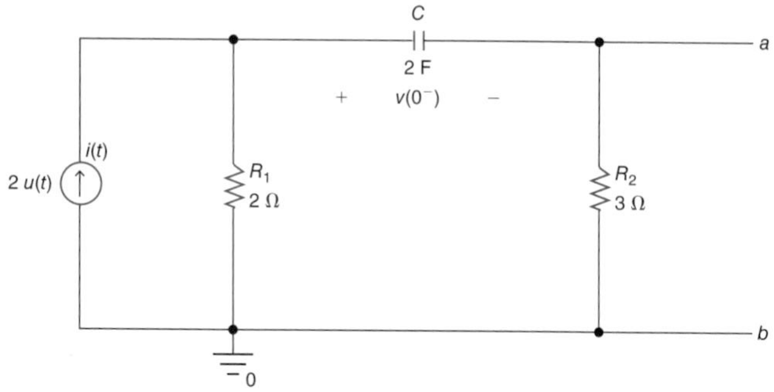

15.23 Find the Thévenin equivalent voltage and the Thévenin equivalent impedance for the circuit shown in Figure P15.23. The initial voltage across the capacitor is $v(0^-) = 2$ V, and the initial current through the inductor is $i(0^-) = 1$ A.

FIGURE P15.23

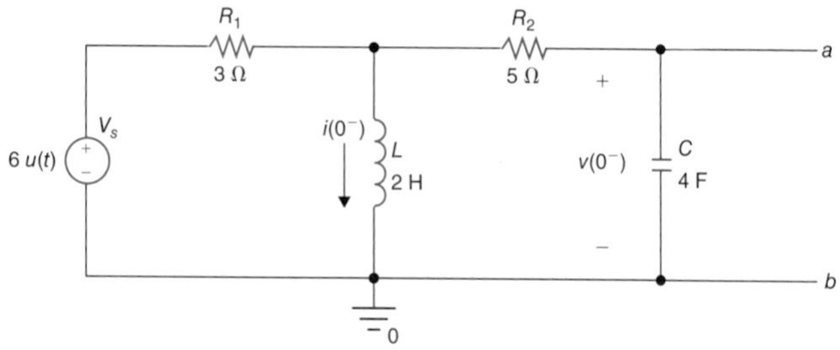

15.24 Find the Thévenin equivalent voltage and the Thévenin equivalent impedance for the circuit shown in Figure P15.24. The initial voltage across the capacitor is $v(0^-) = 1$ V, and the initial current through the inductor is $i(0^-) = 1$ A.

FIGURE P15.24

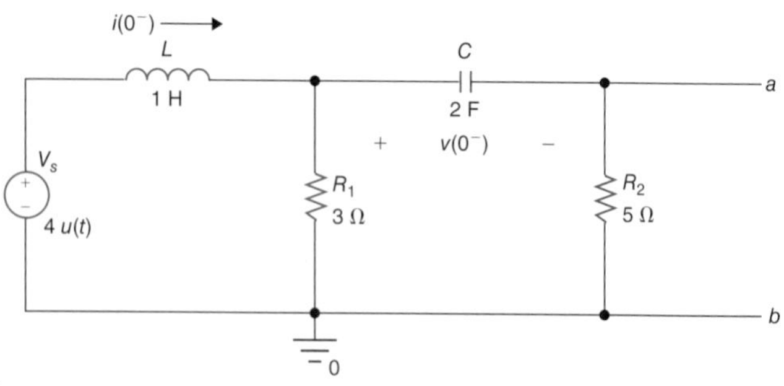

15.25 Find the Thévenin equivalent voltage and the Thévenin equivalent impedance for the circuit shown in Figure P15.25. The initial voltage across the capacitor is $v(0^-) = 2$ V.

FIGURE P15.25

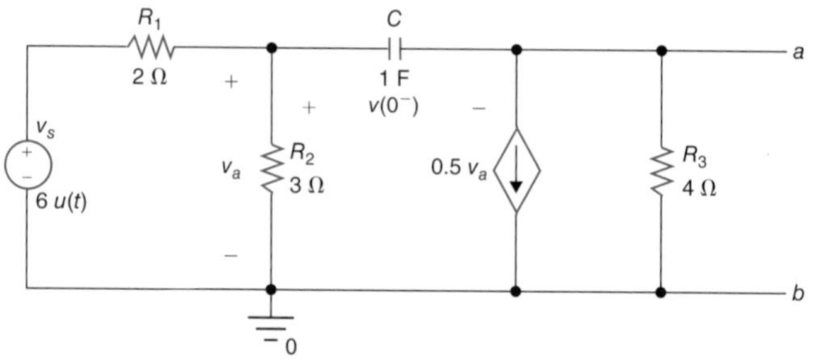

15.26 Find the Thévenin equivalent voltage and the Thévenin equivalent impedance for the circuit shown in Figure P15.26. The initial voltage across the capacitors are $v_1(0^-) = 2$ V and $v_2(0^-) = 1$ V.

FIGURE P15.26

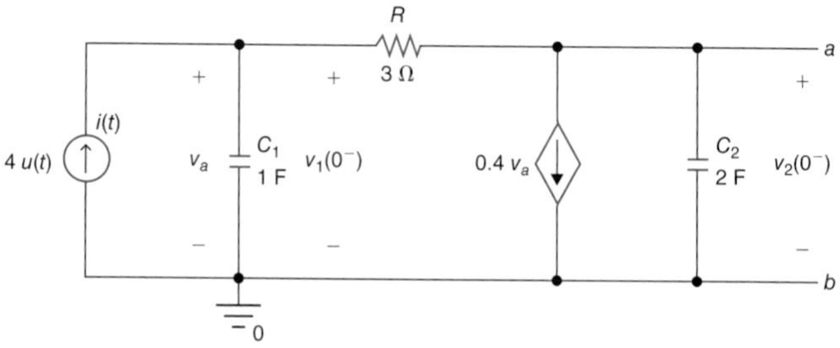

Norton Equivalent Circuit

15.27 Find the Norton equivalent current and the Norton equivalent impedance for the circuit shown in Figure P15.27. The initial voltage across the capacitors are $v_1(0^-) = 1$ V and $v_2(0^-) = 1$ V.

FIGURE P15.27

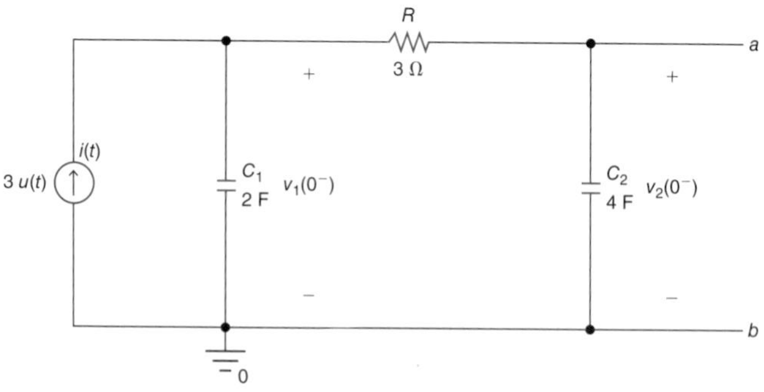

15.28 Find the Norton equivalent current and the Norton equivalent impedance for the circuit shown in Figure P15.28. The initial voltage across the capacitor is $v(0^-) = 2$ V.

FIGURE P15.28

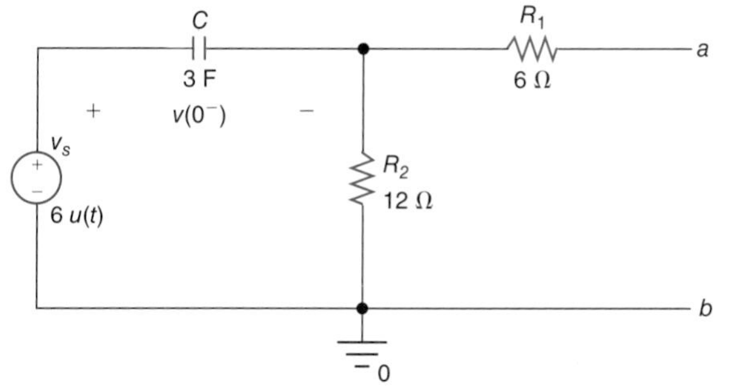

15.29 Find the Norton equivalent current and the Norton equivalent impedance for the circuit shown in Figure P15.29. The initial voltage across the capacitor is $v(0^-) = 1$ V, and the initial current through the inductor is $i(0^-) = 1$ A.

FIGURE P15.29

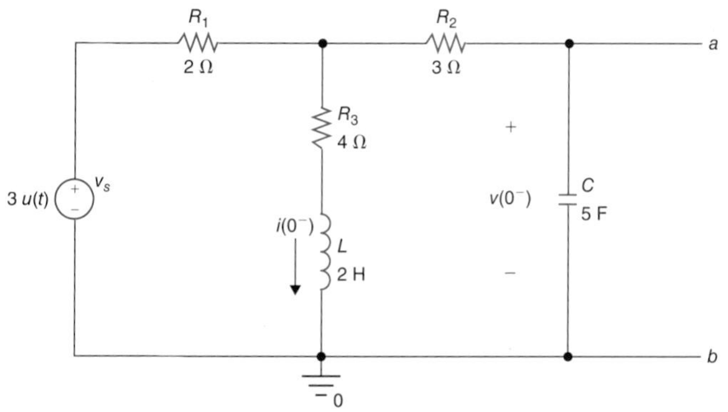

15.30 Find the Norton equivalent current and the Norton equivalent impedance for the circuit shown in Figure P15.30. The initial voltage across the capacitor is $v(0^-) = 1$ V, and the initial current through the inductor is $i(0^-) = 1$ A.

FIGURE P15.30

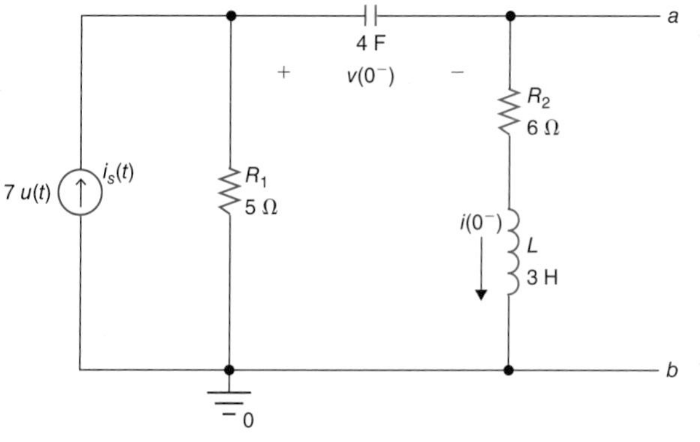

15.31 Find the Norton equivalent current and the Norton equivalent impedance for the circuit shown in Figure P15.31. The initial voltage across the capacitor is $v(0^-) = 1$ V.

FIGURE P15.31

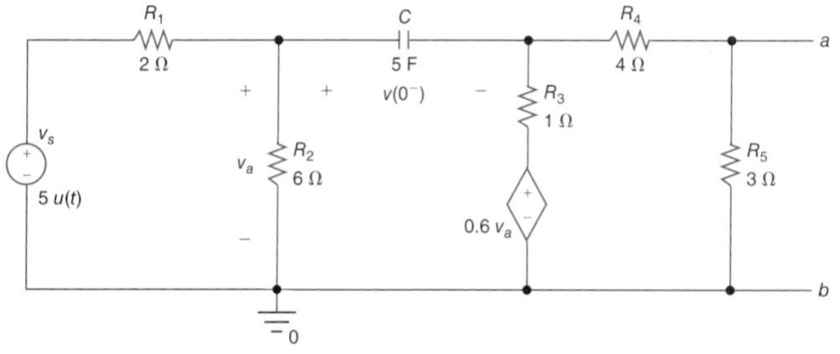

Transfer Function

15.32 In the circuit shown in Figure P15.32, let $R_1 = 2\ \Omega$, $R_2 = 1\ \Omega$, $C_1 = 1$ F, $C_2 = 1$ F, $L = 1$ H. Find the transfer function $H(s) = V_o(s)/V_{in}(s)$.

FIGURE P15.32

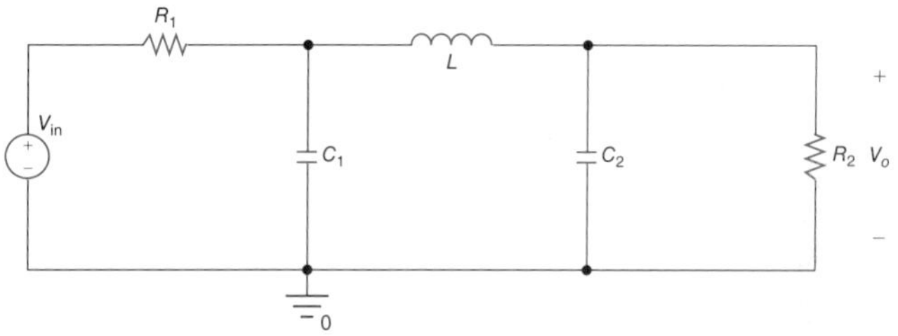

15.33 In the circuit shown in Figure P15.33, let $R_1 = 1\ \Omega$, $R_2 = 1\ \Omega$, $C = 1$ F, $L_1 = 1$ H, $L_2 = 2$ H. Find the transfer function $H(s) = V_o(s)/V_{in}(s)$.

FIGURE P15.33

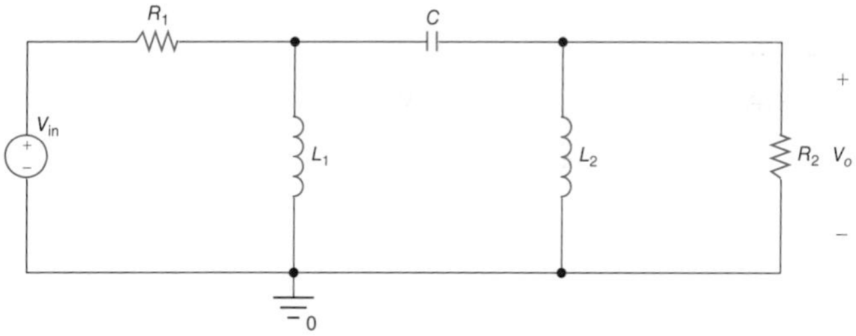

15.34 In the circuit shown in Figure P15.34, let $C_1 = 1$ F, $C_2 = 1$ F, $L = 1$ H, $R = 2\ \Omega$. Find the transfer function $H(s) = V_o(s)/V_{in}(s)$.

FIGURE P15.34

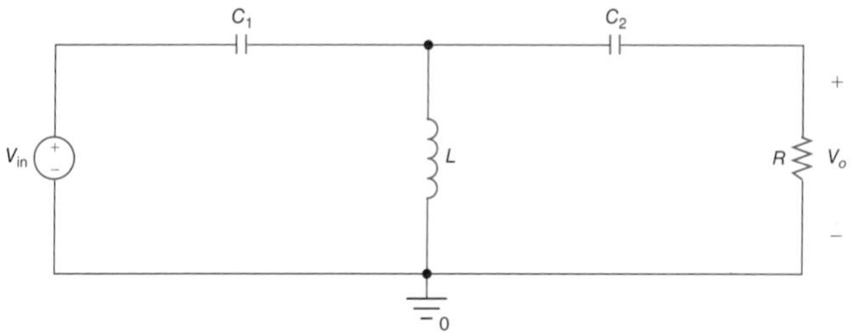

15.35 In the circuit shown in Figure P15.35, let $L_1 = 1$ H, $L_2 = 1$ H, $C = 1$ F, $R = 2\ \Omega$. Find the transfer function $H(s) = V_o(s)/V_{in}(s)$.

FIGURE P15.35

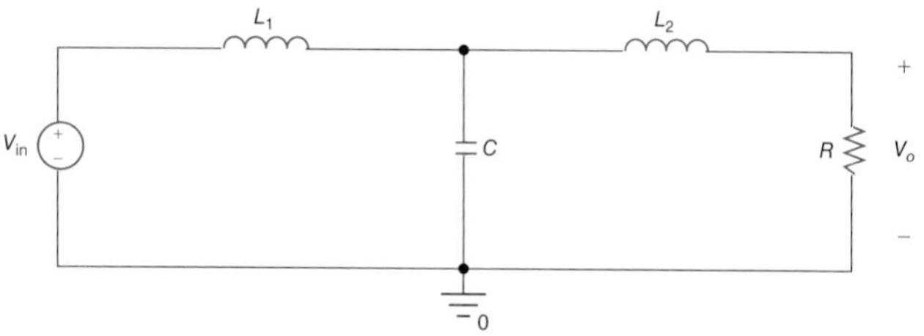

15.36 In the circuit shown in Figure P15.36, let $R_1 = 1\ \Omega$, $R_2 = 2\ \Omega$, $L = 2$ H, $C = 0.1$ F. Find the transfer function $H(s) = V_o(s)/V_{in}(s)$.

FIGURE P15.36

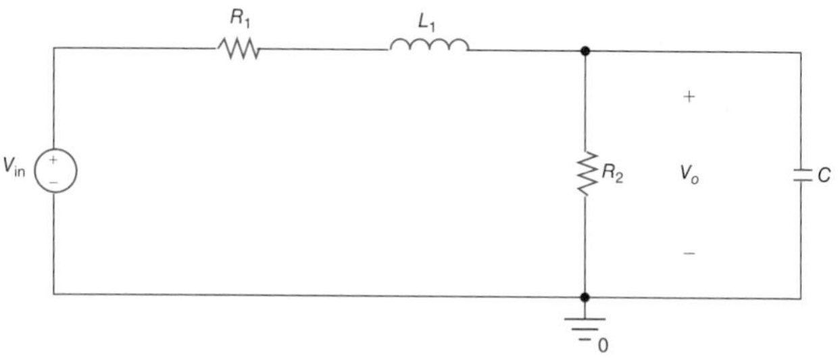

15.37 In the circuit shown in Figure P15.37, find the transfer function $H(s) = V_o(s)/V_{in}(s)$.

FIGURE P15.37

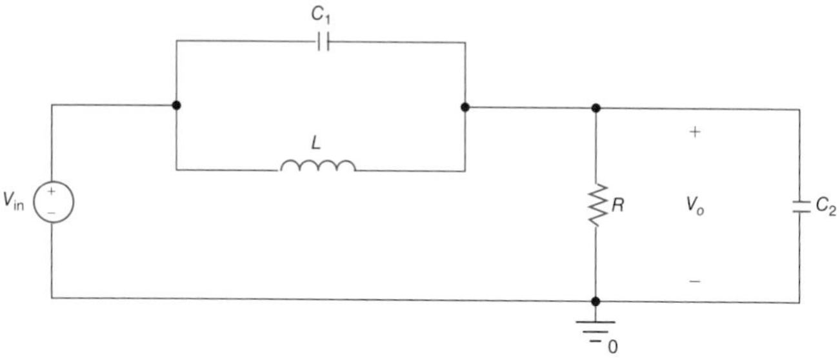

15.38 In the circuit shown in Figure P15.38, let $R_1 = R_2 = 1\ \Omega$, $C_1 = C_2 = 1$ F. Find the transfer function $H(s) = V_o(s)/V_{in}(s)$.

FIGURE P15.38

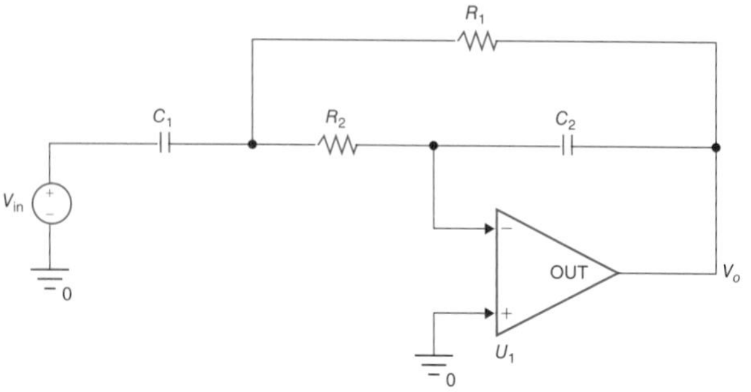

15.39 Find the transfer function $H(s) = V_o(s)/V_{in}(s)$ of the circuit shown in Figure P15.39.

FIGURE P15.39

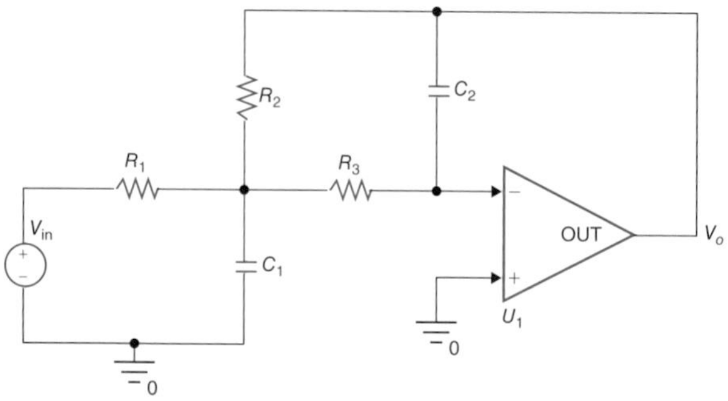

15.40 Find the transfer function $H(s) = V_o(s)/V_{in}(s)$ for the circuit shown in Figure P15.40.

FIGURE P15.40

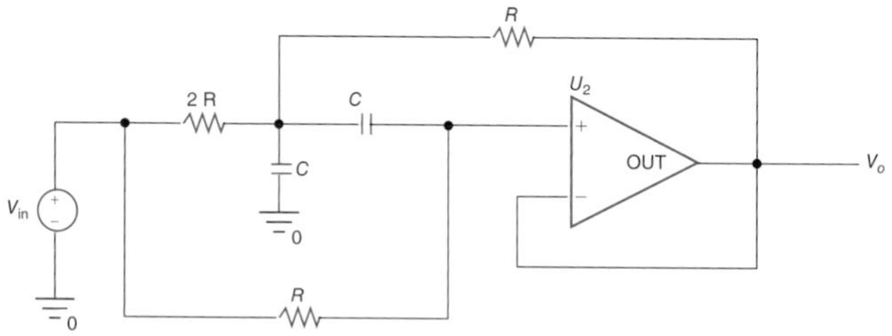

15.41 In the circuit shown in Figure P15.41, $C_1 = C_2 = C_3 = 1$ F. Find the transfer function $H(s) = V_o(s)/V_{in}(s)$ as a function of R_1, R_2.

FIGURE P15.41

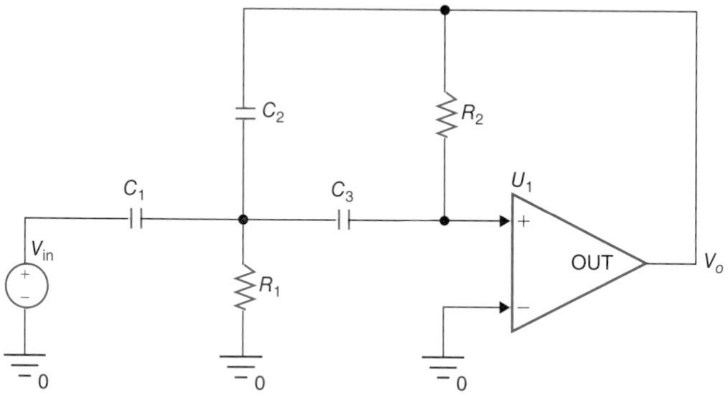

15.42 In the circuit shown in Figure P15.42, let $R_1 = 1\ \Omega$, $R_2 = 1\ \Omega$, $C_1 = 1$ F, $C_2 = 0.5$ F. Find the transfer function $H(s) = V_o(s)/V_{in}(s)$.

FIGURE P15.42

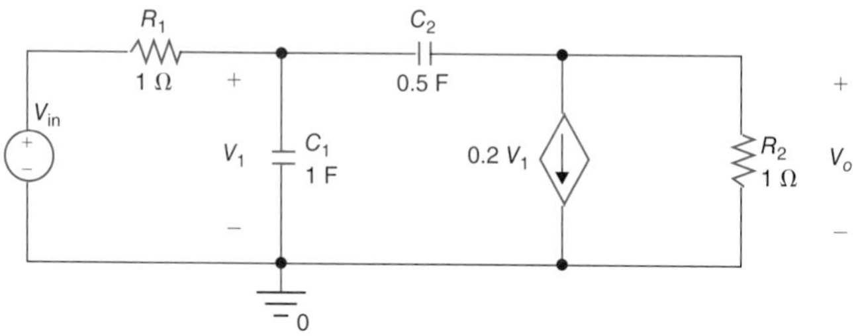

15.43 In the circuit shown in Figure P15.43, let $R_1 = 1\ \Omega$, $R_2 = 0.5\ \Omega$, $R_3 = 1\ \Omega$, $C_1 = 1$, $C_2 = 0.5$ F. Find the transfer function $H(s) = V_o(s)/V_{in}(s)$.

FIGURE P15.43

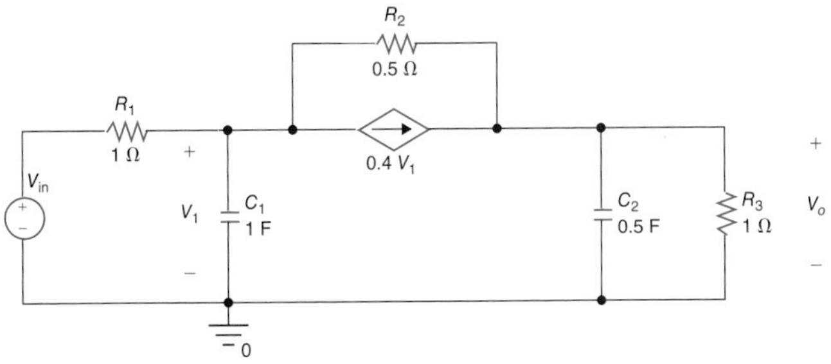

15.44 In the circuit shown in Figure P15.44, let $R_1 = 1\ \Omega$, $R_2 = 0.5\ \Omega$, $R_3 = 1\ \Omega$, $C_1 = 1$ F, $C_2 = 0.5$ F. Find the transfer function $H(s) = V_o(s)/V_{in}(s)$.

FIGURE P15.44

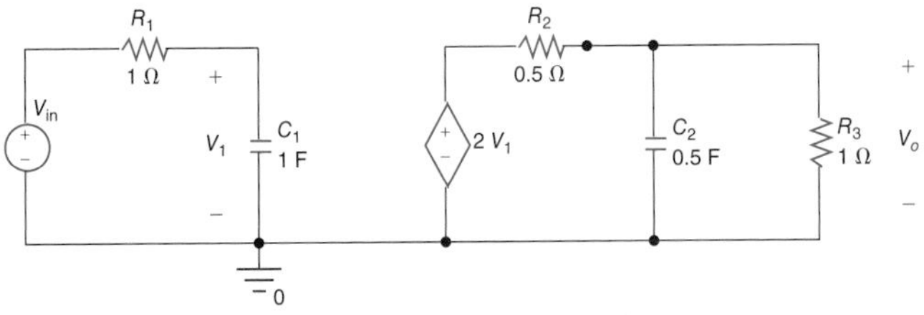

Convolution

15.45 Find the convolution $y(t)$ of $x(t)$ and $h(t)$; that is, find $y(t) = h(t)*x(t)$.

$$x(t) = \begin{cases} 2, & 0 \le t < 5 \\ 0, & \text{otherwise} \end{cases}, \quad h(t) = \begin{cases} 5, & 0 \le t < 3 \\ 0, & \text{otherwise} \end{cases}$$

15.46 Find the convolution $y(t)$ of $x(t)$ and $h(t)$; that is, find $y(t) = h(t)*x(t)$.

$$x(t) = \begin{cases} 2, & -5 \le t < 5 \\ 0, & \text{otherwise} \end{cases}, \quad h(t) = \begin{cases} 5, & -3 \le t < 3 \\ 0, & \text{otherwise} \end{cases}$$

15.47 Find the convolution $y(t)$ of $x(t)$ and $h(t)$; that is, find $y(t) = h(t)*x(t)$.

$$x(t) = \begin{cases} 2, & -2 \le t < 3 \\ 0, & \text{otherwise} \end{cases}, \quad h(t) = \begin{cases} 3, & 5 \le t < 9 \\ 0, & \text{otherwise} \end{cases}$$

15.48 Find the convolution $y(t)$ of $x(t)$ and $h(t)$; that is, find $y(t) = h(t)*x(t)$.

$$x(t) = \begin{cases} 2, & 0 \le t < 5 \\ 0, & \text{otherwise} \end{cases}, \quad h(t) = \begin{cases} 1, & 0 \le t \\ 0, & \text{otherwise} \end{cases}$$

15.49 Find the convolution $y(t)$ of $x(t)$ and $h(t)$; that is, find $y(t) = h(t)*x(t)$.

$$x(t) = \begin{cases} 2, & 3 \le t < 6 \\ 0, & \text{otherwise} \end{cases}, \quad h(t) = \begin{cases} 3, & 5 \le t \\ 0, & \text{otherwise} \end{cases}$$

15.50 Find the convolution $y(t)$ of $x(t)$ and $h(t)$; that is, find $y(t) = h(t)*x(t)$.

$$x(t) = \begin{cases} 2, & 0 \le t < 2 \\ 0, & \text{otherwise} \end{cases}, \quad h(t) = \begin{cases} 2t, & 0 \le t < 5 \\ 0, & \text{otherwise} \end{cases}$$

15.51 Find the convolution $y(t)$ of $x(t)$ and $h(t)$; that is, find $y(t) = h(t)*x(t)$.

$$x(t) = \begin{cases} 2, & 0 \le t < 2 \\ 0, & \text{otherwise} \end{cases},$$

$$h(t) = \begin{cases} -2t + 10, & 0 \le t < 5 \\ 0, & \text{otherwise} \end{cases}$$

15.52 Find the convolution $y(t)$ of $x(t)$ and $h(t)$; that is, find $y(t) = h(t)*x(t)$.

$$x(t) = \begin{cases} 1, & 0 \le t \\ 0, & \text{otherwise} \end{cases}, \quad h(t) = \begin{cases} 2t, & 0 \le t < 5 \\ 0, & \text{otherwise} \end{cases}$$

15.53 Find the convolution $y(t)$ of $x(t)$ and $h(t)$; that is, find $y(t) = h(t)*x(t)$.

$$x(t) = u(t), h(t) = e^{-5t}u(t)$$

15.54 Find the convolution $y(t)$ of $x(t)$ and $h(t)$; that is, find $y(t) = h(t)*x(t)$.

$$x(t) = \begin{cases} 2, & 0 \le t < 3 \\ 0, & \textit{otherwise} \end{cases}, \quad h(t) = e^{-5t}u(t)$$

15.55 Find the convolution $y(t)$ of $x(t)$ and $h(t)$; that is, find $y(t) = h(t)*x(t)$.

$$x(t) = u(t), \quad h(t) = \cos(5t)\, u(t)$$

15.56 Find the convolution $y(t)$ of $x(t)$ and $h(t)$; that is, find $y(t) = h(t)*x(t)$.

$$x(t) = \begin{cases} 2, & 0 \le t < 3 \\ 0, & \text{otherwise} \end{cases}, \quad h(t) = \cos(5t)u(t)$$

15.57 Find the convolution $y(t)$ of $x(t)$ and $h(t)$; that is, find $y(t) = h(t)*x(t)$ for

$$x(t) = u(t), \quad h(t) = e^{-5t}\cos(12t)\, u(t)$$

15.58 **Find the convolution $y(t)$ of $x(t)$ and $h(t)$.**

$$x(t) = \begin{cases} t, & 0 \le t < 6 \\ 0, & \text{otherwise} \end{cases}, \quad h(t) = \begin{cases} t, & 0 \le t < 6 \\ 0, & \text{otherwise} \end{cases}$$

15.59 **Find the convolution $y(t)$ of $x(t)$ and $h(t)$.**

$$x(t) = \begin{cases} -t+1, & 0 \le t < 1 \\ 0, & \text{otherwise} \end{cases}, \quad h(t) = \begin{cases} 1, & 0 \le t < 1 \\ 0, & \text{otherwise} \end{cases}$$

15.60 **Find the convolution $y(t)$ of $x(t)$ and $h(t)$.**

$$x(t) = \begin{cases} 2t, & 0 \le t < 1 \\ -t+3, & 1 \le t < 3, \\ 0, & \text{otherwise} \end{cases} \quad h(t) = u(t)$$

15.61 **Find the convolution $y(t)$ of $x(t)$ and $h(t)$.**

$$x(t) = \begin{cases} t, & 0 \le t < 2 \\ 0, & \text{otherwise} \end{cases}, \quad h(t) = \begin{cases} 1, & 0 \le t < 1 \\ 0, & \text{otherwise} \end{cases}$$

15.62 **Find the convolution $y(t)$ of $h(t)$ and $x(t)$.**

$$h(t) = \begin{cases} 1, & 0 \le t < 3 \\ 0, & \text{otherwise} \end{cases},$$

$$x(t) = \begin{cases} -t+3, & 0 \le t < 3 \\ 0, & \textit{otherwise} \end{cases}$$

15.63 **Find the convolution $y(t)$ of $h(t)$ and $x(t)$.**

$$x(t) = e^{-2t}u(t), \quad h(t) = u(t) - u(t-5)$$

15.64 **Find the convolution $y(t)$ of $x(t)$ and $h(t)$; that is, find $y(t) = h(t)*x(t)$.**

$$x(t) = tu(t), \quad h(t) = (t-2)u(t-2)$$

15.65 **Find the convolution $y(t)$ of $x(t)$ and $h(t)$; that is, find $y(t) = h(t)*x(t)$.**

$$x(t) = tu(t), \quad h(t) = 2\sin(6t)\,u(t)$$

15.66 **Find the convolution $y(t)$ of $h(t)$ and $x(t)$.**

$$x(t) = u(t), \quad h(t) = t^2u(t)$$

15.67 **Use convolution to find $y(t)$ for the circuit shown in Figure P15.67 for the following inputs, assuming that the initial voltage across the capacitor is zero.**

a. $x(t) = \delta(t)$
b. $x(t) = u(t)$
c. $x(t) = e^{-20{,}000t}\,u(t)$

FIGURE P15.67

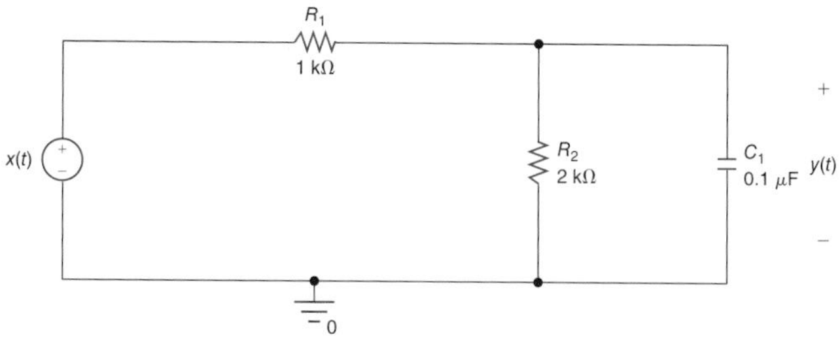

15.68 **Use convolution to find $y(t)$ for the circuit shown in Figure P15.68 for the following inputs, assuming that the initial current across the inductor is zero.**

FIGURE P15.68

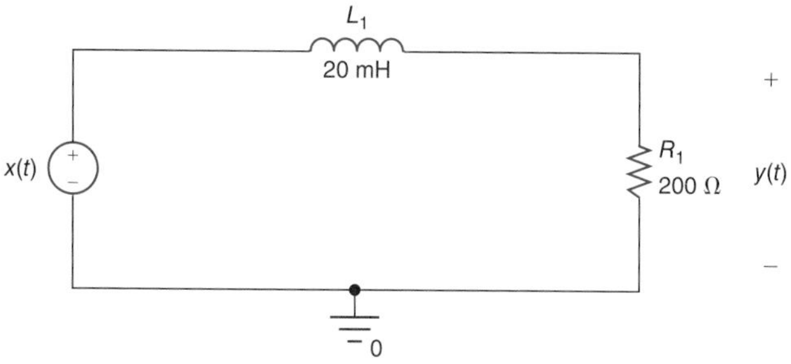

a. $x(t) = \delta(t)$
b. $x(t) = u(t)$
c. $x(t) = e^{-20{,}000t}u(t)$

Bode Diagram

15.69 **Plot the magnitude and phase Bode diagrams of the transfer function given by**

$$H(s) = \frac{10^4}{s + 10^4}$$

15.70 **Plot the magnitude and phase Bode diagrams of the transfer function given by**

$$H(s) = \frac{s}{s + 10^4}$$

15.71 **Plot the magnitude and phase Bode diagrams of the transfer function given by**

$$H(s) = \frac{10^7}{(s + 10^3)(s + 10^4)}$$

15.72 **Plot the magnitude and phase Bode diagrams of the transfer function given by**

$$H(s) = \frac{10^4 s}{(s + 10^3)(s + 10^4)}$$

15.73 Plot the magnitude and phase Bode diagrams of the transfer function given by

$$H(s) = \frac{10s^2}{(s + 10^3)(s + 10^4)}$$

15.74 Plot the magnitude and phase Bode diagrams of the transfer function given by

$$H(s) = \frac{10^8}{(s + 10^4)^2}$$

15.75 Plot the magnitude and phase Bode diagrams of the transfer function given by

$$H(s) = \frac{s^2}{(s + 10^4)^2}$$

15.76 Plot the magnitude and phase Bode diagrams of the transfer function given by

$$H(s) = \frac{10^{11}}{(s + 10^4)^2(s + 10^3)}$$

15.77 Plot the magnitude and phase Bode diagrams of the transfer function given by

$$H(s) = \frac{10^4 s^2}{(s + 10^4)^2(s + 10^3)}$$

15.78 Plot the magnitude and phase Bode diagrams of the transfer function given by

$$H(s) = \frac{s^3}{(s + 10^4)^2(s + 10^3)}$$

15.79 Plot the magnitude and phase Bode diagrams of the transfer function given by

$$H(s) = \frac{10^5(s + 1000)}{(s + 100)^2(s + 10{,}000)}$$

15.80 Plot the magnitude and phase Bode diagrams of the transfer function given by

$$H(s) = \frac{10^4(s + 1000)(s + 10^7)}{s^2(s + 10^4)(s + 10^5)}$$

15.81 Plot the magnitude and phase Bode diagrams of the transfer function given by

$$H(s) = \frac{10^7(s + 1000)^2}{s^2(s + 10^4)(s + 10^5)}$$

15.82 Plot the magnitude and phase Bode diagrams of the transfer function given by

$$H(s) = \frac{10^9(s + 10^4)(s + 10^5)}{(s + 10^3)^2(s + 10^6)^2}$$

15.83 Plot the magnitude and phase Bode diagrams of the transfer function given by

$$H(s) = \frac{10^8 s^2}{(s + 10^3)^2(s + 10^4)^2}$$

15.84 The magnitude Bode diagram of a transfer function $H(s)$ is shown in Figure P15.84. From this, find $H(s)$.

FIGURE P15.84

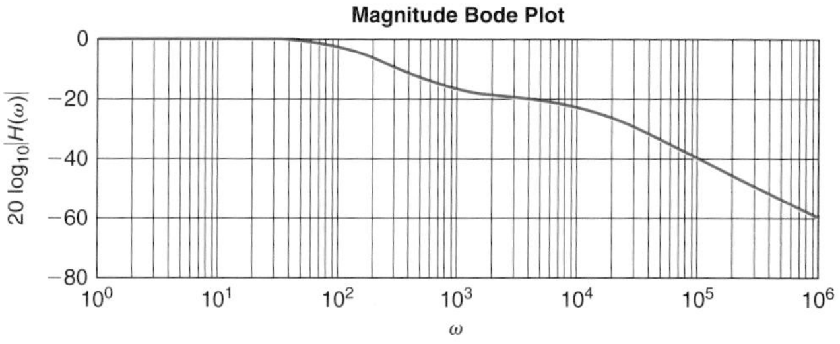

15.85 The magnitude Bode diagram of a transfer function $H(s)$ is shown in Figure P15.85. From this, find $H(s)$.

FIGURE P15.85

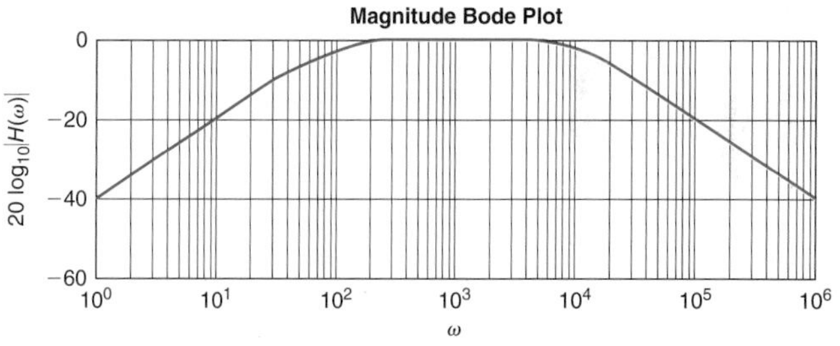

15.86 The transfer function of a linear, time-invariant system is given by

$$H(s) = \frac{10^8}{s^2 + 4000s + 10^8}$$

a. Find the corner frequency ω_0.
b. Find the damping coefficient ζ.
c. Find the magnitude of $H(\omega)$when $\omega = \omega_0$.
d. Find the peak frequency ω_{peak}.
e. Find the magnitude of $H(\omega)$ when $\omega = \omega_{\text{peak}}$.
f. Find the frequency ω_{zc} where the magnitude of $H(\omega)$ is 0 dB.
g. Plot the Bode diagram.

Chapter 16

First- and Second-Order Analog Filters

16.1 Introduction

A *filter* is a circuit that keeps the desired signals and removes unwanted signals. If the desired signals are low-frequency signals, the filter will pass the signals with frequencies below a certain frequency (called the *cutoff frequency*) and block the signals with frequencies above the cutoff frequency. The cutoff frequency is the boundary between the passband and the stopband. This filter is called the *low-pass filter (LPF)*. If the desired signals are high-frequency signals, the filter will pass the signals with frequencies above a certain cutoff frequency and block the signals with frequencies below the cutoff frequency. This filter is called the *high-pass filter (HPF)*. If the desired signals are mid-frequency signals, the filter will pass the signals with frequencies between the low-cutoff frequency and high-cutoff frequency and block the signals with frequencies below the low-cutoff frequency and above the high-cutoff frequency. This filter is called the *bandpass filter (BPF)*. If the unwanted signals are mid-frequency signals, the filter will block the signals with frequencies between the low-cutoff frequency and the high-cutoff frequency and pass the signals with frequencies below the low-cutoff frequency and above the high-cutoff frequency. This filter is called the *bandstop filter (BSF)*.

In this chapter, the analysis and design of the first-order LPF, first-order HPF, second-order LPF, second-order HPF, second-order bandpass filter, and second-order bandstop filter are presented. The first-order and the second-order filters are building blocks for the higher-order filters. For example, a fifth-order LPF can be designed by cascading one first-order LPF and two second-order LPFs. Table 16.1 at the end of the Chapter provides a list of the first-order and the second-order passive filters that use passive elements of R, L, and C. Table 16.2 at the end of the Chapter provides a list of the first-order and the second-order active filters that use op amp, R, and C.

If the cutoff frequency or the corner frequency is one rad/s, the filter is called a normalized filter. For the given transfer function, the component values for the normalized filter are found. Then, the magnitude scaling and the frequency scaling are applied to the normalized component values to obtain the component values for the actual filter we are designing. After the component values are selected, the PSpice or Multisim simulation can be used to verify whether the filter matches the transfer function given.

16.2 Magnitude Scaling and Frequency Scaling

16.2.1 MAGNITUDE SCALING

When the magnitude of the impedance Z is multiplied by a positive constant k_m, the impedance becomes $k_m Z$. This process is called *magnitude scaling*. If $k_m > 1$, the magnitude is scaled up. On the other hand, if $k_m < 1$, the magnitude is scaled down. When the impedances of resistor, inductor, and capacitor are magnitude scaled, the impedances become, respectively,

$$k_m Z_R = k_m R$$

$$k_m Z_L = k_m j\omega L = j\omega(k_m L)$$

$$k_m Z_C = k_m \frac{1}{j\omega C} = \frac{1}{j\omega\left(\frac{C}{k_m}\right)}$$

Thus, to magnitude scale the impedances by k_m, multiply the resistance value by k_m, multiply the inductance value by k_m, and divide the capacitance value by k_m:

$$R_{new} = k_m R_{old} \quad (16.1)$$

$$L_{new} = k_m L_{old} \quad (16.2)$$

$$C_{new} = \frac{C_{old}}{k_m} \quad (16.3)$$

After the magnitude scaling, the ratio of the impedances remains the same. Thus, if the voltage divider rule or the current divider rule can be applied to obtain the transfer function, the transfer function remains the same after the magnitude scaling. Notice that the new values of RC, LC, R/L, L/R are the same as the old values; that is,

$$R_{new}C_{new} = k_m R_{old}\frac{C_{old}}{k_m} = R_{old}C_{old}$$

$$L_{new}C_{new} = k_m L_{old}\frac{C_{old}}{k_m} = L_{old}C_{old}$$

$$\frac{R_{new}}{L_{new}} = \frac{k_m R_{old}}{k_m L_{old}} = \frac{R_{old}}{L_{old}}$$

We can design a circuit using small resistance values and large capacitance values. Then, we can apply magnitude scaling to achieve the component values that are available at lower cost (resistors in the $k\Omega$, capacitors in the $\mu F, nF, pF$).

EXAMPLE 16.1

From the voltage divider rule, the transfer function of an *RC* circuit shown in Figure 16.1 is given by

continued

Example 16.1 continued

FIGURE 16.1

A circuit with $R = 1\ \Omega$ and $C = 1$ F.

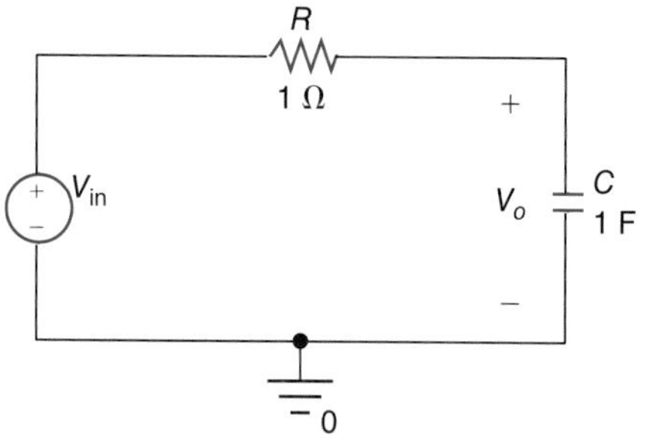

$$H(s) = \frac{V_o(s)}{V_{in}(s)} = \frac{\frac{1}{sC}}{R + \frac{1}{sC}} = \frac{\frac{1}{RC}}{s + \frac{1}{RC}} = \frac{1}{s+1}$$

Find the new values for R and C when the magnitude scale factor is $k_m = 1000$.

The new values of R and C are given by

$$R_{new} = k_m R_{old} = 1000 \times 1\ \Omega = 1\ k\Omega$$

$$C_{new} = \frac{C_{old}}{k_m} = \frac{1\ \text{F}}{1000} = 1\ \text{mF}$$

Notice that $R_{new}C_{new} = 1$ rad/s. Thus, after the magnitude scaling, the transfer function remains the same.

Exercise 16.1

For the circuit shown in Figure 16.1, find the new values for R and C when the magnitude scale factor is $k_m = 10^5$.

Answer:
$R = 100\ k\Omega, \quad C = 10\ \mu F.$

16.2.2 FREQUENCY SCALING

If the frequency is multiplied by k_f without changing the magnitude of the impedance, it is called frequency scaling. For an inductor, to keep the magnitude of the impedance $|j\omega L| = \omega L$ constant after the frequency is scaled by $k_f(\omega \rightarrow k_f\omega)$, the inductance value must be divided by $k_f(L \rightarrow L/k_f)$. Thus, we have

$$Z_L = j\omega L = j(k_f\omega)\left(\frac{L}{k_f}\right)$$

The new value of the inductance is the old value divided by k_f; that is,

$$L_{new} = \frac{L_{old}}{k_f} \tag{16.4}$$

For a capacitor, to keep the magnitude of the impedance $1/|j\omega C| = 1/(\omega C)$ constant after the frequency is scaled by $k_f(\omega \rightarrow k_f\omega)$, the capacitance value must be divided by $k_f(C \rightarrow C/k_f)$. Thus, we have

$$Z_C = \frac{1}{j\omega C} = \frac{1}{j(k_f\omega)\left(\frac{C}{k_f}\right)}$$

The new value of the capacitance is the old value divided by k_f; that is,

$$C_{new} = \frac{C_{old}}{k_f} \tag{16.5}$$

Since the impedance of a resistor is not a function of frequency ω, the frequency scaling has no effect on the resistance value.

$$R_{new} = R_{old} \tag{16.6}$$

The effects of frequency scaling on $1/(RC)$, $1/(LC)$, R/L, and $1/\sqrt{LC}$ are given by

$$\frac{1}{R_{new}C_{new}} = \frac{1}{R_{old}\dfrac{C_{old}}{k_f}} = \frac{k_f}{R_{old}C_{old}}$$

$$\frac{1}{L_{new}C_{new}} = \frac{1}{\dfrac{L_{old}}{k_f}\dfrac{C_{old}}{k_f}} = \frac{k_f^2}{L_{old}C_{old}}$$

$$\frac{R_{new}}{L_{new}} = \frac{R_{old}}{\dfrac{L_{old}}{k_f}} = k_f\frac{R_{old}}{L_{old}}$$

$$\frac{1}{\sqrt{L_{new}C_{new}}} = \frac{1}{\sqrt{\dfrac{L_{old}}{k_f}\dfrac{C_{old}}{k_f}}} = \frac{k_f}{\sqrt{L_{old}C_{old}}}$$

EXAMPLE 16.2

For the circuit shown in Figure 16.1, apply frequency scaling with $k_f = 10{,}000$.

The frequency scaled values of R and C are

$$R_{new} = R_{old} = 1\ \Omega$$

$$C_{new} = \frac{C_{old}}{k_f} = \frac{1\text{ F}}{10{,}000} = 10^{-4}\text{ F} = 100\ \mu F$$

Notice that $R_{new}C_{new} = 10^{-4}$. Thus, the transfer function using the frequency scaled R and C is given by

$$H(s) = \frac{10{,}000}{s + 10{,}000}$$

Exercise 16.2

For the circuit shown in Figure 16.1, apply frequency scaling with $k_f = 40{,}000$.

Answer:
$R_{new} = 1\ \Omega,\quad C_{new} = 25\ \mu F.$

16.2.3 MAGNITUDE AND FREQUENCY SCALING

When both the magnitude scaling by k_m and the frequency scaling by k_f are applied, we can find the new component values by combining Equations (16.1) through (16.3) and Equations (16.4) through (16.6):

$$R_{new} = k_m R_{old} \tag{16.7}$$

$$L_{new} = \frac{k_m}{k_f} L_{old} \tag{16.8}$$

$$C_{new} = \frac{C_{old}}{k_f k_m} \tag{16.9}$$

The effects of magnitude and frequency scaling on $1/(RC), 1/(LC), R/L, 1/\sqrt{LC}$ are given by

$$\frac{1}{R_{new}C_{new}} = \frac{1}{k_m R_{old} \dfrac{1}{k_f k_m} C_{old}} = \frac{k_f}{R_{old}C_{old}}$$

$$\frac{1}{L_{new}C_{new}} = \frac{1}{\dfrac{k_m}{k_f} L_{old} \dfrac{C_{old}}{k_f k_m}} = \frac{k_f^2}{L_{old}C_{old}}$$

$$\frac{R_{new}}{L_{new}} = \frac{k_m R_{old}}{\dfrac{k_m}{k_f} L_{old}} = k_f \frac{R_{old}}{L_{old}}$$

$$\frac{1}{\sqrt{L_{new}C_{new}}} = \frac{1}{\sqrt{\dfrac{k_m}{k_f} L_{old} \dfrac{C_{old}}{k_f k_m}}} = \frac{k_f}{\sqrt{L_{old}C_{old}}}$$

EXAMPLE 16.3

Apply magnitude scaling and frequency scaling with $k_m = 1000$ and $k_f = 10{,}000$ for the circuit shown in Figure 16.1.

The magnitude and frequency scaled values of R and C are

$$R_{new} = k_m R_{old} = 1000 \times 1\ \Omega = 1\ k\Omega$$

$$C_{new} = \frac{C_{old}}{k_f k_m} = \frac{1\text{ F}}{10^4 \times 10^3} = 10^{-7}\text{ F} = 0.1\ \mu F$$

Notice that $R_{new}C_{new} = 10^{-4}$. Thus, the transfer function using the magnitude and frequency scaled R and C is given by

$$H(s) = \frac{10{,}000}{s + 10{,}000}$$

Notice that only the frequency scaling affects the transfer function.

Exercise 16.3

Apply magnitude scaling and frequency scaling with $k_m = 10{,}000$ and $k_f = 20{,}000$ for the circuit shown in Figure 16.1.

Answer:

$R_{new} = k_m R_{old} = 10{,}000 \times 1\ \Omega = 10\ k\Omega.$

$$C_{new} = \frac{C_{old}}{k_f k_m} = \frac{1\text{ F}}{2 \times 10^4 \times 10^4} = 5 \times 10^{-9}\text{ F} = 5\ nF$$

16.3 First-Order LPF

FIGURE 16.2

The pole-zero diagram of the first-order LPF.

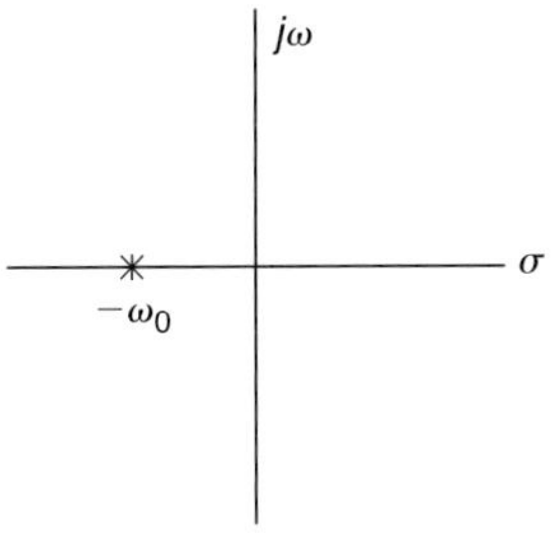

The transfer function of the first-order LPF is given by

$$H(s) = \frac{\omega_0}{s + \omega_0} \tag{16.10}$$

This transfer function has one pole at $s = -\omega_0$ and one zero at $s = \infty$. The pole-zero diagram of this transfer function is shown in Figure 16.2.

The frequency response of this transfer function is given by

$$H(\omega) = H(s)|_{s=j\omega} = \frac{\omega_0}{j\omega + \omega_0} \tag{16.11}$$

The magnitude response is

$$|H(\omega)| = \frac{\omega_0}{\sqrt{\omega^2 + \omega_0^2}} \tag{16.12}$$

and the phase response is

$$\angle H(\omega) = -\tan^{-1}\left(\frac{\omega}{\omega_0}\right) \tag{16.13}$$

Figure 16.3 shows the magnitude response and the phase response. The magnitude response in decibels is given by

$$M(\omega) = 20 \log_{10}(|H(\omega)|) = 20 \log_{10}\frac{\omega_0}{\sqrt{\omega^2 + \omega_0^2}}$$

When $\omega = \omega_0$, the magnitude is

$$|H(\omega_0)| = \frac{\omega_0}{\sqrt{\omega_0^2 + \omega_0^2}} = \frac{1}{\sqrt{2}} = -3.01\text{ dB}$$

The frequency ω_0 is called the 3-dB *cutoff frequency,* or *half-power frequency.*

The first-order LPF can be synthesized using an *RC* circuit, as shown in Figure 16.4. The transfer function of this circuit is given by

$$H(s) = \frac{V_o(s)}{V_{in}(s)} = \frac{\dfrac{1}{sC}}{R + \dfrac{1}{sC}} = \frac{1}{RCs + 1} = \frac{\dfrac{1}{RC}}{s + \dfrac{1}{RC}}$$

FIGURE 16.3

(a) The magnitude response in linear scale, (b) the magnitude response in dB scale, (c) the phase response of the first-order LPF (ω_0 = 1000 rad/s).

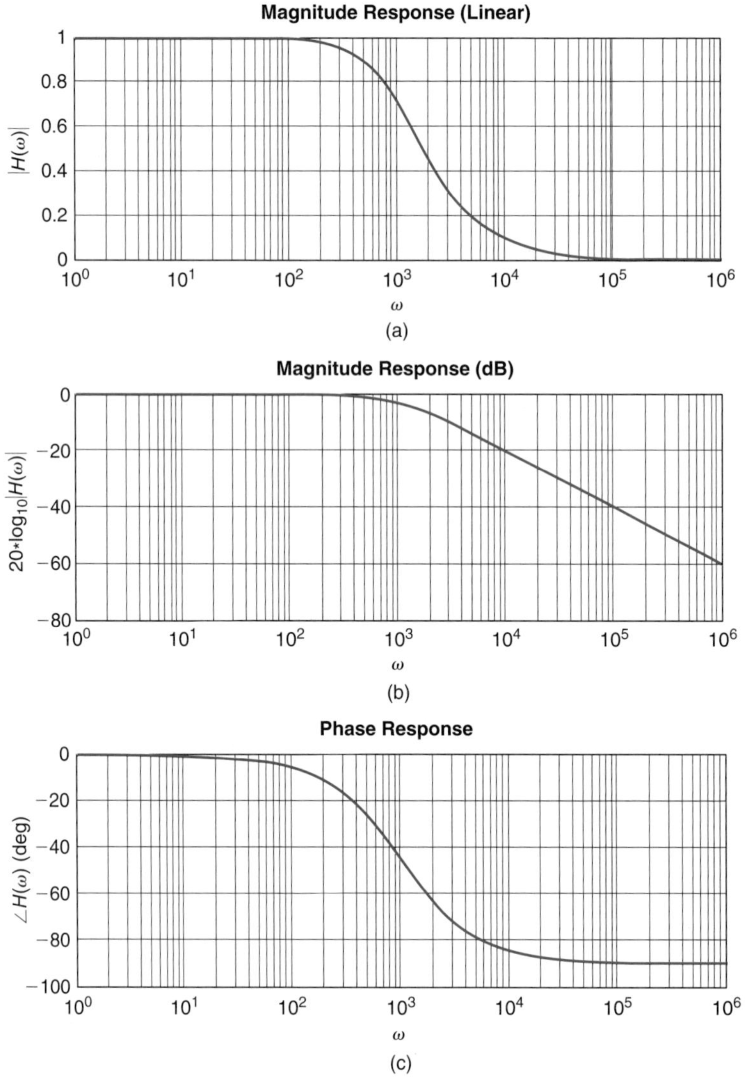

FIGURE 16.4

The first-order LPF using the RC circuit.

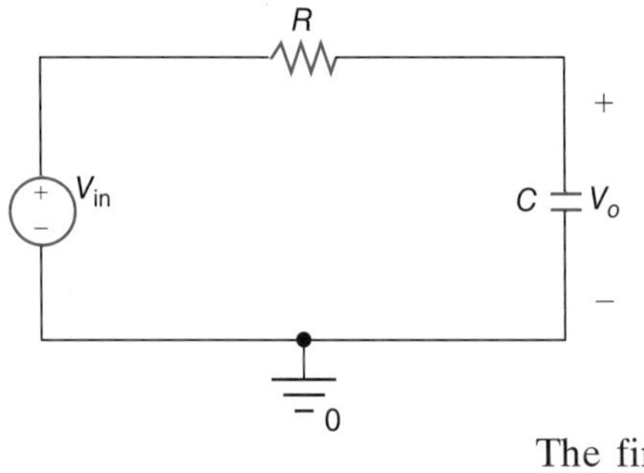

For the normalized LPF, $1/(RC) = 1$ rad/s. Select $R = 1\ \Omega$, $C = 1$ F for the normalized LPF. For the LPF with cutoff frequency ω_0, use frequency scaling $k_f = \omega_0$. The magnitude scaling factor k_m can be chosen for 1000 or 10,000 to get $R = 1\ k\Omega$ or $R = 10\ k\Omega$.

The LR circuit shown in Figure 16.5 can also be used to implement the second-order LPF.

$$H(s) = \frac{R}{sL + R} = \frac{\frac{R}{L}}{s + \frac{R}{L}}$$

The first-order LPF can also be designed using an operational amp, as shown in Figure 16.6.

Summing the currents leaving node 1, we obtain

$$\frac{0 - V_{in}}{R_1} + \frac{0 - V_o}{R_2} + \frac{0 - V_o}{\frac{1}{sC}} = 0$$

FIGURE 16.5

The first-order LPF using an *LR* circuit.

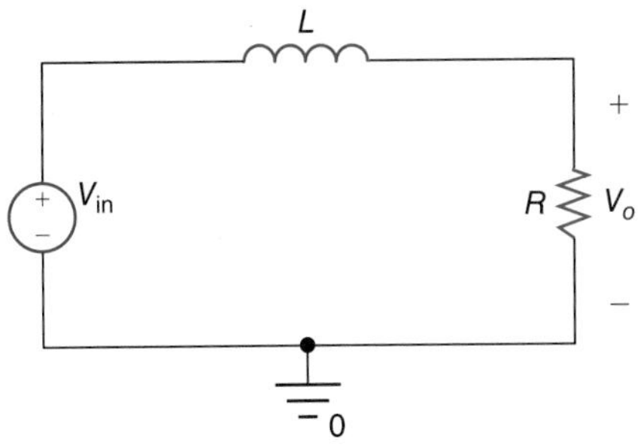

which can be rearranged as

$$\left(sC + \frac{1}{R_2}\right)V_o = \frac{-V_{in}}{R_1}$$

The transfer function of the circuit shown in Figure 16.6 is given by

$$H(s) = \frac{V_o}{V_{in}} = \frac{-\dfrac{1}{R_1}}{sC + \dfrac{1}{R_2}} = -\frac{\dfrac{1}{R_1 C}}{s + \dfrac{1}{R_2 C}} = -\frac{R_2}{R_1}\frac{\dfrac{1}{R_2 C_1}}{s + \dfrac{1}{R_2 C_1}}$$

If $R_2 = R_1$, $H(s)$ becomes

$$H(s) = \frac{V_o(s)}{V_{in}(s)} = -\frac{\dfrac{1}{R_2 C_1}}{s + \dfrac{1}{R_2 C_1}}$$

FIGURE 16.6

The first-order LPF using an operational amp.

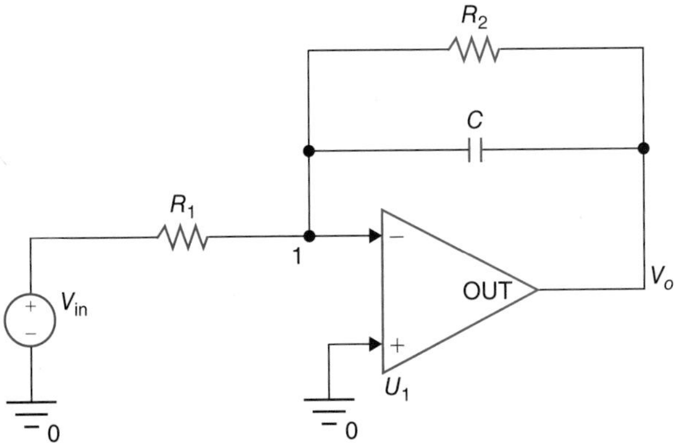

Initially, we can choose $C_1 = 1$ F, $R_2 = 1\ \Omega$. Then frequency scale by $k_f = \omega_0$. The magnitude scaling can be employed to change the component values to more practical values.

Figure 16.7 shows an implementation of the first-order *RC* LPF using an operational amp voltage follower.

Due to a virtual short, the voltage across the capacitor is V_o. Application of the voltage divider rule yields

$$H(s) = \frac{V_o(s)}{V_{in}(s)} = \frac{\dfrac{1}{sC}}{R + \dfrac{1}{sC}} = \frac{1}{RCs + 1} = \frac{\dfrac{1}{RC}}{s + \dfrac{1}{RC}}$$

FIGURE 16.7

The first-order LPF.

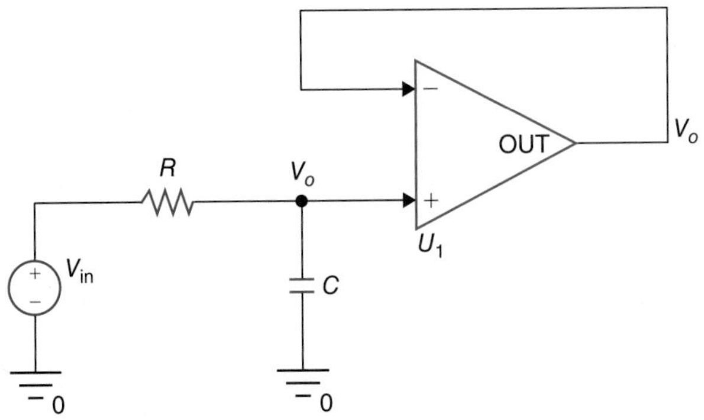

The operational amp provides large input impedance and small output impedance. The first-order section is isolated from other sections in the cascade implementation.

16.4 First-Order HPF

The transfer function of the first-order HPF is given by

$$H(s) = \frac{s}{s + \omega_0} \tag{16.14}$$

This transfer function has one pole at $s = -\omega_0$ and one zero at $s = 0$. The pole-zero diagram of this transfer function is shown in Figure 16.8.

The frequency response of this transfer function is given by

$$H(\omega) = H(s)|_{s=j\omega} = \frac{j\omega}{j\omega + \omega_0} \tag{16.15}$$

FIGURE 16.8

The pole-zero diagram for the first-order HPF.

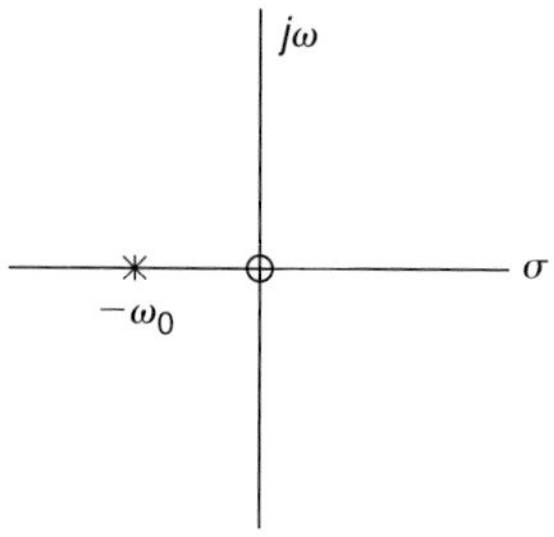

The magnitude response is

$$|H(\omega)| = \frac{\omega}{\sqrt{\omega^2 + \omega_0^2}} \tag{16.16}$$

and the phase response is

$$\angle H(\omega) = \frac{\pi}{2} - \tan^{-1}\left(\frac{\omega}{\omega_0}\right) \tag{16.17}$$

The magnitude response in decibels is given by

$$M(\omega) = 20\log_{10}(|H(\omega)|) = 20\log_{10}\frac{\omega}{\sqrt{\omega^2 + \omega^2}}$$

Figure 16.9 shows the magnitude response (linear scale and dB scale) and the phase response.

FIGURE 16.9

(a) The magnitude response in linear scale, (b) the magnitude response in dB scale, (c) the phase response of the first-order HPF ($\omega_0 = 1000$ rad/s).

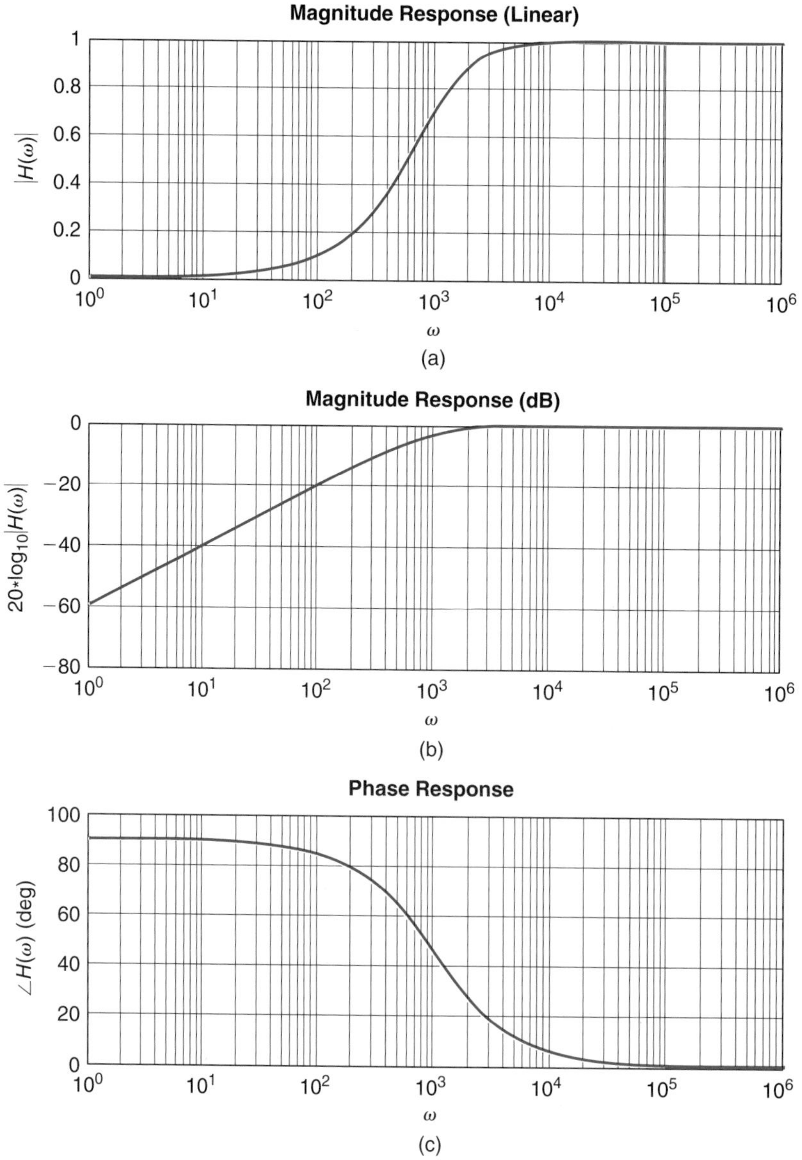

When $\omega = \omega_o$, the magnitude is

$$|H(\omega_o)| = \frac{\omega_0}{\sqrt{\omega_0^2 + \omega_0^2}} = \frac{1}{\sqrt{2}} = -3.01 \text{ dB}$$

The frequency ω_o is called 3-dB *cutoff frequency* or *half-power frequency*.

The first-order HPF can be synthesized using a *CR* circuit, as shown in Figure 16.10. The transfer function of this circuit is given by

FIGURE 16.10

The first-order HPF using a *CR* circuit.

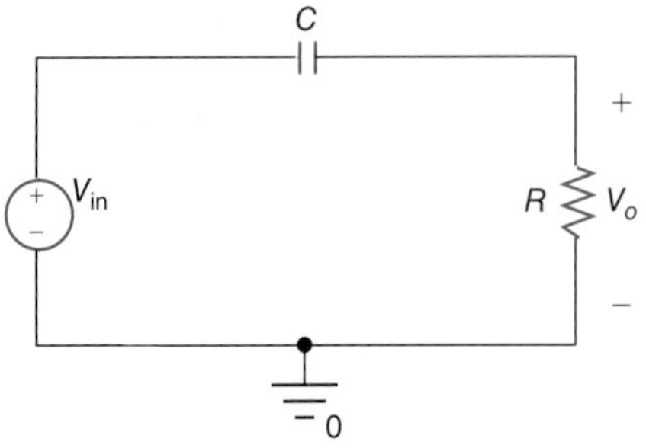

$$H(s) = \frac{V_o(s)}{V_{in}(s)} = \frac{R}{R + \dfrac{1}{sC}} = \frac{RCs}{RCs + 1} = \frac{s}{s + \dfrac{1}{RC}}$$

Initially, the component values can be selected as

$$C = 1 \text{ F}, \quad R = 1\ \Omega.$$

Choose the frequency scale to be $k_f = \omega_0$. The magnitude scaling can now be applied to get practical values for R and C.

The first-order HPF can also be designed using the *RL* circuit shown in Figure 16.11. The transfer function of the circuit shown in Figure 16.11 is given by

FIGURE 16.11

The first-order HPF using an *RL* circuit.

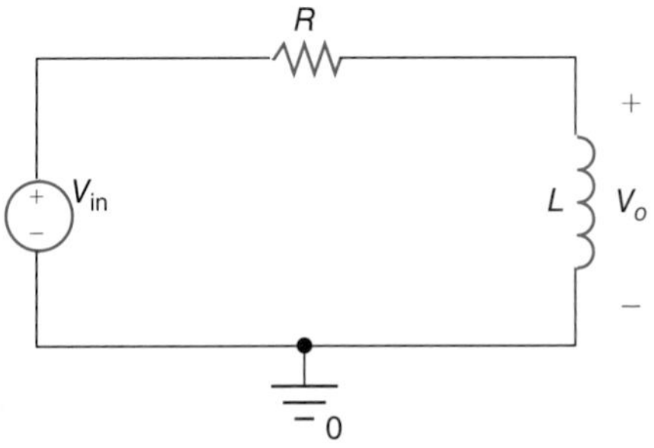

$$H(s) = \frac{sL}{sL + R} = \frac{s}{s + \dfrac{R}{L}}$$

Initially, we can choose $L = 1$ H, $R = 1\ \Omega$. The frequency scaling can be employed to meet the cutoff frequency requirement, and the magnitude scaling can be used to change the component values to more practical values.

The first-order HPF can also be designed using an operational amp circuit shown in Figure 16.12.

FIGURE 16.12

First-order HPF using an operational amp.

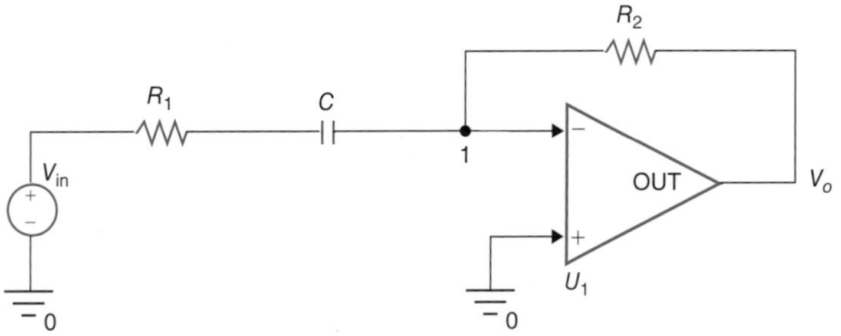

Summing the currents leaving node 1, we obtain

$$\frac{0 - V_{in}}{R_1 + \dfrac{1}{sC}} + \frac{0 - V_o}{R_2} = 0$$

which can be rearranged as

$$\frac{V_o}{R_2} = \frac{-V_{in}}{R_1 + \dfrac{1}{sC}}$$

The transfer function of the circuit shown in Figure 16.12 is given by

$$H(s) = \frac{V_o}{V_{in}} = \frac{-sR_2C}{sR_1C + 1} = -\frac{\frac{R_2}{R_1}s}{s + \frac{1}{R_1C}} = -\frac{R_2}{R_1}\frac{s}{s + \frac{1}{R_1C}}$$

If $R_2 = R_1$, $H(s)$ becomes

$$H(s) = \frac{V_o(s)}{V_{in}(s)} = -\frac{s}{s + \frac{1}{R_1C}}$$

Initially, we can pick $C = 1$ F, $R_1 = 1\ \Omega$. The frequency scaling can be employed to meet the cutoff frequency requirement, and the magnitude scaling can be used to change the component values to more practical values.

Figure 16.13 shows an implementation of the first order *CR* HPF using an operational amp voltage follower.

Due to a virtual short, the voltage across the resistor is V_o. Application of the voltage divider rule yields

$$H(s) = \frac{V_o(s)}{V_{in}(s)} = \frac{R}{R + \frac{1}{sC}} = \frac{RCs}{RCs + 1} = \frac{s}{s + \frac{1}{RC}}$$

The operational amp provides large-input impedance and small-output impedance. The first-order section is isolated from other sections in the cascade implementation.

FIGURE 16.13

The first-order HPF.

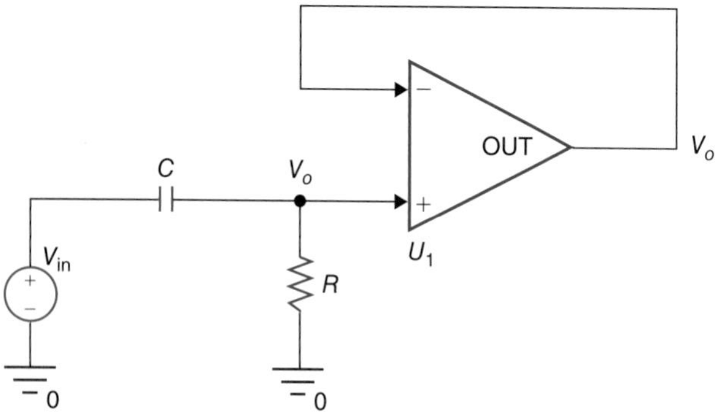

16.5 Second-Order LPF

The transfer function of the second-order LPF is given by

$$H(s) = \frac{\omega_0^2}{s^2 + \frac{\omega_0}{Q}s + \omega_0^2} = \frac{\omega_0^2}{s^2 + 2\zeta\omega_0 s + \omega_0^2} \tag{16.18}$$

Here, ω_0 is called the *corner frequency* (or *center frequency*, or *resonant frequency*) in radians per second, Q is called the *quality factor*, and ζ is called the *damping coefficient*. Notice that

$$\zeta = \frac{1}{2Q} \quad \text{and} \quad Q = \frac{1}{2\zeta}$$

The locations of poles are the roots of $s^2 + \frac{\omega_0}{Q}s + \omega_0^2 = 0$. The pole locations are

$$p_1 = -\frac{\omega_0}{2Q} + \omega_0\sqrt{\frac{1}{4Q^2} - 1} = -\frac{\omega_0}{2Q} + \frac{\omega_0}{2Q}\sqrt{1 - 4Q^2}$$

$$p_2 = -\frac{\omega_0}{2Q} - \omega_0\sqrt{\frac{1}{4Q^2} - 1} = -\frac{\omega_0}{2Q} - \frac{\omega_0}{2Q}\sqrt{1 - 4Q^2} \tag{16.19}$$

For the poles to be complex conjugates,

$$1 - 4Q^2 < 0$$

or $Q > 1/2$. Since $Q = 1/(2\zeta)$, $Q > 1/2$ is equivalent to $0 < \zeta < 1$. When the poles are complex conjugates, we have

$$p_1 = -\frac{\omega_0}{2Q} + j\omega_0\sqrt{1 - \frac{1}{4Q^2}}, \quad p_2 = -\frac{\omega_0}{2Q} - j\omega_0\sqrt{1 - \frac{1}{4Q^2}} \qquad \textbf{(16.20)}$$

FIGURE 16.14

Location of poles as Q is increased from 0 to ∞.

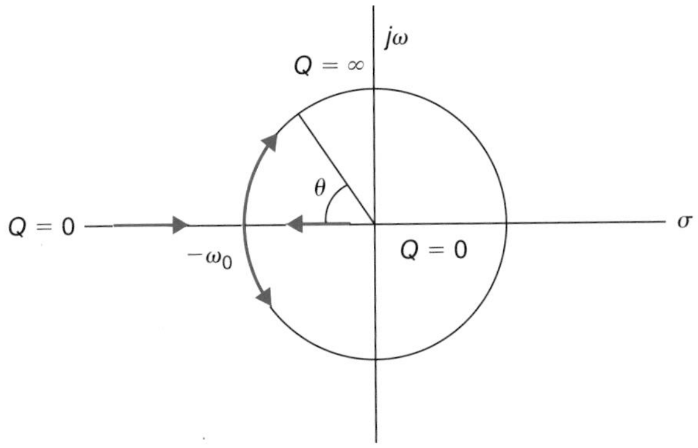

As Q is decreased to zero ($Q \to 0$), from Equation (16.19), we have $p_1 \to 0$ and $p_2 \to -\infty$. As Q is increased from zero to infinity, the two poles at zero and $-\infty$ on the real line move toward $-\omega_0$ and reach $-\omega_0$ when $Q = 1/2$. As Q is increased above 1/2, the poles are on the circle of radius ω_0 and move toward imaginary axis. As $Q \to \infty$, from Equation (16.20), we have $p_1 \to j\omega_0$ and $p_2 \to -j\omega_0$, as shown in Figure 16.14. Notice that $\cos(\theta) = \dfrac{\frac{\omega_0}{2Q}}{\omega_0} = \dfrac{1}{2Q}$. As Q is increased, θ moves toward 90°, and $\cos(\theta)$ approaches zero.

There are two zeros at $s = \infty$. The pole-zero diagram of the second-order LPF is shown in Figure 16.15. The radius from the origin to the poles is ω_0.

FIGURE 16.15

The pole-zero diagram of the second-order LPF.

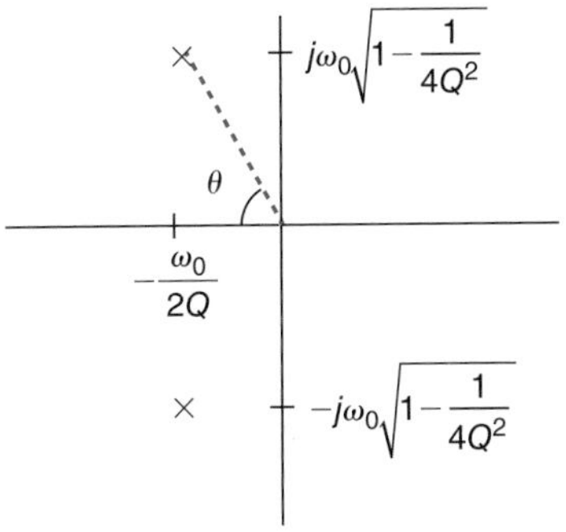

16.5.1 FREQUENCY RESPONSE

The frequency response of the second-order LPF is obtained by substituting $s = j\omega$ in $H(s)$.

$$H(\omega) = H(s)|_{s=j\omega} = \frac{\omega_0^2}{-\omega^2 + \frac{\omega_0}{Q}j\omega + \omega_0^2} \qquad \textbf{(16.21)}$$

16.5.2 MAGNITUDE RESPONSE

The magnitude response of the second-order LPF is given by the absolute value of $H(\omega)$.

$$|H(\omega)| = \frac{\omega_0^2}{\sqrt{(\omega_0^2 - \omega^2)^2 + \left(\frac{\omega_0}{Q}\omega\right)^2}}$$

At $\omega = \omega_0$, the magnitude is $|H(\omega_0)| = Q$. The peak frequency ω_{peak}, where the magnitude is at its maximum, is given by

$$\omega_{peak} = \omega_0\sqrt{1 - \frac{1}{2Q^2}} \qquad \textbf{(16.22)}$$

For the proof, refer to Problem 16.1.

The maximum magnitude at peak frequency is

$$|H(j\omega_{\text{peak}})| = \frac{\omega_0^2}{\sqrt{\left[\omega_0^2 - \omega_0^2\left(1 - \frac{1}{2Q^2}\right)\right]^2 + \frac{1}{Q^2}\omega_0^2\omega_0^2\left(1 - \frac{1}{2Q^2}\right)}} = \frac{Q}{\sqrt{1 - \frac{1}{4Q^2}}} \quad \textbf{(16.23)}$$

The largest frequency for $|H(\omega)| = 1$ (0 dB), called the *zero-crossing frequency*, is given by

$$\omega_{zc} = \omega_0\sqrt{2\left(1 - \frac{1}{2Q^2}\right)} = \sqrt{2}\,\omega_{\text{peak}} \quad \textbf{(16.24)}$$

For the proof, refer to Problem 16.2. The 3-dB cutoff frequency ω_c is given by

$$\omega_c = \omega_0\sqrt{\frac{2 - \frac{1}{Q^2} + \sqrt{\left(2 - \frac{1}{Q^2}\right)^2 + 4}}{2}} \quad \textbf{(16.25)}$$

For the proof, refer to Problem 16.3. The magnitude response of the second-order LPF for $Q = 2$ is shown in Figure 16.16 for $\omega_0 = 1$ rad/s. The magnitudes at frequencies ω_{peak}, ω_0, ω_{zc}, and ω_c are shown in Figure 16.16.

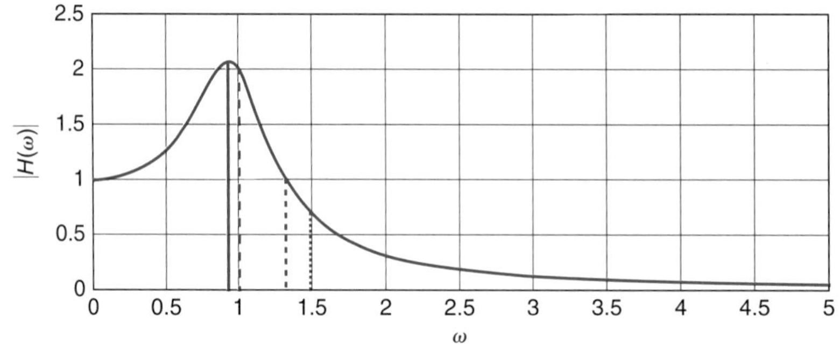

FIGURE 16.16 The magnitude response for $Q = 2$.

16.5.3 PHASE RESPONSE

From Equation (16.21), at $\omega = 0$, we have $H(0) = \frac{\omega_0^2}{\omega_0^2} = 1$. Thus, we get $\angle H(0) = 0$. As ω is increased to infinity ($\omega \rightarrow \infty$), the real part of the denominator $-\omega^2 + \frac{\omega_0}{Q}j\omega + \omega_0^2$ approaches $-\infty$, and the imaginary part of the denominator is small compared to the real part. The angle is in the second quadrant close to the real axis. The angle is $180° - \varepsilon$. Since this is the phase of the denominator, the phase of $H(\omega)$ is $-180°$. As the frequency is scanned from zero to infinity, the phase angle changes from zero to $-180°$.

The phase response of the second-order LPF is obtained from Equation (16.21).

$$\angle H(\omega) = -\tan^{-1}\left(\frac{\frac{\omega_o}{Q}\omega}{\omega_o^2 - \omega^2}\right) \quad \textbf{(16.26)}$$

Each finite zero contributes 90° to phase at $\omega = \infty$. Each finite pole contributes $-90°$ phase at $\omega = \infty$. Ignoring the zeros at infinity, the two poles of the second-order LPF contribute $-180°$ to phase at $\omega = \infty$. The phase response of the second-order LPF for $Q = 2$ is shown in Figure 16.17 for $\omega_0 = 1$ rad/s.

FIGURE 16.17

Phase response for $Q = 2$.

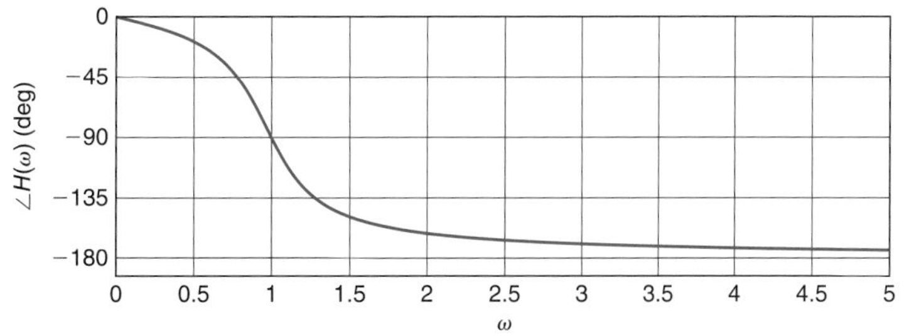

FIGURE 16.18

The second-order LPF using series RLC circuit.

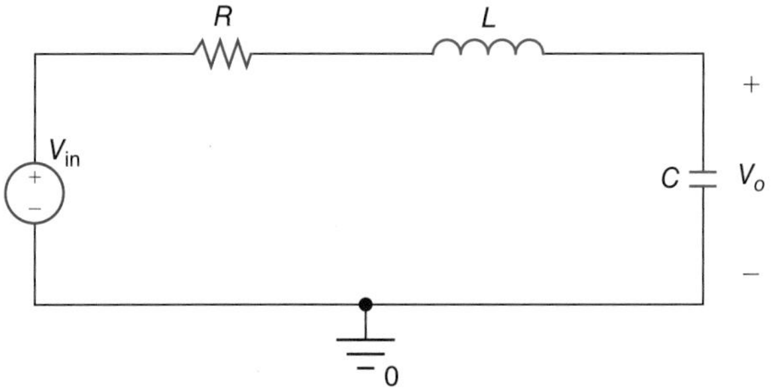

16.5.4 SERIES *RLC* LPF

The second-order LPF can be designed using the series RLC circuit shown in Figure 16.18.

The transfer function of the circuit shown in Figure 16.18 is given by the voltage divider rule

$$H(s) = \frac{V_o(s)}{V_{in}(s)} = \frac{\frac{1}{sC}}{sL + R + \frac{1}{sC}} = \frac{1}{LCs^2 + RCs + 1} = \frac{\frac{1}{LC}}{s^2 + \frac{R}{L}s + \frac{1}{LC}} \tag{16.27}$$

Thus, the corner frequency and the quality factor are expressed using circuit element values as

$$\omega_o = \frac{1}{\sqrt{LC}} \tag{16.28}$$

and

$$Q = \frac{1}{R}\sqrt{\frac{L}{C}} \tag{16.29}$$

respectively. For a normalized LPF, we set $\omega_0 = 1$ rad/s. Then, the transfer function becomes

$$H(s) = \frac{V_o(s)}{V_{in}(s)} = \frac{\omega_0^2}{s^2 + \frac{\omega_0}{Q}s + \omega_0^2} = \frac{1}{s^2 + \frac{1}{Q}s + 1} \tag{16.30}$$

To get $\omega_0 = 1$ rad/s, let us choose $L = 1$ H, and $C = 1$ F. Then, $R = 1/Q$. Apply magnitude scaling k_m and frequency scaling $k_f = \omega_0$ to obtain the scaled values for R, L, and C. After scaling, the transfer function becomes the one given in Equation (16.18).

EXAMPLE 16.4

Design a LPF with the transfer function

$$H(s) = \frac{9 \times 10^8}{s^2 + 1.5 \times 10^4 s + 9 \times 10^8}$$

continued

Example 16.4 continued

using the series RLC circuit. Assume that $k_m = 1000$.

Let $a_1 = 1.5 \times 10^4$ and $a_0 = 9 \times 10^8$. Then, we have

$$\omega_0 = \sqrt{a_0} = \sqrt{9 \times 10^8} = 3 \times 10^4 \text{ rad/s.}$$

Since

$$a_1 = \frac{\omega_0}{Q} = 1.5 \times 10^4, \text{ we get}$$

$$Q = \frac{\omega_0}{a_1} = \frac{3 \times 10^4}{1.5 \times 10^4} = 2$$

Component values for normalized filter:

$$R = 1/Q = 0.5\ \Omega$$
$$L = 1\text{ H}$$
$$C = 1\text{ F}$$

Let $k_m = 1000$ and $k_f = \omega_0 = 3 \times 10^4$. Then, the component values after scaling are given by

$$R = k_m \times 0.5 = 500\ \Omega$$
$$L = 1 \times k_m/k_f = 33.3333\text{ mH}$$
$$C = 1/(k_m k_f) = 0.03333\ \mu F$$

Exercise 16.4

Design a LPF with transfer function $H(s)$ given by

$$H(s) = \frac{9 \times 10^7}{s^2 + 9500s + 9 \times 10^7}$$

using series RLC circuit. Assume that $k_m = 1000$.

Answer:
$R = 1001.39\ \Omega$, $L = 105.41$ mH, $C = 0.10541\ \mu F$.

FIGURE 16.19

The second-order LPF using a parallel RLC circuit.

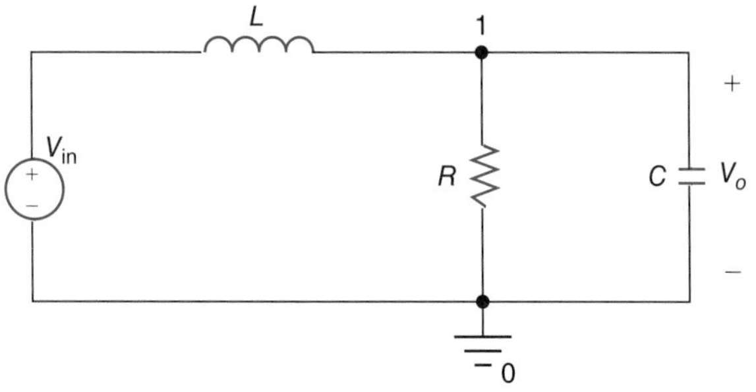

16.5.5 PARALLEL *RLC* LPF

The second-order LPF can be designed using a parallel RLC circuit shown in Figure 16.19.

Summing the currents leaving node 1, we obtain

$$\frac{V_o(s) - V_{in}(s)}{sL} + \frac{V_o(s)}{R} + \frac{V_o(s)}{\frac{1}{sC}} = 0$$

Solving for $V_o(s)/V_{in}(s)$, we get

$$H(s) = \frac{V_o(s)}{V_{in}(s)} = \frac{\frac{1}{sL}}{sC + \frac{1}{R} + \frac{1}{sL}} = \frac{1}{LCs^2 + \frac{L}{R}s + 1} = \frac{\frac{1}{LC}}{s^2 + \frac{1}{RC}s + \frac{1}{LC}} \quad \textbf{(16.31)}$$

Thus, the corner frequency is given by

$$\omega_o = \frac{1}{\sqrt{LC}} \quad \textbf{(16.32)}$$

and the quality factor is given by

$$Q = R\sqrt{\frac{C}{L}} \quad \textbf{(16.33)}$$

The normalized LPF is obtained by setting $\omega_0 = 1$ rad/s. To get $\omega_0 = 1$ rad/s, let us choose $L = 1$ H, and $C = 1$ F. Then, $R = Q$. Apply the magnitude scaling k_m and the frequency scaling $k_f = \omega_0$ to obtain the values for R, L, and C. After scaling, the transfer function becomes the one given by Equation (16.18).

EXAMPLE 16.5

Design a LPF with transfer function

$$H(s) = \frac{25 \times 10^8}{s^2 + 10^4 s + 25 \times 10^8}$$

using a parallel RLC circuit. Assume that $k_m = 1000$.

Let $a_1 = 1 \times 10^4$ and $a_0 = 25 \times 10^8$. Then, we have

$$\omega_0 = \sqrt{a_0} = \sqrt{25 \times 10^8} = 5 \times 10^4 \text{ rad/s}$$

Since

$$a_1 = \frac{\omega_0}{Q} = 1 \times 10^4$$

we get

$$Q = \frac{\omega_0}{a_1} = \frac{5 \times 10^4}{1 \times 10^4} = 5$$

Component values for the normalized filter:

$$R = Q = 5\ \Omega$$
$$L = 1\text{ H}$$
$$C = 1\ F$$

continued

Example 16.5 continued

Let $k_m = 1000$ and $k_f = \omega_0 = 5 \times 10^4$. Then, the component values after scaling are given by

$$R = k_m \times 5 = 5\ k\Omega$$
$$L = 1 \times k_m/k_f = 20\ \text{mH}$$
$$C = 1/(k_m k_f) = 0.02\ \mu F$$

Exercise 16.5

Design an LPF with transfer function

$$H(s) = \frac{9 \times 10^7}{s^2 + 8500s + 9 \times 10^7}$$

using a parallel *RLC* circuit. Assume that $k_m = 1000$.

Answer:
$R = 1116.098\ \Omega$, $L = 105.4093$ mH, $C = 0.1054\ \mu F$.

16.5.6 SALLEN-KEY CIRCUIT FOR THE SECOND-ORDER LPF

The second-order LPF can also be designed using a circuit with an operational amp, as shown in Figure 16.20. This circuit is called a *Sallen-Key circuit.*

FIGURE 16.20

The Sallen-Key circuit for the second-order LPF.

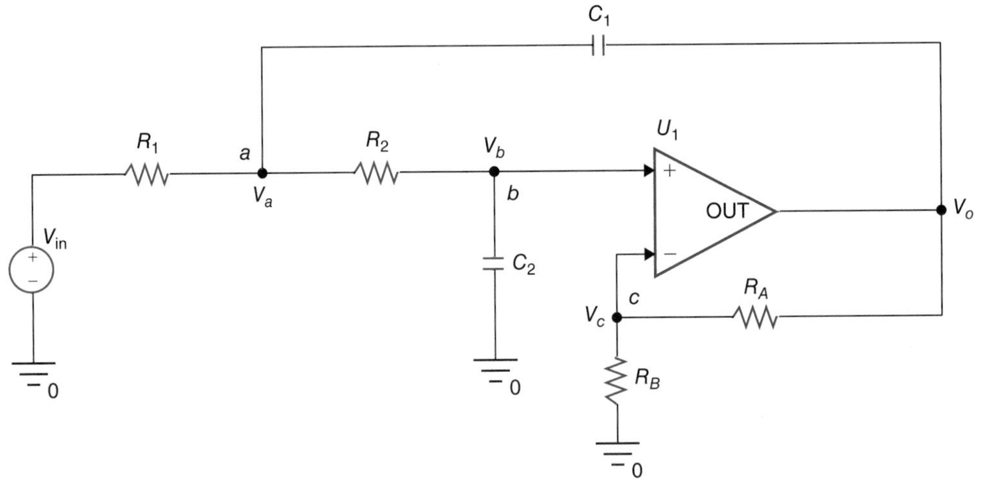

Since the current flowing into the negative input of the operational amp is zero, the sum of currents leaving node *c* is given by

$$\frac{V_c}{R_B} + \frac{V_c - V_o}{R_A} = 0 \quad (16.34)$$

which can be rearranged as

$$\left(\frac{1}{R_B} + \frac{1}{R_A}\right)V_c = \frac{V_o}{R_A}$$

Solving for V_c, we obtain

$$V_c = \frac{\frac{1}{R_A}}{\frac{1}{R_B} + \frac{1}{R_A}} V_o = \frac{R_B}{R_A + R_B} V_o \tag{16.35}$$

Alternatively, the voltage divider rule can be used to get Equation (16.35). Let

$$K = \frac{R_A + R_B}{R_B} = 1 + \frac{R_A}{R_B} \tag{16.36}$$

Then, Equation (16.35) can be rewritten as

$$V_c = \frac{R_B}{R_A + R_B} V_o = \frac{1}{\frac{R_A + R_B}{R_B}} V_o = \frac{V_o}{K} \tag{16.37}$$

Since the current flowing into the positive input of the operational amp is zero, the sum of currents leaving node b is given by

$$\frac{V_b}{\frac{1}{sC_2}} + \frac{V_b - V_a}{R_2} = 0 \tag{16.38}$$

which can be rearranged as

$$\left(sC_2 + \frac{1}{R_2}\right) V_b = \frac{V_a}{R_2}$$

Solving for V_a, we obtain

$$V_a = (sR_2C_2 + 1)V_b \tag{16.39}$$

Due to a virtual short, the voltage V_b is equal to V_c. Thus

$$V_b = V_c = \frac{V_o}{K} \tag{16.40}$$

Substitution of Equation (16.40) into Equation (16.39) yields

$$V_a = (sR_2C_2 + 1)V_b = (sR_2C_2 + 1)\frac{V_o}{K} = \frac{1}{K}(sR_2C_2 + 1)V_o \tag{16.41}$$

Summing the currents leaving node a, we obtain

$$\frac{V_a - V_{in}}{R_1} + \frac{V_a - V_o}{\frac{1}{sC_1}} + \frac{V_a - V_b}{R_2} = 0 \tag{16.42}$$

which can be rearranged as

$$\left(\frac{1}{R_1} + sC_1 + \frac{1}{R_2}\right) V_a - sC_1 V_o - \frac{V_b}{R_2} = \frac{V_{in}}{R_1} \tag{16.43}$$

Substitution of Equations (16.40) and (16.41) into Equation (16.43) yields

$$\left(\frac{1}{R_1}+sC_1+\frac{1}{R_2}\right)\frac{1}{K}(sR_2C_2+1)V_o-sC_1V_o-\frac{1}{R_2}\frac{V_o}{K}=\frac{V_{in}}{R_1} \tag{16.44}$$

The transfer function is given by

$$H(s)=\frac{V_o}{V_{in}}=\frac{\dfrac{1}{R_1}}{\left(\dfrac{1}{R_1}+sC_1+\dfrac{1}{R_2}\right)\dfrac{1}{K}(sR_2C_2+1)-sC_1-\dfrac{1}{R_2K}} \tag{16.45}$$

Multiplication by R_1R_2K yields

$$\begin{aligned}H(s)=\frac{V_o}{V_{in}}&=\frac{R_2K}{(R_2+sR_1R_2C_1+R_1)(sR_2C_2+1)-sKR_1R_2C_1-R_1}\\&=\frac{R_2K}{R_1R_2^2C_1C_2s^2+(R_2^2C_2+R_1R_2C_2+R_1R_2C_1-KR_1R_2C_1)s+R_2}\end{aligned}$$

Dividing by $R_1R_2^2C_1C_2$ for every term, we obtain the transfer function $H(s)$, given by

$$H(s)=\frac{\dfrac{K}{R_1R_2C_1C_2}}{s^2+\left(\dfrac{1}{R_1C_1}+\dfrac{1}{R_2C_1}+\dfrac{1-K}{R_2C_2}\right)s+\dfrac{1}{R_1R_2C_1C_2}} \tag{16.46}$$

To simplify the design, we can use either the equal R, equal C method or the unity gain method.

16.5.7 EQUAL *R*, EQUAL *C* METHOD

Let $R = R_1 = R_2$ and $C = C_1 = C_2$. Then, the transfer function becomes

$$H(s)=\frac{K\left(\dfrac{1}{RC}\right)^2}{s^2+\dfrac{3-K}{RC}s+\left(\dfrac{1}{RC}\right)^2} \tag{16.47}$$

Thus, the corner frequency is given by

$$\omega_0=\frac{1}{RC} \tag{16.48}$$

Since the coefficient of s term is

$$\frac{\omega_0}{Q}=\frac{3-K}{RC}=(3-K)\omega_0$$

the Q value is given by

$$Q=\frac{1}{3-K} \tag{16.49}$$

Solving for K, we obtain

$$K = 3 - \frac{1}{Q} \tag{16.50}$$

From Equations (16.36) and (16.50), we obtain

$$\frac{R_A}{R_B} = K - 1 = 2 - \frac{1}{Q} \tag{16.51}$$

If s is set to zero in Equation (16.47), we obtain

$$H(0) = K \tag{16.52}$$

Thus, the gain of the filter at $\omega = 0$ is K.

16.5.8 NORMALIZED FILTER

For the normalized filter, we set $\omega_0 = 1$ rad/s. From Equation (16.48), we select

$$R = R_1 = R_2 = 1\ \Omega$$
$$C = C_1 = C_2 = 1\ \text{F}.$$

Let $R_B = 1\ \Omega$. Then, from Equation (16.51), we obtain

$$R_A = K - 1 = 2 - \frac{1}{Q}\ \Omega.$$

Use the frequency scale ($k_f = \omega_0$) and the magnitude scale to get the desired values. Choose k_m for the appropriate resistor and capacitor values.

The gain of the filter at $\omega = 0$ can be changed from K to 1 by replacing V_{in} and R_1 by the circuit shown in Figure 16.21.

FIGURE 16.21

A voltage divider circuit.

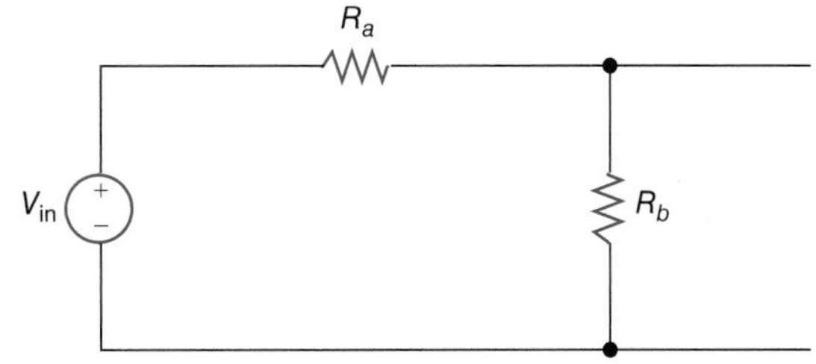

The Thévenin resistance of this circuit must be equal to R_1, and the Thévenin voltage must be V_{in}/K. Thus, we have

$$\frac{R_a R_b}{R_a + R_b} = R_1 = 1 \tag{16.53}$$

$$\frac{R_b}{R_a + R_b} = \frac{1}{K} \tag{16.54}$$

Dividing Equation (16.53) by Equation (16.54), we have

$$R_a = K \tag{16.55}$$

From Equation (16.54), we get

$$R_b = \frac{K}{K-1} \tag{16.56}$$

Substituting K from Equation (16.50), we obtain

$$R_a = 3 - \frac{1}{Q} \tag{16.57}$$

$$R_b = \frac{3Q-1}{2Q-1} \tag{16.58}$$

Figure 16.22 shows the gain-equalized circuit for the second-order LPF.

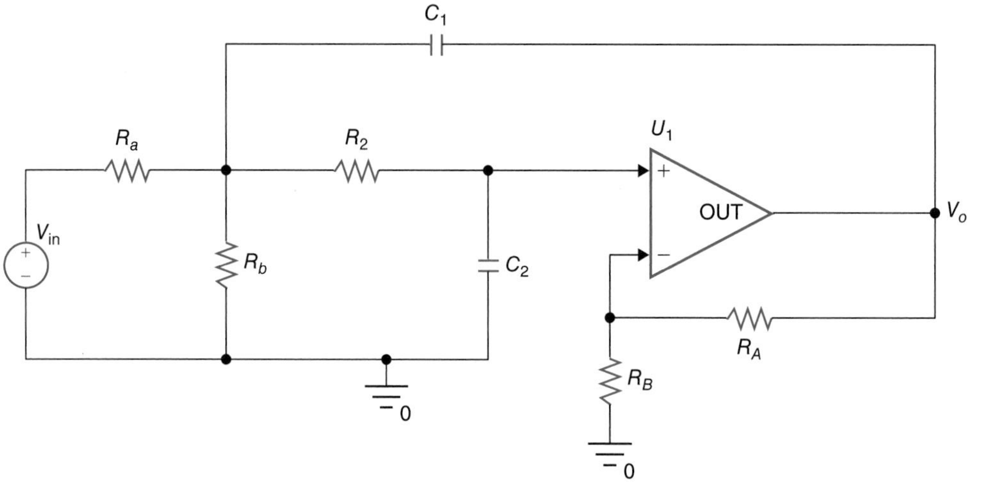

FIGURE 16.22

A Sallen-Key circuit with gain set to 1.

The normalized component values for the circuit shown in Figure 16.22 are given by

$$C = C_1 = C_2 = 1\text{ F},\quad R_2 = 1\ \Omega,\quad R_B = 1\ \Omega,\quad R_A = 2 - \frac{1}{Q},$$
$$R_a = 3 - \frac{1}{Q},\quad R_b = \frac{3Q-1}{2Q-1} \tag{16.59}$$

Use the frequency scale ($k_f = \omega_0$) and the magnitude scale to get the desired values. Choose k_m for the appropriate resistor and capacitor values.

EXAMPLE 16.6

Design a LPF with transfer function

$$H(s) = \frac{93{,}876{,}690.1780}{s^2 + 13{,}702.3130s + 93{,}876{,}690.1780}$$

using a Sallen-Key circuit in the equal R, equal C method. Assume that $k_m = 1000$.

continued

Example 16.6 continued

Let $a_1 = 13{,}702.3130$ and $a_0 = 93{,}876{,}690.1780$. Then, the transfer function can be written as

$$H(s) = \frac{93{,}876{,}690.1780}{s^2 + 13{,}702.3130s + 93{,}876{,}690.1780} = \frac{a_0}{s^2 + a_1 s + a_0} = \frac{\omega_0^2}{s^2 + \frac{\omega_0}{Q}s + \omega_0^2}$$

The corner frequency ω_0 is given by

$$\omega_0 = \sqrt{a_0} = \sqrt{93{,}876{,}690.1780} = 9688.9984 \text{ rad/s}$$

The Q value is given by

$$Q = \frac{\omega_0}{a_1} = \frac{9688.9984}{13{,}702.3130} = 0.7071$$

The normalized component values are given by

$$C_1 = 1\text{ F},\quad C_2 = 1\text{ F},\quad R_2 = 1\ \Omega,\quad R_B = 1\ \Omega,\quad R_A = 2 - 1/Q = 0.5858\ \Omega$$
$$R_a = 3 - 1/Q = 1.5858\ \Omega,\quad R_b = (3Q-1)/(2Q-1) = 2.7071\ \Omega$$

Application of the magnitude scaling ($k_m = 1000$) and the frequency scaling ($k_f = \omega_0 = 9688.9984$) yields

$$C_1 = \frac{1\text{ F}}{k_m \times k_f} = \frac{1}{1000 \times 9688.9984}\text{F} = 0.1032\ \mu F$$

$$C_2 = \frac{1\text{ F}}{k_m \times k_f} = \frac{1}{1000 \times 9688.9984}\text{F} = 0.1032\ \mu F$$

$$R_2 = 1 \times 1000\ \Omega = 1\ k\Omega$$
$$R_B = 1 \times 1000\ \Omega = 1\ k\Omega$$
$$R_A = 2 - 1/Q = 0.5858 \times 1000\ \Omega = 585.8\ \Omega$$
$$R_a = 3 - 1/Q = 1.5858 \times 1000\ \Omega = 1.5858\ k\Omega$$
$$R_b = (3Q-1)/(2Q-1) = 2.7071 \times 1000\ \Omega = 2.7071\ k\Omega$$

Exercise 16.6

Design a lowpass filter with $\omega_0 = 2\pi1000 = 6283.1853$ rad/s and $Q = 2$ using a Sallen-Key circuit in the equal R, equal C method. Assume that $k_m = 1000$.

The normalized component values are given by

$$C_1 = 1\text{ F},\quad C_2 = 1\text{ F},\quad R_2 = 1\ \Omega,\quad R_B = 1\ \Omega,\quad R_A = 2 - 1/Q = 1.5\ \Omega,$$
$$R_a = 3 - 1/Q = 2.5\ \Omega,\quad R_b = (3Q-1)/(2Q-1) = 1.6667\ \Omega$$

Choose $k_f = 2\pi1000 = 6283.1853$ and $k_m = 1000$. The scaled component values are given by

$$C_1 = C_2 = \frac{1}{k_m k_f}\text{F} = 1.592 \times 10^{-7}\text{F} = 0.1592\ \mu F$$

$$R_2 = R_B = 1\ k\Omega,\quad R_A = 1.5\ k\Omega,\quad R_a = 2.5\ k\Omega,\quad R_b = 1.67\ k\Omega.$$

EXAMPLE 16.7

The transfer function of the fourth-order LPF is given by

$$H(s) = \frac{77{,}585{,}411.64}{s^2 + 16{,}275.5388s + 77{,}585{,}411.64} \times \frac{77{,}585{,}411.64}{s^2 + 6741.5489s + 77{,}585{,}411.64}$$

Design this filter by cascading two Sallen-Key second-order lowpass filters using the equal R, equal C method.

a. Find the values of $R_a, R_b, R_2, R_A, R_B, C_1, C_2$ when $k_m = 1000$ for the first second-order section.
b. Find the values of $R_a, R_b, R_2, R_A, R_B, C_1, C_2$ when $k_m = 1000$ for the second second-order section.

a. Let $a_1 = 16{,}275.5388$ and $a_2 = 77{,}585{,}411.64$. Then, the transfer function can be written as

$$H(s) = \frac{77{,}585{,}411.64}{s^2 + 16{,}275.5388s + 77{,}585{,}411.64} = \frac{a_0}{s^2 + a_1 s + a_0} = \frac{\omega_0^2}{s^2 + \frac{\omega_0}{Q}s + \omega_0^2}$$

The corner frequency ω_0 is given by

$$\omega_0 = \sqrt{a_0} = \sqrt{77{,}585{,}411.64} = 8808.2582 \text{ rad/s}$$

The Q value is given by

$$Q = \frac{\omega_0}{a_1} = \frac{8808.2582}{16{,}275.5388} = 0.5412$$

The normalized component values are given by

$$C_1 = 1\text{ F},\quad C_2 = 1\text{ F},\quad R_2 = 1\ \Omega,\quad R_B = 1\ \Omega,\quad R_A = 2 - 1/Q = 0.15224\ \Omega$$
$$R_a = 3 - 1/Q = 1.1522\ \Omega,\quad R_b = (3Q - 1)/(2Q - 1) = 7.5685\ \Omega$$

Application of the magnitude scaling ($k_m = 1000$) and the frequency scaling ($k_f = \omega_0 = 8808.2582$) yields

$$C_1 = \frac{1\text{ F}}{k_m \times k_f} = \frac{1}{1000 \times 8808.2582}\text{F} = 0.1135\ \mu F$$

$$C_2 = \frac{1\text{ F}}{k_m \times k_f} = \frac{1}{1000 \times 8808.2582}\text{F} = 0.1135\ \mu F$$

$$R_2 = 1 \times 1000\ \Omega = 1\ k\Omega$$
$$R_B = 1 \times 1000\ \Omega = 1\ k\Omega$$
$$R_A = 2 - 1/Q = 0.15224 \times 1000\ \Omega = 152.24\ \Omega$$
$$R_a = 3 - 1/Q = 1.1522 \times 1000\ \Omega = 1.1522\ k\Omega$$
$$R_b = (3Q - 1)/(2Q - 1) = 7.5685 \times 1000\ \Omega = 7.5685\ k\Omega$$

continued

Example 16.7 continued

b. Let $a_1 = 6741.5489$ and $a_2 = 77{,}585{,}411.64$. Then, the transfer function can be written as

$$H(s) = \frac{77{,}585{,}411.64}{s^2 + 6741.5489s + 77{,}585{,}411.64} = \frac{a_0}{s^2 + a_1 s + a_0} = \frac{\omega_0^2}{s^2 + \frac{\omega_0}{Q}s + \omega_0^2}$$

The corner frequency ω_0 is given by

$$\omega_0 = \sqrt{a_0} = \sqrt{77{,}585{,}411.64} = 8808.2582 \text{ rad/s}$$

The Q value is given by

$$Q = \frac{\omega_0}{a_1} = \frac{8808.2582}{6741.5489} = 1.3066$$

The normalized component values are given by

$$C_1 = 1\text{ F},\quad C_2 = 1\text{ F},\quad R_2 = 1\ \Omega,\quad R_B = 1\ \Omega,\quad R_A = 2 - 1/Q = 1.2346\ \Omega$$
$$R_a = 3 - 1/Q = 2.2346\ \Omega,\quad R_b = (3Q - 1)/(2Q - 1) = 1.81\ \Omega$$

Application of the magnitude scaling ($k_m = 1000$) and the frequency scaling ($k_f = \omega_0 = 8808.2582$) yields

$$C_1 = \frac{1\text{ F}}{k_m \times k_f} = \frac{1}{1000 \times 8808.2582}\text{F} = 0.1135\ \mu F$$

$$C_2 = \frac{1\text{ F}}{k_m \times k_f} = \frac{1}{1000 \times 8808.2582}\text{F} = 0.1135\ \mu F$$

$$R_2 = 1 \times 1000\ \Omega = 1\ k\Omega$$
$$R_B = 1 \times 1000\ \Omega = 1\ k\Omega$$
$$R_A = 2 - 1/Q = 1.2346 \times 1000\ \Omega = 1.2346\ k\Omega$$
$$R_a = 3 - 1/Q = 2.2346 \times 1000\ \Omega = 2.2346\ k\Omega$$
$$R_b = (3Q - 1)/(2Q - 1) = 1.81 \times 1000\ \Omega = 1.81\ k\Omega$$

Exercise 16.7

The transfer function of the fourth-order LPF is given by

$$H(s) = \frac{139{,}656{,}236}{s^2 + 21{,}836.1218s + 139{,}656{,}236} \times \frac{139{,}656{,}236}{s^2 + 9044.8178s + 139{,}656{,}236}$$

Design this filter by cascading two Sallen-Key second-order lowpass filters using the equal R, equal C method.

a. Find the values of R_a, R_b, R_2, R_A, R_B, C_1, and C_2 when $k_m = 1000$ for the first second-order section.

b. Find the values of R_a, R_b, R_2, R_A, R_B, C_1, and C_2 when $k_m = 1000$ for the second second-order section.

continued

Exercise 16.7 continued

Answer:

a. $C_1 = 84.62\ nF$, $C_2 = 84.62\ nF$, $R_2 = 1\ k\Omega$, $R_B = 1\ k\Omega$, $R_A = 152.24\ \Omega$, $R_a = 1.1522\ k\Omega$, $R_b = 7.5685\ k\Omega$.

b. $C_1 = 84.62\ nF$, $C_2 = 84.62\ nF$, $R_2 = 1\ k\Omega$, $R_B = 1\ k\Omega$, $R_A = 1.2346\ k\Omega$, $R_a = 2.2346\ k\Omega$, $R_b = 1.81\ k\Omega$.

16.5.9 UNITY GAIN METHOD

Set the gain at 1; that is, $K = 1$. Then, from Equation (16.51), we have $R_A = 0$, and R_B does not exist (open circuit). The circuit reduces to the one shown in Figure 16.23. Let

$$R = R_1 = R_2$$

FIGURE 16.23

A unity gain Sallen-Key circuit for the second-order LPF.

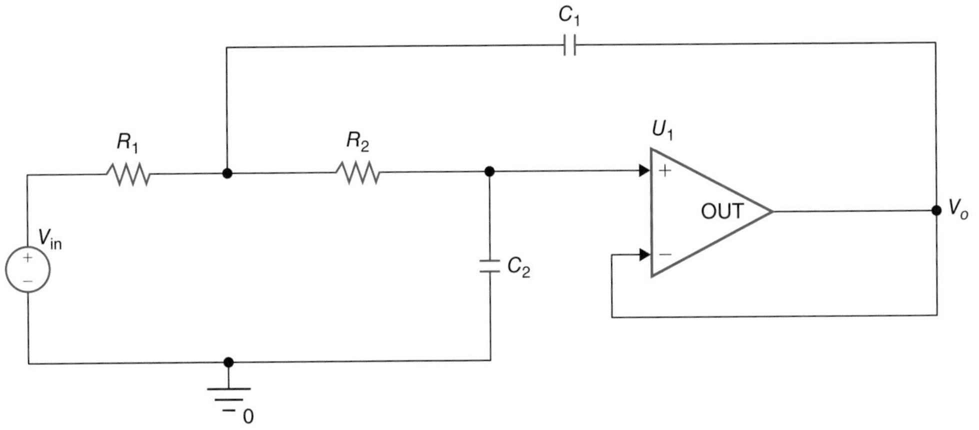

Then, Equation (16.46) becomes

$$H(s) = \frac{\omega_0^2}{s^2 + \frac{\omega_0}{Q}s + \omega_0^2} = \frac{\frac{1}{R^2C_1C_2}}{s^2 + \frac{2}{RC_1}s + \frac{1}{R^2C_1C_2}} \quad \textbf{(16.60)}$$

Comparing the constant terms in the denominator, we obtain

$$\omega_0^2 = \frac{1}{R^2C_1C_2}$$

Thus, the corner frequency is given by

$$\omega_0 = \frac{1}{R\sqrt{C_1C_2}} \quad \textbf{(16.61)}$$

Comparing the coefficients of s terms in the denominator, we obtain

$$\frac{\omega_0}{Q} = \frac{2}{RC_1}$$

Thus, the Q value is given by

$$Q = \frac{RC_1}{2}\omega_0 = \frac{RC_1}{2} \times \frac{1}{R\sqrt{C_1C_2}} = \frac{1}{2}\sqrt{\frac{C_1}{C_2}} \qquad (16.62)$$

For normalized filter, set $\omega_0 = 1$ rad/s. We choose $R = R_1 = R_2 = 1\ \Omega$. Then, Equation (16.60) becomes

$$H(s) = \frac{1}{s^2 + \frac{1}{Q}s + 1} = \frac{\frac{1}{C_1C_2}}{s^2 + \frac{2}{C_1}s + \frac{1}{C_1C_2}} \qquad (16.63)$$

Comparison of the coefficients of s term in the denominator yields

$$\frac{1}{Q} = \frac{2}{C_1}$$

Thus, we have

$$C_1 = 2Q \qquad (16.64)$$

Substituting this into $\frac{1}{C_1C_2} = 1$, we get

$$C_2 = \frac{1}{2Q} \qquad (16.65)$$

In summary, the normalized component values are given by

$$R_1 = 1\ \Omega, \quad R_2 = 1\ \Omega, \quad C_1 = 2Q\ \text{F}, \quad C_2 = \frac{1}{2Q}\ \text{F}$$

EXAMPLE 16.8

Design a LPF with transfer function

$$H(s) = \frac{1.786214 \times 10^8}{s^2 + 18{,}900.8672s + 1.786214 \times 10^8}$$

using the Sallen-Key circuit in the unity gain method. Assume $k_m = 1000$.

Let $a_1 = 18{,}900.8672$ and $a_0 = 1.786214 \times 10^8$. Then, the transfer function can be written as

$$H(s) = \frac{1.786214 \times 10^8}{s^2 + 18{,}900.8672s + 1.786214 \times 10^8} = \frac{a_0}{s^2 + a_1s + a_0} = \frac{\omega_0^2}{s^2 + \frac{\omega_0}{Q}s + \omega_0^2}$$

The corner frequency ω_0 is given by

$$\omega_0 = \sqrt{a_0} = \sqrt{1.786214 \times 10^8} = 13{,}364.9313 \text{ rad/s}$$

continued

Example 16.8 continued

The Q value is given by

$$Q = \frac{\omega_0}{a_1} = \frac{13{,}364.9313}{18{,}900.8672} = 0.7071$$

The normalized component values are given by

$$R_1 = 1\ \Omega, \quad R_2 = 1\ \Omega, \quad C_1 = 2Q\ \text{F} = 1.4142\ \text{F}, \quad C_2 = 1/(2Q)\ \text{F} = 0.7071\ \text{F}$$

Application of the magnitude scaling ($k_m = 1000$) and the frequency scaling ($k_f = \omega_0 = 13{,}364.9313$) yields

$$R_1 = 1\ k\Omega, \quad R_2 = 1\ k\Omega$$

$$C_1 = \frac{1.4142\ \text{F}}{k_m \times k_f} = \frac{1.4142}{1000 \times 13{,}364.9313}\text{F} = 0.105815\ \mu F$$

$$C_2 = \frac{0.7071\ \text{F}}{k_m \times k_f} = \frac{0.7071}{1000 \times 13{,}364.9313}\text{F} = 52.9076\ nF$$

Exercise 16.8

Design a LPF with $\omega_0 = 2\pi 1000 = 6283.1853$ rad/s and $Q = 2$ using the Sallen-Key unity gain method. Assume that $k_m = 1000$.

$$R_1 = R_2 = 1\ \Omega$$
$$C_1 = 2Q = 4\ \text{F}$$
$$C_2 = \frac{1}{2Q} = \frac{1}{4}\text{F}$$

Choose $k_f = 2\pi 1000 = 6283.1853$ and $k_m = 1000$.

$$C_1 = \frac{4}{k_m k_f} = 6.366 \times 10^{-7} = 0.6366\ \mu F$$

$$C_2 = \frac{1}{4k_m k_f} = 3.979 \times 10^{-8} = 0.03979\ \mu F$$

$$R_1 = R_2 = 1\,k\Omega$$

16.6 Second-Order HPF Design

The transfer function of the second-order HPF is given by

$$H(s) = \frac{s^2}{s^2 + \dfrac{\omega_0}{Q}s + \omega_0^2} = \frac{s^2}{s^2 + 2\zeta\omega_0 s + \omega_0^2} \tag{16.66}$$

Here, ω_0 is called the *corner frequency*, Q is called the *quality factor*, and ζ is called the *damping coefficient*. Notice that

$$\zeta = \frac{1}{2Q}, \quad Q = \frac{1}{2\zeta}$$

FIGURE 16.24

The pole-zero diagram of the second-order HPF.

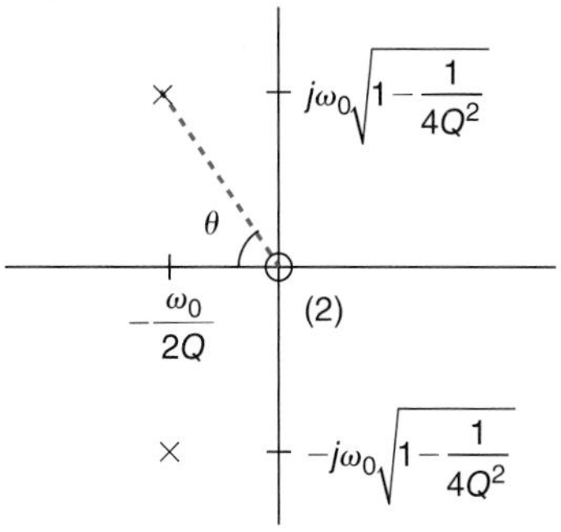

Since the denominator polynomial of $H(s)$ is identical to the one for the second-order LPF, the locations of the poles are the same as those of the second-order LPF given by Equations (16.19) and (16.20). There are two zeros at $s = 0$. The pole-zero diagram of the second-order HPF is shown in Figure 16.24. The distance from the origin to the two complex conjugate poles is ω_0.

16.6.1 FREQUENCY RESPONSE

The frequency response of the second-order HPF is obtained by substituting $s = j\omega$ in $H(s)$.

$$H(\omega) = H(s)|_{s=j\omega} = \frac{-\omega^2}{-\omega^2 + \frac{\omega_0}{Q}j\omega + \omega_0^2} \tag{16.67}$$

16.6.2 MAGNITUDE RESPONSE

The magnitude response of the second-order HPF is given by the absolute value of $H(\omega)$. Thus, we have

$$|H(\omega)| = \frac{\omega^2}{\sqrt{(\omega_0^2 - \omega^2)^2 + \left(\frac{\omega_0}{Q}\omega\right)^2}}$$

At $\omega = \omega_0$, $|H(j\omega_0)| = Q$.

The frequency ω_{peak} where maximum magnitude occurs is given by

$$\omega_{\text{peak}} = \sqrt{\frac{1}{1 - \frac{1}{2Q^2}}}\,\omega_0 = \sqrt{\frac{2\omega_0^2 Q^2}{2Q^2 - 1}} = \frac{\sqrt{2}\omega_0 Q}{\sqrt{2Q^2 - 1}} \tag{16.68}$$

For the proof, refer to Problem 16.19. As Q increases, ω_{peak} approaches ω_0. The maximum magnitude at peak frequency is

$$|H(\omega_{\text{peak}})| = \frac{Q}{\sqrt{1 - \frac{1}{4Q^2}}} \tag{16.69}$$

The frequency for $|H(j\omega)| = 1$ (0 dB), called the *zero-crossing frequency,* is found to be

$$\omega_{zc} = \frac{\omega_0}{\sqrt{2 - \frac{1}{Q^2}}} \tag{16.70}$$

For the proof, refer to Problem 16.20.

The 3-dB cutoff frequency ω_c is given by

$$\omega_c = \omega_0 \frac{\sqrt{-2Q^2 + 1 + \sqrt{8Q^4 - 4Q^2 + 1}}}{\sqrt{2}\,Q} \tag{16.71}$$

For the proof, refer to Problem 16.21. The magnitude response of the second-order HPF for $Q = 2$ is shown in Figure 16.25 for $\omega_0 = 1$ rad/s. The magnitudes at frequencies ω_{peak}, ω_0, ω_{zc}, and ω_c are shown in Figure 16.25.

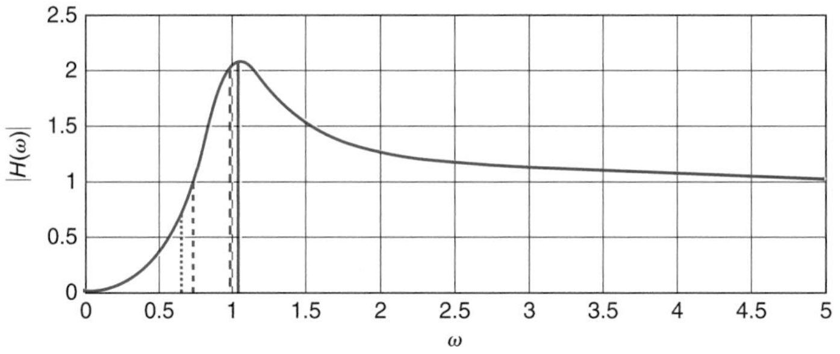

FIGURE 16.25

The magnitude response for $Q = 2$.

16.6.3 PHASE RESPONSE

The phase response of the second-order HPF is obtained from Equation 16.67:

$$\angle H(\omega) = 180° - \tan^{-1}\left(\frac{\frac{\omega_0}{Q}\omega}{\omega_0^2 - \omega^2}\right) \tag{16.72}$$

At $\omega = 0^+$, the phase is 180° due to the minus sign in the numerator. Each finite zero contributes 90° to phase at $\omega = \infty$. Each finite pole contributes $-90°$ phase at $\omega = \infty$. Since there are two zeros and two poles, the phase at $\omega = \infty$ is zero. The phase response of the second-order HPF for $Q = 2$ is shown in Figure 16.26 for $\omega_0 = 1$ rad/s.

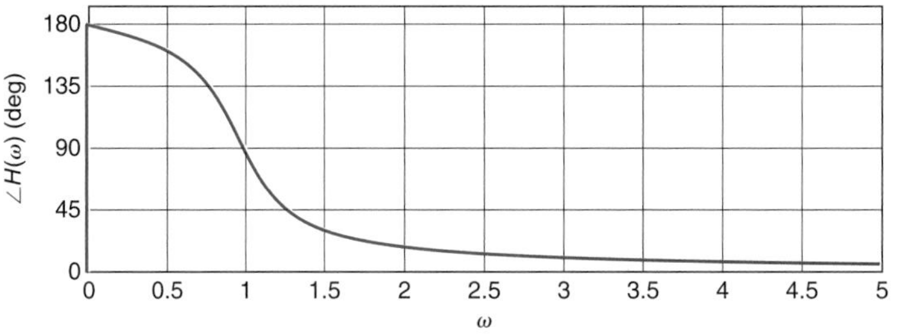

FIGURE 16.26

Phase response for $Q = 2$.

FIGURE 16.27

Implementation of the second-order HPF using the series RLC circuit.

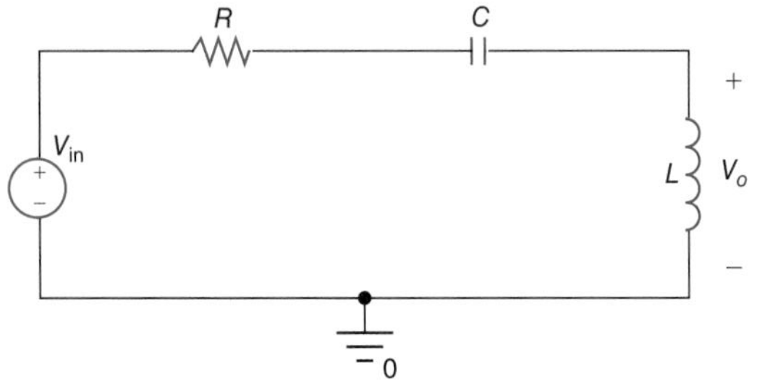

16.6.4 SERIES *RLC* HPF

The second-order HPF can be designed using a series RLC circuit shown in Figure 16.27.

From the voltage divider rule, the transfer function of the circuit shown in Figure 16.27 is

$$H(s) = \frac{V_o(s)}{V_{in}(s)} = \frac{sL}{sL + R + \frac{1}{sC}} = \frac{LCs^2}{LCs^2 + RCs + 1} = \frac{s^2}{s^2 + \frac{R}{L}s + \frac{1}{LC}} \tag{16.73}$$

Thus, the corner frequency is given by

$$\omega_0 = \frac{1}{\sqrt{LC}} \tag{16.74}$$

and the quality factor is given by

$$Q = \frac{1}{R}\sqrt{\frac{L}{C}} \tag{16.75}$$

For a normalized HPF, we set $\omega_0 = 1$ rad/s. Then, the transfer function becomes

$$H(s) = \frac{V_o(s)}{V_{in}(s)} = \frac{s^2}{s^2 + \frac{\omega_0}{Q}s + \omega_0^2} = \frac{s^2}{s^2 + \frac{1}{Q}s + 1} \tag{16.76}$$

To get $\omega_0 = 1$ rad/s, let us choose $L = 1$ H, and $C = 1$ F. Then, $R = 1/Q\ \Omega$. Apply the magnitude scaling k_m and the frequency scaling $k_f = \omega_0$ to obtain the scaled values for R, L, and C. After scaling, the transfer function becomes the one given by Equation (16.66).

EXAMPLE 16.9

Design an HPF with the transfer function

$$H(s) = \frac{s^2}{s^2 + 4 \times 10^4 s + 4 \times 10^8}$$

using a series RLC circuit. Assume that $k_m = 1000$.

Let $a_1 = 4 \times 10^4$ and $a_0 = 4 \times 10^8$. Then, we have

$$\omega_0 = \sqrt{a_0} = \sqrt{4 \times 10^8} = 2 \times 10^4 \text{ rad/s}$$

Since

$a_1 = \omega_0/Q = 4 \times 10^4$, we get

$$Q = \frac{\omega_0}{a_1} = \frac{2 \times 10^4}{4 \times 10^4} = \frac{1}{2}$$

Component values for a normalized filter:

$$R = 1/Q = 2\ \Omega$$
$$L = 1\text{ H}$$
$$C = 1\text{ F}$$

Let $k_m = 1000$ and $k_f = \omega_0 = 2 \times 10^4$. Then, the component values after scaling are given by

$$R = k_m \times 2 = 2\ k\Omega$$
$$L = 1 \times k_m/k_f = 50\text{ mH}$$
$$C = 1/(k_m k_f) = 0.05\ \mu F$$

Exercise 16.9

Design an HPF with the transfer function

$$H(s) = \frac{s^2}{s^2 + 8 \times 10^3 s + 9 \times 10^8}$$

using a series RLC circuit. Assume that $k_m = 1000$.

Answer:
$R = 266.67\ \Omega$, $L = 33.33$ mH, $C = 0.03333\ \mu F$.

16.6.5 PARALLEL *RLC* HPF

The second-order HPF can be designed using a parallel RLC circuit shown in Figure 16.28.

FIGURE 16.28

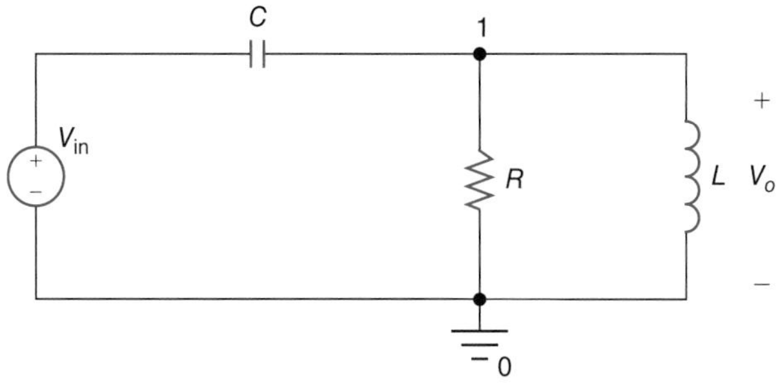

Implementation of the second-order HPF using the parallel RLC circuit.

Summing the currents leaving node 1, we obtain

$$\frac{V_o(s) - V_{in}(s)}{\frac{1}{sC}} + \frac{V_o(s)}{R} + \frac{V_o(s)}{sL} = 0$$

Solving for $V_o(s)/V_{in}(s)$, we get

$$H(s) = \frac{V_o(s)}{V_{in}(s)} = \frac{sC}{sC + \frac{1}{R} + \frac{1}{sL}} = \frac{LCs^2}{LCs^2 + \frac{L}{R}s + 1} = \frac{s^2}{s^2 + \frac{1}{RC}s + \frac{1}{LC}} \tag{16.77}$$

Thus, the corner frequency is given by

$$\omega_0 = \frac{1}{\sqrt{LC}} \tag{16.78}$$

and the quality factor is given by

$$Q = R\sqrt{\frac{C}{L}} \tag{16.79}$$

The normalized HPF is obtained by setting $\omega_0 = 1$ rad/s. To get $\omega_0 = 1$ rad/s, let us choose $L = 1$ H, and $C = 1$ F. Then, $R = Q$. Apply the magnitude scaling k_m and the frequency scaling $k_f = \omega_0$ to obtain the values for R, L, and C. After scaling, the transfer function becomes the one given by Equation (16.66).

EXAMPLE 16.10

Design an HPF with transfer function

$$H(s) = \frac{s^2}{s^2 + 2 \times 10^5 s + 16 \times 10^8}$$

using the parallel RLC circuit. Assume that $k_m = 1000$.

Let $a_1 = 2 \times 10^5$ and $a_0 = 16 \times 10^8$. Then, we have

$$\omega_0 = \sqrt{a_0} = \sqrt{16 \times 10^8} = 4 \times 10^4 \text{ rad/s}$$

Since

$a_1 = \omega_0/Q = 2 \times 10^5$, we get

$$Q = \frac{\omega_0}{a_1} = \frac{4 \times 10^4}{2 \times 10^5} = \frac{1}{5}$$

Component values for the normalized filter:

$$R = Q = 0.2\ \Omega$$
$$L = 1\text{ H}$$
$$C = 1\text{ F}$$

Let $k_m = 1000$ and $k_f = \omega_0 = 4 \times 10^4$. Then, the component values after scaling are given by

$$R = k_m \times 0.2 = 200\ \Omega$$
$$L = 1 \times k_m/k_f = 25\text{ mH}$$
$$C = 1/(k_m k_f) = 0.025\ \mu F$$

Exercise 16.10

Design an HPF with the transfer function

$$H(s) = \frac{s^2}{s^2 + 9000s + 8 \times 10^7}$$

using a parallel RLC circuit. Assume that $k_m = 1000$.

Answer:
$R = 993.81\ \Omega$, $L = 111.8$ mH, $C = 0.1118\ \mu F$.

16.6.6 SALLEN-KEY CIRCUIT FOR THE SECOND-ORDER HPF

The second-order Sallen-Key HPF shown in Figure 16.29 is obtained when the resistors R_1 and R_2 are swapped with the capacitors C_1 and C_2 in the second-order Sallen-Key LPF shown in Figure 16.20.

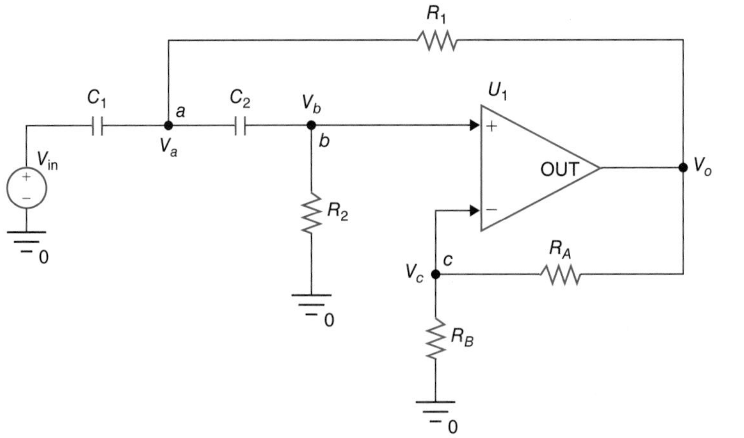

FIGURE 16.29
The second-order Sallen-Key HPF.

Since the current flowing into the negative input of the operational amp is zero, the sum of currents leaving node c is given by

$$\frac{V_c}{R_B} + \frac{V_c - V_o}{R_A} = 0 \tag{16.80}$$

which can be rearranged as

$$\left(\frac{1}{R_B} + \frac{1}{R_A}\right)V_c = \frac{V_o}{R_A}$$

Solving for V_c, we obtain

$$V_c = \frac{\frac{1}{R_A}}{\frac{1}{R_B} + \frac{1}{R_A}}V_o = \frac{R_B}{R_A + R_B}V_o \tag{16.81}$$

Alternatively, the voltage divider rule can be used to get Equation (16.81). Let

$$K = \frac{R_A + R_B}{R_B} = 1 + \frac{R_A}{R_B} \tag{16.82}$$

Then, Equation (16.81) can be rewritten as

$$V_c = \frac{R_B}{R_A + R_B}V_o = \frac{1}{\frac{R_A + R_B}{R_B}}V_o = \frac{V_o}{K} \tag{16.83}$$

Since the current flowing into the positive input of the operational amp is zero, the sum of currents leaving node b is given by

$$\frac{V_b}{R_2} + \frac{V_b - V_a}{\frac{1}{sC_2}} = 0 \tag{16.84}$$

which can be rearranged as

$$\left(sC_2 + \frac{1}{R_2}\right)V_b = sC_2V_a$$

Solving for V_a, we obtain

$$V_a = \frac{sC_2 + \dfrac{1}{R_2}}{sC_2} V_b = \frac{sR_2C_2 + 1}{sR_2C_2} V_b \tag{16.85}$$

Due to a virtual short, voltage V_b is equal to V_c. Thus

$$V_b = V_c = \frac{V_o}{K} \tag{16.86}$$

Substitution of Equation (16.86) into Equation (16.85) yields

$$V_a = \frac{sR_2C_2 + 1}{sR_2C_2} \times \frac{V_o}{K} \tag{16.87}$$

Summing the currents leaving node a, we obtain

$$\frac{V_a - V_{in}}{\dfrac{1}{sC_1}} + \frac{V_a - V_o}{R_1} + \frac{V_a - V_b}{\dfrac{1}{sC_2}} = 0 \tag{16.88}$$

which can be rearranged as

$$\left(sC_1 + \frac{1}{R_1} + sC_2\right)V_a - \frac{1}{R_1}V_o - sC_2V_b = sC_1V_{in} \tag{16.89}$$

Substitution of Equations (16.86) and (16.87) into Equation (16.89) yields

$$\left(sC_1 + \frac{1}{R_1} + sC_2\right)\frac{sR_2C_2 + 1}{sR_2C_2}\frac{V_o}{K} - \frac{1}{R_1}V_o - sC_2\frac{V_o}{K} = sC_1V_{in} \tag{16.90}$$

The transfer function is given by

$$H(s) = \frac{V_o}{V_{in}} = \frac{sC_1}{\left(sC_1 + \dfrac{1}{R_1} + sC_2\right)\dfrac{sR_2C_2 + 1}{sR_2C_2}\dfrac{1}{K} - \dfrac{1}{R_1} - sC_2\dfrac{1}{K}} \tag{16.91}$$

Multiplication by $sR_1R_2C_2K$ yields

$$H(s) = \frac{V_o}{V_{in}} = \frac{s^2R_1R_2C_1C_2K}{(sR_1C_1 + 1 + sR_1C_2)(sR_2C_2 + 1) - sR_2C_2K - s^2R_1R_2C_2^2}$$

$$= \frac{s^2R_1R_2C_1C_2K}{sR_1C_1 + 1 + sR_1C_2 + s^2R_1R_2C_1C_2 + sR_2C_2 + s^2R_1R_2C_2^2 - sR_2C_2K - s^2R_1R_2C_2^2}$$

Dividing by $R_1R_2C_1C_2$, we obtain

$$H(s) = \frac{Ks^2}{s^2 + \left(\dfrac{1-K}{R_1C_1} + \dfrac{1}{R_2C_2} + \dfrac{1}{R_2C_1}\right)s + \dfrac{1}{R_1R_2C_1C_2}} \tag{16.92}$$

The corner frequency is given by

$$\omega_0 = \frac{1}{\sqrt{R_1R_2C_1C_2}} \tag{16.93}$$

The Q value is given by

$$Q = \frac{1}{(1-K)\sqrt{\frac{R_2C_2}{R_1C_1}} + \sqrt{\frac{R_1C_1}{R_2C_2}} + \sqrt{\frac{R_1C_2}{R_2C_1}}} \tag{16.94}$$

To simplify the design, we can use either the equal R and equal C method or the unity gain method.

16.6.7 EQUAL R AND EQUAL C METHOD

In the equal R, equal C method, we choose

$$R = R_1 = R_2$$

and

$$C = C_1 = C_2$$

Then, the transfer function given by Equation (16.92) becomes

$$H(s) = \frac{Ks^2}{s^2 + \frac{3-K}{RC}s + \left(\frac{1}{RC}\right)^2} = \frac{Ks^2}{s^2 + \frac{\omega_0}{Q}s + \omega_0^2} \tag{16.95}$$

The corner frequency is given by

$$\omega_0 = \frac{1}{RC} \tag{16.96}$$

Comparison of the coefficients of s terms in the denominator of Equation (16.95) reveals that

$$Q = \frac{1}{3-K} \tag{16.97}$$

Solving Equation (16.97) for K, we obtain

$$K = 3 - \frac{1}{Q} \tag{16.98}$$

From Equations (16.82) and (16.98), we obtain

$$\frac{R_A}{R_B} = K - 1 = 2 - \frac{1}{Q} \tag{16.99}$$

If s is set to infinity in Equation (16.95), we obtain

$$H(\infty) = K \tag{16.100}$$

Thus, the gain of the filter at $\omega = \infty$ is K. We can use a voltage divider circuit shown in Figure 16.30 to decrease the gain to one. Let $R_C = K - 1$, $R_D = 1$. Then, the voltage across R_D, which is the voltage at the output of the operational amp, is given by

$$V_o = \frac{R_D}{R_C + R_D}V_{in} = \frac{1}{K-1+1}V_{in} = \frac{1}{K}V_{in}$$

FIGURE 16.30

Voltage divider circuit to reduce gain.

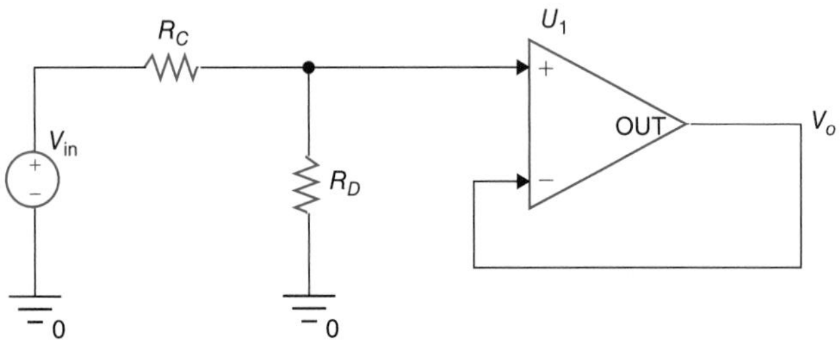

16.6.8 NORMALIZATION

For a normalized filter, we set $\omega_0 = 1$ rad/s. From Equation (16.96), we select

$$R = R_1 = R_2 = 1\ \Omega$$
$$C = C_1 = C_2 = 1\ \text{F}$$

Let $R_B = 1\ \Omega$. Then, from Equation (16.99), we obtain

$$R_A = 2 - \frac{1}{Q}$$

Use the frequency scale ($k_f = \omega_0$) and the magnitude scale to get the desired component values. Choose k_m for the appropriate resistor and capacitor values.

EXAMPLE 16.11

Design an HPF with transfer function

$$H(s) = \frac{s^2}{s^2 + 15{,}515.0995s + 1.2035916 \times 10^8}$$

using a Sallen-Key circuit in the equal R, equal C method. Assume that $k_m = 1000$.

Let $a_1 = 15{,}515.0995$ and $a_0 = 1.2035916 \times 10^8$. Then, the transfer function can be written as

$$H(s) = \frac{s^2}{s^2 + 15{,}515.0995s + 1.2035916 \times 10^8} = \frac{s^2}{s^2 + a_1 s + a_0} = \frac{s^2}{s^2 + \dfrac{\omega_0}{Q}s + \omega_0^2}$$

The corner frequency ω_0 is given by

$$\omega_0 = \sqrt{a_0} = \sqrt{1.2035916 \times 10^8} = 10{,}970.8321 \text{ rad/s}$$

The Q value is given by

$$Q = \frac{\omega_0}{a_1} = \frac{10{,}970.8321}{15{,}515.0995} = 0.7071$$

The normalized component values are given by

$$R = R_1 = R_2 = 1\ \Omega, \quad R_B = 1\ \Omega$$
$$C = C_1 = C_2 = 1\ \text{F}$$
$$R_A = 2 - 1/Q = 0.5858\ \Omega$$

Application of the magnitude scaling ($k_m = 1000$) and the frequency scaling ($k_f = \omega_0 = 10{,}970.8321$) yields

$$C_1 = \frac{1\ \text{F}}{k_m \times k_f} = \frac{1}{1000 \times 10{,}970.8321}\ \text{F} = 91.1508\ nF$$

continued

Example 16.11 continued

$$C_2 = \frac{1\text{ F}}{k_m \times k_f} = \frac{1}{1000 \times 10{,}970.8321}\text{ F} = 91.1508\ nF$$

$$R = R_1 = R_2 = 1\ k\Omega,\quad R_B = 1\ k\Omega,\quad R_A = 585.7865\ \Omega$$

Exercise 16.11

Design an HPF with $\omega_0 = 2\pi 1000$ and $Q = 2$. Assume that $k_m = 1000$.

$$R_1 = R_2 = R_B = 1\ \Omega,\quad C = C_1 = C_2 = 1\text{ F}.$$

$$R_A = 2 - \frac{1}{Q} = 2 - \frac{1}{2} = 1.5\ \Omega$$

Choose $k_f = 2\pi 1000 = 6283.1853$ and $k_m = 1000$.

$$R_1 = R_2 = R_B = 1\ k\Omega$$

$$R_A = 1.5\ k\Omega$$

$$C_1 = C_2 = \frac{1}{k_m k_f} = 1.592 \times 10^{-7} = 0.1592\ \mu F$$

16.6.9 UNITY GAIN METHOD

For the unity gain method, we set $K = 1$. Then, from Equation (16.82), we have

$$\frac{R_A}{R_B} = K - 1 = 0$$

Thus, $R_A = 0$, and R_B does not exist. The output is connected to the inverting input of the operating amp, as shown in Figure 16.31. This is a voltage follower circuit. Let

$$C = C_1 = C_2$$

Then, Equation (16.92) becomes

$$H(s) = \frac{s^2}{s^2 + \dfrac{2}{R_2 C}s + \dfrac{1}{R_1 R_2 C^2}}$$

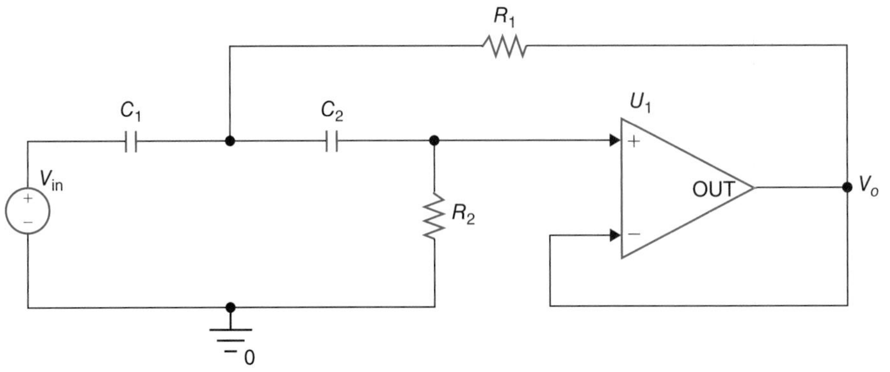

FIGURE 16.31

The Sallen-Key high-pass circuit with the unity gain method.

The corner frequency is given by

$$\omega_0 = \frac{1}{C\sqrt{R_1 R_2}}$$

and the Q value is given by

$$Q = \frac{1}{2}\sqrt{\frac{R_2}{R_1}}$$

Squaring both sides, we obtain

$$4Q^2 = \frac{R_2}{R_1} \tag{16.101}$$

16.6.10 NORMALIZATION

For normalization, we set $\omega_0 = 1$. Let us choose

$$C = C_1 = C_2 = 1\text{ F}$$

Then, the transfer function becomes

$$H(s) = \frac{s^2}{s^2 + \frac{2}{R_2}s + \frac{1}{R_1 R_2}} = \frac{s^2}{s^2 + \frac{1}{Q}s + 1}$$

Comparison of the coefficients of the s-term yields

$$\frac{2}{R_2} = \frac{1}{Q}$$

Thus, we have

$$R_2 = 2Q \tag{16.102}$$

Comparison of the constant term yields

$$R_1 R_2 = 1$$

Thus,

$$R_1 = \frac{1}{R_2} = \frac{1}{2Q} \tag{16.103}$$

EXAMPLE 16.12

Design an HPF with the transfer function

continued

Example 16.12 continued

$$H(s) = \frac{s^2}{s^2 + 11{,}487.9013s + 6.5986 \times 10^7}$$

using a Sallen-Key circuit with the unity gain method. Assume that $k_m = 1000$.

Let $a_1 = 11{,}487.9013$ and $a_0 = 6.5986 \times 10^7$. Then, the transfer function can be written as

$$H(s) = \frac{s^2}{s^2 + 11{,}487.9013s + 6.5986 \times 10^7} = \frac{s^2}{s^2 + a_1 s + a_0} = \frac{s^2}{s^2 + \frac{\omega_0}{Q}s + \omega_0^2}$$

The corner frequency ω_0 is given by

$$\omega_0 = \sqrt{a_0} = \sqrt{6.5986 \times 10^7} = 8123.1729 \text{ rad/s}$$

The Q value is given by

$$Q = \frac{\omega_0}{a_1} = \frac{8123.1729}{11{,}487.9013} = 0.7071$$

The normalized component values are given by

$$R_1 = 1/(2Q) = 0.7071\ \Omega, \quad R_2 = 2Q = 1.4142\ \Omega$$
$$C = C_1 = C_2 = 1\text{ F}$$

Application of the magnitude scaling ($k_m = 1000$) and the frequency scaling ($k_f = \omega_0 = 8123.1729$) yields

$$R_1 = 707.1068\ \Omega, \quad R_2 = 1.4142\ k\Omega$$

$$C_1 = \frac{1\text{ F}}{k_m \times k_f} = \frac{1}{1000 \times 8123.1729}\text{ F} = 0.1231\ \mu F$$

$$C_2 = \frac{1\text{ F}}{k_m \times k_f} = \frac{1}{1000 \times 8123.1729}\text{ F} = 0.1231\ \mu F$$

Exercise 16.12

Design an HPF with $\omega_0 = 2\pi 1000$ and $Q = 2$ using the unity gain method. Assume that $k_m = 1000$.

$$R_1 = \frac{1}{2Q} = \frac{1}{4}\Omega, \quad R_2 = 2Q = 4\ \Omega, \quad C_1 = C_2 = 1\text{ F}$$

Choose $k_f = 2\pi 1000$ and $k_m = 1000$. The scaled component values are given by

$$R_1 = 250\ \Omega, \quad R_2 = 4\ k\Omega$$

$$C_1 = C_2 = \frac{1}{k_m k_f} = 1.592 \times 10^{-7} = 0.1592\ \mu F$$

16.7 Second-Order Bandpass Filter Design

The transfer function of the second-order bandpass filter is given by

$$H(s) = \frac{\frac{\omega_0}{Q}s}{s^2 + \frac{\omega_0}{Q}s + \omega_0^2} = \frac{2\zeta\omega_0 s}{s^2 + 2\zeta\omega_0 s + \omega_0^2} \tag{16.104}$$

Here, ω_0 is called the *corner frequency*, Q is called the *quality factor*, and ζ is called the *damping coefficient*. Notice that

$$\zeta = \frac{1}{2Q}, \quad Q = \frac{1}{2\zeta}$$

Since the denominator polynomial of $H(s)$ is identical to the one for the second-order LPF, the locations of the poles are the same as those of the second-order LPF given by Equations (16.19) and (16.20). There is one zero at $s = 0$ and one zero at $s = \infty$. The pole-zero diagram of the second-order bandpass filter is shown in Figure 16.32. The distance from the origin to the two complex conjugate poles is ω_0.

FIGURE 16.32

The pole-zero diagram of the second-order bandpass filter.

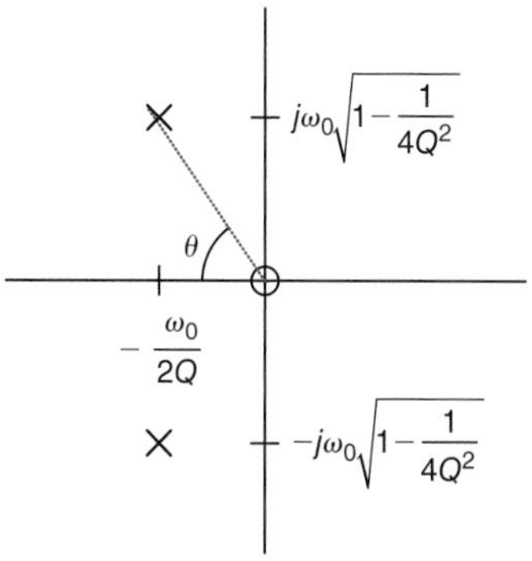

16.7.1 FREQUENCY RESPONSE

The frequency response of the second-order bandpass filter is obtained by substituting $s = j\omega$ in $H(s)$.

$$H(\omega) = H(s)|_{s=j\omega} = \frac{j\frac{\omega_0}{Q}\omega}{-\omega^2 + j\frac{\omega_0}{Q}\omega + \omega_0^2} \tag{16.105}$$

16.7.2 MAGNITUDE RESPONSE

The magnitude response of the second-order bandpass filter is given by the absolute value of $H(\omega)$:

$$|H(\omega)| = \frac{\frac{\omega_0}{Q}\omega}{\sqrt{(\omega_0^2 - \omega^2)^2 + \left(\frac{\omega_0}{Q}\omega\right)^2}}$$

At $\omega = \omega_0$, $|H(\omega_0)| = 1$. The magnitude squared of the second-order bandpass filter is given by

$$|H(\omega)|^2 = \frac{\frac{\omega_0^2}{Q^2}\omega^2}{(\omega_0^2 - \omega^2)^2 + \left(\frac{\omega_0}{Q}\omega\right)^2}$$

The frequency at which the magnitude squared is at its maximum can be found by solving

$$\frac{d}{d\omega}|H(\omega)|^2 = \frac{d}{d\omega}\frac{\frac{\omega_0^2}{Q^2}\omega^2}{(\omega_0^2 - \omega^2)^2 + \left(\frac{\omega_0}{Q}\omega\right)^2} = 0$$

Application of $\frac{d}{dx}\left(\frac{u}{v}\right) = \frac{v\frac{du}{dx} - u\frac{dv}{dx}}{v^2}$ yields

$$\frac{d}{d\omega}|H(\omega)|^2 = \frac{\frac{\omega_0^2}{Q^2}2\omega\left[(\omega_0^2 - \omega^2)^2 + \left(\frac{\omega_0}{Q}\omega\right)^2\right] - \frac{\omega_0^2}{Q^2}\omega^2\left[2(\omega_0^2 - \omega^2)(-2\omega) + \frac{\omega_0^2}{Q^2}2\omega\right]}{\left[(\omega_0^2 - \omega^2)^2 + \left(\frac{\omega_0}{Q}\omega\right)^2\right]^2} = 0$$

Simplification results in

$$2\omega(\omega^4 - \omega_0^4) = 2\omega(\omega^2 - \omega_0^2)(\omega^2 + \omega_0^2) = 2\omega(\omega - \omega_0)(\omega + \omega_0)(\omega - j\omega_0)(\omega + j\omega_0) = 0$$

The solution is given by $\omega = 0, \omega_0, -\omega_0, j\omega_0, -j\omega_0$. We choose $\omega_{peak} = \omega_0$. The magnitude response peaks at $\omega = \omega_0$. The maximum magnitude at $\omega = \omega_0$ is 1.

The 3-dB cutoff frequencies are found by solving

$$\frac{\frac{\omega_0}{Q}\omega}{\sqrt{(\omega_0^2 - \omega^2)^2 + \left(\frac{\omega_0}{Q}\omega\right)^2}} = \frac{1}{\sqrt{2}} \tag{16.106}$$

Squaring Equation (16.106), we obtain

$$2\frac{\omega_0^2}{Q^2}\omega^2 = (\omega_0^2 - \omega^2)^2 + \left(\frac{\omega_0}{Q}\omega\right)^2 \tag{16.107}$$

Equation (16.107) can be written as

$$(\omega_0^2 - \omega^2)^2 = \frac{\omega_0^2}{Q^2}\omega^2 \tag{16.108}$$

Taking the square root on both sides of Equation (16.108), we obtain

$$\omega_0^2 - \omega^2 = \pm\frac{\omega_0}{Q}\omega \tag{16.109}$$

The two cases ($\pm$) from Equation (16.109) can be written as

$$\omega^2 + \frac{\omega_0}{Q}\omega - \omega_0^2 = 0 \tag{16.110}$$

$$\omega^2 - \frac{\omega_0}{Q}\omega - \omega_0^2 = 0 \tag{16.111}$$

The roots of Equation (16.110) are given by

$$\omega = -\frac{\omega_0}{2Q} \pm \frac{\omega_0}{2Q}\sqrt{4Q^2 + 1}$$

Taking the positive frequency, we get

$$\omega_1 = -\frac{\omega_0}{2Q} + \frac{\omega_0}{2Q}\sqrt{4Q^2 + 1} \tag{16.112}$$

The roots of Equation (16.111) are given by

$$\omega = \frac{\omega_0}{2Q} \pm \frac{\omega_0}{2Q}\sqrt{4Q^2+1}$$

Taking the positive frequency, we get

$$\omega_2 = \frac{\omega_0}{2Q} + \frac{\omega_0}{2Q}\sqrt{4Q^2+1} \tag{16.113}$$

The difference between ω_2 and ω_1 is defined as the 3-dB bandwidth. Thus, the 3-dB bandwidth of the second-order bandpass filter is given by

$$\omega_{3\text{dB}} = \omega_2 - \omega_1 = \frac{\omega_0}{Q} \tag{16.114}$$

Notice that the 3-dB bandwidth is the coefficient of s term in the denominator polynomial. If Q is increased, the bandwidth is decreased. If Q is decreased, the bandwidth is increased. The magnitude response of the second-order bandpass filter for different values of Q is shown in Figure 16.33 for $\omega_0 = 1$ rad/s. The bandwidth is the width of the magnitude response between the points that intersect with the horizontal line at 0.7071.

FIGURE 16.33

The magnitude response for different Q values.

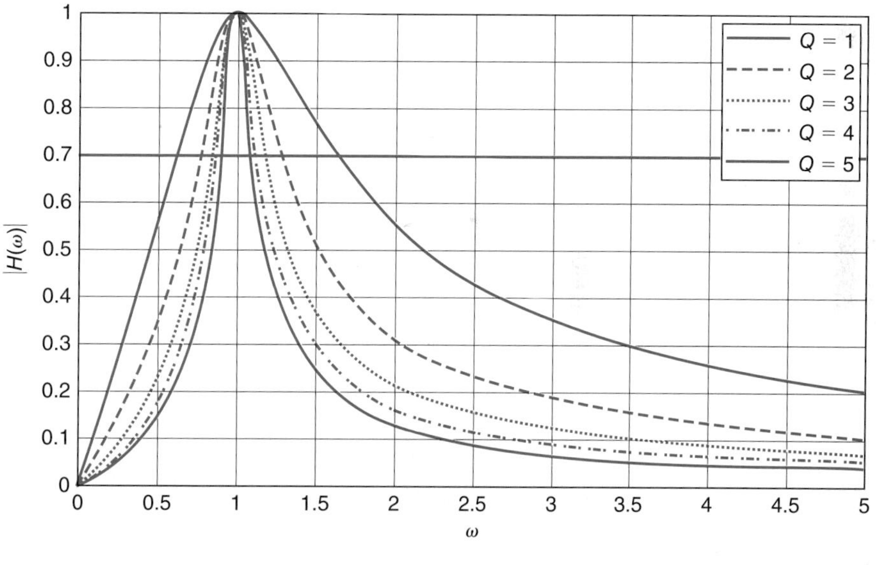

Notice that

$$\omega_1\omega_2 = \left(-\frac{\omega_0}{2Q} + \frac{\omega_0}{2Q}\sqrt{4Q^2+1}\right)\left(\frac{\omega_0}{2Q} + \frac{\omega_0}{2Q}\sqrt{4Q^2+1}\right) = \omega_0^2 \tag{16.115}$$

The corner frequency ω_0 is the geometric mean of ω_1 and ω_2. The bandpass transfer function given by Equation (16.104) can be written as

$$H(s) = \frac{\dfrac{\omega_0}{Q}s}{s^2 + \dfrac{\omega_0}{Q}s + \omega_0^2} = \frac{(\omega_2 - \omega_1)s}{s^2 + (\omega_2 - \omega_1)s + \omega_1\omega_2} \tag{16.116}$$

16.7.3 PHASE RESPONSE

The phase response of the second-order bandpass filter is obtained from Equation (16.105):

$$\angle H(\omega) = 90° - \tan^{-1}\left(\frac{\frac{\omega_0}{Q}\omega}{\omega_0^2 - \omega^2}\right) \tag{16.117}$$

The 90° in the phase response is due to $j = e^{j90°}$ in the numerator. Each finite zero contributes 90° to phase at $\omega = \infty$. Each finite pole contributes $-90°$ phase at $\omega = \infty$. At $\omega = 0^+$, $\angle H(\omega) = 90°$ and at $\omega = \infty$, $\angle H(\omega) = -90°$. The phase response of the second-order bandpass filter for different values of Q is shown in Figure 16.34 for $\omega_0 = 1$ rad/s.

FIGURE 16.34

Phase response for different Q values.

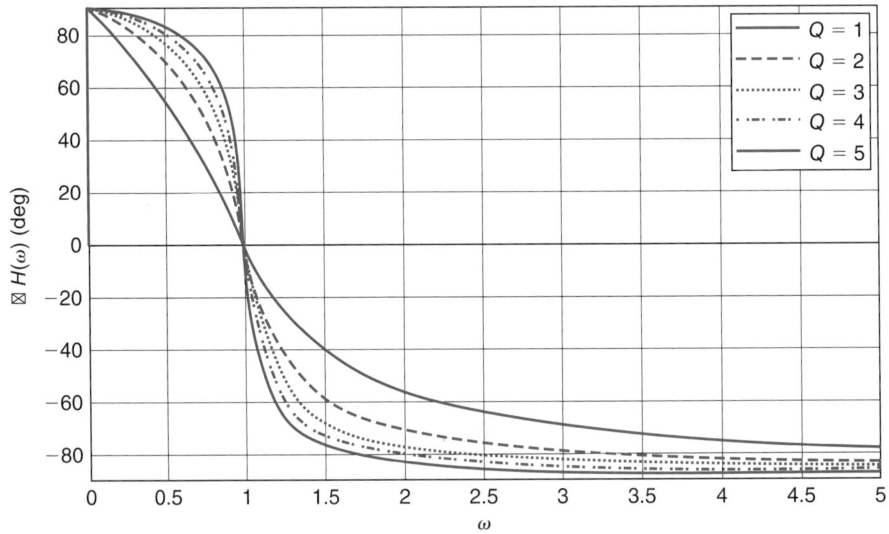

16.7.4 SERIES *RLC* BANDPASS FILTER

The second-order bandpass filter can be implemented using a series RLC circuit shown in Figure 16.35.

Application of the voltage divider rule to the circuit shown in Figure 16.35 yields

$$H(s) = \frac{V_o(s)}{V_{in}(s)} = \frac{R}{sL + R + \frac{1}{sC}} = \frac{RCs}{LCs^2 + RCs + 1} = \frac{\frac{R}{L}s}{s^2 + \frac{R}{L}s + \frac{1}{LC}} \tag{16.118}$$

Comparison of Equations (16.104) and (16.118) reveals that the corner frequency is given by

$$\omega_o = \frac{1}{\sqrt{LC}} \tag{16.119}$$

and the Q value is given by

$$Q = \frac{1}{R}\sqrt{\frac{L}{C}} \tag{16.120}$$

FIGURE 16.35

Second-order bandpass filter.

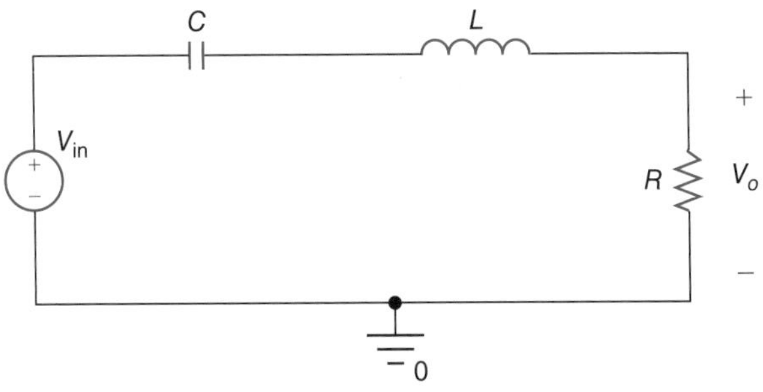

The 3-dB bandwidth is the coefficient of the s term in the denominator. Thus, we have

$$\omega_{3\text{dB}} = R/L \tag{16.121}$$

For the normalized bandpass filter, we set $\omega_0 = 1$. Then, the transfer function becomes

$$H(s) = \frac{V_o(s)}{V_{in}(s)} = \frac{\frac{\omega_0}{Q}s}{s^2 + \frac{\omega_0}{Q}s + \omega_0^2} = \frac{\frac{1}{Q}s}{s^2 + \frac{1}{Q}s + 1}$$

To get $\omega_0 = 1$ rad/s, let us choose $L = 1$ H, and $C = 1$ F. Then, $R = 1/Q\ \Omega$. Apply the magnitude scaling k_m and the frequency scaling $k_f = \omega_0$ to obtain the scaled values for R, L, and C. After scaling, the transfer function becomes the one given in Equation (16.104).

EXAMPLE 16.13

Design a bandpass filter with the transfer function

$$H(s) = \frac{4 \times 10^4 s}{s^2 + 4 \times 10^4 s + 36 \times 10^8}$$

using the series RLC circuit. Assume that $k_m = 1000$.

Let $a_1 = 4 \times 10^4$ and $a_0 = 36 \times 10^8$. Then, we have

$$\omega_0 = \sqrt{a_0} = \sqrt{36 \times 10^8} = 6 \times 10^4 \text{ rad/s}$$

Since

$a_1 = \omega_0/Q = 4 \times 10^4$, we get

$$Q = \frac{\omega_0}{a_1} = \frac{6 \times 10^4}{4 \times 10^4} = 1.5$$

Component values for the normalized filter are:

$$R = 1/Q = 2/3\ \Omega$$
$$L = 1\text{ H}$$
$$C = 1\text{ F}$$

Let $k_m = 1000$ and $k_f = \omega_0 = 6 \times 10^4$. Then, the component values after scaling are given by

$$R = k_m \times 0.6667 = 666.6667\ \Omega$$
$$L = 1 \times k_m/k_f = 16.6667\text{ mH}$$
$$C = 1/(k_m k_f) = 0.01667\ \mu F$$

Exercise 16.13

Design a bandpass filter with the transfer function $H(s)$ given by

$$H(s) = \frac{2 \times 10^4 s}{s^2 + 2 \times 10^4 s + 5 \times 10^8}$$

using the series RLC circuit. Assume that $k_m = 1000$.

Answer:
$R = 894.4272\ \Omega$, $L = 44.7214$ mH, $C = 0.04472\ \mu F$.

16.7.5 PARALLEL *RLC* BANDPASS FILTER

The second-order bandpass filter can be designed using the parallel RLC circuit shown in Figure 16.36.

Summing the currents leaving node 1, we obtain

$$\frac{V_o(s) - V_{in}(s)}{R} + \frac{V_o(s)}{sL} + \frac{V_o(s)}{\frac{1}{sC}} = 0$$

Solving for $V_o(s)/V_{in}(s)$, we get

$$H(s) = \frac{V_o(s)}{V_{in}(s)} = \frac{\frac{1}{R}}{sC + \frac{1}{R} + \frac{1}{sL}} = \frac{\frac{L}{R}s}{LCs^2 + \frac{L}{R}s + 1} = \frac{\frac{1}{RC}s}{s^2 + \frac{1}{RC}s + \frac{1}{LC}} \qquad \textbf{(16.122)}$$

FIGURE 16.36

A parallel RLC bandpass filter.

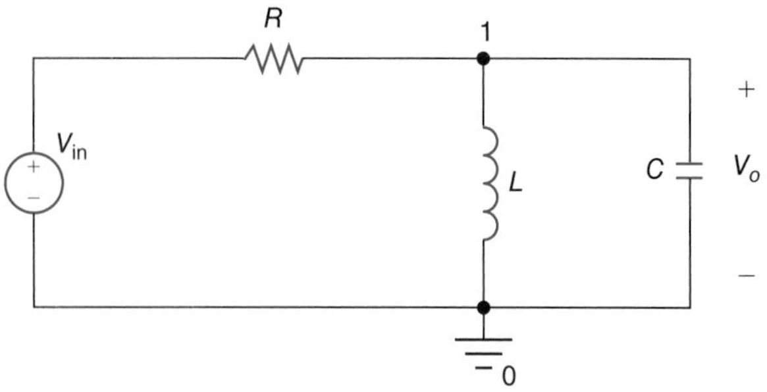

Thus, the corner frequency and the quality factor are expressed using circuit element values as

$$\omega_o = \frac{1}{\sqrt{LC}} \qquad \textbf{(16.123)}$$

and

$$Q = R\sqrt{\frac{C}{L}} \qquad \textbf{(16.124)}$$

The 3-dB bandwidth is the coefficient of the s term in the denominator. Thus, we have

$$\omega_{3dB} = \frac{1}{RC} \qquad \textbf{(16.125)}$$

The normalized bandpass filter is obtained by setting $\omega_0 = 1$. Thus, we have

$$H(s) = \frac{V_o(s)}{V_{in}(s)} = \frac{\frac{\omega_0}{Q}s}{s^2 + \frac{\omega_0}{Q}s + \omega_0^2} = \frac{\frac{1}{Q}s}{s^2 + \frac{1}{Q}s + 1}$$

To get $\omega_0 = 1$ rad/s, let us choose $L = 1$ H, and $C = 1$ F. Then, $R = Q$. Apply the magnitude scaling k_m and the frequency scaling $k_f = \omega_0$ to obtain the values for R, L, and C. After scaling, the transfer function becomes the one given in Equation (16.104).

EXAMPLE 16.14

Design a bandpass filter with the transfer function

$$H(s) = \frac{2 \times 10^4 s}{s^2 + 2 \times 10^4 s + 36 \times 10^8}$$

using a parallel RLC circuit. Assume that $k_m = 1000$.

Let $a_1 = 2 \times 10^4$ and $a_0 = 36 \times 10^8$. Then, we have

$$\omega_0 = \sqrt{a_0} = \sqrt{36 \times 10^8} = 6 \times 10^4 \text{ rad/s}$$

Since

$$a_1 = \omega_0/Q = 2 \times 10^4, \text{ we get}$$

$$Q = \frac{\omega_0}{a_1} = \frac{6 \times 10^4}{2 \times 10^4} = 3$$

Component values for a normalized filter are:

$$R = Q = 3\ \Omega$$
$$L = 1\text{ H}$$
$$C = 1\text{ F}$$

Let $k_m = 1000$ and $k_f = \omega_0 = 6 \times 10^4$. Then, the component values after scaling are given by

$$R = k_m \times 3 = 3\ k\Omega$$
$$L = 1 \times k_m/k_f = 16.6667 \text{ mH}$$
$$C = 1/(k_m k_f) = 0.01667\ \mu F$$

Exercise 16.14

Design a bandpass filter with the transfer function

$$H(s) = \frac{7800s}{s^2 + 7800s + 7 \times 10^8}$$

using a parallel RLC circuit. Assume that $k_m = 1000$.

Answer:

$R = 3391.9889\ \Omega$, $L = 37.7964$ mH, $C = 0.0378\ \mu F$.

16.7.6 SALLEN-KEY CIRCUIT FOR THE SECOND-ORDER BANDPASS FILTER

The Sallen-Key circuit for the second-order bandpass filter is shown in Figure 16.37.

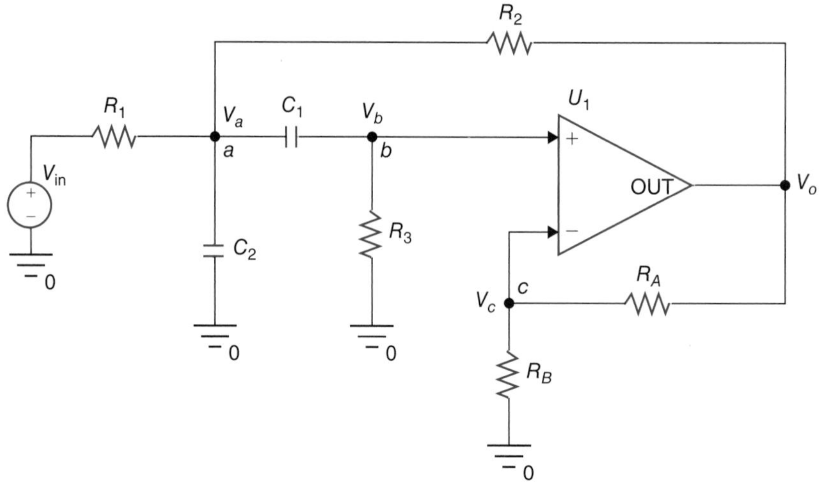

FIGURE 16.37

Sallen-Key second-order bandpass filter.

Since the current flowing into the negative input of the operational amp is zero, the sum of currents leaving node c is given by

$$\frac{V_c}{R_B} + \frac{V_c - V_o}{R_A} = 0 \tag{16.126}$$

which can be rearranged as

$$\left(\frac{1}{R_B} + \frac{1}{R_A}\right)V_c = \frac{V_o}{R_A}$$

Solving for V_c, we obtain

$$V_c = \frac{\dfrac{1}{R_A}}{\dfrac{1}{R_B} + \dfrac{1}{R_A}} V_o = \frac{R_B}{R_A + R_B} V_o \tag{16.127}$$

Alternatively, the voltage divider rule can be used to get Equation (16.127). Let

$$K = \frac{R_A + R_B}{R_B} = 1 + \frac{R_A}{R_B} \tag{16.128}$$

Then, Equation (16.127) can be rewritten as

$$V_c = \frac{R_B}{R_A + R_B} V_o = \frac{1}{\dfrac{R_A + R_B}{R_B}} V_o = \frac{V_o}{K} \tag{16.129}$$

Since the current flowing into the positive input of the operational amp is zero, the sum of currents leaving node b is given by

$$\frac{V_b - V_a}{\dfrac{1}{sC_1}} + \frac{V_b}{R_3} = 0 \tag{16.130}$$

which can be rearranged as

$$\left(sC_1 + \frac{1}{R_3}\right)V_b = sC_1V_a$$

Solving for V_a, we obtain

$$V_a = \left(1 + \frac{1}{sC_1R_3}\right)V_b \tag{16.131}$$

Due to virtual short, the voltage V_b is equal to V_c. Thus,

$$V_b = V_c = \frac{V_o}{K} \tag{16.132}$$

Thus, Equation (16.131) can be rewritten as

$$V_a = \left(1 + \frac{1}{sC_1R_3}\right)\frac{V_o}{K} = \frac{sC_1R_3 + 1}{sC_1R_3} \times \frac{V_o}{K} \tag{16.133}$$

Summing the currents leaving node a, we obtain

$$\frac{V_a - V_{in}}{R_1} + \frac{V_a - V_o}{R_2} + \frac{V_a - V_b}{\frac{1}{sC_1}} + \frac{V_a}{\frac{1}{sC_2}} = 0 \tag{16.134}$$

which can be rearranged as

$$\left(sC_1 + sC_2 + \frac{1}{R_1} + \frac{1}{R_2}\right)V_a - \frac{V_o}{R_2} - sC_1V_b = \frac{V_{in}}{R_1} \tag{16.135}$$

Substituting Equations (16.132) and (16.133) into Equation (16.135), we obtain

$$\left(sC_1 + sC_2 + \frac{1}{R_1} + \frac{1}{R_2}\right)\left(\frac{sC_1R_3 + 1}{sC_1R_3}\right)\frac{V_o}{K} - \frac{V_o}{R_2} - sC_1\frac{V_o}{K} = \frac{V_{in}}{R_1}$$

Taking the ratio of V_o to V_{in}, we obtain

$$H(s) = \frac{V_o}{V_{in}} = \frac{\frac{K}{R_1}}{\left(sC_1 + sC_2 + \frac{1}{R_1} + \frac{1}{R_2}\right)\left(\frac{sC_1R_3 + 1}{sC_1R_3}\right) - sC_1 - \frac{K}{R_2}}$$

$$= \frac{\frac{K}{R_1}sC_1R_3}{\left(sC_1 + sC_2 + \frac{1}{R_1} + \frac{1}{R_2}\right)(sC_1R_3 + 1) - s^2C_1^2R_3 - \frac{K}{R_2}sC_1R_3}$$

$$= \frac{\frac{K}{R_1}sC_1R_3}{s^2C_1C_2R_3 + \frac{sC_1R_3}{R_1} + \frac{sC_1R_3}{R_2} + sC_1 + sC_2 + \frac{1}{R_1} + \frac{1}{R_2} - \frac{K}{R_2}sC_1R_3}$$

$$= \frac{\frac{K}{R_1C_2}s}{s^2 + \left(\frac{1}{R_1C_2} + \frac{1}{R_3C_2} + \frac{1}{R_3C_1} + \frac{1 - K}{R_2C_2}\right)s + \frac{1}{R_1R_3C_1C_2} + \frac{1}{R_2R_3C_1C_2}}$$

Thus, the transfer function $H(s)$ is given by

$$H(s) = \frac{\frac{K}{R_1C_2}s}{s^2 + \left(\frac{1}{R_1C_2} + \frac{1}{R_3C_2} + \frac{1}{R_3C_1} + \frac{1-K}{R_2C_2}\right)s + \frac{1}{R_1R_3C_1C_2} + \frac{1}{R_2R_3C_1C_2}} \quad \textbf{(16.136)}$$

To simplify the design, we can use either the equal R, equal C method or the unity gain method.

16.7.7 EQUAL *R*, EQUAL *C* METHOD

In the equal R, equal C method, we choose

$$R = R_1 = R_2 = R_3$$

and

$$C = C_1 = C_2$$

Then, the transfer function given by Equation (16.136) reduces to

$$H(s) = \frac{\frac{K}{RC}s}{s^2 + \frac{4-K}{RC}s + \frac{2}{(RC)^2}} \quad \textbf{(16.137)}$$

The corner frequency is given by

$$\omega_0 = \frac{\sqrt{2}}{RC} \quad \textbf{(16.138)}$$

Since the coefficient of the s term is

$$\frac{\omega_0}{Q} = \frac{4-K}{RC} = (4-K)\frac{\omega_0}{\sqrt{2}}$$

the Q value is given by

$$Q = \frac{\sqrt{2}}{4-K} \quad \textbf{(16.139)}$$

Solving for K, we obtain

$$K = 4 - \frac{\sqrt{2}}{Q} \quad \textbf{(16.140)}$$

From Equations (16.128) and (16.140), we obtain

$$\frac{R_A}{R_B} = K - 1 = 3 - \frac{\sqrt{2}}{Q} \quad \textbf{(16.141)}$$

If s is set to $j\omega_0 = j\frac{\sqrt{2}}{RC}$ in Equation (16.137), we obtain

$$H(\omega_0) = \frac{K}{4-K} \quad \textbf{(16.142)}$$

Thus, the gain of the filter at $\omega = \omega_0$ is $K/(4 - K)$. This is the maximum value of $|H(\omega)|$.

16.7.8 NORMALIZATION

For the normalized filter, we set $\omega_0 = 1$ rad/s. We select

$$R = R_1 = R_2 = R_3 = 1\ \Omega$$

Then, Equation (16.136) becomes

$$H(s) = \frac{\frac{K}{C}s}{s^2 + \frac{4-K}{C}s + \frac{2}{C^2}} \tag{16.143}$$

For $\omega_0 = 1$ rad/s, we have

$$\omega_0^2 = \frac{2}{C^2} = 1$$

Thus, the C value is given by

$$C = \sqrt{2} \tag{16.144}$$

and ω_0 is given by

$$\omega_0 = \frac{\sqrt{2}}{C} \tag{16.145}$$

Let $R_B = 1$. Then, from Equation (16.141), we obtain

$$R_A = K - 1 = 3 - \frac{\sqrt{2}}{Q} \tag{16.146}$$

Use the frequency scale ($k_f = \omega_0$) and the magnitude scale to get the desired values. Choose k_m for the appropriate resistor and capacitor values. The gain at $s = \pm j\omega_0$ is given by $K/(4 - K)$. To make the gain equal to 1 at the resonant frequency, we replace the source V_{in} and R_1 by a source V_{in} and a voltage divider consisting of R_a and R_b, as shown in Figure 16.38.

FIGURE 16.38

A Sallen-Key bandpass filter circuit with gain adjustment.

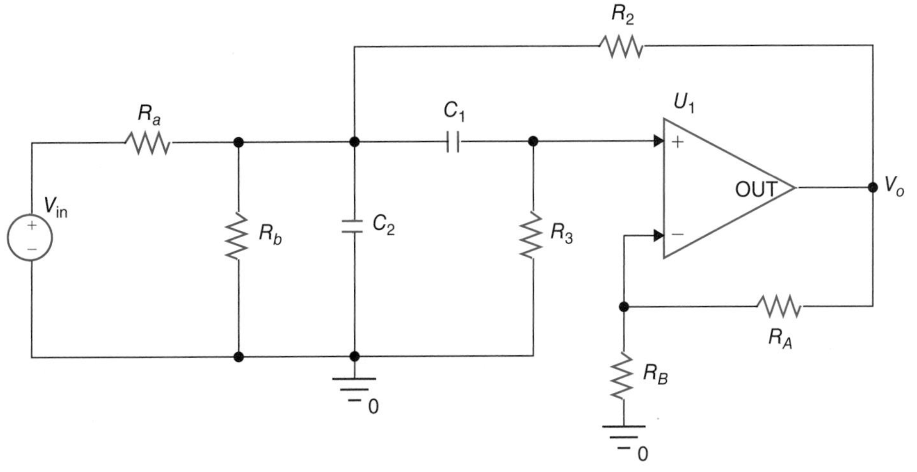

The values of R_a and R_b are found from

$$\frac{R_b}{R_a + R_b} = \frac{4 - K}{K} \tag{16.147}$$

and

$$\frac{R_a R_b}{R_a + R_b} = 1 \tag{16.148}$$

Dividing Equation (16.147) from Equation (16.148), we obtain

$$R_a = \frac{K}{4 - K} \tag{16.149}$$

Substitution of Equation (16.149) into Equation (16.147) yields

$$R_b = \frac{K}{2K - 4} \tag{16.150}$$

The gain adjustment by R_a, R_b makes the maximum gain of each second-order section to be 1.

EXAMPLE 16.15

Design a bandpass filter with the transfer function

$$H(s) = \frac{9165.5892s}{s^2 + 9165.5892s + 2.321331 \times 10^8}$$

using the Sallen-Key circuit shown in Figure 16.38. Assume that $k_m = 1000$.

Let $a_1 = 9165.5892$ and $a_0 = 2.321331 \times 10^8$. Then, we have

$$\omega_0 = \sqrt{a_0} = \sqrt{2.321331 \times 10^8} = 15{,}235.9147 \text{ rad/s}$$

Since

$a_1 = \omega_0/Q = 9165.5892$, we get

$$Q = \frac{\omega_0}{a_1} = \frac{15{,}235.9147}{9165.5892} = 1.6622952$$

Component values for the normalized filter are:

$$K = 4 - \frac{\sqrt{2}}{Q} = 3.14924$$

$$R = R_2 = R_3 = 1\ \Omega$$

$$C = C_1 = C_2 = \sqrt{2} = 1.4142 \text{ F}$$

$$R_B = 1\ \Omega, \quad R_A = K - 1 = 3 - \frac{\sqrt{2}}{Q} = 2.14924\ \Omega, \quad R_a = \frac{K}{4 - K} = 3.70168\ \Omega$$

$$R_b = \frac{K}{2K - 4} = 1.37014\ \Omega$$

continued

Example 16.15 continued

Let $k_m = 1000$ and $k_f = \omega_0 = 15{,}235.9147$. Then, the component values after scaling are given by

$$R = R_2 = R_3 = 1\ k\Omega$$
$$C = C_1 = C_2 = 92.82105\ nF$$
$$R_B = 1\ k\Omega, \quad R_A = 2.14924\ k\Omega, \quad R_a = 3.70168\ k\Omega, \quad R_b = 1.37014\ k\Omega$$

Exercise 16.15

Design a bandpass filter with the transfer function

$$H(s) = \frac{8192.0687s}{s^2 + 8192.0687s + 2.321331 \times 10^8}$$

using the Sallen-Key circuit shown in Figure 16.38. Assume that $k_m = 1000$.

Answer:

$R = R_1 = R_2 = R_3 = 1\ k\Omega$

$C = C_1 = C_2 = 92.82105\ nF$

$R_B = 1\ k\Omega, \quad R_A = 2.2396\ k\Omega, \quad R_a = 4.2604\ k\Omega, \quad R_b = 1.3067\ k\Omega.$

16.7.9 DELIYANNIS-FRIEND CIRCUIT

The Deliyannis-Friend circuit for the second-order bandpass filter is shown in Figure 16.39.

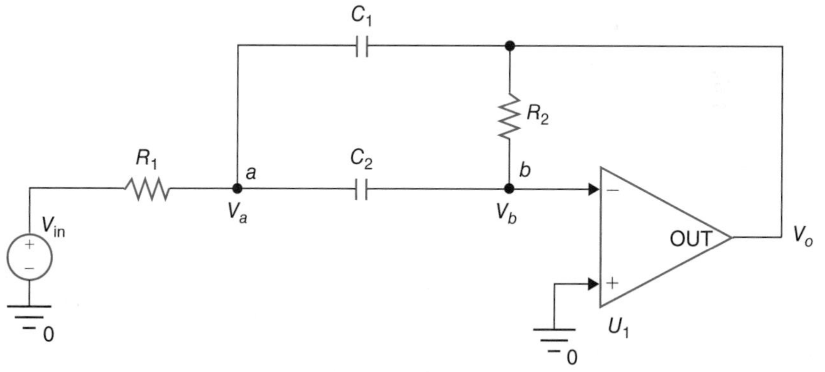

FIGURE 16.39

The Deliyannis-Friend circuit for the second-order bandpass filter.

Notice that $V_b = 0$ due to a virtual short.
Summing the currents leaving node b, we get

$$\frac{0 - V_a}{\frac{1}{sC_2}} + \frac{0 - V_o}{R_2} = 0 \tag{16.151}$$

Solving Equation (16.151) for V_a, we have

$$V_a = -\frac{V_o}{R_2 C_2 s} \tag{16.152}$$

Summing the currents leaving node a, we obtain

$$\frac{V_a - V_{in}}{R_1} + \frac{V_a - V_o}{\frac{1}{sC_1}} + \frac{V_a - 0}{\frac{1}{sC_2}} = 0 \tag{16.153}$$

which can be rearranged as

$$\left(\frac{1}{R_1} + sC_1 + sC_2\right)V_a - sC_1V_o = \frac{V_{in}}{R_1} \tag{16.154}$$

Substitution of Equation (16.152) into Equation (16.154) yields

$$H(s) = \frac{V_o}{V_{in}} = -\frac{\frac{1}{R_1C_1}s}{s^2 + \left(\frac{1}{R_2C_2} + \frac{1}{R_2C_1}\right)s + \frac{1}{R_1R_2C_1C_2}} \tag{16.155}$$

To simplify the design, we apply the equal C method. Let

$$C = C_1 = C_2$$

Then, Equation (16.155) becomes

$$H(s) = \frac{V_o}{V_{in}} = -\frac{\frac{1}{R_1C}s}{s^2 + \frac{2}{R_2C}s + \frac{1}{R_1R_2C^2}} \tag{16.156}$$

16.7.10 NORMALIZATION

For the normalized filter, we set $\omega_0 = 1$ rad/s. Let $R_1 = 1\ \Omega$. Then, Equation (16.156) becomes

$$H(S) = \frac{V_o}{V_{in}} = -\frac{\frac{1}{C}S}{S^2 + \frac{2}{R_2C}S + \frac{1}{R_2C^2}} = -\frac{\frac{1}{C}S}{S^2 + \frac{1}{Q}S + 1} \tag{16.157}$$

Comparing the coefficients of the S term and constant term in the denominator, we obtain

$$\frac{2}{R_2C} = \frac{1}{Q} \tag{16.158}$$

and

$$\frac{1}{R_2C^2} = 1 \tag{16.159}$$

Dividing Equation (16.159) from Equation (16.158), we get

$$2C = \frac{1}{Q}$$

Thus, we have

$$C = \frac{1}{2Q} \tag{16.160}$$

Substitution of Equation (16.160) into Equation (16.159) yields

$$R_2 = 4Q^2 \tag{16.161}$$

The maximum gain of the normalized bandpass filter at $S = j1$ is given by

$$|H(j1)| = \left| -\frac{\frac{1}{C}j1}{-1 + \frac{1}{Q}j1 + 1} \right| = \left| -\frac{Q}{C} \right| = 2Q^2$$

To make the gain of the bandpass filter at ω_0 to be 1, replace the V_{in} and R_1 combination by V_{in}, R_a, R_b, as shown in Figure 16.40. We want the voltage across R_b to be $\frac{V_{in}}{2Q^2}$. Also, the parallel combination of R_a, R_b should be $R_1 = 1$. Thus,

$$\frac{R_b}{R_a + R_b} = \frac{1}{2Q^2} \tag{16.162}$$

$$\frac{R_a R_b}{R_a + R_b} = 1 \tag{16.163}$$

From Equation (16.163), we have $R_a + R_b = R_a R_b$. Substituting this into Equation (16.162), we have

$$R_a = 2Q^2 \tag{16.164}$$

From Equation (16.162), we have

$$R_b = \frac{2Q^2}{2Q^2 - 1} \tag{16.165}$$

FIGURE 16.40

Delyiannis-Friend circuit with a voltage divider.

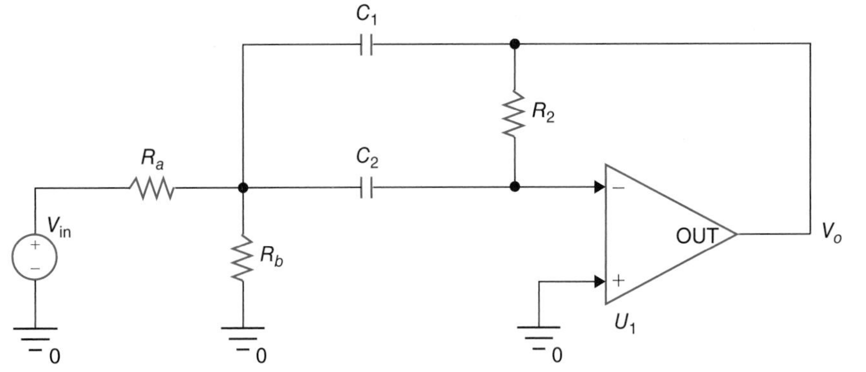

The completed normalized bandpass filter is shown in Figure 16.41 with component values.

Use the frequency scale and the magnitude scale to get the desired values. The frequency scale factor is $k_f = \omega_0$. Choose k_m for the appropriate resistor values. Figure 16.42 shows the Deliyannis-Friend circuit with the scaled component values.

FIGURE 16.41

The Deliyannis-Friend circuit with normalized component values.

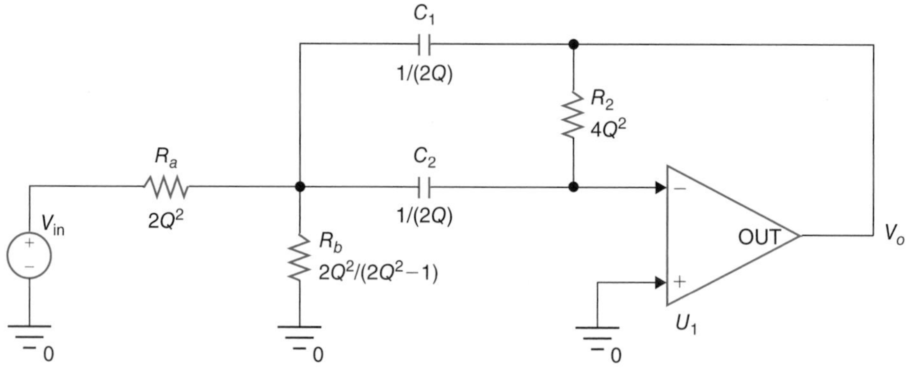

FIGURE 16.42

The Deliyannis-Friend circuit with scaled component values.

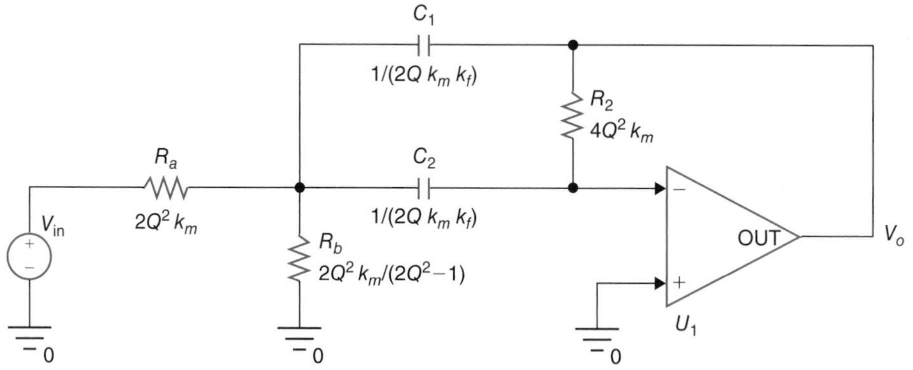

EXAMPLE 16.16

Design a bandpass filter with transfer function

$$H(s) = \frac{14{,}691.5035s}{s^2 + 14{,}691.5035s + 2.1003 \times 10^8}$$

using the Deliyannis-Friend circuit shown in Figure 16.40. Assume that $k_m = 1000$. Let $a_1 = 14{,}691.5035$ and $a_0 = 2.1003 \times 10^8$. Then, we have

$$\omega_0 = \sqrt{a_0} = \sqrt{2.1003 \times 10^8} = 14{,}492.4118 \text{ rad/s}$$

Since

$a_1 = \omega_0/Q = 14{,}691.5035$, we get

$$Q = \frac{\omega_0}{a_1} = \frac{14{,}492.2456}{14{,}691.5035} = 0.98645$$

Component values for the normalized filter are:

$$C = C_1 = C_2 = \frac{1}{2Q} = 0.5069 \text{ F}$$

$$R_2 = 4Q^2 = 3.8923\ \Omega$$

continued

Example 16.16 continued

$$R_a = 2Q^2 = 1.9462\ \Omega$$

$$R_b = \frac{2Q^2}{2Q^2 - 1} = 2.0569\ \Omega$$

Let $k_m = 1000$ and $k_f = \omega_0 = 14{,}492.4118$. Then, the component values after scaling are given by

$$C = C_1 = C_2 = 34.9748\ nF$$
$$R_2 = 3.8923\ k\Omega$$
$$R_a = 1.9462\ k\Omega$$
$$R_b = 2.0569\ k\Omega$$

Exercise 16.16

Design a bandpass filter with the transfer function

$$H(s) = \frac{8192.0687s}{s^2 + 8192.0687s + 2.321331 \times 10^8}$$

using the Deliyannis-Friend circuit shown in Figure 16.40. Assume that $k_m = 1000$.

Answer:
$C = C_1 = C_2 = 17.6452\ nF$, $R_2 = 13.836\ k\Omega$, $R_a = 6.918\ k\Omega$, $R_b = 1.169\ k\Omega$.

16.8 Second-Order Bandstop Filter Design

The transfer function of the second-order bandstop filter is given by

$$H(s) = \frac{s^2 + \omega_z^2}{s^2 + \frac{\omega_0}{Q}s + \omega_0^2} = \frac{s^2 + \omega_z^2}{s^2 + 2\zeta\omega_0 s + \omega_0^2} \tag{16.166}$$

Here, ω_0 is called the *corner frequency*, Q is called the *quality factor*, and ζ is called the *damping coefficient*. Notice that

$$\zeta = \frac{1}{2Q}, \quad Q = \frac{1}{2\zeta}$$

Since the denominator polynomial of $H(s)$ is identical to the one for the second-order LPF, the locations of the poles are the same as those of the second-order LPF given by Equations (16.19) and (16.20). The zeros are located at

$$z_1 = j\omega_z, \quad z_2 = -j\omega_z$$

The null frequency ω_z can be smaller than the corner frequency ω_0 ($\omega_z < \omega_0$), equal to the corner frequency ($\omega_z = \omega_0$), or larger than the corner frequency ($\omega_z > \omega_0$). The pole-zero diagram of the second-order bandstop filter is shown in Figure 16.43. The distance from the origin to the two complex conjugate poles is ω_0.

FIGURE 16.43

The pole-zero diagram of the second-order bandstop filter.

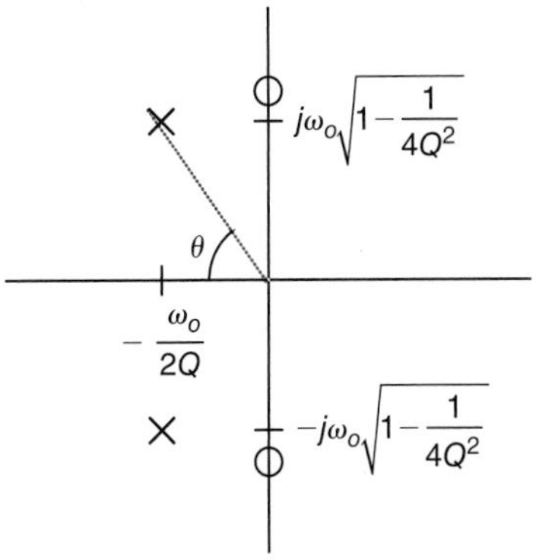

16.8.1 FREQUENCY RESPONSE

The frequency response of the second-order bandstop filter is obtained by substituting $s = j\omega$ in $H(s)$.

$$H(\omega) = H(s)|_{s=j\omega} = \frac{-\omega^2 + \omega_z^2}{-\omega^2 + j\dfrac{\omega_0}{Q}\omega + \omega_0^2} \tag{16.167}$$

16.8.2 MAGNITUDE RESPONSE

The magnitude of the second-order bandstop filter is given by the absolute value of $H(\omega)$:

$$|H(\omega)| = \frac{|\omega_z^2 - \omega^2|}{\sqrt{(\omega_0^2 - \omega^2)^2 + \left(\dfrac{\omega_0}{Q}\omega\right)^2}}$$

At $\omega = \omega_z, |H(\omega_z)| = 0$, and at $\omega = 0, |H(0)| = \dfrac{\omega_z^2}{\omega_0^2}$ and at $\omega = \infty, |H(\infty)| = 1$.

If $\omega_z = \omega_0$, the 3-dB cutoff frequencies are found by solving

$$\frac{|-\omega^2 + \omega_0^2|}{\sqrt{(\omega_0^2 - \omega^2)^2 + \left(\dfrac{\omega_0}{Q}\omega\right)^2}} = \frac{1}{\sqrt{2}} \tag{16.168}$$

Squaring Equation (16.168), we obtain

$$2(\omega_0^2 - \omega^2)^2 = (\omega_0^2 - \omega^2)^2 + \left(\frac{\omega_0}{Q}\omega\right)^2 \tag{16.169}$$

Equation (16.169) can be written as

$$(\omega_0^2 - \omega^2)^2 = \frac{\omega_0^2}{Q^2}\omega^2 \tag{16.170}$$

Taking the square root on both sides of Equation (16.170), we obtain

$$\omega_0^2 - \omega^2 = \pm\frac{\omega_0}{Q}\omega \tag{16.171}$$

The two cases ($\pm$) from Equation (16.171) can be written as

$$\omega^2 + \frac{\omega_0}{Q}\omega - \omega_0^2 = 0 \tag{16.172}$$

$$\omega^2 - \frac{\omega_0}{Q}\omega - \omega_0^2 = 0 \tag{16.173}$$

The roots of Equation (16.172) are given by

$$\omega = -\frac{\omega_0}{2Q} \pm \frac{\omega_0}{2Q}\sqrt{4Q^2 + 1}$$

Taking the positive frequency, we get

$$\omega_1 = -\frac{\omega_0}{2Q} + \frac{\omega_0}{2Q}\sqrt{4Q^2 + 1} \tag{16.174}$$

The roots of Equation (16.173) are given by

$$\omega = \frac{\omega_0}{2Q} \pm \frac{\omega_0}{2Q}\sqrt{4Q^2 + 1}$$

Taking the positive frequency, we get

$$\omega_2 = \frac{\omega_0}{2Q} + \frac{\omega_0}{2Q}\sqrt{4Q^2 + 1} \tag{16.175}$$

The difference between ω_2 and ω_1 is defined as the 3-dB bandwidth. Thus, the 3-dB bandwidth of the second-order bandstop filter is given by

$$\omega_{3\text{dB}} = \omega_2 - \omega_1 = \frac{\omega_0}{Q} \tag{16.176}$$

Notice that the 3-dB bandwidth is the coefficient of the s term in the denominator polynomial.

If Q is increased, the bandwidth is decreased. If Q is decreased, the bandwidth is increased. The magnitude response of the second-order bandstop filter for different values of Q is shown in Figure 16.44 for $\omega_0 = 1$ rad/s and $\omega_z = \omega_0$. The bandwidth is the width of the magnitude response between the points that intersect with the horizontal line at 0.7071. Unlike the bandpass filter, the 3-dB bandwidth of the bandstop filter measures the width of the stopband that is 3-dB lower than the value of the magnitude at $\omega = 0$.

FIGURE 16.44

The magnitude response for different Q values for $\omega_z = \omega_0$.

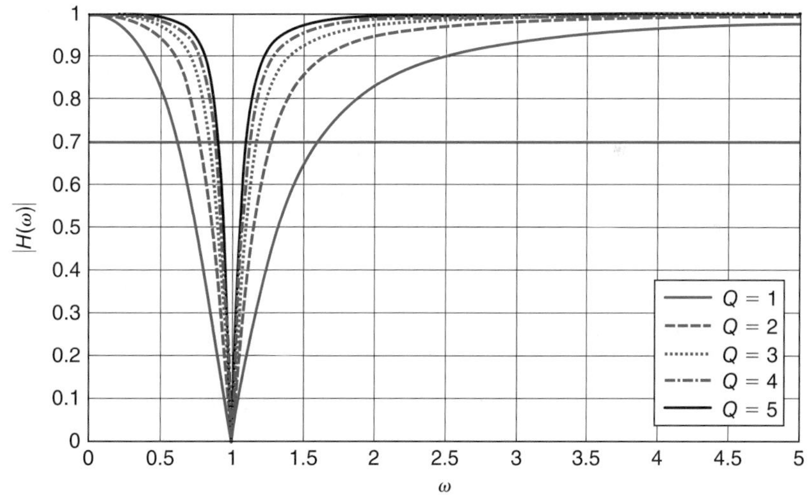

Notice that

$$\omega_1\omega_2 = \left(-\frac{\omega_0}{2Q} + \frac{\omega_0}{2Q}\sqrt{4Q^2 + 1}\right)\left(\frac{\omega_0}{2Q} + \frac{\omega_0}{2Q}\sqrt{4Q^2 + 1}\right) = \omega_0^2 \tag{16.177}$$

The frequency ω_0 where the magnitude is zero (called the *null frequency*) is the geometric mean of ω_1 and ω_2. The bandstop transfer function given by Equation (16.166) for $\omega_z = \omega_0$ can be written as

$$H(s) = \frac{s^2 + \omega_0^2}{s^2 + \dfrac{\omega_0}{Q}s + \omega_0^2} = \frac{s^2 + \omega_1\omega_2}{s^2 + (\omega_2 - \omega_1)s + \omega_1\omega_2} \tag{16.178}$$

16.8.3 PHASE RESPONSE

The phase response of the second-order bandstop filter is obtained from Equation (16.167). The phase of $-\omega^2 + \omega_z^2$ is zero for $\omega \le \omega_z$ and is 180° for $\omega \ge \omega_0$. Thus, the phase response of the second-order bandstop filter is given by

$$\angle H(j\omega) = \begin{cases} -\tan^{-1}\left(\dfrac{\dfrac{\omega_o}{Q}\omega}{\omega_o^2 - \omega^2}\right), & \omega \le \omega_z \\ 180^\circ - \tan^{-1}\left(\dfrac{\dfrac{\omega_o}{Q}\omega}{\omega_o^2 - \omega^2}\right), & \omega > \omega_z \end{cases} \tag{16.179}$$

Each finite zero contributes 90° to phase at $\omega = \infty$. Each finite pole contributes $-90°$ phase at $\omega = \infty$. At $\omega = 0$, $\angle H(\omega) = 0$ and at $\omega = \infty$, $\angle H(j\omega) = 0$. At $\omega = \omega_z$, the phase changes from negative to positive, as shown in Figure 16.45 for $\omega_z = 1$ rad/s and $\omega_z = \omega_0$.

FIGURE 16.45

Phase response for different Q values for $\omega_z = \omega_0$.

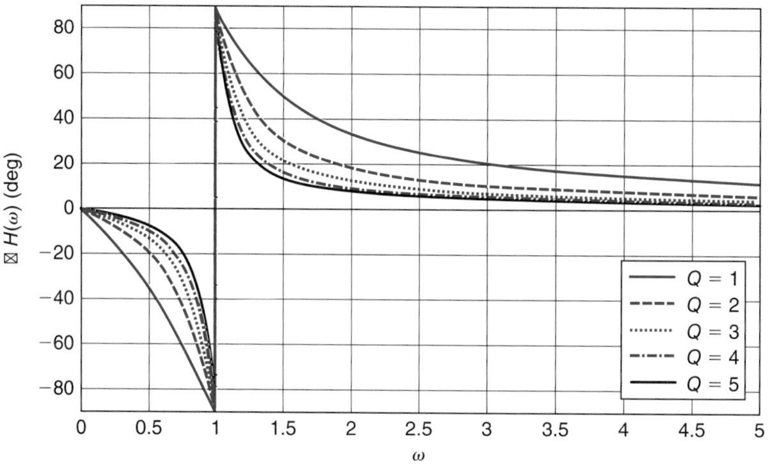

FIGURE 16.46

Second-order bandstop filter.

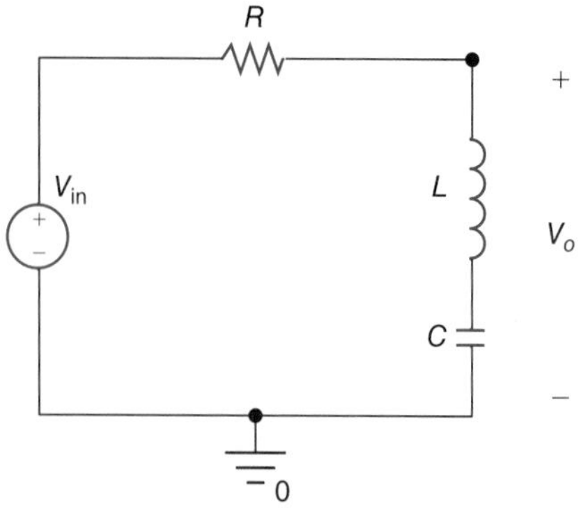

16.8.4 SERIES *RLC* BANDSTOP FILTER

The second-order bandstop filter can be implemented using a series RLC circuit shown in Figure 16.46.

Applying the voltage divider rule, we obtain the transfer function of the circuit shown in Figure 16.46.

$$H(s) = \frac{sL + \dfrac{1}{sC}}{sL + R + \dfrac{1}{sC}} = \frac{s^2 + \dfrac{1}{LC}}{s^2 + \dfrac{R}{L}s + \dfrac{1}{LC}} \tag{16.180}$$

Comparison of Equations (16.166) and (16.180) reveals that the corner frequency and the null frequency are given by

$$\omega_0 = \omega_z = \frac{1}{\sqrt{LC}} \tag{16.181}$$

and the Q value is given by

$$Q = \frac{1}{R}\sqrt{\frac{L}{C}} \tag{16.182}$$

The 3-dB bandwidth is the coefficient of the s term in the denominator. Thus, we have

$$\omega_{3\text{dB}} = R/L \tag{16.183}$$

For the normalized bandstop filter, we set $\omega_0 = 1$ rad/s. Then, the transfer function becomes

$$H(s) = \frac{V_o(s)}{V_{in}(s)} = \frac{s^2 + \omega_0^2}{s^2 + \frac{\omega_0}{Q}s + \omega_0^2} = \frac{s^2 + 1}{s^2 + \frac{1}{Q}s + 1}$$

To get $\omega_0 = 1$ rad/s, let us choose $L = 1$ H, and $C = 1$ F. Then, $R = 1/Q$. Apply the magnitude scaling k_m and the frequency scaling $k_f = \omega_0$ to obtain the scaled values for R, L, and C. After scaling, the transfer function becomes

$$H(s) = \frac{s^2 + \omega_0^2}{s^2 + \frac{\omega_0}{Q}s + \omega_0^2}$$

EXAMPLE 16.17

Design a bandstop filter with the transfer function

$$H(s) = \frac{s^2 + 64 \times 10^8}{s^2 + 2 \times 10^4 s + 64 \times 10^8}$$

using a series RLC circuit. Assume that $k_m = 1000$.

Let $a_1 = 2 \times 10^4$ and $a_0 = 64 \times 10^8$. Then, we have

$$\omega_0 = \sqrt{a_0} = \sqrt{64 \times 10^8} = 8 \times 10^4 \text{ rad/s}$$

Since

$a_1 = \omega_0/Q = 2 \times 10^4$, we get

$$Q = \frac{\omega_0}{a_1} = \frac{8 \times 10^4}{2 \times 10^4} = 4$$

Component values for the normalized filter:

$$R = 1/Q = 0.25\ \Omega$$
$$L = 1\text{ H}$$
$$C = 1\text{ F}$$

Let $k_m = 1000$ and $k_f = \omega_0 = 8 \times 10^4$. Then, the component values after scaling are given by

$$R = k_m \times 0.25 = 250\ \Omega$$
$$L = 1 \times k_m/k_f = 12.5\text{ mH}$$
$$C = 1/(k_m k_f) = 0.0125\ \mu F$$

Exercise 16.17

Design a bandstop filter with the transfer function

$$H(s) = \frac{s^2 + 49 \times 10^8}{s^2 + 4 \times 10^4 s + 49 \times 10^8}$$

using the series RLC circuit. Assume that $k_m = 1000$.

Answer:
$R = 571.4286\ \Omega$, $L = 14.2857$ mH, $C = 0.01429\ \mu F$.

16.8.5 PARALLEL *RLC* BANDSTOP FILTER

The second-order bandstop filter can be designed using the parallel RLC circuit shown in Figure 16.47.

Summing the currents leaving node 1, we obtain

$$\frac{V_o(s) - V_{in}(s)}{sL} + \frac{V_o(s) - V_{in}(s)}{\frac{1}{sC}} + \frac{V_o(s)}{R} = 0$$

Solving for $V_o(s)/V_{in}(s)$, we get

$$H(s) = \frac{V_o(s)}{V_{in}(s)} = \frac{sC + \frac{1}{sL}}{sC + \frac{1}{R} + \frac{1}{sL}} = \frac{LCs^2 + 1}{LCs^2 + \frac{L}{R}s + 1} = \frac{s^2 + \frac{1}{LC}}{s^2 + \frac{1}{RC}s + \frac{1}{LC}} \tag{16.184}$$

Thus, the corner frequency and the null frequency are given by

$$\omega_o = \omega_z = \frac{1}{\sqrt{LC}} \tag{16.185}$$

The quality factor is given by

$$Q = R\sqrt{\frac{C}{L}} \tag{16.186}$$

The 3-dB bandwidth is the coefficient of the s term in the denominator. Thus, we have

$$\omega_{3\text{dB}} = \frac{1}{RC} \tag{16.187}$$

The normalized bandstop filter is obtained by setting $\omega_0 = 1$ rad/s. Thus, we have

$$H(s) = \frac{V_o(s)}{V_{in}(s)} = \frac{s^2 + 1}{s^2 + \frac{1}{Q}s + 1}$$

FIGURE 16.47

Parallel RLC bandstop filter.

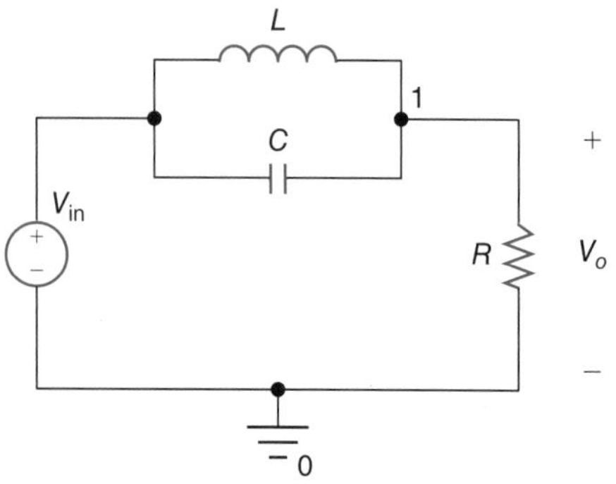

To get $\omega_0 = 1$ rad/s, let us choose $L = 1$ H, and $C = 1$ F. Then, $R = Q$. Apply the magnitude scaling k_m and the frequency scaling $k_f = \omega_0$ to obtain the values for R, L, and C. After scaling, the transfer function becomes

$$H(s) = \frac{s^2 + \omega_0^2}{s^2 + \frac{\omega_0}{Q}s + \omega_0^2}$$

EXAMPLE 16.18

Design a bandstop filter with the transfer function

$$H(s) = \frac{s^2 + 64 \times 10^8}{s^2 + 4 \times 10^4 s + 64 \times 10^8}$$

using the parallel RLC circuit. Assume that $k_m = 1000$.

Let $a_1 = 4 \times 10^4$ and $a_0 = 64 \times 10^8$. Then, we have

$$\omega_0 = \sqrt{a_0} = \sqrt{64 \times 10^8} = 8 \times 10^4 \text{ rad/s}$$

Since

$a_1 = \omega_0/Q = 4 \times 10^4$, we get

$$Q = \frac{\omega_0}{a_1} = \frac{8 \times 10^4}{4 \times 10^4} = 2$$

Component values for the normalized filter:

$$R = Q = 2\ \Omega$$
$$L = 1\text{ H}$$
$$C = 1\text{ F}$$

Let $k_m = 1000$ and $k_f = \omega_0 = 8 \times 10^4$. Then, the component values after scaling are given by

$$R = k_m \times 2 = 2\ k\Omega$$
$$L = 1 \times k_m/k_f = 12.5\text{ mH}$$
$$C = 1/(k_m k_f) = 0.0125\ \mu F$$

Exercise 16.18

Design a bandstop filter with the transfer function

$$H(s) = \frac{s^2 + 81 \times 10^8}{s^2 + 4 \times 10^4 s + 81 \times 10^8}$$

using the parallel RLC circuit. Assume that $k_m = 1000$.

Answer:
$R = 2250\ \Omega$, $L = 11.1111$ mH, $C = 0.01111\ \mu F$.

16.8.6 SALLEN-KEY CIRCUIT FOR THE SECOND-ORDER BANDSTOP FILTER

In general, the transfer function of the second-order bandstop filter is given by

$$H(s) = \frac{V_o(s)}{V_{in}(s)} = \frac{G(s^2 + \omega_z^2)}{s^2 + \dfrac{\omega_0}{Q}s + \omega_0^2} \tag{16.188}$$

where G is the gain of the filter. At $s = j\omega = 0$, $H(0) = \dfrac{G\omega_z^2}{\omega_0^2}$. At $s = j\omega = \infty$, $H(\infty) = G$.

The null frequency ω_z can be greater than the corner frequency ($\omega_z > \omega_0$), equal to the corner frequency ($\omega_z = \omega_0$), or less than the corner frequency ($\omega_z < \omega_0$). If $\omega_0 \geq \omega_z$ (case 1), the circuit shown in Figure 16.48 can be used to implement the second-order bandstop filter. If $\omega_0 \leq \omega_z$ (case 2), the circuit shown in Figure 16.49 can be used to implement the second-order bandstop filter[1]. If R_4 is replaced by C_4 in the circuit shown in Figure 16.48, it changes from case 1 to case 2. If $\omega_z = \omega_0$, either the circuit shown in Figure 16.48 or the circuit shown in Figure 16.49 can be used.

FIGURE 16.48

The circuit for case 1 ($\omega_0 \geq \omega_z$).

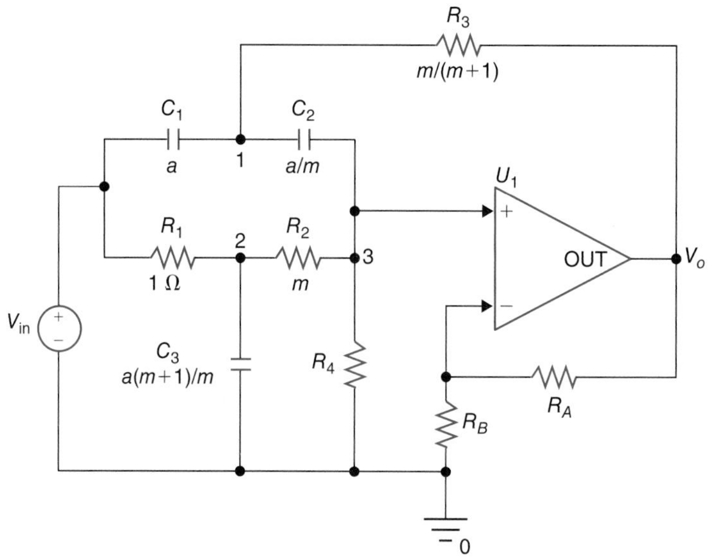

The transfer function of the circuit shown in Figure 16.48 is given by

$$H(s) = \frac{G(s^2 + b_0)}{s^2 + a_1 s + a_0} = \frac{G(s^2 + \omega_z^2)}{s^2 + \dfrac{\omega_0}{Q}s + \omega_0^2} \tag{16.189}$$

Notice that

$$a_0 = \omega_0^2, \quad a_1 = \frac{\omega_0}{Q} \tag{16.190}$$

Thus, we have

$$\omega_0 = \sqrt{a_0}, \quad Q = \frac{\omega_0}{a_1}, \quad \frac{1}{Q} = \frac{a_1}{\omega_0} \tag{16.191}$$

[1]L. P. Huelsman and P. E. Allen, *Introduction to the Theory and Design of Active Filters*, McGraw-Hill, 1980.

For the normalized filter, we set $\omega_0 = 1$ rad/s. The transfer function of the normalized bandstop filter can be written as

$$H(S) = \frac{G(S^2 + B_0)}{S^2 + A_1 S + A_0} = \frac{G(S^2 + B_0)}{S^2 + \frac{1}{Q}s + 1} \tag{16.192}$$

where

$$A_0 = 1, \quad A_1 = \frac{1}{Q} = \frac{a_1}{\omega_0}, \quad B_0 = \frac{\omega_z^2}{\omega_0^2} = \frac{b_0}{a_0} \tag{16.193}$$

The normalized transfer function $H(S)$ using the component values shown in Figure 16.48 is given by

$$H(S) = \frac{K\left(S^2 + \frac{1}{a^2}\right)}{S^2 + \frac{m+1}{a}\left(\frac{1}{R} + \frac{2-K}{m}\right)S + \frac{1 + \frac{m+1}{R}}{a^2}} \tag{16.194}$$

where $R = R_4$ and

$$K = 1 + \frac{R_A}{R_B}$$

The transfer function given by Equation (16.194) can be verified using MATLAB:

```
clear all;
syms Vin a m R K Vo s V1 V2
[V1,V2,Vo]=solve((V1-Vin)*s*a+(V1-Vo)/(m/(m+1))+(V1-Vo/K)*s*a/m==0,...
(V2-Vin)/1+V2*s*a*(m+1)/m+(V2-Vo/K)/m==0,...
(Vo/K-V1)*s*a/m+(Vo/K-V2)/m+Vo/K/R==0,V1,V2,Vo)
Vo=simplify(Vo)
Vo=subs(Vo,Vin,1)
pretty(Vo)
```

The constant terms of the numerator of Equations (16.192) and (16.194) reveal that

$$B_0 = \frac{1}{a^2}$$

Solving for a, we obtain

$$a = \frac{1}{\sqrt{B_0}} \tag{16.195}$$

The constant terms of the denominator of Equations (16.192) and (16.194) reveal that

$$\frac{1 + \frac{m+1}{R}}{a^2} = \left(1 + \frac{m+1}{R}\right)\frac{1}{a^2} = \left(1 + \frac{m+1}{R}\right)B_0 = A_0$$

Solving for R, we obtain

$$R = \frac{m+1}{\frac{A_0}{B_0} - 1} \tag{16.196}$$

The coefficients of the S terms of the denominator of Equations (16.192) and (16.194) reveal that

$$\frac{m+1}{a}\left(\frac{1}{R} + \frac{2-K}{m}\right) = A_1$$

Solving for K, we obtain

$$K = 2 + \frac{m}{m+1}\left(\frac{A_0}{B_0} - 1 - \frac{A_1}{\sqrt{B_0}}\right) \tag{16.197}$$

The gain of the filter at $S = \infty$ is

$$H(\infty) = G = K \tag{16.198}$$

To make the gain equal to 1 at $S = \infty$, we can use a voltage divider with a gain of $1/G$. We choose $R_D = 1$, $R_C = G - 1$ in the circuit shown in Figure 16.51. The normalized component values are given by

$$R_1 = 1,\quad R_2 = m,\quad R_3 = m/(1+m),\quad R_4 = R,\quad R_B = 1,\quad R_A = K - 1.$$
$$C_1 = a,\quad C_2 = a/m,\quad C_3 = a(m+1)/m$$

where a, R, and K are given by Equations (16.195), (16.196), and (16.197), respectively.

If we choose $m = 1$, we have

$$R_1 = 1,\quad R_2 = 1,\quad R_3 = 0.5,\quad R_4 = R,\quad R_B = 1,\quad R_A = K - 1$$
$$C_1 = a,\quad C_2 = a,\quad C_3 = 2a$$
$$G = K > 1,\quad R_D = 1,\quad R_C = K - 1$$

Magnitude scale by k_m (= 1000) and frequency scale by $k_f = \omega_0$ to get the scaled component values.

When $a_0 = b_0$ ($A_0 = B_0$, $\omega_0 = \omega_z$), $R \rightarrow \infty$ (open circuit). Remove R_4 (= R). Let $m = 1$. Then, $R_1 = R_2 = 1$, $R_3 = 0.5$, $C_1 = C_2 = a$, $C_3 = 2a$, and $H(s)$ becomes

$$H(s) = \frac{K(s^2 + b_0)}{s^2 + a_1 s + a_0} = \frac{K(s^2 + \omega_z^2)}{s^2 + \frac{\omega_0}{Q}s + \omega_0^2} = \frac{K\left(s^2 + \frac{1}{a^2}\right)}{s^2 + \frac{4-2K}{a}s + \frac{1}{a^2}} \tag{16.199}$$

The quality factor is given by

$$Q = \frac{1}{4 - 2K}$$

Solving for K, we obtain

$$K = 2 - \frac{1}{2Q}$$

If $a = 1$, Equation (16.199) becomes the normalized transfer function given by

$$H(S) = \frac{K(S^2 + 1)}{S^2 + (4 - 2K)S + 1}$$

If $\omega_0 \leq \omega_z$, it is case 2. In this case, resistor R_4 is replaced by capacitor C_4, as shown in Figure 16.49.

FIGURE 16.49

The circuit for case 2 ($\omega_0 \leq \omega_z$).

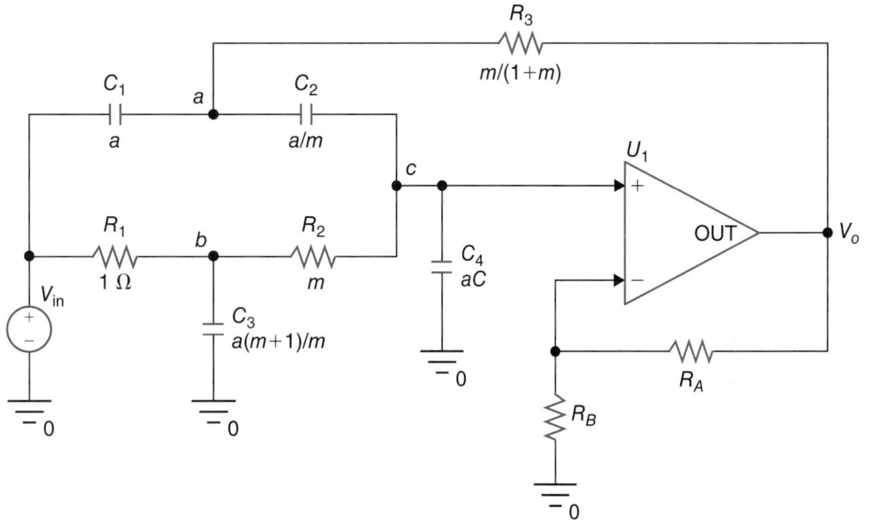

The transfer function of the circuit shown in Figure 16.49 is given by

$$H(S) = \frac{\dfrac{K}{(m+1)C+1}\left(S^2 + \dfrac{1}{a^2}\right)}{S^2 + \dfrac{(m+1)\left(C + \dfrac{2-K}{m}\right)}{a[(m+1)C+1]}S + \dfrac{1}{a^2[(m+1)C+1]}} \tag{16.200}$$

The transfer function given by Equation (16.200) can be verified using MATLAB:

```
clear all;
syms Vin a m R K Vo s C V1 V2
[V1,V2,Vo]=solve((V1-Vin)*s*a+(V1-Vo)/(m/(m+1))+(V1-Vo/K)*s*a/m==0,...
(V2-Vin)/1+V2*s*a*(m+1)/m+(V2-Vo/K)/m==0,...
(Vo/K-V1)*s*a/m+(Vo/K-V2)/m+Vo*s*a*C/K==0,V1,V2,Vo)
Vo=subs(Vo,Vin,1)
Vo=simplify(Vo)
pretty(Vo)
```

The constant terms of the numerator of Equations (16.192) and (16.200) reveal that

$$B_0 = \frac{1}{a^2}$$

Solving for a, we obtain

$$a = \frac{1}{\sqrt{B_0}} \tag{16.201}$$

The constant terms of the denominator of Equations (16.192) and (16.200) reveal that

$$\frac{1}{a^2[(m+1)C+1]} = A_0 \tag{16.202}$$

Solving for C, we obtain

$$C = \frac{\frac{B_0}{A_0} - 1}{m+1} \tag{16.203}$$

The coefficients of the S terms of the denominator of Equations (16.192) and (16.200) reveal that

$$\frac{(m+1)\left(C + \frac{2-K}{m}\right)}{a[(m+1)C+1]} = A_1$$

Solving for K, we obtain

$$K = 2 + \frac{m}{m+1}\left(\frac{B_0}{A_0} - 1 - \frac{A_1\sqrt{B_0}}{A_0}\right) \tag{16.204}$$

The gain of the filter at $S = \infty$ $(j\Omega = \infty)$ is given by, from Equations (16.200) and (16.202),

$$G = \frac{K}{(m+1)C+1} = \frac{K}{\frac{B_0}{A_0}} = \frac{A_0}{B_0}K \tag{16.205}$$

To make the gain equal to 1 at $S = \infty$, we choose $R_D = 1$, $R_C = G - 1$ in the circuit shown in Figure 16.51. The component values of the normalized bandstop filter are given by

$$R_1 = 1,\quad R_2 = m,\quad R_3 = m/(1+m),\quad R_B = 1,\quad R_A = K - 1$$

$$C_1 = a,\quad C_2 = a/m,\quad C_3 = a(m+1)/m,\quad C = \frac{\frac{B_0}{A_0} - 1}{m+1},\quad C_4 = aC$$

where a, C, and K are given by Equations (16.201), (16.203), and (16.204), respectively.

If $m = 1$, the normalized component values become

$$R_1 = 1,\quad R_2 = 1,\quad R_3 = 0.5,\quad R_B = 1,\quad R_A = K - 1$$

$$C_1 = a,\quad C_2 = a,\quad C_3 = 2a,\quad C = 0.5\left(\frac{B_0}{A_0} - 1\right),\quad C_4 = aC$$

Magnitude scale by k_m $(= 1000)$ and frequency scale by $k_f = \omega_0$ to get scaled component values.

When $a_0 = b_0$ $(A_0 = B_0, \omega_0 = \omega_z)$, $C = 0$ and $(1/(sC) = \infty)$ (open circuit). Remove C_4. Let $m = 1$. Then, $R_1 = R_2 = 1$, $R_3 = 0.5$, $C_1 = C_2 = a$, $C_3 = 2a$, and $H(s)$ becomes

$$H(s) = \frac{G(s^2 + b_0)}{s^2 + a_1 s + a_0} = \frac{G(s^2 + \omega_z^2)}{s^2 + \frac{\omega_0}{Q}s + \omega_0^2} = \frac{K\left(s^2 + \frac{1}{a^2}\right)}{s^2 + \frac{4-2K}{a}s + \frac{1}{a^2}} \tag{16.206}$$

The quality factor is given by

$$Q = \frac{1}{4 - 2K}$$

FIGURE 16.50

Gain control ($G < 1$).

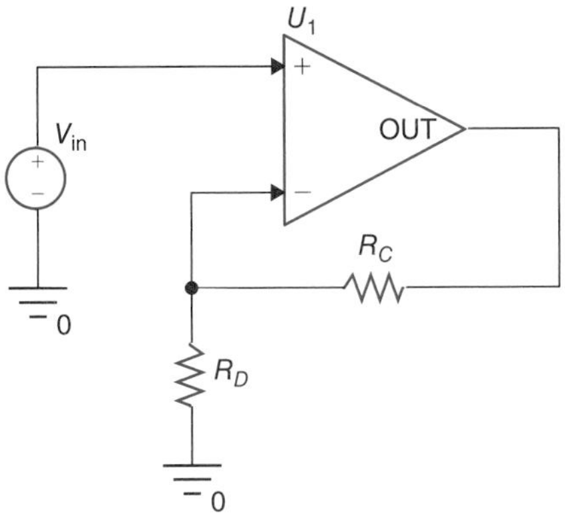

FIGURE 16.51

Gain control ($G > 1$).

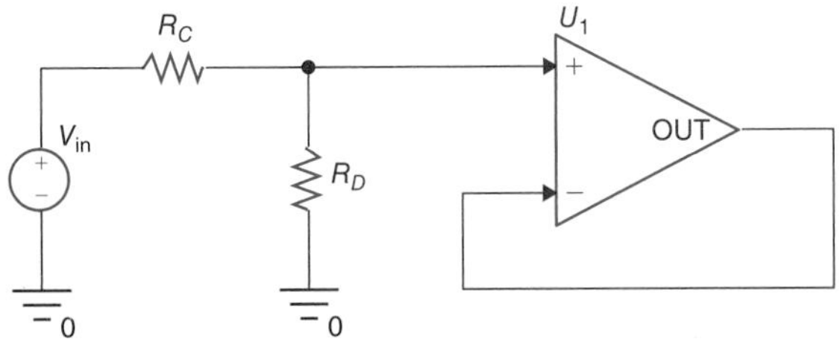

If $a = 1$, Equation (16.206) becomes the transfer function of the normalized bandstop filter given by

$$H(S) = \frac{K(s^2 + 1)}{s^2 + (4 - 2K)s + 1} \qquad \textbf{(16.207)}$$

The normalized component values are given by

$$R_1 = 1,\quad R_2 = m,\quad R_3 = m/(1 + m),\quad R_B = 1,\quad R_A = K - 1$$
$$C_1 = a,\quad C_2 = a/m,\quad C_3 = a(m + 1)/m,\quad C_4 = aC$$

If $m = 1$, we have

$$R_1 = 1,\quad R_2 = 1,\quad R_3 = 0.5,\quad R_B = 1,\quad R_A = K - 1$$
$$C_1 = a,\quad C_2 = a,\quad C_3 = 2a,\quad C_4 = aC$$

Magnitude scale by k_m (= 1000) and frequency scale by $k_f = \omega_0$ to get the scaled component values.

If there is more than one section, multiply the gains from all the sections to get the single gain G_0.

If the gain is less than 1 ($G_0 < 1$), take the inverse of G_0 to get $Gi = 1/G_0 > 1$. Use the noninverting amplifier shown in Figure 16.50 to increase the gain to 1. $R_D = 1, R_C = Gi - 1$.

If the gain is greater than 1 ($G_0 > 1$), use the voltage divider circuit shown in Figure 16.51 to decrease the gain to 1. $R_C = G_0 - 1, R_D = 1$.

EXAMPLE 16.19

Design a bandstop filter with the transfer function

$$H(s) = \frac{s^2 + 1.57913670 \times 10^8}{s^2 + 11{,}087.0258s + 4.41940645 \times 10^8} = \frac{s^2 + b_0}{s^2 + a_1 s + a_0} = \frac{s^2 + \omega_z^2}{s^2 + \dfrac{\omega_0}{Q}s + \omega_0^2}$$

Assume that $k_m = 1000$ and $m = 1$.

$$b_0 = \omega_z^2 = 1.57913670 \times 10^8$$
$$\omega_z = \sqrt{b_0} = 12{,}566.3706 \text{ rad/s}$$
$$a_0 = \omega_0^2 = 4.41940645 \times 10^8$$
$$\omega_0 = \sqrt{a_0} = 21{,}022.3844 \text{ rad/s}$$
$$a_1 = 11{,}087.0258$$

continued

Example 16.19 continued

Since $\omega_0 > \omega_z$, this is case 1. From Equation (16.191), we have

$$Q = \frac{\omega_0}{a_1} = 1.8961$$

$$\frac{1}{Q} = \frac{a_1}{\omega_0} = 0.5274$$

From Equation (16.193), we have

$$A_0 = 1$$

$$A_1 = \frac{1}{Q} = \frac{a_1}{\omega_0} = 0.5274$$

$$B_0 = \frac{\omega_z^2}{\omega_0^2} = \frac{b_0}{a_0} = 0.3573$$

From Equation (16.195), we have

$$a = \frac{1}{\sqrt{B_0}} = 1.6729$$

From Equation (16.196), we have

$$R = \frac{m+1}{\frac{A_0}{B_0} - 1} = 1.112$$

From Equation (16.197), we have

$$K = 2 + \frac{m}{m+1}\left(\frac{A_0}{B_0} - 1 - \frac{A_1}{\sqrt{B_0}}\right) = 2.4582$$

Let $m = 1$. Then, we have

$$R_1 = 1\ \Omega,\quad R_2 = 1\ \Omega,\quad R_3 = 0.5\ \Omega,\quad R_4 = R = 1.112\ \Omega,\quad R_B = 1\ \Omega,$$
$$R_A = K - 1 = 1.4582\ \Omega$$

$$C_1 = a = 1.6729\text{ F},\quad C_2 = a = 1.6729\text{ F},\quad C_3 = 2a = 3.3458\text{ F},\quad R_D = 1\ \Omega,$$
$$R_C = K - 1 = 1.4582\ \Omega$$

Let $k_m = 1000$, $k_f = \omega_0 = 21{,}022.3844$ rad/s. After magnitude and frequency scaling, we obtain

$$R_1 = 1\ k\Omega,\quad R_2 = 1\ k\Omega,\quad R_3 = 500\ \Omega,\quad R_4 = 1.112\ k\Omega,\quad R_B = 1\ k\Omega,$$
$$R_A = 1.4582\ k\Omega$$

$$C_1 = 79.5775\ nF,\quad C_2 = 79.5775\ nF,\quad C_3 = 2a = 0.1592\ \mu F,\quad R_D = 1\ k\Omega,$$
$$R_C = 1.4582\ k\Omega$$

The PSpice schematic for the circuit is shown in Figure 16.52, and the magnitude response is shown in Figure 16.53.

continued

Example 16.19 continued

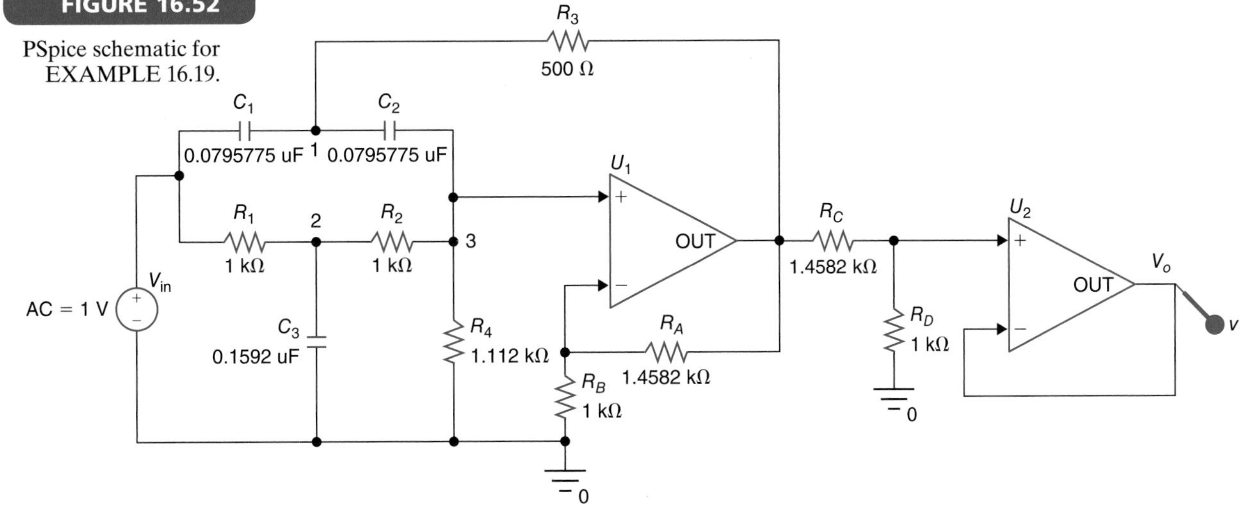

FIGURE 16.52

PSpice schematic for EXAMPLE 16.19.

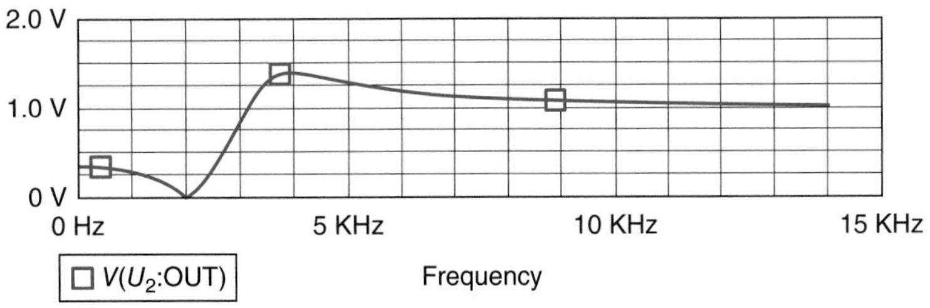

FIGURE 16.53

Magnitude response of the circuit shown in Figure 16.52.

Exercise 16.19

Design a bandstop filter with the transfer function

$$H(s) = \frac{s^2 + 4.41940645 \times 10^8}{s^2 + 11{,}087.0258s + 4.41940645 \times 10^8} = \frac{s^2 + b_0}{s^2 + a_1 s + a_0} = \frac{s^2 + \omega_z^2}{s^2 + \dfrac{\omega_0}{Q}s + \omega_0^2}$$

using the circuit shown in Figure 16.48. Assume that $k_m = 1000$ and $m = 1$.

Answer:

$\omega_z = \omega_0 = 21{,}022.4844$ rad/s, $a_1 = 11{,}087.0258$, $Q = \dfrac{\omega_0}{a_1} = 1.8961$, $\dfrac{1}{Q} = \dfrac{a_1}{\omega_0} = 0.5274$,

$A_0 = 1$, $A_1 = \dfrac{1}{Q} = \dfrac{a_1}{\omega_0} = 0.5274$, $B_0 = \dfrac{\omega_z^2}{\omega_0^2} = \dfrac{b_0}{a_0} = 1$, $a = \dfrac{1}{\sqrt{B_0}} = 1$, $R = \dfrac{m+1}{\dfrac{A_0}{B_0} - 1} = \infty$

(Remove R_4), $K = 2 + \dfrac{m}{m+1}\left(\dfrac{A_0}{B_0} - 1 - \dfrac{A_1}{\sqrt{B_0}}\right) = 2 - \dfrac{m}{m+1}A_1 = 1.7363$, $m = 1$,

$R_1 = 1\ \Omega$, $R_2 = 1\ \Omega$, $R_3 = 0.5\ \Omega$, $R_B = 1\ \Omega$, $R_A = K - 1 = 0.7363\ \Omega$, $C_1 = a = 1$ F,
$C_2 = a = 1$ F, $C_3 = 2a = 2$ F, $R_D = 1\ \Omega$, $R_C = K - 1 = 0.7363\ \Omega$.

continued

Exercise 16.19 continued

Scaling by $k_m = 1000$, $k_f = \omega_0 = 21{,}022.4844$ rad/s, we obtain

$$R_1 = 1\,k\Omega,\quad R_2 = 1\,k\Omega,\quad R_3 = 500\,\Omega,\quad R_B = 1\,k\Omega,\quad R_A = K - 1 = 736.3\,\Omega,$$
$$C_1 = a = 47.5681\,nF,\quad C_2 = a = 47.5681\,nF,\quad C_3 = 2a = 95.1367\,nF,\quad R_D = 1\,k\Omega,$$
$$R_C = 0.7363\,k\Omega$$

The PSpice schematic for the circuit is shown in Figure 16.54, and the magnitude response is shown in Figure 16.55.

FIGURE 16.54

PSpice schematic for EXERCISE 16.19.

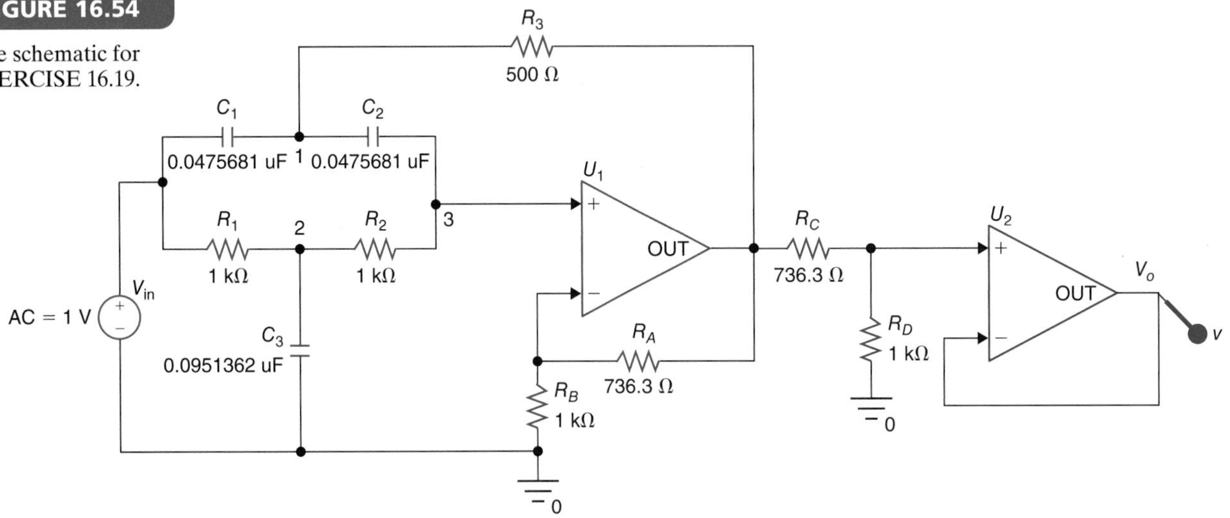

FIGURE 16.55

Magnitude response of the circuit shown in Figure 16.54.

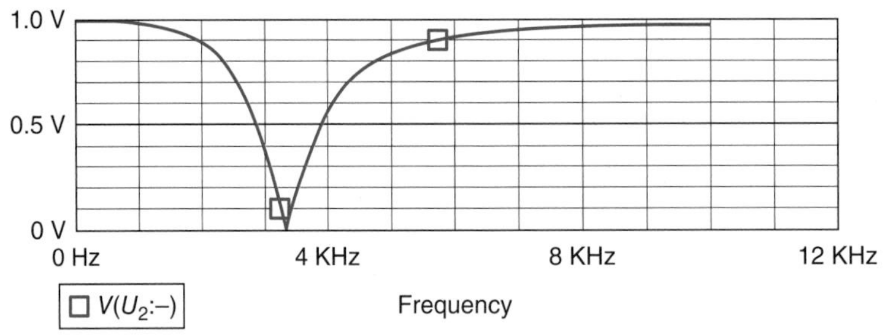

EXAMPLE 16.20

Design a bandstop filter with the transfer function

$$H(s) = \frac{s^2 + 1.57913670 \times 10^8}{s^2 + 3961.602s + 5.64255123 \times 10^7}$$

Assume that $k_m = 1000$ and $m = 1$.

$$b_0 = \omega_z^2 = 1.57913670 \times 10^8$$
$$\omega_z = \sqrt{b_0} = 12{,}566.3706 \text{ rad/s}$$
$$a_0 = \omega_0^2 = 5.64255123 \times 10^7$$

continued

Example 16.20 continued

$$\omega_0 = \sqrt{a_0} = 7511.6917 \text{ rad/s}$$
$$a_1 = 3961.602$$

Since $\omega_0 < \omega_z$, this is case 2. From Equation (16.191), we have

$$Q = \frac{\omega_0}{a_1} = 1.8961$$

$$\frac{1}{Q} = \frac{a_1}{\omega_0} = 0.5274$$

From Equation (16.193), we have

$$A_0 = 1$$

$$A_1 = \frac{1}{Q} = \frac{a_1}{\omega_0} = 0.5274$$

$$B_0 = \frac{\omega_z^2}{\omega_0^2} = \frac{b_0}{a_0} = 2.7986$$

From Equation (16.201), we have

$$a = \frac{1}{\sqrt{B_0}} = 0.5978$$

From Equation (16.203), we have

$$C = \frac{\dfrac{B_0}{A_0} - 1}{m + 1} = 0.8993$$

From Equation (16.204), we have

$$K = 2 + \frac{m}{m+1}\left(\frac{B_0}{A_0} - 1 - \frac{A_1\sqrt{B_0}}{A_0}\right) = 2.4582$$

The gain is given by

$$G = \frac{A_0}{B_0}K = 0.8784$$

The normalized component values are

$$R_1 = 1\ \Omega,\quad R_2 = 1\ \Omega,\quad R_3 = 0.5\ \Omega,\quad R_B = 1\ \Omega,\quad R_A = K - 1 = 1.4582\ \Omega,$$
$$C_1 = a = 0.5978\text{ F},\quad C_2 = a = 0.5978\text{ F},\quad C_3 = 2a = 1.1955\text{ F},\quad C_4 = aC = 0.5376\text{ F}$$

After the magnitude scale by $k_m = 1000$ and the frequency scale by $k_f = \text{w}_0 =$ 7511.6917 rad/s, we obtain

$$R_1 = 1\ k\Omega,\quad R_2 = 1\ k\Omega,\quad R_3 = 0.5\ k\Omega,\quad R_B = 1\ k\Omega,\quad R_A = K - 1 = 1.4582\ k\Omega,$$
$$C_1 = 79.5774\ nF,\quad C_2 = 79.5774\ nF,\quad C_3 = 0.1592\ \mu F,\quad C_4 = 71.5648\ nF$$

continued

Example 16.20 continued

The inverse of gain is 1.1385. Thus, $R_C = 0.1385\ k\Omega$, $R_D = 1\ k\Omega$.

The PSpice schematic for the circuit is shown in Figure 16.56, and the magnitude response is shown in Figure 16.57.

FIGURE 16.56

The PSpice schematic for EXAMPLE 16.20.

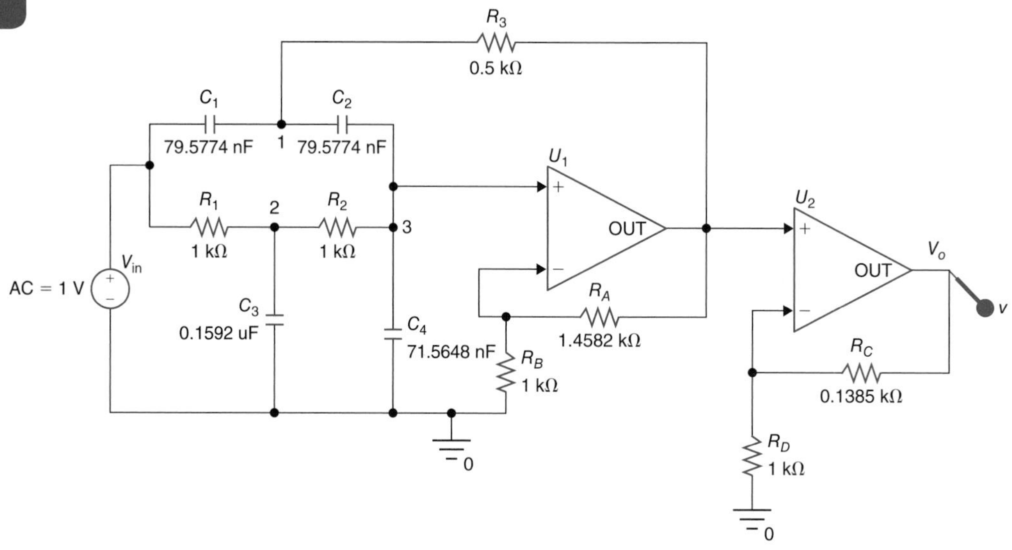

FIGURE 16.57

The magnitude response of the circuit shown in Figure 16.56.

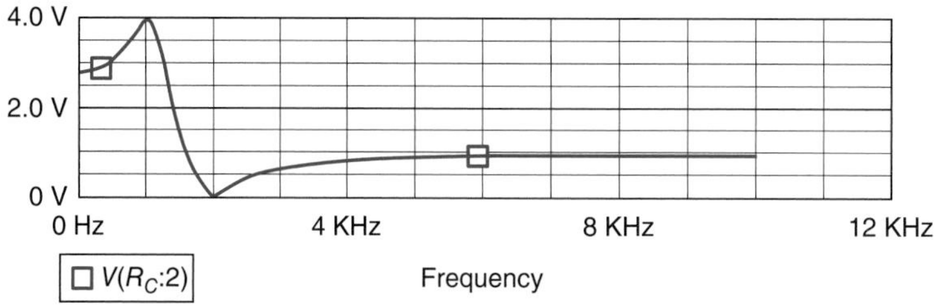

Exercise 16.20

Design a bandstop filter with the transfer function

$$H(s) = \frac{s^2 + 5.64255123 \times 10^7}{s^2 + 3961.602s + 5.64255123 \times 10^7} = \frac{s^2 + b_0}{s^2 + a_1 s + a_0} = \frac{s^2 + \omega_z^2}{s^2 + \dfrac{\omega_0}{Q}s + \omega_0^2}$$

using the circuit shown in Figure 16.49. Assume that $k_m = 1000$ and $m = 1$.

Answer:

$\omega_z = \omega_0 = 7511.6917$ rad/s, $a_1 = 3961.602$, $Q = \dfrac{\omega_0}{a_1} = 1.8961$, $\dfrac{1}{Q} = \dfrac{a_1}{\omega_0} = 0.5274$,

$A_0 = 1$, $A_1 = \dfrac{1}{Q} = \dfrac{a_1}{\omega_0} = 0.5274$, $B_0 = \dfrac{\omega_z^2}{\omega_0^2} = \dfrac{b_0}{a_0} = 1$, $a = \dfrac{1}{\sqrt{B_0}} = 1$, $C = \dfrac{\dfrac{B_0}{A_0} - 1}{m+1} = 0$

(Remove C_4), $K = 2 + \dfrac{m}{m+1}\left(\dfrac{B_0}{A_0} - 1 - \dfrac{A_1\sqrt{B_0}}{A_0}\right) = 2 - \dfrac{m}{m+1}A_1 = 1.7363$, $R_1 = 1\ \Omega$,

$R_2 = 1\ \Omega$, $R_3 = 0.5\ \Omega$, $R_B = 1\ \Omega$, $R_A = K - 1 = 0.7363\ \Omega$, $C_1 = a = 1$ F,

$C_2 = a = 1$ F, $C_3 = 2a = 2$ F, $R_D = 1\ \Omega$, $R_C = K - 1 = 0.7363\ \Omega$.

continued

Exercise 16.20 continued

Scaling by $k_m = 1000$, $k_f = \omega_0 = 7511.6917$ rad/s, we obtain

$$R_1 = 1\ k\Omega,\quad R_2 = 1\ k\Omega,\quad R_3 = 500\ \Omega,\quad R_B = 1\ k\Omega,\quad R_A = K - 1 = 736.3\ \Omega,$$
$$C_1 = 0.13313\ \mu F,\ \ C_2 = 0.13313\ \mu F,\ \ C_3 = 0.26625\ \mu F,\ \ R_D = 1\ k\Omega,\ \ R_C = 736.3\ \Omega$$

The PSpice schematic for the circuit is shown in Figure 16.58, and the magnitude response is shown in Figure 16.59.

FIGURE 16.58

The PSpice schematic for EXERCISE 16.20.

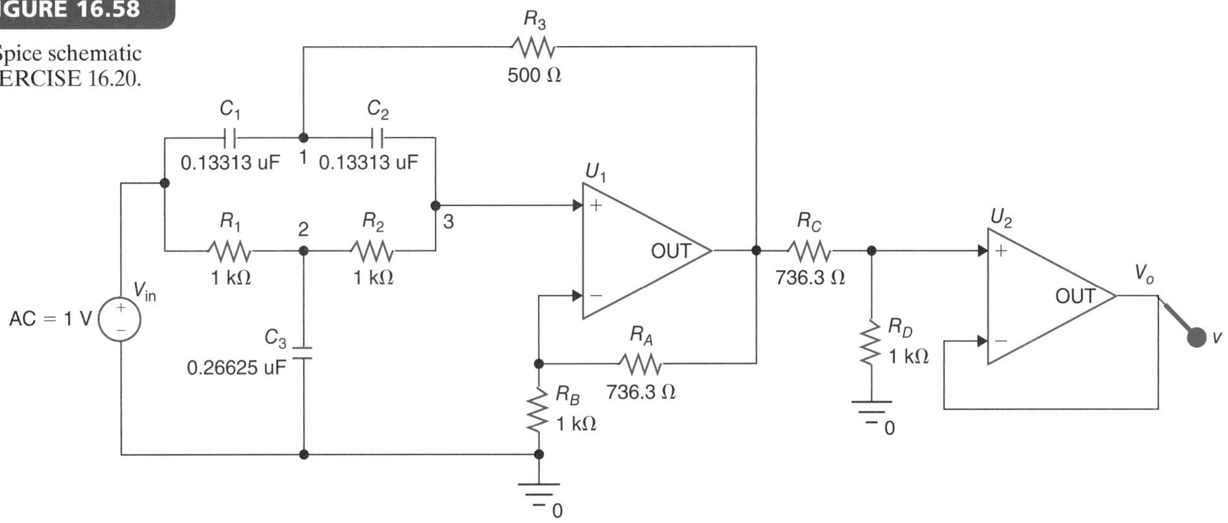

FIGURE 16.59

The magnitude response of the circuit shown in Figure 16.57.

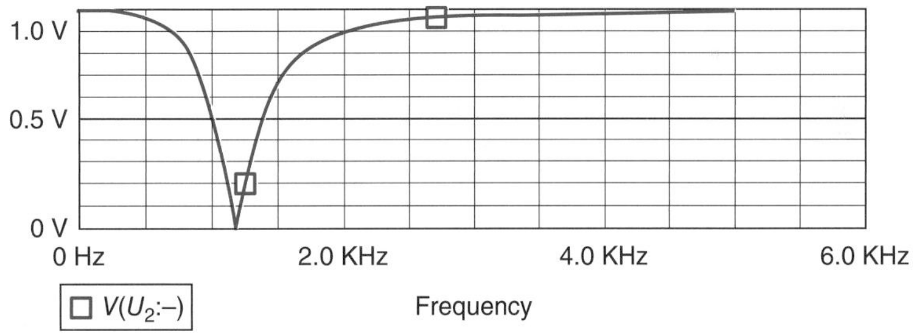

16.9 Simulink

As shown in Chapter 15, the linear analysis tool of Simulink can be used to plot the step response and the impulse response of a system. In this chapter, we are interested in plotting the Bode diagram of a system using the linear analysis tool.

As an example, consider a system with the transfer function

$$H(s) = \frac{9165.5892s}{s^2 + 9165.5892s + 2.321331 \times 10^8}$$

This is the transfer function used in EXAMPLE 16.15 earlier in this chapter. The Simulink model for this transfer function is shown in Figure 16.60.

FIGURE 16.60

A Simulink model for the transfer function.

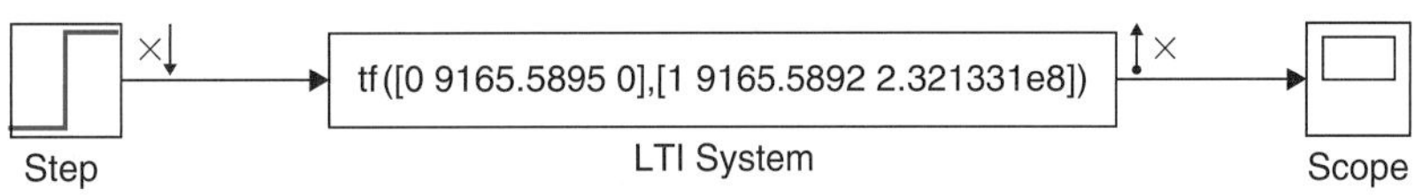

Right-click on the wire connecting the Step block to Transfer Fcn block. Select the open-loop input point. An arrow appears, as shown in Figure 16.60. Similarly, right-click on the wire connecting Transfer Fcn and Scope, and select open-loop output point on. The linear analysis is on the system between the input and the output points. Select Linear Analysis to open the linear analysis tool window. Select New Bode, and click on Linearize to get the Bode diagram shown in Figure 16.61.

FIGURE 16.61

Bode diagram.

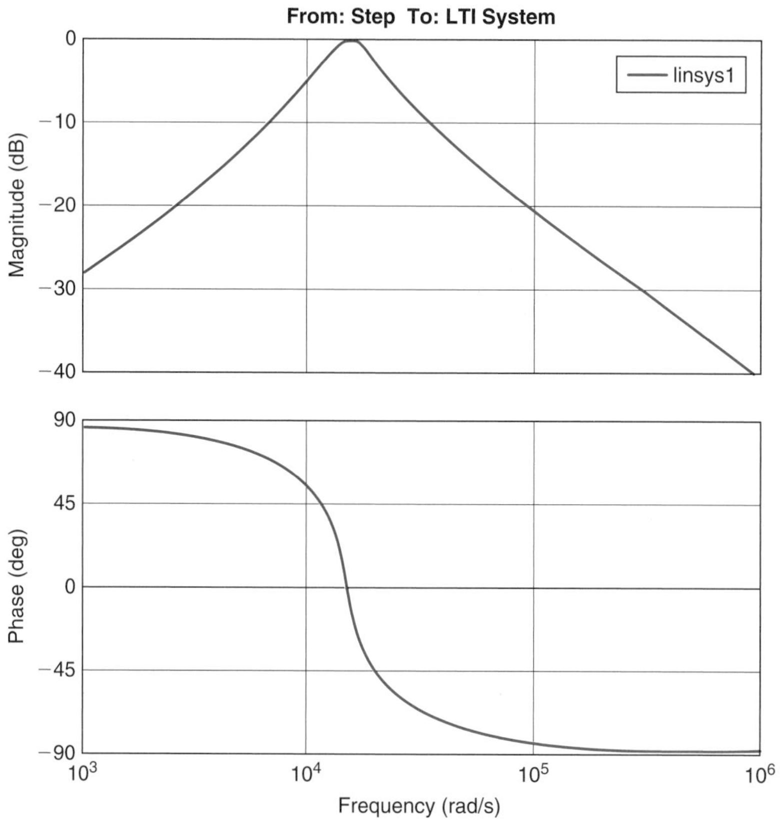

SUMMARY

In this chapter, the analysis and the design of first-order and second-order filters are discussed. The filters can be implemented using passive elements (R, L, C) only or a combination of an operational amp and resistors and capacitors. When a filter implementation includes an operational amp, the filter is called an active filter. The second-order Sallen-Key circuits provide the basic building blocks for the active filters.

The Sallen-Key circuits are used to implement the analog filters designed in Chapter 17. When the order is greater than 2, second-order sections of the Sallen-Key circuits are cascaded. For the low-pass filter and the high-pass filter, if the order of the filter is odd, one of the sections is the first-order filter.

For each second-order or first-order filter, we design the normalized filter, which has a cutoff frequency of 1 rad/s. Then, the component values for the normalized filter are found. These normalized component values are scaled (both magnitude and frequency) to obtain the final component values. The filter can be simulated using PSpice to verify that the designed filter satisfies the specifications.

TABLE 16.1 Passive Filters

First-Order LPF

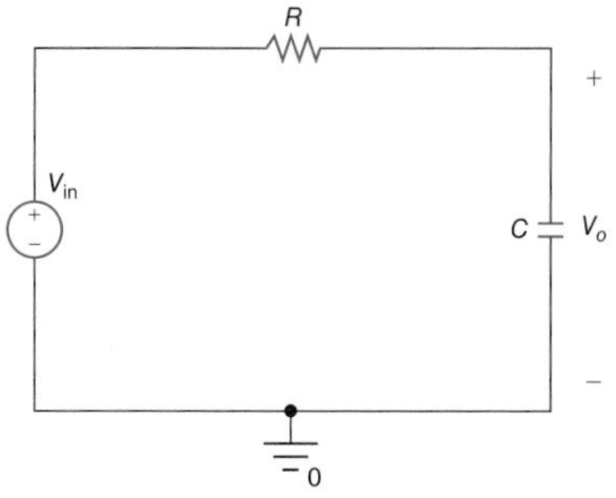

$$H(s) = \frac{\frac{1}{RC}}{s + \frac{1}{RC}} = \frac{\omega_o}{s + \omega_o}, \quad \omega_o = \frac{1}{RC}$$

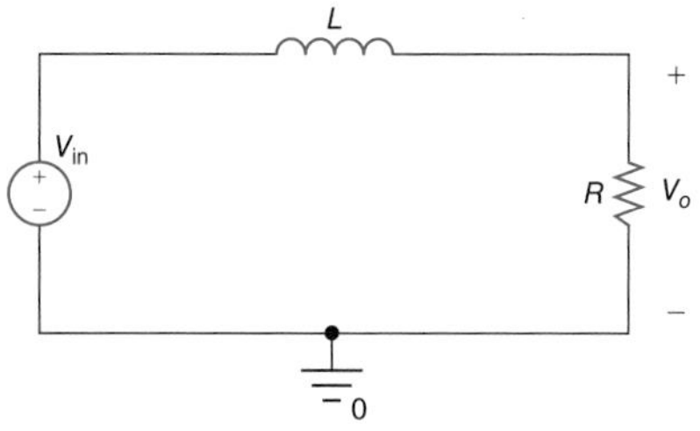

$$H(s) = \frac{R}{sL + R} = \frac{\frac{R}{L}}{s + \frac{R}{L}} = \frac{\omega_o}{s + \omega_o}, \quad \omega_o = \frac{R}{L}$$

First-Order HPF

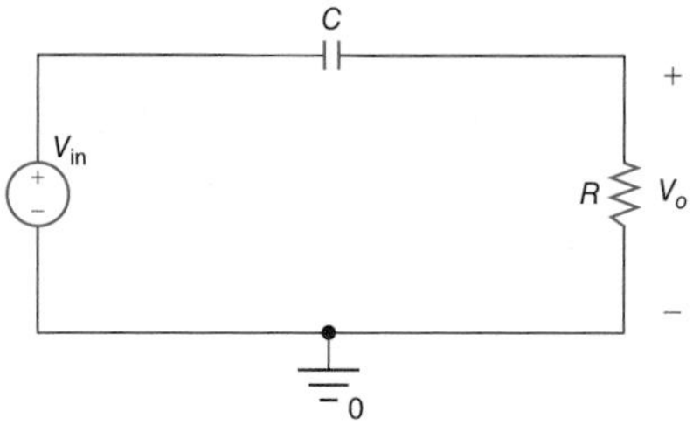

$$H(s) = \frac{s}{s + \frac{1}{RC}} = \frac{s}{s + \omega_o}, \quad \omega_o = \frac{1}{RC}$$

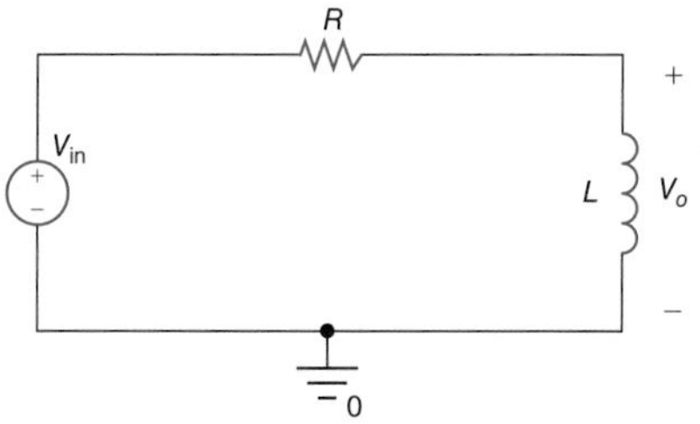

$$H(s) = \frac{sL}{sL + R} = \frac{s}{s + \frac{R}{L}} = \frac{s}{s + \omega_o}, \quad \omega_o = \frac{R}{L}$$

Second-Order LPF

Series *RLC*

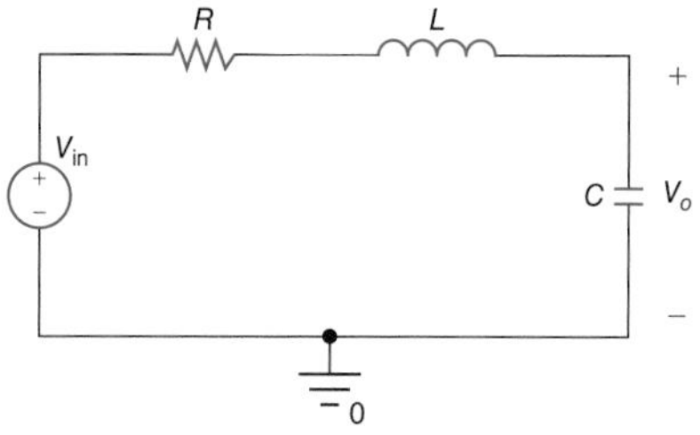

$$H(s) = \frac{\frac{1}{LC}}{s^2 + \frac{R}{L}s + \frac{1}{LC}} = \frac{\omega_o^2}{s^2 + \frac{\omega_o}{Q}s + \omega_0^2}$$

$$\omega_o = \frac{1}{\sqrt{LC}}, \quad Q = \frac{1}{R}\sqrt{\frac{L}{C}}$$

Poles: $s = -\frac{\omega_0}{2Q} \pm j\omega_0\sqrt{1 - \frac{1}{4Q^2}} = -\frac{R}{2L}$

$\pm j\frac{1}{\sqrt{LC}}\sqrt{1 - \frac{R^2C}{4L}}$, Zeros: $s = \infty$, $s = \infty$

$L = 1\text{ H}, \quad C = 1\text{ F}, \quad R = 1/Q\ \Omega$

$k_f = \omega_0, k_m$

Parallel *RLC*

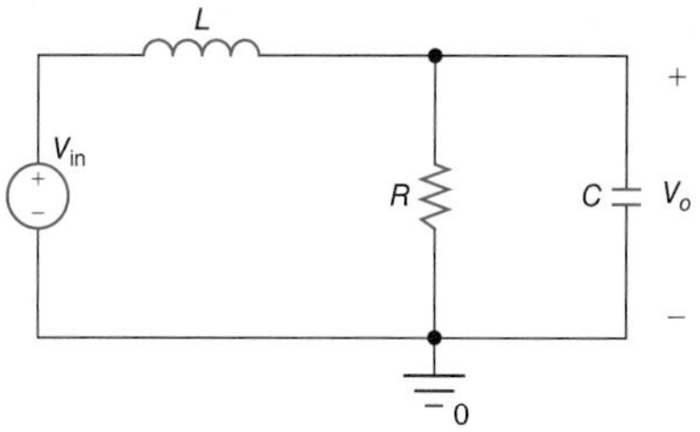

$$H(s) = \frac{\frac{1}{LC}}{s^2 + \frac{1}{RC}s + \frac{1}{LC}} = \frac{\omega_o^2}{s^2 + \frac{\omega_o}{Q}s + \omega_o^2}$$

$$\omega_o = \frac{1}{\sqrt{LC}}, \quad Q = R\sqrt{\frac{C}{L}}$$

continued

Table 16.1 continued

TABLE 16.1

Poles: $s = -\frac{\omega_0}{2Q} \pm j\omega_0\sqrt{1 - \frac{1}{4Q^2}} = -\frac{1}{2RC}$

$\pm j\frac{1}{\sqrt{LC}}\sqrt{1 - \frac{L}{4R^2C}}$, Zeros: $s = \infty$, $s = \infty$

$L = 1\text{ H}$, $C = 1\text{ F}$, $R = Q\ \Omega$

$k_f = \omega_0, k_m$

Second-Order HPF

Series *RLC*

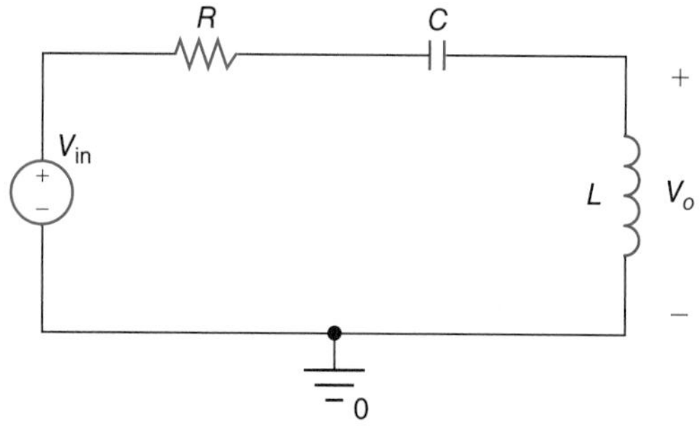

$$H(s) = \frac{s^2}{s^2 + \frac{R}{L}s + \frac{1}{LC}} = \frac{s^2}{s^2 + \frac{\omega_o}{Q}s + \omega_o^2}$$

$$\omega_o = \frac{1}{\sqrt{LC}}, \quad Q = \frac{1}{R}\sqrt{\frac{L}{C}}$$

Poles: $s = -\frac{\omega_0}{2Q} \pm j\omega_0\sqrt{1 - \frac{1}{4Q^2}} = -\frac{R}{2L}$

$\pm j\frac{1}{\sqrt{LC}}\sqrt{1 - \frac{R^2C}{4L}}$, Zeros: $s = 0$, $s = 0$

$L = 1\text{ H}$, $C = 1\text{ F}$, $R = 1/Q\ \Omega$

$k_f = \omega_0, k_m$

Parallel *RLC*

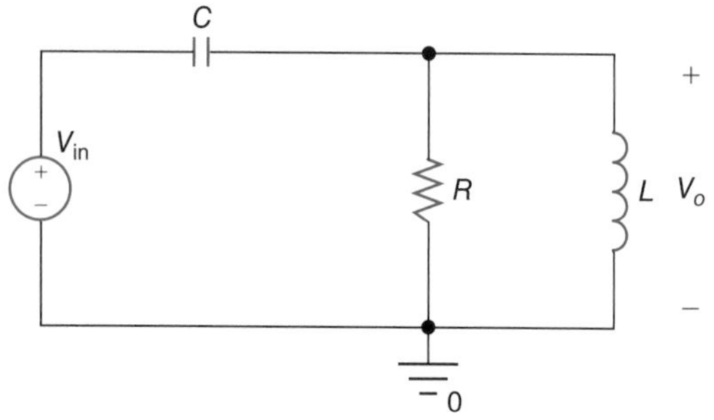

$$H(s) = \frac{s^2}{s^2 + \frac{1}{RC}s + \frac{1}{LC}} = \frac{s^2}{s^2 + \frac{\omega_o}{Q}s + \omega_o^2}$$

$$\omega_o = \frac{1}{\sqrt{LC}}, \quad Q = R\sqrt{\frac{C}{L}}$$

Poles: $s = -\frac{\omega_0}{2Q} \pm j\omega_0\sqrt{1 - \frac{1}{4Q^2}} = -\frac{1}{2RC}$

$\pm j\frac{1}{\sqrt{LC}}\sqrt{1 - \frac{L}{4R^2C}}$, Zeros: $s = 0$, $s = 0$

$L = 1\text{ H}$, $C = 1\text{ F}$, $R = Q\ \Omega$

$k_f = \omega_0, k_m$

Second-Order Bandpass Filter

Series *RLC*

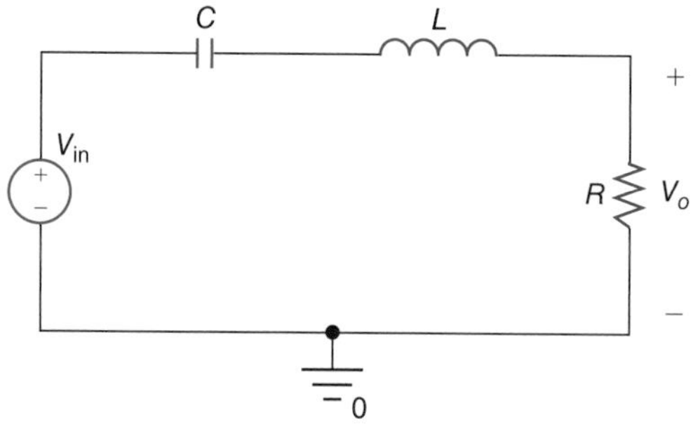

$$H(s) = \frac{\frac{R}{L}s}{s^2 + \frac{R}{L}s + \frac{1}{LC}} = \frac{\frac{\omega_o}{Q}s}{s^2 + \frac{\omega_o}{Q}s + \omega_o^2}$$

$$\omega_o = \frac{1}{\sqrt{LC}}, \quad Q = \frac{1}{R}\sqrt{\frac{L}{C}}$$

Poles: $s = -\frac{\omega_0}{2Q} \pm j\omega_0\sqrt{1 - \frac{1}{4Q^2}} = -\frac{R}{2L}$

$\pm j\frac{1}{\sqrt{LC}}\sqrt{1 - \frac{R^2C}{4L}}$, Zeros: $s = 0$, $s = \infty$

3-dB cutoff frequencies:

$$\omega_1 = -\frac{\omega_0}{2Q} + \frac{\omega_0}{2Q}\sqrt{4Q^2 + 1},$$

$$\omega_2 = \frac{\omega_0}{2Q} + \frac{\omega_0}{2Q}\sqrt{4Q^2 + 1}$$

continued

Table 16.1 continued

TABLE 16.1

3-dB bandwidth: $\omega_{3\text{dB}} = \omega_2 - \omega_1 = \dfrac{\omega_0}{Q}$

$$L = 1\,\text{H}, \quad C = 1\,\text{F}, \quad R = 1/Q\,\Omega$$
$$k_f = \omega_0, k_m$$

Parallel *RLC*

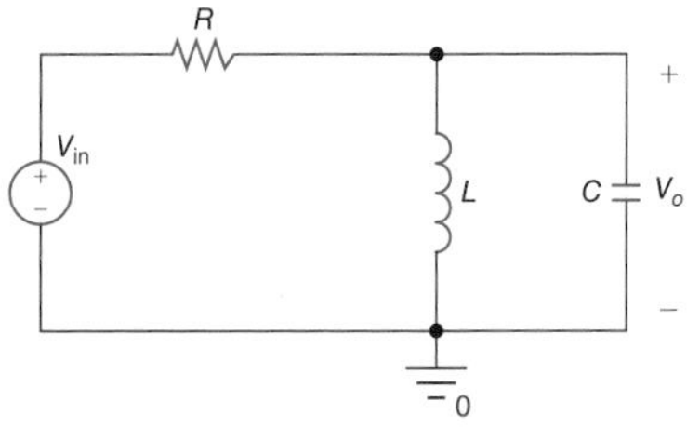

$$H(s) = \frac{\frac{1}{RC}s}{s^2 + \frac{1}{RC}s + \frac{1}{LC}} = \frac{\frac{\omega_o}{Q}s}{s^2 + \frac{\omega_o}{Q}s + \omega_o^2}$$

$$\omega_o = \frac{1}{\sqrt{LC}}, \quad Q = R\sqrt{\frac{C}{L}}$$

Poles: $s = -\dfrac{\omega_0}{2Q} \pm j\omega_0\sqrt{1 - \dfrac{1}{4Q^2}} = -\dfrac{1}{2RC} \pm j\dfrac{1}{\sqrt{LC}}\sqrt{1 - \dfrac{L}{4R^2C}}$, Zeros: $s = 0, \quad s = \infty$

3-dB cutoff frequencies:

$$\omega_1 = -\frac{\omega_0}{2Q} + \frac{\omega_0}{2Q}\sqrt{4Q^2 + 1},$$
$$\omega_2 = \frac{\omega_0}{2Q} + \frac{\omega_0}{2Q}\sqrt{4Q^2 + 1}$$

3-dB bandwidth: $\omega_{3\text{dB}} = \omega_2 - \omega_1 = \dfrac{\omega_0}{Q}$

$$L = 1\,\text{H}, \quad C = 1\,\text{F}, \quad R = Q\,\Omega$$
$$k_f = \omega_0, k_m$$

Second-Order Bandstop Filter

Series *RLC*

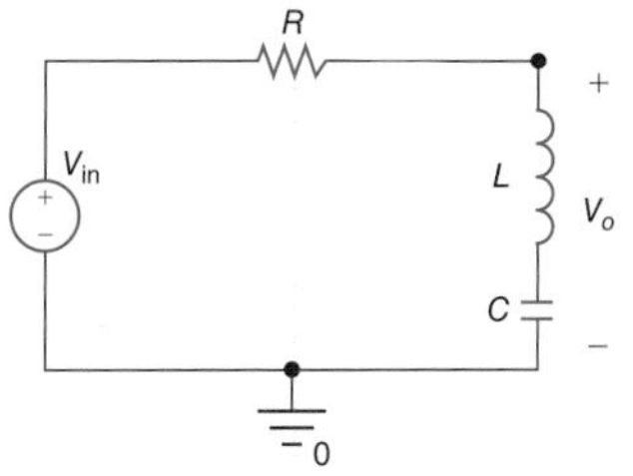

$$H(s) = \frac{s^2 + \frac{1}{LC}}{s^2 + \frac{R}{L}s + \frac{1}{LC}} = \frac{s^2 + \omega_o^2}{s^2 + \frac{\omega_o}{Q}s + \omega_o^2}$$

$$\omega_o = \frac{1}{\sqrt{LC}}, \quad Q = \frac{1}{R}\sqrt{\frac{L}{C}}$$

Poles: $s = -\dfrac{\omega_0}{2Q} \pm j\omega_0\sqrt{1 - \dfrac{1}{4Q^2}} = -\dfrac{R}{2L} \pm j\dfrac{1}{\sqrt{LC}}\sqrt{1 - \dfrac{R^2C}{4L}}$, Zeros: $s = j\dfrac{1}{\sqrt{LC}}$, $s = -j\dfrac{1}{\sqrt{LC}}$

$$L = 1\,\text{H}, \quad C = 1\,\text{F}, \quad R = 1/Q\,\Omega$$
$$k_f = \omega_0, k_m$$

Parallel *RLC*

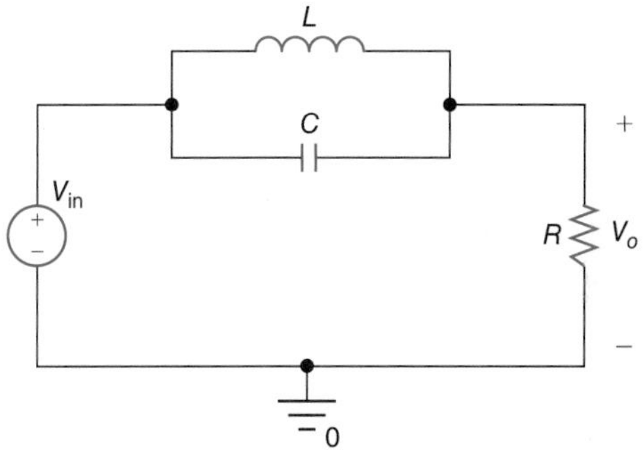

$$H(s) = \frac{s^2 + \frac{1}{LC}}{s^2 + \frac{1}{RC}s + \frac{1}{LC}} = \frac{s^2 + \omega_o^2}{s^2 + \frac{\omega_o}{Q}s + \omega_o^2}$$

$$\omega_o = \frac{1}{\sqrt{LC}}, \quad Q = R\sqrt{\frac{C}{L}}$$

Poles: $s = -\dfrac{\omega_0}{2Q} \pm j\omega_0\sqrt{1 - \dfrac{1}{4Q^2}} = -\dfrac{1}{2RC} \pm j\dfrac{1}{\sqrt{LC}}\sqrt{1 - \dfrac{L}{4R^2C}}$, Zeros: $s = j\dfrac{1}{\sqrt{LC}}$, $s = -j\dfrac{1}{\sqrt{LC}}$

$$L = 1\,\text{H}, \quad C = 1\,\text{F}, \quad R = Q\,\Omega$$
$$k_f = \omega_0, k_m$$

TABLE 16.2 Active Filters

First-Order LPF

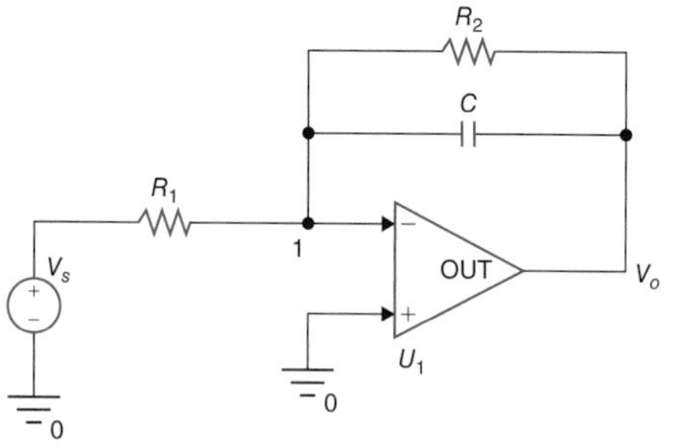

$$H(s) = \frac{V_o(s)}{V_{in}(s)} = -\frac{R_2}{R_1}\frac{\dfrac{1}{R_2C}}{s + \dfrac{1}{R_2C}}.$$

$$\text{If } R_1 = R_2, \quad H(s) = -\frac{\dfrac{1}{R_2C}}{s + \dfrac{1}{R_2C}}$$

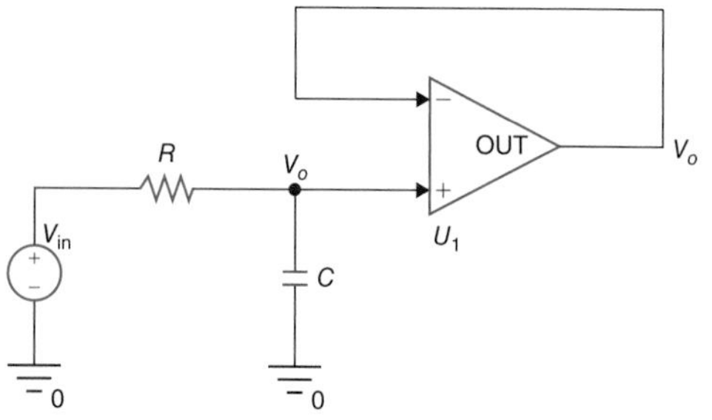

$$H(s) = \frac{V_o(s)}{V_{in}(s)} = \frac{\dfrac{1}{RC}}{s + \dfrac{1}{RC}} = \frac{\omega_o}{s + \omega_o}, \quad \omega_o = \frac{1}{RC}$$

First-Order HPF

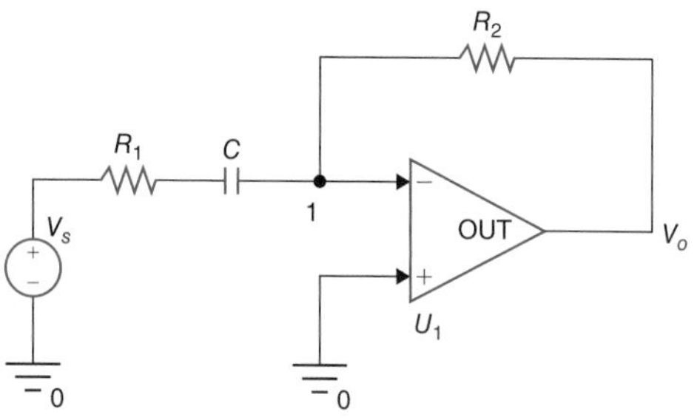

$$H(s) = \frac{V_o(s)}{V_{in}(s)} = -\frac{R_2}{R_1}\frac{s}{s + \dfrac{1}{R_1C}}.$$

$$\text{If } R_1 = R_2, \quad H(s) = -\frac{s}{s + \dfrac{1}{R_1C}}$$

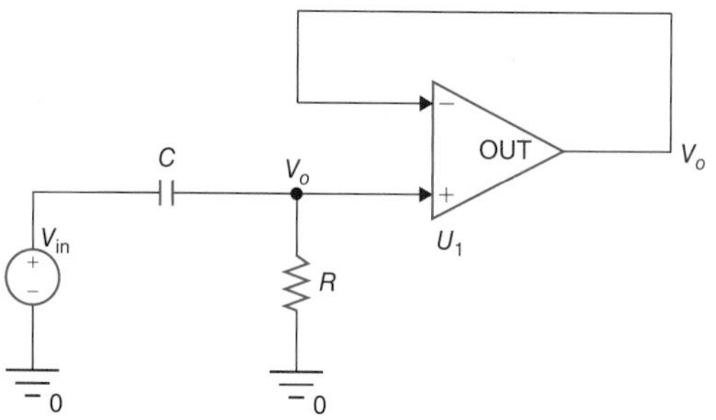

$$H(s) = \frac{V_o(s)}{V_{in}(s)} = \frac{s}{s + \dfrac{1}{RC}} = \frac{s}{s + \omega_o}, \quad \omega_o = \frac{1}{RC}$$

Second-Order LPF (Sallen-Key)

Equal *R* Equal *C*

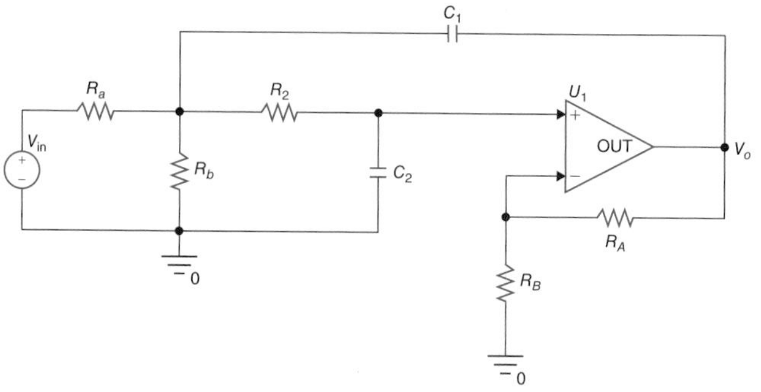

$R_2 = 1, \quad C_1 = C_2 = 1, \quad R_B = 1, \quad K = 3 - 1/Q,$
$R_A = 2 - 1/Q, \quad R_a = 3 - 1/Q,$
$R_b = (3Q - 1)/(2Q - 1)$
$k_f = \omega_0, k_m$

Unity Gain

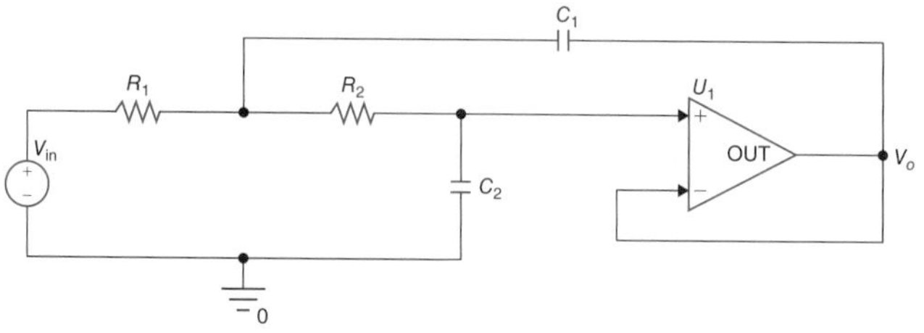

$R_1 = 1, \quad R_2 = 1, \quad C_1 = 2Q, \quad C_2 = 1/(2Q)$
$k_f = \omega_0, k_m$

continued

Table 16.2 continued

TABLE 16.2

Second-Order HPF (Sallen-Key)

Equal R Equal C

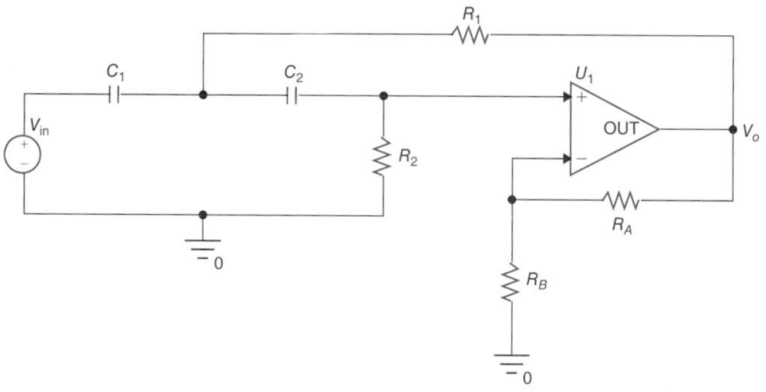

$R_1 = R_2 = 1, \quad C_1 = C_2 = 1, \quad R_B = 1,$
$R_A = 2 - 1/Q, \quad K = 3 - 1/Q$

$k_f = \omega_0, k_m$

Unity Gain

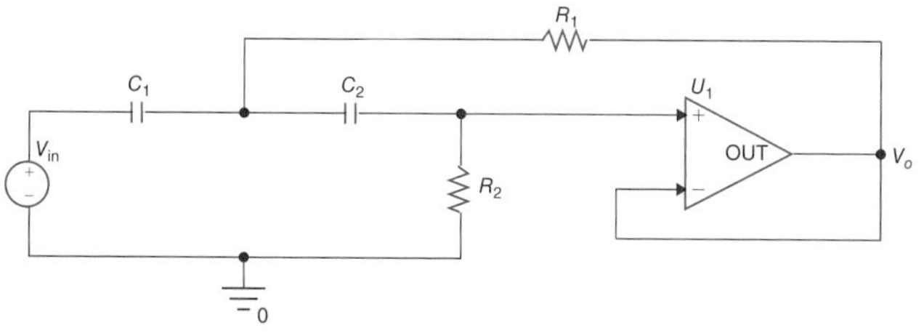

$R_1 = 1/(2Q), \quad R_2 = 2Q, \quad C_1 = 1,$
$C_2 = 1, \quad K = 1$

$k_f = \omega_0, k_m$

Second-Order Bandpass Filter (Sallen-Key)

Equal R Equal C

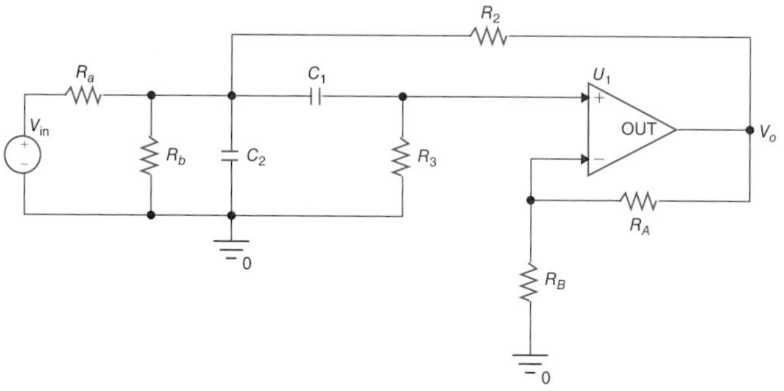

$R_2 = R_3 = 1, \quad C_1 = C_2 = \sqrt{2}, \quad R_B = 1,$
$K = 4 - \sqrt{2}/Q, \quad R_A = K - 1,$
$R_a = K/(4 - K),$

$$R_b = \frac{K}{2K - 4}$$

$k_f = \omega_0, k_m$

Delyiannis-Friend Circuit

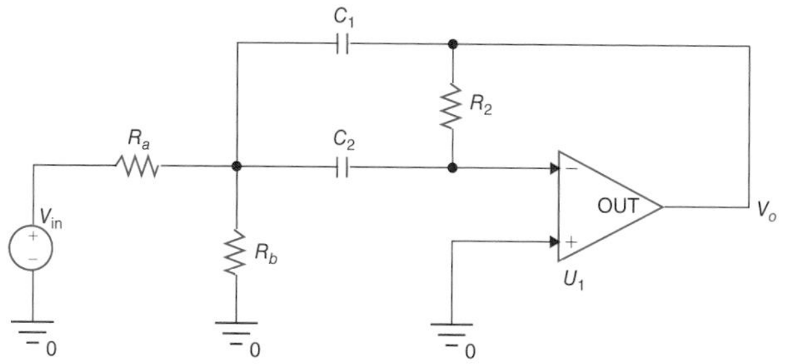

$R_a = 2Q^2, \quad R_b = 2Q^2/(2Q^2 - 1), \quad R_2 = 4Q^2,$
$C_1 = 1/(2Q), \quad C_2 = 1/(2Q)$

$k_f = \omega_0, k_m$

Second-Order Bandstop Filter

$k_f = \omega_0, k_m$

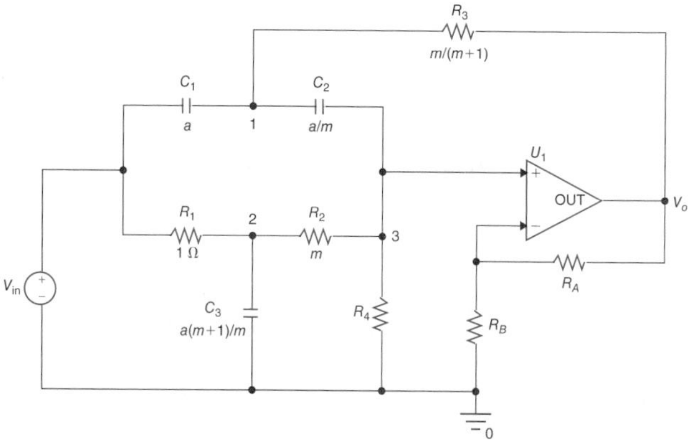

The circuit for case 1 ($\omega_0 \geq \omega_z$). For $\omega_0 = \omega_z$, remove R_4.

If $m = 1$, we have

$R_1 = 1, \quad R_2 = 1, \quad R_3 = 0.5, \quad R_4 = R,$
$R_B = 1, \quad R_A = K - 1.$

$C_1 = a, \quad C_2 = a, \quad C_3 = 2a.$

$$a = \frac{1}{\sqrt{B_0}}, \quad K = 2 + \frac{m}{m+1}\left(\frac{A_0}{B_0} - 1 - \frac{A_1}{\sqrt{B_0}}\right),$$

$$R = \frac{m+1}{\dfrac{A_0}{B_0} - 1}, \quad G = K.$$

continued

Table 16.2 continued

TABLE 16.2

$$H(s) = \frac{K(s^2 + b_0)}{s^2 + a_1 s + a_0} = \frac{K(s^2 + \omega_z^2)}{s^2 + \frac{\omega_0}{Q}s + \omega_0^2},$$

$$a_0 = \omega_0^2,\ a_1 = \frac{\omega_0}{Q},\ \omega_0 = \sqrt{a_0},\ Q = \frac{\omega_0}{a_1},\ \frac{1}{Q} = \frac{a_1}{\omega_0},$$

$$H(S) = \frac{K(S^2 + B_0)}{S^2 + A_1 S + A_0} = \frac{K(S^2 + B_0)}{S^2 + \frac{1}{Q}s + 1},$$

$$A_0 = 1,\ \ A_1 = \frac{1}{Q} = \frac{a_1}{\omega_0},\ \ B_0 = \frac{\omega_z^2}{\omega_0^2} = \frac{b_0}{a_0},$$

$$K = 1 + \frac{R_A}{R_B}.$$

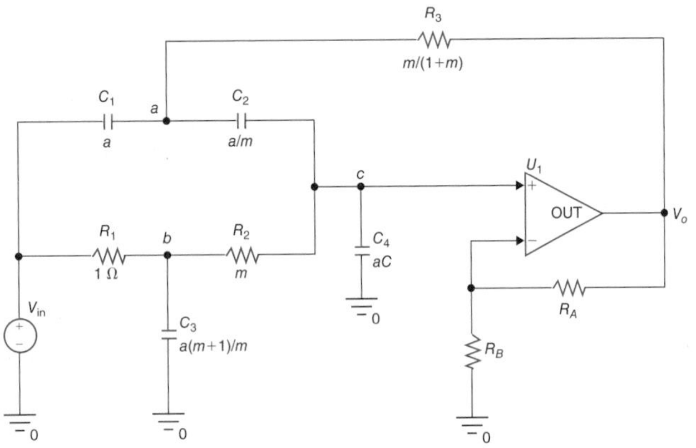

The circuit for case 2 ($\omega_0 \le \omega_z$). Remove C_4 for $\omega_0 = \omega_z$.

$$a = \frac{1}{\sqrt{B_0}},\quad C = \frac{\frac{B_0}{A_0} - 1}{m + 1},\quad C_4 = aC = a\frac{\frac{B_0}{A_0} - 1}{m + 1},$$

$$K = 2 + \frac{m}{m+1}\left(\frac{B_0}{A_0} - 1 - \frac{A_1\sqrt{B_0}}{A_0}\right),$$

$$G = \frac{A_0}{B_0}K.$$

If $m = 1$, we have

$R_1 = 1,\ \ R_2 = 1,\ \ R_3 = 0.5,\ \ R_B = 1,$
$R_A = K - 1.$

$C_1 = a,\ \ C_2 = a,\ \ C_3 = 2a,\ \ C_4 = aC.$

Gain Control

If the required gain G is less than 1, use the voltage divider circuit shown below. If K is the gain of the filter, choose $G = 1/K$ to make the combined gain equal to 1.

$$G = \frac{R_2}{R_1 + R_2}$$

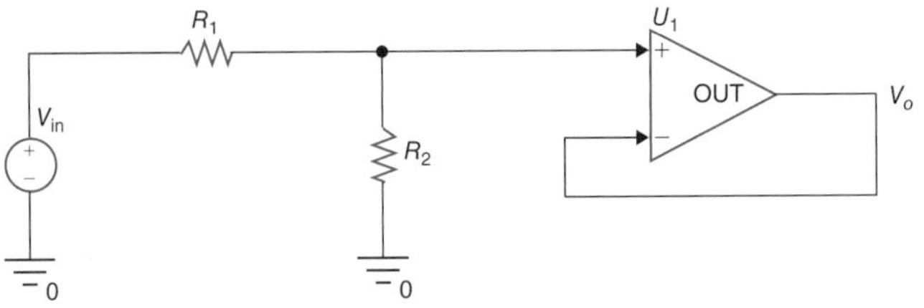

If the required gain G is greater than 1, use the noninverting amplifier circuit shown below. If K is the gain of the filter, choose $G = 1/K$ to make the combined gain equal to 1.

$$G = \left(1 + \frac{R_1}{R_2}\right)V_{in}$$

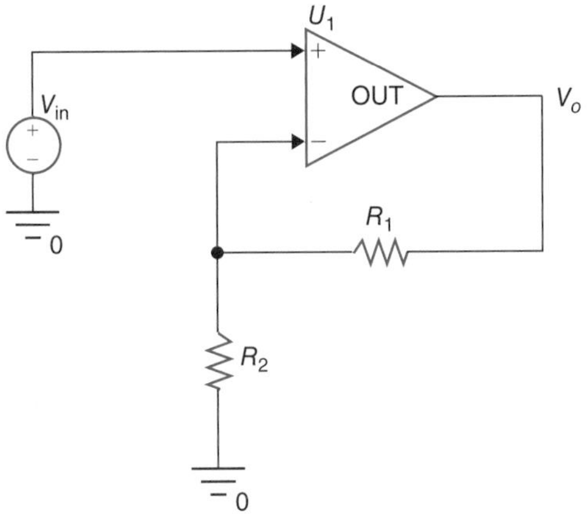

PROBLEMS

LPF Design

16.1 **Prove that the peak frequency for the second-order LPF with the frequency response given by Equation (16.21) is**

$$\omega_{\text{peak}} = \omega_0\sqrt{1 - \frac{1}{2Q^2}}$$

16.2 **Prove that the zero-crossing frequency ω_{zc} for the second-order LPF with the frequency response given by Equation (16.21) is**

$$\omega_{zc} = \omega_0\sqrt{2\left(1 - \frac{1}{2Q^2}\right)}$$

16.3 **Prove that the 3-dB cutoff frequency ω_c for the second-order LPF with the frequency response given by Equation (16.21) is**

$$\omega_c = \omega_0\sqrt{\frac{2 - \frac{1}{Q^2} + \sqrt{\left(2 - \frac{1}{Q^2}\right)^2 + 4}}{2}}$$

16.4 **Let $R = 1\ k\Omega, L = 10$ mH, and $C = 0.1\ \mu F$ for the circuit shown in Figure P16.4.**

FIGURE P16.4

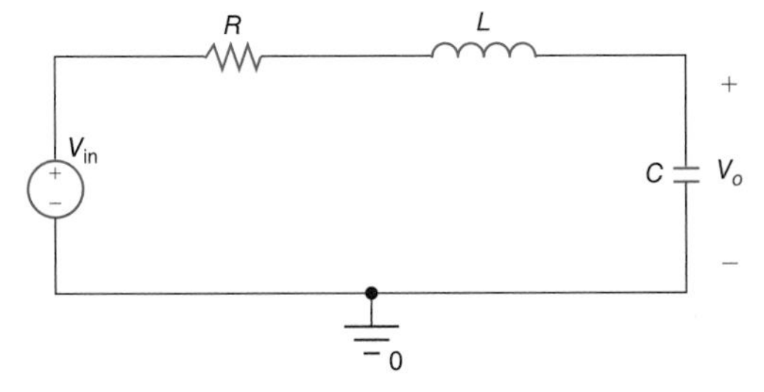

a. Find the transfer function $H(s)$.
b. Find ω_0.
c. Find Q.
d. Find the 3-dB cutoff frequency.
e. Plot the magnitude response $|H(\omega)|$ and the phase response $\angle H(\omega)$ in linear scale.
f. Plot the magnitude response $|H(\omega)|$ in dB scale.

16.5 **Design an LPF with the transfer function**

$$H(s) = \frac{56{,}984{,}067.9}{s^2 + 10{,}675.586s + 56{,}984{,}067.9}$$

using a series RLC circuit. Assume that $k_m = 1000$.

16.6 **Let $R = 10\ k\Omega, L = 50$ mH, and $C = 0.01\ \mu F$ for the circuit shown in Figure P16.16.**

FIGURE P16.6

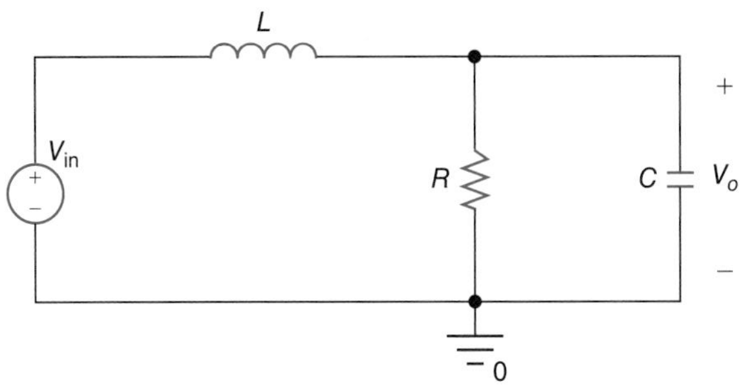

a. Find the transfer function $H(s)$.
b. Find ω_0.
c. Find Q.
d. Find the 3-dB cutoff frequency.
e. Plot the magnitude response $|H(\omega)|$ and the phase response $\angle H(\omega)$ in linear scale.
f. Plot the magnitude response $|H(\omega)|$ in dB scale.

16.7 **Design an LPF with the transfer function**

$$H(s) = \frac{51{,}620{,}410.4}{s^2 + 10{,}160.749s + 51{,}620{,}410.4}$$

using a parallel RLC circuit. Assume that $k_m = 1000$.

16.8 **Design an LPF with the transfer function**

$$H(s) = \frac{1.11721 \times 10^8}{s^2 + 14{,}948s + 1.11721 \times 10^8}$$

using a parallel RLC circuit. Assume that $k_m = 2000$.

a. Find ω_o and Q.
b. Find the component values R, L, C for the normalized filter.
c. Find the component values R, L, C after the frequency scale and the magnitude scale.

16.9 **Let $R_1 = 10\ k\Omega, R_2 = 10\ k\Omega, C_1 = 0.01\ \mu F$, $C_2 = 0.01\ \mu F$ for the circuit shown in Figure P16.9.**

FIGURE P16.9

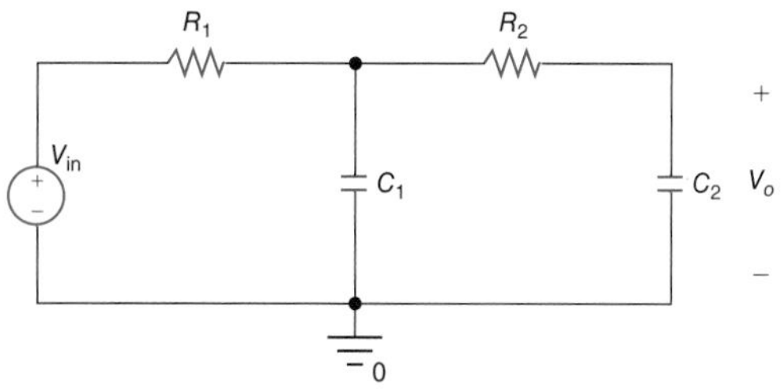

a. Find the transfer function $H(s)$.
b. Find ω_0.
c. Find Q.
d. Find the 3-dB cutoff frequency.
e. Plot the magnitude response $|H(\omega)|$ and the phase response $\angle H(\omega)$.

16.10 **Let $R_1 = 10\ k\Omega$, $R_2 = 10\ k\Omega$, $R_3 = 10\ k\Omega$, $C_1 = 0.01\ \mu F$, $C_2 = 0.01\ \mu F$, $C_3 = 0.01\ \mu F$ for the circuit shown in Figure P16.10.**

FIGURE P16.10

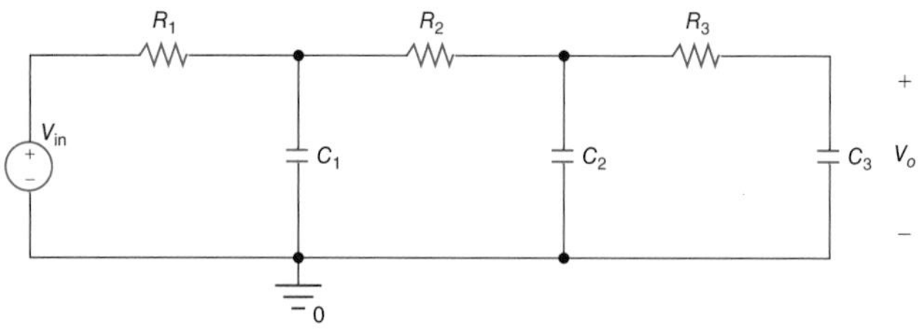

a. Find the transfer function $H(s)$.
b. Find the 3-dB cutoff frequency.
c. Plot the magnitude response $|H(\omega)|$ and the phase response $\angle H(\omega)$ in linear scale.
d. Plot the magnitude response $|H(\omega)|$ in dB scale.

16.11 **Design a second-order LPF with $Q = 3.5$ and $\omega_0 = 2\pi 2500$ using the unity gain circuit shown in Figure P16.11. Find the values of R_1, R_2, C_1, and C_2 when $k_m = 1000$.**

FIGURE P16.11

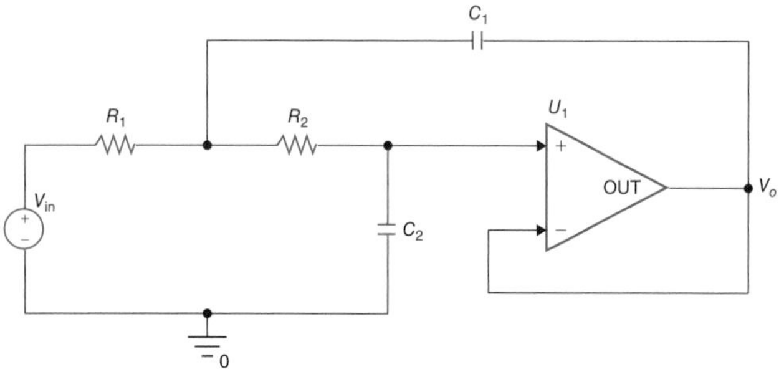

16.12 **The transfer function of a filter is given by**

$$H(s) = \frac{\omega_0^2}{s^2 + \dfrac{\omega_0}{Q}s + \omega_0^2} = \frac{4.9 \times 10^7}{s^2 + 2800s + 4.9 \times 10^7}$$

a. Find ω_o and Q.
b. This filter is implemented using a Sallen-Key circuit with the unity gain configuration, as shown in Figure P16.12. Find the component values R_1, R_2, C_1, and C_2 when $k_m = 10{,}000$, $k_f = \omega_0$.

FIGURE P16.12

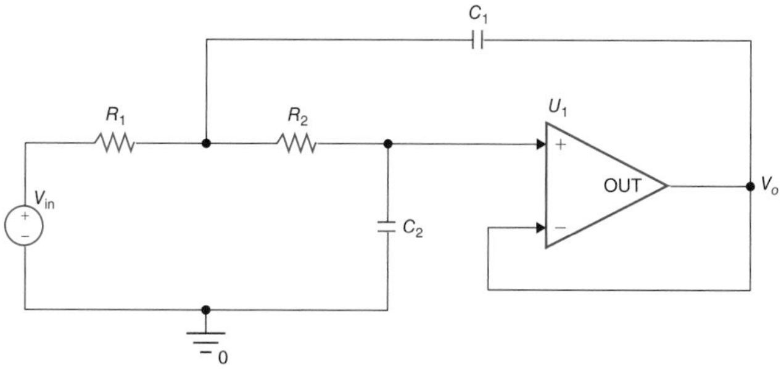

16.13 **Design a fourth-order LPF using two unity gain Sallen-Key circuits. The first second-order section has $Q = 1.3066$ and $\omega_0 = 6718.86$ rad/s, and the second second-order section has $Q = 0.5412$ and $\omega_0 = 6718.86$ rad/s. Assume that $k_m = 1000$.**

16.14 **Design a fifth-order LPF using two unity gain Sallen-Key circuits and one first-order circuit. The first second-order section has $Q = 1.5$ and $\omega_0 = 6718.86$ rad/s, the second second-order section has $Q = 0.5$ and $\omega_0 = 6718.86$ rad/s, and the first-order section has $\omega_0 = 6718.86$ rad/s. Assume that $k_m = 1000$.**

16.15 **Design an active LPF with the transfer function**

$$H(s) = \frac{10{,}297.33}{s + 10{,}297.33} \times \frac{1.06035 \times 10^8}{s^2 + 16{,}661.44s + 1.06035 \times 10^8} \times \frac{1.06035 \times 10^8}{s^2 + 6364.1s + 1.06035 \times 10^8}$$

using the unity gain method. Assume that $k_m = 1000$.

a. Find ω_o.
b. Find Q for each second-order section.

c. Find all the component values for the normalized filter.
d. Find all the component values after the frequency scale and the magnitude scale.

16.16 The transfer function of a fourth-order LPF is given by

$$H(s) = \frac{44{,}457{,}526}{s^2 + 12{,}320s + 44{,}457{,}526} \times \frac{44{,}457{,}526}{s^2 + 5103s + 44{,}457{,}526}$$

This filter can be implemented by cascading two second-order LPFs. Each second-order section is implemented using a Sallen-Key circuit with the equal R, equal C design with gain of 1 at $\omega = 0$.

a. Find ω_0 and Q for filter #1 with the transfer function $H(s) = \dfrac{44{,}457{,}526}{s^2 + 12{,}320s + 44{,}457{,}526}$.
b. Find the normalized component values (all R and C) for filter #1.
c. Find the component values for filter #1 after magnitude scaling and frequency scaling. $k_m = 1000$.
d. Find ω_0 and Q for filter #2 with the transfer function $H(s) = \dfrac{44{,}457{,}526}{s^2 + 5103s + 44{,}457{,}526}$.
e. Find the normalized component values (all R and C) for filter #2.
f. Find the component values for filter #2 after magnitude scaling and frequency scaling. $k_m = 1000$.

16.17 Design an active LPF with the transfer function

$$H(s) = \frac{8732.4802}{s + 8732.4802} \times \frac{7.6256 \times 10^7}{s^2 + 14{,}129.45s + 7.6256 \times 10^7} \times \frac{7.6256 \times 10^7}{s^2 + 5396.97s + 7.6256 \times 10^7}$$

using the equal R, equal C method with a gain of 1 for each first-order and second-order sections. $k_m = 2000$.

16.18 Design a second-order LPF with $Q = 2.5$ and $\omega_0 = 2\pi 2000 = 12{,}566.3706$ rad/s using the equal R, equal C circuit shown in Figure P16.18. R_a and R_b are chosen to have a gain of 1 at $\omega = 0$. Find the values of R_a, R_b, R_2, R_A, R_B, C_1, and C_2 when $k_m = 1000$.

FIGURE P16.18

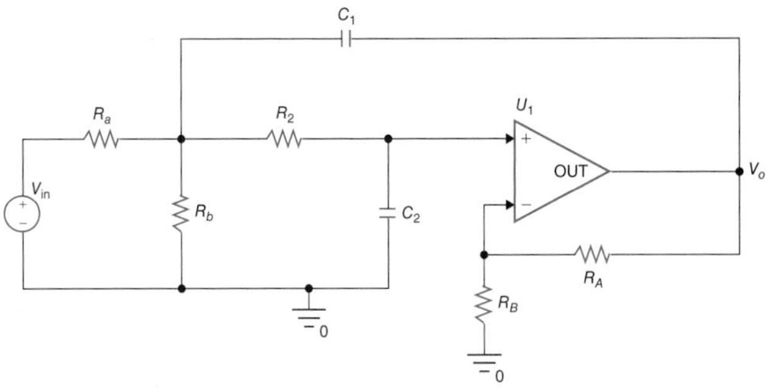

HPF Design

16.19 Prove that the peak frequency for the second-order HPF with the frequency response given by Equation (16.67) is

$$\omega_{\text{peak}} = \frac{1}{\sqrt{1 - \dfrac{1}{2Q^2}}}\omega_0 = \sqrt{\frac{2\omega_0^2 Q^2}{2Q^2 - 1}} = \frac{\sqrt{2}\omega_0 Q}{\sqrt{2Q^2 - 1}}$$

16.20 Prove that the zero-crossing frequency ω_{zc} for the second-order HPF with the frequency response given by Equation (16.67) is

$$\omega_{zc} = \frac{\omega_0}{\sqrt{2 - \dfrac{1}{Q^2}}}$$

16.21 Prove that the 3-dB cutoff frequency ω_c for the second-order HPF with the frequency response given by Equation (16.67) is

$$\omega_c = \omega_0 \frac{\sqrt{-2Q^2 + 1 + \sqrt{8Q^4 - 4Q^2 + 1}}}{\sqrt{2}\,Q}$$

16.22 Let $R = 1\ k\Omega$, $L = 100$ mH, and $C = 0.01\ \mu F$ for the circuit shown in Figure P16.22.

FIGURE P16.22

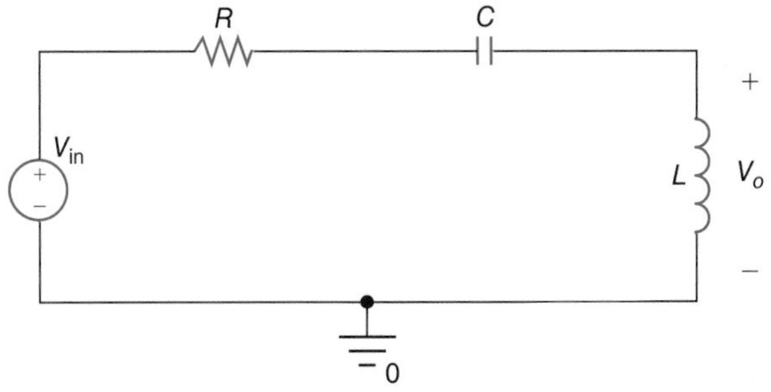

a. Find the transfer function $H(s)$.
b. Find ω_0.
c. Find Q.

d. Find the 3-dB cutoff frequency.
e. Plot the magnitude response $|H(\omega)|$ and the phase response $\angle H(\omega)$ in linear scale.
f. Plot the magnitude response $|H(\omega)|$ in dB scale.

16.23 Design an HPF with the transfer function

$$H(s) = \frac{s^2}{s^2 + 17{,}699.1877s + 156{,}630{,}621.85}$$

using a series *RLC* circuit. Assume that $k_m = 1000$.

16.24 Let $R = 2\ k\Omega$, $L = 50$ mH, and $C = 0.01\ \mu F$ for the circuit shown in Figure P16.24.

FIGURE P16.24

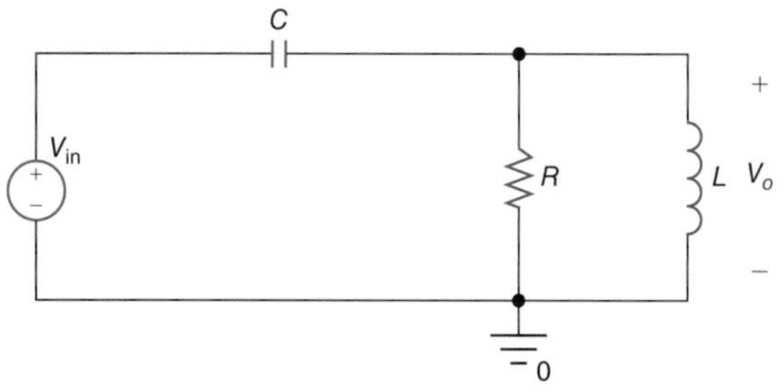

a. Find the transfer function $H(s)$.
b. Find ω_0.
c. Find Q.
d. Find the 3-dB cutoff frequency.
e. Plot the magnitude response $|H(\omega)|$ and the phase response $\angle H(\omega)$ in linear scale.
f. Plot the magnitude response $|H(\omega)|$ in dB scale.

16.25 Design an HPF with the transfer function

$$H(s) = \frac{s^2}{s^2 + 18{,}261.308s + 166{,}737{,}676.2}$$

using a parallel *RLC* circuit. $k_m = 1000$.

16.26 The transfer function of a filter is given by

$$H(s) = \frac{s^2}{s^2 + \frac{\omega_o}{Q}s + \omega_o^2} = \frac{s^2}{s^2 + 5000s + 1.44 \times 10^8}$$

a. Find ω_0 and Q.
b. This filter is implemented using a Sallen-Key circuit with the unity gain configuration, as shown in Figure P16.26. Find the component values R_1, R_2, C_1, and C_2 when $k_m = 10{,}000$, $k_f = \omega_0$.

FIGURE P16.26

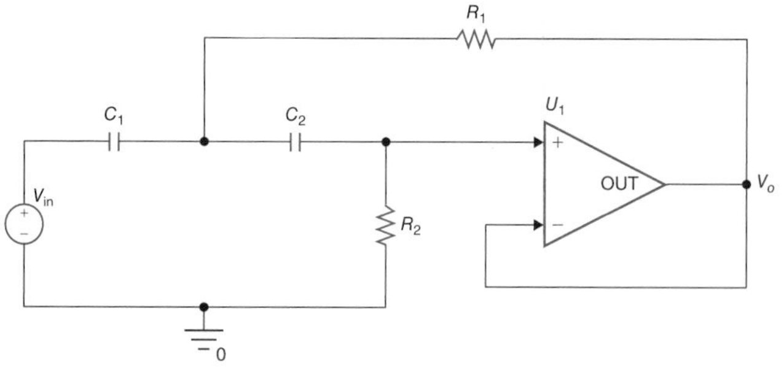

16.27 Design a second-order HPF with $Q = 4.5$ and $\omega_0 = 2\pi 5000 = 31{,}415.9265$ rad/s using the unity gain circuit shown in Figure P16.27. Find the values of R_1, R_2, C_1, and C_2 when $k_m = 1000$.

FIGURE P16.27

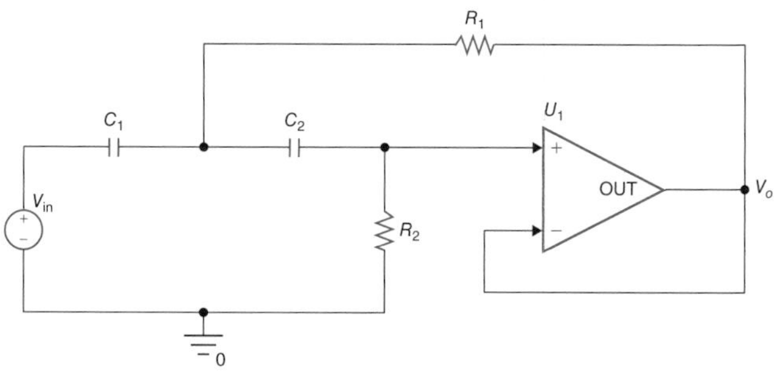

16.28 Design a fourth-order HPF using two unity gain Sallen-Key circuits. The first second-order section has $Q = 1.3066$ and $\omega_0 = 10{,}576.38$ rad/s, and the second second-order section has $Q = 0.5412$ and $\omega_0 = 10{,}576.38$ rad/s. Assume that $k_m = 1000$.

16.29 Design a fourth-order HPF using two unity gain Sallen-Key circuits. The transfer function of this filter is given by

$$H(s) = \frac{s^2}{s^2 + 27{,}272s + 4.5781 \times 10^8} \times \frac{s^2}{s^2 + 3199s + 1.2966 \times 10^8}$$

Assume that $k_m = 1000$.

a. Find ω_0 and Q for each second-order section.
b. For each second-order section, find the normalized component values.
c. For each second-order section, find the scaled component values.

16.30 The transfer function of a filter is given by

$$H(s) = \frac{s^2}{s^2 + \frac{\omega_o}{Q}s + \omega_o^2} = \frac{s^2}{s^2 + 3000s + 1.44 \times 10^8}$$

a. Find ω_0 and Q.
b. This filter is implemented using a Sallen-Key circuit with the equal R, equal C configuration as shown in Figure P16.30. Find the normalized component values.
c. Find the scaled component values when $k_m = 1000$.

FIGURE P16.30

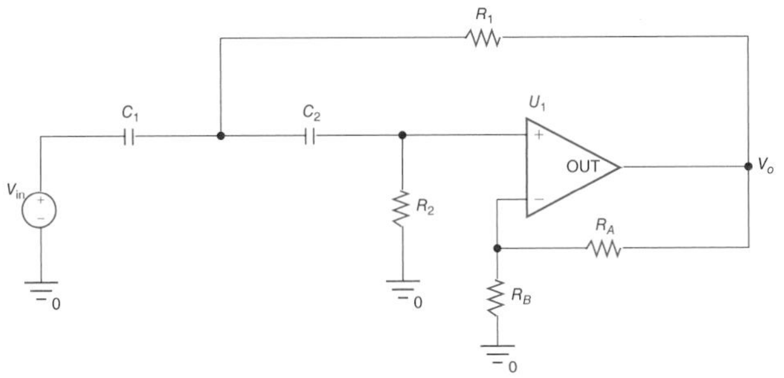

16.31 **Design a second-order HPF with $Q = 1.5$ and $\omega_0 = 2\pi 3000 = 18{,}849.5559$ rad/s using the equal R, equal C circuit shown in Figure P16.31. Find the values of R_1, R_2, R_A, R_B, C_1, C_2, and K when $k_m = 1000$.**

FIGURE P16.31

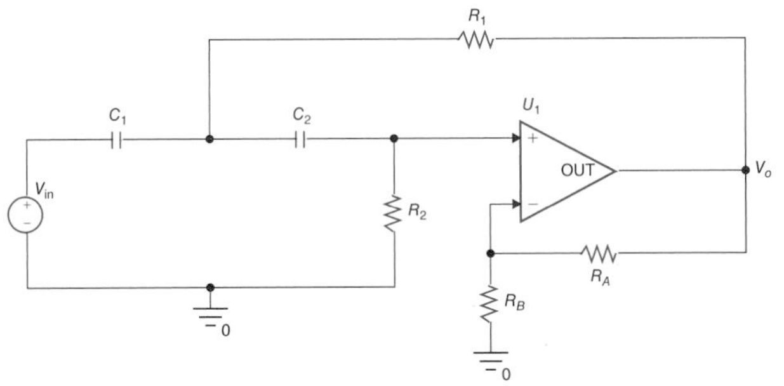

16.32 **Find the transfer function $H(s) = V_o(s)/V_{in}(s)$ of the circuit shown in Figure P16.32.**

FIGURE P16.32

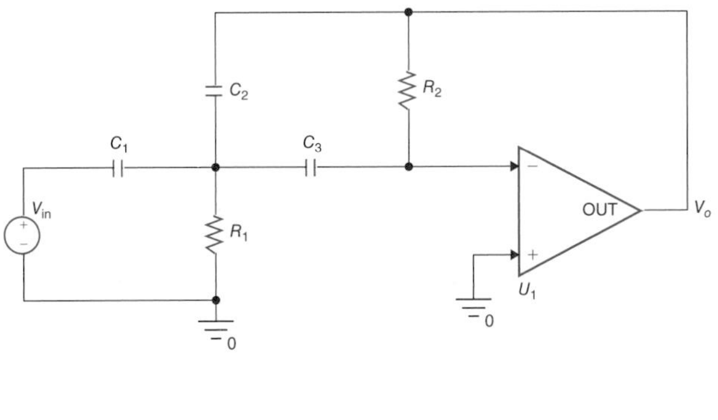

Bandpass Filter Design

16.33 **Let $R = 1\ k\Omega$, $L = 10$ mH, and $C = 0.01\ \mu F$ for the circuit shown in Figure P16.33.**

FIGURE P16.33

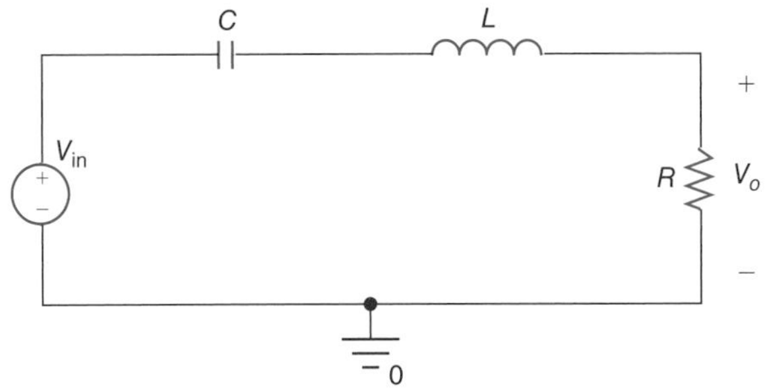

a. Find the transfer function $H(s)$.
b. Find ω_0.
c. Find Q.
d. Find the 3-dB cutoff frequencies ω_1 and ω_2 and the 3-dB bandwidth.
e. Plot the magnitude response $|H(\omega)|$ and the phase response $\angle H(\omega)$.

16.34 **Design a bandpass filter with the transfer function**

$$H(s) = \frac{6710s}{s^2 + 6710s + 208{,}149{,}957}$$

using a series RLC circuit. Assume that $k_m = 1000$.

16.35 **Let $R = 50\ k\Omega$, $L = 10$ mH, and $C = 0.001\ \mu F$ for the circuit shown in Figure P16.35.**

FIGURE P16.35

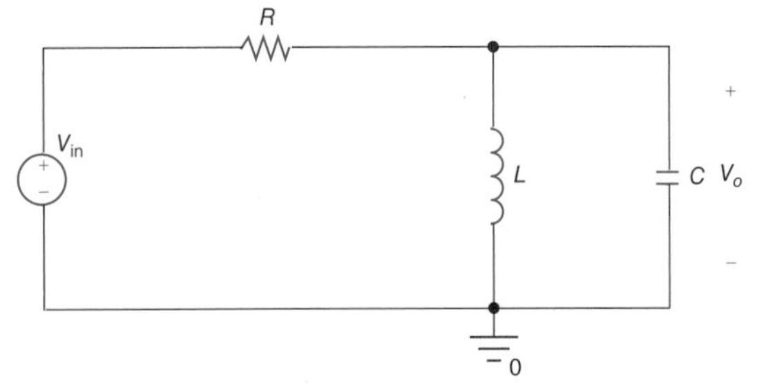

a. Find the transfer function $H(s)$.
b. Find ω_0.
c. Find Q.
d. Find the 3-dB cutoff frequencies ω_1 and ω_2 and the 3-dB bandwidth.
e. Plot the magnitude response $|H(\omega)|$ and the phase response $\angle H(\omega)$.

16.36 Design a bandpass filter with the transfer function

$$H(s) = \frac{7394s}{s^2 + 7394s + 1.8308116 \times 10^8}$$

using a parallel *RLC* circuit. Assume that $k_m = 5000$.

a. Find ω_0 and Q.
b. Find the component values R, L, C for the normalized filter.
c. Find the component values R, L, and C after the frequency scale and the magnitude scale.

16.37 Design a bandpass filter with the transfer function

$$H(s) = \frac{12{,}390.94s}{s^2 + 12{,}390.94s + 1.0782 \times 10^9}$$

using a parallel *RLC* circuit. $k_m = 1000$.

16.38 The transfer function of a filter is given by

$$H(s) = \frac{-\dfrac{\omega_o}{Q}s}{s^2 + \dfrac{\omega_o}{Q}s + \omega_o^2} = \frac{-6200s}{s^2 + 6200s + 2.25 \times 10^8}$$

a. Find ω_0 and Q.
b. This filter is implemented using the circuit shown in Figure P16.38. Find the component values R_2, R_a, R_b, C_1, and C_2 when $k_m = 1000$ and $k_f = \omega_0$.

FIGURE P16.38

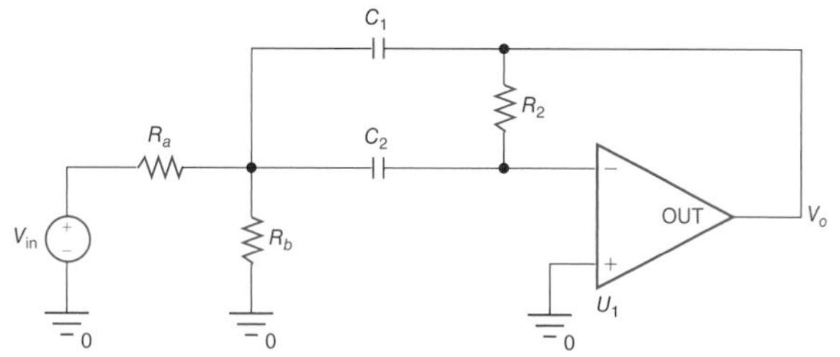

16.39 Design a fourth-order BPF using a Deliyannis-Friend circuit with the gain adjusted to 1 using R_a and R_b for each second-order section. The transfer function of this filter is given by

$$H(s) = \frac{4072s}{s^2 + 4072s + 3.4143 \times 10^8} \times \frac{2825s}{s^2 + 2825s + 1.6433 \times 10^8}$$

Assume that $k_m = 1000$.

16.40 A fourth-order bandpass filter with the transfer function

$$H(s) = \frac{4242s}{s^2 + 4242s + 169{,}778{,}736} \times \frac{5919s}{s^2 + 5919s + 330{,}475{,}051}$$

is implemented using the circuit shown in Figure P16.40 (two sections).

FIGURE P16.40

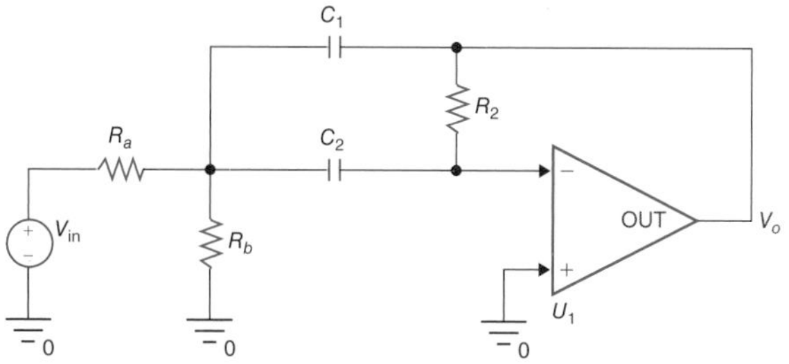

a. Find ω_0 and Q value for both second-order sections.
b. Find the normalized values of R_a, R_b, R_2, C_1, and C_2 for both second-order sections.
c. Let the magnitude scale factor be $k_m = 1000$. Find the scaled (both magnitude and frequency) values of R_a, R_b, R_2, C_1, and C_2 for both sections.

16.41 Design a second-order bandpass filter with $Q = 2.5$ and $\omega_0 = 2\pi 2700 = 16{,}964.6003$ rad/s using a equal *C* circuit (Deliyannis-Friend circuit) shown in Figure P16.41. R_a and R_b are chosen to have a gain of 1 at $\omega = \omega_0$. Find the values of R_2, R_a, R_b, C_1, and C_2 when $k_m = 1000$.

FIGURE P16.41

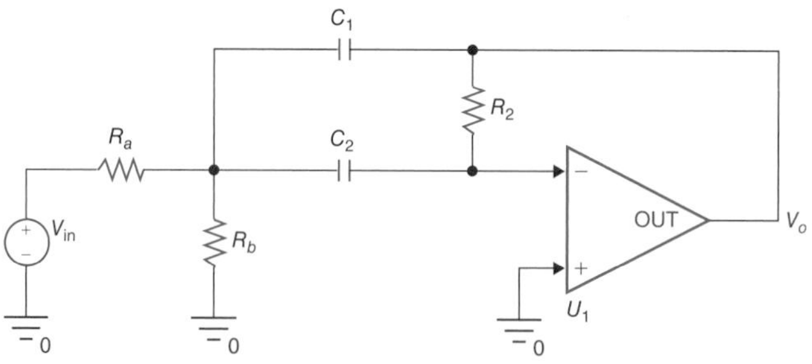

16.42 Design a fourth-order BPF using two unity gain Deliyannis-Friend circuits. The first second-order section has $Q = 3.072$ and $\omega_0 = 13{,}030$ rad/s, and the second second-order section has $Q = 3.072$ and $\omega_0 = 18{,}179$ rad/s. Assume that $k_m = 1000$.

16.43 **Design a second-order bandpass filter with $Q = 2.3$ and $\omega_0 = 2\pi 2000 = 12{,}566.3706$ rad/s using the equal R, equal C Sallen-Key circuit shown in Figure P16.43. R_a and R_b are chosen to have a gain of 1 at $\omega = \omega_0$. Find the values of R_a, R_b, R_2, R_3, R_A, R_B, C_1, and C_2 when $k_m = 1000$.**

FIGURE P16.43

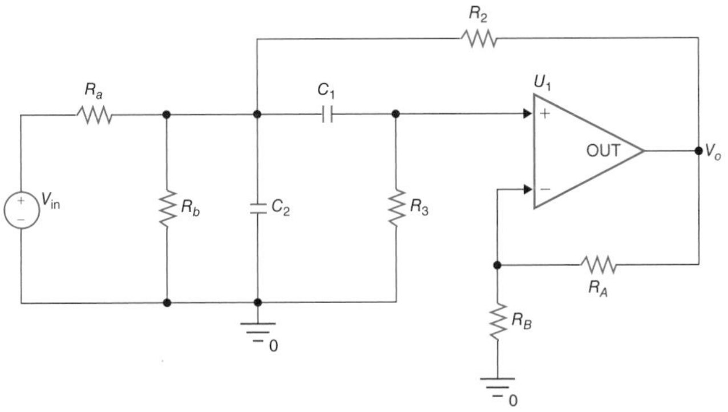

16.44 **Design a fourth-order BPF using two Sallen-Key circuits shown in Figure P16.43. The first second-order section has $Q = 3.072$ and $\omega_0 = 13{,}030$ rad/s, and the second second-order section has $Q = 3.072$ and $\omega_0 = 18{,}179$ rad/s. Assume that $k_m = 1000$.**

Bandstop Filter Design

16.45 **Let $R = 1\ k\Omega$, $L = 10$ mH, $C = 0.01\ \mu F$ for the circuit shown in Figure P16.45.**

FIGURE P16.45

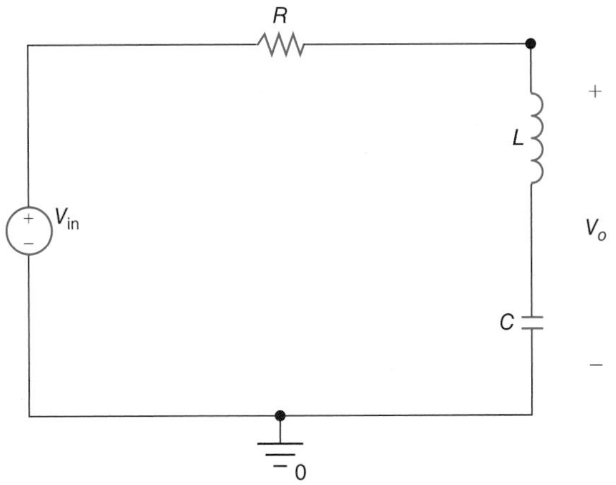

a. Find the transfer function $H(s)$.
b. Find ω_0.
c. Find Q.
d. Find the 3-dB cutoff frequencies ω_1 and ω_2 and the 3-dB bandwidth of the stopband.
e. Plot the magnitude response $|H(\omega)|$ and the phase response $\angle H(\omega)$.

16.46 **Design a bandstop filter with the transfer function**

$$H(s) = \frac{s^2 + 2.273957 \times 10^8}{s^2 + 13{,}592s + 2.273957 \times 10^8}$$

using a series RLC circuit. Assume that $k_m = 10{,}000$.

a. Find ω_0 and Q.
b. Find the component values R, L, and C for the normalized filter.
c. Find the component values R, L, and C after the frequency scale and the magnitude scale.

16.47 **Design a bandstop filter with the transfer function**

$$H(s) = \frac{s^2 + 204{,}103{,}419}{s^2 + 20{,}736.3s + 204{,}103{,}419}$$

using a series RLC circuit. Assume that $k_m = 1000$.

16.48 **Let $R = 5\ k\Omega$, $L = 15$ mH, $C = 0.001\ \mu F$ for the circuit shown in Figure P16.48.**

FIGURE P16.48

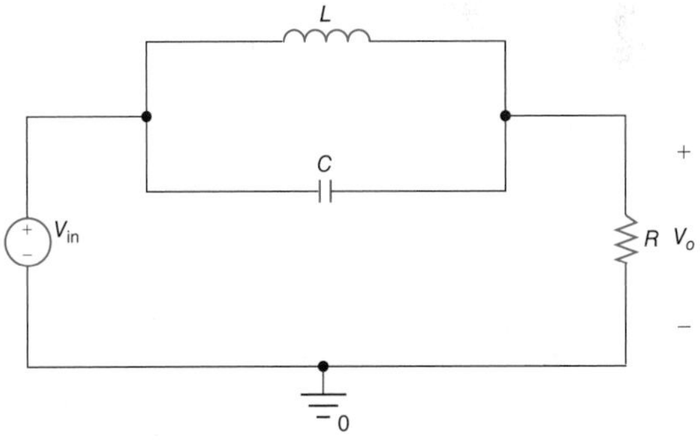

a. Find the transfer function $H(s)$.
b. Find ω_0.
c. Find Q.
d. Find the 3-dB cutoff frequencies ω_1 and ω_2 and the 3-dB bandwidth of the stopband.
e. Plot the magnitude response $|H(\omega)|$ and the phase response $\angle H(\omega)$.

16.49 **The transfer function of a filter is given by**

$$H(s) = \frac{s^2 + \omega_0^2}{s^2 + \dfrac{\omega_0}{Q}s + \omega_0^2} = \frac{s^2 + 9 \times 10^8}{s^2 + 15{,}500s + 9 \times 10^8}$$

FIGURE P16.49

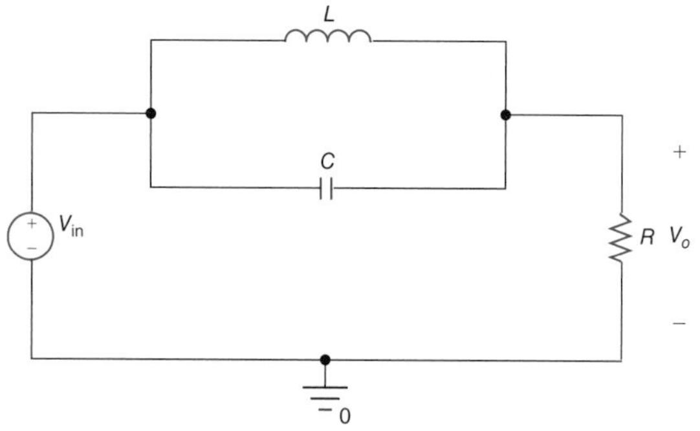

a. Find ω_0 and Q.
b. This filter is implemented using the circuit shown in Figure P16.49. Find the component values R, L, and C when $k_m = 1000$, $k_f = \omega_0$ assuming that the normalized value of $C = 1$ F.

16.50 **Design a bandstop filter with the transfer function**

$$H(s) = \frac{s^2 + 227{,}395{,}685}{s^2 + 13{,}591.81s + 227{,}395{,}685}$$

using a parallel *RLC* circuit. Assume that $k_m = 1000$.

16.51 **Design a second-order bandstop filter with $Q = 2.6$ and $\omega_0 = 2\pi 2700$ using the circuit shown in Figure P16.51. Find the values of R_A, R_B, R, and C when $k_m = 1000$.**

FIGURE P16.51

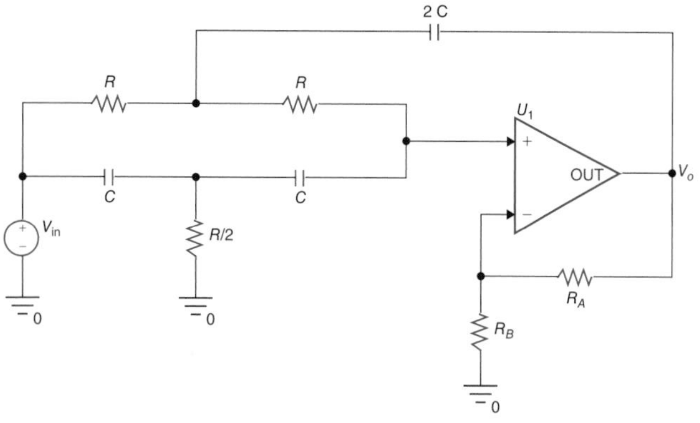

16.52 **Design a fourth-order bandstop filter. The first second-order section has $Q = 3.072$ and $\omega_0 = 13{,}030$ rad/s, and the second second-order section has $Q = 3.072$ and $\omega_0 = 18{,}179$ rad/s. Assume that $k_m = 1000$.**

General Filters

16.53 **In the circuit shown in Figure P16.53, let $R_1 = R_2 = 1\ k\Omega$ and $C_1 = C_2 = 0.1\ \mu F$.**

FIGURE P16.53

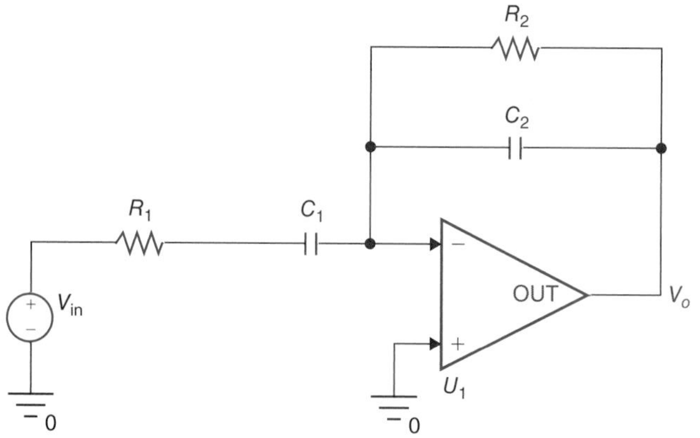

a. Find the transfer function $H(s) = V_o(s)/V_{in}(s)$.
b. What is the value of ω_0 and Q for this circuit?
c. What type of filter (LPF, HPF, BPF, or BSF) is this?

16.54 **In the circuit shown in Figure P16.54, let $R_1 = R_2 = 1\ \Omega$.**

FIGURE P16.54

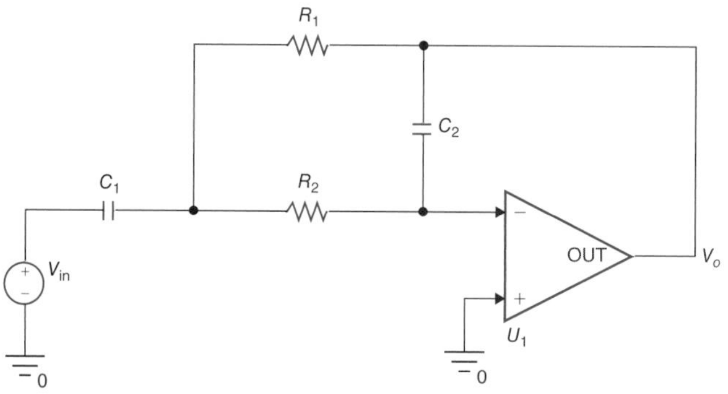

a. Find the transfer function $H(s) = V_o(s)/V_{in}(s)$.
b. Represent C_1 and C_2 as a function of Q if the normalized denominator polynomial is $s^2 + (1/Q)\,s + 1$.
c. Find the numerical values of C_1 and C_2 when $Q = 1.5$ for the normalized transfer function.
d. What type of filter (LPF, HPF, BPF, or BSF) is this?

16.55 **In the circuit shown in Figure P16.55, let $R_1 = R_2 = 2\ k\Omega$ and $C_1 = C_2 = 0.01\ \mu F$.**

FIGURE P16.55

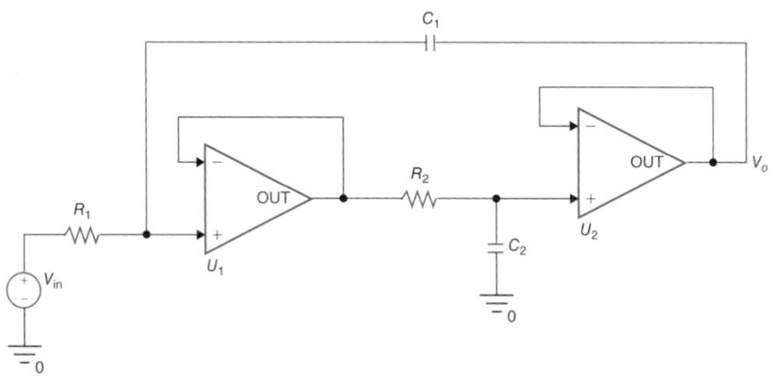

a. Find the transfer function $H(s) = V_o(s)/V_{in}(s)$.
b. What is the value of ω_0 and Q for this circuit?
c. What type of filter (LPF, HPF, BPF, or BSF) is this?

16.56 **In the circuit shown in Figure P16.56, the component vales are given by**

$$R_1 = 2\ \Omega,\quad R_2 = 1\ \Omega,\quad R_3 = 1\ \Omega,$$
$$C_1 = 1\ \text{F},\quad C_2 = 1\ \text{F}$$

FIGURE P16.56

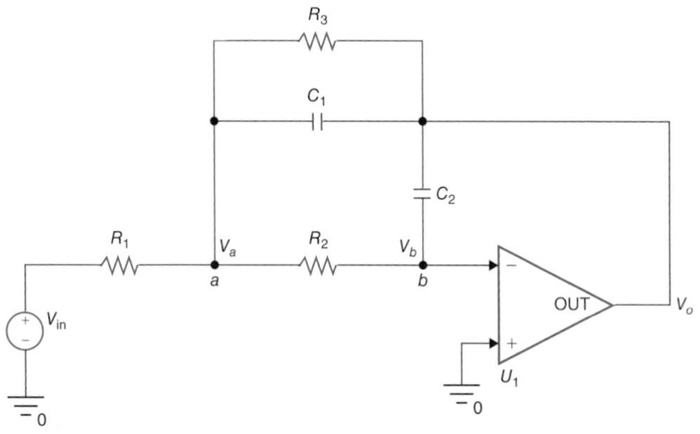

a. Write a node equation at node b by summing the currents leaving node b, and represent V_a as a function of V_o.
b. Write a node equation at node a by summing the currents leaving node a.
c. Find the transfer function $H(s) = V_o(s)/V_{in}(s)$.
d. What type of filter (LPF, HPF, BPF, or BSF) is this?

16.57 **In the circuit shown in Figure P16.57, the component values are given by**

$$R_1 = 1\ \Omega,\quad R_2 = 1\ \Omega,\quad R_3 = 1\ \Omega,$$
$$C_1 = 1\ \text{F},\quad C_2 = 0.5\ \text{F},\quad C_3 = 0.5\ \text{F},$$
$$R_A = 1\ \Omega, \text{and}\quad R_B = 1\ \Omega$$

FIGURE P16.57

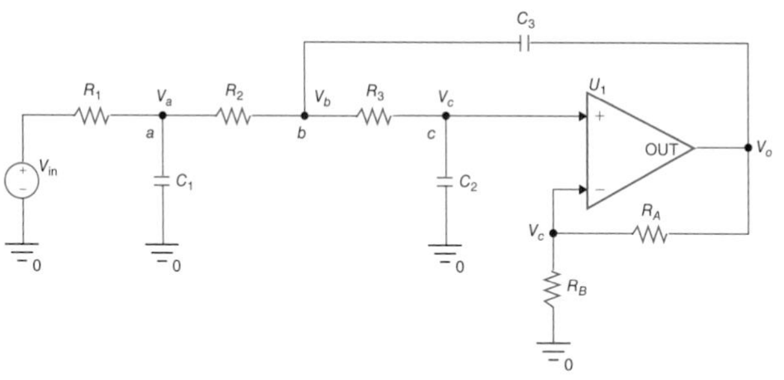

a. Find the transfer function $H(s) = V_o(s)/V_{in}(s)$.
b. What type of filter (LPF, HPF, BPF, or BSF) is this?

16.58 **In the circuit shown in Figure P16.58, $R_1 = R_2 = R_3 = R_4 = R_5 = R_6 = 1\ \Omega$.**

FIGURE P16.58

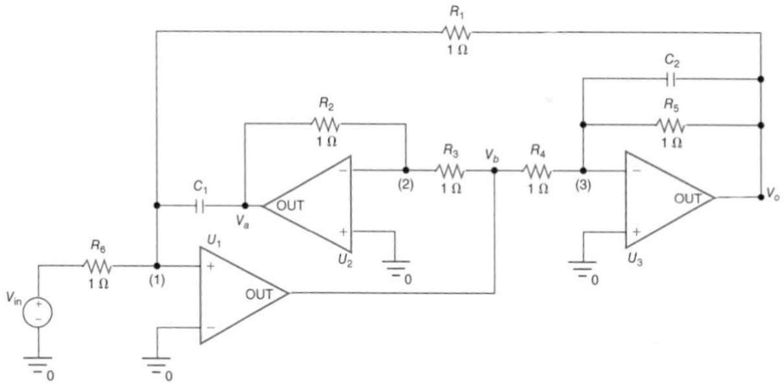

a. Write a node equation at node (3) by summing the currents away from node (3). Solve the node equation for V_b; that is, represent V_b as a function of V_o.
b. Write a node equation at node (2) by summing the currents away from node (2). Solve the node equation for V_a, and represent V_a as a function of V_o by using the result from part (a).
c. Write a node equation at node (1) by summing the currents away from node (1). Substitute the results from (a) and (b), and find the transfer function $H(s) = V_o(s)/V_{in}(s)$.
d. Represent C_1 and C_2 as a function of Q if the normalized denominator polynomial is $s^2 + (1/Q)\, s + 1$.
e. Find the numerical values of C_1 and C_2 when $Q = 1.8$ for the normalized transfer function.

16.59 **In the circuit shown in Figure P16.59, let $R_1 = 2\ \Omega, R_2 = 0.1\ \Omega, C = 0.5$ F, $L = 1$ H.**

FIGURE P16.59

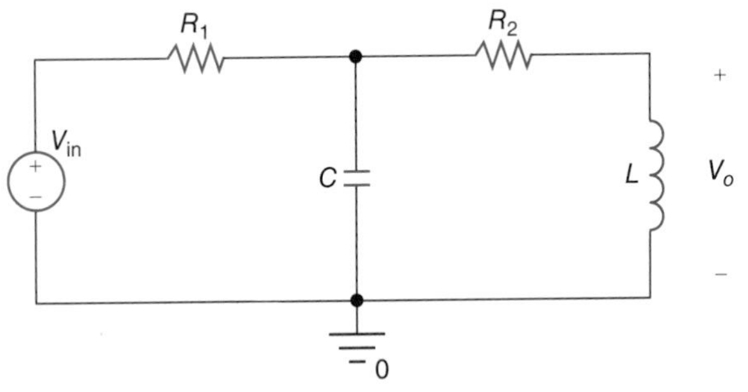

a. Find the transfer function $H(s) = V_o(s)/V_{in}(s)$.
b. What is the value of ω_0 and Q for this circuit?
c. What type of filter (LPF, HPF, BPF, or BSF) is this?

16.60 **For the circuit shown in Figure P16.60,**

FIGURE P16.60

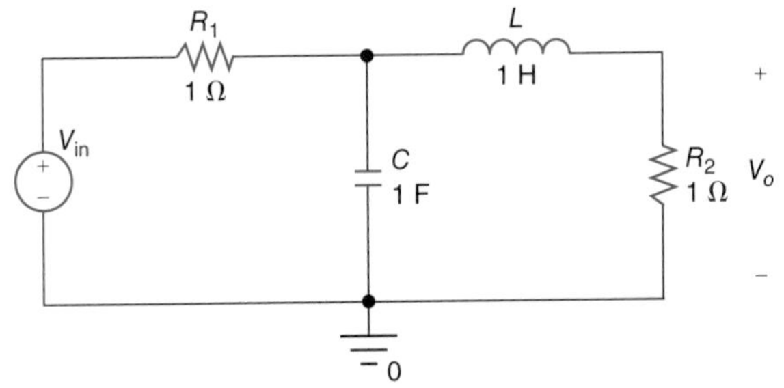

a. Find the transfer function $H(s) = V_o(s)/V_{in}(s)$.
b. Evaluate $H(s)$ for $s = j0$ and $s = j\infty$.
c. What type (LPF, HPF, BPF, or BSF) of filter is this circuit?

16.61 **Let $R_1 = 1\ \Omega, R_2 = 1\ \Omega, L = 0.5$ H, $C_1 = 1$ F, $C_2 = 1$ F in the circuit shown in Figure P16.61.**

FIGURE P16.61

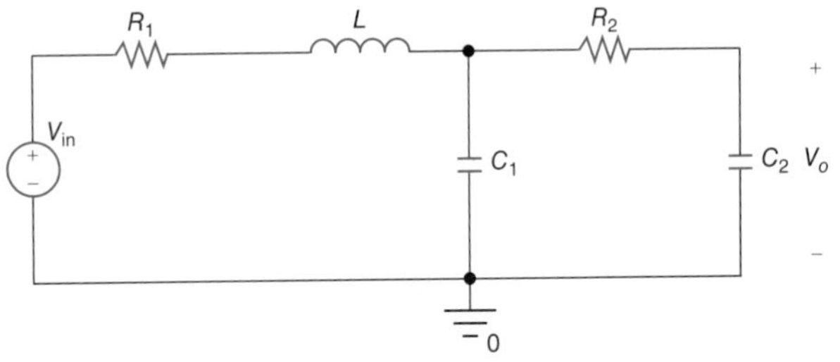

a. Find the transfer function $H(s) = V_o(s)/V_{in}(s)$.
b. What type of filter (LPF, HPF, BPF, BSF) is this?

16.62 **In the circuit shown in Figure P16.62, $C_1 = C_2 = 1$ F and $R_1 = R_2 = R_3 = 1\ \Omega$ and k ≠ 1.**

FIGURE P16.62

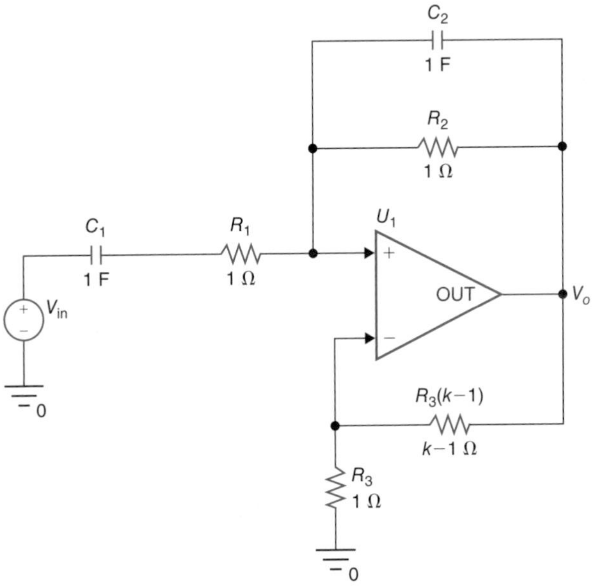

Find the transfer function $H(s) = V_o(s)/V_{in}(s)$. What type of filter (LPF, HPF, BPF, or BSF) is this circuit?

16.63 **Let $R_1 = R_2 = R_3 = R_4 = 1\ \Omega, C_1 = C_2 = 1$ F in the circuit shown in Figure P16.63. Find the transfer function $H(s) = V_o(s)/V_{in}(s)$. What type of filter (LPF, HPF, BPF, or BSF) is this?**

FIGURE P16.63

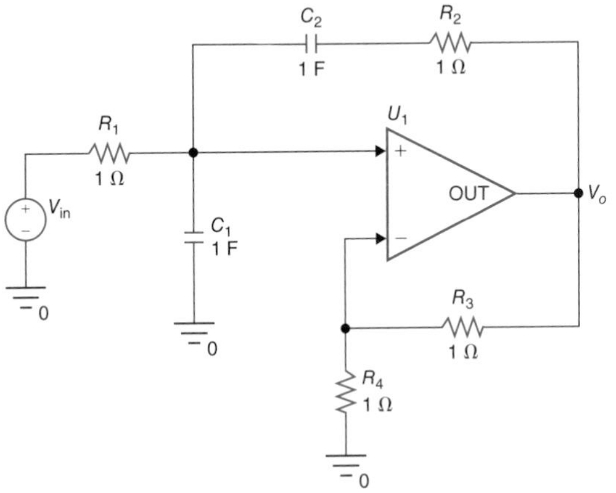

16.64 **The transfer function of an analog filter is given by**

$$H(s) = \frac{s + 16.66}{s^2 + 4.2s + 16.66}$$

a. Find the poles and the zeros of $H(s)$, and plot the pole-zero diagram.
b. Find $H(j0)$ and $H(j\infty)$, and plot the magnitude response of the filter; that is, $|H(\omega)|$.
c. Plot the phase response of the filter; that is, $\angle H(\omega)$.
d. What type of filter (LPF, HPF, BPF, or BSF) is $H(s)$?

16.65 For the biquad circuit shown in Figure P16.65, find

FIGURE P16.65

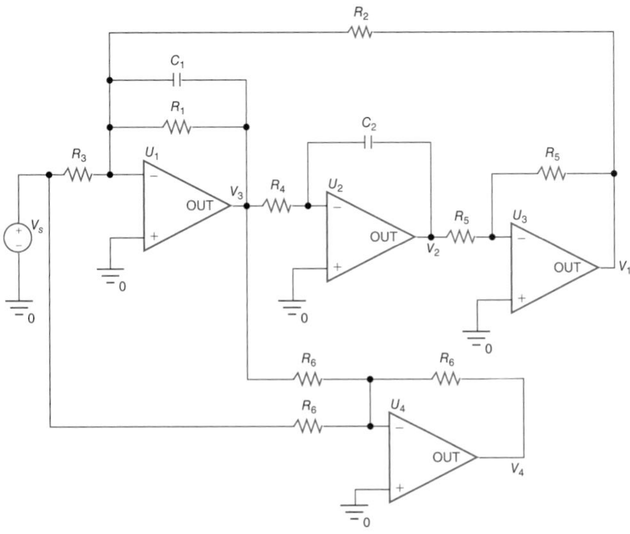

a. $V_1(s)/V_s(s)$
b. $V_3(s)/V_s(s)$
c. $V_4(s)/V_s(s)$

Assume that $R_1 = R_3$ for this question.

16.66 For the biquad circuit shown in Figure P16.66, find $V_4(s)/V_s(s)$. Assume that $R_1 = R_3$ and $R_2 = R_3$.

FIGURE P16.66

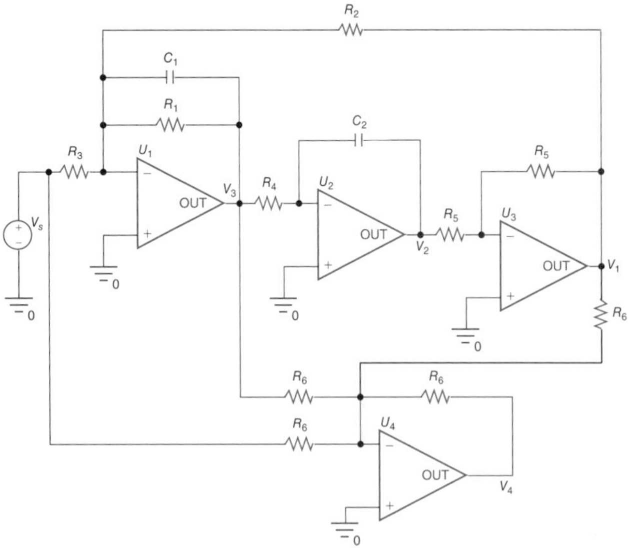

Chapter 17

Analog Filter Design

17.1 Introduction

In this chapter, the design and the implementation of analog filters are presented. The types of filters discussed are low-pass filter (LPF), high-pass filter (HPF), bandpass filter, and bandstop filter.

The specifications of filters are the performance criteria for the filters. The specifications include cutoff frequencies, maximum attenuation in the passband, and minimum attenuation in the stopband. If we are designing an LPF, we are given four numbers. These are the passband cutoff frequency, the stopband cutoff frequency, the maximum attenuation in the passband, and the minimum attenuation in the stopband. For the LPF, the passband refers to a band of frequencies from 0 hertz to the passband cutoff frequency f_p, and the stopband refers to a frequency band from the stopband cutoff frequency f_s to infinity ($f_p < f_s$). The passband cutoff frequency refers to the largest frequency of the passband. Below this frequency, the attenuation cannot exceed the maximum attenuation in the passband, A_p. The stopband cutoff frequency refers to the smallest frequency of the stopband. Above this frequency, the attenuation cannot be smaller than the minimum attenuation in the stopband, A_s.

The normalized LPF means that the passband cutoff frequency of an LPF is 1 rad/s. Regardless of the type of filter we design, we can first design a normalized LPF, whose specifications are derived from the filter we are designing, and then transform it to LPF, HPF, bandpass filter, and bandstop filter. The transformation can be achieved by transforming the poles and the zeros of the normalized LPF.

Once the poles, the zeros, and the transfer function are obtained, the filter can be implemented using the circuits discussed in Chapter 16. The Sallen-Key circuits can be used to implement the filters using an operational amp. The component values for the normalized filter are first found. Then, magnitude scaling and frequency scaling are applied to get the final component values. Once all the component values are selected, the circuit can be simulated in PSpice or Multisim to verify the design.

17.2 Analog Butterworth LPF Design

The **attenuation**, also called **loss**, is the opposite of gain and is expressed in decibels. When gain is large, attenuation is small, and when gain is small, attenuation is large. The attenuation A in decibels is defined as

$$A = -20 \log_{10}(G)$$

where G is gain. Table 17.1 shows the gain and attenuation in decibels.

TABLE 17.1

Gain vs. Attenuation

Gain	Attenuation (dB)
10	−20
1	0
0.1	20
0.01	40
0.001	60
0.0001	80
0.00001	100

Figure 17.1(a) shows the specifications of an LPF in gain, and Figure 17.1(b) shows the specifications of an LPF in attenuation in decibels.

The notations used in the specifications are as follows:

$\omega_p = 2\pi f_p$ = passband cutoff frequency in rad/s.
$\omega_s = 2\pi f_s$ = stopband cutoff frequency in rad/s.
G_p = minimum gain in the passband. Maximum gain in the passband is 1.
G_s = maximum gain in the stopband. Minimum gain in the stopband is 0.
δ_p = passband tolerance.
δ_s = stopband tolerance.
A_p = maximum attenuation in decibels in the passband.
A_s = minimum attenuation in decibels in the stopband.
ε_p = passband ripple parameter.
ε_s = stopband ripple parameter.

FIGURE 17.1

Specifications of LPF (a) in gain and (b) in attenuation (dB).

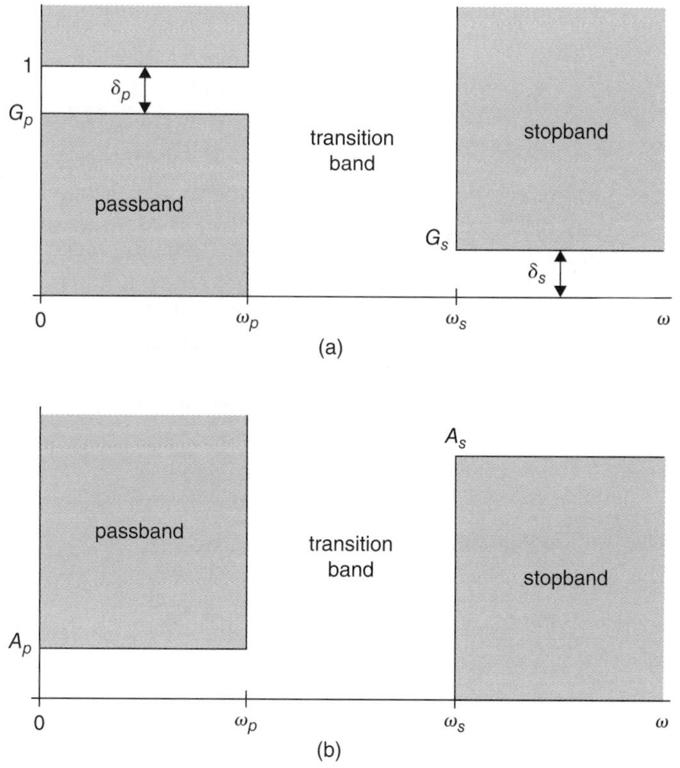

The parameters are related through the following equations:

$$G_p = 1 - \delta_p$$

$$G_s = \delta_s$$

$$A_p = -20 \log_{10} G_p = -20 \log_{10}(1 - \delta_p)$$

$$A_s = -20 \log_{10} G_s = -20 \log_{10} \delta_s$$

$$G_p = 1 - \delta_p = 10^{-\frac{A_p}{20}}$$

$$G_s = \delta_s = 10^{-\frac{A_s}{20}}$$

$$\delta_p = 1 - 10^{-\frac{A_p}{20}}$$

The filter design procedure can be broken down to several steps, including backward transformation, finding the order, adjusting the parameters to assign excess tolerance, finding normalized low-pass poles and zeros, forward transformation, finding the transfer function and component values of the second-order and first-order sections, magnitude and frequency scaling, implementing the filter, and verifying the design.

17.2.1 BACKWARD TRANSFORMATION

Regardless of the filter type (low-pass, high-pass, bandpass, or bandstop) we are designing, we initially design a normalized LPF and transform the filter back to the filter we are designing. The transformation of specifications from the original filter to the normalized LPF is called backward transformation. Designing the normalized LPF first and then transforming to the original filter simplifies the filter design procedure. The passband cutoff frequency of the normalized LPF is 1 rad/s. Ω is used for radian frequency for the normalized filter, and S is used for the complex variable for the normalized LPFs. ω and s are used for the frequency-transformed filters. If we are designing an LPF, the backward transformation is trivial. The normalized cutoff frequencies are obtained by dividing the cutoff frequencies of the original LPF by ω_p, which is the passband cutoff frequency. Thus, we have

$$\text{Normalized frequency: } \Omega = \frac{\omega}{\omega_p}$$

$$\text{Passband cutoff frequency: } \Omega_p = \frac{\omega_p}{\omega_p} = 1 \text{ rad/s}$$

$$\text{Stopband cutoff frequency: } \Omega_s = \frac{\omega_s}{\omega_p} \text{ rad/s}$$

Figure 17.2 shows the specifications for the normalized LPF. There are no changes in gain and attenuation specifications.

FIGURE 17.2

Specifications for the normalized LPF (a) in gain (b) in attenuation (dB).

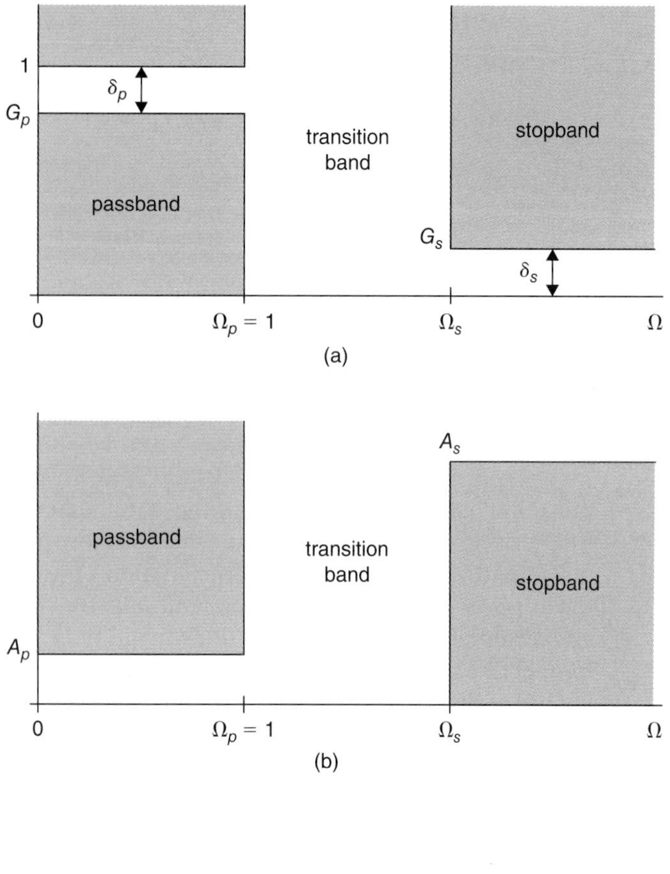

17.2.2 FINDING THE ORDER OF THE NORMALIZED LPF

The magnitude squared of the nth-order normalized Butterworth LPF is given by

$$|H_n(j\Omega)|^2 = \frac{1}{1 + \varepsilon_p^2 \Omega^{2n}} \tag{17.1}$$

where the passband ripple parameter ε_p is related to passband tolerance δ_p, passband gain G_p, and maximum attenuation in the passband A_p (refer to Equation (17.8)). If δ_p is decreased toward 0, G_p is increased toward 1, A_p is decreased toward 0, and ε_p is decreased toward 0. The magnitude response of the nth-order normalized Butterworth LPF is given by

$$|H_n(j\Omega)| = \frac{1}{\sqrt{1 + \varepsilon_p^2 \Omega^{2n}}} \tag{17.2}$$

The gain in decibels is given by

$$G = 20\log_{10}|H_n(j\Omega)| = 10\log_{10}\left[\frac{1}{1 + \varepsilon_p^2 \Omega^{2n}}\right]$$
$$= -10\log_{10}\left[1 + \varepsilon_p^2 \Omega^{2n}\right] \text{dB} \tag{17.3}$$

The attenuation (also called loss) A is defined as

$$A = -G \text{ dB} \tag{17.4}$$

Thus, the attenuation in decibels of the nth-order normalized Butterworth LPF is

$$\begin{aligned} A &= -20\log_{10}\frac{1}{\sqrt{1+\varepsilon_p^2\Omega^{2n}}} = -20\log_{10}\left(1+\varepsilon_p^2\Omega^{2n}\right)^{-\frac{1}{2}} \\ &= 10\log_{10}\left[1+\varepsilon_p^2\Omega^{2n}\right]\text{ dB} \end{aligned} \tag{17.5}$$

At the passband cutoff frequency $\Omega = \Omega_p = 1$ rad/s, from Equation (17.2), the gain is given by

$$|H_n(j\Omega_p)| = \frac{1}{\sqrt{1+\varepsilon_p^2}} \tag{17.6}$$

The attenuation in decibels at the passband cutoff frequency $\Omega = \Omega_p = 1$ rad/s is given by

$$A_p = -20\log_{10}|H(j\Omega_p)| = -20\log_{10}\frac{1}{\sqrt{1+\varepsilon_p^2}} = 10\log_{10}(1+\varepsilon_p^2) \tag{17.7}$$

Solving for the passband ripple parameter ε_p in Equation (17.7), we obtain

$$\varepsilon_p = \sqrt{10^{\frac{A_p}{10}}-1} = \sqrt{\frac{1}{G_p^2}-1} = \sqrt{\frac{1}{(1-\delta_p)^2}-1} \tag{17.8}$$

At the stopband cutoff frequency $\Omega = \Omega_s$ rad/s, from Equation (17.2), the gain is given by

$$|H_n(j\Omega_s)| = \frac{1}{\sqrt{1+\varepsilon_p^2\Omega_s^{2n}}} = G_s = \delta_s = \frac{1}{\sqrt{1+\varepsilon_s^2}} \tag{17.9}$$

where

$$\varepsilon_p^2\Omega_s^{2n} = \varepsilon_s^2 \quad \text{or} \quad \varepsilon_s = \varepsilon_p\,\Omega_s^n \tag{17.10}$$

Here, ε_s is called the *stopband ripple parameter*. Notice that

$$\Omega_s^n = \frac{\varepsilon_s}{\varepsilon_p} \tag{17.11}$$

The attenuation in decibels at the stopband cutoff frequency $\Omega = \Omega_s$ rad/s, is given by

$$\begin{aligned} A_s &= -20\log_{10}|H_n(j\Omega_s)| = -20\log_{10}\frac{1}{\sqrt{1+\varepsilon_p^2\Omega_s^{2n}}} \\ &= 10\log_{10}\left(1+\varepsilon_p^2\Omega_s^{2n}\right) \end{aligned} \tag{17.12}$$

Since $\varepsilon_p^2\Omega_s^{2n} = \varepsilon_s^2$ from Equation (17.10), Equation (17.12) becomes

$$A_s = -20\log_{10}|H_n(j\Omega_s)| = -20\log_{10}\frac{1}{\sqrt{1+\varepsilon_s^2}} = 10\log_{10}(1+\varepsilon_s^2) \tag{17.13}$$

Solving Equation (17.13) for the stopband ripple parameter ε_s, we have

$$\varepsilon_s = \sqrt{10^{\frac{A_s}{10}}-1} \tag{17.14}$$

Substituting Equations (17.8) and (17.14) into Equation (17.11), we obtain

$$\Omega_s^n = \frac{\sqrt{10^{\frac{A_s}{10}} - 1}}{\sqrt{10^{\frac{A_p}{10}} - 1}} \tag{17.15}$$

Taking the log on both sides of Equation (17.15), we obtain

$$n = \frac{\log\left(\dfrac{10^{0.1A_s} - 1}{10^{0.1A_p} - 1}\right)}{2\log(\Omega_s)} \tag{17.16}$$

The base of the log can be 10 or e. This equation can be used to find the order n of a filter given A_p, A_s, ω_p, and ω_s (notice that $\Omega_p = 1, \Omega_s = \omega_s/\omega_p$). Since the order n of a filter must be an integer, the n value must be rounded up. Thus, we have

$$n = \left\lceil \frac{\log\left(\dfrac{10^{0.1A_s} - 1}{10^{0.1A_p} - 1}\right)}{2\log(\Omega_s)} \right\rceil = ceil\left[\frac{\log\left(\dfrac{10^{0.1A_s} - 1}{10^{0.1A_p} - 1}\right)}{2\log(\Omega_s)}\right] \tag{17.17}$$

Since the order is rounded up, the actual filter exceeds the specifications unless n is an integer before rounding up. The excess tolerance can be assigned to stopband or passband. If the excess tolerance is assigned to the stopband, the resulting filter will exceed the specifications in the stopband (attenuation is greater than A_s dB at ω_s) and will meet the specifications exactly in the passband (attenuation is A_p dB at ω_p). On the other hand, if the excess tolerance is assigned to the passband, the resulting filter will exceed the specifications in the passband (attenuation is smaller than A_p dB at ω_p) and will meet the specifications exactly in the stopband (attenuation is A_s dB at ω_s). If the excess tolerance is assigned to the stopband, the passband attenuation A_p is fixed at the value given in the specifications. Substituting the rounded-up n value from Equation (17.17) into Equation (17.15) and solving for A_s, we obtain new A_s, denoted by A_{s_1}, given by

$$A_{s_1} = 10\log_{10}\left[\Omega_s^{2n}\left(10^{\frac{A_p}{10}} - 1\right) + 1\right] \tag{17.18}$$

Since A_p is fixed, there is no change to the passband ripple parameter ε_p given by Equation (17.8).

If the excess tolerance is assigned to the passband, the stopband attenuation A_s is fixed at the value given in the specifications. Substituting the rounded-up n value from Equation (17.17) into Equation (17.15) and solving for A_p, we obtain new A_p, denoted by A_{p_1}, given by

$$A_{p_1} = 10\log_{10}\left(\frac{10^{\frac{A_s}{10}} - 1}{\Omega_s^{2n}} + 1\right) \tag{17.19}$$

The passband ripple parameter ε_p can be recalculated using Equation (17.8):

$$\varepsilon_p = \sqrt{10^{\frac{A_{p_1}}{10}} - 1} \tag{17.20}$$

17.2.3 FINDING THE POLE LOCATIONS

The magnitude squared of the nth-order normalized Butterworth LPF is given by

$$|H_n(j\Omega)|^2 = H_n(j\Omega)H_n(-j\Omega) = \frac{1}{1 + \varepsilon_p^2\Omega^{2n}} \tag{17.21}$$

Let S be the normalized complex variable. Then, $S = j\Omega$, and $\Omega = S/j$. Substituting these into Equation (17.21), we obtain

$$H_n(S)H_n(-S) = \frac{1}{1 + \varepsilon_p^2\left(\frac{S}{j}\right)^{2n}} = \frac{1}{1 + (-1)^n \varepsilon_p^2 S^{2n}} \tag{17.22}$$

The poles of $H_n(S)H_n(-S)$ are the roots of the denominator polynomial

$$1 + (-1)^n \varepsilon_p^2 S^{2n} = 0 \tag{17.23}$$

When n is odd, this equation becomes

$$1 - \varepsilon_p^2 S^{2n} = 0 \quad \text{or} \quad S^{2n} = \frac{1}{\varepsilon_p^2} \quad \text{or} \quad S^{2n} = \frac{1}{\varepsilon_p^2}e^{j2\pi k} \tag{17.24}$$

Thus the $2n$ poles are located at

$$S = \frac{1}{\varepsilon_p^{\frac{1}{n}}}e^{j\frac{\pi k}{n}} = \Omega_c e^{j\frac{\pi k}{n}}, \quad k = 0, 1, 2, \ldots, 2n - 1 \tag{17.25}$$

where

$$\Omega_c = \frac{1}{\varepsilon_p^{\frac{1}{n}}} \tag{17.26}$$

Notice that at $\Omega = \Omega_c$, the attenuation is, from Equation (17.5),

$$A = 10\log_{10}(2) = 3.01\,\text{dB} \tag{17.27}$$

At $\Omega = \Omega_c$, the attenuation is 3 dB. The radian frequency Ω_c is called the *3-dB cutoff frequency* of the normalized LPF.

The roots are located on the circle of radius Ω_c in the complex S-plane. The roots are separated by π/n radians on the unit circle. When $k = 0$, $S = \Omega_c$. Thus there is a pole at $(\Omega_c, 0)$. In rectangular coordinate, the poles are located at

$$S_k = \Omega_c\cos\left(\frac{\pi k}{n}\right) + j\Omega_c\sin\left(\frac{\pi k}{n}\right), \quad k = 0,1,2,\ldots,2n - 1 \tag{17.28}$$

For a stable filter, we select the n poles on the left half of the S plane. There is one real pole at

$$S = -\Omega_c \tag{17.29}$$

and $(n-1)/2$ complex conjugate poles at

$$S_k = \Omega_c e^{\pm j\frac{\pi k}{n}} = \Omega_c \cos\left(\frac{\pi k}{n}\right) \pm j\Omega_c \sin\left(\frac{\pi k}{n}\right),$$
$$k = \frac{n+1}{2}, \frac{n+3}{2}, \ldots, n-1 \tag{17.30}$$

Let $k' = k - \frac{n+1}{2}$. Then, $k = k' + \frac{n+1}{2}$. Equation (17.30) becomes

$$S_{k'} = \Omega_c e^{\pm j\frac{\pi\left(k'+\frac{n+1}{2}\right)}{n}} = \Omega_c \cos\left[\frac{\pi\left(k'+\frac{n+1}{2}\right)}{n}\right] \pm j\Omega_c \sin\left[\frac{\pi\left(k'+\frac{n+1}{2}\right)}{n}\right],$$
$$k' = 0, 1, \ldots, \frac{n-3}{2}$$

Changing k' back to k, we have

$$S_k = \Omega_c e^{\pm j\frac{\pi\left(k+\frac{n+1}{2}\right)}{n}} = \Omega_c \cos\left[\frac{\pi\left(k+\frac{n+1}{2}\right)}{n}\right] \pm j\Omega_c \sin\left[\frac{\pi\left(k+\frac{n+1}{2}\right)}{n}\right],$$
$$k = 0,1, \ldots, \frac{n-3}{2} \tag{17.31)}$$

When n is even, Equation (17.23) becomes

$$1 + \varepsilon_p^2 S^{2n} = 0 \tag{17.32}$$

or

$$S^{2n} = -\frac{1}{\varepsilon_p^2} = \frac{1}{\varepsilon_p^2} e^{j\pi(2k-1)}, \quad k = 1, 2, 3, \ldots, 2n \tag{17.33}$$

Thus, the $2n$ roots are located at

$$S = \frac{1}{\varepsilon_p^{\frac{1}{n}}} e^{j\frac{\pi(2k-1)}{2n}} = \Omega_c e^{j\frac{\pi(2k-1)}{2n}}, \quad k = 1, 2, 3, \ldots, 2n \tag{17.34}$$

The roots are located on the circle of radius Ω_c in the complex S-plane. The roots are separated by π/n radians on the circle of radius Ω_c. When $k = 1$,

$$S = \Omega_c e^{j\frac{\pi}{2n}} \tag{17.35}$$

and when $k = 2n$,

$$S = \Omega_c e^{j\frac{\pi(4n-1)}{2n}} = \Omega_c e^{j\left(2\pi - \frac{\pi}{2n}\right)} = \Omega_c e^{j\left(-\frac{\pi}{2n}\right)} \tag{17.36}$$

Thus there are poles on a circle of radius Ω_c at angles $\pi/2n$ and $-\pi/2n$ from the real axis. In Cartesian coordinates, the poles are located at

$$S_k = \Omega_c \cos\left[\frac{\pi(2k-1)}{2n}\right] + j\Omega_c \sin\left[\frac{\pi(2k-1)}{2n}\right], \quad k = 1,2,3, \ldots, 2n \tag{17.37}$$

For a stable filter, we select the n poles on the left half of the S plane. There are $n/2$ complex conjugate poles at

$$S_k = \Omega_c e^{\pm j\frac{\pi(2k-1)}{2n}} = \Omega_c \cos\left[\frac{\pi(2k-1)}{2n}\right] \pm j\Omega_c \sin\left(\frac{\pi(2k-1)}{2n}\right), \quad k = \frac{n}{2}+1, \frac{n}{2}+2, \ldots, n \tag{17.38}$$

Let $k' = k - \frac{n}{2} - 1$. Then, $k = k' + \frac{n}{2} + 1$. The above equation becomes

$$S_{k9} = \Omega_c e^{\pm j\frac{\pi(2k'+n+1)}{2n}} = \Omega_c \cos\left(\frac{\pi(2k'+n+1)}{2n}\right) \pm j\Omega_c \sin\left(\frac{\pi(2k'+n+1)}{2n}\right), \quad k' = 0, 1, \ldots, \frac{n}{2}-1$$

Changing the index k' back to k, we have

$$S_k = \Omega_c e^{\pm j\frac{\pi(2k+n+1)}{2n}} = \Omega_c \cos\left(\frac{\pi(2k+n+1)}{2n}\right) \pm j\Omega_c \sin\left(\frac{\pi(2k+n+1)}{2n}\right), \quad k = 0, 1, \ldots, \frac{n}{2}-1 \tag{17.39}$$

Since Ω_c is multiplied everywhere, we can find the poles assuming $\Omega_c = 1$ (poles are on the unit circle, normalized half-power frequency $\Omega_c = 1$ rad/s), and then later multiply these poles by Ω_c.

There are n zeros at $S = \infty$.

In summary, the design procedure for the Butterworth LPF is as follows.

1. Change the Cutoff Frequencies from Hertz to Radians

If the cutoff frequencies are given in hertz, change the cutoff frequencies to radians.

$$\omega_p = 2\pi f_p, \quad \omega_s = 2\pi f_s \tag{17.40}$$

2. Backward Transformation

Given the passband cutoff frequency ω_p, stopband cutoff frequency ω_s, maximum attenuation (in decibels) in the passband A_p, and minimum attenuation (in decibels) in the stopband A_s, find the cutoff frequencies for the normalized LPF by dividing by ω_p:

$$\text{Passband cutoff frequency for the normalized LPF} = \Omega_p = 1 \text{ rad/s} \tag{17.41}$$

$$\text{Stopband cutoff frequency for the normalized LPF} = \Omega_s = \omega_s/\omega_p \text{ rad/s} \tag{17.42}$$

Notice that $\omega_s/\omega_p = f_s/f_p > 1$

3. Order

Find the order n of the normalized LPF using

$$n = ceil\left[\frac{\log\left(\frac{10^{0.1A_s}-1}{10^{0.1A_p}-1}\right)}{2\log(\Omega_s)}\right] \tag{17.43}$$

4. Recalculation of the Parameters

If excess tolerance is assigned to the stopband, we have new A_s given by

$$A_{s_1} = 10\log_{10}\left[\Omega_s^{2n}\left(10^{\frac{A_p}{10}} - 1\right) + 1\right] \tag{17.44}$$

The value of A_p remains the same. The passband ripple parameter is given by

$$\varepsilon_p = \sqrt{10^{\frac{A_p}{10}} - 1} \tag{17.45}$$

The 3-dB cutoff frequency is given by

$$\Omega_c = \frac{1}{\varepsilon_p^{\frac{1}{n}}} \tag{17.46}$$

If excess tolerance is assigned to the passband, we have a new A_p, given by

$$A_{p_1} = 10\log_{10}\left(\frac{10^{\frac{A_s}{10}} - 1}{\Omega_s^{2n}} + 1\right) \tag{17.47}$$

The passband ripple parameter ε_p can be recalculated using

$$\varepsilon_{p_1} = \sqrt{10^{\frac{A_{p_1}}{10}} - 1} \tag{17.48}$$

The 3-dB cutoff frequency can be recalculated from

$$\Omega_{c_1} = \frac{1}{\varepsilon_{p_1}^{\frac{1}{n}}} \tag{17.49}$$

5. Normalized Low-Pass Poles and Zeros

Find poles and zeros assuming $\Omega_c = 1$ rad/s

n odd:

There is one pole at $S = -1$ and $(n-1)/2$ complex conjugate poles at

$$S_k = e^{\pm j\frac{\pi(2k+n+1)}{2n}} = \cos\left[\frac{\pi(2k+n+1)}{2n}\right] \pm j\sin\left[\frac{\pi(2k+n+1)}{2n}\right], \quad k = 0, 1, \ldots, \frac{n-3}{2} \tag{17.50}$$

n even:

There are $n/2$ complex conjugate poles at

$$S_k = e^{\pm j\frac{\pi(2k+n+1)}{2n}} = \cos\left[\frac{\pi(2k+n+1)}{2n}\right] \pm j\sin\left[\frac{\pi(2k+n+1)}{2n}\right], \quad k = 0, 1, \ldots, \frac{n}{2} - 1 \tag{17.51}$$

There are n zeros at $S = \infty$.

6. Transfer Function of the Normalized LPF

Convert the poles and the zeros of the normalized LPF into polynomials, and write the transfer function as a product of the first-order and the second-order sections. If the order of the filter is odd, there is one first-order section and $(n - 1)/2$ second-order sections. If the order of the filter is even, there are $n/2$ second-order sections. The numerator is a constant equal to the constant term of the denominator for each first- or second-order term. Also, find the Q value of each second-order section.

7. Scaled Poles and Zeros

Multiply the poles and the zeros from step 5 by Ω_c. There are n zeros at $S = \infty$.

8. Frequency Transformation

Trasnsform the LPF with the passband cutoff frequency $\Omega_p = 1$ rad/s to the LPF with the passband cutoff frequency ω_p by multiplying the poles and the zeros from step 7 above by ω_p. Let s be the location of poles and zeros after the transformation, and let S be the location of poles and zeros before the transformation. Then, we have

$$s = \omega_p S \tag{17.52}$$

There are n zeros at $s = \infty$. Notice that steps 7 and 8 can be combined into a single step by multiplying the poles and the zeros from step 5 by

$$\omega_0 = \omega_p \Omega_c \tag{17.53}$$

9. Transfer Function

Convert the poles and the zeros of the frequency-transfromed LPF into polynomials, and write the transfer function as a product of the first-order and the second-order sections. If the order of the filter is odd, there is one first-order section and $(n - 1)/2$ second-order sections. If the order of the filter is even, there are $n/2$ second-order sections. The numerator is a constant equal to the constant term of the denominator for each first- or second-order term.

10. Verify

Calculate the attenuation at ω_p and ω_s, and compare with the specifications. If the excess tolerance is assigned to the stopband, we have

$$-20 \log_{10}|H(j\omega_p)| = A_p, \quad -20 \log_{10}|H(j\omega_s)| \geq A_s \tag{17.54}$$

If the excess tolerance is assigned to the passband, we have

$$-20 \log_{10}|H(j\omega_p)| \leq A_p, \quad -20 \log_{10}|H(j\omega_s)| = A_s \tag{17.55}$$

EXAMPLE 17.1

Design a Butterworth LPF with the following specifications:

Passband cutoff frequency $= f_p = 1$ kHz
Stopband cutoff frequency $= f_s = 2$ kHz
Maximum attenuation in the passband $= A_p = 1$ dB
Minimum attenuation in the stopband $= A_s = 20$ dB
Excess tolerance = stopband

Refer to Section 17.6 for a MATLAB script for this EXAMPLE.

continued

Example 17.1 continued

1. Change the Cutoff Frequencies from Hertz to Radians

$$\omega_p = 2\pi 1000 \text{ rad/s} = 6283.1853 \text{ rad/s}$$
$$\omega_s = 2\pi 2000 \text{ rad/s} = 12{,}566.3706 \text{ rad/s}$$

2. Backward Transformation

The cutoff frequencies of the normalized LPF are given by

$$\Omega_p = 1 \text{ rad/s}, \Omega_s = \omega_s/\omega_p = 2 \text{ rad/s}$$

3. Order

The order of the filter is

$$n = \left\lceil \frac{\log\left(\frac{10^{0.1A_s} - 1}{10^{0.1A_p} - 1}\right)}{2\log(\Omega_s)} \right\rceil = \left\lceil \frac{\log\left(\frac{10^{0.1\times 20} - 1}{10^{0.1\times 1} - 1}\right)}{2\log(2)} \right\rceil = ceil\,[4.29] = 5$$

4. Recalculation of the Parameters

Since the excess tolerance is assigned in the stopband, the new attenuation in the stopband is given by

$$A_{s_1} = 10\log_{10}\left[\Omega_s^{2n}\left(10^{\frac{A_p}{10}} - 1\right) + 1\right] = 24.2511 \text{ dB}$$

The attenuation in the stopband is 4.2511 dB larger than the minimum required attenuation of 20 dB. The value of ε_p is given by

$$\varepsilon_p = \sqrt{10^{\frac{A_p}{10}} - 1} = \sqrt{10^{0.1} - 1} = 0.50885$$

The 3-dB cutoff frequency (scale factor) Ω_c is given by

$$\Omega_c = \frac{1}{\varepsilon_p^{\frac{1}{n}}} = 1.1447$$

5. Normalized Low-Pass Poles and Zeros

Since n is odd, assuming $\Omega_c = 1$, there is one pole at $S = -1$ and two pairs of normalized complex conjugate poles at

$$S_k = e^{\pm j\frac{\pi(2k+n+1)}{2n}} = \cos\left(\frac{\pi(2k+n+1)}{2n}\right) \pm j\sin\left(\frac{\pi(2k+n+1)}{2n}\right), \quad k = 0,1$$

Thus, the pole locations ($\Omega_c = 1$) are

-1.0
$-0.3090 \pm j0.9511$
$-0.8090 \pm j0.5878$

There are five zeros at $S = \infty$.

continued

Example 17.1 continued

6. Transfer Function of the Normalized LPF

The transfer function of the normalized LPF ($\Omega_c = 1$) is given by

$$H(S) = \frac{1}{(S+1)(S^2+0.6180S+1)(S^2+1.6180S+1)}$$

$$= \frac{1}{S+1} \times \frac{1}{S^2+0.6180S+1} \times \frac{1}{S^2+1.6180S+1}$$

Notice that each pair of conjugate poles generates one second-order term. The coefficient of the S term is the $-2\ \mathrm{Re}(S)$, and the constant term is the product of the conjugate poles

$$-2\ \mathrm{Re}(-0.3090 \pm j0.9511) = 0.6180$$

$$-2\ \mathrm{Re}(-0.8090 \pm j0.5878) = 1.6180$$

$$(-0.3090 + j0.9511) \times (-0.3090 - j0.9511) = 1$$

$$(-0.8090 + j0.5878) \times (-0.8090 - j0.5878) = 1$$

For the first second-order section

$$H(S) = \frac{1}{S^2 + \frac{1}{Q}S + 1} = \frac{1}{S^2+0.6180S+1}$$

the Q value is $Q = 1/0.6180 = 1.6180$. For the second second-order section

$$H(S) = \frac{1}{S^2 + \frac{1}{Q}S + 1} = \frac{1}{S^2+1.6180S+1}$$

the Q value is $Q = 1/1.618 = 0.6180$

7. Scaled Poles and Zeros

Multiplying the poles by Ω_c, we obtain the low-pass poles

-1.1447

$-0.3537 \pm j1.0887$

$-0.9260 \pm j0.6728$

There are five zeros at $S = \infty$

8. Frequency Transformation

The frequency transformed poles are obtained by multiplying the poles given in step 7 by $\omega_p = 2\pi 1000$. The frequency transformed poles are given by

-7192.2107

$-2222.5153 \pm j6840.1988$

$-5818.6207 \pm j4227.4754$

There are five zeros at $s = \infty$. Notice that $\omega_0 = \Omega_c \omega_p = 7192.2107$ rad/s

continued

Example 17.1 continued

9. Transfer Function

The transfer function of the LPF is

$$H(s) = \frac{7192.21^5}{(s + 7192.21)(s^2 + 4445.03s + 7192.21^2)(s^2 + 11{,}637.24s + 7192.21^2)}$$

$$= \frac{7192.21}{s + 7192.21} \times \frac{51{,}727{,}894.51}{s^2 + 4445.03s + 51{,}727{,}894.51}$$

$$\times \frac{51{,}727{,}894.51}{s^2 + 11{,}637.24s + 51{,}727{,}894.51}$$

Second-order term	Corner frequency (ω_0)	Q value
$s^2 + 4445.03s + 51{,}727{,}894.51$	7192.2107 rad/s	1.618
$s^2 + 11{,}637.24s + 51{,}727{,}894.51$	7192.2107 rad/s	0.618

10. Verify

$$-20\log_{10}|H(j\omega_p)| = 1\,\text{dB}, \quad -20\log_{10}|H(j\omega_s)| = 24.2511\,\text{dB}$$

Exercise 17.1

Design a Butterworth LPF with the following specifications:

Passband cutoff frequency = f_p = 1.2 kHz
Stopband cutoff frequency = f_s = 2.3 kHz
Maximum attenuation in the passband = A_p = 0.6 dB
Minimum attenuation in the stopband = A_s = 12 dB
Excess tolerance = passband

Answer:

ω_p = 7539.8224 rad/s, ω_s = 14,451.3262, $n = 4$, A_{p_1} = 0.3404, ε_p = 0.2855, Ω_c = 1.3680.

Normalized poles: $-0.9239 \pm j0.3827, -0.3827 \pm j0.9239$

Normalized zeros: $\infty, \infty, \infty, \infty$

Normalized transfer function

$$H_n(S) = \frac{1}{S^2 + 1.8478S + 1} \times \frac{1}{S^2 + 0.7654S + 1}$$

Q values: 0.5412, 1.3066

$$\omega_0 = \Omega_c\omega_p = 10{,}314.4435 \text{ rad/s}$$

Frequency-transformed LPF transfer function

$$H(s) = \frac{1.0639 \times 10^8}{s^2 + 19{,}058.6065s + 1.0639 \times 10^8} \times \frac{1.0639 \times 10^8}{s^2 + 7894.3333s + 1.0639 \times 10^8}$$

$$-20\log_{10}|H(j\omega_p)| = 0.3404 \text{ dB}, \quad -20\log_{10}|H(j\omega_s)| = 12 \text{ dB}$$

EXAMPLE 17.2

Implement the filter designed in EXAMPLE 17.1 using the Sallen-Key unity gain method. Assume that $k_m = 1000$.

$$k_f = \omega_0 = \Omega_c \omega_p = 7192.2107 \text{ rad/s}$$

For transfer function $H(S) = \dfrac{1}{S+1}$

Normalized component values: $R = 1\ \Omega$, $C = 1$ F
Scaled component values: $R = 1\ k\Omega$, $C = 1/(k_m k_f)$ F $= 0.139\ \mu$F

For transfer function $H(S) = \dfrac{1}{S^2 + \dfrac{1}{Q}S + 1} = \dfrac{1}{S^2 + 0.6180S + 1}$

The Q value is $Q = 1/0.6180 = 1.6180$

Normalized component values:

$$R_1 = 1\ \Omega,\quad R_2 = 1\ \Omega,\quad C_1 = 2Q \text{ F} = 3.2361 \text{ F},\quad C_2 = 1/(2Q) \text{ F} = 0.3090 \text{ F}$$

Scaled component values:

$$R_1 = 1\ k\Omega,\quad R_2 = 1\ k\Omega,\quad C_1 = 2Q \text{ F} = 3.2361/(k_m k_f) \text{ F} = 0.4499\ \mu\text{F},$$
$$C_2 = 1/(2Q) \text{ F} = 0.3090/(k_m k_f) \text{ F} = 0.04297\ \mu\text{F}$$

For transfer function $H(S) = \dfrac{1}{S^2 + \dfrac{1}{Q}S + 1} = \dfrac{1}{S^2 + 1.6180S + 1}$

The Q value is $Q = 1/1.6180 = 0.6180$

Normalized component values:

$$R_1 = 1\ \Omega,\quad R_2 = 1\ \Omega,\quad C_1 = 2Q \text{ F} = 1.2361 \text{ F},\quad C_2 = 1/(2Q) \text{ F} = 0.8090 \text{ F}$$

Scaled component values:

$$R_1 = 1\ k\Omega,\quad R_2 = 1\ k\Omega,\quad C_1 = 2Q \text{ F} = 1.2361/(k_m k_f) \text{ F} = 0.1719\ \mu\text{F},$$
$$C_2 = 1/(2Q) \text{ F} = 0.8090/(k_m k_f) \text{ F} = 0.1125\ \mu\text{F}$$

FIGURE 17.3

PSpice schematic for Sallen-Key unity gain circuit.

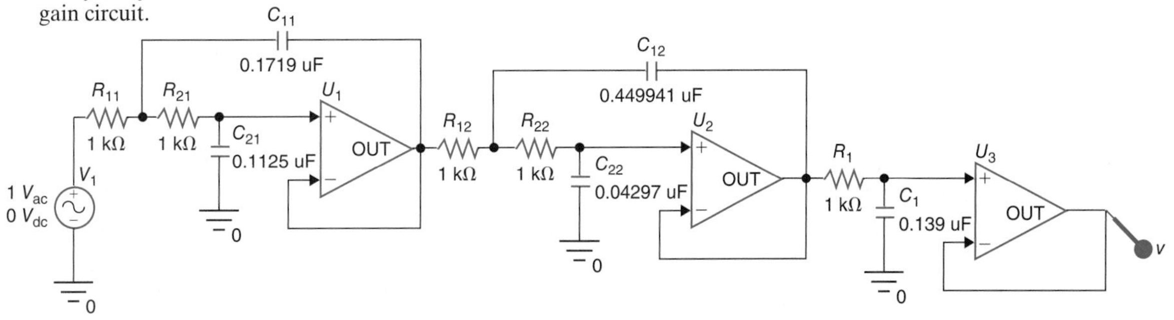

continued

Example 17.2 continued

FIGURE 17.4

Magnitude response for the circuit shown in Figure 17.3.

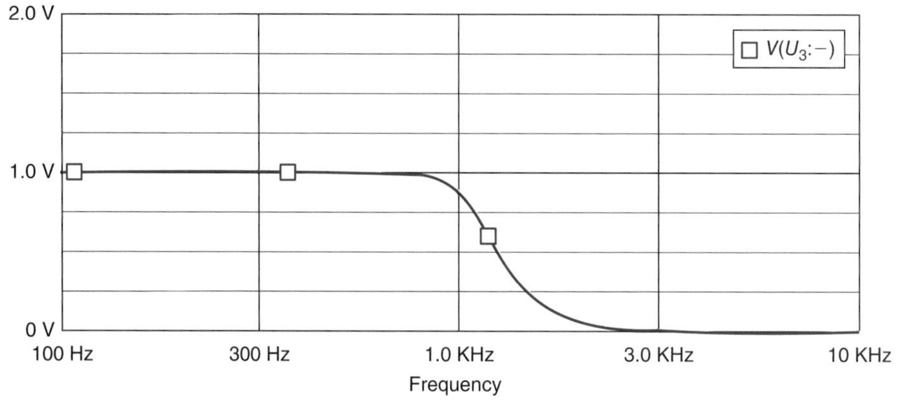

Exercise 17.2

Implement the filter designed in EXERCISE 17.1 using the Sallen-Key equal *R*, equal *C* method. Assume $k_m = 1000$.

$$\omega_0 = \Omega_c \omega_p = 10{,}314.4435 \text{ rad/s}$$

For transfer function $H(S) = \dfrac{1}{S^2 + \dfrac{1}{Q}S + 1} = \dfrac{1}{S^2 + 1.8478S + 1}$

The Q value is $Q = 1/1.8478 = 0.5412$

Normalized component values

$$R_2 = 1\ \Omega,\quad R_B = 1\ \Omega,\quad R_A = 0.1522\ \Omega,\quad R_a = 1.1522\ \Omega,\quad R_b = 7.5685\ \Omega$$
$$C_1 = 1\text{ F},\quad C_2 = 1\text{ F}$$

Scaled component values

$$R_2 = 1\ k\Omega,\quad R_B = 1\ k\Omega,\quad R_A = 152.2\ \Omega,\quad R_a = 1.1522\ k\Omega,\quad R_b = 7.5685\ k\Omega$$
$$C_1 = 96.9514\ n\text{F},\quad C_2 = 96.9514\ n\text{F}$$

For transfer function $H(S) = \dfrac{1}{S^2 + \dfrac{1}{Q}S + 1} = \dfrac{1}{S^2 + 0.7654S + 1}$

The Q value is $Q = 1/0.7654 = 1.3066$

$$R_2 = 1\ \Omega,\quad R_B = 1\ \Omega,\quad R_A = 1.2346\ \Omega,\quad R_a = 2.2346\ \Omega,\quad R_b = 1.81\ \Omega$$
$$C_1 = 1\text{ F},\quad C_2 = 1\text{ F}$$

Scaled component values:

$$R_2 = 1\ k\Omega,\quad R_B = 1\ k\Omega,\quad R_A = 1.2346\ k\Omega,\quad R_a = 2.2346\ k\Omega,\quad R_b = 1.81\ k\Omega$$
$$C_1 = 96.9514\ n\text{F},\quad C_2 = 96.9514\ n\text{F}$$

continued

Exercise 17.2 continued

The PSpice simulation is shown in Figure 17.5, and the magnitude response is shown in Figure 17.6.

FIGURE 17.5

PSpice circuit for EXERCISE 17.2.

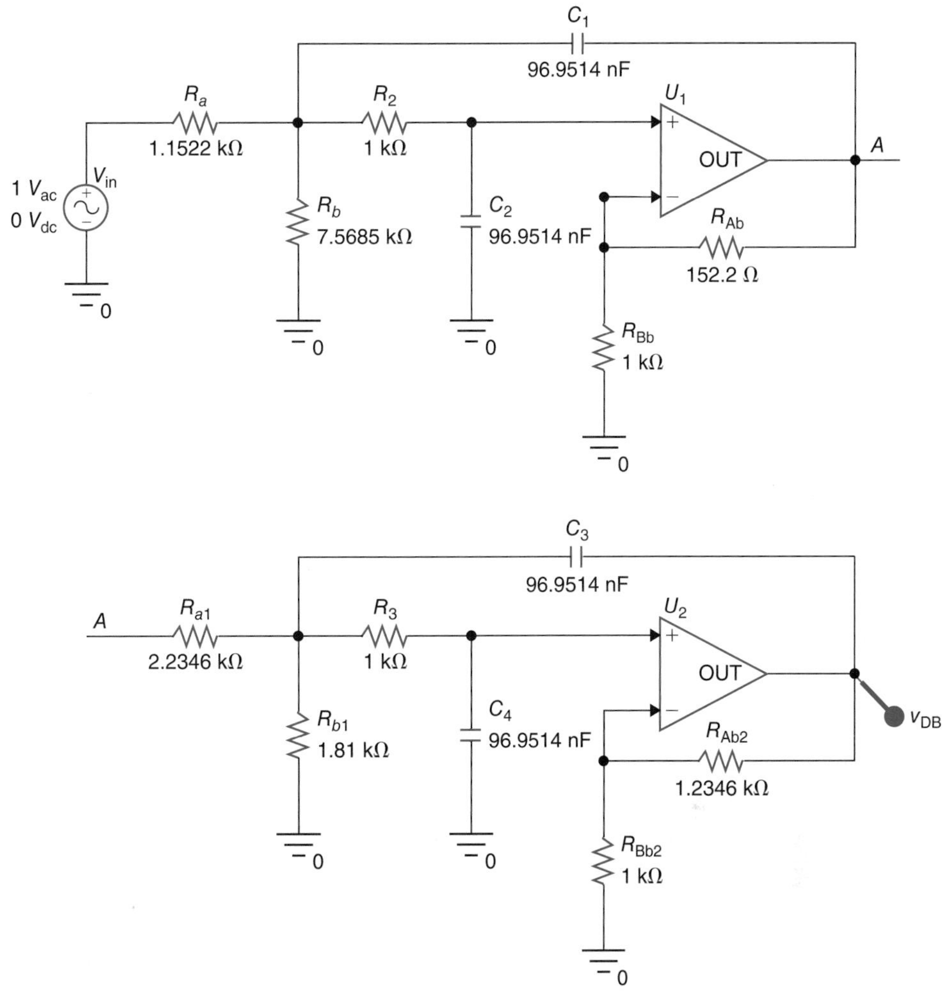

FIGURE 17.6

Magnitude response in decibels for the circuit shown in Figure 17.5.

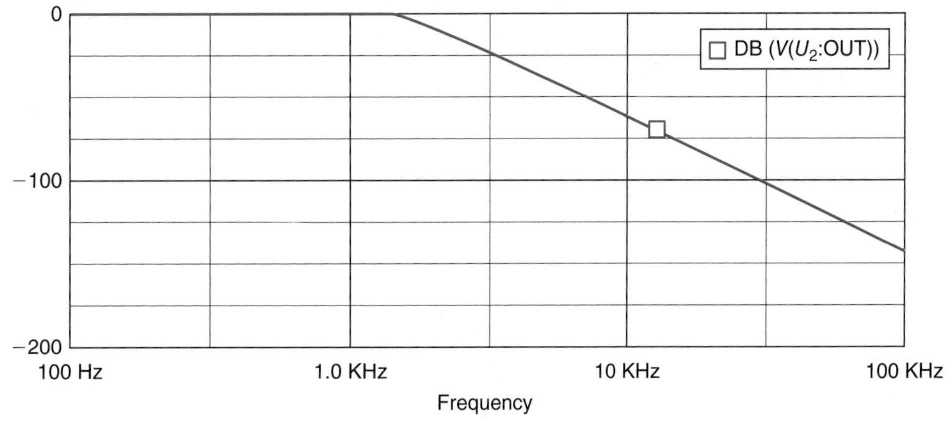

17.3 Analog Butterworth HPF Design

In this section, we are designing an analog Butterworth HPF that satisfies the specifications shown in Figure 17.7.

FIGURE 17.7

Specifications for the Butterworth HPF.

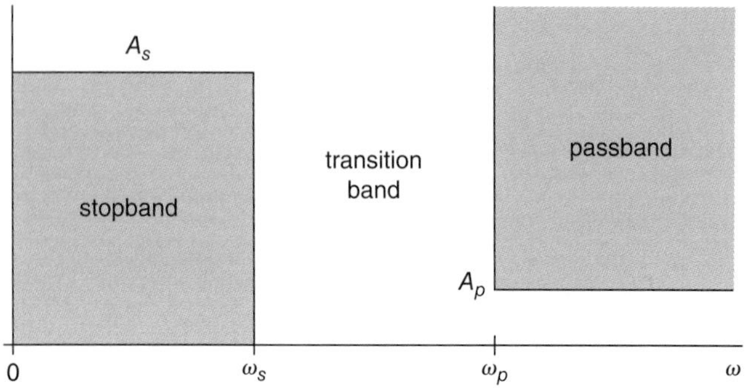

Here, A_p is the maximum attenuation (loss) in decibels in the passband, A_s is the minimum attenuation (loss) in decibels in the stopband, ω_p is the passband cutoff frequency, and ω_s is the stopband cutoff frequency. Notice that $\omega_p > \omega_s$.

The HPF is designed by first transforming the specifications for the HPF to the specifications for the normalized LPF. Then, the order and the poles and the zeros of the normalized LPF are found. These poles and zeros are transformed to high-pass poles and zeros. From the poles and the zeros, the transfer function of the HPF can be found. To verify this procedure, let us consider the first-order normalized LPF. The first-order normalized ($\Omega_p = 1$ rad/s) LPF has a transfer function

$$H(S) = \frac{1}{S+1}$$

This filter has a cutoff frequency of 1 rad/s. There is one pole at $S = -1$ and one zero at $S = \infty$. This filter can be transformed to an HPF with cutoff frequency ω_p by replacing S by ω_p/s; that is

$$H(s) = H(S)\Big|_{S=\frac{\omega_p}{s}} = \frac{1}{S+1}\Big|_{S=\frac{\omega_p}{s}} = \frac{1}{\dfrac{\omega_p}{s}+1} = \frac{s}{s+\omega_p}$$

The HPF has a pole at $s = -\omega_p$ and zero at $s = 0$. Solving $S = \dfrac{\omega_p}{s}$ for s, we obtain

$$s = \frac{\omega_p}{S} \qquad \textbf{(17.56)}$$

The high-pass poles and zeros are obtained by taking the inverse of the normalized low-pass poles and zeros and then multiplying by ω_p. Equation (17.56) can be used to transform normalized low-pass poles and zeros to high-pass poles and zeros. For the current example, the normalized low-pass pole at $S = -1$ is transformed to a high-pass pole at

$$s = \frac{\omega_p}{-1} = -\omega_p$$

and the normalized low-pass zero at $S = \infty$ is transformed to a high-pass zero at

$$s = \frac{\omega_p}{\infty} = 0$$

The transformation of the specifications from the HPF to the normalized LPF is called *backward transformation*. The specifications for the normalized LPF are shown in Figure 17.8. The passband cutoff

FIGURE 17.8

Specifications for the normalized LPF.

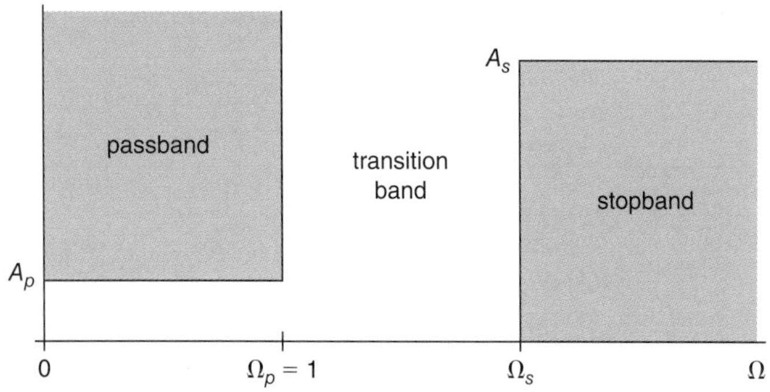

frequency is $\Omega_p = 1$ rad/s, and the stopband cutoff frequency is given by

$$\Omega_s = \frac{\omega_p}{\omega_s} \tag{17.57}$$

where $\omega_p > \omega_s$.

The design procedure for the Butterworth HPF is as follows:

1. Change the Cutoff Frequencies from Hertz to Radians

If the cutoff frequencies are given in hertz, change the cutoff frequencies to radians.

$$\omega_p = 2\pi f_p, \quad \omega_s = 2\pi f_s \tag{17.58}$$

2. Backward Transformation

Given the passband cutoff frequency ω_p, stopband cutoff frequency ω_s ($\omega_s < \omega_p$), maximum attenuation (in decibels) in the passband A_p, and minimum attenuation (in decibels) in the stopband A_s of the HPF, find the cutoff frequencies of the normalized LPF using

$$\text{Passband cutoff frequency for the normalized LPF} = \Omega_p = 1 \text{ rad/s} \tag{17.59}$$

$$\text{Stopband cutoff frequency for the normalized LPF} = \Omega_s = \omega_p/\omega_s \text{ rad/s} \tag{17.60}$$

Notice that $\omega_p/\omega_s = f_p/f_s > 1$.

3. Order

Find the order n of the normalized LPF using

$$n = \left\lceil \frac{\log\left(\dfrac{10^{0.1A_s} - 1}{10^{0.1A_p} - 1}\right)}{2\log\left(\dfrac{\Omega_s}{\Omega_p}\right)} \right\rceil = ceil\left[\frac{\log\left(\dfrac{10^{0.1A_s} - 1}{10^{0.1A_p} - 1}\right)}{2\log(\Omega_s)}\right] \tag{17.61}$$

4. Recalculation of the Parameters

If excess tolerance is assigned to the stopband, we have a new A_s given by

$$A_{s_1} = 10\log_{10}\left[\Omega_s^{2n}\left(10^{\frac{A_p}{10}} - 1\right) + 1\right] \tag{17.62}$$

The value of A_p remains the same. The passband ripple parameter is given by

$$\varepsilon_p = \sqrt{10^{\frac{A_p}{10}} - 1} \tag{17.63}$$

The 3-dB cutoff frequency is given by

$$\Omega_c = \frac{1}{\varepsilon_p^{\frac{1}{n}}} \tag{17.64}$$

If excess tolerance is assigned to the passband, we have a new A_p given by

$$A_{p_1} = 10\log_{10}\left(\frac{10^{\frac{A_s}{10}} - 1}{\Omega_s^{2n}} + 1\right) \tag{17.65}$$

The passband ripple parameter ε_p can be recalculated using

$$\varepsilon_{p_1} = \sqrt{10^{\frac{A_{p_1}}{10}} - 1} \tag{17.66}$$

The 3-dB cutoff frequency can be recalculated from

$$\Omega_{c_1} = \frac{1}{\varepsilon_{p_1}^{\frac{1}{n}}} \tag{17.67}$$

5. Normalized Low-Pass Poles and Zeros

Find the normalized ($\Omega_c = 1$ rad/s) poles and zeros.

n odd:
There is one pole at $S = -1$ and $(n - 1)/2$ complex conjugate poles at

$$S_k = e^{\pm j\frac{\pi(2k+n+1)}{2n}} = \cos\left(\frac{\pi(2k + n + 1)}{2n}\right) \pm j\sin\left(\frac{\pi(2k + n + 1)}{2n}\right),$$
$$k = 0, 1, \ldots, \frac{n-3}{2} \tag{17.68}$$

n even:
There are $n/2$ complex conjugate poles at

$$S_k = e^{\pm j\frac{\pi(2k+n+1)}{2n}} = \cos\left(\frac{\pi(2k + n + 1)}{2n}\right) \pm j\sin\left(\frac{\pi(2k + n + 1)}{2n}\right),$$
$$k = 0, 1, \ldots, \frac{n}{2} - 1 \tag{17.69}$$

There are n zeros at $S = \infty$.

6. Transfer Function of the Normalized LPF

Convert the poles and the zeros of the normalized LPF into polynomials, and write the transfer function as a product of the first-order and the second-order sections. If the order of the filter is odd, there is one first-order section and $(n - 1)/2$ second-order sections. If the order of the filter is even, there are $n/2$ second-order sections. The numerator is a constant equal to the constant term of the denominator for each first- or second-order term. Also, find the Q value of each second-order section.

7. Scaled Poles and Zeros

Multiply the poles and the zeros of the normalized LPF from step 5 by Ω_c. There are n zeros at $S = \infty$.

8. Frequency Transformation

The low-pass poles and zeros are transformed to high-pass poles and zeros. The high-pass poles and zeros are obtained by taking the inverse of $\Omega_c S$ and multiplying by ω_p; that is,

$$s = \frac{\omega_p}{\Omega_c S} = \frac{\omega_0}{S} \tag{17.70}$$

where S is the normalized ($\Omega_p = 1$) low-pass poles and

$$\omega_0 = \frac{\omega_p}{\Omega_c} \tag{17.71}$$

Also, n zeros at $S = \infty$ are transformed to n zeros at $s = 0$.

9. Transfer Function

Convert the poles and the zeros of the frequency-transformed HPF found in step 8 into polynomials, and write the transfer function as a product of the first-order and the second-order sections. If the order of the filter is odd, there is one first-order section and $(n - 1)/2$

second-order sections. If the order of the filter is even, there are $n/2$ second-order sections. The numerator polynomial for the first-order section is s, and the numerator polynomial for each second-order section is s^2.

10. Verify

Calculate the attenuation at ω_p and ω_s, and compare with the specifications. If the excess tolerance is assigned to the stopband, we have

$$-20\log_{10}|H(j\omega_p)| = A_p, \quad -20\log_{10}|H(j\omega_s)| \geq A_s \tag{17.72}$$

If the excess tolerance is assigned to the passband, we have

$$-20\log_{10}|H(j\omega_p)| \leq A_p, \quad -20\log_{10}|H(j\omega_s)| = A_s \tag{17.73}$$

EXAMPLE 17.3

Design a Butterworth HPF with the following specifications:

Passband cutoff frequency = 2.5 kHz
Stopband cutoff frequency = 1.5 kHz
Maximum attenuation in the passband = 1 dB
Minimum attenuation in the stopband = 15 dB
Excess tolerance = stopband

1. Change the Cutoff Frequencies from Hertz to Radians

$$\omega_p = 2\pi f_p = 15{,}707.9633 \text{ rad/s}, \quad \omega_s = 2\pi f_s = 9424.7780 \text{ rad/s}$$

2. Backward Transformation

The cutoff frequencies of the normalized LPF are given by

$$\Omega_p = 1 \text{ rad/s}, \Omega_s = \omega_p/\omega_s = 1.6667$$

3. Order

The order of the normalized LPF is

$$n = ceil\left\lceil \frac{\log\left(\frac{10^{0.1\times15} - 1}{10^{0.1\times1} - 1}\right)}{2\log\left(\frac{2\pi 2500}{2\pi 1500}\right)} \right\rceil = ceil(4.67) = 5$$

4. Recalculation of the Parameters

Since the excess tolerance is assigned to the stopband, we have

$$A_{s_1} = 10\log_{10}\left[\Omega_s^{2n}\left(10^{\frac{A_p}{10}} - 1\right) + 1\right] = 16.4169 \text{ dB}$$

$$\varepsilon_p = \sqrt{10^{\frac{A_p}{10}} - 1} = 0.5088$$

$$\Omega_c = \frac{1}{\varepsilon_p^{\frac{1}{n}}} = 1.1447$$

continued

Example 17.3 continued

5. Normalized Low-Pass Poles and Zeros

Since n is odd, there is one pole at $S = -1$ and $(n-1)/2 = 2$ complex conjugate poles at

$$S_k = e^{\pm j\frac{\pi(2k+n+1)}{2n}} = \cos\left(\frac{\pi(2k+n+1)}{2n}\right) \pm j\sin\left(\frac{\pi(2k+n+1)}{2n}\right),\ k = 0, 1$$

Thus, the pole locations ($\Omega_c = 1$) are

$$-1.0$$
$$-0.3090 \pm j0.9511$$
$$-0.8090 \pm j0.5878$$

There are five zeros at $S = \infty$

6. Transfer Function of the Normalized LPF ($\Omega_c = 1$)

$$H(S) = \frac{1}{(S+1)(S^2+0.6180S+1)(S^2+1.6180S+1)}$$
$$= \frac{1}{S+1} \times \frac{1}{S^2+0.6180S+1} \times \frac{1}{S^2+1.6180S+1}$$

The Q values are 1.618 and 0.618.

7. Scaled Poles and Zeros

Multiplying the poles by Ω_c, we obtain the low-pass poles:

$$-1.1447$$
$$-0.3537 \pm j1.0887$$
$$-0.9261 \pm j0.6728$$

There are five zeros at $S = \infty$.

8. Frequency Transformation

$$\omega_0 = \frac{\omega_p}{\Omega_c} = 13{,}722.6297 \text{ rad/s}$$

$$s = \frac{\omega_p}{\Omega_c S} = \frac{\omega_0}{S}$$

where S is the normalized ($\Omega_p = 1$) low-pass poles from step 5. The frequency transformed high-pass poles are given by

$$s = -13{,}722.6297$$
$$s = -11{,}101.8406 \pm j8065.9594$$
$$s = -4240.5258 \pm j13{,}050.9964$$

Five zeros at $S = \infty$ are transformed to five zeros at $s = 0$:

$$s = 0,\quad s = 0,\quad s = 0,\quad s = 0,\quad s = 0$$

continued

Example 17.3 continued

9. Transfer Function

The transfer function of the HPF is given by

$$H(s) = \frac{s^5}{(s + 13{,}722.63)(s^2 + 22{,}203.68s + 13{,}722.63^2)(s^2 + 8481.05s + 13{,}722.63^2)}$$

$$= \frac{s}{s + 13{,}722.63} \times \frac{s^2}{s^2 + 22{,}203.68s + 1.88311 \times 10^8}$$

$$\times \frac{s^2}{s^2 + 8184.0.5s + 1.88311 \times 10^8}$$

Second-order term	Corner frequency (ω_0)	Q value
$s^2 + 22{,}203.68s + 1.88311 \times 10^8$	13,722.6297 rad/s	0.618
$s^2 + 8184.0.5s + 1.88311 \times 10^8$	13,722.6297 rad/s	1.618

10. Verify

$$-20\log_{10}|H(j\omega_p)| = 1\,\text{dB}, \quad -20\log_{10}|H(j\omega_s)| = 16.4169\,\text{dB}$$

Exercise 17.3

Design a Butterworth HPF with the following specifications:

Passband cutoff frequency = 3 kHz
Stopband cutoff frequency = 1.5 kHz
Maximum attenuation in the passband = 0.8 dB
Minimum attenuation in the stopband = 15 dB
Excess tolerance = passband

Answer:

$\omega_p = 18{,}849.5559$ rad/s, $\omega_s = 9424.7780$ rad/s $\Omega_p = 1$ rad/s, $\Omega_s = 2$ rad/s,
$n = \text{ceil}\,(3.62) = 4$, $A_{p_1} = 0.4907$ dB, $\varepsilon_p = 0.3459$, $\Omega_c = 1.3040$

Normalized poles: $-0.9239 \pm j0.3827, -0.3827 \pm j0.9239$. Four zeros at $S = \infty$.

Normalized transfer function:

$$H_n(S) = \frac{1}{S^2 + 1.8478S + 1} \times \frac{1}{S^2 + 0.7654S + 1}$$

Q values: 0.5412, 1.3066

$$\omega_0 = \omega_p/\Omega_c = 14{,}455.2863 \text{ rad/s}$$

High-pass poles:

$$s = -13{,}354.9432 \pm j5531.7986, \quad -5531.7986 \pm j13{,}354.9432$$

Four zeros at $s = 0$

continued

Exercise 17.3 continued

Frequency-transformed HPF transfer function:

$$H(s) = \frac{s^2}{s^2 + 26{,}709.8863s + 2.0896 \times 10^8} \times \frac{s^2}{s^2 + 11{,}063.5972s + 2.0896 \times 10^8}$$

$$-20\log_{10}|H(j\omega_p)| = 0.4907\,\text{dB}, \quad -20\log_{10}|H(j\omega_s)| = 15\,\text{dB}$$

EXAMPLE 17.4

Implement the filter designed in EXAMPLE 17.3 using the Sallen-Key unity gain method. Assume $k_m = 1000$.

$$k_f = \omega_0 = \omega_p/\Omega_c = 13{,}722.6297 \text{ rad/s}$$

For transfer function $H(S) = \dfrac{1}{S+1}$

Normalized component values: $R = 1\ \Omega, C = 1$ F

Scaled component values: $R = 1\ k\Omega, C = 1/(k_m k_f)$ F $= 72.87233\ n$F

For transfer function $H(S) = \dfrac{1}{S^2 + \dfrac{1}{Q}S + 1} = \dfrac{1}{S^2 + 0.6180S + 1}$

The Q value is $Q = 1/0.6180 = 1.6180$

Normalized component values

$$R_1 = 1/(2Q) = 0.309\ \Omega, \quad R_2 = 2Q = 3.2361\ \Omega, \quad C_1 = 1\text{ F}, \quad C_2 = 1\text{ F}$$

Scaled component values

$$R_1 = 309\ \Omega, \quad R_2 = 3.2361\ k\Omega, \quad C_1 = 1/(k_m k_f)\text{ F} = 72.87233\ n\text{F}$$
$$C_2 = 1/(k_m k_f)\text{ F} = 72.87233\ n\text{F}$$

For transfer function $H(S) = \dfrac{1}{S^2 + \dfrac{1}{Q}S + 1} = \dfrac{1}{S^2 + 1.6180S + 1}$

The Q value is $Q = 1/1.6180 = 0.6180$

Normalized component values

$$R_1 = 1/(2Q) = 0.809\ \Omega, \quad R_2 = 2Q = 1.2361\ \Omega, \quad C_1 = 1\text{ F}, \quad C_2 = 1\text{ F}$$

Scaled component values

$$R_1 = 809\ \Omega, \quad R_2 = 1.2361\ k\Omega, \quad C_1 = 1/(k_m k_f)\text{ F} = 72.87233\ n\text{F}$$
$$C_2 = 1/(k_m k_f)\text{ F} = 72.87233\ n\text{F}$$

The PSpice circuit is shown in Figure 17.9, and the magnitude response is shown in Figure 17.10.

continued

Example 17.4 continued

FIGURE 17.9

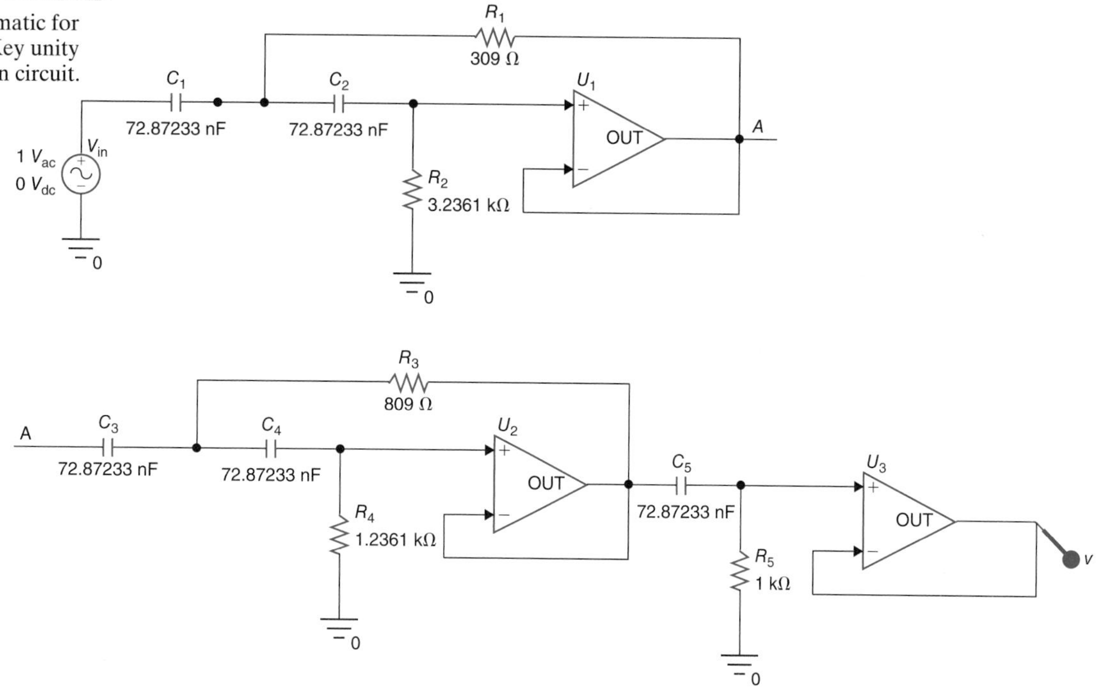

PSpice schematic for the Sallen-Key unity gain circuit.

FIGURE 17.10

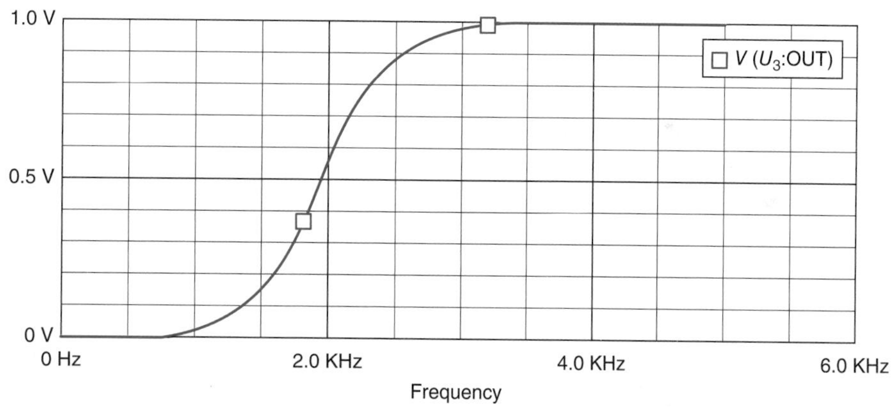

Magnitude response for the circuit shown in Figure 17.9.

Exercise 17.4

Implement the filter designed in EXERCISE 17.3 using the Sallen-Key equal *R*, equal *C* method. Assume $k_m = 1000$.

$$\omega_0 = \omega_p/\Omega_c = 14{,}455.2863 \text{ rad/s}$$

For transfer function $H(S) = \dfrac{1}{S^2 + \dfrac{1}{Q}S + 1} = \dfrac{1}{S^2 + 1.8478S + 1}$

The Q value is $Q = 1/1.8478 = 0.5412$

Normalized component values

$$R_1 = 1\ \Omega, \quad R_2 = 1\ \Omega, \quad R_B = 1\ \Omega, \quad R_A = 0.1522\ \Omega, \quad C_1 = 1\text{ F}, \quad C_2 = 1\text{ F}$$

continued

Exercise 17.4 continued

Scaled component values

$$R_1 = 1\,k\Omega,\quad R_2 = 1\,k\Omega,\quad R_B = 1\,k\Omega,\quad R_A = 152.241\,\Omega,\quad C_1 = 69.1788\,n\text{F},\quad C_2 = 69.1788\,n\text{F}$$

For transfer function $H(S) = \dfrac{1}{S^2 + \dfrac{1}{Q}S + 1} = \dfrac{1}{S^2 + 0.7654S + 1}$

The Q value is $Q = 1/0.7654 = 1.3066$

Normalized component values

$$R_1 = 1\,\Omega,\quad R_2 = 1\,\Omega,\quad R_B = 1\,\Omega,\quad R_A = 1.2346\,\Omega,\quad C_1 = 1\,\text{F},\quad C_2 = 1\,\text{F}$$

Scaled component values

$$R_1 = 1\,k\Omega,\quad R_2 = 1\,k\Omega,\quad R_B = 1\,k\Omega,\quad R_A = 1.2346\,k\Omega,\quad C_1 = 69.1788\,n\text{F},\quad C_2 = 69.1788\,n\text{F}$$

The PSpice simulation is shown in Figure 17.11, and the magnitude response is shown in Figure 17.12.

FIGURE 17.11

PSpice circuit for EXERCISE 17.4.

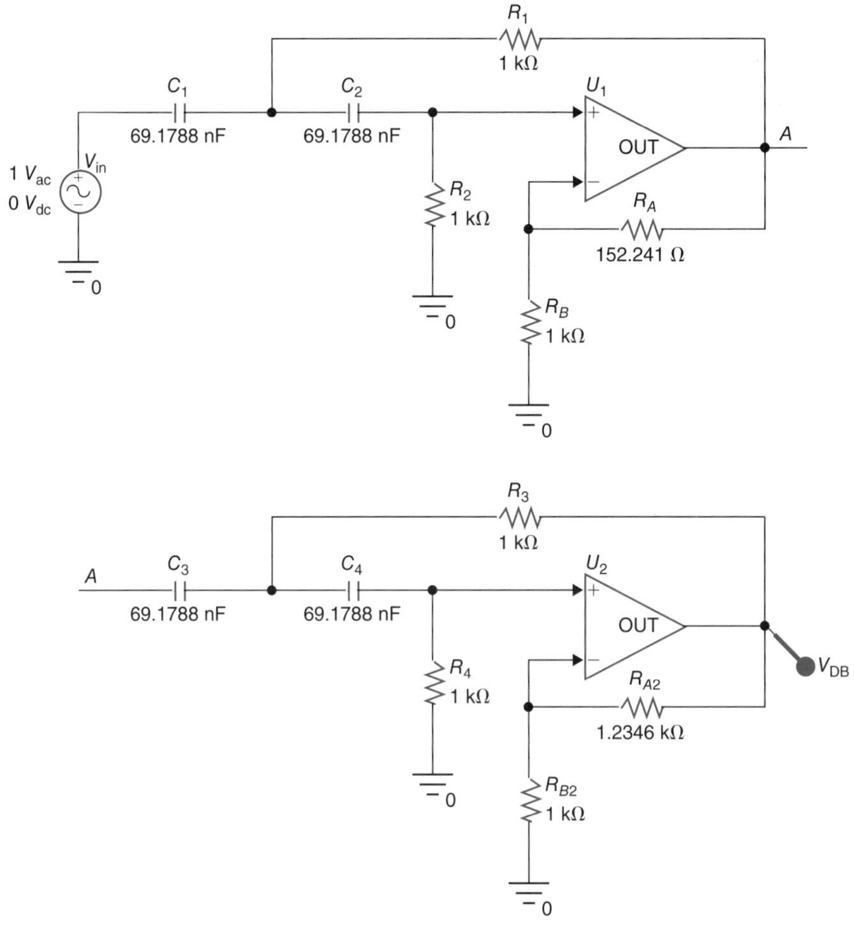

continued

Exercise 17.4 continued

FIGURE 17.12

Magnitude response in decibels for the circuit shown in Figure 17.11.

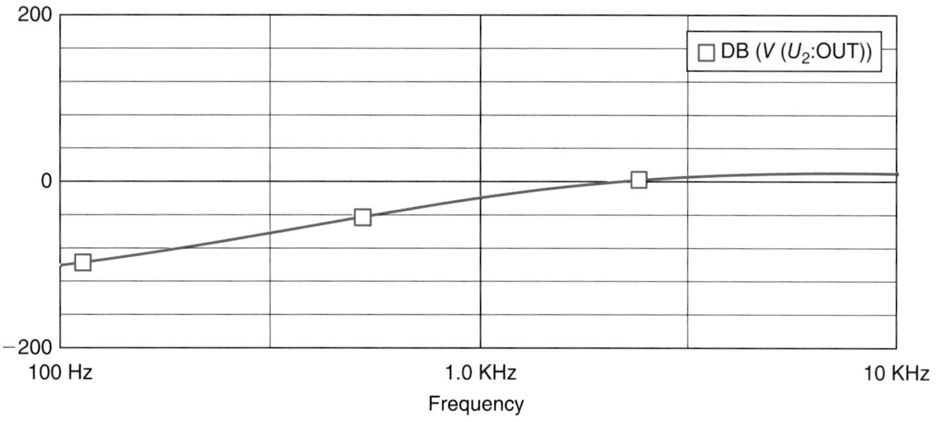

17.4 Analog Butterworth Bandpass Filter Design

The specifications for the bandpass filter are:

Maximum attenuation in the passband = A_p dB
Minimum attenuation in the stopband = A_s dB
Lower stopband cutoff frequency = f_1 Hz
Lower passband cutoff frequency = f_2 Hz
Upper passband cutoff frequency = f_3 Hz
Upper stopband cutoff frequency = f_4 Hz

These specifications are shown in Figure 17.13 with cutoff frequencies in rad/s instead of hertz.

Notice that $\omega_1 < \omega_2 < \omega_3 < \omega_4$. The minimum losses for the two stopbands are assumed to be the same. Assume that the passband cutoff frequencies ω_2 and ω_3 are fixed, but the stopband cutoff frequencies ω_1 and ω_4 can be adjusted.

The bandpass filter is designed by first transforming the specifications for the bandpass filter to the specifications for the normalized LPF. Then, the order and the poles and zeros of the normalized LPF are found. These poles and zeros are transformed into bandpass poles and zeros. From the poles and the zeros, the transfer function of the bandpass filter can be found. To verify this procedure, let us consider the first-order normalized LPF. The first-order normalized ($\Omega_p = 1$ rad/s) LPF has a transfer function

$$H(S) = \frac{1}{S+1}$$

This filter has a cutoff frequency of 1 rad/s. There is one pole at $S = -1$ and one zero at $S = \infty$. This filter can be transformed to a bandpass filter by replacing S by

$$S = \frac{s^2 + \omega_2\omega_3}{(\omega_3 - \omega_2)s} \tag{17.74}$$

FIGURE 17.13

Specifications for the bandpass filter.

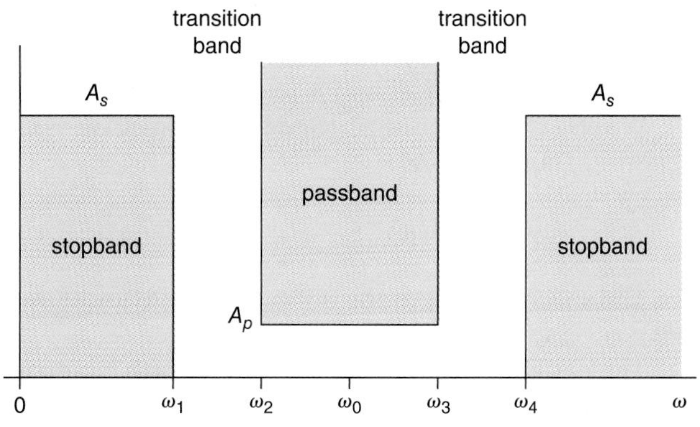

Thus, we have

$$H(s) = H(S)\Big|_{S=\frac{s^2+\omega_2\omega_3}{(\omega_3-\omega_2)s}} = \frac{1}{S+1}\Big|_{S=\frac{s^2+\omega_2\omega_3}{(\omega_3-\omega_2)s}} = \frac{1}{\frac{s^2+\omega_2\omega_3}{(\omega_3-\omega_2)s}+1}$$
$$= \frac{(\omega_3-\omega_2)s}{s^2+(\omega_3-\omega_2)s+\omega_2\omega_3} \quad \textbf{(17.75)}$$

The transfer function given by Equation (17.75) represents a bandpass filter with 3-dB bandwidth of

$$\omega_{3dB} = \omega_3 - \omega_2 \quad \textbf{(17.76)}$$

and corner frequency of

$$\omega_0 = \sqrt{\omega_2\omega_3} \quad \textbf{(17.77)}$$

The location of poles for the bandpass filter can be found by solving $S = \frac{s^2+\omega_2\omega_3}{(\omega_3-\omega_2)s}$ for s. Thus, we obtain

$$s^2 - S(\omega_3-\omega_2)s + \omega_2\omega_3 = 0 \quad \textbf{(17.78)}$$

The two roots of Equation (17.78) are given by

$$s = \frac{S(\omega_3-\omega_2) + \sqrt{S^2(\omega_3-\omega_2)^2 - 4\omega_2\omega_3}}{2} \quad \textbf{(17.79)}$$

$$s = \frac{S(\omega_3-\omega_2) - \sqrt{S^2(\omega_3-\omega_2)^2 - 4\omega_2\omega_3}}{2} \quad \textbf{(17.80)}$$

The stopband cutoff frequency Ω_s of the normalized LPF can be found from Equation (17.74), which relates normalized LPF and bandpass filter. Substitution of $S = j\Omega_{s_1}$ and $s = j\omega_1$ in Equation (17.74) yields

$$\Omega_{s_1} = \frac{\omega_1^2 - \omega_2\omega_3}{(\omega_3-\omega_2)\omega_1} \quad \textbf{(17.81)}$$

Substitution of $S = j\Omega_{s_2}$ and $s = j\omega_4$ in Equation (17.74) yields

$$\Omega_{s_2} = \frac{\omega_4^2 - \omega_2\omega_3}{(\omega_3-\omega_2)\omega_4} \quad \textbf{(17.82)}$$

We choose smaller of $|\Omega_{s_1}|$ and $|\Omega_{s_2}|$ to satisfy the specifications. Thus, we have

$$\Omega_s = \min(|\Omega_{s_1}|, |\Omega_{s_2}|) \quad \textbf{(17.83)}$$

The specifications of the normalized LPF are shown in Figure 17.14.

The design procedure for the Butterworth bandpass filter is as follows.

FIGURE 17.14

Specifications for the normalized LPF.

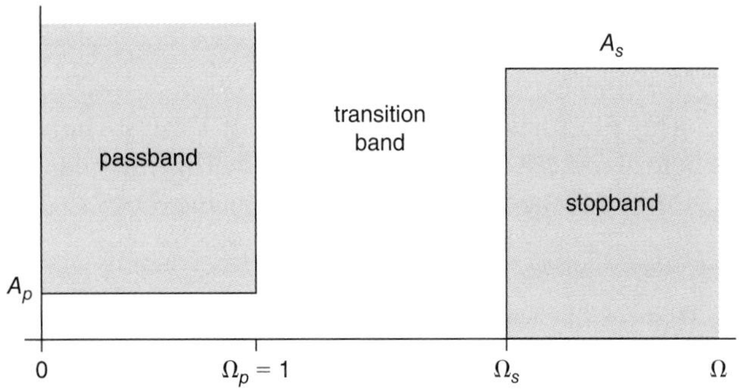

1. Change the Cutoff Frequencies from Hertz to Radians

If the cutoff frequencies are given in hertz, change the cutoff frequencies to rad/s.

$$\omega_1 = 2\pi f_1, \quad \omega_2 = 2\pi f_2, \quad \omega_3 = 2\pi f_3, \quad \omega_4 = 2\pi f_4 \tag{17.84}$$

2. Backward Transformation

Find the cutoff frequencies of the normalized LPF using

$$\text{Passband cutoff frequency for the normalized LPF} = \Omega_p = 1 \text{ rad/s} \tag{17.85}$$

$$\text{Stopband cutoff frequency for the normalized LPF} = \Omega_s = \min(|\Omega_{s_1}|, |\Omega_{s_2}|) \text{ rad/s} \tag{17.86}$$

where

$$\Omega_{s_1} = \frac{\omega_1^2 - \omega_2\omega_3}{(\omega_3 - \omega_2)\omega_1} \tag{17.87}$$

$$\Omega_{s_2} = \frac{\omega_4^2 - \omega_2\omega_3}{(\omega_3 - \omega_2)\omega_4} \tag{17.88}$$

3. Order

Find the order n of the normalized LPF using

$$n = \left\lceil \frac{\log\left(\dfrac{10^{0.1A_s} - 1}{10^{0.1A_p} - 1}\right)}{2\log\left(\dfrac{\Omega_s}{\Omega_p}\right)} \right\rceil = ceil\left[\frac{\log\left(\dfrac{10^{0.1A_s} - 1}{10^{0.1A_p} - 1}\right)}{2\log(\Omega_s)}\right] \tag{17.89}$$

4. Recalculation of the Parameters

If excess tolerance is assigned to the stopband, we have new A_s given by

$$A_{s_1} = 10\log_{10}\left[\Omega_s^{2n}\left(10^{\frac{A_p}{10}} - 1\right) + 1\right] \tag{17.90}$$

The value of A_p remains the same. The passband ripple parameter is given by

$$\varepsilon_p = \sqrt{10^{\frac{A_p}{10}} - 1} \tag{17.91}$$

The 3-dB cutoff frequency is given by

$$\Omega_c = \frac{1}{\varepsilon_p^{\frac{1}{n}}} \tag{17.92}$$

If excess tolerance is assigned to the passband, we have new A_p given by

$$A_{p_1} = 10\log_{10}\left(\frac{10^{\frac{A_s}{10}} - 1}{\Omega_s^{2n}} + 1\right) \tag{17.93}$$

The passband ripple parameter ε_p can be recalculated using

$$\varepsilon_{p_1} = \sqrt{10^{\frac{A_p}{10}} - 1} \tag{17.94}$$

The 3-dB cutoff frequency can be recalculated from

$$\Omega_{c_1} = \frac{1}{\varepsilon_{p_1}^{\frac{1}{n}}} \tag{17.95}$$

5. Normalized Lowpass Poles and Zeros

Find the normalized ($\Omega_c = 1$ rad/s) poles and zeros.

n odd:
There is one pole at $S = -1$ and $(n - 1)/2$ complex conjugate poles at

$$S_k = e^{\pm j\frac{\pi(2k+n+1)}{2n}} = \cos\left(\frac{\pi(2k + n + 1)}{2n}\right) \pm j\sin\left(\frac{\pi(2k + n + 1)}{2n}\right),\quad k = 0, 1, \ldots, \frac{n-3}{2} \tag{17.96}$$

n even:
There are $n/2$ complex conjugate poles at

$$S_k = e^{\pm j\frac{\pi(2k+n+1)}{2n}} = \cos\left(\frac{\pi(2k + n + 1)}{2n}\right) \pm j\sin\left(\frac{\pi(2k + n + 1)}{2n}\right),\quad k = 0, 1, \ldots, \frac{n}{2} - 1 \tag{17.97}$$

There are n zeros at $S = \infty$

6. Transfer Function of the Normalized LPF

Convert the poles and the zeros of the normalized LPF into polynomials, and write the transfer function as a product of the first-order and the second-order sections. If the order of the filter is odd, there is one first-order section and $(n-1)/2$ second-order sections. If the order of the filter is even, there are $n/2$ second-order sections. The numerator is a constant equal to the constant term of the denominator for each first- or second-order term.

7. Scaled Poles and Zeros

Multiply the poles and the zeros of the normalized LPF from step 5 by Ω_c. There are n zeros at $S = \infty$.

8. Frequency Transformation

The low-pass poles and zeros are transformed to bandpass poles and zeros. The bandpass poles are obtained from

$$s = \frac{S(\omega_3 - \omega_2) + \sqrt{S^2(\omega_3 - \omega_2)^2 - 4\omega_2\omega_3}}{2} \tag{17.98}$$

$$s = \frac{S(\omega_3 - \omega_2) - \sqrt{S^2(\omega_3 - \omega_2)^2 - 4\omega_2\omega_3}}{2} \tag{17.99}$$

Also, n zeros at $S = \infty$ are transformed to n zeros at $s = 0$ and n zeros at $s = \infty$.

9. Transfer Function

Convert the poles and the zeros of the frequency-transformed bandpass filter into polynomials, and write the transfer function as a product of n second-order sections. For each second-order section, find the corner frequency ω_0, Q value and the constant term b_0. Find the single b_0 by multiplying all b_0's.

10. Verify

Calculate the attenuation at the cutoff frequencies, and compare with the specifications. If $|\Omega_{s_1}| < |\Omega_{s_2}|$, the frequency ω_4 could be lowered to

$$\omega_{4a} = \frac{\omega_2\omega_3}{\omega_1}$$

The losses at ω_1 and ω_{4a} are the same. The loss at ω_4 will be greater than the loss at ω_{4a}. If $|\Omega_{s_1}| > |\Omega_{s_2}|$, the frequency ω_1 could be increased to

$$\omega_{1a} = \frac{\omega_2\omega_3}{\omega_4}$$

The losses at ω_{1a} and ω_4 are the same.

If the excess tolerance is assigned to the stopband, we have

$$-20\log_{10}|H(j\omega_2)| = A_p, \quad -20\log_{10}|H(j\omega_3)| = A_p \tag{17.100}$$

$$-20\log_{10}|H(j\omega_1)| \geq A_s, \quad -20\log_{10}|H(j\omega_4)| \geq A_s \tag{17.101}$$

If the excess tolerance is assigned to the passband, we have

$$-20\log_{10}|H(j\omega_2)| \leq A_p, \quad -20\log_{10}|H(j\omega_3)| \leq A_p \tag{17.102}$$

If $|\Omega_{s_1}| < |\Omega_{s_2}|$,

$$-20\log_{10}|H(j\omega_1)| = A_s \quad \text{and} \quad -20\log_{10}|H(j\omega_{4a})| = A_s \tag{17.103}$$

If $|\Omega_{s_1}| > |\Omega_{s_2}|$,

$$-20\log_{10}|H(j\omega_{1a})| = A_s \quad \text{and} \quad -20\log_{10}|H(j\omega_4)| = A_s \tag{17.104}$$

EXAMPLE 17.5

Design a Butterworth bandpass filter with the following specifications:

Maximum loss in the passband = A_p = 2 dB
Minimum loss in the stopband = A_s = 17 dB
Lower stopband cutoff frequency = f_1 = 1500 Hz
Lower passband cutoff frequency = f_2 = 2000 Hz
Upper passband cutoff frequency = f_3 = 3000 Hz
Upper stopband cutoff frequency = f_4 = 4500 Hz
Excess tolerance = stopband

continued

Example 17.5 continued

1. Change the Cutoff Frequencies from Hertz to Radians

The cutoff frequencies in rad/s are

$$\omega_1 = 2\pi f_1 = 9424.7780 \text{ rad/s}$$
$$\omega_2 = 2\pi f_2 = 12{,}566.3706 \text{ rad/s}$$
$$\omega_3 = 2\pi f_3 = 18{,}849.5559 \text{ rad/s}$$
$$\omega_4 = 2\pi f_4 = 28{,}274.3339 \text{ rad/s}$$

2. Backward Transformation

$$\Omega_{s_1} = \frac{\omega_2\omega_3 - \omega_1^2}{(\omega_3 - \omega_2)\omega_1} = 2.5 \text{ rad/s}$$

$$\Omega_{s_2} = \frac{\omega_4^2 - \omega_2\omega_3}{(\omega_3 - \omega_2)\omega_4} = 3.1667 \text{ rad/s}$$

$$\Omega_p = 1 \text{ rad/s}$$

$$\Omega_s = \min(|\Omega_{s_1}|, |\Omega_{s_2}|) = 2.5 \text{ rad/s}$$

3. Order

$$n = ceil\left[\frac{\log\left(\dfrac{10^{0.1A_s} - 1}{10^{0.1A_p} - 1}\right)}{2\log(\Omega_s)}\right] = ceil\left[\frac{\log\left(\dfrac{10^{1.7} - 1}{10^{0.2} - 1}\right)}{2\log(2.5)}\right] = ceil(2.42) = 3$$

4. Recalculation of the Parameters

Since the excess tolerance is assigned to the stopband, we have

$$A_{s_1} = 10\log_{10}\left[\Omega_s^{2n}\left(10^{\frac{A_p}{10}} - 1\right) + 1\right] = 21.5775 \text{ dB}$$

The value of A_p remains the same. The ripple parameter is given by

$$\varepsilon_p = \sqrt{10^{\frac{A_p}{10}} - 1} = 0.7648$$

The 3-dB cutoff frequency is given by

$$\Omega_c = \frac{1}{\varepsilon_p^{\frac{1}{n}}} = 1.0935$$

5. Normalized Lowpass Poles and Zeros

Pole locations

$$S = -1$$

$$S = e^{\pm j\frac{\pi(3+1)}{2\times 3}} = \cos\left(\frac{\pi(3+1)}{2\times 3}\right) \pm j\sin\left(\frac{\pi(3+1)}{2\times 3}\right) = \cos\left(\frac{2\pi}{3}\right) \pm j\sin\left(\frac{2\pi}{3}\right)$$
$$= -0.5 \pm j0.8660$$

There are 3 zeros at $S = \infty$.

continued

Example 17.5 continued

6. Transfer Function of the Normalized LPF

$$H(s) = \frac{1}{S+1} \times \frac{1}{(S+0.5-j0.866)(S+0.5+j0.866)} = \frac{1}{S+1} \times \frac{1}{S^2+S+1}$$

7. Scaled Poles and Zeros

$$S = \Omega_c S$$
$$-1.0935$$
$$-0.5468 \pm j0.9470$$

8. Frequency Transformation

$$s = -3435 \pm j15{,}002.2967$$
$$s = -2046 \pm j18{,}559.6668$$
$$s = -1390 \pm j12{,}609.4735$$

There are three zeros at $s = 0$ and three zeros at $s = \infty$.

9. Transfer Function

$$H(s) = \frac{6870.6914s}{s^2 + 6870.6914s + 236{,}870{,}506} \times \frac{4091.1537s}{s^2 + 4091.1537s + 348{,}645{,}616} \times \frac{2779.5377s}{s^2 + 2779.5377s + 160{,}930{,}280}$$

Second-order term	Corner frequency	Q value	b_0
$s^2 + 6870.6914s + 236{,}870{,}506$	15,390.5980	2.24	1
$s^2 + 4091.1537s + 348{,}645{,}616$	18,672.0544	4.564	2.0375
$s^2 + 2779.5377s + 160{,}930{,}280$	12,685.8299	4.564	2.0375

The center frequency of the passband is

$$\omega_0 = \sqrt{\omega_2 \omega_3} = 15{,}390.5980 \text{ rad/s}$$

The constant term b_0 can be found by evaluating the gain at $\omega = \omega_0 = 15{,}390.5980$ rad/s for each second-order section and taking the inverse. The single b_0 is the product of all b_0's.

$$\text{Single } b_0 = 1 \times 2.0375 \times 2.0375 = 4.1513$$

Thus, the transfer function of the bandpass filter becomes

$$H(s) = 4.1513 \times \frac{6870.6914s}{s^2 + 6870.6914s + 236{,}870{,}506} \times \frac{4091.1537s}{s^2 + 4091.1537s + 348{,}645{,}616} \times \frac{2779.5377s}{s^2 + 2779.5377s + 160{,}930{,}280}$$

continued

Example 17.5 continued

10. Verify

$$-20\log_{10}|H(j\omega_2)| = 2\,\text{dB},\ -20\log_{10}|H(j\omega_3)| = 2\,\text{dB}$$

$$-20\log_{10}|H(j\omega_1)| = 21.5775\,\text{dB},\ -20\log_{10}|H(j\omega_{4a})| = 21.5775\,\text{dB},$$

$$-20\log_{10}|H(j\omega_4)| = 27.7143\,\text{dB}$$

Exercise 17.5

Design a Butterworth bandpass filter with the following specifications:

Maximum loss in the passband = A_p = 1 dB
Minimum loss in the stopband = A_s = 10 dB
Lower stopband cutoff frequency = f_1 = 1000 Hz
Lower passband cutoff frequency = f_2 = 2000 Hz
Upper passband cutoff frequency = f_3 = 3000 Hz
Upper stopband cutoff frequency = f_4 = 4000 Hz
Excess tolerance = passband

Answer:

$\omega_1 = 6283.1853$ rad/s, $\omega_2 = 12{,}566.3706$ rad/s, $\omega_3 = 18{,}849.5559$ rad/s,
$\omega_4 = 25{,}132.7412$ rad/s, $\omega_{1a} = 9424.778$ rad/s

$\Omega_p = 1$ rad/s, $\Omega_s = 2.5$ rad/s, $n = \text{ceil}(1.94) = 2$, $A_{p_1} = 0.9005$ dB, $\varepsilon_p = 0.48$,
$\Omega_c = 1.4434$.

Normalized ($\Omega_c = 1$) low-pass poles and zeros: $S = -0.7071 \pm j0.7071$, two zeros at $S = \infty$.

Transfer function of normalized LPF:

$$H(S) = \frac{1}{S^2 + 1.4142S + 1}$$

Multiplication of poles by Ω_c: $S = \Omega_c S$

$S = -1.0206 \pm j1.0206$, two zeros at ∞.

Bandpass poles and zeros:

$s = -3873.7422 \pm j18{,}611.4349$, $s = -2539.0070 \pm j12{,}198.6958$, two zeros at $s = 0$ and two zeros at $s = \infty$

Transfer function of bandpass filter:

$$H(s) = \frac{7747.4843s}{s^2 + 7747.4843s + 361{,}391{,}389} \times \frac{5078.0140s}{s^2 + 5078.0140s + 155{,}254{,}492}$$

Second-order term	Corner frequency	Q value	b_0
$s^2 + 7747.4843s + 361{,}391{,}389$	19,010.2969	2.4537	1.4459
$s^2 + 5078.0140s + 155{,}254{,}492$	12,460.1160	2.4537	1.4459

continued

Exercise 17.5 continued

Single $b_0 = 2.0906$

The center frequency of the passband is

$$\omega_0 = \sqrt{\omega_2\omega_3} = 15{,}390.5980 \text{ rad/s}$$
$$-20\log_{10}|H(j\omega_2)| = 0.9 \text{ dB}, \; -20\log_{10}|H(j\omega_3)| = 0.9\,\text{dB}$$
$$-20\log_{10}|H(j\omega_1)| = 21.6137 \text{ dB}, \;\; -20\log_{10}|H(j\omega_{1a})| = 10 \text{ dB},$$
$$-20\log_{10}|H(j\omega_4)| = 10 \text{ dB}$$

EXAMPLE 17.6

Implement the filter designed in EXAMPLE 17.5 using the Deliyannis-Friend circuit. Assume $k_m = 1000$.

For the first second-order section, $Q = 2.24$. Thus, the normalized component values are given by

$$C = C_1 = C_2 = \frac{1}{2Q} = 0.2232 \text{ F}, \quad R_2 = 4Q^2 = 20.0711\ \Omega, \quad R_a = 2Q^2 = 10.0355\ \Omega$$
$$R_b = \frac{2Q^2}{2Q^2 - 1} = 1.1107\ \Omega$$

Application of the scaling by $k_m = 1000$ and $k_f = 15{,}390.5980$ yields

$$C_1 = C_2 = 14.5031\ n\text{F}, \quad R_2 = 20.0711\ k\Omega, \quad R_a = 10.0355\ k\Omega, \quad R_b = 1.1107\ k\Omega$$

For the second second-order section, $Q = 4.564$. Thus, the normalized component values are given by

$$C = C_1 = C_2 = \frac{1}{2Q} = 0.1096 \text{ F}, \quad R_2 = 4Q^2 = 83.3206\ \Omega, \quad R_a = 2Q^2 = 41.6603\ \Omega$$
$$R_b = \frac{2Q^2}{2Q^2 - 1} = 1.0246\ \Omega$$

Application of the scaling by $k_m = 1000$ and $k_f = 18{,}672.0544$ yields

$$C_1 = C_2 = 5.8672\ n\text{F}, \quad R_2 = 83.3206\ k\Omega, \quad R_a = 41.6603\ k\Omega, \quad R_b = 1.0246\ k\Omega$$

For the third second-order section, $Q = 4.564$. Thus, the normalized component values are given by

$$C = C_1 = C_2 = \frac{1}{2Q} = 0.1096 \text{ F}, \quad R_2 = 4Q^2 = 83.3206\ \Omega, \quad R_a = 2Q^2 = 41.6603\ \Omega,$$
$$R_b = \frac{2Q^2}{2Q^2 - 1} = 1.0246\ \Omega$$

Application of the scaling by $k_m = 1000$ and $k_f = 12{,}685.8299$ yields

$$C_1 = C_2 = 8.6358\ n\text{F}, \quad R_2 = 83.3206\ k\Omega, \quad R_a = 41.6603\ k\Omega, \quad R_b = 1.0246\ k\Omega$$
$$R_D = 1\ k\Omega, R_C = (\text{Single } b_0 - 1)\ k\Omega = (4.1513 - 1)\ k\Omega = 3.1513\ k\Omega$$

continued

Example 17.6 continued

FIGURE 17.15

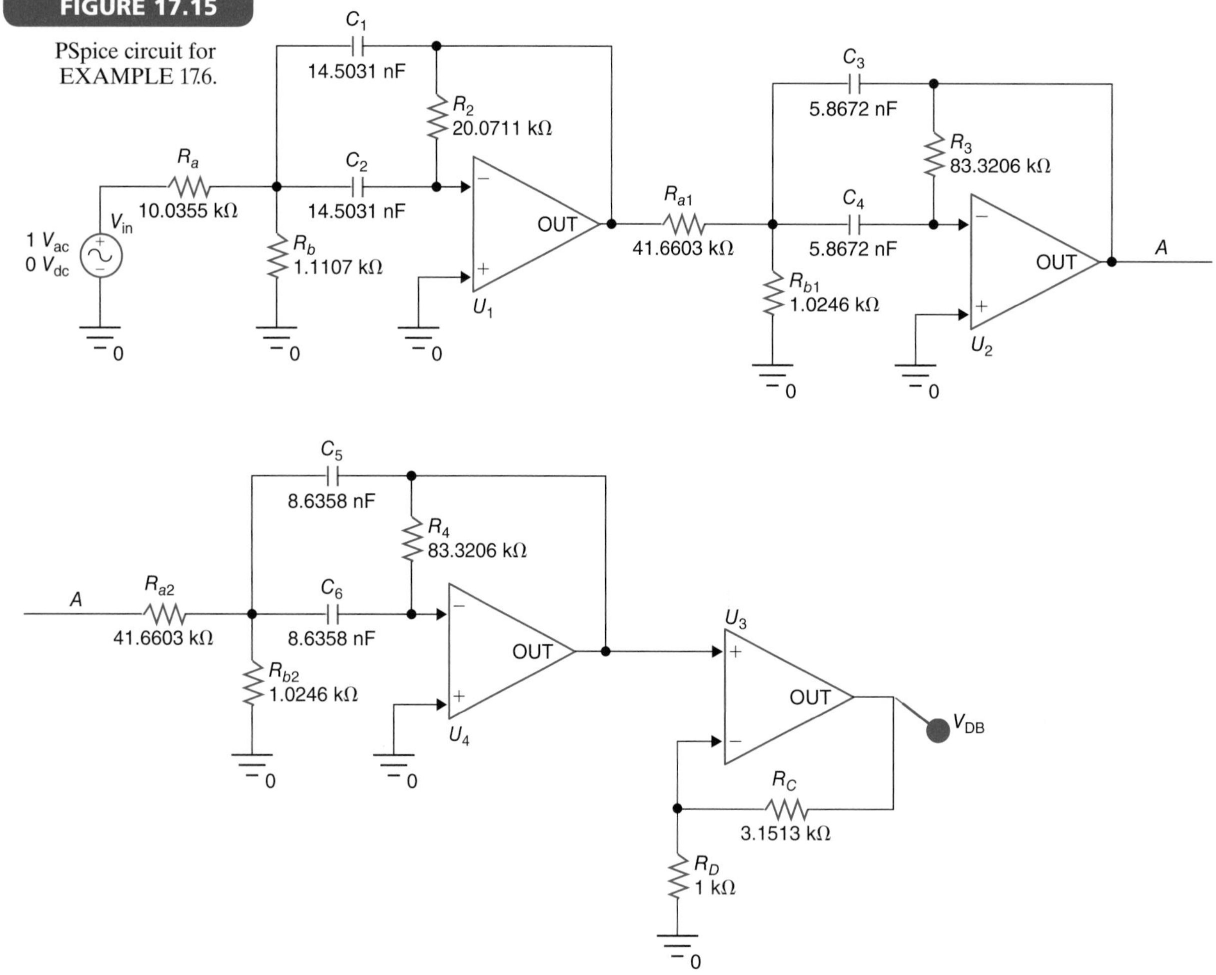

PSpice circuit for EXAMPLE 17.6.

FIGURE 17.16

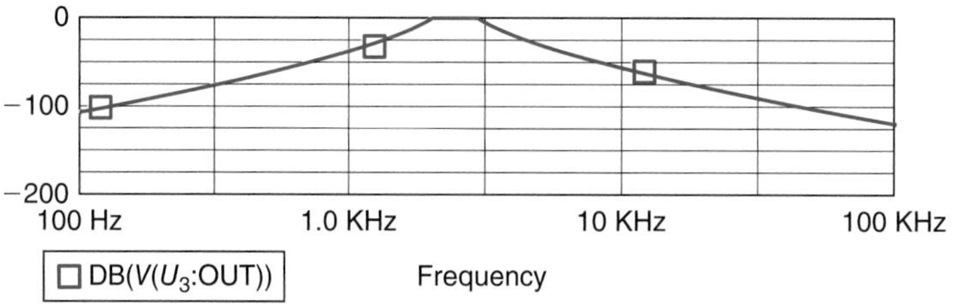

The magnitude response of the circuit shown in Figure 17.15.

Exercise 17.6

Implement the filter designed in EXERCISE 17.5 using the Sallen-Key circuit with the equal *R*, equal *C* method. Assume $k_m = 1000$.

First second-order section

$$\frac{7747.4843s}{s^2 + 7747.4843s + 361{,}391{,}389}$$

$k_f = 19{,}010.2969$ rad/s, $Q = 2.4537$

continued

Exercise 17.6 continued

Normalized component values:

$R_2 = 1\ \Omega,\quad R_3 = 1\ \Omega,\quad R_a = 5.9402\ \Omega,\quad R_b = 1.2024\ \Omega,\quad R_A = 2.4236\ \Omega,$
$R_B = 1\ \Omega$

$C_1 = 1.4142\text{ F},\quad C_2 = 1.4142\text{ F}$

Frequency scaled component values:

$R_2 = 1\ k\Omega,\quad R_3 = 1\ k\Omega,\quad R_a = 5.9402\ k\Omega,\quad R_b = 1.2024\ k\Omega,\quad R_A = 2.4236\ k\Omega,$
$R_B = 1\ k\Omega$

$C_1 = 74.392\ n\text{F},\quad C_2 = 74.392\ n\text{F}$

Second second-order section

$$\frac{5078.0140s}{s^2 + 5078.0140s + 155{,}254{,}492}$$

$k_f = 12{,}460.1160$ rad/s, $Q = 2.4537$

Normalized component values:

$R_2 = 1\ \Omega,\quad R_3 = 1\ \Omega,\quad R_a = 5.9402\ \Omega,\quad R_b = 1.2024\ \Omega,\quad R_A = 2.4236\ \Omega,$
$R_B = 1\ \Omega$

$C_1 = 1.4142\text{ F},\quad C_2 = 1.4142\text{ F}$

Frequency-scaled component values:

$R_2 = 1\ k\Omega,\quad R_3 = 1\ k\Omega,\quad R_a = 5.9402\ k\Omega,\quad R_b = 1.2024\ k\Omega,\quad R_A = 2.4236\ k\Omega,$
$R_B = 1\ k\Omega$

$C_1 = 0.1135\ \mu\text{F}, C_2 = 0.1135\ \mu\text{F}$

$R_D = 1\ k\Omega,\quad R_C = (\text{Single } b_0 - 1)\ k\Omega = (2.0906 - 1)\ k\Omega = 1.0906\ k\Omega$

FIGURE 17.17

PSpice circuit for EXERCISE 17.6.

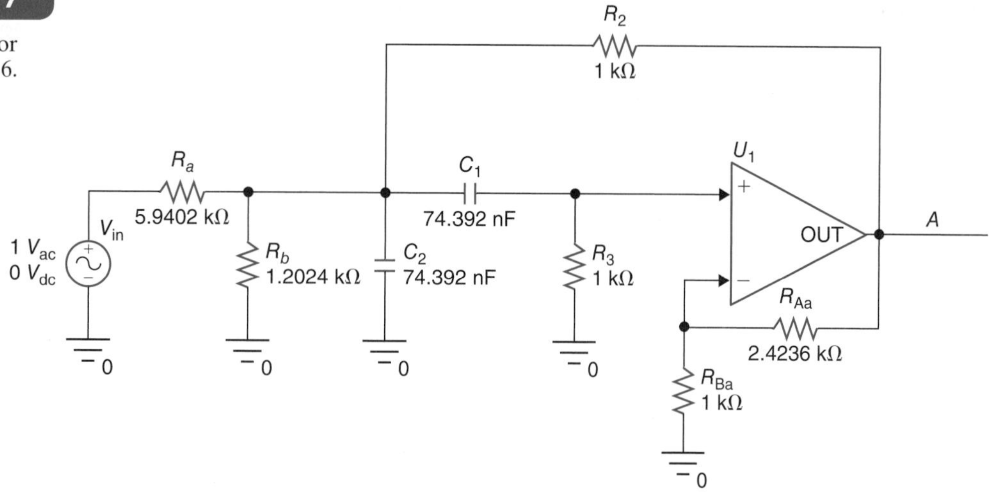

continued

Exercise 17.6 continued

FIGURE 17.17

continued

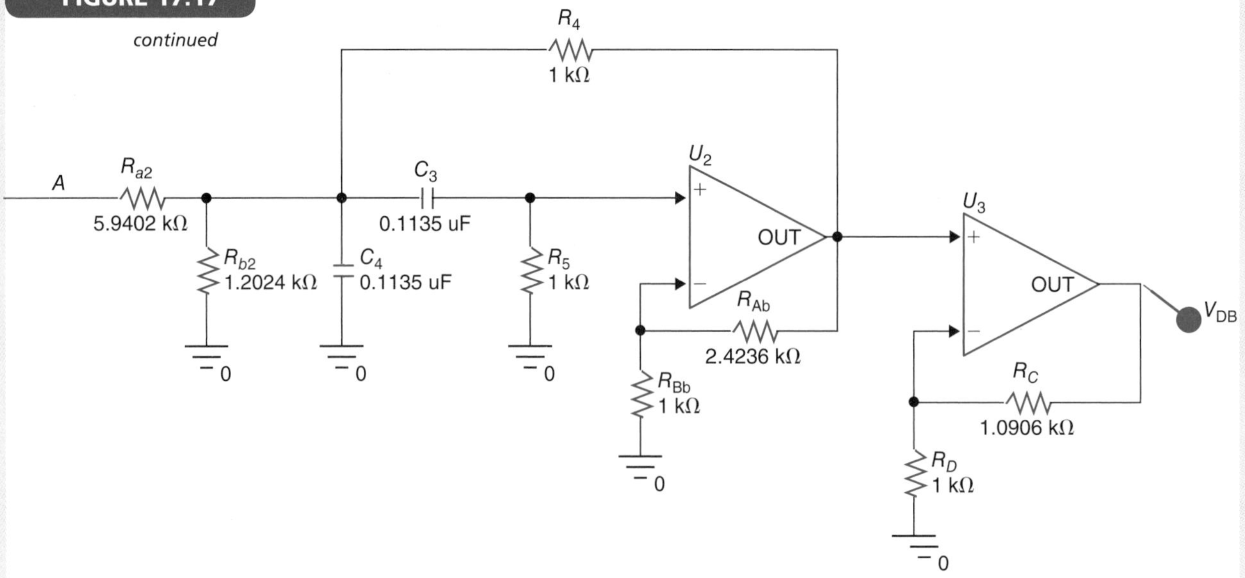

FIGURE 17.18

The magnitude response of the circuit shown in Figure 17.17.

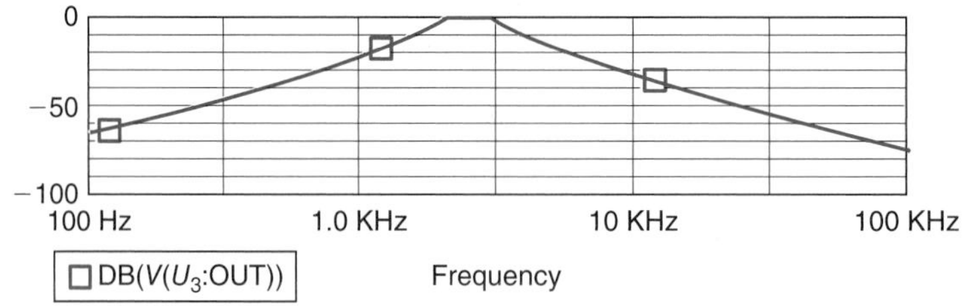

17.5 Analog Butterworth Bandstop Filter Design

FIGURE 17.19

Specifications for the bandstop filter.

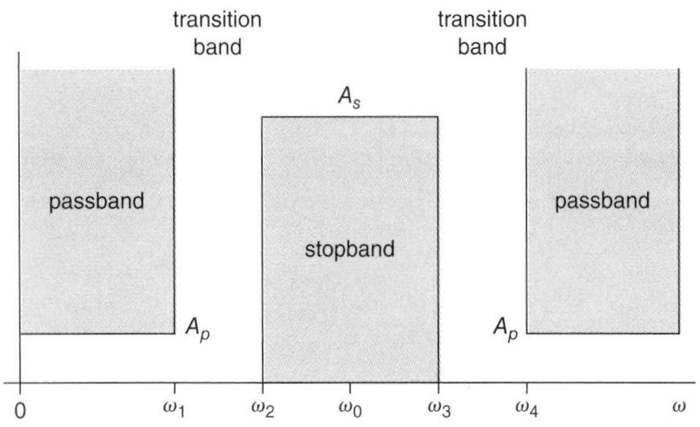

The specifications for the bandstop filter are:

Maximum attenuation in the passband = A_p dB
Minimum attenuation in the stopband = A_s dB
Lower passband cutoff frequency = f_1 Hz
Lower stopband cutoff frequency = f_2 Hz
Upper stopband cutoff frequency = f_3 Hz
Upper passband cutoff frequency = f_4 Hz

These specifications are shown in Figure 17.19 with cutoff frequencies in rad/s instead of hertz.

Notice that $\omega_1 < \omega_2 < \omega_3 < \omega_4$. The maximum attenuations for the two passbands are assumed to be equal. Assume that the passband cutoff frequencies ω_1 and ω_4 are fixed, but the stopband cutoff frequencies ω_2 and ω_3 can be adjusted.

The bandstop filter is designed by first transforming the specifications for the bandstop filter to the specifications for the normalized LPF. Then, the order and the poles and the zeros of the normalized LPF are found. These poles and zeros are transformed to bandstop poles and zeros. From the poles and the zeros, the transfer function of the bandstop filter can be found. To verify this procedure, let us consider the first-order normalized LPF. The first-order normalized ($\Omega_p = 1$ rad/s) LPF has a transfer function

$$H(S) = \frac{1}{S+1}$$

This filter has a cutoff frequency of 1 rad/s. There is one pole at $S = -1$ and one zero at $S = \infty$. This filter can be transformed to a bandstop filter by replacing S by

$$S = \frac{(\omega_4 - \omega_1)s}{s^2 + \omega_1\omega_4} \tag{17.105}$$

Thus, we have

$$H(s) = H(S)\Big|_{S=\frac{(\omega_4-\omega_1)s}{s^2+\omega_1\omega_4}} = \frac{1}{S+1}\Big|_{S=\frac{(\omega_4-\omega_1)s}{s^2+\omega_1\omega_4}} = \frac{1}{\dfrac{(\omega_4 - \omega_1)s}{s^2 + \omega_1\omega_4} + 1}$$
$$= \frac{s^2 + \omega_1\omega_4}{s^2 + (\omega_4 - \omega_1)s + \omega_1\omega_4} \tag{17.106}$$

The transfer function given by Equation (17.106) represents a bandstop filter with a 3-dB bandwidth of

$$\omega_{3dB} = \omega_4 - \omega_1 \tag{17.107}$$

and corner frequency of

$$\omega_0 = \sqrt{\omega_1\omega_4} \tag{17.108}$$

The location of poles for the bandstop filter can be found by solving $S = \dfrac{(\omega_4 - \omega_1)s}{s^2 + \omega_1\omega_4}$ for s. Thus, we obtain

$$Ss^2 - (\omega_4 - \omega_1)s + S\omega_1\omega_4 = 0 \tag{17.109}$$

The two roots of Equation (17.109) are given by

$$s = \frac{(\omega_4 - \omega_1) + \sqrt{(\omega_4 - \omega_1)^2 - 4S^2\omega_1\omega_4}}{2S} \tag{17.110}$$

$$s = \frac{(\omega_4 - \omega_1) - \sqrt{(\omega_4 - \omega_1)^2 - 4S^2\omega_1\omega_4}}{2S} \tag{17.111}$$

The stopband cutoff frequency Ω_s of the normalized LPF can be found from Equation (17.105), which relates to normalized LPF and bandstop filter. Substitution of $S = j\Omega_{s_1}$ and $s = j\omega_2$ in Equation (17.105) yields

$$\Omega_{s_1} = \frac{(\omega_4 - \omega_1) \cdot \omega_2}{\omega_1\omega_4 - \omega_2^2} \tag{17.112}$$

Substitution of $S = j\Omega_{s_2}$ and $s = j\omega_3$ in Equation (17.105) yields

$$\Omega_{s_2} = \frac{(\omega_4 - \omega_1) \cdot \omega_3}{\omega_3^2 - \omega_1\omega_4} \tag{17.113}$$

FIGURE 17.20

Specifications for the normalized LPF.

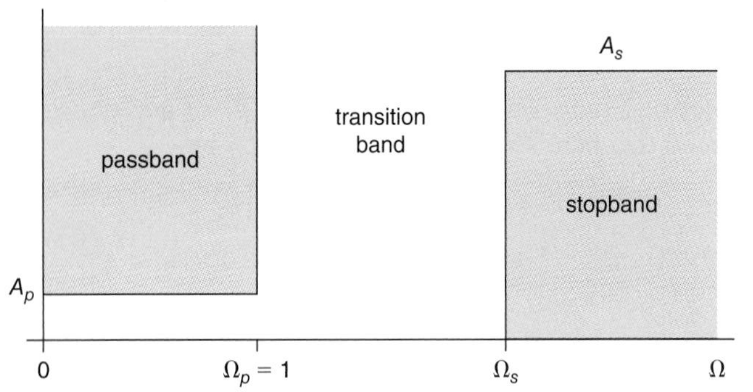

We choose the smaller of $|\Omega_{s_1}|$ and $|\Omega_{s_2}|$ to satisfy the specifications. Thus, we have

$$\Omega_s = \min(|\Omega_{s_1}|, |\Omega_{s_2}|). \quad \textbf{(17.114)}$$

The specifications of the normalized LPF are shown in Figure 17.20.

The design procedure for the Butterworth bandstop filter is as follows.

1. Change the Cutoff Frequencies from Hertz to Radians

If the cutoff frequencies are given in hertz, change the cutoff frequencies to radians.

$$\omega_1 = 2\pi f_1, \quad \omega_2 = 2\pi f_2, \quad \omega_3 = 2\pi f_3, \quad \omega_4 = 2\pi f_4 \quad \textbf{(17.115)}$$

2. Backward Transformation

Find the cutoff frequencies of the normalized LPF using

$$\text{Passband cutoff frequency for the normalized LPF} = \Omega_p = 1 \text{ rad/s} \quad \textbf{(17.116)}$$

$$\begin{aligned}&\text{Stopband cutoff frequency for the normalized LPF} = \Omega_s \\ &= \min(|\Omega_{s_1}|,|\Omega_{s_2}|) \text{ rad/s}\end{aligned} \quad \textbf{(17.117)}$$

where

$$\Omega_{s_1} = \frac{(\omega_4 - \omega_1)\cdot\omega_2}{\omega_1\omega_4 - \omega_2^2} \quad \textbf{(17.118)}$$

$$\Omega_{s_2} = \frac{(\omega_4 - \omega_1)\cdot\omega_3}{\omega_3^2 - \omega_1\omega_4} \quad \textbf{(17.119)}$$

3. Order

Find the order n of the normalized LPF using

$$n = \left\lceil \frac{\log\left(\frac{10^{0.1A_s} - 1}{10^{0.1A_p} - 1}\right)}{2\log\left(\frac{\Omega_s}{\Omega_p}\right)} \right\rceil = ceil\left[\frac{\log\left(\frac{10^{0.1A_s} - 1}{10^{0.1A_p} - 1}\right)}{2\log(\Omega_s)}\right] \quad \textbf{(17.120)}$$

4. Recalculation of the Parameters

If excess tolerance is assigned to the stopband, we have a new A_s, given by

$$A_{s_1} = 10\log_{10}\left[\Omega_s^{2n}\left(10^{\frac{A_p}{10}} - 1\right) + 1\right] \quad \textbf{(17.121)}$$

The value of A_p remains the same. The passband ripple parameter is given by

$$\varepsilon_p = \sqrt{10^{\frac{A_p}{10}} - 1} \quad \textbf{(17.122)}$$

The 3-dB cutoff frequency is given by

$$\Omega_c = \frac{1}{\varepsilon_p^{\frac{1}{n}}} \tag{17.123}$$

If excess tolerance is assigned to the passband, we have a new A_p, given by

$$A_{p_1} = 10 \log_{10}\left(\frac{10^{\frac{A_s}{10}} - 1}{\Omega_s^{2n}} + 1\right) \tag{17.124}$$

The passband ripple parameter ε_p can be recalculated using

$$\varepsilon_{p_1} = \sqrt{10^{\frac{A_{p_1}}{10}} - 1} \tag{17.125}$$

The 3-dB cutoff frequency can be recalculated from

$$\Omega_{c_1} = \frac{1}{\varepsilon_{p_1}^{\frac{1}{n}}} \tag{17.126}$$

5. Normalized Lowpass Poles and Zeros

Find the normalized ($\Omega_c = 1$ rad/s) poles and zeros.

n odd:
One pole at $S = -1$ and $(n-1)/2$ complex conjugate poles at

$$S_k = e^{\pm j\frac{\pi(2k+n+1)}{2n}} = \cos\left(\frac{\pi(2k+n+1)}{2n}\right) \pm j\sin\left(\frac{\pi(2k+n+1)}{2n}\right), \quad k = 0, 1, \ldots, \frac{n-3}{2} \tag{17.127}$$

n even:
There are $n/2$ complex conjugate poles at

$$S_k = e^{\pm j\frac{\pi(2k+n+1)}{2n}} = \cos\left(\frac{\pi(2k+n+1)}{2n}\right) \pm j\sin\left(\frac{\pi(2k+n+1)}{2n}\right), \quad k = 0, 1, \ldots, \frac{n}{2} - 1 \tag{17.128}$$

There are n zeros at $S = \infty$.

6. Transfer Function of the Normalized LPF

Convert the poles and the zeros of the normalized LPF into polynomials, and write the transfer function as a product of the first-order and the second-order sections. If the order of the filter is odd, there is one first-order section and $(n - 1)/2$ second-order sections. If the order of the filter is even, there are $n/2$ second-order sections. The numerator is a constant equal to the constant term of the denominator for each first- or second-order term. Also, find the Q value of each second-order section.

7. Scaled Poles and Zeros

Multiply the poles and the zeros of the normalized LPF from step 5 by Ω_c. There are n zeros at $S = \infty$.

8. Frequency Transformation

The lowpass poles and zeros are transformed to bandstop poles and zeros. The bandstop poles are obtained from

$$s = \frac{(\omega_4 - \omega_1) + \sqrt{(\omega_4 - \omega_1)^2 - 4S^2\omega_1\omega_4}}{2S} \quad (17.129)$$

$$s = \frac{(\omega_4 - \omega_1) - \sqrt{(\omega_4 - \omega_1)^2 - 4S^2\omega_1\omega_4}}{2S} \quad (17.130)$$

Also, n zeros at $S = \infty$ are transformed to n zeros at $s = j\sqrt{\omega_1\omega_4}$ and n zeros at $s = -j\sqrt{\omega_1\omega_4}$.

9. Transfer Function

Convert the poles and the zeros into polynomials, and write the transfer function as a product of n second-order sections.

10. Verify

Calculate the attenuation at the cutoff frequencies, and compare with the specifications. If $|\Omega_{s_1}| < |\Omega_{s_2}|$, the frequency ω_3 could be increased to

$$\omega_{3a} = \frac{\omega_1\omega_4}{\omega_2}$$

The losses at ω_2 and ω_{3a} are the same. The loss at ω_3 will be greater than the loss at ω_{3a}. If $|\Omega_{s_1}| > |\Omega_{s_2}|$, the frequency ω_2 could be decreased to

$$\omega_{2a} = \frac{\omega_1\omega_4}{\omega_3}$$

The losses at ω_{2a} and ω_3 are the same. The loss at ω_2 will be greater than the loss at ω_{2a}. If the excess tolerance is assigned to the stopband, we have

$$-20\log_{10}|H(j\omega_1)| = A_p, \quad -20\log_{10}|H(j\omega_4)| = A_p \quad (17.131)$$

$$-20\log_{10}|H(j\omega_2)| \geq A_s, \quad -20\log_{10}|H(j\omega_3)| \geq A_s \quad (17.132)$$

If the excess tolerance is assigned to the passband, we have

$$-20\log_{10}|H(j\omega_1)| \leq A_p, \quad -20\log_{10}|H(j\omega_4)| \leq A_p \quad (17.133)$$

If $|\Omega_{s_1}| < |\Omega_{s_2}|$,

$$-20\log_{10}|H(j\omega_2)| = A_s \quad \text{and} \quad -20\log_{10}|H(j\omega_{3a})| = A_s \quad (17.134)$$

If $|\Omega_{s_1}| > |\Omega_{s_2}|$,

$$-20\log_{10}|H(j\omega_{2a})| = A_s \quad \text{and} \quad -20\log_{10}|H(j\omega_3)| = A_s \quad (17.135)$$

EXAMPLE 17.7

Design a Butterworth bandstop filter with the following specifications:

Maximum loss in the passband = $A_p = 1$ dB
Minimum loss in the stopband = $A_s = 18$ dB
Lower passband cutoff frequency = $f_1 = 1200$ Hz

continued

Example 17.7 continued

Lower stopband cutoff frequency $= f_2 = 2000$ Hz
Upper stopband cutoff frequency $= f_3 = 3000$ Hz
Upper passband cutoff frequency $= f_4 = 4500$ Hz
Excess tolerance = stopband

1. Change the Cutoff Frequencies from Hertz to Radians

$$\omega_1 = 2\pi f_1 = 7539.8224 \text{ rad/s}$$
$$\omega_2 = 2\pi f_2 = 12{,}566.3706 \text{ rad/s}$$
$$\omega_3 = 2\pi f_3 = 18{,}849.5559 \text{ rad/s}$$
$$\omega_4 = 2\pi f_4 = 28{,}274.3339 \text{ rad/s}$$

2. Backward Transformation

$$\Omega_p = 1 \text{ rad/s}$$
$$\Omega_{s_1} = \frac{\omega_2(\omega_4 - \omega_1)}{\omega_1\omega_4 - \omega_2^2} = 4.7143$$
$$\Omega_{s_2} = \frac{\omega_3(\omega_4 - \omega_1)}{\omega_3^2 - \omega_1\omega_4} = 2.75$$
$$\Omega_s = \min(|\Omega_{s_1}|, |\Omega_{s_2}|) = 2.75$$

Since $|\Omega_{s_1}| > |\Omega_{s_2}|$, we calculate ω_{2a} given by

$$\omega_{2a} = \frac{\omega_1\omega_4}{\omega_3} = 11{,}309.7336$$

3. Order

The order of the normalized LPF is

$$n = \left\lceil \frac{\log\left(\frac{10^{0.1A_s} - 1}{10^{0.1A_p} - 1}\right)}{2\log(\Omega_s)} \right\rceil = \left\lceil \frac{\log\left(\frac{10^{0.1\times18} - 1}{10^{0.1\times1} - 1}\right)}{2\log(2.75)} \right\rceil = ceil\,[2.71] = 3$$

4. Recalculation of the Parameters

Since the excess tolerance is assigned to the stopband, we have

$$A_{s_1} = 10\log_{10}\left[\Omega_s^{2n}\left(10^{\frac{A_p}{10}} - 1\right) + 1\right] = 20.5303 \text{ dB}$$

The value of A_p remains the same. The passband ripple parameter is given by

$$\varepsilon_p = \sqrt{10^{\frac{A_p}{10}} - 1} = 0.5088$$

The 3-dB cutoff frequency is given by

$$\Omega_c = \frac{1}{\varepsilon_p^{\frac{1}{n}}} = 1.2526$$

continued

Example 17.7 continued

5. Normalized Lowpass Poles and Zeros

Pole locations

$$S = -1$$
$$S = -0.5 \pm j0.8660$$

There are three zeros at $S = \infty$.

6. Transfer Function of the Normalized LPF

$$H(s) = \frac{1}{S+1} \times \frac{1}{S^2 + S + 1}$$

7. Scaled Poles and Zeros

$$S = \Omega_c S$$
$$-1.2526$$
$$-0.6263 \pm j1.0848$$

8. Frequency Transformation

$$s = -8276.7453 \pm j12{,}028.2560$$
$$s = -6010.9209 \pm j23{,}009.0253$$
$$s = -2265.8244 \pm j8673.2819$$

There are six zeros:

$$sz = \pm j14{,}600.8032$$
$$sz = \pm j14{,}600.8032$$
$$sz = \pm j14{,}600.8032$$

9. Transfer Function

$$H(s) = \frac{s^2 + 2.131835 \times 10^8}{s^2 + 16{,}553.4906s + 2.131835 \times 10^8} \times \frac{s^2 + 2.131835 \times 10^8}{s^2 + 12{,}021.8418s + 5.655464 \times 10^8} \times \frac{s^2 + 2.131835 \times 10^8}{s^2 + 4531.6488s + 8.035978 \times 10^7}$$

Second-order term	ω_0	Q value	B_0	A_1
$s^2 + 16{,}553.4906s + 2.131835 \times 10^8$	14,600.8032	0.882	1	1.1337
$s^2 + 12{,}021.8418s + 5.655464 \times 10^8$	23,781.2198	1.9782	0.377	0.5055
$s^2 + 4531.6488s + 8.035978 \times 10^7$	8964.3616	1.9782	2.6529	0.5055

Second-order term	K	Gain	Case
$s^2 + 16{,}553.4906s + 2.131835 \times 10^8$	1.4331	1.4331	1 or 2
$s^2 + 12{,}021.8418s + 5.655464 \times 10^8$	2.4147	2.4147	1
$s^2 + 4531.6488s + 8.035978 \times 10^7$	2.4147	0.9102	2

continued

Example 17.7 continued

Refer to Chapter 16 for the definitions of B_0, A_1, K, gain, and case. The total gain is $G_0 = 1.4331 \times 2.4147 \times 0.8102 = 3.15$. Thus, $R_D = 1\ k\Omega, R_C = 2.15\ k\Omega$.

The center frequency of the stopband is

$$\omega_0 = \sqrt{\omega_1\omega_4} = 14{,}600.8032 \text{ rad/s}$$

$\omega_z = 14{,}600.8032$ rad/s for all three second-order sections.

10. Verify

$$-20\log_{10}|H(j\omega_1)| = 1\text{ dB}, \quad -20\log_{10}|H(j\omega_4)| = 1\text{ dB}$$

$$-20\log_{10}|H(j\omega_{2a})| = 20.53\text{ dB}, \quad -20\log_{10}|H(j\omega_2)| = 34.54\text{ dB},$$

$$-20\log_{10}|H(j\omega_3)| = 20.53\text{ dB}$$

Exercise 17.7

Design a Butterworth bandstop filter with the following specifications:

Maximum loss in the passband = $A_p = 0.8$ dB
Minimum loss in the stopband = $A_s = 4$ dB
Lower passband cutoff frequency = $f_1 = 1000$ Hz
Lower stopband cutoff frequency = $f_2 = 2000$ Hz
Upper stopband cutoff frequency = $f_3 = 3000$ Hz
Upper passband cutoff frequency = $f_4 = 4000$ Hz
Excess tolerance = passband

Answers:

$\omega_1 = 2\pi f_1 = 6283.1853$ rad/s
$\omega_2 = 2\pi f_2 = 12{,}566.3706$ rad/s
$\omega_3 = 2\pi f_3 = 18{,}849.5559$ rad/s
$\omega_4 = 2\pi f_4 = 25{,}132.7412$ rad/s
$\omega_{2a} = 8377.5804$ rad/s.

$$\Omega_p = 1 \text{ rad/s}, \quad \Omega_s = 1.8 \text{ rad/s}, \quad n = \text{ceil}(1.71) = 2, \quad A_{p_1} = 0.5843 \text{ dB},$$

$$\varepsilon_p = 0.3795, \quad \Omega_c = 1.6233 \text{ rad/s}$$

Normalized ($\Omega_c = 1$) LPF poles and zeros:
Poles: $S = -0.7071 \pm j0.7071$, two zeros at $S = \infty$
Transfer function of normalized LPF

$$H(S) = \frac{1}{S^2 + 1.4142S + 1}$$

Multiplication of poles by Ω_c: $S = \Omega_c S$

$$S = -1.1478 \pm j1.1478, \quad \text{two zeros at } S = \infty$$

Bandstop poles and zeros:

$$s = -5439.2556 \pm j16{,}742.4304, \quad s = -2771.6945 \pm j8531.4803$$

$$sz = \pm j12{,}566.3706, \quad sz = \pm j12{,}566.3706$$

continued

Exercise 17.7 continued

Transfer function of bandpass filter:

$$H(s) = \frac{s^2 + 1.5791367 \times 10^8}{s^2 + 10{,}878.5112s + 3.098945 \times 10^8} \times \frac{s^2 + 1.5791367 \times 10^8}{s^2 + 5543.389s + 8.04684468 \times 10^7}$$

Second-order term	Corner frequency	Q value	K	b_0
$s^2 + 10{,}878.5112s + 3.098945 \times 10^8$	17,603.82	1.618	2.0484	1.96
$s^2 + 5543.389s + 8.04684468 \times 10^7$	8970.4207	1.618	2.0484	0.51

Single $b_0 = 1$
The center frequency of the passband is

$$\omega_0 = \sqrt{\omega_1 \omega_4} = 12{,}566.3706 \text{ rad/s}$$

$$\omega_z = 12{,}566.3705 \text{ rad/s}$$

$$-20\log_{10}|H(j\omega_1)| = 0.5843\,\text{dB}, \quad -20\log_{10}|H(j\omega_4)| = 0.5843\,\text{dB}$$

$$-20\log_{10}|H(j\omega_{2a})| = 4\,\text{dB}, \quad -20\log_{10}|H(j\omega_3)| = 4\,\text{dB}$$

EXAMPLE 17.8

Implement the filter designed in EXAMPLE 17.7 using the Sallen-Key circuit for the bandstop filter. Assume $k_m = 1000$.

Transfer function:

$$\frac{s^2 + 2.131835 \times 10^8}{s^2 + 16{,}553.4906s + 2.131835 \times 10^8}$$

$\omega_0 = k_f = 14{,}600.8032$ rad/s, $\omega_z = 14{,}600.8032$ rad/s, $Q = 0.882$, $A_0 = 1$, $B_0 = 1$, $A_1 = 1.1337$. Since $\omega_z = \omega_0$, this is case 1 or case 2.

Normalized component values:

$$m = 1, \quad a = 1, \quad K = 1.4331, \quad R_1 = 1\ \Omega, \quad R_2 = 1\ \Omega, \quad R_3 = 0.5\ \Omega, \quad C_1 = 1\text{ F},$$
$$C_2 = 1\text{ F}, \quad C_3 = 2\text{ F}, \quad R_B = 1\ \Omega, \quad R_A = 0.4331\ \Omega.$$

Scaled component values:

$$R_1 = 1\ k\Omega, \quad R_2 = 1\ k\Omega, \quad R_3 = 500\ \Omega, \quad C_1 = 68.4894\ n\text{F}, \quad C_2 = 68.4894\ n\text{F},$$
$$C_3 = 0.137\ \mu\text{F}, \quad R_B = 1\ k\Omega, \quad R_A = 433.1\ \Omega$$

Transfer function:

$$\frac{s^2 + 2.131835 \times 10^8}{s^2 + 12{,}021.8418s + 5.655464 \times 10^8}$$

$\omega_0 = k_f = 23{,}781.2198$ rad/s, $\omega_z = 14{,}600.8032$ rad/s, $Q = 1.9782$, $A_0 = 1$, $B_0 = 0.377$, $A_1 = 0.5055$. Since $\omega_z < \omega_0$, this is case 1.

continued

Example 17.8 continued

Normalized component values:

$$m = 1, \quad a = 1.6288, \quad K = 2.4147, \quad R_1 = 1\ \Omega, \quad R_2 = 1\ \Omega, \quad R_3 = 0.5\ \Omega,$$
$$R_4 = 1.21\ \Omega, \quad C_1 = 1.6288\ \text{F}, \quad C_2 = 1.6288\ \text{F}, \quad C_3 = 3.2575\ \text{F}, \quad R_B = 1\ \Omega,$$
$$R_A = 1.4147\ \Omega$$

Scaled component values:

$$R_1 = 1\ k\Omega, \quad R_2 = 1\ k\Omega, \quad R_3 = 500\ \Omega, \quad R_4 = 1.21\ k\Omega, \quad C_1 = 68.4894\ n\text{F},$$
$$C_2 = 68.4894\ n\text{F}, \quad C_3 = 0.137\ \mu\text{F}, \quad R_B = 1\ k\Omega, \quad R_A = 1.4147\ k\Omega$$

Transfer function:

$$\frac{s^2 + 2.131835 \times 10^8}{s^2 + 4531.6488s + 8.035978 \times 10^7}$$

$$\omega_0 = k_f = 8964.3616\ \text{rad/s}, \quad \omega_z = 14{,}600.8032\ \text{rad/s}, \quad Q = 1.9782, \quad A_0 = 1,$$
$B_0 = 2.6529, \quad A_1 = 0.5055$. Since $\omega_z > \omega_0$, this is case 2.

Normalized component values:

$$m = 1, \quad a = 0.614, \quad K = 2.4147, \quad R_1 = 1\ \Omega, \quad R_2 = 1\ \Omega, \quad R_3 = 0.5\ \Omega,$$
$$R_4 = 2.0781\ \Omega, \quad C_1 = 0.614\ \text{F}, \quad C_2 = 0.614\ \text{F}, \quad C_3 = 1.2279\ \text{F}, \quad C_4 = 0.5074\ \text{F},$$
$$R_B = 1\ \Omega, \quad R_A = 1.4147\ \Omega$$

Scaled component values:

$$R_1 = 1\ k\Omega, \quad R_2 = 1\ k\Omega, \quad R_3 = 500\ \Omega, \quad C_1 = 68.4894\ n\text{F}, \quad C_2 = 68.4894\ n\text{F},$$
$$C_3 = 0.137\ \mu\text{F}, \quad C_4 = 56.6018\ n\text{F}, \quad R_B = 1\ k\Omega, \quad R_A = 1.4147\ k\Omega$$

The total gain is 3.15. $R_D = 1\ k\Omega$, $R_C = 2.15\ k\Omega$. The implementation of the design using the Sallen-Key circuit is shown in Figure 17.21, and the magnitude response of the circuit is shown in Figure 17.22.

FIGURE 17.21

PSpice simulation of the circuit for EXAMPLE 17.8.

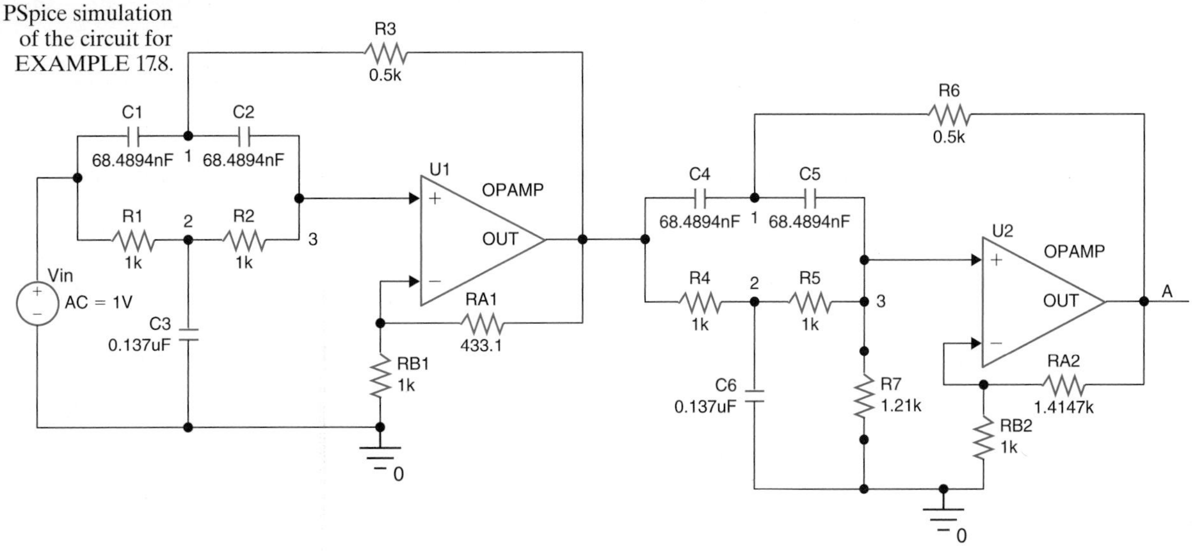

continued

Example 17.8 continued

FIGURE 17.21

continued

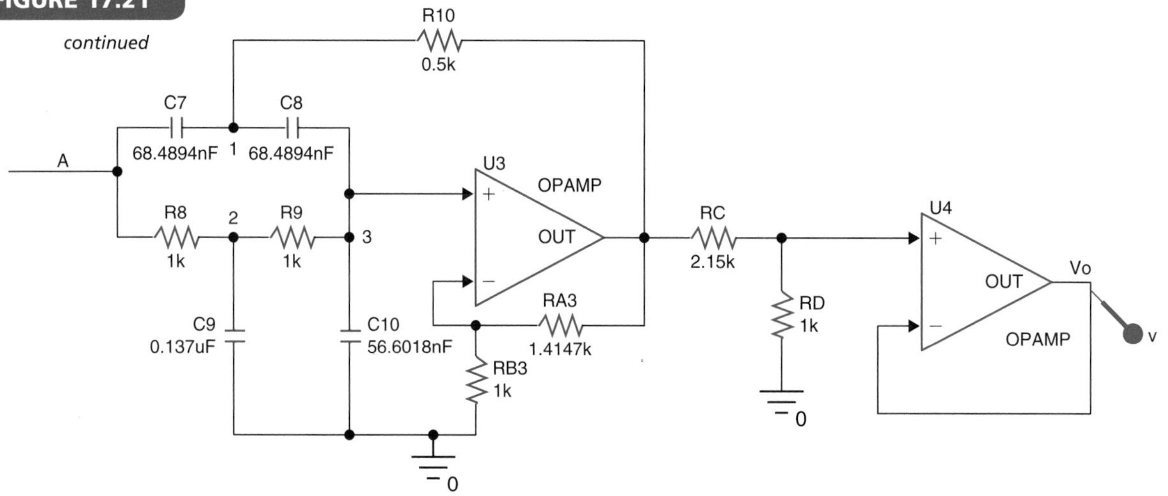

FIGURE 17.22

Magnitude response of the circuit shown in Figure 17.21.

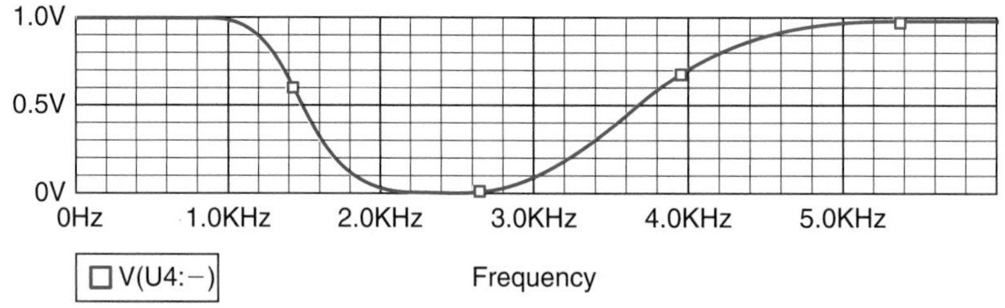

Exercise 17.8

Implement the filter designed in EXERCISE 17.7 using the Sallen-Key circuit. Assume $k_m = 1000$.

Transfer function:

$$\frac{s^2 + 1.5791367 \times 10^8}{s^2 + 10{,}878.5112s + 3.098945 \times 10^8}$$

$\omega_0 = k_f = 17{,}603.82$ rad/s, $\omega_z = 12{,}566.3706$ rad/s, $Q = 1.6182$, $A_0 = 1$, $B_0 = 0.5096$, $A_1 = 0.618$. Since $\omega_z < \omega_0$, this is case 1.

Normalized component values:

$m = 1$, $a = 1.4009$, $K = 2.0484$, $R_1 = 1\ \Omega$, $R_2 = 1\ \Omega$, $R_3 = 0.5\ \Omega$, $R_4 = 2.0781\ \Omega$, $C_1 = 1.4009$ F, $C_2 = 1.4009$ F, $C_3 = 2.8017$ F, $R_B = 1\ \Omega$, $R_A = 1.0484\ \Omega$

Scaled component values:

$R_1 = 1\ k\Omega$, $R_2 = 1\ k\Omega$, $R_3 = 500\ \Omega$, $R_4 = 2.0781\ k\Omega$, $C_1 = 79.5775\ n$F, $C_2 = 79.5775\ n$F, $C_3 = 0.1592\ \mu$F, $R_B = 1\ k\Omega$, $R_A = 1.0484\ k\Omega$

continued

Exercise 17.8 continued

Transfer function:

$$\frac{s^2 + 1.5791367 \times 10^8}{s^2 + 5543.389s + 8.04684468 \times 10^7}$$

$\omega_0 = k_f = 8970.4207$ rad/s, $\omega_z = 12{,}566.3706$ rad/s, $Q = 1.6182$, $A_0 = 1$, $B_0 = 1.9624$, $A_1 = 0.618$. Since $\omega_z > \omega_0$, this is case 2.

Normalized component values:

$m = 1$, $a = 0.7138$, $K = 2.0484$, $R_1 = 1\ \Omega$, $R_2 = 1\ \Omega$, $R_3 = 0.5\ \Omega$, $C_1 = 0.7138$ F, $C_2 = 0.7138$ F, $C_3 = 1.4277$ F, $C_4 = 0.3435$ F, $R_B = 1\ \Omega$, $R_A = 1.0484\ \Omega$

Scaled component values:

$R_1 = 1\ k\Omega$, $R_2 = 1\ k\Omega$, $R_3 = 500\ \Omega$, $C_1 = 79.5775\ n$F, $C_2 = 79.5775\ n$F, $C_3 = 0.1592\ \mu$F, $C_4 = 38.2939\ n$F, $R_B = 1\ k\Omega$, $R_A = 1.0484\ k\Omega$

Total gain = 2.138. $R_D = 1\ k\Omega$, $R_C = 1.138\ k\Omega$. The implementation of the design using the Sallen-Key circuit is shown in Figure 17.23, and the magnitude response of the circuit is shown in Figure 17.24.

FIGURE 17.23

PSpice simulation of the circuit for EXERCISE 17.8.

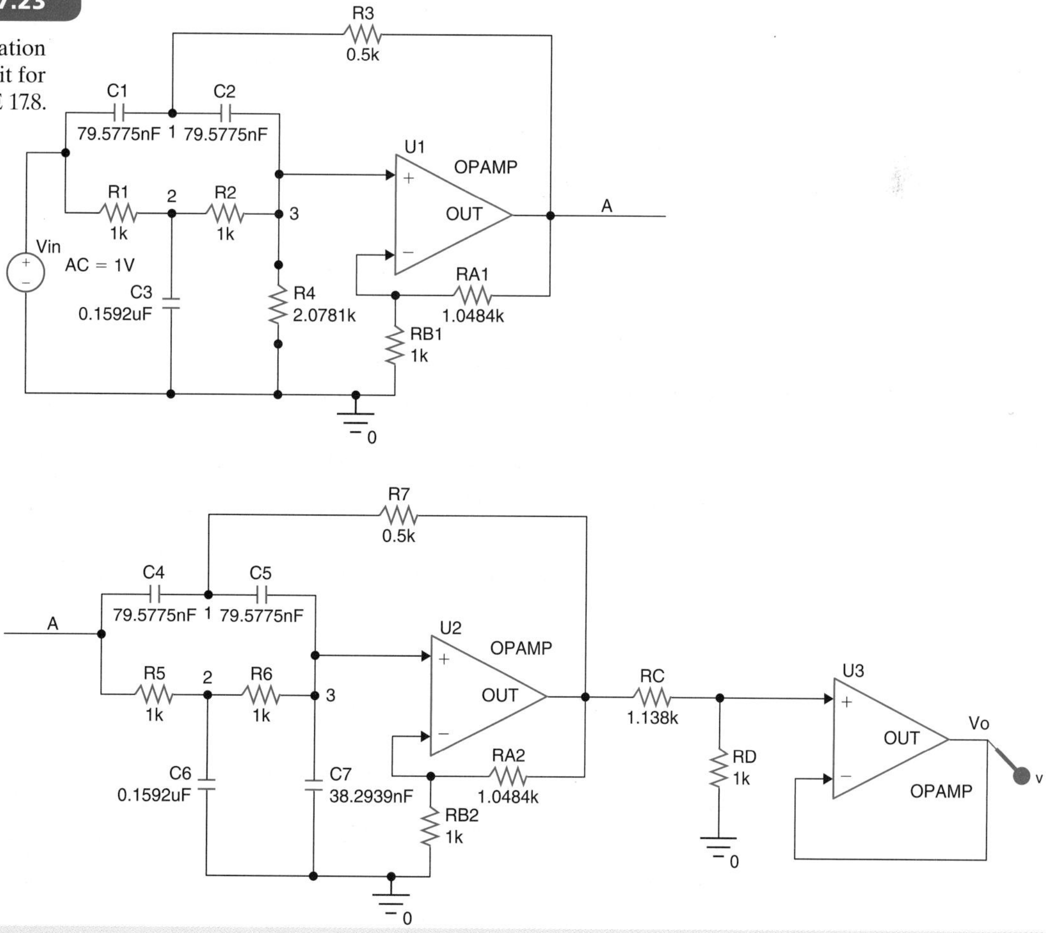

continued

Exercise 17.8 continued

FIGURE 17.24

Magnitude response of the circuit shown in Figure 17.23.

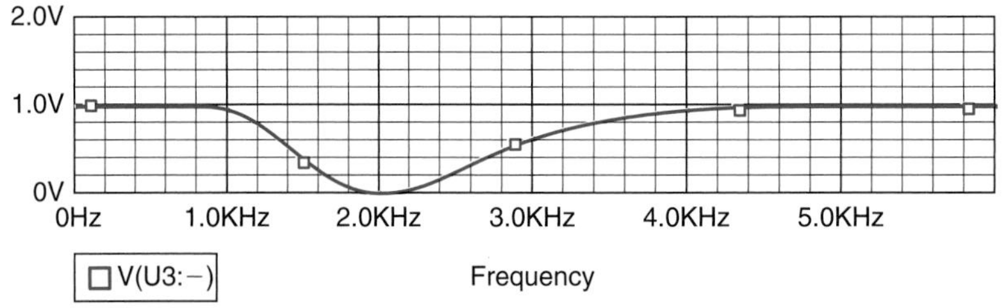

17.6 Analog Chebyshev Type 1 LPF Design

The Chebyshev type 1 approximation introduces ripples in the passband and does not introduce ripples in the stopband. The magnitude-squared response of the normalized ($\Omega_p = 1$ rad/s) Chebyshev type 1 LPF is given by

$$|H_n(j\Omega)|^2 = \frac{1}{1 + \varepsilon_p^2 C_n^2(\Omega)} \tag{17.136}$$

where $C_n(\Omega)$ is an nth-order Chebyshev polynomial, and ε_p is the passband ripple parameter. The magnitude response of the normalized ($\Omega_p = 1$ rad/s) Chebyshev type 1 LPF is given by

$$|H_n(j\Omega)| = \frac{1}{\sqrt{1 + \varepsilon_p^2 C_n^2(\Omega)}} \tag{17.137}$$

The nth-order Chebyshev polynomial can be written as

$$C_n(\Omega) = \begin{cases} \cos(n\cos^{-1}(\Omega)), & |\Omega| \le 1 \\ \cosh(n\cosh^{-1}(\Omega)), & |\Omega| \ge 1 \end{cases} \tag{17.138}$$

Notice that arc cosine is defined as $\cos^{-1}(x) = \frac{\pi}{2} + j\ln(jx + \sqrt{1 - x^2})$. Also, notice that

$$C_0(\Omega) = \cos(0\cos^{-1}(\Omega)) = \cos(0) = 1$$
$$C_1(\Omega) = \cos(1\cos^{-1}(\Omega)) = \cos(\cos^{-1}(\Omega)) = \Omega$$

Let $\Omega = \cos(\phi)$. Then, $\phi = \cos^{-1}(\Omega)$ and $C_n(\Omega) = \cos(n\phi)$. For order $n - 1$ and $n + 1$, we have

$$C_{n-1}(\Omega) = \cos[(n-1)\phi] = \cos(n\phi - \phi) = \cos(n\phi)\cos(\phi) + \sin(n\phi)\sin(\phi)$$
$$C_{n+1}(\Omega) = \cos[(n+1)\phi] = \cos(n\phi + \phi) = \cos(n\phi)\cos(\phi) - \sin(n\phi)\sin(\phi),$$

respectively. Adding these two equations, we get

$$C_{n+1}(\Omega) = 2\Omega C_n(\Omega) - C_{n-1}(\Omega) \text{ for } n \ge 1 \tag{17.139}$$

Equation (17.139) can be used to generate the Chebyshev polynomials recursively. The Chebyshev polynomial $C_n(\Omega)$ for $n = 1, 2, 3, 4, 5$ is shown in Figure 17.25.

FIGURE 17.25 Chebyshev polynomials for $1 \leq n \leq 5$.

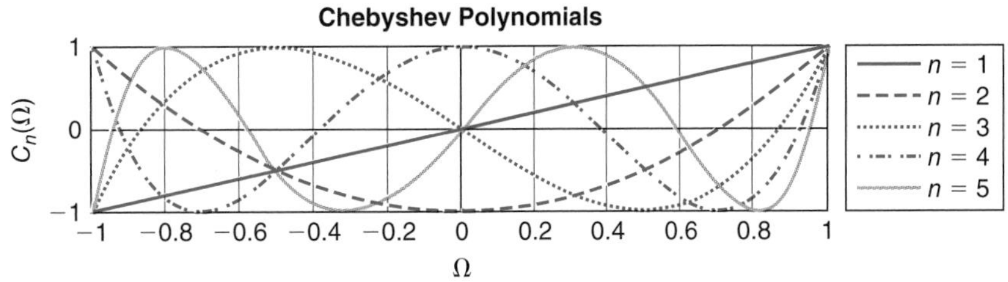

When n is odd, $C_n(0) = 0$. On the other hand, when n is even, $C_n(0) = 1$ or -1. When $n = 2, 6, 10, 14, \ldots$ $C_n(0) = -1$ and when $n = 4, 8, 12, 16, \ldots, C_n(0) = 1$. Thus, when n is odd, we have

$$|H_n(j0)|^2 = \frac{1}{1 + \varepsilon_p^2 C_n^2(0)} = 1$$

and

$$|H_n(j0)| = 1$$

When n is even, we have

$$|H_n(j0)|^2 = \frac{1}{1 + \varepsilon_p^2 C_n^2(0)} = \frac{1}{1 + \varepsilon_p^2}$$

and

$$|H_n(j0)| = \frac{1}{\sqrt{1 + \varepsilon_p^2 C_n^2(0)}} = \frac{1}{\sqrt{1 + \varepsilon_p^2}} \tag{17.140}$$

The attenuation (or loss) in decibels for the Chebyshev type 1 response is defined as

$$\begin{aligned} A &= -20 \log_{10}(|H_n(j\Omega)|) = -20 \log_{10} \frac{1}{\sqrt{1 + \varepsilon_p^2 C_n^2(\Omega)}} \\ &= 10 \log_{10}[1 + \varepsilon_p^2 C_n^2(\Omega)] \text{ dB} \end{aligned} \tag{17.141}$$

Notice that $C_n(1) = \cos(n \cos^{-1}(1)) = \cos(0) = 1$. Thus, the magnitude at the passband cutoff frequency $\Omega = \Omega_p = 1$ rad/s is given by

$$|H_n(j1)| = \frac{1}{\sqrt{1 + \varepsilon_p^2 C_n^2(1)}} = \frac{1}{\sqrt{1 + \varepsilon_p^2}} \tag{17.142}$$

and the attenuation in decibels is given by

$$A_p = 10 \log_{10}[1 + \varepsilon_p^2 C_n^2(1)] = 10 \log_{10}[1 + \varepsilon_p^2] \text{ dB} \tag{17.143}$$

Solving for the passband ripple parameter ε_p, we obtain

$$\varepsilon_p = \sqrt{10^{\frac{A_p}{10}} - 1} = \sqrt{\frac{1}{G_p^2} - 1} = \sqrt{\frac{1}{(1 - \delta_p)^2} - 1} \tag{17.144}$$

At the stopband cutoff frequency $\Omega = \Omega_s$ rad/s, from Equation (17.137), the magnitude is given by

$$|H_n(j\Omega_s)| = \frac{1}{\sqrt{1+\varepsilon_p^2 C_n^2(\Omega_s)}} = \frac{1}{\sqrt{1+\varepsilon_p^2 \cosh^2(n\cosh^{-1}(\Omega_s))}} = G_s = \delta_s = \frac{1}{\sqrt{1+\varepsilon_s^2}} \quad (17.145)$$

where

$$\varepsilon_p^2 \cosh^2(n\cosh^{-1}(\Omega_s)) = \varepsilon_s^2 \quad \text{or} \quad \varepsilon_s = \varepsilon_p \cosh(n\cosh^{-1}(\Omega_s)) \quad (17.146)$$

ε_s is called the *stopband ripple parameter.* Notice that

$$\cosh(n\cosh^{-1}(\Omega_s)) = \frac{\varepsilon_s}{\varepsilon_p} \quad (17.147)$$

The attenuation (in decibels) at the stopband cutoff frequency $\Omega = \Omega_s$ rad/s is given by

$$A_s = -20\log_{10}|H_n(j\Omega_s)| = -20\log_{10}\frac{1}{\sqrt{1+\varepsilon_p^2 \cosh^2(n\cosh^{-1}(\Omega_s))}} = 10\log_{10}(1+\varepsilon_p^2\cosh^2(n\cosh^{-1}(\Omega_s))) \quad (17.148)$$

Since $\varepsilon_p^2 \cosh^2(n\cosh^{-1}(\Omega_s)) = \varepsilon_s^2$ from Equation (17.146), Equation (17.148) becomes

$$A_s = -20\log_{10}|H_n(j\Omega_s)| = -20\log_{10}\frac{1}{\sqrt{1+\varepsilon_s^2}} = 10\log_{10}(1+\varepsilon_s^2) \quad (17.149)$$

Solving Equation (17.149) for ε_s, we have

$$\varepsilon_s = \sqrt{10^{\frac{A_s}{10}} - 1} \quad (17.150)$$

Substituting Equations (17.144) and (17.150) into Equation (17.147), we obtain

$$\cosh(n\cosh^{-1}(\Omega_s)) = \frac{\sqrt{10^{\frac{A_s}{10}} - 1}}{\sqrt{10^{\frac{A_p}{10}} - 1}} \quad (17.151)$$

Taking the arc cosh (acosh) on both sides of Equation (17.151), we obtain

$$n = \frac{\text{acosh}\left(\sqrt{\frac{10^{0.1A_s}-1}{10^{0.1A_p}-1}}\right)}{\text{acosh}(\Omega_s)} \quad (17.152)$$

This equation can be used to find the order n of a filter given $A_p, A_s, \omega_p, \omega_s$ (notice that $\Omega_p = 1, \Omega_s = \omega_s/\omega_p$). Since the order n of a filter must be an integer, the n value must be rounded up. Thus, we have

$$n = \left\lceil \frac{\text{acosh}\left(\sqrt{\frac{10^{0.1A_s}-1}{10^{0.1A_p}-1}}\right)}{\text{acosh}(\Omega_s)} \right\rceil = ceil\left(\frac{\text{acosh}\left(\sqrt{\frac{10^{0.1A_s}-1}{10^{0.1A_p}-1}}\right)}{\text{acosh}(\Omega_s)}\right) \quad (17.153)$$

Notice that

$$\cosh^{-1}(x) = \ln(x + \sqrt{x^2 - 1})$$
$$\sinh^{-1}(x) = \ln(x + \sqrt{x^2 + 1})$$

Since the order is rounded up, the actual filter exceeds the specifications unless n is an integer before rounding up. The excess tolerance can be assigned to the stopband or the passband. If the excess tolerance is assigned to the stopband, the resulting filter will exceed the specifications in the stopband (attenuation is greater than A_s dB at ω_s) and will meet the specifications exactly in the passband (attenuation is A_p dB at ω_p). On the other hand, if the excess tolerance is assigned to the passband, the resulting filter will exceed the specifications in the passband (attenuation is smaller than A_p dB at ω_p) and will meet the specifications exactly in the stopband (attenuation is A_s dB at ω_s). If the excess tolerance is assigned to the stopband, the passband attenuation A_p is fixed at the value given in the specifications. Substituting the rounded-up n value from Equation (17.153) into Equation (17.151) and solving for A_s, we obtain new A_s, denoted by A_{s_1}, given by

$$A_{s_1} = 10\log_{10}[1 + (10^{0.1A_p} - 1)\cosh^2(n\,\text{acosh}(\Omega_s))] \qquad \textbf{(17.154)}$$

Since A_p is fixed, there is no change to the passband ripple parameter ε_p given by Equation (17.144).

If the excess tolerance is assigned to the passband, the stopband attenuation A_s is fixed at the value given in the specifications. Substituting the rounded-up n value from Equation (17.153) into Equation (17.151) and solving for A_p, we obtain new A_p, denoted by A_{p_1}, given by

$$A_{p_1} = 10\log_{10}\left[1 + \frac{(10^{0.1A_s} - 1)}{\cosh^2(n\,\text{acosh}(\Omega_s))}\right] \qquad \textbf{(17.155)}$$

The passband ripple parameter ε_p can be recalculated using Equation (17.144)

$$\varepsilon_p = \sqrt{10^{\frac{A_p}{10}} - 1} \qquad \textbf{(17.156)}$$

Chebyshev Type 1 Pole Locations

Let a parameter a be

$$a = \frac{1}{n}\sinh^{-1}\left(\frac{1}{\varepsilon_p}\right) = \frac{1}{n}\sinh^{-1}\left(\frac{1}{\sqrt{10^{0.1A_p} - 1}}\right) \qquad \textbf{(17.157)}$$

It can be shown that[1,2,3] for n odd, there is one pole at $S = -\sinh(a)$ and $(n - 1)/2$ complex conjugate poles at

$$S_k = -\sinh(a)\sin\left(\frac{\pi(2k+1)}{2n}\right) \pm j\cosh(a)\cos\left(\frac{\pi(2k+1)}{2n}\right),$$
$$k = 0, 1, 2, \ldots, \frac{n-3}{2} \qquad \textbf{(17.158)}$$

[1] M. E. Van Valkenburg, *Analog Filter Design*, Holt, Rinehart and Winston, 1982.
[2] T. W. Parks and C. S. Burrus, *Digital Filter Design*, Wiley, 1987.
[3] Sophocles J. Orfanidis, *Introduction to Signal Processing*, Prentice Hall, 1996.

For n even, there are $n/2$ complex conjugate poles at

$$S_k = -\sinh(a)\sin\left(\frac{\pi(2k+1)}{2n}\right) \pm j\cosh(a)\cos\left(\frac{\pi(2k+1)}{2n}\right),$$
$$k = 0,1,2,\ldots,\frac{n-2}{2} \quad (17.159)$$

In summary, the design procedure for the Chebyshev type 1 LPF is as follows.

1. Change the Cutoff Frequencies from Hertz to Radians

If the cutoff frequencies are given in hertz, change the cutoff frequencies to radians.

$$\omega_p = 2\pi f_p, \quad \omega_s = 2\pi f_s \quad (17.160)$$

2. Backward Transformation

Given the passband cutoff frequency ω_p, stopband cutoff frequency ω_s, maximum attenuation (in decibels) in the passband A_p, and minimum attenuation (in decibels) in the stopband A_s, find the cutoff frequencies for the normalized LPF by dividing by ω_p.

$$\text{Passband cutoff frequency for the normalized LPF} = \Omega_p = 1 \text{ rad/s} \quad (17.161)$$

$$\text{Stopband cutoff frequency for the normalized LPF} = \Omega_s = \omega_s/\omega_p \text{ rad/s} \quad (17.162)$$

Notice that $\omega_s/\omega_p = f_s/f_p > 1$.

3. Order

Find the order n of the normalized LPF using

$$n = ceil\left(\frac{\text{acosh}\left(\sqrt{\frac{10^{0.1A_s}-1}{10^{0.1A_p}-1}}\right)}{\text{acosh}(\Omega_s)}\right) \quad (17.163)$$

4. Recalculation of the Parameters

If excess tolerance is assigned to the stopband, we have new A_s given by

$$A_{s_1} = 10\log_{10}[1 + (10^{0.1A_p} - 1)\cosh^2(n\,\text{acosh}(\Omega_s))] \quad (17.164)$$

The value of A_p remains the same. The passband ripple parameter is given by

$$\varepsilon_p = \sqrt{10^{\frac{A_p}{10}} - 1} \quad (17.165)$$

If excess tolerance is assigned to the passband, we have new A_p given by

$$A_{p_1} = 10\log_{10}\left[1 + \frac{(10^{0.1A_s} - 1)}{\cosh^2(n\,\text{acosh}(\Omega_s))}\right] \quad (17.166)$$

The passband ripple parameter ε_p can be recalculated using

$$\varepsilon_{p_1} = \sqrt{10^{\frac{A_{p_1}}{10}} - 1} \quad (17.167)$$

5. Normalized Lowpass Poles and Zeros

$$a = \frac{1}{n}\sinh^{-1}\left(\frac{1}{\sqrt{10^{0.1A_p} - 1}}\right)$$

n odd:

There is one pole at $S = -\sinh(a)$ and $(n - 1)/2$ complex conjugate poles at

$$S_k = -\sinh(a)\sin\left(\frac{\pi(2k+1)}{2n}\right) \pm j\cosh(a)\cos\left(\frac{\pi(2k+1)}{2n}\right),$$
$$k = 0, 1, 2, \ldots, \frac{n-3}{2} \quad \textbf{(17.168)}$$

n even:

There are $n/2$ complex conjugate poles at

$$S_k = -\sinh(a)\sin\left(\frac{\pi(2k+1)}{2n}\right) \pm j\cosh(a)\cos\left(\frac{\pi(2k+1)}{2n}\right),$$
$$k = 0, 1, 2, \ldots, \frac{n-2}{2} \quad \textbf{(17.169)}$$

There are n zeros at $S = \infty$.

6. Transfer Function of the Normalized LPF

Convert the poles and the zeros of the normalized LPF into polynomials, and write the transfer function as a product of the first-order and the second-order sections. If the order of the filter is odd, there is one first-order section and $(n - 1)/2$ second-order sections. If the order of the filter is even, there are $n/2$ second-order sections. The numerator is a constant equal to the constant term of the denominator for each first- or second-order term. Also, find the Q value of each second-order section.

7. Frequency Transformation

Transform the LPF with the passband cutoff frequency $\Omega_p = 1$ rad/s to the LPF with the passband cutoff frequency ω_p by multiplying the poles and the zeros from step 5 above by ω_p. Let s be the location of poles and zeros after the transformation, and let S be the location of poles and zeros before the transformation. Then, we have

$$s = \omega_p S \quad \textbf{(17.170)}$$

There are n zeros at $s = \infty$.

8. Transfer Function

Convert the poles and the zeros of the frequency-transformed LPF into polynomials, and write the transfer function as a product of the first-order and the second-order sections. If the order of the filter is odd, there is one first-order section and $(n - 1)/2$ second-order sections. If the order of the filter is even, there are $n/2$ second-order sections. The numerator is a constant equal to the constant term of the denominator for each first- or second-order term.

9. Constant Term b_0

If n is odd, $C_n(0) = 0$. On the other hand, if n is even, $C_n(0) = 1$ or -1. When $n = 2, 6, 10, 14, \ldots, C_n(0) = -1$ and when $n = 4, 8, 12, 16, \ldots, C_n(0) = 1$. Thus, if n is odd, we have

$$|H_n(j0)|^2 = \frac{1}{1 + \varepsilon_p^2 C_n^2(0)} = 1$$

and

$$|H_n(j0)| = 1$$

Since the gain is 1 at $\Omega = 0$, there is no need to adjust the gain. Thus, the constant term is given by $b_0 = 1$ when n is odd.

If n is even, we have

$$|H_n(j0)|^2 = \frac{1}{1 + \varepsilon_p^2 C_n^2(0)} = \frac{1}{1 + \varepsilon_p^2}$$

and

$$|H_n(j0)| = \frac{1}{\sqrt{1 + \varepsilon_p^2 C_n^2(0)}} = \frac{1}{\sqrt{1 + \varepsilon_p^2}}$$

Notice that $C_n(1) = \cos(n \cos^{-1}(1)) = \cos(0) = 1$. Thus, at the normalized cutoff frequency ($\Omega = \Omega_p = 1$), the magnitude response of the Chebyshev filter is

$$|H_n(j1)| = \frac{1}{\sqrt{1 + \varepsilon^2 C_n^2(1)}} = \frac{1}{\sqrt{1 + \varepsilon_p^2}}$$

Since $A_p = 10 \log_{10}(1 + \varepsilon_p^2)$,

$$|H_n(j0)| = \frac{1}{\sqrt{1 + \varepsilon_p^2 C_n^2(0)}} = \frac{1}{\sqrt{1 + \varepsilon_p^2}} = \frac{1}{10^{0.05A_p}} = 10^{-0.05A_p}$$

To make the gain to be $10^{-0.05A_p}$ at $\Omega = 0$, we multiply by

$$b_0 = 10^{-0.05A_p} \tag{17.171}$$

when n is even.

10. Verify

Calculate the attenuation at ω_p and ω_s, and compare with the specifications. If the excess tolerance is assigned to the stopband, we have

$$-20 \log_{10}|H(j\omega_p)| = A_p, \quad -20 \log_{10}|H(j\omega_s)| \geq A_s \tag{17.172}$$

If the excess tolerance is assigned to the passband, we have

$$-20 \log_{10}|H(j\omega_p)| \leq A_p, \quad -20 \log_{10}|H(j\omega_s)| = A_s \tag{17.173}$$

EXAMPLE 17.9

Design a Chebyshev type 1 LPF with the following specifications:

Passband cutoff frequency = f_p = 1 kHz
Stopband cutoff frequency = f_s = 1.5 kHz
Maximum attenuation in the passband = A_p = 0.5 dB
Minimum attenuation in the stopband = A_s = 25 dB
Excess tolerance = stopband

continued

Example 17.9 continued

1. Change the Cutoff Frequencies from Hertz to Radians

$$\omega_p = 2\pi f_p = 2\pi 1000 = 6283.1853$$
$$\omega_s = 2\pi f_s = 2\pi 1500 = 9424.7780$$

2. Backward Transformation

The cutoff frequencies of the normalized LPF are given by

$$\Omega_p = 1 \text{ rad/s}, \quad \Omega_s = \omega_s/\omega_p = 1.5 \text{ rad/s}$$

3. Order

The order of the filter is

$$n = \left\lceil \frac{\operatorname{acosh}\left(\sqrt{\dfrac{10^{0.1A_s} - 1}{10^{0.1A_p} - 1}}\right)}{\operatorname{acosh}(\Omega_s)} \right\rceil = ceil\left(\frac{\operatorname{acosh}\left(\sqrt{\dfrac{10^{0.1A_s} - 1}{10^{0.1A_p} - 1}}\right)}{\operatorname{acosh}(\Omega_s)} \right) = ceil(4.802) = 5$$

4. Recalculation of the Parameters

Since the excess tolerance is assigned in the stopband, the new attenuation in the stopband is given by

$$A_{s_1} = 10\log_{10}[1 + (10^{0.1A_p} - 1)\cosh^2(n \operatorname{acosh}(\Omega_s))] = 26.6512 \text{ dB}$$

The attenuation in the stopband is 1.6511 dB larger than the minimum required attenuation of 25 dB. The value of ε_p is given by

$$\varepsilon_p = \sqrt{10^{\frac{A_p}{10}} - 1} = \sqrt{10^{0.05} - 1} = 0.3493$$

5. Normalized Lowpass Poles and Zeros

Constant a

$$a = \frac{1}{n}\sinh^{-1}\left(\frac{1}{\varepsilon}\right) = \frac{1}{n}\sinh^{-1}\left(\frac{1}{\sqrt{10^{0.1A_{p_2}} - 1}}\right) = 0.3548$$

Since n is odd, there is one pole at $S = -\sinh(a)$ and two pairs of normalized complex conjugate poles at

$$S_k = -\sinh(a)\sin\left(\frac{\pi(2k+1)}{2n}\right) \pm j\cosh(a)\cos\left(\frac{\pi(2k+1)}{2n}\right), \quad k = 0,1$$

Thus, the pole locations are:

-0.3623
$-0.2931 \pm j0.6252$
$-0.1120 \pm j1.0116$

There are five zeros at $S = \infty$.

continued

Example 17.9 continued

6. Transfer Function of the Normalized LPF

The transfer function of the normalized LPF is given by

$$H(S) = \frac{0.3623}{S + 0.3623} \times \frac{0.4768}{S^2 + 0.5862S + 0.4768} \times \frac{1.0358}{S^2 + 0.2239S + 1.0358}$$

7. Frequency Transformation

The frequency-transformed poles are obtained by multiplying the poles given in step 5 by $\omega_p = 2\pi 1000$. The frequency-transformed poles are given by

-2276.5213

$-1841.7445 \pm j3928.1019$

$-703.4838 \pm j6355.8024$

There are five zeros at $s = \infty$.

8. Transfer Function

The transfer function of the LPF is

$$H(s) = \frac{2276.52}{s + 2276.52} \times \frac{1.8822 \times 10^7}{s^2 + 3683.49s + 1.8822 \times 10^7}$$
$$\times \frac{4.0891 \times 10^7}{s^2 + 1406.97s + 4.0891 \times 10^7}$$

Second-order term	Corner frequency (ω_0)	Q value
$s^2 + 3683.48s + 1.8822 \times 10^7$	4338.4337 rad/s	1.1778
$s^2 + 1406.97s + 4.0891 \times 10^7$	6394.6160 rad/s	4.5450

9. Constant Term b_0

Since n is odd in this example, $b_0 = 1$.

10. Verify

$$-20 \log_{10}|H(j\omega_p)| = 0.5\,\text{dB}, \quad -20 \log_{10}|H(j\omega_s)| = 26.6512\,\text{dB}$$

Exercise 17.9

Design a Chebyshev type 1 LPF with the following specifications:

Passband cutoff frequency = f_p = 1.2 kHz
Stopband cutoff frequency = f_s = 1.9 kHz
Maximum attenuation in the passband = A_p = 0.8 dB
Minimum attenuation in the stopband = A_s = 20 dB
Excess tolerance = passband

Answer:

$\omega_p = 7539.8224$ rad/s, $\omega_s = 11{,}938.0521$, $\Omega_p = 1$ rad/s, $\Omega_s = 1.5833$ rad/s, $n = 4$, $A_{p_1} = 0.42$, $\varepsilon_p = 0.3187$.

Normalized poles: $-0.4456 \pm j0.4249$, $-0.1846 \pm j1.0257$. Four zeros at ∞.
Normalized transfer function:

$$H_n(S) = \frac{0.3790}{S^2 + 0.8911S + 0.3790} \times \frac{1.0861}{S^2 + 0.3691S + 1.0861}$$

continued

Exercise 17.9 continued

Frequency-transformed LPF transfer function:

$$H(s) = \frac{2.1548 \times 10^7}{s^2 + 6718.9228s + 2.1548 \times 10^7} \times \frac{6.1746 \times 10^7}{s^2 + 2783.069s + 6.1746 \times 10^7}$$

$$-20\log_{10}|H(j\omega_p)| = 0.42\,\text{dB}, \quad -20\log_{10}|H(j\omega_s)| = 20\,\text{dB}$$

EXAMPLE 17.10

Implement the filter designed in EXAMPLE 17.9 using the Sallen-Key unity gain method. Assume $k_m = 1000$.

For transfer function: $H(s) = \dfrac{2276.52}{s + 2276.52}$

$k_f = 2276.52$ rad/s

Normalized component values: $R = 1\ \Omega, \quad C = 1\ \text{F}$
Scaled component values: $R = 1\ k\Omega, \quad C = 1/(k_m k_f)\ \text{F} = 0.4392667\ \mu\text{F}$
For the transfer function:

$$H(s) = \frac{1.8822 \times 10^7}{s^2 + 3683.49s + 1.8822 \times 10^7} = \frac{\omega_0^2}{s^2 + \dfrac{\omega_0}{Q}s + \omega_0^2}$$

Corner frequency $= \omega_0 = \sqrt{1.8822 \times 10^7} = 4338.4337$ rad/s

Q value $= \omega_0/3683.49 = 1.1778$
$k_f = \omega_0 = 4338.4337$ rad/s

Normalized component values:

$$R_1 = 1\ \Omega, \quad R_2 = 1\ \Omega, \quad C_1 = 2Q\ \text{F} = 2.3556\ \text{F}, \quad C_2 = 1/(2Q)\ \text{F} = 0.4245\ \text{F}$$

Scaled component values:

$$R_1 = 1\ k\Omega, \quad R_2 = 1\ k\Omega, \quad C_1 = 2.3556/(k_m k_f)\ \text{F} = 0.543\ \mu\text{F}$$
$$C_2 = 0.4245/(k_m k_f)\ \text{F} = 0.0979\ \mu\text{F}$$

For transfer function:

$$H(s) = \frac{4.0891 \times 10^7}{s^2 + 1406.97s + 4.0891 \times 10^7} = \frac{\omega_0^2}{s^2 + \dfrac{\omega_0}{Q}s + \omega_0^2}$$

Corner frequency $= \omega_0 = \sqrt{4.0891 \times 10^7} = 6394.6160$ rad/s

Q value $= \omega_0/1406.97 = 4.5450$
$k_f = \omega_0 = 6394.6160$ rad/s

Normalized component values:

$$R_1 = 1\ \Omega, \quad R_2 = 1\ \Omega, \quad C_1 = 2Q\ \text{F} = 9.09\ \text{F}, \quad C_2 = 1/(2Q)\ \text{F} = 0.11\ \text{F}$$

continued

Example 17.10 continued

Scaled component values:

$$R_1 = 1\ k\Omega, \quad R_2 = 1\ k\Omega, \quad C_1 = 9.09/(k_m k_f)\ \text{F} = 1.4215\ \mu\text{F}$$
$$C_2 = 0.11/(k_m k_f)\ \text{F} = 0.0172\ \mu\text{F}$$

The PSpice simulation is shown in Figure 17.26, and the magnitude response is shown in Figure 17.27.

FIGURE 17.26

PSpice schematic for the Sallen-Key unity gain circuit.

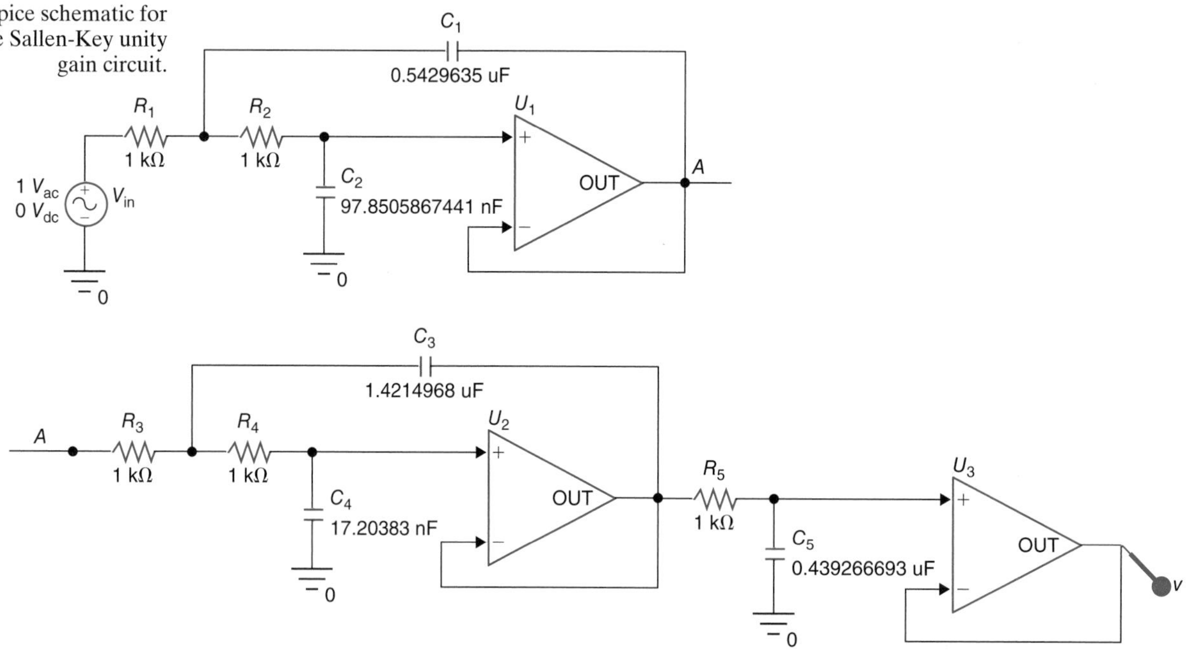

FIGURE 17.27

Magnitude response for the circuit shown in Figure 17.26.

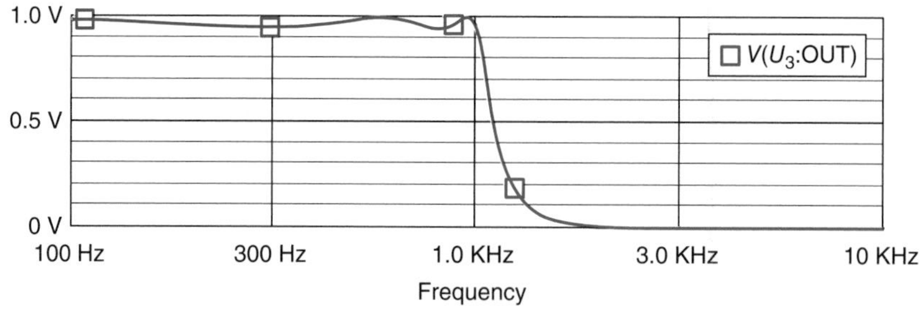

Exercise 17.10

Implement the filter designed in EXERCISE 17.9 using the Sallen-Key equal *R*, equal *C* method. Assume $k_m = 1000$.

For transfer function

$$H(s) = \frac{2.1548 \times 10^7}{s^2 + 6718.9228s + 2.1548 \times 10^7} = \frac{\omega_0^2}{s^2 + \dfrac{\omega_0}{Q}s + \omega_0^2}$$

continued

Exercise 17.10 continued

Corner frequency $= \omega_0 = \sqrt{2.1548 \times 10^7} = 4641.948$ rad/s

Q value $= \omega_0/6718.92 = 0.6909$

$k_f = \omega_0 = 4641.948$ rad/s

Normalized component values:

$R_2 = 1\ \Omega,\quad R_B = 1\ \Omega,\quad R_A = 0.5526\ \Omega,\quad R_a = 1.5526\ \Omega,\quad R_b = 2.8097\ \Omega$

$C_1 = 1\text{ F},\quad C_2 = 1\text{ F}$

Scaled component values:

$R_2 = 1\ k\Omega,\quad R_B = 1\ k\Omega,\quad R_A = 552.56\ \Omega,\quad R_a = 1.5526\ k\Omega,\quad R_b = 2.8097\ k\Omega$

$C_1 = 0.2154\ \mu\text{F},\quad C_2 = 0.2154\ \mu\text{F}$

For transfer function:

$$H(s) = \frac{6.1746 \times 10^7}{s^2 + 2783.069s + 6.1746 \times 10^7} = \frac{\omega_0^2}{s^2 + \dfrac{\omega_0}{Q}s + \omega_0^2}$$

Corner frequency $= \omega_0 = \sqrt{6.1746 \times 10^7} = 7857.8584$ rad/s

Q value $= \omega_0/2783.069 = 2.82345$

$k_f = \omega_0 = 7857.8584$ rad/s

Normalized component values:

$R_2 = 1\ \Omega,\quad R_B = 1\ \Omega,\quad R_A = 1.6458\ \Omega,\quad R_a = 2.6458\ \Omega,\quad R_b = 1.6076\ \Omega$

$C_1 = 1\text{ F},\quad C_2 = 1\text{ F}$

Scaled component values

$R_2 = 1\ k\Omega,\quad R_B = 1\ k\Omega,\quad R_A = 1.6458\ k\Omega,\quad R_a = 2.6458\ k\Omega,\quad R_b = 1.6076\ k\Omega$

$C_1 = 0.1273\ \mu\text{F},\quad C_2 = 0.1273\ \mu\text{F}$

Since the order is even ($n = 4$), the constant term is given by

$b_0 = 10^{-0.05A_p} = 0.95279$

A voltage divider can be used to provide this gain. The PSpice simulation is shown in Figure 17.28, and the magnitude response is shown in Figure 17.29.

continued

Exercise 17.10 continued

FIGURE 17.28

PSpice circuit for EXERCISE 17.10.

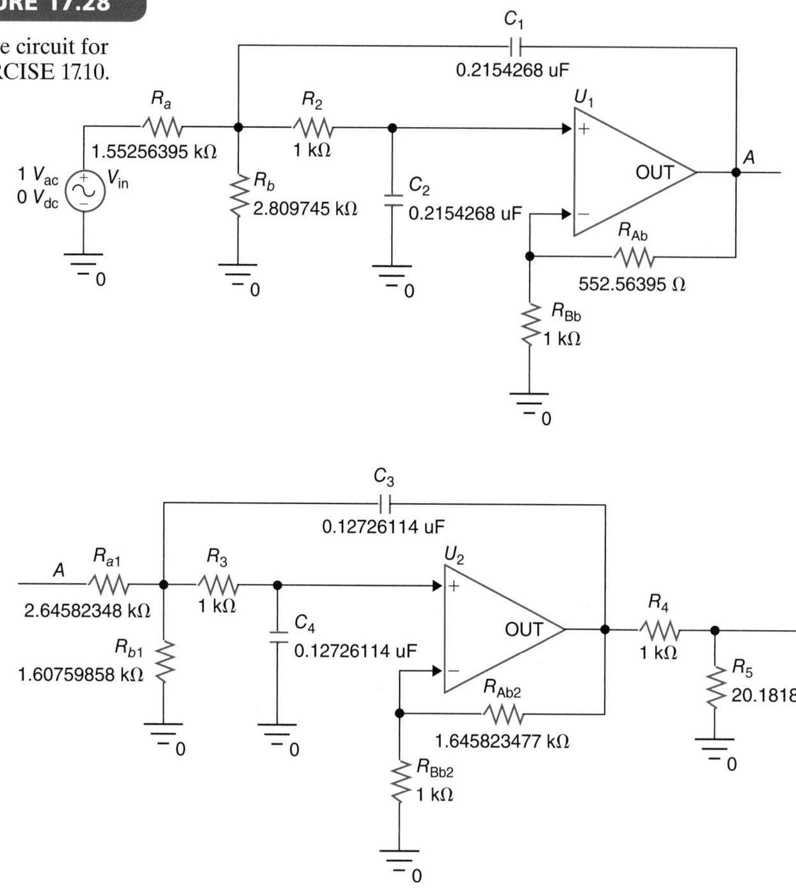

FIGURE 17.29

Magnitude response in decibels for the circuit shown in Figure 17.28.

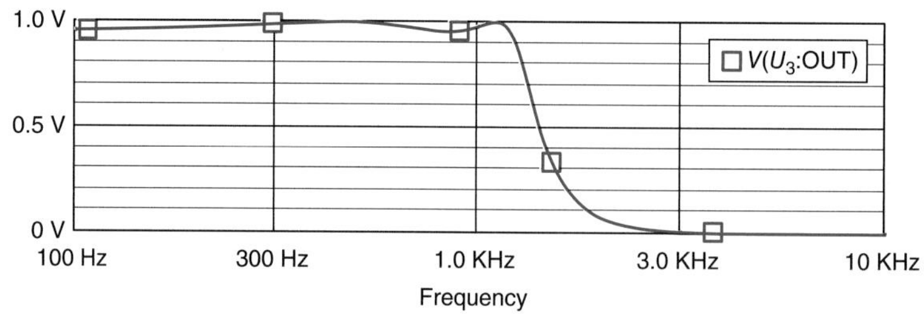

17.7 Analog Chebyshev Type 2 LPF Design

The Chebyshev type 2 approximation introduces ripples into the stopband and does not introduce ripples into the passband. The magnitude-squared response of the normalized ($\Omega_p = 1$ rad/s) Chebyshev type 2 LPF is given by

$$|H_n(j\Omega)|^2 = \frac{1}{1 + \dfrac{\varepsilon_s^2}{C_n^2\left(\dfrac{\Omega_s}{\Omega}\right)}} = \frac{C_n^2\left(\dfrac{\Omega_s}{\Omega}\right)}{C_n^2\left(\dfrac{\Omega_s}{\Omega}\right) + \varepsilon_s^2} \tag{17.174}$$

where $C_n(\Omega)$ is an nth-order Chebyshev polynomial defined by Equation (17.138), and ε_s is the stopband ripple parameter. The magnitude response of the normalized ($\Omega_p = 1$ rad/s) Chebyshev type 2 LPF is given by

$$|H_n(j\Omega)| = \frac{1}{\sqrt{1 + \dfrac{\varepsilon_s^2}{C_n^2\left(\dfrac{\Omega_s}{\Omega}\right)}}} \tag{17.175}$$

When $\Omega = 0$, $\Omega_s/\Omega = \infty$, and $C_n(\Omega_s/\Omega) = \infty$. Thus, we have

$$|H_n(j0)|^2 = \frac{1}{1 + \dfrac{\varepsilon_s^2}{C_n^2\left(\dfrac{\Omega_s}{0}\right)}} = \frac{1}{1 + \dfrac{\varepsilon_s^2}{\infty}} = 1$$

and

$$|H_n(j0)| = 1$$

The attenuation (or loss) in decibels for the Chebyshev type 2 response is defined as

$$A = -20\log_{10}(|H_n(j\Omega)|) = -20\log_{10}\frac{1}{\sqrt{1+\dfrac{\varepsilon_s^2}{C_n^2\left(\dfrac{\Omega_s}{\Omega}\right)}}} = 10\log_{10}\left[1+\frac{\varepsilon_s^2}{C_n^2\left(\dfrac{\Omega_s}{\Omega}\right)}\right] \text{dB}$$

Thus, the attenuation (in decibels) at the passband cutoff frequency $\Omega = \Omega_p = 1$ rad/s is given by

$$|H_n(j1)| = \frac{1}{\sqrt{1 + \dfrac{\varepsilon_s^2}{C_n^2\left(\dfrac{\Omega_s}{1}\right)}}} = \frac{1}{\sqrt{1 + \dfrac{\varepsilon_s^2}{C_n^2(\Omega_s)}}} = \frac{1}{\sqrt{1 + \dfrac{\varepsilon_s^2}{\cosh^2(n\cosh^{-1}(\Omega_s))}}} = \frac{1}{\sqrt{1+\varepsilon_p^2}} \tag{17.176}$$

where

$$\varepsilon_p^2\cosh^2(n\cosh^{-1}(\Omega_s)) = \varepsilon_s^2 \quad \text{or} \quad \varepsilon_s = \varepsilon_p\cosh(n\cosh^{-1}(\Omega_s)) \tag{17.177}$$

Notice that

$$\cosh(n\cosh^{-1}(\Omega_s)) = \frac{\varepsilon_s}{\varepsilon_p} \tag{17.178}$$

and the attenuation (in decibels) at the passband cutoff frequency is given by

$$A_p = 10\log_{10}[1 + \varepsilon_p^2]\ \text{dB}$$

Solving for ε_p, we obtain

$$\varepsilon_p = \sqrt{10^{\frac{A_p}{10}} - 1} = \sqrt{\frac{1}{G_p^2} - 1} = \sqrt{\frac{1}{(1-\delta_p)^2} - 1} \quad (17.179)$$

Notice that $C_n(1) = \cos(n\cos^{-1}(1)) = \cos(0) = 1$. At the stopband cutoff frequency $\Omega = \Omega_s$ rad/s from Equation (17.175), the magnitude is given by

$$|H_n(j\Omega_s)| = \frac{1}{\sqrt{1 + \dfrac{\varepsilon_s^2}{C_n^2\left(\dfrac{\Omega_s}{\Omega_s}\right)}}} = \frac{1}{\sqrt{1 + \dfrac{\varepsilon_s^2}{C_n^2(1)}}} = \frac{1}{\sqrt{1+\varepsilon_s^2}} \quad (17.180)$$

The attenuation in decibels at the stopband cutoff frequency $\Omega = \Omega_s$ rad/s is given by

$$A_s = -20\log_{10}|H_n(j\Omega_s)| = -20\log_{10}\frac{1}{\sqrt{1+\varepsilon_s^2}} = 10\log_{10}(1+\varepsilon_s^2) \quad (17.181)$$

Solving Equation (17.181) for ε_s, we have

$$\varepsilon_s = \sqrt{10^{\frac{A_s}{10}} - 1} \quad (17.182)$$

Substituting Equations (17.179) and (17.182) into Equation (17.178), we obtain

$$\cosh(n\cosh^{-1}(\Omega_s)) = \frac{\sqrt{10^{\frac{A_s}{10}} - 1}}{\sqrt{10^{\frac{A_p}{10}} - 1}} \quad (17.183)$$

Taking the arc cosh (acosh) on both sides of Equation (17.183), we obtain

$$n = \frac{\operatorname{acosh}\left(\sqrt{\dfrac{10^{0.1A_s} - 1}{10^{0.1A_p} - 1}}\right)}{\operatorname{acosh}(\Omega_s)} \quad (17.184)$$

This equation can be used to find the order n of a filter given $A_p, A_s, \omega_p, \omega_s$ (notice that $\Omega_p = 1, \Omega_s = \omega_s/\omega_p$). Since the order n of a filter must be an integer, the n value must be rounded up. Thus, we have

$$n = \left\lceil \frac{\operatorname{acosh}\left(\sqrt{\dfrac{10^{0.1A_s} - 1}{10^{0.1A_p} - 1}}\right)}{\operatorname{acosh}(\Omega_s)} \right\rceil = ceil\left(\frac{\operatorname{acosh}\left(\sqrt{\dfrac{10^{0.1A_s} - 1}{10^{0.1A_p} - 1}}\right)}{\operatorname{acosh}(\Omega_s)}\right) \quad (17.185)$$

Notice that

$$\cosh^{-1}(x) = \ln(x + \sqrt{x^2 - 1})$$

$$\sinh^{-1}(x) = \ln(x + \sqrt{x^2 + 1})$$

Since the order is rounded up, the actual filter exceeds the specifications unless n is an integer before rounding up. The excess tolerance can be assigned to the stopband or the passband. If the excess tolerance is assigned to the stopband, the resulting filter will exceed the specifications in the stopband (attenuation is greater than A_s dB at ω_s) and will meet the specifications exactly in the passband (attenuation is A_p dB at ω_p). On the other hand, if the excess tolerance is assigned to the passband, the resulting filter will exceed the specifications in the passband (attenuation is smaller than A_p dB at ω_p) and will meet the specifications exactly in the stopband (attenuation is A_s dB at ω_s). If the excess tolerance is assigned to the stopband, the passband attenuation A_p is fixed at the value given in the specifications. Substituting the rounded-up n value from Equation (17.185) into Equation (17.183) and solving for A_s, we obtain new A_s, denoted by A_{s_1}, given by

$$A_{s_1} = 10 \log_{10}[1 + (10^{0.1A_p} - 1) \cosh^2(n \operatorname{acosh}(\Omega_s))] \tag{17.186}$$

Since A_p is fixed, there is no change to the passband ripple parameter ε_p given by Equation (17.179).

If the excess tolerance is assigned to the passband, the stopband attenuation A_s is fixed at the value given in the specifications. Substituting the rounded-up n value from Equation (17.185) into Equation (17.183) and solving for A_p, we obtain new A_p, denoted by A_{p_1}, given by

$$A_{p_1} = 10 \log_{10}\left[1 + \frac{(10^{0.1A_s} - 1)}{\cosh^2(n \operatorname{acosh}(\Omega_s))}\right] \tag{17.187}$$

The passband ripple parameter ε_p can be recalculated using Equation (17.179).

$$\varepsilon_p = \sqrt{10^{\frac{A_p}{10}} - 1} \tag{17.188}$$

Chebyshev Type 2 Pole and Zero Locations

Let a parameter a be

$$a = \frac{1}{n}\sinh^{-1}(\sqrt{10^{0.1A_s} - 1}) \tag{17.189}$$

It can be shown (see the references given in Section 17.6) that for n odd, there is one real pole at $S = \dfrac{-1}{\sinh(a)}$ and $(n - 1)/2$ pairs of complex conjugate poles at

$$S_k = \frac{-\Omega_s}{\sinh(a)\sin\left(\frac{\pi}{2n}(2k+1)\right) \pm j\cosh(a)\cos\left(\frac{\pi}{2n}(2k+1)\right)}, \quad k = 0, 1, \ldots, \frac{n-3}{2} \tag{17.190}$$

If n is even, there are $(n/2)$ pairs of complex conjugate poles at

$$S_k = \frac{-\Omega_s}{\sinh(a)\sin\left(\frac{\pi}{2n}(2k+1)\right) \pm j\cosh(a)\cos\left(\frac{\pi}{2n}(2k+1)\right)}, \quad k = 0, 1, \ldots, \frac{n-2}{2} \tag{17.191}$$

In summary, the design procedure for the Chebyshev type 2 LPF is as follows.

1. Change the Cutoff Frequencies from Hertz to Radians

If the cutoff frequencies are given in hertz, change the cutoff frequencies to radians.

$$\omega_p = 2\pi f_p, \quad \omega_s = 2\pi f_s \tag{17.192}$$

2. Backward Transformation

Given the passband cutoff frequency ω_p, stopband cutoff frequency ω_s, maximum attenuation (in decibels) in the passband A_p, and minimum attenuation (in decibels) in the stopband A_s, find the cutoff frequencies for the normalized LPF by dividing by ω_p.

$$\text{Passband cutoff frequency for the normalized LPF} = \Omega_p = 1 \text{ rad/s} \quad (17.193)$$

$$\text{Stopband cutoff frequency for the normalized LPF} = \Omega_s = \omega_s/\omega_p \text{ rad/s} \quad (17.194)$$

Notice that $\omega_s/\omega_p = f_s/f_p > 1$.

3. Order

Find the order n of the normalized LPF using

$$n = ceil\left(\frac{\operatorname{acosh}\left(\sqrt{\dfrac{10^{0.1A_s}-1}{10^{0.1A_p}-1}}\right)}{\operatorname{acosh}(\Omega_s)}\right) \quad (17.195)$$

4. Recalculation of the Parameters

If excess tolerance is assigned to the stopband, we have new A_s given by

$$A_{s_1} = 10\log_{10}[1 + (10^{0.1A_p} - 1)\cosh^2(n\operatorname{acosh}(\Omega_s))] \quad (17.196)$$

The value of A_p remains the same. The passband ripple parameter is given by

$$\varepsilon_p = \sqrt{10^{\frac{A_p}{10}} - 1} \quad (17.197)$$

If excess tolerance is assigned to the passband, we have new A_p given by

$$A_{p_1} = 10\log_{10}\left[1 + \frac{(10^{0.1A_s} - 1)}{\cosh^2(n\operatorname{acosh}(\Omega_s))}\right] \quad (17.198)$$

The passband ripple parameter ε_p can be recalculated using

$$\varepsilon_{p_1} = \sqrt{10^{\frac{A_{p_1}}{10}} - 1} \quad (17.199)$$

5. Normalized Lowpass Poles and Zeros ($\Omega_s = 1$)

For simplicity, we set $\Omega_s = 1$ rad/s for normalized poles and zeros. The parameter a is given by

$$a = \frac{1}{n}\sinh^{-1}\left(\sqrt{10^{0.1A_s} - 1}\right)$$

n odd:

There is one pole at $S = \dfrac{-1}{\sinh(a)}$ and $(n-1)/2$ complex conjugate poles at

$$S_k = \frac{-1}{\sinh(a)\sin\left(\dfrac{2k+1}{2n}\pi\right) \pm j\cosh(a)\cos\left(\dfrac{2k+1}{2n}\pi\right)},$$
$$k = 0, 1, 2, \ldots, \frac{n-3}{2} \quad (17.200)$$

and $(n-1)$ zeros are located at

$$S_k = \frac{\pm j}{\cos\left(\frac{\pi(2k+1)}{2n}\right)} = \pm j\sec\left(\frac{\pi(2k+1)}{2n}\right), \quad k = 0, 1, 2, \ldots, \left\lfloor \frac{n-2}{2} \right\rfloor \tag{17.201}$$

n even:

There are $n/2$ complex conjugate poles at

$$S_k = -\frac{1}{\sinh(a)\sin\left(\frac{2k+1}{2n}\pi\right) \pm j\cosh(a)\cos\left(\frac{2k+1}{2n}\pi\right)}, \quad k = 0, 1, 2, \ldots, \frac{n-2}{2} \tag{17.202}$$

and n zeros at

$$S_k = \frac{\pm j}{\cos\left(\frac{\pi(2k+1)}{2n}\right)} = \pm j\sec\left(\frac{\pi(2k+1)}{2n}\right), \quad k = 0, 1, 2, \ldots, \frac{n-2}{2} \tag{17.203}$$

6. Transfer Function of the Normalized LPF

Convert the poles and the zeros of the normalized LPF into polynomials, and write the transfer function as a product of the first-order and the second-order sections. If the order of the filter is odd, there is one first-order section and $(n-1)/2$ second-order sections. If the order of the filter is even, there are $n/2$ second-order sections.

7. Frequency Transformation

Transform the poles and the zeros of the normalized LPF to the poles and zeros of the frequency-transformed LPF by multiplying the poles and the zeros from step 5 above by ω_s. Let s be the location of poles and zeros after the transformation, and let S be the location of poles and zeros before the transformation. Then, we have

$$s = \omega_s S \tag{17.204}$$

There is one zero at $s = \infty$ when n is odd.

8. Transfer Function

Convert the poles and the zeros of the frequency-transformed LPF into polynomials, and write the transfer function as a product of the first-order and the second-order sections. If the order of the filter is odd, there is one first-order section and $(n-1)/2$ second-order sections. If the order of the filter is even, there are $n/2$ second-order sections.

9. Verify

Calculate the attenuation at ω_p and ω_s, and compare with the specifications. If the excess tolerance is assigned to stopband, we have

$$-20\log_{10}|H(j\omega_p)| = A_p, \quad -20\log_{10}|H(j\omega_s)| \geq A_s \tag{17.205}$$

If the excess tolerance is assigned to passband, we have

$$-20\log_{10}|H(j\omega_p)| \le A_p, \quad -20\log_{10}|H(j\omega_s)| = A_s \tag{17.206}$$

EXAMPLE 17.11

Design a Chebyshev type 2 LPF with the following specifications:

Passband cutoff frequency = f_p = 2 kHz
Stopband cutoff frequency = f_s = 3 kHz
Maximum attenuation in the passband = A_p = 2 dB
Minimum attenuation in the stopband = A_s = 30 dB
Excess tolerance = stopband

1. Change the Cutoff Frequencies from Hertz to Radians

$$\omega_p = 2\pi f_p = 2\pi 2000 = 12{,}566.3706 \text{ rad/s}$$
$$\omega_s = 2\pi f_s = 2\pi 3000 = 18{,}849.5559 \text{ rad/s}$$

2. Backward Transformation

The cutoff frequencies of the normalized LPF are given by

$$\Omega_p = 1 \text{ rad/s}, \Omega_s = \omega_s/\omega_p = 1.5 \text{ rad/s}$$

3. Order

The order of the filter is

$$n = \left\lceil \frac{\operatorname{acosh}\left(\sqrt{\frac{10^{0.1A_s}-1}{10^{0.1A_p}-1}}\right)}{\operatorname{acosh}(\Omega_s)} \right\rceil = ceil\left(\frac{\operatorname{acosh}\left(\sqrt{\frac{10^{0.1\times 30}-1}{10^{0.1\times 2}-1}}\right)}{\operatorname{acosh}(1.5)}\right) = ceil(4.5869) = 5$$

4. Recalculation of the Parameters

Since the excess tolerance is assigned in the stopband, the new attenuation in the stopband is given by

$$A_{s_1} = 10\log_{10}[1 + (10^{0.1A_p} - 1)\cosh^2(n\operatorname{acosh}(\Omega_s))] = 33.4502 \text{ dB}$$

The attenuation in the stopband is 3.4502 dB larger than the minimum required attenuation of 30 dB. The value of ε_p is given by

$$\varepsilon_p = \sqrt{10^{\frac{A_p}{10}} - 1} = \sqrt{10^{0.2} - 1} = 0.7648$$

5. Normalized Lowpass Poles and Zeros

Constant a

$$a = \frac{1}{n}\sinh^{-1}(\sqrt{10^{0.1A_s} = 1}) = 0.9088$$

continued

Example 17.11 continued

Since n is odd, there is one pole at $S = -1/\sinh(a)$ and two pairs of normalized complex conjugate poles at

$$S_k = \frac{-1}{\sinh(a)\sin\left(\frac{2k+1}{2n}\pi\right) \pm j\cosh(a)\cos\left(\frac{2k+1}{2n}\pi\right)}, \quad k = 0,1$$

Thus, the pole locations are

–0.9623
–0.5898 ± j0.5947
–0.1618 ± j0.6912

The zero locations are found from

$$S_k = \frac{\pm j}{\cos\left(\frac{\pi(2k+1)}{2n}\right)} = \pm j\sec\left(\frac{\pi(2k+1)}{2n}\right), \quad k = 0, 1$$

The zero locations are

∞
± j1.7013
± j1.0515

6. Transfer Function of the Normalized LPF

The transfer function of the normalized LPF is given by

$$H(S) = \frac{0.9623}{S + 0.9623} \times \frac{S^2 + 2.8944}{S^2 + 1.1796S + 0.7015} \times \frac{S^2 + 1.1056}{S^2 + 0.3236S + 0.5039}$$

7. Frequency Transformation

The frequency-transformed poles and zeros are obtained by multiplying the poles and the zeros given in step 5 by $\omega_s = 2\pi 3000$. The frequency-transformed poles are given by

–18,138.4115
–11,117.6032 ± j11,209.7728
–3050.3051 ± j13,028.4229

The frequency-transformed zeros are given by

∞
± j32,068.78
± j19,819.596

8. Transfer Function

The transfer function of the LPF is

$$H(s) = \frac{18{,}138.41}{s + 18{,}138.41} \times \frac{s^2 + 1.0284 \times 10^9}{s^2 + 22{,}235.21s + 2.4926 \times 10^8}$$
$$\times \frac{s^2 + 3.9282 \times 10^8}{s^2 + 6100.61s + 1.79044 \times 10^8}$$

continued

Example 17.11 continued

Second-order term	Corner frequency (ω_0)	Q value
$s^2 + 22{,}235.21s + 2.4926 \times 10^8$	15,787.9734 rad/s	0.71
$s^2 + 6100.61s + 1.79044 \times 10^8$	13,380.7386 rad/s	2.1933

9. Constant Term b_0

The product of the gains at $\omega = 0$ ($s = j\omega = 0$) is given by

$$g_0 = \frac{18{,}138.41}{18{,}138.41} \times \frac{1.0284 \times 10^9}{2.4926 \times 10^8} \times \frac{3.9282 \times 10^8}{1.79044 \times 10^8} = 1 \times 4.1258 \times 2.194 = 9.05194$$

To make the gain equal to 1 at $\omega = 0$, we take the inverse of gain for each term and multiply them to obtain the single gain denoted by b_0. The single gain b_0 is given by

$$b_0 = \frac{1}{1} \times \frac{1}{4.1258} \times \frac{1}{2.194} = 1 \times 0.2424 \times 0.4558 = 0.1105$$

10. Verify

$$-20\log_{10}|H(j\omega_p)| = 2\,\text{dB}, \quad -20\log_{10}|H(j\omega_s)| = 33.4502\,\text{dB}$$

Exercise 17.11

Design a Chebyshev type 2 LPF with the following specifications:

Passband cutoff frequency $= f_p = 1$ kHz
Stopband cutoff frequency $= f_s = 2$ kHz
Maximum attenuation in the passband $= A_p = 1$ dB
Minimum attenuation in the stopband $= A_s = 30$ dB
Excess tolerance = passband

Answer:
$\omega_p = 6283.1853$ rad/s, $\omega_s = 12{,}566.3706$, $\Omega_p = 1$ rad/s, $\Omega_s = 2$ rad/s, $n = 4$, $A_{p_1} = 0.4382$, $\varepsilon_p = 0.5088$.

Normalized poles: $-0.6836 \pm j0.3646, -0.1988 \pm j0.6180$.
Normalized zeros: $\pm j2.6131, \pm j1.0824$.
Normalized LPF transfer function

$$H(S) = \frac{S^2 + 6.8284}{S^2 + 1.3673S + 0.6003} \times \frac{S^2 + 1.1716}{S^2 + 0.3976S + 0.4214}$$

Frequency-transformed LPF transfer function

$$H(s) = \frac{s^2 + 1.0783 \times 10^9}{s^2 + 17{,}181.4585s + 9.4797 \times 10^7} \times \frac{s^2 + 1.8501 \times 10^8}{s^2 + 4996.0601s + 6.6548 \times 10^7}$$

$$-20\log_{10}|H(j\omega_p)| = 0.4382\text{ dB}, \quad -20\log_{10}|H(j\omega_s)| = 30\,\text{dB}$$

EXAMPLE 17.12

Implement the filter designed in EXAMPLE 17.11 using the Sallen-Key circuit for the bandstop filter. Assume $k_m = 1000$.

For transfer function: $H(s) = \dfrac{18{,}138.41}{s + 18{,}138.41}$

$k_f = 18{,}138.41$ rad/s

Normalized component values: $R = 1\ \Omega, C = 1$ F
Scaled component values: $R = 1\ k\Omega, C = 1/(k_m k_f)$ F $= 55.13161936\ n$F
Transfer function:

$$H(s) = \frac{s^2 + 1.0284 \times 10^9}{s^2 + 22{,}235.21s + 2.4926 \times 10^8} = \frac{s^2 + b_0}{s^2 + a_1 s + a_0} = \frac{s^2 + \omega_z^2}{s^2 + \dfrac{\omega_0}{Q}s + \omega_0^2}$$

$$b_0 = \omega_z^2 = 1.0284 \times 10^9$$

$$\omega_z = \sqrt{b_0} = 32{,}068.78 \text{ rad/s}$$

$$a_0 = \omega_0^2 = 2.4926 \times 10^8$$

$$\omega_0 = \sqrt{a_0} = 15{,}787.9734 \text{ rad/s}$$

$$a_1 = 22{,}235.21$$

Since $\omega_0 < \omega_z$, this is case 2. From Equation (16.191), we have

$$Q = \frac{\omega_0}{a_1} = 0.710044$$

$$\frac{1}{Q} = \frac{a_1}{\omega_0} = 1.4084$$

From Equation (16.193), we have

$$A_0 = 1$$

$$A_1 = \frac{1}{Q} = \frac{a_1}{\omega_0} = 1.4084$$

$$B_0 = \frac{\omega_z^2}{\omega_0^2} = \frac{b_0}{a_0} = 4.1258$$

From Equation (16.201), we have

$$a = \frac{1}{\sqrt{B_0}} = 0.4923$$

From Equation (16.203), we have

$$C = \frac{\dfrac{B_0}{A_0} - 1}{m + 1} = 1.5629$$

continued

Example 17.12 continued

From Equation (16.204), we have

$$K = 2 + \frac{m}{m+1}\left(\frac{B_0}{A_0} - 1 - \frac{A_1\sqrt{B_0}}{A_0}\right) = 2.1326$$

The gain at $S = 0$ is given by

$$G = \frac{K}{(m+1)C+1} \times \frac{A_0}{B_0} = 2.1326$$

The normalized component values are

$$R_1 = 1\ \Omega,\quad R_2 = 1\ \Omega,\quad R_3 = 0.5\ \Omega,\quad R_B = 1\ \Omega,\quad R_A = K - 1 = 1.1326\ \Omega$$
$$C_1 = a = 0.4923\text{ F},\quad C_2 = a = 0.4923\text{ F},\quad C_3 = 2a = 0.9846\text{ F},\quad C_4 = aC = 0.7694\text{ F}$$

After magnitude scale by $k_m = 1000$ and frequency scale by $k_f = \omega_0 =$ 15,787.9734 rad/s, we obtain

$$R_1 = 1\ k\Omega,\quad R_2 = 1\ k\Omega,\quad R_3 = 0.5\ k\Omega,\quad R_B = 1\ k\Omega,\quad R_A = K - 1 = 1.1326\ k\Omega$$
$$C_1 = 31.183\ n\text{F},\quad C_2 = 31.183\ n\text{F},\quad C_3 = 62.366\ n\text{F},\quad C_4 = 48.7365\ n\text{F}$$

Transfer function:

$$H(s) = \frac{s^2 + 3.9282 \times 10^8}{s^2 + 6100.61s + 1.79044 \times 10^8} = \frac{s^2 + b_0}{s^2 + a_1 s + a_0} = \frac{s^2 + \omega_z^2}{s^2 + \dfrac{\omega_0}{Q}s + \omega_0^2}$$

$$b_0 = \omega_z^2 = 3.9282 \times 10^8$$
$$\omega_z = \sqrt{b_0} = 19{,}819.596 \text{ rad/s}$$
$$a_0 = \omega_0^2 = 1.79044 \times 10^8$$
$$\omega_0 = \sqrt{a_0} = 13{,}380.7386 \text{ rad/s}$$
$$a_1 = 6100.61$$

Since $\omega_0 < \omega_z$, this is case 2. From Equation (16.191), we have

$$Q = \frac{\omega_0}{a_1} = 2.193344$$

$$\frac{1}{Q} = \frac{a_1}{\omega_0} = 0.4559$$

From Equation (16.193), we have

$$A_0 = 1$$

$$A_1 = \frac{1}{Q} = \frac{a_1}{\omega_0} = 0.4559$$

$$B_0 = \frac{\omega_z^2}{\omega_0^2} = \frac{b_0}{a_0} = 2.194$$

continued

Example 17.12 continued

From Equation (16.201), we have

$$a = \frac{1}{\sqrt{B_0}} = 0.6751$$

From Equation (16.203), we have

$$C = \frac{\frac{B_0}{A_0} - 1}{m + 1} = 0.597$$

From Equation (16.204), we have

$$K = 2 + \frac{m}{m+1}\left(\frac{B_0}{A_0} - 1 - \frac{A_1\sqrt{B_0}}{A_0}\right) = 2.2593$$

The gain at $S = 0$ is given by

$$G = \frac{K}{(m+1)C + 1} \times \frac{A_0}{B_0} = 2.2593$$

The product of the second-order gains is given by

$$G_p = 2.1326 \times 2.2593 = 4.8182$$

To make the gain equal to 0 at $\omega = 0$, we need to reduce the gain to 1. The inverse of G_p is given by

$$G_{pi} = 1/G_p = 0.2075$$

A voltage-divider circuit can be used to reduce the gain to 1. Let $R_C = 1\ \Omega$ in the circuit shown in Figure 17.1. Then, R_D is given by

$$R_D = \frac{G_{pi}}{1 - G_{pi}} = 0.2619\ \Omega$$

The normalized component values are

$$R_1 = 1\ \Omega,\quad R_2 = 1\ \Omega,\quad R_3 = 0.5\ \Omega,\quad R_B = 1\ \Omega,\quad R_A = K - 1 = 1.2593\ \Omega$$
$$C_1 = a = 0.6751\ \text{F},\quad C_2 = a = 0.6751\ \text{F},\quad C_3 = 2a = 1.3503\ \text{F},$$
$$C_4 = aC = 0.403038\ \text{F}$$

After magnitude scale by $k_m = 1000$ and frequency scale by $k_f = \omega_0 =$ 13,380.7386 rad/s, we obtain

$$R_1 = 1\ k\Omega,\quad R_2 = 1\ k\Omega,\quad R_3 = 0.5\ k\Omega,\quad R_B = 1\ k\Omega,\quad R_A = K - 1 = 1.2593\ k\Omega,$$
$$R_C = 1\ k\Omega,\quad R_D = 261.9\ \Omega,\quad C_1 = 50.4551\ n\text{F},\quad C_2 = 50.4551\ n\text{F},$$
$$C_3 = 0.10091\ \mu\text{F},\quad C_4 = 30.1208\ n\text{F}$$

The PSpice simulation is shown in Figure 17.30, and the magnitude response is shown in Figure 17.31.

continued

Example 17.12 continued

FIGURE 17.30

PSpice schematic for the Sallen-Key circuit for BSF.

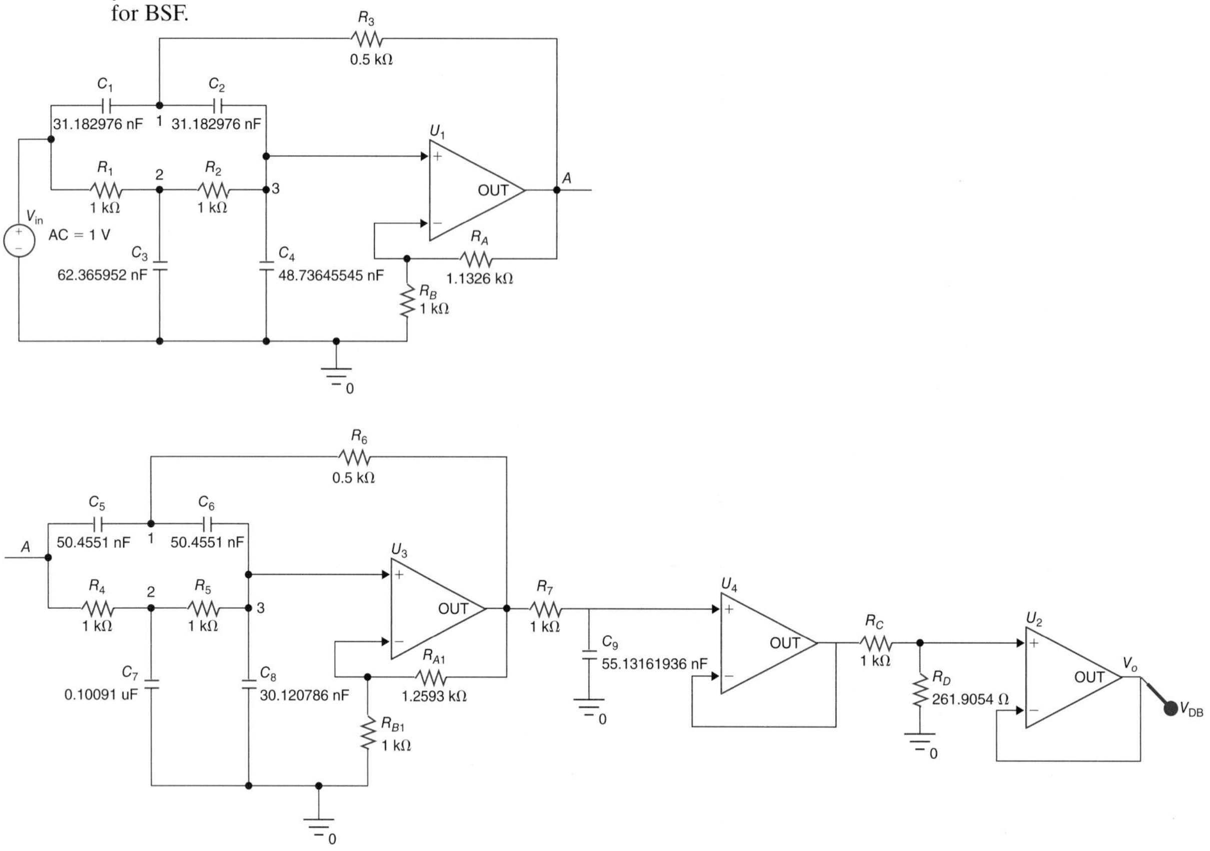

FIGURE 17.31

Magnitude response for the circuit shown in Figure 17.30.

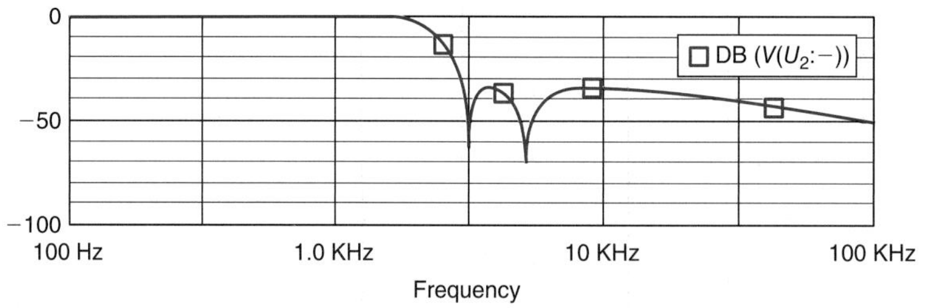

Exercise 17.12

Implement the filter designed in EXERCISE 17.11 using the Sallen-Key circuit for the bandstop filter. Assume $k_m = 1000$.

Transfer function:

$$H(s) = \frac{s^2 + 1.0783 \times 10^9}{s^2 + 17{,}181.4585s + 9.4797 \times 10^7} = \frac{s^2 + b_0}{s^2 + a_1 s + a_0} = \frac{s^2 + \omega_z^2}{s^2 + \dfrac{\omega_0}{Q}s + \omega_0^2}$$

continued

Exercise 17.12 continued

$$b_0 = \omega_z^2 = 1.0783 \times 10^9$$
$$\omega_z = \sqrt{b_0} = 32{,}837.5089 \text{ rad/s}$$
$$a_0 = \omega_0^2 = 9.4797 \times 10^7$$
$$\omega_0 = \sqrt{a_0} = 9736.358 \text{ rad/s}$$
$$a_1 = 17{,}181.4585$$

Since $\omega_0 < \omega_z$, this is case 2. From Equation (16.191), we have

$$Q = \frac{\omega_0}{a_1} = 0.5667$$

$$\frac{1}{Q} = \frac{a_1}{\omega_0} = 1.7647$$

From Equation (16.193), we have

$$A_0 = 1$$

$$A_1 = \frac{1}{Q} = \frac{a_1}{\omega_0} = 1.7647$$

$$B_0 = \frac{\omega_z^2}{\omega_0^2} = \frac{b_0}{a_0} = 11.3749$$

From Equation (16.201), we have

$$a = \frac{1}{\sqrt{B_0}} = 0.2965$$

From Equation (16.203), we have

$$C = \frac{\dfrac{B_0}{A_0} - 1}{m + 1} = 5.18745$$

From Equation (16.204), we have

$$K = 2 + \frac{m}{m+1}\left(\frac{B_0}{A_0} - 1 - \frac{A_1\sqrt{B_0}}{A_0}\right) = 4.2116$$

The gain at $S = 0$ is given by

$$G = \frac{K}{(m+1)C + 1} \times \frac{A_0}{B_0} = 4.2116$$

The normalized component values are

$$R_1 = 1\ \Omega,\quad R_2 = 1\ \Omega,\quad R_3 = 0.5\ \Omega,\quad R_B = 1\ \Omega,\quad R_A = K - 1 = 3.2116\ \Omega$$
$$C_1 = a = 0.2965\text{ F},\quad C_2 = a = 0.2965\text{ F},\quad C_3 = 2a = 0.5930\text{ F},\quad C_4 = aC = 1.5381\text{ F}$$

continued

Exercise 17.12 continued

After magnitude scale by $k_m = 1000$ and frequency scale by $k_f = \omega_0 = 9736.358$ rad/s, we obtain

$$R_1 = 1\ k\Omega,\quad R_2 = 1\ k\Omega,\quad R_3 = 0.5\ k\Omega,\quad R_B = 1\ k\Omega,\quad R_A = K - 1 = 3.2116\ k\Omega$$
$$C_1 = 30.453\ n\text{F},\quad C_2 = 30.453\ n\text{F},\quad C_3 = 60.906\ n\text{F},\quad C_4 = 0.158\ \mu\text{F}$$

Transfer function:

$$H(s) = \frac{s^2 + 1.8501 \times 10^8}{s^2 + 4996.0601s + 6.6548 \times 10^7} = \frac{s^2 + b_0}{s^2 + a_1 s + a_0} = \frac{s^2 + \omega_z^2}{s^2 + \dfrac{\omega_0}{Q}s + \omega_0^2}$$

$$b_0 = \omega_z^2 = 1.8501 \times 10^8$$
$$\omega_z = \sqrt{b_0} = 13{,}601.7415 \text{ rad/s}$$
$$a_0 = \omega_0^2 = 6.6548 \times 10^7$$
$$\omega_0 = \sqrt{a_0} = 8157.7086 \text{ rad/s}$$
$$a_1 = 4996.0601$$

Since $\omega_0 < \omega_z$, this is case 2. From Equation (16.191), we have

$$Q = \frac{\omega_0}{a_1} = 1.6328$$
$$\frac{1}{Q} = \frac{a_1}{\omega_0} = 0.6124$$

From Equation (16.193), we have

$$A_0 = 1$$
$$A_1 = \frac{1}{Q} = \frac{a_1}{\omega_0} = 0.6124$$
$$B_0 = \frac{\omega_z^2}{\omega_0^2} = \frac{b_0}{a_0} = 2.7801$$

From Equation (16.201), we have

$$a = \frac{1}{\sqrt{B_0}} = 0.5998$$

From Equation (16.203), we have

$$C = \frac{\dfrac{B_0}{A_0} - 1}{m + 1} = 0.89$$

From Equation (16.204), we have

$$K = 2 + \frac{m}{m+1}\left(\frac{B_0}{A_0} - 1 - \frac{A_1\sqrt{B_0}}{A_0}\right) = 2.3795$$

continued

Exercise 17.12 continued

The gain at $S = 0$ is given by

$$G = \frac{K}{(m+1)C+1} \times \frac{A_0}{B_0} = 2.3794$$

The product of all the gains is given by

$$G_p = 4.2116 \times 2.3794 = 10.0214$$

To make the gain equal to 0 at $\omega = 0$, we need to reduce the gain to 1. The inverse of G_p is given by

$$G_{pi} = 1/G_p = 0.09979$$

A voltage-divider circuit can be used to reduce the gain to 1. Let $R_C = 1\ \Omega$ in the circuit shown in Figure 17.1. Then, R_D is given by

$$R_D = \frac{G_{pi}}{1 - G_{pi}} = 0.110848\ \Omega$$

The normalized component values are

$$R_1 = 1\ \Omega,\quad R_2 = 1\ \Omega,\quad R_3 = 0.5\ \Omega,\quad R_B = 1\ \Omega,\quad R_A = K - 1 = 1.3795\ \Omega$$
$$C_1 = a = 0.5998\text{ F},\quad C_2 = a = 0.5998\text{ F},\quad C_3 = 2a = 1.1995\text{ F},\quad C_4 = aC = 0.5338\text{ F}$$

After magnitude scale by $k_m = 1000$ and frequency scale by $k_f = \omega_0 = 8157.7086$ rad/s, we obtain

$$R_1 = 1\ k\Omega,\quad R_2 = 1\ k\Omega,\quad R_3 = 0.5\ k\Omega,\quad R_B = 1\ k\Omega,\quad R_A = K - 1 = 1.3795\ k\Omega$$
$$C_1 = 73.52\ nF,\quad C_2 = 73.52\ nF,\quad C_3 = 0.147\ \mu F,\quad C_4 = 65.4346\ nF,\quad R_C = 1\ k\Omega,$$
$$R_D = 110.848\ \Omega$$

The PSpice simulation is shown in Figure 17.32, and the magnitude response is shown in Figure 17.33.

FIGURE 17.32

PSpice circuit for EXERCISE 17.12.

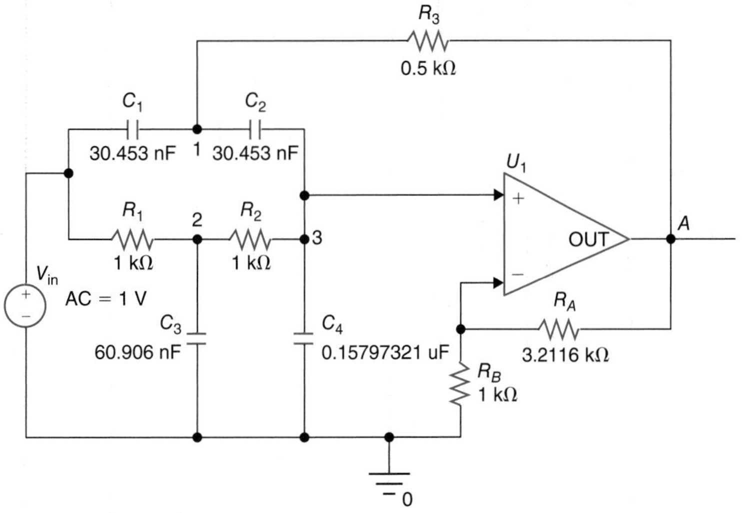

continued

Exercise 17.12 continued

FIGURE 17.32

continued

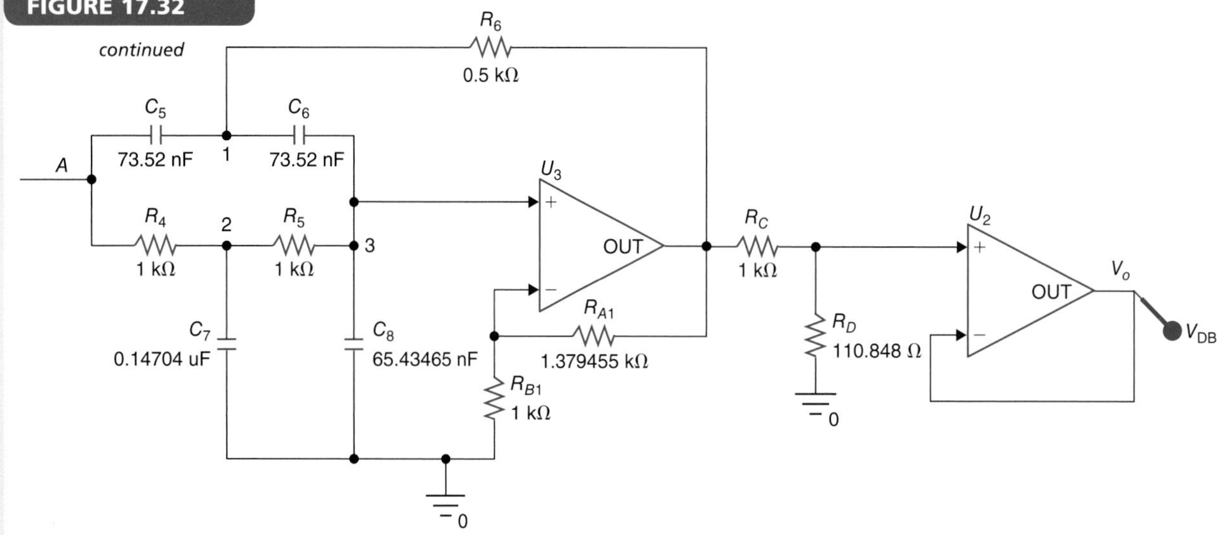

FIGURE 17.33

Magnitude response in decibels for the circuit shown in Figure 17.32.

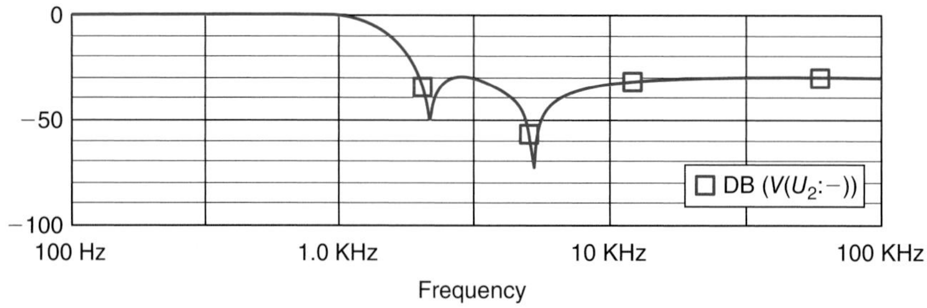

17.8 MATLAB

MATLAB is a useful tool in filter design. The MATLAB script for the low-pass filter design of EXAMPLE 17.1 is shown below as a reference. This script can be modified to other designs.

```
% Analog Butterworth Lowpass Filter Design.
% EXAMPLE 17.1
clear all;close all;format long;
Ap=1;As=20;fp=1000;fs=2000;excess='stop'; %excess='pass'
%
disp('Analog Butterworth Lowpass Filter Design')
disp(['Maximum attenuation in the passband, Ap = ', num2str(Ap), ' dB']);
disp(['Minimum attenuation in the stopband, As = ', num2str(As), ' dB']);
disp(['Passband cutoff frequency, fp = ', num2str(fp), ' Hz']);
disp(['Stopband cutoff frequency, fs = ', num2str(fs), ' Hz']);
disp(['Excess tolerance assigned in ', excess, 'band']);
% f (Hz) to w (rad/s)
disp('Cutoff frequencies in rad/s');
wp=2*pi*fp
ws=2*pi*fs
% Backward Transformation
disp('Cutoff frequencies of normalized LPF');
```

```
Wp=1
Ws=ws/wp
% Calculate Order of Normalized LPF
disp('Order n');
n=log10((10^(0.1*As)-1)/(10^(0.1*Ap)-1))/(2*log10(Ws))
n=ceil(n)
% Recalculation of Ap or As, Calculation of ep, Wc
disp('Ap1, As1, epsilon, Wc');
Ap1=Ap;
As1=As;
if excess == lower('stop')
    As1=10*log10(Ws^(2*n)*(10^(0.1*Ap1)-1)+1);
    epsilon=sqrt(10^(0.1*As1)-1)/Ws^n;
    Wc=1/epsilon^(1/n);
else
    Ap1=10*log10((10^(0.1*As1)-1)/Ws^(2*n)+1);
    epsilon=sqrt(10^(0.1*Ap1)-1);
    Wc=1/epsilon^(1/n);
end
Ap1
As1
epsilon
Wc
% Normalized Lowpass Poles and Zeros
disp('Normalized lowpass poles (S) and zeros (SZ');
k=0:ceil((n-3)/2);
Sa=exp(j*pi*(2*k+n+1)/(2*n));
Sa=[Sa,conj(Sa)]
Sa=cplxpair(Sa)
Sb=reshape(Sa,[2,ceil((n-1)/2)])
S=Sa;
if rem(n,2)==1 S=[-1,Sa]; end
S
SZ=inf*ones(1,n)
% Transfer Function of Normalized LPF
DenSOS=[ones(1,ceil((n-1)/2));-sum(Sb);prod(Sb)];
NumSOS=[zeros(1,ceil((n-1)/2));zeros(1,ceil((n-1)/2));DenSOS(3,:)];
Qn=1./DenSOS(2,:);
if rem(n,2)==1
    DenSOS=[[0 1 1]',DenSOS];
    NumSOS=[[0 0 1]',NumSOS];
end
disp('Transfer function of normalized LPF');
DenSOS
NumSOS
Qn
TwoQ=2*Qn
InvTwoQ=1./(2*Qn)
W0=DenSOS(3,:)
% Scale S by Wc
disp('Sc = Wc*wp');
Sc=Wc*S
% Frequency Transformation from Normalized LPF to LPF
disp('Frequency transformed poles (s) and zeros (sz)');
s=wp*Sc
sz=inf*ones(1,n)
wo=Wc*wp
% Transfer Function of Frequency Transformed LPF
denSOS=[DenSOS(1,:);wo*DenSOS(2,:);wo^2*DenSOS(3,:)];
numSOS=[NumSOS(1,:);wo*NumSOS(2,:);wo^2*NumSOS(3,:)];
```

```
if rem(n,2)==1
    denSOS(2,1)=1;denSOS(3,1)=wo;numSOS(3,1)=wo;
end
disp('Transfer function of frequency transformed LPF');
denSOS
numSOS
% Plots
% Normalized Lowpass Filter (Wc = 1) Pole Locations
x=-1:0.01:1;ty=sqrt(1-x.^2);
figure(1);
plot(x,ty,':b',x,-ty,':b');hold on;
plot(real(S),imag(S),'bx','MarkerSize',12); grid on;
xlabel('Real Part');ylabel('Imaginary Part');axis square;
title('Normalized Lowpass Pole Locations')
% Magnitude Response (linear) of Normalized LPF
W=0:Wp/100:5*Wp;
Wa=[-W.^2;j*W;ones(1,length(W))];
Sd=DenSOS.'*Wa;
if n <=2
    HL=NumSOS(3,:)./Sd;
else
    HL=prod(DenSOS(3,:)')./prod(Sd);
end
figure(2);
plot(W,abs(HL));grid on;
xlabel('\Omega (rad)');ylabel('|H(\Omega)|');
title('Magnitude Response (Linear) of Normalized LPF');
% Magnitude Response (dB) of Normalized LPF
figure(3);
plot(W,20*log10(abs(HL)));grid on;
xlabel('\Omega (rad)');ylabel('20 log_1_0(|H(\Omega)|)');
title('Magnitude Response (dB) of Normalized LPF');
% Phase Response (deg) of Normalized LPF
figure(4);
plot(W,unwrap(angle(HL))*180/pi);grid on;
xlabel('\Omega (rad)');ylabel('\angleH(\Omega) deg');
title('Phase Response of Normalized LPF');
% Frequency Transformed LPF Pole Locations
x=-wo:0.01*wo:wo;ty=sqrt(wo^2-x.^2);
figure(5);
plot(x,ty,':b',x,-ty,':b');hold on;
plot(real(s),imag(s),'bx','MarkerSize',12); grid on;
xlabel('Real Part');ylabel('Imaginary Part');axis square;
title('Frequency Transformed Lowpass Pole Locations')
% Magnitude Response (linear) of Frequency Transformed LPF
w=0:wo/100:5*wo;
wa=[-w.^2;j*w;ones(1,length(w))];
sd=denSOS.'*wa;
if n <=2
    H=numSOS(3,:)./sd;
else
    H=prod(denSOS(3,:)')./prod(sd);
end
figure(6);
plot(w,abs(H));grid on;
xlabel('\omega (rad/s)');ylabel('|H(\omega)|');
title('Magnitude Response (Linear) of Frequency Transformed LPF');
% Magnitude Response (dB) of Frequency Transformed LPF
figure(7);
plot(w,20*log10(abs(H)));grid on;
```

```
xlabel('\omega (rad/s)');ylabel('20 log_1_0(|H(\omega)|)');
title('Magnitude Response (dB) of Frequency Transformed LPF');
% Phase Response (deg) of Frequency Transformed LPF
figure(8);
plot(w,unwrap(angle(H))*180/pi);grid on;
xlabel('\omega (rad/s)');ylabel('\angleH(\omega) deg');
title('Phase Response of Frequency Transformed LPF');
% Verification
disp('Verification');
w1=[wp,ws];wa1=[-w1.^2;j*w1;ones(1,length(w1))];
sd1=denSOS.'*wa1;
if n <=2
    H1=numSOS(3,:)./sd1;
else
    H1=prod(denSOS(3,:)')./prod(sd1);
end
Ap2=-20*log10(abs(H1(1)))
As2=-20*log10(abs(H1(2)))
```

SUMMARY

In this chapter, the design and the implementation of analog filters using the Butterworth approximation are discussed. The types of filters discussed in this chapter are low-pass filter (LPF), high-pass filter (HPF), bandpass filter, and bandstop filter. To simplify the design, regardless of the type of filter we are designing, the specifications are transformed to those of the normalized LPF. The order and the locations of poles and zeros of the normalized LPF are found. These poles and zeros are transformed to those of the original filter by frequency transformation. From the poles and the zeros, we can find the first-order and the second-order transfer functions. The first-order and the second-order sections can be implemented using Sallen-Key circuits and cascaded to get the final design. The circuit can be simulated by PSpice to compare the specifications and the results from the circuit.

The approximations for filter design include the Butterworth approximation, Chebyshev type 1 approximation, Chebyshev type 2 approximation, and elliptic approximation. In the Butterworth approximation, there are no ripples in the passband and no ripples in the stopband. In the Chebyshev type 1 approximation, there are ripples in the passband and no ripples in the stopband. In the Chebyshev type 2 approximation, there are ripples in the stopband and no ripples in the passband. In the elliptic approximation, there are ripples in the passband and stopband. The Chebyshev types 1 and 2 and the elliptic approximation provide a smaller order n for the same specifications than the Butterworth approximation, but the Butterworth approximation provides a phase response that is more close to linear than the other approximations.

The design examples given in this chapter are Butterworth approximation for LPF, HPF, BPF, and BSF design; Chebyshev type 1 approximation for LPF design; and Chebyshev type 2 approximation for LPF design. These examples can be applied to other designs with modifications.

PROBLEMS

Butterworth LPF Design

17.1 **Design a Butterworth LPF with the following specifications:**

Passband cutoff frequency $= f_p = 910$ Hz.

Stopband cutoff frequency $= f_s = 1950$ Hz.

Maximum attenuation in the passband (dB) $= A_p = 0.8$ dB.

Minimum attenuation in the stopband (dB) $= A_s = 5$ dB.

`excess='pass'`

a. Find Ω_s and order n of the normalized LPF. Also, find the recalculated values of $A_{s_1}, A_{p_1}, \varepsilon_{p_1}, \Omega_{c_1}$.
b. Find the poles S and zeros SZ of the normalized LPF (do not multiply by Ω_c).
c. Find the transfer function of the normalized LPF, and find Q values of the second-order sections.
d. Find the normalized component values for the equal R, equal C Sallen-Key circuit with gain adjusted to 1 using R_a and R_b for each second-order section.

e. Find the scaled ($k_m = 1000, k_f = \Omega_c\omega_p$) component values for the equal R, equal C Sallen-Key circuit with gain adjusted to 1 using R_a and R_b for each second-order section.

17.2 Design a Butterworth LPF with the following specifications:

Passband cutoff frequency = f_p = 900 Hz.

Stopband cutoff frequency = f_s = 2100 Hz.

Maximum attenuation in the passband (dB) = A_p = 0.85 dB.

Minimum attenuation in the stopband (dB) = A_s = 5.5 dB.

`excess='pass'`

a. Find Ω_s and order n of the normalized LPF. Also, find the recalculated values of $A_{s_1}, A_{p_1}, \varepsilon_{p_1}, \Omega_{c_1}$.
b. Find the poles S and the zeros SZ of the normalized LPF (do not multiply by Ω_c).
c. Find the transfer function of the normalized LPF, and find Q values of the second-order sections.
d. Find the normalized component values for the unity gain Sallen-Key circuit.
e. Find the scaled ($k_m = 1000, k_f = \Omega_c\omega_p$) component values for the unity gain Sallen-Key circuit.

17.3 Design a Butterworth LPF with the following specifications:

Passband cutoff frequency = f_p = 950 Hz.

Stopband cutoff frequency = f_s = 2250 Hz.

Maximum attenuation in the passband (dB) = A_p = 0.9 dB.

Minimum attenuation in the stopband (dB) = A_s = 5.5 dB.

`excess='stop'`

a. Find Ω_s and order n of the normalized LPF. Also, find the recalculated values of $A_{s_1}, A_{p_1}, \varepsilon_{p_1}, \Omega_{c_1}$.
b. Find the poles S and the zeros SZ of the normalized LPF (do not multiply by Ω_c).
c. Find the transfer function of the normalized LPF, and find Q values of the second-order sections.
d. Find the normalized component values for the equal R, equal C Sallen-Key circuit with gain adjusted to 1 using R_a and R_b for each second-order section.
e. Find the scaled ($k_m = 1000, k_f = \Omega_c\omega_p$) component values for the equal R, equal C Sallen-Key circuit with gain adjusted to 1 using R_a and R_b for each second-order section.

17.4 Design a Butterworth LPF with the following specifications:

Passband cutoff frequency = f_p = 960 Hz.

Stopband cutoff frequency = f_s = 2350 Hz.

Maximum attenuation in the passband (dB) = A_p = 0.95 dB.

Minimum attenuation in the stopband (dB) = A_s = 6.5 dB.

`excess='stop'`

a. Find Ω_s and order n of the normalized LPF. Also, find the recalculated values of $A_{s_1}, A_{p_1}, \varepsilon_{p_1}, \Omega_{c_1}$.
b. Find the poles S and the zeros SZ of the normalized LPF (do not multiply by Ω_c).
c. Find the transfer function of the normalized LPF, and find Q values of the second-order sections.
d. Find the normalized component values for the unity gain Sallen-Key circuit.
e. Find the scaled ($k_m = 1000, k_f = \Omega_c\omega_p$) component values for the unity gain Sallen-Key circuit.

17.5 Design a Butterworth LPF with the following specifications:

Passband cutoff frequency = f_p = 900 Hz.

Stopband cutoff frequency = f_s = 2100 Hz.

Maximum attenuation in the passband (dB) = A_p = 0.8 dB.

Minimum attenuation in the stopband (dB) = A_s = 12 dB.

`excess='pass'`

a. Find Ω_s and order n of the normalized LPF. Also, find the recalculated values of $A_{s_1}, A_{p_1}, \varepsilon_{p_1}, \Omega_{c_1}$.
b. Find the poles S and the zeros SZ of the normalized LPF (do not multiply by Ω_c).
c. Find the transfer function of the normalized LPF, and find Q values of the second-order sections.
d. Find the normalized component values for the unity gain Sallen-Key circuit.
e. Find the scaled ($k_m = 1000, k_f = \Omega_c\omega_p$) component values for the unity gain Sallen-Key circuit.

17.6 **Design a Butterworth LPF with the following specifications:**

Passband cutoff frequency $= f_p = 900$ Hz.

Stopband cutoff frequency $= f_s = 1900$ Hz.

Maximum attenuation in the passband (dB) $= A_p = 0.8$ dB.

Minimum attenuation in the stopband (dB) $= A_s = 10$ dB.

`excess='pass'`

a. Find Ω_s and order n of the normalized LPF. Also, find the recalculated values of A_{s_1}, A_{p_1}, ε_{p_1}, Ω_{c_1}.
b. Find the poles S and the zeros SZ of the normalized LPF (do not multiply by Ω_c).
c. Find the transfer function of the normalized LPF, and find Q values of the second-order sections.
d. Find the normalized component values for the equal R, equal C Sallen-Key circuit with gain adjusted to 1 using R_a and R_b for each second-order section.
e. Find the scaled ($k_m = 1000$, $k_f = \Omega_c\omega_p$) component values for the equal R, equal C Sallen-Key circuit with gain adjusted to 1 using R_a and R_b for each second-order section.

17.7 **Design a Butterworth LPF with the following specifications:**

Passband cutoff frequency $= f_p = 900$ Hz.

Stopband cutoff frequency $= f_s = 2200$ Hz.

Maximum attenuation in the passband (dB) $= A_p = 0.8$ dB.

Minimum attenuation in the stopband (dB) $= A_s = 11$ dB.

`excess='stop'`

a. Find Ω_s and order n of the normalized LPF. Also, find the recalculated values of A_{s_1}, A_{p_1}, ε_{p_1}, Ω_{c_1}.
b. Find the poles S and the zeros SZ of the normalized LPF (do not multiply by Ω_c).
c. Find the transfer function of the normalized LPF, and find Q values of the second-order sections.
d. Find the normalized component values for the equal R, equal C Sallen-Key circuit with gain adjusted to 1 using R_a and R_b for each second-order section.
e. Find the scaled ($k_m = 1000$, $k_f = \Omega_c\omega_p$) component values for the equal R, equal C Sallen-Key circuit with gain adjusted to 1 using R_a and R_b for each second-order section.

17.8 **Design a Butterworth LPF with the following specifications:**

Passband cutoff frequency $= f_p = 900$ Hz.

Stopband cutoff frequency $= f_s = 2300$ Hz.

Maximum attenuation in the passband (dB) $= A_p = 0.95$ dB.

Minimum attenuation in the stopband (dB) $= A_s = 14.5$ dB.

`excess='stop'`

a. Find Ω_s and order n of the normalized LPF. Also, find the recalculated values of A_{s_1}, A_{p_1}, ε_{p_1}, Ω_{c_1}.
b. Find the poles S and the zeros SZ of the normalized LPF (do not multiply by Ω_c).
c. Find the transfer function of the normalized LPF, and find Q values of the second-order sections.
d. Find the normalized component values for the unity gain Sallen-Key circuit.
e. Find the scaled ($k_m = 1000$, $k_f = \Omega_c\omega_p$) component values for the unity gain Sallen-Key circuit.

17.9 **Design a Butterworth LPF with the following specifications:**

Passband cutoff frequency $= f_p = 1100$ Hz.

Stopband cutoff frequency $= f_s = 2200$ Hz.

Maximum attenuation in the passband (dB) $= A_p = 1.1$ dB.

Minimum attenuation in the stopband (dB) $= A_s = 15$ dB.

`excess='stop'`

a. Find Ω_s and order n of the normalized LPF. Also, find the recalculated values of A_{s_1}, A_{p_1}, ε_{p_1}, Ω_{c_1}.
b. Find the poles S and the zeros SZ of the normalized LPF (do not multiply by Ω_c).
c. Find the transfer function of the normalized LPF, and find Q values of the second-order sections.
d. Find the normalized component values for the equal R, equal C Sallen-Key circuit with gain adjusted to 1 using R_a and R_b for each second-order section.
e. Find the scaled ($k_m = 1000$, $k_f = \Omega_c\omega_p$) component values for the equal R, equal C Sallen-Key circuit with gain adjusted to 1 using R_a and R_b for each second-order section.

17.10 **Design a Butterworth LPF with the following specifications:**

Passband cutoff frequency = f_p = 1250 Hz.

Stopband cutoff frequency = f_s = 2350 Hz.

Maximum attenuation in the passband (dB) = A_p = 1.2 dB.

Minimum attenuation in the stopband (dB) = A_s = 15.5 dB.

`excess='pass'`

a. Find Ω_s and order n of the normalized LPF. Also, find the recalculated values of A_{s_1}, A_{p_1}, ε_{p_1}, Ω_{c_1}.
b. Find the poles S and the zeros SZ of the normalized LPF (do not multiply by Ω_c).
c. Find the transfer function of the normalized LPF, and find Q values of the second-order sections.
d. Find the normalized component values for the equal R, equal C Sallen-Key circuit with gain adjusted to 1 using R_a and R_b for each second-order section.
e. Find the scaled ($k_m = 1000$, $k_f = \Omega_c\omega_p$) component values for the equal R, equal C Sallen-Key circuit with gain adjusted to 1 using R_a and R_b for each second-order section.

17.11 **Design a Butterworth LPF with the following specifications:**

Passband cutoff frequency = f_p = 1150 Hz.

Stopband cutoff frequency = f_s = 2250 Hz.

Maximum attenuation in the passband (dB) = A_p = 1.15 dB.

Minimum attenuation in the stopband (dB) = A_s = 16 dB.

`excess='stop'`

a. Find Ω_s and order n of the normalized LPF. Also, find the recalculated values of A_{s_1}, A_{p_1}, ε_{p_1}, Ω_{c_1}.
b. Find the poles S and the zeros SZ of the normalized LPF (do not multiply by Ω_c).
c. Find the transfer function of the normalized LPF, and find Q values of the second-order sections.
d. Find the normalized component values for the unity gain Sallen-Key circuit.
e. Find the scaled ($k_m = 1000$, $k_f = \Omega_c\omega_p$) component values for the unity gain Sallen-Key circuit.

17.12 **Design a Butterworth LPF with the following specifications:**

Passband cutoff frequency = f_p = 1250 Hz.

Stopband cutoff frequency = f_s = 2450 Hz.

Maximum attenuation in the passband (dB) = A_p = 1.5 dB.

Minimum attenuation in the stopband (dB) = A_s = 14.8 dB.

`excess='pass'`

a. Find Ω_s and order n of the normalized LPF. Also, find the recalculated values of A_{s_1}, A_{p_1}, ε_{p_1}, Ω_{c_1}.
b. Find the poles S and the zeros SZ of the normalized LPF (do not multiply by Ω_c).
c. Find the transfer function of the normalized LPF, and find Q values of the second-order sections.
d. Find the normalized component values for the unity gain Sallen-Key circuit.
e. Find the scaled ($k_m = 1000$, $k_f = \Omega_c\omega_p$) component values for the unity gain Sallen-Key circuit.

17.13 **Design a Butterworth LPF with the following specifications:**

Passband cutoff frequency = f_p = 1250 Hz.

Stopband cutoff frequency = f_s = 2450 Hz.

Maximum attenuation in the passband (dB) = A_p = 1.5 dB.

Minimum attenuation in the stopband (dB) = A_s = 21.8 dB.

`excess='stop'`

a. Find Ω_s and order n of the normalized LPF. Also, find the recalculated values of A_{s_1}, A_{p_1}, ε_{p_1}, Ω_{c_1}.
b. Find the poles S and the zeros SZ of the normalized LPF (do not multiply by Ω_c).
c. Find the transfer function of the normalized LPF, and find Q values of the second-order sections.
d. Find the normalized component values for the equal R, equal C Sallen-Key circuit with gain adjusted to 1 using R_a and R_b for each second-order section.
e. Find the scaled ($k_m = 1000$, $k_f = \Omega_c\omega_p$) component values for the equal R, equal C Sallen-Key circuit with gain adjusted to 1 using R_a and R_b for each second-order section.

17.14 **Design an analog Butterworth LPF with the following specifications:**

Passband cutoff frequency = f_p = 1150 Hz.

Stopband cutoff frequency = f_s = 2150 Hz.

Maximum attenuation in the passband (dB) = A_p = 0.5 dB.

Minimum attenuation in the stopband (dB) = A_s = 25 dB.

`excess='stop'`

a. Find ω_p, Ω_s, order n, recalculated values of $A_{p_1}, A_{s_1}, \varepsilon_{p_1}, \Omega_{c_1}$.
b. Find the poles S and the zeros SZ of the normalized LPF (do not multiply by Ω_c).
c. Find the transfer function of the normalized LPF, and find Q values of the second-order sections.
d. Find the normalized component values for the equal R, equal C Sallen-Key circuit with gain adjusted to 1 using R_a and R_b for each second-order section.
e. Find the scaled ($k_m = 1000, k_f = \Omega_c\omega_p$) component values for the equal R, equal C Sallen-Key circuit with gain adjusted to 1 using R_a and R_b for each second-order section.

Butterworth High-Pass Filter Design

17.15 **Design a Butterworth HPF with the following specifications:**

Passband cutoff frequency = f_p = 2650 Hz.

Stopband cutoff frequency = f_s = 1250 Hz.

Maximum attenuation in the passband (dB) = A_p = 0.81 dB.

Minimum attenuation in the stopband (dB) = A_s = 6 dB.

`excess='pass'`

a. Find Ω_s and order n of the normalized LPF. Also, find the recalculated values of $A_{s_1}, A_{p_1}, \varepsilon_{p_1}, \Omega_{c_1}$.
b. Find the poles S and the zeros SZ of the normalized LPF (do not multiply by Ω_c).
c. Find the transfer function of the normalized LPF, and find Q values of the second-order sections.
d. Find the normalized component values for the equal R, equal C Sallen-Key circuit.
e. Find the scaled ($k_m = 1000, k_f = \omega_p/\Omega_c$) component values for the equal R, equal C Sallen-Key circuit.

17.16 **Design a Butterworth HPF with the following specifications:**

Passband cutoff frequency = f_p = 2750 Hz.

Stopband cutoff frequency = f_s = 1150 Hz.

Maximum attenuation in the passband (dB) = A_p = 0.85 dB.

Minimum attenuation in the stopband (dB) = A_s = 7.5 dB.

`excess='stop'`

a. Find Ω_s and order n of the normalized LPF. Also, find the recalculated values of $A_{s_1}, A_{p_1}, \varepsilon_{p_1}, \Omega_{c_1}$.
b. Find the poles S and the zeros SZ of the normalized LPF (do not multiply by Ω_c).
c. Find the transfer function of the normalized LPF, and find Q values of the second-order sections.
d. Find the normalized component values for the equal R, equal C Sallen-Key circuit.
e. Find the scaled ($k_m = 1000, k_f = \omega_p/\Omega_c$) component values for the equal R, equal C Sallen-Key circuit.

17.17 **Design a Butterworth HPF with the following specifications:**

Passband cutoff frequency = f_p = 2550 Hz.

Stopband cutoff frequency = f_s = 1250 Hz.

Maximum attenuation in the passband (dB) = A_p = 0.83 dB.

Minimum attenuation in the stopband (dB) = A_s = 6.5 dB.

`excess='pass'`

a. Find Ω_s and order n of the normalized LPF. Also, find the recalculated values of $A_{s_1}, A_{p_1}, \varepsilon_{p_1}, \Omega_{c_1}$.
b. Find the poles S and the zeros SZ of the normalized LPF (do not multiply by Ω_c).
c. Find the transfer function of the normalized LPF, and find Q values of the second-order sections.
d. Find the normalized component values for the unity gain Sallen-Key circuit.
e. Find the scaled ($k_m = 1000, k_f = \omega_p/\Omega_c$) component values for the unity gain Sallen-Key circuit.

17.18 **Design a Butterworth HPF with the following specifications:**

Passband cutoff frequency = f_p = 3050 Hz.

Stopband cutoff frequency = f_s = 1150 Hz.

Maximum attenuation in the passband (dB) = A_p = 0.95 dB.

Minimum attenuation in the stopband (dB) = A_s = 8.5 dB.

`excess='stop'`

a. Find Ω_s and order n of the normalized LPF. Also, find the recalculated values of $A_{s_1}, A_{p_1}, \varepsilon_{p_1}, \Omega_{c_1}$.
b. Find the poles S and the zeros SZ of the normalized LPF (do not multiply by Ω_c).
c. Find the transfer function of the normalized LPF, and find Q values of the second-order sections.
d. Find the normalized component values for the unity gain Sallen-Key circuit.
e. Find the scaled ($k_m = 1000$, $k_f = \omega_p/\Omega_c$) component values for the unity gain Sallen-Key circuit.

17.19 Design a Butterworth HPF with the following specifications:

Passband cutoff frequency = f_p = 2550 Hz.

Stopband cutoff frequency = f_s = 1250 Hz.

Maximum attenuation in the passband (dB) = A_p = 0.8 dB.

Minimum attenuation in the stopband (dB) = A_s = 11 dB.

`excess='pass'`

a. Find Ω_s and order n of the normalized LPF. Also, find the recalculated values of $A_{s_1}, A_{p_1}, \varepsilon_{p_1}, \Omega_{c_1}$.
b. Find the poles S and the zeros SZ of the normalized LPF (do not multiply by Ω_c).
c. Find the transfer function of the normalized LPF, and find Q values of the second-order sections.
d. Find the normalized component values for the equal R, equal C Sallen-Key circuit.
e. Find the scaled ($k_m = 1000$, $k_f = \omega_p/\Omega_c$) component values for the equal R, equal C Sallen-Key circuit.

17.20 Design a Butterworth HPF with the following specifications:

Passband cutoff frequency = f_p = 2750 Hz.

Stopband cutoff frequency = f_s = 1150 Hz.

Maximum attenuation in the passband (dB) = A_p = 0.8 dB.

Minimum attenuation in the stopband (dB) = A_s = 10.5 dB.

`excess='stop'`

a. Find Ω_s and order n of the normalized LPF. Also, find the recalculated values of $A_{s_1}, A_{p_1}, \varepsilon_{p_1}, \Omega_{c_1}$.
b. Find the poles S and the zeros SZ of the normalized LPF (do not multiply by Ω_c).
c. Find the transfer function of the normalized LPF, and find Q values of the second-order sections.
d. Find the normalized component values for the equal R, equal C Sallen-Key circuit.
e. Find the scaled ($k_m = 1000$, $k_f = \omega_p/\Omega_c$) component values for the equal R, equal C Sallen-Key circuit.

17.21 Design a Butterworth HPF with the following specifications:

Passband cutoff frequency = f_p = 2650 Hz.

Stopband cutoff frequency = f_s = 1250 Hz.

Maximum attenuation in the passband (dB) = A_p = 0.8 dB.

Minimum attenuation in the stopband (dB) = A_s = 10 dB.

`excess='pass'`

a. Find Ω_s and order n of the normalized LPF. Also, find the recalculated values of $A_{s_1}, A_{p_1}, \varepsilon_{p_1}, \Omega_{c_1}$.
b. Find the poles S and the zeros SZ of the normalized LPF (do not multiply by Ω_c).
c. Find the transfer function of the normalized LPF, and find Q values of the second-order sections.
d. Find the normalized component values for the unity gain Sallen-Key circuit.
e. Find the scaled ($k_m = 1000$, $k_f = \omega_p/\Omega_c$) component values for the unity gain Sallen-Key circuit.

17.22 Design a Butterworth HPF with the following specifications:

Passband cutoff frequency = f_p = 2950 Hz.

Stopband cutoff frequency = f_s = 1150 Hz.

Maximum attenuation in the passband (dB) = A_p = 0.85 dB.

Minimum attenuation in the stopband (dB) = A_s = 12.5 dB.

`excess='stop'`

a. Find Ω_s and order n of the normalized LPF. Also, find the recalculated values of $A_{s_1}, A_{p_1}, \varepsilon_{p_1}, \Omega_{c_1}$.
b. Find the poles S and the zeros SZ of the normalized LPF (do not multiply by Ω_c).

c. Find the transfer function of the normalized LPF, and find Q values of the second-order sections.
d. Find the normalized component values for the unity gain Sallen-Key circuit.
e. Find the scaled ($k_m = 1000, k_f = \omega_p/\Omega_c$) component values for the unity gain Sallen-Key circuit.

17.23 **Design a Butterworth HPF with the following specifications:**

Passband cutoff frequency $= f_p = 2550$ Hz.

Stopband cutoff frequency $= f_s = 1250$ Hz.

Maximum attenuation in the passband (dB) $= A_p = 1.05$ dB.

Minimum attenuation in the stopband (dB) $= A_s = 14$ dB.

`excess='pass'`

a. Find Ω_s and order n of the normalized LPF. Also, find the recalculated values of $A_{s_1}, A_{p_1}, \varepsilon_{p_1}, \Omega_{c_1}$.
b. Find the poles S and the zeros SZ of the normalized LPF (do not multiply by Ω_c).
c. Find the transfer function of the normalized LPF, and find Q values of the second-order sections.
d. Find the normalized component values for the equal R, equal C Sallen-Key circuit.
e. Find the scaled ($k_m = 1000, k_f = \omega_p/\Omega_c$) component values for the equal R, equal C Sallen-Key circuit.

17.24 **Design a Butterworth HPF with the following specifications:**

Passband cutoff frequency $= f_p = 2650$ Hz.

Stopband cutoff frequency $= f_s = 1250$ Hz.

Maximum attenuation in the passband (dB) $= A_p = 1.1$ dB.

Minimum attenuation in the stopband (dB) $= A_s = 15$ dB.

`excess='pass'`

a. Find Ω_s and order n of the normalized LPF. Also, find the recalculated values of $A_{s_1}, A_{p_1}, \varepsilon_{p_1}, \Omega_{c_1}$.
b. Find the poles S and the zeros SZ of the normalized LPF (do not multiply by Ω_c).
c. Find the transfer function of the normalized LPF, and find Q values of the second-order sections.
d. Find the normalized component values for the unity gain Sallen-Key circuit.
e. Find the scaled ($k_m = 1000, k_f = \omega_p/\Omega_c$) component values for the unity gain Sallen-Key circuit.

17.25 **Design a Butterworth HPF with the following specifications:**

Passband cutoff frequency $= f_p = 2850$ Hz.

Stopband cutoff frequency $= f_s = 1450$ Hz.

Maximum attenuation in the passband (dB) $= A_p = 1.2$ dB.

Minimum attenuation in the stopband (dB) $= A_s = 16.5$ dB.

`excess='stop'`

a. Find Ω_s and order n of the normalized LPF. Also, find the recalculated values of $A_{s_1}, A_{p_1}, \varepsilon_{p_1}, \Omega_{c_1}$.
b. Find the poles S and the zeros SZ of the normalized LPF (do not multiply by Ω_c).
c. Find the transfer function of the normalized LPF, and find Q values of the second-order sections.
d. Find the normalized component values for the unity gain Sallen-Key circuit.
e. Find the scaled ($k_m = 1000, k_f = \omega_p/\Omega_c$) component values for the unity gain Sallen-Key circuit.

17.26 **Design a Butterworth HPF with the following specifications:**

Passband cutoff frequency $= f_p = 2750$ Hz.

Stopband cutoff frequency $= f_s = 1350$ Hz.

Maximum attenuation in the passband (dB) $= A_p = 1.15$ dB.

Minimum attenuation in the stopband (dB) $= A_s = 15.5$ dB.

`excess='stop'`

a. Find Ω_s and order n of the normalized LPF. Also, find the recalculated values of $A_{s_1}, A_{p_1}, \varepsilon_{p_1}, \Omega_{c_1}$.
b. Find the poles S and the zeros SZ of the normalized LPF (do not multiply by Ω_c).
c. Find the transfer function of the normalized LPF, and find Q values of the second-order sections.
d. Find the normalized component values for the equal R, equal C Sallen-Key circuit.
e. Find the scaled ($k_m = 1000, k_f = \omega_p/\Omega_c$) component values for the equal R, equal C Sallen-Key circuit.

17.27 Design a Butterworth HPF with the following specifications:

Passband cutoff frequency = f_p = 2850 Hz.

Stopband cutoff frequency = f_s = 1550 Hz.

Maximum attenuation in the passband (dB) = A_p = 1.2 dB.

Minimum attenuation in the stopband (dB) = A_s = 18.5 dB.

`excess='stop'`

a. Find Ω_s and order n of the normalized LPF. Also, find the recalculated values of $A_{s_1}, A_{p_1}, \varepsilon_{p_1}, \Omega_{c_1}$.
b. Find the poles S and the zeros SZ of the normalized LPF (do not multiply by Ω_c).
c. Find the transfer function of the normalized LPF, and find Q values of the second-order sections.
d. Find the normalized component values for the equal R, equal C Sallen-Key circuit.
e. Find the scaled ($k_m = 1000, k_f = \omega_p/\Omega_c$) component values for the equal R, equal C Sallen-Key circuit.

17.28 Design an analog Butterworth HPF with the following specifications:

Passband cutoff frequency = f_p = 2850 Hz.

Stopband cutoff frequency = f_s = 1650 Hz.

Maximum attenuation in the passband (dB) = A_p = 1.2 dB.

Minimum attenuation in the stopband (dB) = A_s = 20.5 dB.

`excess='stop'`

a. Find ω_p, Ω_s, order n, recalculated values of $A_{p_1}, A_{s_1}, \varepsilon_{p_1}, \Omega_{c_1}$.
b. Find the poles S and the zeros SZ of the normalized LPF (do not multiply by Ω_c).
c. Find the transfer function of the normalized LPF, and find Q values of the second-order sections.
d. Find the normalized component values for the equal R, equal C Sallen-Key circuit.
e. Find the scaled ($k_m = 1000, k_f = \omega_p/\Omega_c$) component values for the equal R, equal C Sallen-Key circuit.

Butterworth Bandpass Filter Design

17.29 Design a Butterworth bandpass filter with the following specifications:

Lower stopband cutoff frequency = f_1 = 1150 Hz.

Lower passband cutoff frequency = f_2 = 2250 Hz.

Upper passband cutoff frequency = f_3 = 3260 Hz.

Upper stopband cutoff frequency = f_4 = 4150 Hz.

Maximum attenuation in the passband (dB) = A_p = 0.9 dB.

Minimum attenuation in the stopband (dB) = A_s = 3 dB.

`excess='pass'`

k_m = 1000

a. Find Ω_s and order n of the normalized LPF. Also, find the recalculated values of $A_{s_1}, A_{p_1}, \varepsilon_{p_1}, \Omega_{c_1}$.
b. Find the poles S and the zeros SZ of the normalized LPF before multiplication by Ω_c.
c. Find the transfer function of the frequency-transformed bandpass filter, and find Q values and ω_0 values of the second-order sections.
d. Find the normalized component values for the Sallen-Key circuit for each second-order section.
e. Find the scaled component values for the Sallen-Key circuit for each second-order section.

17.30 Design a Butterworth bandpass filter with the following specifications:

Lower stopband cutoff frequency = f_1 = 1150 Hz.

Lower passband cutoff frequency = f_2 = 2350 Hz.

Upper passband cutoff frequency = f_3 = 3160 Hz.

Upper stopband cutoff frequency = f_4 = 4150 Hz.

Maximum attenuation in the passband (dB) = A_p = 0.9 dB.

Minimum attenuation in the stopband (dB) = A_s = 2.8 dB.

`excess='pass'`

k_m = 1000

a. Find Ω_s and order n of the normalized LPF. Also, find the recalculated values of $A_{s_1}, A_{p_1}, \varepsilon_{p_1}, \Omega_{c_1}$.
b. Find the poles S and the zeros SZ of the normalized LPF before multiplication by Ω_c.
c. Find the transfer function of the frequency-transformed bandpass filter, and find Q values and ω_0 values of the second-order sections.
d. Find the normalized component values for the Deliyannis-Friend circuit.
e. Find the scaled component values for the Deliyannis-Friend circuit.

17.31 **Design a Butterworth bandpass filter with the following specifications:**

Lower stopband cutoff frequency = f_1 = 1150 Hz.

Lower passband cutoff frequency = f_2 = 2350 Hz.

Upper passband cutoff frequency = f_3 = 3160 Hz.

Upper stopband cutoff frequency = f_4 = 4150 Hz.

Maximum attenuation in the passband (dB) = A_p = 0.9 dB.

Minimum attenuation in the stopband (dB) = A_s = 2.7 dB.

`excess='stop'`

k_m = `1000`

a. Find Ω_s and order n of the normalized LPF. Also, find the recalculated values of $A_{p_1}, A_{s_1}, \varepsilon_{p_1}, \Omega_{c_1}$.
b. Find the poles S and the zeros SZ of the normalized LPF before multiplication by Ω_c.
c. Find the transfer function of the frequency-transformed bandpass filter, and find Q values and ω_0 values of the second-order sections.
d. Find the normalized component values for the Deliyannis-Friend circuit.
e. Find the scaled component values for the Deliyannis-Friend circuit.

17.32 **Design a Butterworth bandpass filter with the following specifications:**

Lower stopband cutoff frequency = f_1 = 1150 Hz.

Lower passband cutoff frequency = f_2 = 2450 Hz.

Upper passband cutoff frequency = f_3 = 3060 Hz.

Upper stopband cutoff frequency = f_4 = 4150 Hz.

Maximum attenuation in the passband (dB) = A_p = 0.9 dB.

Minimum attenuation in the stopband (dB) = A_s = 2.6 dB.

`excess='stop'`

a. Find Ω_s and order n of the normalized LPF. Also, find the recalculated values of $A_{p_1}, A_{s_1}, \varepsilon_{p_1}, \Omega_{c_1}$.
b. Find the poles S and the zeros SZ of the normalized LPF before multiplication by Ω_c.
c. Find the transfer function of the frequency-transformed bandpass filter, and find Q values and ω_0 values of the second-order sections.
d. Find the normalized component values for the Sallen-Key circuit for each second-order section.
e. Find the scaled component values for the Sallen-Key circuit for each second-order section.

17.33 **Design a Butterworth bandpass filter with the following specifications:**

Lower stopband cutoff frequency = f_1 = 1150 Hz.

Lower passband cutoff frequency = f_2 = 2150 Hz.

Upper passband cutoff frequency = f_3 = 3360 Hz.

Upper stopband cutoff frequency = f_4 = 4150 Hz.

Maximum attenuation in the passband (dB) = A_p = 0.95 dB.

Minimum attenuation in the stopband (dB) = A_s = 5 dB.

`excess='pass'`

k_m = `1000`

a. Find Ω_s and order n of the normalized LPF. Also, find the recalculated values of $A_{p_1}, A_{s_1}, \varepsilon_{p_1}, \Omega_{c_1}$.
b. Find the poles S and the zeros SZ of the normalized LPF before multiplication by Ω_c.
c. Find the transfer function of the frequency-transformed bandpass filter, and find Q values and ω_0 values of the second-order sections.
d. Find the normalized component values for the Sallen-Key circuit for each second-order section.
e. Find the scaled component values for the Sallen-Key circuit for each second-order section.

17.34 **Design a Butterworth bandpass filter with the following specifications:**

Lower stopband cutoff frequency = f_1 = 1150 Hz.

Lower passband cutoff frequency = f_2 = 2250 Hz.

Upper passband cutoff frequency = f_3 = 3260 Hz.

Upper stopband cutoff frequency = f_4 = 4150 Hz.

Maximum attenuation in the passband (dB) = A_p = 0.96 dB.

Minimum attenuation in the stopband (dB) = A_s = 5.2 dB.

`excess='pass'`

k_m = `1000`

a. Find Ω_s and order n of the normalized LPF. Also, find the recalculated values of $A_{p_1}, A_{s_1}, \varepsilon_{p_1}, \Omega_{c_1}$.
b. Find the poles S and the zeros SZ of the normalized LPF before multiplication by Ω_c.
c. Find the transfer function of the frequency-transformed bandpass filter, and find Q values and ω_0 values of the second-order sections.
d. Find the normalized component values for the Deliyannis-Friend circuit.

e. Find the scaled component values for the Deliyannis-Friend circuit.

17.35 Design a Butterworth bandpass filter with the following specifications:

Lower stopband cutoff frequency = f_1 = 1250 Hz.

Lower passband cutoff frequency = f_2 = 2250 Hz.

Upper passband cutoff frequency = f_3 = 3360 Hz.

Upper stopband cutoff frequency = f_4 = 4150 Hz.

Maximum attenuation in the passband (dB) = A_p = 1.1 dB.

Minimum attenuation in the stopband (dB) = A_s = 6 dB.

`excess='stop'`

k_m = 1000

a. Find Ω_s and order n of the normalized LPF. Also, find the recalculated values of A_{p_1}, A_{s_1}, ε_{p_1}, Ω_{c_1}.
b. Find the poles S and the zeros SZ of the normalized LPF before multiplication by Ω_c.
c. Find the transfer function of the frequency-transformed bandpass filter, and find Q values and ω_0 values of the second-order sections.
d. Find the normalized component values for the Deliyannis-Friend circuit.
e. Find the scaled component values for the Deliyannis-Friend circuit.

17.36 Design a Butterworth bandpass filter with the following specifications:

Lower stopband cutoff frequency = f_1 = 1250 Hz.

Lower passband cutoff frequency = f_2 = 2350 Hz.

Upper passband cutoff frequency = f_3 = 3260 Hz.

Upper stopband cutoff frequency = f_4 = 4150 Hz.

Maximum attenuation in the passband (dB) = A_p = 1.05 dB.

Minimum attenuation in the stopband (dB) = A_s = 5.6 dB.

`excess='stop'`

k_m = 1000

a. Find Ω_s and order n of the normalized LPF. Also, find the recalculated values of A_{p_1}, A_{s_1}, ε_{p_1}, Ω_{c_1}.
b. Find the poles S and the zeros SZ of the normalized LPF before multiplication by Ω_c.
c. Find the transfer function of the frequency-transformed bandpass filter, and find Q values and ω_0 values of the second-order sections.
d. Find the normalized component values for the Sallen-Key circuit for each second-order section.
e. Find the scaled component values for the Sallen-Key circuit for each second-order section.

17.37 Design a Butterworth bandpass filter with the following specifications:

Lower stopband cutoff frequency = f_1 = 1250 Hz.

Lower passband cutoff frequency = f_2 = 2450 Hz.

Upper passband cutoff frequency = f_3 = 3260 Hz.

Upper stopband cutoff frequency = f_4 = 4150 Hz.

Maximum attenuation in the passband (dB) = A_p = 1.15 dB.

Minimum attenuation in the stopband (dB) = A_s = 15 dB.

`excess='stop'`

k_m = 1000

a. Find Ω_s and order n of the normalized LPF. Also, find the recalculated values of A_{p_1}, A_{s_1}, ε_{p_1}, Ω_{c_1}.
b. Find the poles S and the zeros SZ of the normalized LPF before multiplication by Ω_c.
c. Find the transfer function of the frequency-transformed bandpass filter, and find Q values and ω_0 values of the second-order sections.
d. Find the normalized component values for the Sallen-Key circuit for each second-order section.
e. Find the scaled component values for the Sallen-Key circuit for each second-order section.

17.38 Design an analog Butterworth bandpass filter with the following specifications:

Lower stopband cutoff frequency = f_1 = 1250 Hz.

Lower passband cutoff frequency = f_2 = 2350 Hz.

Upper passband cutoff frequency = f_3 = 3360 Hz.

Upper stopband cutoff frequency = f_4 = 4150 Hz.

Maximum attenuation in the passband (dB) = A_p = 1.1 dB.

Minimum attenuation in the stopband (dB) = A_s = 21 dB.

`excess='stop'`

k_m = 1000

a. Find Ω_s and order n of the normalized LPF. Also, find the recalculated values of A_{p_1}, A_{s_1}, ε_{p_1}, Ω_{c_1}.

b. Find the poles S and the zeros SZ of the normalized LPF before multiplication by Ω_c.
c. Find the transfer function of the frequency-transformed bandpass filter, and find Q values and ω_0 values of the second-order sections.
d. Find the normalized component values for the Deliyannis-Friend circuit.
e. Find the scaled component values for the Deliyannis-Friend circuit.

Butterworth Bandstop Filter Design

17.39 Design a Butterworth bandstop filter with the following specifications:

Lower passband cutoff frequency = f_1 = 1150 Hz.

Lower stopband cutoff frequency = f_2 = 2350 Hz.

Upper stopband cutoff frequency = f_3 = 3050 Hz.

Upper passband cutoff frequency = f_4 = 4150 Hz.

Maximum attenuation in the passband (dB) = A_p = 0.9 dB.

Minimum attenuation in the stopband (dB) = A_s = 2.7 dB.

`excess='pass'`

k_m = `1000`

a. Find Ω_s and order n of the normalized LPF. Also, find the recalculated values of A_{p_1}, A_{s_1}, ε_{p_1}, Ω_{c_1}.
b. Find the poles S and the zeros SZ of the normalized LPF before multiplication by Ω_c.
c. Find the transfer function of the frequency-transformed bandstop filter, and find Q values and ω_0 values of the second-order sections.
d. Find the normalized component values for the Sallen-Key circuit for each second-order section.
e. Find the scaled component values for the Sallen-Key circuit for each second-order section.

17.40 Design a Butterworth bandstop filter with the following specifications:

Lower passband cutoff frequency = f_1 = 1150 Hz.

Lower stopband cutoff frequency = f_2 = 2250 Hz.

Upper stopband cutoff frequency = f_3 = 3150 Hz.

Upper passband cutoff frequency = f_4 = 4150 Hz.

Maximum attenuation in the passband (dB) = A_p = 0.85 dB.

Minimum attenuation in the stopband (dB) = A_s = 2.2 dB.

`excess='pass'`

k_m = `1000`

a. Find Ω_s and order n of the normalized LPF. Also, find the recalculated values of A_{p_1}, A_{s_1}, ε_{p_1}, Ω_{c_1}.
b. Find the poles S and the zeros SZ of the normalized LPF before multiplication by Ω_c.
c. Find the transfer function of the frequency-transformed bandstop filter, and find Q values and ω_0 values of the second-order sections.
d. Find the normalized component values for the Sallen-Key circuit for each second-order section.
e. Find the scaled component values for the Sallen-Key circuit for each second-order section.

17.41 Design a Butterworth bandstop filter with the following specifications:

Lower passband cutoff frequency = f_1 = 1150 Hz.

Lower stopband cutoff frequency = f_2 = 2350 Hz.

Upper stopband cutoff frequency = f_3 = 3050 Hz.

Upper passband cutoff frequency = f_4 = 4150 Hz.

Maximum attenuation in the passband (dB) = A_p = 0.75 dB.

Minimum attenuation in the stopband (dB) = A_s = 2.1 dB.

`excess='stop'`

a. Find Ω_s and order n of the normalized LPF. Also, find the recalculated values of A_{p_1}, A_{s_1}, ε_{p_1}, Ω_{c_1}.
b. Find the poles S and the zeros SZ of the normalized LPF before multiplication by Ω_c.
c. Find the transfer function of the frequency-transformed bandstop filter, and find Q values and ω_0 values of the second-order sections.
d. Find the normalized component values for the Sallen-Key circuit for each second-order section.
e. Find the scaled component values for the Sallen-Key circuit for each second-order section.

17.42 **Design a Butterworth bandstop filter with the following specifications:**

Lower passband cutoff frequency = f_1 = 1050 Hz.

Lower stopband cutoff frequency = f_2 = 2250 Hz.

Upper stopband cutoff frequency = f_3 = 3150 Hz.

Upper passband cutoff frequency = f_4 = 4150 Hz.

Maximum attenuation in the passband (dB) = A_p = 0.8 dB.

Minimum attenuation in the stopband (dB) = A_s = 2.0 dB.

`excess='stop'`

k_m = 1000

a. Find Ω_s and order n of the normalized LPF. Also, find the recalculated values of A_{p_1}, A_{s_1}, ε_{p_1}, Ω_{c_1}.
b. Find the poles S and the zeros SZ of the normalized LPF before multiplication by Ω_c.
c. Find the transfer function of the frequency-transformed bandstop filter, and find Q values and ω_0 values of the second-order sections.
d. Find the normalized component values for the Sallen-Key circuit for each second-order section.
e. Find the scaled component values for the Sallen-Key circuit for each second-order section.

17.43 **Design a Butterworth bandstop filter with the following specifications:**

Lower passband cutoff frequency = f_1 = 1150 Hz.

Lower stopband cutoff frequency = f_2 = 2150 Hz.

Upper stopband cutoff frequency = f_3 = 3250 Hz.

Upper passband cutoff frequency = f_4 = 4150 Hz.

Maximum attenuation in the passband (dB) = A_p = 0.83 dB.

Minimum attenuation in the stopband (dB) = A_s = 3.1 dB.

`excess='stop'`

k_m = 1000

a. Find Ω_s and order n of the normalized LPF. Also, find the recalculated values of A_{p_1}, A_{s_1}, ε_{p_1}, Ω_{c_1}.
b. Find the poles S and the zeros SZ of the normalized LPF before multiplication by Ω_c.
c. Find the transfer function of the frequency-transformed bandstop filter, and find Q values and ω_0 values of the second-order sections.
d. Find the normalized component values for the Sallen-Key circuit for each second-order section.
e. Find the scaled component values for the Sallen-Key circuit for each second-order section.

17.44 **Design a Butterworth bandstop filter with the following specifications:**

Lower passband cutoff frequency = f_1 = 1050 Hz.

Lower stopband cutoff frequency = f_2 = 2150 Hz.

Upper stopband cutoff frequency = f_3 = 3150 Hz.

Upper passband cutoff frequency = f_4 = 4150 Hz.

Maximum attenuation in the passband (dB) = A_p = 0.85 dB.

Minimum attenuation in the stopband (dB) = A_s = 3.3 dB.

`excess='stop'`

k_m = 1000

a. Find Ω_s and order n of the normalized LPF. Also, find the recalculated values of A_{p_1}, A_{s_1}, ε_{p_1}, Ω_{c_1}.
b. Find the poles S and the zeros SZ of the normalized LPF before multiplication by Ω_c.
c. Find the transfer function of the frequency-transformed bandstop filter, and find Q values and ω_0 values of the second-order sections.
d. Find the normalized component values for the Sallen-Key circuit for each second-order section.
e. Find the scaled component values for the Sallen-Key circuit for each second-order section.

17.45 **Design a Butterworth bandstop filter with the following specifications:**

Lower passband cutoff frequency = f_1 = 1150 Hz.

Lower stopband cutoff frequency = f_2 = 2250 Hz.

Upper stopband cutoff frequency = f_3 = 3050 Hz.

Upper passband cutoff frequency = f_4 = 4150 Hz.

Maximum attenuation in the passband (dB) = A_p = 0.87 dB.

Minimum attenuation in the stopband (dB) = A_s = 3.5 dB.

`excess='pass'`

k_m = 1000

a. Find Ω_s and order n of the normalized LPF. Also, find the recalculated values of A_{p_1}, A_{s_1}, ε_{p_1}, Ω_{c_1}.

b. Find the poles S and the zeros SZ of the normalized LPF before multiplication by Ω_c.
c. Find the transfer function of the frequency-transformed bandstop filter, and find Q values and ω_0 values of the second-order sections.
d. Find the normalized component values for the Sallen-Key circuit for each second-order section.
e. Find the scaled component values for the Sallen-Key circuit for each second-order section.

17.46 **Design a Butterworth bandstop filter with the following specifications:**

Lower passband cutoff frequency = f_1 = 1050 Hz.

Lower stopband cutoff frequency = f_2 = 2350 Hz.

Upper stopband cutoff frequency = f_3 = 3150 Hz.

Upper passband cutoff frequency = f_4 = 4150 Hz.

Maximum attenuation in the passband (dB) = A_p = 0.92 dB.

Minimum attenuation in the stopband (dB) = A_s = 3.8 dB.

`excess='pass'`

k_m = 1000

a. Find Ω_s and order n of the normalized LPF. Also, find the recalculated values of A_{p_1}, A_{s_1}, ε_{p_1}, Ω_{c_1}.
b. Find the poles S and the zeros SZ of the normalized LPF before multiplication by Ω_c.
c. Find the transfer function of the frequency-transformed bandstop filter, and find Q values and ω_0 values of the second-order sections.
d. Find the normalized component values for the Sallen-Key circuit for each second-order section.
e. Find the scaled component values for the Sallen-Key circuit for each second-order section.

17.47 **Design a Butterworth bandstop filter with the following specifications:**

Lower passband cutoff frequency = f_1 = 1150 Hz.

Lower stopband cutoff frequency = f_2 = 2150 Hz.

Upper stopband cutoff frequency = f_3 = 3250 Hz.

Upper passband cutoff frequency = f_4 = 4150 Hz.

Maximum attenuation in the passband (dB) = A_p = 0.83 dB.

Minimum attenuation in the stopband (dB) = A_s = 5.1 dB.

`excess='stop'`

k_m = 1000

a. Find Ω_s and order n of the normalized LPF. Also, find the recalculated values of A_{p_1}, A_{s_1}, ε_{p_1}, Ω_{c_1}.
b. Find the poles S and the zeros SZ of the normalized LPF before multiplication by Ω_c.
c. Find the transfer function of the frequency-transformed bandstop filter, and find Q values and ω_0 values of the second-order sections.
d. Find the normalized component values for the Sallen-Key circuit for each second-order section.
e. Find the scaled component values for the Sallen-Key circuit for each second-order section.

17.48 **Design an analog Butterworth bandstop filter with the following specifications:**

Lower passband cutoff frequency = f_1 = 1250 Hz.

Lower stopband cutoff frequency = f_2 = 2250 Hz.

Upper stopband cutoff frequency = f_3 = 3150 Hz.

Upper passband cutoff frequency = f_4 = 4150 Hz.

Maximum attenuation in the passband (dB) = A_p = 1.05 dB.

Minimum attenuation in the stopband (dB) = A_s = 15 dB.

`excess='stop'`

k_m = 1000

a. Find Ω_s and order n of the normalized LPF. Also, find the recalculated values of A_{p_1}, A_{s_1}, ε_{p_1}, Ω_{c_1}.
b. Find the poles S and the zeros SZ of the normalized LPF before multiplication by Ω_c.
c. Find the transfer function of the frequency-transformed bandstop filter, and find Q values and ω_0 values of the second-order sections.
d. Find the normalized component values for the Sallen-Key circuit for each second-order section.
e. Find the scaled component values for the Sallen-Key circuit for each second-order section.

Chebyshev Type 1 LPF Design

17.49 **Design a Chebyshev type 1 LPF with the following specifications:**

Passband cutoff frequency = f_p = 1250 Hz.

Stopband cutoff frequency = f_s = 2550 Hz.

Maximum attenuation in the passband (dB) = A_p = 1.15 dB.

Minimum attenuation in the stopband (dB) = A_s =9.3 dB.

`excess='stop'`

a. Find Ω_s and order n of the normalized LPF. Also, find the recalculated values of A_{s_1}, A_{p_1}.
b. Find the poles S and the zeros SZ of the normalized LPF.
c. Find the transfer function of the LPF, and find Q values of the second-order sections.
d. Find the normalized component values for the equal R, equal C Sallen-Key circuit with gain adjusted to 1 using R_a and R_b for each second-order section.
e. Find the scaled (k_m = 1000, $k_f = \omega_0$) component values for the equal R, equal C Sallen-Key circuit with gain adjusted to 1 using R_a and R_b for each second-order section.

17.50 **Design a Chebyshev type 1 LPF with the following specifications:**

Passband cutoff frequency = f_p = 1150 Hz.

Stopband cutoff frequency = f_s = 2350 Hz.

Maximum attenuation in the passband (dB) = A_p = 1.25 dB.

Minimum attenuation in the stopband (dB) = A_s =15.3 dB.

`excess='stop'`

a. Find Ω_s and order n of the normalized LPF. Also, find the recalculated values of A_{s_1}, A_{p_1}.
b. Find the poles S and the zeros SZ of the normalized LPF.
c. Find the transfer function of the LPF, and find Q values of the second-order sections.
d. Find the normalized component values for the equal R, equal C Sallen-Key circuit with gain adjusted to 1 using R_a and R_b for each second-order section.
e. Find the scaled (k_m = 1000, $k_f = \omega_0$) component values for the equal R, equal C Sallen-Key circuit with gain adjusted to 1 using R_a and R_b for each second-order section.

Chebyshev Type 2 LPF Design

17.51 **Design a Chebyshev type 2 LPF with the following specifications:**

Passband cutoff frequency = f_p = 1250 Hz.

Stopband cutoff frequency = f_s = 2650 Hz.

Maximum attenuation in the passband (dB) = A_p = 1.25 dB.

Minimum attenuation in the stopband (dB) = A_s = 8.7 dB.

`excess='stop'`

a. Find Ω_s and order n of the normalized LPF. Also, find the recalculated values of A_{s_1}, $A_{p_1}, \varepsilon_{p_1}$.
b. Find the poles S and the zeros SZ of the normalized LPF.
c. Find the transfer function of the LPF, and find Q values of the second-order sections.
d. Find the normalized component values for the Sallen-Key circuit for the bandstop filter.
e. Find the scaled (k_m = 1000, $k_f = \omega_0$) component values for the Sallen-Key circuit for the bandstop filter.

17.52 **Design a Chebyshev type 2 LPF with the following specifications:**

Passband cutoff frequency = f_p = 1450 Hz.

Stopband cutoff frequency = f_s = 2250 Hz.

Maximum attenuation in the passband (dB) = A_p = 1.2 dB.

Minimum attenuation in the stopband (dB) = A_s =20.5 dB.

`excess='stop'`

a. Find Ω_s and order n of the normalized LPF. Also, find the recalculated values of $A_{s_1}, A_{p_1}, \varepsilon_{p_1}$.
b. Find the poles S and the zeros SZ of the normalized LPF.
c. Find the transfer function of the LPF, and find Q values of the second-order sections.
d. Find the normalized component values for the Sallen-Key circuit for the bandstop filter.
e. Find the scaled (k_m = 1000, $k_f = \omega_0$) component values for the Sallen-Key circuit for the bandstop filter.

Chapter 18

Fourier Series

18.1 Introduction

If a sinusoidal signal is applied to a circuit, we can find the steady-state response of the circuit using the phasor transformed circuit, as shown in Chapter 10. This process is a lot simpler than solving differential equations to get the steady-state response. This simpler method can be applied to the nonsinusoidal input signal if the input signal can be expressed as a sum of sines and cosines. Fourier series representation makes it possible to represent a signal as a sum of sines and cosines.

The mathematical foundation of a Fourier series is the representation of signals as a linear combination of a set of orthogonal functions. In Fourier series representation, the set of orthogonal functions consists of sines and cosines. There are many other sets of orthogonal functions. Some of these other sets of orthogonal functions may be useful in applications such as data compression, noise elimination, and pattern recognition.

The Fourier series provides analysis of systems in the frequency domain. The frequency domain analysis of signals and systems makes it possible to understand and design various devices and systems such as communication systems.

18.2 Signal Representation Using Orthogonal Functions

18.2.1 ORTHOGONAL FUNCTIONS

Two lines are orthogonal if they are perpendicular at the point of itersection. Two vectors x_1 and x_2 are orthogonal if the inner product (dot product) is zero; that is, $x_1^H x_2 = 0$, where H denotes conjugate transpose. For example, if $x_1 = [1\ 1]^T$ and $x_2 = [1\ -1]^T$, the inner product is zero. Thus, x_1 and x_2 are orthogonal.

The **inner product** of two functions (signals) $\phi_0(t)$ and $\phi_1(t)$, denoted by $\langle \phi_0, \phi_1 \rangle$, over the interval $t_0 < t < t_1$ (the interval is denoted by (t_0, t_1)) is defined as

$$\langle \phi_0, \phi_1 \rangle = \int_{t_0}^{t_1} \phi_0(t)\phi_1^*(t)dt \quad \textbf{(18.1)}$$

where $\phi_1^*(t)$ is the complex conjugate of $\phi_1(t)$. Note that if $\phi_1(t)$ is real, $\phi_1^*(t) = \phi_1(t)$.

Two functions $\phi_0(t)$ and $\phi_1(t)$ are **orthogonal** over the interval (t_0, t_1) if the inner product is zero; that is,

$$\langle \phi_0, \phi_1 \rangle = \int_{t_0}^{t_1} \phi_0(t)\phi_1^*(t)dt = \int_{t_0}^{t_1} \phi_0^*(t)\phi_1(t)dt = 0 \quad \textbf{(18.2)}$$

The inner product of $\phi_0(t)$ and $\phi_0(t)$ is the energy μ_0 of $\phi_0(t)$; that is,

$$\mu_0 = \langle \phi_0, \phi_0 \rangle = \int_{t_0}^{t_1} \phi_0(t)\phi_0^*(t)dt = \int_{t_0}^{t_1} |\phi_0(t)|^2 dt \quad \textbf{(18.3)}$$

Similarly, the energy of $\phi_1(t)$ is given by

$$\mu_1 = \langle \phi_1, \phi_1 \rangle = \int_{t_0}^{t_1} \phi_1(t)\phi_1^*(t)dt = \int_{t_0}^{t_1} |\phi_1(t)|^2 dt \quad \textbf{(18.4)}$$

If two signals $\phi_0(t)$ and $\phi_0(t)$ are orthogonal and the energy of each signal is 1 ($\mu_0 = 1$ and $\mu_1 = 1$), they are called **orthonormal**, meaning orthogonal and normal.

EXAMPLE 18.1

Two signals

$$\phi_0(t) = rect\left(\frac{t - \frac{1}{2}}{1}\right) = \begin{cases} 1, & 0 \le t < 1 \\ 0, & elsewhere \end{cases}$$

and

$$\phi_1(t) = rect\left(\frac{t - \frac{1}{4}}{\frac{1}{2}}\right) - rect\left(\frac{t - \frac{3}{4}}{\frac{1}{2}}\right) = \begin{cases} 1, & 0 \le t < \frac{1}{2} \\ -1, & \frac{1}{2} \le t < 1 \\ 0, & elsewhere \end{cases}$$

are shown in Figure 18.1. Find the inner product of $\phi_0(t)$ and $\phi_1(t)$ over the interval $(0, 1)$, and state whether they are orthogonal or not. Also find the energy of $\phi_0(t)$ and $\phi_1(t)$.

continued

Example 18.1 continued

FIGURE 18.1

Two signals $\phi_0(t)$ and $\phi_1(t)$.

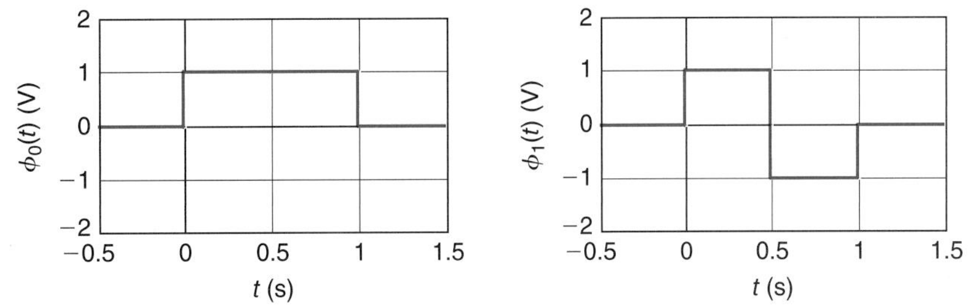

The product of $\phi_0(t)$ and $\phi_1(t)$ is $\phi_1(t)$.

$$\int_{t_0}^{t_1} \phi_0(t)\phi_1^*(t)dt = \int_{t_0}^{t_1} \phi_0(t)\phi_1(t)dt = \int_0^{0.5} 1 \times 1\,dt + \int_{0.5}^{1} 1 \times (-1)\,dt = \frac{1}{2} - \frac{1}{2} = 0$$

Therefore, $\phi_0(t)$ and $\phi_1(t)$ are orthogonal over the interval $(0, 1)$. The energy of $\phi_0(t)$ is

$$\mu_0 = \int_0^1 |\phi_0(t)|^2 dt = \int_0^1 1^2 dt = 1$$

The energy of $\phi_1(t)$ is given by

$$\mu_1 = \int_0^1 |\phi_1(t)|^2 dt = \int_0^1 1^2 dt = 1$$

Thus, $\phi_0(t)$ and $\phi_1(t)$ are orthonormal over the interval $(0, 1)$.

Exercise 18.1

Let

$$\phi_0(t) = \begin{cases} 1, & 0 \le t < 0.5 \\ -1, & 0.5 \le t < 1, \\ 0, & elsewhere \end{cases} \quad \phi_1(t) = \begin{cases} 1, & 0 \le t < 0.25 \\ -1, & 0.25 \le t < 0.75 \\ 1, & 0.75 \le t < 1 \\ 0, & elsewhere \end{cases}$$

Find the inner product of $\phi_0(t)$ and $\phi_1(t)$ over the interval $(0, 1)$, and state whether they are orthogonal or not.

Answer:
$\langle \phi_0, \phi_1 \rangle = 0$. $\phi_0(t)$ and $\phi_1(t)$ are orthogonal over the interval $(0, 1)$.

EXAMPLE 18.2

Let

$$\phi_0(t) = \begin{cases} \cos\left(\dfrac{2\pi t}{T_0}\right), & 0 \le t \le T_0 \\ 0, & otherwise \end{cases}$$

and

$$\phi_1(t) = \begin{cases} \sin\left(\dfrac{2\pi t}{T_0}\right), & 0 \le t \le T_0 \\ 0, & otherwise \end{cases}$$

Show that $\phi_0(t)$ and $\phi_1(t)$ are orthogonal over the interval $(0, T_0)$. Also, find the energy of $\phi_0(t)$ and $\phi_1(t)$.

Figure 18.2 shows $\phi_0(t)$ and $\phi_1(t)$.

FIGURE 18.2

Two signals $\phi_0(t)$ and $\phi_1(t)$.

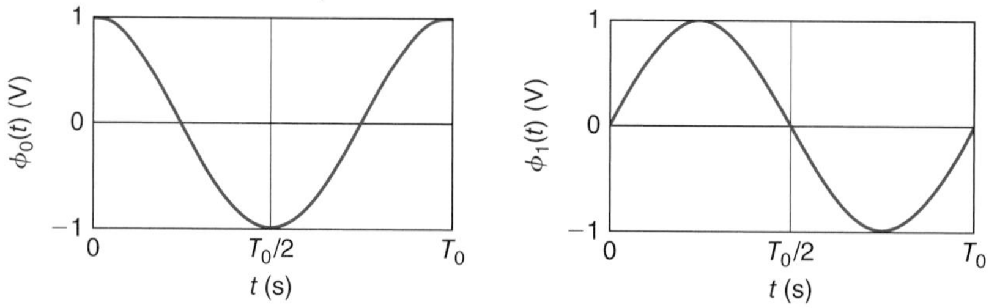

The product of $\phi_0(t)$ and $\phi_1(t)$ is a sine wave with amplitude 1/2 and period $T_0/2$. The integral of this product from 0 to T_0 is zero because the area under two cycles of sine wave is equal to zero. Evaluating the integral, we have

$$\int_0^{T_0} \sin\left(\frac{2\pi t}{T_0}\right)\cos\left(\frac{2\pi t}{T_0}\right)dt = \frac{1}{2}\int_0^{T_0} \sin\left(2\frac{2\pi t}{T_0}\right)dt = \frac{1}{2}\frac{-\cos\left(2\dfrac{2\pi t}{T_0}\right)\Bigg|_0^{T_0}}{2\dfrac{2\pi}{T_0}} = \frac{-1+1}{\dfrac{8\pi}{T_0}} = 0$$

Therefore, $\phi_0(t)$ and $\phi_1(t)$ are orthogonal over $(0, T_0)$. The energy of $\phi_0(t)$ is given by

$$\mu_0 = \int_0^{T_0} |\phi_0(t)|^2 dt = \int_0^{T_0} \cos^2\left(\frac{2\pi}{T_0}t\right)dt = \int_0^{T_0}\left[\frac{1}{2} + \frac{1}{2}\cos\left(\frac{4\pi}{T_0}t\right)\right]dt = \frac{T_0}{2} + \frac{1}{2}\frac{\sin\left(\dfrac{4\pi}{T_0}t\right)\Bigg|_0^{T_0}}{\dfrac{4\pi}{T_0}}$$

$$= \frac{T_0}{2} + \frac{1}{2}\frac{\sin\left(\dfrac{4\pi}{T_0}T_0\right) - \sin\left(\dfrac{4\pi}{T_0}0\right)}{\dfrac{4\pi}{T_0}} = \frac{T_0}{2} + \frac{1}{2}\frac{0-0}{\dfrac{4\pi}{T_0}} = \frac{T_0}{2}$$

continued

Example 18.2 continued

The energy of $\phi_1(t)$ is given by

$$\mu_1 = \int_0^{T_0} |\phi_1(t)|^2 dt = \int_0^{T_0} \sin^2\left(\frac{2\pi}{T_0}t\right)dt = \int_0^{T_0}\left[\frac{1}{2} - \frac{1}{2}\cos\left(\frac{4\pi}{T_0}t\right)\right]dt$$

$$= \frac{T_0}{2} - \frac{1}{2}\frac{\sin\left(\frac{4\pi}{T_0}t\right)\Big|_0^{T_0}}{\frac{4\pi}{T_0}} = \frac{T_0}{2}$$

Exercise 18.2

Let

$$\phi_0(t) = \begin{cases} \cos\left(\frac{\pi t}{T_0}\right), & 0 \le t \le T_0 \\ 0, & otherwise \end{cases}$$

and

$$\phi_1(t) = \begin{cases} \sin\left(\frac{\pi t}{T_0}\right), & 0 \le t \le T_0 \\ 0, & otherwise \end{cases}$$

Find the inner product of $\phi_0(t)$ and $\phi_1(t)$ over the interval $(0, T_0)$, and state whether they are orthogonal or not.

Answer:
$\langle \phi_0, \phi_1 \rangle = 0$. $\phi_0(t)$ and $\phi_1(t)$ are orthogonal over the interval $(0, T_0)$.

The definition of orthogonal functions can be generalized to more than two functions.

A set of N functions $[\phi_0(t), \phi_1(t), \ldots, \phi_{N-1}(t)]$, where N could be infinite, is orthogonal over the interval (t_0, t_1) if

$$\langle \phi_m, \phi_n \rangle = \int_{t_0}^{t_1} \phi_m(t)\phi_n^*(t)dt = \begin{cases} 0, & if\, m \ne n \\ \mu_m, & if\, m = n \end{cases} \tag{18.5}$$

If $m = n$, the inner product of $\phi_m(t)$ and $\phi_m(t)$, $<\phi_m, \phi_m>$, represents the energy of $\phi_m(t)$, μ_m, over the interval (t_0, t_1); that is,

$$\langle \phi_m, \phi_n \rangle = \int_{t_0}^{t_1} \phi_m(t)\phi_m^*(t)dt = \int_{t_0}^{t_1} |\phi_m(t)|^2 dt = \mu_m \tag{18.6}$$

If $\mu_m = 1$ (energy is 1) for all m, the set of functions is called orthogonal and normal, in short, orthonormal on the interval (t_0, t_1). An orthogonal set $[\phi_0(t), \phi_1(t), \ldots, \phi_{N-1}(t)]$ is complete if there is no other function that is orthogonal to each of the functions in the set.

EXAMPLE 18.3

The signals $\phi_0(t)$ and $\phi_1(t)$ discussed in EXAMPLE 18.1 are orthogonal and normal. The number of orthogonal signals can be increased by applying scaling and translation to $\phi_1(t)$. The signal $\phi_1(t)$ is called a wavelet, which means a small wave. The number of orthonormal basis functions can be increased to four by defining $\phi_2(t)$ and $\phi_3(t)$ as

$$\phi_2(t) = \sqrt{2}\phi_1(2t) = \sqrt{2}\phi_1\left(\frac{t}{\frac{1}{2}}\right) = \sqrt{2}rect\left(\frac{t-\frac{1}{8}}{\frac{1}{4}}\right) - \sqrt{2}rect\left(\frac{t-\frac{3}{8}}{\frac{1}{4}}\right)$$

$$= \begin{cases} \sqrt{2}, & 0 \le t < \frac{1}{4} \\ -\sqrt{2}, & \frac{1}{4} \le t < \frac{1}{2} \\ 0, & otherwise \end{cases}$$

$$\phi_3(t) = \sqrt{2}\phi_1(2t-1) = \sqrt{2}\phi_1\left(\frac{t-\frac{1}{2}}{\frac{1}{2}}\right) = \sqrt{2}rect\left(\frac{t-\frac{5}{8}}{\frac{1}{4}}\right) - \sqrt{2}rect\left(\frac{t-\frac{7}{8}}{\frac{1}{4}}\right)$$

$$= \begin{cases} \sqrt{2}, & \frac{1}{2} \le t < \frac{3}{4} \\ -\sqrt{2}, & \frac{3}{4} \le t < 1 \\ 0, & otherwise \end{cases}$$

Plot $\phi_0(t), \phi_1(t), \phi_2(t)$, and $\phi_3(t)$, and verify that these four signals are orthonormal.
Figure 18.3 shows $\phi_0(t), \phi_1(t), \phi_2(t)$, and $\phi_3(t)$.
The energy of $\phi_0(t), \phi_1(t), \phi_2(t)$, and $\phi_3(t)$ are given, respectively, as

$$\mu_0 = \int_0^1 \phi_0(t)^2 dt = \int_0^1 1^2 dt = 1$$

$$\mu_1 = \int_0^1 \phi_1(t)^2 dt = \int_0^{\frac{1}{2}} 1^2 dt + \int_{\frac{1}{2}}^1 (-1)^2 dt = 1$$

$$\mu_2 = \int_0^1 \phi_2(t)^2 dt = \int_0^{\frac{1}{4}} (\sqrt{2})^2 dt + \int_{\frac{1}{4}}^{\frac{1}{2}} (-\sqrt{2})^2 dt = 1$$

$$\mu_3 = \int_0^1 \phi_3(t)^2 dt = \int_{\frac{1}{2}}^{\frac{3}{4}} (\sqrt{2})^2 dt + \int_{\frac{3}{4}}^1 (-\sqrt{2})^2 dt = 1$$

continued

Example 18.3 continued

FIGURE 18.3

Four orthonormal signals $\phi_0(t)$, $\phi_1(t)$, $\phi_2(t)$, $\phi_3(t)$.

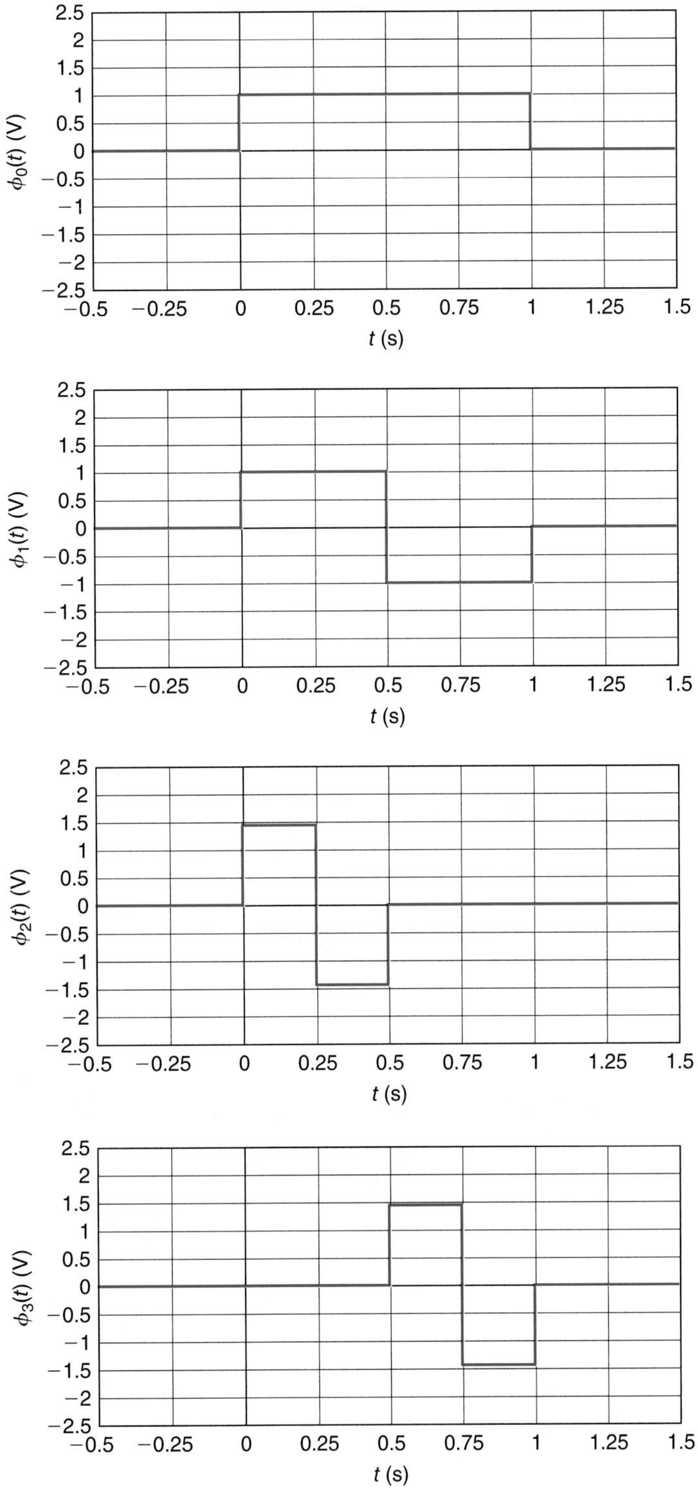

The inner products are given by

$$\int_0^1 \phi_0(t)\phi_1(t)dt = \int_0^1 \phi_1(t)dt = \int_0^{\frac{1}{2}} 1dt + \int_{\frac{1}{2}}^1 (-1)dt = 0$$

continued

Example 18.3 continued

$$\int_0^1 \phi_0(t)\phi_2(t)dt = \int_0^1 \phi_2(t)dt = \int_0^{\frac{1}{4}} (\sqrt{2})dt + \int_{\frac{1}{4}}^{\frac{1}{2}} (-\sqrt{2})dt = 0$$

$$\int_0^1 \phi_0(t)\phi_3(t)dt = \int_0^1 \phi_3(t)dt = \int_{\frac{1}{2}}^{\frac{3}{4}} (\sqrt{2})dt + \int_{\frac{3}{4}}^{1} (-\sqrt{2})dt = 0$$

$$\int_0^1 \phi_1(t)\phi_2(t)dt = \int_0^{\frac{1}{2}} \phi_2(t)dt = \int_0^{\frac{1}{4}} (\sqrt{2})dt + \int_{\frac{1}{4}}^{\frac{1}{2}} (-\sqrt{2})dt = 0$$

$$\int_0^1 \phi_1(t)\phi_3(t)dt = -\int_0^1 \phi_3(t)dt = -\int_{\frac{1}{2}}^{\frac{3}{4}} (\sqrt{2})dt - \int_{\frac{3}{4}}^{1} (-\sqrt{2})dt = 0$$

$$\int_0^1 \phi_2(t)\phi_3(t)dt = \int_0^{\frac{1}{2}} 0dt + \int_{\frac{1}{2}}^{1} 0dt = 0$$

Since all the inner products are zero, the signals $\phi_0(t), \phi_1(t), \phi_2(t)$, and $\phi_3(t)$ are orthogonal. Also, since the energy is one for each of the four signals, the set is orthonormal. The number of basis functions can be increased to eight from four by scaling and translating $\phi_1(t)$. By continuing this process, the number of basis functions can be increased to $N = 2^n$ where n is a positive integer. The signals $\phi_0(t), \phi_1(t), \phi_2(t), \phi_3(t), \ldots, \phi_{N-1}(t)$ are called **Haar wavelets**. In general, the signals $\phi_2(t), \phi_3(t), \ldots, \phi_{N-1}(t)$ can be obtained by scaling and translating $\phi_1(t)$ using

$$\phi_{m,k}(t) = (\sqrt{2})^m \phi_1(2^m t - k), \quad m = 1,2,\ldots,n-1, \quad k = 0,1,2,\ldots, 2^m - 1 \qquad \textbf{(18.7)}$$

Exercise 18.3

Find the eight basis functions ($N = 8 = 2^3$) using Equation (18.7), and plot them.

Answer:

$$\phi_2(t) = \phi_{1,0}(t) = (\sqrt{2})^1 \phi_1(2^1 t - 0) = \sqrt{2}\phi_1(2t) = \sqrt{2}\phi_1\left(\frac{t}{\frac{1}{2}}\right)$$

$$= \begin{cases} \sqrt{2}, & 0 \le t < \frac{1}{4} \\ -\sqrt{2}, & \frac{1}{4} \le t < \frac{1}{2} \\ 0, & otherwise \end{cases}$$

continued

Exercise 18.3 continued

$$\phi_3(t) = \phi_{1,1}(t) = (\sqrt{2})^1\phi_1(2^1t - 1) = \sqrt{2}\phi_1(2t-1) = \sqrt{2}\phi_1\left(\frac{t-\frac{1}{2}}{\frac{1}{2}}\right)$$

$$= \begin{cases} \sqrt{2}, & \frac{1}{2} \le t < \frac{3}{4} \\ -\sqrt{2}, & \frac{3}{4} \le t < 1 \\ 0, & otherwise \end{cases}$$

$$\phi_4(t) = \phi_{2,0}(t) = (\sqrt{2})^2\phi_1(2^2t - 0) = 2\phi_1(4t) = 2\phi_1\left(\frac{t}{\frac{1}{4}}\right) = \begin{cases} 2, & 0 \le t < \frac{1}{8} \\ -2, & \frac{1}{8} \le t < \frac{1}{4} \\ 0, & otherwise \end{cases}$$

$$\phi_5(t) = \phi_{2,1}(t) = (\sqrt{2})^2\phi_1(2^2t - 1) = 2\phi_1(4t-1) = 2\phi_1\left(\frac{t-\frac{1}{4}}{\frac{1}{4}}\right) = \begin{cases} 2, & \frac{1}{4} \le t < \frac{3}{8} \\ -2, & \frac{3}{8} \le t < \frac{1}{2} \\ 0, & otherwise \end{cases}$$

$$\phi_6(t) = \phi_{2,2}(t) = (\sqrt{2})^2\phi_1(2^2t - 2) = 2\phi_1(4t-2) = 2\phi_1\left(\frac{t-\frac{2}{4}}{\frac{1}{4}}\right)$$

$$= \begin{cases} 2, & \frac{1}{2} \le t < \frac{5}{8} \\ -2, & \frac{5}{8} \le t < \frac{3}{4} \\ 0, & otherwise \end{cases}$$

$$\phi_7(t) = \phi_{2,3}(t) = (\sqrt{2})^2\phi_1(2^2t - 3) = 2\phi_1(4t-3) = 2\phi_1\left(\frac{t-\frac{3}{4}}{\frac{1}{4}}\right)$$

$$= \begin{cases} 2, & \frac{3}{4} \le t < \frac{7}{8} \\ -2, & \frac{7}{8} \le t < 1 \\ 0, & otherwise \end{cases}$$

continued

Exercise 18.3 continued

The basis functions $\phi_0(t)$, $\phi_1(t)$, $\phi_2(t)$, $\phi_3(t)$, $\phi_4(t)$, $\phi_5(t)$, $\phi_6(t)$, and $\phi_7(t)$ are plotted in Figure 18.4.

FIGURE 18.4

Eight orthonormal signals: $\phi_0(t)$, $\phi_1(t)$, $\phi_2(t)$, $\phi_3(t)$, $\phi_4(t)$, $\phi_5(t)$, $\phi_6(t)$, and $\phi_7(t)$.

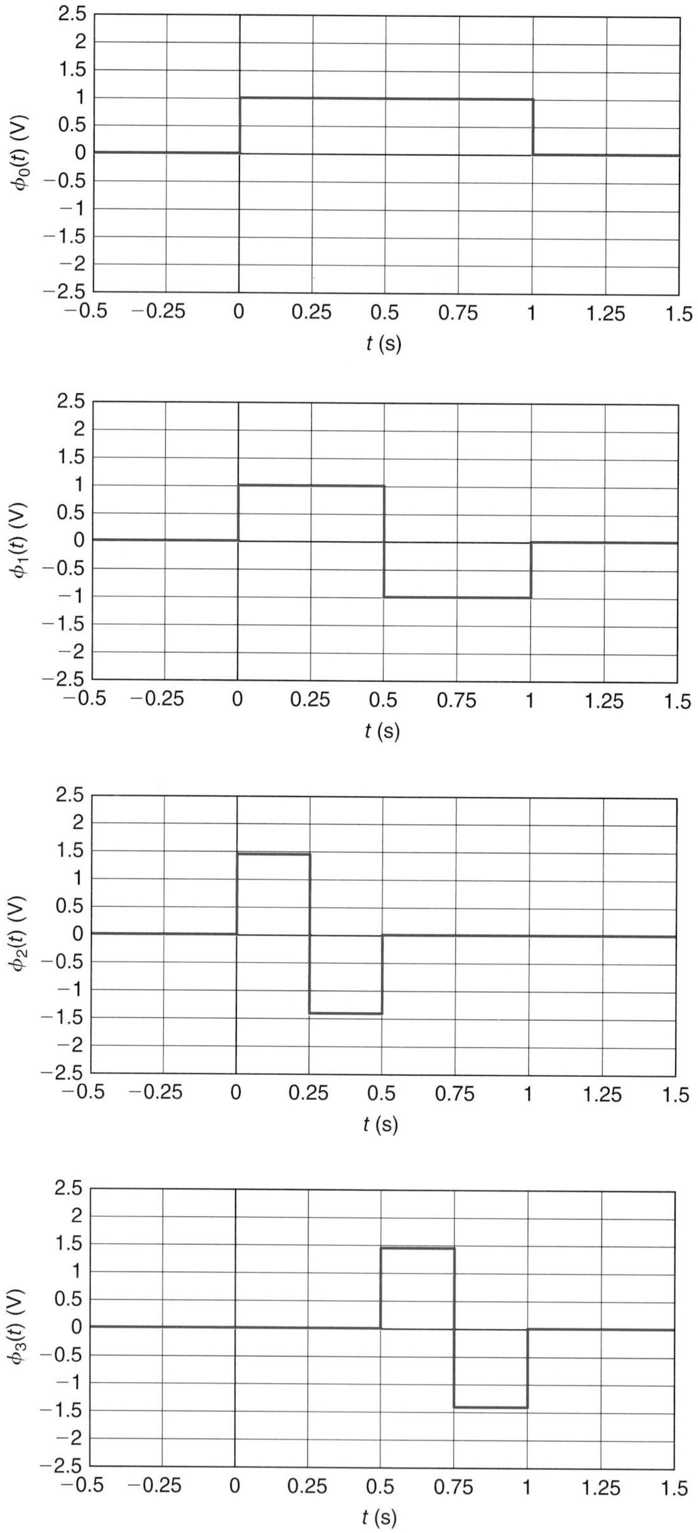

continued

Exercise 18.3 continued

FIGURE 18.4

continued

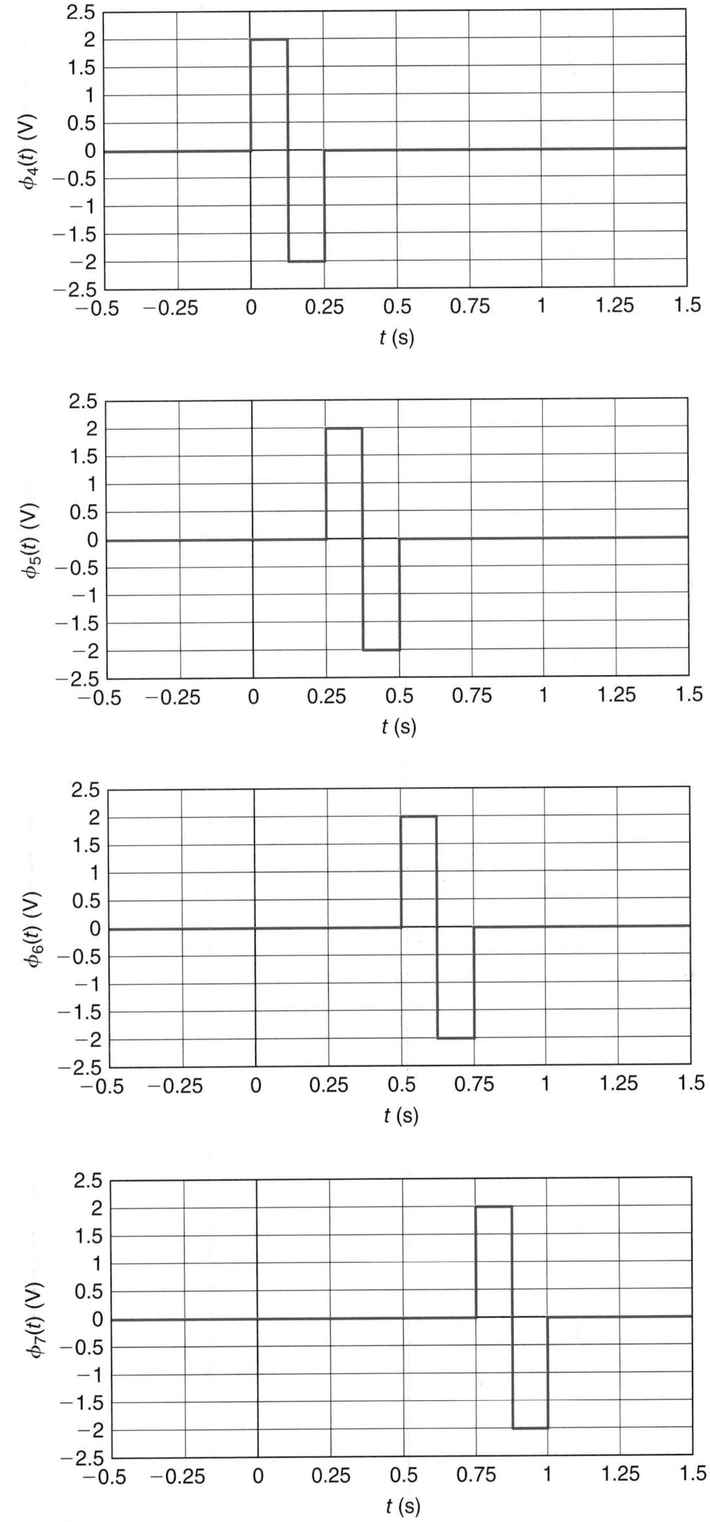

18.2.2 REPRESENTATION OF AN ARBITRARY SIGNAL BY ORTHOGONAL FUNCTIONS

An arbitrary signal $f(t)$ can be approximated over the interval (t_0, t_1) by a linear combination of an arbitrary set of orthogonal functions $[\phi_0(t), \phi_1(t), \ldots, \phi_{N-1}(t)]$ if the energy of the signal over the interval (t_0, t_1) is finite; that is,

$$E_f = \int_{t_0}^{t_1} |f(t)|^2 dt < \infty$$

Let $\hat{f}(t)$ be the approximation. Then

$$\hat{f}(t) = a_0\phi_0(t) + a_1\phi_1(t) + \cdots + a_{N-1}\phi_{N-1}(t) = \sum_{n=0}^{N-1} a_n\phi_n(t) \qquad \textbf{(18.8)}$$

The set of orthogonal functions $[\phi_0(t), \phi_1(t), \ldots, \phi_{N-1}(t)]$ is called basis functions because it forms the basis for the representation given by Equation (18.8). The difference between the original signal $f(t)$ and its approximation $\hat{f}(t)$ is defined as the error signal:

$$e(t) = f(t) - \hat{f}(t) = f(t) - \sum_{n=0}^{N-1} a_n\phi_n(t) \qquad \textbf{(18.9)}$$

The coefficients $a_0, a_1, \ldots, a_{N-1}$ in the linear combination are determined to minimize the energy of the error signal $e(t)$ over the interval (t_0, t_1), given by

$$\begin{aligned}
E_e &= \int_{t_0}^{t_1} |e(t)|^2 dt = \int_{t_0}^{t_1} \left| f(t) - \sum_{n=0}^{N-1} a_n\phi_n(t) \right|^2 dt \\
&= \int_{t_0}^{t_1} \left[f(t) - \sum_{n=0}^{N-1} a_n\phi_n(t) \right] \left[f(t) - \sum_{n=0}^{N-1} a_n\phi_n(t) \right]^* dt \\
&= \int_{t_0}^{t_1} \left[f(t) - \sum_{n=0}^{N-1} a_n\phi_n(t) \right] \left[f^*(t) - \sum_{n=0}^{N-1} a_n^*\phi_n^*(t) \right] dt \\
&= \int_{t_0}^{t_1} f(t)f^*(t)dt - \int_{t_0}^{t_1} f(t)\sum_{n=0}^{N-1} a_n^*\phi_n^*(t)dt - \sum_{n=0}^{N-1} a_n \int_{t_0}^{t_1} f^*(t)\phi_n(t)dt \\
&\quad + \sum_{n=0}^{N-1}\sum_{m=0}^{N-1} a_n a_m^* \int_{t_0}^{t_1} \phi_n(t)\phi_m^*(t)dt \\
&= \int_{t_0}^{t_1} |f(t)|^2 dt - \sum_{n=0}^{N-1} a_n^* \int_{t_0}^{t_1} f(t)\phi_n^*(t)dt - \sum_{n=0}^{N-1} a_n \int_{t_0}^{t_1} f^*(t)\phi_n(t)dt + \sum_{n=0}^{N-1} \mu_n a_n a_n^*
\end{aligned} \qquad \textbf{(18.10)}$$

To find the coefficient a_n that minimizes the energy of the error signal, we take the partial differential of E_e with respect to a_n and set the result to zero. Before we take the partial differential, we will review the partial differential of complex variables. Let a complex variable be z_n. Then, it can be represented as

$$z_n = x_n + jy_n$$

where x_n and y_n are real variables. Then, $z_n^* = x_n - jy_n$ and

$$z_n z_n^* = (x_n + jy_n)(x_n - jy_n) = x_n^2 + y_n^2$$

$$\frac{\partial(z_n z_n^*)}{\partial z_n} = \frac{\partial(z_n z_n^*)}{\partial x_n} + j\frac{\partial(z_n z_n^*)}{\partial y_n} = \frac{\partial(x_n^2 + y_n^2)}{\partial x_n} + j\frac{\partial(x_n^2 + y_n^2)}{\partial y_n} = 2x_n + j2y_n = 2z_n$$

$$\frac{\partial(z_n)}{\partial z_n} = \frac{\partial(z_n)}{\partial x_n} + j\frac{\partial(z_n)}{\partial y_n} = \frac{\partial(x_n + jy_n)}{\partial x_n} + j\frac{\partial(x_n + jy_n)}{\partial y_n} = 1 + j^2 = 1 - 1 = 0$$

$$\frac{\partial(z_n^*)}{\partial z_n} = \frac{\partial(z_n^*)}{\partial x_n} + j\frac{\partial(z_n^*)}{\partial y_n} = \frac{\partial(x_n - jy_n)}{\partial x_n} + j\frac{\partial(x_n - jy_n)}{\partial y_n} = 1 - j^2 = 1 + 1 = 2$$

Taking the partial differential of E_e with respect to a_n and set it equal to zero, we have

$$\frac{\partial E_e}{\partial a_n} = \frac{\partial}{\partial a_n}\int_{t_0}^{t_1} |f(t)|^2 dt - \frac{\partial}{\partial a_n}\sum_{n=0}^{N-1} a_n^* \int_{t_0}^{t_1} f(t)\phi_n^*(t)dt - \frac{\partial}{\partial a_n}\sum_{n=0}^{N-1} a_n \int_{t_0}^{t_1} f^*(t)\phi_n(t)dt + \frac{\partial}{\partial a_n}\sum_{n=0}^{N-1} \mu_n a_n a_n^* = 0 \quad \textbf{(18.11)}$$

The first term on the right side is zero because the integral is not a function of a_n. In the second term on the right side, there is only one a_n^* in the summation, and since $\frac{\partial(a_n^*)}{\partial a_n} = 2$, this term becomes $-2\int_{t_0}^{t_1} f(t)\phi_n^*(t)dt$. In the third term on the right side, there is only one a_n in the summation, and since $\frac{\partial(a_n)}{\partial a_n} = 0$, this term is zero. In the fourth term on the right side, there is only one $a_n a_n^*$ in the summation, and since $\frac{\partial(a_n a_n^*)}{\partial a_n} = 2a_n$, this term becomes $2\mu_n a_n$. Thus, Equation (18.11) becomes

$$\frac{\partial E_e}{\partial a_n} = -2\int_{t_0}^{t_1} f(t)\phi_n^*(t)dt + 2\mu_n a_n = 0 \quad \textbf{(18.12)}$$

Solving Equation (18.12) for a_n, we obtain

$$a_n = \frac{1}{\mu_n}\int_{t_0}^{t_1} f(t)\phi_n^*(t)dt \quad \textbf{(18.13)}$$

Equation (18.13) indicates that the coefficient a_n that minimizes the energy of the error signal is obtained by multiplying the signal $f(t)$ by the orthonormal basis function $\phi_n^*(t)$ and integrates the product for the interval (t_0, t_1) and divides by the energy of $\phi_n(t)$. If the two signals are similar in shape, we have large magnitude for a_n, and if they do not possess similarity (for example, one is changing rapidly and the other is changing slowly), the magnitude of a_n is small.

Taking the complex conjugate of Equation (18.13), we obtain

$$a_n^* = \frac{1}{\mu_n}\int_{t_0}^{t_1} f^*(t)\phi_n(t)dt \quad \textbf{(18.14)}$$

Substitution of Equations (18.13) and (18.14) into Equation (18.10) yields

$$\begin{aligned} E_e &= \int_{t_0}^{t_1} |f(t)|^2 dt - \sum_{n=0}^{N-1} \mu_n a_n^* a_n - \sum_{n=0}^{N-1} \mu_n a_n a_n^* + \sum_{n=0}^{N-1} \mu_n a_n a_n^* \\ &= \int_{t_0}^{t_1} |f(t)|^2 dt - \sum_{n=0}^{N-1} \mu_n |a_n|^2 \end{aligned} \tag{18.15}$$

The first term on the right side of Equation (18.15) is the energy of $f(t)$, and the second term on the right side is the energy of the approximation of $f(t)$; that is, $\hat{f}(t)$. The proof that the energy of $\hat{f}(t)$ is given by $\sum_{n=0}^{N-1} \mu_n |a_n|^2$ is shown below.

$$\begin{aligned} E_{\hat{f}} &= \int_{t_0}^{t_1} \hat{f}(t)\hat{f}^*(t)dt = \int_{t_0}^{t_1} \sum_{n=0}^{N-1} a_n \phi_n(t) \sum_{m=0}^{N-1} a_m^* \phi_m^*(t)dt \\ &= \sum_{n=0}^{N-1} \sum_{m=0}^{N-1} a_n a_m^* \int_{t_0}^{t_1} \phi_n(t)\phi_m^*(t)dt = \sum_{n=0}^{N-1} \mu_n |a_n|^2 \end{aligned} \tag{18.16}$$

If $f(t)$ has a finite energy in the interval (t_0, t_1); that is,

$$E_f = \int_{t_0}^{t_1} |f(t)|^2 dt < \infty$$

then, $f(t)$ is said to be a **square-integrable** function. If $f(t)$ is square-integrable, then the first term on the right side of Equation (18.15), which is the energy of $f(t)$, is fixed and is independent of N. As N (the number of basis functions in the orthogonal set) is increased, the second term on the right side of Equation (18.15) (summation) is increased or stays the same because each term in the summation is nonnegative. The sum is a bounded non-decreasing sequence. Thus, as N approaches infinity, the sum converges to the energy of $f(t)$, and the energy of the error approaches zero. Thus, from Equation (18.9), we have

$$\lim_{N\to\infty} \int_{t_0}^{t_1} \left| f(t) - \sum_{n=0}^{N-1} a_n \phi_n(t) \right|^2 dt = 0 \tag{18.17}$$

This equation indicates that, as $N \to \infty$, the approximation $\hat{f}(t) = \sum_{n=0}^{N-1} a_n \phi_n(t)$ converges to $f(t)$. Thus, we have

$$f(t) = \sum_{n=0}^{\infty} a_n \phi_n(t) \tag{18.18}$$

As $N \to \infty$, $E_e \to 0$, from Equation (18.15), we have

$$\sum_{n=0}^{\infty} \mu_n |a_n|^2 = \int_{t_0}^{t_1} |f(t)|^2 dt \tag{18.19}$$

If the set of orthogonal functions is also normal, then $\mu_n = 1$ for all n, and Equation (18.19) becomes

$$\sum_{n=0}^{\infty} |a_n|^2 = \int_{t_0}^{t_1} |f(t)|^2 dt \tag{18.20}$$

Equations (18.19) and (18.20) are known as **Parseval's formula**. Parseval's formula says that the energy of the signal over the interval (t_0, t_1) can be obtained as a sum of the squares of the coefficients (times the energy of the orthogonal signals) in the expansion.

EXAMPLE 18.4

Let a signal $f(t)$ be

$$f(t) = \begin{cases} 2, & 0 \le t < \dfrac{3}{4} \\ -3, & \dfrac{3}{4} \le t \le 1 \\ 0, & otherwise \end{cases}$$

This signal is represented as a linear combination of orthonormal signals $\phi_0(t)$ and $\phi_1(t)$ shown in Figure 18.1; that is,

$$\hat{f}(t) = a_0\phi_0(t) + a_1\phi_1(t)$$

a. Find the coefficients a_0 and a_1.
b. Plot the approximation $\hat{f}(t)$.
c. Plot the error signal $e(t) = f(t) - \hat{f}(t)$.
d. Find the energy of $f(t)$ and $e(t)$.

Figure 18.5 shows signals $\phi_0(t)$, $\phi_1(t)$, and $f(t)$. As shown earlier, the energy of $\phi_0(t)$ is 1, and the energy $\phi_1(t)$ is 1; that is, $\mu_0 = 1, \mu_1 = 1$.

a.

FIGURE 18.5

Signals $\phi_0(t)$, $\phi_1(t)$, and $f(t)$.

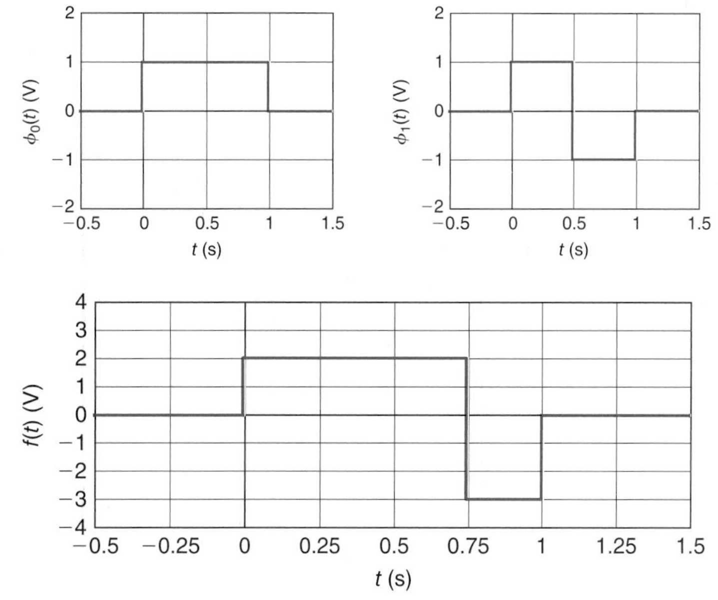

continued

Example 18.4 continued

From Equation (18.13), we obtain

$$a_0 = \frac{1}{\mu_0}\int_0^1 f(t)\phi_0(t)dt = \frac{1}{1}\int_0^1 f(t)dt = 2 \times \frac{3}{4} + (-3) \times \frac{1}{4} = \frac{3}{4} = 0.75$$

$$a_1 = \frac{1}{\mu_1}\int_0^1 f(t)\phi_1(t)dt = \frac{1}{1}\int_0^{\frac{1}{2}} 2 \times 1dt + \frac{1}{1}\int_{\frac{1}{2}}^{\frac{3}{4}} 2 \times (-1)dt + \frac{1}{1}\int_{\frac{3}{4}}^{1} (-3) \times (-1)dt$$

$$= 2 \times \frac{1}{2} + (-2) \times \frac{1}{4} + 3 \times \frac{1}{4} = \frac{5}{4} = 1.25$$

b. The approximation is given by

$$\hat{f}(t) = a_0\phi_0(t) + a_1\phi_1(t) = 0.75\phi_0(t) + 1.25\phi_1(t) = \begin{cases} 2, & 0 \le t < 0.5 \\ -0.5, & 0.5 \le t < 1 \\ 0, & \textit{elsewhere} \end{cases}$$

and is shown in Figure 18.6.

FIGURE 18.6

Approximation $\hat{f}(t)$.

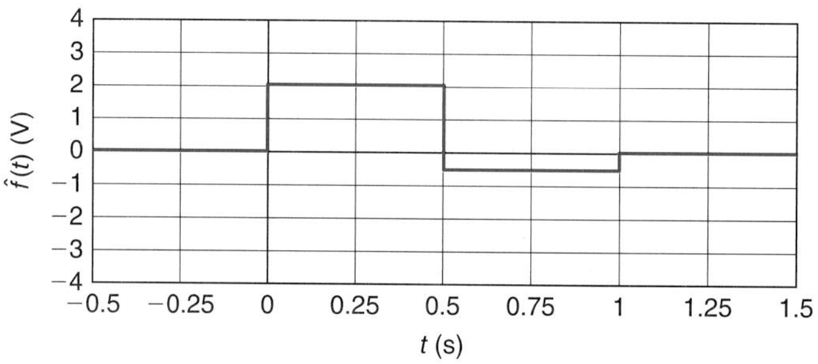

c. The error signal in the representation is given by

$$e(t) = f(t) - \hat{f}(t) = \begin{cases} 2.5, & 0.5 \le t < 0.75 \\ -2.5, & 0.75 \le t < 1 \\ 0, & \textit{elsewhere} \end{cases}$$

and is shown in Figure 18.7.

FIGURE 18.7

Error signal $e(t)$.

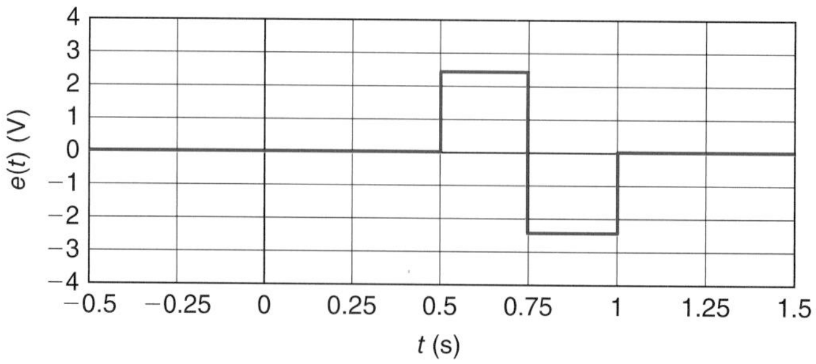

continued

Example 18.4 continued

d. The energy of $f(t)$ is given by

$$E_f = \int_0^1 f(t)^2 dt = \int_0^{\frac{3}{4}} 2^2 dt + \int_{\frac{3}{4}}^1 (-3)^2 dt = 3 + 2.25 = 5.25 \text{ J}$$

The energy of error signal $e(t)$ is

$$E_e = \int_0^1 e(t)^2 dt = \int_{\frac{1}{2}}^{\frac{3}{4}} (2.5)^2 dt + \int_{\frac{3}{4}}^1 (-2.5)^2 dt = 1.5625 + 1.5625 = 3.125 \text{ J}$$

The difference between the energy of $f(t)$ and the energy of the approximation $\hat{f}(t)$ is the energy of error signal

$$E_f - (\mu_0 a_0^2 + \mu_1 a_1^2) = 5.25 - (1 \times 0.75^2 + 1 \times 1.25^2) = 3.125 \text{ J}$$

Exercise 18.4

Let a signal $f(t)$ be

$$f(t) = \begin{cases} -2, & 0 \le t < \frac{1}{4} \\ 1, & \frac{1}{4} \le t \le 1 \\ 0, & otherwise \end{cases}$$

This signal is represented as a linear combination of orthonormal signals $\phi_0(t)$ and $\phi_1(t)$ shown in Figure 18.1; that is, $\hat{f}(t) = a_0\phi_0(t) + a_1\phi_1(t)$. Find the coefficents a_0 and a_1.

Answer:
$a_0 = 0.25, \quad a_1 = -0.75.$

EXAMPLE 18.5

Repeat Example 18.4 with the orthonormal signals $\phi_0(t)$, $\phi_1(t)$, $\phi_2(t)$, and $\phi_3(t)$ shown in Figure 18.3.

The energy of each of the orthonomal basis functions is 1; that is, $\mu_n = 1$ for $0 \le n \le 3$. The coefficients are given by

$$a_0 = \frac{1}{\mu_0} \int_0^1 f(t)\phi_0(t) dt = 0.75$$

$$a_1 = \frac{1}{\mu_1} \int_0^1 f(t)\phi_1(t) dt = 1.25$$

continued

Example 18.5 continued

$$a_2 = \frac{1}{\mu_2}\int_0^1 f(t)\phi_2(t)dt = \frac{1}{1}\int_0^{\frac{1}{4}} 2 \times \sqrt{2}dt + \frac{1}{1}\int_{\frac{1}{4}}^{\frac{1}{2}} 2 \times (-\sqrt{2})dt = 0$$

$$a_3 = \frac{1}{\mu_3}\int_0^1 f(t)\phi_3(t)dt = \frac{1}{1}\int_{0.5}^{0.75} 2 \times \sqrt{2}dt + \frac{1}{1}\int_{0.75}^{1} (-3) \times (-\sqrt{2})dt = \frac{5\sqrt{2}}{4} = 1.7678$$

Thus, the approximation is given by

$$\begin{aligned}\hat{f}(t) &= a_0\phi_0(t) + a_1\phi_1(t) + a_2\phi_2(t) + a_3\phi_3(t) \\ &= 0.75\phi_0(t) + 1.25\phi_1(t) + 1.7678\phi_3(t) = f(t)\end{aligned}$$

Notice that the approximation is identical to $f(t)$. This can be verified by plotting $\hat{f}(t)$. The error signal is zero in this case.

$$e(t) = f(t) - \hat{f}(t) = 0$$

The energy of $f(t)$ is given by

$$E_f = \int_0^1 f(t)^2 dt = \int_0^{\frac{3}{4}} 2^2 dt + \int_{\frac{3}{4}}^{1} (-3)^2 dt = 3 + 2.25 = 5.25 \text{ J}$$

The energy of error signal is zero:

$$E_e = \int_0^1 e^2(t)dt = 0$$

The difference between the energy of $f(t)$ and and the energy of approximation is also zero:

$$E_f - (\mu_0 a_0^2 + \mu_1 a_1^2 + \mu_2 a_2^2 + \mu_3 a_3^2) = 0$$

Exercise 18.5

Let a signal $f(t)$ be

$$f(t) = \begin{cases} -4, & 0 \le t < \frac{3}{8} \\ 2, & \frac{3}{8} \le t \le 1 \\ 0, & otherwise \end{cases}$$

This signal is represented as a linear combination of orthonormal signals $\phi_0(t)$, $\phi_1(t)$, $\phi_2(t)$, and $\phi_3(t)$ shown in Figure 18.3; that is, $\hat{f}(t) = a_0\phi_0(t) + a_1\phi_1(t) + a_2\phi_2(t) + a_3\phi_3(t)$. Find the coefficents a_0, a_1, a_2, and a_3.

Answer:

$a_0 = -0.25, \quad a_1 = -2.25, \quad a_2 = -3\sqrt{2}/4 = -1.0607, \quad a_3 = 0.$

EXAMPLE 18.6

Repeat Example 18.4 with the orthogonal signals $\phi_0(t)$ and $\phi_1(t)$ shown in Figure 18.2.

a. $T_0 = 1\text{s}$

$$\phi_0(t) = \begin{cases} \cos\left(\dfrac{2\pi t}{T_0}\right), & 0 \le t \le T_0 \\ 0, & otherwise \end{cases}$$

and

$$\phi_1(t) = \begin{cases} \sin\left(\dfrac{2\pi t}{T_0}\right), & 0 \le t \le T_0 \\ 0, & otherwise \end{cases}$$

$$\mu_0 = T_0/2 = 0.5\text{ J}, \quad \mu_1 = T_0/2 = 0.5\text{ J}$$

$$f(t) = \begin{cases} 2, & 0 \le t < \dfrac{3}{4} \\ -3, & \dfrac{3}{4} \le t \le 1 \\ 0, & otherwise \end{cases}$$

$$a_0 = \frac{1}{\mu_0}\int_0^1 f(t)\phi_0(t)dt = \frac{1}{0.5}\int_0^{0.75} 2\cos(2\pi t)dt + \frac{1}{0.5}\int_{0.75}^{1} (-3)\cos(2\pi t)dt = -1.592$$

$$a_1 = \frac{1}{\mu_1}\int_0^1 f(t)\phi_1(t)dt = \frac{1}{0.5}\int_0^{0.75} 2\sin(2\pi t)dt + \frac{1}{0.5}\int_{0.75}^{1} (-3)\sin(2\pi t)dt = 1.592$$

FIGURE 18.8

Approximation $\hat{f}(t)$.

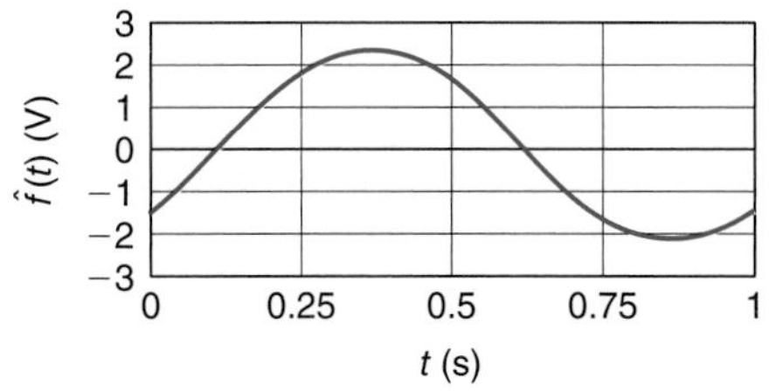

FIGURE 18.9

Error signal $e(t)$.

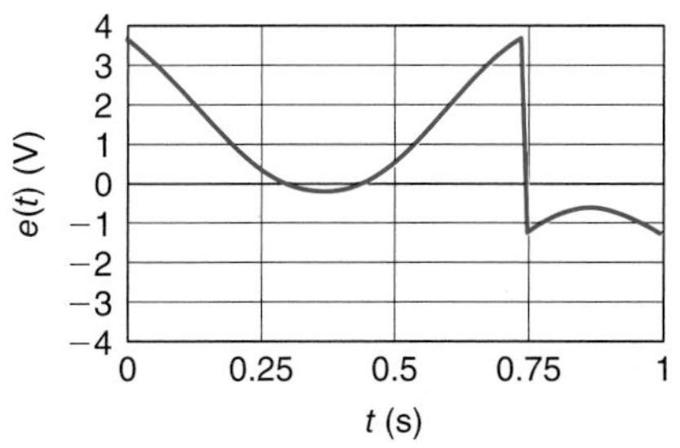

b. The approximation is given by

$$\hat{f}(t) = a_0\phi_0(t) + a_1\phi_1(t) = -1.5915\cos(2\pi t) + 1.5915\sin(2\pi t)$$

and is plotted in Figure 18.8.

c. The error signal is shown in Figure 18.9.

$$e(t) = f(t) - \hat{f}(t)$$

The large error is the result of dissimilarity between the orthogonal signals $\phi_0(t)$, $\phi_1(t)$, and $f(t)$.

continued

Example 18.6 continued

d. The energy of $f(t)$, $e(t)$, and $\hat{f}(t)$ are given, respectively, by

$$E_f = \int_0^1 f(t)^2 dt = \int_0^{\frac{3}{4}} 2^2 dt + \int_{\frac{3}{4}}^1 (-3)^2 dt = 3 + 2.25 = 5.25 \text{ J}$$

$$E_e = \int_0^1 e^2(t) dt = 2.717 \text{ J}$$

$$E_{\hat{f}} = \mu_0 a_0^2 + \mu_1 a_1^2 = 2.533 \text{ J}$$

Exercise 18.6

Let a signal $f(t)$ be

$$f(t) = \cos\left(\frac{2\pi t}{T_0} + \frac{\pi}{4}\right), \quad 0 \le t \le T_0$$

This signal is represented as a linear combination of orthonormal signals $\phi_0(t)$ and $\phi_1(t)$ shown in Figure 18.2; that is, $\hat{f}(t) = a_0\phi_0(t) + a_1\phi_1(t)$. Find coefficents a_0 and a_1.

Answer:
$a_0 = 1/\sqrt{2} = 0.7071, \quad a_1 = -1/\sqrt{2} = -0.7071.$

18.2.3 TRIGONOMETRIC FOURIER SERIES

A set of functions

$$[1, \cos(\omega_0 t), \cos(2\omega_0 t), \ldots, \sin(\omega_0 t), \sin(2\omega_0 t), \ldots]$$

is orthogonal (the proof is given here) and complete over the interval $(t_0, t_0 + T_0)$, where

$$\omega_0 = \frac{2\pi}{T_0} = 2\pi f_0$$

is the **fundamental frequency** in rad/s, and $f_0 = 1/T_0$ is the fundamental frequency in hertz. Notice that t_0 is arbitrary. The most common choices for t_0 are either 0 or $-T_0/2$. If $t_0 = 0$, the interval is $(0, T_0)$, and if $t_0 = -T_0/2$, the interval is $(-T_0/2, T_0/2)$. Since the set of the signals is orthogonal, a square-integrable function $f(t)$ can be represented over the interval $(t_0, t_0 + T_0)$ as follows:

$$f(t) = a_0 + \sum_{n=1}^{\infty}[a_n \cos(n\omega_0 t) + b_n \sin(n\omega_0 t)], \quad t_0 \le t \le t_0 + T_0 \tag{18.21}$$

The coefficients a_0, a_n, and b_n can be computed from Equation (18.13) as

$$a_0 = \frac{\int_{t_0}^{t_0+T_0} f(t) dt}{\int_{t_0}^{t_0+T_0} 1^2 dt} = \frac{1}{T_0}\int_{t_0}^{t_0+T_0} f(t) dt \tag{18.22}$$

$$a_n = \frac{\int_{t_0}^{t_0+T_0} f(t)\cos(n\omega_0 t)dt}{\int_{t_0}^{t_0+T_0} \cos^2(n\omega_0 t)dt} = \frac{2}{T_0}\int_{t_0}^{t_0+T_0} f(t)\cos(n\omega_0 t)dt, \quad n = 1,2,3,\ldots \tag{18.23}$$

$$b_n = \frac{\int_{t_0}^{t_0+T_0} f(t)\sin(n\omega_0 t)dt}{\int_{t_0}^{t_0+T_0} \sin^2(n\omega_0 t)dt} = \frac{2}{T_0}\int_{t_0}^{t_1+T_0} f(t)\sin(n\omega_0 t)dt, \quad n = 1,2,3,\ldots \tag{18.24}$$

If $f(t)$ is periodic, the representation given by Equation (18.21) is valid for all t; that is, $-\infty < t < \infty$. If $f(t)$ is nonperiodic, the representation given by Equation (18.21) is valid for $t_0 < t < t_0 + T_0$. Unless mentioned otherwise, we assume $f(t)$ is periodic with period T_0.

The representation given by Equation (18.21) is called trigonometric Fourier series expansion, and a_0, a_n, b_n are called trigonometric Fourier coefficients. The coefficient a_0 is given by the integral of the signal $f(t)$ for one period divided by the period T_0. This is the **average value** of the signal or the **dc offset**. The coefficient a_0 is called the **dc component**. If there is no dc offset in the waveform, a_0 is zero. The coefficient a_n measures the correlation (similarity) between $f(t)$ and $\cos(n\omega_0 t)$, and the coefficient b_n measures the correlation (similarity) between $f(t)$ and $\sin(n\omega_0 t)$.

18.2.4 PROOF OF ORTHOGONALITY

We are presenting the proof that a set of sinusoidal signals

$$\left[1,\ \sin\left(\frac{2\pi}{T_0}t\right),\ \sin\left(2\frac{2\pi}{T_0}t\right),\ \sin\left(3\frac{2\pi}{T_0}t\right),\ldots,\right.$$
$$\left.\cos\left(\frac{2\pi}{T_0}t\right),\ \cos\left(2\frac{2\pi}{T_0}t\right),\ \cos\left(3\frac{2\pi}{T_0}t\right),\ldots\right]$$

is orthogonal over $(t_0, t_0 + T_0)$, where t_0 is arbitrary.

The inner product of 1 and $\sin(n2\pi t/T_0)$ is given by

$$\int_{t_0}^{t_0+T_0} \sin\left(\frac{n2\pi t}{T_0}\right)dt = \frac{-\cos\left(\frac{n2\pi t}{T_0}\right)\Big|_{t_0}^{t_0+T_0}}{\frac{n2\pi}{T_0}} = \frac{-\cos\left(\frac{n2\pi(t_0+T_0)}{T_0}\right)+\cos\left(\frac{n2\pi t_0}{T_0}\right)}{\frac{n2\pi}{T_0}}$$

$$= \frac{-\cos\left(\frac{n2\pi t_0}{T_0}+2\pi n\right)+\cos\left(\frac{n2\pi t_0}{T_0}\right)}{\frac{n2\pi}{T_0}} = \frac{-\cos\left(\frac{n2\pi t_0}{T_0}\right)+\cos\left(\frac{n2\pi t_0}{T_0}\right)}{\frac{n2\pi}{T_0}} = 0$$

Similarly, the inner product of 1 and $\cos(n2\pi t/T_0)$ is zero for $n = 1, 2, 3, \ldots$

$$\int_{t_0}^{t_0+T_0} \cos\left(\frac{n2\pi t}{T_0}\right)dt = \frac{\sin\left(\frac{n2\pi t}{T_0}\right)\Big|_{t_0}^{t_0+T_0}}{\frac{n2\pi}{T_0}} = \frac{\sin\left(\frac{n2\pi(t_0+T_0)}{T_0}\right) - \sin\left(\frac{n2\pi t_0}{T_0}\right)}{\frac{n2\pi}{T_0}}$$

$$= \frac{\sin\left(\frac{n2\pi t_0}{T_0} + 2\pi n\right) - \sin\left(\frac{n2\pi t_0}{T_0}\right)}{\frac{n2\pi}{T_0}} = \frac{\sin\left(\frac{n2\pi t_0}{T_0}\right) - \sin\left(\frac{n2\pi t_0}{T_0}\right)}{\frac{n2\pi}{T_0}} = 0$$

For $n \neq m$, using the trigonometric identities

$$\sin(\alpha)\sin(\beta) = \frac{1}{2}\cos(\alpha - \beta) - \frac{1}{2}\cos(\alpha + \beta)$$

$$\cos(\alpha)\cos(\beta) = \frac{1}{2}\cos(\alpha + \beta) + \frac{1}{2}\cos(\alpha - \beta)$$

$$\sin(\alpha)\cos(\beta) = \frac{1}{2}\sin(\alpha + \beta) + \frac{1}{2}\sin(\alpha - \beta)$$

we can show that the inner product of $\sin(n2\pi t/T_0)$ and $\sin(m2\pi t/T_0)$, the inner product of $\cos(n2\pi t/T_0)$ and $\cos(m2\pi t/T_0)$, the inner product of $\sin(n2\pi t/T_0)$ and $\cos(m2\pi t/T_0)$ are all zero assuming $m \neq n$. The inner product of $\sin(n2\pi t/T_0)$ and $\sin(m2\pi t/T_0)$ is given by

$$\int_{t_0}^{t_0+T_0} \sin\left(\frac{n2\pi t}{T_0}\right)\sin\left(\frac{m2\pi t}{T_0}\right)dt = \frac{1}{2}\int_{t_0}^{t_0+T_0} \cos\left(\frac{(n-m)2\pi t}{T_0}\right)dt - \frac{1}{2}\int_{t_0}^{t_0+T_0} \cos\left(\frac{(n+m)2\pi t}{T_0}\right)dt$$

$$= \frac{1}{2}\frac{\sin\left(\frac{(n-m)2\pi t}{T_0}\right)\Big|_{t_0}^{t_0+T_0}}{\frac{(n-m)2\pi}{T_0}} - \frac{1}{2}\frac{\sin\left(\frac{(n+m)2\pi t}{T_0}\right)\Big|_{t_0}^{t_0+T_0}}{\frac{(n+m)2\pi}{T_0}}$$

$$= \frac{1}{2}\frac{\sin\left(\frac{(n-m)2\pi(t_0+T_0)}{T_0}\right) - \sin\left(\frac{(n-m)2\pi t_0}{T_0}\right)}{\frac{(n-m)2\pi}{T_0}}$$

$$- \frac{1}{2}\frac{\sin\left(\frac{(n+m)2\pi(t_0+T_0)}{T_0}\right) - \sin\left(\frac{(n+m)2\pi t_0}{T_0}\right)}{\frac{(n+m)2\pi}{T_0}}$$

$$= \frac{1}{2}\frac{\sin\left((n-m)2\pi + \frac{(n-m)2\pi t_0}{T_0}\right) - \sin\left(\frac{(n-m)2\pi t_0}{T_0}\right)}{\frac{(n-m)2\pi}{T_0}}$$

$$- \frac{1}{2}\frac{\sin\left((n+m)2\pi + \frac{(n+m)2\pi t_0}{T_0}\right) - \sin\left(\frac{(n+m)2\pi t_0}{T_0}\right)}{\frac{(n+m)2\pi}{T_0}}$$

$$= \frac{1}{2}\frac{\sin\left(\frac{(n-m)2\pi t_0}{T_0}\right) - \sin\left(\frac{(n-m)2\pi t_0}{T_0}\right)}{\frac{(n-m)2\pi}{T_0}} - \frac{1}{2}\frac{\sin\left(\frac{(n+m)2\pi t_0}{T_0}\right) - \sin\left(\frac{(n+m)2\pi t_0}{T_0}\right)}{\frac{(n+m)2\pi}{T_0}} = 0$$

The inner product of $\cos(n2\pi t/T_0)$ and $\cos(m2\pi t/T_0)$ is given by

$$\int_{t_0}^{t_0+T_0} \cos\left(\frac{n2\pi t}{T_0}\right)\cos\left(\frac{m2\pi t}{T_0}\right)dt = \frac{1}{2}\int_{t_0}^{t_0+T_0}\cos\left(\frac{(n+m)2\pi t}{T_0}\right)dt + \frac{1}{2}\int_{t_0}^{t_0+T_0}\cos\left(\frac{(n-m)2\pi t}{T_0}\right)dt$$

$$= \frac{1}{2}\frac{\sin\left(\frac{(n+m)2\pi t}{T_0}\right)\Big|_{t_0}^{t_0+T_0}}{\frac{(n+m)2\pi}{T_0}} + \frac{1}{2}\frac{\sin\left(\frac{(n-m)2\pi t}{T_0}\right)\Big|_{t_0}^{t_0+T_0}}{\frac{(n-m)2\pi}{T_0}} = 0$$

The inner product of $\sin(n2\pi t/T_0)$ and $\cos(m2\pi t/T_0)$ is given by

$$\int_{t_0}^{t_0+T_0} \sin\left(\frac{n2\pi t}{T_0}\right)\cos\left(\frac{m2\pi t}{T_0}\right)dt = \frac{1}{2}\int_{t_0}^{t_0+T_0}\sin\left(\frac{(n+m)2\pi t}{T_0}\right)dt + \frac{1}{2}\int_{t_0}^{t_0+T_0}\sin\left(\frac{(n-m)2\pi t}{T_0}\right)dt$$

$$= \frac{1}{2}\frac{-\cos\left(\frac{(n+m)2\pi t}{T_0}\right)\Big|_{t_0}^{t_0+T_0}}{\frac{(n+m)2\pi}{T_0}} + \frac{1}{2}\frac{-\cos\left(\frac{(n-m)2\pi t}{T_0}\right)\Big|_{t_0}^{t_0+T_0}}{\frac{(n-m)2\pi}{T_0}} = 0$$

Thus, all pairs of inner product of the set of signals are zero. The energy of 1 is

$$\int_{t_0}^{t_0+T_0} 1^2 dt = T_0$$

The energy of $\cos\left(\frac{n2\pi t}{T_0}\right)$, $n = 1, 2, 3, \ldots$, is

$$\int_{t_0}^{t_0+T_0} \cos^2\left(\frac{n2\pi t}{T_0}\right)dt = \int_{t_0}^{t_0+T_0}\left[\frac{1}{2} + \frac{1}{2}\cos\left(\frac{2n2\pi t}{T_0}\right)\right]dt = \frac{T_0}{2}$$

The energy of $\sin\left(\frac{n2\pi t}{T_0}\right), n = 1, 2, 3, \ldots,$ is

$$\int_{t_0}^{t_0+T_0} \sin^2\left(\frac{n2\pi t}{T_0}\right)dt = \int_{t_0}^{t_0+T_0}\left[\frac{1}{2} - \frac{1}{2}\cos\left(\frac{2n2\pi t}{T_0}\right)\right]dt = \frac{T_0}{2}$$

This concludes the proof.

18.2.5 EXPONENTIAL FOURIER SERIES

A set of complex exponential functions

$$e^{jn\omega_0 t}, \quad n = 0, \pm 1, \pm 2, \pm 3, \ldots$$

is orthogonal and complete over the interval $(t_0, t_0 + T_0)$, where

$$\omega_0 = \frac{2\pi}{T_0} = 2\pi f_0$$

is the **fundamental frequency** in rad/s, and $f_0 = 1/T_0$ is the fundamental frequency in hertz, and t_0 is arbitrary. The most common choices for t_0 are either 0 or $-T_0/2$. If $t_0 = 0$, the interval is $(0, T_0)$, and if $t_0 = -T_0/2$, the interval is $(-T_0/2, T_0/2)$. Since the set of the signals is orthogonal, a square-integrable function $f(t)$ can be represented over the interval $(t_0, t_0 + T_0)$ as follows

$$f(t) = \sum_{n=-\infty}^{\infty} F_n e^{jn\omega_0 t}, \quad t_0 \le t \le t_0 + T_0 \tag{18.25}$$

The coefficients F_0 and $F_n, n = \pm 1, \pm 2, \pm 3, \ldots,$ can be obtained from Equation (18.13) as

$$F_0 = \frac{\int_{t_1}^{t_0+T_0} f(t)e^0 dt}{\int_{t_0}^{t_0+T_0} 1^2 dt} = \frac{1}{T_0}\int_{t_0}^{t_0+T_0} f(t)dt \tag{18.26}$$

$$F_n = \frac{\int_{t_1}^{t_0+T_0} f(t)e^{-jn\omega_0 t} dt}{\int_{t_0}^{t_0+T_0} e^{jn\omega_0 t}e^{-jn\omega_0 t} dt} = \frac{1}{T_0}\int_{t_0}^{t_0+T_0} f(t)e^{-jn\omega_0 t}dt, \quad n = \pm 1, \pm 2, \pm 3, \ldots \tag{18.27}$$

If $f(t)$ is periodic, the representation given by Equation (18.25) is valid for all t; that is, $-\infty < t < \infty$. If $f(t)$ is nonperiodic, the representation given by Equation (18.25) is valid for $t_0 < t < t_0 + T_0$. Unless mentioned otherwise, we assume that $f(t)$ is periodic with period T_0.

The representation given by Equation (18.25) is called *exponential Fourier series expansion,* and F_n is called an *exponential Fourier coefficient*. Coefficient F_0 is given by the integral of the signal $f(t)$ for one period divided by the period T_0. This is the **average value** of the signal or the **dc offset**. The coefficient F_0 is called the **dc component**. If there is no dc offset in the waveform, F_0 is zero. Notice that $F_0 = a_0$. The coefficient F_n measures the correlation (similarity) between $f(t)$ and $e^{jn\omega_0 t}$.

18.2.6 PROOF OF ORTHOGONALITY

The proof that a set of exponential signals

$$\phi_n(t) = e^{jn\frac{2\pi}{T_0}t}, \quad n = 0, \pm 1, \pm 2, \pm 3, \ldots,$$

is orthogonal over $(t_0, t_0 + T_0)$, where t_0 is arbitrary, is presented.

If $m \neq n, m - n \neq 0$, and the inner product of $\phi_m(t)$ and $\phi_n(t)$ is given by

$$\langle \phi_m, \phi_n \rangle = \int_{t_0}^{t_0+T_0} \phi_m(t)\phi_n^*(t)dt = \int_{t_0}^{t_0+T_0} e^{jm\frac{2\pi}{T_0}t} e^{-jn\frac{2\pi}{T_0}t} dt = \int_{t_0}^{t_0+T_0} e^{j(m-n)\frac{2\pi}{T_0}t} dt = \frac{e^{j(m-n)\frac{2\pi}{T_0}t}\Big|_{t_0}^{t_0+T_0}}{j(m-n)\frac{2\pi}{T_0}}$$

$$= \frac{e^{j(m-n)\frac{2\pi}{T_0}(t_0+T_0)} - e^{j(m-n)\frac{2\pi}{T_0}t_0}}{j(m-n)\frac{2\pi}{T_0}} = \frac{e^{j(m-n)\frac{2\pi}{T_0}t_0}(e^{j(m-n)\frac{2\pi}{T_0}T_0} - 1)}{j(m-n)\frac{2\pi}{T_0}} = \frac{e^{j(m-n)\frac{2\pi}{T_0}t_0}(1-1)}{j(m-n)\frac{2\pi}{T_0}} = 0$$

If $m = n, m - n = 0$, and the inner product of $\phi_m(t)$ and $\phi_m(t)$ is given by

$$\langle \phi_m, \phi_m \rangle = \int_{t_0}^{t_0+T_0} \phi_m(t)\phi_m^*(t)dt = \int_{t_0}^{t_0+T_0} e^{jm\frac{2\pi}{T_0}t} e^{-jm\frac{2\pi}{T_0}t} dt = \int_{t_0}^{t_0+T_0} 1dt = T_0$$

Notice that the inner product $\langle \phi_m, \phi_m \rangle$ is the energy of $\phi_m(t)$. Thus, the energy of each basis function in the set is T_0.

The trigonometric Fourier series is dicussed in detail in Section 18.3, and the exponential Fourier series is discussed in detail in Section 18.5.

18.3 Trigonometric Fourier Series

As shown in Section 18.2 earlier in this chapter, a square-integrable function $f(t)$ can be represented over the interval $(t_0, t_0 + T_0)$ by

$$f(t) = a_0 + \sum_{n=1}^{\infty}\left[a_n \cos\left(n\frac{2\pi}{T_0}t\right) + b_n \sin\left(n\frac{2\pi}{T_0}t\right)\right] \tag{18.28}$$

The coefficients a_0, a_n, and b_n are given by

$$a_0 = \frac{1}{T_0}\int_{t_0}^{t_0+T_0} f(t)dt \tag{18.29}$$

$$a_n = \frac{2}{T_0}\int_{t_0}^{t_0+T_0} f(t)\cos\left(n\frac{2\pi}{T_0}t\right)dt, \quad n = 1,2,3,\ldots \tag{18.30}$$

$$b_n = \frac{2}{T_0}\int_{t_0}^{t_1+T_0} f(t)\sin\left(n\frac{2\pi}{T_0}t\right)dt, \quad n = 1,2,3,\ldots \tag{18.31}$$

Notice that $\frac{2\pi}{T_0} = \omega_0 = 2\pi f_0$ is the fundamental frequency in rad/s, and $f_0 = 1/T_0$ is the fundamental frequency in hertz. If $f(t)$ is periodic, the representation given by Equation (18.28) is valid for all t; that is, $-\infty < t < \infty$. If $f(t)$ is nonperiodic, the representation given by Equation (18.28) is valid for $t_0 < t < t_0 + T_0$. Unless mentioned otherwise, we assume that $f(t)$ is periodic. Nonperiodic signals can be represented by Fourier transform, as explained in Chapter 19.

The representation of a signal by sum of sines and cosines given by Equation (18.28) is called the **trigonometric Fourier series** in honor of Jean-Baptiste Joseph Fourier (1768–1830). The coefficients $a_0, a_n, b_n, n = 1, 2, 3, \ldots$ are called **trigonometric Fourier coefficients**.

The lower limit of integration in Equations (8.29) through (8.31) is given by t_0. The value of the lower limit t_0 is arbitrary. As long as the interval of integration is one period, T_0, the starting point of the integration can be any value. Depending on the shape of the signal $f(t)$, we can choose appropriate value for t_0. The most common choices are $t_0 = 0$ or $t_0 = -T_0/2$. If $t_0 = 0$ is chosen, Equations (18.29) through (18.31) become

$$a_0 = \frac{1}{T_0}\int_0^{T_0} f(t)dt \tag{18.32}$$

$$a_n = \frac{2}{T_0}\int_0^{T_0} f(t)\cos\left(n\frac{2\pi}{T_0}t\right)dt, \quad n = 1,2,3,\ldots \tag{18.33}$$

$$b_n = \frac{2}{T_0}\int_0^{T_0} f(t)\sin\left(n\frac{2\pi}{T_0}t\right)dt, \quad n = 1,2,3,\ldots \tag{18.34}$$

If $t_0 = -T_0/2$ is chosen, Equations (18.29) through (18.31) become

$$a_0 = \frac{1}{T_0}\int_{-\frac{T_0}{2}}^{\frac{T_0}{2}} f(t)dt \tag{18.35}$$

$$a_n = \frac{2}{T_0}\int_{-\frac{T_0}{2}}^{\frac{T_0}{2}} f(t)\cos\left(n\frac{2\pi}{T_0}t\right)dt, \quad n = 1,2,3,\ldots \tag{18.36}$$

$$b_n = \frac{2}{T_0}\int_{-\frac{T_0}{2}}^{\frac{T_0}{2}} f(t)\sin\left(n\frac{2\pi}{T_0}t\right)dt, \quad n = 1,2,3,\ldots \tag{18.37}$$

If an even function is multiplied by another even function, the resulting function is also an even function. If an odd function is multiplied by an even function, the resulting function is an odd function. If an odd function is multiplied by another odd function, the resulting function is an even function. In summary, we have

(even function) × (even function) = (even function)
(odd function) × (even function) = (odd function)
(even function) × (odd function) = (odd function)
(odd function) × (odd function) = (even function)

Notice that $\cos\left(n\frac{2\pi}{T_0}t\right)$ is an even function of t, and $\sin\left(n\frac{2\pi}{T_0}t\right)$ is an odd function of t.

If $f(t)$ possesses **even symmetry** so that $f(-t) = f(t)$, then $f(t)\cos(n\omega_0 t)$ is an even function of t, and $f(t)\sin(n\omega_0 t)$ is an odd function of t. Thus, if $f(t)$ is even, Equations (18.35) through (18.37) simplify to

$$a_0 = \frac{2}{T_0}\int_0^{\frac{T_0}{2}} f(t)dt \tag{18.38}$$

$$a_n = \frac{4}{T_0}\int_0^{\frac{T_0}{2}} f(t)\cos\left(n\frac{2\pi}{T_0}t\right)dt, \quad n = 1, 2, 3, \ldots \tag{18.39}$$

$$b_n = \frac{2}{T_0}\int_{-\frac{T_0}{2}}^{\frac{T_0}{2}} f(t)\sin\left(n\frac{2\pi}{T_0}t\right)dt = 0, \quad n = 1, 2, 3, \ldots \tag{18.40}$$

If $f(t)$ possesses **odd symmetry** so that $f(-t) = -f(t)$, then $f(t)\cos(n\omega_0 t)$ is an odd function of t, and $f(t)\sin(n\omega_0 t)$ is an even function of t. Thus, if $f(t)$ is odd, Equations (18.35) through (18.37) simplify to

$$a_0 = \frac{1}{T_0}\int_{-\frac{T_0}{2}}^{\frac{T_0}{2}} f(t)dt = 0 \tag{18.41}$$

$$a_n = \frac{2}{T_0}\int_{-\frac{T_0}{2}}^{\frac{T_0}{2}} f(t)\cos\left(n\frac{2\pi}{T_0}t\right)dt = 0, \quad n = 1, 2, 3, \ldots \tag{18.42}$$

$$b_n = \frac{4}{T_0}\int_0^{\frac{T_0}{2}} f(t)\sin\left(n\frac{2\pi}{T_0}t\right)dt, \quad n = 1, 2, 3, \ldots \tag{18.43}$$

The coefficient a_0 is given by the integral of the signal $f(t)$ for one period divided by the period T_0. This is the average value of the signal $f(t)$ or dc offset. The coefficient a_0 is called the **dc component**. If there is no dc offset in the waveform, a_0 is zero.

The first term in the Fourier sum given by Equation (18.28) is

$$a_1\cos\left(\frac{2\pi}{T_0}t\right) + b_1\sin\left(\frac{2\pi}{T_0}t\right) = a_1\cos(\omega_0 t) + b_1\sin(\omega_0 t)$$

The radian frequency of the cosine and the sine is given by $\omega_0 = 2\pi/T_0$ rad/s. This frequency is called the **fundamental frequency** because all the other terms in the Fourier sum have frequencies that are integer multiples of ω_0. In terms of hertz (Hz), the fundamental frequency is given by $f_0 = 1/T_0$ Hz. The coefficients a_1 and b_1 are called fundamental components of the Fourier coefficients. The coefficient a_1 is a measure of similarity between the signal $f(t)$ and $\cos(2\pi t/T_0)$. Similarly, the coefficient b_1 is a measure of similarity between the signal $f(t)$ and $\sin(2\pi t/T_0)$. If $f(t)$ is very similar in shape with $\cos(2\pi t/T_0)$, the coefficient a_1 is large. Otherwise, a_1 is small. The second term in the sum is given by

$$a_2 \cos\left(2\frac{2\pi}{T_0}t\right) + b_2 \sin\left(2\frac{2\pi}{T_0}t\right) = a_2 \cos(2\omega_0 t) + b_2 \sin(2\omega_0 t)$$

The radian frequency of the cosine and the sine is given by $2\omega_0 = 4\pi/T_0$ rad/s. This frequency is called the **second harmonic frequency** because it is twice the fundamental frequency ω_0. In terms of hertz (Hz), the second harmonic frequency is given by $2f_0 = 2/T_0$ Hz. The coefficients a_2 and b_2 are called the second harmonic component of the Fourier coefficients. The coefficient a_2 is a measure of similarity between the signal $f(t)$ and $\cos(4\pi t/T_0)$. Similarly, the coefficient b_2 is a measure of similarity between the signal $f(t)$ and $\sin(4\pi t/T_0)$. If $f(t)$ is very similar in shape with $\sin(4\pi t/T_0)$, the coefficient b_2 is large. Otherwise, b_2 is small. Similarly, we can interpret all the rest of the terms in the Fourier sum.

18.3.1 TRIGONOMETRIC FOURIER SERIES USING COSINES ONLY

The two terms $a_n \cos(n\omega_0 t)$ and $b_n \sin(n\omega_0 t)$ in the Fourier series representation

$$f(t) = a_0 + \sum_{n=1}^{\infty}[a_n \cos(n\omega_0 t) + b_n \sin(n\omega_0 t)]$$

can be rewritten as

$$f(t) = a_0 + \sum_{n=1}^{\infty}\sqrt{a_n^2 + b_n^2}\left[\frac{a_n}{\sqrt{a_n^2 + b_n^2}}\cos(n\omega_0 t) - \frac{-b_n}{\sqrt{a_n^2 + b_n^2}}\sin(n\omega_0 t)\right]$$

Let

$$c_0 = a_0 \quad (18.44)$$

$$c_n = \sqrt{a_n^2 + b_n^2} \quad (18.45)$$

$$\theta_n = \tan^{-1}\left(\frac{-b_n}{a_n}\right) = -\tan^{-1}\left(\frac{b_n}{a_n}\right) \quad (18.46)$$

Refer to Figure 18.10 for the definitions of c_n and θ_n. Using the definitions of c_n and θ_n, we can rewrite $f(t)$ as

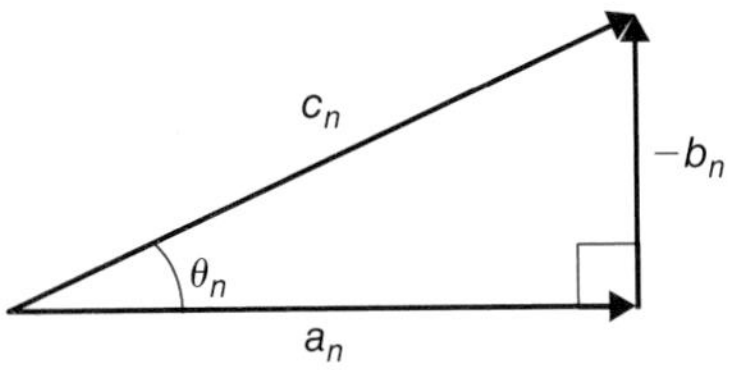

FIGURE 18.10

Definition of c_n and θ_n.

$$\begin{aligned} f(t) &= c_0 + \sum_{n=1}^{\infty} c_n\left[\frac{a_n}{c_n}\cos(n\omega_0 t) - \frac{-b_n}{c_n}\sin(n\omega_0 t)\right] \\ &= c_0 + \sum_{n=1}^{\infty} c_n[\cos(\theta_n)\cos(n\omega_0 t) - \sin(\theta_n)\sin(n\omega_0 t)] \\ &= c_0 + \sum_{n=1}^{\infty} c_n \cos(n\omega_0 t + \theta_n) \end{aligned} \quad (18.47)$$

Notice that $\theta_0 = \angle c_0 = \angle a_0$
If $a_0 \geq 0$, $\angle a_0 = 0$, and if $a_0 < 0$, $|a_0| = -a_0$, $\angle a_0 = \pi = 180°$ or $\angle a_0 = -\pi = -180°$.
If $b_n = 0$ for $n \geq 1$, we have

$$f(t) = a_0 + \sum_{n=1}^{\infty} a_n \cos(n\omega_0 t) = c_0 + \sum_{n=1}^{\infty} c_n \cos(n\omega_0 t + \theta_n)$$

The magnitude is given by

$$c_n = |a_n|$$

and the phase is given by

$$\theta_n = \begin{cases} 0, & \text{if } a_n \geq 0 \\ \pi, & \text{if } a_n < 0 \end{cases} \quad \text{or} \quad \theta_n = \begin{cases} 0, & \text{if } a_n \geq 0 \\ 180°, & \text{if } a_n < 0 \end{cases}$$

Notice that $\theta_n = \pi$ (180°) is equivalent to $\theta_n = -\pi$ (−180°).
If $a_n = 0$ for $n \geq 1$

$$f(t) = a_0 + \sum_{n=1}^{\infty} b_n \sin(n\omega_0 t) = a_0 + \sum_{n=1}^{\infty} b_n \cos(n\omega_0 t - 90°) = c_0 + \sum_{n=1}^{\infty} c_n \cos(n\omega_0 t + \theta_n)$$

The magnitude is given by

$$c_n = |b_n|$$

and the phase is given by

$$\theta_n = \begin{cases} \dfrac{-\pi}{2}, & \text{if } b_n \geq 0 \\ \dfrac{\pi}{2}, & \text{if } b_n < 0 \end{cases} \quad \text{or} \quad \theta_n = \begin{cases} -90°, & \text{if } b_n \geq 0 \\ 90°, & \text{if } b_n < 0 \end{cases}$$

If $a_n > 0$ and $b_n \neq 0$,

$$\theta_n = -\tan^{-1}\left(\frac{b_n}{a_n}\right)$$

If $a_n < 0$ and $b_n > 0$,

$$\theta_n = -\pi - \tan^{-1}\left(\frac{b_n}{a_n}\right)$$

If $a_n < 0$ and $b_n < 0$,

$$\theta_n = \pi - \tan^{-1}\left(\frac{b_n}{a_n}\right)$$

Table 18.3, at the end of Chapter 18, shows Fourier coefficients and Fourier series representation of common signals.

18.3.2 ONE-SIDED MAGNITUDE SPECTRUM AND ONE-SIDED PHASE SPECTRUM

If the magnitudes $|c_0|, c_1, c_2, c_3, \ldots$ are plotted against frequencies, in radians/s, $0, \omega_0 = 2\pi/T_0, 2\omega_0 = 4\pi/T_0, 3\omega_0 = 6\pi/T_0, \ldots$ (or in hertz, $0, f_0 = 1/T_0, 2f_0 = 2/T_0, 3f_0 = 3/T_0, \ldots$), we obtain what is referred to as a *one-sided magnitude spectrum.* The magnitudes $|c_0|, c_1, c_2, c_3 \ldots$ are nonnegative. The one-sided magnitude spectrum shows the magnitudes of the dc component, fundamental frequency component, second harmonic frequency component, third harmonic frequency component, etc. The magnitude spectrum reveals the relative strength of the various harmonic frequency components of the given signal. If the signal changes slowly with time (smooth signal or low-pass signal), the magnitudes of lower frequency components are stronger than those of higher frequency components. If the signal changes rapidly with time (such as bandpass signals), the magnitudes of certain harmonic components are higher than others. On the other hand, if the phases $\theta_0 = \angle c_0$, $\theta_1, \theta_2, \theta_3, \ldots$ are plotted against frequencies, in radians, $0, 2\pi/T_0, 4\pi/T_0, 6\pi/T_0, \ldots$ (or, in hertz, $0, f_0, 2f_0, 3f_0, \ldots$), we obtain what is referred to as a *one-sided phase spectrum*. The one-sided phase spectrum shows the phases of the dc component, fundamental frequency component, second harmonic frequency component, third harmonic frequency component, etc. The phase spectrum reveals the phase values of the various harmonic frequency components of the given signal.

Consider a rectangular pulse train with amplitude A, period T_0, and pulse width τ, as shown in Figure 18.11.

FIGURE 18.11

A rectangular pulse train.

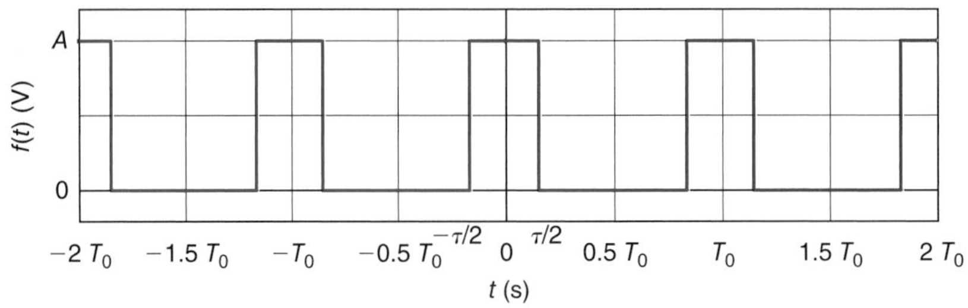

The duty cycle of a pulse train is defined as the ratio of pulse width to period. Thus, we have

$$d = \frac{\tau}{T_0} \tag{18.48}$$

The dc component of $f(t)$ shown in Figure 18.11 is given by

$$a_0 = \frac{1}{T_0}\int_{-\frac{T_0}{2}}^{\frac{T_0}{2}} f(t)dt = \frac{1}{T_0}\int_{\frac{-\tau}{2}}^{\frac{\tau}{2}} Adt = \frac{A}{T_0}[t]_{-\frac{\tau}{2}}^{\frac{\tau}{2}} = \frac{A\tau}{T_0} = Ad \tag{18.49}$$

Notice that the dc component is given by the area of the pulse, $A\tau$, divided by the period T_0. Since $f(t)$ possesses even symmetry, from Equation (18.40), we have

$$b_n = 0, \quad n = 1, 2, 3, \ldots \tag{18.50}$$

The Fourier coefficient a_n is given by, from Equation (18.39)

$$\begin{aligned} a_n &= \frac{4}{T_0}\int_0^{\frac{T_0}{2}} f(t)\cos\left(n\frac{2\pi}{T_0}t\right)dt = \frac{4}{T_0}\int_0^{\frac{\tau}{2}} A\cos\left(n\frac{2\pi}{T_0}t\right)dt = \frac{4A\sin\left(n\frac{2\pi}{T_0}t\right)\Big|_0^{\frac{\tau}{2}}}{T_0 n\frac{2\pi}{T_0}} \\ &= \frac{4A\sin\left(n\frac{2\pi}{T_0}\frac{\tau}{2}\right)}{T_0 n\frac{2\pi}{T_0}} = \frac{2A\sin\left(\pi n\frac{\tau}{T_0}\right)}{\pi n} = \frac{2A\frac{\tau}{T_0}\sin\left(\pi n\frac{\tau}{T_0}\right)}{\pi n\frac{\tau}{T_0}} \\ &= 2A\frac{\tau}{T_0}\operatorname{sinc}\left(n\frac{\tau}{T_0}\right) = 2Ad\operatorname{sinc}(nd), \quad n = 1, 2, 3, \ldots \end{aligned} \tag{18.51}$$

FIGURE 18.12

Plot of sinc(x).

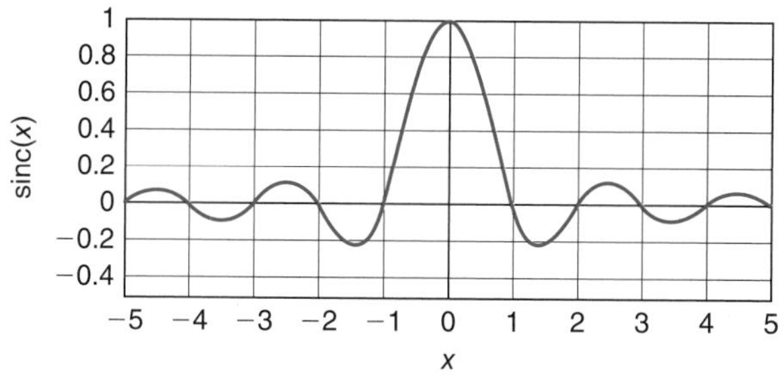

where the sinc function is defined as

$$\text{sinc}(x) = \frac{\sin(\pi x)}{\pi x} \quad \textbf{(18.52)}$$

Figure 18.12 shows sinc(x) as a function of x. From L'Hospital's rule, we have

$$\begin{aligned} \lim_{x\to 0} \text{sinc}(x) &= \lim_{x\to 0} \frac{\sin(\pi x)}{\pi x} \\ &= \lim_{x\to 0} \frac{\pi \cos(\pi x)}{\pi} = 1 \end{aligned} \quad \textbf{(18.53)}$$

For integer values of x other than 0, sinc(x) = 0. Notice that sinc function oscillates between positive and negative and decays toward zero as $|x|$ is increased from zero to infinity, $\sim 1/(\pi|x|)$.

The trigonometric Fourier series representation is given by

$$f(t) = Ad + \sum_{n=1}^{\infty} 2Ad \,\text{sinc}(nd) \cos\left(n\frac{2\pi}{T_0}t\right) \quad \textbf{(18.54)}$$

or

$$f(t) = c_0 + \sum_{n=1}^{\infty} c_n \cos(n\omega_0 t + \theta_n) = Ad + \sum_{n=1}^{\infty} |2Ad \,\text{sinc}(nd)| \cos(n\omega_0 t + \theta_n) \quad \textbf{(18.55)}$$

Notice that

$$c_0 = \text{Ad} \quad \textbf{(18.56)}$$

$$c_n = |\text{2Ad sinc(nd)}| \quad \textbf{(18.57)}$$

$$\theta_n = \begin{cases} 0, & \textit{if } Ad \,\text{sinc}(nd) \geq 0 \\ 180°, & \textit{if } Ad \,\text{sinc}(nd) < 0 \end{cases} \quad \textbf{(18.58)}$$

When Ad sinc(nd) is negative, the phase angle is 180° (or −180°) because Ad sinc(nd) is in the negative real axis. The spacings between the spectral lines is $f_0 = 1/T_0$ Hz, and the zero crossing points are multiples of $1/\tau$. Each $1/\tau$ spacing between the neighboring zero crossing points is called a lobe. There are $(1/\tau)/(1/T_0) = T_0/\tau$ spectral lines per lobe.

EXAMPLE 18.7

Let $A = 1$ V, $T_0 = 1$ ms, $\tau = 0.5$ ms for the rectangular pulse train shown in Figure 18.11. This pulse train is shown in Figure 18.13.

a. Find f_0, ω_0, d, a_0, a_n, b_n, c_n, and θ_n.
b. Evaluate a_n, c_n, and θ_n for $0 \leq n \leq 10$, and plot as a function of frequency.
c. Approximate $f(t)$ by the dc component and the first ten harmonics and plot the approximation.
d. Write a MATLAB script to answer (a)–(c).

continued

Example 18.7 continued

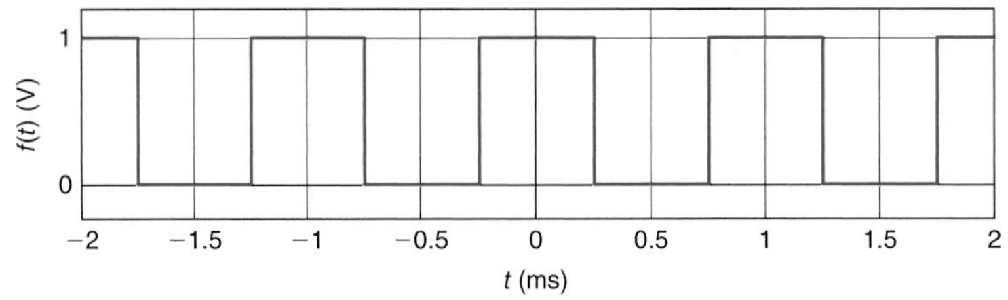

FIGURE 18.13

A rectangular pulse train with duty cycle 0.5.

a. $f_0 = \dfrac{1}{T_0} = \dfrac{1}{0.001s} = 1000\text{ Hz} = 1\text{ kHz}$

$$\omega_0 = \frac{2\pi}{T_0} = \frac{2\pi}{0.001s} = 2\pi \times 1000\text{ rad/s} = 6283.1853\text{ rad/s}$$

$$d = \frac{\tau}{T_0} = \frac{0.5\text{ ms}}{1\text{ ms}} = 0.5$$

$$a_0 = Ad = 1 \times \frac{1}{2} = 0.5$$

$$a_n = 2Ad\,\text{sinc}(nd) = 2 \times 1 \times 0.5 \times \text{sinc}(n0.5) = \text{sinc}(0.5n), \quad n = 1, 2, 3, \ldots$$

$$b_n = 0, \quad n = 1, 2, 3, \ldots$$

$$c_0 = a_0 = 0.5$$

$$c_n = \sqrt{a_n^2 + b_n^2} = |a_n| = |\text{sinc}(0.5n)|$$

$$\theta_n = \begin{cases} 0, & a_n \geq 0 \\ 180°, & a_n < 0 \end{cases} = \begin{cases} 0, & n = 0, 1, 2, 4, 5, 6, 8, 9, 10, 12, 13, 14, \ldots \\ 180°, & n = 3, 7, 11, 15, \ldots \end{cases}$$

b.

n:	0	1	2	3	4	5	6	7	8	9	10
a_n:	0.5	0.6366	0	-0.2122	0	0.1273	0	-0.09095	0	0.07074	0
c_n:	0.5,	0.6366	0	0.2122	0	0.1273	0	0.09095	0	0.07074	0
θ_n:	0	0	0	180	0	0	0	180	0	0	0

The spectrums a_n, c_n, θ_n are shown in Figure 18.14.

FIGURE 18.14

Plot of a_n, c_n, and θ_n.

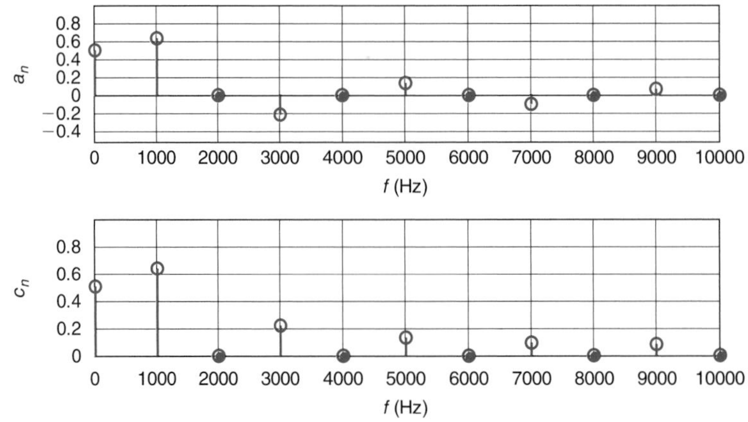

continued

Example 18.7 continued

FIGURE 18.14

continued

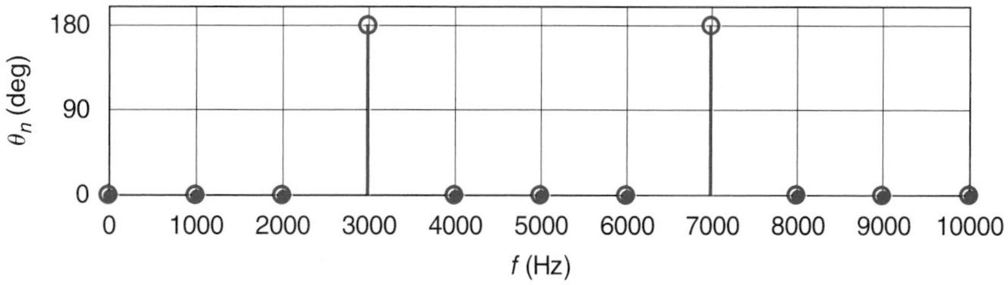

c. $\hat{f}(t) = 0.5 + 0.6366\cos(2\pi 1000t) - 0.2122\cos(2\pi 3000t) + 0.1273\cos(2\pi 5000t) - 0.09095\cos(2\pi 7000t) + 0.07074\cos(2\pi 9000t)$

or

$\hat{f}(t) = 0.5 + 0.6366\cos(2\pi 1000t) + 0.2122\cos(2\pi 3000t + \pi) + 0.1273\cos(2\pi 5000t) + 0.09095\cos(2\pi 7000t + \pi) + 0.07074\cos(2\pi 9000t)$

The approximation $\hat{f}(t)$ shown in Figure 18.15 shows overshoot (ringing) when the original pulse train changes level from 0 V to 1 V and 1 V to 0 V. This phenomenon is called the **Gibbs phenomenon**. Figure 18.16 shows the approximation $\hat{f}(t)$ as more harmonics are added.

FIGURE 18.15

Plot of $f(t)$ and $\hat{f}(t)$.

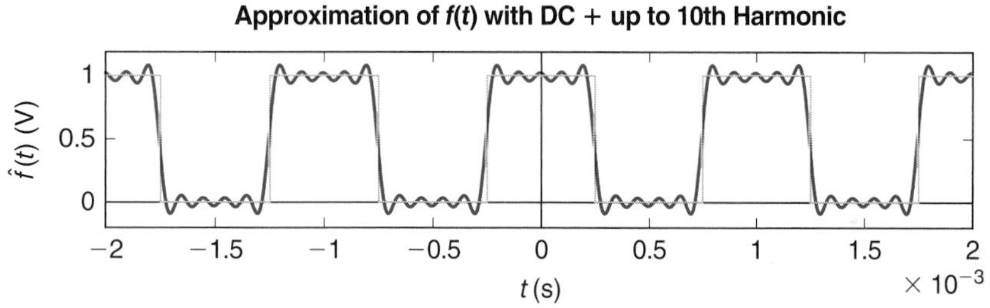

FIGURE 18.16

Approximation of rectangular pulse train by Fourier series as more harmonics are added.

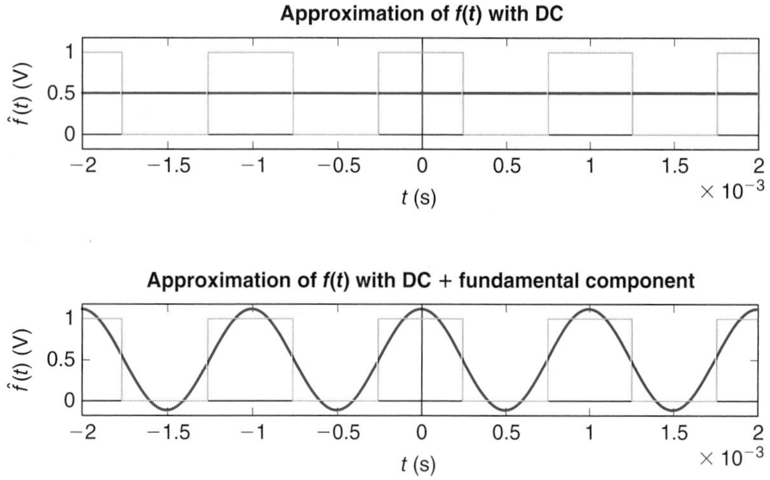

continued

Example 18.7 continued

FIGURE 18.16

continued

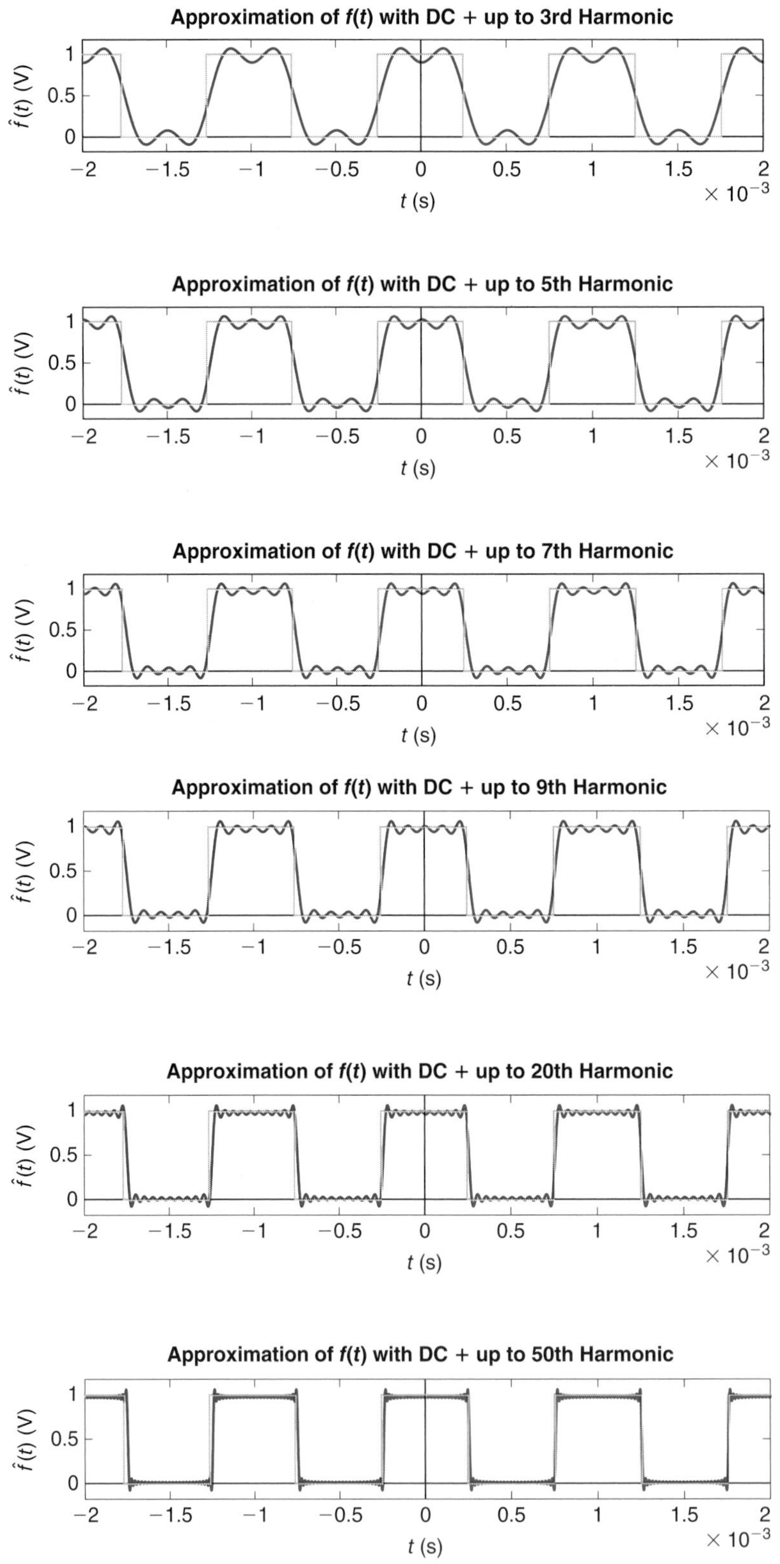

continued

Example 18.7 continued

FIGURE 18.16

continued

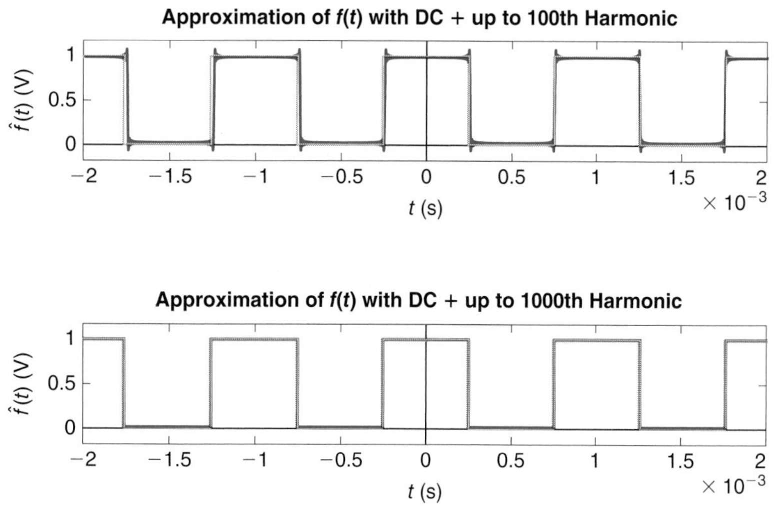

d. A MATLAB script is shown here:

```
% EXAMPLE 18.7
clear all;format short;
A=1;T0=1e-3;tau=0.5e-3;N=10;M1=2;M2=1000;Th=1e-10;
f0=1/T0
w0=2*pi/T0
d=tau/T0
a0=A*tau/T0
n1=1:N;
n=0:N;
a=2*A*d*sinc(n1*d);
a=[a0,a]
b=zeros(1,N+1)
c=sqrt(a.^2+b.^2)
theta=zeros(1,N+1);
for i=1:length(a)
    if (abs(a(i))<=Th)&&(abs(b(i))<=Th)
        theta(i)=0;
    elseif (a(i)<-Th)&&(abs(b(i))<=Th)
        theta(i)=pi;
    elseif (a(i)>Th)&&(abs(b(i))<=Th)
        theta(i)=0;
    elseif (abs(a(i))<Th)&&(b(i)>Th)
        theta(i)=-pi/2;
    elseif (abs(a(i))<Th)&&(b(i)<-Th)
        theta(i)=pi/2;
    else
        theta(i)=-atan2(b(i),a(i));
    end
end
theta
thetad=theta*180/pi
f1=f0*n;
figure(1)
stem(f1,a,'LineWidth',2);grid;xlabel('f (Hz)');ylabel('a_n');
```

continued

Example 18.7 continued

```
set(gca,'YTick',-0.4:0.2:0.8)
set(gca,'YTickLabel',-0.4:0.2:0.8)
axis([0,N*f0,-0.4,0.8])
figure(2)
stem(f1,c,'LineWidth',2);grid;xlabel('f (Hz)');ylabel('c_n');
set(gca,'YTick',0:0.2:0.8)
set(gca,'YTickLabel',0:0.2:0.8)
axis([0,N*f0,0,0.8])
figure(3)
stem(f1,thetad,'LineWidth',2);grid;
xlabel('f (Hz)');ylabel('\theta_n (deg)');
set(gca,'YTick',0:90:180)
set(gca,'YTickLabel',0:90:180)
axis([0,N*f0,0,180])
%
t=-M1*T0:T0/M2:M1*T0;
f0a=0.5*(1+square(2*pi*(t+0.25*T0)/T0,50));
M=10;
f=a(1);
for i=2:M+1
    f=f+a(i)*cos(2*pi*(i-1)*t/T0)+b(i)*sin(2*pi*(i-1)*t/T0);
end
figure(4)
plot(t,f,'LineWidth',2);grid;hold on;
plot(t,f0a,'r-');xlabel('t');ylabel('f^\wedge(t)');hold off;
axis([-M1*T0,M1*T0,-0.2*A,1.2*A]);
title(['Approximation of f(t) with DC + up to ',num2str(M),'th Harmonic']);
```

Exercise 18.7

Let $A = 1$ V, $T_0 = 1$ ms, $\tau = 0.25$ ms for the rectangular pulse train shown in Figure 18.11. The pulse train is shown in Figure 18.17.

a. Find f_0, ω_0, d, a_0, a_n, b_n, c_n, and θ_n.
b. Evaluate a_n, c_n, and θ_n for $0 \le n \le 10$, and plot as a function of frequency.
c. Approximate $f(t)$ by the dc component and the first 10 harmonics, and plot the approximation.

FIGURE 18.17

A rectangular pulse train with duty cycle 0.25.

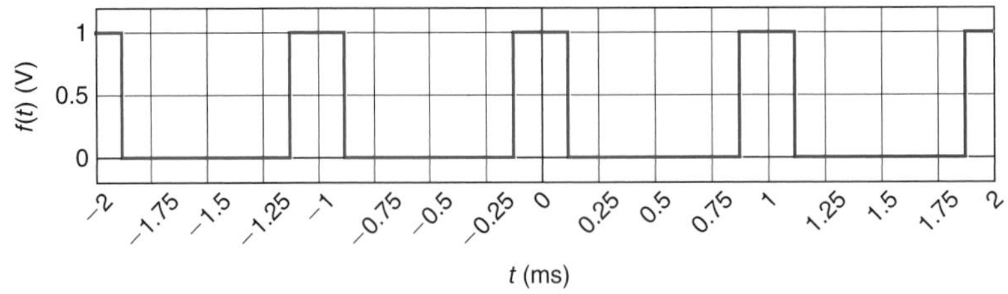

a. $f_0 = \dfrac{1}{T_0} = \dfrac{1}{0.001s} = 1000 \text{ Hz} = 1 \text{ kHz}$

continued

Exercise 18.7 continued

$$\omega_0 = \frac{2\pi}{T_0} = \frac{2\pi}{0.001s} = 2\pi \times 1000 \text{ rad/s} = 6283.1853 \text{ rad/s}$$

$$d = \frac{\tau}{T_0} = \frac{0.25 \text{ ms}}{1 \text{ ms}} = 0.25$$

$$a_n = 2Ad\,\text{sinc}(nd) = 2 \times 1 \times 0.25 \times \text{sinc}(n0.25)$$
$$= 0.5 \times \text{sinc}(0.25n), \quad n = 1, 2, 3, \ldots$$

$$b_n = 0, \quad n = 1, 2, 3, \ldots$$

$$c_n = \sqrt{a_n^2 + b_n^2} = |a_n| = 0.5|\text{sinc}(0.25n)|$$

$$\theta_n = \begin{cases} 0, & a_n \geq 0 \\ 180°, & a_n < 0 \end{cases} = \begin{cases} 0, & n = 0, 1, 2, 3, 4, 8, 9, 10, 11, 12, 16, 17, \ldots \\ 180°, & n = 5, 6, 7, 13, 14, 15, \ldots \end{cases}$$

b.

n:	0	1	2	3	4	5	6	7	8	9	10
a_n:	0.25	0.4502	0.3183	0.1501	0	0.09003	0.1061	-0.06431	0	0.05002	0.06366
c_n:	0.25	0.4502	0.3183	0.1501	0	0.09003	0.1061	0.06431	0	0.05002	0.06366
θ_n:	0	0	0	0	0	180	180	180	0	0	0

The spectrums a_n, c_n, θ_n are shown in Figure 18.18.

FIGURE 18.18

Plot of a_n, c_n, and θ_n.

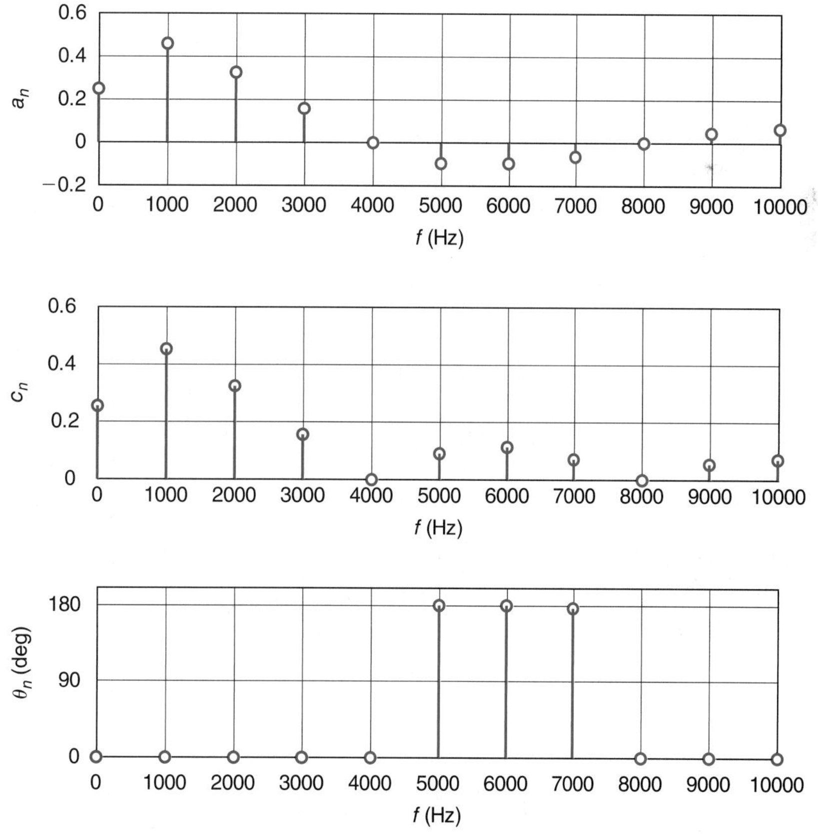

continued

Exercise 18.7 continued

c. $\hat{f}(t) = 0.25 + 0.4502\cos(2\pi 1000t) + 0.3183\cos(2\pi 2000t) + 0.1501\cos(2\pi 3000t) - 0.09003\cos(2\pi 5000t) - 0.1061\cos(2\pi 6000t) - 0.06431\cos(2\pi 7000t) + 0.05002\cos(2\pi 9000t) + 0.06366\cos(2\pi 10{,}000t)$

or

$\hat{f}(t) = 0.25 + 0.4502\cos(2\pi 1000t) + 0.3183\cos(2\pi 2000t) + 0.1501\cos(2\pi 3000t) + 0.09003\cos(2\pi 5000t + \pi) + 0.1061\cos(2\pi 6000t + \pi) + 0.06431\cos(2\pi 7000t + \pi) + 0.05002\cos(2\pi 9000t) + 0.06366\cos(2\pi 10{,}000t)$

The approximation is shown in Figure 18.19.

FIGURE 18.19

Plot of $f(t)$ and $\hat{f}(t)$.

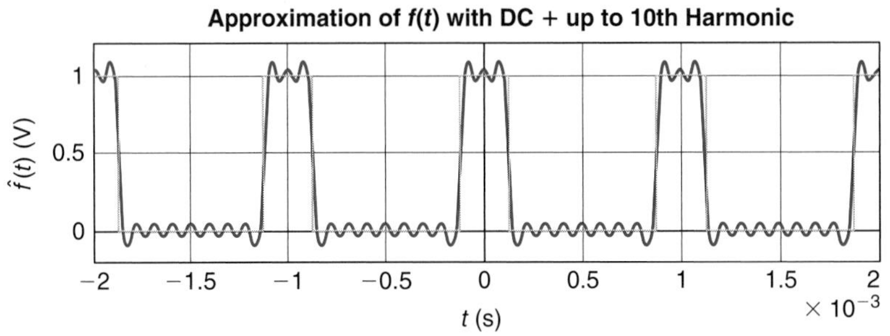

18.3.3 DC LEVEL

The Fourier series representation of a dc signal $f(t) = A_{dc}$ is the signal itself, since $a_0 = A_{dc}$ (the average value of $f(t)$), $a_n = 0$ for all n, and $b_n = 0$ for all n because the area under integer number of cycles of sine and cosine is zero. If a signal can be written as a sum of dc and some other signal such as $g(t) = A_{dc} + f(t)$, then the Fourier coefficient for $g(t)$ is identical to those of $f(t)$ except the dc component, which is given by the dc component of $f(t)$ plus A_{dc}. In certain cases, it is convenient to find the Fourier coefficient of $f(t)$ rather than those of $g(t)$, or the Fourier coefficient of $f(t)$ is already known. Then the Fourier coefficient of $g(t)$ is identical to those of $f(t)$ except the dc component. In conclusion, adding a dc bias to a signal affects the dc component of the Fourier series representation, but it has no effects on the other components. The dc component of $g(t)$ can also be found by finding the average value of the signal; that is, by integrating the signal for one period and then dividing by the period T_0:

$$a_0 = \frac{1}{T_0}\int_0^{T_0} g(t)dt$$

EXAMPLE 18.8

Find the dc component of $f(t)$ shown in Figure 18.20.

Since the dc component is the average value of $f(t)$, we find the area under $f(t)$ for one period and divide by the period. The period is $T_0 = 10$ s. From -8 s to -6 s, the area under $f(t)$ is

$$0.6 \times 2 = 1.2$$

continued

Example 18.8 continued

FIGURE 18.20

Waveform for EXAMPLE 18.8.

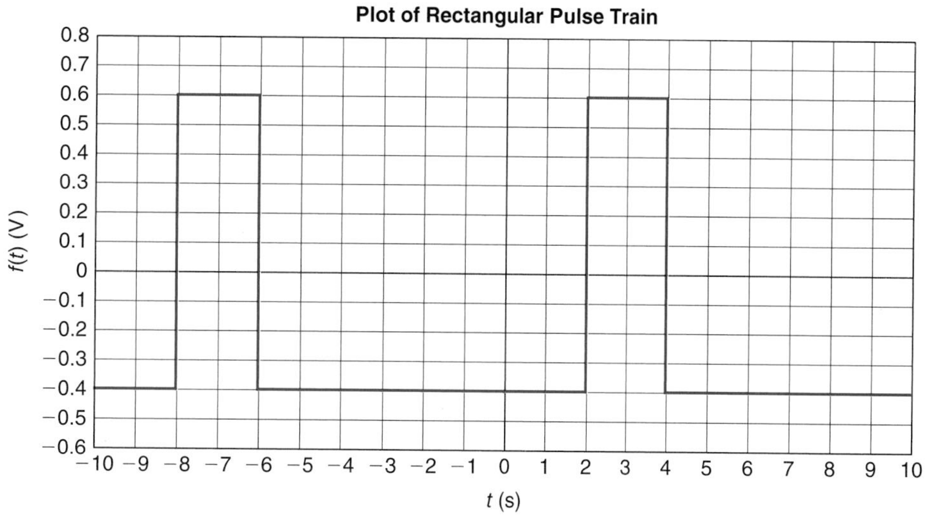

From −6s to 2s, the area under $f(t)$ is

$$-0.4 \times 8 = -3.2$$

Thus, the dc component is given by

$$a_0 = c_0 = F_0 = \frac{1.2 - 3.2}{10} = -0.2 \text{ V}$$

Exercise 18.8

Find the dc component of $f(t)$ shown in Figure 18.21.

FIGURE 18.21

Waveform for EXERCISE 18.8.

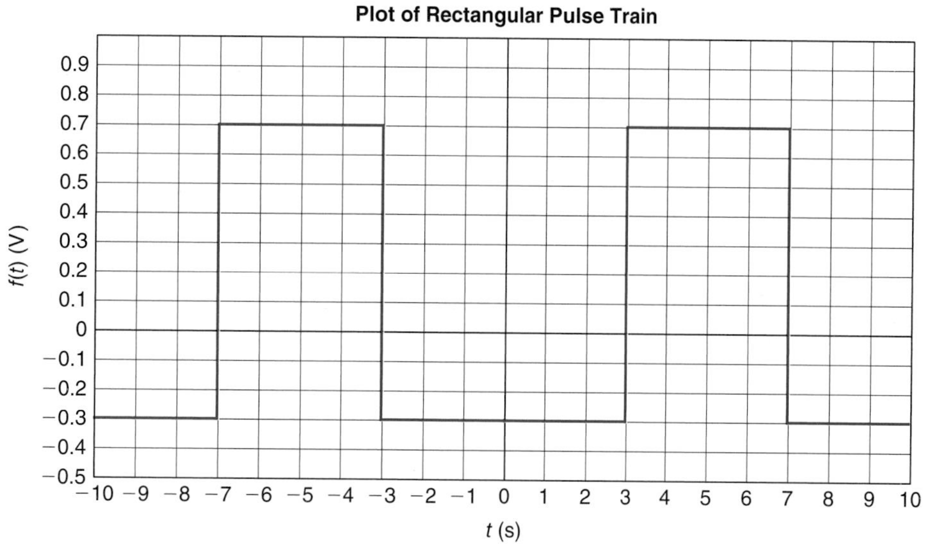

Answer:

$a_0 = c_0 = F_0 = 0.1$ V.

18.3.4 TIME SHIFTING

The Fourier coefficients of time-shifted signal $f(t - t_d)$ can be found from the Fourier coefficient of $f(t)$. The Fourier series representation of $f(t)$ is given by

$$f(t) = a_0 + \sum_{n=1}^{\infty}\left[a_n \cos\left(n\frac{2\pi}{T_0}t\right) + b_n \sin\left(n\frac{2\pi}{T_0}t\right)\right] = c_0 + \sum_{n=1}^{\infty} c_n \cos\left(n\frac{2\pi}{T_0}t + \theta_n\right)$$

Let $g(t) = f(t - t_d)$. Let the Fourier series representation of $g(t)$ be

$$g(t) = ga_0 + \sum_{n=1}^{\infty}\left[ga_n \cos\left(n\frac{2\pi}{T_0}t\right) + gb_n \sin\left(n\frac{2\pi}{T_0}t\right)\right]$$

Then, we have

$$\begin{aligned}
g(t) = f(t - t_d) &= a_0 + \sum_{n=1}^{\infty}\left[a_n \cos\left(n\frac{2\pi}{T_0}(t - t_d)\right) + b_n \sin\left(n\frac{2\pi}{T_0}(t - t_d)\right)\right] \\
&= a_0 + \sum_{n=1}^{\infty}\left[a_n \cos\left(n\frac{2\pi}{T_0}t_d\right)\cos\left(n\frac{2\pi}{T_0}t\right) + a_n \sin\left(n\frac{2\pi}{T_0}t_d\right)\sin\left(n\frac{2\pi}{T_0}t\right)\right] \\
&\quad + \sum_{n=1}^{\infty}\left[b_n \cos\left(n\frac{2\pi}{T_0}t_d\right)\sin\left(n\frac{2\pi}{T_0}t\right) - b_n \sin\left(n\frac{2\pi}{T_0}t_d\right)\cos\left(n\frac{2\pi}{T_0}t\right)\right] \\
&= a_0 + \sum_{n=1}^{\infty}\left[a_n \cos\left(n\frac{2\pi}{T_0}t_d\right) - b_n \sin\left(n\frac{2\pi}{T_0}t_d\right)\right]\cos\left(n\frac{2\pi}{T_0}t\right) \\
&\quad + \sum_{n=1}^{\infty}\left[a_n \sin\left(n\frac{2\pi}{T_0}t_d\right) + b_n \cos\left(n\frac{2\pi}{T_0}t_d\right)\right]\sin\left(n\frac{2\pi}{T_0}t\right)
\end{aligned}$$

Therefore, we have

$$ga_0 = a_0 \tag{18.59}$$

$$ga_n = a_n \cos\left(n\frac{2\pi}{T_0}t_d\right) - b_n \sin\left(n\frac{2\pi}{T_0}t_d\right) \tag{18.60}$$

$$gb_n = a_n \sin\left(n\frac{2\pi}{T_0}t_d\right) + b_n \cos\left(n\frac{2\pi}{T_0}t_d\right) \tag{18.61}$$

Using the cosine-only representation, we obtain

$$\begin{aligned}
g(t) = f(t - t_d) &= c_0 + \sum_{n=1}^{\infty} c_n \cos\left(n\frac{2\pi}{T_0}(t - t_d) + \theta_n\right) \\
&= c_0 + \sum_{n=1}^{\infty} c_n \cos\left(n\frac{2\pi}{T_0}t + \theta_n - n\frac{2\pi}{T_0}t_d\right)
\end{aligned}$$

The magnitude of $g(t)$ is the same as that of $f(t)$, and the phase of $g(t)$ is given by

$$g\theta_n = \theta_n - n\omega_0 t_d \tag{18.62}$$

The time shifting introduces the linear phase term $-n\omega_0 t_d$. Notice that the time shifting does not change the dc component.

EXAMPLE 18.9

Find the Fourier coefficients for the rectangular pulse train with amplitude A, period T_0, and pulse width τ shown in Figure 18.22.

FIGURE 18.22

A rectangular pulse train with time shift of $\tau/2$ to the right.

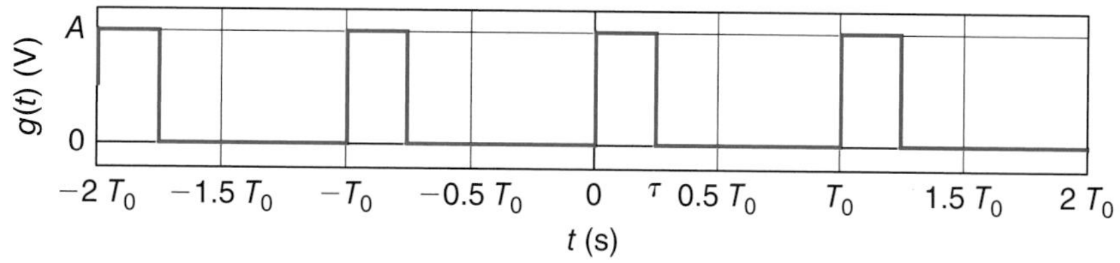

The rectangular pulse train shown in Figure 18.22 is the time shift of the rectangular pulse train shown in Figure 18.11 by $\tau/2$ to the right; that is, $g(t) = f(t - \tau/2)$. The Fourier coefficients of the rectangular pulse train shown in Figure 18.11 are given in Equations (8.49) through (8.51), and Equations (8.56) through (8.58).

Let the Fourier series representation of $g(t)$ be

$$g(t) = ga_0 + \sum_{n=1}^{\infty}\left[ga_n \cos\left(n\frac{2\pi}{T_0}t\right) + gb_n \sin\left(n\frac{2\pi}{T_0}t\right)\right]$$

$$= gc_0 + \sum_{n=1}^{\infty} gc_n \cos\left(n\frac{2\pi}{T_0}t + g\theta_n\right)$$

Then, from Equations (18.59) through (18.62), since $b_n = 0$, we have

$$ga_0 = gc_0 = a_0 = \text{Ad} \tag{18.63}$$

$$ga_n = a_n \cos\left(n\frac{2\pi}{T_0}t_d\right) - b_n \sin\left(n\frac{2\pi}{T_0}t_d\right)$$
$$= a_n \cos\left(n\frac{2\pi}{T_0}\frac{\tau}{2}\right) = 2Ad\,\text{sinc}(nd)\cos(n\pi d) \tag{18.64}$$

$$gb_n = a_n \sin\left(n\frac{2\pi}{T_0}t_d\right) + b_n \cos\left(n\frac{2\pi}{T_0}t_d\right)$$
$$= a_n \sin\left(n\frac{2\pi}{T_0}\frac{\tau}{2}\right) = 2Ad\,\text{sinc}(nd)\sin(n\pi d) \tag{18.65}$$

$$gc_n = \sqrt{ga_n^2 + gb_n^2} = |2Ad\,\text{sinc}(nd)| = c_n \tag{18.66}$$

$$g\theta_n = \theta_n - n\pi d \tag{18.67}$$

Exercise 18.9

The rectangular pulse train shown in Figure 18.23 has amplitude $A = 1$ V, period $T_0 = 1$ ms, and pulse width $\tau = 250\ \mu$s. Find the Fourier coefficients, and plot the magnitude spectrum and the phase spectrum for $0 \le f \le 20f_0$. Plot the approximation, including the dc component and the first 10 harmonics.

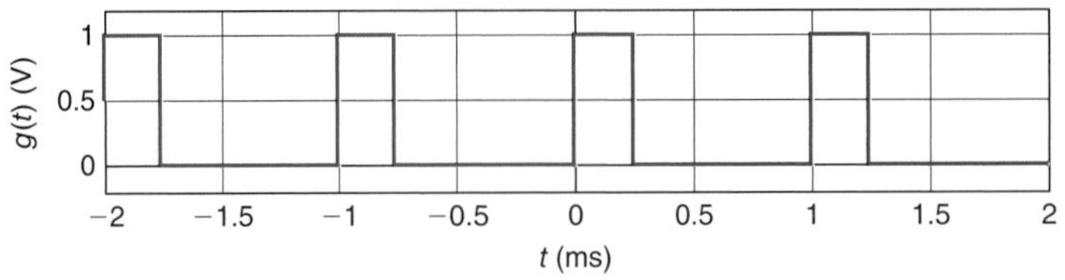

FIGURE 18.23 A rectangular pulse train with $A = 1$ V, $T_0 = 1$ ms, $\tau = 250\ \mu$s.

Answers:

$$g(t) = ga_0 + \sum_{n=1}^{\infty}\left[ga_n \cos\left(n\frac{2\pi}{T_0}t\right) + gb_n \sin\left(n\frac{2\pi}{T_0}t\right)\right] = gc_0 + \sum_{n=1}^{\infty} gc_n \cos\left(n\frac{2\pi}{T_0}t + g\theta_n\right)$$

$A = 1\text{ V},\ \ f_0 = 1/T_0 = 1\text{ kHz},\ \ d = \tau/T_0 = 1/4 = 0.25$
$ga_0 = a_0 = \text{Ad} = 1 \times 0.25 = 0.25.$
$ga_n = 2\text{Ad sinc(nd)} \cos(n\pi d) = 0.5\ \text{sinc}(n/4)\cos(n\pi/4)$
$gb_n = 2\text{Ad sinc(nd)} \sin(n\pi d) = 0.5\ \text{sinc}(n/4)\sin(n\pi/4)$
$gc_n = |0.5\ \text{sinc}(n/4)|$
$g\theta_n = -n\pi/4\text{ rad} = -45n\text{ deg if sinc}(n/4) \ge 0$
$g\theta_n = \pi - n\pi/4\text{ rad} = 180 - 45n\text{ deg if sinc}(n/4) < 0$

Figure 18.24 shows the magnitude spectrum gc_n, the phase spectrum $g\theta_n$, and approximation.

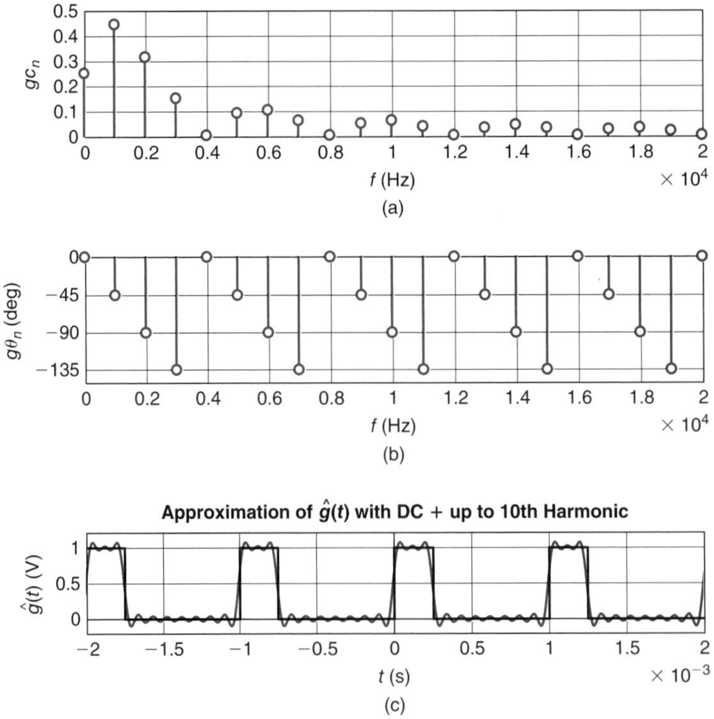

FIGURE 18.24 (a) Magnitude spectrum. (b) Phase spectrum. (c) Approximation.

continued

Exercise 18.9 continued

MATLAB

```
% Exercise 18.9
clear all;format short;
A=1;T0=1e-3;tau=0.25e-3;N=20;td=0.125e-3;
f0=1/T0
w0=2*pi/T0
d=tau/T0
ga0=A*tau/T0
n1=1:N;
n=0:N;
ga=2*A*d*sinc(n1*d).*cos(n1*w0*td);
gb=2*A*d*sinc(n1*d).*sin(n1*w0*td);
ga=[ga0,ga]
gb=[0,gb]
gc=sqrt(ga.^2+gb.^2)
Th=1e-10;
theta=zeros(1,N+1);
for i=1:length(ga)
    if (abs(ga(i))<=Th)&&(abs(gb(i))<=Th)
        gtheta(i)=0;
    elseif (ga(i)<-Th)&&(abs(gb(i))<=Th)
        gtheta(i)=pi;
    elseif (ga(i)>Th)&&(abs(gb(i))<=Th)
        gtheta(i)=0;
    elseif (abs(ga(i))<Th)&&(gb(i)>Th)
        gtheta(i)=-pi/2;
    elseif (abs(ga(i))<Th)&&(gb(i)<-Th)
        gtheta(i)=pi/2;
    else
        gtheta(i)=-atan2(gb(i),ga(i));
    end
end
gtheta
gthetad=gtheta*180/pi
f=f0*n;
figure(1)
stem(f,gc,'LineWidth',2);grid;
xlabel('f (Hz)');ylabel('gc_n');
set(gca,'YTick',0:0.1:0.5)
set(gca,'YTickLabel',0:0.1:0.5)
axis([0,N*f0,0,0.5])
figure(2)
stem(f,gthetad,'LineWidth',2);grid;
xlabel('f (Hz)');ylabel('g\theta_n (deg)');
set(gca,'YTick',-180:45:0)
set(gca,'YTickLabel',-180:45:0)
%
M1=1000;
t1=-2*T0:T0/M1:2*T0;
g0=0.5*(1+square(2*pi*(t1+0*T0)/T0,100*d));
M=10;
g2=ga(1);
for i=2:M+1
    g2=g2+ga(i)*cos(2*pi*(i-1)*t1/T0)+gb(i)*sin(2*pi*(i-1)*t1/T0);
end
figure(3)
```

continued

Exercise 18.9 continued
MATLAB continued

```
plot(t1,g2,'LineWidth',2);grid;hold on;
plot(t1,g0,'r-');xlabel('t');ylabel('g^\wedge(t)');hold off;
axis([-2*T0,2*T0,-0.2*A,1.2*A]);
title(['Approximation of g(t) with DC + up to ',num2str(M),'th Harmonic']);
```

18.3.5 TRIANGULAR PULSE TRAIN

A triangular pulse train is shown in Figure 18.25.

FIGURE 18.25

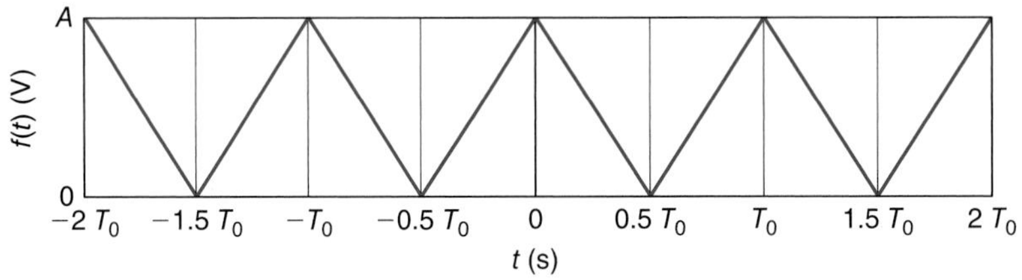

A triangular pulse train.

The equation of $f(t)$ is $-2At/T_0 + A$ for $0 \le t < T_0/2$. Since $f(t)$ possesses even symmetry, we have

$$b_n = 0, \quad n = 1, 2, 3, \ldots$$

Since the average value of $f(t)$ is $A/2$, the dc component is $A/2$. To verify this, from Equation (18.38), we have

$$a_0 = \frac{2}{T_0}\int_0^{\frac{T_0}{2}}\left(\frac{-2A}{T_0}t + A\right)dt = \frac{2}{T_0}\left(\int_0^{\frac{T_0}{2}}\frac{-2A}{T_0}tdt + \int_0^{\frac{T_0}{2}}Adt\right) = \frac{2}{T_0}\left(\frac{-2A}{T_0}\frac{\frac{T_0^2}{4}}{2} + A\frac{T_0}{2}\right)$$
$$= 2\left(\frac{-A}{4} + \frac{2A}{4}\right) = \frac{A}{2} \tag{18.68}$$

The coefficient a_n can be obtained from Equation (18.39)

$$a_n = \frac{4}{T_0}\int_0^{\frac{T_0}{2}}\left(\frac{-2A}{T_0}t + A\right)\cos\left(n\frac{2\pi}{T_0}t\right)dt = \frac{4}{T_0}\frac{-2A}{T_0}\int_0^{\frac{T_0}{2}}t\cos\left(n\frac{2\pi}{T_0}t\right)dt$$
$$+\frac{4A}{T_0}\int_0^{\frac{T_0}{2}}\cos\left(n\frac{2\pi}{T_0}t\right)dt$$

The second integral is zero since

$$\int_0^{\frac{T_0}{2}}\cos\left(n\frac{2\pi}{T_0}t\right)dt = \frac{\sin\left(n\frac{2\pi}{T_0}t\right)\Big|_0^{\frac{T_0}{2}}}{n\frac{2\pi}{T_0}} = \frac{\sin\left(n\frac{2\pi}{T_0}\frac{T_0}{2}\right) - \sin\left(n\frac{2\pi}{T_0}0\right)}{n\frac{2\pi}{T_0}} = \frac{\sin(n\pi)}{n\frac{2\pi}{T_0}} = 0$$

Application of the integral formula $\int t\cos(at)\,dt = \frac{1}{a^2}[\cos(at) + at\sin(at)]$ to the first integral yields

$$a_n = \frac{-8A}{T_0^2}\frac{1}{\left(n\frac{2\pi}{T_0}\right)^2}\left[\cos\left(n\frac{2\pi}{T_0}t\right) + n\frac{2\pi}{T_0}t\sin\left(n\frac{2\pi}{T_0}t\right)\right]_0^{\frac{T_0}{2}} = \frac{-8A}{T_0^2}\frac{[\cos(n\pi) - 1]}{\left(n\frac{2\pi}{T_0}\right)^2}$$

$$= \frac{2A[1 - \cos(n\pi)]}{\pi^2 n^2} = \begin{cases} \frac{4A}{\pi^2 n^2}, & n = 1, 3, 5, \ldots \\ 0, & n = 2, 4, 6, \ldots \end{cases} \quad \textbf{(18.69)}$$

Alternatively, applying $1 - \cos(2\alpha) = 2\sin^2(\alpha)$, we can write a_n as

$$a_n = \frac{2A2\sin^2\left(\frac{n\pi}{2}\right)}{\pi^2 n^2} = \frac{A\sin^2\left(\frac{n\pi}{2}\right)}{\left(\frac{n\pi}{2}\right)^2} = A\,\mathrm{sinc}^2\left(\frac{n}{2}\right) \quad \textbf{(18.70)}$$

Since $a_n \geq 0$ for all n, we have

$$c_n = a_n = \frac{2A2\sin^2\left(\frac{n\pi}{2}\right)}{\pi^2 n^2} = \frac{A\sin^2\left(\frac{n\pi}{2}\right)}{\left(\frac{n\pi}{2}\right)^2} = A\,\mathrm{sinc}^2\left(\frac{n}{2}\right)$$

$$= \begin{cases} \frac{4A}{\pi^2 n^2}, & n = 1,3,5, \ldots \\ 0, & n = 2,4,6, \ldots \end{cases} \quad \textbf{(18.71)}$$

and

$$\theta_n = 0, \quad n = 1, 2, 3, \ldots \quad \textbf{(18.72)}$$

The Fourier series representation of the triangular pulse train shown in Figure 18.25 is given by

$$f(t) = \frac{A}{2} + \sum_{n=1}^{\infty}\frac{4A}{(2n-1)^2\pi^2}\cos\left((2n-1)\frac{2\pi}{T_0}t\right) \quad \textbf{(18.73)}$$

Alternatively, from Equations (18.71) and (18.72), the Fourier series representation can be written as

$$f(t) = \frac{A}{2} + \sum_{n=1}^{\infty} A\,\mathrm{sinc}^2\left(\frac{n}{2}\right)\cos\left(n\frac{2\pi}{T_0}t\right) \quad \textbf{(18.74)}$$

MATLAB

```
clear all;
syms A T0 t
syms n integer
assume (n > 0)
f=-2*A*t/T0+A;
a0=(2/T0)*int(f,t,0,T0/2)
a=(4/T0)*int(f*cos(n*2*pi*t/T0),t,0,T0/2)
```

continued

MATLAB continued

```
Answer:
a0 =
A/2
a =
(4*A*sin((pi*n)/2)^2)/(pi^2*n^2)
```

$$a_n = 4A\frac{\sin^2\left(\pi\frac{n}{2}\right)}{\pi^2 n^2} = A\frac{\sin^2\left(\pi\frac{n}{2}\right)}{\frac{\pi^2 n^2}{4}} = A\left[\frac{\sin\left(\pi\frac{n}{2}\right)}{\pi\frac{n}{2}}\right]^2 = A\,\text{sinc}^2\left(\frac{n}{2}\right)$$

EXAMPLE 18.10

Find the trigonometric Fourier coefficients of the triangular pulse train $g(t)$ shown in Figure 18.26, and represent it by Fourier series.

FIGURE 18.26

A triangular pulse train.

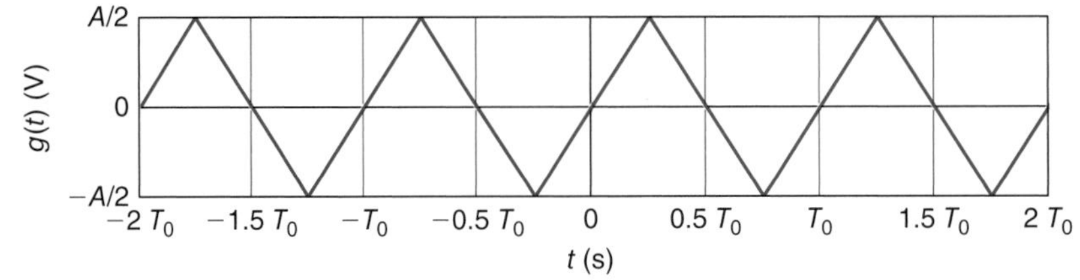

Since $g(t)$ possesses odd symmetry, we have

$$a_0 = 0$$
$$a_n = 0 \text{ for } n = 1, 2, 3, 4, 5, \ldots$$

The trigonometric pulse train shown in Figure 18.26 is similar to the one shown in Figure 18.25. The differences are the dc offset and the time shift of $T_0/4$ to the right. Since the average value of $g(t)$ is zero, the dc component of $g(t)$ is zero. Since the Fourier series representation of the triangular pulse train shown in Figure 18.25 is given by Equation (18.74), the Fourier series representation of the triangular pulse train shown in Figure 18.26 is given by

$$\begin{aligned} g(t) &= \sum_{n=1}^{\infty} A\,\text{sinc}^2\left(\frac{n}{2}\right)\cos\left(n\frac{2\pi}{T_0}\left(t-\frac{T_0}{4}\right)\right) \\ &= \sum_{n=1}^{\infty} A\,\text{sinc}^2\left(\frac{n}{2}\right)\cos\left(n\frac{2\pi}{T_0}t-\frac{n\pi}{2}\right) \end{aligned} \tag{18.75}$$

Notice that $gc_n = A\,\text{sinc}^2(n/2)$ and $g\theta_n = -n\pi/2$.
Expanding the cosine, we obtain

$$\begin{aligned} &A\text{sinc}^2\left(\frac{n}{2}\right)\cos\left(n\frac{2\pi}{T_0}t-\frac{n\pi}{2}\right) \\ &= A\text{sinc}^2\left(\frac{n}{2}\right)\cos\left(\frac{n\pi}{2}\right)\cos\left(n\frac{2\pi}{T_0}t\right) + A\text{sinc}^2\left(\frac{n}{2}\right)\sin\left(\frac{n\pi}{2}\right)\sin\left(n\frac{2\pi}{T_0}t\right) \end{aligned}$$

continued

Example 18.10 continued

$$= A\frac{\sin\left(\frac{n\pi}{2}\right)\sin\left(\frac{n\pi}{2}\right)\cos\left(\frac{n\pi}{2}\right)}{\left(\frac{n\pi}{2}\right)^2}\cos\left(n\frac{2\pi}{T_0}t\right) + A\frac{\sin\left(\frac{n\pi}{2}\right)\sin\left(\frac{n\pi}{2}\right)\sin\left(\frac{n\pi}{2}\right)}{\left(\frac{n\pi}{2}\right)^2}\sin\left(n\frac{2\pi}{T_0}t\right) \quad \textbf{(18.76)}$$

Since

$$\sin\left(\frac{n\pi}{2}\right)\cos\left(\frac{n\pi}{2}\right) = \frac{1}{2}\sin(n\pi) = 0$$

the first term of Equation (18.76) is zero. This is expected because $g(t)$ possesses odd symmetry. Notice that $\sin\left(\frac{n\pi}{2}\right) = 0$ for even values of n, and $\sin\left(\frac{n\pi}{2}\right) = 1$ for $n = 1, 5, 9, \ldots$ and $\sin\left(\frac{n\pi}{2}\right) = -1$ for $n = 3, 7, 11, \ldots$.

Thus, we conclude that the coefficient b_n for $g(t)$ is given by

$$b_n = A\text{sinc}^2\left(\frac{n}{2}\right)\sin\left(\frac{n\pi}{2}\right) = \begin{cases} \frac{4A}{n^2\pi^2}\sin\left(\frac{n\pi}{2}\right), & n\,odd \\ 0, & n\,even \end{cases} = \begin{cases} \frac{4A}{n^2\pi^2}, & n = 1, 5, 9, \ldots \\ \frac{-4A}{n^2\pi^2}, & n = 3, 7, 11, \ldots \\ 0, & n\,even \end{cases}$$

The Fourier series representation of $g(t)$ is

$$g(t) = \sum_{n=1}^{\infty} A\,\text{sinc}^2\left(\frac{n}{2}\right)\sin\left(\frac{n\pi}{2}\right)\sin\left(n\frac{2\pi}{T_0}t\right) \quad \textbf{(18.77)}$$

or

$$g(t) = \sum_{n=1}^{\infty}\frac{(-1)^{n+1}4A}{(2n-1)^2\pi^2}\sin\left((2n-1)\frac{2\pi}{T_0}t\right) \quad \textbf{(18.78)}$$

Exercise 18.10

Find the trigonometric Fourier coefficients of the periodic triangular waveform $h(t)$ shown in Figure 18.27, and represent $h(t)$ by its Fourier series.

FIGURE 18.27

A triangular pulse train.

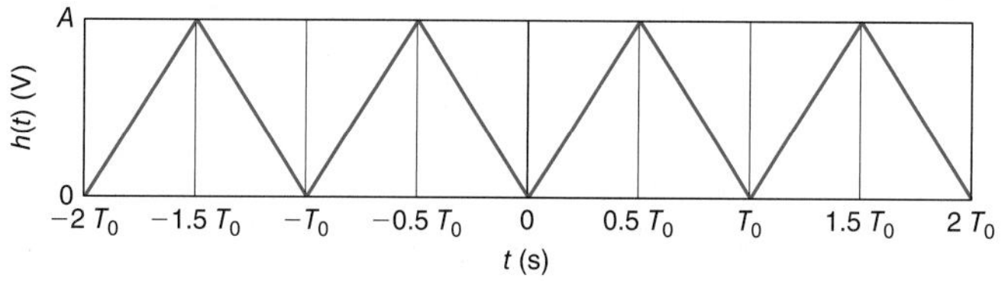

continued

Exercise 18.10 continued

Answers:

$h(t) = f\left(t - \frac{T_0}{2}\right)$. $f(t)$ is shown in Figure 18.25. $h(t)$ is an even function of t.

$$a_0 = A/2,\quad b_n = 0,\quad a_n = A\,\text{sinc}^2\left(\frac{n}{2}\right)\cos(n\pi).$$

$$\begin{aligned} h(t) &= \frac{A}{2} + \sum_{n=1}^{\infty} A\,\text{sinc}^2\left(\frac{n}{2}\right)\cos\left(n\frac{2\pi}{T_0}\left(t - \frac{T_0}{2}\right)\right) \\ &= \frac{A}{2} + \sum_{n=1}^{\infty} A\,\text{sinc}^2\left(\frac{n}{2}\right)\cos\left(n\frac{2\pi}{T_0}t - n\pi\right) \\ &= \frac{A}{2} + \sum_{n=1}^{\infty} A\,\text{sinc}^2\left(\frac{n}{2}\right)\cos(n\pi)\cos\left(n\frac{2\pi}{T_0}t\right) \\ &= \frac{A}{2} + \sum_{n=1}^{\infty} (-1)^n A\,\text{sinc}^2\left(\frac{n}{2}\right)\cos\left(n\frac{2\pi}{T_0}t\right) \end{aligned} \tag{18.79}$$

18.3.6 SAWTOOTH PULSE TRAIN

A sawtooth waveform is shown in Figure 18.28.

FIGURE 18.28

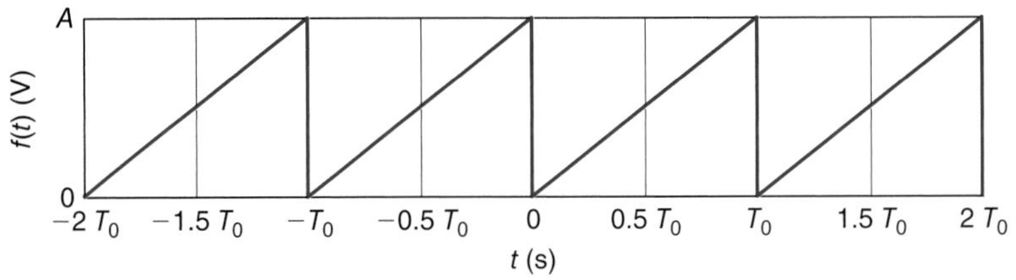

A sawtooth waveform.

The equation of the one period of $f(t)$ for $0 \le t < T_0$ is given by

$$f(t) = \begin{cases} \frac{A}{T_0}t, & 0 \le t < T_0 \\ 0, & \textit{otherwise} \end{cases}$$

The average value of $f(t)$ is $A/2$. Thus, the dc component is $a_0 = A/2$. We can verify this coefficient from Equation (18.32)

$$a_0 = \frac{1}{T_0}\int_0^{T_0} f(t)dt = \frac{1}{T_0}\int_0^{T_0} \frac{At}{T_0}dt = \frac{A}{T_0^2}\int_0^{T_0} t\,dt = \frac{A}{T_0^2}\frac{T_0^2}{2} = \frac{A}{2} \tag{18.80}$$

The Fourier coefficient a_n is given by, from Equation (18.33)

$$a_n = \frac{2}{T_0}\int_0^{T_0} f(t)\cos\left(n\frac{2\pi}{T_0}t\right)dt = \frac{2}{T_0}\int_0^{T_0} \frac{At}{T_0}\cos\left(n\frac{2\pi}{T_0}t\right)dt = \frac{2A}{T_0^2}\int_0^{T_0} t\cos\left(n\frac{2\pi}{T_0}t\right)dt$$

Application of the integral formula $\int t\cos(at)\,dt = \frac{1}{a^2}[\cos(at) + at\sin(at)]$ yields

$$a_n = \frac{2A}{T_0^2}\frac{1}{\left(n\frac{2\pi}{T_0}\right)^2}\left[\cos\left(n\frac{2\pi}{T_0}t\right) + n\frac{2\pi}{T_0}t\sin\left(n\frac{2\pi}{T_0}t\right)\right]_0^{T_0}$$

$$= \frac{2A}{T_0^2}\frac{1}{\left(n\frac{2\pi}{T_0}\right)^2}\left[\cos\left(n\frac{2\pi}{T_0}T_0\right) + n\frac{2\pi}{T_0}T_0\sin\left(n\frac{2\pi}{T_0}T_0\right) - 1\right]$$

$$= \frac{A}{2\pi^2n^2}[\cos(2\pi n) + 2\pi n\sin(2\pi n) - 1] = 0, \quad n = 1, 2, 3, \ldots$$

The Fourier coefficient b_n is given by, from Equation (18.34)

$$b_n = \frac{2}{T_0}\int_0^{T_0} f(t)\sin\left(n\frac{2\pi}{T_0}t\right)dt = \frac{2}{T_0}\int_0^{T_0}\frac{At}{T_0}\sin\left(n\frac{2\pi}{T_0}t\right)dt = \frac{2A}{T_0^2}\int_0^{T_0} t\sin\left(n\frac{2\pi}{T_0}t\right)dt$$

Application of the integral formula $\int t\sin(at)\,dt = \frac{1}{a^2}[\sin(at) - at\cos(at)]$ yields

$$b_n = \frac{2A}{T_0^2}\frac{1}{\left(n\frac{2\pi}{T_0}\right)^2}\left[\sin\left(n\frac{2\pi}{T_0}t\right) - n\frac{2\pi}{T_0}t\cos\left(n\frac{2\pi}{T_0}t\right)\right]_0^{T_0}$$

$$= \frac{2A}{T_0^2}\frac{1}{\left(n\frac{2\pi}{T_0}\right)^2}\left[\sin\left(n\frac{2\pi}{T_0}T_0\right) - n\frac{2\pi}{T_0}T_0\cos\left(n\frac{2\pi}{T_0}T_0\right)\right] \quad \textbf{(18.81)}$$

$$= \frac{A}{2\pi^2n^2}[\sin(2\pi n) - 2\pi n\cos(2\pi n)] = -\frac{A}{\pi n}, \quad n = 1, 2, 3, \ldots$$

$$f(t) = \frac{A}{2} + \sum_{n=1}^{\infty}\frac{-A}{\pi n}\sin\left(n\frac{2\pi}{T_0}t\right) \quad \textbf{(18.82)}$$

Equation (18.82) can be rearranged as

$$f(t) = \frac{A}{2} + \sum_{n=1}^{\infty}\frac{A}{\pi n}\cos\left(n\frac{2\pi}{T_0}t - \frac{\pi}{2} + \pi\right) = \frac{A}{2} + \sum_{n=1}^{\infty}\frac{A}{\pi n}\cos\left(n\frac{2\pi}{T_0}t + \frac{\pi}{2}\right) \quad \textbf{(18.83)}$$

From Equation (18.83), we can find the Fourier coefficients c_n and θ_n:

$$c_0 = a_0 = \frac{A}{2}$$

$$c_n = \frac{A}{\pi n}$$

$$\theta_n = \pi/2 \text{ rad} = 90°, \quad n = 1, 2, 3, \ldots$$

MATLAB

```
clear all;
syms A T0 t n
assume(n,'integer')
assume(n > 0)
f=A*t/T0;
a0=(1/T0)*int(f,t,0,T0)
a=(2/T0)*int(f*cos(n*2*pi*t/T0),t,0,T0)
b=(2/T0)*int(f*sin(n*2*pi*t/T0),t,0,T0)

Answer:
a0 =
A/2
a =
-(A*(2*sin(pi*n)^2 - 2*pi*n*sin(2*pi*n)))/(2*pi^2*n^2)
b =
(2*A*(sin(2*pi*n)/4 - (pi*n*cos(2*pi*n))/2))/(pi^2*n^2)
```

Since $\sin(\pi n) = 0$ and $\sin(2\pi n) = 0$, $a_n = 0$. Since $\sin(2\pi n) = 0$ and $\cos(2\pi n) = 1$, $b_n = -A/(\pi n)$.

EXAMPLE 18.11

Find the trigonometric Fourier coefficients of the sawtooth waveform shown in Figure 18.29. Represent *f*(*t*) by a Fourier series.

FIGURE 18.29

A sawtooth waveform.

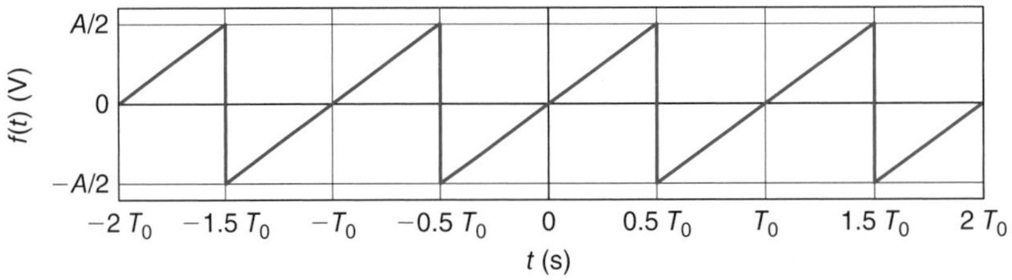

The equation of $f(t)$ from 0 to $T_0/2$ is At/T_0. Since $f(t)$ possesses odd symmetry, we have

$$a_0 = 0$$
$$a_n = 0, \quad n = 1, 2, 3, \ldots$$

Using the integral formula $\int t\sin(at)\,dt = \frac{1}{a^2}[\sin(at) - at\cos(at)]$, we obtain

$$b_n = \frac{4}{T_0}\int_0^{\frac{T_0}{2}} f(t)\sin\left(n\frac{2\pi}{T_0}t\right)dt = \frac{4}{T_0}\int_0^{\frac{T_0}{2}} \frac{A}{T_0}t\sin\left(n\frac{2\pi}{T_0}t\right)dt$$

$$= \frac{4A}{T_0^2}\frac{1}{\left(n\frac{2\pi}{T_0}\right)^2}\left[\sin\left(n\frac{2\pi}{T_0}t\right) - n\frac{2\pi}{T_0}t\cos\left(n\frac{2\pi}{T_0}t\right)\right]_0^{\frac{T_0}{2}}$$

continued

Example 18.11 continued

$$= \frac{A}{n^2\pi^2}\left[\sin\left(n\frac{2\pi}{T_0}\frac{T_0}{2}\right) - n\frac{2\pi}{T_0}\frac{T_0}{2}\cos\left(n\frac{2\pi}{T_0}\frac{T_0}{2}\right)\right]$$

$$= \frac{A}{n^2\pi^2}[-n\pi\cos(n\pi)] = \frac{-A\cos(n\pi)}{n\pi} = \frac{-A(-1)^n}{n\pi} \tag{18.84}$$

$$= \frac{A(-1)^{n+1}}{n\pi} = \begin{cases} \frac{A}{n\pi}, & n = 1, 3, 5, \ldots \\ \frac{-A}{n\pi}, & n = 2, 4, 6, \ldots \end{cases}$$

The Fourier series representation of $f(t)$ is given by

$$f(t) = \sum_{n=1}^{\infty} \frac{(-1)^{n+1}A}{\pi n} \sin\left(n\frac{2\pi}{T_0}t\right) \tag{18.85}$$

Exercise 18.11

Find the trigonometric Fourier coefficients of the periodic waveform shown in Figure 18.30. Represent $f(t)$ by its Fourier series.

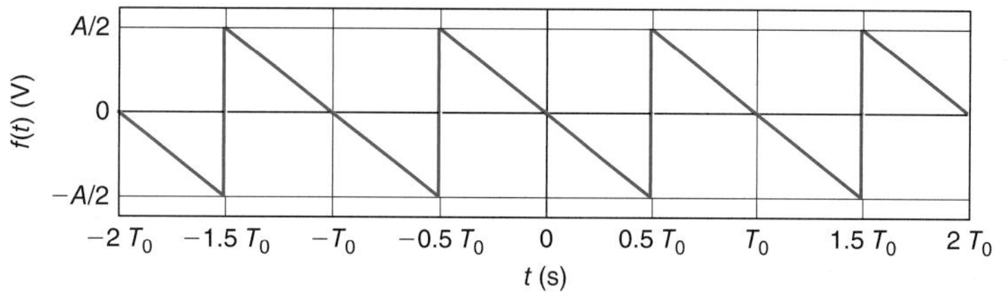

FIGURE 18.30 A sawtooth wave.

Answer:

$a_0 = 0, \quad a_n = 0, \quad n = 1, 2, 3, \ldots,$

$$b_n = \frac{A(-1)^n}{n\pi} = \begin{cases} \frac{-A}{n\pi}, & n = 1, 3, 5, \ldots \\ \frac{A}{n\pi}, & n = 2, 4, 6, \ldots \end{cases}$$

$$f(t) = \sum_{n=1}^{\infty} \frac{(-1)^n A}{\pi n} \sin\left(n\frac{2\pi}{T_0}t\right).$$

18.3.7 RECTIFIED COSINE

A half-wave rectified cosine waveform is shown in Figure 18.31.

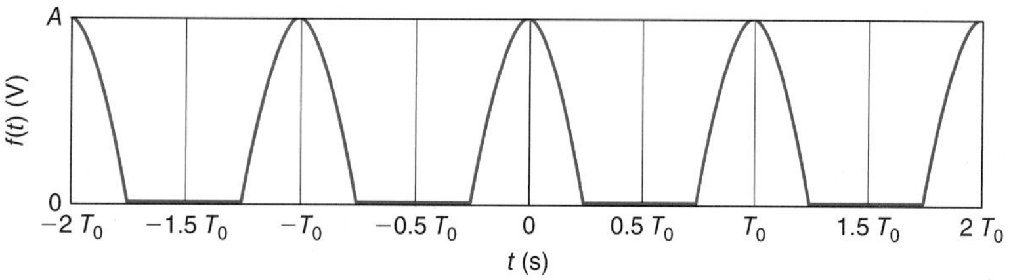

FIGURE 18.31 A half-wave rectified cosine.

The equation of $f(t)$ is $A\cos(2\pi t/T_0)$ for $-T_0/4 \le t < T_0/4$. Since $f(t)$ possesses even symmetry, we have

$$b_n = 0, \quad n = 1, 2, 3, \dots$$

The dc component is obtained from Equation (18.38)

$$a_0 = \frac{2}{T_0}\int_0^{\frac{T_0}{2}} f(t)dt = \frac{2}{T_0}\int_0^{\frac{T_0}{4}} A\cos\left(\frac{2\pi}{T_0}t\right)dt = \frac{2A}{T_0}\frac{\sin\left(\frac{2\pi}{T_0}t\right)\Big|_0^{\frac{T_0}{4}}}{\frac{2\pi}{T_0}} = \frac{A}{\pi} \tag{18.86}$$

From Equation (18.39) and using the trigonometric identity

$$\cos(\alpha)\cos(\beta) = (1/2)\left[\cos(\alpha+\beta) + \cos(\alpha-\beta)\right]$$

we can find the coefficients a_n

$$a_n = \frac{4}{T_0}\int_0^{\frac{T_0}{2}} f(t)\cos\left(n\frac{2\pi}{T_0}t\right)dt = \frac{4}{T_0}\int_0^{\frac{T_0}{4}} A\cos\left(\frac{2\pi}{T_0}t\right)\cos\left(n\frac{2\pi}{T_0}t\right)dt$$

$$a_n = \frac{2A}{T_0}\int_0^{\frac{T_0}{4}}\cos\left((n+1)\frac{2\pi}{T_0}t\right)dt + \frac{2A}{T_0}\int_0^{\frac{T_0}{4}}\cos\left((n-1)\frac{2\pi}{T_0}t\right)dt \tag{18.87}$$

If $n = 1$, Equation (18.87) becomes

$$a_1 = \frac{2A}{T_0}\frac{\sin\left((1+1)\frac{2\pi}{T_0}\frac{T_0}{4}\right)}{(1+1)\frac{2\pi}{T_0}} + \frac{2A}{T_0}\frac{T_0}{4} = \frac{A}{2}$$

If $n > 1$, Equation (18.87) becomes

$$a_n = \frac{2A}{T_0}\frac{\sin\left((n+1)\frac{2\pi}{T_0}\frac{T_0}{4}\right)}{(n+1)\frac{2\pi}{T_0}} + \frac{2A}{T_0}\frac{\sin\left((n-1)\frac{2\pi}{T_0}\frac{T_0}{4}\right)}{(n-1)\frac{2\pi}{T_0}}$$

$$= \frac{A}{\pi}\frac{\sin\left((n+1)\frac{\pi}{2}\right)}{(n+1)} + \frac{A}{\pi}\frac{\sin\left((n-1)\frac{\pi}{2}\right)}{(n-1)}$$

$$= \frac{A}{\pi}\frac{\cos\left(\frac{n\pi}{2}\right)}{(n+1)} - \frac{A}{\pi}\frac{\cos\left(\frac{n\pi}{2}\right)}{(n-1)} = \frac{A}{\pi}\frac{-2\cos\left(\frac{n\pi}{2}\right)}{n^2-1}$$

$$= \frac{-2A\cos\left(\frac{n\pi}{2}\right)}{\pi(n^2-1)}, \quad n = 2, 3, 4, 5, \dots$$

Thus, we have

$$a_n = \frac{2A}{\pi}\frac{-\cos\left(\frac{n\pi}{2}\right)}{n^2-1} = \begin{cases} 0, & n = 3, 5, 7, 9, \ldots \\ \dfrac{2A}{\pi(n^2-1)}, & n = 2, 6, 10, \ldots \\ \dfrac{-2A}{\pi(n^2-1)}, & n = 4, 8, 12, \ldots \end{cases} \quad \textbf{(18.88)}$$

Since $b_n = 0$, $c_n = |a_n|$ and $\theta_n = \angle a_n$. Therefore,

$$c_n = \begin{cases} \dfrac{A}{\pi}, & n = 0 \\ \dfrac{A}{2}, & n = 1 \\ 0, & n = 3, 5, 7, 9, \ldots \\ \dfrac{2A}{\pi(n^2-1)}, & n = 2, 4, 6, \ldots \end{cases}, \quad \textbf{(18.89)}$$

$$\theta_n = \begin{cases} 0, & n = 0, 1, 2, 3, 5, 6, 7, 9, 10, \ldots \\ \pi, & n = 4, 8, 12, \ldots \end{cases}$$

$$f(t) = \frac{A}{\pi} + \frac{A}{2}\cos\left(\frac{2\pi}{T_0}t\right) + \sum_{n=1}^{\infty}\frac{(-1)^{n-1}2A}{\pi(4n^2-1)}\cos\left(2n\frac{2\pi}{T_0}t\right) \quad \textbf{(18.90)}$$

MATLAB

```
clear all;
syms A T0 t
syms n integer
assume (n > 0)
f=A*cos(2*pi*t/T0);
a0=(1/T0)*int(f,t,-T0/4,T0/4)
a=(4/T0)*int(f*cos(n*2*pi*t/T0),t,0,T0/4)

Answer:
a0 =
A/pi
a =
piecewise([n == 1, A/2], [n ~= 1, -(2*A*cos((pi*n)/2))/(pi*(n^2 - 1))])
```

The answers match those given by Equations (18.86) and (18.88).
A full-wave rectified cosine waveform is shown in Figure 18.32.

FIGURE 18.32

A full-wave rectified cosine wave.

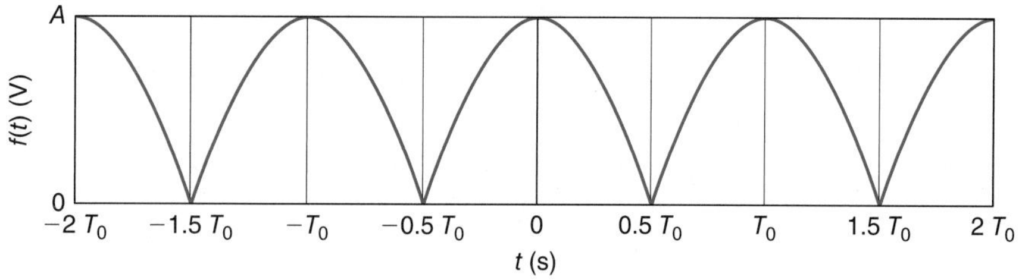

The equation of $f(t)$ is $A\cos(\pi t/T_0)$ for $-T_0/2 \le t < T_0/2$. Since $f(t)$ possesses even symmetry, we have

$$b_n = 0, \quad n = 1, 2, 3, \ldots$$

The dc component is obtained from Equation (18.38)

$$a_0 = \frac{2}{T_0}\int_0^{\frac{T_0}{2}} f(t)dt = \frac{2}{T_0}\int_0^{\frac{T_0}{2}} A\cos\left(\frac{\pi}{T_0}t\right)dt = \frac{2A}{T_0}\frac{\sin\left(\frac{\pi}{T_0}t\right)\Big|_0^{\frac{T_0}{2}}}{\frac{\pi}{T_0}} = \frac{2A}{\pi} \tag{18.91}$$

Notice that the average value of the full-wave rectified cosine is twice the average value of the half-wave rectified cosine because the number of half cycles of cosine per given time is doubled. Using Equation (8.39) and the trigonometric identity

$$\cos(\alpha)\cos(\beta) = (1/2)\left[\cos(\alpha + \beta) + \cos(\alpha - \beta)\right]$$

we have

$$\begin{aligned} a_n &= \frac{4}{T_0}\int_0^{\frac{T_0}{2}} f(t)\cos\left(n\frac{2\pi}{T_0}t\right)dt = \frac{4}{T_0}\int_0^{\frac{T_0}{2}} A\cos\left(\frac{\pi}{T_0}t\right)\cos\left(n\frac{2\pi}{T_0}t\right)dt \\ &= \frac{2A}{T_0}\int_0^{\frac{T_0}{2}} \cos\left((2n+1)\frac{\pi}{T_0}t\right)dt + \frac{2A}{T_0}\int_0^{\frac{T_0}{2}} \cos\left((2n-1)\frac{\pi}{T_0}t\right)dt \\ &= \frac{2A}{T_0}\frac{\sin\left((2n+1)\frac{\pi}{T_0}\frac{T_0}{2}\right)}{(2n+1)\frac{\pi}{T_0}} + \frac{2A}{T_0}\frac{\sin\left((2n-1)\frac{\pi}{T_0}\frac{T_0}{2}\right)}{(2n-1)\frac{\pi}{T_0}} \\ &= \frac{2A}{\pi}\frac{\sin\left((2n+1)\frac{\pi}{2}\right)}{(2n+1)} + \frac{2A}{\pi}\frac{\sin\left((2n-1)\frac{\pi}{2}\right)}{(2n-1)} \\ &= \frac{2A}{\pi}\frac{\cos(n\pi)}{(2n+1)} - \frac{2A}{\pi}\frac{\cos(n\pi)}{(2n-1)} = \frac{-4A}{\pi}\frac{\cos(n\pi)}{4n^2-1} = \frac{(-1)^{n+1}4A}{\pi(4n^2-1)} \end{aligned} \tag{18.92}$$

Since $b_n = 0$, $c_n = |a_n|$ and $\theta_n = \angle a_n$. Therefore,

$$c_n = \frac{4A}{\pi(4n^2-1)}, \quad n = 1, 2, 3, \ldots \tag{18.93}$$

$$\theta_n = \begin{cases} 0, & n = 0, 1, 3, 5, \ldots \\ 180°, & n = 2, 4, 6, \ldots \end{cases} \tag{18.94}$$

The Fourier series representation of the full-wave rectified cosine is given by

$$f(t) = \frac{2A}{\pi} + \sum_{n=1}^{\infty}\frac{(-1)^{n+1}4A}{\pi(4n^2-1)}\cos\left(n\frac{2\pi}{T_0}t\right) \tag{18.95}$$

MATLAB

```
clear all;
syms A T0 t
syms n integer
f=A*cos(pi*t/T0);
a0=(2/T0)*int(f,t,0,T0/2)
a=(4/T0)*int(f*cos(n*2*pi*t/T0),t,0,T0/2)

Answer:
a0 =
(2*A)/pi
a =
-(4*A*cos(pi*n))/(pi*(4*n^2 - 1))
```

The answers match those given by Equations (18.91) and (18.92).

18.3.8 RECTIFIED SINE

A half-wave rectified sine waveform is shown in Figure 18.33.

FIGURE 18.33

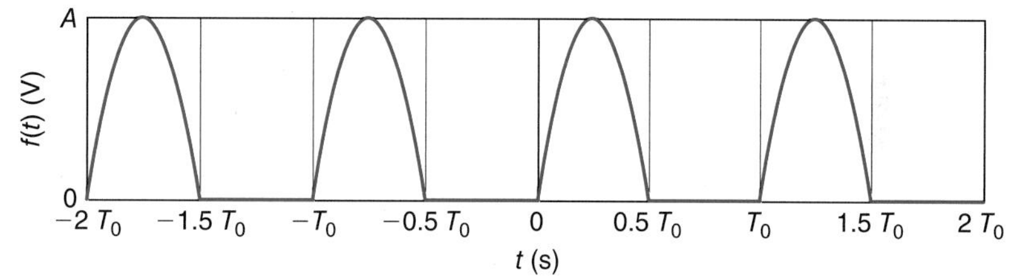

A half-wave rectified sine.

For $0 \le t < T_0/2$, the equation of the waveform is $A\sin(2\pi t/T_0)$ for $0 \le t < T_0/2$. The dc component is obtained from Equation (18.32)

$$a_0 = \frac{1}{T_0}\int_0^{T_0} f(t)dt = \frac{1}{T_0}\int_0^{\frac{T_0}{2}} A\sin\left(\frac{2\pi}{T_0}t\right)dt = \frac{A}{T_0}\frac{-\cos\left(\frac{2\pi}{T_0}t\right)\Big|_0^{\frac{T_0}{2}}}{\frac{2\pi}{T_0}} = \frac{A}{\pi} \tag{18.96}$$

From Equation (18.33) and using the trigonometric identity

$$\sin(\alpha)\cos(\beta) = (1/2)\left[\sin(\alpha+\beta) + \sin(\alpha-\beta)\right]$$

we can find the coefficient a_n:

$$\begin{aligned} a_n &= \frac{2}{T_0}\int_0^{T_0} f(t)\cos\left(n\frac{2\pi}{T_0}t\right)dt = \frac{2A}{T_0}\int_0^{\frac{T_0}{2}} \sin\left(\frac{2\pi}{T_0}t\right)\cos\left(n\frac{2\pi}{T_0}t\right)dt \\ &= \frac{A}{T_0}\int_0^{\frac{T_0}{2}} \sin\left((1+n)\frac{2\pi}{T_0}t\right)dt + \frac{A}{T_0}\int_0^{\frac{T_0}{2}} \sin\left((1-n)\frac{2\pi}{T_0}t\right)dt \end{aligned} \tag{18.97}$$

If $n = 1$, Equation (18.97) becomes

$$a_1 = \frac{A}{T_0}\frac{-\cos\left((1+1)\frac{2\pi}{T_0}\frac{T_0}{2}\right)+1}{(1+1)\frac{2\pi}{T_0}} + 0 = 0$$

For $n > 1$, Equation (18.97) becomes

$$a_n = \frac{A}{T_0}\frac{-\cos\left((1+n)\frac{2\pi}{T_0}t\right)\Big|_0^{\frac{T_0}{2}}}{(1+n)\frac{2\pi}{T_0}} + \frac{A}{T_0}\frac{-\cos\left((1-n)\frac{2\pi}{T_0}t\right)\Big|_0^{\frac{T_0}{2}}}{(1-n)\frac{2\pi}{T_0}}$$
$$= \frac{A}{T_0}\frac{1-\cos\left((1+n)\frac{2\pi}{T_0}\frac{T_0}{2}\right)}{(1+n)\frac{2\pi}{T_0}} + \frac{A}{T_0}\frac{1-\cos\left((1-n)\frac{2\pi}{T_0}\frac{T_0}{2}\right)}{(1-n)\frac{2\pi}{T_0}} \tag{18.98}$$
$$= \frac{A}{T_0}\frac{1-\cos[(1+n)\pi]}{(1+n)\frac{2\pi}{T_0}} + \frac{A}{T_0}\frac{1-\cos[(1-n)\pi]}{(1-n)\frac{2\pi}{T_0}}$$
$$= \frac{A}{\pi}\frac{1+\cos(n\pi)}{1-n^2} = \begin{cases} 0, & n = 3, 5, 7, \ldots \\ \frac{-2A}{\pi(n^2-1)}, & n = 2, 4, 6, \ldots \end{cases}$$

From Equation (18.34) and using the trigonometric identity

$$\sin(\alpha)\sin(\beta) = (1/2)\left[\cos(\alpha-\beta) - \cos(\alpha+\beta)\right]$$

we can find the coefficients b_n:

$$b_n = \frac{2}{T_0}\int_0^{T_0} f(t)\sin(n\omega_0 t)dt = \frac{2}{T_0}\int_0^{\frac{T_0}{2}} A\sin\left(\frac{2\pi}{T_0}t\right)\sin\left(n\frac{2\pi}{T_0}t\right)dt$$
$$= \frac{A}{T_0}\int_0^{\frac{T_0}{2}}\cos\left(\frac{2\pi}{T_0}(1-n)t\right)dt - \frac{A}{T_0}\int_0^{\frac{T_0}{2}}\cos\left(\frac{2\pi}{T_0}(1+n)t\right)dt \tag{18.99}$$

If $n = 1$, Equation (18.99) becomes

$$b_1 = \frac{A}{T_0}\int_0^{\frac{T_0}{2}} 1dt - \frac{A}{T_0}\frac{\sin\left(\frac{2\pi}{T_0}(1+1)t\right)\Big|_0^{\frac{T_0}{2}}}{\frac{2\pi}{T_0}(1+1)} = \frac{A}{2} - 0 = \frac{A}{2} \tag{18.100}$$

If $n > 1$, Equation (18.99) becomes

$$b_n = \frac{A}{T_0}\frac{\sin\left(\frac{2\pi}{T_0}(1-n)t\right)\Big|_0^{\frac{T_0}{2}}}{\frac{2\pi}{T_0}(1-n)} - \frac{A}{T_0}\frac{\sin\left(\frac{2\pi}{T_0}(1+n)t\right)\Big|_0^{\frac{T_0}{2}}}{\frac{2\pi}{T_0}(1+n)}$$
$$= \frac{A}{T_0}\frac{\sin\left(\frac{2\pi}{T_0}(1-n)\frac{T_0}{2}\right)}{\frac{2\pi}{T_0}(1-n)} - \frac{A}{T_0}\frac{\sin\left(\frac{2\pi}{T_0}(1+n)\frac{T_0}{2}\right)}{\frac{2\pi}{T_0}(1+n)}$$
$$= \frac{A}{2\pi}\frac{\sin(\pi(1-n))}{(1-n)} - \frac{A}{2\pi}\frac{\sin(\pi(1+n))}{(1+n)}$$
$$= 0, \quad n = 2,3,4,\ldots$$

The coefficients c_n and θ_n are given by

$$c_n = \begin{cases} \dfrac{A}{\pi}, & n = 0 \\ \dfrac{A}{2}, & n = 1 \\ 0, & n = 3, 5, 7, 9, \ldots \\ \dfrac{2A}{\pi(n^2-1)}, & n = 2, 4, 6, \ldots \end{cases}, \quad \theta_n = \begin{cases} 0, & n = 0, 1, 3, 5, 7, 9, \ldots \\ -90°, & n = 1 \\ 180°, & n = 2, 4, 6, 8, \ldots \end{cases} \tag{18.101}$$

The Fourier series representation is given by

$$f(t) = \frac{A}{\pi} + \frac{A}{2}\sin\left(\frac{2\pi}{T_0}t\right) + \sum_{n=1}^{\infty} \frac{-2A}{\pi(4n^2-1)}\cos\left(2n\frac{2\pi}{T_0}t\right) \tag{18.102}$$

MATLAB

```
clear all;
syms A T0 t n
assume(n,'integer')
f=A*sin(2*pi*t/T0);
a0=(1/T0)*int(f,t,0,T0/2)
assume(n > 0)
a=(2/T0)*int(f*cos(n*2*pi*t/T0),t,0,T0/2)
assume(n == 1)
b=(2/T0)*int(f*sin(n*2*pi*t/T0),t,0,T0/2)
assume(n > 1)
b=(2/T0)*int(f*sin(n*2*pi*t/T0),t,0,T0/2)

Answer:
a0 =
A/pi
a =
piecewise([n == 1, 0], [n ~= 1, -(A*(cos(pi*n) + 1))/(pi*(n^2 - 1))])
b =
A/2
b =
-(A*sin(pi*n))/(pi*(n^2 - 1))
```

Since $\sin(\pi n) = 0$, $b_n = 0$ for $n > 1$. For $n = 1$, $b_1 = A/2$. The answers for a_0 and a_n match those given by Equations (18.96) and (18.98).

A full-wave rectified sine waveform shown in Figure 18.34.

FIGURE 18.34

Full-wave rectified sine wave.

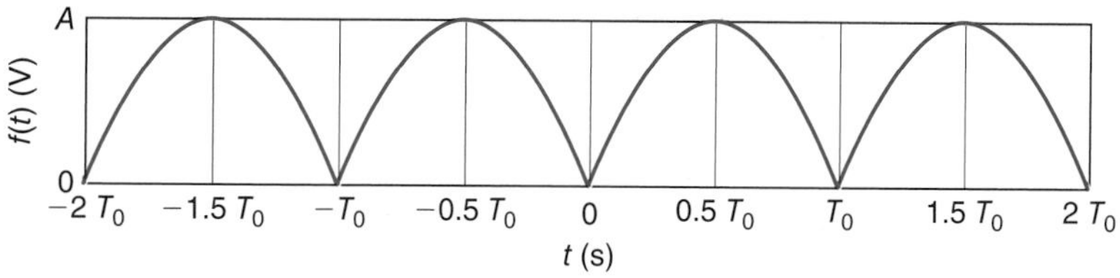

For $0 \le t < T_0$, the equation of the waveform is $A\sin(\pi t/T_0)$. Since $f(t)$ possesses even symmetry, we have

$$b_n = 0, \quad n = 1, 2, 3, \ldots$$

The dc component is obtained from Equation (18.38)

$$a_0 = \frac{2}{T_0}\int_0^{\frac{T_0}{2}} f(t)dt = \frac{2}{T_0}\int_0^{\frac{T_0}{2}} A\sin\left(\frac{\pi}{T_0}t\right)dt = \frac{2A}{T_0}\frac{-\cos\left(\frac{\pi}{T_0}t\right)\Big|_0^{\frac{T_0}{2}}}{\frac{\pi}{T_0}} = \frac{2A}{\pi} \quad \textbf{(18.103)}$$

From Equation (18.39) and using the trigonometric identity

$$\sin(\alpha)\cos(\beta) = (1/2)\left[\sin(\alpha+\beta) + \sin(\alpha-\beta)\right]$$

we can find the coefficients a_n:

$$\begin{aligned}
a_n &= \frac{4}{T_0}\int_0^{\frac{T_0}{2}} f(t)\cos\left(n\frac{2\pi}{T_0}t\right)dt = \frac{4A}{T_0}\int_0^{\frac{T_0}{2}} \sin\left(\frac{\pi}{T_0}t\right)\cos\left(n\frac{2\pi}{T_0}t\right)dt \\
&= \frac{2A}{T_0}\int_0^{\frac{T_0}{2}} \sin\left((1+2n)\frac{\pi}{T_0}t\right)dt + \frac{2A}{T_0}\int_0^{\frac{T_0}{2}} \sin\left((1-2n)\frac{\pi}{T_0}t\right)dt \\
&= \frac{2A}{T_0}\frac{-\cos\left((1+2n)\frac{\pi}{T_0}t\right)\Big|_0^{\frac{T_0}{2}}}{(1+2n)\frac{\pi}{T_0}} + \frac{2A}{T_0}\frac{-\cos\left((1-2n)\frac{\pi}{T_0}t\right)\Big|_0^{\frac{T_0}{2}}}{(1-2n)\frac{\pi}{T_0}} \\
&= \frac{2A}{T_0}\frac{1-\cos\left((1+2n)\frac{\pi}{T_0}\frac{T_0}{2}\right)}{(1+2n)\frac{\pi}{T_0}} + \frac{2A}{T_0}\frac{1-\cos\left((1-2n)\frac{\pi}{T_0}\frac{T_0}{2}\right)}{(1-2n)\frac{\pi}{T_0}} \\
&= \frac{2A}{T_0}\frac{1-\cos\left[(1+2n)\frac{\pi}{2}\right]}{(1+2n)\frac{\pi}{T_0}} + \frac{2A}{T_0}\frac{1-\cos\left[(1-2n)\frac{\pi}{2}\right]}{(1-2n)\frac{\pi}{T_0}} \\
&= \frac{2A}{\pi}\left[\frac{1}{(1+2n)} + \frac{1}{(1-2n)}\right] = \frac{-4A}{\pi(4n^2-1)}, \quad n = 1, 2, 3, \ldots
\end{aligned} \quad \textbf{(18.104)}$$

$$c_n = \frac{4A}{\pi(4n^2-1)}, \quad n = 1, 2, 3, \ldots \quad \textbf{(18.105)}$$

$$\theta_n = \angle a_n = 180°, \quad n = 1, 2, 3, \ldots \quad \textbf{(18.106)}$$

The Fourier series representation of $f(t)$ is given by

$$f(t) = \frac{2A}{\pi} - \sum_{n=1}^{\infty}\frac{4A}{(4n^2-1)\pi}\cos\left(n\frac{2\pi}{T_0}t\right) \quad \textbf{(18.107)}$$

MATLAB

```
clear all;
syms A T0 t
syms n integer
f=A*sin(pi*t/T0);
a0=(2/T0)*int(f,t,0,T0/2)
a=(4/T0)*int(f*cos(n*2*pi*t/T0),t,0,T0/2)
```

continued

MATLAB continued

```
Answer:
a0 =
(2*A)/pi
a =
(4*A*(2*n*sin(pi*n) - 1))/(pi*(4*n^2 - 1))
```

Since $\sin(\pi n) = 0$, these answers match those given by Equations (18.103) and (18.104).

18.3.9 AVERAGE POWER OF PERIODIC SIGNALS

If an arbitrary voltage signal $f(t)$ is applied to a resistor with resistance R, the instantaneous power on the resistor is $|f(t)|^2/R$. If this instantaneous power is averaged, we obtain the average power P dissipated on the resistor

$$P = \lim_{T\to\infty} \frac{1}{T}\int_{-\frac{T}{2}}^{\frac{T}{2}} \frac{|f(t)|^2}{R}dt \tag{18.108}$$

If $f(t)$ is periodic with period T_0, the average power is calculated by averaging the instantaneous power over one period:

$$P = \frac{1}{T_0}\int_{-\frac{T_0}{2}}^{\frac{T_0}{2}} \frac{|f(t)|^2}{R}dt \tag{18.109}$$

The **average power** is usually defined as the power dissipated on a 1-Ω resistor. Setting $R = 1\ \Omega$ in Equation (18.109), we obtain

$$P = \frac{1}{T_0}\int_{-\frac{T_0}{2}}^{\frac{T_0}{2}} |f(t)|^2 dt \tag{18.110}$$

If the trigonometric Fourier series representation

$$f(t) = a_0 + \sum_{n=1}^{\infty}[a_n\cos(n\omega_0 t) + b_n\sin(n\omega_0 t)] = c_0 + \sum_{n=1}^{\infty} c_n\cos(n\omega_0 t + \theta_n)$$

is substituted into Equation (18.110), we get

$$P = \frac{1}{T_0}\int_{-\frac{T_0}{2}}^{\frac{T_0}{2}} |f(t)|^2 dt = a_0^2 + \sum_{n=1}^{\infty}\frac{a_n^2 + b_n^2}{2} = c_0^2 + \sum_{n=1}^{\infty}\frac{c_n^2}{2} \tag{18.111}$$

since all the cross-product terms in the expansion of $|f(t)|^2 = f(t) \times f^*(t) = f^2(t)$ vanish because of the orthogonality of the signals $\cos(2\pi nt/T_0)$ and $\sin(2\pi nt/T_0)$ for $n = 1, 2, 3, \ldots$; that is,

$$\frac{1}{T_0}\int_{-\frac{T_0}{2}}^{\frac{T_0}{2}} \sin\left(2\pi\frac{n}{T_0}t\right)\sin\left(2\pi\frac{m}{T_0}t\right)dt = \begin{cases}\frac{1}{2} & for\ n = m\\ 0 & otherwise\end{cases}$$

$$\frac{1}{T_0}\int_{-\frac{T_0}{2}}^{\frac{T_0}{2}} \cos\left(2\pi\frac{n}{T_0}t\right)\cos\left(2\pi\frac{m}{T_0}t\right)dt = \begin{cases}\frac{1}{2} & for\ n = m\\ 0 & otherwise\end{cases}$$

$$\frac{1}{T_0}\int_{-\frac{T_0}{2}}^{\frac{T_0}{2}} \sin\left(2\pi\frac{n}{T_0}t\right)\cos\left(2\pi\frac{m}{T_0}t\right)dt = 0$$

$$\frac{1}{T_0}\int_{-\frac{T_0}{2}}^{\frac{T_0}{2}} \cos\left(2\pi\frac{n}{T_0}t\right)dt = 0$$

$$\frac{1}{T_0}\int_{-\frac{T_0}{2}}^{\frac{T_0}{2}} \sin\left(2\pi\frac{n}{T_0}t\right)dt = 0$$

Equation (18.111) indicates that the total average power can be broken down into powers contained in various harmonic components. The power on the dc component is

$$P_0 = a_0^2 = c_0^2 \tag{18.112}$$

The power on the fundamental component is

$$P_1 = \frac{a_1^2 + b_1^2}{2} = \frac{c_1^2}{2} \tag{18.113}$$

and the power contained in the second harmonic component is

$$P_2 = \frac{a_2^2 + b_2^2}{2} = \frac{c_2^2}{2} \tag{18.114}$$

etc. The plot of $P_0, P_1, P_2, \ldots$ as a function of harmonic frequencies $nf_0 = n/T_0$, $n = 0, 1, 2, 3, \ldots$, is referred to as the **power spectral density (PSD)**, since it shows relative strength of powers of various harmonic frequency components as a function of frequency. Equation (18.111) is called **Parseval's theorem**.

The **mean square** value of a signal is defined as the average of the square of the signal. The average power of a periodic signal is also the mean square value of the periodic signal. The square root of the mean square value is called the **root mean square (rms)** value of the signal. The rms value of a periodic signal with period T_0 is given by

$$F_{rms} = \sqrt{P} = \sqrt{\frac{1}{T_0}\int_{-T_0/2}^{T_0/2} |f(t)|^2 dt} = \sqrt{a_0^2 + \sum_{n=1}^{\infty}\frac{a_n^2 + b_n^2}{2}} = \sqrt{a_0^2 + \sum_{n=1}^{\infty}\left(\frac{a_n}{\sqrt{2}}\right)^2 + \sum_{n=1}^{\infty}\left(\frac{b_n}{\sqrt{2}}\right)^2}$$
$$= \sqrt{c_0^n + \sum_{n=1}^{\infty}\frac{c_n^2}{2}} = \sqrt{c_0^n + \sum_{n=1}^{\infty}\left(\frac{c_n}{\sqrt{2}}\right)^2} \tag{18.115}$$

The rms value of a periodic signal is the square root of the sum of powers of the dc component, the fundamental frequency component, the second harmonic frequency component, and all the rest of the frequency components. The power of each harmonic frequency component can be obtained by squaring the rms value of the harmonic frequency component. The power of the nth harmonic frequency component $\frac{c_n^2}{2}$ is the square of the rms value $c_n/\sqrt{2}$ of the nth harmonic component

$$c_n \cos\left(2\pi\frac{n}{T_0}t + \theta_n\right)$$

EXAMPLE 18.12

For the rectangular pulse train shown in Figure 18.13, find the average power of *f(t)*. Also, find the power on the first ten harmonics, including the dc component. What percentage of power is in the first ten harmonics, including the dc component?

The average power of $f(t)$ is

$$P = \frac{1}{T_0}\int_{-\frac{T_0}{2}}^{\frac{T_0}{2}} |f(t)|^2 dt = \frac{1}{1\text{ ms}}\int_{-0.25\text{ ms}}^{0.25\text{ ms}} 1^2 dt = \frac{1}{2}\text{ W}$$

From Example 18.7, we obtain the following powers:

$$P_0 = a_0^2 = c_0^2 = 0.5^2 = 0.25\text{ W}$$

$$P_1 = \frac{a_1^2 + b_1^2}{2} = \frac{c_1^2}{2} = 0.5 \times 0.6366^2 = 0.2026\text{ W}$$

$$P_3 = \frac{a_3^2 + b_3^2}{2} = \frac{c_3^2}{2} = 0.5 \times 0.2122^2 = 0.0225\text{ W}$$

$$P_5 = \frac{a_5^2 + b_5^2}{2} = \frac{c_5^2}{2} = 0.5 \times 0.1273^2 = 0.0081\text{ W}$$

$$P_7 = \frac{a_7^2 + b_7^2}{2} = \frac{c_7^2}{2} = 0.5 \times 0.09095^2 = 0.0041\text{ W}$$

$$P_9 = \frac{a_9^2 + b_9^2}{2} = \frac{c_9^2}{2} = 0.5 \times 0.07074^2 = 0.0025\text{ W}$$

$$P_2 = 0,\quad P_4 = 0,\quad P_6 = 0,\quad P_8 = 0,\quad P_{10} = 0$$

The sum of P_0 to P_{10} is 0.4899 W, which is 97.98% of 0.5 W.

Exercise 18.12

Let $A = 1$ V and $T_0 = 1$ ms for the triangular pulse train shown in Figure 18.25. Find the average power of *f(t)*. Also, find the power on the first 10 harmonics, including the dc component. What percentage of power is in the first 10 harmonics, including the dc component? What is the rms value of *f(t)*?

Answer:

$$P = \frac{2}{T_0}\int_0^{\frac{T_0}{2}} \left(-\frac{2A}{T_0}t + A\right)^2 dt = \frac{2}{T_0}\int_0^{\frac{T_0}{2}} \left(\frac{4A^2}{T_0^2}t^2 - \frac{4A^2}{T_0}t + A^2\right)^2 dt = \frac{A^2}{3} = \frac{1}{3}\text{ W}$$

$$a_0 = 0.5,\quad b_n = 0, \text{from Equation (18.70): } a_n = A\,\text{sinc}^2\!\left(\frac{n}{2}\right) = \text{sinc}^2\!\left(\frac{n}{2}\right)$$

continued

Exercise 18.12 continued

n:	0	1	2	3	4	5	6	7	8	9	10
a_n:	0.5	0.4053	0	0.0450	0	0.0162	0	0.0083	0	0.0050	0
c_n:	0.5	0.4053	0	0.0450	0	0.0162	0	0.0083	0	0.0050	0
θ_n:	0	0	0	0	0	0	0	0	0	0	0

$P_0 = 0.25, P_1 = 0.08213$ W, $P_2 = 0$ W, $P_3 = 0.001014$ W, $P_4 = 0$ W, $P_5 = 0.0001314$ W, $P_6 = 0$ W, $P_7 = 0.00003421$ W, $P_8 = 0$ W, $P_9 = 0.00001252$ W, $P_{10} = 0$ W

The sum of P_0 to P_{10} is 0.333319910054502 W, which is 99.996% of 1/3 W.

The rms amplitude is $1/\sqrt{3} = 0.5774$ V.

18.3.10 HALF-WAVE SYMMETRY

A periodic signal $f(t)$ with period T_0 is said to possess **half-wave symmetry** if we shift the signal by half a period to the right (or to the left), and flipped with respect to the time axis, we get the original signal. Mathematically, $f(t)$ is half-wave symmetry if

$$f(t) = -f\left(t - \frac{T_0}{2}\right) = -f\left(t + \frac{T_0}{2}\right)$$

If $f(t)$ possesses **half–wave symmetry**, we have

$$a_0 = 0 \tag{18.116}$$

$$a_n = 0, \quad \text{for } n \text{ even } (n = 2, 4, 6, \ldots)$$

$$a_n = \frac{4}{T_0}\int_0^{\frac{T_0}{2}} f(t)\cos\left(n\frac{2\pi}{T_0}t\right)dt = 0, \quad n = 1, 3, 5, \ldots \tag{18.117}$$

$$b_n = 0, \quad \text{for } n \text{ even } (n = 2, 4, 6, \ldots)$$

$$b_n = \frac{4}{T_0}\int_0^{\frac{T_0}{2}} f(t)\sin\left(n\frac{2\pi}{T_0}t\right)dt, \quad n = 1, 3, 5, \ldots \tag{18.118}$$

If $f(t)$ possesses **both even symmetry and half-wave symmetry**, the Fourier coefficients are given by

$$a_0 = 0 \tag{18.119}$$

$$a_n = 0, \quad \text{for } n \text{ even } (n = 2, 4, 6, \ldots)$$

$$a_n = \frac{4}{T_0}\int_0^{\frac{T_0}{2}} f(t)\cos\left(n\frac{2\pi}{T_0}t\right)dt = 0, \quad n = 1, 3, 5, \ldots \tag{18.120}$$

$$b_n = 0, \quad n = 1, 2, 3, \ldots \tag{18.121}$$

If $f(t)$ possesses **both odd symmetry and half-wave symmetry**, the Fourier coefficients are given by

$$a_0 = \frac{1}{T_0}\int_{-\frac{T_0}{2}}^{\frac{T_0}{2}} f(t)\,dt = 0 \tag{18.122}$$

$$a_n = \frac{2}{T_0}\int_{-\frac{T_0}{2}}^{\frac{T_0}{2}} f(t)\cos\left(n\frac{2\pi}{T_0}t\right)dt = 0, \quad n = 1, 2, 3, \ldots \tag{18.123}$$

$b_n = 0, \quad \text{for } n \text{ even } (n = 2, 4, 6, \ldots)$

$$b_n = \frac{4}{T_0}\int_{0}^{\frac{T_0}{2}} f(t)\sin\left(n\frac{2\pi}{T_0}t\right)dt, \quad n = 1, 3, 5, \ldots \tag{18.124}$$

The proof for Equations (18.116) through (18.118) is given here.

$$a_0 = \frac{1}{T_0}\int_{-\frac{T_0}{2}}^{\frac{T_0}{2}} f(t)dt = \frac{1}{T_0}\int_{-\frac{T_0}{2}}^{0} f(t)dt + \frac{1}{T_0}\int_{0}^{\frac{T_0}{2}} f(t)dt = -\frac{1}{T_0}\int_{-\frac{T_0}{2}}^{0} f\left(t - \frac{T_0}{2}\right)dt + \frac{1}{T_0}\int_{0}^{\frac{T_0}{2}} f(t)dt$$

Let us apply the change of variable $t' = t + T_0/2$ to the first integral. Then $t = t' - T_0/2$, $dt = dt'$. When $t = -T_0/2$, $t' = 0$, and when $t = 0$, $t' = T_0/2$. Thus, a_0 becomes

$$a_0 = -\frac{1}{T_0}\int_{0}^{\frac{T_0}{2}} f(t' - T_0)dt' + \frac{1}{T_0}\int_{0}^{\frac{T_0}{2}} f(t)dt$$

Since $f(t)$ is periodic, $f(t - T_0) = f(t)$. Thus we have

$$a_0 = -\frac{1}{T_0}\int_{0}^{\frac{T_0}{2}} f(t')dt' + \frac{1}{T_0}\int_{0}^{\frac{T_0}{2}} f(t)dt = 0$$

For a_n, we have

$$a_n = \frac{2}{T_0}\int_{-\frac{T_0}{2}}^{\frac{T_0}{2}} f(t)\cos(n\omega_0 t)dt = \frac{2}{T_0}\int_{-\frac{T_0}{2}}^{0} f(t)\cos(n\omega_0 t)dt + \frac{2}{T_0}\int_{0}^{\frac{T_0}{2}} f(t)\cos(n\omega_0 t)dt$$

$$= -\frac{2}{T_0}\int_{-\frac{T_0}{2}}^{0} f\left(t - \frac{T_0}{2}\right)\cos(n\omega_0 t)dt + \frac{2}{T_0}\int_{0}^{\frac{T_0}{2}} f(t)\cos(n\omega_0 t)dt$$

Let us apply the change of variable $t' = t + T_0/2$ to the first integral. Then $t = t' - T_0/2$, $dt = dt'$. When $t = -T_0/2$, $t' = 0$, and when $t = 0$, $t' = T_0/2$. Thus

$$a_n = -\frac{2}{T_0}\int_{0}^{\frac{T_0}{2}} f(t' - T_0)\cos\left[n\omega_0\left(t' - \frac{T_0}{2}\right)\right]dt' + \frac{2}{T_0}\int_{0}^{\frac{T_0}{2}} f(t)\cos(n\omega_0 t)dt$$

Since

$$\cos\left[n\omega_0\left(t' - \frac{T_0}{2}\right)\right] = \cos(n\omega_0 t')\cos\left(n\omega_0\frac{T_0}{2}\right) + \sin(n\omega_0 t')\sin\left(n\omega_0\frac{T_0}{2}\right)$$

$$= \cos(n\omega_0 t')\cos\left(n\frac{2\pi}{T_0}\frac{T_0}{2}\right) + \sin(n\omega_0 t')\sin\left(n\frac{2\pi}{T_0}\frac{T_0}{2}\right)$$

$$= \cos(n\omega_0 t')\cos(n\pi) + \sin(n\omega_0 t')\sin(n\pi) = (-1)^n\cos(n\omega_0 t')$$

and $f(t' - T_0) = f(t')$, we get

$$a_n = -\frac{2(-1)^n}{T_0}\int_0^{\frac{T_0}{2}} f(t')\cos(n\omega_0 t')dt' + \frac{2}{T_0}\int_0^{\frac{T_0}{2}} f(t)\cos(n\omega_0 t)dt$$

$$= \begin{cases} 0, & n\ even \\ \dfrac{4}{T_0}\displaystyle\int_0^{\frac{T_0}{2}} f(t)\cos(n\omega_0 t)dt, & n\ odd \end{cases}$$

Similarly, for b_n, we have

$$b_n = \frac{2}{T_0}\int_{-\frac{T_0}{2}}^{\frac{T_0}{2}} f(t)\sin(n\omega_0 t)dt = \frac{2}{T_0}\int_{-\frac{T_0}{2}}^{0} f(t)\sin(n\omega_0 t)dt + \frac{2}{T_0}\int_0^{\frac{T_0}{2}} f(t)\sin(n\omega_0 t)dt$$

$$= -\frac{2}{T_0}\int_{-\frac{T_0}{2}}^{0} f\left(t - \frac{T_0}{2}\right)\sin(n\omega_0 t)dt + \frac{2}{T_0}\int_0^{\frac{T_0}{2}} f(t)\sin(n\omega_0 t)dt$$

Let us apply the change of variable $t' = t + T_0/2$ to the first integral. Then $t = t' - T_0/2$, $dt = dt'$. When $t = -T_0/2, t' = 0$, and when $t = 0, t' = T_0/2$. Thus

$$b_n = -\frac{2}{T_0}\int_0^{\frac{T_0}{2}} f(t' - T_0)\sin\left[n\omega_0\left(t' - \frac{T_0}{2}\right)\right]dt' + \frac{2}{T_0}\int_0^{\frac{T_0}{2}} f(t)\sin(n\omega_0 t)dt$$

Since

$$\sin\left[n\omega_0\left(t' - \frac{T_0}{2}\right)\right] = \sin(n\omega_0 t')\cos\left(n\omega_0\frac{T_0}{2}\right) - \cos(n\omega_0 t')\sin\left(n\omega_0\frac{T_0}{2}\right)$$

$$= \sin(n\omega_0 t')\cos\left(n\frac{2\pi}{T_0}\frac{T_0}{2}\right) - \cos(n\omega_0 t')\sin\left(n\frac{2\pi}{T_0}\frac{T_0}{2}\right)$$

$$= \sin(n\omega_0 t')\cos(n\pi) - \cos(n\omega_0 t')\sin(n\pi) = (-1)^n\sin(n\omega_0 t')$$

and $f(t' - T_0) = f(t')$, we get

$$b_n = -\frac{2(-1)^n}{T_0}\int_0^{\frac{T_0}{2}} f(t')\sin(n\omega_0 t')dt' + \frac{2}{T_0}\int_0^{\frac{T_0}{2}} f(t)\sin(n\omega_0 t)dt$$

$$= \begin{cases} 0, & n\ even \\ \dfrac{4}{T_0}\displaystyle\int_0^{\frac{T_0}{2}} f(t)\sin(n\omega_0 t)dt, & n\ odd \end{cases}$$

EXAMPLE 18.13

Find the trigonometric Fourier coefficients of the rectangular pulse train shown in Figure 18.35. Represent *f*(*t*) by its Fourier series.

FIGURE 18.35

A rectangular pulse train.

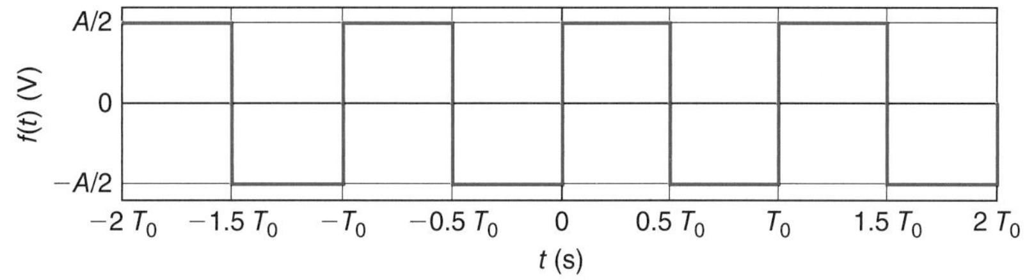

Since $f(t)$ possesses odd symmetry and half-wave symmetry, we have

$$a_0 = 0$$
$$a_n = 0, \quad n = 1, 2, 3, \ldots$$
$$b_n = 0, \quad n = 2, 4, 6, \ldots$$

For $n = 1, 3, 5, \ldots, b_n$ is given by

$$b_n = \frac{4}{T_0}\int_0^{\frac{T_0}{2}} f(t)\sin\left(n\frac{2\pi}{T_0}t\right)dt = \frac{4}{T_0}\int_0^{\frac{T_0}{2}} \frac{A}{2}\sin\left(n\frac{2\pi}{T_0}t\right)dt = \frac{2A}{T_0}\frac{-\cos\left(n\frac{2\pi}{T_0}t\right)\Big|_0^{\frac{T_0}{2}}}{n\frac{2\pi}{T_0}}$$

$$= \frac{2A}{T_0}\frac{-\cos\left(n\frac{2\pi}{T_0}\frac{T_0}{2}\right)+1}{n\frac{2\pi}{T_0}} = \frac{A}{n\pi}[1 - \cos(n\pi)] = \begin{cases} 0, & n = 2, 4, 6, \ldots \\ \frac{2A}{n\pi}, & n = 1, 3, 5, \ldots \end{cases}$$

The Fourier series representation of $f(t)$ is given by

$$f(t) = \sum_{\substack{n=1 \\ n=odd}}^{\infty} \frac{2A}{n\pi}\sin\left(\frac{n2\pi t}{T_0}\right) = \sum_{n=1}^{\infty}\frac{2A}{(2n-1)\pi}\sin\left(\frac{(2n-1)2\pi t}{T_0}\right)$$

Exercise 18.13

Find the trigonometric Fourier coefficients of the rectangular pulse train shown in Figure 18.36. Represent *f*(*t*) by its Fourier series.

continued

Exercise 18.13 continued

FIGURE 18.36

A rectangular pulse train.

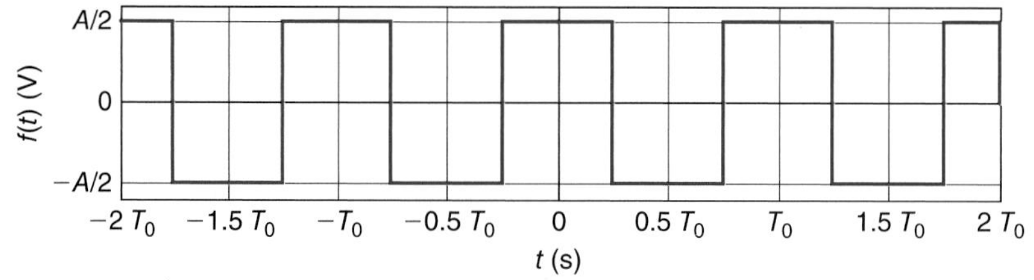

Answer:
$f(t)$ possesses both even symmetry and half-wave symmetry.

$a_0 = 0,\quad b_n = 0,\quad n = 1, 2, 3, \ldots,\quad a_n = 0,\quad n = 2, 4, 6, \ldots$ For $n =$ odd,

$$a_n = \frac{2A}{n\pi}\sin\left(\frac{n\pi}{2}\right),\quad n = 1, 3, 5, \ldots$$

18.4 Solving Circuit Problems Using Trigonometric Fourier Series

The steady-state response of a linear, time-invariant (LTI) system with transfer function $H(\omega)$ to a sinusoidal input signal (cosine)

$$x(t) = A\cos(\omega_0 t) \tag{18.125}$$

is given by

$$y(t) = A|H(\omega_0)|\cos(\omega_0 t + \angle H(\omega_0)) \tag{18.126}$$

Similarly, the steady-state response of an LTI system with a transfer function $H(\omega)$ to a sinusoidal input signal (sine)

$$x(t) = A\sin(\omega_0 t) \tag{18.127}$$

is given by

$$y(t) = A|H(\omega_0)|\sin(\omega_0 t + \angle H(\omega_0)) \tag{18.128}$$

If the frequency of the sinusoidal input signal is changed from ω_0 to $n\omega_0$, Equation (18.125) becomes

$$x(t) = A\cos(n\omega_0 t) \tag{18.129}$$

The steady-state response (output) of an LTI system to the input signal given by Equation (18.129) is

$$y(t) = A|H(n\omega_0)|\cos(n\omega_0 t + \angle H(n\omega_0)) \tag{18.130}$$

Similarly, for the input signal

$$x(t) = A\sin(n\omega_0 t) \tag{18.131}$$

the steady-state response is given by

$$y(t) = A|H(n\omega_0)| \sin(n\omega_0 t + \angle H(n\omega_0)) \tag{18.132}$$

If a signal $f(t)$ can be represented by the trigonometric Fourier series, it can be written as

$$f(t) = a_0 + \sum_{n=1}^{\infty}[a_n \cos(n\omega_0 t) + b_n \sin(n\omega_0 t)] \tag{18.133}$$

where the coefficients a_0, a_n, and b_n are given by Equations (18.29) through (18.31). Suppose that the signal $f(t)$ given by Equation (18.133) is applied to an LTI system with a transfer function $H(\omega)$. Then, from the superposition principle of an LTI system, the steady-state response of the system is the sum of the steady-state response of each term in Equation (18.133). The value of the transfer function for $\omega = 0$ is $H(0)$. Thus, the steady-state response of the system to the dc component a_0 is $a_0H(0)$. From Equation (18.130), the steady-state response of the system to the input $a_n \cos(n\omega_0 t)$ is $a_n|H(n\omega_0)| \cos(n\omega_0 t + \angle H(n\omega_0))$. From Equation (18.132), the steady-state response of the system to the input $b_n \sin(n\omega_0 t)$ is $b_n|H(n\omega_0)| \sin(n\omega_0 t + \angle H(n\omega_0))$. Thus, from the superposition principle, the steady-state response of an LTI system to the input given by Equation (18.133) is

$$\begin{aligned} y(t) = a_0H(0) + \sum_{n=1}^{\infty}[a_n|H(n\omega_0)| \cos(n\omega_0 t + \angle H(n\omega_0)) \\ + b_n|H(n\omega_0)| \sin(n\omega_0 t + \angle H(n\omega_0))] \end{aligned} \tag{18.134}$$

The trigonometric Fourier series representation given by Equation (18.133) can be rewritten as

$$f(t) = c_0 + \sum_{n=1}^{\infty} c_n \cos(n\omega_0 t + \theta_n) \tag{18.135}$$

where c_0, c_n, and θ_n are given by Equations (18.44) through (18.46). The steady-state response of an LTI system with the transfer function $H(\omega)$ to the input signal $f(t)$ given by Equation (18.135) is

$$y(t) = c_0H(0) + \sum_{n=1}^{\infty} c_n|H(n\omega_0)| \cos(n\omega_0 t + \theta_n + \angle H(n\omega_0)) \tag{18.136}$$

If $f(t)$ is periodic with period T_0, the steady-state response $y(t)$ is also periodic with period T_0. The signal $y(t)$ can be represented by a trigonometric Fourier series as

$$\begin{aligned} y(t) &= ya_0 + \sum_{n=1}^{\infty}[ya_n \cos(n\omega_0 t) + yb_n \sin(n\omega_0 t)] \\ &= yc_0 + \sum_{n=1}^{\infty} yc_n \cos(n\omega_0 t + y\theta_n) \end{aligned} \tag{18.137}$$

where

$$\begin{aligned} ya_0 &= yc_0 = a_0H(0) = c_0H(0) \\ ya_n &= |H(n\omega_0)|[a_n \cos(\angle H(n\omega_0)) + b_n \sin(\angle H(n\omega_0))] \\ yb_n &= |H(n\omega_0)|[-a_n \sin(\angle H(n\omega_0)) + b_n \cos(\angle H(n\omega_0))] \\ yc_n &= c_n|H(n\omega_0)| \\ y\theta_n &= \theta_n + \angle H(n\omega_0) \end{aligned}$$

EXAMPLE 18.14

A rectangular pulse train with amplitude $A = 1$ V, period $T_0 = 1$ ms, and pulse width $\tau = 0.5$ ms, shown in Figure 18.37, is applied to an RC circuit with $C = 0.1\ \mu$F and $R = 1\ k\Omega$, shown in Figure 18.38.

a. Find f_0, ω_0, d, a_0, a_n, b_n, c_n, θ_n, $H(\omega)$, $H(n\omega_0)$, $|H(n\omega_0)|$, $\angle H(n\omega_0)$, yc_n, and $y\theta_n$.
b. Evaluate a_n, c_n, θ_n, $|H(n\omega_0)|$, $\angle H(n\omega_0)$, yc_n, and $y\theta_n$ for $0 \le n \le 10$, and plot them.
c. Approximate $f(t)$ by the dc component and the first ten harmonics, and plot the approximation.
d. Approximate $y(t)$ by the dc component and the first ten harmonics, and plot the approximation.

FIGURE 18.37

A rectangular pulse train.

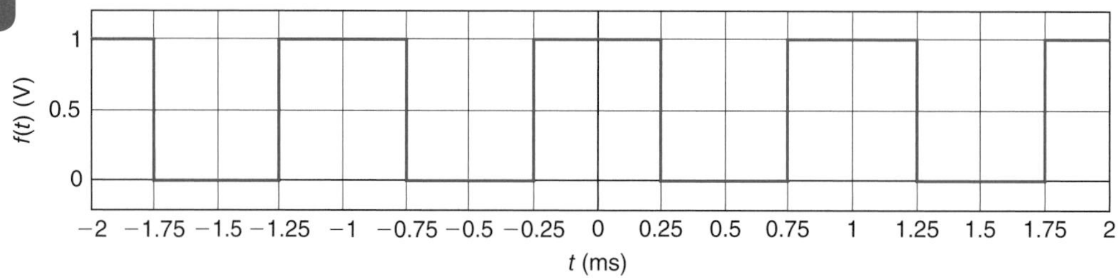

FIGURE 18.38

RC circuit.

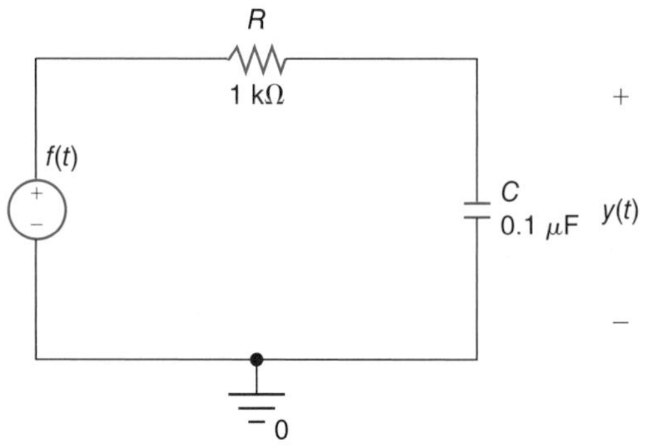

a. $$f_0 = \frac{1}{T_0} = \frac{1}{0.001s} = 1000 \text{ Hz} = 1 \text{ kHz}$$

$$\omega_0 = \frac{2\pi}{T_0} = \frac{2\pi}{0.001s} = 2\pi \times 1000 \text{ rad/s} = 6283.1853 \text{ rad/s}$$

$$d = \frac{\tau}{T_0} = \frac{0.5 \text{ ms}}{1 \text{ ms}} = 0.5$$

$$a_0 = Ad = 1 \times \frac{1}{2} = 0.5$$

$$a_n = 2Ad \operatorname{sinc}(nd) = 2 \times 1 \times 0.5 \times \operatorname{sinc}(n0.5)$$
$$= \operatorname{sinc}(0.5n), \quad n = 1, 2, 3, \ldots$$

$$b_n = 0, \quad n = 1, 2, 3, \ldots$$

$$c_0 = a_0 = 0.5$$

$$c_n = \sqrt{a_n^2 + b_n^2} = |a_n| = |\operatorname{sinc}(0.5n)|, \quad n = 1, 2, 3, \ldots$$

$$\theta_n = \begin{cases} 0, & a_n \ge 0 \\ 180°, & a_n < 0 \end{cases} = \begin{cases} 0, & n = 0, 1, 2, 4, 5, 6, 8, 9, 10, 12, 13, 14, \ldots \\ 180°, & n = 3, 7, 11, 15, \ldots \end{cases}$$

$$H(s) = \frac{\dfrac{1}{sC}}{R + \dfrac{1}{sC}} = \frac{1}{sRC + 1} = \frac{\dfrac{1}{RC}}{s + \dfrac{1}{RC}} = \frac{10{,}000}{s + 10{,}000}$$

continued

Example 18.14 continued

$$H(\omega) = H(s)|_{s=j\omega} = \frac{\frac{1}{RC}}{j\omega + \frac{1}{RC}} = \frac{\frac{1}{RC}}{j\omega + \frac{1}{RC}} = \frac{10{,}000}{j\omega + 10{,}000}$$

$$H(n\omega_0) = H(\omega)|_{\omega=n\omega_0} = \frac{10{,}000}{jn\omega_0 + 10{,}000} = \frac{10{,}000}{jn6283.1853 + 10{,}000} = \frac{1.59155}{jn + 1.59155}$$

$$|H(n\omega_0)| = \frac{1.59155}{|jn + 1.59155|} = \frac{1.59155}{\sqrt{n^2 + 1.59155^2}}$$

$$\angle H(n\omega_0) = -\angle(jn + 1.59155) = -\tan^{-1}\left(\frac{n}{1.59155}\right)$$

$$yc_n = c_n|H(n\omega_0)| = \frac{1.59155|\text{sinc}(0.5n)|}{\sqrt{n^2 + 1.59155^2}}$$

$$y\theta_n = \theta_n + \angle H(n\omega_0) = \theta_n - \tan^{-1}\left(\frac{n}{1.59155}\right)$$

b.

n:	0	1	2	3	4	5	6	7	8	9	10
a_n:	0.5	0.6366	0	-0.2122	0	0.1273	0	-0.09095	0	0.07074	0
c_n:	0.5,	0.6366	0	0.2122	0	0.1273	0	0.09095	0	0.07074	0
θ_n (rad)	0	0	0	π	0	0	0	π	0	0	0
θ_n (deg)	0	0	0	180	0	0	0	180	0	0	0
|H|	1	0.8467	0.6227	0.4686	0.3697	0.3033	0.2564	0.2217	0.1951	0.1741	0.1572
∠H (rad)	0	-0.561	-0.899	-1.083	-1.192	-1.263	-1.312	-1.347	-1.374	-1.396	-1.413
∠H (deg)	0	-32.14	-51.49	-62.05	-68.30	-72.34	-75.14	-77.19	-78.75	-79.97	-80.96
yc	0.5	0.539	0	0.0995	0	0.0386	0	0.0202	0	0.0123	0
yθ (rad)	0	-0.561	-0.899	2.0586	-1.192	-1.263	-1.312	1.794	-1.374	-1.396	-1.413
yθ (deg)	0	-32.14	-51.49	117.95	-68.30	-72.34	-75.14	102.81	-78.75	-79.97	-80.96

c. $f(t) = 0.5 + 0.6266\cos(2\pi 1000t) - 0.2122\cos(2\pi 3000t) + 0.1273\cos(2\pi 5000t) - 0.09095\cos(2\pi 7000t) + 0.07074\cos(2\pi 9000t)$

d. $y(t) = 0.5 + 0.539\cos(2\pi 1000t - 32.13°) + 0.0995\cos(2\pi 3000t + 117.95°) + 0.0386\cos(2\pi 5000t - 72.34°) + 0.0202\cos(2\pi 7000t + 102.81°) + 0.0123\cos(2\pi 9000t - 79.97°)$

The plots are shown in Figure 18.39.

FIGURE 18.39

Plot of (a) a_n; (b) c_n; (c) θ_n; (d) $|H(n\omega_0)|$; (e) $\angle H(n\omega_0)$; (f) yc_n; (g) $y\theta_n$; (h) $f(t)$ and $\hat{f}(t)$; and (i) $f(t)$ and $\hat{y}(t)$.

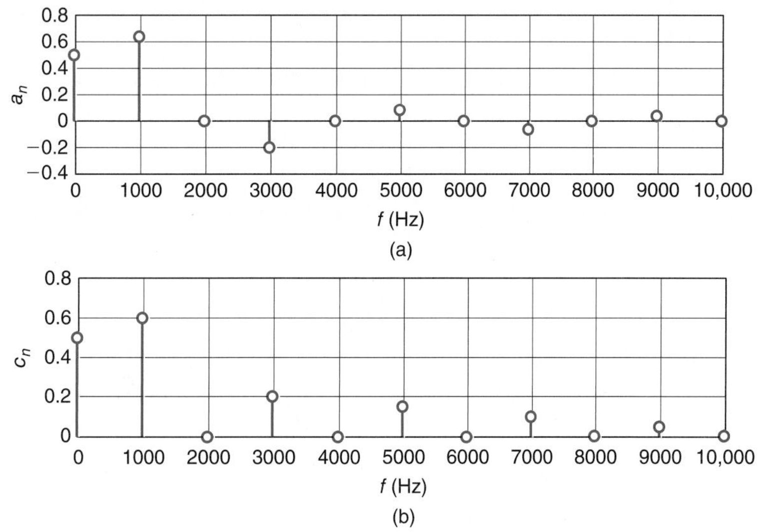

continued

Example 18.14 continued

FIGURE 18.39

continued

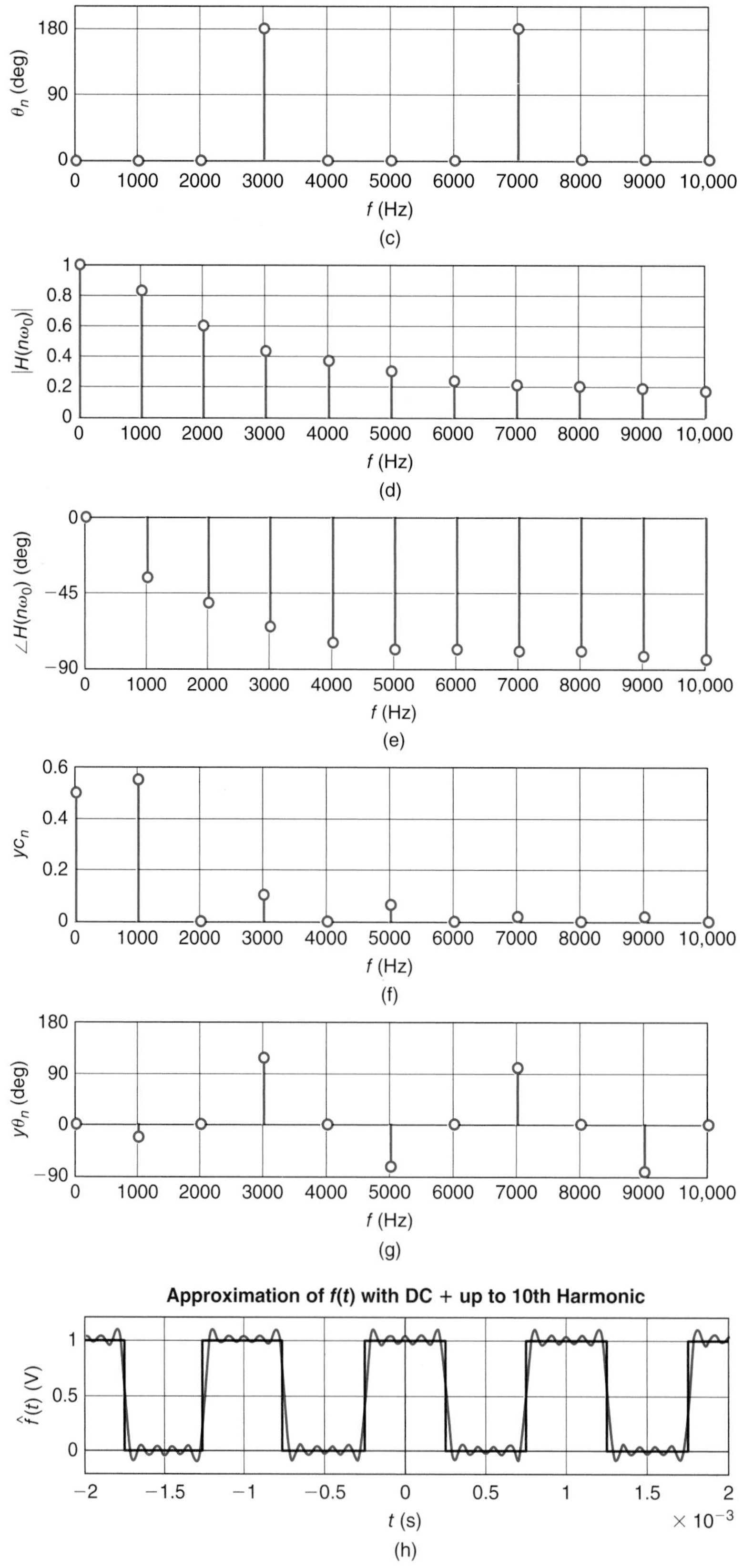

continued

Example 18.14 continued

FIGURE 18.39

continued

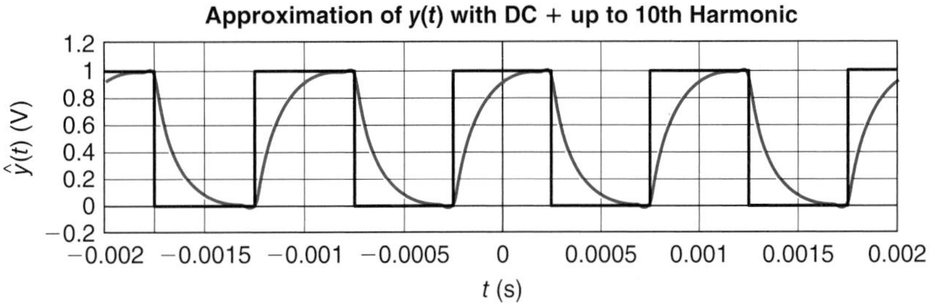

(i)

MATLAB

```
% EXAMPLE 18.14
clear all;format short;
A=1;T0=1e-3;tau=0.5e-3;N=10;R=1000;C=0.1e-6;M1=2;M2=1000;Th=1e-10;
f0=1/T0
w0=2*pi/T0
d=tau/T0
a0=A*tau/T0
n1=1:N;
n=0:N;
a=2*A*d*sinc(n1*d);
a=[a0,a]
b=zeros(1,N+1)
c=sqrt(a.^2+b.^2)
theta=zeros(1,N+1);
for i=1:length(a)
    if (abs(a(i))<=Th)&&(abs(b(i))<=Th)
        theta(i)=0;
    elseif (a(i)<-Th)&&(abs(b(i))<=Th)
        theta(i)=pi;
    elseif (a(i)>Th)&&(abs(b(i))<=Th)
        theta(i)=0;
    elseif (abs(a(i))<Th)&&(b(i)>Th)
        theta(i)=-pi/2;
    elseif (abs(a(i))<Th)&&(b(i)<-Th)
        theta(i)=pi/2;
    else
        theta(i)=-atan2(b(i),a(i));
    end
end
theta
thetad=theta*180/pi
f1=f0*n;
%
H=(1/(R*C))./(j*n*w0+(1/(R*C))) %TF (one-sided) H(nw0)
Hmag=abs(H)                     %abs(H(nwo))
Hphase=angle(H)                 %angle(H(nw0))
Hphased=Hphase*180/pi           %angle(H(nw0)) in degrees
ya=a.*H                         %Trig coeffs an of output y(t)
yc=c.*Hmag                      %Trig coeffs cn of the output y(t)
yp=(theta+Hphase).*(yc>=Th)     %phase ytheta(n) of y(t)
ypd=yp*180/pi                   %phase ytheta(n) in degrees
t=-M1*T0:T0/M2:M1*T0;
f0a=0.5*(1+square(2*pi*(t+0.25*T0)/T0,50));
```

continued

Example 18.14 continued

MATLAB continued

```
M=10;
f=a(1);
for i=2:M+1
    f=f+a(i)*cos(2*pi*(i-1)*t/T0)+b(i)*sin(2*pi*(i-1)*t/T0);
end
y=yc(1);
for i=2:M+1
    y=y+yc(i)*cos(2*pi*(i-1)*t/T0+yp(i));
end
%
figure(1)
stem(f1,a,'LineWidth',2);grid;xlabel('f (Hz)');ylabel('a_n');
set(gca,'YTick',-0.4:0.2:0.8)
set(gca,'YTickLabel',-0.4:0.2:0.8)
axis([0,N*f0,-0.4,0.8])
figure(2)
stem(f1,c,'LineWidth',2);grid;xlabel('f (Hz)');ylabel('c_n');
set(gca,'YTick',0:0.2:0.8)
set(gca,'YTickLabel',0:0.2:0.8)
axis([0,N*f0,0,0.8])
figure(3)
stem(f1,thetad,'LineWidth',2);grid;
xlabel('f (Hz)');ylabel('\theta_n (deg)');
set(gca,'YTick',0:90:180)
set(gca,'YTickLabel',0:90:180)
figure(4)
plot(t,f,'LineWidth',2);grid;hold on;
plot(t,f0a,'r-');xlabel('t');ylabel('f^\wedge(t)');hold off;
axis([-M1*T0,M1*T0,-0.2*A,1.2*A]);
title(['Approximation of f(t) with dc + up to ',num2str(M),'th Harmonic']);
figure(5)
stem(f1,Hmag,'LineWidth',2);grid;
xlabel('f (Hz)');ylabel('|H(n\omega_0)|');
set(gca,'YTick',0:0.2:1)
set(gca,'YTickLabel',0:0.2:1)
figure(6)
stem(f1,Hphased,'LineWidth',2);grid;
xlabel('f (Hz)');ylabel('\angleH(n\omega_0)(deg)');
set(gca,'YTick',-90:45:0)
set(gca,'YTickLabel',-90:45:0)
axis([0,N*f0,-90,0])
figure(7)
stem(f1,yc,'LineWidth',2);grid;xlabel('f (Hz)');ylabel('yc_n');
set(gca,'YTick',-0.4:0.2:0.8)
set(gca,'YTickLabel',-0.4:0.2:0.8)
figure(8)
stem(f1,ypd,'LineWidth',2);grid;
xlabel('f (Hz)');ylabel('y\theta_n (deg)');
set(gca,'YTick',-180:90:180)
set(gca,'YTickLabel',-180:90:180)
figure(9)
plot(t,y,'LineWidth',2);grid;hold on;
plot(t,f0a,'r-');xlabel('t');ylabel('y^\wedge(t)');hold off;
axis([-M1*T0,M1*T0,-0.2*A,1.2*A]);
title(['Approximation of y(t) with dc + up to ',num2str(M),'th Harmonic']);
set(gca,'XTick',-M1*T0:T0/2:M1*T0)
```

continued

Example 18.14 continued

MATLAB continued

```
set(gca,'XTickLabel',-M1*T0:T0/2:M1*T0)
set(gca,'YTick',-0.2*A:0.2*A:1.2*A)
set(gca,'YTickLabel',-0.2*A:0.2*A:1.2*A)
axis([-M1*T0,M1*T0,-0.2*A,1.2*A]);
```

Exercise 18.14

Represent the input $f(t)$ and output $y(t)$ by a Fourier series up to seventh harmonic for the circuit shown in Figure 18.38 when the input is a triangular pulse train with amplitude 1 V, period 1 ms, as shown in Figure 18.40.

FIGURE 18.40

Triangular pulse train.

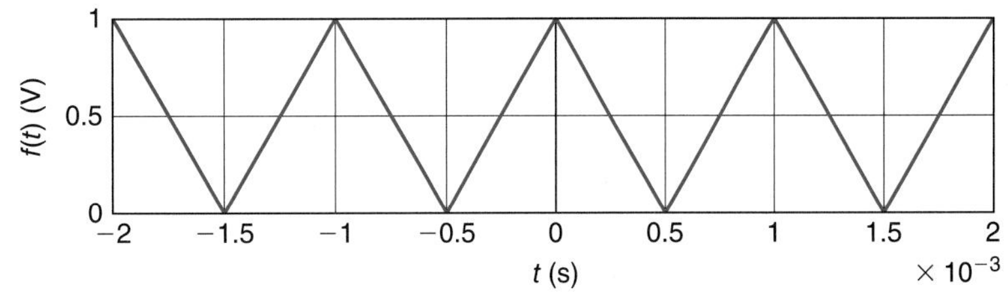

Answer:

$f(t) = 0.5 + 0.4053\cos(2\pi 1000t) + 0.045\cos(2\pi 3000t) + 0.0162\cos(2\pi 5000t) + 0.0083\cos(2\pi 7000t)$ V

$y(t) = 0.5 + 0.3432\cos(2\pi 1000t - 0.561) + 0.0211\cos(2\pi 3000t - 1.083) + 0.0049\cos(2\pi 5000t - 1.2626) + 0.0018\cos(2\pi 7000t - 1.3472)$ V. The phase is in radians.

EXAMPLE 18.15

A sawtooth pulse train $f(t)$ with amplitude 1 V and period 1 ms, shown in Figure 18.41, is applied to a series *RLC* circuit shown in Figure 18.42. Represent $f(t)$ and $y(t)$ by Fourier series up to the fifth harmonic.

FIGURE 18.41

Sawtooth pulse train.

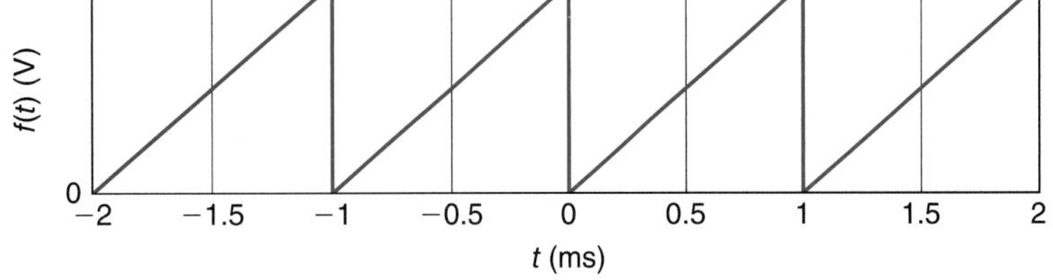

From Equations (18.80) through (18.82), we obtain

$$a_0 = 0.5,\quad a_n = 0 \text{ for } n = 1, 2, 3, \ldots,\quad b_n = -A/(\pi n) = -1/(\pi n),$$
$$n = 1, 2, 3, \ldots$$

continued

Example 18.15 continued

FIGURE 18.42

Series *RLC* circuit for EXAMPLE 18.15.

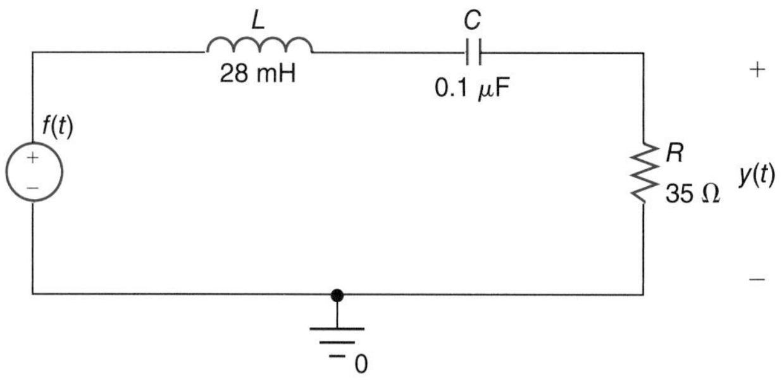

$$f(t) = \frac{A}{2} + \sum_{n=1}^{\infty} \frac{-A}{\pi n} \sin\left(n\frac{2\pi}{T_0}t\right)$$
$$= \frac{1}{2} + \sum_{n=1}^{\infty} \frac{-1}{\pi n} \sin(n2\pi 1000t)$$

The Fourier series up to the fifth harmonic is given by

$$f(t) = 0.5 - 0.3183\sin(2\pi 1000t) - 0.1592\sin(2\pi 2000t) - 0.1061\sin(2\pi 3000t) - 0.07958\sin(2\pi 4000t) - 0.06366\sin(2\pi 5000t)$$

Application of the voltage divider rule to the circuit shown in Figure 18.40 yields

$$H(s) = \frac{\frac{R}{L}s}{s^2 + \frac{R}{L}s + \frac{1}{LC}} = \frac{1250s}{s^2 + 1250s + 3.5714 \times 10^8}$$

Substitution of $s = jn\omega_0 = jn2\pi 1000 = jn6283.2$ into $H(s)$ yields

$$H(jn\omega_0) = \frac{1250jn6283.2}{-n^2 3.9478 \times 10^7 + 1250jn6283.2 + 3.5714 \times 10^8}$$

Evaluating $|H(jn\omega_0)|$ and $\angle H(jn\omega_0)$ for $n = 0, 1, 2, 3, 4, 5$, we obtain

n	0	1	2	3	4	5
$\|H(jn\omega_0)\|$	0	0.0247	0.0786	0.997	0.1137	0.06223
$\angle H(jn\omega_0)$	0	88.5837°	85.4919°	4.4583°	-83.4713°	-86.4322°

The voltage across the resistor is given by

$$y(t) = -0.007868\sin(2\pi 1000t + 88.5837°) - 0.01251\sin(2\pi 2000t + 85.4919°) - 0.1058\sin(2\pi 3000t + 4.4583°) - 0.009048\sin(2\pi 4000t - 83.4713°) - 0.003962\sin(2\pi 5000t - 86.4322°)\text{ V}$$

MATLAB

```
clear all;format long;
A=1;T0=1e-3;R=35;L=28e-3;C=0.1e-6;M=5;
n1=1:M;n=0:M;
f0=1/T0
w0=2*pi*f0
b=[A/2,-1./(n1*pi)] %A/2 represents dc component
H=R/L*j*n*w0./((j*n*w0).^2+R/L*j*n*w0+1/(L*C))
Hmag=abs(H)
Hphase=angle(H)*180/pi
ybamp=b.*Hmag
```

Exercise 18.15

A sawtooth pulse train $f(t)$ with amplitude 1 V and period 1 ms, shown in Figure 18.41, is applied to a series *RLC* circuit shown in Figure 18.43. Represent $y(t)$ by its Fourier series up to the fifth harmonic.

FIGURE 18.43

Series RLC circuit for EXERCISE 18.15.

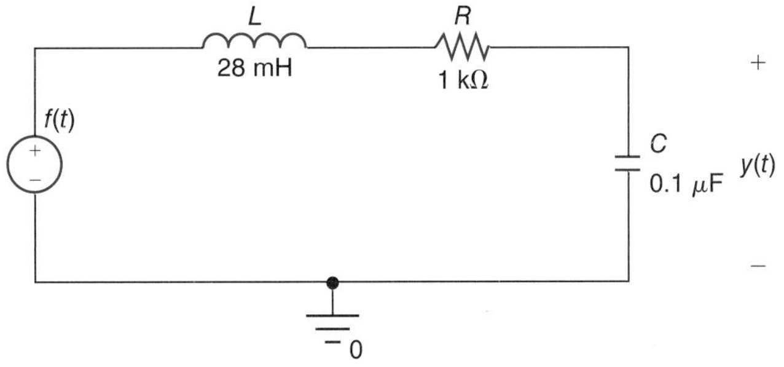

Answer:
$y(t) = 0.5 - 0.2923 \sin(2\pi 1000t - 35.2375°) - 0.1158 \sin(2\pi 2000t - 66.0628°) - 0.05629 \sin(2\pi 3000t - 89.8436°) - 0.03028 \sin(2\pi 4000t - 107.0052°) - 0.01767 \sin(2\pi 5000t - 119.3071°)$ V.

18.5 Exponential Fourier Series

As shown in Section 18.2 earlier in this chapter, a set of complex exponential functions

$$e^{jn\omega_0 t}, \quad n = 0, \pm 1, \pm 2, \pm 3, \ldots \tag{18.138}$$

is orthogonal and complete on the interval $(t_0, t_0 + T_0)$, where

$$\omega_0 = \frac{2\pi}{T_0} = 2\pi f_0 \tag{18.139}$$

is the fundamental frequency in radians/s ($f_0 = 1/T_0$ is the fundamental frequency in Hz), and t_0 is arbitrary. Therefore, a square-integrable function $f(t)$ can be represented as a linear combination of this set of complete orthogonal functions in the interval $(t_0, t_0 + T_0)$:

$$f(t) = \sum_{n=-\infty}^{\infty} F_n e^{jn\omega_0 t}, \quad t_0 \le t \le t_0 + T_0 \tag{18.140}$$

The coefficients F_0 and $F_n, n = \pm 1, \pm 2, \pm 3, \ldots$ can be obtained from Equations (18.26) and (18.27), respectively,

$$F_0 = \frac{1}{T_0} \int_{t_0}^{t_0+T_0} f(t)dt \tag{18.141}$$

$$F_n = \frac{1}{T_0} \int_{t_0}^{t_0+T_0} f(t)e^{-jn\omega_0 t}dt, \quad n = \pm 1, \pm 2, \pm 3, \ldots \tag{18.142}$$

If $f(t)$ is periodic, the representation given by Equation (18.140) is valid for all t; that is, $-\infty < t < \infty$. If $f(t)$ is nonperiodic, the representation given by Equation (18.140) is valid for $t_0 < t < t_0 + T_0$. Unless mentioned otherwise, we assume that $f(t)$ is periodic. Nonperiodic signals can be represented by the Fourier transform, as explained in Chapter 19.

The representation of a signal by the sum of complex exponential signals given by Equation (18.140) is called an **exponential Fourier series.** The coefficients $F_n, n = 0, \pm 1, \pm 2, \pm 3, \ldots$ are called **exponential Fourier coefficients.** In general, the coefficients F_n are complex numbers unless certain symmetry exits in $f(t)$.

The lower limit of integration in Equations (18.141) and (18.142) is given by t_0. The value of the lower limit t_0 is arbitrary. As long as the interval of integration is one period, T_0, the starting point of the integratrion can be any value. Depending on the shape of the signal $f(t)$, we can choose the appropriate value for t_0. Most common choices are $t_0 = 0$ or $t_0 = -T_0/2$. If $t_0 = 0$ is chosen, Equations (18.141) and (18.142) become, respectively,

$$F_0 = \frac{1}{T_0}\int_0^{T_0} f(t)dt \tag{18.143}$$

$$F_n = \frac{1}{T_0}\int_0^{T_0} f(t)e^{-jn\frac{2\pi}{T_0}t}dt, \quad n = \pm 1, \pm 2, \pm 3, \ldots \tag{18.144}$$

If $t_0 = -T_0/2$ is chosen, Equations (18.141) and (18.142) become, respectively,

$$F_0 = \frac{1}{T_0}\int_{-\frac{T_0}{2}}^{\frac{T_0}{2}} f(t)dt \tag{18.145}$$

$$F_n = \frac{1}{T_0}\int_{-\frac{T_0}{2}}^{\frac{T_0}{2}} f(t)e^{-jn\frac{2\pi}{T_0}t}dt, \quad n = \pm 1, \pm 2, \pm 3, \ldots \tag{18.146}$$

The exponential Fourier series expansion can be interpreted as follows. A square-integrable function $f(t)$ can be represented as a sum of complex exponential signals whose frequencies are harmonically related. The dc component F_0 represents a point P_0 in the real axis of a complex plane. A typical term $F_n e^{jn\omega_0 t}$ represents a point P_n in the complex plane that revolves around P_0 in a circular path at a constant speed of $n\omega_0$ radians/s (or nf_0 revolutions per second). The radius of the circular path on which the point P_n travels is given by $|F_n|$. At $t = 0$, the point P_n is at $|F_n| \angle F_n$. In other words, the point P_n starts at an angle $\angle F_n$ with respect to the positive real axis. Thus $\sum_{n=-\infty}^{\infty} F_n e^{jn\omega_0 t}$ represents the sum of an infinite number of points around the point P_0 revolving at different speeds and starting points. Positive n values represent points revolving in the counterclockwise direction, while negative n values represent points revolving in the clockwise direction. For each point revolving in the counterclockwise direction, there is a point revolving in the clockwise direction with same speed. For example, for $F_n e^{jn\omega_0 t}$, there is $F_{-n} e^{-jn\omega_0 t}$ which starts at $F_{-n} = F_n^*$ and revolves in the opposite direction. They cross each other twice per revolution (cycle). When the conjugate pair $F_n e^{jn\omega_0 t}$ and $F_{-n} e^{-jn\omega_0 t}$ are added, we have $2|F_n| \cos(n\omega_0 t + \angle F_n)$, which is a real signal. Notice that the sum of the conjugate pair is twice the projection of either point to the real axis. The reason for this is that the projections of P_n and P_{-n} to the imaginary axis cancel each other out, but the projections of P_n and P_{-n} to the real axis are added. If the speed of the point revolving in the counterclockwise direction is defined as a positive frequency, the speed of the point revolving in the clockwise direction is defined as a negative frequency.

The complex exponential Fourier coefficients given by Equation (18.146) can be rewritten as

$$F_n = \frac{1}{T_0}\int_{-\frac{T_0}{2}}^{\frac{T_0}{2}} f(t)\cos\left(n\frac{2\pi}{T_0}t\right)dt - j\frac{1}{T_0}\int_{-\frac{T_0}{2}}^{\frac{T_0}{2}} f(t)\sin\left(n\frac{2\pi}{T_0}t\right)dt$$
$$= \frac{a_n}{2} - j\frac{b_n}{2}, \quad n = 1, 2, 3, \ldots \tag{18.147}$$

Notice that

$$\text{Re}(F_n) = \frac{a_n}{2}, \quad \text{Im}(F_n) = -\frac{b_n}{2}$$

If $f(t)$ possesses **even symmetry**, $f(t)\cos(n\omega_0 t)$ is an even function of t, and $f(t)\sin(n\omega_0 t)$ is an odd function of t. Thus, the second integral of Equation (18.147) vanishes, and the first integral is two times the integral from 0 to $T_0/2$, and F_n reduces to

$$F_n = \frac{2}{T_0}\int_0^{\frac{T_0}{2}} f(t)\cos\left(n\frac{2\pi}{T_0}t\right)dt = \frac{a_n}{2}, \quad n = 1, 2, 3, \ldots \tag{18.148}$$

$$b_n = -2\,\text{Im}(F_n) = 0, \quad n = 1, 2, 3, \ldots$$

Notice that since $\cos\left(-n\frac{2\pi}{T_0}t\right) = \cos\left(n\frac{2\pi}{T_0}t\right)$

$$F_{-n} = F_n = \frac{a_n}{2}, \quad n = 1, 2, 3, \ldots \tag{18.149}$$

If $f(t)$ possesses **even symmetry**, we have

$$a_n = 2F_n, \quad n = 1, 2, 3, \ldots$$
$$b_n = 0, \quad n = 1, 2, 3, \ldots$$
$$|a_n| = 2|F_n|, \quad n = 1, 2, 3, \ldots$$
$$\angle a_n = \angle F_n, \quad n = 1, 2, 3, \ldots$$
$$c_n = |a_n| = 2\,|F_n|, \quad n = 1, 2, 3, \ldots$$
$$\theta_n = \angle F_n, \quad n = 1, 2, 3, \ldots$$

Thus, F_n is real and possesses even symmetry with respect to n (or frequency $n\omega_0$ or nf_0). Notice that the dc component is given by

$$F_0 = \frac{2}{T_0}\int_0^{\frac{T_0}{2}} f(t)dt$$

If $f(t)$ possesses **odd symmetry**, $f(t)\cos(n\omega_0 t)$ is an odd function of t, and $f(t)\sin(n\omega_0 t)$ is an even function of t. Thus, the first integral of Equation (18.147) vanishes, and the second integral is two times the integral from 0 to $T_0/2$, and F_n reduces to

$$F_n = -j\frac{2}{T_0}\int_0^{\frac{T_0}{2}} f(t)\sin\left(n\frac{2\pi}{T_0}t\right)dt = -j\frac{b_n}{2} \tag{18.150}$$

Notice that

$$F_{-n} = -F_n = j\frac{b_n}{2}, \quad n = 1, 2, 3, \ldots \tag{18.151}$$

Thus, F_n is purely imaginary and possesses odd symmetry with respect to n (or frequency $n\omega_0$ or nf_0). Here, odd symmetry $F_{-n} = -F_n$ implies

$$|F_{-n}| = |F_n|, \quad n = 1, 2, 3, \ldots \tag{18.152}$$

and

$$\angle F_{-n} = -\angle F_n, \quad n = 1, 2, 3, \ldots \tag{18.153}$$

Notice that the dc component is 0; that is,

$$F_0 = 0$$

If $f(t)$ possesses **odd symmetry**, we have

$$a_n = 0, \quad n = 1,2,3,\ldots$$
$$b_n = j2F_n, \quad n = 1, 2, 3, \ldots$$
$$|b_n| = 2|F_n|, \quad n = 1, 2, 3, \ldots$$
$$\angle b_n = \angle F_n + 90°, \quad n = 1, 2, 3, \ldots$$
$$c_n = 2|F_n|, \quad n = 1,2,3,\ldots$$
$$\theta_n = \angle F_n, \quad n = 0,1,2,\ldots$$

18.5.1 CONVERSION OF FOURIER COEFFICIENTS

The dc component F_0 given by Equation (18.145) is the same as the dc component in the trigonometric Fourier series given by Equation (18.35). Thus, we have

$$F_0 = a_0 = c_0 = \frac{1}{T_0}\int_{-\frac{T_0}{2}}^{\frac{T_0}{2}} f(t)dt \tag{18.154}$$

The exponential Fourier coefficients given by Equation (18.146) can be rewritten as

$$F_n = \frac{1}{T_0}\int_{-\frac{T_0}{2}}^{\frac{T_0}{2}} f(t)\cos\left(n\frac{2\pi}{T_0}t\right)dt - j\frac{1}{T_0}\int_{-\frac{T_0}{2}}^{\frac{T_0}{2}} f(t)\sin\left(n\frac{2\pi}{T_0}t\right)dt, \quad n = 1, 2, 3, \ldots \tag{18.155}$$

Comparing Equation (18.155) with Equations (18.36) and (18.37), we conclude that

$$F_n = \frac{a_n}{2} - j\frac{b_n}{2} = |F_n|e^{j\angle F_n}, \quad n = 1, 2, 3, \ldots \tag{18.156}$$

Setting $n = -n$ and noting that $\cos(-n\omega_0 t) = \cos(n\omega_0 t)$ and $\sin(-n\omega_0 t) = -\sin(n\omega_0 t)$ in Equation (18.155), we have

$$F_{-n} = \frac{1}{T_0}\int_{-\frac{T_0}{2}}^{\frac{T_0}{2}} f(t)\cos\left(n\frac{2\pi}{T_0}t\right)dt + j\frac{1}{T_0}\int_{-\frac{T_0}{2}}^{\frac{T_0}{2}} f(t)\sin\left(n\frac{2\pi}{T_0}t\right)dt, \quad n = 1, 2, 3, \ldots \tag{18.157}$$

Comparing Equation (18.157) with Equations (18.36) and (18.37), we have

$$F_{-n} = \frac{a_n}{2} + j\frac{b_n}{2} = F_n^* = |F_n|e^{-j\angle F_n}, \quad n = 1, 2, 3, \ldots \tag{18.158}$$

Adding Equations (18.156) and (18.158), we have

$$\begin{aligned} a_n &= 2\text{Re}[F_n] = F_n + F_{-n} = |F_n|e^{j\angle F_n} + |F_n|e^{-j\angle F_n} \\ &= 2|F_n|\cos(\angle F_n), \quad n = 1, 2, 3, \ldots \end{aligned} \tag{18.159}$$

Subtracting Equation (18.156) from Equation (18.158) and dividing by j, we have

$$\begin{aligned} b_n &= -2\text{Im}[F_n] = j[F_n - F_{-n}] = \frac{F_{-n} - F_n}{j} = \frac{|F_n|e^{-j\angle F_n} - |F_n|e^{j\angle F_n}}{j} \\ &= -2|F_n|\sin(\angle F_n), \quad n = 1, 2, 3, \ldots \end{aligned} \tag{18.160}$$

The coefficients c_n can be found from Equations (18.159) and (18.160)

$$\begin{aligned} &c_n = \sqrt{a_n^2 + b_n^2} = \sqrt{4|F_n|^2\cos^2(\angle F_n) + 4|F_n|^2\sin^2(\angle F_n)} = 2|F_n|, \\ &n = 1, 2, 3, \ldots \end{aligned} \tag{18.161}$$

From Equations (18.159) and (18.160), we have

$$\begin{aligned} \theta_n &= -\tan^{-1}\left(\frac{b_n}{a_n}\right) = -\tan^{-1}\left(\frac{-2|F_n|\sin(\angle F_n)}{2|F_n|\cos(\angle F_n)}\right) \\ &= \tan^{-1}(\tan(\angle F_n)) = \angle F_n, \quad n = 1, 2, 3, \ldots \end{aligned} \tag{18.162}$$

From Equation (18.161), we have

$$|F_n| = |F_{-n}| = \frac{c_n}{2}, \quad n = 1, 2, 3, \ldots \tag{18.163}$$

From Equation (18.162) we have

$$\angle F_n = \theta_n, \quad n = 1, 2, 3, \ldots \tag{18.164}$$

From Equation (18.158) we have

$$\angle F_{-n} = -\angle F_n = -\theta_n, \quad n = 1, 2, 3, \ldots \tag{18.165}$$

Table 18.3 at the end of Chapter 18 shows Fourier coefficients and Fourier series representation of common signals.

18.5.2 TWO-SIDED MAGNITUDE SPECTRUM AND TWO-SIDED PHASE SPECTRUM

If the magnitudes (or radii)..., $|F_{-3}|, |F_{-2}|, |F_{-1}|, |F_0|, |F_1|, |F_2|, |F_3|, \ldots$ are plotted against frequencies (direction and speed of revolutions) in radians..., $-3\omega_0$, $-2\omega_0$, $-\omega_0$, 0, ω_0, $2\omega_0$, $3\omega_0$,...(or in hertz..., $-3f_0$, $-2f_0$, $-f_0$, 0, f_0, $2f_0$, $3f_0$,...), we obtain what is referred to as a **two-sided magnitude spectrum**. The two-sided magnitude spectrum shows the magnitudes of the dc component, the fundamental components, the second harmonic components, the third harmonic components, etc. The magnitude spectrum reveals the relative strength of the various magnitude components of the given signal. If the signal changes slowly with time (smooth signal or lowpass signal), the magnitudes of lower frequency components are stronger than those of higher frequency components. If the signal changes fastly with time (such as a bandpass signal), the magnitudes of certain components are higher than others. On the other hand, if the phases (initial phase with respect to the real axis) ..., $\angle F_{-3}, \angle F_{-2}, \angle F_{-1}, \angle F_0, \angle F_1, \angle F_2, \angle F_3, \ldots$ are plotted against frequencies (direction and speed of revolutions) in radians..., $-3\omega_0,\ -2\omega_0,\ -\omega_0,\ 0,\ \omega_0,\ 2\omega_0, 3\omega_0, \ldots$(or in hertz ..., $-3f_0, -2f_0, -f_0, 0, f_0, 2f_0, 3f_0, \ldots$), we obtain what is referred to as a **two-sided phase spectrum**. The two-sided phase spectrum shows the phases of the dc component, the fundamental components, the second harmonic components, the third harmonic components, etc. The phase spectrum reveals the relative values of the various phase components of the given signal.

Notice that the dc component is the same whether the Fourier series representation is trigonometric or exponential, since the dc component is defined as the average value of the signal. Since the two-sided magnitude spectrum is an even function of frequency for real signal $f(t)$, the one-sided magnitude spectrum is split in half and placed at positive and negative frequencies. Since the two-sided phase spectrum of a real signal $f(t)$ is an odd function of frequency, the two-sided phase spectrum for positive frequencies is identical to that of a one-sided phase spectrum. The two-sided phase spectrums for negative frequencies have opposite polarity to those for positive frequencies.

A rectangular pulse train with amplitude A and pulse width τ is shown in Figure 18.44. Notice that the pulse train is centered at $t = 0$.

FIGURE 18.44

A rectangular pulse train.

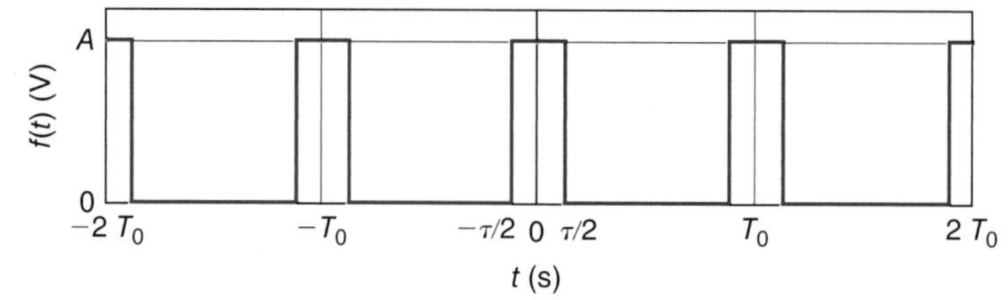

The dc component is given by

$$F_0 = \frac{1}{T_0}\int_{-\frac{T_0}{2}}^{\frac{T_0}{2}} f(t)dt = \frac{1}{T_0}\int_{-\frac{\tau}{2}}^{\frac{\tau}{2}} Adt = \frac{A}{T_0}t\Big|_{-\frac{\tau}{2}}^{\frac{\tau}{2}} = \frac{A\tau}{T_0} = Ad \tag{18.166}$$

where d is the duty cycle defined as the ratio of the pulse width τ to the period T_0; that is,

$$d = \frac{\tau}{T_0} \tag{18.167}$$

The complex exponential Fourier coefficients are given by

$$\begin{aligned} F_n &= \frac{1}{T_0}\int_{-\frac{T_0}{2}}^{\frac{T_0}{2}} f(t)e^{-jn\frac{2\pi}{T_0}t}dt = \frac{1}{T_0}\int_{-\frac{\tau}{2}}^{\frac{\tau}{2}} Ae^{-jn\frac{2\pi}{T_0}t}dt = \frac{A}{T_0}\frac{e^{-jn\frac{2\pi}{T_0}t}\Big|_{-\frac{\tau}{2}}^{\frac{\tau}{2}}}{-jn\frac{2\pi}{T_0}} \\ &= \frac{A}{T_0}\frac{e^{-jn\frac{2\pi}{T_0}\frac{\tau}{2}} - e^{jn\frac{2\pi}{T_0}\frac{\tau}{2}}}{-jn\frac{2\pi}{T_0}} = \frac{A}{T_0}\frac{e^{jn\frac{2\pi}{T_0}\frac{\tau}{2}} - e^{-jn\frac{2\pi}{T_0}\frac{\tau}{2}}}{jn\frac{2\pi}{T_0}} = \frac{A\tau}{T_0}\frac{\frac{e^{jn\frac{\pi}{T_0}\tau} - e^{-jn\frac{\pi}{T_0}\tau}}{2j}}{n\frac{\pi}{T_0}\tau} \\ &= \frac{A\tau}{T_0}\frac{\sin\left(n\frac{\pi}{T_0}\tau\right)}{n\frac{\pi}{T_0}\tau} = \frac{A\tau}{T_0}\frac{\sin\left(\pi n\frac{\tau}{T_0}\right)}{\pi n\frac{\tau}{T_0}} \\ &= Ad\,\mathrm{sinc}(nd), \quad n = 0, \pm 1, \pm 2, \pm 3, \ldots \end{aligned} \tag{18.168}$$

where the sinc function is defined as

$$\text{sinc}(x) = \frac{\sin(\pi x)}{\pi x} \tag{18.169}$$

Equation (18.168) includes the dc component. Notice that we can also find F_n from

$$a_n = 2\text{Ad sinc(nd)}$$

$$F_n = \frac{a_n}{2} = Ad\,\text{sinc}(nd)$$

EXAMPLE 18.16

A rectangular pulse train shown in Figure 18.45 has $A = 1$ V, $T_0 = 1$ ms, $\tau = 0.5$ ms. The pulse train is centered at $t = 0$.

a. Find $f_0, \omega_0, d, F_0, F_n$.
b. Evaluate F_0, F_n for $-10 \le n \le 10$, and plot $F_n, |F_n|, \angle F_n$ as a function of frequency.
c. Approximate *f*(*t*) by the dc component and the first 10 harmonics (both + and −), and plot the approximation.

FIGURE 18.45

A rectangular pulse train with $A = 1$ V, $T_0 = 1$ ms, $d = 1/2$.

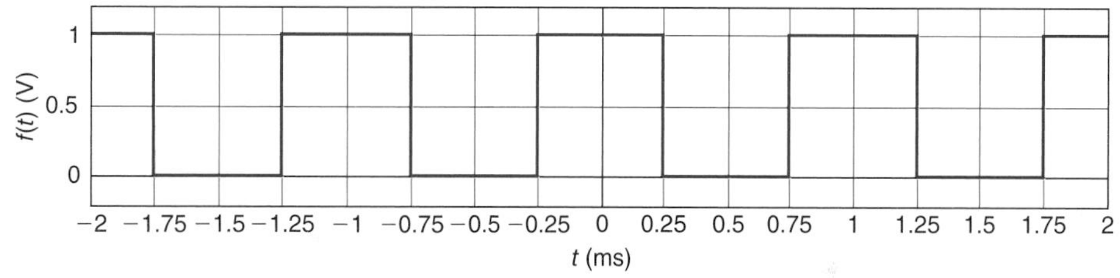

a. $f_0 = \dfrac{1}{T_0} = \dfrac{1}{0.001s} = 1000 \text{ Hz} = 1 \text{ kHz}$

$$\omega_0 = \frac{2\pi}{T_0} = \frac{2\pi}{0.001s} = 2\pi \times 1000 \text{ rad/s} = 6283.1853 \text{ rad/s}$$

$$d = \frac{\tau}{T_0} = \frac{0.5 \text{ ms}}{1 \text{ ms}} = 0.5$$

$$F_0 = a_0 = c_0 = Ad = 1 \times \frac{1}{2} = 0.5$$

$$F_n = Ad\,\text{sinc}(nd) = 1 \times 0.5 \times \text{sinc}(n0.5) = 0.5 \times \text{sinc}(0.5n),$$
$$n = 0, \pm1, \pm2, \pm3, \ldots$$

$$|F_n| = |0.5 \times \text{sinc}(0.5n)|$$

$$\angle F_n = \begin{cases} 0, & F_n \ge 0 \\ 180°, & F_n < 0 \end{cases}$$

$$= \begin{cases} 0, & n = 0, \pm1, \pm2, \pm4, \pm5, \pm6, \pm8, \pm9, \pm10, \pm12, \pm13, \pm14, \ldots \\ 180°, & n = \pm3, \pm7, \pm11, \pm15, \ldots \end{cases}$$

continued

Example 18.16 continued

b.

n:	0	±1	±2	±3	±4	±5	±6	±7	±8	±9	±10
F_n:	0.5	0.3183	0	-0.1061	0	0.0637	0	-0.0455	0	0.0354	0
$\|F_n\|$:	0.5	0.3183	0	0.1061	0	0.0637	0	0.0455	0	0.0354	0
θ_n:	0	0	0	180	0	0	0	180	0	0	0

The plot of F_n, $|F_n|$, and $\angle F_n$ are shown in Figure 18.46.

FIGURE 18.46

Plot of F_n, $|F_n|$, $\angle F_n$.

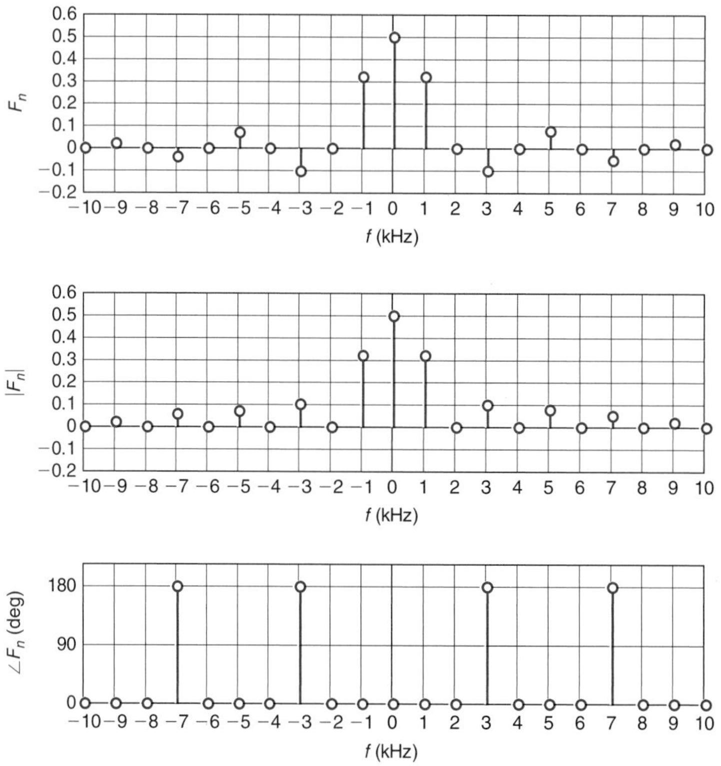

c. $$\hat{f}(t) = 0.5 + 0.3183\exp(j2\pi1000t) + 0.3183\exp(-j2\pi1000t)$$
$$- 0.1061\exp(j2\pi3000t) - 0.1061\exp(-j2\pi3000t)$$
$$+ 0.0637\exp(j2\pi5000t) + 0.0637\exp(-j2\pi5000t)$$
$$- 0.0455\exp(j2\pi7000t) - 0.0455\exp(-j2\pi7000t)$$
$$+ 0.0354\exp(j2\pi9000t) + 0.0354\exp(-j2\pi9000t)$$

Plot of $f(t)$ and $\hat{f}(t)$ is shown in Figure 18.47.

FIGURE 18.47

Plot of $f(t)$ (black) and $\hat{f}(t)$ (blue).

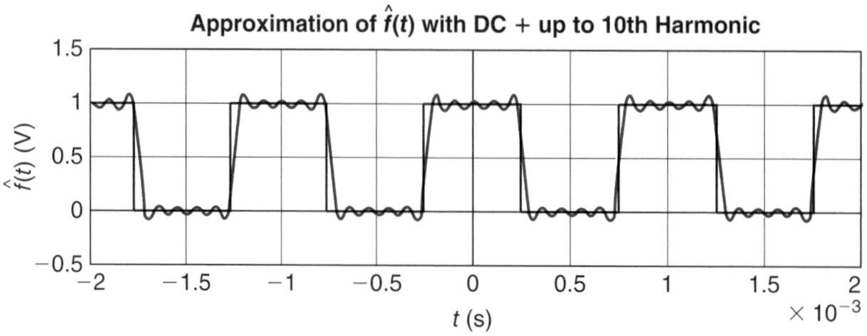

continued

Example 18.16 continued

The ringing at the amplitude changes is called the *Gibbs phenomenon.*

MATLAB

```
% EXAMPLE 18.16
clear all;format long;
A=1;T0=1e-3;tau=0.5e-3;N=10;Th=1e-10;M1=2;M2=1000;
f0=1/T0
w0=2*pi/T0
d=tau/T0
a0=A*tau/T0
n=-N:N;
f2=n*f0;
F=A*d*sinc(n*d)
Fmag=abs(F)
Fphase=angle(F).*(abs(F)>Th)
Fphased=Fphase*180/pi
figure(1)
stem(f2,Fmag,'LineWidth',2);grid;
xlabel('f (kHz)');ylabel('|F_n|');
set(gca,'XTick',-N*f0:f0:N*f0)
set(gca,'XTickLabel',-N:N)
set(gca,'YTick',0:0.1:0.6)
set(gca,'YTickLabel',0:0.1:0.6)
axis([-N*f0,N*f0,0,0.6]);
figure(2)
stem(f2,Fphased,'LineWidth',2);grid;
xlabel('f (kHz)');ylabel('\angleF_n (deg)');
set(gca,'XTick',-N*f0:f0:N*f0)
set(gca,'XTickLabel',-N:N)
set(gca,'YTick',0:90:180)
set(gca,'YTickLabel',0:90:180)
axis([-N*f0,N*f0,0,200]);
t=-M1*T0:T0/M2:M1*T0;
f=F(N+1);
for i=1:N
    f=f+F(N+1+i)*exp(2j*pi*(i)*t/T0)+F(N+1-i)*exp(2j*pi*(-i)*t/T0);
end
f0a=0.5*(1+square(2*pi*(t+0.25*T0)/T0,50));
figure(3)
plot(t,f,'LineWidth',2);grid;hold on;
plot(t,f0a,'r-');xlabel('t');ylabel('f^\wedge(t)');hold off;
title(['Approximation of f(t) with DC + up to ',num2str(N),'th Harmonic']);
axis([-M1*T0,M1*T0,-0.2*A,1.2*A]);
```

Exercise 18.16

A rectangular pulse train shown in Figure 18.48 has $A = 1$ V, $T_0 = 1$ ms, $\tau = 0.25$ ms. The pulse train is centered at $t = 0$.

a. Find f_0, ω_0, d, F_0, F_n.
b. Evaluate F_0, F_n for $-10 \le n \le 10$, and plot F_n, $|F_n|$, $\angle F_n$ as a function of frequency.
c. Approximate $f(t)$ by the dc component and the first tenth harmonics (both + and −), and plot the approximation.

continued

Exercise 18.16 continued

FIGURE 18.48

A rectangular pulse train with $A = 1$, $T_0 = 1$ ms, $d = 1/4$.

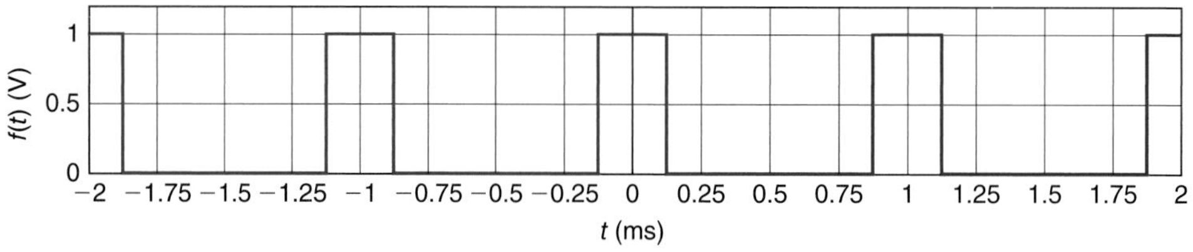

Answer:

a. $f_0 = \frac{1}{T_0} = \frac{1}{0.001s} = 1000 \text{ Hz} = 1 \text{ kHz}$

$$\omega_0 = \frac{2\pi}{T_0} = \frac{2\pi}{0.001s} = 2\pi \times 1000 \text{ rad/s} = 6283.1853 \text{ rad/s}$$

$$d = \frac{\tau}{T_0} = \frac{0.25 \text{ ms}}{1 \text{ ms}} = 0.25$$

$$F_0 = a_0 = c_0 = Ad = 1 \times \frac{1}{4} = 0.25$$

$F_n = Ad \operatorname{sinc}(nd) = 1 \times 0.25 \times \operatorname{sinc}(n0.25) = 0.25 \times \operatorname{sinc}(0.25n)$,
$n = 0, \pm1, \pm2, \pm3, \ldots$

$|F_n| = |0.25 \times \operatorname{sinc}(0.25n)|$

$$\angle F_n = \begin{cases} 0, & F_n \geq 0 \\ 180°, & F_n < 0 \end{cases}$$

$$= \begin{cases} 0, & n = 0, \pm1, \pm2, \pm3, \pm4, \pm8, \pm9, \pm10, \pm11, \pm12, \ldots \\ 180°, & n = \pm5, \pm6, \pm7, \pm13, \ldots \end{cases}$$

b.

n:	0	±1	±2	±3	±4	±5	±6	±7	±8	±9	±10
F_n:	0.25	0.2251	0.1592	0.075	0	-0.045	-0.053	-0.032	0	0.025	0.032
$\|F_n\|$:	0.25	0.2251	0.1592	0.075	0	0.045	0.053	0.032	0	0.025	0.032
θ_n:	0	0	0	0	0	-180	-180	-180	0	0	0

The plot of F_n, $|F_n|$, and $\angle F_n$ are shown in Figure 18.49.

FIGURE 18.49

Plot of F_n, $|F_n|$, and $\angle F_n$.

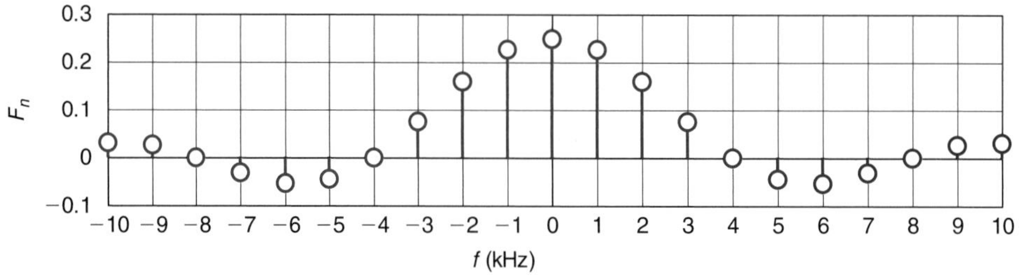

continued

Exercise 18.16 continued

FIGURE 18.49

continued

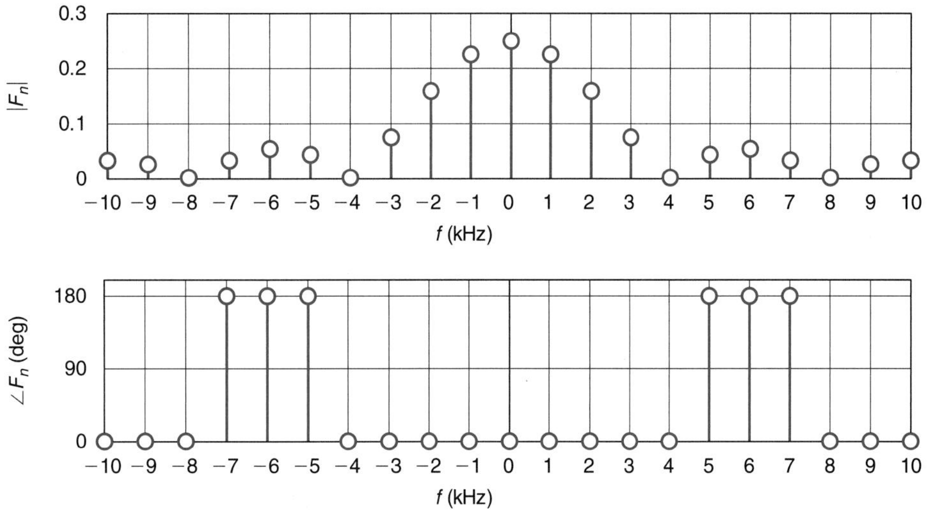

(c)
$$
\begin{aligned}
\hat{f}(t) = 0.25 &+ 0.2251 \exp(j2\pi 1000t) + 0.2251 \exp(-j2\pi 1000t) \\
&+ 0.1592 \exp(j2\pi 2000t) + 0.1592 \exp(-j2\pi 2000t) \\
&+ 0.07503 \exp(j2\pi 3000t) + 0.07503 \exp(-j2\pi 3000t) \\
&- 0.04502 \exp(j2\pi 5000t) - 0.04502 \exp(-j2\pi 5000t) \\
&- 0.05305 \exp(j2\pi 6000t) - 0.05305 \exp(-j2\pi 6000t) \\
&- 0.03215 \exp(j2\pi 7000t) - 0.03215 \exp(-j2\pi 7000t) \\
&+ 0.02501 \exp(j2\pi 9000t) + 0.02501 \exp(-j2\pi 9000t) \\
&+ 0.03183 \exp(j2\pi 10000t) + 0.03183 \exp(-j2\pi 10000t)
\end{aligned}
$$

Plot of $f(t)$ and $\hat{f}(t)$ is shown in Figure 18.50.

FIGURE 18.50

Plot of $f(t)$ (black) and $\hat{f}(t)$ (blue).

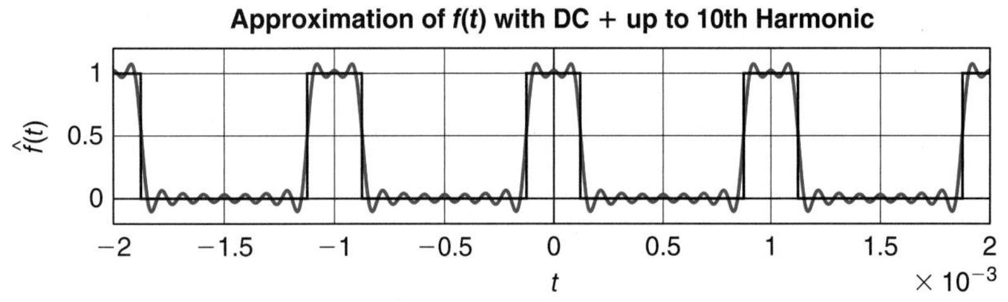

18.5.3 TRIANGULAR PULSE TRAIN

A triangular pulse train is shown in Figure 18.51.

FIGURE 18.51

A triangular pulse train.

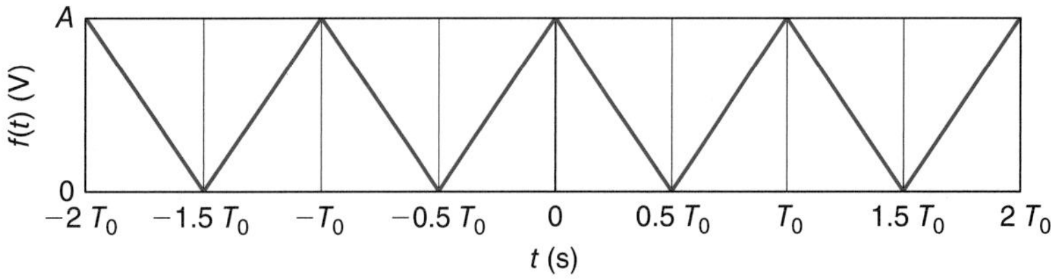

The equation of $f(t)$ for $-T_0/2 \le t < T_0/2$ is given by

$$f(t) = \begin{cases} \dfrac{2A}{T_0}t + A, & \dfrac{-T_0}{2} \le t < 0 \\ \dfrac{-2A}{T_0}t + A, & 0 \le t < \dfrac{T_0}{2} \end{cases}$$

The signal $f(t)$ possesses even symmetry. The dc component is given by

$$\begin{aligned} F_0 &= \frac{2}{T_0}\int_0^{\frac{T_0}{2}} f(t)dt = \frac{2}{T_0}\int_0^{\frac{T_0}{2}}\left(\frac{-2A}{T_0}t + A\right)dt = \frac{-4A}{T_0^2}\int_0^{\frac{T_0}{2}} tdt + \frac{2A}{T_0}\int_0^{\frac{T_0}{2}} 1dt \\ &= \frac{-4A}{T_0^2}\left[\frac{t^2}{2}\right]_0^{\frac{T_0}{2}} + \frac{2A}{T_0}[t]_0^{\frac{T_0}{2}} = \frac{-4A}{T_0^2} \times \frac{T_0^2}{8} + A = \frac{A}{2} \end{aligned} \quad \textbf{(18.170)}$$

The coefficients F_n can be obtained from Equation (18.146):

$$\begin{aligned} F_n &= \frac{1}{T_0}\int_{-\frac{T_0}{2}}^{0}\left(\frac{2A}{T_0}t + A\right)e^{-jn\frac{2\pi}{T_0}t}dt + \frac{1}{T_0}\int_0^{\frac{T_0}{2}}\left(\frac{-2A}{T_0}t + A\right)e^{-jn\frac{2\pi}{T_0}t}dt \\ &= \frac{2A}{T_0^2}\int_{-\frac{T_0}{2}}^{0} te^{-jn\frac{2\pi}{T_0}t}dt - \frac{2A}{T_0^2}\int_0^{\frac{T_0}{2}} te^{-jn\frac{2\pi}{T_0}t}dt + \frac{A}{T_0}\int_{-\frac{T_0}{2}}^{\frac{T_0}{2}} e^{-jn\frac{2\pi}{T_0}t}dt \end{aligned}$$

The third integral is equal to 0:

$$\frac{A}{T_0}\int_{-\frac{T_0}{2}}^{\frac{T_0}{2}} e^{-jn\frac{2\pi}{T_0}t}dt = \frac{A}{T_0}\frac{e^{-jn\frac{2\pi}{T_0}\frac{T_0}{2}} - e^{jn\frac{2\pi}{T_0}\frac{T_0}{2}}}{-jn\dfrac{2\pi}{T_0}} = \frac{A}{T_0}\frac{e^{jn\pi} - e^{-jn\pi}}{jn\dfrac{2\pi}{T_0}} = \frac{A\sin(n\pi)}{n\pi} = 0$$

Application of the integral formula $\int te^{at}dt = \dfrac{e^{at}(at-1)}{a^2}$ to the remaining integrals yields

$$\begin{aligned} F_n &= \frac{2A}{T_0^2}\left[\frac{e^{-jn\frac{2\pi}{T_0}t}\left(-jn\dfrac{2\pi}{T_0}t - 1\right)\Big|_{-\frac{T_0}{2}}^{0} - e^{-jn\frac{2\pi}{T_0}t}\left(-jn\dfrac{2\pi}{T_0}t - 1\right)\Big|_0^{\frac{T_0}{2}}}{\left(-jn\dfrac{2\pi}{T_0}\right)^2}\right] \\ &= 2A\left[\frac{-1 - e^{jn\frac{2\pi}{T_0}\frac{T_0}{2}}\left(jn\dfrac{2\pi}{T_0}\dfrac{T_0}{2} - 1\right) - \left[e^{-jn\frac{2\pi}{T_0}\frac{T_0}{2}}\left(-jn\dfrac{2\pi}{T_0}\dfrac{T_0}{2} - 1\right) - (-1)\right]}{-(2\pi n)^2}\right] \\ &= 2A\left[\frac{-1 - e^{jn\pi}(jn\pi - 1) - [e^{-jn\pi}(-jn\pi - 1) - (-1)]}{-(2\pi n)^2}\right] \\ &= 2A\left[\frac{-1 - jn\pi e^{jn\pi} + e^{jn\pi} + jn\pi e^{-jn\pi} + e^{-jn\pi} - 1}{-(2\pi n)^2}\right] \end{aligned}$$

$$= 2A\left[\frac{-2 + e^{jn\pi} + e^{-jn\pi} - jn\pi(e^{jn\pi} - e^{-jn\pi})}{-(2\pi n)^2}\right]$$

$$= 2A\left[\frac{-2 + 2\cos(n\pi) - jn\pi 2j\sin(n\pi)}{-(2\pi n)^2}\right]$$

$$= 4A\frac{1-\cos(n\pi)}{(2\pi n)^2} = 4A\frac{2\sin^2\left(\frac{n\pi}{2}\right)}{(2\pi n)^2} = \frac{A}{2}\frac{\sin^2\left(\frac{n\pi}{2}\right)}{\left(\frac{n\pi}{2}\right)^2} = \frac{A}{2}\text{sinc}^2\left(\frac{n}{2}\right)$$

Thus, the complex exponential Fourier coefficients for the triangular pulse train shown in Figure 18.51 is given by

$$F_n = \frac{A}{2}\text{sinc}^2\left(\frac{n}{2}\right) \tag{18.171}$$

If A is positive, F_n is greater than or equal to zero for all n; that is, $F_n \geq 0$. Thus, the phase response is zero for all n; that is, $\angle F_n = 0$.

Since $f(t)$ is even, F_n can be evaluated using Equation (18.148):

$$F_n = \frac{2}{T_0}\int_0^{\frac{T_0}{2}} f(t)\cos\left(n\frac{2\pi}{T_0}t\right)dt = \frac{2}{T_0}\int_0^{\frac{T_0}{2}}\left(\frac{-2A}{T_0}t + A\right)\cos\left(n\frac{2\pi}{T_0}t\right)dt$$

MATLAB can be used to evaluate this integral:

```
clear all;
syms A T0 t n
f=A-2*A/T0*t;
F=(2/T0)*int(f*cos(n*2*pi*t/T0),t,0,T0/2)

Answer:
F =
(2*A*sin((pi*n)/2)^2)/(pi^2*n^2)
```

The answer can be rewritten as

$$F_n = 2A\frac{\sin^2\left(\pi\frac{n}{2}\right)}{\pi^2 n^2} = \frac{2A}{4}\frac{\sin^2\left(\pi\frac{n}{2}\right)}{\frac{\pi^2 n^2}{4}} = \frac{A}{2}\left[\frac{\sin\left(\pi\frac{n}{2}\right)}{\pi\frac{n}{2}}\right]^2 = \frac{A}{2}\text{sinc}^2\left(\frac{n}{2}\right)$$

EXAMPLE 18.17

Find the exponential Fourier coefficients F_n for the triangular pulse train shown in Figure 18.52.

continued

Example 18.17 continued

FIGURE 18.52

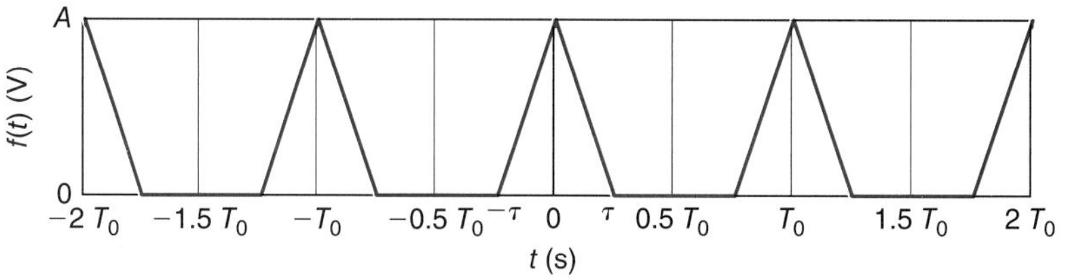

A triangular pulse train.

The equation of $f(t)$ for $0 \le t < \tau$ is given by

$$f(t) = \frac{-A}{\tau}t + A$$

The signal $f(t)$ possesses even symmetry. The dc component is given by

$$F_0 = \frac{2}{T_0}\int_0^{\frac{T_0}{2}} f(t)dt = \frac{2}{T_0}\int_0^{\tau}\left(\frac{-A}{\tau}t + A\right)dt = \frac{-2A}{T_0\tau}\int_0^{\tau} t dt + \frac{2A}{T_0}\int_0^{\tau} 1 dt$$
$$= \frac{-2A}{T_0\tau}\left[\frac{t^2}{2}\right]_0^{\tau} + \frac{2A}{T_0}[t]_0^{\tau} = \frac{-2A}{T_0\tau} \times \frac{\tau^2}{2} + \frac{2A\tau}{T_0} = \frac{A\tau}{T_0} = Ad \quad \textbf{(18.172)}$$

where d is defined as

$$d = \frac{\tau}{T_0} \quad \textbf{(18.173)}$$

Since $f(t)$ possesses even symmetry, from Equation (18.148), F_n is given by

$$F_n = \frac{2}{T_0}\int_0^{\frac{T_0}{2}} f(t)\cos\left(n\frac{2\pi}{T_0}t\right)dt = \frac{2}{T_0}\int_0^{\tau}\left(\frac{-A}{\tau}t + A\right)\cos\left(n\frac{2\pi}{T_0}t\right)dt$$
$$= \frac{-2A}{T_0\tau}\int_0^{\tau} t\cos\left(n\frac{2\pi}{T_0}t\right)dt + \frac{2A}{T_0}\int_0^{\tau}\cos\left(n\frac{2\pi}{T_0}t\right)dt$$

Application of the integral formula

$$\int t\cos(at)\,dt = \frac{1}{a^2}[\cos(at) + at\sin(at)]$$

results in

$$F_n = \frac{-2A}{T_0\tau}\,\frac{\left[\cos\left(n\frac{2\pi}{T_0}t\right) + n\frac{2\pi}{T_0}t\sin\left(n\frac{2\pi}{T_0}t\right)\right]_0^{\tau}}{\left(n\frac{2\pi}{T_0}\right)^2} + \frac{2A}{T_0}\,\frac{\left[\cos\left(n\frac{2\pi}{T_0}t\right)\right]_0^{\tau}}{n\frac{2\pi}{T_0}}$$
$$= \frac{-2A}{T_0\tau_0}\,\frac{\cos\left(n\frac{2\pi}{T_0}\tau\right) + n\frac{2\pi}{T_0}\tau\sin\left(n\frac{2\pi}{T_0}\tau\right) - 1}{\left(n\frac{2\pi}{T_0}\right)^2} + \frac{2A}{T_0}\,\frac{\sin\left(n\frac{2\pi}{T_0}\tau\right)}{n\frac{2\pi}{T_0}}$$

continued

Example 18.17 continued

$$= \frac{2A}{T_0\tau} \frac{1 - \cos\left(n\frac{2\pi}{T_0}\tau\right)}{\left(n\frac{2\pi}{T_0}\right)^2}$$

Application of $1 - \cos(2\theta) = 2\sin^2(\theta)$ results in

$$F_n = \frac{2A}{T_0\tau} \frac{2\sin^2\left(n\frac{\pi}{T_0}\tau\right)}{\left(n\frac{2\pi}{T_0}\right)^2} = \frac{A\tau}{T_0} \frac{\sin^2\left(n\frac{\pi}{T_0}\tau\right)}{\left(n\frac{\pi}{T_0}\tau\right)^2}$$
$$= \frac{A\tau}{T_0}\left[\frac{\sin\left(\pi n\frac{\tau}{T_0}\right)}{\pi n\frac{\tau}{T_0}}\right]^2 = Ad\,\text{sinc}^2(nd) \tag{18.174}$$

If $\tau = T_0/2$, Equations (18.172) through (18.174) become

$$d = \frac{\tau}{T_0} = \frac{1}{2} \tag{18.175}$$

$$F_0 = Ad = \frac{A}{2} \tag{18.176}$$

$$F_n = \frac{A}{2}\text{sinc}^2\left(\frac{n}{2}\right) \tag{18.177}$$

Exercise 18.17

Let $A = 1$ V and $T_0 = 1$ ms for the triangular pulse train shown in Figure 18.51. The waveform is shown in Figure 18.53. Find the exponential Fourier coefficients F_n for $-5 \le n \le 5$, and represent $f(t)$ by the dc component and up to the fifth harmonic components.

FIGURE 18.53

A triangular pulse train.

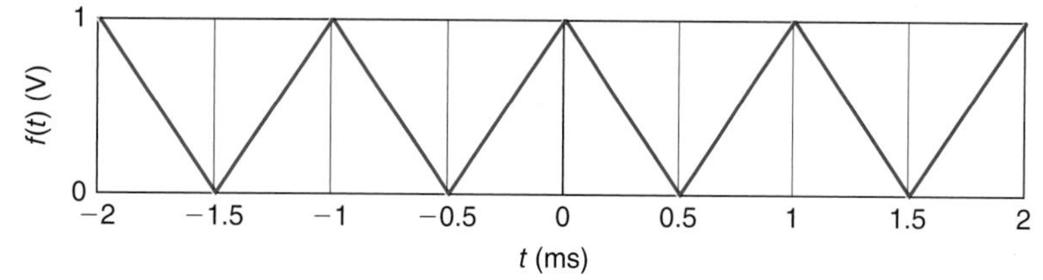

Answer:

$$F_n = \frac{1}{2}\text{sinc}^2\left(\frac{n}{2}\right).$$

n:	-5	-4	-3	-2	-1	0	1	2	3	4	5
F_n:	0.0081	0	0.02252	0	0.2026	0.5	0.2026	0	0.02252	0	0.0081

continued

Exercise 18.17 continued

$$f(t) = 0.5 + 0.2026(e^{j2\pi 1000t} + e^{-j2\pi 1000t}) + 0.02252(e^{j2\pi 3000t} + e^{-j2\pi 3000t}) + 0.0081(e^{j2\pi 5000t} + e^{-j2\pi 5000t}) = 0.5 + 0.4053\cos(2\pi 1000t) + 0.045\cos(2\pi 3000t) + 0.01621\cos(2\pi 5000t)$$

18.5.4 SAWTOOTH PULSE TRAIN

A sawtooth waveform is shown in Figure 18.54.

FIGURE 18.54

A sawtooth pulse train.

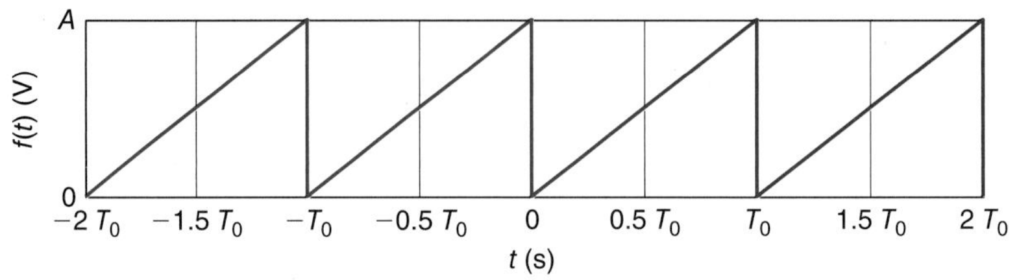

The equation of $f(t)$ for $0 \le t < T_0$ is given by $f(t) = \frac{A}{T_0}t$. The dc component is given by, from Equation (8.143),

$$F_0 = \frac{1}{T_0}\int_0^{T_0} f(t)dt = \frac{1}{T_0}\int_0^{T_0} \frac{A}{T_0}tdt = \frac{A}{T_0^2}\int_0^{T_0} tdt = \frac{A}{T_0^2}\frac{t^2|_0^{T_0}}{2} = \frac{A}{T_0^2}\frac{T_0^2}{2} = \frac{A}{2} \tag{18.178}$$

Using the integral formula $\int te^{at}\,dt = \frac{1}{a^2}e^{at}(at - 1)$ and Equation (8.144), we obtain

$$\begin{aligned} F_n &= \frac{1}{T_0}\int_0^{T_0} f(t)e^{-jn\frac{2\pi}{T_0}t}dt = \frac{1}{T_0}\int_0^{T_0}\frac{A}{T_0}te^{-jn\frac{2\pi}{T_0}t}dt = \frac{A}{T_0^2}\frac{\left[e^{-jn\frac{2\pi}{T_0}t}\left(-jn\frac{2\pi}{T_0}t - 1\right)\right]_0^{T_0}}{\left(-jn\frac{2\pi}{T_0}\right)^2} \\ &= \frac{A}{T_0^2}\frac{e^{-jn\frac{2\pi}{T_0}T_0}\left(-jn\frac{2\pi}{T_0}T_0 - 1\right) - (-1)}{-\left(n\frac{2\pi}{T_0}\right)^2} = \frac{A}{T_0^2}\frac{-jn\frac{2\pi}{T_0}T_0 - 1 - (-1)}{-\left(n\frac{2\pi}{T_0}\right)^2} \\ &= A\frac{jn2\pi}{(n2\pi)^2} = \frac{jA}{2\pi n}, \quad n \ne 0 \end{aligned} \tag{18.179}$$

Notice that $\angle F_n = 90°$ for $n > 0$, and $\angle F_n = -90°$ for $n < 0$.

MATLAB

```
clear all;
syms A T0 t
syms n integer
f=A*t/T0;
F=(1/T0)*int(f*exp(-j*n*2*pi*t/T0),t,0,T0)
F=simplify(F)

Answer:
F =
-A*(1/(4*pi^2*n^2) - (exp(-pi*n*2i)*(pi*n*2i + 1))/(4*pi^2*n^2))
F =
(A*1i)/(2*pi*n)
```

$$F_n = -A\left(\frac{1}{4\pi^2 n^2} - \frac{e^{-j2\pi n}(j2\pi n + 1)}{4\pi^2 n^2}\right) = -A\frac{1 - j2\pi n - 1}{4\pi^2 n^2} = \frac{jA}{2\pi n}$$

EXAMPLE 18.18

Find the exponential Fourier coefficients for the sawtooth waveform shown in Figure 18.55.

FIGURE 18.55

A sawtooth waveform.

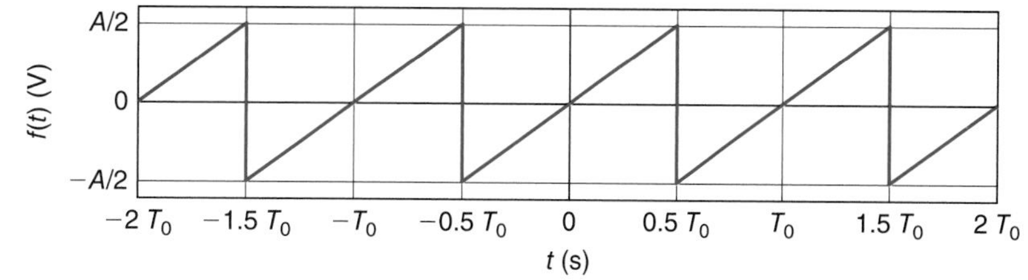

The equation of $f(t)$ for $0 \le t < T_0/2$ is given by $f(t) = \frac{A}{T_0}t$. Since $f(t)$ possesses odd symmetry, the dc component is 0; that is,

$$F_0 = 0 \quad \textbf{(17.180)}$$

Using the integral formula $\int t\sin(at)\,dt = \frac{1}{a^2}[\sin(at) - at\cos(at)]$ and Equation (18.150), we obtain

$$\begin{aligned}
F_n &= -j\frac{2}{T_0}\int_0^{\frac{T_0}{2}} f(t)\sin\left(n\frac{2\pi}{T_0}t\right)dt = -j\frac{2}{T_0}\int_0^{\frac{T_0}{2}} \frac{A}{T_0}t\sin\left(n\frac{2\pi}{T_0}t\right)dt \\
&= -j\frac{2A}{T_0^2}\frac{\left[\sin\left(n\frac{2\pi}{T_0}t\right) - n\frac{2\pi}{T_0}t\cos\left(n\frac{2\pi}{T_0}t\right)\right]_0^{\frac{T_0}{2}}}{\left(n\frac{2\pi}{T_0}\right)^2} \\
&= -j\frac{2A}{T_0^2}\frac{\sin\left(n\frac{2\pi}{T_0}\frac{T_0}{2}\right) - n\frac{2\pi}{T_0}\frac{T_0}{2}\cos\left(n\frac{2\pi}{T_0}\frac{T_0}{2}\right)}{\left(n\frac{2\pi}{T_0}\right)^2} \\
&= -j\frac{2A}{T_0^2}\frac{\sin(n\pi) - n\pi\cos(n\pi)}{\left(n\frac{2\pi}{T_0}\right)^2} = j\frac{(-1)^n A}{2n\pi}
\end{aligned} \quad \textbf{(17.181)}$$

Exercise 18.18

Find the exponential Fourier coefficients for the waveform shown in Figure 18.56. The equation of $f(t)$ for $0 \le t < \tau/2$ is given by $f(t) = \frac{A}{\tau}t$.

continued

Exercise 18.18 continued

FIGURE 18.56

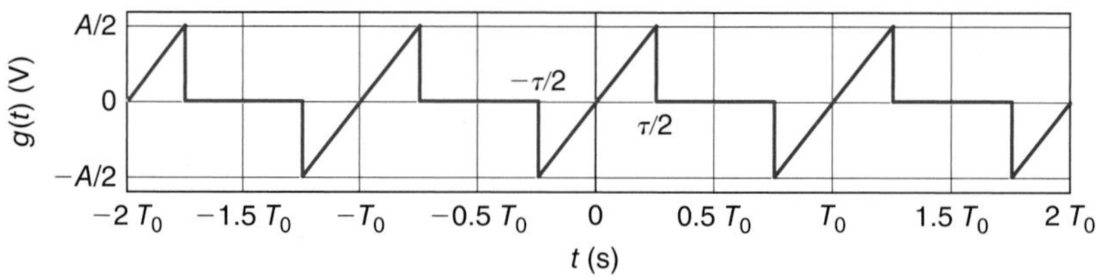

Waveform for EXERCISE 18.18.

Answer:

$F_0 = 0,$

$$F_n = j\frac{2A}{T_0\tau}\frac{n\pi\frac{\tau}{T_0}\cos\left(n\pi\frac{\tau}{T_0}\right) - \sin\left(n\pi\frac{\tau}{T_0}\right)}{\left(n\frac{2\pi}{T_0}\right)^2} \quad \textbf{(18.182)}$$

18.5.5 RECTIFIED COSINE

A half-wave rectified cosine waveform is shown in Figure 18.57.

FIGURE 18.57

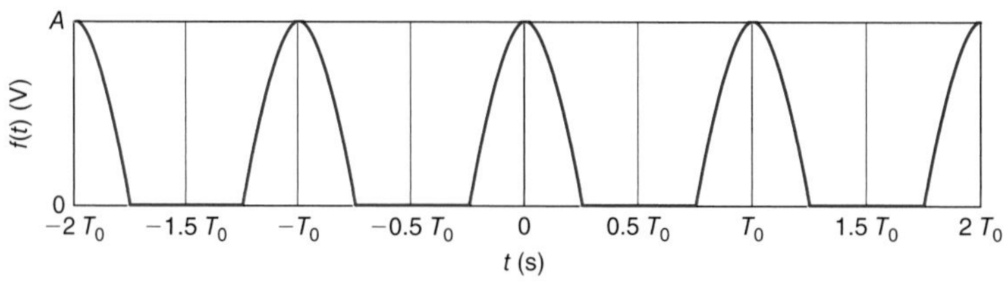

A half-wave rectified cosine.

The equation of $f(t)$ is $A\cos(2\pi t/T_0)$ for $-T_0/4 \leq t < T_0/4$. Since $f(t)$ possesses even symmetry, we have

$$F_0 = \frac{2}{T_0}\int_0^{\frac{T_0}{2}} f(t)dt = \frac{2}{T_0}\int_0^{\frac{T_0}{4}} A\cos\left(\frac{2\pi}{T_0}t\right)dt = \frac{2A}{T_0}\frac{\sin\left(\frac{2\pi}{T_0}t\right)\Big|_0^{\frac{T_0}{2}}}{\frac{2\pi}{T_0}} = \frac{A}{\pi}$$

If $n = 1$, we have

$$F_1 = \frac{A}{T_0}\int_{-\frac{T_0}{4}}^{\frac{T_0}{4}} \cos\left(\frac{2\pi t}{T_0}\right)e^{-j\frac{2\pi}{T_0}t}dt = \frac{A}{T_0}\int_{-\frac{T_0}{4}}^{\frac{T_0}{4}} \cos\left(\frac{2\pi t}{T_0}\right)\left[\cos\left(\frac{2\pi t}{T_0}\right) - j\sin\left(\frac{2\pi t}{T_0}\right)\right]dt$$

$$= \frac{A}{T_0}\int_{-\frac{T_0}{4}}^{\frac{T_0}{4}} \cos^2\left(\frac{2\pi t}{T_0}\right)dt - j\frac{A}{T_0}\int_{-\frac{T_0}{4}}^{\frac{T_0}{4}} \sin\left(\frac{2\pi t}{T_0}\right)\cos\left(\frac{2\pi t}{T_0}\right)dt$$

$$= \frac{A}{T_0}\int_{-\frac{T_0}{4}}^{\frac{T_0}{4}} \frac{1 + \cos\left(\frac{4\pi t}{T_0}\right)}{2}dt - j\frac{A}{2T_0}\int_{-\frac{T_0}{4}}^{\frac{T_0}{4}} \sin\left(\frac{4\pi t}{T_0}\right)dt = \frac{A}{4}$$

For $n = -1$, we have

$$F_{-1} = \frac{A}{4}$$

If $n \neq 1$, from Equation (8.146) and integral formula

$$\int e^{at}\cos(bt)dt = \frac{e^{at}[a\cos(bt) + b\sin(bt)]}{a^2 + b^2}, \text{ we have}$$

$$F_n = \frac{A}{T_0}\int_{-\frac{T_0}{4}}^{\frac{T_0}{4}} \cos\left(\frac{2\pi t}{T_0}\right) e^{-jn\frac{2\pi}{T_0}t}\,dt = \frac{A}{T_0}\frac{e^{-jn\frac{2\pi}{T_0}t}\left[-jn\frac{2\pi}{T_0}\cos\left(\frac{2\pi t}{T_0}\right) + \frac{2\pi}{T_0}\sin\left(\frac{2\pi t}{T_0}\right)\right]\Bigg|_{-\frac{T_0}{4}}^{\frac{T_0}{4}}}{\left(-jn\frac{2\pi}{T_0}\right)^2 + \left(\frac{2\pi}{T_0}\right)^2}$$

$$= \frac{A}{T_0\left(\frac{2\pi}{T_0}\right)^2(1-n^2)}\left\{\begin{matrix} e^{-jn\frac{2\pi}{T_0}\frac{T_0}{4}}\left[-jn\frac{2\pi}{T_0}\cos\left(\frac{2\pi\frac{T_0}{4}}{T_0}\right) + \frac{2\pi}{T_0}\sin\left(\frac{2\pi\frac{T_0}{4}}{T_0}\right)\right] \\ -e^{-jn\frac{2\pi}{T_0}\frac{-T_0}{4}}\left[-jn\frac{2\pi}{T_0}\cos\left(\frac{2\pi\frac{-T_0}{4}}{T_0}\right) + \frac{2\pi}{T_0}\sin\left(\frac{2\pi\frac{-T_0}{4}}{T_0}\right)\right]\end{matrix}\right\}$$

$$= \frac{A}{T_0\left(\frac{2\pi}{T_0}\right)^2(1-n^2)}\left\{\begin{matrix} e^{-j\frac{n\pi}{2}}\left[-jn\frac{2\pi}{T_0}\cos\left(\frac{\pi}{2}\right) + \frac{2\pi}{T_0}\sin\left(\frac{\pi}{2}\right)\right] \\ -e^{j\frac{n\pi}{2}}\left[jn\frac{2\pi}{T_0}\cos\left(\frac{-\pi}{2}\right) + \frac{2\pi}{T_0}\sin\left(\frac{-\pi}{2}\right)\right]\end{matrix}\right\}$$

$$= \frac{A}{T_0\left(\frac{2\pi}{T_0}\right)^2(1-n^2)}\left[\frac{2\pi}{T_0}e^{-j\frac{n\pi}{2}} + \frac{2\pi}{T_0}e^{j\frac{n\pi}{2}}\right] = \frac{A\cos\left(\frac{n\pi}{2}\right)}{\pi(1-n^2)} = \frac{-A\cos\left(\frac{n\pi}{2}\right)}{\pi(n^2-1)}, \quad n \neq 1$$

Thus, we have

$$F_n = \frac{-A\cos\left(\frac{n\pi}{2}\right)}{\pi(n^2-1)} = \begin{cases} 0, & n = 3, 5, 7, 9, \ldots \\ \frac{A}{\pi(n^2-1)}, & n = 2, 6, 10, \ldots \\ \frac{-A}{\pi(n^2-1)}, & n = 4, 8, 12, \ldots \end{cases} \tag{18.183}$$

$$F_1 = \frac{A}{4}, \quad F_{-1} = \frac{A}{4}$$

Since $f(t)$ is even, F_n can be evaluated using Equation (18.148)

$$F_n = \frac{2}{T_0}\int_0^{\frac{T_0}{2}} f(t)\cos\left(n\frac{2\pi}{T_0}t\right)dt = \frac{2}{T_0}\int_0^{\frac{T_0}{4}} A\cos\left(\frac{2\pi}{T_0}t\right)\cos\left(n\frac{2\pi}{T_0}t\right)dt$$

MATLAB can be used to evaluate this integral as follows:

```
clear all;
syms A T0 t n
f=A*cos(2*pi*t/T0);
F=(2/T0)*int(f*cos(n*2*pi*t/T0),t,0,T0/4)
assume (n, 'integer')
F=(2/T0)*int(f*cos(n*2*pi*t/T0),t,0,T0/4)
```

```
Answer:
F =
piecewise([n in {-1, 1}, A/4], [~n in {-1, 1}, -(A*cos((pi*n)/2))/(pi*(n^2 - 1))])
F =
piecewise([n in {-1, 1}, A/4], [~n in {-1, 1}, -((-1)^(n/2)*A*((-1)^n + 1))/(2*pi*(n^2 - 1))])
```

The answer matches the one given by Equation (18.183).
A full-wave rectified cosine waveform is shown in Figure 18.58.

FIGURE 18.58

A full-wave rectified cosine waveform.

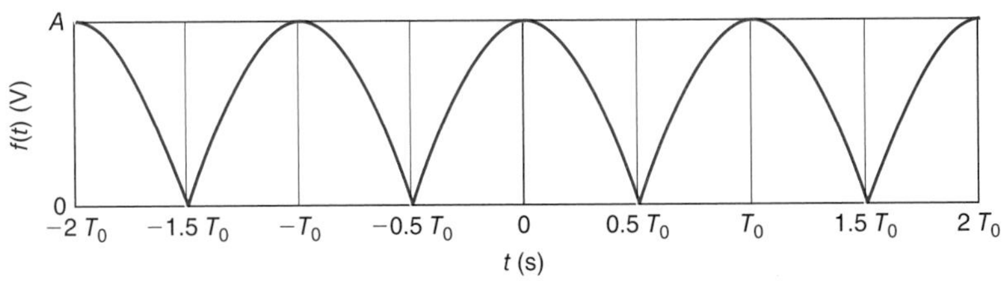

The equation of $f(t)$ is $A\cos(\pi t/T_0)$ for $-T_0/2 \le t < T_0/2$. Since $f(t)$ possesses even symmetry, we have

$$F_0 = \frac{2}{T_0}\int_0^{\frac{T_0}{2}} f(t)dt = \frac{2}{T_0}\int_0^{\frac{T_0}{2}} A\cos\left(\frac{\pi}{T_0}t\right)dt = \frac{2A}{T_0}\frac{\sin\left(\frac{\pi}{T_0}t\right)\Big|_0^{\frac{T_0}{2}}}{\frac{\pi}{T_0}} = \frac{2A}{\pi}$$

From Equation (8.141) and integral formula

$$\int e^{at}\cos(bt)dt = \frac{e^{at}[a\cos(bt) + b\sin(bt)]}{a^2 + b^2}, \text{ we have}$$

$$F_n = \frac{A}{T_0}\int_{-\frac{T_0}{2}}^{\frac{T_0}{2}} \cos\left(\frac{\pi t}{T_0}\right)e^{-jn\frac{2\pi}{T_0}t}dt = \frac{A}{T_0}\frac{e^{-jn\frac{2\pi}{T_0}t}\left[-jn\frac{2\pi}{T_0}\cos\left(\frac{\pi t}{T_0}\right) + \frac{\pi}{T_0}\sin\left(\frac{\pi t}{T_0}\right)\right]\Bigg|_{-\frac{T_0}{2}}^{\frac{T_0}{2}}}{\left(-jn\frac{2\pi}{T_0}\right)^2 + \left(\frac{\pi}{T_0}\right)^2}$$

$$= \frac{A}{T_0\left(\frac{\pi}{T_0}\right)^2(1-4n^2)}\left\{\begin{array}{l} e^{-jn\frac{2\pi}{T_0}\frac{T_0}{2}}\left[-jn\frac{2\pi}{T_0}\cos\left(\frac{\pi\frac{T_0}{2}}{T_0}\right) + \frac{\pi}{T_0}\sin\left(\frac{\pi\frac{T_0}{2}}{T_0}\right)\right] \\ -e^{-jn\frac{2\pi}{T_0}\frac{-T_0}{2}}\left[-jn\frac{2\pi}{T_0}\cos\left(\frac{\pi\frac{-T_0}{2}}{T_0}\right) + \frac{\pi}{T_0}\sin\left(\frac{\pi\frac{-T_0}{2}}{T_0}\right)\right] \end{array}\right\}$$

$$= \frac{A}{T_0\left(\frac{\pi}{T_0}\right)^2(1-4n^2)}\left\{\begin{array}{l} e^{-jn\pi}\left[-jn\frac{2\pi}{T_0}\cos\left(\frac{\pi}{2}\right)+\frac{\pi}{T_0}\sin\left(\frac{\pi}{2}\right)\right] \\ -e^{jn\pi}\left[jn\frac{2\pi}{T_0}\cos\left(\frac{-\pi}{2}\right)+\frac{\pi}{T_0}\sin\left(\frac{-\pi}{2}\right)\right]\end{array}\right\}$$

$$= \frac{A}{T_0\left(\frac{\pi}{T_0}\right)^2(1-4n^2)}\left[\frac{\pi}{T_0}e^{-jn\pi}+\frac{\pi}{T_0}e^{jn\pi}\right] = \frac{2A\cos(n\pi)}{\pi(1-4n^2)} \quad \textbf{(18.184)}$$

$$= \frac{-2A\cos(n\pi)}{\pi(4n^2-1)} = \frac{(-1)^{n+1}2A}{\pi(4n^2-1)}$$

Since $f(t)$ is even, F_n can be evaluated using Equation (18.148):

$$F_n = \frac{2}{T_0}\int_0^{\frac{T_0}{2}} f(t)\cos\left(n\frac{2\pi}{T_0}t\right)dt = \frac{2}{T_0}\int_0^{\frac{T_0}{2}} A\cos\left(\frac{\pi}{T_0}t\right)\cos\left(n\frac{2\pi}{T_0}t\right)dt$$

MATLAB can be used to evaluate this integral:

```
clear all;
syms A T0 t n
assume (n, 'integer')
f=A*cos(pi*t/T0);
F=(2/T0)*int(f*cos(n*2*pi*t/T0),t,0,T0/2)

Answer:
F =
-(2*A*cos(pi*n))/(pi*(4*n^2 - 1))
```

The answer matches the one given by Equation (18.184).

18.5.6 RECTIFIED SINE

A half-wave rectified sine waveform is shown in Figure 18.59.

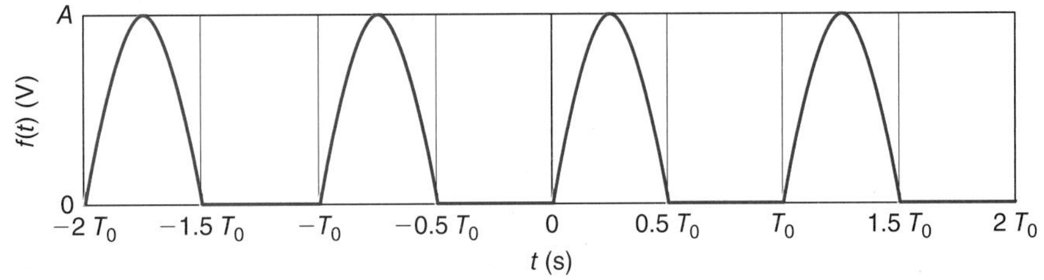

FIGURE 18.59 A half-wave rectified sine.

The equation of $f(t)$ is $A\sin(2\pi t/T_0)$ for $0 \le t < T_0/2$. The dc component is obtained from Equation (18.138)

$$F_0 = \frac{1}{T_0}\int_0^{\frac{T_0}{2}} f(t)dt = \frac{1}{T_0}\int_0^{\frac{T_0}{2}} A\sin\left(\frac{2\pi}{T_0}t\right)dt = \frac{A}{T_0}\,\frac{-\cos\left(\frac{2\pi}{T_0}t\right)\Big|_0^{\frac{T_0}{2}}}{\frac{2\pi}{T_0}}$$

$$= A\frac{-\cos\left(\frac{2\pi}{T_0}\frac{T_0}{2}\right) + 1}{2\pi} = \frac{A}{\pi}$$

For $n \neq 1$, from Equation (18.139), we have

$$F_n = \frac{1}{T_0}\int_0^{\frac{T_0}{2}} A \sin\left(\frac{2\pi}{T_0}t\right)e^{-jn\frac{2\pi}{T_0}t}dt \tag{18.185}$$

If $n = 1$, we have

$$F_1 = \frac{A}{T_0}\int_0^{\frac{T_0}{2}} \sin\left(\frac{2\pi t}{T_0}\right)e^{-j\frac{2\pi}{T_0}t}dt = \frac{A}{T_0}\int_0^{\frac{T_0}{2}} \sin\left(\frac{2\pi t}{T_0}\right)\left[\cos\left(\frac{2\pi t}{T_0}\right) - j\sin\left(\frac{2\pi t}{T_0}\right)\right]dt$$

$$= \frac{A}{T_0}\int_0^{\frac{T_0}{2}} \sin\left(\frac{2\pi t}{T_0}\right)\cos\left(\frac{2\pi t}{T_0}\right)dt - j\frac{A}{T_0}\int_0^{\frac{T_0}{2}} \sin^2\left(\frac{2\pi t}{T_0}\right)dt$$

$$= \frac{A}{2T_0}\int_0^{\frac{T_0}{2}} \sin\left(\frac{4\pi t}{T_0}\right)dt - j\frac{A}{T_0}\int_0^{\frac{T_0}{2}} \frac{1 - \cos\left(\frac{4\pi t}{T_0}\right)}{2}dt$$

$$= \frac{-A}{2T_0}\frac{\cos\left(\frac{4\pi t}{T_0}\right)\Big|_0^{\frac{T_0}{2}}}{\frac{4\pi}{T_0}} - j\frac{A}{4} + j\frac{A}{2T_0}\frac{\sin\left(\frac{4\pi t}{T_0}\right)\Big|_0^{\frac{T_0}{2}}}{\frac{4\pi}{T_0}} = -j\frac{A}{4}$$

For $n = -1$, we have

$$F_{-1} = \frac{jA}{4}$$

If $n \neq 1$, from Equation (8.181) and integral formula

$$\int e^{at}\sin(bt)dt = \frac{e^{at}[a\sin(bt) - b\cos(bt)]}{a^2 + b^2}, \text{ we have}$$

$$F_n = \frac{A}{T_0}\int_0^{\frac{T_0}{2}} \sin\left(\frac{2\pi t}{T_0}\right)e^{-jn\frac{2\pi}{T_0}t}dt = \frac{A}{T_0}\frac{e^{-jn\frac{2\pi}{T_0}t}\left[-jn\frac{2\pi}{T_0}\sin\left(\frac{2\pi t}{T_0}\right) - \frac{2\pi}{T_0}\cos\left(\frac{2\pi t}{T_0}\right)\right]\Big|_0^{\frac{T_0}{2}}}{\left(-jn\frac{2\pi}{T_0}\right)^2 + \left(\frac{2\pi}{T_0}\right)^2}$$

$$= \frac{A}{T_0\left(\frac{2\pi}{T_0}\right)^2(1-n^2)}\left\{\begin{aligned}&e^{-jn\frac{2\pi}{T_0}\frac{T_0}{2}}\left[-jn\frac{2\pi}{T_0}\sin\left(\frac{2\pi\frac{T_0}{2}}{T_0}\right) - \frac{2\pi}{T_0}\cos\left(\frac{2\pi\frac{T_0}{2}}{T_0}\right)\right]\\&-e^{-jn\frac{2\pi}{T_0}0}\left[-jn\frac{2\pi}{T_0}\sin\left(\frac{2\pi 0}{T_0}\right) - \frac{2\pi}{T_0}\cos\left(\frac{2\pi 0}{T_0}\right)\right]\end{aligned}\right\}$$

$$= \frac{A}{T_0\left(\frac{2\pi}{T_0}\right)^2(1-n^2)}\left[e^{-jn\pi}\left[-\frac{2\pi}{T_0}\cos(\pi)\right]+\frac{2\pi}{T_0}\right] = \frac{A[1+e^{-jn\pi}]}{2\pi(1-n^2)} = \frac{A[1+\cos(n\pi)]}{2\pi(1-n^2)}$$

(18.186)

$$= \frac{-A[1+\cos(n\pi)]}{2\pi(n^2-1)} = \frac{-A\cos^2\left(\frac{n\pi}{2}\right)}{\pi(n^2-1)} = \begin{cases} 0, & n = \pm3, \pm5, \pm7, \ldots \\ \frac{-A}{\pi(n^2-1)}, & n = \pm2, \pm4, \pm6, \ldots \end{cases}$$

MATLAB

```
clear all;
syms A T0 t
syms n integer
f=A*sin(2*pi*t/T0);
F=(1/T0)*int(f*exp(-j*n*2*pi*t/T0),t,0,T0/2)
F=simplify(F)

Answer:
F =
-(A*(exp(-pi*n*1i) + 1))/(2*pi*(n - 1)*(n + 1))
F =
-(A*((-1)^n + 1))/(2*pi*(n^2 - 1))
```

$$F_n = \frac{-A[e^{-j\pi n}+1]}{2\pi(n^2-1)} = \frac{-A[\cos(n\pi)+1]}{2\pi(n^2-1)} = \frac{-A\cos^2\left(\frac{n\pi}{2}\right)}{\pi(n^2-1)}$$

$$= \begin{cases} 0, & n = \pm3, \pm5, \pm7, \ldots \\ \frac{-A}{\pi(n^2-1)}, & n = \pm2, \pm4, \pm6, \ldots \end{cases}$$

Using L'Hospital's rule, we have $F_1 = -jA/4$ and $F_{-1} = jA/4$.
A full-wave rectified sine waveform is shown in Figure 18.60.

FIGURE 18.60

Full-wave rectified sine wave.

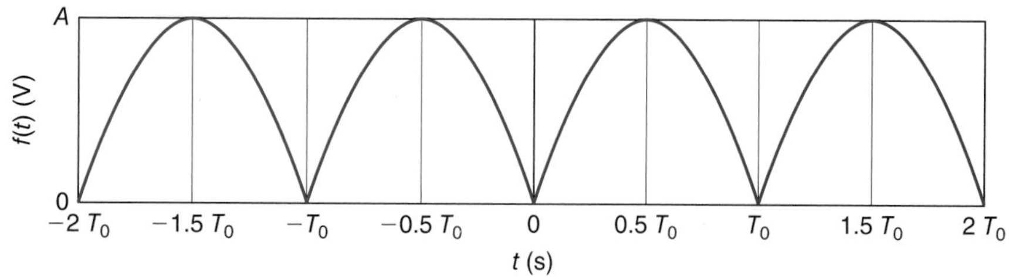

For $0 \le t < T_0$, the equation of the waveform is $A\sin(\pi t/T_0)$. Since $f(t)$ possesses even symmetry, we have

$$F_0 = \frac{2}{T_0}\int_0^{\frac{T_0}{2}} f(t)dt = \frac{2}{T_0}\int_0^{\frac{T_0}{2}} A\sin\left(\frac{\pi}{T_0}t\right)dt$$

$$= \frac{2A}{T_0}\frac{-\cos\left(\frac{\pi}{T_0}t\right)\Big|_0^{\frac{T_0}{2}}}{\frac{\pi}{T_0}} = \frac{2A}{T_0}\frac{-\cos\left(\frac{\pi}{T_0}\frac{T_0}{2}\right)+1}{\frac{\pi}{T_0}} = \frac{2A}{\pi}$$

From Equation (8.141) and integral formula

$$\int e^{at}\sin(bt)dt = \frac{e^{at}[a\sin(bt) - b\cos(bt)]}{a^2 + b^2}, \text{ we have}$$

$$F_n = \frac{A}{T_0}\int_0^{T_0}\sin\left(\frac{\pi t}{T_0}\right)e^{-jn\frac{2\pi}{T_0}t}dt = \frac{A}{T_0}\frac{e^{-jn\frac{2\pi}{T_0}t}\left[-jn\frac{2\pi}{T_0}\sin\left(\frac{\pi t}{T_0}\right) - \frac{\pi}{T_0}\cos\left(\frac{\pi t}{T_0}\right)\right]\Bigg|_0^{T_0}}{\left(-jn\frac{2\pi}{T_0}\right)^2 + \left(\frac{\pi}{T_0}\right)^2}$$

$$= \frac{A}{T_0\left(\frac{\pi}{T_0}\right)^2(1-4n^2)}\left\{\begin{matrix} e^{-jn\frac{2\pi}{T_0}T_0}\left[-jn\frac{2\pi}{T_0}\sin\left(\frac{\pi T_0}{T_0}\right) - \frac{\pi}{T_0}\cos\left(\frac{\pi T_0}{T_0}\right)\right] \\ -e^{-jn\frac{2\pi}{T_0}0}\left[-jn\frac{2\pi}{T_0}\sin\left(\frac{\pi 0}{T_0}\right) - \frac{\pi}{T_0}\cos\left(\frac{2\pi 0}{T_0}\right)\right]\end{matrix}\right\} \quad \textbf{(18.187)}$$

$$= \frac{A}{T_0\left(\frac{\pi}{T_0}\right)^2(1-4n^2)}\left[e^{-jn2\pi}\left(-\frac{\pi}{T_0}\cos(\pi)\right) + \frac{\pi}{T_0}\right] = \frac{-2A}{\pi(4n^2-1)},$$

$n = 0, \pm1, \pm2, \ldots$

Since $f(t)$ is even, F_n can be evaluated using Equation (18.148):

$$F_n = \frac{2}{T_0}\int_0^{\frac{T_0}{2}} f(t)\cos\left(n\frac{2\pi}{T_0}t\right)dt = \frac{2}{T_0}\int_0^{\frac{T_0}{2}} A\sin\left(\frac{\pi}{T_0}t\right)\cos\left(n\frac{2\pi}{T_0}t\right)dt$$

MATLAB can be used to evaluate this integral:

```
clear all;
syms A T0 t n
assume (n, 'integer')
f=A*sin(pi*t/T0);
F=(2/T0)*int(f*cos(n*2*pi*t/T0),t,0,T0/2)

Answer:
F =
(2*A*(2*n*sin(pi*n) - 1))/(pi*(4*n^2 - 1))
```

Since $\sin(\pi n) = 0$, the answer matches the one given by Equation (18.187).

18.5.7 AVERAGE POWER OF PERIODIC SIGNALS

If the exponential Fourier series representation

$$f(t) = \sum_{n=-\infty}^{\infty} F_n e^{jn\omega_0 t}$$

is substituted into the equation for average power

$$P = \frac{1}{T_0}\int_{-\frac{T_0}{2}}^{\frac{T_0}{2}} |f(t)|^2 dt$$

we get

$$P = \frac{1}{T_0}\int_{-\frac{T_0}{2}}^{\frac{T_0}{2}} |f(t)|^2 dt = \sum_{n=-\infty}^{\infty} |F_n|^2 = |F_0|^2 + \sum_{n=1}^{\infty} 2|F_n|^2 \quad \textbf{(18.188)}$$

since all the cross-product terms in the expansion of $|f(t)|^2 = f(t) \times f^*(t)$ are zero because of the orthonormality of the exponential terms. Including the average power expressed in terms of trogonometric Fourier coefficients, we have

$$\begin{aligned} P &= \frac{1}{T_0}\int_{-\frac{T_0}{2}}^{\frac{T_0}{2}} |f(t)|^2 dt = a_0^2 + \sum_{n=1}^{\infty}\left(\frac{a_n^2}{2} + \frac{b_n^2}{2}\right) \\ &= c_0^2 + \sum_{n=1}^{\infty}\frac{c_n^2}{2} = \sum_{n=-\infty}^{\infty} |F_n|^2 = |F_0|^2 + \sum_{n=1}^{\infty} 2|F_n|^2 \end{aligned} \quad \textbf{(18.189)}$$

This equation indicates that the total average power P can be broken down into power contained in the dc component, $P_0 = |F_0|^2$, power contained in fundamental frequency $f_0 = 1/T_0$, $P_1 = |F_1|^2$, power contained in frequency $-f_0 = -1/T_0$, $P_{-1} = |F_{-1}|^2 = |F_1|^2$, power contained in frequency $2f_0$, $P_2 = |F_2|^2$, power contained in frequency $-2f_0$, $P_{-2} = |F_{-2}|^2 = |F_2|^2$, etc. If the various power components are plotted with respect to frequencies, we obtain the power spectral density. The power spectral density shows the relative strength of power at various harmonic frequency components. Equation (18.189) is called **Parseval's theorem**.

The **mean square value** of a signal is defined as the average of the square of the signal. The average power of the periodic signal is also the mean square value of the periodic signal. The square root of the mean square value is called the **root mean square (rms)** value of the signal. The rms value of a periodic signal with period T_0 is given by

$$\begin{aligned} F_{rms} &= \sqrt{P} = \sqrt{\frac{1}{T_0}\int_{-T_0/2}^{T_0/2} |f(t)|^2 dt} = \sqrt{a_0^2 + \sum_{n=1}^{\infty}\frac{a_n^2 + b_n^2}{2}} \\ &= \sqrt{a_0^2 + \sum_{n=1}^{\infty}\left(\frac{a_n}{\sqrt{2}}\right)^2 + \sum_{n=1}^{\infty}\left(\frac{b_n}{\sqrt{2}}\right)^2} = \sqrt{c_0^2 + \sum_{n=1}^{\infty}\frac{c_n^2}{2}} \\ &= \sqrt{c_0^n + \sum_{n=1}^{\infty}\left(\frac{c_n}{\sqrt{2}}\right)^2} = \sqrt{|F_0|^2 + \sum_{n=1}^{\infty} 2|F_n|^2} = \sqrt{|F_0|^2 + \sum_{n=1}^{\infty} 4\left(\frac{|F_n|}{\sqrt{2}}\right)^2} \end{aligned} \quad \textbf{(18.190)}$$

18.6 Properties of Exponential Fourier Coefficients

We can gain some insights on the relation between the time domain signals and the Fourier coefficients by examining several properties of Fourier coefficients. Application of these properties facilitates the calculation of Fourier coefficients. The signal and its Fourier coefficients are denoted by $f(t) \leftrightarrow F_n$.

18.6.1 DC LEVEL

Let F_n be the exponential Fourier coefficients of a periodic signal $f(t)$. Then, the Fourier coefficients G_n of a signal $g(t) = f(t) + c$, where c is a dc signal, are given by

$$G_n = \begin{cases} F_0 + c, & n = 0 \\ F_n, & n \neq 0 \end{cases}$$

In other words, adding a dc level to a signal $f(t)$ changes the dc component F_0, but it does not change any other Fourier coefficients $F_n, n \neq 0$.

18.6.2 LINEARITY PROPERTY (SUPERPOSITION PRINCIPLE)

Let F_n be the Fourier coefficients of a periodic signal $f(t)$ with period T_0, and let G_n be the Fourier coefficients of another periodic signal $g(t)$ with the same period T_0. Then, the Fourier coefficients of $af(t) + bg(t)$ are $aF_n + bG_n$.

Proof

$$\frac{1}{T_0}\int_{t_0}^{t_0+T_0}[af(t) + bg(t)]e^{-jn\omega_0 t}dt = \frac{a}{T_0}\int_{t_0}^{t_0+T_0} f(t)e^{-jn\omega_0 t}dt + \frac{b}{T_0}\int_{t_0}^{t_0+T_0} g(t)e^{-jn\omega_0 t}dt$$
$$= aF_n + bG_n \qquad (18.191)$$

18.6.3 TIME-SHIFTING PROPERTY

Let F_n be the Fourier coefficients of a periodic signal $f(t)$ with period T_0, and let $g(t) = f(t - t_d)$ where t_d is a constant. Then

$$G_n = F_n e^{-jn\omega_0 t_d}, \quad \omega_0 = \frac{2\pi}{T_0} \qquad (18.192)$$

Since $F_n = |F_n|e^{j\angle F_n}$, we have

$$G_n = F_n e^{-jn\omega_0 t_d} = |F_n|e^{j\angle F_n}e^{-jn\omega_0 t_d} = |F_n|e^{j(\angle F_n - n\omega_0 t_d)} \qquad (18.193)$$

Therefore

$$|G_n| = |F_n| \qquad (18.194)$$

and

$$\angle G_n = \angle F_n - n\omega_0 t_d \qquad (18.195)$$

Thus, we conclude that the time shifting does not change the magnitude spectrum, but it introduces a linear phase term $-n\omega_0 t_d$. The slope $-t_d$ of this linear phase is proportional to the amount of delay.

Proof

The Fourier coefficients of $g(t) = f(t - t_d)$ are given by

$$G_n = \frac{1}{T_0}\int_{t_0}^{t_0+T_0} f(t - t_d)e^{-jn\omega_0 t}dt$$

Let $t' = t - t_d$. Then, $t = t' + t_d$ and $dt = dt'$. When $t = t_0$, $t' = t_0 - t_d$ and when $t = t_0 + T_0, t' = t_0 + T_0 - t_d$. Thus, we have

$$G_n = \frac{1}{T_0}\int_{t_0}^{t_0+T_0} f(t - t_d)e^{-jn\omega_0 t}dt = \frac{1}{T_0}\int_{t_0-t_d}^{t_0+T_0-t_d} f(t')e^{-jn\omega_0(t'+t_d)}dt'$$
$$= e^{-jn\omega_0 t_d}\frac{1}{T_0}\int_{t_0-t_d}^{t_0+T_0-t_d} f(t')e^{-jn\omega_0 t'}dt' = e^{-jn\omega_0 t_d}F_n$$

EXAMPLE 18.19

Let $A = 1$ V, $T_0 = 1$ ms, and $\tau = 250\ \mu$s for the rectangular pulse train $f(t)$ shown in Figure 18.61(a). The rectangular pulse train $g(t)$ shown in Figure 18.61(b) is $f(t)$ shifted to the right by $\tau/2$. Find the exponential Fourier coefficients F_n and G_n, and plot the magnitude spectrum and phase spectrum for both.

FIGURE 18.61

(a) A rectangular pulse train centered at $t = 0$. (b) A rectangular pulse train centered at $t = \tau/2$.

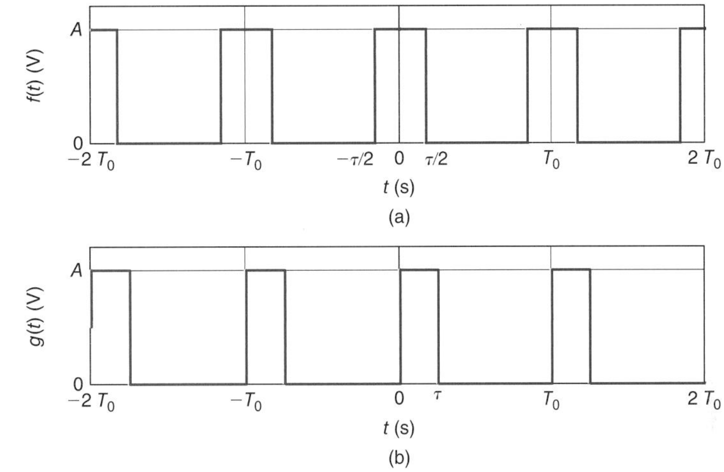

The duty cycle of the pulse train $f(t)$ is

$$d = \frac{\tau}{T_0} = \frac{0.25 \text{ ms}}{1 \text{ ms}} = \frac{1}{4} = 0.25$$

The exponential Fourier coefficients of a rectangular pulse train $f(t)$ are

$$F_n = \text{Ad sinc(nd)} = 1 \times 0.25 \times \text{sinc}(0.25n) = 0.25 \times \text{sinc}(0.25n) \quad \text{for}$$
$$n = 0, \pm 1, \pm 2, \pm 3, \ldots$$

Since $g(t) = f(t - 0.125 \times 10^{-3})$, the Fourier coefficients of $g(t)$ are given by

$$G_n = F_n e^{-jn\frac{2\pi}{T_0}t_d} = Ad\,\text{sinc}\,(nd)e^{-jn\frac{2\pi}{T_0}t_d} = 1 \times \frac{1}{4}\text{sinc}\left(\frac{n}{4}\right)e^{-jn\frac{2\pi}{1\times 10^{-3}}0.125\times 10^{-3}}$$

$$= \frac{1}{4}\text{sinc}\left(\frac{n}{4}\right)e^{-j\frac{n\pi}{4}}$$

$$|G_n| = |F_n| = \frac{1}{4}\left|\text{sinc}\left(\frac{n}{4}\right)\right|$$

$$\angle G_n = \angle F_n - \frac{n\pi}{4}$$

Notice that

$$\angle F_n = \begin{cases} 0, & \frac{1}{4}\text{sinc}\left(\frac{n}{4}\right) \geq 0 \\ 180°, & \frac{1}{4}\text{sinc}\left(\frac{n}{4}\right) < 0 \end{cases}$$

continued

Example 18.19 continued

Figure 18.62 shows F_n, $|F_n|$, $\angle F_n$, $|G_n|$, and $\angle G_n$.

FIGURE 18.62

Plot of F_n, $|F_n|$, $\angle F_n$, $|G_n|$, and $\angle G_n$.

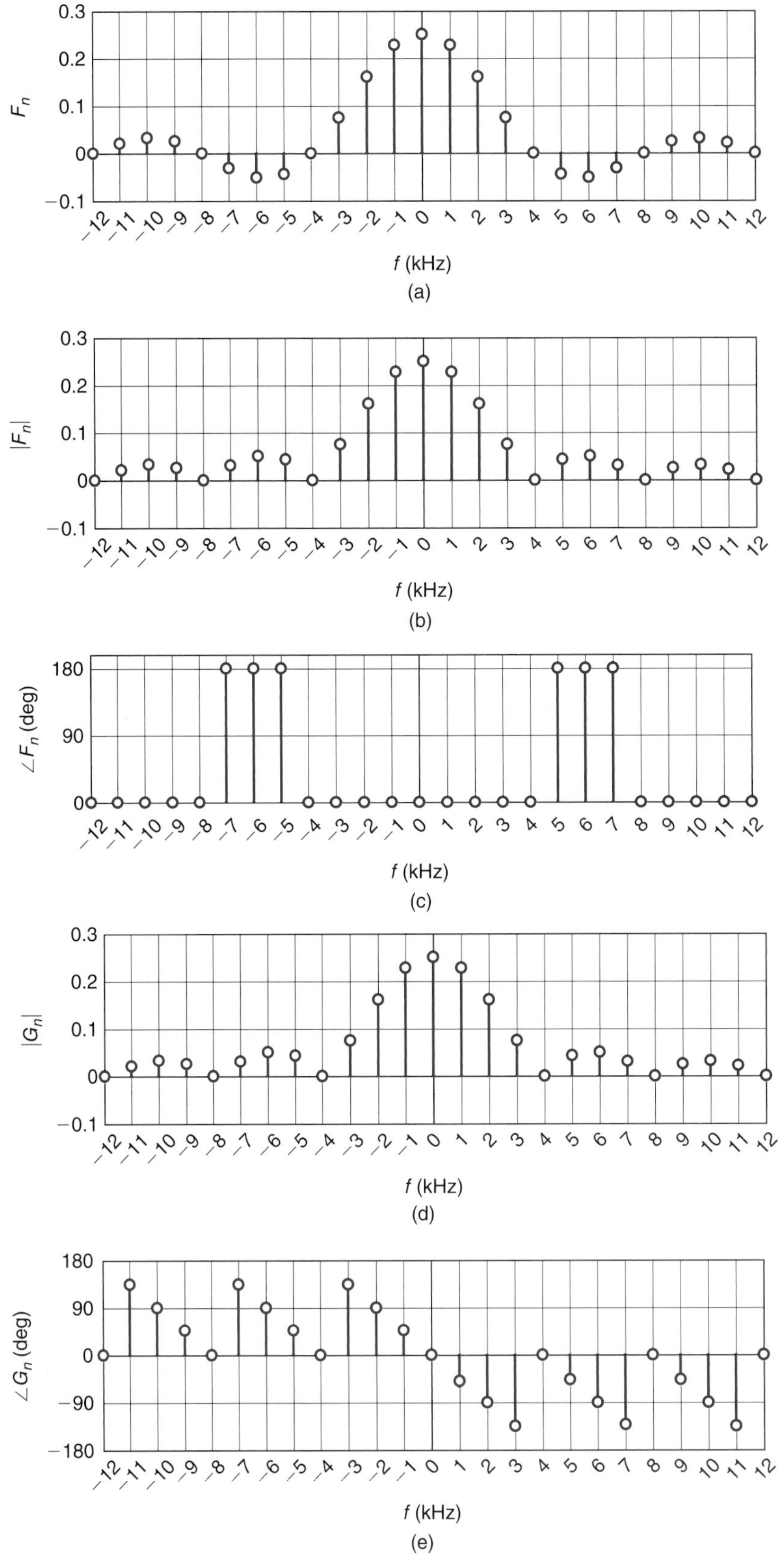

continued

Example 18.19 continued

MATLAB

```
% EXAMPLE 18.19
clear all;format long;
A=1;T0=1e-3;tau=0.25e-3;N=12;Th=1e-10;M1=2;M2=1000;td=0.125e-3;
f0=1/T0
w0=2*pi/T0
d=tau/T0
a0=A*tau/T0
n=-N:N;
f2=n*f0;
F=A*d*sinc(n*d)
Fmag=abs(F)
Fphase=angle(F).*(abs(F)>Th)
Fphased=Fphase*180/pi
G=A*d*sinc(n*d).*exp(-j*n*w0*td)
Gmag=abs(G)
Gphase=angle(G).*(abs(G)>Th)
Gphased=Gphase*180/pi
figure(1)
stem(f2,F,'LineWidth',2);grid;
xlabel('f (kHz)');ylabel('|F_n|');
set(gca,'XTick',-N*f0:f0:N*f0)
set(gca,'XTickLabel',-N:N)
set(gca,'YTick',-0.1:0.1:0.3)
set(gca,'YTickLabel',-0.1:0.1:0.3)
axis([-N*f0,N*f0,-0.1,0.3]);
figure(2)
stem(f2,Fmag,'LineWidth',2);grid;
xlabel('f (kHz)');ylabel('|F_n|');
set(gca,'XTick',-N*f0:f0:N*f0)
set(gca,'XTickLabel',-N:N)
set(gca,'YTick',0:0.1:0.3)
set(gca,'YTickLabel',0:0.1:0.3)
axis([-N*f0,N*f0,0,0.3]);
figure(3)
stem(f2,Fphased,'LineWidth',2);grid;
xlabel('f (kHz)');ylabel('\angleF_n (deg)');
set(gca,'XTick',-N*f0:f0:N*f0)
set(gca,'XTickLabel',-N:N)
set(gca,'YTick',0:90:180)
set(gca,'YTickLabel',0:90:180)
axis([-N*f0,N*f0,0,200]);
figure(4)
stem(f2,Gmag,'LineWidth',2);grid;
xlabel('f (kHz)');ylabel('|G_n|');
set(gca,'XTick',-N*f0:f0:N*f0)
set(gca,'XTickLabel',-N:N)
set(gca,'YTick',0:0.05:0.3)
set(gca,'YTickLabel',0:0.05:0.3)
axis([-N*f0,N*f0,0,0.3]);
figure(5)
stem(f2,Gphased,'LineWidth',2);grid;
xlabel('f (kHz)');ylabel('\angleG_n (deg)');
set(gca,'XTick',-N*f0:f0:N*f0)
set(gca,'XTickLabel',-N:N)
set(gca,'YTick',-180:45:180)
set(gca,'YTickLabel',-180:45:180)
axis([-N*f0,N*f0,-180,180]);
t=-M1*T0:T0/M2:M1*T0;
g=G(N+1);
for i=1:N
```

continued

Example 18.19 continued
MATLAB continued

```
    g=g+G(N+1+i)*exp(2j*pi*(i)*t/T0)+G(N+1-i)*exp(2j*pi*(-i)*t/T0);
end
g0a=0.5*(1+square(2*pi*(t+0*T0)/T0,25));
figure(6)
plot(t,g,'LineWidth',2);grid;hold on;
plot(t,g0a,'r-');xlabel('t');ylabel('g^\wedge(t)');hold off;
title(['Approximation of g(t) with DC + Upto ',num2str(N),'th Harmonic']);
axis([-M1*T0,M1*T0,-0.2*A,1.2*A]);
```

Exercise 18.19

Find the exponential and trigonometric Fourier coefficients of $f(t)$ shown in Figure 18.63.

FIGURE 18.63

A square wave for EXERCISE 18.19.

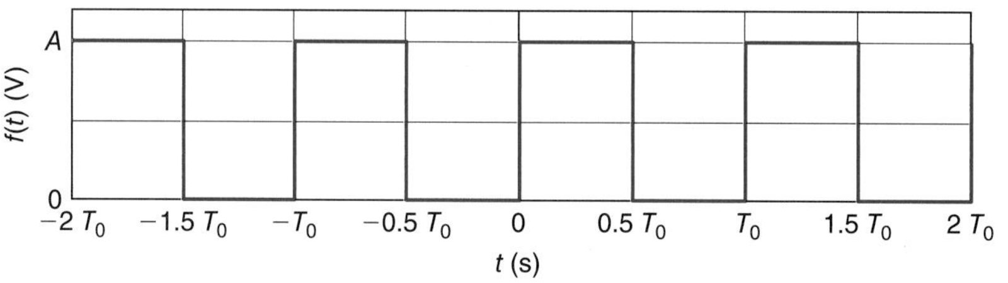

The dc component is $F_0 = A/2$, and the duty cycle is $d = 1/2$. The time delay is $0.25T_0$. Thus, we have

$$F_n = \frac{A}{2}\operatorname{sinc}\left(\frac{n}{2}\right)e^{-jn\frac{2\pi}{T_0}\frac{T_0}{4}} = \frac{A}{2}\operatorname{sinc}\left(\frac{n}{2}\right)e^{-j\frac{n\pi}{2}} = \frac{A}{2}\frac{\sin\left(\frac{n\pi}{2}\right)\left[\cos\left(\frac{n\pi}{2}\right) - j\sin\left(\frac{n\pi}{2}\right)\right]}{\frac{n\pi}{2}}$$

$$= \frac{A}{2}\frac{-j\sin^2\left(\frac{n\pi}{2}\right)}{\frac{n\pi}{2}} = -jA\frac{\sin^2\left(\frac{n\pi}{2}\right)}{n\pi} = \begin{cases} 0, & n = even \\ \frac{-jA}{n\pi}, & n = odd \end{cases}$$

$$a_0 = F_0 = A/2$$

$$a_n = 2\,\mathrm{Re}[F_n] = 0$$

$$b_n = -2\,\mathrm{Im}[F_n] = \begin{cases} 0, & n = even \\ \frac{2A}{n\pi}, & n = odd \end{cases}$$

EXAMPLE 18.20

Find the exponential Fourier coefficients of $f(t)$, $g(t)$, and $h(t)$ shown in Figure 18.64.

continued

Example 18.20 continued

The Fourier coefficients of a triangular pulse train $f(t)$ shown in Figure 18.64(a) are given by

$$F_n = \frac{A}{2}\text{sinc}^2\left(\frac{n}{2}\right)$$

When $A/2$ is subtracted from $f(t)$, a new triangular pulse train $g(t)$ shown in Figure 18.64(b) is obtained. Thus, $g(t) = f(t) - A/2$. Since only the dc component is reduced by $A/2$, the Fourier coefficient G_n is given by

$$G_n = \begin{cases} 0, & n = 0 \\ \frac{A}{2}\text{sinc}^2\left(\frac{n}{2}\right), & n \neq 0 \end{cases}$$

When $g(t)$ is shifted by $T_0/4$ to the right, another triangular pulse train shown in Figure 18.64(c) is obtained. Applying the time-shifting property, we obtain

$$H_n = \begin{cases} 0, & n = 0 \\ \frac{A}{2}\text{sinc}^2\left(\frac{n}{2}\right)e^{-jn\frac{2\pi}{T_0}\frac{T_0}{4}} = \frac{A}{2}\text{sinc}^2\left(\frac{n}{2}\right)e^{-j\frac{n\pi}{2}}, & n \neq 0 \end{cases}$$

FIGURE 18.64

Triangular pulse trains for EXAMPLE 18.20.

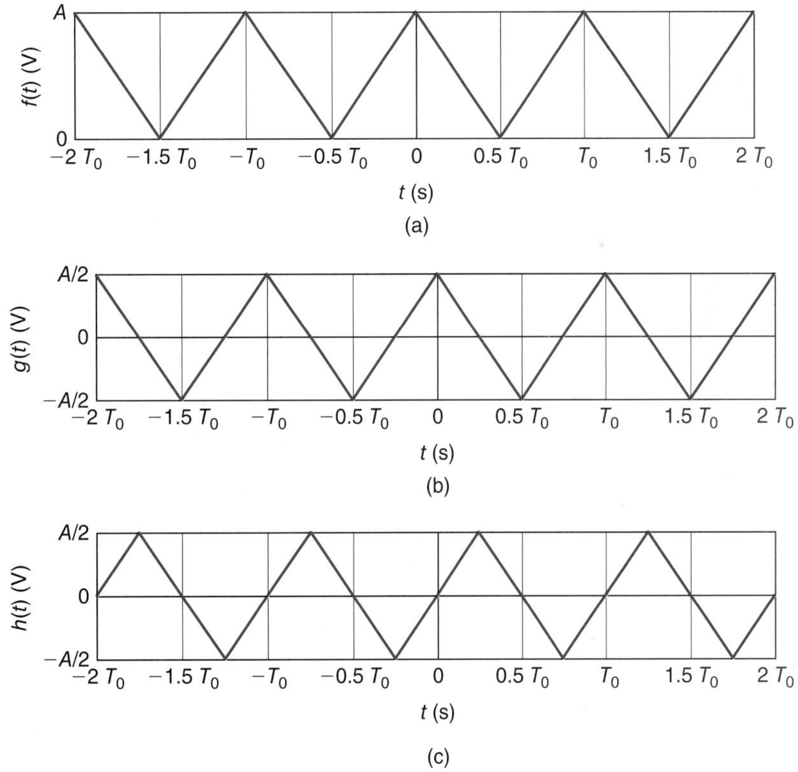

Exercise 18.20

Find the exponential and trigonometric Fourier coefficients of $f(t)$ shown in Figure 18.65.

continued

Exercise 18.20 continued

FIGURE 18.65

A triangular pulse train.

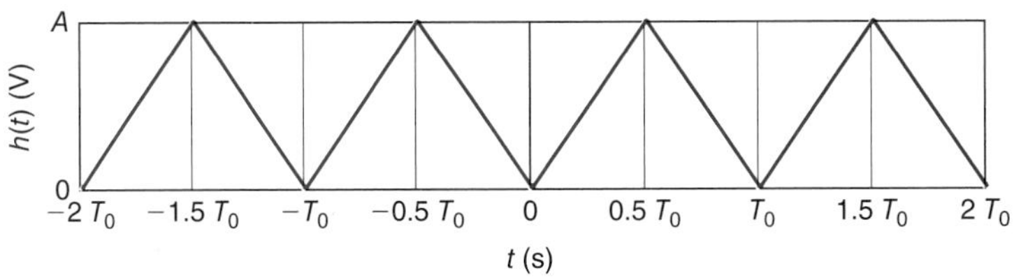

The dc component is $F_0 = A/2$. The time delay is $0.5T_0$ from the triangular pulse train shown in Figure 18.64(a). Thus, we have

$$F_n = \frac{A}{2}\operatorname{sinc}^2\left(\frac{n}{2}\right)e^{-jn\frac{2\pi}{T_0}\frac{T_0}{2}} = \frac{A}{2}\operatorname{sinc}^2\left(\frac{n}{2}\right)e^{-jn\pi} = \frac{A}{2}\frac{\sin^2\left(\frac{n\pi}{2}\right)[\cos(n\pi) - j\sin(n\pi)]}{\left(\frac{n\pi}{2}\right)^2}$$

$$= \frac{A}{2}\frac{\sin^2\left(\frac{n\pi}{2}\right)\cos(n\pi)}{\left(\frac{n\pi}{2}\right)^2} = \frac{2A}{n^2\pi^2}(-1)^n\sin^2\left(\frac{n\pi}{2}\right) = \begin{cases} 0, & n = even \\ \dfrac{-2A}{n^2\pi^2}, & n = odd \end{cases}$$

$$a_0 = F_0 = A/2$$

$$a_n = 2\,\mathrm{Re}[F_n] = \begin{cases} 0, & n = even \\ \dfrac{-4A}{n^2\pi^2}, & n = odd \end{cases}$$

$$b_n = -2\,\mathrm{Im}[F_n] = 0$$

18.6.4 TIME REVERSAL PROPERTY

Let F_n be the Fourier coefficients of a periodic signal $f(t)$ with period T_0. Then, the Fourier coefficients of $f(-t)$ are given by

$$F_{-n} = F_n^* \tag{18.196}$$

Proof

The Fourier coefficients of $f(-t)$ are given by

$$f(-t) \leftrightarrow \frac{1}{T_0}\int_{t_0}^{t_0+T_0} f(-t)e^{-jn\omega_0 t}dt$$

Let $t' = -t$. Then, $t = -t'$ and $dt = -dt'$. When $t = t_0$, $t' = -t_0$. When $t = t_0 + T_0$, $t' = -t_0 - T_0$. Thus, we have

$$\frac{1}{T_0}\int_{t_0}^{t_0+T_0} f(-t)e^{-jn\omega_0 t}dt = \frac{1}{T_0}\int_{-t_0}^{-t_0-T_0} f(t')e^{-j(-n)\omega_0 t'}(-dt')$$

$$= \frac{1}{T_0}\int_{-t_0-T_0}^{-t_0} f(t')e^{-j(-n)\omega_0 t'}dt' = F_{-n} = F_n^*$$

18.6.5 TIME DIFFERENTIATION PROPERTY

Let F_n be the Fourier coefficients of a periodic signal $f(t)$ with period T_0. Then, the Fourier coefficients of the first derivative $f'(t)$ are given by $jn\omega_0 F_n$; that is,

$$f'(t) \leftrightarrow jn\omega_0 F_n \tag{18.197}$$

Proof

Taking the derivative on both sides of

$$f(t) = \sum_{n=-\infty}^{\infty} F_n e^{jn\omega_0 t}$$

we obtain

$$f'(t) = \sum_{n=-\infty}^{\infty} jn\omega_0 F_n e^{jn\omega_0 t}$$

Thus, the Fourier coefficients of $f'(t)$ are $jn\omega_0 F_n$.

18.6.6 CONVOLUTION PROPERTY

Let F_n be the Fourier coefficients of a periodic signal $f(t)$ with period T_0, let G_n be the Fourier coefficients of a periodic signal $g(t)$ with period T_0, and let Z_n be the Fourier coefficients of a periodic signal $z(t)$ with period T_0, which is the convolution of $f(t)$ and $g(t)$. Then, we have

$$Z_n = T_0 F_n G_n \tag{18.198}$$

Proof

The Fourier coefficients of

$$z(t) = \int_{t_0}^{t_0+T_0} f(\lambda) g(t-\lambda) d\lambda$$

are given by

$$Z_n = \frac{1}{T_0} \int_{t_0}^{t_0+T_0} \left[\int_{t_0}^{t_0+T_0} f(\lambda) g(t-\lambda) d\lambda \right] e^{-jn\omega_0 t} dt$$

Let $t' = t - \lambda$. Then, $t = t' + \lambda$ and $dt = dt'$, and Z_n becomes

$$Z_n = \frac{1}{T_0} \int_{t_0}^{t_0+T_0} f(\lambda) e^{-jn\omega_0 \lambda} d\lambda \int_{t_0}^{t_0+T_0} g(t') e^{-jn\omega_0 t'} dt' = T_0 F_n G_n.$$

18.7 Solving Circuit Problems Using Exponential Fourier Series

If a complex exponential signal

$$x(t) = Ae^{j\omega_0 t} \tag{18.199}$$

is applied to a linear, time invariant system with the transfer function $H(s)$, the steady-state output of the system is given by

$$y(t) = AH(j\omega_0)e^{j\omega_0 t} \tag{18.200}$$

Since $H(j\omega_0)$ is complex, it can be represented as $|H(j\omega_0)|e^{j\angle H(j\omega_0)}$. Thus, the output can also be expressed as

$$\begin{aligned} y(t) &= AH(j\omega_0)e^{j\omega_0 t} = A|H(j\omega_0)|e^{j\angle H(j\omega_0)}e^{j\omega_0 t} \\ &= A|H(j\omega_0)|e^{j(\omega_0 t+\angle H(j\omega_0))} \end{aligned} \tag{18.201}$$

If the frequency of the input signal is changed from ω_0 to $n\omega_0$, the steady-state output signal is given by

$$\begin{aligned} y(t) &= AH(jn\omega_0)e^{jn\omega_0 t} = A|H(jn\omega_0)|e^{j\angle H(jn\omega_0)}e^{jn\omega_0 t} \\ &= A|H(jn\omega_0)|e^{j(n\omega_0 t+\angle H(jn\omega_0))} \end{aligned} \tag{18.202}$$

If a signal $f(t)$ with exponential Fourier series representation

$$f(t) = \sum_{n=-\infty}^{\infty} F_n e^{jn\omega_0 t} \tag{18.203}$$

is applied to a linear, time invariant system with the transfer function $H(s)$, from the superposition principle and Equation (18.202), the steady-state response of the system is given by

$$y(t) = \sum_{n=-\infty}^{\infty} F_n H(jn\omega_0)e^{jn\omega_0 t} \tag{18.204}$$

If $f(t)$ is periodic with period T_0, the steady-state response $y(t)$ is also periodic with period T_0. Let the Fourier series representation of $y(t)$ be

$$y(t) = \sum_{n=-\infty}^{\infty} Y_n e^{jn\omega_0 t} \tag{18.205}$$

Then, from Equations (18.204) and (18.205), the Fourier coefficients Y_n are given by

$$Y_n = F_n H(jn\omega_0) = |F_n|\,|H(jn\omega_0)|e^{j(\angle F_n+\angle H(jn\omega_0))} \tag{18.206}$$

From Equation (18.158), we have

$$F_{-n} = F_n^* = |F_n|e^{-j\angle F_n}, \quad n = 1, 2, 3, \ldots \tag{18.207}$$

We can show that

$$H(-jn\omega_0) = H^*(jn\omega_0) = |H(jn\omega_0)|e^{-j\angle H(jn\omega_0)} \tag{18.208}$$

From Equations (18.206) through (18.208), we have

$$Y_{-n} = F_{-n}H(-jn\omega_0) = |F_n|\,|H(jn\omega_0)|e^{-j(\angle F_n+\angle H(jn\omega_0))} \tag{18.209}$$

From Equations (8.206) and (8.209), we have

$$|Y_n| = |Y_{-n}| = |F_n|\,|H(jn\omega_0)|, \quad n = 1, 2, 3, \ldots \tag{18.210}$$

$$\angle Y_n = -\angle Y_{-n} = \angle F_n H(jn\omega_0) = \angle F_n + \angle H(jn\omega_0), \quad n = 1, 2, 3, \ldots \tag{18.211}$$

According to Equation (18.210), the magnitude spectrum at the output of the system is an even function of frequency. The magnitude of the nth harmonic frequency component at the output of the system is given by the product of the magnitude of the nth harmonic frequency component at the input of the system and the magnitude response of the system at the nth harmonic frequency. According to Equation (18.211), the phase spectrum at the output of the system is an odd function of frequency. The phase of the nth harmonic frequency component at the output of the system is given by the sum of the phase of the nth harmonic frequency component at the input of the system and the phase response of the system at the nth harmonic frequency.

Using Equations (18.207) and (18.208), we can rewrite Equation (18.204) as

$$\begin{aligned} y(t) &= F_0H(0) + \sum_{n=1}^{\infty}[F_nH(jn\omega_0)e^{jn\omega_0 t} + F_{-n}H(-jn\omega_0)e^{-jn\omega_0 t}] \\ &= F_0H(0) + \sum_{n=1}^{\infty}[|F_n||H(jn\omega_0)|e^{j(n\omega_0 t+\angle F_n+\angle H(jn\omega_0))} \\ &\quad + |F_n||H(jn\omega_0)|e^{-j(n\omega_0 t+\angle F_n+\angle H(jn\omega_0))}] \\ &= F_0H(0) + \sum_{n=1}^{\infty}2|F_n||H(jn\omega_0)|\cos[n\omega_0 t + \angle F_n + \angle H(jn\omega_0)] \end{aligned} \tag{18.212}$$

If $y(t)$ is represented by a trigonometric Fourier series using cosines only, it can be written as

$$y(t) = e_0 + \sum_{n=1}^{\infty} e_n \cos(n\omega_0 t + \phi_n) \tag{18.213}$$

From Equations (18.212) and (18.213), we have

$$e_0 = F_0H(0) \tag{18.214}$$

$$e_n = 2|F_n||H(jn\omega_0)| \tag{18.215}$$

$$\phi_n = \angle F_n + \angle H(jn\omega_0) \tag{18.216}$$

EXAMPLE 18.21

A rectangular pulse train with amplitude $A = 1$ V, period $T_0 = 1$ ms, and pulse width $\tau = 0.5$ ms, shown in Figure 18.66, is applied to an RC circuit with $C = 0.1\ \mu$F and $R = 800\ \Omega$, shown in Figure 18.67.

- **a.** Find $f_0, \omega_0, d, F_n, |F_n|, \angle F_n, H(\omega), H(n\omega_0), |H(n\omega_0)|, \angle H(n\omega_0), |Y_n|$, and $\angle Y_n$.
- **b.** Evaluate $|F_n|, \angle F_n, |H(n\omega_0)|, \angle H(n\omega_0), |Y_n|$, and $\angle Y_n$ for $0 \le n \le 10$, and plot them.
- **c.** Approximate $f(t)$ by the dc component and the first 10 harmonics, and plot the approximation.
- **d.** Approximate $y(t)$ by the dc component and the first 10 harmonics, and plot the approximation.

FIGURE 18.66

A rectangular pulse train.

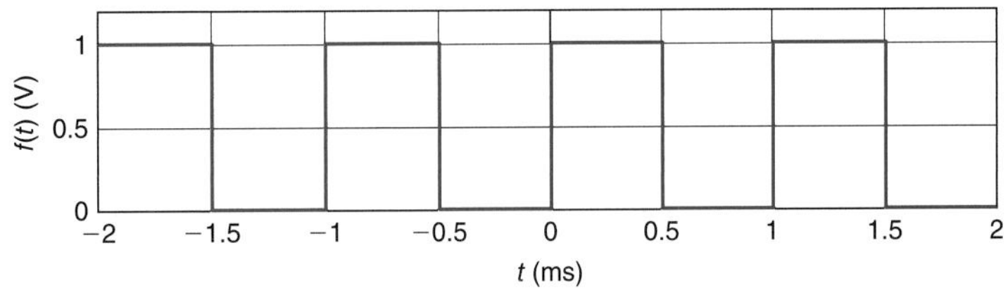

continued

Example 18.21 continued

FIGURE 18.67

RC circuit.

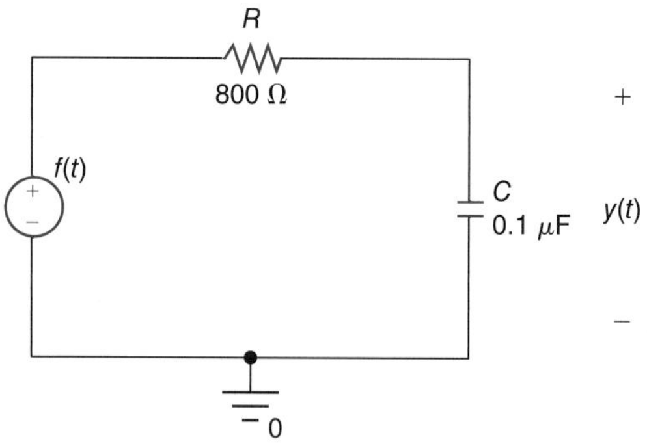

a. $$f_0 = \frac{1}{T_0} = \frac{1}{0.001 \text{ s}} = 1000 \text{ Hz} = 1 \text{ kHz}$$

$$\omega_0 = \frac{2\pi}{T_0} = \frac{2\pi}{0.001 \text{ s}} = 2\pi \times 1000 \text{ rad/s} = 6283.1853 \text{ rad/s}$$

$$d = \frac{\tau}{T_0} = \frac{0.5 \text{ ms}}{1 \text{ ms}} = 0.5$$

$$F_0 = Ad = 1 \times \frac{1}{2} = 0.5$$

$$F_n = \frac{1}{2}\operatorname{sinc}\left(\frac{n}{2}\right)e^{-jn\frac{2\pi}{1 \text{ ms}}0.25 \text{ ms}} = \frac{1}{2}\operatorname{sinc}\left(\frac{n}{2}\right)e^{-j\frac{n\pi}{2}}$$

$$|F_n| = \left|\frac{1}{2}\operatorname{sinc}\left(\frac{n}{2}\right)\right|$$

$$\angle F_n = \begin{cases} -\dfrac{n\pi}{2} + \pi, & \textit{if}\operatorname{sinc}\left(\dfrac{n}{2}\right) < 0 \\ -\dfrac{n\pi}{2}, & \textit{if}\operatorname{sinc}\left(\dfrac{n}{2}\right) \geq 0 \end{cases}$$

$$H(s) = \frac{\dfrac{1}{sC}}{R + \dfrac{1}{sC}} = \frac{1}{sRC + 1} = \frac{\dfrac{1}{RC}}{s + \dfrac{1}{RC}} = \frac{12{,}500}{s + 12{,}500}$$

$$H(\omega) = H(s)|_{s=j\omega} = \frac{\dfrac{1}{RC}}{j\omega + \dfrac{1}{RC}} = \frac{\dfrac{1}{RC}}{j\omega + \dfrac{1}{RC}} = \frac{12{,}500}{j\omega + 12{,}500}$$

$$H(n\omega_0) = H(\omega)|_{\omega=n\omega_0} = \frac{12{,}500}{jn\omega_0 + 12{,}500} = \frac{12{,}500}{jn6283.1853 + 12{,}500} = \frac{1.9894}{jn + 1.9894}$$

$$|H(n\omega_0)| = \frac{1.9894}{|jn + 1.9894|} = \frac{1.9894}{\sqrt{n^2 + 1.9894^2}}$$

$$\angle H(n\omega_0) = -\angle(jn + 1.9894) = -\tan^{-1}\left(\frac{n}{1.9894}\right)$$

$$Y_n = F_n H(n\omega_0)$$

$$|Y_n| = |F_n|\,|H(n\omega_0)|$$

$$\angle Y_n = \angle F_n + \angle H(n\omega_0|)|$$

continued

Example 18.21 continued

b.

n:	0	1	2	3	4	5	6	7	8	9	10
$\|F_n\|$:	0.5	0.3183	0	0.1061	0	0.0637	0	0.0455	0	0.0354	0
$\angle F_n$(r):	0	-1.571	0	-1.571	0	-1.571	0	-1.571	0	-1.571	0
$\angle F_n$(d):	0	-90	0	-90	0	-90	0	-90	0	-90	0
$\|H\|$	1	0.8935	0.7052	0.5527	0.4453	0.3697	0.3147	0.2734	0.2413	0.2158	0.1951
$\angle H$(rad)	0	-0.466	-0.788	-0.985	-1.109	-1.192	-1.251	-1.294	-1.327	-1.353	-1.374
$\angle H$(deg)	0	-36.69	-45.15	-56.45	-63.56	-68.30	-71.66	-74.13	-76.03	-77.54	-78.75
$\|Y_n\|$:	0.5	0.2844	0	0.0586	0	0.0235	0	0.0124	0	0.0076	0
$\angle Y_n$(r):	0	-2.037	-0.788	-2.556	-1.109	-2.763	-1.251	-2.865	-1.327	-2.924	-1.374
$\angle Y_n$(d):	0	-116.7	-45.15	-146.4	-63.56	-158.3	-71.66	-164.1	-76.03	-167.5	-78.75

The magnitude for the negative n values are identical to the magnitude for the positive n values, and the phase for the negative n values have opposite sign of the phase for the positive n values.

c. $f(t) = 0.5 + 0.3183\,[e^{j(2\pi 1000t-\pi/2)} + e^{-j(2\pi 1000t-\pi/2)}] + 0.1061[e^{j(2\pi 3000t-\pi/2)} + e^{-j(2\pi 3000t-\pi/2)}] + 0.0637\,[e^{j(2\pi 5000t-\pi/2)} + e^{-j(2\pi 5000t-\pi/2)}] + 0.0455[e^{j(2\pi 7000t-\pi/2)} + e^{-j(2\pi 7000t-\pi/2)}] + 0.0354\,[e^{j(2\pi 9000t-\pi/2)} + e^{-j(2\pi 9000t-\pi/2)}]$ V. The phase is in radians.

d. $y(t) = 0.5 + 0.2844\,[e^{j(2\pi 1000t-2.037)} + e^{-j(2\pi 1000t-2.037)}] + 0.0586\,[e^{j(2\pi 3000t-2.556)} + e^{-j(2\pi 3000t-2.556)}] + 0.0235\,[e^{j(2\pi 5000t-2.763)} + e^{-j(2\pi 5000t-2.763)}] + 0.0124\,[e^{j(2\pi 7000t-2.865)} + e^{-j(2\pi 7000t-2.865)}] + 0.0076\,[e^{j(2\pi 9000t-2.924)} + e^{-j(2\pi 9000t-2.924)}]$ V. The phase is in radians.

The plots are shown in Figure 18.68.

FIGURE 18.68

Plot of (a) $|F_n|$ (b) $\angle F_n$ (c) $|H(n\omega_0)|$ (d) $\angle H(n\omega_o)$ (e) $|Y_n|$ (f) $\angle Y_n$(g) $f(t)$ and $\hat{f}(t)$ (h) $f(t)$ and $\hat{y}(t)$.

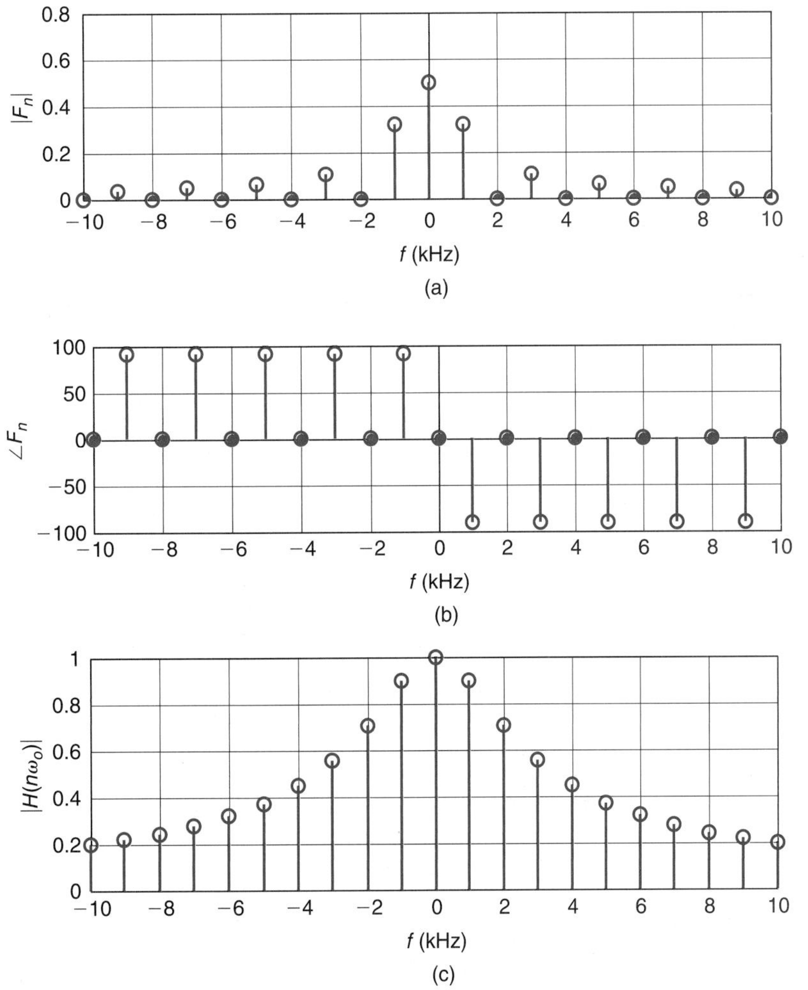

continued

Example 18.21 continued

FIGURE 18.68

continued

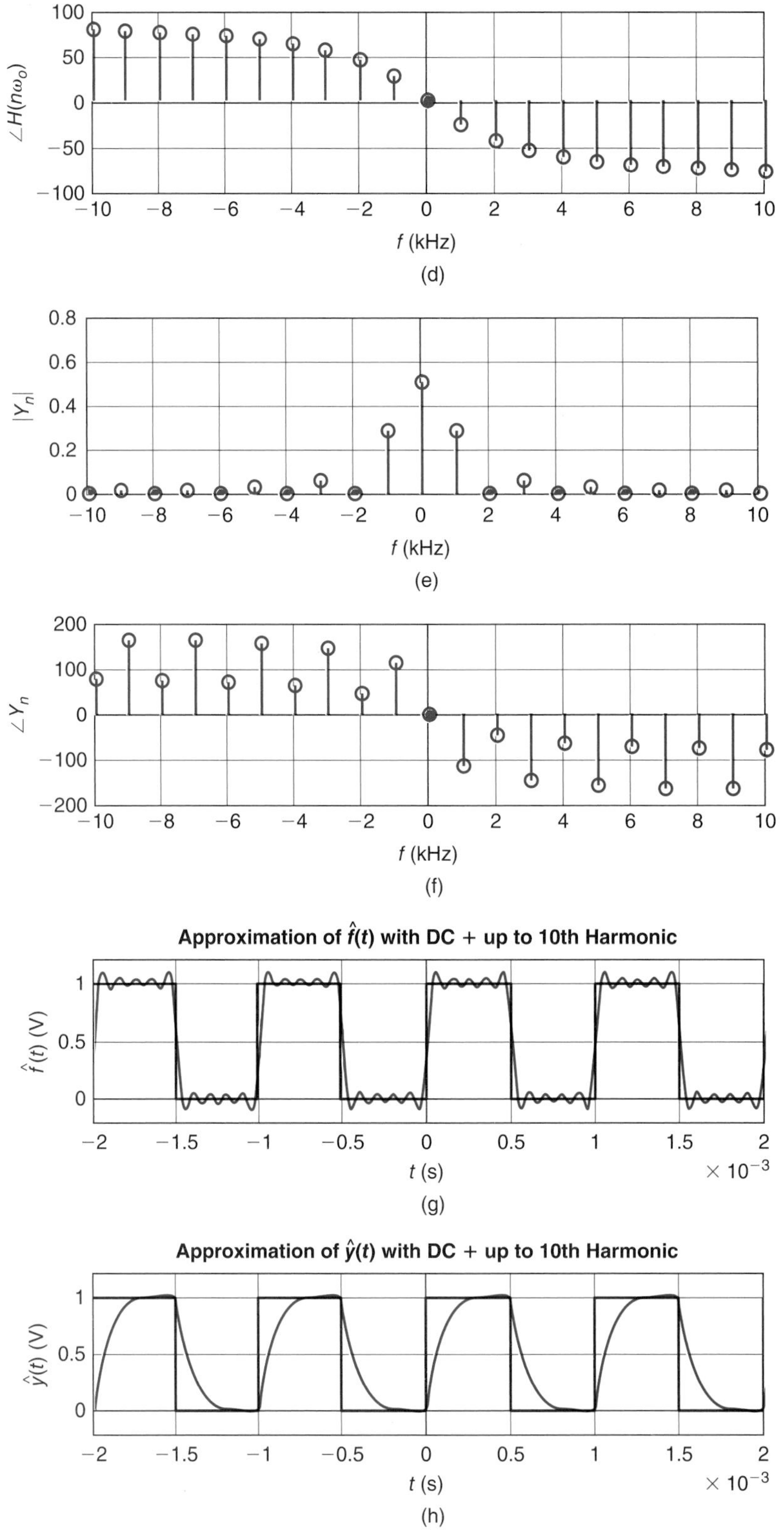

continued

Example 18.21 continued

MATLAB

```
% EXAMPLE 18.21
clear all;format long;
A=1;T0=1e-3;tau=0.5e-3;N=10;Th=1e-10;M1=2;M2=1000;td=0.25e-3;
R=800;C=0.1e-6;
f0=1/T0
w0=2*pi/T0
d=tau/T0
a0=A*tau/T0
n=-N:N;
f2=n*f0;
F=A*d*sinc(n*d).*exp(-j*n*w0*td)
Fmag=abs(F)
Fphase=angle(F).*(abs(F)>Th)
Fphased=Fphase*180/pi
%
H=(1/(R*C))./(j*n*w0+(1/(R*C)))
Hmag=abs(H)
Hphase=angle(H)
Hphased=Hphase*180/pi
Y=F.*H
Ymag=abs(Y)
YPhase=Fphase+Hphase
Yphase=YPhase.*(abs(Y)>Th)
Yphased=Yphase*180/pi
%
t=-M1*T0:T0/M2:M1*T0;
f=F(N+1);
for i=1:N
    f=f+F(N+1+i)*exp(2j*pi*(i)*t/T0)+F(N+1-i)*exp(2j*pi*(-i)*t/T0);
end
y=Y(N+1);
for i=1:N
    y=y+Y(N+1+i)*exp(2j*pi*(i)*t/T0)+Y(N+1-i)*exp(2j*pi*(-i)*t/T0);
end
f0a=0.5*(1+square(2*pi*(t+0*T0)/T0,50));
%
figure(1)
stem(f2,Fmag,'LineWidth',2);grid;
xlabel('f (kHz)');ylabel('|F_n|');
set(gca,'XTick',-N*f0:f0:N*f0)
set(gca,'XTickLabel',-N:N)
set(gca,'YTick',0:0.1:0.5)
set(gca,'YTickLabel',0:0.1:0.5)
axis([-N*f0,N*f0,0,0.5]);
figure(2)
stem(f2,Fphased,'LineWidth',2);grid;
xlabel('f (kHz)');ylabel('\angleF_n (deg)');
set(gca,'XTick',-N*f0:f0:N*f0)
set(gca,'XTickLabel',-N:N)
set(gca,'YTick',-90:90:90)
set(gca,'YTickLabel',-90:90:90)
axis([-N*f0,N*f0,-90,90]);
```

continued

Example 18.21 continued

MATLAB continued

```
figure(3)
stem(f2,Hmag,'LineWidth',2);grid;
xlabel('f (kHz)');ylabel('|H_n|');
set(gca,'XTick',-N*f0:f0:N*f0)
set(gca,'XTickLabel',-N:N)
set(gca,'YTick',0:0.2:1)
set(gca,'YTickLabel',0:0.2:1)
axis([-N*f0,N*f0,0,1]);
figure(4)
stem(f2,Hphased,'LineWidth',2);grid;
xlabel('f (kHz)');ylabel('\angleH_n (deg)');
set(gca,'XTick',-N*f0:f0:N*f0)
set(gca,'XTickLabel',-N:N)
set(gca,'YTick',-90:90:90)
set(gca,'YTickLabel',-90:90:90)
axis([-N*f0,N*f0,-90,90]);
figure(5)
stem(f2,Ymag,'LineWidth',2);grid;
xlabel('f (kHz)');ylabel('|Y_n|');
set(gca,'XTick',-N*f0:f0:N*f0)
set(gca,'XTickLabel',-N:N)
set(gca,'YTick',0:0.1:0.5)
set(gca,'YTickLabel',0:0.1:0.5)
axis([-N*f0,N*f0,0,0.5]);
figure(6)
stem(f2,Yphased,'LineWidth',2);grid;
xlabel('f (kHz)');ylabel('\angleY_n (deg)');
set(gca,'XTick',-N*f0:f0:N*f0)
set(gca,'XTickLabel',-N:N)
set(gca,'YTick',-180:90:180)
set(gca,'YTickLabel',-180:90:180)
axis([-N*f0,N*f0,-180,180]);
figure(7)
plot(t,f,'LineWidth',2);grid;hold on;
plot(t,f0a,'r-');xlabel('t');ylabel('f^\wedge(t)');hold off;
title(['Approximation of f(t) with DC + up to ',num2str(N),'th Harmonic']);
axis([-M1*T0,M1*T0,-0.2*A,1.2*A]);
figure(8)
plot(t,y,'LineWidth',2);grid;hold on;
plot(t,f0a,'r-');xlabel('t');ylabel('y^\wedge(t)');hold off;
title(['Approximation of y(t) with DC + up to ',num2str(N),'th Harmonic']);
axis([-M1*T0,M1*T0,-0.2*A,1.2*A]);
```

Exercise 18.21

Represent the input $f(t)$ and output $y(t)$ by Fourier series up to the third harmonic for the circuit shown in Figure 18.67 when the input is a triangular pulse train with amplitude 1 V, period 1 ms, as shown in Figure 18.69.

continued

Exercise 18.21 continued

FIGURE 18.69

Triangular pulse train.

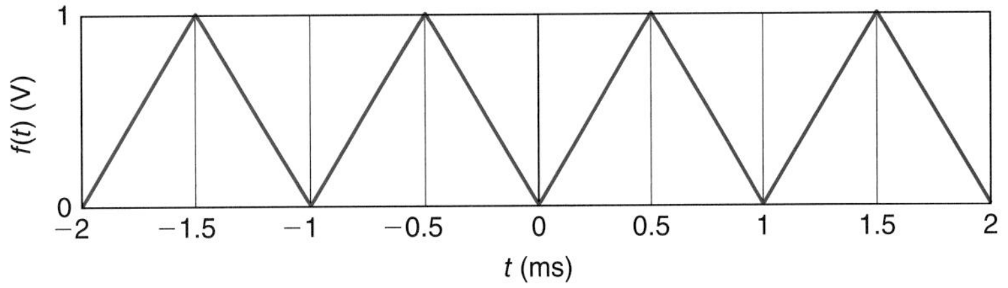

Answer:

$f(t) = 0.5 + 0.2026[e^{j(2\pi 1000t-\pi)} + e^{-j(2\pi 1000t-\pi)}] + 0.02252[e^{j(2\pi 3000t-\pi)} + e^{-j(2\pi 3000t-\pi)}]$ V

The phase is in radians.

$y(t) = 0.5 + 0.1811[e^{j(2\pi 1000t-3.6074)} + e^{-j(2\pi 1000t-3.6074)}] + 0.01244[e^{j(2\pi 1000t-0.7880)} + e^{-j(2\pi 1000t-0.7880)}]$ V

The phase is in radians.

18.8 PSpice and Simulink

FIGURE 18.70

An *RC* circuit with square wave as an input.

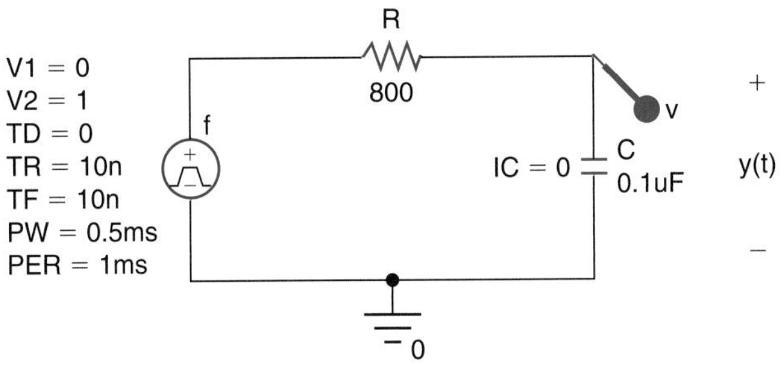

PSpice provides Fourier coefficients c_n. An *RC* circuit with a rectangular pulse train as input is shown in Figure 18.70. This is the circuit discussed in Example 18.21. The square pulse can be obtained by entering VPULSE in the place part. The input wave is shown in Figure 18.71. Clicking the FFT (fast Fourier transform) button results in the one-sided magnitude spectrum shown in Figure 18.72.

FIGURE 18.71

Input signal.

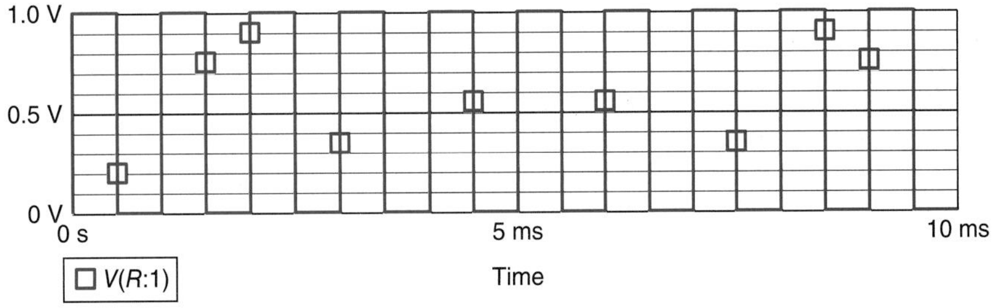

FIGURE 18.72

FFT of the input pulse train.

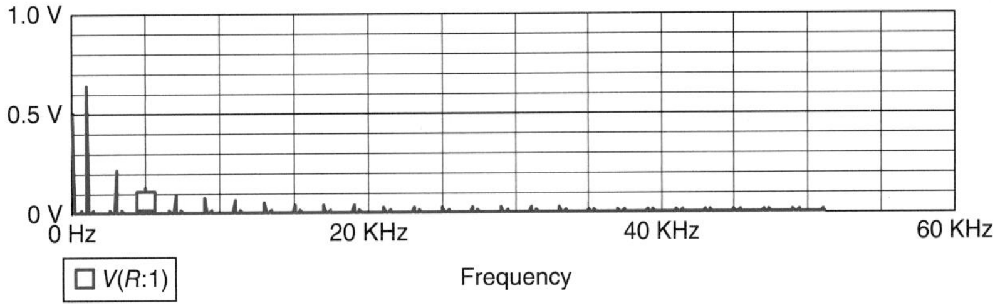

To check the accuracy, we can get the data by selecting Edit → Select All → Copy, and paste in Word. Table 18.1 shows the Fourier coefficients from the PSpice and the correct value. If more periods are displayed in the time domain, the accuracy improves.

TABLE 18.1

FFT from PSpice and Exact Value.

Frequency	V(R:1)	Correct Value
0	0.500488738306234	0.5
1000	0.637246058058807	0.6366
2000	3.84080225904469e-008	0
3000	0.212425968970605	0.2122
4000	3.29053845998752e-008	0
5000	0.127468334807191	0.1273
6000	2.598793402319e-008	0
7000	0.0910624899880495	0.09095
8000	2.00663796499394e-008	0
9000	0.0708405789378011	0.0707
10,000	1.64211290436882e-008	0

The voltage across the capacitor is shown in Figure 18.73. The waveform looks very similar to the one shown in Figure 18.68. The FFT of the voltage across the capacitor is shown in Figure 18.74. The values of the FFT are shown in Table 18.2.

FIGURE 18.73

The voltage across the capacitor.

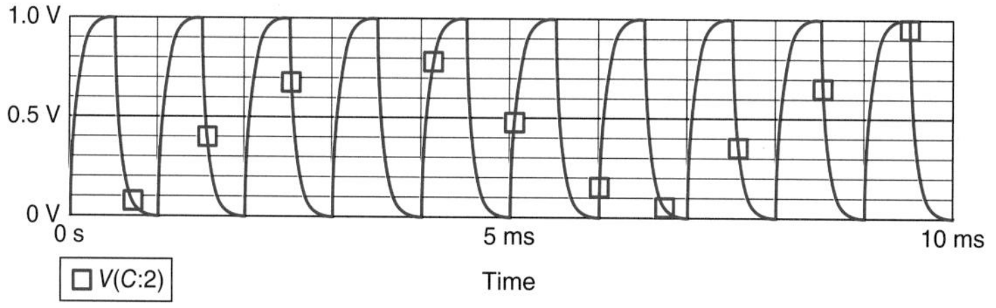

FIGURE 18.74

FFT of the output signal.

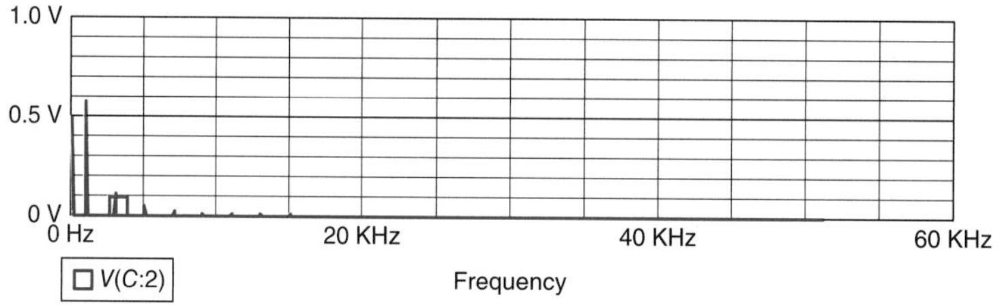

TABLE 18.2

FFT from PSpice and Exact Value

Frequency	V(C:2)	Correct Value
0	0.498983852562746	0.5
1000	0.573989436422028	0.5688
2000	0.00281660546666292O	0
3000	0.115413011378013	0.1173
4000	0.00189608896386068	0
5000	0.0450179299421487	0.04707
6000	0.0012679628159854	0
7000	0.0247169523308761	0.02486
8000	0.00173508018228611	0
9000	0.0143175896761499	0.01527
10,000	0.0015083394959239	0

The exponential Fourier coefficients F_n of a periodic signal $f(t)$ can be approximated from the fast Fourier transform (FFT) of the sampled signal $f(k)$. When the sampling rate is large compared to the fundamental frequency f_0 of the periodic signal $f(t)$, the approximation is very good for $n << N/2$. Let $F(n)$ be the FFT of $f(k)$. Then, the Fourier coefficients F_n can be approximated by

$$F_n \approx \frac{1}{N} F(n)$$

where N is the length of the FFT.

The Simulink model shown in Figure 18.75 can be used to find the Fourier coefficients of a rectangular pulse train with amplitude $A = 1$ V, period $T_0 = 1$ s, and duty cycle $d = 1/2$. Figure 18.76 shows the magnitude response $|F_n|$.

FIGURE 18.75

Simulink model to calculate F_n.

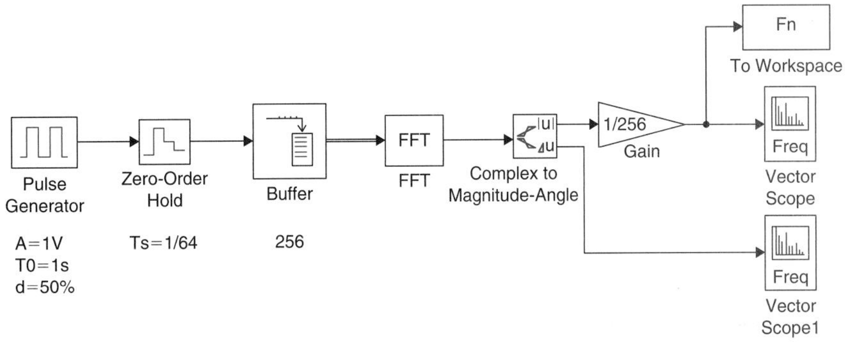

FIGURE 18.76

Magnitude response $|F_n|$.

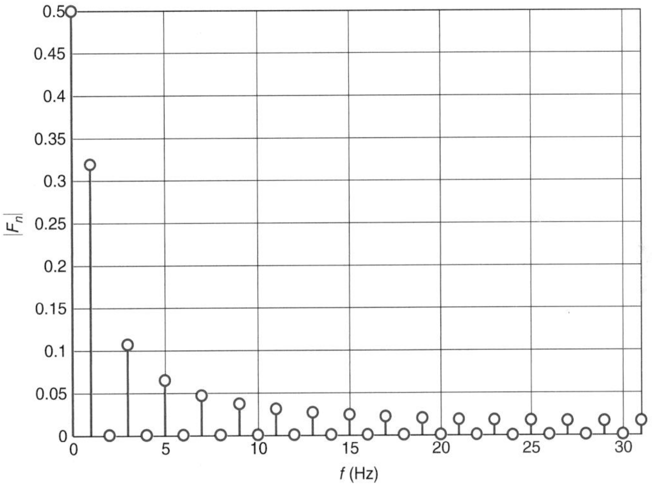

The dc component and the first ten harmonics and the exact values are:

n	From Simulink	Exact Values
0	0.500000000000000	0.5
1	0.318437753860877	0.318309886183791
2	0	0
3	0.106487763077147	0.106103295394597
4	0	0
5	0.064305656561044	0.063661977236758
6	0	0
7	0.046380144259818	0.045472840883399
8	0	0
9	0.036544997919778	0.035367765131532
10	0	0

As n value is increased, the error increases.

EXAMPLE 18.22

Use the Simulink model shown in Figure 18.77 to find the Fourier coefficients of a sawtooth pulse train with amplitude $A = 1$ V, period $T_0 = 1$s. Plot the magnitude spectrum of F_n. Calculate the exact values of F_n for $0 \le n \le 10$. $F_0 = A/2, F_n = jA/(2\pi n)$. Compare the values of exact F_n and approximation obtained from Simulink.

FIGURE 18.77

A Simulink mode to calculate F_n for sawtooth.

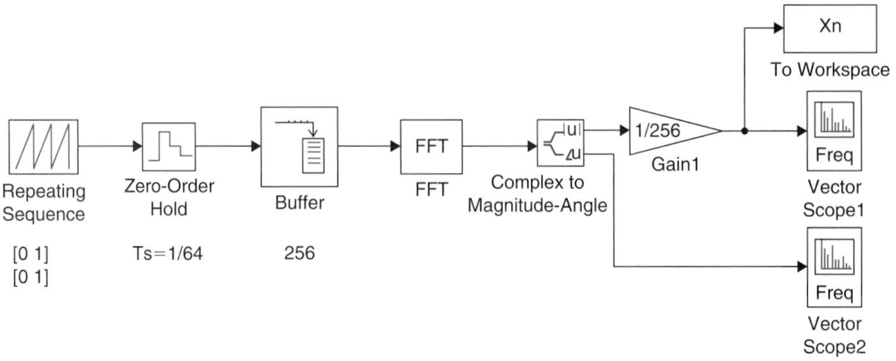

Figure 18.78 shows the magnitude response $|F_n|$.

FIGURE 18.78

Magnitude response $|F_n|$.

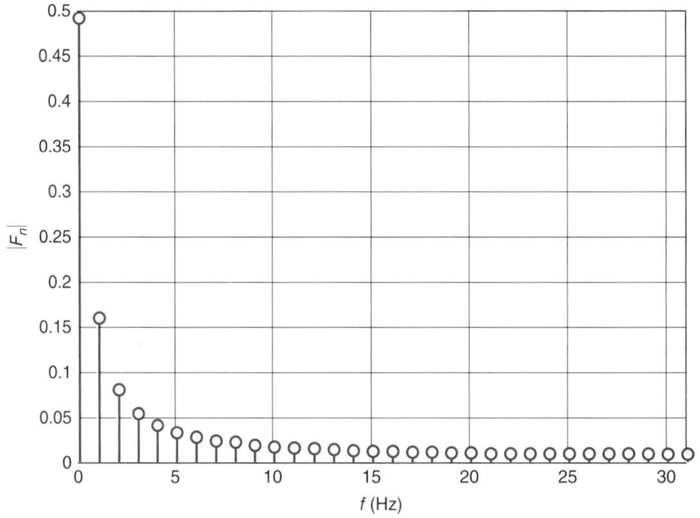

n	From Simulink	Exact Values
0	0.492187500000000	0.5
1	0.159218876930438	0.159154943091895
2	0.079705447167018	0.079577471545948
3	0.053243881538574	0.053051647697298
4	0.040045553870961	0.039788735772974
5	0.032152828280522	0.031830988618379
6	0.026913235909974	0.026525823848649
7	0.023190072129909	0.022736420441699
8	0.020415046326193	0.019894367886487
9	0.018272498959889	0.017683882565766
10	0.016573088843599	0.015915494309190

SUMMARY

In this chapter, the orthogonal functions are introduced. If a set of orthogonal functions is available, arbitrary signals can be represented as a linear combination of the orthogonal functions. The error between the original signal and the approximation by the orthogonal signals is decreased as the number of orthogonal signals is increased.

One of the complete set of orthogonal signals is the sines and the cosines that are harmonically related. The representation of periodic signals as a linear combination of sines and cosines is called a Fourier series representation. Instead of sines and cosines, a set of complex exponential signals can be used as a basis functions. The coefficients in the representations are called Fourier coefficients. The plot of spectrums shows the characteristic of the signals in the frequency domain.

One of the applications of the Fourier series is to find the response of a circuit to a periodic signal. The periodic input signal is first represented by its Fourier series. Next, we find the magnitude response and the phase response of the circuit from the transfer function. Finally, the magnitude and the phase of the output signal are found by multiplying the magnitudes and adding the phases.

TABLE 18.3 Fourier Series Representation of Common Signals

Trigonometric Fourier Series Representation

$$f(t) = a_0 + \sum_{n=1}^{\infty}[a_n \cos(n\omega_0 t) + b_n \sin(n\omega_0 t)]$$

$$f(t) = c_0 + \sum_{n=1}^{\infty} c_n \cos(n\omega_0 t + \theta_n)$$

$$a_0 = \frac{1}{T_0}\int_{t_0}^{t_0+T_0} f(t)dt$$

$$a_n = \frac{2}{T_0}\int_{t_0}^{t_0+T_0} f(t)\cos(n\omega_0 t)dt, \quad n = 1, 2, 3\ldots$$

$$b_n = \frac{2}{T_0}\int_{t_0}^{t_1+T_0} f(t)\sin(n\omega_0 t)dt, \quad n = 1, 2, 3\ldots$$

$T_0 =$ period.
$\omega_0 = 2\pi/T_0 =$ fundamental frequency in rad/s.
$f_0 = 1/T_0 =$ fundamental frequency in hertz.
t_0 is arbitrary. A common choice for t_0 is 0 or $-T_0/2$.

If $f(t)$ possesses even symmetry, the coefficients become

$$a_0 = \frac{1}{T_0}\int_{-\frac{T_0}{2}}^{\frac{T_0}{2}} f(t)dt = \frac{2}{T_0}\int_{0}^{\frac{T_0}{2}} f(t)dt$$

$$a_n = \frac{4}{T_0}\int_{0}^{\frac{T_0}{2}} f(t)\cos\left(n\frac{2\pi}{T_0}t\right)dt, \quad n = 1, 2, 3, \ldots$$

$$b_n = 0, \quad n = 1, 2, 3, \ldots$$

If $f(t)$ possesses odd symmetry, the coefficients become

$$a_0 = 0 \quad a_n = 0, \quad n = 1, 2, 3, \ldots$$

$$b_n = \frac{4}{T_0}\int_{0}^{\frac{T_0}{2}} f(t)\sin\left(n\frac{2\pi}{T_0}t\right)dt, \quad n = 1, 2, 3\ldots$$

Exponential Fourier Series Representation

$$f(t) = \sum_{n=-\infty}^{\infty} F_n e^{jn\omega_0 t}, \quad t_0 \le t \le t_0 + T_0$$

$$F_0 = \frac{1}{T_0}\int_{t_0}^{t_0+T_0} f(t)dt$$

$$F_n = \frac{1}{T_0}\int_{t_0}^{t_0+T_0} f(t)e^{-jn\omega_0 t}dt, \quad n = \pm 1, \pm 2, \pm 3, \ldots$$

If $f(t)$ possesses even symmetry, we have

$$F_n = \frac{2}{T_0}\int_{0}^{\frac{T_0}{2}} f(t)\cos(n\omega_0 t)dt = \frac{a_n}{2} = F_{-n}$$

If $f(t)$ possesses odd symmetry, we have

$$F_n = -j\frac{2}{T_0}\int_{0}^{\frac{T_0}{2}} f(t)\sin(n\omega_0 t)dt = -j\frac{b_n}{2} = -F_{-n}$$

continued

Table 18.3 continued

TABLE 18.3

T_0 = period.
$\omega_0 = 2\pi/T_0$ = fundamental frequency in rad/s.
$f_0 = 1/T_0$ = fundamental frequency in Hz.
t_0 is arbitrary. A common choice for t_0 is 0 or $T_0/2$.

Conversion of Fourier Coefficients

$$F_0 = a_0 = c_0$$

$$F_n = \frac{a_n}{2} - j\frac{b_n}{2} = |F_n|e^{j\angle F_n}, \quad n = 1, 2, 3, \ldots.$$

$$F_{-n} = \frac{a_n}{2} + j\frac{b_n}{2} = F_n^* = |F_n|e^{-j\angle F_n}, \quad n = 1, 2, 3, \ldots$$

$$a_n = 2\,\mathrm{Re}[F_n] = F_n + F_{-n} = 2|F_n|\cos(\angle F_n), \quad n = 1, 2, 3, \ldots$$

$$b_n = -2\,\mathrm{Im}[F_n] = j[F_n - F_{-n}] = -2|F_n|\sin(\angle F_n), \quad n = 1, 2, 3, \ldots$$

$$c_n = \sqrt{a_n^2 + b_n^2} = 2|F_n|, \quad n = 1, 2, 3, \ldots$$

$$\theta_n = -\tan^{-1}\left(\frac{b_n}{a_n}\right) = \angle F_n, \quad n = 1, 2, 3, \ldots$$

If $c_0 \geq 0$ ($a_0 \geq 0$), $\theta_0 = 0$. If $c_0 < 0$ ($a_0 < 0$), $\theta_0 = \pi$.
If $a_n = 0$ and $b_n = 0$, $\theta_n = 0$.
If $a_n \geq 0$ and $b_n = 0$, $\theta_n = 0$. If $a_n < 0$ and $b_n = 0$, $\theta_n = \pi$.
If $a_n = 0$ and $b_n > 0$, $\theta_n = -\pi/2$. If $a_n = 0$ and $b_n < 0$, $\theta_n = \pi/2$.
If $a_n > 0$ and $|b_n| > 0$, $\theta_n = -\tan^{-1}(b_n/a_n)$.
If $a_n < 0$ and $b_n > 0$, $\theta_n = -\pi - \tan^{-1}(b_n/a_n)$.
If $a_n < 0$ and $b_n < 0$, $\theta_n = \pi - \tan^{-1}(b_n/a_n)$.

General Rectangular Pulse Train

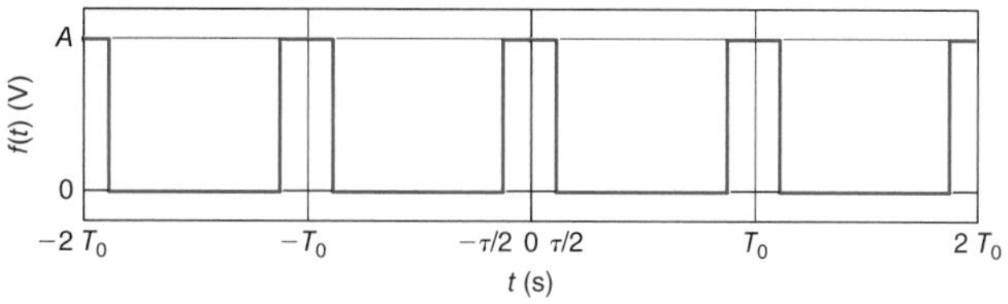

One period of $f(t)$:

$$f(t) = \begin{cases} A, & \frac{-\tau}{2} \leq t < \frac{\tau}{2} \\ 0, & otherwise \end{cases}$$

τ = pulse width ($-\tau/2$ to $\tau/2$)
d = duty cycle = τ/T_0
ω_0 = fundamental frequency = $2\pi/T_0$
$\mathrm{sinc}(x) = \sin(\pi x)/(\pi x)$

$$F_n = \frac{A\tau}{T_0}\mathrm{sinc}(nd) = Ad\,\mathrm{sinc}(nd) = \begin{cases} Ad, & n = 0 \\ A\frac{\sin(n\pi d)}{n\pi}, & n \neq 0 \end{cases}$$

$$F_0 = a_0 = c_0 = Ad = \frac{A\tau}{T_0}, \quad \text{for } n \geq 1,$$

$$a_n = 2Ad\,\mathrm{sinc}(nd) = \frac{2A}{n\pi}\sin\left(\frac{n\pi\tau}{T_0}\right), \quad b_n = 0$$

$$c_0 = Ad = \frac{A\tau}{T_0}, \quad \textit{for } n \geq 1,\ c_n = |2Ad\,\mathrm{sinc}(nd)|,$$

$$\theta_n = \begin{cases} \pi, & 2Ad\,\mathrm{sinc}(nd) < 0 \\ 0, & otherwise \end{cases}$$

$$f(t) = Ad + \sum_{n=1}^{\infty} 2Ad\,\mathrm{sinc}(nd)\cos(n\omega_0 t) = \frac{A\tau}{T_0} + \sum_{n=1}^{\infty}\frac{2A}{n\pi}\sin\left(\frac{n\pi\tau}{T_0}\right)\cos\left(n\frac{2\pi}{T_0}t\right)$$

Rectangular Pulse Train with Duty Cycle 1/2

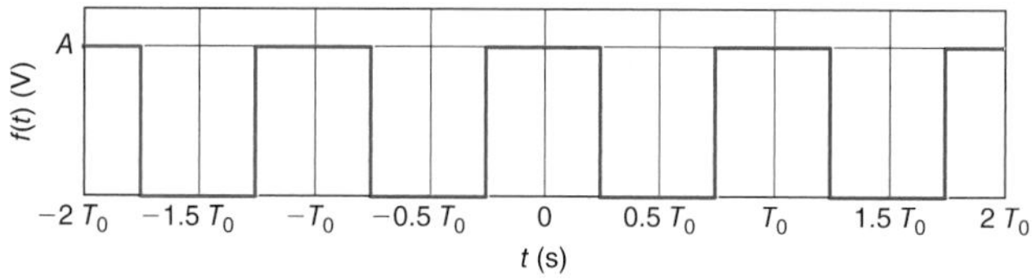

If $\tau = T_0/2$, $d = 1/2$, and the equations given here become

$$a_0 = \frac{A}{2}, \quad a_n = \frac{2A}{n\pi}\sin\left(\frac{n\pi}{2}\right) = \begin{cases} 0, & n = even \\ \frac{2A}{n\pi}\sin\left(\frac{n\pi}{2}\right), & n = odd \end{cases}$$

$$f(t) = \frac{A}{2} + \sum_{n=1}^{\infty} A\,\mathrm{sinc}\left(\frac{n}{2}\right)\cos\left(n\frac{2\pi}{T_0}t\right) = \frac{A}{2} + \sum_{n=1}^{\infty}\frac{(-1)^{n+1}2A}{(2n-1)\pi}\cos\left((2n-1)\frac{2\pi}{T_0}t\right)$$

continued

Table 18.3 continued

TABLE 18.3

Rectangular Pulse Train with Duty Cycle 1/2 and Zero DC Offset

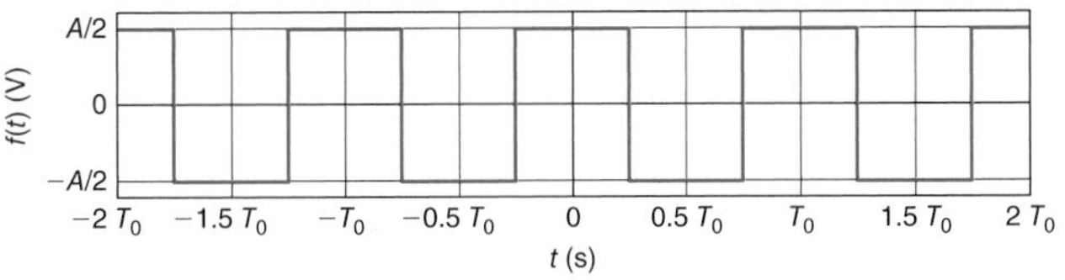

$$a_0 = 0,$$

$$a_n = \frac{2A}{n\pi}\sin\left(\frac{n\pi}{2}\right) = \begin{cases} 0, & n = even \\ \dfrac{2A}{n\pi}\sin\left(\dfrac{n\pi}{2}\right), & n = odd \end{cases}$$

$$f(t) = \sum_{n=1}^{\infty} A\,\text{sinc}\left(\frac{n}{2}\right)\cos\left(\frac{n2\pi t}{T_0}\right)$$

$$= \sum_{n=1}^{\infty}\frac{(-1)^{n+1}2A}{(2n-1)\pi}\cos\left(\frac{(2n-1)2\pi t}{T_0}\right)$$

Rectangular Pulse Train with Duty Cycle 1/2 and Time Shift of $T_0/4$

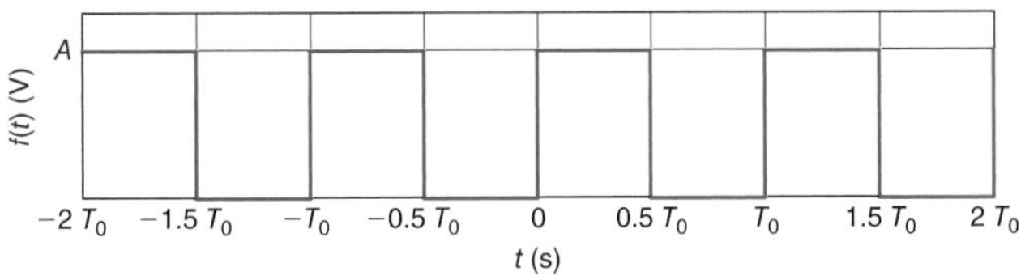

$$F_0 = \frac{A}{2},\quad F_n = \begin{cases} 0, & n = even \\ \dfrac{-jA}{n\pi}, & n = odd \end{cases}$$

$$a_0 = \frac{A}{2},\quad a_n = 0,\quad n = 1, 2, 3, \ldots$$

$$b_n = \frac{A}{n\pi}[1 - \cos(n\pi)] = \begin{cases} 0, & n = even \\ \dfrac{2A}{n\pi}, & n = odd \end{cases}$$

$$f(t) = \frac{A}{2} + \sum_{\substack{n=1 \\ n=odd}}^{\infty}\frac{2A}{n\pi}\sin\left(\frac{n2\pi t}{T_0}\right)$$

$$= \frac{A}{2} + \sum_{n=1}^{\infty}\frac{2A}{(2n-1)\pi}\sin\left(\frac{(2n-1)2\pi t}{T_0}\right)$$

Rectangular Pulse Train with Duty Cycle 1/2, Time Shift $T_0/4$, and DC Offset 0

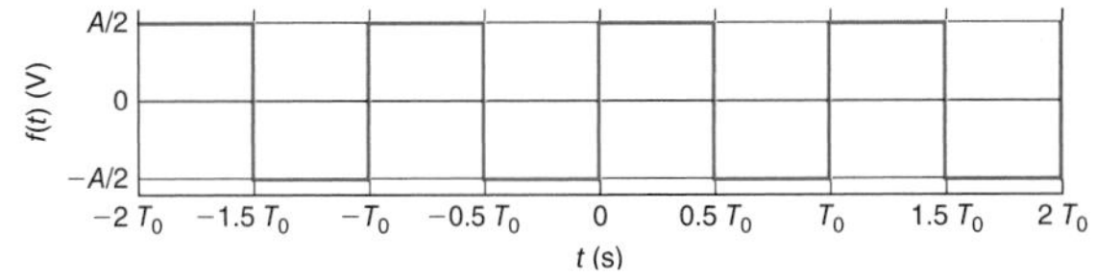

$$a_0 = 0,\quad a_n = 0,\quad n = 1, 2, 3, \ldots$$

$$b_n = \frac{A}{n\pi}[1 - \cos(n\pi)] = \begin{cases} 0, & n = even \\ \dfrac{2A}{n\pi}, & n = odd \end{cases}$$

$$f(t) = \sum_{n=1}^{\infty}\frac{2A}{(2n-1)\pi}\sin\left(\frac{(2n-1)2\pi t}{T_0}\right)$$

Rectangular Pulse Train with Time Shift $\tau/2$

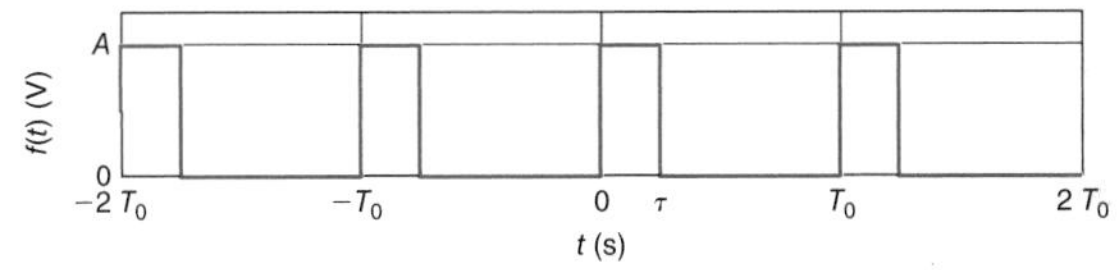

$$d = \frac{\tau}{T_0} = \text{duty cycle}$$

$$F_n = Ad\,\text{sinc}(nd)e^{-jn\frac{2\pi}{T_0}t_0}$$

$$= Ad\,\text{sinc}(nd)e^{-jn\frac{2\pi}{T_0}\frac{\tau}{2}} = Ad\,\text{sinc}(nd)e^{-jn\pi d}$$

$$F_n = Ad\,\text{sinc}(nd)\cos(n\pi d) - jAd\,\text{sinc}(nd)\sin(n\pi d)$$

$$a_0 = Ad,\ \text{for } n \geq 1, a_n = 2Ad\,\text{sinc}(nd)\cos(n\pi d)$$

$$= \frac{A}{\pi n}\sin\left(\frac{2\pi n\tau}{T_0}\right)$$

$$b_n = 2Ad\,\text{sinc}(nd)\sin(n\pi d) = \frac{2A}{\pi n}\sin^2\left(\frac{n\pi\tau}{T_0}\right)$$

$$c_0 = Ad,\ \text{for } n \geq 1, c_n = 2|Ad\,\text{sinc}(nd)|,$$

$$\theta_n = \angle F_n = \begin{cases} \dfrac{-n\pi\tau}{T_0}, & \text{sinc}\left(\dfrac{n\tau}{T_0}\right) \geq 0 \\ \dfrac{-n\pi\tau}{T_0} + \pi, & \text{sinc}\left(\dfrac{n\tau}{T_0}\right) < 0 \end{cases}$$

$$f(t) = \frac{A\tau}{T_0} + \sum_{n=1}^{\infty}\frac{A}{n\pi}\sin\left(\frac{2\pi n\tau}{T_0}\right)\cos\left(n\frac{2\pi}{T_0}t\right)$$

$$+ \sum_{n=1}^{\infty}\frac{2A}{n\pi}\sin^2\left(\frac{n\pi\tau}{T_0}\right)\sin\left(n\frac{2\pi}{T_0}t\right)$$

continued

Table 18.3 continued

TABLE 18.3

Rectangular Pulse Train with General Time Shift t_d

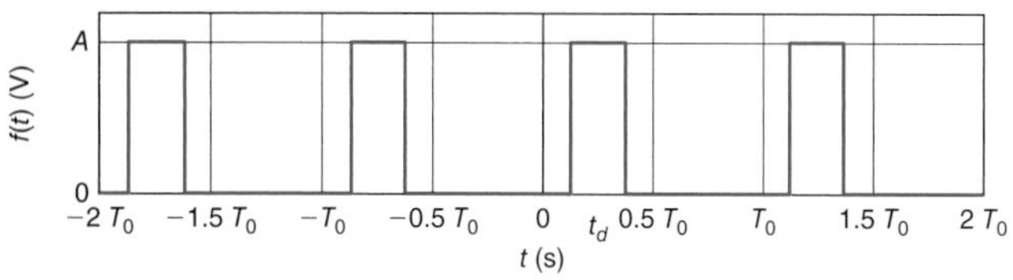

$$d = \frac{\tau}{T_0} = \text{duty cycle}$$

$$F_n = Ad\,\text{sinc}(nd)e^{-jn\frac{2\pi}{T_0}t_d} = Ad\,\text{sinc}(nd)\cos\left(n\frac{2\pi}{T_0}t_d\right)$$

$$-jAd\,\text{sinc}(nd)\sin\left(n\frac{2\pi}{T_0}t_d\right)$$

$$a_0 = Ad$$

$$\text{For } n \geq 1, \quad a_n = 2Ad\,\text{sinc}(nd)\cos\left(n\frac{2\pi}{T_0}t_d\right) = \frac{2A}{\pi n}\sin\left(\frac{n\pi\tau}{T_0}\right)\cos\left(n\frac{2\pi}{T_0}t_d\right)$$

$$b_n = 2Ad\,\text{sinc}(nd)\sin\left(n\frac{2\pi}{T_0}t_d\right) = \frac{2A}{\pi n}\sin\left(\frac{n\pi\tau}{T_0}\right)\sin\left(n\frac{2\pi}{T_0}t_d\right)$$

$$c_0 = Ad, \quad \text{for } n \geq 1, \quad c_n = 2|Ad\,\text{sinc}(nd)|,$$

$$\theta_n = \angle F_n = \begin{cases} -n\frac{2\pi}{T_0}t_d, & \text{sinc}\left(\frac{n\tau}{T_0}\right) \geq 0 \\ -n\frac{2\pi}{T_0}t_d + \pi, & \text{sinc}\left(\frac{n\tau}{T_0}\right) < 0 \end{cases}$$

$$f(t) = \frac{A\tau}{T_0} + \sum_{n=1}^{\infty}\frac{2A}{\pi n}\sin\left(\frac{n\pi\tau}{T_0}\right)\cos\left(n\frac{2\pi}{T_0}t_d\right)\cos\left(\frac{n2\pi}{T_0}t\right) + \sum_{n=1}^{\infty}\frac{2A}{n\pi}\sin\left(\frac{n\pi\tau}{T_0}\right)\cos\left(n\frac{2\pi}{T_0}t_d\right)\sin\left(\frac{n2\pi}{T_0}t\right)$$

General Triangular Pulse Train

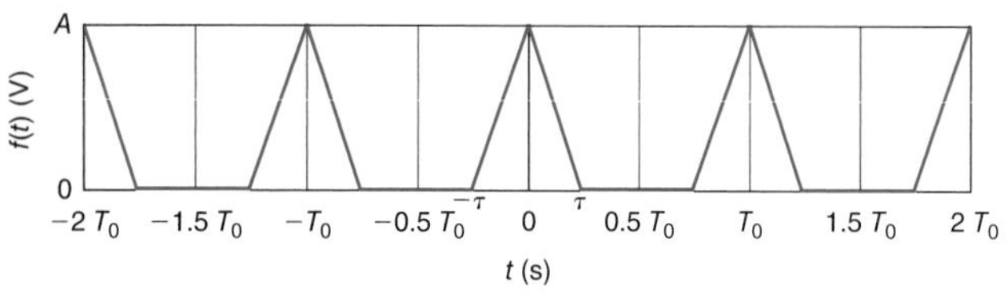

One period of $f(t)$:

$$f(t) = \begin{cases} A\frac{t}{\tau} + A, & -\tau \leq t < 0 \\ -A\frac{t}{\tau} + A, & 0 \leq t < \tau \\ 0, & \textit{otherwise} \end{cases}$$

τ = half of the base of the triangle $(-\tau \leq t \leq \tau)$

$d = \tau/T_0$.

ω_0 = fundamental frequency $= 2\pi/T_0$

$F_n = Ad\,\text{sinc}^2(nd)$

$F_0 = a_0 = c_0 = Ad,$

$a_n = c_n = 2F_n = 2Ad\,\text{sinc}^2(nd), \quad b_n = 0$

$$f(t) = \frac{A\tau}{T_0} + \sum_{n=1}^{\infty}\frac{2A\tau}{T_0}\text{sinc}^2\left(\frac{n\tau}{T_0}\right)\cos\left(n\frac{2\pi}{T_0}t\right)$$

Triangular Pulse Train with $\tau = T_0/2$

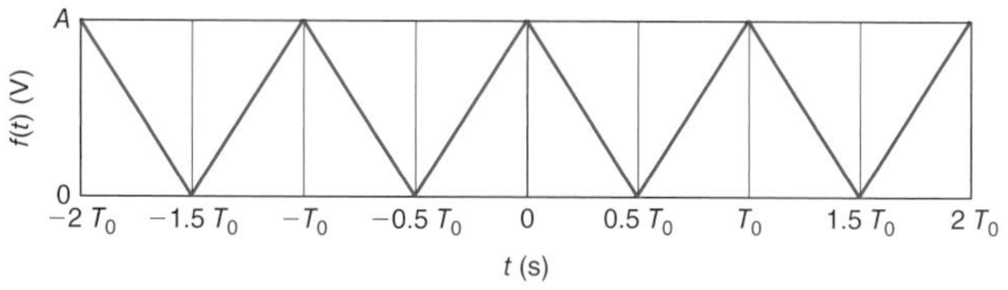

$$F_n = \frac{A}{2}\text{sinc}^2\left(\frac{n}{2}\right) = \begin{cases} \frac{A}{2}, & n = 0 \\ 0, & n = \textit{even} \\ \frac{2A}{n^2\pi^2}, & n = \textit{odd} \end{cases}$$

$$F_0 = a_0 = c_0 = \frac{A}{2},$$

$$\text{For } n \geq 1, \quad a_n = c_n = 2F_n = A\,\text{sinc}^2\left(\frac{n}{2}\right) = \begin{cases} 0, & n = \textit{even} \\ \frac{4A}{n^2\pi^2}, & n = \textit{odd} \end{cases}, \quad b_n = 0$$

$$f(t) = \frac{A}{2} + \sum_{n=1}^{\infty}\frac{4A}{(2n-1)^2\pi^2}\cos\left((2n-1)\frac{2\pi}{T_0}t\right)$$

continued

Table 18.3 continued

TABLE 18.3

Triangular Pulse Train with $\tau = T_0/2$, Time Shift $T_0/2$

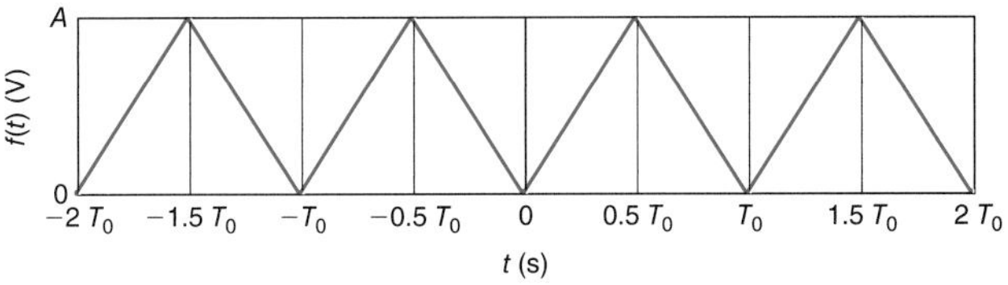

$$F_n = \frac{A}{2}\text{sinc}^2\left(\frac{n}{2}\right)e^{-jn\frac{2\pi}{T_0}\frac{T_0}{2}} = \frac{A}{2}\text{sinc}^2\left(\frac{n}{2}\right)e^{-jn\pi}$$

$$= \frac{A}{2}\text{sinc}^2\left(\frac{n}{2}\right)\cos(n\pi) = \begin{cases} \frac{A}{2}, & n = 0 \\ 0, & n = even \\ \frac{-2A}{n^2\pi^2}, & n = odd \end{cases}$$

$$F_0 = a_0 = c_0 = \frac{A}{2}$$

$$\text{For } n \geq 1,\quad a_n = c_n = 2F_n = A\,\text{sinc}^2\left(\frac{n}{2}\right)\cos(n\pi)$$

$$= \begin{cases} 0, & n = even \\ \frac{-4A}{n^2\pi^2}, & n = odd \end{cases},\quad b_n = 0$$

$$f(t) = \frac{A}{2} + \sum_{n=1}^{\infty}\frac{-4A}{(2n-1)^2\pi^2}\cos\left((2n-1)\frac{2\pi}{T_0}t\right)$$

Triangular Pulse Train with $\tau = T_0/2$, DC Offset 0

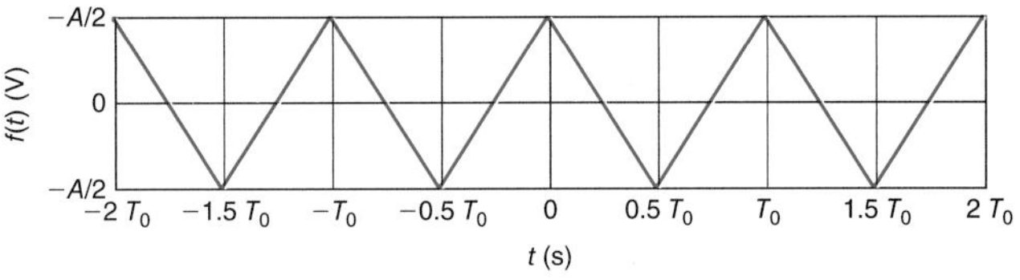

$$F_0 = 0,\quad F_n = \frac{A}{2}\text{sinc}^2\left(\frac{n}{2}\right) = \begin{cases} 0, & n = even \\ \frac{2A}{n^2\pi^2}, & n = odd \end{cases}$$

$$F_0 = a_0 = c_0 = 0$$

$$\text{For } n \geq 1,\quad a_n = c_n = 2F_n = A\,\text{sinc}^2\left(\frac{n}{2}\right)$$

$$= \begin{cases} 0, & n = even \\ \frac{4A}{n^2\pi^2}, & n = odd \end{cases},\quad b_n = 0$$

$$f(t) = \sum_{n=1}^{\infty}\frac{4A}{(2n-1)^2\pi^2}\cos\left((2n-1)\frac{2\pi}{T_0}t\right)$$

Triangular Pulse Train with $\tau = T_0/2$, DC Offset 0, Time Shift $T_0/4$

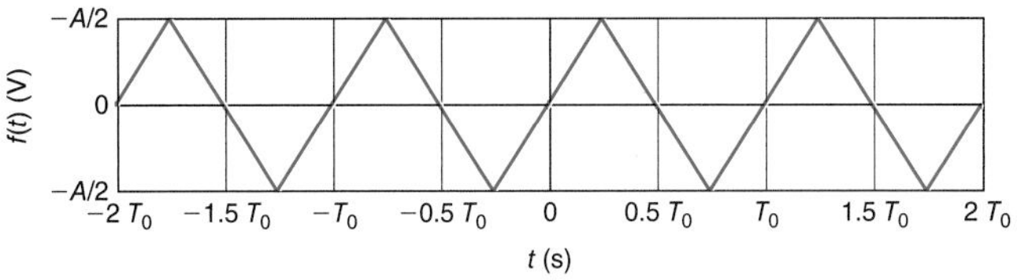

$$F_0 = 0,$$

$$F_n = \frac{A}{2}\text{sinc}^2\left(\frac{n}{2}\right)e^{-jn\frac{2\pi}{T_0}\frac{T_0}{4}} = \frac{A}{2}\text{sinc}^2\left(\frac{n}{2}\right)e^{-j\frac{n\pi}{2}}$$

$$F_n = \frac{A}{2}\text{sinc}^2\left(\frac{n}{2}\right)\cos\left(\frac{n\pi}{2}\right) - j\frac{A}{2}\text{sinc}^2\left(\frac{n}{2}\right)\sin\left(\frac{n\pi}{2}\right)$$

$F_0 = a_0 = c_0 = 0$. For $n \geq 1, a_n = 0$,

$$b_n = A\,\text{sinc}^2\left(\frac{n}{2}\right)\sin\left(\frac{n\pi}{2}\right) = \begin{cases} 0, & n = even \\ \frac{4A}{n^2\pi^2}, & n = 1,5,9,\ldots \\ \frac{-4A}{n^2\pi^2}, & n = 3,7,11,\ldots \end{cases}$$

$$c_n = A\,\text{sinc}^2\left(\frac{n}{2}\right) = \begin{cases} 0, & n = even \\ \frac{4A}{n^2\pi^2}, & n = odd \end{cases},$$

$$\theta_n = \begin{cases} 0, & n = even \\ \frac{-\pi}{2}, & n = 1,5,9,\ldots \\ \frac{\pi}{2}, & n = 3,7,11,\ldots \end{cases}$$

$$f(t) = \sum_{n=1}^{\infty}\frac{(-1)^{n+1}4A}{(2n-1)^2\pi^2}\sin\left((2n-1)\frac{2\pi}{T_0}t\right)$$

Half-Wave Rectified Cosine

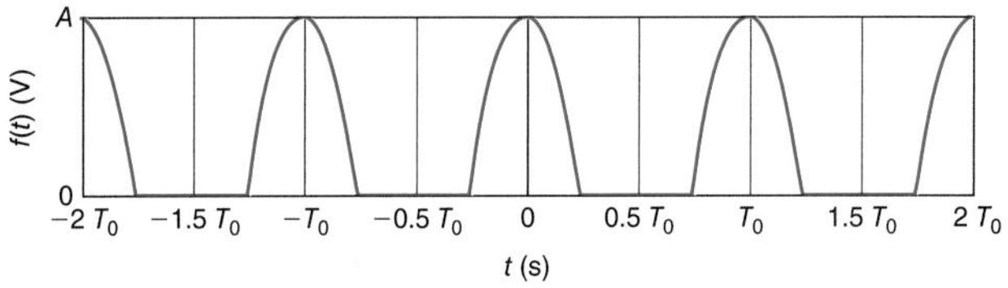

continued

Table 18.3 continued

TABLE 18.3

One period of $f(t)$:

$$f(t) = \begin{cases} A\cos\left(\dfrac{2\pi t}{T_0}\right), & \dfrac{-T_0}{4} \le t < \dfrac{T_0}{4} \\ 0, & otherwise \end{cases}$$

$$F_0 = \frac{A}{\pi}, \quad F_1 = \frac{A}{4}, \quad F_{-1} = \frac{A}{4},$$

$$F_n = \frac{A}{\pi}\frac{\cos\left(\dfrac{n\pi}{2}\right)}{1-n^2}, \quad n \ne \pm 1$$

$$F_0 = a_0 = c_0 = \frac{A}{\pi}, \quad a_1 = c_1 = \frac{A}{2},$$

$$a_n = \frac{2A}{\pi}\frac{\cos\left(\dfrac{n\pi}{2}\right)}{1-n^2} = \begin{cases} 0, & n = 3,5,7,9,\ldots \\ \dfrac{2A}{\pi(n^2-1)}, & n = 2,6,10,\ldots \\ \dfrac{-2A}{\pi(n^2-1)}, & n = 4,8,12,\ldots \end{cases},$$

$$b_n = 0$$

$$c_n = \begin{cases} \dfrac{A}{\pi}, & n = 0 \\ \dfrac{A}{2}, & n = 1 \\ 0, & n = 3,5,7,9,\ldots \\ \dfrac{2A}{\pi(n^2-1)}, & n = 2,4,6,\ldots \end{cases},$$

$$\theta_n = \begin{cases} 0, & n = 0,1,2,3,5,6,7,9,10,\ldots \\ \pi, & n = 4,8,12,\ldots \end{cases}$$

$$f(t) = \frac{A}{\pi} + \frac{A}{2}\cos\left(\frac{2\pi}{T_0}t\right) + \sum_{n=1}^{\infty}\frac{(-1)^{n-1}2A}{\pi(4n^2-1)}\cos\left(2n\frac{2\pi}{T_0}t\right)$$

Half-Wave Rectified Sine

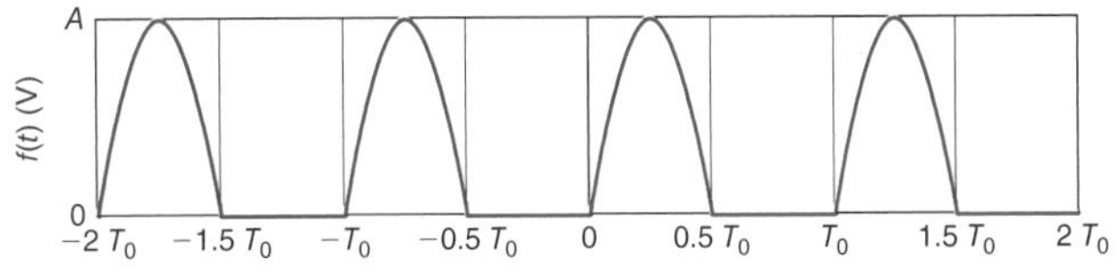

One period of $f(t)$:

$$f(t) = \begin{cases} A\sin\left(\dfrac{2\pi t}{T_0}\right), & 0 \le t < \dfrac{T_0}{2} \\ 0, & otherwise \end{cases}$$

$$F_0 = \frac{A}{\pi}, \quad F_1 = \frac{-jA}{4}, \quad F_{-1} = \frac{jA}{4}$$

$$F_n = \frac{A}{\pi}\frac{\cos^2\left(\dfrac{n\pi}{2}\right)}{1-n^2} = \begin{cases} 0, & n = \pm 3, \pm 5, \pm 7,\ldots \\ \dfrac{A}{\pi}\dfrac{1}{1-n^2}, & n = \pm 2, \pm 4, \pm 6,\ldots \end{cases}$$

$$F_0 = a_0 = c_0 = \frac{A}{\pi}, \quad a_1 = 0,$$

$$a_n = \frac{2A}{\pi}\frac{\cos^2\left(\dfrac{n\pi}{2}\right)}{1-n^2} = \begin{cases} 0, & n = 3,5,7,9,\ldots \\ \dfrac{-2A}{\pi(n^2-1)}, & n = 2,4,6,8,\ldots \end{cases}$$

$$b_1 = \frac{A}{2}, \quad b_n = 0, \quad n = 2,3,4,5,\ldots$$

$$c_n = \begin{cases} \dfrac{A}{\pi}, & n = 0 \\ \dfrac{A}{2}, & n = 1 \\ 0, & n = 3,5,7,9,\ldots \\ \dfrac{2A}{\pi(n^2-1)}, & n = 2,4,6,\ldots \end{cases},$$

$$\theta_n = \begin{cases} 0, & n = 0,1,3,5,7,9,\ldots \\ \pi, & n = 2,4,6,8,\ldots \end{cases}$$

$$f(t) = \frac{A}{\pi} + \frac{A}{2}\sin\left(\frac{2\pi}{T_0}t\right) + \sum_{n=1}^{\infty}\frac{-2A}{\pi(4n^2-1)}\cos\left(2n\frac{2\pi}{T_0}t\right)$$

Full-Wave Rectified Cosine

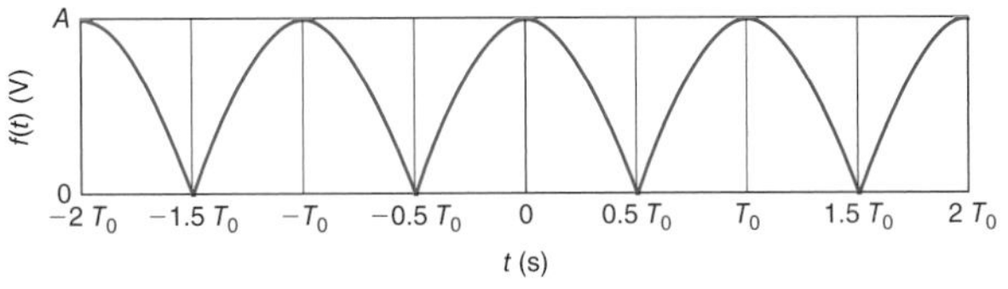

One period of $f(t)$:

$$f(t) = \begin{cases} A\cos\left(\dfrac{\pi t}{T_0}\right), & \dfrac{-T_0}{2} \le t < \dfrac{T_0}{2} \\ 0, & otherwise \end{cases}$$

$$F_n = \frac{2A}{\pi}\frac{\cos(n\pi)}{1-4n^2} = \frac{2A}{\pi}\frac{(-1)^n}{1-4n^2} = \frac{(-1)^{n+1}2A}{\pi(4n^2-1)}$$

continued

Table 18.3 continued

TABLE 18.3

$$F_0 = a_0 = c_0 = \frac{2A}{\pi},$$

$$a_n = \frac{4A}{\pi}\frac{(-1)^n}{1-4n^2}, \quad n = 1, 2, 3, \ldots,$$

$$b_n = 0, \quad n = 1, 2, 3, \ldots$$

$$c_n = \frac{4A}{\pi(4n^2-1)}, \quad n = 1, 2, 3, \ldots,$$

$$\theta_n = \begin{cases} 0, & n = 0, 1, 3, 5, \ldots \\ \pi, & n = 2, 4, 6, \ldots \end{cases}$$

$$f(t) = \frac{2A}{\pi} + \sum_{n=1}^{\infty} \frac{(-1)^{n+1}4A}{\pi(4n^2-1)} \cos\left(n\frac{2\pi}{T_0}t\right)$$

Full-Wave Rectified Sine

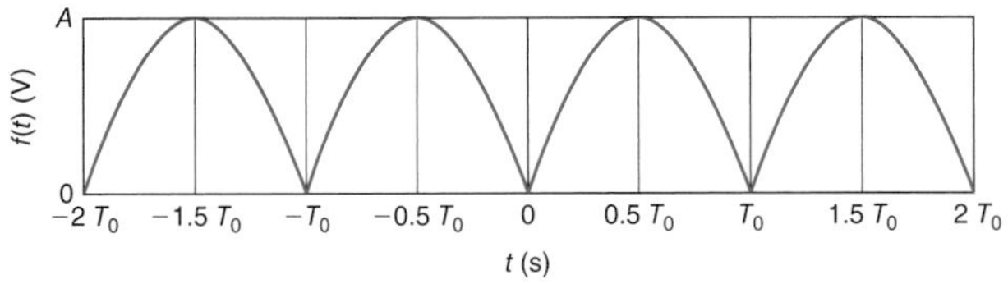

One period of $f(t)$:

$$f(t) = \begin{cases} A\sin\left(\frac{\pi t}{T_0}\right), & 0 \le t < T_0 \\ 0, & otherwise \end{cases}$$

$$F_n = \frac{2A}{\pi}\frac{\cos^2(n\pi)}{1-4n^2} = \frac{2A}{\pi}\frac{1}{1-4n^2}$$

$$F_0 = a_0 = c_0 = \frac{2A}{\pi},$$

$$a_n = \frac{-4A}{\pi(4n^2-1)}, \quad n = 1, 2, 3, \ldots,$$

$$b_n = 0, \quad n = 1, 2, 3, \ldots$$

$$c_n = \frac{4A}{\pi(4n^2-1)}, \quad n = 1, 2, 3, \ldots,$$

$$\theta_n = \begin{cases} 0, & n = 0 \\ \pi, & n = 1, 2, 3, \ldots \end{cases}$$

$$f(t) = \frac{2A}{\pi} + \sum_{n=1}^{\infty} \frac{-4A}{\pi(4n^2-1)} \cos\left(n\frac{2\pi}{T_0}t\right)$$

Sawtooth

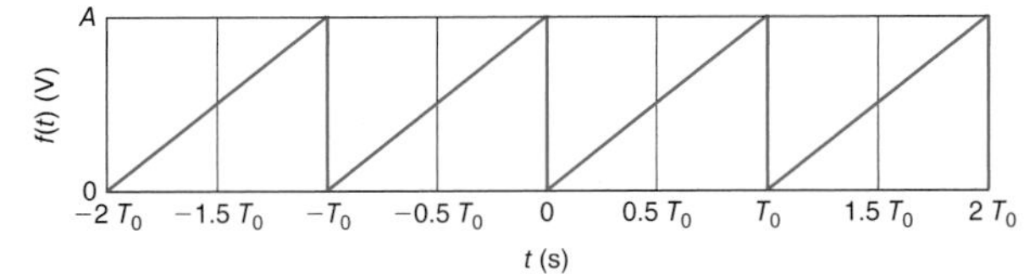

One period of $f(t)$:

$$f(t) = \begin{cases} A\frac{t}{T_0}, & 0 \le t < T_0 \\ 0, & otherwise \end{cases}$$

$$F_0 = \frac{A}{2}, \quad F_n = \frac{jA}{2\pi n}, \quad n \ne 0$$

$$F_0 = a_0 = c_0 = \frac{A}{2}, \quad a_n = 0, \quad n = 1, 2, 3, \ldots,$$

$$b_n = \frac{-A}{\pi n}, \quad n = 1, 2, 3, \ldots$$

$$c_n = \frac{A}{\pi n}, \quad \theta_n = \pi/2, \quad n = 1, 2, 3, \ldots$$

$$f(t) = \frac{A}{2} + \sum_{n=1}^{\infty} \frac{-A}{\pi n} \sin\left(n\frac{2\pi}{T_0}t\right)$$

Sawtooth with Time Shift $T_0/2$, DC Offset 0

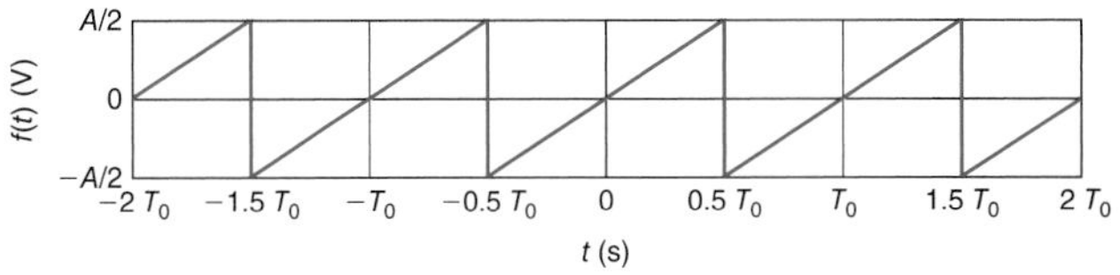

One period of $f(t)$:

$$f(t) = \begin{cases} A\frac{t}{T_0}, & -\frac{T_0}{2} \le t < \frac{T_0}{2} \\ 0, & otherwise \end{cases}$$

$$F_0 = 0, \quad F_n = \frac{jA\cos(n\pi)}{2\pi n}, \quad n \ne 0$$

$$F_0 = a_0 = c_0 = 0, \quad a_n = 0, \quad n = 1, 2, 3, \ldots,$$

$$b_n = \frac{(-1)^{n+1}A}{\pi n}, \quad n = 1, 2, 3, \ldots$$

$$c_n = \frac{A}{\pi n}, \quad \theta_n = \begin{cases} 0, & n = 0 \\ \frac{-\pi}{2}, & n = 1,3,5, \ldots \\ \frac{\pi}{2}, & n = 2,4,6, \ldots \end{cases}$$

$$f(t) = \sum_{n=1}^{\infty} \frac{(-1)^{n+1}A}{\pi n} \sin\left(n\frac{2\pi}{T_0}t\right)$$

continued

Table 18.3 continued

TABLE 18.3

Exponential Decay

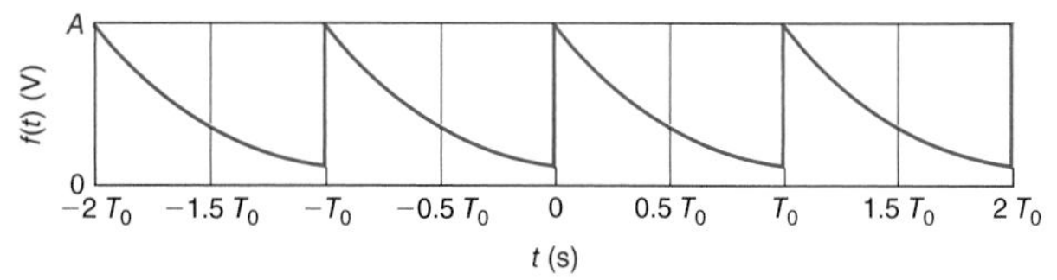

One period of $f(t)$:

$$f(t) = \begin{cases} Ae^{-Kt}, & 0 \leq t < T_0 \\ 0, & otherwise \end{cases}$$

$$F_n = \frac{A}{T_0} \frac{1 - e^{-KT_0}}{K + jn\frac{2\pi}{T_0}}$$

$$F_0 = a_0 = c_0 = \frac{A(1 - e^{-KT_0})}{KT_0},$$

$$a_n = \frac{2AKT_0(1 - e^{-KT_0})}{K^2T_0^2 + 4\pi^2n^2},$$

$$b_n = \frac{4\pi An(1 - e^{-KT_0})}{K^2T_0^2 + 4\pi^2n^2}$$

$$f(t) = \frac{A(1 - e^{-KT_0})}{KT_0} + \sum_{n=1}^{\infty}\left[\frac{2AKT_0(1 - e^{-KT_0})}{K^2T_0^2 + 4\pi^2n^2} \cos\left(n\frac{2\pi}{T_0}t\right) + \frac{4\pi An(1 - e^{-KT_0})}{K^2T_0^2 + 4\pi^2n^2} \sin\left(n\frac{2\pi}{T_0}t\right)\right]$$

PROBLEMS

Orthogonal Functions

18.1 **Let a signal $f(t)$ be**

$$f(t) = \begin{cases} -5, & 0 \leq t < \frac{1}{4} \\ 2, & \frac{1}{4} \leq t \leq 1 \\ 0, & otherwise \end{cases}$$

This signal is represented as a linear combination of orthonormal signals $\phi_0(t)$ and $\phi_1(t)$, as shown in Figure P18.1; that is,

$$\hat{f}(t) = a_0\phi_0(t) + a_1\phi_1(t)$$

FIGURE P18.1

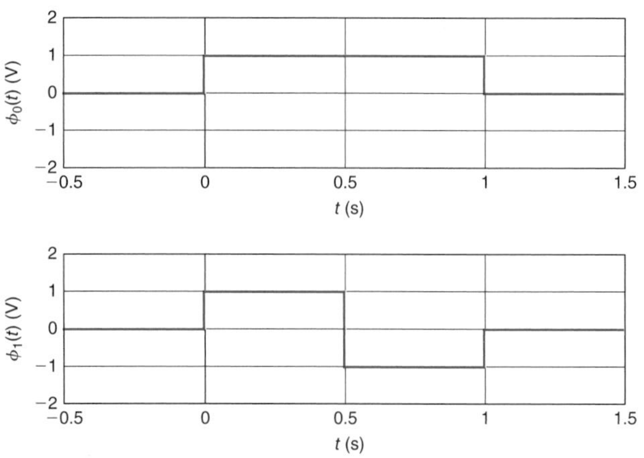

a. Find the coefficients a_0 and a_1.
b. Plot the approximation $\hat{f}(t)$.
c. Plot the error signal $e(t) = f(t) - \hat{f}(t)$.
d. Find the energy of $f(t)$ and $e(t)$.

18.2 **Let a signal $f(t)$ be**

$$f(t) = \begin{cases} 3, & 0 \leq t < \frac{3}{8} \\ -2, & \frac{3}{8} \leq t \leq 1 \\ 0, & otherwise \end{cases}$$

This signal is represented as a linear combination of orthonormal signals $\phi_0(t)$ and $\phi_1(t)$ shown in Figure P18.1; that is,

$$\hat{f}(t) = a_0\phi_0(t) + a_1\phi_1(t)$$

a. Find the coefficients a_0 and a_1.
b. Plot the approximation $\hat{f}(t)$.
c. Plot the error signal $e(t) = f(t) - \hat{f}(t)$.

18.3 Let

$$\phi_0(t) = \begin{cases} 1, & 0 \leq t < 1 \\ 0, & otherwise \end{cases}, \quad \phi_1(t) = \begin{cases} 1, & 0 \leq t < \frac{1}{2} \\ -1, & \frac{1}{2} \leq t < 1 \\ 0, & otherwise \end{cases},$$

$$\phi_2(t) = \begin{cases} \sqrt{2}, & 0 \leq t < \frac{1}{4} \\ -\sqrt{2}, & \frac{1}{4} \leq t < \frac{1}{2} \\ 0, & otherwise \end{cases}$$

$$\phi_3(t) = \begin{cases} \sqrt{2}, & \frac{1}{2} \le t < \frac{3}{4} \\ -\sqrt{2}, & \frac{3}{4} \le t < 1, \\ 0, & otherwise \end{cases}$$

$$f(t) = \begin{cases} 3, & 0 \le t < 0.3 \\ -2, & 0.3 \le t < 1 \\ 0, & otherwise \end{cases}$$

a. Let the approximation be $\hat{f}(t) = a_0\phi_0(t) + a_1\phi_1(t) + a_2\phi_2(t) + a_3\phi_3(t)$. Find a_0, a_1, a_2, a_3 to minimize the energy of the error signal $e(t) = f(t) - \hat{f}(t)$.
b. Plot the approximation $\hat{f}(t) = a_0\phi_0(t) + a_1\phi_1(t) + a_2\phi_2(t) + a_3\phi_3(t)$ for $0 \le t < 1$.
c. Plot the error signal $e(t) = f(t) - \hat{f}(t)$ for $0 \le t < 1$.

18.4 Let

$$\phi_0(t) = \begin{cases} 1, & 0 \le t < 1 \\ 0, & otherwise \end{cases}, \quad \phi_1(t) = \begin{cases} 1, & 0 \le t < \frac{1}{2} \\ -1, & \frac{1}{2} \le t < 1, \\ 0, & otherwise \end{cases}$$

$$\phi_2(t) = \begin{cases} \sqrt{2}, & 0 \le t < \frac{1}{4} \\ -\sqrt{2}, & \frac{1}{4} \le t < \frac{1}{2} \\ 0, & otherwise \end{cases}$$

$$\phi_3(t) = \begin{cases} \sqrt{2}, & \frac{1}{2} \le t < \frac{3}{4} \\ -\sqrt{2}, & \frac{3}{4} \le t < 1, \\ 0, & otherwise \end{cases}$$

$$f(t) = \begin{cases} 2, & 0 \le t < 0.6 \\ -4, & 0.6 \le t < 1 \\ 0, & otherwise \end{cases}$$

a. Let the approximation be $\hat{f}(t) = a_0\phi_0(t) + a_1\phi_1(t) + a_2\phi_2(t) + a_3\phi_3(t)$. Find a_0, a_1, a_2, a_3 to minimize the energy of the error signal $e(t) = f(t) - \hat{f}(t)$.
b. Plot the approximation $\hat{f}(t) = a_0\phi_0(t) + a_1\phi_1(t) + a_2\phi_2(t) + a_3\phi_3(t)$ for $0 \le t < 1$.
c. Plot the error signal $e(t) = f(t) - \hat{f}(t)$ for $0 \le t < 1$.

18.5 The signals $f(t)$, $\phi_0(t)$, and $\phi_1(t)$ are given by

$$f(t) = \begin{cases} 2t, & 0 \le t < \frac{1}{2} \\ 2 - 2t, & \frac{1}{2} \le t < 1 \\ 0, & otherwise \end{cases}$$

$$\phi_0(t) = \begin{cases} 1, & 0 \le t < 1 \\ 0, & otherwise \end{cases}$$

$$\phi_1(t) = \begin{cases} 1, & 0 \le t < \frac{1}{2} \\ -1, & \frac{1}{2} \le t < 1 \\ 0, & otherwise \end{cases}$$

a. Let $\hat{f}(t) = a_0\phi_0(t) + a_1\phi_1(t)$. Find a_0 and a_1 to minimize the energy of the error signal $e(t) = f(t) - \hat{f}(t)$.
b. Plot the approximation $\hat{f}(t) = a_0\phi_0(t) + + a_1\phi_1(t)$ for $0 \le t < 1$.
c. Plot the error signal $e(t) = f(t) - \hat{f}(t)$ for $0 \le t < 1$.

18.6 The signals $f(t)$, $\phi_0(t)$, and $\phi_1(t)$ are given by

$$f(t) = \begin{cases} \sin(\pi t), & 0 \le t < 1 \\ 0, & otherwise \end{cases}$$

$$\phi_0(t) = \begin{cases} 1, & 0 \le t < 1 \\ 0, & otherwise \end{cases}$$

$$\phi_1(t) = \begin{cases} 1, & 0 \le t < \frac{1}{2} \\ -1, & \frac{1}{2} \le t < 1 \\ 0, & otherwise \end{cases}$$

a. Let $\hat{f}(t) = a_0\phi_0(t) + a_1\phi_1(t)$. Find a_0 and a_1 to minimize the energy of the error signal $e(t) = f(t) - \hat{f}(t)$.
b. Plot the approximation $\hat{f}(t) = a_0\phi_0(t) + a_1\phi_1(t)$ for $0 \le t < 1$.
c. Plot the error signal $e(t) = f(t) - \hat{f}(t)$ for $0 \le t < 1$.

18.7 The signals $f(t)$, $\phi_0(t)$, and $\phi_1(t)$ are given by

$$f(t) = \begin{cases} \cos(\pi t), & 0 \le t < 1 \\ 0, & otherwise \end{cases}$$

$$\phi_0(t) = \begin{cases} 1, & 0 \le t < 1 \\ 0, & otherwise \end{cases}$$

$$\phi_1(t) = \begin{cases} 1, & 0 \le t < \frac{1}{2} \\ -1, & \frac{1}{2} \le t < 1 \\ 0, & otherwise \end{cases}$$

a. Let $\hat{f}(t) = a_0\phi_0(t) + a_1\phi_1(t)$. Find a_0 and a_1 to minimize the energy of the error signal $e(t) = f(t) - \hat{f}(t)$.
b. Plot the approximation $\hat{f}(t) = a_0\phi_0(t) + a_1\phi_1(t)$ for $0 \le t < 1$.
c. Plot the error signal $e(t) = f(t) - \hat{f}(t)$ for $0 \le t < 1$.

18.8 Given

$$f(t) = \begin{cases} 2t, & 0 \le t < \frac{1}{2} \\ 2 - 2t, & \frac{1}{2} \le t < 1 \\ 0, & otherwise \end{cases}$$

$$\phi_0(t) = \begin{cases} 1, & 0 \le t < 1 \\ 0, & otherwise \end{cases}$$

$$\phi_1(t) = \begin{cases} 1, & 0 \le t < \frac{1}{2} \\ -1, & \frac{1}{2} \le t < 1 \\ 0, & otherwise \end{cases}$$

$$\phi_2(t) = \begin{cases} \sqrt{2}, & 0 \le t < \frac{1}{4} \\ -\sqrt{2}, & \frac{1}{4} \le t < \frac{1}{2} \\ 0, & otherwise \end{cases}$$

$$\phi_3(t) = \begin{cases} \sqrt{2}, & \frac{1}{2} \le t < \frac{3}{4} \\ -\sqrt{2}, & \frac{3}{4} \le t < 1 \\ 0, & otherwise \end{cases}$$

a. Let $\hat{f}(t) = a_0\phi_0(t) + a_1\phi_1(t) + a_2\phi_2(t) + a_3\phi_3(t)$. Find a_0, a_1, a_2, a_3 to minimize the energy of the error signal $e(t) = f(t) - \hat{f}(t)$.
b. Plot the approximation $\hat{f}(t) = a_0\phi_0(t) + a_1\phi_1(t) + a_2\phi_2(t) + a_3\phi_3(t)$ for $0 \le t < 1$.
c. Plot the error signal $e(t) = f(t) - \hat{f}(t)$ for $0 \le t < 1$.

18.9 Given

$$f(t) = \begin{cases} \sin(\pi t), & 0 \le t < 1 \\ 0, & otherwise \end{cases}$$

$$\phi_0(t) = \begin{cases} 1, & 0 \le t < 1 \\ 0, & otherwise \end{cases}$$

$$\phi_1(t) = \begin{cases} 1, & 0 \le t < \frac{1}{2} \\ -1, & \frac{1}{2} \le t < 1 \\ 0, & otherwise \end{cases}$$

$$\phi_2(t) = \begin{cases} \sqrt{2}, & 0 \le t < \frac{1}{4} \\ -\sqrt{2}, & \frac{1}{4} \le t < \frac{1}{2} \\ 0, & otherwise \end{cases}$$

$$\phi_3(t) = \begin{cases} \sqrt{2}, & \frac{1}{2} \le t < \frac{3}{4} \\ -\sqrt{2}, & \frac{3}{4} \le t < 1 \\ 0, & otherwise \end{cases}$$

$$\phi_4(t) = \begin{cases} 2, & 0 \le t < \frac{1}{8} \\ -2, & \frac{1}{8} \le t < \frac{1}{4} \\ 0, & otherwise \end{cases}$$

$$\phi_5(t) = \begin{cases} 2, & \frac{1}{4} \le t < \frac{3}{8} \\ -2, & \frac{3}{8} \le t < \frac{1}{2} \\ 0, & otherwise \end{cases}$$

$$\phi_6(t) = \begin{cases} 2, & \frac{1}{2} \le t < \frac{5}{8} \\ -2, & \frac{5}{8} \le t < \frac{3}{4} \\ 0, & otherwise \end{cases}$$

$$\phi_7(t) = \begin{cases} 2, & \frac{3}{4} \le t < \frac{7}{8} \\ -2, & \frac{7}{8} \le t < 1 \\ 0, & otherwise \end{cases}$$

a. Let $\hat{f}(t) = a_0\phi_0(t) + a_1\phi_1(t) + a_2\phi_2(t) + a_3\phi_3(t) + a_4\phi_4(t) + a_5\phi_5(t) + a_6\phi_6(t) + a_7\phi_7(t)$. Find $a_0, a_1, a_2, a_3, a_4, a_5, a_6$, and a_7 to minimize the energy of the error signal $e(t) = f(t) - \hat{f}(t)$.
b. Plot the approximation $\hat{f}(t) = a_0\phi_0(t) + a_1\phi_1(t) + a_2\phi_2(t) + a_3\phi_3(t) + a_4\phi_4(t) + a_5\phi_5(t) + a_6\phi_6(t) + a_7\phi_7(t)$ for $0 \le t < 1$.
c. Plot the error signal $e(t) = f(t) - \hat{f}(t)$ for $0 \le t < 1$.

18.10 Given

$$f(t) = \begin{cases} \cos(\pi t), & 0 \le t < 1 \\ 0, & otherwise \end{cases}$$

$$\phi_0(t) = \begin{cases} 1, & 0 \le t < 1 \\ 0, & otherwise \end{cases}$$

$$\phi_1(t) = \begin{cases} 1, & 0 \le t < \frac{1}{2} \\ -1, & \frac{1}{2} \le t < 1 \\ 0, & otherwise \end{cases}$$

$$\phi_2(t) = \begin{cases} \sqrt{2}, & 0 \le t < \frac{1}{4} \\ -\sqrt{2}, & \frac{1}{4} \le t < \frac{1}{2} \\ 0, & otherwise \end{cases}$$

$$\phi_3(t) = \begin{cases} \sqrt{2}, & \frac{1}{2} \le t < \frac{3}{4} \\ -\sqrt{2}, & \frac{3}{4} \le t < 1 \\ 0, & otherwise \end{cases}$$

$$\phi_4(t) = \begin{cases} 2, & 0 \le t < \frac{1}{8} \\ -2, & \frac{1}{8} \le t < \frac{1}{4} \\ 0, & otherwise \end{cases}$$

$$\phi_5(t) = \begin{cases} 2, & \frac{1}{4} \le t < \frac{3}{8} \\ -2, & \frac{3}{8} \le t < \frac{1}{2} \\ 0, & otherwise \end{cases}$$

$$\phi_6(t) = \begin{cases} 2, & \frac{1}{2} \le t < \frac{5}{8} \\ -2, & \frac{5}{8} \le t < \frac{3}{4} \\ 0, & otherwise \end{cases}$$

$$\phi_7(t) = \begin{cases} 2, & \frac{3}{4} \le t < \frac{7}{8} \\ -2, & \frac{7}{8} \le t < 1 \\ 0, & otherwise \end{cases}$$

a. Let $\hat{f}(t) = a_0\phi_0(t) + a_1\phi_1(t) + a_2\phi_2(t) + a_3\phi_3(t) + a_4\phi_4(t) + a_5\phi_5(t) + a_6\phi_6(t) + a_7\phi_7(t)$. Find $a_0, a_1, a_2, a_3, a_4, a_5, a_6, a_7$ to minimize the energy of the error signal $e(t) = f(t) - \hat{f}(t)$.
b. Plot the approximation $\hat{f}(t) = a_0\phi_0(t) + a_1\phi_1(t) + a_2\phi_2(t) + a_3\phi_3(t) + a_4\phi_4(t) + a_5\phi_5(t) + a_6\phi_6(t) + a_7\phi_7(t)$ for $0 \le t < 1$.
c. Plot the error signal $e(t) = f(t) - \hat{f}(t)$ for $0 \le t < 1$.

Trigonometric Fourier Series

18.11 **A periodic rectangular pulse train $f(t)$, as shown in Figure P8.11, has amplitude $A = 1$ V, pulse width $\tau = 1/3$ ms, and period $T_0 = 1$ ms. The pulse is centered at $t = 0$.**

FIGURE P8.11

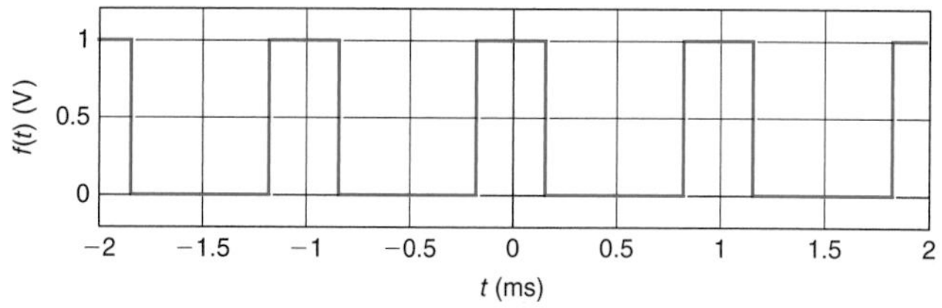

a. Find the fundamental frequency f_0 of this signal.
b. Find the duty cycle d of this signal.
c. Find the expression of the trigonometric Fourier coefficients a_0, a_n, b_n, F_n.
d. Find the numerical values of a_0, and a_n for $1 \le n \le 6$.
e. Plot the one-sided magnitude spectrum for $0 \le f \le 6f_0$.
f. Plot the one-sided phase spectrum for $0 \le f \le 6f_0$.
g. Find the expression of approximation $\hat{f}(t)$ consisting of the dc component, the fundamental component, the second harmonic component, the third harmonic component,

and the fourth harmonic component of its trigonometric Fourier series (sum of cosines).

18.12 **A rectangular pulse train $f(t)$ with period $T_0 = 1$ ms, peak-to-peak amplitude $A = 1$ V, and pulse width $\tau = 0.6$ ms is shown in Figure P18.12.**

FIGURE P18.12

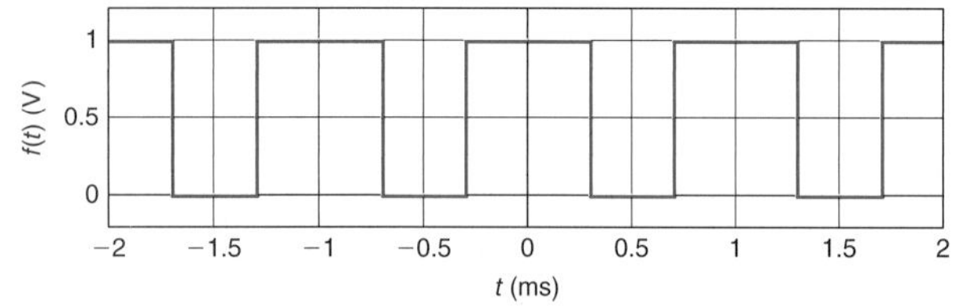

a. Find the fundamental frequency ω_0 in rad/s and f_0 in hertz.
b. Find the duty cycle d.
c. Find the trigonometric Fourier coefficients a_0, a_n, and b_n.
d. Find the trigonometric Fourier coefficients c_n and θ_n.
e. Plot c_n and θ_n as a function of frequency.
f. Find the approximation $\hat{f}(t)$ consisting of the dc component and up to the fifth harmonic, and plot the approximation $\hat{f}(t)$.

18.13 **A rectangular pulse train $f(t)$ with period $T_0 = 1$ ms, peak-to-peak amplitude $A = 2$ V, and pulse width $\tau = 0.5$ ms is shown in Figure P18.13.**

FIGURE P18.13

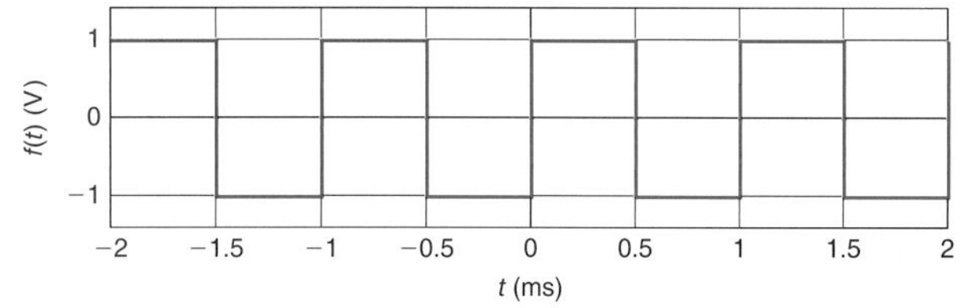

a. Find the fundamental frequency ω_0 in rad/s and f_0 in hertz.
b. Find the duty cycle d.
c. Find the trigonometric Fourier coefficients a_0, a_n, and b_n.
d. Find the trigonometric Fourier coefficients c_n and θ_n.
e. Plot c_n and θ_n as a function of frequency.
f. Find the approximation $\hat{f}(t)$ consisting of the dc component and up to the fifth harmonic, and plot the approximation $\hat{f}(t)$.

18.14 **A rectangular pulse train $f(t)$ with period $T_0 = 1$ ms, peak-to-peak amplitude $A = 1$ V, and pulse width $\tau = 0.25$ ms is shown in Figure P18.14.**

FIGURE P18.14

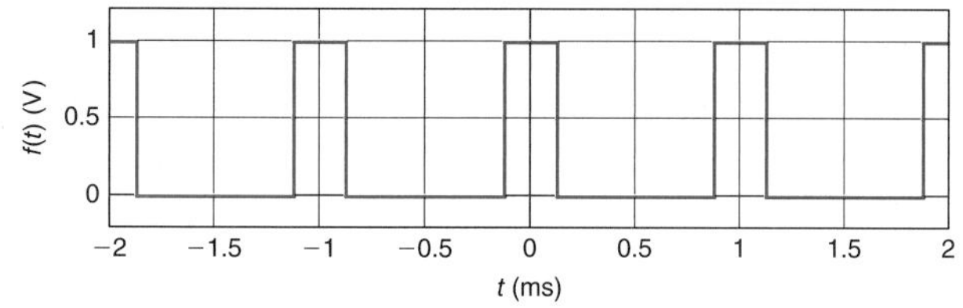

a. Find the fundamental frequency ω_0 in rad/s and f_0 in hertz.
b. Find the duty cycle d.
c. Find the trigonometric Fourier coefficients a_0, a_n, and b_n.
d. Find the trigonometric Fourier coefficients c_n and θ_n.
e. Plot c_n and θ_n as a function of frequency.
f. Find the approximation $\hat{f}(t)$ consisting of the dc component and up to the fifth harmonic, and plot the approximation $\hat{f}(t)$.

18.15 **A rectangular pulse train $f(t)$ with period $T_0 = 1$ ms, peak-to-peak amplitude $A = 2$ V, and pulse width $\tau = 0.25$ ms is shown in Figure P18.15.**

FIGURE P18.15

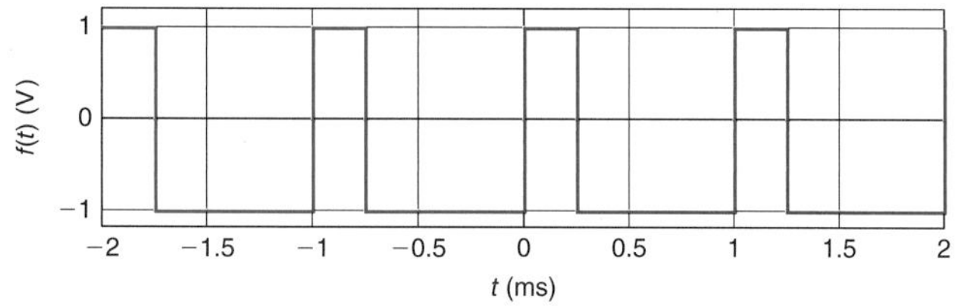

a. Find the fundamental frequency ω_0 in rad/s and f_0 in hertz.
b. Find the duty cycle d.
c. Find the trigonometric Fourier coefficients a_0, a_n, and b_n.
d. Find the trigonometric Fourier coefficients c_n and θ_n.
e. Plot c_n and θ_n as a function of frequency.
f. Find the approximation $\hat{f}(t)$ consisting of the dc component and up to the fifth harmonic, and plot the approximation $\hat{f}(t)$.

18.16 **A rectangular pulse train $f(t)$ with period $T_0 = 1$ ms, peak-to-peak amplitude $A = 1$ V, and pulse width $\tau = 0.2$ ms is shown in Figure P18.16.**

FIGURE P18.16

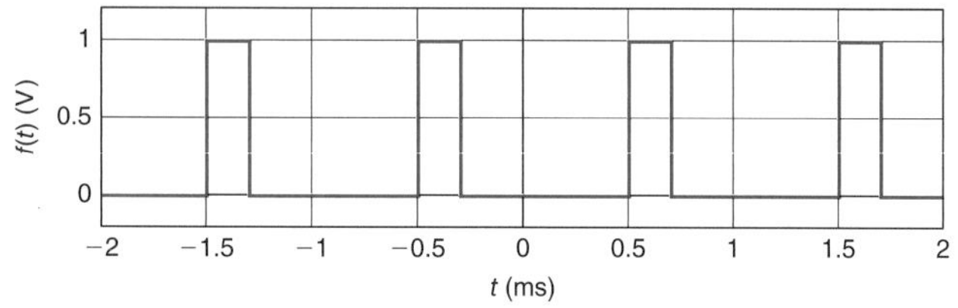

a. Find the fundamental frequency ω_0 in rad/s and f_0 in hertz.
b. Find the duty cycle d.
c. Find the trigonometric Fourier coefficients a_0, a_n, and b_n.
d. Find the trigonometric Fourier coefficients c_n and θ_n.
e. Plot c_n and θ_n as a function of frequency.
f. Find the approximation $\hat{f}(t)$ consisting of the dc component and up to the fifth harmonic, and plot the approximation $\hat{f}(t)$.

18.17 **A rectangular pulse train $f(t)$ with period $T_0 = 1$ ms, peak-to-peak amplitude $A = 2$ V, and pulse width $\tau = 0.2$ ms is shown in Figure P18.17. Notice that for $0 \le t < T_0, f(t)$ is given by**

$$f(t) = \begin{cases} \dfrac{-A}{2}, & 0 \le t < t_0 \\ \dfrac{A}{2}, & t_0 \le t < t_0 + \tau \\ \dfrac{-A}{2}, & t_0 + \tau \le t < T_0 \end{cases}$$

where $t_0 = 0.3$ ms is the time delay.

FIGURE P18.17

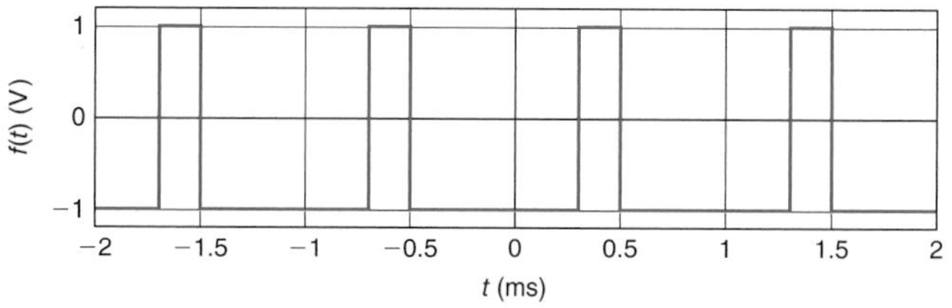

a. Find the fundamental frequency ω_0 in rad/s and f_0 in hertz.
b. Find the duty cycle d.
c. Find the trigonometric Fourier coefficients a_0, a_n, and b_n.
d. Find the trigonometric Fourier coefficients c_n and θ_n.
e. Plot c_n and θ_n as a function of frequency.
f. Find the approximation $\hat{f}(t)$ consisting of the dc component and up to the fifth harmonic, and plot the approximation $\hat{f}(t)$.

18.18 **A periodic rectangular pulse train $f(t)$ is shown in Figure P18.18. The period is $T_0 = 10$ s. Find the trigonometric Fourier coefficients a_0, a_n, and b_n.**

FIGURE P18.18

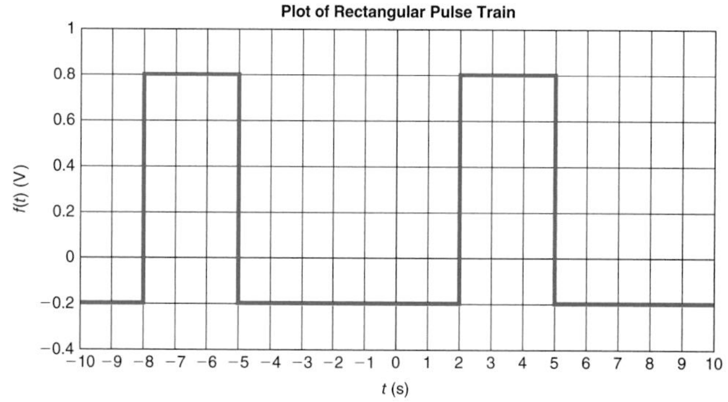

18.19 **A periodic signal $f(t)$ is shown in Figure P18.19. The period is $T_0 = 5$ s. Find the trigonometric Fourier coefficients a_0, a_n, and b_n.**

FIGURE P18.19.

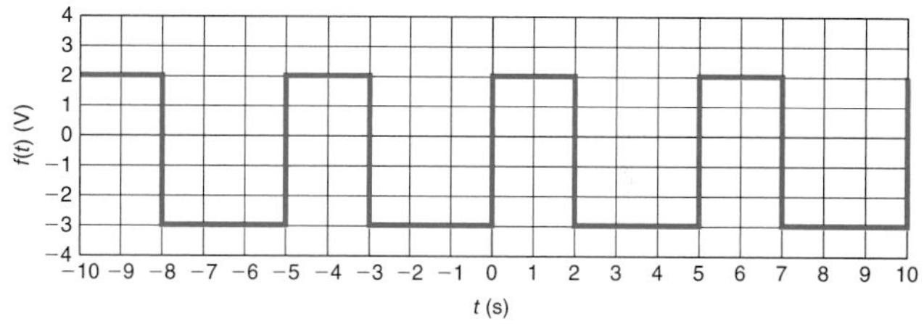

18.20 **A triangular pulse train $f(t)$ with period $T_0 = 100$ μs and peak-to-peak amplitude $A = 1$ V is shown in Figure P18.20.**

FIGURE P18.20

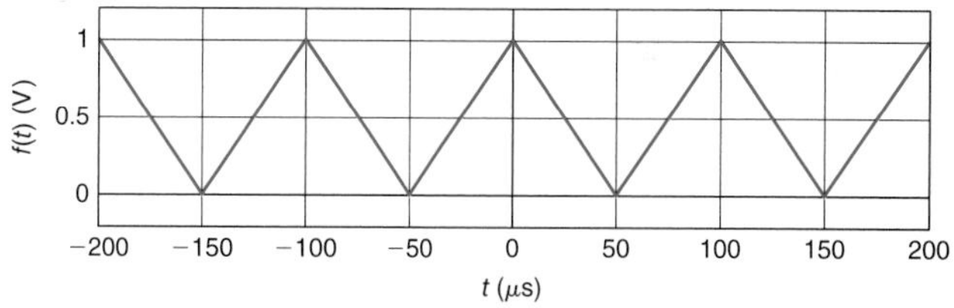

a. Find the fundamental frequency ω_0 in rad/s and f_0 in hertz.
b. Find the trigonometric Fourier coefficients a_0, a_n, and b_n.
c. Find the trigonometric Fourier coefficients c_n and θ_n.
d. Plot c_n and θ_n as a function of frequency.
e. Find the approximation $\hat{f}(t)$ consisting of the dc component and up to the fifth harmonic, and plot the approximation $\hat{f}(t)$.

18.21 **A triangular pulse train $f(t)$ with period $T_0 = 100$ μs and peak-to-peak amplitude $A = 2$ V is shown in Figure P18.21.**

FIGURE P18.21

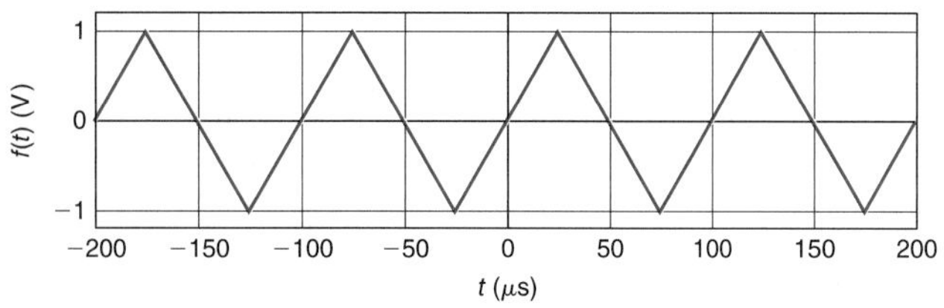

a. Find the fundamental frequency ω_0 in rad/s and f_0 in Hz.
b. Find the trigonometric Fourier coefficients a_0, a_n, and b_n.
c. Find the trigonometric Fourier coefficients c_n and θ_n.
d. Plot c_n and θ_n as a function of frequency.
e. Plot the approximation $\hat{f}(t)$ consisting of the dc component and up to the fifth harmonic, and plot the approximation $\hat{f}(t)$.

18.22 **A triangular pulse train $f(t)$ with period $T_0 = 10\ \mu s$ and peak-to-peak amplitude $A = 1$ V is shown in Figure P18.22.**

FIGURE P18.22

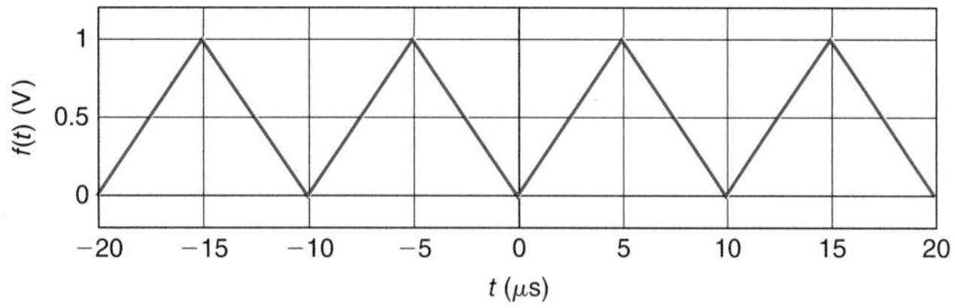

a. Find the fundamental frequency ω_0 in rad/s and f_0 in hertz.
b. Find the trigonometric Fourier coefficients a_0, a_n, and b_n.
c. Find the trigonometric Fourier coefficients c_n and θ_n.
d. Plot c_n and θ_n as a function of frequency.
e. Find the approximation $\hat{f}(t)$ consisting of the dc component and up to the fifth harmonic, and plot the approximation $\hat{f}(t)$.

18.23 **A triangular pulse train $f(t)$ with period $T_0 = 10\ \mu s$ and peak-to-peak amplitude $A = 2$ V is shown in Figure P18.23.**

FIGURE P18.23

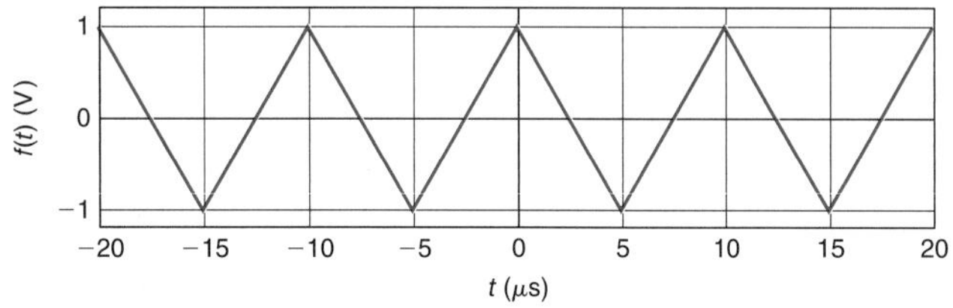

a. Find the fundamental frequency ω_0 in rad/s and f_0 in hertz.
b. Find the trigonometric Fourier coefficients a_0, a_n, and b_n.
c. Find the trigonometric Fourier coefficients c_n and θ_n.
d. Plot c_n and θ_n as a function of frequency.
e. Find the complex exponential Fourier coefficients F_n.
f. Plot the magnitude $|F_n|$ and phase $\angle F_n$ as a function of frequency.
g. Find the approximation $\hat{f}(t)$ consisting of the dc component and up to the fifth harmonic, and plot the approximation $\hat{f}(t)$.

18.24 **A sawtooth pulse train $f(t)$ with period $T_0 = 1\ \mu s$ and peak-to-peak amplitude $A = 1$ V is shown in Figure P18.24.**

FIGURE P18.24

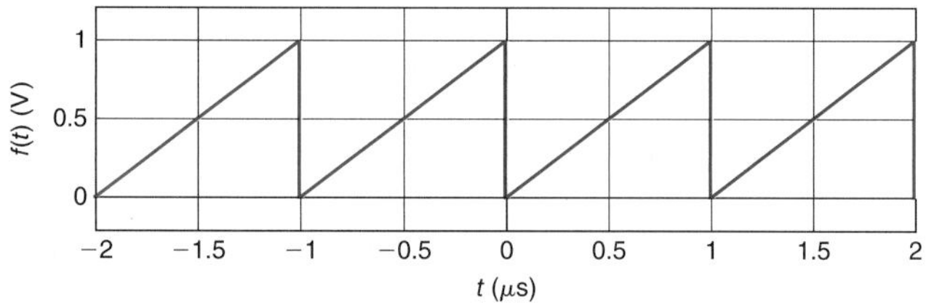

a. Find the fundamental frequency ω_0 in rad/s and f_0 in hertz.
b. Find the trigonometric Fourier coefficients a_0, a_n, and b_n.
c. Find the trigonometric Fourier coefficients c_n and θ_n.
d. Plot c_n and θ_n as a function of frequency.
e. Find the approximation $\hat{f}(t)$ consisting of the dc component and up to the fifth harmonic, and plot the approximation $\hat{f}(t)$.

18.25 **A sawtooth pulse train $f(t)$ with period $T_0 = 1\ \mu s$, peak-to-peak amplitude $A = 2$ V, is shown in Figure P18.25.**

FIGURE P18.25

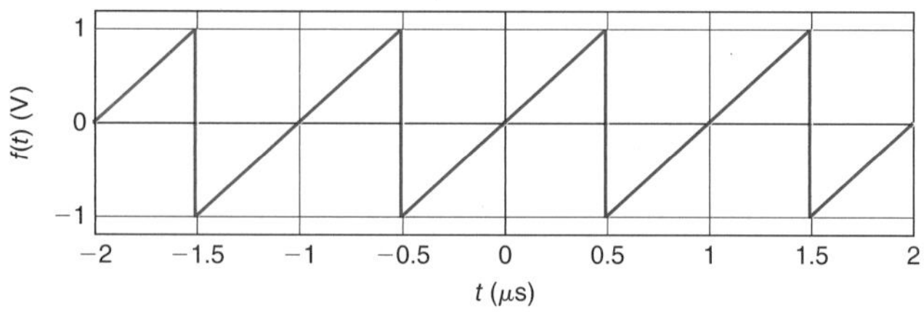

a. Find the fundamental frequency ω_0 in rad/s and f_0 in hertz.
b. Find the trigonometric Fourier coefficients a_0, a_n, and b_n.

c. Find the trigonometric Fourier coefficients c_n and θ_n.
d. Plot c_n and θ_n as a function of frequency.
e. Find the approximation $\hat{f}(t)$ consisting of the dc component and up to the fifth harmonic, and plot the approximation $\hat{f}(t)$.

18.26 **A sawtooth pulse train $f(t)$ with period $T_0 = 1\ \mu s$ and peak-to-peak amplitude $A = 1$ V is shown in Figure P18.26.**

FIGURE P18.26

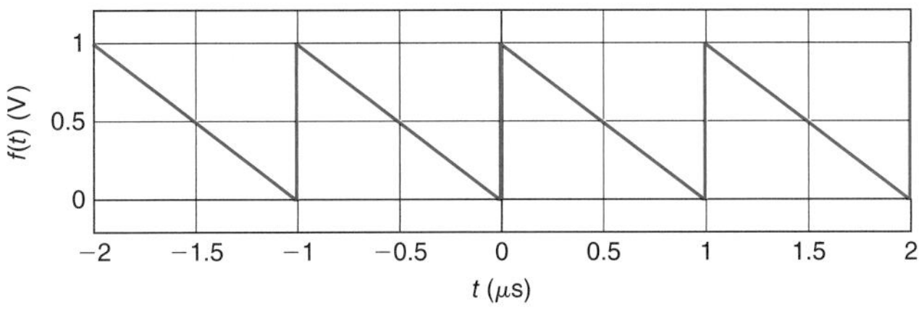

a. Find the fundamental frequency ω_0 in rad/s and f_0 in hertz.
b. Find the trigonometric Fourier coefficients a_0, a_n, and b_n.
c. Find the trigonometric Fourier coefficients c_n and θ_n.
d. Plot c_n and θ_n as a function of frequency.
e. Find the approximation $\hat{f}(t)$ consisting of the dc component and up to the fifth harmonic, and plot the approximation $\hat{f}(t)$.

18.27 **A half-wave rectified sine waveform $f(t)$ with period $T_0 = 1\ \mu s$ and peak-to-peak amplitude $A = 1$ V is shown in Figure P18.27. Notice that the equation of $f(t)$ for $0 \le t < T_0$ is given by $f(t) = A\sin(2\pi t/T_0)$ for $0 \le t < T_0/2$, and 0 for $T_0/2 \le t < T_0$.**

FIGURE P18.27

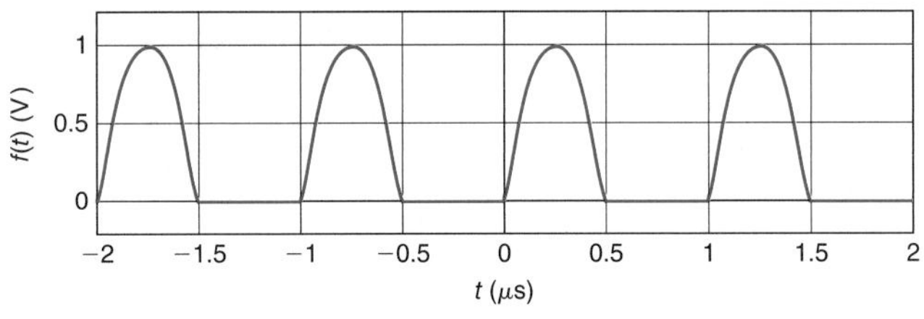

a. Find the fundamental frequency ω_0 in rad/s and f_0 in hertz.
b. Find the trigonometric Fourier coefficients a_0, a_n, and b_n.
c. Find the trigonometric Fourier coefficients c_n and θ_n.
d. Plot c_n and θ_n as a function of frequency.
e. Find the approximation $\hat{f}(t)$ consisting of the dc component and up to the fifth harmonic, and plot the approximation $\hat{f}(t)$.

18.28 **A half-wave rectified cosine waveform $f(t)$ with period $T_0 = 1\ \mu s$ and peak-to-peak amplitude $A = 1$ V is shown in Figure P18.28. Notice that the equation of $f(t)$ for $-T_0/2 \le t < T_0/2$ is given by $f(t) = A\cos(2\pi t/T_0)$ for $-T_0/4 \le t < T_0/4$, and 0 elsewhere.**

FIGURE P18.28

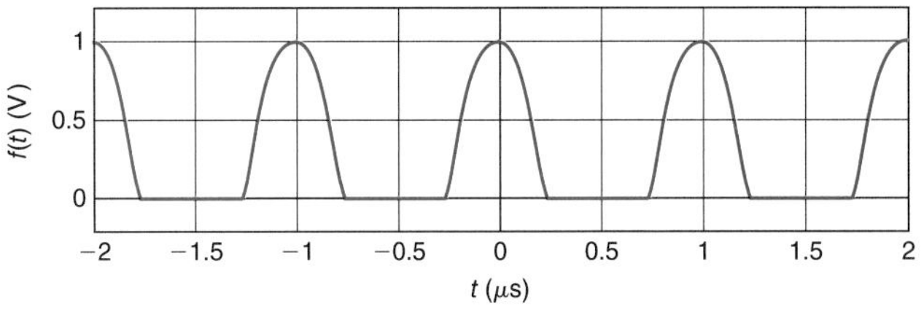

a. Find the fundamental frequency ω_0 in rad/s and f_0 in hertz.
b. Find the trigonometric Fourier coefficients a_0, a_n, and b_n.
c. Find the trigonometric Fourier coefficients c_n and θ_n.
d. Plot c_n and θ_n as a function of frequency.
e. Find the approximation $\hat{f}(t)$ consisting of the dc component and up to the fifth harmonic, and plot the approximation $\hat{f}(t)$.

18.29 **A full-wave rectified sine waveform $f(t)$ with period $T_0 = 1$ ms and peak-to-peak amplitude $A = 1$ V is shown in Figure P18.29. Notice that the equation of $f(t)$ for $0 \le t < T_0$ is given by $f(t) = A\sin(\pi t/T_0)$.**

FIGURE P18.29

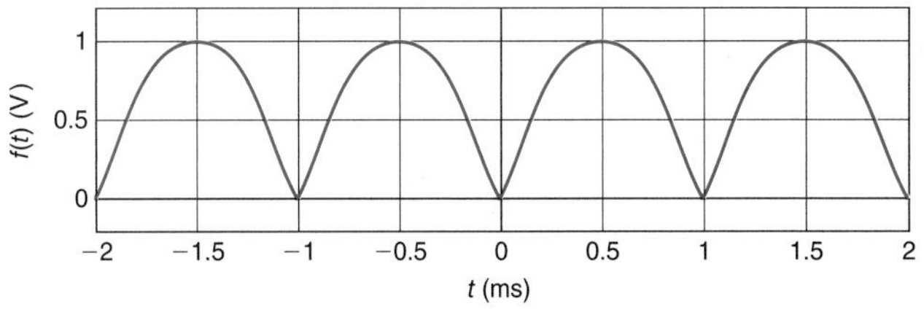

a. Find the fundamental frequency ω_0 in rad/s and f_0 in hertz.
b. Find the trigonometric Fourier coefficients a_0, a_n, and b_n.
c. Find the trigonometric Fourier coefficients c_n and θ_n.
d. Plot c_n and θ_n as a function of frequency.
e. Find the approximation $\hat{f}(t)$ consisting of the dc component and up to the fifth harmonic, and plot the approximation $\hat{f}(t)$.

18.30 **A full-wave rectified cosine waveform $f(t)$ with period $T_0 = 1$ ms and peak-to-peak amplitude $A = 1$ V is shown in Figure P18.30. Notice that**

the equation of $f(t)$ for $-T_0/2 \le t < T_0/2$ is given by $f(t) = A\cos(\pi t/T_0)$.

FIGURE P18.30

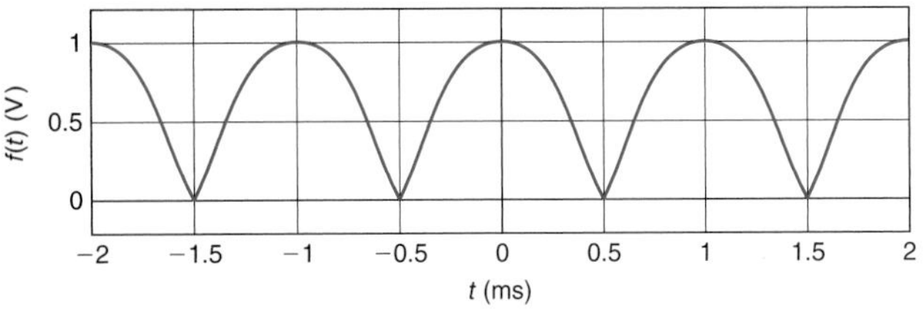

a. Find the fundamental frequency ω_0 in rad/s and f_0 in hertz.
b. Find the trigonometric Fourier coefficients a_0, a_n, and b_n.
c. Find the trigonometric Fourier coefficients c_n and θ_n.
d. Plot c_n and θ_n as a function of frequency.
e. Find the approximation $\hat{f}(t)$ consisting of the dc component and up to the fifth harmonic, and plot the approximation $\hat{f}(t)$.

18.31 **An exponential decay waveform $f(t)$ is shown in Figure P18.31. One period of the waveform is given by**

$$f_1(t) = \begin{cases} Ae^{-Kt}, & 0 \le t < T_0 \\ 0, & otherwise \end{cases}$$

Assume that $A = 1$ V, $K = 2$, and $T_0 = 1$ s.

FIGURE P18.31

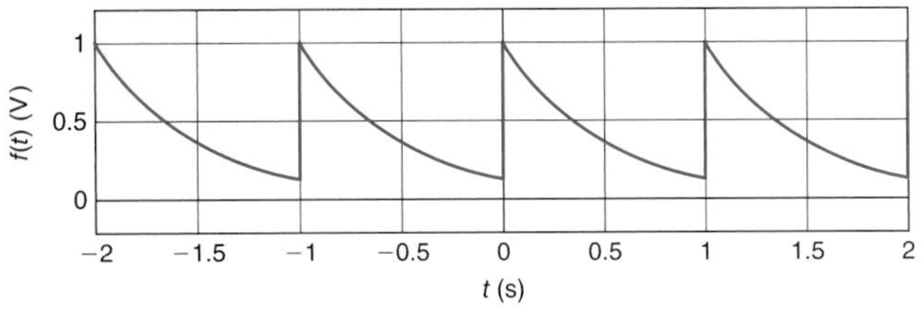

a. Find the fundamental frequency ω_0 in rad/s and f_0 in hertz.
b. Find the trigonometric Fourier coefficients a_0, a_n, and b_n.
c. Find the trigonometric Fourier coefficients c_n and θ_n.
d. Plot c_n and θ_n as a function of frequency.
e. Find the approximation $\hat{f}(t)$ consisting of the dc component and up to the fifth harmonic, and plot the approximation $\hat{f}(t)$.

18.32 **A waveform $f(t)$ with period $T_0 = 2$ ms and peak-to-peak amplitude $A = 1$ V is shown in Figure P18.32. Notice that the equation of $f(t)$ for $0 \le t < T_0$ is given by $f(t) = At/T_1$ for $0 \le t < T_1$, and 0 for $T_1 \le t < T_0$.**

FIGURE P18.32

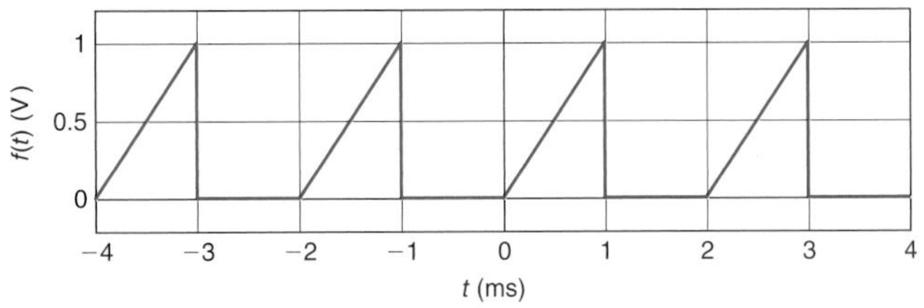

Assume that $T_0 = 2$ ms, $T_1 = 1$ ms.

a. Find the fundamental frequency ω_0 in rad/s and f_0 in hertz.
b. Find the trigonometric Fourier coefficients a_0, a_n, and b_n.
c. Find the trigonometric Fourier coefficients c_n and θ_n.
d. Plot c_n and θ_n as a function of frequency.
e. Find the approximation $\hat{f}(t)$ consisting of the dc component and up to the fifth harmonic, and plot the approximation $\hat{f}(t)$.

Solving Circuit Problems Using Trigonometric Fourier Series

18.33 **A rectangular pulse train with amplitude 1 V (High = 1 V, Low = 0 V), period (T_0) 1 ms, pulse width 0.25 ms, and delay 0.125 ms, as shown in Figure P18.33(a), is applied to a series RC circuit with $R = 800\ \Omega$, $C = 0.1\ \mu$F shown in Figure P18.33(b). Let $f(t)$ be the rectangular pulse train, and let $y(t)$ be the voltage across the capacitor.**

FIGURE P18.33

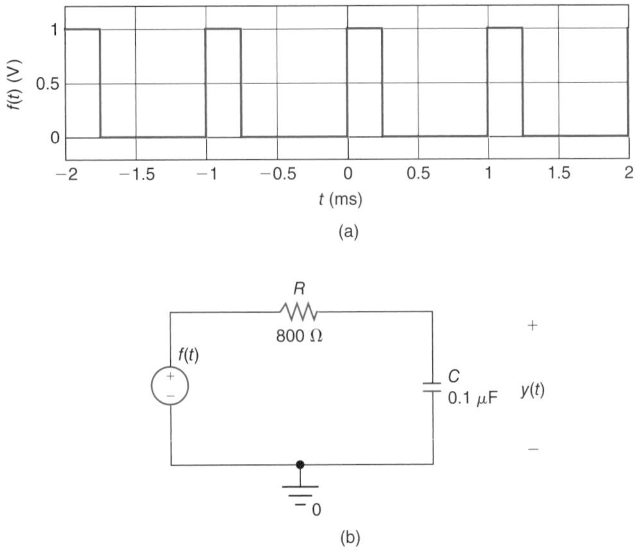

a. Find the Fourier coefficients of the rectangular pulse train; that is, find c_0, c_n, and θ_n.
b. Find the transfer function $H(\omega) = Y(\omega)/F(\omega)$ of the circuit, and also find $H(n\omega_0)$ where $\omega_0 = 2\pi/T_0$.
c. Find the Fourier coefficients of the output signal; that is, find yc_0, yc_n, and $y\theta_n$.
d. Represent $y(t)$ by its trigonometric Fourier series.
e. Approximate $y(t)$ by adding the dc component and up to the fifth harmonic and plot.

18.34 **A rectangular pulse train with amplitude 1 V (High = 1 V, Low = 0 V), period (T_0) 1 ms, pulse width 0.5 ms, and delay 0.25 ms, as shown in Figure P18.34(a), is applied to the series CR circuit with $R = 1\ k\Omega$, $C = 0.05\ \mu F$ shown in Figure P18.34(b). Let $f(t)$ be the rectangular pulse train, and let $y(t)$ be the voltage across the resistor.**

FIGURE P18.34

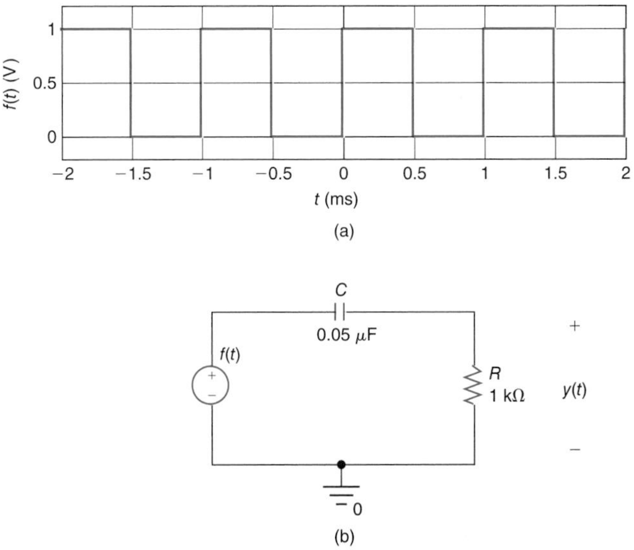

(a)

(b)

a. Find the Fourier coefficients of the rectangular pulse train; that is, find c_0, c_n, and θ_n.
b. Find the transfer function $H(\omega) = Y(\omega)/F(\omega)$ of the circuit, and also find $H(n\omega_0)$ where $\omega_0 = 2\pi/T_0$.
c. Find the Fourier coefficients of the output signal; that is, find yc_0, yc_n, and $y\theta_n$.
d. Approximate $y(t)$ by adding the dc component and up to the fifth harmonic and plot.

18.35 **A sawtooth waveform with amplitude 1 V (High = 1 V, Low = 0 V), period (T_0) 1 ms, as shown in Figure P18.35(a), is applied to the series *RLC* circuit shown in Figure P18.35(b), with $R = 1\ k\Omega$, $L = 20$ mH, and $C = 0.1\ \mu F$. Let $f(t)$ be the sawtooth waveform, and let $y(t)$ be the voltage across the capacitor.**

FIGURE P18.35

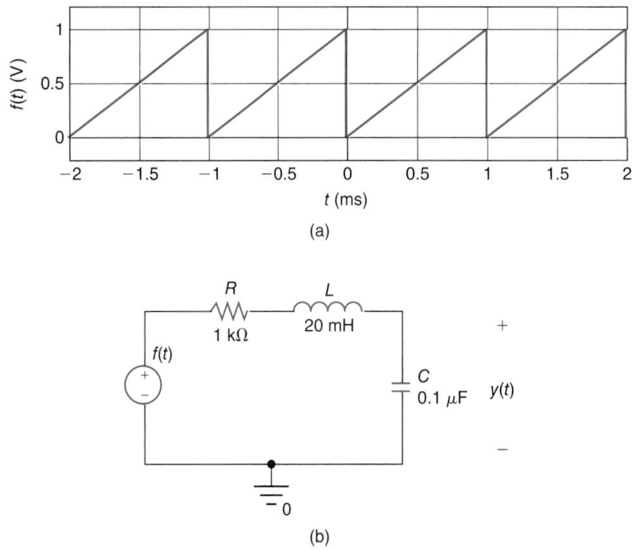

(a)

(b)

a. Find the Fourier coefficients of the sawtooth waveform; that is, find c_0, c_n, and θ_n.
b. Find the transfer function $H(\omega) = Y(\omega)/F(\omega)$ of the circuit, and also find $H(n\omega_0)$ where $\omega_0 = 2\pi/T_0$.
c. Find the Fourier coefficients of the output signal; that is, find yc_0, yc_n, and $y\theta_n$.
d. Approximate $y(t)$ by adding the dc component and up to the fifth harmonic and plot.

18.36 **A sawtooth waveform with amplitude 1 V (High = 1 V, Low = 0 V), period (T_0) 1 ms, as shown in Figure P18.36(a), is applied to the series *RLC* circuit shown in Figure P18.36(b), with $R = 20\ \Omega$, $L = 28$ mH, $C = 0.1\ \mu F$. Let $f(t)$ be the sawtooth waveform, and let $y(t)$ be the voltage across the resistor.**

a. Find the Fourier coefficients of the sawtooth waveform; that is, find c_0, c_n, and θ_n.
b. Find the transfer function $H(\omega) = Y(\omega)/F(\omega)$ of the circuit, and also find $H(n\omega_0)$ where $\omega_0 = 2\pi/T_0$.
c. Find the Fourier coefficients of the output signal; that is, find yc_0, yc_n, and $y\theta_n$.
d. Approximate $y(t)$ by adding the dc component and up to the fifth harmonic and plot.

FIGURE P18.36

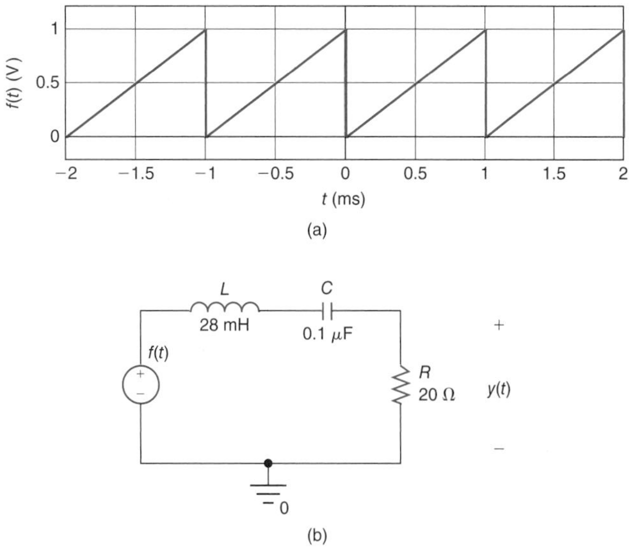

18.37 **A sawtooth waveform with amplitude 1 V (High = 1 V, Low = 0 V), period (T_0) 1 ms, as shown in Figure P18.37(a), is applied to the series *RLC* circuit shown in Figure P18.37(b), with $R = 1\ k\Omega$, $L = 20$ mH, $C = 0.1\ \mu$F. Let $f(t)$ be the sawtooth waveform, and let $y(t)$ be the voltage across the inductor.**

FIGURE P18.37

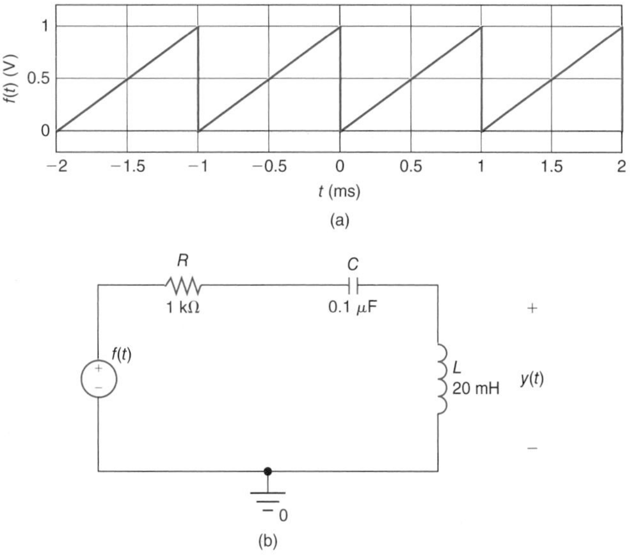

a. Find the Fourier coefficients of the sawtooth waveform; that is, find c_0, c_n, and θ_n.
b. Find the transfer function $H(\omega) = Y(\omega)/F(\omega)$ of the circuit, and also find $H(n\omega_0)$ where $\omega_0 = 2\pi/T_0$.
c. Find the Fourier coefficients of the output signal; that is, find yc_0, yc_n, and $y\theta_n$.
d. Approximate $y(t)$ by adding the dc component and up to the fifth harmonic and plot.

18.38 **A full-wave rectified cosine waveform with amplitude 1 V (High = 1 V, Low = 0 V), period (T_0) 1 ms, as shown in Figure P18.38(a), is applied to the circuit shown in Figure P18.38(b). Let $f(t)$ be the full-wave rectified cosine waveform, and let $y(t)$ be the voltage at the output of the operational amp.**

FIGURE P18.38

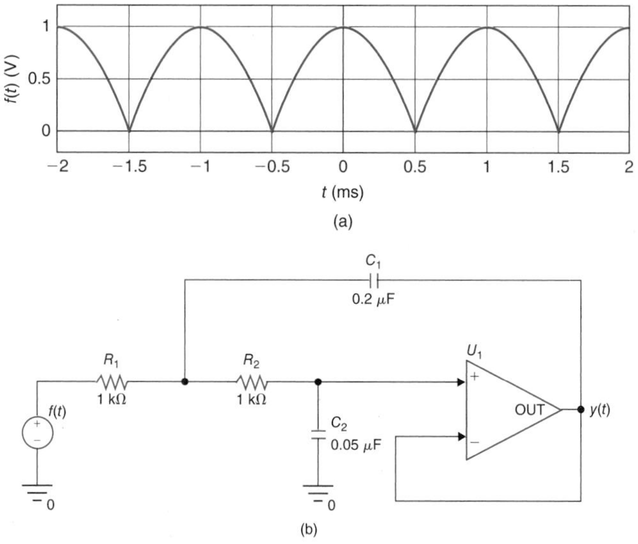

a. Find the Fourier coefficients of the full-wave rectified cosine waveform; that is, find c_0, c_n, and θ_n.
b. Find the transfer function $H(\omega) = Y(\omega)/F(\omega)$ of the circuit, and also find $H(n\omega_0)$ where $\omega_0 = 2\pi/T_0$.
c. Find the Fourier coefficients of the output signal; that is, find yc_0, yc_n, and $y\theta_n$.
d. Approximate $y(t)$ by adding the dc component and up to the fifth harmonic and plot.

18.39 **A half-wave rectified cosine waveform with amplitude 1 V (High = 1 V, Low = 0 V), period (T_0) 1 ms, as shown in Figure P18.39(a), is applied to the circuit shown in Figure P18.39(b). Let $f(t)$ be the full-wave rectified cosine waveform, and let $y(t)$ be the voltage at the output of the operational amp.**

a. Find the Fourier coefficients of the half-wave rectified cosine waveform, that is, find c_0, c_n, and θ_n.
b. Find the transfer function $H(\omega) = Y(\omega)/F(\omega)$ of the circuit, and also find $H(n\omega_0)$ where $\omega_0 = 2\pi/T_0$.
c. Find the Fourier coefficients of the output signal; that is, find yc_0, yc_n, and $y\theta_n$.
d. Approximate $y(t)$ by adding the dc component and up to the fifth harmonic and plot.

FIGURE P18.39

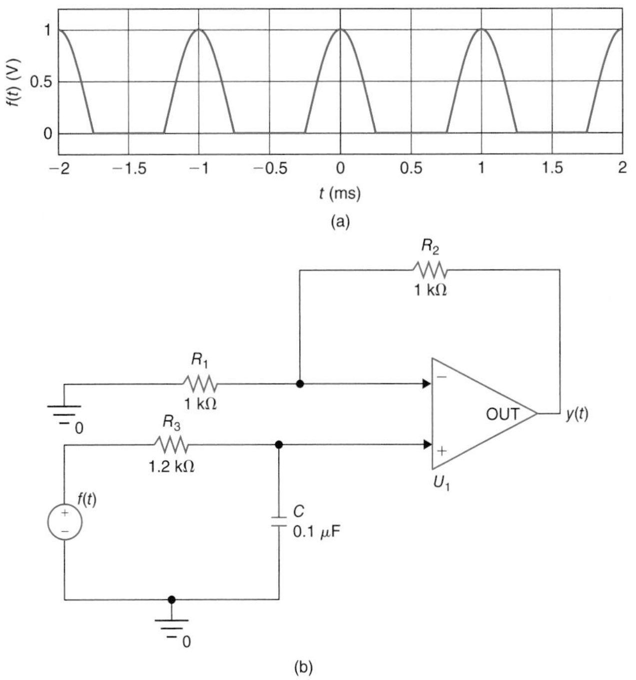

Exponential Fourier Series

18.40 A periodic rectangular pulse train $f(t)$ is shown in Figure P18.40. The period is $T_0 = 1\ \mu s$, the pulse width is $\tau = 0.75\ \mu s$, the peak-to-peak amplitude is 4 V (max $= -2$ V, min $= -6$ V), and the center of the pulses are at $-0.125\ \mu s \pm nT_0$ s ($n =$ integer). Find the expression of the exponential Fourier coefficients F_n.

FIGURE P18.40

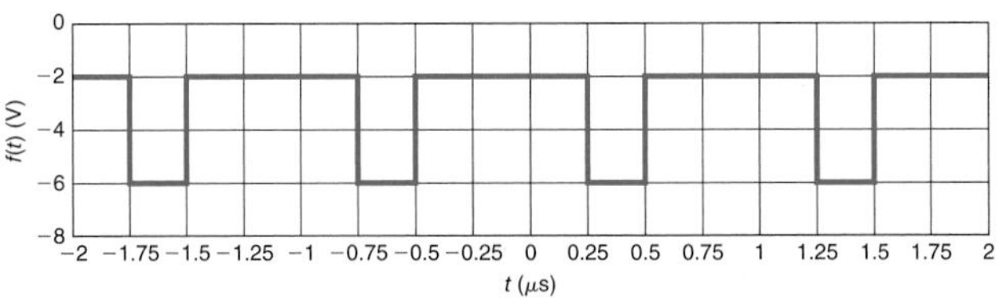

18.41 A periodic rectangular pulse train $f(t)$ is shown in Figure P18.41. The period is $T_0 = 2\ \mu s$, the pulse width is $\tau = 0.8\ \mu s$, the peak-to-peak amplitude is 1 V (max $= 1$ V, min $= 0$ V), and the center of the pulses are at $1.4\ \mu s \pm nT_0$ s ($n =$ integer). Find the expression of the exponential Fourier coefficients F_n.

FIGURE P18.41

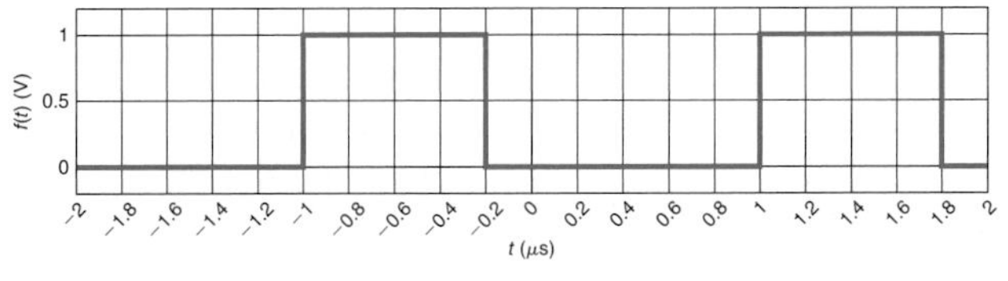

18.42 A periodic rectangular pulse train $f(t)$ is shown in Figure P18.42. The period is $T_0 = 1$ ms, the pulse width is $\tau = 0.4$ ms, the peak-to-peak amplitude is 1 V (max $= 1$ V, min $= 0$ V), and the center of the pulses are at 0.2 ms $\pm nT_0$ s ($n =$ integer). Find the expression of the exponential Fourier coefficients F_n.

FIGURE P18.42

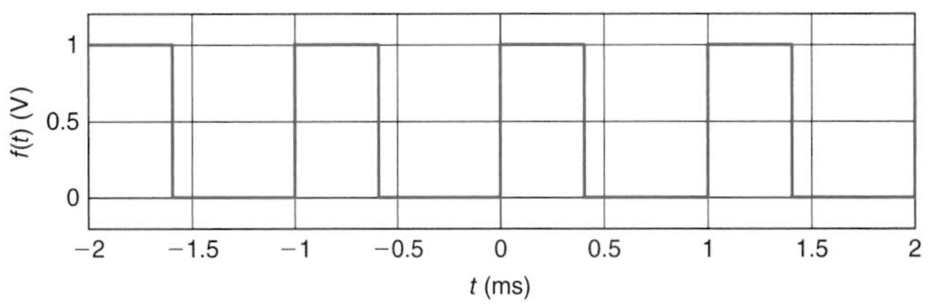

18.43 Find the exponential Fourier coefficients F_n for the signal shown in Figure P18.43.

FIGURE P18.43

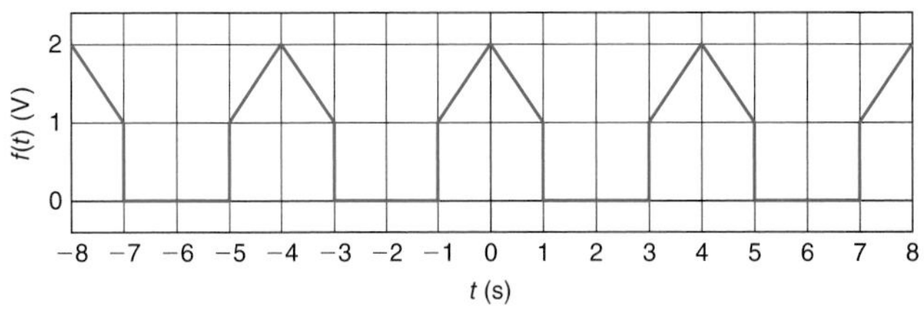

18.44 Find the exponential Fourier coefficients F_n for the signal shown in Figure P18.44.

FIGURE P18.44

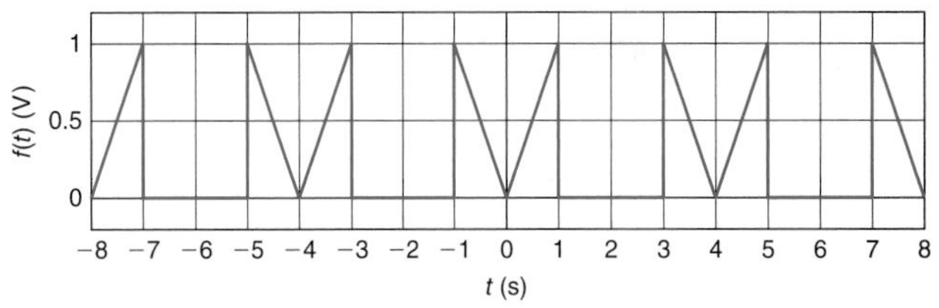

18.45 Find the exponential Fourier coefficients F_n for the signal shown in Figure P18.45.

FIGURE P18.45

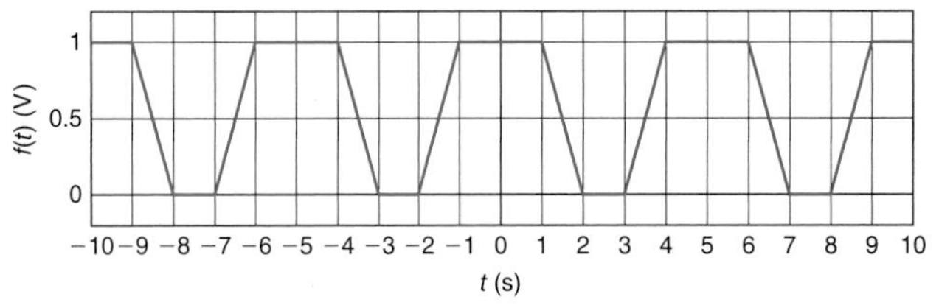

18.46 Find the exponential Fourier coefficients F_n for the signal shown in Figure P18.46.

FIGURE P18.46

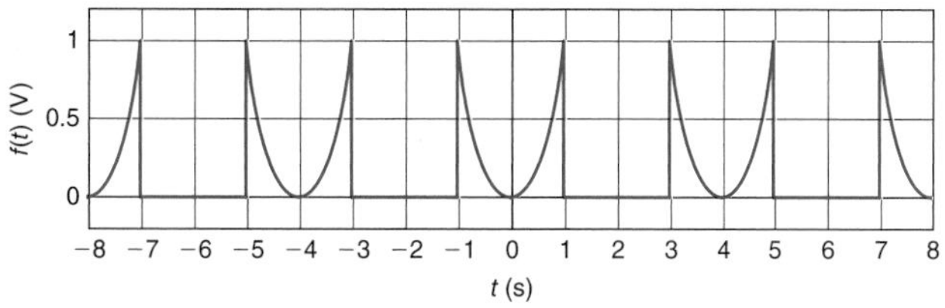

18.47 **Find the exponential Fourier coefficients F_n for the signal shown in Figure P18.47.**

$$f(t) = 1 - t^2 \text{ and } |t| \le 1$$

FIGURE P18.47

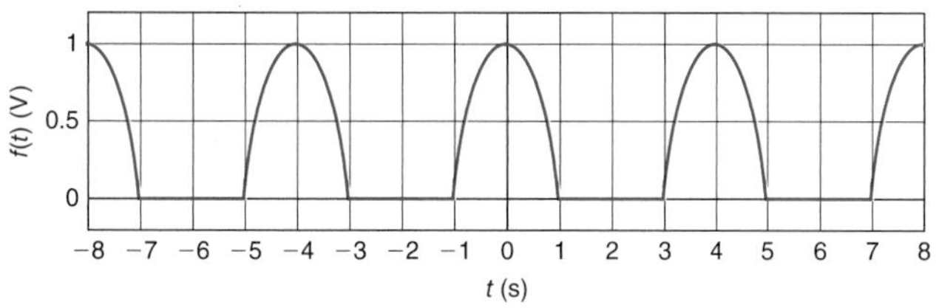

18.48 **A double exponential decay waveform $f(t)$ with period $T_0 = 1$s, amplitude $A = 1$ V, is shown in Figure P18.48. One period of a periodic signal $f(t)$ is given by**

$$f(t) = Ae^{-K|t|}, \quad -\frac{T_0}{2} \le t < \frac{T_0}{2}$$

Assume that $K = 4$. Find the exponential Fourier coefficients F_n.

FIGURE P18.48

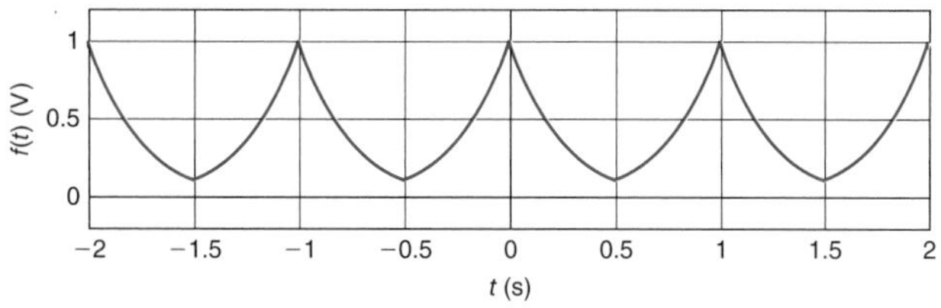

18.49 **A periodic rectangular pulse train $f(t)$ is shown in Figure P18.49. Find the expression of the exponential Fourier coefficients F_n. Plot the two-sided magnitude spectrum and the two-sided phase spectrum for $-10f_0 \le f \le 10f_0$. Find the approximation $\hat{f}(t)$ consisting of the dc component and up to the fifth harmonic, and plot the approximation $\hat{f}(t)$.**

FIGURE P18.49

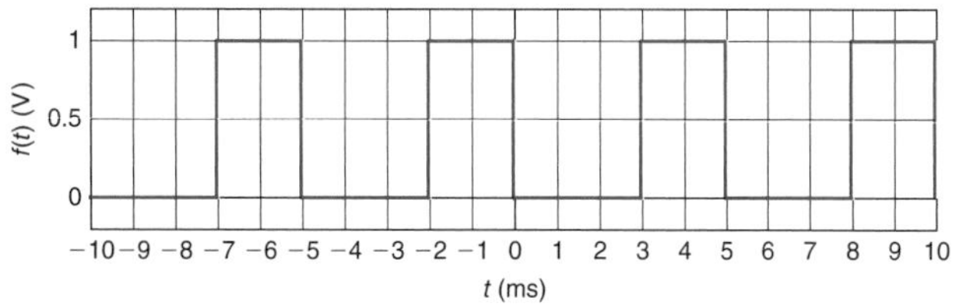

18.50 **A periodic triangular pulse train $f(t)$ is shown in Figure P18.50. Find the expression of the exponential Fourier coefficients F_n. Plot the two-sided magnitude spectrum and the two-sided phase spectrum for $-10f_0 \le f \le 10f_0$. Find the approximation $\hat{f}(t)$ consisting of the dc component and up to the fifth harmonic, and plot the approximation $\hat{f}(t)$.**

FIGURE P18.50

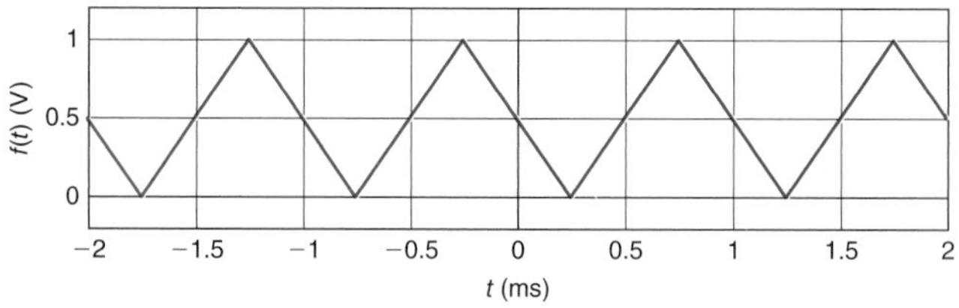

Solving Circuit Problems Using Exponential Fourier Series

18.51 **A rectangular pulse train with amplitude 1 V (High = 1 V, Low = 0 V), period (T_0) 1 ms, and pulse width 0.5 ms, as shown in Figure P18.51(a), is applied to the LR circuit shown in Figure P18.51(b), with $L = 60$ mH, $R = 1000\ \Omega$. Let $f(t)$ be the rectangular pulse train, and let $y(t)$ be the voltage across the resistor.**

FIGURE P18.51
(a) Rectangular pulse train.
(b) *LR* circuit.

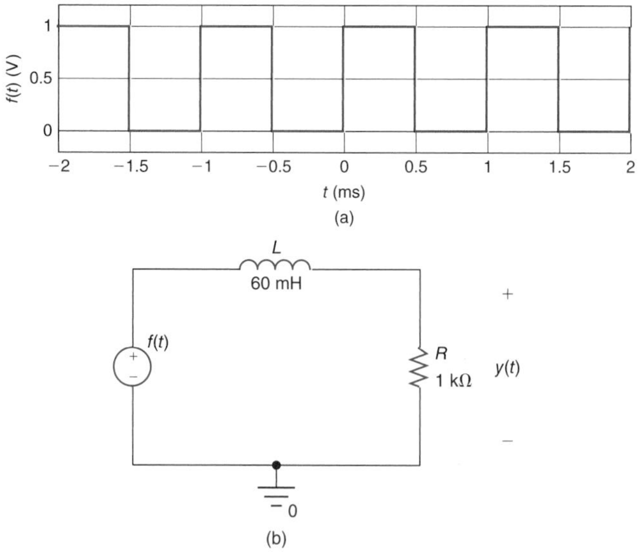

a. Find the exponential Fourier coefficients F_n of the rectangular pulse train.
b. Find the transfer function $H(\omega) = Y(\omega)/F(\omega)$, and also find $H(n\omega_0)$ where $\omega_0 = 2\pi/T_0$.
c. Find the exponential Fourier coefficients of the output signal; that is, find Y_n.

d. Approximate $y(t)$ by the dc component and the first five harmonics. Plot $y(t)$.

18.52 A full-wave rectified sine wave with amplitude 1 V and period 1 ms, as shown in Figure P18.52(a), is applied to the *RL* circuit shown in Figure P18.52(b), with $L = 50$ mH, $R = 1000\ \Omega$. Let $f(t)$ be the full-wave rectified sine wave, and let $y(t)$ be the voltage across the inductor.

FIGURE P18.52

(a) Full-wave rectified sine wave.
(b) *RL* circuit.

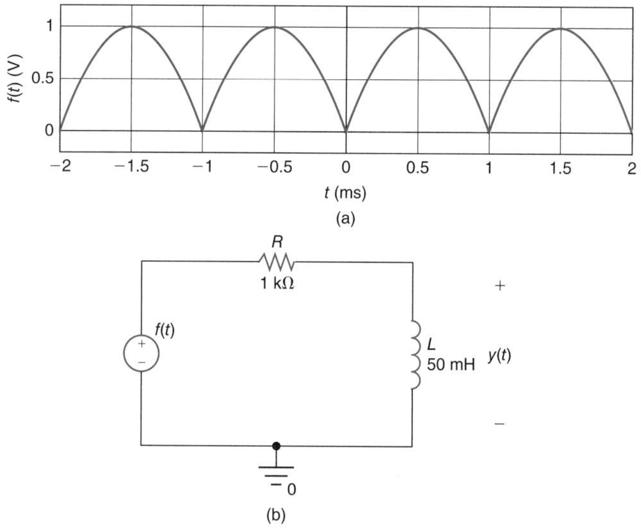

a. Find the exponential Fourier coefficients F_n of the full-wave rectified sine wave.
b. Find the transfer function $H(\omega) = Y(\omega)/F(\omega)$, and also find $H(n\omega_0)$ where $\omega_0 = 2\pi/T_0$.
c. Find the exponential Fourier coefficients of the output signal; that is, find Y_n.
d. Approximate $y(t)$ by the dc component and the first seven harmonics.

18.53 A sawtooth wave with amplitude 1 V and period 1 ms, as shown in Figure P18.53(a), is applied to the parallel *RLC* circuit shown in Figure P18.53(b) with $R = 100\ k\Omega$, $L = 70.3$ mH, C = 0.01 μF. Let $f(t)$ be the sawtooth wave, and let $y(t)$ be the voltage across the inductor.

a. Find the exponential Fourier coefficients F_n of the sawtooth wave.
b. Find the transfer function $H(\omega) = Y(\omega)/F(\omega)$, and also find $H(n\omega_0)$ where $\omega_0 = 2\pi/T_0$.
c. Find the exponential Fourier coefficients of the output signal; that is, find Y_n.
d. Approximate $y(t)$ by the dc component and the first two harmonics. Plot $y(t)$, including dc and the first ten harmonics.

FIGURE P18.53

(a) Sawtooth wave.
(b) Parallel *RLC* circuit.

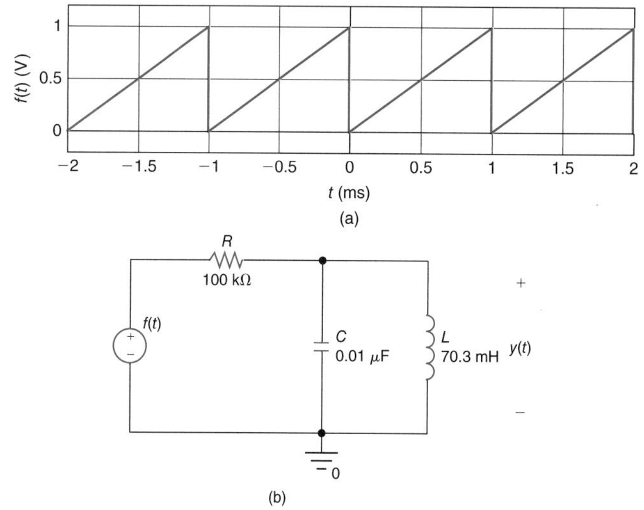

18.54 A full-wave rectified cosine wave with amplitude 1 V and period 1 ms, as shown in Figure P18.54(a), is applied to the parallel *RLC* circuit shown in Figure P18.54(b), with $R = 10\ k\Omega$, $L = 500$ mH, $C = 1\ \mu$F. Let $f(t)$ be the full-wave rectified cosine wave, and let $y(t)$ be the voltage across the capacitor.

FIGURE P18.54

(a) Full-wave rectified sine wave.
(b) Parallel *RLC* circuit.

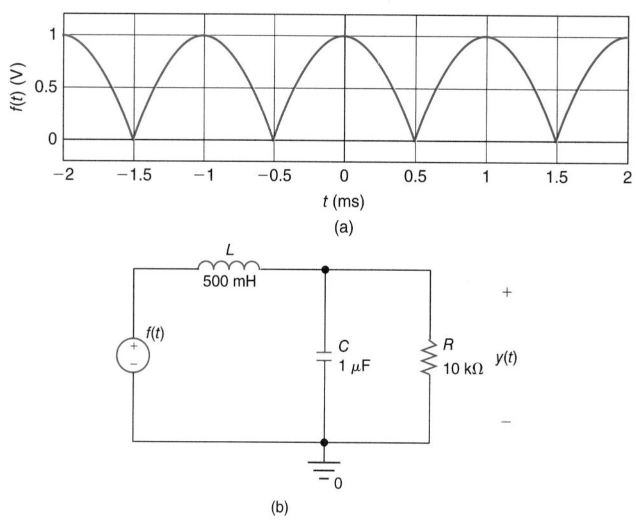

a. Find the exponential Fourier coefficients F_n of the full-wave rectified cosine wave.

b. Find the transfer function $H(\omega) = Y(\omega)/F(\omega)$, and also find $H(n\omega_0)$ where $\omega_0 = 2\pi/T_0$.
c. Find the exponential Fourier coefficients of the output signal; that is, find Y_n.
d. Approximate $y(t)$ by the dc component and the first two harmonics. Plot $y(t)$, including dc and the first ten harmonics.

18.55 **A triangular pulse train with amplitude 1 V and period 1 ms, as shown in Figure P18.55(a), is applied to the parallel *RLC* circuit shown in Figure P18.53(b), with $R = 10\ k\Omega$, $L = 253$ mH, $C = 0.1\ \mu$F. Let $f(t)$ be the triangular pulse train, and let $y(t)$ be the voltage across the resistor.**

FIGURE P18.55

(a) Triangular pulse train.
(b) Parallel *RLC* circuit.

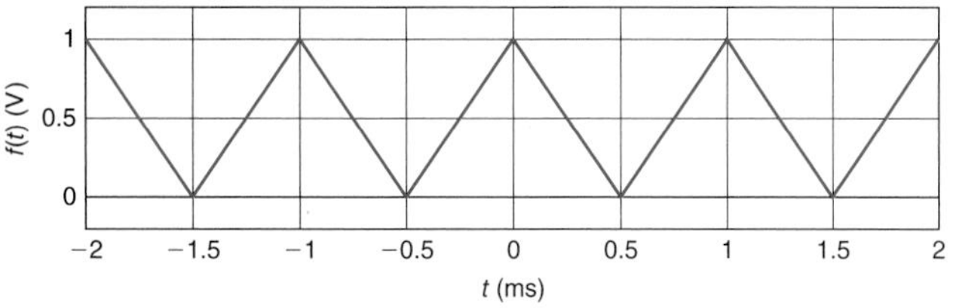

(a)

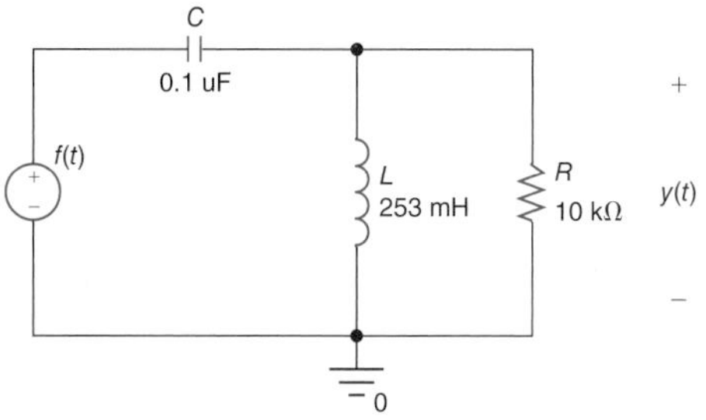

(b)

a. Find the exponential Fourier coefficients F_n of the triangular pulse train.
b. Find the transfer function $H(\omega) = Y(\omega)/F(\omega)$, and also find $H(n\omega_0)$ where $\omega_0 = 2\pi/T_0$.
c. Find the exponential Fourier coefficients of the output signal; that is, find Y_n.
d. Approximate $y(t)$ by the dc component and the first two harmonics. Plot $y(t)$, including dc and the first ten harmonics.

Chapter 19

Fourier Transform

19.1 Introduction

The Fourier series is a representation of a signal as a sum of the dc component and sinusoids with frequencies that are harmonics to the frequency of the fundamental component. The frequency of the fundamental component is inversely related to the period of the signal. The Fourier series is primarily used for periodic signals. If the period of the periodic signal is increased to infinity, the periodic signal becomes a nonperiodic signal, but the Fourier coefficients will be zero due to division by the period in the evaluation of the coefficients. To get around this problem, we can multiply the period to the Fourier coefficients and then increase the period to infinity. As the period is increased to infinity, the distance between the spectral lines is decreased to zero, resulting in a continuous spectrum instead of discrete spectral lines.

In this chapter, the definition of the Fourier transform is discussed along with the various properties of the Fourier transform. Properties of the Fourier transform include time-shifting property, frequency-shifting property, modulation property, duality property, multiplication property, and convolution property. These properties provide tools to analyze signals and systems in the frequency domain.

19.2 Definition of Fourier Transform

Let us consider a rectangular pulse train shown in Figure 19.1.

FIGURE 19.1

A rectangular pulse train.

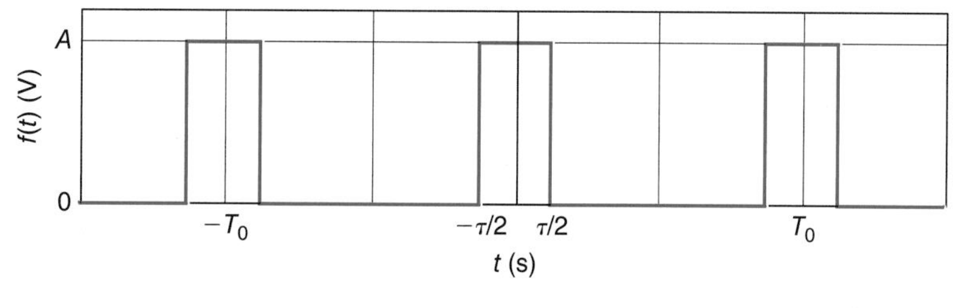

As shown in Chapter 18, the complex exponential Fourier coefficients of this pulse train are given by

$$F_n = Ad\,\text{sinc}(nd) = \frac{A\tau}{T_0}\text{sinc}\left(\frac{n\tau}{T_0}\right) = \frac{A\tau}{T_0}\text{sinc}\left(\frac{n\frac{2\pi}{T_0}\tau}{2\pi}\right) = \frac{A\tau}{T_0}\text{sinc}\left(\frac{n\omega_0\tau}{2\pi}\right) \quad \textbf{(19.1)}$$

where

$$\text{sinc}(x) = \frac{\sin(\pi x)}{\pi x} \quad \textbf{(19.2)}$$

and

$$\omega_0 = \frac{2\pi}{T_0} \quad \textbf{(19.3)}$$

is the fundamental frequency of the pulse train in radians per second. The Fourier coefficients are plotted in Figure 19.2.

FIGURE 19.2

Fourier coefficients of a rectangular pulse train.

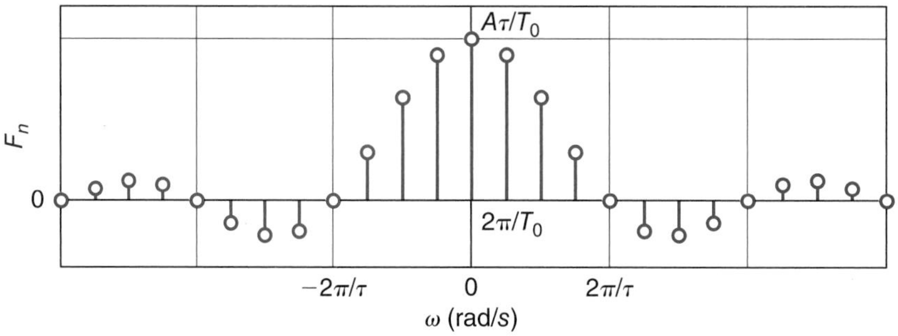

As shown in Chapter 18, the spacing between the spectral lines is $2\pi/T_0$, and the zero-crossing points occur at integer (except 0) multiples of $2\pi/\tau$. The discrete spectral lines can be viewed as samples of a continuous curve

$$E(\omega) = \frac{A\tau}{T_0}\text{sinc}\left(\frac{\omega\tau}{2\pi}\right) \quad \textbf{(19.4)}$$

called the **envelope** at an interval of ω_0. With amplitude A and pulse width τ fixed, as the period T_0 is increased to infinity, the spacing between the spectral lines $\omega_0 = 2\pi/T_0$ is decreased to 0, and the discrete spectral lines merge together, but because of $A\tau/T_0$ in front of the sinc function, the amplitude is decreased to 0, and F_n approaches 0 for all n. In the time domain, as the period T_0 is increased to infinity, a periodic pulse train reduces to a single pulse. Since the amplitude of the spectrum is so small, a frequency domain description of the signal is impossible if F_n is used. To get around this problem, we multiply T_0 on both sides of Equation (19.1) to get

$$F_n T_0 = A\tau\,\text{sinc}\left(\frac{n\omega_0\tau}{2\pi}\right) \quad \textbf{(19.5)}$$

In this case, as the period T_0 is increased to infinity, the spacing between the spectral lines $2\pi/T_0$ decreases to 0, but the envelope $T_0\,E(\omega) = A\tau\,\text{sinc}(\omega\tau/2\pi)$ does not change. The discrete spectral lines are merged to form a continuous spectrum $A\tau\,\text{sinc}(\omega\tau/2\pi)$. This continuous spectrum is defined to be the Fourier transform of a single pulse at the origin shown in Figure 19.3 and is denoted by $F(\omega)$. Thus, the Fourier transform of a single pulse $f(t)$ is

$$F(\omega) = A\tau \operatorname{sinc}\left(\frac{\omega\tau}{2\pi}\right) \qquad (19.6)$$

which is shown in Figure 19.4.

FIGURE 19.3

A rectangular pulse.

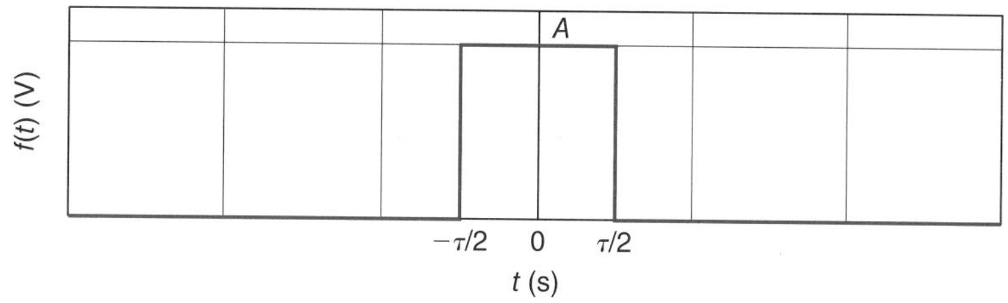

FIGURE 19.4

The Fourier transform of a rectangular pulse.

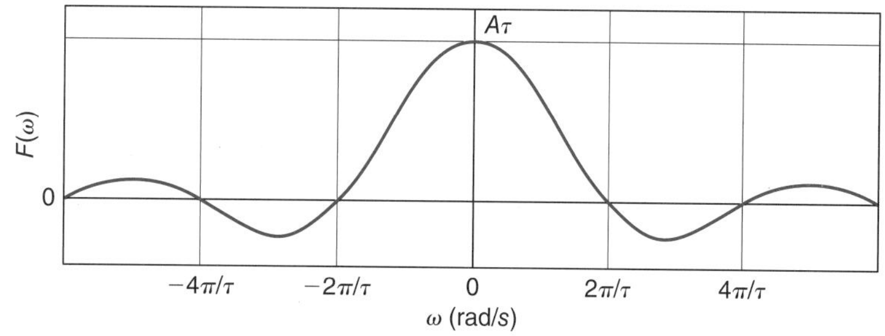

Comparing Equations (19.5) and (19.6), or by observing Figures 19.2 and 19.4, we can see that $F_n T_0$ are samples of $F(\omega)$ at $n\omega_0$; that is,

$$F_n T_0 = F(\omega)\big|_{\omega=n\omega_0} = F(n\omega_0) \qquad (19.7)$$

or

$$F_n = \frac{1}{T_0} F(\omega)\bigg|_{\omega=n\omega_0} = \frac{F(n\omega_0)}{T_0}, \quad \omega_0 = \frac{2\pi}{T_0} \qquad (19.8)$$

From the definition of the exponential Fourier coefficients given by

$$F_n = \frac{1}{T_0} \int_{-\frac{T_0}{2}}^{\frac{T_0}{2}} f(t) e^{-jn\omega_0 t} dt$$

we have

$$F_n T_0 = \int_{-\frac{T_0}{2}}^{\frac{T_0}{2}} f(t) e^{-jn\omega_0 t} dt \qquad (19.9)$$

As T_0 is increased to infinity, $n\omega_0$ becomes ω, and $F_n T_0$ given by Equation (19.9) becomes the Fourier transform (FT)

$$F(\omega) = \lim_{T_0 \to \infty} F_n T_0 = \lim_{T_0 \to \infty} \int_{-\frac{T_0}{2}}^{\frac{T_0}{2}} f(t) e^{-jn\omega_0 t} dt = \int_{-\infty}^{\infty} f(t) e^{-j\omega t} dt \qquad (19.10)$$

The exponential Fourier series representation of $f(t)$ given by

$$f(t) = \sum_{n=-\infty}^{\infty} F_n e^{jn\omega_0 t}, \quad \omega_0 = \frac{2\pi}{T_0} \tag{19.11}$$

can be rewritten as

$$\begin{aligned} f(t) &= \sum_{n=-\infty}^{\infty} \frac{F(n\omega_0)}{T_0} e^{jn\omega_0 t} = \frac{1}{2\pi} \sum_{n=-\infty}^{\infty} F(n\omega_0) e^{jn\omega_0 t} \frac{2\pi}{T_0} \\ &= \frac{1}{2\pi} \sum_{n=-\infty}^{\infty} F(n\omega_0) e^{jn\omega_0 t} \omega_0 \end{aligned} \tag{19.12}$$

As T_0 approaches infinity, $n\omega_0$ becomes ω, ω_0 becomes $d\omega$, and the summation becomes integral. Thus, $f(t)$ can be written as

$$f(t) = \frac{1}{2\pi} \int_{-\infty}^{\infty} F(\omega) e^{j\omega t} d\omega \tag{19.13}$$

which is called the inverse Fourier transform (IFT). Equation (19.13) indicates that a function $f(t)$ can be expanded as a sum of infinitely many sinusoidal oscillations of frequencies ω, varying from $-\infty$ to ∞. The weighting function in this expansion is given by $F(\omega)d\omega$. The Fourier transform $F(\omega)$ indicates the spectral content of the signal $f(t)$. If $f(t)$ is changing slowly with time, the spectrum $F(\omega)$ is concentrated near dc (0 hertz), and if $f(t)$ contains oscillations around certain frequency ω_c, the spectrum $F(\omega)$ is concentrated near ω_c. The Fourier transform in terms of frequency in hertz, f, is obtained by replacing ω by $2\pi f$. Thus, we have

$$F(f) = \int_{-\infty}^{\infty} f(t) e^{-j2\pi f t} dt \tag{19.14}$$

and the corresponding inverse Fourier transform is

$$f(t) = \int_{-\infty}^{\infty} F(f) e^{j2\pi f t} df \tag{19.15}$$

Notice the absence of $1/(2\pi)$ in this equation. This is due to $\omega = 2\pi f$.

The Fourier transform given by Equation (19.10) is the two-sided Laplace transform with $\sigma = 0$. As shown in Chapter 14, the two-sided Laplace transform is defined as

$$F(s) = \int_{-\infty}^{\infty} f(t) e^{-st} dt$$

Substituting the complex variable $s = \sigma + j\omega$, we obtain

$$F(\sigma + j\omega) = \int_{-\infty}^{\infty} f(t) e^{-\sigma t} e^{-j\omega t} dt$$

Setting $\sigma = 0$, we have

$$F(j\omega) = \int_{-\infty}^{\infty} f(t) e^{-j\omega t} dt$$

Evaluation of the Laplace transform on the imaginary axis ($s = j\omega$) yields Fourier transform.

19.2.1 SYMMETRIES

Applying Euler's rule, $e^{-j\omega t} = \cos(\omega t) - j\sin(\omega t)$, to the definition of the Fourier transform given by Equation (19.10), we obtain

$$F(\omega) = \int_{-\infty}^{\infty} f(t)e^{-j\omega t}\,dt = \int_{-\infty}^{\infty} f(t)\cos(\omega t)dt - j\int_{-\infty}^{\infty} f(t)\sin(\omega t)dt \quad \textbf{(19.16)}$$

If $f(t)$ is real, the real part of $F(\omega)$ is

$$F_R(\omega) = \int_{-\infty}^{\infty} f(t)\cos(\omega t)dt \quad \textbf{(19.17)}$$

and the imaginary part of $F(\omega)$ is

$$F_I(\omega) = -\int_{-\infty}^{\infty} f(t)\sin(\omega t)dt \quad \textbf{(19.18)}$$

The Fourier transform $F(\omega)$ can be written as

$$F(\omega) = |F(\omega)|e^{j\angle F(\omega)} = F_R(\omega) + jF_I(\omega) \quad \textbf{(19.19)}$$

If $f(t)$ is real, the magnitude of $F(\omega)$ is given by

$$|F(\omega)| = \sqrt{|F_R(\omega)|^2 + |F_I(\omega)|^2} \quad \textbf{(19.20)}$$

and the phase of $F(\omega)$ is given by

$$\angle F(\omega) = \tan^{-1}\left[\frac{F_I(\omega)}{F_R(\omega)}\right] \quad \textbf{(19.21)}$$

If $f(t)$ possesses **even symmetry**, $f(t)\cos(\omega t)$ is an even function of t, and $f(t)\sin(\omega t)$ is an odd function of t. Thus, $F_I(\omega)$ is equal to 0, and $F_R(\omega)$ is two times the integral from 0 to infinity. Therefore, the Fourier transform $F(\omega)$ reduces to

$$F(\omega) = 2\int_{0}^{\infty} f(t)\cos(\omega t)dt \quad \textbf{(19.22)}$$

which is a real function of ω.

If $f(t)$ possesses **odd symmetry**, $f(t)\cos(\omega t)$ is an odd function of t, and $f(t)\sin(\omega t)$ is an even function of t. Thus, $F_R(\omega)$ is equal to 0, and $F_I(\omega)$ is obtained by multiplying two times the integral from 0 to infinity and $-j$. Therefore, the Fourier transform $F(\omega)$ reduces to

$$F(\omega) = -j2\int_{0}^{\infty} f(t)\sin(\omega t)dt \quad \textbf{(19.23)}$$

which is purely imaginary.

EXAMPLE 19.1

Determine the Fourier transform of a rectangular pulse shown in Figure 19.3, and plot $F(\omega)$, $|F(\omega)|$, and $\angle F(\omega)$.

$$F(\omega) = \int_{-\frac{\tau}{2}}^{\frac{\tau}{2}} A e^{-j\omega t} dt = A \frac{e^{-j\omega t}\Big|_{-\frac{\tau}{2}}^{\frac{\tau}{2}}}{-j\omega} = A \frac{e^{-j\omega\frac{\tau}{2}} - e^{j\omega\frac{\tau}{2}}}{-j\omega} = A\tau \frac{e^{j\omega\frac{\tau}{2}} - e^{-j\omega\frac{\tau}{2}}}{2j\omega\frac{\tau}{2}}$$

$$= A\tau \frac{\sin\left(\frac{\omega\tau}{2}\right)}{\frac{\omega\tau}{2}} = A\tau \operatorname{sinc}\left(\frac{\omega\tau}{2\pi}\right) = \begin{cases} \frac{2A}{\omega}\sin\left(\frac{\omega\tau}{2}\right), & \omega \neq 0 \\ A\tau, & \omega = 0 \end{cases} \quad \textbf{(19.24)}$$

MATLAB

```
syms t w A tau
F=int(A*exp(-j*w*t),t,-tau/2,tau/2)
Answer:
F =
(2*A*sin((tau*w)/2))/w
```

If f is used instead of ω, this result becomes

$$F(f) = A\tau \operatorname{sinc}(f\tau) \quad \textbf{(19.25)}$$

The Fourier transform $F(\omega)$, its magnitude spectrum $|F(\omega)|$, and the phase spectrum $\angle F(\omega)$ are shown in Figure 19.5. The zero-crossing points are integer multiples of $2\pi/\tau$. The spectrum from $-2\pi/\tau$ to $2\pi/\tau$ is called the *main lobe*, the spectrums from $2\pi/\tau$ to $4\pi/\tau$ and $-2\pi/\tau$ to $-4\pi/\tau$ are called the *first-side lobes*, the spectrums from $4\pi/\tau$ to $6\pi/\tau$ and $-4\pi/\tau$ to $-6\pi/\tau$ are called the *second-side lobes*, etc. Since $F(\omega) \neq 0$ for all ω except zero-crossing points, the bandwidth of $F(\omega)$ is infinity, but as the frequency ω approaches infinity, $F(\omega)$ approaches 0. Thus, the bandwidth can be defined as the first zero-crossing frequency; that is,

$$\text{BW} \approx 2\pi/\tau \quad \textbf{(19.26)}$$

This bandwidth is called the *first zero-crossing bandwidth*. If a more accurate bandwidth is needed, then we define the bandwidth as the second zero-crossing frequency or the third zero-crossing frequency.

FIGURE 19.5

Fourier transform, magnitude spectrum, and phase spectrum of a rectangular pulse.

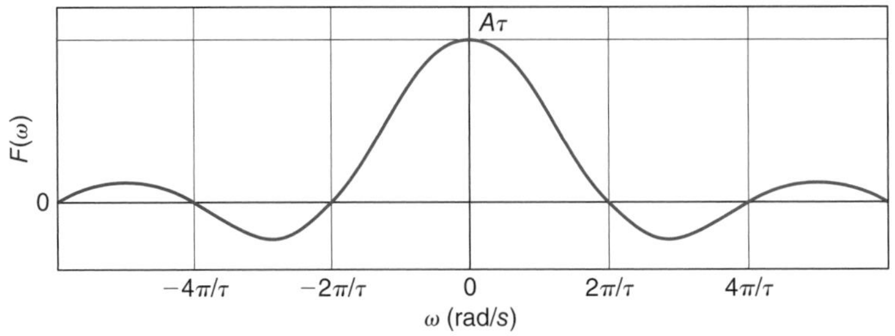

continued

Example 19.1 continued

FIGURE 19.5

continued

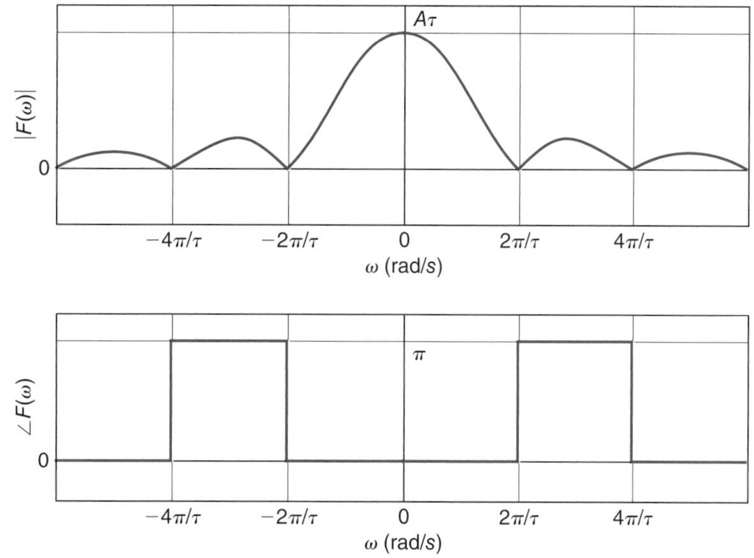

The Fourier transform $F(f)$, the magnitude spectrum $|F(f)| = |A\tau \operatorname{sinc}(f\tau)|$, and the phase spectrum $\angle F(f)$ are plotted in Figure 19.6.

FIGURE 19.6

Fourier transform, magnitude spectrum, and phase spectrum of a rectangular pulse.

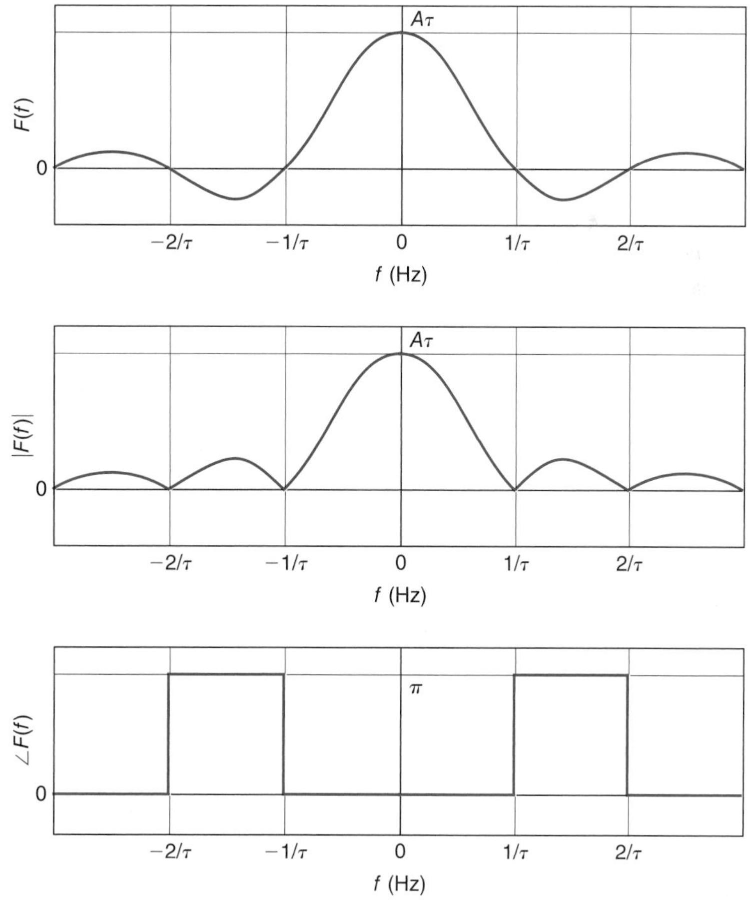

Exercise 19.1

Find the Fourier transform of $f(t) = u(t-1) - u(t-3)$.

Answer:

$$F(\omega) = \int_{1}^{3} 1e^{-j\omega t}\,dt = \frac{e^{-j\omega 3} - e^{-j\omega 1}}{-j\omega} = 2e^{-j\omega 2}\frac{e^{j\omega 1} - e^{-j\omega 1}}{2j\omega} = 2e^{-j\omega 2}\frac{\sin(\omega)}{\omega} = 2e^{-j\omega 2}\,\text{sinc}\left(\frac{\omega}{\pi}\right).$$

EXAMPLE 19.2

Determine the Fourier transform of the Dirac delta function $\delta(t)$.

$$F[\delta(t)] = \int_{-\infty}^{\infty} \delta(t)e^{-j\omega t}\,dt = e^{-j\omega 0} = 1 \quad \textbf{(19.27)}$$

where use has been made of a sifting property of a Dirac delta function

$$\int_{-\infty}^{\infty} \delta(t-a)f(t)dt = f(a) \quad \textbf{(19.28)}$$

Alternatively, since the Dirac delta function is defined as a rectangular pulse with amplitude $1/\tau$ and pulse width τ as the pulse width τ is decreased to 0; that is,

$$\delta(t) = \lim_{\tau \to 0} \frac{1}{\tau}\,rect\left(\frac{t}{\tau}\right) \quad \textbf{(19.29)}$$

the Fourier transform of the Dirac delta function is defined as the Fourier transform of a rectangular pulse with amplitude $1/\tau$ and pulse width τ as the pulse width τ is decreased to 0; that is,

$$F[\delta(t)] = \lim_{\tau \to 0} F\left[\frac{1}{\tau}\,rect\left(\frac{t}{\tau}\right)\right] = \lim_{\tau \to 0} \text{sinc}\left(\frac{\omega\tau}{2\pi}\right) = 1 \quad \textbf{(19.30)}$$

This result indicates that the Fourier transform of the Dirac delta function $\delta(t)$ is 1 for $-\infty \leq f \leq \infty$. The Dirac delta dunction $\delta(t)$ contains all the frequency components from 0 hertz to infinite frequency with equal weights. The Dirac delta function changes very rapidly in the time domain, since it changes instantly from 0 to infinity and back to 0. This fast change in the time domain translates into strong high-frequency components in the frequency domain.

Exercise 19.2

Find the Fourier transform of $f(t) = \delta(t-5)$.

Answer:

$$F[\delta(t-5)] = \int_{-\infty}^{\infty} \delta(t-5)e^{-j\omega t}\,dt = e^{-j\omega 5}.$$

19.2.2 FINDING FOURIER TRANSFORM FROM FOURIER COEFFICIENTS

Let $F(\omega)$ be the Fourier transform of a signal $f(t)$. Let F_n be the exponential Fourier coefficients of the periodic signal formed by repeating $f(t)$ every T_0. Then, the Fourier transform $F(\omega)$ (or $F(f)$) can be found from the exponential Fourier coefficients F_n by multiplying F_n by the period T_0 and replacing $n\omega_0$ by ω (or nf_0 by f); that is,

$$F(\omega) = T_0 F_n \text{ with } n\omega_0 \text{ replaced by } \omega \tag{19.31}$$

or

$$F(f) = T_0 F_n \text{ with } nf_0 \text{ replaced by } f \tag{19.32}$$

In terms of the envelope $E(\omega)$ of the Fourier coefficients F_n, the Fourier transform $F(\omega)$ is given by

$$F(\omega) = T_0\, E(\omega) \tag{19.33}$$

or

$$F(f) = T_0\, E(f) \tag{19.34}$$

where $E(f)$ is the envelope of F_n obtained by replacing nf_0 by f.

EXAMPLE 19.3

Find the Fourier transform of $f(t) = A\,\text{tri}(t/\tau)$.

As shown in Chapter 18, the exponential Fourier coefficients of a triangular pulse train is given by

$$F_n = \frac{A\tau}{T_0}\text{sinc}^2\left(\frac{n\omega_0\tau}{2\pi}\right) = \frac{A\tau}{T_0}\text{sinc}^2(nf_0\tau) \tag{19.35}$$

Multiplying F_n by T_0 and replacing $n\omega_0$ by ω, we get

$$F(\omega) = A\tau\ \text{sinc}^2\left(\frac{\omega\tau}{2\pi}\right) \tag{19.36}$$

Alternatively, multiplying F_n by T_0 and replacing nf_0 by f, we get

$$F(f) = A\tau\,\text{sinc}^2(f\tau) \tag{19.37}$$

Exercise 19.3

Find the Fourier transform of

$$f(t) = \begin{cases} A\cos\left(\dfrac{\pi t}{T_0}\right), & |t| \le \dfrac{T_0}{2} \\ 0, & \textit{elsewhere} \end{cases}$$

continued

Exercise 19.3 continued

As shown in Chapter 18, the exponential Fourier coefficients of a full-wave rectified cosine waveform is given by

$$F_n = \frac{2A\cos(n\pi)}{\pi(1-4n^2)} = \frac{T_0 2A\cos(n\pi)}{\pi T_0(1-4n^2)} = \frac{T_0 2A\cos\left(n\frac{2\pi}{T_0}\frac{T_0}{2}\right)}{\pi T_0\left(1-4n\frac{2\pi}{T_0}n\frac{2\pi}{T_0}\frac{T_0^2}{4\pi^2}\right)}$$
$$= \frac{T_0 2A\cos\left(n\omega_0\frac{T_0}{2}\right)}{\pi T_0\left(1-4n^2\omega_0^2\frac{T_0^2}{4\pi^2}\right)} \qquad \textbf{(19.38)}$$

Replacing $n\omega_0$ by ω and multiplying by T_0, we obtain

$$F(\omega) = \frac{2AT_0}{\pi}\frac{\cos\left(\omega\frac{T_0}{2}\right)}{1-\left(\frac{\omega T_0}{\pi}\right)^2} \qquad \textbf{(19.39)}$$

Replacing ω by $2\pi f$, we get

$$F(f) = \frac{2AT_0}{\pi}\frac{\cos(f\pi T_0)}{1-(f2T_0)^2} \qquad \textbf{(19.40)}$$

19.3 Properties of Fourier Transform

We present several important properties of the Fourier transform in this section. In many occasions, the evaluation of the Fourier transform is simplified by applying the appropriate properties. Also, we can gain some insights into the relation between the time domain and the frequency domain description of the signals. We will use F to denote the Fourier transform and F^{-1} to denote the inverse Fourier transform. Thus, $F[f(t)] = F(\omega)$, and $F^{-1}[F(\omega)] = f(t)$. Sometimes double arrows are used to indicate Fourier transform pairs. Thus, $f(t) \leftrightarrow F(\omega)$. Table 19.1 shows the properties of the Fourier transform, and Table 19.2 shows the Fourier transform pairs for common signals.

TABLE 19.1
Properties of the Fourier Transform

Property	$f(t)$	$F(\omega)$	$F(f)$
Linearity	$a_1f_1(t) + a_2f_2(t)$	$a_1F_1(\omega) + a_2F_2(\omega)$	$a_1F_1(f) + a_2F_2(f)$
Time shifting	$f(t-t_d)$	$e^{-j\omega t_d}F(\omega)$	$e^{-j2\pi f t_d}F(f)$
Time scaling	$f(ct)$	$\frac{1}{\lvert c\rvert}F\left(\frac{\omega}{c}\right)$	$\frac{1}{\lvert c\rvert}F\left(\frac{f}{c}\right)$
Symmetry (duality)	$F(t)$	$2\pi f(-\omega)$	$f(-f)$
Time reversal	$f(-t)$	$F(-\omega)$	$F(-f) = F^*(f)$

continued

Table 19.1 continued

TABLE 19.1

Property	$f(t)$	$F(\omega)$	$F(f)$
Frequency shifting	$f(t)e^{j\omega_c t}$	$F(\omega - \omega_c)$	$F(f - f_c)$
Modulation	$f(t)\cos(\omega_c t)$	$\frac{1}{2}F(\omega - \omega_c) + \frac{1}{2}F(\omega + \omega_c)$	$\frac{1}{2}F(f - f_c) + \frac{1}{2}F(f + f_c)$
Time differentiation	$\frac{df(t)}{dt}$	$j\omega F(\omega)$	$j2\pi fF(f)$
	$\frac{d^n f(t)}{dt^n}$	$(j\omega)^n F(\omega)$	$(j2\pi f)^n F(f)$
Frequency differentiation	$tf(t)$	$j\frac{dF(\omega)}{d\omega}$	$j\frac{dF(f)}{2\pi df}$
	$t^n f(t)$	$j^n\frac{d^n F(\omega)}{d\omega^n}$	$\frac{d^n F(f)}{(-j2\pi)^n df^n}$
Conjugate	$f^*(t)$	$F^*(-\omega)$	$F^*(-f)$
Integration	$\int_{-\infty}^{t} f(\lambda)d\lambda$	$\frac{1}{j\omega}F(\omega) + \pi F(0)\delta(\omega)$	$\frac{1}{j2\pi f}F(f) + \frac{1}{2}F(0)\delta(f)$
Convolution	$\int_{-\infty}^{\infty} h(\lambda)f(t-\lambda)d\lambda$	$H(\omega)F(\omega)$	$H(f)F(f)$
Multiplication	$f_1(t)f_2(t)$	$\frac{1}{2\pi}\int_{-\infty}^{\infty} F_1(v)F_2(\omega - v)dv$	$\int_{-\infty}^{\infty} F_1(v)F_2(f - v)dv$
Parseval's theorem	$\int_{-\infty}^{\infty} \lvert f(t)\rvert^2 dt$	$\frac{1}{2\pi}\int_{-\infty}^{\infty} \lvert F(\omega)\rvert^2 d\omega$	$\int_{-\infty}^{\infty} \lvert F(f)\rvert^2 df$
Plancherel theorem	$\int_{-\infty}^{\infty} f(t)g^*(t)dt$	$\frac{1}{2\pi}\int_{-\infty}^{\infty} F(\omega)G^*(\omega)d\omega$	$\int_{-\infty}^{\infty} F(f)G^*(f)df$

TABLE 19.2

Fourier Transform Pairs ($a > 0$)

$f(t)$	$F(\omega)$	$F(f)$
$rect\left(\frac{t}{a}\right) = \begin{cases} 1, & \frac{-a}{2} < t < \frac{a}{2} \\ 0, & otherwise \end{cases}$	$a\,\text{sinc}\left(\frac{\omega a}{2\pi}\right)$	$a\,\text{sinc}(fa)$
$tri\left(\frac{t}{a}\right) = \begin{cases} 1 + \frac{t}{a}, & -a \le t < 0 \\ 1 - \frac{t}{a}, & 0 \le t < a \\ 0, & otherwise \end{cases}$	$a\,\text{sinc}^2\left(\frac{\omega a}{2\pi}\right)$	$a\,\text{sinc}^2(fa)$

continued

Table 19.2 continued

TABLE 19.2

$f(t)$	$F(\omega)$	$F(f)$
$tri\left(\frac{t}{\frac{T}{2}}\right) = \begin{cases} 1 + \frac{t}{\frac{T}{2}}, & -\frac{T}{2} \le t < 0 \\ 1 - \frac{t}{\frac{T}{2}}, & 0 \le t < \frac{T}{2} \\ 0, & otherwise \end{cases}$	$\frac{T}{2}\,\text{sinc}^2\left(\frac{\omega T}{4\pi}\right)$	$\frac{T}{2}\,\text{sinc}^2\left(f\frac{T}{2}\right)$
$e^{-at}u(t)$	$\frac{1}{j\omega + a}$	$\frac{1}{j2\pi f + a}$
$e^{at}u(-t)$	$\frac{1}{-j\omega + a}$	$\frac{1}{-j2\pi f + a}$
$e^{-a\|t\|}$	$\frac{2a}{\omega^2 + a^2}$	$\frac{2a}{4\pi^2 f^2 + a^2}$
$\frac{1}{t^2 + a^2}$	$\frac{\pi}{a}e^{-a\|\omega\|}$	$\frac{\pi}{a}e^{-2\pi a\|f\|}$
$e^{-at}u(t) - e^{at}u(-t)$	$\frac{-2j\omega}{\omega^2 + a^2}$	$\frac{-j4\pi f}{4\pi^2 f^2 + a^2}$
$\delta(t)$	1	1
1	$2\pi\delta(\omega)$	$\delta(f)$
$u(t)$	$\pi\delta(\omega) + \frac{1}{j\omega}$	$\frac{1}{2}\delta(f) + \frac{1}{j2\pi f}$
$tu(t)$	$\frac{\pi}{j\omega}\delta(\omega) + \frac{1}{(j\omega)^2}$	$\frac{1}{j4\pi f}\delta(f) + \frac{1}{(j2\pi f)^2}$
$te^{-at}u(t)$	$\frac{1}{(j\omega + a)^2}$	$\frac{1}{(j2\pi f + a)^2}$
$\cos(\omega_c t) = \cos(2\pi f_c t)$	$\pi[\delta(\omega - \omega_c) + \delta(\omega + \omega_c)]$	$\frac{1}{2}[\delta(f - f_c) + \delta(f + f_c)]$
$\sin(\omega_c t) = \sin(2\pi f_c t)$	$-j\pi[\delta(\omega - \omega_c) - \delta(\omega + \omega_c)]$	$\frac{-j}{2}[\delta(f - f_c) - \delta(f + f_c)]$
$e^{-at}u(t)\cos(\omega_c t)$	$\frac{j\omega + a}{(j\omega + a)^2 + \omega_c^2}$	$\frac{j2\pi f + a}{(j2\pi f + a)^2 + (2\pi f_c)^2}$
$e^{-at}u(t)\sin(\omega_c t)$	$\frac{\omega_c}{(j\omega + a)^2 + \omega_c^2}$	$\frac{2\pi f_c}{(j2\pi f + a)^2 + (2\pi f_c)^2}$
$sgn(t) = \begin{cases} -1, & t < 0 \\ 0, & t = 0 \\ 1, & t > 0 \end{cases}$	$\frac{2}{j\omega}$	$\frac{1}{j\pi f}$
$\text{sinc}(ct) = \frac{\sin(\pi ct)}{\pi ct}$	$\frac{1}{c}rect\left(\frac{\omega}{2\pi c}\right)$	$\frac{1}{c}rect\left(\frac{f}{c}\right)$

continued

Table 19.2 continued

TABLE 19.2

$f(t)$	$F(\omega)$	$F(f)$
$\text{sinc}^2(ct) = \dfrac{\sin^2(\pi ct)}{(\pi ct)^2}$	$\dfrac{1}{c} tri\left(\dfrac{\omega}{2\pi c}\right)$	$\dfrac{1}{c} tri\left(\dfrac{f}{c}\right)$
$\cos\left(\dfrac{\pi t}{a}\right) rect\left(\dfrac{t}{a}\right)$	$\dfrac{2a}{\pi} \dfrac{\cos\left(\dfrac{\omega a}{2}\right)}{1 - \left(\dfrac{\omega a}{\pi}\right)^2}$	$\dfrac{2a}{\pi} \dfrac{\cos(\pi a f)}{1 - (2af)^2}$
$\dfrac{1}{2}\left[1 + \cos\left(\dfrac{\pi t}{a}\right)\right] rect\left(\dfrac{t}{2a}\right)$	$a \dfrac{\sin(\omega a)}{\omega a\left[1 - \left(\dfrac{\omega a}{\pi}\right)^2\right]}$	$a \dfrac{\sin(2\pi f a)}{2\pi f a[1 - (2af)^2]}$
$\dfrac{1}{\sqrt{2\pi}\sigma} e^{-\frac{t^2}{2\sigma^2}}$	$e^{-\frac{\sigma^2\omega^2}{2}}$	$e^{-2(\sigma\pi f)^2}$
$\sum_{k=-\infty}^{\infty} \delta(t - kT_s)$	$\sum_{n=-\infty}^{\infty} \dfrac{2\pi}{T_s}\delta\left(\omega - n\dfrac{2\pi}{T_s}\right)$	$\sum_{n=-\infty}^{\infty} \dfrac{1}{T_s}\delta\left(f - \dfrac{n}{T_s}\right)$

19.3.1 LINEARITY PROPERTY (SUPERPOSITION PRINCIPLE)

If $F[f_1(t)] = F_1(\omega)$ and $F[f_2(t)] = F_2(\omega)$, then

$$F[a_1 f_1(t) + a_2 f_2(t)] = a_1 F_1(\omega) + a_2 F_2(\omega) \quad \textbf{(19.41)}$$

Proof

$$F[a_1 f_1(t) + a_2 f_2(t)] = \int_{-\infty}^{\infty} [a_1 f_1(t) + a_2 f_2(t)] e^{-j\omega t}\, dt$$

$$= a_1 \int_{-\infty}^{\infty} f_1(t) e^{-j\omega t}\, dt + a_2 \int_{-\infty}^{\infty} f_2(t) e^{-j\omega t}\, dt = a_1 F_1(\omega) + a_2 F_2(\omega)$$

The linearity property says that the Fourier transform of a linear combination of signals is given by the linear combination of the Fourier transforms of the signals. The linearity property is also called the superposition principle.

19.3.2 TIME-SHIFTING PROPERTY

If $F[f(t)] = F(\omega)$, then

$$F[f(t - t_d)] = F(\omega) e^{-j\omega t_d} \quad \textbf{(19.42)}$$

Proof

Let $t' = t - t_d$. Then $t = t' + t_d$ and $dt = dt'$.

$$F[f(t - t_d)] = \int_{-\infty}^{\infty} f(t - t_d) e^{-j\omega t}\, dt = \int_{-\infty}^{\infty} f(t') e^{-j\omega(t' + t_d)}\, dt' = e^{-j\omega t_d} \int_{-\infty}^{\infty} f(t') e^{-j\omega t'}\, dt' = F(\omega) e^{-j\omega t_d}$$

Let $g(t) = f(t - t_d)$. Then, this theorem states that

$$G(\omega) = F(\omega)e^{-j\omega t_d} = |F(\omega)|e^{j\angle F(\omega)}e^{-j\omega t_d} = |F(\omega)|e^{j(\angle F(\omega) - \omega t_d)}$$

Thus

$$|G(\omega)| = |F(\omega)|$$

and

$$\angle G(\omega) = \angle F(\omega) - \omega t_d$$

Therefore, the magnitude spectrum of the time-shifted signal is the same as the magnitude spectrum of the original signal, and the phase spectrum of the time-shifted signal is given by the sum of the phase spectrum of the original signal and a linear phase term $-\omega t_d$. In other words, a time shift in the time domain introduces a linear phase term in the frequency domain. Notice that the signal $f(t - t_d)$ has the same shape as $f(t)$, but it is shifted by t_d relative to $f(t)$. Thus, time shifting is not considered as a distortion. The slope of the linear phase term $-\omega t_d$ is given by

$$\frac{d(-\omega t_d)}{d\omega} = -t_d$$

Delaying $f(t)$ by t_d introduces a linear phase with a slope of $-t_d$.

EXAMPLE 19.4

Given a rectangular pulse with amplitude A and width τ shown in Figure 19.7, determine the Fourier transform $F(\omega)$, and plot $|F(\omega)|$ and $\angle F(\omega)$.

FIGURE 19.7

A rectangular pulse.

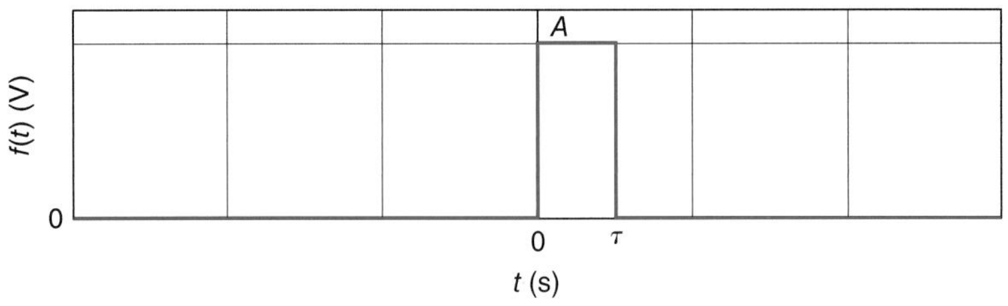

The rectangular pulse shown in Figure 19.7 is the rectangular pulse shown in Figure 19.3 shifted to the right by $\tau/2$. Thus, the Fourier transform of the rectangular pulse shown in Figure 19.7 is given by

$$F(\omega) = A\tau \text{ sinc}\left(\frac{\omega\tau}{2\pi}\right)e^{-j\omega\frac{\tau}{2}}$$

Thus, we have

$$|F(\omega)| = \left|A\tau \text{ sinc}\left(\frac{\omega\tau}{2\pi}\right)\right|$$

continued

Example 19.4 continued

and

$$\angle F(\omega) = \begin{cases} -\omega\dfrac{\tau}{2}, & A\tau\,\mathrm{sinc}\left(\dfrac{\omega\tau}{2\pi}\right) \geq 0 \\ \pi - \omega\dfrac{\tau}{2}, & A\tau\,\mathrm{sinc}\left(\dfrac{\omega\tau}{2\pi}\right) < 0 \end{cases}$$

The magnitude spectrum $|F(\omega)|$ is shown in Figure 19.8(a), and the phase spectrum $\angle F(\omega)$ is shown in Figure 19.8(b).

FIGURE 19.8

(a) Magnitude spectrum and (b) phase spectrum of $F(\omega)$.

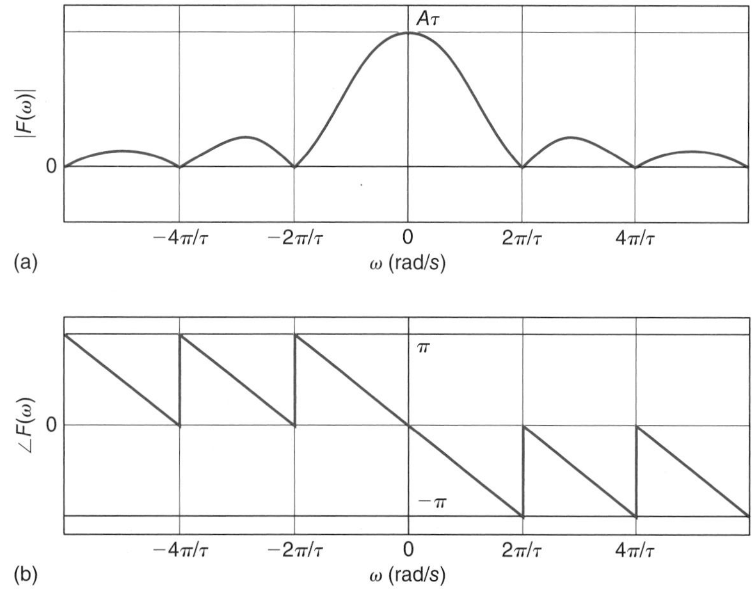

Exercise 19.4

Determine the Fourier transform of the signal shown in Figure 19.9.

FIGURE 19.9

Waveform for EXERCISE 19.4

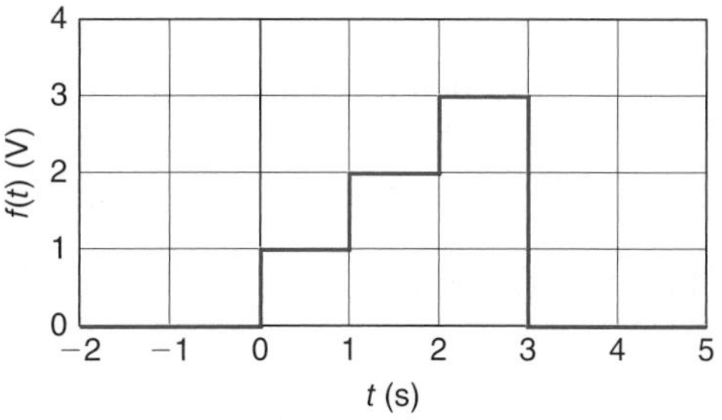

Answer:

$$F(\omega) = \mathrm{sinc}\left(\frac{\omega}{2\pi}\right)(e^{-j\omega 0.5} + 2e^{-j\omega 1.5} + 3e^{-j\omega 2.5}).$$

19.3.3 TIME-SCALING PROPERTY

If $F[f(t)] = F(\omega)$, then

$$F[f(ct)] = \frac{1}{|c|}F\left(\frac{\omega}{c}\right) \qquad (19.43)$$

Proof

Let $t' = ct$. Then, $t = t'/c$ and $dt = dt'/c$. If $c > 0$, then

$$F[f(ct)] = \int_{-\infty}^{\infty} f(ct)e^{-j\omega t}dt = \int_{-\infty}^{\infty} f(t')e^{-j\left(\frac{\omega}{c}\right)t'}\left(\frac{1}{c}\right)dt' = \frac{1}{c}F\left(\frac{\omega}{c}\right)$$

If $c < 0$, then

$$F[f(ct)] = \int_{-\infty}^{\infty} f(ct)e^{-j\omega t}dt = \int_{\infty}^{-\infty} f(t')e^{-j\left(\frac{\omega}{c}\right)t'}\left(\frac{1}{c}\right)dt' = \frac{1}{|c|}\int_{-\infty}^{\infty} f(t')e^{-j\left(\frac{\omega}{c}\right)t'}dt' = \frac{1}{|c|}F\left(\frac{\omega}{c}\right)$$

Combining these two results, we obtain the desired result.

The signal $f(at)$, $a > 1$, is a squeezed version of $f(t)$, and $f(at)$, $a < 1$, is an expanded version of $f(t)$. When the signal is squeezed in the time domain, the spectrum in the frequency domain is expanded. Thus, the spectrum is spread out farther into higher frequencies. On the other hand, when the signal is expanded in the time domain, the spectrum in the frequency domain is squeezed into lower frequencies.

EXAMPLE 19.5

If $F[f(t)] = F(\omega)$, then what is the Fourier transform of the following signals?

a. $f(t/5)$ **b.** $f(-5t)$ **c.** $f[-5(t-2)]$

a. $F\left[f\left(\frac{t}{5}\right)\right] = 5F(5\omega)$ **b.** $F[f(-5t)] = \frac{1}{5}F\left(-\frac{\omega}{5}\right)$ **c.** $F\{f[-5(t-2)]\} = \frac{1}{5}F\left(-\frac{\omega}{5}\right)e^{-j\omega 2}$

Notice that the time-shifting property and the time-scaling property are utilized in (c). Apply the time-shifting property last.

Exercise 19.5

If $F[f(t)] = F(\omega)$, then what is the Fourier transform of the following signals?

a. $f(-t/3)$ **b.** $f\left(\frac{-2t-6}{5}\right)$ **c.** $f\left(\frac{-3t+12}{7}\right)$

Answer:

a. $3F(-3\omega)$ **b.** $2.5F(-2.5\omega)\,e^{i\omega 3}$ **c.** $\frac{7}{3}F\left(-\frac{7}{3}\omega\right)e^{-j\omega 4}$

EXAMPLE 19.6

Let $f_1(t) = rect(t)$, $f_2(t) = rect(5t)$ and $f_3(t) = rect(t/5)$. Determine the Fourier transforms of these signals. Plot these three signals and their Fourier transforms.

$$F_1(\omega) = \text{sinc}\left(\frac{\omega}{2\pi}\right)$$

$$F_2(\omega) = \frac{1}{5}F_1\left(\frac{\omega}{5}\right) = \frac{1}{5}\text{sinc}\left(\frac{\omega}{10\pi}\right)$$

$$F_3(\omega) = 5F_1(5\omega) = 5\,\text{sinc}\left(\frac{\omega}{0.4\pi}\right)$$

Figure 19.10 shows the three signals in the time domain and their Fourier transforms in the frequency domain.

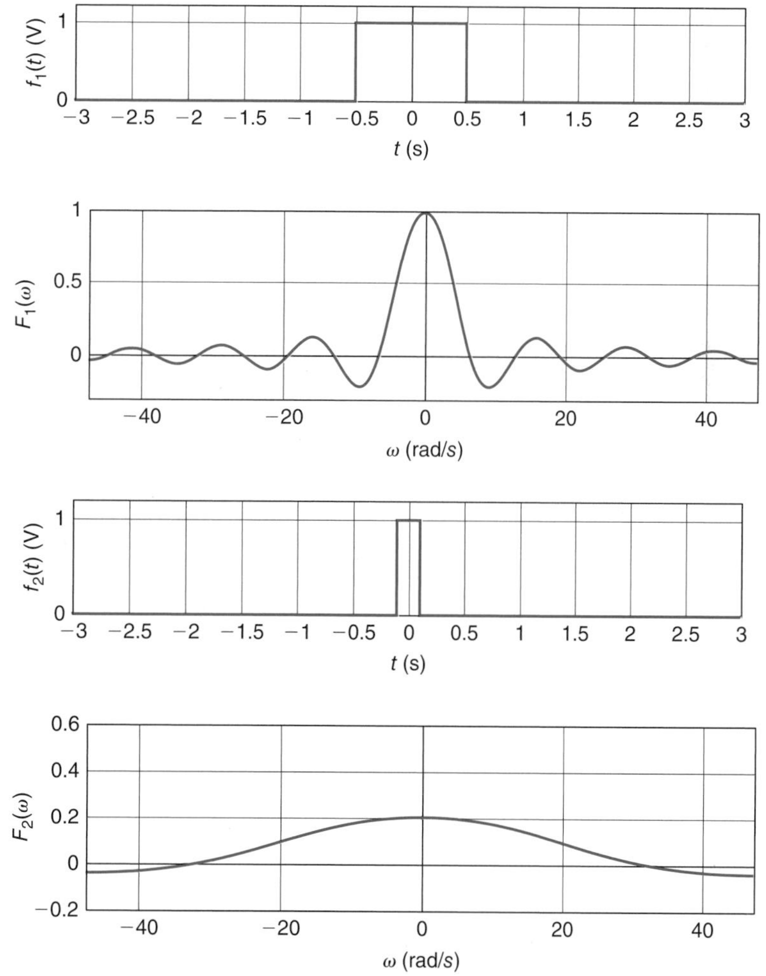

FIGURE 19.10

Rectangular pulse in the time domain and in the frequency domain for three different pulse widths.

continued

Example 19.6 continued

FIGURE 19.10

continued

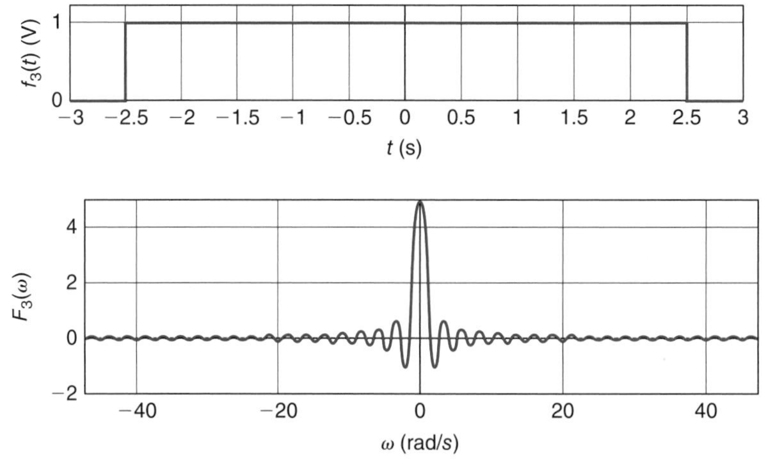

The first zero-crossing bandwidth of the pulse *rect*(t) with pulse width 1 s is 2π rad/s, while the first zero-crossing bandwidth of the pulse *rect*($5t$) with pulse width $1/5$ s is 10π rad/s. As the pulse width is reduced by a factor of 5, the bandwidth is increased by a factor of 5. The first zero-crossing bandwidth of the pulse *rect*($t/5$) with pulse width 5 s is 0.4π rad/s. As the pulse width is increased by a factor of 5, the bandwidth is decreased by a factor of 5. Notice that the pulse with the smallest pulse width in the time domain corresponds with the widest spread in the frequency domain, and the largest pulse width in the time domain corresponds to the spectrum with the narrowest spread in the frequency domain. Thus, there exists an inverse relation between the pulse width in the time domain and the spread (bandwidth) in the frequency domain. In general, more rapid variation in the time domain corresponds to the higher high-frequency contents. This is called the inverse time-frequency relation.

Exercise 19.6

Find the Fourier transform of *tri*($2t$) and *tri*($t/2$).

Answer:

$$F\left[tri\left(\frac{t}{a}\right)\right] = a\,\text{sinc}^2\left(\frac{\omega a}{2\pi}\right),\quad F\left[tri\left(\frac{t}{\frac{1}{2}}\right)\right] = \frac{1}{2}\,\text{sinc}^2\left(\frac{\omega}{4\pi}\right),$$

$$F\left[tri\left(\frac{t}{2}\right)\right] = 2\,\text{sinc}^2\left(\frac{\omega 2}{2\pi}\right) = 2\,\text{sinc}^2\left(\frac{\omega}{\pi}\right).$$

19.3.4 SYMMETRY PROPERTY (DUALITY PROPERTY)

If $F[f(t)] = F(\omega)$, then

$$F[F(t)] = 2\pi f(-\omega) \qquad \textbf{(19.44)}$$

or

$$F[F(t)] = f(-f) \qquad \textbf{(19.45)}$$

Proof

Since t and ω are arbitrary variables in the inverse Fourier transform

$$f(t) = \frac{1}{2\pi}\int_{-\infty}^{\infty} F(\omega)e^{j\omega t}\,d\omega$$

we can replace ω with t and t with $-\omega$ to get

$$f(-\omega) = \frac{1}{2\pi}\int_{-\infty}^{\infty} F(t)e^{-j\omega t}dt$$

Therefore,

$$F[F(t)] = \int_{-\infty}^{\infty} F(t)e^{-j\omega t}dt = 2\pi f(-\omega)$$

Similarly, if we replace f with t and t with $-f$ in the inverse Fourier transform

$$f(t) = \int_{-\infty}^{\infty} F(f)e^{j2\pi ft}\,df$$

we obtain

$$f(-f) = \int_{-\infty}^{\infty} F(t)e^{-j2\pi ft}\,dt$$

Therefore,

$$F[F(t)] = f(-f)$$

If the frequency domain function $F(\omega)$, which is the Fourier transform of $f(t)$, is made to be the time function $F(t)$, then the Fourier transform of $F(t)$ is $f(-f)$. Notice that $f(-f)$ is the folded (flipped) function of $f(f)$, which is the original time function $f(t)$ in the frequency domain. This is a role reversal between $f(t)$ and $F(\omega)$.

EXAMPLE 19.7

It has been shown that $F[\delta(t)] = 1$. Applying the symmetry property, we obtain $F[1] = 2\pi\delta(-\omega) = 2\pi\delta(\omega)$, since $\delta(\omega)$ is an even function of ω. In terms of f in hertz, we have $F[1] = \delta(f)$. Figure 19.11 illustrates the Fourier transform pairs $F[\delta(t)] = 1$ and $F[1] = \delta(f)$.

FIGURE 19.11

$F[\delta(t)] = 1$ and $F[1] = \delta(f)$.

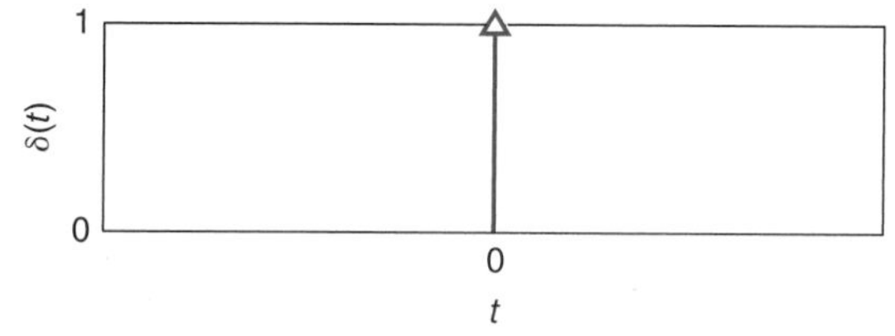

continued

Example 19.7 continued

FIGURE 19.11

continued

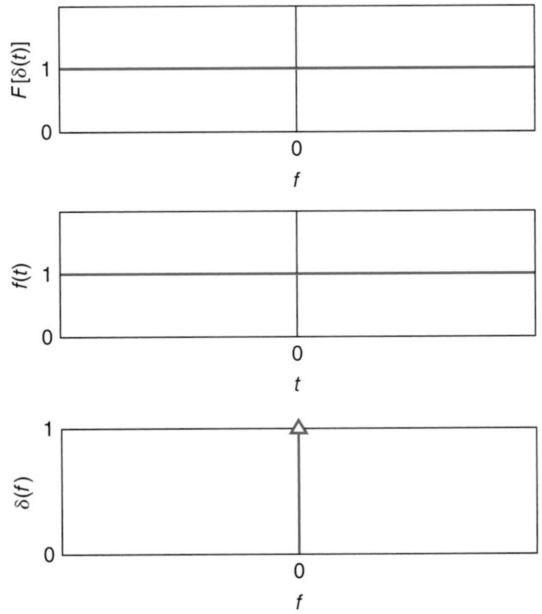

According to this example, an impulse $\delta(t)$ contains all the frequency components from minus infinity to plus infinity with equal weights. The signal $\delta(t)$ instantaneously changes from 0 to plus infinity and back to 0. This indicates a strong presence of high-frequency components based on the inverse time-frequency relation. Obviously, a dc signal has only the zero-frequency component.

Exercise 19.7

If $F\left[\cos\left(\frac{\pi t}{a}\right) rect\left(\frac{t}{a}\right)\right] = \frac{2a}{\pi}\frac{\cos(\pi af)}{1-(2af)^2}$, what is the Fourier transform of $\frac{2a}{\pi}\frac{\cos(\pi at)}{1-(2at)^2}$?

Answer:

$$F\left[\frac{2a}{\pi}\frac{\cos(\pi at)}{1-(2at)^2}\right] = \cos\left(\frac{\pi f}{a}\right) rect\left(\frac{f}{a}\right).$$

EXAMPLE 19.8

Determine the Fourier transform of sinc(ct).

Applying the symmetry property to the Fourier transform pair

$$F\left[rect\left(\frac{t}{\tau}\right)\right] = \tau\,\text{sinc}\left(\frac{\omega\tau}{2\pi}\right),$$

we obtain

$$F\left[\tau\,\text{sinc}\left(\frac{t\tau}{2\pi}\right)\right] = 2\pi\,rect\left(\frac{-\omega}{\tau}\right) = 2\pi\,rect\left(\frac{\omega}{\tau}\right)$$

continued

Example 19.8 continued

since $rect(\omega/\tau)$ is an even function of ω. Let $\tau/(2\pi) = c$. Then $\tau = 2\pi c$.

$$F[2\pi c \text{ sinc}(ct)] = 2\pi \, rect\left(\frac{\omega}{2\pi c}\right)$$

Therefore

$$F[\text{sinc}(ct)] = \frac{1}{c} rect\left(\frac{\omega}{2\pi c}\right)$$

In terms of f (in hertz), we have

$$F[\text{sinc}(ct)] = \frac{1}{c} rect\left(\frac{f}{c}\right)$$

Figure 19.12 depicts the two Fourier transform pairs discussed in this example.

FIGURE 19.12

The two Fourier transform pairs in EXAMPLE 19.8.

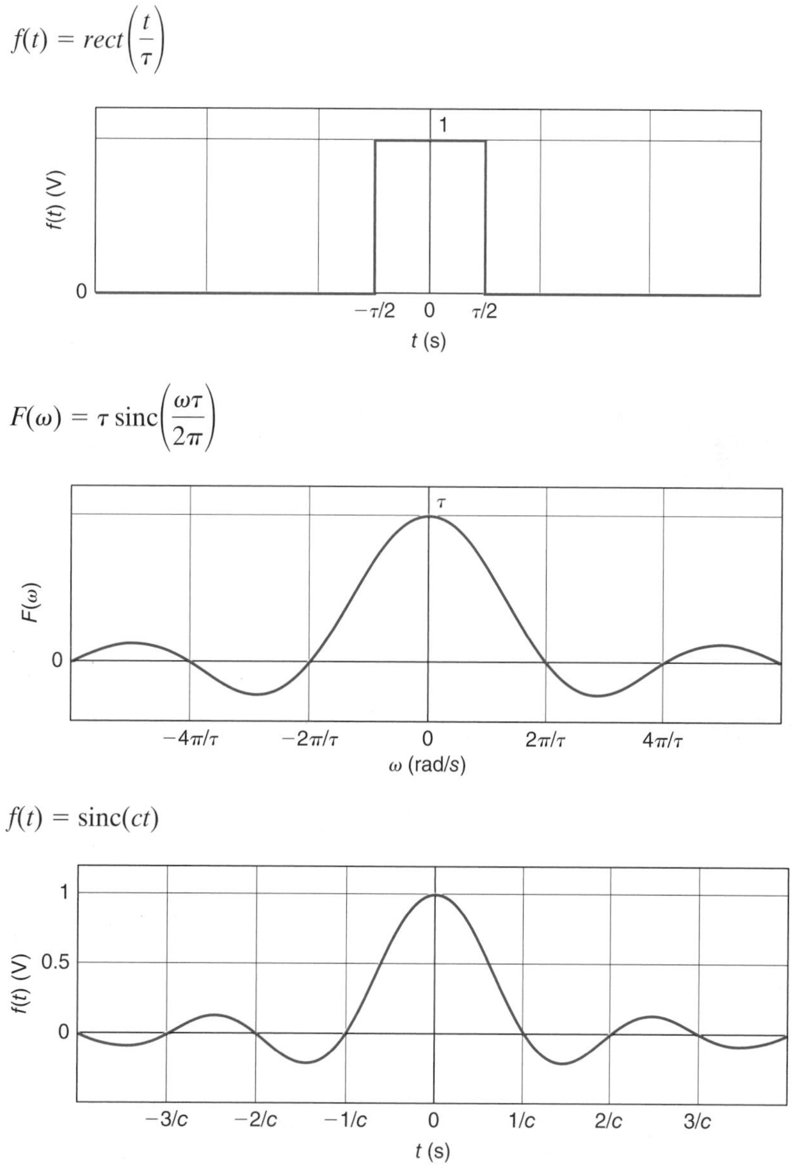

continued

Example 19.8 continued

FIGURE 19.12

continued

$$F(\omega) = \frac{1}{c} rect\left(\frac{\omega}{2\pi c}\right)$$

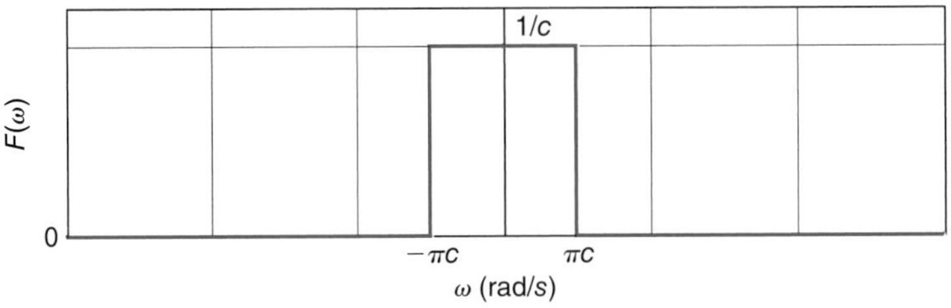

$$F\left[rect\left(\frac{t}{\tau}\right)\right] = \tau \operatorname{sinc}\left(\frac{\omega\tau}{2\pi}\right), \quad F[\operatorname{sinc}(ct)] = \frac{1}{c} rect\left(\frac{\omega}{2\pi c}\right)$$

We all know that a rectangular pulse in the time domain results in a sinc-shaped Fourier transform in the frequency domain. This example shows that a sinc pulse in the time domain gives us a rectangular-shaped Fourier transform in the frequency domain. Also notice that time-limited signals such as a rectangular pulse in the time domain are not band-limited, since the bandwidth of sinc is infinity. On the other hand, band-limited signals such as the rectangular-shaped spectrum are not time-limited, since a sinc pulse in the time domain extends from negative infinity to positive infinity. If a rectangular pulse is used to transmit binary data, the required bandwidth is infinity to receive the same pulse at the receiver. On the other hand, if the sinc pulse is used to transmit binary data, the required bandwidth is half the data rate.

Exercise 19.8

Find the Fourier transform for $\operatorname{sinc}^2(ct)$.

Answer:

$$F\left[tri\left(\frac{t}{c}\right)\right] = c \operatorname{sinc}^2\left(\frac{\omega c}{2\pi}\right) = c \operatorname{sinc}^2(fc), \quad F[\operatorname{sinc}^2(ct)] = \frac{1}{c} tri\left(\frac{\omega}{2\pi c}\right) = \frac{1}{c} tri\left(\frac{f}{c}\right).$$

19.3.5 TIME-REVERSAL PROPERTY

If $F[f(t)] = F(\omega)$, then $F[f(-t)] = F(-\omega)$.

Proof

Let $-t = t'$. Then, $t = -t'$ and $dt = -dt'$.

$$F[f(-t)] = \int_{-\infty}^{\infty} f(-t)e^{-j\omega t}dt = -\int_{\infty}^{-\infty} f(t')e^{-j(-\omega)t'}dt' = \int_{-\infty}^{\infty} f(t')e^{-j(-\omega)t'}dt' = F(-\omega)$$

If $f(t)$ is folded (flipped) with respect to the ordinate, the Fourier transform of the folded signal $f(-t)$ is given by $F(-\omega)$.

EXAMPLE 19.9

Determine the Fourier transform of

a. $f_1(t) = e^{-at}u(t)$ **b.** $f_2(t) = e^{at}u(-t)$ **c.** $f_3(t) = e^{-a|t|}$

d. $f_1(t) = e^{-at}u(t) - e^{at}u(-t)$ **e.** $sgn(t)$ **f.** $u(t)$

shown in Figure 19.13. Assume $a > 0$.

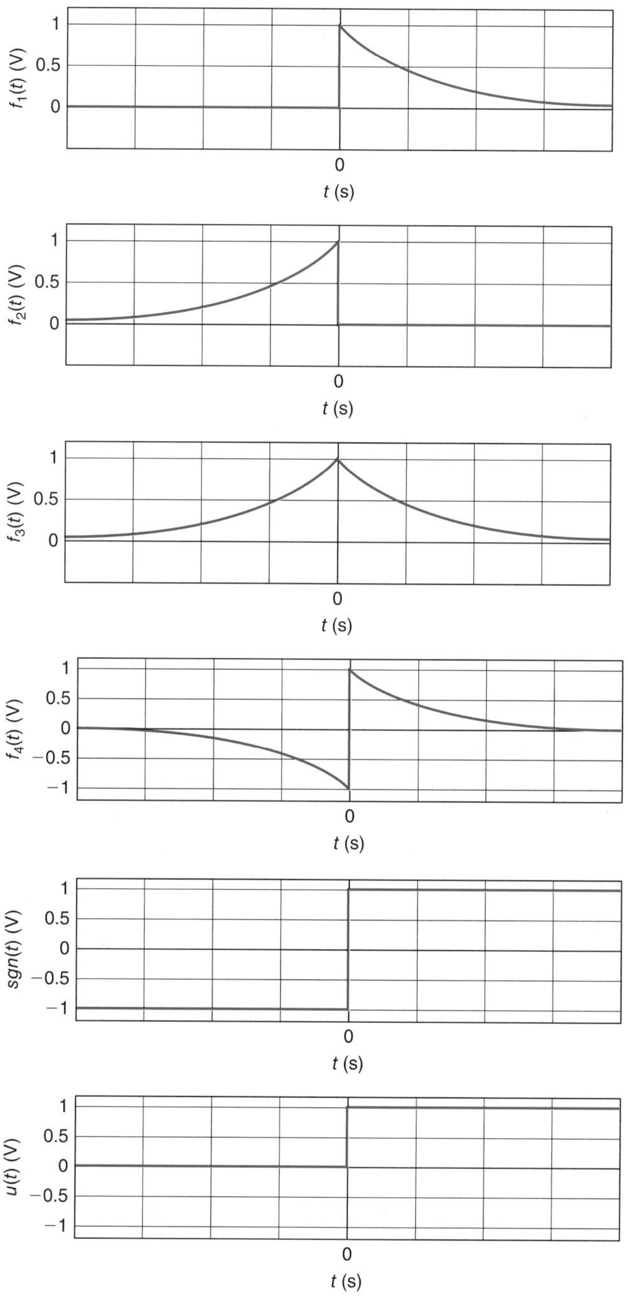

FIGURE 19.13

Waveforms for EXAMPLE 19.9

a. The Fourier transform of $f_1(t)$ is given by

$$F_1(\omega) = \int_0^\infty e^{-at}e^{-j\omega t}dt = \int_0^\infty e^{-(j\omega+a)t}dt = \frac{1}{j\omega + a}$$

b. Since $f_2(t) = f_1(-t)$

$$F_2(\omega) = F_1(-\omega) = \frac{1}{-j\omega + a}$$

c. Since $f_3(t) = f_1(t) + f_2(t)$

$$F_3(\omega) = F_1(\omega) + F_2(\omega) = \frac{1}{j\omega + a} + \frac{1}{-j\omega + a}$$

$$= \frac{2a}{\omega^2 + a^2}$$

d. Since $f_4(t) = f_1(t) - f_2(t)$

$$F_4(\omega) = F_1(\omega) - F_2(\omega) = \frac{1}{j\omega + a} - \frac{1}{-j\omega + a}$$

$$= \frac{-j2\omega}{\omega^2 + a^2}$$

e. Notice that the signum function $sgn(t)$, defined by

$$sgn(t) = \begin{cases} 1, & t > 0 \\ 0, & t = 0 \\ -1, & t < 0 \end{cases}$$

is obtained from the waveform $f_4(t)$ by letting $a \to 0$. Therefore

$$F[sgn(t)] = \lim_{a\to 0}[F_4(\omega)] = \frac{2}{j\omega}$$

continued

Example 19.9 continued

f. Since $u(t) = 0.5[1 + sgn(t)]$

$$F[u(t)] = F[0.5] + F[0.5\,sgn(t)] = \pi\delta(\omega) + \frac{1}{j\omega}$$

Although it is not possible to find the Fourier transform of $u(t)$ directly from

$$F[u(t)] = \int_{-\infty}^{\infty} u(t)e^{-j\omega t}dt = \int_{0}^{\infty} e^{-j\omega t}dt$$

we found $F[u(t)]$ using $F[0.5]$ and $F[sgn(t)]$.

Exercise 19.9

Find the Fourier transform for $u(-t)$.

Answer:

$$F[\delta(-t)] = \pi\delta(\omega) - \frac{1}{j\omega}.$$

19.3.6 FREQUENCY-SHIFTING PROPERTY

If $F[f(t)] = F(\omega)$, then

$$F[f(t)e^{j\omega_c t}] = F(\omega - \omega_c) \tag{19.46}$$

$$F[f(t)e^{j\omega_c t}] = F(f - f_c) \tag{19.47}$$

Proof

$$F[f(t)e^{j\omega_c t}] = \int_{-\infty}^{\infty} f(t)e^{j\omega_c t}e^{-j\omega t}dt = \int_{-\infty}^{\infty} f(t)e^{-j(\omega-\omega_c)t}dt = F(\omega - \omega_c)$$

The frequency-shifting property says that multiplication of a signal $f(t)$ by $e^{j\omega_c t}$ translates the entire spectrum $F(\omega)$ by ω_c.

EXAMPLE 19.10

Find and plot the Fourier transform of

a. $f(t) = e^{-5|t|}$

b. $g(t) = e^{-5|t|}e^{j60t}$

a. $F(\omega) = \dfrac{2 \times 5}{\omega^2 + 5^2} = \dfrac{10}{\omega^2 + 25}$. $F(\omega)$ is shown in Figure 19.14.

continued

Example 19.10 continued

FIGURE 19.14

Plot of $F(\omega)$.

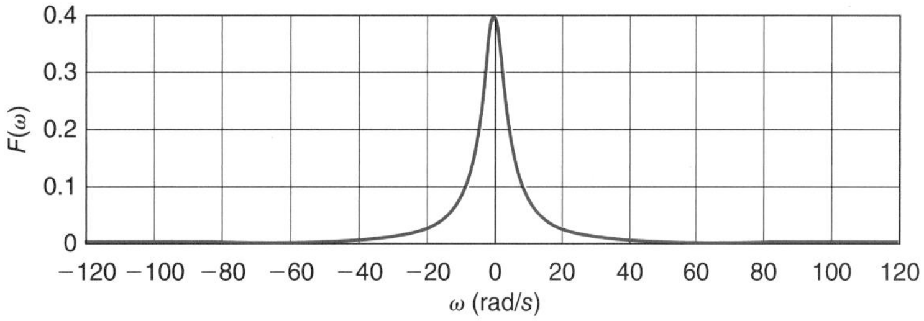

b. $G(\omega) = F(\omega - 60) = \dfrac{10}{(\omega - 60)^2 + 25}$. $G(\omega)$ is shown in Figure 19.15.

FIGURE 19.15

Plot of $G(\omega)$.

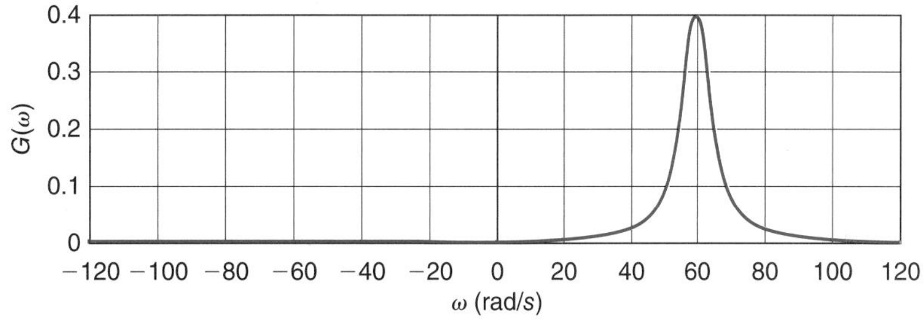

In this plot, the spectrum $G(\omega) = F(\omega - \omega_c)$ is not symmetrical with respect to $\omega = 0$. This can happen since $f(t)e^{j\omega_c t}$ is a complex signal. The magnitude spectrum of real signal $f(t)$ possesses even symmetry, $|F(-\omega)| = |F(\omega)|$, and the phase spectrum of real signal $f(t)$ posesses odd symmetry, $\angle F(-\omega) = -\angle F(\omega)$.

Exercise 19.10

Find the Fourier transform of $f(t) = rect(t)e^{j30t}$.

Answer:

$$F(\omega) = \text{sinc}\left(\frac{\omega - 30}{2\pi}\right).$$

EXAMPLE 19.11

Determine the Fourier transform of (a) $\cos(\omega_c t)$ and (b) $\sin(\omega_c t)$.

a. $F[\cos(\omega_c t)] = F\left[\dfrac{1}{2}e^{j\omega_c t}\right] + F\left[\dfrac{1}{2}e^{-j\omega_c t}\right] = \pi[\delta(\omega - \omega_c) + \delta(\omega + \omega_c)]$

where we applied the frequency-shifting property to $F[1] = 2\pi\delta(\omega)$.
Since $F[1] = \delta(f)$, in terms of f (in hertz)

$$F[\cos(\omega_c t)] = F\left[\frac{1}{2}e^{j\omega_c t}\right] + F\left[\frac{1}{2}e^{-j\omega_c t}\right] = \frac{1}{2}[\delta(f - f_c) + \delta(f + f_c)]$$

continued

Example 19.11 continued

This equation indicates that the spectrum of a single cosine of frequency f_c consists of two impulses with the weight one-half each at f_c and $-f_c$, confirming the fact that $\cos(\omega_c t)$ has only one frequency component. The magnitude spectrum of $F[\cos(\omega_c t)]$ is plotted in Figure 19.16. Notice that the phase spectrum is zero everywhere.

FIGURE 19.16

The magnitude spectrum of $\cos(2\pi f_c t)$.

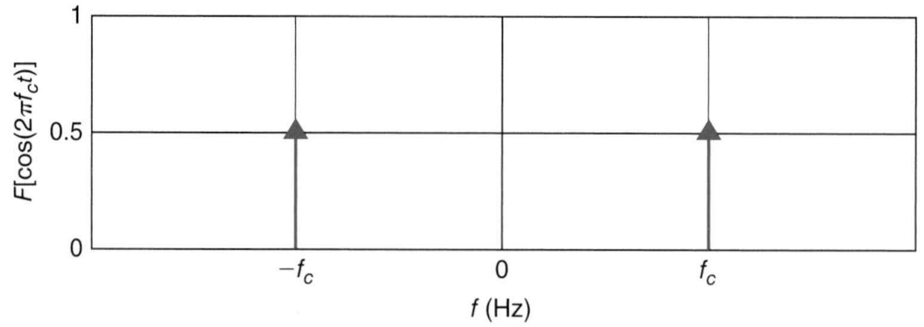

b. $F[\sin(\omega_c t)] = F\left[\frac{1}{2j}e^{j\omega_c t}\right] - F\left[\frac{1}{2j}e^{-j\omega_c t}\right] = -j\pi\,[\delta(\omega - \omega_c) - \delta(\omega + \omega_c)]$

In terms of f (in hertz)

$$F[\sin(\omega_c t)] = F\left[\frac{1}{2j}e^{j\omega_c t}\right] - F\left[\frac{1}{2j}e^{-j\omega_c t}\right] = -\frac{j}{2}[\delta(f - f_c) - \delta(f + f_c)]$$

Notice that $F[\sin(\omega_c t)]$ has the same magnitude spectrum as $F[\cos(\omega_c t)]$ shown in Figure 19.16, but there exists a negative 90° phase offset in the phase spectrum. The phase spectrum of $F[\sin(\omega_c t)]$ is shown in Figure 19.17.

FIGURE 19.17

The phase spectrum of $\sin(2\pi f_c t)$.

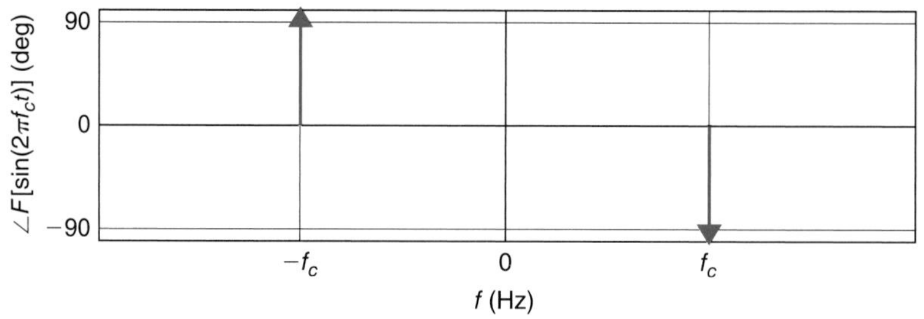

Exercise 19.11

Find the Fourier transform of

a. $\cos(\omega_c t + \phi)$ **b.** $\sin(\omega_c t + \phi)$

a. $F[\cos(\omega_c t + \phi)] = F\left[\frac{1}{2}e^{j\phi}e^{j\omega_c t}\right] + F\left[\frac{1}{2}e^{-j\phi}e^{-j\omega_c t}\right] = \pi\left[\delta(\omega - \omega_c)e^{j\phi} + \delta(\omega + \omega_c)e^{-j\phi}\right]$

continued

Exercise 19.11 continued

or in terms of frequency f (in hertz)

$$F[\cos(\omega_c t + \phi)] = F\left[\frac{1}{2}e^{j\phi}e^{j\omega_c t}\right] + F\left[\frac{1}{2}e^{-j\phi}e^{-j\omega_c t}\right] = \frac{1}{2}\left[\delta(f - f_c)e^{j\phi} + \delta(f + f_c)e^{-j\phi}\right]$$

b. $$F[\sin(\omega_c t + \phi)] = F\left[\frac{1}{2j}e^{j\phi}e^{j\omega_c t}\right] - F\left[\frac{1}{2}e^{-j\phi}e^{-j\omega_c t}\right] = -j\pi\left[\delta(\omega - \omega_c)e^{j\phi} - \delta(\omega + \omega_c)e^{-j\phi}\right]$$

or in terms of frequency f (in hertz)

$$F[\sin(\omega_c t + \phi)] = F\left[\frac{1}{2j}e^{j\phi}e^{j\omega_c t}\right] - F\left[\frac{1}{2j}e^{-j\phi}e^{-j\omega_c t}\right] = \frac{-j}{2}\left[\delta(f - f_c)e^{j\phi} - \delta(f + f_c)e^{-j\phi}\right]$$

Notice that $F[\cos(\omega_c t + \phi)]$ and $F[\sin(\omega_c t + \phi)]$ have the same magnitude spectrum as $F[\cos(\omega_c t)]$, shown in Figure 19.16. The phase spectrum of $F[\cos(\omega_c t + \phi)]$ is shown in Figure 19.18.

FIGURE 19.18

Phase spectrum of $\cos(\omega_c t + \phi)$.

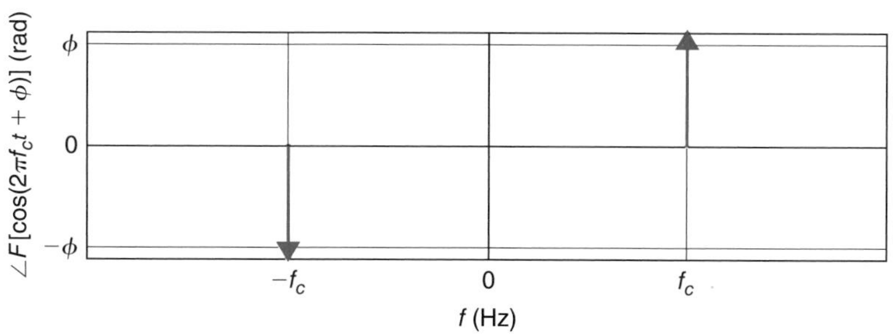

and the phase spectrum of $F[\sin(\omega_c t + \phi)]$ is shown in Figure 19.19.

FIGURE 19.19

Phase spectrum of $\sin(\omega_c t + \phi)$.

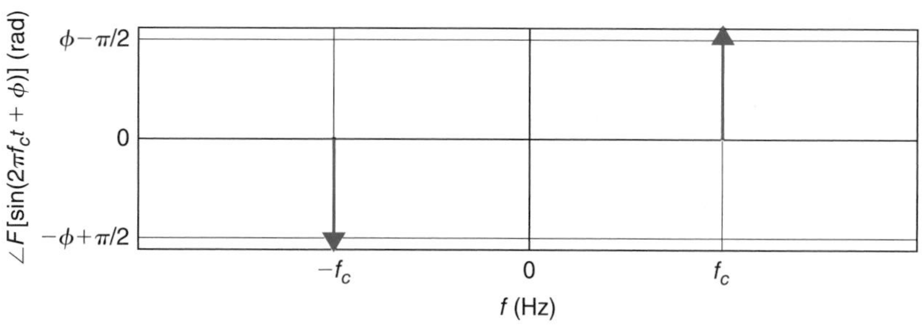

19.3.7 MODULATION PROPERTY

If $F[f(t)] = F(\omega)$, then

$$F[f(t)\cos(\omega_c t)] = \frac{1}{2}F(\omega - \omega_c) + \frac{1}{2}F(\omega + \omega_c) \tag{19.48}$$

$$F[f(t)\cos(\omega_c t)] = \frac{1}{2}F(f - f_c) + \frac{1}{2}F(f + f_c) \tag{19.49}$$

Proof

$$F[f(t)\cos(\omega_c t)] = F\left[\frac{1}{2}f(t)e^{j\omega_c t}\right] + F\left[\frac{1}{2}f(t)e^{-j\omega_c t}\right] = \frac{1}{2}F(\omega - \omega_c) + \frac{1}{2}F(\omega + \omega_c)$$

When $f(t)$ is multiplied by $\cos(\omega_c t)$, the entire spectrum $F(\omega)$ is shifted to ω_c and $-\omega_c$ with the amplitude reduced to half.

EXAMPLE 19.12

Find and plot the Fourier transform of

a. $f(t) = e^{-5|t|}$ **b.** $g(t) = e^{-5|t|}\cos(60t)$

and find the 3-dB bandwidth of $f(t)$ and $g(t)$.

a. $F(\omega) = \dfrac{2 \times 5}{\omega^2 + 5^2} = \dfrac{10}{\omega^2 + 25}$

$F(\omega)$ is shown in Figure 19.20.

FIGURE 19.20

Plot of $F(\omega)$.

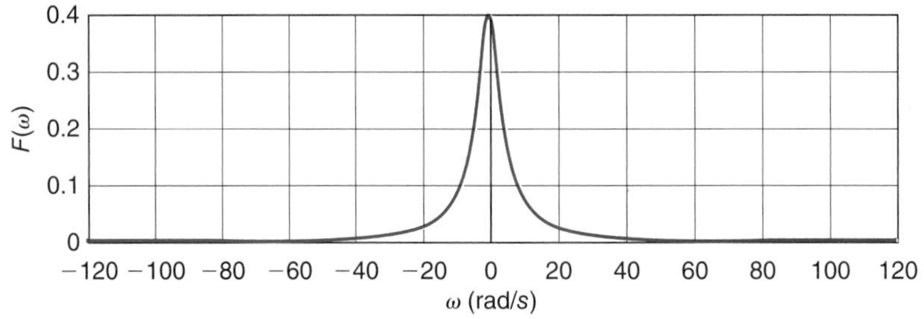

b. $G(\omega) = \dfrac{1}{2}F(\omega - 60) + \dfrac{1}{2}F(\omega + 60) = \dfrac{5}{(\omega - 60)^2 + 25} + \dfrac{5}{(\omega + 60)^2 + 25}$

$G(\omega)$ is shown in Figure 19.21.

FIGURE 19.21

Plot of $G(\omega)$.

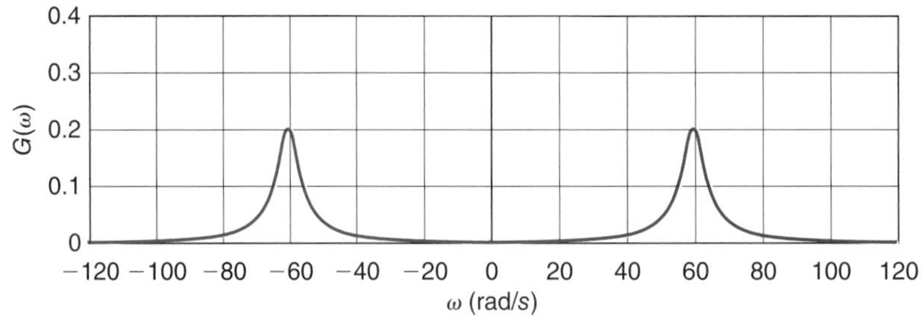

continued

Example 19.12 continued

For $f(t)$, the 3-dB bandwidth is found by solving the following equation for ω

$$\frac{10}{\omega^2 + 25} = \frac{0.4}{\sqrt{2}}$$

$$\omega_{3dBf} = 5\sqrt{\sqrt{2} - 1} = 3.218 \text{ rad/s}$$

For $g(t)$, the 3-dB bandwidth is two times the 3-dB bandwidth of $f(t)$. Thus, we have

$$\omega_{3dBg} = 2 \times \omega_{3dBf} = 6.4359 \text{ rad/s}$$

Exercise 19.12

Find the Fourier transform of $f(t) = 15e^{-10t}u(t) \times \cos(20t)$.

Answer:

$$F(\omega) = \frac{7.5}{j(\omega - 20) + 10} + \frac{7.5}{j(\omega + 20) + 10}.$$

EXAMPLE 19.13

Find the Fourier transform of

a. $f(t) = A\,rect\left(\frac{t}{\tau}\right)$ **b.** $g(t) = A\,rect\left(\frac{t}{\tau}\right)\cos(\omega_c t)$

and plot $F(\omega)$ and $G(\omega)$ for $A = 1$ V, $\tau = 2\pi/100 = 62.831853$ ms, $\omega_c = 600$ rad/s

a. $F(\omega) = A\tau \text{ sinc}\left(\frac{\omega\tau}{2\pi}\right)$

$F(\omega)$ is shown in Figure 19.22.

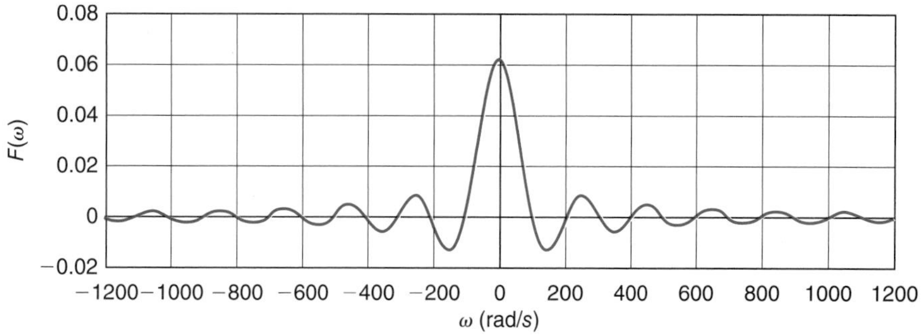

FIGURE 19.22 Plot of $F(\omega)$.

continued

Example 19.13 continued

b. $$G(\omega) = \frac{A\tau}{2}\text{sinc}\left[\frac{(\omega - \omega_c)\tau}{2\pi}\right] + \frac{A\tau}{2}\text{sinc}\left[\frac{(\omega + \omega_c)\tau}{2\pi}\right]$$

$G(\omega)$ is shown in Figure 19.23.

FIGURE 19.23

Plot of $G(\omega)$.

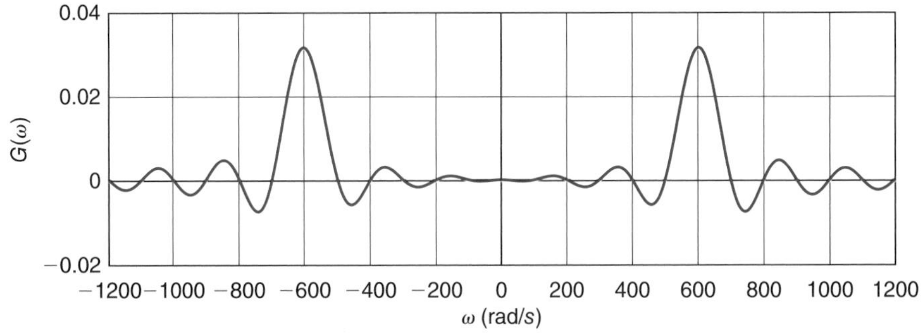

Exercise 19.13

Find the Fourier transform of $f(t) = tri(t)\cos(30t)$.

Answer:

$$F(\omega) = \frac{1}{2}\text{sinc}^2\left(\frac{\omega - 30}{2\pi}\right) + \frac{1}{2}\text{sinc}^2\left(\frac{\omega + 30}{2\pi}\right).$$

19.3.8 TIME-DIFFERENTIATION PROPERTY

If $F[f(t)] = F(\omega)$, then

$$F\left[\frac{df(t)}{dt}\right] = j\omega F(\omega) \tag{19.50}$$

In general

$$F\left[\frac{d^n f(t)}{dt^n}\right] = (j\omega)^n F(\omega) \tag{19.51}$$

Proof

Taking the time derivative of the inverse Fourier transform

$$f(t) = \frac{1}{2\pi}\int_{-\infty}^{\infty} F(\omega)e^{j\omega t}\,d\omega$$

we get

$$\frac{df(t)}{dt} = \frac{1}{2\pi}\int_{-\infty}^{\infty} j\omega F(\omega)e^{j\omega t}d\omega$$

Therefore

$$F\left[\frac{df(t)}{dt}\right] = j\omega F(\omega)$$

Repeating this procedure for n times, we obtain the general formula. Time differentiation introduces a 90° phase shift, emphasizes (amplifies) the high-frequency components, and attenuates the low-frequency components, since $F(\omega)$ is multiplied by $j\omega$. Thus, time differentiation can be considered as a high-pass operation. The transfer function of time differentiation is given by

$$H(\omega) = j\omega$$

EXAMPLE 19.14

Use the time-differentiation property to find the Fourier transform of the triangular pulse $f(t) = A\, tri\left(\frac{t}{\tau}\right)$ shown in Figure 19.24, and plot the Fourier transform.

FIGURE 19.24

A triangular pulse and its derivative.

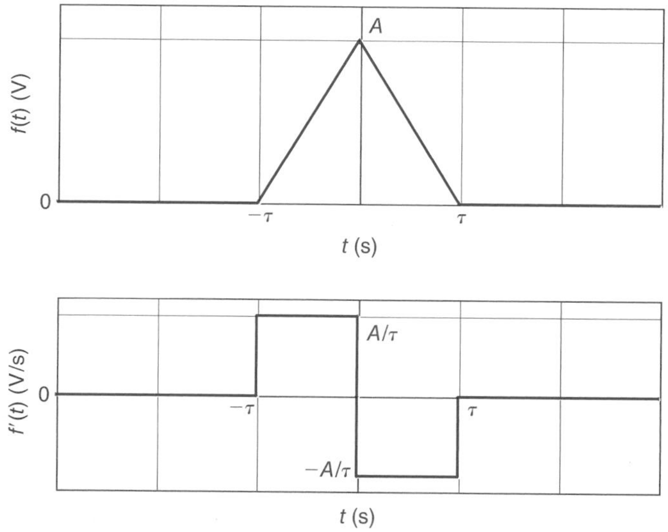

The derivative of $f(t)$ is shown in Figure 19.24. From this figure, it is obvious that

$$j\omega F(\omega) = A \operatorname{sinc}\left(\frac{\omega\tau}{2\pi}\right)\left(e^{j\omega\frac{\tau}{2}} - e^{-j\omega\frac{\tau}{2}}\right)$$

$$= A \operatorname{sinc}\left(\frac{\omega\tau}{2\pi}\right) 2j \sin\left(\frac{\omega\tau}{2}\right)$$

Thus,

$$F(\omega) = \frac{A\tau \operatorname{sinc}\left(\frac{\omega\tau}{2\pi}\right)\sin\left(\frac{\omega\tau}{2}\right)}{\frac{\omega\tau}{2}}$$

$$= A\tau \operatorname{sinc}^2\left(\frac{\omega\tau}{2\pi}\right)$$

which is shown in Figure 19.25.

Since $F(\omega) \geq 0$ for all ω, the phase angle of $F(\omega)$ is 0 for all ω. The side-lobe levels are smaller than those of $\operatorname{sinc}\left(\frac{\omega\tau}{2\pi}\right)$ because the triangular pulse does not have discontinuous points like the rectangular pulse. To see the side lobes better, we can plot the magnitude in the decibel scale given by $20 \log_{10}|H(\omega)|$ instead of $|H(\omega)|$. Figure 19.26 shows the magnitude for a rectangular pulse and a triangular pulse in dB.

FIGURE 19.25

The Fourier transform of a triangular pulse.

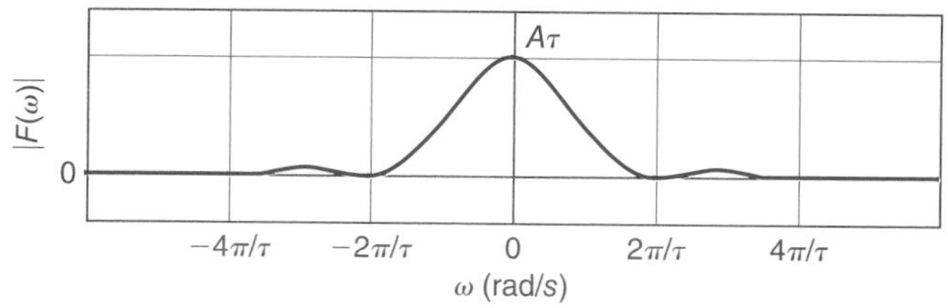

continued

Example 19.14 continued

FIGURE 19.26

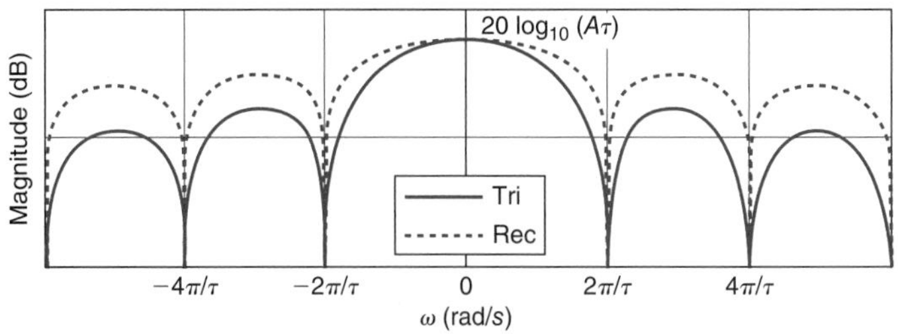

Magnitude spectrum of a rectangular pulse and a triangular pulse.

Exercise 19.14

Use the time-differentiation property to find the Fourier transform of $f(t)$ shown in Figure 19.27

Answer:

$$j\omega F(\omega) = A\,\mathrm{sinc}\left(\frac{\omega\tau}{2\pi}\right)e^{-j\omega\frac{\tau}{2}} - Ae^{-j\omega\tau} \qquad F(\omega) = \frac{A}{j\omega}e^{-j\omega\tau}\left[\mathrm{sinc}\left(\frac{\omega\tau}{2\pi}\right)e^{j\omega\frac{\tau}{2}} - 1\right].$$

FIGURE 19.27

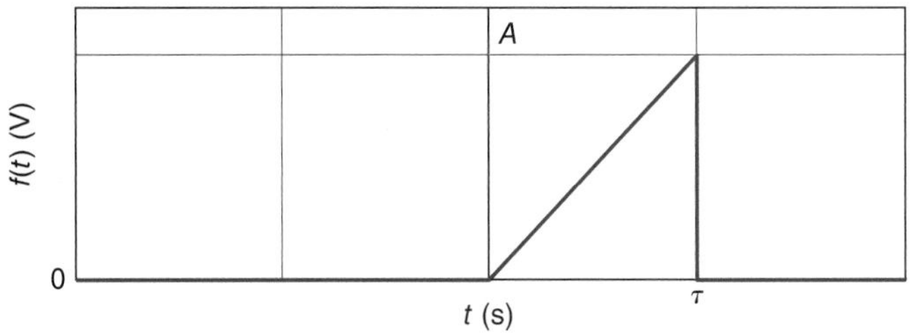

Waveform for EXERCISE 19.14.

FIGURE 19.28

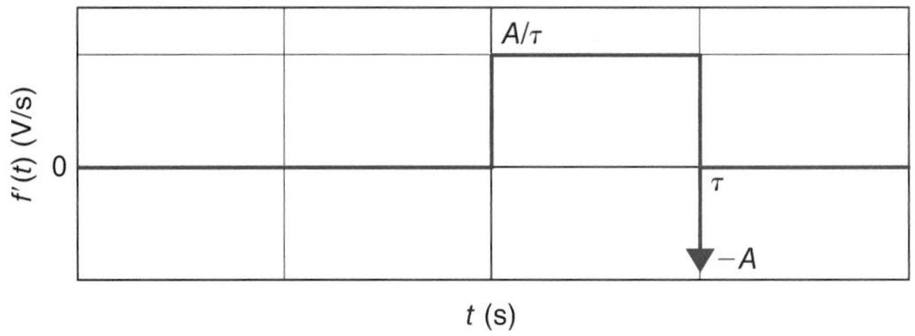

A signal $df(t)/dt$.

EXAMPLE 19.15

Use the time-differentiation property to find the Fourier transform of the ramp function $f(t) = t\,u(t)$.

Since $f'(t) = u(t)$ and $F[u(t)] = \pi\delta(\omega) + 1/(j\omega)$, we have

$$j\omega F(\omega) = \pi\delta(\omega) + \frac{1}{j\omega}$$

Therefore

$$F[tu(t)] = \frac{\pi}{j\omega}\delta(\omega) + \frac{1}{(j\omega)^2}$$

Exercise 19.15

Find the Fourier transform of $\delta'(t)$ and $\delta^{(n)}(t)$.

Answer:
$F[\delta'(t)] = j\omega, \quad F[\delta^{(n)}(t)] = (j\omega)^n.$

19.3.9 FREQUENCY-DIFFERENTIATION PROPERTY

If $F[f(t)] = F(\omega)$, then

$$F[tf(t)] = j\frac{dF(\omega)}{d\omega} \tag{19.52}$$

In general,

$$F[t^n f(t)] = j^n \frac{d^n F(\omega)}{d\omega^n} \tag{19.53}$$

Proof

Taking the derivative of the definition of the Fourier transform

$$F(\omega) = \int_{-\infty}^{\infty} f(t)e^{-j\omega t}\,dt$$

with respect to ω, we get

$$\frac{dF(\omega)}{d\omega} = \int_{-\infty}^{\infty} (-jt)f(t)e^{-j\omega t}\,dt$$

Therefore,

$$F[-jtf(t)] = \frac{dF(\omega)}{d\omega}$$

or

$$F[tf(t)] = j\frac{dF(\omega)}{d\omega}$$

If this procedure is repeated n times, the general formula is obtained. The frequency-differentiation property is the converse to the time-differentiation property.

EXAMPLE 19.16

Determine the Fourier transform of $f(t) = te^{-at}u(t)$, $a > 0$.

continued

Example 19.16 continued

It has been shown that

$$F[e^{-at}u(t)] = \frac{1}{j\omega + a}$$

Therefore,

$$F[te^{-at}u(t)] = j\frac{d}{d\omega}\left[\frac{1}{j\omega + a}\right] = \frac{1}{(j\omega + a)^2}$$

Exercise 19.16

Find the Fourier transform of $f(t) = t^2 e^{-at}u(t), a > 0$.

Answer:

$$F(\omega) = \frac{2}{(j\omega + a)^3}.$$

19.3.10 CONJUGATE PROPERTY

If $F[f(t)] = F(\omega)$, then

$$F[f^*(t)] = F^*(-\omega) \quad \textbf{(19.54)}$$

Proof

$$F[f^*(t)] = \int_{-\infty}^{\infty} f^*(t)e^{-j\omega t}dt = \left[\int_{-\infty}^{\infty} f(t)e^{-j(-\omega)t}dt\right]^* = F^*(-\omega)$$

Applying the time-reversal property, we get

$$F[f^*(-t)] = F^*[-(-\omega)] = F^*(\omega)$$

If $f(t)$ is real, $f^*(t) = f(t)$. Since $F[f(t)] = F(\omega)$ and $F[f^*(t)] = F^*(-\omega)$, we have

$$F(\omega) = F^*(-\omega)$$

or

$$F^*(\omega) = F(-\omega)$$

Furthermore, if $f(t)$ is real and even so that $f^*(-t) = f(t)$, then

$$F^*(\omega) = F(\omega)$$

Therefore, $F(\omega)$ is even.

19.3.11 INTEGRATION PROPERTY

If $F[f(t)] = F(\omega)$, then

$$F\left[\int_{-\infty}^{t} f(\lambda)d\lambda\right] = \frac{1}{j\omega}F(\omega) + \pi F(0)\delta(\omega) \quad \textbf{(19.55)}$$

If $F(0) = 0$, the integration property reduces to

$$F\left[\int_{-\infty}^{t} f(\lambda)d\lambda\right] = \frac{1}{j\omega}F(\omega) = \frac{-j}{\omega}F(\omega)$$

Integration introduces a -90° phase shift, emphasizes the low-frequency components, and attenuates the high-frequency components. Thus, integration has the role of the low-pass filter (LPF) in smoothing out signals.

Proof

Note that

$$\int_{-\infty}^{t} f(\lambda)d\lambda = \int_{-\infty}^{\infty} f(\lambda)u(t-\lambda)d\lambda$$

since $u(t-\lambda)$ is 1 for $\lambda < t$ and is 0 for $\lambda > t$, as shown in Figure 19.29.

FIGURE 19.29

Plot of $u(t-\lambda)$.

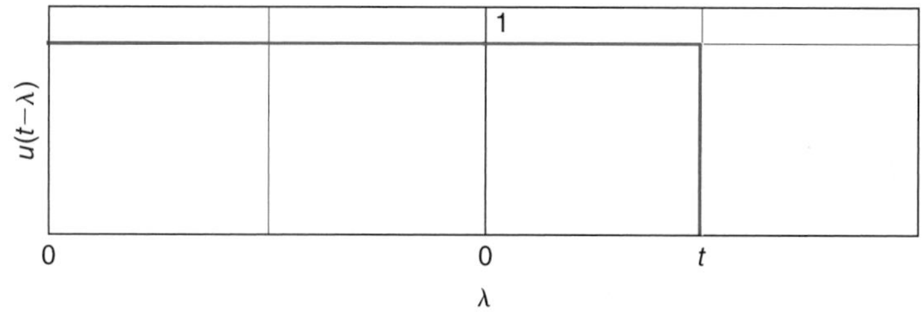

Therefore

$$F\left[\int_{-\infty}^{t} f(\lambda)d\lambda\right] = \int_{-\infty}^{\infty}\left[\int_{-\infty}^{\infty} f(\lambda)u(t-\lambda)d\lambda\right]e^{-j\omega t}dt$$

Interchanging the order of integration, we obtain

$$F\left[\int_{-\infty}^{t} f(\lambda)d\lambda\right] = \int_{-\infty}^{\infty} f(\lambda)\left[\int_{-\infty}^{\infty} u(t-\lambda)e^{-j\omega t}dt\right]d\lambda$$

As a function of t, $u(t-\lambda)$ is a unit step function delayed by λ. Applying the time-shifting property to

$$F[u(t)] = \frac{1}{j\omega} + \pi\delta(\omega)$$

we get

$$F[u(t-\lambda)] = \left[\frac{1}{j\omega} + \pi\delta(\omega)\right]e^{-j\omega\lambda}$$

Therefore,

$$F\left[\int_{-\infty}^{t} f(\lambda)d\lambda\right] = \int_{-\infty}^{\infty} f(\lambda)\left[\frac{1}{j\omega} + \pi\delta(\omega)\right]e^{-j\omega\lambda}d\lambda = \left[\frac{1}{j\omega} + \pi\delta(\omega)\right]\int_{-\infty}^{\infty} f(\lambda)e^{-j\omega\lambda}d\lambda$$

$$= \left[\frac{1}{j\omega} + \pi\delta(\omega)\right]F(\omega) = \frac{F(\omega)}{j\omega} + \pi F(0)\delta(\omega)$$

since $F(\omega)\delta(\omega) = F(0)\delta(\omega)$

19.3.12 CONVOLUTION PROPERTY

If $F[x(t)] = X(\omega), F[h(t)] = H(\omega), F[y(t)] = Y(\omega)$, and

$$y(t) = h(t)*x(t) = \int_{-\infty}^{\infty} h(\lambda)x(t-\lambda)d\lambda \tag{19.56}$$

then

$$Y(\omega) = H(\omega)X(\omega) \tag{19.57}$$

Proof

$$Y(\omega) = \int_{-\infty}^{\infty}\left[\int_{-\infty}^{\infty} h(\lambda)x(t-\lambda)d\lambda\right]e^{-j\omega t}dt$$

Interchanging the order of integration, we obtain

$$Y(\omega) = \int_{-\infty}^{\infty} h(\lambda)\left[\int_{-\infty}^{\infty} x(t-\lambda)e^{-j\omega t}dt\right]d\lambda$$

From the time-shifting property, the integral inside the bracket is $X(\omega)e^{-j\omega\lambda}$. Therefore,

$$Y(\omega) = \int_{-\infty}^{\infty} h(\lambda)X(\omega)e^{-j\omega\lambda}d\lambda = X(\omega)\int_{-\infty}^{\infty} h(\lambda)e^{-j\omega\lambda}d\lambda = X(\omega)H(\omega) = H(\omega)X(\omega)$$

The convolution property says that the convolution in the time domain corresponds to the multiplication in the frequency domain. If a system is linear and time-invariant (LTI), the output of the system is given by the convolution of the input and the impulse response of the system. This means that the Fourier transform of the output of the LTI system shown in Figure 19.30 is given by the product of the Fourier transform of the input signal and the transfer function, which is the Fourier transform of the impulse response of the system. Thus, the output of the system can be obtained by taking the inverse Fourier transform of the output Fourier transform.

FIGURE 19.30

Block diagram for an LTI system.

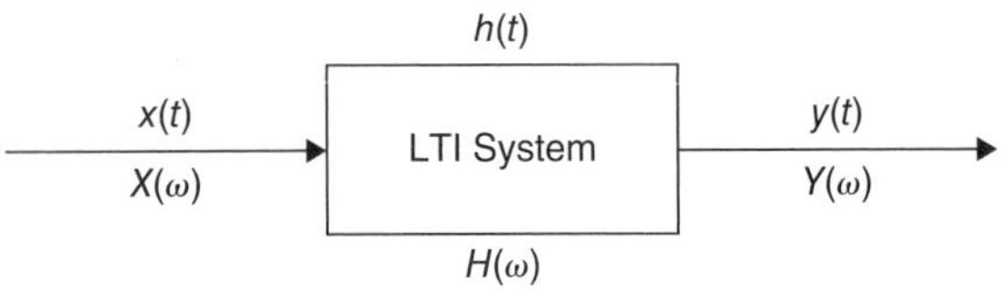

EXAMPLE 19.17

Use the convolution property to find the Fourier transform of the triangular pulse $f(t) = A\, tri\left(\frac{t}{\tau}\right)$.

Since

$$f(t) = \sqrt{\frac{A}{\tau}} rect\left(\frac{t}{\tau}\right) * \sqrt{\frac{A}{\tau}} rect\left(\frac{t}{\tau}\right)$$

we obtain

$$F(\omega) = \left[\sqrt{\frac{A}{\tau}}\tau \operatorname{sinc}\left(\frac{\omega\tau}{2\pi}\right)\right]^2 = A\tau \operatorname{sinc}^2\left(\frac{\omega\tau}{2\pi}\right)$$

which is the same as that obtained in EXAMPLE 19.14 by using the time-differentiation property.

Exercise 19.17

Find the Fourier transform of $f(t) = tri(t) * tri(t)$.

Answer:

$F(\omega) = \operatorname{sinc}^4\left(\frac{\omega}{2\pi}\right).$

EXAMPLE 19.18

Use the convolution property to find the convolution of $f_1(t) = e^{-2t}u(t)$ and $f_2(t) = e^{5t}u(-t)$ shown here.

FIGURE 19.31

Waveforms $f_1(t)$ and $f_2(t)$.

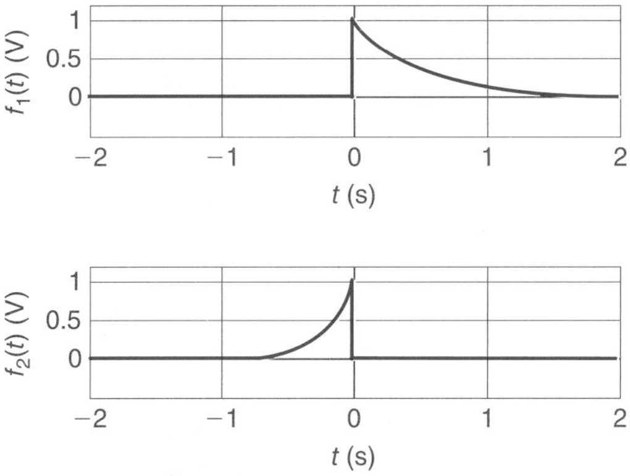

It has been shown in EXAMPLE 19.9 that

$$F_1(\omega) = \frac{1}{j\omega + 2} \quad \text{and} \quad F_2(\omega) = \frac{1}{-j\omega + 5}$$

Therefore,

$$F(\omega) = \frac{1}{(j\omega + 2)(-j\omega + 5)}$$

The convolution $f(t)$ can be found by taking the inverse Fourier transform of $F(\omega)$; that is,

$$f(t) = F^{-1}[F(\omega)]$$

continued

Example 19.18 continued

To facilitate the procedure of taking the inverse Fourier transform, we expand $F(\omega)$ in partial fraction as

$$\frac{1}{(j\omega + 2)(-j\omega + 5)} = \frac{A}{j\omega + 2} + \frac{B}{-j\omega + 5}$$

Clearing the fractions on the right side, we obtain

$$\frac{1}{(j\omega + 2)(-j\omega + 5)} = \frac{(-A + B)j\omega + (5A + 2B)}{(j\omega + 2)(-j\omega + 5)}$$

Comparing the coefficients in the numerator, we get

$$-A + B = 0, \quad 5A + 2B = 1$$

Therefore, $A = B = 1/7$. The convolution is given by

$$f(t) = F^{-1}\left[\frac{\frac{1}{7}}{j\omega + 2} + \frac{\frac{1}{7}}{-j\omega + 5}\right] = \frac{1}{7}e^{-2t}u(t) + \frac{1}{7}e^{5t}u(-t)$$

Exercise 19.18

Find the Fourier transform of $f(t) = e^{-2t}u(t) * e^{-3t}u(t)$.

Answer:
$f(t) = (e^{-2t} - e^{-3t})u(t)$.

EXAMPLE 19.19

A signal $x(t) = u(t)$ is applied to an LTI system with the impulse response $h(t) = e^{-2t}u(t)$. Find the output $y(t)$ of the system.

The Fourier transform of $u(t)$ is given by

$$X(\omega) = F[u(t)] = \pi\delta(\omega) + \frac{1}{j\omega}$$

The transfer function of the LTI system is given by

$$H(\omega) = F[h(t)] = \frac{1}{j\omega + 2}$$

continued

Example 19.19 continued

The Fourier transform $Y(\omega)$ of the output $y(t)$ is the product of $X(\omega)$ and $H(\omega)$; that is,

$$Y(\omega) = X(\omega)H(\omega) = \left[\pi\delta(\omega) + \frac{1}{j\omega}\right]\frac{1}{j\omega + 2} = \frac{\pi\delta(\omega)}{j\omega + 2} + \frac{1}{j\omega(j\omega + 2)}$$

$$= \frac{\pi}{2}\delta(\omega) + \frac{\frac{1}{2}}{j\omega} - \frac{\frac{1}{2}}{j\omega + 2} = \frac{1}{2}\left[\pi\delta(\omega) + \frac{1}{j\omega}\right] - \frac{1}{2}\frac{1}{j\omega + 2}$$

Taking the inverse Fourier transform, we obtain

$$y(t) = \frac{1}{2}(1 - e^{-2t})u(t)$$

Exercise 19.19

A signal $x(t) = u(t) - u(t - 2)$ is applied to an LTI system with the impulse response $h(t) = e^{-2t}u(t)$. Find the output $y(t)$ of the system.

Answer:

$$y(t) = \frac{1}{2}(1 - e^{-2t})u(t) - \frac{1}{2}(1 - e^{-2(t-2)})u(t - 2).$$

19.3.13 MULTIPLICATION PROPERTY

If $F[f_1(t)] = F_1(\omega)$ and $F[f_2(t)] = F_2(\omega)$, then

$$F[f_1(t)f_2(t)] = \frac{1}{2\pi}F_1(\omega) * F_1(\omega) = \frac{1}{2\pi}\int_{-\infty}^{\infty} F_1(v)F_2(\omega - v)dv \tag{19.58}$$

or

$$F[f_1(t)f_2(t)] = F_1(f) * F_1(f) = \int_{-\infty}^{\infty} F_1(v)F_2(f - v)dv \tag{19.59}$$

In general,

$$F[f_1(t)f_2(t)\ldots f_n(t)] = \frac{1}{(2\pi)^{n-1}}F_1(\omega) * F_1(\omega) * \ldots * F_n(\omega) \tag{19.60}$$

or

$$F[f_1(t)f_2(t)\ldots f_n(t)] = F_1(f) * F_1(f) * \ldots * F_n(f) \tag{19.61}$$

Proof

$$F[f_1(t)f_2(t)] = \int_{-\infty}^{\infty} f_1(t)f_2(t)e^{-j\omega t}dt$$

Replacing $f_1(t)$ by

$$f_1(t) = \frac{1}{2\pi}\int_{-\infty}^{\infty} F_1(v)e^{jvt}\,dv$$

we obtain

$$F[f_1(t)f_2(t)] = \int_{-\infty}^{\infty}\left[\frac{1}{2\pi}\int_{-\infty}^{\infty} F_1(v)e^{jvt}\,dv\right]f_2(t)e^{-j\omega t}\,dt$$

Interchanging the order of integration, we get

$$F[f_1(t)f_2(t)] = \frac{1}{2\pi}\int_{-\infty}^{\infty} F_1(v)\left[\int_{-\infty}^{\infty} f_2(t)e^{-j(\omega-v)t}dt\right]dv = \frac{1}{2\pi}\int_{-\infty}^{\infty} F_1(v)F_2(\omega - v)dv$$

$$= \frac{1}{2\pi}F_1(\omega) * F_2(\omega)$$

Repeating this procedure n times, we obtain the general formula. The multiplication property is the converse to the convolution property; that is, the multiplication of two signals in the time domain results in the convolution in the frequency domain.

EXAMPLE 19.20

Consider a nonlinear system whose output $y(t)$ is given by the power series of the input $x(t)$:

$$y(t) = a_0 + a_1x(t) + a_2x^2(t) + a_3x^3(t)$$

Express the Fourier transform of the output as a function of the Fourier transform of the input.

$$Y(\omega) = a_0 2\pi\delta(\omega) + a_1X(\omega) + \frac{a_2}{2\pi}X(\omega) * X(\omega) + \frac{a_3}{(2\pi)^2}X(\omega) * X(\omega) * X(\omega)$$

Exercise 19.20

Let $f_1(t) = 2\cos(2\pi \times 1000t)$ and $f_2(t) = 2\cos(2\pi \times 10{,}000t)$. Find the Fourier transform of $f(t) = f_1(t)f_2(t)$.

Answer:
$F(f) = \delta(f - 11{,}000) + \delta(f - 9000) + \delta(f + 9000) + \delta(f + 11{,}000)$.

19.4 Fourier Transform of Periodic Signals

If a signal $f(t)$ is periodic with period T_0, it is not square-integrable. Thus, its Fourier transform $F(\omega)$ cannot be obtained directly from the definition

$$F(\omega) = \int_{-\infty}^{\infty} f(t)e^{-j\omega t}\,dt$$

since the integral does not converge, but $f(t)$ can be expanded in the Fourier series as

$$f(t) = \sum_{n=-\infty}^{\infty} F_n e^{jn\omega_0 t}$$

where $\omega_0 = \dfrac{2\pi}{T_0} = 2\pi f_0$ is the fundamental frequency in radians/s, and F_n is the exponential Fourier coefficients given by

$$F_n = \frac{1}{T_0}\int_{t_0}^{t_0+T_0} f(t)e^{-jn\omega_0 t}\,dt$$

Application of

$$F[e^{jn\omega_0 t}] = 2\pi\delta(\omega - n\omega_0)$$

to

$$f(t) = \sum_{n=-\infty}^{\infty} F_n e^{jn\omega_0 t}$$

gives us

$$F(\omega) = \sum_{n=-\infty}^{\infty} 2\pi F_n \delta(\omega - n\omega_0) \qquad \textbf{(19.62)}$$

which is the Fourier transform of the periodic signal $f(t)$. If frequency f (in hertz) is used instead of ω, this Fourier transform becomes

$$F(f) = \sum_{n=-\infty}^{\infty} F_n \delta(f - nf_0) \qquad \textbf{(19.63)}$$

This equation states that the Fourier transform of a periodic signal consists of a train of impulses separated by f_0 Hz, and the strength (area) of the impulse at nf_0 is given by the nth Fourier coefficient F_n. Thus, the plot of $F(f)$ is exactly the same as that of F_n if each line spectrum is interpreted as an impulse.

EXAMPLE 19.21

Find and sketch the Fourier transform of a rectangular pulse train with amplitude $A = 1$ V, pulse width $\tau = 1/4$ ms, and period $T_0 = 1$ ms.

continued

Example 19.21 continued

The Fourier transform is

$$F(f) = \sum_{n=-\infty}^{\infty} F_n \delta(f - nf_0)$$

with

$$F_n = \frac{1}{4}\text{sinc}\left(\frac{n}{4}\right)$$

The Fourier transform $F(f)$ is shown in Figure 19.32.

FIGURE 19.32

The Fourier transform of a rectangular pulse train.

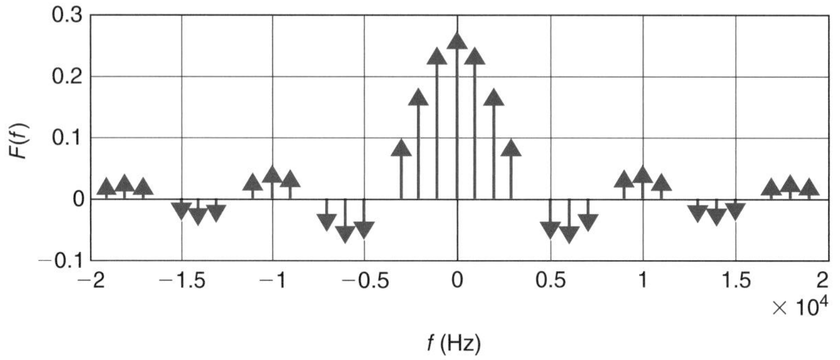

Exercise 19.21

Find the Fourier transform of a triangular pulse train with amplitude $A = 1$ V and period $T_0 = 1$ ms.

Answer:

$$F(f) = \sum_{n=-\infty}^{\infty} \frac{1}{2}\text{sinc}^2\left(\frac{n}{2}\right)\delta(f - n1000).$$

19.4.1 FOURIER SERIES AND FOURIER TRANSFORM OF IMPULSE TRAIN

Consider a rectangular pulse train with amplitude $1/\tau$, period T_0, and pulse width τ shown in Figure 19.33.

FIGURE 19.33

A rectangular pulse train.

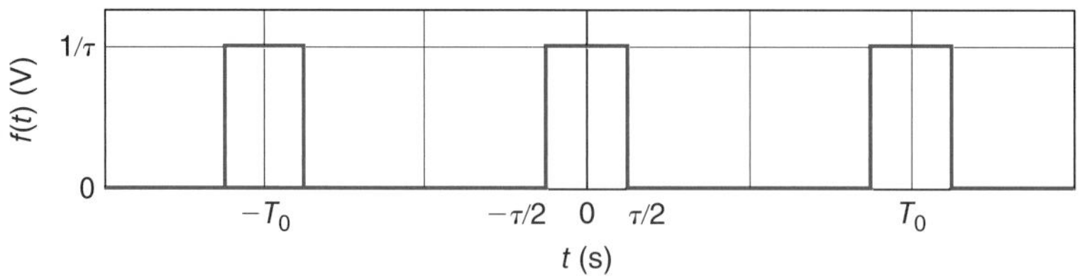

As shown earlier in this chapter, the Fourier series representation of this signal is

$$f(t) = \sum_{n=-\infty}^{\infty} \frac{1}{\tau}\frac{\tau}{T_0}\text{sinc}\left(\frac{n\tau}{T_0}\right)e^{jn\frac{2\pi}{T_0}t} = \sum_{n=-\infty}^{\infty} \frac{1}{T_0}\text{sinc}\left(\frac{n\tau}{T_0}\right)e^{jn\frac{2\pi}{T_0}t}$$

The spectrum of the Fourier coefficient F_n is plotted in Figure 19.34 for duty cycle of 1/4 and period $T_0 = 1$ ms.

FIGURE 19.34

The Fourier coefficients of a rectangular pulse train.

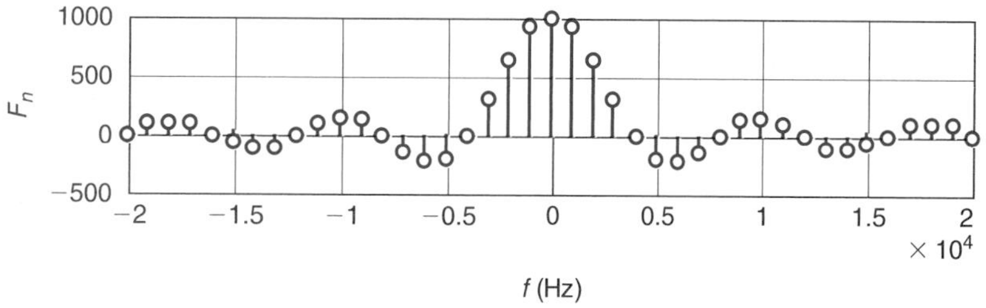

Let us now decrease the pulse width τ to 0 while maintaining the area of each pulse at 1. Then, each pulse becomes an impulse, and the rectangular pulse train becomes the impulse train $f_\delta(t)$ shown in Figure 19.35.

FIGURE 19.35

Impulse train.

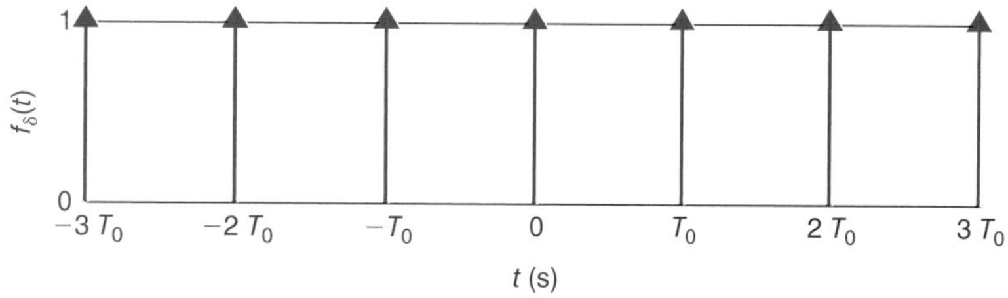

The Fourier coefficient of this impulse train is given by

$$F_{\delta_n} = \lim_{\tau \to 0} \frac{1}{T_0} \operatorname{sinc}\left(\frac{n\tau}{T_0}\right) = \frac{1}{T_0} = f_0 \text{ for all } n$$

which is plotted in Figure 19.36.

FIGURE 19.36

The Fourier coefficients of an impulse train.

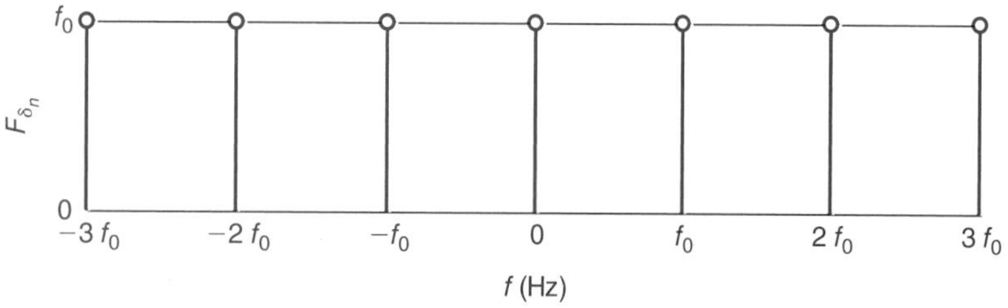

Therefore, the Fourier series representation of the impulse train is

$$f_\delta(t) = \sum_{n=-\infty}^{\infty} \delta(t - nT_0) = \sum_{n=-\infty}^{\infty} \frac{1}{T_0} e^{jn\frac{2\pi}{T_0}t} = \frac{1}{T_0}\left[1 + 2\sum_{n=1}^{\infty} \cos\left(n\frac{2\pi}{T_0}t\right)\right] \quad \textbf{(19.64)}$$

The Fourier transform of the impulse train is obtained by applying the Fourier transform pair

$$F[e^{jn\omega_0 t}] = 2\pi\delta(\omega - n\omega_0)$$

to the Fourier series representation given by Equation (19.64). Notice that $\omega_0 = 2\pi/T_0$.

$$F_\delta(\omega) = \sum_{n=-\infty}^{\infty} \frac{1}{T_0} 2\pi\delta(\omega - n\omega_0) \tag{19.65}$$

or

$$F_\delta(f) = \sum_{n=-\infty}^{\infty} f_0\delta(f - nf_0) \tag{19.66}$$

which is plotted in Figure 19.37.

FIGURE 19.37

The Fourier transform of an impulse train.

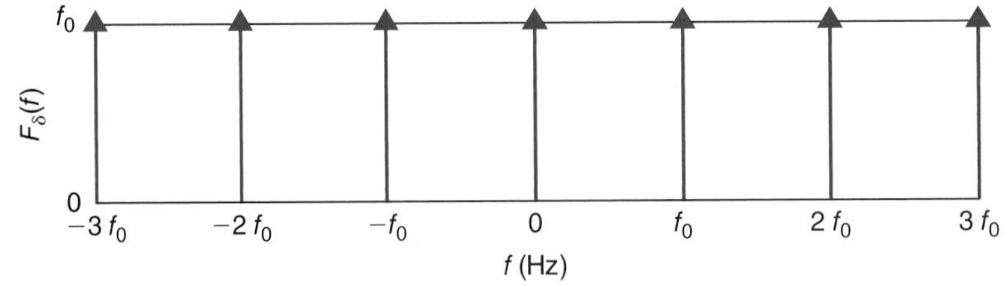

Notice that $F_\delta(f)$ is periodic with period f_0. Thus, it can be expanded in the Fourier series as

$$F_\delta(f) = \sum_{n=-\infty}^{\infty} e^{jn2\pi fT_0} = 1 + 2\sum_{n=1}^{\infty} \cos(n2\pi fT_0)$$

Since $F_\delta(f)$ can be written as

$$F_\delta(f) = \sum_{n=-\infty}^{\infty} f_0\delta(f - nf_0) = \sum_{n=-\infty}^{\infty} \frac{1}{T_0}\delta\left(f - \frac{n}{T_0}\right)$$

we have the following identity:

$$\sum_{n=-\infty}^{\infty} \frac{1}{T_0}\delta\left(f - \frac{n}{T_0}\right) = \sum_{n=-\infty}^{\infty} e^{jn2\pi fT_0} = 1 + 2\sum_{n=1}^{\infty} \cos(n2\pi fT_0) \tag{19.67}$$

If a signal $f(t)$ with the Fourier transform $F(\omega)$ is multiplied by an impulse train $f_\delta(t)$, we obtain

$$f_s(t) = f(t)f_\delta(t) = \sum_{n=-\infty}^{\infty} f(nT_0)\delta(t - nT_0) = \sum_{n=-\infty}^{\infty} \frac{1}{T_0} f(t)e^{jn\frac{2\pi}{T_0}t}$$

The Fourier transform of the product $f_s(t)$ is given by

$$F_s(\omega) = \sum_{n=-\infty}^{\infty} f(nT_0)e^{-j\omega nT_0} = \sum_{n=-\infty}^{\infty} \frac{1}{T_0} 2\pi F\left(\omega - n\frac{2\pi}{T_0}\right) \tag{19.68}$$

In f (in hertz), we have

$$F_s(f) = \sum_{n=-\infty}^{\infty} f(nT_0)e^{-j2\pi fnT_0} = \sum_{n=-\infty}^{\infty} \frac{1}{T_0} F(f - nf_0) \tag{19.69}$$

EXAMPLE 19.22

A signal $f(t) = 2\cos(2\pi 2000t)$ is multiplied by an impulse train with period 0.1 ms. Find the Fourier transform of the product $f_s(t)$, and plot it.

The Fourier transform of $f(t)$ is given by

$$F(f) = \delta(f - 2000) + \delta(f + 2000)$$

The Fourier transform of the product is given by

$$F_s(f) = \sum_{n=-\infty}^{\infty} \frac{1}{T_0} F(f - nf_0)$$

$$= \sum_{n=-\infty}^{\infty} [10{,}000\,\delta(f - n10{,}000 - 2000) + 10{,}000\,\delta(f - n10{,}000 + 2000)]$$

The spectrum of the product of a sinusoid and an impulse train is shown in Figure 19.38.

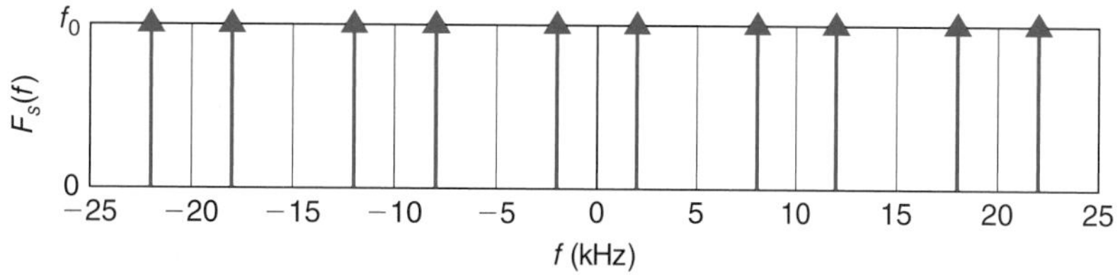

FIGURE 19.38

The spectrum of the product of a sinusoid and an impulse train.

Exercise 19.22

A signal $f(t) = 2\cos(2\pi 1000t) + 4\cos(2\pi 2000t)$ is multiplied by an inpulse train with period 0.1 ms. Find the Fourier transform of the product.

Answer:

$$F(f) = \delta(f - 1000) + \delta(f + 1000) + 2\delta(f - 2000) + 2\delta(f + 2000)$$

$$F_s(f) = \sum_{n=-\infty}^{\infty} \frac{1}{T_0} F(f - nf_0)$$

$$= \sum_{n=-\infty}^{\infty} [10{,}000\,\delta(f - n10{,}000 - 1000) + 10{,}000\,\delta(f - n10{,}000 + 1000)$$

$$+ 20{,}000\,\delta(f - n10{,}000 - 2000) + 20{,}000\,\delta(f - n10{,}000 + 2000)].$$

19.5 Parseval's Theorem

If $F[f(t)] = F(\omega)$, then the total normalized (based on a 1-Ω resistor) energy E of $f(t)$ is given by

$$E = \int_{-\infty}^{\infty} |f(t)|^2 dt = \frac{1}{2\pi}\int_{-\infty}^{\infty} |F(\omega)|^2 d\omega = \int_{-\infty}^{\infty} |F(f)|^2 df \qquad (19.70)$$

Proof

Taking the conjugate on both sides of the inverse Fourier transform

$$f(t) = \frac{1}{2\pi}\int_{-\infty}^{\infty} F(\omega)e^{j\omega t}d\omega$$

we obtain

$$f^*(t) = \frac{1}{2\pi}\int_{-\infty}^{\infty} F^*(\omega)e^{-j\omega t}d\omega$$

Thus, the energy of $f(t)$ becomes

$$E = \int_{-\infty}^{\infty} |f(t)|^2 dt = \int_{-\infty}^{\infty} f(t)f^*(t)dt = \int_{-\infty}^{\infty} f(t)\left[\frac{1}{2\pi}\int_{-\infty}^{\infty} F^*(\omega)e^{-j\omega t}d\omega\right]dt \quad \textbf{(19.71)}$$

Interchanging the order of integration, we obtain

$$\begin{aligned} E &= \frac{1}{2\pi}\int_{-\infty}^{\infty}\left[\int_{-\infty}^{\infty} f(t)e^{-j\omega t}dt\right]F^*(\omega)d\omega = \frac{1}{2\pi}\int_{-\infty}^{\infty} F(\omega)F^*(\omega)d\omega \\ &= \frac{1}{2\pi}\int_{-\infty}^{\infty} |F(\omega)|^2 d\omega = \int_{-\infty}^{\infty} |F(f)|^2 df \end{aligned} \quad \textbf{(19.72)}$$

The integral of $|F(\omega)|^2$ from $-\infty$ to $+\infty$ divided by 2π gives us the total energy of the signal. Thus, $|F(\omega)|^2$ can be interpreted as the energy density in J/rad/s. This quantity is called the **energy spectral density (ESD)**. Equations (19.71) and (19.72) can be written as

$$\int_{-\infty}^{\infty} f(t)f^*(t)dt = \frac{1}{2\pi}\int_{-\infty}^{\infty} F(\omega)F^*(\omega)d\omega \quad \textbf{(19.73)}$$

If Parseval's theorem, given by Equation (19.73), is generalized by replacing $f^*(t)$ by $g^*(t)$, we obtain

$$\int_{-\infty}^{\infty} f(t)g^*(t)dt = \frac{1}{2\pi}\int_{-\infty}^{\infty} F(\omega)G^*(\omega)d\omega \quad \textbf{(19.74)}$$

Equation (19.74) is called the *Plancherel theorem*. The proof is as follows:

$$\begin{aligned} \int_{-\infty}^{\infty} f(t)g^*(t)dt &= \int_{-\infty}^{\infty} f(t)\left[\frac{1}{2\pi}\int_{-\infty}^{\infty} G^*(\omega)e^{-j\omega t}d\omega\right] = \frac{1}{2\pi}\int_{-\infty}^{\infty}\left[\int_{-\infty}^{\infty} f(t)e^{-j\omega t}d\omega\right]G^*(\omega)d\omega \\ &= \frac{1}{2\pi}\int_{-\infty}^{\infty} F(\omega)G^*(\omega)d\omega \end{aligned}$$

If ω is changed to f (in hertz), the Plancherel theorem becomes

$$\int_{-\infty}^{\infty} f(t)g^*(t)dt = \int_{-\infty}^{\infty} F(f)G^*(f)df \tag{19.75}$$

EXAMPLE 19.23

Given $f(t) = 15\,e^{-6t}u(t)$, compute the total energy of $f(t)$ in the time domain and in the frequency domain. What percentage of this total energy lies in $-2 \le \omega \le 2$?

In the time domain, the total energy is given by

$$E = \int_{-\infty}^{\infty} |f(t)|^2 dt = \int_{0}^{\infty} 15^2 e^{-12t} dt = 18.75\ (Joules)$$

In the frequency domain, we have

$$E = \frac{1}{2\pi}\int_{-\infty}^{\infty} |F(\omega)|^2 d\omega = \frac{15^2}{2\pi}\int_{-\infty}^{\infty} \frac{1}{\omega^2 + 6^2} d\omega = \frac{15^2}{\pi}\int_{0}^{\infty} \frac{1}{\omega^2 + 6^2} d\omega$$

since the integrand is even. Now, we apply the integral formula

$$\int \frac{dx}{x^2 + a^2} = \frac{1}{a}\tan^{-1}\left(\frac{x}{a}\right)$$

to get

$$E = \left(\frac{15^2}{\pi}\right)\left(\frac{1}{6}\right)\left[\tan^{-1}\left(\frac{\omega}{6}\right)\right]_0^{\infty} = 18.75\ (Joules)$$

Let E_1 be the energy contained in $-2 \le \omega \le 2$. Then

$$E_1 = \frac{15^2}{2\pi}\int_{-2}^{2} \frac{1}{\omega^2 + 6^2} d\omega = 3.8406\ (Joules)$$

Thus, 20.48% of the total energy lies in $-2 \le \omega \le 2$.

Exercise 19.23

Find the energy that lies in $-2 \le \omega \le 2$ for $f(t) = \text{sinc}^2(t)$.

Answer:
3.3634 J.

EXAMPLE 19.24

A signal $x(t) = e^{-2t}u(t)$ is applied to a circuit shown in Figure 19.39. Let $R = 1\ \Omega$, $C = 0.2$ F.

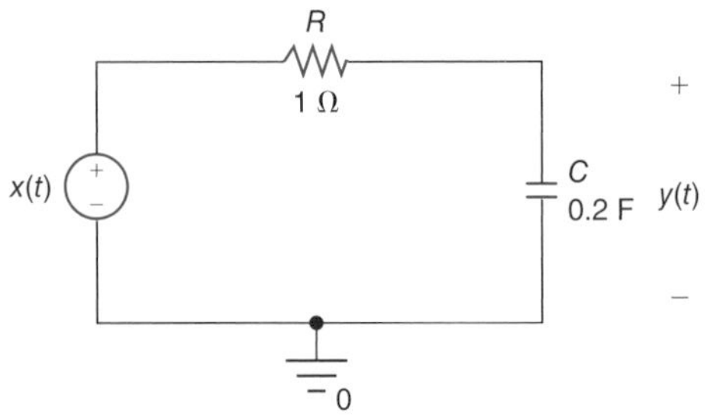

FIGURE 19.39

Circuit for EXAMPLE 19.24.

a. Find the total energy E_x contained in the signal $x(t)$.
b. Find the Fourier transform $X(\omega)$ of the input signal $x(t)$, and find the energy of $x(t)$ using $X(\omega)$.
c. Find the transfer function $H(s) = Y(s)/X(s)$ of the circuit shown in Figure 19.39. Find $H(\omega) = H(s)|_{s=j\omega}$.
d. Find the Fourier transform of the $y(t)$ given by $Y(\omega) = H(\omega)X(\omega)$.
e. Find $y(t)$ from $Y(\omega)$ using partial fraction expansion.
f. Find the energy of $y(t)$ using $y(t)$.
g. Find $|Y(\omega)|^2$.
h. Represent $|Y(\omega)|^2$ by partial fraction expansion.
i. Find the total energy E_y contained in the output $y(t)$ using $Y(\omega)$.
j. Find the energy contained in the input $x(t)$ between $-2 \le \omega \le 2$.
k. Find the energy contained in the output $y(t)$ between $-2 \le \omega \le 2$.

a. $$E_x = \int_{-\infty}^{\infty} x^2(t)dt = \int_0^{\infty} e^{-4t}dt = \frac{1}{4} = 0.25 \text{ J}$$

b. $$X(\omega) = \frac{1}{j\omega + 2}$$

$$E_x = \frac{1}{2\pi}\int_{-\infty}^{\infty} |X(\omega)|^2 d\omega = \frac{1}{2\pi}\int_{-\infty}^{\infty} \frac{1}{\omega^2 + 2^2}d\omega = \frac{2}{2\pi}\int_0^{\infty} \frac{1}{\omega^2 + 2^2}d\omega = \frac{2}{2\pi}\frac{1}{2}\tan^{-1}\left(\frac{\omega}{2}\right)\Bigg|_0^{\infty}$$

$$= \frac{2}{2\pi}\frac{1}{2}\left[\tan^{-1}\left(\frac{\infty}{2}\right) - \tan^{-1}\left(\frac{0}{2}\right)\right] = \frac{2}{2\pi}\frac{1}{2}\frac{\pi}{2} = \frac{1}{4} = 0.25 \text{ J}$$

c. $$H(s) = \frac{Y(s)}{X(s)} = \frac{\frac{1}{sC}}{R + \frac{1}{sC}} = \frac{\frac{1}{RC}}{s + \frac{1}{RC}} = \frac{5}{s+5}, \quad H(\omega) = \frac{5}{j\omega + 5}$$

d. $$Y(\omega) = H(\omega)X(\omega) = \frac{5}{(j\omega + 2)(j\omega + 5)}$$

e. $$Y(\omega) = \frac{5}{(j\omega + 2)(j\omega + 5)} = \frac{A}{j\omega + 2} + \frac{B}{j\omega + 5} = \frac{(A+B)j\omega + 5A + 2B}{(j\omega + 2)(j\omega + 5)}$$

$A + B = 0, \quad 5A + 2B = 5$

$B = -A, \quad 3A = 5, \quad A = 5/3, \quad B = -5/3.$

$$Y(\omega) = \frac{\frac{5}{3}}{j\omega + 2} + \frac{\frac{-5}{3}}{j\omega + 5}$$

$$y(t) = \left(\frac{5}{3}e^{-2t} - \frac{5}{3}e^{-5t}\right)u(t)$$

continued

Example 19.24 continued

f. Find the energy of $y(t)$ using $y(t)$.

$$E_y = \int_0^\infty y^2(t)dt = \int_0^\infty \left(\frac{25}{9}e^{-4t} + \frac{25}{9}e^{-10t} - \frac{50}{9}e^{-7t}\right)dt$$

$$= \frac{25}{9}\left(\frac{1}{4} + \frac{1}{10} - \frac{2}{7}\right) = \frac{5}{28} = 0.1786 \text{ J}$$

g. Find $|Y(\omega)|^2$.

$$|Y(\omega)|^2 = \frac{25}{|j\omega + 2|^2|j\omega + 5|^2} = \frac{25}{(\omega^2 + 2^2)(\omega^2 + 5^2)}$$

h. $$|Y(\omega)|^2 = \frac{A}{\omega^2 + 2^2} + \frac{B}{\omega^2 + 5^2} = \frac{(A + B)\omega^2 + 25A + 4B}{(\omega^2 + 2^2)(\omega^2 + 5^2)}$$

$A + B = 0$, $25A + 4B = 25$, $A = 25/21 = 1.1905$, $B = -25/21 = -1.1905$.

i. Find the total energy E_y contained in the output $y(t)$ using $Y(\omega)$.

$$E_y = \frac{1}{2\pi}\int_{-\infty}^{\infty} |Y(\omega)|^2 d\omega = \frac{1}{2\pi}\int_{-\infty}^{\infty} \frac{\frac{25}{21}}{\omega^2 + 2^2}d\omega + \frac{1}{2\pi}\int_{-\infty}^{\infty} \frac{\frac{-25}{21}}{\omega^2 + 5^2}d\omega$$

$$= \frac{2}{2\pi}\frac{25}{21}\int_0^\infty \frac{1}{\omega^2 + 2^2}d\omega + \frac{2}{2\pi}\left(\frac{-25}{21}\right)\int_0^\infty \frac{1}{\omega^2 + 5^2}d\omega$$

Application of the integral formula

$$\int \frac{dx}{x^2 + a^2} = \frac{1}{a}\tan^{-1}\left(\frac{x}{a}\right)$$

yields

$$E_y = \frac{2}{2\pi}\frac{25}{21}\frac{1}{2}\tan^{-1}\left(\frac{\omega}{2}\right)\Big|_0^\infty + \frac{2}{2\pi}\left(\frac{-25}{21}\right)\frac{1}{5}\tan^{-1}\left(\frac{\omega}{5}\right)\Big|_0^\infty$$

$$= \frac{2}{2\pi}\frac{25}{21}\frac{1}{2}\frac{\pi}{2} - \frac{2}{2\pi}\frac{25}{21}\frac{1}{5}\frac{\pi}{2} = \frac{25}{84} - \frac{25}{210} = \frac{5}{28} = 0.1786 \text{ J}$$

j. $$E_2 = \frac{1}{2\pi}\int_{-2}^{2} |X(\omega)|^2 d\omega = \frac{1}{2\pi}\int_{-2}^{2} \frac{1}{\omega^2 + 2^2}d\omega = \frac{2}{2\pi}\int_0^2 \frac{1}{\omega^2 + 2^2}d\omega$$

$$= \frac{2}{2\pi}\frac{1}{2}\tan^{-1}\left(\frac{\omega}{2}\right)\Big|_0^2 = \frac{1}{2\pi}\tan^{-1}\left(\frac{2}{2}\right) = 0.125 \text{ J}$$

continued

Example 19.24 continued

k. $$E_y = \frac{2}{2\pi}\frac{25}{21}\frac{1}{2}\tan^{-1}\left(\frac{\omega}{2}\right)\Bigg|_0^2 + \frac{2}{2\pi}\left(\frac{-25}{21}\right)\frac{1}{5}\tan^{-1}\left(\frac{\omega}{5}\right)\Bigg|_0^2$$

$$= \frac{25}{42\pi}\tan^{-1}\left(\frac{2}{2}\right) - \frac{5}{21\pi}\tan^{-1}\left(\frac{2}{5}\right) = 0.12 \text{ J}$$

MATLAB

```
% EXAMPLE 19.24
clear all;
a=2;R=1;C=0.2;
syms t s w
x=exp(-a*t)
Ex=int(x^2,t,0,inf)
Ex=vpa(Ex,8)
wc=1/(R*C)
H=wc/(s+wc)
h=ilaplace(H)
Ha=subs(H,s,j*w)
X=laplace(x)
Xa=fourier(x*heaviside(t))
Exa=(1/pi)*int(abs(Xa)^2,w,0,inf)
Exa=vpa(Exa,7)
Y=X*H
y=ilaplace(Y)
Ya=Ha*Xa
Ey=int(y^2,t,0,inf)
Ey=double(Ey)
Eya=(1/pi)*int(abs(Ya)^2,w,0,inf)
Eya=vpa(Eya,8)
z=x-y
Ez=int(z^2,t,0,inf)
Ez=double(Ez)
Exb=Ez+Ey
Ex1=(1/(2*pi))*int(abs(Xa)^2,w,-2,2)
Ex1=vpa(Ex1,8)
Ey1=(1/(2*pi))*int(abs(Ya)^2,w,-2,2)
Ey1=vpa(Ey1,8)
```

Exercise 19.24

Find the energy of the output $y(t)$ for the circuit shown in Figure 19.40 if $x(t) = \text{sinc}(t)$.

FIGURE 19.40

Circuit for EXERCISE 19.24.

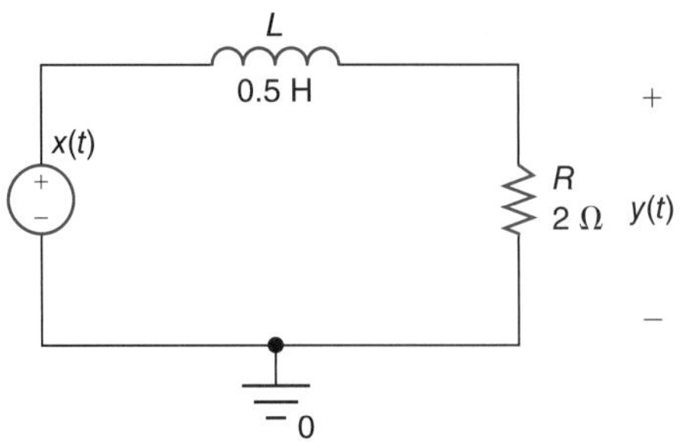

Answer:
$E_y = 0.8477$ J.

EXAMPLE 19.25

Find the energy of $f(t) = A\ \mathrm{sinc}(ct)$.

The Fourier transform of $f(t)$ is given by

$$F(f) = \frac{A}{c} rect\left(\frac{f}{c}\right)$$

Application of Parseval's theorem given by Equation (19.70) yields

$$E = \int_{-\infty}^{\infty} |F(f)|^2 df = \int_{-\infty}^{\infty} \frac{A^2}{c^2} rect\left(\frac{f}{c}\right) df = \frac{A^2}{c^2} \times c = \frac{A^2}{c}\ \mathrm{J}$$

Exercise 19.25

Find the energy of $f(t) = A\ \mathrm{sinc}^2(ct)$.

Answer:

$$F(f) = \frac{A}{c} tri\left(\frac{f}{c}\right), \quad E = \frac{2A^2}{3c}\ \mathrm{J}.$$

19.6 Simulink

The Fourier transform of a continuous time signal $f(t)$ is

$$F(\omega) = \int_{-\infty}^{\infty} f(t) e^{-j\omega t} dt$$

This integral can be computed numerically as

$$\int_{-\infty}^{\infty} f(t) e^{-j\omega t} dt \approx \sum_{k=-\infty}^{\infty} f(kT_s) e^{-j\omega k T_s} T_s$$

Meanwhile, when $f(t)$ is sampled at a rate of $f_s = 1/T_s$ samples/s, the spectrum of the sampled waveform is

$$F_s(\omega) = \frac{1}{T_s} \sum_{k=-\infty}^{\infty} F(\omega - k\omega_s) = \sum_{k=-\infty}^{\infty} f(kT_s) e^{-j\omega k T_s}$$

and the discrete-time Fourier transform of the discrete-time signal $f(k)$ is

$$F(\theta) = \sum_{k=-\infty}^{\infty} f(k) e^{-j\theta k}$$

Comparing the last two equations, we have

$$F(\theta) = F_s(\omega)|_{\omega=\frac{\theta}{T_s}}$$

and

$$F_s(\omega) = F(\theta)|_{\theta=\omega T_s}$$

If $f(t)$ is band-limited to B Hz, and if the sampling rate is greater than $2B$ samples/s,

$$F(\omega) = T_s F(\theta)|_{\theta=\omega T_s}, \quad |\omega| \leq 2\pi B$$

and

$$F(\theta) = \frac{1}{T_s} F(\omega)|_{\omega=\frac{\theta}{T_s}}, \quad |\theta| \leq \pi$$

If $f(t)$ is time-limited, then the bandwidth of $f(t)$ is infinite and aliasing error results. In this case, these two equations hold only for $\omega \ll \omega_s/2$.

In general, the evaluation of $F(\theta)$ for every value of θ is not possible, since there are infinite values of θ for $0 \leq \theta \leq 2\pi$. Evaluation of $F(\theta)$for N values of θ equally spaced on the unit circle may be sufficient to get accurate characterization of $F(\theta)$. Let these N samples be $F(n)$. Then, we have

$$F(n) = F(\theta)|_{\theta=n\frac{2\pi}{N}} = \sum_{k=0}^{N-1} f(k)e^{-j\frac{2\pi}{N}kn} = \sum_{k=0}^{N-1} f(k)W_N^{kn}$$

where W_N is the nth root of unity given by

$$W_N = e^{-j\frac{2\pi}{N}} = (e^{-j2\pi})^{\frac{1}{N}} = (1)^{\frac{1}{N}}$$

The N samples of $F(\theta)$, $F(n)$, is called the *discrete Fourier transform* (DFT). The fast Fourier transform (FFT) is an efficient calculation of DFT. In conclusion, $F(\omega)$ can be approximated by

$$F(\omega) = T_s F(\theta)|_{\theta=\omega T_s} = T_s F(\theta)|_{\theta=n\frac{2\pi}{N}T_s} = T_s F\left(n\frac{2\pi}{N}T_s\right) = T_s F\left(n\frac{2\pi}{N}\right)$$

EXAMPLE 19.26

Use the Simulink model shown in Figure 19.41 to plot the magnitude of the Fourier transform of a rectangular pulse with amplitude $A = 1$ V and pulse width $\tau = 1$ s.

FIGURE 19.41

Simulink model to find the Fourier transform of a rectangular pulse.

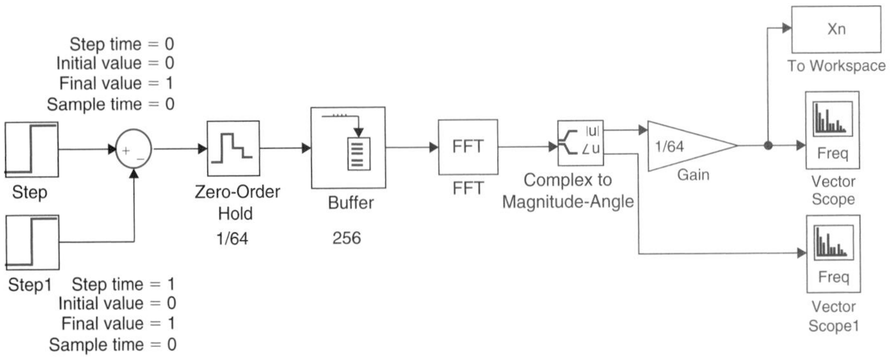

continued

Example 19.26 continued

The magnitude spectrum is shown in Figure 19.42.

FIGURE 19.42

Magnitude spectrum of a rectangular pulse.

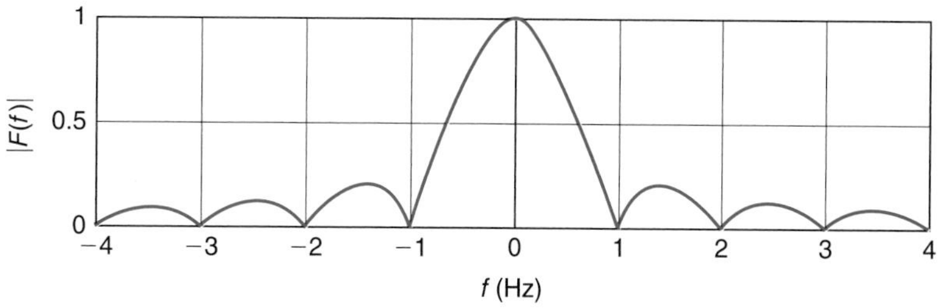

Exercise 19.26

Use the Simulink model shown in Figure 19.43 to plot the magnitude spectrum of the Fourier transform of a triangular pulse with amplitude $A = 1$ V and $\tau = 1$.

FIGURE 19.43

Simulink model to find the Fourier transform of a triangular pulse.

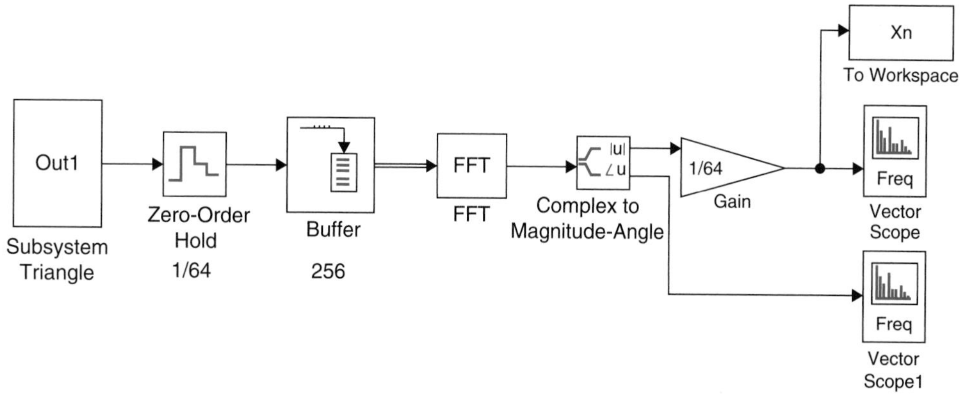

The subsystem to generate the triangular pulse is shown in Figure 19.44.

FIGURE 19.44

Subsystem to generate a triangular pulse.

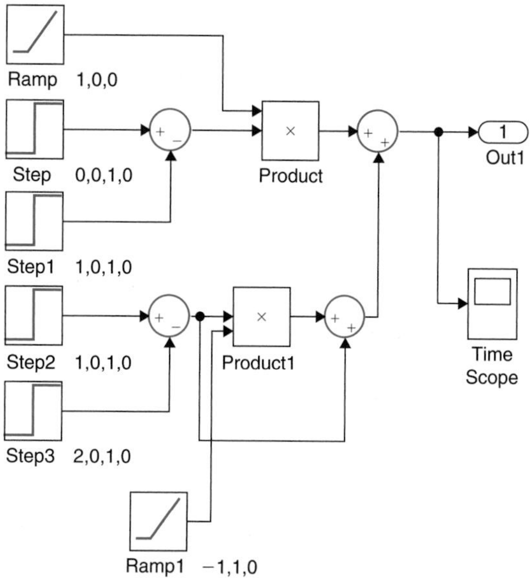

continued

Exercise 19.26 continued

FIGURE 19.45

Magnitude spectrum of a triangular pulse.

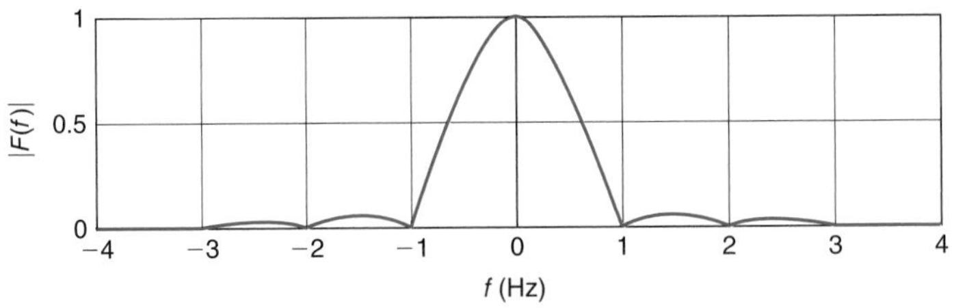

SUMMARY

The Fourier transform provides the means to analyze signals in the frequency domain. In this chapter, the definition of the Fourier transform, along with the properties of the Fourier transform, are presented. The properties discussed in this chapter include time shifting, time scaling, time reversal, time differentiation, duality, frequency scaling, modulation, convolution, and multiplication. These properties relate the characteristics of the signals in the time domain and the frequency domain.

Parseval's theorem says the total energy of a signal can be evaluated in the time domain or in the frequency domain. In the frequency domain, the total energy of a signal is the integral of the energy spectral density. The energy spectral density shows the energy as a function of frequency.

PROBLEMS

Definition

19.1 **Find the Fourier transform of**

a. $f_1(t) = e^{-3t}u(t)$
b. $f_2(t) = e^{3t}u(-t)$
c. $f_3(t) = e^{-3|t|}$

Sifting Property of Impulse

19.2 **Use the sifting property of the Dirac delta function to evaluate the following integrals.**

a. $$\int_{-\infty}^{\infty} \frac{1}{5t}\sin\left(\frac{\pi t}{2}\right)e^{-2t}\delta(t-1)\,dt$$

b. $$\int_{-\infty}^{\infty} e^{j2\pi t}\delta\left(t-\frac{1}{8}\right)dt$$

Time-Shifting Property

19.3 **Find the Fourier transform of $f(t) = \delta(t+5)$.**

19.4 **Find the Fourier transform of $f(t) = u(t-7)$.**

19.5 **If $F[f(t)] = F(\omega)$, what is the Fourier transform of $g(t)$ as a function of $F(\omega)$?**

$$g(t) = 3f(t+7)$$

19.6 **If $F[f(t)] = F(\omega)$, what is the Fourier transform of $g(t)$ as a function of $F(\omega)$?**

$$g(t) = (t-5)f(3-t)$$

19.7 **Let $f(t) = 2\,rect(t/4)$. Find the Fourier transform of the following signals.**

a. $f(t)$
b. $f(t-10)$

Time-Scaling Property

19.8 **If $F[f(t)] = F(\omega)$, what is the Fourier transform of $g(t)$ as a function of $F(\omega)$?**

$$g(t) = 9f\left(\frac{t-7}{5}\right)$$

19.9 **Let $f(t) = 5\,\mathrm{sinc}(5t)$. Find the Fourier transform of the following signals.**

a. $g(t) = f(t/2)$
b. $g(t) = f(4t)$

Symmetry Property

19.10 **Apply the symmetry property (duality property) to**

$$F[sgn(t)] = 2/(j\omega)$$

to find the Fourier transform of $1/(\pi t)$.

19.11 **Apply the symmetry property (duality property) to**

$$F[e^{-a|t|}] = \frac{2a}{\omega^2 + a^2}$$

to find the Fourier transform of $1/(t^2 + a^2)$.

Frequency Shifting and Modulation Property

19.12 **Find the Fourier transform of**

$$f(t) = 3\cos(2\pi 1000t + \pi/3)$$

19.13 **Find the Fourier transform of**

$$f(t) = 8\sin(2\pi 20{,}000t - 3\pi/4)$$

19.14 **If $F[f(t)] = F(\omega)$, what is the Fourier transform of $g(t)$ as a function of $F(\omega)$?**

$$g(t) = -f\left(-\frac{t+20}{12}\right)e^{j1000t}$$

19.15 **If $F[f(t)] = F(\omega)$, what is the Fourier transform of $g(t)$ as a function of $F(\omega)$?**

$$g(t) = f\left(\frac{t-10}{5}\right)\cos(50t)$$

19.16 **Find and plot the Fourier transform of**

a. $f(t) = e^{-5|t|}$
b. $g(t) = e^{-5|t|}e^{j50t}$
c. $h(t) = e^{-5|t|}\cos(50t)$

19.17 **Find and plot the Fourier transform of**

a. $f_1(t) = \text{sinc}(1000t)$
b. $f_2(t) = \text{sinc}(1000t)e^{j10{,}000t}$
c. $f_3(t) = \text{sinc}(1000t)\cos(10{,}000t)$

19.18 **Find and plot the Fourier transform of**

a. $f_1(t) = \text{sinc}^2(1000t)$
b. $f_2(t) = \text{sinc}^2(1000t)e^{j10{,}000t}$
c. $f_3(t) = \text{sinc}^2(1000t)\cos(10{,}000t)$

19.19 **Let $f(t) = 2\,rect(t/4)$. Find the Fourier transform of the following signals.**

a. $f(t)$
b. $g(t) = f(t)e^{j2\pi 5t}$
c. $h(t) = f(t)\cos(2\pi 5t)$

19.20 **The Fourier transform $F(\omega)$ of a signal $f(t)$ is given by**

$$F(\omega) = \begin{cases} 8, & -6 \le \omega \le -2 \\ 8, & 2 \le \omega \le 6 \\ 0, & otherwise \end{cases}$$

Find and sketch $f(t)$.

19.21 **Let $f(t) = \dfrac{1}{t^2 + a^2}$ and $a = 1$.**

a. Find and plot the Fourier transform $G(\omega)$ of $g(t) = f(t)e^{j20t}$
b. Find and plot the Fourier transform $H(\omega)$ of $h(t) = f(t)\cos(20t)$

19.22 **Sketch the spectrums $X(f)$ and $Y(f)$ in the system shown in Figure P19.22.**

FIGURE P19.22

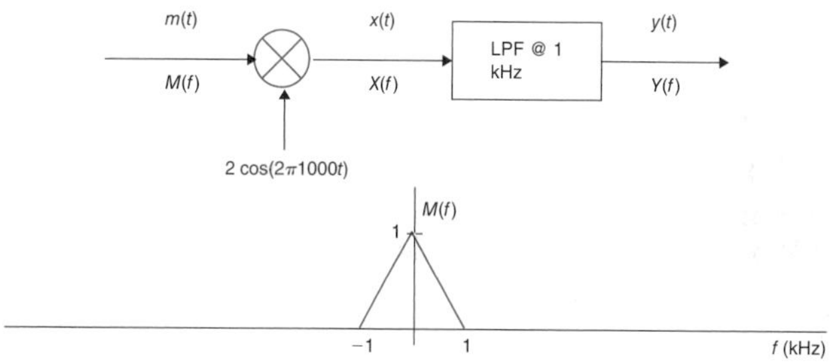

19.23 **Sketch the spectrums $Y(f)$ and $Z(f)$ in the system shown in Figure P19.23.**

FIGURE P19.23

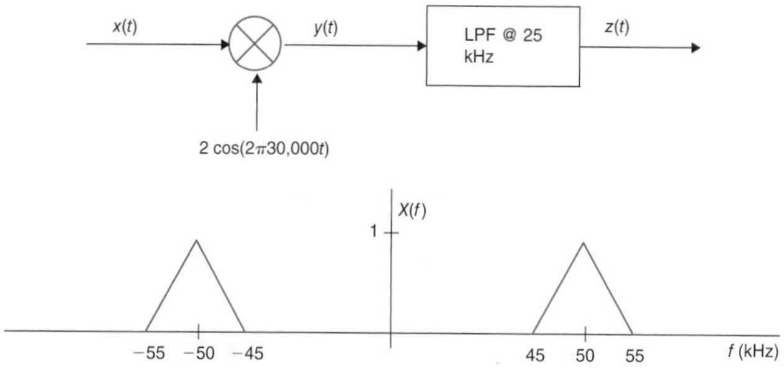

19.24 A message $m(t) = 24\cos(2\pi 100t) + 16\cos(2\pi 200t)$ is applied to the system shown in Figure P19.24. Assume that

$$H(f) = \begin{cases} 1, & 1000 \le f \le 1500 \\ 1, & -1500 \le f \le -1000 \\ 0, & elsewhere \end{cases}$$

a. Determine $x(t)$ and sketch its spectrum.
b. Determine $y(t)$ and sketch its spectrum.

FIGURE P19.24

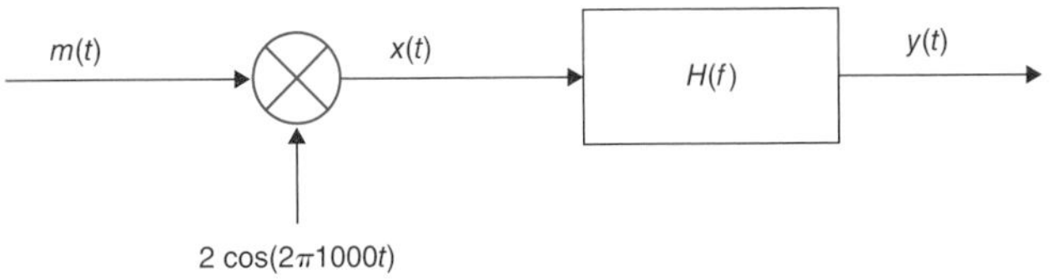

19.25 The system shown in Figure P19.25 can be used to send two messages on one carrier. If $m_1(t) = \cos(2\pi 1000t)$ and $m_2(t) = \cos(2\pi 4000t)$, find the output $y(t)$.

FIGURE P19.25

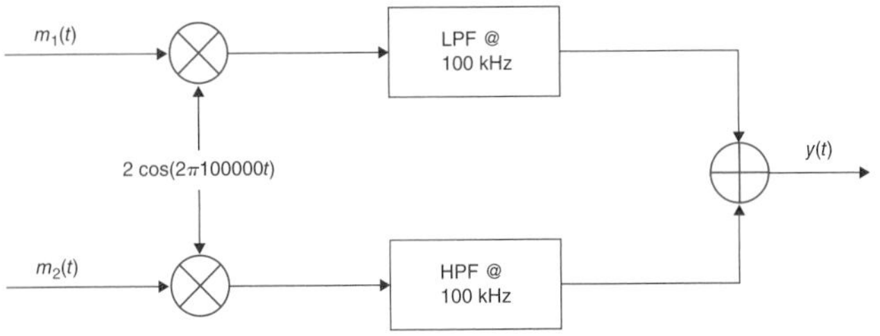

19.26 Determine the signal $f(t)$ with the Fourier transform $F(\omega)$ shown in Figure P19.26.

FIGURE P19.26

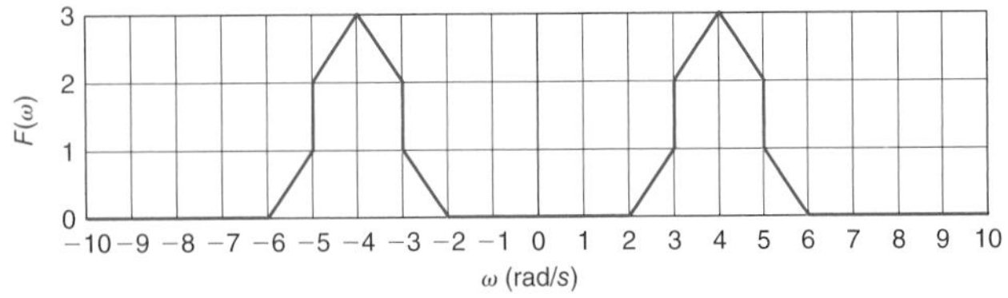

19.27 A signal $m(t) = 2\cos(2\pi 1000t)$ is multiplied by $c(t) = 10\cos(2\pi 10{,}000t)$. Let the resulting signal be $s(t)$. Thus, $s(t) = m(t)c(t)$. Plot $s(t)$ in the time domain. Find and plot the Fourier transform of $s(t)$, that is, $S(f) = F[s(t)]$.

19.28 A signal $m(t) = 1 + 0.9\cos(2\pi 1000t)$ is multiplied by $c(t) = 10\cos(2\pi 10{,}000t)$. Let the resulting signal be $s(t)$. Thus, $s(t) = m(t)c(t)$. Plot $s(t)$ in the time domain. Find and plot the Fourier transform of $s(t)$, that is, $S(f) = F[s(t)]$.

Time-Differentiation Property

19.29 Use the time-differentiation property to find the Fourier transform $F(\omega)$ of $f(t)$, as shown in Figure P19.29

FIGURE P19.29

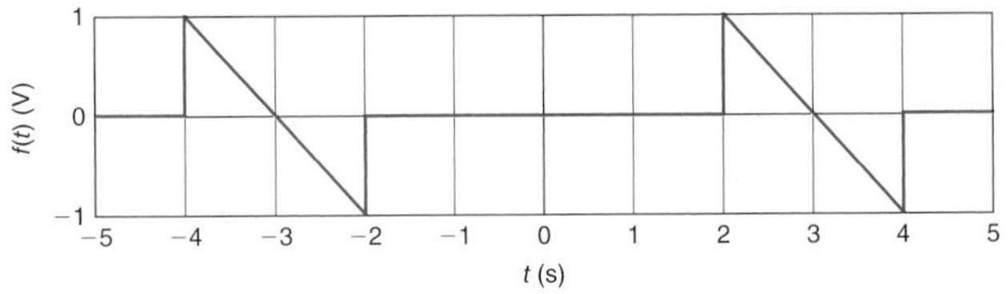

19.30 Use the time-differentiation property to find the Fourier transform of

$$f(t) = \begin{cases} t+1, & 0 \le t < 1 \\ -t+1, & -1 \le t < 0 \\ 0, & otherwise \end{cases}$$

19.31 Use the time-differentiation property to find the Fourier transform of

$$f(t) = \begin{cases} t, & 0 \le t < 1 \\ -\frac{3}{2}t + \frac{9}{2}, & 1 \le t < 3 \\ 0, & otherwise \end{cases}$$

Frequency-Differentiation Property

19.32 If $F[f(t)] = F(\omega)$, what is the Fourier transform of $g(t)$ as a function of $F(\omega)$?

$$g(t) = 3tf\left(\frac{t+2}{8}\right)$$

19.33 Use the frequency-differentiation property to find the Fourier transform of

a. $f_1(t) = te^{-5t}u(t)$
b. $f_2(t) = t^2e^{-5t}u(t)$
c. $f_3(t) = t^3e^{-5t}u(t)$

Convolution Property

19.34 If $F[f(t)] = F(\omega)$, what is the Fourier transform of $g(t)$ as a function of $F(\omega)$?

$$g(t) = 3tf\left(\frac{t+5}{6}\right) * \frac{df(t)}{dt}$$

19.35 If $F[f(t)] = F(\omega)$, what is the Fourier transform of $g(t)$ as a function of $F(\omega)$?

$$g(t) = f\left(\frac{t}{3}\right) * F(-5t)$$

19.36 Use the convolution property of the Fourier transform to find the convolution of $f_1(t)$ and $f_2(t)$.

$$f_1(t) = e^{-2t}u(t), \quad f_2(t) = e^{-5t}u(t)$$

19.37 Use the convolution property of the Fourier transform to find the convolution of $f_1(t)$ and $f_2(t)$.

$$f_1(t) = e^{-3t}u(t), \quad f_2(t) = e^{6t}u(t)$$

19.38 Use Fourier transform to find the following convolution.

$$f(t) = \text{sinc}(2t) * \text{sinc}(2t)$$

19.39 Use the Fourier transform to find the following convolution.

$$f(t) = \text{sinc}(2t) * \text{sinc}(4t)$$

19.40 Use the Fourier transform to find the following convolution.

$$f(t) = \text{sinc}(2t) * \text{sinc}^2(t)$$

19.41 Use the Fourier transform to find the following convolution.

$$f(t) = \text{sinc}(2t) * \text{sinc}^2(2t)$$

19.42 The impulse response of a linear, time-invariant system is given by $u(t)$.

Sketch the output $y(t)$ of the system if the input is given by $f(t) = \delta(t) + \delta(t-1) + \delta(t-2) - 5\delta(t-3) + 3\delta(t-4) - \delta(t-5)$

19.43 The transfer function of an ideal low-pass filter is given by $H(\omega) = rect[\omega/(40\pi)]$.

What is the output $y(t)$ of the filter when $\delta(t)$ is applied at the input?

19.44 The transfer function of an ideal low-pass filter is given by $H(\omega) = rect[\omega/(40\pi)]$.

What is the output $y(t)$ of the filter when 10 sinc$(10t)$ is applied at the input?

Multiplication Property

19.45 If $F[f(t)] = F(\omega)$, what is the Fourier transform of $g(t)$ as a function of $F(\omega)$?

$$g(t) = f^3(t) + 4f^2(t) - 9f(t) + 15$$

19.46 Let $F(\omega) = rect\left(\frac{\omega}{5}\right)$.

a. Plot $F(\omega)$
b. Find and plot $f(t)$
c. Find and plot $f^2(t)$
d. Find and plot $F[f^2(t)]$

19.47 A signal $f(t)$ is band-limited to B Hz. What is the bandwidth of $y(t) = f^3(t)$?

19.48 The Fourier transform of $f(t)$ is shown in Figure P19.48. Find the Fourier transform of $f^2(t)$.

FIGURE P19.48

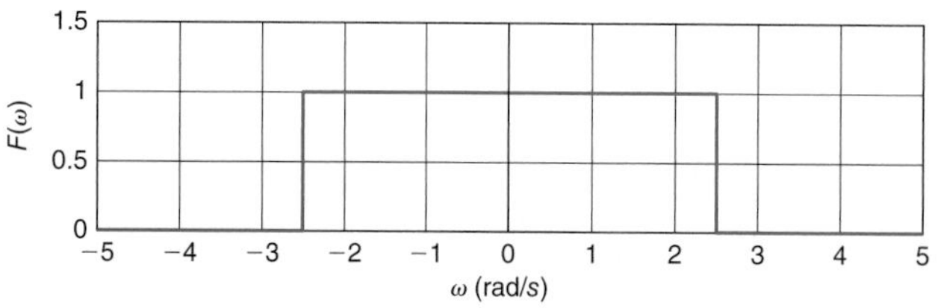

19.49 Find the Fourier transform $F(\omega)$ of $f(t) = \text{sinc}^2[5(t-6)]$.

Parseval's Theorem

19.50 The magnitude of the Fourier transform of a signal $f(t)$ is given by

$|F(\omega)| = 2\ tri(\omega/2)$. Find the energy of $f(t)$

19.51 A signal $x(t) = e^{-5t}u(t)$ is applied to a circuit shown in Figure P19.51. $R = 1\ \Omega$, $C = 0.1$ F.

FIGURE P19.51

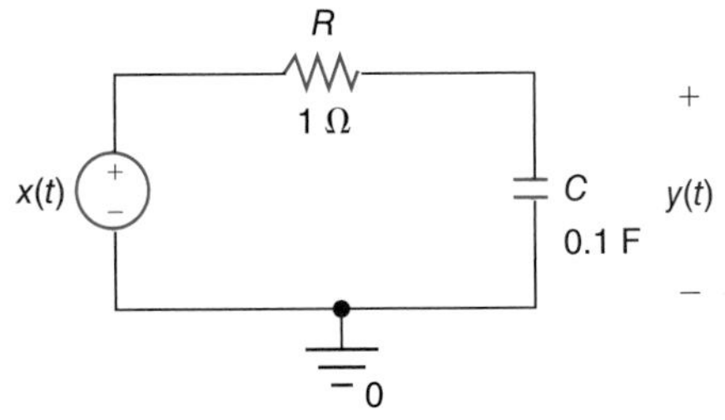

a. Find the total energy E_x contained in the signal $x(t)$.
b. Find the Fourier transform $X(\omega)$ of the input signal $x(t)$.
c. Find the transfer function $H(s) = Y(s)/X(s)$ of the circuit shown in Figure P19.51. Find $H(\omega) = H(s)|_{s=j\omega}$.
d. Find the Fourier transform of the $y(t)$ given by $Y(\omega) = H(\omega)X(\omega)$.
e. Find $|Y(\omega)|^2$.
f. Represent $|Y(\omega)|^2$ by partial fraction expansion; that is, find A and B in the following equation:

$$|Y(\omega)|^2 = \frac{A}{\omega^2 + 5^2} + \frac{B}{\omega^2 + 10^2}$$

g. Find the total energy E_y contained in the output $y(t)$.
h. Find the energy contained in the input $x(t)$ between $-2 \le \omega \le 2$.
i. Find the energy contained in the output $y(t)$ between $-2 \le \omega \le 2$.

19.52 A signal $x(t) = e^{-3t}u(t)$ is applied to a circuit shown in Figure P19.52. Let $R = 1\ \Omega$, $L = 0.25\ H$.

FIGURE P19.52

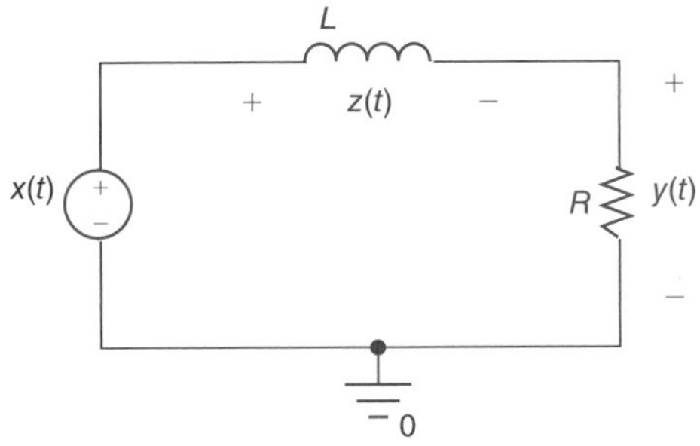

a. Find the total energy E_x contained in the signal $x(t)$.
b. Find the transfer function $H(s) = Y(s)/X(s)$ of the circuit shown in Figure P19.52. Find $H(\omega) = H(s)|_{s=j\omega}$ and the Fourier transform $X(\omega)$ of $x(t)$.
c. Find $y(t)$ from $Y(\omega) = H(\omega)X(\omega)$ using partial fraction expansion.
d. Find the energy E_y of $y(t)$.
e. Find the energy contained in the output $y(t)$ between $-1 \le \omega \le 1$.
f. Find $z(t)$ and energy E_z of $z(t)$.

Chapter 20

Two-Port Circuits

20.1 Introduction

If a circuit is connected to a pair of ports, it is called a *two-port circuit* (*two-port network*). The port on the left side is called port 1, and the port on the right side is called port 2. An input signal to the circuit is applied through port 1, and an output signal from the circuit is taken from port 2. We assume that the two-port circuit does not contain independent sources. There are four variables in the two-port circuit. These are voltage V_1 and current I_1 at port 1 and voltage V_2 and current I_2 at port 2. Two of the four variables are independent, and the other two variables are dependent.

Depending on which two of the four parameters are selected as independent variables, we have six different representations of the circuit. The coefficients of the representation are called *parameters*. The six parameters are z-parameters, y-parameters, h-parameters, g-parameters, $ABCD$ parameters, and b-parameters. The z-parameters are ratios of voltage to current with units of ohms, and they are called *impedance parameters*. The y-parameters are ratios of current to voltage with units of siemens, and they are called *admittance parameters*. The h-parameters represent input impedance, reverse voltage gain, forward current gain, and output admittance, and they are called *hybrid parameters*. The g-parameters represent output impedance, forward voltage gain, reverse current gain, and input admittance, and they are called *inverse hybrid parameters*. The $ABCD$ parameters represent voltage ratio, transfer impedance, transfer admittance, and current ratio, and they are called *transmission parameters*. The b-parameters represent voltage gain, transfer impedance, transfer admittance, and current gain, and they are called *inverse transmission parameters*.

If one of the six parameters is known, the other five parameters are obtained by applying the conversion formulas. The two-port circuits can be interconnected. The interconnections can be in cascade, series, parallel, series-parallel, and parallel-series.

20.2 Two-Port Circuit

FIGURE 20.1

One-port circuit.

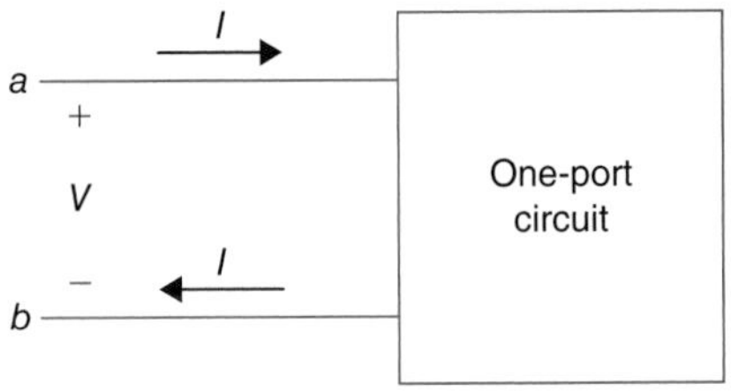

Figure 20.1 shows a one-port circuit (or one-port network). The circuit is connected to the outside world with a pair of terminals *a* and *b*. The pair of terminals, such as *a* and *b* shown in Figure 20.1, is called a *port*. The current *I* entering the circuit from terminal *a* is the same current exiting the circuit at terminal *b*. The loop current in the loop formed by the external circuit and the one-port circuit is *I*.

If a circuit (network) is connected to a pair of ports, as shown in Figure 20.2, it is called a *two-port circuit* (*two-port network*). The port on the left side is called port 1, and the port on the right side is called port 2. The voltage across terminals in port 1 is V_1, and the voltage across terminals in port 2 is V_2. The loop current for port 1 is I_1, and the loop current for port 2 is I_2, as shown in Figure 20.2. An input signal to the circuit is applied through port 1, and an output signal from the circuit is taken from port 2. We assume the two-port circuit does not contain independent sources. Two of the four variables are independent. The other two variables are dependent.

FIGURE 20.2

Two-port circuit.

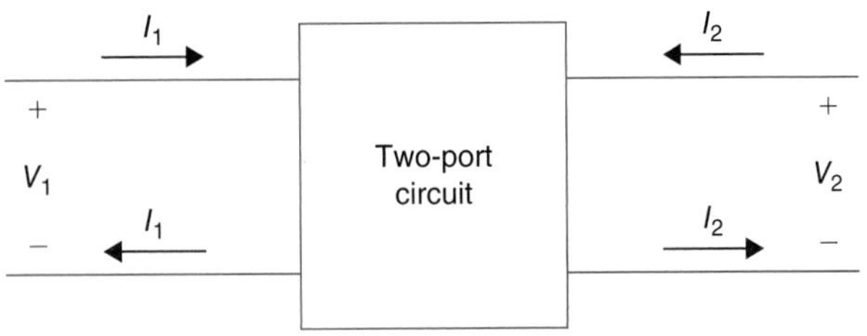

20.2.1 *z*-PARAMETERS (IMPEDANCE PARAMETERS)

If I_1 and I_2 are independent variables, then V_1 and V_2 are dependent variables. The dependent variables are written as linear combinations of independent variables as

$$V_1 = z_{11}I_1 + z_{12}I_2 \quad \textbf{(20.1)}$$

$$V_2 = z_{21}I_1 + z_{22}I_2 \quad \textbf{(20.2)}$$

In matrix notation, we have

$$\begin{bmatrix} V_1 \\ V_2 \end{bmatrix} = \begin{bmatrix} z_{11} & z_{12} \\ z_{21} & z_{22} \end{bmatrix} \begin{bmatrix} I_1 \\ I_2 \end{bmatrix} \quad \textbf{(20.3)}$$

This equation can be written as

$$\mathbf{v} = \mathbf{Z}\mathbf{i} \quad \textbf{(20.4)}$$

where

$$\mathbf{v} = \begin{bmatrix} V_1 \\ V_2 \end{bmatrix}, \quad \mathbf{Z} = \begin{bmatrix} z_{11} & z_{12} \\ z_{21} & z_{22} \end{bmatrix}, \quad \mathbf{i} = \begin{bmatrix} I_1 \\ I_2 \end{bmatrix} \quad \textbf{(20.5)}$$

Coefficient z_{11} can be found by taking the ratio of V_1 over I_1 when port 2 is open-circuited ($I_2 = 0$)

$$z_{11} = \left.\frac{V_1}{I_1}\right|_{I_2=0} \quad \textbf{(20.6)}$$

Coefficient z_{11} is the input impedance with port 2 open-circuited. Similarly, coefficients z_{12}, z_{21}, and z_{22} are found by taking the ratio of voltage to current while one of the ports is open circuited

$$z_{12} = \left.\frac{V_1}{I_2}\right|_{I_1=0} \quad \textbf{(20.7)}$$

$$z_{21} = \left.\frac{V_2}{I_1}\right|_{I_2=0} \quad \textbf{(20.8)}$$

$$z_{22} = \left.\frac{V_2}{I_2}\right|_{I_1=0} \tag{20.9}$$

Coefficient z_{12} is the transfer impedance with port 1 open-circuited, z_{21} is the transfer impedance with port 2 open-circuited, and z_{22} is the output impedance with port 1 open circuited. Coefficients z_{11}, z_{12}, z_{21}, and z_{22} are measured in ohms (V/A). To find the parameters z_{11} and z_{21}, with port 2 open-circuited ($I_2 = 0$), apply voltage V_1 (or current I_1) at port 1, as shown in Figure 20.3, and measure I_1 and V_2. The parameter z_{11} is the ratio of V_1 to I_1:

$$z_{11} = \frac{V_1}{I_1} \tag{20.10}$$

and the parameter z_{21} is the ratio of V_2 to I_1:

$$z_{21} = \frac{V_2}{I_1} \tag{20.11}$$

To find the parameters z_{12} and z_{22}, with port 1 open-circuited ($I_1 = 0$), apply voltage V_2 (or current I_2) at port 2, as shown in Figure 20.4, and measure I_2 and V_1. The parameter z_{12} is the ratio of V_1 to I_2:

$$z_{12} = \frac{V_1}{I_2} \tag{20.12}$$

and the parameter z_{22} is the ratio of V_2 to I_2:

$$z_{22} = \frac{V_2}{I_2} \tag{20.13}$$

The equivalent circuit for the z-parameters based on Equations (20.1) and (20.2) is shown in Figure 20.5.

FIGURE 20.3

The circuit to measure z_{11} and z_{21}.

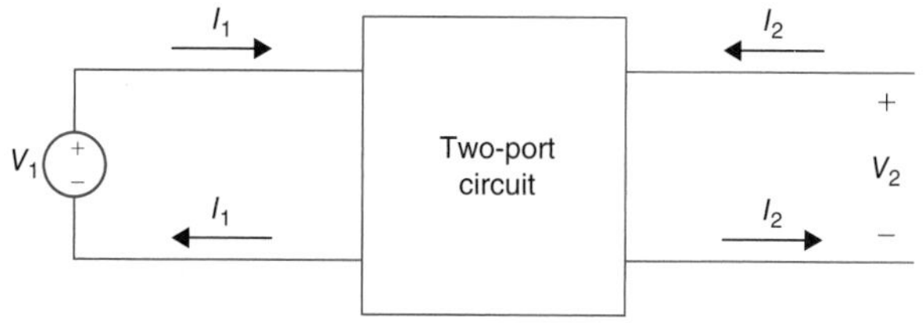

FIGURE 20.4

The circuit to measure z_{12} and z_{22}.

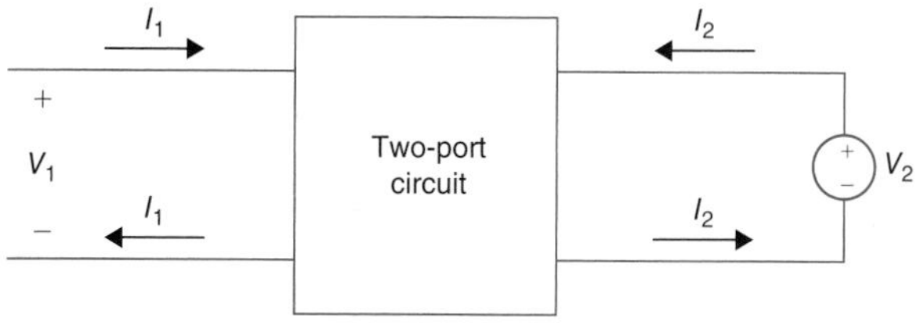

FIGURE 20.5

The equivalent circuit for the z-parameters based on Equations (20.1) and (20.2).

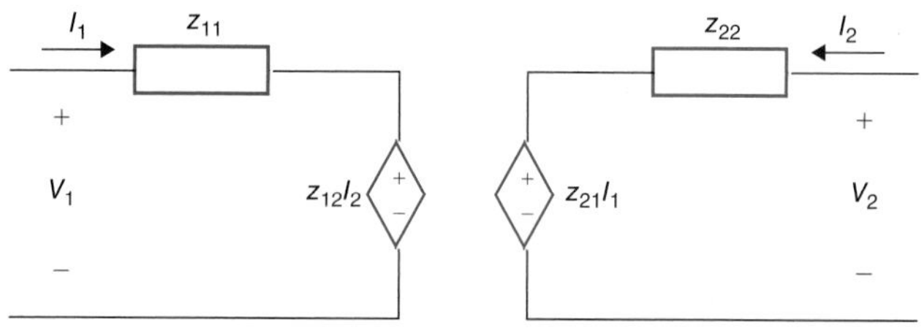

Figure 20.6(a) shows the equivalent circuit with port 2 open-circuited, and port 1 is driven by a current source with current I. Since current I_2 is zero, the voltage drop across z_{22} is zero. Thus, the voltage at port 2 is

$$V_2 = z_{21}I \tag{20.14}$$

FIGURE 20.6

(a) The current source is at port 1. (b) The current source is at port 2.

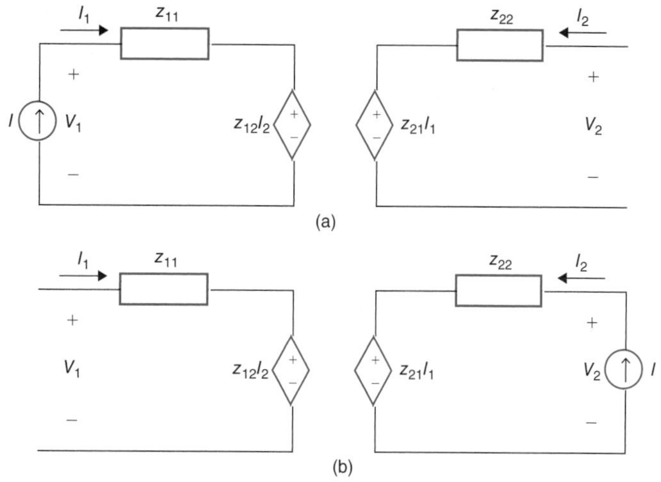

FIGURE 20.7

Equivalent circuit for the reciprocal two-port circuit.

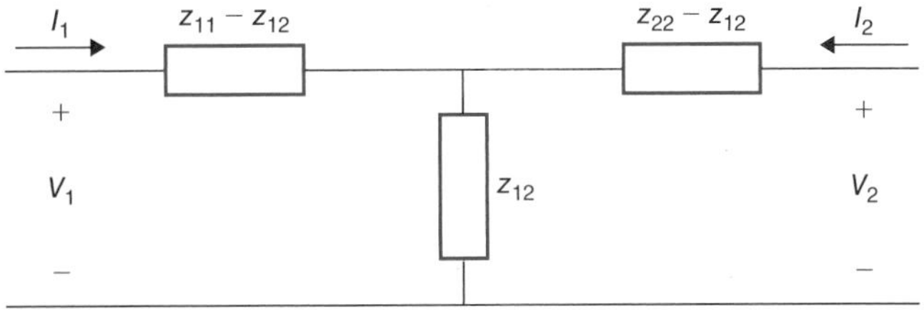

Figure 20.6(b) shows the equivalent circuit with port 1 open-circuited, and port 2 is driven by a current source with current I. Since the current I_1 is zero, the voltage drop across z_{11} is zero. Thus, the voltage at port 1 is

$$V_1 = z_{12}I \tag{20.15}$$

A two-port circuit is said to be **reciprocal** if $V_2 = V_1$ in Equations (20.14) and (20.15). In other words, a two-port circuit is reciprocal if the input and the output are reversed, and the voltage at the open port remains the same for the same driving current. If a two-port circuit is reciprocal, from Equations (20.14) and (20.15), we have

$$z_{21} = z_{12} \tag{20.16}$$

If a circuit consists of R, L, C, and M only, the circuit is reciprocal. If a circuit has dependent sources, in general, it is not reciprocal.

If a two-port circuit is reciprocal, Equations (20.1) and (20.2) can be rewritten as

$$V_1 = z_{11}I_1 - z_{12}I_1 + z_{12}I_1 + z_{12}I_2 = (z_{11} - z_{12})I_1 + z_{12}(I_1 + I_2) \tag{20.17}$$

$$V_2 = z_{12}I_1 + z_{12}I_2 + z_{22}I_2 - z_{12}I_2 = z_{12}(I_1 + I_2) + (z_{22} - z_{12})I_2 \tag{20.18}$$

The equivalent circuit based on Equations (20.17) and (20.18) is shown in Figure 20.7. If $z_{11} = z_{22}$, the two-port circuit is called **symmetrical**.

EXAMPLE 20.1

Find the z-parameters for the circuit shown in Figure 20.8.

FIGURE 20.8

Circuit for EXAMPLE 20.1.

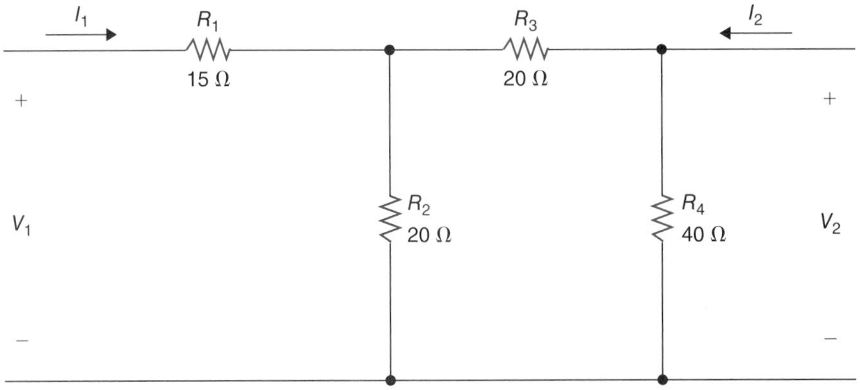

Since the circuit consists of only resistors, it is reciprocal. To find z_{11} and z_{21}, we apply V_1 at port 1 with port 2 open-circuited, as shown in Figure 20.9.

continued

Example 20.1 continued

FIGURE 20.9

Circuit to find z_{11} and z_{21}.

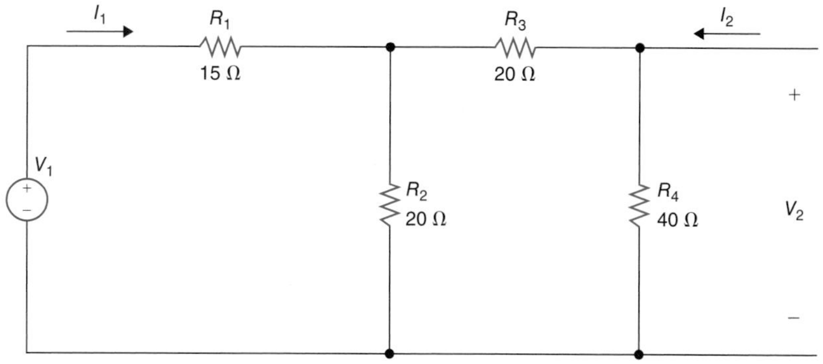

Since port 2 is open-circuited, $I_2 = 0$. The equivalent impedance of R_1, R_2, R_3, and R_4 is

$$R_{\text{eq1}} = R_1 + [R_2 \| (R_3 + R_4)] = 15 + \frac{20 \times (20 + 40)}{20 + (20 + 40)} = 15 + 15 = 30\,\Omega$$

The current I_1 is given by

$$I_1 = \frac{V_1}{R_{\text{eq1}}} = \frac{V_1}{30}$$

From Equation (20.10), the parameter z_{11} is given by

$$z_{11} = \frac{V_1}{I_1} = \frac{V_1}{\frac{V_1}{30}} = 30\,\Omega$$

Application of the voltage divider rule yields

$$V_2 = V_1 \times \frac{15}{30} \times \frac{40}{60} = \frac{V_1}{3}$$

From Equation (20.11), the parameter z_{21} is given by

$$z_{21} = \frac{V_2}{I_1} = \frac{\frac{V_1}{3}}{\frac{V_1}{30}} = 10\,\Omega$$

To find z_{12} and z_{22}, we apply V_2 at port 2 with port 1 open-circuited, as shown in Figure 20.10.

FIGURE 20.10

Circuit to find z_{12} and z_{22}.

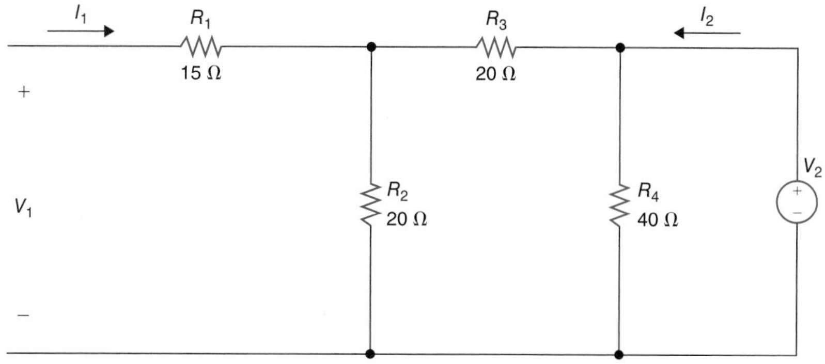

continued

Example 20.1 continued

Since port 1 is open-circuited, $I_1 = 0$. The equivalent impedance of R_4, R_3, and R_2 is

$$R_{eq2} = R_4 \| (R_3 + R_2) = \frac{R_4 \times (R_3 + R_2)}{R_4 + (R_3 + R_2)} = \frac{40 \times 40}{40 + 40}\ \Omega = 20\ \Omega$$

The current I_2 is given by

$$I_2 = \frac{V_2}{R_{eq2}} = \frac{V_2}{20}$$

From Equation (20.13), the parameter z_{22} is given by

$$z_{22} = \frac{V_2}{I_2} = \frac{V_2}{\frac{V_2}{20}} = 20\ \Omega$$

Application of the voltage divider rule yields

$$V_1 = V_2 \times \frac{R_2}{R_2 + R_3} = V_2 \times \frac{20}{20 + 20} = \frac{V_2}{2}$$

From Equation (20.12), the parameter z_{12} is given by

$$z_{12} = \frac{V_1}{I_2} = \frac{\frac{V_2}{2}}{\frac{V_2}{20}} = 10\ \Omega$$

Since $z_{12} = z_{21}$, the circuit is reciprocal as expected.

Exercise 20.1

Find the z-parameters for the circuit shown in Figure 20.11.

FIGURE 20.11

Circuit for EXERCISE 20.1.

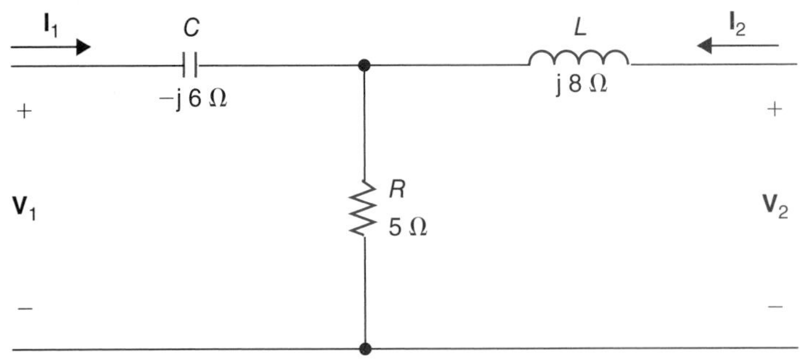

Answer:

$z_{11} = 5 - j6\ \Omega$, $z_{12} = 5\ \Omega$, $z_{21} = 5\ \Omega$, $z_{22} = 5 + j8\ \Omega$.

EXAMPLE 20.2

Find the z-parameters for the coupled coils shown in Figure 20.12.

FIGURE 20.12

Coupled coils.

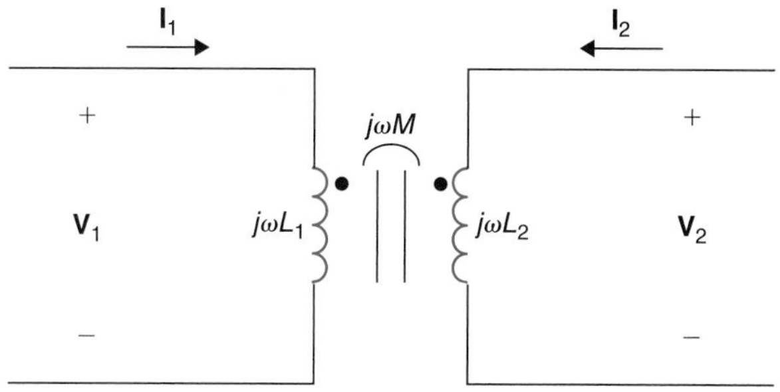

The mesh equation on the mesh 1 (left side) is given by

$$\mathbf{V}_1 = j\omega L_1 \mathbf{I}_1 + j\omega M \mathbf{I}_2 \tag{20.19}$$

The mesh equation on the mesh 2 (right side) is given by

$$\mathbf{V}_2 = j\omega M \mathbf{I}_1 + j\omega L_2 \mathbf{I}_2 \tag{20.20}$$

Comparison of Equations (20.19) and (20.20) with Equations (20.1) and (20.2) reveals that

$$z_{11} = j\omega L_1, \quad z_{12} = j\omega M, \quad z_{21} = j\omega M, \quad z_{22} = j\omega L_2 \tag{20.21}$$

EXAMPLE 20.3

Use z-parameters to find $\mathbf{I}_1$, $\mathbf{I}_2$, $\mathbf{V}_1$, $\mathbf{V}_2$ for the circuit shown in Figure 20.13.

FIGURE 20.13

Circuit for EXAMPLE 20.3.

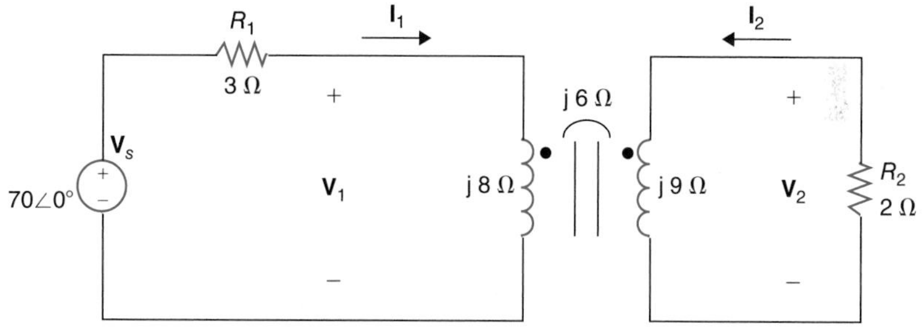

From Equation (20.21), the z-parameters are given by

$$z_{11} = j8, \quad z_{12} = j6, \quad z_{21} = j6, \quad z_{22} = j9 \tag{20.22}$$

Equations (20.19) and (20.20) become, respectively,

$$\mathbf{V}_1 = j8\,\mathbf{I}_1 + j6\,\mathbf{I}_2 \tag{20.23}$$

and

$$\mathbf{V}_2 = j6\,\mathbf{I}_1 + j9\,\mathbf{I}_2 \tag{20.24}$$

Notice that

$$\mathbf{V}_2 = -R_2\mathbf{I}_2 = -2\mathbf{I}_2 \tag{20.25}$$

continued

Example 20.3 continued

$$\mathbf{V}_1 = 70 - R_1\mathbf{I}_1 = 70 - 3\mathbf{I}_1 \quad (20.26)$$

Substitution of Equation (20.25) into Equation (20.24) yields

$$-2\mathbf{I}_2 = j6\,\mathbf{I}_1 + j9\,\mathbf{I}_2$$

which can be rearranged as

$$j6\,\mathbf{I}_1 + (2 + j9)\,\mathbf{I}_2 = 0 \quad (20.27)$$

Substitution of Equation (20.26) into Equation (20.23) results in

$$70 - 3\,\mathbf{I}_1 = j8\,\mathbf{I}_1 + j6\,\mathbf{I}_2$$

which can be rearranged as

$$(3 + j8)\,\mathbf{I}_1 + j6\,\mathbf{I}_2 = 70 \quad (20.28)$$

Application of Cramer's rule to Equations (20.27) and (20.28) yields

$$\mathbf{I}_1 = \frac{\begin{vmatrix} 0 & 2 + j9 \\ 70 & j6 \end{vmatrix}}{\begin{vmatrix} j6 & 2 + j9 \\ 3 + j8 & j6 \end{vmatrix}} = \frac{-140 - j630}{-36 + 72 - 6 - j43} = \frac{-140 - j630}{30 - j43} = 8.3267 - j9.0651$$

$$= 12.3089\angle -47.4313° \text{ A}$$

$$\mathbf{I}_2 = \frac{\begin{vmatrix} j6 & 0 \\ 3 + j8 & 70 \end{vmatrix}}{\begin{vmatrix} j6 & 2 + j9 \\ 3 + j8 & j6 \end{vmatrix}} = \frac{j420}{-36 + 72 - 6 - j43} = \frac{j420}{30 - j43} = -6.5697 + j4.5835$$

$$= 8.0105\angle 145.0975° \text{ A}$$

From Equation (20.25), we get

$$\mathbf{V}_2 = R_2\mathbf{I}_2 = -2\mathbf{I}_2 = 13.1393 - j9.1670 = 16.0211\angle -34.9025° \text{ V}$$

From Equation (20.26), we get

$$\mathbf{V}_1 = 70 - R_1\mathbf{I}_1 = 70 - 3\mathbf{I}_1 = 45.0200 + j27.1953 = 52.5965\angle 31.1350° \text{ V}$$

20.2.2 *y*-PARAMETERS (ADMITTANCE PARAMETERS)

If V_1 and V_2 are independent variables, then I_1 and I_2 are dependent variables. The dependent variables are written as linear combinations of independent variables as

$$I_1 = y_{11}V_1 + y_{12}V_2 \quad (20.29)$$

$$I_2 = y_{21}V_1 + y_{22}V_2 \quad (20.30)$$

In matrix notation, we have

$$\begin{bmatrix} I_1 \\ I_2 \end{bmatrix} = \begin{bmatrix} y_{11} & y_{12} \\ y_{21} & y_{22} \end{bmatrix} \begin{bmatrix} V_1 \\ V_2 \end{bmatrix} \quad (20.31)$$

This equation can be written as

$$i = Yv \tag{20.32}$$

where

$$i = \begin{bmatrix} I_1 \\ I_2 \end{bmatrix}, \quad Y = \begin{bmatrix} y_{11} & y_{12} \\ y_{21} & y_{22} \end{bmatrix}, \quad v = \begin{bmatrix} V_1 \\ V_2 \end{bmatrix} \tag{20.33}$$

Coefficient y_{11} can be measured by taking the ratio of I_1 over V_1 when port 2 is short-circuited ($V_2 = 0$)

$$y_{11} = \left.\frac{I_1}{V_1}\right|_{V_2=0} \tag{20.34}$$

Coefficient y_{11} is the input admittance with port 2 short-circuited. Similarly, coefficients y_{12}, y_{21}, and y_{22} are found by taking the ratio of current to voltage while one of the ports is short-circuited:

$$y_{12} = \left.\frac{I_1}{V_2}\right|_{V_1=0} \tag{20.35}$$

$$y_{21} = \left.\frac{I_2}{V_1}\right|_{V_2=0} \tag{20.36}$$

$$y_{22} = \left.\frac{I_2}{V_2}\right|_{V_1=0} \tag{20.37}$$

Coefficient y_{12} is the transfer admittance with port 1 short-circuited, y_{21} is the transfer admittance with port 2 short-circuited, and y_{22} is the output admittance with port 1 short-circuited. Coefficients y_{11}, y_{12}, y_{21}, and y_{22} have unit of siemens (A/V). Notice that if V is replaced by I, and I is replaced by V in the right side of Equations (20.6) through (20.9), the z-parameters become y-parameters. The impedance parameters and admittance parameters are called *immittance parameters*. The equivalent circuit for the y-parameters based on Equations (20.29) and (20.30) is shown in Figure 20.14.

If a two-port circuit is reciprocal, we have

$$y_{12} = y_{21} \tag{20.38}$$

If a two-port circuit is reciprocal, Equations (20.29) and (20.30) can be written as

$$\begin{aligned} I_1 &= y_{11}V_1 + y_{12}V_1 - y_{12}V_1 + y_{12}V_2 \\ &= (y_{11} + y_{12})V_1 - y_{12}(V_1 - V_2) \end{aligned} \tag{20.39}$$

$$\begin{aligned} I_2 &= y_{12}V_1 - y_{12}V_2 + y_{12}V_2 + y_{22}V_2 \\ &= -y_{12}(V_2 - V_1) + (y_{12} + y_{22})V_2 \end{aligned} \tag{20.40}$$

The equivalent circuit based on Equations (20.39) and (20.40) is shown in Figure 20.15.

FIGURE 20.14

The equivalent circuit for the y-parameters.

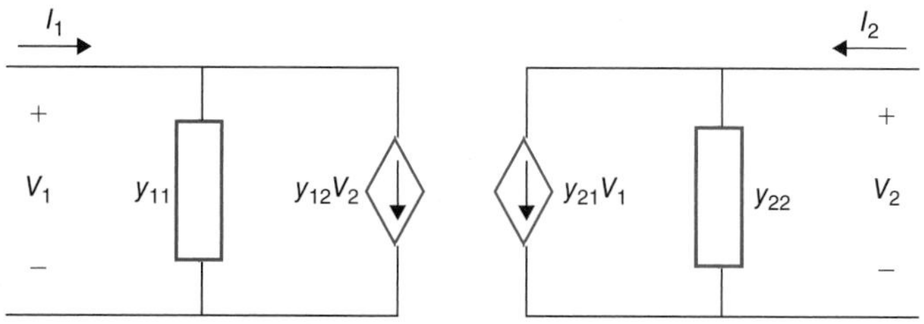

FIGURE 20.15

The equivalent circuit for the reciprocal y-parameters.

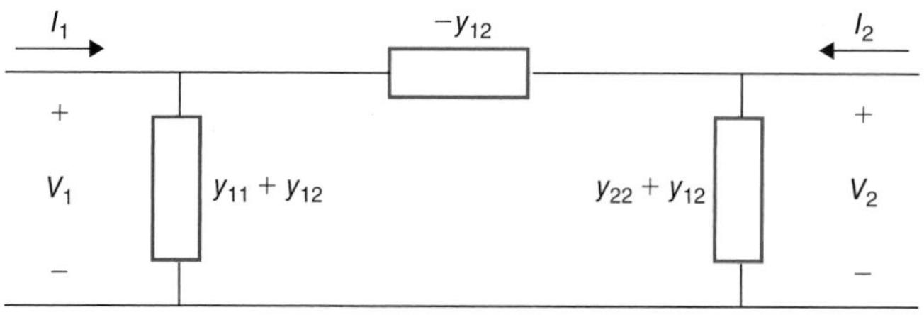

EXAMPLE 20.4

Find the y-parameters for the circuit shown in Figure 20.16.

FIGURE 20.16

Circuit for EXAMPLE 20.4.

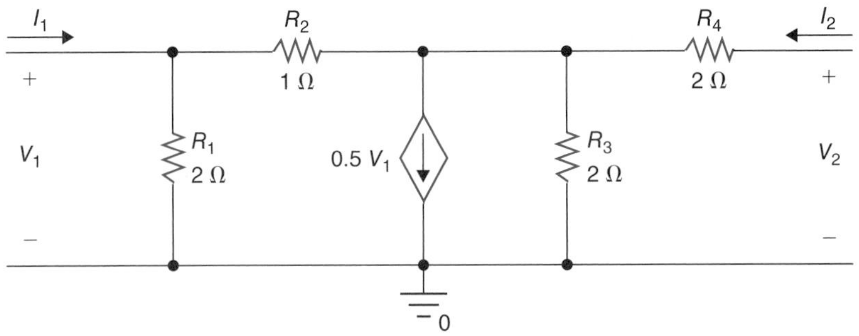

Since the circuit contains a dependent source, it is not reciprocal. To find y_{11} and y_{21}, we apply V_1 at port 1 with V_2 short-circuited, as shown in Figure 20.17.

FIGURE 20.17

Circuit to find y_{11} and y_{21}.

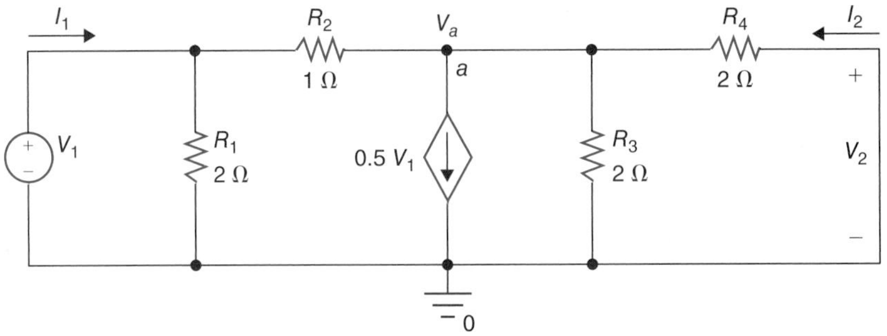

Since port 2 is short-circuited, $V_2 = 0$. Summing the currents leaving node a, we obtain

$$\frac{V_a - V_1}{1} + 0.5V_1 + \frac{V_a}{2} + \frac{V_a}{2} = 0$$

Multiplication by 2 yields

$$2V_a - 2V_1 + V_1 + V_a + V_a = 0$$

which can be simplified to

$$4V_a = V_1$$

Thus, we obtain

$$V_a = 0.25V_1$$

The current I_1 is given by

$$I_1 = \frac{V_1}{2} + \frac{V_1 - V_a}{1} = \frac{V_1}{2} + \frac{3V_1}{4} = \frac{5V_1}{4} = 1.25V_1$$

continued

Example 20.4 continued

From Equation (20.34), the parameter y_{11} is given by

$$y_{11} = \frac{I_1}{V_1} = \frac{1.25V_1}{V_1} = 1.25 \text{ S}$$

The current I_2 is given by

$$I_2 = -\frac{V_a}{2} = -\frac{\frac{V_1}{4}}{2} = -\frac{1}{8}V_1$$

From Equation (20.36), the parameter y_{21} is given by

$$y_{21} = \frac{I_2}{V_1} = \frac{\frac{-1}{8}V_1}{V_1} = \frac{-1}{8} \text{ S} = -0.125 \text{ S}$$

To find y_{12} and y_{22}, we apply V_2 at port 2 and short-circuit V_1, as shown in Figure 20.18.

FIGURE 20.18

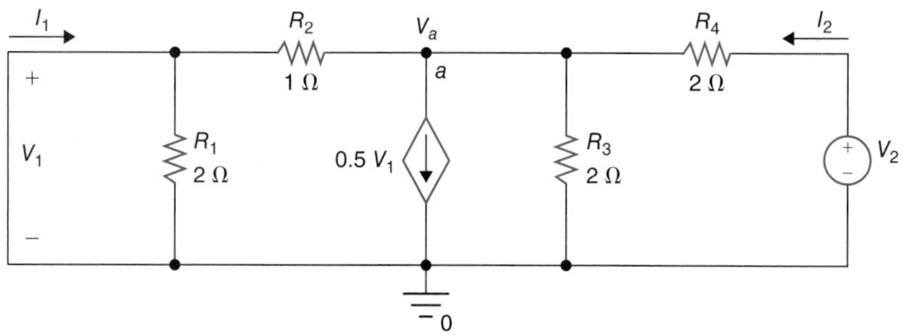

Circuit to find y_{12} and y_{22}.

Since port 1 is short-circuited, $V_1 = 0$. There is no current through the voltage-controlled current source. Summing the currents leaving node *a*, we obtain

$$\frac{V_a}{1} + \frac{V_a}{2} + \frac{V_a - V_2}{2} = 0$$

Multiplication by 2 yields

$$2V_a + V_a + V_a - V_2 = 0$$

which can be simplified to

$$4V_a = V_2$$

Thus, we obtain

$$V_a = 0.25V_2$$

The current I_2 is given by

$$I_2 = \frac{V_2 - V_a}{2} = \frac{3V_2}{8} = 0.375V_2$$

From Equation (20.37), the parameter y_{22} is given by

$$y_{22} = \frac{I_2}{V_2} = \frac{0.375V_2}{V_2} = 0.375 \text{ S}$$

continued

Example 20.4 continued

The current I_1 is given by

$$I_1 = -\frac{V_a}{1} = -\frac{\frac{V_2}{4}}{1} = -\frac{1}{4}V_2$$

From Equation (20.35), the parameter y_{12} is given by

$$y_{12} = \frac{I_1}{V_2} = \frac{\frac{-1}{4}V_2}{V_2} = \frac{-1}{4}\,\text{S} = -0.25\ \text{S}$$

Since $y_{12} \neq y_{21}$, the circuit is not reciprocal.

EXAMPLE 20.5

A voltage source with voltage of 1 V and a load resistor with resistance 2 Ω are added to the circuit shown in Figure 20.16. The resulting circuit is shown in Figure 20.19. Use the *y*-parameters obtained in EXAMPLE 20.4 to find I_2, V_2 and I_1.

FIGURE 20.19

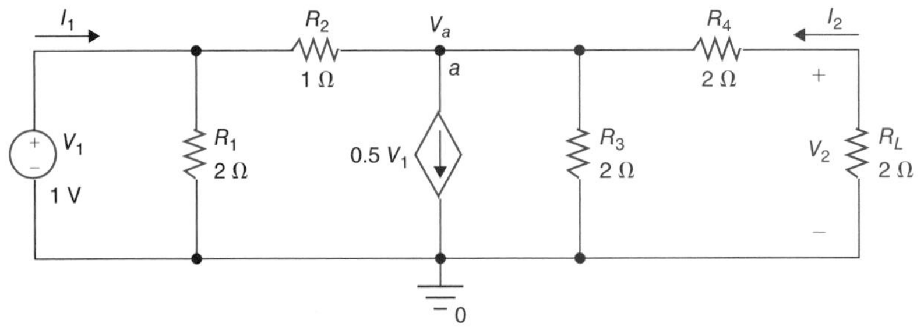

Circuit for EXAMPLE 20.5.

From Equations (20.29) and (20.30) and EXAMPLE 20.4, we obtain

$$I_1 = y_{11}V_1 + y_{12}V_2 = 1.25V_1 - 0.25V_2 \quad \textbf{(20.41)}$$

$$I_2 = y_{21}V_1 + y_{22}V_2 = -0.125V_1 + 0.375V_2 \quad \textbf{(20.42)}$$

Substitution of $V_2 = -R_LI_2 = -2I_2$ into Equation (20.42) yields

$$I_2 + 0.75I_2 = -0.125V_1$$

Solving for I_2, we obtain

$$I_2 = \frac{-0.125}{1.75}V_1 = \frac{-0.125}{1.75} \times 1 = -0.07143\ \text{A}$$

Voltage V_2 is given by

$$V_2 = -2I_2 = 0.1429\ \text{V}$$

From Equation (20.41), we get

$$I_1 = 1.25V_1 - 0.25V_2 = 1.2143\ \text{A}$$

Exercise 20.2

Find the y-parameters for the circuit shown in Figure 20.20.

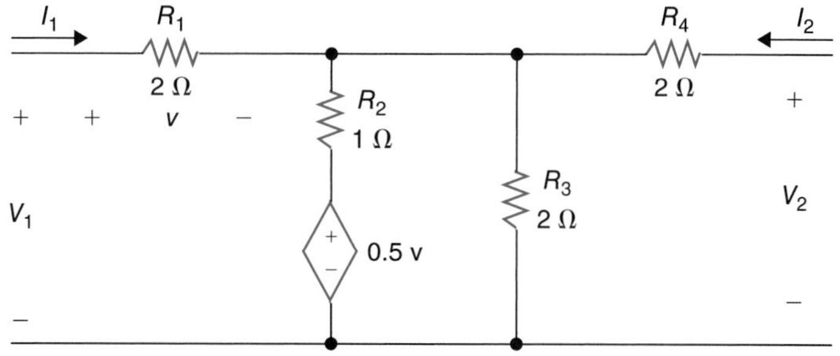

FIGURE 20.20

Circuit for EXERCISE 20.2.

Answer:

$y_{11} = 1/3$ S, $y_{12} = -1/12$ S, $y_{21} = -1/12$ S, $y_{22} = 5/12$ S.

EXAMPLE 20.6

Find the y-parameters for the circuit shown in Figure 20.21.

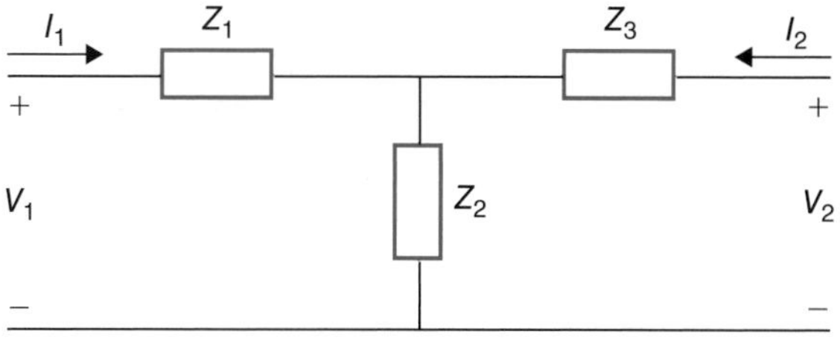

FIGURE 20.21

Circuit for EXAMPLE 20.6.

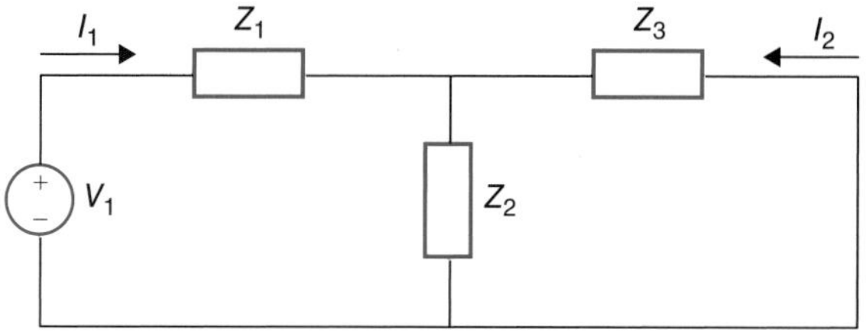

FIGURE 20.22

Circuit with V_2 short-circuited.

To find y_{11} and y_{21}, V_2 is short-circuited, as shown in Figure 20.22.

The equivalent impedance seen from V_1 is given by

$$Z_a = Z_1 + (Z_2 \| Z_3) = Z_1 + \frac{Z_2 Z_3}{Z_2 + Z_3}$$

$$= \frac{Z_1 Z_2 + Z_1 Z_3 + Z_2 Z_3}{Z_2 + Z_3}$$

The current I_1 is given by

$$I_1 = \frac{V_1}{Z_a} = \frac{Z_2 + Z_3}{Z_1 Z_2 + Z_1 Z_3 + Z_2 Z_3} V_1$$

Thus, y_{11} is given by

$$y_{11} = \frac{I_1}{V_1} = \frac{Z_2 + Z_3}{Z_1 Z_2 + Z_1 Z_3 + Z_2 Z_3} \tag{20.43}$$

Application of the current divider rule yields

$$I_2 = -I_1 \times \frac{Z_2}{Z_2 + Z_3} = -\frac{Z_2}{Z_1 Z_2 + Z_1 Z_3 + Z_2 Z_3} V_1$$

Dividing by V_1, we obtain

$$y_{21} = \frac{I_2}{V_1} = \frac{-Z_2}{Z_1 Z_2 + Z_1 Z_3 + Z_2 Z_3} \tag{20.44}$$

continued

Example 20.6 continued

FIGURE 20.23

Circuit with V_1 short-circuited.

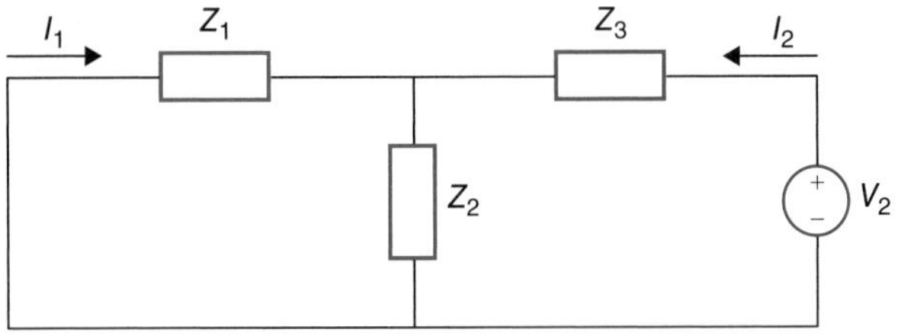

To find y_{22} and y_{12}, V_1 is short-circuited, as shown in Figure 20.23.

The equivalent impedance seen from V_2 is given by

$$Z_b = Z_3 + (Z_1 \| Z_2) = Z_3 + \frac{Z_1 Z_2}{Z_1 + Z_2}$$
$$= \frac{Z_1 Z_2 + Z_1 Z_3 + Z_2 Z_3}{Z_1 + Z_2}$$

The current I_2 is given by

$$I_2 = \frac{V_2}{Z_b} = \frac{Z_1 + Z_2}{Z_1 Z_2 + Z_1 Z_3 + Z_2 Z_3} V_2$$

Thus, y_{22} is given by

$$y_{22} = \frac{I_2}{V_2} = \frac{Z_1 + Z_2}{Z_1 Z_2 + Z_1 Z_3 + Z_2 Z_3} \tag{20.45}$$

Application of the current divider rule yields

$$I_1 = -I_2 \times \frac{Z_2}{Z_1 + Z_2} = -\frac{Z_2}{Z_1 Z_2 + Z_1 Z_3 + Z_2 Z_3} V_2$$

Dividing by V_2, we obtain

$$y_{12} = \frac{I_1}{V_2} = \frac{-Z_2}{Z_1 Z_2 + Z_1 Z_3 + Z_2 Z_3} \tag{20.46}$$

Notice that the circuit is reciprocal.

Exercise 20.3

Let $Z_1 = 2\ \Omega$, $Z_2 = 3\ \Omega$, and $Z_3 = 5\ \Omega$ for the circuit shown in Figure 20.21. Find the y-parameters for the circuit shown in Figure 20.21.

Answer:

$y_{11} = 8/31 = 0.2581$ S, $y_{12} = -3/31 = -0.09677$ S, $y_{21} = -3/31 = -0.09677$ S, $y_{22} = 5/31 = 0.1613$ S.

20.2.3 *h*-PARAMETERS (HYBRID PARAMETERS)

If I_1 and V_2 are independent variables, then V_1 and I_2 are dependent variables. The dependent variables are written as linear combinations of independent variables as

$$V_1 = h_{11} I_1 + h_{12} V_2 \tag{20.47}$$

$$I_2 = h_{21} I_1 + h_{22} V_2 \tag{20.48}$$

In matrix notation, we have

$$\begin{bmatrix} V_1 \\ I_2 \end{bmatrix} = \begin{bmatrix} h_{11} & h_{12} \\ h_{21} & h_{22} \end{bmatrix} \begin{bmatrix} I_1 \\ V_2 \end{bmatrix} \tag{20.49}$$

This equation can be written as

$$\boldsymbol{d}_h = \boldsymbol{H}\boldsymbol{n}_h$$

where

$$\boldsymbol{d}_h = \begin{bmatrix} V_1 \\ I_2 \end{bmatrix}, \quad \boldsymbol{H} = \begin{bmatrix} h_{11} & h_{12} \\ h_{21} & h_{22} \end{bmatrix}, \quad \boldsymbol{n}_h = \begin{bmatrix} I_1 \\ V_2 \end{bmatrix}$$

Coefficient h_{11} can be measured by taking the ratio of V_1 over I_1 when port 2 is short-circuited ($V_2 = 0$)

$$h_{11} = \left.\frac{V_1}{I_1}\right|_{V_2=0} \tag{20.50}$$

Coefficient h_{11} is the input impedance with port 2 short-circuited. The other coefficients h_{12}, h_{21}, and h_{22} are given by

$$h_{12} = \left.\frac{V_1}{V_2}\right|_{I_1=0} \tag{20.51}$$

$$h_{21} = \left.\frac{I_2}{I_1}\right|_{V_2=0} \tag{20.52}$$

$$h_{22} = \left.\frac{I_2}{V_2}\right|_{I_1=0} \tag{20.53}$$

FIGURE 20.24

The equivalent circuit for the h-parameters.

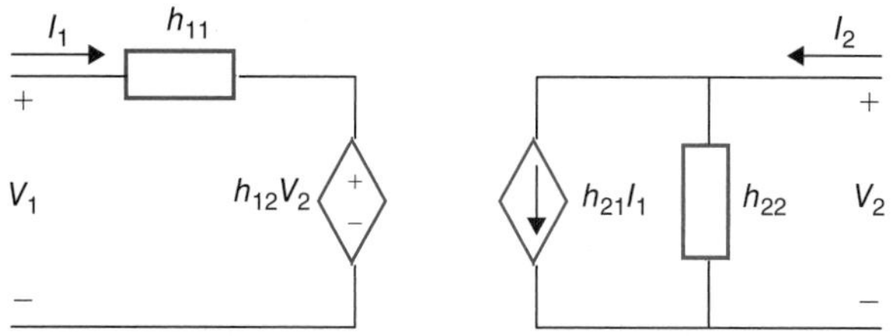

Coefficient h_{12} is the reverse voltage gain with port 1 open-circuited, h_{21} is the forward current gain with port 2 short-circuited, and h_{22} is the output admittance with port 1 open-circuited. Coefficient h_{11} has a unit of ohm (V/A), coefficient h_{12} is unitless (V/V), coefficient h_{21} is unitless (A/A), and coefficient h_{22} has a unit of siemens (A/V). If a two-port circuit is reciprocal, $h_{12} = -h_{21}$. The equivalent circuit for the h-parameters based on Equations (20.47) and (20.48) is shown in Figure 20.24.

EXAMPLE 20.7

Find the h-parameters for the circuit shown in Figure 20.25.

To find h_{11} and h_{21}, port 2 is short-circuited, and voltage V_1 is applied at port 1, as shown in Figure 20.26.

Summing the currents leaving node a, we obtain

$$\frac{V_a - V_1}{1} + \frac{V_a}{s} + \frac{V_a}{1} = 0$$

which can be written as

$$\left(2 + \frac{1}{s}\right)V_a = V_1$$

continued

Example 20.7 continued

FIGURE 20.25

Circuit for EXAMPLE 20.7.

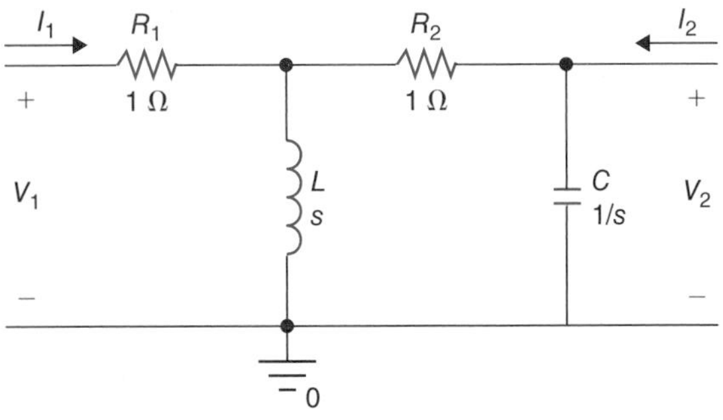

FIGURE 20.26

Circuit used to find h_{11} and h_{21}.

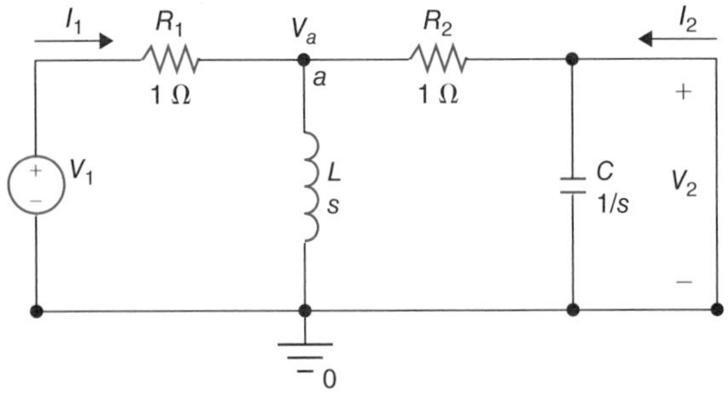

FIGURE 20.27

Circuit used to find h_{22} and h_{12}.

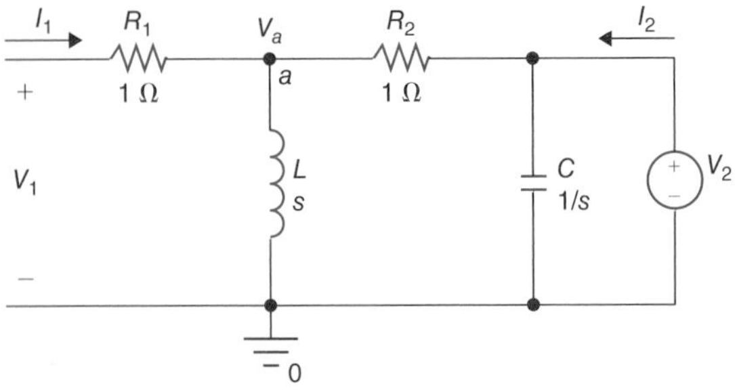

Solving for V_a, we get

$$V_a = \frac{s}{2s+1}V_1$$

The current I_1 is the current through R_1. Thus we have

$$I_1 = \frac{V_1 - V_a}{1} = \frac{V_1 - \dfrac{s}{2s+1}V_1}{1} = \frac{s+1}{2s+1}V_1$$

Thus, we obtain

$$h_{11} = \frac{V_1}{I_1} = \frac{2s+1}{s+1}$$

Current I_2 is given by

$$I_2 = \frac{-V_a}{R_2} = \frac{-\dfrac{s}{2s+1}V_1}{1} = -\frac{s}{2s+1}V_1$$

And h_{21} is given by

$$h_{21} = \frac{I_2}{I_1} = \frac{-\dfrac{s}{2s+1}V_1}{\dfrac{s+1}{2s+1}V_1} = \frac{-s}{s+1}$$

To find h_{22} and h_{12}, port 1 is open-circuited, and voltage V_2 is applied at port 2, as shown in Figure 20.27.

Application of the voltage divider rule on the series connection of R_2 and L yields

$$V_a = V_2 \times \frac{s}{s+1} = \frac{s}{s+1}V_2 = V_1$$

Current I_2 is given by

$$I_2 = \frac{V_2}{\dfrac{1}{s}} + \frac{V_2}{s+1} = \left(s + \frac{1}{s+1}\right)V_2 = \frac{s^2+s+1}{s+1}V_2$$

Therefore, h_{22} is given by

$$h_{22} = \frac{I_2}{V_2} = \frac{s^2+s+1}{s+1}$$

Since $V_1 = V_a$, h_{12} is given by

$$h_{12} = \frac{V_1}{V_2} = \frac{\dfrac{s}{s+1}V_2}{V_2} = \frac{s}{s+1}$$

Notice that $h_{12} = -\,h_{21}$ due to reciprocity.

Exercise 20.4

Find the h-parameters for the circuit shown in Figure 20.28.

FIGURE 20.28

Circuit for EXERCISE 20.4.

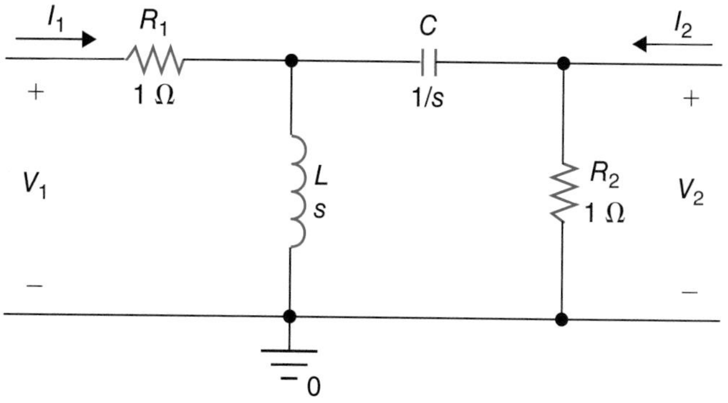

Answer:

$$h_{11} = \frac{s^2 + s + 1}{s^2 + 1}, \quad h_{21} = -\frac{s^2}{s^2 + 1}, \quad h_{22} = \frac{s^2 + s + 1}{s^2 + 1}, \quad h_{12} = \frac{s^2}{s^2 + 1}.$$

20.2.4 g-PARAMETERS (INVERSE HYBRID PARAMETERS)

If V_1 and I_2 are independent variables, then I_1 and V_2 are dependent variables. The dependent variables are written as linear combinations of independent variables as

$$I_1 = g_{11}V_1 + g_{12}I_2 \tag{20.54}$$

$$V_2 = g_{21}V_1 + g_{22}I_2 \tag{20.55}$$

In matrix notation, we have

$$\begin{bmatrix} I_1 \\ V_2 \end{bmatrix} = \begin{bmatrix} g_{11} & g_{12} \\ g_{21} & g_{22} \end{bmatrix} \begin{bmatrix} V_1 \\ I_2 \end{bmatrix} \tag{20.56}$$

This equation can be written as

$$\boldsymbol{d}_g = \boldsymbol{G}\boldsymbol{n}_g$$

where

$$\boldsymbol{d}_g = \begin{bmatrix} I_1 \\ V_2 \end{bmatrix}, \quad \boldsymbol{G} = \begin{bmatrix} g_{11} & g_{12} \\ g_{21} & g_{22} \end{bmatrix}, \quad \boldsymbol{n}_g = \begin{bmatrix} V_1 \\ I_2 \end{bmatrix}$$

Coefficient g_{11} can be measured by taking the ratio of I_1 over V_1 when port 2 is open-circuited ($I_2 = 0$)

$$g_{11} = \left.\frac{I_1}{V_1}\right|_{I_2=0} \tag{20.57}$$

Coefficient g_{11} is the input admittance with port 2 open-circuited. Coefficients g_{12}, g_{21}, and g_{22} are given by

$$g_{12} = \left.\frac{I_1}{I_2}\right|_{V_1=0} \tag{20.58}$$

$$g_{21} = \left.\frac{V_2}{V_1}\right|_{I_2=0} \tag{20.59}$$

$$g_{22} = \left.\frac{V_2}{I_2}\right|_{V_1=0} \tag{20.60}$$

FIGURE 20.29

The equivalent circuit for the *g*-parameters.

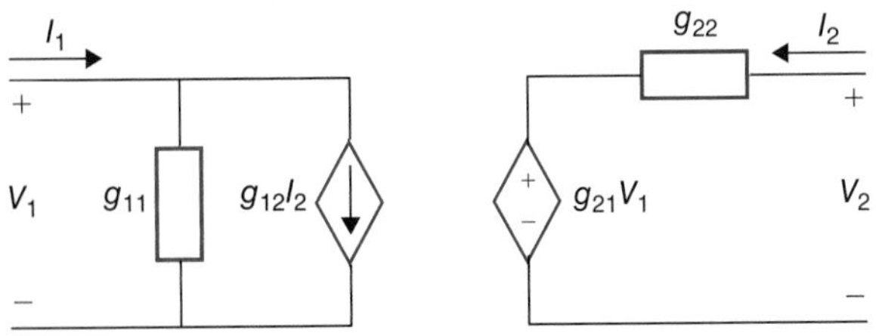

Coefficient g_{12} is the reverse current gain with port 1 short-circuited, g_{21} is the forward voltage gain with port 2 open-circuited, and g_{22} is the output impedance with port 1 short-circuited. Coefficient g_{11} has a unit of siemens (A/V), coefficient g_{12} is unitless (A/A), coefficient g_{21} is unitless (V/V), and coefficient g_{22} has a unit of ohm (V/A). Notice that if V is replaced by I, and I is replaced by V in the right side of Equations (20.50) through (20.53), the *h*-parameters become *g*-parameters. The equivalent circuit for the *g*-parameters based on Equations (20.54) and (20.55) is shown in Figure 20.29.

EXAMPLE 20.8

Find the *g*-parameters for the two-port circuit shown in Figure 20.30.

To find g_{11} and g_{21}, port 2 is open-circuited, and voltage V_1 is applied at port 1, as shown in Figure 20.31.

Since the current leaving node *a* through the voltage-controlled current source (VCCS) is returning back to node *a*, it has no effect on the net current leaving node *a*. Summing the currents leaving node *a*, we obtain

FIGURE 20.30

Circuit for EXAMPLE 20.8.

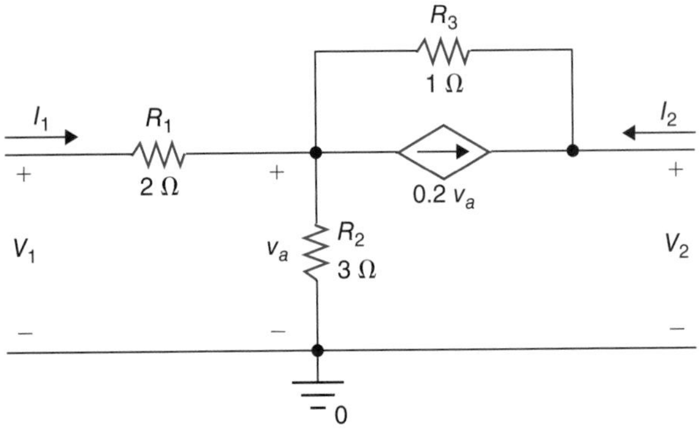

$$\frac{v_a - V_1}{2} + \frac{v_a}{3} = 0$$

Multiplication by 6 yields

$$3v_a - 3V_1 + 2v_a = 0$$

Solving v_a, we obtain

$$v_a = 0.6V_1$$

Current I_1 is given by

$$I_1 = \frac{V_1 - v_a}{2} = \frac{1 - 0.6}{2}V_1 = 0.2\,V_1$$

continued

Example 20.8 continued

FIGURE 20.31

Circuit with port 2 open-circuited.

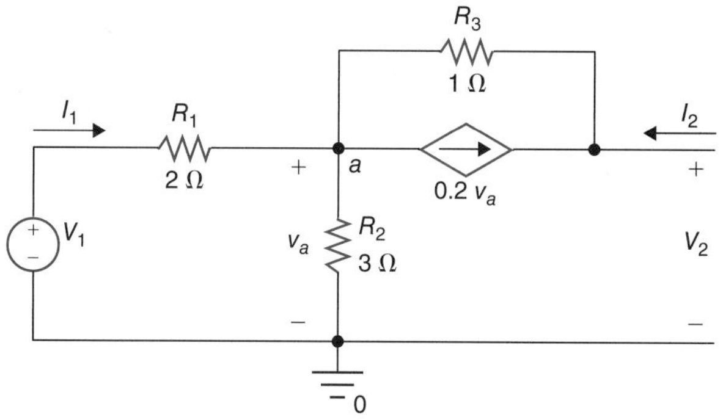

FIGURE 20.32

Circuit with port 1 short-circuited.

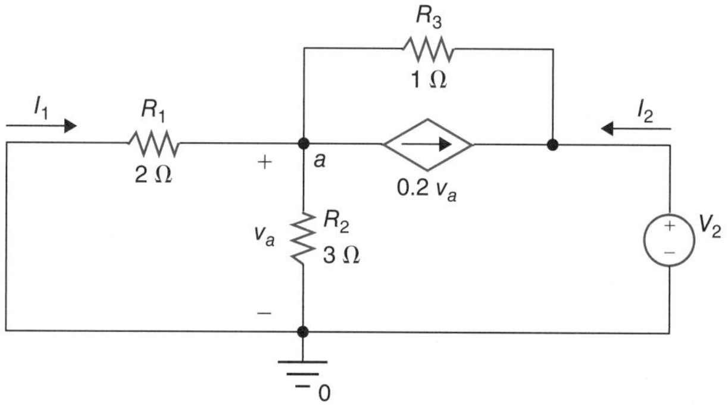

Thus, we have

$$g_{11} = \frac{I_1}{V_1} = 0.2\ \text{S}$$

Voltage V_2 is the sum of v_a and the voltage across R_3. Thus

$$V_2 = v_a + 1 \times 0.2v_a = 1.2v_a = 0.72\ V_1$$

Coefficient g_{21} is given by

$$g_{21} = \frac{V_2}{V_1} = 0.72$$

To find g_{12} and g_{22}, port 1 is short-circuited, and voltage V_2 is applied at port 2, as shown in Figure 20.32.

Summing the currents leaving node a, we obtain

$$\frac{v_a}{2} + \frac{v_a - V_2}{1} + 0.2v_a + \frac{v_a}{3} = 0$$

Multiplication by 6 yields

$$3v_a + 6v_a - 6V_2 + 1.2v_a + 2v_a = 0$$

which can be simplified to

$$12.2v_a = 6V_2$$

Thus,

$$v_a = \frac{6}{12.2}V_2 = 0.4918\,V_2$$

Current I_2 is the sum of $-0.2v_a$ and the current through R_3 is

$$I_2 = -0.2v_a + \frac{V_2 - v_a}{1} = V_2 - 1.2v_a = V_2 - 0.5902V_2 = 0.4098V_2$$

Parameter g_{22} is given by

$$g_{22} = \frac{V_2}{I_2} = \frac{1}{0.4098} = 2.44\ \Omega$$

Current I_1 is given by

$$I_1 = \frac{-v_a}{R_1} = \frac{-0.4918}{2}V_2 = -0.2549V_2$$

Parameter g_{12} is given by

$$g_{12} = \frac{I_1}{I_2} = \frac{-0.2549V_2}{0.4098V_2} = -0.6$$

EXAMPLE 20.9

A voltage source with voltage of 1 V and a load resistor with resistance 2 Ω are added to the circuit shown in Figure 20.30. The resulting circuit is shown in Figure 20.33. Use the *g*-parameters obtained in EXAMPLE 20.8 to find I_2, V_2, and I_1.

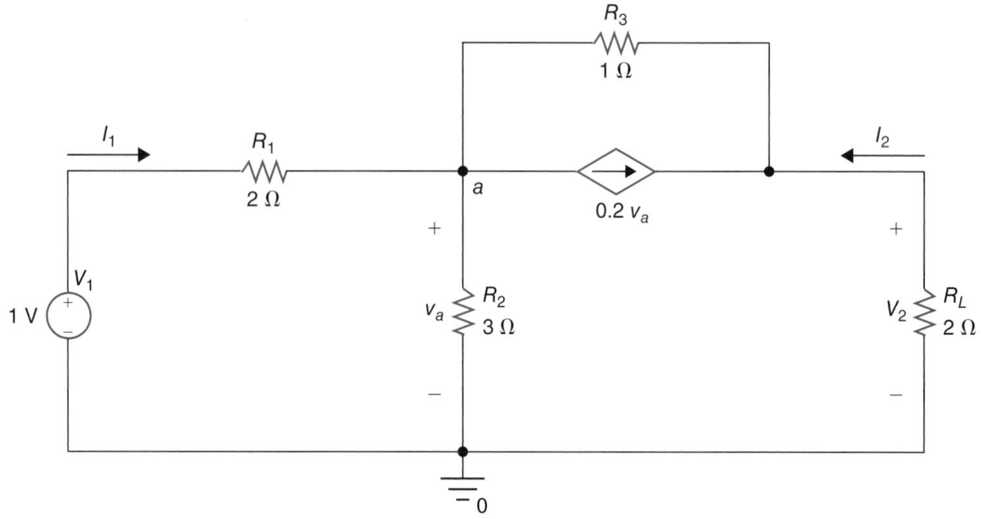

FIGURE 20.33

Circuit for EXAMPLE 20.9.

From Equations (20.54) and (20.55) and EXAMPLE 20.8, we obtain

$$I_1 = g_{11}V_1 + g_{12}I_2 = 0.2V_1 - 0.6I_2 \quad \textbf{(20.61)}$$

$$V_2 = g_{21}V_1 + g_{22}I_2 = 0.72V_1 + 2.44I_2 \quad \textbf{(20.62)}$$

Substitution of $V_2 = -R_LI_2 = -2I_2$ into Equation (20.62) yields

$$-2I_2 = 0.72V_1 + 2.44I_2$$

Solving for I_2, we obtain

$$I_2 = \frac{-0.72}{4.44}V_1 = \frac{-0.72}{4.44} \times 1 = -0.1622\,\text{A}$$

Voltage V_2 is given by

$$V_2 = -2I_2 = 0.3243\,\text{V}$$

From Equation (20.61), we can get I_1:

$$I_1 = 0.2V_1 - 0.6I_2 = 0.2 \times 1 - 0.6 \times (-0.1622) = 0.2973\,\text{A}$$

Exercise 20.5

Find the g-parameters for the circuit shown in Figure 20.34.

FIGURE 20.34

Circuit for EXERCISE 20.5.

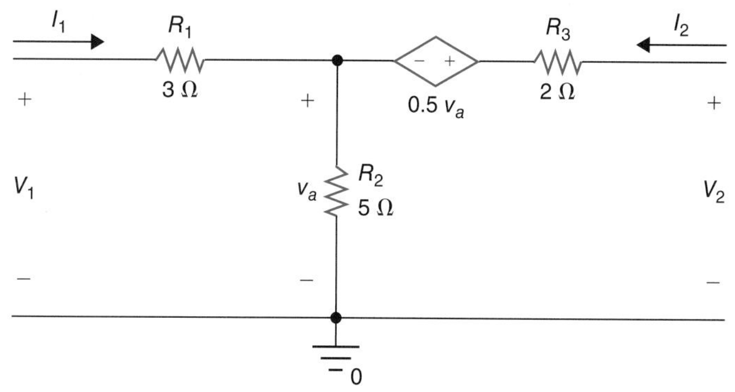

Answer:

$g_{11} = 0.125, \quad g_{21} = 0.9375, \quad g_{22} = 4.8125, \quad g_{12} = -0.625.$

20.2.5 *ABCD*-PARAMETERS (TRANSMISSION PARAMETERS, *a*-PARAMETERS)

If V_2 and I_2 are independent variables, then V_1 and I_1 are dependent variables. The dependent variables are written as linear combinations of independent variables as

$$V_1 = AV_2 - BI_2 \tag{20.63}$$

$$I_1 = CV_2 - DI_2 \tag{20.64}$$

In matrix notation, we have

$$\begin{bmatrix} V_1 \\ I_1 \end{bmatrix} = \begin{bmatrix} A & B \\ C & D \end{bmatrix} \begin{bmatrix} V_2 \\ -I_2 \end{bmatrix} \tag{20.65}$$

This equation can be written as

$$\boldsymbol{d}_a = \boldsymbol{T}\boldsymbol{n}_a$$

where

$$\boldsymbol{d}_a = \begin{bmatrix} V_1 \\ I_1 \end{bmatrix}, \quad \boldsymbol{T} = \begin{bmatrix} A & B \\ C & D \end{bmatrix}, \quad \boldsymbol{n}_a = \begin{bmatrix} V_2 \\ -I_2 \end{bmatrix}$$

Coefficient A can be found by taking the ratio of V_1 over V_2 when port 2 is open-circuited ($I_2 = 0$):

$$A = \left.\frac{V_1}{V_2}\right|_{I_2=0} \tag{20.66}$$

Coefficient A is the voltage ratio with port 2 open-circuited. Coefficients B, C, and D are given by

$$B = \left.\frac{V_1}{-I_2}\right|_{V_2=0} \quad \textbf{(20.67)}$$

$$C = \left.\frac{I_1}{V_2}\right|_{I_2=0} \quad \textbf{(20.68)}$$

$$D = \left.\frac{I_1}{-I_2}\right|_{V_2=0} \quad \textbf{(20.69)}$$

Coefficient B is the negative transfer impedance with port 2 short-circuited, C is the transfer admittance with port 2 open-circuited, and D is the negative current ratio with port 2 short-circuited. Coefficient A is unitless (V/V), coefficient B has a unit of ohm (V/A), coefficient C has a unit of siemens (A/V), and coefficient D is unitless (A/A). The a-parameters refer to defining parameters as $a_{11} = A, a_{12} = B, a_{21} = C, a_{22} = D$.

EXAMPLE 20.10

Find the *ABCD* parameters for the circuits shown in Figure 20.35.

FIGURE 20.35

Circuit for EXAMPLE 20.10.

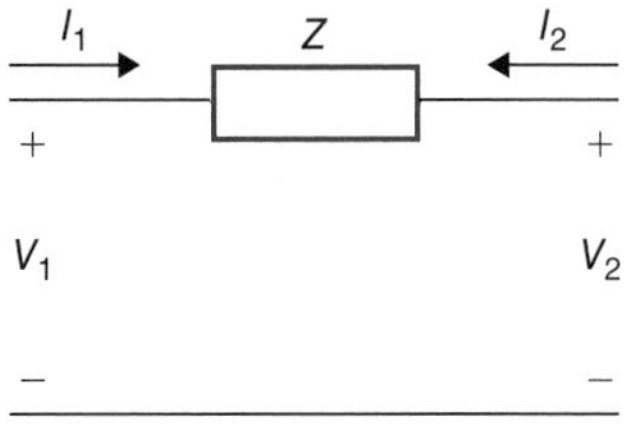

For a series impedance shown in Figure 20.35, if port 2 is open-circuited ($I_2 = 0$), there is no closed path in the circuit, and currents I_2 and I_1 are 0. The voltage drop across Z is 0. Thus, $V_2 = V_1$. Coefficients A and C are given, respectively, by

$$A = \left.\frac{V_1}{V_2}\right|_{I_2=0} = \frac{V_1}{V_1} = 1$$

$$C = \left.\frac{I_1}{V_2}\right|_{I_2=0} = \frac{0}{V_1} = 0$$

If port 2 is short-circuited ($V_2 = 0$), the voltage drop across Z is V_1. Thus, $I_1 = V_1/Z = -I_2$ and $V_1/(-I_2) = Z$. Coefficients B and D are given, respectively, as

$$B = -\left.\frac{V_1}{I_2}\right|_{V_2=0} = \frac{V_1}{\frac{V_1}{Z}} = Z$$

$$D = -\left.\frac{I_1}{I_2}\right|_{I_2=0} = -\frac{I_1}{-I_1} = 1$$

Putting the parameters in matrix form, we have

$$\boldsymbol{T} = \begin{bmatrix} A & B \\ C & D \end{bmatrix} = \begin{bmatrix} 1 & Z \\ 0 & 1 \end{bmatrix} \quad \textbf{(20.70)}$$

EXAMPLE 20.11

Find the *ABCD* parameters for the circuits shown in Figure 20.36.

FIGURE 20.36

Circuit for EXAMPLE 20.11.

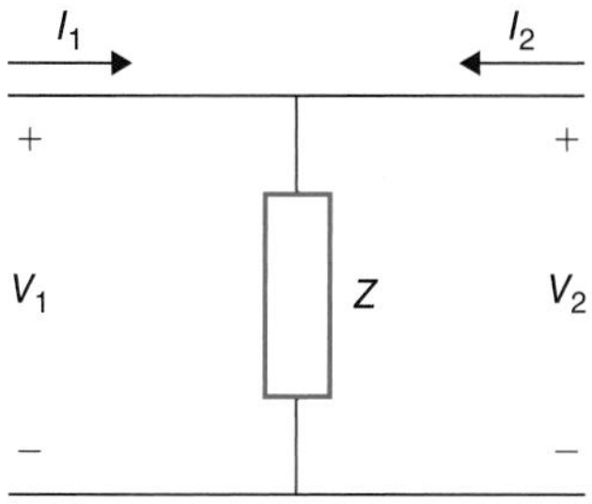

For a parallel impedance shown in Figure 20.36, if port 2 is open-circuited ($I_2 = 0$), the voltage drop across Z is V_1, which is also V_2. The current through Z is $I_1 = V_1/Z$. Coefficients A and C are given, respectively, by

$$A = \left.\frac{V_1}{V_2}\right|_{I_2=0} = \frac{V_1}{V_1} = 1$$

$$C = \left.\frac{I_1}{V_2}\right|_{I_2=0} = \frac{\frac{V_1}{Z}}{V_1} = \frac{1}{Z}$$

If port 2 is short-circuited ($V_2 = 0$), the voltage drop across Z is 0. Thus, $V_1 = V_2 = 0$. Also, since there is no current through Z, we have $I_1 = -I_2$. Coefficients B and D are given, respectively, as

$$B = \left.\frac{V_1}{-I_2}\right|_{V_2=0} = \frac{0}{-I_2} = 0$$

$$D = \left.\frac{I_1}{-I_2}\right|_{I_2=0} = \frac{I_1}{I_1} = 1$$

Putting the parameters in matrix form, we have

$$\boldsymbol{T} = \begin{bmatrix} A & B \\ C & D \end{bmatrix} = \begin{bmatrix} 1 & 0 \\ \frac{1}{Z} & 1 \end{bmatrix} \tag{20.71}$$

EXAMPLE 20.12

Find the *ABCD* parameters for the circuit shown in Figure 20.37.

FIGURE 20.37

Circuit for EXAMPLE 20.12.

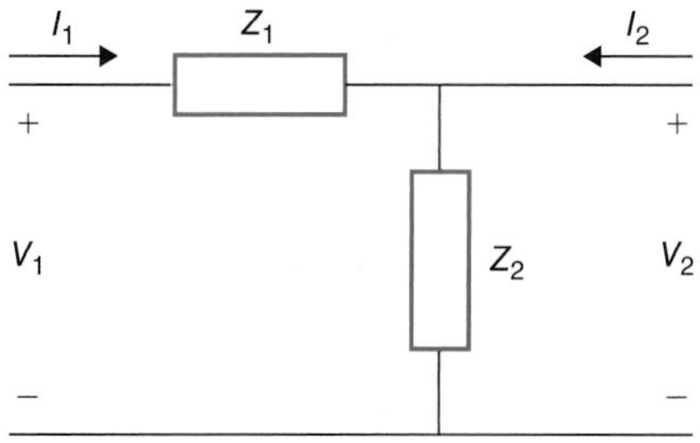

For the circuit shown in Figure 20.37, if port 2 is open-circuited ($I_2 = 0$), from the voltage divider rule, voltage V_2 is given by

$$V_2 = \frac{Z_2}{Z_1 + Z_2} V_1$$

Current I_1 is given by

$$I_1 = \frac{V_1}{Z_1 + Z_2}$$

continued

Example 20.12 continued

Coefficients A and C are given, respectively, by

$$A = \left.\frac{V_1}{V_2}\right|_{I_2=0} = \frac{Z_1 + Z_2}{Z_2} = 1 + \frac{Z_1}{Z_2}$$

$$C = \left.\frac{I_1}{V_2}\right|_{I_2=0} = \frac{\dfrac{V_1}{Z_1 + Z_2}}{\dfrac{Z_2}{Z_1 + Z_2}V_1} = \frac{1}{Z_2}$$

If port 2 is short-circuited ($V_2 = 0$), the voltage drop across Z_2 is 0. Current I_1 is given by

$$I_1 = \frac{V_1}{Z_1} = -I_2$$

Coefficients B and D are given, respectively, by

$$B = \left.\frac{V_1}{-I_2}\right|_{V_2=0} = \frac{V_1}{\dfrac{V_1}{Z_1}} = Z_1$$

$$D = \left.\frac{I_1}{-I_2}\right|_{I_2=0} = -\frac{I_1}{I_1} = 1$$

Putting the parameters in matrix form, we have

$$\boldsymbol{T} = \begin{bmatrix} A & B \\ C & D \end{bmatrix} = \begin{bmatrix} 1 + \dfrac{Z_1}{Z_2} & Z_1 \\ \dfrac{1}{Z_2} & 1 \end{bmatrix} \tag{20.72}$$

The circuit shown in Figure 20.37 can be interpreted as a cascade of series impedance Z_1 and parallel impedance Z_2. Matrix $\boldsymbol{T}$ can be obtained by multiplying the matrices of each section:

$$\boldsymbol{T} = \begin{bmatrix} A & B \\ C & D \end{bmatrix} = \begin{bmatrix} 1 & Z_1 \\ 0 & 1 \end{bmatrix} \times \begin{bmatrix} 1 & 0 \\ \dfrac{1}{Z_2} & 1 \end{bmatrix} = \begin{bmatrix} 1 + \dfrac{Z_1}{Z_2} & Z_1 \\ \dfrac{1}{Z_2} & 1 \end{bmatrix}$$

which matches direct calculation. The cascade connection of two circuits is discussed in more detail in Section 20.4.1, later in this chapter. The determinant of the product of the two square matrices of equal size is the product of the determinants of each matrix, that is

$$\det(EF) = \det(E)\det(F)$$

The determinant of the T matrix, given by Equation (20.72), is given by

$$\det(T) = \Delta_T = AD - BC = \left(1 + \frac{Z_1}{Z_2}\right) \times 1 - \frac{Z_1}{Z_2} = 1$$

$$\det\left(\begin{bmatrix} 1 & Z_1 \\ 0 & 1 \end{bmatrix}\right) \times \det\left(\begin{bmatrix} 1 & 0 \\ \dfrac{1}{Z_2} & 1 \end{bmatrix}\right) = 1 \times 1 = 1$$

EXAMPLE 20.13

Find the *ABCD* parameters for the circuits shown in Figure 20.38.

FIGURE 20.38

Circuit for EXAMPLE 20.13.

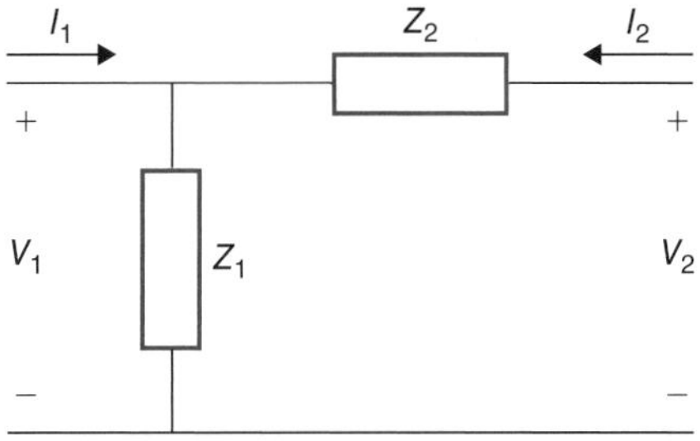

For the circuit shown in Figure 20.38, if port 2 is open-circuited ($I_2 = 0$), we have $V_2 = V_1$. The current I_1 is given by

$$I_1 = \frac{V_1}{Z_1}$$

Coefficients A and C are given, respectively, by

$$A = \left.\frac{V_1}{V_2}\right|_{I_2=0} = 1$$

$$C = \left.\frac{I_1}{V_2}\right|_{I_2=0} = \frac{\frac{V_1}{Z_1}}{V_1} = \frac{1}{Z_1}$$

If port 2 is short-circuited ($V_2 = 0$), the current I_1 is given by

$$I_1 = \frac{V_1}{Z_1} + \frac{V_1}{Z_2} = \left(\frac{1}{Z_1} + \frac{1}{Z_2}\right)V_1$$

and the current $-I_2$ is given by

$$-I_2 = \frac{V_1}{Z_2}$$

Coefficients B and D are given, respectively, as

$$B = \left.\frac{V_1}{-I_2}\right|_{V_2=0} = \frac{V_1}{\frac{V_1}{Z_2}} = Z_2$$

$$D = \left.\frac{I_1}{-I_2}\right|_{I_2=0} = \frac{\left(\frac{1}{Z_1} + \frac{1}{Z_2}\right)V_1}{\frac{V_1}{Z_2}} = 1 + \frac{Z_2}{Z_1}$$

Putting the parameters in matrix form, we have

$$\boldsymbol{T} = \begin{bmatrix} A & B \\ C & D \end{bmatrix} = \begin{bmatrix} 1 & Z_2 \\ \frac{1}{Z_1} & 1+\frac{Z_2}{Z_1} \end{bmatrix} \tag{20.73}$$

The circuit shown in Figure 20.38 can be interpreted as a cascade of parallel impedance Z_1 and series impedance Z_2. Matrix $\boldsymbol{T}$ can be obtained by multiplying the matrices of each section:

$$\boldsymbol{T} = \begin{bmatrix} A & B \\ C & D \end{bmatrix} = \begin{bmatrix} 1 & 0 \\ \frac{1}{Z_1} & 1 \end{bmatrix} \times \begin{bmatrix} 1 & Z_2 \\ 0 & 1 \end{bmatrix} = \begin{bmatrix} 1 & Z_2 \\ \frac{1}{Z_1} & 1+\frac{Z_2}{Z_1} \end{bmatrix}$$

continued

Example 20.13 continued

which matches direct calculation. Notice that

$$\det(T) = \Delta_T = AD - BC = \left(1 + \frac{Z_2}{Z_1}\right) \times 1 - \frac{Z_2}{Z_1} = 1$$

EXAMPLE 20.14

Find the *ABCD* parameters for the *T* circuit shown in Figure 20.39.

The circuit shown in Figure 20.39 can be interpreted as a cascade of the circuit shown in Figure 20.37 and the circuit shown in Figure 20.35. Matrix $\boldsymbol{T}$ can be obtained by multiplying the matrices of each section:

FIGURE 20.39

Circuit for EXAMPLE 20.14.

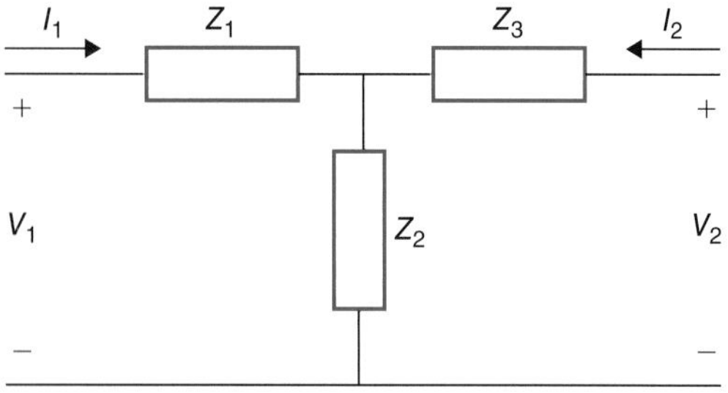

$$\boldsymbol{T} = \begin{bmatrix} A & B \\ C & D \end{bmatrix} = \begin{bmatrix} 1 + \dfrac{Z_1}{Z_2} & Z_1 \\ \dfrac{1}{Z_2} & 1 \end{bmatrix} \times \begin{bmatrix} 1 & Z_3 \\ 0 & 1 \end{bmatrix} = \begin{bmatrix} 1 + \dfrac{Z_1}{Z_2} & Z_1 + Z_3 + \dfrac{Z_1 Z_3}{Z_2} \\ \dfrac{1}{Z_2} & 1 + \dfrac{Z_3}{Z_2} \end{bmatrix} \tag{20.74}$$

which matches direct calculation. Notice that

$$\det(T) = \Delta_T = AD - BC = 1$$

EXAMPLE 20.15

Find the *ABCD* parameters for the Pi circuit shown in Figure 20.40.

The circuit shown in Figure 20.40 can be interpreted as a cascade of the circuit shown in Figure 20.38 and the circuit shown in Figure 20.36. Matrix $\boldsymbol{T}$ can be obtained by multiplying the matrices of each section:

FIGURE 20.40

Circuit for EXAMPLE 20.15.

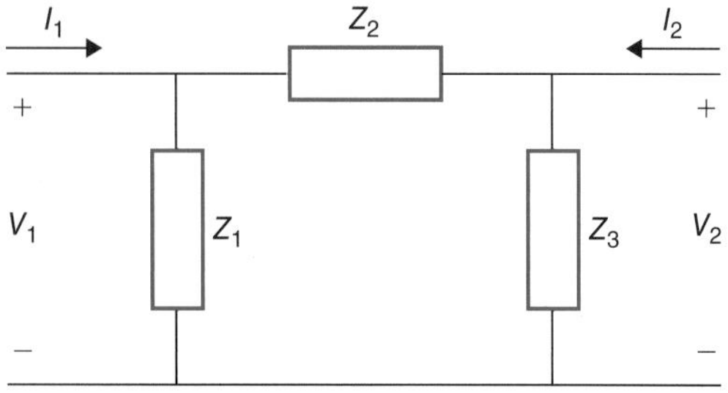

$$\boldsymbol{T} = \begin{bmatrix} A & B \\ C & D \end{bmatrix} = \begin{bmatrix} 1 & Z_2 \\ \dfrac{1}{Z_1} & 1 + \dfrac{Z_2}{Z_1} \end{bmatrix} \times \begin{bmatrix} 1 & 0 \\ \dfrac{1}{Z_3} & 1 \end{bmatrix} = \begin{bmatrix} \dfrac{Z_2 + Z_3}{Z_3} & Z_2 \\ \dfrac{Z_1 + Z_2 + Z_3}{Z_1 Z_3} & \dfrac{Z_1 + Z_2}{Z_1} \end{bmatrix} \tag{20.75}$$

which matches direct calculation. Notice that

$$\det(T) = \Delta_T = AD - BC = 1$$

EXAMPLE 20.16

Find the *ABCD* parameters for the circuits shown in Figure 20.41.

FIGURE 20.41

Circuit for EXAMPLE 20.16.

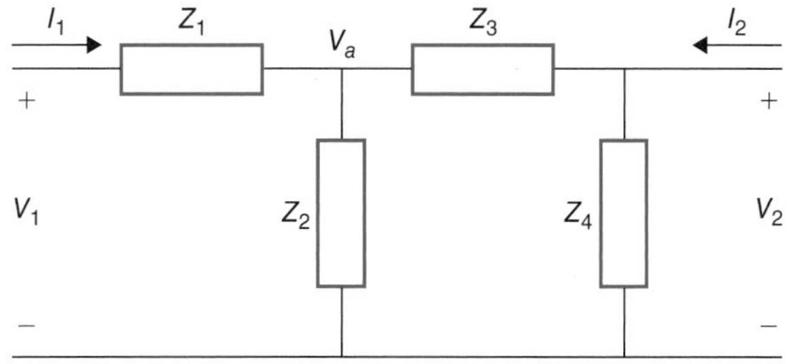

The circuit shown in Figure 20.41 can be interpreted as a cascade of the circuit shown in Figure 20.37 and the circuit shown in Figure 20.37. Matrix $\boldsymbol{T}$ can be obtained by multiplying the matrices of each section:

$$\boldsymbol{T} = \begin{bmatrix} A & B \\ C & D \end{bmatrix} = \begin{bmatrix} 1 + \dfrac{Z_1}{Z_2} & Z_1 \\ \dfrac{1}{Z_2} & 1 \end{bmatrix} \times \begin{bmatrix} 1 + \dfrac{Z_3}{Z_4} & Z_3 \\ \dfrac{1}{Z_4} & 1 \end{bmatrix}$$

$$= \begin{bmatrix} 1 + \dfrac{Z_1}{Z_2} + \dfrac{Z_1}{Z_4} + \dfrac{Z_3}{Z_4} + \dfrac{Z_1 Z_3}{Z_2 Z_4} & Z_1 + Z_3 + \dfrac{Z_1 Z_3}{Z_2} \\ \dfrac{1}{Z_2} + \dfrac{1}{Z_4} + \dfrac{Z_3}{Z_2 Z_4} & 1 + \dfrac{Z_3}{Z_2} \end{bmatrix} \quad (20.76)$$

$$= \begin{bmatrix} \dfrac{Z_1Z_2 + Z_1Z_3 + Z_1Z_4 + Z_2Z_3 + Z_2Z_4}{Z_2Z_4} & \dfrac{Z_1Z_2 + Z_1Z_3 + Z_2Z_3}{Z_2} \\ \dfrac{Z_2 + Z_3 + Z_4}{Z_2Z_4} & \dfrac{Z_2 + Z_3}{Z_2} \end{bmatrix}$$

Notice that

$$\det(T) = \Delta_T = AD - BC = 1$$

EXAMPLE 20.17

Find the transfer function $H(s) = V_2(s)/V_1(s)$ for the circuit shown in Figure 20.42.

The circuit shown in Figure 20.42 has the same shape as the circuit shown in Figure 20.41 with $Z_1 = R_1, Z_2 = 1/(sC_1), Z_3 = R_2, Z_4 = 1/(sC_2)$.

From Equation (20.66), we have

$$A = \left.\frac{V_1}{V_2}\right|_{I_2=0}$$

continued

Example 20.17 continued

The transfer function is the inverse of A. Thus, from Equation (20.76), we obtain

FIGURE 20.42

Circuit for EXAMPLE 20.17.

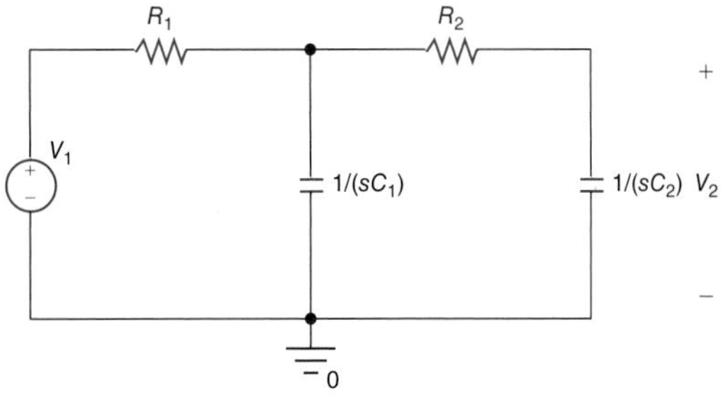

$$H(s)=\frac{V_2}{V_1}=\frac{1}{A}=\frac{Z_2Z_4}{Z_1Z_2+Z_1Z_3+Z_1Z_4+Z_2Z_3+Z_2Z_4}$$
$$=\frac{\frac{1}{sC_1}\frac{1}{sC_2}}{R_1\frac{1}{sC_1}+R_1R_2+R_1\frac{1}{sC_2}+\frac{1}{sC_1}R_2+\frac{1}{sC_1}\frac{1}{sC_2}}$$
$$=\frac{1}{R_1sC_2+R_1R_2sC_1sC_2+R_1sC_1+R_2sC_2+1}$$
$$=\frac{\frac{1}{R_1R_2C_1C_2}}{s^2+\left(\frac{1}{R_1C_1}+\frac{1}{R_2C_1}+\frac{1}{R_2C_2}\right)s+\frac{1}{R_1R_2C_1C_2}} \tag{20.77}$$

Exercise 20.6

Find the transfer function $H(s)=V_2(s)/V_1(s)$ for the circuit shown in Figure 20.43.

FIGURE 20.43

Circuit for EXERCISE 20.6.

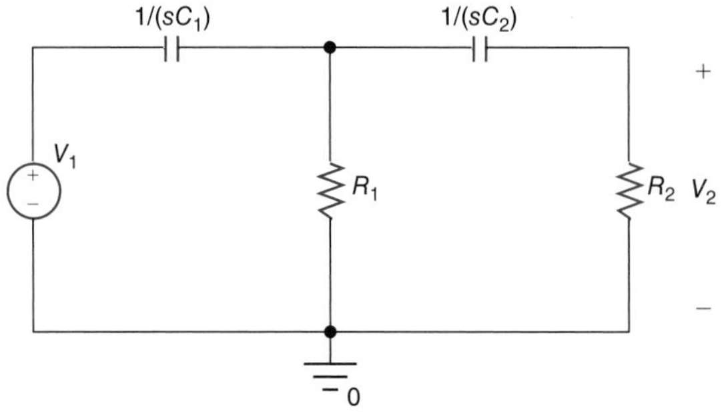

Answer:

$Z_1=1/(sC_1),\quad Z_2=R_1,\quad Z_3=1/(sC_2),\quad Z_4=R_2.$

$$H(s)=\frac{V_2(s)}{V_1(s)}=\frac{s^2}{s^2+\left(\frac{1}{R_1C_1}+\frac{1}{R_2C_1}+\frac{1}{R_2C_2}\right)s+\frac{1}{R_1R_2C_1C_2}} \tag{20.78}$$

EXAMPLE 20.18

Find the *ABCD*-parameters for an ideal transformer shown in Figure 20.44.

For an ideal transformer, we have (for more information, refer to Section 13.7 in Chapter 13)

$$V_1=\frac{1}{n}V_2 \tag{20.79}$$

continued

Example 20.18 continued

FIGURE 20.44

An ideal transformer ($n = N_2/N_1$).

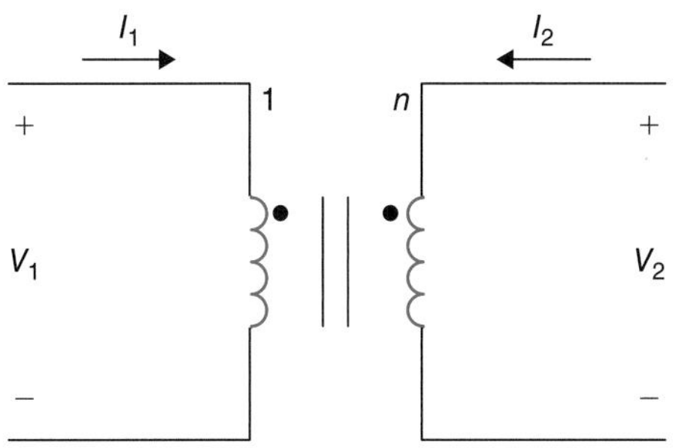

and

$$I_1 = n(-I_2) \tag{20.80}$$

Equations (20.79) and (20.80) can be put in matrix form as

$$\begin{bmatrix} V_1 \\ I_1 \end{bmatrix} = \begin{bmatrix} \frac{1}{n} & 0 \\ 0 & n \end{bmatrix} \begin{bmatrix} V_2 \\ -I_2 \end{bmatrix} \tag{20.81}$$

Comparing Equation (20.81) with Equation (20.65), we conclude that

$$A = 1/n, \quad B = 0, \quad C = 0, \quad D = n$$

EXAMPLE 20.19

Find the input impedance $Z_{in} = V_1/I_1$ in the circuit shown in Figure 20.45.

FIGURE 20.45

Circuit for EXAMPLE 20.19.

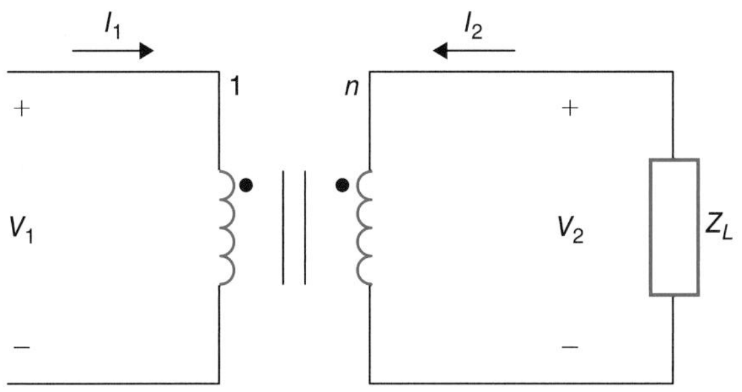

Voltage V_2 is given by

$$V_2 = -I_2 Z_L$$

which can be rearranged as

$$-I_2 = \frac{V_2}{Z_L} \tag{20.82}$$

Substitution of Equation (20.82) into Equation (20.81) yields

$$\begin{bmatrix} V_1 \\ I_1 \end{bmatrix} = \begin{bmatrix} \frac{1}{n} & 0 \\ 0 & n \end{bmatrix} \begin{bmatrix} V_2 \\ \frac{V_2}{Z_L} \end{bmatrix} = \begin{bmatrix} \frac{1}{n} V_2 \\ \frac{n V_2}{Z_L} \end{bmatrix} \tag{20.83}$$

The input impedance is given by

$$Z_{in} = \frac{V_1}{I_1} = \frac{\frac{1}{n} V_2}{\frac{n V_2}{Z_L}} = \frac{Z_L}{n^2} \tag{20.84}$$

20.2.6 INVERSE TRANSMISSION PARAMETERS (*b*-PARAMETERS)

If V_1 and I_1 are independent variables, then V_2 and I_2 are dependent variables. The dependent variables are written as linear combinations of independent variables as

$$V_2 = b_{11} V_1 - b_{12} I_1 \tag{20.85}$$

$$I_2 = b_{21}V_1 - b_{22}I_1 \tag{20.86}$$

In matrix notation, we have

$$\begin{bmatrix} V_2 \\ I_2 \end{bmatrix} = \begin{bmatrix} b_{11} & b_{12} \\ b_{21} & b_{22} \end{bmatrix} \begin{bmatrix} V_1 \\ -I_1 \end{bmatrix} \tag{20.87}$$

This equation can be written as

$$\boldsymbol{d}_b = \boldsymbol{B}\boldsymbol{n}_b$$

where

$$\boldsymbol{d}_b = \begin{bmatrix} V_2 \\ I_2 \end{bmatrix}, \quad \boldsymbol{B} = \begin{bmatrix} b_{11} & b_{12} \\ b_{21} & b_{22} \end{bmatrix}, \quad \boldsymbol{n}_b = \begin{bmatrix} V_1 \\ -I_1 \end{bmatrix}$$

Coefficient b_{11} can be measured by taking the ratio of V_2 over V_1 when port 1 is open-circuited ($I_1 = 0$)

$$b_{11} = \left.\frac{V_2}{V_1}\right|_{I_1=0} \tag{20.88}$$

Coefficient b_{11} is the voltage gain with port 1 open-circuited. Coefficients b_{12}, b_{21}, and b_{22} are given by

$$b_{12} = \left.\frac{V_2}{-I_1}\right|_{V_1=0} \tag{20.89}$$

$$b_{21} = \left.\frac{I_2}{V_1}\right|_{I_1=0} \tag{20.90}$$

$$b_{22} = \left.\frac{I_2}{-I_1}\right|_{V_1=0} \tag{20.91}$$

Coefficient b_{12} is the negative transfer impedance with port 1 short-circuited, b_{21} is the transfer admittance with port 1 open-circuited, and b_{22} is the negative current gain with port 1 short-circuited. Coefficient b_{11} is unitless (V/V), coefficient b_{12} has a unit of ohm (V/A), coefficient b_{21} has a unit of siemens (A/V), and coefficient b_{22} is unitless (A/A).

If $\Delta_T = AD - BC = 1$ for the $ABCD$-parameters, the b-parameters can be obtained from the $ABCD$-parameters through the conversion given by

$$b_{11} = D, \quad b_{12} = B, \quad b_{21} = C, \quad b_{22} = A$$

If $\Delta_b = b_{11}b_{22} - b_{12}b_{21} = 1$ for the b-parameters, the $ABCD$-parameters can be obtained from the b-parameters through the conversion given by

$$A = b_{22}, \quad B = b_{12}, \quad C = b_{21}, \quad D = b_{11}$$

The conversion of parameters is discussed in Section 20.3.

EXAMPLE 20.20

Find the b-parameters for the circuit shown in Figure 20.46.

FIGURE 20.46

Circuit for EXAMPLE 20.20.

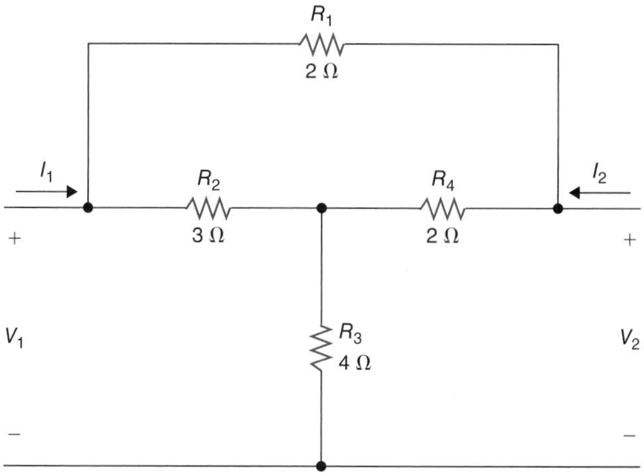

FIGURE 20.47

Circuit with port 1 open-circuited.

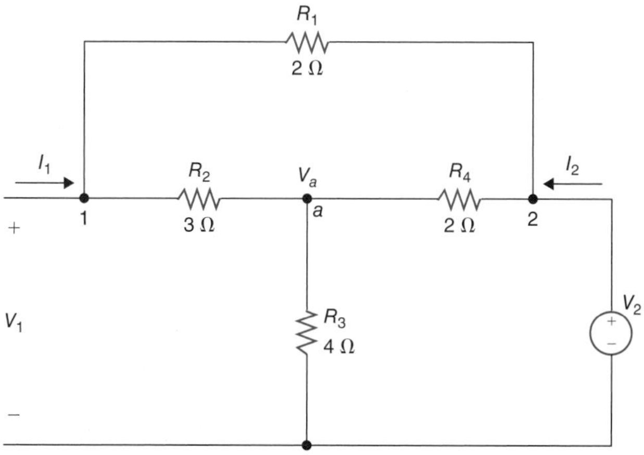

To find b_{11} and b_{21}, port 1 is open-circuited ($I_1 = 0$), and V_2 is applied at port 2, as shown in Figure 20.47.

Summing the currents leaving node 1, we obtain

$$\frac{V_1 - V_2}{2} + \frac{V_1 - V_a}{3} = 0$$

Multiplication by 6 yields

$$3V_1 - 3V_2 + 2V_1 - 2V_a = 0$$

which can be simplified to

$$5V_1 - 2V_a = 3V_2 \qquad \textbf{(20.92)}$$

Summing the currents leaving node a, we obtain

$$\frac{V_a - V_1}{3} + \frac{V_a}{4} + \frac{V_a - V_2}{2} = 0$$

Multiplication by 12 yields

$$4V_a - 4V_1 + 3V_a + 6V_a - 6V_2 = 0$$

which can be simplified to

$$-4V_1 + 13V_a = 6V_2 \qquad \textbf{(20.93)}$$

Application of Cramer's rule to Equations (20.92) and (20.93) yields

$$V_1 = \frac{\begin{vmatrix} 3 & -2 \\ 6 & 13 \end{vmatrix}}{\begin{vmatrix} 5 & -2 \\ -4 & 13 \end{vmatrix}} V_2 = \frac{39 + 12}{65 - 8} V_2 = \frac{51}{57} V_2$$

$$= \frac{17}{19} V_2 = 0.8947 V_2$$

$$V_a = \frac{\begin{vmatrix} 5 & 3 \\ -4 & 6 \end{vmatrix}}{\begin{vmatrix} 5 & -2 \\ -4 & 13 \end{vmatrix}} V_2 = \frac{30 + 12}{65 - 8} V_2 = \frac{42}{57} V_2 = \frac{14}{19} V_2 = 0.7368 V_2$$

Current I_2 is given by

$$I_2 = \frac{V_2 - V_1}{R_1} + \frac{V_2 - V_a}{R_4} = \frac{V_2 - \frac{17}{19}V_2}{2} + \frac{V_2 - \frac{14}{19}V_2}{2} = \frac{7}{38} V_2 = 0.1842 V_2$$

continued

Example 20.20 continued

Parameters b_{11} and b_{21} are given by

$$b_{11} = \left.\frac{V_2}{V_1}\right|_{I_1=0} = \frac{V_2}{\frac{17}{19}V_2} = \frac{19}{17} = 1.117647$$

$$b_{21} = \left.\frac{I_2}{V_1}\right|_{I_1=0} = \frac{\frac{7}{38}V_2}{\frac{17}{19}V_2} = \frac{7}{34} = 0.20588 \text{ S}$$

To find b_{12} and b_{22}, port 1 is short-circuited ($V_1 = 0$), and V_2 is applied at port 2, as shown in Figure 20.48.

FIGURE 20.48

Circuit with port 1 short-circuited.

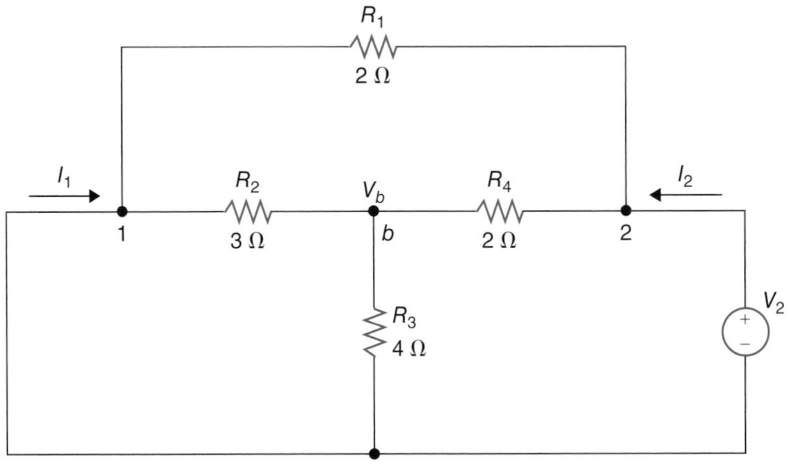

Summing the currents leaving node b, we obtain

$$\frac{V_b - 0}{3} + \frac{V_b}{4} + \frac{V_b - V_2}{2} = 0$$

Multiplication by 12 yields

$$4V_b + 3V_b + 6V_b - 6V_2 = 0$$

which can be simplified to

$$13V_b = 6V_2$$

Thus, we have

$$V_b = \frac{6}{13}V_2 = 0.46154V_2$$

Current I_2 is given by

$$I_2 = \frac{V_2}{R_1} + \frac{V_2 - V_b}{R_4} = \frac{V_2}{2} + \frac{V_2 - \frac{6}{13}V_2}{2} = \frac{20}{26}V_2 = 0.76923V_2$$

continued

Example 20.20 continued

Current I_1 is given by

$$I_1 = \frac{0 - V_2}{R_1} + \frac{0 - V_b}{R_2} = \frac{-V_2}{2} + \frac{0 - \frac{6}{13}V_2}{3} = -\frac{51}{78}V_2 = -0.653846V_2$$

$$b_{12} = \left.\frac{V_2}{-I_1}\right|_{V_1=0} = 1.529412\ \Omega$$

$$b_{22} = \left.\frac{I_2}{-I_1}\right|_{V_1=0} = 1.176471$$

20.3 Conversion of Parameters

Given a circuit with any one of the two-port parameters, we can find the other parameters by using conversion formulas. Depending on which two of the four variables (V_1, V_2, I_1, and I_2) are chosen to be independent variables (the other two are dependent variables), we have six different two-port parameters. Converting one parameter to another parameter is equivalent to solving the two equations for the given parameters for the new independent variables. Due to similarities in the conversions, only the conversion of z-parameters to all the other parameters is presented in detail. For other conversions, the results are given without proof.

20.3.1 CONVERSION OF z-PARAMETERS TO ALL THE OTHER PARAMETERS

For convenience, the voltage-current relations of the z-parameter are repeated here.

$$V_1 = z_{11}I_1 + z_{12}I_2 \quad (20.94)$$

$$V_2 = z_{21}I_1 + z_{22}I_2 \quad (20.95)$$

In matrix notation, we have

$$\begin{bmatrix} V_1 \\ V_2 \end{bmatrix} = \begin{bmatrix} z_{11} & z_{12} \\ z_{21} & z_{22} \end{bmatrix} \begin{bmatrix} I_1 \\ I_2 \end{bmatrix} \quad (20.96)$$

20.3.2 CONVERSION OF z-PARAMETERS TO y-PARAMETERS

The equations for the y-parameters are

$$I_1 = y_{11}V_1 + y_{12}V_2 \quad (20.97)$$

$$I_2 = y_{21}V_1 + y_{22}V_2 \quad (20.98)$$

In matrix notation, Equations (20.97) and (20.98) become

$$\begin{bmatrix} I_1 \\ I_2 \end{bmatrix} = \begin{bmatrix} y_{11} & y_{12} \\ y_{21} & y_{22} \end{bmatrix} \begin{bmatrix} V_1 \\ V_2 \end{bmatrix} \quad (20.99)$$

Equations (20.94) and (20.95) can be converted to Equations (20.97) and (20.98) by solving Equations (20.94) and (20.95) for I_1 and I_2. Currents I_1 and I_2 can be found by solving Equation (20.96) for I_1 and I_2. Multiplying the inverse of the Z matrix on both sides of Equation (20.96), we have

$$\begin{bmatrix} z_{11} & z_{12} \\ z_{21} & z_{22} \end{bmatrix}^{-1} \begin{bmatrix} V_1 \\ V_2 \end{bmatrix} = \begin{bmatrix} z_{11} & z_{12} \\ z_{21} & z_{22} \end{bmatrix}^{-1} \begin{bmatrix} z_{11} & z_{12} \\ z_{21} & z_{22} \end{bmatrix} \begin{bmatrix} I_1 \\ I_2 \end{bmatrix} \tag{20.100}$$

Since the product of the inverse of Z-matrix and the Z-matrix is an identity matrix, we have

$$\begin{bmatrix} I_1 \\ I_2 \end{bmatrix} = \begin{bmatrix} z_{11} & z_{12} \\ z_{21} & z_{22} \end{bmatrix}^{-1} \begin{bmatrix} V_1 \\ V_2 \end{bmatrix} \tag{20.101}$$

Comparing Equations (20.99) and (20.101), we conclude that

$$\begin{bmatrix} y_{11} & y_{12} \\ y_{21} & y_{22} \end{bmatrix} = \begin{bmatrix} z_{11} & z_{12} \\ z_{21} & z_{22} \end{bmatrix}^{-1} \tag{20.102}$$

The inverse of the Z-matrix is given by

$$\begin{bmatrix} y_{11} & y_{12} \\ y_{21} & y_{22} \end{bmatrix} = \begin{bmatrix} z_{11} & z_{12} \\ z_{21} & z_{22} \end{bmatrix}^{-1} = \frac{Adj(z)}{Det(z)} = \frac{\begin{bmatrix} z_{22} & -z_{12} \\ -z_{21} & z_{11} \end{bmatrix}}{z_{11}z_{22} - z_{12}z_{21}} = \begin{bmatrix} \dfrac{z_{22}}{\Delta_z} & \dfrac{-z_{12}}{\Delta_z} \\ \dfrac{-z_{21}}{\Delta_z} & \dfrac{z_{11}}{\Delta_z} \end{bmatrix} \tag{20.103}$$

where

$$\Delta_z = \det(\mathbf{Z}) = z_{11}z_{22} - z_{12}z_{21} \tag{20.104}$$

From Equation (20.103), we conclude that

$$y_{11} = \frac{z_{22}}{\Delta_z}, \quad y_{12} = \frac{-z_{12}}{\Delta_z}, \quad y_{21} = \frac{-z_{21}}{\Delta_z}, \quad y_{22} = \frac{z_{11}}{\Delta_z} \tag{20.105}$$

The numerator of the inverse is given by the classical adjoint matrix Adj(z). The classical adjoint matrix can be found using cofactors of the matrix. The cofactor of an element at row i and column j is the determinant of the matrix after deleting row i and column j of the matrix. The classical adjoint matrix is obtained by transposing the matrix of cofactors after adjusting the polarity of each element. For the element at row i and column j, the polarity is $(-1)^{i+j}$.

20.3.3 CONVERSION OF *z*-PARAMETERS TO *ABCD* PARAMETERS

The equations for $ABCD$ parameters are

$$V_1 = AV_2 - BI_2 \tag{20.106}$$

$$I_1 = CV_2 - DI_2 \tag{20.107}$$

From Equation (20.95), we have

$$I_1 = \frac{1}{z_{21}}V_2 - \frac{z_{22}}{z_{21}}I_2 \tag{20.108}$$

Comparison of Equations (20.107) and (20.108) reveals that

$$C = \frac{1}{z_{21}}, \quad D = \frac{z_{22}}{z_{21}} \tag{20.109}$$

Substitution of Equation (20.108) into Equation (20.94) results in

$$V_1 = z_{11}\left(\frac{1}{z_{21}}V_2 - \frac{z_{22}}{z_{21}}I_2\right) + z_{12}I_2 = \frac{z_{11}}{z_{21}}V_2 - \frac{z_{11}z_{22} - z_{12}z_{21}}{z_{21}}I_2 \tag{20.110}$$

Comparing Equations (20.106) and (20.110) provides us

$$A = \frac{z_{11}}{z_{21}}, \quad B = \frac{z_{11}z_{22} - z_{12}z_{21}}{z_{21}} = \frac{\Delta_z}{z_{21}} \tag{20.111}$$

20.3.4 CONVERSION OF *z*-PARAMETERS TO *b*-PARAMETERS

The equations for b-parameters are

$$V_2 = b_{11}V_1 - b_{12}I_1 \tag{20.112}$$

$$I_2 = b_{21}V_1 - b_{22}I_1 \tag{20.113}$$

From Equation (20.94), we have

$$I_2 = \frac{1}{z_{12}}V_1 - \frac{z_{11}}{z_{12}}I_1 \tag{20.114}$$

Comparison of Equations (20.113) and (20.114) reveals that

$$b_{21} = \frac{1}{z_{12}}, \quad b_{22} = \frac{z_{11}}{z_{12}} \tag{20.115}$$

Substitution of Equation (20.114) into Equation (20.95) results in

$$V_2 = z_{21}I_1 + z_{22}\left(\frac{1}{z_{12}}V_1 - \frac{z_{11}}{z_{12}}I_1\right) = \frac{z_{22}}{z_{12}}V_1 - \frac{z_{11}z_{22} - z_{12}z_{21}}{z_{12}}I_1 \tag{20.116}$$

Comparing Equations (20.112) and (20.116) provides us

$$b_{11} = \frac{z_{22}}{z_{12}}, \quad b_{12} = \frac{z_{11}z_{22} - z_{12}z_{21}}{z_{12}} = \frac{\Delta_z}{z_{12}} \tag{20.117}$$

20.3.5 CONVERSION OF *z*-PARAMETERS TO *h*-PARAMETERS

The equations for h-parameters are

$$V_1 = h_{11}I_1 + h_{12}V_2 \tag{20.118}$$

$$I_2 = h_{21}I_1 + h_{22}V_2 \tag{20.119}$$

From Equation (20.95), we have

$$I_2 = -\frac{z_{21}}{z_{22}}I_1 + \frac{1}{z_{22}}V_2 \tag{20.120}$$

Comparison of Equations (20.119) and (20.120) reveals that

$$h_{21} = -\frac{z_{21}}{z_{22}}, \quad h_{22} = \frac{1}{z_{22}} \tag{20.121}$$

Substitution of Equation (20.120) into Equation (20.94) results in

$$V_1 = z_{11}I_1 + z_{12}\left(-\frac{z_{21}}{z_{22}}I_1 + \frac{1}{z_{22}}V_2\right) = \frac{z_{11}z_{22} - z_{12}z_{21}}{z_{22}}I_1 + \frac{z_{12}}{z_{22}}V_2 \tag{20.122}$$

Comparing Equations (20.118) and (20.122) provides us

$$h_{11} = \frac{z_{11}z_{22} - z_{12}z_{21}}{z_{22}} = \frac{\Delta_z}{z_{22}}, \quad h_{12} = \frac{z_{12}}{z_{22}} \tag{20.123}$$

20.3.6 CONVERSION OF z-PARAMETERS TO g-PARAMETERS

The equations for g-parameters are

$$I_1 = g_{11}V_1 + g_{12}I_2 \tag{20.124}$$

$$V_2 = g_{21}V_1 + g_{22}I_2 \tag{20.125}$$

From Equation (20.94), we have

$$I_1 = \frac{1}{z_{11}}V_1 - \frac{z_{12}}{z_{11}}I_2 \tag{20.126}$$

Comparison of Equations (20.124) and (20.126) reveals that

$$g_{11} = \frac{1}{z_{11}}, \quad g_{12} = -\frac{z_{12}}{z_{11}} \tag{20.127}$$

Substitution of Equation (20.126) into Equation (20.95) results in

$$V_2 = z_{21}\left(\frac{1}{z_{11}}V_1 - \frac{z_{12}}{z_{11}}I_2\right) + z_{22}I_2 = \frac{z_{21}}{z_{11}}V_1 + \frac{z_{11}z_{22} - z_{12}z_{21}}{z_{11}}I_2 \tag{20.128}$$

Comparing Equations (20.125) and (20.128) provides the following:

$$g_{21} = \frac{z_{21}}{z_{11}}, \quad g_{22} = \frac{z_{11}z_{22} - z_{12}z_{21}}{z_{11}} = \frac{\Delta_z}{z_{11}} \tag{20.129}$$

The conversion of z-parameters to all other parameters are summarized here:

$$\Delta_z = \det(z) = z_{11}z_{22} - z_{12}z_{21} \tag{20.130}$$

Conversion of z-parameters to y-parameters:

$$y_{11} = \frac{z_{22}}{\Delta_z}, \quad y_{12} = \frac{-z_{12}}{\Delta_z}, \quad y_{21} = \frac{-z_{21}}{\Delta_z}, \quad y_{22} = \frac{z_{11}}{\Delta_z} \tag{20.131}$$

Conversion of z-parameters to $ABCD$ parameters:

$$A = \frac{z_{11}}{z_{21}}, \quad B = \frac{z_{11}z_{22} - z_{12}z_{21}}{z_{21}} = \frac{\Delta z}{z_{21}}, \quad C = \frac{1}{z_{21}}, \quad D = \frac{z_{22}}{z_{21}} \tag{20.132}$$

Conversion of z-parameters to b-parameters:

$$b_{11} = \frac{z_{22}}{z_{12}}, \quad b_{12} = \frac{z_{11}z_{22} - z_{12}z_{21}}{z_{12}} = \frac{\Delta z}{z_{12}}, \quad b_{21} = \frac{1}{z_{12}}, \quad b_{22} = \frac{z_{11}}{z_{12}} \tag{20.133}$$

Conversion of z-parameters to h-parameters:

$$h_{11} = \frac{z_{11}z_{22} - z_{12}z_{21}}{z_{22}} = \frac{\Delta z}{z_{22}}, \quad h_{12} = \frac{z_{12}}{z_{22}}, \quad h_{21} = -\frac{z_{21}}{z_{22}}, \quad h_{22} = \frac{1}{z_{22}} \tag{20.134}$$

Conversion of z-parameters to g-parameters

$$g_{11} = \frac{1}{z_{11}}, \quad g_{12} = -\frac{z_{12}}{z_{11}}, \quad g_{21} = \frac{z_{21}}{z_{11}}, \quad g_{22} = \frac{z_{11}z_{22} - z_{12}z_{21}}{z_{11}} = \frac{\Delta z}{z_{11}} \tag{20.135}$$

20.3.7 CONVERSION OF *y*-PARAMETERS TO ALL THE OTHER PARAMETERS

The equations for the y-parameters are

$$I_1 = y_{11}V_1 + y_{12}V_2 \tag{20.136}$$

$$I_2 = y_{21}V_1 + y_{22}V_2 \tag{20.137}$$

In matrix notation, we have

$$\begin{bmatrix} I_1 \\ I_2 \end{bmatrix} = \begin{bmatrix} y_{11} & y_{12} \\ y_{21} & y_{22} \end{bmatrix} \begin{bmatrix} V_1 \\ V_2 \end{bmatrix} \tag{20.138}$$

$$\Delta_y = \det(y) = y_{11}y_{22} - y_{12}y_{21} \tag{20.139}$$

Conversion of y-parameters to z-parameters:

$$z_{11} = \frac{y_{22}}{\Delta_y}, \quad z_{12} = \frac{-y_{12}}{\Delta_y}, \quad z_{21} = \frac{-y_{21}}{\Delta_y}, \quad z_{22} = \frac{y_{11}}{\Delta_y} \tag{20.140}$$

Conversion of y-parameters to $ABCD$ parameters:

$$A = -\frac{y_{22}}{y_{21}}, \quad B = -\frac{1}{y_{21}}, \quad C = -\frac{y_{11}y_{22} - y_{12}y_{21}}{y_{21}} = \frac{-\Delta_y}{y_{21}}, \quad D = -\frac{y_{11}}{y_{21}} \tag{20.141}$$

Conversion of y-parameters to b-parameters:

$$b_{11} = -\frac{y_{11}}{y_{12}}, \quad b_{12} = -\frac{1}{y_{12}}, \quad b_{21} = -\frac{y_{11}y_{22} - y_{12}y_{21}}{y_{12}} = -\frac{\Delta_y}{y_{12}}, \quad b_{22} = -\frac{y_{22}}{y_{12}} \tag{20.142}$$

Conversion of y-parameters to h-parameters:

$$h_{11} = \frac{1}{y_{11}}, \quad h_{12} = -\frac{y_{12}}{y_{11}}, \quad h_{21} = \frac{y_{21}}{y_{11}}, \quad h_{22} = \frac{y_{11}y_{22} - y_{12}y_{21}}{y_{11}} = \frac{\Delta_y}{y_{11}} \tag{20.143}$$

Conversion of y-parameters to g-parameters:

$$g_{11} = \frac{y_{11}y_{22} - y_{12}y_{21}}{y_{22}} = \frac{\Delta_y}{y_{22}}, \quad g_{12} = \frac{y_{12}}{y_{22}}, \quad g_{21} = -\frac{y_{21}}{y_{22}}, \quad g_{22} = \frac{1}{y_{22}} \tag{20.144}$$

20.3.8 CONVERSION OF *h*-PARAMETERS TO ALL THE OTHER PARAMETERS

The equations for the h-parameters are

$$V_1 = h_{11}I_1 + h_{12}V_2 \tag{20.145}$$

$$I_2 = h_{21}I_1 + h_{22}V_2 \tag{20.146}$$

In matrix notation, we have

$$\begin{bmatrix} V_1 \\ I_2 \end{bmatrix} = \begin{bmatrix} h_{11} & h_{12} \\ h_{21} & h_{22} \end{bmatrix} \begin{bmatrix} I_1 \\ V_2 \end{bmatrix} \tag{20.147}$$

$$\Delta_h = h_{11}h_{22} - h_{12}h_{21} \tag{20.148}$$

Conversion of h-parameters to z-parameters:

$$z_{11} = \frac{h_{11}h_{22} - h_{12}h_{21}}{h_{22}} = \frac{\Delta_h}{h_{22}}, \quad z_{12} = \frac{h_{12}}{h_{22}}, \quad z_{21} = -\frac{h_{21}}{h_{22}}, \quad z_{22} = \frac{1}{h_{22}} \tag{20.149}$$

Conversion of h-parameters to y-parameters:

$$y_{11} = \frac{1}{h_{11}}, \quad y_{12} = -\frac{h_{12}}{h_{11}}, \quad y_{21} = \frac{h_{21}}{h_{11}}, \quad y_{22} = \frac{h_{11}h_{22} - h_{12}h_{21}}{h_{11}} = \frac{\Delta_h}{h_{11}} \tag{20.150}$$

Conversion of h-parameters to $ABCD$ parameters:

$$A = -\frac{h_{11}h_{22} - h_{12}h_{21}}{h_{21}} = -\frac{\Delta_h}{h_{21}}, \quad B = -\frac{h_{11}}{h_{21}}, \quad C = -\frac{h_{22}}{h_{21}}, \quad D = -\frac{1}{h_{21}} \tag{20.151}$$

Conversion of h-parameters to b-parameters:

$$b_{11} = \frac{1}{h_{12}}, \quad b_{12} = \frac{h_{11}}{h_{12}}, \quad b_{21} = \frac{h_{22}}{h_{12}}, \quad b_{22} = \frac{h_{11}h_{22} - h_{12}h_{21}}{h_{12}} = \frac{\Delta_h}{h_{12}} \tag{20.152}$$

Conversion of h-parameters to g-parameters:

$$g_{11} = \frac{h_{22}}{\Delta_h}, \quad g_{12} = \frac{-h_{12}}{\Delta_h}, \quad g_{21} = \frac{-h_{21}}{\Delta_h}, \quad g_{22} = \frac{h_{11}}{\Delta_h} \tag{20.153}$$

20.3.9 CONVERSION OF *g*-PARAMETERS TO ALL THE OTHER PARAMETERS

The equations for the g-parameters are

$$I_1 = g_{11}V_1 + g_{12}I_2 \tag{20.154}$$

$$V_2 = g_{21}V_1 + g_{22}I_2 \tag{20.155}$$

In matrix notation, we have

$$\begin{bmatrix} I_1 \\ V_2 \end{bmatrix} = \begin{bmatrix} g_{11} & g_{12} \\ g_{21} & g_{22} \end{bmatrix} \begin{bmatrix} V_1 \\ I_2 \end{bmatrix} \tag{20.156}$$

$$\Delta_g = g_{11}g_{22} - g_{12}g_{21} \tag{20.157}$$

Conversion of g-parameters to z-parameters:

$$z_{11} = \frac{1}{g_{11}}, \quad z_{12} = -\frac{g_{12}}{g_{11}}, \quad z_{21} = \frac{g_{21}}{g_{11}}, \quad z_{22} = \frac{g_{11}g_{22} - g_{12}g_{21}}{g_{11}} = \frac{\Delta_g}{g_{11}} \tag{20.158}$$

Conversion of g-parameters to y-parameters:

$$y_{11} = \frac{g_{11}g_{22} - g_{12}g_{21}}{g_{22}} = \frac{\Delta_g}{g_{22}}, \quad y_{12} = \frac{g_{12}}{g_{22}}, \quad y_{21} = -\frac{g_{21}}{g_{22}}, \quad y_{22} = \frac{1}{g_{22}} \tag{20.159}$$

Conversion of g-parameters to $ABCD$ parameters:

$$a_{11} = \frac{1}{g_{21}}, \quad a_{12} = \frac{g_{22}}{g_{21}}, \quad a_{21} = \frac{g_{11}}{g_{21}}, \quad a_{22} = \frac{g_{11}g_{22} - g_{12}g_{21}}{g_{21}} = \frac{\Delta_g}{g_{21}} \tag{20.160}$$

Conversion of g-parameters to b-parameters:

$$b_{11} = -\frac{g_{11}g_{22} - g_{12}g_{21}}{g_{12}} = \frac{-\Delta_g}{g_{12}}, \quad b_{12} = \frac{-g_{22}}{g_{12}}, \quad b_{21} = -\frac{g_{11}}{g_{12}}, \quad b_{22} = -\frac{1}{g_{12}} \tag{20.161}$$

Conversion of g-parameters to h-parameters:

$$h_{11} = \frac{g_{22}}{\Delta_g}, \quad h_{12} = \frac{-g_{12}}{\Delta_g}, \quad h_{21} = \frac{-g_{21}}{\Delta_g}, \quad h_{22} = \frac{g_{11}}{\Delta_g} \tag{20.162}$$

20.3.10 CONVERSION OF *ABCD* PARAMETERS TO ALL THE OTHER PARAMETERS

The equations for the a-parameters are

$$V_1 = AV_2 - BI_2 \tag{20.163}$$

$$I_1 = CV_2 - DI_2 \tag{20.164}$$

In matrix notation, we have

$$\begin{bmatrix} V_1 \\ I_1 \end{bmatrix} = \begin{bmatrix} A & B \\ C & D \end{bmatrix} \begin{bmatrix} V_2 \\ -I_2 \end{bmatrix} \tag{20.165}$$

$$\Delta_T = AD - BC \tag{20.166}$$

Conversion of $ABCD$ parameters to z-parameters:

$$z_{11} = \frac{A}{C}, \quad z_{12} = \frac{AD - BC}{C} = \frac{\Delta_T}{C}, \quad z_{21} = \frac{1}{C}, \quad z_{22} = \frac{D}{C} \tag{20.167}$$

Conversion of $ABCD$ parameters to y-parameters:

$$y_{11} = \frac{D}{B}, \quad y_{12} = -\frac{AD - BC}{B} = -\frac{\Delta_T}{B}, \quad y_{21} = -\frac{1}{B}, \quad y_{22} = \frac{A}{B} \tag{20.168}$$

Conversion of $ABCD$ parameters to b-parameters:

$$b_{11} = \frac{D}{\Delta_T}, \quad b_{12} = \frac{B}{\Delta_T}, \quad b_{21} = \frac{C}{\Delta_T}, \quad b_{22} = \frac{A}{\Delta_T} \tag{20.169}$$

If $\Delta_T = 1$, Equation (20.167) becomes

$$b_{11} = D, \quad b_{12} = B, \quad b_{21} = C, \quad b_{22} = A$$

Conversion of $ABCD$ parameters to h-parameters:

$$h_{11} = \frac{B}{D}, \quad h_{12} = \frac{AD - BC}{D} = \frac{\Delta_T}{D}, \quad h_{21} = -\frac{1}{D}, \quad h_{22} = \frac{C}{D} \tag{20.170}$$

Conversion of $ABCD$ parameters to g-parameters:

$$g_{11} = \frac{C}{A}, \quad g_{12} = -\frac{AD - BC}{A} = \frac{-\Delta_T}{A}, \quad g_{21} = \frac{1}{A}, \quad g_{22} = \frac{B}{A} \tag{20.171}$$

20.3.11 CONVERSION OF *b*-PARAMETERS TO ALL THE OTHER PARAMETERS

The equations for the b-parameters are

$$V_2 = b_{11}V_1 - b_{12}I_1 \tag{20.172}$$

$$I_2 = b_{21}V_1 - b_{22}I_1 \tag{20.173}$$

In matrix notation, we have

$$\begin{bmatrix} V_2 \\ I_2 \end{bmatrix} = \begin{bmatrix} b_{11} & b_{12} \\ b_{21} & b_{22} \end{bmatrix} \begin{bmatrix} V_1 \\ -I_1 \end{bmatrix} \tag{20.174}$$

$$\Delta_b = b_{11}b_{22} - b_{12}b_{21} \tag{20.175}$$

Conversion of b-parameters to z-parameters:

$$z_{11} = \frac{b_{22}}{b_{21}}, \quad z_{12} = \frac{1}{b_{21}}, \quad z_{21} = \frac{b_{11}b_{22} - b_{12}b_{21}}{b_{21}} = \frac{\Delta_b}{b_{21}}, \quad z_{22} = \frac{b_{11}}{b_{21}} \tag{20.176}$$

Conversion of b-parameters to y-parameters:

$$y_{11} = \frac{b_{11}}{b_{12}}, \quad y_{12} = -\frac{1}{b_{12}}, \quad y_{21} = -\frac{b_{11}b_{22} - b_{12}b_{21}}{b_{12}} = \frac{-\Delta_b}{b_{12}}, \quad y_{22} = \frac{b_{22}}{b_{12}} \tag{20.177}$$

Conversion of b-parameters to $ABCD$ parameters:

$$A = \frac{b_{22}}{\Delta_b}, \quad B = \frac{b_{12}}{\Delta_b}, \quad C = \frac{b_{21}}{\Delta_b}, \quad D = \frac{b_{11}}{\Delta_b} \tag{20.178}$$

If $\Delta_b = 1$, Equation (20.176) becomes

$$A = b_{22}, \quad B = b_{12}, \quad C = b_{21}, \quad D = b_{11}$$

Conversion of b-parameters to h-parameters:

$$h_{11} = \frac{b_{12}}{b_{11}}, \quad h_{12} = \frac{1}{b_{11}}, \quad h_{21} = -\frac{b_{11}b_{22} - b_{12}b_{21}}{b_{11}} = -\frac{\Delta_b}{b_{11}}, \quad h_{22} = \frac{b_{21}}{b_{11}} \tag{20.179}$$

Conversion of b-parameters to g-parameters:

$$g_{11} = \frac{b_{21}}{b_{22}}, \quad g_{12} = -\frac{1}{b_{22}}, \quad g_{21} = \frac{b_{11}b_{22} - b_{12}b_{21}}{b_{22}} = \frac{\Delta_b}{b_{22}}, \quad g_{22} = \frac{b_{12}}{b_{22}} \tag{20.180}$$

EXAMPLE 20.21

Let the z-parameters be

$$z_{11} = 1, \quad z_{12} = 2, \quad z_{21} = 3, \quad z_{22} = 4$$

Convert the z-parameters to all other parameters.
A MATLAB function **z2all**, shown here, can find all the other parameters:

```
function [ w ] = z2all( z )
%Input is z-parameters, z=[z11 z12 z21 z22]
%Output has 6 rows: row1=z,row2=y;row3=T;row4=b,row5=h,row6=g.
D=z(1)*z(4)-z(2)*z(3);
y=[z(4),-z(2),-z(3),z(1)]/D;
T=[z(1),D,1,z(4)]/z(3);
b=[z(4),D,1,z(1)]/z(2);
h=[D,z(2),-z(3),1]/z(4);
g=[1,-z(2),z(3),D]/z(1);
w=[z;y;T;b;h;g];
disp('row1=[z11,z12,z21,z22],row2=y,row3=[A,B,C,D],row4=b,row5=h,row6=g')
end
```

To run the function **z2all**, in the command window, enter

```
>> param1=z2all([1 2 3 4])
```

The answer shown here is printed.

```
row1=[z11,z12,z21,z22],row2=y,row3=[A,B,C,D],row4=b,row5=h,row6=g
param1 =
   1.000000000000000   2.000000000000000   3.000000000000000   4.000000000000000
  -2.000000000000000   1.000000000000000   1.500000000000000  -0.500000000000000
   0.333333333333333  -0.666666666666667   0.333333333333333   1.333333333333333
   2.000000000000000  -1.000000000000000   0.500000000000000   0.500000000000000
  -0.500000000000000   0.500000000000000  -0.750000000000000   0.250000000000000
   1.000000000000000  -2.000000000000000   3.000000000000000  -2.000000000000000
```

Exercise 20.7

Let the y-parameters be

$$y_{11} = -2, \quad y_{12} = 1, \quad y_{21} = 1.5, \quad y_{22} = -0.5$$

Convert the y-parameters to all other parameters.

Answer:

Create a function **y2all.m** shown here and save in the current folder.

```
function [ w ] = y2all( y )
%Input is g-parameters, y=[y11 y12 y21 y22]
%Output has 6 rows: row1=z,row2=y;row3=T;row4=b,row5=h,row6=g.
D=y(1)*y(4)-y(2)*y(3);
z=[y(4),-y(2),-y(3),y(1)]/D;
T=[-y(4),-1,-D,-y(1)]/y(3);
b=[-y(1),-1,-D,-y(4)]/y(2);
h=[1,-y(2),y(3),D]/y(1);
g=[D,y(2),-y(3),1]/y(4);
w=[z;y;T;b;h;g];
```

continued

Exercise 20.7 continued

```
disp('row1=[z11,z12,z21,z22],row2=y,row3=[A,B,C,D],row4=b,row5=h,row6=g')
end
```

To run the function **y2all**, in the command window, enter

```
>> param2=y2all([-2 1 1.5 -0.5])
```

The answer shown here is printed.

```
row1=[z11,z12,z21,z22],row2=y,row3=[A,B,C,D],row4=b,row5=h,row6=g
param2 =
   1.000000000000000   2.000000000000000   3.000000000000000   4.000000000000000
  -2.000000000000000   1.000000000000000   1.500000000000000  -0.500000000000000
   0.333333333333333  -0.666666666666667   0.333333333333333   1.333333333333333
   2.000000000000000  -1.000000000000000   0.500000000000000   0.500000000000000
  -0.500000000000000   0.500000000000000  -0.750000000000000   0.250000000000000
   1.000000000000000  -2.000000000000000   3.000000000000000  -2.000000000000000
```

EXAMPLE 20.22

Convert the *ABCD* parameters for the *T* circuit given by Equation (20.74) to *z*-parameters and *y*-parameters.

A MATLAB script given here, can be utilized for the conversion:

```
%ABCD to z and y for T circuit
clear all;
syms Z1 Z2 Z3 A B C D
A=(Z1+Z2)/Z2
B=(Z1*Z2+Z1*Z3+Z2*Z3)/Z2
C=1/Z2
D=(Z2+Z3)/Z2
z11=A/C
z12=1/C
Z21=1/C
z22=D/C
y11=D/B
y12=-1/B
y21=-1/B
y22=A/B

Answer:
z11 =
Z1 + Z2
z12 =
Z2
Z21 =
Z2
z22 =
Z2 + Z3
y11 =
(Z2 + Z3)/(Z1*Z2 + Z1*Z3 + Z2*Z3)
y12 =
-Z2/(Z1*Z2 + Z1*Z3 + Z2*Z3)
y21 =
-Z2/(Z1*Z2 + Z1*Z3 + Z2*Z3)
y22 =
(Z1 + Z2)/(Z1*Z2 + Z1*Z3 + Z2*Z3)
```

EXAMPLE 20.23

Convert the *ABCD* parameters for the Pi circuit given by Equation (20.75) to *z*-parameters and *y*-parameters.

```
%ABCD to z  and y for Pi circuit
clear all;
syms Z1 Z2 Z3 A B C D
A=(Z2+Z3)/Z3
B=Z2
C=(Z1+Z2+Z3)/(Z1*Z3)
D=(Z1+Z2)/Z1
z11=A/C
z12=1/C
Z21=1/C
z22=D/C
y11=D/B
y12=-1/B
y21=-1/B
y22=A/B

Answer:

z11 =
(Z1*(Z2 + Z3))/(Z1 + Z2 + Z3)
z12 =
(Z1*Z3)/(Z1 + Z2 + Z3)
Z21 =
(Z1*Z3)/(Z1 + Z2 + Z3)
z22 =
(Z3*(Z1 + Z2))/(Z1 + Z2 + Z3)
y11 =
(Z1 + Z2)/(Z1*Z2)
y12 =
-1/Z2
y21 =
-1/Z2
y22 =
(Z2 + Z3)/(Z2*Z3)
```

The rest of the parameter conversion functions are given here:

```
function [ w ] = h2all( h )
%Input is h-parameters, h=[h11 h12 h21 h22]
%Output has 6 rows: row1=z,row2=y;row3=T;row4=b,row5=h,row6=g.
D=h(1)*h(4)-h(2)*h(3);
z=[D,h(2),-h(3),1]/h(4);
y=[1,-h(2),h(3),D]/h(1);
T=[-D,-h(1),-h(4),-1]/h(3);
b=[1,h(1),h(4),D]/h(2);
g=[h(4),-h(2),-h(3),h(1)]/D;
w=[z;y;T;b;h;g];
disp('row1=[z11,z12,z21,z22],row2=y,row3=[A,B,C,D],row4=b,row5=h,row6=g')
end
```

continued

Example 20.23 continued

```
function [ w ] = g2all( g )
%Input is g-parameters, g=[g11 g12 g21 g22]
%Output has 6 rows: row1=z,row2=y;row3=T;row4=b,row5=h,row6=g.
D=g(1)*g(4)-g(2)*g(3);
z=[1,-g(2),g(3),D]/g(1);
y=[D,g(2),-g(3),1]/g(4);
T=[1,g(4),g(1),D]/g(3);
b=[-D,-g(4),-g(1),-1]/g(2);
h=[g(4),-g(2),-g(3),g(1)]/D;
w=[z;y;T;b;h;g];
disp('row1=[z11,z12,z21,z22],row2=y,row3=[A,B,C,D],row4=b,row5=h,row6=g')
end

function [ w ] = T2all( T )
%Input is g-parameters, T=[T11 T12 T21 T22]
%Output has 6 rows: row1=z,row2=y;row3=T;row4=b,row5=h,row6=g.
D=T(1)*T(4)-T(2)*T(3);
z=[T(1),D,1,T(4)]/T(3);
y=[T(4),-D,-1,T(1)]/T(2);
b=[T(4),T(2),T(3),T(1)]/D;
h=[T(2),D,-1,T(3)]/T(4);
g=[T(3),-D,1,T(2)]/T(1);
w=[z;y;T;b;h;g];
disp('row1=[z11,z12,z21,z22],row2=y,row3=[A,B,C,D],row4=b,row5=h,row6=g')
end

function [ w ] = b2all( b )
%Input is g-parameters, b=[b11 b12 b21 b22]
%Output has 6 rows: row1=z,row2=y;row3=T;row4=b,row5=h,row6=g.
D=b(1)*b(4)-b(2)*b(3);
z=[b(4),1,D,b(1)]/b(3);
y=[b(1),-1,-D,b(4)]/b(2);
T=[b(4),b(2),b(3),b(1)]/D;
h=[b(2),1,-D,b(3)]/b(1);
g=[b(3),-1,D,b(2)]/b(4);
w=[z;y;T;b;h;g];
disp('row1=[z11,z12,z21,z22],row2=y,row3=[A,B,C,D],row4=b,row5=h,row6=g')
end
```

20.4 Interconnection of Two-Port Circuits

Two 2-port networks can be interconnected in cascade, in series, in parallel, in series-parallel, and in parallel series. We are interested in finding the parameters of the equivalent 2-port network as a function of parameters of two 2-port networks interconnected in cascade, in series, in parallel, in series-parallel, and in parallel series.

20.4.1 CASCADE CONNECTION

Figure 20.49 shows two 2-port networks connected in cascade. When two 2-port networks are connected in cascade, the matrix of the *ABCD* parameters of the equivalent 2-port network shown in Figure 20.50 is the product of the matrices of the *ABCD* parameters of the two networks. If other than *ABCD* parameters are given for the original two ports,

FIGURE 20.49

Two 2-port networks connected in cascade.

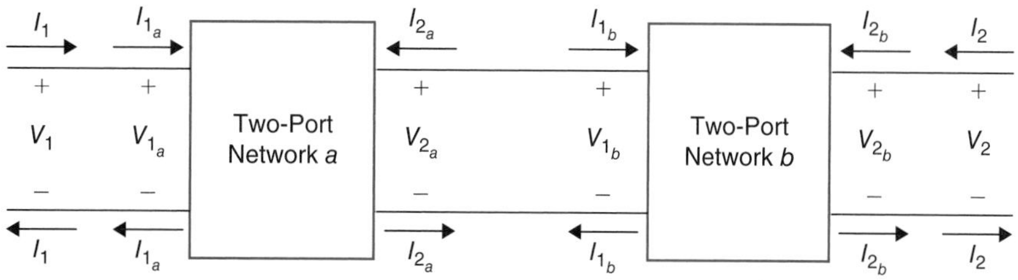

FIGURE 20.50

The equivalent two-port network.

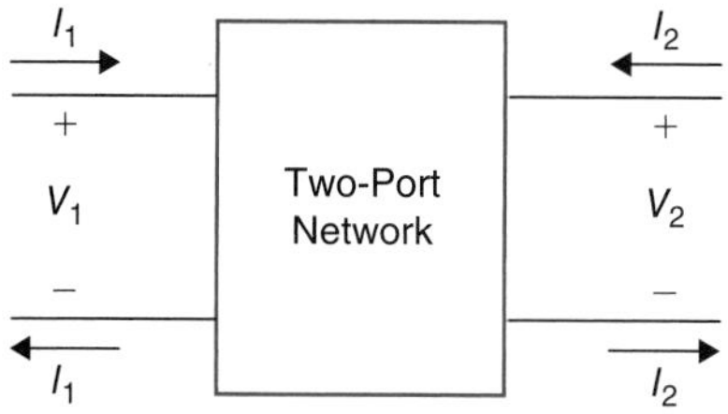

parameter conversion can be used to transform the given parameters to *ABCD* parameters. Once the *ABCD* parameters of the equivalent circuit are found, all the other parameters can be found by parameter conversion.

For the first two-port network, from Equation (20.65), we have

$$\begin{bmatrix} V_{1_a} \\ I_{1_a} \end{bmatrix} = \begin{bmatrix} A_a & B_a \\ C_a & D_a \end{bmatrix} \begin{bmatrix} V_{2_a} \\ -I_{2_a} \end{bmatrix} \tag{20.181}$$

Similarly, for the second two-port network, we have

$$\begin{bmatrix} V_{1_b} \\ I_{1_b} \end{bmatrix} = \begin{bmatrix} A_b & B_b \\ C_b & D_b \end{bmatrix} \begin{bmatrix} V_{2_b} \\ -I_{2_b} \end{bmatrix} \tag{20.182}$$

Since $V_{2_a} = V_{1_b}$ and $-I_{2_a} = I_{1_b}$, from Equation (20.182), we have

$$\begin{bmatrix} V_{2_a} \\ -I_{2_a} \end{bmatrix} = \begin{bmatrix} V_{1_b} \\ I_{1_b} \end{bmatrix} = \begin{bmatrix} A_b & B_b \\ C_b & D_b \end{bmatrix} \begin{bmatrix} V_{2_b} \\ -I_{2_b} \end{bmatrix} \tag{20.183}$$

Substituting Equation (20.183) into Equation (20.181), we obtain

$$\begin{bmatrix} V_{1_a} \\ I_{1_a} \end{bmatrix} = \begin{bmatrix} A_a & B_a \\ C_a & D_a \end{bmatrix} \begin{bmatrix} V_{2_a} \\ -I_{2_a} \end{bmatrix} = \begin{bmatrix} A_a & B_a \\ C_a & D_a \end{bmatrix} \begin{bmatrix} A_b & B_b \\ C_b & D_b \end{bmatrix} \begin{bmatrix} V_{2_b} \\ -I_{2_b} \end{bmatrix} \tag{20.184}$$

According to Equation (20.184), when two sections of *ABCD* parameters are cascaded, the combined *ABCD* parameters are obtained by multiplying the ***T*** matrices. Let

$$\boldsymbol{T} = \begin{bmatrix} A & B \\ C & D \end{bmatrix} = \begin{bmatrix} A_a & B_a \\ C_a & D_a \end{bmatrix} \begin{bmatrix} A_b & B_b \\ C_b & D_b \end{bmatrix} = \boldsymbol{T}_a \boldsymbol{T}_b \tag{20.185}$$

Since $V_{1_a} = V_1, I_{1_a} = I_1, V_{2_b} = V_2, I_{2_b} = I_2$, Equation (20.184) can be rewritten as

$$\begin{bmatrix} V_1 \\ I_1 \end{bmatrix} = \begin{bmatrix} A & B \\ C & D \end{bmatrix} \begin{bmatrix} V_2 \\ -I_2 \end{bmatrix} \tag{20.186}$$

Equation (20.186) represents the equivalent circuit shown in Figure 20.50 with *ABCD* parameters.

EXAMPLE 20.24

Find the *ABCD* parameters for the circuit shown in Figure 20.51 by cascading the circuit consisting of R_1 and C_1 and the circuit consisting of C_2 and R_2.

continued

Example 20.24 continued

FIGURE 20.51

Circuit for EXAMPLE 20.24.

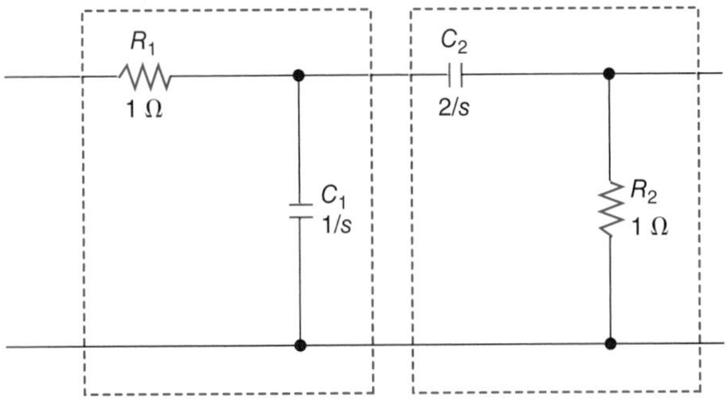

The $ABCD$ parameters for the circuit shown in Figure 20.37 are given by, from Equation (20.72),

$$\boldsymbol{T} = \begin{bmatrix} A & B \\ C & D \end{bmatrix} = \begin{bmatrix} 1 + \dfrac{Z_1}{Z_2} & Z_1 \\ \dfrac{1}{Z_2} & 1 \end{bmatrix}$$

where Z_1 is the impedance of the horizontal component, and Z_2 is the impedance of the vertical component. For the circuit consisting of R_1 and C_1, $Z_1 = 1$, and $Z_2 = 1/s$. Thus, the $ABCD$ parameters are given by

$$\boldsymbol{T}_a = \begin{bmatrix} A_a & B_a \\ C_a & D_a \end{bmatrix} = \begin{bmatrix} 1+s & 1 \\ s & 1 \end{bmatrix}$$

For the circuit consisting of C_2 and R_2, $Z_1 = 2/s$, and $Z_2 = 1$. Thus, the $ABCD$ parameters are given by

$$\boldsymbol{T}_b = \begin{bmatrix} A_b & B_b \\ C_b & D_b \end{bmatrix} = \begin{bmatrix} \dfrac{s+2}{s} & \dfrac{2}{s} \\ 1 & 1 \end{bmatrix}$$

Since the two circuits are cascaded, the $ABCD$ parameters for the circuit shown in Figure 20.51 is given by

$$\boldsymbol{T} = \boldsymbol{T}_a\boldsymbol{T}_b = \begin{bmatrix} 1+s & 1 \\ s & 1 \end{bmatrix}\begin{bmatrix} \dfrac{s+2}{s} & \dfrac{2}{s} \\ 1 & 1 \end{bmatrix} = \begin{bmatrix} \dfrac{s^2+4s+2}{s} & \dfrac{3s+2}{s} \\ s+3 & 3 \end{bmatrix}$$

The transfer function of the circuit shown in Figure 20.51 is the inverse of parameter A. Thus, we have

$$H(s) = \frac{V_2}{V_1} = \frac{1}{A} = \frac{s}{s^2+4s+2}$$

20.4.2 SERIES CONNECTION

Figure 20.52 shows two 2-port networks connected in series. When two 2-port networks are connected in series, the matrix of the z-parameters of the equivalent 2-port network shown in Figure 20.50 is the sum of the matrices of the z-parameters of the two networks.

From Equation (20.3), the first 2-port is described by the z-parameters as

$$\begin{bmatrix} V_{1_a} \\ V_{2_a} \end{bmatrix} = \begin{bmatrix} z_{11_a} & z_{12_a} \\ z_{21_a} & z_{22_a} \end{bmatrix}\begin{bmatrix} I_{1_a} \\ I_{2_a} \end{bmatrix} \tag{20.187}$$

Similarly, the second 2-port is described by the z-parameters as

$$\begin{bmatrix} V_{1_b} \\ V_{2_b} \end{bmatrix} = \begin{bmatrix} z_{11_b} & z_{12_b} \\ z_{21_b} & z_{22_b} \end{bmatrix}\begin{bmatrix} I_{1_b} \\ I_{2_b} \end{bmatrix} \tag{20.188}$$

FIGURE 20.52

Two 2-port networks connected in series.

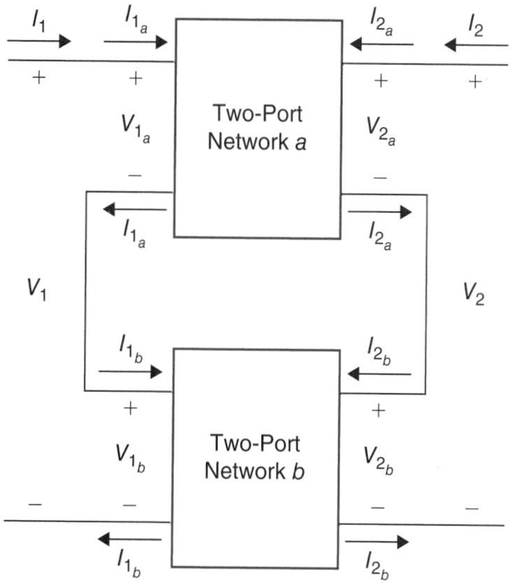

The equivalent circuit shown in Figure 20.50 is described by

$$\begin{bmatrix} V_1 \\ V_2 \end{bmatrix} = \begin{bmatrix} z_{11} & z_{12} \\ z_{21} & z_{22} \end{bmatrix} \begin{bmatrix} I_1 \\ I_2 \end{bmatrix} \tag{20.189}$$

Since $I_1 = I_{1_a} = I_{1_b}, I_2 = I_{2_a} = I_{2_b}, V_1 = V_{1_a} + V_{1_b}$, and $V_2 = V_{2_a} + V_{2_b}$, we have

$$\begin{aligned} \begin{bmatrix} V_1 \\ V_2 \end{bmatrix} &= \begin{bmatrix} V_{1_a} \\ V_{2_a} \end{bmatrix} + \begin{bmatrix} V_{1_b} \\ V_{2_b} \end{bmatrix} = \begin{bmatrix} z_{11_a} & z_{12_a} \\ z_{21_a} & z_{22_a} \end{bmatrix} \begin{bmatrix} I_{1_a} \\ I_{2_a} \end{bmatrix} + \begin{bmatrix} z_{11_b} & z_{12_b} \\ z_{21_b} & z_{22_b} \end{bmatrix} \begin{bmatrix} I_{1_b} \\ I_{2_b} \end{bmatrix} \\ &= \begin{bmatrix} z_{11_a} & z_{12_a} \\ z_{21_a} & z_{22_a} \end{bmatrix} \begin{bmatrix} I_1 \\ I_2 \end{bmatrix} + \begin{bmatrix} z_{11_b} & z_{12_b} \\ z_{21_b} & z_{22_b} \end{bmatrix} \begin{bmatrix} I_1 \\ I_2 \end{bmatrix} \\ &= \left(\begin{bmatrix} z_{11_a} & z_{12_a} \\ z_{21_a} & z_{22_a} \end{bmatrix} + \begin{bmatrix} z_{11_b} & z_{12_b} \\ z_{21_b} & z_{22_b} \end{bmatrix} \right) \begin{bmatrix} I_1 \\ I_2 \end{bmatrix} \end{aligned} \tag{20.190}$$

Comparing Equations (20.189) and (20.190), we conclude that

$$\begin{aligned} \begin{bmatrix} z_{11} & z_{12} \\ z_{21} & z_{22} \end{bmatrix} &= \begin{bmatrix} z_{11_a} & z_{12_a} \\ z_{21_a} & z_{22_a} \end{bmatrix} + \begin{bmatrix} z_{11_b} & z_{12_b} \\ z_{21_b} & z_{22_b} \end{bmatrix} \\ &= \begin{bmatrix} z_{11_a} + z_{11_b} & z_{12_a} + z_{12_b} \\ z_{21_a} + z_{21_b} & z_{22_a} + z_{22_b} \end{bmatrix} \end{aligned} \tag{20.191}$$

or

$$\mathbf{Z} = \mathbf{Z}_a + \mathbf{Z}_b \tag{20.192}$$

According to Equation (20.192), when two sections of z-parameters are connected in series, the equivalent **Z** matrix is obtained by adding the matrices $\mathbf{Z}_a$ and $\mathbf{Z}_b$.

EXAMPLE 20.25

Find the z-parameters for the circuit shown in Figure 20.53 by adding the z-parameters of Section 1 and Section 2. Section 1 consists of R_1, R_2, R_3, and R_4, and Section 2 consists of the single resistor R_5.

FIGURE 20.53

Series connection of two circuits.

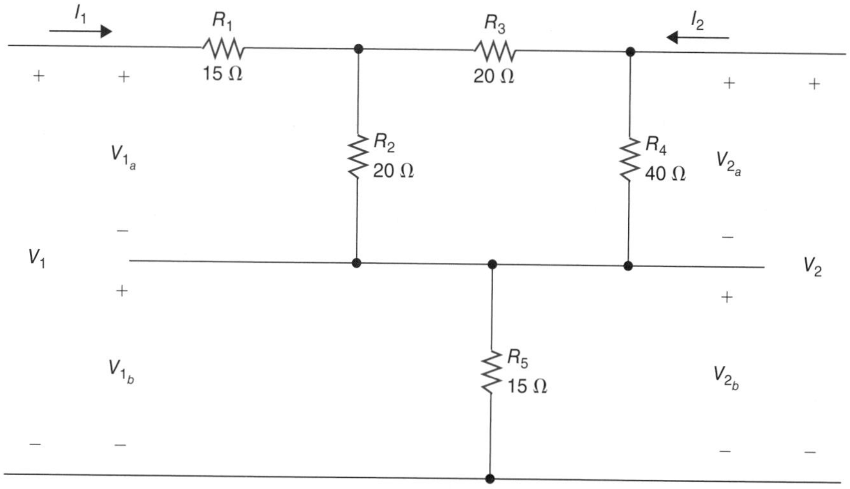

continued

Example 20.25 continued

As shown in EXAMPLE 20.1, the z-parameters of Section 1 consisting of R_1, R_2, R_3, and R_4 are given by

$$z_{11} = 30\ \Omega, \quad z_{12} = 10\ \Omega, \quad z_{21} = 10\ \Omega, \quad z_{22} = 20\ \Omega$$

The matrix of z-parameters is given by

$$\mathbf{Z}_a = \begin{bmatrix} z_{11_a} & z_{12_a} \\ z_{21_a} & z_{22_a} \end{bmatrix} = \begin{bmatrix} 30 & 10 \\ 10 & 20 \end{bmatrix}$$

Section 2, consisting of R_5, is shown in Figure 20.54.

FIGURE 20.54

Circuit consisting of R_5.

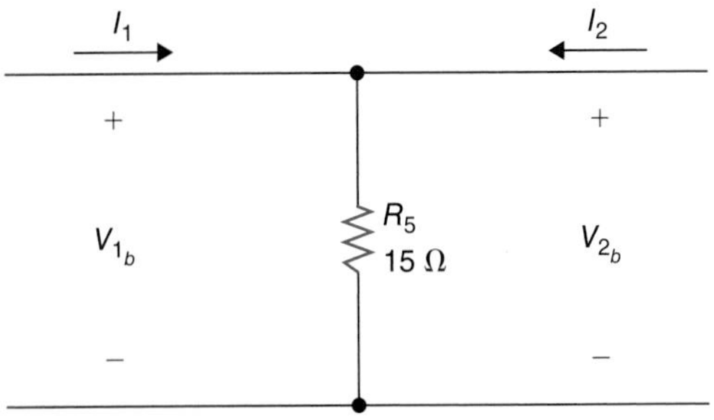

To find z_{11} and z_{21}, we apply 1 V at the input port with the output port open-circuited ($I_2 = 0$), as shown in Figure 20.55.

FIGURE 20.55

Circuit to measure z_{11} and z_{21}.

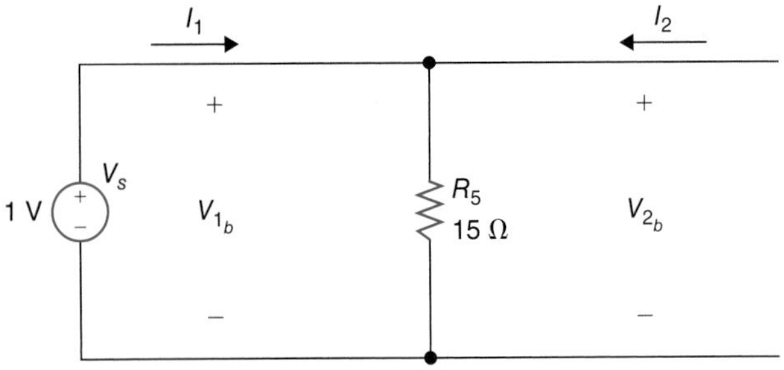

Notice that $V_{1_b} = V_{2_b} = V_s = 1$ V. The parameter z_{11} is given by

$$z_{11} = \frac{V_{1_b}}{I_1} = \frac{1\text{ V}}{\dfrac{1\text{ V}}{15\ \Omega}} = 15\ \Omega$$

The parameter z_{21} is given by

$$z_{21} = \frac{V_{2_b}}{I_1} = \frac{1\text{ V}}{\dfrac{1\text{ V}}{15\ \Omega}} = 15\ \Omega$$

To find z_{22} and z_{12}, we apply 1 V at the output port with the input port open-circuited ($I_2 = 0$), as shown in Figure 20.56.

FIGURE 20.56

Circuit to measure z_{22} and z_{12}.

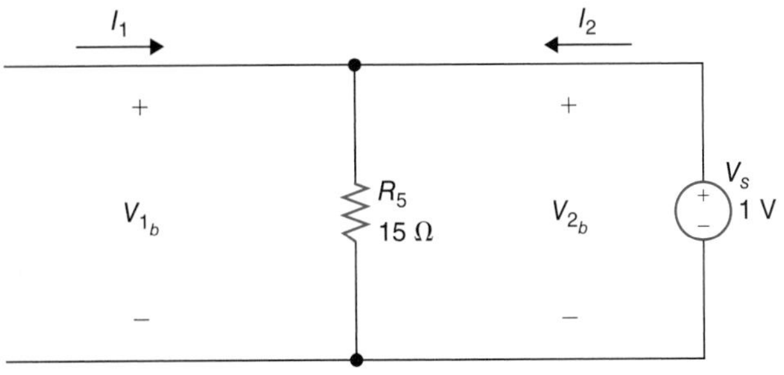

Notice that $V_{1_b} = V_{2_b} = V_s = 1$ V. Parameter z_{22} is given by

$$z_{22} = \frac{V_{2_b}}{I_2} = \frac{1\text{ V}}{\dfrac{1\text{ V}}{15\ \Omega}} = 15\ \Omega$$

Parameter z_{12} is given by

$$z_{12} = \frac{V_{1_b}}{I_2} = \frac{1\text{ V}}{\dfrac{1\text{ V}}{15\ \Omega}} = 15\ \Omega$$

The matrix of z-parameters is given by

$$Z_b = \begin{bmatrix} z_{11_b} & z_{12_b} \\ z_{21_b} & z_{22_b} \end{bmatrix} = \begin{bmatrix} 15 & 15 \\ 15 & 15 \end{bmatrix}$$

The matrix of z-parameters for the circuit shown in Figure 20.53 is given by

$$\mathbf{Z} = \begin{bmatrix} z_{11} & z_{12} \\ z_{21} & z_{22} \end{bmatrix} = \mathbf{Z}_a + \mathbf{Z}_b = \begin{bmatrix} 45 & 25 \\ 25 & 35 \end{bmatrix}$$

FIGURE 20.57

Two 2-port networks connected in parallel.

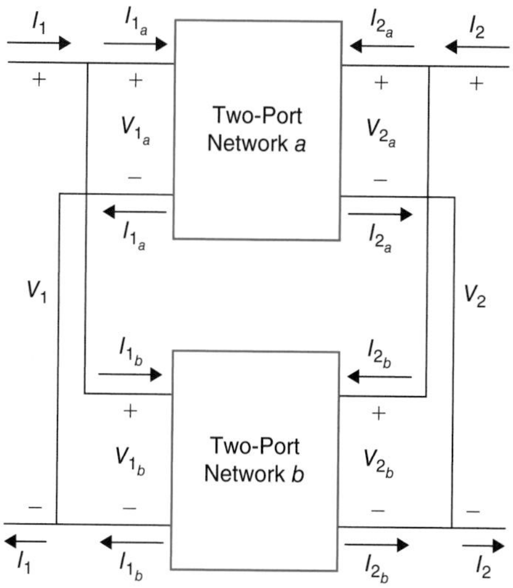

20.4.3 PARALLEL CONNECTION

Figure 20.57 shows two 2-port networks connected in parallel. When two 2-port networks are connected in parallel, the matrix of the y-parameters of the equivalent 2-port network shown in Figure 20.50 is the sum of the matrices of the y-parameters of the two networks. From Equation (20.31), the first 2-port network is described by y-parameters as

$$\begin{bmatrix} I_{1_a} \\ I_{2_a} \end{bmatrix} = \begin{bmatrix} y_{11_a} & y_{12_a} \\ y_{21_a} & y_{22_a} \end{bmatrix} \begin{bmatrix} V_{1_a} \\ V_{2_a} \end{bmatrix} \quad (20.193)$$

Similarly, the second two-port network is described by y-parameters as

$$\begin{bmatrix} I_{1_b} \\ I_{2_b} \end{bmatrix} = \begin{bmatrix} y_{11_b} & y_{12_b} \\ y_{21_b} & y_{22_b} \end{bmatrix} \begin{bmatrix} V_{1_b} \\ V_{2_b} \end{bmatrix} \quad (20.194)$$

The equivalent circuit shown in Figure 20.50 can be described by y-parameters as

$$\begin{bmatrix} I_1 \\ I_2 \end{bmatrix} = \begin{bmatrix} y_{11} & y_{12} \\ y_{21} & y_{22} \end{bmatrix} \begin{bmatrix} V_1 \\ V_2 \end{bmatrix} \quad (20.195)$$

Since $V_1 = V_{1_a} = V_{1_b}$, $V_2 = V_{2_a} = V_{2_b}$, $I_1 = I_{1_a} + I_{1_b}$, $I_2 = I_{2_a} + I_{2_b}$, we have

$$\begin{aligned} \begin{bmatrix} I_1 \\ I_2 \end{bmatrix} &= \begin{bmatrix} I_{1_a} \\ I_{2_a} \end{bmatrix} + \begin{bmatrix} I_{1_b} \\ I_{2_b} \end{bmatrix} = \begin{bmatrix} y_{11_a} & y_{12_a} \\ y_{21_a} & y_{22_a} \end{bmatrix} \begin{bmatrix} V_{1_a} \\ V_{2_a} \end{bmatrix} + \begin{bmatrix} y_{11_b} & y_{12_b} \\ y_{21_b} & y_{22_b} \end{bmatrix} \begin{bmatrix} V_{1_b} \\ V_{2_b} \end{bmatrix} \\ &= \left(\begin{bmatrix} y_{11_a} & y_{12_a} \\ y_{21_a} & y_{22_a} \end{bmatrix} + \begin{bmatrix} y_{11_b} & y_{12_b} \\ y_{21_b} & y_{22_b} \end{bmatrix} \right) \begin{bmatrix} V_1 \\ V_2 \end{bmatrix} = \begin{bmatrix} y_{11_a} + y_{11_b} & y_{12_a} + y_{12_b} \\ y_{21_a} + y_{21_b} & y_{22_a} + y_{22_b} \end{bmatrix} \begin{bmatrix} V_1 \\ V_2 \end{bmatrix} \end{aligned} \quad (20.196)$$

From Equations (20.195) and (20.196), we have

$$\begin{bmatrix} y_{11} & y_{12} \\ y_{21} & y_{22} \end{bmatrix} = \begin{bmatrix} y_{11_a} + y_{11_b} & y_{12_a} + y_{12_b} \\ y_{21_a} + y_{21_b} & y_{22_a} + y_{22_b} \end{bmatrix} \quad (20.197)$$

$$\boldsymbol{Y} = \boldsymbol{Y}_a + \boldsymbol{Y}_b \quad (20.198)$$

According to Equation (20.198), when two sections of y-parameters are connected in parallel, the combined $\boldsymbol{Y}$ matrix is obtained by adding $\boldsymbol{Y}_a$ and $\boldsymbol{Y}_b$.

EXAMPLE 20.26

Find the y-parameters for the twin T circuit shown in Figure 20.58.

The y-parameters for the T circuit are derived in EXAMPLE 20.6. For the T circuit shown in Figure 20.21, the y-parameters are given by

$$y_{11} = \frac{I_1}{V_1} = \frac{Z_2 + Z_3}{Z_1Z_2 + Z_1Z_3 + Z_2Z_3}$$

$$y_{21} = \frac{I_2}{V_1} = \frac{-Z_2}{Z_1Z_2 + Z_1Z_3 + Z_2Z_3}$$

continued

Example 20.26 continued

FIGURE 20.58

Twin *T* circuit.

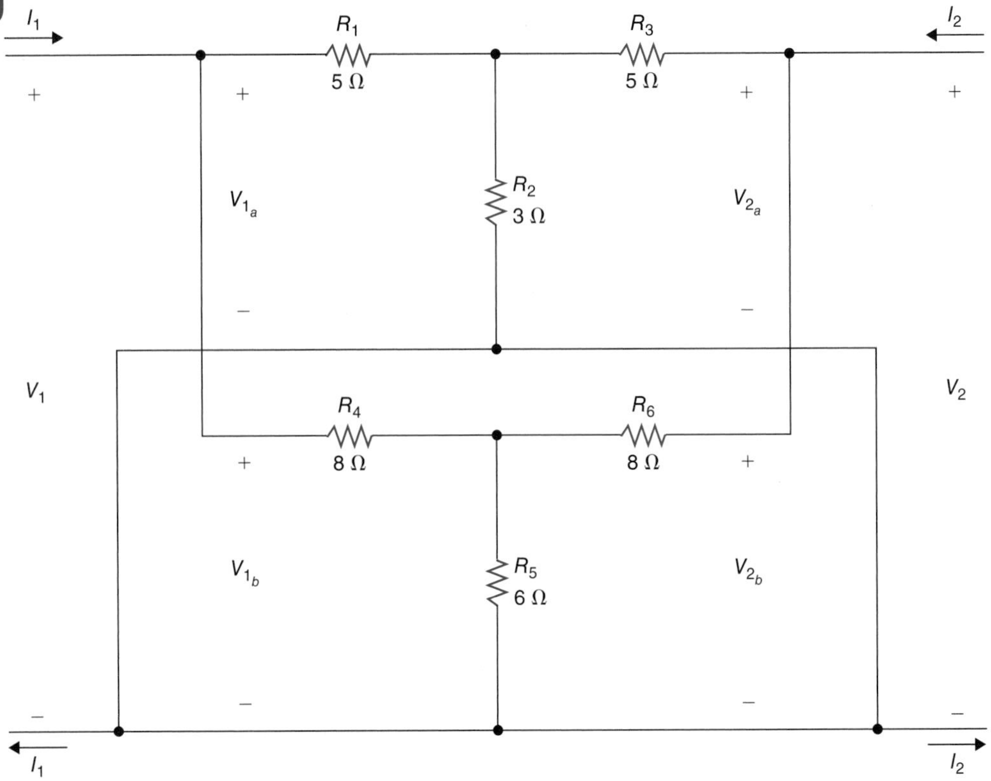

$$y_{22} = \frac{I_2}{V_2} = \frac{Z_1 + Z_2}{Z_1Z_2 + Z_1Z_3 + Z_2Z_3}$$

$$y_{12} = \frac{I_1}{V_2} = \frac{-Z_2}{Z_1Z_2 + Z_1Z_3 + Z_2Z_3}$$

For the *T* circuit consisting of R_1, R_2, and R_3, the *y*-parameters are given by

$$y_{11} = \frac{R_2 + R_3}{R_1R_2 + R_1R_3 + R_2R_3} = \frac{8}{55} = 0.1455$$

$$y_{21} = \frac{-R_2}{R_1R_2 + R_1R_3 + R_2R_3} = \frac{-3}{55} = -0.054545$$

$$y_{22} = \frac{R_1 + R_2}{R_1R_2 + R_1R_3 + R_2R_3} = \frac{8}{55} = 0.1455$$

$$y_{12} = \frac{-R_2}{R_1R_2 + R_1R_3 + R_2R_3} = \frac{-3}{55} = -0.054545$$

The matrix of *y*-parameters is given by

$$Y_a = \begin{bmatrix} y_{11_a} & y_{12_a} \\ y_{21_a} & y_{22_a} \end{bmatrix} = \begin{bmatrix} 0.1455 & -0.054545 \\ -0.054545 & 0.1455 \end{bmatrix}$$

continued

Example 20.26 continued

For the T circuit consisting of R_4, R_5, and R_6, the y-parameters are given by

$$y_{11} = \frac{R_5 + R_6}{R_4R_5 + R_4R_6 + R_5R_6} = \frac{14}{160} = 0.0875$$

$$y_{21} = \frac{-R_5}{R_4R_5 + R_4R_6 + R_5R_6} = \frac{-6}{160} = -0.0375$$

$$y_{22} = \frac{R_4 + R_5}{R_4R_5 + R_4R_6 + R_5R_6} = \frac{14}{160} = 0.0875$$

$$y_{12} = \frac{-R_5}{R_4R_5 + R_4R_6 + R_5R_6} = \frac{-6}{160} = -0.0375$$

The matrix of y-parameters is given by

$$\boldsymbol{Y_b} = \begin{bmatrix} y_{11_b} & y_{12_b} \\ y_{21_b} & y_{22_b} \end{bmatrix} = \begin{bmatrix} 0.0875 & -0.0375 \\ -0.0375 & 0.0875 \end{bmatrix}$$

The matrix of y-parameters for the circuit shown in Figure 20.58 is given by

$$\boldsymbol{Y} = \boldsymbol{Y_a} + \boldsymbol{Y_b} = \begin{bmatrix} 0.2330 & -0.092045 \\ -0.092045 & 0.2330 \end{bmatrix}$$

FIGURE 20.59

Two 2-port networks connected in series-parallel.

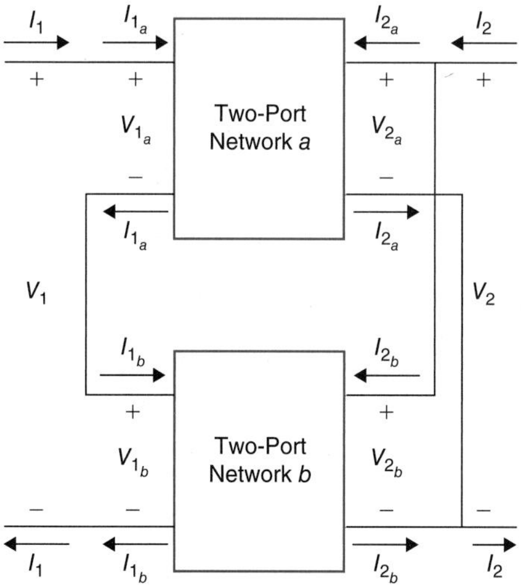

20.4.4 SERIES-PARALLEL CONNECTION

Figure 20.59 shows two 2-port networks connected in series-parallel. When two 2-port networks are connected in series-parallel, the matrix of the h-parameters of the equivalent 2-port network shown in Figure 20.50 is the sum of the matrices of the h-parameters of the two networks. From Equation (20.49), the first 2-port network is described by h-parameters as

$$\begin{bmatrix} V_{1_a} \\ I_{2_a} \end{bmatrix} = \begin{bmatrix} h_{11_a} & h_{12_a} \\ h_{21_a} & h_{22_a} \end{bmatrix} \begin{bmatrix} I_{1_a} \\ V_{2_a} \end{bmatrix} \tag{20.199}$$

Similarly, the second 2-port network is described by h-parameters as

$$\begin{bmatrix} V_{1_b} \\ I_{2_b} \end{bmatrix} = \begin{bmatrix} h_{11_b} & h_{12_b} \\ h_{21_b} & h_{22_b} \end{bmatrix} \begin{bmatrix} I_{1_b} \\ V_{2_b} \end{bmatrix} \tag{20.200}$$

The equivalent circuit shown in Figure 20.50 can be described by h-parameters as

$$\begin{bmatrix} V_1 \\ I_2 \end{bmatrix} = \begin{bmatrix} h_{11} & h_{12} \\ h_{21} & h_{22} \end{bmatrix} \begin{bmatrix} I_1 \\ V_2 \end{bmatrix} \tag{20.201}$$

Since $V_1 = V_{1_a} + V_{1_b}$, $V_2 = V_{2_a} = V_{2_b}$, $I_1 = I_{1_a} = I_{1_b}$, $I_2 = I_{2_a} + I_{2_b}$, we have

$$\begin{aligned} \begin{bmatrix} V_1 \\ I_2 \end{bmatrix} &= \begin{bmatrix} V_{1_a} \\ I_{2_a} \end{bmatrix} + \begin{bmatrix} V_{1_b} \\ I_{2_b} \end{bmatrix} = \begin{bmatrix} h_{11_a} & h_{12_a} \\ h_{21_a} & h_{22_a} \end{bmatrix} \begin{bmatrix} I_{1_a} \\ V_{2_a} \end{bmatrix} + \begin{bmatrix} h_{11_b} & h_{12_b} \\ h_{21_b} & h_{22_b} \end{bmatrix} \begin{bmatrix} I_{1_b} \\ V_{2_b} \end{bmatrix} \\ &= \left(\begin{bmatrix} h_{11_a} & h_{12_a} \\ h_{21_a} & h_{22_a} \end{bmatrix} + \begin{bmatrix} h_{11_b} & h_{12_b} \\ h_{21_b} & h_{22_b} \end{bmatrix} \right) \begin{bmatrix} I_1 \\ V_2 \end{bmatrix} = \begin{bmatrix} h_{11_a} + h_{11_b} & h_{12_a} + h_{12_b} \\ h_{21_a} + h_{21_b} & h_{22_a} + h_{22_b} \end{bmatrix} \begin{bmatrix} I_1 \\ V_2 \end{bmatrix} \end{aligned} \tag{20.202}$$

From Equations (20.201) and (20.202), we have

$$\begin{bmatrix} h_{11} & h_{12} \\ h_{21} & h_{22} \end{bmatrix} = \begin{bmatrix} h_{11_a} + h_{11_b} & h_{12_a} + h_{12_b} \\ h_{21_a} + h_{21_b} & h_{22_a} + h_{22_b} \end{bmatrix} \quad \textbf{(20.203)}$$

$$\boldsymbol{H} = \boldsymbol{H_a} + \boldsymbol{H_b} \quad \textbf{(20.204)}$$

According to Equation (20.204), when two sections of h-parameters are connected in series-parallel, the combined $\boldsymbol{H}$ matrix is obtained by adding $\boldsymbol{H_a}$ and $\boldsymbol{H_b}$.

FIGURE 20.60

Two 2-port networks connected in parallel series.

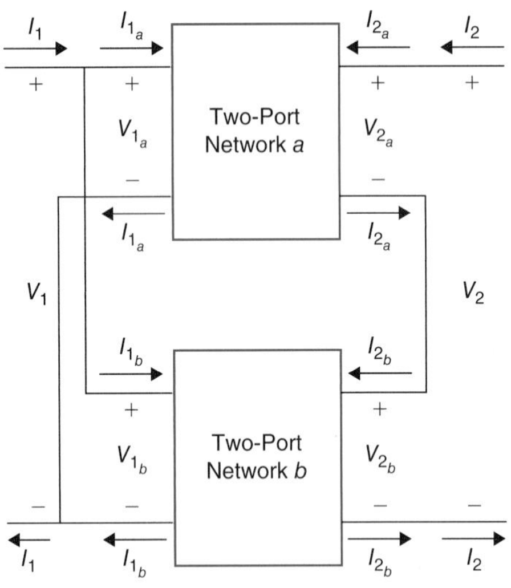

20.4.5 PARALLEL-SERIES CONNECTION

Figure 20.60 shows two 2-port networks connected in parallel-series. When two 2-port networks are connected in parallel-series, the matrix of the g-parameters of the equivalent 2-port network shown in Figure 20.50 is the sum of the matrices of the g-parameters of the two networks. From Equation (20.56), the first 2-port network is described by g-parameters as

$$\begin{bmatrix} I_{1_a} \\ V_{2_a} \end{bmatrix} = \begin{bmatrix} g_{11_a} & g_{12_a} \\ g_{21_a} & g_{22_a} \end{bmatrix} \begin{bmatrix} V_{1_a} \\ I_{2_a} \end{bmatrix} \quad \textbf{(20.205)}$$

Similarly, the second two-port network is described by g-parameters as

$$\begin{bmatrix} I_{1_b} \\ V_{2_b} \end{bmatrix} = \begin{bmatrix} g_{11_b} & g_{12_b} \\ g_{21_b} & g_{22_b} \end{bmatrix} \begin{bmatrix} V_{1_b} \\ I_{2_b} \end{bmatrix} \quad \textbf{(20.206)}$$

The equivalent circuit shown in Figure 20.50 can be described by g-parameters as

$$\begin{bmatrix} I_1 \\ V_2 \end{bmatrix} = \begin{bmatrix} g_{11} & g_{12} \\ g_{21} & g_{22} \end{bmatrix} \begin{bmatrix} V_1 \\ I_2 \end{bmatrix} \quad \textbf{(20.207)}$$

Since $V_1 = V_{1_a} = V_{1_b}, V_2 = V_{2_a} + V_{2_b}, I_1 = I_{1_a} + I_{1_b}, I_2 = I_{2_a} = I_{2_b}$, we have

$$\begin{aligned} \begin{bmatrix} I_1 \\ V_2 \end{bmatrix} &= \begin{bmatrix} I_{1_a} \\ V_{2_a} \end{bmatrix} + \begin{bmatrix} I_{1_b} \\ V_{2_b} \end{bmatrix} = \begin{bmatrix} g_{11_a} & g_{12_a} \\ g_{21_a} & g_{22_a} \end{bmatrix} \begin{bmatrix} V_{1_a} \\ I_{2_a} \end{bmatrix} + \begin{bmatrix} g_{11_b} & g_{12_b} \\ g_{21_b} & g_{22_b} \end{bmatrix} \begin{bmatrix} V_{1_b} \\ I_{2_b} \end{bmatrix} \\ &= \left(\begin{bmatrix} g_{11_a} & g_{12_a} \\ g_{21_a} & g_{22_a} \end{bmatrix} + \begin{bmatrix} g_{11_b} & g_{12_b} \\ g_{21_b} & g_{22_b} \end{bmatrix} \right) \begin{bmatrix} V_1 \\ I_2 \end{bmatrix} = \begin{bmatrix} g_{11_a} + g_{11_b} & g_{12_a} + g_{12_b} \\ g_{21_a} + g_{21_b} & g_{22_a} + g_{22_b} \end{bmatrix} \begin{bmatrix} V_1 \\ I_2 \end{bmatrix} \end{aligned} \quad \textbf{(20.208)}$$

From Equations (20.207) and (20.208), we have

$$\begin{bmatrix} g_{11} & g_{12} \\ g_{21} & g_{22} \end{bmatrix} = \begin{bmatrix} g_{11_a} + g_{11_b} & g_{12_a} + g_{12_b} \\ g_{21_a} + g_{21_b} & g_{22_a} + g_{22_b} \end{bmatrix} \quad \textbf{(20.209)}$$

$$\boldsymbol{G} = \boldsymbol{G_a} + \boldsymbol{G_b} \quad \textbf{(20.210)}$$

According to Equation (20.210), when two sections of g-parameters are connected in parallel-series, the combined $\boldsymbol{G}$ matrix is obtained by adding $\boldsymbol{G_a}$ and $\boldsymbol{G_b}$.

20.4.6 CASCADE CONNECTION FOR *b*-PARAMETERS

Figure 20.49 shows two 2-port networks connected in cascade. When two 2-port networks are connected in cascade, the matrix of the b-parameters of the equivalent 2-port network shown in Figure 20.50 is the product of the matrices of the b-parameters of the two networks. From Equation (20.87), the first 2-port network is described by b-parameters as

$$\begin{bmatrix} V_{2_a} \\ I_{2_a} \end{bmatrix} = \begin{bmatrix} b_{11_a} & b_{12_a} \\ b_{21_a} & b_{22_a} \end{bmatrix} \begin{bmatrix} V_{1_a} \\ -I_{1_a} \end{bmatrix} \tag{20.211}$$

Similarly, the second two-port network is described by b-parameters as

$$\begin{bmatrix} V_{2_b} \\ I_{2_b} \end{bmatrix} = \begin{bmatrix} b_{11_b} & b_{12_b} \\ b_{21_b} & b_{22_b} \end{bmatrix} \begin{bmatrix} V_{1_b} \\ -I_{1_b} \end{bmatrix} \tag{20.212}$$

Since $V_{2_a} = V_{1_b}$ and $-I_{2_a} = I_{1_b}$, from Equation (20.211), we have

$$\begin{bmatrix} V_{1_b} \\ -I_{1_b} \end{bmatrix} = \begin{bmatrix} V_{2_a} \\ I_{2_a} \end{bmatrix} = \begin{bmatrix} b_{11_a} & b_{12_a} \\ b_{21_a} & b_{22_a} \end{bmatrix} \begin{bmatrix} V_{1_a} \\ -I_{1_a} \end{bmatrix} \tag{20.213}$$

Substituting Equation (20.213) into Equation (20.212), we have

$$\begin{bmatrix} V_{2_b} \\ I_{2_b} \end{bmatrix} = \begin{bmatrix} b_{11_b} & b_{12_b} \\ b_{21_b} & b_{22_b} \end{bmatrix} \begin{bmatrix} V_{1_b} \\ -I_{1_b} \end{bmatrix} = \begin{bmatrix} b_{11_b} & b_{12_b} \\ b_{21_b} & b_{22_b} \end{bmatrix} \begin{bmatrix} b_{11_a} & b_{12_a} \\ b_{21_a} & b_{22_a} \end{bmatrix} \begin{bmatrix} V_{1_a} \\ -I_{1_a} \end{bmatrix} \tag{20.214}$$

According to Equation (20.214), when two sections of b-parameters are cascaded, the combined $\boldsymbol{B}$ matrix is obtained by multiplying the $\boldsymbol{B}$ matrices; that is,

$$\begin{bmatrix} b_{11} & b_{12} \\ b_{21} & b_{22} \end{bmatrix} = \begin{bmatrix} b_{11_b} & b_{12_b} \\ b_{21_b} & b_{22_b} \end{bmatrix} \begin{bmatrix} b_{11_a} & b_{12_a} \\ b_{21_a} & b_{22_a} \end{bmatrix} \tag{20.215}$$

or

$$\boldsymbol{B} = \boldsymbol{B}_b \boldsymbol{B}_a \tag{20.216}$$

20.5 PSpice and Simulink

The two-port parameters can be measured by building a circuit in PSpice or Simulink. As an example, let us measure the y-parameters of T circuit shown in Figure 20.61.

In EXERCISE 20.3, we found that

FIGURE 20.61

T circuit.

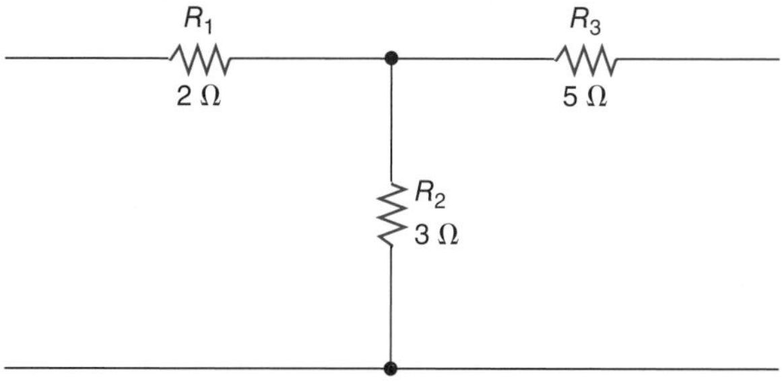

$$y_{11} = 8/31 = 0.2581\ \text{S}, \quad y_{12} = -3/31 = -0.09677\ \text{S},$$

$$y_{21} = -3/31 = -0.09677\ \text{S}, \quad y_{22} = 5/31 = 0.1613\ \text{S}$$

To measure y_{11} and y_{21}, V_2 is short-circuited, and 1 V is applied at the input ($V_1 = 1$ V), as shown in Figure 20.62. The current from the positive terminal of voltage source is $I_1 = 258.1$ mA. Thus, y_{11} is given by

$$y_{11} = \left.\frac{I_1}{V_1}\right|_{V_2=0} = \frac{258.1\,\text{mA}}{1\,\text{V}} = 0.2581\,\text{S}$$

FIGURE 20.62

$V_1 = 1$ V and $V_2 = 0$ V.

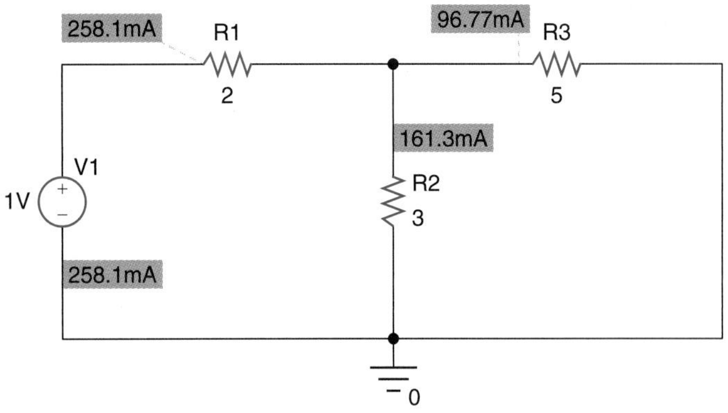

The current through R_3 (→) at the output is 96.77 mA. Since I_2 is defined as the current into the circuit from the output port, $I_2 = -96.77$ mA. Thus, we have

$$y_{21} = \left.\frac{I_2}{V_1}\right|_{V_2=0} = \frac{-96.77\text{ mA}}{1\text{ V}} = -0.09677\text{ S}$$

To measure y_{22} and y_{12}, V_1 is short-circuited, and 1 V is applied at the output ($V_2 = 1$ V), as shown in Figure 20.63. The current from the positive terminal of voltage source is $I_2 = 161.3$ mA. Thus, y_{22} is given by

$$y_{22} = \left.\frac{I_2}{V_2}\right|_{V_1=0} = \frac{161.3\text{ mA}}{1\text{ V}} = 0.1613\text{ S}$$

FIGURE 20.63

$V_1 = 0$ V and $V_2 = 1$ V.

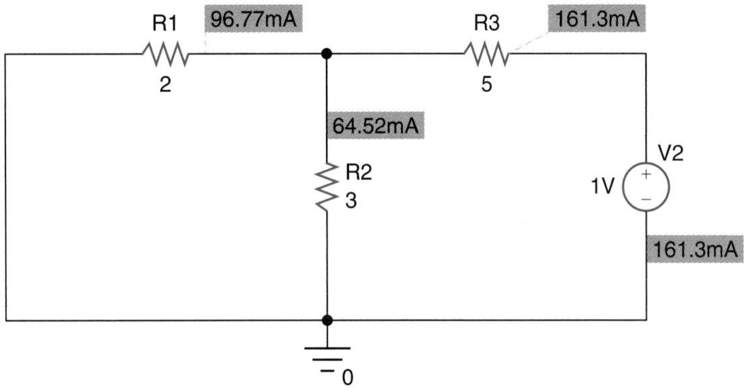

The current through R_1 (←) at the input is 96.77 mA. Since I_1 is defined as the current into the circuit from the input port, $I_1 = -96.77$ mA. Thus, we have

$$y_{12} = \left.\frac{I_1}{V_2}\right|_{V_1=0} = \frac{-96.77\text{ mA}}{1\text{ V}} = -0.09677\text{ S}$$

The values of the y-parameters from PSpice match those from direct calculation.

A Simulink model to measure y-parameters is shown in Figure 20.64. The input voltage is set at 1 V, and the output voltage is set at 0 V. Double-clicking Manual Switch changes voltages from 1 V to 0 V and 0 V to 1 V. The same circuit can be used to measure all four y-parameters by changing the switch positions.

FIGURE 20.64

(a) Simulink model to measure y_{11} and y_{21}. (b) Simulink model to measure y_{22} and y_{12}.

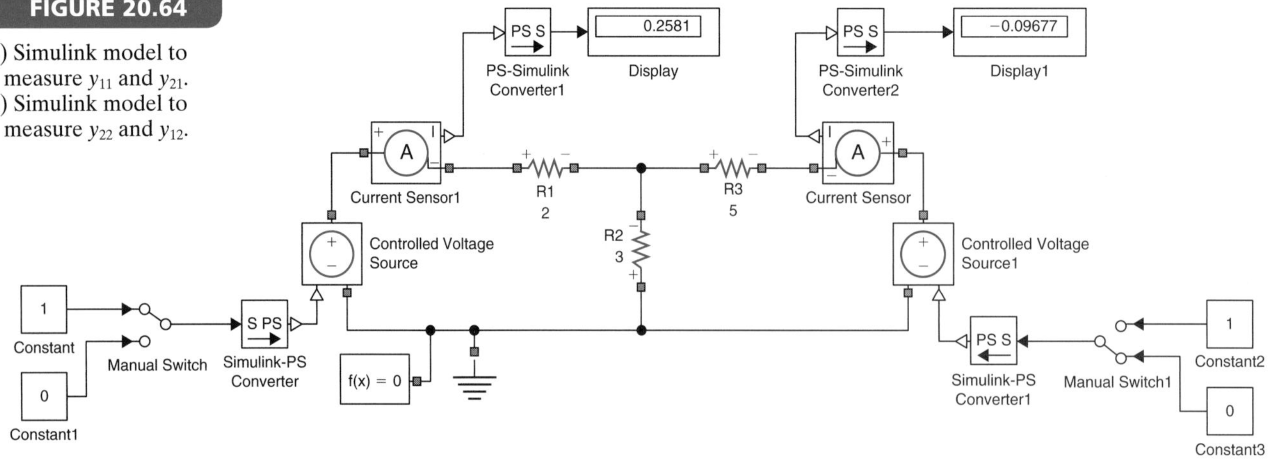

(a)

continued

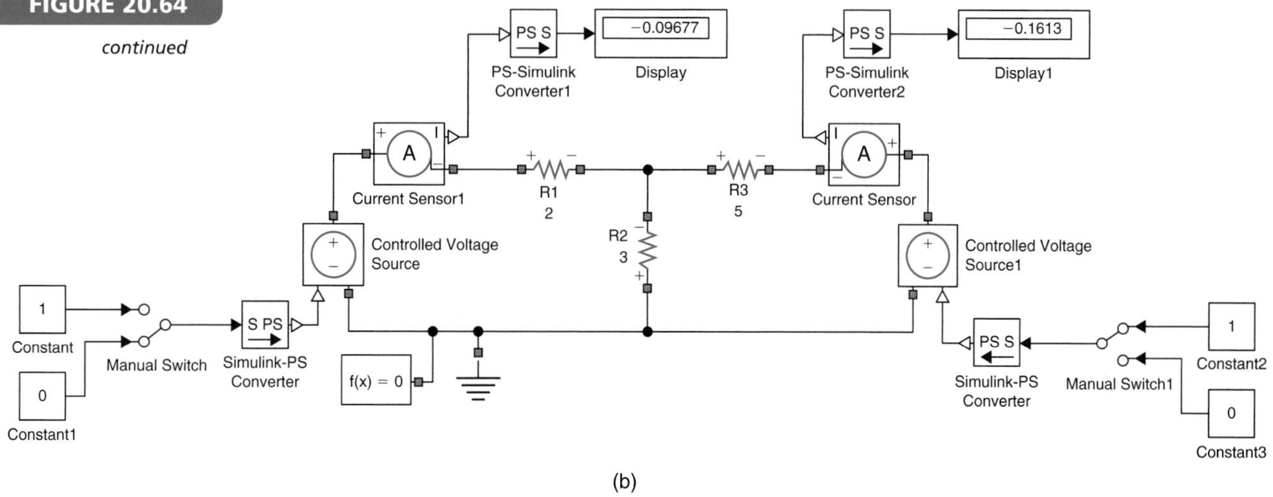

FIGURE 20.64

continued

The g-parameters for the circuit shown in Figure 20.30 can be measured in PSpice or Simulink. The g-parameters found in EXAMPLE 20.8 are

$$g_{11} = 0.2\text{ S},\quad g_{21} = 0.72,\quad g_{22} = 2.44\ \Omega,\quad g_{12} = -0.6$$

The PSpice simulation results are shown in in Figure 20.65. The g-parameters match those from direct calculation.

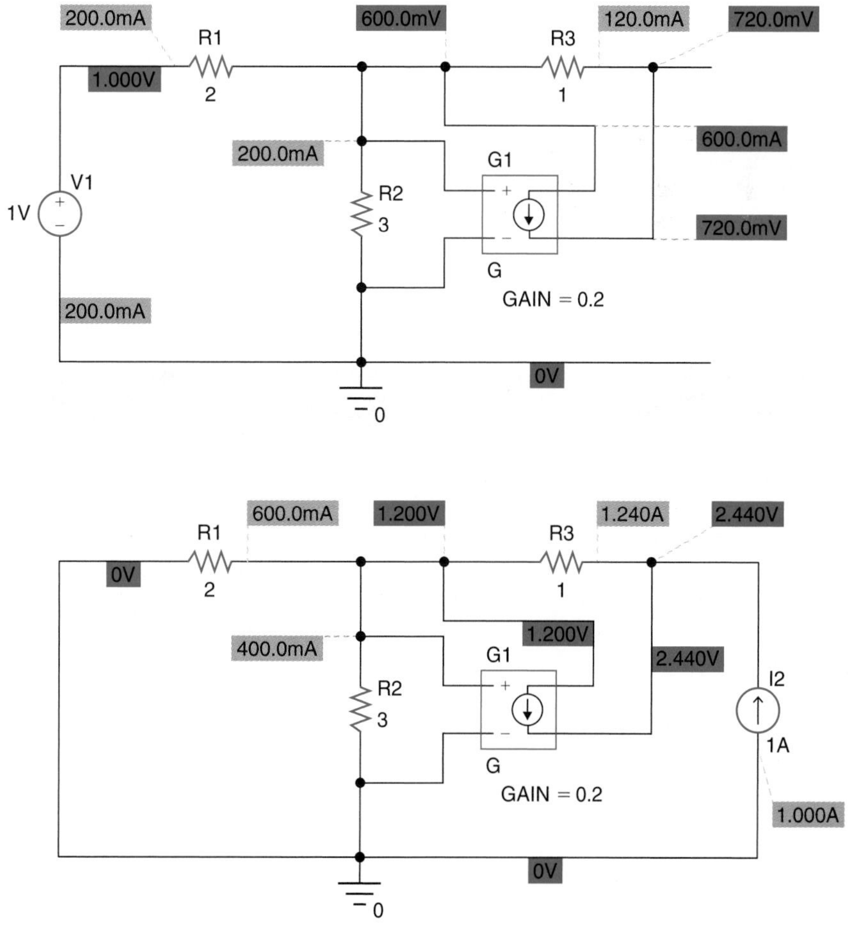

FIGURE 20.65

PSpice circuits to measure g-parameters.

The Simulink model to measure the *g*-parameters is shown in Figure 20.66.

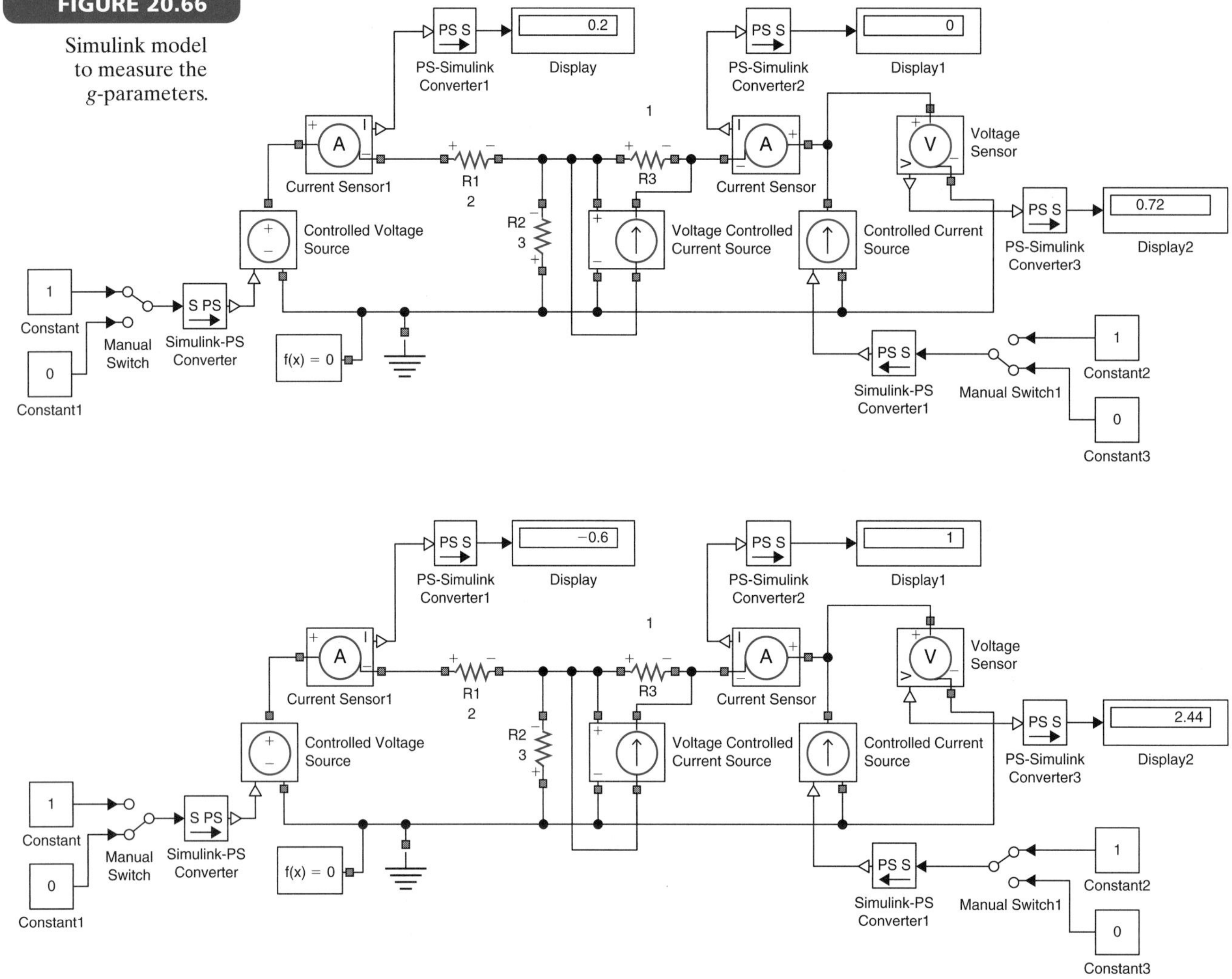

FIGURE 20.66 Simulink model to measure the *g*-parameters.

SUMMARY

The two-port circuits can be characterized in six different ways depending on which two of the four variables (V_1, I_1, V_2, I_2) are independent variables and the other two are dependent variables. The coefficients of the independent variables are called parameters. The six different parameters are *z*-parameters, *y*-parameters, *h*-parameters, *g*-parameters, *ABCD*-parameters, and *b*-parameters.

Given one parameter, all the rest of the parameters can be found by simple conversion formulas. For example, if *z*-parameters are known from the measurements, the other five parameters are obtained by conversion.

The two-port circuits can be interconnected. Depending on how the two port circuits are interconnected, we can find one of the six parameters by adding the matrices of the parameters of each section or multiplying the matrices of the parameters of each section. For example, if the two-port circuits are connected in series, the *z*-parameters of the interconnected circuit is the sum of the *z*-parameters of each section.

PROBLEMS

Parameters

20.1 **Find the z-parameters for the circuit shown in Figure P20.1.**

FIGURE P20.1

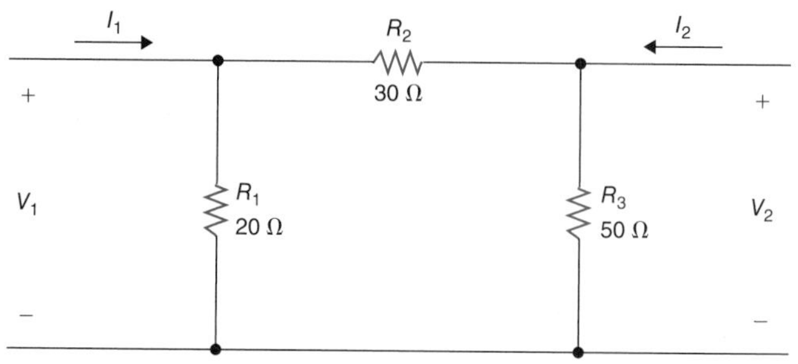

20.2 **Find the z-parameters for the circuit shown in Figure P20.2.**

FIGURE P20.2

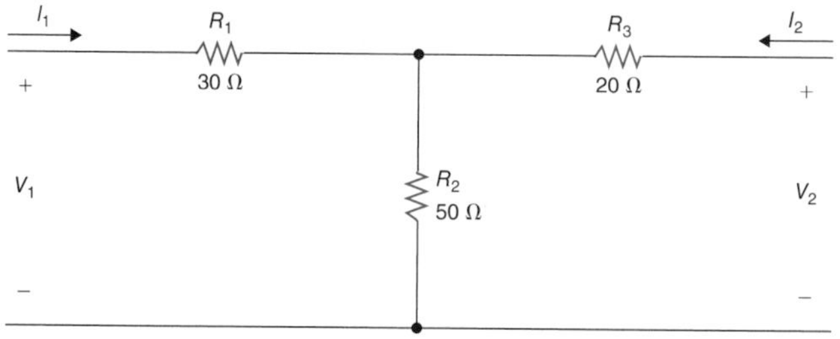

20.3 **Find the z-parameters for the circuit shown in Figure P20.3.**

FIGURE P20.3

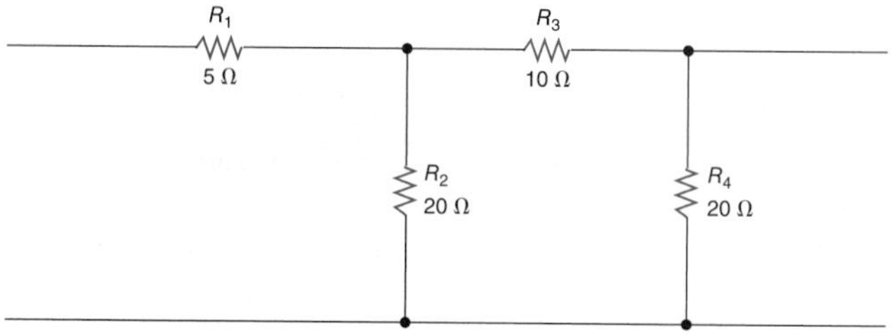

20.4 **Find the z-parameters for the circuit shown in Figure P20.4.**

FIGURE P20.4

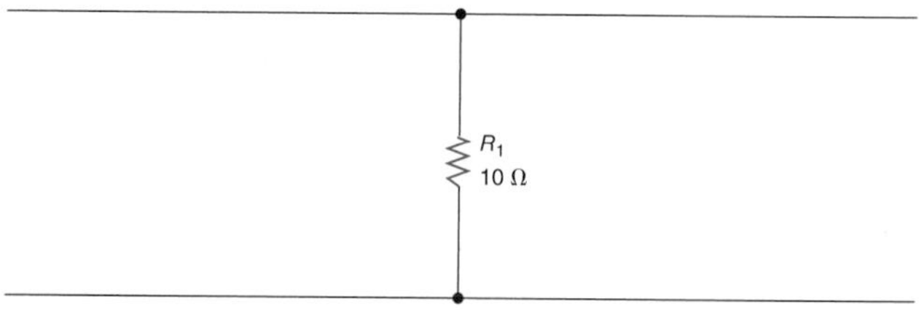

20.5 **Find the z-parameters for the circuit shown in Figure P20.5.**

FIGURE P20.5

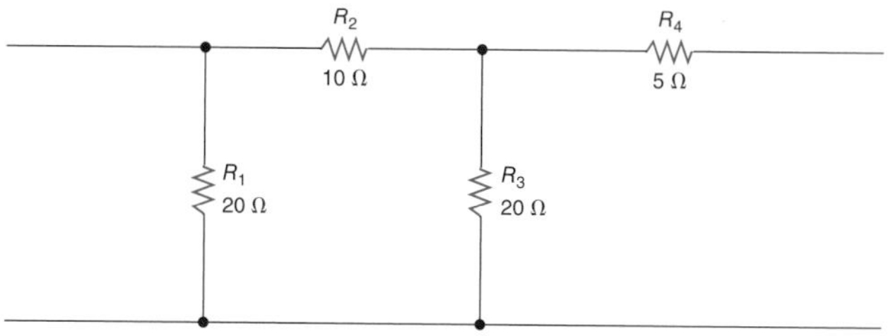

20.6 **Find the z-parameters for the coupled coils shown in Figure P20.6.**

FIGURE P20.6

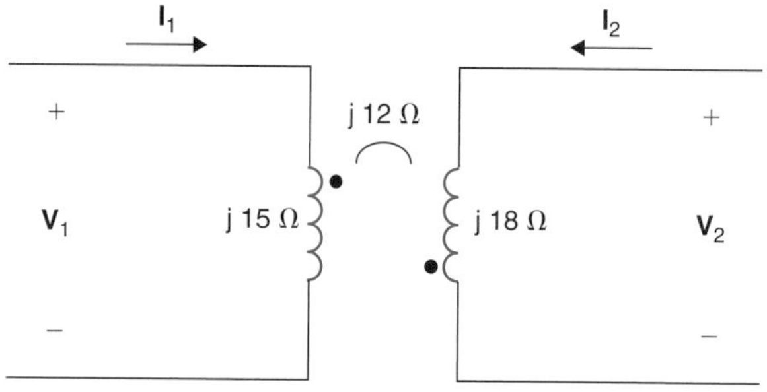

20.7 **Find the z-parameters for the coupled coils shown in Figure P20.7.**

FIGURE P20.7

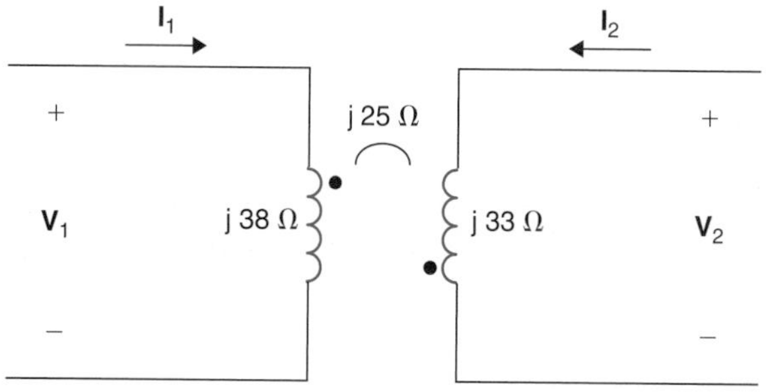

20.8 **Find the h-parameters for the circuit shown in Figure P20.8.**

FIGURE P20.8

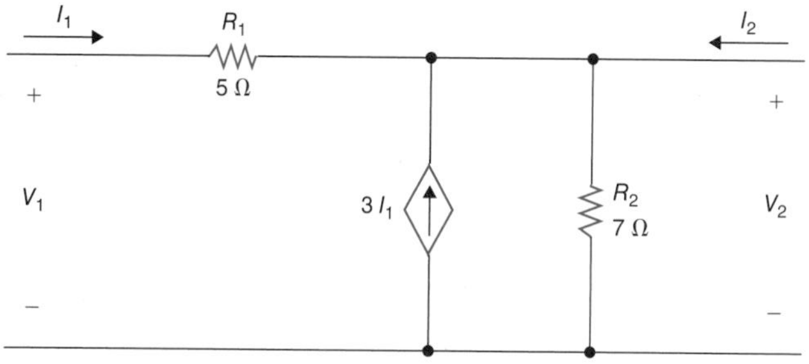

20.9 Find the g-parameters for the circuit shown in Figure P20.9.

FIGURE P20.9

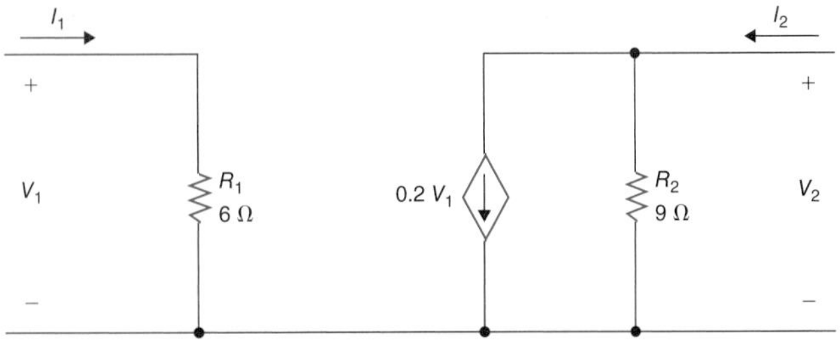

20.10 Find the $ABCD$ parameters for the circuit shown in Figure P20.10.

FIGURE P20.10

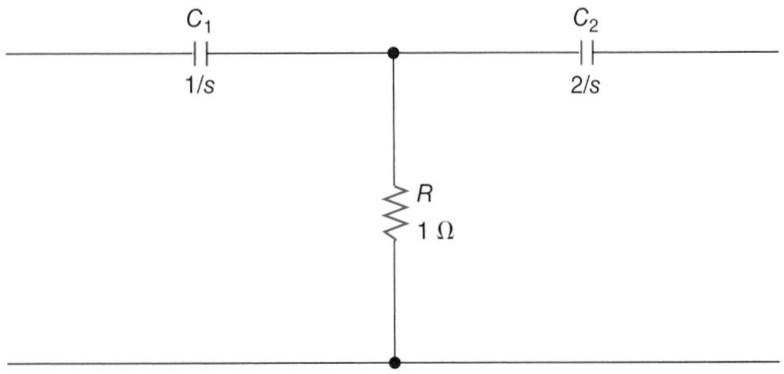

20.11 Find the $ABCD$ parameters for the circuit shown in Figure P20.11.

FIGURE P20.11

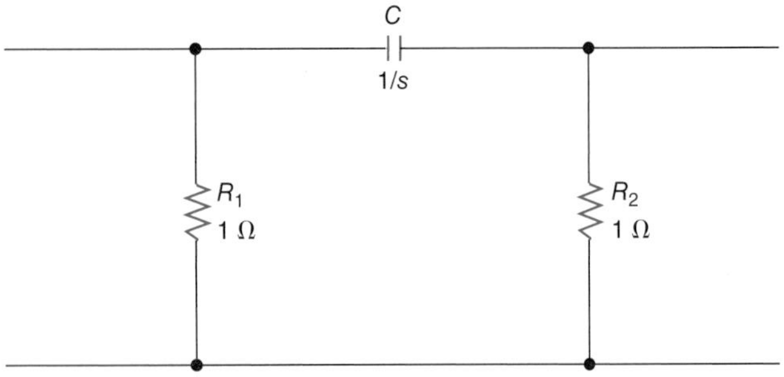

20.12 Find the $ABCD$ parameters for the circuit shown in Figure P20.12.

FIGURE P20.12

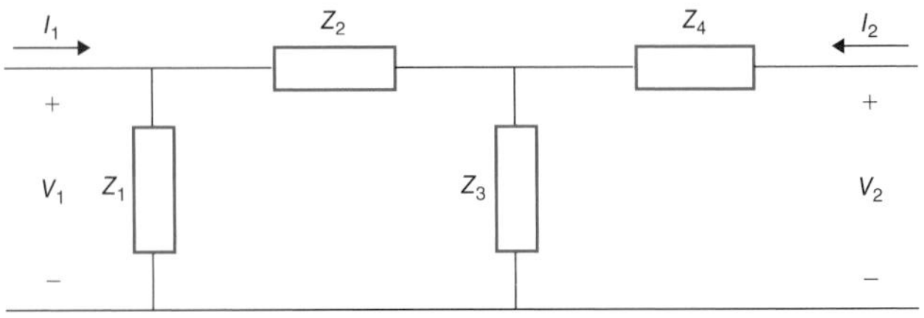

20.13 Find the transfer function $H(s) = V_2(s)/V_1(s)$ for the circuit shown in Figure P20.13.

FIGURE P20.13

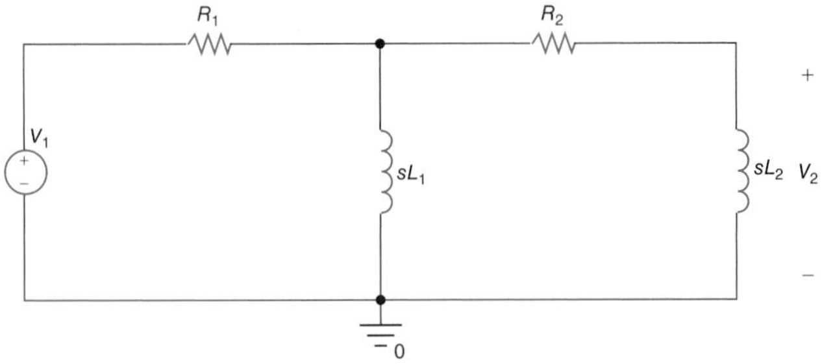

20.14 Find the transfer function $H(s) = V_2(s)/V_1(s)$ for the circuit shown in Figure P20.14.

FIGURE P20.14

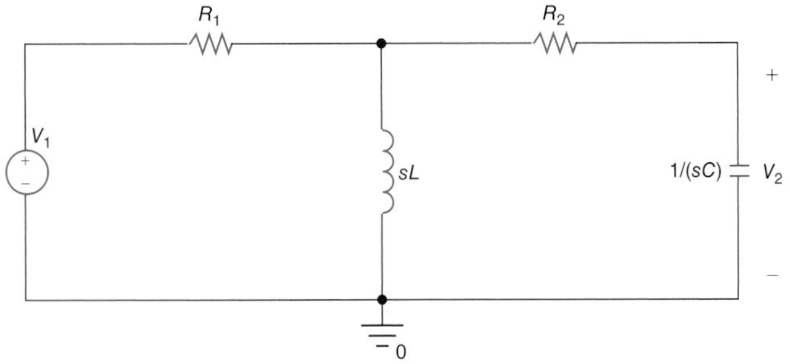

20.15 Find the b-parameters for the circuit shown in Figure P20.15.

FIGURE P20.15

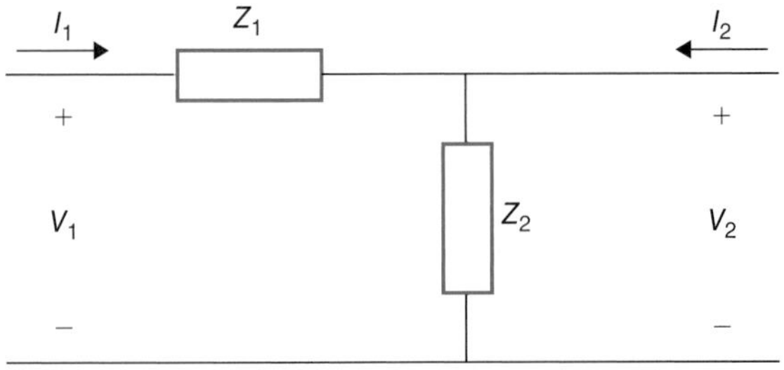

20.16 Find the b-parameters for the circuit shown in Figure P20.16.

FIGURE P20.16

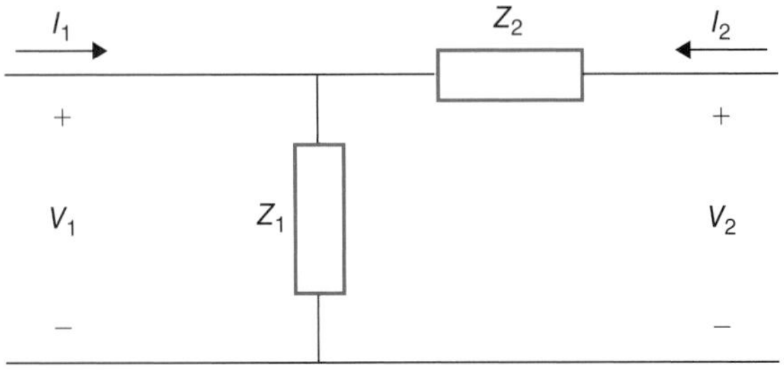

Conversion of Parameters

20.17 **Convert the *ABCD* parameters for the circuit shown in Figure P20.17 to *h*-parameters.**

FIGURE P20.17

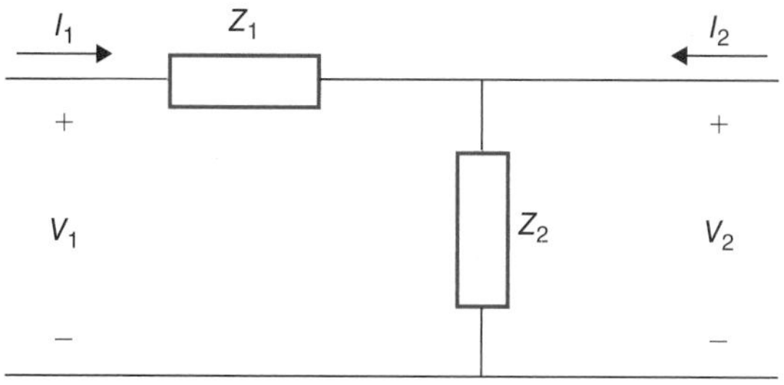

20.18 **Convert the *ABCD* parameters for the circuit shown in Figure P20.18 to *g*-parameters.**

FIGURE P20.18

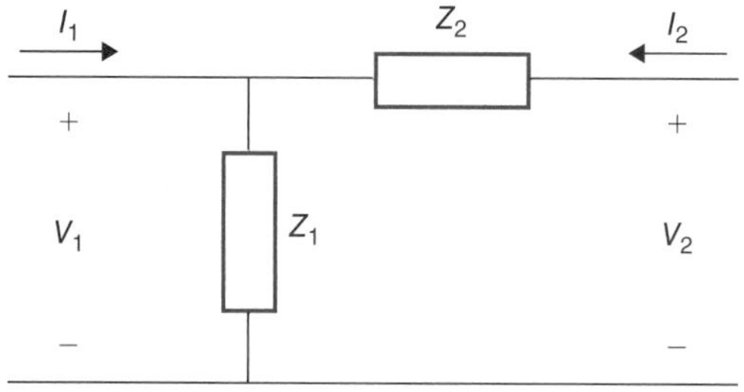

Interconnection of Parameters

20.19 **When the circuits shown in Figure P20.10 and Figure P20.11 are cascaded, we obtain the circuit shown in Figure P20.19. Find the *ABCD* parameters for the circuit shown in Figure P20.19.**

FIGURE P20.19

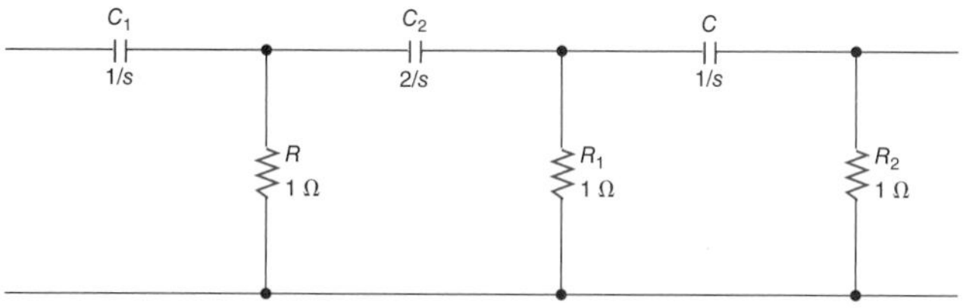

20.20 **When the circuits shown in Figure P20.3 and Figure P20.4 are connected in series, we obtain the circuit shown in Figure P20.20. Find the *z*-parameters for the circuit shown in Figure P20.20.**

FIGURE P20.20

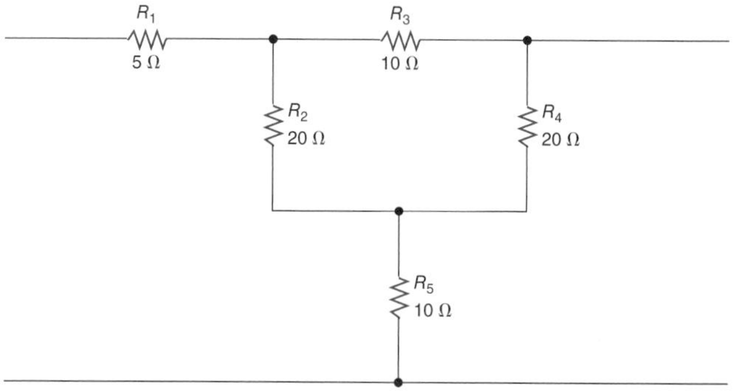

20.21 **When the circuits shown in Figure P20.5 and Figure P20.4 are connected in series, we obtain the circuit shown in Figure P20.21. Find the *z*-parameters for the circuit shown in Figure P20.21.**

FIGURE P20.21

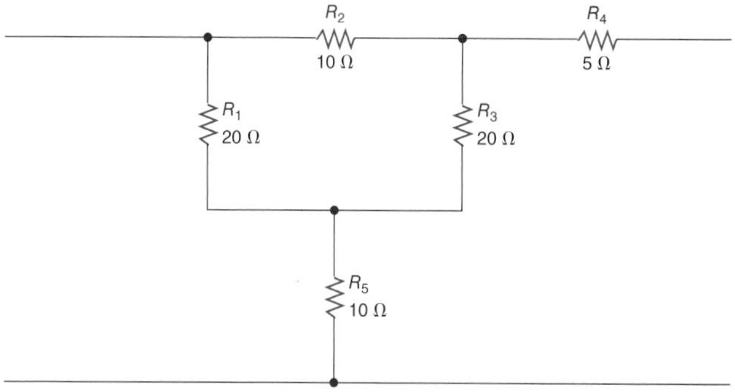

20.22 **Find the *y*-parameters for the bridged-*T* circuit shown in Figure P20.22 using the parallel connection of *T* circuit (R_1, R_2, and R_3) and R_4.**

FIGURE P20.22

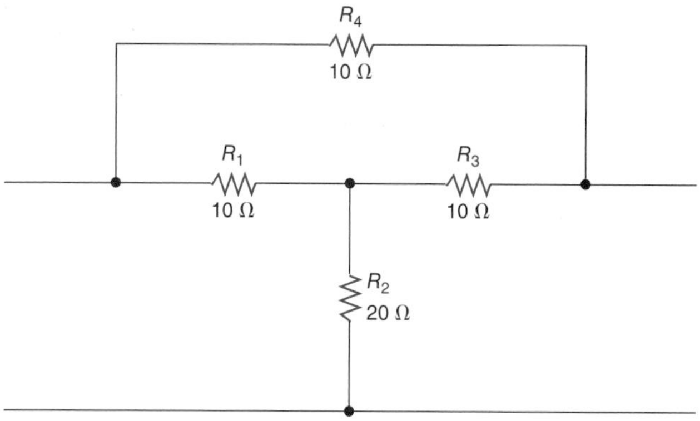

20.23 Find the y-parameters for the twin-T circuit shown in Figure P20.23 using the parallel connection of T circuit (R_1, R_2, and C_3) and another T circuit (C_1, C_2, and R_3).

FIGURE P20.23

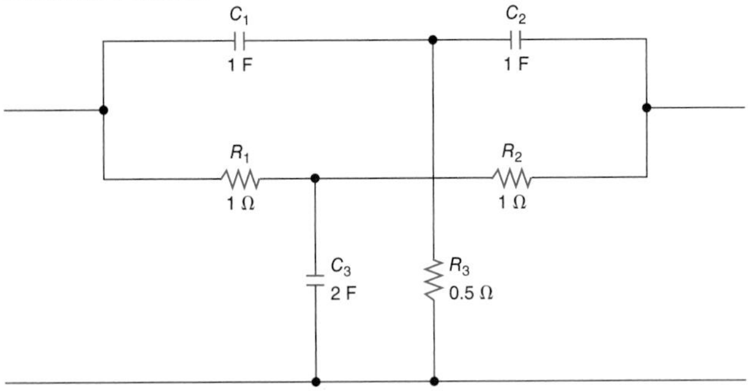

20.24 Find the y-parameters for the circuit shown in Figure P20.24 using the parallel connection of pi circuit (R_1, R_2, and R_3) and T circuit (R_4, R_5, and R_6).

FIGURE P20.24

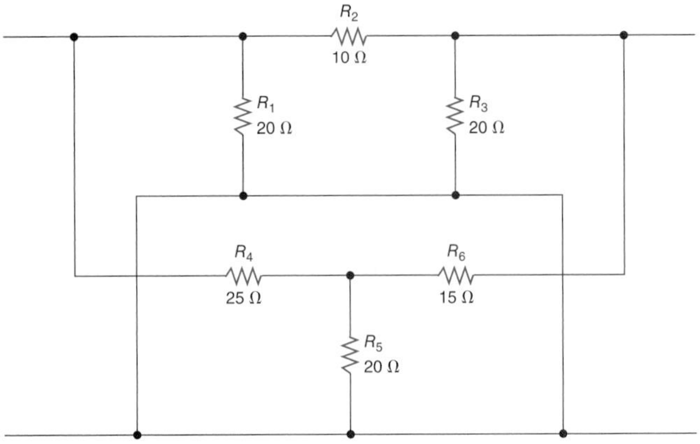

Answers to Odd-Numbered Questions

CHAPTER 1

1.1 $i(t) = \begin{cases} 2 \text{ mA}, & t \geq 0 \\ 0 \text{ mA}, & t < 0 \end{cases}$

1.3 $i(t) = \begin{cases} 24e^{-0.003t} \text{ mA}, & t \geq 0 \\ 0 \text{ mA}, & t < 0 \end{cases}$

1.5 $i(t) = \begin{cases} 50.2655 \cos(2\pi \times 1000t) \text{ mA}, & 1 \geq 0 \\ 0 \text{ mA}, & t < 0 \end{cases}$

1.7 $Q = 25 \times 10^{-3}$ C

1.9 $Q = 9.4925$ C

1.11 $Q = 22.2817$ C

1.13 $P = VI = 2 \text{ V} \times (-3 \text{ A}) = -6$ W, delivering power

1.15 $P = VI = (-12 \text{ V}) \times (-10 \text{ A}) = 120$ mW, absorbing power

1.17 $p(t) = 62.5 \sin(2\pi 2000t)$ mW

1.19 $p(t) = [12 + 12 \cos(2\pi 200t)]$ mW

1.21

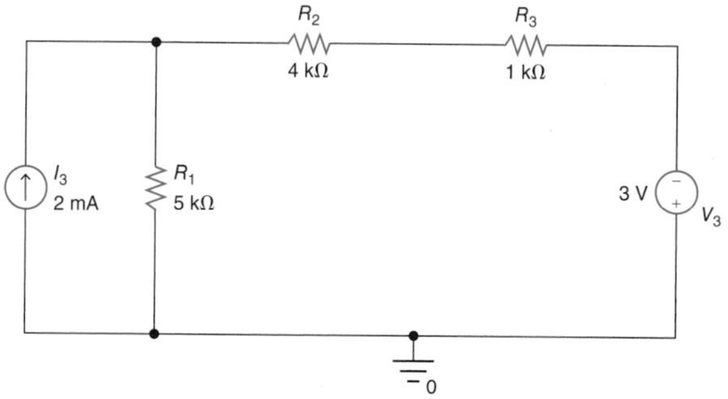

1.23

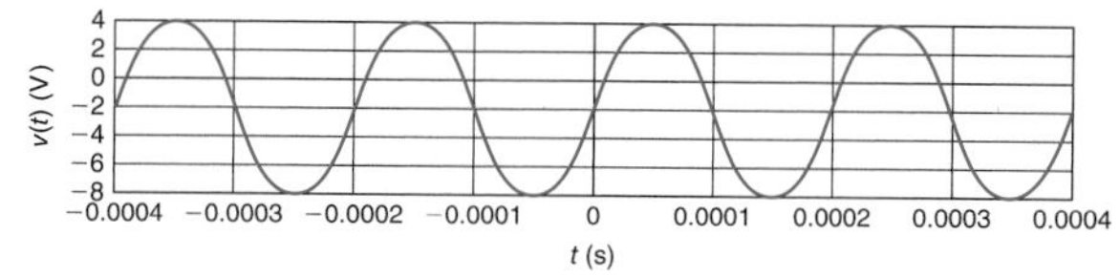

1.25 The voltage across the VCVS: $0.5\, v_a = 0.5 \times 1.2908 \text{ V} = 0.6454$ V
The current through the VCCS: $0.001\, v_a = 0.001 \text{ (A/V)} \times 1.2908 \text{ V} = 1.2908$ mA

1.27

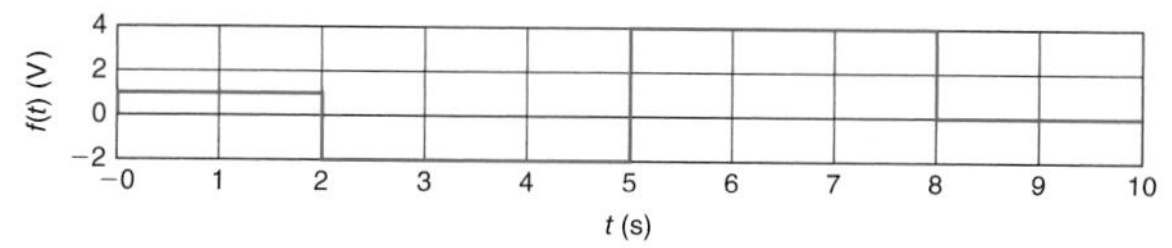

1.29

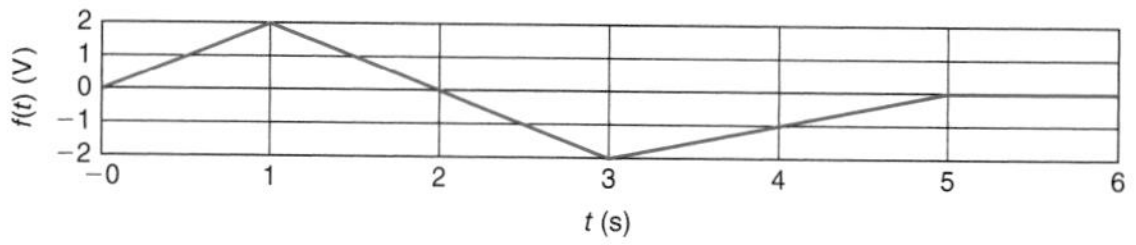

1.31

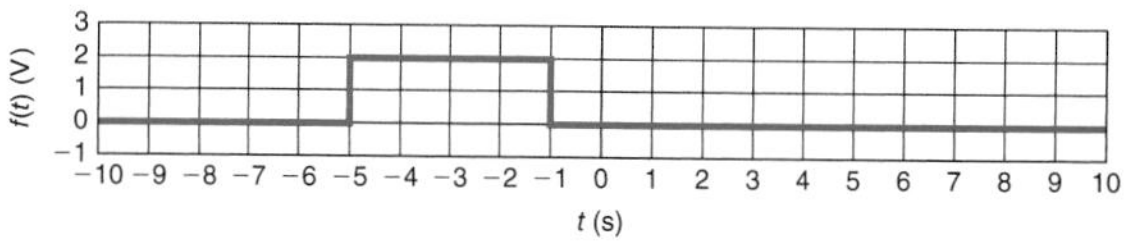

CHAPTER 2

2.1 $I_1 = 2$ mA, $I_2 = 1$ mA

2.3 $I_1 = 0.6$ mA, $I_2 = 0.4$ mA, $I_3 = 0.6667$ mA, $I_4 = 0.2$ mA, $I_5 = 0.1333$ mA

2.5 $V_o = 3.6$ V, $I_3 = 0.6$ mA

2.7 $V_o = 3.5$ V, $R_2 = 30\ k\Omega$

2.9 $P_{I_s} = -4.8$ mW (delivered), $P_{R_1} = 2.88$ mW (absorbed), $P_{R_2} = 1.92$ mW (absorbed)

2.11 $i = 1$ mA, $I_1 = 2$ mA, $I_2 = 1$ mA, $I_3 = 2$ mA, $I_4 = 1$ mA, $I_5 = 2$ mA

2.13 $I = 0.55556$ mA, $V_1 = 1.6667$ V, $V_2 = 2.7778$ V, $V_3 = 2.2222$ V, $V_4 = 1.1111$ V, $V_5 = 2.2222$ V

2.15 $V_2 = 6$ V, $I_3 = 2$ mA, $I_1 = 5$ mA, $V_1 = 5$ V

2.17 $V_2 = 6$ V, $I_3 = 2$ mA, $I_2 = 3$ mA, $I_1 = 2$ mA, $V_1 = 9$ V

2.19 $I_3 = 2$ mA, $V_3 = 4$ V, $V_2 = 9$ V, $I_2 = 3$ mA, $I_1 = 5$ mA, $V_1 = 5$ V

2.21 $I_2 = 5$ mA, $I_4 = 3$ mA, $I_5 = 12$ mA

2.23 $I_1 = 1$ A, $I_2 = 3$ A, $I_3 = -6$ A, $I_4 = -7$ A, and $I_5 = -12$ A

2.25 $V_{R_2} = 5$ V, $V_{R_4} = 15$ V, $V_{R_5} = 10$ V

2.27 $I_5 = 6$ mA, $I_1 = 2$ mA, $I_3 = 4$ mA, $V_4 = 10$ V, $I_4 = 2$ mA, $I_2 = 6$ mA

2.29 $V_1 = 5$ V, $I_2 = 4$ mA, $V_2 = 3$ V, $I_3 = 3$ mA, $I_4 = 1$ mA

2.31 $V_1 = -3$ V, $V_2 = 4$ V, $V_5 = 1$ V, $I_1 = 4$ A, $I_4 = 5$ A

2.33 $R_{eq} = 15\ k\Omega$

2.35 $R_{eq} = 30\ k\Omega$

2.37 $R_{eq} = 60\ k\Omega$

2.39 $R_{eq} = 10\ k\Omega$

2.41 $R_{eq} = 2.8772\ k\Omega$

2.43 $R_{eq} = 15\ k\Omega$

2.45 $R_{eq} = 2.4\ k\Omega$

2.47 $V_1 = 5$ V, $V_2 = 15$ V

2.49 $V_1 = 11.4$ V, $V_2 = 18.6$ V
2.51 $V_1 = 36$ V, $V_2 = 7.3636$ V
2.53 $V_1 = 6.4$ V, $V_2 = 3.84$ V
2.55 $V_1 = 27.5$ V, $V_2 = 20$ V
2.57 $V_1 = 9$ V, $V_2 = 4$ V, $V_3 = 3.2$ V, $I_1 = 2$ mA, $I_2 = 1$ mA, $I_3 = 1$ mA, $I_4 = 1.6$ mA, $I_5 = 0.4$ mA
2.59 $I_{R_1} = 12$ mA, $I_{R_2} = 8$ mA, $I_{R_3} = 6$ mA
2.61 $I_1 = 0.6$ mA, $I_2 = 1.4$ mA, $I_3 = 0.84$ mA, $I_4 = 0.56$ mA, $V_1 = 4.2$ V, $V_2 = 3.36$ V
2.63 $I_1 = 2$ mA, $I_2 = 1$ mA, $I_3 = 1$ mA, $V_1 = 20$ V
2.65 $I_1 = 4.8$ mA, $I_2 = 2.4$ mA, $I_3 = 2.4$ mA, $I_4 = 4.32$ mA, $I_5 = 2.88$ mA, $V_1 = 0.864$ V, $V_2 = 0.432$ V
2.67 $I = 1.3333$ mA
2.69 $R_{eq} = 12.1685\ k\Omega$
2.71 $V_1 = 5$ V, $V_2 = 6.5$ V

CHAPTER 3

3.1 $V_1 = 6$ V
3.3 $V_1 = 12$ V
3.5 $V_1 = 9$ V, $I_1 = 1.5$ mA, $I_2 = 3$ mA
3.7 $V_1 = 5$ V, $V_2 = 3$ V
3.9 $V_1 = 4$ V, $V_2 = 7.5$ V
3.11 $V_1 = 2.8689$ V, $V_2 = 1.3934$ V
3.13 $V_1 = 22.7273$ V, $V_2 = 9.0909$ V
3.15 $V_1 = 2.5276$ V, $V_2 = 4.1811$ V
3.17 $V_1 = 4.2$ V, $V_2 = 4.4$ V, $I_1 = -0.2$ mA, $I_2 = -0.9$ mA
3.19 $V_1 = 25$ V, $V_2 = 20$ V, $V_3 = 10$ V
3.21 $V_1 = 16$ V, $V_2 = 8$ V, $V_3 = 6$ V
3.23 $V_1 = 1.6364$ V
3.25 $V_1 = 1.25$ V, $V_2 = 3.5$ V
3.27 $V_1 = 5.1429$ V, $V_2 = 9.7143$ V
3.29 $V_1 = 2.0339$ V, $V_2 = 3.2464$ V
3.31 $V_1 = 1.2245$ V, $V_2 = 6.36735$ V
3.33 $V_1 = 3.2987$ V, $V_2 = 6.2987$ V
3.35 $V_1 = 1.4220$ V, $V_2 = 2.4220$ V
3.37 $V_1 = 12$ V, $V_2 = 8$ V, $V_3 = 15$ V
3.39 $V_1 = 20$ V, $V_2 = 15$ V, $V_3 = 8$ V, $V_4 = 12$ V
3.41 $V_1 = 2.10325$ V, $V_2 = 2.94455$ V
3.43 $V_1 = 8$ V, $V_2 = 6$ V, $V_3 = 4$ V
3.45 $V_1 = 10$ V, $V_2 = 4$ V, $V_3 = 16$ V
3.47 $V_1 = 13.8462$ V, $V_2 = 24.2308$ V
3.49 $V_1 = 4.35294$ V, $V_2 = 3.0588$ V
3.51 $I_1 = 1$ mA, $I_2 = -1$ mA, $V_1 = 2$ V
3.53 $I_1 = 2$ mA, $I_2 = 0.85246$ mA, $I_3 = 0.88525$ mA, $V_1 = 7.8689$ V, $V_2 = 5.5738$ V, $V_3 = 5.3115$ V
3.55 $I_1 = -2$ mA, $I_2 = -4$ mA, $I_3 = -5$ mA, $V_1 = 5$ V, $V_2 = 9$ V
3.57 $I_1 = 2$ mA, $I_2 = 1$ mA, $I_3 = 6$ mA, $V_1 = 8$ V, $V_2 = 5$ V, $V_3 = 3$ V
3.59 $I_1 = 0.71429$ mA, $I_2 = 0.2381$ mA, $V_1 = 1.4286$ V, $V_2 = 0.4762$ V
3.61 $I_1 = 1.3095$ mA, $I_2 = 0.7143$ mA, $V_1 = 2.3810$ V, $V_2 = 3.5714$ V
3.63 $v_o = 0.07634$ V
3.65 $I_1 = 4$ mA, $I_2 = -1$ mA, $I_3 = 7$ mA, $I_4 = 5$ mA, $V_1 = 15$ V, $V_2 = 10$ V, $V_3 = 20$ V
3.67 $I_1 = -5$ mA, $I_2 = -10$ mA, $I_3 = 0$ mA, $I_4 = -10$ mA, $V_1 = 5$ V, $V_2 = 20$ V, $V_3 = 10$ V, $V_4 = 15$ V
3.69 $I_1 = 0.2707$ mA, $I_2 = 2.1654$ mA, $I_3 = 1.6917$ mA, $V_1 = 1.4211$ V, $V_2 = 8.4586$ V
3.71 $I_1 = -4$ mA, $I_2 = -2$ mA, $I_3 = -5$ mA, $I_4 = -7$ mA, $V_1 = 10$ V, $V_2 = 5$ V, $V_3 = 15$ V
3.73 $I_1 = 5.5556$ mA, $I_2 = 4.4444$ mA, $I_3 = 16.6667$ mA, $I_4 = 5.5556$ mA, $V_1 = 14.4444$ V, $V_2 = 8.8889$ V, $V_3 = 3.3333$ V

CHAPTER 4

4.1 $V_1 = 3$ V
4.3 $V_1 = 5.3571$ V
4.5 $V_1 = 9.5$ V
4.7 $V_1 = 4$ V, $V_2 = 2$ V
4.9 $V_1 = 8$ V, $V_2 = 15$ V
4.11 $V_1 = 4$ V, $V_2 = 6$ V
4.13 $V_1 = 6.5$ V, $V_2 = -1$ V
4.15 $V_1 = 4$ V, $V_2 = 8$ V, $V_3 = 10$ V
4.17 $I_1 = 0.4444$ mA, $I_2 = 0.3333$ mA
4.19 $V_o = 5.5556$ V
4.21 $V_o = 5.4066$ V
4.23 $V_o = 0.5672$ V
4.25 $V_1 = 9.1364$ V, $V_2 = 2.4545$ V
4.27 $V_1 = 10.08$ V, $V_2 = 18$ V, $V_3 = 12$ V
4.29 $V_1 = 10$ V, $V_2 = 30$ V, $V_3 = 16$ V
4.31 $V_{th} = 3$ V, $R_{th} = 2.4\ k\Omega$
4.33 $V_{th} = 6$ V, $R_{th} = 2\ k\Omega$
4.35 $V_{th} = 5.6571$ V, $R_{th} = 6\ k\Omega$
4.37 $V_{th} = 14$ V, $R_{th} = 6\ \Omega$
4.39 $V_{th} = 40.7595$ V, $R_{th} = 8.8608\ k\Omega$
4.41 $V_{th} = 15$ V, $R_{th} = 10\ k\Omega$
4.43 $V_{th} = -0.9$ V, $R_{th} = 1.2\ k\Omega$
4.45 $V_{th} = 15$ V, $R_{th} = 2.3864\ k\Omega$
4.47 $V_{th} = 4$ V, $R_{th} = 1\ \Omega$
4.49 $V_{th} = 3.3191$ V, $R_{th} = 808.5106\ \Omega$
4.51 $V_{th} = -8.5714$ V, $R_{th} = 1.4286\ k\Omega$
4.53 $V_{th} = 7.425$ V, $R_{th} = 2.1\ k\Omega$
4.55 $V_{th} = 3$ V, $R_{th} = 722.1154\ \Omega$
4.57 $V_{th} = 10.5$ V, $R_{th} = 1\ \Omega$
4.59 $V_{th} = 7.2$ V, $R_{th} = 5\ \Omega$
4.61 $V_{th} = 1.3333$ V, $R_{th} = 944.4444\ \Omega$
4.63 $I_n = 2$ mA, $R_n = 3.3333\ k\Omega$
4.65 $I_n = 5$ mA, $R_n = 7\ k\Omega$
4.67 $I_n = 1.1489$ mA, $R_n = 3.4815\ k\Omega$
4.69 $I_n = 1.2037$ mA, $R_n = 43.2\ k\Omega$
4.71 $I_n = 4.4516$ mA, $R_n = 449.2754\ \Omega$
4.73 $I_n = 0.6667$ mA, $R_n = 7.5\ k\Omega$
4.75 $I_n = 4.8$ mA, $R_n = 416.6667\ \Omega$
4.77 $I_n = 1.1104$ mA, $R_n = 3.7531\ k\Omega$
4.79 $V_{th} = 8$ V, $R_L = R_{th} = 20\ k\Omega$, $P_L = 0.8$ mW
4.81 $V_{th} = 5$ V, $R_L = R_{th} = 777.7778\ \Omega$, $P_L = 8.0357$ mW

4.83 $V_{th} = 2\,V, R_L = R_{th} = 188.8889\,\Omega, P_L = 5.2941\,mW$
4.85 $V_{th} = 6.6667\,V, R_L = R_{th} = 6.6667\,\Omega, P_L = 1.6667\,W$
4.87 $V_{th} = 8\,V, R_L = R_{th} = 4\,k\Omega, P_L = 4\,mW$
4.89 $V_{th} = 12\,V, R_L = R_{th} = 2.25\,k\Omega, P_L = 16\,mW$

CHAPTER 5

5.1 $V_o = -3\,V$
5.3 $V_o = -2.3\,V$
5.5 $V_o = 0.8\,V$
5.7 $V_o = 1.636364\,V$
5.9 $V_o = 0.75\,V$
5.11 $R_2 = 4.5\,k\Omega$

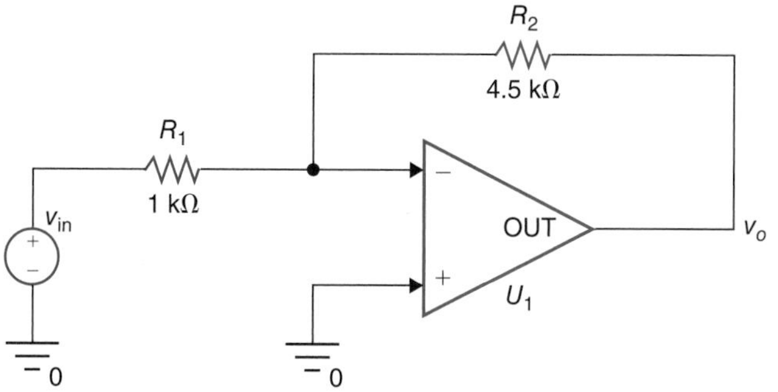

5.13 $R_2 = 3\,k\Omega$

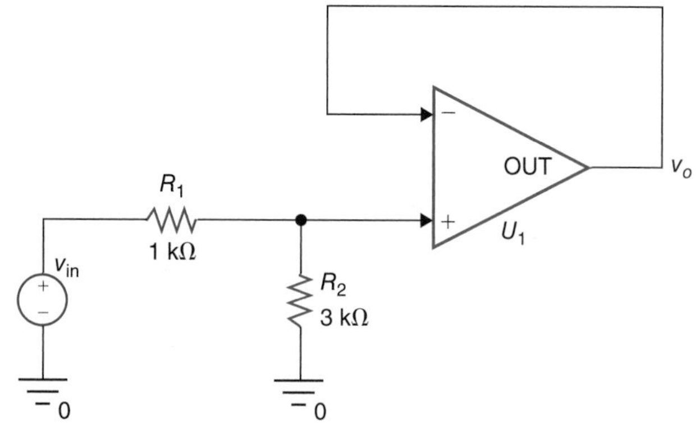

5.15 $V_o = 4\,V$
5.17 $V_a = 3\,V, V_b = 5\,V, V_c = 6\,V, V_o = 11\,V$
5.19 $V_o = 9.3\,V$
5.21

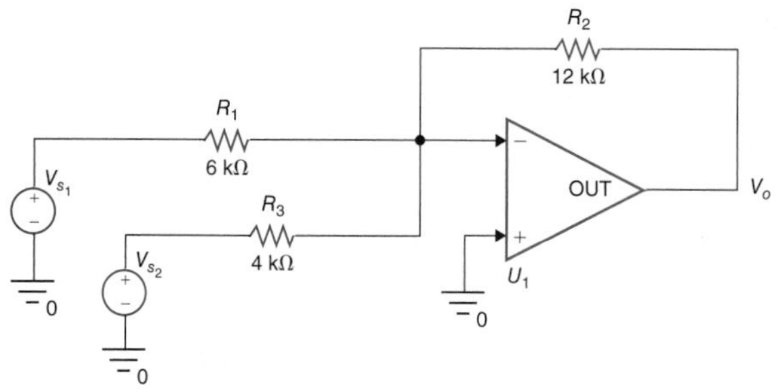

5.23 One solution is to use the circuit shown in Figure 5.38 with three inputs ($n = 3$).
$R = 1\,k\Omega, R_1 = 2\,R = 2\,k\Omega, R_2 = 4\,R = 4\,k\Omega,$
$R_3 = 8\,R = 8\,k\Omega, R_4 = R_5 = R = 1\,k\Omega,$
$R_6 = R(1/2 + 1/4 + 1/8) = (7/8)\,R = 875\,\Omega.$

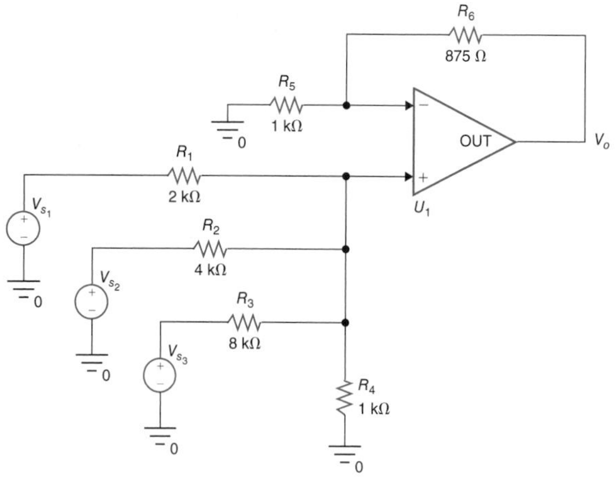

5.25

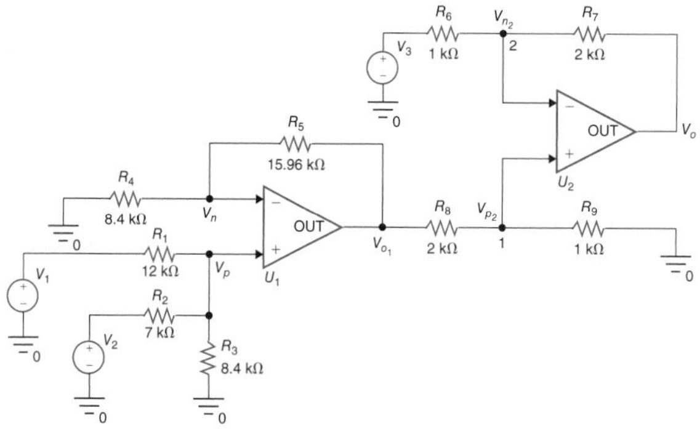

5.27 a. $V_{o_1} = V_s(R_1 + R_2)/R_2$
b. $V_{o_2} = -V_s R_3/R_2$
c. $V_{o_1} = 5\,V, V_{o_2} = -5\,V$
d. $R_1 = 9\,k\Omega, R_2 = 1\,k\Omega, R_3 = 5\,k\Omega$
5.29 $V_o = -6\,V$
5.31 $I_o = -2\,mA$
5.33 a. $I_o = \dfrac{R_1 R_4}{R_1 R_3 R_4 + R_1 R_4 R_5 - R_2 R_3 R_5} V_s$
b. $I_o = 0.5\,mA$
c. $I_o = 0.403846\,mA$
5.35

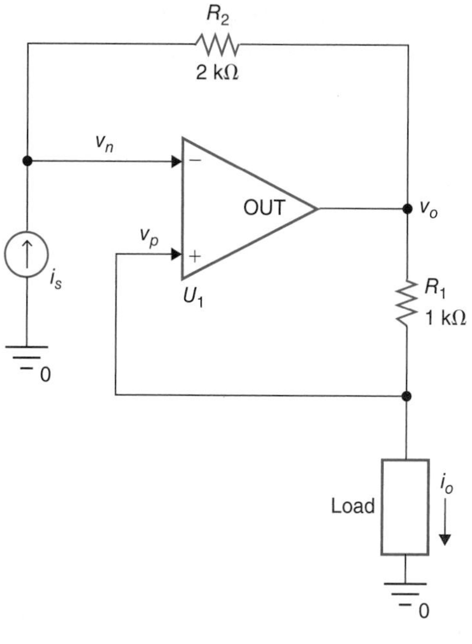

5.37

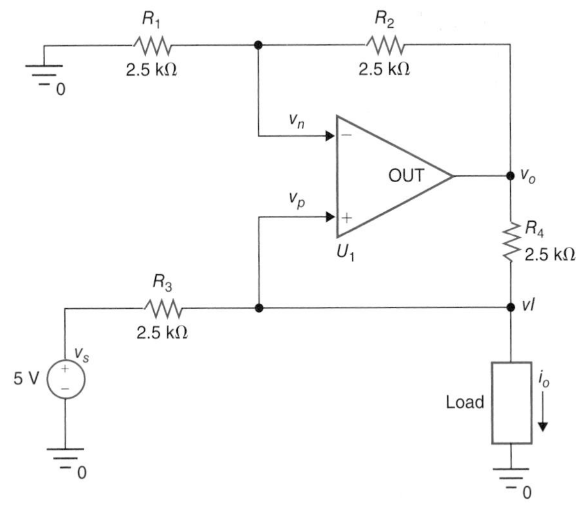

5.39 $V_o = -2.9986676$ V

5.41 $R_{out} = 9.8731\ \Omega$

5.43 $R_{in} = 1028.8462\ \Omega$

5.45 $V_o = 4.9972475$ V

5.47 $R_{out} = 13.5301\ \Omega$

5.49 $R_{in} = 51.75\ k\Omega$

CHAPTER 6

6.1 $i(t) = 2$ mA

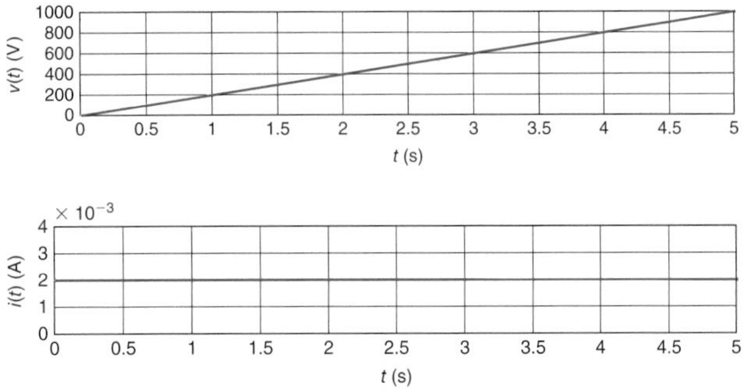

6.3 $i(t) = -0.1257 \sin(2\pi 1000t)$ mA

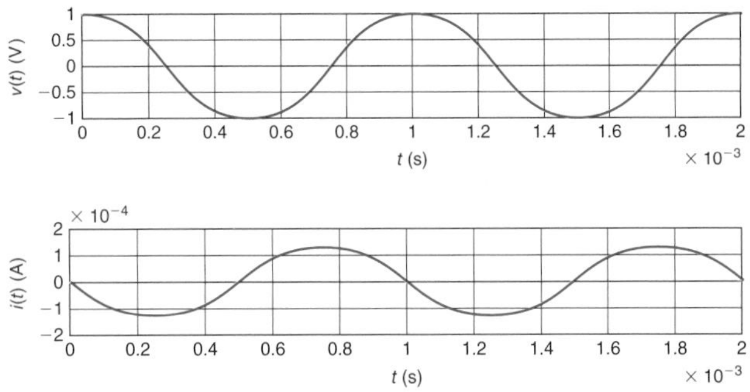

6.5 $i(t) = e^{-5000t}[-10\cos(2\pi 8000t) - 100\sin(2\pi 8000t)]\ \mu$A

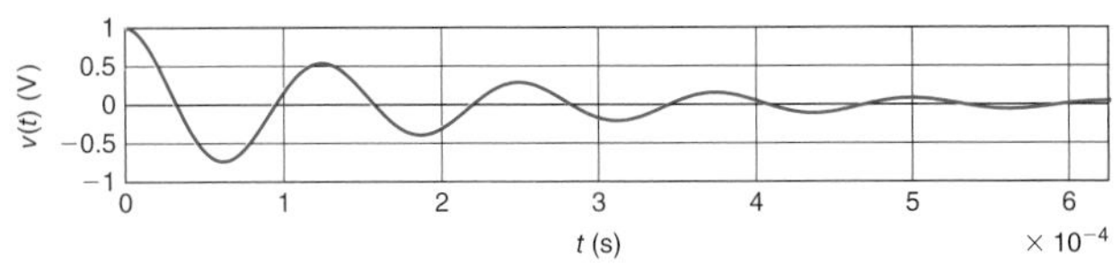

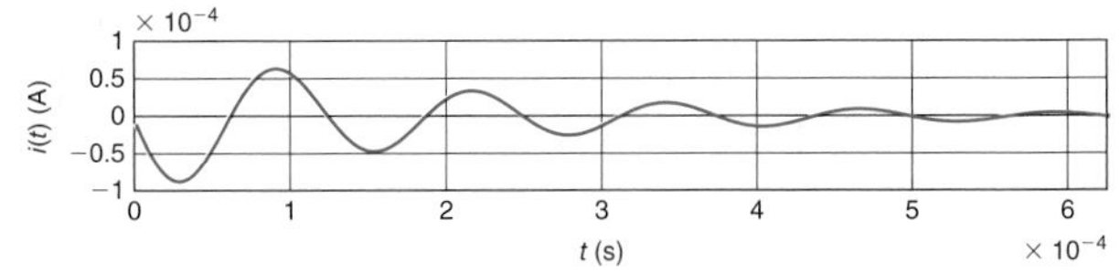

6.7 $i(t) = -C \times 2\pi \times 2000 \times \sin(2\pi \times 2000t) = -0.6283 \times 10^{-3} \times \sin(2\pi \times 2000t)$

$p(t) = -0.6283 \times 10^{-3} \times \sin(2\pi \times 2000t) \times \cos(2\pi \times 2000t) = -\pi \times 10^{-4} \sin(2\pi \times 4000t)$

$w(t) = 0.5 \times 5 \times 10^{-8} \times \cos^2(2\pi \times 2000t)$

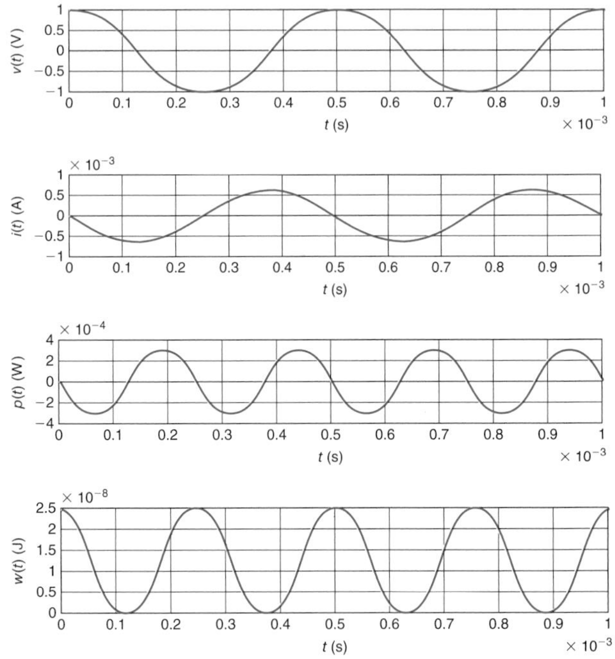

6.9

$$i_C(t) = \begin{cases} 23.3, & 0 \le t < 1 \\ -46.6, & 1 \le t < 2 \\ 23.3, & 2 \le t < 4 \\ -11.65, & 4 \le t < 6 \\ 0, & otherwise \end{cases} \text{ mA}$$

$$p_C(t) = \begin{cases} 2.33t, & 0 \le t < 1 \\ 9.32t - 13.98, & 1 \le t < 2 \\ 2.33t - 6.99, & 2 \le t < 4 \\ 0.5825t - 3.495, & 4 \le t < 6 \\ 0, & otherwise \end{cases} \text{ W}$$

$$w_C(t) = 0.29125 \begin{cases} 1.165t^2, & 0 \le t < 1 \\ 4.66t^2 - 13.98t + 10.485, & 1 \le t < 2 \\ 1.165t^2 - 6.99t + 10.485, & 2 \le t < 4 \\ 4.66t^2 - 3.495t + 10.485, & 4 \le t < 6 \\ 0, & otherwise \end{cases} \text{ J}$$

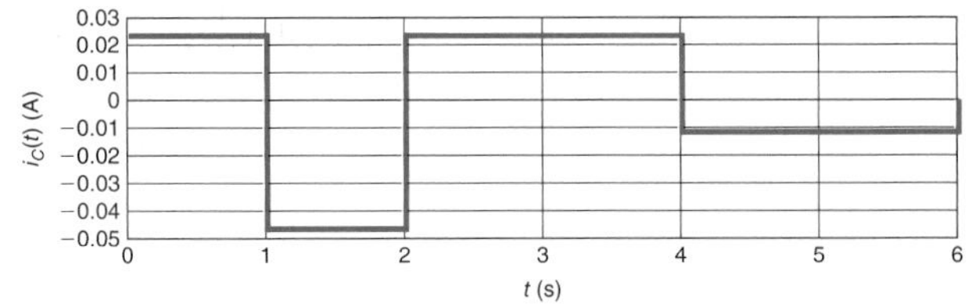

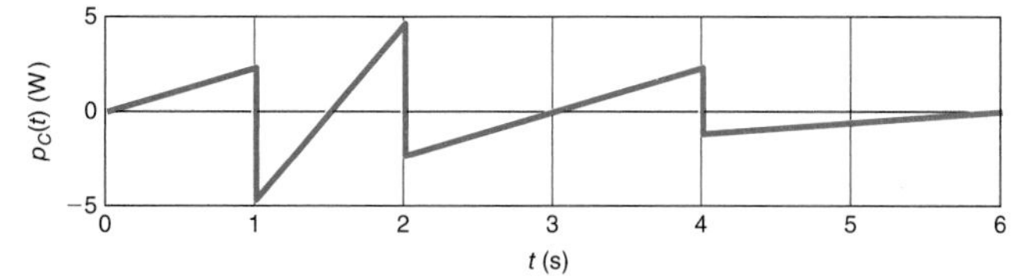

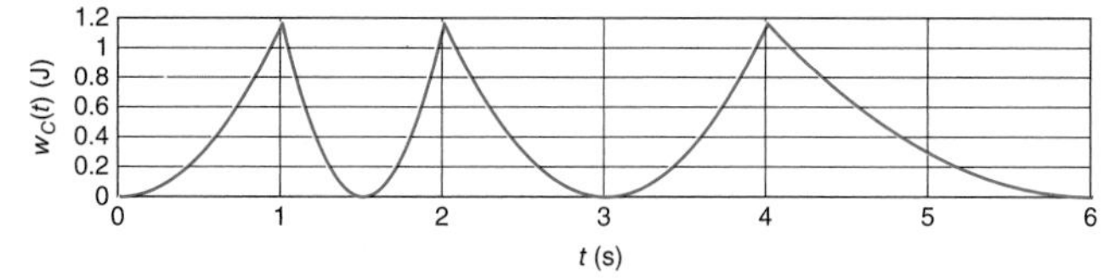

Power is delivered: $1 \le t < 1.5, 2 \le t < 3, 4 \le t < 6$.

6.11 $v_C(t) = 80\,(1 - \exp(-500t))\,u(t)$ V

$p_C(t) = v_C(t)i_C(t) = 0.08[\exp(-500t) - \exp(-1000t)]\,u(t)$ W

$w_C(t) = 0.0000000125 \times 80^2[1 - \exp(-500t)]^2 = 8 \times 10^{-5}[1 - \exp(-500t)]^2\,u(t)$ J

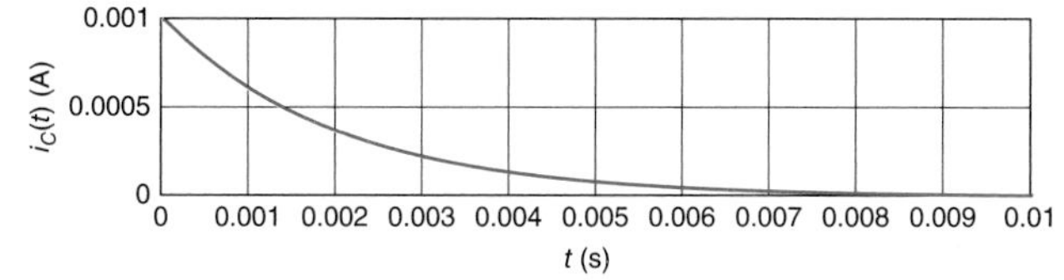

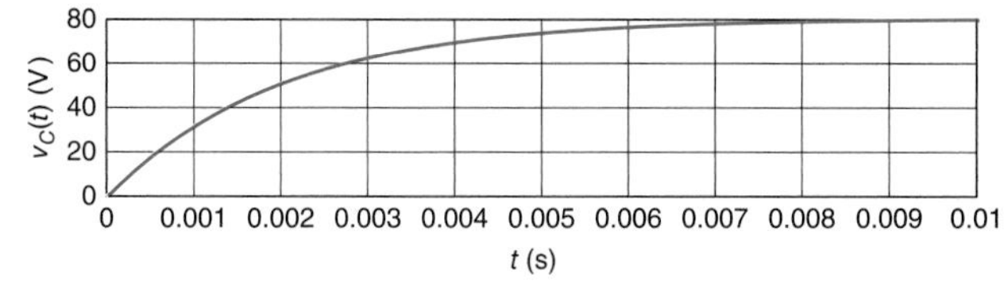

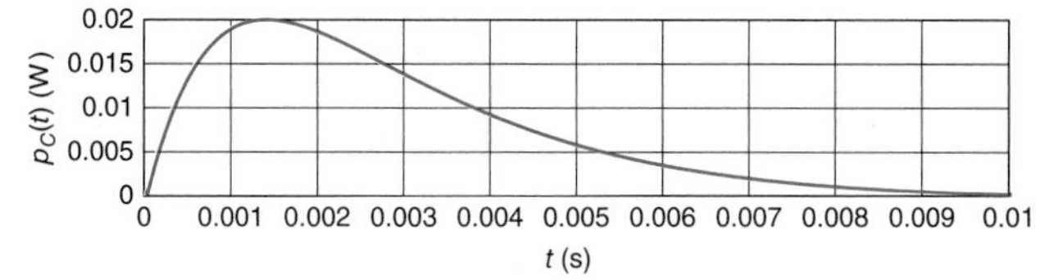

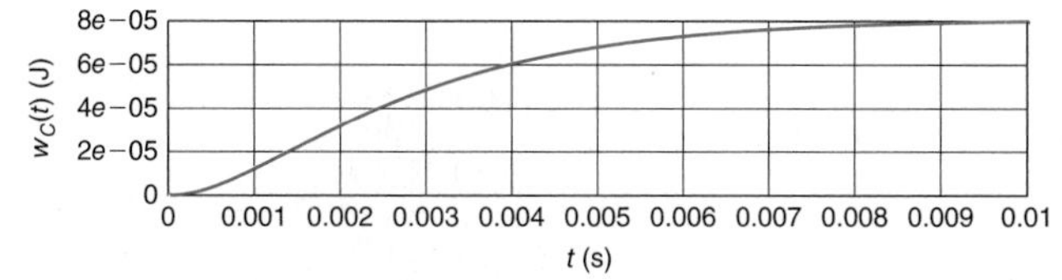

6.13 $v_C(t) = 0.04\,[\sin(2500t)]^2\,u(t)$ V

$p(t) = [\sin(2500t)^2 \sin(5000t)]/25{,}000 = 4 \times 10^{-5} \times \sin(5000t)[\sin(2500t)]^2\,u(t)$ W

$w(t) = 0.000000008\,[\sin(2500.0t)]^4 = 8 \times 10^{-9} \times [\sin(2500t)]^4\,u(t)$ J

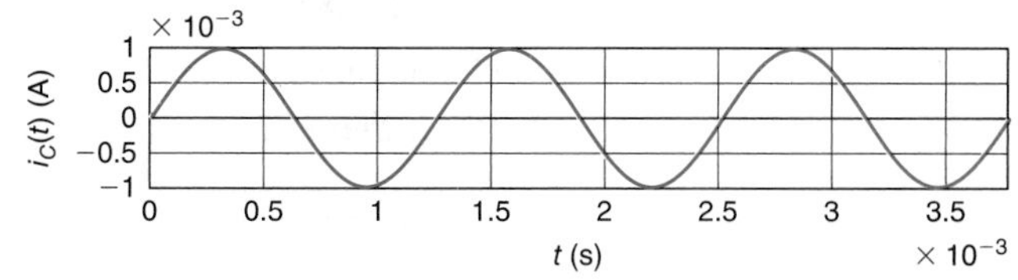

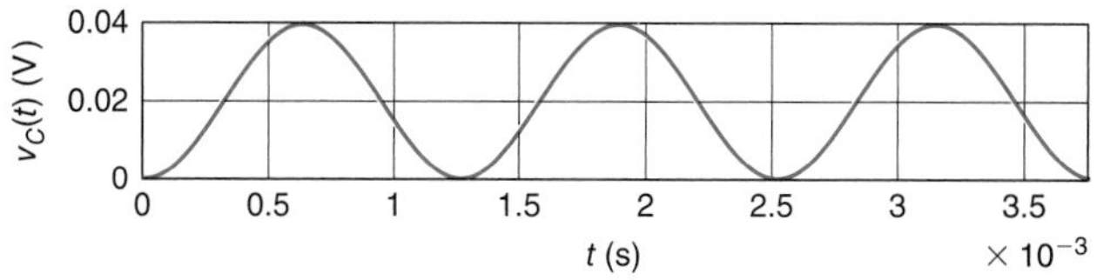

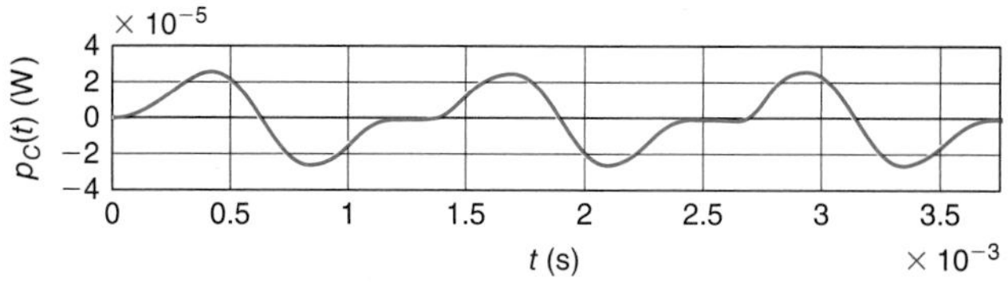

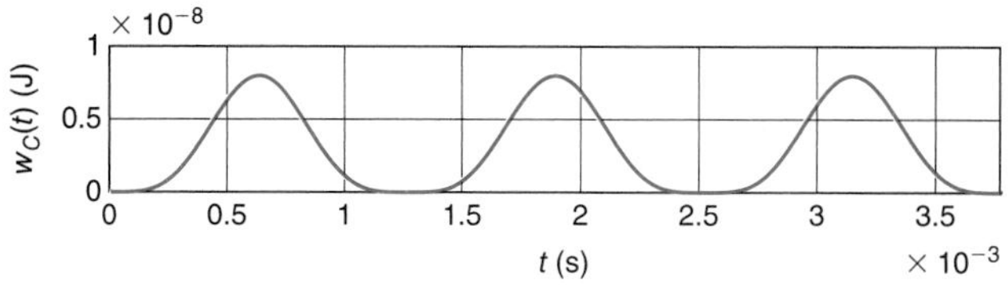

6.15
$$v_C(t) = \begin{cases} 3.3333t^2, & 0 \le t < 1 \\ 6.6667t - 3.3333, & 1 \le t < 3 \\ -3.3333t^2 + 26.6667t - 33.3333, & 3 \le t < 4 \\ 20, & 4 \le t \end{cases} \text{ V}$$

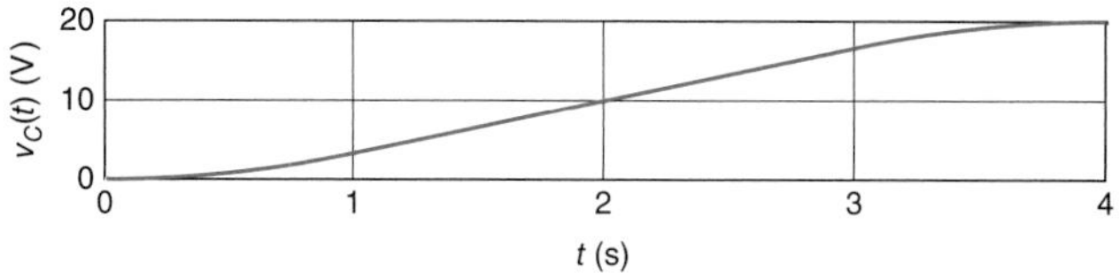

6.17
$$i_C(t) = \begin{cases} 1.08, & 0 \le t < 1 \\ -2.16, & 1 \le t < 2 \\ 2.16, & 2 \le t < 3 \\ -0.54, & 3 \le t < 5 \\ 0, & \textit{otherwise} \end{cases} \text{ mA}$$

$$p_C(t) = \begin{cases} 0.00432t, & 0 \le t < 1 \\ 0.01728t - 0.02592, & 1 \le t < 2 \\ 0.01728t - 0.0432, & 2 \le t < 3 \\ 0.00108t - 0.0054, & 3 \le t < 5 \\ 0, & \textit{otherwise} \end{cases} \text{ W}$$

$$w_C(t) = \begin{cases} 0.00216t^2, & 0 \le t < 1 \\ 0.00864t^2 - 0.02592t + 0.01944, & 1 \le t < 2 \\ 0.00864t^2 - 0.0432t + 0.054, & 2 \le t < 3 \\ 0.00054t^2 - 0.0054t + 0.0135, & 3 \le t < 5 \\ 0, & \textit{otherwise} \end{cases} \text{ J}$$

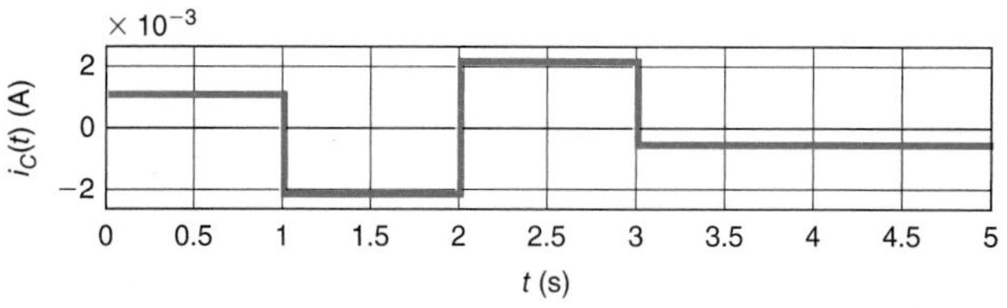

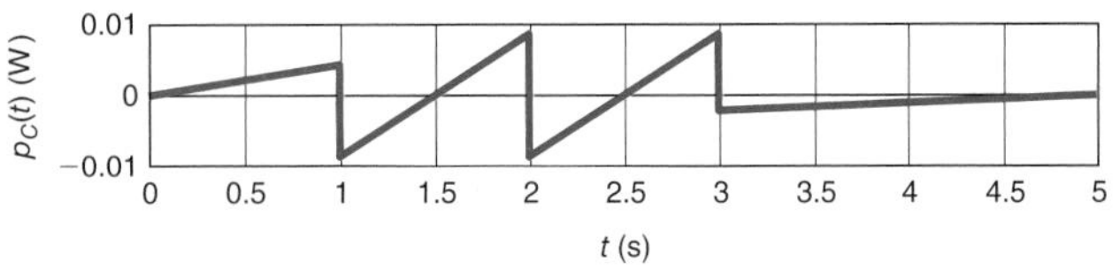

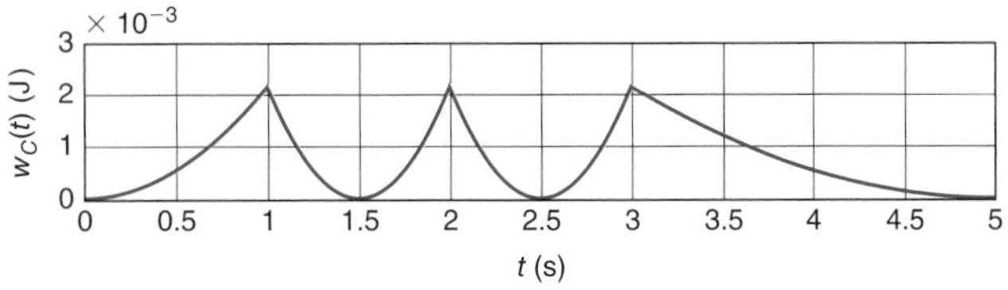

6.19 $$i_C(t) = \begin{cases} 0.014, & 0 \le t < 2 \\ -0.028, & 2 \le t < 4 \\ 0.014, & 4 \le t < 6 \\ 0, & otherwise \end{cases} \text{ A}$$

$$p_C(t) = \begin{cases} 0.56t, & 0 \le t < 2 \\ 2.24t - 6.72, & 2 \le t < 4 \\ 0.56t - 3.36, & 4 \le t < 6 \\ 0, & otherwise \end{cases} \text{ W}$$

$$w_C(t) = \begin{cases} 0.28t^2, & 0 \le t < 2 \\ 1.12t^2 - 6.72t + 10.08, & 2 \le t < 4 \\ 0.28t^2 - 3.36t + 10.08, & 4 \le t < 6 \\ 0, & otherwise \end{cases} \text{ J}$$

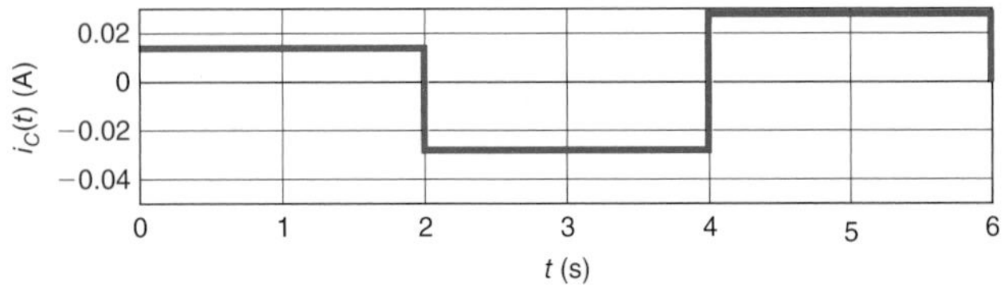

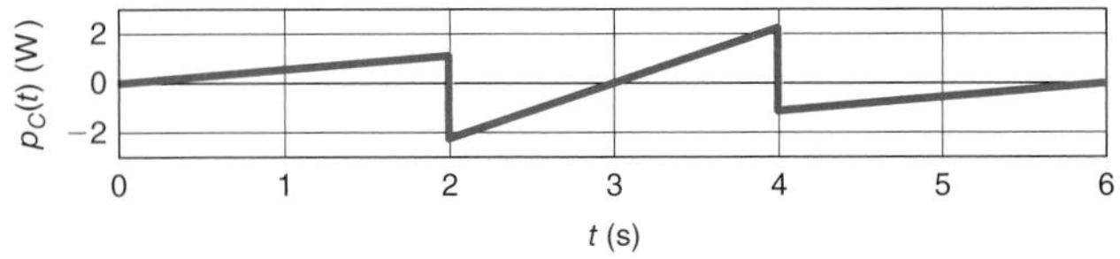

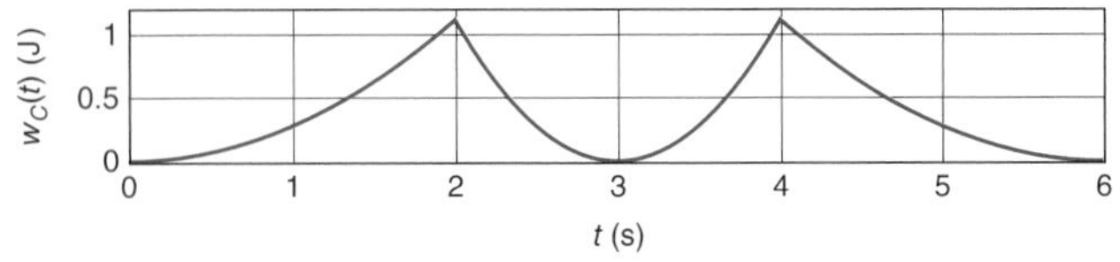

6.21 $$i_C(t) = \begin{cases} 0.0099, & 0 \le t < 2 \\ -0.0198, & 2 \le t < 4 \\ 0.0099, & 4 \le t < 6 \\ 0, & otherwise \end{cases} \text{ A}$$

$$p_C(t) = \begin{cases} 0.297t, & 0 \le t < 2 \\ 1.188t - 3.564, & 2 \le t < 4 \\ 0.297t - 1.782, & 4 \le t < 6 \\ 0, & otherwise \end{cases} \text{ W}$$

$$w_C(t) = \begin{cases} 0.1485t^2, & 0 \le t < 2 \\ 0.594t^2 - 3.564t + 5.346, & 2 \le t < 4 \\ 0.1485t^2 - 1.782t + 5.346, & 4 \le t < 6 \\ 0, & otherwise \end{cases} \text{ J}$$

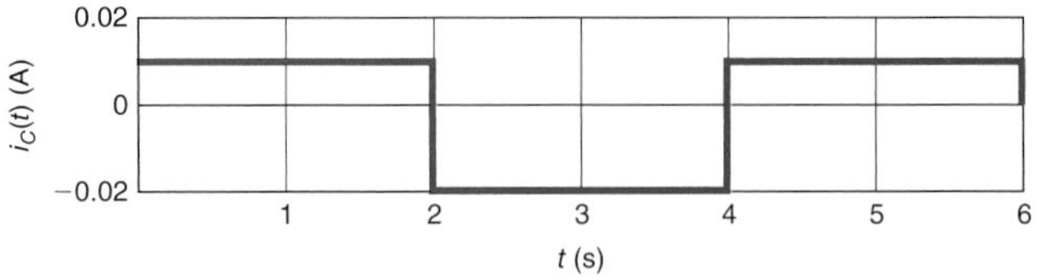

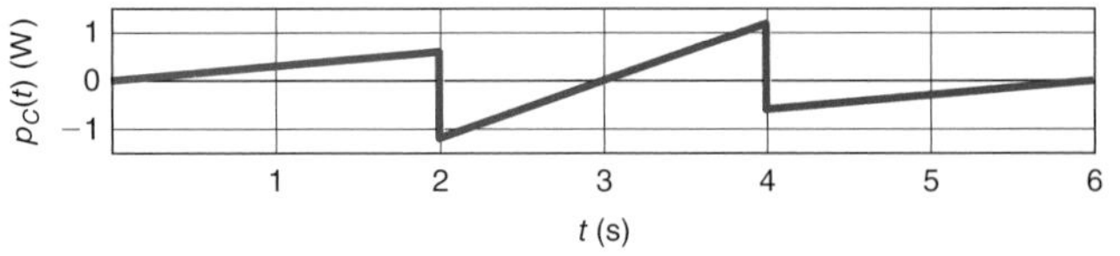

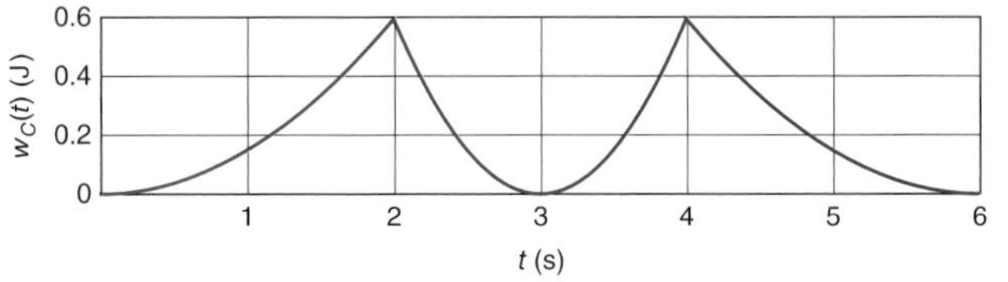

6.23 $C_{eq} = 6.3\ \mu F$
6.25 $C_{eq} = 2.2634\ nF$
6.27 $C_{eq} = 9.8652\ nF$
6.29 $C_{eq} = 3.6364\ nF$

6.31 $v_L(t) = L\dfrac{di_L(t)}{dt} = 0.0012 \times (-2000) \times e^{-2000t}$

$u(t) = -2.4e^{-2000t}\,u(t)$ V

$p(t) = i_L(t)v_L(t) = -2.4e^{-4000t}u(t)$ W

$w(t) = \dfrac{1}{2}Li_L^2(t) = 0.0006e^{-4000t}u(t) = 0.6e^{-4000t}u(t)$ mJ

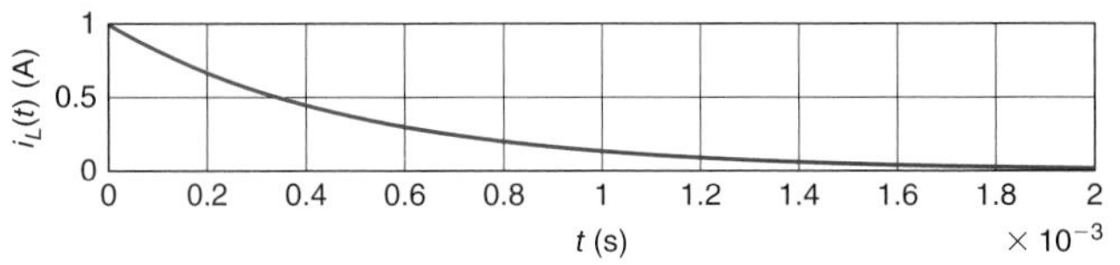

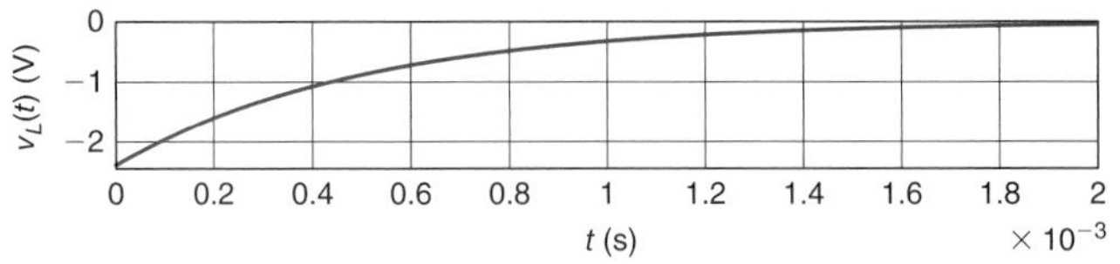

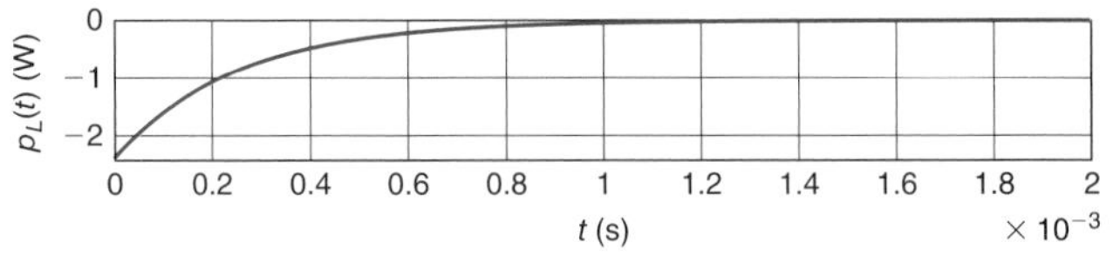

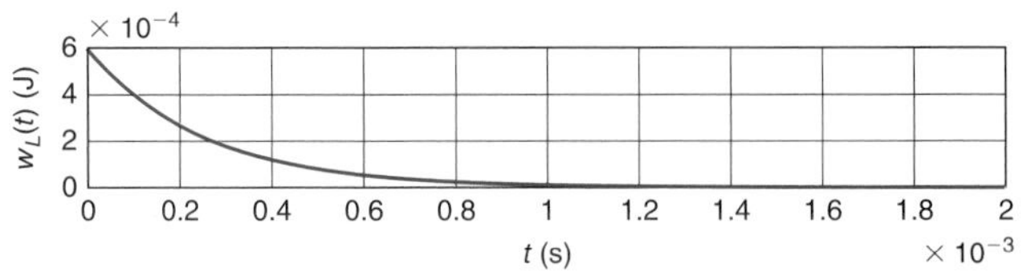

6.33 $v_L(t) = L\dfrac{di_L(t)}{dt} = 0.00027 \times (2\pi 5000) \times \cos(2\pi 5000t)u(t) = 8.4823\cos(2\pi 5000t)u(t)$ V

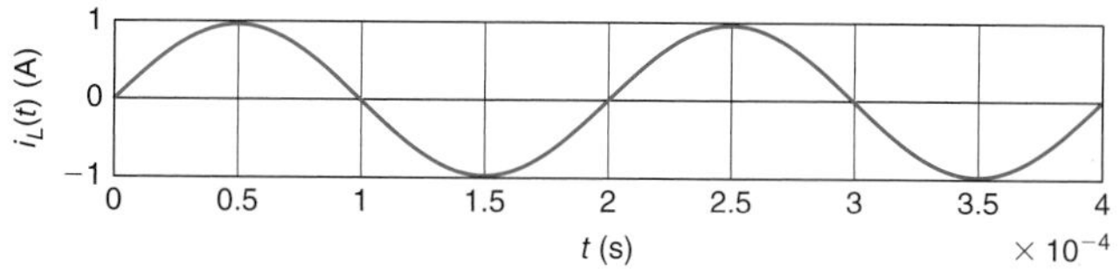

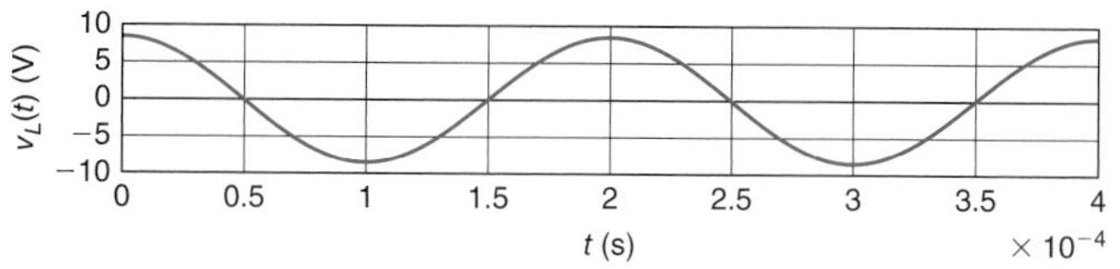

6.35 $v_L(t) = -0.044 \times \exp(-2000t) \times \sin(2\pi \times 1000t) + 0.1382 \times \exp(-2000t) \times \cos(2\pi \times 1000t)$ V

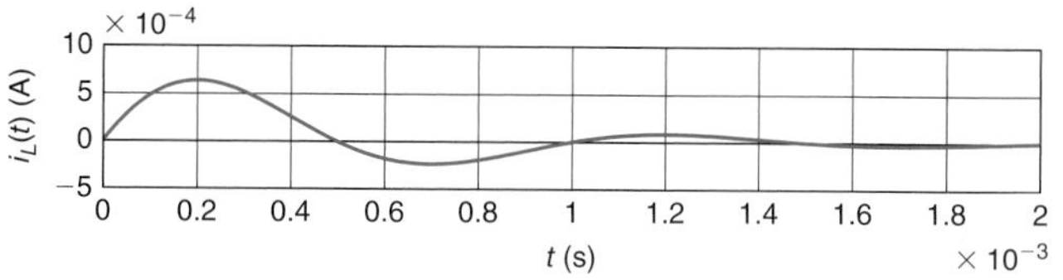

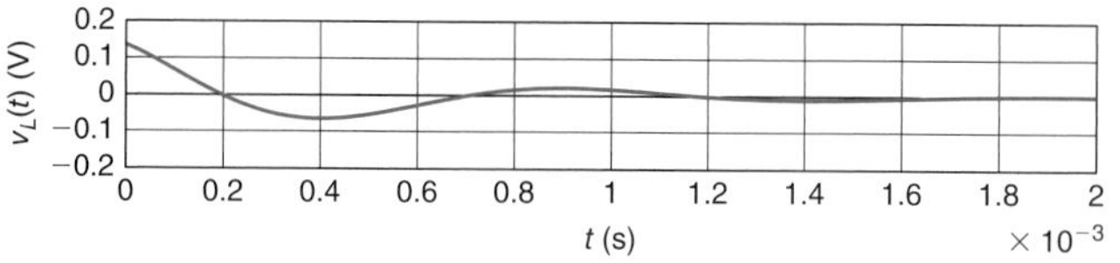

6.37

$$v_L(t) = \begin{cases} -0.56, & 0 \le t < 1 \\ 1.12, & 1 \le t < 2 \\ -0.56, & 2 \le t < 3 \\ 0, & otherwise \end{cases} \text{ V}$$

$$p_L(t) = \begin{cases} 5.6t, & 0 \le t < 1 \\ 22.4t - 33.6, & 1 \le t < 2 \\ 5.6t - 16.8, & 2 \le t < 3 \\ 0, & otherwise \end{cases} \text{ W}$$

$$w(t) = \begin{cases} 2.8t^2, & 0 \le t < 1 \\ 11.2t^2 - 33.6t + 25.2, & 1 \le t < 2 \\ 2.8t^2 - 16.8t + 25.2, & 2 \le t < 3 \\ 0, & otherwise \end{cases} \text{ J}$$

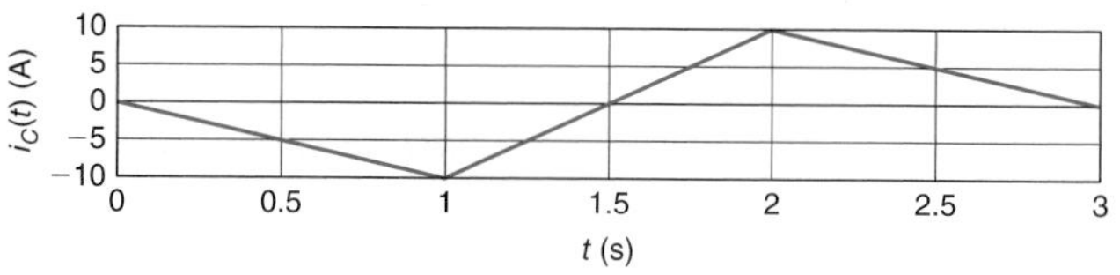

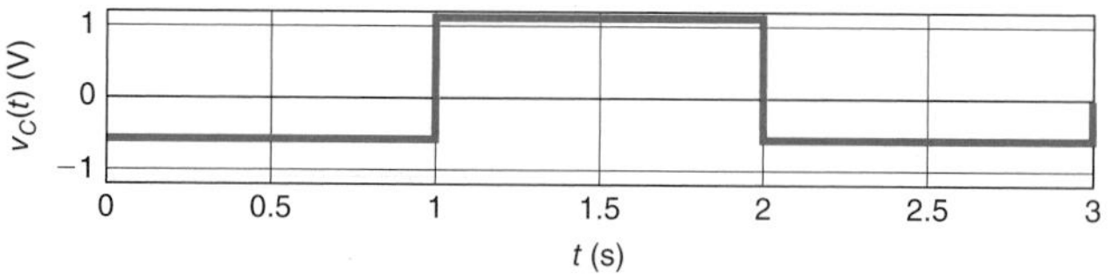

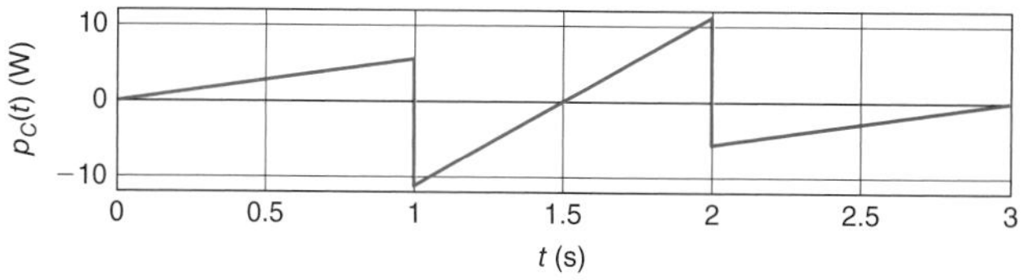

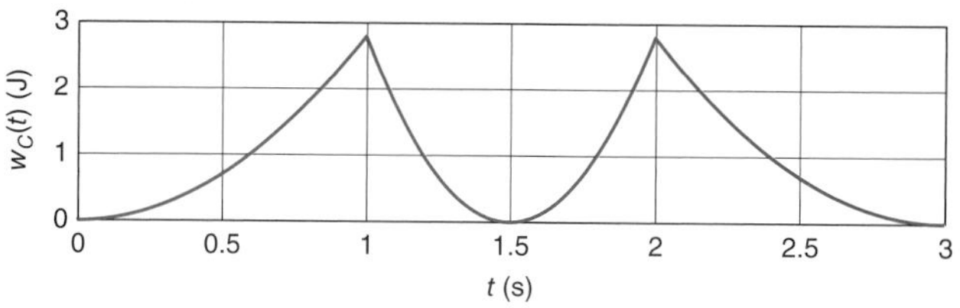

6.39 $i_L(t) = 11.1111\,(1 - \exp(-5t))\,u(t)$ A

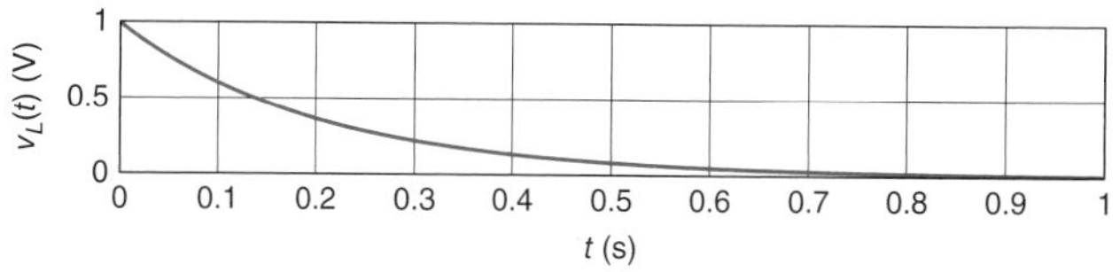

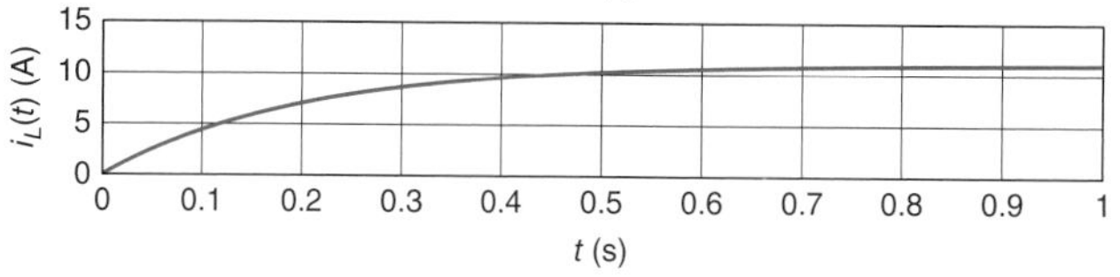

6.41 $i_L(t) = 2.5[\sin(25t)]^2\,u(t)$ A, $p(t) = 2.5[\sin(25t)]^2 \sin(50t)u(t)$ W, $w(t) = 0.05[\sin(25t)]^4\,u(t)$ J

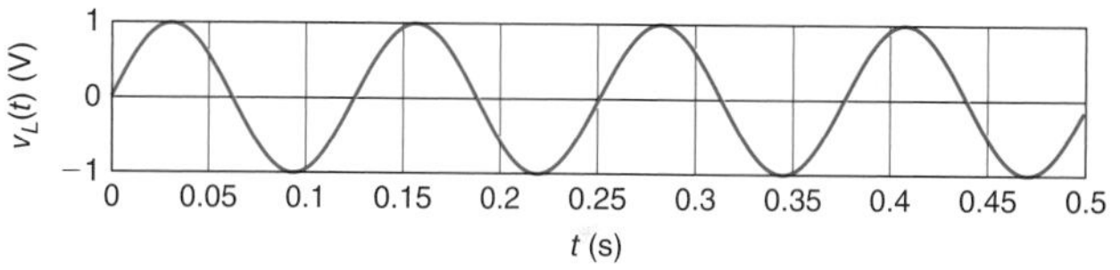

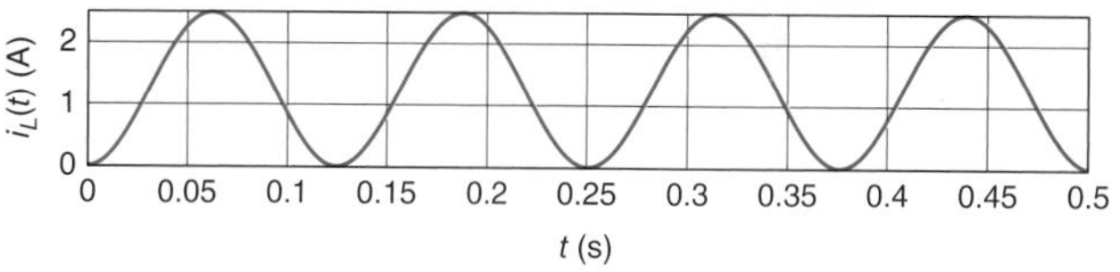

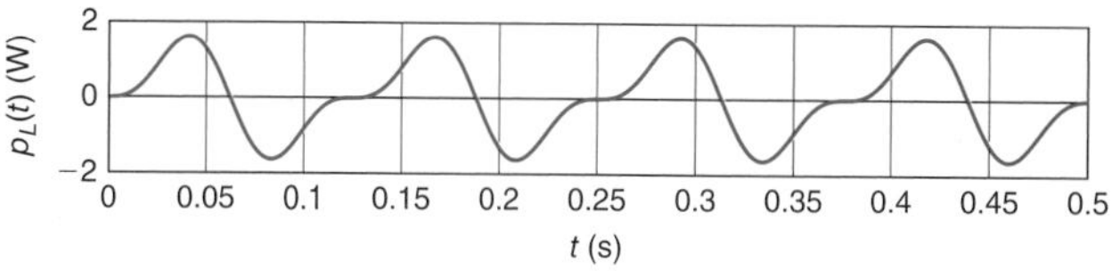

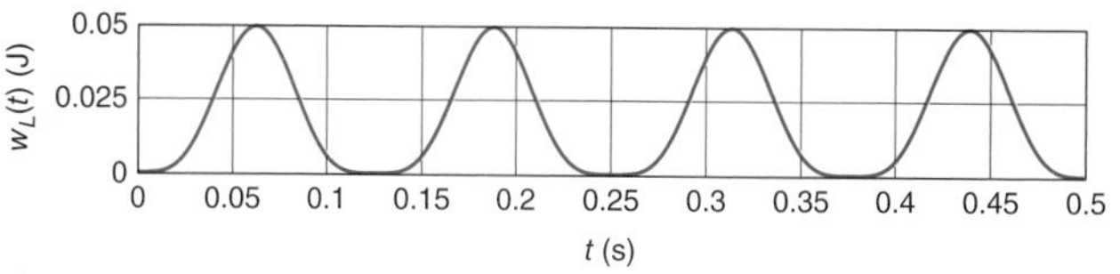

6.43

$$i_L(t) = \begin{cases} 2.272727t^2, & 0 \le t < 2 \\ -4.545455t^2 + 27.27273t - 27.27273, & 2 \le t < 3 \\ 13.6364, & 3 \le t \\ 0, & elsewhere \end{cases} \text{ A}$$

$$p_L(t) = \begin{cases} 3.4091t^3, & 0 \le t < 2 \\ 13.6364t^3 - 122.7273t^2 + 327.2727t - 245.4545, & 2 \le t < 3 \text{ W} \\ 0, & elsewhere \end{cases}$$

$$w_L(t)=\begin{cases}0.8523t^4, & 0\le t<2\\ 3.4091t^4-40.9091t^3+163.6364t^2-245.4545t+122.7273, & 2\le t<3\\ 30.6818, & 3\le t\\ 0, & elsewhere\end{cases}\ \text{J}$$

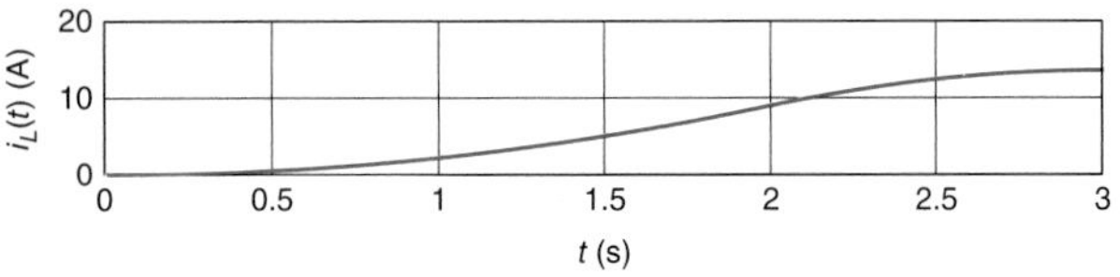

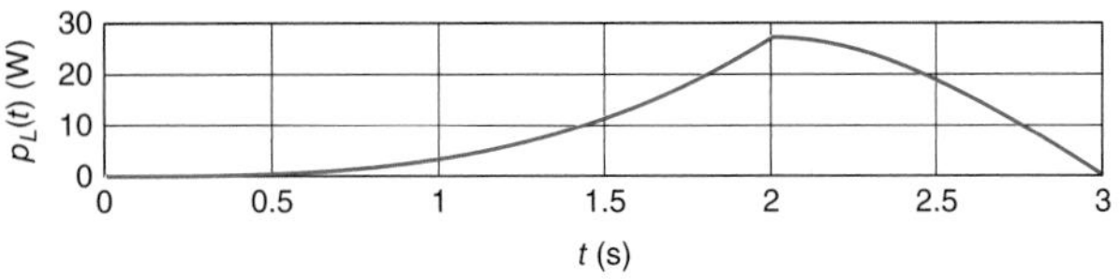

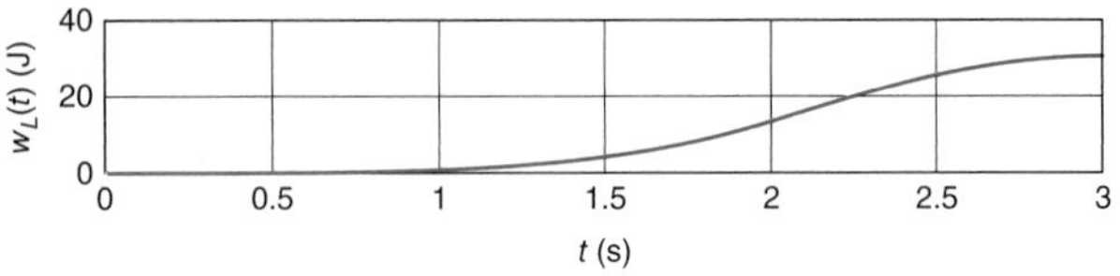

6.45 $L_{eq} = 132.2433$ mH
6.47 $L_{eq} = 21.1111\ \mu$H

CHAPTER 7

7.1 a. $v(t) = V_0 e^{-\frac{t}{RC}} = V_0 e^{-\frac{t}{\tau}} = 5e^{-\frac{t}{0.0015}} u(t)$ V
b. $i(t) = -1.6667e^{-\frac{t}{0.0015}} u(t)$ mA

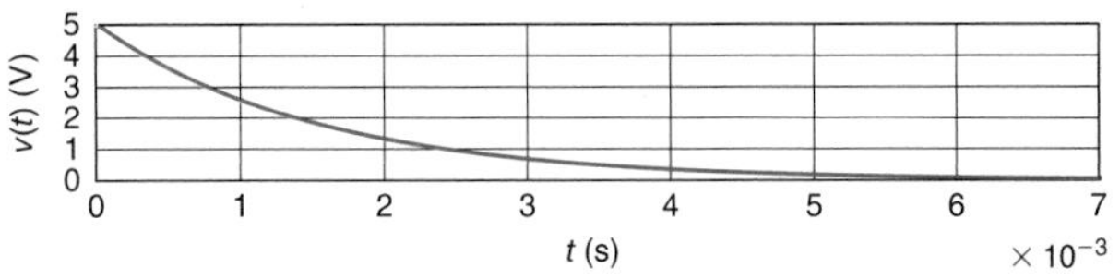

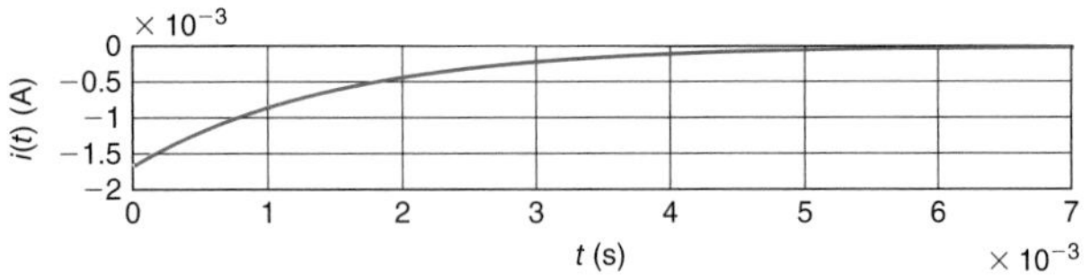

7.3 a. $v(t) = 5e^{\frac{-t}{0.036}} u(t)$ V $= 5e^{-27.7778t} u(t)$ V
b. $v_o(t) = 3.3333e^{\frac{-t}{0.036}} u(t)$ V

7.5 a. $v(t) = 5e^{\frac{-t}{0.2}} u(t)$ V $= 5e^{-5t} u(t)$ V
b. $v_o(t) = 2.25e^{\frac{-t}{0.2}} u(t)$ V

7.7 a. $v(t) = V_o e^{-\frac{t}{RC}} = V_o e^{-\frac{t}{\tau}} = 12e^{-\frac{t}{0.0006}} u(t)$ V
b. $i(t) = -0.6e^{-\frac{t}{0.6\times10^{-3}}} u(t)$ mA

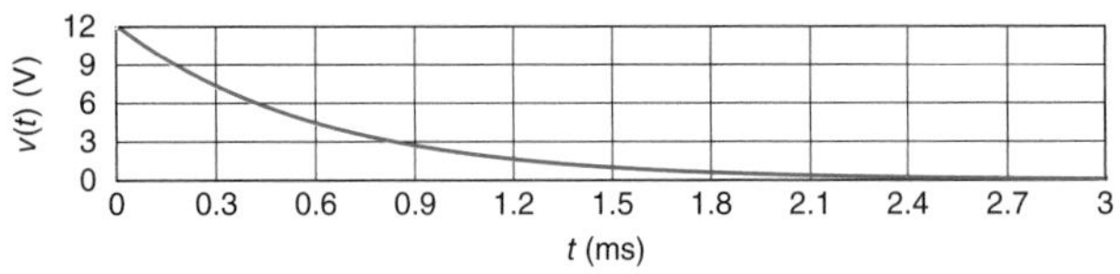

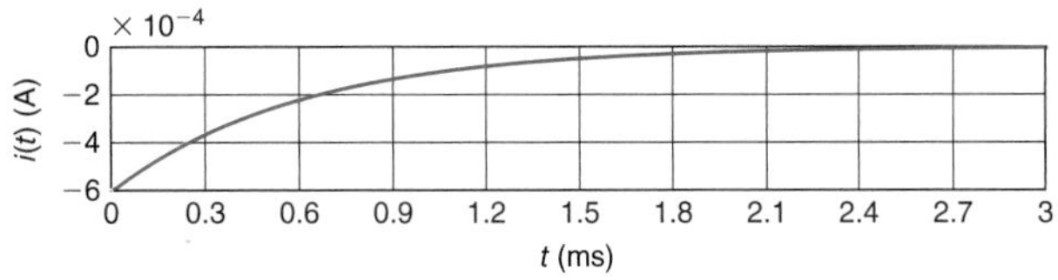

7.9 $R_{th} = 16/11\ k\Omega = 1.4545\ k\Omega, \tau = 1.4545$ ms,
$v(t) = 8e^{-\frac{t}{16/11\times10^{-3}}} = 8e^{-\frac{t}{1.4545\times10^{-3}}} u(t)$

7.11 a. $v(t) = (10 - 7e^{-\frac{t}{0.004}})\, u(t)$,
b. $i(t) = 87.5e^{-\frac{t}{0.004}} u(t)\ \mu$A

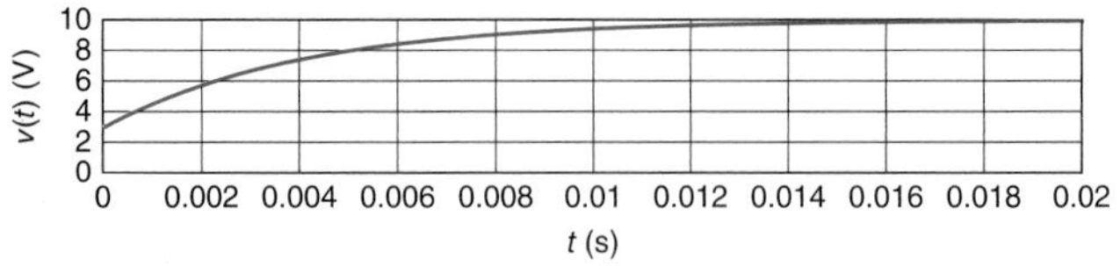

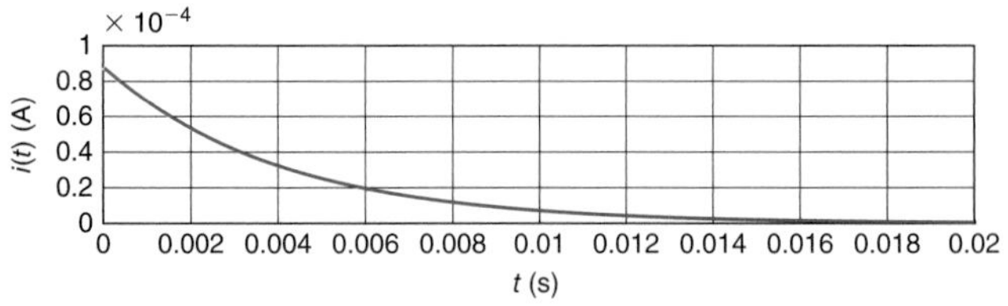

7.13 $v(t) = (7 - 5e^{-10t})\, u(t)$ V
7.15 $v(t) = (12 - 9e^{-50t})\, u(t)$ V
7.17 a. $v(t) = (10 - 6e^{-\frac{t}{3\times10^{-3}}})u(t)$ V,
b. $i(t) = 1.2e^{-\frac{t}{0.003}} u(t)$ mA

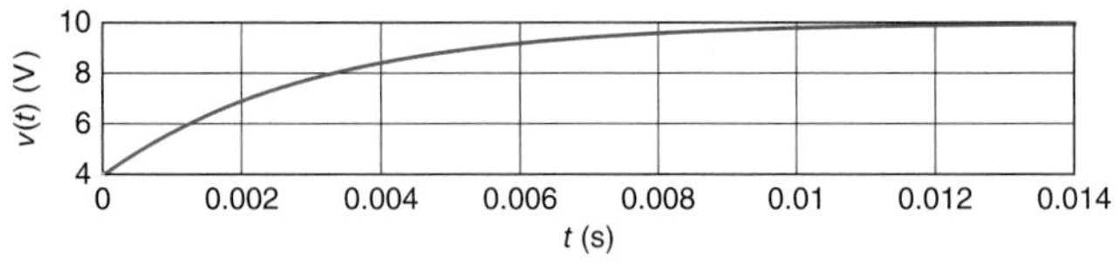

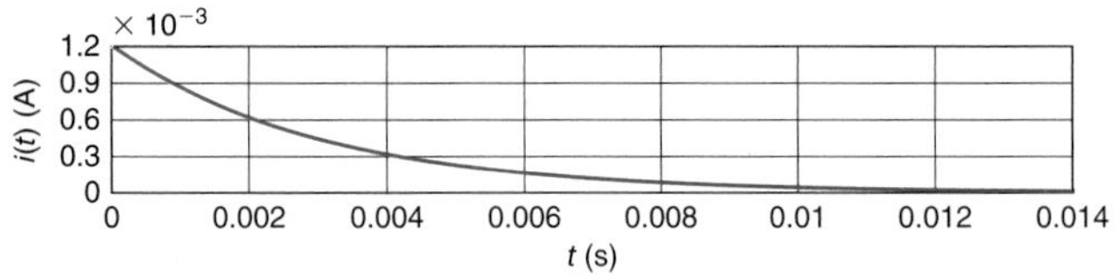

7.19 $v(t) = (9.375 - 6.375e^{-20t})\, u(t)$ V

7.21 a. $V_0 = 15$ V
b. $v(t) = 15e^{-\frac{t}{2\times10^{-3}}}$ V for $0 \le t < 4$ ms. At $t = 4$ ms, $v(t) = 15e^{-\frac{4\times10^{-3}}{2\times10^{-3}}} = 2.03$ V
c. $v(t) = (4 - 1.97e^{-\frac{t-4\times10^{-3}}{1\times10^{-3}}})u(t - 4\text{ ms})$ V for 4 ms $\le t$
d.

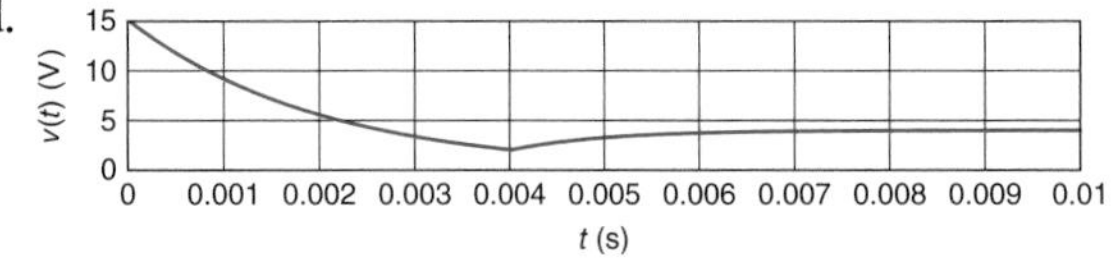

7.23 a. $V_0 = 8$ V
b. $v(\infty) = 24$ V
c. $R_{eq} = 10/3\ k\Omega = 3.3333\ k\Omega$
d. $v(t) = (24 - 16e^{-60t})\, u(t)$ V

7.25 a. $V_0 = 16$ V
b. $v(t) = V_0 e^{-\frac{t}{\tau_1}} u(t) = 16e^{-\frac{t}{0.112}} [u(t) - u(t - 0.12)]$
c. $v(0.12) = 16e^{-\frac{0.12}{0.112}} = 5.4803$ V
d. $v(t) = (8 - 2.5197e^{-\frac{t-0.12}{0.08}})u(t - 0.12)$ V

7.27 a. $i(t) = 5e^{-8000t} u(t)$ mA
b. $i_1(t) = -4e^{-8000t} u(t)$ mA

7.29 a. $i(t) = 1e^{-\frac{t}{1.5\times10^{-5}}} u(t)$ mA
b. $v(t) = -20e^{-\frac{t}{1.5\times10^{-5}}} u(t)$ V

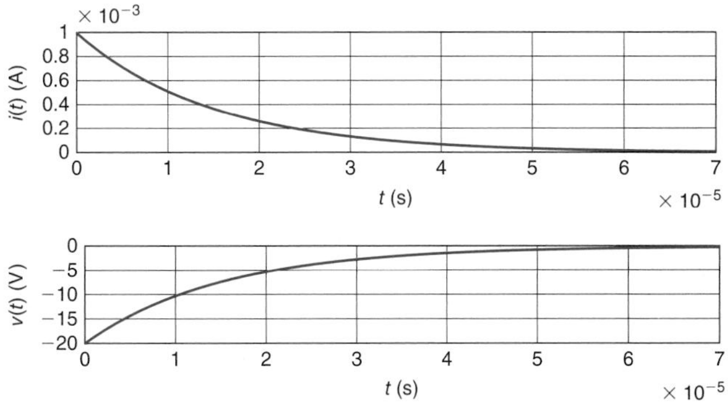

7.31 a. $I_0 = 0.6$ mA
b. $R_{eq} = 15\ k\Omega$
c. $\tau = L/R = 2 \times 10^{-5}s$
d. $i(t) = 0.6e^{-\frac{t}{2\times10^{-5}}}u(t)$ mA $= 0.6e^{-50{,}000t}u(t)$ mA

7.33 $R_{th} = 2.3846\ k\Omega, \tau = 4.1935 \times 10^{-8}s,$
$i(t) = 1 \times e^{-\frac{t}{4.1935\times10^{-8}}}u(t)$ mA

7.35 a. $I_0 = 0.9$ mA
b. $R_{eq} = 6.8966\ k\Omega$
c. $\tau = 6.525 \times 10^{-5}s = 65.25\ \mu s$
d. $i(t) = 0.9e^{-\frac{t}{6.525\times10^{-5}}}u(t)$ mA $= 0.9e^{-15{,}325.6705t}u(t)$ mA

7.37 $i(t) = (6 - 5e^{-\frac{t}{1\times10^{-4}}})u(t)$ mA

7.39 $i(t) = (2 - 1.5e^{-\frac{t}{2\times10^{-4}}})u(t)$ mA

7.41 $i(t) = (1.25 + 0.75e^{-\frac{t}{6.25\times10^{-9}}})u(t)$ mA
$v(t) = -6e^{-\frac{t}{6.25\times10^{-9}}}u(t)$ V

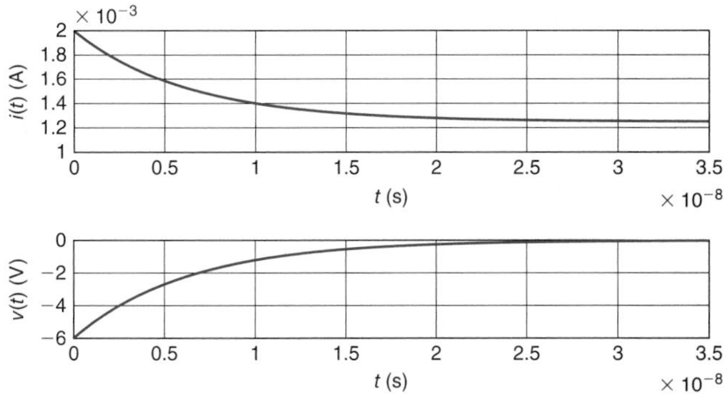

7.43 a. $i(\infty) = 11$ mA
b. $R_{eq} = 2.4\ k\Omega$
c. $\tau = 10^{-4}s$
d. $i(t) = (11 - 7e^{-\frac{t}{10^{-4}}})u(t)$ mA

7.45 a. $I_0 = 3$ mA
b. $i(t) = 3e^{-\frac{t}{10^{-5}}}$ mA. At $t = 10\ \mu s$,
$i(t) = 3e^{-\frac{10^{-5}}{10^{-5}}} = 1.10364$ mA
c. $i(t) = [4 - 2.89636e^{-50{,}000(t-10\times10^{-6})}]$
$\times\ u(t - 10 \times 10^{-6})$ mA
d.

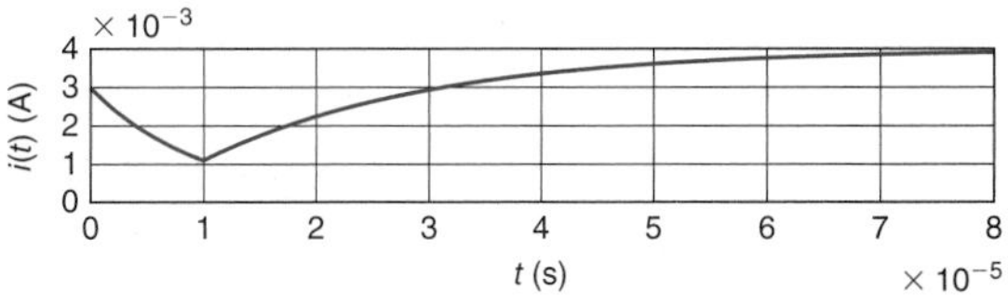

7.47 a. $i(\infty) = 2.6$ mA
b. $R_{eq} = 4\ k\Omega$
c. $\tau = 7.5 \times 10^{-5}s$
d. $i(t) = (2.6 - 1.6e^{-\frac{t}{7.5\times10^{-5}}})u(t)$ mA

7.49 a. $\frac{dv}{dt} + \frac{1}{RC}v = \frac{V_s}{RC}, \quad \frac{dv}{dt} + \frac{1}{0.5}v = \frac{10}{0.5}, \quad v(0) = 3$
b. $v(t) = (10 - 7e^{-\frac{t}{0.5}})u(t)$ V

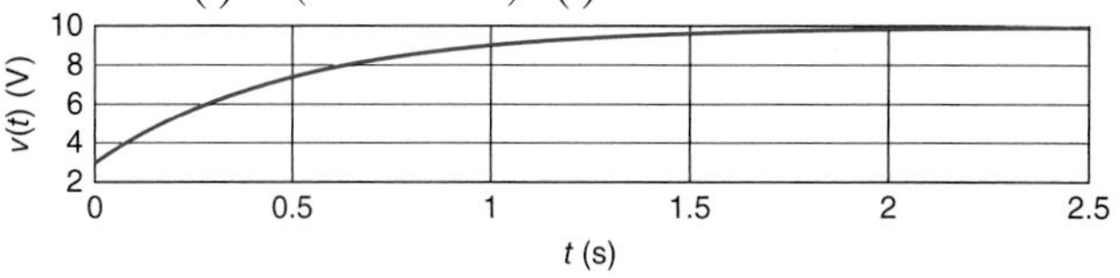

c.

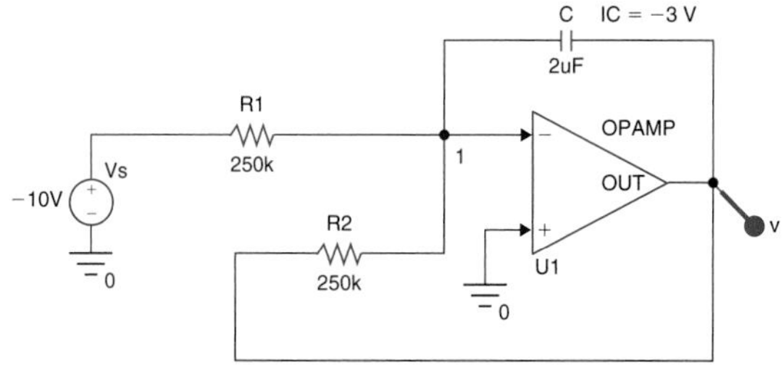

d.

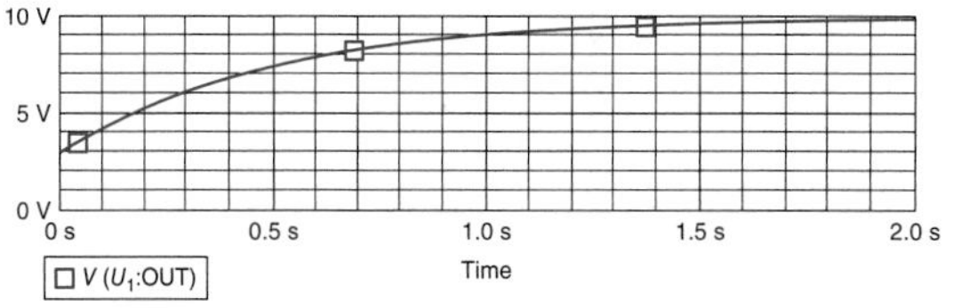

CHAPTER 8

8.1 a. $d^2v/dt^2 + 6000\ dv/dt + 8 \times 10^6 v = 0$
b. $\alpha = 3000$ Np/s, $\omega_0 = 2828.4271$ rad/s
c. $s^2 + 6000s + 8 \times 10^6 = 0$
d. $s_1 = -2000, s_2 = -4000$
e. $v(t) = (13e^{-2000t} - 9e^{-4000t})\ u(t)$ V
f. $i(t) = (-13e^{-2000t} + 18e^{-4000t})\ u(t)$ mA

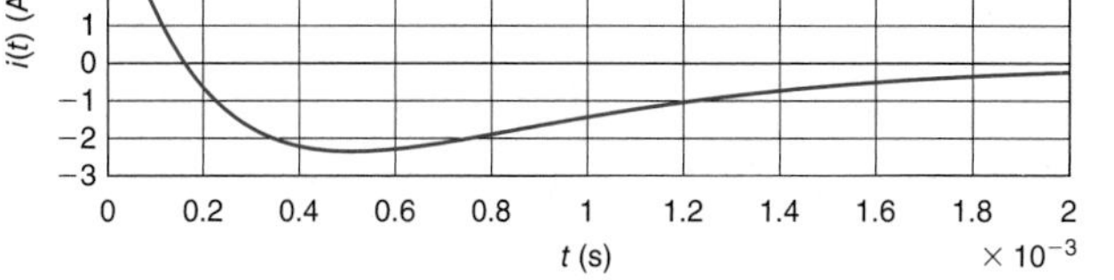

8.3 a. $d^2v/dt^2 + 5000\ dv/dt + 6.25 \times 10^7 v = 0$
b. $\alpha = 2500$ Np/s, $\omega_0 = 7905.69415$ rad/s
c. $s^2 + 5000s + 6.25 \times 10^7 = 0$
d. $s_1 = -2500 + j7500, s_2 = -2500 - j7500$
e. $v(t) = e^{-2500t}$ [7 cos(7500t) + 8.583333 sin(7500t)] V

f. $i(t) = e^{-2500t}\,[3\cos(7500t) - 4.73333\sin(7500t)]\,u(t)$ mA

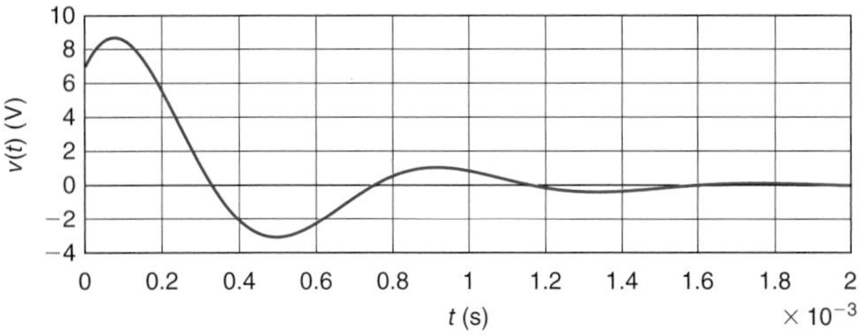

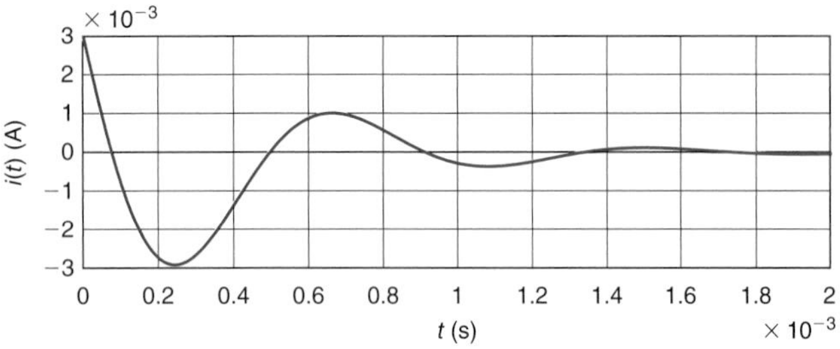

8.5 a. $d^2i/dt^2 + 3600\,di/dt + 2.88 \times 10^6 i = 0$

b. $\alpha = 1800$ Np/s, $\omega_0 = 1697.05627$ rad/s

c. $s^2 + 3600s + 2.88 \times 10^6 = 0$

d. $s_1 = -1200, s_2 = -2400$

e. $i(t) = [-51e^{-1200t} + 54e^{-2400t}]u(t)$ mA

f. $v_2(t) = (5.3125e^{-1200t} - 2.8125e^{-2400t})u(t)$ V

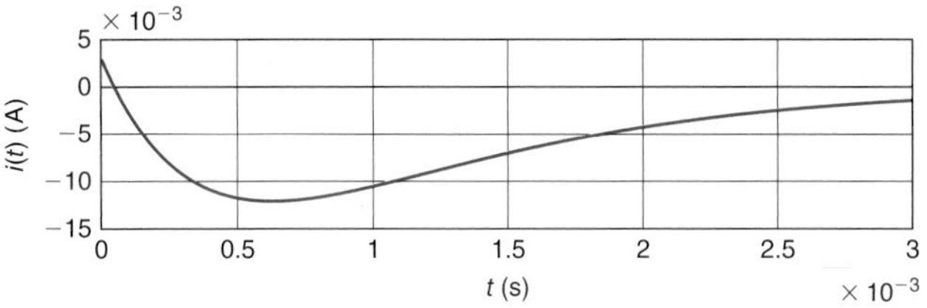

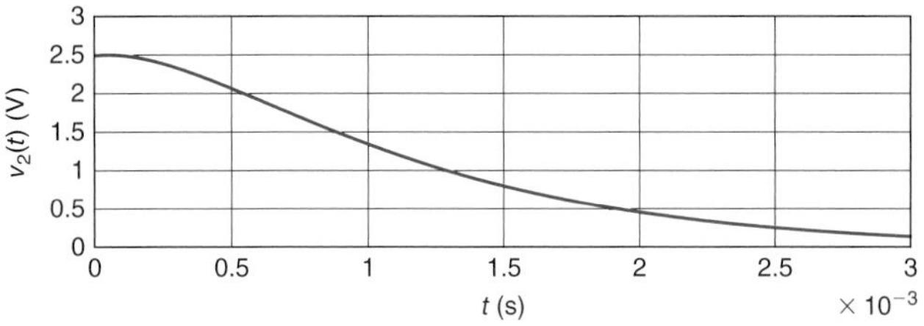

8.7 a. $d^2v/dt^2 + 3100\,dv/dt + 2.4 \times 10^6 v = 0$

b. $\alpha = 1550$ Np/s, $\omega_0 = 1549.1933$ rad/s

c. $s^2 + 3100s + 2.4 \times 10^6 = (s + 1500)(s + 1600) = 0$

d. $s_1 = -1500, s_2 = -1600$

e. $v(t) = [69.04348e^{-1500t} - 65.54348e^{-1600t}]u(t)$ V

f. $i(t) = [482.4e^{-1500t}\ 476.4e^{-1600t}]u(t)$ A

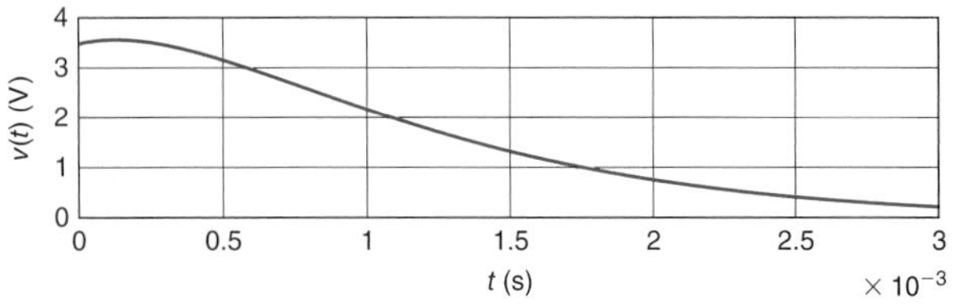

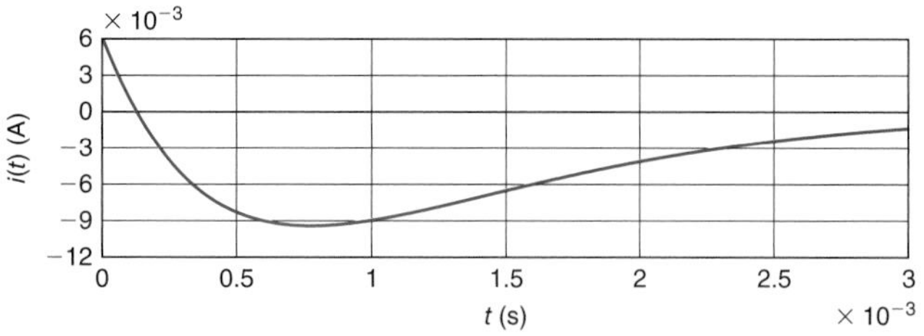

8.9 a. $d^2v/dt^2 + 1.75dv/dt + 0.75v = 0$

b. $\alpha = 0.875$ Np/s, $\omega_0 = 0.8660254$ rad/s

c. $s^2 + 1.75s + 0.75 = 0$

d. $s_1 = -0.75, s_2 = -1$

e. $v(t) = [14e^{-0.75t} - 9e^{-t}]u(t)$ V

f. $i(t) = [-7e^{-0.75t} + 9e^{-t}]u(t)$ A

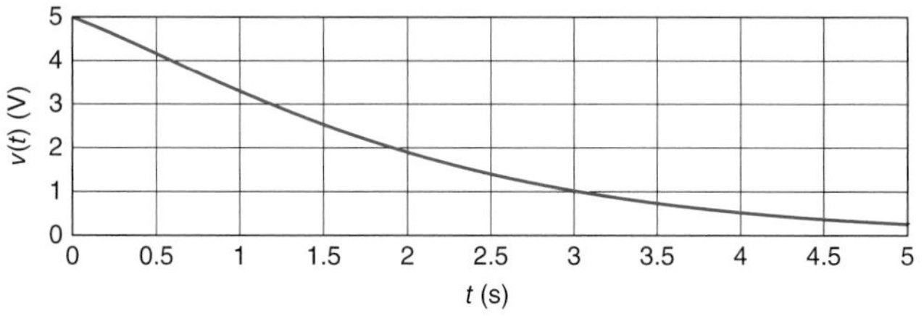

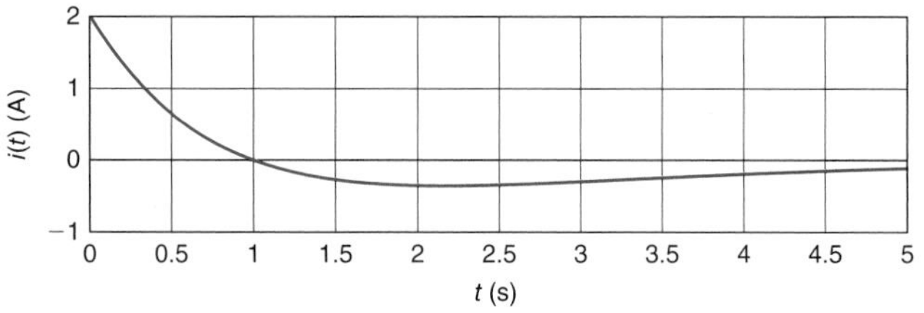

8.11 $I_0 = 2$ A, $v(0) = 1$ V, $\dfrac{dv(0)}{dt} = -8$ V/s, $s^2 + 2.5s + 4 = 0$, $\alpha = 1.25$, $\beta = 1.56125$, $v(t) = e^{-1.25t}[\cos(1.56125t) - 4.32346\sin(1.56125t)]u(t)$ V

8.13 $v(t) = e^{-t}\,[4\cos(1.391941t) - 12.2132 \times \sin(1.391941t)]u(t)$ V

8.15 a. $\alpha = 10{,}000$ Np/s

b. $\omega_0 = 10{,}000$ rad/s

c. $\dfrac{d^2i(t)}{dt^2} + \dfrac{1}{RC}\dfrac{di(t)}{dt} + \dfrac{1}{LC}i(t) = 0$

d. $s^2 + 20{,}000s + 1 \times 10^8 = 0$

e. $s_1 = -10{,}000, s_2 = -10{,}000$

f. $i(t) = (0.004e^{-10{,}000t} + 280te^{-10{,}000t})\,u(t)$ A

g. $v(t) = (12e^{-10{,}000t} - 140{,}000te^{-10{,}000t})\,u(t)$ V

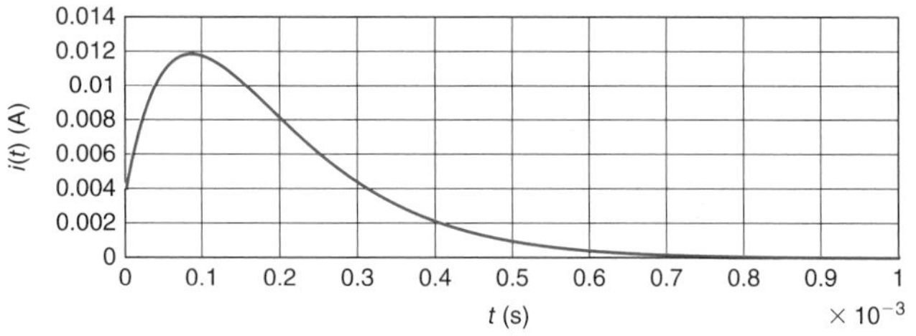

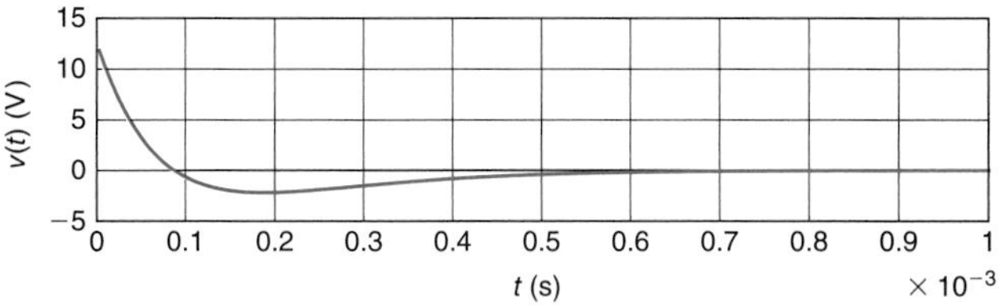

8.17 $i(t) = e^{-0.5t}[2\cos(1.9365t) + 0.5164\sin(1.9365t)]u(t)$ A

8.19 $i(t) = e^{-1.8333t}[2\cos(1.6245t) + 2.2572\sin(1.6245t)]u(t)$ A

8.21 $i(t) = e^{-0.625t}[3\cos(2.1469t) + 0.8733\sin(2.1469t)]u(t)$ A

8.23 a. $\alpha = 17{,}500$ Np/s

b. $\omega_0 = 15{,}811.3883$ rad/s

c. $\dfrac{d^2v(t)}{dt^2} + \dfrac{R}{L}\dfrac{dv(t)}{dt} + \dfrac{1}{LC}v(t) = \dfrac{V_s}{LC}$

d. $s^2 + 35{,}000s + 2.5 \times 10^8 = 0$

e. $s_1 = -10{,}000,\ s_2 = -25{,}000$

f. $v(\infty) = 10$ V

g. $v(t) = (10 - 10e^{-10{,}000t} + 2e^{-25{,}000t})\,u(t)$ V

h. $i(t) = (8e^{-10{,}000t} - 4e^{-25{,}000t})\,u(t)$ mA

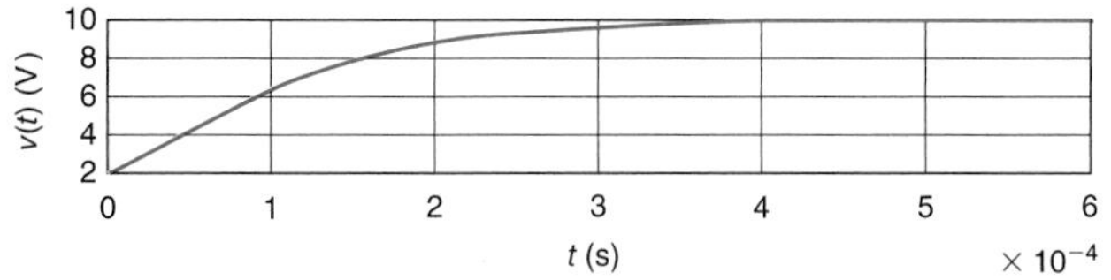

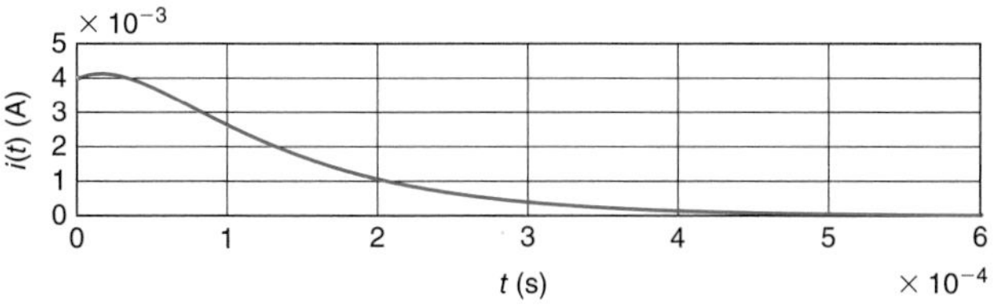

8.25 a. $\alpha = 20{,}000$ Np/s

b. $\omega_0 = 44{,}721.35955$ rad/s

c. $\dfrac{d^2i(t)}{dt^2} + \dfrac{R}{L}\dfrac{dv(t)}{dt} + \dfrac{1}{LC}v(t) = \dfrac{V_s}{LC}$

d. $s^2 + 40{,}000s + 2 \times 10^9 = 0$

e. $s_1 = -20{,}000 + j40{,}000,\ s_2 = -20{,}000 - j40{,}000$

f. $v(\infty) = 15$ V

g. $v(t) = [15 - 10e^{-20{,}000t}\cos(40{,}000t) + 15e^{-20{,}000t}\sin(40{,}000t)]\,u(t)$ V

h. $i(t) = [8e^{-20{,}000t}\cos(40{,}000t) + e^{-20{,}000t} \times \sin(40{,}000t)]\,u(t)$ mA

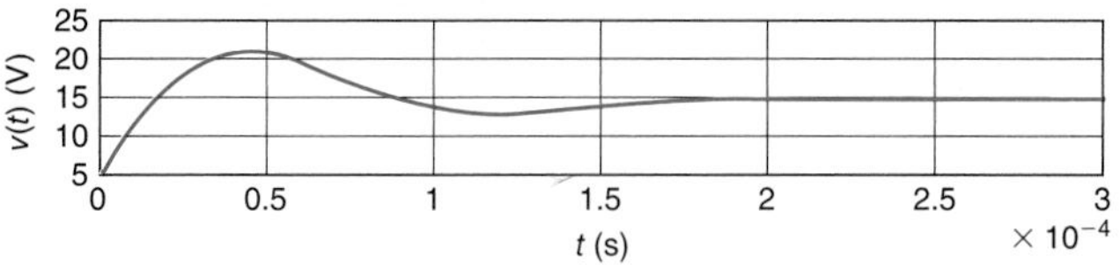

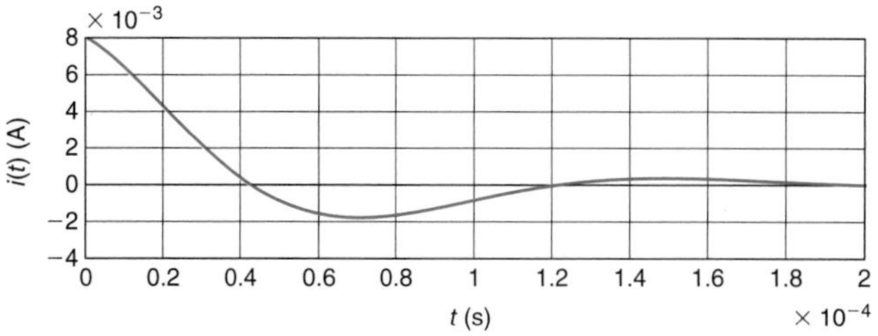

8.27 a. $v(t) = [16 - 12e^{-3t}\cos(4t) - 6.5\,e^{-3t}\sin(4t)]\,u(t)$

b. $i(t) = e^{-3t}[2\cos(4t) + 13.5\sin(4t)]\,u(t)$

8.29 $v(t) = \{4.5 + e^{-1.75t}[-8.5\cos(2.6339t) + 1.9458 \times \sin(2.6339t)]\}u(t)$ V

8.31 a. $\alpha = 50{,}000$ Np/s

b. $\omega_0 = 50{,}000$ rad/s

c. $\dfrac{d^2i(t)}{dt^2} + \dfrac{1}{RC}\dfrac{di(t)}{dt} + \dfrac{1}{LC}i(t) = \dfrac{I_s}{LC}$

d. $s^2 + 1 \times 10^5 s + 2.5 \times 10^9 = 0$

e. $s_1 = -50{,}000,\ s_2 = -50{,}000$

f. $i(\infty) = 8$ mA

g $i(t) = (0.008 - 0.006e^{-50{,}000t} - 237.5te^{-50{,}000t})u(t)$ A

h. $v(t) = L\dfrac{di(t)}{dt} = (5e^{-50{,}000t} + 950{,}000te^{-50{,}000t})u(t)$ V

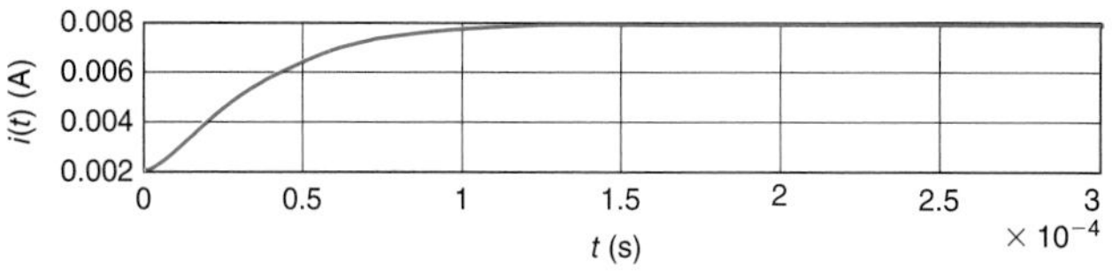

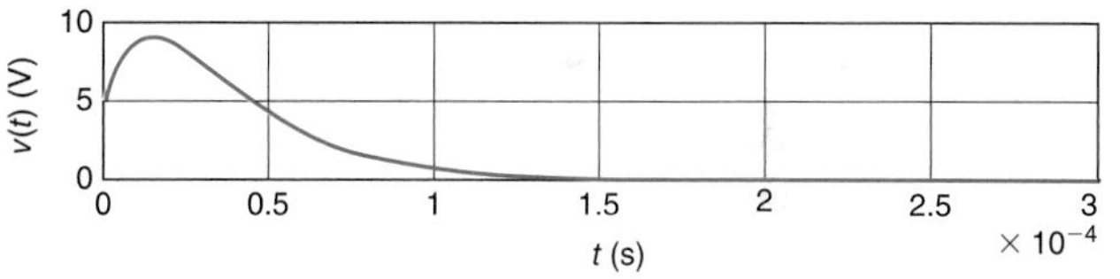

8.33 a. $i(t) = \{6 + e^{-2t}[-4\cos(4t) + 4.25\sin(4t)]\}u(t)$ A

b. $v(t) = e^{-2t}[5\cos(4t) + 1.5\sin(4t)]u(t)$ V

8.35 $i(t) = \{1.3333 + e^{-3t}[-0.583333\cos(1.7321t) - 1.0104\sin(1.7321t)]\}\,u(t)$ A

8.37 a. $I_0 = 2$ A, $V_0 = 2$ V

b. $-20 + 3\,i(t) + 0.5\,di(t)/dt + v(t) = 0$

c. $i(t) = 0.5\,dv(t)/dt + v(t)/1$

d. $d^2v(t)/dt^2 + 8\,dv(t)/dt + 16\,v(t) = 80$

e. $v(t) = [5 - 12te^{-4t} - 3e^{-4t}]\,u(t)$ V

f. $i(t) = [5 + 12te^{-4t} - 3e^{-4t}]\,u(t)$ A

8.39 a. $v(t) - 20 + i(t) + 0.5\,dv(t)/dt = 0$

b. $v(t) = 4\,i(t) + 0.5\,di(t)/dt$

c. $d^2i(t)/dt^2 + 10\,di(t)/dt + 20\,i(t) = 80$

d. $i(t) = [4 + 1.4069e^{-7.2361t} - 4.4069e^{-2.7639t}]u(t)$

8.41 a. $v(t) - 12 + 0.2\,dv(t)/dt + i(t) = 0$

b. $v(t) = 2\,i(t) + di(t)/dt,\ di(0)/dt = 1$ A/s

c. $d^2i(t)/dt^2 + 7\,di(t)/dt + 15\,i(t) = 60$

d. $i(\infty) = 4$ A

e. $i(t) = \{4 + e^{-3.5t}[-3\cos(1.6583t) - 5.7287\sin(1.6583t)]\}u(t)$ A

8.43 a. $(di(t)/dt - 4)/2 + i(t) + (di(t)/dt - v(t))/4 = 0$, $di(0)/dt = 2$
b. $(v(t) - di(t)/dt)/4 + v(t)/1 + dv(t)/dt = 0$
c. $d^2i(t)/dt^2 + 2.5\,di(t)/dt + (5/3)\,i(t) = 10/3$, $s = -1.25 \pm j0.32275$
d. $i(t) = \{2 + e^{-1.25t}[-\cos(0.32275t) + 2.3238 \times \sin(0.32275t)]\}\,u(t)$ A

8.45 a. $(di_1(t) - 9)/1 + i_1(t) + i_2(t) = 0$, $di_1(0)/dt = 6$
b. $di_1(t)/dt = di_2(t)/dt + 2i_2(t)$
c. $d^2i_1(t)/dt^2 + 4\,di_1(t)/dt + 2i_1(t) = 18$
d. $i_1(t) = (9 - 0.6716e^{-3.4142t} - 6.3284e^{-0.5858t}) \times u(t)$ A

8.47 a. $I_0 = 1$ A, $V_0 = 2$ V
b. $(di(t)/dt - 4)/4 + i(t) + (di(t)/dt - v(t))/1 = 0$, $di(0)/dt = 1.6$
c. $(v(t) - 4)/2 + (v(t) - di(t)/dt)/1 + v(t)/2 + dv(t)/dt = 0$
d. $d^2i(t)/dt^2 + 2\,di(t)/dt + 1.6\,i(t) = 3.2$
e. $i(t) = \{2 + e^{-t}[-\cos(0.7746t) + 0.7746 \times \sin(0.7746t)]\}\,u(t)$ A

8.49 a. $i(0) = 1$ A, $v(0) = -8$ V
b. $V_{th} = 20$ V, $R_{th} = 4\ \Omega$
c. $d^2v(t)/dt^2 + 4\,dv(t)/dt + 3\,v(t) = 60$
d. $v(\infty) = 60/3 = 20$ V
e. $s^2 + 4s + 3 = 0$
f. $s = -3, s = -1$
g. $v(t) = [20 + 9.5e^{-3t} - 37.5e^{-t}]\,u(t)$ V

8.51 a. $V_0 = 4.8$ V, $I_0 = 1.2$ A
b. $i(t) + 0.1\,dv(t)/dt + v(t)/1 - 15 = 0$
c. $v(t) = 4\,i(t) + 0.2\,di(t)/dt$, $di(0)/dt = 0$
d. $d^2i(t)/dt^2 + 30\,di(t)/dt + 250\,i(t) = 750$
e. $i(t) = \{3 + e^{-15t}[-1.8\cos(5t) - 5.4\sin(5t)]\} \times u(t)$ A

CHAPTER 9

9.1 a. 7 V
b. 14 V
c. 0 V
d. $7/\sqrt{2} = 4.9497$ V

9.3 a. −135°
b. 187.5 μs
c. $\theta_i(t) = 2\pi 2000t - 3\pi/4$

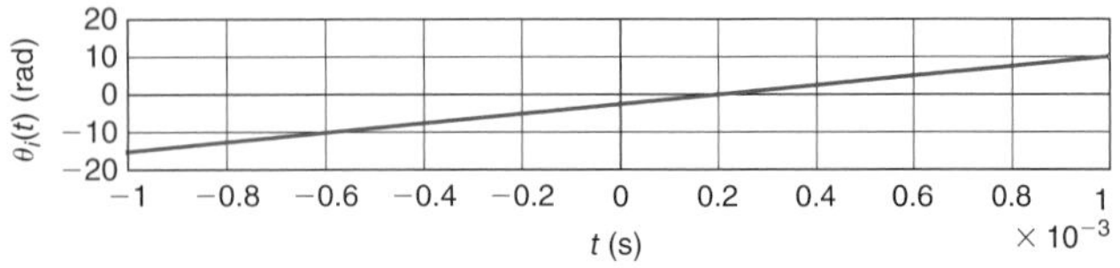

d.

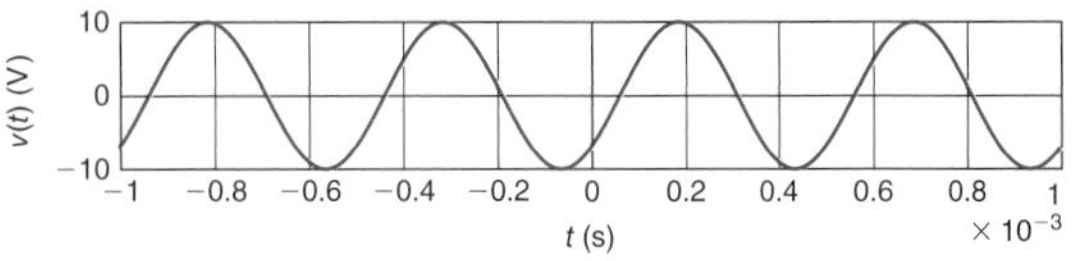

9.5

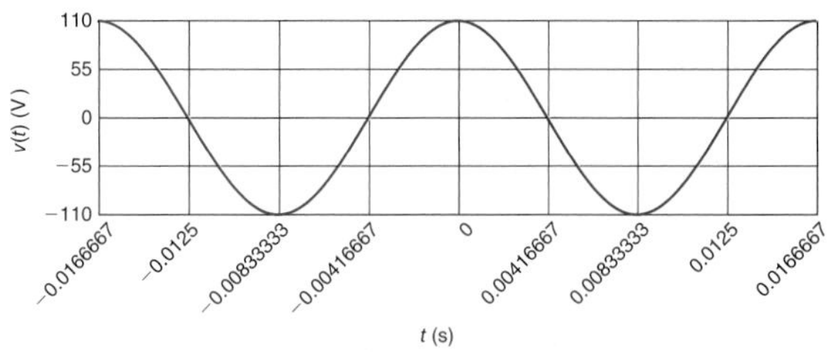

9.7

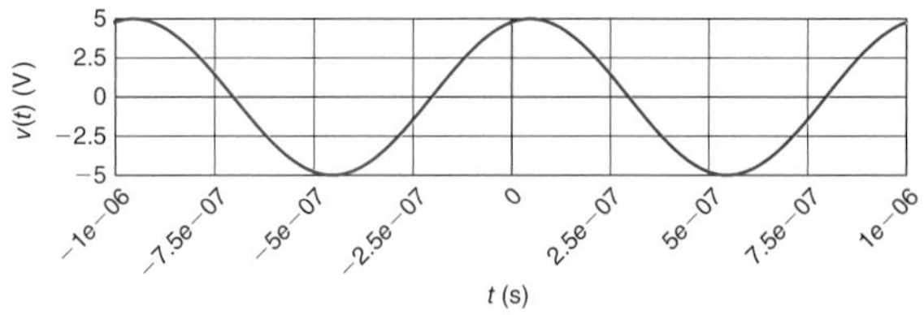

9.9

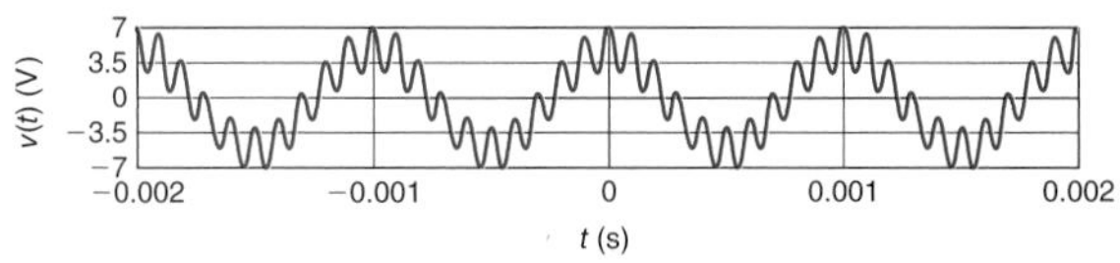

9.11

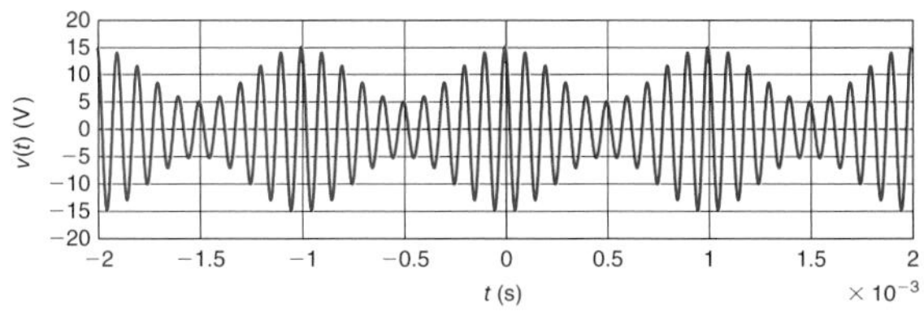

9.13 $V_{rms} = A/\sqrt{2}$
9.15 $V_{rms} = A/\sqrt{3}$
9.17 $\mathbf{V} = 110e^{j120°} = 110\angle 120°$

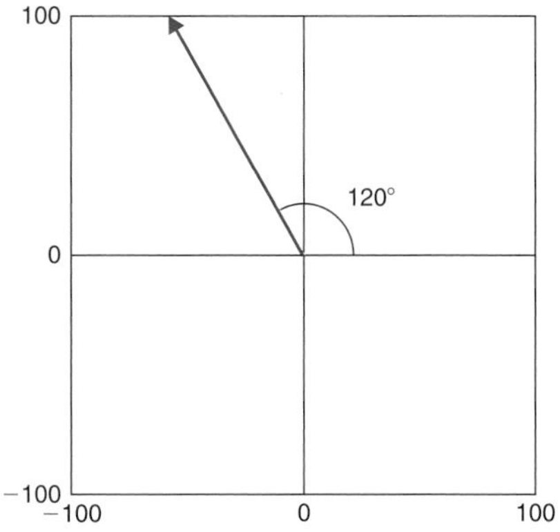

9.19 $\mathbf{I} = 10e^{j120°} = 10\angle -120°$

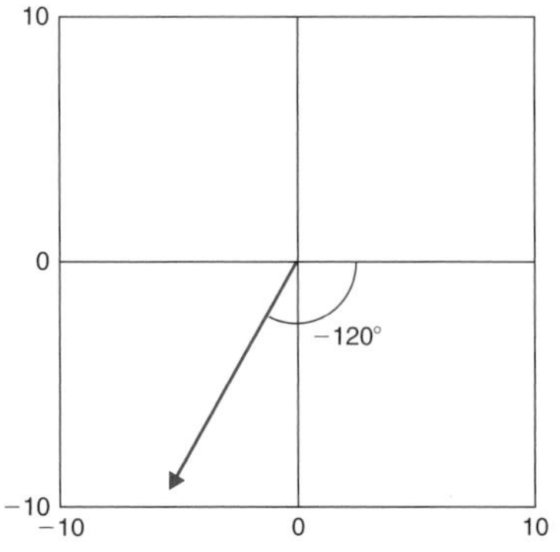

9.21 a. $v(t) = 120\cos(2\pi 60t + 120°)$ V
b. $v(t) = 220\cos(2\pi 60t - 150°)$ V

9.23 a. $\mathbf{V} = 22.3607\angle 63.4349°$ V
b. $\mathbf{V} = 58.3095\angle 120.9638°$ V
c. $\mathbf{V} = 223.6068\angle -153.4349°$ V
d. $\mathbf{V} = 583.0952\angle -30.9638°$ V

9.25 a. $\mathbf{V} = 55 + j95.2628$ V
b. $\mathbf{V} = -190.5256 - j110$ V
c. $\mathbf{V} = -415.6922 - j240$ V
d. $\mathbf{V} = 762.1024 - j440$ V

9.27 a. $\mathbf{I} = 4.8366\angle 41.9325°$ A, $i(t) = 4.8366\cos(2\pi 60 + 41.9325°)$ A
b. $\mathbf{I} = 7.8102\angle 86.3295°$ A, $i(t) = 7.8102\cos(2\pi 60 + 86.3295°)$ A
c. $\mathbf{I} = 42\angle -90°$ A, $i(t) = 42\cos(2\pi 60 - 90°)$ A
d. $\mathbf{I} = 1.6667\angle -60°$ A, $i(t) = 1.6667\cos(2\pi 60 - 60°)$ A

9.29 a. $v(t) = 435.8899\cos(2\pi 60t - 173.4132°)$ V
b. $v(t) = 45.8258\cos(2\pi 60t + 79.1066°)$ V

9.31 a. $V_s = 120\angle 60° = 60 + j103.9230$ V, $Z_R = 55\ \Omega$, $Z_L = j\omega L = j47.1239\ \Omega$

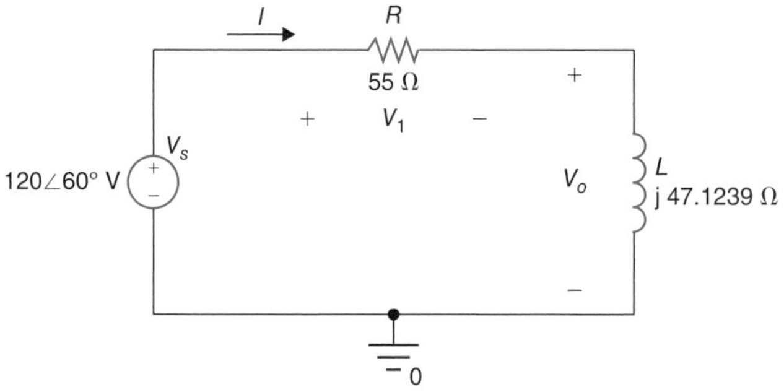

b. $I = 1.6568\angle 19.4101°$ A, $V_1 = 91.1263\angle 19.4101°$ V, $V_o = 78.0768\angle 109.4101°$ V
c. $i(t) = 1.6568\cos(2\pi 60t + 19.4101°)$ A, $v_1(t) = 91.1263\cos(2\pi 60t + 19.4101°)$ V, $v_o(t) = 78.0768\cos(2\pi 60t + 109.4101°)$ V.

9.33 a. $Z_L = j35.8142\ \Omega$

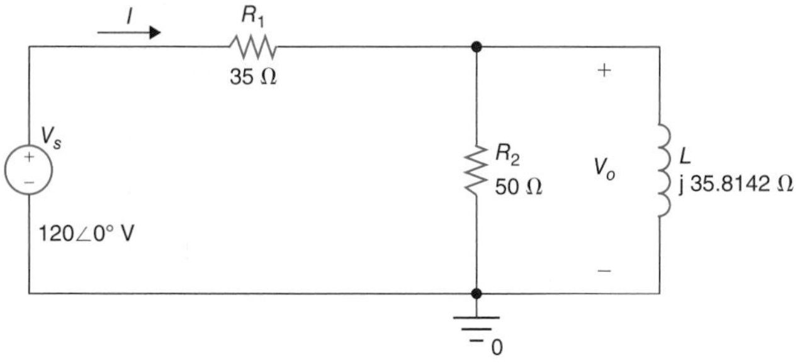

b. $I = 2.1019\angle -24.4936°$ A, $V_o = 61.1970\angle 29.8930°$ V
c. $i(t) = 2.1019\cos(2\pi 60t - 24.4936°)$ A, $v_o(t) = 61.1970\cos(2\pi 60t + 29.8930°)$ V

9.35 a. $Z_L = j32.0442\ \Omega$, $Z_C = -j58.9463\ \Omega$

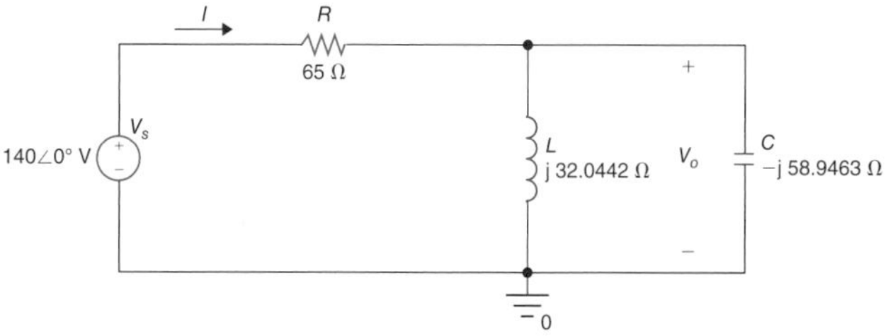

b. $I = 1.4632\angle -47.2081°$ A, $V_o = 102.7357\angle 42.7919°$ V
c. $i(t) = 1.4632\cos(2\pi 60t - 47.2081°)$ A, $v_o(t) = 102.7357\cos(2\pi 60t + 42.7919°)$ V

9.37 $V_o = 12.6723\angle -9.7276°$ V

9.39 $V_o = 21.8413\angle -41.6427°$ V

CHAPTER 10

10.1

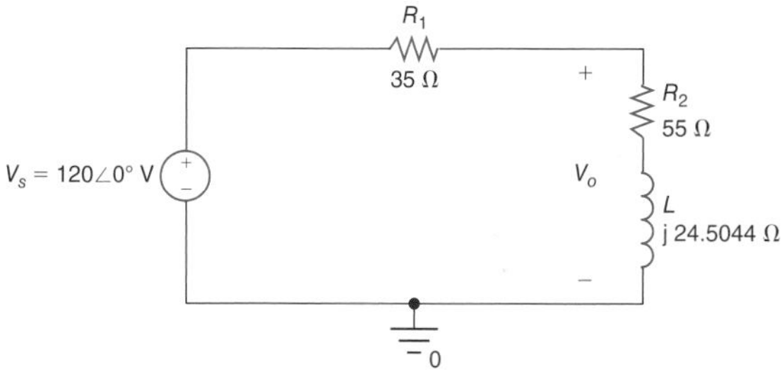

$\mathbf{V_o} = 77.4626\angle 8.7838°$ V, $v_o(t) = 77.4626\cos(2\pi 60t + 8.7838°)$ V

10.3

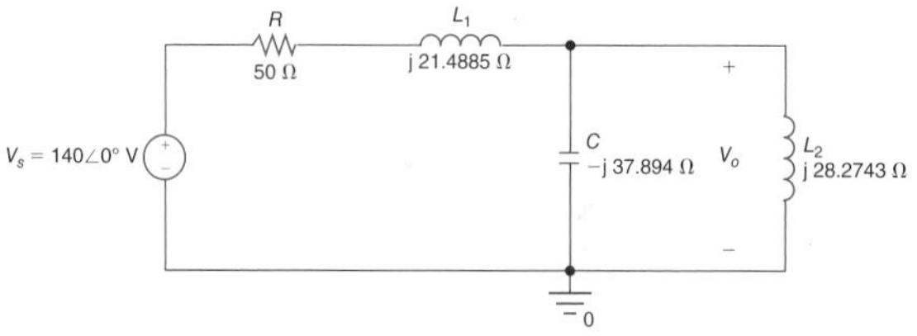

$\mathbf{V_o} = 109.838\angle 20.6221°$ V, $v_o(t) = 109.838 \times \cos(2\pi 60t + 20.6221°)$ V

10.5

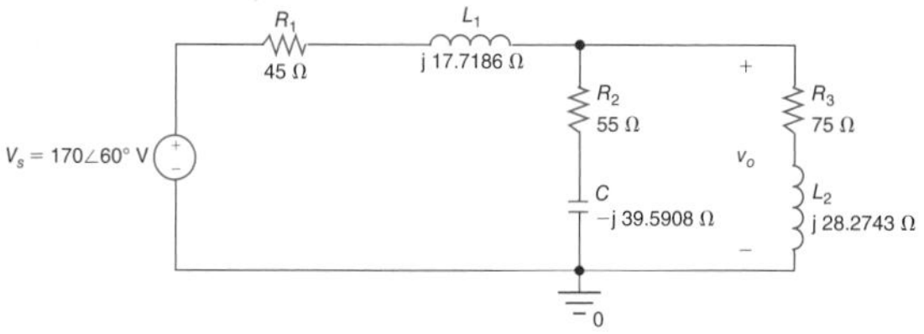

$\mathbf{V_o} = 81.707\angle 42.9818°$ V, $v_o(t) = 81.707 \times \cos(2\pi 60t + 42.9818°)$ V

10.7 $\mathbf{V_o} = 52.0161\angle 48.8925°$ V

10.9

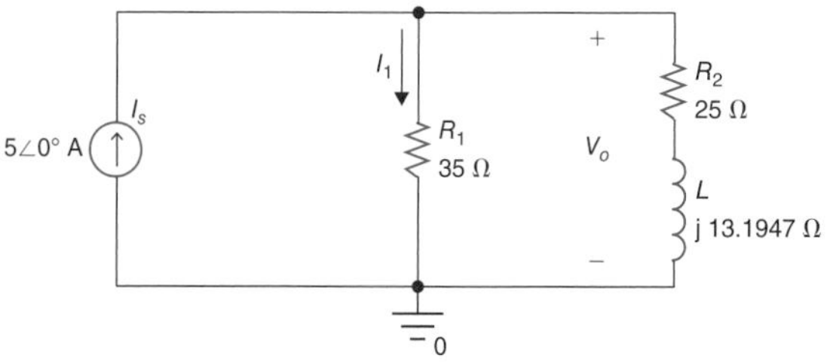

$I_1 = 2.3007\angle 15.422°$ A, $V_o = 80.5252\angle 15.422°$ V, $v_o(t) = 80.5252\cos(2\pi 60t + 15.422°)$ V

10.11

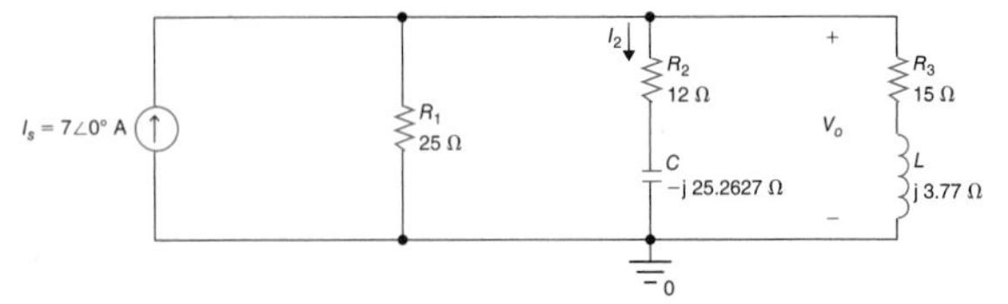

$I_2 = 2.09973\angle 56.6172°$ A, $V_o = 58.7249\angle -7.9746°$ V, $v_o(t) = 58.7249\cos(2\pi 60t - 7.9746°)$ V

10.13 $I_3 = 2.2575\angle -91.232°$ A, $V_o = 18.0603\angle -91.232°$ V

10.15 $I_3 = 7.6935\angle 65.285°$ A, $V_o = 46.1612\angle -24.715°$ V

10.17 $V_o = 67.2124\angle 12.9465°$ V

10.19 $V_o = 50.4598\angle 17.9125°$ V

10.21 $V_1 = 57.5965\angle 5.44°$ V, $V_2 = 25.7579\angle -57.9946°$ V

10.23 $V_1 = 43.6318\angle 48.2843°$ V, $V_2 = 50.1655\angle 42.4718°$ V

10.25 $V_1 = 3.123\angle -8°$ V, $V_2 = 6.9466\ \angle 115.5151°$ V

10.27 $V_1 = 8.9289\angle -3.9568°$ V, $V_2 = 3.0511\angle 4.965°$ V, $V_3 = 5.9762\angle -1.688°$ V

10.29 $I_1 = 0.7531\angle 30.4518°$ A, $I_2 = 3\angle 0°$ A, $I_3 = 1.2064\angle 15.4334°$ A, $V_1 = 5.5948\angle -9.9128°$ V, $V_2 = 6.032\angle 15.4334°$ V

10.31 $I_1 = 1.07968\angle -22.04883°$ A, $I_2 = 0.7212\angle 75.8946°$ A, $V_1 = 8.0395\angle 5.7869°$ V, $V_2 = 2.937\angle 54.5437°$ V

10.33 $V_1 = 28.7348\angle 73.3008°$ V, $V_2 = 40\angle 56.6015°$ V

10.35 $V_1 = 43.4936\angle 31.8796°$ V, $V_2 = 25.4748\angle 68.7495°$ V

10.37 $V_1 = 34.602\angle -5.0772°$ V, $V_2 = 38.3955\angle 21.3874°$ V

10.39 $V_o = 11.8329\angle 75.5595°$ V

10.41 $V_o = 7.1261\angle 105.2551°$ V

10.43 $V_{th} = 11.5383\angle -105.9454°$ V, $Z_{th} = 5.5283 + j1.8491 = 5.8293\angle 18.4936°\ \Omega$

10.45 $V_{th} = 32.8356\angle -22.8337°$ V, $Z_{th} = 6.1321 + j1.8715 = 6.4114\angle 16.9719°\ \Omega$

10.47 $V_{th} = 2.1649\angle 74.8355°$ V, $Z_{th} = 3.5476 - j2.083 = 4.1139\angle -30.4196°\ \Omega$

10.49 $I_n = 2.095\angle 16.2202°$ V, $Z_n = 6.6765 - j7.2059 = 9.8234\angle -47.184°\ \Omega$

10.51 $I_n = 0.5233\angle -24.3045°$ V, $Z_n = 9.332 - j3.5609 = 9.9883\angle -20.8858°\ \Omega$

10.53 $H(\omega) = \dfrac{1.6667}{j\omega + 2.5}$, LPF

10.55 $H(\omega) = \dfrac{60{,}000\, j\omega + 3.3333 \times 10^8}{(j\omega)^2 + 60{,}000\, j\omega + 3.3333 \times 10^8}$, LPF

10.57 $H(\omega) = \dfrac{(j\omega)^2 + 10{,}000\, j\omega}{(j\omega)^2 + 10{,}000\, j\omega + 1.6667 \times 10^8}$, HPF

10.59 $H(\omega) = \dfrac{(j\omega)^2}{(j\omega)^2 + 3.5\, j\omega + 7.5}$, HPF

10.61 $H(\omega) = \dfrac{(j\omega)^2}{(j\omega)^2 + 22.5\, j\omega + 10}$, HPF

10.63 $H(\omega) = \dfrac{j\omega + 2}{j\omega + 1}$

CHAPTER 11

11.1 $\mathbf{V} = 120\angle 120°$, $\mathbf{I} = 5\angle 60°$, $\mathbf{S} = 150 + j259.8076 = 300\angle 60°$ VA, $|\mathbf{S}| = 300$ VA, $P = 150$ W, $Q = 259.8076$ VAR, $pf = \cos(60°) = 0.5$

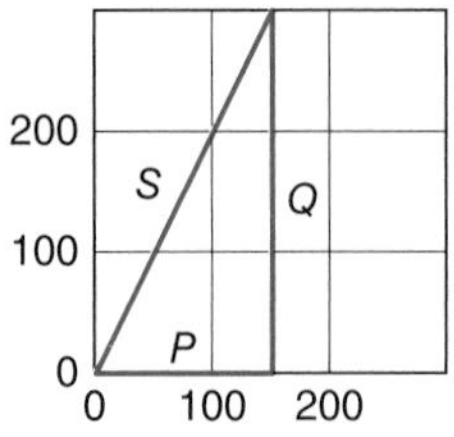

11.3 $\mathbf{V} = 5\angle 60°$, $\mathbf{I} = 2\angle 30°$, $\mathbf{S} = 4.33 + j2.5 = 5\angle 30°$ VA, $|\mathbf{S}| = 5$ VA, $P = 4.33$ W, $Q = 2.5$ VAR, $pf = \cos(30°) = 0.866$

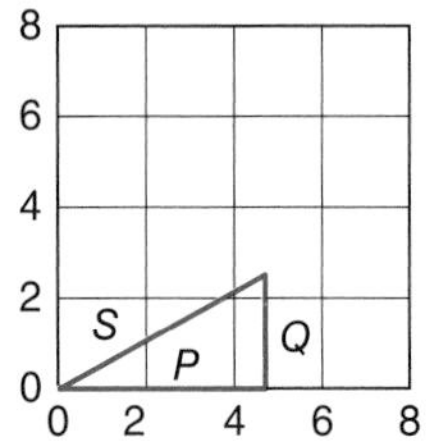

11.5 $\mathbf{V} = 80\angle 135°$, $\mathbf{I} = 2\angle 120°$, $\mathbf{S} = 77.2741 + j20.7055 = 80\angle 15°$ VA, $|\mathbf{S}| = 80$ VA, $P = 77.2741$ W, $Q = 20.7055$ VAR, $pf = \cos(15°) = 0.9659$

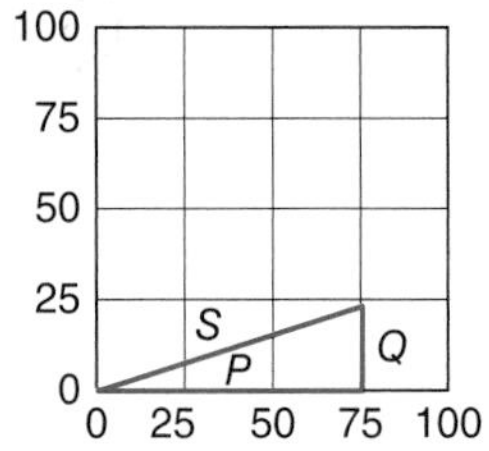

11.7 $Z = 76.8148 + j31.1587\ \Omega$, $\mathbf{S} = 698.6845 + j283.4104 = 753.9771\angle 22.0792°$ VA, $|\mathbf{S}| = 753.9771$ VA, $P = 698.6845$ W, $Q = 283.4104$ VAR, $pf = 0.9267$

11.9 $Z = 116.6133 + j36.3414\ \Omega$, $\mathbf{S} = 2915.3332 + j908.5346 = 3053.6213\angle 17.3091°$ VA, $|\mathbf{S}| = 3053.6213$ VA, $P = 2915.3332$ W, $Q = 908.5346$ VAR, $pf = 0.9547$

11.11 $Z = 70.4694 - j52.1610\ \Omega$, $\mathbf{S} = 3452.9987 - j2555.8908 = 4296.0188\angle -36.5086°$ VA, $|\mathbf{S}| = 4296.0188$ VA, $P = 3452.9987$ W, $Q = -2555.8908$ VAR, $pf = 0.8038$

11.13 $Z = 47.4816 + j6.6126\ \Omega$, $pf = 0.9904$, $R = 47.4816\ \Omega$, L = 17.5405 mH

11.15 $|\mathbf{S}| = 9322.4086$ VA, $P = 8147.7852$ W, $\mathbf{S} = 8147.7852 + j4530$ V

11.17 $\mathbf{S} = 25{,}513.6 - j8203.4270$ VA, $P = 25{,}513.6$ W, $Q = -8203.4270$ VAR

11.19 $|\mathbf{S}| = 4420.4597$ VA, $P = 4186.1753$ W, $\mathbf{S} = 4186.1753 + j1420$ V, $Z = 47.3237 + j16.05274\ \Omega$, $R = 47.3237\ \Omega$, $X = 16.05274\ \Omega$, $L = 42.5812$ mH

11.21 $P = 20{,}984.3$ W, $Q = -7546.4663$ VAR, $\mathbf{S} = 20{,}984.3 - j7546.4663$ VA, $Z = 76.1515 - j27.3859\ \Omega$, $R = 76.1515\ \Omega$, $X = -27.3859\ \Omega$, $C = 96.8593\ \mu F$

11.23 a. $Z_{R_1} = 20\ \Omega$, $Z_L = j37.6991\ \Omega$, $Z_{R_2} = 200\ \Omega$, $Z_C = -j265.2582\ \Omega$, $I_{R_1} = 0.7563\angle 51.6128°$ A, $V_{R_1} = 15.1261\angle 51.6128°$ V, $V_L = 28.5120\angle 141.6128°$ V, $V_{R_2} = 120.7773\angle 14.5972°$ V
b. $P_{R_1} = 11.4399$ W, $P_{R_2} = 72.9358$ W
c. $S_w = 11.4399 + j21.5637$ VA, $S_2 = 72.9358 - j54.9923\ \Omega$, $pf_w = 0.4686$, $pf_2 = 0.7985$

11.25 a. $I_1 = 7.3333 - j12.7017$ A
b. $I_2 = 1.012 + j6.3225$ A
c. $I_3 = 8.3453 - j6.3792$ A
d. $V_s = 516.7238\angle -39.7194°$ V
e. $S_1 = 2758.4837 - j6620.3608$ VA, $S_L = j9680$ VA, $S_2 = 2664.8471 - j3279.8118$ VA
f. $S_{V_s} = -5423.3307 + j220.172549$, $S_{V_s} + S_1 + S_L + S_2 = 0 + j0$

11.27 $S_{R_1} = 73.1927$ VA, $S_{R_2} = 156.8415$ VA, $S_{L_1} = j35.4767$ VA, $S_{L_2} = j51.2442$ VA, $S_{V_s} = -230.0343 - j86.7209$ VA, Sum $= 0 + j0$

11.29 $S_{R_1} = 264.2234$ VA, $S_{R_2} = 63.1399$ VA, $S_{L_1} = j109.5708$ VA, $S_{L_2} = j260.9672$ VA, $S_C = -26.5847$ VA, $S_{V_s} = -327.3633 - j343.9533$ VA, Sum $= 0 + j0$

11.31 $S_{R_1} = 603.1725$ VA, $S_{R_2} = 378.2718$ VA, $S_{L_1} = j94.0535$ VA, $S_{L_2} = j160.4307$ VA, $S_{C_1} = -539.5161$ VA, $S_{C_2} = -518.2285$ VA, $S_{Vs} = -981.4443 + j803.2603$ VA, Sum $= 0 + j0$

11.33 $Z_{\text{Load}} = 7.68 + j5.24$, $P_{\max} = 75$ W

11.35 $Z_{\text{Load}} = 41.432 - j12.376$, $P_{\max} = 89.2836$ W

11.37 $pf_1 = 0.8362$, $C = 11.5046\ \mu F$

11.39 $C = 14.0257\ \mu F$

11.41 $C = 73.4163\ \mu F$

CHAPTER 12

12.1 $V_{an} = 460\angle 0°$ V, $V_{bn} = 460\angle -120°$ V, $V_{cn} = 460\angle 120°$ V, $V_{ab} = 796.7434\angle 30°$ V, $V_{bc} = 796.7434\angle -90°$ V, $V_{ca} = 796.7434\angle 150°$ V

12.3 $V_{bn} = 250\angle -150°$ V, $V_{an} = 250\angle -30°$ V, $V_{ab} = 433.0127\angle 0°$ V

12.5 $V_{ab} = 168.6783\angle 32.5°$ V, $V_{bc} = 168.6783\angle -82.5°$ V, $V_{ca} = 181.2616\angle 155°$ V

12.7 $I_A = 2.8398\angle -30.2564°$ A, $V_{AN} = 236.7136\angle 0°$ V, $V_{AB} = 410\angle 30°$ V

12.9 $V_{an} = 248.2606\angle 0°$ V, $V_{bn} = 248.2606\angle -120°$ V, $V_{cn} = 248.2606\angle 120°$ V, $I_A = 2.5175\angle -30.4655°$ A, $I_B = 2.5175\angle -150.4655°$ A, $I_C = 2.5175\angle 89.5345°$ A, $V_{AN} = 195.9259\angle -2.9127°$ V, $V_{BN} = 195.9259\angle -122.9127°$ V, $V_{CN} = 195.9259\angle 117.0873°$ V, $V_{AB} = 339.3536\angle 27.0873°$ V

12.11 $I_A = 3.065\angle -15.1011°$ A, $I_B = 2.9704\angle -142.4226°$ A, $I_C = 2.8985\angle 90.382°$ A, $V_{AN} = 153.2488\angle -15.1011°$ V, $V_{BN} = 148.5202\angle -132.4226°$ V, $V_{CN} = 144.9270\angle 110.382°$ V, $V_{AB} = 257.7535\angle 15.6914°$ V, $V_{BC} = 250.473\angle -101.4486°$ V, $V_{CA} = 265.0906\angle 138.4642°$ V

12.13 $I_A = 3.065\angle -15.1011°$ A, $I_B = 3.065\angle -145.1011°$ A, $I_C = 3.065\angle 84.8989°$ A, $V_{AN} = 153.2488\angle -15.1011°$ V, $V_{BN} = 153.2488\angle -145.1011°$ V, $V_{CN} = 153.2488\angle 84.8989°$ V, $V_{AB} = 277.7812\angle 9.8989°$ V, $V_{BC} = 277.7812\angle -120.1011°$ V, $V_{CA} = 234.7908\angle 124.8989°$ V

12.15 $I_A = 18.9697\angle -30.1414°$ A, $I_B = 18.9697\angle -150.1414°$ A, $I_C = 18.9697\angle 89.8586°$ A, $I_{AB} = 10.9521\angle -0.1414°$ A, $I_{BC} = 10.9521\angle -120.1414°$ A, $I_{CA} = 10.9521\angle 119.8586°$ A

12.17 $I_A = 26.2717\angle -14.8976°$ A, $I_B = 26.5590\angle -138.7585°$ A, $I_C = 24.8604\angle 102.8877°$ A, $I_{AB} = 16.4027\angle 11.7299°$ A, $I_{BC} = 14.7035\angle -105.4242°$ A, $I_{CA} = 13.7407\angle 132.7576°$ A

12.19 $I_{AB} = 9.6896\angle -26.5651°$ A, $I_{BC} = 9.6896\angle -146.5651°$ A, $I_{CA} = 9.6896\angle 93.4349°$ A, $I_A = 16.7829\angle -56.5651°$ A, $I_B = 16.7829\angle -176.5651°$ A, $I_C = 16.7829\angle 63.4349°$ A, $S_L = 6760 + j3380 = 7557.9097\angle 26.5651°$ VA

12.21 $I_A = 8.5863\angle -50.9597°$ A, $I_B = 10.3118\angle -167.1008°$ A, $I_C = 10.1014\angle 62.6345°$ A, $I_{AB} = 5.8738\angle -9.1554°$ A, $I_{BC} = 5.3441\angle -142.7262°$ A, $I_{CA} = 5.7477\angle 86.1027°$ A, $S_L = 13240.1229 + j3927.0079 = 13810.2225\angle 16.5203°$ VA

12.23 $V_{ab} = 450\angle 0°$ V, $V_{bc} = 450\angle -120°$ V, $V_{ca} = 450\angle 120°$ V, $V_{an} = 259.8076\angle -30°$ V, $V_{bn} = 259.8076\angle -150°$ V, $V_{cn} = 259.8076\angle -90°$ V, $I_A = 8.2782\angle -60.6507°$ A, $I_B = 8.2782\angle 179.3493°$ A, $I_C = 8.2782\angle 59.3493°$ A, $S_L = 5550.7614 + j3289.3401 = 6452.1865\angle 30.6507°$ VA

12.25 $V_{ab} = 470\angle 0°$ V, $V_{bc} = 470\angle -120°$ V, $V_{ca} = 470\angle 120°$ V, $V_{an} = 271.3546\angle -30°$ V, $V_{bn} = 271.3546\angle -150°$ V, $V_{cn} = 271.3546\angle 90°$ V, $I_A = 6.3467\angle -40.7843°$ A, $I_B = 6.2246\angle -167.6386°$ A, $I_C = 6.1469\angle 65.5654°$ A, $V_{AN} = 190.4015\angle -40.7843°$ V, $V_{BN} = 186.7388\angle -157.6386°$ V, $V_{CN} = 184.4084\angle 85.5654°$ V, $S_1 = 1208.4245 = 1208.4245\ \angle 0°$ VA, $S_2 = 1144.7199 + j201.8450 = 1162.3791\angle 10°$ VA, $S_3 = 1065.1868 + j387.6963 = 1133.5481\angle 20°$ VA

CHAPTER 13

13.1 $M = 18.8496$ mH

13.3 $M = 3.06495$ mH

13.5
$$v_1(t) = L_1\frac{di_1(t)}{dt} + M\frac{di_2(t)}{dt},$$
$$v_2(t) = M\frac{di_1(t)}{dt} + L_2\frac{di_2(t)}{dt}$$

13.7 $v_1(t) = L_1\frac{di_1(t)}{dt} - M\frac{di_2(t)}{dt}$,

$v_2(t) = M\frac{di_1(t)}{dt} - L_2\frac{di_2(t)}{dt}$

13.9 $L = \frac{L_1L_2 - M^2}{L_1 + L_2 + 2M}$

13.11 $L = \frac{L_1L_2L_3 - L_1M^2 - L_2M^2 - L_3M^2 - 2M^3}{L_1L_2 + L_1L_3 + L_2L_3 + 3M^2 + 2L_1M + 2L_2M + 2L_3M}$

13.13 $I_1 = 3.2089 - j1.7357 = 3.6482\angle -28.4094°$ A, $I_2 = -2.1149 + j0.2917 = 2.1350\angle 172.1467°$ A, $V_1 = 8.9557 + j8.6785 = 12.4708\angle 44.0996°$ V, $V_2 = 6.3448 - j0.8751 = 6.4049\angle -7.8533°$ V

13.15 $V_1 = 10.9431 + j7.9556 = 13.5293\angle 36.0169°$ V, $V_2 = 1.5756 + j7.5324 = 7.6954\angle 78.1853°$ V, $I_1 = 2.6771 - j2.7576 = 3.8434\angle -45.8491°$ A, $I_2 = 2.0268 - j1.4007 = 2.4637\angle -34.6483°$ A

13.17 $I_1 = 5.5369 + j4.6710 = 7.2440\angle 40.1518°$ A, $I_2 = -1.6052 - j2.5092 = 2.9787\angle -122.6082°$ A, $I_3 = -0.3501 - j3.1476 = 3.1670\angle -96.3463°$ A

13.19 $Z_a = Z_s + R_1 + j\omega L_1$, $Z_b = j\omega L_2 + R_2$,

$V_{th} = \frac{j\omega M}{Z_a}V_s$, $Z_{th} = Z_b + \frac{(\omega M)^2}{Z_a}$

13.21 $V_{th} = 33.6763 + j64.6122$, $Z_{th} = 14.1176 + j19.5294$, $V_o = 24.1616 + j11.6199 = 26.8105\angle 25.6840°$

13.23 $V_1 = 11.6827 + j5.5515 = 12.9346\angle 25.4167°$ V, $V_2 = 8.5624 + j1.7186 = 8.7332\angle 11.3492°$ V, $I_1 = 1.4686 - j1.9915 = 2.4744\angle -53.5932°$ A, $I_2 = 0.7881 - j0.3763 = 0.8733\angle -25.5207°$ A

13.25 $n = 1/9$

13.27 $R_{in} = 4.5\ k\Omega$

13.29 $R_{in} = 1.6926\ k\Omega$

13.31 $V_1 = -0.3937 + j0.7330 = 0.8321\angle 118.2374°$ V, $V_2 = -1.9683 + j3.6652 = 4.1603\angle 118.2374°$ V, $V_3 = -2.362 + j4.3982 = 4.9923\angle 118.2374°$ V, $I_1 = 0.2036 + j4.2760 = 4.2809\angle 87.2737°$ A, $I_2 = 0.04072 + j0.8552 = 0.8562\angle 87.2737°$ A, $I_a = 0.2443 + j5.1312 = 5.137\angle 87.2737°$ A

13.33 $V_1 = 14.8147 - j0.2389 = 14.8166\angle -0.9240°$ V, $V_2 = 2.1164 - j0.03414 = 2.1167\angle -0.9240°$ V, $V_3 = 12.6983 - j0.2048 = 12.7\angle -0.924°$ V, $I_1 = 0.03706 + j0.04779 = 0.06048\angle 52.2061°$ A, $I_2 = 0.2224 + j0.2867 = 0.3629\angle 52.2061°$ A, $I_a = 0.2594 + j0.3345 = 0.4233\angle 52.2061°$ A

CHAPTER 14

14.1 $F(s) = 7e^{-5s}$

14.3 $F(s) = 5/s$

14.5 $F(s) = e^{-3s}/s^2 + 5e^{-3s}/s = (5s + 1)e^{-3s}/s^2$

14.7 $F(s) = 4/(s + 2) + 3/(s + 5) = (7s + 26)/(s^2 + 7s + 10)$

14.9 $F(s) = (6e)e^{-5s}/(s + 2) = 16.3097e^{-5s}/(s + 2)$

14.11 $F(s) = e^{-4}/(s + 2)$

14.13 $F(s) = 1/(s - \ln(5)) = 1/(s - 1.6094)$

14.15 $F(s) = 5/(s^2 + 6s - 16)$

14.17 $F(s) = \frac{-3.5s + 19.8109}{s^2 + 6s + 34}$

14.19 $F(s) = -3\left[\frac{-j(s + j5 + 3)^3 + j(s - j5 + 3)^3}{((s + 3)^2 + 25)^3}\right]$

$-5.1962\left[\frac{(s + j5 + 3)^3 + (s - j5 + 3)^3}{((s + 3)^2 + 25)^3}\right]$

14.21 $F(s) = \frac{1 - e^{-s} - e^{-4s} + e^{-5s}}{s^2}$

14.23 $F(s) = \frac{-1 + 3e^{-s} - 3e^{-2s} + e^{-3s}}{s^2}$

14.25 $F(s) = \frac{15}{s^2 + 9}$, $G(s) = \frac{-135}{s^2 + 9}$

14.27 $f(t) = 3e^{-3t}u(t) - e^{-4t}u(t)$

14.29 $f(t) = [1.5e^{-2t} - 2e^{-3t} + 0.5e^{-4t}]u(t)$

14.31 $f(t) = [2t - 0.3333 + 0.3333e^{-3t}]u(t)$

14.33 $f(t) = 2\,e^{-2t}\cos(5t)\,u(t) + e^{-2t}\sin(5t)\,u(t)$

14.35 $f(t) = 0.758621e^{-t}u(t) - 0.758621e^{-3t}\cos(5t) \times u(t) + 0.096552e^{-3t}\sin(5t)\,u(t)$

14.37 $f(t) = \{4.5e^{-3t} - e^{-4t}[1.5\cos(5t) + 3.1\sin(5t)]\}\,u(t)$

14.39 $f(t) = \delta''(t) + 6\delta'(t) - 97\delta(t) - 715e^{-5t}u(t) + 1684e^{-7t}u(t)$

14.41 $f(t) = 10t^2e^{-2t}u(t) - 34te^{-2t}u(t) + 39e^{-2t}u(t) - 39e^{-3t}u(t)$

14.43 $f(t) = e^{-4(t-3)}\cos(5(t - 3))\,u(t - 3) + 0.4e^{-4(t-3)}\sin(5(t - 3))\,u(t - 3)$

14.45 $f(t) = -3te^{-t}u(t) + 6e^{-t}u(t) - 9te^{-3t}u(t) - 6e^{-3t}u(t)$

14.47 $f(0^+) = 3, f(\infty) = 6$

14.49 $f(0^+) = 2, f(\infty) = 8$

CHAPTER 15

15.1 $v(t) = (3 + e^{-2t})\,u(t)$ V

15.3 $v_o(t) = \{5 - 4e^{-8333.3333t}[\cos(7993.0525t) + 1.1468\sin(7993.0525t)]\}u(t)$ V

15.5 $v_o(t) = [0.9091 - 0.4064e^{-1.6217t} - 0.5027e^{-0.6783t}]\,u(t)$ V

15.7 $v_o(t) = [-1.05882e^{-3.3333t} + 7.05882e^{-0.5t}]u(t)$ V

15.9 $v_o(t) = [4e^{-0.32t}\sin(0.24t)]u(t)$ V

15.11 $v_o(t) = [2.5 + 1.2780561e^{-1.4534t} - 1.7781e^{-0.7733t} \times \cos(1.4677t) - 3.7592e^{-0.7733t}\sin(1.4677t)]u(t)$ V

15.13 $v_o(t) = [36e^{-2t} - 33.3333e^{-1.6667t}]\,u(t)$ V

15.15 $v_o(t) = [-1.5e^{-0.625t}\cos(1.0532687t) - 0.534052e^{-0.625t} \times \sin(1.0532687t)]u(t)$ V

15.17 $v_o(t) = [2 + 2e^{-0.333333t}\cos(1.37436854t) - 2.91043e^{-0.333333t}\sin(1.37436854t)]\,u(t)$ V

15.19 $v_o(t) = [2.5e^{-t} - 0.5e^{-3t}]u(t)$ V

15.21 $g(t) = [0.6667e^{-20{,}000t}]u(t)$ V

15.23 $V_{th}(s) = \frac{2s + 1.96875}{s^2 + 0.96875s + 0.046875}$,

$Z_{th}(s) = \frac{0.25s + 0.234375}{s^2 + 0.96875s + 0.046875}$

15.25 $V_{th}(s) = \dfrac{0.2105s - 0.9474}{s^2 + 0.1316s}$,

$Z_{th}(s) = \dfrac{0.6316s + 0.5263}{s + 0.1316}$

15.27 $I_n(s) = \dfrac{4s^2 + s + 0.5}{s^2 + 0.16667s}$,

$Z_n(s) = \dfrac{6s + 1}{24s^2 + 6s} = \dfrac{0.25s + 0.0416667}{s^2 + 0.25s}$

15.29 $I_n(s) = \dfrac{5s^2 + 13.2s + 1.2}{s^2 + 2.6s}$,

$Z_n(s) = \dfrac{0.2s + 0.52}{s^2 + 2.64s + 0.12}$

15.31 $I_n(s) = \dfrac{0.462s + 0.05696}{s^2 + 0.1266s}$, $Z_n(s) = \dfrac{1.866s + 0.2362}{s + 0.126}$

15.33 $H(s) = \dfrac{0.5s^3}{s^3 + 1.25s^2 + 0.75s + 0.25}$

15.35 $H(s) = \dfrac{2}{s^3 + 2s^2 + 2s + 2}$

15.37 $H(s) = \dfrac{\left(\dfrac{C_1}{C_1 + C_2}\right)s^2 + \dfrac{1}{L(C_1 + C_2)}}{s^2 + \dfrac{1}{R(C_1 + C_2)}s + \dfrac{1}{L(C_1 + C_2)}}$

15.39 $H(s) = \dfrac{\dfrac{-1}{R_1R_3C_1C_2}}{s^2 + \left(\dfrac{1}{R_1C_1} + \dfrac{1}{R_2C_1} + \dfrac{1}{R_3C_1}\right)s + \dfrac{1}{R_2R_3C_1C_2}}$

15.41 $H(s) = \dfrac{-s^2}{s^2 + \dfrac{3}{R_2}s + \dfrac{1}{R_1R_2}}$

15.43 $H(s) = \dfrac{4.8}{s^2 + 9.4s + 10.8}$

15.45 $y(t) = \begin{cases} 10t, & 0 \le t < 3 \\ 30, & 3 \le t < 5 \\ -10t + 80, & 5 \le t < 8 \\ 0, & \textit{elsewhere} \end{cases}$

15.47 $y(t) = \begin{cases} 6t - 18, & 3 \le t < 7 \\ 24, & 7 \le t < 8 \\ -6t + 72, & 8 \le t < 12 \\ 0, & \textit{elsewhere} \end{cases}$

15.49 $y(t) = \begin{cases} 0, & t < 8 \\ 6t - 48, & 8 \le t < 11 \\ 18, & 11 \le t \end{cases}$

15.51 $y(t) = \begin{cases} 0, & t < 0 \\ -2t^2 + 20t, & 0 \le t < 2 \\ -8t + 48, & 2 \le t < 5 \\ 2(t - 7)^2, & 5 \le t < 7 \\ 0, & 7 \le t \end{cases}$

15.53 $y(t) = 0.2(1 - e^{-5t})\,u(t)$

15.55 $y(t) = 0.2\sin(5t)\,u(t)$

15.57 $y(t) = \{0.029586 - 0.029586e^{-5t}[\cos(12t) - 2.4 \times \sin(12t)]\}\,u(t)$

15.59 $y(t) = \begin{cases} 0, & t < 0 \\ \dfrac{-1}{2}t^2 + t, & 0 \le t < 1 \\ \dfrac{1}{2}(t - 2)^2, & 1 \le t < 2 \\ 0, & 2 \le t \end{cases}$

15.61 $y(t) = \begin{cases} 0, & t < 0 \\ 0.5t^2, & 0 \le t < 1 \\ t - 0.5, & 1 \le t < 2 \\ 2 - 0.5(t - 1)^2, & 2 \le t < 3 \\ 0, & 3 \le t \end{cases}$

15.63 $y(t) = \begin{cases} 0, & t < 0 \\ 0.5(1 - e^{-2t}), & 0 \le t < 5 \\ \dfrac{e^{10} - 1}{2}e^{-2t}, & 5 \le t \end{cases}$

15.65 $y(t) = [(1/3)t - (1/18)\sin(6t)]\,u(t)$

15.67 $H(s) = 10{,}000/(s + 15{,}000)$

a. $X(s) = 1$, $Y(s) = H(s)X(s) = H(s)$, $y(t) = 10{,}000e^{-15{,}000t}u(t)$

b. $X(s) = 1/s$, $Y(s) = H(s)X(s) = 10{,}000/[s(s + 15{,}000)]$, $y(t) = 0.6667\,(1 - e^{-15{,}000t})\,u(t)$

c. $X(s) = 1/(s + 20{,}000)$, $Y(s) = H(s)X(s) = 10{,}000/[(s + 20{,}000)(s + 15{,}000)]$, $y(t) = 2\,(e^{-15{,}000t} - e^{-20{,}000t})\,u(t)$

15.69

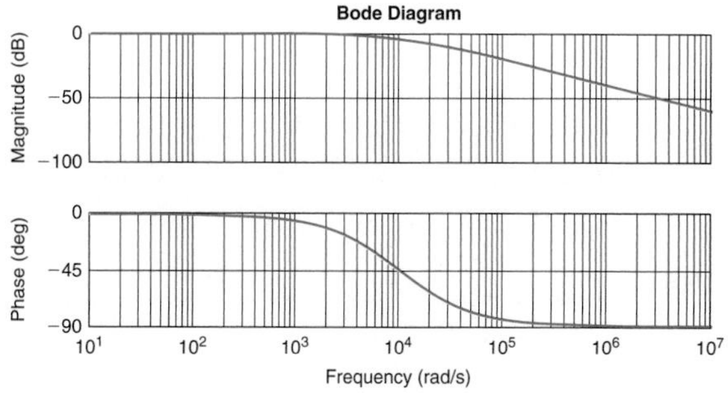

15.71

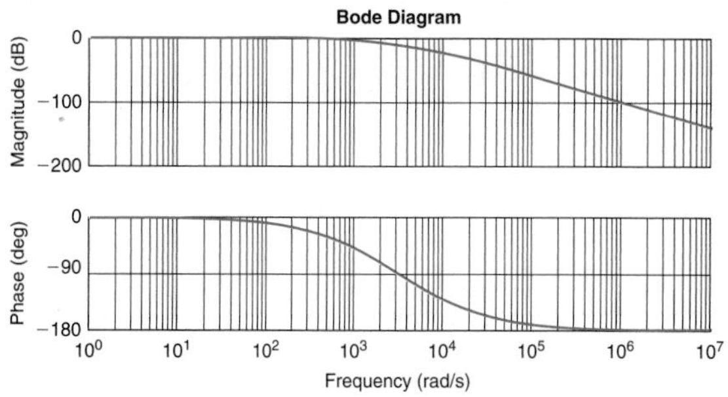

15.73

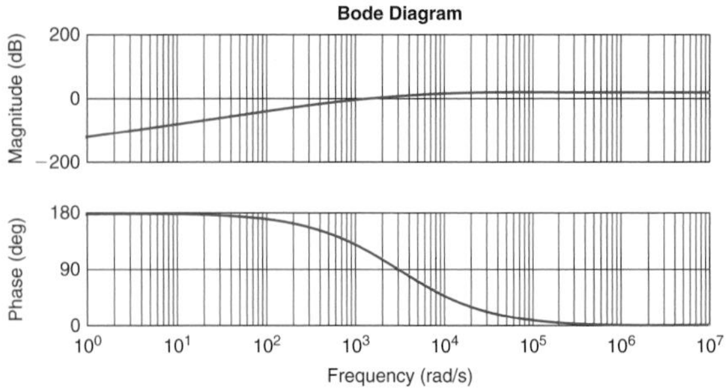

15.75

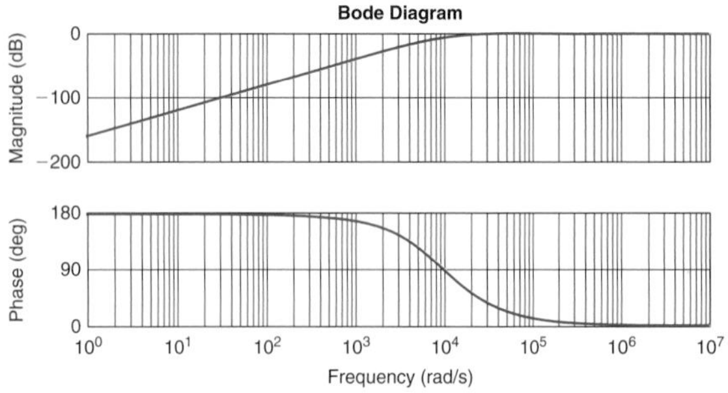

15.77

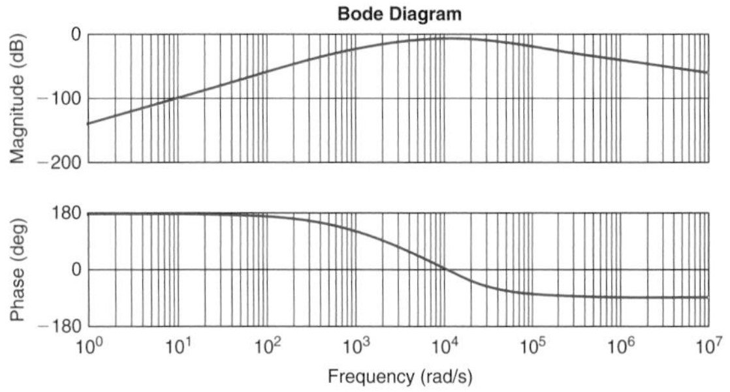

15.79

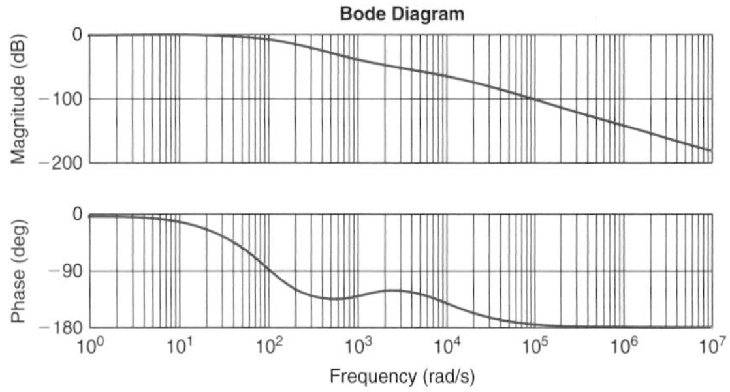

15.81

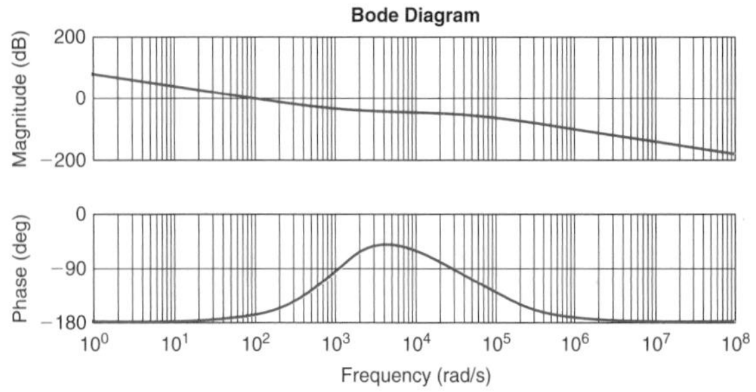

15.83

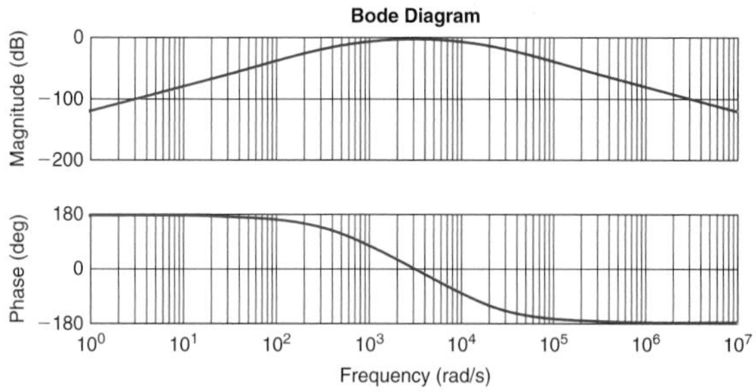

15.85 $H(s) = \dfrac{10^4 s}{(s + 100)(s + 10{,}000)}$

CHAPTER 16

16.5 $R = 1.4142\ k\Omega, L = 132.4718$ mH, $C = 0.13247\ \mu F$

16.7 $R = 707.1068\ \Omega, L = 139.184$ mH, $C = 0.139184\ \mu F$

16.9 a. $10^8/(s^2 + 30{,}000s + 10^8)$

b. $\omega_0 = 10{,}000$ rad/s

c. $Q = 1/3$

d. $\omega_c = 3742.3915$ rad/s

e.

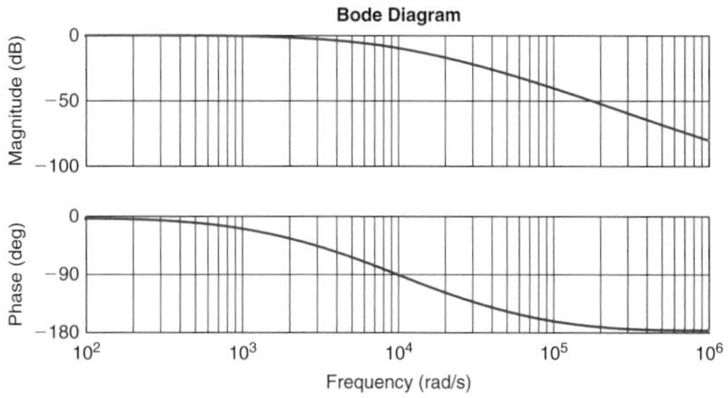

16.11 $R_1 = 1\ k\Omega, R_2 = 1\ k\Omega, C_1 = 0.4456\ \mu F,$
$C_2 = 9.0946\ nF$

16.13 First Second-Order Section: $R_1 = 1\ k\Omega, R_2 = 1\ k\Omega,$
$C_1 = 0.3889\ \mu F, C_2 = 56.955\ nF$
Second Second-Order Section: $R_1 = 1\ k\Omega,$
$R_2 = 1\ k\Omega, C_1 = 0.1611\ \mu F, C_2 = 0.1375\ \mu F$

16.15 First-Order Section: $R = 1\ k\Omega, C = 97.1126\ nF$
First Second-Order Section: $R_1 = 1\ k\Omega, R_2 = 1\ k\Omega,$
$C_1 = 0.12\ \mu F, C_2 = 78.5658\ nF$
Second Second-Order Section: $R_1 = 1\ k\Omega,$
$R_2 = 1\ k\Omega, C_1 = 0.3143\ \mu F, C_2 = 30.0094\ nF$

16.17 First-Order Section: $R = 1\ k\Omega, C = 0.1145\ \mu F,$
First Second-Order Section: $R_2 = 1\ k\Omega, R_B = 1\ k\Omega,$
$R_A = 381.9638\ \Omega, R_a = 1.382\ k\Omega, R_b = 3.618\ k\Omega,$
$C_1 = 0.1145\ \mu F, C_2 = 0.1145\ \mu F$
Second Second-Order Section: $R_2 = 1\ k\Omega,$
$R_B = 1\ k\Omega, R_A = 1.382\ k\Omega, R_a = 2.382\ k\Omega,$
$R_b = 1.7236\ k\Omega, C_1 = 0.1145\ \mu F, C_2 = 0.1145\ \mu F$

16.23 $R = 1.4142\ k\Omega$, L = 79.9027 mH, $C = 79.9027\ nF$

16.25 $R = 707.1068\ \Omega$, L = 77.4432 mH, $C = 77.4432\ nF$

16.27 $R_1 = 111.1111\ \Omega, R_2 = 9\ k\Omega, C_1 = 31.831\ nF,$
$C_2 = 31.831\ nF$

16.29 First Second-Order Section:
a. $\omega_0 = 21396.4950$ rad/s, $Q = 0.7846$

b. $R_1 = 0.6373\ \Omega, R_2 = 1.5691\ \Omega, C_1 = 1\ \text{F}, C_2 = 1\ \text{F}$
c. $R_1 = 637.3006\ \Omega, R_2 = 1.5691\ k\Omega, C_1 = 46.7366\ nF$, $C_2 = 46.7366\ nF$

Second Second-Order Section:

a. $\omega_0 = 11386.8345$ rad/s, $Q = 3.5595$
b. $R_1 = 0.1404692\ \Omega, R_2 = 7.119\ \Omega, C_1 = 1\ \text{F}$, $C_2 = 1\ \text{F}$
c. $R_1 = 140.4692\ \Omega, R_2 = 7.119\ k\Omega, C_1 = 87.8207\ nF$, $C_2 = 87.8207\ nF$

16.31 $R_1 = 1\ k\Omega, R_2 = 1\ k\Omega, R_B = 1\ k\Omega, R_A = 1.3333\ k\Omega$, $C_1 = 53.0516\ nF, C_2 = 53.0516\ nF$

16.33 a. $H(s) = 10^5 s/(s^2 + 10^5 s + 10^{10})$
b. $\omega_0 = 10^5$
c. $Q = 1$
d. $\omega_1 = 61803.4$ rad/s, $\omega_2 = 161803.4$ rad/s, $\omega_{3\text{dB}} = 10^5$
e.

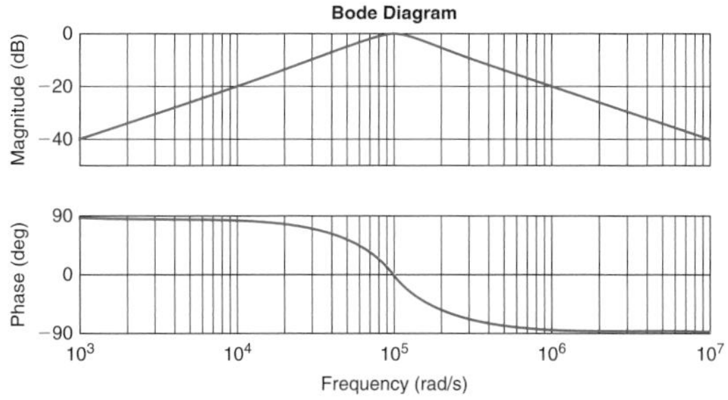

16.35 a. $H(s) = 20{,}000s/(s^2 + 20{,}000s + 10^{11})$
b. $\omega_0 = 3.16228 \times 10^5$
c. $Q = 15.8114$
d. $\omega_1 = 306385.84$ rad/s, $\omega_2 = 32{,}6385.84$ rad/s, $\omega_{3\text{dB}} = 20{,}000$
e.

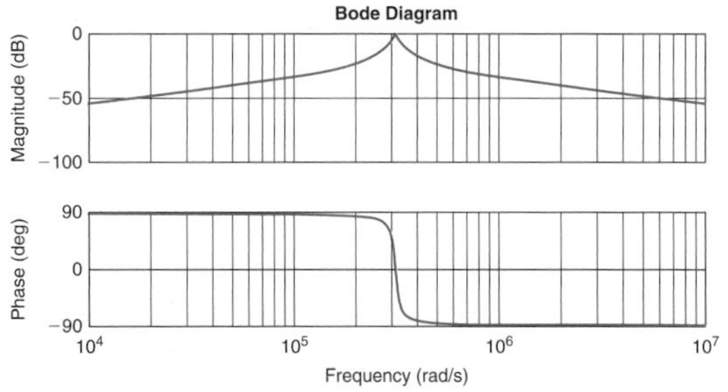

16.37 $R = 2.65\ k\Omega, L = 30.4544$ mH, $C = 30.4544\ nF$

16.39 First Second-Order Section: $R_2 = 82.3657\ k\Omega$, $R_a = 41.1828\ k\Omega, R_b = 1.0249\ k\Omega, C = C_1 = C_2 = 5.9632\ nF$
Second Second-Order Section: $R_2 = 82.3657\ k\Omega$, $R_a = 41.1822\ k\Omega, R_b = 1.0249\ k\Omega, C = C_1 = C_2 = 8.5955\ nF$

16.41 $R_2 = 25\ k\Omega, R_a = 12.5\ k\Omega, R_b = 1.08696\ k\Omega$, $C = C_1 = C_2 = 11.7893\ nF$

16.43 $R_2 = R_3 = R_B = 1\ k\Omega, R_A = 2.3851\ k\Omega$, $R_a = 5.5054\ k\Omega, R_b = 1.222\ k\Omega, C = C_1 = C_2 = 0.1125\ \mu F$

16.45 a. $H(s) = (s^2 + 10^{10})/(s^2 + 10^5 s + 10^{10})$
b. $\omega_0 = 10^5$
c. $Q = 1$
d. $\omega_1 = 61803.4$ rad/s, $\omega_2 = 161803.4$ rad/s, $\omega_{3\text{dB}} = 10^5$
e.

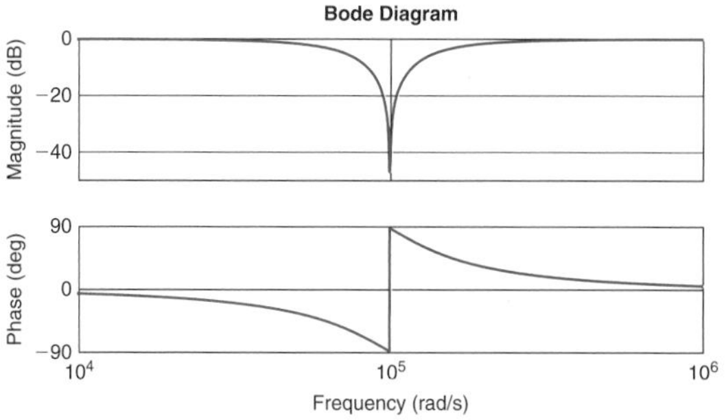

16.47 $R = 1.4515\ k\Omega, L = 69.9963$ mH, $C = 69.9963\ nF$

16.49 a. $\omega_0 = 30{,}000$ rad/s, $Q = 1.9355$
b. $R = 1.9355\ k\Omega, L = 33.3333$ mH, $C = 33.3333\ nF$

16.51 $R = 1\ k\Omega, R/2 = 500\ \Omega, R_B = 1\ k\Omega$, $R_A = 807.6923\ \Omega, C = 58.9463\ nF, 2C = 0.1179\ \mu F$, $R_D = 1\ k\Omega, R_C = 807.6923\ \Omega$

16.53 a. $H(s) = -10{,}000\text{s}/(s^2 + 20{,}000\text{s} + 10^8)$
b. $\omega_0 = 10{,}000, Q = 0.5$
c. BPF

16.55 a. $H(s) = 2.5 \times 10^9/(s^2 + 50{,}000\text{s} + 2.5 \times 10^9)$
b. $\omega_0 = 50{,}000, Q = 1$
c. LPF

16.57 a. $H(s) = 8/(s^3 + 4s^2 + 6s + 4)$
b. LPF

16.59 a. $H(s) = s/(s^2 + 1.1s + 2.1)$
b. $\omega_0 = 1.4491, Q = 1.3174$
c. BPF

16.61 a. $H(s) = 2/(s^3 + 4s^2 + 6s + 2)$
b. LPF

16.63 $H(s) = (2s + 2)/(s^2 + s + 1)$, LPF

16.65 a. $$\frac{V_1(s)}{V_s(s)} = \frac{\dfrac{-1}{R_3R_4C_1C_2}}{s^2 + \dfrac{1}{R_1C_1}s + \dfrac{1}{R_2R_4C_1C_2}}$$

b. $$\frac{V_3(s)}{V_s(s)} = \frac{\dfrac{-1}{R_3C_1}s}{s^2 + \dfrac{1}{R_1C_1}s + \dfrac{1}{R_2R_4C_1C_2}}$$

c. $$\frac{V_4(s)}{V_s(s)} = \frac{s^2 + \dfrac{1}{R_2C_4C_1C_2}s}{s^2 + \dfrac{1}{R_1C_1}s + \dfrac{1}{R_2R_4C_1C_2}}$$

CHAPTER 17

17.1 a. $\Omega_s = 2.1429$ rad/s, $n = 2, A_{s_1} = 5$ dB, $A_{p_1} = 0.424$ dB, $\varepsilon_{p_1} = 0.3202, \Omega_{c_1} = 1.7671$ rad/s
b. $S = -0.7071 \pm j0.7071, SZ = \infty, SZ = \infty$
c. $H(S) = 1/(S^2 + 1.4142S + 1), Q = 0.7071$

d. $R_2 = 1\ \Omega, R_B = 1\ \Omega, R_A = 0.5858\ \Omega,$
$R_a = 1.5858\ \Omega, R_b = 2.7071\ \Omega, C_1 = 1$ F,
$C_2 = 1$ F
e. $R_2 = 1\ k\Omega, R_B = 1\ k\Omega, R_A = 585.7864\ \Omega,$
$R_a = 1.5858\ k\Omega, R_b = 2.7071\ k\Omega,$
$C_1 = 98.9722\ nF, C_2 = 98.9722\ nF$

17.3 a. $\Omega_s = 2.3684$ rad/s, $n = 2, A_{p_1} = 0.9$ dB,
$A_{s_1} = 9.1622$ dB, $\varepsilon_{p_1} = 0.4799, \Omega_{c_1} = 1.4436$ rad/s
b. $S = -0.7071 \pm j0.7071, SZ = \infty, SZ = \infty$
c. $H(S) = 1/(S^2 + 1.4142S + 1), Q = 0.7071$
d. $R_2 = 1\ \Omega, R_B = 1\ \Omega, R_A = 0.5858\ \Omega, R_a =$
$1.5858\ \Omega, R_b = 2.7071\ \Omega, C_1 = 1$ F, $C_2 = 1$ F
e. $R_2 = 1\ k\Omega, R_B = 1\ k\Omega, R_A = 585.7864\ \Omega,$
$R_a = 1.5858\ k\Omega, R_b = 2.7071\ k\Omega,$
$C_1 = 0.1161\ \mu F, C_2 = 0.1161\ \mu F$

17.5 a. $\Omega_s = 2.3333$ rad/s, $n = 3, A_{p_1} = 0.3823$ dB,
$A_{s_1} = 12$ dB, $\varepsilon_{p_1} = 0.3033, \Omega_{c_1} = 1.4883$ rad/s
b. $S = -1, S = -0.5 \pm j0.866, SZ = \infty, SZ = \infty,$
$SZ = \infty$
c. $H(S) = 1/[(S + 1)(S^2 + S + 1)], Q = 1$
d. Second-Order Section:
$R_1 = 1\ \Omega, R_2 = 1\ \Omega, C_1 = 2$ F, $C_2 = 0.5$ F
First-Order Section:
$R = 1\ \Omega, C = 1$ F
e. Second-Order Section:
$R_1 = 1\ k\Omega, R_2 = 1\ k\Omega, C_1 = 0.2376\ \mu F,$
$C_2 = 0.05941\ \mu F$
First-Order Section: $R = 1\ k\Omega, C = 0.1188\ \mu F$

17.7 a. $\Omega_s = 2.4444$ rad/s, $n = 3, A_{p_1} = 0.8$ dB, $A_{s_1} =$
16.4495 dB, $\varepsilon_{p_1} = 0.4497, \Omega_{c_1} = 1.3052$ rad/s
b. $S = -1, S = -0.5 \pm j0.866, SZ = \infty, SZ = \infty,$
$SZ = \infty$
c. $H(S) = 1/[(S + 1)(S^2 + S + 1)], Q = 1$
d. Second-Order Section: $R_2 = 1\ \Omega, R_A = 1\ \Omega,$
$R_B = 1\ \Omega, R_a = 2\ \Omega, R_b = 2\ \Omega, C_1 = 1$ F, $C_2 = 1$ F
First-Order Section: $R = 1\ \Omega, C = 1$ F
e. Second-Order Section: $R_2 = 1\ k\Omega, R_A = 1\ k\Omega,$
$R_B = 1\ k\Omega, R_a = 2\ k\Omega, R_b = 2\ k\Omega, C_1 = 0.1355\ \mu F,$
$C_2 = 0.1355\ \mu F$
First-Order Section: $R = 1\ k\Omega, C = 0.1355\ \mu F$

17.9 a. $\Omega_s = 2$ rad/s, $n = 4, A_{p_1} = 1.1$ dB, $A_{s_1} = 18.7385$ dB,
$\varepsilon_{p_1} = 0.5369, \Omega_{c_1} = 1.1682$ rad/s
b. $S = -0.9239 \pm j0.3827, S = -0.3827 \pm j0.9239,$
$SZ = \infty, SZ = \infty, SZ = \infty, SZ = \infty$
c. $H(S) = 1/[(S^2 + 1.8478S + 1)(S^2 + 0.7654S$
$+ 1)], Q = 0.5412, Q = 1.3066$
d. Second-Order Section 1: $R_2 = 1\ \Omega,$
$R_A = 0.1522\ \Omega, R_B = 1\ \Omega, R_a = 1.1522\ \Omega,$
$R_b = 7.5685\ \Omega, C_1 = 1$ F, $C_2 = 1$ F
Second-Order Section 2: $R_2 = 1\ \Omega,$
$R_A = 1.2346\ \Omega, R_B = 1\ \Omega, R_a = 2.2346\ \Omega,$
$R_b = 1.81\ \Omega, C_1 = 1$ F, $C_2 = 1$ F
e. Second-Order Section 1: $R_2 = 1\ k\Omega,$
$R_A = 152.241\ \Omega, R_B = 1\ k\Omega, R_a = 1.1522\ k\Omega,$
$R_b = 7.5685\ k\Omega, C_1 = 0.1239\ \mu F, C_2 = 0.1239\ \mu F$
Second-Order Section 2: $R_2 = 1\ k\Omega,$
$R_A = 1.2346\ k\Omega, R_B = 1\ k\Omega, R_a = 2.2346\ k\Omega,$
$R_b = 1.81\ k\Omega, C_1 = 0.1239\ \mu F, C_2 = 0.1239\ \mu F$

17.11 a. $\Omega_s = 1.9565$ rad/s, $n = 4, A_{p_1} = 1.15$ dB, $A_{s_1} =$
18.2018 dB, $\varepsilon_{p_1} = 0.5506, \Omega_{c_1} = 1.1609$ rad/s
b. $S = -0.9239 \pm j0.3827, S = -0.3827 \pm j0.9239,$
$SZ = \infty, SZ = \infty, SZ = \infty, SZ = \infty$
c. $H(S) = 1/[(S^2 + 1.8478S + 1)(S^2 + 0.7654S + 1)],$
$Q = 0.5412, Q = 1.3066$
d. Second-Order Section 1: $R_1 = 1\ \Omega, R_2 = 1\ \Omega,$
$C_1 = 1.0824$ F, $C_2 = 0.9239$ F
Second-Order Section 2: $R_1 = 1\ \Omega, R_2 = 1\ \Omega,$
$C_1 = 2.6131$ F, $C_2 = 0.3827$ F
e. Second-Order Section 1: $R_1 = 1\ k\Omega, R_2 = 1\ k\Omega,$
$C_1 = 0.129\ \mu F, C_2 = 0.1101\ \mu F$
Second-Order Section 2: $R_1 = 1\ k\Omega, R_2 = 1\ k\Omega,$
$C_1 = 0.3115\ \mu F, C_2 = 0.04562\ \mu F$

17.13 a. $\Omega_s = 1.96$ rad/s, $n = 5, A_{p_1} = 1.5$ dB,
$A_{s_1} = 25.3928$ dB, $\varepsilon_{p_1} = 0.6423, \Omega_{c_1} = 1.0926$ rad/s
b. $S = -1, S = -0.8090 \pm j0.5878, S = -0.3090 \pm$
$j0.9511, SZ = \infty, SZ = \infty, SZ = \infty, SZ = \infty,$
$SZ = \infty$
c. $H(S) = 1/[(S + 1)(S^2 + 1.618S + 1)$
$(S^2 + 0.618S + 1)], Q = 0.618, Q = 1.618$
d. Second-Order Section 1: $R_2 = 1\ \Omega, R_A = 0.382\ \Omega,$
$R_B = 1\ \Omega, R_a = 1.382\ \Omega, R_b = 3.618\ \Omega,$
$C_1 = 1$ F, $C_2 = 1$ F
Second-Order Section 2: $R_2 = 1\ \Omega,$
$R_A = 1.382\ \Omega, R_B = 1\ \Omega, R_a = 2.382\ \Omega,$
$R_b = 1.7236\ \Omega, C_1 = 1$ F, $C_2 = 1$ F
First-Order Section: $R = 1\ \Omega, C = 1$ F
e. Second-Order Section 1: $R_2 = 1\ k\Omega,$
$R_A = 381.966\ \Omega, R_B = 1\ k\Omega, R_a = 1.382\ k\Omega,$
$R_b = 3.618\ k\Omega, C_1 = 0.1165\ \mu F, C_2 = 0.1165\ \mu F$
Second-Order Section 2: $R_2 = 1\ k\Omega,$
$R_A = 1.382\ k\Omega, R_B = 1\ k\Omega, R_a = 2.382\ k\Omega,$
$R_b = 1.7236\ k\Omega, C_1 = 0.1165\ \mu F, C_2 = 0.1165\ \mu F$
First-Order Section: $R = 1\ k\Omega, C = 0.1165\ \mu F$

17.15 a. $\Omega_s = 2.12$ rad/s, $n = 2, A_{s_1} = 6$ dB, $A_{p_1} = 0.5978$ dB,
$\varepsilon_{p_1} = 0.3842, \Omega_{c_1} = 1.6134$ rad/s
b. $S = -0.7071 \pm j0.7071, SZ = \infty, SZ = \infty$
c. $H(S) = 1/(S^2 + 1.4142S + 1), Q = 0.7071$
d. $R_1 = 1\ \Omega, R_2 = 1\ \Omega, R_B = 1\ \Omega, R_A = 0.5858\ \Omega,$
$C_1 = 1$ F, $C_2 = 1$ F, $R_D = 1\ \Omega, R_C = 0.5858\ \Omega$
e. $R_1 = 1\ k\Omega, R_2 = 1\ k\Omega, R_B = 1\ k\Omega,$
$R_A = 585.7864\ \Omega, C_1 = 96.8985\ nF,$
$C_2 = 96.8985\ nF, R_D = 1\ k\Omega, R_C = 585.8\ \Omega$

17.17 a. $\Omega_s = 2.04$ rad/s, $n = 2, A_{s_1} = 6.5$ dB,
$A_{p_1} = 0.7925$ dB, $\varepsilon_{p_1} = 0.4474, \Omega_{c_1} = 1.4950$ rad/s
b. $S = -0.7071 \pm j0.7071, SZ = \infty, SZ = \infty$
c. $H(S) = 1/(S^2 + 1.4142S + 1), Q = 0.7071$
d. $R_1 = 0.7071\ \Omega, R_2 = 1.4142\ \Omega, C_1 = 1$ F, $C_2 = 1$ F
e. $R_1 = 707.1\ \Omega, R_2 = 1.4142\ k\Omega, C_1 = 93.3097\ nF,$
$C_2 = 93.3097\ nF$

17.19 a. $\Omega_s = 2.04$ rad/s, $n = 3$, $A_{s_1} = 11$ dB, $A_{p_1} = 0.6476$ dB, $\varepsilon_{p_1} = 0.401$, $\Omega_{c_1} = 1.3561$ rad/s
b. $S = -1, S = -0.5 \pm j0.866, SZ = \infty, SZ = \infty, SZ = \infty$
c. $H(S) = 1/[(S+1)(S^2+S+1)], Q = 1$
d. Second-Order Section: $R_1 = 1\ \Omega, R_2 = 1\ \Omega, R_B = 1\ \Omega, R_A = 1\ \Omega, C_1 = 1$ F, $C_2 = 1$ F, $R_D = 1\ \Omega, R_C = 1\ \Omega$
First-Order Section: $R = 1\ \Omega, C = 1$ F
e. $R_1 = 1\ k\Omega, R_2 = 1\ k\Omega, R_B = 1\ k\Omega, R_A = 1\ k\Omega, C_1 = 84.6384\ nF, C_2 = 84.6384\ nF, R_D = 1\ k\Omega, R_C = 1\ k\Omega$
First-Order Section: $R = 1\ k\Omega, C = 84.6384\ nF$

17.21 a. $\Omega_s = 2.12$ rad/s, $n = 3$, $A_{s_1} = 10$ dB, $A_{p_1} = 0.4105$ dB, $\varepsilon_{p_1} = 0.3149$, $\Omega_{c_1} = 1.4699$ rad/s
b. $S = -1, S = -0.5 \pm j0.866, SZ = \infty, SZ = \infty, SZ = \infty$
c. $H(S) = 1/[(S+1)(S^2+S+1)], Q = 1$
d. Second-Order Section: $R_1 = 0.5\ \Omega, R_2 = 2\ \Omega, C_1 = 1$ F, $C_2 = 1$ F
First-Order Section: $R = 1\ \Omega, C = 1$ F
e. $R_1 = 500\ \Omega, R_2 = 2\ k\Omega, C_1 = 88.2815\ nF, C_2 = 88.2815\ nF$
First-Order Section: $R = 1\ k\Omega, C = 88.2815\ nF$

17.23 a. $\Omega_s = 2.04$ rad/s, $n = 4$, $A_{s_1} = 14$ dB, $A_{p_1} = 0.3359$ dB, $\varepsilon_{p_1} = 0.2836$, $\Omega_{c_1} = 1.3704$ rad/s
b. $S = -0.9239 \pm j0.3827, S = -0.3827 \pm j0.9239, SZ = \infty. SZ = \infty, SZ = \infty, SZ = \infty$
c. $H(S) = 1/[(S^2 + 1.8478S + 1)(S^2 + 0.7654S + 1)], Q = 0.5412, Q = 1.3066$
d. Second-Order Section 1: $R_1 = 1\ \Omega, R_2 = 1\ \Omega, R_B = 1\ \Omega, R_A = 0.1522409\ \Omega, C_1 = 1$ F, $C_2 = 1$ F, $K = 1.1522409, R_D = 1\ \Omega, R_C = K - 1 = 0.1522409\ \Omega$
Second-Order Section 2: $R_1 = 1\ \Omega, R_2 = 1\ \Omega, R_B = 1\ \Omega, R_A = 1.2346\ \Omega, C_1 = 1$ F, $C_2 = 1$ F, $K = 2.2346, R_D = 1\ \Omega, R_C = K - 1 = 1.2346\ \Omega$
e. Second-Order Section 1: $R_1 = 1\ k\Omega, R_2 = 1\ k\Omega, R_B = 1\ k\Omega, R_A = 152.2409\ \Omega, C_1 = 85.5294\ nF, C_2 = 85.5294\ nF, R_D = 1\ k\Omega, R_C = 152.2409\ \Omega$
Second-Order Section 2: $R_1 = 1\ k\Omega, R_2 = 1\ k\Omega, R_B = 1\ k\Omega, R_A = 1.2346\ k\Omega, C_1 = 85.5294\ nF, C_2 = 85.5294\ nF, R_D = 1\ k\Omega, R_C = 1.2346\ k\Omega$
Single$K = 1.1522409 \times 2.2346331 = 2.574836$
$R_D = 1\ \Omega, R_C =$ Single$K - 1 = 1.574836$

17.25 a. $\Omega_s = 1.9655$ rad/s, $n = 4$, $A_{s_1} = 18.5668$ dB, $A_{p_1} = 1.2$ dB, $\varepsilon_{p_1} = 0.5641$, $\Omega_{c_1} = 1.1539$ rad/s
b. $S = -0.9239 \pm j0.3827, S = -0.3827 \pm j0.9239, SZ = \infty. SZ = \infty, SZ = \infty, SZ = \infty$
c. $H(S) = 1/[(S^2 + 1.8478S + 1)(S^2 + 0.7654S + 1)], Q = 0.5412, Q = 1.3066$
d. Second-Order Section 1: $R_1 = 0.9239\ \Omega, R_2 = 1.0824\ \Omega, C_1 = 1$ F, $C_2 = 1$ F
Second-Order Section 2: $R_1 = 0.3827\ \Omega, R_2 = 2.6131\ \Omega, C_1 = 1$ F, $C_2 = 1$ F
e. Second-Order Section 1: $R_1 = 923.8795\ \Omega, R_2 = 1.0824\ k\Omega, C_1 = 64.4359\ nF, C_2 = 64.4359\ nF$
Second-Order Section 2: $R_1 = 382.6834\ \Omega, R_2 = 2.6131\ k\Omega, C_1 = 64.4359\ nF, C_2 = 64.4359\ nF$

17.27 a. $\Omega_s = 1.8387$ rad/s, $n = 5$, $A_{s_1} = 21.5099$ dB, $A_{p_1} = 1.2$ dB, $\varepsilon_{p_1} = 0.5641$, $\Omega_{c_1} = 1.1213$ rad/s
b. $S = -1, S = -0.8090 \pm j0.5878, S = -0.3090 \pm j0.9511, SZ = \infty, SZ = \infty. SZ = \infty, SZ = \infty, SZ = \infty$
c. $H(S) = 1/[(S+1)(S^2 + 1.618S + 1)(S^2 + 0.618S + 1)], Q = 0.618, Q = 1.618$
d. Second-Order Section 1: $R_1 = 1\ \Omega, R_2 = 1\ \Omega, R_B = 1\ \Omega, R_A = 0.382\ \Omega, C_1 = 1$ F, $C_2 = 1$ F, $R_D = 1\ \Omega, R_C = 0.381966\ \Omega$
Second-Order Section 2: $R_1 = 1\ \Omega, R_2 = 1\ \Omega, R_B = 1\ \Omega, R_A = 1.382\ \Omega, C_1 = 1$ F, $C_2 = 1$ F, $R_D = 1\ \Omega, R_C = 1.2346\ \Omega$
First-Order Section: $R_1 = 1\ \Omega, C_1 = 1$ F
Single$K = 3.2918$
$R_D = 1\ \Omega, R_C =$ Single$K - 1 = 2.2918$
e. Second-Order Section 1: $R_1 = 1\ k\Omega, R_2 = 1\ k\Omega, R_B = 1\ k\Omega, R_A = 381.966\ \Omega, C_1 = 62.6178\ nF, C_2 = 62.6178\ nF, R_D = 1\ k\Omega, R_C = 381.966\ \Omega$
Second-Order Section 2: $R_1 = 1\ k\Omega, R_2 = 1\ k\Omega, R_B = 1\ k\Omega, R_A = 1.382\ k\Omega, C_1 = 62.6178\ nF, C_2 = 62.6178\ nF, R_D = 1\ k\Omega, R_C = 1.2346\ k\Omega$
First-Order Section: $R = 1\ k\Omega, C = 62.6178\ nF$
$R_D = 1\ k\Omega, R_C = 2.2918\ k\Omega$

17.29 a. $\Omega_s = 2.3589$ rad/s, $n = 1$, $A_{s_1} = 3$ dB, $A_{p_1} = 0.7146$ dB, $\varepsilon_{p_1} = 0.4229$, $\Omega_{c_1} = 2.3645$ rad/s
b. $S = -1, SZ = \infty$
c. $H(s) = 15{,}005.466s/(s^2 + 15{,}005.466s + 2.8957 \times 10^8), Q = 1.134, \omega_0 = 17{,}016.8797$ rad/s
d. $R_2 = 1\ \Omega, R_3 = 1\ \Omega, R_B = 1\ \Omega, R_A = 1.7529\ \Omega, R_a = 2.2076\ \Omega, R_b = 1.8281\ \Omega, C_1 = 1.4142$ F, $C_2 = 1.4142$ F
e. $R_2 = 1\ k\Omega, R_3 = 1\ k\Omega, R_B = 1\ k\Omega, R_A = 1.7529\ k\Omega, R_a = 2.2076\ k\Omega, R_b = 1.8281\ k\Omega, C_1 = 83.1065\ nF, C_2 = 83.1065\ nF$

17.31 a. $\Omega_s = 2.9142$ rad/s, $n = 1$, $A_{s_1} = 4.7067$ dB, $A_{p_1} = 0.9$ dB, $\varepsilon_{p_1} = 0.4799$, $\Omega_{c_1} = 2.0839$ rad/s
b. $S = -1, SZ = \infty$
c. $H(s) = 10{,}605.896s/(s^2 + 10{,}605.896s + 2.9317 \times 10^8), Q = 1.6144, \omega_0 = 17{,}122.1123$ rad/s
d. $R_2 = 10.4251\ \Omega, R_a = 5.2125\ \Omega, R_b = 1.2374\ \Omega, C_1 = 0.3097$ F, $C_2 = 0.3097$ F
e. $R_2 = 10.4251\ k\Omega, R_a = 5.2125\ k\Omega, R_b = 1.2374\ k\Omega, C_1 = 18.0885\ nF, C_2 = 18.0885\ nF$

17.33 a. $\Omega_s = 1.9911$ rad/s, $n = 2$, $A_{s_1} = 5$ dB, $A_{p_1} = 0.5598$ dB, $\varepsilon_{p_1} = 0.3709$, $\Omega_{c_1} = 1.6420$ rad/s
b. $S = -0.7071 \pm j0.7071, SZ = \infty, SZ = \infty$
c. $H_1(s) = 11{,}128.8842s/(s^2 + 11{,}128.8842s + 4.8638 \times 10^8), Q = 1.9817, \omega_0 = 22{,}053.9332$, $H_2(s) = 6525.5480s/(s^2 + 6525.5480s + 1.6723 \times 10^8), Q = 1.9817, \omega_0 = 12{,}931.5749$ rad/s

d. Second-Order Section 1: $R_2 = 1\ \Omega, R_3 = 1\ \Omega$, $R_B = 1\ \Omega, R_A = 2.2864\ \Omega, R_a = 4.6050\ \Omega$, $R_b = 1.2774\ \Omega, C_1 = 1.4142$ F, $C_2 = 1.4142$ F
Second-Order Section 2: $R_2 = 1\ \Omega, R_3 = 1\ \Omega$, $R_B = 1\ \Omega, R_A = 2.2864\ \Omega, R_a = 4.6050\ \Omega$, $R_b = 1.2774\ \Omega, C_1 = 1.4142$ F, $C_2 = 1.4142$ F
$R_D = 1\ \Omega, R_C =$ Singleb0 $- 1 = 1.1459\ \Omega$
e. Second-Order Section 1: $R_2 = 1\ k\Omega, R_3 = 1\ k\Omega$, $R_B = 1\ k\Omega, R_A = 2.2864\ k\Omega, R_a = 4.6050\ k\Omega$, $R_b = 1.2774\ \Omega, C_1 = 0.06413\ \mu F, C_2 = 0.06413\ \mu F$
Second-Order Section 2: $R_2 = 1\ k\Omega, R_3 = 1\ k\Omega$, $R_B = 1\ k\Omega, R_A = 2.2864\ k\Omega, R_a = 4.6050\ k\Omega$, $R_b = 1.2774\ k\Omega, C_1 = 0.1094\ \mu F, C_2 = 0.1094\ \mu F$
$R_D = 1\ k\Omega, R_C = 1.1459\ k\Omega$

17.35 a. $\Omega_s = 2.0976$ rad/s, $n = 2, A_{s_1} = 8.1823$ dB, $A_{p_1} = 1.1$ dB, $\varepsilon_{p_1} = 0.5369, \Omega_{c_1} = 1.3648$ rad/s
b. $S = -0.7071 \pm j0.7071, SZ = \infty, SZ = \infty$
c. $H_1(s) = 8040.5873s/(s^2 + 8040.5873s + 4.4273 \times 10^8), Q = 2.6169, \omega_0 = 21{,}041.2001, H_2(s) = 5420.3619s/(s^2 + 5420.3619s + 2.012 \times 10^8)$, $Q = 2.6169, \omega_0 = 14{,}184.4018$ rad/s
d. Second-Order Section 1: $R_2 = 27.3921\ \Omega$, $R_a = 13.6961\ \Omega, R_b = 1.0788\ \Omega, C_1 = 0.1911$ F, $C_2 = 0.1911$ F
Second-Order Section 2: $R_2 = 27.3921\ \Omega$, $R_a = 13.6961\ \Omega, R_b = 1.0788\ \Omega, C_1 = 0.1911$ F, $C_2 = 0.1911$ F
$R_D = 1\ \Omega, R_C =$ Singleb0 $- 1 = 1.07876\ \Omega$
e. Second-Order Section 1: $R_2 = 27.3921\ k\Omega$, $R_a = 13.6961\ k\Omega, R_b = 1.0788\ k\Omega, C_1 = 9.0806\ nF, C_2 = 9.0806\ nF$
Second-Order Section 2: $R_2 = 27.3921\ k\Omega$, $R_a = 13.6961\ k\Omega, R_b = 1.0788\ k\Omega$, $C_1 = 13.4703\ nF, C_2 = C_1 = 13.4703\ nF$
$R_D = 1\ k\Omega, R_C = 1.07876\ k\Omega$

17.37 a. $\Omega_s = 2.7474$ rad/s, $n = 3, A_{s_1} = 21.1856$ dB, $A_{p_1} = 1.15$ dB, $\varepsilon_{p_1} = 0.5506, \Omega_{c_1} = 1.2201$ rad/s
b. $S = -1, S = -0.5 \pm j0.866, SZ = \infty, SZ = \infty$, $SZ = \infty$
c. $H_1(s) = 6209.4236s/(s^2 + 6209.4236s + 3.1531 \times 10^8), Q = 2.8597, \omega_0 = 17{,}757.0865, H_2(s) = 3571.2321s/(s^2 + 3571.2321s + 4.2683 \times 10^8)$, $Q = 5.7851, \omega_0 = 20{,}659.8706$ rad/s, $H_3(s) = 2638.1914s/(s^2 + 2638.1914s + 2.3293 \times 10^8)$, $Q = 5.7851, \omega_0 = 15{,}262.1538$ rad/s
d. Second-Order Section 1: $R_2 = 1\ \Omega, R_3 = 1\ \Omega$, $R_B = 1\ \Omega, R_A = 2.5055\ \Omega, R_a = 7.0885\ \Omega$, $R_b = 1.1642\ \Omega, C_1 = 1.4142$ F, $C_2 = 1.4142$ F
Second-Order Section 2: $R_2 = 1\ \Omega, R_3 = 1\ \Omega$, $R_B = 1\ \Omega, R_A = 2.7555\ \Omega, R_a = 15.3627\ \Omega$, $R_b = 1.0696\ \Omega, C_1 = 1.4142$ F, $C_2 = 1.4142$ F
Second-Order Section 3: $R_2 = 1\ \Omega, R_3 = 1\ \Omega$, $R_B = 1\ \Omega, R_A = 2.7555\ \Omega, R_a = 15.3627\ \Omega$, $R_b = 1.0696\ \Omega, C_1 = 1.4142$ F, $C_2 = 1.4142$ F
$R_D = 1\ \Omega, R_C =$ Singleb0 $- 1 = 3.0924\ \Omega$
e. Second-Order Section 1: $R_2 = 1\ k\Omega, R_3 = 1\ k\Omega$, $R_B = 1\ k\Omega, R_A = 2.5055\ k\Omega, R_a = 7.0885\ k\Omega$, $R_b = 1.1642\ k\Omega, C_1 = 79.6422\ nF, C_2 = 79.6422\ nF$
Second-Order Section 2: $R_2 = 1\ k\Omega, R_3 = 1\ k\Omega$, $R_B = 1\ k\Omega, R_A = 2.7555\ k\Omega, R_a = 15.3627\ k\Omega$, $R_b = 1.0696\ k\Omega, C_1 = 68.4522\ nF, C_2 = 68.4522\ nF$
Second-Order Section 3: $R_2 = 1\ k\Omega, R_3 = 1\ k\Omega$, $R_B = 1\ k\Omega, R_A = 2.7555\ k\Omega, R_a = 15.3627\ k\Omega$, $R_b = 1.0696\ k\Omega, C_1 = 92.6615\ nF, C_2 = 92.6615\ nF$
$R_D = 1\ k\Omega, R_C = 3.0924\ k\Omega$

17.39 a. $\Omega_s = 2.02$ rad/s, $n = 1, A_{s_1} = 2.7$ dB, $A_{p_1} = 0.8325$ dB, $\varepsilon_{p_1} = 0.4597, \Omega_{c_1} = 2.1754$ rad/s
b. $S = -1, SZ = \infty$
c. $H(s) = (s^2 + 1.8841 \times 10^8)/(s^2 + 8664.7056s + 1.8841 \times 10^8), Q = 1.5842, \omega_0 = 13{,}726.2795$ rad/s, $\omega_z = 13{,}726.2795$ rad/s
d. $K = 1.6844, R_1 = 1\ \Omega, R_2 = 1\ \Omega, R_3 = 0.5\ \Omega$, $R_B = 1\ \Omega, R_A = 0.6844\ \Omega, C_1 = 1$ F, $C_2 = 1$ F, $C_3 = 2$ F, $R_D = 1\ \Omega, R_C =$ Total Gain $- 1 = 0.6843753\ \Omega$
e. $R_1 = 1\ k\Omega, R_2 = 1\ k\Omega, R_3 = 500\ \Omega, R_B = 1\ k\Omega, R_A = 684.3753\ \Omega, C_1 = 72.853\ nF$, $C_2 = 72.853\ nF, C_3 = 0.1457\ \mu F, R_D = 1\ k\Omega$, $R_C = 684.3753\ \Omega$

17.41 a. $\Omega_s = 2.02$ rad/s, $n = 1, A_{s_1} = 2.4774$ dB, $A_{p_1} = 0.75$ dB, $\varepsilon_{p_1} = 0.4342, \Omega_{c_1} = 2.3034$ rad/s
b. $S = -1, SZ = \infty$
c. $H(s) = (s^2 + 1.8841 \times 10^8)/(s^2 + 8183.8821s + 1.8841 \times 10^8), Q = 1.6772, \omega_0 = 13{,}726.2795$ rad/s, $\omega_z = 13{,}726.2795$ rad/s
d. $K = 1.7019, R_1 = 1\ \Omega, R_2 = 1\ \Omega, R_3 = 0.5\ \Omega$, $R_B = 1\ \Omega, R_A = 0.7019\ \Omega, C_1 = 1$ F, $C_2 = 1$ F, $C_3 = 2$ F, $R_D = 1\ \Omega, R_C =$ Total Gain $- 1 = 0.70189\ \Omega$
e. $R_1 = 1\ k\Omega, R_2 = 1\ k\Omega, R_3 = 500\ \Omega, R_B = 1\ k\Omega$, $R_A = 701.89\ \Omega, C_1 = 72.853\ nF, C_2 = 72.853\ nF$, $C_3 = 0.1457\ \mu F, R_D = 1\ k\Omega, R_C = 701.89\ \Omega$

17.43 a. $\Omega_s = 1.6839$ rad/s, $n = 2, A_{s_1} = 4.303$ dB, $A_{p_1} = 0.83$ dB, $\varepsilon_{p_1} = 0.4589, \Omega_{c_1} = 1.4762$ rad/s
b. $S = -0.7071 \pm j0.7071, SZ = \infty, SZ = \infty$
c. $H_1(s) = (s^2 + 1.8841 \times 10^8)/(s^2 + 11{,}981.9082s + 3.7152 \times 10^8), Q = 1.6087, \omega_0 = 19{,}274.7457$ rad/s, $\omega_z = 13{,}726.2795$ rad/s, $H_2(s) = (s^2 + 1.8841 \times 10^8)/(s^2 + 6076.5118s + 0.9555 \times 10^8), Q = 1.6087$, $\omega_0 = 9775.0057$ rad/s, $\omega_z = 13{,}726.2795$ rad/s
d. Second-Order Section 1: $K = 1.7019, R_1 = 1\ \Omega$, $R_2 = 1\ \Omega, R_3 = 0.5\ \Omega, R_4 = 2.058\ \Omega, R_B = 1\ \Omega$, $R_A = 1.04946\ \Omega, C_1 = 1.4042$ F, $C_2 = 1.4042$ F, $C_3 = 2.8084$ F
Second-Order Section 2: $K = 1.7019, R_1 = 1\ \Omega$, $R_2 = 1\ \Omega, R_3 = 0.5\ \Omega, R_B = 1\ \Omega, R_A = 1.04946\ \Omega$, $C_1 = 0.7121$ F, $C_2 = 0.7121$ F, $C_3 = 1.4243$ F, $C_4 = 0.3460$ F, $R_D = 1\ \Omega, R_C =$ Total Gain $- 1 = 1.1301\ \Omega$

e. Second-Order Section 1: $R_1 = 1\ k\Omega, R_2 = 1\ k\Omega,$ $R_3 = 500\ \Omega, R_4 = 2.058\ k\Omega, R_B = 1\ k\Omega,$ $R_A = 1.04946\ k\Omega, C_1 = 72.853\ nF, C_2 = 72.853\ nF, C_3 = 0.1457\ \mu F$
Second-Order Section 2: $R_1 = 1\ k\Omega, R_2 = 1\ k\Omega,$ $R_3 = 500\ \Omega, R_B = 1\ k\Omega, R_A = 1.04946\ k\Omega,$ $C_1 = 72.853\ nF, C_2 = 72.853\ nF, C_3 = 0.1457\ \mu F,$ $C_4 = 35.4007\ nF$
$R_D = 1\ k\Omega, R_C = 1.1301\ k\Omega$

17.45 a. $\Omega_s = 2.02$ rad/s, $n = 2, A_{s_1} = 3.5$ dB, $A_{p_1} = 0.3117$ dB, $\varepsilon_{p_1} = 0.2728, \Omega_{c_1} = 1.9146$ rad/s
b. $S = -0.7071 \pm j0.7071, SZ = \infty, SZ = \infty$
c. $H_1(s) = (s^2 + 1.8841 \times 10^8)/(s^2 + 8723.2853s + 3.1608 \times 10^8), Q = 2.0381, \omega_0 = 17{,}778.6195$ rad/s, $\omega_z = 13{,}726.2795$ rad/s, $H_2(s) = (s^2 + 1.8841 \times 10^8)/(s^2 + 5199.8363s + 1.1231 \times 10^8),$ $Q = 2.0381, \omega_0 = 10{,}597.6028$ rad/s, $\omega_z = 13{,}726.2795$ rad/s
d. Second-Order Section 1: $K = 2.0210, R_1 = 1\ \Omega,$ $R_2 = 1\ \Omega, R_3 = 0.5\ \Omega, R_4 = 2.9516\ \Omega,$ $R_B = 1\ \Omega, R_A = 1.02105\ \Omega, C_1 = 1.2952$ F, $C_2 = 1.2952$ F, $C_3 = 2.5904$ F
Second-Order Section 2: $K = 2.0210,$ $R_1 = 1\ \Omega, R_2 = 1\ \Omega, R_3 = 0.5\ \Omega, R_B = 1\ \Omega,$ $R_A = 1.02105\ \Omega, C_1 = 0.7721$ F, $C_2 = 0.7721$ F, $C_3 = 1.5441$ F, $C_4 = 0.2616$ F, $R_D = 1\ \Omega,$ $R_C =$ Total Gain $- 1 = 1.4348\ \Omega$
e. Second-Order Section 1: $R_1 = 1\ k\Omega, R_2 = 1\ k\Omega,$ $R_3 = 500\ \Omega, R_4 = 2.9516\ k\Omega, R_B = 1\ k\Omega,$ $R_A = 1.02105\ k\Omega, C_1 = 72.853\ nF, C_2 = 72.853\ nF,$ $C_3 = 0.1457\ \mu F$
Second-Order Section 2: $R_1 = 1\ k\Omega, R_2 = 1\ k\Omega,$ $R_3 = 500\ \Omega, R_B = 1\ k\Omega, R_A = 1.02105\ k\Omega,$ $C_1 = 72.853\ nF, C_2 = 72.853\ nF, C_3 = 0.1457\ \mu F,$ $C_4 = 24.6829\ nF$
$R_D = 1\ k\Omega, R_C = 1.4348\ k\Omega$

17.47 a. $\Omega_s = 1.6839$ rad/s, $n = 3, A_{s_1} = 7.6357$ dB, $A_{p_1} = 0.83$ dB, $\varepsilon_{p_1} = 0.4589, \Omega_{c_1} = 1.2965$ rad/s
b. $S = -1, S = -0.5 \pm j0.866, SZ = \infty, SZ = \infty,$ $SZ = \infty$
c. $H_1(s) = (s^2 + 1.8841 \times 10^8)/(s^2 + 14{,}539.3015s + 1.8841 \times 10^8), Q = 0.9441, \omega_0 = 13{,}726.2795$ rad/s, $\omega_z = 13{,}726.2795$ rad/s, $H_2(s) = (s^2 + 1.8841 \times 10^8)/(s^2 + 10{,}374.8014s + 4.6938 \times 10^8),$ $Q = 2.0882, \omega_0 = 21{,}665.13$ rad/s, $\omega_z = 13{,}726.2795$ rad/s, $H_3(s) = (s^2 + 1.8841 \times 10^8)/(s^2 + 4164.5s + 0.7563 \times 10^8), Q = 2.0882,$ $\omega_0 = 8696.4975$ rad/s, $\omega_z = 13{,}726.2795$ rad/s
d. Second-Order Section 1: $K = 1.4704, R_1 = 1\ \Omega,$ $R_2 = 1\ \Omega, R_3 = 0.5\ \Omega, R_B = 1\ \Omega, R_A = 0.4704\ \Omega,$ $C_1 = 1$ F, $C_2 = 1$ F, $C_3 = 2$ F
Second-Order Section 2: $K = 2.3677, R_1 = 1\ \Omega,$ $R_2 = 1\ \Omega, R_3 = 0.5\ \Omega, R_4 = 1.3412\ \Omega, R_B = 1\ \Omega,$ $R_A = 1.3677\ \Omega, C_1 = 1.5784$ F, $C_2 = 1.5784$ F, $C_3 = 3.1567$ F
Second-Order Section 3: $K = 2.3677, R_1 = 1\ \Omega,$ $R_2 = 1\ \Omega, R_3 = 0.5\ \Omega, R_B = 1\ \Omega, R_A = 1.3667\ \Omega,$ $C_1 = 0.6336$ F, $C_2 = 0.6336$ F, $C_3 = 1.2671$ F, $C_4 = 0.4724$ F, $R_D = 1\ \Omega,$ $R_C =$ Total Gain $- 1 = 2.3088\ \Omega$
e. Second-Order Section 1: $R_1 = 1\ k\Omega, R_2 = 1\ k\Omega,$ $R_3 = 500\ \Omega, R_B = 1\ k\Omega, R_A = 470.3845\ \Omega,$ $C_1 = 72.853\ nF, C_2 = 72.853\ nF, C_3 = 0.1457\ \mu F$
Second-Order Section 2: $R_1 = 1\ k\Omega, R_2 = 1\ k\Omega,$ $R_3 = 500\ \Omega, R_4 = 1.3412\ k\Omega, R_B = 1\ k\Omega,$ $R_A = 1.3677\ k\Omega, C_1 = 72.853\ nF, C_2 = 72.853\ nF,$ $C_3 = 0.1457\ \mu F$
Second-Order Section 3: $R_1 = 1\ k\Omega, R_2 = 1\ k\Omega,$ $R_3 = 500\ \Omega, R_B = 1\ k\Omega, R_A = 1.3677\ k\Omega,$ $C_1 = 72.853\ nF, C_2 = 72.853\ nF, C_3 = 0.1457\ \mu F, C_4 = 54.3209\ nF$
$R_D = 1\ k\Omega, R_C = 2.3088\ k\Omega$

17.49 a. $\Omega_s = 2.04$ rad/s, $n = 2, A_{p_1} = 1.15$ dB, $A_{s_1} = 12.37$ dB
b. $S = -0.518 \pm j0.8765, SZ = \infty, SZ = \infty$
c. $H_1(s) = (6.3945 \times 10^7)/(s^2 + 8136.6867s + 6.3945 \times 10^7), Q = 0.9828, \omega_0 = 7996.5838$ rad/s
d. $R_2 = 1\ \Omega, R_B = 1\ \Omega, R_A = 0.9825\ \Omega, R_a = 1.9825\ \Omega, R_b = 2.0178\ \Omega, C_1 = 1$ F, $C_2 = 1$ F, $R_D = 1\ \Omega, R_C =$ b0i $- 1 = 0.1415633\ \Omega$
e. $R_2 = 1\ k\Omega, R_B = 1\ k\Omega, R_A = 982.4796\ \Omega,$ $R_a = 1.9825\ k\Omega, R_b = 2.0178\ k\Omega, C_1 = 0.1251\ \mu F,$ $C_2 = 0.1251\ \mu F, R_D = 1\ k\Omega, R_C = 141.5633\ \Omega$

17.51 a. $\Omega_s = 2.12$ rad/s, $n = 2, A_{p_1} = 1.25$ dB, $A_{s_1} = 13.4803$ dB, $\varepsilon_p = 0.3949$
b. $S = -0.4086 \pm j0.5067, SZ = \pm j1.4142$
c. $H_1(s) = (s^2 + 5.5447 \times 10^8)/(s^2 + 13{,}606.9062s + 1.1745 \times 10^8), Q = 0.79648, \omega_0 = 10{,}837.6328$ rad/s
d. $R_1 = 1\ \Omega, R_2 = 1\ \Omega, R_3 = 0.5\ \Omega, R_B = 1\ \Omega,$ $R_A = 1.4964\ \Omega, C_1 = 0.46025$ F, $C_2 = 0.46025$ F, $C_3 = 0.9205$ F, $C_4 = 0.8562$ F, $R_D = 1\ \Omega,$ $R_C =$ Single$K - 1 = 1.4964\ \Omega$
e. $R_1 = 1\ k\Omega, R_2 = 1\ k\Omega, R_3 = 500\ \Omega, R_B = 1\ k\Omega,$ $R_A = 1.4964\ k\Omega, C_1 = 42.4678\ nF,$ $C_2 = 42.4678\ nF, C_3 = 84.9355\ nF,$ $C_4 = 79.0063\ nF, R_D = 1\ k\Omega, R_C = 1.4964\ k\Omega$

CHAPTER 18

18.1 a. $a_0 = 0.25, a_1 = -1.75$
b.
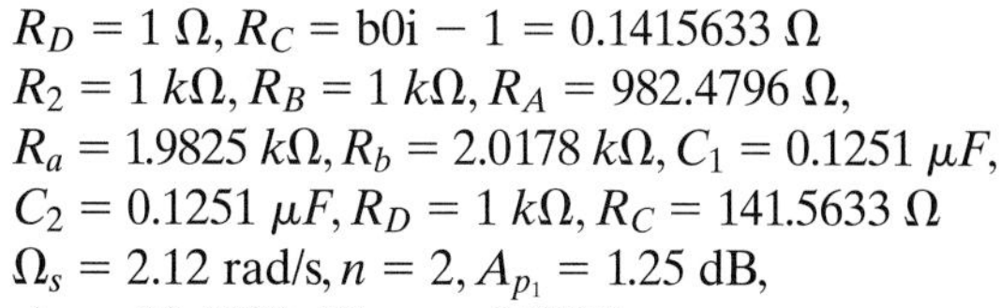

c.
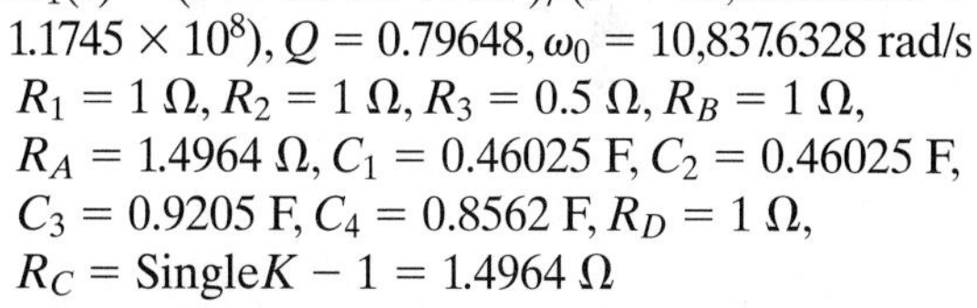

d. Energy of $f(t) = 9.25$ J, Energy of $e(t) = 6.125$ J

18.3 a. $a_0 = -0.5, a_1 = 1.5, a_2 = 1.4142, a_3 = 0$

b.

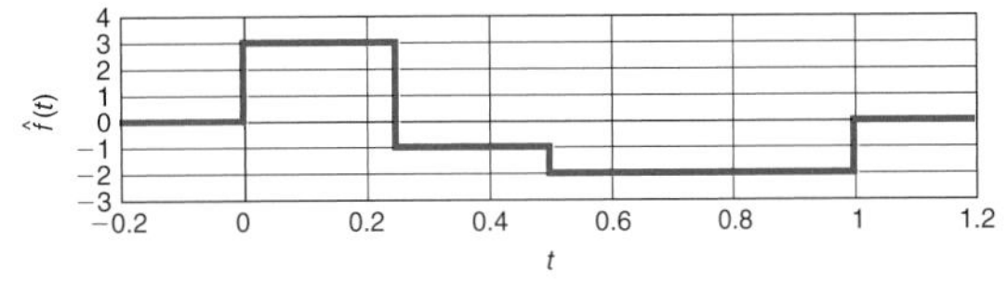

c.

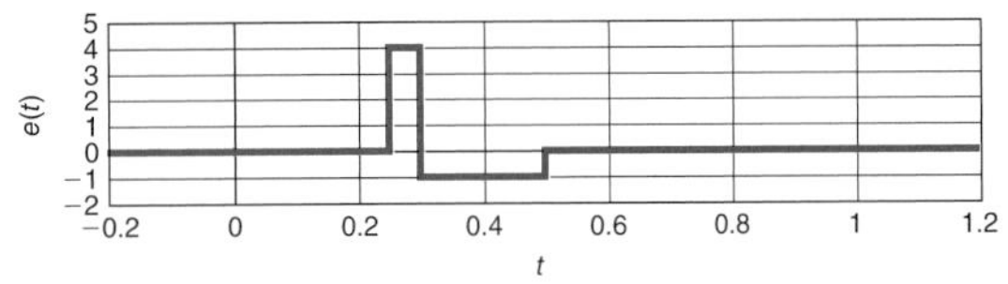

18.5 a. $a_0 = 0.5, a_1 = 0$

b.

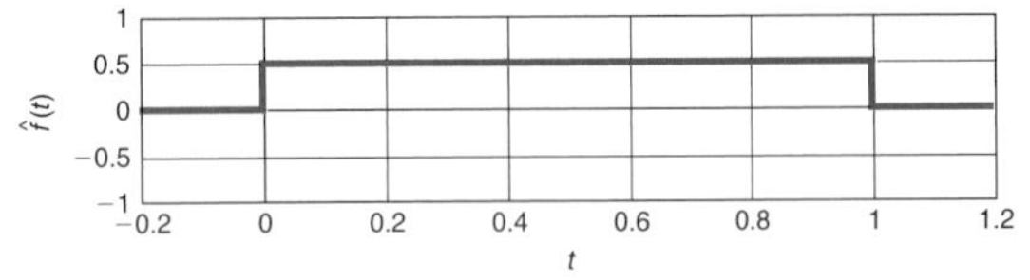

c.

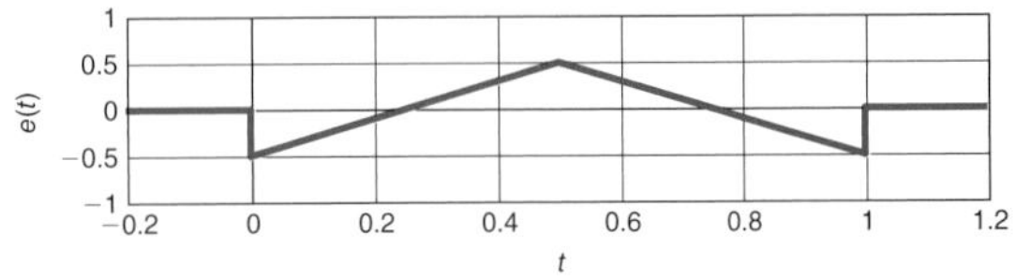

18.7 a. $a_0 = 0, a_1 = 0.63662$

b.

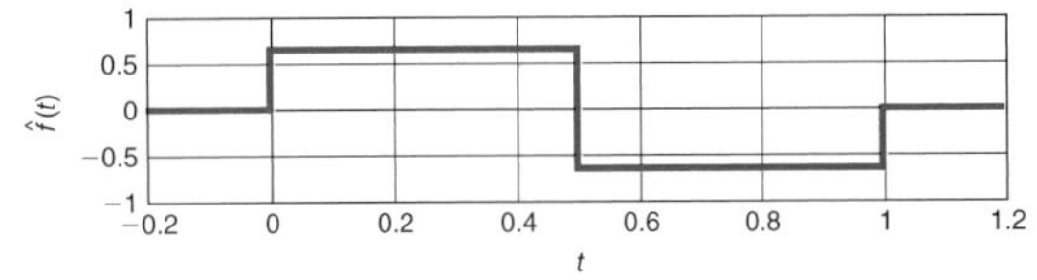

c.

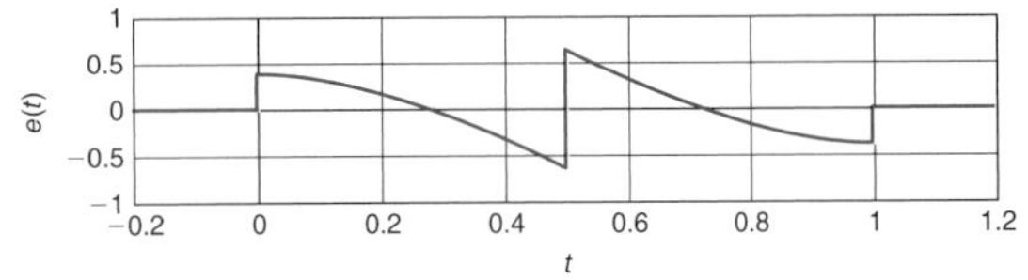

18.9 a. $a_0 = 0.63662, a_1 = 0, a_2 = -0.1865, a_3 = 0.1865, a_4 = -0.089542, a_5 = -0.03709, a_6 = 0.03709, a_7 = 0.089542$

b.

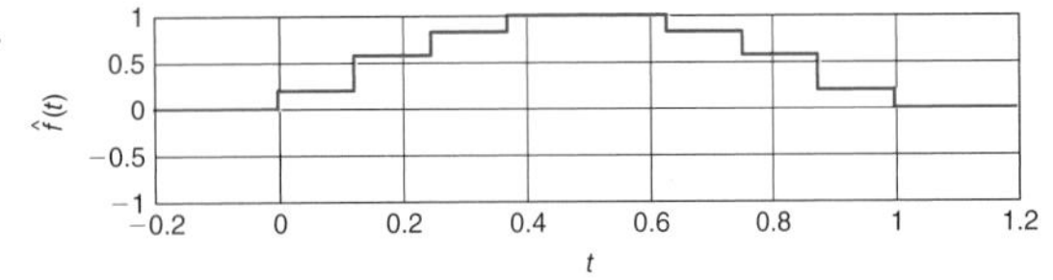

c.

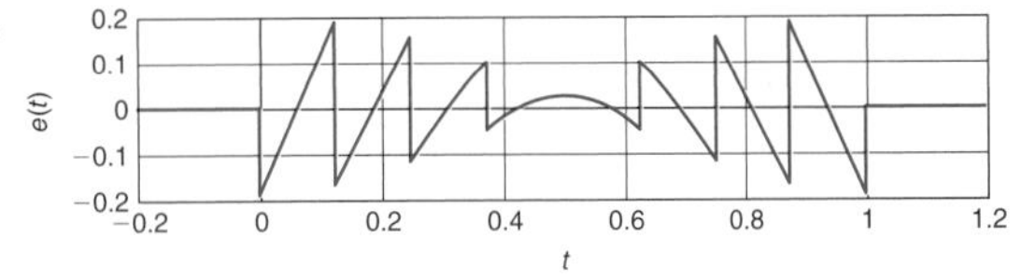

18.11 a. $f_0 = 1$ kHz

b. $d = 1/3$

c. $a_0 = 1/3, a_n = 2Ad\ \text{sinc}(nd) = (2/3)\ \text{sinc}(n/3)$, $b_n = 0, F_n = (1/3)\ \text{sinc}(n/3)$

d. 0.33333, 0.55133, 0.27566, 0, −0.13783, −0.11027, 0

e.

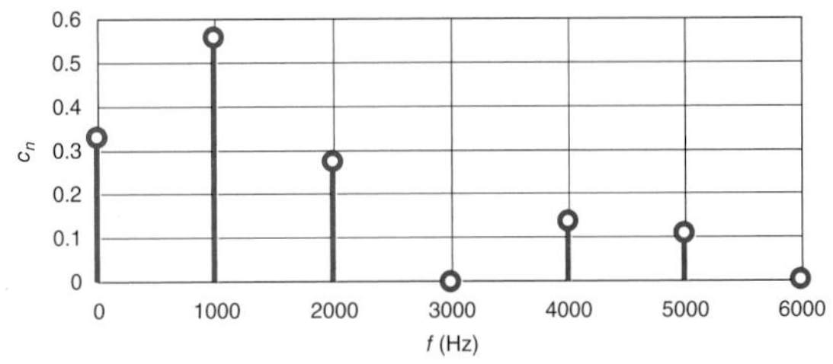

f.

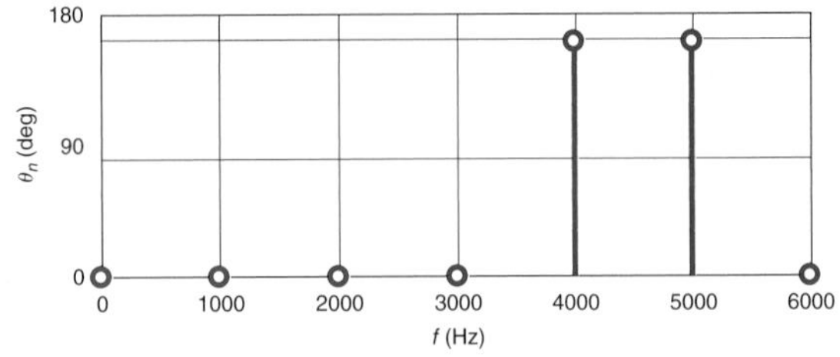

g. $f(t) = 0.33333 + 0.55133\cos(2\pi 1000t) + 0.27566\cos(2\pi 2000t) - 0.13783\cos(2\pi 4000t)$

18.13 a. $\omega_0 = 6283.1853$ rad/s, $f_0 = 1$ kHz

b. $d = 1/2$

c. $a_0 = 0, a_n = 0,$

$$b_n = \text{sinc}\left(\frac{n}{2}\right)\sin\left(\frac{n\pi}{2}\right) = \frac{A}{n\pi}[1 - \cos(n\pi)]$$

$$= \begin{cases} 0, & n = even \\ \dfrac{2A}{n\pi}, & n = odd \end{cases}$$

d. $c_n = b_n, \theta_n = -90°$ when n is odd and zero when n is even.

e.

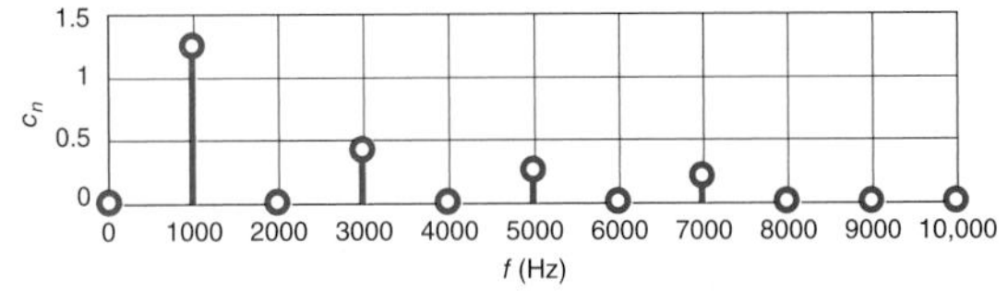

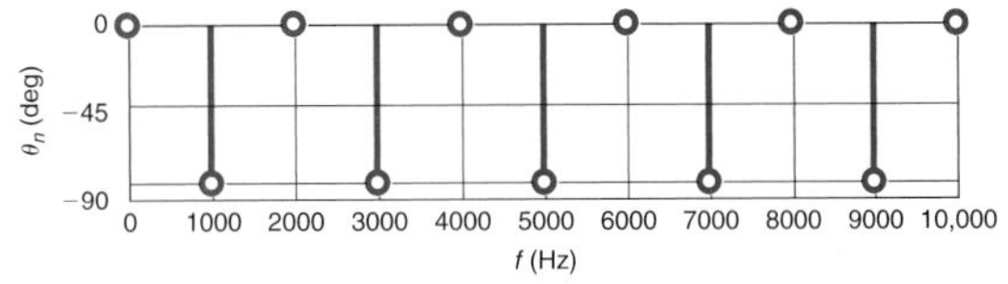

f. $\hat{f}(t) = 1.2732\sin(2\pi 1000t) + 0.4244\sin(2\pi 3000t) + 0.2546\sin(2\pi 5000t)$

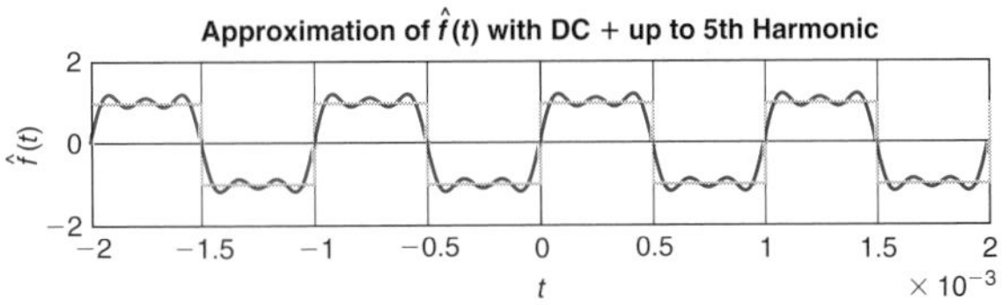

18.15 a. $\omega_0 = 6283.1853$ rad/s, $f_0 = 1$ kHz

b. $d = 1/4$

c. $a_0 = -0.5, a_n = \text{sinc}\left(\frac{n}{4}\right)\cos\left(\frac{n\pi}{4}\right)$,

$$b_n = \text{sinc}\left(\frac{n}{4}\right)\sin\left(\frac{n\pi}{4}\right)$$

d. $c_n = \left|\operatorname{sinc}\left(\frac{n}{4}\right)\right| = \left|\frac{4}{n\pi}\sin\left(\frac{n\pi}{4}\right)\right|, \quad n = 1, 2, 3, \ldots,$

$$\theta_n = \angle F_n = \begin{cases} \frac{-n\pi}{4}, & \operatorname{sinc}\left(\frac{n\tau}{T_0}\right) \geq 0 \\ \frac{-n\pi}{4} + \pi, & \operatorname{sinc}\left(\frac{n\tau}{T_0}\right) < 0 \end{cases}$$

n	0	1	2	3	4	5
a_n	-0.5	0.6366	0	-0.2122	0	0.1273
b_n	0	0.6366	0.6366	0.2122	0	0.1273
c_n	0.5	0.9	0.6366	0.3001	0	0.1801
θ_n (deg)	180	-45	-90	-135	0	-45

n	6	7	8	9	10
a_n	0	-0.09	0	0.07074	0
b_n	0.2122	0.09095	0	0.07074	0.1273
c_n	0.2122	0.1286	0	0.1	0.1273
θ_n (deg)	-90	-135	0	-45	-90

e.

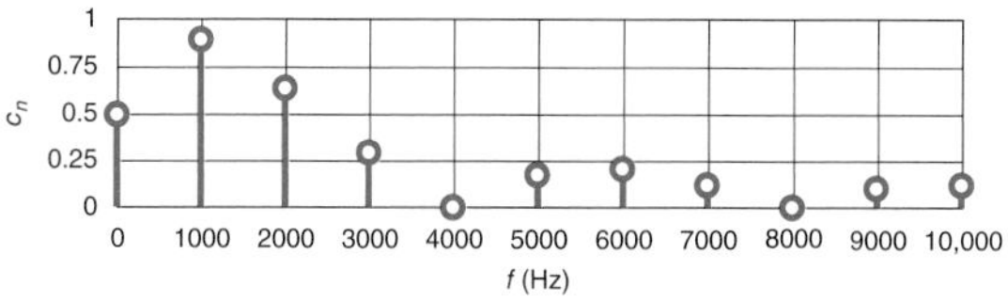

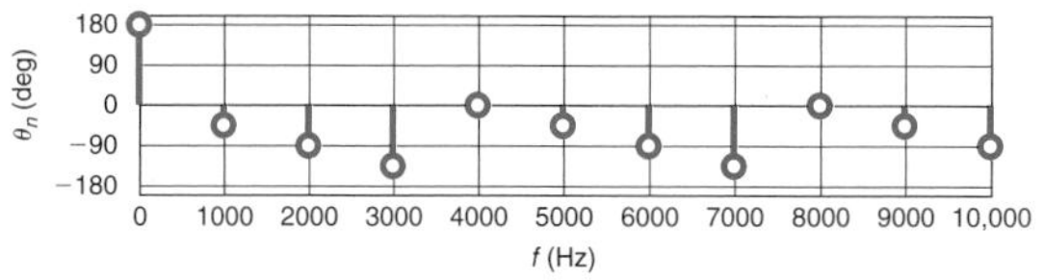

f. $\hat{f}(t) = -0.5 + 0.9003\cos(2\pi 1000t - \pi/4) + 0.6366\cos(2\pi 2000t - \pi/2) + 0.3001 \times \cos(2\pi 3000t - 3\pi/4) + 0.1801 \times \cos(2\pi 5000t - \pi/4)$

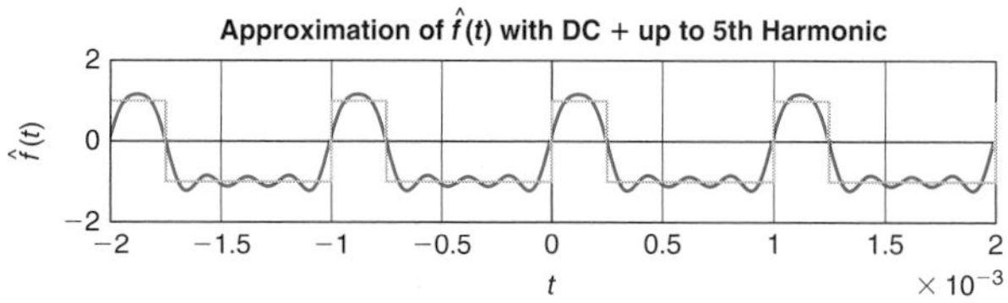

18.17 a. $\omega_0 = 6283.1853$ rad/s, $f_0 = 1$ kHz

b. $d = 1/5$

c. $a_0 = -0.6, a_n = \frac{4}{5}\operatorname{sinc}\left(\frac{n}{5}\right)\cos\left(\frac{4n\pi}{5}\right),$

$$b_n = \frac{4}{5}\operatorname{sinc}\left(\frac{n}{5}\right)\sin\left(\frac{4n\pi}{5}\right)$$

d. $c_n = \left|\frac{4}{5}\operatorname{sinc}\left(\frac{n}{5}\right)\right| = \left|\frac{4}{n\pi}\sin\left(\frac{n\pi}{5}\right)\right|,$

$$\theta_n = \angle F_n = \begin{cases} -\frac{4n\pi}{5}, & \operatorname{sinc}\left(\frac{n\tau}{T_0}\right) \geq 0 \\ -\frac{4n\pi}{5} + \pi, & \operatorname{sinc}\left(\frac{n\tau}{T_0}\right) < 0 \end{cases}$$

e.

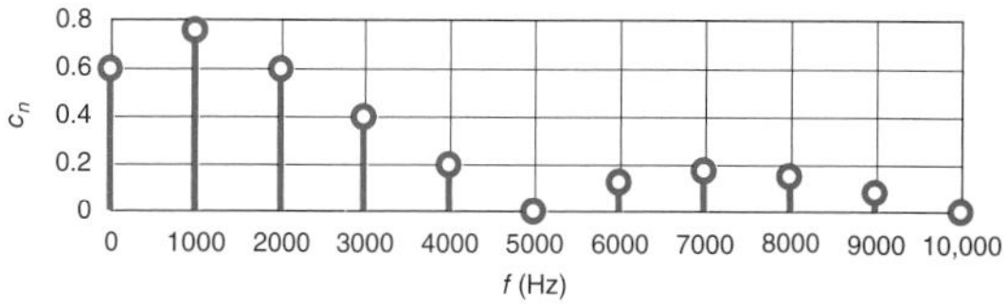

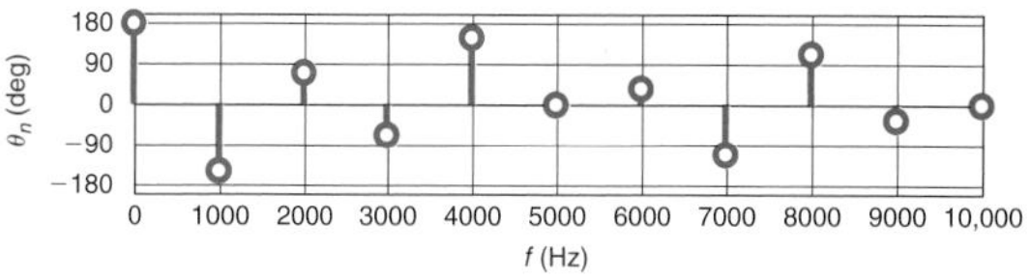

f. $\hat{f}(t) = -0.6 + 0.7484\cos(2\pi 1000t - 144°) + 0.6055\cos(2\pi 2000t + 72°) + 0.4036 \times \cos(2\pi 3000t - 72°) + 0.1871\cos(2\pi 4000t + 144°).$

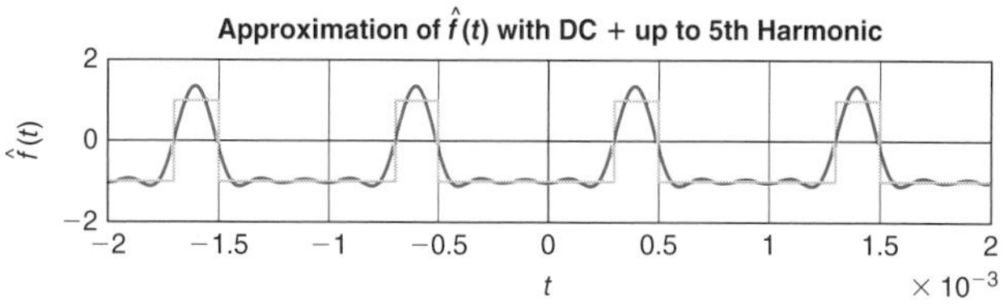

18.19 $a_0 = -1$ V, $f_0 = 1/5 = 0.2$ Hz, $A = 5$ V, $d = 2/5 = 0.4$, $t_d = 1s$.

$$a_n = 2Ad\operatorname{sinc}(nd)\cos\left(n\frac{2\pi}{T_0}t_d\right)$$

$$= 4\operatorname{sinc}\left(\frac{2n}{5}\right)\cos\left(n\frac{2\pi}{5}1\right),$$

$$b_n = 2Ad\operatorname{sinc}(nd)\sin\left(n\frac{2\pi}{T_0}t_d\right)$$

$$= 4\operatorname{sinc}\left(\frac{2n}{5}\right)\sin\left(n\frac{2\pi}{5}1\right)$$

18.21 a. $\omega_0 = 62{,}831.8531$ rad/s, $f_0 = 10$ kHz

b. $a_0 = 0, a_n = 2 \times A \times \frac{1}{2}\operatorname{sinc}^2\left(\frac{n}{2}\right)\cos\left(n\frac{2\pi}{T_0}t_d\right)$

$$= 2\operatorname{sinc}^2\left(\frac{n}{2}\right)\cos\left(\frac{n\pi}{2}\right), b_n = 2\operatorname{sinc}^2\left(\frac{n}{2}\right)\sin\left(\frac{n\pi}{2}\right)$$

c. $c_n = 2\operatorname{sinc}^2\left(\frac{n}{2}\right), \ \theta_n = \angle F_n = -n\frac{2\pi}{T_0}t_d = -\frac{n\pi}{2}$

d.

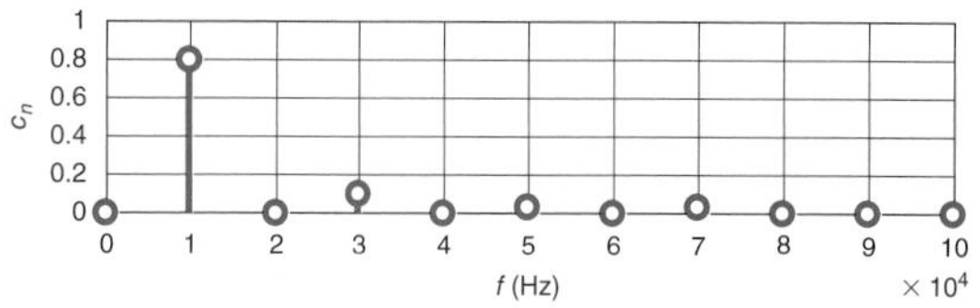

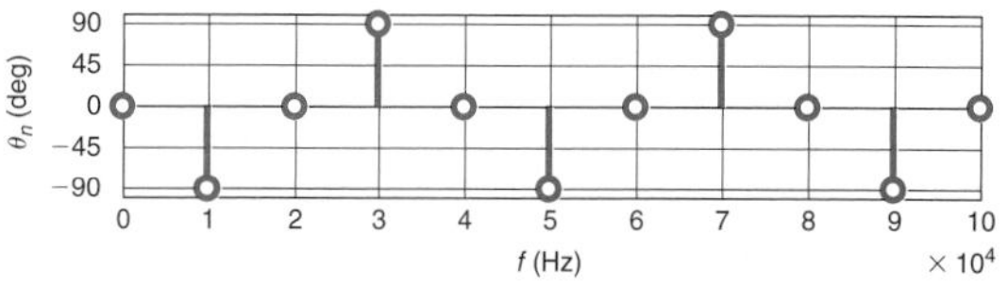

e. $\hat{f}(t) = 0 + 0.8106\cos(2\pi 1000t - \pi/2) + 0.09006\cos(2\pi 2000t + \pi/2) + 0.03242 \times$

$\cos(2\pi 3000t - \pi/2) + 0.01654\cos(2\pi 4000t + \pi/2)$. Angles given in radians.

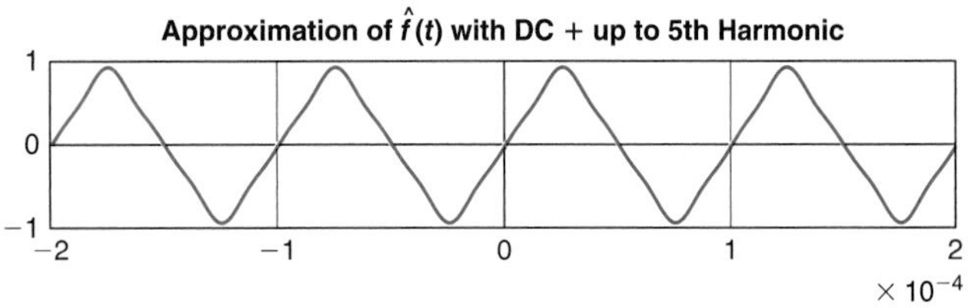

18.23 a. $\omega_0 = 628{,}318.5307$ rad/s, $f_0 = 100$ kHz

b. $a_0 = 0, a_n = 2\,\mathrm{sinc}^2\left(\dfrac{n}{2}\right), b_n = 0$

c. $c_n = a_n, \theta_n = \angle F_n = 0$

d.

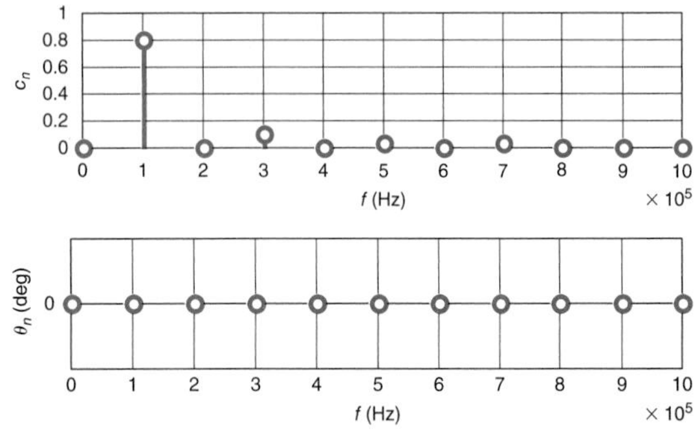

e. $F_0 = 0, F_n = \dfrac{a_n}{2} = \dfrac{1}{2} \times 2\,\mathrm{sinc}^2\left(n\dfrac{1}{2}\right) = \mathrm{sinc}^2\left(n\dfrac{1}{2}\right)$

f.

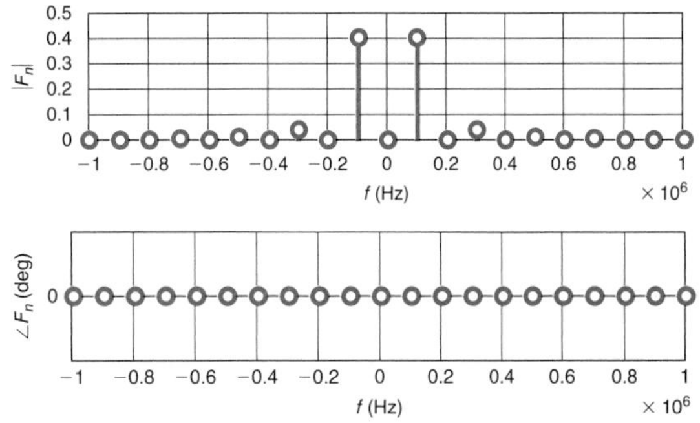

g. $\hat{f}(t) = 0 + 0.8106\cos(2\pi 100{,}000t) + 0.09006 \times \cos(2\pi 300{,}000t) + 0.03242\cos(2\pi 500{,}000t)$

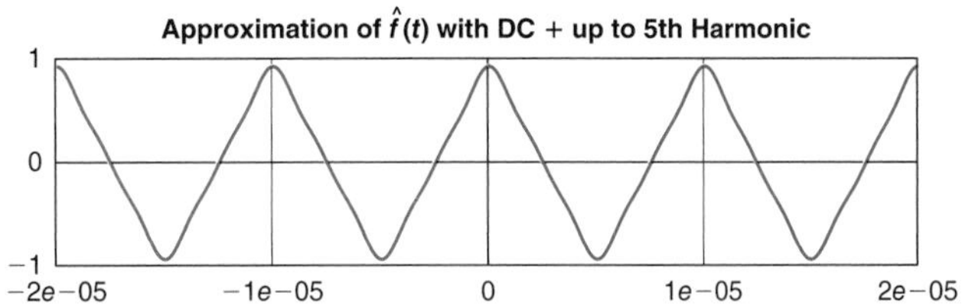

18.25 a. $\omega_0 = 6.2832 \times 10^6$ rad/s, $f_0 = 1$ MHz

b. $A = 2, a_0 = 0, a_n = 0,$

$$b_n = \frac{-2}{\pi n}\cos(n\pi) = \frac{(-1)^{n+1}2}{\pi n}, \quad n = 1, 2, 3, \ldots$$

c. $$c_n = \frac{2}{\pi n}, \theta_n = \begin{cases} 0, & n = 0 \\ \dfrac{-\pi}{2}, & n = 1, 3, 5, \ldots \\ \dfrac{\pi}{2}, & n = 2, 4, 6, \ldots \end{cases}$$

d.

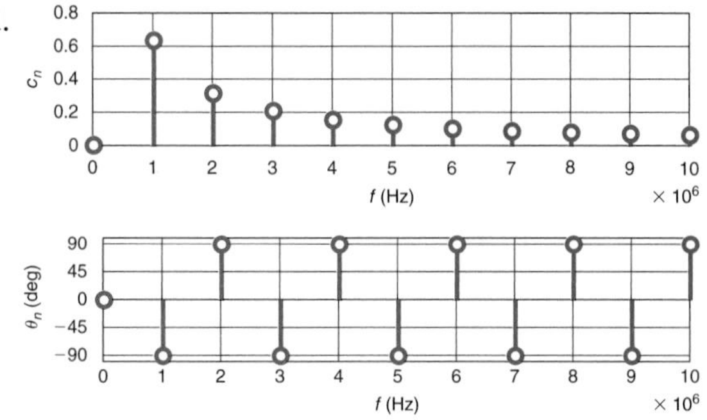

e. $\hat{f}(t) = 0.6366\cos(2\pi 10^6 t - \pi/2) + 0.3183 \times \cos(2\pi 2 \times 10^6 t + \pi/2) + 0.2122\cos(2\pi 3 \times 10^6 t - \pi/2) + 0.1592\cos(2\pi 4 \times 10^6 t + \pi/2) + 0.1273\cos(2\pi 5 \times 10^6 t - \pi/2) + \ldots$. Angles given in radians.

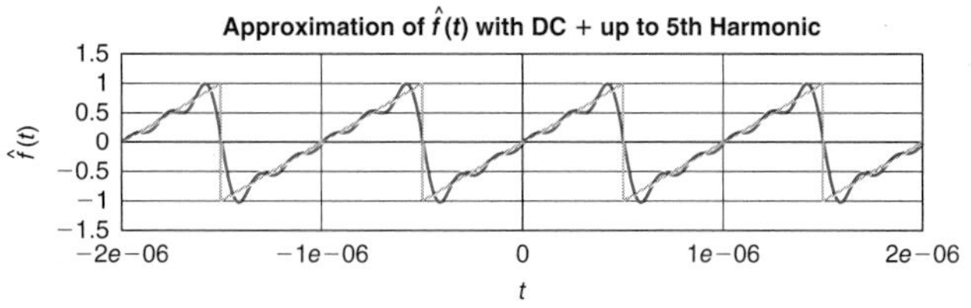

18.27 a. $\omega_0 = 6.2832 \times 10^6$ rad/s, $f_0 = 1$ MHz

b. $a_0 = 1/\pi, a_1 = 0, b_1 = A/2 = 0.5, b_n = 0, n \geq 2,$

$$a_n = \frac{A}{\pi}\frac{1 + \cos(n\pi)}{1 - n^2} = \begin{cases} 0, & n = 3, 5, 7, \ldots \\ \dfrac{-2A}{\pi(n^2 - 1)}, & n = 2, 4, 6, \ldots \end{cases}$$

c. $$c_n = \begin{cases} \dfrac{1}{\pi}, & n = 0 \\ \dfrac{1}{2}, & n = 1 \\ 0, & n = 3,5,7,9, \ldots \\ \dfrac{2}{\pi(n^2 - 1)}, & n = 2,4,6, \ldots \end{cases},$$

$$\theta_n = \begin{cases} 0, & n = 0, 1, 3, 5, 7, 9, \ldots \\ -90^\circ, & n = 1 \\ 180^\circ, & n = 2, 4, 6, 8, \ldots \end{cases}$$

d. 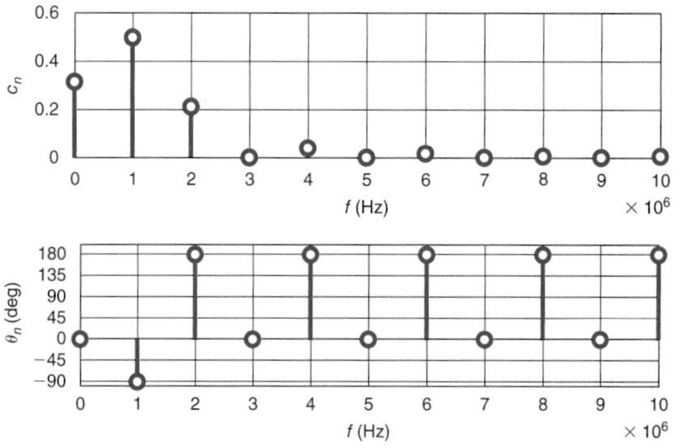

e. $\hat{f}(t) = 0.3183 + 0.5\cos(2\pi 10^6 t - \pi/2) + 0.2122 \times \cos(2\pi 2 \times 10^6 t + \pi) + 0.04244\cos(2\pi 4 \times 10^6 t + \pi) + \ldots$. Angles given in radians.

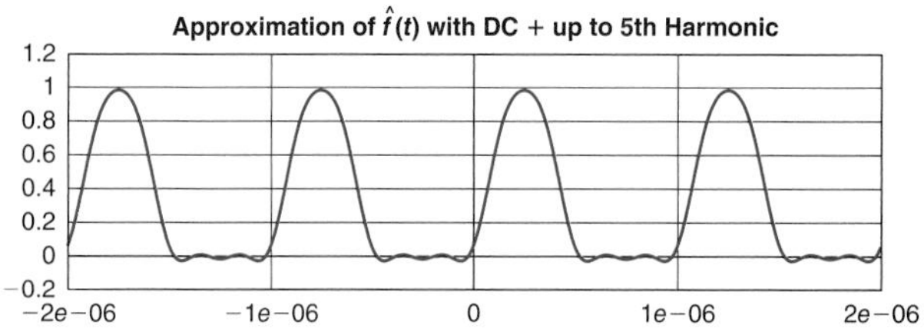

18.29 a. $\omega_0 = 6.2832 \times 10^3$ rad/s, $f_0 = 1$ kHz, $A = 1$ V

b. $a_0 = 2A/\pi = 0.6366$ V,

$$a_n = \frac{-4A}{\pi(4n^2 - 1)}, \quad n = 1, 2, 3, \ldots, b_n = 0$$

c. $$c_n = \frac{4A}{\pi(4n^2 - 1)}, \quad n = 1, 2, 3, \ldots,$$

$$\theta_n = \begin{cases} 0, & n = 0 \\ \pi, & n = 1, 2, 3, \ldots \end{cases}$$

d. 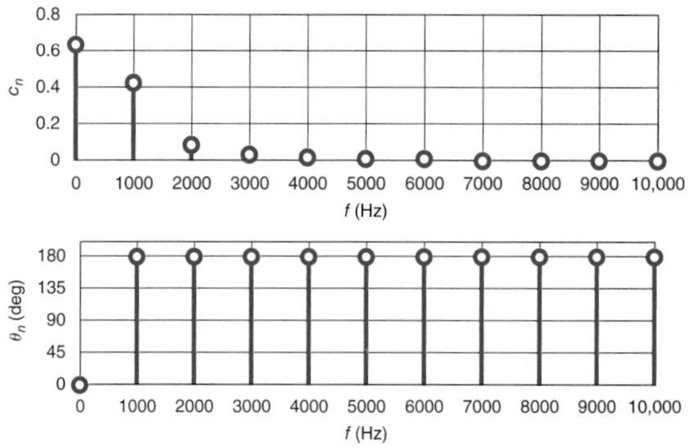

e. $\hat{f}(t) = 0.6366 - 0.4244\cos(2\pi 10^3 t) - 0.08488 \times \cos(2\pi 2 \times 10^3 t) - 0.03638\cos(2\pi 3 \times 10^3 t) - 0.02021\cos(2\pi 4 \times 10^3 t) - 0.01286\cos(2\pi 5 \times 10^3 t) - \ldots$. Angles given in radians.

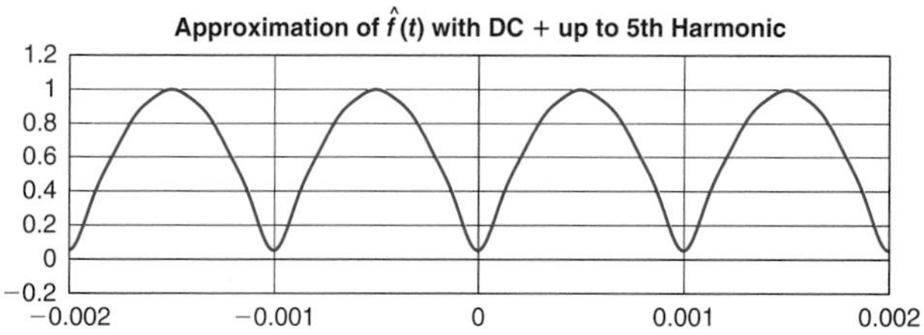

18.31 a. $\omega_0 = 6.2832$ rad/s, $f_0 = 1$ Hz

b. $$F_0 = a_0 = c_0 = \frac{A(1 - e^{-KT_0})}{KT_0} = 0.4323,$$

$$a_n = \frac{2AKT_0(1 - e^{-KT_0})}{K^2T_0^2 + 4\pi^2 n^2}, b_n = \frac{4\pi A n(1 - e^{-KT_0})}{K^2T_0^2 + 4\pi^2 n^2}$$

c. $$c_n = \sqrt{a_n^2 + b_n^2}, \quad \theta_n = -\tan^{-1}\left(\frac{b_n}{a_n}\right)$$

$$= \angle F_n, \quad n = 1, 2, 3, \ldots$$

d. 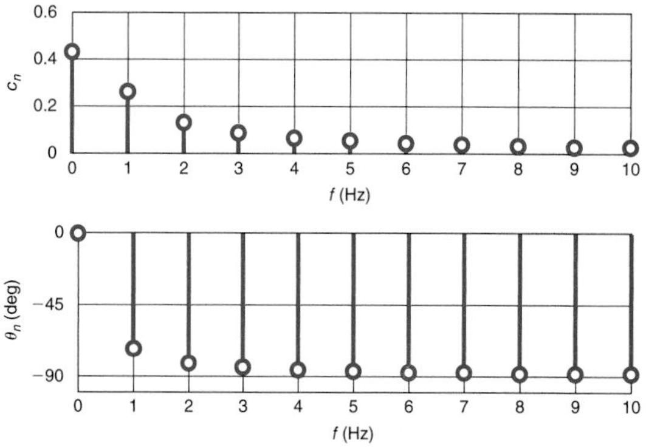

e. $\hat{f}(t) = 0.4323 + 0.2623\cos(2\pi 1t - 1.2626) + 0.1359\cos(2\pi 2t - 1.413) + 0.09123\cos(2\pi 3t - 1.465) + 0.06859\cos(2\pi 4t - 1.4914) + 0.05494\cos(2\pi 5t - 1.5072)$. Angles given in radians.

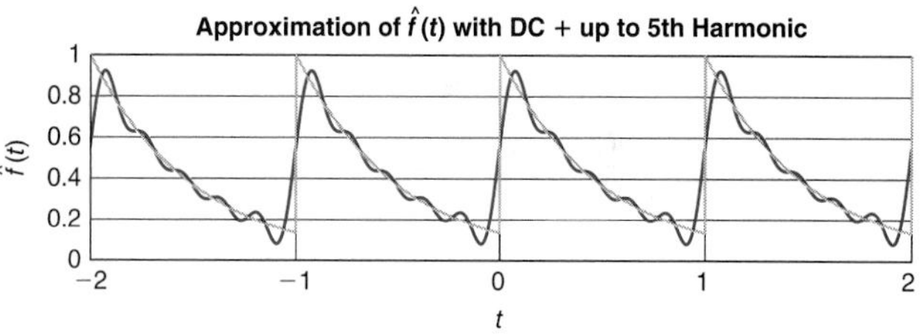

18.33 a. $$a_0 = c_0 = F_0 = \frac{1}{4}\text{ V} = 0.25\text{ V},$$

$$a_n = \frac{1}{4}\text{sinc}\left(\frac{n}{4}\right)\cos\left(\frac{n\pi}{4}\right), b_n = \frac{1}{4}\text{sinc}\left(\frac{n}{4}\right)\sin\left(\frac{n\pi}{4}\right),$$

$$c_n = \left|\frac{1}{4}\text{sinc}\left(\frac{n}{4}\right)\right|$$

b. $H(\omega) = 12{,}500/(j\omega + 12{,}500)$, $H(n\omega_0) = 12{,}500/(jn\omega_0 + 12500)$, $\omega_0 = 6283.1853$ rad/s

c. $yc_n = c_n|H(n\omega_0)|$, $y\theta_n = \theta_n + \angle H(n\omega_0)$

d. $$y(t) = yc_0 + \sum_{n=1}^{\infty} yc_n\cos(n\omega_0 t + \theta y_n)$$

e. $y(t) = 0.25 + 0.40221\cos(2\pi 1000t - 1.2512) + 0.22448\cos(2\pi 2000t - 2.3588) + 0.082929 \times \cos(2\pi 3000t - 3.3414) + 0.03329\cos(2\pi 5000t - 1.9775)$

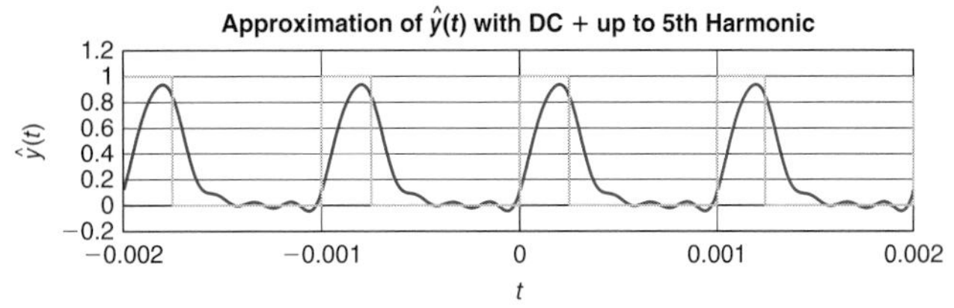

n	0	1	2	3	4	5
c_n	0.25	0.45016	0.31831	0.15005	0	0.090032
θ_n (deg)	0	-45	-90	-135	0	-45
$\lvert H(n\omega_0)\rvert$	1	0.89348	0.70523	0.55267	0.44532	0.3697
$\angle H(n\omega_0)$	0	-26.687	-45.152	-56.45	-63.556	-68.303
cy_n	0.25	0.40221	0.22448	0.082929	0	0.033285
θy_n (deg)	0	-71.687	-135.15	-191.45	0	-113.3

n	6	7	8	9	10
c_n	0.1061	0.064308	0	0.050018	0.063662
θ_n (deg)	-90	-135	0	-45	-90
$\lvert H(n\omega_0)\rvert$	0.31472	0.27338	0.24133	0.21584	0.19512
$\angle H(n\omega_0)$	-71.656	-74.135	-76.035	-77.535	-78.748
cy_n	0.033393	0.017581	0	0.010796	0.012422
θy_n (deg)	-161.66	-209.13	0	-122.54	-168.75

18.35 a. $a_0 = c_0 = F_0 = \dfrac{1}{2}$ V $= 0.5$ V, $c_n = \dfrac{A}{\pi n}$,

$\theta_n = \pi/2, n = 1, 2, 3, \ldots$

b. $H(\omega) = (1/(LC))/((j\omega)^2 + (R/L)j\omega + 1/(LC)) = 5 \times 10^8/(-\omega^2 + 50{,}000 j\omega + 5 \times 10^8), H(n\omega_0) = 5 \times 10^8/(-n^2\omega_0^2 + 50000 jn\omega_0 + 5 \times 10^8)$, $\omega_0 = 6283.1853$ rad/s

c. $yc_n = c_n|H(n\omega_0)|, y\theta_n = \theta_n + \angle H(n\omega_0)$

d. $\hat{y}(t) = 0.5 + 0.2855\cos(2\pi 1000t + 55.699°) + 0.1112\cos(2\pi 2000t + 28.566°) + 0.05564 \times \cos(2\pi 3000t + 8.7282°) + 0.03149\cos(2\pi 4000t - 5.9809°) + 0.01936\cos(2\pi 5000t - 17.224°)$

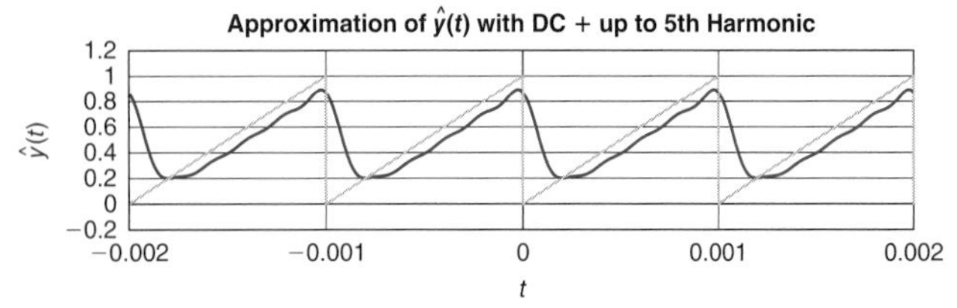

n	0	1	2	3	4	5
c_n	0.5	0.31831	0.15915	0.1061	0.079577	0.063662
θ_n(deg)	0	90	90	90	90	90
$\lvert H(n\omega_0)\rvert$	1	0.8969	0.6989	0.52437	0.39572	0.30404
$\angle H(n\omega_0)$	0	-34.301	-61.434	-81.272	-95.981	-107.22
cy_n	0.5	0.28549	0.11123	0.055638	0.031491	0.019355
θy_n(deg)	0	55.699	28.566	8.7282	-5.9809	-17.224

n	6	7	8	9	10
c_n	0.053052	0.045473	0.039789	0.035368	0.031831
θ_n(deg)	90	90	90	90	90
$\lvert H(n\omega_0)\rvert$	0.23832	0.19043	0.15487	0.12794	0.10719
$\angle H(n\omega_0)$	-116.05	-123.12	-128.88	-133.66	-137.66
cy_n	0.012643	0.00866	0.006162	0.004525	0.003412
θy_n(deg)	-26.046	-33.116	-38.882	-43.655	-47.661

18.37 a. $a_0 = c_0 = F_0 = \dfrac{1}{2}$ V $= 0.5$ V, $c_n = \dfrac{A}{\pi n}$,

$\theta_n = \pi/2, n = 1, 2, 3, \ldots$

b. $H(\omega) = ((j\omega)^2)/((j\omega)^2 + (R/L)j\omega + 1/(LC)) = -\omega^2/(-\omega^2 + 50{,}000 j\omega + 5 \times 10^8), H(n\omega_0) = -n^2\omega_0^2/(-n^2\omega_0^2 + 50{,}000 jn\omega_0 + 5 \times 10^8)$, $\omega_0 = 6283.1853$ rad/s

c. $yc_n = c_n|H(n\omega_0)|, y\theta_n = \theta_n + \angle H(n\omega_0)$

d. $y(t) = 0.02254\cos(2\pi 1000t + 4.1137) + 0.03513 \times \cos(2\pi 2000t + 3.6402) + 0.03954\cos(2\pi 3000t + 3.2939) + 0.03978\cos(2\pi 4000t + 3.0372) + 0.03821\cos(2\pi 5000t + 2.841)$

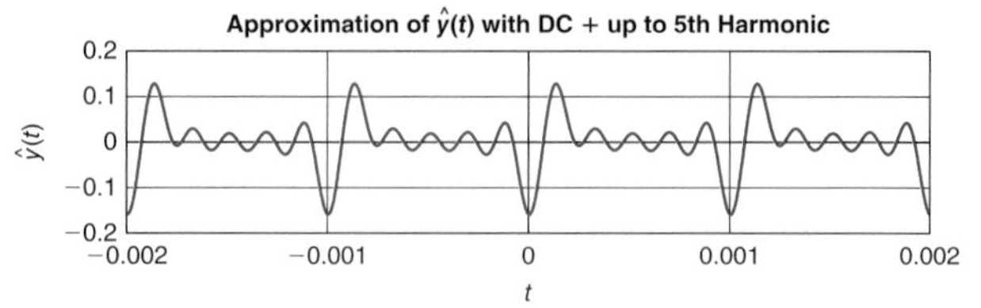

n	0	1	2	3	4	5
c_n	0.5	0.31831	0.15915	0.1061	0.079577	0.063662
θ_n(deg)	0	90	90	90	90	90
$\lvert H(n\omega_0)\rvert$	0	0.07082	0.22073	0.37263	0.49992	0.60014
$\angle H(n\omega_0)$	0	145.7	118.57	98.728	84.019	72.776
cy_n	0	0.02254	0.03513	0.03954	0.03978	0.03821
θy_n(deg)	0	235.7	208.57	188.73	174.02	162.78

n	6	7	8	9	10
c_n	0.053052	0.045473	0.039789	0.035368	0.031831
θ_n(deg)	90	90	90	90	90
$\lvert H(n\omega_0)\rvert$	0.67741	0.73676	0.78258	0.81826	0.84637
$\angle H(n\omega_0)$	63.954	56.884	51.118	46.345	42.339
cy_n	0.03594	0.0335	0.03114	0.02894	0.02694
θy_n(deg)	153.95	146.88	141.12	136.34	132.34

18.39 a. $a_0 = c_0 = F_0 = \dfrac{A}{\pi} = \dfrac{1}{\pi} = 0.3183$ V,

$$c_n = \begin{cases} \dfrac{A}{\pi}, & n = 0 \\ \dfrac{A}{2}, & n = 1 \\ 0, & n = 3, 5, 7, 9, \ldots \\ \dfrac{2A}{\pi(n^2 - 1)}, & n = 2, 4, 6, \ldots \end{cases},$$

$$\theta_n = \begin{cases} 0, & n = 0, 1, 2, 3, 5, 6, 7, 9, 10, \ldots \\ \pi, & n = 4, 8, 12, \ldots \end{cases}$$

b. $H(\omega) = (2/(R_3C))/(j\omega + 1/(R_3C)) = 16{,}666.6667/(j\omega + 8333.3333), H(n\omega_0) = 16{,}666.6667/(jn\omega_0 + 8333.3333)$, $\omega_0 = 6283.1853$ rad/s

c. $yc_n = c_n|H(n\omega_0)|, y\theta_n = \theta_n + \angle H(n\omega_0)$

d. $y(t) = 0.6366 + 0.7985\cos(2\pi 1000t - 0.6460) + 0.2346\cos(2\pi 2000t - 0.9852) + 0.02672 \times \cos(2\pi 4000t + 1.891)$

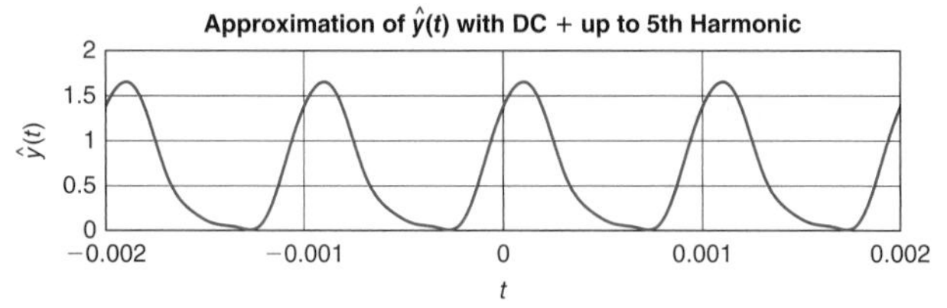

n	0	1	2	3	4	5
c_n	0.31831	0.5	0.21221	0	0.04244	0
θ_n (deg)	0	0	0	0	180	0
$\|H(n\omega_0)\|$	2	1.5969	1.1053	0.80869	0.62945	0.51278
$\angle H(n\omega_0)$	0	-37.016	-56.45	-66.15	-71.656	-75.144
cy_n	0.63662	0.79847	0.23456	0	0.02671	0
θy_n (deg)	0	-37.016	-56.45	0	108.34	0

n	6	7	8	9	10
c_n	0.01819	0	0.01011	0	0.00643
θ_n (deg)	0	0	180	0	0
$\|H(n\omega_0)\|$	0.43168	0.37232	0.32711	0.29158	0.26296
$\angle H(n\omega_0)$	-77.535	-79.271	-80.587	-81.617	-82.445
cy_n	0.00785	0	0.00331	0	0.00169
θy_n (deg)	-77.535	0	99.413	0	-82.445

18.41 $A = 1, d = 0.8/2 = 0.4, t_d = -0.6\ \mu s,$
$F_0 = 0.4\text{ V}, d = 0.4,$

$$F_n = Ad\,\text{sinc}(nd)\,e^{jn\frac{2\pi}{T_0}t_d} = 0.4\,\text{sinc}(0.4n)\,e^{jn\frac{2\pi}{2\mu s}0.6\mu s} = 0.4\,\text{sinc}(0.4n)e^{jn0.6\pi}$$

18.43 $F_n = \dfrac{1}{2}\text{sinc}\left(\dfrac{n}{2}\right) + \dfrac{1}{4}\text{sinc}^2\left(\dfrac{n}{4}\right)$

18.45 $F_n = 0.8\,\text{sinc}^2(0.4n) - 0.2\,\text{sinc}^2(0.2n)$

18.47 $F_0 = 1/3, F_n = \dfrac{8\sin\left(\dfrac{n\pi}{2}\right) - 4\pi n\cos\left(\dfrac{n\pi}{2}\right)}{\pi^3 n^3}$

18.49 $A = 1\text{ V}, T_0 = 5\text{ ms}, f_0 = 200\text{ Hz}, \omega_0 = 1256.6371\text{ rad/s}, d = 0.4, t_d = 1\text{ ms}.$

$$F_n = Ad\,\text{sinc}(nd)\,e^{jn\frac{2\pi}{T_0}t_d} = 0.4\,\text{sinc}(0.4n)\,e^{jn\frac{2\pi}{5ms}1ms} = 0.4\,\text{sinc}(0.4n)e^{jn0.4\pi}$$

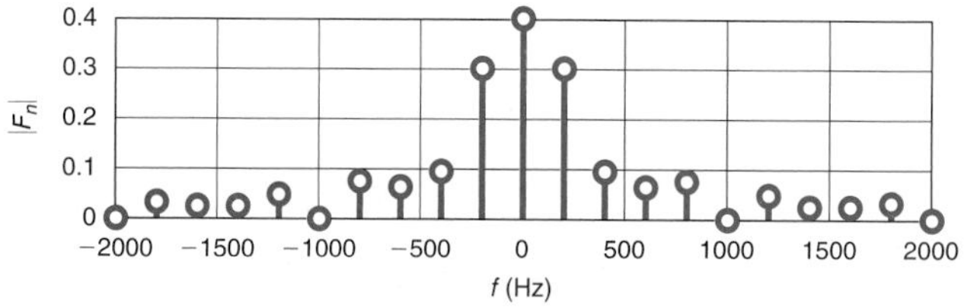

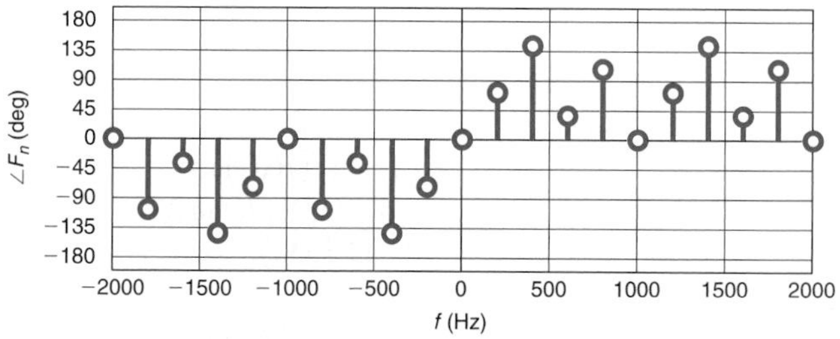

$$\hat{f}(t) = 0.4 + 0.30273[e^{j(2\pi 200t+72°)} + e^{-j(2\pi 200t+72°)}] + 0.093549[e^{j(2\pi 400t+144°)} + e^{-j(2\pi 400t+144°)}] + 0.062366[e^{j(2\pi 600t+36°)} + e^{-j(2\pi 600t+36°)}] + 0.075683[e^{j(2\pi 800t+108°)} + e^{-j(2\pi 800t+108°)}]$$

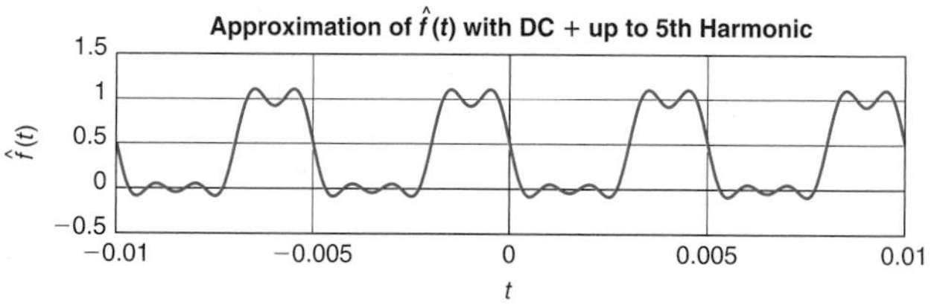

18.51 a. $A = 1\text{ V}, d = 1/2, t_d = 0.25\text{ ms},$

$$F_n = Ad\,\text{sinc}(nd)e^{-jn\frac{2\pi}{T_0}t_d} = 0.5\,\text{sinc}(0.5n)e^{-jn\frac{2\pi}{1ms}0.25ms} = 0.5\,\text{sinc}(0.5n)e^{-jn0.5\pi}$$

b. $H(\omega) = (R/L)/(j\omega + R/L) = 16{,}666.67/(j\omega + 16{,}666.67), H(n\omega_0) = 16{,}666.67/(jn\omega_0 + 16{,}666.67)$

c. $Y_n = F_nH(n\omega_0) = 0.5\,\text{sinc}(0.5n)e^{-jn0.5\pi} \times 16{,}666.67/(jn\omega_0 + 16{,}666.67)$

d. $\hat{y}(t) = 0.5 + 0.29785[e^{j(2\pi 1000t+110.66°)} + e^{-j(2\pi 1000t+110.66°)}] + 0.07028[e^{j(2\pi 3000t-138.52°)} + e^{-j(2\pi 3000t-138.52°)}] + 0.02984[e^{j(2\pi 5000t-152.05°)} + e^{-j(2\pi 5000t-152.05°)}]$

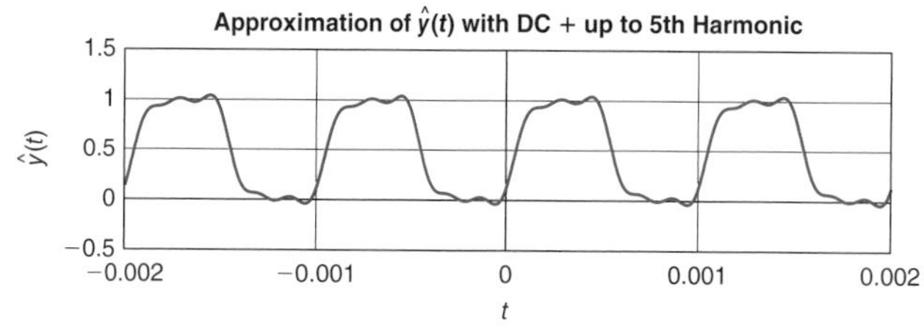

18.53 a. $A = 1\text{ V}, F_0 = \dfrac{A}{2} = 0.5\text{ V}, F_n = \dfrac{jA}{2\pi n}, \quad n \neq 0,$
$\omega_0 = 2\pi 1000\text{ rad/s}$

b. $H(\omega) = (1/(RC))j\omega/((j\omega)^2 + (1/(RC))j\omega + 1/(LC)), 1/(RC) = 1000, 1/(LC) = 1.4225 \times 10^9$

c. $Y_n = F_nH(n\omega_0)$

$$= \frac{j}{2\pi n} \times \frac{1000jn\omega_0}{-n^2\omega_0^2 + 1000jn\omega_0 + 1.4225 \times 10^9}$$

d. $\hat{y}(t) = 0 + 0.000723[e^{j(2\pi 1000t+179.74°)} + e^{-j(2\pi 1000t+179.74°)}] + 0.000791[e^{j(2\pi 2000t+179.43°)} + e^{-j(2\pi 2000t+179.43°)}]$

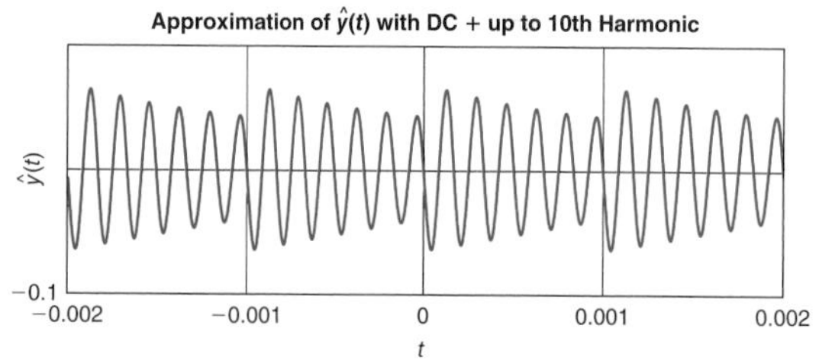

18.55 a. $A = 1\text{ V}, F_0 = \dfrac{A}{2} = 0.5\text{ V}, F_n = \dfrac{A}{2}\text{sinc}^2\left(\dfrac{n}{2}\right),$
$\omega_0 = 2\pi 1000\text{ rad/s}$

b. $H(\omega) = (j\omega)^2/((j\omega)^2 + (1/(RC))j\omega + 1/(LC)), 1/(RC) = 1000, 1/(LC) = 3.9526 \times 10^7$

c. $Y_n = F_nH(n\omega_0)$

$$= \frac{1}{2}\text{sinc}^2\left(\frac{n}{2}\right) \times \frac{-n^2\omega_0^2}{-n^2\omega_0^2 + 1000jn\omega_0 + 3.9526 \times 10^7}$$

d. $\hat{y}(t) = 0 + 1.2732[e^{j(2\pi 1000t+90.431°)} + e^{-j(2\pi 1000t+90.431°)}] + 0.02529[e^{j(2\pi 3000t+3.416°)} + e^{-j(2\pi 3000t+3.416°)}]$

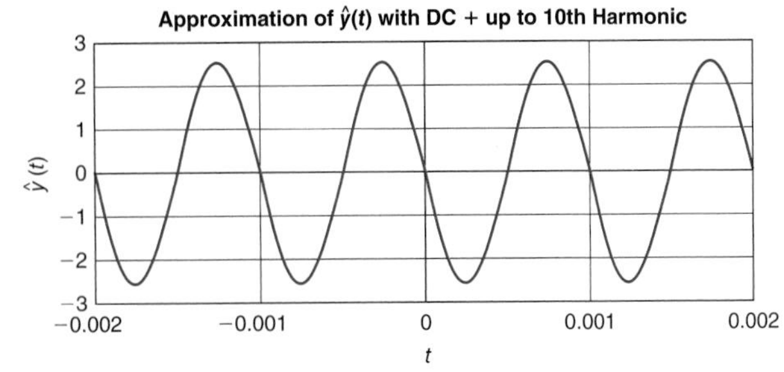

CHAPTER 19

19.1 a. $F_1(\omega) = 1/(j\omega + 3)$,
b. $F_2(\omega) = 1/(-j\omega + 3)$,
c. $F(\omega) = 6/(\omega^2 + 9)$

19.3 $F(\omega) = e^{j\omega 5}$

19.5 $G(\omega) = 3\ F(\omega)e^{j\omega 7}$

19.7 a. $F(\omega) = 8\ \text{sinc}(2\omega/\pi)$
b. $F[f(t - 10)] = 8\ \text{sinc}(2\omega/\pi)e^{-j\omega 10}$

19.9 $F(\omega) = rect\left(\dfrac{\omega}{10\pi}\right)$

a. $G(\omega) = 2F(2\omega) = 2rect\left(\dfrac{\omega}{5\pi}\right)$

b. $G(\omega) = \dfrac{1}{4}F\left(\dfrac{\omega}{4}\right) = \dfrac{1}{4}rect\left(\dfrac{\omega}{40\pi}\right)$

19.11 $F\left[\dfrac{1}{t^2 + a^2}\right] = \dfrac{\pi}{a}e^{-a|\omega|}$

19.13 $F(\omega) = j5.6569\pi[\delta(\omega - 2\pi 20{,}000) - \delta(\omega + 2\pi 20{,}000)] - 5.6569\pi[\delta(\omega - 2\pi 20{,}000) + \delta(\omega + 2\pi 20{,}000)]$

19.15 $G(\omega) = 2.5F[5(\omega - 50)]e^{-j(\omega-50)10} + 2.5F[5(\omega + 50)]e^{-j(\omega+50)10}$

19.17 a. $F_1(\omega) = \dfrac{1}{1000}rect\left(\dfrac{\omega}{2\pi 1000}\right)$,

$F_1(f) = \dfrac{1}{1000}rect\left(\dfrac{f}{1000}\right)$

b. $F_2(\omega) = \dfrac{1}{1000}rect\left(\dfrac{\omega - 10{,}000}{2\pi 1000}\right)$,

$F_2(f) = \dfrac{1}{1000}rect\left(\dfrac{f - \dfrac{10{,}000}{2\pi}}{1000}\right)$

c. $F_3(\omega) = \dfrac{1}{2000}rect\left(\dfrac{\omega - 10{,}000}{2\pi 1000}\right) + \dfrac{1}{2000}rect\left(\dfrac{\omega + 10{,}000}{2\pi 1000}\right)$

$$F_3(f) = \frac{1}{2000}rect\left(\frac{f - \dfrac{10{,}000}{2\pi}}{1000}\right) + \frac{1}{2000}rect\left(\frac{f + \dfrac{10{,}000}{2\pi}}{1000}\right)$$

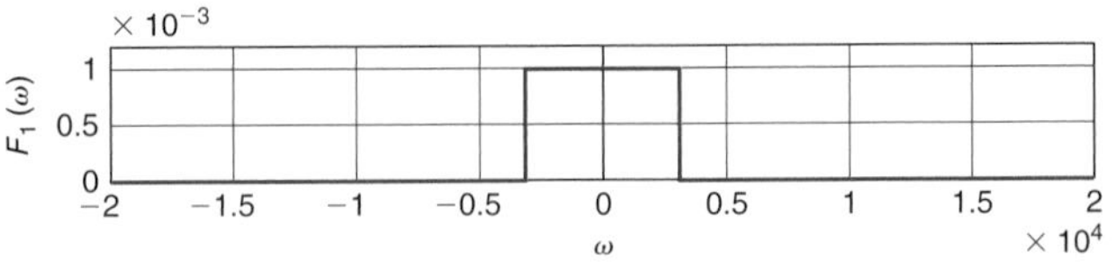

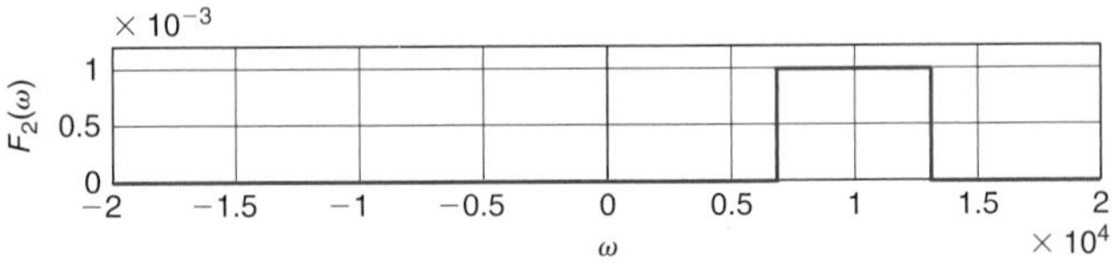

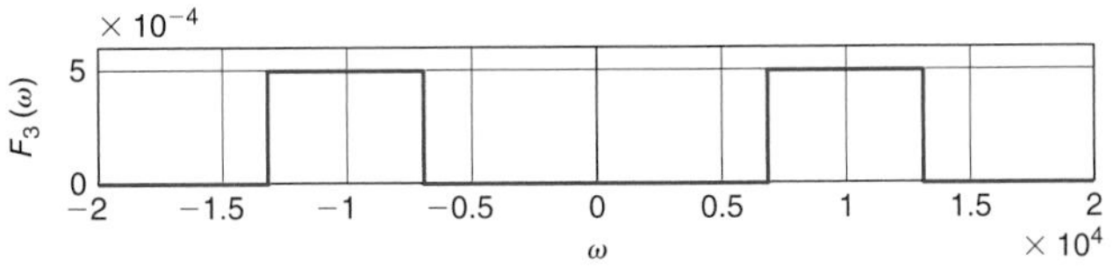

19.19 a. $F(\omega) = 8\ \text{sinc}(2\omega/\pi)$
b. $8\ \text{sinc}(2(\omega - 10\pi)/\pi)$
c. $4\ \text{sinc}(2(\omega - 10\pi)/\pi) + 4\ \text{sinc}(2(\omega + 10\pi)/\pi)$

19.21 a. $F(\omega) = \pi e^{-|\omega|}$
b. $G(\omega) = \pi e^{-|\omega-20|}$
c. $H(\omega) = \dfrac{\pi}{2}e^{-|\omega-20|} + \dfrac{\pi}{2}e^{-|\omega+20|}$

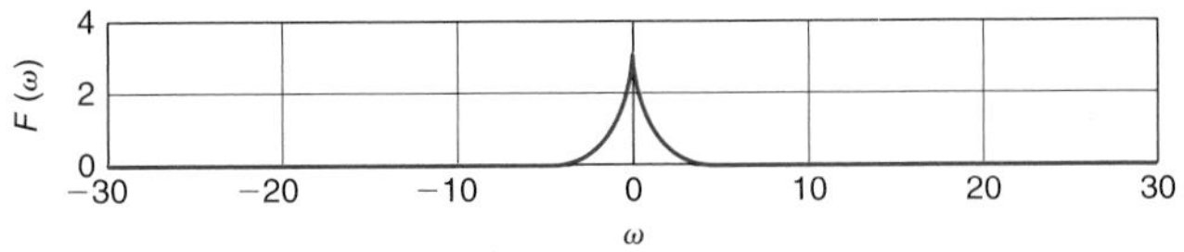

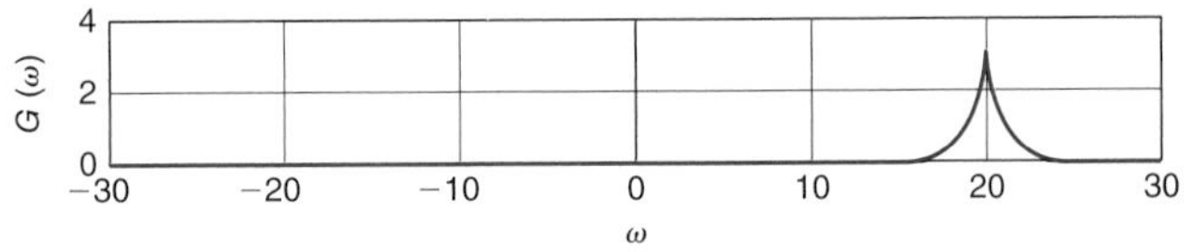

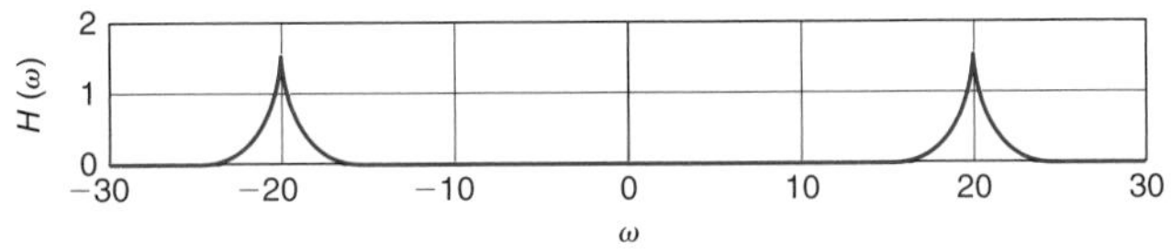

19.23

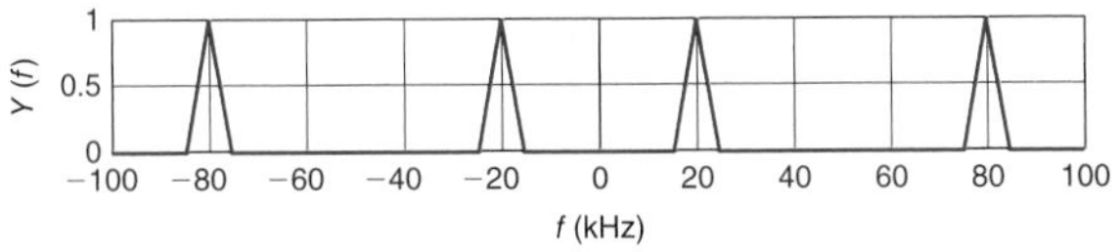

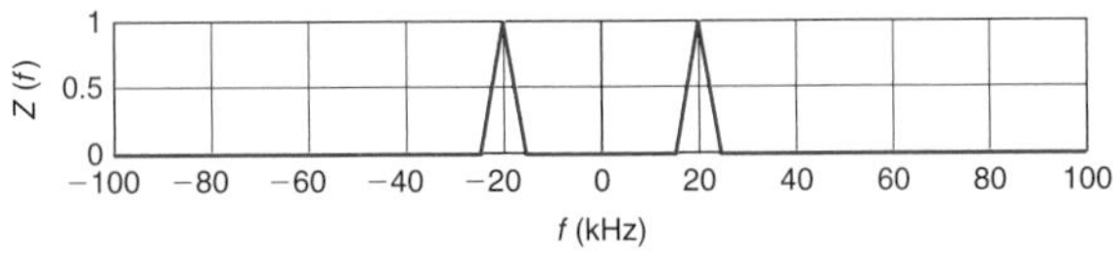

19.25 $y(t) = \cos(2\pi 99{,}000t) + \cos(2\pi 104{,}000t)$.

19.27 $s(t) = m(t)\,c(t) = 2\cos(2\pi 1000t) \times 10 \times \cos(2\pi 10{,}000t) = 10\cos(2\pi 9000t) + 10 \times \cos(2\pi 11{,}000t)$.

$S(f) = 5\delta(f - 9000) + 5\delta(f + 9000) + 5\delta(f - 11{,}000) + 5\delta(f + 11{,}000)$

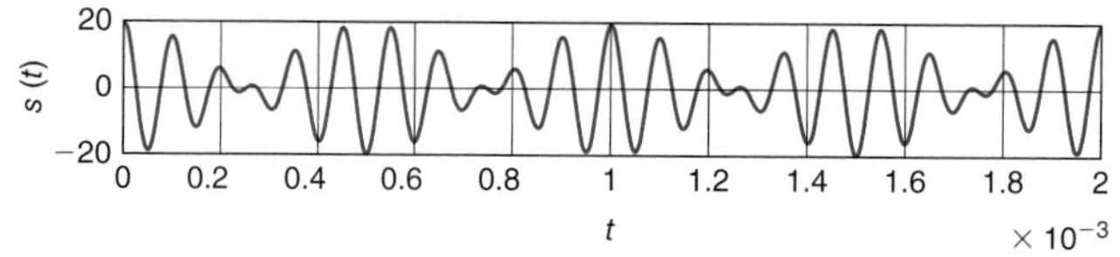

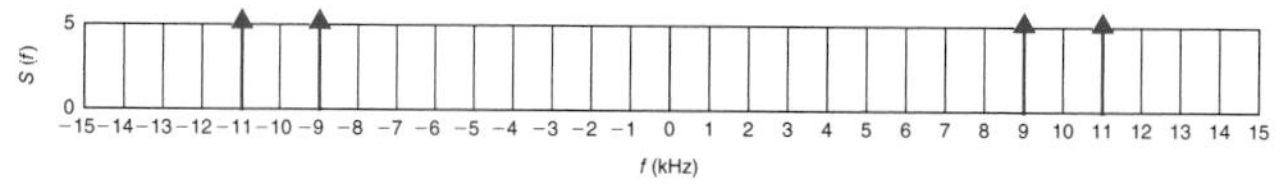

19.29 $$F(\omega) = \frac{-4\,\mathrm{sinc}\left(\frac{\omega}{\pi}\right)\cos(3\omega) + 2\cos(2\omega) + 2\cos(4\omega)}{j\omega}$$

19.31 $$F(\omega) = \frac{\mathrm{sinc}\left(\frac{\omega}{2\pi}\right)e^{-j0.5\omega} - 3\,\mathrm{sinc}\left(\frac{\omega}{\pi}\right)e^{-j2\omega} + 2e^{-j\omega}}{j\omega}$$

19.33 $F[e^{-5t}u(t)] = 1/(j\omega + 5)$

a. $F_1(\omega) = 1/(j\omega + 5)^2$

b. $F_2(\omega) = 2/(j\omega + 5)^3$

c. $F_3(\omega) = 6/(j\omega + 5)^4$

19.35 $$G(\omega) = \frac{6\pi}{5}F(3\omega)f\left(\frac{\omega}{5}\right)$$

19.37 $$f(t) = f_1(t)*f_2(t) = \left[\frac{-1}{9}e^{-3t} + \frac{1}{9}e^{6t}\right]u(t)$$

19.39 $f(t) = (1/4)\,\mathrm{sinc}(2t)$

19.41 $f(t) = (1/4)\,\mathrm{sinc}(2t) + (1/8)\,\mathrm{sinc}^2(t)$

19.43 $y(t) = 20\,\mathrm{sinc}(20t)$

19.45 $$G(\omega) = \frac{1}{(2\pi)^2}F(\omega)*F(\omega)*F(\omega) + \frac{4}{2\pi}F(\omega)*F(\omega) - 9F(\omega) + 30\pi\delta(\omega)$$

$G(f) = F(f)*F(f)*F(f) + 4F(f)*F(f) - 9F(f) + 15\delta(f)$

19.47 $BW = 3B$ Hz

19.49 $$F(\omega) = \frac{1}{5}tri\left(\frac{\omega}{10\pi}\right)e^{-j6\omega}$$

19.51 a. $E_x = 0.1$ J

b. $X(\omega) = 1/(j\omega + 5)$

c. $H(\omega) = 10/(j\omega + 10)$

d. $Y(\omega) = 10/[(j\omega + 5)(j\omega + 10)]$

e. $|Y(\omega)|^2 = 100/[(\omega^2 + 5^2)(\omega^2 + 10^2)]$

f. $|Y(\omega)|^2 = (4/3)/(\omega^2 + 5^2)\text{-}(4/3)/(\omega^2 + 10^2)$

g. $E_y = 1/15$ J $= 0.06667$ J

h. $E_2 = 0.024224$ J

i. $E_y = 0.023921$ J

CHAPTER 20

20.1 $z_{11} = 16\ \Omega, z_{21} = 10\ \Omega, z_{22} = 25\ \Omega, z_{12} = 10\ \Omega$

20.3 $z_{11} = 17\ \Omega, z_{21} = 8\ \Omega, z_{22} = 12\ \Omega, z_{12} = 8\ \Omega$

20.5 $z_{11} = 12\ \Omega, z_{21} = 8\ \Omega, z_{22} = 17\ \Omega, z_{12} = 8\ \Omega$

20.7 $z_{11} = j38\ \Omega, z_{21} = -j25\ \Omega, z_{22} = j33\ \Omega, z_{12} = -j25\ \Omega$

20.9 $g_{11} = 0.16667\ S, g_{21} = -1.8, g_{22} = 9\ \Omega, g_{12} = 0$

20.11 $A = (s + 1)/s, C = (2s + 1)/s\ S, B = 1/s\ \Omega, D = (s + 1)/s$

20.13 $$H(s) = \frac{s^2}{s^2 + \left(\frac{R_1}{L_2} + \frac{R_1}{L_1} + \frac{R_2}{L_2}\right)s + \frac{R_1R_2}{L_1L_2}}$$

20.15 $b_{11} = 1, b_{21} = 1/Z_2\ S, b_{12} = Z_1\ \Omega, b_{22} = 1 + Z_1/Z_2$

20.17 $h_{11} = Z_1\ \Omega, h_{12} = 1, h_{21} = -1, h_{22} = 1/Z_2\ S$

20.19 $$\boldsymbol{T} = \begin{bmatrix} \dfrac{s^3 + 8s^2 + 8s + 2}{s^3} & \dfrac{2(2s^2 + 3s + 1)}{s^3} \\ \dfrac{3s^2 + 6s + 2}{s^2} & \dfrac{s^2 + 4s + 2}{s^2} \end{bmatrix}$$

20.21 $$\boldsymbol{Z} = \begin{bmatrix} 22 & 18 \\ 18 & 27 \end{bmatrix}$$

20.23 $$\boldsymbol{Y} = \begin{bmatrix} \dfrac{0.5s^2 + 2s + 0.5}{s + 1} & \dfrac{-0.5s^2 - 0.5}{s + 1} \\ \dfrac{-0.5s^2 - 0.5}{s + 1} & \dfrac{0.5s^2 + 2s + 0.5}{s + 1} \end{bmatrix}$$

Index

A

C

G

M